PRINCIPLES OF HUMAN ANATOMY

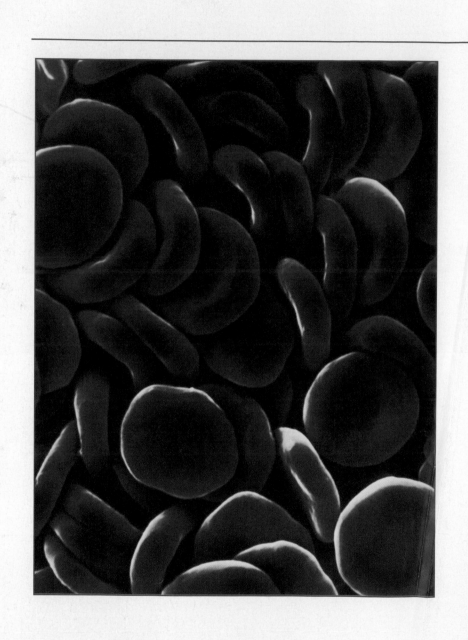

PRINCIPLES OF HUMAN ANATOMY

TENTH EDITION

Gerard J. Tortora

Bergen Community College

WILEY

John Wiley & Sons, Inc.

Executive Editor	Bonnie Roesch
Executive Marketing Manager	Clay Stone
Developmental Editor	Karen Trost
Associate Production Manager	Kelly Tavares
Cover/Text Designer	Karin Gerdes Kincheloe
Art Coordinator	Claudia Durrell
Illustration Editor	Anna Melhorn
Art Studio	Imagineering, Inc.
Photo Researcher	Teri Strafford
Photo Editor	Hilary Newman

Cover Photo Credits: Background photo: ©imagingbody.com. *Left inset*: ©Dennis Kunkel/Phototake.*Right insets* (top to bottom): ©David Becker/Photo Researchers; ©Science Photo Library/Photo Researchers; ©CNRI/Photo Researchers; ©CNRI/Phototake; ©CNRI/Phototake; ©imagingbody.com; ©ISM/Phototake; ©CNRI/Phototake.

This book was set in Janson by Progressive Information Technologies and printed and bound by Von Hoffman Press. The cover was printed by Lehigh Press.

ISBN: 0-471-42081-6
Printed in the United States of America
10 9 8 7 6 5 4 3

ABOUT THE AUTHOR

Jerry Tortora is Professor of Biology and former Biology Coordinator at Bergen Community College in Paramus, New Jersey, where he teaches human anatomy and physiology as well as microbiology. He received his bachelor's degree in biology from Fairleigh Dickinson University and his master's degree in science education from Montclair State College. He is a member of many professional organizations, including the Human Anatomy and Physiology Society (HAPS), the American Society of Microbiology (ASM), American Association for the Advancement of Science (AAAS), National Education Association (NEA), and the Metropolitan Association of College and University Biologists (MACUB).

Above all, Jerry is devoted to his students and their aspirations. In recognition of this commitment, Jerry was the recipient of MACUB's 1992 President's Memorial Award. In 1996, he received a National Institute for Staff and Organizational Development (NISOD) excellent award from the University of Texas and was selected to represent Bergen Community College in a campaign to increase awareness of the contributions of community colleges to higher education.

Jerry is the author of several best-selling science textbooks and laboratory manuals, a calling that often requires an additional 40 hours per week beyond his teaching responsibilities. Nevertheless, he still makes time for four or five weekly aerobic workouts that include biking and running. He also enjoys attending college basketball and professional hockey games and performances at the Metropolitan Opera House.

To my mother,
Angelina M. Tortora
Her love, guidance, faith, support,
and example continue to be
the cornerstone of my personal
and professional life.

G.J.T.

PREFACE

Principles of Human Anatomy, tenth edition, is designed for introductory courses in human anatomy. The highly successful approach of previous editions–to provide students with an accurate, clearly written, and expertly illustrated presentation of the structure of the human body, to offer insights into the connections between structure and function, and to explore the practical and relevant applications of anatomical knowledge to everyday life and career development—has been retained. This new edition enhances and improves upon these strengths, while at the same time offering some new and innovative features to increase student motivation and success.

ORGANIZATION, SPECIAL TOPICS, AND CONTENT IMPROVEMENTS

Like most undergraduate anatomy courses, *Principles of Human Anatomy* follows a systems approach. The first three chapters introduce the basic nomenclature and conventions used to study human anatomy, review the fundamentals of cell biology, and introduce tissues, respectively.

A major organizational change in the tenth edition is the placement of the concepts of development early in the book in Chapter 4. This provides students with the basics of development *before* studying the developmental biology of body systems in their respective chapters.

Within the systems chapters is another major organizational change. The autonomic nervous system, which previously followed the special senses, now follows coverage of the spinal cord and brain to provide students with greater continuity.

VISUALIZING THE HUMAN BODY

A new feature of the tenth edition are 11 perforated pull-outs visualizing the human body. Each inset is devoted to a specific body system and presents a series of images that highlight various ways to visualize the anatomical features of that system. The front of the inset arranges the images in a dramatic visual presentation, suitable as a poster if desired. On the back, the images are repeated, but this time accompanied by a caption and a few labels of key anatomical features where appropriate.

Images common to multiple body systems include diagrams, cadaver photos, CT and MRI scans, radiographs, and micrographs (scanning electron micrographs, transmission electron micrographs, and photomicrographs). Among the images unique to one or a few specific body systems are urograms (urinary organs), PET scans (brain and heart), barium contrast x-rays (gastrointestinal organs), and angiograms (blood vessels). The variety of images presented provides students with an understanding of the many ways to study both normal anatomy and applied anatomy in clinical medicine in the diagnosis of disease.

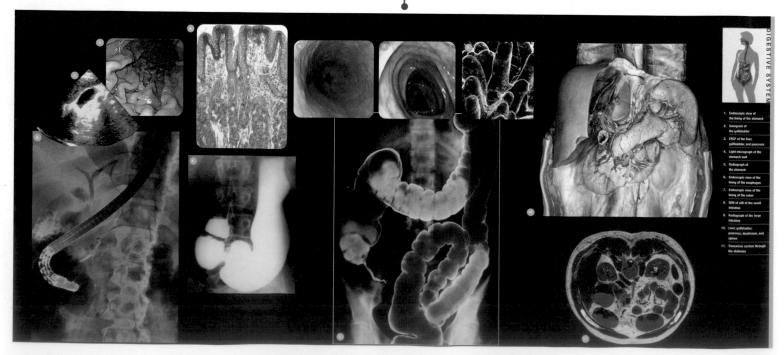

MNEMONICS

At times, students require extra help to learn specific anatomical features of the various body systems. One way to do this is the use of mnemonics, aids to help memory. In this edition, the number of mnemonics has been increased significantly. You are encouraged to not only use the mnemonics provided, but to also substitute your own or even devise new ones. If you would like to share one or more original mnemonics, please send them to me in care of John Wiley & Sons, 111 River Street, Hoboken, NJ 07030, and they will be considered for publication on the companion website, as well as in the next edition of the book along with your name.

DEVELOPMENTAL BIOLOGY

Following an introduction to the concepts of development in Chapter 4, illustrated discussions of developmental biology are found near the conclusion of most body system chapters. These sections, entitled "Development of . . .", are identified with a distinctive "fetus" icon. Placing this coverage at the end of chapters enables students to master the concepts and anatomical terminology they need in order to learn about embryonic and fetal structures.

AGING

Anatomy is not static. As the body ages, its structure and related functions change in subtle and not so subtle ways. Moreover, aging is a professionally relevant topic for the majority of this book's readers, who will go on to careers in health-related fields in which the median age of the client population is steadily advancing. For these reasons, many of the body system chapters explore age-related changes in anatomy and function.

EXERCISE

Physical exercise can produce favorable changes in some anatomical structures, most notably those associated with the musculoskeletal and cardiovascular systems. This information has special relevance to readers embarking on careers in physical education, sports training, and dance. Key chapters include brief discussions of exercise, indicated by a distinctive "running shoe" icon.

DEVELOPMENT OF THE EYES AND EARS

OBJECTIVE

● Describe the development of the eyes and the ears.

Development of the Eyes

The *eyes* begin to develop about 22 days after fertilization when the **ectoderm** of the lateral walls of the prosencephalon (fore-

---

AGING AND THE INTEGUMENTARY SYSTEM

OBJECTIVE

● Describe the effects of aging on the integumentary system.

The pronounced effects of skin aging do not become noticeable until people reach their late forties. Most of the age-related changes occur in the dermis. Collagen fibers in the dermis begin to decrease in number, stiffen, break apart, and disorganize into a shapeless, matted tangle. Elastic fibers lose some of their elas-

---

EXERCISE AND THE HEART

OBJECTIVE

● Explain the relationship between exercise and the heart.

A person's fitness, regardless of level, can be improved at any age with regular exercise. Of the various types of exercise, some are more effective than others for improving the health of the cardiovascular system. **Aerobics,** any activity that works large body muscles for at least 20 minutes, elevates cardiac output and accelerates metabolic rate. Three to five such sessions a week are

Every chapter in the tenth edition of *Principles of Human Anatomy* incorporates a host of improvements to both the text and the art suggested by reviewers, educators, or students. Here are just some of the more noteworthy changes:

- **Chapter 1 An Introduction to the Human Body** includes expanded coverage of radiography, new sections on radionuclide scanning and single-photo-emission computerized tomography (SPECT) scanning, new illustrations of body systems, and new photographs of coronary angiography, intravenous urography, barium contrast x-ray, mammogram, bone density, radionuclide scan, and SPECT scan.

- **Chapter 2 Cells** This fundamental chapter features five new clinical applications; a new section on proteasomes; new illustrations of cell division and meiosis to accompany the section on reproductive cell division, which has been moved to this chapter from the reproductive systems chapter; and an expanded key medical terms list.

- **Chapter 3 Tissues** Among the changes to this chapter are three new clinical applications on basement membranes and disease, chondroitin sulfate, glucosamine, and joint disease, and ligaments and sprains; an expanded discussion of synovial membranes; an expanded key medical terms list; new line art on epithelial cell shapes and arrangements and types of membranes; and several new epithelial tissue photos.

- **Chapter 4 Development** For the first time, the basic concepts of development, including the developmental stages, prenatal diagnostic tests, material changes during pregnancy, exercise and pregnancy, and labor, are considered early in the book, rather than in the last chapter. This provides students with the basics of development *before* studying the developmental biology of the various body systems later in the book.

  Organizational changes were made within the chapter as well. The embryonic period is now discussed chronologically by weeks (1, 2, 3, 4, 5-8) to emphasize the developmental changes that occur in specific time frames.

  New to the chapter are sections dealing with development of lacunar networks, gastrulation, induction, neurulation, development of intraembryonic coelom, development of the cardiovascular system, organogenesis, embryonic folding, and development of pharyngeal arches and pouches. In addition, new clinical applications on stem cell research and therapeutic cloning and neural tube defects have been added, and the discussion of somite development and the key medical terms list have been expanded.

  In addition to the numerous text changes, the bulk of the art has been redrawn and many new pieces have been added. The new art helps to bridge the gap between major developmental events. All new photos of embryos and fetuses have also been added.

- **Chapter 5 The Integumentary System** This important chapter now includes a new section on types of hairs, new clinical applications on tattooing and chemotherapy and hair loss, and an expanded section on development of the integumentary system. There are also new photomicro-

graphs of sebaceous and sudoriferous glands, new photos of malignant melanoma, burns, and pressure ulcers, and new line art on the development of the skin, hair, sebaceous glands, and sudoriferous glands.

- **Chapter 6 Bone Tissue** features a new clinical application on remodeling and orthodontics.

- **Chapter 7 The Skeletal System: The Axial Skeleton** now has a new clinical application on dislocated and separated ribs; a new surface projection illustration of the paranasal sinuses; new illustrations of the position of the hyoid bone, isolated sternum, and abnormal curves of the vertebral column; a new photograph illustrating spina bifida; and the addition of a key medical terms list.

- **Chapter 8 The Skeletal System: The Appendicular Skeleton** contains an expanded version of the section on the development of the skeletal system formerly included in the bone tissue chapter. New line illustrations show a superior view of the clavicle and the development of the cranium.

- **Chapter 9 Joints** includes new clinical applications on cartilage replacement and ankle sprains and fractures, a new section on arthroplasty, and a new exhibit and line art on the ankle joint.

- **Chapter 10 Muscular Tissue** features an expanded discussion of the development of muscle and a new clinical application on the regeneration of heart cells.

- **Chapter 11 The Muscular System** has a new section on the femoral triangle, a new clinical application on benefits of stretching, and a new exhibit and illustration on muscles of the anterior neck.

- **Chapter 12 Surface Anatomy** Several of the photographs in this chapter have been replaced to enhance the presentation of surface anatomy.

- **Chapter 13 The Cardiovascular System: Blood** New clinical applications on induced polycythemia in athletes, bone marrow aspiration and biopsy, and iron overload are featured.

- **Chapter 14 The Cardiovascular System: The Heart** includes expanded discussions of the development of the heart and the development of atherosclerotic plaques, a new clinical application on help for failing hearts, new illustrations on the development of the heart and congenital heart defects, and an expanded key medical terms list.

- **Chapter 15 The Cardiovascular System: Blood Vessels** features an expanded discussion of the development of blood vessels; new clinical information related to sites for taking blood pressure and pulse, and coronary artery bypass grafting; new line art showing generalized views of circulatory routes, pulmonary circulation, and fetal circulation; and an expanded key medical terms list.

- **Chapter 16 The Lymphatic and Immune System** Phonetic pronunciations have been added to names of lymph nodes, and the AIDS discussion has been expanded and updated. In addition, new line art shows components of the lymphatic system, routes of lymph drainage, structure of a lymph node, and principal lymph nodes of the head and neck, upper limbs, lower limbs, abdomen and pelvis, and thorax. Photomicrographs of the thymic corpuscle, lymph nodes, and the spleen have also been added.

- **Chapter 17 Nervous Tissue** includes a completely revised and expanded discussion of neuroglia, along with expanded coverage of neurotransmitters and a revised discussion of neurons and synapses. This chapter also has new clinical applications on neurotoxins and local anesthesia and demyelination, and new illustrations of the structure of a neuron, structural classification of neurons, and neuroglia.

- **Chapter 18 The Spinal Cord and Spinal Nerves** features an expanded version of the section on spinal cord disorders that formerly appeared in the special senses chapter; new sections on spinal cord compression and degenerative diseases of the spinal cord; new clinical applications on the vertebral canal and spinal injuries, spinal nerve root damage, and reflexes and diagnosis; a new key medical terms list; and new illustrations of the gross anatomy of the spinal cord, external anatomy of the spinal cord and spinal nerves, transverse section of the spinal cord, sensory and motor tracts in the spinal cord, patellar reflex, connective tissue coverings of a spinal nerve, and branches of a spinal nerve.

- **Chapter 19 The Brain and Cranial Nerves** includes a new section on brain tumors; new tables on brain development and functional differences between the cerebral hemispheres; new clinical applications on subdural hemorrhage, ataxia, and damage to basla ganglia; an expanded key medical terms list, and numerous new illustrations.

- **Chapter 20 The Autonomic Nervous System** This chapter now directly follows the spinal cord and brain coverage to provide students with continuity of the major parts of the nervous system. It includes an expanded table comparing the sympathetic and parasympathetic divisions of the autonomic nervous system, a new key medical terms list, and new illustrations of motor neuron pathways, sympathetic and parasympathetic divisions of the autonomic nervous system, and ganglia of the autonomic nervous system.

- **Chapter 21 Somatic Senses** This chapter has been retitled to reflect its major content. Included are new clinical applications on amyotrophic lateral sclerosis, damage to basal ganglia, and damage to the cerebellum; an expanded key medical terms list; and new illustrations of proprioceptors, somatic sensory pathways, primary somatosensory and primary motor areas of the cerebrum, and direct motor pathways.

- **Chapter 22 The Special Senses** features new sections on development of the eyes and ears and the effects of aging on the special senses; new clinical applications on taste aversion, LASIK, color blindness and night blindness, and age-related macular disease; an expanded key medical terms list; and new illustrations of the develoment of the eye and ear.

- **Chapter 23 The Endocrine System** includes a new section and photomicrograph on the histology of the anterior pituitary, and new illustrations of the thyroid gland, parathyroid glands, adrenal glands, pancreas, development of the pituitary gland, and endocrine disorders.

- **Chapter 24 The Respiratory System** The section on the development of the respiratory system has been revised and expanded. Also included are a new table on modified respiratory movements; a new clinical application on nasal polyps; new illustrations of the structure of the respiratory system, respiratory structures in the head and neck, pharynx, trachea, bronchial tree, surface anatomy of the lungs, and bronchopulmonary segments; and new photos of the lungs, histology of a bronchiole, alveoli, and nervous control of respiration.

- **Chapter 25 The Digestive System** features new clinical applications on Zollinger-Ellison syndrome, liver biopsy, lactose intolerance, absorption of alcohol, polyps of the colon, and colonoscopy; an expanded key medical terms list; new illustrations of layers of the gastrointestinal tract, salivary glands, liver histology, and large intestine; new photomicrographs of the histology of the small intestine and large intestine; and a new photograph of the stomach.

- **Chapter 26 The Urinary System** includes a new section on kidney dialysis; expanded coverage of development and the effects of aging on the urinary system; new clinical applications on kidney transplants, diuretics, and cystoscopy; new tables on physical characteristics and abnormal constituents of urine; an expanded key medical terms list; and new illustrations of development and a comparison of the female and male urethra.

- **Chapter 27 The Reproductive Systems** has revised sections on the structure of a sperm cell and erectile tissue and erection; expansion of coverage of the effects of aging; a new table comparing oogenesis and follicular development; new clinical applications on testicular injury, ovarian cysts, abnormal vaginal discharge and bleeding, the female athlete triad, and deficiency of 5-alpha-reductase; and new illustrations of the sperm cell and of development of the reproductive systems.

ENHANCEMENTS TO THE ILLUSTRATION PROGRAM

NEW DESIGN

A textbook with beautiful illustrations or photographs on most pages requires a carefully crafted and functional design. The design for the tenth edition has been transformed to assist students in making the most of the text's many features and outstanding art. Each page is carefully laid out to place related text, figures, and tables near one another, minimizing page-turning during the reading of a topic. Red print is used to indicate the first mention of a figure or table. Not only is the reader alerted to refer to the figure or table, but the color print also serves as a place locator for easy return to the narrative.

The distinctive icons incorporated throughout the text signal special features and make them easy to find during review. These include the **key** with the Key Concept Statements; the **question mark** with the applicable questions that enhance every figure; the **stethoscope** indicating a clinical application within a chapter narrative; the **fetus** icon announcing the developmental biology section of a chapter; the **running shoe** highlighting content relevant to exercise, and the icons that indicate the distinctive types of **chapter-ending questions.**

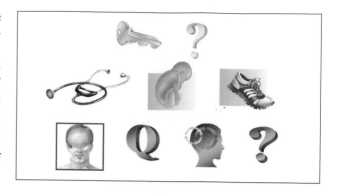

NEW ART

An outstanding illustration program has always been a signature feature of this text. Beautiful artwork, carefully chosen photographs and photomicrographs, and unique pedagogical enhancements all combine to make the visual appeal and usefulness of the illustration program in *Principles of Human Anatomy* distinctive. Continuing in this tradition, you will find exciting new three-dimensional illustrations gracing the pages of nearly every chapter in the text.

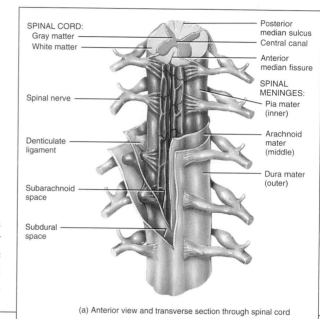

(a) Anterior view and transverse section through spinal cord

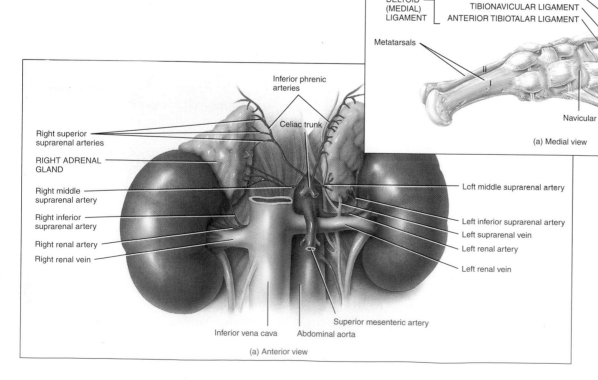

(a) Medial view

(a) Anterior view

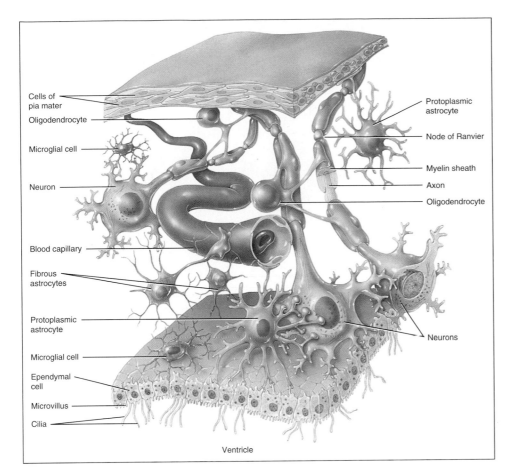

Ventricle

MORE NEW HISTOLOGY-BASED ART

As part of our plan of continuous improvement, many of the anatomical illustrations based on histological preparations have been replaced in this edition, while others are new additions.

NEW PHOTOMICROGRAPHS

Dr. Michael Ross of the University of Florida has once again provided me with beautiful, customized photomicrographs of various tissues of the body. I have always considered Dr. Ross' photos among the best available; their inclusion greatly enhances the illustration program.

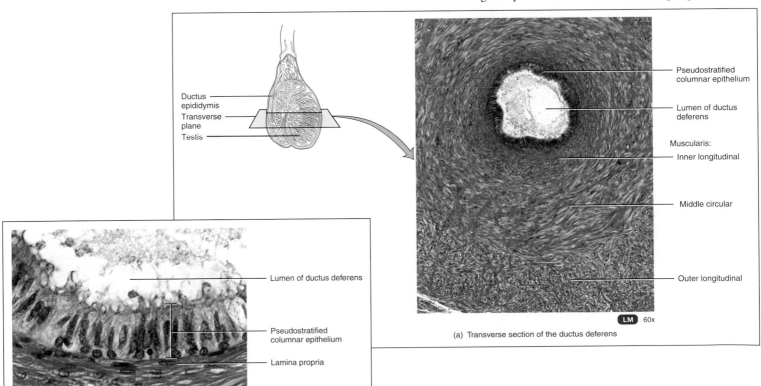

(a) Transverse section of the ductus deferens

(b) Details of the epithelium

CADAVER PHOTOS

An assortment of large, clear cadaver photos are included at strategic points in many chapters. What is more, many anatomy illustrations are keyed to the large cadaver photos included in my companion text, *A Photographic Atlas of the Human Body, Second Edition.*

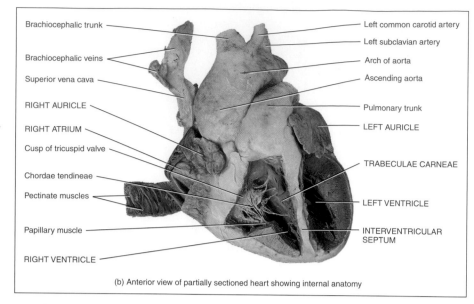

(b) Anterior view of partially sectioned heart showing internal anatomy

ORIENTATION DIAGRAMS

Students sometimes need help figuring out the plane of view of anatomy illustrations—descriptions alone do not always suffice. An orientation diagram explaining the perspective of the view represented in the figure accompanies every major anatomy illustration. There are three types of diagrams: (1) planes used to indicate where certain sections are made when a part of the body is cut;

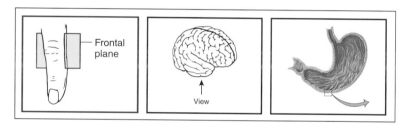

(2) diagrams containing a directional arrow and the word "View" to indicate the direction from which the body part is viewed, e.g. superior, inferior, posterior, anterior; and (3) diagrams with arrows to direct attention to enlarged and detailed parts of illustrations.

KEY CONCEPT STATEMENTS

Included above every figure and denoted by the "key" icon, this feature summarizes an idea that is discussed in the text and demonstrated in the figure. A pedagogical feature unique to our text, these statements help students keep focused on the relevance of the figure to their understanding of specific content.

REVISED FIGURE QUESTIONS

This highly applauded feature asks readers to synthesize verbal and visual information, think critically, and/or draw conclusions about what they see in a figure. Each Figure Question appears under its illustration and is highlighted in this edition by the distinctive "Question Mark" icon. Answers are located at the end of each chapter.

Figure 25.2 Layers of the gastrointestinal tract.

The four layers of the gastrointestinal tract, from deep to superficial, are the mucosa, submucosa, muscularis, and serosa.

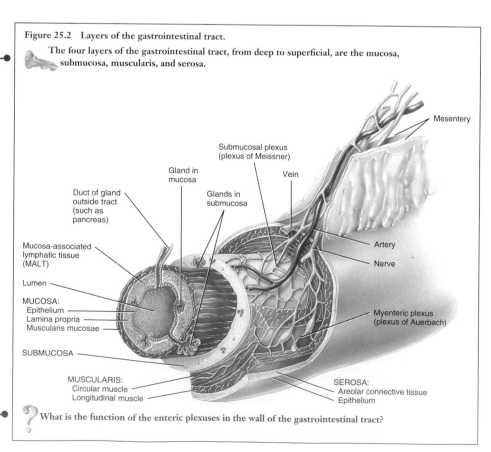

What is the function of the enteric plexuses in the wall of the gastrointestinal tract?

OVERVIEWS OF FUNCTIONAL ANATOMY

This art-related feature gives a brief summary of the functions of the anatomical structure or system depicted, reinforcing the connection between structure and function.

FUNCTIONS
1. Body temperature regulation
2. Reservoir for blood
3. Protection from external environment
4. Cutaneous sensations
5. Excretion and absorption
6. Vitamin D synthesis

Epidermal ridge

Dermal papilla

Capillary loop

Sweat pore

Sebaceous (oil) gland

Corpuscle of touch (Meissner corpuscle)

Arrector pili muscle

Hair follicle

Hair root

Eccrine sweat gland

Apocrine sweat gland

Lamellated (pacinian) corpuscle

Sensory nerve

Adipose tissue

Hair shaft

Free nerve ending

EPIDERMIS

Papillary region

DERMIS

Reticular region

Subcutaneous layer

Blood vessels:
Vein
Artery

ENHANCED CLINICAL AND PRACTICAL RELEVANCE

Principles of anatomy and functional anatomy sometimes are better understood by considering variations from normal. Moreover, most students are naturally curious about how the information in this book or in their anatomy course pertains to a potential career choice, or how it relates to them or family members. Most chapters in the text offer several opportunities for students to make connections between the normal and abnormal.

CHEMOTHERAPY AND HAIR LOSS

Chemotherapy is the treatment of disease, usually cancer, by means of chemical substances or drugs. Chemotherapeutic agents interrupt the life cycle of rapidly dividing cancer cells. Unfortunately, the drugs also affect other rapidly dividing cells in the body, such as the matrix cells of a hair. It is for this reason that individuals undergoing chemotherapy experience hair loss. Since about 15% of the matrix cells of scalp hairs are in the resting stage, these cells are not affected by chemotherapy. Once chemotherapy is stopped, the matrix cells replace lost hair follicles and hair growth resumes. ■

CLINICAL APPLICATIONS

This feature is a perennial favorite among students. A variety of clinical perspectives—many proposed by students—are introduced directly following the discussions to which they relate, highlighted by the distinctive stethoscope icon. Some applications are new to this edition, and all have been reviewed and updated.

APPLICATIONS TO HEALTH

This popular feature at the end of relevant chapters consists of major diseases, disorders, and medical conditions that illustrate departures from normal anatomy and function. These descriptions, which provide answers to many questions that students ask about medical disorders and diseases, have been reviewed and updated; some have been simplified, and others expanded.

KEY MEDICAL TERMS

Also found at the end of most chapters, vocabulary-building glossaries of selected medical terms and conditions with phonetic pronunciations have been fully updated for the tenth edition.

APPLICATIONS TO HEALTH

MULTIPLE SCLEROSIS

Multiple sclerosis (MS) is a disease that causes a progressive destruction of myelin sheaths of neurons in the CNS. It afflicts about 350,000 people in the United States and 2 million people worldwide. It usually appears between the ages of 20 and 40, affecting females twice as often as males. MS is most common in whites, less common in blacks, and rare in Asians. MS is an autoimmune disease—the body's own immune system spearheads the attack. The condition's name describes the anatomical

EPILEPSY

Epilepsy is ch sensory, or psy affects intellig about 1% of abnormal, syn neurons in the ating circuits. send nerve imp noise, o

KEY MEDICAL TERMS ASSOCIATED WITH NERVOUS TISSUE

Guillain-Barré Syndrome (GBS) (GĒ-an ba-RĀ) An acute demyelinating disorder in which macrophages strip myelin from axons in the PNS. It is the most common cause of acute paralysis in North America and Europe and may result from the immune system's response to a bacterial infection. Most patients recover completely or partially, but about 15% remain paralyzed.

Neuropathy (noo-ROP-a-thē; *neuro-*=a nerve; *-pathy*=disease) Any disorder that affects the nervous system but particularly a disorder of a cranial or spinal nerve. An example is *facial neuropathy* (Bell's palsy), a disorder of the facial (VII) nerve.

Neuroblastoma (noor-ō-blas- consists of immature ner commonly in the abdon adrenal glands. Although in infants.

Rabies (RĀ-bēz; *rabi-*=mad, a virus that reaches the C mitted by the bite of a The symptoms are exciter followed by paralysis and

HALLMARK FEATURES

The tenth edition of *Principles of Human Anatomy* builds on the legacy of thoughtfully designed and class-tested pedagogical features that provide a complete learning system for students as they navigate their way through the text and course. Many—such as critical thinking questions and end of chapter quizzes—have been revised to reflect the enhancements to the text and art.

CONTENTS AT A GLANCE

A brief outline opens each chapter of the text to give the student a preview of what will be covered.

STUDENT OBJECTIVES

Student learning objectives are integrated into the body of the chapters, at the beginning of major sections, where they serve as both a guide for section content and a handy review tool.

CHECKPOINTS

Placed at strategic intervals within chapters, these review questions give students the chance to validate their understanding as they study. The numbered checkpoints relate directly back to the student objective at the beginning of the text section, while the answers to those in the exhibits can be found within the exhibit.

HELPFUL TABLES AND EXHIBITS

The utility and readability of the helpful *Tables* included throughout the text has been improved.

In addition, *Exhibits,* self-contained features developed to give students the extra help they need to learn the anatomy of complex body systems—most notably skeletal muscles, articulations, blood vessels, and nerves—have also been improved. Each Exhibit consists of an objective, an overview, a tabular summary of the relevant anatomy, an associated suite of illustrations or photographs, and checkpoint questions. Some Exhibits contain a relevant clinical application as well. Students will find this clear and concise presentation to be the ideal study vehicle for organizing and learning the details of these important systems.

CROSS REFERENCES

This edition features cross-references that guide the reader to specific pages and figures. Most will help students relate new concepts to previously learned material. However, I also acknowledge the really ambitious student by including some cross-references to material that has yet to be considered.

PRONUNCIATIONS AND WORD ROOTS

Students—even the best—generally find it difficult at first to read and pronounce anatomical terms. As an educator in the largest metropolitan region of the United States, I teach and am sympathetic to the needs of the growing ranks of college students who speak English as a second language. Because of these reasons, the features included in *Principles of Human Anatomy* to help students build a working and useful vocabulary are carefully scrutinized during development of each edition. New terms are highlighted in boldface type, and most highlighted terms in the text, tables, and exhibits are accompanied by easy-to-understand pronunciation guides. As a further aid, the most important highlighted terms include word roots, designed to both provide an understanding of the meaning of the term at hand and help students learn the meaning of related terms they may encounter.

STUDY OUTLINE

As always, readers will benefit from the popular end-of-chapter summaries of major topics that are page referenced to the chapter discussions.

SELF QUIZZES

Quizzes at the end of every chapter include fill-in-the-blank, multiple choice, and matching questions. Many of the questions are new to this edition. These quizzes are meant not only for students to test their ability to memorize the facts presented in each chapter, but also to sharpen their critical thinking skills by applying the concepts and processes they have just mastered. Answers to the Self-Quiz items are presented in an appendix at the end of the book.

CRITICAL THINKING APPLICATIONS

Critical Thinking Questions are provided at the end of each chapter. These questions, indicated by a "thinking" head icon, are essay-style problems that encourage students to think about and apply the concepts they have studied in each chapter. The style of these questions ought to make students smile on occasion as well as think! Although many of these questions have no one right answer, suggested answers appear in an appendix at the end of the book.

GLOSSARY

A full glossary of terms with phonetic pronunciations appears at the end of the book. The basic building blocks of medical terminology—**Combining Forms, Word Roots, Prefixes, and Suffixes**—are listed inside the back cover, as is a listing of **Eponyms**, traditional terms that include reference to a person's name, along with the current terminology.

ACKNOWLWDGEMENTS

I wish to especially thank several of my academic colleagues for their helpful contributions to this edition.

- **Frances Frierson, M.D.**, of Valencia Community College provided invaluable assistance with the preparation of the Visualizing the Human Body Insets. Her expertise as both a teacher and a medical doctor helped define the choice of images most appropriate for depicting each body system.

- **Leslie Miller M.S.N.** of Iowa State University provided many insights and suggestions in her review of the clinical material throughout the text. Her practical experience as a registered nurse enabled her to make many helpful contributions along the way.

- **Pam Langley** of New Hampshire Community Technical College contributed to the completion of this edition by updating and revising the Self-Quizzes and Critical Thinking Questions at the end of each chapter. In the process she supplied many new, challenging, and diverse test questions.

- Happily, **Dr. Michael Ross** of the University of Florida once again agreed to provide new histology photographs especially to illustrate our work. I am grateful to all for their fine work in making this an even better edition for students to use.

In addition to the enhancements to the text contributed by those named above, a terrific group of academics provided the insights and materials to for tremendous ancillary support available for this edition. I wish to acknowledge each and thank them for their work - Izak Paul, Mt. Royal College for Instructor's Resources and the creation of WebCT and Blackboard courses; Kathleen Anderson, University of Iowa for the Testbank; Ameed Raoof, University of Michigan for the Illustrated Powerpoint Lecture Slides; Ron Gaines, Cameron University, Jon Jackson, University of North Dakota, Ameed Raoof, University of Michigan, and Brian Wisenden, University of Minnesota for the bank of Practice Quizzes on the Student Companion Website; Tony Yates, Seminole State College for identifying and reviewing web links for the Companion Sites; and Kevin Petti, San Diego Miramar College for his excellent Changing Image essays and activities.

The publication of the tenth edition of **Principles of Human Anatomy** is my second edition of this text with John Wiley and Sons, Inc. and I continue to enjoy collaborating with this enthusiastic, dedicated and talented team of publishing professionals. My thanks to the entire "Tortora Team" - Bonnie Roesch, Executive Editor; Karen Trost, Developmental Editor; Mary O'Sullivan, Project Editor and Supplements Coordinator; Maureen Powers, Program Assistant; Karin Kincheloe, Senior Designer; Claudia Durell, Art Coordinator; Kelly Tavares, Senior Production Editor; Hillary Newman, Photo Editor; Teri Strafford, Photo Researcher; and Clay Stone, Executive Marketing Manager. I appreciate all that you do.

I am always extremely grateful to my colleagues who have reviewed the manuscript and offered numerous suggestions for improvement. The reviewers who have provided their time and expertise to maintain this book's accuracy are noted in the Reviewer list that follows:

Kathleen Andersen
University of Iowa

Frank Baker
Golden West College

Leann Blem
Virginia Commonwealth University

Alphonse Burdi
University of Michigan

Alan Dietsche
University of Rochester

Ron Gaines
Cameron University

David Hammerman
Long Island University

Robert S. Hikida
Ohio University

Grant Hurlburt
California State University, Bakersfield

Stephen K. Itaya
University of South Alabama

Jon Jackso
University of North Dakota

Kelly Johnson
University of Kansas

Lyle Konigsberg
University of Tennessee

Frank T. Logiudice
University of Central Florida

Rebecca Moore Peterson
Pennsylvania State University

Virginia L. Naples
Northern Illinois University

Daniel Olsen
Northern Illinois University

Russell Peterson
Indiana University of Pennsylvania

Catrin Pittack
University of Washington

Ameed Raoof
University of Michigan

Ahnya Redman
Pennsylvania State University

James Strauss
Pennsylvania State University

Barbara Walton
University of Tennessee at Chattanooga

Brian Wisenden
Minnesota State University, Moorhead

David A. Woodman
University of Nebraska-Lincoln

Tony Yates
Seminole State College

As always, my family, colleagues, and friends have supported my writing activities in more ways than I could ever mention here. Their understanding and encouragement will always be appreciated and will never be taken for granted.

Gerard J. Tortora
Department of Science and Health, S229
Bergen Community College
400 Paramus Road
Paramus, NJ 07652

NOTE TO STUDENTS

Your book has a variety of special features that will make your time studying anatomy a more rewarding experience. These have been developed based on feedback from students—like you—who have used previous editions of the text. Below are some hints for using some of these helpful aids. A review of the preface will give you insight, both visually and in narrative, to all of the text's distinctive features.

Begin your study by anticipating what is to be learned from each chapter. A brief, page referenced overview called **Contents at a Glance** begins each chapter. As you begin each narrative section of the chapter, be sure to take note of the **Objectives** at the beginning of the section to help you focus on what is important as you read. At the end of the section, take time to try and answer the **Checkpoint** questions placed there. If you can, then you are ready to move on to the next section. If you experience difficulty answering the questions, you may want to re-read the section before continuing.

OVERVIEW OF BRAIN ORGANIZATION AND BLOOD SUPPLY

OBJECTIVES

- Identify the major parts of the brain.
- Describe how the brain is protected.
- Describe the blood supply of the brain.

CHECKPOINT

1. Compare the sizes and locations of the cerebrum and cerebellum.
2. Describe the locations of the cranial meninges.
3. Explain the blood supply to the brain and the importance of the blood–brain barrier.

Studying the figures (illustrations that include artwork and photographs) in this book is as important as reading the text. To get the most out of the visual parts of this book, use the tools we have added to the figures to help you understand the concepts being presented. Start by reading the **legend,** which explains what the figure is about. Next, study the **key concept statement,** indicated by a "key" icon, which reveals a basic idea portrayed in the figure. Added to many figures you will also find an **orientation diagram** to help you understand the perspective from which you are viewing a particular piece of anatomical art. Finally, at the bottom of each figure you will find a **figure question,** accompanied by a "question mark" icon. If you try to answer these questions as you go along, they will serve as self-checks to help you understand the mater-

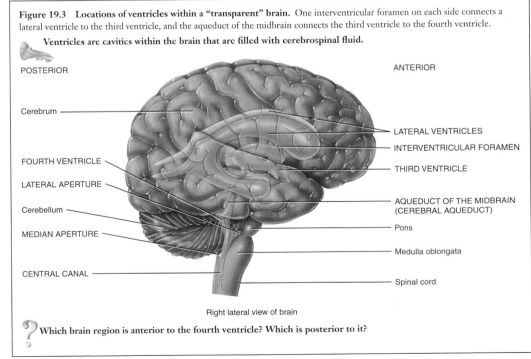

Figure 19.3   Locations of ventricles within a "transparent" brain. One interventricular foramen on each side connects a lateral ventricle to the third ventricle, and the aqueduct of the midbrain connects the third ventricle to the fourth ventricle.

🔑 **Ventricles are cavities within the brain that are filled with cerebrospinal fluid.**

POSTERIOR — ANTERIOR

Cerebrum — LATERAL VENTRICLES — INTERVENTRICULAR FORAMEN — THIRD VENTRICLE

FOURTH VENTRICLE — LATERAL APERTURE — Cerebellum — MEDIAN APERTURE — CENTRAL CANAL

AQUEDUCT OF THE MIDBRAIN (CEREBRAL AQUEDUCT) — Pons — Medulla oblongata — Spinal cord

Right lateral view of brain

❓ **Which brain region is anterior to the fourth ventricle? Which is posterior to it?**

ial. Often it will be possible to answer a question by examining the figure itself. Other questions will encourage you to integrate the knowledge you've gained by carefully reading the text associated with the figure. Still other questions may prompt you to think critically about the topic at hand or predict a consequence in advance of its description in the text. You will find the answer to each figure question at the end of the chapter in which the figure appears.

Learning the complex anatomy and all of the terminology involved for each body system can be a daunting task. For many topics — most notably the skeletal muscles, articulations, blood vessels, and nerves – I have created special **Exhibits** which organize the material into manageable segments. Each Exhibit consists of an objective, an overview, a tabular summary of the relevant anatomy, an associated suite of illustrations or photographs, and checkpoint questions. Some Exhibits contain a relevant clinical application as well.

EXHIBIT 11.4   MUSCLES THAT MOVE THE TONGUE—EXTRINSIC TONGUE MUSCLES (Figure 11.7)

OBJECTIVE

● Describe the origin, insertion, action and innervation of the extrinsic muscles of the tongue.

The tongue is a highly mobile structure that is vital to digestive functions such as mastication, perception of taste, and deglutition (swallowing). It is also important in speech. The tongue's mobility is greatly aided by its suspension from the mandible, styloid process of the temporal bone, and hyoid bone.

The tongue is divided into lateral halves by a median fibrous septum. The septum extends throughout the length of the tongue. Inferiorly, the septum attaches to the hyoid bone. Muscles of the tongue are of two principal types: extrinsic and intrinsic. **Extrinsic tongue muscles** originate outside the tongue and insert into it. They move the entire tongue in various directions, such as anteriorly, posteriorly, and laterally. **Intrinsic tongue muscles** originate and insert within the tongue. These muscles alter the shape of the tongue rather than moving the entire tongue. The extrinsic and intrinsic muscles of the tongue insert into both lateral halves of the tongue.

When you study the extrinsic tongue muscles, you will notice that all of their names end in *glossus*, meaning tongue. You will

● Palatoglossus: the pharyngeal plexus, which contains axons from both the vagus (X) nerve and the accessory (XI) nerve

INTUBATION DURING ANESTHESIA

When general anesthesia is administered during surgery, a total relaxation of the muscles results. Once one of the various types of drugs for anesthesia have been given (especially the paralytic agents), the patient's airway must be protected and the lungs ventilated because the muscles involved with respiration are among those paralyzed. Paralysis of the genioglossus muscle causes the tongue to fall posteriorly, which may obstruct the airway to the lungs. To avoid this, the mandible is either manually thrust forward and held in place (known as the "sniffing position"), or a tube is inserted from the lips through the laryngopharynx (inferior portion of the throat) into the trachea (endotracheal intubation). People can also be intubated nasally. ■

Relating Muscles to Movements

Arrange the muscles in this exhibit according to the following actions on the tongue: (1) depression, (2) elevation, (3) protraction, and (4) retraction. The same muscle may be mentioned more than once.

■ CHECKPOINT

1. When your physician says, "open your mouth, stick out your tongue and say *ahh*" so she can examine the inside of your mouth for possible signs of infection, which muscles do you contract?

EXHIBIT 11.4   MUSCLES THAT MOVE THE TONGUE—EXTRINSIC TONGUE MUSCLES (Figure 11.7)

CONTINUED

Figure 11.7   Muscles that move the tongue.

The extrinsic and intrinsic muscles of the tongue are arranged in both lateral halves of the tongue.

Superior constrictor
Styloid process of temporal bone
Mastoid process of temporal bone
Digastric (posterior belly-cut)
Middle constrictor
Stylohyoid
Stylopharyngeus
HYOGLOSSUS
Hyoid bone
Inferior constrictor
Thyroid cartilage of larynx

STYLOGLOSSUS
PALATOGLOSSUS
Palatine tonsil
Hard palate (cut)
Tongue
GENIOGLOSSUS
Mandible (cut)
GENIOHYOID
Mylohyoid
Intermediate tendon of digastric
Fibrous loop for intermediate tendon of digastric
Thyrohyoid membrane (connects hyoid bone to larynx)

DANK

Right side deep view

? What are the functions of the tongue?

INSERTION	ACTION
Undersurface of tongue and hyoid bone.	Depresses tongue and thrusts it anteriorly (protraction).
Side and undersurface of tongue.	Elevates tongue and draws it posteriorly (retraction).
Side of tongue.	Elevates posterior portion of tongue and draws soft palate down on tongue.
Side of tongue.	Depresses tongue and draws down its sides.

continues

At the end of each chapter are other resources that you will find useful. The **Study Outline** is a concise statement of important topics discussed in the chapter. Page numbers are listed next to key concepts so you can refer easily to specific passages in the text for clarification or amplification. The **Self-quiz Questions** are designed to help you evaluate your understanding of the chapter contents. **Critical Thinking Questions** are word problems that allow you to apply the concepts you have studied in the chapter to specific situations.

STUDY OUTLINE

Overview of Brain Organization and Blood Supply (p. 584)

1. The major parts of the brain are the brain stem, cerebellum, diencephalon, and cerebrum.
2. The brain is protected by cranial bones and the cranial meninges.

visual reflexes. It also contains nuclei associated with cranial nerves III and IV.

4. A large part of the brain stem consists of small areas of gray matter and white matter called the reticular formation, which helps maintain consciousness, causes awakening from sleep, and contributes

Q SELF-QUIZ QUESTIONS

Choose the one best answer to the following questions.

1. The diencephalon consists of the:
   a. insula and hypothalamus.   b. pons and hypothalamus.

a. microglia.   b. oligodendrocytes.   c. neurons.
d. endothelial cells.   e. ependymal cells.

4. Which of the following is not located in the cerebrum?
   a. globus pallidus   b. corpora quadrigemina

ANSWERS TO FIGURE QUESTIONS **631**

CRITICAL THINKING QUESTIONS

1. An elderly relative suffered a stroke and now has difficulty moving her right arm and also has speech problems. What areas of the brain were damaged by the stroke?
2. Wolfgang partied a little too hard one night and passed out drunk in his bathroom at home. He awoke with a lump on his head from

4. Alicia just figured out that her Human Anatomy class actually starts at 9:00 A.M. and not at 9:15 A.M., which has been her arrival time since the beginning of the term. One of the other students remarks that Alicia's "gray matter is pretty thin." Should Alicia thank him?

ANSWERS TO FIGURE QUESTIONS

19.1 The largest part of the brain is the cerebrum.

19.2 From superficial to deep, the three cranial meninges are the dura mater, arachnoid mater, and pia mater.

19.3 The brain stem is anterior to the fourth ventricle, and the cerebellum is posterior to it.

19.4 Cerebrospinal fluid is reabsorbed by the arachnoid villi that project into the dural venous sinuses.

19.5 The medulla oblongata contains the pyramids; the midbrain contains the cerebral peduncles; pons means "bridge."

19.6 Decussation means crossing to the opposite side. The functional consequence of decussation of the pyramids is that one side of the cerebrum controls muscles on the opposite side of the body.

19.7 The cerebral peduncles are the main sites where tracts extend and nerve impulses are conducted between the superior parts of the brain and the inferior parts of the brain and spinal cord.

19.15 The common integrative area integrates interpretation of visual, auditory, and somatic sensations; the motor speech area translates thoughts into speech; the premotor area controls skilled muscular movements; the gustatory areas interpret sensations related to taste; the auditory areas interpret pitch and rhythm; the visual areas interpret shape, color, and movement of objects; the frontal eye field area controls voluntary scanning movements of the eyes.

19.16 Axons in the olfactory tracts terminate in the primary olfactory area in the temporal lobe of the cerebral cortex.

19.17 Most axons in the optic tracts terminate in the primary visual area in the occipital lobe of the cerebral cortex.

19.18 The superior branch of the oculomotor nerve is distributed to the superior rectus muscle; the trochlear nerve is the smallest cranial nerve.

19.19 The trigeminal nerve is the largest cranial nerve.

Throughout the text we have included **Pronunciations** and, sometimes, **Word Roots,** for many terms that may be new to you. These appear in parentheses immediately following the new words, and the pronunciations are repeated in the glossary at the back of the book. Look at the words carefully and say them out loud several times. Learning to pronounce a new word

will help you remember it and make it a useful part of your medical vocabulary. Take a few minutes now to read the following pronunciation key, so it will be familiar as you encounter new words. The key is repeated at the beginning of the Glossary, page G-1.

PRONUNCIATION KEY

1. The most strongly accented syllable appears in capital letters, for example, bilateral (bī-LAT-er-al) and diagnosis (dī-ag-NŌ-sis).

2. If there is a secondary accent, it is noted by a prime ('), for example, physiology (fiz'-ē-OL-ō-jē). Any additional secondary accents are also noted by a prime, for example, decarboxylation (dē'-kar-bok'-si-LĀ-shun).

3. Vowels marked by a line above the letter are pronounced with the long sound, as in the following common words:

   ā as in *māke*   ī as in *īvy*   ū as in *cūte*
   ē as in *bē*   ō as in *pōle*

4. Vowels not marked by a line above the letter are pronounced with the short sound, as in the following words:

   a as in *above* or *at*   i as in *sip*   u as in *bud*
   e as in *bet*   o as in *not*

5. Other vowel sounds are indicated as follows:

   oy as in *oil*   oo as in *root*

   Consonant sounds are pronounced as in the following words:

   b as in *bat*   m as in *mother*
   ch as in *chair*   n as in *no*
   d as in *dog*   p as in *pick*
   f as in *father*   r as in *rib*
   g as in *get*   s as in *so*
   h as in *hat*   t as in *tea*
   j as in *jump*   v as in *very*
   k as in *can*   w as in *welcome*
   ks as in *tax*   z as in *zero*
   kw as in *quit*   zh as in *lesion*
   l as in *let*

BRIEF TABLE OF CONTENTS

CONTENTS

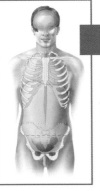

XXI

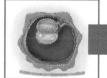

4 DEVELOPMENT 93

5 THE INTEGUMENTARY SYSTEM 121

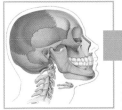
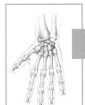

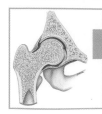

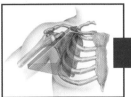

12 SURFACE ANATOMY 381

13 THE CARDIOVASCULAR SYSTEM: BLOOD 405

14 THE CARDIOVASCULAR SYSTEM: THE HEART 424

15 THE CARDIOVASCULAR SYSTEM: BLOOD VESSELS 453

16 THE LYMPHATIC AND IMMUNE SYSTEM 511

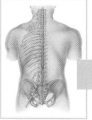
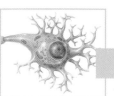

17 NERVOUS TISSUE 536

18 THE SPINAL CORD AND SPINAL NERVES 553

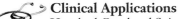

19 THE BRAIN AND CRANIAL NERVES 583

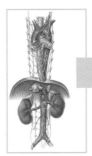

20 THE AUTONOMIC NERVOUS SYSTEM 632

21 SOMATIC SENSES 651

Clinical Applications

Clinical Applications

22 SPECIAL SENSES 671

Introduction 671

Olfaction: Sense of Smell 672

Gustation: Sense of Taste 674

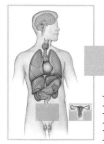

23 THE ENDOCRINE SYSTEM 704

Introduction 704
Endocrine Glands Defined 705
Hormones 705

Hypothalmus and Pituitary Gland 707

Thyroid Gland 711
Parathyroid Glands 713
Adrenal Glands 715

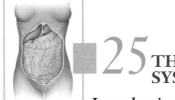

PRINCIPLES OF HUMAN ANATOMY

AN INTRODUCTION TO THE HUMAN BODY

1

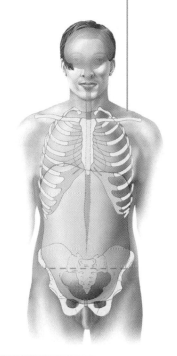

INTRODUCTION   You are about to begin a study of the human body to learn how it is organized and how it functions. In order to understand what happens to the body when it is injured, diseased, or placed under stress, you must know how it is organized and how its different parts normally work. Much of what you study in this chapter will help you visualize the body as anatomists do, and you will learn a basic anatomical vocabulary that will help you talk about the body in a way that is understood by professionals in various fields.

Can you determine what is being demonstrated in this image?

www.wiley.com/college/apcentral

ANATOMY DEFINED

● Define anatomy and physiology, and name several subdisciplines of anatomy.

Anatomy (a-NAT-ō-mē; *ana-*=upward; *-tomy*=to cut) is the study of *structure* and the relationships among structures. It was first studied by **dissection** (dis-SEK-shun; *dis-*=apart; *-section*= act of cutting), the careful cutting apart of body structures to study their relationships. Today, a variety of imaging techniques also contribute to the advancement of anatomical knowledge. We will describe and compare some common imaging techniques in Table 1.4 on pages 18–20. The anatomy of the human body can be studied at various levels of structural organization, ranging from microscopic to macroscopic. These levels and the different methods used to study them provide the basis for the subdisciplines of anatomy, several of which are described in Table 1.1.

Whereas anatomy deals with structures of the body, **physiology** (fiz'-ē-OL-ō-jē; *physio-*=nature; *-logy*=study of) deals with *functions* of body parts—that is, how they work. Because function cannot be completely separated from structure, you will learn how each structure of the body often reflects its functions. For example, the hairs lining the nose filter air that you inhale. The bones of the

skull are tightly joined to protect the brain. The bones of the fingers, by contrast, are more loosely joined to permit various movements. The external ear is shaped in such a way as to collect sound waves, which facilitates hearing. The lungs are filled with millions of air sacs that are so thin that they permit both the movement of oxygen into the blood for use by body cells and the movement of carbon dioxide out of the blood to be exhaled.

1. Which subdisciplines of anatomy would be used when dissecting a cadaver?
2. Give several examples of how the structure and function of the human body are related.

# NONINVASIVE DIAGNOSTIC TECHNIQUES

Several **noninvasive diagnostic techniques** are commonly used by health-care professionals and students to assess certain aspects of body structure and function. In **inspection,** the first diagnostic technique, the examiner observes the body for any changes that deviate from normal. In **palpation** (pal-PĀ-shun; *palpa-*=gently touching) the examiner feels body surfaces with the hands. An example is palpating a blood vessel (called an artery) to find the pulse and measure the heart rate. In **auscultation** (aus-cul-TĀ-shun; *ausculta-*=listening) the examiner listens to body sounds to evaluate the functioning of certain organs, often using a stethoscope to amplify the sounds. An example is auscultation of the lungs during breathing to check for crackling sounds associated with abnormal fluid accumulation in the lungs. In **percussion** (pur-KUSH-un; *percus-*=beat through) the examiner taps on the body surface with the fingertips and listens to the resulting echo. For example, percussion may reveal the abnormal presence of fluid in the lungs or air in the intestines. It is also used to reveal the size, consistency, and position of an underlying structure. ■

LEVELS OF BODY ORGANIZATION

● Describe the levels of structural organization that make up the human body.
● List the 11 systems of the human body, the organs present in each, and their general functions.

The levels of organization of a language—letters of the alphabet, words, sentences, paragraphs, and so on—provide a useful comparison to the levels of organization of the human body. Your exploration of the human body will extend from some of the smallest body structures and their functions to the largest structure—an entire person. From the smallest to the largest size of their components, six levels of organization are relevant to understanding anatomy: the chemical, cellular, tissue, organ, system, and organismal levels of organization (Figure 1.1).

❶ The *chemical level* includes **atoms,** the smallest components of a chemical element that retain the properties of

TABLE 1.1 SELECTED SUBDISCIPLINES OF ANATOMY	
SUBDISCIPLINE	**STUDY OF**
Embryology (em'-brē-OL-ō-jē; *embry-*=embryo; *-logy*=study of)	Structures that emerge from the time of the fertilized egg through the eighth week in utero.
Developmental biology	Structures that emerge from the time of the fertilized egg to the adult form.
Histology (hiss'-TOL-ō-jē; *hist-*=tissue)	Microscopic structure of tissues.
Surface anatomy	Anatomical landmarks on the surface of the body through visualization and palpation.
Gross anatomy	Structures that can be examined without using a microscope.
Systemic anatomy	Structure of specific systems of the body such as the nervous or respiratory systems.
Regional anatomy	Specific regions of the body such as the head or chest.
Radiographic anatomy (rā'-dē-ō-GRAF-ik; *radio-*=ray; *-graphic*=to write)	Body structures that can be visualized with x-rays.
Pathological anatomy (path'-ō-LOJ-i-kal; *path-*=disease)	Structural changes (from gross to microscopic) associated with disease.

the element, and **molecules,** two or more atoms joined together. Certain atoms, such as carbon (C), hydrogen (H), oxygen (O), nitrogen (N), phosphorus (P), and calcium (Ca), are essential for life. Familiar examples of molecules found in the body are deoxyribonucleic acid (DNA), the genetic material passed from one generation to the next; hemoglobin, a protein that carries oxygen in the blood; and glucose, commonly known as blood sugar.

2 Molecules, in turn, combine to form structures at the next level of organization—the *cellular level*. **Cells** are the basic structural and functional units of an organism and are the smallest living units in the human body. Among the many kinds of cells in your body are muscle cells, nerve cells, and blood cells. Shown in Figure 1.1 is a smooth muscle cell, one of three different types of muscle cells in your body. Chapter 2 focuses on the cellular level of organization.

3 The next level of structural organization is the *tissue level*. **Tissues** are groups of cells and the materials surrounding them that work together to perform a particular function. There are just four basic types of tissue in your body: *epithelial tissue, connective tissue, muscular tissue,* and *nervous tissue,* which are described in Chapter 3. Smooth muscle tissue consists of tightly packed smooth muscle cells.

Figure 1.1 Levels of structural organization in the human body.

The levels of structural organization are chemical, cellular, tissue, organ, system, and organismal.

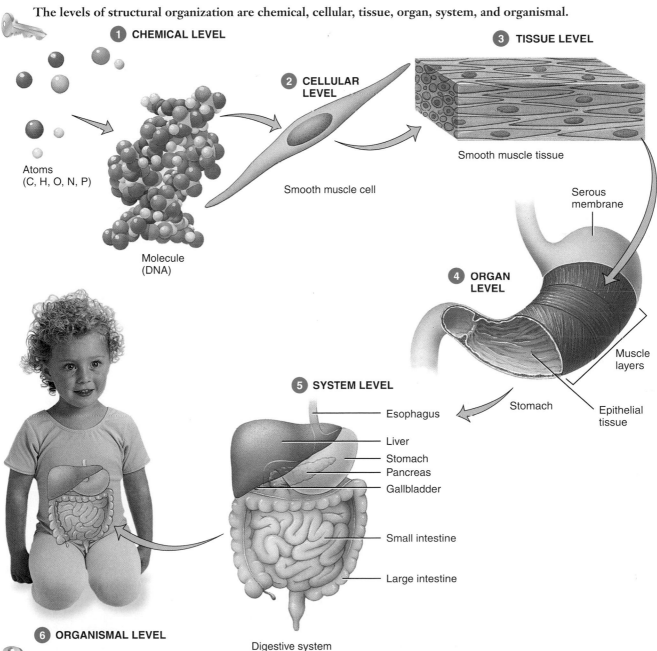

1 CHEMICAL LEVEL

Atoms (C, H, O, N, P)

Molecule (DNA)

2 CELLULAR LEVEL

Smooth muscle cell

3 TISSUE LEVEL

Smooth muscle tissue

Serous membrane

4 ORGAN LEVEL

Muscle layers

Stomach

Epithelial tissue

5 SYSTEM LEVEL

Esophagus
Liver
Stomach
Pancreas
Gallbladder
Small intestine
Large intestine

6 ORGANISMAL LEVEL

Digestive system

Which level of structural organization is composed of two or more different types of tissues that work together to perform a specific function?

④ When different kinds of tissues are joined together, they form the next level of organization, the *organ level*. **Organs** are structures that are composed of two or more different types of tissues; they have specific functions and usually have recognizable shapes. Examples of organs are the stomach, heart, liver, lungs, and brain. Figure 1.1 shows how several tissues make up the stomach. The outer covering is a *serous membrane*, a layer of epithelial tissue and connective tissue that protects the stomach and other organs and reduces friction when the stomach moves and rubs against other organs. Underneath are the *muscle tissue layers* (smooth muscle), which contract to churn and mix food and push it on to the next digestive organ, the small intestine. The innermost lining is an *epithelial tissue layer* that produces fluid and chemicals responsible for digestion in the stomach.

⑤ The next level of structural organization in the body is the *system level*, also called the *organ-system level*. A **system** consists of related organs that have a common function. An example is the digestive system, which breaks down and

TABLE 1.2 THE 11 SYSTEMS OF THE HUMAN BODY	
INTEGUMENTARY SYSTEM	**SKELETAL SYSTEM**

INTEGUMENTARY SYSTEM

Components Skin, and structures derived from it, such as hair, nails, sweat glands, and oil glands.

Functions Protects the body; helps regulate body temperature; eliminates some wastes; helps make vitamin D; and detects sensations such as touch, pain, warmth, and cold.

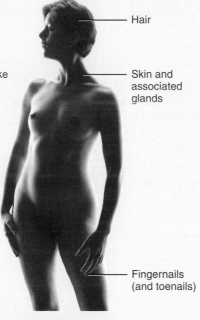

Hair

Skin and associated glands

Fingernails (and toenails)

SKELETAL SYSTEM

Components Bones and joints of the body and their associated cartilages.

Functions Supports and protects the body; provides a surface area for muscle attachment; aids body movements; houses cells that produce blood cells; stores minerals and lipids (fats).

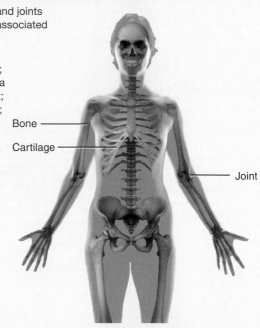

Bone

Cartilage

Joint

MUSCULAR SYSTEM

Components Muscles composed of skeletal muscle tissue, so-named because it is usually attached to bones.

Functions Produces body movements, such as walking; stabilizes body position (posture); generates heat.

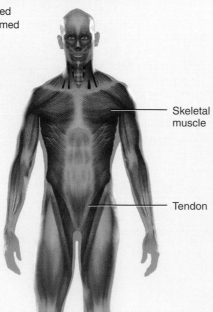

Skeletal muscle

Tendon

CARDIOVASCULAR SYSTEM

Components Blood, heart, and blood vessels.

Functions Heart pumps blood through blood vessels; blood carries oxygen and nutrients to cells and carbon dioxide and wastes away from cells and helps regulate acid–base balance, temperature, and water content of body fluids; blood components help defend against disease and mend damaged blood vessels.

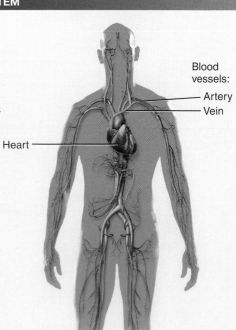

Blood vessels:

Artery

Vein

Heart

absorbs food. Its organs include the mouth, salivary glands, pharynx (throat), esophagus (food tube), stomach, small intestine, large intestine, liver, gallbladder, and pancreas. Sometimes an organ is part of more than one system. The pancreas, for example, is part of both the digestive system and the hormone-producing endocrine system.

6 The final organizational level is the *organismal level.* An **organism** is any living individual, whether human or not. In terms of humans, all the parts of the human body functioning together constitute the total organism—one living person.

In the following chapters, you will study the anatomy and physiology of the body systems. Table 1.2 introduces the components and functions of these systems in the order they are discussed in the book.

CHECKPOINT

3. Define the following terms: atom, molecule, cell, tissue, organ, system, and organism.
4. Referring to Table 1.2, which body systems help eliminate wastes?

LYMPHATIC AND IMMUNE SYSTEM

Components Lymphatic fluid and vessels; also includes spleen, thymus, lymph nodes, and tonsils.

Functions Returns proteins and fluid to blood; carries lipids from gastrointestinal tract to blood; includes structures where lymphocytes that protect against disease-causing organisms mature and proliferate.

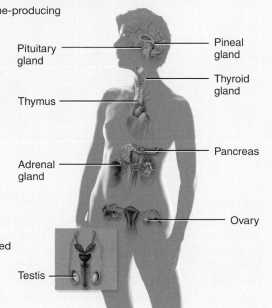

Tonsil
Thymus
Thoracic duct
Spleen
Lymph node
Lymphatic vessel

NERVOUS SYSTEM

Components Brain, spinal cord, nerves, and special sense organs, such as the eye and ear.

Functions Generates action potentials (nerve impulses) to regulate body activities; detects changes in the body's internal and external environment, interprets the changes, and responds by causing muscular contractions or glandular secretions.

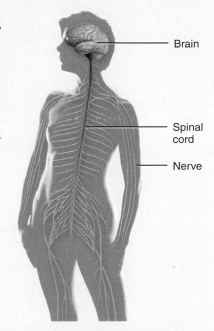

Brain
Spinal cord
Nerve

ENDOCRINE SYSTEM

Components Hormone-producing glands (pineal gland, hypothalamus, pituitary gland, thymus, thyroid gland, parathyroid glands, adrenal glands, pancreas, ovaries, and testes) and hormone-producing cells in several other organs.

Functions Regulates body activities by releasing hormones, which are chemical messengers transported in blood from an endocrine gland to a target organ.

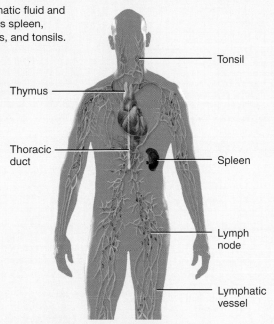

Pituitary gland
Pineal gland
Thyroid gland
Thymus
Pancreas
Adrenal gland
Ovary
Testis

RESPIRATORY SYSTEM

Components Lungs and air passageways such as the pharynx (throat), larynx (voice box), trachea (windpipe), and bronchial tubes leading into and out of them.

Functions Transfers oxygen from inhaled air to blood and carbon dioxide from blood to exhaled air; helps regulate acid–base balance of body fluids; air flowing out of lungs through vocal cords produces sounds.

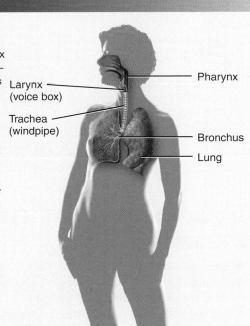

Larynx (voice box)
Trachea (windpipe)
Pharynx
Bronchus
Lung

continues

TABLE 1.2 THE 11 SYSTEMS OF THE HUMAN BODY (continued)

DIGESTIVE SYSTEM

Components
Organs of gastrointestinal tract, a long tube that includes the mouth, pharynx, esophagus, stomach, small and large intestines, and anus; also includes accessory organs that assist in digestive processes, such as the salivary glands, liver, gallbladder, and pancreas.

Functions Achieves physical and chemical breakdown of food; absorbs nutrients; eliminates solid wastes.

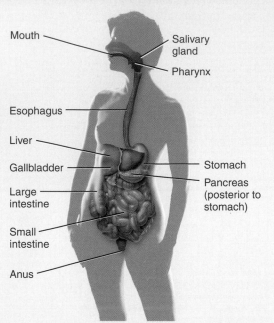

Mouth
Salivary gland
Pharynx
Esophagus
Liver
Gallbladder
Stomach
Pancreas (posterior to stomach)
Large intestine
Small intestine
Anus

URINARY SYSTEM

Components Kidneys, ureters, urinary bladder, and urethra.

Functions Produces, stores, and eliminates urine; eliminates wastes and regulates volume and chemical composition of blood; helps maintain the acid–base balance of body fluids; maintains body's mineral balance; helps regulate production of red blood cells.

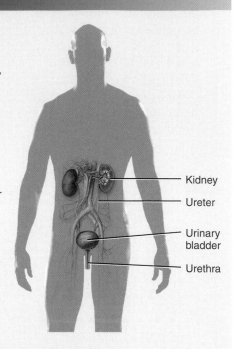

Kidney
Ureter
Urinary bladder
Urethra

REPRODUCTIVE SYSTEM

Components Gonads (testes in males and ovaries in females) and associated organs (uterine tubes, uterus, and vagina in females and epididymis, ductus deferens, and penis in males).

Functions Gonads produce gametes (sperm or oocytes) that unite to form a new organism; gonads also release hormones that regulate reproduction and other body processes; associated organs transport and store gametes.

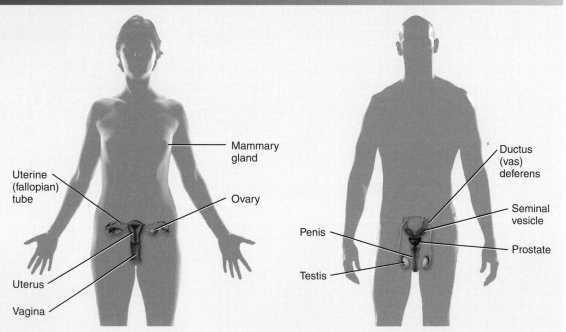

Mammary gland
Uterine (fallopian) tube
Ovary
Uterus
Vagina
Ductus (vas) deferens
Seminal vesicle
Penis
Prostate
Testis

BASIC ANATOMICAL TERMINOLOGY

OBJECTIVES

● Describe the orientation of the body in the anatomical position.
● Relate the common names to the corresponding anatomical descriptive terms for various regions of the human body.
● Define the anatomical planes, sections, and the directional terms used to describe the human body.

Scientists and health-care professionals use a common language of special terms when referring to body structures and their functions. The language of anatomy has precisely defined meanings that allow us to communicate without using unneeded or ambiguous words. For example, is it correct to say, "The wrist is above the fingers?" This might be true if your arms are at your sides. But if you hold your hands up above your head, your fingers would be above your wrists. To prevent this kind of confusion, anatomists developed a standard anatomical position and a special vocabulary for relating body parts to one another.

Anatomical Position

In the study of anatomy, descriptions of any part of the human body assume that the body is in a specific position called the **anatomical position.** In the anatomical position, the subject stands erect facing the observer, with the head level and the eyes facing directly anteriorly. The feet are flat on the floor and directed forward, and the arms are at the sides with the palms turned forward (Figure 1.2). Once the body is in the anatomical position, it is easier to visualize and understand how it is organized into various regions.

In the anatomical position, the body is upright. Two terms describe a reclining body. If the body is lying face down, it is in

Figure 1.2   The anatomical position.   The common names and corresponding anatomical terms (in parentheses) are indicated for specific body regions. For example, the head is the cephalic region.

🔑   **In the anatomical position, the subject stands erect facing the observer with the head level and the eyes facing forward. The feet are flat on the floor and directed forward, and the arms are at the sides with the palms facing forward.**

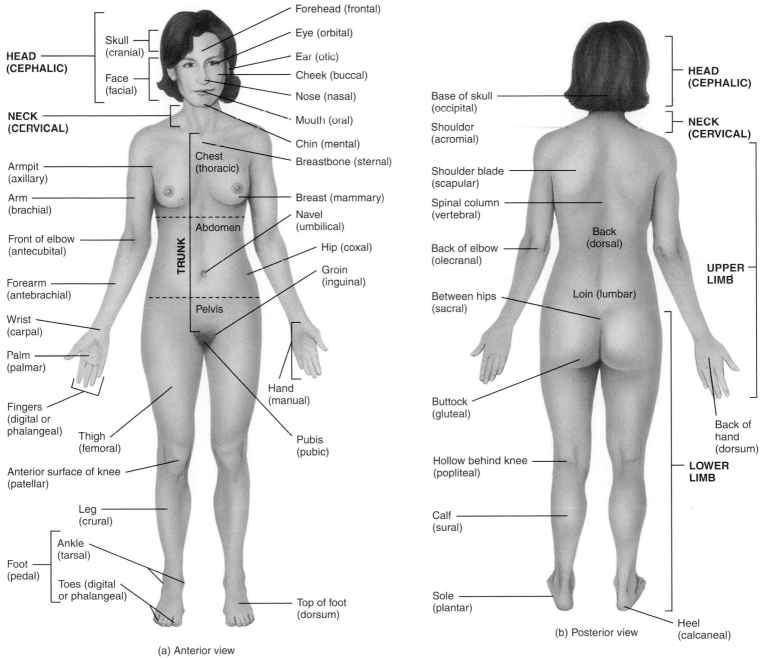

(a) Anterior view

(b) Posterior view

❓**What is the usefulness of defining one standard anatomical position?**

the **prone** position. If the body is lying face up, it is in the **supine** position.

Regional Names

The human body is divided into several major regions that can be identified externally. These are the head, neck, trunk, upper limbs, and lower limbs (see Figure 1.2). The **head** consists of the skull and face. The **skull** encloses and protects the brain, while the **face** is the front portion of the head that includes the eyes, nose, mouth, forehead, cheeks, and chin. The **neck** supports the head and attaches it to the trunk. The **trunk** consists of the chest, abdomen, and pelvis. Each **upper limb (extremity)** is attached to the trunk and consists of the shoulder, armpit, arm (portion of the limb from the shoulder to the elbow), forearm (portion of the limb from the elbow to the wrist), wrist, and hand. Each **lower limb (extremity)** is also attached to the trunk and consists of the buttock, thigh (portion of the limb from the buttock to the knee), leg (portion of the limb from the knee to the ankle), ankle, and foot. The *groin* is the area on the front surface of the body marked by a crease on each side, where the trunk attaches to the thighs.

Figure 1.2 shows the common names of major parts of the body. The corresponding anatomical descriptive form (adjective) for each part appears in parentheses next to the common name. For example, if you receive a tetanus shot in your *buttock*, it is a *gluteal* injection. Why does the descriptive form of a body part look different from its common name? The reason is that the descriptive form is based on a Greek or Latin word or "root" for the same part or area. The Latin word for armpit is *axilla* (ak-SIL-a), for example, and thus one of the nerves passing within the armpit is named the axillary nerve. You will learn more about the word roots of anatomical and physiological terms as you read this book.

Planes and Sections

You will study parts of the body relative to **planes**—imaginary flat surfaces that pass through them (Figure 1.3). A **sagittal plane** (SAJ-i-tal; *sagitt-*=arrow) is a vertical plane that divides the body or organ into right and left sides. More specifically, when such a plane passes through the midline of the body or organ and divides it into *equal* right and left sides, it is called a **midsagittal plane,** or a **median plane.** If the sagittal plane does not pass through the midline but instead divides the body or organ into *unequal* right and left sides, it is called a **parasagittal plane** (*para-*=near). A **frontal,** or **coronal, plane** (kō-RŌ-nal; *corona*=crown) divides the body or organ into front and back portions. A **transverse plane** divides the body or organ into upper and lower portions. A transverse plane may also be termed a *cross-sectional plane* or *horizontal plane*. Sagittal, frontal, and transverse planes are all at right angles to one another. An **oblique plane,** by contrast, passes through the body or organ at an angle between the transverse plane and either a sagittal or frontal plane.

Figure 1.3 Planes through the human body.

Frontal, transverse, sagittal, and oblique planes divide the body in specific ways.

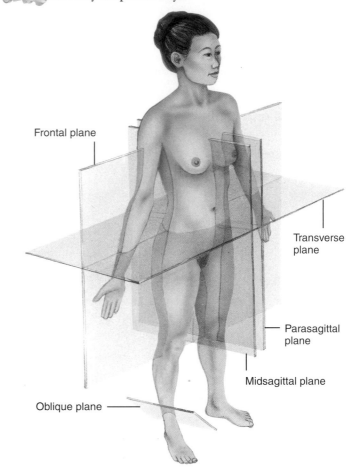

Frontal plane

Transverse plane

Parasagittal plane

Midsagittal plane

Oblique plane

Right anterolateral view

Which plane divides the heart into anterior and posterior portions?

When you study a body region, you often view it in section. A **section** is one flat surface of the three-dimensional structure or a cut along a plane. It is important to know the plane of the section so you can understand the anatomical relationship of one part to another. Figure 1.4 indicates how three different sections—a *transverse section*, a *frontal section*, and a *midsagittal section*—provide different views of the brain.

CHECKPOINT

5. Describe the anatomical position and explain why it is used.
6. Locate each region in Figure 1.2 on your own body, and then identify it by its common name and the corresponding anatomical descriptive form.
7. Which of the planes that divide the body are vertical?
8. What is the difference between a plane and a section?

Figure 1.4 Planes and sections through different parts of the brain. The diagrams (left) show the planes, and the photographs (right) show the resulting sections.
Note: The arrows in the diagrams indicate the direction from which each section is viewed. This aid is used throughout the book to indicate viewing perspective.

Planes divide the body in various ways to produce sections.

(a)

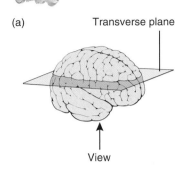

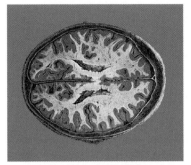

Transverse section

(b)

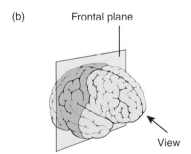

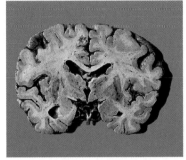

Frontal section

(c)

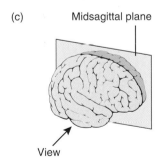

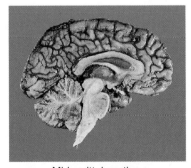

Midsagittal section

 Which plane divides the brain into unequal right and left portions?

Directional Terms

To locate various body structures, anatomists use specific **directional terms,** words that describe the position of one body part relative to another. Several directional terms can be grouped in pairs that have opposite meanings—for example, anterior (front) and posterior (back). Exhibit 1.1 and Figure 1.5 present the principal directional terms.

EXHIBIT 1.1 DIRECTIONAL TERMS (Figure 1.5)

OBJECTIVE

● Define each directional term used to describe the human body.

Overview

Most of the directional terms used to describe the human body can be grouped into pairs that have opposite meanings. For example, **superior** means toward the upper part of the body, whereas **inferior** means toward the lower part of the body. Moreover, it is important to understand that directional terms have relative meanings; they make sense only when used to describe the position of one structure relative to another. For example, your knee is superior to your ankle, even though both are located in the inferior half of the body. Study the directional terms below and the example of how each is used. As you read the examples, refer to Figure 1.5 on page 11 to see the location of each structure.

■ CHECKPOINT

Which directional terms can be used to specify the relationships between (1) the elbow and the shoulder, (2) the left and right shoulders, (3) the sternum and the humerus, and (4) the heart and the diaphragm?

DIRECTIONAL TERM	DEFINITION	EXAMPLE OF USE
Superior (soo-PEER-ē-or) (**cephalic** or **cranial**)	Toward the head, or the upper part of a structure.	The heart is superior to the liver.
Inferior (in-FEER-ē-or) **(caudal)**	Away from the head, or the lower part of a structure.	The stomach is inferior to the lungs.
Anterior (an-TEER-ē-or) **(ventral)***	Nearer to or at the front of the body.	The sternum (breastbone) is anterior to the heart.
Posterior (pos-TEER-ē-or) **(dorsal)***	Nearer to or at the back of the body.	The esophagus is posterior to the trachea (windpipe).
Medial (MĒ-dē-al)	Nearer to the midline.†	The ulna is medial to the radius.
Lateral (LAT-er-al)	Farther from the midline.	The lungs are lateral to the heart.
Intermediate (in′-ter-MĒ-dē-at)	Between two structures.	The transverse colon is intermediate between the ascending colon and descending colon.
Ipsilateral (ip-si-LAT-er-al)	On the same side of the body as another structure.	The gallbladder and ascending colon are ipsilateral.
Contralateral (CON-tra-lat-er-al)	On the opposite side of the body from another structure.	The ascending colon and descending colon are contralateral.
Proximal (PROK-si-mal)	Nearer to the attachment of a limb to the trunk; nearer to the origination of a structure.	The humerus is proximal to the radius.
Distal (DIS-tal)	Farther from the attachment of a limb to the trunk; farther from the origination of a structure.	The phalanges are distal to the carpals.
Superficial (soo′-per-FISH-al)	Toward or on the surface of the body.	The ribs are superficial to the lungs.
Deep (DĒP)	Away from the surface of the body.	The ribs are deep to the skin of the chest and back.

*Ventral refers to the belly side, whereas dorsal refers to the back side. In four-legged animals, anterior=cephalic (toward the head), ventral=inferior, posterior=caudal (toward the tail), and dorsal=superior.

†The midline is an imaginary vertical line that divides the body into equal right and left sides.

Figure 1.5 Directional terms.

Directional terms precisely locate various parts of the body relative to one another.

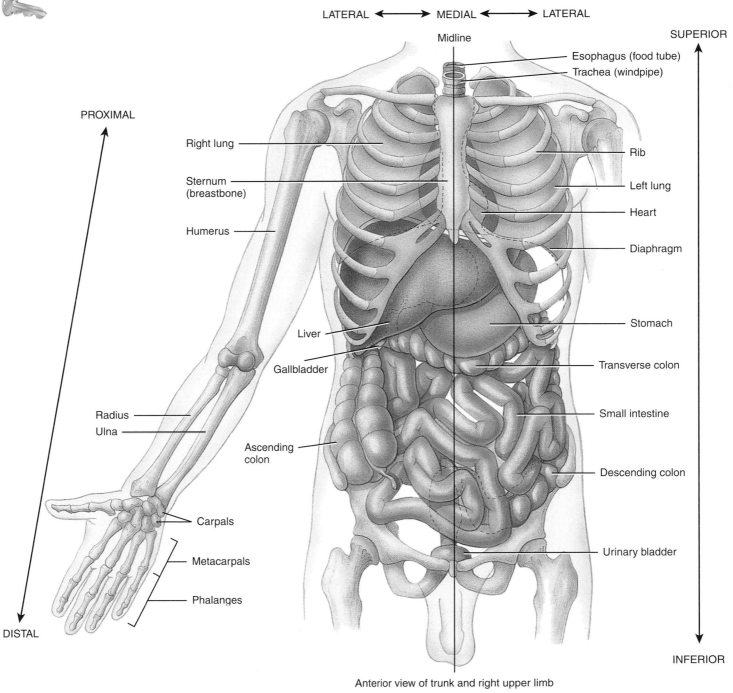

Anterior view of trunk and right upper limb

Is the radius proximal to the humerus? Is the esophagus anterior to the trachea? Are the ribs superficial to the lungs? Is the urinary bladder medial to the ascending colon? Is the sternum lateral to the descending colon?

Body Cavities

OBJECTIVE

- Describe the principal body cavities, the organs they contain, and their associated linings.

Body cavities are spaces within the body that help protect, separate, and support internal organs. Bones, muscles, and ligaments separate the various body cavities from one another. The two principal cavities are the dorsal and ventral body cavities (Figure 1.6).

Dorsal Body Cavity

The **dorsal body cavity** is located near the dorsal (posterior) surface of the body and has two subdivisions, the cranial cavity and the vertebral canal. The **cranial cavity** is formed by the cranial bones and contains the brain. The **vertebral (spinal) canal** is formed by the bones of the vertebral column (backbone) and contains the spinal cord. Three layers of protective tissue, called **meninges** (me-NIN-jēz), line the dorsal body cavity.

Ventral Body Cavity

The other principal body cavity—the **ventral body cavity**—is located on the ventral (anterior) aspect of the body. The ventral body cavity also has two main subdivisions, the superior thoracic and the inferior abdominopelvic cavities. The **diaphragm** (DĪ-a-fram;=partition or wall), the dome-shaped muscle that powers lung expansion during breathing, separates the thoracic cavity from the abdominopelvic cavity. The organs inside the ventral body cavity are termed the **viscera** (VIS-er-a).

Figure 1.6 Body cavities. The dashed lines in (a) and (b) indicate the border between the abdominal and pelvic cavities. (See Tortora, *A Photographic Atlas of the Human Body*, 2e, Figures 6.6 and 11.13.)

The two principal body cavities are the dorsal and ventral body cavities.

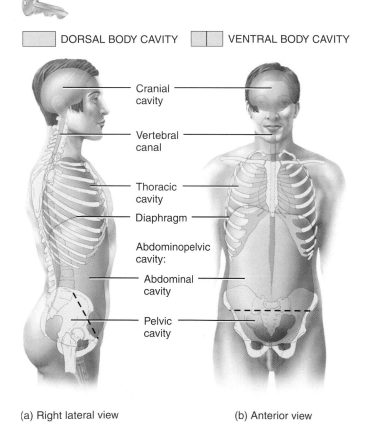

DORSAL BODY CAVITY   VENTRAL BODY CAVITY

- Cranial cavity
- Vertebral canal
- Thoracic cavity
- Diaphragm
- Abdominopelvic cavity:
- Abdominal cavity
- Pelvic cavity

(a) Right lateral view   (b) Anterior view

CAVITY	COMMENTS
DORSAL CAVITY	
Cranial cavity	Formed by cranial bones and contains brain.
Vertebral cavity	Formed by vertebral column and contains spinal cord and the beginnings of spinal nerves.
VENTRAL CAVITY*	
Thoracic cavity	Chest cavity; superior portion of ventral body cavity; contains pleural and pericardial cavities and mediastinum.
Pleural cavity	Each surrounds a lung; the serous membrane of the pleural cavities is the pleura.
Pericardial cavity	Surrounds the heart; the serous membrane of the pericardial cavity is the pericardium.
Mediastinum	Central portion of thoracic cavity between the lungs; extends from sternum to vertebral column and from neck to diaphragm; contains heart, thymus, esophagus, trachea, and several large blood vessels.
Abdominopelvic cavity	Inferior portion of ventral body cavity; subdivided into abdominal and pelvic cavities.
Abdominal cavity	Contains stomach, spleen, liver, gallbladder, small intestine, and most of large intestine; the serous membrane of the abdominal cavity is the peritoneum.
Pelvic cavity	Contains urinary bladder, portions of large intestine, and internal organs of reproduction.

* See figure 1.7 for details of the thoracic cavity

In which cavities are the following organs located: urinary bladder, stomach, heart, small intestine, lungs, internal female reproductive organs, thymus, spleen, liver? Use the following symbols for your response: T=thoracic cavity, A=abdominal cavity, or P=pelvic cavity.

The superior portion of the ventral body cavity is the **thoracic cavity** (thor-AS-ik; *thorac-*=chest), or chest cavity (Figure 1.7). The thoracic cavity is encircled by the ribs, the muscles of the chest, the sternum (breastbone), and the thoracic portion of the vertebral column (backbone). Within the thoracic cavity are three smaller cavities: the **pericardial cavity** (per′-i-KAR-dē-al; *peri-*=around; *-cardial*=heart), a fluid-filled space that surrounds the heart, and two **pleural cavities** (PLOOR-al; *pleur-*=rib or side). Each pleural cavity surrounds one lung and contains a small amount of fluid. The central portion of the thoracic cavity is called the **mediastinum** (mē′-dē-as-TĪ-num; *media-*=middle; *-stinum*=partition). It is located between the lungs and extends from the sternum to the vertebral column, and from the neck to the diaphragm (Figure 1.7). The mediastinum contains all thoracic viscera except the lungs themselves. Among the structures in the mediastinum are the heart, esophagus, trachea, thymus, and several large blood vessels.

Figure 1.7   The thoracic cavity. The dashed lines in (a), (b), and (d) indicate the borders of the mediastinum. Notice that the pericardial cavity surrounds the heart, and that the pleural cavities surround the lungs.

Note: When transverse sections, such as those shown in (b) and (c), are viewed inferiorly (from below) the anterior aspect of the body appears on the top of the illustration and the left side of the body appears on the right side of the illustration. (See Tortora, *A Photographic Atlas of the Human Body*, 2e, Figure 6.6.)

The mediastinum is medial to the lungs; it extends from the sternum to the vertebral column and from the neck to the diaphragm.

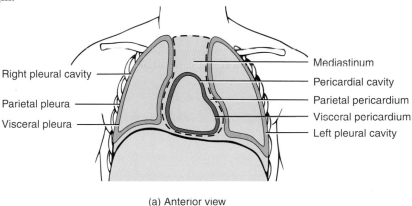

(a) Anterior view

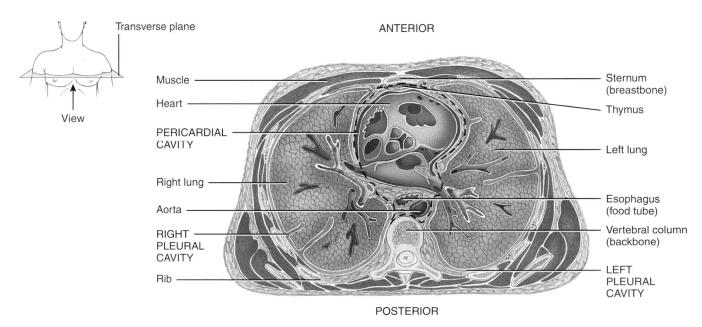

(b) Inferior view of transverse section of thoracic cavity

continues

Figure 1.7 The thoracic cavity (continued)

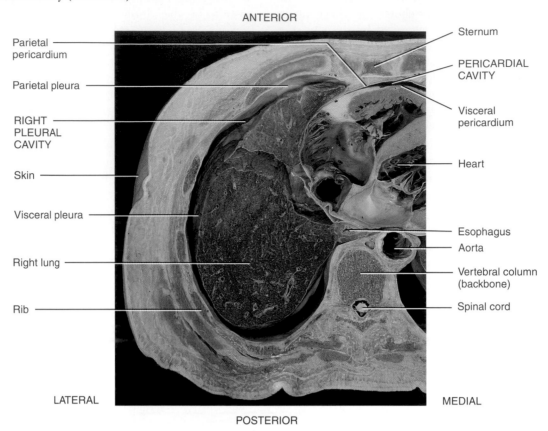

(c) Inferior view of transverse section of thoracic cavity

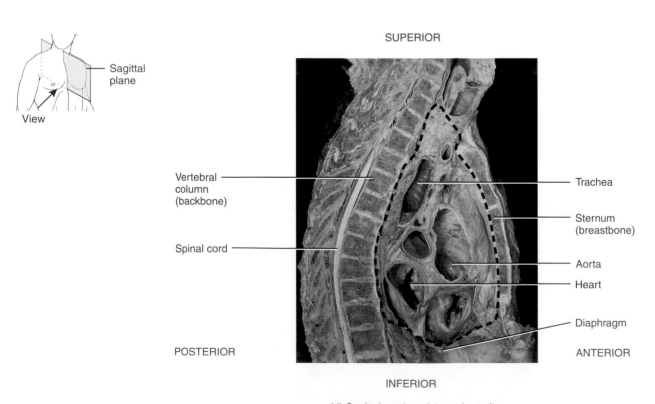

(d) Sagittal section of thoracic cavity

? Which of the following structures are contained in the mediastinum: right lung, heart, esophagus, spinal cord, trachea, rib, thymus, left pleural cavity?

The inferior portion of the ventral body cavity is the **abdominopelvic cavity** (ab-dom′-i-nō-PEL-vik; Figure 1.8), which extends from the diaphragm to the groin and is encircled by the abdominal wall and the bones and muscles of the pelvis. As the name suggests, the abdominopelvic cavity is divided into two portions, even though no wall separates them (Figure 1.8). The superior portion, the **abdominal cavity** (*abdomin-*=belly), contains the stomach, spleen, liver, gallbladder, small intestine, and most of the large intestine. The inferior portion, the **pelvic cavity** (*pelv-*=basin), contains the urinary bladder, portions of the large intestine, and internal organs of the reproductive system.

Figure 1.8   The abdominopelvic cavity. The horizontal dashed black line approximates the point of separation of the abdominal and pelvic cavities. (See Tortora, *A Photographic Atlas of the Human Body*, 2e, Figure 12.2.)

The abdominopelvic cavity extends from the diaphragm to the groin.

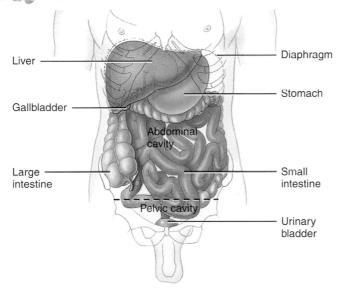

Anterior view

To which body systems do the organs shown here within the abdominal and pelvic cavities belong? (Hint: *Refer to Table 1.2.*)

Thoracic and Abdominal Cavity Membranes

A thin, slippery **serous membrane** covers the viscera within the thoracic and abdominal cavities and also lines the walls of the thorax and abdomen. The parts of a serous membrane are (1) the *parietal layer* (pa-RĪ-e-tal), which lines the walls of the cavities, and (2) the *visceral layer*, which covers and adheres to the viscera within the cavities. Serous fluid between the two layers reduces friction, allowing the viscera to slide somewhat during movements—for example, when the lungs inflate and deflate during breathing.

The serous membrane of the pleural cavities is called the **pleura** (PLOO-ra) (see Figure 1.7a, c). The *visceral pleura* clings to the surface of the lungs, whereas the *parietal pleura* lines the chest wall. In between is the pleural cavity, filled with a small volume of serous fluid. The serous membrane of the pericardial cavity is the **pericardium** (see Figure 1.7a, c). The *visceral pericardium* covers the surface of the heart, whereas the anterior portion of the *parietal pericardium* lines the chest wall. Between them is the pericardial cavity. The **peritoneum** (per-i-tō-NĒ-um) is the serous membrane of the abdominal cavity (see Figure 25.3a on page 000). The *visceral peritoneum* covers the abdominal viscera, whereas the *parietal peritoneum* lines the abdominal wall. Between them is the peritoneal cavity. Most abdominal organs are located in the peritoneal cavity. Some are located behind the parietal peritoneum—they are between it and the posterior abdominal wall. Such organs are said to be *retroperitoneal* (re′-trō-per-i-tō-NĒ-al; *retro*= behind). Examples include the kidneys, adrenal glands, pancreas, duodenum of the small intestine, ascending and descending colons of the large intestine, and portions of the abdominal aorta and inferior vena cava.

A summary of body cavities and their membranes is presented in Figure 1.6 on page 12.

CHECKPOINT

9. What structures separate the various body cavities from one another?
10. What structures are found in the mediastinum?

Abdominopelvic Regions and Quadrants

OBJECTIVE

● Name and describe the abdominopelvic regions and the abdominopelvic quadrants.

To describe the location of the many abdominal and pelvic organs more easily, anatomists and clinicians use two methods of dividing the abdominopelvic cavity into smaller areas. In the first method, two transverse and two vertical lines, aligned like a tick-tack-toe grid, partition this cavity into nine **abdominopelvic regions** (Figure 1.9a). The *subcostal* (top transverse) *line* is drawn just inferior to the rib cage, across the inferior portion of the stomach; the *transtubercular* (bottom transverse) *line* is drawn just inferior to the tops of the hip bones. The left and right *midclavicular* (two vertical) *lines* are drawn through the midpoints of the clavicles (collar bones), just medial to the nipples. The four lines divide the

Figure 1.9 Abdominopelvic regions. (a) The nine regions. (b) The greater omentum has been removed. The greater omentum is a serous membrane that contains fatty tissue which covers some of the abdominal organs. (c) Some anterior organs have been removed, exposing the more posterior structures. The internal reproductive organs in the pelvic cavity are shown in Figures 27.1 on page 836 and 27.11 on page 850.

The nine-region designation is used for anatomical studies.

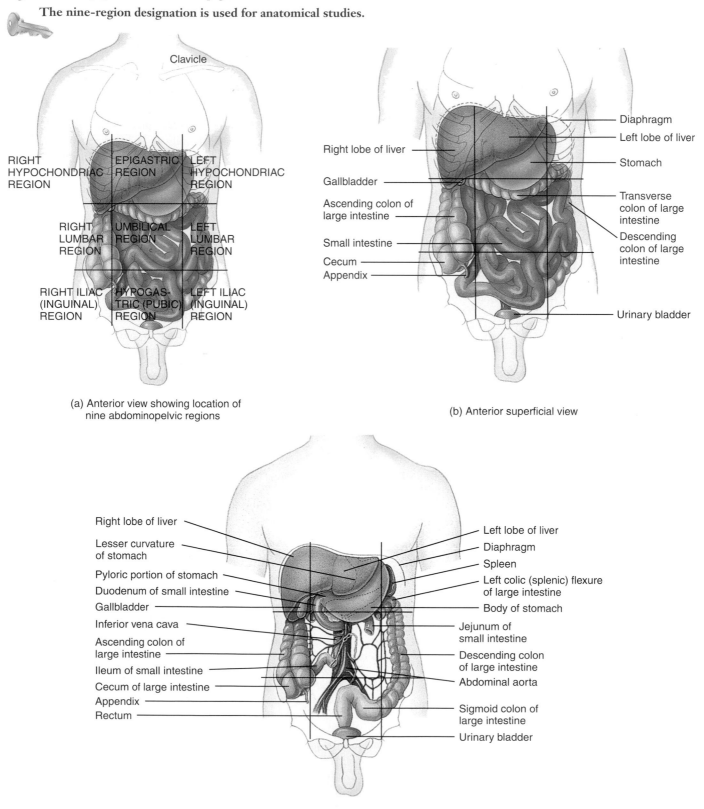

(a) Anterior view showing location of nine abdominopelvic regions

(b) Anterior superficial view

(c) Anterior deep view

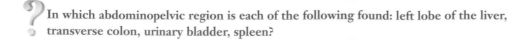

In which abdominopelvic region is each of the following found: left lobe of the liver, transverse colon, urinary bladder, spleen?

TABLE 1.3 REPRESENTATIVE STRUCTURES FOUND IN THE ABDOMINOPELVIC REGIONS

REGION	REPRESENTATIVE STRUCTURES
Right hypochondriac (hī-pō-KON-drē-ak; *hypo*-under; *chondro*=cartilage)	Right lobe of liver, gallbladder, and superior third of right kidney.
Epigastric (ep-i-GAS-trik; *epi*-above; *gaster*=stomach)	Left lobe and medial part of right lobe of liver, pyloric portion and lesser curvature of stomach, superior and descending portions of duodenum, body and superior part of head of pancreas, and right and left adrenal (suprarenal) glands.
Left hypochondriac	Body and fundus of stomach, spleen, left colic (splenic) flexure, superior two-thirds of left kidney, and tail of pancreas.
Right lumbar (*lumbus*=loin)	Superior part of cecum, ascending colon, right colic (hepatic) flexure, inferior lateral portion of right kidney, and part of small intestine.
Umbilical (um-BIL-i-kul; *umbilikus*=navel)	Middle portion of tranverse colon, part of small intestine, and bifurcations (branching) of abdominal aorta and inferior vena cava.
Left lumbar	Descending colon, inferior third of left kidney, and part of small intestine.
Right iliac (inguinal) (IL-ē-ak; iliac refers to superior part of hip bone)	Lower end of cecum, appendix, and part of small intestine.
Hypogastric (pubic)	Urinary bladder when full, small intestine, and part of sigmoid colon.
Left iliac (inguinal)	Junction of descending and sigmoid parts of colon and part of small intestine.

Figure 1.10 Quadrants of the abdominopelvic cavity. The two lines intersect at right angles at the umbilicus (navel).

🔑 **The quadrant designation is used to locate the site of pain, a tumor, or some other abnormality.**

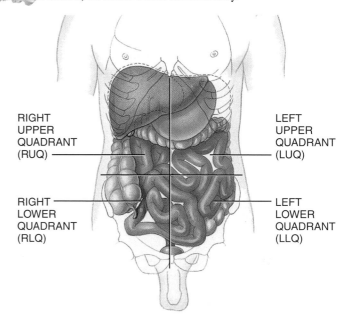

RIGHT UPPER QUADRANT (RUQ)

LEFT UPPER QUADRANT (LUQ)

RIGHT LOWER QUADRANT (RLQ)

LEFT LOWER QUADRANT (LLQ)

Anterior view

❓ **In which quadrant would the pain from appendicitis (inflammation of the appendix) be felt?**

abdominopelvic cavity into a larger middle section and smaller left and right sections. The names of the nine abdominopelvic regions are **right hypochondriac, epigastric, left hypochondriac, right lumbar, umbilical, left lumbar, right inguinal (iliac), hypogastric (pubic),** and **left inguinal (iliac).** Note which organs and parts of organs are in the different regions by carefully examining Figure 1.9b–c and Table 1.3. Figure 1.9b–c illustrates successively deeper views of the abdominopelvic contents in their respective regions. Although some parts of the body in these illustrations are most likely unfamiliar to you at this point, they will be discussed in detail in later chapters.

The second method is simpler and divides the abdominopelvic cavity into **quadrants** (KWOD-rantz; *quad-*=one-fourth), as shown in Figure 1.10. In this method, a transverse plane and a midsagittal plane are passed through the **umbilicus** (um-bi-LĪ-kus; *umbilic-*=navel) or *belly button*. The names of the abdominopelvic quadrants are **right upper quadrant (RUQ), left upper quadrant (LUQ), right lower**

quadrant (RLQ), and **left lower quadrant (LLQ).** Whereas the nine-region division is more widely used for anatomical studies, quadrants are more commonly used by clinicians for describing the site of abdominopelvic pain, tumor, or other abnormality.

CHECKPOINT

11. Locate the abdominopelvic regions and the abdominopelvic quadrants on yourself, and list some of the organs found in each.

AUTOPSY

An **autopsy** (AW-top-sē; *auto-*=self; *-opsy*=to see) is a postmortem (after-death) examination of the body and dissection of its internal organs to confirm or determine the cause of death. An autopsy can uncover the existence of diseases not detected during life, determine the extent of injuries, and explain how those injuries may have contributed to a person's death. It also may support the accuracy of diagnostic tests, establish the beneficial and adverse effects of drugs, reveal the effects of environmental influences on the body, provide more information about a disease, assist in the accumulation of statistical data, and educate health-care students. Moreover, an autopsy can reveal conditions that may affect offspring or siblings (such as congenital heart defects). An autopsy may be legally required, such as in the course of a criminal investigation, or to resolve disputes between beneficiaries and insurance companies about the cause of death. ■

MEDICAL IMAGING

OBJECTIVE

● Describe the principles and importance of medical imaging procedures in the evaluation of organ functions and the diagnosis of disease.

Various kinds of **medical imaging** procedures allow visualization of structures inside our bodies and are increasingly helpful for precise diagnosis of a wide range of anatomical and physiological disorders. The grandparent of all medical imaging tech-

TABLE 1.4   COMMON MEDICAL IMAGING PROCEDURES
RADIOGRAPHY

Procedure A single barrage of x-rays passes through the body, producing an image of interior structures on x-ray-sensitive film. The resulting two-dimensional image is a *radiograph* (RĀ-dē-ō-graf), commonly called an *x-ray.*

Comments Radiographs are relatively inexpensive, quick, and simple to perform, and usually provide sufficient information for diagnosis. X-rays do not easily pass through dense structures so bones appear white. Hollow structures, such as the lungs, appear black. Structures of intermediate density, such as skin, fat, and muscle, appear as varying shades of gray.

At low doses, x-rays are useful for examining soft tissues such as the breast (*mammography*) and bone density (*bone densitometry).*

It is necessary to use a substance called a contrast medium to make hollow or fluid-filled structures visible in radiographs. X-rays make structures that contain contrast media appear white. The medium may be introduced by injection, orally, or rectally depending on the structure to be imaged. Contrast x-rays are used to image blood vessels (*angiography*), the urinary system (*intravenous urography*), and the gastrointestinal tract (*barium contrast x-ray*).

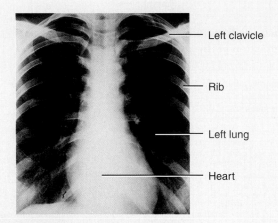

Radiograph of the thorax in anterior view
— Left clavicle — Rib — Left lung — Heart

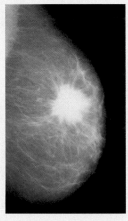

Mammogram of a female breast showing a cancerous tumor (white mass with uneven border)

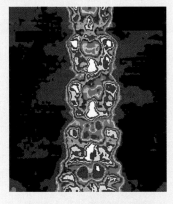

Bone densiometry scan of the lumbar spine in anterior view

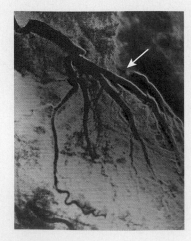

Angiogram of an adult human heart showing a blockage in a coronary artery (arrow)

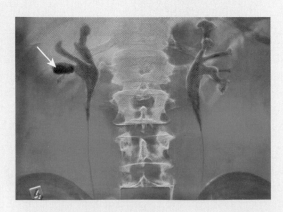

Intravenous urogram showing a kidney stone (arrow) in the right ureter

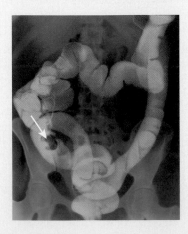

Barium contrast x-ray showing a cancer of the ascending colon (arrow)

niques is conventional radiography (x-rays), in medical use since the late 1940s. The newer imaging technologies not only contribute to diagnosis of disease, but they also are advancing our understanding of normal physiology. Table 1.4 describes some commonly used medical imaging techniques. In addition, throughout the book you will find three-page, folded inserts that contain various ways for imaging the systems of the human body, including medical imaging. They are designed to show you the many ways that the anatomy of the body can be visualized and studied.

CHECKPOINT

12. Which forms of medical imaging use radiation? Which do not?
13. Which imaging technique best reveals the physiology of a structure?
14. Which imaging technique would best show a broken bone?

MAGNETIC RESONANCE IMAGING (MRI)

Procedure The body is exposed to a high-energy magnetic field, which causes protons (small positive particles within atoms, such as hydrogen) in body fluids and tissues to arrange themselves in relation to the field. Then a pulse of radiowaves "reads" these ion patterns, and a color-coded image is assembled on a video monitor. The result is a two- or three-dimensional blueprint of cellular chemistry.

Comments Relatively safe, but can't be used on patients with metal in their bodies. Shows fine details for soft tissues but not for bones. Most useful for differentiating between normal and abnormal tissues. Used to detect tumors and artery-clogging fatty plaques, reveal brain abnormalities, measure blood flow, and detect a variety of musculoskeletal, liver, and kidney disorders.

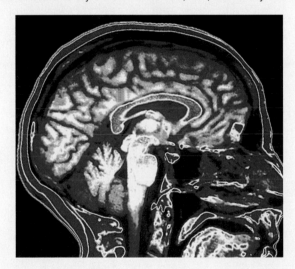

Magnetic resonance image of the brain in sagittal section

COMPUTED TOMOGRAPHY (CT)

[formerly called computerized axial tomography (CAT) scanning]

Procedure Computer-assisted radiography in which an x-ray beam traces an arc at multiple angles around a section of the body. The resulting transverse section of the body, called a *CT scan,* is reproduced on a video monitor.

Comments Visualizes soft tissues and organs with much more detail than conventional radiographs. Differing tissue densities show up as various shades of gray. Multiple scans can be assembled to build three-dimensional views of structures.

ANTERIOR

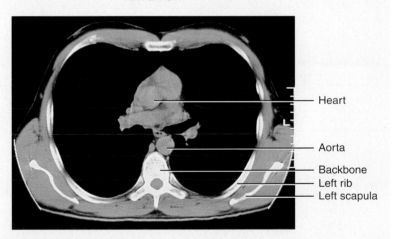

POSTERIOR

Computed tomography scan of the thorax in inferior view

continues

TABLE 1.4 COMMON MEDICAL IMAGING PROCEDURES

ULTRASOUND SCANNING

Procedure High-frequency sound waves produced by a handheld wand reflect off body tissues and are detected by the same instrument. The image, which may be still or moving, is called a *sonogram* (SON-ō-gram) and is reproduced on a video monitor.

Comments Safe, noninvasive, painless, and uses no dyes. Most commonly used to visualize the fetus during pregnancy. Also used to observe the size, location, and actions of organs and blood flow through blood vessels (*doppler ultrasound*).

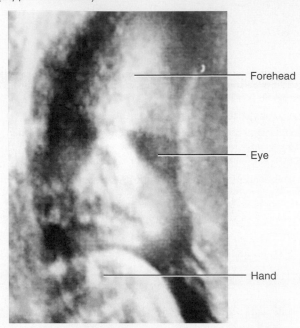

— Forehead

— Eye

— Hand

Sonogram of a fetus (Courtesy of Andrew Joseph Tortora and Damaris Soler)

POSITRON EMISSION TOMOGRAPHY (PET)

Procedure A substance that emits positrons (positively charged particles) is injected into the body, where it is taken up by tissues. The collision of positrons with negatively charged electrons in body tissues produces gamma rays (similar to x-rays) that are detected by gamma cameras positioned around the subject. A computer receives signals from the gamma cameras and constructs a *PET* scan image, displayed in color on a video monitor. The PET scan shows where the injected substance is being used in the body. In the PET scan images shown here, the black and blue colors indicate minimal activity, whereas the red, orange, yellow, and white colors indicate areas of increasingly greater activity.

Comments Used to study the physiology of body structures, such as metabolism in the brain or heart.

ANTERIOR

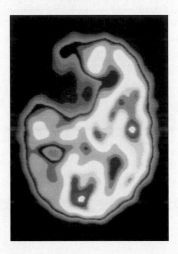

POSTERIOR

Positron emission tomography scan of a transverse section of the brain (darkened area at upper left indicates where a stroke has occurred)

RADIONUCLIDE SCANNING

Procedure A *radionuclide* (radioactive substance) is introduced intravenously into the body and carried by the blood to the tissue to be imaged. Gamma rays emitted by the radionuclide are detected by a gamma camera outside the subject and fed into a computer. The computer constructs a *radionuclide image* and displays it in color on a video monitor. Areas of intense color take up a lot of the radionuclide and represent high tissue activity; areas of less intense color take up smaller amounts of the radionuclide and represent low tissue activity. *Single-photo-emission computerized tomography (SPECT) scanning* is a specialized type of radionuclide scanning that is especially useful for studying the brain, heart, lungs, and liver.

Comments Used to study activity of a tissue or organ, such as the heart, thyroid gland, and kidneys.

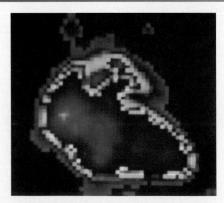

Radionuclide (nuclear) scan of a normal human heart

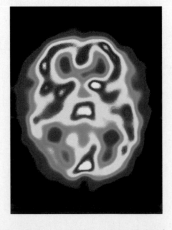

Single-photon-emission computerized tomography (SPECT) scan of a transverse section of the brain (green area at lower left indicates a migraine attack)

MEASURING THE HUMAN BODY

An important aspect of describing the body and understanding how it works is *measurement*—what the dimensions of an organ are, how much it weighs, how long it takes for a physiological event to occur. Such measurements also have clinical importance, for example, in determining how much of a given medication should be administered. As you will see, measurements involving time, weight, temperature, size, and volume are a routine part of your studies in a medical science program.

Whenever you come across a measurement in this text, the measurement will be given in metric units. The metric system is the standard used in the sciences. To help you compare the metric unit to a familiar unit, the approximate U.S. equivalent will also be given in parentheses directly after the metric unit. For example, you might be told that the length of a particular part of the body is 2.54 cm (1 in.).

To help you understand the correlation between the metric system and the U.S. system of measurement, there are five exhibits located in Appendix A for your reference.

STUDY OUTLINE

Anatomy Defined (p. 2)

1. Anatomy is the science of body structures and the relationships among structures; physiology is the science of body functions.
2. Some subdisciplines of anatomy are embryology, developmental biology, histology, surface anatomy, gross anatomy, systemic anatomy, regional anatomy, radiographic anatomy, and pathological anatomy (see Table 1.1 on page 2).

Levels of Body Organization (p. 2)

1. The human body consists of six levels of structural organization: chemical, cellular, tissue, organ, system, and organismal.
2. Cells are the basic structural and functional living units of an organism and the smallest living units in the human body.
3. Tissues are groups of cells and the materials surrounding them that work together to perform a particular function.
4. Organs are composed of two or more different types of tissues; they have specific functions and usually have recognizable shapes.
5. Systems consist of related organs that have a common function.
6. An organism is any living individual.
7. Table 1.2 on pages 4–6 introduces the 11 systems of the human organism: the integumentary, skeletal, muscular, nervous, endocrine, cardiovascular, lymphatic and immune, respiratory, digestive, urinary, and reproductive systems.

Basic Anatomical Terminology (p. 6)

Anatomical Position (p. 7)

1. Descriptions of any region of the body assume the body is in the anatomical position, in which the subject stands erect facing the observer, with the head level and the eyes facing directly forward. The feet are flat on the floor and directed forward, and the arms are at the sides, with the palms turned forward.
2. A body lying face down is prone, whereas a body lying face up is supine.

Regional Names (p. 8)

1. Regional names are terms given to specific regions of the body. The principal regions are the head, neck, trunk, upper limbs, and lower limbs.
2. Within the regions, specific body parts have common names and are specified by corresponding anatomical terms. Examples are chest (thoracic), nose (nasal), and wrist (carpal).

Planes and Sections (p. 8)

1. Planes are imaginary flat surfaces that are used to divide the body or organs into definite areas. A midsagittal plane divides the body or organ into equal right and left sides; a parasagittal plane divides the body or organ into unequal right and left sides; a frontal plane divides the body or organ into anterior and posterior portions; a transverse plane divides the body or organ into superior and inferior portions; and an oblique plane passes through the body or organ at an angle between a transverse plane and either a midsagittal, parasagittal, or frontal plane.
2. Sections are flat surfaces of three-dimensional structures or cuts along a plane. They are named according to the plane along which the cut is made and include transverse, frontal, and sagittal sections.

Directional Terms (p. 19)

1. Directional terms indicate the relationship of one part of the body to another.
2. Exhibit 1.1 on pages 10–11 summarizes commonly used directional terms.

Body Cavities (p. 12)

1. Spaces in the body that help protect, separate, and support internal organs are called body cavities.
2. The dorsal and ventral cavities are the two principal body cavities.
3. The dorsal body cavity is subdivided into the cranial cavity, which contains the brain, and the vertebral canal, which contains the spinal cord. The meninges are protective tissues that line the dorsal body cavity.
4. The ventral body cavity is separated by the diaphragm into a superior thoracic cavity and an inferior abdominopelvic cavity. Viscera are organs within the ventral body cavity. A serous membrane lines the wall of the cavity and adheres to the viscera.
5. The thoracic cavity is subdivided into three smaller cavities: a pericardial cavity, which contains the heart, and two pleural cavities, each of which contains one lung.
6. The central portion of the thoracic cavity is the mediastinum. It is located between the pleural cavities and extends from the sternum to the vertebral column and from the neck to the diaphragm. It contains all thoracic viscera except the lungs.
7. The abdominopelvic cavity is divided into a superior abdominal and an inferior pelvic cavity.

8. Viscera of the abdominal cavity include the stomach, spleen, liver, gallbladder, small intestine, and most of the large intestine.

9. Viscera of the pelvic cavity include the urinary bladder, portions of the large intestine, and internal organs of the reproductive system.

10. Serous membranes line the walls of the thoracic and abdominal cavities and cover the organs within them. They include the pleura, associated with the lungs; the pericardium, associated with the heart; and the peritoneum, associated with the abdominal cavity.

11. Figure 1.6 on page 12 summarizes body cavities and their membranes.

Abdominopelvic Regions and Quadrants (p. 15)

1. To describe the location of organs easily, the abdominopelvic cavity may be divided into nine regions by drawing four imaginary lines (left midclavicular, right midclavicular, subcostal, and transtubercular).

2. The names of the abdominopelvic regions are right hypochondriac, epigastric, left hypochondriac, right lumbar, umbilical, left lumbar, right iliac (inguinal), hypogastric (pubic), and left iliac (inguinal).

3. To locate the site of an abdominopelvic abnormality in clinical studies, the abdominopelvic cavity may be divided into quadrants by passing imaginary horizontal and vertical lines through the umbilicus.

4. The names of the abdominopelvic quadrants are right upper quadrant (RUQ), left upper quadrant (LUQ), right lower quadrant (RLQ), and left lower quadrant (LLQ).

Medical Imaging (p. 18)

1. Medical imaging techniques allow visualization of internal structures to diagnose abnormal anatomy and deviations from normal physiology.

2. Table 1.4 on pages 18–20 summarizes several medical imaging techniques.

Measuring the Human Body (p. 21)

1. Measurements involving time, weight, temperature, size, and volume are used in clinical situations.

2. Measurements in this book are given in metric units followed by U.S. equivalents in parentheses.

 SELF-QUIZ QUESTIONS

Choose the one best answer to the following questions.

1. Which word describes the location of the lungs with reference to the liver?
   a. anterior   b. inferior   c. distal
   d. medial   e. superior

2. Which is located most inferiorly?
   a. abdominal cavity   b. pleural cavity   c. mediastinum
   d. cranial cavity   e. pelvic cavity

3. The spleen, tonsils, and thymus are all organs in which system?
   a. nervous   b. lymphatic and immune   c. cardiovascular
   d. digestive   e. endocrine

4. A transverse plane is:
   a. the same as a horizontal plane.
   b. perpendicular to a sagittal plane.
   c. one that divides the body into right and left sides.
   d. Both A and B are correct.
   e. Only C is correct.

5. Which of the following statements about the function of the respiratory system is *not* true?
   a. It supplies oxygen.
   b. It eliminates carbon dioxide.
   c. It helps regulate acid–base balance in the body.
   d. It filters blood.
   e. It includes the lungs.

6. The body in the anatomical position:
   a. is standing with upper limbs parallel to the floor.
   b. is prone.
   c. has the palms turned anteriorly.
   d. has the head turned to the right.
   e. Both A and C are correct..

Complete the following.

7. The four basic types of tissues in the body are _____, _____, _____, and _____.

8. The serous membrane lining the abdominal cavity is the _____.

9. The abdominal region directly superior to the umbilical region is the _____.

10. A _____ plane divides the body into unequal right and left sections.

11. Body structures known as _____ are composed of two or more different tissues, have specific functions, and usually have recognizable shapes.

12. Because the liver and ascending colon are both located on the right side of the abdomen, they could be described as _____-lateral.

13. In terms of position, the spleen is _____ to the diaphragm and _____ to the stomach.

Are the following statements true or false?

14. The joints of the body are part of the muscular system.

15. "Tissue" is a higher level of structural organization than "organ."

16. The parietal layer of a serous membrane adheres to the organs of the thoracic and abdmonial cavities.

17. The diaphragm divides the ventral cavity into abdominal and pelvic portions.

18. The mediastinum is part of the thoracic cavity.

Matching:

19. Match the following common and anatomical terms:

— **(a)** front of elbow   **(1)** gluteal
— **(b)** arm   **(2)** inguinal
— **(c)** shoulder   **(3)** cervical
— **(d)** buttock   **(4)** plantar
— **(e)** groin   **(5)** antecubital
— **(f)** thigh   **(6)** brachial
— **(g)** neck   **(7)** acromial
— **(h)** sole   **(8)** femoral

20. Match the following:

— **(a)** superficial   **(1)** toward the front of the body
— **(b)** deep   **(2)** point on a limb nearer to the trunk
— **(c)** anterior (ventral)   **(3)** toward the surface of the body
— **(d)** posterior (dorsal)   **(4)** nearer the midsagittal plane
— **(e)** medial   **(5)** at the back of the body
— **(f)** lateral   **(6)** point on a limb farther from the trunk
— **(g)** proximal
— **(h)** distal   **(7)** farther from the midsagittal plane
   (8) away from the surface of the body

CRITICAL THINKING QUESTIONS

1. The surgeon was describing an experimental procedure to view the surface of the kidney. "We'll insert a scope without penetrating the abdominal cavity." How is this possible?

2. Taylor was going for the record for the longest upside-down hang from the monkey bars. She didn't make it and she may have broken her arm. The emergency room technician would like an x-ray film of her arm in anatomical position. Use the proper anatomical terms to describe the position of Taylor's arm in the x-ray.

3. An alien landed in your backyard, abducted your cat, and flew off. Being an observant student of anatomy, you later described the alien's appearance to the FBI as follows: "It had 2 caudal exten-

sions, 6 bilateral extremities, 4 axillae, and 1 oral orifice in place of an umbilicus." What did the alien look like?

4. Your anatomy professor displays an MRI scan that shows a parasagittal section of the torso taken through the midpoint of the left mammary gland. Name five organs you would expect to see in this image. (You may consult a general photo of the human body.)

5. Mikhail has been diagnosed with a ruptured appendix, which has allowed bacteria from his intestinal tract to infect his peritoneum. The doctors are very concerned. Why do they consider this condition (peritonitis) to be so dangerous?

ANSWERS TO FIGURE QUESTIONS

1.1 Organs are composed of two or more different types of tissues that work together to perform a specific function.

1.2 Having one standard anatomical position allows directional terms to be clearly defined, so that any body part can be described in relation to any other part.

1.3 The frontal plane divides the heart into anterior and posterior portions.

1.4 The parasagittal plane divides the brain into unequal right and left portions.

1.5 No, No, Yes, Yes, No

1.6 P, A, T, A, T, P, T, A, A

1.7 Structures in the mediastinum include the heart, esophagus, trachea, and thymus.

1.8 The illustrated abdominal cavity organs all belong to the digestive system (liver, gallbladder, stomach, appendix, small intestine, and most of the large intestine). Illustrated pelvic cavity organs belong to the urinary system (the urinary bladder) and the digestive system (part of the large intestine).

1.9 The liver is mostly in the epigastric region; the transverse colon is in the umbilical region; the urinary bladder is in the hypogastric region; the spleen is in the left hypochondriac region.

1.10 The pain associated with appendicitis would be felt in the right lower quadrant (RLQ).

2

CELLS

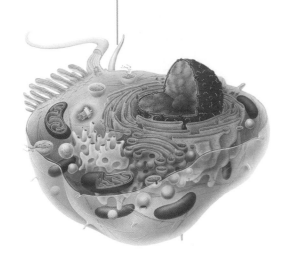

INTRODUCTION **Cells** are the basic, living, structural and functional units of the body. There are about 200 different types of cells in the body. Cells are composed of characteristic parts, whose coordinated functioning enables each cell type to fulfill a unique biochemical or structural role. **Cell biology** is the study of cellular structure and function. As you study the various parts of a cell and their relationships to each other, you will learn that cell structure and function are interdependent and inseparable. Cells perform numerous chemical reactions to create life processes. One way that they do this is by compartmentalization, the isolation of specific kinds of chemical reactions within specialized structures inside the cell. The isolated reactions are coordinated with one another to maintain life in a cell, tissue, organ, system, and organism.

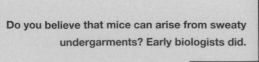

Do you believe that mice can arise from sweaty undergarments? Early biologists did.

www.wiley.com/college/apcentral

A GENERALIZED CELL

● Name and describe the three principal parts of a cell.

Figure 2.1 is a cell that is a composite of many different cell types in the body. Most cells have many of the features shown in this diagram, but not necessarily all of them. For ease of study, we can divide a cell into three principal parts: the plasma membrane, cytoplasm, and nucleus.

● The **plasma membrane** forms the cell's sturdy outer surface, separating the cell's internal environment from the environment outside the cell. The plasma membrane is a selective barrier that orchestrates the flow of materials into and out of a cell to establish and maintain the appropriate environment for normal cellular activities. It also plays a key role in communication among cells and between cells and their external environment.

● The **cytoplasm** (SĪ-tō-plazm; -*plasm*=formed or molded) is all the cellular contents between the plasma membrane and the nucleus. This compartment has two components: the cytosol and organelles. **Cytosol** (SĪ-tō-sol) is the fluid portion of cytoplasm that consists mostly of water plus dissolved solutes and suspended particles. Within the cytosol are **organelles** (or-ga-NELZ=little organs), highly organized subcellular structures, each having a characteristic shape and specific functions. Examples are the cytoskeleton, ribosomes, endoplasmic reticulum, Golgi complex, lysosomes, peroxisomes, and mitochondria.

● The **nucleus** (NOO-klē-us=nut kernel) is a large organelle that houses most of a cell's DNA. Within the nucleus are the **chromosomes** (*chromo-*=colored), each of which consists of a single molecule of DNA associated with several proteins. A chromosome contains thousands of hereditary units called **genes** that control most aspects of cellular structure and function.

1. What are the general features of the three main parts of a cell?

Figure 2.1   Generalized body cell.

The cell is the basic, living, structural and functional unit of the body.

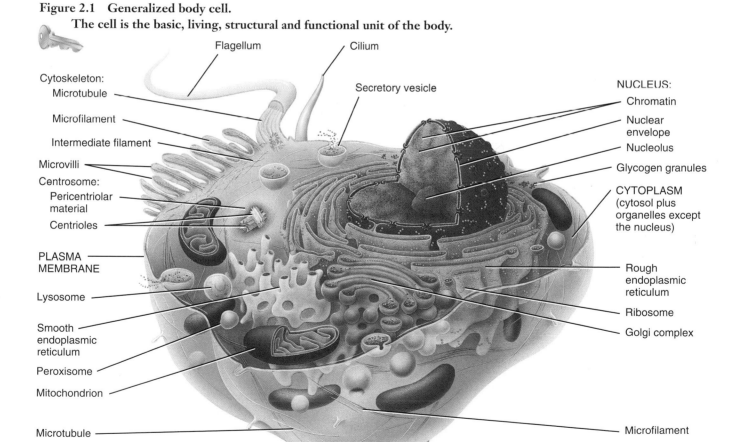

Sectional view

What are the three principal parts of a cell?

THE PLASMA MEMBRANE

OBJECTIVES

- Describe the structure and functions of the plasma membrane.
- Describe the processes that transport substances across the plasma membrane.

The **plasma membrane** is a flexible yet sturdy barrier that surrounds and contains the cytoplasm of a cell. The *fluid mosaic model* describes its structure: According to this model, the molecular arrangement of the plasma membrane resembles an ever-moving sea of lipids that contains a "mosaic" of many different proteins (Figure 2.2). The proteins may float freely like icebergs, be moored at specific locations, or be moved through the lipid sea. The lipids act as a barrier to the entry or exit of various substances, while some of the proteins in the plasma membrane act as "gatekeepers" that regulate the passage of other molecules and ions (charged particles) into and out of the cell.

Structure of the Membrane

The basic structural framework of the plasma membrane is the **lipid bilayer,** two back-to-back layers made up of three types of lipid molecules—phospholipids, cholesterol, and glycolipids (Figure 2.2). About 75% of the membrane lipids are *phospholipids*, lipids that contain phosphate groups. About 20% of plasma membrane lipids are *cholesterol* molecules, which are interspersed among the other lipids in both layers of the membrane. *Glycolipids*, which are lipids attached to carbohydrates (*glyco-*=carbohydrate), account for the other 5%.

Membrane proteins are divided into two categories—integral and peripheral—according to whether or not they are firmly embedded in the membrane (Figure 2.2). **Integral proteins** extend into or through the lipid bilayer and are firmly embedded in it. Most integral proteins are **transmembrane proteins,** which means that they span the entire lipid bilayer and protrude into both the cytosol and extracellular fluid. **Peripheral proteins,** by contrast, are not as firmly embedded in the membrane and are attached to membrane lipids or integral proteins at the inner or outer surface of the membrane.

Many membrane proteins are **glycoproteins,** proteins with carbohydrate groups attached to the ends that protrude into the extracellular fluid. The carbohydrate portions of glycolipids and glycoproteins form an extensive sugary coat called the **glycocalyx** (glī-kō-KAL-iks). The composition of the glycocalyx acts like a molecular "signature" that enables cells to recognize one another. For example, a white blood cell's ability to detect a "foreign" glycocalyx is one basis of the immune response that

Figure 2.2   Structure of the plasma membrane.

  The basic framework of the plasma membrane is the lipid bilayer.

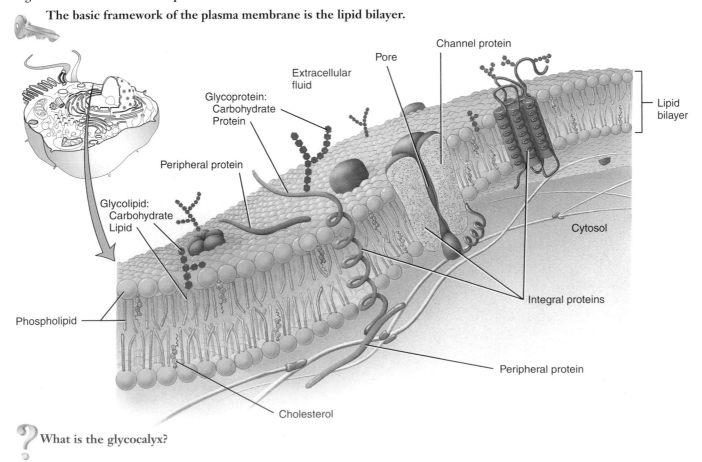

What is the glycocalyx?

helps us destroy invading organisms. In addition, the glycocalyx enables cells to adhere to one another in some tissues, and it protects cells from being digested by enzymes in the extracellular fluid. The chemical properties of the glycocalyx attract a film of fluid to the surface of many cells. This action makes red blood cells slippery as they flow through narrow blood vessels and protects cells that line the airways and the gastrointestinal tract from drying out.

Functions of Membrane Proteins

Generally, the types of lipids in cellular membranes vary only slightly from one membrane to another. In contrast, the membranes of different cells and various intracellular organelles have remarkably different assortments of proteins. Their proteins determine many of the functions that membranes can perform. Some integral membrane proteins are **ion channels** that have a *pore* or hole through which specific ions, such as potassium ions (K^+), can flow into or out of the cell. Most ion channels are *selective;* they allow only a single type of ion to pass through. Other membrane proteins act as **transporters,** which selectively move a substance from one side of the membrane to the other. Integral proteins called **receptors** serve as cellular recognition sites. Each type of receptor recognizes and binds a specific type of molecule. For instance, insulin receptors bind the hormone insulin. A specific molecule that binds to a receptor is called a **ligand** (LĪ-gand; *liga*=tied) of that receptor. Some integral and peripheral proteins are **enzymes** that catalyze specific chemical reactions at the inside or outside surface of the cell. Membrane glycoproteins and glycolipids often are **cell-identity markers.** They may enable a cell to recognize other cells of the same kind during tissue formation or to recognize and respond to potentially dangerous foreign cells. The ABO blood type markers are one example of cell identity markers. When you receive a blood transfusion, the blood type must be compatible with your own. Integral and peripheral proteins may serve as **linkers,** which anchor proteins in the plasma membranes of neighboring cells to one another or to protein filaments inside and outside the cell.

Membrane Permeability

A membrane is *permeable* to things that can pass through it and *impermeable* to those that cannot. Although plasma membranes are not completely permeable to any substance, they do permit some substances to pass more readily than others. This property of membranes is called **selective permeability.**

The lipid bilayer portion of the membrane is permeable to some molecules such as oxygen, carbon dioxide, and steroids, but is impermeable to ions and molecules such as glucose. It is also permeable to water. Transmembrane proteins that act as channels and transporters increase the plasma membrane's permeability to a variety of small- and medium-sized charged substances (including ions) that cannot cross the lipid bilayer. These proteins are very selective—each one helps only a specific molecule or ion to cross the membrane. Macromolecules, such as proteins, cannot pass

through the plasma membrane except by endocytosis and exocytosis (discussed later in this chapter).

2. What is the composition of the lipid bilayer?
3. Distinguish integral from peripheral proteins.
4. What are the major functions of membrane proteins?

Transport Across the Plasma Membrane

Before discussing how materials move into and out of cells, we will describe the locations of the various fluids through which the substances move. Fluid within cells is called **intracellular** (*intra-*=within, inside) **fluid (ICF)** or **cytosol.** Fluid outside body cells is called **extracellular** (*extra-*=outside) **fluid (ECF)** and is found in several locations. (1) The ECF filling the microscopic spaces between the cells of tissues is called **interstitial** (in′-ter-STISH-al) **fluid** (*inter-*=between), or *intercellular fluid.* (2) The ECF in blood vessels is termed **plasma** (Figure 2.3); in lymphatic vessels it is called **lymph.** Among the substances in extracellular fluid are gases, nutrients, and ions—all needed for the maintenance of life.

Extracellular fluid circulates through the blood and lymphatic vessels and into the spaces between the tissue cells. Thus, it is in constant motion throughout the body. Essentially, all body cells are surrounded by the same fluid environment. The

Figure 2.3 Body fluids. Intracellular fluid (ICF) is the fluid within cells. Extracellular fluid (ECF) is found outside cells, in blood vessels as plasma, lymphatic vessels as lymph, and between tissue cells as interstitial fluid.

Plasma membranes regulate fluid movements from one compartment to another.

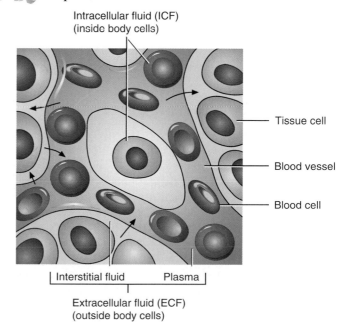

Intracellular fluid (ICF)
(inside body cells)

Tissue cell

Blood vessel

Blood cell

Interstitial fluid   Plasma

Extracellular fluid (ECF)
(outside body cells)

What is another name for intracellular fluid?

movement of substances across a plasma membrane and across membranes within cells is essential to the life of the cell and the organism. Certain substances—oxygen, for example—must move into the cell to support life, whereas waste materials or harmful substances must be moved out. Plasma membranes regulate the movements of such materials.

Several processes are involved in the movement of substances across plasma membranes. These include passive processes, active processes, and transport in vesicles. In **passive processes,** substances move across plasma membranes without the use of energy from the breakdown of ATP (adenosine triphosphate) by the cell. ATP is a molecule that stores energy for various cellular uses. The movement of substances in passive processes involves the *kinetic energy* (the energy of motion) of individual molecules or ions. The substances move on their own down a concentration gradient—that is, from an area where their concentration is higher to an area where their concentration is lower. The substances may also be forced across the plasma membrane by pressure from an area where the pressure is high to an area where it is low. In **active processes,** the cell uses some of its stored energy (from the breakdown of ATP) in moving the substance (usually ions) across the membrane since the substance typically moves against a concentration gradient—that is, from an area where its concentration is low to an area where its concentration is high. In addition, transporters in the membrane are involved in the movement of the ions.

Besides passive and active processes, materials can enter and leave cells in vesicles, spherical membranous sacs formed from pieces of the plasma membrane. Even large particles, such as whole bacteria and red blood cells, and macromolecules, such as polysaccharides and proteins, may enter and leave cells in this manner.

Passive Processes

The passive processes considered here are simple diffusion, facilitated diffusion, osmosis, and filtration.

SIMPLE DIFFUSION **Simple diffusion** (di-FŪ-zhun; *diffus-*= spreading) is a *net* (greater) movement of a substance from a region of higher concentration to a region of lower concentration—that is, the substance moves from an area where there is more of it to an area where there is less of it. A good example of diffusion in the body is the movement of oxygen from the blood into the body cells and the movement of carbon dioxide from the cells back into the blood. This movement ensures that cells receive adequate amounts of oxygen and eliminate carbon dioxide as part of their normal metabolism.

FACILITATED DIFFUSION **Facilitated diffusion** is accomplished with the assistance of integral membrane proteins that function as transporters. In this process, some large molecules and molecules that are insoluble in lipids can still pass through the plasma membrane. Among these are various sugars, especially glucose. In facilitated diffusion, glucose binds to a specific transporter protein on one side of the plasma membrane, the transporter undergoes a change in shape, and, as a result, glucose is released on the opposite side of the membrane.

OSMOSIS **Osmosis** (oz-MŌ-sis) is the net movement of water molecules through a selectively permeable membrane from an area of higher water concentration to an area of lower water concentration. The water molecules pass through pores (holes) in integral proteins in the membrane and between neighboring phospholipid molecules, and the net movement continues until equilibrium is reached. Water moves between various compartments of the body by osmosis.

FILTRATION **Filtration** involves the movement of solvents such as water and dissolved substances such as sugar across a selectively permeable membrane by gravity or mechanical pressure, usually hydrostatic (water) pressure. In the body, the driving pressure is usually not water pressure but blood pressure generated by the pumping action of the heart. Filtration is a very important process in which water and nutrients in the blood are pushed into interstitial fluid for use by body cells. It is also the primary process in the initial stage of urine formation.

Active Process

The most important active process in the human body is active transport.

ACTIVE TRANSPORT The process by which substances, usually ions, are transported across plasma membranes with the expenditure of energy by the cell, typically from an area of lower concentration to an area of higher concentration, is called **active transport.** In active transport the substance being moved makes contact with a specific site on a transporter protein. Then the ATP splits, and the energy from the breakdown of ATP causes a change in the shape of the transporter protein that expels the substance on the opposite side of the membrane. Active transport is vitally important in maintaining ion concentrations both in body cells and in extracellular fluids. For example, before a nerve cell can conduct a nerve impulse, the concentration of potassium ions (K^+) must be considerably higher inside the nerve cell than outside and the concentration of sodium ions (Na^+) must be higher outside than inside.

Transport in Vesicles

A **vesicle** (VES-i-kul), as noted earlier, is a small, spherical, membranous sac formed by budding off from an existing membrane that transports substances from one structure to another within cells, takes in substances from extracellular fluid, or releases substances into extracellular fluid. In **endocytosis** (*endo-*=within), materials move into a cell in a vesicle formed from the plasma membrane. In **exocytosis** (*exo-*=out), materials move out of a cell by the fusion of vesicles with the plasma membrane. Both endocytosis and exocytosis require cellular energy supplied by the breakdown of ATP.

ENDOCYTOSIS Here we consider three types of endocytosis: receptor-mediated endocytosis, phagocytosis, and bulk-phase endocytosis.

Receptor-mediated endocytosis is a highly selective type of endocytosis by which cells take up specific ligands. (Recall that ligands are molecules that bind to specific receptors.)

A vesicle forms after a receptor protein in the plasma membrane recognizes and binds to a particular particle in the extracellular fluid. For instance, cells take up cholesterol contained in low-density lipoproteins (LDLs), transferrin (an iron-transporting protein in the blood), some vitamins, antibodies, and certain hormones by receptor-mediated endocytosis. One variation of receptor-mediated endocytosis occurs as follows (Figure 2.4):

1 *Binding.* On the extracellular side of the plasma membrane, a ligand binds to a specific receptor in the plasma membrane to form a receptor-ligand complex. The receptors are integral membrane proteins that are concentrated in regions of the plasma membrane, called *clathrin-coated pits.* Here, a protein called clathrin attaches to the membrane on its cytoplasmic side. Many clathrin molecules come together, forming a basketlike structure around the receptor-ligand complexes that causes the membrane to invaginate (fold inward).

2 *Vesicle formation.* The invaginated edges of the membrane around the clathrin-coated pit fuse and a small piece of the membrane pinches off. The resulting vesicle, known as a *clathrin-coated vesicle,* contains the receptor-ligand complexes.

3 *Uncoating.* Almost immediately after it is formed, the clathrin-coated vesicle loses its clathrin coat to become an *uncoated vesicle.* Clathrin molecules either return to the inner surface of the plasma membrane or help form coats on other vesicles inside the cell.

4 *Fusion with endosome.* The uncoated vesicle quickly fuses with a vesicle known as an *endosome.* Within an endosome, the ligands separate from their receptors.

5 *Recycling of receptors to plasma membrane.* Most of the receptors accumulate in elongated protrusions of the endosome. These pinch off, forming a *transport vesicle* that contains the receptors and returns them to the plasma membrane.

6 *Degradation in lysosome.* Other transport vesicles, which contain the ligands, bud off the endosome and soon fuse with a *lysosome.* Within lysosomes are many digestive enzymes. Certain enzymes break down the ligand. These smaller molecules then leave the lysosome.

VIRUSES AND RECEPTOR-MEDIATED ENDOCYTOSIS

Although receptor-mediated endocytosis normally imports needed materials, some viruses are able to use this mechanism to enter and infect body cells. For example, the human immunodeficiency virus (HIV), which causes acquired immunodeficiency syndrome (AIDS), can attach to a receptor called CD4. This receptor is present in the plasma membrane of white blood cells called helper T cells. After binding to CD4, HIV enters the helper T cell via receptor-mediated endocytosis. ■

Phagocytosis (fag'-ō-sī-TŌ-sis; *phago-*=to eat) is a form of endocytosis in which the cell engulfs large solid particles, such as worn-out cells, whole bacteria, or viruses (Figure 2.5). Only a few body cells, termed **phagocytes,** are able to carry out phagocytosis. Two main types of phagocytes are *macrophages,* located in many body tissues, and *neutrophils,* a type of white blood cell. Phagocytosis begins when the particle binds to a plasma membrane receptor, causing the cell to extend **pseudopods** (SOO-dō-pods; *pseudo-*=false; *-pods*=feet), which are projections of its plasma membrane and cytoplasm. Pseudopods surround the particle outside the cell, and the membranes fuse to form a vesicle called a *phagosome,* which enters the cytoplasm. The phagosome fuses with one or more cellular organelles called lysosomes, and lysosomal enzymes break down the ingested material. In most

Figure 2.4   Receptor-mediated endocytosis.

Receptor-mediated endocytosis imports materials that are needed by cells.

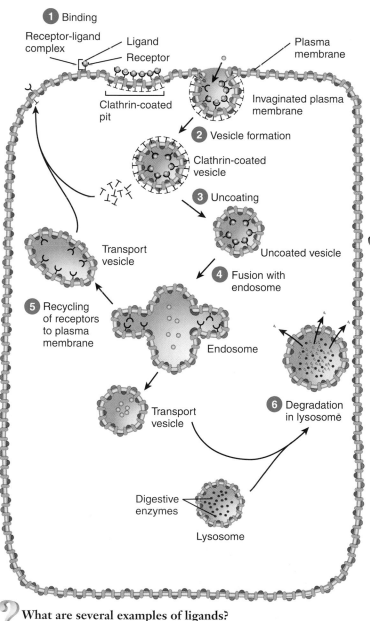

1 Binding
Receptor-ligand complex
Ligand
Receptor
Plasma membrane
Clathrin-coated pit
Invaginated plasma membrane
2 Vesicle formation
Clathrin-coated vesicle
3 Uncoating
Uncoated vesicle
4 Fusion with endosome
Transport vesicle
5 Recycling of receptors to plasma membrane
Endosome
Transport vesicle
6 Degradation in lysosome
Digestive enzymes
Lysosome

What are several examples of ligands?

Figure 2.5 Phagocytosis.

Phagocytosis is a vital defense mechanism that helps protect the body from disease.

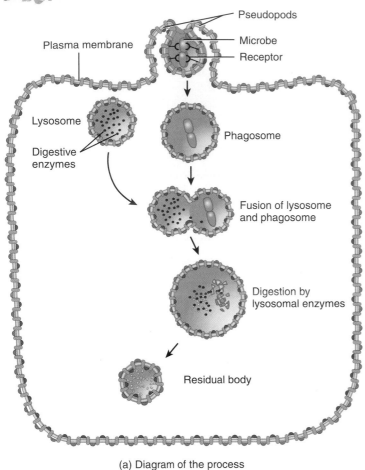

(a) Diagram of the process

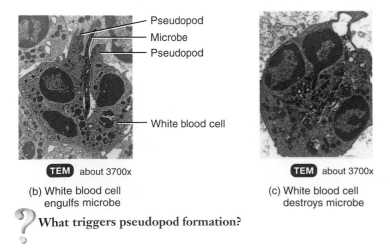

TEM about 3700x

(b) White blood cell engulfs microbe

TEM about 3700x

(c) White blood cell destroys microbe

What triggers pseudopod formation?

cases, any undigested materials in the phagosome remain indefinitely in a vesicle called a *residual body*. The process of phagocytosis is a vital defense mechanism that helps protect the body from disease. Through phagocytosis, macrophages dispose of billions of aged, worn-out red blood cells every day and neutrophils help rid the body of invading microorganisms.

Most body cells carry out **bulk-phase endocytosis (pinocytosis),** a form of endocytosis in which tiny droplets of extracellular fluid are taken up. No receptor proteins are involved; all solutes dissolved in the extracellular fluid are brought into the cell. During bulk-phase endocytosis, the plasma membrane folds inward and forms a vesicle containing a droplet of extracellular fluid. The vesicle detaches or "pinches off" from the plasma membrane and enters the cytosol. Within the cell, the vesicle fuses with a lysosome, where enzymes degrade the engulfed solutes. The resulting smaller molecules, such as amino acids and fatty acids, leave the lysosome to be used elsewhere in the cell.

EXOCYTOSIS In contrast with endocytosis, which brings materials into a cell, **exocytosis** releases materials from a cell. All cells carry out exocytosis, but it is especially important in two types of cells: (1) Secretory cells that liberate digestive enzymes, hormones, mucus, or other secretions; (2) nerve cells that release substances called *neurotransmitters*. In some cases, wastes are also released by exocytosis. During exocytosis, membrane-enclosed vesicles called *secretory vesicles* form inside the cell, fuse with the plasma membrane, and release their contents into the extracellular fluid.

Table 2.1 summarizes the processes by which materials move into and out of cells.

CHECKPOINT

5. Distinguish a passive process from an active process.
6. Why are simple diffusion, facilitated diffusion, osmosis, filtration, and active transport important to the body?
7. Describe each type of transport in vesicles. How is each important to the body?

CYTOPLASM

OBJECTIVE

● Describe the structure and function of cytoplasm, cytosol, and organelles.

Cytoplasm has two components: (1) the cytosol and (2) a variety of organelles, tiny structures that perform different functions in the cell.

Cytosol

The **cytosol** (intracellular fluid), the fluid portion of the cytoplasm that surrounds organelles (see Figure 2.1), constitutes about 55% of total cell volume. Although it varies in composition and consistency from one part of a cell to another, cytosol is 75–90% water plus dissolved and suspended components. Among these are various ions, glucose, amino acids, fatty acids, proteins, lipids, ATP, and waste products. Also present are various organic molecules that aggregate into masses for storage. These aggregations may appear and disappear at various times in the life of a cell. Examples include *lipid droplets* that contain

TABLE 2.1 PROCESSES BY WHICH SUBSTANCES MOVE ACROSS PLASMA MEMBRANES

PROCESS	DESCRIPTION
Passive processes	Substances move down a concentration gradient from an area of higher to lower concentration or pressure; cell does not expend energy by the breakdown of ATP.
Simple diffusion	Net movement of substances from an area of higher to lower concentration until an equilibrium is reached.
Facilitated diffusion	Movement of larger molecules across a selectively permeable membrane with the assistance of integral membrane proteins that serve as transporters.
Osmosis	Net movement of water molecules across a selectively permeable membrane from an area of higher to lower concentration of water until an equilibrium is reached.
Filtration	Movement of solvents (such as water) and solutes (such as glucose) across a selectively permeable membrane as a result of gravity or hydrostatic (water) pressure from an area of higher to lower pressure.
Active process	Substances move against a concentration gradient from an area of lower to higher concentration; cell must expend energy released by the breakdown of ATP.
Active transport	Movement of substances, usually ions, across a selectively permeable membrane from a region of lower to higher concentration by an interaction with transporter proteins in the membrane.
Transport in vesicles	Movement of substances into or out of a cell in vesicles that bud from the plasma membrane; requires energy supplied by the breakdown of ATP.
Endocytosis	Movement of substances into a cell in vesicles.
Receptor-mediated endocytosis	Ligand-receptor complexes trigger infolding of a clathrin-coated pit that forms a vesicle containing ligands.
Phagocytosis	"Cell eating"; movement of a solid particle into a cell after pseudopods engulf it to form a phagosome.
Bulk-phase endocytosis	"Cell drinking"; movement of extracellular fluid into a cell by infolding of plasma membrane, forming a vesicle.
Exocytosis	Movement of substances out of a cell in secretory vesicles that fuse with the plasma membrane and release their contents into the extracellular fluid.

triglycerides and clusters of glycogen molecules called *glycogen granules* (see Figure 2.1).

The cytosol is the site of many chemical reactions required for a cell's existence. For example, enzymes in cytosol catalyze numerous chemical reactions. As a result of these reactions, energy is released and captured to drive cellular activities. In addition, these reactions provide the building blocks for maintaining cell structure, function, and growth.

Organelles

As noted previously, **organelles** are specialized structures that have characteristic shapes and that perform specific functions in cellular growth, maintenance, and reproduction. Despite the many chemical reactions occurring in a cell at the same time, there is little interference among reactions because they occur in different organelles. Each type of organelle is a functional compartment with its own set of enzymes where specific physiological processes take place. Moreover, organelles often cooperate to maintain homeostasis. The numbers and types of organelles vary in different cells, depending on the function of the cell. Although the nucleus is an organelle, it is discussed separately because of its special importance in directing the life of a cell.

The Cytoskeleton

The **cytoskeleton** is a network of protein filaments that extend throughout the cytosol (Figure 2.6). The cytoskeleton provides a structural framework for the cell, serving as a scaffold that helps to determine a cell's shape and organizes the cellular contents. The cytoskeleton is also responsible for cell movements, including the internal transport of organelles and some chemicals, the movement of chromosomes during cell division, and the movement of whole cells such as phagocytes. In order of their increasing diameter, components of the cytoskeleton are microfilaments, intermediate filaments, and microtubules.

MICROFILAMENTS These are the thinnest elements of the cytoskeleton. Microfilaments are concentrated at the periphery of a cell (Figure 2.6a). They are composed of the protein *actin* and have two general functions: movement and mechanical support. With respect to movement, microfilaments are involved in muscle contraction, cell division, and cell locomotion, such as occurs during the migration of embryonic cells during development, the invasion of tissues by white blood cells to fight infection, or the migration of skin cells during wound healing.

Microfilaments provide much of the mechanical support that is responsible for the basic strength and shapes of cells. They anchor the cytoskeleton to integral proteins in the plasma membrane. Microfilaments also provide mechanical support for cell extensions called **microvilli** (*micro-*=small; *-villi*=tufts of hair), which are nonmotile, microscopic fingerlike projections of the plasma membrane. Within a microvillus is a core of parallel microfilaments that supports it and attaches it to the cytoskeleton (Figure 2.6a). Microvilli are abundant on the surfaces of cells involved in absorption, such as the epithelial cells that line the small intestine. Some microfilaments extend beyond the plasma membrane and help cells attach to one another or to extracellular material.

Figure 2.6 Cytoskeleton.

The cytoskeleton is a network of three kinds of protein filaments that extend throughout the cytoplasm: microfilaments, intermediate filaments, and microtubules.

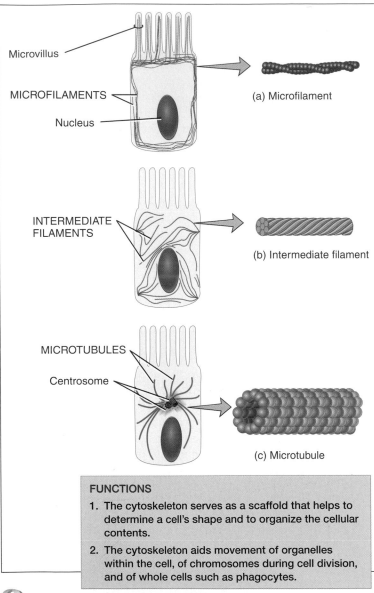

(a) Microfilament

(b) Intermediate filament

(c) Microtubule

Microvillus

MICROFILAMENTS

Nucleus

INTERMEDIATE FILAMENTS

MICROTUBULES

Centrosome

FUNCTIONS

1. The cytoskeleton serves as a scaffold that helps to determine a cell's shape and to organize the cellular contents.

2. The cytoskeleton aids movement of organelles within the cell, of chromosomes during cell division, and of whole cells such as phagocytes.

Which cytoskeletal component helps form the structure of centrioles, cilia, and flagella?

INTERMEDIATE FILAMENTS As their name suggests, intermediate filaments are thicker than microfilaments but thinner than microtubules (Figure 2.6b). Several different proteins can compose intermediate filaments, which are exceptionally strong. They are found in parts of cells subject to mechanical stress and also help anchor organelles such as the nucleus and help attach cells to one another.

MICROTUBULES The largest of the cytoskeletal components, microtubules are long, unbranched hollow tubes composed mainly of a protein called *tubulin*. The centrosome (discussed shortly) serves as the initiation site for the assembly of microtubules. The microtubules grow outward from the centrosome toward the periphery of the cell (Figure 2.6c). Microtubules help determine cell shape and function in the intracellular transport of organelles, such as secretory vesicles, and the migration of chromosomes during cell division. They also participate in the movement of specialized cell projections such as cilia and flagella.

Centrosome

The **centrosome,** located near the nucleus, consists of the pericentriolar material and centrioles (Figure 2.7a). The **pericentriolar material** (per′-ē-sen′-trē-Ō-lar) is a region of the cytosol composed of a dense network of small protein fibers (tubulins). This region is the organizing center for the growth of the mitotic spindle, which plays a critical role in cell division, and for microtubule formation in nondividing cells. Within the pericentriolar material is a pair of cylindrical structures called **centrioles,** each of which is composed of nine clusters of three microtubules (triplets) arranged in a circular pattern, an arrangement called a *9+0 array* (Figure 2.7b). The 9 refers to the nine clusters of microtubules, and the 0 refers to the absence of microtubules in the center. The long axis of one centriole is at a right angle to the long axis of the other (Figure 2.7a and c).

Cilia and Flagella

Microtubules are the dominant structural and functional components of cilia and flagella, both of which are motile projections of the cell surface. In the human body, cells that are firmly anchored in place use cilia to move fluids across their surfaces; motile cells, such as sperm cells, use flagella to propel themselves through a liquid medium.

Cilia (SIL-ē-a=eyelashes) are numerous, short, hairlike projections that extend from the surface of the cell (see Figure 2.1). Each cilium (the singular form) contains a core of microtubules surrounded by plasma membrane (Figure 2.8). The microtubules are arranged with one pair in the center surrounded by nine clusters of two microtubules (doublets), an arrangement called a *9+2 array*. Each cilium is anchored to a *basal body* just below the surface of the plasma membrane. A basal body is similar in structure to a centriole. In fact, basal bodies and centrioles are considered to be two different functional manifestations of the same structure. The function of a basal body is to initiate the assembly of cilia and flagella. The coordinated movement of numerous cilia on the surface of a cell ensures the steady movement of fluid over the cell's surface. Many cells of the respiratory tract, for example, have hundreds of cilia that help sweep foreign

Figure 2.7 Centrosome.

Centrioles are composed of clusters of microtubules.

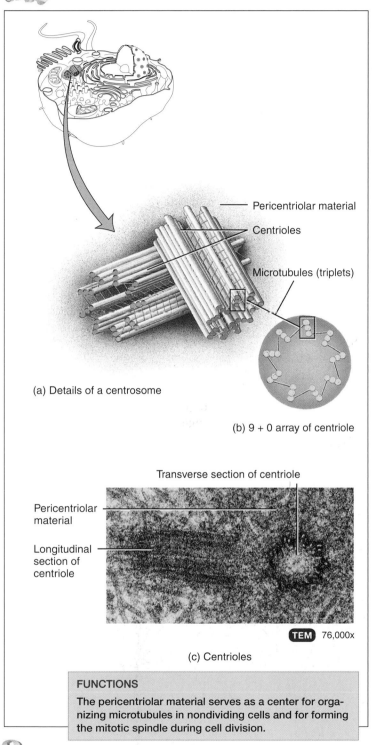

Pericentriolar material

Centrioles

Microtubules (triplets)

(a) Details of a centrosome

(b) 9 + 0 array of centriole

Transverse section of centriole

Pericentriolar material

Longitudinal section of centriole

TEM 76,000x

(c) Centrioles

FUNCTIONS

The pericentriolar material serves as a center for organizing microtubules in nondividing cells and for forming the mitotic spindle during cell division.

If you observed that a cell did not have a centrosome, what could you predict about its capacity for cell division?

particles trapped in mucus away from the lungs. Their movement is paralyzed by nicotine in cigarette smoke. For this reason, smokers cough often to remove foreign particles from their airways. Cells that line the uterine (fallopian) tubes also have cilia that sweep oocytes (egg cells) toward the uterus.

Flagella (fla-JEL-a; singular is *flagellum*=whip) are similar in structure to cilia but are much longer (see Figure 2.1). Flagella usually move an entire cell. The only example of a flagellum in the human body is a sperm cell's tail, which propels the sperm toward its rendezvous with an ovum.

Ribosomes

Ribosomes (RĪ-bō-sōms; *-somes*=bodies) are sites of protein synthesis. These tiny organelles are packages of **ribosomal RNA (rRNA)** and many ribosomal proteins. Ribosomes are so named because of their high content of *ribonucleic* acid.

Figure 2.8 Cilia and flagella.

Whereas centrioles have a 9+0 array of microtubules, cilia and flagella have a 9+2 array.

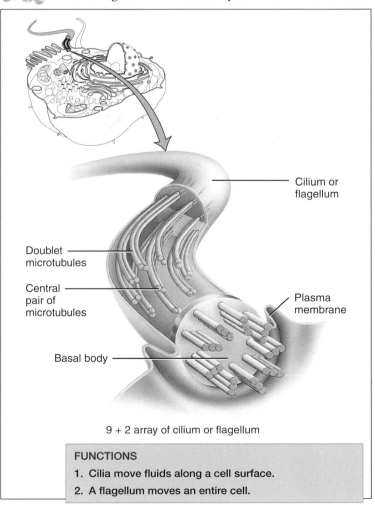

Cilium or flagellum

Doublet microtubules

Central pair of microtubules

Plasma membrane

Basal body

9 + 2 array of cilium or flagellum

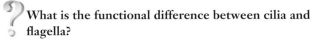

FUNCTIONS

1. Cilia move fluids along a cell surface.
2. A flagellum moves an entire cell.

What is the functional difference between cilia and flagella?

Structurally, a ribosome consists of two subunits, one about half the size of the other (Figure 2.9). The two subunits are made separately in the nucleolus, a spherical body inside the nucleus (see page 41). Once produced, they exit the nucleus and join together in the cytosol where they become functional.

Some ribosomes, called *free ribosomes*, are unattached to any structure in the cytoplasm. Primarily, free ribosomes synthesize proteins used *inside* the cell. Other ribosomes, called *membrane-bound ribosomes*, attach to the nuclear membrane and to an extensively folded membrane called the endoplasmic reticulum (see Figure 2.10b). These ribosomes synthesize proteins destined for specific organelles, insertion in the plasma membrane, or for export from the cell. Ribosomes are also located within mitochondria, where they synthesize mitochondrial proteins. Sometimes 10–20 ribosomes occur in a stringlike arrangement called a *polyribosome*.

Endoplasmic Reticulum

The **endoplasmic reticulum** (en′-dō-PLAS-mik re-TIK-ū-lum; *-plasmic*=cytoplasm; *reticulum*=network), or **ER,** is a network of membranes in the form of flattened sacs or tubules (Figure 2.10). The ER extends from the nuclear envelope (membrane around the nucleus), to which it is connected, throughout the cytoplasm. The ER is so extensive that it consti-

tutes more than half of the membranous surfaces within the cytoplasm of most cells.

Cells contain two distinct forms of ER, which differ in structure and function. **Rough ER** is continuous with the nuclear membrane and usually is folded into a series of flattened sacs. The outer surface of rough ER is studded with ribosomes,

Figure 2.9   Ribosomes.

Ribosomes are the sites of protein synthesis.

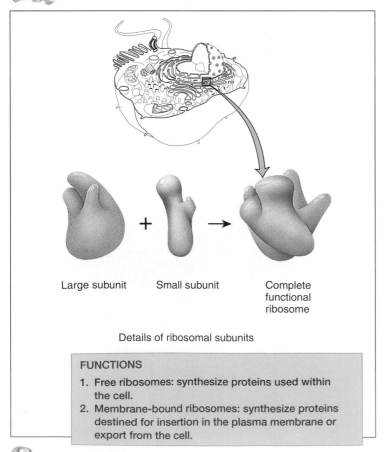

Large subunit   Small subunit   Complete functional ribosome

Details of ribosomal subunits

FUNCTIONS
1. Free ribosomes: synthesize proteins used within the cell.
2. Membrane-bound ribosomes: synthesize proteins destined for insertion in the plasma membrane or export from the cell.

Where are subunits of ribosomes synthesized and assembled?

Figure 2.10   Endoplasmic reticulum.

The endoplasmic reticulum is a network of membrane-enclosed sacs or tubules that extend throughout the cytoplasm and connect to the nuclear envelope.

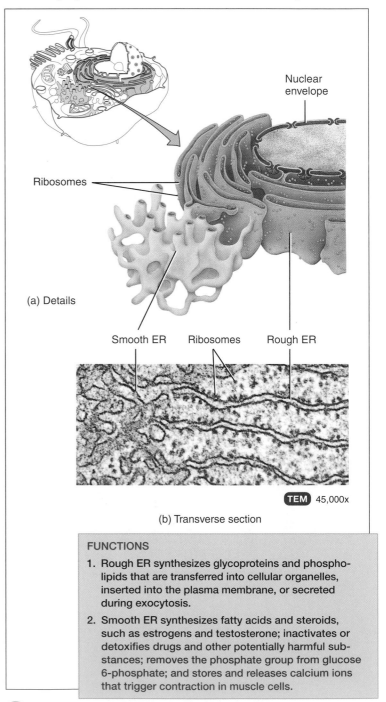

Nuclear envelope

Ribosomes

(a) Details

Smooth ER   Ribosomes   Rough ER

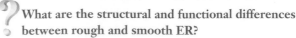

TEM   45,000x

(b) Transverse section

FUNCTIONS
1. Rough ER synthesizes glycoproteins and phospholipids that are transferred into cellular organelles, inserted into the plasma membrane, or secreted during exocytosis.
2. Smooth ER synthesizes fatty acids and steroids, such as estrogens and testosterone; inactivates or detoxifies drugs and other potentially harmful substances; removes the phosphate group from glucose 6-phosphate; and stores and releases calcium ions that trigger contraction in muscle cells.

What are the structural and functional differences between rough and smooth ER?

the sites of protein synthesis. Proteins synthesized by ribosomes attached to rough ER enter spaces within the ER for processing and sorting. In some cases, enzymes attach the proteins to carbohydrates to form glycoproteins. In other cases, enzymes attach the proteins to phospholipids, also synthesized by rough ER. These molecules may be incorporated into the membranes of organelles, inserted into the plasma membrane or secreted via exocytosis. Thus rough ER produces secretory proteins, membrane proteins, and many organellar proteins.

Smooth ER extends from the rough ER to form a network of membrane tubules (Figure 2.10). Unlike rough ER, smooth ER does not have ribosomes on the outer surface of its membrane. However, smooth ER contains unique enzymes that make it functionally more diverse than rough ER. Although it does not synthesize proteins, smooth ER does synthesize fatty acids and steroids, such as estrogens and testosterone. In liver cells, enzymes of the smooth ER help release glucose into the bloodstream and inactivate or detoxify lipid-soluble drugs and other potentially harmful substances, for example, alcohol. In liver, kidney, and intestinal cells a smooth ER enzyme removes the phosphate group from glucose 6-phosphate, which allows the "free" glucose to enter the bloodstream. In muscle cells, calcium ions that trigger contraction are released from the sarcoplasmic reticulum, a form of smooth ER.

SMOOTH ER AND DRUG TOLERANCE

One of the functions of smooth ER, as noted earlier, is to detoxify certain drugs. Individuals who repeatedly take such drugs, for example, the sedative phenobarbital, develop changes in the smooth ER in their liver cells. Prolonged administration of phenobarbital results in increased tolerance to the drug so that the same dose no longer produces the same degree of sedation. This is due to an increase in the amount of smooth ER and its enzymes to further protect the cell from the toxic effects of the drug. As the amount of smooth ER increases, higher and higher dosages are needed to achieve the original effect of the drug. ■

Golgi Complex

Most of the proteins synthesized by ribosomes attached to rough ER are ultimately transported to other regions of the cell. The first step in the transport pathway is through an organelle called the **Golgi complex** (GOL-jē). It consists of 3–20 flattened, membranous sacs with bulging edges, called **cisternae** (sis-TER-nē = cavities; singular is *cisterna*), that resemble a stack of pita bread (Figure 2.11). The cisternae are often curved, giving the Golgi complex a cuplike shape. Most cells have several Golgi com-

Figure 2.11   Golgi complex.

The opposite faces of a Golgi complex differ in size, shape, content, and enzymatic activity.

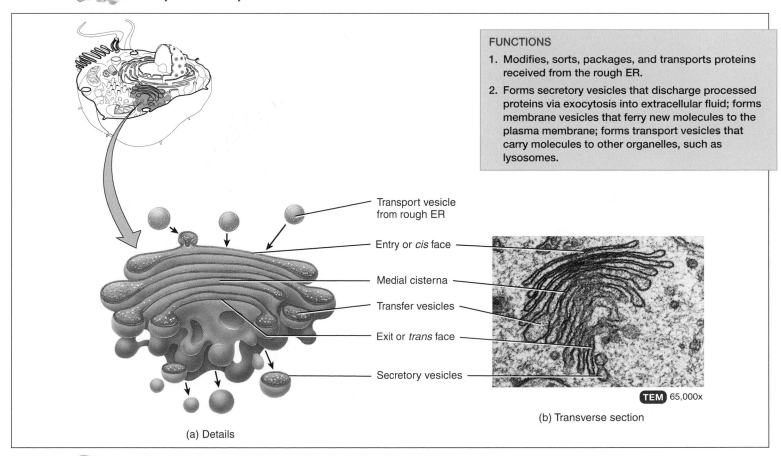

FUNCTIONS

1. Modifies, sorts, packages, and transports proteins received from the rough ER.

2. Forms secretory vesicles that discharge processed proteins via exocytosis into extracellular fluid; forms membrane vesicles that ferry new molecules to the plasma membrane; forms transport vesicles that carry molecules to other organelles, such as lysosomes.

Transport vesicle from rough ER

Entry or *cis* face

Medial cisterna

Transfer vesicles

Exit or *trans* face

Secretory vesicles

TEM 65,000x

(b) Transverse section

(a) Details

How do the entry and exit faces differ in function?

plexes. The Golgi complex is more extensive in cells that secrete proteins into the extracellular fluid, for example, pancreatic cells. This fact is a clue to the organelle's role in the cell.

The cisternae at the opposite ends of a Golgi complex differ from each other in size, shape, content, enzymatic activity, and types of vesicles (described shortly). The convex **entry,** or *cis* **face,** is composed of cisternae that face the rough ER. The concave **exit,** or *trans* **face,** is composed of cisternae that face the plasma membrane. Cisternae between the entry and exit faces are called **medial cisternae.** Transport vesicles (described shortly) from the ER merge to form the entry face. From the entry face, the cisternae are thought to mature, in turn becoming medial and then exit cisternae.

The entry face, medial cisternae, and exit face of the Golgi complex each contain different enzymes that permit them to modify, sort, and package proteins for transport to different destinations. For example, the entry face receives and modifies proteins produced by the rough ER. The medial cisternae add sugars to proteins and lipids to form glycoproteins and glycolipids, and they add proteins to lipids to form lipoproteins. The exit face modifies the molecules further and then sorts and packages them for transport to their destinations.

Proteins arriving at, passing through, and exiting the Golgi complex do so through maturation of the cisternae and exchanges that occur via transfer vesicles (Figure 2.12).

1 Proteins synthesized by ribosomes on the rough ER are surrounded by a portion of the ER membrane, which eventually buds from the membrane surface to form **transport vesicles.**

2 The transport vesicles move toward the entry face of the Golgi complex.

3 Fusion of several transport vesicles creates the entry face of the Golgi complex and releases proteins into its lumen.

4 The proteins move from the entry face into one or more medial cisternae. Enzymes in the medial cisternae modify the proteins to form glycoproteins, glycolipids, and lipoproteins. **Transfer vesicles** that bud from the edges of the cisternae move specific enzymes back toward the entry face and also move some partially modified proteins forward toward the exit face.

5 The products of the medial cisternae move into the lumen of the exit face.

6 Within the exit face cisterna, the products are further modified and are sorted and packaged.

7 Some of the processed proteins leave the exit face in **secretory vesicles.** These vesicles deliver the proteins to the plasma membrane, where they are discharged by exocytosis

Figure 2.12   Processing and packaging of synthesized proteins by the Golgi complex.

All proteins exported from the cell are processed in the Golgi complex.

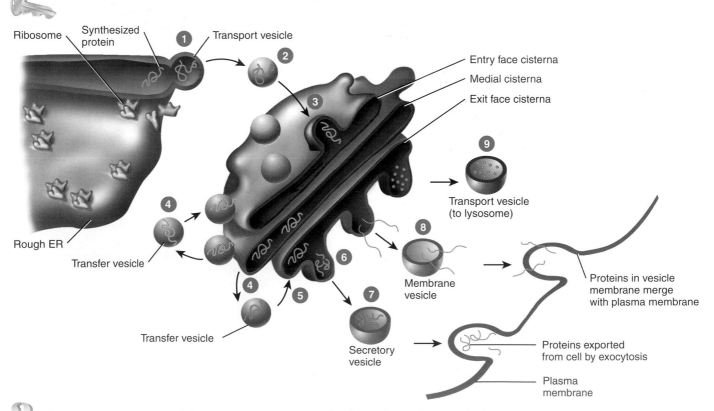

? **What are the three general destinations for proteins that leave the Golgi complex?**

into the extracellular fluid. For example, certain pancreatic cells release the hormone insulin in this way.

8 Other processed proteins leave the exit face in **membrane vesicles** that deliver their contents to the plasma membrane for incorporation into the membrane. In doing so, the Golgi complex adds new segments of plasma membrane as existing segments are lost and modifies the number and distribution of membrane molecules.

9 Finally, some processed proteins leave the exit face in transport vesicles that will carry the proteins to another cellular destination. For instance, transport vesicles ferry digestive enzymes to lysosomes, whose structure and functions are discussed next.

Lysosomes

Lysosomes (LĪ-sō-sōms; *lyso-*=dissolving; *-somes*=bodies) are membrane-enclosed vesicles that form from the Golgi complex (Figure 2.13). Inside can be as many as 60 kinds of powerful digestive enzymes that are capable of breaking down a wide variety of molecules. Lysosomes fuse with vesicles formed during endocytosis and the lysosomal enzymes break down the contents of the vesicles. Proteins in the lysosomal membrane allow the final products of digestion, such as sugars, fatty acids, and amino acids, to be transported into the cytosol. In a similar way, lysosomes in phagocytes can break down and destroy microbes, such as bacteria and viruses.

Lysosomal enzymes also recycle the cell's own structures. A lysosome can engulf another organelle, digest it, and return the digested components to the cytosol for reuse. In this way, old organelles are continually replaced. The process by which worn-out organelles are digested is called **autophagy** (aw-TOF-a-jē; *auto-*=self; *-phagy*=eating). In autophagy, the organelle to be digested is enclosed by a membrane derived from the ER to create a vesicle called an **autophagosome;** the vesicle then fuses with a lysosome. In this way, a human liver cell, for example, recycles about half its cytoplasmic contents every week. Lysosomal enzymes may also destroy the cell that contains them, a process known as **autolysis** (aw-TOL-i-sis). Autolysis occurs in some pathological conditions and also is responsible for the tissue deterioration that occurs immediately after death.

Although most of the digestive processes involving lysosomal enzymes occur within a cell, there are some instances in which the enzymes operate in extracellular digestion. One example is the release of lysosomal enzymes during fertilization. Enzymes from lysosomes in the head of a sperm cell help the sperm cell penetrate the surface of the ovum.

TAY-SACHS DISEASE

Some disorders are caused by faulty or absent lysosomal enzymes. For instance, **Tay-Sachs disease,** which most often affects children of Ashkenazi (eastern European Jewish) descent, is an inherited condition characterized by the absence of a single

Figure 2.13 Lysosomes.

🔑 Lysosomes arise in the Golgi complex and store several kinds of powerful digestive enzymes.

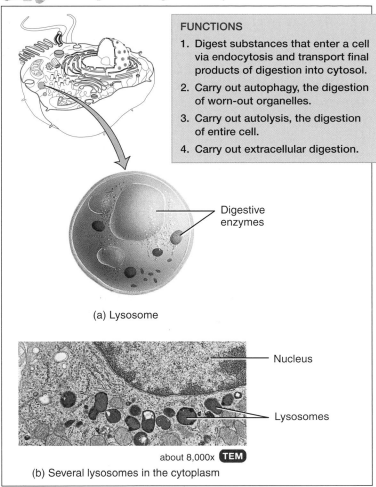

FUNCTIONS

1. Digest substances that enter a cell via endocytosis and transport final products of digestion into cytosol.
2. Carry out autophagy, the digestion of worn-out organelles.
3. Carry out autolysis, the digestion of entire cell.
4. Carry out extracellular digestion.

Digestive enzymes

(a) Lysosome

Nucleus

Lysosomes

about 8,000x **TEM**

(b) Several lysosomes in the cytoplasm

❓ What is the name of the process by which worn-out organelles are digested by lysosomes?

lysosomal enzyme called Hex A. This enzyme normally breaks down a membrane glycolipid called ganglioside G_{M2} that is especially prevalent in nerve cells. As the excess ganglioside G_{M2} accumulates, the nerve cells function less efficiently. Children with Tay-Sachs disease typically experience seizures and muscle rigidity. They gradually become blind, demented, and uncoordinated and usually die before the age of 5. Tests can now reveal whether an adult is a carrier of the defective gene. ∎

Peroxisomes

Another group of organelles similar in structure to lysosomes, but smaller, are called **peroxisomes** (pe-ROKS-i-sōms; *peroxi-*= peroxide; *-somes*=bodies; see Figure 2.1). Peroxisomes contain several enzymes called oxidases that can oxidize (remove hydrogen atoms from) various organic substances. For example, amino acids and fatty acids are oxidized in peroxisomes as part of normal metabolism. In addition, enzymes in peroxisomes oxidize toxic substances, such as alcohol. Thus, peroxisomes are very abundant in the liver, where detoxification of alcohol and other damaging substances takes place. A byproduct of the oxidation reactions is hydrogen peroxide (H_2O_2), a potentially toxic compound. However, peroxisomes also contain an enzyme called *catalase*, which decomposes H_2O_2. Because the generation and degradation of H_2O_2 occurs within the same organelle, peroxisomes protect other parts of the cell from the toxic effects of H_2O_2. New peroxisomes form by budding off from preexisting ones.

Proteasomes

Although lysosomes degrade proteins delivered to them in vesicles, cytosolic proteins also require disposal at certain times in the life of a cell. Continuous destruction of unneeded, damaged, or faulty proteins is the function of tiny structures called **proteasomes** (PRŌ-tē-a-sōmes=protein bodies). For example, proteins that are part of metabolic pathways are degraded after they have accomplished their function. Such protein destruction halts the pathway once the appropriate response has been achieved. A typical body cell contains many thousands of proteasomes, in both the cytosol and the nucleus. They were discovered only recently because they are far too small to discern under the light microscope and do not show up well in electron micrographs. Proteasomes were so-named because they contain myriad *proteases*, enzymes that cut proteins into small peptides. Once the enzymes of a proteasome have chopped up a protein into smaller chunks, other enzymes then break down the peptides into amino acids, which can be recycled into new proteins.

PROTEASOMES AND DISEASE

Proteasomes are thought to be a factor in several diseases. For instance, people who have **cystic fibrosis** produce a misshapen membrane transporter protein whose mission normally is to help pump chloride ions (Cl^-) out of certain cells. Proteosomes degrade the defective transporter before it can reach the plasma membrane. The result is an imbalance in the transport of fluid and ions across the plasma membrane that causes the buildup of thick mucus outside certain types of cells. The accumulated mucus clogs the airways in the lungs, causing breathing difficulty, and prevents proper secretion of digestive enzymes by the pancreas, causing digestive problems. Disease could also result from failure of proteasomes to degrade abnormal proteins. For example, clumps of misfolded proteins accumulate in brain cells of people with Parkinson's disease and Alzheimer's disease. Discovering why the proteasomes fail to clear these abnormal proteins is a goal of ongoing research. ∎

Mitochondria

Because of their function in generating ATP, **mitochondria** (mī-tō-KON-drē-a; *mito-*=thread; *-chondria*=granules) are the "powerhouses" of the cell. A cell may have as few as a hundred or as many as several thousand mitochondria (singular is *mitochondrion*). Physiologically active cells, such as those found in the muscles, liver, and kidneys, have a large number of mitochondria because they use ATP at a high rate. Mitochondria are usually located in a cell where the energy need is greatest, such as between the contractile proteins in muscle cells.

A mitochondrion consists of an **outer mitochondrial membrane** and an **inner mitochondrial membrane** with a small fluid-filled space between them (Figure 2.14). Both membranes are similar in structure to the plasma membrane. The inner mitochondrial membrane contains a series of folds called **cristae** (KRIS-tē=ridges). The large central fluid-filled cavity of a mitochondrion, enclosed by the inner mitochondrial membrane, is the **matrix.** The elaborate folds of the cristae provide an enormous surface area for the chemical reactions that are part of the aerobic phase of *cellular respiration.* These reactions produce most of a cell's ATP and the enzymes that catalyze them are located on the cristae and in the matrix.

Like peroxisomes, mitochondria self-replicate, a process that occurs during times of increased cellular energy demand or before cell division. Each mitochondrion has multiple identical copies of a circular DNA molecule that contains 37 genes. The mitochondrial genes along with the genes in the cell's nucleus control the production of proteins that build mitochondrial components. Because ribosomes are also present in the mitochondrial matrix, some protein synthesis occurs inside mitochondria.

Although the nucleus of each somatic cell contains genes from both your mother and father, mitochondrial genes usually are inherited only from your mother. The head of a sperm (the part that penetrates and fertilizes an ovum) normally lacks most organelles, such as mitochondria, ribosomes, endoplasmic reticulum, and Golgi complexes.

MITOCHONDRIAL MYOPATHIES

Mitochondrial myopathies (mī-OP-a-thēs; *myo-*=muscle; *path-*=disease) are rare inherited muscle disorders due to faulty

Figure 2.14 Mitochondria.

 Within mitochondria, chemical reactions of aerobic cellular respiration generate ATP.

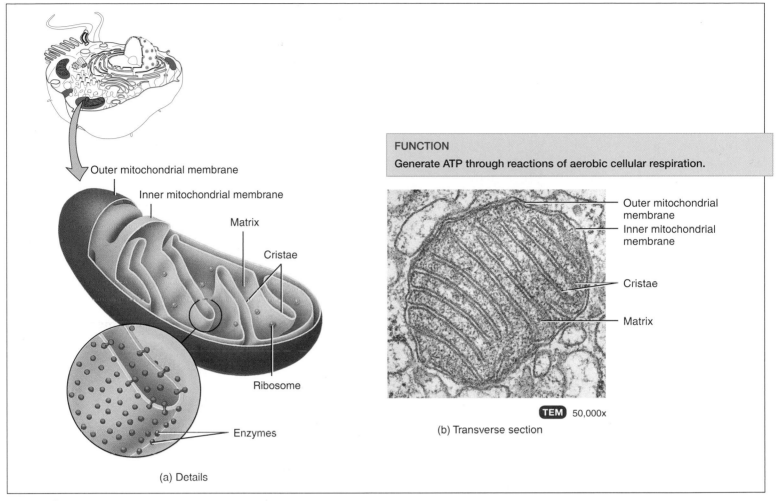

Outer mitochondrial membrane

Inner mitochondrial membrane

Matrix

Cristae

Ribosome

Enzymes

(a) Details

FUNCTION

Generate ATP through reactions of aerobic cellular respiration.

Outer mitochondrial membrane

Inner mitochondrial membrane

Cristae

Matrix

TEM 50,000x

(b) Transverse section

? How do the cristae of a mitochondrion contribute to its ATP-producing function?

mitochondrial genes. One such example is *ophthalmoplegia* (of-thal-mō-PLĒ-jē-a; *ophthalmo-*=eye; *-plegia*=blow or strike), weakness or paralysis of eye muscles. ■

CHECKPOINT

8. What does cytoplasm have that cytosol lacks?
9. Which organelles are surrounded by a membrane and which are not?
10. Which organelles contribute to synthesizing protein hormones and packaging them into secretory vesicles?
11. What happens on the cristae and in the matrix of mitochondria?

NUCLEUS

● Describe the structure and functions of the nucleus.

The **nucleus** is a spherical or oval-shaped structure that is usually the most prominent feature of a cell (Figure 2.15). Most body cells have a single nucleus, although some, such as mature red blood cells, have none. In contrast, skeletal muscle cells and a few other cells have several nuclei. A double membrane called the **nuclear envelope** separates the nucleus from the cytoplasm. Both layers of the nuclear envelope are lipid

Figure 2.15 Nucleus.

The nucleus contains most of a cell's genes, which are located on chromosomes.

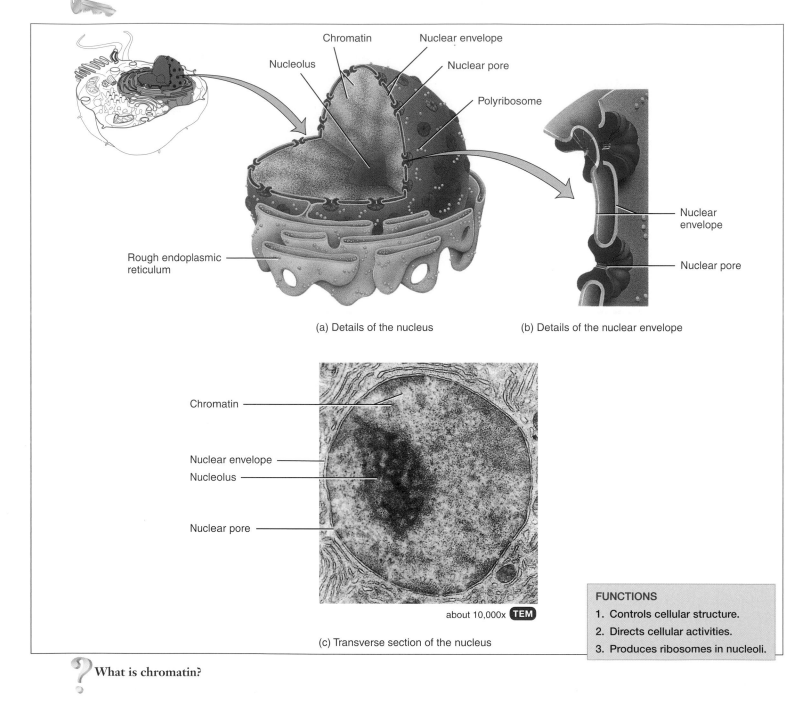

(a) Details of the nucleus

(b) Details of the nuclear envelope

about 10,000x **TEM**

(c) Transverse section of the nucleus

FUNCTIONS
1. Controls cellular structure.
2. Directs cellular activities.
3. Produces ribosomes in nucleoli.

What is chromatin?

bilayers similar to the plasma membrane. The outer membrane of the nuclear envelope is continuous with rough ER and resembles it in structure. The nuclear envelope is perforated by numerous openings called **nuclear pores** (Figure 2.15c). Each pore consists of a circular arrangement of proteins that surrounds a large central channel. This channel is about 10 times larger than the pore of a channel protein in the plasma membrane.

Nuclear pores control the movement of substances between the nucleus and the cytoplasm. Small molecules and ions diffuse passively through the pores. Most large molecules, however, such as RNAs and proteins, cannot pass through the nuclear pores by diffusion. Instead, their passage involves an active transport process in which the molecules are recognized and selectively transported through the nuclear pore into or out of the nucleus. For example, proteins needed for nuclear functions move from the cytosol into the nucleus whereas RNA molecules move from the nucleus into the cytosol.

Inside the nucleus are one or more spherical bodies called **nucleoli** (noo′-KLĒ-ō-lī; singular is *nucleolus*) that function in producing ribosomes. Each nucleolus is a cluster of protein, DNA, and RNA that is not enclosed by a membrane. Nucleoli are the sites of synthesis of one type of RNA and its assembly with proteins into ribosomal subunits. Nucleoli are quite prominent in cells that synthesize large amounts of protein, such as muscle and liver cells. Nucleoli disperse and disappear during cell division and reorganize once new cells are formed.

Within the nucleus are most of the cell's hereditary units, called **genes,** which control cellular structure and direct cellular activities. Genes are arranged in single file along **chromosomes** (*chromo-*—colored). Human somatic (body) cells have 46 chromosomes, 23 inherited from each parent. Each chromosome is a long molecule of DNA that is coiled together with several proteins (Figure 2.16). This complex of DNA, proteins, and some RNA is called **chromatin.** The total genetic information carried in a cell or an organism is its **genome.**

In cells that are not dividing, the chromatin appears as a diffuse, granular mass. Electron micrographs reveal that chromatin has a beads-on-a-string structure. Each bead is a **nucleosome** and consists of double-stranded DNA wrapped twice around a core of eight proteins called **histones,** which help organize the coiling and folding of DNA. The string between the beads is **linker DNA,** which holds adjacent nucleosomes together. Another histone promotes coiling of nucleosomes into a larger-diameter **chromatin fiber,** which then folds into large loops. In cells that are not dividing, this is how DNA is packed. Just before cell division takes place, however, the DNA replicates (duplicates) and the loops condense even more, forming a pair of **chromatids.** As you will see shortly, during cell division a pair of chromatids constitutes a chromosome.

GENOMICS

In the last decade of the twentieth century, the genomes of humans, mice, fruit flies, and more than 50 microbes were sequenced. As a result, research in the field of **genomics,** the study of the relationships between the genome and the biological functions of an organism, has flourished. The Human Genome Project began in June 1990 as an effort to sequence all of the nearly 3.2 billion nucleotides of our genome. The project was completed in April 2003. Scientists now know that the total number of genes in the human genome is about 30,000, far fewer than the 100,000 previously predicted to exist. Information regarding the human genome and the effects of environment on the genome aim to

Figure 2.16   Packing of DNA into a chromosome in a dividing cell. When packing is complete, two identical DNAs and their histones form a pair of chromatids, held together by a centromere.

A chromosome is a highly coiled and folded DNA molecule that is combined with protein molecules.

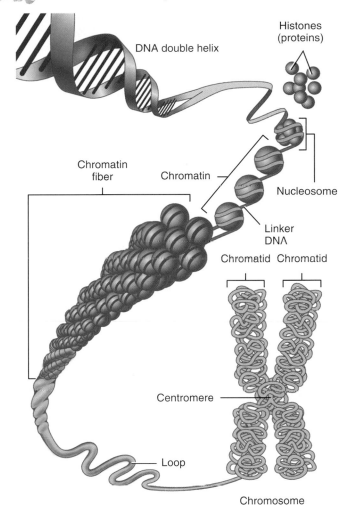

What are the components of a nucleosome?

identify the specific genes that play a role in genetic diseases and how the genes produce various diseases. Genomic medicine also aims to design new drugs to treat genetic diseases and to provide screening tests to indicate susceptibility to certain genetic diseases. It is hoped that physicians may be able to provide counseling and treatment that is more effective for disorders that have significant genetic components, for instance, hypertension (high blood pressure), obesity, diabetes, and cancer. ■

The main parts of a cell and their functions are summarized in Table 2.2.

CHECKPOINT

12. How is DNA packed into the nucleus?

TABLE 2.2   CELL PARTS AND THEIR FUNCTIONS

PART	STRUCTURE	FUNCTIONS
Plasma membrane	Fluid-mosaic lipid bilayer (phospholipids, cholesterol, and glycolipids) studded with proteins; surrounds cytoplasm.	Protects cellular contents; makes contact with other cells; contains channels, transporters, receptors,enzymes, cell-identity markers, and linker proteins; mediates the entry and exit of substances.
Cytoplasm	Cellular contents between the plasma membrane and nucleus — cytosol and organelles.	Site of all intracellular activities except those occurring in the nucleus.
Cytosol	Composed of water, solutes, suspended particles, lipid droplets, and glycogen granules	Medium in which many of cell's chemical reactions occur.
Organelles	Specialized structures with characteristic shapes.	Each organelle has one or more specific functions.
Cytoskeleton	Network of three types of protein filaments: microfilaments, intermediate filaments, and microtubules.	Maintains shape and general organization of cellular contents; responsible for cellular movements.
Centrosome	A pair of centrioles plus pericentriolar material.	The pericentriolar material contains tubulins, which are used for growth of the mitotic spindle and microtubule formation.
Cilia and flagella	Motile cell surface projections that contain 20 microtubules and a basal body.	Cilia move fluids over a cell's surface; flagella move an entire cell.
Ribosome	Composed of two subunits containing ribosomal RNA and proteins; may be free in cytosol or attached to rough ER.	Protein synthesis.
Endoplasmic reticulum (ER)	Membranous network of flattened sacs or tubules. Rough ER is covered by ribosomes and is attached to nuclear envelope; smooth ER lacks ribosomes.	Rough ER synthesizes glycoproteins and phospholipids that are transferred to cellular organelles, inserted into the plasma membrane, or secreted during exocytosis. Smooth ER synthesizes fatty acids and steroids; inactivates or detoxifies drugs; removes phosphate group from glucose 6-phosphate; and stores and releases calcium ions in muscle cells.
Golgi complex	Consists of 3–20 flattened membranous sacs called cisternae; structurally and functionally divided into entry (*cis*) face, medial cisternae, and exit (*trans*) face.	Entry (*cis*) face accepts proteins from rough ER; medial cisternae form glycoproteins, glycolipids, and lipoproteins; exit (*trans*) face modifies the molecules further, then sorts and packages them for transport to their destinations.
Lysosome	Vesicle formed from Golgi complex; contains digestive enzymes.	Fuses with and digests contents of endosomes, pinocytic vesicles, and phagosomes and transports final products of digestion into cytosol; digests worn-out organelles (autophagy), entire cells (autolysis), and extracellular materials.
Peroxisome	Vesicle containing oxidases (oxidative enzymes) and catalase (decomposes hydrogen peroxide); new peroxisomes bud from preexisting ones.	Oxidizes amino acids and fatty acids; detoxifies harmful substances, such as alcohol; produces hydrogen peroxide.
Proteasome	Tiny structure that contains proteases (proteolytic enzymes).	Degrades unneeded, damaged, or faulty proteins by cutting them into small peptides.
Mitochondrion	Consists of outer and inner mitochondrial membranes, cristae, and matrix; new mitochodria form from preexisting ones.	Site of aerobic cellular respiration reactions that produce most of a cell's ATP.
Nucleus	Consists of nuclear envelope with pores, nucleoli, and chromosomes, which exist as a tangled mass of chromatin in interphase cells.	Contains genes, which control cellular structure and direct most cellular functions.

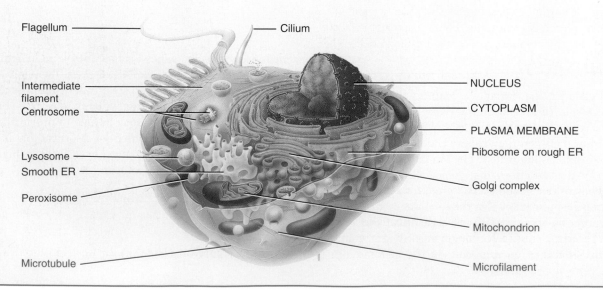

CELL DIVISION

- Discuss the stages, events, and significance of somatic and reproductive cell division.

As cells become damaged diseased, or worn out, they are replaced by **cell division,** the process whereby cells reproduce themselves. Cell division also occurs during growth of tissues. The two types of cell division—somatic cell division and reproductive cell division—accomplish different goals for the organism.

A **somatic cell** (*soma*=body) is any cell of the body other than a germ cell, that is, a gamete (sperm or oocyte) or any precursor cell destined to become a gamete. In **somatic cell division,** a cell undergoes a nuclear division called **mitosis** and a cytoplasmic division called **cytokinesis** to produce two identical cells, each having the same number and kind of chromosomes as the original cell. Somatic cell division replaces dead or injured cells and adds new ones for tissue growth.

Reproductive cell division is the mechanism that produces gametes, the cells needed to form the next generation of sexually reproducing organisms. This process consists of a special two-step division called **meiosis,** in which the number of chromosomes in the nucleus is reduced by half.

Somatic Cell Division

The **cell cycle** is an orderly sequence of events in which a somatic cell duplicates its contents and divides in two. Human somatic cells, such as those in the brain, stomach, and kidneys, contain 23 pairs of chromosomes, or a total of 46 chromosomes. One member of each pair is inherited from each parent. The two chromosomes that make up each pair are called **homologous chromosomes** (hō-MOL-ō-gus; *homo-*=same) or **homologs;** they contain similar genes arranged in the same (or almost the same) order. When examined under a light microscope, homologous chromosomes generally look very similar. The exception to this rule is one pair of chromosomes called the **sex chromosomes,** designated X and Y. In females the homologous pair of sex chromosomes consists of two large X chromosomes; in males the pair consists of an X and a much smaller Y chromosome. Because somatic cells contain two sets of chromosomes, they are called **diploid cells** (DIP-loid; *dipl-*=double; *-oid*=form), symbolized **2n.**

When a cell reproduces, it must replicate (duplicate) all its chromosomes so that its genes may be passed on to the next generation of cells. The cell cycle consists of two major periods: interphase, when a cell is not dividing, and the mitotic (M) phase, when a cell is dividing (Figure 2.17).

Interphase

During **interphase** the cell replicates its DNA. It also produces additional organelles and cytosolic components in anticipation of cell division. Interphase is a state of high metabolic activity, and during this time the cell does most of its growing. Interphase consists of three phases: G_1, S, and G_2 (Figure 2.17). The S stands for *synthesis* of DNA. Because the G-phases are

periods when there is no activity related to DNA duplication, they are thought of as *gaps* or interruptions in DNA duplication.

The **G_1 phase** is the interval between the mitotic phase and the S phase. During G_1, the cell is metabolically active; it duplicates most of its organelles and cytosolic components but not its DNA. Replication of centrosomes also begins in the G_1 phase. Virtually all the cellular activities described in this chapter happen during G_1. For a cell with a total cell cycle time of 24 hours, G_1 lasts 8 to 10 hours. The duration of this phase, however, is quite variable. It is very short in many embryonic cells or cancer cells.

The **S phase** is the interval between G_1 and G_2 and lasts about 8 hours. During the S phase, DNA replication occurs. As a result, the two identical cells formed during cell division will have the same genetic material. The centrosome also replicates during the S phase. The **G_2 phase** is the interval between the S phase and the mitotic phase. It lasts 4 to 6 hours. During G_2, cell growth continues and enzymes and other proteins are synthesized in preparation for cell division and replication of centrosomes is completed. Cells that remain in G_1 for a very long time, perhaps destined never to divide again, are said to be in the **G_0 state.** For example, most nerve cells are in this state. Once a cell enters the S phase, however, it is committed to go through cell division.

When DNA replicates during the S phase, its helical structure partially uncoils, and the two strands separate at the points

Figure 2.17 The cell cycle for a typical body cell with a total cell cycle time of 24 hours. Not illustrated is cytokinesis, division of the cytoplasm, which occurs during late anaphase or early telophase of the mitotic phase.

In a complete cell cycle, a cell duplicates its contents and divides into two identical cells.

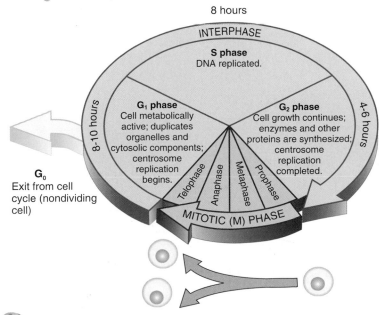

? In which phase of the cell cycle does DNA replication occur?

where hydrogen bonds connect base pairs (Figure 2.18). Each exposed base of the old DNA strand then pairs with the complementary base of a newly synthesized nucleotide. A new DNA strand takes shape as chemical bonds form between neighboring nucleotides. The uncoiling and complementary base pairing continues until each of the two original DNA strands is joined with a newly formed complementary DNA strand. The original DNA molecule has become two identical DNA molecules.

A microscopic view of a cell during interphase shows a clearly defined nuclear envelope, a nucleolus, and a tangled mass of chro-

Figure 2.18 Replication of DNA. The two strands of the double helix separate by breaking the hydrogen bonds (shown as dotted lines) between nucleotides. New, complementary nucleotides attach at the proper sites, and a new strand of DNA is synthesized alongside each of the original strands. Arrows indicate hydrogen bonds forming again between pairs of bases.

Replication doubles the amount of DNA.

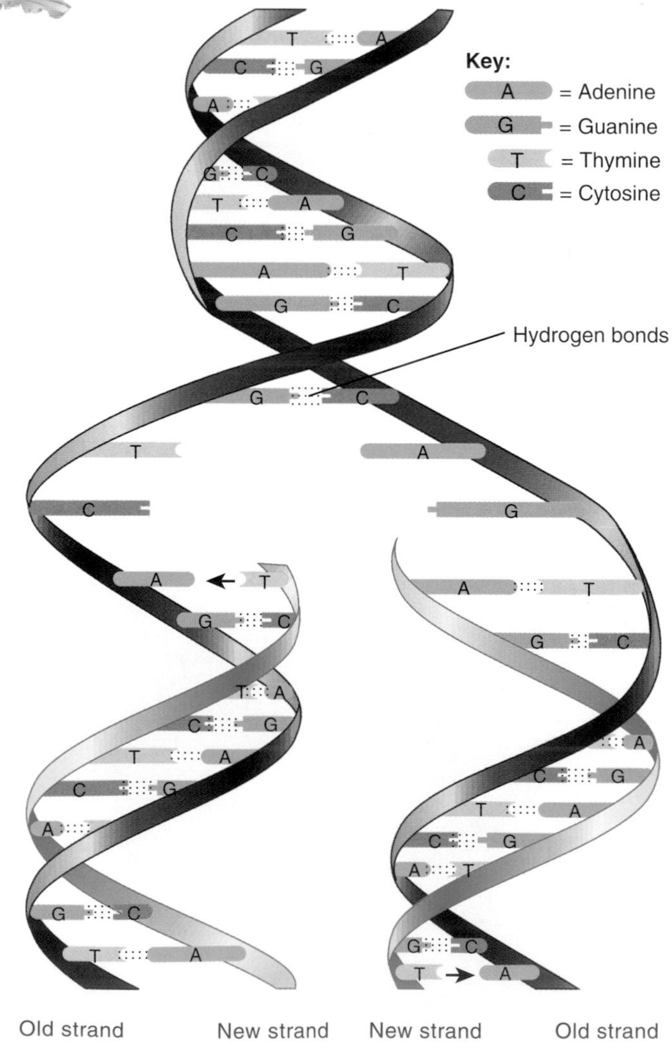

Key:
A = Adenine
G = Guanine
T = Thymine
C = Cytosine

Hydrogen bonds

Old strand   New strand   New strand   Old strand

Why is it crucial that DNA replication occurs prior to cytokinesis in somatic cell division?

matin (Figure 2.19a). Once a cell completes its activities during the G_1, S, and G_2 phase of interphase, the mitotic phase begins.

Mitotic Phase

The **mitotic (M) phase** of the cell cycle consists of a nuclear division, or mitosis, and a cytoplasmic division, or cytokinesis. The events that occur during mitosis and cytokinesis are plainly visible under a microscope because chromatin condenses into discrete chromosomes.

NUCLEAR DIVISION: MITOSIS **Mitosis** is the distribution of the two sets of chromosomes, one set into each of two separate nuclei. The process results in the *exact* partitioning of genetic information. For convenience, biologists divide the process into four stages: prophase, metaphase, anaphase, and telophase. However, mitosis is a continuous process; one stage merges imperceptibly into the next.

1. **Prophase.** During early prophase, the chromatin fibers condense and shorten into chromosomes that are visible under the light microscope (Figure 2.19b). The condensation process may prevent entangling of the long DNA strands as they move during mitosis. Because DNA replication took place during the S phase of interphase, each prophase chromosome consists of a pair of identical, double-stranded **chromatids.** A constricted region called a **centromere** holds the chromatid pair together. At the outside of each centromere is a protein complex known as the **kinetochore** (ki-NET-ō-kor). Later in prophase, tubulins in the pericentriolar material of the centrosomes start to form the **mitotic spindle,** a football-shaped assembly of microtubules that attach to the kinetochore (Figure 2.19b). As the microtubules lengthen, they push the centrosomes to the poles (ends) of the cell so that the spindle extends from pole to pole. The spindle is responsible for the separation of chromatids to opposite poles of the cell. Then, the nucleolus and nuclear envelope break down.

2. **Metaphase.** During metaphase, the microtubules align the centromeres of the chromatid pairs at the exact center of the mitotic spindle (Figure 2.19c). This midpoint region is called the **metaphase plate.**

3. **Anaphase.** During anaphase, the centromeres split, separating the two members of each chromatid pair, which move toward opposite poles of the cell (Figure 2.19d). Once separated, the chromatids are termed chromosomes. As the chromosomes are pulled by the microtubules during anaphase, they appear V-shaped because the centromeres lead the way, dragging the trailing arms of the chromosomes toward the pole.

4. **Telophase.** The final stage of mitosis, telophase, begins after chromosomal movement stops (Figure 2.19e). The identical sets of chromosomes, now at opposite poles of the cell, uncoil and revert to the threadlike chromatin form. A nuclear envelope forms around each chromatin mass, nucleoli reappear in the two identical nuclei, and the mitotic spindle breaks up.

Figure 2.19 Cell division: mitosis and cytokinesis. Begin the sequence at ❶ at the top of the figure and read clockwise until you complete the process.

🔑 **In somatic cell division, a single diploid cell divides to produce two identical diploid cells.**

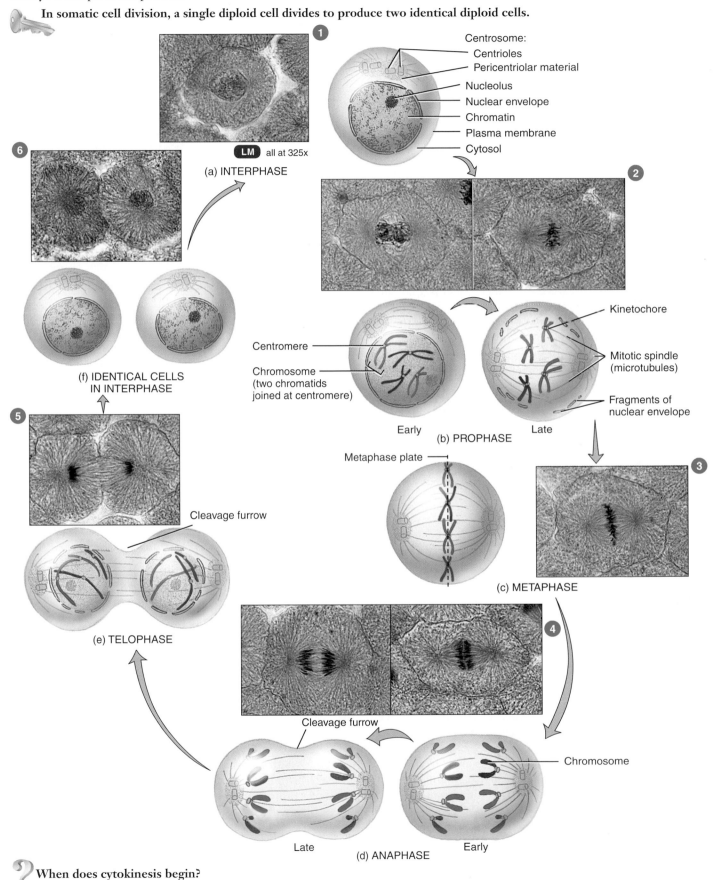

❶

Centrosome:
Centrioles
Pericentriolar material
Nucleolus
Nuclear envelope
Chromatin
Plasma membrane
Cytosol

LM all at 325x

(a) INTERPHASE

❻

❷

Kinetochore

Mitotic spindle (microtubules)

Fragments of nuclear envelope

Centromere

Chromosome (two chromatids joined at centromere)

(f) IDENTICAL CELLS IN INTERPHASE

Early   Late

(b) PROPHASE

❺

Metaphase plate

❸

Cleavage furrow

(c) METAPHASE

(e) TELOPHASE

❹

Chromosome

Cleavage furrow

Late   Early

(d) ANAPHASE

❓ **When does cytokinesis begin?**

CYTOPLASMIC DIVISION: CYTOKINESIS Division of a cell's cytoplasm and organelles into two identical cells is called **cytokinesis** (si′-tō-ki-NĒ-sis; -*kinesis*=motion). This process begins in late anaphase or early telophase with the formation of a **cleavage furrow,** a slight indentation of the plasma membrane. The cleavage furrow usually appears midway between the centrosomes and extends around the periphery of the cell (Figure 2.19d and e). Actin microfilaments that lie just inside the plasma membrane form a *contractile ring* that pulls the plasma membrane progressively inward. The ring constricts the center of the cell, like tightening a belt around the waist, and ultimately pinches it in two. Because the plane of the cleavage furrow is always perpendicular to the mitotic spindle, the two sets of chromosomes end up in separate cells. When cytokinesis is complete, interphase begins (Figure 2.19f).

THE MITOTIC SPINDLE AND CANCER

One of the distinguishing features of cancer cells is that their division is uncontrolled. The mass of cells resulting from this division is called a neoplasm or tumor. One of the ways to treat cancer is by chemotherapy, the use of anticancer drugs. Some of these drugs inhibit the formation of the mitotic spindle and cell division stops. Unfortunately, anticancer drugs also kill all types of rapidly dividing cells in the body, causing side effects such as nausea, diarrhea, hair loss, fatigue, and decreased resistance to disease. ■

Considering the cell cycle in its entirety, the sequence of events is

$$G_1 \longrightarrow S \text{ phase} \longrightarrow G_2 \text{ phase} \longrightarrow \text{mitosis} \longrightarrow \text{cytokinesis}$$

Table 2.3 summarizes the events of the cell cycle in somatic cells.

Control of Cell Destiny

A cell has three possible destinies—to remain alive and functioning without dividing, to grow and divide, or to die. Homeostasis is maintained when there is a balance between cell proliferation and cell death. There are signals that tell a cell when to exist in the G_0 phase, when to divide, and when to die.

Cellular death is also regulated. Throughout the lifetime of an organism, certain cells undergo **apoptosis** (ap-ō-TŌ-sis=a falling off), an orderly, genetically programmed death. In apoptosis, a triggering agent from either outside or inside the cell causes "cell-suicide" genes to produce enzymes that damage the cell in several ways, including disrupting its cytoskeleton and nucleus. As a result, the cell shrinks and pulls away from neighboring cells. The DNA within the nucleus fragments, and the cytoplasm shrinks, although the plasma membrane remains intact. Phagocytes in the vicinity then ingest the dying cell. Apoptosis is especially useful because it removes unneeded cells during development before birth, for example, the separation of webbed digits during fetal development. It continues to occur after birth to regulate the number of cells in a tissue and eliminate potentially dangerous cells such as cancer cells.

Apoptosis, a normal type of cell death, contrasts with **necrosis** (ne-KRŌ-sis=death), a pathological type of cell death that results from tissue injury. In necrosis, many adjacent cells swell, burst, and spill their cytoplasm into the interstitial fluid. The cellular debris usually stimulates an inflammatory response by the immune system, which does not occur in apoptosis.

TUMOR-SUPPRESSOR GENES

Abnormalities in genes that regulate the cell cycle or apoptosis are associated with many diseases. For example, some cancers are caused by damage to genes called **tumor-suppressor genes,** which produce proteins that normally inhibit cell division. Loss or alteration of a tumor-suppressor gene called *p53* on chromosome 17 is the most common genetic change leading to a wide variety of tumors, including breast and colon cancers. The normal p53 protein arrests cells in the G_1 phase, which prevents cell division. Normal p53 protein also assists in repair of damaged DNA and induces apoptosis in the cells where DNA repair was not successful. For this reason, the p53 gene is nicknamed "the guardian angel of the genome." ■

TABLE 2.3	EVENTS OF THE SOMATIC CELL CYCLE
PHASE	**ACTIVITY**
Interphase	Period between cell divisions; chromosomes not visible under light microscope.
G₁ phase	Metabolically active cell duplicates organelles and cytosolic components; replication of centrosomes begins.
S phase	Replication of DNA and centrosomes.
G₂ phase	Cell growth, enzyme and protein synthesis continues; replication of centrosomes complete.
Mitotic phase	Cell produces identical cells with identical chromosomes; chromosomes visible under light microscope.
Mitosis	Nuclear division; distribution of two sets of chromosomes into separate nuclei.
Prophase	Chromatin fibers condense into paired chromatids; nucleolus and nuclear envelope disappear; each centrosome moves to an opposite pole of the cell.
Metaphase	Centromeres of chromatid pairs line up at metaphase plate.
Anaphase	Centromeres split; identical sets of chromosomes move to opposite poles of cell.
Telophase	Nuclear envelopes and nucleoli reappear; chromosomes resume chromatin form; mitotic spindle disappears.
Cytokinesis	Cytoplasmic division; contractile ring forms cleavage furrow around center of cell, dividing cytoplasm into separate and equal portions.

Reproductive Cell Division

In sexual reproduction, each new organism is the result of the union of two different gametes (fertilization), one produced by each parent. If gametes had the same number of chromosomes as somatic cells, the number of chromosomes would double at fertilization. **Meiosis** (mī-Ō-sis; *mei-*=lessening; *-osis*=condition of), the reproductive cell division that occurs in the gonads (ovaries and testes), produces gametes in which the number of chromosomes is reduced by half. As a result, gametes contain a single set of 23 chromosomes and thus are **haploid (*n*) cells** (HAP-loyd; *hapl-*=single). Fertilization then restores the diploid number of chromosomes.

Meiosis

Unlike mitosis, which is complete after a single round, meiosis occurs in two successive stages: **meiosis I** and **meiosis II**. During the interphase that precedes meiosis I, the chromosomes of the diploid starting cell replicate. As a result of replication, each chromosome consists of two genetically identical (sister) chromatids, which are attached at their centromeres. This replication of chromosomes is similar to the one that precedes mitosis in somatic cell division.

MEIOSIS I   Meiosis I, which begins once chromosomal replication is complete, consists of four phases: prophase I, metaphase I, anaphase I, and telophase I (Figure 2.20a). Prophase I is an extended phase in which the chromosomes shorten and thicken, the nuclear envelope and nucleoli disappear, and the mitotic spindle forms. Two events that are not seen in mitotic prophase occur during prophase I of meiosis. First, the two sister chromatids of each pair of homologous chromosomes pair off, an event called **synapsis** (Figure 2.20b). The resulting four chromatids form a structure called a **tetrad**. Second, parts of the chromatids of two homologous chromosomes may be exchanged with one another. Such an exchange between parts of genetically different (nonsister) chromatids is termed **crossing-over** (Figure 2.20b). This process, among others, permits an exchange of genes between chromatids of homologous chromosomes. Due to crossing-over, the resulting cells are genetically unlike each other and genetically unlike the parent cell that produced them. Crossing-over results in **genetic recombination**—that is, the formation of new combinations of genes—and accounts for part of the great genetic variation among humans and other organisms that form gametes via meiosis.

In metaphase 1, the tetrads formed by homologous pairs of chromosomes line up along the metaphase plate of the cell, with homologous chromosomes side by side (Figure 2.20a). During anaphase I, the members of each homologous pair of chromosomes separate as they are pulled to opposite poles of the cell by the kinetochore microtubules attached to the centromeres. The paired chromatids, held by a centromere, remain together. (Recall that during mitotic anaphase, the centromeres split and the sister chromatids separate.) Telophase I and cytokinesis of meiosis are similar to telophase and cytokinesis of mitosis. The net effect of meiosis I is that each resulting cell contains the haploid number of chromosomes because it contains only one member of each pair of the homologous chromosomes present in the parent cell.

MEIOSIS II   The second stage of meiosis, meiosis II, also consists of four phases: prophase II, metaphase II, anaphase II, and telophase II (Figure 2.20d). These phases are similar to those that occur during mitosis; the centromeres split, and the sister chromatids separate and move toward opposite poles of the cell.

In summary, meiosis I begins with a diploid starting cell and ends with two cells, each with the haploid number of chromosomes. During meiosis II, each of the two haploid cells formed during meiosis I divides, and the net result is four haploid gametes that are genetically different from the original diploid starting cell.

Figure 2.21 on page 49 compares the events of meiosis and mitosis.

CHECKPOINT

13. Distinguish between the somatic and reproductive types of cell division. Why is each important?
14. Define interphase. When does DNA replicate?
15. What are the major events of each stage of the mitotic phase?
16. How are apoptosis and necrosis similar and different?
17. How are haploid (*n*) and diploid (*2n*) cells different?
18. What are homologous chromosomes?

CELLULAR DIVERSITY

OBJECTIVE

● Describe how cells differ in size and shape.

The body of an average human adult is composed of nearly 100 trillion cells. All of these cells can be classified into about 200 different cell types. Cells vary considerably in size. High-powered microscopes are needed to see the smallest cells of the body. The largest cell, a single oocyte, is barely visible to the unaided eye. The sizes of cells are measured in units called *micrometers*. One micrometer (μm) is equal to 1 one-millionth of a meter, or 10^{-6}m (1/25,000 of an inch). Whereas a red blood cell has a diameter of 8 μm, an oocyte has a diameter of about 140 μm.

The shapes of cells also vary considerably (Figure 2.22 on page 50). They may be round, oval, flat, cuboidal, columnar, elongated, star-shaped, cylindrical, or disc-shaped. A cell's shape is related to its function in the body. For example, a sperm cell has a long whip-like tail (flagellum) that it uses for locomotion. The disc shape of a red blood cell gives it a large surface area that enhances its ability to pass oxygen to other cells. The long, spindle shape of a relaxed smooth muscle cell shortens as it contracts. This change in shape allows groups of smooth muscle cells to narrow or widen the passage for blood flowing through blood vessels. In this way, they regulate blood flow through various tissues. Some cells contain microvilli, which greatly increase their surface area. Microvilli are common in the epithelial cells that line the small intestine, where the large surface area speeds the absorption of digested food.

Figure 2.20 Meiosis, reproductive cell division. Details of events are discussed in the text.

In reproductive cell division, a single diploid cell undergoes meiosis I and meiosis II to produce four haploid gametes that are genetically different from the diploid cell that produced them.

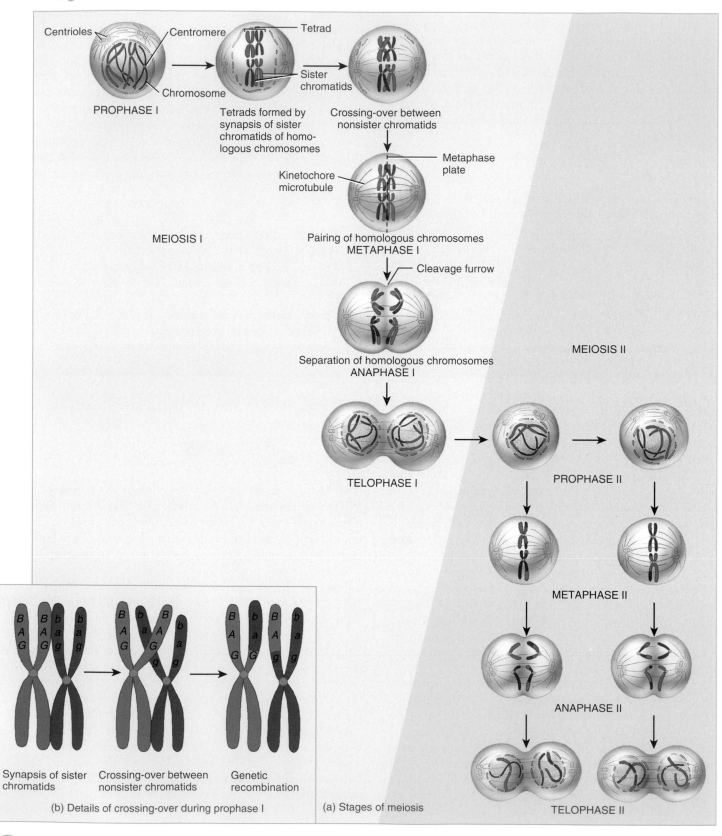

Centrioles Centromere Tetrad

PROPHASE I

Chromosome

Tetrads formed by synapsis of sister chromatids of homologous chromosomes

Sister chromatids

Crossing-over between nonsister chromatids

MEIOSIS I

Kinetochore microtubule

Metaphase plate

Pairing of homologous chromosomes
METAPHASE I

Cleavage furrow

Separation of homologous chromosomes
ANAPHASE I

MEIOSIS II

TELOPHASE I

PROPHASE II

METAPHASE II

ANAPHASE II

TELOPHASE II

Synapsis of sister chromatids

Crossing-over between nonsister chromatids

Genetic recombination

(b) Details of crossing-over during prophase I

(a) Stages of meiosis

How does crossing-over affect the genetic content of the four haploid gametes?

Figure 2.21   Comparison between mitosis (left) and meiosis (right) in which the parent cell has two pairs of homologous chromosomes.

The phases of meiosis II and mitosis are similar.

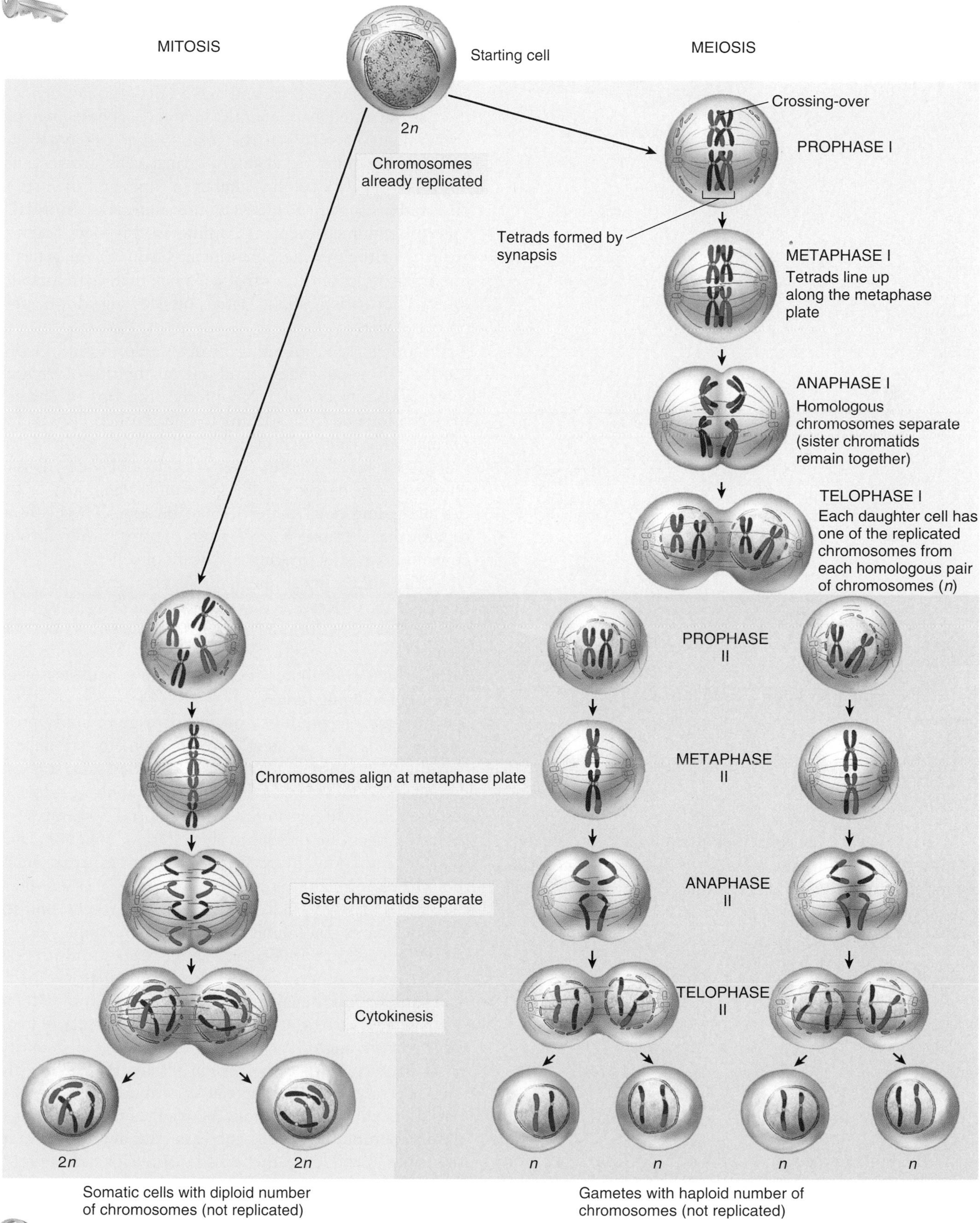

MITOSIS

MEIOSIS

Starting cell

2n

Chromosomes already replicated

Crossing-over

PROPHASE I

Tetrads formed by synapsis

METAPHASE I
Tetrads line up along the metaphase plate

ANAPHASE I
Homologous chromosomes separate (sister chromatids remain together)

TELOPHASE I
Each daughter cell has one of the replicated chromosomes from each homologous pair of chromosomes (n)

PROPHASE II

Chromosomes align at metaphase plate

METAPHASE II

Sister chromatids separate

ANAPHASE II

Cytokinesis

TELOPHASE II

2n   2n   n   n   n   n

Somatic cells with diploid number of chromosomes (not replicated)

Gametes with haploid number of chromosomes (not replicated)

How does anaphase I of meiosis differ from anaphase of mitosis and anaphase II of meiosis?

Figure 2.22 Diverse shapes and sizes of human cells. The relative difference in size between the smallest and largest cells of the human body is actually much greater than shown here.

🔑 **The nearly 100 trillion cells in an average adult can be classified into about 200 different cell types.**

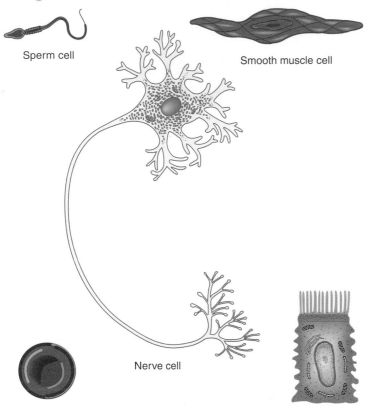

Sperm cell

Smooth muscle cell

Nerve cell

Red blood cell

Epithelial cell

❓ **Why are sperm the only body cells that need a flagellum?**

Nerve cells have long extensions that permit them to conduct nerve impulses over great distances. As you will see in the next several chapters, cellular diversity also permits organization of cells into more complex tissues and organs.

CHECKPOINT

19. How is cell shape related to function?

AGING AND CELLS

OBJECTIVE

• Describe the cellular changes that occur with aging.

Aging is a normal process accompanied by a progressive alteration of the body's homeostatic adaptive responses. It produces observable changes in structure and function and increases vulnerability

to environmental stress and disease. The specialized branch of medicine that deals with the medical problems and care of elderly persons is **geriatrics** (jer′-ē-AT-riks; *ger-*=old age; *-iatrics*= medicine). **Gerontology** (jer′-on-TOL-ō-jē) is the scientific study of the process and problems associated with aging.

Although many millions of new cells normally are produced each minute, several kinds of cells in the body—heart muscle cells, skeletal muscle cells, and nerve cells—do not divide because they are arrested permanently in the G_0 phase (see page 43). Experiments have shown that many other cell types have only a limited capability to divide. Normal cells grown outside the body divide only a certain number of times and then stop. These observations suggest that cessation of mitosis is a normal, genetically programmed event. According to this view, "aging genes" are part of the genetic blueprint at birth. These genes have an important function in normal cells but their activities slow over time. They bring about aging by slowing down or halting processes vital to life.

Another aspect of aging involves **telomeres** (TĒ-lō-merz), specific DNA sequences found only at the tips of each chromosome. These pieces of DNA protect the tips of chromosomes from erosion and from sticking to one another. However, in most normal body cells each cycle of cell division shortens the telomeres. Eventually, after many cycles of cell division, the telomeres can be completely gone and even some of the functional chromosomal material may be lost. These observations suggest that erosion of DNA from the tips of our chromosomes contributes greatly to aging and death of cells.

Glucose, the most abundant sugar in the body, plays a role in the aging process. It is haphazardly added to proteins inside and outside cells, forming irreversible cross-links between adjacent protein molecules. With advancing age, more cross-links form, which contributes to the stiffening and loss of elasticity that occur in aging tissues.

Free radicals produce oxidative damage in lipids, proteins, or nucleic acids by "stealing" an electron to accompany their unpaired electrons. Some effects are wrinkled skin, stiff joints, and hardened arteries. Normal cellular metabolism—for example, aerobic cellular respiration in mitochondria—produces some free radicals. Others are present in air pollution, radiation, and certain foods we eat. Naturally occurring enzymes in peroxisomes and in the cytosol normally dispose of free radicals. Certain dietary substances, such as vitamin E, vitamin C, betacarotene, and selenium, are antioxidants that inhibit free radical formation.

Whereas some theories of aging explain the process at the cellular level, others concentrate on regulatory mechanisms operating within the entire organism. For example, the immune system may start to attack the body's own cells. This *autoimmune response* might be caused by changes in cell-identity markers at the surface of cells (certain plasma membrane glycoproteins and glycolipids) that cause antibodies to attach to and mark the cell for destruction. As changes in the proteins on the plasma membrane of cells increase, the autoimmune response intensifies, producing the well-known signs of aging. In the chapters that follow, we will discuss the effects of aging on each body system.

PROGERIA AND WERNER SYNDROME

Progeria (prō-JER-ē-a) is a noninherited disease characterized by normal development in the first year of life followed by rapid aging. The condition is expressed by dry and wrinkled skin, total baldness, and birdlike facial features. Death usually occurs around age 13. Progeria is caused by a genetic defect in which telomeres are considerably shorter than usual.

Werner syndrome is a rare, inherited disease that causes a rapid acceleration of aging, usually while the person is only in his or her twenties. It is characterized by wrinkling of the skin, graying of the hair and baldness, cataracts, muscular atrophy, and a tendency to develop diabetes mellitus, cancer, and cardiovascular disease. Most afflicted individuals die before age 50. Recently, the gene that causes Werner syndrome has been identified. Researchers hope to use this information to gain insight into the mechanisms of aging. ■

CHECKPOINT

20. What is one reason that some tissues become stiffer as they age?

APPLICATIONS TO HEALTH

CANCER

Cancer is a group of diseases characterized by uncontrolled or abnormal cell proliferation. When cells in a part of the body divide without control, the excess tissue that develops is called a **tumor** or **neoplasm** (NĒ-ō-plazm; *neo-*=new). The study of tumors is called **oncology** (on-KOL-ō-jē; *onco-*=swelling or mass). Tumors may be cancerous and often fatal, or they may be harmless. A cancerous neoplasm is called a **malignant tumor or malignancy.** One property of most malignant tumors is their ability to undergo **metastasis** (me-TAS-ta-sis), the spread of cancerous cells to other parts of the body. A **benign tumor** is a neoplasm that does not metastasize. An example is a wart. A benign tumor may be removed surgically if it interferes with normal body function or becomes disfiguring.

Growth and Spread of Cancer

Cells of malignant tumors duplicate rapidly and continuously. As malignant cells invade surrounding tissues, they often trigger **angiogenesis,** the growth of new networks of blood vessels. Proteins that stimulate angiogenesis in tumors are called **tumor angiogenesis factors (TAFs).** However, inhibitors of angiogenesis may also be present. Thus, the formation of new blood vessels can occur either by overproduction of TAFs or by the lack of naturally occurring angiogenesis inhibitors. As the cancer grows, it begins to compete with normal tissues for space and nutrients. Eventually, the normal tissue decreases in size and dies. Some malignant cells may detach from the initial (primary) tumor and invade a body cavity or enter the blood or lymph, then circulate to and invade other body tissues, establishing secondary tumors. The pain associated with cancer develops when the tumor presses on nerves, blocks a passageway in an organ so that secretions build up pressure, or as a result of dying tissues or organs.

Causes of Cancer

Several factors may trigger a normal cell to lose control and become cancerous. One cause is environmental agents: substances in the air we breathe, the water we drink, and the food we eat. A chemical agent or radiation that produces cancer is called a **carcinogen** (car-SIN-ō-jen). Carcinogens induce **mutations,** permanent changes in the DNA base sequence of a gene. The World Health Organization estimates that carcinogens are associated with 60–90% of all human cancers. Examples of carcinogens are hydrocarbons found in cigarette tar, radon gas from the earth, and ultraviolet (UV) radiation in sunlight.

Intensive research efforts are now directed toward studying cancer-causing genes, or **oncogenes** (ON-kō-jēnz). These genes, when inappropriately activated, have the ability to transform a normal cell into a cancerous cell. Most oncogenes derive from normal genes called **proto-oncogenes** that regulate growth and development. The proto-oncogene undergoes some change that either causes it to produce an abnormal product or disrupts its control. It may be expressed inappropriately or make its products in excessive amounts or at the wrong time. Some oncogenes cause excessive production of growth factors, chemicals that stimulate cell growth. Others may trigger changes in a cell-surface receptor, causing it to send signals as though it were being activated by a growth factor. As a result, the growth pattern of the cell becomes abnormal.

Proto-oncogenes in every cell carry out normal cellular functions until a malignant change occurs. It appears that some proto-oncogenes are activated to oncogenes by mutations in which the DNA of the proto-oncogene is altered. Other proto-oncogenes are activated by a rearrangement of the chromosomes so that segments of DNA are exchanged. Rearrangement activates proto-oncogenes by placing them near genes that enhance their activity.

Some cancers have a viral origin. Viruses are tiny packages of nucleic acids, either RNA or DNA, that can reproduce only while inside the cells they infect. Some viruses, termed **oncogenic viruses,** cause cancer by stimulating abnormal proliferation of cells. For instance, the *human papillomavirus (HPV)* causes virtually all cervical cancers in women. The virus produces a protein that causes proteasomes to destroy the p53 protein, (a protein that normally suppresses unregulated cell division.) In the absence of this suppressor protein, cells proliferate without control.

Carcinogenesis: A Multistep Process

Carcinogenesis (kar′-si-nō-JEN-e-sis), the process by which cancer develops, is a multistep process in which as many as 10 distinct mutations may have to accumulate in a cell before it becomes cancerous. The progression of genetic changes leading to cancer is best understood for colon (colorectal) cancer. Such cancers, as well as lung and breast cancer, take years or decades to develop. In colon cancer, the tumor begins as an area of increased cell proliferation that results from one mutation. This growth then progresses to abnormal, but noncancerous, growths called adenomas. After two or three additional mutations, a mutation of the tumor-suppressor gene *p53* occurs and a carcinoma develops. The fact that so many mutations are needed for a cancer to develop indicates that cell growth is normally controlled with many sets of checks and balances.

Treatment of Cancer

Many cancers are removed surgically. However, when cancer is widely distributed throughout the body or exists in organs such as the brain whose functioning would be greatly harmed by surgery, chemotherapy and radiation therapy may be used instead. Sometimes surgery, chemotherapy, and radiation therapy are used in combination. Chemotherapy involves administering drugs that cause death of cancerous cells. Radiation therapy breaks chromosomes, thus blocking cell division. Because cancerous cells divide rapidly, they are more vulnerable to the destructive effects of chemotherapy and radiation therapy than are normal cells. Unfortunately for the patients, hair follicle cells, red bone marrow cells, and cells lining the gastrointestinal tract also are rapidly dividing. Hence, the side effects of chemotherapy and radiation therapy include hair loss due to death of hair follicle cells, vomiting and nausea due to death of cells lining the stomach and intestines, and susceptibility to infection due to slowed production of white blood cells in red bone marrow.

Treating cancer is difficult because it is not a single disease and because the cells in a single tumor population rarely behave all in the same way. Although most cancers are thought to derive from a single abnormal cell, by the time a tumor reaches a clinically detectable size, it may contain a diverse population of abnormal cells. For example, some cancerous cells metastasize readily, and others do not. Some are sensitive to chemotherapy drugs and some are drug resistant. Because of differences in drug resistance, a single chemotherapeutic agent may destroy susceptible cells but permit resistant cells to proliferate.

KEY MEDICAL TERMS ASSOCIATED WITH CELLS

NOTE TO THE STUDENT

Most chapters in this text are followed by a glossary of key medical terms that include both normal and pathological conditions. You should familiarize yourself with the terms because they will play an essential role in your medical vocabulary.

Some of these disorders, as well as disorders discussed in the text, are referred to as local or systemic. A *local disease* is one that affects one part or a limited area of the body. A *systemic disease* affects either the entire body or several parts.

The science that deals with why, when, and where diseases occur and how they are transmitted in a human community is known as **epidemiology** (ep′-i-dē-mē-OL-ō-jē; *epidemios-*=prevalent; *-logos*=study of). The science that deals with the effects and uses of drugs in the treatment of disease is called **pharmacology** (far′-ma-KOL-ō-jē; *pharmakon-*=medicine; *-logos*=study of).

Anaplasia (an′-a-PLĀ-zē-a; *an-*=not; *-plasia*=to shape) The loss of tissue differentiation and function that is characteristic of most malignancies.

Atrophy (AT-rō-fē; *a-*=without; *-trophy*=nourishment) A decrease in the size of cells, with a subsequent decrease in the size of the affected tissue or organ; wasting away.

Dysplasia (dis-PLĀ-zē-a; *dys-*=abnormal) Alteration in the size, shape, and organization of cells due to chronic irritation or inflammation; may progress to neoplasia (tumor formation, usually malignant) or revert to normal if the irritation is removed.

Hyperplasia (hī-per-PLĀ-zē-a; *hyper-*=over) Increase in the number of cells of a tissue due to an increase in the frequency of cell division.

Hypertrophy (hī-PER-trō-fē) Increase in the size of cells without cell division.

Karyotype (KAR-ē-ō-tīp; *karyo-*=nucleus) The chromosomal characteristics of an individual presented as a systematic arrangement of pairs of metaphase chromosomes arrayed in descending order of size and according to the position of the centromere; useful in judging whether chromosomes are normal in number and structure.

Metaplasia (met′-a-PLĀ-zē-a; *meta-*=change) The transformation of one type of cell into another.

Progeny (PROJ-e-nē; *pro-*=forward; *-geny*=production) Offspring or descendants.

Tumor marker A substance introduced into circulation by tumor tissue that indicates the presence of a tumor as well as the specific type. Tumor markers may be used to screen, diagnose, assess prognosis, evaluate a response to treatment, and monitor for recurrence of cancer.

STUDY OUTLINE

Introduction (p. 24)

1. A cell is the basic, living, structural and functional unit of the body.
2. Cell biology is the scientific study of cellular structure and function.

A Generalized Cell (p. 25)

1. Figure 2.1 shows a cell that is a composite of many different cells in the body.
2. The principal parts of a cell are the plasma membrane; the cytoplasm, which consists of cytosol and organelles; and the nucleus.

The Plasma Membrane (p. 26)

Structure of the Membrane (p. 26)

1. The plasma membrane surrounds and contains the cytoplasm of a cell.
2. According to the fluid mosaic model, the membrane is a mosaic of proteins floating like icebergs in a bilayer sea of lipids.
3. The lipid bilayer consists of two back-to-back layers of phospholipids, cholesterol, and glycolipids.
4. Integral proteins extend into or through the plasma membrane and are firmly embedded in the membrane.
5. Many integral proteins are glycoproteins that have sugar groups attached to the ends that face the extracellular fluid. Together with glycolipids, the glycoproteins form a glycocalyx on the extracellular surface of cells.
6. Peripheral proteins are not as firmly embedded in the plasma membrane and are attached to membrane lipid or integral proteins and the inner or outer surface of the membrane.

Functions of Membrane Proteins (p. 27)

1. Membrane proteins have a variety of functions. Ion channels and transporters are membrane proteins that help specific solutes across the membrane; receptors serve as cellular recognition sites; and linkers anchor proteins in the plasma membranes to filaments inside and outside the cell. Some membrane proteins are enzymes and others are cell-identity markers.
2. The membrane's selective permeability permits some substances to pass more readily than others. The lipid bilayer is permeable to water and molecules such as oxygen, carbon dioxide, and steroids. Channels and transporters increase the plasma membrane's permeability to small- and medium-sized substances, including ions, that cannot cross the lipid bilayer.

Transport Across the Plasma Membrane (p. 27)

1. Passive processes depend on the concentration of substances and their kinetic energy.
2. Simple diffusion is the net movement of a substance from an area of higher concentration to an area of lower concentration until an equilibrium is reached.
3. In facilitated diffusion, certain molecules, such as glucose, move through the membrane with the help of integral membrane proteins.
4. Osmosis is the movement of water through a selectively permeable membrane from an area of higher water concentration to an area of lower water concentration.
5. Filtration is the movement of water and dissolved substances across a membrane due to gravity or hydrostatic pressure.
6. Active processes depend on the use of ATP by the cell.

7. Active transport is the movement of a substance (usually ions) across a cell membrane from lower to higher concentration using energy derived from ATP and a carrier protein.
8. Transport in vesicles includes both endocytosis and exocytosis.
9. Receptor-mediated endocytosis is the selective uptake of large molecules and particles (ligands) that bind to specific receptors in membrane areas called clathrin-coated pits.
10. Phagocytosis is the ingestion of solid particles. It is an important process used by some white blood cells to destroy bacteria that enter the body.
11. Bulk-phase endocytosis is the ingestion of extracellular fluid. In this process, the fluid becomes surrounded by a pinocytic vesicle.
12. Exocytosis involves movement of secretory or waste products out of a cell by fusion of vesicles with the plasma membrane.

Cytoplasm (p. 30)

1. Cytoplasm is all the cellular contents between the plasma membrane and the nucleus.
2. Cytoplasm consists of cytosol and organelles.

Cytosol (p. 30)

1. Cytosol (intracellular fluid) is the fluid portion of cytoplasm.
2. Cytosol contains mostly water, plus ions, glucose, amino acids, fatty acids, proteins, lipids, ATP, and waste products.
3. Cytosol is the site of many chemical reactions required for a cell's existence.

Organelles (p. 31)

1. Organelles are specialized structures with characteristic shapes and specific functions.
2. The cytoskeleton is a network of several kinds of protein filaments that extend throughout the cytoplasm. Components of the cytoskeleton are microfilaments, intermediate filaments, and microtubules. The cytoskeleton provides a structural framework for the cell and is responsible for cell movements.
3. The centrosome consists of pericentriolar material and centrioles. The pericentriolar material organizes microtubules in nondividing cells and the mitotic spindle in dividing cells.
4. Cilia and flagella are motile projections of the cell surface. Cilia move fluids along the cell surface. Flagella move an entire cell.
5. Ribosomes, composed of ribosomal RNA and ribosomal proteins, consist of two subunits made in the nucleus. Free ribosomes are not attached to any cytoplasmic structure; membrane-bound ribosomes are attached to endoplasmic reticulum. Ribosomes are sites of protein synthesis.
6. Endoplasmic reticulum (ER) is a network of membranes in the form of flattened sacs or tubules; it extends from the nuclear envelope throughout the cytoplasm.
7. Rough ER is covered by ribosomes, and its primary function is protein synthesis. It also forms glycoproteins and attaches proteins to phospholipids, which it also synthesizes.
8. Smooth ER lacks ribosomes. It synthesizes fatty acids and steroids; releases glucose from the liver into the bloodstream; inactivates or detoxifies drugs and other potentially harmful substances; and releases calcium ions that trigger contraction in muscle cells.
9. The Golgi complex consists of flattened sacs called cisternae that differ in size, shape, content, enzymatic activity, and types of vesi-

cles. Entry face cisternae face the rough ER, exit face cisternae face the plasma membrane, and medial cisternae are between the two. The Golgi complex receives synthesized products from the rough ER and modifies, sorts, packages, and transports them to different destinations.

10. Some processed proteins leave the cell in secretory vesicles, some leave in membrane vesicles and are incorporated into the plasma membrane, and some leave in transport vesicles and enter lysosomes.

11. Lysosomes are membrane-enclosed vesicles that form from the Golgi complex and contain digestive enzymes. Lysosomes function in digestion of worn-out organelles (autophagy), digestion of a host cell (autolysis), and extracellular digestion.

12. Peroxisomes are similar to lysosomes, but smaller. Peroxisomes oxidize various organic substances such as amino acids, fatty acids, and toxic substances and, in the process, produce hydrogen peroxide. The hydrogen peroxide produced from oxidation is degraded by an enzyme in peroxisomes called catalase.

13. Proteasomes continually degrade unneeded, damaged, or faulty proteins. Their proteases cut proteins into small peptides.

14. Mitochondria consist of a smooth outer membrane, an inner membrane containing cristae, and a large central cavity filled with a fluid called the matrix. Mitochondria are called the "powerhouses" of the cell because they produce ATP.

Nucleus (p. 39)

1. The nucleus consists of a double nuclear envelope; nuclear pores, which control the movement of substances between the nucleus and cytoplasm; nucleoli, which produce ribosomes; and genes arranged on chromosomes. Genes control cellular structure and most cellular functions.

2. Human somatic cells have 46 chromosomes, 23 inherited from each parent. The total genetic information carried in a cell or an organism is its genome.

Cell Division (p. 43)

1. Cell division is the process by which cells reproduce themselves. It consists of nuclear division (mitosis or meiosis) and cytoplasmic division (cytokinesis).

2. Cell division that replaces cells or adds new ones is called somatic cell division and involves a nuclear division called mitosis plus cytokinesis.

3. Cell division that results in the production of gametes (sperm and ova) is called reproductive cell division and consists of a nuclear division called meiosis plus cytokinesis.

Somatic Cell Division (p. 43)

1. Human somatic cells contain 23 pairs of chromosomes (homologous chromosomes) and are diploid ($2n$).

2. Before the mitotic phase, the DNA molecules, or chromosomes, replicate themselves so that identical chromosomes can be passed on to the next generation of cells.

3. A cell that is between divisions and is carrying on every life process except division is said to be in interphase, which consists of three phases: G_1, S, and G_2.

4. During the G_1 phase, the cell duplicates its organelles and cytosolic components and centrosome replication begins; during the S phase, DNA replication occurs; during the G_2 phase, enzymes and other proteins are synthesized and centrosome replication is complete.

5. Mitosis is the splitting of the cell chromosomes and the distribution of the two identical sets of chromosomes into separate and equal nuclei; mitosis consists of prophase, metaphase, anaphase, and telophase.

6. Cytokinesis usually begins in late anaphase and ends in telophase.

7. A cleavage furrow forms at the cell's metaphase plate and progresses inward, pinching in through the cell to form two separate portions of cytoplasm.

Control of Cell Destiny (p. 46)

1. A cell can either remain alive and functioning without dividing, grow and divide, or die.

2. Apoptosis is programmed cell death, a normal type of cell death. It first occurs during embryological development and continues for the lifetime of an organism.

3. Certain genes regulate both cell division and apoptosis. Abnormalities in these genes are associated with a wide variety of diseases and disorders.

Reproductive Cell Division (p. 47)

1. In sexual reproduction, each new organism is the result of the union of gametes produced in the ovaries and testes.

2. Gametes contain a single set of chromosomes (23) and thus are haploid (n).

3. Meiosis is the process that produces haploid gametes; it consists of two successive nuclear divisions called meiosis I and meiosis II.

4. During meiosis I, homologous chromosomes undergo synapsis (pairing) and crossing-over; the net result is two haploid cells that are genetically unlike each other and unlike the starting diploid cell that produced them.

5. During meiosis II, the two haploid cells divide to form four haploid cells.

Cellular Diversity (p. 47)

1. The almost 200 cell types in the body vary considerably in size and shape.

2. The sizes of cells are measured in micrometers. One micrometer equals 10^{-6}m (1/25,000 of an inch).

3. A cell's shape is related to its function.

Aging and Cells (p. 50)

1. Aging is a normal process accompanied by progressive alteration of the body's homeostatic adaptive responses.

2. Many theories of aging have been proposed, including genetically programmed cessation of cell division, the buildup of free radicals, and an intensified autoimmune response.

Q ■ SELF-QUIZ QUESTIONS

Choose the one best answer to the following questions.

1. The fluid portion of the cytoplasm is called the:
   a. cytosol   b. interstitial fluid   c. plasma
   d. cisternae   e. cristae

2. Respiratory gases move between the lungs and blood by means of:
   a. simple diffusion   b. osmosis   c. active transport
   d. phagocytosis   e. bulk-phase endocytosis

3. Which of the following is *not* a passive form of transport?
   a. phagocytosis   b. osmosis   c. simple diffusion
   d. facilitated diffusion   e. filtration

4. The glycocalyx is a surface coat on cells that:
   a. aids the movement of red blood cells through small blood vessels.
   b. consists of carbohydrate portions of membrane glycolipids and glycoproteins.
   c. serves as a recognition signal for other body cells.
   d. facilitates the adherence of cells to each other in some tissues.
   e. All of the above are correct.

5. Cytokinesis is:
   a. a passive form of transmembrane transport.
   b. how macrophages engulf bacteria.
   c. the separation of chromatid pairs during metaphase.
   d. the division of the cytoplasm during somatic cell division.
   e. the development of flagella on sperm.

6. As a result of somatic cell division, each cell produced has:
   a. half as many chromosomes as the original cell.
   b. twice as many chromosomes as the original cell.
   c. exactly the same number of chromosomes as the original cell.
   d. one-quarter as many chromosomes as the original cell.
   e. none of the above.

7. The process by which worn-out cells or whole bacteria are engulfed and destroyed is called:
   a. exocytosis   b. hemolysis   c. phagocytosis
   d. autolysis   e. autophagy

8. Division of the nucleus during somatic cell division is called:
   a. apoptosis   b. mitosis   c. meiosis
   d. metastasis   e. metaphase

Complete the following.

9. Solvents move down a pressure gradient across a selectively permeable membrane by a passive process called _____.

10. Reproductive cell division consists of a special two-step process of division called _____.

11. Clathrin-coated pits participate in a membrane transport process called _____.

12. The plasma membrane is composed of two main chemical components: a bilayer of _____ and protein.

13. Specific molecules that bind to receptors are called _____.

14. The extracellular fluid filling the microscopic spaces between the cells of tissues is called _____.

15. The net movement of water molecules through a selectively permeable membrane from an area of higher water concentration to an area of lower water concentration is called _____.

Are the following statements true or false?

16. Cholesterol molecules make up the largest percentage of lipid molecules in the plasma membrane.

17. Peripheral proteins span the entire lipid bilayer.

18. Facilitated diffusion is considered an active form of membrane transport because it requires the assistance of a transporter protein.

Matching

19. Match the following:

___ **(a)** directs cellular activities by means of genes located here

___ **(b)** sites of protein synthesis; may occur attached to ER or scattered freely in cytoplasm

___ **(c)** system of membranous channels providing pathways for transport within cell and surface areas for chemical reactions; may be rough or smooth

___ **(d)** stacks of cisternae; involved in packaging and secretion of glycolipids, glycoproteins, and lipoproteins

___ **(e)** may release enzymes that lead to autolysis of the cell

___ **(f)** similar to lysosomes, but smaller; contain the enzyme catalase

___ **(g)** cristae-containing structures, called "powerhouses of the cell" because ATP production occurs within them

___ **(h)** part of cytoskeleton; peripherally located in cytoplasm; involved with cell migration and contraction

___ **(i)** part of cytoskeleton; give shape to cell; found in flagella, cilia, centrioles, and mitotic spindle

___ **(j)** helps organize mitotic spindle used in cell division

___ **(k)** long, hairlike structures that help move an entire cell

___ **(l)** short, hairlike structures that move particles over cell surface

(1) centrosome
(2) cilia
(3) endoplasmic reticulum
(4) flagella
(5) lysosomes
(6) Golgi complex
(7) microfilaments
(8) mitochondria
(9) nucleus
(10) microtubules
(11) ribosomes
(12) peroxisomes

20. Match the following:

___ **(a)** time during mitosis which chromosomes move toward opposite poles of the cell

___ **(b)** the result of reproductive cell division

___ **(c)** time during mitosis which chromosomes uncoil into chromatin, the mitotic spindle breaks up, and the nuclear envelope appears

___ **(d)** spherical bodies that function in producing ribosomes

___ **(e)** time during interphase when a cell is duplicating organelles and cytosolic components, but not DNA

___ **(f)** cells that contain 23 pairs of chromosomes

___ **(g)** members of a chromosome pair, one inherited from each parent

___ **(h)** time during mitosis when the nuclear envelope disappears, chromatin thickens into distinct chromosomes, and mitotic spindle forms

___ **(i)** proteins that form the core around which DNA wraps in chromatin

___ **(j)** time during interphase when DNA replication occurs

___ **(k)** the genetically programmed death of certain cells

___ **(l)** time during mitosis which chromatids line up on the equator of the mitotic spindle

(1) S phase
(2) anaphase
(3) metaphase
(4) prophase
(5) telophase
(6) homologs
(7) gametes
(8) diploid cells
(9) apoptosis
(10) G_1 phase
(11) histones
(12) nucleoli

CRITICAL THINKING QUESTIONS

1. Jim had some blood work done recently. When his white blood cells were counted, the test indicated that 95% of them were neutrophils; the normal percentage of neutrophils in a white blood cell count is 75%. What does the increased percentage of neutrophils in Jim's blood suggest?

2. You may inherit your brown eyes from either your mother or your father but some traits can only be passed down from mother to child. "Maternal inheritance" is due to genetic material that is not located in the nucleus. Can you suggest an explanation?

3. In the "old days," intravenous solutions did not come "prepackaged" and had to be mixed at the hospital. Maureen wasn't very good at arithmetic and misplaced her decimal point when calculating how much glucose to add to the intravenous solution for her patient. Instead of making a 0.9% saline solution, she made a 9.0% saline solution. Using your knowledge of osmosis, predict what would happen to her patient's blood cells if she injected this solution into her patient's bloodstream.

4. A child was brought to the emergency room after eating rat poison containing arsenic. Arsenic kills rats by blocking the function of the mitochondria. What effect would the poison have on the child's body functions?

5. Imagine that a new chemotherapy agent has been discovered that disrupts microtubules in cancerous cells but leaves normal cells unaffected. What effect would this agent have on the cancer cells?

ANSWERS TO FIGURE QUESTIONS

2.1 The three principal parts of the cell are the plasma membrane, cytoplasm, and nucleus.

2.2 The glycocalyx is a coat on the extracellular surface of the plasma membrane that is composed of the carbohydrate portions of membrane glycolipids and glycoproteins.

2.3 Another name for intracellular fluid is cytosol.

2.4 Cholesterol, iron, vitamins, and hormones are examples of ligands.

2.5 The binding of particles to a plasma membrane receptor triggers pseudopod formation.

2.6 Microtubules help form the structure of centrioles, cilia, and flagella.

2.7 A cell without a centrosome probably would not be able to undergo cell division.

2.8 Cilia move fluids across cell surfaces, whereas flagella move entire cells.

2.9 Large and small ribosomal subunits are synthesized in the nucleolus in the nucleus and then join together in the cytoplasm.

2.10 Rough ER has attached ribosomes, whereas smooth ER does not. Rough ER synthesizes proteins that will be exported from the cell; smooth ER is associated with lipid synthesis and other metabolic reactions.

2.11 The entry face receives and modifies proteins from rough ER while the exit face modifies, sorts, packages, and transports molecules.

2.12 Some proteins are discharged from the cell by exocytosis, some are incorporated into the plasma membrane, and some occupy storage vesicles that become lysosomes.

2.13 Digestion of worn-out organelles by lysosomes is called autophagy.

2.14 Mitochondrial cristae increase the surface area available for chemical reactions and contain the enzymes needed for ATP production.

2.15 Chromatin is a complex of DNA, proteins, and some RNA.

2.16 A nucleosome is a double-stranded molecule of DNA wrapped twice around a core of eight histones (proteins).

2.17 Chromosomes replicate during the S phase.

2.18 DNA replication occurs before cytokinesis so that each of the new cells will have a complete set of genes.

2.19 Cytokinesis usually starts in late anaphase or early telophase.

2.20 The result of crossing-over is that the cells that follow are genetically unlike each other and genetically unlike the starting cell that produced them.

2.21 During anaphase I of meiosis, the paired chromatids are held together by a centromere and do not separate. During anaphase II of meiosis, the centromeres split and the paired chromatids separate.

2.22 Sperm, which use the flagella for locomotion, are the only body cells required to move considerable distances.

TISSUES 3

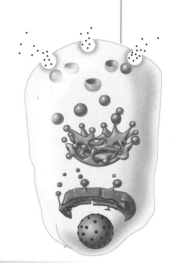

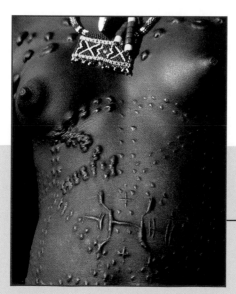

INTRODUCTION   As you learned in Chapter 2, a cell is a complex collection of compartments, each of which carries out a host of biochemical reactions that make life possible. However, a cell seldom functions as an isolated unit in the body. Instead, cells usually work together in groups called tissues. A **tissue** is a group of similar cells that usually have a common embryonic origin and function together to carry out specialized activities. The structure and properties of a specific tissue are influenced by factors such as the nature of the extracellular material that surrounds the tissue cells and the connections between the cells that compose the tissue. Tissues may be hard, semisolid, or even liquid in their consistency, a range exemplified by bone, fat, and blood. In addition, tissues vary tremendously with respect to the kinds of cells present, how the cells are arranged, and the types of fibers present, if any.

Do you have a scar as the result of a healed wound? Do you try to conceal it? Would it surprise you to know that some cultures actually induce scarring as a form of body art and expression?

www.wiley.com/college/apcentral

Histology (hiss'-TOL-ō-jē; *hist-*=tissue; *-ology*=study of) is the science that deals with the study of tissues. A **pathologist** (pa-THOL-ō-gist; *patho-*=disease) is a physician who specializes in laboratory studies of cells and tissues to help other physicians make accurate diagnoses. One of the principal functions of a pathologist is to examine tissues for any changes that might indicate disease.

TYPES OF TISSUES AND THEIR ORIGINS

OBJECTIVE

- Name the four basic types of tissues that make up the human body, and state the characteristics of each.

Body tissues can be classified into four basic types according to their function and structure:

1. **Epithelial tissue** covers body surfaces and lines hollow organs, body cavities, and ducts. It also forms glands.
2. **Connective tissue** protects and supports the body and its organs. Various types of connective tissue bind organs together, store energy reserves as fat, and help provide immunity to disease-causing organisms.
3. **Muscular tissue** generates the physical force needed to make body structures move.
4. **Nervous tissue** detects changes in a variety of conditions inside and outside the body and responds by generating nerve impulses. The nervous tissue in the brain helps to maintain homeostasis.

Epithelial tissue and connective tissue, except for bone tissue and blood, are discussed in detail in this chapter. The general features of bone tissue and blood will be introduced here, but their detailed discussion is presented in Chapters 6 and 13, respectively. Similarly, the structure and function of muscular tissue and nervous tissue are examined in detail in Chapters 10 and 17, respectively.

As you will see in the next chapter, tissues of the body develop from three **primary germ layers,** which are called the **ectoderm, endoderm,** and **mesoderm,** the first tissues formed in a human embryo. Epithelial tissues develop from all three primary germ layers. All connective tissues and most muscle tissues derive from mesoderm. Nervous tissue develops from ectoderm. (Figure 4-7b on page 103 illustrates the primary germ layers and Table 4-1 on page 104 provides a list of structures derived from the primary germ layers.)

Normally, most cells within a tissue remain anchored to other cells, to basement membranes (described shortly), and to connective tissues. A few cells, such as phagocytes, move freely through the body, searching for invaders. Before birth, however, many cells migrate extensively as part of the growth and development process.

BIOPSY

A **biopsy** (BĪ-op-sē; *bio-*=life; *-opsy*=to view) is the removal of a sample of living tissue for microscopic examination. This procedure is used to help diagnose many disorders, especially cancer, and to discover the cause of unexplained infections and inflammations. Both normal and potentially diseased tissues are removed for purposes of comparison. Once the tissue sample is removed, either surgically or through a needle and syringe, it may be preserved, stained to highlight special properties, or cut into thin sections for microscopic observation. Sometimes a biopsy is conducted while a patient is anesthetized during surgery to help a physician determine the most appropriate treatment. For example, if a biopsy of thyroid tissue reveals malignant cells, the surgeon can immediately proceed with the most appropriate procedure. ■

CHECKPOINT

1. Define a tissue.
2. What are the four basic types of human tissue?

CELL JUNCTIONS

OBJECTIVE

- Describe the structure and functions of the five main types of cell junctions.

Most epithelial cells and some muscle and nerve cells are tightly joined into functional units. **Cell junctions** are contact points between the plasma membranes of tissue cells. Depending on their structure, cell junctions may serve one of three functions: (1) forming seals between cells, like a "zip-lock" at the top of a plastic storage bag, (2) anchoring cells to one another or to extracellular material, or (3) providing channels that allow ions and molecules to pass from cell to cell within a tissue. Here we consider the five most important types of cell junctions: tight junctions, adherens junctions, desmosomes, hemidesmosomes, and gap junctions (Figure 3.1).

Tight Junctions

Tight junctions consist of weblike strands of transmembrane proteins that fuse the outer surfaces of adjacent plasma membranes together (Figure 3.1a). Cells of epithelial tissues that line the stomach, intestines, and urinary bladder have many tight junctions. They retard the passage of substances between cells and thus prevent the contents of these organs from leaking into the blood or surrounding tissues.

Adherens Junctions

Adherens junctions (ad-HER-ens) contain **plaque,** a dense layer of proteins on the inside of the plasma membrane that attaches to both cytoskeleton proteins and membrane proteins (Figure 3.1b). Actin microfilaments extend from the plaque into the cell's cytosol. Transmembrane glycoproteins called **cadherins**

Figure 3.1 Cell junctions.

Most epithelial cells and some muscle and nerve cells contain cell junctions.

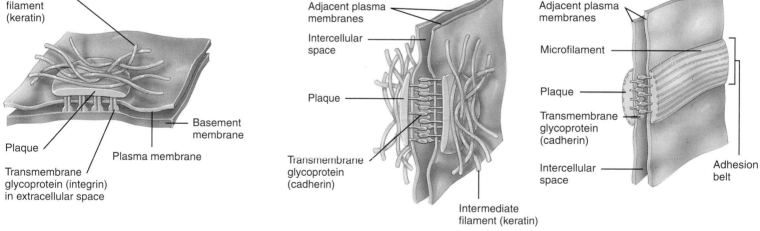

(c) Gap junction

(a) Tight junction

(d) Hemidesmosome

(c) Desmosome

(b) Adherens junction

Which type of junction functions in communication between adjacent cells?

insert into the plaque from the opposite side. Each cadherin partially crosses the intercellular space (the space between the cells) and connects to cadherins of the adjacent cell, thus attaching the two cells. In epithelial cells, adherens junctions often form extensive zones called *adhesion belts* that encircle the cell. Adherens junctions help epithelial surfaces resist separation during various contractile processes, for example, in the intestines as food moves through them for processing.

Desmosomes

Like adherens junctions, **desmosomes** (DEZ-mō-sōms; *desmo-* = band) contain plaque and have transmembrane glycoproteins (cadherins) that extend into the intercellular space between adjacent cell membranes and attach cells to one another (Figure 3.1c). Intermediate filaments (made of keratin) extend from desmosomes on one side of the cell across the cytosol to desmosomes on the opposite side of the cell. This structural arrangement contributes to the stability of the cells and tissue. Desmosomes are spotweld-like junctions and are common among the cells that make up the epidermis (the outermost layer of the skin) and between cardiac muscle cells in the heart. Desmosomes prevent epidermal cells from separating under tension and cardiac muscle cells from pulling apart during contraction.

Hemidesmosomes

Hemidesmosomes (*hemi-*=half) resemble desmosomes but they lack links to adjacent cells. They look like half of a desmosome, thus the name (Figure 3.1d). However, the transmembrane glycoproteins in hemidesmosomes are **integrins** rather than cadherins. On the inside of the plasma membrane, integrins attach to intermediate filaments made of the protein keratin. On the outside of the plasma membrane, the integrins attach to the protein *laminin*, which is present in the basement membrane (discussed shortly). Thus, hemidesmosomes anchor cells to the basement membrane.

Gap Junctions

At **gap junctions**, membrane proteins called **connexins** form tiny fluid-filled tunnels called **connexons** that connect neighboring cells (Figure 3.1e). The plasma membranes of gap junctions are not fused together as in tight junctions but are separated by a very narrow intercellular gap (space). Ions and small molecules can diffuse through connexons from the cytosol of one cell to another. Gap junctions allow the cells in a tissue to communicate with each other. In a developing embryo, some of the chemical and electrical signals that regulate growth and cell differentiation travel via gap junctions. Moreover, gap junctions enable nerve or muscle impulses to spread rapidly among cells, a process that is crucial for the normal operation of some parts of the nervous system and for the contraction of muscle in the heart and gastrointestinal tract.

CHECKPOINT

3. Which type of cell junctions allows cellular communication?
4. Which types of cell junctions are found in epithelial tissues?

EPITHELIAL TISSUE

OBJECTIVES

● Describe the general features of epithelial tissues.
● List the location, structure, and function of each different type of epithelium.

An **epithelial tissue** (ep-i-THĒ-lē-al), or **epithelium** (plural is *epithelia*), consists of cells arranged in continuous sheets, in either single or multiple layers. Because the cells are closely packed and are held tightly together by many cell junctions, there is little intercellular space between adjacent plasma membranes.

The various surfaces of epithelial cells often differ in structure and have specialized functions. The **apical (free) surface** of an epithelial cell faces the body surface, a body cavity, the lumen

(interior space) of an internal organ, or a tubular duct that receives secretions from the cells (Figure 3.2). Apical surfaces may contain cilia or microvilli. The **lateral surfaces** of an epithelial cell face the adjacent cells on either side. As you have seen in Figure 3.1, lateral surfaces may contain tight junctions, adherens junctions, desmosomes, and/or gap junctions. The **basal surface** of an epithelial cell is opposite the apical surface and adheres to extracellular materials. Hemidesmosomes in the basal surfaces of epithelial cells anchor the epithelium to the basement membrane. In subsequent discussions about epithelia with multiple layers the term *apical layer* refers to the most superficial layer of cells whereas the *basal layer* is the deepest layer of cells.

The **basement membrane** is a thin extracellular layer that commonly consists of two layers, the basal lamina and reticular lamina. The *basal lamina* (*lamina*=thin layer) is closer to the epithelial cells and is secreted by them. It contains proteins such as collagen (described shortly) and laminin, as well as glycoproteins and proteoglycans (also described shortly). The laminin molecules in the basal lamina adhere to integrins in hemidesmosomes and thus attach epithelial cells to the basement membrane (see Figure 3.1d). The *reticular lamina* is closer to the underlying connective tissue and contains fibrous proteins produced by connective tissue cells called fibroblasts. Besides their function of attaching to and supporting the overlying epithelial tissue, basement membranes have other roles. They form a surface along which epithelial cells migrate during growth or wound healing, they restrict passage of larger molecules between epithelium and connective tissue, and they participate in filtration of blood in the kidneys.

Figure 3.2   Surfaces of epithelial cells and the structure and location of the basement membrane.

The basement membrane is found between epithelium and connective tissue.

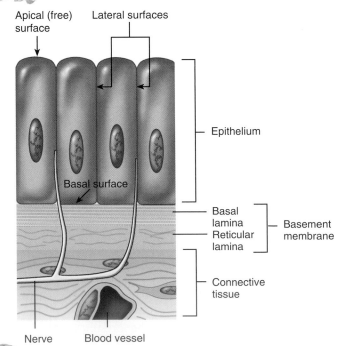

What are the functions of the basement membrane?

BASEMENT MEMBRANES AND DISEASE

Under certain conditions, basement membranes become markedly thickened, due to increased production of collagen and laminin. In untreated cases of diabetes mellitus, the basement membrane of small blood vessels (capillaries) thickens, especially those in the eyes and kidneys. As a result the blood vessels do not function properly and blindness and kidney failure may result. ■

Epithelial tissue is **avascular** (*a-*=without; *vascular*=vessel); that is, it lacks its own blood supply. The blood vessels that bring in nutrients and remove wastes are located in the adjacent connective tissue. Exchange of substances between epithelium and connective tissue occurs by diffusion. Although epithelial tissue is avascular, it has a nerve supply.

Because epithelial tissue forms boundaries between the body's organs, or between the body and the external environment, it is repeatedly subjected to physical stress and injury. A high rate of cell division, however, allows epithelial tissue to constantly renew and repair itself by sloughing off dead or injured cells and replacing them with new ones. Epithelial tissue plays many different roles in the body, the most important of which are protection, filtration, secretion, absorption, and excretion. In addition, epithelial tissue combines with nervous tissue to form special organs for smell, hearing, vision, and touch.

Epithelial tissue may be divided into two types. **Covering and lining epithelium** forms the outer covering of the skin and some internal organs. It also forms the inner lining of blood vessels, ducts, and body cavities, and the interior of the respiratory, digestive, urinary, and reproductive systems. **Glandular epithelium**

constitutes the secreting portion of glands such as the thyroid gland, adrenal glands, and sweat glands.

Covering and Lining Epithelium

The types of covering and lining epithelial tissue are classified according to two characteristics: the arrangement of cells into layers and the shapes of the cells (Figure 3.3).

1. ***Arrangement of cells in layers.*** The cells of covering and lining epithelia are arranged in one or more layers depending on the functions the epithelium performs:

   a. *Simple epithelium* is a single layer of cells that functions in diffusion, osmosis, filtration, secretion (the production and release of substances such as mucus, sweat, or enzymes), and absorption (the intake of fluids or other substances by cells).

   b. *Stratified* (*stratum*=layer) *epithelium* consists of two or more layers of cells that protect underlying tissues in locations where there is considerable wear and tear.

   Pseudostratified epithelium appears to have multiple layers of cells because the cell nuclei lie at different levels and not all cells reach the apical surface. Cells that do extend to the apical surface may contain cilia; others (goblet cells) secrete mucus. Pseudostratified epithelium is actually a simple epithelium and all cells rest on the basement membrane.

2. ***Cell shapes.***

   a. *Squamous cells* (SKWĀ-mus=flat) are arranged like floor tiles and are thin, which allows for the rapid movement of substances through them.

Figure 3.3 Cell shapes and arrangement of layers for covering and lining epithelium.

🔑 Cell shapes and arrangement of layers are the bases for classifying covering and lining epithelium.

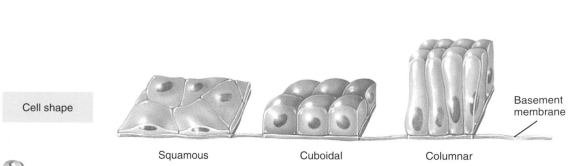

Which cell shape is best adapted for the rapid movement of substances from one cell to another?

b. *Cuboidal cells* are as tall as they are wide and are shaped like cubes or hexagons. They may have microvilli at their apical surface and function in either secretion or absorption.

c. *Columnar cells* are much taller than they are wide and protect underlying tissues. Their apical surface may have cilia or microvilli, and they often are specialized for secretion and absorption.

d. *Transitional cells* change shape, from cuboidal to flat and back, as organs stretch (distend) to a larger size then collapse to a smaller size.

Considering the arrangements of layers and the cell shapes, the types of covering and lining epithelia are:

I. Simple epithelium
   A. Simple squamous epithelium
   B. Simple cuboidal epithelium
   C. Simple columnar epithelium (nonciliated and ciliated)
   D. Pseudostratified columnar epithelium (nonciliated and ciliated)

II. Stratified epithelium
   A. Stratified squamous epithelium (keratinized and nonkeratinized)*
   B. Stratified cuboidal epithelium*
   C. Stratified columnar epithelium*
   D. Transitional epithelium

Each of these covering and lining epithelial tissues is described in the following sections and illustrated in Table 3.1. The tables throughout this chapter contain figures consisting of a photomicrograph, a corresponding diagram, and an inset that identifies a major location of the tissue in the body. Along with the illustrations are descriptions, locations, and functions of the tissues.

Simple Epithelium

SIMPLE SQUAMOUS EPITHELIUM This tissue consists of a single layer of flat cells that resembles a tiled floor when viewed from

———————

* This classification is based on the shape of the cells in the apical layer.

TABLE 3.1 EPITHELIAL TISSUES

COVERING AND LINING EPITHELIUM

A. Simple squamous epithelium

Description Single layer of flat cells; centrally located nucleus.

Location Lines heart, blood vessels, lymphatic vessels, air sacs of lungs, glomerular (Bowman's) capsule of kidneys, and inner surface of the tympanic membrane (eardrum); forms epithelial layer of serous membranes, such as the peritoneum.

Function Filtration, diffusion, osmosis, and secretion in serous membranes.

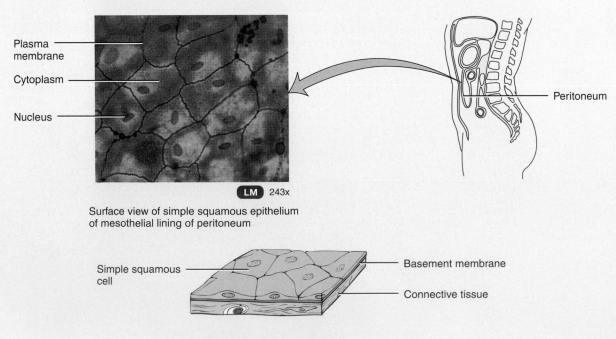

Plasma membrane
Cytoplasm
Nucleus
Peritoneum

LM 243x

Surface view of simple squamous epithelium of mesothelial lining of peritoneum

Simple squamous cell
Basement membrane
Connective tissue

Simple squamous epithelium

their apical surface (Table 3.1A). The nucleus of each cell is a flattened oval or sphere and is centrally located. Simple squamous epithelium is present at sites where the processes of filtration, such as blood filtration in the kidneys, or diffusion, for instance, diffusion of oxygen into blood vessels of the lungs, are occurring. It is not found in body areas that are subject to wear and tear.

The simple squamous epithelium that lines the heart, blood vessels, and lymphatic vessels is known as **endothelium** (*endo-*= within; *-thelium*=covering); the type that forms the epithelial layer of serous membranes (such as the peritoneum) is called **mesothelium** (*meso-*=middle). Unlike other epithelial tissues, which arise from embryonic ectoderm or endoderm, endothelium and mesothelium both are derived from embryonic mesoderm.

SIMPLE CUBOIDAL EPITHELIUM The cuboidal shape of the cells in this tissue (Table 3.1B) is obvious when the tissue is sectioned and viewed from the side. Cell nuclei are usually round and centrally located. Simple cuboidal epithelium performs the functions of secretion and absorption.

SIMPLE COLUMNAR EPITHELIUM When viewed from the side, the cells appear rectangular with oval nuclei near the base of the cells. Simple columnar epithelium exists in two forms: nonciliated simple columnar epithelium and ciliated simple columnar epithelium.

Nonciliated simple columnar epithelium contains two types of cells—columnar epithelial cells with microvilli at their apical surface and goblet cells (Table 3.1C). **Microvilli** are microscopic fingerlike cytoplasmic projections that increase the surface area of the plasma membrane (see Figure 2.1 on page 25). Their presence increases the rate of absorption by the cell. **Goblet cells** are modified columnar epithelial cells that secrete mucus, a slightly sticky fluid, at their apical surfaces. Before it is released, mucus accumulates in the upper portion of the cell, causing that area to bulge out. The whole cell then resembles a goblet or wine glass. Secreted mucus serves as a lubricant for the linings of the digestive, respiratory, reproductive, and most of the urinary tracts. Mucus also helps prevent destruction of the stomach lining by acidic gastric juice secreted by the stomach.

Ciliated simple columnar epithelium contains columnar epithelial cells with cilia at their apical surface (Table 3.1D). In certain parts of the airways of the upper respiratory tract, goblet cells are interspersed among ciliated columnar cells. Mucus secreted by the goblet cells forms a film over the airway surface that traps

B. Simple cuboidal epithelium

Description Single layer of cube-shaped cells; centrally located nucleus.

Location Covers surface of ovary, lines anterior surface of capsule of the lens of the eye, forms the pigmented epithelium at the posterior surface of the eye, lines kidney tubules and smaller ducts of many glands, and makes up the secreting portion of some glands such as the thyroid gland and the ducts of some glands such as the pancreas.

Function Secretion and absorption.

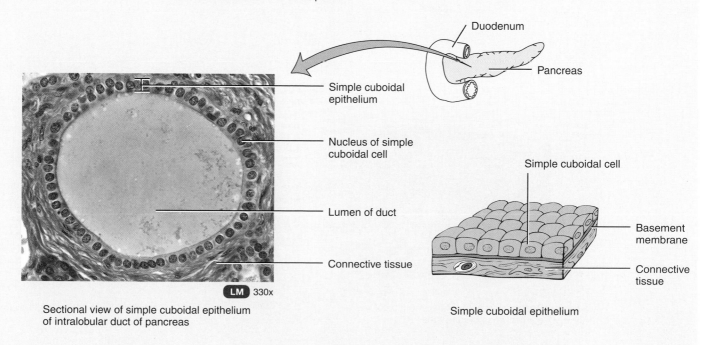

Duodenum

Pancreas

Simple cuboidal epithelium

Nucleus of simple cuboidal cell

Lumen of duct

Connective tissue

LM 330x

Sectional view of simple cuboidal epithelium of intralobular duct of pancreas

Simple cuboidal cell

Basement membrane

Connective tissue

Simple cuboidal epithelium

continues

TABLE 3.1   EPITHELIAL TISSUES (continued)

COVERING AND LINING EPITHELIUM

C. Nonciliated simple columnar epithelium

Description   Single layer of nonciliated rectangular cells with nuclei near base of cells; contains goblet cells and absorptive cell with microvilli in some locations.

Location   Lines the gastrointestinal tract from the stomach to the anus, ducts of many glands, and gallbladder.

Function   Secretion and absorption.

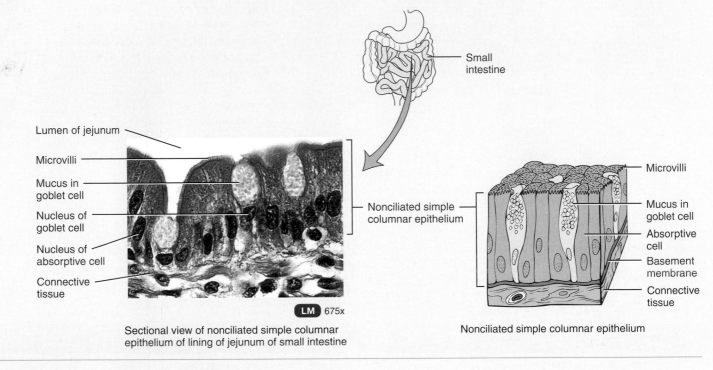

Small intestine

Lumen of jejunum

Microvilli

Mucus in goblet cell

Nucleus of goblet cell

Nucleus of absorptive cell

Connective tissue

Nonciliated simple columnar epithelium

LM 675x

Sectional view of nonciliated simple columnar epithelium of lining of jejunum of small intestine

Microvilli

Mucus in goblet cell

Absorptive cell

Basement membrane

Connective tissue

Nonciliated simple columnar epithelium

D. Ciliated simple columnar epithelium

Description   Single layer of ciliated rectangular cells with nuclei near base of cells; contains goblet cells in some locations.

Location   Lines a few portions of upper respiratory tract, uterine (fallopian) tubes, uterus, some paranasal sinuses, and central canal of spinal cord.

Function   Moves mucus and other substances by ciliary action.

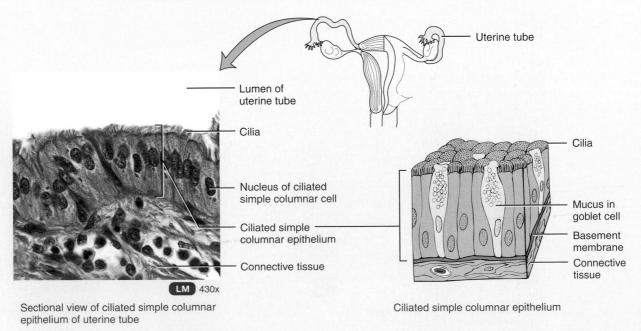

Uterine tube

Lumen of uterine tube

Cilia

Nucleus of ciliated simple columnar cell

Ciliated simple columnar epithelium

Connective tissue

LM 430x

Sectional view of ciliated simple columnar epithelium of uterine tube

Cilia

Mucus in goblet cell

Basement membrane

Connective tissue

Ciliated simple columnar epithelium

E. Pseudostratified columnar epithelium

Description Not a true stratified tissue; nuclei of cells are at different levels; all cells are attached to basement membrane, but not all reach the apical surface.

Location Pseudostratified ciliated columnar epithelium lines the airways of most of upper respiratory tract; pseudostratified nonciliated columnar epithelium lines larger ducts of many glands, epididymis, and part of male urethra.

Function Secretion and movement of mucus by ciliary action.

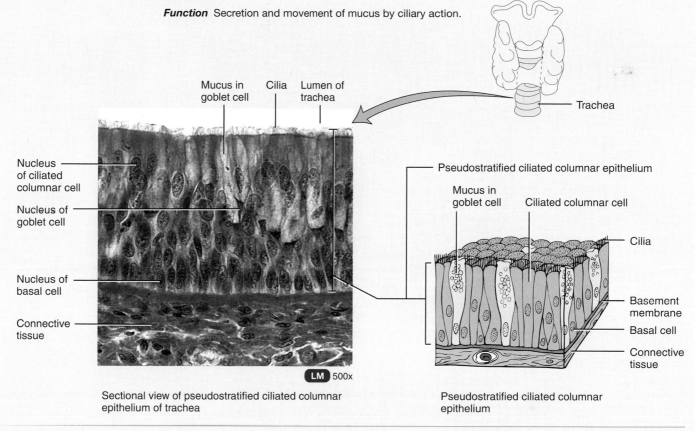

Sectional view of pseudostratified ciliated columnar epithelium of trachea

Pseudostratified ciliated columnar epithelium

F. Stratified squamous epithelium

Description Several layers of cells; cuboidal to columnar shape in deep layers; squamous cells form the apical layer and several layers deep to it; cells from the basal layer replace surface cells as they are lost.

Location Keratinized variety forms superficial layer of skin; nonkeratinized variety lines wet surfaces, such as lining of the mouth, esophagus, part of epiglottis, and vagina, and covers the tongue.

Function Protection.

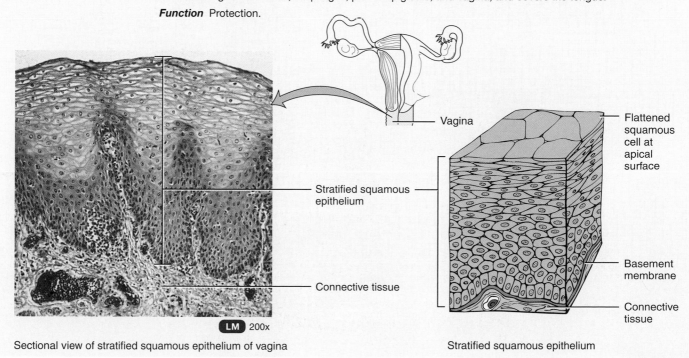

Sectional view of stratified squamous epithelium of vagina

Stratified squamous epithelium

continues

TABLE 3.1   EPITHELIAL TISSUES (continued)

COVERING AND LINING EPITHELIUM

G. Stratified cuboidal epithelium

Description Two or more layers of cells in which the cells in the apical layer are cube-shaped.

Location Ducts of adult sweat glands and esophageal glands and part of male urethra.

Function Protection and limited secretion and absorption.

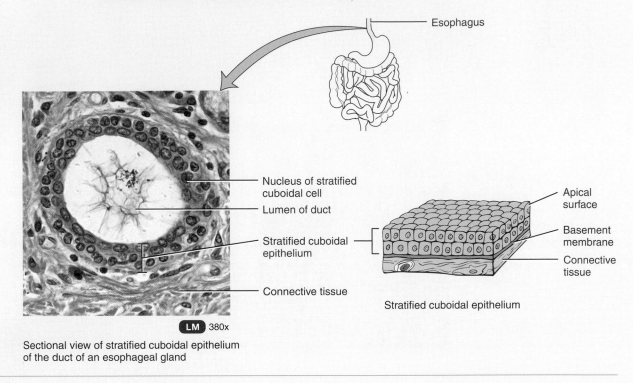

Sectional view of stratified cuboidal epithelium
of the duct of an esophageal gland

H. Stratified columnar epithelium

Description Several layers of irregularly shaped cells; columnar cells are only in the apical layer.

Location Lines part of urethra, large excretory ducts of some glands, such as esophageal glands, small areas in anal mucous membrane, and part of the conjunctiva of the eye.

Function Protection and secretion.

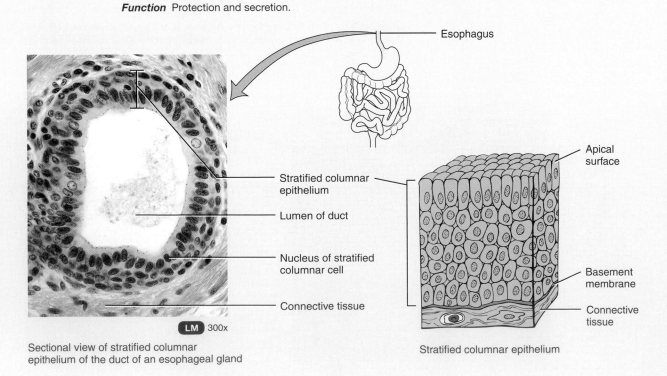

Sectional view of stratified columnar
epithelium of the duct of an esophageal gland

I. Transitional epithelium

Description Appearance is variable (transitional); shape of cells in apical layer ranges from squamous (when stretched) to cuboidal (when relaxed).

Location Lines urinary bladder and portions of ureters and urethra.

Function Permits distention.

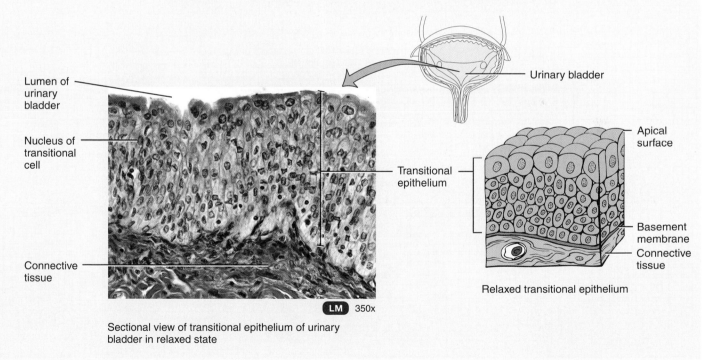

Lumen of urinary bladder

Nucleus of transitional cell

Connective tissue

LM 350x

Sectional view of transitional epithelium of urinary bladder in relaxed state

Urinary bladder

Transitional epithelium

Apical surface

Basement membrane

Connective tissue

Relaxed transitional epithelium

continues

TABLE 3.1 EPITHELIAL TISSUES (continued)
GLANDULAR EPITHELIUM

J. Endocrine glands

Description Secretory products (hormones) diffuse into blood after passing through interstitial fluid.

Location Examples include pituitary gland at base of brain, pineal gland in brain, thyroid and parathyroid glands near larynx (voice box), adrenal glands superior to kidneys, pancreas near stomach, ovaries in pelvic cavity, testes in scrotum, and thymus in thoracic cavity.

Function Produce hormones that regulate various body activities.

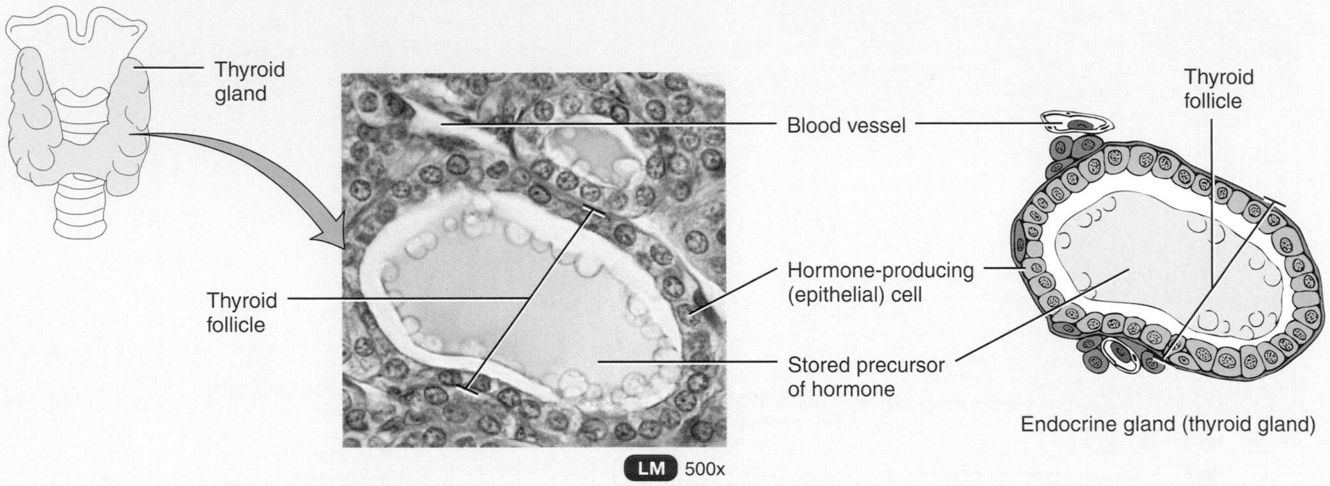

Thyroid gland
Blood vessel
Thyroid follicle
Thyroid follicle
Hormone-producing (epithelial) cell
Stored precursor of hormone
Endocrine gland (thyroid gland)

LM 500x

Sectional view of endocrine gland (thyroid gland)

K. Exocrine Glands

Description Secretory products released into ducts.

Location Sweat, oil, and earwax glands of the skin; digestive glands such as salivary glands, which secrete into mouth cavity, and pancreas, which secretes into the small intestine.

Function Produce substances such as sweat, oil, earwax, saliva, or digestive enzymes.

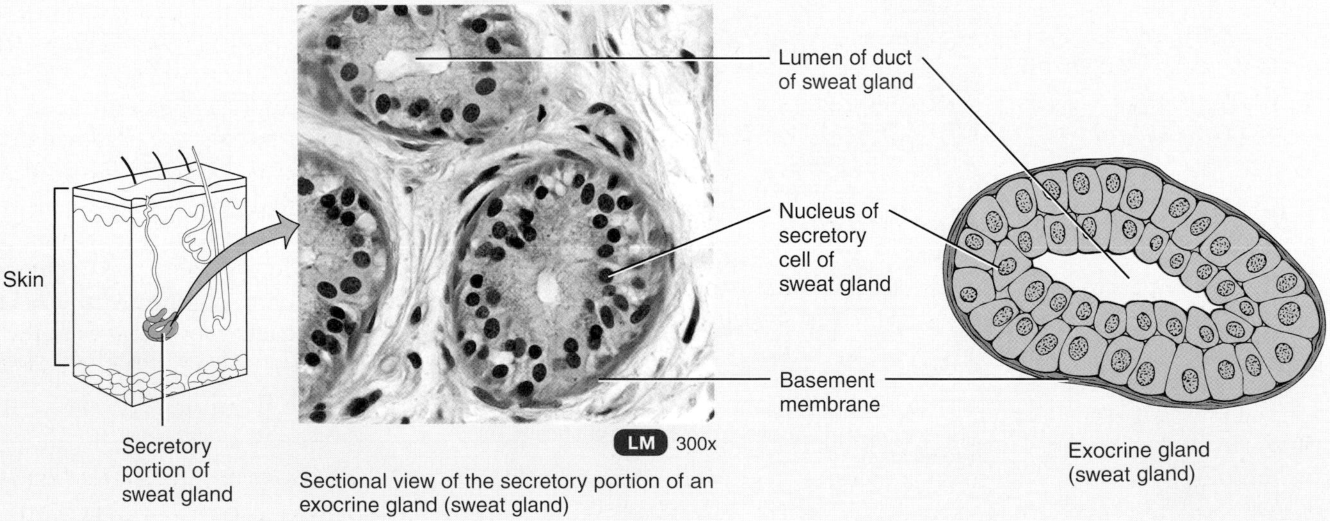

Skin
Secretory portion of sweat gland
Lumen of duct of sweat gland
Nucleus of secretory cell of sweat gland
Basement membrane
Exocrine gland (sweat gland)

LM 300x

Sectional view of the secretory portion of an exocrine gland (sweat gland)

inhaled foreign particles. The cilia beat in unison, moving the mucus and any foreign particles toward the throat, where it can be coughed up and swallowed or spit out. Coughing and sneezing speed up the movement of cilia and mucus. Cilia also help move oocytes expelled from the ovaries through the uterine (fallopian) tubes into the uterus.

PSEUDOSTRATIFIED COLUMNAR EPITHELIUM As noted earlier, this tissue appears to have several layers because the nuclei of the cells are at various depths (Table 3.1E). Even though all the cells are attached to the basement membrane in a single layer, some cells do not extend to the surface. When viewed from the side, these features give the false impression of a multilayered tissue—thus the name *pseudo*stratified epithelium (*pseudo-*=false). In *pseudostratified ciliated columnar epithelium,* the cells that extend to the surface either secrete mucus (goblet cells) or bear cilia. The mucus traps foreign particles and the cilia sweep away mucus for eventual elimination from the body. *Pseudostratified nonciliated columnar epithelium* contains cells without cilia and lacks goblet cells.

Stratified Epithelium

In contrast to simple epithelium, stratified epithelium has two or more layers of cells. Thus, it is more durable and can better protect underlying tissues. Some cells of stratified epithelia also produce secretions. The name of the specific kind of stratified epithelium depends on the shape of the cells in the apical layer.

STRATIFIED SQUAMOUS EPITHELIUM Cells in the apical layer of this type of epithelium are flat, whereas in the deep layers, cells vary in shape from cuboidal to columnar (Table 3.1F). The basal (deepest) cells continually undergo cell division. As new cells grow, the cells of the basal layer are pushed upward toward the apical layer. As they move farther from the deeper layers and from their blood supply in the underlying connective tissue, they become dehydrated, shrunken, and harder and then die. At the apical layer, the cells lose their cell junctions and are sloughed off, but they are replaced as new cells continually emerge from the basal cells.

Stratified squamous epithelium exists in both keratinized and nonkeratinized forms. In *keratinized stratified squamous epithelium,* the apical layer and several layers deep to it are partially dehydrated and contain a layer of **keratin,** a tough, fibrous protein that helps protect the skin and underlying tissues from heat, microbes, and chemicals. *Nonkeratinized stratified squamous epithelium* does not contain keratin in the apical layer and several layers deep to it and remains moist. Stratified squamous epithelium forms the first line of defense against microbes.

PAPANICOLAOU TEST

A **Papanicolaou test** (pa-pa-NI-kō-lō), **Pap test,** or **Pap smear** involves collecting and microscopically examining epithelial cells that have sloughed off the apical layer of a tissue. A very common type of Pap test involves examining the cells from the nonkeratinized stratified squamous epithelium of the cervix and vagina. This type of Pap test is performed mainly to detect early changes in the cells of the female reproductive system that may indicate cancer or a precancerous condition. An annual Pap test is recommended for all women as part of a routine pelvic exam. ■

STRATIFIED CUBOIDAL EPITHELIUM This fairly rare type of epithelium sometimes consists of more than two layers of cells (Table 3.1G). Cells in the apical layer are cuboidal. The function of stratified cuboidal epithelium is mainly protective but it also functions in limited secretion and absorption.

STRATIFIED COLUMNAR EPITHELIUM Like stratified cuboidal epithelium, this type of tissue also is uncommon. Usually the basal layer consists of shortened, irregularly shaped cells; only the apical layer has cells that are columnar in form (Table 3.1H). This type of epithelium functions in protection and secretion.

TRANSITIONAL EPITHELIUM This kind of stratified epithelium is present only in the urinary system and has a variable appearance. In its relaxed (unstretched) state (Table 3.1I), transitional epithelium looks similar to stratified cuboidal epithelium, except that the cells in the apical layer tend to be large and rounded. As the tissue is stretched, its cells become flatter, giving the appearance of stratified squamous epithelium. Because of its elasticity, transitional epithelium is ideal for lining hollow structures that are subjected to expansion from within, such as the urinary bladder. It allows organs to stretch to hold a variable amount of fluid without rupturing.

Glandular Epithelium

The function of glandular epithelium is secretion, which is accomplished by glandular cells that often lie in clusters deep to the covering and lining epithelium. A **gland** may consist of a single cell or a group of cells that secrete substances into ducts (tubes), onto a surface, or into the blood. All glands of the body are classified as either endocrine or exocrine.

The secretions of **endocrine glands** (*endo-*=within; *-crine*= secretion) (Table 3.1J) enter the interstitial fluid and then diffuse directly into the bloodstream without flowing through a duct. These secretions, called *hormones*, regulate many metabolic and physiological activities to maintain homeostasis. The pituitary, thyroid, and adrenal glands are examples of endocrine glands. Endocrine glands will be described in detail in Chapter 23.

Exocrine glands (*exo-*=outside) (Table 3.1K) secrete their products into ducts that empty onto the surface of a covering and lining epithelium. Thus, the product of an exocrine gland is released at the skin surface or into the lumen of a hollow organ. The secretions of exocrine glands include mucus, sweat, oil, earwax, saliva, and digestive enzymes. Examples of exocrine glands are sudoriferous (sweat) glands, which produce sweat to help lower body temperature, and salivary glands, which secrete saliva. Saliva contains mucus and digestive enzymes among other substances. As you will see later, some glands of the body, such as the pancreas, ovaries, and testes, are mixed glands that contain both endocrine and exocrine tissue.

Structural Classification of Exocrine Glands

Exocrine glands are classified into unicellular and multicellular types. **Unicellular glands** are single-celled. Goblet cells are unicellular exocrine glands that secrete mucus directly onto the apical surface of a lining epithelium rather than into ducts. Most glands are **multicellular glands,** composed of many cells that form a distinctive microscopic structure or macroscopic organ. Examples are sudoriferous, sebaceous (oil), and salivary glands.

Multicellular glands are categorized according to two criteria: by whether the ducts of the glands are branched or unbranched and by the shape of the secretory portions of the gland

(Figure 3.4). If the duct of the gland does not branch, it is a **simple gland.** If the duct branches, it is a **compound gland.** Glands with tubular secretory parts are **tubular glands;** those with more rounded secretory portions are **acinar glands** (AS-i-nar; *acin-=* berry). **Tubuloacinar glands** have both tubular and more rounded secretory parts.

Combinations of the degree of duct branching and the shape of the secretory part are the criteria for the following structural classification scheme for multicellular exocrine glands:

I. *Simple glands*

A. *Simple tubular.* Tubular secretory part is straight and attaches to a single unbranched duct. Example: glands in the large intestine.

B. *Simple branched tubular.* Tubular secretory part is branched and attaches to a single unbranched duct. Example: gastric glands.

C. *Simple coiled tubular.* Tubular secretory part is coiled and attaches to a single unbranched duct. Example: sweat glands.

D. *Simple acinar.* Secretory portion is rounded and attaches to a single unbranched duct. Example: glands of the penile urethra.

E. *Simple branched acinar.* Rounded secretory part is branched and attaches to a single unbranched duct. Example: sebaceous glands.

II. *Compound glands*

A. *Compound tubular.* Secretory portion is tubular and attaches to a branched duct. Example: bulbourethral (Cowper's) glands.

B. *Compound acinar.* Secretory portion is rounded and attaches to a branched duct. Example: mammary glands.

C. *Compound tubuloacinar.* Secretory portion is both tubular and rounded and attaches to a branched duct. Example: acinar glands of the pancreas.

Functional Classification of Exocrine Glands

The functional classification of exocrine glands is based on how their secretion is released. In **merocrine glands** (MER-ō-krin;

Figure 3.4   Multicellular exocrine glands. Pink represents the secretory portion; lavender represents the duct.

Structural classification of multicellular exocrine glands is based on the branching pattern of the duct and the shape of the secreting portion.

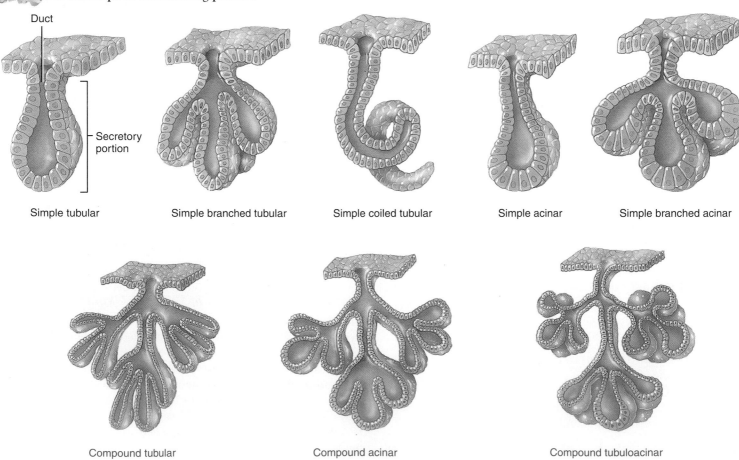

Simple tubular   Simple branched tubular   Simple coiled tubular   Simple acinar   Simple branched acinar

Compound tubular   Compound acinar   Compound tubuloacinar

How do simple multicellular glands differ from compound ones?

mero-=a part), the secretion is synthesized on ribosomes attached to rough ER; processed, sorted, and packaged by the Golgi complex; and released from the cell in secretory vesicles via exocytosis (Figure 3.5a). Most exocrine glands of the body are merocrine glands. Examples are the salivary glands and pancreas. **Apocrine glands** (AP-ō-krin; *apo-*=from) accumulate their secretory product at the apical surface of the secreting cell. Then, that portion of the cell pinches off from the rest of the cell to release the secretion (Figure 3.5b). The remaining part of the cell repairs itself and repeats the process. Electron micrographic studies have called into

question whether humans have apocrine glands. What were once thought to be apocrine glands—for example, mammary glands that secrete milk—are probably merocrine glands. The cells of **holocrine glands** (HŌ-lō-krin; *holo-*=entire) accumulate a secretory product in their cytosol. As the secretory cell matures, it ruptures and becomes the secretory product (Figure 3.5c). The sloughed off cell is replaced by a new cell. One example of a holocrine gland is a sebaceous gland of the skin.

CHECKPOINT

5. Describe the various layering arrangements and cell shapes of epithelium.

6. What characteristics are common to all epithelial tissues?

7. How is the structure of the following kinds of epithelium related to their functions: simple squamous, simple cuboidal, simple columnar (nonciliated and ciliated), pseudostratified columnar (ciliated and nonciliated), stratified squamous (keratinized and nonkeratinized), stratified cuboidal, stratified columnar and transitional?

8. Where are endothelium and mesothelium located?

9. What distinguishes endocrine glands and exocrine glands? Name and give examples of the three functional classes of exocrine glands.

Figure 3.5 Functional classification of multicellular exocrine glands.

The functional classification of exocrine glands is based on whether a secretion is a product of a cell or consists of an entire or a partial glandular cell.

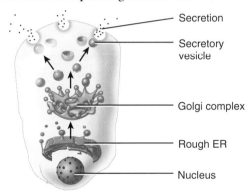

(a) Merocrine secretion

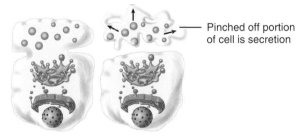

(b) Apocrine secretion

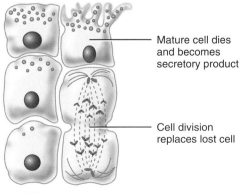

(c) Holocrine secretion

? **What class of glands are sebaceous (oil) glands? Salivary glands?**

CONNECTIVE TISSUE

OBJECTIVES

● Describe the general features of connective tissue.
● Describe the structure, location, and function of the various types of connective tissues.

Connective tissue is one of the most abundant and widely distributed tissues in the body. In its various forms, connective tissue has a variety of functions. It binds together, supports, and strengthens other body tissues; protects and insulates internal organs; compartmentalizes structures such as skeletal muscles; is the major transport system within the body (blood, a fluid connective tissue); is the major site of stored energy reserves (adipose, or fat, tissue); and is the main site of immune responses.

General Features of Connective Tissue

Connective tissue consists of two basic elements: cells and extracellular matrix. A connective tissue's **extracellular matrix** is the material located between its widely spaced cells. The extracellular matrix consists of protein fibers and ground substance, the material between the cells and the fibers. It is usually secreted by the connective tissue cells and determines the tissue's qualities. For instance, in cartilage, the extracellular matrix is firm but pliable. In bone, by contrast, the extracellular matrix is hard and not pliable.

In contrast to epithelia, connective tissues do not usually occur on body surfaces, such as the covering or lining of internal organs, the lining of body cavities, or the external surface of the body. However, a type of connective tissue called areolar connective tissue lines joint cavities. Also, unlike epithelia, connective tissues usually are highly vascular; that is, they have a rich blood supply. Exceptions include cartilage, which is avascular, and tendons, which have a scanty blood supply. Except for cartilage, connective tissues, like epithelia, have a nerve supply.

Connective Tissue Cells

Mesodermal embryonic cells called mesenchymal cells give rise to the cells of connective tissue. Each major type of connective tissue contains an immature class of cells whose name ends in -*blast*, which means "to bud or sprout." These immature cells are called *fibroblasts* in loose and dense connective tissue, *chondroblasts* in cartilage, and *osteoblasts* in bone. Blast cells retain the capacity for cell division and secrete the matrix that is characteristic of the tissue. In cartilage and bone, once the matrix is produced, the fibroblasts differentiate into mature cells whose names end in -*cyte*, namely chondrocytes and osteocytes. Mature cells have reduced capacity for cell division and matrix formation and are mostly involved in maintaining the matrix.

The types of cells present in connective tissues vary according to the type of tissue and include the following (Figure 3.6):

1. **Fibroblasts** (FĪ-brō-blasts; *fibro-*=fibers) are large, flat cells with branching processes. They are present in several connective tissues, where they usually are the most numerous connective tissue cells. Fibroblasts migrate through the connective tissue, secreting the fibers and ground substance of the matrix.

2. **Macrophages** (MAK-rō-fā-jez; *macro-*=large; -*phages*= eaters) develop from monocytes, a type of white blood cell. Macrophages have an irregular shape with short branching projections and are capable of engulfing bacteria and cellular debris by phagocytosis. Some are fixed macrophages, which means they reside in a particular tissue. Examples are alveolar macrophages in the lungs or spleen macrophages in the spleen. Others are wandering macrophages, which roam the tissues and gather at sites of infection or inflammation.

3. **Plasma cells** are small cells that develop from a type of white blood cell called a B lymphocyte. Plasma cells secrete antibodies, proteins that attack or neutralize foreign substances in the body. Thus, plasma cells are an important part of the body's immune system. Although they are found in many places in the body, most plasma cells reside in connective tissues, especially in the gastrointestinal and respiratory tracts. They are also abundant in the salivary glands, lymph nodes, and red bone marrow.

4. **Mast cells** are abundant alongside the blood vessels that supply connective tissue. They produce histamine, a chemical that dilates small blood vessels as part of the inflammatory response, the body's reaction to injury or infection.

5. **Adipocytes,** also called fat cells or adipose cells, are connective tissue cells that store triglycerides (fats). They are found deep to the skin and around organs such as the heart and kidneys.

6. **White blood cells** are not found in significant numbers in normal connective tissue. However, in response to certain conditions they migrate from blood into connective tissues. For example, neutrophils gather at sites of infection, and eosinophils migrate to sites of parasitic invasions and allergic responses.

Connective Tissue Extracellular Matrix

Each type of connective tissue has unique properties, based on the specific extracellular matrix materials between the cells. The extracellular matrix consists of ground substance and fibres. The ground substance may be fluid, semifluid, gelatinous, or calcified.

Figure 3.6 Representative cells and fibers present in connective tissue.

🔑 **Fibroblasts are usually the most numerous connective tissue cells.**

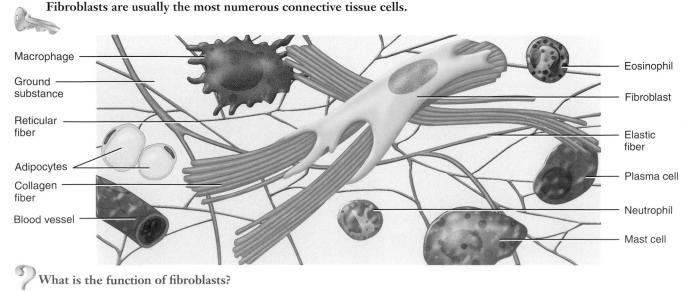

Macrophage
Ground substance
Reticular fiber
Adipocytes
Collagen fiber
Blood vessel

Eosinophil
Fibroblast
Elastic fiber
Plasma cell
Neutrophil
Mast cell

❓ **What is the function of fibroblasts?**

Ground Substance

As mentioned earlier, the **ground substance** is the component of a connective tissue between the cells and fibers. The ground substance supports cells, binds them together, stores water, and provides a medium through which substances are exchanged between the blood and cells. It plays an active role in how tissues develop, migrate, proliferate, and change shape, and in how they carry out their metabolic functions.

Ground substance contains water and an assortment of large organic molecules, many of which are complex combinations of polysaccharides and proteins. The polysaccharides include hyaluronic acid, chondroitin sulfate, dermatan sulfate, and keratan sulfate. Collectively, they are referred to as **glycosaminoglycans** (glī-kos-a-mē′-nō-GLĪ-kans) or **GAGs.** Except for hyaluronic acid, the GAGs are associated with proteins called **proteoglycans** (prō′-tē-ō-GLĪ-kans). The proteoglycans form a core protein and the GAGs project from the protein like the bristles of a brush. One of the most important properties of GAGs is that they trap water, making the ground substance more jellylike.

Hyaluronic acid (hī-a-loo-RON-ik) is a viscous, slippery substance that binds cells together, lubricates joints, and helps maintain the shape of the eyeballs. White blood cells, sperm cells, and some bacteria produce *hyaluronidase,* an enzyme that breaks apart hyaluronic acid, thus causing the ground substance of connective tissue to become more liquid. The ability to produce hyaluronidase helps white blood cells move more easily through connective tissues to reach sites of infection and aids penetration of an oocyte by a sperm cell during fertilization. It may also allow bacteria to spread more rapidly through connective tissues. **Chondroitin sulfate** (kon-DROY-tin) provides support and adhesiveness in cartilage, bone, skin, and blood vessels. The skin, tendons, blood vessels, and heart valves contain **dermatan sulfate,** whereas bone, cartilage, and the cornea of the eye contain **keratan sulfate.** Also present in the ground substance are **adhesion proteins,** which are responsible for linking components of the ground substance to each other and to the surfaces of cells. The main adhesion protein of connective tissue is **fibronectin,** which binds to both collagen fibers (discussed shortly) and ground substance thereby linking them together. It also attaches cells to the ground substance.

CHONDROITIN SULFATE, GLUCOSAMINE, AND JOINT DISEASE

In recent years, chondroitin sulfate and glucosamine (a proteoglycan) have been used as nutritional supplements either alone or in combination to promote and maintain the structure and function of joint cartilage, to provide pain relief from osteoarthritis, and to reduce joint inflammation. Although these supplements have benefited some individuals, more research is needed to determine how they act and why some people benefit and others do not. ■

Fibers

Three types of **fibers** are embedded in the extracellular matrix between the cells: collagen fibers, elastic fibers, and reticular fibers. They function to strengthen and support connective tissues.

Collagen fibers (*colla*=glue) are very strong and resist pulling forces, but they are not stiff, which promotes tissue flexibility. Different types of collagen fibers in various tissues have slightly different properties. For example, the collagen fibers found in cartilage attract more water molecules than do the collagen fibers in bone, which gives cartilage a more cushioning consistency. Collagen fibers often occur in bundles lying parallel to one another (Figure 3.6). The bundle arrangement affords great strength. Chemically, collagen fibers consist of the protein *collagen.* This is the most abundant protein in your body, representing about 25% of the total protein. Collagen fibers are found in most types of connective tissues, especially bone, cartilage, tendons (attach muscle to bone), and ligaments (attach bone to bone).

LIGAMENTS AND SPRAINS

Despite the strength of ligaments, they may be stressed beyond their normal capacity. This results in a **sprain,** a stretched or torn ligament. The ankle joint is most frequently strained. Because of their poor blood supply, the healing of even partially torn ligaments is a very slow process; completely torn ligaments require surgical repair. ■

Elastic fibers, which are smaller in diameter than collagen fibers, branch and join together to form a network within a tissue. An elastic fiber consists of molecules of the protein elastin surrounded by a glycoprotein named fibrillin, which strengthens and stabilizes elastic fibers. Because of their unique molecular structure, elastic fibers are strong but can be stretched up to 150% of their relaxed length without breaking. Equally important, elastic fibers have the ability to return to their original shape after being stretched, a property called elasticity. Elastic fibers are plentiful in skin, blood vessel walls, and lung tissue.

Reticular fibers (*reticul-*=net), consisting of *collagen* arranged in fine bundles and a coating of glycoprotein, provide support in the walls of blood vessels and form a network around the cells in some tissues, for instance, areolar connective tissue, adipose tissue, and smooth muscle tissue. Produced by fibroblasts, reticular fibers are much thinner than collagen fibers and form branching networks. Like collagen fibers, reticular fibers provide support and strength. Reticular fibers are plentiful in reticular connective tissue, which forms the **stroma** (=bed or covering) or supporting framework of many soft organs, such as the spleen and lymph nodes. These fibers also help form the basement membrane.

MARFAN SYNDROME

Marfan syndrome (MAR-fan) is an inherited disorder caused by a defective fibrillin gene. The result is abnormal development of elastic fibers. Tissues rich in elastic fibers are malformed or weakened. Structures affected most seriously are the covering layer of bones (periosteum), the ligament that suspends the lens of the eye, and the walls of the large arteries. People with Marfan syndrome tend to be tall and have disproportionately long arms, legs, fingers, and toes. A common symptom is blurred vision caused by displacement of the lens of the eye. The most life-threatening complication of Marfan syndrome is weakening of the aorta (the main artery that emerges from the heart), which can suddenly tear or burst. ■

Classification of Connective Tissues

Classifying connective tissues can be challenging because of the diversity of cells and extracellular matrix present, and the differences in their relative proportions. Thus, the grouping of connective tissues into categories is not always clear-cut. We will classify them as follows:

I. Embryonic connective tissue

   A. Mesenchyme

   B. Mucous connective tissue

II. Mature connective tissue

   A. Loose connective tissue

      1. Areolar connective tissue

      2. Adipose tissue

      3. Reticular connective tissue

   B. Dense connective tissue

      1. Dense regular connective tissue

      2. Dense irregular connective tissue

      3. Elastic connective tissue

   C. Cartilage

      1. Hyaline cartilage

      2. Fibrocartilage

      3. Elastic cartilage

   D. Bone tissue

   E. Liquid connective tissue

      1. Blood tissue

      2. Lymph

Note that our classification scheme has two major subclasses of connective tissue: embryonic and mature. **Embryonic connective tissue** is present primarily in the *embryo*, the developing human from fertilization through the first two months of pregnancy, and in the *fetus*, the developing human from the third month of pregnancy to birth.

One example of embryonic connective tissue found almost exclusively in the embryo is **mesenchyme** (MEZ-en-kīm), the tissue from which all other connective tissues eventually arise (Table 3.2A). Mesenchyme is composed of irregularly shaped cells, a semifluid ground substance, and delicate reticular fibers. Another kind of embryonic tissue is **mucous connective tissue (Wharton's jelly),** found mainly in the umbilical cord of the fetus. Mucous connective tissue is a form of mesenchyme that

contains widely scattered fibroblasts, a more viscous jellylike ground substance, and collagen fibers (Table 3.2B).

The second major subclass of connective tissue, **mature connective tissue,** is present in the newborn. Its cells arise from mesenchyme. Mature connective tissue is of several types, which we explore next.

Types of Mature Connective Tissue

The five types of mature connective tissue are (1) loose connective tissue, (2) dense connective tissue, (3) cartilage, (4) bone tissue, and (5) liquid connective tissue (blood tissue and lymph). We now examine each in detail.

Loose Connective Tissue

In **loose connective tissue** the fibers are loosely intertwined and many cells are present. The types of loose connective tissue are areolar connective tissue, adipose tissue, and reticular connective tissue.

AREOLAR CONNECTIVE TISSUE One of the most widely distributed connective tissues in the body is areolar connective tissue (a-RĒ-ō-lar; *areol-*=a small space). It contains several kinds of cells, including fibroblasts, macrophages, plasma cells, mast cells, adipocytes, and a few white blood cells (Table 3.3A). All three types of fibers—collagen, elastic, and reticular—are arranged randomly throughout the tissue. The ground substance contains hyaluronic acid, chondroitin sulfate, dermatan sulfate, and keratan sulfate. Combined with adipose tissue, areolar connective tissue forms the *subcutaneous layer*, the layer of tissue that attaches the skin to underlying tissues and organs.

ADIPOSE TISSUE Adipose tissue is a loose connective tissue in which the cells, called **adipocytes** (*adipo-*=fat), are specialized for storage of triglycerides (Table 3.3B). Adipocytes are derived from fibroblasts. Because the cell fills up with a single, large triglyceride droplet, the cytoplasm and nucleus are pushed to the periphery of the cell. Adipose tissue is found wherever areolar connective tissue is located. Adipose tissue is a good insulator and can therefore reduce heat loss through the skin. It is a major energy reserve and generally supports and protects various organs. As the amount of adipose tissue increases with weight gain, new blood vessels form. Thus, an obese person has many more miles of blood vessels than does a lean person, a situation that can cause high blood pressure.

Most adipose tissue in adults is **white adipose tissue,** the type just described. Another type, called **brown adipose tissue (BAT),** obtains its darker color from a very rich blood supply and numerous mitochondria, which contain colored pigments that participate in aerobic cellular respiration. Although BAT is widespread in the fetus and infant; in adults only small amounts are present. BAT generates considerable heat and probably helps to maintain body temperature in the newborn. The heat generated by the many mitochondria is carried away to other body tissues by the extensive blood supply.

TABLE 3.2 EMBRYONIC CONNECTIVE TISSUES

A. Mesenchyme

Description Consists of irregularly shaped mesenchymal cells embedded in a semifluid ground substance that contains reticular fibers.

Location Under skin and along developing bones of embryo; some mesenchymal cells are found in adult connective tissue, especially along blood vessels.

Function Forms all other kinds of connective tissue.

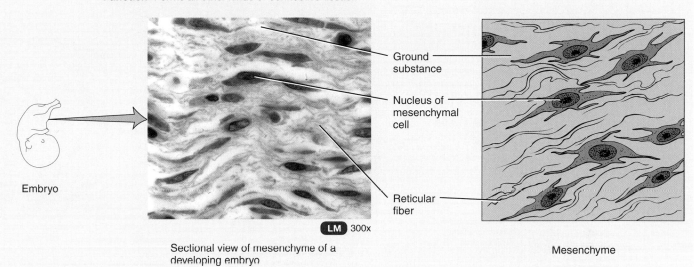

Embryo

Ground substance

Nucleus of mesenchymal cell

Reticular fiber

LM 300x

Sectional view of mesenchyme of a developing embryo

Mesenchyme

B. Mucous connective tissue

Description Consists of widely scattered fibroblasts embedded in a viscous, jellylike ground substance that contains fine collagen fibers.

Location Umbilical cord of fetus.

Function Support.

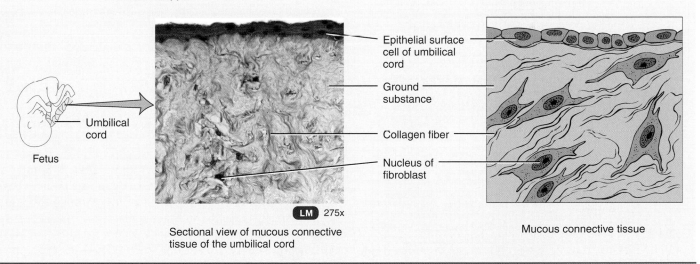

Umbilical cord

Fetus

Epithelial surface cell of umbilical cord

Ground substance

Collagen fiber

Nucleus of fibroblast

LM 275x

Sectional view of mucous connective tissue of the umbilical cord

Mucous connective tissue

LIPOSUCTION

A surgical procedure, called **liposuction** (*lip-*=fat) or **suction lipectomy** (*-ectomy*=to cut out), involves suctioning out small amounts of adipose tissue from various areas of the body. The technique can be used as a body-contouring procedure in regions such as the thighs, buttocks, arms, breasts, and abdomen. Postsurgical complications that may develop include fat emboli (clots), infection, fluid depletion, injury to internal structures, and severe postoperative pain. ■

RETICULAR CONNECTIVE TISSUE Reticular connective tissue consists of fine interlacing reticular fibers and reticular cells (Table 3.3C). Reticular connective tissue forms the stroma (supporting framework) of the liver, spleen, and lymph nodes and helps bind together smooth muscle cells. Additionally, retic-

TABLE 3.3 MATURE CONNECTIVE TISSUES
LOOSE CONNECTIVE TISSUE

A. Areolar connective tissue

Description Consists of fibers (collagen, elastic, and reticular) and several kinds of cells (fibroblasts, macrophages, plasma cells, adipocytes, and mast cells) embedded in a semifluid ground substance.

Location Subcutaneous layer deep to skin; papillary (superficial) region of dermis of skin; lamina propria of mucous membranes; and around blood vessels, nerves, and body organs.

Function Strength, elasticity, and support.

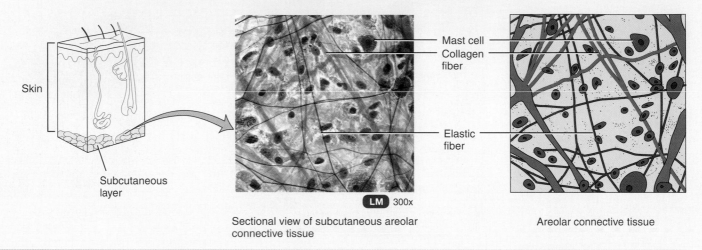

Skin

Subcutaneous layer

Mast cell
Collagen fiber

Elastic fiber

LM 300x

Sectional view of subcutaneous areolar connective tissue

Areolar connective tissue

B. Adipose tissue

Description Consists of adipocytes, cells specialized to store triglycerides (fats) as a large centrally located droplet; nucleus and cytoplasm are peripherally located.

Location Subcutaneous layer deep to skin, around heart and kidneys, yellow bone marrow, and padding around joints and behind eyeball in eye socket.

Function Reduces heat loss through skin, serves as an energy reserve, supports, and protects. In newborns, brown adipose tissue generates considerable heat that helps maintain proper body temperature.

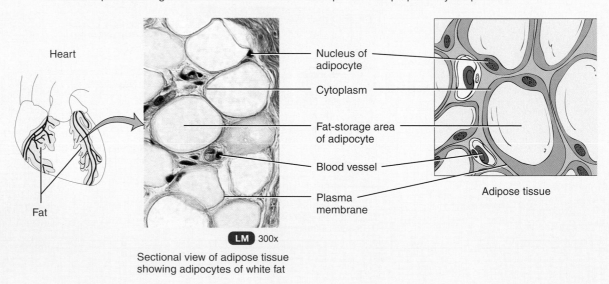

Heart

Fat

Nucleus of adipocyte

Cytoplasm

Fat-storage area of adipocyte

Blood vessel

Plasma membrane

Adipose tissue

LM 300x

Sectional view of adipose tissue showing adipocytes of white fat

ular fibers in the spleen filter blood and remove worn-out blood cells, and reticular fibers in lymph nodes filter lymph and remove bacteria.

Dense Connective Tissue

Dense connective tissue contains more numerous, thicker, and denser fibers but considerably fewer cells than loose con-

nective tissue. There are three types: dense regular connective tissue, dense irregular connective tissue, and elastic connective tissue.

DENSE REGULAR CONNECTIVE TISSUE In this tissue, bundles of collagen fibers are *regularly* arranged in parallel patterns that provide the tissue with great strength (Table 3.3D). The tissue withstands pulling along the axis of the fibers.

C. Reticular connective tissue

Description Consists of a network of interlacing reticular fibers and reticular cells.

Location Stroma (supporting framework) of liver, spleen, lymph nodes; red bone marrow, which gives rise to blood cells; reticular lamina of the basement membrane; and around blood vessels and muscles.

Function Forms stroma of organs; binds together smooth muscle tissue cells; filters and removes worn-out blood cells in the spleen and microbes in lymph nodes.

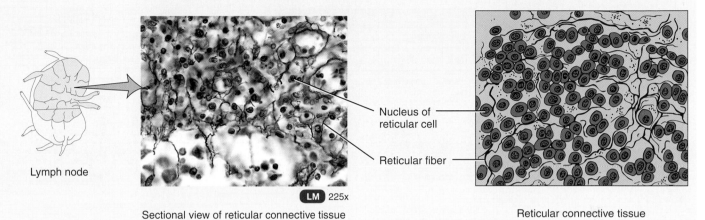

Lymph node

Nucleus of reticular cell

Reticular fiber

LM 225x

Sectional view of reticular connective tissue of a lymph node

Reticular connective tissue

DENSE CONNECTIVE TISSUE

D. Dense regular connective tissue

Description Matrix looks shiny white; consists mainly of collagen fibers arranged in bundles; fibroblasts present in rows between bundles.

Location Forms tendons (attach muscle to bone), most ligaments (attach bone to bone), and aponeuroses (sheetlike tendons that attach muscle to muscle or muscle to bone).

Function Provides strong attachment between various structures.

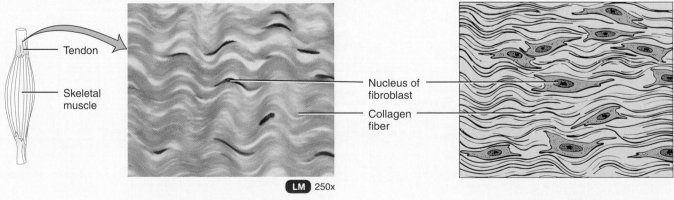

Tendon

Skeletal muscle

Nucleus of fibroblast

Collagen fiber

LM 250x

Sectional view of dense regular connective tissue of a tendon

Dense regular connective tissue

continues

Fibroblasts, which produce the fibers and ground substance, appear in rows between the fibers. The tissue is silvery white and tough, yet somewhat pliable. Examples are tendons and most ligaments.

DENSE IRREGULAR CONNECTIVE TISSUE This tissue contains collagen fibers that are packed more closely together than in loose connective tissue and are usually *irregularly* arranged (Table 3.3E). It is found in parts of the body where pulling forces are exerted in various directions. The tissue often occurs in sheets, such as in the dermis of the skin, which is deep to the epidermis, or the pericardium around the heart. Heart valves, the perichondrium (the membrane surrounding cartilage), and the periosteum (the membrane surrounding bone) are dense irregular connective tissues, although they have a fairly orderly arrangement of their collagen fibers.

TABLE 3.3   MATURE CONNECTIVE TISSUES (continued)
DENSE CONNECTIVE TISSUE

E. Dense irregular connective tissue

Description   Consists predominantly of collagen fibers, randomly arranged, and a few fibroblasts.

Location   Fasciae (tissue beneath skin and around muscles and other organs), reticular (deeper) region of dermis of skin, periosteum of bone, perichondrium of cartilage, joint capsules, membrane capsules around various organs (kidneys, liver, testes, lymph nodes), pericardium of the heart, and heart valves.

Function   Provides strength.

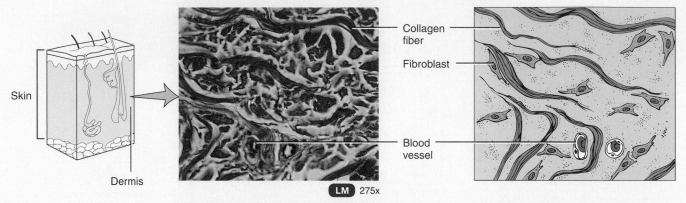

Sectional view of dense irregular connective tissue of reticular region of dermis

Dense irregular connective tissue

F. Elastic connective tissue

Description   Consists of predominantly freely branching elastic fibers; fibroblasts present in spaces between fibers.

Location   Lung tissue, walls of elastic arteries, trachea, bronchial tubes, true vocal cords, suspensory ligament of penis, and ligaments between vertebrae.

Function   Allows stretching of various organs.

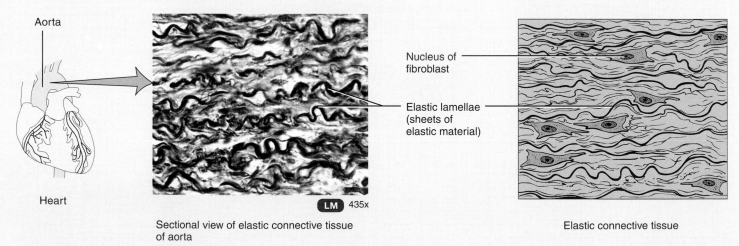

Sectional view of elastic connective tissue of aorta

Elastic connective tissue

ELASTIC CONNECTIVE TISSUE   Branching elastic fibers predominate in elastic connective tissue (Table 3.3F), giving the unstained tissue a yellowish color. Fibroblasts are present in the spaces between the fibers. Elastic connective tissue is quite strong and can recoil to its original shape after being stretched. Elasticity is important to the normal functioning of lung tissue, which recoils as you exhale, and elastic arteries, whose recoil between heartbeats helps maintain blood flow.

Cartilage

Cartilage consists of a dense network of collagen fibers and elastic fibers firmly embedded in chondroitin sulfate, a gel-like component of the ground substance. Cartilage can endure considerably more stress than loose and dense connective tissues. Whereas the strength of cartilage is due to its collagen fibers, its resilience (ability to assume its original shape after deformation) is due to chondroitin sulfate.

CARTILAGE

G. Hyaline cartilage

Description Consists of a bluish-white, shiny ground substance with fine collagen fibers and many chondrocytes; most abundant type of cartilage.

Location Ends of long bones, anterior ends of ribs, nose, parts of larynx, trachea, bronchi, bronchial tubes, and embryonic and fetal skeleton.

Function Provides smooth surfaces for movement at joints, as well as flexibility and support.

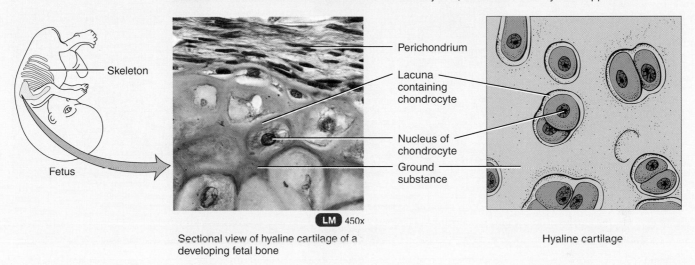

Skeleton

Fetus

Perichondrium

Lacuna containing chondrocyte

Nucleus of chondrocyte

Ground substance

LM 450x

Sectional view of hyaline cartilage of a developing fetal bone

Hyaline cartilage

H. Fibrocartilage

Description Consists of chondrocytes scattered among bundles of collagen fibers within the extracellular matrix.

Location Pubic symphysis (point where hip bones join anteriorly), intervertebral discs (discs between vertebrae), menisci (cartilage pads) of knee, and portions of tendons that insert into cartilage.

Function Support and fusion.

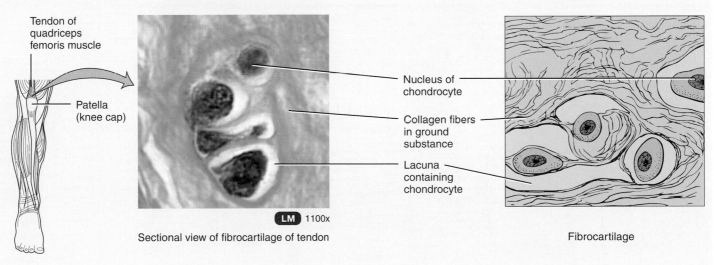

Tendon of quadriceps femoris muscle

Patella (knee cap)

Nucleus of chondrocyte

Collagen fibers in ground substance

Lacuna containing chondrocyte

LM 1100x

Sectional view of fibrocartilage of tendon

Fibrocartilage

continues

The cells of mature cartilage, called **chondrocytes** (KON-drō-sīts; *chondro-*=cartilage), occur singly or in groups within spaces called **lacunae** (la-KOO-nē= little lakes; singular is *lacuna*) in the extracellular matrix. A membrane of dense irregular connective tissue called the **perichondrium** (per'-i-KON-drē-um; *peri-*=around) covers the surface of most cartilage. Unlike other connective tissues, cartilage has no blood vessels or nerves, except in the perichondrium. Since cartilage has no blood supply, it heals poorly following an injury. Three kinds of cartilage are recognized: hyaline cartilage, fibrocartilage, and elastic cartilage.

HYALINE CARTILAGE This type of cartilage contains a resilient gel as its ground substance and appears in the body as a bluish-white, shiny substance. The fine collagen fibers are not visible with ordinary staining techniques, and prominent chondrocytes are found in lacunae (Table 3.3G). Most hyaline cartilage is sur-

TABLE 3.3   **MATURE CONNECTIVE TISSUES (continued)**
CARTILAGE

I. Elastic cartilage

Description Consists of chondrocytes located in a threadlike network of elastic fibers within the extracellular matrix.

Location Lid on top of larynx (epiglottis), part of external ear (auricle), and auditory (eustachian) tubes.

Function Gives support and maintains shape.

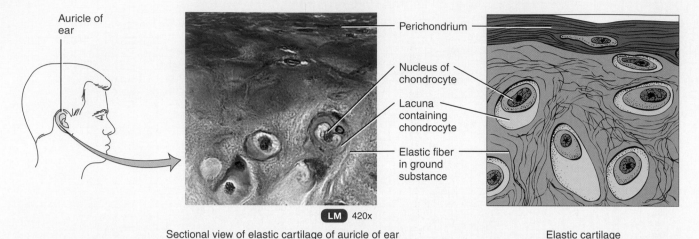

Sectional view of elastic cartilage of auricle of ear

Elastic cartilage

BONE TISSUE

J. Bone Tissue

Description Compact bone tissue consists of osteons (haversian systems) that contain lamellae, lacunae, osteocytes, canaliculi, and central (haversian) canals. By contrast, spongy bone tissue (see Figure 6.3 on page 149) consists of thin columns called trabeculae; spaces between trabeculae are filled with red bone marrow.

Location Both compact and spongy bone tissue make up the various parts of bones of the body.

Function Support, protection, storage; houses blood-forming tissue; serves as levers that act together with muscle tissue to enable movement.

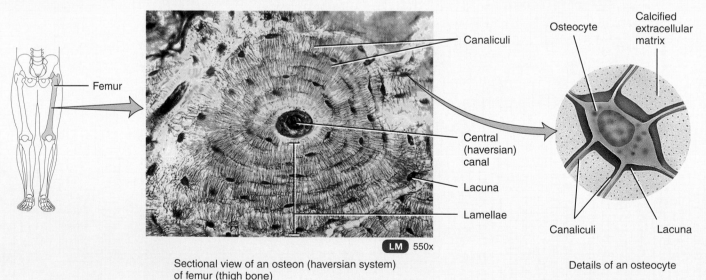

Sectional view of an osteon (haversian system) of femur (thigh bone)

Details of an osteocyte

rounded by a perichondrium. The exceptions are the articular cartilage in joints and at the epiphyseal plates, the regions where bones lengthen as a person grows. Hyaline cartilage is the most abundant cartilage in the body. It affords flexibility and support and, at joints, reduces friction and absorbs shock. Hyaline cartilage is the weakest of the three types of cartilage.

FIBROCARTILAGE Chondrocytes are scattered among clearly visible bundles of collagen fibers within the extracellular matrix

of this type of cartilage (Table 3.3H). Fibrocartilage lacks a perichondrium. This tissue combines strength and rigidity and is the strongest of the three types of cartilage. One location of fibrocartilage is the intervertebral discs, the disk-shaped material between the vertebrae (backbones).

ELASTIC CARTILAGE In this tissue, chondrocytes are located within a threadlike network of elastic fibers within the extracellular matrix (Table 3.3I). A perichondrium is present. Elastic

LIQUID CONNECTIVE TISSUE

K. Blood Tissue

Description Consists of blood plasma and formed elements: red blood cells (erythrocytes), white blood cells (leukocytes), and platelets (thrombocytes).

Location Within blood vessels (arteries, arterioles, capillaries, venules, and veins) and within the chambers of the heart.

Function Red blood cells transport oxygen and some carbon dioxide; white blood cells carry on phagocytosis and are involved in allergic reactions and immune system responses; platelets are essential for the clotting of blood.

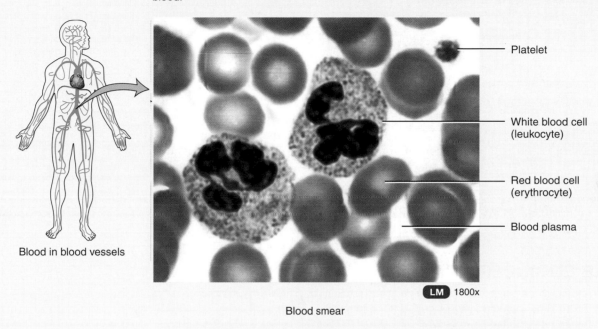

Platelet

White blood cell (leukocyte)

Red blood cell (erythrocyte)

Blood plasma

LM 1800x

Blood smear

Blood in blood vessels

Red blood cells

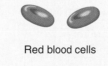

White blood cells

Platelets

cartilage provides strength and elasticity and maintains the shape of certain structures, such as the external ear.

Bone Tissue

Cartilage, joints, and bones make up the skeletal system. The skeletal system supports soft tissues, protects delicate structures, and works with skeletal muscles to generate movement. Bones store calcium and phosphorus; house red bone marrow, which produces blood cells; and contain yellow bone marrow, a storage site for triglycerides. Bones are organs composed of several different connective tissues, including **bone** or **osseous tissue** (OS-ē-us), the periosteum, red and yellow bone marrow, and the endosteum (a membrane that lines a space within bone that stores yellow bone marrow). Bone tissue is classified as either compact or spongy, depending on how its matrix and cells are organized.

The basic unit of **compact bone** is an **osteon** or **haversian system** (Table 3.3J). Each osteon has four parts.

1. The **lamellae** (la-MEL-lē = little plates) are concentric rings of extracellular matrix that consist of mineral salts (mostly calcium and phosphates), which give bone its hardness, and collagen fibers, which give bone its strength. The lamellae are responsible for the compact nature of this type of bone tissue.

2. **Lacunae** are small spaces between lamellae that contain mature bone cells called **osteocytes.**

3. Projecting from the lacunae are **canaliculi** (KAN-a-lik-ū-lī = little canals), networks of minute canals containing the processes of osteocytes. Canaliculi provide routes for nutrients to reach osteocytes and for wastes to leave them.

4. A **central (haversian) canal** contains blood vessels and nerves.

Spongy bone lacks osteons. Rather, it consists of columns of bone called **trabeculae** (tra-BEK-ū-lē = little beams), which contain lamellae, osteocytes, lacunae, and canaliculi. Spaces between lamellae are filled with red bone marrow. Chapter 6 presents bone tissue histology in more detail.

TISSUE ENGINEERING

The technology of **tissue engineering** has allowed scientists to grow new tissues in the laboratory for replacement of damaged tissues in the body. Tissue engineers have already developed laboratory-grown versions of skin and cartilage. In the procedure, scaffolding beds of biodegradable synthetic materials or collagen are used as substrates that permit body cells such as skin cells or cartilage cells to be cultured. As the cells divide and assemble, the scaffolding degrades, and the new, permanent tissue is then implanted in the patient. Other structures being developed by tissue engineers include bones, tendons, heart valves, bone marrow, and intestines. Work is also under way to develop insulin-producing cells for diabetics, dopamine-producing cells for Parkinson's disease patients, and even entire livers and kidneys. ■

Liquid Connective Tissue

In a liquid connective tissue, the cells are suspended in a liquid extracellular matrix. The two types of liquid connective tissues are blood tissue and lymph.

BLOOD TISSUE **Blood tissue** (or simply **blood**) is a connective tissue with a liquid extracellular matrix called **blood plasma,** a pale yellow fluid that consists mostly of water with a wide variety of dissolved substances—nutrients, wastes, enzymes, plasma proteins, hormones, respiratory gases, and ions (Table 3.3K). Suspended in the blood plasma are formed elements—red blood cells (erythrocytes), white blood cells (leukocytes), and platelets (thrombocytes). **Red blood cells** transport oxygen to body cells and remove some carbon dioxide from them. **White blood cells** are involved in phagocytosis, immunity, and allergic reactions. **Platelets** participate in blood clotting. The details of blood are considered in Chapter 13.

LYMPH **Lymph** is the extracellular fluid that flows in lymphatic vessels. It is a connective tissue that consists of several types of cells in a clear liquid extracellular matrix that is similar to blood plasma but with much less protein. In the lymph are various cells and chemicals, whose composition varies from one part of the body to another. For example, lymph leaving lymph nodes includes many lymphocytes, a type of white blood cell, whereas lymph from the small intestine has a high content of newly absorbed dietary lipids. The details of lymph are considered in Chapter 16.

10. In what ways do connective tissues differ from epithelia?
11. What are the features of the cells, ground substance, and fibers that make up connective tissue?
12. How are connective tissues classified? List the various types.
13. Describe how the structures of the following connective tissues are related to their functions: areolar connective tissue, adipose tissue, reticular connective tissue, dense regular connective tissue, dense irregular connective tissue, elastic connective tissue, hyaline cartilage, fibrocartilage, elastic cartilage, bone tissue, blood tissue, and lymph?

MEMBRANES

OBJECTIVES

● Define a membrane.
● Describe the classification of membranes.

Membranes are flat sheets of pliable tissue that cover or line a part of the body. The combination of an epithelial layer and an underlying connective tissue layer constitutes an **epithelial membrane.** The principal epithelial membranes of the body are mucous membranes, serous membranes, and the cutaneous membrane, or skin. Another kind of membrane, a **synovial membrane,** lines joints and contains connective tissue but no epithelium.

Epithelial Membranes
Mucous Membranes

A **mucous membrane** or **mucosa** (mū-KŌ-sa) lines a body cavity that opens directly to the exterior. Mucous membranes line the entire digestive, respiratory, and reproductive tracts, and much of the urinary tract. They consist of both a lining layer of epithelium and an underlying layer of connective tissue (Figure 3.7a).

The epithelial layer of a mucous membrane is an important feature of the body's defense mechanisms because it is a barrier that microbes and other pathogens have difficulty penetrating. Usually, tight junctions connect the cells, so materials cannot leak in between them. Goblet cells and other cells of the epithelial layer of a mucous membrane secrete mucus, and this slippery fluid prevents the cavities from drying out. It also traps particles in the respiratory passageways and lubricates food as it moves through the gastrointestinal tract. In addition, the epithelial layer secretes some of the enzymes needed for digestion and is the site of food and fluid absorption in the gastrointestinal tract.

Figure 3.7 Membranes.

A membrane is a flat sheet of pliable tissue that covers or lines a part of the body.

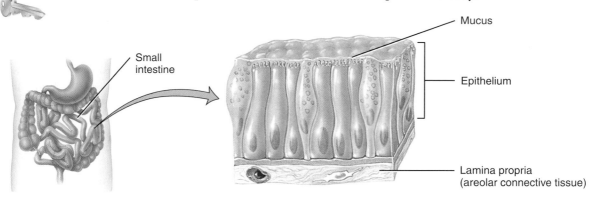

- Small intestine
- Mucus
- Epithelium
- Lamina propria (areolar connective tissue)

(a) Mucous membrane

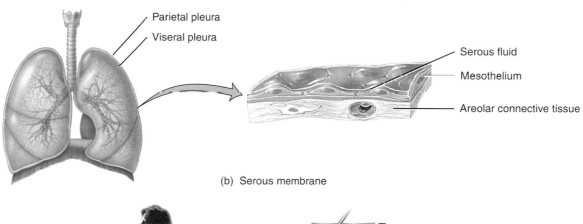

- Parietal pleura
- Viseral pleura
- Serous fluid
- Mesothelium
- Areolar connective tissue

(b) Serous membrane

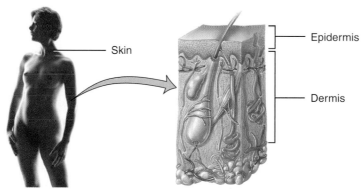

- Skin
- Epidermis
- Dermis

(c) Skin

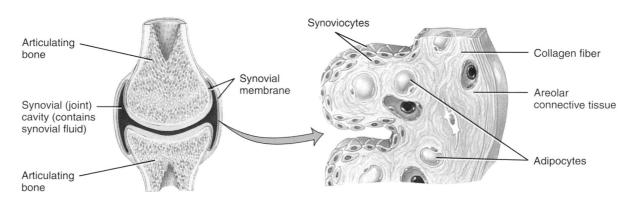

- Synoviocytes
- Articulating bone
- Collagen fiber
- Synovial membrane
- Synovial (joint) cavity (contains synovial fluid)
- Areolar connective tissue
- Articulating bone
- Adipocytes

(d) Synovial membrane

What is an epithelial membrane?

The epithelia of mucous membranes vary greatly in different parts of the body. For example, the epithelium of the mucous membrane of the small intestine is nonciliated simple columnar (see Table 3.1C), whereas the epithelium of the large airways to the lungs is pseudostratified ciliated columnar (see Table 31.E).

The connective tissue layer of a mucous membrane is areolar connective tissue and is called the **lamina propria** (LAM-i-na PRŌ-prē-a). The lamina propria is so named because it belongs to the mucous membrane (*propria*=one's own). The lamina propria supports the epithelium, binds it to the underlying structures, and allows some flexibility of the membrane. It also holds blood vessels in place and protects underlying muscles from abrasion or puncture. Oxygen and nutrients diffuse from the lamina propria to the epithelium covering it whereas carbon dioxide and wastes diffuse in the opposite direction.

Serous Membranes

A **serous membrane** (*serous*=watery) or **serosa** lines a body cavity that does not open directly to the exterior, and it covers the organs that lie within the cavity. Serous membranes consist of areolar connective tissue covered by mesothelium (simple squamous epithelium) (Figure 3.7b). Serous membranes have two layers: The layer attached to the cavity wall is called the **parietal layer** (pa-RĪ-e-tal; *pariet-*=wall); the layer that covers and attaches to the organs inside the cavity is the **visceral layer** (*viscer-*=body organ). The mesothelium of a serous membrane secretes **serous fluid,** a watery lubricating fluid that allows organs to glide easily over one another or to slide against the walls of cavities.

The serous membrane lining the thoracic cavity and covering the lungs is the **pleura.** The serous membrane lining the heart cavity and covering the heart is the **pericardium.** The serous membrane lining the abdominal cavity and covering the abdominal organs is the **peritoneum.**

Cutaneous Membrane

The **cutaneous membrane** or skin covers the surface of the body and consists of a superficial portion called the *epidermis* and a deeper portion called the *dermis* (Figure 3.7c). The epidermis consists of keratinized stratified squamous epithelium, which protects underlying tissues. The dermis is composed of connective tissue (areolar connective tissue and dense irregular connective tissue). Details of the cutaneous membrane are presented in Chapter 5.

Synovial Membranes

Synovial membranes (sin-Ō-vē-al; *syn-*=together, referring here to a place where bones come together) line the cavities of freely movable joints. Like serous membranes, synovial membranes line structures that do not open to the exterior. Unlike mucous, serous, and cutaneous membranes, they lack epithelium and are therefore not epithelial membranes. Synovial membranes are composed of a discontinuous layer of cells called **synoviocytes** (si-NŌ-vē-ō-sītes), which are closer to the synovial cavity (space between the bones) and a layer of connective tissue deep to the synoviocytes (Figure 3.7d). The synoviocytes are separated by the connective tissue and are of two types: Some are macrophage-like cells that are phagocytic. Others resemble fibroblasts. These cells secrete some of the components of synovial fluid. The connective tissue layer is usually composed of areolar connective tissue and a variable amount of adipocytes, elastic fibers, and collagen fibers. **Synovial fluid** lubricates and nourishes the cartilage covering the bones at movable joints and contains macrophages that remove microbes and debris from the joint cavity. Synovial membranes associated with joints are called *articular synovial membranes*. Other synovial membranes line cushioning sacs, called *bursae*, and *tendon sheaths*, for example, in the hands and feet, thus easing the movement of muscle tendons.

CHECKPOINT

14. Define the following kinds of membranes: mucous, serous, cutaneous, and synovial.
15. Where is each type of membrane located in the body? What are their functions?

MUSCULAR TISSUE

OBJECTIVES

- Describe the general features of muscular tissue.
- Contrast the structure, location, and mode of control of skeletal, cardiac, and smooth muscle tissue.

Muscular tissue consists of elongated cells called *muscle fibers* that can use ATP to generate force. As a result, muscular tissue produces body movements, maintains posture, and generates heat. It also affords protection. Based on its location and certain structural and functional features, muscular tissue is classified into three types: skeletal, cardiac, and smooth (Table 3.4).

Skeletal muscle tissue is named for its location—usually attached to the bones of the skeleton (Table 3.4A). It is also *striated*; the fibers contain alternating light and dark bands called *striations* that are visible under a light microscope. Skeletal muscle is *voluntary* because it can be made to contract or relax by conscious control. A single skeletal muscle fiber is very long (up to 30–40 cm [about 12–16 in.] in your longest muscles). A muscle fiber is roughly cylindrical in shape, and has many nuclei located at the periphery of the cell. Within a whole muscle, the individual muscle fibers are parallel to each other.

Cardiac muscle tissue forms most of the wall of the heart (Table 3.4B). Like skeletal muscle, it is striated. However, unlike skeletal muscle tissue, it is *involuntary*; its contraction is not consciously controlled. Cardiac muscle fibers are branched and usually have only one centrally located nucleus, although an occasional cell has two nuclei. They attach end to end to each other by transverse thickenings of the plasma membrane called **intercalated discs** (*intercalat-*=to insert between), which contain both desmosomes and gap junctions. Intercalated discs are unique to cardiac muscle. The desmosomes strengthen the tissue and hold the fibers together during their vigorous contractions. The gap junctions provide a route for quick conduction of muscle action potentials throughout the heart.

TABLE 3.4 MUSCULAR TISSUES

A. Skeletal muscle tissue

Description Long, cylindrical, striated fibers with many peripherally located nuclei; voluntary control.

Location Usually attached to bones by tendons.

Function Motion, posture, heat production, and protection.

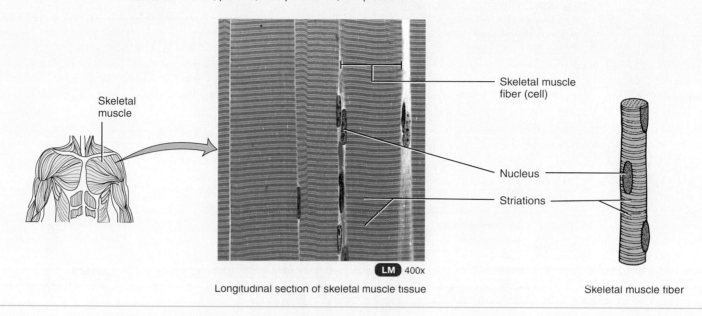

LM 400x

Longitudinal section of skeletal muscle tissue

Skeletal muscle fiber

B. Cardiac muscle tissue

Description Branched striated fibers with one or two centrally located nuclei; contains intercalated discs; involuntary control.

Location Heart wall.

Function Pumps blood to all parts of the body.

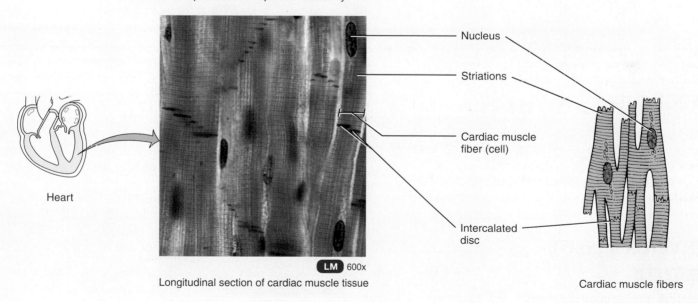

LM 600x

Longitudinal section of cardiac muscle tissue

Cardiac muscle fibers

continues

Smooth muscle tissue is located in the walls of hollow internal structures such as blood vessels, airways to the lungs, the stomach, intestines, gallbladder, and urinary bladder (Table 3.4C). Its contraction helps constrict or narrow the lumen of blood vessels, physically break down and move food along the gastrointestinal tract, move fluids through the body, and eliminate wastes. Smooth muscle fibers are usually *involuntary*, and they are *nonstriated* (lack striations), hence the term *smooth*. A smooth muscle fiber is small, thickest in the middle, and tapering at each end. It contains a single, centrally located nucleus. Gap junctions connect many individual fibers in some smooth muscle tissues, for example, in the wall of the intestines. Such muscle tissues can produce powerful contractions as many muscle fibers contract in unison. In other locations such as the

TABLE 3.4	MUSCULAR TISSUES (continued)
C. Smooth muscle tissue	**Description** Spindle-shaped (thickest in middle and tapering at both ends), nonstriated fibers with one centrally located nucleus; involuntary control. **Location** Iris of the eyes and walls of hollow internal structures such as blood vessels, airways to the lungs, stomach, intestines, gallbladder, urinary bladder, and uterus. **Function** Motion (constriction of blood vessels and airways, propulsion of foods through gastrointestinal tract, contraction of urinary bladder and gallbladder).

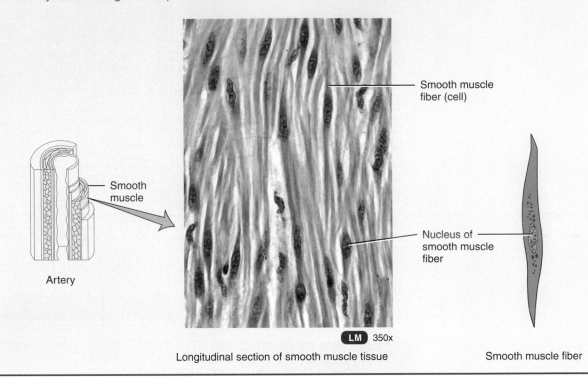

Smooth muscle fiber (cell)

Smooth muscle

Nucleus of smooth muscle fiber

Artery

LM 350x

Longitudinal section of smooth muscle tissue

Smooth muscle fiber

---

iris of the eye, smooth muscle fibers contract individually, like skeletal muscle fibers, because gap junctions are absent. Chapter 10 provides a detailed discussion of muscle tissue.

CHECKPOINT

16. Which muscles are striated and which are smooth?

17. Which types of muscular tissue have gap junctions?

NERVOUS TISSUE

OBJECTIVE

• Describe the structural features and functions of nervous tissue.

Despite the awesome complexity of the nervous system, it consists of only two principal types of cells: neurons and neuroglia. **Neurons** (*neur-*=nerve, nervous tissue, nervous system) or nerve cells, are sensitive to various stimuli. They convert stimuli into nerve impulses (action potentials) and conduct these impulses to other neurons, to muscle tissue, or to glands. Most neurons consist of three basic parts: a cell body and two kinds of cell processes—dendrites and axons (Table 3.5). The **cell body** contains the nucleus and other organelles. **Dendrites** (*dendr-*=tree)

are tapering, highly branched, and usually short cell processes. They are the major receiving or input portion of a neuron. The **axon** (*axo-*=axis) of a neuron is a single, thin, cylindrical process that may be very long. It is the output portion of a neuron, conducting nerve impulses toward another neuron or to some other tissue.

Even though **neuroglia** (noo-RŌG-lē-a; *-glia*=glue) do not generate or conduct nerve impulses, these cells do have many important supportive functions (see page 543). The detailed structure and function of neurons and neuroglia are considered in Chapter 17.

CHECKPOINT

18. What are the functions of the dendrites, cell body, and axon of a neuron?

AGING AND TISSUES

OBJECTIVE

• Describe the effects of aging on tissues.

Generally, tissues heal faster and leave less obvious scars in the young than in the aged. In fact, surgery performed on fetuses

TABLE 3.5 NERVOUS TISSUE

Description Consists of neurons (nerve cells) and neuroglia. Neurons consist of a cell body and processes extending from the cell body (multiple dendrites and a single axon). Neuroglia do not generate or conduct nerve impulses but have other important supporting functions.

Location Nervous system.

Functions Exhibits sensitivity to various types of stimuli, converts stimuli into nerve impulses (action potentials), and conducts nerve impulses to other neurons, muscle fibers, or glands.

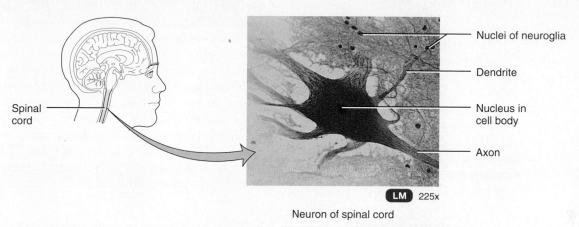

Spinal cord

Nuclei of neuroglia

Dendrite

Nucleus in cell body

Axon

LM 225x

Neuron of spinal cord

leaves no scars. The younger body is generally in a better nutritional state, its tissues have a better blood supply, and its cells have a higher metabolic rate. Thus, cells can synthesize needed materials and divide more quickly. The extracellular components of tissues also change with age. Glucose, the most abundant sugar in the body, plays a role in the aging process. Glucose is haphazardly added to proteins inside and outside cells, forming irreversible cross-links between adjacent protein molecules. With advancing age, more cross-links form, which contributes to the stiffening and loss of elasticity that occur in aging tissues. Collagen fibers, responsible for the strength of tendons, increase in number and change in quality with aging. These changes in

the collagen of arterial walls are as much responsible for their loss of extensibility as are the deposits associated with atherosclerosis, the deposition of fatty materials in arterial walls. Elastin, another extracellular component, is responsible for the elasticity of blood vessels and skin. It thickens, fragments, and acquires a greater affinity for calcium with age—changes that may also be associated with the development of atherosclerosis.

CHECKPOINT

19. What common changes occur in epithelial and connective tissues with aging?

APPLICATIONS TO HEALTH

Disorders of epithelial tissues are mainly specific to individual organs, for example, peptic ulcer disease (PUD), which erodes the epithelial lining of the stomach or small intestine. For this reason, epithelial disorders are described together with the relevant body system throughout the text. The most prevalent disorders of connective tissues are **autoimmune diseases**—diseases in which antibodies produced by the immune system fail to distinguish what is foreign from what is self and attack the body's own tissues. One of the most common autoimmune disorders is rheumatoid arthritis, which attacks the synovial membranes of joints. Because connective tissue is one of the most abundant and widely distributed of the four main types of tissues, its disorders often affect multiple body systems. Common disorders of muscle

tissue and nervous tissue are described at the ends of Chapters 10 and 16, respectively.

SJÖGREN'S SYNDROME

Sjögren's syndrome (SHŌ-grenz) is a common autoimmune disorder that causes inflammation and destruction of exocrine glands, especially the lacrimal (tear) glands and salivary glands. Signs include dryness of the mucous membranes in the eyes, mouth, and nose and salivary gland enlargement. Systemic effects include fatigue, arthritis, difficulty in swallowing, pancreatitis (inflammation of the pancreas), pleuritis (inflammation of the pleurae of the lungs), and muscle and joint pain: The disorder affects females more than males by a ratio of 9 to 1. About 20% of older adults experience some signs of Sjögren's. Treatment is supportive, including using artificial tears to moisten the eyes, sipping fluids, chewing sugarless gum, and using a saliva substitute to moisten the mouth.

SYSTEMIC LUPUS ERYTHEMATOSUS

Systemic lupus erythematosus (er-i-thē-ma-TŌ-sus), **SLE**, or simply lupus, is a chronic inflammatory disease of connective tissue occurring mostly in nonwhite women during their child-bearing years. It is an autoimmune disease that can cause tissue damage in every body system. The disease, which can range from a mild condition in most patients to a rapidly fatal disease, is marked by periods of exacerbation and remission. The prevalence of SLE is about 1 in 2000 persons, with females more likely to be affected than males by a ratio of 8 or 9 to 1.

Although the cause of SLE is not known, genetic, environmental, and hormonal factors are implicated. The genetic component is suggested by studies of twins and family history. Environmental factors include viruses, bacteria, chemicals, drugs, exposure to excessive sunlight, and emotional stress. With regard to hormones, sex hormones, such as estrogens, may trigger SLE.

Signs and symptoms of SLE include painful joints, low-grade fever, fatigue, mouth ulcers, weight loss, enlarged lymph nodes and spleen, sensitivity to sunlight, rapid loss of large amounts of scalp hair, and anorexia. A distinguishing feature of lupus is an eruption across the bridge of the nose and cheeks called a "butterfly rash." Other skin lesions may occur, including blistering and ulceration. The erosive nature of some SLE skin lesions was thought to resemble the damage inflicted by the bite of a wolf—thus, the term *lupus* (=wolf). The most serious complications of the disease involve inflammation of the kidneys, liver, spleen, lungs, heart, brain, and gastrointestinal tract. Because there is no cure for SLE, treatment is supportive, including anti-inflammatory drugs, such as aspirin, and immunosuppressive drugs.

KEY MEDICAL TERMS ASSOCIATED WITH TISSUES

Atrophy (AT-rō-fē; *a-*=without; *-trophy*=nourishment) A decrease in the size of cells, with a subsequent decrease in the size of the affected tissue or organ.

Hypertrophy (hī-PER-trō-fē; *hyper-*=above or excessive) Increase in the size of a tissue because its cells enlarge without undergoing cell division.

Polyp (POL-ip; *polys-*=many; *-pous*=foot) Any mass of tissue that projects upward or outward from a mucous membrane of a hollow structure such as the nose, uterus, colon, or urinary bladder. May be benign or malignant.

Tissue rejection An immune response of the body directed at foreign proteins in a transplanted tissue or organ; immunosuppressive drugs, such as cyclosporine, have largely overcome tissue rejection in heart-, kidney-, and liver-transplant patients.

Tissue transplantation The replacement of a diseased or injured tissue or organ. The most successful transplants involve use of a persons's own tissues or those from an identical twin.

Xenotransplantation (zen′-ō-trans-plan-TĀ-shun; *xeno-*= strange, foreign) The replacement of a diseased or injured tissue or organ with cells or tissues from an animal. Porcine (from pigs) and bovine (from cows) heart valves are used for some heart-valve replacement surgeries.

STUDY OUTLINE

Types of Tissues and Their Origins (p. 58)

1. A tissue is a group of similar cells that usually has a similar embryological origin and is specialized for a particular function.
2. The various tissues of the body are classified into four basic types: epithelial, connective, muscular, and nervous.
3. All tissues of the body develop from three primary germ layers, the first tissues that form in a human embryo: ectoderm, mesoderm, and endoderm.

Cell Junctions (p. 58)

1. Cell junctions are points of contact between adjacent plasma membranes.
2. Tight junctions form fluid-tight seals between cells: adherens junctions, desmosomes, and hemidesmosomes anchor cells to one another or to the basement membrane, and gap junctions permit electrical and chemical signals to pass between cells.

Epithelial Tissue (p. 60)

1. The subtypes of epithelia include covering and lining epithelia and glandular epithelia.
2. An epithelium consists mostly of cells with little extracellular material between adjacent plasma membranes. The apical, lateral, and basal surfaces of epithelial cells are modified in various ways to carry out specific functions. Epithelium is arranged in sheets and attached to a basement membrane. Although it is avascular, it has a nerve supply. Epithelia are derived from all three primary germ layers and have a high capacity for renewal.
3. Epithelial layers are simple (one layer) or stratified (several layers). The cell shapes may be squamous (flat), cuboidal (cubelike), columnar (rectangular), or transitional (variable).
4. Simple squamous epithelium consists of a single layer of flat cells (Table 3.1A). It is found in parts of the body where filtration or diffusion are priority processes. One type, endothelium, lines the heart and blood vessels. Another type, mesothelium, forms the

serous membranes that line the thoracic and abdominal cavities and cover the organs within them.

5. Simple cuboidal epithelium consists of a single layer of cube-shaped cells that function in secretion and absorption (Table 3.1B). It is found covering the ovaries, in the kidneys and eyes, and lining some glandular ducts.

6. Nonciliated simple columnar epithelium consists of a single layer of nonciliated rectangular cells (Table 3.1C). It lines most of the gastrointestinal tract. Specialized cells containing microvilli perform absorption. Goblet cells secrete mucus.

7. Ciliated simple columnar epithelium consists of a single layer of ciliated rectangular cells (Table 3.1D). It is found in a few portions of the upper respiratory tract, where it moves foreign particles trapped in mucus out of the respiratory tract.

8. Pseudostratified columnar epithelium has only one layer but gives the appearance of many (Table 3.1E). A ciliated variety contains goblet cells and lines most of the upper respiratory tract; a nonciliated variety has no goblet cells and lines ducts of many glands, the epididymis, and part of the male urethra.

9. Stratified squamous epithelium consists of several layers of cells: Cells of the apical layer and several layers deep to it are flat (Table 3.1F). A nonkeratinized variety lines the mouth: A keratinized variety forms the epidermis, the most superficial layer of the skin.

10. Stratified cuboidal epithelium consists of several layers of cells: Cells at the apical layer are cube-shaped (Table 3.1G). It is found in adult sweat glands and a portion of the male urethra.

11. Stratified columnar epithelium consists of several layers of cells: Cells of the apical layer have a columnar shape (3.1H). It is found in a portion of the male urethra and large excretory ducts of some glands.

12. Transitional epithelium consists of several layers of cells whose appearance varies with the degree of stretching (Table 3.1I). It lines the urinary bladder.

13. A gland is a single cell or a group of epithelial cells adapted for secretion.

14. Endocrine glands secrete hormones into interstitial fluid and then into the blood (Table 3.1J).

15. Exocrine glands (mucous, sweat, oil, and digestive glands) secrete into ducts or directly onto a free surface (Table 3.1K).

16. Structural classification of exocrine glands includes unicellular and multicellular glands.

17. Functional classification of exocrine glands includes holocrine, apocrine, and merocrine glands.

Connective Tissue (p. 71)

1. Connective tissue is one of the most abundant body tissues.

2. Connective tissue consists of relatively few cells and an abundant extracellular matrix of ground substance and fibers: It does not usually occur on free surfaces, has a nerve supply (except for cartilage), and is highly vascular (except for cartilage, tendons, and ligaments).

3. Cells in connective tissue are derived from mesenchymal cells.

4. Cell types include fibroblasts (secrete matrix), macrophages (perform phagocytosis), plasma cells (secrete antibodies), mast cells (produce histamine), adipocytes (store fat), and white blood cells (migrate from blood in response to infections).

5. The ground substance and fibers make up the extracellular matrix.

6. The ground substance supports and binds cells together, provides a medium for the exchange of materials, stores water, and is active in influencing cell functions.

7. Substances found in the ground substance include water and polysaccharides such as hyaluronic acid, chondroitin sulfate, dermatan sulfate, and keratan sulfate (glycosaminoglycans). Also present are proteoglycans and adhesion proteins.

8. The fibers in the extracellular matrix provide strength and support and are of three types: (a) Collagen fibers (composed of collagen) are found in large amounts in bone, tendons, and ligaments; (b) elastic fibers (composed of elastin, fibrillin, and other glycoprotein) are found in skin, blood vessel walls, and lungs; (c) reticular fibers (composed of collagen and glycoprotein) are found around fat cells, nerve fibers, and skeletal and smooth muscle cells.

9. The two major subclasses of connective tissue are embryonic connective tissue (found in the embryo and fetus) and mature connective tissue (present in the newborn).

10. The embryonic connective tissues are mesenchyme, which forms all other connective tissues (Table 3.2A), and mucous connective tissue, found in the umbilical cord of the fetus, where it gives support (Table 3.2B).

11. Mature connective tissue differentiates from mesenchyme. It is subdivided into several kinds: loose or dense connective tissue, cartilage, bone tissue, and liquid connective tissue.

12. Loose connective tissue includes areolar connective tissue, adipose tissue and reticular connective tissue.

13. Areolar connective tissue consists of the three types of fibers, several types of cells, and a semifluid ground substance (Table 3.3A). It is found in the subcutaneous layer, in mucous membranes, and around blood vessels, nerves, and body organs.

14. Adipose tissue consists of adipocytes, which store triglycerides (Table 3.3B). It is found in the subcutaneous layer, around organs, and in yellow bone marrow. Brown adipose tissue (BAT) generates heat.

15. Reticular connective tissue consists of reticular fibers and reticular cells and is found in the liver, spleen, and lymph nodes (Table 3.3C).

16. Dense connective tissue includes dense regular connective tissue, dense irregular connective tissue, and elastic connective tissue.

17. Dense regular connective tissue consists of parallel bundles of collagen fibers and fibroblasts (Table 3.3D). It forms tendons, most ligaments, and aponeuroses.

18. Dense irregular connective tissue consists of usually randomly arranged collagen fibers and a few fibroblasts (Table 3.3E). It is found in fasciae, the dermis of skin, and membrane capsules around organs.

19. Elastic connective tissue consists of branching elastic fibers and fibroblasts (Table 3.3F). It is found in the walls of large arteries, lungs, trachea, and bronchial tubes.

20. Cartilage contains chondrocytes and has a rubbery matrix (chondroitin sulfate) containing collagen and elastic fibers.

21. Hyaline cartilage is found in the embryonic skeleton, at the ends of bones, in the nose, and in respiratory structures (Table 3.3G). It is flexible, allows movement, and provides support.

22. Fibrocartilage is found in the pubic symphysis, intervertebral discs, and menisci (cartilage pads) of the knee joint (Table 3.3H).

23. Elastic cartilage maintains the shape of organs such as the epiglottis of the larynx, auditory (eustachian) tubes, and external ear (Table 3.3I).

24. Bone or osseous tissue consists of a matrix of mineral salts and collagen fibers that contribute to the hardness of bone, and osteocytes that are located in lacunae (Table 3.3J). It supports, protects, provides a surface area for muscle attachment, helps provide movement, stores minerals, and houses blood-forming tissue.

25. Blood tissue is a liquid connective tissue that consists of blood plasma and formed elements—red blood cells, white blood cells, and platelets (Table 3.3K). Its cells function to transport oxygen and carbon dioxide, carry on phagocytosis, participate in allergic reactions, provide immunity, and bring about blood clotting.

26. Lymph is also a liquid connective tissue—it is the extracellular fluid that flows in lymphatic vessels. Lymph is a clear fluid similar to blood plasma but with less protein.

Membranes (p. 82)

1. An epithelial membrane consists of an epithelial layer overlying a connective tissue layer. Examples are mucous, serous, and cutaneous membranes.

2. Mucous membranes line cavities that open to the exterior, such as the gastrointestinal tract.

3. Serous membranes line closed cavities (pleura, pericardium, peritoneum) and cover the organs in the cavities. These membranes consist of parietal and visceral layers.

4. Synovial membranes line joint cavities, bursae, and tendon sheaths and consist of areolar connective tissue instead of epithelium.

Muscular Tissue (p. 84)

1. Muscular tissue consists of fibers that are specialized for contraction. It provides motion, maintenance of posture, heat production, and protection.

2. Skeletal muscle tissue is attached to bones and is striated, and its action is voluntary (Table 3.4A).

3. Cardiac muscle tissue forms most of the heart wall and is striated, and its action is involuntary (Table 3.4B).

4. Smooth muscle tissue is found in the walls of hollow internal structures (blood vessels and viscera) and is nonstriated, and its action is involuntary (Table 3.4C).

Nervous Tissue (p. 86)

1. The nervous system is composed of neurons (nerve cells) and neuroglia (protective and supporting cells) (Table 3.5).

2. Neurons are sensitive to stimuli, convert stimuli into nerve impulses, and conduct nerve impulses.

3. Most neurons consist of a cell body and two types of processes, dendrites and axons.

Aging and Tissues (p. 86)

1. Tissues heal faster and leave less obvious scars in the young than in the aged; surgery performed on fetuses leaves no scars.

2. The extracellular components of tissues, such as collagen and elastic fibers, also change with age.

SELF-QUIZ QUESTIONS

Choose the one best answer to the following questions.

1. A group of similar cells that has the same embryological origin and operates together to perform a specialized activity is called a(n):
   a. organ.   b. tissue.   c. system.   d. organelle.   e. organism.

2. Which of the following is an avascular tissue?
   a. bone   b. fibrocartilage   c. stratified squamous epithelium
   d. Both b and c are correct.
   e. All three types of tissue are avascular.

3. The two layers of a basement membrane are the:
   a. apical and basal layers.   b. parietal and visceral layers.
   c. lamina propria and epithelial layers.
   d. avascular and vascular layers.
   e. basal lamina and reticular lamina.

4. Mesenchyme is:
   a. part of the extracellular matrix of connective tissues.
   b. the embryonic connective tissue from which all other connective tissues arise.
   c. a mucus-producing cell.
   d. a type of fiber found in connective tissues.
   e. the covering of bone and cartilage.

5. The connective tissue best designed to resist pulling forces in various directions is:
   a. fibrocartilage.   b. bone.   c. dense regular connective tissue.
   d. dense irregular connective tissue.
   e. areolar connective tissue.

6. Supporting cells in nervous tissue are called:
   a. neuroglia   b. neurons   c. cadherins
   d. mesenchyme   e. synoviocytes

7. Which statement best describes connective tissue?
   a. usually contains large amount of extracellular matrix
   b. always arranged in a single layer of cells
   c. primarily concerned with secretion
   d. cells always very closely packed together
   e. is avascular

8. Which of the following is *voluntary* and *striated*?
   a. skeletal muscle tissue
   b. cardiac muscle tissue
   c. smooth muscle tissue
   d. both skeletal and cardiac muscle tissue
   e. These terms do not apply to muscle tissue.

Complete the following.

9. The muscle tissue that forms most of the wall of the heart is _____ muscle tissue.

10. A gland that secretes its product into a duct is referred to as a(n) _____ gland.

11. Membrane proteins called connexins form tunnels called connexons in the type of cell junction known as a(n) _____.

12. The three types of fibers that strengthen and support the extracellular matrix of connective tissues are _____ fibers, _____ fibers, and _____ fibers.

13. Hollow organs belonging to systems that do not open to the outside of the body are lined with a _____ membrane.
14. The connective tissue layer of a mucous membrane is called the _____.

Are the following statements true or false?

15. Blood and lymph are types of connective tissues.
16. Tight junctions are common between epithelial cells lining the stomach or urinary bladder where they prevent fluid from passing between the lining cells.
17. Hyaline cartilage provides flexibility and acts as a shock absorber in joints and epiphyseal plates.
18. Desmosomes connect lateral surfaces of adjacent cells to one another, and hemidesmosomes connect the basal surface of a cell to a basement membrane.

Matching

19. Match the following:

___ **(a)** a mature bone cell; resides in a space called a lacuna

___ **(b)** large phagocytic cell derived from a monocyte; engulfs bacteria and cleans up debris; wandering or fixed

___ **(c)** derived from a fibroblast; specialized for triglyceride storage

___ **(d)** a mucus-producing cell

___ **(e)** large, branched cell; secretes the extracellular matrix of connective tissues

___ **(f)** located along the walls of blood vessels; secrete histamine, which dilates blood vessels during inflammation

___ **(g)** cartilage cell; resides in a space called a lacuna

___ **(h)** derived from a B lymphocyte; produces antibodies as part of an immune response

(1) fibroblast
(2) mast cell
(3) plasma cell
(4) osteocyte
(5) adipocyte
(6) chondrocyte
(7) macrophage
(8) goblet cell

20. Match the following:

___ **(a)** lines inner surface of the stomach and intestine

___ **(b)** avascular connective tissue that occurs as three main types

___ **(c)** lines urinary bladder, ureters, and urethra

___ **(d)** forms fasciae and dermis of skin

___ **(e)** has cells that are specialized for triglyceride storage

___ **(f)** all cells attached to the basement membrane, but not all cells reach the surface of tissue

___ **(g)** subcutaneous tissue containing several kinds of cells and all three fiber types

___ **(h)** lines the mouth; present on the outer surface of the skin

___ **(i)** lines air sacs of the lungs; suited for diffusion of gases

___ **(j)** strong tissue found in the lungs; can recoil back to its original shape after being stretched

(1) dense irregular connective tissue
(2) adipose tissue
(3) simple columnar epithelium
(4) stratified squamous epithelium
(5) cartilage
(6) areolar connective tissue
(7) transitional epithelium
(8) pseudostratified columnar epithelium
(9) elastic connective tissue
(10) simple squamous epithelium

 CRITICAL THINKING QUESTIONS

1. Mike has had a series of respiratory tract infections this winter. His doctor has just prescribed a mucus-thinning drug. Using your knowledge of the structure of the mucous membrane lining the respiratory tract, how do you think this type of drug will help Mike get better?

2. You've gone out to eat at your favorite fast-food joint: General Bellisimo's Fried Chicken. A health-food zealot waves a chicken drumstick and declares "This is all fat!" Using your knowledge of tissues, defend the contents of the chicken leg.

3. Janelle has been an anorexic for several years. As a result of her chronically low daily caloric intake, her adipocytes are storing little

or no triglycerides. What structural problems might Janelle suffer as a result?

4. The neighborhood kids are walking around with common pins and sewing needles stuck into their fingertips. There is no visible bleeding. What type of tissue have they pierced? How do you know?

5. Sometime during your daily activities you come in contact with a colony of bacteria that gets onto your skin. You have no cuts in the area, and your skin is in good condition. However, these microbes manage to penetrate your tissues and get into your blood. What specific structural barriers would the bacteria have to overcome to accomplish this?

ANSWERS TO FIGURE QUESTIONS

3.1 Gap junctions allow the spread of electrical and chemical signals between adjacent cells.

3.2 The basement membrane provides physical support for epithelium, serves as a filter in the kidneys, and guides cell migration during development and tissue repair.

3.3 Substances would move most rapidly through squamous cells.

3.4 Simple multicellular exocrine glands have a nonbranched duct; compound multicellular exocrine glands have a branched duct.

3.5 Sebaceous (oil) glands are holocrine glands, and salivary glands are merocrine glands.

3.6 Fibroblasts secrete the fibers and ground substance of the extracellular matrix.

3.7 An epithelial membrane is a membrane that consists of an epithelial layer and an underlying connective tissue layer.

DEVELOPMENT

4

I N T R O D U C T I O N   Before we examine the first body system in Chapter 5 (The Integumentary System), we will take a look at how body systems develop. Knowledge of their origins will enhance your understanding of the different systems of the human body. Later in the book, you will learn more about development in the context of the various body systems.

Sexual reproduction is the process by which organisms produce off-spring by making sex cells (immature eggs) called **gametes** (GAM-ēts=spouses). Male gametes are called **sperm (spermatozoa)** and female gametes are called **secondary oocytes**. The organs that produce gametes are called gonads. They are the testes in the male and the ovaries in the female. The details of sperm formation in the testes and secondary oocyte formation in the ovaries are discussed in Chapter 27. Once sperm and a secondary oocyte have developed and the sperm have been deposited in the female reproductive tract, pregnancy can occur.

Examine this image and see if you notice anything unusual. In which direction is time being depicted? What is the symbolism of the flowers? Count the number of women and guess why this number was chosen by the artist.

www.wiley.com/college/apcentral

Pregnancy is a sequence of events that begins with fertilization, proceeds to implantation, embryonic development, and fetal development, and normally ends with birth about 38 weeks later, or 40 weeks after the last menstrual period.

Developmental biology is the study of the sequence of events from the fertilization of a secondary oocyte by a sperm cell to the formation of an adult organism. From fertilization through the eighth week of development, a stage called the **embryonic period,** the developing human is called an **embryo** (*em-*=into;*-bryo*= grow). **Embryology** (em-brē-OL-ō-jē) is the study of development from the fertilized egg through the eighth week. The **fetal period** begins at week nine and continues until birth. During this time, the developing human is called a **fetus** (FĒ-tus=offspring).

Prenatal development (prē-NĀ-tal; *pre-*=before; *natal* = birth) is the time from fertilization to birth and includes both the embryonic and fetal periods. Prenatal development is divided into periods of three calendar months each, called **trimesters.**

1. The **first trimester** is the most critical stage of development during which the rudiments of all the major organs systems appear, and also during which the developing organism is the most vulnerable to the effects of drugs, radiation, and microbes.

2. The **second trimester** is characterized by the nearly complete development of organ systems. By the end of this stage, the fetus assumes distinctively human features.

3. The **third trimester** represents a period of rapid fetal growth. During the early stages of this period, most of the organ systems are becoming fully functional.

In this chapter, we focus on the developmental sequence from fertilization through implantation and embryonic and fetal development.

Among the female reproductive organs involved in the major events of the embryonic and fetal periods are the *ovaries*, paired organs in the superior portion of the pelvic cavity on either side of the uterus (Figure 4.1). During the reproductive years of a female, they produce secondary oocytes and discharge them into the pelvic cavity each month, a process called *ovulation*. The *uterine (fallopian) tubes*, which extend laterally from the uterus, provide a route for sperm to reach a secondary oocyte and transport a fertilized (or unfertilized) oocyte to the uterus. The *uterus*, an inverted pear-shaped organ in the pelvic cavity, consists of a superior portion (fundus), middle portion (body), and inferior portion (cervix). The interior of the body is called the *uterine cavity*, whereas the interior of the cervix is known as the *cervical canal*. Part of the *endometrium*, the lining of the uterus, is shed during menstruation. Deep to the endometrium is a muscle layer called the *myometrium*. The uterus is the site of implantation of a fertilized ovum, development of the fetus during pregnancy, and labor. The cervix of the uterus connects to the *vagina*, a canal that opens to the exterior, which serves as the receptacle for the penis during sexual intercourse, the outlet for the menstrual flow, and the passageway for childbirth.

EMBRYONIC PERIOD

● Explain the major developmental events that occur during the embryonic period.

First Week of Development

The first week of development is characterized by several significant events including fertilization, cleavage of the zygote, blastocyst formation, and implantation.

Fertilization

During **fertilization** (fer-til-i-ZĀ-shun; *fertil-*=fruitful) the genetic material from a haploid sperm cell and a haploid secondary oocyte merges into a single diploid nucleus. Of approximately 200 million sperm introduced into the vagina, fewer than 2 million (1%) reach the cervix of the uterus and only about 200 reach the secondary oocyte. Fertilization normally occurs in the uterine (fallopian) tube within 12 to 24 hours after ovulation. Sperm can remain viable for about 48 hours after deposition in the vagina, although a secondary oocyte is viable for only about 24 hours after ovulation. Thus, pregnancy is *most likely* to occur if intercourse takes place during a 3-day window—from 2 days before ovulation to 1 day after ovulation.

Sperm swim from the vagina into the cervical canal propelled by the whiplike movements of their tails (flagella). The passage of sperm through the rest of the uterus and then into the uterine tube results mainly from contractions of the walls of these organs. Sperm that reach the vicinity of the oocyte within minutes after ejaculation *are not capable* of fertilizing it until about seven hours later. During this time in the female reproductive tract, mostly in the uterine tube, sperm undergo **capacitation** (ka-pas′-i-TĀ-shun; *capacit-*=capable of), a series of functional changes that cause the sperm's tail to beat even more vigorously and prepare its plasma membrane to fuse with the oocyte's plasma membrane. During capacitation, sperm are acted upon by secretions in the female reproductive tract that result in the removal of cholesterol, glycoproteins, and proteins from the plasma membrane around the head of the sperm cell.

For fertilization to occur, a sperm cell first must penetrate the **corona radiata** (kō-RŌ-na=crown; rā-dē-A-ta=to shine), the cells that surround the secondary oocyte, and then the **zona pellucida** (ZŌ-na=zone; pe-LOO-si-da=allowing passage of light), the clear glycoprotein layer between the corona radiata and the oocyte's plasma membrane (Figure 4.2a). One of the glycoproteins in the zona pellucida acts as a sperm receptor. Its binding to specific membrane proteins in the sperm head triggers the **acrosomal reaction,** the release of the contents of the acrosome. The acrosome is a helmetlike structure that covers the head of a sperm and contains several enzymes (see Figure 27.6 on page 842). The acrosomal enzymes digest a path through the zona pellucida as the lashing sperm tail pushes the sperm cell onward. Although many sperm undergo acrosomal reactions, only the first sperm cell to penetrate the entire zona

Figure 4.1 Uterus and associated structures. The left side of the uterine tube, uterus, and vagina have been sectioned.

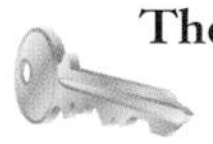

The ovaries produce female gametes called secondary oocytes.

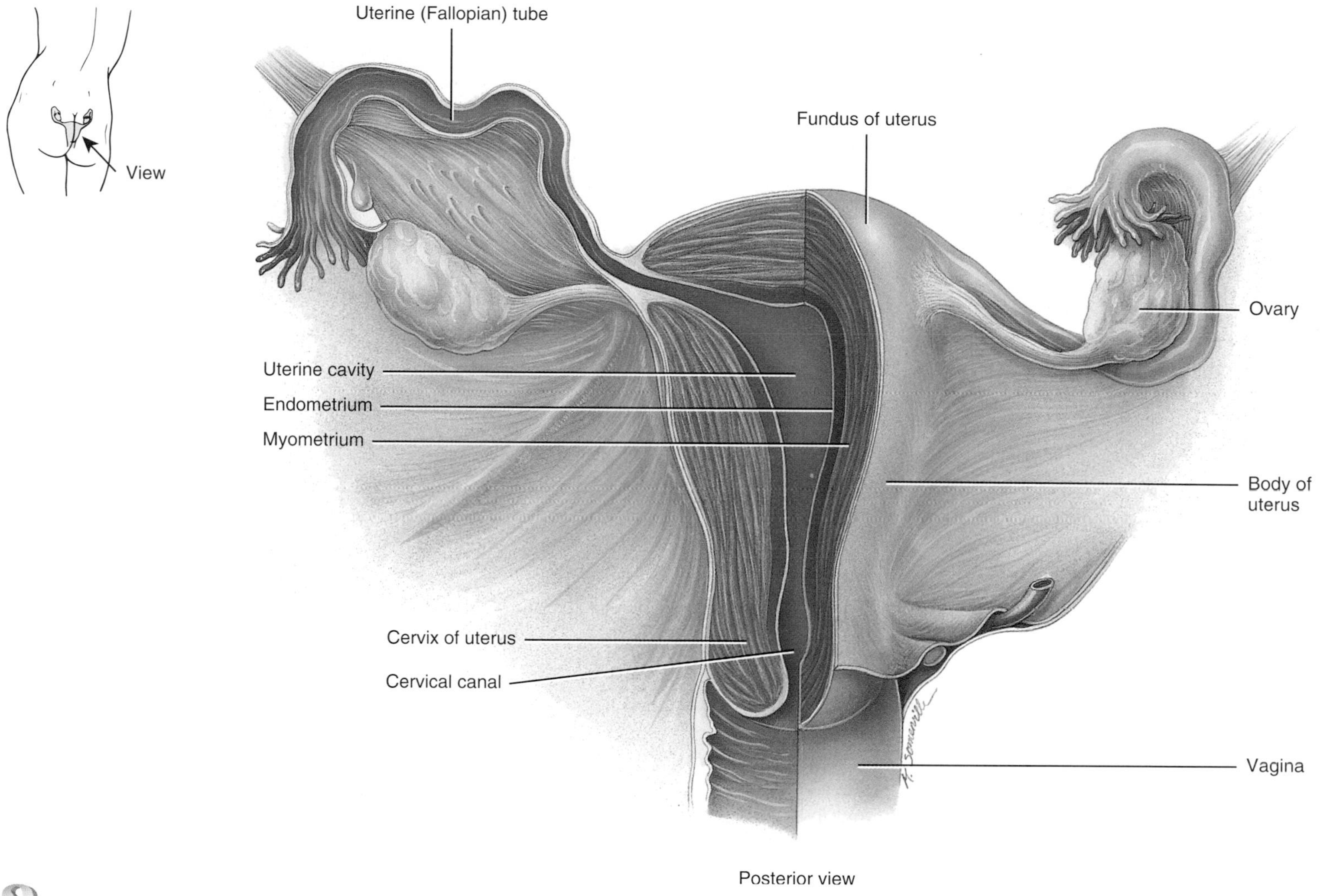

Uterine (Fallopian) tube

Fundus of uterus

Ovary

Uterine cavity

Endometrium

Myometrium

Body of uterus

Cervix of uterus

Cervical canal

Vagina

Posterior view

? **What are the functions of the uterine tubes?**

pellucida and reach the oocyte's plasma membrane fuses with the oocyte. The fusion of a sperm with a secondary oocyte, called **syngamy** (*syn-*=coming together; *-gamy*=marriage), sets in motion events that block **polyspermy,** fertilization by more than one sperm cell.

Once a sperm cell enters a secondary oocyte, the oocyte first must complete meiosis II. It divides into a larger ovum (mature egg) and a smaller second polar body that fragments and disintegrates (see Figure 27.15 on page 855). The nucleus in the head of the sperm develops into the **male pronucleus,** and the nucleus in the fertilized ovum develops into the **female pronucleus** (Figure 4.2c). After the male and female pronuclei form, they fuse, producing a single diploid nucleus that contains 23 chromosomes from each pronucleus. Fertilization has taken place. Thus, the fusion of the haploid (*n*) pronuclei restores the diploid number (2*n*) of 46

chromosomes. The fertilized ovum now is called a **zygote** (ZĪ-gōt; *zygon*=yolk).

Dizygotic (fraternal) twins are produced from the independent release of two secondary oocytes and the subsequent fertilization of each by different sperm. They are the same age and in the uterus at the same time, but they are genetically as dissimilar as are any other siblings. Dizygotic twins may or may not be the same sex. Because **monozygotic (identical) twins** develop from a single fertilized ovum, they contain exactly the same genetic material and are always the same sex. Monozygotic twins arise from separation of the developing zygote into two embryos, which occurs within 8 days of fertilization 99% of the time. Separations that occur later than 8 days are likely to produce **conjoined twins,** a situation in which the twins are joined together and share some body structures.

Figure 4.2 Selected structures and events in fertilization. (a) A sperm cell penetrating the corona radiata and zona pellucida around a secondary oocyte. (b) A sperm cell in contact with a secondary oocyte. (c) Male and female pronuclei.

During fertilization, genetic material from a sperm cell and a secondary oocyte merge to form a single diploid nucleus.

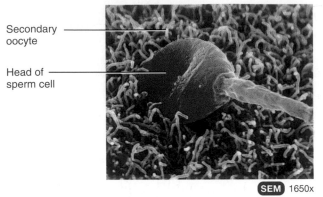

Secondary oocyte

Head of sperm cell

SEM 1650x

(b) Sperm cell in contact with a secondary oocyte

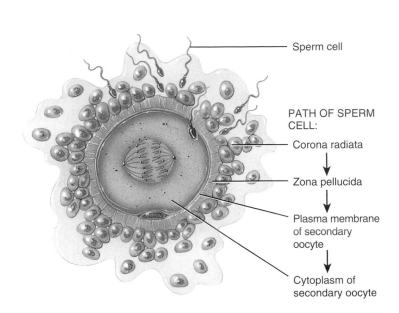

Sperm cell

PATH OF SPERM CELL:

Corona radiata

Zona pellucida

Plasma membrane of secondary oocyte

Cytoplasm of secondary oocyte

(a) Sperm cell penetrating a secondary oocyte

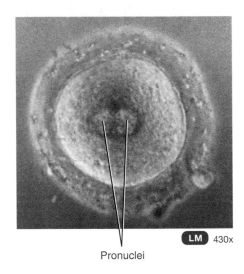

LM 430x

Pronuclei

(c) Male and female pronuclei

What is capacitation?

Cleavage of the Zygote

After fertilization, rapid mitotic cell divisions of the zygote called **cleavage** (KLĒV-ij) take place (Figure 4.3). The first division of the zygote begins about 24 hours after fertilization and is completed about 6 hours later. Each succeeding division takes slightly less time. By the second day after fertilization, the second cleavage is completed and there are four cells (Figure 4.3b). By the end of the third day, there are 16 cells. The progressively smaller cells produced by cleavage are called **blastomeres** (BLAS-tō-mērz; *blasto-*=germ or sprout; *-meres*=parts). Successive cleavages eventually produce a solid sphere of cells called the **morula** (MOR-ū-la; *morula*=mulberry). The morula is still surrounded by the zona pellucida and is about the same size as the original zygote (Figure 4.3c).

Blastocyst Formation

By the end of the fourth day, the number of cells in the morula increases as it continues to move through the uterine tube toward the uterine cavity. When the morula enters the uterine cavity on day 4 or 5, a glycogen-rich secretion from the glands of the endometrium passes into the uterine cavity and enters the morula through the zona pellucida. This secretion, called **uterine milk,** along with nutrients stored in the cytoplasm of the blastomeres of the morula, provides nourishment for the developing morula. At the 32-cell stage, the fluid enters the morula, collects between the blastomeres, and reorganizes them around a large fluid-filled cavity called the **blastocyst cavity** (BLAS-tō-sist; *blasto-*=germ or sprout; *-cyst*=bag) (Figure 4.3e). With the formation of this cavity, the developing mass is then called the **blastocyst.** Though it now has hundreds of cells, the blastocyst is still about the same size as the original zygote.

Further rearrangement of the blastomeres results in the formation of two distinct structures: the inner cell mass and trophoblast (Figure 4.3e). The **inner cell mass** is located internally and eventually develops into the embryo. The **trophoblast** (TRŌF-ō-blast; *tropho-*=develop or nourish) is an outer superficial layer of cells that forms the wall of the blastocyst. It will

Figure 4.3 Cleavage and the formation of the morula and blastocyst.

🔑 **Cleavage refers to the early, rapid mitotic divisions of a zygote.**

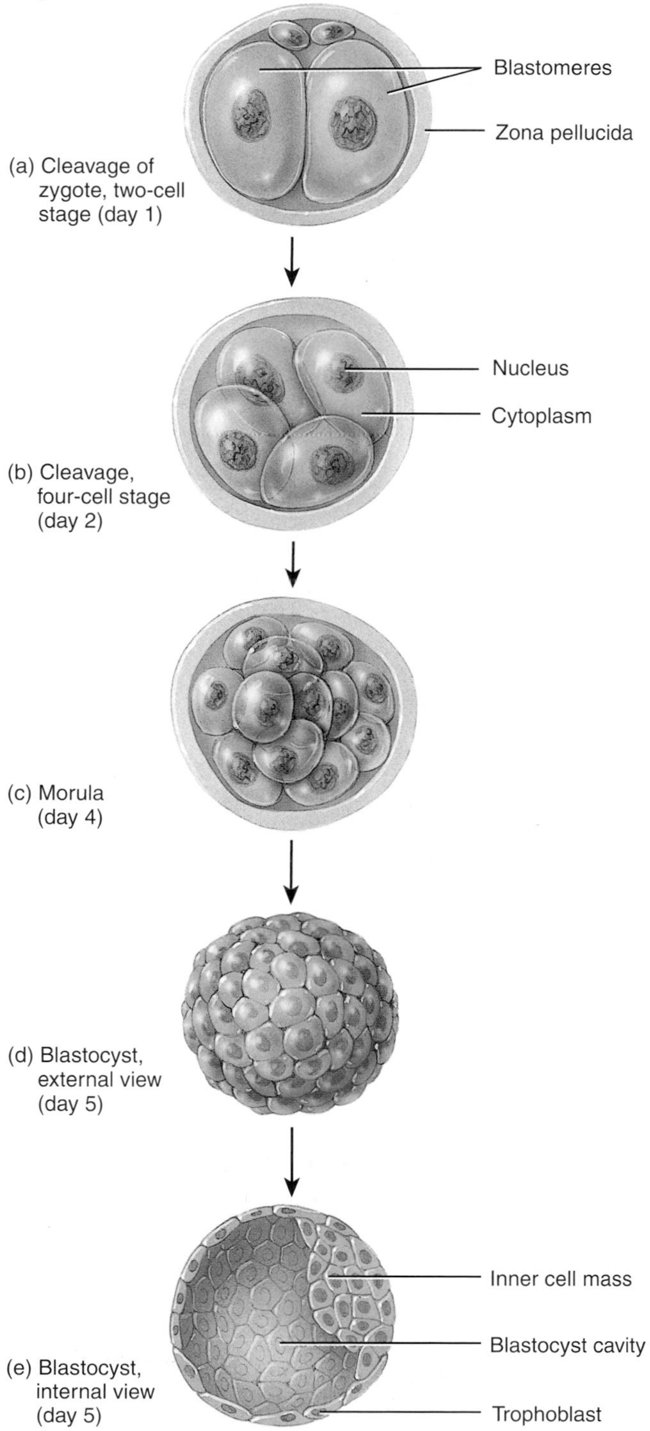

(a) Cleavage of zygote, two-cell stage (day 1)

Blastomeres

Zona pellucida

(b) Cleavage, four-cell stage (day 2)

Nucleus

Cytoplasm

(c) Morula (day 4)

(d) Blastocyst, external view (day 5)

(e) Blastocyst, internal view (day 5)

Inner cell mass

Blastocyst cavity

Trophoblast

❓ **What is the histological difference between a morula and a blastocyst?**

ultimately develop into the fetal portion of the placenta, the site of exchange of nutrients and wastes between the mother and fetus. On about the fifth day after fertilization, the blastocyst hatches from the zona pellucida by digesting a hole in it with an enzyme, and then the blastocyst squeezes through the hole. Shedding of the zona pellucida is necessary in order to permit implantation.

🩺 STEM CELL RESEARCH AND THERAPEUTIC CLONING

Stem cells are unspecialized cells that have the ability to divide for indefinite periods and give rise to specialized cells. In the context of human development, a zygote (fertilized ovum) is a stem cell. Because it has the potential to form an entire organism, a zygote is known as a *totipotent stem cell* (tō-TIP-ō-tent; *totus-*=whole; *-potentia*=power). Inner cell mass cells, by contrast, can give rise to many (but not all) different types of cells. They are called *pluripotent stem cells* (ploo-RIP-ō-tent; *plur-*=several). Later, pluripotent stem cells can undergo further specialization into cells that have a specific function. These cells are called *multipotent stem cells* (mul-TIP-ō-tent). Examples include keratinocytes that produce new skin cells, myeloid and lymphoid stem cells that develop into blood cells, and spermatogonia that give rise to sperm.

Pluripotent stem cells currently used in research are derived from (1) the inner cell mass of embryos in the blastocyst stage that were destined to be used for infertility treatments but were not needed and from (2) nonliving fetuses terminated during the first three months of pregnancy. On October 13, 2001, researchers reported cloning of the first human embryo to grow cells to treat human diseases. **Therapeutic cloning** is envisioned as a procedure in which the genetic material of a patient with a particular disease is used to create pluripotent stem cells to treat the disease. Using the principles of therapeutic cloning, scientists hope to make an embryo clone of a patient, remove the pluripotent stem cells from the embryo, and then use them to grow tissues to treat particular diseases. For example, they might be used to generate cells and tissues for transplantation to treat conditions such as cancer, Parkinson's and Alzheimer's disease, spinal cord injury, diabetes, heart disease, stroke, burns, birth defects, osteoarthritis, and rheumatoid arthritis. Presumably, the tissues would not be rejected since they would contain the patient's own genetic material. ■

Implantation

About 6 days after fertilization, the blastocyst loosely attaches to the endometrium, a process called **implantation** (Figure 4.4). As the blastocyst implants, it orients with the inner cell mass toward the endometrium (Figure 4.4b). About 7 days after fertilization, the blastocyst attaches to the endometrium more firmly, endometrial glands in the vicinity enlarge, and the endometrium becomes more vascularized (forms new blood vessels).

Following implantation, the endometrium is known as the **decidua** (de-SID-ū-a=falling off). The decidua separates from the endometrium after the fetus is delivered much as it does

Figure 4.4 Relation of a blastocyst to the endometrium of the uterus at the time of implantation.

Implantation, the attachment of a blastocyst to the endometrium, occurs about 6 days after fertilization.

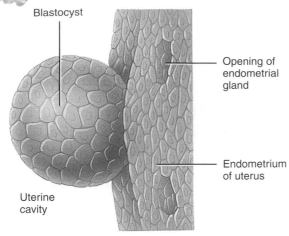

(a) External view of blastocyst, about 6 days after fertilization

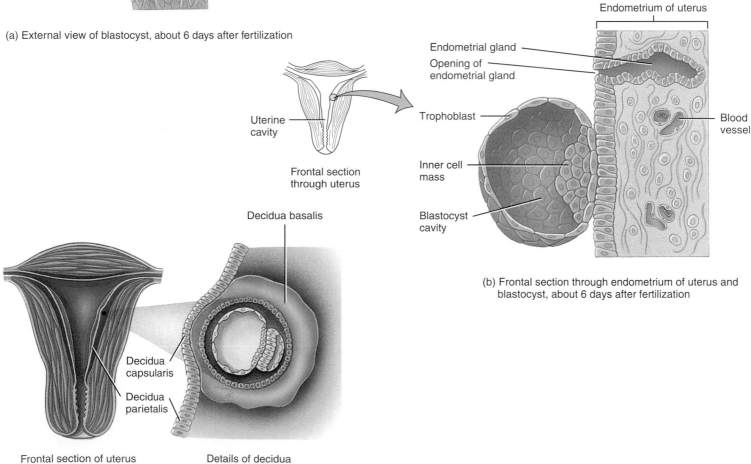

Frontal section through uterus

(b) Frontal section through endometrium of uterus and blastocyst, about 6 days after fertilization

Frontal section of uterus

Details of decidua

(c) Regions of the decidua

How does the blastocyst merge with and burrow into the endometrium?

in normal menstruation. Different regions of the decidua are named based on their positions relative to the site of the implanted blastocyst (Figure 4.4c). The **decidua basalis** is the portion of the endometrium between the embryo and the stratum basalis of the uterus; it provides large amounts of glycogen and lipids for the developing embryo and fetus and later becomes the maternal part of the placenta. The **decidua capsularis** is the portion of the endometrium that is located between the embryo and the uterine cavity. The **decidua parietalis** (par-ri-e-TAL-is) is the remaining modified endometrium that lines the noninvolved areas of the rest of the uterus. As the embryo and later the fetus enlarges, the decidua capsularis bulges into the uterine cavity and fuses with the decidua parietalis, thereby obliterating the uterine cavity. By about 27 weeks, the decidua capsularis degenerates and disappears.

The major events associated with the first week of development are summarized in Figure 4.5.

ECTOPIC PREGNANCY

Ectopic pregnancy (ek-TOP-ik; *ec-*=out of; *-topic*=place) is the development of an embryo or fetus outside the uterine cavity. An ectopic pregnancy usually occurs when movement of the fertilized ovum through the uterine tube is impaired. Situations that impair movement include scarring due to a prior tubal infection, decreased movement of the uterine tube smooth muscle, or abnormal tubal anatomy. Although the most common site of ectopic pregnancy is the uterine tube, ectopic pregnancies may also occur in the ovary, abdominal cavity, or uterine cervix. Compared to nonsmokers, women who smoke are twice as likely to have an ectopic pregnancy because nicotine in cigarette smoke paralyzes the cilia in the lining of the uterine tube (as it does those in the respiratory airways). Scars from pelvic inflammatory disease, previous uterine tube surgery, and previous ectopic pregnancy may also hinder movement of the fertilized ovum.

The signs and symptoms of ectopic pregnancy include one or two missed menstrual cycles followed by bleeding and acute abdominal and pelvic pain. Unless removed, the developing embryo can rupture the uterine tube, often resulting in death of the mother. Treatment options include surgery or the use of a cancer drug called methotrexate, which causes embryonic cells to stop dividing and eventually disappear. ■

CHECKPOINT

1. Where does fertilization normally occur?
2. What is a morula, and how is it formed?
3. Describe the layers of a blastocyst, and their eventual fates.
4. When, where, and how does implantation occur?
5. On what basis are the three regions of the decidua named?

Second Week of Development
Development of the Trophoblast

About 8 days after fertilization, the trophoblast develops into two layers in the region of contact between the blastocyst and endometrium. These layers are a **syncytiotrophoblast** (sin-sīt′-ē-ō-TRŌF-ō-blast) that contains no distinct cell boundaries, and a **cytotrophoblast** (sī-tō-TRŌF-ō-blast) between the inner

Figure 4.5   Summary of events associated with the first week of development.

Fertilization usually occurs in the uterine tube.

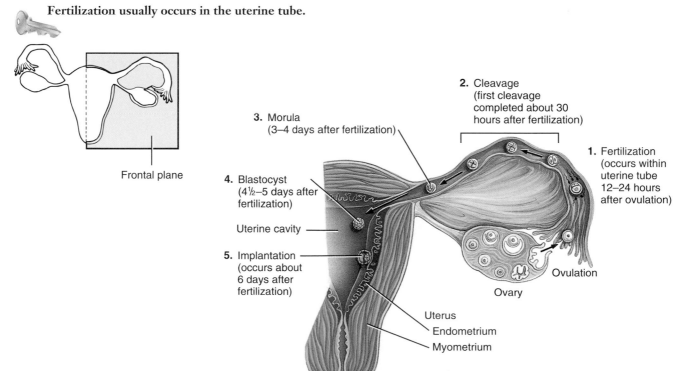

3. Morula
(3–4 days after fertilization)

2. Cleavage
(first cleavage completed about 30 hours after fertilization)

1. Fertilization
(occurs within uterine tube 12–24 hours after ovulation)

Frontal plane

4. Blastocyst
(4½–5 days after fertilization)

Uterine cavity

5. Implantation
(occurs about 6 days after fertilization)

Ovulation

Ovary

Uterus
Endometrium
Myometrium

Frontal section through uterus, uterine tube, and ovary

 In which portion of the uterus does implantation usually occur?

cell mass and syncytiotrophoblast that is composed of distinct cells (Figure 4.6a). The two layers of trophoblast become part of the chorion (one of the fetal membranes) as they undergo further growth (see Figure 4.11a inset). During implantation, the syncytiotrophoblast secretes enzymes that enable the blastocyst to penetrate the uterine lining by digesting and liquefying the endometrial cells. Eventually, the blastocyst becomes buried in the endometrium and inner one-third of the myometrium (mus-

cle layer of the uterus). Another secretion of the trophoblast is human chorionic gonadotropin (hCG), which rescues the corpus luteum from degeneration and sustains its secretion of progesterone and estrogens. These hormones maintain the uterine lining in a secretory state and thereby prevent menstruation. Peak secretion of hCG occurs about the ninth week of pregnancy at which time the placenta is fully developed and produces the progesterone and estrogens that continue to sustain the preg-

Figure 4.6 Principal events of the second week of development.

🔑 **About 8 days after fertilization, the trophoblast develops into a syncytiotrophoblast and a cytotrophoblast; the inner cell mass develops into a hypoblast and epiblast (bilaminar embryonic disc).**

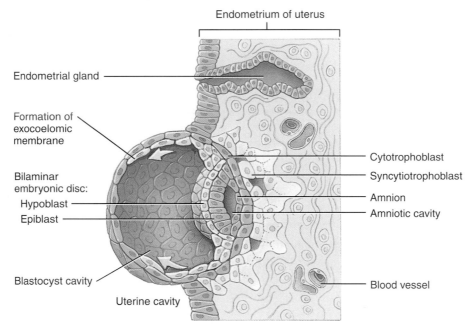

(a) Frontal section through endometrium of uterus showing blastocyst, about 8 days after fertilization

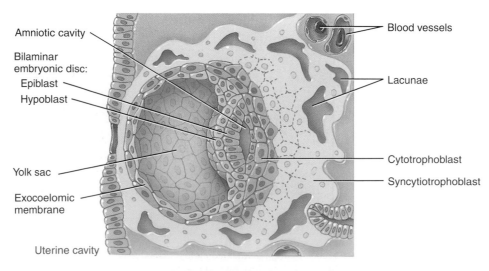

(b) Frontal section through endometrium of uterus showing blastocyst, about 9 days after fertilization

nancy. The presence of hCG in maternal blood or urine is an indicator of pregnancy and is the hormone detected by home pregnancy tests.

Development of the Bilaminar Embryonic Disc

Cells of the inner cell mass also differentiate into two layers around 8 days after fertilization: a **hypoblast (primitive endoderm)** and **epiblast (primitive ectoderm)** (Figure 4.6a). Cells of the hypoblast and epiblast together form a flat disc referred to as the **bilaminar embryonic disc** (bī-LAM-in-ar= two-layered). In addition, a small cavity appears within the epiblast and eventually enlarges to form the **amniotic cavity** (am-nē-OT-ik; *amnio-*=lamb).

Development of the Amnion

As the amniotic cavity enlarges, a thin protective membrane called the **amnion** (AM-nē-on) develops from the epiblast (Figure 4.6a). Whereas the amnion forms the roof of the amniotic cavity, the epiblast forms the floor. Initially, the amnion overlies only the bilaminar embryonic disc. However, as the embryo grows, the amnion eventually surrounds the entire embryo (see Figure 4.11a inset), creating the amniotic cavity that becomes filled with **amniotic fluid.** Most amniotic fluid is initially derived from maternal blood. Later, the fetus contributes to the fluid by excreting urine into the amniotic cavity. Amniotic fluid serves as a shock absorber for the fetus, helps regulate fetal body temperature, helps prevent drying out, and prevents adhesions between the skin of the fetus and surrounding tissues. The amnion usually ruptures just before birth; it and its fluid constitute the "bag of waters." Embryonic cells are normally sloughed off into amniotic fluid. They can be examined in a procedure called amniocentesis, which involves withdrawing some of the amniotic fluid that bathes the developing fetus and analyzing the fetal cells and dissolved substances (see page 117).

Development of the Yolk Sac

Also on the eighth day after fertilization, cells of the hypoblast migrate and cover the inner surface of the blastocyst wall (Figure 4.6a). The migrating cells form a thin membrane called the **exocoelomic membrane** (ek′-sō-sē-LŌ-mik; *exo-*=outside; *-koilos*=space). Together with the hypoblast, the exocoelomic membrane forms the wall of the **yolk sac,** formerly called the blastocyst cavity (Figure 4.6b). As a result, the bilaminar embryonic disc is now positioned between the amniotic cavity and yolk sac.

Since human embryos receive their nutrients from the endometrium, the yolk sac is relatively empty, small, and decreases in size as development progresses (see Figure 4.11a inset). Nevertheless, the yolk sac has several important functions in humans. It supplies nutrients to the embryo during the second and third weeks of development, is the source of blood cells from the third through sixth weeks, contains the first cells (primordial germ cells) that will eventually migrate into the developing gonads and there will differentiate into the primitive germ cells that will form gametes, forms part of the gut (gastrointestinal tract), functions as a shock absorber, and helps prevent drying out of the embryo.

Development of Sinusoids

On the ninth day after fertilization, the blastocyst becomes completely embedded in the endometrium. As the syncytiotrophoblast expands into the endometrium and around the yolk sac, small spaces called **lacunae** (la-KOO-nē=little lakes) develop within it (Figure 4.6b).

By the twelfth day of development, the lacunae fuse to form larger, interconnecting spaces called **lacunar networks** (Figure 4.6c). Endometrial capillaries (microscopic blood vessels) around the developing embryo expand and are referred to as **sinusoids.** As the syncytiotrophoblast erodes some of the sinu-

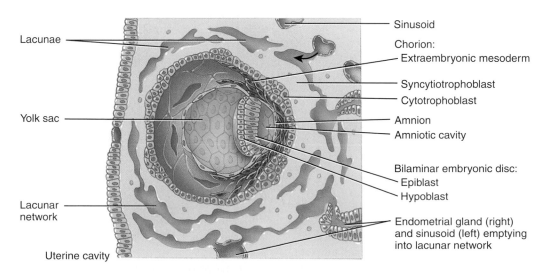

Lacunae

Yolk sac

Lacunar network

Uterine cavity

Sinusoid

Chorion:
Extraembryonic mesoderm

Syncytiotrophoblast

Cytotrophoblast

Amnion

Amniotic cavity

Bilaminar embryonic disc:
Epiblast
Hypoblast

Endometrial gland (right) and sinusoid (left) emptying into lacunar network

(c) Frontal section through endometrium of uterus showing blastocyst, about 12 days after fertilization

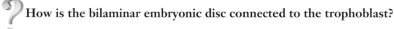

How is the bilaminar embryonic disc connected to the trophoblast?

soids and endometrial glands, maternal blood and glandular secretions enter the lacunar networks, which serve as both a rich source of materials for embryonic nutrition and a disposal site for the embryo's wastes.

Development of the Extraembryonic Coelom

About the twelfth day after fertilization, the **extraembryonic mesoderm** develops. These mesodermal cells are derived from the yolk sac and form a connective tissue (mesenchyme) around the amnion and yolk sac (Figure 4.6c). Soon a number of large cavities develop in the extraembryonic mesoderm, which then fuse to form a single, larger cavity called the **extraembryonic coelom** (SĒ-Lōm).

Development of the Chorion

The extraembryonic mesoderm and the two layers of the trophoblast (the cytotrophoblast and the syncytiotrophoblast) together form the **chorion** (KOR-ē-on=membrane) (Figure 4.6c). It surrounds the embryo and, later, the fetus (see Figure 4.11a inset). Eventually the chorion becomes the principal embryonic part of the placenta, the structure for exchange of materials between mother and fetus. The chorion also protects the embryo and fetus from the immune responses of the mother and produces human chorionic gonadotropin (hCG).

The inner layer of the chorion eventually fuses with the amnion. With the development of the chorion, the extraembryonic coelom is now referred to as the **chorionic cavity.** By the end of the second week of development, the bilaminar embryonic disc becomes connected to the trophoblast by a band of extraembryonic mesoderm called the **connecting (body) stalk** (see Figure 4.7, inset), the future umbilical cord.

CHECKPOINT

6. What are the functions of the trophoblast?
7. How is the bilaminar embryonic disc formed?
8. Describe the formation of the amnion, yolk sac, and chorion and explain their functions.
9. Why are sinusoids important during embryonic development?

Third Week of Development

The third week of development begins a six-week period of rapid embryonic development and differentiation. During the third week, the three primary germ layers are established, which lays the groundwork for organ development in weeks four through eight.

Gastrulation

The first major event of the third week of development is called **gastrulation** (gas′-troo-LĀ-shun) and occurs about 15 days after fertilization. In this process, the bilaminar (two-layered) embryonic disc, consisting of epiblast and hypoblast, is transformed into a **trilaminar** (three-layered) **embryonic disc** consisting of three primary germ layers: the ectoderm, mesoderm, and endoderm. The primary germ layers are the major embryonic tissues from which the various tissues and organs of the body develop.

Gastrulation involves the rearrangement and migration of cells from the epiblast. The first evidence of gastrulation is the formation of the **primitive streak,** a faint groove on the dorsal surface of the epiblast that elongates from the posterior to the anterior part of the embryo (Figure 4.7a). The primitive streak clearly establishes the head and tail ends of the embryo, as well as its right and left sides. At the head end of the primitive streak a small group of epiblastic cells forms a rounded structure called the **primitive node.**

Following formation of the primitive streak, cells of the epiblast move inward below the primitive streak and detach from the epiblast (Figure 4.7b). This inward movement is called **invagination** (in-vaj-i-NĀ-shun). Once the cells have invaginated, some of them displace the hypoblast, forming the **endoderm** (*endo-*=inside; *-derm*=skin). Other cells remain between the epiblast and newly formed endoderm to form the **mesoderm** (*meso-*=middle). Cells remaining in the epiblast then form the **ectoderm** (*ecto-*=outside). Table 4.1 on page 104 provides details about the fates of these primary germ layers; coverage is also included in later chapters in the context of the various body systems.

About 16 days after fertilization, mesodermal cells from the primitive node migrate toward the head end of the embryo and form a hollow tube of cells in the midline called the **notochordal process** (nō-tō-KOR-dal) (Figure 4.8 on page 104). By days 22–24, the notochordal process becomes a solid cylinder of cells called the **notochord** (nō-tō-KORD; *noto-*=back; *-chord*=cord) (see Figure 4.9a). This structure plays an extremely important role in **induction** (in-DUK-shun), the process by which one tissue (*inducing tissue*) stimulates the development of an adjacent unspecialized tissue (*responding tissue*) into a specialized one. An inducing tissue usually produces a chemical substance that influences the responding tissue. The notochord induces certain mesodermal cells to develop into parts of vertebrae (back bones) and contributes to the formation of intervertebral discs between vertebrae (see Figure 7.17 on page 187).

During the third week of development, two faint depressions appear on the dorsal surface of the embryo. The structure closer to the head end is called the **oropharyngeal membrane** (or-ō-fa-RIN-jē-al; *oro-*=mouth; *-pharyngeal*=pertaining to the pharynx) (Figure 4.8a, b). It breaks down during the fourth week to connect the mouth cavity to the pharynx (throat) and the remainder of the gastrointestinal tract. The structure closer to the tail end is called the **cloacal membrane** (klō-Ā-kul=sewer), which degenerates in the seventh week to form the openings of the anus and urinary and reproductive tracts.

When the cloacal membrane appears, the wall of the yolk sac forms a small vascularized outpouching called the **allantois** (a-LAN-tō-is; *allant-*=sausage) that extends into the connecting stalk (Figure 4.8b). In most other mammals, the allantois is used for gas exchange and waste removal. Because of the role of the human placenta in these activities, the allantois is not a prominent structure in humans (see Figure 4.11a, inset). Nevertheless, it does function in early formation of blood and blood vessels and it is associated with the development of the urinary bladder.

Figure 4.7   Gastrulation.

Gastrulation involves the rearrangement and migration of cells from the epiblast.

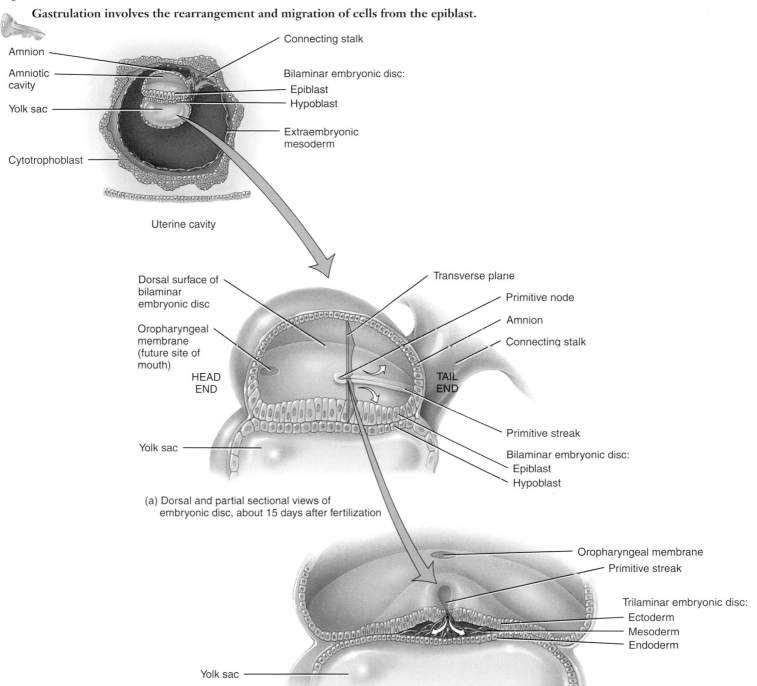

(a) Dorsal and partial sectional views of embryonic disc, about 15 days after fertilization

(b) Transverse section of trilaminar embryonic disc, about 16 days after fertilization

What is the significance of gastrulation?

TABLE 4.1 STRUCTURES PRODUCED BY THE THREE PRIMARY GERM LAYERS*

ENDODERM	MESODERM	ECTODERM
Epithelial lining of gastrointestinal tract (except the oral cavity and anal canal) and the epithelium of its glands.	All skeletal and cardiac muscle tissue and most smooth muscle tissue.	All nervous tissue.
Epithelial lining of urinary bladder, gallbladder, and liver.	Cartilage, bone, and other connective tissues.	Epidermis of skin.
Epithelial lining of pharynx, auditory (eustachian) tubes, tonsils, larynx, trachea, bronchi, and lungs.	Blood, red bone marrow, and lymphatic tissue.	Hair follicles, arrector pili muscles, nails, epithelium of skin glands (sebaceous and sudoriferous), and mammary glands.
Epithelium of thyroid gland, parathyroid glands, pancreas, and thymus.	Endothelium of blood vessels and lymphatic vessels.	Lens, cornea, and internal eye muscles.
Epithelial lining of prostate and bulbourethral (Cowper's) glands, vagina, vestibule, urethra, and associated glands such as the greater (Bartholin's) vestibular and lesser vestibular glands.	Dermis of skin.	Internal and external ear.
	Fibrous tunic and vascular tunic of eye.	Neuroepithelium of sense organs.
	Middle ear.	Epithelium of oral cavity, nasal cavity, paranasal sinuses, salivary glands, and anal canal.
	Mesothelium of thoracic, abdominal, and pelvic body cavities.	Epithelium of pineal gland, pituitary gland, and adrenal medullae.
	Epithelium of kidneys and ureters.	
	Epithelium of adrenal cortex.	
	Epithelium of gonads and genital ducts.	

* Discussed in detail in later chapters.

Figure 4.8 Development of the notochordal process.

The notochordal process develops from the primitive node and later becomes the notochord.

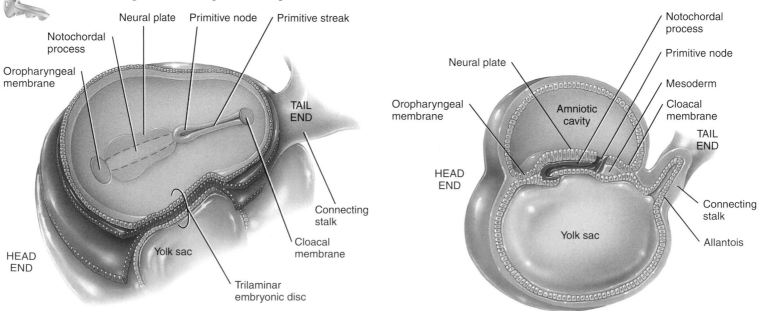

(a) Dorsal and partial sectional views of trilaminar embryonic disc, about 16 days after fertilization

(b) Sagittal section of trilaminar embryonic disc, about 16 days after fertilization

What is the significance of the notochord?

Neurulation

In addition to inducing mesodermal cells to develop into parts of vertebrae, the notochord also induces ectodermal cells over it to form the **neural plate** (Figure 4.9a). (Also see Figure 19.26 on page 624) By the end of the third week, the lateral edges of the neural plate become more elevated and form the **neural fold** (Figure 4.9b). The depressed region between the folds is called the **neural groove** (Figure 4.9c). Generally, the neural folds approach each other and fuse, thus converting the neural plate into a **neural tube** (Figure 4.9d). Neural tube cells then develop into the brain and spinal cord. The process by which the neural plate, neural folds, and neural tube form is called **neurulation** (noor-oo-LĀ-shun).

Figure 4.9 Neurulation and the development of somites.

Neurulation is the process by which the neural plate, neural folds, and neural tube form.

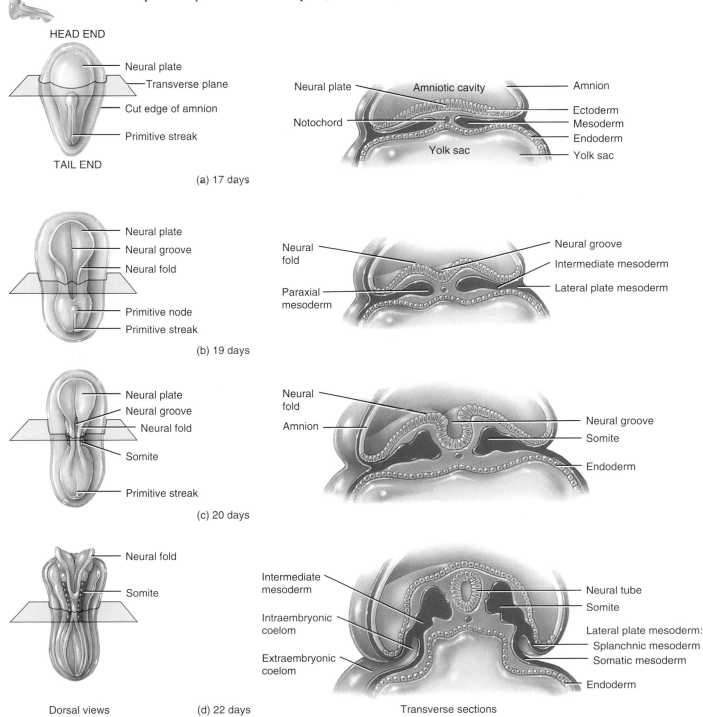

Dorsal views   (d) 22 days   Transverse sections

Which structures develop from the neural tube and somites?

As the neural tube forms, some of the ectodermal cells from the tube migrate to form several layers of cells called the **neural crest** (see Figure 19.26 on page 624). Neural crest cells give rise to nerves, the meninges (coverings) of the brain and spinal cord, and several skeletal and muscular components of the head.

At about four weeks after fertilization, the head end of the neural tube develops into three enlarged areas called **primary brain vesicles** (see Figure 19.27 on page 625), from which the brain develops. The parts of the brain that develop from the various brain vesicles are described on page 625 in Chapter 19.

ANENCEPHALY

Neural tube defects (NTDs) are caused by arrest of the normal development and closure of the neural tube. These include spina bifida (discussed on page 199) and **anencephaly** (an′-en-SEPH-a-lē; *an-*=without; *encephal*=brain). In anencephaly, the cranial bones fail to develop and certain parts of the brain remain in contact with amniotic fluid and degenerate. Usually, a part of the brain that controls vital functions such as breathing and regulation of the heart is also affected. Infants with anencephaly are stillborn or die within a few days after birth. The condition occurs about once in every 1000 births and is 2 to 4 times more common in female infants than males. ■

Development of Somites

By about the seventeenth day after fertilization, the mesoderm adjacent to the notochord and neural tube forms paired longitudinal columns of **paraxial mesoderm** (par-AK-sē-al; *para-*=near) (Figure 4.9b). The mesoderm lateral to the paraxial mesoderm forms paired cylindrical masses called **intermediate mesoderm.** The mesoderm lateral to the intermediate mesoderm consists of a pair of flattened sheets called **lateral plate mesoderm.** The paraxial mesoderm soon segments into a series of paired, cube-shaped structures called **somites** (SŌ-mīts=little bodies). By the end of the fifth week, 42–44 pairs of somites are present. The number of somites that develop over a given period can be correlated to the approximate age of the embryo.

Each somite differentiates into three regions: a **myotome,** a **dermatome,** and a **sclerotome** (see Figure 10.10 on page 282). The myotomes develop into the skeletal muscles of the neck, trunk, and limbs; the dermatomes form connective tissue, including the dermis of the skin; and the sclerotomes give rise to the vertebrae and ribs.

Development of the Intraembryonic Coelom

In the third week of development, small spaces appear in the lateral plate mesoderm. These spaces soon merge to form a larger cavity called the **intraembryonic coelom** (SĒ-lom=cavity). This cavity splits the lateral plate mesoderm into two parts called the splanchnic mesoderm and somatic mesoderm (Figure 4.9d). During the second month of development, the intraembryonic coelom divides into the pericardial, pleural, and peritoneal cavities. **Splanchnic mesoderm** (SPLANGK-nik=visceral) forms the heart and the visceral layer of the serous pericardium, blood vessels, the smooth muscle and connective tissues of the respiratory and digestive organs, and the visceral layer of the serous membrane of the pleurae and peritoneum. **Somatic mesoderm** (sō-MAT-ik; *soma-*=body) gives rise to the bones, ligaments, and dermis of the limbs and the parietal layer of the serous membrane of the pericardium, pleurae, and peritoneum.

Development of the Cardiovascular System

At the beginning of the third week, **angiogenesis** (an-jē-ō-JEN-e-sis; *angio-*=vessel; *-genesis*=production), the formation of blood vessels, begins in the extraembryonic mesoderm of the yolk sac, connecting stalk, and chorion. This early development is necessary because there is insufficient yolk in the yolk sac and ovum to provide adequate nutrition for the rapidly developing embryo. Angiogenesis is initiated when mesodermal cells differentiate into **hemangioblasts.** These then develop into cells called **angioblasts,** which aggregate to form isolated masses of cells referred to as **blood islands** (see Figure 15.18 on page 505). As the blood islands grow and fuse, they soon form an extensive system of blood vessels throughout the embryo.

About 3 weeks after fertilization, blood cells and blood plasma begin to develop *outside* the embryo from hemangioblasts in the blood vessels in the walls of the yolk sac, allantois, and chorion. These then develop into **pluripotent stem cells** that form blood cells. Blood formation begins *within* the embryo at about the fifth week in the liver and the twelfth week in the spleen, red bone marrow, and thymus.

The heart forms from splanchnic mesoderm in the head end of the embryo on day 18 or 19 after fertilization. This region of mesodermal cells is called the **cardiogenic area** (kar-dē-ō-JEN-ik; *cardio-*=heart; *-genic*=producing). In response to induction signals from the underlying endoderm, these mesodermal cells ultimately form a pair of **endocardial tubes** (see Figure 14.12 on page 443). The tubes then fuse to form a single **primitive heart tube.** By the end of the third week, the primitive heart tube bends on itself, becomes S-shaped, and begins to beat. It then joins blood vessels in other parts of the embryo, connecting stalk, chorion, and yolk sac to form a primitive cardiovascular system.

Development of the Chorionic Villi and Placenta

By the end of the second week of development, **chorionic villi** (ko-rē-ON-ik VIL-ī) begin to develop. These fingerlike projections consist of chorion (syncytiotrophoblast surrounded by cytotrophoblast) (Figure 4.10a). By the end of the third week, blood capillaries develop in the chorionic villi (Figure 4.10b). Blood vessels in the chorionic villi connect to the embryonic heart by way of the umbilical arteries and umbilical vein (Figure 4.10c). As a result, maternal and fetal blood vessels are in close proximity. Note, however, that maternal and fetal blood vessels do not join, and the blood they carry does not normally mix. Instead, oxygen and nutrients in the blood of the mother's **intervillous spaces,** which are the spaces between chorionic villi, diffuse across the cell membranes into the capillaries of the villi. Waste products such as carbon dioxide diffuse in the opposite direction.

Figure 4.10 Development of chorionic villi.

Blood vessels in chorionic villi connect to the embryonic heart via the umbilical arteries and umbilical vein.

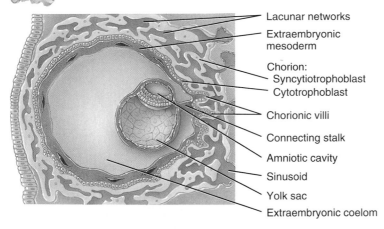

- Lacunar networks
- Extraembryonic mesoderm
- Chorion:
  - Syncytiotrophoblast
  - Cytotrophoblast
- Chorionic villi
- Connecting stalk
- Amniotic cavity
- Sinusoid
- Yolk sac
- Extraembryonic coelom

(a) Frontal section through uterus showing blastocyst, about 13 days after fertilization

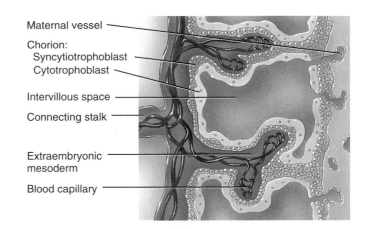

- Maternal vessel
- Chorion:
  - Syncytiotrophoblast
  - Cytotrophoblast
- Intervillous space
- Connecting stalk
- Extraembryonic mesoderm
- Blood capillary

(b) Details of two chorionic villi, about 21 days after fertilization

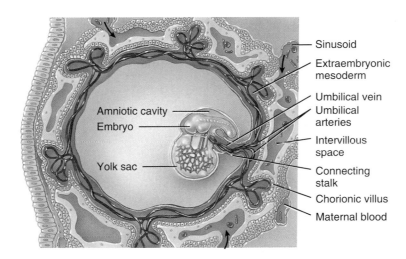

- Amniotic cavity
- Embryo
- Yolk sac

- Sinusoid
- Extraembryonic mesoderm
- Umbilical vein
- Umbilical arteries
- Intervillous space
- Connecting stalk
- Chorionic villus
- Maternal blood

(c) Frontal section through uterus showing an embryo and its vascular supply, about 21 days after fertilization

 Why is development of chorionic villi important?

Placentation (plas′-en-TĀ-shun) is the process of forming the **placenta** (pla-SEN-ta=flat cake), the site of exchange of nutrients and wastes between the mother and fetus. The placenta also produces hormones needed to sustain the pregnancy. The placenta is a unique structure because it develops from two separate individuals, the mother and the fetus.

By the beginning of the twelfth week, the placenta has two distinct parts: (1) the fetal portion formed by the chorionic villi of the chorion and (2) the maternal portion formed by the decidua basalis of the endometrium (Figure 4.11a). When fully developed, the placenta is shaped like a pancake (Figure 4.11b).

Functionally, the placenta allows oxygen and nutrients to diffuse from maternal blood into fetal blood while carbon dioxide and wastes diffuse from fetal blood into maternal blood. The placenta also is a protective barrier because most microorganisms cannot pass through it. However, certain viruses, such as those that cause AIDS, German measles, chickenpox, measles, encephalitis, and poliomyelitis, can cross the placenta. Many drugs, alcohol, and some substances that can cause birth defects also pass freely through the placenta. The placenta stores nutrients such as carbohydrates, proteins, calcium, and iron, which are released into fetal circulation as required.

Figure 4.11 Placenta and umbilical cord.

The placenta is formed by the chorionic villi of the embryo and the decidua basalis of the endometrium of the mother.

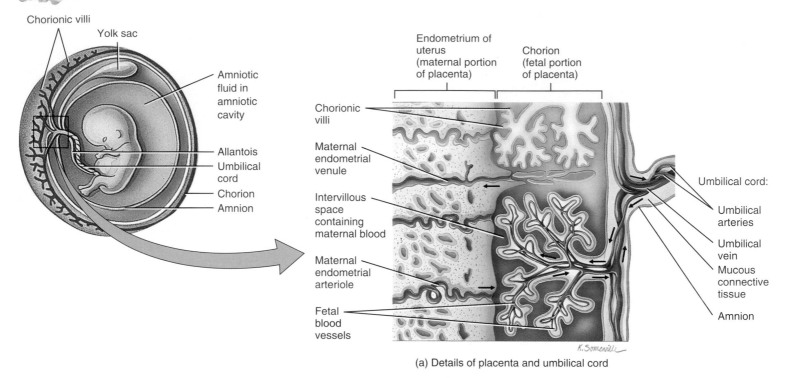

(a) Details of placenta and umbilical cord

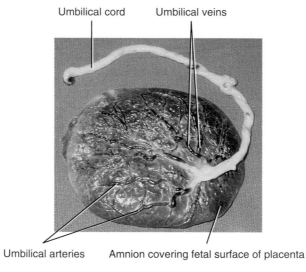

(b) Fetal aspect of placenta

What is the function of the placenta?

The actual connection between the placenta and embryo, and later the fetus, is through the **umbilical cord** (um-BIL-i-kul=navel), which develops from the connecting stalk. The umbilical cord consists of two umbilical arteries that carry deoxygenated fetal blood to the placenta, one umbilical vein that carries oxygenated maternal blood into the fetus, and supporting mucous connective tissue called **Wharton's jelly** derived from the allantois. A layer of amnion surrounds the entire umbilical cord and gives it a shiny appearance (Figure 4.11b).

CHECKPOINT

10. What is the significance of gastrulation?
11. How do the three primary germ layers form? Why are they important?
12. What is the function of the notochord?
13. Describe how neurulation occurs. Why is it important?
14. What are the functions of somites?
15. How does the cardiovascular system develop?
16. How does the placenta form and what is its importance?

Fourth Week of Development

The fourth through eighth weeks of development are very significant in embryonic development because all major organs appear during this time. The term **organogenesis** (or′-ga-nō-JEN-e-sis) refers to the formation of body organs and systems. By the end of the eighth week, all the major body systems have begun to develop, although their functions for the most part are minimal. Organogenesis requires the presence of blood vessels to supply developing organs with oxygen and other nutrients. However, recent studies suggest that blood vessels play a significant role in organogenesis even before blood begins to flow within them. The endothelial cells of blood vessels apparently provide some type of developmental signal, either a secreted substance or a direct cell-to-cell interaction, that is necessary for organogenesis.

During the fourth week after fertilization, the embryo undergoes dramatic changes in shape and size, nearly tripling its size. It is essentially converted from a flat, two-dimensional trilaminar embryonic disc to a three-dimensional cylinder, a process called **embryonic folding** (Figure 4.12). The cylinder consists of endoderm in the center (gut), ectoderm on the outside (epidermis), and mesoderm in between. The main force responsible for embryonic folding is the different rates of growth of various parts of the embryo, especially the rapid longitudinal growth of the nervous system (neural tube). Folding in the median plane produces a **head fold** and a **tail fold,** while folding in the horizontal plane results in the two **lateral folds.** Overall, due to the foldings, the embryo curves into a C-shape.

The head fold brings the developing heart and mouth into their eventual adult positions. The tail fold brings the developing anus into its eventual adult position. The lateral folds form as the lateral margins of the trilaminar embryonic disc bend ventrally. As they move toward the midline, the lateral folds incorporate the dorsal part of the yolk sac into the embryo as the **primitive gut,** the forerunner of the gastrointestinal tract (Figure 4.12b). The primitive gut differentiates into an anterior **foregut,** an intermediate **midgut,** and a posterior **hindgut** (Figure 4.12c). The fates of the foregut, midgut, and hindgut are described on page 799. Recall that the oropharyngeal membrane is located in the head end of the embryo (see Figure 4.8). It separates the future pharyngeal (throat) region of the foregut from the **stomodeum** (stō-mō-DĒ-um; *stomo-*=mouth), the future oral (mouth) cavity. Because of head folding, the oropharyngeal membrane moves downward and the foregut and stomodeum move closer to their final positions. When the oropharyngeal membrane ruptures during the fourth week, the pharyngeal region of the pharynx and stomodeum are brought into contact with each other.

In a developing embryo, the last part of the hindgut expands into a cavity called the **cloaca** (see Figure 26.13a on page 827). On the outside of an embryo is a small cavity in the tail region called the **proctodeum** (prok-tō-DĒ-um; *procto-*=anus) (Figure 4.12c). Separating the cloaca from the proctodeum is the **cloacal membrane** (see Figure 4.8). During embryonic development, the cloaca divides into a ventral urogenital sinus and a dorsal anorectal canal. As a result of tail folding, the cloacal membrane moves downward and the urogenital sinus, anorectal canal, and proctodeum move closer to their final positions. When the cloacal membrane ruptures during the seventh week of development, the urogenital and anal openings are created.

Along with embryonic folding, development of somites and the neural tube (previously described) occur during the fourth week of development. In addition, eventually five pairs of **pharyngeal (branchial) arches** develop on each side of the future head and neck regions (Figure 4.13 on page 111). These five paired structures begin their development on the twenty-second day after fertilization and form swellings on the surface of the embryo. Each pharyngeal arch consists of an outer covering of ectoderm, an inner covering of endoderm, and mesoderm in between. The pharyngeal arches also contain an artery, a cranial nerve, cartilage, and muscle tissue. Also on the outside of the embryo are a series of grooves between the pharyngeal arches called **pharyngeal clefts,** which separate the pharyngeal arches (Figure 4.13a). Simultaneous with the development of the pharyngeal arches and clefts, four distinct pairs of **pharyngeal (branchial) pouches** develop inside the embryo (Figure 4.13b). The pharyngeal pouches are endoderm-lined, balloon-like outgrowths of the primitive pharynx, the most cranial part of the foregut.

Together, the pharyngeal arches, clefts, and pouches give rise to the structures of the head and neck. The first sign of a developing ear is a thickened area of ectoderm, the **otic placode** (future internal ear), which can be distinguished about 22 days after fertilization (Figure 4.13a). A thickened area of ectoderm called the **lens placode** (Figure 4.13a), which will become the eye, also appears at this time.

By the middle of the fourth week, the upper limbs begin their development as outgrowths of mesoderm covered by ectoderm called **upper limb buds** (see Figure 8.18a on page 224). By the end of the fourth week, the **lower limb buds** develop. The heart also forms a distinct projection on the ventral surface of the embryo called the **heart prominence** (see Figure 8.18b), and a **tail** becomes visible (see Figure 8.18a).

Fifth Through Eighth Weeks of Development

During the fifth week of development, there is a very rapid development of the brain and thus growth of the head is considerable. By the end of the sixth week, the head grows even larger relative to the trunk, and the limbs show substantial development (see Figure 8.18b on page 225). In addition, the neck and trunk begin to straighten, and the heart is now four-chambered. By the seventh week, the various regions of the limbs become distinct and the beginnings of digits appear (see Figure 8.18c on page 225). At the start of the eighth week, the final week of the embryonic period, the digits of the hands are short and webbed and the tail is still visible, but shorter. In addition, the eyes are open and the auricles of the ears are visible (see Figure 8.18b on page 225). By the end of the eighth week, all regions of limbs are apparent; the digits are distinct and no longer webbed due to removal of cells via apoptosis. Also, the eyelids come together and may fuse, the tail disappears, and the external genitals begin to differentiate. The embryo now has clearly human characteristics.

Figure 4.12 Embryonic folding.

Embryonic folding converts the two-dimensional trilaminar embryonic disc into a three-dimensional cylinder.

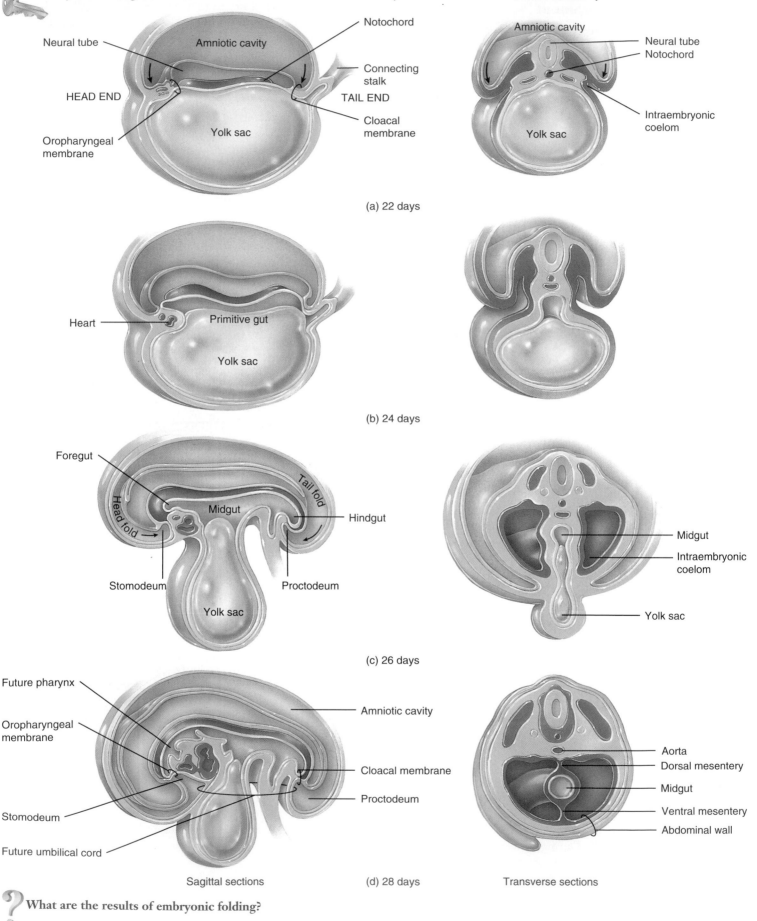

(a) 22 days

(b) 24 days

(c) 26 days

Sagittal sections (d) 28 days Transverse sections

What are the results of embryonic folding?

Figure 4.13 Development of pharyngeal arches, pharyngeal clefts, and pharyngeal pouches.

The five pairs of pharyngeal pouches consist of ectoderm, mesoderm, and endoderm and contain blood vessels, cranial nerves, cartilage, and musclar tissue.

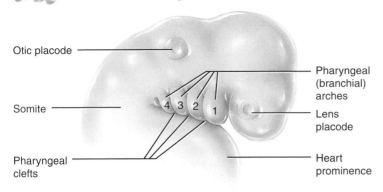

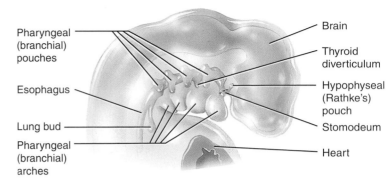

(a) External view, about 28-day embryo (b) Sagittal section, about 28-day embryo

Why are pharyngeal arches, clefts, and pouches important?

CHECKPOINT

17. How does embryonic folding occur?
18. How does the primitive gut form and what is its significance?
19. What is the origin of the structures of the head and neck?
20. What are limb buds?
21. What changes occur in the limbs during the second half of the embryonic period?

FETAL PERIOD

OBJECTIVE

● Describe the major events of the fetal period.

During the fetal period, tissues and organs that developed during the embryonic period grow and differentiate. Very few new structures appear during the fetal period, but the rate of body growth is remarkable, especially during the second half of intrauterine life. For example, during the last two and one half months of intrauterine life, half of the full-term weight is added. At the beginning of the fetal period, the head is half the length of the body. By the end of the fetal period, the head size is only one-quarter the length of the body. During the same period, the limbs also increase in size from one-eighth to one-half the fetal length. The fetus is also less vulnerable to the damaging effects of drugs, radiation, and microbes than it was as an embryo.

A summary of the major developmental events of the embryonic and fetal period is presented in Table 4.2 and illustrated in Figure 4.14 on page 113-114.

Throughout the text we will discuss the developmental biology of the various body systems in their respective chapters. The following list of these sections is presented for your reference.

Integumentary System (page 135)
Skeletal System (page 225)
Muscular System (page 282)
Heart (page 442)
Blood and Blood Vessels (page 505)
Lymphatic System (page 528)
Nervous System (page 624)
Endocrine System (page 721)
Respiratory System (page 758)
Digestive System (page 764)
Urinary System (page 808)
Reproductive Systems (page 835)

CHECKPOINT

22. What are the general developmental trends during the fetal period?
23. Using Table 4.2 as a guide, select any one body structure in weeks 9 through 12 and trace its development through the remainder of the fetal period.

TABLE 4.2   SUMMARY OF CHANGES DURING EMBRYONIC AND FETAL DEVELOPMENT

TIME	APPROXIMATE SIZE AND WEIGHT	REPRESENTATIVE CHANGES
EMBRYONIC PERIOD 1–4 weeks	0.6 cm (3/16 in.)	Primary germ layers and notochord develop. Neurulation occurs. Primary brain vesicles, somites, and intraembryonic coelom develop. Blood vessel formation begins and blood forms in yolk sac, allantois, and chorion. Heart forms and begins to beat. Chorionic villi develop and placental formation begins. The embryo folds. The primitive gut, pharyngeal arches, and limb buds develop. Eyes and ears begin to develop, tail forms, and body systems begin to form.
5–8 weeks	3 cm (1.25 in.) 1 g (1/30 oz)	Limbs become distinct and digits appear. Heart becomes four-chambered. Eyes are far apart and eyelids are fused. Nose develops and is flat. Face is more human-like. Bone formation begins. Blood cells start to form in liver. External genitals begin to differentiate. Tail disappears. Major blood vessels form. Many internal organs continue to develop.
FETAL PERIOD 9–12 weeks	7.5 cm (3 in.) 30 g (1 oz)	Head constitutes about half the length of the fetal body, and fetal length nearly doubles. Brain continues to enlarge. Face is broad, with eyes fully developed, closed, and widely separated. Nose develops a bridge. External ears develop and are low set. Bone formation continues. Upper limbs almost reach final relative length but lower limbs are not quite as well developed. Heartbeat can be detected. Gender is distinguishable from external genitals. Urine secreted by fetus is added to amniotic fluid. Red bone marrow, thymus, and spleen participate in blood cell formation. Fetus begins to move, but its movements cannot be felt yet by the mother. Body systems continue to develop.
13–16 weeks	18 cm (6.5–7 in.) 100 g (4 oz)	Head is relatively smaller than rest of body. Eyes move medially to their final positions, and ears move to their final positions on the sides of the head. Lower limbs lengthen. Fetus appears even more human-like. Rapid development of body systems occurs.
17–20 weeks	25–30 cm (10–12 in.) 200–450 g (0.5–1 lb)	Head is more proportionate to rest of body. Eyebrows and head hair are visible. Growth slows but lower limbs continue to lengthen. Vernix caseosa (fatty secretions of oil glands and dead epithelial cells) and lanugo (delicate fetal hair) cover fetus. Brown fat forms and is the site of heat production. Fetal movements are commonly felt by mother (quickening).
21–25 weeks	27–35 cm (11–14 in.) 550–800 g (1.25–1.5 lb)	Head becomes even more proportionate to rest of body. Weight gain is substantial, and skin is pink and wrinkled.
26–29 weeks	32–42 cm (13–17 in.) 110–1350 g (2.5–3 lb)	Head and body are more proportionate and eyes are open. Toenails are visible. Body fat is 3.5% of total body mass and additional subcutaneous fat smoothes out some wrinkles. Testes begin to descend toward scrotum at 28 to 32 weeks. Red bone marrow is major site of blood cell production. Many fetuses born prematurely during this period survive if given intensive care because lungs can provide adequate ventilation and central nervous system is developed enough to control breathing and body temperature.
30–34 weeks	41–45 cm (16.5–18 in.) 2000–2300 g (4.5–5 lb)	Skin is pink and smooth. Fetus assumes upside down position. Body fat is 8% of total body mass. Fetuses 33 weeks and older usually survive if born prematurely.
35–38 weeks	50 cm (20 in.) 3200–3400 g (7–7.5 lb)	By 38 weeks circumference of fetal abdomen is greater than that of head. Skin is usually bluish-pink, and growth slows as birth approaches. Body fat is 16% of total body mass. Testes are usually in scrotum in full-term male infants. Even after birth, an infant is not completely developed; an additional year is required, especially for complete development of the nervous system.

| 4 | 8 | 12 | 16 | 20 | 24 | 28 | 32 | 36 | (weeks) |

Figure 4.14 **Summary of representative developmental events of the embryonic and fetal periods.**
The embryos and fetuses are not shown at their actual sizes.

Development during the fetal period is mostly concerned with the growth and differentiation of tissues and organs formed during the embryonic period.

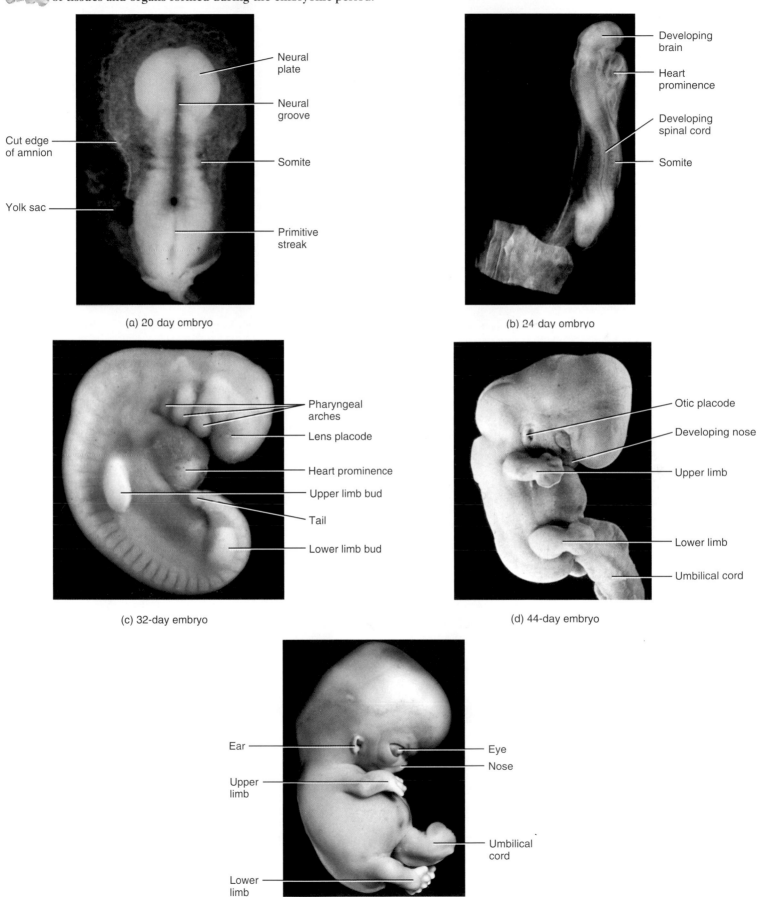

(a) 20-day embryo

- Neural plate
- Neural groove
- Cut edge of amnion
- Somite
- Yolk sac
- Primitive streak

(b) 24-day embryo

- Developing brain
- Heart prominence
- Developing spinal cord
- Somite

(c) 32-day embryo

- Pharyngeal arches
- Lens placode
- Heart prominence
- Upper limb bud
- Tail
- Lower limb bud

(d) 44-day embryo

- Otic placode
- Developing nose
- Upper limb
- Lower limb
- Umbilical cord

(e) 52-day embryo

- Ear
- Eye
- Nose
- Upper limb
- Umbilical cord
- Lower limb

continues

Figure 4.14 (continued)

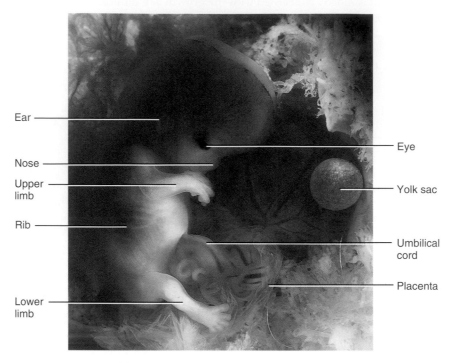

Ear

Nose

Upper
limb

Rib

Lower
limb

Eye

Yolk sac

Umbilical
cord

Placenta

(f) Ten-week fetus

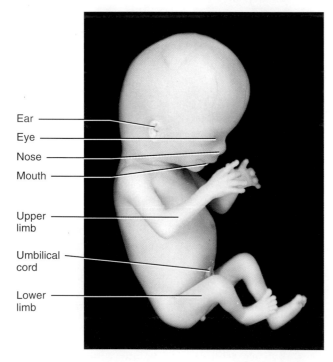

Ear

Eye

Nose

Mouth

Upper
limb

Umbilical
cord

Lower
limb

(g) Thirteen-week fetus

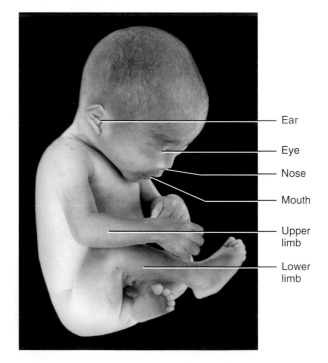

Ear

Eye

Nose

Mouth

Upper
limb

Lower
limb

(h) Twenty-six-week fetus

How much weight is gained by a fetus during the last two and half months of intrauterine life?

MATERNAL CHANGES DURING PREGNANCY

● Describe the effects of pregnancy on various body systems.

By about the end of the third month of pregnancy, the uterus occupies most of the pelvic cavity. As the fetus continues to grow, the uterus extends higher into the abdominal cavity. Toward the end of a full-term pregnancy, the uterus fills almost the entire abdominal cavity. It pushes the maternal intestines, liver, and stomach superiorly, elevates the diaphragm, and widens the thoracic cavity. Pressure on the stomach may force the stomach contents superiorly into the esophagus, resulting in heartburn. In the pelvic cavity, compression of the ureters and urinary bladder occurs.

Besides the anatomical changes associated with pregnancy, pregnancy-induced physiological changes also occur, including weight gain due to the fetus, amniotic fluid, the placenta, uterine enlargement, and increased total body water; increased storage of nutrients; marked breast enlargement in preparation for milk secretion and ejection; and lower back pain due to stress on the lower backbone.

Several changes occur in the maternal cardiovascular system. For example, heart rate increases and blood volume increases, mostly during the second half of pregnancy. These increases are necessary to meet the additional demands of the fetus for nutrients and oxygen. The total volume of air inhaled and exhaled per minute also increases to meet the added oxygen demands of the fetus.

With regard to the digestive activities, movements of the gastrointestinal tract, pregnant women experience an increase in appetite. A general decrease in gastrointestinal tract movement can cause constipation, and produce nausea, vomiting, and heartburn.

Pressure on the urinary bladder by the enlarging uterus can produce urinary symptoms, such as increased frequency and urgency of urination, and incontinence (inability to retain urine). An increase in blood flow through the kidneys allows faster elimination of the extra wastes produced by the fetus.

Changes in the skin during pregnancy are more apparent in some women than in others. Included are increased pigmentation around the eyes and cheekbones in a masklike pattern and in the circular area around the nipples of the breasts. Some females also exhibit a dark, vertical line along the lower abdomen. Stretch marks over the abdomen can occur as the uterus enlarges, and hair loss increases.

Changes in the reproductive system include swelling of the external genitals, as well as increased blood supply, and increased pliability of the vagina. The uterus increases from its nonpregnant mass of 60–80 g to 900–1200 g at term due to growth of muscle fibers in the myometrium.

24. What structural and functional changes occur in the mother during pregnancy?

LABOR

● Explain the events associated with the three stages of labor.

Obstetrics (ob-STET-riks; *obstetrix*=midwife) is the branch of medicine that deals with the management of pregnancy, labor, and the **neonatal period,** the first 28 days after birth. **Labor** is the process by which the fetus is expelled from the uterus through the vagina. **Parturition** (par′-toor-ISH-un; *parturit-*=childbirth) also means giving birth.

Uterine contractions occur in waves that start at the top of the uterus and move downward, eventually expelling the fetus. **True labor** begins when uterine contractions occur at regular intervals, usually producing pain. As the interval between contractions shortens, the contractions intensify. Another symptom of true labor in some women is localization of pain in the back that is intensified by walking. The reliable indicator of true labor is dilation (expansion) of the cervix and the "show," a discharge of a blood-containing mucus that appears in the cervical canal during labor. In **false labor,** pain is felt in the abdomen at irregular intervals, but it does not intensify and walking does not alter it significantly. There is no "show" and no cervical dilation.

True labor can be divided into three stages (Figure 4.15):

❶ Stage of dilation. The time from the onset of labor to the complete dilation of the cervix is the **stage of dilation.** This stage, which typically lasts 6–12 hours, features regular contractions of the uterus, usually a rupturing of the amniotic sac, and complete dilation (to 10 cm) of the cervix. If the amniotic sac does not rupture spontaneously, it is ruptured intentionally.

❷ Stage of expulsion. The time (10 minutes to several hours) from complete cervical dilation to delivery of the baby is the **stage of expulsion.**

❸ Placental stage. The time (5–30 minutes or more) after delivery until the placenta or "afterbirth" is expelled by powerful uterine contractions is the **placental stage.** These contractions also constrict blood vessels that were torn during delivery, thereby reducing the likelihood of hemorrhage.

As a rule, labor lasts longer with first babies, typically about 14 hours. For women who have previously given birth, the average duration of labor is about 8 hours—although the time varies enormously among births.

Delivery of a physiologically immature baby carries certain risks. A **premature infant** or "preemie" is generally considered a baby who weighs less than 2500 g (5.5 lb) at birth. Poor prenatal care, drug abuse, history of a previous premature delivery, and mother's age below 16 or above 35 increase the chance of premature delivery. The body of a premature infant is not yet ready to sustain some critical functions, and thus its survival is uncertain without medical intervention.

After the birth of the baby, the umbilical cord is tied off and then severed, leaving the baby on its own. The small portion

Figure 4.15 Stages of true labor.

The term parturition refers to birth.

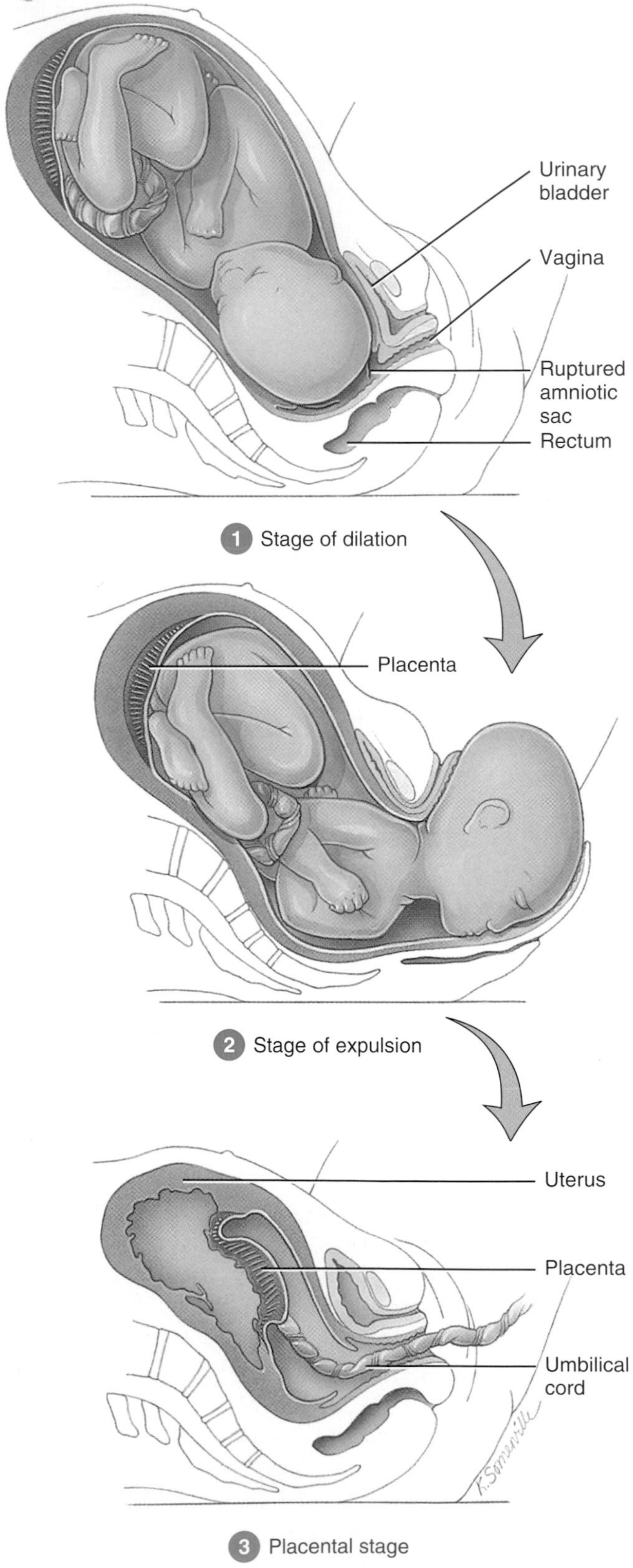

1 Stage of dilation

2 Stage of expulsion

3 Placental stage

What event marks the beginning of the stage of expulsion?

(about an inch) of the cord that remains attached to the infant begins to wither and falls off, usually within 12 to 15 days after birth. The area where the cord was attached becomes covered by a thin layer of skin, and scar tissue forms. The scar is the **umbilicus** (navel). Pharmaceutical companies use human placentas as a source of hormones, drugs, and blood; portions of placentas are also used for burn coverage. The placental and umbilical cord veins can also be used in blood vessel grafts, and cord blood can be frozen to provide a future source of pluripotent stem cells, for example, to repopulate red bone marrow following radiotherapy for cancer.

About 7% of pregnant women do not deliver by 2 weeks after their due date. Such cases carry an increased risk of brain damage to the fetus, and even fetal death, due to inadequate supplies of oxygen and nutrients from an aging placenta. Post-term deliveries may be facilitated by inducing labor, initiated by administration of oxytocin (Pitocin), or by surgical delivery (cesarean section).

DYSTOCIA AND CESAREAN SECTION

Dystocia (dis-TŌ-sē-a; *dys-*=painful or difficult; *toc-*= birth), or difficult labor, may result either from an abnormal position (presentation) of the fetus or a birth canal of inadequate size to permit vaginal delivery. In a **breech presentation,** for example, the fetal buttocks or lower limbs, rather than the head, enter the birth canal first; this occurs most often in premature births. If fetal or maternal distress prevents a vaginal birth, the baby may be delivered surgically through an abdominal incision. A low, horizontal cut is made through the abdominal wall and lower portion of the uterus, through which the baby and placenta are removed. Even though it is popularly associated with the birth of Julius Caesar, the true reason this procedure is termed a **cesarean section (C-section)** is because it was described in Roman Law, *lex cesarea*, about 600 years before Julius Caesar was born. Even a history of multiple C-sections need not exclude a pregnant woman from attempting a vaginal delivery. ■

CHECKPOINT

25. What is the difference between false labor and true labor?
26. What happens during the stage of dilation, the stage of expulsion, and the placental stage of true labor?

APPLICATIONS TO HEALTH

INFERTILITY

Female infertility, or the inability to conceive, occurs in about 10% of all women of reproductive age in the United States. Female infertility may be caused by ovarian disease, obstruction of the uterine tubes, or conditions in which the uterus is not adequately prepared to receive a fertilized ovum. **Male infertility (sterility)** is an inability to fertilize a secondary oocyte; it does not imply erectile dysfunction (impotence). Male fertility requires production of adequate quantities of viable, normal sperm by the testes, unobstructed transport of sperm though the ducts, and satisfactory deposition of sperm in the vagina. The sperm-producing structures of the testes are sensitive to many factors—x-rays, infections, toxins, malnutrition, and higher-than-normal scrotal temperatures—that may cause degenerative changes and produce male sterility.

To begin and maintain a normal reproductive cycle, a female must have a minimum amount of body fat. Even a moderate deficiency of fat—10% to 15% below normal weight for height—may delay the onset of menstruation, inhibit ovulation during the reproductive cycle, or cause amenorrhea (cessation of menstruation). Both dieting and intensive exercise may reduce body fat below the minimum amount and lead to infertility that is reversible, if weight gain or reduction of intensive exercise or both occurs.

Studies of very obese women indicate that they, like very lean ones, experience problems with amenorrhea and infertility. Males also experience reproductive problems in response to undernutrition and weight loss. For example, they produce less prostatic fluid and reduced numbers of sperm having decreased motility.

DOWN'S SYNDROME

Down's syndrome is a disorder that most often results during meiosis when an extra chromosome passes to one of the gametes. Most of the time the extra chromosome comes from the mother, a not-too-surprising finding given that all her oocytes began meiosis when she herself was a fetus. They may have been exposed to chromosome-damaging chemicals and radiation for years. (Sperm, by contrast, usually are less than 10 weeks old at the time they fertilize a secondary oocyte.) The chance of conceiving a baby with this syndrome, which is less than 1 in 3000 for women under age 30, increases to 1 in 300 in the 35 to 39 age group, and to 1 in 9 at age 48.

Down's syndrome is characterized by mental retardation; retarded physical development (short stature and stubby fingers), distinctive facial structures (large tongue, flat profile, broad skull, slanting eyes, and round head), as well as malformations of the heart, ears, hands, and feet. Sexual maturity is rarely attained.

KEY MEDICAL TERMS ASSOCIATED WITH DEVELOPMENT

Amniocentesis (am′-nē-ō-sen-TĒ-sis; *amnio-*=amnion; *-centesis*= puncture to remove fluid) A prenatal diagnostic test in which some of the amniotic fluid that bathes the developing fetus is withdrawn and the fetal cells and dissolved substances are analyzed. It is used to test for the presence of certain genetic disorders, such as Down's syndrome, hemophilia, Tay-Sachs disease, sickle-cell disease, and certain muscular dystrophies. The test is usually done at 14–18 weeks of gestation. It carries about a 0.5% chance of spontaneous abortion.

Chorionic villi sampling (CVS) (ko-rē-ON-ik VIL-ī) A prenatal diagnostic test in which a catheter is guided through the vagina and cervix of the uterus and then advanced to the chorionic villi and about 30 milligrams of tissue are suctioned out and prepared for chromosomal analysis. Alternatively, the chorionic villi can be sampled by inserting a needle through the abdominal cavity, as performed in amniocentesis. CVS can be performed as early as eight weeks of gestation, and test results are available in only a few days. CVS carries a 1–2% chance of spontaneous abortion after the procedure.

Fetal alcohol syndrome (FAS) A specific pattern of fetal malformation due to intrauterine exposure to alcohol. FAS is one of the most common causes of mental retardation and the most common preventable cause of birth defects in the United States. The symptoms of FAS may include slow growth before and after birth, characteristic facial features (short palpebral fissures, a thin upper lip, and sunken nasal bridge), defective heart and other organs, malformed limbs, genital abnormalities, and central nervous system damage. Behavioral problems, such as hyperactivity, extreme nervousness, reduced ability to concentrate, and an inability to appreciate cause-and-effect relationships, are common.

Fetal surgery Surgical procedure to repair and correct certain fetal disorders prior to childbirth; also called **in-utero surgery.**

Lethal gene (LĒ-thal JĒN; *lethum*=death) A gene that, when expressed, results in death either in the embryonic state or shortly after birth.

Maternal alpha-fetoprotein (AFP) test A prenatal diagnostic test in which the mother's blood is analyzed for the presence of AFP, a protein synthesized in the fetus that passes into the maternal circulation. The highest levels of AFP normally occur during weeks 12 through 15 of pregnancy. A high level of AFP after week 16 usually indicates that the fetus has a neural tube defect, such as spina bifida or anencephaly (absence of a cerebrum). Because the test is 95% accurate, it is now recommended that all pregnant women be tested.

Mutation (mū-TĀ-shun) Any change in the sequence of bases in a DNA molecule resulting in a permanent alteration of an inherited trait.

Puerperal fever (pū-ER-per-al; *puer*=child) An infectious disease of childbirth, also called **puerperal sepsis** and

childbed fever. The disease, which results from an infection originating in the birth canal, affects the endometrium. It may spread to other pelvic structures and lead to septicemia.

Teratogen (TER-a-tō-jēn; *terato-*=monster; *-gen*=creating) Any agent or influence that causes developmental defects in the embryo. Examples include alcohol, pesticides, industrial chemicals, antibiotics, thalidomide, LSD, and cocaine.

 ## STUDY OUTLINE

Embryonic Period (p. 94)

1. Pregnancy is a sequence of events that begins with fertilization, and proceeds to implantation, embryonic development, and fetal development. It normally ends in birth.

2. During fertilization a sperm cell penetrates a secondary oocyte and their pronuclei unite. Penetration of the zona pellucida is facilitated by enzymes in the sperm's acrosome. The resulting cell is a zygote.

3. Normally, only one sperm cell fertilizes a secondary oocyte.

4. Early rapid cell division of a zygote is called cleavage, and the cells produced by cleavage are called blastomeres. The solid sphere of cells produced by cleavage is a morula.

5. The morula develops into a blastocyst, a hollow ball of cells differentiated into a trophoblast and an inner cell mass.

6. The attachment of a blastocyst to the endometrium is termed implantation; it occurs as a result of enzymatic degradation of the endometrium.

7. The trophoblast develops into the syncytiotrophoblast and cytotrophoblast, both of which become part of the chorion.

8. The inner cell mass differentiates into hypoblast and epiblast, the bilaminar (two-layered) embryonic disc.

9. The amnion is a thin protective membrane that develops from the cytotrophoblast.

10. The exocoelomic membrane and hypoblast form the yolk sac, which transfers nutrients to the embryo, forms blood cells, produces primordial germ cells, and forms part of the gut.

11. Erosion of sinusoids and endometrial glands provides blood and secretions, which enter lacunar networks to supply nutrition to and remove wastes from the embryo.

12. The extraembryonic coelom forms within extraembryonic mesoderm.

13. The extraembryonic mesoderm and trophoblast form the chorion, the principal embryonic part of the placenta.

14. The third week of development is characterized by gastrulation, the conversion of the bilaminar disc into a trilaminar (three-layered) embryonic disc consisting of ectoderm, mesoderm, and endoderm.

15. The first evidence of gastrulation is formation of the primitive streak and then the primitive node, notochordal process, and notochord.

16. The three primary germ layers form all tissues and organs of the developing organism. Table 4.1 on page 104 summarizes the structures that develop from the primary germ layers.

17. Also during the third week, the oropharyngeal and cloacal membranes form. The wall of the yolk sac forms a small vascularized outpouching called the allantois, which functions in blood formation and development of the urinary bladder.

18. The process by which the neural plate, neural folds, and neural tube form is called neurulation. The brain and spinal cord develop from the neural tube.

19. Paraxial mesoderm segments to form somites from which skeletal muscles of the neck, trunk, and limbs develop. Somites also form connective tissues and vertebrae.

20. Blood vessel formation, called angiogenesis, begins in mesodermal cells called angioblasts.

21. The heart develops from mesodermal cells called the cardiogenic area. By the end of the third week, the primitive heart beats and circulates blood.

22. Chorionic villi, projections of the chorion, connect to the embryonic heart so that maternal and fetal blood vessels are brought into close proximity. Thus, nutrients and wastes are exchanged between maternal and fetal blood.

23. Placentation refers to formation of the placenta, the site of exchange of nutrients and wastes between the mother and fetus. The placenta also functions as a protective barrier, stores nutrients, and produces several hormones to maintain pregnancy.

24. The actual connection between the placenta and embryo (and later the fetus) is the umbilical cord.

25. Organogenesis refers to the formation of body organs and systems and occurs during the fourth week of development.

26. Conversion of the flat, two-dimensional trilaminar embryonic disc to the three-dimensional cylinder occurs by a process called embryonic folding.

27. Embryonic folding brings various organs into their final adult positions and helps form the gastrointestinal tract.

28. Pharyngeal arches, clefts, and pouches give rise to the structures of the head and neck.

29. By the end of the fourth week, upper and lower limb buds develop and by the end of the eighth week the embryo has clearly human features.

Fetal Period (p. 111)

1. The fetal period is primarily concerned with the growth and differentiation of tissues and organs that developed during the embryonic period.

2. The rate of body growth is remarkable, especially during the ninth and sixteenth weeks.

3. The principal changes associated with embryonic and fetal growth are summarized in Table 4.2 on page 112.

Maternal Changes During Pregnancy (p. 115)

1. During pregnancy, several anatomical and physiological changes occur in the mother.

2. The uterus nearly fills the abdominal cavity toward the end of full-term pregnancy, pushing the viscera out of their normal position.

3. Physiological changes include weight gain, increased skin pigmentation in certain areas, and various alterations in the cardiovascular, respiratory, digestive, urinary, and reproductive systems.

Labor (p. 115)

1. Labor is the process by which the fetus is expelled from the uterus through the vagina to the outside.

2. True labor involves dilation of the cervix, expulsion of the fetus, and delivery of the placenta.

 SELF-QUIZ QUESTIONS

Choose the one best answer to the following questions.

1. The embryonic period includes:

   a. the first and second trimesters.
   b. weeks 9–38 of development.
   c. the time between ovulation and fertilization.
   d. the first 8 weeks of development.
   e. the full 38 weeks of development.

2. Approximately how long after fertilization does implantation of an embryo usually occur?

   a. 3 weeks   b. about 6 days   c. 1 day
   d. about 3 days   e. 30 minutes

3. Dizygotic (fraternal) twins result from:

   a. two secondary oocytes and two sperm
   b. two secondary oocytes and one sperm.
   c. one secondary oocyte and two sperm.
   d. one secondary oocyte and one sperm.
   e. one secondary oocyte and three sperm.

4. A series of functional changes that cause a sperm's tail to beat more vigorously and prepare its plasma membrane to fuse with the oocyte's plasma membrane is called:

   a. fertilization.   b. implantation.   c. capacitation.
   d. syngamy.   e. cleavage.

5. Chemicals from the notochord induce the development of:

   a. the heart.
   b. parts of the vertebral column and the neural plate.
   c. openings to the urinary and reproductive tracts.
   d. serous membranes.
   e. the maternal portion of the placenta.

6. Oxygen and nutrients from the maternal blood must pass through which of the following structures before entering the fetal blood?

   a. umbilical vein   b. umbilical artery   c. decidua capsularis
   d. intervillous spaces   e. allantois

Complete the following.

7. Fusion of a sperm with a secondary oocyte is called _____.

8. The primary germ layers are the _____, the _____, and the _____.

9. The primary germ layers form by a process called _____, which is the major event in the third week of development.

10. _____, which is derived initially from maternal blood, serves as a shock absorber for the fetus and helps regulate fetal body temperature.

11. The process of blood vessel formation is called _____.

12. Following implantation, the endometrium of the uterus becomes known as the _____.

13. The trophoblast secretes a hormone called _____, which maintains the corpus luteum.

Are the following statements true or false?

14. Organ systems are almost completely developed by the end of the first trimester.

15. The chorionic villi form the maternal portion of the placenta.

16. Development of the placenta occurs during the third month of pregnancy.

17. The primary germ layers, which give rise to all tissues and organs, are present at the time of implantation.

18. The umbilical cord contains two umbilical arteries that carry oxygenated blood to the fetus, and one umbilical vein that carries fetal blood to the placenta.

19. Fertilization normally occurs in the vagina.

Matching

20. Match the following (answers are used more than once).

    ___ (a) epithelial lining of all gastrointestinal, respiratory, and genitourinary tracts except near openings to the exterior of the body

    ___ (b) epidermis of skin, epithelial lining of entrances to the body (such as mouth, nose, and anus), hair, nails

    ___ (c) all of the skeletal system (bone, cartilage, and other connective tissues)

    ___ (d) muscle (skeletal, cardiac, and most smooth)

    ___ (e) blood and all blood and lymphatic vessels

    ___ (f) entire nervous system

    ___ (g) thyroid, parathyroid, thymus, and pancreas

    (1) develop(s) from ectoderm
    (2) develop(s) from endoderm
    (3) develop(s) from mesoderm

21. Match the following:

    ___ (a) induces the formation of the neural plate

    ___ (b) forms the heart

    ___ (c) develops into the brain and spinal cord

    ___ (d) the initial result of gastrulation; establishes head and tail ends of the embryo

    ___ (e) forms bones and ligaments of the limbs

    ___ (f) degenerates to form the openings of the anus and the urinary and reproductive tracts

    ___ (g) forms the meninges (coverings) of the brain and spinal cord

    (1) primitive streak
    (2) notochord
    (3) cloacal membrane
    (4) neural tube
    (5) neural crest
    (6) splanchnic mesoderm
    (7) somatic mesoderm

22. Match the following:

____ **(a)** the structure that implants into the uterine wall

____ **(b)** the result of the fusion of the haploid pronuclei

____ **(c)** the layer of the trophoblast that secretes enzymes that liquefy endometrial cells during implantation and eventually develops into the chorion

____ **(d)** the form of the female gamete that is fertilized by a sperm cell

____ **(e)** the portion of the blastocyst that eventually develops into the embryo

____ **(f)** with the decidua basalis, forms the placenta

____ **(g)** the amnion eventually develops from this structure

____ **(h)** the form of the female gamete that is the result of meiosis II following fertilization

____ **(i)** with the exocoelomic membrane, forms the wall of the yolk sac

____ **(j)** the solid sphere of blastomeres that results from cleavage of the zygote

(1) secondary oocyte
(2) ovum
(3) zygote
(4) morula
(5) blastocyst
(6) inner cell mass
(7) syncytiotrophoblast
(8) chorionic villi
(9) hypoblast
(10) epiblast

CRITICAL THINKING QUESTIONS

1. Your neighbors from down the street put up a sign to announce the birth of their twins—a girl and a boy. Your next-door neighbor says, "Oh, how sweet! I wonder if they're identical?" Without even seeing the twins, what can you tell her?

2. Some disorders of the nervous system may cause particular skin problems. For example, a person with a type of nervous system tumor may show coffee-colored spots on the skin. How does the structure of the nervous system relate to that of the skin?

3. Josefina, in the last 2 weeks of her first pregnancy, anxiously called her doctor to ask if she should leave for the hospital. She was experiencing irregular "labor" pains which were thankfully eased by walking. She had no other signs to report. The doctor told Josefina to stay home—it wasn't time yet. Why did the doctor tell Josefina to stay home? How did he know it "wasn't time"?

4. Infection of the mother by certain viruses is known to result in severe birth defects in the child. If there is no mixing of maternal and fetal blood in the placenta, how can a viral infection in the mother cause problems in the child? At what point during the pregnancy do you think the greatest risk to the child would occur?

5. Larry's wife Elena is eight months pregnant. He has been fussing at her lately about the increase in their weekly grocery bill "even though there are still only two of us," and he has complained that they have been "going through toilet paper and antacids like crazy lately." When he wakes up in the morning, he discovers that Elena has stolen his extra pillow to prop herself up higher. What are some of the specific anatomical and physiological changes Elena has been experiencing that might explain these lifestyle changes?

ANSWERS TO FIGURE QUESTIONS

4.1 The uterine tubes provide a route for sperm to reach a secondary oocyte and transport a fertilized (or unfertilized) oocyte to the uterus.

4.2 Capacitation is the series of changes in sperm after they have been deposited in the female reproductive tract that enable them to fertilize a secondary oocyte.

4.3 A morula is a solid ball of cells, whereas a blastocyst consists of a rim of cells (trophoblast) surrounding a cavity (blastocyst cavity) and an inner cell mass.

4.4 The blastocyst secretes digestive enzymes that eat away the endometrial lining at the site of implantation.

4.5 Implantation usually occurs in the posterior portion of the fundus or body of the uterus.

4.6 The bilaminar embryonic disc is attached to the trophoblast by the connecting stalk.

4.7 Gastrulation converts a two-dimensional bilaminar embryonic disc into a two-dimensional trilaminar embryonic disc.

4.8 The notochord induces mesodermal cells to develop into vertebral bodies and forms a portion of the intervertebral discs.

4.9 The neural tube forms the brain and spinal cord; somites develop into skeletal muscles, connective tissue, and the vertebrae.

4.10 Chorionic villi help to bring the fetal and maternal blood vessels into close proximity.

4.11 The placenta functions in the exchange of materials between the fetus and the mother and as a protective barrier against many microbes, and stores nutrients.

4.12 Because of embryo folding, the embryo curves into a C-shape, various organs are brought into their eventual adult positions, and the primitive gut is formed.

4.13 Pharyngeal arches, clefts, and pouches give rise to structures of the head and neck.

4.14 During this time, fetal weight doubles.

4.15 Complete dilation of the cervix marks the onset of the stage of expulsion.

THE INTEGUMENTARY SYSTEM

5

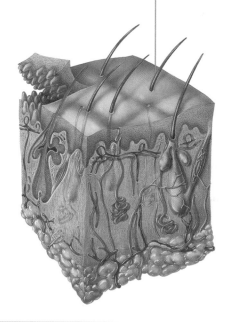

INTRODUCTION Skin and its accessory structures—hair and nails, along with various glands, muscles, and nerves—make up the **integumentary system** (in-teg-ū-MEN-tar-ē; *inte-=* whole; *-gument*=body covering). The integumentary system helps protect the body, helps maintain a constant body temperature, and provides sensory information about the surrounding environment. Of all the body's organs, none is more easily inspected or more exposed to infection, disease, and injury than the skin. Although the skin's location makes it vulnerable to damage from trauma, sunlight, microbes, and pollutants in the environment, it has protective features that ward off such damage. Because of its visibility, skin reflects our emotions and some aspects of normal physiology, as evidenced by frowning, blushing, and sweating. Changes in skin color may indicate homeostatic imbalances in the body. For example, a bluish skin color indicates

Has someone you know used the skin on their body as a canvas?
Have you? What do you think about the role of tattoos as an
expression of culture?

www.wiley.com/college/apcentral

hypoxia (oxygen deficiency at the tissue level) and is one sign of heart failure as well as other disorders. Abnormal skin eruptions or rashes such as chickenpox, cold sores, or measles may reveal systemic infections or diseases of internal organs. Other conditions may involve just the skin itself, such as warts, age spots, or pimples. So important is the skin to one's image that many people spend a great deal of time and money to restore it to a more normal or youthful appearance. **Dermatology** (der'-ma-TOL-ō-jē; *dermato-*=skin; *-logy*=study of) is the medical specialty for the diagnosis and treatment of integumentary system disorders.

STRUCTURE OF THE SKIN

OBJECTIVES

• Describe the layers of the epidermis and the cells that compose them.

• Compare the composition of the papillary and reticular regions of the dermis.

• Explain the basis for different skin colors.

The **skin** or **cutaneous membrane** covers the external surface of the body. It is the largest organ of the body in surface area and

Figure 5.1   Components of the integumentary system. The skin consists of a superficial, thin epidermis and a deep, thicker dermis. Deep to the skin is the subcutaneous layer, which attaches the dermis to underlying organs and tissues.

The integumentary system includes the skin and its accessory structures—hair, nails, and skin glands—along with associated muscles and nerves.

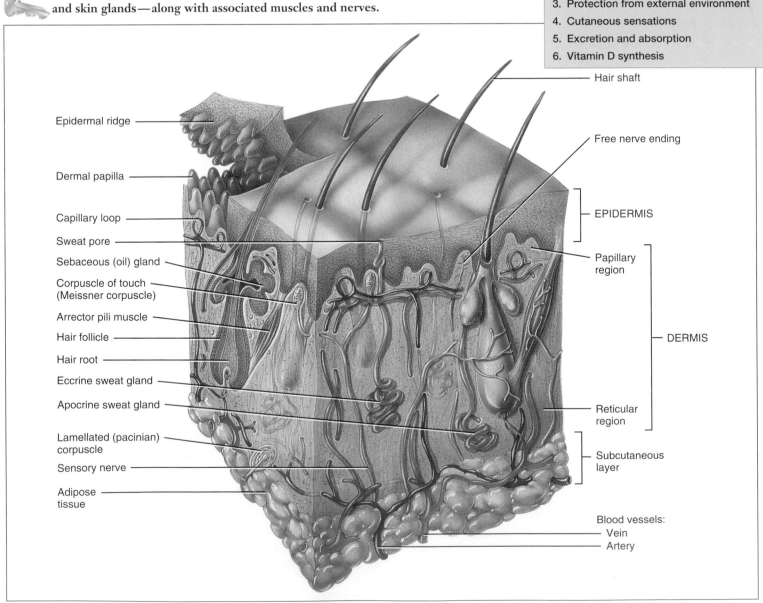

FUNCTIONS
1. Body temperature regulation
2. Reservoir for blood
3. Protection from external environment
4. Cutaneous sensations
5. Excretion and absorption
6. Vitamin D synthesis

? **What types of tissues make up the epidermis and the dermis?**

weight. In adults, the skin covers an area of about 2 square meters (22 square feet) and weighs 4.5–5 kg (10–11 lb), about 16% of total body weight. It ranges in thickness from 0.5 mm (.02 in.) on the eyelids to 4.0 mm (.16 in.) on the heels. However, over most of the body it is 1–2 mm (.04–.08 in.) thick. Structurally, the skin consists of two principal parts (Figure 5.1). The superficial, thinner portion, which is composed of *epithelial tissue*, is the **epidermis** (ep′-i-DERM-is; *epi-*=above). The deeper, thicker, *connective tissue* part is the **dermis.**

Deep to the dermis, and not part of the skin, is the **subcutaneous (subQ) layer,** or **hypodermis** (*hypo-*=below). This layer consists of areolar and adipose tissues. Fibers that extend from the dermis anchor the skin to the subcutaneous layer, which, in turn, attaches to underlying tissues and organs. The subcutaneous layer serves as a storage depot for fat and contains large blood vessels that supply the skin. This region (and sometimes the dermis) also contains nerve endings called **lamellated (pacinian) corpuscles** (pa-SIN-ē-an) that are sensitive to pressure (Figure 5.1).

Epidermis

The **epidermis** is composed of keratinized stratified squamous epithelium that contains four principal types of cells: keratinocytes, melanocytes, Langerhans cells, and Merkel cells (Figure 5.2). About 90% of epidermal cells are **keratinocytes** (ker-a-TIN-ō-sīts; *keratino-*=hornlike; *-cytes*=cells) (Figure 5.2a), which are arranged in four or five layers. These cells produce **keratin,** a tough, fibrous protein that helps protect the skin and underlying tissues from heat, microbes, and chemicals. Keratinocytes also produce lamellar granules, which release a waterproofing sealant.

About 8% of the epidermal cells are **melanocytes** (MEL-a-nō-sīts; *melano-*=black), which develop from neural crest cells (ectoderm) of a developing embryo. Melanocytes (Figure 4.2b) produce the pigment melanin. Their long, slender projections extend between the keratinocytes and transfer melanin granules to them. **Melanin** is a brown-black pigment that contributes to skin color and absorbs damaging ultraviolet (UV) light. Once inside keratinocytes, the melanin granules cluster to form a protective veil over the nucleus on the side toward the skin surface. In this way they shield the nuclear DNA from UV light. Although keratinocytes gain some protection from melanin granules, melanocytes themselves are particularly susceptible to damage by UV light.

Langerhans cells (LANG-er-hans) arise from red bone marrow and migrate to the epidermis (Figure 5.2c), where they constitute a small proportion of the epidermal cells. They participate in immune responses mounted against microbes that invade the skin, and they are easily damaged by UV light.

Merkel cells are the least numerous of the epidermal cells. They are located in the deepest layer of the epidermis, where they contact the flattened process of a sensory neuron (nerve cell), a structure called a **tactile (Merkel) disc** (Figure 5.2d). Merkel cells and tactile discs function in the sensation of touch.

Several distinct layers of keratinocytes in various stages of development form the epidermis (Figure 5.3). In most regions of the body the epidermis has four strata or layers—stratum

Figure 5.2 Types of cells in the epidermis. Besides keratinocytes, the epidermis contains melanocytes, which produce the pigment melanin; Langerhans cells, which participate in immune responses; and Merkel cells, which function in the sensation of touch.

Most of the epidermis consists of keratinocytes, which produce the protein keratin (protects underlying tissues) and lamellar granules (contains a waterproof sealant).

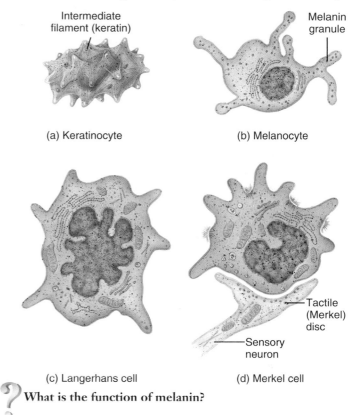

(a) Keratinocyte

(b) Melanocyte

(c) Langerhans cell

(d) Merkel cell

What is the function of melanin?

basale, stratum spinosum, stratum granulosum, and a thin stratum corneum. This is called thin skin. Where exposure to friction is greatest, as in the fingertips, palms, and soles, the epidermis has five layers—stratum basale, stratum spinosum, stratum granulosum, stratum lucidum, and a thick stratum corneum. This is called thick skin. The details of thin and thick skin are discussed later in the chapter.

Stratum Basale

The deepest layer of the epidermis, the **stratum basale** (ba-SA-lē; *basal-*=base), is composed of a single row of cuboidal or columnar keratinocytes, some of which are *stem cells* that undergo cell division to continually produce new keratinocytes. The nuclei of these keratinocytes are large, and their cytoplasm contains many ribosomes, a small Golgi complex, a few mitochondria, and some rough endoplasmic reticulum. The cytoskeleton within keratinocytes of the stratum basale includes scattered intermediate filaments called *tonofilaments*. Tonofilaments are composed of a protein that forms keratin in the more superficial epidermal layers. Tonofilaments attach to desmosomes, which bind cells of the stratum basale to each other and to the cells of the adjacent stratum

Figure 5.3 Layers of the epidermis.

The epidermis consists of keratinized stratified squamous epithelium.

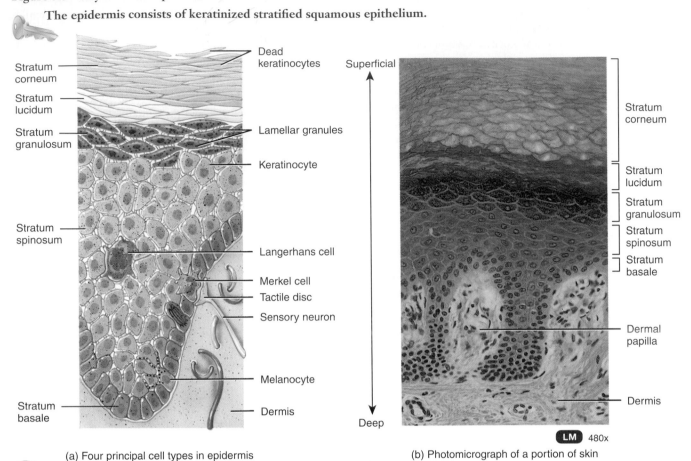

(a) Four principal cell types in epidermis

(b) Photomicrograph of a portion of skin

Which epidermal layer includes stem cells that continually undergo cell division?

spinosum, and to hemidesmosomes, which bind the keratinocytes to the basement membrane between the epidermis and the dermis. Keratin also protects deeper layers from injury. Melanocytes, Langerhans cells, and Merkel cells (with their associated tactile discs) are scattered among the keratinocytes of the basal layer. The stratum basale is sometimes referred to as the **stratum germinativum** (jer′-mi-na-TĒ-vum; *germ-*=sprout) to indicate its role in forming new cells.

SKIN GRAFTS

New skin cannot regenerate if an injury destroys the stratum basale and its stem cells. Skin wounds of this magnitude require skin grafts in order to heal. A **skin graft** involves covering the wound with a patch of healthy skin taken from a donor site. To avoid tissue rejection, the transplanted skin is usually taken from the same individual (*autograft*) or an identical twin (*isograft*). If skin damage is so extensive that an autograft would cause harm, a self-donation procedure called *autologous skin transplantation* (aw-TOL-ō-gus) may be used. In this procedure, performed most often for severely burned patients, small amounts of an individual's epidermis are removed, and the keratinocytes are cultured in the laboratory to produce thin sheets of skin. The new skin is transplanted back to the patient so that it covers the burn wound and generates a permanent skin. Also available as skin grafts for wound coverage are products developed from the foreskins of circumcised infants (Apligraft and Transite) that are grown in the laboratory. ■

Stratum Spinosum

Superficial to the stratum basale is the **stratum spinosum** (spi-NŌ-sum; *spinos-*=thornlike), where 8–10 layers of polyhedral (many-sided) keratinocytes fit closely together. Cells in the more superficial portions of this layer become somewhat flattened. These keratinocytes have the same organelles as cells of the stratum basale, and some cells in this layer retain their ability to divide. When cells of the stratum spinosum are prepared for microscopic examination, they shrink and pull apart such that they appear to be covered with thornlike spines (Figure 5.3a), although they are rounded and larger in living tissue. At each spinelike projection, bundles of tonofilaments insert into desmosomes, which tightly join the cells to one another. This arrangement provides both strength and flexibility to the skin. Projections of both Langerhans cells and melanocytes also appear in this layer.

Stratum Granulosum

At about the middle of the epidermis, the **stratum granulosum** (gran-ū-LŌ-sum; *granulos-*=little grains) consists of three to five layers of flattened keratinocytes that are undergoing apoptosis. (Recall from Chapter 2 that apoptosis is an orderly, genetically programmed cell death in which the nucleus fragments before the cells die.) The nuclei and other organelles of these cells begin to degenerate, and tonofilaments become more apparent. A distinctive feature of cells in this layer is the presence of darkly staining granules of a protein called **keratohyalin** (ker′-a-tō-HĪ-a-lin), which converts the tonofilaments into keratin. Also present in the keratinocytes are membrane-enclosed **lamellar granules,** which release a lipid-rich secretion. This secretion fills the spaces between cells of the stratum granulosum, stratum lucidum, and stratum corneum. The lipid-rich secretion acts as a water-repellent sealant, retarding loss of body fluids and entry of foreign materials. As their nuclei break down during apoptosis, the keratinocytes of the stratum granulosum can no longer carry on vital metabolic reactions, and they die. Thus, the stratum granulosum marks the transition between the deeper, metabolically active strata and the dead cells of the more superficial strata.

Stratum Lucidum

The **stratum lucidum** (LOO-si-dum; *lucid-*=clear) is present only in the thick skin of the fingertips, palms, and soles. It consists of three to five layers of clear, flat, dead keratinocytes that contain large amounts of keratin and thickened plasma membranes.

Stratum Corneum

The **stratum corneum** (COR-nē-um; *corne-*=hard or hooflike) consists of 25–30 layers of dead, flat keratinocytes. These cells are continuously shed and replaced by cells from the deeper strata. The interior of the cells contains mostly keratin. Between the cells are lipids from lamellar granules. The stratum corneum thus protects against injury and microbes and serves as an effective water-repellent barrier.

Keratinization and Growth of the Epidermis

Newly formed cells in the stratum basale are pushed slowly to the surface. As the cells move from one epidermal layer to the next, they accumulate more and more keratin in a process called **keratinization** (ker-a-tin-i-ZĀ-shun). Then they undergo apoptosis. Eventually the keratinized cells slough off and are replaced by underlying cells that, in turn, become keratinized. The whole process by which cells form in the stratum basale, rise to the surface, become keratinized, and slough off takes about four weeks in an average epidermis of 0.1 mm (.04 in.) thickness. The rate of cell division in the stratum basale increases when the outer layers of the epidermis are stripped away, as occurs in abrasions and burns. The mechanisms that regulate this remarkable growth are not well understood, but hormonelike proteins such as **epidermal growth factor (EGF)** play a role. An excessive

| TABLE 5.1 | SUMMARY OF EPIDERMAL STRATA |
STRATUM	DESCRIPTION
Stratum basale	Deepest layer, composed of a single row of cuboidal or columnar keratinocytes that contain scattered tonofilaments (intermediate filaments); stem cells undergo cell division to produce new keratinocytes; melanocytes, Langerhans cells, and Merkel cells associated with tactile discs are scattered among the keratinocytes.
Stratum spinosum	Eight to ten rows of many-sided keratinocytes with bundles of tonofilaments; includes projections of melanocytes and Langerhans cells.
Stratum granulosum	Three to five rows of flattened keratinocytes, in which organelles are beginning to degenerate; cells contain the protein keratohyalin, which converts tonofilaments into keratin, and lamellar granules, which release a lipid-rich, water-repellent secretion.
Stratum lucidum	Present only in skin of fingertips, palms, and soles; consists of three to five rows of clear, flat, dead keratinocytes with large amounts of keratin.
Stratum corneum	Twenty-five to thirty rows of dead, flat keratinocytes that contain mostly keratin.

amount of keratinized cells shed from the skin of the scalp is called *dandruff.*

Table 5.1 presents a summary of the distinctive features of the epidermal strata.

PSORIASIS

Psoriasis is a common and chronic skin disorder in which keratinocytes divide and move more quickly than normal from the stratum basale to the stratum corneum. They are shed prematurely in as little as 7–10 days. The immature keratinocytes make an abnormal keratin that forms flaky, silvery scales at the skin surface, most often on the knees, elbows, and scalp (dandruff). Effective treatments—various topical ointments and UV phototherapy—suppress cell division, decrease the rate of cell growth, or inhibit keratinization. ■

CHECKPOINT

1. How does the subcutaneous layer relate to the skin?
2. List the distinctive features of the epidermal layers from deepest to most superficial.

Dermis

The second, deeper part of the skin, the **dermis,** is composed mainly of connective tissue containing collagen and elastic fibers. The few cells present in the dermis include fibroblasts, macrophages, and some adipocytes. Blood vessels, nerves, glands, and hair follicles are embedded in dermal tissue. Based on its tissue structure, the dermis can be divided into a superficial papillary region and a deeper reticular region.

The **papillary region** makes up about one-fifth of the thickness of the total layer (see Figure 5.1). It consists of areolar connective tissue containing fine elastic fibers. Its surface area is greatly increased by small, fingerlike projections called **dermal papillae** (pa-PIL-ē=nipples). These nipple-shaped structures indent the epidermis and some contain **capillary loops** (blood capillaries). Other dermal papillae contain touch receptors called **corpuscles of touch,** or **Meissner corpuscles,** which contain sensitive nerve endings. Also present in the dermal papillae are **free nerve endings,** dendrites that lack any apparent structural specialization. Different free nerve endings initiate signals that produce sensations of warmth, coolness, pain, tickling, and itching.

The deeper portion of the dermis is called the **reticular region** (*reticul-*=netlike). It is attached to the subcutaneous layer and consists of dense irregular connective tissue containing bundles of collagen and some coarse elastic fibers. The bundles of collagen fibers in the reticular region interlace in a netlike manner. Spaces between the fibers are occupied by a few adipose cells, hair follicles, nerves, sebaceous (oil) glands, and sudoriferous (sweat) glands.

The combination of collagen and elastic fibers in the reticular region provides the skin with strength, **extensibility** (ability to stretch), and **elasticity** (ability to return to its original shape after stretching). The extensibility of skin can readily be seen in pregnancy and obesity. Small tears that occur in the dermis caused by extreme stretching produce **striae** (STRĪ-ē; *striae*=streaks), or stretch marks, which are visible as red or silvery white streaks on the skin surface.

Tattooing is a permanent coloration of the skin in which a foreign pigment is deposited with a needle into the dermis. It is believed by many researchers that the practice originated in Ancient Egypt between 4000 and 2000 B.C. Today, tattooing is performed in one form or another by nearly all peoples of the world. Although tattoos can be removed by lasers, the removal procedure requires a considerable investment in time and money and can be quite painful.

LINES OF CLEAVAGE AND SURGERY

In certain regions of the body, collagen fibers tend to orient more in one direction than another. **Lines of cleavage (tension lines)** in the skin indicate the predominant direction of underlying collagen fibers (Figure 5.4). The lines are especially evident on the palmar surfaces of the fingers, where they are arranged parallel to the long axis of the digits. Knowledge of lines of cleavage is especially important to plastic surgeons. For example, a surgical incision running parallel to the collagen fibers will heal with only a fine scar. A surgical incision made across the

Figure 5.4 Lines of cleavage.

Lines of cleavage in the skin indicate the predominant direction of underlying collagen fibers in the reticular region.

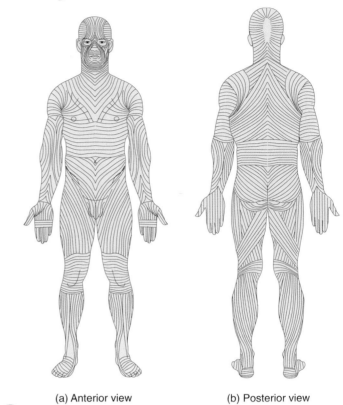

(a) Anterior view   (b) Posterior view

Why are lines of cleavage clinically important?

rows of fibers disrupts the collagen, and the wound tends to gape open and heal in a broad, thick scar. ■

The surfaces of the palms, fingers, soles, and toes are marked by series of ridges and grooves. They appear either as straight lines or as a pattern of loops and whorls, as on the tips of the digits. These **epidermal ridges** develop during the third month of fetal development as the epidermis projects downward into the dermis between the dermal papillae of the papillary region (see Figure 5.1). The ridges increase the surface area of the epidermis and thus function to increase the grip of the hand or foot by increasing friction. Because the ducts of sweat glands open on the tops of the epidermal ridges as sweat pores, the sweat and ridges form fingerprints (or footprints) when a smooth object is touched. The epidermal ridge pattern is genetically determined and is unique for each individual. Normally, the ridge pattern does not change during life, except to enlarge, and thus can serve as the basis for identification through fingerprints or footprints. The study of the pattern of epidermal ridges that is concerned with the identification and classification of fingerprints is called **dermatoglyphics** (der′-ma-tō-GLIF-iks; *glyphe*=carved work).

TABLE 5.2	SUMMARY OF PAPILLARY AND RETICULAR REGIONS OF THE DERMIS
REGION	**DESCRIPTION**
Papillary region	The superficial portion of the dermis (about one-fifth); consists of areolar connective tissue with elastic fibers; contains dermal papillae that house capillaries, corpuscles of touch, and free nerve endings.
Reticular region	The deeper portion of the dermis (about four-fifths); consists of dense irregular connective tissue with bundles of collagen and some coarse elastic fibers. Spaces between fibers contain some adipose cells, hair follicles, nerves, sebaceous glands, and sudoriferous glands.

Table 5.2 summarizes the structural features of the papillary and reticular regions of the dermis.

CHECKPOINT

3. Compare the structure and functions of the epidermis and dermis.
4. Compare the composition of the papillary and reticular regions of the dermis.
5. How are epidermal ridges formed?

The Structural Basis of Skin Color

Melanin, carotene, and hemoglobin are three pigments that give skin a wide variety of colors. The amount of **melanin,** which is located mostly in the epidermis, causes the skin's color to vary from pale yellow to red to brown to black. Melanocytes are most plentiful in the mucous membranes, penis, nipples of the breasts and the area just around the nipples (areolae), face, and limbs. Because the *number* of melanocytes is about the same in all races, differences in skin color are due mainly to the *amount of pigment* the melanocytes produce and transfer to keratinocytes. In some people, melanin tends to accumulate in patches called *freckles*. As one grows older, *age (liver) spots* may develop. These are flat skin patches that look like freckles and range in color from light brown to black. Like freckles, age spots are accumulations of melanin.

Melanocytes synthesize melanin from the amino acid *tyrosine* in the presence of an enzyme called *tyrosinase*. Synthesis occurs in an organelle called a **melanosome.** Exposure to UV light increases the enzymatic activity within melanosomes and leads to increased melanin production. Both the amount and darkness of melanin increase, which gives the skin a tanned appearance and further protects the body against UV radiation. Melanin absorbs UV radiation and prevents damage to DNA in epidermal cells. Melanin also neutralizes free radicals that form in the skin following damage by UV radiation. Thus, within limits, melanin serves a protective function. As you will see, however, repeatedly exposing the skin to UV light may cause skin cancer. A tan is lost when the melanin-containing keratinocytes are shed from the stratum corneum.

Carotene (KAR-ō-tēn; *carot-*=carrot) is a yellow-orange pigment that gives egg yolk and carrots their color. It is the precursor of vitamin A, which is used to synthesize pigments needed for vision. Carotene is found in the stratum corneum and fatty areas of the dermis and subcutaneous layer. When little melanin or carotene are present, the epidermis appears translucent. Thus, the skin of European-American people appears pink to red, depending on the amount and oxygen content of the blood moving through capillaries in the dermis. The red color is due to **hemoglobin,** the oxygen-carrying pigment in red blood cells.

Albinism (AL-bin-izm; *albin-*=white) is the inherited inability of an individual to produce melanin. Most **albinos** (al-BĪ-nōs), people affected by albinism, have melanocytes that are unable to synthesize tyrosinase. Melanin is missing from their hair, eyes, and skin. In another condition, called **vitiligo** (vit-i-LĪ-gō), the partial or complete loss of melanocytes from patches of skin produces irregular white spots. The loss of melanocytes may be related to an immune system malfunction in which antibodies attack the melanocytes.

There are enormous variations in skin color among different populations. Individuals who live in the tropics (near the equator) where the sun's rays are intense have evolved darkly pigmented skin; people who inhabit the subtropics and temperate regions have moderately pigmented skin and have the ability to tan; and individuals who live in regions near the poles are very light-skinned and burn easily.

⚕ SKIN COLOR AS A DIAGNOSTIC CLUE

The color of skin and mucous membranes can provide clues for diagnosing certain conditions. When blood is not picking up an adequate amount of oxygen in the lungs, such as in someone who has stopped breathing, the mucous membranes, nail beds, and skin appear bluish or **cyanotic** (sī-an-OT-ic; *cyan-*=blue). This occurs because hemoglobin that is depleted of oxygen looks deep, purplish blue. **Jaundice** (JON-dis; *jaund-*=yellow) is due to a buildup of the yellow pigment bilirubin in the blood. This condition gives a yellowish appearance to the skin and the whites of the eyes. Jaundice usually indicates liver disease. **Erythema** (er-e-THĒ-ma; *eryth-*=red), redness of the skin, is caused by engorgement of capillaries in the dermis with blood due to skin injury, exposure to heat, infection, inflammation, or allergic reactions. All skin color changes are observed most readily in people with lighter-colored skin and may be more difficult to discern in people with darker skin. **Pallor** (PAL-or), or paleness of the skin, may occur in conditions such as shock and anemia. ∎

CHECKPOINT

6. What are the three pigments in the skin and how do they contribute to skin color?
7. How do albinism and vitiligo differ?

ACCESSORY STRUCTURES OF THE SKIN

OBJECTIVE

• Compare the structure, distribution, and functions of hair, skin glands, and nails.

Accessory structures of the skin—hair, skin glands, and nails—develop from the embryonic epidermis. They have a host of important functions, among them the following: hair and nails protect the body, and sweat glands help regulate body temperature.

Hair

Hairs, or *pili* (PI-lē), are present on most skin surfaces except the palms, palmar surfaces of the fingers, soles, and plantar surfaces of the toes. In adults, hair is usually most heavily distributed across the scalp, over the brows of the eyes, and around the external genitalia. Genetic and hormonal influences largely determine the thickness and pattern of distribution of hair.

Although the protection it offers is limited, hair on the head guards the scalp from injury and the sun's rays. It also decreases heat loss from the scalp. Eyebrows and eyelashes protect the eyes from foreign particles, as does hair in the nostrils and in the external ear canal. Touch receptors associated with hair follicles (hair root plexuses) are activated whenever a hair is even slightly moved. Thus, hairs function in sensing light touch.

Anatomy of a Hair

Each hair is composed of columns of dead, keratinized cells bonded together by extracellular proteins. The **shaft** is the superficial portion of the hair, most of which projects from the surface of the skin (Figure 5.5a).

The transverse-sectional shape of the shafts of hairs varies in relation to race. Straight hair is round in transverse-section; wavy hair is oval in transverse-section; and curly hair is kidney shaped in transverse-section. The **root** is the portion of the hair deep to the shaft that penetrates into the dermis, and sometimes into the subcutaneous layer. The shaft and root both consist of three concentric layers (Figure 5.5c, d). The inner *medulla* is composed of two or three rows of polyhedral cells containing pigment granules and air spaces. The middle *cortex* forms the major part of the shaft and consists of elongated cells that contain pigment granules in dark hair but mostly air in gray or white hair. The *cuticle* of the hair, the outermost layer, consists of a single layer of thin, flat cells that are the most heavily keratinized. Cuticle cells are arranged like shingles on the side of a house, with their free edges pointing toward the free end of the hair (Figure 5.5b).

Surrounding the root of the hair is the **hair follicle,** which is made up of an external root sheath and an internal root sheath, together referred to as an **epithelial root sheath** (Figure 5.5c, d). The *external root sheath* is a downward continuation of the epidermis. Near the surface of the skin, it contains all the epidermal layers. At the base of the hair follicle, the external root sheath contains only the stratum basale. The *internal root sheath* is produced by the matrix (described shortly) and forms a cellular tubular sheath of epithelium between the external root sheath and the hair. The dense dermis surrounding the hair follicle is called the **dermal root sheath.**

The base of each hair follicle is an onion-shaped structure, the **bulb** (Figure 5.5c). This structure houses a nipple-shaped indentation, the **papilla of the hair,** which contains areolar connective tissue and many blood vessels that nourish the growing hair follicle. The bulb also contains a germinal layer of cells called the **matrix.** Matrix cells arise from the stratum basale, the site of cell division. Hence, matrix cells are responsible for the growth of existing hairs, and they produce new hairs when old hairs are shed. This replacement process occurs within the same follicle. Matrix cells also give rise to the cells of the internal root sheath.

HAIR REMOVAL

A substance that removes superfluous hair is called a **depilatory.** It dissolves the protein in the hair shaft, turning it into a gelatinous mass that can be wiped away. Because the hair root is not affected, regrowth of the hair occurs. In **electrolysis,** the hair matrix is destroyed by an electric current so that the hair cannot regrow. **Laser treatments** may also be used to remove hair. ∎

Sebaceous (oil) glands (discussed shortly) and a bundle of smooth muscle cells are also associated with hairs (Figure 5.5a). The smooth muscle is called **arrector pili** (a-REK-tor PI-lē; *arrect-*=to raise). It extends from the superficial dermis of the skin to the connective tissue sheath around the hair follicle. In its normal position, hair emerges at an angle to the surface of the skin. Under physiologic or emotional stress, such as cold or fright, autonomic nerve endings stimulate the arrector pili muscles to contract, which pulls the hair shafts perpendicular to the skin surface. This action causes "goose bumps" or "gooseflesh" because the skin around the shaft forms slight elevations.

Surrounding each hair follicle are dendrites of neurons, called **hair root plexuses,** that are sensitive to touch (Figure 5.5a). The hair root plexuses generate nerve impulses if their hair shaft is moved, for example, when an insect bumps into it when crawling across your skin.

Hair Growth

Each hair follicle goes through a growth cycle, which consists of a growth stage and a resting stage. During the **growth stage,** cells of the matrix differentiate, keratinize, and die. This process forms the root sheath and hair shaft. As new cells are added at the base of the hair root, the hair grows longer. In time, the growth of the hair stops and the **resting stage** begins. After the resting stage, a new growth cycle begins. The old hair root falls out or is pushed out of the hair follicle, and a new hair begins to grow in its place. Scalp hair grows for 2–6 years and rests for about 3 months. At any time, about 85% of scalp hairs are in the growth stage. Visible hair is dead, but until the hair is pushed out of its follicle by a new hair, portions of the root within the scalp are alive.

Figure 5.5 Hair.

Hairs are growths of epidermis composed of dead, keratinized cells.

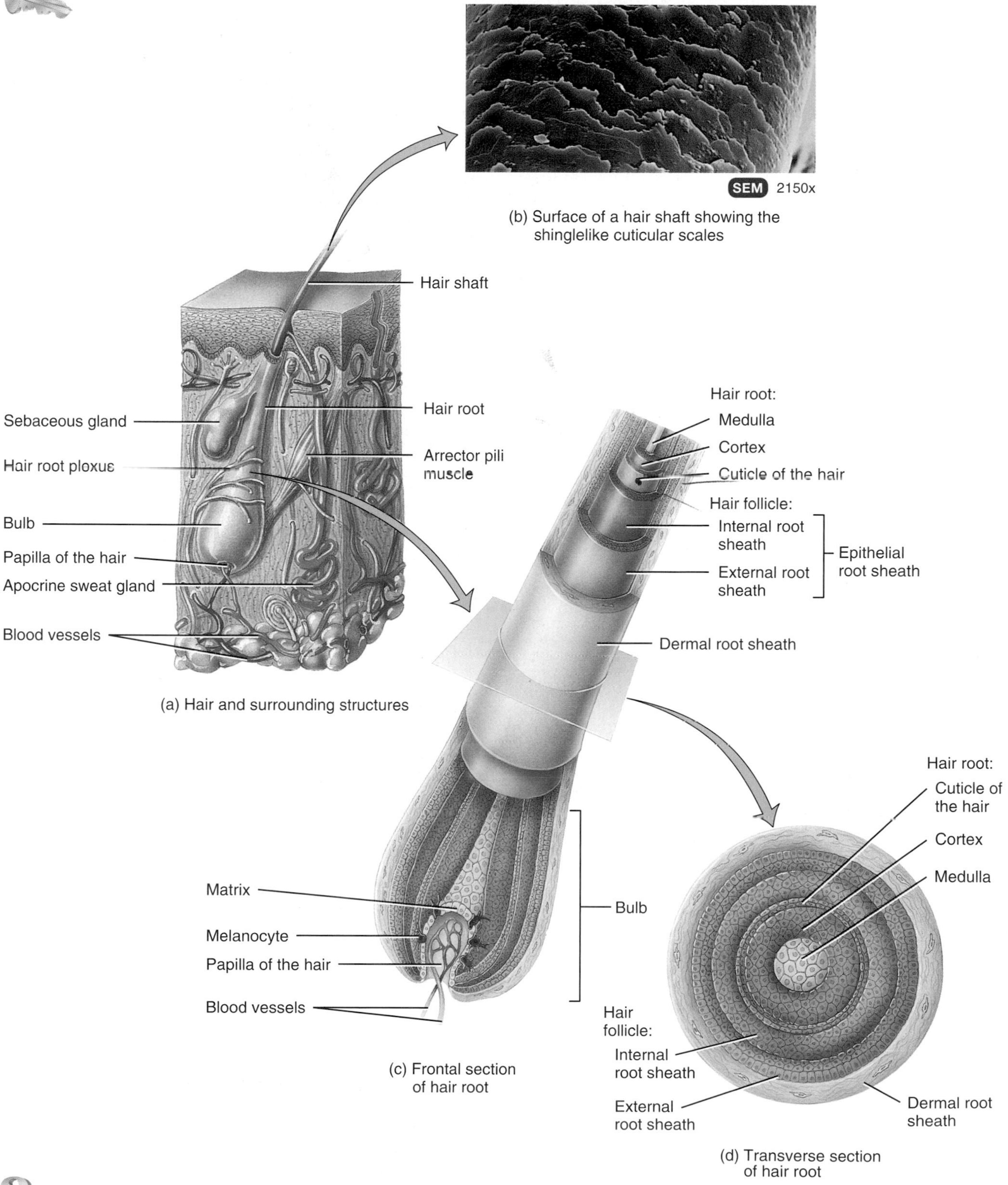

SEM 2150x

(b) Surface of a hair shaft showing the shinglelike cuticular scales

Hair shaft

Hair root

Arrector pili muscle

Sebaceous gland

Hair root plexus

Bulb

Papilla of the hair

Apocrine sweat gland

Blood vessels

(a) Hair and surrounding structures

Hair root:
Medulla
Cortex
Cuticle of the hair

Hair follicle:
Internal root sheath
External root sheath

Epithelial root sheath

Dermal root sheath

Matrix

Melanocyte

Papilla of the hair

Blood vessels

Bulb

Hair follicle:
Internal root sheath
External root sheath

Dermal root sheath

(c) Frontal section of hair root

Hair root:
Cuticle of the hair
Cortex
Medulla

(d) Transverse section of hair root

Why does it hurt when you pluck a hair but not when you have a haircut?

Normal hair loss in an adult scalp is about 70–100 hairs per day. Both the rate of growth and the replacement cycle can be altered by illness, diet, high fever, surgery, blood loss, severe emotional stress, and gender. Rapid weight-loss diets that severely restrict calories or protein increase hair loss. An increase in the rate of hair shedding can also occur with certain drugs, after radiation therapies for cancer, and for 3–4 months after childbirth. **Alopecia** (al′-o-PĒ-shē-a), the partial or complete lack of hair, may result from genetic factors, aging, endocrine disorders, chemotherapy for cancer, or skin disease.

CHEMOTHERAPY AND HAIR LOSS

Chemotherapy is the treatment of disease, usually cancer, by means of chemical substances or drugs. Chemotherapeutic agents interrupt the life cycle of rapidly dividing cancer cells. Unfortunately, the drugs also affect other rapidly dividing cells in the body, such as the matrix cells of a hair. It is for this reason that individuals undergoing chemotherapy experience hair loss. Since about 15% of the matrix cells of scalp hairs are in the resting stage, these cells are not affected by chemotherapy. Once chemotherapy is stopped, the matrix cells replace lost hair follicles and hair growth resumes. ■

Types of Hairs

Hair follicles develop between the ninth and twelfth weeks after fertilization as downgrowths of the stratum basale of the epidermis into the dermis (see Figure 5.8e). Usually by the fifth month of development, the follicles produce very fine, nonpigmented hairs called **lanugo** (la-NOO-gō = wool or down) that cover the body of the fetus. This hair is shed before birth, except in the scalp, eyebrows, and eyelashes. A few months after birth, slightly thicker hairs replace these downy hairs. Over the remainder of the body of an infant, a new growth of short, fine hair occurs. These hairs are known as **vellus hairs** (VEL-us = fleece), commonly called "peach fuzz." In response to hormones (androgens) secreted at puberty, coarse pigmented and frequently curly hair develops in the axillae (armpits) and pubic region. In males, these hairs also appear on the face and other parts of the body. The hairs that develop at puberty, together with those of the head, eyebrows, and eyelashes, are called **terminal hairs.** About 95% of body hair on males is terminal hair (5% vellus hair), whereas about 35% of body hair on females is terminal (65% vellus hair).

Hair Color

The color of hair is due primarily to the amount and type of melanin in its keratinized cells. Melanin is synthesized by melanocytes scattered in the matrix of the bulb and passes into cells of the cortex and medulla (Figure 5.5c). Dark-colored hair contains mostly true melanin, whereas blond and red hair contain variants of melanin in which there is iron and more sulfur. Graying hair occurs because of a progressive decline in tyrosinase (an enzyme required for the production of melanin), whereas white hair results from accumulation of air bubbles in the medullary shaft.

HAIR AND HORMONES

Occasionally, a tumor of the adrenal glands, testes, or ovaries produces an excessive amount of androgens. The result in females or prepubertal males is **hirsutism** (HER-soo-tizm; *hirsut-* = shaggy), a condition of excessive body hair, especially on the face, trunk, and limbs. Surprisingly, androgens also must be present for occurrence of the most common form of baldness, **androgenic alopecia** or **male-pattern baldness.** In genetically predisposed adults, androgens inhibit hair growth. In men, hair loss is most obvious at the temples and crown. Women are more likely to have thinning of hair on top of the head. The first drug approved for enhancing scalp hair growth was minoxidil (Rogaine). It causes vasodilation (widening of blood vessels), thus increasing circulation. In about a third of the people who try it, minoxidil improves hair growth, causing scalp follicles to enlarge and lengthening the growth cycle. For many, however, the hair growth is meager. Minoxidil does not help people who are already bald. ■

Skin Glands

Recall from Chapter 3 that glands are single or groups of epithelial cells that secrete a substance. Several kinds of exocrine glands are associated with the skin: sebaceous (oil) glands, sudoriferous (sweat) glands, ceruminous glands, and mammary glands. Mammary glands, which are specialized sudoriferous glands that secrete milk, are discussed in Chapter 27 along with the female reproductive system.

Sebaceous Glands

Sebaceous glands (se-BĀ-shus; *sebace-* = greasy), or **oil glands,** are simple, branched acinar glands. With few exceptions, they are connected to hair follicles (see Figures 5.6a and 5.1 and 5.5a). The secreting portion of a sebaceous gland lies in the dermis and usually opens into the neck of the hair follicle. In other locations, such as the lips, glans penis, labia minora, and tarsal glands of the eyelids, sebaceous glands open directly onto the surface of the skin. Sebaceous glands, which vary in size and shape, are found in the skin over all regions of the body except the palms and soles. They are small in most areas of the trunk and limbs, but large in the skin of the breasts, face, neck, and upper chest.

Sebaceous glands secrete an oily substance called **sebum** (SĒ-bum), which is a mixture of fats, cholesterol, proteins, and inorganic salts. Sebum coats the surface of hairs and helps keep them from drying and becoming brittle. Sebum also prevents excessive evaporation of water from the skin, keeps the skin soft and pliable, and inhibits the growth of certain bacteria.

ACNE

Acne is an inflammation of sebaceous glands that usually begins at puberty, when the sebaceous glands grow in size and increase their production of sebum. Androgens from the testes, ovaries, and adrenal glands play the greatest role in stimulating sebaceous glands. Acne occurs predominantly in sebaceous follicles that have been colonized by bacteria, some of which thrive in the lipid-rich sebum. The infection may cause a cyst or sac of

Figure 5.6   Histology of skin glands.

Whereas sebaceous glands are simple branched acinar glands, eccrine and apocrine sweat glands are simple coiled tubular glands.

Hair follicle

Sebaceous gland

LM 45x

(a) Sebaceous gland

Secretory portion

Duct

LM 200x

(b) Eccrine sweat gland

Sebaceous gland

Eccrine sweat gland

Apocrine sweat gland

Location of skin glands

Duct

Secretory portion

LM 120x

(c) Apocrine sweat gland

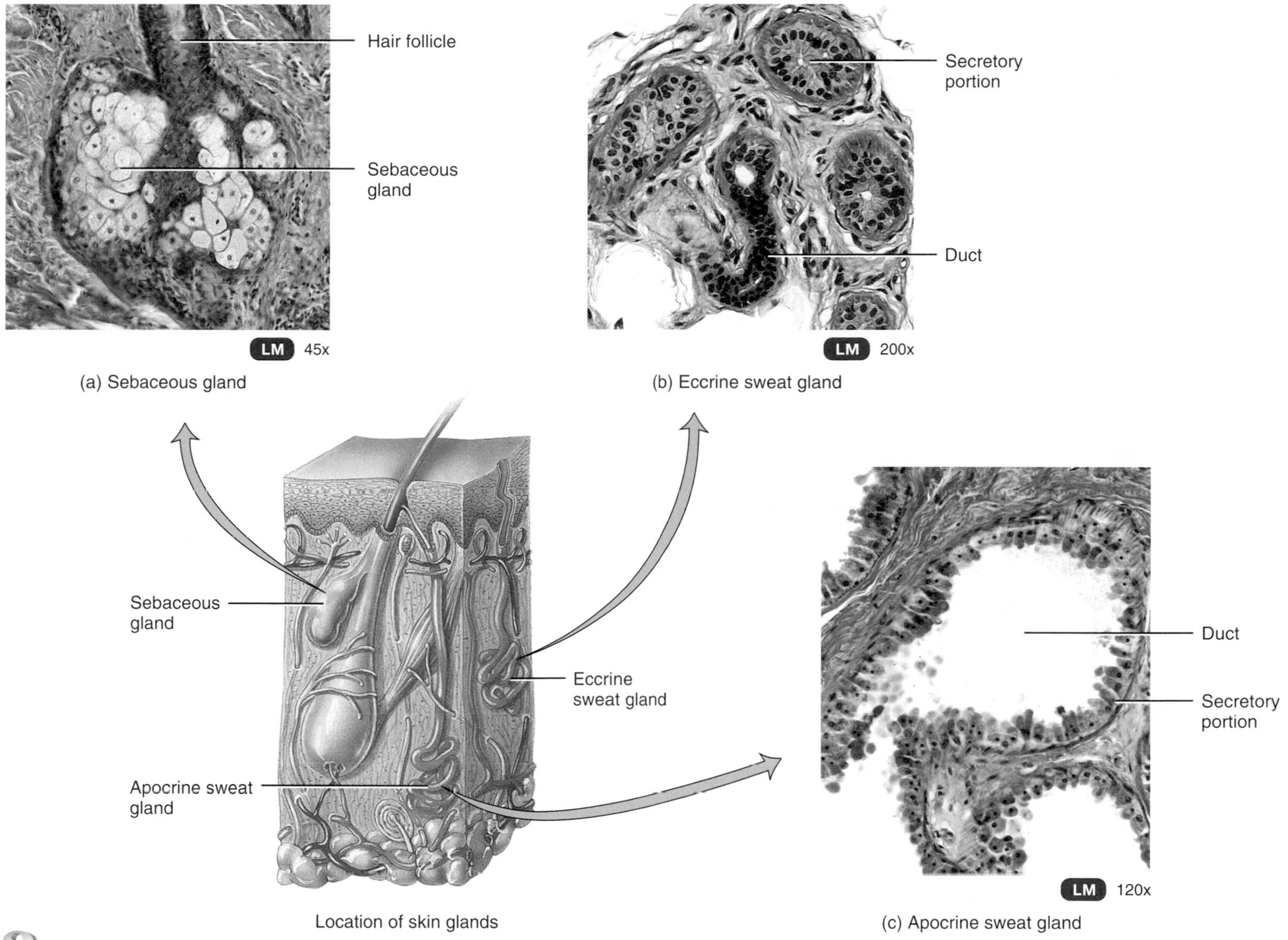

? What is the main function of eccrine sweat glands?

connective tissue cells to form, which can destroy and displace epidermal cells. This condition, called **cystic acne,** can permanently scar the epidermis.   ■

Sudoriferous Glands

There are three to four million **sweat glands,** or **sudoriferous glands** (sū-dor-IF-er-us; *sudori-*=sweat; *-ferous*=bearing). The cells of sweat glands release their secretions by exocytosis and empty them onto the skin surface through pores or into hair follicles. Depending on their structure, location, and type of secretion, the glands are classified as either eccrine or apocrine.

Eccrine sweat glands (*eccrine*=secreting outwardly) are simple, coiled tubular glands and are much more common than apocrine sweat glands (Figure 5.6b). They are distributed throughout the skin, except for the margins of the lips, nail beds

of the fingers and toes, glans penis, glans clitoris, labia minora, and eardrums. Eccrine sweat glands are most numerous in the skin of the forehead, palms, and soles; their density can be as high as 450 per square centimeter (3000 per square inch) in the palms. The secretory portion of eccrine sweat glands is located mostly in the deep dermis (sometimes in the upper subcutaneous layer). The excretory duct projects through the dermis and epidermis and ends as a pore at the surface of the epidermis (see also Figure 5.1).

The sweat produced by eccrine sweat glands (about 600 mL per day) consists of water, ions (mostly Na^+ and Cl^-), urea, uric acid, ammonia, amino acids, glucose, and lactic acid. The main function of eccrine gland sweat is to help regulate body temperature through evaporation. As sweat evaporates, large quantities of heat energy leave the body surface. Sweat, or perspiration, usually occurs first on the forehead and scalp, extends to the face and the rest of the body, and occurs last on the palms and soles. Under conditions of emotional stress, however, the palms, soles, and axillae are the first surfaces to sweat. Eccrine sweat also plays a small role in eliminating wastes such as urea, uric acid, and ammonia. Sweat that evaporates from the skin before it is perceived as moisture is referred to as **insensible perspiration.** Sweat that is excreted in larger amounts and is perceived as moisture on the skin is called **sensible perspiration.**

Apocrine sweat glands are also simple, coiled tubular glands. They are found mainly in the skin of the axilla (armpit), groin, areolae (pigmented areas around the nipples) of the breasts, and bearded regions of the face in adult males (Figure 5.6c). These glands were once thought to release their secretions in an apocrine manner—by pinching off a portion of the cell. We now know, however, that their secretion is released by exocytosis, which is characteristic of the release of secretions by merocrine glands (see page 70). Nevertheless, the term *apocrine* is still used. The secretory portion of these sweat glands is located mostly in the subcutaneous layer, and the excretory duct opens into hair follicles (see also Figure 5.1). Their secretory product is slightly viscous compared to eccrine secretions and contains the same components as eccrine sweat plus lipids and proteins. In women, cells of apocrine sweat glands enlarge about the time of ovulation and shrink during menstruation. Whereas eccrine sweat glands start to function soon after birth, apocrine sweat glands do not begin to function until puberty. Apocrine sweat glands are stimulated during emotional stress and sexual excitement; these secretions are commonly known as a "cold sweat."

Table 5.3 presents a comparison of eccrine and apocrine sweat glands.

Ceruminous Glands

Modified sweat glands in the external ear, called **ceruminous glands** (se-RŪ-mi-nus; *cer-*=wax), produce a waxy secretion. The secretory portions of ceruminous glands lie in the subcutaneous layer, deep to sebaceous glands. Their excretory ducts open either directly onto the surface of the external auditory canal (ear canal) or into ducts of sebaceous glands. The combined secretion of the ceruminous and sebaceous glands is called **cerumen,** or earwax. Cerumen, together with hairs in the external auditory canal, provides a sticky barrier that prevents the entrance of foreign bodies.

IMPACTED CERUMEN

Some people produce an abnormally large amount of cerumen in the external auditory canal. The cerumen may accumulate until it becomes impacted (firmly wedged), which prevents sound waves from reaching the eardrum. The treatment for **impacted cerumen** is usually periodic ear irrigation with enzymes to dissolve the wax or removal of wax with a blunt instrument by trained medical personnel. The use of cotton-tipped swabs or sharp objects is not recommended for this purpose because they may push the cerumen farther into the external auditory canal and damage the eardrum. ■

Nails

Nails are plates of tightly packed, hard, keratinized epidermal cells. The cells form a clear, solid covering over the dorsal surfaces of the distal portions of the digits. Each nail consists of a nail body, a free edge, and a nail root (Figure 5.7). The **nail body** is the portion of the nail that is visible, the **free edge** is the part that may extend past the distal end of the digit, and the **nail root** is the portion that is buried in a fold of skin. Most of the nail body appears pink because of blood flowing through underlying capillaries. The free edge is white because there are no underlying capillaries. The whitish, crescent-shaped area of the proximal end of the nail body is called the **lunula** (LOO-noo-la=little moon). It appears whitish because the vascular tissue underneath does not show through due to the thickened stratum basale in the area. Beneath the free edge is a thickened region of stratum corneum called the **hyponychium** (hī-′pō-

TABLE 5.3	COMPARISON OF ECCRINE AND APOCRINE SWEAT GLANDS	
FEATURE	**ECCRINE SWEAT GLANDS**	**APOCRINE SWEAT GLANDS**
Distribution	Throughout skin, except for margins of lips, nail beds, glans penis and clitoris, labia minora, and eardrums.	Skin of the axilla, groin, areolae, and bearded regions of the face.
Location of secretory portion	Mostly in deep dermis.	Mostly in subcutaneous layer.
Termination of excretory duct	Surface of epidermis.	Hair follicle.
Secretion	Less viscous; consists of water, ions (Na⁺, Cl⁻), urea, uric acid, ammonia, amino acids, glucose, and lactic acid.	More viscous; consists of the same components as eccrine sweat glands plus lipids and proteins.
Functions	Regulation of body temperature and waste removal.	Stimulated during emotional stress and sexual excitement.
Onset of Function	Soon after birth.	Puberty.

Figure 5.7 **Nails.** Shown is a fingernail.

Nail cells arise by transformation of superficial cells of the nail matrix into nail cells.

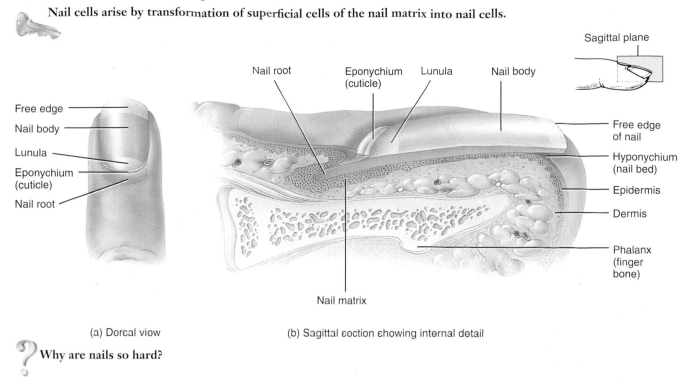

(a) Dorsal view

(b) Sagittal section showing internal detail

? Why are nails so hard?

NIK-ē-um; *hypo-*=below; *-onych*=nail), which secures the nail to the fingertip. The **eponychium** (ep′-ō-NIK-ē-um; *ep-*=above) or **cuticle** is a narrow band of epidermis that extends from and adheres to the margin (lateral border) of the nail wall. It occupies the proximal border of the nail and consists of stratum corneum.

The epithelium deep to the nail root is known as the **nail matrix.** The matrix cells divide mitotically to produce growth. Nail growth occurs by the transformation of superficial matrix cells into nail cells. In the process, the harder outer layer is pushed forward over the stratum basale. The growth rate of nails is determined by the rate at which matrix cells divide, which is influenced by factors such as a person's age, health, and nutritional status. Nail growth also varies according to the season, the time of day, and environmental temperature. The average growth in the length of fingernails is about 1 mm (.04 in.) per week. The growth rate is somewhat slower in toenails. The longer the digit, the faster the nail grows.

Functionally, nails help us grasp and manipulate small objects in various ways, provide protection against trauma to the ends of the digits, and allow us to scratch various parts of the body.

CHECKPOINT

8. Describe the structure of a hair. What produces "goose bumps"?
9. Contrast the locations and functions of sebaceous (oil) glands, sudoriferous (sweat) glands, and ceruminous glands.
10. Describe the principal parts of a nail.

TYPES OF SKIN

OBJECTIVE

• Compare structural and functional differences in thin and thick skin.

Although the skin over the entire body is similar in structure, there are quite a few local variations related to thickness of the epidermis, strength, flexibility, degree of keratinization, distribution and type of hair, density and types of glands, pigmentation, vascularity (blood supply), and innervation (nerve supply). On the basis of certain structural and functional properties, we recognize two major types of skin: thin (hairy) and thick (hairless).

Thin skin covers all parts of the body except for the palms, palmar surfaces of the digits, and the soles. Its epidermis is thin, just 0.10–0.15 mm (.004–.006 in.). A distinct stratum lucidum is lacking, and the strata spinosum and corneum are relatively thin. Thin skin has lower, broader, and fewer dermal papillae than thick skin and thus lacks epidermal ridges; it has hair follicles, arrector pili muscles, and sebaceous (oil) glands, but it has fewer sweat glands than thick skin. Finally, thin skin has a sparser distribution of sensory receptors than thick skin.

Thick skin covers the palms, palmar surfaces of the digits, and soles. Its epidermis is relatively thick, 0.6–4.5 mm (.024–.18 in.), and features a distinct stratum lucidum as well as thicker strata spinosum and corneum. The dermal papillae of thick skin are higher, narrower, and more numerous than those in thin skin, and thus thick skin has epidermal ridges. Thick skin

TABLE 5.4	COMPARISON OF THIN AND THICK SKIN	
FEATURE	**THIN SKIN**	**THICK SKIN**
Distribution	All parts of the body except palms, palmar surface of digits, and soles.	Palms, palmar surface of digits, and soles.
Epidermal thickness	0.10–0.15 mm (0.004–0.006 in.).	0.6–4.5 mm (0.024–0.18 in.).
Epidermal strata	Stratum lucidum essentially lacking; thinner strata spinosum and corneum.	Thick strata lucidum, spinosum, and corneum.
Epidermal ridges	Lacking due to poorly developed and fewer dermal papillae.	Present due to well-developed and more numerous dermal papillae.
Hair follicles and arrector pili muscles	Present.	Absent.
Sebaceous glands	Present.	Absent.
Sudoriferous glands	Fewer.	More numerous.
Sensory receptors	Sparser.	Denser.

lacks hair follicles, arrector pili muscles, and sebaceous glands, and has more sweat glands than thin skin. Also, sensory receptors are more densely clustered in thick skin.

Table 5.4 presents a summary of the features of thin and thick skin.

CHECKPOINT

11. What criteria are used to distinguish thin and thick skin?

FUNCTIONS OF THE SKIN

OBJECTIVE

● Describe how the skin contributes to body temperature regulation, protection, sensation, excretion and absorption, and synthesis of vitamin D.

Now that you have a basic understanding of the structure of the skin, you can better appreciate its many functions. Among the numerous functions of the integumentary system (mainly the skin) are the following:

1. ***Regulation of body temperature.*** In response to high environmental temperature or strenuous exercise, the evaporation of sweat from the skin surface helps lower an elevated body temperature to normal. In response to low environmental temperature, production of sweat is decreased, which helps conserve heat.

2. ***Blood reservoir.*** The dermis houses an extensive network of blood vessels that carry 8–10% of the total blood flow in a resting adult. For this reason, the skin acts as a *blood reservoir*. During moderate exercise, the flow of blood through the skin increases, which increases the amount of heat radiated from the body. During very strenuous exercise, however, skin blood vessels constrict (narrow) somewhat, diverting more blood to contracting skeletal muscles and to the heart. Because of this shunting of blood away from the skin, however, less heat is lost from the skin, and body temperature tends to rise.

3. ***Protection.*** The skin covers the body and provides protection in various ways. Keratin in the skin protects underlying tissues from microbes, abrasion, heat, and chemicals and the tightly interlocked keratinocytes resist invasion by microbes. Lipids released by lamellar granules retard evaporation of water from the skin surface, thus protecting the body from dehydration; they also retard entry of water across the skin surface during showers and swims. The oily sebum from the sebaceous glands protects skin and hairs from drying out and contains bactericidal chemicals that kill surface bacteria. The pigment melanin provides some protection against the damaging effects of UV light. Protective functions that are immunological in nature are carried out by epidermal Langerhans cells, which alert the immune system to the presence of potentially harmful microbial invaders, and by macrophages in the dermis, which phagocytize bacteria and viruses that manage to penetrate the skin surface.

4. ***Cutaneous sensations.*** *Cutaneous sensations* are those that arise in the skin. These include tactile sensations—touch, pressure, vibration, and tickling—as well as thermal sensations such as warmth and coolness. Another cutaneous sensation, pain, usually is an indication of impending or actual tissue damage. Some of the wide variety of abundantly distributed nerve endings and receptors in the skin include the tactile discs of the epidermis, the corpuscles of touch in the dermis, and hair root plexuses around each hair follicle. Chapter 21 provides more details on the topic of cutaneous sensations.

5. ***Excretion and absorption.*** The skin plays minor roles in *excretion*, the elimination of substances from the body, and *absorption*, the passage of materials from the external environment into body cells. Besides removing water and heat (by evaporation), sweat also is the vehicle for excretion of small amounts of salts, carbon dioxide, and two organic products of protein breakdown—ammonia and urea. The absorption of water-soluble substances through the skin is negligible, but certain lipid-soluble materials do penetrate the skin. These include fat-soluble vitamins (A, D, E, and K) and oxygen and carbon dioxide gases. Toxic materials that can be absorbed through the skin include organic solvents such as acetone (in some nail polish removers) and carbon tetrachloride (dry-cleaning fluid); salts of heavy metals such as lead, mercury, and arsenic; and the toxins in poison ivy and poison oak.

6. ***Synthesis of vitamin D.*** What is commonly called vitamin D is actually a group of closely related compounds. Synthesis of vitamin D requires activation of a precursor molecule in the skin by UV rays in sunlight. Enzymes in the liver and kidneys then modify the activated molecule, finally producing *calcitriol*, the most active form of vitamin D. Calcitriol aids in the absorption of calcium in foods from the gastrointestinal tract into the blood.

TRANSDERMAL DRUG ADMINISTRATION

Most drugs are either absorbed into the body through the digestive system or injected into subcutaneous tissue or muscle. An alternative route, **transdermal drug administration,** enables a drug contained within an adhesive skin patch to pass across the epidermis and into the blood vessels of the dermis. The drug is released at a controlled rate over one to several days. Because the major barrier to penetration of most drugs is the stratum corneum, transdermal absorption is most rapid in regions of the skin where this layer is thin, such as the scrotum, face, and scalp. A growing number of drugs are available for transdermal administration. These drugs include nitroglycerin, for prevention of angina pectoris (chest pain associated with heart disease); scopolamine, for motion sickness; estradiol, used for estrogen-replacement therapy during menopause; ethinyl estradiol and norelgestromin in contraceptive patches; nicotine, to help people stop smoking; and fentanyl, used to relieve severe pain, for example, in cancer patients. ■

CHECKPOINT

12. In what two ways does the skin help regulate body temperature?
13. How does the skin serve as a protective barrier?
14. What sensations arise from stimulation of neurons in the skin?
15. What types of molecules can penetrate the stratum corneum?

BLOOD SUPPLY OF THE INTEGUMENTARY SYSTEM

OBJECTIVE

● Describe the blood supply of the integumentary system.

Although the epidermis is avascular, the dermis is well supplied with blood (see Figure 5.1). The arteries supplying the dermis are generally derived from branches of arteries supplying skeletal muscles in a particular region. Some arteries supply the skin directly. One plexus (network) of arteries, the **cutaneous plexus,** is located at the junction of the dermis and subcutaneous layer and sends branches that supply the sebaceous (oil) and sudoriferous (sweat) glands, the deep portions of hair follicles, and adipose tissue. The **papillary plexus,** formed at the level of the papillary region, sends branches that supply the capillary loops in the dermal papillae, sebaceous (oil) glands, and the superficial portion of hair follicles. The arterial plexuses are accompanied by venous plexuses that drain blood from the dermis into larger subcutaneous veins.

CHECKPOINT

16. How do the cutaneous and papillary plexus differ in distribution?

DEVELOPMENT OF THE INTEGUMENTARY SYSTEM

OBJECTIVE

● Describe the development of the epidermis, its accessory structures, and the dermis.

The *epidermis* is derived from the **ectoderm,** which covers the surface of the embryo. Initially, at about the fourth week after fertilization, the epidermis consists of only a single layer of ectodermal cells (Figure 5.8a). At the beginning of the seventh week the single layer, called the **basal layer,** divides and forms a superficial protective layer of flattened cells called the **periderm** (Figure 5.8b).

The peridermal cells are continuously sloughed off and by the fifth month of development, secretions from sebaceous glands mix with them and hairs to form a fatty substance called **vernix caseosa** (VER-niks KĀ-sē-ō-sa; *vernix*=varnish, *caseosa*=cheese). This substance covers and protects the skin of the fetus from the constant exposure to the amniotic fluid in which it is bathed. In addition, the vernix caseosa facilitates the birth of the fetus because of its slippery nature and protects the skin from being damaged by the nails.

By about eleven weeks, the basal layer forms an intermediate layer of cells (Figure 5.8c). Proliferation of the basal cells eventually forms all layers of the epidermis, which are present at birth (Figure 5.8d). *Epidermal ridges* form along with the epidermal layers (Figure 5.8c). By about the eleventh week, cells from the neural crest (see Figure 19.26b on page 624) migrate into the dermis and differentiate into **melanoblasts** (Figure 5.8c). These cells soon enter the epidermis and differentiate into **melanocytes.** Later in the first trimester of pregnancy, *Langerhans cells*, which arise from red bone marrow, invade the epidermis. *Merkel cells*, whose origin is unknown, appear in the epidermis in the fourth to sixth months.

The *dermis* arises from **lateral plate mesoderm** (see Figure 4.9b on page 105) and the **dermatome of somites** (see Figure 10.10 on page 282). The mesoderm from these two sources is located deep to the surface ectoderm. The mesoderm gives rise to a loosely organized embryonic connective tissue called **mesenchyme** (MEZ-en-kīm; see Figure 5.8a). By eleven weeks, the mesenchymal cells differentiate into fibroblasts and begin to form collagen and elastic fibers. As the epidermal ridges form, parts of the superficial dermis project into the epidermis and develop into the *dermal papillae*, which contain capillary loops, corpuscles of touch, and free nerve endings (Figure 5.8c).

Figure 5.8   Development of the integumentary system.

Whereas the epidermis develops from ectoderm, the dermis develops from mesoderm.

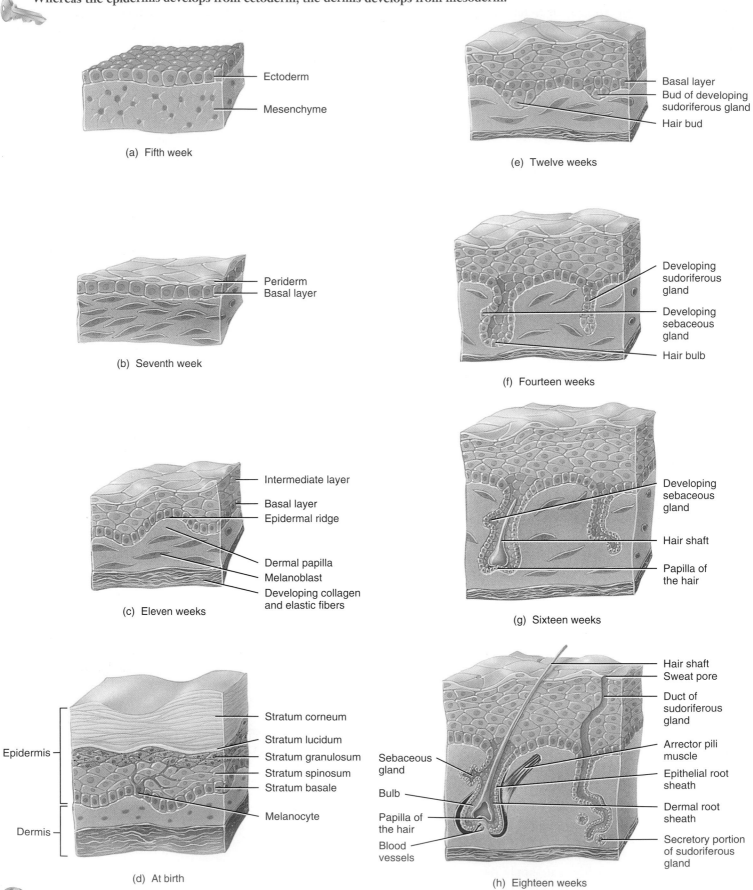

(a) Fifth week
- Ectoderm
- Mesenchyme

(b) Seventh week
- Periderm
- Basal layer

(c) Eleven weeks
- Intermediate layer
- Basal layer
- Epidermal ridge
- Dermal papilla
- Melanoblast
- Developing collagen and elastic fibers

(d) At birth
- Epidermis
- Stratum corneum
- Stratum lucidum
- Stratum granulosum
- Stratum spinosum
- Stratum basale
- Melanocyte
- Dermis

(e) Twelve weeks
- Basal layer
- Bud of developing sudoriferous gland
- Hair bud

(f) Fourteen weeks
- Developing sudoriferous gland
- Developing sebaceous gland
- Hair bulb

(g) Sixteen weeks
- Developing sebaceous gland
- Hair shaft
- Papilla of the hair

(h) Eighteen weeks
- Hair shaft
- Sweat pore
- Duct of sudoriferous gland
- Arrector pili muscle
- Epithelial root sheath
- Dermal root sheath
- Secretory portion of sudoriferous gland
- Sebaceous gland
- Bulb
- Papilla of the hair
- Blood vessels

What is the composition of vernix caseosa?

Hair follicles develop between the ninth and twelfth weeks as downgrowths of the basal layer of the epidermis into the deeper dermis. The downgrowths are called **hair buds** (Figure 5.8e). As the hair buds penetrate deeper into the dermis, their distal ends become club-shaped and are called **hair bulbs** (Figure 5.8f). Invaginations of the hair bulbs, called *papillae of the hair*, fill with mesoderm in which blood vessels and nerve endings develop (Figure 5.8g). Cells in the center of a hair bulb develop into the *matrix*, which forms the *hair*, whereas the peripheral cells of the hair bulb form the *epithelial root sheath* (Figure 5.8h). Mesenchyme in the surrounding dermis develops into the *dermal root sheath* and *arrector pili muscle* (Figure 5.8h). By the fifth month, the hair follicles produce lanugo (delicate fetal hair; see page 130). It is produced first on the head and then on other parts of the body, and is usually shed prior to birth.

Most *sebaceous (oil) glands* develop as outgrowths from the sides of hair follicles at about four months and remain connected to the follicles (Figure 5.8f). Most *sudoriferous (sweat)* glands are derived from downgrowths (buds) of the stratum basale of the epidermis into the dermis (Figure 5.8e). As the buds penetrate into the dermis, the proximal portion forms the duct of the sweat gland and the distal portion coils and forms the secretory portion of the gland (Figure 5.8h). Sweat glands appear at about five months on the palms and soles and a little later in other regions.

Nails are developed at about 10 weeks. Initially they consist of a thick layer of epithelium called the **primary nail field.** The nail itself is keratinized epithelium and grows distally from its base. It is not until the ninth month that the nails actually reach the tips of the digits.

CHECKPOINT

17. Which structures develop as downgrowths of the basal layer of epidermal cells?

AGING AND THE INTEGUMENTARY SYSTEM

OBJECTIVE

- Describe the effects of aging on the integumentary system.

The pronounced effects of skin aging do not become noticeable until people reach their late forties. Most of the age-related changes occur in the dermis. Collagen fibers in the dermis begin to decrease in number, stiffen, break apart, and disorganize into a shapeless, matted tangle. Elastic fibers lose some of their elasticity, thicken into clumps, and fray, an effect that is greatly accelerated in the skin of smokers. Fibroblasts, which produce both collagen and elastic fibers, decrease in number. As a result, the skin forms the characteristic crevices and furrows known as wrinkles.

With further aging, Langerhans cells dwindle in number and macrophages become less-efficient phagocytes, thus decreasing the skin's immune responsiveness. Moreover, decreased size of sebaceous glands leads to dry and broken skin that is more susceptible to infection. Production of sweat diminishes, which probably contributes to the increased incidence of heat stroke in the elderly. There is a decrease in the number of functioning melanocytes, resulting in gray hair and atypical skin pigmentation. An increase in the size of some melanocytes produces pigmented blotching (age spots). Walls of blood vessels in the dermis become thicker and less permeable, and subcutaneous adipose tissue is lost. Aged skin (especially the dermis) is thinner than young skin, and the migration of cells from the basal layer to the epidermal surface slows considerably. With the onset of old age, skin heals poorly and becomes more susceptible to pathological conditions such as skin cancer and pressure ulcers. **Rosacea** (rō-ZĀ-she-a=rosy) is a skin condition that affects mostly fair-skinned adults between ages 30–60. It is characterized by redness, tiny pimples, and noticeable blood vessels, usually in the central area of the face.

Growth of nails and hair slows during the second and third decades of life. The nails also may become more brittle with age, often due to dehydration or repeated use of cuticle remover or nail polish.

PHOTODAMAGE AND PHOTOSENSITIVITY REACTIONS

Although basking in the warmth of the sun may feel good, it is not a healthy practice. Both the longer-wavelength ultraviolet A (UVA) rays and the shorter-wavelength ultraviolet B (UVB) rays cause **photodamage** of the skin. Light-skinned and dark-skinned individuals alike experience the effect of acute overexposure to UVB rays—sunburn. Even if sunburn does not occur, the UVB rays can damage DNA, producing genetic mutations in epidermal cells. In turn, the mutations can cause skin cancer. As UVA rays penetrate to the dermis, they produce oxygen free radicals that disrupt collagen and elastic fibers in the extracellular matrix. This is the main reason for the severe wrinkling that occurs in those who spend a great deal of time in the sun without protection. Besides causing photodamage to the skin, exposure to direct sunlight may also produce **photosensitivity reactions** in some individuals taking certain oral medications, such as erythromycin, or herbal supplements, such as St. John's wort. The reactions may include redness, peeling, hives, blisters, and even shock. ∎

CHECKPOINT

18. What factors contribute to the susceptibility of aging skin to infection?

Applications to Health

SKIN CANCER

Excessive exposure to the sun causes virtually all of the 1 million cases of **skin cancer** diagnosed in the United States annually. There are three common forms of skin cancer. **Basal cell carcinomas** account for about 78% of all skin cancers. The tumors arise from cells in the stratum basale of the epidermis and rarely metastasize. **Squamous cell carcinomas**, which account for about 20% of all skin cancers, arise from squamous cells of the epidermis, and they have a variable tendency to metastasize. Most arise from preexisting lesions of damaged tissue in sun-exposed skin. Basal and squamous cell carcinomas are together known as *non-melanoma skin cancer* and are 50% more common in males than in females. **Malignant melanomas** arise from melanocytes and account for about 2% of all skin cancers. The American Academy of Dermatology estimates that the lifetime risk of developing melanoma is currently 1 in 75, double the risk only 15 years ago. In part, this increase is due to depletion of the ozone layer, which absorbs some UV light high in the atmosphere. But the main reason for the increase is that more people are spending more time in the sun. Malignant melanomas metastasize rapidly and can kill a person within months of diagnosis.

The key to successful treatment of malignant melanoma is early detection. The early warning signs of malignant melanoma are identified by the acronym ABCD (Figure 5.9) *A* is for *asymmetry;* malignant melanomas tend to lack symmetry. *B* is for *border;* malignant melanomas have notched, indented, scalloped, or indistinct borders. *C* is for *color;* malignant melanomas have uneven coloration and may contain several colors. *D* is for *diameter;* ordinary moles tend to be smaller than 6 mm (0.25 in.), about the size of a pencil eraser. Once a malignant melanoma has the characteristics of A, B, and C, it is usually larger than 6 mm.

Among the risk factors for skin cancer are the following:

Figure 5.9 Comparison of a normal nevus (mole) and a malignant melanoma.

 Excessive exposure to the sun accounts for almost all cases of skin cancer.

(a) Normal nevus (mole)   (b) Malignant melanoma

? **Which is the most common type of skin cancer?**

1. *Skin type.* Individuals with light-colored skin who never tan but always burn are at high risk.

2. *Sun exposure.* People who live in areas with many days of sunlight per year and at high altitudes (where ultraviolet light is more intense) have a higher risk of developing skin cancer. Likewise, people who engage in outdoor occupations and those who have suffered three or more severe sunburns have a higher risk.

3. *Family history.* Skin cancer rates are higher in some families than in others.

4. *Age.* Older people are more prone to skin cancer owing to longer total exposure to sunlight.

5. *Immunological status.* Individuals who are immunosuppressed have a higher incidence of skin cancer.

BURNS

A **burn** is tissue damage caused by excessive heat, electricity, radioactivity, or corrosive chemicals that destroy (denature) the proteins in the skin cells. Burns destroy some of the skin's important contributions to homeostasis—protection against microbial invasion and desiccation, and regulation of body temperature.

Burns are graded according to their severity. A *first-degree burn* involves only the epidermis (Figure 5.10a). It is characterized by mild pain and erythema (redness) but no blisters. Skin functions remain intact. The pain and damage caused by a first-degree burn may be lessened by immediately flushing it with cold water. Generally, a first-degree burn will heal in about 3–6 days and may be accompanied by flaking or peeling. One example of a first-degree burn is a mild sunburn.

A *second-degree burn* destroys a portion of the epidermis and possibly parts of the dermis (Figure 5.10b). Some skin functions are lost. In a second-degree burn, redness, blister formation, edema, and pain result. (Blister formation is caused by separation of the epidermis from the dermis due to the accumulation of tissue fluid between the layers.) Associated structures, such as hair follicles, sebaceous glands, and sweat glands, usually are not injured. If there is no infection, second-degree burns heal without skin grafting in about 3–4 weeks, but scarring may result. First- and second-degree burns are collectively referred to as *partial-thickness burns.*

A *third-degree burn*, or *full-thickness burn*, destroys a portion of the epidermis, the underlying dermis, and associated structures (Figure 5.10c). Most skin functions are lost. Such burns vary in appearance from marble-white to mahogany colored to charred, dry wounds. There is marked edema, and the burned region is numb because sensory nerve endings have been destroyed. Regeneration occurs slowly, and much granulation tissue forms before being covered by epithelium. Skin grafting may be required to promote healing and to minimize scarring. The injury to the skin tissues directly in contact with the damag-

Figure 5.10 Burns.

A burn is tissue damage caused by agents that destroy the proteins in skin cells.

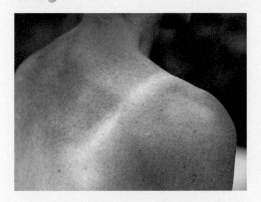 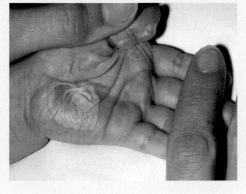 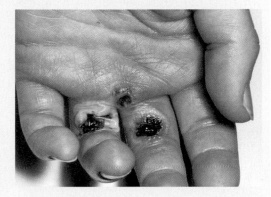

(a) First-degree burn (sunburn)   (b) Second-degree burn (note the blister)   (c) Third-degree burn

What factors determine the seriousness of a burn?

ing agent is the *local effect* of a burn. Generally, however, the *systemic effects* of a major burn are a greater threat to life. The systemic effects of a burn may include (1) a large loss of water, plasma, and plasma proteins, which causes shock; (2) bacterial infection; (3) reduced circulation of blood; (4) decreased production of urine; and (5) diminished immune responses.

The seriousness of a burn is determined by its depth and extent of area involved, as well as the person's age and general health. According to the American Burn Association's classification of burn injury, a major burn includes third-degree burns over 10% of body surface area; or second-degree burns over 25% of body surface area; or any third-degree burns on the face, hands, feet, or *perineum* (per-i-NĒ-um, which includes the anal and urogenital regions). When the burn area exceeds 70%, more than half the victims die.

Many people who have been burned in fires also inhale smoke. If the smoke is unusually hot or dense or if inhalation is prolonged, serious problems can develop. The hot smoke can damage the trachea (windpipe), causing its lining to swell. As the swelling narrows the trachea, airflow into the lungs is obstructed. Further, small airways inside the lungs can also narrow, producing wheezing and shortness of breath. A person who has inhaled smoke is given oxygen through a face mask, and a tube may be inserted into the trachea to assist breathing.

PRESSURE ULCERS

Pressure ulcers, also known as *decubitus ulcers* (dē-KŪ-bi-tus), or *bedsores*, are a shedding of epithelium caused by a constant deficiency of blood flow to tissues (Figure 5.11). Typically the affected tissue overlies a bony projection that has been subjected to prolonged pressure against an object

such as a bed, cast, or splint. If the pressure is relieved in a few hours, redness occurs but no lasting tissue damage results. Blistering of the affected area may indicate superficial damage, whereas a reddish-blue discoloration may indicate deep tissue damage. Prolonged pressure results in tissue ulceration. Small breaks in the epidermis become infected, and the sensitive subcutaneous layer and deeper tissues are damaged. Eventually, the tissue dies. Pressure ulcers occur most often in patients who are bedridden. With proper care, pressure ulcers are preventable.

Figure 5.11 Pressure ulcers.

A pressure ulcer is a shedding of epithelium caused by a constant deficiency of blood flow to tissues.

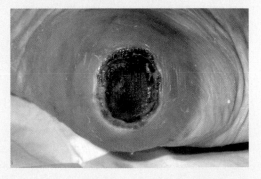

Pressure ulcer on heel

What parts of the body are usually affected by pressure ulcers?

KEY MEDICAL TERMS ASSOCIATED WITH THE INTEGUMENTARY SYSTEM

Abrasion (a-BRĀ-shun; *ab-*=away; *-rasion*=scraped) An area where skin has been scraped away.

Athlete's foot A superficial fungus infection of the skin of the foot.

Blister A collection of serous fluid within the epidermis or between the epidermis and dermis, due to short-term but severe friction. The term *bulla* (BUL-a) refers to a large blister.

Callus (KAL-lus=hard skin) An area of hardened and thickened skin that is usually seen in palms and soles and is due to persistent pressure and friction.

Cold sore A lesion, usually in the oral mucous membrane, caused by Type 1 herpes simplex virus (HSV) transmitted by oral or respiratory routes. The virus remains dormant until triggered by factors such as ultraviolet light, hormonal changes, and emotional stress. Also called a *fever blister*.

Comedo (KOM-ē-dō; *comedo*=to eat up) A collection of sebaceous material and dead cells in the hair follicle and excretory duct of the sebaceous (oil) gland. Usually found over the face, chest, and back, and more commonly during adolescence. Also called *blackhead*.

Contact dermatitis (der-ma-TĪ-tis; *dermat-*=skin; *-itis*=inflammation of) Inflammation of the skin characterized by redness, itching, and swelling and caused by exposure of the skin to chemicals that bring about an allergic reaction, such as poison ivy toxin.

Contusion (kon-TOO-shun; *contundere*=to bruise) Condition in which tissue deep to the skin is damaged, but the epidermis is not broken.

Corn A painful conical thickening of the stratum corneum of the epidermis found principally over toe joints and between the toes, often caused by friction or pressure. Corns may be hard or soft, depending on their location. Hard corns are usually found over toe joints, and soft corns are usually found between the fourth and fifth toes.

Cutaneous anthrax One of the three forms of anthrax that accounts for 95% of all naturally occurring anthrax. Endospores (spores) introduced at the site of a cut or abrasion produce a lesion that turns into a black scab and falls off in a week or two. The disease is characterized by low-grade fever, malaise (general feeling of discomfort), and enlarged lymph nodes. Without antibiotic treatment the mortality rate is as high as 20%.

Cyst (SIST; *cyst*=sac containing fluid) A sac with a distinct connective tissue wall, containing a fluid or other material.

Eczema (EK-ze-ma, *ekzeo-*=to boil over) An inflammation of the skin characterized by patches of red, blistering, dry, extremely itchy skin. It occurs mostly in skin creases in the wrists, backs of the kness, and front of the elbows. It typically begins in infancy, and many children outgrow the condition. The cause is unknown but is linked to genetics and allergies.

Frostbite Local destruction of skin and subcutaneous tissue on exposed surfaces as a result of extreme cold. In mild cases, the skin is blue and swollen and there is slight pain. In severe cases there is considerable swelling, some bleeding, no pain, and blistering. If untreated, gangrene may develop. Frostbite is treated by rapid rewarming.

Furuncle (FŪ-rung-kul) A boil; an abscess resulting from infection of a hair follicle. Usually caused by *Staphylococcus aureus*.

Hemangioma (he-man′-jē-Ō-ma; *hem-*=blood; *-angi-*=blood vessel; *-oma*=tumor) Localized tumor of the skin and subcutaneous layer that results from an abnormal increase in blood vessels. One type is a *portwine stain*, a flat, pink, red, or purple lesion present at birth, usually at the nape of the neck.

Hives Condition of the skin marked by reddened elevated patches that are often itchy. Most commonly caused by infections, physical trauma, medications, emotional stress, food additives, and certain food allergies. Also called *urticaria* (ūr-ti-KAR-ē-a).

Impetigo (im′-pe-TĪ-gō) Superficial skin infection caused by *Staphylococcus* bacteria; most common in children.

Intradermal (in-tra-DER-mal; *intra-*=within) Within the skin. Also called *intracutaneous*.

Keloid (KĒ-loid; *kelis*=tumor) An elevated, irregular darkened area of excess scar tissue caused by collagen formation during healing. It extends beyond the original injury and is tender and frequently painful. It occurs in the dermis and underlying subcutaneous tissue, usually after trauma, surgery, a burn, or severe acne; more common in people of African descent.

Keratosis (ker′-a-TŌ-sis; *kera-*=horn) Formation of a hardened growth of epidermal tissue, such as a *solar keratosis*, a premalignant lesion of the sun-exposed skin of the face and hands.

Laceration (las-er-Ā-shun; *lacer-*=torn) An irregular tear of the skin.

Nevus (NĒ-vus) A round, flat, or raised area of pigmented skin that may be present at birth or may develop later. Varies in color from yellow-brown to black. Also called a *mole* or *birthmark*.

Papule (PAP-ūl; *papula*=pimple) A small, round skin elevation less than 1 cm in diameter. One example is a pimple.

Pruritus (proo-RĪ-tus; *pruri-*=to itch) Itching, one of the most common dermatological disorders. It may be caused by skin disorders (infections), systemic disorders (cancer, kidney failure), psychogenic factors (emotional stress), or allergic reactions.

Topical In reference to a medication, applied to the skin surface rather than ingested or injected.

Wart Mass produced by uncontrolled growth of epithelial skin cells; caused by a papilloma virus. Most warts are noncancerous.

STUDY OUTLINE

Structure of the Skin (p. 122)

1. The integumentary system consists of the skin and its accessory structures—hair, nails, glands, muscles, and nerves.

2. The skin is the largest organ of the body in surface area and weight. The principal parts of the skin are the epidermis (superficial) and dermis (deep).

3. The subcutaneous layer (hypodermis) is deep to the dermis and not part of the skin. It anchors the dermis to underlying tissues and organs, and it contains lamellated (pacinian) corpuscles.

4. The types of cells in the epidermis are keratinocytes, melanocytes, Langerhans cells, and Merkel cells.

5. The epidermal layers, from deep to superficial, are the stratum basale, stratum spinosum, stratum granulosum, stratum lucidum (in thick skin only), and stratum corneum (see Table 5.1). Stem cells in the stratum basale undergo continuous cell division, producing keratinocytes for the other layers.

6. The dermis consists of papillary and reticular regions. The papillary region is composed of areolar connective tissue containing fine elastic fibers, dermal papillae, and Meissner corpuscles. The reticular region is composed of dense irregular connective tissue containing interlaced collagen and coarse elastic fibers, adipose tissue, hair follicles, nerves, sebaceous (oil) glands, and ducts of sudoriferous (sweat) glands.

7. Epidermal ridges provide the basis for fingerprints and footprints.

8. The color of skin is due to melanin, carotene, and hemoglobin.

Accessory Structures of the Skin (p. 128)

1. Accessory structures of the skin—hair, skin glands, and nails—develop from the embryonic epidermis.

2. A hair consists of a shaft, most of which is superficial to the surface, a root that penetrates the dermis and sometimes the subcutaneous layer, and a hair follicle.

3. Associated with each hair follicle is a sebaceous (oil) gland, an arrector pili muscle, and a hair root plexus.

4. New hairs develop from division of matrix cells in the bulb; hair replacement and growth occur in a cyclic pattern consisting of alternating growth and resting stages.

5. Hairs offer a limited amount of protection from the sun, heat loss, and entry of foreign particles into the eyes, nose, and ears. They also function in sensing light touch.

6. Sebaceous (oil) glands are usually connected to hair follicles; they are absent in the palms and soles. Sebaceous glands produce sebum, which moistens hairs and waterproofs the skin. Clogged sebaceous glands may produce acne.

7. There are two types of sudoriferous (sweat) glands: eccrine and apocrine. Eccrine sweat glands have an extensive distribution; their ducts terminate at pores at the surface of the epidermis. Apocrine sweat glands are limited to the skin of the axilla, groin, and areolae;

their ducts open into hair follicles. They begin functioning at puberty and are stimulated during emotional stress and sexual excitement. Mammary glands are specialized sudoriferous glands that secrete milk.

8. Ceruminous glands are modified sudoriferous glands that secrete cerumen (ear wax). They are found in the external auditory canal (ear canal).

9. Lanugo of the fetus is shed before birth, except in the scalp, eyebrows, and eyelashes. Most body hair on males is terminal (coarse, pigmented); most body hair on females is vellus (fine).

10. Nails are hard, keratinized epidermal cells over the dorsal surfaces of the distal portions of the digits.

11. The principal parts of a nail are the nail body, free edge, nail root, lunula, eponychium, and matrix. Cell division of the matrix cells produces new nails.

Types of Skin (p. 133)

1. Thin skin covers all parts of the body except for the palms, palmar surfaces of the digits, and the soles.

2. Thick skin covers the palms, palmar surfaces of the digits, and soles.

Functions of the Skin (p. 134)

1. Skin functions include body temperature regulation, protection, sensation, excretion and absorption, and synthesis of vitamin D.

2. The skin participates in thermoregulation by liberating sweat at its surface and functions as a blood reservoir.

3. The skin provides physical, chemical, and biological barriers that help protect the body.

4. Cutaneous sensations include touch, hot and cold, and pain.

Blood Supply of the Integumentary System (p. 135)

1. The epidermis is avascular.

2. The dermis is supplied by the cutaneous and papillary plexuses.

Development of the Integumentary System (p. 135)

1. The epidermis develops from the embryonic ectoderm, and the accessory structures of the skin (hair, nails, and skin glands) are epidermal derivatives.

2. The dermis is derived from mesodermal cells.

Aging and the Integumentary System (p. 137)

1. Most effects of aging begin to occur when people reach their late forties.

2. Among the effects of aging are wrinkling, loss of subcutaneous fat, atrophy of sebaceous glands, and decrease in the number of melanocytes and Langerhans cells.

SELF-QUIZ QUESTIONS

Choose the one best answer to the following questions.

1. Which of the following statements about the function of skin is *not* true?
   a. It helps regulate body temperature.
   b. It protects against bacterial invasion and dehydration.
   c. It absorbs water and salts.
   d. It participates in the synthesis of vitamin D.
   e. It detects stimuli related to temperature and pain.

2. The layer of the skin from which new epidermal cells are derived is the:
   a. stratum corneum   b. stratum basale   c. stratum lucidum
   d. stratum granulosum   e. stratum spinosum

3. The substance that prevents excessive evaporation of water from the skin, keeps the skin soft and pliable, and inhibits the growth of bacteria is:
   a. sebum.   b. keratin.   c. cerumen.
   d. sweat.   e. carotene.

4. The subcutaneous layer consists mostly of:
   a. melanin.   b. simple squamous epithelial tissue.
   c. keratin.   d. areolar and adipose tissue.   e. smooth muscle.

5. Tyrosinase is an enzyme necessary for the production of:
   a. melanin.   b. vitamin D.   c. sweat.
   d. carotene.   e. sebum.

6. Lanugo is:
   a. a water-repelling protein in the epidermis.
   b. a type of skin pigment.
   c. the cuticle of a nail.
   d. nonpigmented hairs covering the body of the fetus.
   e. hairs that appear at puberty in response to hormone changes.

Complete the following.

7. _____ glands, which secrete earwax, are located in the lining of the external auditory meatus.

8. The two principal parts of the skin are the superficial _____, and the deeper _____.

9. Sweat glands are also known as _____ glands.

Are the following statements true or false?

10. Hair follicles and sebaceous glands are found only in areas covered by thick skin.

11. The scientific name for the cuticle of a nail is the lunula.

12. The dermis receives its blood supply from two arterial networks or plexuses: the cutaneous plexus and the epidermal plexus.

13. Eccrine sweat glands are more common than apocrine sweat glands.

14. The dermis consists of two regions; the papillary region is superficial, and the reticular region is deep.

15. Racial differences in skin color are mainly due to the number of melanocytes in the epidermis.

Matching

16. Match the following:
   ___ **(a)** assist in immune responses; easily damaged by UV light
   ___ **(b)** produce pigment that shields cell nuclei from UV light
   ___ **(c)** located in subcutaneous tissue; sensitive to pressure
   ___ **(d)** most abundant epidermal cells
   ___ **(e)** touch receptor cells in the stratum basale layer
   ___ **(f)** tactile receptors located in dermal papillae

   (1) melanocytes
   (2) keratinocytes
   (3) Merkel cells
   (4) Langerhans cells
   (5) Meissner corpuscles (corpuscles of touch)
   (6) lamellated (pacinian) corpuscles

17. Match the following:
   ___ **(a)** helps control the rate of cell division in the stratum basale
   ___ **(b)** lipid-rich secretion of keratinocytes that fills spaces between the layers of the epidermis and that acts as a water-repellant sealant
   ___ **(c)** reddish brown/black pigment that neutralizes free radicals in the skin following damage by UV radiation
   ___ **(d)** a yellow-orange pigment found in the stratum corneum, dermis, and subcutaneous layer
   ___ **(e)** tough, fibrous protective protein in the epidermis
   ___ **(f)** fatty substance that helps prevent water evaporation from skin and inhibits bacterial growth

   (1) melanin
   (2) keratin
   (3) lamellar granules
   (4) epidermal growth factor
   (5) carotene
   (6) sebum

CRITICAL THINKING QUESTIONS

1. Your 65-year-old aunt has spent every sunny day at the beach for as long as you can remember. Her skin looks a lot like a comfortable lounge chair—brown and wrinkled. Recently her dermatologist removed a suspicious growth from the skin of her face. What would you suspect is the problem and its likely cause?

2. Lindsay is partying with his friends on a cold winter night. The outside temperature is below freezing, but they are not wearing their coats when they step outside to smoke. They say they don't need their coats because they are drinking alcoholic beverages, which they say keeps them warm enough. You have learned that alcohol dilates the blood vessels in the skin. Do you think Lindsay and his friends have chosen an effective way to control their body temperature? Why or why not?

3. Felicity is very concerned about her appearance. She heard that the skin should be protected against sun exposure and dryness to prevent wrinkles. Felicity's new plan for perfect skin is to only go out at night and to wear a waterproof covering over her entire body. Comment on Felicity's beauty plan.

4. Your nephew has been learning about cells in science class and now refuses to take a bath. He asks, "If all cells have a semipermeable membrane, and my skin is made out of cells, then won't I swell up and pop when I take a bath?" Explain this dilemma to your nephew before he starts attracting flies.

5. People always say, "It's not the heat; it's the humidity," when complaining about summer weather. Why do you think people feel hotter when it's 95 degrees Fahrenheit and 95% humidity than when it's 95 degrees and 30% humidity? (Humidity refers to the amount of water vapor in the air.)

ANSWERS TO FIGURE QUESTIONS

5.1 The epidermis is composed of epithelial tissue, whereas the dermis is made up of connective tissue.

5.2 Melanin protects the DNA of the nucleus of keratinocytes from damage from UV light.

5.3 The stratum basale is the layer of the epidermis that contains stem cells.

5.4 Lines of cleavage are clinically important because surgical incisions parallel to them leave only fine scars, whereas a surgical incision made across lines of cleavage will tend to heal in a broad, thick scar.

5.5 Plucking a hair stimulates hair root plexuses in the dermis, some of which are sensitive to pain. Because the cells of a hair shaft are already dead and the hair shaft lacks nerves, cutting hair is not painful.

5.6 The main function of eccrine sweat glands is to help regulate body temperature through evaporation.

5.7 Nails are hard because they are composed of tightly packed, keratinized epidermal cells.

5.8 Vernix caseosa consists of secretions from sebaceous glands, sloughed off peridermal cells, and hairs.

5.9 Basal cell carcinoma is the most common type of skin cancer.

5.10 The seriousness of a burn is determined by the depth and extent of the area involved, the individual's age, and general health.

5.11 Pressure ulcers typically develop in tissues that overlie bony projections subjected to pressure, such as the shoulders, hips, buttocks, heels, and ankles.

6

BONE TISSUE

INTRODUCTION   A bone is made up of several different tissues working together: bone or osseous tissue, cartilage, dense connective tissues, epithelium, adipose tissue, and nervous tissue. For this reason, each individual bone is considered an organ. Bone tissue is a complex and dynamic living tissue. It continually engages in a process called remodeling—building new bone tissue and breaking down old bone tissue. The entire framework of bones and their cartilages constitute the **skeletal system**. This chapter will survey the various components of bones so you can understand how bones form, how they age, and how exercise affects the density and strength of bones. The study of bone structure and the treatment of bone disorders is referred to as **osteology** (os-tē-OL-ō-jē; *osteo-*=bone; *-logy*=study of).

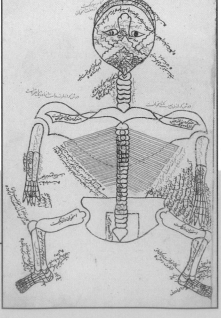

Can you guess from which culture and era
this skeletal image originates?

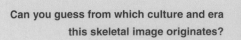
www.wiley.com/college/apcentral

FUNCTIONS OF THE SKELETAL SYSTEM

OBJECTIVE

● Describe the six main functions of the skeletal system.

Bone tissue and the skeletal system perform several basic functions:

1. *Support.* The skeleton serves as the structural framework for the body by supporting soft tissues and providing attachment points for the tendons of most skeletal muscles.
2. *Protection.* The skeleton protects many internal organs from injury. For example, cranial bones protect the brain, backbones protect the spinal cord, and the rib cage protects the heart and lungs.
3. *Assistance in movement.* Because skeletal muscles attach to bones, when muscles contract, they pull on bones. Together, bones and muscles produce movement.
4. *Mineral storage and release.* Bone tissue stores several minerals, especially calcium and phosphorus, which contribute to the strength of the bone. On demand bones can release minerals into the bloodstream to maintain critical mineral balances and to distribute minerals to other organs.
5. *Blood cell production.* Within certain bones, a connective tissue called **red bone marrow** produces red blood cells, white blood cells, and platelets, a process called **hemopoiesis** (hēm-ō-poy-Ē-sis; *hemo-*=blood; *-poiesis*=making). Red bone marrow consists of developing blood cells, adipocytes, fibroblasts, and macrophages within a network of reticular fibers. It is present in developing bones of the fetus and in some adult bones, such as the pelvis, ribs, breastbone, backbones, skull, and ends of the arm bones and thighbones.
6. *Triglyceride storage.* **Yellow bone marrow** consists mainly of adipose cells, which store triglycerides, and a few blood cells. Triglycerides stored in the adipose cells of yellow bone marrow are a potential chemical energy reserve. In the newborn, all bone marrow is red and is involved in hemopoiesis. With increasing age, much of the bone marrow changes from red to yellow.

CHECKPOINT

1. What types of tissues make up the skeletal system?
2. How do red and yellow bone marrow differ in composition and function?

ANATOMY OF A BONE

OBJECTIVE

● Describe the parts of a long bone.

We will now examine the structure of bone at both the macroscopic and microscopic levels. The structure of a bone may be analyzed by considering the parts of a long bone such as the humerus (the arm bone) or the femur (the thigh bone)

(Figure 6.1). A *long bone* is one that has greater length than width. A typical long bone consists of the following parts:

1. The **diaphysis** (dī-AF-i-sis=growing between) is the bone's shaft, or body—the long, cylindrical, main portion of the bone.
2. The **epiphyses** (e-PIF-i-sēz=growing own; singular is *epiphysis*) are the distal and proximal ends of the bone.
3. The **metaphyses** (me-TAF-i-sēz; *meta-*=between; singular is *metaphysis*) are the regions in a mature bone where the diaphysis joins the epiphyses. In a growing bone, each metaphysis includes an **epiphyseal plate** (ep′-i-FIZ-ē-al), a layer of hyaline cartilage that allows the diaphysis of the bone to grow in length (described later in the chapter). When bone growth in length stops, the cartilage in the epiphyseal plate is replaced by bone and the resulting bony structure is known as the **epiphyseal line.**
4. The **articular cartilage** is a thin layer of hyaline cartilage covering the part of the epiphysis where the bone forms an articulation (joint) with another bone. Articular cartilage reduces friction and absorbs shock at freely movable joints. Because articular cartilage lacks a perichondrium, repair of damage is limited.
5. The **periosteum** (per′-ē-OS-tē-um; *peri-*=around) is a tough sheath of dense irregular connective tissue that surrounds the bone surface wherever it is not covered by articular cartilage. The periosteum contains bone-forming cells that enable bone to grow in width, but not in length. It also protects the bone, assists in fracture repair, helps nourish bone tissue, and serves as an attachment point for ligaments and tendons. The periosteum is attached to the underlying bone by **perforating (Sharpey's) fibers,** thick bundles of collagen that extend from the periosteum into the bone matrix.
6. The **medullary cavity** (MED-ū-lar′-ē; *medulla-*=marrow, pith), or **marrow cavity,** is the space within the diaphysis that contains fatty yellow bone marrow in adults.
7. The **endosteum** (end-OS-tē-um; *endo-*=within) is a thin membrane that lines the medullary cavity. It contains a single layer of bone-forming cells and a small amount of connective tissue.

CHECKPOINT

3. Diagram the parts of a long bone, and list the functions of each part.

HISTOLOGY OF BONE TISSUE

OBJECTIVE

● Describe the histological features of bone tissue.

Like other connective tissues, **bone,** or **osseous tissue** (OS-ē-us), contains an abundant extracellular matrix that surrounds widely separated cells. The extracellular matrix is about 25% water, 25% collagen fibers, and 50% crystallized mineral salts.

Figure 6.1 Parts of a long bone. The spongy bone tissue of the epiphysis and metaphysis contains red bone marrow, whereas the medullary cavity of the diaphysis contains yellow bone marrow (in adults).

🔑 **A long bone is covered by articular cartilage at its proximal and distal epiphyses and by periosteum around the diaphysis.**

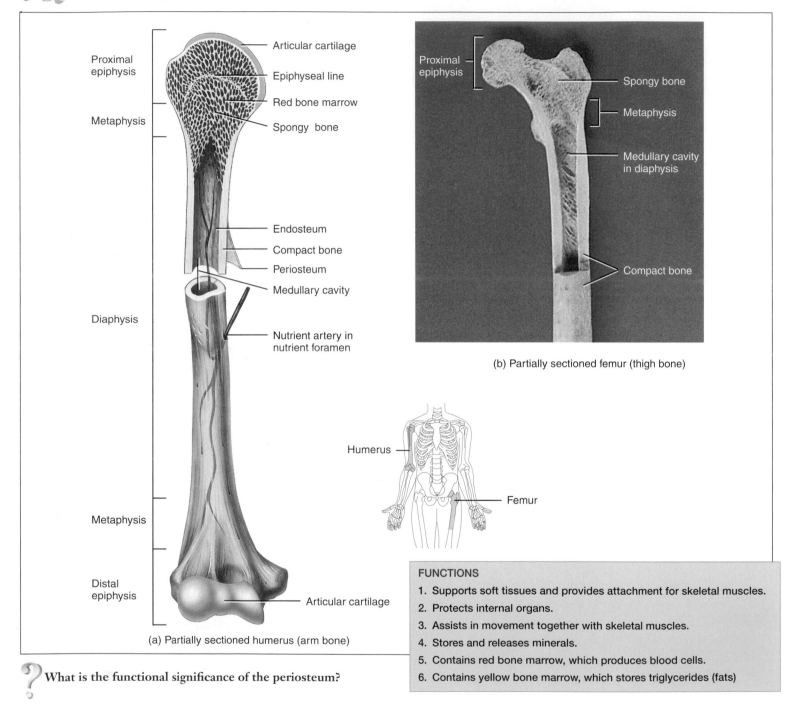

Proximal epiphysis

Articular cartilage

Epiphyseal line

Metaphysis

Red bone marrow

Spongy bone

Endosteum

Compact bone

Periosteum

Diaphysis

Medullary cavity

Nutrient artery in nutrient foramen

Metaphysis

Distal epiphysis

Articular cartilage

(a) Partially sectioned humerus (arm bone)

Proximal epiphysis

Spongy bone

Metaphysis

Medullary cavity in diaphysis

Compact bone

(b) Partially sectioned femur (thigh bone)

Humerus

Femur

FUNCTIONS

1. Supports soft tissues and provides attachment for skeletal muscles.
2. Protects internal organs.
3. Assists in movement together with skeletal muscles.
4. Stores and releases minerals.
5. Contains red bone marrow, which produces blood cells.
6. Contains yellow bone marrow, which stores triglycerides (fats)

❓ **What is the functional significance of the periosteum?**

The abundant inorganic mineral salts are mainly *hydroxyapatite* (calcium phosphate and calcium carbonate). In addition, the extracellular matrix of bone includes small amounts of magnesium hydroxide, fluoride, and sulfate. As these mineral salts are deposited in the framework formed by the collagen fibers of the extracellular matrix, they crystallize and the tissue hardens. This process of **calcification** (kal′-si-fi-KĀ-shun) is initiated by osteoblasts, the bone-building cells.

It was once thought that calcification simply occurred when enough mineral salts were present to form crystals. Now, however, we know that the process occurs only in the presence of collagen fibers. Mineral salts begin to crystallize in the microscopic spaces between collagen fibers. After the spaces are filled, mineral crystals accumulate around the collagen fibers. The combination of crystallized salts and collagen fibers is responsible for the hardness that is characteristic of bone.

Although a bone's *hardness* depends on the crystallized inorganic mineral salts, a bone's *flexibility* depends on its collagen fibers. Like reinforcing metal rods in concrete, collagen fibers and other organic molecules provide *tensile strength*, which is resistance to being stretched or torn apart. Soaking a bone in an acidic solution, such as vinegar, dissolves its mineral salts, causing the bone to become rubbery and flexible. As you will see shortly, bone cells called osteoclasts secrete enzymes and acids that break down the extracellular matrix of bone. Four types of cells are present in bone tissue: osteogenic cells, osteoblasts, osteocytes, and osteoclasts (Figure 6.2).

1. **Osteogenic cells** (os′-tē-ō-JEN-ik; *-genic*=producing) are unspecialized stem cells derived from mesenchyme, the tissue from which all connective tissues are formed. They are the only bone cells to undergo cell division; the resulting cells develop into osteblasts. Osteogenic cells are found along the inner portion of the periosteum, in the endosteum, and in the canals within bone that contain blood vessels.

2. **Osteoblasts** (OS-tē-ō-blasts′; *-blasts*=buds or sprouts) are bone-building cells. They synthesize and secrete collagen fibers and other organic components needed to build the extracellular matrix of bone tissue, and they initiate calcification. As osteoblasts surround themselves with extracellular matrix, they become trapped in their secretions and become osteocytes. (Note: *Blasts* in bone or any other connective tissue secrete extracellular matrix.)

3. **Osteocytes** (OS-tē-ō-sīts′; *-cytes*=cells), mature bone cells, are the main cells in bone tissue and maintain its daily metabolism, such as the exchange of nutrients and wastes with the blood. Like osteoblasts, osteocytes do not undergo cell division. (Note: *Cytes* in bone or any other tissue maintain the tissue.)

4. **Osteoclasts** (OS-tē-ō-clasts′; *-clast*=break) are huge cells derived from the fusion of as many as 50 monocytes (a type of white blood cell) and are concentrated in the endosteum. On the side of the cell that faces the bone surface, the osteoclast's plasma membrane is deeply folded into a *ruffled border*. Here the cell releases powerful lysosomal enzymes and acids that digest the protein and mineral components of the underlying extracellular matrix of bone. This breakdown of the extracellular matrix of bone, termed **resorption** (rē-SORP-shun), is part of the normal development, growth, maintenance, and repair of bone. (Note: *Clasts* in bone break down extracellular matrix.)

Bone is not completely solid but has many small spaces between its cells and extracellular matrix components. Some spaces are channels for blood vessels that supply bone cells with nutrients. Other spaces are storage areas for red bone marrow. Depending on the size and distribution of the spaces, the regions of a bone may be categorized as compact or spongy (see Figure 6.1). Overall, about 80% of the skeleton is compact bone and 20% is spongy bone.

Compact Bone Tissue

Compact bone tissue contains few spaces. It forms the external layer of all bones and makes up the bulk of the diaphyses of long

Figure 6.2   Types of cells in bone tissue.

Osteogenic cells undergo cell division and develop into osteoblasts, which secrete bone extracellular matrix.

| Osteogenic cell (develops into an osteoblast) | Osteoblast (forms bone matrix) | Osteocyte (maintains bone tissue) | Osteoclast (functions in resorption, the breakdown of bone matrix) |

Ruffled border

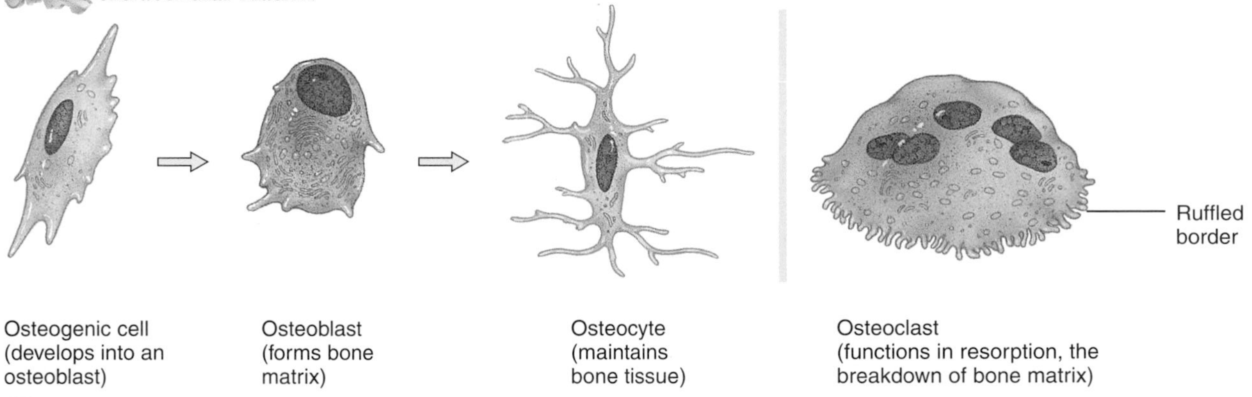

? Why is bone resorption important?

bones. Compact bone tissue provides protection and support and resists the stresses produced by weight and movement.

Compact bone tissue is arranged in units called **osteons** or **haversian systems** (ha-VER-shun) (Figure 6.3a). Blood vessels, lymphatic vessels, and nerves from the periosteum penetrate the compact bone through transverse **perforating** or **Volkmann's canals** (FOLK-mans). The vessels and nerves of the perforating canals connect with those of the medullary cavity, periosteum, and **central** or **haversian canals.** The central canals run longitudinally through the bone. Around the canals are **concentric lamellae** (la-MEL-ē)—rings of hard, calcified extracellular matrix. Between the lamellae are small spaces called **lacunae** (la-KOO-nē = little lakes; singular is *lacuna*), which contain osteocytes. Radiating in all directions from the lacunae are tiny **canaliculi** (kan′-a-LIK-ū-lī = small channels), which are filled with extracellular fluid. Inside the canaliculi are slender fingerlike processes of osteocytes (see inset at right of Figure 6.3a). Neighboring osteocytes communicate via gap junctions. The canaliculi connect lacunae with one another and with the central canals. Thus, an intricate, miniature system of interconnected canals throughout the bone provides many routes for nutrients and oxygen to reach the osteocytes and for wastes to diffuse away. This is very important because diffusion through the lamellae is extremely slow.

Osteons in compact bone tissue are aligned in the same direction along lines of stress. In the shaft, for example, they are parallel to the long axis of the bone. As a result, the shaft of a long bone resists bending or fracturing even when considerable force is applied from either end. Compact bone tissue tends to be thickest in those parts of a bone where stresses are applied in relatively few directions. The lines of stress in a bone are not static. They change as a person learns to walk and in response to repeated strenuous physical activity, such as occurs when a person undertakes weight training. The lines of stress in a bone also can change because of fractures or physical deformity. Thus, the organization of osteons changes over time in response to the physical demands placed on the skeleton.

The areas between osteons contain **interstitial lamellae,** which also have lacunae with osteocytes and canaliculi. Interstitial lamellae are fragments of older osteons that have been partially destroyed during bone rebuilding or growth. Lamellae that encircle the bone just beneath the periosteum or encircle the medullary cavity are called **circumferential lamellae.**

Spongy Bone Tissue

In contrast to compact bone tissue, **spongy bone tissue** does not contain osteons (Figure 6.3b, c). It consists of lamellae that are arranged in an irregular lattice of thin columns of bone called **trabeculae** (tra-BEK-ū-lē = little beams; singular is *trabecula*). The macroscopic spaces between the trabeculae of some bones are filled with red bone marrow, which produces blood cells. Within each trabecula are osteocytes that lie in lacunae. Radiating from the lacunae are canaliculi. Because osteocytes are located on the superficial surfaces of trabeculae, they receive nourishment directly from the blood circulating through the medullary cavities.

Spongy bone tissue makes up most of the bone tissue of short, flat, and irregularly shaped bones; most of the epiphyses of long bones; and a narrow rim around the medullary cavity of the diaphysis of long bones.

At first glance, the trabeculae of spongy bone tissue may appear to be much less well organized than the osteons of compact bone tissue. However, the trabeculae of spongy bone tissue are precisely oriented along lines of stress, a characteristic that helps bones resist stresses and transfer force without breaking. Spongy bone tissue tends to be located where bones are not heavily stressed or where stresses are applied from many directions.

Spongy bone tissue is different from compact bone tissue in two respects. First, spongy bone tissue is light, which reduces the overall weight of a bone so that it moves more readily when pulled by a skeletal muscle. Second, the trabeculae of spongy bone tissue support and protect the red bone marrow. Spongy bone in the hip bones, ribs, breastbone, backbones, and the ends of long bones is where red bone marrow is stored and where hemopoiesis occurs in adults.

BONE SCAN

A **bone scan** is a diagnostic procedure that takes advantage of the fact that bone is living tissue. A small amount of a radioactive tracer compound that is readily absorbed by bone is injected intravenously. The degree of uptake of the tracer is related to the amount of blood flow to the bone. A scanning device measures the radiation emitted from the bones, and the information is translated into a photograph or diagram that can be read like an x-ray. Normal bone tissue is identified by a consistent gray color throughout because of its uniform uptake of the radioactive tracer. Darker or lighter areas, however, may indicate bone abnormalities. Darker areas, called "hot spots," are areas of increased metabolism that absorb more of the radioactive tracer. Hot spots may indicate bone cancer, abnormal healing of fractures, or abnormal bone growth. Lighter areas, called "cold spots," are areas of decreased metabolism that absorb less of the radioactive tracer. Cold spots may indicate problems such as degenerative bone disease, decalcified bone, fractures, bone infections, Paget's disease, and rheumatoid arthritis. A bone scan not only detects abnormalities 3–6 months sooner than standard x-ray procedures, it exposes the patient to less radiation. A bone scan is the standard test for bone density screening for females. ■

CHECKPOINT

4. Why is bone considered a connective tissue?
5. Describe the four types of cells in bone tissue.
6. What is the composition of the matrix of bone tissue?
7. Distinguish between spongy and compact bone tissue in terms of microscopic appearance, location, and function.

Figure 6.3   Histology of compact and spongy bone. (a) Sections through the diaphysis of a long bone, from the surrounding periosteum on the right, to compact bone in the middle, to spongy bone and the medullary cavity on the left. The inset at the upper right shows an osteocyte in a lacuna. (b and c) Details of spongy bone.

 Osteocytes lie in lacunae arranged in concentric circles around a central canal in compact bone, and in lacunae arranged irregularly in trabeculae in spongy bone.

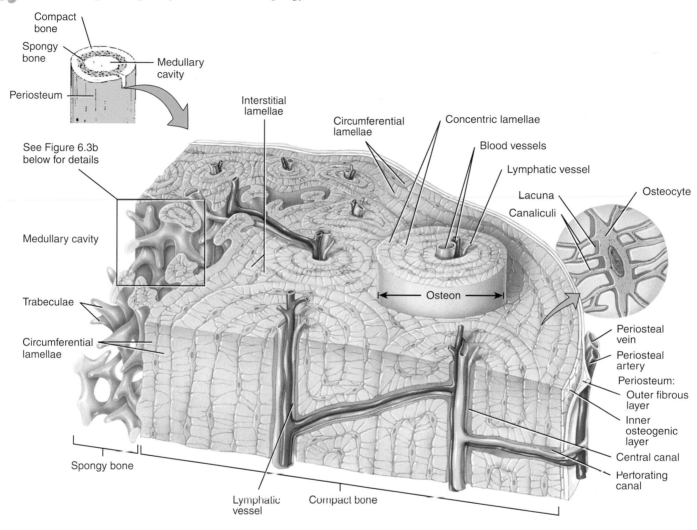

(a) Osteons (haversian systems) in compact bone and trabeculae in spongy bone

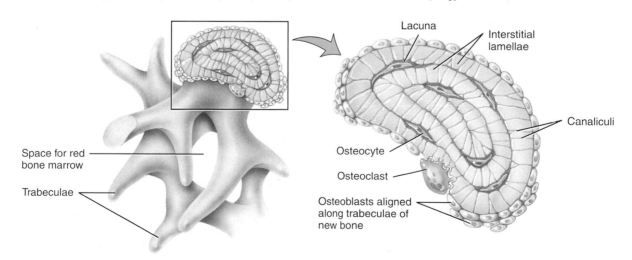

(b) Enlarged aspect of spongy bone trabeculae   (c) Details of a section of a trabecula

As people age, some central (haversian) canals may become blocked. What effect would this have on the osteocytes?

BLOOD AND NERVE SUPPLY OF BONE

• Describe the blood and nerve supply of bone.

Bone is richly supplied with blood. Blood vessels, which are especially abundant in portions of bone containing red bone marrow, pass into bones from the periosteum. We will consider the blood supply of a long bone such as the mature tibia (shin bone) shown in Figure 6.4.

Periosteal arteries (per-ē-OS-tē-al) accompanied by nerves enter the diaphysis through numerous perforating (Volkmann's) canals and supply the periosteum and outer part of the compact bone (see Figure 6.3a). Near the center of the diaphysis, a large **nutrient artery** enters the compact bone through a hole called the **nutrient foramen** (Figure 6.4). On entering the medullary cavity, the nutrient artery divides into proximal and distal branches that supply both the inner part of compact bone tissue of the diaphysis and the spongy bone tissue and red marrow as far as the epiphyseal plates (or lines). Some bones, like the tibia, have only one nutrient artery; others like the femur (thigh bone) have several. The ends of long bones are supplied by the metaphyseal and epiphyseal arteries, which arise from arteries that supply the associated joint. The **metaphyseal**

arteries (met-a-FIZ-ē-al) enter the metaphyses of a long bone and, together with the nutrient artery, supply the red bone marrow and bone tissue of the metaphyses. The **epiphyseal arteries** (ep′-i-FIZ-ē-al) enter the epiphyses of a long bone and supply the red bone marrow and bone tissue of the epiphyses.

Veins that carry blood away from long bones are evident in three places: (1) One or two **nutrient veins** accompany the nutrient artery in the diaphysis; (2) numerous **epiphyseal veins** and **metaphyseal veins** exit with their respective arteries in the epiphyses; and (3) many small **periosteal veins** exit with their respective arteries in the periosteum.

Nerves accompany the blood vessels that supply bones. The periosteum is rich in sensory nerves, some of which carry pain sensations. These nerves are especially sensitive to tearing or tension, which explains the severe pain resulting from a fracture or a bone tumor. For the same reason there is some pain associated with a bone marrow needle biopsy. In this procedure, a needle is inserted into the middle of the bone to withdraw a sample of red bone marrow to examine it microscopically for conditions such as leukemias, metastatic neoplasms, lymphoma, Hodgkin's disease, and aplastic anemia. As the needle penetrates the periosteum, pain is felt. Once it passes through, there is little pain.

8. Explain the location and roles of the nutrient arteries, nutrient foramina, epiphyseal arteries, and periosteal arteries.

Figure 6.4   Blood supply of a mature long bone, the tibia (shinbone).

Bone is richly supplied with blood vessels.

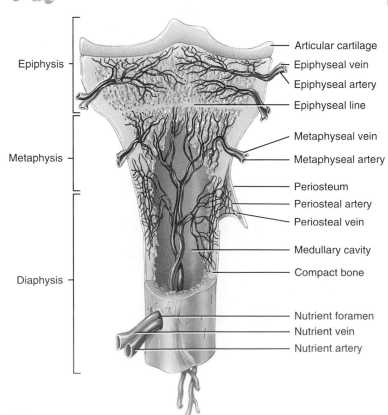

Epiphysis
- Articular cartilage
- Epiphyseal vein
- Epiphyseal artery
- Epiphyseal line

Metaphysis
- Metaphyseal vein
- Metaphyseal artery

- Periosteum
- Periosteal artery
- Periosteal vein

Diaphysis
- Medullary cavity
- Compact bone

- Nutrient foramen
- Nutrient vein
- Nutrient artery

Where do periosteal arteries enter bone tissue?

BONE FORMATION

• Describe the steps involved in intramembranous and endochondral ossification.

The process by which bone forms is called **ossification** (os′-i-fi-KĀ-shun; *ossi-*=bone; *-fication*=making) or **osteogenesis** (os′-tē-ō-JEN-e-sis). The "skeleton" of a human embryo is composed of mesenchymal cells, which are shaped like bones and are the sites where ossification occurs. These "bones" provide the template for subsequent ossification, which begins during the sixth week of embryonic development and follows one of two patterns.

The two methods of bone formation involve the replacement of a preexisting connective tissue with bone. Both methods of ossification do not lead to differences in the structure of mature bones, but are simply different methods of bone development. In the first type of ossification, called **intramembranous ossification** (in′-tra-MEM-bra-nus; *intra-*=within; *-membran*=membrane), bone forms directly within mesenchyme arranged in sheet-like layers that resemble membranes. In the second type, **endochondral ossification** (en′-dō-KON-dral; *endo-*=within; *-chondral*=cartilage), bone forms within hyaline cartilage that develops from mesenchyme.

Intramembranous Ossification

Intramembranous ossification is the simpler of the two methods of bone formation. The flat bones of the skull and mandible (lower jawbone) are formed in this way. Also, the "soft spots" that help the fetal skull pass through the birth canal later harden as they undergo intramembranous ossification, which occurs as follows (Figure 6.5):

❶ *Development of the center of ossification.* At the site where the bone will develop, mesenchymal cells cluster together and differentiate, first into osteogenic cells and then into osteoblasts. (Recall that *mesenchyme* is the tissue from which all other connective tissues arise.) The site of such a cluster is called a **center of ossification.** Osteoblasts secrete the organic matrix of bone until they are surrounded by it.

❷ *Calcification.* Then secretion of extracellular matrix stops and the cells, now called osteocytes, lie in lacunae and extend their narrow cytoplasmic processes into canaliculi

that radiate in all directions. Within a few days, calcium and other mineral salts are deposited and the extracellular matrix hardens or calcifies.

❸ *Formation of trabeculae.* As the extracellular matrix forms, it develops into trabeculae that fuse with one another to form spongy bone. Blood vessels grow into the spaces between the trabeculae and the mesenchyme along the surface of the newly formed bone. Connective tissue that is associated with the blood vessels in the trabeculae differentiates into red bone marrow.

❹ *Development of the periosteum.* At the periphery of the bone, the mesenchyme condenses and develops into the periosteum. Eventually, a thin layer of compact bone replaces the surface layers of the spongy bone, but spongy

Figure 6.5 Intramembranous ossification. Illustrations 1 and 2 show a smaller field of vision at higher magnification than illustrations 3 and 4. Refer to this figure as you read the corresponding numbered paragraphs in the text.

Intramembranous ossification involves the formation of bone within mesenchyme arranged in sheet-like layers that resemble membranes.

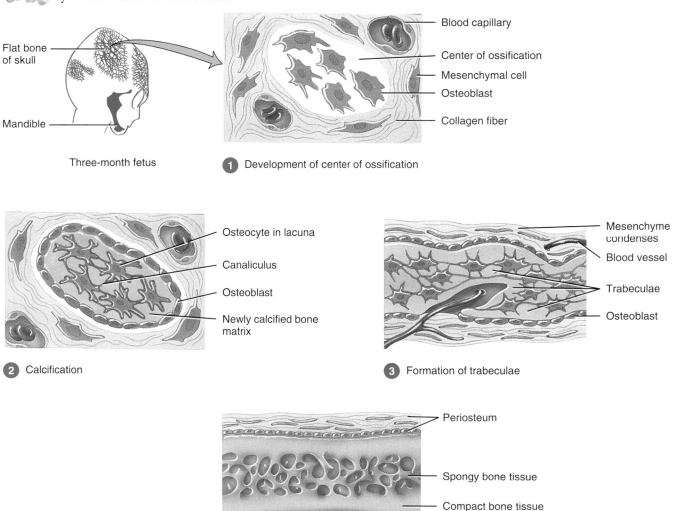

❶ Development of center of ossification

❷ Calcification

❸ Formation of trabeculae

❹ Development of the periosteum

 Which bones of the body develop by intramembranous ossification?

bone remains in the center. Much of the newly formed bone is remodeled (destroyed and reformed) as the bone is transformed into its adult size and shape.

Endochondral Ossification

The replacement of cartilage by bone is called **endochondral ossification.** Although most bones of the body are formed in this way, the process is best observed in a long bone. It proceeds as follows (Figure 6.6):

1 *Development of the cartilage model.* At the site where the bone is going to form, mesenchymal cells crowd together in the shape of the future bone, and then develop into chondroblasts. The chondroblasts secrete cartilage matrix, producing a **cartilage model** consisting of hyaline cartilage. A membrane called the **perichondrium** (per-i-KON-drē-um) develops around the cartilage model.

2 *Growth of the cartilage model.* Once chondroblasts become deeply buried in the extracellular matrix of cartilage, they are called chondrocytes. The cartilage model grows in length by continual cell division of chondrocytes accompanied by further secretion of the cartilage extracellular matrix. This type of growth is termed **interstitial growth** and results in an increase in length. In contrast, growth of the cartilage in thickness is due mainly to the addition of more extracellular matrix material to the periphery of the model by new chondroblasts that develop from the perichondrium. This growth pattern in which extracellular matrix is deposited on the cartilage surface is called **appositional growth** (a-pō-ZISH-i-nal).

As the cartilage model continues to grow, chondrocytes in its mid-region hypertrophy (increase in size). Some hypertrophied cells burst and release their contents, which increases the pH of the surrounding extracellular matrix. This change in pH triggers calcification. Other chondrocytes within the calcifying cartilage die because nutrients can no longer diffuse quickly enough through the extracellular matrix. As chondrocytes die, lacunae form and eventually merge into small cavities.

3 *Development of the primary ossification center.* Primary ossification proceeds *inward* from the external surface of the bone. A nutrient artery penetrates the perichondrium and the calcifying cartilage model through a nutrient foramen in the mid-region of the cartilage model, stimulating osteogenic cells in the perichondrium to differentiate into osteoblasts. Deep to the perichondrium the osteoblasts secrete a thin shell of compact bone called the **periosteal bone collar.** Once the perichondrium starts to form bone, it is known as the **periosteum.** Near the middle of the model, periosteal capillaries grow into the disintegrating calcified cartilage. Upon growing into the cartilage model, the capillaries induce growth of a **primary ossification center,** a region where bone tissue will replace most of the cartilage. Osteoblasts then begin to deposit bone extracellular matrix over the remnants of calcified cartilage, forming spongy bone trabeculae. As the ossification center grows toward the ends of the

bone, osteoclasts break down some of the newly formed spongy bone trabeculae. This activity leaves a cavity, the medullary (marrow) cavity, in the core of the model. The medullary cavity then fills with red bone marrow.

4 *Development of the secondary ossification centers.* The diaphysis (shaft), which was once a solid mass of hyaline cartilage, is replaced by compact bone, the core of which contains a red bone marrow–filled medullary cavity. When branches of the epiphyseal artery enter the epiphyses, **secondary ossification centers** develop, usually around the time of birth. Bone formation is similar to that in primary ossification centers. One difference, however, is that spongy bone remains in the interior of the epiphyses (no medullary cavities are formed there). Secondary ossification proceeds *outward* from the center of the epiphysis toward the outer surface of the bone.

5 *Formation of articular cartilage and the epiphyseal plate.* The hyaline cartilage that covers the epiphyses becomes the articular cartilage. Prior to adulthood, hyaline cartilage remains between the diaphysis and epiphysis as the **epiphyseal plate,** which is responsible for the lengthwise growth of long bones.

CHECKPOINT

9. What are the major events of intramembranous ossification and endochondral ossification and how are they different?

BONE GROWTH

OBJECTIVE

● Describe how bones grow in length and thickness.

During childhood, bones throughout the body grow in width by appositional growth, and long bones lengthen by the addition of bone material on the diaphyseal side of the epiphyseal plate.

Growth in Length

To understand how a bone grows in length, you need to know about the structure of the epiphyseal plate. The **epiphyseal plate** (Figure 6.7 on page 154) (ep′-i-FIZ-ē-al), a layer of hyaline cartilage in the metaphysis of a growing bone, consists of four zones (Figure 6.7b).

1. *Zone of resting cartilage.* This layer is nearest the epiphysis and consists of small, scattered chondrocytes. The cells do not function in bone growth (thus the term *resting*). Instead, they anchor the epiphyseal plate to the bone of the epiphysis.

2. *Zone of proliferating cartilage.* Slightly larger chondrocytes in this zone are arranged like stacks of coins. The chondrocytes divide to replace those that die at the diaphyseal side of the epiphyseal plate.

3. *Zone of hypertrophic cartilage* (hī-per-TRŌ-fic). In this layer, the chondrocytes are even larger and remain arranged in columns. The chondrocytes accumulate glycogen in their cytoplasm, and the extracellular matrix between lacunae

Figure 6.6 Endochondral ossification.

During endochondral ossification, bone gradually replaces a cartilage model.

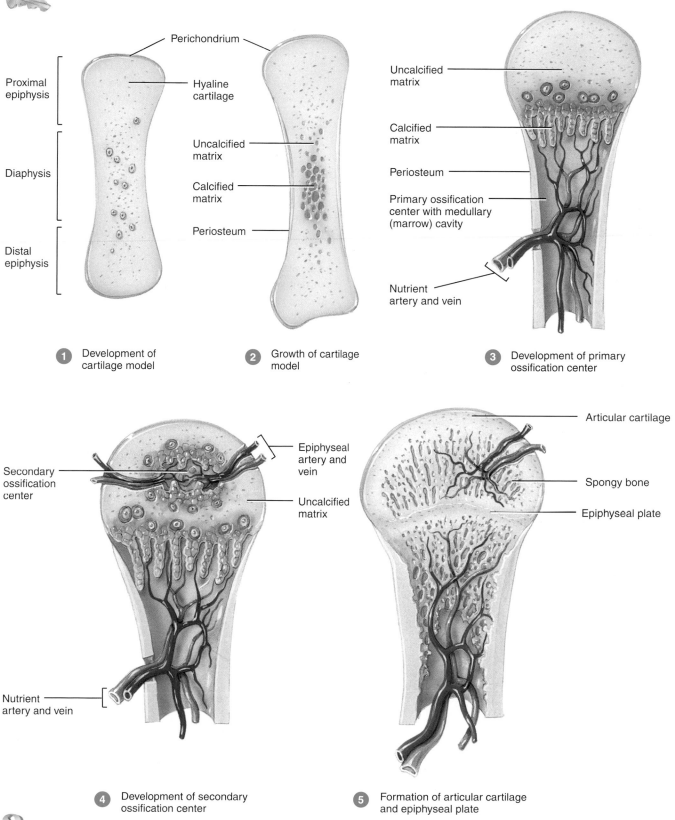

Proximal epiphysis

Diaphysis

Distal epiphysis

Perichondrium

Hyaline cartilage

Uncalcified matrix

Calcified matrix

Periosteum

1 Development of cartilage model

2 Growth of cartilage model

Uncalcified matrix

Calcified matrix

Periosteum

Primary ossification center with medullary (marrow) cavity

Nutrient artery and vein

3 Development of primary ossification center

Secondary ossification center

Epiphyseal artery and vein

Uncalcified matrix

Nutrient artery and vein

4 Development of secondary ossification center

Articular cartilage

Spongy bone

Epiphyseal plate

5 Formation of articular cartilage and epiphyseal plate

If radiographs of an 18-year-old basketball star player show clear epiphyseal plates but no epiphyseal lines, is she likely to grow taller?

Figure 6.7 **The epiphyseal plate is a layer of hyaline cartilage in the metaphysis of a growing bone.** The epiphyseal plate appears as a dark band between whiter calcified areas in the radiograph shown in part (a).

🔑 **The epiphyseal plate allows the diaphysis of a bone to increase in length.**

(a) Radiograph showing the epiphyseal plate of the femur of a 3-year-old

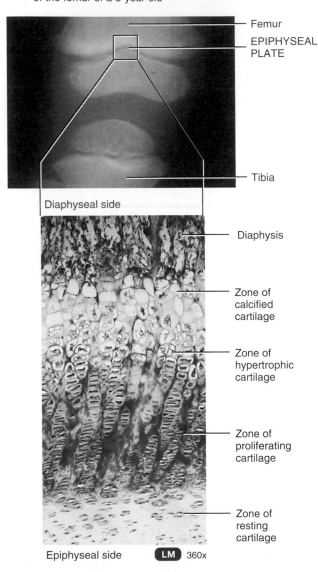

(b) Histology of the epiphyseal plate

❓ **What activities of the epiphyseal plate account for the lengthwise growth of the diaphysis?**

narrows in response to the growth of lacunae. The lengthening of the diaphysis is the result of cell divisions in the zone of proliferating cartilage and maturation of the cells in the zone of hypertrophic cartilage.

4. ***Zone of calcified cartilage.*** The final zone of the epiphyseal plate is only a few cells thick and consists mostly of dead chondrocytes because the extracellular matrix around them has calcified. The calcified cartilage is dissolved by osteoclasts, and the area is invaded by osteoblasts and capillaries from the diaphysis. The osteoblasts lay down bone extracellular matrix, replacing the calcified cartilage. As a result, the diaphyseal border of the epiphyseal plate is firmly cemented to the bone of the diaphysis.

The activity of the epiphyseal plate is the only way that the diaphysis can increase in length. As a bone grows, chondrocytes proliferate on the epiphyseal side of the plate. New chondrocytes cover older ones, which are then destroyed by calcification. Thus, the cartilage is replaced by bone on the diaphyseal side of the plate. In this way the thickness of the epiphyseal plate remains relatively constant, but the bone on the diaphyseal side increases in length.

At about age 18 in females and 21 in males, the epiphyseal plates close; that is, the epiphyseal cartilage cells stop dividing, and bone replaces all the cartilage. The epiphyseal plate fades, leaving a bony structure called the **epiphyseal line.** The appearance of the epiphyseal line signifies that the bone has stopped growing in length. The clavicle is the last bone to stop growing. If a bone fracture damages the epiphyseal plate, the fractured bone may be shorter than normal once adult stature is reached. This is because damage to cartilage, which is avascular, accelerates closure of the epiphyseal plate, and growth in length of the bone is inhibited.

Growth in Thickness

Unlike cartilage, which can thicken by both interstitial and appositional growth, bone can grow in thickness (diameter) only by **appositional growth** (Figure 6.8):

1️⃣ At the bone surface, periosteal cells differentiate into osteoblasts, which secrete the collagen fibers and other organic molecules that form bone extracellular matrix. The osteoblasts become surrounded by extracellular matrix and develop into osteocytes. This process forms bone ridges on either side of a periosteal blood vessel. The ridges slowly enlarge and create a groove for the periosteal blood vessel.

2️⃣ Eventually, the ridges fold together and fuse, and the groove becomes a tunnel that encloses the blood vessel. The former periosteum now becomes the endosteum that lines the tunnel.

3️⃣ Osteoblasts in the endosteum deposit bone extracellular matrix, forming new concentric lamellae. The formation of additional concentric lamellae proceeds inward toward the periosteal blood vessel. In this way, the tunnel fills in, and a new osteon is created.

Figure 6.8 Bone growth in thickness.

As new bone tissue is deposited on the outer surface of bone by osteoblasts, the bone tissue lining the medullary cavity is destroyed by osteoclasts in the endosteum.

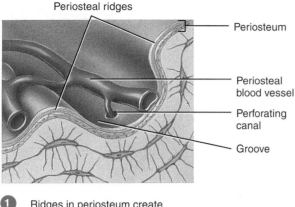

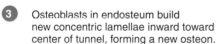

Ridges in periosteum create groove for periosteal blood vessel.

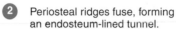

Periosteal ridges fuse, forming an endosteum-lined tunnel.

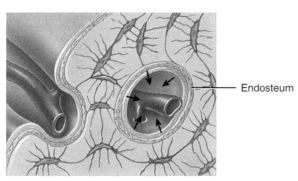

Osteoblasts in endosteum build new concentric lamellae inward toward center of tunnel, forming a new osteon.

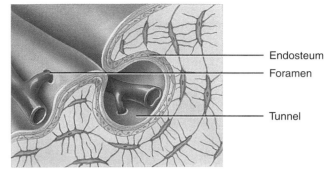

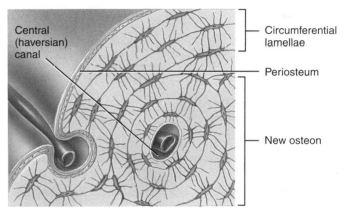

Bone grows outward as osteoblasts in periosteum build new circumferential lamellae. Osteon formation repeats as new periosteal ridges fold over blood vessels.

How does the medullary cavity enlarge during growth in thickness?

As an osteon is forming, osteoblasts under the periosteum deposit new circumferential lamellae, further increasing the thickness of the bone. As additional periosteal blood vessels become enclosed as in step ❶, the growth process continues.

Recall that as new bone tissue is being deposited on the outer surface of bone, the bone tissue lining the medullary cavity is destroyed by osteoclasts in the endosteum. In this way, the medullary cavity enlarges as the bone increases in thickness.

CHECKPOINT

10. Describe the zones of the epiphyseal plate and their functions.
11. Explain how bone grows in length and thickness.
12. What is the significance of the epiphyseal line?

BONE REMODELING

OBJECTIVE

● Describe the processes involved in bone remodeling.

Bone, like skin, forms before birth but continually renews itself thereafter. **Bone remodeling** is the ongoing replacement of old bone tissue by new bone tissue. It involves **bone resorption,** the removal of minerals and collagen (destruction of extracellular matrix) by osteoclasts and the deposition of minerals and collagen (building of extracellular matrix) by osteoblasts. At any given time, about 5% of the total bone mass in the body is remodeled. The renewal rate for compact bone tissue is about 4% per year and for spongy bone tissue is about 20% per year. Remodeling also takes place at different rates in various body regions. The distal portion of the thighbone (femur) is replaced about every four months. By contrast, bone in certain areas of the shaft of the femur will not be completely replaced during an individual's life. Even after bones have reached their adult shapes and sizes, old bone is continually destroyed and new bone tissue is formed in its place. Remodeling also removes injured bone, replacing it with new bone tissue.

Remodeling has several other advantages. Since the strength of bone is related to the degree to which it is stressed, if newly formed bone is subjected to heavy loads, it will be stronger than the old bone. Also, the shape of a bone can be altered for proper support based on certain stress patterns as it remodels. Finally, new bone is less brittle and stronger than old bone.

REMODELING AND ORTHODONTICS

Orthodontics (or-thō-DON-tiks) is the branch of dentistry concerned with the prevention and correction of poorly aligned teeth. The movement of teeth by braces places a stress on the bone that forms the sockets of the teeth. In response to the stress, osteoclasts and osteoblasts remodel the sockets and reshape them so that the teeth can be aligned properly. ■

A delicate balance exists between the actions of osteoclasts and osteoblasts. Should too much new tissue be formed, the bones become abnormally thick and heavy. If too much mineral material is deposited in the bone, the surplus may form thick bumps, called *spurs*, on the bone that interfere with movement at joints. A loss of too much calcium or tissue weakens the bones, and they may break, as occurs in osteoporosis, or they may become too flexible, as in rickets and osteomalacia. (For more on these disorders, see the Applications to Health section at the end of the chapter.) Abnormal acceleration of the remodeling process results in a condition called Paget's disease in which bone deformity results, especially in the bones of the limbs, pelvis, lower vertebrae, and skull. The newly formed bone becomes hard and brittle and fractures easily.

Normal bone growth in the young and bone replacement in the adult depend on the presence of (1) several minerals: calcium, phosphorus, fluoride, magnesium, iron, and manganese; (2) several vitamins: A, B12, C, D, and K; and (3) several hormones: human growth hormone, sex hormones (estrogens and testosterone), thyroid hormones, calcitonin, parathyroid hormone, and insulin.

CHECKPOINT

13. Define bone remodeling, and describe the roles of osteoblasts and osteoclasts in the process.

FRACTURE AND REPAIR OF BONE

A **fracture** is any break in a bone. Fractures are named according to their severity, the shape or position of the fracture line, or even the physician who first described them. Among the common kinds of fractures are the following (Figure 6.9):

Figure 6.9   Types of bone fractures.

A fracture is any break in a bone.

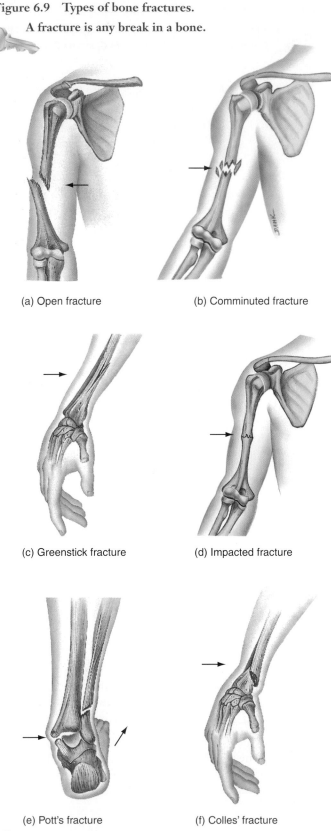

(a) Open fracture

(b) Comminuted fracture

(c) Greenstick fracture

(d) Impacted fracture

(e) Pott's fracture

(f) Colles' fracture

What is a fibrocartilaginous callus?

- **Open (compound) fracture:** The broken ends of the bone protrude through the skin (Figure 6.9a). Conversely, a **closed (simple) fracture** does not break the skin.

- **Comminuted fracture** (KOM-i-noo-ted; *com-*=together; *-minuted*=crumbled): The bone splinters at the site of impact, and smaller bone fragments lie between the two main fragments (Figure 6.9b).

- **Greenstick fracture:** A partial fracture in which one side of the bone is broken and the other side bends; occurs only in children, whose bones are not yet fully ossified and contain more organic material than inorganic material (Figure 6.9c).

- **Impacted fracture:** One end of the fractured bone is forcefully driven into the interior of the other (Figure 6.9d).

- **Pott's fracture:** A fracture of the distal end of the lateral leg bone (fibula), with serious injury of the distal tibial articulation (Figure 6.9e).

- **Colles' fracture** (KOL-ez): A fracture of the distal end of the lateral forearm bone (radius) in which the distal fragment is displaced posteriorly (Figure 6.9f).

In some cases, a bone may fracture without visibly breaking. For example, a **stress fracture** is a series of microscopic fissures in bone that forms without any evidence of injury to other tissues. In healthy adults, stress fractures result from repeated, strenuous activities such as running, jumping, or aerobic dancing. Stress fractures also result from disease processes that disrupt normal bone calcification, such as osteoporosis (discussed later). About 25% of stress fractures involve the tibia. Although standard x-ray images often fail to reveal the presence of stress fractures, they show up clearly in a bone scan.

The following steps occur in the repair of a bone fracture (Figure 6.10):

1 **Formation of fracture hematoma.** As a result of the fracture, blood vessels crossing the fracture line are broken. These include vessels in the periosteum, osteons (haversian systems), medullary cavity, and perforating canals. As blood leaks from the torn ends of the vessels, it forms a clot around the site of the fracture. This clot, called a fracture **hematoma** (hē′-ma-TŌ-ma; *hemat-*=blood; *-oma*=tumor), usually forms 6–8 hours after the injury. Because the circulation of blood stops when the fracture hematoma forms, bone cells at the fracture site die. The hematoma serves as a focus for the influx of cells that follows. Swelling and inflammation occur in response to dead bone cells, producing additional cellular debris. Blood capillaries grow into the blood clot, and phagocytes (neutrophils and macrophages) and osteoclasts begin to remove the dead or damaged tissue in and around the fracture hematoma. This stage may last up to several weeks.

2 **Fibrocartilaginous callus formation.** The infiltration of new blood capillaries into the fracture hematoma helps organize it into an actively growing connective tissue called a **procallus.** Next, fibroblasts from the periosteum and osteogenic cells from the periosteum, endosteum, and red bone marrow invade the procallus. The fibroblasts produce collagen fibers,

which help to connect the broken ends of the bones. Phagocytes continue to remove cellular debris. Osteogenic cells develop into chondroblasts in areas of avascular healthy bone tissue and begin to produce fibrocartilage. Eventually, the procallus is transformed into a **fibrocartilaginous callus** (fi-brō-kar-ti-LAJ-i-nus), a mass of repair tissue that bridges the broken ends of the bones. The stage of the fibrocartilaginous callus lasts about 3 weeks.

Figure 6.10 Steps involved in repair of a bone fracture.

Bone heals more rapidly than cartilage because its blood supply is more plentiful.

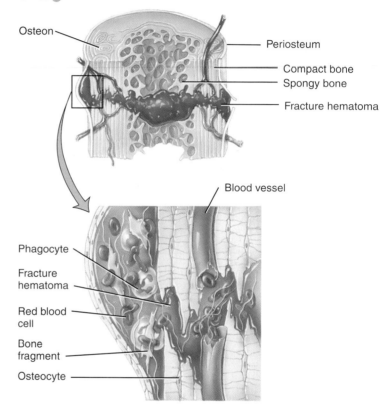

1 Formation of fracture hematoma

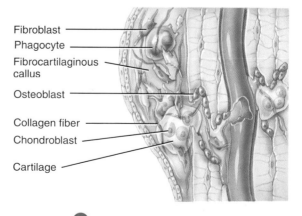

2 Fibrocartilaginous callus fomation

Figure 6.10 (continued)

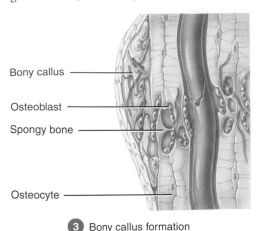

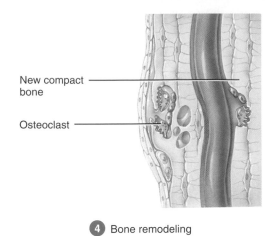

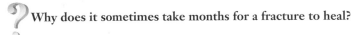

③ Bony callus formation

④ Bone remodeling

? Why does it sometimes take months for a fracture to heal?

③ **Bony callus formation.** In areas closer to well-vascularized healthy bone tissue, osteogenic cells develop into osteoblasts, which begin to produce spongy bone trabeculae. The trabeculae join living and dead portions of the original bone fragments. In time, the fibrocartilage is converted to spongy bone, and the callus is then referred to as a **bony callus.** The stage of the bony callus lasts about 3–4 months.

④ **Bone remodeling.** The final phase of fracture repair is bone remodeling of the callus. Dead portions of the original fragments of broken bone are gradually resorbed by osteoclasts. Compact bone replaces spongy bone around the periphery of the fracture. Sometimes, the repair process is so thorough that the fracture line is undetectable, even in a radiograph (x-ray). However, a thickened area on the surface of the bone may remain as evidence of a healed fracture.

Although bone has a generous blood supply, healing sometimes takes months. The calcium and phosphorus needed to strengthen and harden new bone are deposited only gradually, and bone cells generally grow and reproduce slowly. Moreover, the blood supply in a fractured bone may be temporarily disrupted, which helps to explain the difficulty in the healing of severely fractured bones.

TREATMENTS FOR FRACTURES

Treatments for fractures vary according to the person's age, the type of fracture, and the bone involved. The ultimate goals of fracture treatment are the anatomic realignment of the bone fragments, immobilization to maintain realignment, and restoration of function. For bones to unite, the fractured ends must be brought into alignment, a process called **reduction,** and they must be immobilized during the time required for the fracture to heal. In **closed reduction,** the fractured ends of a bone are brought into alignment by manual manipulation, and the skin remains intact. In **open reduction,** the fractured ends of a bone

are brought into alignment by a surgical procedure in which internal fixation devices such as screws, plates, pins, rods, and wires are used. Following reduction, a fractured bone may be kept immobilized by a cast, sling, splint, elastic bandage, external fixation device, or a combination of these devices. ∎

CHECKPOINT

14. Define a fracture and outline the four steps involved in fracture repair.

EXERCISE AND BONE TISSUE

OBJECTIVE

• Describe how exercise and mechanical stress affect bone tissue.

Within limits, bone tissue has the ability to alter its strength in response to changes in mechanical stress. When placed under stress, bone tissue adapts by becoming stronger through increased deposition of mineral salts and production of collagen fibers. Another effect of stress is to increase the production of calcitonin, a hormone that inhibits bone resorption. Without mechanical stress, bone does not remodel normally because resorption outstrips bone formation. Removal of mechanical stress weakens bone through loss of bone minerals and decreased numbers of collagen fibers.

The main mechanical stresses on bone result from the contraction of skeletal muscles and the pull of gravity. If a person is bedridden or has a fractured bone in a cast, the strength of the unstressed bones diminishes. Astronauts subjected to the microgravity of space also lose bone mass. In both cases, bone loss can be dramatic—as much as 1% per week. Bones of athletes, which are repetitively and highly stressed, become notably thicker than those of nonathletes. Weight-bearing activities, such as walking or moderate weight lifting, help build and retain bone mass. In recent years it has become increasingly clear that adolescents and young adults should engage in regular weight-bearing exercise prior to the closure of the epiphyseal plates. This helps to build total bone mass prior to its inevitable reduction with aging. Even elderly people can strengthen their bones by engaging in weight-bearing exercise.

CHECKPOINT

15. Explain the types of mechanical stresses that may be used to strengthen bone tissue.

AGING AND BONE TISSUE

OBJECTIVE

● Describe the effects of aging on bone tissue.

From birth through adolescence, more bone tissue is produced than is lost during bone remodeling. In young adults the rates of bone deposition and resorption are about the same. As the level of sex steroids diminishes during middle age, especially in women after menopause, a decrease in bone mass occurs because bone resorption outpaces bone deposition. In old age, loss of bone through resorption occurs more rapidly than bone gain. Because women's bones generally are smaller and less massive than men's bones to begin with, loss of bone mass in old age typically has a greater adverse effect in women. These factors contribute to a higher incidence of osteoporosis in females.

There are two principal effects of aging on bone tissue: loss of bone mass and brittleness. Loss of bone mass results from demineralization (dē-min′-er-al-i-ZĀ-shun), the loss of calcium and other minerals from bone extracellular matrix. This loss usually begins after age 30 in females, accelerates greatly around age 45 as levels of estrogens decrease, and continues until as much as 30% of the calcium in bones is lost by age 70. Once bone loss begins in females, about 8% of bone mass is lost every 10 years. In males, calcium loss typically does not begin until after age 60, and about 3% of bone mass is lost every 10 years. The loss of calcium from bones is one of the problems in osteoporosis (described shortly in the Applications to Health section of the chapter).

The second principal effect of aging on the skeletal system, brittleness, results from a decreased rate of protein synthesis. Recall that the organic part of bone matrix, mainly collagen fibers, gives bone its tensile strength. The loss of tensile strength causes the bones to become very brittle and susceptible to fractures. In some elderly people, collagen fiber synthesis slows, in part, due to diminished production of human growth hormone. In addition to increasing the susceptibility to fractures, loss of bone mass also leads to deformity, pain, stiffness, loss of height, and loss of teeth.

CHECKPOINT

16. What is demineralization, and how does it affect the functioning of bone?
17. What changes occur in the organic part of bone extracellular matrix with aging?

APPLICATIONS TO HEALTH

OSTEOPOROSIS

Osteoporosis (os′-tē-ō-pō-RŌ-sis; *por*=passageway; *-osis*=condition) is literally a condition of porous bones (Figure 6.11). The basic problem is that bone resorption outpaces bone deposition. In large part this is due to depletion of calcium from the body—more calcium is lost in urine, feces, and sweat than is absorbed from the diet. Bone mass becomes so depleted that bones fracture, often spontaneously, under the mechanical stresses of everyday living. For example, a hip fracture might result from simply sitting down too quickly. In the United States, osteoporosis causes more than a million fractures a year, mainly in the hip, wrist, and vertebrae. Osteoporosis afflicts the entire skeletal system. In addition to fractures, osteoporosis causes shrinkage of vertebrae, height loss, hunched backs, and bone pain.

Thirty million people in the United States suffer from osteoporosis. The disorder primarily affects middle-aged and elderly people, 80% of them women. Older women suffer from osteoporosis more often than men for two reasons: Women's bones are less massive than men's bones, and production of estrogens in women declines dramatically at menopause, whereas production of the main androgen, testosterone, in older men wanes gradually and only slightly. Estrogens and testos-

Figure 6.11 Comparison of spongy bone tissue from (a) a normal young adult and (b) a person with osteoporosis. Notice the weakened trabeculae in (b). Compact bone tissue is similarly affected by osteoporosis.

In osteoporosis, bone resorption outpaces bone formation, so bone mass decreases.

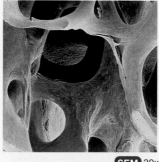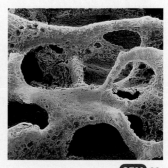

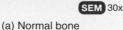
(a) Normal bone   (b) Osteoporotic bone

If you wanted to develop a drug to lessen the effects of osteoporosis, would you look for a chemical that inhibits the activity of osteoblasts, or that of osteoclasts?

terone stimulate osteoblast activity and synthesis of bone matrix. Besides gender, risk factors for developing osteoporosis include a family history of the disease, European or Asian ancestry, thin or small body build, an inactive lifestyle, cigarette smoking, a diet low in calcium and vitamin D, more than two alcoholic drinks a day, and the use of certain medications.

In postmenopausal women, treatment of osteoporosis may include estrogen replacement therapy (ERT; low doses of estrogens) or hormone replacement therapy (HRT; a combination of estrogens and progesterone, another sex steroid). Although such treatments help combat osteoporosis, they may increase a woman's risk of breast cancer. The drug Raloxifene (Evista) mimics the beneficial effects of estrogens on bone without increasing the risk of breast cancer. Another drug that may be used is the nonhormone drug Alendronate (Fosamax), which blocks resorption of bone by osteoclasts.

Perhaps more important than treatment is prevention. Adequate calcium intake and weight-bearing exercise may be more beneficial to a woman than drugs and calcium supplements when she is older.

RICKETS AND OSTEOMALACIA

Rickets and **osteomalacia** (os′-tē-ō-ma-LĀ-shē-a; *-malacia=* softness) are disorders in which bones fail to calcify. Although the organic extracellular matrix is still produced, calcium salts are not deposited, and the bones become "soft" or rubbery and easily deformed. Rickets affects the growing bones of children. Because new bone formed at the epiphyseal plates fails to ossify, bowed legs and deformities of the skull, rib cage, and pelvis are common. In osteomalacia, sometimes called "adult rickets," new bone formed during remodeling fails to calcify. This disorder causes varying degrees of pain and tenderness in bones, especially in the hip and leg. Bone fractures also result from minor trauma.

KEY MEDICAL TERMS ASSOCIATED WITH BONE TISSUE

Osteoarthritis (os′-tē-ō -ar-THRĪ-tis; *arthr*=joint) The degeneration of articular cartilage such that the bony ends touch; the resulting friction of bone against bone worsens the condition. Usually associated with the elderly.

Osteogenic sarcoma (os′-tē-ō-JEN-ik sar-KŌ-ma; *sarcoma=* connective tissue tumor) Bone cancer that primarily affects osteoblasts and occurs most often in teenagers during their growth spurt; the most common sites are the metaphyses of the thighbone (femur), shinbone (tibia), and arm bone (humerus). Metastases occur most often in lungs; treatment consists of multidrug chemotherapy and removal of the malignant growth, or amputation of the limb.

Osteomyelitis (os′-tē-ō-mī-el-Ī-tis) An infection of bone characterized by high fever, sweating, chills, pain, and nausea, pus formation, edema, and warmth over the affected bone and rigid overlying muscles. It is often caused by bacteria, usually *Staphylococcus aureus*. The bacteria may reach the bone from outside the body (through open fractures, penetrating wounds, or orthopedic surgical procedures); from other sites of infection in the body (abscessed teeth, burn infections, urinary tract infections, or upper respiratory infections) via the blood; and from adjacent soft tissue infections (as occurs in diabetes mellitus).

Osteopenia (os′-tē-ō-PĒ-nē-a; *-penia*=poverty) Reduced bone mass due to a decrease in the rate of bone synthesis to a level insufficient to compensate for normal bone resorption; any decrease in bone mass below normal. An example is osteoporosis.

STUDY OUTLINE

Introduction (p. 144)

1. A bone is made up of several different tissues: bone, or osseous tissue, cartilage, dense connective tissues, epithelium, various blood-forming tissues, adipose tissue, and nervous tissue.
2. The entire framework of bones and their cartilages constitute the skeletal system.

Functions of the Skeletal System (p. 145)

1. The skeletal system functions in support, protection, movement, mineral storage and release, blood cell production, and triglyceride storage.

Anatomy of a Bone (p. 145)

1. Parts of a typical long bone are the diaphysis (shaft), proximal and distal epiphyses (ends), metaphyses, articular cartilage, periosteum, medullary (marrow) cavity, and endosteum.

Histology of Bone Tissue (p. 145)

1. Bone tissue consists of widely separated cells surrounded by large amounts of extracellular matrix.
2. The four principal types of cells in bone tissue are osteogenic cells, osteoblasts, osteocytes, and osteoclasts.
3. The extracellular matrix of bone contains abundant mineral salts (mostly hydroxyapatite) and collagen fibers.
4. Compact bone tissue consists of osteons (haversian systems) with little space between them.
5. Compact bone tissue lies over spongy bone tissue in the epiphyses and makes up most of the bone tissue of the diaphysis. Functionally, compact bone tissue protects, supports, and resists stress.
6. Spongy bone tissue does not contain osteons. It consists of trabeculae surrounding spaces filled with red bone marrow.

7. Spongy bone tissue forms most of the structure of short, flat, and irregular bones, and the epiphyses of long bones. Functionally, spongy bone tissue trabeculae offer resistance along lines of stress, support and protect red bone marrow, and make bones lighter for easier movement.

Blood and Nerve Supply of Bone (p. 150)

1. Long bones are supplied by periosteal, nutrient, and epiphyseal arteries; veins accompany the arteries.

2. Nerves accompany blood vessels in bone; the periosteum is rich in sensory neurons.

Bone Formation (p. 150)

1. Bone forms by a process called ossification (osteogenesis), which begins when mesenchymal cells are transformed into osteogenic cells. These undergo cell division and give rise to cells that differentiate into osteoblasts and osteoclasts.

2. Ossification begins during the sixth or seventh week of embryonic development. The two types of ossification, intramembranous and endochondral, involve the replacement of a preexisting connective tissue with bone.

3. Intramembranous ossification refers to bone formation directly within mesenchyme arranged in sheet-like layers that resemble membranes.

4. Endochondral ossification refers to bone formation within hyaline cartilage that develops from mesenchyme. The primary ossification center of a long bone is in the diaphysis. Cartilage degenerates, leaving cavities that merge to form the medullary cavity. Osteoblasts lay down bone. Next, ossification occurs in the epiphyses, where bone replaces cartilage, except for the epiphyseal plate.

Bone Growth (p. 152)

1. The epiphyseal plate consists of four zones: resting cartilage, proliferating cartilage, hypertrophic cartilage, and calcified cartilage.

2. Because of the activity at the epiphyseal plate, the diaphysis of a bone increases in length.

3. Bone grows in width as a result of the addition of new bone tissue by periosteal osteoblasts around the outer surface of the bone (appositional growth).

Bone Remodeling (p. 155)

1. Bone remodeling is the replacement of old bone tissue by new bone tissue.

2. Old bone tissue is constantly destroyed by osteoclasts, whereas new bone is constructed by osteoblasts.

3. Remodeling requires minerals (calcium, phosphorus, fluoride, magnesium, iron, and manganese), vitamins (A, B12, C, D, and K), and hormones (human growth hormone, sex hormones, insulin, thyroid hormones, parathyroid hormone, and calcitonin).

Fracture and Repair of Bone (p. 156)

1. A fracture is any break in a bone.

2. Fracture repair involves formation of a fracture hematoma, a fibrocartilaginous callus, and a bony callus, followed by bone remodeling.

3. Types of fractures include closed (simple), open (compound), comminuted, greenstick, impacted, stress, Pott's, and Colles'.

Exercise and Bone Tissue (p. 158)

1. Mechanical stress increases bone strength by increasing deposition of mineral salts and production of collagen fibers.

2. Removal of mechanical stress weakens bone through demineralization and collagen fiber reduction.

Aging and Bone Tissue (p. 159)

1. The principal effect of aging is a loss of calcium from bones, which may result in osteoporosis.

2. Another effect is a decreased production of extracellular matrix proteins (mostly collagen fibers), which makes bones more brittle and thus more susceptible to fracture.

Q SELF-QUIZ QUESTIONS

Choose the one best answer to the following questions:

1. Which of the following is *not* considered a function of bone tissue and the skeletal system?

   a. attachment site of muscles, tendons, and ligaments
   b. storage of calcium and phosphorus
   c. formation of blood cells
   d. support and protection of soft organs and tissues
   e. synthesis of vitamin D

2. Yellow bone marrow consists mainly of:

   a. adipose cells   b. hemopoietic tissue   c. calcium salts
   d. collagen   e. elastic cartilage

3. Haversian canals contain:

   a. high concentrations of calcium salts   b. blood vessels
   c. osteocytes   d. articular cartilage   e. the epiphyseal plate

4. The thin columns of bone that form an irregular lattice in spongy bone are called:

   a. osteons   b. canaliculi   c. lacunae
   d. Volkmann's canals   e. trabeculae

5. Periosteal arteries enter a bone through:

   a. haversian canals   b. lacunae   c. canaliculi
   d. Volkmann's canals   e. trabeculae

Complete the following:

6. The periosteum is made of _____ tissue.

7. Most bones of the body are formed from cartilage by the process of _____ ossification.

8. The process by which mineral salts crystallize around collagen fibers in bone is called _____.

9. Osteocytes are located in spaces called _____.

10. The medullary cavity in the long bone of an adult contains _____.

11. The cells responsible for the resorption (destruction) of bone tissue are _____.

12. The process by which a bone grows in width is called _____ growth.

Are the following statements true or false?

13. The flat bones of the skull develop by intramembranous ossification.

14. Canaliculi are microscopic canals running longitudinally through bone.

15. Bones do not have a nerve supply.

16. A fracture hematoma is a mass of fibrocartilage that bridges the broken ends of the bones.

17. The organic extracellular matrix of bone tissue is secreted by osteoblasts.

Matching

18. Match the following.

____ **(a)** thin layer of hyaline cartilage at end of long bone

____ **(b)** region of mature bone where diaphysis joins epiphysis

____ **(c)** covering over bone to which ligaments and tendons attach

____ **(d)** a layer of hyaline cartilage in the metaphysis of a growing bone

____ **(e)** lining of the medullary cavity; contains osteoprogenitor cells and osteoblasts

(1) articular cartilage
(2) endosteum
(3) fibrous periosteum
(4) metaphysis
(5) epiphyseal plate

CRITICAL THINKING QUESTIONS

1. Lynda, a petite 55-year-old couch potato who smokes heavily, wants to lose 50 pounds before her next class reunion. Her diet consists mostly of diet soda and crackers. Explain the effects of her age and lifestyle on her bone composition.

2. In a popular children's book, the bones in the hero's arm are magically removed by accident. (Don't worry—they grow back overnight.) What would be the results if your skeleton were to magically disappear?

3. Aunt Edith is 95 years old today. She comments that she's been getting shorter every year and soon she'll fade out all together. What's happening to Aunt Edith?

4. Astronaut John Glenn exercised every day while in space, yet he and the other astronauts experienced weakness upon their return to earth. Why?

5. Chantal was concerned that her new baby was a "cone head" when she saw him for the first time. Later on, after carefully following the advice to always lay the baby on his back for sleeping, she became concerned that the back of his head was getting flat. If bone is hard, why does the baby's head keep changing shape?

ANSWERS TO FIGURE QUESTIONS

6.1 The periosteum is essential for growth in bone thickness, bone repair, and bone nutrition. It also serves as a point of attachment for ligaments and tendons.

6.2 Bone resorption is necessary for the development, growth, maintenance, and repair of bone.

6.3 The central (haversian) canals are the main blood supply to the osteocytes of an osteon (haversian system), so their blockage would lead to death of the osteocytes.

6.4 Periosteal arteries enter bone tissue through perforating (Volkmann's) canals.

6.5 Flat bones of the skull and mandible (lower jawbone) develop by intramembranous ossification.

6.6 Yes, she probably will grow taller. The absence of epiphyseal lines, indications of growth zones that have ceased to function, indicates that the bone is still lengthening.

6.7 The lengthwise growth of the diaphysis is caused by cell divisions in the zone of proliferating cartilage and maturation of the cells in the zone of hypertrophic cartilage.

6.8 The medullary cavity enlarges by activity of the osteoclasts in the endosteum.

6.9 A fibrocartilaginous callus is a mass of fibrocartilaginous repair tissue that bridges the ends of broken bones.

6.10 Healing of bone fractures can take months because calcium and phosphorus deposition is a slow process, and bone cells generally grow and reproduce slowly.

6.11 A drug that inhibits the activity of osteoclasts might lessen the effects of osteoporosis.

THE SKELETAL SYSTEM: THE AXIAL SKELETON

CONTENTS AT A GLANCE

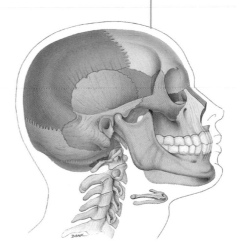

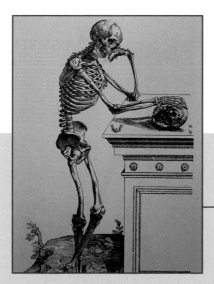

INTRODUCTION Without bones, you would be unable to perform movements such as walking or grasping. The slightest blow to your head or chest could damage your brain or heart. It would even be impossible for you to chew food. Because the skeletal system forms the framework of the body, a familiarity with the names, shapes, and positions of individual bones will help you locate other organ systems. For example, the radial artery, the site where pulse is usually taken, is named for its proximity to the radius, the lateral bone of the forearm. The frontal lobe of the brain lies deep to the frontal (forehead) bone. The tibialis anterior muscle is located near the anterior surface of the tibia (shinbone). The ulnar nerve is named for its proximity to the ulna, the medial bone of the forearm. Parts of certain bones also serve to locate structures within the skull and to outline the lungs and heart and abdominal and pelvic organs.

What do you think of when you see a picture of a skeleton? What words come to mind—symbol of death or foundation of our bodies?

www.wiley.com/college/apcentral

Movements such as throwing a ball, biking, and walking require an interaction between bones and muscles. To understand how muscles produce different movements, you will learn where the muscles attach to individual bones and the types of joints acted on by the contracting muscles. The bones, muscles, and joints together form an integrated system called the **musculoskeletal system.** The branch of medical science that is concerned with the prevention or correction of disorders of the musculoskeletal system is called **orthopedics** (or´-thō-PĒ -diks; *ortho-*=correct; *pedi*=child).

DIVISIONS OF THE SKELETAL SYSTEM

OBJECTIVE

• Describe how the skeleton is divided into axial and appendicular divisions.

The adult human skeleton consists of 206 named bones. The skeletons of infants and children have more than 206 bones because some of their bones, such as the hip bones and certain bones of the backbone fuse later in life. There are two principal divisions in the adult skeleton: the **axial skeleton** and the **appendicular skeleton.** Refer to Figure 7.1 to see how the two divisions join to form the complete skeleton. The bones of the axial skeleton are shown in blue. The longitudinal **axis,** or center, of the human body is a straight line that runs through the body's center of gravity. This imaginary line extends through the head and down to the space between the feet. The axial skeleton consists of the bones that lie around the axis: skull bones, auditory ossicles (ear bones), hyoid bone, ribs, breastbone, and bones of the backbone. Although the auditory ossicles are not considered part of the axial or appendicular skeleton, but rather as a separate group of bones, they are placed with the axial skeleton for convenience. The middle portion of each ear contains three auditory ossicles held together by a series of ligaments. The auditory ossicles vibrate in response to sound waves that strike the eardrum and have a key role in the mechanism of hearing. This is described in detail in Chapter 22. The appendicular skeleton contains the bones of the **upper** and **lower limbs (extremities),** plus the bones called **girdles** that connect the limbs to the axial skeleton. Table 7.1 on page 166 presents the standard grouping of the 80 bones of the axial skeleton and the 126 bones of the appendicular skeleton.

We will organize our study of the skeletal system around the two divisions of the skeleton, with emphasis on the interrelationships of the many bones of the body. In this chapter we focus on the axial skeleton, looking first at the skull and then at the bones of the vertebral column (backbone) and the chest. In Chapter 8 we explore the appendicular skeleton, examining in turn the bones of the pectoral (shoulder) girdle and upper limbs, and then the pelvic (hip) girdle and the lower limbs. But before we examine the skull we direct our attention to some rather general characteristics of bones.

CHECKPOINT

1. List the bones that make up the axial and appendicular divisions of the skeleton.

TYPES OF BONES

OBJECTIVE

• Classify bones on the basis of shape and location.

Almost all the bones of the body can be classified into five principal types on the basis of shape: long, short, flat, irregular, and sesamoid (Figure 7.2 on page 166). As you learned in Chapter 6, **long bones** have greater length than width and consist of a shaft and a variable number of extremities (ends). They are slightly curved for strength. A curved bone absorbs the stress of the body's weight at several different points so that the stress is evenly distributed. If such bones were straight, the weight of the body would be unevenly distributed and the bone would easily fracture. Long bones consist mostly of *compact bone tissue*, which is dense and has few spaces, but they also contain considerable amounts of *spongy bone tissue*, which has larger spaces (see Figure 6.3 on page 149). Long bones include those in the thigh (femur), leg (tibia and fibula), toes (phalanges), arm (humerus), forearm (ulna and radius), and fingers (phalanges).

Short bones are somewhat cube-shaped and nearly equal in length and width. They consist of spongy bone except at the surface, where there is a thin layer of compact bone. Examples of short bones are the wrist or carpal bones (except for the pisiform, which is a sesamoid bone) and the ankle or tarsal bones (except for the calcaneus, which is an irregular bone).

Flat bones are generally thin and composed of two nearly parallel plates of compact bone enclosing a layer of spongy bone. The layers of compact bone are called external and internal *tables*, whereas the spongy bone is referred to as *diploe* (DIP-lō-ē). Flat bones afford considerable protection and provide extensive areas for muscle attachment. Flat bones include the cranial bones, which protect the brain; the breastbone (sternum) and ribs, which protect organs in the thorax; and the shoulder blades (scapulae).

Irregular bones have complex shapes and cannot be grouped into any of the three categories just described. They also vary in the amount of spongy and compact bone present. Such bones include the backbones (vertebrae), certain facial bones, and the aforementioned heel bone (calcaneus).

Sesamoid bones (meaning shaped like a sesame seed) develop in certain tendons where there is considerable friction, tension, and physical stress. They are not always completely ossified and measure only a few millimeters in diameter except for the two patellae (kneecaps), the largest of the sesamoid bones. Sesamoid bones vary in number from person to person except for the patellae, which are located in the quadriceps

Figure 7.1 **Divisions of the skeletal system.** The axial skeleton is indicated in blue, the appendicular skeleton in yellow. Note the position of the hyoid bone in Figure 7.4a. (See Tortora, *A Photographic Atlas of the Human Body*, 2e, Figure 3.1.)

🔑 **The adult human skeleton consists of 206 bones grouped into axial and appendicular divisions.**

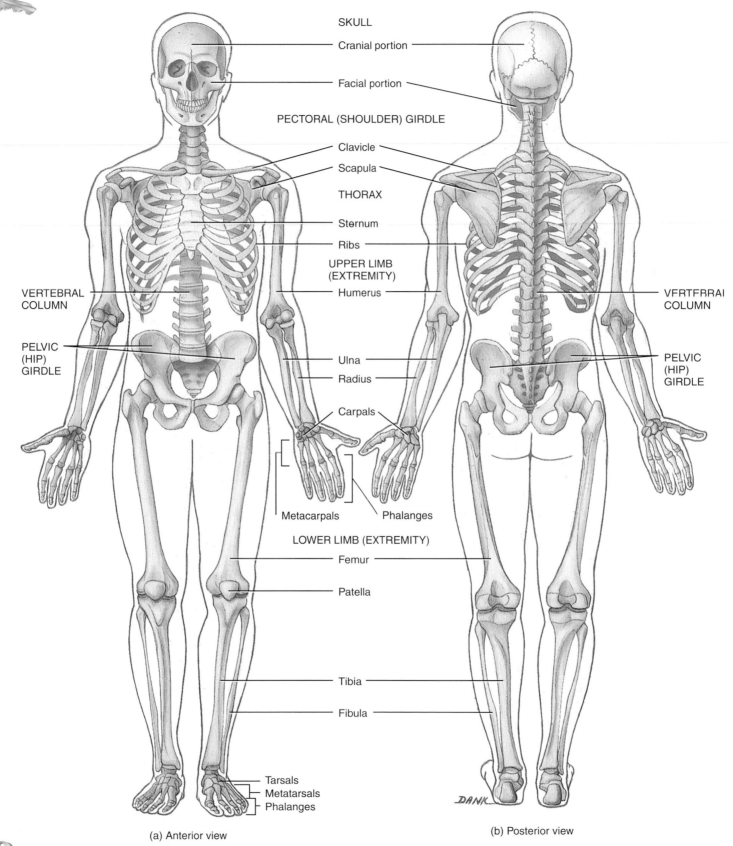

(a) Anterior view

(b) Posterior view

❓ **Which of the following structures are part of the axial skeleton, and which are part of the appendicular skeleton? Skull, clavicle, vertebral column, shoulder girdle, humerus, pelvic girdle, and femur.**

TABLE 7.1 THE BONES OF THE ADULT SKELETAL SYSTEM

DIVISION OF THE SKELETON	STRUCTURE	NUMBER OF BONES	DIVISION OF THE SKELETON	STRUCTURE	NUMBER OF BONES
Axial Skeleton 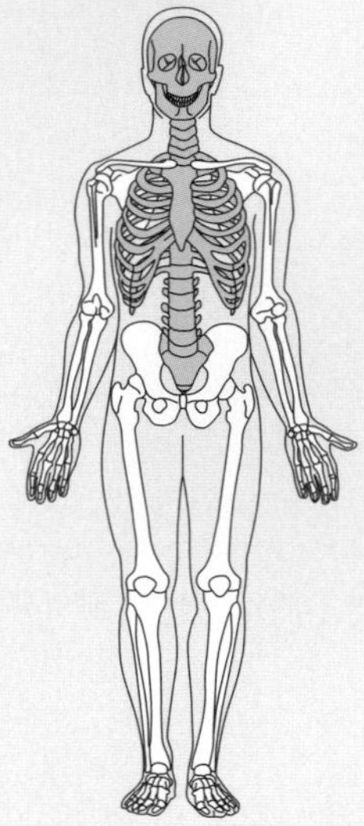	Skull		**Appendicular Skeleton** 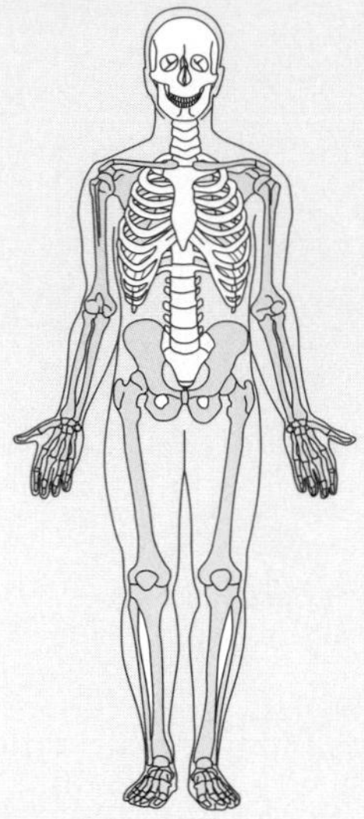	Pectoral (shoulder) girdles	
	Cranium	8		*Clavicle*	2
	Face	14		*Scapula*	2
	Hyoid	1		Upper limbs (extremities)	
	Auditory ossicles	6		*Humerus*	2
				Ulna	2
	Vertebral column	26		*Radius*	2
				Carpals	16
	Thorax			*Metacarpals*	10
	Sternum	1		*Phalanges*	28
	Ribs	24		Pelvic (hip) girdle	
		Subtotal=80		*Hip, pelvic, or coxal bone*	2
				Lower limbs (extremities)	
				Femur	2
				Patella	2
				Fibula	2
				Tibia	2
				Tarsals	14
				Metatarsals	10
				Phalanges	28
					Subtotal=126
					TOTAL=206

Figure 7.2 Types of bones based on shape. The bones are not drawn to scale.

The shapes of bones largely determine their functions.

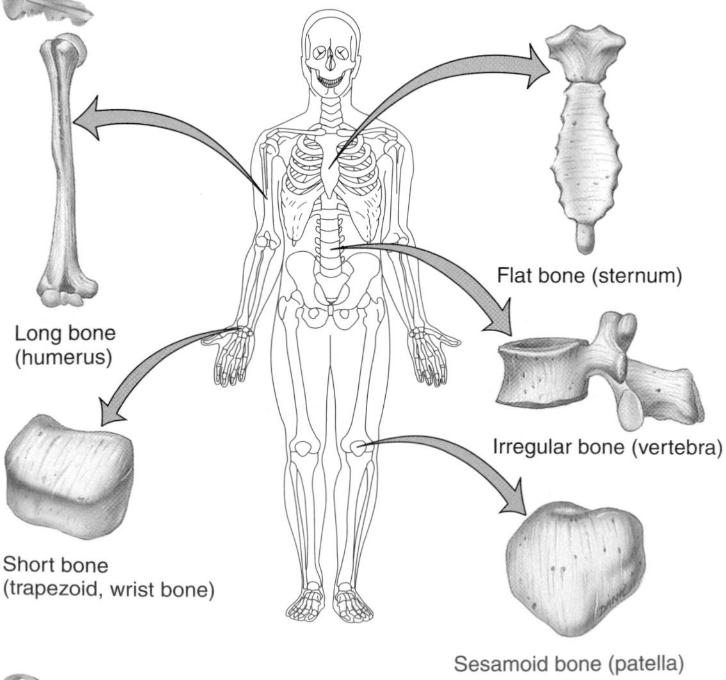

Long bone (humerus)

Flat bone (sternum)

Irregular bone (vertebra)

Short bone (trapezoid, wrist bone)

Sesamoid bone (patella)

? **Which type of bone primarily protects and provides a large surface area for muscle attachment?**

femoris tendon (see Figure 11.24a, b on page 368) and are normally present in all individuals. Functionally, sesamoid bones protect tendons from excessive wear and tear, and they often change the direction of pull of a tendon, which improves the mechanical advantage at a joint.

In the upper limbs, sesamoid bones usually occur only in the joints of the palmar surface of the hands. Two frequently encountered sesamoid bones are in the tendons of the adductor pollicis and flexor pollicis brevis muscles at the metacarpophalangeal joint of the thumb (see Figure 8.8a on page 211). In the lower limbs, aside from the patellae, two constant sesamoid bones occur on the plantar surface of the foot in the tendons of the flexor hallucis brevis muscle at the metatarsophalangeal joint of the great (big) toe (see Figure 8.16b on page 223).

An additional type of bone is not included in this classification by shape, but instead is classified by location. **Sutural** (SOŌ-chur-al; *sutura*=seam) or **Wormian bones** (named after a Danish anatomist, O. Worm, 1588–1654) are small bones located within the sutures (joints) of certain cranial bones (see Figure 7.6). The number of sutural bones varies greatly from person to person.

During active growth of the skeleton, red bone marrow is progressively replaced by yellow bone marrow in most of the long bones. In adults, red bone marrow is restricted to flat bones such as the ribs, sternum (breastbone), and skull; irregular bones such as vertebrae (backbones) and hip bones; long bones such as the proximal epiphyses of the femur (thigh bone) and humerus (arm bone); and some short bones.

2. Give examples of long, short, flat, and irregular bones.

BONE SURFACE MARKINGS

OBJECTIVE

• Describe the principal surface markings on bones and the functions of each.

Bones have characteristic **surface markings,** structural features adapted for specific functions. Most are not present at birth but develop later in response to certain forces and are most prominent during adult life. In response to tension on a bone surface where tendons, ligaments, aponeuroses, and fasciae pull on the periosteum of bone, new bone is deposited, resulting in raised or roughened areas. Conversely, compression on a bone surface results in a depression.

There are two major types of surface markings: (1) *depressions and openings,* which form joints or allow the passage of soft tissues (such as blood vessels and nerves), and (2) *processes,* projections or outgrowths that either help form joints or serve as attachment points for connective tissue (such as ligaments and tendons). Table 7.2 describes the various surface markings and provides examples of each.

MEDICAL ANTHROPOLOGY

Skeletal remains may persist for thousands of years after a person has died. They make it possible to trace patterns of disease and nutrition, evaluate the effects of certain social and economic changes, and deduce patterns of reproduction and mortality. Skeletal remains also may reveal an individual's sex, age, height, and race.

Many disorders can leave permanent effects on skeletal material. Three of the many common causes of bone pathologies are malnutrition, tumors, and infections. Each may cause specific changes in bone that permit diagnosis hundreds or even thousands of years after a person's death. ■

CHECKPOINT

3. List and describe several bone surface markings, and give an example of each. Check your list against Table 7.2.

TABLE 7.2 BONE SURFACE MARKINGS

MARKING	DESCRIPTION	EXAMPLE
Depressions and Openings: Sites allowing the passage of soft tissue (nerves, blood vessels, ligaments, tendons) or formation of joints		
Fissure (FISH-ur)	Narrow slit between adjacent parts of bones through which blood vessels or nerves pass.	Superior orbital fissure of the sphenoid bone (Figure 7.12).
Foramen (fō-RĀ-men; plural is *foramina*)	Opening (*foramen*=hole) through which blood vessels, nerves, or ligaments pass.	Optic foramen of the sphenoid bone (Figure 7.12).
Fossa (FOS-a)	Shallow depression (*fossa*=trench).	Coronoid fossa of the humerus (Figure 8.5a).
Sulcus (SUL-kus)	Furrow (*sulcus*=groove) along a bone surface that accommodates a blood vessel, nerve, or tendon.	Intertubercular sulcus of the humerus (Figure 8.5a).
Meatus (mē-Ā-tus)	Tubelike opening (*meatus*=passageway).	External auditory meatus of the temporal bone (Figure 7.4a).
Processes: Projections or outgrowths on bone that form joints or attachment points for connective tissue, such as ligaments and tendons. *Processes that form joints:*		
Condyle (KON-dīl)	Large, round protuberance (*condylus*=knuckle) at the end of a bone.	Lateral condyle of the femur (Figure 8.13b).
Facet	Smooth flat articular surface.	Superior articular facet of a vertebra (Figure 7.18c).
Head	Rounded articular projection supported on the neck (constricted portion) of a bone.	Head of the femur (Figure 8.13c).
Processes that form attachment points for connective tissue:		
Crest	Prominent ridge or elongated projection.	Iliac crest of the hip bone (Figure 8.10c).
Epicondyle	Projection above (*epi-*=above) a condyle.	Medial epicondyle of the femur (Figure 8.13a).
Line	Long, narrow ridge or border (less prominent than a crest).	Linea aspera of the femur (Figure 8.13b).
Spinous process	Sharp, slender projection.	Spinous process of a vertebra (Figure 7.18d).
Trochanter (trō-KAN-ter)	Very large projection.	Greater trochanter of the femur (Figure 8.13b).
Tubercle (TOO-ber-kul)	Small, rounded projection (*tuber-*=knob).	Greater tubercle of the humerus (Figure 8.5a).
Tuberosity	Large, rounded, usually roughened projection.	Ischial tuberosity of the hip bone (Figure 8.10c).

SKULL

OBJECTIVES

- Name the cranial and facial bones and indicate the number of each.
- Describe the following unique features of the skull: sutures, paranasal sinuses, and fontanels.

The **skull,** which contains 22 bones, rests on the superior end of the vertebral column (backbone). It includes two sets of bones: cranial bones and facial bones (Table 7.3). The **cranial bones** (*crani-*=brain case) form the cranial cavity, which encloses and protects the brain. The eight cranial bones are the frontal bone, two parietal bones, two temporal bones, the occipital bone, the sphenoid bone, and the ethmoid bone. Fourteen **facial bones** form the face: two nasal bones, two maxillae (or maxillas), two zygomatic bones, the mandible, two lacrimal bones, two palatine bones, two inferior nasal conchae, and the vomer. Figures 7.3 through 7.14 illustrate these bones from different viewing directions.

TABLE: 7.3 SUMMARY OF BONES OF THE ADULT SKULL	
CRANIAL BONES	**FACIAL BONES**
Frontal (1)	Nasal (2)
Parietal (2)	Maxillae (2)
Temporal (2)	Zygomatic (2)
Occipital (1)	Mandible (1)
Sphenoid (1)	Lacrimal (2)
Ethmoid (1)	Palatine (2)
Inferior nasal conchae (2)	
Vomer (1)	

The numbers in parentheses indicate how many of each bone are present.

General Features and Functions

Besides forming the large cranial cavity, the skull also forms several smaller cavities, including the nasal cavity and orbits

Figure 7.3   Skull.

The skull consists of cranial bones and facial bones.

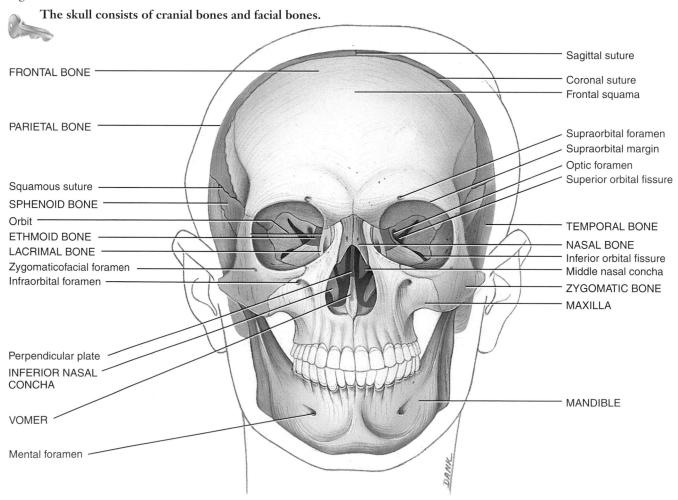

FRONTAL BONE

PARIETAL BONE

Squamous suture
SPHENOID BONE
Orbit
ETHMOID BONE
LACRIMAL BONE
Zygomaticofacial foramen
Infraorbital foramen

Perpendicular plate
INFERIOR NASAL CONCHA

VOMER

Mental foramen

Sagittal suture
Coronal suture
Frontal squama

Supraorbital foramen
Supraorbital margin
Optic foramen
Superior orbital fissure

TEMPORAL BONE
NASAL BONE
Inferior orbital fissure
Middle nasal concha
ZYGOMATIC BONE
MAXILLA

MANDIBLE

(a) Anterior view

(eye sockets), which open to the exterior. Certain skull bones also contain cavities called paranasal sinuses that are lined with mucous membranes and open into the nasal cavity. Also within the skull are small cavities that house the structures involved in hearing and equilibrium.

Other than the auditory ossicles, which are involved in hearing and are located within the temporal bones, the mandible is the only movable bone of the skull. Most of the skull bones are held together by immovable joints called sutures, which are especially noticeable on the outer surface of the skull.

The skull has numerous surface markings, such as foramina and fissures through which blood vessels and nerves pass. You will learn the names of important skull bone surface markings as the various bones are described.

The cranial bones have other functions in addition to protecting the brain. Their inner surfaces attach to membranes (meninges) that stabilize the positions of the brain, blood vessels, and nerves. The outer surfaces of cranial bones provide large areas of attachment for muscles that move various parts of the head. The bones also provide attachment for some muscles that are involved in producing facial expressions. Besides forming the framework of the face, the facial bones protect and provide support for the entrances to the digestive and respiratory systems. Together, the cranial and facial bones protect and support the delicate special sense organs for vision, taste, smell, hearing, and equilibrium (balance).

Cranial Bones

Frontal Bone

The **frontal bone** forms the forehead (the anterior part of the cranium), the roofs of the *orbits* (eye sockets), and most of the anterior part of the cranial floor (Figure 7.3). Soon after birth the left and right sides of the frontal bone are united by the *metopic suture*, which usually disappears by age 6–8.

If you examine the anterior view of the skull in Figure 7.3, you will note the *frontal squama*, a scalelike plate of bone that forms the forehead. It gradually slopes inferiorly from the coronal suture on top of the skull, then angles abruptly and becomes almost vertical. Superior to the orbits the frontal bone thickens,

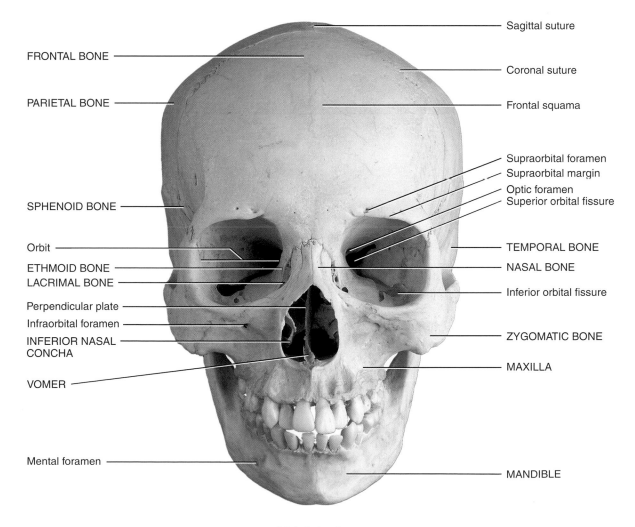

(b) Anterior view

 Which of the bones shown here are cranial bones?

forming the *supraorbital margin* (*supra-*=above; *-orbital*=wheel rut). From this margin the frontal bone extends posteriorly to form the roof of the orbit and part of the floor of the cranial cavity. Within the supraorbital margin, slightly medial to its midpoint, is a hole called the *supraorbital foramen* through which the supraorbital nerve and artery pass. Sometimes this foramen is incomplete and is called the *supraorbital notch*. The *frontal sinuses* lie deep to the frontal squama.

BLACK EYE

Just superior to the supraorbital margin is a sharp ridge. A blow to the ridge often fractures the frontal bone or lacerates the skin over it, resulting in bleeding. Bruising of the skin over the ridge causes tissue fluid and blood to accumulate in the surrounding connective tissue. The resulting swelling and discoloration is called a **black eye. ■**

Parietal Bones

The two **parietal bones** (pa-RĪ-e-tal; *pariet-*=wall) form the greater portion of the sides and roof of the cranial cavity (Figure 7.4). The internal surfaces of the parietal bones contain many protrusions and depressions that accommodate the blood vessels supplying the dura mater, the superficial membrane (meninx) covering the brain.

Temporal Bones

The two **temporal bones** (*tempor-*=temple) form the inferior lateral aspects of the cranium and part of the cranial floor. In the lateral view of the skull (Figure 7.4), note the *temporal squama*, the thin, flat portion of the temporal bone that forms the anterior and superior part of the temple. Projecting from the inferior portion of the temporal squama is the *zygomatic process*,

Figure 7.4 Skull. Although the hyoid bone is not part of the skull, it is included in the illustration for reference.

🔑 **The zygomatic arch is formed by the zygomatic process of the temporal bone and the temporal process of the zygomatic bone.**

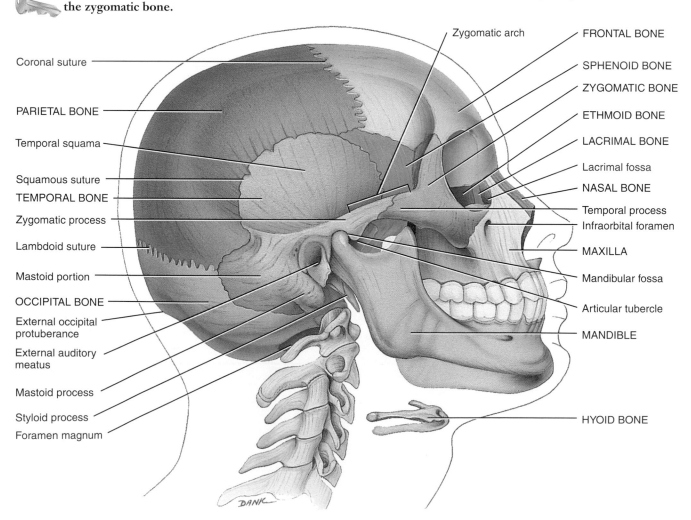

(a) Right lateral view

which articulates (forms a joint) with the temporal process of the zygomatic (cheek) bone. Together, the zygomatic process of the temporal bone and the temporal process of the zygomatic bone form the *zygomatic arch*.

On the inferior posterior surface of the zygomatic process of the temporal bone is a socket called the *mandibular fossa*. Anterior to the mandibular fossa is a rounded elevation, the *articular tubercle* (Figure 7.4). The mandibular fossa and articular tubercle articulate with the mandible (lower jawbone) to form the *temporomandibular joint (TMJ)*.

Located posteriorly on the temporal bone is the *mastoid portion* (*mastid*=breast-shaped) (Figure 7.4). The mastoid portion is located posterior and inferior to the *external auditory meatus* (*meatus*=passageway), or ear canal, which directs sound waves into the ear. In the adult, this portion of the bone contains several *mastoid "air cells."* These tiny air-filled compartments are separated from the brain by thin bony partitions. **Mastoiditis**

(inflammation of the mastoid air cells) can spread an infection to the middle ear and to the brain.

The *mastoid process* is a rounded projection of the mastoid portion of the temporal bone posterior to the external auditory meatus. It serves as a point of attachment for several neck muscles (Figure 7.4). The *internal auditory meatus* (Figure 7.5) is the opening through which the facial (VII) and vestibulocochlear (VIII) cranial nerves pass. The *styloid process* (*styl-*=stake or pole) projects inferiorly from the inferior surface of the temporal bone and serves as a point of attachment for muscles and ligaments of the tongue and neck (see Figure 7.4). Between the styloid process and the mastoid process is the *stylomastoid foramen*, through which the facial (VII) nerve and stylomastoid artery pass (see Figure 7.7).

At the floor of the cranial cavity (see Figure 7.8a) is the *petrous portion* (*petrous*=rock) of the temporal bone. This portion is triangular and located at the base of the skull between the

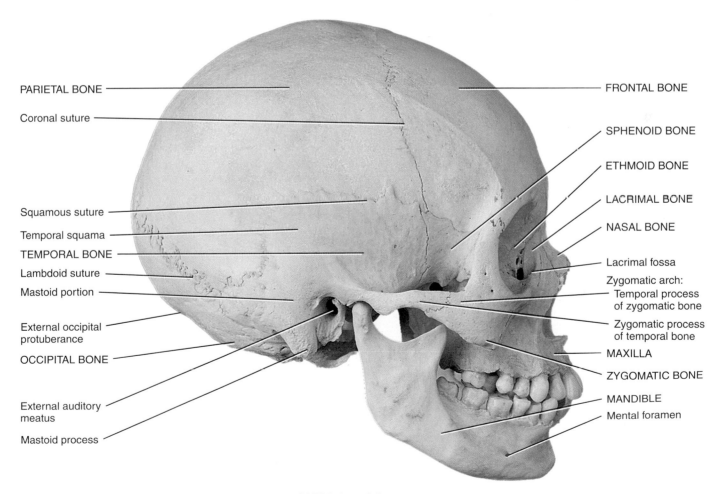

PARIETAL BONE

Coronal suture

Squamous suture

Temporal squama

TEMPORAL BONE

Lambdoid suture

Mastoid portion

External occipital protuberance

OCCIPITAL BONE

External auditory meatus

Mastoid process

FRONTAL BONE

SPHENOID BONE

ETHMOID BONE

LACRIMAL BONE

NASAL BONE

Lacrimal fossa

Zygomatic arch:
Temporal process of zygomatic bone

Zygomatic process of temporal bone

MAXILLA

ZYGOMATIC BONE

MANDIBLE

Mental foramen

(b) Right lateral view

 What are the major bones on either side of the squamous suture, the lambdoid suture, and the coronal suture?

Figure 7.5 Skull.

The cranial bones are the frontal, parietal, temporal, occipital, sphenoid, and ethmoid bones. The facial bones are the nasal bone, maxillae, zygomatic bones, lacrimal bones, palatine bones, mandible, and vomer.

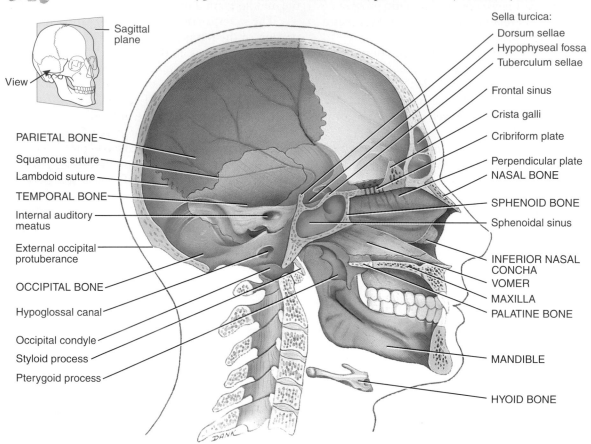

Sagittal plane

View

Sella turcica:
Dorsum sellae
Hypophyseal fossa
Tuberculum sellae

Frontal sinus

Crista galli

Cribriform plate

Perpendicular plate
NASAL BONE

SPHENOID BONE

Sphenoidal sinus

INFERIOR NASAL CONCHA
VOMER
MAXILLA
PALATINE BONE

MANDIBLE

HYOID BONE

PARIETAL BONE

Squamous suture

Lambdoid suture

TEMPORAL BONE

Internal auditory meatus

External occipital protuberance

OCCIPITAL BONE

Hypoglossal canal

Occipital condyle

Styloid process

Pterygoid process

(a) Medial view of sagittal section

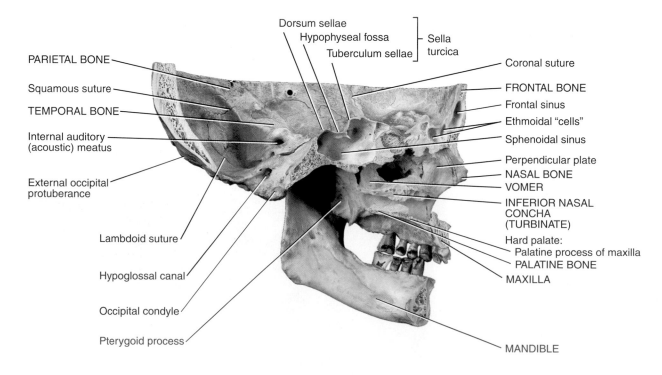

Dorsum sellae
Hypophyseal fossa   Sella
Tuberculum sellae   turcica

PARIETAL BONE

Squamous suture

TEMPORAL BONE

Internal auditory (acoustic) meatus

External occipital protuberance

Lambdoid suture

Hypoglossal canal

Occipital condyle

Pterygoid process

Coronal suture

FRONTAL BONE

Frontal sinus

Ethmoidal "cells"

Sphenoidal sinus

Perpendicular plate
NASAL BONE
VOMER

INFERIOR NASAL CONCHA (TURBINATE)

Hard palate:
Palatine process of maxilla
PALATINE BONE

MAXILLA

MANDIBLE

(b) Medial view of sagittal section

With which bones does the temporal bone articulate?

sphenoid and occipital bones. The petrous portion houses the internal ear and the middle ear, structures involved in hearing and equilibrium. It also contains the *carotid foramen*, through which the carotid artery passes (see Figure 7.7). Posterior to the carotid foramen and anterior to the occipital bone is the *jugular foramen*, a passageway for the jugular vein.

Occipital Bone

The **occipital bone** (ok-SIP-i-tal; *occipit-*=back of head) forms the posterior part and most of the base of the cranium (Figure 7.6; also see Figure 7.4). The *foramen magnum* (=large hole) is in the inferior part of the bone. Within this foramen, the medulla oblongata (inferior part of the brain) connects with the spinal cord. The vertebral and spinal arteries also pass through this foramen. The *occipital condyles* are oval processes with convex surfaces, one on either side of the foramen magnum (see Figure 7.7). They articulate with depressions on the first cervical vertebra (atlas) to form the *atlanto-occipital joints.* Superior to each occipital condyle on the inferior surface of the skull is the *hypoglossal canal* (*hypo-*=under;*-glossal*=tongue), through which the hypoglossal (XII) nerve and a branch of the ascending pharyngeal artery pass (see Figure 7.5).

The *external occipital protuberance* is a prominent midline projection on the posterior surface of the bone just superior to the foramen magnum. You may be able to feel this structure as a definite bump on the back of your head, just above your neck (see Figure 7.4). A large fibrous, elastic ligament, the *ligamentum nuchae* (*nucha-*=nape of neck), which helps support the head, extends from the external occipital protuberance to the seventh

Figure 7.6   Skull. The sutures are exaggerated for emphasis. (See Tortora, *A Photographic Atlas of the Human Body*, Figure 3.5.)

> The occipital bone forms most of the posterior and inferior portions of the cranium.

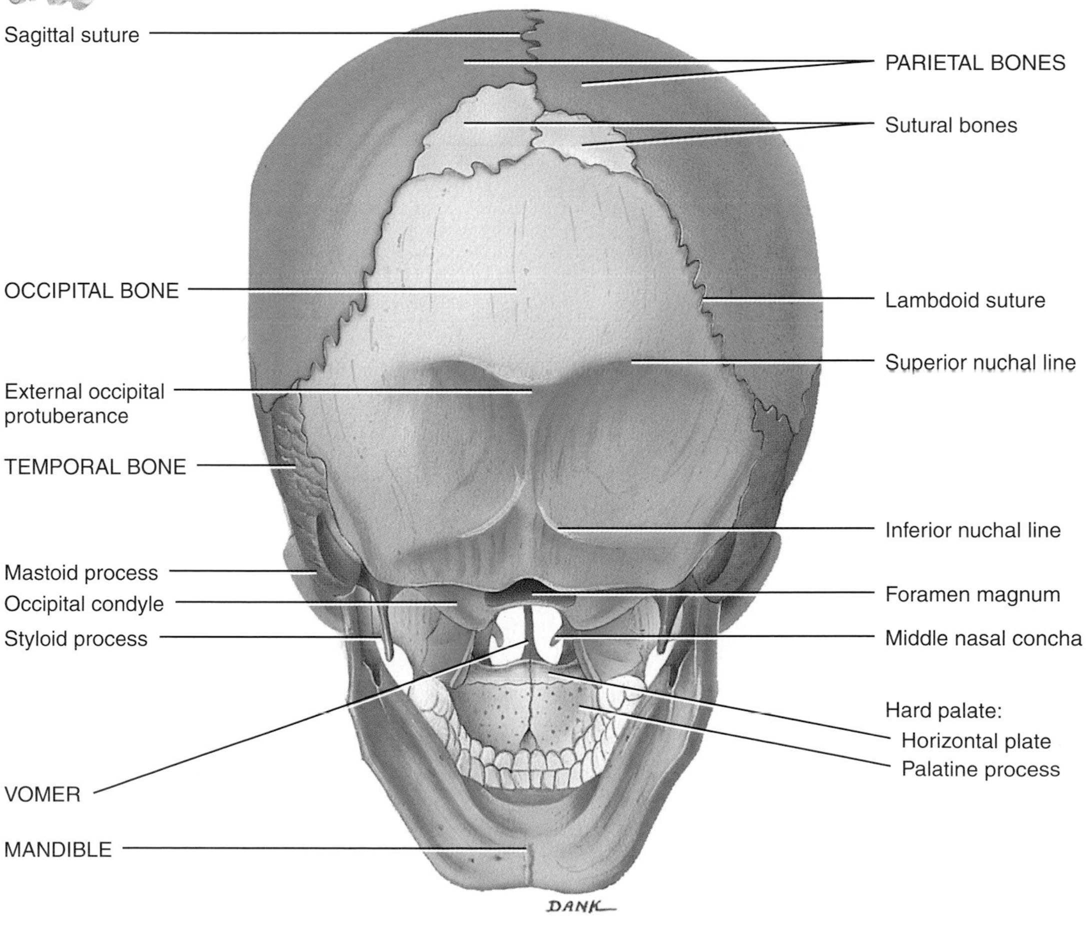

Posterior view

? Which bones form the posterior, lateral portion of the cranium?

cervical vertebra. Extending laterally from the protuberance are two curved lines, the *superior nuchal lines*, and below these are two *inferior nuchal lines*, which are areas of muscle attachment (Figure 7.7). It is possible to view the parts of the occipital bone, as well as surrounding structures, in the inferior view of the skull in Figure 7.7.

Sphenoid Bone

The **sphenoid bone** (SFĒ-noyd=wedge-shaped) lies at the middle part of the base of the skull (Figure 7.7 and Figure 7.8 on pages 176–177). This bone is called the keystone of the cranial floor because it articulates with all the other cranial bones, holding them together. Viewing the floor of the cranium superiorly (Figure 7.8a), note the sphenoid articulations: anteriorly with the frontal bone, laterally with the temporal bones, and posteriorly with the occipital bone. It lies posterior and slightly superior to the nasal cavity and forms part of the floor, sidewalls, and rear wall of the orbit (see Figure 7.12).

The shape of the sphenoid resembles a bat with outstretched wings (see Figure 7.8b). The *body* of the sphenoid is the cube-like medial portion between the ethmoid and occipital bones. It contains the *sphenoidal sinuses*, which drain into the nasal cavity (see

Figure 7.7 Skull. The mandible (lower jawbone) has been removed.

🔑 **The occipital condyles of the occipital bone articulate with the first cervical vertebra to form the atlanto-occipital joints.**

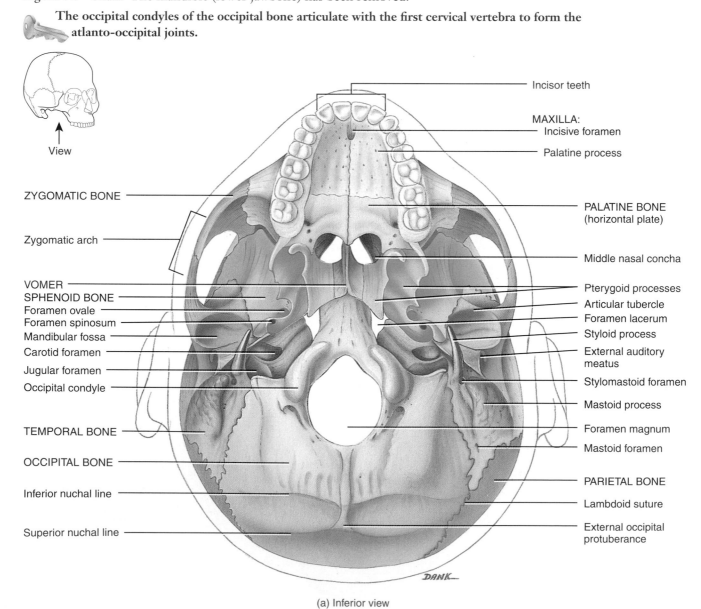

View

Incisor teeth

MAXILLA:
Incisive foramen
Palatine process

ZYGOMATIC BONE

PALATINE BONE
(horizontal plate)

Zygomatic arch

Middle nasal concha

VOMER
SPHENOID BONE
Foramen ovale
Foramen spinosum
Mandibular fossa
Carotid foramen
Jugular foramen
Occipital condyle

Pterygoid processes
Articular tubercle
Foramen lacerum
Styloid process
External auditory meatus
Stylomastoid foramen

TEMPORAL BONE

Mastoid process

Foramen magnum
Mastoid foramen

OCCIPITAL BONE

PARIETAL BONE

Inferior nuchal line

Lambdoid suture

Superior nuchal line

External occipital protuberance

DANK

(a) Inferior view

Figure 7.13). The *sella turcica* (SEL-a TUR-si-ka; *sella*=saddle; *turcica*=Turkish) is a bony saddle-shaped structure on the superior surface of the body of the sphenoid (Figure 7.8a). The anterior part of the sella turcica, which forms the horn of the saddle, is a ridge called the *tuberculum sellae*. The seat of the saddle is a depression, the *hypophyseal fossa* (hī-pō-FIZ-ē-al), which contains the pituitary gland. The posterior part of the sella turcica, which forms the back of the saddle, is another ridge called the *dorsum sellae*.

The *greater wings* of the sphenoid project laterally from the body, forming the anterolateral floor of the cranium. The greater wings also form part of the lateral wall of the skull just

anterior to the temporal bone and can be viewed externally. The *lesser wings*, which are smaller than the greater wings, form a ridge of bone anterior and superior to the greater wings. They form part of the floor of the cranium and the posterior part of the orbit of the eye.

Between the body and lesser wing just anterior to the sella turcica is the *optic foramen* (*optic*=eye), through which the optic (II) nerve and ophthalmic artery pass. Lateral to the body between the greater and lesser wings is a somewhat triangular slit called the *superior orbital fissure*. This fissure may also be seen in the anterior view of the orbit in Figure 7.12.

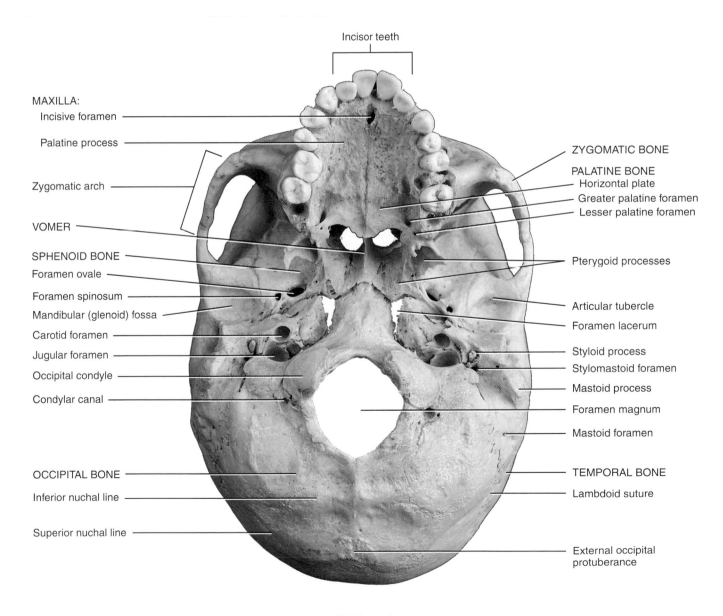

Incisor teeth

MAXILLA:
Incisive foramen
Palatine process

Zygomatic arch

VOMER

SPHENOID BONE
Foramen ovale
Foramen spinosum
Mandibular (glenoid) fossa
Carotid foramen
Jugular foramen
Occipital condyle
Condylar canal

OCCIPITAL BONE
Inferior nuchal line

Superior nuchal line

ZYGOMATIC BONE
PALATINE BONE
Horizontal plate
Greater palatine foramen
Lesser palatine foramen

Pterygoid processes

Articular tubercle
Foramen lacerum
Styloid process
Stylomastoid foramen
Mastoid process
Foramen magnum
Mastoid foramen

TEMPORAL BONE
Lambdoid suture

External occipital protuberance

(b) Inferior view

What parts of the nervous system join together within the foramen magnum?

Figure 7.8 **Sphenoid bone.** (See Tortora, *A Photographic Atlas of the Human Body*, 2e, Figures 3.8 and 3.9.)

The sphenoid bone is called the keystone of the cranial floor because it articulates with all other cranial bones, holding them together.

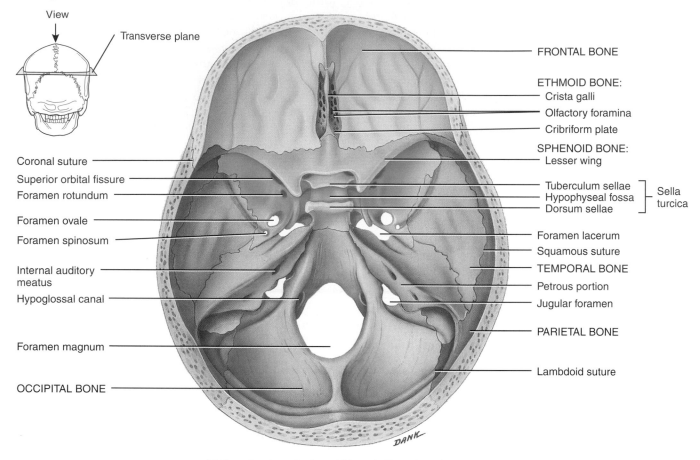

(a) Superior view of sphenoid bone in floor of cranium

In Figures 7.7 and 7.8b you can see the *pterygoid processes* (TER-i-goyd=winglike) extending from the inferior part of the sphenoid bone. These structures project inferiorly from the points where the body and greater wings unite and form the lateral posterior region of the nasal cavity. Some of the muscles that move the mandible attach to the pterygoid processes. At the base of the lateral pterygoid process in the greater wing is the *foramen ovale* (=oval), an opening for the mandibular branch of the trigeminal (V) nerve. Another foramen, the *foramen spinosum* (=resembling a spine), lies at the posterior angle of the sphenoid and transmits the middle meningeal blood vessels. The *foramen lacerum* (=lacerated) is bounded anteriorly by the sphenoid bone and medially by the sphenoid and occipital bones. The foramen is covered in part by a layer of fibrocartilage in living subjects. It transmits a branch of the ascending pharyngeal artery. Another foramen associated with the sphenoid bone is

the *foramen rotundum* (=round) located at the junction of the anterior and medial parts of the sphenoid bone. Through it passes the maxillary branch of the trigeminal (V) nerve.

Ethmoid Bone

The **ethmoid bone** (ETH-moyd;=like a sieve) is a light, spongelike bone located on the midline in the anterior part of the cranial floor medial to the orbits (Figure 7.9 on page 178). It is anterior to the sphenoid and posterior to the nasal bones. The ethmoid bone forms (1) part of the anterior portion of the cranial floor; (2) the medial wall of the orbits; (3) the superior portion of the nasal septum, a partition that divides the nasal cavity into right and left sides; and (4) most of the superior sidewalls of the nasal cavity. The ethmoid is a major superior supporting structure of the nasal cavity.

The *lateral masses* of the ethmoid bone compose most of the wall between the nasal cavity and the orbits. They contain 3 to

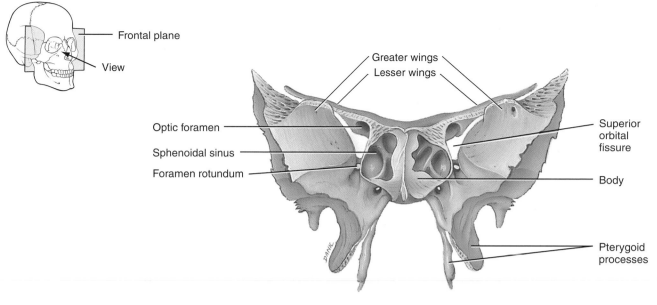

Frontal plane

View

Greater wings
Lesser wings

Optic foramen

Sphenoidal sinus

Foramen rotundum

Superior orbital fissure

Body

Pterygoid processes

(b) Anterior view of sphenoid bone

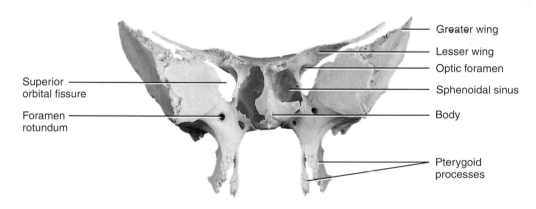

Superior orbital fissure

Foramen rotundum

Greater wing

Lesser wing

Optic foramen

Sphenoidal sinus

Body

Pterygoid processes

(c) Anterior view

Name the bones that articulate with the sphenoid bone, starting at the crista galli of the ethmoid bone and going in a clockwise direction.

18 air spaces, or cells. The ethmoidal cells together form the *ethmoidal sinuses* (see Figure 7.13). The *perpendicular plate* forms the superior portion of the nasal septum (see Figure 7.14). The *cribriform plate* (*cribri-*=sieve) lies in the anterior floor of the cranium and forms the roof of the nasal cavity. The cribriform plate contains the *olfactory foramina* (*olfact-*=smell), through which the olfactory (I) nerve passes. Projecting superiorly from the cribriform plate is a sharp triangular process called the *crista galli* (*crista*=crest; *galli*=cock). This structure serves as a point of attachment for the membranes that cover the brain.

The lateral masses of the ethmoid bone contain two thin, scroll-shaped projections lateral to the nasal septum. These are called the *superior nasal concha* (KONG-ka=shell) or *turbinate* and the *middle nasal concha (turbinate)*. The plural form is *conchae* (KONG-kē). A third pair of conchae, the *inferior nasal conchae*, are separate bones (discussed shortly). The con-

chae increase the vascular and mucous membrane surface area in the nasal cavities, which warms and moistens inhaled air before passing into the lungs. The conchae also cause inhaled air to swirl; the result is that many inhaled particles strike and become trapped in the mucus that lines the nasal passageways. The action of the conchae helps cleanse inhaled air before it passes into the rest of the respiratory tract. Air striking the mucous lining of the conchae is also warmed and moistened. The superior nasal conchae also participate in the sense of smell.

Facial Bones

The shape of the face changes dramatically during the first two years after birth. The brain and cranial bones expand, the teeth form and erupt, and the paranasal sinuses increase in size. Growth of the face ceases at about 16 years of age. The 14 facial

Figure 7.9 Ethmoid bone. (See Tortora, *A Photographic Atlas of the Human Body*, 2e, Figure 3.10.)

The ethmoid bone forms part of the anterior portion of the cranial floor, the medial wall of the orbits, the superior portions of the nasal septum, and most of the sidewalls of the nasal cavity.

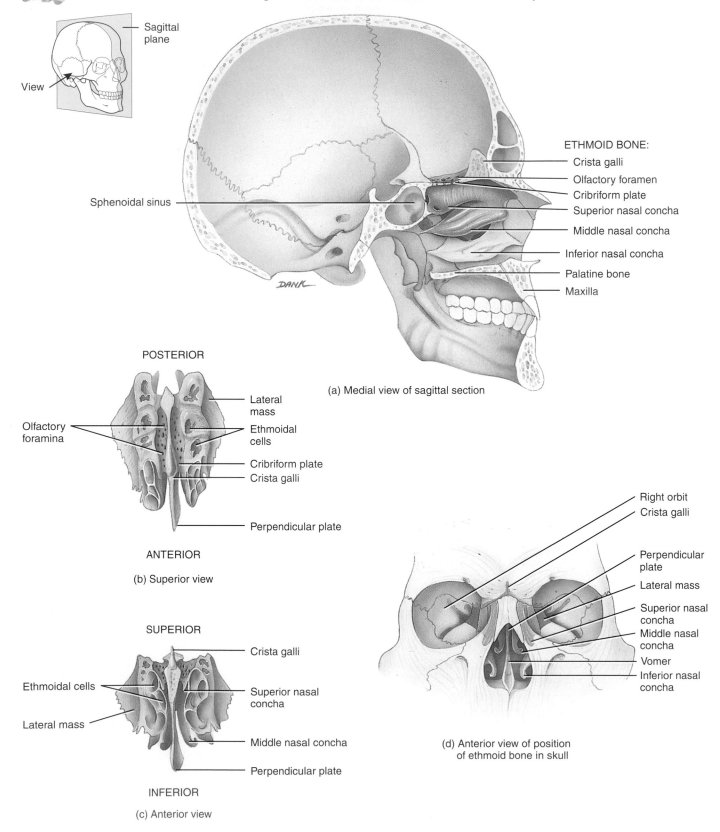

Sagittal plane

View

ETHMOID BONE:

Crista galli

Olfactory foramen

Cribriform plate

Superior nasal concha

Sphenoidal sinus

Middle nasal concha

Inferior nasal concha

Palatine bone

Maxilla

(a) Medial view of sagittal section

POSTERIOR

Lateral mass

Olfactory foramina

Ethmoidal cells

Cribriform plate

Crista galli

Perpendicular plate

ANTERIOR

(b) Superior view

SUPERIOR

Crista galli

Ethmoidal cells

Superior nasal concha

Lateral mass

Middle nasal concha

Perpendicular plate

INFERIOR

(c) Anterior view

Right orbit

Crista galli

Perpendicular plate

Lateral mass

Superior nasal concha

Middle nasal concha

Vomer

Inferior nasal concha

(d) Anterior view of position of ethmoid bone in skull

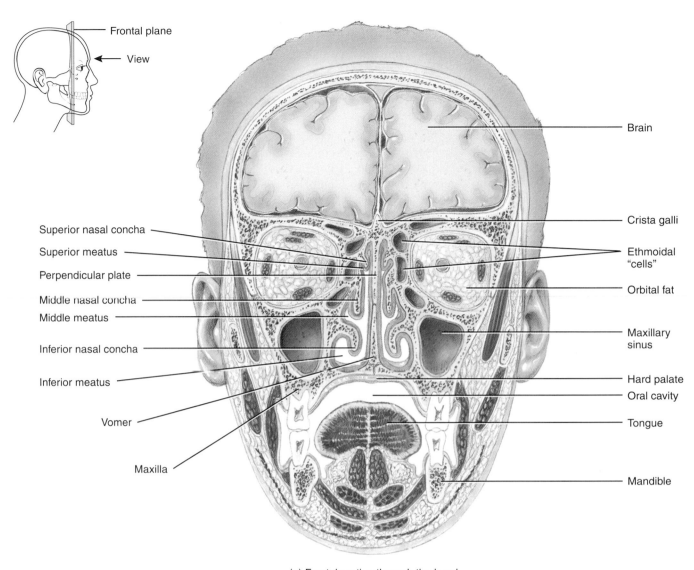

Frontal plane

View

Brain

Crista galli

Superior nasal concha

Superior meatus

Ethmoidal "cells"

Perpendicular plate

Orbital fat

Middle nasal concha

Middle meatus

Maxillary sinus

Inferior nasal concha

Inferior meatus

Hard palate

Oral cavity

Vomer

Tongue

Maxilla

Mandible

(e) Frontal section through the head

? **What part of the ethmoid bone forms the superior part of the nasal septum? The medial walls of the orbits?**

bones include two nasal bones, two maxillae (or maxillas), two zygomatic bones, the mandible, two lacrimal bones, two palatine bones, two inferior nasal conchae, and the vomer.

Nasal Bones

The paired **nasal bones** meet at the midline (see Figure 7.3) and form part of the bridge of the nose. The major structural portion of the nose consists of cartilage.

Maxillae

The paired **maxillae** (mak-SIL-ē = jawbones; singular is *maxilla*) unite to form the upper jawbone. They articulate with every bone of the face except the mandible, or lower jawbone (see Figures 7.4 and 7.7). The maxillae form part of the floors of the orbits, part of the lateral walls and floor of the nasal cavity, and most of the hard palate. The hard palate is a bony partition

formed by the palatine processes of the maxillae and horizontal plates of the palatine bones that forms the roof of the mouth.

Each maxilla contains a large *maxillary sinus* that empties into the nasal cavity (see Figure 7.11). The *alveolar process* (al-VĒ-ō-lar; *alveol-*=small cavity) is an arch that contains the *alveoli* (sockets) for the maxillary (upper) teeth. The *palatine process* is a horizontal projection of the maxilla that forms the anterior three-quarters of the hard palate. The union and fusion of the maxillary bones is normally completed before birth.

The *infraorbital foramen* (*infra-*=below; *orbital*=orbit), which can be seen in the anterior view of the skull in Figure 7.3, is an opening in the maxilla inferior to the orbit. Through it passes the infraorbital nerve and blood vessels and a branch of the maxillary division of the trigeminal (V) nerve. Another prominent foramen in the maxilla is the *incisive foramen* (=incisor teeth) just posterior to the incisor teeth (see Figure

7.7). It transmits branches of the greater palatine blood vessels and nasopalatine nerve. A final structure associated with the maxilla and sphenoid bone is the *inferior orbital fissure*, which is located between the greater wing of the sphenoid and the maxilla (see Figure 7.12).

CLEFT PALATE AND CLEFT LIP

Usually the palatine processes of the maxillary bones unite during weeks 10 to 12 of embryonic development. Failure to do so can result in one type of **cleft palate.** The condition may also involve incomplete fusion of the horizontal plates of the palatine bones (see Figure 7.7). Another form of this condition, called **cleft lip,** involves a split in the upper lip. Cleft lip and cleft palate often occur together. Depending on the extent and position of the cleft, speech and swallowing may be affected. In addition, children with cleft palate tend to have many ear infections that can lead to hearing loss. Facial and oral surgeons recommend closure of cleft lip during the first few weeks following birth, and surgical results are excellent. Repair of cleft palate typically is done between 12 and 18 months of age, ideally before the child begins to talk. Speech therapy may be needed, because the palate is important for pronouncing consonants, and orthodontic therapy may be needed to align the teeth. Again, results are usually excellent. Folic acid (one of the B vitamins) supplementation during pregnancy decreases the incidence of cleft palate and cleft lip. ■

Zygomatic Bones

The two **zygomatic bones** (*zygo-*=yokelike), commonly called cheekbones, form the prominences of the cheeks and part of the lateral wall and floor of each orbit (see Figure 7.12). They articulate with the frontal, maxilla, sphenoid, and temporal bones.

The *temporal process* of the zygomatic bone projects posteriorly and articulates with the zygomatic process of the temporal bone to form the *zygomatic arch* (see Figure 7.4b). A foramen associated with the zygomatic bone is the *zygomaticofacial foramen* near the center of the bone (Figure 7.3a). It transmits the zygomaticofacial nerve and vessels.

Lacrimal Bones

The paired **lacrimal bones** (LAK-ri-mal; *lacrim-*=teardrops) are thin and roughly resemble a fingernail in size and shape (see Figures 7.3, 7.4, and 7.12). These bones, the smallest bones of the face, are posterior and lateral to the nasal bones and form a part of the medial wall of each orbit. The lacrimal bones each contain a *lacrimal fossa*, a vertical tunnel formed with the maxilla, that houses the lacrimal sac, a structure that gathers tears and passes them into the nasal cavity (see Figure 7.12).

Palatine Bones

The two L-shaped **palatine bones** (PAL-a-tīn) form the posterior portion of the hard palate, part of the floor, and lateral wall of the nasal cavity, and a small portion of the floors of the orbits (see Figures 7.7 and 7.12). The posterior portion of the hard palate, which separates the nasal cavity from the oral cavity, is formed by the *horizontal plates* of the palatine bones (see Figures 7.6 and 7.7).

Inferior Nasal Conchae

The two **inferior nasal conchae** are inferior to the middle nasal conchae of the ethmoid bone (see Figures 7.3 and 7.9a). These scroll-like bones form a part of the inferior lateral wall of the nasal cavity and project into the nasal cavity. The inferior nasal conchae are separate bones; they are not part of the ethmoid bone. All three pairs of nasal conchae help swirl and filter air before it passes into the lungs. However, only the superior nasal conchae of the ethmoid bone are involved in the sense of smell.

Vomer

The **vomer** (VŌ-mer=plowshare) is a roughly triangular bone on the floor of the nasal cavity that articulates superiorly with the perpendicular plate of the ethmoid bone and inferiorly with both the maxillae and palatine bones along the midline (see Figures 7.3, 7.7, and 7.11). It is one of the components of the nasal septum, the partition that divides the nasal cavity into right and left sides.

Mandible

The **mandible** (*mand-*=to chew), or lower jawbone, is the largest, strongest facial bone (Figure 7.10). It is the only movable skull bone (other than the auditory ossicles). In the lateral view, you can see that the mandible consists of a curved, horizontal portion, the *body*, and two perpendicular portions, the *rami* (RĀ-mī=branches). The *angle* of the mandible is the area where each *ramus* (singular form) meets the body. Each ramus

Figure 7.10 Mandible.

The mandible is the largest and strongest facial bone.

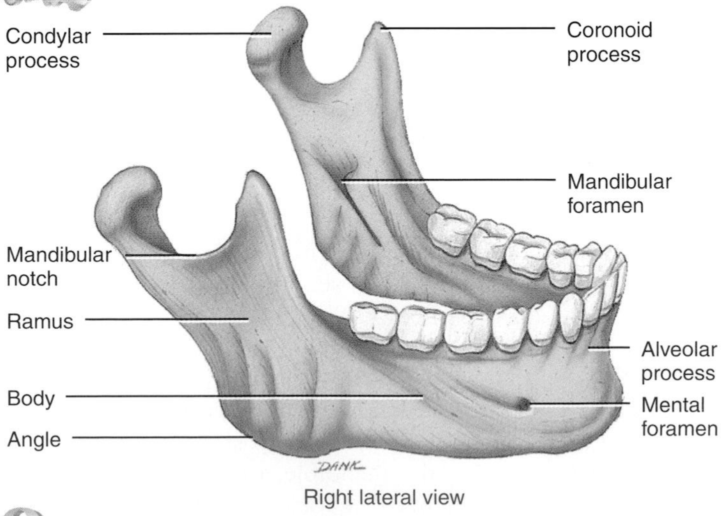

Right lateral view

? **What is the distinctive functional feature of the mandible among all the skull bones?**

has a posterior *condylar process* (KON-di-lar) that articulates with the mandibular fossa and articular tubercle of the temporal bone (see Figure 7.4) to form the **temporomandibular joint (TMJ).** It also has an anterior *coronoid process* (KOR-ō-noyd) to which the temporalis muscle attaches. The depression between the coronoid and condylar processes is called the *mandibular notch*. The *alveolar process* is an arch containing the *alveoli* (sockets) for the mandibular (lower) teeth.

The *mental foramen* (ment-=chin) is approximately inferior to the second premolar tooth. It is near this foramen that dentists reach the mental nerve when injecting anesthetics. Another foramen associated with the mandible is the *mandibular foramen* on the medial surface of each ramus, another site often used by dentists to inject anesthetics. The mandibular foramen is the beginning of the *mandibular canal*, which runs obliquely in the ramus and anteriorly to the body. Through the canal pass the inferior alveolar nerves and blood vessels, which are distributed to the mandibular teeth.

TEMPOROMANDIBULAR JOINT SYNDROME

One problem associated with the temporomandibular joint is **temporomandibular joint (TMJ) syndrome.** It is characterized by dull pain around the ear, tenderness of the jaw muscles, a clicking or popping noise when opening or closing the mouth, limited or abnormal opening of the mouth, headache, tooth sensitivity, and abnormal wearing of the teeth. TMJ syndrome can be caused by improperly aligned teeth, grinding or clenching the teeth, trauma to the head and neck, or arthritis. Treatment may involve applying moist heat or ice, eating soft foods, taking pain relievers such as aspirin, muscle retraining, adjusting or reshaping the teeth, orthodontic treatment, or surgery. ■

Nasal Septum

The inside of the nose, called the nasal cavity, is divided into right and left sides by a vertical partition called the **nasal septum.** The three components of the nasal septum are the vomer, septal cartilage, and the perpendicular plate of the ethmoid bone (Figure 7.11). The anterior border of the vomer articulates with the septal cartilage, which is hyaline cartilage, to form the anterior portion of the septum. The superior border of the vomer articulates with the perpendicular plate of the ethmoid bone to form the remainder of the nasal septum. The term "broken nose," in most cases, refers to damage to the septal cartilage rather than the nasal bones themselves.

DEVIATED NASAL SEPTUM

A **deviated nasal septum** is one that is deflected laterally from the midline of the nose. The deviation usually occurs at the junction of the vomer bone with the septal cartilage. Septal deviations may occur due to developmental abnormality or trauma. If the deviation is severe, it may entirely block the nasal passageway. Even a partial blockage may lead to infection. If inflammation occurs, it may cause nasal congestion, blockage of the paranasal sinus openings, chronic sinusitis, headache, and nosebleeds. The condition usually can be corrected or improved surgically. ■

Figure 7.11 Nasal septum. (See Tortora, *A Photographic Atlas of the Human Body*, 2e, Figure 3.4.)

The structures that form the nasal septum are the perpendicular plate of the ethmoid bone, the vomer, and septal cartilage.

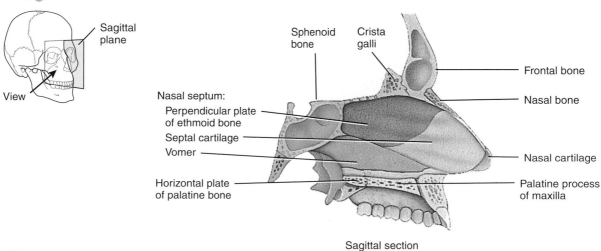

Sagittal section

What is the function of the nasal septum?

Orbits

Seven bones of the skull join to form each **orbit** (eye socket), which contains the eyeball and associated structures (Figure 7.12). The three cranial bones of the orbit are the frontal, sphenoid, and ethmoid; the four facial bones are the palatine, zygomatic, lacrimal, and maxilla. Each pyramid-shaped orbit has four regions that converge posteriorly:

1. Parts of the frontal and sphenoid bones comprise the *roof* of the orbit.

2. Parts of the zygomatic and sphenoid bones form the *lateral wall* of the orbit.

3. Parts of the maxilla, zygomatic, and palatine bones make up the *floor* of the orbit.

4. Parts of the maxilla, lacrimal, ethmoid, and sphenoid bones form the *medial wall* of the orbit.

    Associated with each orbit are five openings:

1. The *optic foramen* is at the junction of the roof and medial wall.

2. The *superior orbital fissure* is at the superior lateral angle of the apex.

3. The *inferior orbital fissure* is at the junction of the lateral wall and floor.

4. The *supraorbital foramen* is on the medial side of the supraorbital margin of the frontal bone.

5. The *lacrimal fossa* is in the lacrimal bone.

Foramina

We mentioned most of the **foramina** (openings for blood vessels, nerves, or ligaments) of the skull in the descriptions of the cranial and facial bones that they penetrate. As preparation for studying other systems of the body, especially the nervous and cardiovascular systems, these foramina and the structures passing through them are listed in Table 7.4. For your convenience and for future reference, the foramina are listed alphabetically.

Unique Features of the Skull

The skull exhibits several unique features not seen in other bones of the body. These include sutures, paranasal sinuses, and fontanels.

Sutures

A **suture** (SOO-chur=seam) is an immovable joint in an adult that is found only between skull bones and that holds most skull bones together. Sutures in the skulls of infants and children often are movable. The names of many sutures reflect the bones they unite. For example, the frontozygomatic suture is between the frontal bone and the zygomatic bone. Similarly, the sphenoparietal suture is between the sphenoid bone and the parietal bone. In other cases, however, the names of sutures are not so obvious. Of the many sutures found in the skull, we will identify only four prominent ones:

1. The **coronal suture** (kō-RŌ-nal; *coron-*=crown) unites the frontal bone and both parietal bones (see Figure 7.4).

Figure 7.12   Details of the orbit (eye socket). (See Tortora, *A Photographic Atlas of the Human Body*, 2e, Figure 3.11.)

The orbit is a pyramid-shaped structure that contains the eyeball and associated structures.

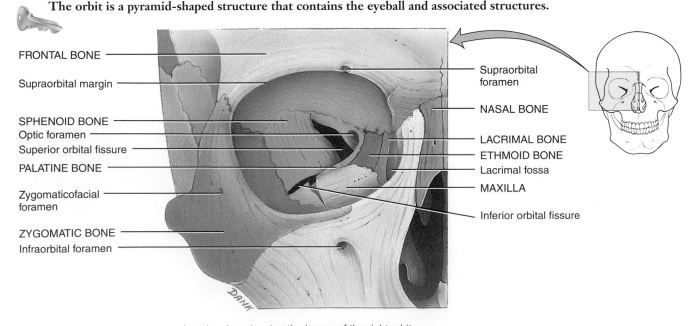

Anterior view showing the bones of the right orbit

 Which seven bones form the orbit?

2. The **sagittal suture** (SAJ-i-tal; *sagitt-*=arrow) unites the two parietal bones on the superior midline of the skull (see Figure 7.6). The sagittal suture is so named because in the infant, before the bones of the skull are firmly united, the suture and the fontanels (soft spots) associated with it resemble an arrow.

3. The **lambdoid suture** (LAM-doyd) unites the two parietal bones to the occipital bone. This suture is so named because of its resemblance to the Greek letter lambda (Λ), as can be seen in Figure 7.6. Sutural bones may occur within the sagittal and lambdoid sutures.

4. The **squamous sutures** (SKWĀ-mus; *squam-*=flat) unite the parietal and temporal bones on the lateral aspects of the skull (see Figure 7.4).

Paranasal Sinuses

The **paranasal sinuses** (*para-*=beside) are cavities in certain cranial and facial bones near the nasal cavity. They are most evi-

---

TABLE 7.4   PRINCIPAL FORAMINA OF THE SKULL

FORAMEN	LOCATION	STRUCTURES PASSING THROUGH
Carotid (relating to carotid artery in neck)	Petrous portion of temporal bone (Figure 7.7a).	Internal carotid artery and sympathetic nerves for eyes.
Hypoglossal (*hypo-*=under; *glossus*=tongue)	Superior to base of occipital condyles (Figure 7.8a).	Cranial nerve XII (hypoglossal) and branch of ascending pharyngeal artery.
Incisive (*incisive*= pertaining to incisor teeth)	Posterior to incisor teeth in maxilla (Figure 7.7a).	Branches of greater palatine blood vessels and nasopalatine nerve.
Infraorbital (*infra*=below)	Inferior to orbit in maxilla (Figure 7.12).	Infraorbital nerve and blood vessels and a branch of the maxillary division of cranial nerve V (trigeminal).
Jugular (*jugular*=throat)	Posterior to carotid canal between petrous portion of temporal bone and occipital bone (Figure 7.8a).	Internal jugular vein, cranial nerves IX (glossopharyngeal), X (vagus), and XI (accessory).
Lacerum (*lacerum*=lacerated)	Bounded anteriorly by sphenoid bone, posteriorly by petrous portion of temporal bone, and medially by the sphenoid bone and occipital bone (Figure 7.8a).	Branch of ascending pharyngeal artery in palatine bones.
Magnum (*magnum*=large)	Occipital bone (Figure 7.7a).	Medulla oblongata and its membranes (meninges), cranial nerve XI (accessory), and vertebral and spinal arteries.
Mandibular (*mand*=to chew)	Medial surface of ramus of mandible (Figure 7.10).	Inferior alveolar nerve and blood vessels.
Mastoid (=breast-shaped)	Posterior border of mastoid process of temporal bone (Figure 7.7a).	Emissary vein to transverse sinus and branch of occipital artery to dura mater.
Mental (*ment-*=chin)	Inferior to second premolar tooth in mandible (Figure 7.10).	Mental nerve and vessels.
Olfactory (*olfact*=to smell)	Cribriform plate of ethmoid bone (Figure 7.8a).	Cranial nerve I (olfactory).
Optic (=eye)	Between superior and inferior portions of small wing of sphenoid bone (Figure 7.12).	Cranial nerve II (optic) and ophthalmic artery.
Ovale (*ovale*=oval)	Greater wing of sphenoid bone (Figure 7.8a).	Mandibular branch of cranial nerve V (trigeminal).
Rotundum (=round)	Junction of anterior and medial parts of sphenoid bone (Figure 7.8a).	Maxillary branch of cranial nerve V (trigeminal).
Spinosum (=resembling a spine)	Posterior angle of sphenoid bone (Figure 7.8a).	Middle meningeal blood vessels.
Stylomastoid (*stylo*=stake or pole)	Between styloid and mastoid processes of temporal bone (Figure 7.7a).	Cranial nerve VII (facial) and stylomastoid artery.
Supraorbital (*supra-*=above)	Supraorbital margin of orbit in frontal bone (Figure 7.12).	Supraorbital nerve and artery.

dent in a sagittal section of the skull (Figure 7.13). The paranasal sinuses are lined with mucous membranes that are continuous with the lining of the nasal cavity. Skull bones con-

taining the paranasal sinuses are the frontal, sphenoid, ethmoid, and maxillary. Besides producing mucus, the paranasal sinuses serve as resonating chambers for sound as we speak or sing.

Figure 7.13 Paranasal sinuses. (a and b) Location of the paranasal sinuses. (c) Openings of the paranasal sinuses into the nasal cavity. Portions of the conchae have been sectioned. (See Tortora, *A Photographic Atlas of the Human Body*, Figure 3.4.)

Paranasal sinuses are mucous membrane-lined spaces in the frontal, sphenoid, ethmoid, and maxillary bones that connect to the nasal cavity.

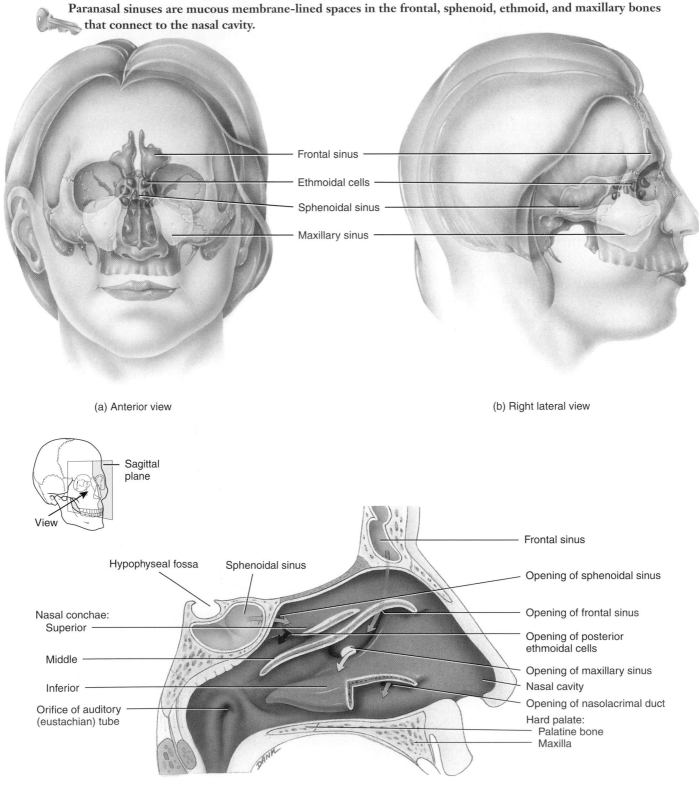

(a) Anterior view

(b) Right lateral view

(c) Sagittal section

What are the functions of the paranasal sinuses?

SINUSITIS

Secretions produced by the mucous membranes of the paranasal sinuses drain into the nasal cavity. An inflammation of the membranes due to an allergic reaction or infection is called **sinusitis.** If the membranes swell enough to block drainage into the nasal cavity, fluid pressure builds up in the paranasal sinuses, and a sinus headache results. A severely deviated nasal septum or nasal polyps, growths that can be removed surgically, may also cause chronic sinusitis. ■

Fontanels

The skeleton of a newly formed embryo consists of cartilage or mesenchyme arranged in sheet-like layers that resemble membranes shaped like bones. Gradually, ossification occurs—bone replaces the cartilage and mesenchyme. At birth, mesenchyme-filled spaces called **fontanels** (fon-ta-NELZ=little fountains) are present between the cranial bones (Figure 7.14). Commonly called "soft spots," fontanels are areas of unossified mesenchyme. Eventually, they will be replaced with bone by intramembranous ossification and become sutures. Functionally, the fontanels provide some flexibility to the fetal skull. They allow the skull to change shape as it passes through the birth canal and permit rapid growth of the brain during infancy. Although an infant may have many fontanels at birth, the form and location of six are fairly constant:

1. The unpaired **anterior fontanel,** located at the midline between the two parietal bones and the frontal bone, is roughly diamond-shaped and is the largest fontanel. It usually closes 18 to 24 months after birth.

2. The unpaired **posterior fontanel** is located at the midline between the two parietal bones and the occipital bone. Because it is much smaller than the anterior fontanel, it generally closes about 2 months after birth.

3. The paired **anterolateral fontanels,** located laterally between the frontal, parietal, temporal, and sphenoid bones, are small and irregular in shape. Normally, they close about 3 months after birth.

4. The paired **posterolateral fontanels,** located laterally between the parietal, occipital, and temporal bones, are irregularly shaped. They begin to close 1 to 2 months after birth, but closure is generally not complete until 12 months.

The amount of closure in fontanels helps a physician gauge the degree of brain development. In addition, the anterior fontanel serves as a landmark for withdrawal of blood for analysis from the superior sagittal sinus (a large vein on the midline surface of the brain).

Cranial Fossae

The floor of the cranium contains three distinct levels, from anterior to posterior, called **cranial fossae** (Figure 7.15). The fossae contain depressions for the various brain convolutions, grooves for cranial blood vessels, and numerous foramina. From anterior to posterior, they are named the anterior cranial fossa, middle cranial fossa, and posterior cranial fossa. The highest

Figure 7.14   Fontanels at birth. (See Tortora, *A Photographic Atlas of the Human Body*, 2e, Figure 3.12.)

Fontanels are mesenchyme-filled spaces between cranial bones that are present at birth.

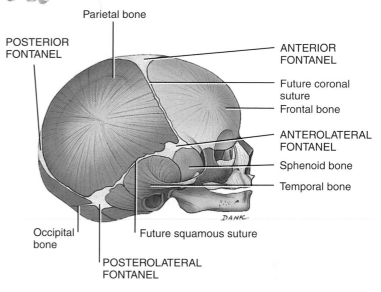

Right lateral view

Which fontanel is bordered by four different skull bones?

Figure 7.15   Cranial fossae.

Cranial fossae are levels in the cranial floor that contain depressions for brain convolutions, grooves for blood vessels, and numerous foramina.

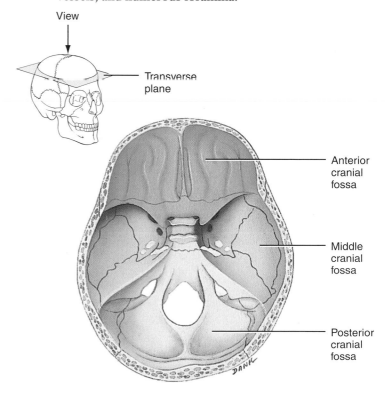

Superior view of floor of cranium

Which cranial fossa is the largest?

level, the *anterior cranial fossa*, is formed largely by the portion of the frontal bone that constitutes the roof of the orbits and nasal cavity, the crista galli and cribriform plate of the ethmoid bone, and the lesser wings and part of the body of the sphenoid bone. This fossa houses the frontal lobes of the cerebral hemispheres of the brain. The rough surface of the frontal bone can lead to tearing of the frontal lobes of the cerebral hemispheres during head trauma. The *middle cranial fossa* is inferior and posterior to the anterior cranial fossa. It is shaped like a butterfly with a small median portion and two expanded lateral portions. The median portion is formed by part of the body of the sphenoid bone, and the lateral portions are formed by the greater wings of the sphenoid bone, temporal squama, and parietal bone. The middle cranial fossa cradles the temporal lobes of the cerebral hemispheres. The last fossa, at the most inferior level, is the *posterior cranial fossa*, the largest of the fossae. It is formed largely by the occipital bone and the petrous and mastoid portions of the temporal bone. It is a very deep fossa that accommodates the cerebellum, pons, and medulla oblongata of the brain.

CHECKPOINT

4. Describe the general features of the skull.
5. What bones constitute the orbit?
6. What structures make up the nasal septum?
7. Define the following: foramen, suture, paranasal sinus, and fontanel.
8. Name the cranial fossae from inferior to superior.

HYOID BONE

OBJECTIVE

● Describe the relationship of the hyoid bone to the skull.

The single **hyoid bone** (=U-shaped) is a unique component of the axial skeleton because it does not articulate with any other bone. Rather, it is suspended from the styloid processes of the temporal bones by ligaments and muscles (see Figure 11.7 on page 308). Located in the anterior neck between the mandible and larynx (Figure 7.16a), the hyoid bone supports the tongue, providing attachment sites for some tongue muscles and for muscles of the neck and pharynx. The hyoid bone consists of a horizontal *body* and paired projections called the *lesser horns* and the *greater horns* (Figure 7.16 b, c). Muscles and ligaments attach to these paired projections.

The hyoid bone, as well as cartilages of the larynx and trachea, are often fractured during strangulation. As a result, they are carefully examined at autopsy when strangulation is suspected.

CHECKPOINT

9. What are the functions of the hyoid bone?

VERTEBRAL COLUMN

OBJECTIVE

● Identify the regions and normal curves of the vertebral column and describe its structural and functional features.

Figure 7.16 Hyoid bone. (See Tortora, *A Photographic Atlas of the Human Body*, Figure 3.13.)

The hyoid bone supports the tongue, providing attachment sites for muscles of the tongue, neck, and pharynx.

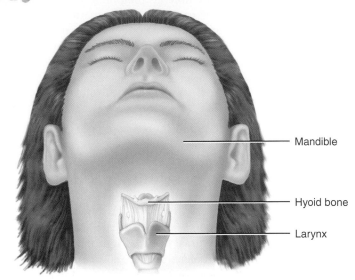

(a) Position of hyoid

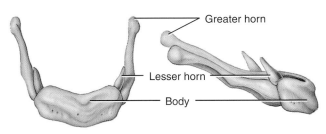

(b) Anterior view   (c) Right lateral view

In what way is the hyoid bone different from all the other bones of the axial skeleton?

The **vertebral column,** also called the *spine* or *backbone*, makes up about two-fifths of the total height of the body and is composed of a series of bones called **vertebrae** (VER-te-brē; singular is *vertebra*). The vertebral column consists of bone and connective tissue; the spinal cord that it surrounds and protects consists of nervous tissue. The length of the column is about 71 cm (28 in.) in an average adult male and about 61 cm (24 in.) in an average adult female. The vertebral column functions as a strong, flexible rod with elements that can move forward, move backward, move sideways, and rotate. It encloses and protects the spinal cord, supports the head, and serves as a point of attachment for the ribs, pelvic girdle, and muscles of the back.

The total number of vertebrae during early development is 33. Then, several vertebrae in the sacral and coccygeal regions fuse. As a result, the adult vertebral column, also called the spinal column, typically contains 26 vertebrae (Figure 7.17a). These are distributed as follows:

● 7 **cervical vertebrae** (*cervic-*=neck) are in the neck region.
● 12 **thoracic vertebrae** (*thorax*=chest) are posterior to the thoracic cavity.

Figure 7.17 Vertebral column. The numbers in parentheses in (a) indicate the number of vertebrae in each region. In (d), the relative size of the disc has been enlarged for emphasis. A "window" has been cut in the annulus fibrosus so that the nucleus pulposus can be seen. (See Tortora, *A Photographic Atlas of the Human Body*, Figure 3.15.)

The adult vertebral column typically contains 26 vertebrae.

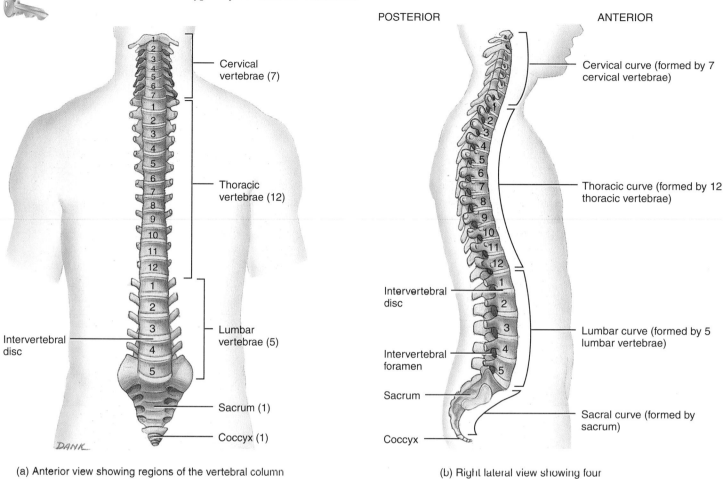

(a) Anterior view showing regions of the vertebral column

(b) Right lateral view showing four normal curves

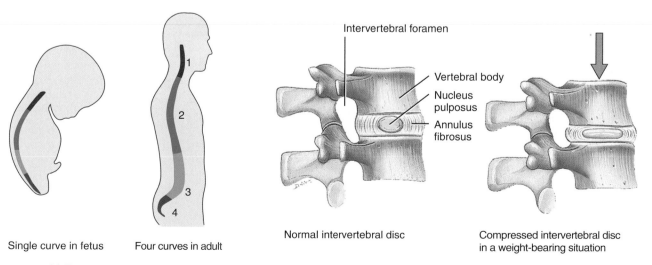

Single curve in fetus   Four curves in adult

(c) Fetal and adult curves

Normal intervertebral disc

Compressed intervertebral disc in a weight-bearing situation

(d) Intervertebral disc

Which curves of the adult vertebral column are concave (relative to the anterior side of the body)?

- 5 **lumbar vertebrae** (*lumb-*=loin) support the lower back.
- 1 **sacrum** (SĀ-krum=sacred bone) consists of five fused **sacral vertebrae.**
- 1 **coccyx** (KOK-siks=cuckoo, because the shape resembles the bill of a cuckoo bird), consists of four fused **coccygeal vertebrae** (kok-SIJ-ē-al).

Whereas the cervical, thoracic, and lumbar vertebrae are movable, the sacrum and coccyx are not. We will discuss each of these regions in detail shortly.

Normal Curves of the Vertebral Column

When viewed from the side, the adult vertebral column shows four slight bends called **normal curves** (Figure 7.17b). Relative to the front of the body, the *cervical* and *lumbar curves* are convex (bulging out), whereas the *thoracic* and *sacral curves* are concave (cupping in).

The curves of the vertebral column increase its strength, help maintain balance in the upright position, absorb shocks during walking, and help protect the vertebrae from fracture.

In the fetus, there is only a single anteriorly concave curve (Figure 7.17c). At about the third month after birth, when an infant begins to hold its head erect, the cervical curve develops. Later, when the child sits up, stands, and walks, the lumbar curve develops. The thoracic (2) and sacral (4) curves are called *primary curves* because they form first during fetal development. The cervical (1) and lumbar (3) curves are known as *secondary curves* because they begin to form later, several months after birth. All curves are fully developed by age 10. However, secondary curves may be progressively lost in old age.

Various conditions may exaggerate the normal curves of the vertebral column, or the column may acquire a lateral bend, resulting in **abnormal curves** of the vertebral column. Three such abnormal curves—kyphosis, lordosis, and scoliosis—are described in the Applications to Health section on page 199.

Figure 7.18   Structure of a typical vertebra, as illustrated by a thoracic vertebra. In (b), only one spinal nerve has been included, and it has been extended beyond the intervertebral foramen for clarity. The sympathetic chain is part of the autonomic nervous system (see Figure 20.5 on page 639).

A vertebra consists of a body, a vertebral arch, and several processes.

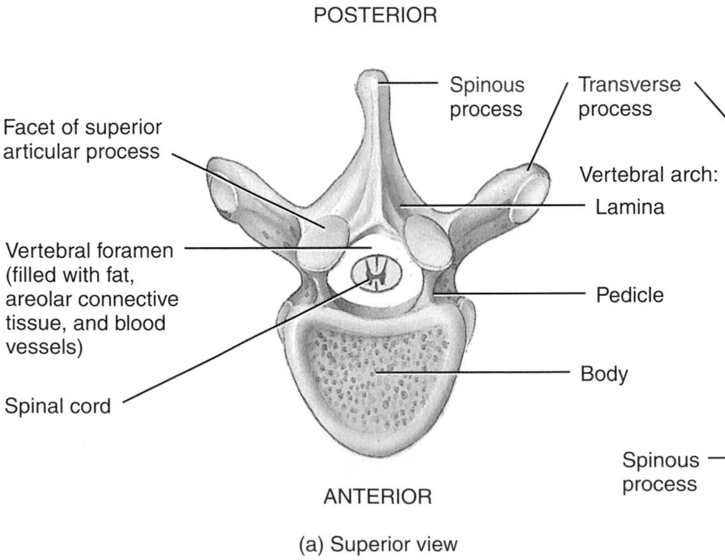

(a) Superior view

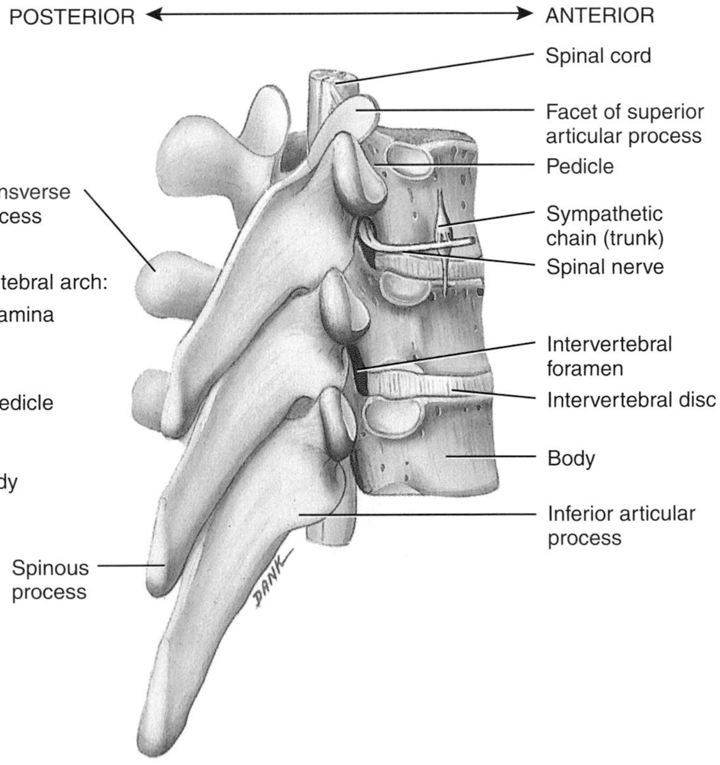

(b) Right posterolateral view of articulated vertebrae

Intervertebral Discs

Between the bodies of adjacent vertebrae from the second cervical vertebra to the sacrum are **intervertebral discs** (Figure 7.17d). Each disc has an outer fibrous ring consisting of fibrocartilage called the *annulus fibrosus* (*annulus*=ringlike) and an inner soft, pulpy, highly elastic substance called the *nucleus pulposus* (*pulposus*=pulplike). The discs form strong joints, permit various movements of the vertebral column, and absorb vertical shock. Under compression, they flatten and broaden; with age, the nucleus pulposus hardens and becomes less elastic. Narrowing of the discs and compression of vertebrae results in a decrease in height with age.

Parts of a Typical Vertebra

Even though vertebrae in different regions of the spinal column vary in size, shape, and detail, they are similar enough that we can discuss the structures (and the functions) of a typical vertebra (Figure 7.18). Vertebrae typically consist of a body, a vertebral arch, and several processes.

Body

The **body** is the thick, disc-shaped anterior portion that is the weight-bearing part of a vertebra. Its superior and inferior surfaces are roughened for the attachment of cartilaginous intervertebral discs. The anterior and lateral surfaces contain nutrient foramina, openings for blood vessels that deliver nutrients and oxygen and remove carbon dioxide and wastes from bone tissue.

Vertebral Arch

The **vertebral arch** extends posteriorly from the body of the vertebra and together with the body of the vertebra surrounds the spinal cord. Two short, thick processes, the *pedicles* (PED-i-kuls=little feet), form the vertebral arch. The pedicles project posteriorly from the body to unite with the laminae. The *laminae*

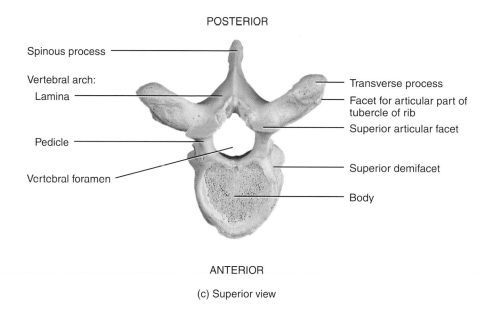

(c) Superior view

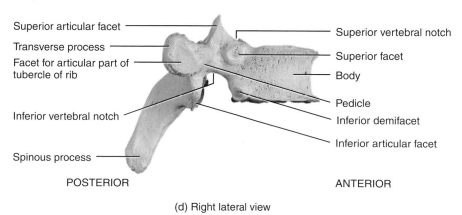

(d) Right lateral view

What are the functions of the vertebral and intervertebral foramina?

(LAM-i-nē =thin layers) are the flat parts that join to form the posterior portion of the vertebral arch. The *vertebral foramen* lies between the vertebral arch and body and contains the spinal cord, adipose tissue, areolar connective tissue, and blood vessels. Collectively, the vertebral foramina of all vertebrae form the *vertebral (spinal) canal*. The pedicles exhibit superior and inferior indentations called *vertebral notches*. When the vertebral notches are stacked on top of one another, they form an opening between adjoining vertebrae on both sides of the column. Each opening, called an *intervertebral foramen*, permits the passage of a single spinal nerve.

Processes

Seven **processes** arise from the vertebral arch. At the point where a lamina and pedicle join, a *transverse process* extends laterally on each side. A single *spinous process (spine)* projects posteriorly from the junction of the laminae. These three processes serve as points of attachment for muscles. The remaining four processes form joints with other vertebrae above or below. The two *superior articular processes* of a vertebra articulate (form joints) with the two inferior articular processes of the vertebra immediately superior to them. In turn, the two *inferior articular processes* of that vertebra articulate with the two superior articular

processes of the vertebra immediately inferior to them, and so on.

The articulating surfaces of the articular processes are referred to as *facets* (=little faces), and are covered with hyaline cartilage. The articulations formed between the bodies and articular facets of successive vertebrae are called *intervertebral joints*.

Regions of the Vertebral Column

We turn now to the five regions of the vertebral column, beginning superiorly and moving inferiorly. Note that vertebrae in each region are numbered in sequence, from superior to inferior.

Cervical Region

The bodies of **cervical vertebrae** (C1–C7) are smaller than those of thoracic vertebrae (Figure 7.19a). The vertebral arches, however, are larger. All cervical vertebrae have three foramina: one vertebral foramen and two transverse foramina (Figure 7.19d). The vertebral foramina of cervical vertebrae are the largest in the spinal column because they house the cervical enlargement of the spinal cord. Each cervical transverse process contains a *transverse foramen* through which the vertebral artery and its accompanying vein and nerve fibers pass. The spinous

Figure 7.19 Cervical vertebrae.

The cervical vertebrae are found in the neck region.

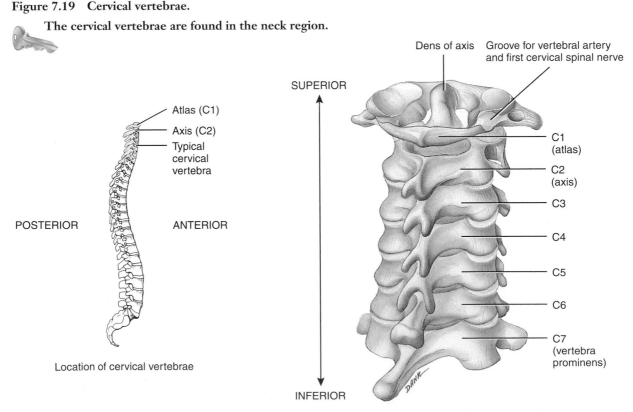

Atlas (C1)
Axis (C2)
Typical cervical vertebra

POSTERIOR　　ANTERIOR

Location of cervical vertebrae

SUPERIOR

INFERIOR

Dens of axis　Groove for vertebral artery and first cervical spinal nerve

C1 (atlas)
C2 (axis)
C3
C4
C5
C6
C7 (vertebra prominens)

(a) Posterior view of articulated cervical vertebrae

processes of C2 through C6 are often *bifid*—that is, split into two parts (Figure 7.19a, d).

The first two cervical vertebrae differ considerably from the others. Named after the mythological Atlas, who supported the world on his shoulders, the first cervical vertebra (C1), the **atlas,** supports the head (Figure 7.19a, b). The atlas is a ring of bone with *anterior* and *posterior arches* and large *lateral masses.* It lacks a body and a spinous process. The superior surfaces of the lateral masses, called *superior articular facets,* are concave and articulate with the occipital condyles of the occipital bone to form the paired *atlanto-occipital joints.* These articulations permit the movement seen when moving the head to signify "yes." The inferior surfaces of the lateral masses, the *inferior articular facets,* articulate with the second cervical vertebra. The transverse processes and transverse foramina of the atlas are quite large.

The second cervical vertebra (C2), called the **axis** (Figure 7.19a, c), does have a body. A peglike process called the *dens* (=tooth) or *odontoid process* projects superiorly through the anterior portion of the vertebral foramen of the atlas. The dens makes a pivot on which the atlas and head rotate. This arrangement permits side-to-side movement of the head, as when you move your head to signify "no." The articulation formed between the anterior arch of the atlas and dens of the axis, and

between their articular facets, is called the *atlanto-axial joint.* In some instances of trauma, the dens of the axis may be driven into the medulla oblongata of the brain. This type of injury is the usual cause of death from whiplash injuries.

The third through sixth cervical vertebrae (C3–C6), represented by the vertebra in Figure 7.19d, correspond to the structural pattern of the typical cervical vertebra previously described. The seventh cervical vertebra (C7), called the *vertebra prominens,* is somewhat different (Figure 7.19a). It has a single large spinous process that may be seen and felt at the base of the neck.

Thoracic Region

Thoracic vertebrae (T1–T12; Figure 7.20) are considerably larger and stronger than cervical vertebrae. In addition, the spinous processes on T1 and T2 are long, laterally flattened, and directed inferiorly. In contrast, the spinous processes on T11 and T12 are shorter, broader, and directed more posteriorly. Compared to cervical vertebrae, thoracic vertebrae also have longer and larger transverse processes.

The most distinguishing feature of thoracic vertebrae is that they articulate with the ribs. Except for T11 and T12, the transverse processes have facets for articulating with the *tubercles* of the ribs. The bodies of thoracic vertebrae also have either facets

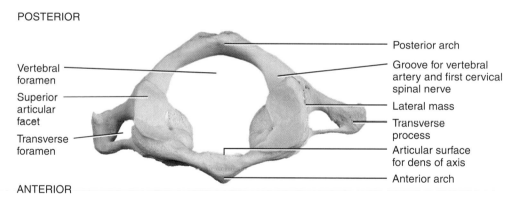

POSTERIOR

Vertebral foramen

Superior articular facet

Transverse foramen

ANTERIOR

Posterior arch

Groove for vertebral artery and first cervical spinal nerve

Lateral mass

Transverse process

Articular surface for dens of axis

Anterior arch

(b) Superior view of atlas (C1)

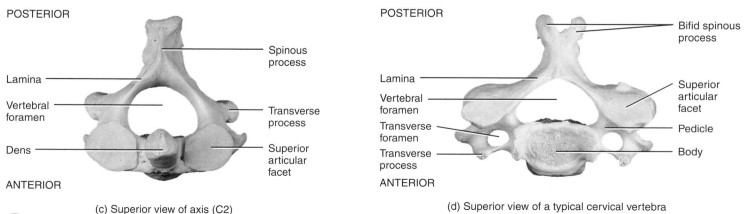

POSTERIOR

Lamina

Vertebral foramen

Dens

ANTERIOR

Spinous process

Transverse process

Superior articular facet

(c) Superior view of axis (C2)

POSTERIOR

Lamina

Vertebral foramen

Transverse foramen

Transverse process

ANTERIOR

Bifid spinous process

Superior articular facet

Pedicle

Body

(d) Superior view of a typical cervical vertebra

 Which bones permit the movement of the head to signify "no"?

Figure 7.20   Thoracic vertebrae.

The thoracic vertebrae are found in the chest region and articulate with the ribs.

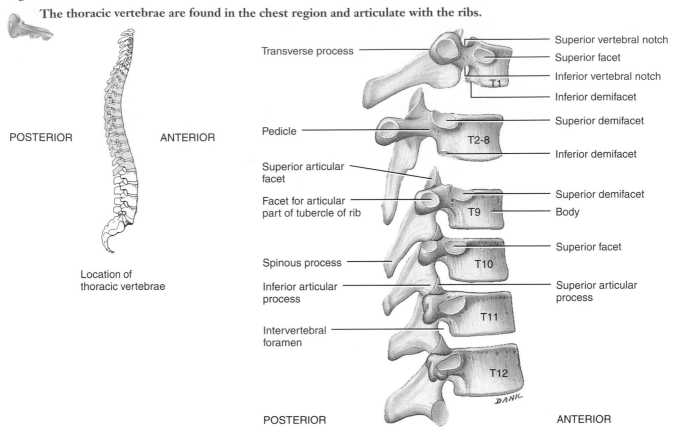

POSTERIOR   ANTERIOR

Location of
thoracic vertebrae

Right lateral view of several articulated thoracic vertebrae

Which parts of thoracic vertebrae articulate with the ribs?

or demifacets (half facets) for articulation with the *heads* of the ribs. The articulations between the thoracic vertebrae and ribs are called *vertebrocostal joints.* As you can see in Figure 7.20, T1 has a superior facet and an inferior demifacet, one on each side of the vertebral body. T2–T8 have a superior and inferior demifacet, one on each side of the vertebral body. T9 has a superior demifacet on each side of the vertebral body, and T10–T12 have a superior facet on each side of the vertebral body. Movements of the thoracic region are limited by thin intervertebral discs and by the attachment of the ribs to the sternum.

Lumbar Region

The **lumbar vertebrae** (L1–L5) are the largest and strongest in the vertebral column (Figure 7.21) because the amount of body weight supported by the vertebrae increases toward the inferior end of the backbone. Their various projections are short and thick. The superior articular processes are directed medially instead of superiorly, and the inferior articular processes are directed laterally instead of inferiorly. The spinous processes are quadrilateral in shape, thick and broad, and project nearly straight posteriorly. The spinous processes are well-adapted for the attachment of the large back muscles.

A summary of the major structural differences among cervical, thoracic, and lumbar vertebrae is presented in Table 7.5 on page 194.

Sacrum

The **sacrum** is a triangular bone formed by the union of five sacral vertebrae (S1–S5), indicated in Figure 7.22a on page 194. The sacral vertebrae begin to fuse in individuals between 16 and 18 years of age, a process usually completed by age 30. The sacrum serves as a strong foundation for the pelvic girdle. It is positioned at the posterior portion of the pelvic cavity medial to the two hip bones. The female sacrum is shorter, wider, and more curved between S2 and S3 than the male sacrum (see Table 8.1 on page 217).

The concave anterior side of the sacrum faces the pelvic cavity. It is smooth and contains four *transverse lines (ridges)* that mark the joining of the sacral vertebral bodies (Figure 7.22a). At the ends of these lines are four pairs of *anterior sacral foramina.* The lateral portion of the superior surface of the sacrum contains a smooth surface called the *sacral ala* (5 wing), which is formed by the fused transverse processes of the first sacral vertebra (S1).

The convex, posterior surface of the sacrum contains a *median sacral crest,* which is the fused spinous processes of the

Figure 7.21   Lumbar vertebrae.

Lumbar vertebrae are found in the lower back.

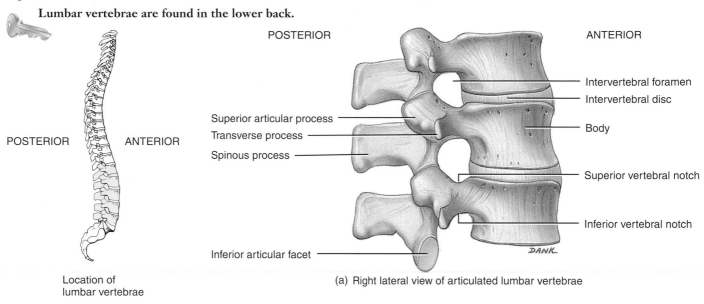

POSTERIOR   ANTERIOR

Intervertebral foramen
Intervertebral disc

Superior articular process
Transverse process   Body
Spinous process

Superior vertebral notch

Inferior vertebral notch

Inferior articular facet

DANK

POSTERIOR   ANTERIOR

Location of
lumbar vertebrae

(a) Right lateral view of articulated lumbar vertebrae

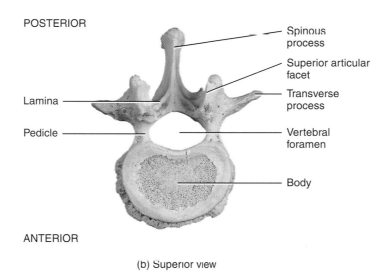

POSTERIOR

Spinous
process

Superior articular
facet

Lamina

Transverse
process

Pedicle

Vertebral
foramen

Body

ANTERIOR

(b) Superior view

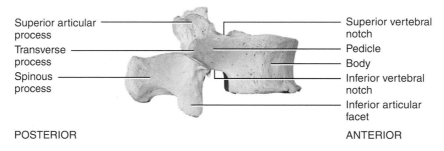

Superior articular
process   Superior vertebral
                                                   notch
Transverse
process   Pedicle
Spinous   Body
process   Inferior vertebral
                                                   notch
                                                   Inferior articular
                                                   facet

POSTERIOR   ANTERIOR

(c) Right lateral view

Why are the lumbar vertebrae the largest and strongest in the vertebral column?

CHARACTERISTIC	CERVICAL	THORACIC	LUMBAR
Overall structure			
Size	Small	Larger	Largest
Foramina	One vertebral and two transverse	One vertebral	One vertebral
Spinous processes	Slender and often bifid (C2–C6)	Long and fairly thick (most project inferiorly)	Short and blunt (project posteriorly rather than inferiorly)
Transverse processes	Small	Fairly large	Large and blunt
Articular facets for ribs	Absent	Present	Absent
Direction of articular facets			
Superior	Posterosuperior	Posterolateral	Medial
Inferior	Anteroinferior	Anteromedial	Lateral
Size of intervertebral discs	Thick relative to size of vertebral bodies	Thin relative to vertebral bodies	Massive

Figure 7.22 **Sacrum and coccyx.** (See Tortora, *A Photographic Atlas of the Human Body*, 2e, Figure 3.19.)

The sacrum is formed by the union of five sacral vertebrae, and the coccyx is formed by the union of usually four coccygeal vertebrae.

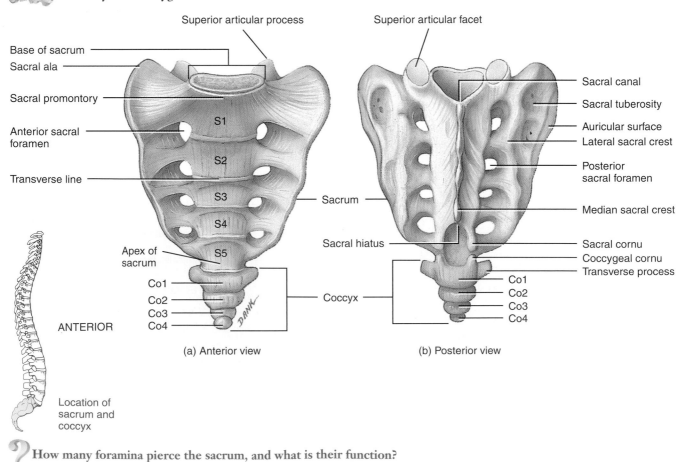

(a) Anterior view

(b) Posterior view

ANTERIOR

Location of sacrum and coccyx

How many foramina pierce the sacrum, and what is their function?

upper sacral vertebrae; a *lateral sacral crest*, which is the fused transverse processes of the sacral vertebrae; and four pairs of *posterior sacral foramina* (Figure 7.22b). These foramina connect with the anterior sacral foramina to allow passage of nerves and blood vessels. The *sacral canal* is a continuation of the vertebral canal. The laminae of the fifth sacral vertebra, and sometimes the fourth, fail to meet. This leaves an inferior entrance to the vertebral canal called the *sacral hiatus* (hī-Ā-tus=opening). On either side of the sacral hiatus are the *sacral cornua* (KOR-noo-a; *cornu*=horn), the inferior articular processes of the fifth sacral vertebra. They are connected by ligaments to the coccyx.

The narrow inferior portion of the sacrum is known as the *apex*. The broad superior portion of the sacrum is called the *base*. The anteriorly projecting border of the base, called the *sacral promontory* (PROM-on-tō-rē), is one of the points used for measurements of the pelvis. On both lateral surfaces the sacrum has a large ear-shaped *auricular surface* that articulates with the ilium of each hip bone to form the *sacroiliac joint*. Posterior to the auricular surface is a roughened surface, the *sacral tuberosity*, that contains depressions for the attachment of ligaments. The sacral tuberosity is another surface of the sacrum that unites with the hip bones to form the sacroiliac joints. The *superior articular processes* of the sacrum articulate with the fifth lumbar vertebra, and the base of the sacrum articulates with the body of the fifth lumbar vertebra, to form the *lumbosacral joint*.

Coccyx

The **coccyx** is also triangular in shape and is usually formed by the fusion of four coccygeal vertebrae, indicated in Figure 7.22 as Co1–Co4. The coccygeal vertebrae fuse when a person is between 20 and 30 years of age. The dorsal surface of the body of the coccyx contains two long *coccygeal cornua* that are connected by ligaments to the sacral cornua. The coccygeal cornua are the pedicles and superior articular processes of the first coccygeal vertebra. On the lateral surfaces of the coccyx are a series of *transverse processes*, the first pair being the largest. The coccyx articulates superiorly with the apex of the sacrum. In females, the coccyx points inferiorly; in males, it points anteriorly (see Table 8.1 on page 217).

CAUDAL ANESTHESIA

Anesthetic agents that act on the sacral and coccygeal nerves are sometimes injected through the sacral hiatus, a procedure called **caudal anesthesia** or **epidural block.** The procedure is used most often to relieve pain during labor and to provide anesthesia to the perineal area. Because the sacral hiatus is between the sacral cornua, the cornua are important bony landmarks for locating the hiatus. Anesthetic agents also may be injected through the posterior sacral foramina. ■

CHECKPOINT

10. What are the functions of the vertebral column?
11. When do the secondary vertebral curves develop?
12. What are the principal distinguishing characteristics of the bones of the various regions of the vertebral column?

THORAX

OBJECTIVE

● Identify the bones of the thorax.

The term **thorax** refers to the entire chest. The skeletal part of the thorax, the **thoracic cage,** is a bony enclosure formed by the sternum, costal cartilages, ribs, and the bodies of the thoracic vertebrae (Figure 7.23). The thoracic cage is narrower at its superior end and broader at its inferior end and is flattened from front to back. It encloses and protects the organs in the thoracic and superior abdominal cavities and provides support for the bones of the shoulder girdle and upper limbs.

Sternum

The **sternum,** or breastbone, is a flat, narrow bone located in the center of the anterior thoracic wall that measures about 15 cm (6 in.) in length and consists of three parts (Figure 7.23). The superior part is the **manubrium** (ma-NOO-brē-um=handle-like); the middle and largest part is the **body;** and the inferior, smallest part is the **xiphoid process** (ZĪ-foyd=sword-shaped). The segments of the sternum typically fuse by age 25 and the points of fusion are marked by transverse ridges.

The junction of the manubrium and body forms the *sternal angle*. The manubrium has a depression on its superior surface, the *suprasternal notch*. Lateral to the suprasternal notch are *clavicular notches* that articulate with the medial ends of the clavicles to form the *sternoclavicular joints*. The manubrium also articulates with the costal cartilages of the first and second ribs to form the *sternocostal joints*.

The body of the sternum articulates directly or indirectly with the costal cartilages of the second through tenth ribs. The xiphoid process consists of hyaline cartilage during infancy and childhood and does not ossify completely until about age 40. No ribs are attached to it, but the xiphoid process provides attachment for some abdominal muscles. Incorrect positioning of the hands of a rescuer during cardiopulmonary resuscitation (CPR) may fracture the xiphoid process, driving it into internal organs. During thoracic surgery, the sternum may be split along the midline and the halves spread apart to allow surgeons access to structures in the thoracic cavity such as the thymus, heart, and great vessels of the heart. After the surgery, the halves of the sternum are held together with wire sutures.

Ribs

Twelve pairs of **ribs** give structural support to the sides of the thoracic cavity (Figure 7.23b). The ribs increase in length from the first through seventh, then decrease in length to the twelfth rib. Each articulates posteriorly with its corresponding thoracic vertebra.

The first through seventh pairs of ribs have a direct anterior attachment to the sternum by a strip of hyaline cartilage called *costal cartilage* (cost-=rib). The costal cartilages contribute to the elasticity of the thoracic cage and prevent various blows to the chest from fracturing the sternum and/or ribs. The ribs that

Figure 7.23   Skeleton of the thorax.

The bones of the thorax enclose and protect organs in the thoracic cavity and upper abdominal cavity.

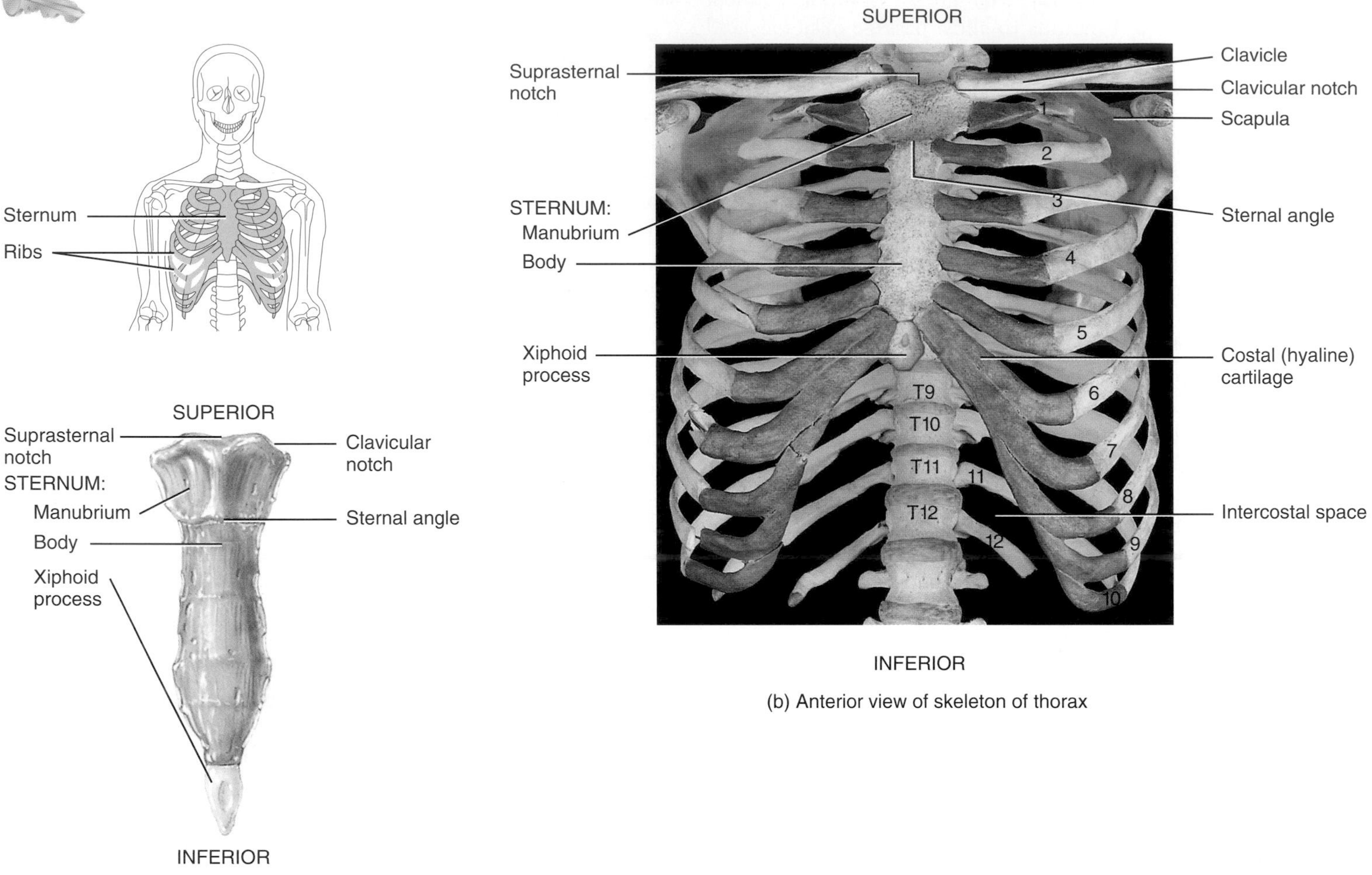

(a) Anterior view of sternum

(b) Anterior view of skeleton of thorax

Which ribs are true ribs, false ribs, and floating ribs?

have costal cartilages and attach directly to the sternum are called *true (vertebrosternal) ribs*. The remaining five pairs of ribs are termed *false ribs* because their costal cartilages either attach indirectly to the sternum or do not attach to the sternum at all. The cartilages of the eighth, ninth, and tenth pairs of ribs attach to one another and then to the cartilages of the seventh pair of ribs. These false ribs are called *vertebrochondral ribs*. The eleventh and twelfth pairs of ribs are false ribs designated as *floating (vertebral) ribs* because the costal cartilage at their anterior ends does not attach to the sternum at all. These ribs attach only posteriorly to the thoracic vertebrae. Inflammation of one or more costal cartilages, called *costochondritis*, is characterized by local tenderness and pain in the anterior chest wall that may radiate. The symptoms mime the chest pain associated with a heart attack (angina pectoris).

Figure 7.24a shows the parts of a typical (third through ninth) rib. The *head* is a projection at the posterior end of the rib. The facet of the head fits into either a facet on the body of one vertebra or into the demifacets of two adjoining vertebrae to form *vertebrocostal joints*. The *neck* is a constricted portion just

lateral to the head. A knoblike structure on the posterior surface where the neck joins the body is called a *tubercle* (TOO-ber-kul). The *nonarticular part* of the tubercle attaches to a ligament (lateral costotransverse ligament) that attaches the transverse process of a vertebra to the nonarticular part of the corresponding rib. The *articular part* of the tubercle articulates with the facet of a transverse process of the inferior of the two vertebrae to which the head of the rib is connected (Figure 7.24c). These articulations also form vertebrocostal joints. The *body (shaft)* is the main part of the rib. A short distance beyond the tubercle, an abrupt change in the curvature of the shaft occurs. This point is called the *costal angle*. The inner surface of the rib has a *costal groove* that protects blood vessels and a small nerve.

In summary, the posterior portion of the rib is connected to a thoracic vertebra by its head and the articular part of a tubercle. The facet of the head fits into a facet on the body of one vertebra or into the demifacets of two adjoining vertebrae. The articular part of the tubercle articulates with the facet of the transverse process of the vertebra.

Figure 7.24   The structure of ribs. Each rib has a head, a neck, and a body. The facets and the articular part of the tubercle of a rib articulate with a thoracic vertebra. (See Tortora, *A Photographic Atlas of the Human Body*, 2e, Figure 3.21.)

Each rib articulates posteriorly with its corresponding thoracic vertebra.

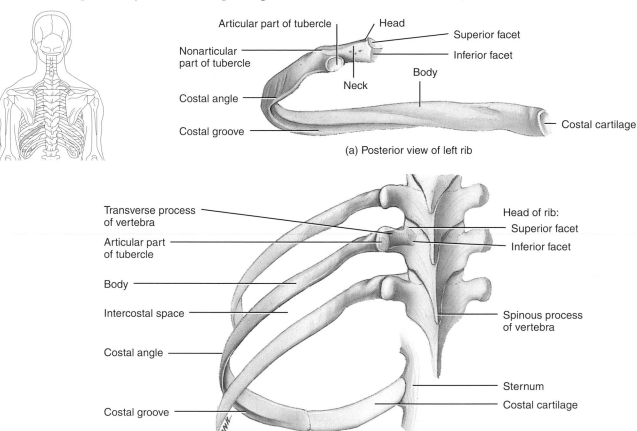

(a) Posterior view of left rib

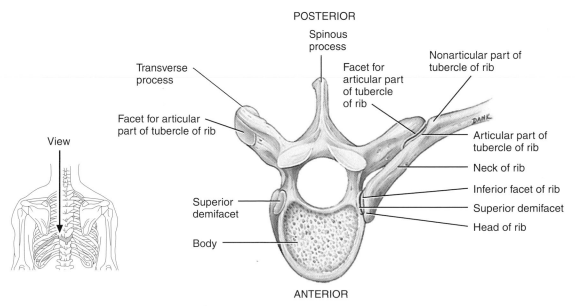

(b) Posterior view of left ribs articulated with thoracic vertebrae and sternum

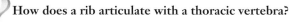

(c) Superior view of left rib articulated with thoracic vertebra

How does a rib articulate with a thoracic vertebra?

If you examine Figure 7.23b, you will notice that the first rib is the shortest, broadest, and most sharply curved. The first rib is an important landmark because of its close relationship to the nerves of the *brachial plexus* (the entire nerve supply of the shoulder and upper limb), two major blood vessels, the subclavian artery and vein, and two skeletal muscles, the anterior and medial scalene muscles. The superior surface of the first rib has two shallow grooves, one for the subclavian vein and one for the subclavian artery and inferior trunk of the brachial plexus. The second rib is thinner, less curved, and considerably longer than the first. Unlike the paired facets of the typical ribs, the tenth rib has a single articular facet on its head. The eleventh and twelfth ribs also have single articular facets on their heads, but no necks, tubercles, or costal angles.

Spaces between ribs, called *intercostal spaces*, are occupied by intercostal muscles, blood vessels, and nerves. Surgical access to the lungs or other structures in the thoracic cavity is commonly obtained through an intercostal space. Special rib retractors are used to create a wide separation between ribs. The costal cartilages are sufficiently elastic in younger individuals to permit considerable bending without breaking.

Structures passing between the thoracic cavity and the neck pass through an opening called the **superior thoracic aperture.** Among these structures are the trachea, esophagus, nerves, and blood vessels that supply and drain the head, neck, and upper limbs. The aperture is bordered by the first thoracic vertebra (posteriorly), the first pair of ribs and their cartilages, and the superior border of the manubrium of the sternum. Structures passing between the thoracic cavity and abdominal cavity pass through the **inferior thoracic aperture.** Through this large opening, which is closed by the diaphragm, pass structures such as the esophagus, nerves, and large blood vessels. This aperture is bordered by the twelfth thoracic vertebra (posteriorly), the eleventh and twelfth pair of ribs, the costal cartilages of ribs 7 through 10, and the joint between the body and xiphoid process of the sternum (anteriorly).

RIB FRACTURES, DISLOCATIONS, AND SEPARATIONS

Rib fractures are the most common chest injuries, and they usually result from direct blows, most often from impact with a steering wheel, falls, and crushing injuries to the chest. Ribs tend to break at the point where the greatest force is applied, but they may also break at their weakest point—the site of greatest curvature, which is just anterior to the costal angle. The middle ribs are the most commonly fractured. In some cases, fractured ribs may puncture the heart, great vessels of the heart, lungs, trachea, bronchi, esophagus, spleen, liver, and kidneys. Rib fractures are usually quite painful.

Dislocated ribs, which are common in body contact sports, involve displacement of a costal cartilage from the sternum, with resulting pain, especially during deep inhalations.

Separated ribs involve displacement of a rib and its costal cartilage; as a result, a rib may move superiorly, overriding the rib above and causing severe pain. ■

CHECKPOINT

13. What bones form the skeleton of the thorax?
14. What are the functions of the bones of the thorax?
15. How are ribs classified?

APPLICATIONS TO HEALTH

HERNIATED (SLIPPED) DISC

In their function as shock absorbers, intervertebral discs are constantly being compressed. If the anterior and posterior ligaments of the discs become injured or weakened, the pressure developed in the nucleus pulposus may be great enough to rupture the surrounding fibrocartilage (annulus fibrosus). If this occurs, the nucleus pulposus may herniate (protrude) posteriorly or into one of the adjacent vertebral bodies (Figure 7.25). This condition is called a **herniated (slipped) disc.** Because the lumbar region bears much of the weight of the body, and is the region of the most flexing and bending, herniated discs most often occur in the lumbar area.

Frequently, the nucleus pulposus slips posteriorly toward the spinal cord and spinal nerves. This movement exerts pressure on the spinal nerves, causing acute pain. If the roots of the sciatic nerve, which passes from the spinal cord to the foot, are compressed, the pain radiates down the posterior thigh, through the calf, and occasionally into the foot. If pressure is exerted on the spinal cord itself, some of its neurons may be destroyed. Treatment options include bed rest, medications for pain, physical therapy and exercises, and traction. A person with a herniated disc

Figure 7.25   Herniated (slipped) disc.

Most often the nucleus pulposus herniates posteriorly.

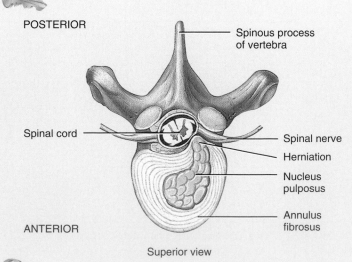

POSTERIOR

Spinous process of vertebra

Spinal cord

Spinal nerve

Herniation

Nucleus pulposus

ANTERIOR

Annulus fibrosus

Superior view

Why do most herniated discs occur in the lumbar region?

may also undergo a *laminectomy*, a procedure in which parts of the laminae of the vertebra and intervertebral disc are removed to relieve pressure on the nerves.

ABNORMAL CURVES OF THE VERTEBRAL COLUMN

Various conditions may exaggerate the normal curves of the vertebral column, or the column may acquire a lateral bend, resulting in **abnormal curves** of the vertebral column.

Scoliosis (skō-lē-Ō-sis; *scolio-*=crooked) is a lateral bending of the vertebral column, usually in the thoracic region (Figure 7.26a). This is the most common of the abnormal curves. It may result from *congenitally* (present at birth) malformed vertebrae, chronic sciatica, paralysis of muscles on one side of the vertebral column, poor posture, or one leg being shorter than the other.

Kyphosis (kī-FŌ-sis; *kyphos-*=hump) is an exaggeration of the thoracic curve of the vertebral column (Figure 7.26b). In tuberculosis of the spine, vertebral bodies may partially collapse, causing an acute angular bending of the vertebral column. In the elderly, degeneration of the intervertebral discs leads to kyphosis. Kyphosis may also be caused by rickets and poor posture. It is also common in females with advanced osteoporosis. The term *round-shouldered* is an expression for mild kyphosis.

Lordosis (lor-DŌ-sis; *lord-*=bent backward), sometimes called *hollow back*, is an exaggeration of the lumbar curve of the vertebral column (Figure 7.26c). It may result from increased weight of the abdomen as in pregnancy, or extreme obesity, poor posture, rickets, osteoporosis, or tuberculosis of the spine.

Figure 7.26 Abnormal curves of the vertebral column.

An abnormal curve is the result of an exaggerated normal curve.

Scoliosis   Kyphosis   Lordosis

Which abnormal curve is common in individuals with advanced osteoporosis?

SPINA BIFIDA

Spina bifida (SPĪ-na BIF-i-da) is a congenital defect of the vertebral column in which laminae of L5 and/or S1 fail to develop normally and unite at the midline. The least serious form is called *spina bifida occulta*. It occurs in L5 or S1 and produces no symptoms. The only evidence of its presence is a small dimple with a tuft of hair in the overlying skin. Several types of spina bifida involve pro-

trusion of the meninges (membranes) and/or spinal cord through the defect in the laminae and are collectively termed *spina bifida cystica* because of the presence of a cystlike sac protruding from the backbone (Figure 7.27). If the sac contains the meninges from the spinal cord and cerebrospinal fluid, the condition is called *spina bifida with meningocele* (me-NING-gō-sēl). If the spinal cord and/or its nerve roots are in the sac, the condition is called *spina bifida with meningomyelocele* (me-ning-gō-MĪ-e-lō-sēl). The larger the cyst and the number of neural structures it contains, the more serious the neurological problems. In severe cases, there may be partial or complete paralysis, partial or complete loss of urinary bladder and bowel control, and the absence of reflexes. An increased risk of spina bifida is associated with low levels of a B vitamin called folic acid during pregnancy. Spina bifida may be diagnosed prenatally by a test of the mother's blood for a substance produced by the fetus called alphafetoprotein, by sonography or by amniocentesis (withdrawal of amniotic fluid for analysis) (see page 117).

Figure 7.27 Spina bifida.

Spina bifida is caused by a failure of laminae to unite at the midline.

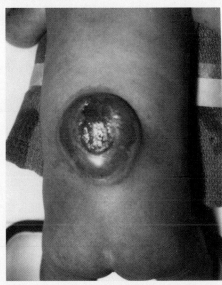

Deficiency of which B vitamin is linked to spina bifida?

FRACTURES OF THE VERTEBRAL COLUMN

Fractures of the vertebral column often involve C1, C2, C4–T7, and T12–L2. Cervical or lumbar fractures usually result from a flexion-compression type of injury such as might be sustained in landing on the feet or buttocks after a fall or having a weight fall on the shoulders. Cervical vertebrae may be fractured or dislodged by a fall on the head with acute flexion of the neck, as might happen on diving into shallow water or being thrown from a horse. Dislocation may result from the sudden forward-then-backward jerk ("whiplash") that may occur in an automobile crash. Spinal cord or spinal nerve damage may occur as a result of fractures of the vertebral column.

KEY MEDICAL TERMS ASSOCIATED WITH THE AXIAL SKELETON

Craniostenosis (krā-nē-ō-sten-Ō-sis; *cranio-*=skull; *-stenosis*=narrowing) Premature closure of one or more cranial sutures during the first 18 to 20 months of life, resulting in a distorted skull. Premature closure of the sagittal suture produces a long narrow skull, whereas premature closure of the coronal suture results in a broad skull. Premature closure of all sutures restricts brain growth and development; surgery is necessary to prevent brain damage.

Craniotomy (krā-nē-OT-ō-mē; *cranio-*=skull; *-tone*=cutting) Surgical procedure in which part of the cranium is removed. It may be performed to remove a blood clot, a brain tumor, or a sample of brain tissue for biopsy.

Laminectomy (lam′-i-NEK-tō-mē; *lamina-*=layer) Surgical procedure to remove a vertebral lamina. It may be performed to access the vertebral canal and relieve the symptoms of a herniated disc.

Lumbar spine stenosis (*sten-*=narrowed) Narrowing of the spinal canal in the lumbar part of the vertebral column, due to hypertrophy of surrounding bone or soft tissues. It may be caused by arthritic changes in the intervertebral discs and is a common cause of back and leg pain.

Spinal fusion (FŪ-zhun) Surgical procedure in which two or more vertebrae of the vertebral column are stabilized with a bone graft or synthetic device. It may be performed to treat a fracture of a vertebra or following removal of a herniated disc.

Whiplash injury Injury to the neck region due to severe hyperextension (backward tilting) of the head followed by severe hyperflexion (forward tilting) of the head, usually associated with a rear-end automobile collision. Symptoms are related to stretching and tearing of ligaments and muscles, vertebral fractures, and herniated vertebral discs.

STUDY OUTLINE

Introduction (p. 163)

1. Bones protect soft body parts and make movement possible; they also serve as landmarks for locating parts of other body systems.
2. The musculoskeletal system is composed of the bones, joints, and muscles working together.

Divisions of the Skeletal System (p. 164)

1. The axial skeleton consists of bones arranged along the longitudinal axis. The parts of the axial skeleton are the skull, auditory ossicles (ear bones), hyoid bone, vertebral column, sternum, and ribs.
2. The appendicular skeleton consists of the bones of the girdles and the upper and lower limbs (extremities). The parts of the appendicular skeleton are the pectoral (shoulder) girdles, bones of the upper limbs, pelvic (hip) girdles, and bones of the lower limbs.

Types of Bones (p. 164)

1. On the basis of shape, bones are classified as long, short, flat, irregular, or sesamoid. Sesamoid bones develop in tendons or ligaments.
2. Sutural bones are found within the sutures of certain cranial bones.

Bone Surface Markings (p. 167)

1. Surface markings are structural features visible on the surfaces of bones.
2. Each marking—whether a depression, an opening, or a process—is structured for a specific function, such as joint formation, muscle attachment, or passage of nerves and blood vessels (see Table 7.2 on page 167).

Skull (p. 168)

1. The 22 bones of the skull include cranial bones and facial bones.
2. The eight cranial bones include the frontal, parietal (2), temporal (2), occipital, sphenoid, and ethmoid.

3. The 14 facial bones are the nasal (2), maxillae (2), zygomatic (2), lacrimal (2), palatine (2), inferior nasal conchae (2), vomer, and mandible.
4. The nasal septum consists of the vomer, perpendicular plate of the ethmoid, and septal cartilage. The nasal septum divides the nasal cavity into left and right sides.
5. Seven skull bones form each of the orbits (eye sockets).
6. The foramina of the skull bones provide passages for nerves and blood vessels (Table 7.4 on page 183).
7. Sutures are immovable joints that connect most bones of the skull. Examples are the coronal, sagittal, lambdoid, and squamous sutures.
8. Paranasal sinuses are cavities in bones of the skull lined with mucous membranes that communicate with the nasal cavity. The frontal, sphenoid, and ethmoid bones and the maxillae contain paranasal sinuses.
9. Fontanels are mesenchyme-filled spaces between the cranial bones of fetuses and infants. The major fontanels are the anterior, posterior, anterolaterals (2), and posterolaterals (2). After birth, the fontanels fill in with bone and become sutures.
10. The three cranial fossae contain depressions for brain convolutions, grooves for cranial blood vessels, and numerous foramina.

Hyoid Bone (p. 186)

1. The hyoid bone is a U-shaped bone that does not articulate with any other bone.
2. It supports the tongue and provides attachment for some tongue muscles and for some muscles of the throat and neck.

Vertebral Column (p. 186)

1. The vertebral column, sternum, and ribs constitute the skeleton of the body's trunk.
2. The 26 bones of the adult vertebral column are the cervical vertebrae

(7), the thoracic vertebrae (12), the lumbar vertebrae (5), the sacrum (5 fused vertebrae), and the coccyx (usually 4 fused vertebrae).

3. The adult vertebral column contains four normal curves (cervical, thoracic, lumbar, and sacral) that give strength, support, and balance.

4. The vertebrae are similar in structure, each usually consisting of a body, vertebral arch, and seven processes. Vertebrae in the different regions of the column vary in size, shape, and detail.

Thorax (p. 195)

1. The thoracic skeleton consists of the sternum, ribs and costal cartilages, and thoracic vertebrae.

2. The thoracic cage protects vital organs in the chest area and upper abdomen.

Q SELF-QUIZ QUESTIONS

Choose the one best answer to the following questions.

1. A foramen is:
   a. a cavity within a bone. b. a depression.
   c. a hole for blood vessels and nerves. d. a ridge.
   e. a site for muscle attachment.

2. Normal skulls have two of each of the following *except*:
   a. parietal bones. b. nasal bones. c. temporal bones.
   d. sphenoid bones. e. maxillae.

3. Tears pass into the nasal cavity through a tunnel formed in part by the:
   a. nasal bone. b. lacrimal bone. c. vomer.
   d. Both a and b are correct. e. a, b, and c are all correct.

4. Which of the following bones is correctly matched with the process?
   a. zygomatic bone – temporal process
   b. maxilla – pterygoid process
   c. vomer – mandibular process
   d. temporal – occipital process
   e. sphenoid – coronoid process

5. All of the following articulate with the maxilla except the:
   a. nasal bone. b. frontal bone. c. palatine bone.
   d. mandible. e. zygomatic bone.

6. Which of the following is *not* a component of the orbit?
   a. temporal. b. ethmoid. c. zygomatic. d. maxilla.
   e. lacrimal.

7. The nasal septum is formed by parts of the:
   a. ethmoid bone b. nasal bone c. vomer
   d. palatine bone e. Both a and c are correct.

8. Which of the following does *not* contain a paranasal sinus?
   a. zygomatic b. sphenoid c. ethmoid d. frontal
   e. maxilla

9. The skeleton of the thorax:
   a. is formed by 12 pairs of ribs and costal cartilages, the sternum, and 12 thoracic vertebrae.
   b. protects the internal chest organs, as well as the liver.
   c. is narrower at its superior end.
   d. aids in supporting the bones of the shoulder girdle.
   e. is described by all of the above.

10. The meninges (coverings) of the brain attach to the crista galli, which is part of the:
    a. frontal bone. b. temporal bone. c. sphenoid bone.
    d. parietal bone. e. ethmoid bone.

Complete the following.

11. The first cervical vertebra is the _____, and the second cervical vertebra is the _____.

12. According to shape classification, phalanges are _____ bones, carpals are _____ bones, and ribs are _____ bones.

13. Bones located in sutures are called sutural or _____ bones.

14. The only bone of the axial skeleton that does not articulate with any other bone is the _____.

Are the following statements true or false?

15. The opening in the occipital bone through which the medulla oblongata connects with the spinal cord is the foramen magnum.

16. The sagittal suture joins the parietal bones.

17. Sesamoid bones protect tendons from wear and tear and may improve the mechanical advantage at a joint by altering the direction of pull of a tendon.

18. The sacral promontory projects anteriorly from the apex of the sacrum.

19. The tubercle of a rib articulates with demifacets on the bodies of adjacent vertebrae.

Matching

20. Match the following (answers may be used more than once):

    ___ (a) mandibular fossa   (1) zygomatic
    ___ (b) optic foramen   (2) frontal
    ___ (c) superior nasal concha   (3) axis
    ___ (d) superior and inferior nuchal lines (4) ethmoid
    ___ (e) horizontal plate   (5) maxilla
    ___ (f) pterygoid process   (6) temporal
    ___ (g) coronoid process   (7) sternum
    ___ (h) mastoid process   (8) occipital
    ___ (i) temporal process   (9) palatine
    ___ (j) supraorbital margin   (10) mandible
    ___ (k) olfactory foramina   (11) sphenoid
    ___ (l) sella turcica
    ___ (m) mental foramen
    ___ (n) external auditory meatus
    ___ (o) perpendicular plate
    ___ (p) palatine process
    ___ (q) xiphoid process
    ___ (r) odontoid process

CRITICAL THINKING QUESTIONS

1. While investigating her new baby brother, 4-year-old Pattee finds a soft spot on the baby's skull and announces that the baby needs to go back because "it's not finished yet." Explain the presence of soft spots in the infant.

2. Dave has a tumor on his pituitary gland that requires surgery. The operation will require exceptional surgical skill both to reach the gland and to avoid serious, possibly fatal, complications. Describe the location of the pituitary gland and the associated skeletal (and other) anatomical features that make such an operation so delicate.

3. The ad reads "New Postureperfect mattress! Keeps spine perfectly straight—just like when you were born! A straight spine means a great sleep!" Would you buy a mattress from this company? Explain.

4. Trina slipped on the ice, fell backwards, and hit her head on the ice. She's okay now, but her doctor said she could have died. Why might a serious blow to the occipital bone be fatal?

5. John had complaints of severe headaches and a yellow-green nasal discharge. His physician ordered x-rays of his head. John was surprised to see several fuzzy-looking holes on the x-ray. Why does John have holes in his head?

ANSWERS TO FIGURE QUESTIONS

7.1 The skull and vertebral column are part of the axial skeleton. The clavicle, shoulder girdle, humerus, pelvic girdle, and femur are part of the appendicular skeleton.

7.2 Flat bones protect and provide a large surface area for muscle attachment.

7.3 The frontal, parietal, sphenoid, ethmoid, and temporal bones are cranial bones.

7.4 The parietal and temporal bones are on either side of the squamous suture. The parietal and occipital bones are on either side of the lambdoid suture. The parietal and frontal bones surround the coronal suture.

7.5 The temporal bone articulates with the parietal, sphenoid, zygomatic, and occipital bones.

7.6 The parietal bones form the posterior, lateral portion of the cranium.

7.7 The medulla oblongata of the brain connects with the spinal cord in the foramen magnum.

7.8 From the crista galli of the ethmoid bone, the sphenoid articulates with the frontal, parietal, temporal, occipital, temporal, parietal, and frontal bones, ending again at the crista galli of the ethmoid bone.

7.9 The perpendicular plate of the ethmoid bone forms the superior part of the nasal septum, and the lateral masses compose most of the medial walls of the orbits.

7.10 The mandible is the only movable skull bone, other than the auditory ossicles.

7.11 The nasal septum divides the nasal cavity into right and left sides.

7.12 Bones forming the orbit are the frontal, sphenoid, zygomatic, maxilla, lacrimal, ethmoid, and palatine.

7.13 The paranasal sinuses produce mucus and serve as resonating chambers for vocalization.

7.14 The anterolateral fontanel is bordered by four different skull bones.

7.15 The posterior cranial fossa is the largest.

7.16 The hyoid bone does not articulate with any other bone.

7.17 The thoracic and sacral curves of the vertebral column are concave relative to the anterior of the body.

7.18 The vertebral foramina enclose the spinal cord, whereas the intervertebral foramina provide spaces for spinal nerves to exit the vertebral column.

7.19 The atlas moving on the axis permits movement of the head to signify "no."

7.20 The facets and demifacets on the bodies of the thoracic vertebrae articulate with the heads of the ribs, and the facets on the transverse processes of these vertebrae articulate with the tubercles of the ribs.

7.21 The lumbar vertebrae are the largest and strongest in the body because the amount of weight supported by vertebrae increases toward the inferior end of the vertebral column.

7.22 There are four pairs of sacral foramina, for a total of eight. Each anterior sacral foramen joins a posterior sacral foramen at the intervertebral foramen. Nerves and blood vessels pass through these tunnels in the bone.

7.23 Pairs 1–7 are the true ribs, pairs 8–12 are the false ribs, and pairs 11 and 12 are the floating ribs.

7.24 The facet on the head of a rib fits into a facet on the body of a vertebra, and the articular part of the tubercle of a rib articulates with the facet of the transverse process of a vertebra.

7.25 Most herniated discs occur in the lumbar region because it bears most of the body weight and most flexing and bending occurs there.

7.26 Kyphosis is common in individuals with advanced osteoporosis.

7.27 Deficiency of folic acid is associated with spina bifida.

THE SKELETAL SYSTEM: THE APPENDICULAR SKELETON

8

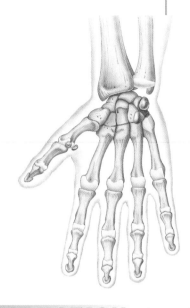

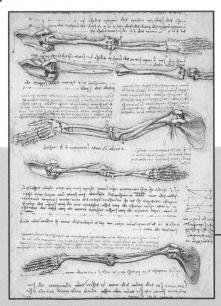

INTRODUCTION As noted in Chapter 7, the two main divisions of the skeletal system are the axial skeleton and the appendicular skeleton. Whereas the general function of the axial skeleton is the protection of internal organs, the primary function of the appendicular skeleton, the focus of this chapter, is movement. The appendicular skeleton includes the bones that make up the upper and lower limbs as well as the bones, arranged in formations called girdles, that attach the limbs to the axial skeleton. As you progress through this chapter, you will see how the bones of the appendicular skeleton are connected with one another and with skeletal muscles, making possible a wide array of movements. This arrangement permits us to walk, write, use a computer, dance, swim, and play a musical instrument.

Most of us know Leonardo da Vinci for his fine art masterpieces such as the Mona Lisa and The Last Supper. Did you know that Leonardo was also an accomplished anatomist? Have you ever seen any of his anatomical renderings?

www.wiley.com/college/apcentral

PECTORAL (SHOULDER) GIRDLE

OBJECTIVE

• Identify the bones of the pectoral (shoulder) girdle and their principal markings.

The **pectoral** (PEK-tō-ral) or **shoulder girdles** attach the bones of the upper limbs to the axial skeleton (Figure 8.1). Each of the two pectoral girdles consists of a clavicle and a scapula. The clavicle is the anterior bone and articulates with the manubrium of the sternum at the *sternoclavicular joint*. The scapula is the posterior bone and articulates with the clavicle at the *acromioclavicular joint* and with the humerus at the *glenohumeral (shoulder) joint*. The pectoral girdles do not articulate with the vertebral column and are held in position instead by complex muscle attachments.

Clavicle

Each slender, S-shaped **clavicle** (KLAV-i-kul=key), or *collarbone*, lies horizontally across the anterior part of the thorax superior to the first rib (Figure 8.2). The medial half of the clavicle is convex anteriorly, whereas the lateral half is concave anteriorly. The medial end of the clavicle, called the *sternal end*, is rounded and articulates with the manubrium of the sternum to form the *sternoclavicular joint*. The broad, flat, lateral end, the *acromial end* (a-KRŌ-mē-al), articulates with the acromion of the scapula. This joint is called the *acromioclavicular joint* (Figure 8.1). The *conoid tubercle* (KŌ-noyd=conelike) on the inferior surface of the lateral end of the bone is a point of attachment for the conoid ligament. As its name implies, the *impression for the costoclavicular ligament* on the inferior surface of the sternal end is a point of attachment for the costoclavicular ligament.

FRACTURED CLAVICLE

The clavicle transmits mechanical force from the upper limb to the trunk. If the force transmitted to the clavicle is excessive, as in falling on one's outstretched arm, a **fractured clavicle** may result. The clavicle is one of the most frequently broken bones in the body. Because the junction of the clavicle's two curves is its weakest point, the clavicular midregion is the most frequent fracture site. Even in the absence of fracture, compression of the clavicle as a result of automobile accidents involving the use of shoulder harness seatbelts often causes damage to the median nerve, which lies between the clavicle and the second rib. A fractured clavicle is usually treated with a regular sling to keep the arm from moving outward. ■

Figure 8.1 Right pectoral (shoulder) girdle. (See Tortora, *A Photographic Atlas of the Human Body*, 2e, *Figure 3.1.*)

The clavicle is the anterior bone of the pectoral girdle and the scapula is the posterior bone.

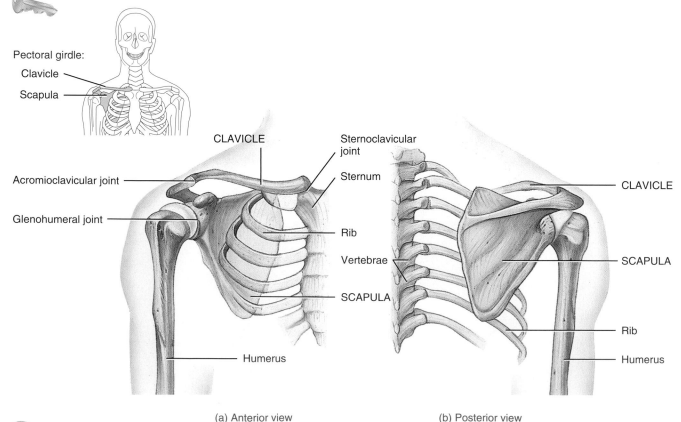

Pectoral girdle:
Clavicle
Scapula

CLAVICLE
Sternoclavicular joint
Sternum
Acromioclavicular joint
Glenohumeral joint
Rib
Vertebrae
SCAPULA
Humerus

CLAVICLE
SCAPULA
Rib
Humerus

(a) Anterior view

(b) Posterior view

What is the function of the pectoral girdles?

Figure 8.2   Right clavicle.

The clavicle articulates medially with the manubrium of the sternum and laterally with the acromion of the scapula.

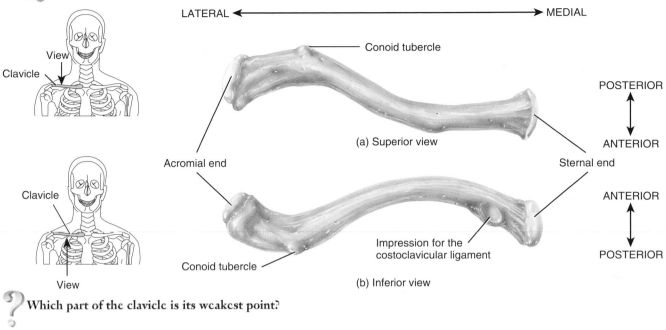

LATERAL ⟵   ⟶ MEDIAL

Conoid tubercle

POSTERIOR

ANTERIOR

(a) Superior view

Acromial end

Sternal end

ANTERIOR

POSTERIOR

Impression for the costoclavicular ligament

Conoid tubercle

(b) Inferior view

Which part of the clavicle is its weakest point?

Scapula

Each **scapula** (SCAP-ū-la; plural is *scapulae*), or *shoulder blade*, is a large, triangular, flat bone situated in the superior part of the posterior thorax between the levels of the second and seventh ribs (Figure 8.3). A prominent ridge called the *spine* runs diagonally across the posterior surface of the flattened, triangular *body*

Figure 8.3   Right scapula (shoulder blade). (See Tortora, *A Photographic Atlas of the Human Body*, 2e, Figure 3.22.)

The glenoid cavity of the scapula articulates with the head of the humerus to form the shoulder joint.

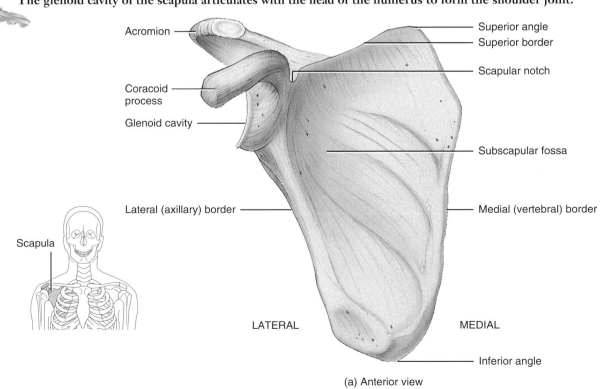

Acromion

Coracoid process

Glenoid cavity

Lateral (axillary) border

Scapula

LATERAL

Superior angle
Superior border

Scapular notch

Subscapular fossa

Medial (vertebral) border

MEDIAL

Inferior angle

(a) Anterior view

continues

Figure 8.3 (continued)

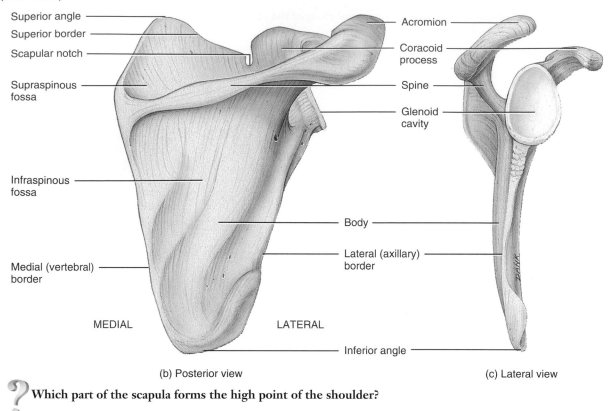

(b) Posterior view

(c) Lateral view

? **Which part of the scapula forms the high point of the shoulder?**

of the scapula (Figure 8.3b). The lateral end of the spine projects as a flattened, expanded process called the *acromion* (a-KRŌ-mē-on; *acrom-*=topmost), easily felt as the high point of the shoulder. Tailors measure the length of the upper limb from the acromion. The acromion articulates with the acromial end of the clavicle to form the *acromioclavicular joint.* Inferior to the acromion is a shallow depression, the *glenoid cavity,* that accepts the head of the humerus (arm bone) to form the *glenohumeral joint* (see Figure 8.1).

The thin edge of the scapula closer to the vertebral column is called the *medial (vertebral) border.* The medial borders of the scapulae lie about 5 cm (2 in.) from the vertebral column. The thick edge of the scapula closer to the arm is called the *lateral (axillary) border.* The medial and lateral borders join at the *inferior angle.* The superior edge of the scapula, called the *superior border,* joins the vertebral border at the *superior angle.* The *scapular notch* is a prominent indentation along the superior border through which the suprascapular nerve passes.

At the lateral end of the superior border of the scapula is a projection of the anterior surface called the *coracoid process* (KOR-a-koyd=like a crow's beak), to which the tendons of muscles attach. Superior and inferior to the spine are two fossae: the *supraspinous fossa* (sū-pra-SPĪ-nus) and the *infraspinous fossa*

(in-fra-SPĪ-nus), respectively. Both serve as surfaces of attachment for the tendons of the supraspinatus and infraspinatus muscles of the shoulder. On the anterior surface is a slightly hollowed-out area called the *subscapular fossa,* also a surface of attachment for the tendons of shoulder muscles.

CHECKPOINT

1. Which bones or parts of bones of the pectoral girdle form the sternoclavicular, acromioclavicular, and glenohumeral joints?

UPPER LIMB (EXTREMITY)

OBJECTIVE

● Identify the bones of the upper limb and their principal markings.

Each **upper limb (extremity)** has 30 bones in three locations— (1) the humerus in the arm; (2) the ulna and radius in the forearm; and (3) the 8 carpals in the carpus (wrist), the 5 metacarpals in the metacarpus (palm), and the 14 phalanges (bones of the digits) in the hand (Figure 8.4).

Figure 8.4   Right upper limb.

🔑 **Each upper limb consists of a humerus, ulna, radius, carpals, metacarpals, and phalanges.**

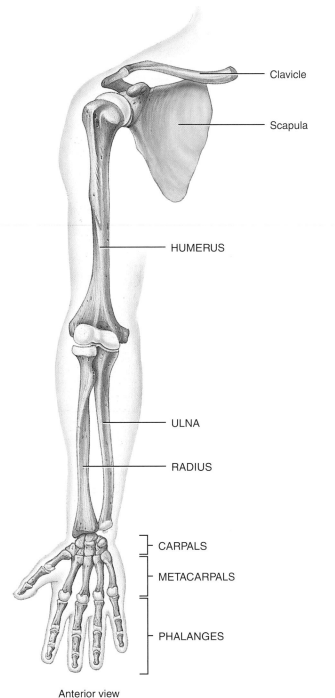

Clavicle

Scapula

HUMERUS

ULNA

RADIUS

CARPALS

METACARPALS

PHALANGES

Anterior view

❓ **How many bones make up each upper limb?**

Humerus

The **humerus** (HŪ-mer-us), or arm bone, is the longest and largest bone of the upper limb (Figure 8.5). It articulates proximally with the scapula and distally at the elbow with both the ulna and the radius.

The proximal end of the humerus features a rounded *head* that articulates with the glenoid cavity of the scapula to form the *glenohumeral joint*. Distal to the head is the *anatomical neck*, which is visible as an oblique groove. The *greater tubercle* is a lateral projection distal to the anatomical neck. It is the most laterally palpable bony landmark of the shoulder region. The *lesser tubercle* projects anteriorly. Between both tubercles runs an *intertubercular sulcus*. The *surgical neck* is a constriction in the humerus just distal to the tubercles, where the head tapers to the shaft; it is so named because fractures often occur here.

The *body (shaft)* of the humerus is roughly cylindrical at its proximal end, but it gradually becomes triangular until it is flattened and broad at its distal end. Laterally, at the middle portion of the shaft, there is a roughened, V-shaped area called the *deltoid tuberosity*. This area serves as a point of attachment for the tendons of the deltoid muscle.

Several prominent features are evident at the distal end of the humerus. The *capitulum* (ka-PIT-ū-lum; *capit-*=head) is a rounded knob on the lateral aspect of the bone that articulates with the head of the radius. The *radial fossa* is an anterior depression that receives the head of the radius when the forearm is flexed (bent). The *trochlea* (TRŌK-lē-a), located medial to the capitulum, is a spool-shaped surface that articulates with the ulna. The *coronoid fossa* (KOR-ō-noyd=crown-shaped) is an anterior depression that receives the coronoid process of the ulna when the forearm is flexed. The *olecranon fossa* (ō-LEK-ra-non=elbow) is a posterior depression that receives the olecranon of the ulna when the forearm is extended (straightened). The *medial epicondyle* and *lateral epicondyle* are rough projections on either side of the distal end of the humerus to which the tendons of most muscles of the forearm are attached. The ulnar nerve may be easily palpated by rolling a finger over the skin surface above the posterior surface of the medial epicondyle.

Ulna and Radius

The **ulna** is located on the medial aspect (the little-finger side) of the forearm and is longer than the radius (Figure 8.6 on page 209). It is sometimes convenient to use an aid to help remember information that may be unfamiliar. Such an aid is called a *mnemonic device* (nē-MON-ik=memory). One such mnemonic to help you remember the location of the ulna in relation to the hand is "p.u." (the *p*inky is on the *u*lna side). At the proximal end of the ulna (Figure 8.6b) is the *olecranon*, which forms the prominence of the elbow. The *coronoid process* (Figure 8.6a) is an anterior projection that, together with the olecranon, receives the trochlea of the humerus. The *trochlear notch* is a large curved area between the olecranon and coronoid process that forms

Figure 8.5 **Right humerus in relation to the scapula, ulna, and radius.** (See Tortora, *A Photographic Atlas of the Human Body*, 2e, Figure 3.23.)

The humerus is the longest and largest bone of the upper limb.

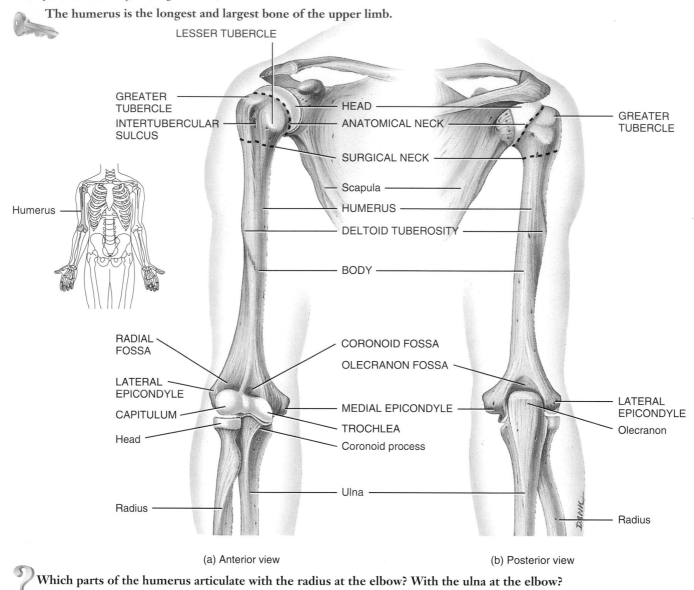

(a) Anterior view

(b) Posterior view

Which parts of the humerus articulate with the radius at the elbow? With the ulna at the elbow?

part of the elbow joint (see Figure 8.7b). Just inferior to the coronoid process is the *ulnar tuberosity*. The distal end of the ulna consists of a *head* that is separated from the wrist by a disc of fibrocartilage. A *styloid process* is located on the posterior side of the ulna's distal end.

The **radius** is located on the lateral aspect (thumb side) of the forearm (see Figure 8.6). The proximal end of the radius has a disc-shaped *head* that articulates with the capitulum of the humerus and the radial notch of the ulna. Inferior to the head is the constricted *neck*. A roughened area inferior to the neck on the medial side, called the *radial tuberosity*, is a point of attachment for the tendons of the biceps brachii muscle. The shaft of the radius widens distally to form a *styloid process* on the lateral side. Fracture of the distal end of the radius is the most common fracture in adults older than 50 years.

The ulna and radius articulate with the humerus at the *elbow joint*. The articulation occurs in two places: where the head of the radius articulates with the capitulum of the humerus (Figure 8.7a on page 210), and where the trochlear notch of the ulna receives the trochlea of the humerus (Figure 8.7b).

The ulna and the radius connect with one another at three sites. First, a broad, flat, fibrous connective tissue called the *interosseous membrane* (in′-ter-OS-ē-us; *inter-*=between, *osse-*= bone) joins the shafts of the two bones. This membrane also provides a site of attachment for some tendons of deep skeletal muscles of the forearm. The ulna and radius articulate at their proximal and distal ends. Proximally, the head of the radius articulates with the ulna's *radial notch*, a depression that is lateral and inferior to the trochlear notch (Figure 8.7b). This articulation is the *proximal radioulnar joint*. Distally, the head of the ulna

Figure 8.6 **Right ulna and radius in relation to the humerus and carpals.** (See Tortora, *A Photographic Atlas of the Human Body*, 2e, Figure 3.24.)

In the forearm, the longer ulna is on the medial side, whereas the shorter radius is on the lateral side.

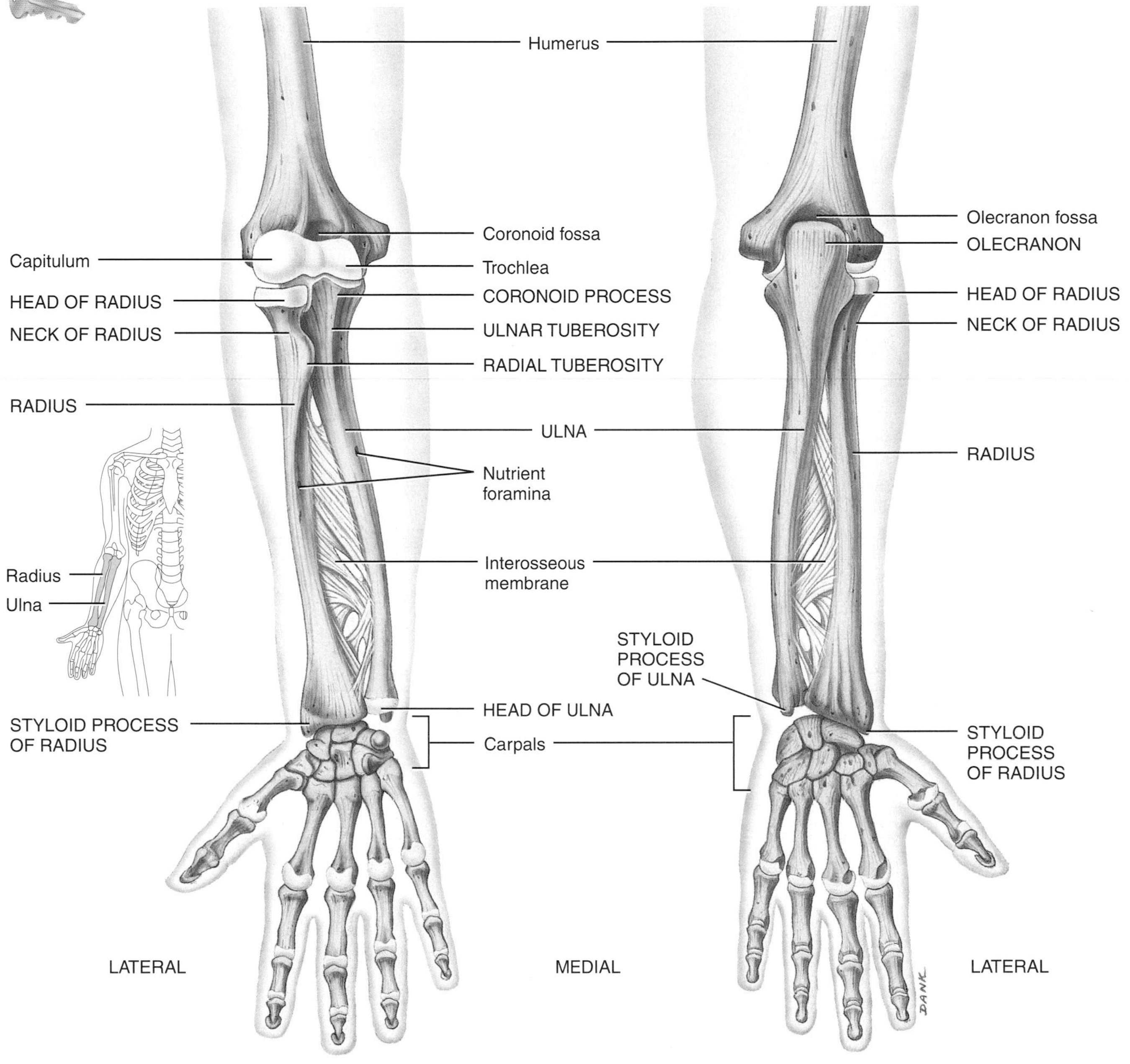

Humerus

Coronoid fossa

Capitulum

Trochlea

HEAD OF RADIUS

CORONOID PROCESS

NECK OF RADIUS

ULNAR TUBEROSITY

RADIAL TUBEROSITY

RADIUS

ULNA

Nutrient foramina

Radius

Ulna

Interosseous membrane

STYLOID PROCESS OF RADIUS

HEAD OF ULNA

Carpals

Olecranon fossa

OLECRANON

HEAD OF RADIUS

NECK OF RADIUS

RADIUS

STYLOID PROCESS OF ULNA

STYLOID PROCESS OF RADIUS

LATERAL

MEDIAL

LATERAL

(a) Anterior view

(b) Posterior view

What part of the ulna is called the "elbow"?

articulates with the *ulnar notch* of the radius (Figure 8.7c). This articulation is the *distal radioulnar joint*. Finally, the distal end of the radius articulates with three bones of the wrist—the lunate, the scaphoid, and the triquetrum—to form the *radiocarpal (wrist) joint*.

Carpals, Metacarpals, and Phalanges

The **carpus** (wrist) is the proximal region of the hand and consists of eight small bones, the **carpals**, joined to one another by ligaments (Figure 8.8). Articulations between carpal bones are called *intercarpal joints*. The carpals are arranged in two transverse rows of four bones each. Their names reflect their shapes. The carpals in the proximal row, from lateral to medial, are the **scaphoid** (SKAF-oid=boatlike), **lunate** (LOŌ-nāt=moon-shaped), **triquetrum** (trī-KWĒ-trum=three-cornered), and **pisiform** (PIS-i-form=pea-shaped). The carpals in the distal row, from lateral to medial, are the **trapezium** (tra-PĒ-zē-um=four-sided figure with no two sides parallel), **trapezoid** (TRAP-e-zoid=four-sided figure with two sides parallel), **capitate** (KAP-i-tāt=head-shaped), and **hamate** (HAM-āt=hooked).

Figure 8.7 Articulations formed by the ulna and radius. (a) Elbow joint. (b) Joint surfaces at proximal end of the ulna. (c) Joint surfaces at distal ends of radius and ulna. The ulna and radius are also attached by the interosseous membrane.

🔑 **The elbow joint is formed by two articulations: (1) the trochlear notch of the ulna with the trochlea of the humerus and (2) the head of the radius with the capitulum of the humerus.**

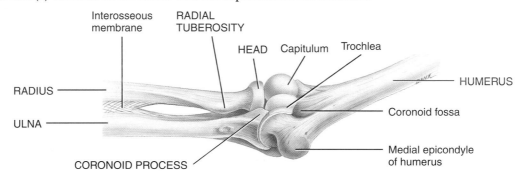

(a) Medial view in relation to humerus

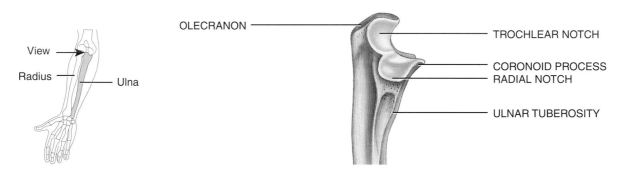

(b) Lateral view of proximal end of ulna

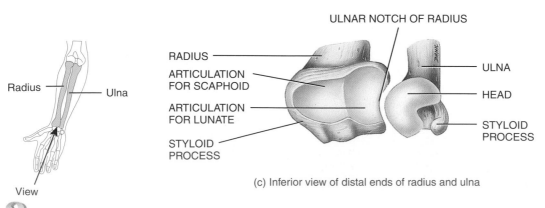

(c) Inferior view of distal ends of radius and ulna

❓ **How many joints are formed between the radius and ulna?**

Figure 8.8   Right wrist and hand in relation to the ulna and radius.

The skeleton of the hand consists of the proximal carpals, the intermediate metacarpals, and the distal phalanges.

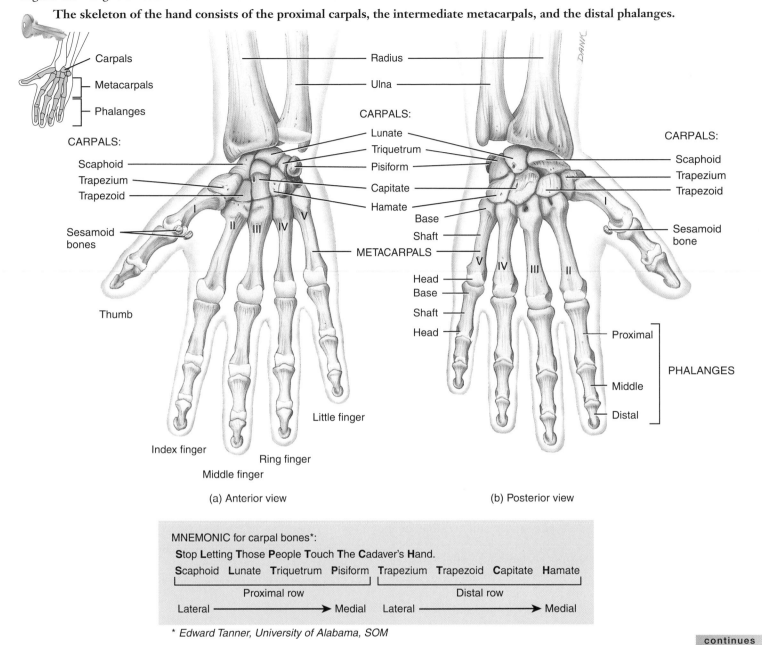

(a) Anterior view

(b) Posterior view

MNEMONIC for carpal bones*:

Stop **L**etting **T**hose **P**eople **T**ouch **T**he **C**adaver's **H**and.

Scaphoid **L**unate **T**riquetrum **P**isiform   **T**rapezium **T**rapezoid **C**apitate **H**amate

Proximal row	Distal row
Lateral ⟶ Medial	Lateral ⟶ Medial

Edward Tanner, University of Alabama, SOM

continues

The capitate is the largest carpal bone; its rounded projection, the head, articulates with the lunate. The hamate is named for a large hook-shaped projection on its anterior surface. In about 70% of carpal fractures, only the scaphoid is broken. This is because the force of a fall on an outstretched hand is transmitted from the capitate through the scaphoid to the radius.

The concave space formed by the pisiform and hamate (on the ulnar side), and the scaphoid and trapezium (on the radial side), plus the *flexor retinaculum* (fibrous bands of deep fascia) is the **carpal tunnel.** The long flexor tendons of the digits and thumb and the median nerve pass through the carpal tunnel. Narrowing of the carpal tunnel may give rise to a condition called carpal tunnel syndrome (described on page 348).

A mnemonic for learning the names of the carpal bones is shown in Figure 8.8. The first letters of the carpal bones from lateral to medial (proximal row, then distal row) correspond to the first letter of each word in the mnemonic.

The **metacarpus** (*meta-*=beyond), or palm, is the intermediate region of the hand and consists of five bones called **metacarpals.** Each metacarpal bone consists of a proximal *base*, an intermediate *shaft*, and a distal *head* (Figure 8.8b). The metacarpal bones are numbered I to V (or 1–5), starting with the thumb, from lateral to medial. The bases articulate with the distal row of carpal bones to form the *carpometacarpal joints.* The heads articulate with the proximal phalanges to form the *metacarpophalangeal joints.* The heads of the metacarpals

Figure 8.8   (continued)

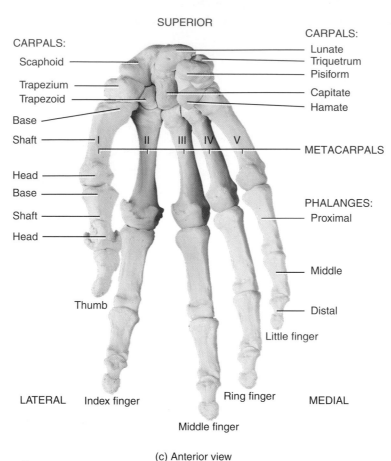

(c) Anterior view

? **Which is the most frequently fractured wrist bone?**

are commonly called "knuckles" and are readily visible in a clenched fist.

The **phalanges** (fa-LAN-jēz; *phalan-*=a battle line), or bones of the digits, make up the distal region of the hand. There are 14 phalanges in the five digits of each hand and, like the metacarpals, the digits are numbered I to V (or 1–5), beginning with the thumb, from lateral to medial. A single bone of a digit is referred to as a *phalanx* (FĀ-lanks). Each phalanx consists of a proximal *base*, an intermediate *shaft*, and a distal *head*. The thumb (*pollex*) has two phalanges, and there are three phalanges in each of the other four digits. In order from the thumb, these other four digits are commonly referred to as the index finger, middle finger, ring finger, and little finger. The first row of phalanges, the *proximal row*, articulates with the metacarpal bones and second row of phalanges. The second row of phalanges, the *middle row*, articulates with the proximal row and the third row. The third row of phalanges, the *distal row*, articulates with the middle row. The thumb has no middle phalanx. Joints between phalanges are called *interphalangeal joints.*

CHECKPOINT

2. Name the bones that form the upper limb, from proximal to distal.
3. Describe the joints of the upper limb.

PELVIC (HIP) GIRDLE

OBJECTIVE

• Identify the bones of the pelvic girdle and their principal markings.

The **pelvic (hip) girdle** consists of the two **hip bones**, also called **coxal bones** (KOK-sal; *cox-*=hip) (Figure 8.9). The hip bones unite anteriorly at a joint called the **pubic symphysis** (PŪ-bik SIM-fi-sis). They unite posteriorly with the sacrum at the *sacroiliac joints.* The complete ring composed of the hip bones, pubic symphysis, and sacrum forms a deep, basinlike

Figure 8.9 Bony pelvis. Shown here is the female bony pelvis. (See Tortora, *A Photographic Atlas of the Human Body*, 2e, Figure 3.27.)

The hip bones are united anteriorly at the pubic symphysis and posteriorly at the sacrum to form the bony pelvis.

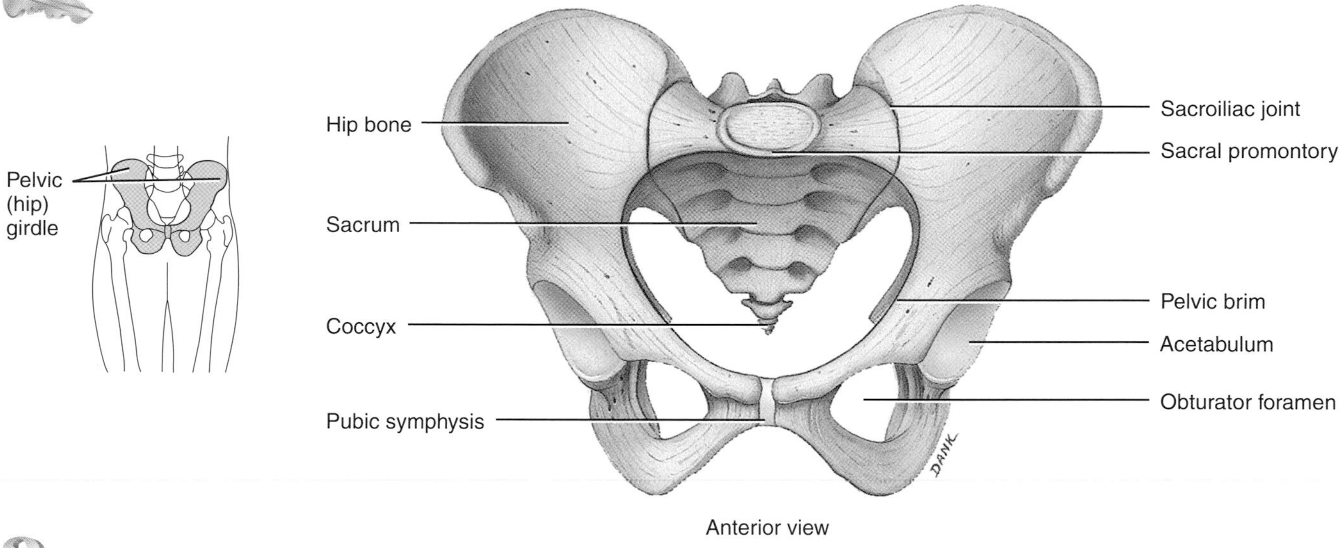

Pelvic (hip) girdle

Hip bone
Sacrum
Coccyx
Pubic symphysis

Sacroiliac joint
Sacral promontory
Pelvic brim
Acetabulum
Obturator foramen

Anterior view

What are the functions of the bony pelvis?

structure called the **bony pelvis** (*pelv-*=basin). The plural is *pelves* (PEL-vēz) or *pelvises*. Functionally, the bony pelvis provides a strong and stable support for the vertebral column and pelvic organs. The pelvic girdle of the bony pelvis also accepts the bones of the lower limbs, connecting them to the axial skeleton.

Each of the two hip bones of a newborn consists of three bones separated by cartilage: a superior *ilium*, an inferior and anterior *pubis*, and an inferior and posterior *ischium*. By age 23, the three separate bones fuse together (Figure 8.10a). Although the hip bones function as single bones, anatomists commonly discuss them as though they still consisted of three bones.

Ilium

The **ilium** (IL-ē-um=flank) is the largest of the three components of the hip bone (Figure 8.10b, c). A superior *ala* (=wing) and an inferior *body*, which enters into the formation of the *acetabulum*, the socket for the head of the femur, comprise the ilium. Its superior border, the *iliac crest*, ends anteriorly in a blunt *anterior superior iliac spine*. Bruising of the anterior superior iliac spine and associated soft tissues, such as occurs in body contact sports, is called a **hip pointer.** Below this spine is the *anterior inferior iliac spine*. Posteriorly, the iliac crest ends in a sharp *posterior superior iliac spine*. Below this spine is the *posterior inferior iliac spine*. The spines serve as points of attachment for the tendons of the muscles of the trunk, hip, and thighs. Below

the posterior inferior iliac spine is the *greater sciatic notch* (sī-AT-ik), through which the sciatic nerve, the longest nerve in the body, passes.

The medial surface of the ilium contains the *iliac fossa*, a concavity where the tendon of the iliacus muscle attaches. Posterior to this fossa are the *iliac tuberosity*, a point of attachment for the sacroiliac ligament, and the *auricular surface* (*auric-*=ear-shaped), which articulates with the sacrum to form the *sacroiliac joint* (see Figure 8.9). Projecting anteriorly and inferiorly from the auricular surface is a ridge called the *arcuate line* (AR-kū-āt; *arc-*=bow).

The other conspicuous markings of the ilium are three arched lines on its lateral surface called the *posterior gluteal line* (*glut-*=buttock), the *anterior gluteal line*, and the *inferior gluteal line*. The tendons of the gluteal muscles attach to the ilium between these lines.

BONE MARROW EXAMINATION

Sometimes a sample of red bone marrow must be obtained in order to diagnose certain blood disorders, such as leukemia and severe anemias. **Bone marrow examination** may involve *bone marrow aspiration* (withdrawal of a small amount of red bone marrow with a fine needle and syringe) or a *bone marrow biopsy* (removal of a core of red bone marrow with a larger needle). Both types of samples are usually taken from the iliac crest of

Figure 8.10 Right hip bone. The lines of fusion of the ilium, ischium, and pubis depicted in (a) are not always visible in an adult. (See Tortora, *A Photographic Atlas of the Human Body, 2e,* Figure 3.26.)

🔑 **The acetabulum is the socket formed where the three parts of the hip bone converge.**

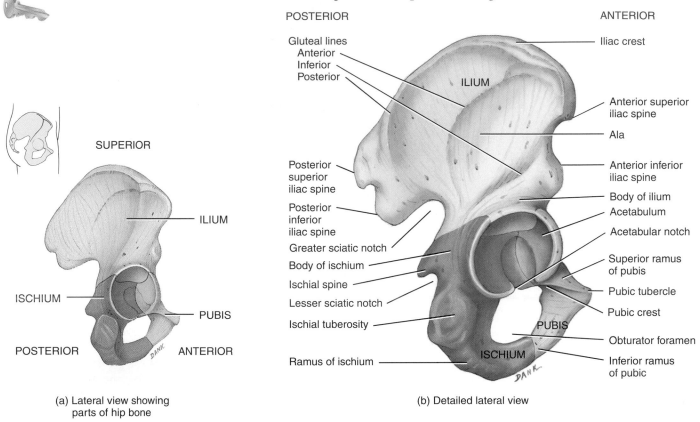

POSTERIOR   ANTERIOR

Gluteal lines
  Anterior
  Inferior
  Posterior
ILIUM
Iliac crest
Anterior superior iliac spine
Ala
Anterior inferior iliac spine
Posterior superior iliac spine
Body of ilium
Acetabulum
Posterior inferior iliac spine
Acetabular notch
Greater sciatic notch
Superior ramus of pubis
Body of ischium
Pubic tubercle
Ischial spine
Pubic crest
Lesser sciatic notch
PUBIS
Ischial tuberosity
Obturator foramen
ISCHIUM
Ramus of ischium
Inferior ramus of pubic

(b) Detailed lateral view

SUPERIOR

ILIUM

ISCHIUM

PUBIS

POSTERIOR   ANTERIOR

(a) Lateral view showing parts of hip bone

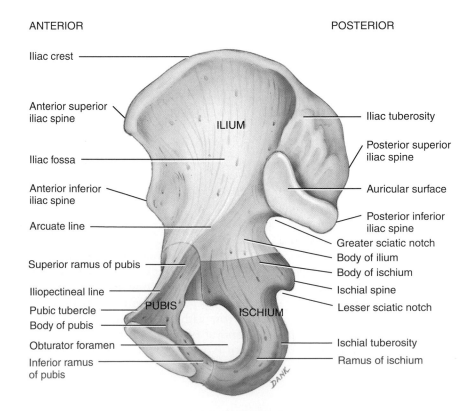

ANTERIOR   POSTERIOR

Iliac crest
ILIUM
Anterior superior iliac spine
Iliac tuberosity
Iliac fossa
Posterior superior iliac spine
Anterior inferior iliac spine
Auricular surface
Arcuate line
Posterior inferior iliac spine
Greater sciatic notch
Superior ramus of pubis
Body of ilium
Body of ischium
Iliopectineal line
Ischial spine
Pubic tubercle
PUBIS
Lesser sciatic notch
Body of pubis
ISCHIUM
Obturator foramen
Ischial tuberosity
Inferior ramus of pubis
Ramus of ischium

(c) Detailed medial view

❓ **Which part of the hip bone articulates with the femur? With the sacrum?**

the hip bone, although aspirates are sometimes taken from the sternum. In young children, bone marrow samples are taken from a vertebra or tibia (shin bone). ■

Ischium

The **ischium** (IS-kē-um=hip), the inferior, posterior portion of the hip bone (Figure 8.10b, c), is comprised of a superior *body* and an inferior *ramus* (ram-=branch; plural is *rami*). The ramus joins the pubis. Features of the ischium include the prominent *ischial spine*, a *lesser sciatic notch* below the spine, and a rough and thickened *ischial tuberosity*. This prominent tuberosity may hurt someone's thigh when you sit on their lap. Together, the ramus and the pubis surround the *obturator foramen* (OB-too-rā-tōr; *obtur-*=closed up), the largest foramen in the skeleton. The foramen is so-named because, even though blood vessels and nerves pass through it, it is nearly completely closed by the fibrous *obturator membrane*.

Pubis

The **pubis** or **os pubis**, meaning pubic bone, is the anterior and inferior part of the hip bone (Figure 8.10b, c). A *superior ramus*, an *inferior ramus*, and a *body* between the rami comprise the pubis. The anterior border of the body is the *pubic crest*, and at its lateral end is a projection called the *pubic tubercle*. This tubercle is the beginning of a raised line, the *iliopectineal line* (il-ē-ō-pek-TIN-ē-al), which extends superiorly and laterally along the superior ramus to merge with the arcuate line of the ilium. These lines, as you will see shortly, are important landmarks for distinguishing the superior and inferior portions of the bony pelvis.

The *pubic symphysis* is the joint between the two hip bones (see Figure 8.9). It consists of a disc of fibrocartilage. Inferior to this joint, the inferior rami of the two pubic bones converge to form the *pubic arch*. In the later stages of pregnancy, the hormone relaxin (produced by the ovaries and placenta) increases the flexibility of the pubic symphysis to ease delivery of the baby. Weakening of the joint, together with an already compromised center of gravity due to an enlarged uterus, also alters the gait during pregnancy.

The *acetabulum* (as-e-TAB-ū-lum=vinegar cup) is a deep fossa formed by the ilium, ischium, and pubis. It functions as the socket that accepts the rounded head of the femur. Together, the acetabulum and the femoral head form the *hip (coxal) joint*. On the inferior side of the acetabulum is a deep indentation, the *acetabular notch*. It forms a foramen through which blood vessels and nerves pass, and it serves as a point of attachment for ligaments of the femur (for example, the ligament of the head of the femur).

False and True Pelves

The bony pelvis is divided into superior and inferior portions by a boundary called the *pelvic brim* (Figure 8.11a). You can trace the pelvic brim by following the landmarks around parts of the hip bones to form the outline of an oblique plane. Beginning posteriorly at the *sacral promontory* of the sacrum, trace laterally and inferiorly along the *arcuate lines* of the ilium. Continue inferiorly along the *iliopectineal lines* of the pubis. Finally, trace anteriorly to the superior portion of the *pubic symphysis*. Together, these points form an oblique plane that is higher in the back than in the front. The circumference of this plane is the pelvic brim.

Figure 8.11   True and false pelves. Shown here is the female pelvis. For simplicity, in part (a) the landmarks of the pelvic brim are shown only on the left side of the body, and the outline of the pelvic brim is shown only on the right side. The entire pelvic brim is shown in Figure 8.9. (See Tortora, *A Photographic Atlas of the Human Body*, Figure 3.27.)

🔑 **The true and false pelves are separated by the pelvic brim.**

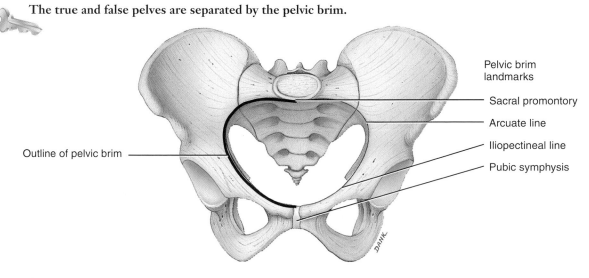

Pelvic brim landmarks

Sacral promontory

Arcuate line

Iliopectineal line

Pubic symphysis

Outline of pelvic brim

(a) Anterior view of borders of pelvic brim

continues

Figure 8.11   (continued)

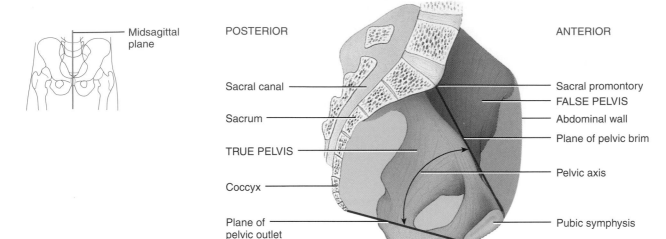

(b) Midsagittal section indicating locations of true and false pelves

What is the significance of the pelvic axis?

The portion of the bony pelvis superior to the pelvic brim is the **false (greater) pelvis** (Figure 8.11b). It is bordered by the lumbar vertebrae posteriorly, the upper portions of the hip bones laterally, and the abdominal wall anteriorly. The space enclosed by the false pelvis is part of the abdomen; it does not contain pelvic organs, except for the urinary bladder (when it is full) and the uterus during pregnancy.

The portion of the bony pelvis inferior to the pelvic brim is the **true (lesser) pelvis** (Figure 8.11b). It is bounded by the sacrum and coccyx posteriorly, inferior portions of the ilium and ischium laterally, and the pubic bones anteriorly. The true pelvis surrounds the pelvic cavity (see Figure 1.6 on page 12). The superior opening of the true pelvis, bordered by the pelvic brim, is called the *pelvic inlet;* the inferior opening of the true pelvis is the *pelvic outlet.* The *pelvic axis* is an imaginary line that curves through the true pelvis from the central point of the plane of the pelvic inlet to the central point of the plane of the pelvic outlet. During childbirth the pelvic axis is the route taken by the baby's head as it descends through the pelvis.

 **PELVIMETRY**

Pelvimetry is the measurement of the size of the inlet and outlet of the birth canal, which may be done by ultrasonography or physical examination. Measurement of the pelvic cavity in pregnant females is important because the fetus must pass through the narrower opening of the pelvis at birth. ■

CHECKPOINT

4. Distinguish between the false and true pelves.

COMPARISON OF FEMALE AND MALE PELVES

OBJECTIVE

● Compare the principal structural differences between female and male pelves.

In the following discussion it is assumed that the male and female are comparable in age and physical stature. Generally, the bones of a male are larger and heavier than those of a female and have larger surface markings. Sex-related differences in the features of bones are readily apparent when comparing the female and male pelves. Most of the structural differences in the pelves are adaptations to the requirements of pregnancy and childbirth. The female's pelvis is wider and shallower than the male's. Consequently, there is more space in the true pelvis of the female, especially in the pelvic inlet and pelvic outlet, which accommodate the passage of the infant's head at birth. Other significant structural differences between pelves of females and males are listed and illustrated in Table 8.1.

CHECKPOINT

5. Why are structural differences between female and male pelves important?

TABLE 8.1   COMPARISON OF FEMALE AND MALE PELVES

POINT OF COMPARISON	FEMALE	MALE
General structure	Light and thin.	Heavy and thick.
False (greater) pelvis	Shallow.	Deep.
Pelvic brim (inlet)	Larger and more oval.	Smaller and heart-shaped.
Acetabulum	Small and faces anteriorly.	Large and faces laterally.
Obturator foramen	Oval.	Round.
Pubic arch	Greater than 90° angle.	Less than 90° angle.

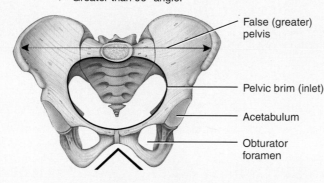

False (greater) pelvis

Pelvic brim (inlet)

Acetabulum

Obturator foramen

Pubic arch (greater than 90°)

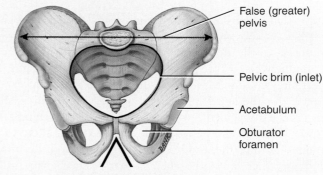

False (greater) pelvis

Pelvic brim (inlet)

Acetabulum

Obturator foramen

Pubic arch (less than 90°)

Anterior views

POINT OF COMPARISON	FEMALE	MALE
Iliac crest	Less curved.	More curved.
Ilium	Less vertical.	More vertical.
Greater sciatic notch	Wide.	Narrow.
Coccyx	More movable and more curved anteriorly.	Less movable and less curved anteriorly.
Sacrum	Shorter, wider (see anterior views), and more curved anteriorly.	Longer, narrower (see anterior views), and less curved anteriorly.

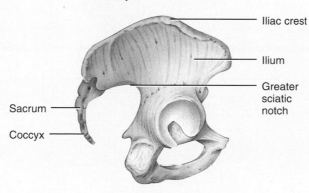

Iliac crest

Ilium

Greater sciatic notch

Sacrum

Coccyx

Iliac crest

Ilium

Greater sciatic notch

Sacrum

Coccyx

Right lateral views

POINT OF COMPARISON	FEMALE	MALE
Pelvic outlet	Wider.	Narrower.
Ischial tuberosity	Shorter, farther apart, and more laterally projecting.	Longer, closer together, and more medially projecting.

Ischial tuberosity   Pelvic outlet

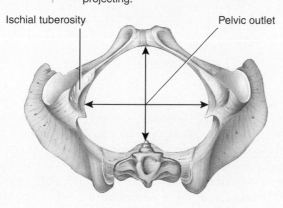

Ischial tuberosity   Pelvic outlet

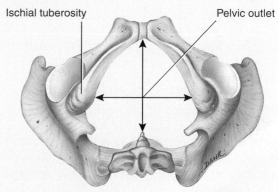

Inferior views

COMPARISON OF PECTORAL AND PELVIC GIRDLES

● Describe the structural differences between the pectoral and pelvic girdles.

Now that we've studied the structures of the pectoral and pelvic girdles, we can note some of their significant differences. The pectoral girdle does not directly articulate with the vertebral column, whereas the pelvic girdle directly articulates with the vertebral column via the sacroiliac joint. The sockets (glenoid fossae) for the upper limbs in the pectoral girdle are shallow and maximize movement, whereas the sockets (acetabula) for the lower limbs in the pelvic girdle are deep and allow less movement. Overall, the structure of the pectoral girdle offers more mobility than strength, whereas that of the pelvic girdle offers more strength than mobility.

6. What is the overall difference between the pectoral and pelvic girdles?

LOWER LIMB (EXTREMITY)

● Identify the bones of the lower limb and their principal markings.

Each **lower limb (extremity)** has 30 bones in four locations—(1) the femur in the thigh; (2) the patella (kneecap); (3) the tibia and fibula in the leg; and (4) the 7 tarsals in the tarsus (ankle), the 5 metatarsals in the metatarsus, and the 14 phalanges (bones of the digits) in the foot (Figure 8.12).

Femur

The **femur**, or thigh bone, is the longest, heaviest, and strongest bone in the body (Figure 8.13). Its proximal end articulates with the acetabulum of the hip bone. Its distal end articulates with the tibia and patella. The *body (shaft)* of the femur angles medially and, as a result, the knee joints are closer to the midline. The angle is greater in females because the female pelvis is broader.

The proximal end of the femur consists of a rounded *head* that articulates with the acetabulum of the hip bone to form the *hip (coxal) joint*. The head contains a small centered depression (pit) called the *fovea capitis* (FŌ-vē-a CAP-i-tis; *fovea*=pit; *capitis*=of the head). The ligament of the head of the femur connects the fovea capitis of the femur to the acetabulum of the hip bone. The *neck* of the femur is a constricted region distal to the head. The *greater trochanter* (trō-KAN-ter) and *lesser trochanter* are projections that serve as points of attachment for the tendons of some of the thigh and buttock muscles. The greater trochanter is the prominence felt and seen anterior to the hollow on the side of the hip. It is a landmark commonly used to locate the site for intramuscular injections into the lateral surface of the thigh. The lesser trochanter is inferior and

Right lower limb.

Each lower limb consists of a femur, patella (kneecap), tibia, fibula, tarsals (ankle bones), metatarsals, and phalanges (bones of the digits).

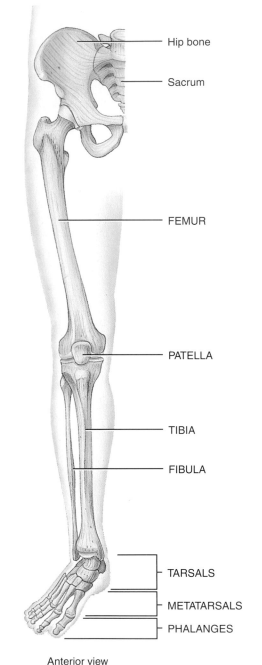

Anterior view

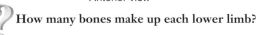 **How many bones make up each lower limb?**

medial to the greater trochanter. Between the anterior surfaces of the trochanters is a narrow *intertrochanteric line* (Figure 8.13a). A ridge called the *intertrochanteric crest* appears between the posterior surfaces of the trochanters (Figure 8.13b).

Inferior to the intertrochanteric crest on the posterior surface of the body of the femur is a vertical ridge called the *gluteal tuberosity*. It blends into another vertical ridge called the

Figure 8.13 Right femur in relation to the hip bone, patella, tibia, and fibula. (See Tortora, *A Photographic Atlas of the Human Body*, Figure 3.28.)

The acetabulum of the hip bone and head of the femur articulate to form the hip joint.

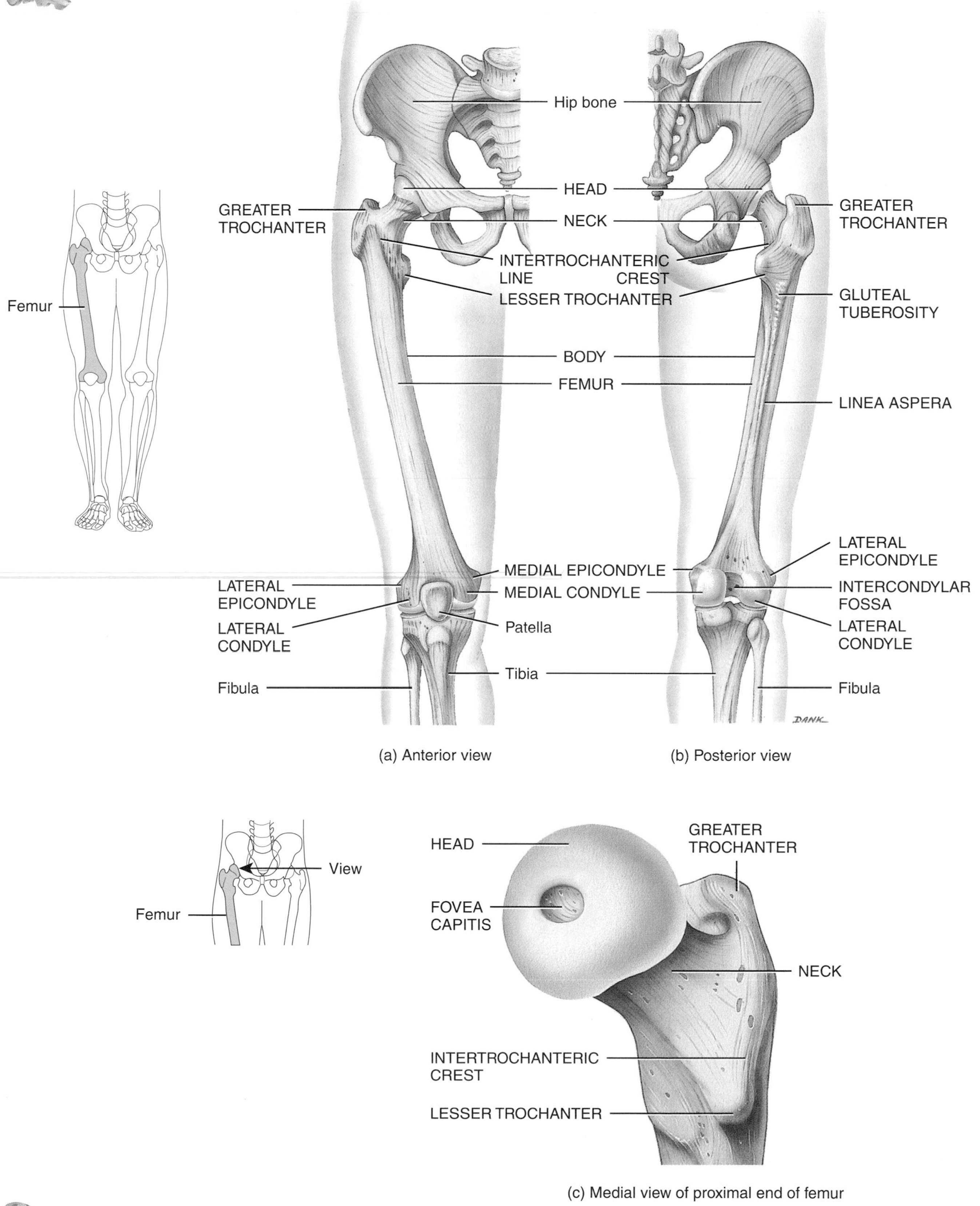

Femur

GREATER TROCHANTER

Hip bone

HEAD

NECK

INTERTROCHANTERIC LINE   CREST

LESSER TROCHANTER

BODY

FEMUR

MEDIAL EPICONDYLE

MEDIAL CONDYLE

LATERAL EPICONDYLE

LATERAL CONDYLE

Patella

Fibula

Tibia

GREATER TROCHANTER

GLUTEAL TUBEROSITY

LINEA ASPERA

LATERAL EPICONDYLE

INTERCONDYLAR FOSSA

LATERAL CONDYLE

Fibula

(a) Anterior view   (b) Posterior view

Femur

View

HEAD

FOVEA CAPITIS

INTERTROCHANTERIC CREST

LESSER TROCHANTER

GREATER TROCHANTER

NECK

(c) Medial view of proximal end of femur

Why is the angle of convergence of the femurs greater in females than males?

linea aspera (LIN-ē-a AS-per-a; *asper*=rough). Both ridges serve as attachment points for the tendons of several thigh muscles.

The distal end of the femur expands to include the *medial condyle* and the *lateral condyle*. These articulate with the medial and lateral condyles of the tibia. Superior to the condyles are the *medial epicondyle* and the *lateral epicondyle*, to which ligaments of the knee joint attach. A depressed area between the condyles on the posterior surface is called the *intercondylar fossa* (in-ter-KON-di-lar). The *patellar surface* is located between the condyles on the anterior surface.

Patella

The **patella** (=little dish), or kneecap, is a small, triangular bone located anterior to the knee joint (Figure 8.14). It is a sesamoid bone that develops in the tendon of the quadriceps femoris muscle. The broad superior end of the patella is called the *base*. The pointed inferior end is the *apex*. The posterior surface contains two *articular facets*, one for the medial condyle of the femur and the other for the lateral condyle of the femur. The patellar ligament attaches the patella to the tibial tuberosity. The *patellofemoral joint*, between the posterior surface of the patella and the patellar surface of the femur, is the intermediate component of the *tibiofemoral (knee) joint*. The functions of the patella are to increase the leverage of the tendon of the quadriceps femoris muscle, to maintain the position of the tendon when the knee is bent (flexed), and to protect the knee joint.

PATELLOFEMORAL STRESS SYNDROME

Patellofemoral stress syndrome ("runner's knee") is one of the most common problems runners experience. During normal flexion and extension of the knee, the patella tracks (glides) superiorly and inferiorly in the groove between the femoral condyles. In patellofemoral stress syndrome, normal tracking does not occur; instead, the patella tracks laterally as well as superiorly and inferiorly, and the increased pressure on the joint causes aching or tenderness around or under the patella. The pain typically occurs after a person has been sitting for awhile, especially after exercise. It is worsened by squatting or walking down stairs. One cause of runner's knee is constantly walking, running, or jogging on the same side of the road. Because roads slope down on the sides, the knee that is closer to the center of the road endures greater mechanical stress because it does not fully extend during a stride. Other predisposing factors include having knock-knees, running on hills, and running long distances.   ■

Tibia and Fibula

The **tibia**, or shin bone, is the larger, medial, weight-bearing bone of the leg (Figure 8.15). The tibia articulates at its proximal end with the femur and fibula and at its distal end with the fibula and the talus bone of the ankle. The tibia and fibula, like the ulna and radius, are connected by an interosseous membrane.

The proximal end of the tibia is expanded into a *lateral condyle* and a *medial condyle*. These articulate with the condyles of the femur to form the lateral and medial *tibiofemoral (knee) joints*. The inferior surface of the lateral condyle articulates with the head of the fibula. The slightly concave condyles are separated by an upward projection called the *intercondylar eminence* (Figure 8.15b). The *tibial tuberosity* on the anterior surface is a point of attachment for the patellar ligament. Inferior to and continuous with the tibial tuberosity is a sharp ridge that can be felt below the skin and is known as the *anterior border (crest)* or shin.

The medial surface of the distal end of the tibia forms the *medial malleolus* (mal-LĒ-ō-lus=hammer). This structure articulates with the talus of the ankle and forms the prominence that can be felt on the medial surface of the ankle. The *fibular notch* (Figure 8.15c) articulates with the distal end of the fibula to

Figure 8.14   Right patella.

The patella articulates with the lateral and medial condyles of the femur.

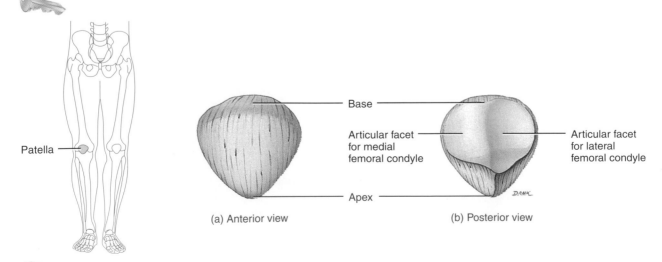

Patella

Base

Articular facet for medial femoral condyle

Articular facet for lateral femoral condyle

Apex

(a) Anterior view

(b) Posterior view

The patella is classified as which type of bone? Why?

Figure 8.15 **Right tibia and fibula in relation to the femur, patella, and talus.** (See Tortora, *A Photographic Atlas of the Human Body*, 2e, Figure 3.30.)

The tibia articulates with the femur and fibula proximally, and with the fibula and talus distally.

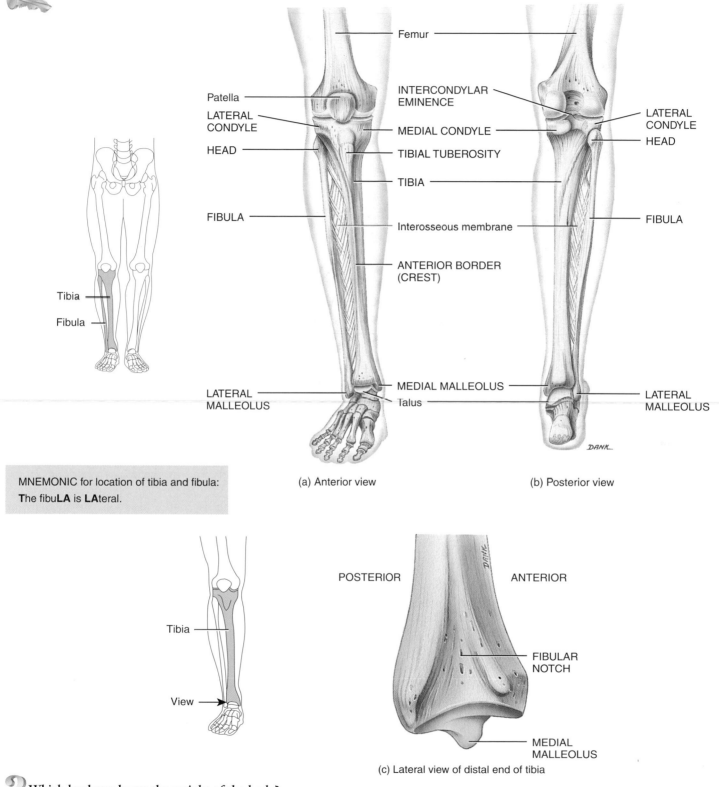

Femur

Patella

LATERAL CONDYLE

HEAD

FIBULA

LATERAL MALLEOLUS

INTERCONDYLAR EMINENCE

MEDIAL CONDYLE

TIBIAL TUBEROSITY

TIBIA

Interosseous membrane

ANTERIOR BORDER (CREST)

MEDIAL MALLEOLUS

Talus

LATERAL CONDYLE

HEAD

FIBULA

LATERAL MALLEOLUS

Tibia

Fibula

DANK

(a) Anterior view

(b) Posterior view

MNEMONIC for location of tibia and fibula:
The fibu**LA** is **LA**teral.

Tibia

View

POSTERIOR   ANTERIOR

DANK

FIBULAR NOTCH

MEDIAL MALLEOLUS

(c) Lateral view of distal end of tibia

Which leg bone bears the weight of the body?

form the *distal tibiofibular joint*. Of all the long bones of the body, the tibia is the most frequently fractured and is also the most frequent site of an open (compound) fracture.

The **fibula** is parallel and lateral to the tibia, but is considerably smaller. The *head* of the fibula, the proximal end, articulates with the inferior surface of the lateral condyle of the tibia below the level of the knee joint to form the *proximal tibiofibular joint*. The distal end has a projection called the *lateral malleolus* that articulates with the talus of the ankle. This forms the prominence on the lateral surface of the ankle. As noted previously, the fibula also articulates with the tibia at the fibular notch.

BONE GRAFTING

Bone grafting generally consists of taking a piece of bone, along with its periosteum and nutrient artery, from one part of the body to replace missing bone in another part of the body. The transplanted bone restores the blood supply to the transplanted site and healing occurs as in a fracture. The fibula is a common source of bone for grafting and, even after a piece of the fibula has been removed, walking, running, and jumping can be normal. ■

Tarsals, Metatarsals, and Phalanges

The **tarsus** (ankle) is the proximal region of the foot and consists of seven **tarsal bones** (Figure 8.16). They include the **talus** (TĀ-lus=ankle bone) and **calcaneus** (kal-KĀ-nē-us= heel), located in the posterior part of the foot. The calcaneus is the largest and strongest tarsal bone. The anterior tarsal bones are the **navicular** (=like a little boat), three **cuneiform bones** (=wedge-shaped) called the **third (lateral), second (intermediate)**, and **first (medial) cuneiforms**, and the **cuboid** (=cube-shaped). Joints between tarsal bones are called *intertarsal joints*. The talus, the most superior tarsal bone, is the only bone of the foot that articulates with the fibula and tibia. It articulates on one side with the medial malleolus of the tibia and on the other side with the lateral malleolus of the fibula. These articulations form the *talocrural (ankle) joint*. During walking, the talus transmits about half the weight of the body to the calcaneus. The remainder is transmitted to the other tarsal bones.

The **metatarsus** is the intermediate region of the foot and consists of five **metatarsal bones** numbered I to V (or 1–5) from the medial to lateral position (Figure 8.16). Like the metacarpals of the palm, each metatarsal consists of a proximal *base*, an intermediate *shaft*, and a distal *head*. The metatarsals articulate proximally with the first, second, and third cuneiform bones and with the cuboid to form the *tarsometatarsal joints*. Distally, they articulate with the proximal row of phalanges to form the *metatarsophalangeal joints*. The first metatarsal is thicker than the others because it bears more weight.

FRACTURES OF THE METATARSALS

Fractures of the metatarsals occur when a heavy object falls on the foot or when a heavy object rolls over the foot. Such frac-

tures are also common among dancers, especially female ballet dancers. If a ballet dancer is on the tip of her toes and loses her balance, the full body weight is placed on the metatarsals, thus fracturing them. ■

The **phalanges** comprise the distal component of the foot and resemble those of the hand both in number and arrangement. The toes are numbered I to V (or 1–5) beginning with the great toe, from medial to lateral. Each *phalanx* (singular) consists of a proximal *base*, an intermediate *shaft*, and a distal *head*. The great or big toe (*hallux*), has two large, heavy phalanges called proximal and distal phalanges. The other four toes each have three phalanges—proximal, middle, and distal. Joints between phalanges of the foot, like those of the hand, are called *interphalangeal joints*.

Arches of the Foot

The bones of the foot are arranged in two **arches** (Figure 8.17 on page 224). The arches enable the foot to support the weight of the body, provide an ideal distribution of body weight over the soft and hard tissues of the foot, and provide leverage when walking. The arches are not rigid; they yield as weight is applied and spring back when the weight is lifted, thus helping to absorb shocks. Usually, the arches are fully developed by the time children reach age 12 or 13.

The **longitudinal arch** has two parts, both of which consist of tarsal and metatarsal bones arranged to form an arch from the anterior to the posterior part of the foot. The *medial part* of the longitudinal arch originates at the calcaneus. It rises to the talus and descends through the navicular, the three cuneiforms, and the heads of the three medial metatarsals. The *lateral part* of the longitudinal arch also begins at the calcaneus. It rises at the cuboid and descends to the heads of the two lateral metatarsals.

The **transverse arch** is found between the medial and lateral aspects of the foot and is formed by the navicular, three cuneiforms, and the bases of the five metatarsals.

As noted earlier, one function of the arches is to distribute body weight over the soft and hard tissues of the body. Normally, the ball of the foot carries about 40 percent of the weight and the heel carries about 60 percent. The ball of the foot is the padded portion of the sole superficial to heads of the metatarsals. When a person wears high-heeled shoes, however, the distribution of weight changes so that the ball of the foot may carry up to 80 percent and the heel 20 percent. As a result, the fat pads at the ball of the foot are damaged, joint pain develops, and structural changes in bones may occur.

FLATFOOT AND CLAWFOOT

The bones composing the arches are held in position by ligaments and tendons. If these ligaments and tendons are weakened, the height of the medial longitudinal arch may decrease or "fall." The result is **flatfoot**, the causes of which include excessive weight, postural abnormalities, weakened supporting tissues, and genetic predisposition. A custom-designed arch support often is prescribed to treat flatfoot.

Figure 8.16 Right foot.

The skeleton of the foot consists of the proximal tarsals, the intermediate metatarsals, and the distal phalanges.

Superior view

Tarsals
Metatarsals
Phalanges

Inferior view

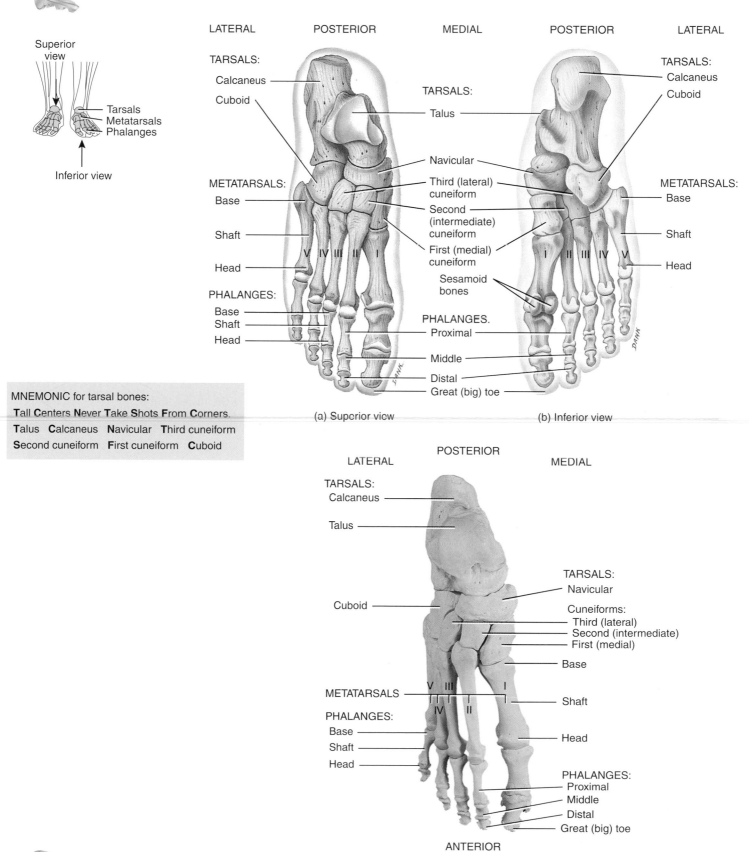

LATERAL   POSTERIOR   MEDIAL   POSTERIOR   LATERAL

TARSALS:
Calcaneus
Cuboid

TARSALS:
Talus

Navicular
Third (lateral) cuneiform
Second (intermediate) cuneiform
First (medial) cuneiform

METATARSALS:
Base

Shaft

Head

PHALANGES:
Base
Shaft
Head

V IV III II I

TARSALS:
Calcaneus
Cuboid

METATARSALS:
Base

Shaft

Head

I II III IV V

Sesamoid bones

PHALANGES.
Proximal

Middle

Distal

Great (big) toe

(a) Superior view

(b) Inferior view

MNEMONIC for tarsal bones:

Tall **C**enters **N**ever **T**ake **S**hots **F**rom **C**orners.

Talus **C**alcaneus **N**avicular **T**hird cuneiform
Second cuneiform **F**irst cuneiform **C**uboid

POSTERIOR

LATERAL   MEDIAL

TARSALS:
Calcaneus

Talus

TARSALS:
Navicular

Cuneiforms:
Third (lateral)
Second (intermediate)
First (medial)

Base

Shaft

Cuboid

METATARSALS

PHALANGES:
Base
Shaft
Head

V III
IV II
I

Head

PHALANGES:
Proximal
Middle
Distal
Great (big) toe

ANTERIOR

Which tarsal bone articulates with the tibia and fibula?

Figure 8.17 Arches of the right foot.

Arches help the foot support and distribute the weight of the body and provide leverage during walking.

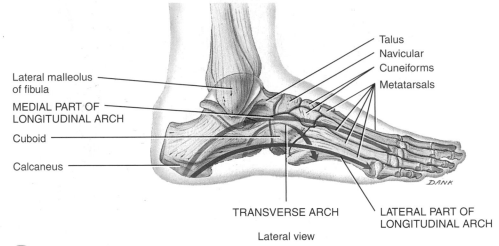

Lateral view

What structural feature of the arches allows them to absorb shocks?

Clawfoot is a condition in which the medial longitudinal arch is abnormally elevated. It is often caused by muscle deformities, such as may occur in diabetics whose neurological lesions lead to atrophy of muscles of the foot. ■

CHECKPOINT

7. Name the bones that form the lower limb, from proximal to distal.
8. Describe the joints of the lower limb.
9. What are the functions of the arches of the foot?

Figure 8.18 Development of the skeletal system. Bones that develop from the cartilaginous neurocranium are indicated light blue; from the cartilaginous viscerocranium in dark blue; from the membranous neurocranium in dark red; and from the membranous viscerocranium in light red.

After the limb buds develop, endochondral ossification of the limb bones begins by the end of the eighth embryonic week.

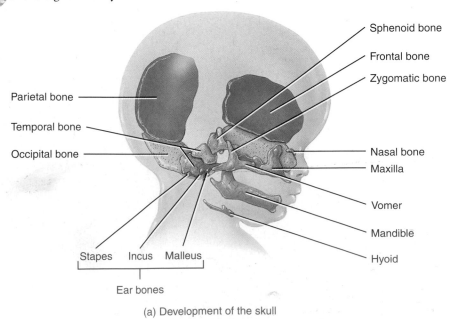

(a) Development of the skull

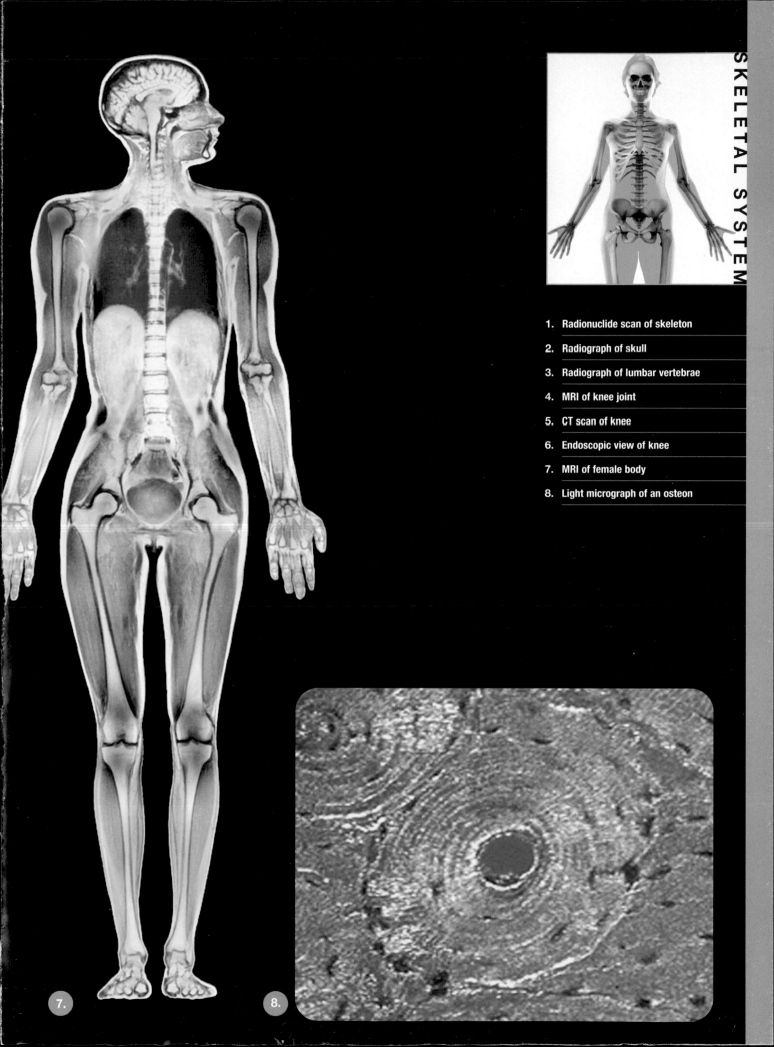

1. Radionuclide scan of skeleton

2. Radiograph of skull

3. Radiograph of lumbar vertebrae

4. MRI of knee joint

5. CT scan of knee

6. Endoscopic view of knee

7. MRI of female body

8. Light micrograph of an osteon

7.

8.

SKELETAL SYSTEM

1. **Radionuclide scan of skeleton.** Radionuclide scan of the entire skeleton in anterior view. Areas of intense color represent high tissue activity.

2. **Radiograph of skull.** Radiograph of the skull and some cervical vertebrae in lateral view.

3. **Radiograph of lumbar vertebrae.** Radiograph of lumbar vertebrae in anterior view with their associated intervertebral discs.

4. **MRI of knee joint.** Magnetic Resonance Image (MRI) of the knee joint in sagittal section.

5. **CT scan of knee.** Computerized Tomography (CT) scan of the knee in sagittal section.

6. **Endoscopic view of knee.** Endoscopic view of the inside of a knee showing an articular disc (pad of fibrocartilage).

7. **MRI of female body.** Magnetic Resonance Image (MRI) of the entire body of a female in frontal section (the head is in sagittal section).

8. **Light micrograph of an osteon.** Light micrograph of an osteon (haversian system).

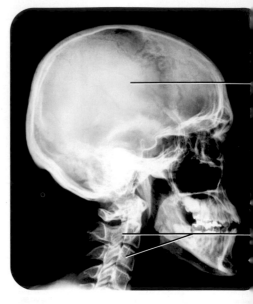

2.

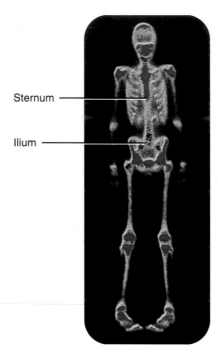

Sternum

Ilium

1.

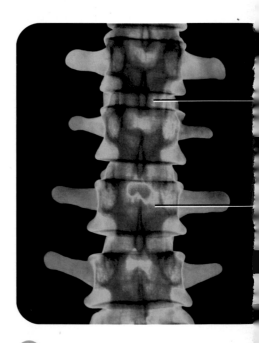

3.

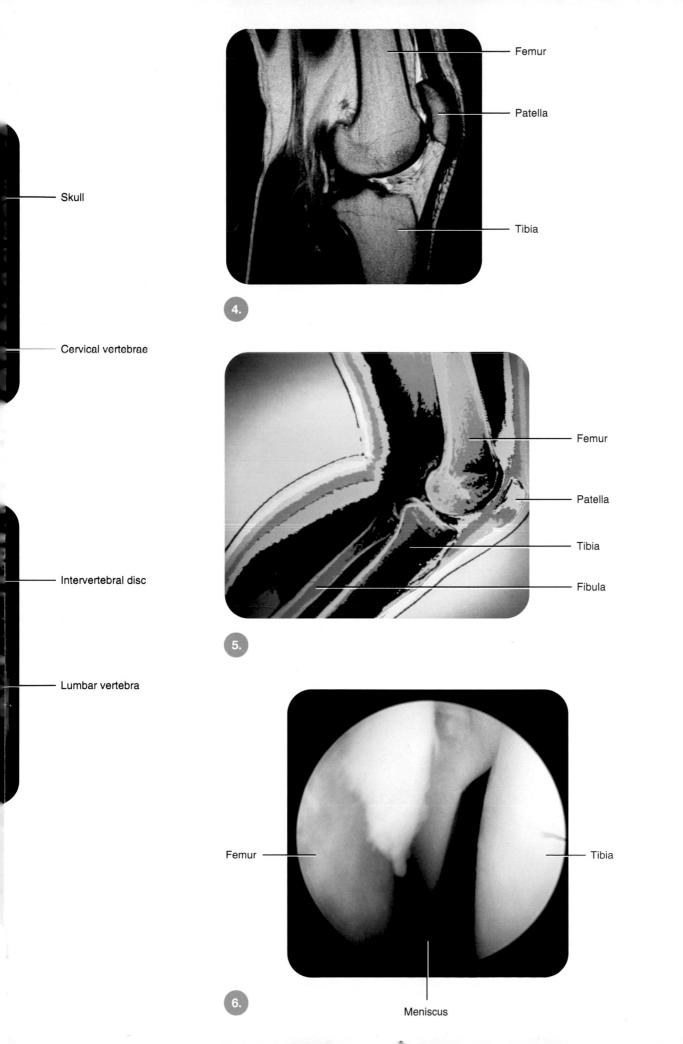

Skull

Cervical vertebrae

Intervertebral disc

Lumbar vertebra

Femur

Patella

Tibia

4.

Femur

Patella

Tibia

Fibula

5.

7.

Femur

Tibia

Meniscus

6.

4.

5.

6.

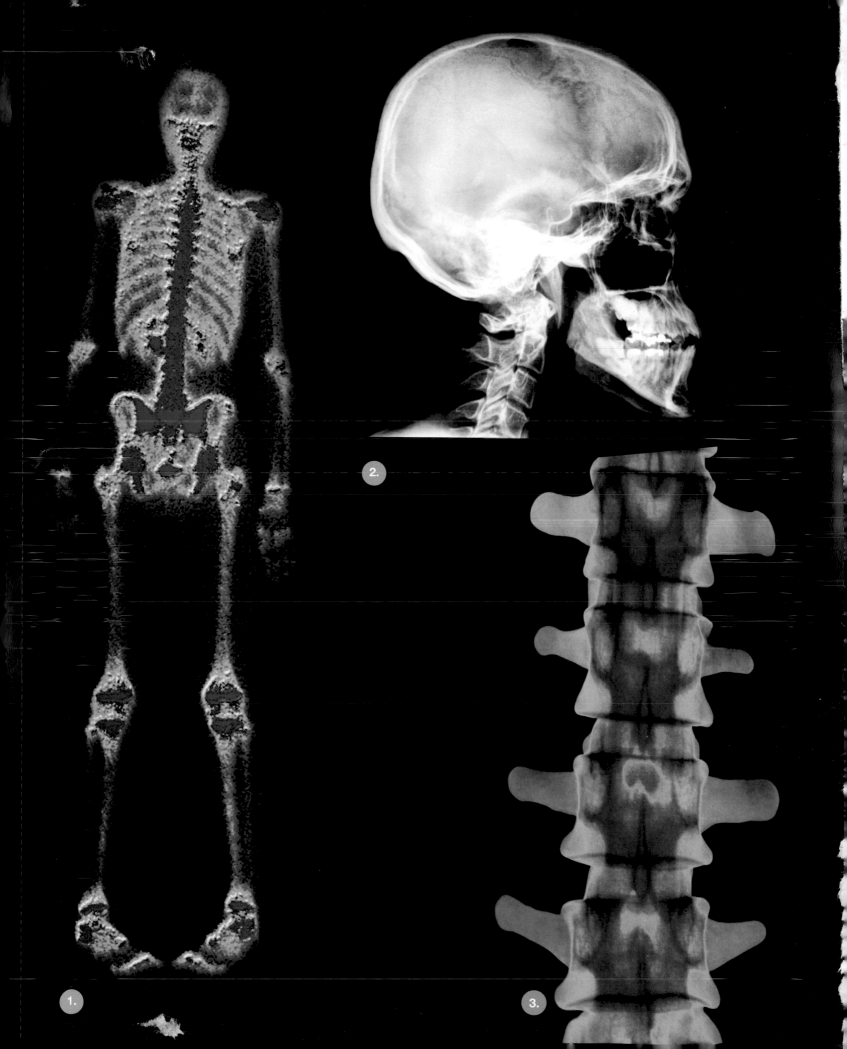

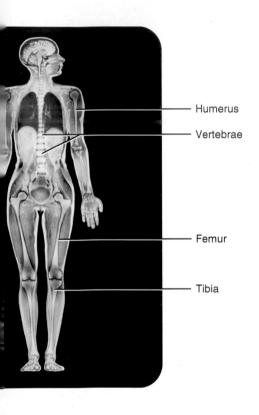

Humerus

Vertebrae

Femur

Tibia

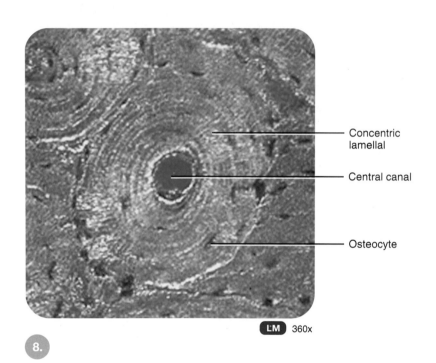

Concentric
lamellal

Central canal

Osteocyte

LM 360x

8.

DEVELOPMENT OF THE SKELETAL SYSTEM

● Describe the development of the skeletal system.

All skeletal tissue arises from *mesenchymal cells*, connective tissue cells derived from **mesoderm.** The mesenchymal cells condense and form models of bone in areas where the bones themselves will ultimately form. In some cases, the bones form directly within the mesenchyme (intramembranous ossification; see Figure 6.5 on page 151). In other cases, the bones form within hyaline cartilage that develops from mesenchyme (endochondral ossification; see Figure 6.6 on page 153).

The *skull* begins development during the fourth week after fertilization. It develops from mesenchyme around the developing brain and consists of two major portions: **neurocranium,** which forms the bones of the skull, and **viscerocranium,** which forms the bones of the face (Figure 8.18a). The neurocranium is divided into two parts called the **cartilagenous neurocranium** and **membranous neurocranium.** The cartilaginous neurocranium consists of hyaline cartilage developed from mesenchyme at the base of the developing skull. It later undergoes endochondral ossification to form the *flat bones at the base of the skull.* The membranous neurocranium consists of mesenchyme and later undergoes intramembranous ossification to form the *flat bones that make up the roof and sides of the skull.* During fetal life and infancy the flat bones are separated by membrane-filled spaces called fontanels (see Figure 7.14 on page 185). The viscerocranium, like the neurocranium, is divided into two parts: **cartilaginous viscerocranium** and **membranous viscerocranium.** The cartilaginous viscerocranium is derived from the cartilage of the first two pharyngeal (branchial) arches (see Figure 4.13 on page 111). Endochondral ossification of these cartilages forms

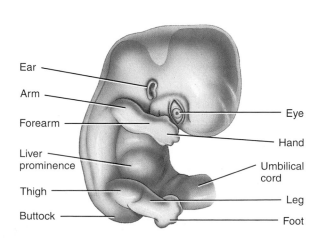

(b) Four-week embryo showing development of limb buds

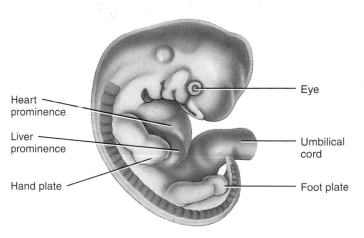

(c) Six-week embryo showing development of hand and foot plates

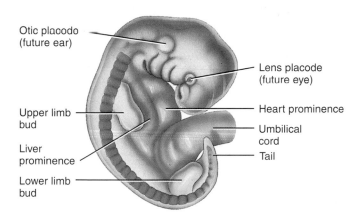

(d) Seven-week embryo showing development of arm, forearm, and hand in upper limb bud and thigh, leg, and foot in lower limb bud

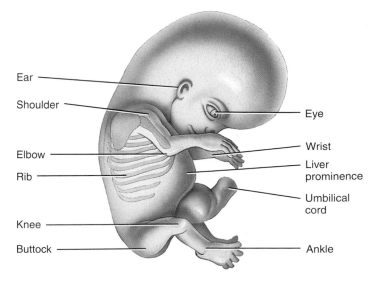

(e) Eight-week embryo in which limb buds have developed into upper and lower limbs

Which of the three basic embryonic tissues—ectoderm, mesoderm, and endoderm—gives rise to the skeletal system?

the *ear bones* and *hyoid bone*. The membranous viscerocardium is derived from mesenchyme in the first pharyngeal arch and, following intramembranous ossification, forms the facial bones.

Vertebrae are derived from the sclerotomes of somites (see Figure 10.10 on page 282). Mesenchymal cells from these regions surround the notochord (see Figure 10.10) at about four weeks after fertilization. The notochord induces the mesenchymal cells to form the *vertebral bodies*. Between the vertebral bodies, the notochord induces mesenchymal cells to form the *nucleus pulposus* of an intervertebral disc and surrounding mesenchymal cells form the *annulus fibrosus* of an intervertebral disc. As development continues, other parts of a vertebra form and the *vertebral arch* surrounds the spinal cord (failure of the vertebral arch to develop properly results in a condition called spina bifida; see page 199 in Chapter 7). In the thoracic region, processes from the vertebrae develop into the *ribs*. The *sternum* develops from mesoderm in the ventral body wall.

The *skeleton of the limbs* is derived from somatic mesoderm of the lateral plate mesoderm (see Figure 4.9d on page 105). During the middle of the fourth week after fertilization, the upper limbs appear as small elevations at the sides of the trunk called **upper limb buds** (Figure 8.18b). About 2 days later, the **lower limb buds** appear. The limb buds consist of **mesenchyme** covered by **ectoderm.** At this point, a mesenchymal skeleton exists in the limbs; some of the masses of mesoderm surrounding the developing bones will become the skeletal muscles of the limbs.

By the sixth week, the limb buds develop a constriction around the middle portion. The constriction produces flattened distal segments of the upper buds called **hand plates** and distal segments of the lower buds called **foot plates** (Figure 8.18c). These plates represent the beginnings of the hands and feet, respectively. At this stage of limb development, a cartilaginous skeleton formed from mesenchyme is present. By the seventh week (Figure 8.18d), the *arm, forearm,* and *hand* are evident in the upper limb bud, and the *thigh, leg,* and *foot* appear in the lower limb bud. By the eighth week (Figure 8.18e), as the shoulder, elbow, and wrist areas become apparent, the upper limb bud is appropriately called the upper limb, and the lower limb bud is now the lower limb.

Endochondral ossification of the limb bones begins by the end of the eighth week after fertilization. By the twelfth week, primary ossification centers are present in most of the limb bones. Most secondary ossification centers appear after birth.

CHECKPOINT

10. When and how do the limbs develop?

APPLICATIONS TO HEALTH

HIP FRACTURE

Although any region of the hip girdle may fracture, the term **hip fracture** most commonly applies to a break in the bones associated with the hip joint—the head, neck, or trochanteric regions of the femur, or the bones that form the acetabulum. In the United States, 300,000–500,000 people sustain hip fractures each year. The incidence of hip fractures is increasing, due in part to longer life spans. Decreases in bone mass due to osteoporosis (which occurs more often in females), along with an increased tendency to fall, predispose elderly people to hip fractures.

Hip fractures often require surgical treatment, the goal of which is to repair and stabilize the fracture, increase mobility, and decrease pain. Sometimes the repair is accomplished by using surgical pins, screws, nails, and plates to secure the head of the femur. In severe hip fractures, the femoral head or the acetabulum of the hip bone may be replaced by prostheses (artificial devices). The procedure of replacing either the femoral head or the acetabulum is *hemiarthroplasty* (hem-ē-AR-thrō-plas-tē; *hemi-*=one half; *arthro*=joint; *-plasty*=molding). Replacement of both the femoral head and acetabulum is *total hip arthroplasty*. The acetabular prosthesis is made of plastic, whereas the femoral prosthesis is metal; both are designed to withstand a high degree of stress. The prostheses are attached to healthy portions of bone with acrylic cement and screws (see Figure 9.17 on page 260).

KEY MEDICAL TERMS ASSOCIATED WITH THE APPENDICULAR SKELETON

Clubfoot or *talipes equinovarus* (*-pes*=foot; *equino-*=horse) An inherited deformity in which the foot is twisted inferiorly and medially, and the angle of the arch is increased; occurs in 1 of every 1000 births. Treatment consists of manipulating the arch to a normal curvature by casts or adhesive tape, usually soon after birth. Corrective shoes or surgery may also be required.

Genu valgum (JĒ-noo VAL-gum; *genu-*=knee; *valgum*=bent outward) A deformity in which the knees are abnormally close together and the space between the ankles is increased due to a lateral angulation of the tibia.

Genu varum (JĒ-noo VAR-um; *varum*=bent toward the midline) A deformity in which the knees are abnormally seperated and the lower limbs are bowed medially. Also called **bowleg.**

Hallux valgus (HAL-uks VAL-gus; *hallux*=great toe) Angulation of the great toe away from the midline of the body, typically caused by wearing tightly fitting shoes. Involves lateral deviation of the proximal phalanx of the great toe and medial displacement of metatarsal I. Also called a **bunion.**

STUDY OUTLINE

Pectoral (Shoulder) Girdle (p. 204)

1. Each pectoral (shoulder) girdle consists of a clavicle and scapula.
2. Each pectoral girdle attaches an upper limb to the axial skeleton.

Upper Limb (Extremity) (p. 206)

1. Each of the two upper limbs (extremities) contains 30 bones.
2. The bones of each upper limb include the humerus, the ulna, the radius, the carpals, the metacarpals, and the phalanges.

Pelvic (Hip) Girdle (p. 212)

1. The pelvic (hip) girdle consists of two hip bones.
2. Each hip bone consists of three fused bones: the ilium, pubis, and ischium.
3. The hip bones, sacrum, and pubic symphysis form the bony pelvis. It supports the vertebral column and pelvic viscera and attaches the lower limbs to the axial skeleton.
4. The true pelvis is separated from the false pelvis by the pelvic brim.

Comparison of Female and Male Pelves (p. 216)

1. Bones of males are generally larger and heavier than bones of females, with more prominent markings for muscle attachment.

2. The female pelvis is adapted for pregnancy and childbirth. Gender-related differences in pelvic structure are listed and illustrated in Table 8.1 on page 217.

Comparison of Pectoral and Pelvic Girdles (p. 218)

1. The pectoral girdle does not directly articulate with the vertebral column; the pelvic girdle does.
2. The glenoid fossae of the scapulae are shallow and maximize movement; the acetabula of the hip bones are deep and allow less movement.

Lower Limb (Extremity) (p. 218)

1. Each of the two lower limbs (extremities) contains 30 bones.
2. The bones of each lower limb include the femur, the patella, the tibia, the fibula, the tarsals, the metatarsals, and the phalanges.
3. The bones of the foot are arranged in two arches, the longitudinal arch and the transverse arch, to provide support and leverage.

Development of the Skeletal System (p. 225)

1. Bone forms from mesoderm by intramembranous or endochondral ossification.
2. Limbs develop from limb buds, which consist of mesoderm and ectoderm.

SELF-QUIZ QUESTIONS

Choose the one best answer to the following questions.

1. The olecranon is at the proximal end of the:
   a. humerus.   b. radius.   c. ulna.
   d. tibia.   e. scapula.

2. The tibia articulates distally with the:
   a. femur.   b. fibula.   c. talus.
   d. cuboid.   e. Both b and c are correct.

3. Which structures are on the posterior surface of the upper limb?
   a. radial fossa and radial notch
   b. trochlea and capitulum
   c. coronoid process and coronoid fossa
   d. olecranon process and olecranon fossa
   e. lesser tubercle and intertubercular sulcus

4. The bones of the pectoral girdle:
   a. articulate with the sternum anteriorly and the vertebrae posteriorly.
   b. include both clavicles, both scapulae, and the manubrium of the sternum.
   c. are considered to be part of the axial skeleton.
   d. articulate with the head of the humerus, forming the shoulder joint.
   e. are described by none of the above.

5. The anatomical name for the socket into which the humerus fits is the:
   a. acetabulum.   b. coronoid fossa.
   c. glenoid cavity.   d. supraspinous fossa.
   e. iliac fossa.

6. All of the following are tarsal bones *except* the:
   a. cuboid.   b. triquetrum.   c. navicular.
   d. first cuneiform.   e. third cuneiform.

7. Which of the following articulates with the sacrum?
   a. ilium   b. ischium   c. pubisd.   d. Both a and b are correct.
   e. All three bones articulate with the sacrum.

8. The greater trochanter is a large bony prominence located:
   a. on the proximal part of the humerus.
   b. on the proximal part of the femur.
   c. near the tibial tuberosity.
   d. on the ilium.
   e. on the posterior surface of the scapula.

9. Which of the following statements regarding the male pelvis is *not* true?
   a. The bones are heavier and thicker than in the female.
   b. The male pelvis is narrow and deep.
   c. The coccyx is less curved anteriorly.
   d. The pelvic outlet is narrower than in the female.
   e. None of the above statements is true.

10. Which of the following is not a carpal bone?
   a. hamate.   b. cuboid.   c. pisiform.
   d. trapezium.   e. scaphoid.

Complete the following.

11. The greater sciatic notch is an indentation seen on the _____.
12. The sesamoid bone that forms in the tendon of the quadriceps muscle is the _____.

13. The lesser trochanter is on the _____ surface of the _____.

14. The portion of the pelvis superior to the pelvic brim is the _____ pelvis.

15. The _____ of the _____ fits into a depression called the acetabulum.

Are the following statements true or false?

16. The metatarsals are proximal to the tarsals.

17. The hamate is a carpal bone with a hook-shaped projection on its anterior surface.

18. The acromial extremity is located at the lateral end of the clavicle.

19. The capitulum of the humerus articulates with the styloid process of the radius.

Matching

20. Match the following bony landmarks and bones. Answers may be used more than once.

___ **(a)** fibular notch	**(1)** radius
___ **(b)** acromion	**(2)** femur
___ **(c)** coracoid process	**(3)** tibia
___ **(d)** lateral malleolus	**(4)** humerus
___ **(e)** deltoid tuberosity	**(5)** scapula
___ **(f)** radial tuberosity	**(6)** ischium
___ **(g)** anterior inferior iliac spine	**(7)** ulna
___ **(h)** linea aspera	**(8)** fibula
___ **(i)** ischial tuberosity	**(9)** ilium
___ **(j)** coracoid process	**(10)** clavicle
___ **(k)** medial malleolus	
___ **(l)** conoid tubercle	
___ **(m)** glenoid cavity	
___ **(n)** coronoid fossa	
___ **(o)** obturator foramen	
___ **(p)** gluteal tuberosity	
___ **(q)** trochlear notch	

CRITICAL THINKING QUESTIONS

1. The *Local News* reported that farmer Bob Ramsey caught his hand in a piece of machinery on Tuesday. He lost the lateral two fingers of his left hand. Science reporter Kent Clark, a high school junior, reports that Farmer Ramsey has 3 remaining phalanges. Is Kent correct or is he anatomy-challenged?

2. Rose had flat feet as a child and was told to take up ballet dancing to correct the condition. Now Rose has hallux valgus and trouble with her talocrural joint, but at least her flat feet are cured. Explain how Rose's problems relate to each other.

3. On the coast of Alaska, archaeologists have unearthed an ancient burial site that contains the skeletal remains of some Native Americans alongside some very large kayaks (a type of paddled boat). The humerus bones show unusually pronounced deltoid tuberosities. Why/How might these projections have grown so large?

4. Amy was visiting her Grandmother Amelia in the hospital. Grandmother said she had a broken hip but when Amy peeked at the chart, it said she had a fractured femur. Explain the discrepancy. Meanwhile, Grandpa Jeremiah is complaining that his "sacroiliac is acting up," making it painful for him to stand or sit for long or to walk very far. Does Grandpa really know what he's talking about? Explain.

5. Derrick, the high school gym teacher, jogs on the same banked high school track at the same time and in the same direction every day. Recently he has been experiencing an ache in one knee, especially after he sits and reads the newspaper following his jog. Nothing else in his routine has changed. What could be the problem with his knee?

ANSWERS TO FIGURE QUESTIONS

8.1 The pectoral girdles attach the upper limbs to the axial skeleton.

8.2 The weakest part of the clavicle is its midregion at the junction of the two curves.

8.3 The acromion forms the high point of the shoulder.

8.4 Each upper limb has 30 bones.

8.5 The radius articulates at the elbow with the capitulum and radial fossa of the humerus. The ulna articulates at the elbow with the trochlea, coronoid fossa, and olecranon fossa of the humerus.

8.6 The olecranon is the "elbow" part of the ulna.

8.7 The radius and ulna form the proximal and distal radioulnar joints. Their shafts are also connected by the interosseous membrane.

8.8 The scaphoid is the most frequently fractured wrist bone.

8.9 The bony pelvis attaches the lower limbs to the axial skeleton and supports the backbone and pelvic viscera.

8.10 The femur articulates with the acetabulum of the hip bone; the sacrum articulates with the auricular surface of the hip bone.

8.11 The pelvic axis is the course taken by a baby's head as it descends through the pelvis during childbirth.

8.12 Each lower limb has 30 bones.

8.13 The angle of convergence of the femurs is greater in females than males because the female pelvis is broader.

8.14 The patella is the largest sesamoid bone in the body. It is classified as a sesamoid bone because it develops in a tendon (the tendon of the quadriceps femoris muscle of the thigh).

8.15 The tibia is the weight-bearing bone of the leg.

8.16 The talus is the only tarsal bone that articulates with the tibia and the fibula.

8.17 Because the arches are not rigid, they yield when weight is applied and spring back when weight is lifted, allowing them to absorb the shock of walking.

8.18 The skeletal system arises from embryonic mesoderm.

JOINTS

9

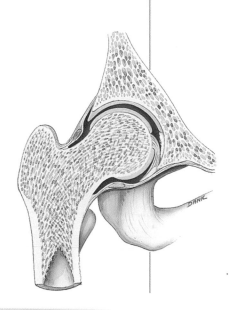

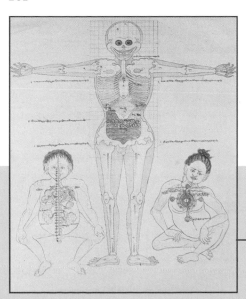

INTRODUCTION Bones are too rigid to bend without being damaged. Fortunately, flexible connective tissues form joints that hold bones together while still permitting, in most cases, some degree of movement. A **joint**, also called an **articulation** (ar-tik-ū-LĀ-shun) or **arthrosis** (ar-THRŌ-sis), is a point of contact between two bones, between bone and cartilage, or between bone and teeth. When we say one bone *articulates* with another bone, we mean that the bones form a joint. Because most movements of the body occur at joints, you can appreciate their importance. Imagine how a cast over your knee joint makes walking difficult, or how a splint on a finger limits your ability to manipulate small objects. The scientific study of joints is termed **arthrology** (ar-THROL-ō-jē; *arthr-*=joint; *-ology*=study of). The study of motion of the human body is called **kinesiology** (ki-nē-sē′-OL-ō-jē; *kinesi-*=*movement*).

We often think of medicine as a scientific search for the truth. It's easy to forget that, like art or politics, medicine is also an expression of culture.

www.wiley.com/college/apcentral

JOINT CLASSIFICATIONS

- Describe the structural and functional classifications of joints.

Joints are classified structurally based on their anatomical characteristics and functionally based on the type of movement they permit.

The structural classification of joints is based on (1) the presence or absence of a space between the articulating bones, called a synovial cavity, and (2) the type of connective tissue that binds the bones together. Structurally, joints are classified as one of the following types:

- **Fibrous joints** (FĪ-brus): The bones are held together by fibrous connective tissue that is rich in collagen fibers and there is no synovial cavity.
- **Cartilaginous joints** (kar-ti-LAJ-i-nus): The bones are held together by cartilage and there is no synovial cavity.
- **Synovial joints** (sī-NŌ-vē-al; *syn-*=together): The bones forming the joint have a synovial cavity and are united by the dense irregular connective tissue of an articular capsule, and often by accessory ligaments.

The functional classification of joints relates to the degree of movement they permit. Functionally, joints are classified as one of the following types:

- **Synarthrosis** (sin′-ar-THRŌ-sis): An immovable joint. The plural is *synarthroses.*
- **Amphiarthrosis** (am′-fē-ar-THRŌ-sis; *amphi-*=on both sides): A slightly movable joint. The plural is *amphiarthroses.*
- **Diarthrosis** (dī-ar-THRŌ-sis=movable joint): A freely movable joint. The plural is *diarthroses.* All diarthroses are synovial joints. They have a variety of shapes and permit several different types of movements (discussed later).

The following sections present the joints of the body according to their structural classifications. As we examine the structure of each type of joint, we will also explore its functional attributes.

1. On what basis are joints classified?

FIBROUS JOINTS

- Describe the structure and functions of the three types of fibrous joints.

As previously noted, **fibrous joints** lack a synovial cavity, and the articulating bones are held together very closely by fibrous connective tissue. These joints, which permit little or no movement, include sutures, syndesmoses, and gomphoses.

Sutures

A **suture** (SOO-chur; *sutur-*=seam) is a fibrous joint composed of a thin layer of dense fibrous connective tissue. Such joints are found only in the skull. An example is the coronal suture between the parietal and frontal bones (Figure 9.1a). The irregular, interlocking edges of sutures give them added strength and decrease their chance of fracturing. Because a suture is immovable, it is classified functionally as a synarthrosis.

Some sutures, although present during growth of the skull, are replaced by bone in the adult. Such a suture is called a **synostosis** (sin′-os-TŌ-sis; *os-*=bone), or bony joint—a joint in which there is a complete fusion of bone across the suture line. For example, the frontal bone grows in halves that grow together across a suture line. Usually they are completely fused by age 6 and the suture becomes obscure. If the suture persists beyond age 6, it is called a **metopic suture** (me-TŌ-pik; *metopon*=forehead). A synostosis is also classified functionally as a synarthrosis.

Syndesmoses

A **syndesmosis** (sin′-dez-MŌ-sis; *syndesmo-*=band or ligament) is a fibrous joint in which there is a greater distance between the articulating bones and more dense fibrous connective tissue than in a suture. The fibrous connective tissue can be arranged either as a bundle (ligament) or as a sheet (interosseous membrane) (Figure 9.1b). One example of a syndesmosis is the distal tibiofibular joint, where the anterior tibiofibular ligament connects the tibia and fibula. Another example is the interosseous membrane between the parallel borders of the tibia and fibula. Because it permits slight movement, a syndesmosis is classified functionally as an amphiarthrosis.

Gomphoses

A **gomphosis** (gom-FŌ-sis; *gompho-*=a bolt or nail) or *dentoalveolar joint* is a type of fibrous joint in which a cone-shaped peg fits into a socket. The only examples of gomphoses are the articulations between the roots of the teeth and their sockets (alveoli) in the maxillae and mandible (Figure 9.1c). The dense fibrous connective tissue between a tooth and its socket is the periodontal ligament (membrane). A gomphosis is classified functionally as a synarthrosis, an immovable joint. Inflammation and degeneration of the gums, periodontal ligament, and bone is called *periodontal disease.*

2. Which fibrous joints are classified as synarthroses? Which are amphiarthroses?

CARTILAGINOUS JOINTS

- Describe the structure and functions of the two types of cartilaginous joints.

Like a fibrous joint, a **cartilaginous joint** lacks a synovial cavity and allows little or no movement. Here the articulating bones

Figure 9.1   Fibrous joints.

At a fibrous joint the bones are held together by fibrous connective tissue.

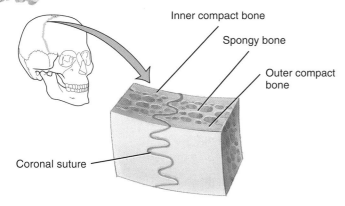

(a) Suture between skull bones

Inner compact bone
Spongy bone
Outer compact bone
Coronal suture

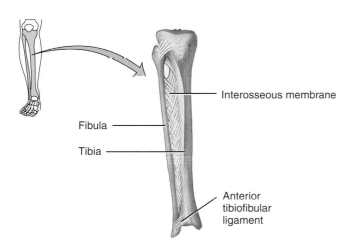

(b) Syndesmoses between tibia and fibula

Interosseous membrane
Fibula
Tibia
Anterior tibiofibular ligament

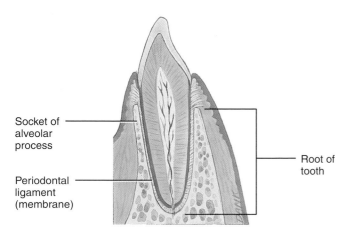

(c) Gomphosis between tooth and socket of alveolar process

Socket of alveolar process
Periodontal ligament (membrane)
Root of tooth

Functionally, why are sutures classified as synarthroses, and syndesmoses as amphiarthroses?

are tightly connected, either by hyaline cartilage or by fibrocartilage (see Table 3.3 H on page 79). The two types of cartilaginous joints are synchondroses and symphyses.

Synchondroses

A **synchondrosis** (sin′-kon-DRŌ-sis; *chondro-*=cartilage) is a cartilaginous joint in which the connecting material is hyaline cartilage. An example of a synchondrosis is the epiphyseal plate that connects the epiphysis and diaphysis of a growing bone (Figure 9.2a). A photomicrograph of the epiphyseal plate is shown in Figure 6.7 on page 154. Functionally, a synchondrosis is a synarthrosis. When bone elongation ceases, bone replaces the hyaline cartilage, and the synchondrosis becomes a synostosis, a bony joint. Another example of a synchondrosis is the joint between the first rib and the manubrium of the sternum, which also ossifies during adult life and becomes an immovable synostosis (see Figure 7.23 on page 196).

Symphyses

A **symphysis** (SIM-fi-sis=growing together) is a cartilaginous joint in which the ends of the articulating bones are covered with hyaline cartilage, but the bones are connected by a broad,

Figure 9.2   Cartilaginous joints.

At a cartilaginous joint the bones are held together by cartilage.

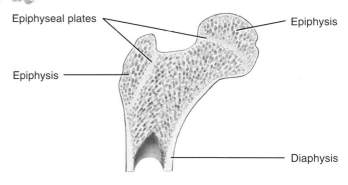

Epiphyseal plates
Epiphysis
Epiphysis
Diaphysis

(a) Synchondrosis

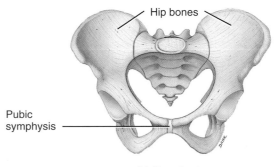

Hip bones
Pubic symphysis

(b) Symphysis

What is the structural difference between a synchondrosis and a symphysis?

flat disc of fibrocartilage. All symphyses occur in the midline of the body. The pubic symphysis between the anterior surfaces of the hip bones is one example of a symphysis (Figure 9.2b). This type of joint is also found at the junction of the manubrium and body of the sternum (see Figure 7.23 on page 196) and at the intervertebral joints between the bodies of vertebrae (see Figure 7.21a on page 193). A portion of the intervertebral disc is made up of fibrocartilage. A symphysis is classified as an amphiarthrosis.

CHECKPOINT

3. Which cartilaginous joints are synarthroses? Which are amphiarthroses?

SYNOVIAL JOINTS

OBJECTIVES

- Describe the structure of synovial joints.
- Describe the six subtypes of synovial joints.
- Describe the structure and function of bursae and tendon sheaths.

Structure of Synovial Joints

Synovial joints (si-NŌ-vē-al) have certain characteristics that distinguish them from other joints. The unique characteristic of a synovial joint is the presence of a space called a **synovial (joint) cavity** between the articulating bones (Figure 9.3). The synovial cavity allows a joint to be freely moveable. Hence, all synovial joints are classified functionally as diarthroses. The bones at a synovial joint are covered by a layer of hyaline cartilage called **articular cartilage.** The cartilage covers the articulating surfaces of the bones with a smooth, slippery surface but does not bind them together. Articular cartilage reduces friction between bones in the joint during movement and helps to absorb shock.

CARTILAGE REPLACEMENT

Several methods are available to replace damaged or diseased articular cartilage to restore normal function. In **cartilage transplantation,** chondrocytes (cartilage cells) from healthy cartilage are removed from a person's joint and sent to a laboratory and grown in a culture to produce many more cells. The newly grown cells are then placed in the damaged joint. Currently this procedure is used only in the knee joint and for individuals who have small areas of arthritis. Another method, which has been used in animal studies, involves replacing eroded cartilage with **synthetic materials,** such as polyurethane. Once the damaged cartilage is scraped away, liquid polyurethane is applied. Once it hardens it forms a smooth surface. Researchers are also using **stem cells** to try to make new cartilage. The idea is to insert stem cells into existing, damaged cartilage where they will produce new cartilage. ∎

Articular Capsule

A sleevelike **articular capsule** surrounds a synovial joint, encloses the synovial cavity, and unites the articulating bones. The articular capsule is composed of two layers, an outer fibrous capsule and an inner synovial membrane (Figure 9.3). The **fibrous capsule** usually consists of dense, irregular connective tissue (mostly collagen fibers) that attaches to the periosteum of the articulating bones. The flexibility of the fibrous capsule permits considerable movement at a joint while its great tensile strength (resistance to stretching) helps prevent the bones from dislocating. The fibers of some fibrous capsules are arranged in parallel bundles that are highly adapted for resisting strains. Such fiber bundles, called **ligaments** (liga-=bound or tied), are often designated by individual names. The strength of ligaments is one of the principal mechanical factors that holds bones close together in a synovial joint. The inner layer of the articular capsule, the **synovial membrane,** is composed of areolar connective tissue with elastic fibers. At many synovial joints, the synovial membrane includes accumulations of adipose tissue called **articular fat pads.** An example is the *infrapatellar fat pad* in the knee (see Figure 9.15c and e).

A "double-jointed" person does not really have extra joints. Individuals who are "double-jointed" have greater flexibility in their articular capsules and ligaments; the resulting increase in range of motion allows them to entertain fellow partygoers with activities such as touching their thumbs to their wrists and putting their ankles or elbows behind their necks. Unfortunately, such flexible joints are less structurally stable and are more easily dislocated.

Synovial Fluid

The synovial membrane secretes **synovial fluid** (ov-=egg), which forms a thin film over the surfaces within the articular capsule. This viscous, clear or pale yellow fluid was named for its similarity in appearance and consistency to uncooked egg white (albumin). Synovial fluid consists of hyaluronic acid secreted by fibroblast-like cells in the synovial membrane and interstitial fluid filtered from blood plasma. Its functions include reducing friction by lubricating the joint, absorbing shocks, and supplying oxygen and nutrients to and removing carbon dioxide and metabolic wastes from the chondrocytes within articular cartilage. (Recall that cartilage is an avascular tissue, so it does not have blood vessels to perform the latter function.) Synovial fluid also contains phagocytic cells that remove microbes and the debris that results from normal wear and tear in the joint. When a synovial joint is immobile for a time, the fluid becomes quite viscous (gel-like), but as joint movement increases, the fluid becomes less viscous. One of the benefits of warming up before exercise is that it stimulates the production and secretion of synovial fluid; more fluid means less stress on the joints during exercise.

We are all familiar with the cracking sounds heard as certain joints move, or the popping sounds that arise when a person pulls on the fingers to crack their knuckles. According to one theory, when the synovial cavity expands, the pressure of the synovial fluid decreases, creating a partial vacuum. The suction draws carbon dioxide and oxygen out of blood vessels in the synovial membrane, forming bubbles in the fluid. When the bubbles burst, the cracking or popping sound is heard.

◆ ASPIRATION OF SYNOVIAL FLUID

As a result of various injuries or diseases, there may be a buildup of synovial fluid in a joint cavity, resulting in pain and decreased mobility. In order to relieve the pressure and ease the pain, **aspiration** (as´-pi-RĀ-shun) **of synovial fluid** may be necessary. In this procedure, a needle is inserted into the joint cavity and the fluid is withdrawn into a syringe; the fluid is then analyzed for diagnostic purposes. For example, the fluid may contain bacteria, which confirms a diagnosis of infection, or it may contain urea crystals, which confirms a diagnosis of gout. It is also possible to inject medication into a joint cavity; anti–

Figure 9.3   Structure of a typical synovial joint. Note the two layers of the articular capsule—the fibrous capsule and the synovial membrane. Synovial fluid fills the joint cavity between the synovial membrane and the articular cartilage.

The distinguishing feature of a synovial joint is the synovial cavity between the articulating bones.

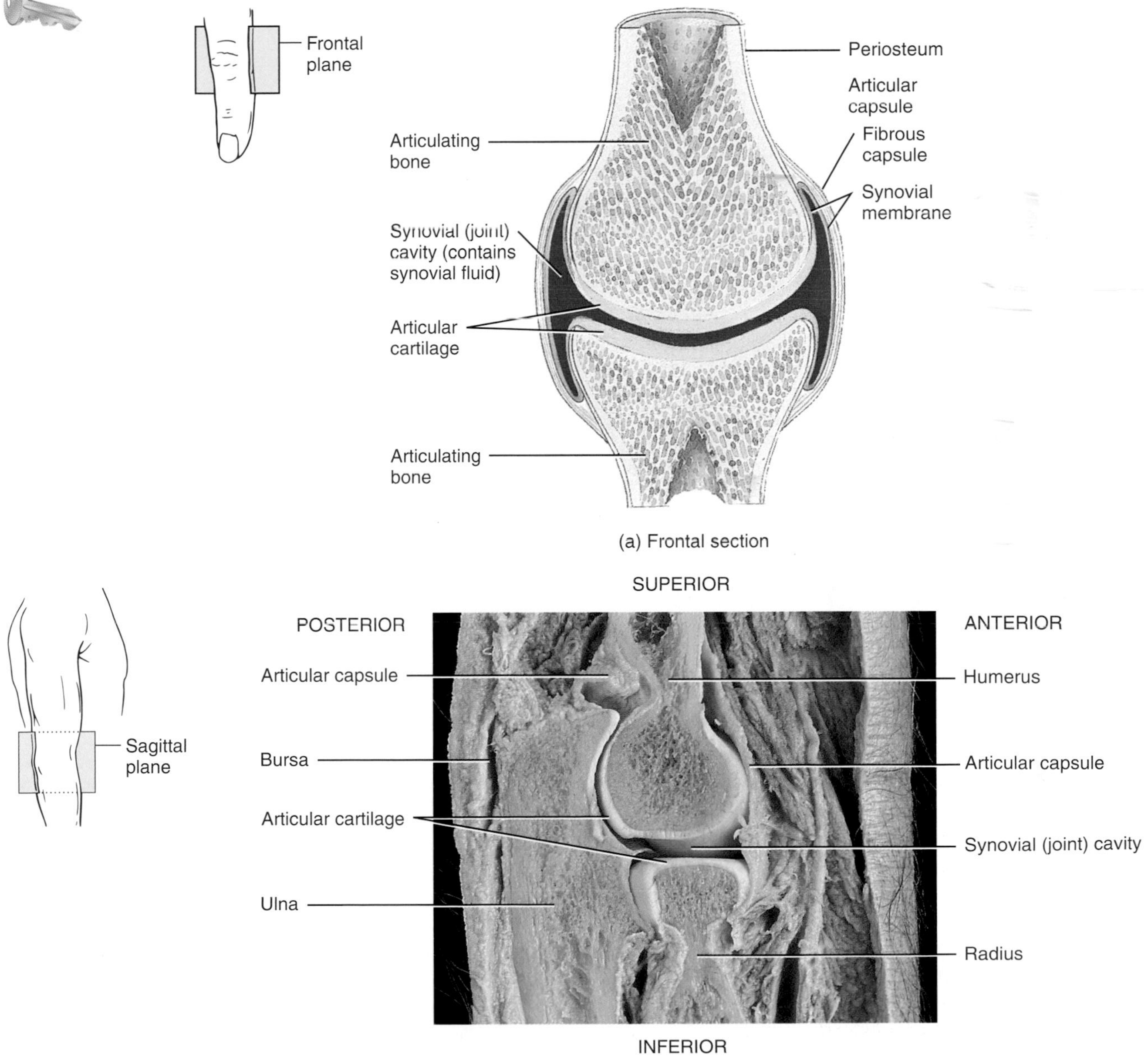

(a) Frontal section

(b) Sagittal section of right elbow joint

? What is the functional classification of synovial joints?

inflammatory drugs such as cortisol are used to decrease joint swelling caused by retention of fluid from the blood that entered the joint. ■

Accessory Ligaments and Articular Discs

Many synovial joints also contain **accessory ligaments** called extracapsular ligaments and intracapsular ligaments. *Extracapsular ligaments* lie outside the articular capsule. Examples are the fibular and tibial collateral ligaments of the knee joint (see Figure 9.15d and f). *Intracapsular ligaments* occur within the articular capsule but are excluded from the synovial cavity by folds of the synovial membrane. Examples are the anterior and posterior cruciate ligaments of the knee joint (see Figure 9.15d and f).

Inside some synovial joints, such as the knee, pads of fibrocartilage lie between the articular surfaces of the bones and are attached to the fibrous capsule. These pads are called **articular discs** or **menisci** (me-NIS-sī or me-NIS-kī; the singular is *meniscus*). Figures 9.15d and f depict the lateral and medial menisci in the knee joint. The discs usually subdivide the synovial cavity into two separate spaces. This separation can allow separate movements to occur in each space. As you will see later, separate movements also occur in the respective compartments of the temporomandibular joint (TMJ) (see page 245). By modifying the shape of the joint surfaces of the articulating bones, articular discs allow two bones of different shapes to fit together more tightly. Articular discs also help to maintain the stability of the joint and direct the flow of synovial fluid to the areas of greatest friction.

TORN CARTILAGE AND ARTHROSCOPY

The tearing of articular discs (menisci) in the knee, commonly called **torn cartilage,** occurs often among athletes. Such damaged cartilage will begin to wear and may precipitate arthritis unless surgically removed (meniscectomy). Surgical repair of the torn cartilage may be assisted by **arthroscopy** (ar-THROS-kō-pē; *-scopy*=observation). This minimally invasive procedure involves examination of the interior of a joint, usually the knee, with an *arthroscope,* a lighted pencil-thin instrument used for visualization. Arthroscopy is used to determine the nature and extent of damage following knee injury and to monitor the progression of disease and the effects of therapy. In addition, the insertion of surgical instruments also enables a physician to help remove torn cartilage and repair damaged cruciate ligaments in the knee; to help obtain tissue samples for analysis; and to help perform surgery on other joints, such as the shoulder, elbow, ankle, and wrist. ■

Nerve and Blood Supply

The nerves that supply a joint are the same as those that supply the skeletal muscles that move the joint. Synovial joints contain many nerve endings that are distributed to the articular capsule and associated ligaments. Some of the nerve endings convey information about pain from the joint to the spinal cord and brain for process-

ing. Other nerve endings are responsive to the degree of movement and stretch at a joint. This information is also relayed to the spinal cord and brain, which may respond by sending impulses through different nerves to the muscles to adjust body movements.

Although many of the components of synovial joints are avascular, arteries in the vicinity send out numerous branches that penetrate the ligaments and articular capsule to deliver oxygen and nutrients. Veins remove carbon dioxide and wastes from the joints. The arterial branches from several different arteries typically join together around a joint before penetrating the articular capsule. The chondrocytes of articular cartilage of a synovial joint receive oxygen and nutrients from synovial fluid derived from blood, whereas all other joint tissues are supplied directly by arteries. Carbon dioxide and wastes pass from chondrocytes of articular cartilage into synovial fluid and then into veins; carbon dioxide and wastes from all other joint structures pass directly into veins.

SPRAIN AND STRAIN

A **sprain** is the forcible wrenching or twisting of a joint that stretches or tears its ligaments but does not dislocate the bones. It occurs when the ligaments are stressed beyond their normal capacity. Sprains also may damage surrounding blood vessels, muscles, tendons, or nerves. Severe sprains may be so painful that the joint cannot be moved. There is considerable swelling, which results from hemorrhage of ruptured blood vessels. The ankle joint is most often sprained; the lower back is another frequent location. A **strain** is a stretched or partially torn muscle. It often occurs when a muscle contracts suddenly and powerfully—for example, in the leg muscles of sprinters when they accelerate quickly. ■

Bursae and Tendon Sheaths

The various movements of the body create friction between moving parts. Saclike structures called **bursae** (BER-sē= purses singular is *bursa*) are strategically situated to alleviate friction in some joints, such as the shoulder and knee joints (see Figures 9.12 and 9.15c and e). Bursae are not strictly parts of synovial joints, but they do resemble joint capsules because their walls consist of connective tissue lined by a synovial membrane. They are also filled with a small amount of fluid similar to synovial fluid. Bursae are located between the skin and bone, tendons and bones, muscles and bones, and ligaments and bones. The fluid-filled bursal sacs cushion the movement of these body parts over one another.

Structures called tendon sheaths also reduce friction at joints. **Tendon sheaths** are tubelike bursae that wrap around tendons that experience considerable friction. This occurs where tendons pass through synovial cavities, such as the tendon of the biceps brachii muscle at the shoulder joint (see Figure 9.12c). Tendon sheaths are also found at the wrist and ankle, where many tendons come together in a confined space (see Figure 11.24 on page 368), and in the fingers and toes, where there is a great deal of movement (see Figure 11.20 on page 350).

BURSITIS

An acute or chronic inflammation of a bursa is called **bursitis.** It is usually caused by irritation from repeated, excessive exertion of a joint. The condition may also be caused by trauma, an acute or chronic infection (including syphilis and tuberculosis), or rheumatoid arthritis (described on page 260). Symptoms include pain, swelling, tenderness, and limited movement. Treatment may include oral anti-inflammatory agents and injections of cortisol-like steroids. ■

Types of Synovial Joints

Although all synovial joints are similar in structure, the shapes of the articulating surfaces vary and thus various types of movements are possible. Accordingly, synovial joints are divided into six subtypes: planar, hinge, pivot, condyloid, saddle, and ball-and-socket.

Planar Joints

The articulating surfaces of bones in a **planar joint** are flat or slightly curved (Figure 9.4a). Planar joints primarily permit side-to-side and back-and-forth gliding movements (described shortly). These joints are said to be *nonaxial* because the motion they allow does not occur around an axis or along a plane. Some examples of planar joints are the intercarpal joints (between carpal bones at the wrist); intertarsal joints (between tarsal bones at the ankle); sternoclavicular joints (between the manubrium of the sternum and the clavicle); acromioclavicular joints (between the acromion of the scapula and the clavicle); sternocostal joints (between the sternum and ends of the costal cartilages at the tips of the second through seventh pairs of ribs); and vertebrocostal joints (between the heads and tubercles of ribs and transverse processes of thoracic vertebrae). X-ray films made during wrist and ankle movements reveal some rotation of the small carpal and tarsal bones in addition to their predominant gliding movements.

Hinge Joints

In a **hinge joint,** the convex surface of one bone fits into the concave surface of another bone (Figure 9.4b). As the name implies, hinge joints produce an angular, opening-and-closing motion like that of a hinged door. In most joint movements, one bone remains in a fixed position while the other moves around an axis. Hinge joints are said to be *monaxial (uniaxial)* because they typically allow motion around a single axis. Examples of hinge joints are the knee, elbow, ankle, and interphalangeal joints.

Pivot Joints

In a **pivot joint,** the rounded or pointed surface of one bone articulates with a ring formed partly by another bone and partly by a ligament (Figure 9.4c). A pivot joint is monaxial because it allows rotation around its own longitudinal axis only. Examples of pivot joints are the atlanto-axial joint, in which the atlas rotates around the axis and permits the head to turn from side-to-side as in signifying "no" (see Figure 9.9a), and the radioulnar joints that enable the palms to turn anteriorly and posteriorly (see Figure 9.10h).

Figure 9.4   Subtypes of synovial joints. For each subtype, a drawing of the actual joint and a simplified diagram are shown.

🔑 **Synovial joints are classified into subtypes on the basis of the shapes of the articulating bone surfaces.**

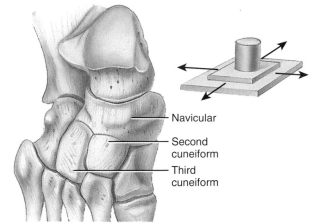

(a) Planar joint between the navicular and second and third cuneiforms of the tarsus in the foot

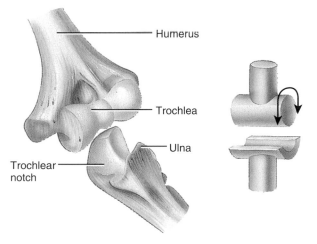

(b) Hinge joint between trochlea of humerus and trochlear notch of ulna at the elbow

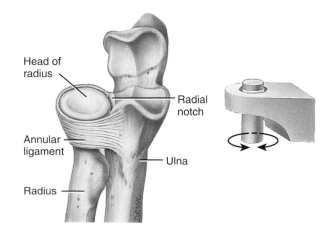

(c) Pivot joint between head of radius and radial notch of ulna

continues

Figure 9.4 (continued)

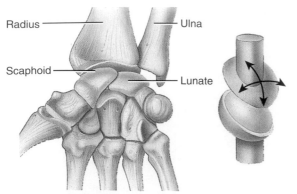

(d) Condyloid joint between radius and scaphoid and lunate bones of the carpus (wrist)

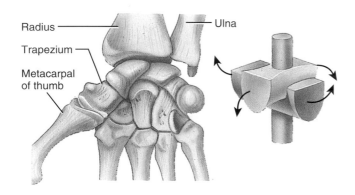

(e) Saddle joint between trapezium of carpus (wrist) and metacarpal of thumb

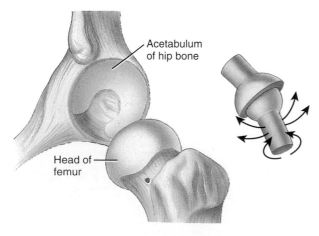

(f) Ball-and-socket joint between head of the femur and acetabulum of the hip bone

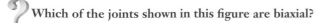

? Which of the joints shown in this figure are biaxial?

Condyloid Joints

In a **condyloid joint** (KON-di-loyd; *condyl-*=knuckle), also called an *ellipsoidal joint*, the convex oval-shaped projection of one bone fits into the oval-shaped depression of another bone (Figure 9.4d). A condyloid joint is *biaxial* because the movement it permits is around two axes. Notice that you can move your index finger both up-and-down and from side-to-side. Examples are the wrist and the metacarpophalangeal joints of the second through fifth digits.

Saddle Joints

In a **saddle joint,** the articular surface of one bone is saddle-shaped, and the articular surface of the other bone fits into the "saddle" as a sitting rider would (Figure 9.4e). A saddle joint is a modified condyloid joint in which the movement is somewhat freer. Saddle joints are *biaxial*, producing side-to-side and up-and-down movements. An example of a saddle joint is the carpometacarpal joint between the trapezium of the carpus and metacarpal of the thumb.

Ball-and-Socket Joints

A **ball-and-socket joint** consists of the ball-like surface of one bone fitting into a cuplike depression of another bone (Figure 9.4f). Such joints are *multiaxial (polyaxial)* because they permit movement around three axes plus all directions in between. Examples of functional ball-and-socket joints are the shoulder and hip joints. At the shoulder joint the head of the humerus fits into the glenoid cavity of the scapula. At the hip joint the head of the femur fits into the acetabulum of the hip bone.

Table 9.1 summarizes the structural and functional categories of joints.

CHECKPOINT

4. How does the structure of synovial joints classify them as diarthroses?
5. What are the functions of articular cartilage, synovial fluid, and articular discs?
6. What types of sensations are perceived at joints, and from what sources do joint components receive their nourishment?
7. In what ways are bursae similar to joint capsules? How do they differ?
8. Which types of joints are nonaxial, monaxial, biaxial, and multiaxial?

TYPES OF MOVEMENTS AT SYNOVIAL JOINTS

OBJECTIVE

● Describe the types of movements that can occur at synovial joints.

Anatomists, physical therapists, and kinesiologists use specific terminology to designate movements that can occur at synovial joints. These precise terms may indicate the form of motion, the direction of movement, or the relationship of one body part to

TABLE 9.1   SUMMARY OF STRUCTURAL AND FUNCTIONAL CLASSIFICATIONS OF JOINTS

STRUCTURAL CLASSIFICATION	DESCRIPTION	FUNCTIONAL	EXAMPLE
Fibrous	Articulating bones held together by fibrous connective tissue; no synovial cavity.		
Suture	Articulating bones united by a thin layer of dense fibrous connective tissue, found between bones of the skull. With age, some sutures are replaced by a synostosis, in which the bones fuse across the former suture.	Synarthrosis (immovable).	Frontal suture.
Syndesmosis	Articulating bones united by dense fibrous connective tissue, either a ligament or an interosseous membrane.	Amphiarthrosis (slightly movable).	Distal tibiofibular joint.
Gomphosis	Articulating bones united by periodontal ligament; cone-shaped peg fits into a socket.	Synarthrosis.	Roots of teeth in alveoli (sockets) of maxillae and mandible.
Cartilaginous	Articulating bones united by cartilage; no synovial cavity.		
Synchondrosis	Connecting material is hyaline cartilage; becomes a synostosis when bone elongation ceases.	Synarthrosis.	Epiphyseal plate at joint between the diaphysis and epiphysis of a long bone.
Symphysis	Connecting material is a broad, flat disc of fibrocartilage.	Amphiarthrosis.	Intervertebral joints and pubic symphysis.
Synovial	Characterized by a synovial cavity, articular cartilage, and an articular capsule; may contain accessory ligaments, articular discs, and bursae.		
Planar	Articulated surfaces are flat or slightly curved.	Nonaxial diarthrosis (freely movable); gliding motion.	Intercarpal, intertarsal, sternocostal (between sternum and the 2nd–7th pairs of ribs), vertebrocostal joints.
Hinge	Convex surface fits into a concave surface.	Monaxial diarthrosis; angular motion.	Elbow, ankle, and interphalangeal joints.
Pivot	Rounded or pointed surface fits into a ring formed partly by bone and partly by a ligament.	Monaxial diarthrosis; rotation.	Atlanto-axial and radioulnar joints.
Condyloid	Oval-shaped projection fits into an oval-shaped depression.	Biaxial diarthrosis; angular motion.	Radiocarpal and metacarpophalangeal joints.
Saddle	Articular surface of one bone is saddle-shaped, and the articular surface of the other bone "sits" in the saddle.	Biaxial diarthrosis; angular motion.	Carpometacarpal joint between trapezium and thumb.
Ball-and-socket	Ball-like surface fits into a cuplike depression.	Multiaxial diarthrosis; angular motion and rotation.	Shoulder and hip joints.

another during movement. Movements at synovial joints are grouped into four main categories: (1) gliding, (2) angular movements, (3) rotation, and (4) special movements. This last category includes movements that occur only at certain joints.

Gliding

Gliding is a simple movement in which relatively flat bone surfaces move back-and-forth and from side-to-side with respect to one another (Figure 9.5). There is no significant alteration of the angle between the bones. Gliding movements are limited in range due to the structure of the articular capsule and associated ligaments and bones. Gliding occurs at planar joints (see Figure 9.4a).

Angular Movements

In **angular movements,** there is an increase or a decrease in the angle between articulating bones. The principal angular move-

Figure 9.5   Gliding movements at synovial joints.

Gliding motion consists of side-to-side and back-and-forth movements.

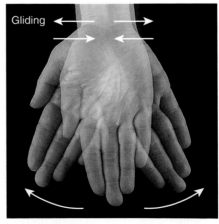

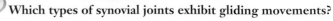

Which types of synovial joints exhibit gliding movements?

Figure 9.6   Angular movements at synovial joints—flexion, extension, hyperextension, and lateral flexion.

In angular movements, there is an increase or decrease in the angle between articulating bones.

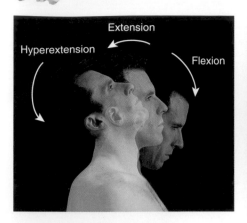

(a) Atlanto-occipital and cervical intervertebral joints

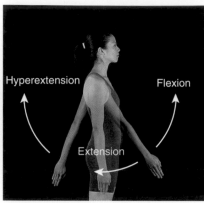

(b) Shoulder joint

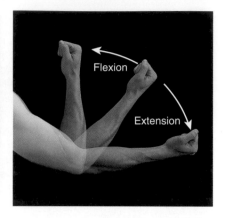

(c) Elbow joint

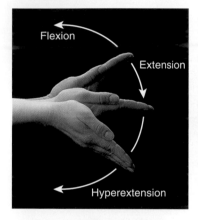

(d) Wrist joint

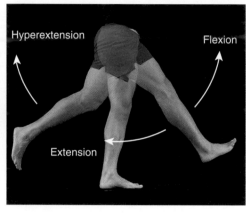

(e) Hip joint

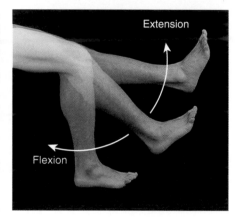

(f) Knee joint

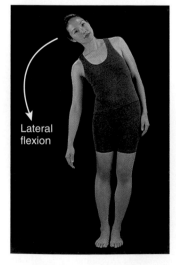

(g) Intervertebral joints

? **What two examples of flexion do not occur in the sagittal plane?**

ments are flexion, extension, lateral extension, hyperextension, abduction, adduction, and circumduction. These movements are always discussed with respect to the body in the anatomical position.

Flexion, Extension, Lateral Flexion, and Hyperextension

Flexion and extension are opposite movements. In **flexion** (FLEK-shun; *flex-*=to bend) there is a decrease in the angle between articulating bones; in **extension** (eks-TEN-shun; *exten-*=to stretch out) there is an increase in the angle between articulating bones, often to restore a part of the body to the anatomical position after it has been flexed (Figure 9.6). Both movements usually occur along the sagittal plane. Hinge, pivot, condyloid, saddle, and ball-and-socket joints all permit flexion and extension (see Figure 9.4b–f).

All of the following are examples of flexion:

- Bending the head toward the chest at the atlanto-occipital joint between the atlas (the first vertebra) and the occipital bone of the skull, and at the cervical intervertebral joints between the cervical vertebrae (Figure 9.6a)
- Bending the trunk forward at the intervertebral joints
- Moving the humerus forward at the shoulder joint, as in swinging the arms forward while walking (Figure 9.6b)
- Moving the forearm toward the arm at the elbow joint between the humerus, ulna, and radius (Figure 9.6c)

Figure 9.7 **Angular movements at synovial joints—abduction and adduction.**

Abduction and adduction usually occur along the frontal plane.

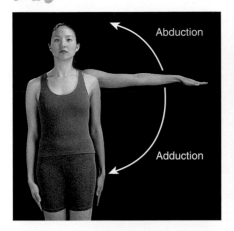

(a) Shoulder joint

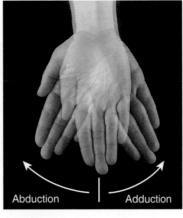

(b) Wrist joint

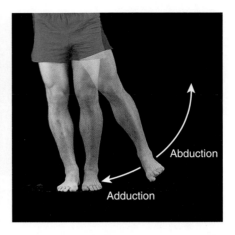

(c) Hip joint

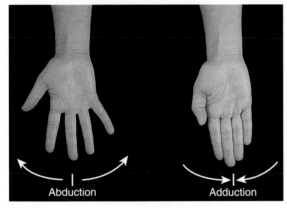

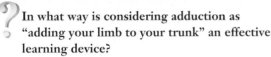

In what way is considering adduction as "adding your limb to your trunk" an effective learning device?

- Moving the palm toward the forearm at the wrist or radiocarpal joint between the radius and carpals (Figure 9.6d)
- Bending of the digits of the hand or feet at the interphalangeal joints between phalanges
- Moving the femur forward at the hip joint between the femur and hip bone, as in walking (Figure 9.6e)
- Moving the leg toward the thigh at the knee or tibiofemoral joint between the tibia, femur, and patella, as occurs when bending the knee (Figure 9.6f)

Although flexion and extension usually occur along the sagittal plane, there are a few exceptions. For example, flexion of the thumb involves movement of the thumb medially across the palm at the carpometacarpal joint between the trapezium and metacarpal of the thumb, as in touching the thumb to the opposite side of the palm (see Figure 11.20f on page 351). Another example is movement of the trunk sideways to the right or left at the waist as in a sidebend. This movement, which occurs along the frontal plane and involves the intervertebral joints, is called **lateral flexion** (Figure 9.6g).

Continuation of extension beyond the anatomical position is called **hyperextension** (*hyper-*=beyond or excessive). Examples of hyperextension include:

- Bending the head backward at the atlanto-occipital and cervical intervertebral joints (Figure 8.6a)
- Bending the trunk backward at the intervertebral joints as in a backbend
- Moving the humerus backward at the shoulder joint, as in swinging the arms backward while walking (Figure 9.6b)
- Moving the palm backward at the wrist joint (Figure 9.6d)
- Moving the femur backward at the hip joint, as in walking (Figure 9.6e)

Hyperextension of hinge joints, such as the elbow, interphalangeal, and knee joints, is usually prevented by factors such as the arrangement of ligaments and the anatomical alignment of the bones.

Abduction, Adduction, and Circumduction

Abduction (ab-DUK-shun; *ab-*=away; *-duct*=to lead) is the movement of a bone away from the midline, whereas **adduction** (ad-DUK-shun; *ad-*=toward) is the movement of a bone toward the midline. Both movements usually occur along the frontal plane. Condyloid, saddle, and ball-and-socket joints permit abduction and adduction. Examples of abduction include moving the humerus laterally at the shoulder joint, moving the palm laterally at the wrist joint, and moving the femur laterally at the hip joint (Figure 9.7a–c). The movement that returns each of these body parts to the anatomical position is adduction (Figure 9.7a–c).

With respect to the digits, the midline of the body is not used as a point of reference for abduction and adduction. In abduction of the fingers (but not the thumb), an imaginary line is drawn through the longitudinal axis of the middle (longest)

Figure 9.8 Angular movements at synovial joints—circumduction.

 Circumduction is the movement of the distal end of a body part in a circle.

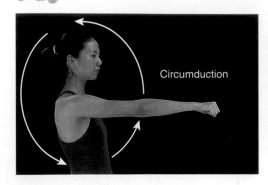 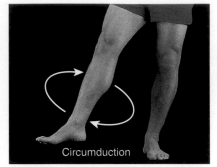

(a) Shoulder joint   (b) Hip joint

Which movements in continuous sequence produce circumduction?

finger, and the fingers move away (spread out) from the middle finger (Figure 9.7d). In abduction of the thumb, the thumb moves away from the palm in the sagittal plane (see Figure 11.20f on page 351). Abduction of the toes is relative to an imaginary line drawn through the second toe. Adduction of the fingers and toes involves returning them to the anatomical position. Adduction of the thumb moves the thumb toward the palm in the sagittal plane (see Figure 11.20f).

Circumduction (ser-kum-DUK-shun; *circ-*=circle) is movement of the distal end of a body part in a circle (Figure 9.8). Circumduction occurs as a result of a continuous sequence of flexion, abduction, extension, and adduction. Examples of circumduction are moving the humerus in a circle at the shoulder joint (Figure 9.8a), moving the hand in a circle at the wrist joint, moving the thumb in a circle at the carpometacarpal joint, moving the fingers in a circle at the metacarpophalangeal joints (between the metacarpals and phalanges), and moving the femur in a circle at the hip joint (Figure 9.8b). Both the shoulder and

hip joints permit circumduction, but its components, flexion, abduction, extension, and adduction, are more limited in the hip joints than the shoulder joints due to the tension on certain ligaments and muscles (see Exhibits 9.2 and 9.4).

Rotation

In **rotation** (rō-TĀ-shun; *rota-*=revolve), a bone revolves around its own longitudinal axis. Pivot and ball-and-socket joints permit rotation. One example is turning the head from side-to-side at the atlanto-axial joint, as in signifying "no" (Figure 9.9a). Another is turning the trunk from side-to-side at the intervertebral joints while keeping the hips and lower limbs in the anatomical position. In the limbs, rotation is defined relative to the midline, and specific qualifying terms are used. If the anterior surface of a bone of the limb is turned toward the midline, the movement is called *medial (internal) rotation*. You can medially rotate the humerus at the shoulder joint as follows:

Figure 9.9 Rotation at synovial joints.

In rotation, a bone revolves around its own longitudinal axis.

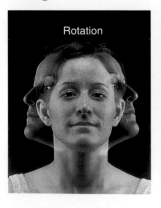 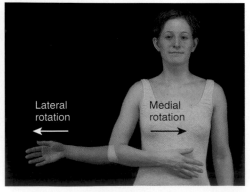 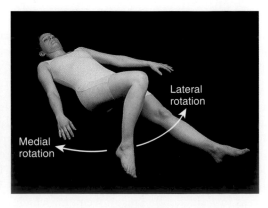

(a) Atlanto-axial joint   (b) Shoulder joint   (c) Hip joint

How do medial and lateral rotation differ?

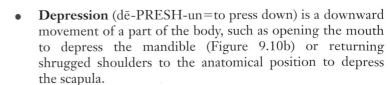

Starting in the anatomical position, flex your elbow and then draw your palm across the chest (Figure 9.9b). Medial rotation of the forearm at the radioulnar joints (between the radius and ulna) involves turning the palm medially from the anatomical position (see Figure 9.10h). You can medially rotate the femur at the hip joint as follows: Lie on your back, bend your knee, and then move your leg and foot laterally from the midline. Although you are moving your leg and foot laterally, the femur is rotating medially (Figure 9.9c). Medial rotation of the leg at the knee joint can be produced by sitting on a chair, bending your knee, raising your lower limb off the floor, and turning your toes medially. If the anterior surface of the bone of a limb is turned away from the midline, the movement is called *lateral (external) rotation* (see Figure 9.9b, c).

Special Movements

As noted previously, special movements occur only at certain joints. They include elevation, depression, protraction, retraction, inversion, eversion, dorsiflexion, plantar flexion, supination, pronation, and opposition (Figure 9.10):

- **Elevation** (el-e-VĀ-shun=to lift up) is an upward movement of a part of the body, such as closing the mouth at the temporomandibular joint (between the mandible and temporal bone) to elevate the mandible (Figure 9.10a) or shrugging the shoulders at the acromioclavicular joint to elevate the scapula. Other bones that may be elevated (or depressed) include the hyoid, clavicle, and ribs.

- **Depression** (dē-PRESH-un=to press down) is a downward movement of a part of the body, such as opening the mouth to depress the mandible (Figure 9.10b) or returning shrugged shoulders to the anatomical position to depress the scapula.

- **Protraction** (prō-TRAK-shun=to draw forth) is a movement of a part of the body anteriorly in the transverse plane. You can protract your mandible at the temporomandibular joint by thrusting it outward (Figure 9.10c) or protract your clavicles at the acromioclavicular and sternoclavicular joints by crossing your arms.

- **Retraction** (rē-TRAK-shun=to draw back) is a movement of a protracted part of the body back to the anatomical position (Figure 9.10d).

- **Inversion** (in-VER-zhun=to turn inward) is movement of the soles of the feet medially at the intertarsal joints (between the tarsals) (Figure 9.10e). Physical therapists also refer to inversion of the feet as *supination*.

- **Eversion** (ē-VER-zhun=to turn outward) is a movement of the soles laterally at the intertarsal joints (Figure 9.10f). Physical therapists also refer to eversion of the feet as *pronation*.

- **Dorsiflexion** (dor-si-FLEK-shun) refers to bending of the foot at the ankle or talocrural joint (between the tibia, fibula, and talus) in the direction of the dorsum (superior surface) (Figure 9.10g). Dorsiflexion occurs when you stand on your heels.

Figure 9.10 Special movements at synovial joints.

Special movements occur only at certain synovial joints.

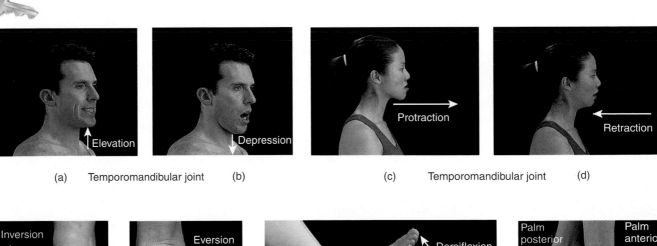
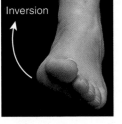
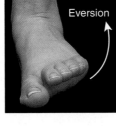
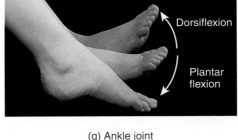
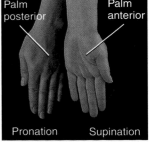

(a)   Temporomandibular joint   (b)   (c)   Temporomandibular joint   (d)

(e)   Intertarsal joints   (f)   (g) Ankle joint

(h) Radioulnar joint

What movement of the shoulder girdle is involved in bringing the arms forward until the elbows touch?

- **Plantar flexion** involves bending of the foot at the ankle joint in the direction of the plantar or inferior surface (Figure 9.10g), as when you elevate your body by standing on your toes.
- **Supination** (soo-pi-NĀ-shun) is a movement of the forearm at the proximal and distal radioulnar joints in which the palm is turned anteriorly (Figure 9.10h). This position of the palms is one of the defining features of the anatomical position.
- **Pronation** (prō-NĀ-shun) is a movement of the forearm at the proximal and distal radioulnar joints in which the distal end of the radius crosses over the distal end of the ulna and the palm is turned posteriorly (Figure 9.10h).

- **Opposition** (op-ō-ZISH-un) is the movement of the thumb at the carpometacarpal joint (between the trapezium and metacarpal of the thumb) in which the thumb moves across the palm to touch the tips of the fingers on the same hand (see Figure 11.20f on page 351). This is the distinctive digital movement that gives humans and other primates the ability to grasp and manipulate objects very precisely.

A summary of the movements that occur at synovial joints is presented in Table 9.2.

CHECKPOINT

9. Alone or with a partner, demonstrate each movement listed in Table 9.2.

TABLE 9.2 SUMMARY OF MOVEMENTS AT SYNOVIAL JOINTS

MOVEMENT	DESCRIPTION
Gliding	Movement of relatively flat bone surfaces back-and-forth and from side-to-side over one another; little change in the angle between bones.
Angular	Increase or decrease in the angle between bones.
Flexion	Decrease in the angle between articulating bones, usually in the sagittal plane.
Lateral flexion	Movement of the trunk in the frontal plane.
Extension	Increase in the angle between articulating bones, usually in the sagittal plane.
Hyperextension	Extension beyond the anatomical position.
Abduction	Movement of a bone away from the midline, usually in the frontal plane.
Adduction	Movement of a bone toward the midline, usually in the frontal plane.
Circumduction	Flexion, abduction, extension, and adduction in succession, in which the distal end of a body part moves in a circle.
Rotation	Movement of a bone around its longitudinal axis; in the limbs, it may be medial (toward midline) or lateral (away from midline).
Special	Occurs at specific joints.
Elevation	Superior movement of a body part.
Depression	Inferior movement of a body part.
Protraction	Anterior movement of a body part in the transverse plane.
Retraction	Posterior movement of a body part in the transverse plane.
Eversion	Lateral movement of the soles so that they face away from each other.
Inversion	Medial movement of the soles so that they face each other.
Dorsiflexion	Bending the foot in the direction of the dorsum (superior surface).
Plantar flexion	Bending the foot in the direction of the plantar surface (sole).
Supination	Movement of the forearm that turns the palm anteriorly.
Pronation	Movement of the forearm that turns the palm posteriorly.
Opposition	Movement of the thumb across the palm to touch fingertips on the same hand.

SELECTED JOINTS OF THE BODY

In Chapters 7 and 8 we discussed the major bones and their markings. In this chapter we have examined how joints are classified according to their structure and function, and we have introduced the movements that occur at joints. Table 9.3 (selected joints of the axial skeleton) and Table 9.4 (selected joints of the appendicular skeleton) will help you integrate the information you have learned in all three chapters. These tables list some of the major joints of the body according to their articular components (the bones that enter into their formation), their structural and functional classification, and the type(s) of movement that occurs at each joint.

Next we examine in detail six selected joints of the body in a series of exhibits. Each exhibit considers a specific synovial joint and contains (1) a definition—a description of the type of joint and the bones that form the joint; (2) the anatomical components—a description of the major connecting ligaments, articular disc, articular capsule, and other distinguishing features of the joint; and (3) the joint's possible movements. Each exhibit also refers you to a figure that illustrates the joint. The joints described are the temporomandibular joint (TMJ), shoulder (humeroscapular or glenohumeral) joint, elbow joint, hip (coxal) joint, knee (tibiofemoral) joint, and ankle (talocrural) joint. Because these joints are described in detail in Exhibits 9.1 through 9.6, they are not included in Tables 9.3 and 9.4.

TABLE 9.3   SELECTED JOINTS OF THE AXIAL SKELETON

JOINT	ARTICULAR COMPONENTS	CLASSIFICATION	MOVEMENTS
Suture	Between skull bones.	*Structural:* fibrous. *Functional:* synarthrosis.	None.
Atlanto-occipital	Between superior articular facets of atlas and occipital condyles of occipital bone.	*Structural:* synovial (condyloid). *Functional:* diarthrosis.	Flexion and extension of head and slight lateral flexion of head to either side.
Atlanto-axial	(1) Between dens of axis and anterior arch of atlas and (2) between lateral masses of atlas and axis.	*Structural:* synovial (pivot) between dens and anterior arch, and synovial (planar) between lateral masses. *Functional:* diarthrosis.	Rotation of head.
Intervertebral	(1) Between vertebral bodies and (2) between vertebral arches.	*Structural:* cartilaginous (symphysis) between vertebral bodies, and synovial (planar) between vertebral arches. *Functional:* amphiarthrosis between vertebral bodies, and diarthrosis between vertebral arches.	Flexion, extension, lateral flexion, and rotation of vertebral column.
Vertebrocostal	(1) Between facets of heads of ribs and facets of bodies of adjacent thoracic vertebrae and (2) between articular part of tubercles of ribs and facets of transverse processes of thoracic vertebrae.	*Structural:* synovial (planar). *Functional:* diarthrosis.	Slight gliding.
Sternocostal	Between sternum and first seven pairs of ribs.	*Structural:* cartilaginous (synchondrosis) between sternum and first pair of ribs, and synovial (planar) between sternum and second through seventh pairs of ribs. *Functional:* synarthrosis between sternum and first pair of ribs, and diarthrosis between sternum and second through seventh pairs of ribs.	None between sternum and first pair of ribs; slight gliding between sternum and second through seventh pairs of ribs.
Lumbosacral	(1) Between body of fifth lumbar vertebra and base of sacrum and (2) between inferior articular facets of fifth lumbar vertebra and superior articular facets of first vertebra of sacrum.	*Structural:* cartilaginous (symphysis) between body and base, and synovial (planar) between articular facets. *Functional:* amphiarthrosis between body and base, and diarthrosis between articular facets.	Flexion, extension, lateral flexion, and rotation of vertebral column.

TABLE 9.4   SELECTED JOINTS OF THE APPENDICULAR SKELETON

JOINT	ARTICULAR COMPONENTS	CLASSIFICATION	MOVEMENTS
Sternoclavicular	Between sternal end of clavicle, manubrium of sternum, and first costal cartilage.	*Structural:* synovial (planar and pivot). *Functional:* diarthrosis.	Gliding, with limited movements in nearly every direction.
Acromioclavicular	Between acromion of scapula and acromial end of clavicle.	*Structural:* synovial (planar). *Functional:* diarthrosis.	Gliding and rotation of scapula on clavicle.
Radioulnar	Proximal radioulnar joint between head of radius and radial notch of ulna; distal radioulnar joint between ulnar notch of radius and head of ulna.	*Structural:* synovial (pivot). *Functional:* diarthrosis.	Rotation of forearm.
Wrist (radiocarpal)	Between distal end of radius and scaphoid, lunate, and triquetrum of carpus.	*Structural:* synovial (condyloid). *Functional:* diarthrosis.	Flexion, extension, abduction, adduction, circumduction, and slight hyperextension of wrist.
Intercarpal	Between proximal row of carpal bones, distal row carpal bones, and between both rows of carpal bones (midcarpal joints).	*Structural:* synovial (planar), except for hamate, scaphoid, and lunate (midcarpal) joint, which is synovial (saddle). *Functional:* diarthrosis.	Gliding plus flexion, extension abduction, adduction, and slight rotation at midcarpal joints.
Carpometacarpal	Carpometacarpal joint of thumb between trapezium of carpus and first metacarpal; carpometacarpal joints of remaining digits formed between carpus and second through fifth metacarpals.	*Structural:* synovial (saddle) at thumb and synovial (planar) at remaining digits. *Functional:* diarthrosis.	Flexion, extension, abduction, adduction, and circumduction at thumb, and gliding at remaining digits.
Metacarpophalangeal and metatarsophalangeal	Between heads of metacarpals (or metatarsals) and bases of proximal phalanges.	*Structural:* synovial (condyloid). *Functional:* diarthrosis.	Flexion, extension, abduction, adduction, and circumduction of phalanges.
Interphalangeal	Between heads of phalanges and bases of more distal phalanges.	*Structural:* synovial (hinge). *Functional:* diarthrosis.	Flexion and extension of phalanges.
Sacroiliac	Between auricular surfaces of sacrum and ilia of hip bones.	*Structural:* synovial (planar). *Functional:* diarthrosis.	Slight gliding (even more so during pregnancy).
Pubic symphysis	Between anterior surfaces of hip bones.	*Structural:* cartilaginous (symphysis) *Functional:* amphiarthrosis.	Slight movements (even more so during pregnancy).
Tibiofibular	Proximal tibiofibular joint between lateral condyle of tibia and head of fibula; distal tibiofibular joint between distal end of fibula and fibular notch of tibia.	*Structural:* synovial (planar) at proximal joint, and fibrous (syndesmosis) at distal joint. *Functional:* diarthrosis at proximal joint, and amphiarthrosis at distal joint.	Slight gliding at proximal joint, and slight rotation of fibula during dorsiflexion of foot.
Intertarsal	Subtalar joint between talus and calcaneus of tarsus; talocalcaneonavicular joint between talus and calcaneus and navicular of tarsus; calcaneocuboid joint between calcaneus and cuboid of tarsus.	*Structural:* synovial (planar) at subtalar and calcaneocuboid joints, and synovial at talocalcaneonavicular joint. *Functional:* diarthrosis.	Inversion and eversion of foot.
Tarsometatarsal	Between three cuneiforms of tarsus and bases of five metatarsal bones.	*Structural:* synovial (planar). *Functional:* diarthrosis.	Slight gliding.

EXHIBIT 9.1   TEMPOROMANDIBULAR JOINT OR TMJ (Figure 9.11)

OBJECTIVE

● Describe the anatomical components of the temporomandibular joint, and explain the movements that can occur at this joint.

Definition

The **temporomandibular joint** is a combined hinge and planar joint formed by the condylar process of the mandible and the mandibular fossa and articular tubercle of the temporal bone. The temporomandibular joint is the only movable joint between skull bones; all other skull joints are sutures and therefore immovable.

Anatomical Components

1. *Articular disc (meniscus).* Fibrocartilage disc that separates the joint cavity into superior and inferior compartments, each with a synovial membrane (Figure 9.11c).
2. *Articular capsule.* Thin, fairly loose envelope around the circumference of the joint (Figure 9.11a).
3. *Lateral ligament.* Two short bands on the lateral surface of the articular capsule that extend inferiorly and posteriorly from the inferior border and tubercle of the zygomatic process of the temporal bone to the lateral and posterior aspect of the neck of the mandible. The lateral ligament is covered by the parotid gland and helps prevent displacement of the mandible (Figure 9.11a).
4. *Sphenomandibular ligament.* Thin band that extends inferiorly and anteriorly from the spine of the sphenoid bone to the ramus of the mandible (Figure 9.11b).
5. *Stylomandibular ligament.* Thickened band of deep cervical fascia that extends from the styloid process of the temporal bone to the inferior and posterior border of the ramus of the mandible. This ligament separates the parotid gland from the submandibular gland (Figure 9.11a, b).

Movements

In the temporomandibular joint, only the mandible moves because the maxilla is firmly anchored to other bones of the skull by sutures. Accordingly, the mandible may function in depression (jaw opening) and elevation (jaw closing), which occurs in the inferior compartment, and protraction, retraction, lateral displacement, and slight rotation, which occur in the superior compartment (see Figure 9.10a–d).

DISLOCATED MANDIBLE

A **dislocation** (dis′-lo-KĀ-shun; *dis-*=apart) or *luxation* (luks-Ā-shun; *luxatio*=dislocation) is the displacement of a bone from a joint with tearing of ligaments, tendons, and articular capsules. It is usually caused by a blow or fall, although unusual physical effort may be a factor. For example, if the condylar processes of the mandible pass anterior to the articular tubercles when you yawn or take a large bite, a dislocated mandible (anterior displacement) may occur. When the mandible is displaced in this manner, the mouth remains wide open and the person is unable to close it. This may be corrected by pressing the thumbs downward on the lower molar teeth and pushing the mandible backward. Other causes of a dislocated mandible include a lateral blow to the chin when the mouth is open and a fracture of the mandible.

■ CHECKPOINT

1. Why is the temporomandibular joint unique among joints of the skull?

Figure 9.11   Right temporomandibular joint (TMJ).

The TMJ is the only movable joint between skull bones.

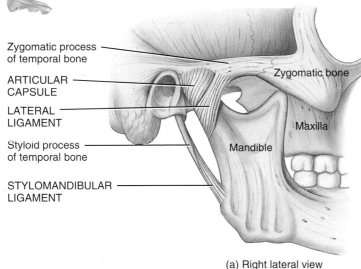

(a) Right lateral view

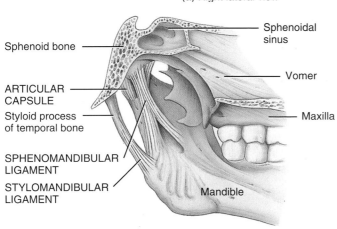

(b) Medial view

Which ligament prevents displacement of the mandible?

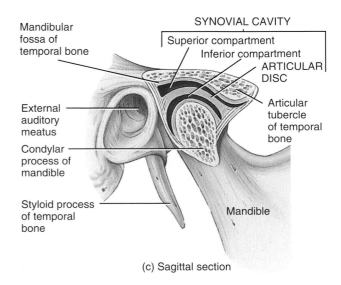

(c) Sagittal section

● Describe the anatomical components of the shoulder joint and the movements that can occur at this joint.

Definition

The **shoulder joint** is a ball-and-socket joint formed by the head of the humerus and the glenoid cavity of the scapula. It also is referred to as the *humeroscapular* or *glenohumeral joint.*

Anatomical Components

1. *Articular capsule.* Thin, loose sac that completely envelops the joint and extends from the glenoid cavity to the anatomical neck of the humerus. The inferior part of the capsule is its weakest area (Figure 9.12a, c).

2. *Coracohumeral ligament.* Strong, broad ligament that strengthens the superior part of the articular capsule and extends from the coracoid process of the scapula to the greater tubercle of the humerus (Figure 9.12a).

3. *Glenohumeral ligaments.* Three thickenings of the articular capsule over the anterior surface of the joint. They extend from the glenoid cavity to the lesser tubercle and anatomical neck of the humerus. These ligaments are often indistinct or absent and provide only minimal strength (Figure 9.12a, b).

4. *Transverse humeral ligament.* Narrow sheet extending from the greater tubercle to the lesser tubercle of the humerus (Figure 9.12a).

5. *Glenoid labrum.* Narrow rim of fibrocartilage around the edge of the glenoid cavity. It slightly deepens and enlarges the glenoid cavity (Figure 9.12b, c).

6. *Bursae.* Four bursae (see page 234) are associated with the shoulder joint. They are the *subscapular bursa* (Figure 9.12a), *subdeltoid bursa, subacromial bursa* (Figure 9.12a–c), and *subcoracoid bursa.*

Movements

The shoulder joint allows flexion, extension, abduction, adduction, medial rotation, lateral rotation, and circumduction of the arm (see Figures 9.6–9.9). It has more freedom of movement than any other joint of the body. This freedom results from the looseness of the articular capsule and shallowness of the glenoid cavity in relation to the large size of the head of the humerus.

Although the ligaments of the shoulder joint strengthen it to some extent, most of the strength results from the muscles that surround the joint, especially the *rotator cuff muscles.* These muscles (supraspinatus, infraspinatus, teres minor, and subscapularis) join the scapula to the humerus (see also Figure 11.17 on pages 334–335). The tendons of the rotator cuff muscles encircle the joint (except for the inferior portion) and fuse with the articular capsule. The rotator cuff muscles work as a group to hold the head of the humerus in the glenoid cavity.

Figure 9.12   Right shoulder (humeroscapular or glenohumeral) joint. (See Tortora, *A Photographic Atlas of the Human Body,* 2e, Figure 4.1.)

Most of the stability of the shoulder joints results from the arrangement of the rotator cuff muscles.

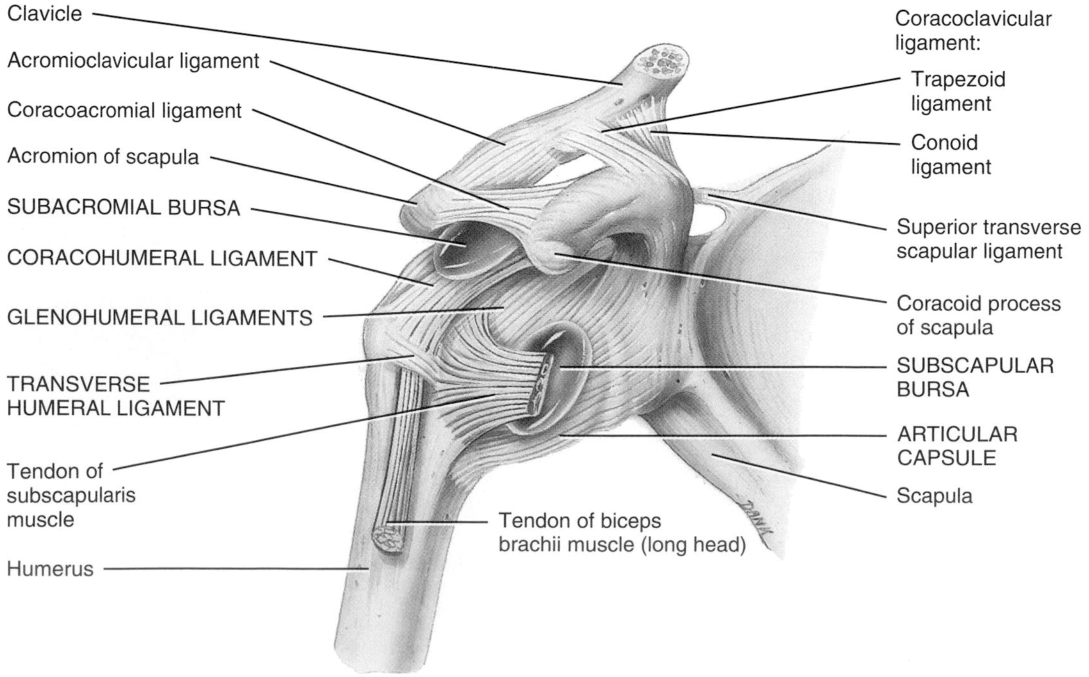

(a) Anterior view

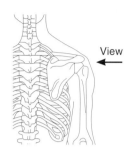

SUPERIOR

Acromion of scapula ———————— Coracoacromial ligament

SUBACROMIAL BURSA ———————— Tendon of supraspinatus muscle

Tendon of biceps brachii muscle (long head) ———— CORACOHUMERAL LIGAMENT

Coracoid process of scapula

Tendon of infraspinatus muscle ———— Tendon of subscapularis muscle

Glenoid cavity ————

ARTICULAR CAPSULE ———— GLENOHUMERAL LIGAMENTS

Tendon of teres minor muscle ———— GLENOID LABRUM

POSTERIOR   ANTERIOR

(b) Lateral view (opened)

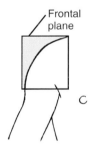

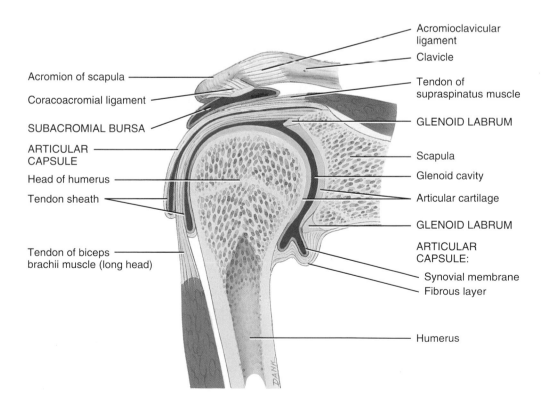

Acromioclavicular ligament

Clavicle

Acromion of scapula ————

Tendon of supraspinatus muscle

Coracoacromial ligament ————

GLENOID LABRUM

SUBACROMIAL BURSA ————

ARTICULAR CAPSULE ————

Scapula

Head of humerus ————

Glenoid cavity

Tendon sheath ————

Articular cartilage

GLENOID LABRUM

ARTICULAR CAPSULE:

Tendon of biceps brachii muscle (long head) ————

Synovial membrane

Fibrous layer

Humerus

(c) Frontal section

continues

247

EXHIBIT 9.2   SHOULDER JOINT (Figure 9.12)

continued

ROTATOR CUFF INJURY AND DISLOCATED AND SEPARATED SHOULDER

Rotator cuff injury is common among baseball pitchers, volleyball players, racket sports players, and swimmers due to shoulder movements that involve vigorous circumduction. It also occurs as a result of wear and tear, trauma, and repetitive motions in certain jobs, such as placing items on a shelf above your head. Most often, there is tearing of the supraspinatus muscle tendon of the rotator cuff. This tendon is especially predisposed to wear-and-tear because of its location between the head of the humerus and acromion of the scapula, which compresses the tendon during shoulder movements.

The joint most commonly dislocated in adults is the shoulder joint because its socket is quite shallow and the bones are held together mainly by supporting muscles. Usually in a **dislocated shoulder,** the head of the humerus becomes displaced inferiorly, where the articular capsule is least protected. Dislocations of the mandible, elbow, fingers, knee, or hip are less common.

A **separated shoulder** refers to an injury of the acromioclavicular joint, a joint formed by the acromion of the scapula and the acromial end of the clavicle. This condition is usually the result of forceful trauma to the joint, as when the shoulder strikes the ground in a fall.

■ **CHECKPOINT**

1. Which tendons at the shoulder joint of a baseball pitcher are most likely to be torn due to excessive circumduction?

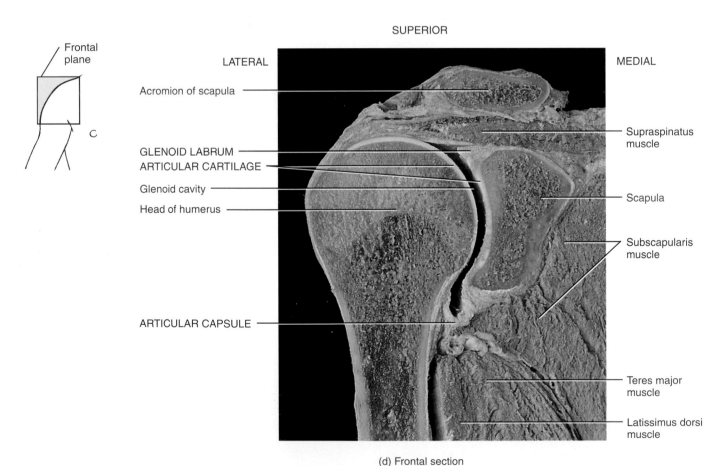

Frontal plane

SUPERIOR

LATERAL

MEDIAL

Acromion of scapula

Supraspinatus muscle

GLENOID LABRUM

ARTICULAR CARTILAGE

Glenoid cavity

Scapula

Head of humerus

Subscapularis muscle

ARTICULAR CAPSULE

Teres major muscle

Latissimus dorsi muscle

(d) Frontal section

? Why does the shoulder joint have more freedom of movement than any other joint of the body?

EXHIBIT 9.3 ELBOW JOINT (Figure 9.13)

OBJECTIVE

● Describe the anatomical components of the elbow joint and the movements that can occur at this joint.

Definition

The **elbow joint** is a hinge joint formed by the trochlea of the humerus, the trochlear notch of the ulna, and the head of the radius.

Anatomical Components

1. **Articular capsule.** The anterior part of the articular capsule covers the anterior part of the elbow joint, from the radial and coronoid fossae of the humerus to the coronoid process of the ulna and the annular ligament of the radius. The posterior part extends from the capitulum, olecranon fossa, and lateral epicondyle of the humerus to the annular ligament of the radius,

the olecranon of the ulna, and the ulna posterior to the radial notch (Figure 9.13a, b).

2. **Ulnar collateral ligament.** Thick, triangular ligament that extends from the medial epicondyle of the humerus to the coronoid process and olecranon of the ulna (Figure 9.13a).

3. **Radial collateral ligament.** Strong, triangular ligament that extends from the lateral epicondyle of the humerus to the annular ligament of the radius and the radial notch of the ulna (Figure 9.13b).

Movements

The elbow joint allows flexion and extension of the forearm (see Figure 9.6c).

Figure 9.13 Right elbow joint. (See Tortora, *A Photographic Atlas of the Human Body*, 2e, Figure 4.3. See also Figure 9.3b in this text.)

The elbow joint is formed by parts of three bones: humerus, ulna, and radius.

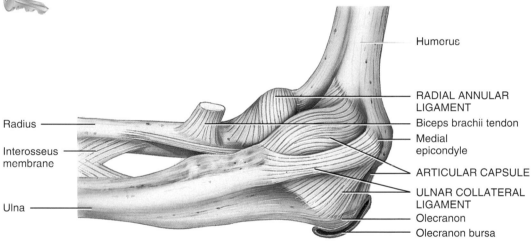

(a) Medial aspect

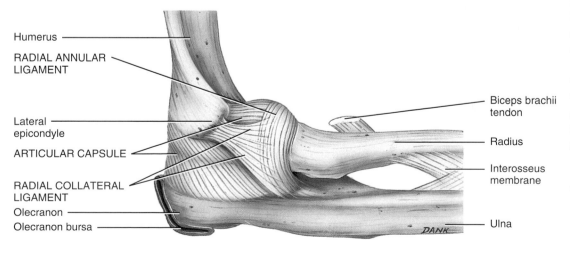

(b) Lateral aspect

  Which movements are possible at a hinge joint?

TENNIS ELBOW, LITTLE-LEAGUE ELBOW, AND DISLOCATION OF THE RADIAL HEAD

Tennis elbow most commonly refers to pain at or near the lateral epicondyle of the humerus, usually caused by an improperly executed backhand. The extensor muscles strain or sprain, resulting in pain. **Little-league elbow** typically develops as a result of a heavy pitching schedule and/or a schedule that involves throwing curve balls, especially among youngsters. In this disorder, the elbow may enlarge, fragment, or separate.

A **dislocation of the radial head** is the most common upper limb dislocation in children. In this injury, the head of the radius slides past or ruptures the radial annular ligament, a ligament that forms a collar around the head of the radius at the proximal radioulnar joint. Dislocation is most apt to occur when a strong pull is applied to the forearm while it is extended and supinated, for instance while swinging a child around with outstretched arms.

■ **CHECKPOINT**

1. At the elbow joint, which ligaments connect (a) the humerus and the ulna, and (b) the humerus and the radius?

EXHIBIT 9.4   HIP JOINT (Figure 9.14)

- Describe the anatomical components of the hip joint and the movements that can occur at this joint.

Definition

The **hip joint** *(coxal joint)* is a ball-and-socket joint formed by the head of the femur and the acetabulum of the hip bone.

Anatomical Components

1. ***Articular capsule.*** Very dense and strong capsule that extends from the rim of the acetabulum to the neck of the femur (Figure 9.14b, d). One of the strongest structures of the body, the capsule consists of circular and longitudinal fibers. The circular fibers, called the *zona orbicularis,* form a collar around the neck of the femur. Accessory ligaments known as the iliofemoral ligament, pubofemoral ligament, and ischiofemoral ligament reinforce the longitudinal fibers.

2. ***Iliofemoral ligament.*** Thickened portion of the articular capsule that extends from the anterior inferior iliac spine of the hip bone to the intertrochanteric line of the femur (Figure 9.14a, c).

3. ***Pubofemoral ligament.*** Thickened portion of the articular capsule that extends from the pubic part of the rim of the acetabulum to the neck of the femur (Figure 9.14a).

4. ***Ischiofemoral ligament.*** Thickened portion of the articular capsule that extends from the ischial wall of the acetabulum to the neck of the femur (Figure 9.14c).

5. ***Ligament of the head of the femur.*** Flat, triangular band that extends from the fossa of the acetabulum to the fovea capitis of the head of the femur (Figure 9.14b, d).

6. ***Acetabular labrum.*** Fibrocartilage rim attached to the margin of the acetabulum that enhances the depth of the acetabulum. Because the diameter of the acetabular rim is smaller than that of the head of the femur, dislocation of the femur is rare (Figure 9.14b, d).

7. ***Transverse ligament of the acetabulum.*** Strong ligament that crosses over the acetabular notch. It supports part of the acetabular labrum and is connected with the ligament of the head of the femur and the articular capsule (Figure 9.14b).

Movements

The hip joint allows flexion, extension, abduction, adduction, circumduction, medial rotation, and lateral rotation of thigh (see Figures 9.6–9.9). The extreme stability of the hip joint is related to the very strong articular capsule and its accessory ligaments, the manner in which the femur fits into the acetabulum, and the mus-

Figure 9.14   Right hip (coxal) joint.

The articular capsule of the hip joint is one of the strongest structures in the body.

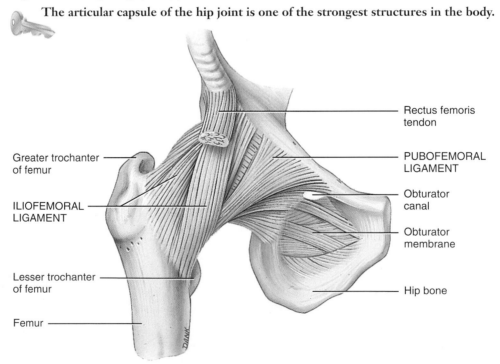

Rectus femoris tendon

Greater trochanter of femur

PUBOFEMORAL LIGAMENT

Obturator canal

ILIOFEMORAL LIGAMENT

Obturator membrane

Lesser trochanter of femur

Hip bone

Femur

(a) Anterior view

cles surrounding the joint. Although the shoulder and hip joints are both ball-and-socket joints, the hip joints do not have as wide a range of motion. Flexion is limited by the anterior surface of the thigh coming into contact with the anterior abdominal wall when the knee is flexed and by tension of the hamstring muscles when the knee is extended. Extension is limited by tension of the iliofemoral, pubofemoral, and ischiofemoral ligaments. Abduction is limited by the tension of the pubofemoral ligament, and adduction is limited by contact with the opposite limb and tension in the ligament of the head of the femur. Medial rotation is limited by the tension in the ischiofemoral ligament, and lateral rotation is limited by tension in the iliofemoral and pubofemoral ligaments.

■ CHECKPOINT

1. What factors limit the degree of flexion and abduction at the hip joint?

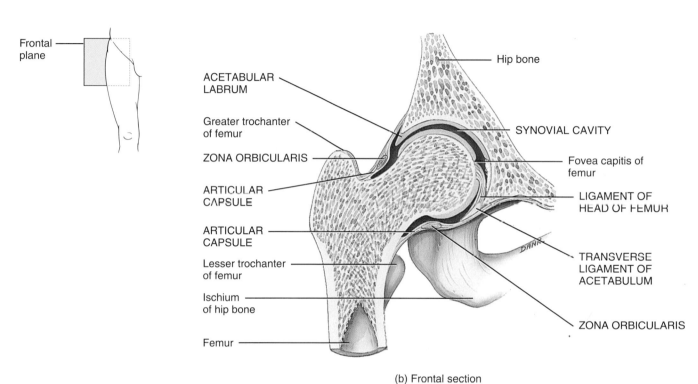

(b) Frontal section

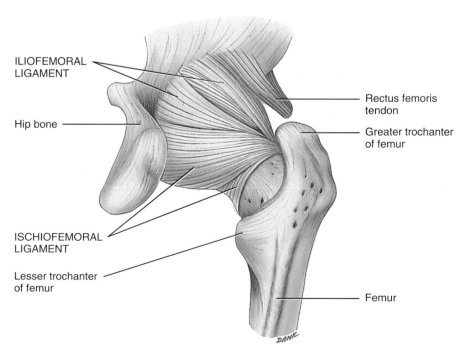

(c) Posterior view

continues

EXHIBIT 9.4   HIP JOINT (Figure 9.14)

continued

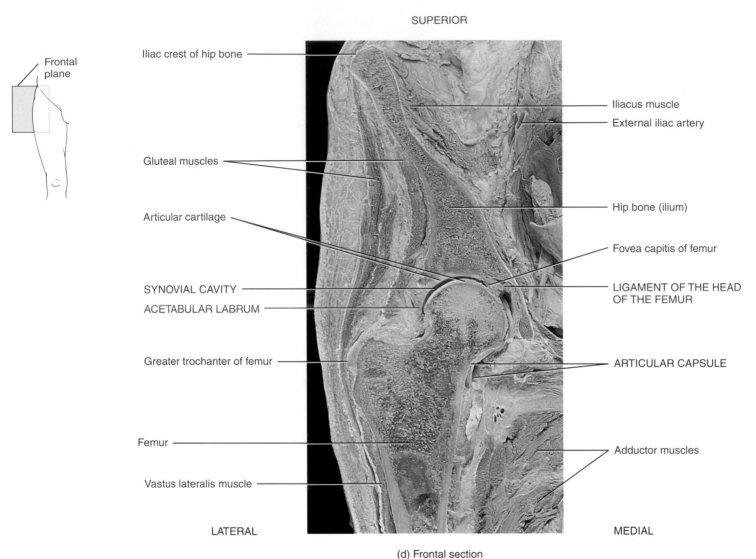

Frontal plane

SUPERIOR

Iliac crest of hip bone

Iliacus muscle

External iliac artery

Gluteal muscles

Hip bone (ilium)

Articular cartilage

Fovea capitis of femur

SYNOVIAL CAVITY

LIGAMENT OF THE HEAD OF THE FEMUR

ACETABULAR LABRUM

Greater trochanter of femur

ARTICULAR CAPSULE

Femur

Adductor muscles

Vastus lateralis muscle

LATERAL

MEDIAL

(d) Frontal section

? Which ligaments limit the degree of extension that is possible at the hip joint?

EXHIBIT 9.5   KNEE JOINT (Figure 9.15)

• Describe the main anatomical components of the knee joint and explain the movements that can occur at this joint.

Definition

The **knee joint** (tibiofemoral joint) is the largest and most complex joint of the body, actually consisting of three joints within a single synovial cavity:

1. Laterally is a tibiofemoral joint, between the lateral condyle of the femur, lateral meniscus, and lateral condyle of the tibia. It is a modified hinge joint.

2. Medially is a second tibiofemoral joint, between the medial condyle of the femur, medial meniscus, and medial condyle of the tibia. It is also a modified hinge joint.

3. An intermediate patellofemoral joint, between the patella and the patellar surface of the femur, is a planar joint.

Anatomical Components

1. **Articular capsule.** No complete, independent capsule unites the bones. The ligamentous sheath surrounding the joint consists mostly of muscle tendons or their expansions (Figure 9.15a, b). There are, however, some capsular fibers connecting the articulating bones.

2. **Medial and lateral patellar retinacula.** Fused tendons of insertion of the quadriceps femoris muscle and the fascia lata (deep fascia of thigh) that strengthen the anterior surface of the joint (Figure 9.15a).

3. **Patellar ligament.** Continuation of the common tendon of insertion of the quadriceps femoris muscle that extends from the patella to the tibial tuberosity. This ligament also strengthens the anterior surface of the joint. The posterior surface of the ligament is separated from the synovial membrane of the joint by an infrapatellar fat pad (Figure 9.15c, e).

4. **Oblique popliteal ligament.** Broad, flat ligament that extends from the intercondylar fossa of the femur to the head of the tibia (Figure 9.15b). The tendon of the semimembranosus muscle is superficial to this ligament and passes from the medial condyle of the tibia to the lateral condyle of the femur. The ligament and tendon strengthen the posterior surface of the joint.

5. **Arcuate popliteal ligament.** Extends from the lateral condyle of the femur to the styloid process of the head of the fibula. It strengthens the lower lateral part of the posterior surface of the joint (Figure 9.15b).

6. **Tibial collateral ligament.** Broad, flat ligament on the medial surface of the joint that extends from the medial condyle of the femur to the medial condyle of the tibia (Figure 9.15a, b, d, f). Tendons of the sartorius, gracilis, and semitendinosus muscles, all of which strengthen the medial aspect of the joint, cross the ligament. Because the tibial collateral ligament is firmly attached to the medial meniscus, tearing of the ligament frequently results in tearing of the meniscus and damage to the anterior cruciate ligament, described under number 8a.

7. **Fibular collateral ligament.** Strong, rounded ligament on the lateral surface of the joint that extends from the lateral condyle of the femur to the lateral side of the head of the fibula (Figure 9.15 a, b, d, f). It strengthens the lateral aspect of the joint. The ligament is covered by the tendon of the biceps femoris muscle. The tendon of the popliteal muscle is deep to the ligament.

8. **Intracapsular ligaments.** Ligaments within the capsule that connect the tibia and femur. The anterior and posterior cruciate ligaments (KROO-shē-āt=shaped like a cross) are named based on their origins relative to the intercondylar area of the tibia. Following their originations, they cross on their way to their destinations on the femur.

   a. *Anterior cruciate ligament (ACL).* Extends posteriorly and laterally from a point *anterior* to the intercondylar area of the tibia to the posterior part of the medial surface of the lateral condyle of the femur (Figure 9.15d, f). The ACL limits hyperextension of the knee and prevents the anterior sliding of the tibia on the femur. This ligament is stretched or torn in about 70% of all serious knee injuries.

   b. *Posterior cruciate ligament (PCL).* Extends anteriorly and medially from a depression on the *posterior* intercondylar area of the tibia and lateral meniscus to the anterior part of the lateral surface of the medial condyle of the femur (Figure 9.15d, f). The PCL prevents the posterior sliding of the tibia on the femur, especially when the knee is flexed. This is very important when walking down stairs or a steep incline.

9. **Articular discs (menisci).** Two fibrocartilage discs between the tibial and femoral condyles that help compensate for the irregular shapes of the bones and circulate synovial fluid.

   a. *Medial meniscus.* Semicircular piece of fibrocartilage (C-shaped). Its anterior end is attached to the anterior intercondylar fossa of the tibia, anterior to the anterior cruciate ligament. Its posterior end is attached to the posterior intercondylar fossa of the tibia between the attachments of the posterior cruciate ligament and lateral meniscus (Figure 9.15d–f).

   b. *Lateral meniscus.* Nearly circular piece of fibrocartilage (approaches an incomplete O in shape). Its anterior end is attached anterior to the intercondylar eminence of the tibia, and lateral and posterior to the anterior cruciate ligament. Its posterior end is attached posterior to the intercondylar eminence of the tibia, and anterior to the posterior end of the medial meniscus (Figure 9.15c, d, f). The medial and lateral menisci are connected to each other by the *transverse ligament* (Figure 9.15d) and to the margins of the head of tibia by the *coronary ligaments* (not illustrated).

10. The more important *bursae* of the knee include the following:

    a. *Prepatellar bursa* between the patella and skin (Figure 9.15c)

    b. *Infrapatellar bursa* between superior part of tibia and patellar ligament (Figure 9.15a, c, e)

    c. *Suprapatellar bursa* between inferior part of femur and deep surface of quadriceps femoris muscle (Figure 9.15a, c, e)

continues

EXHIBIT 9.5   KNEE JOINT (Figure 9.15)

continued

Figure 9.15   Right knee (tibiofemoral) joint. (See Tortora, *A Photographic Atlas of the Human Body*, Figure 4.7.)

The knee joint is the largest and most complex joint in the body.

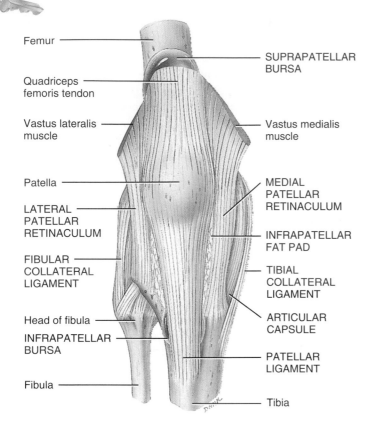

(a) Anterior superficial view

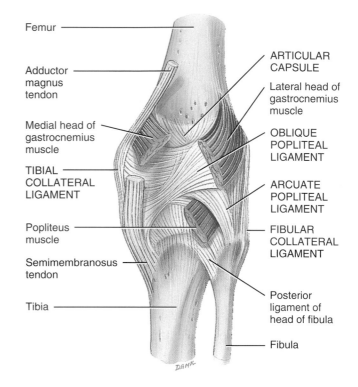

(b) Posterior deep view

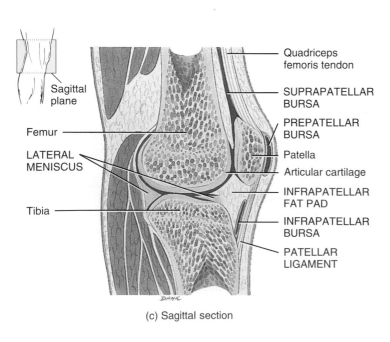

(c) Sagittal section

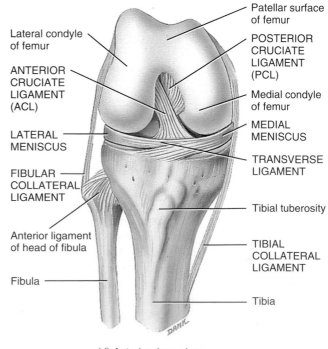

(d) Anterior deep view

Movements

The knee joint allows flexion, extension, slight medial rotation, and lateral rotation of the leg in the flexed position (see Figures 9.6f and 9.9c).

KNEE INJURIES

The knee joint is the joint most vulnerable to damage because it is a mobile, weight-bearing joint and its stability depends almost entirely on its associated ligaments and muscles. Further, there is no correspondence of the articulating bones. A **swollen knee** may occur immediately or hours after an injury. Immediate swelling is due to escape of blood from damaged blood vessels adjacent to areas involving rupture of the anterior cruciate ligament, damage to synovial membranes, torn menisci, fractures, or collateral ligament sprains. Delayed swelling is due to excessive production of synovial fluid, a condition commonly referred to as "water on the knee." A common type of knee injury in football is **rupture of the tibial collateral ligaments,** often associated with tearing of the anterior cruciate ligament and medial meniscus (torn cartilage). Usually, a hard blow to the lateral side of the knee while the foot is fixed on the ground causes the damage. A **dislocated knee** refers to the

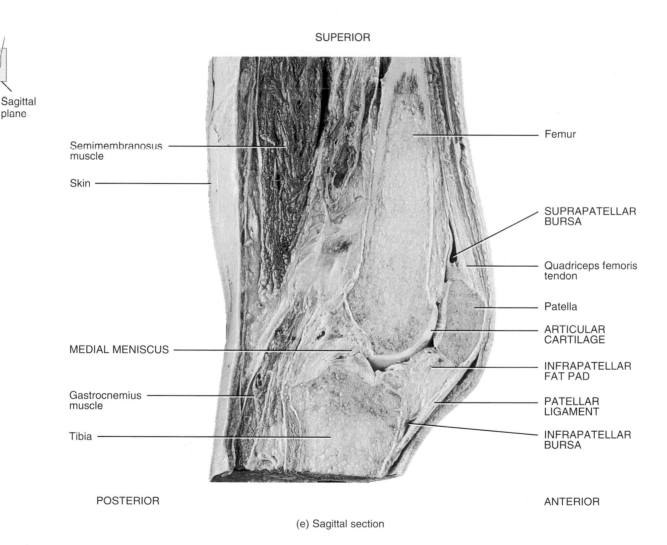

SUPERIOR

Semimembranosus muscle

Skin

MEDIAL MENISCUS

Gastrocnemius muscle

Tibia

Femur

SUPRAPATELLAR BURSA

Quadriceps femoris tendon

Patella

ARTICULAR CARTILAGE

INFRAPATELLAR FAT PAD

PATELLAR LIGAMENT

INFRAPATELLAR BURSA

Sagittal plane

POSTERIOR

ANTERIOR

(e) Sagittal section

continues

EXHIBIT 9.5   KNEE JOINT (Figure 9.15)

continued

displacement of the tibia relative to the femur. The most common type is dislocation anteriorly, resulting from hyperextension of the knee. A frequent consequence of a dislocated knee is damage to the popliteal artery.

■ **CHECKPOINT**

1. What are the opposing functions of the anterior and posterior cruciate ligaments?

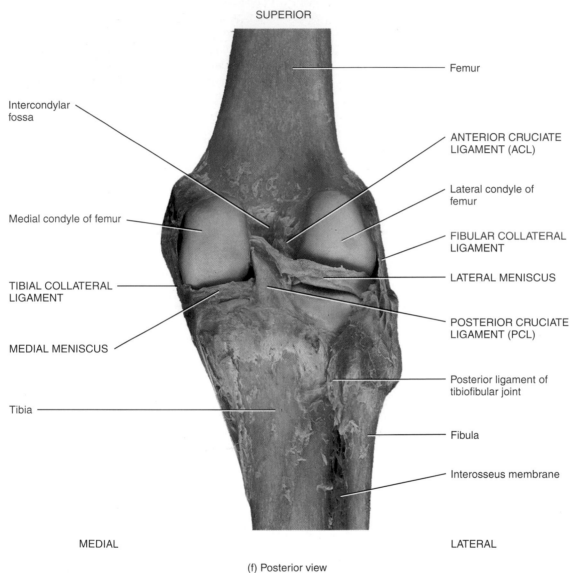

SUPERIOR

Femur

Intercondylar fossa

ANTERIOR CRUCIATE LIGAMENT (ACL)

Lateral condyle of femur

Medial condyle of femur

FIBULAR COLLATERAL LIGAMENT

LATERAL MENISCUS

TIBIAL COLLATERAL LIGAMENT

POSTERIOR CRUCIATE LIGAMENT (PCL)

MEDIAL MENISCUS

Posterior ligament of tibiofibular joint

Tibia

Fibula

Interosseus membrane

MEDIAL

LATERAL

(f) Posterior view

What movement occurs at the knee joint when the quadriceps femoris (anterior thigh) muscles contract?

EXHIBIT 9.6   ANKLE JOINT (Figure 9.16)

- Describe the anatomical components of the ankle joint and explain the movements that can occur at this joint.

Definition

The **ankle joint** *(talocrural joint)* is a hinge joint formed by (1) the distal end of the tibia and its medial malleolus with the talus and (2) the lateral malleolus of the fibula with the talus. It is a strong and stable joint due to the shapes of the articulating bones, the strength of its ligaments, and the tendons that surround it.

Anatomically, the ankle is the region that extends from the distal region of the leg to the proximal region of the foot and contains the ankle joint. In this transition region, there is a change in orientation from a vertical position of the bones and muscles and associated structures in the leg to a horizontal position of the structures in the foot. As a result, there is a turning anteriorly of the tendons, blood vessels, and nerves in the ankle as they enter the foot.

As the structures pass from the leg into the foot at the ankle from a vertical to a horizontal orientation, they are anchored by thickenings of deep fascia called **retinacula** (ret-i-NAK-yoo-la; *retineo*=to hold back). Two principal retinacula are the superior and inferior extensor retinacula (see Figure 11.24 on page 368).

Anatomical Components

1. ***Articular capsule.*** Completely surrounds the joint and is attached superiorly to the tibia and fibula and inferiorly to the talus. The capsule is thin (and weak) anteriorly and posteriorly to permit dorsiflexion and plantar flexion.

2. ***Deltoid (medial) ligament.*** Strong, flat, triangular ligament that extends from the medial malleolus to the talus, navicular, and calcaneus of the tarsus. It is divisible into superficial and deep parts. The superficial components from anterior to posterior are the *tibionavicular ligament, tibiocalcaneal ligament,* and *posterior tibiotalar ligament.* The deep component is the *anterior tibiotalar ligament* (Figure 9.16a).

3. ***Lateral ligament.*** Not as strong as the deltoid ligament. It extends from the lateral malleolus to the talus and calcaneus and is divisible into three components: *anterior talofibular ligament, posterior talofibular ligament, and calcaneofibular ligament* (Figure 9.16b).

Movements

The ankle joint permits dorsiflexion and plantar flexion (see Figure 9.10g).

Figure 9.16   Right ankle (talocrural) joint.

The strength and stability of the ankle joint are due to the shapes of the articulating bone, the strength of its ligaments, and the tendons that surround it.

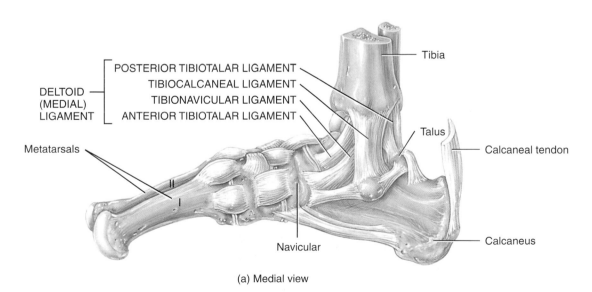

(a) Medial view

continues

EXHIBIT 9.6 ANKLE JOINT (Figure 9.16)

continued

ANKLE SPRAINS AND FRACTURES

The ankle is the most frequently injured major joint in the body. **Ankle sprains** are the most common ankle injuries and often occur in sports that involve running and jumping. *Sprains of the lateral ankle* occur more frequently than those of the medial ankle and are usually caused by excessive inversion (supination) of the foot with plantar flexion of the ankle. As a result, the weaker lateral ligament is partially torn and there is considerable pain and local swelling. Less common *sprains of the medial ankle* occur as a result of excessive eversion (pronation). In the process, the deltoid ligament may be torn but usually the ligament does not tear due to its great strength; instead it may pull off the tip of the medial malleolus. Ankle sprains are treated with RICE: rest, ice, compression, and elevation. Severe sprains may require cast immobilization or surgery.

A **fracture** of the distal end of the leg that involves both the medial and lateral malleoli is called a *Pott's fracture.* Such a fracture may result in dislocation of the talus. Sometimes only lateral malleolar fractures occur; even less frequently, only medial malleolar fractures occur.

■ CHECKPOINT

1. Why do sprains of the lateral ankle occur more frequently than sprains of the medial ankle?

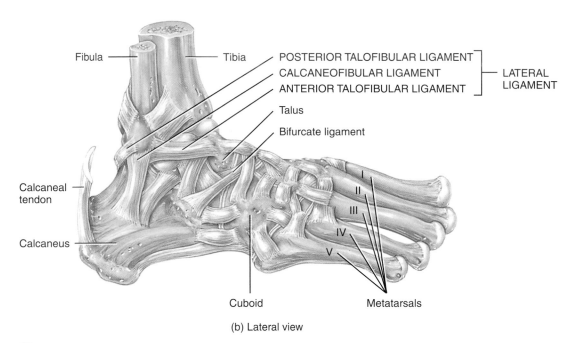

(b) Lateral view

? Why are sprains of the lateral ankle more common than sprains of the medial ankle?

FACTORS AFFECTING CONTACT AND RANGE OF MOTION AT SYNOVIAL JOINTS

OBJECTIVE

• Describe six factors that influence the type of movement and range of motion possible at a synovial joint.

The articular surfaces of synovial joints contact one another and determine the type and range of motion that is possible. **Range of motion (ROM)** refers to the range, measured in degrees of a circle, through which the bones of a joint can be moved. The following factors contribute to keeping the articular surfaces in contact and affect range of motion:

1. *Structure or shape of the articulating bones.* The structure or shape of the articulating bones determines how closely they can fit together. The articular surfaces of some bones have a complementary relationship with one another. This spatial relationship is very obvious at the hip joint, where the head of the femur articulates with the acetabulum of the hip bone. An interlocking fit allows rotational movement.

2. *Strength and tension (tautness) of the joint ligaments.* The different components of a fibrous capsule are tense or taut only when the joint is in certain positions. Tense ligaments not only restrict the range of motion but also direct the movement of the articulating bones with respect to each other. In the knee joint, for example, the anterior cruciate ligament is taut and the posterior cruciate ligament is loose when the knee is straightened, and the reverse occurs when the knee is bent.

3. *Arrangement and tension of the muscles.* Muscle tension reinforces the restraint placed on a joint by its ligaments, and thus restricts movement. A good example of the effect of muscle tension on a joint is seen at the hip joint. When the thigh is fixed with the knee extended, the movement is restricted by the tension of the hamstring muscles on the posterior surface of the thigh. But if the knee is flexed, the tension on the hamstring muscles is lessened, and the thigh can be raised farther.

4. *Contact of soft parts.* The point at which one body surface contacts another may limit mobility. For example, if you bend your arm at the elbow, it can move no further after the anterior surface of the forearm meets with and presses against the biceps brachii muscle of the arm. Joint movement may also be restricted by the presence of adipose tissue.

5. *Hormones.* Joint flexibility may also be affected by hormones. For example, relaxin, a hormone produced by the placenta and ovaries, increases the flexibility of the fibrocartilage of the pubic symphysis and loosens the ligaments between the sacrum, hip bone, and coccyx toward the end of pregnancy. These changes permit expansion of the pelvic outlet, which assists in delivery of the baby.

6. *Disuse.* Movement at a joint may be restricted if a joint has not been used for an extended period. For example, if an elbow joint is immobilized by a cast, range of motion at the joint may be limited for a time after the cast is removed. Disuse may also result in decreased amounts of synovial fluid, diminished flexibility of ligaments and tendons, and **muscular atrophy,** a reduction in size or wasting of a muscle.

CHECKPOINT

10. How do the strength and tension of ligaments determine range of motion?

AGING AND JOINTS

OBJECTIVE

• Explain the effects of aging on joints.

Aging usually results in decreased production of synovial fluid in joints. In addition, the articular cartilage becomes thinner with age, and ligaments shorten and lose some of their flexibility. The effects of aging on joints, which vary considerably from one person to another, are influenced by genetic factors and by wear and tear. Although degenerative changes in joints may begin in those as young as 20 years of age, most changes do not occur until much later. By age 80, almost everyone develops some type of degeneration in the knees, elbows, hips, and shoulders. It is also common for elderly individuals to develop degenerative changes in the vertebral column, resulting in a hunched-over posture and pressure on nerve roots. One type of arthritis, called osteoarthritis (see Applications to Health on page 261), is at least partially age-related. Nearly everyone over age 70 has evidence of some osteoarthritic changes. Stretching and aerobic exercises that attempt to maintain full range of motion are very important in minimizing the effects of aging. They help to maintain the effective functioning of ligaments, tendons, muscles, synovial fluid, and articular cartilage.

CHECKPOINT

11. Which joints show evidence of degeneration in nearly all individuals as aging progresses?

ARTHROPLASTY

OBJECTIVE

• Explain how a total hip replacement is performed.

Joints that have been severely damaged by a disease such as arthritis, or by an injury, may be replaced surgically with artificial joints in a procedure referred to as **arthroplasty** (AR-thrō-plas′-tē; *arthr-*=joint; *-plasty*=plastic repair of). Although most joints in the body can undergo arthroplasty, the ones most commonly replaced are the hips, knees, and shoulders. During the procedure, the ends of the damaged bones are removed and the artificial components are fixed in place. Artificial joint components are made of metal, ceramic, or plastic. The goals of arthroplasty are to relieve pain and increase range of motion.

Thousands of partial hip replacements, involving only the femur, are performed annually. In a total hip replacement, the procedure involves both the acetabulum and head of the femur (Figure 9.17). The damaged portions of the acetabulum and head of the femur are replaced by prefabricated prostheses (artificial devices). The acetabulum is shaped to accept the new socket, the head of the femur is removed, and the center of the bone is shaped to fit the femoral component. The acetabular component consists of polyethylene, whereas the femoral component is composed of cobalt-chrome, titanium alloys, or stainless steel. These materials are designed to withstand a high degree of stress. Once the appropriate acetabular and femoral components are selected, they are attached to the healthy portion of bone with acrylic cement, which forms an interlocking mechanical bond. Researchers are continually seeking to improve the strength of the cement and devise ways to stimulate bone growth around the implanted area.

CHECKPOINT

12. Which joints of the body most commonly undergo arthroplasty?

Figure 9.17 Total hip replacement.

In a total hip replacement, damaged portions of the acetabulum and head of the femur are replaced by prostheses.

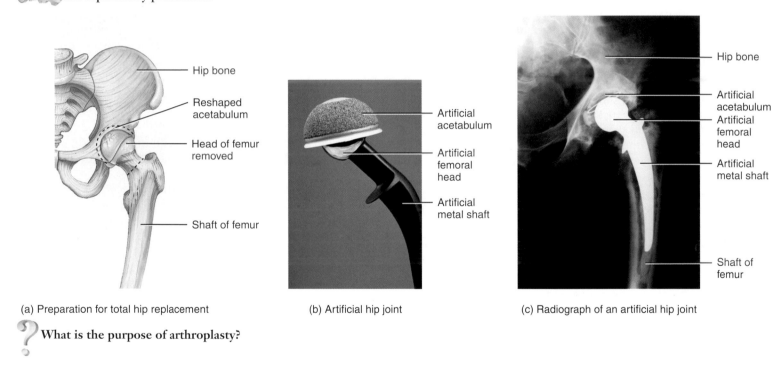

(a) Preparation for total hip replacement

(b) Artificial hip joint

(c) Radiograph of an artificial hip joint

What is the purpose of arthroplasty?

APPLICATIONS TO HEALTH

RHEUMATISM AND ARTHRITIS

Rheumatism (ROO-ma-tizm) is any painful disorder of the supporting structures of the body—bones, ligaments, tendons, or muscles—that is not caused by infection or injury. Arthritis is a form of rheumatism in which the joints are swollen, stiff, and painful. It afflicts about 40 million people in the United States.

RHEUMATOID ARTHRITIS

Rheumatoid arthritis (RA) is an autoimmune disease in which the immune system of the body attacks its own tissues—in this case, its own cartilage and joint linings. RA is characterized by inflammation of the joint, which causes swelling, pain, and loss of function. Usually, this form of arthritis occurs bilaterally: If one wrist is affected, the other is also likely to be affected, although often not to the same degree.

The primary symptom of RA is inflammation of the synovial membrane. If untreated, the membrane thickens, and synovial fluid accumulates. The resulting pressure causes pain and tenderness. The membrane then produces an abnormal granulation tissue, called *pannus*, that adheres to the surface of the articular cartilage and sometimes erodes the cartilage completely. When the cartilage is destroyed, fibrous tissue joins the exposed bone ends. The fibrous tissue ossifies and fuses the joint so that it becomes immovable—the ultimate crippling effect of rheumatoid arthritis. The growth of the granulation tissue causes the distortion of the fingers that characterizes hands of RA sufferers.

OSTEOARTHRITIS

Osteoarthritis (OA) (os′-tē-ō-ar-THRĪ-tis) is a degenerative joint disease in which joint cartilage is gradually lost. It results from a combination of aging, irritation of the joints, and wear and abrasion. Commonly known as "wear-and-tear" arthritis, osteoarthritis is the leading cause of disability in older persons.

Osteoarthritis is a progressive disorder of synovial joints, particularly weight-bearing joints. Articular cartilage deteriorates and new bone forms in the subchondral areas and at the margins of the joint. The cartilage slowly degenerates, and as the bone ends become exposed, spurs (small bumps) of new osseous tissue are deposited on them. These spurs decrease the space of the joint cavity and restrict joint movement. Unlike rheumatoid arthritis, osteoarthritis affects mainly the articular cartilage, although the synovial membrane often becomes inflamed late in the disease. A major distinction between osteoarthritis and rheumatoid arthritis is that osteoarthritis first afflicts the larger joints (knees, hips), whereas rheumatoid arthritis first strikes smaller joints.

GOUTY ARTHRITIS

Uric acid (a substance that gives urine its name) is a waste product produced during the metabolism of nucleic acid (DNA and RNA) subunits. A person who suffers from **gout** (GOWT) either produces excessive amounts of uric acid or is not able to excrete as much as normal. The result is a buildup of uric acid in the blood. This excess acid then reacts with sodium to form a salt called sodium urate. Crystals of this salt accumulate in soft tissues such as the kidneys and in the cartilage of the ears and joints.

In **gouty arthritis,** sodium urate crystals are deposited in the soft tissues of the joints. Gout most often affects the joints of the feet, especially at the base of the great toe. The crystals irritate and erode the cartilage, causing inflammation, swelling, and acute pain. Eventually, the crystals destroy all joint tissues. If the disorder is untreated, the ends of the articulating bones fuse, and the joint becomes immovable. Treatment consists of pain relief (ibuprofen, naproxen, colchicine, and cortisone) followed by administration of allopurinol to keep uric acid levels low so that crystals do not form.

LYME DISEASE

A spiral-shaped bacterium called *Borrelia burgdorferi* causes **Lyme disease,** named for the town of Lyme, Connecticut, where it was first reported in 1975. The bacteria are transmitted to humans mainly by deer ticks (*Ixodes dammini*). These ticks are so small that their bites often go unnoticed. Within a few weeks of the tick bite, a rash may appear at the site. Although the rash often resembles a bull's eye target, there are many variations, and some people never develop a rash. Other symptoms include joint stiffness, fever and chills, headache, stiff neck, nausea, and low back pain. In advanced stages of the disease, arthritis is the main complication. It usually afflicts the larger joints such as the knee, ankle, hip, elbow, or wrist. Antibiotics are generally effective against Lyme disease, especially if they are given promptly. However, some symptoms may linger for years.

ANKYLOSING SPONDYLITIS

Ankylosing spondylitis (ang′-ki-LŌ-sing spon′-di-LĪ-tis; *ankyle*=stiff; *spondyl*=vertebra) is an inflammatory disease that affects joints between vertebrae (intervertebral) and between the sacrum and hip bone (sacroiliac joint). The cause is unknown. The disease is more common in males and has its onset between the ages of 20 and 40 years. It is characterized by pain and stiffness in the hips and lower back that progress upward along the backbone. Inflammation can lead to *ankylosis* (severe or complete loss of movement at a joint) and *kyphosis* (hunchback). Treatment consists of anti-inflammatory drugs, heat, massage, and supervised exercise.

KEY MEDICAL TERMS ASSOCIATED WITH JOINTS

Arthralgia (ar-THRAL-jē-a; *arthr-*=joint; *-algia*=pain) Pain in a joint.

Bursectomy (bur-SEK-tō-mē; *-ectomy*=removal of) Removal of a bursa.

Chondritis (kon-DRĪ-tis; *chondr-*=cartilage) Inflammation of cartilage.

Subluxation (sub-luks-Ā-shun) A partial or incomplete dislocation.

Synovitis (sin′-ō-VĪ-tis) Inflammation of a synovial membrane in a joint.

STUDY OUTLINE

Introduction (p. 229)

1. A joint (articulation or arthrosis) is a point of contact between two bones, between bone and cartilage, or between bone and teeth.

2. A joint's structure may permit no movement, slight movement, or free movement.

Joint Classifications (p. 230)

1. Structural classification is based on the presence or absence of a synovial cavity and the type of connecting tissue. Structurally, joints are classified as fibrous, cartilaginous, or synovial.

2. Functional classification of joints is based on the degree of movement permitted. Joints may be synarthroses (immovable), amphiarthroses (slightly movable), or diarthroses (freely movable).

Fibrous Joints (p. 230)

1. The bones of fibrous joints are held together by fibrous connective tissue.

2. These joints include immovable sutures (found between skull bones), slightly movable syndesmoses (such as the distal tibiofibular joint), and immovable gomphoses (roots of teeth in the sockets in the mandible and maxilla).

Cartilaginous Joints (p. 230)

1. The bones of cartilaginous joints are held together by cartilage.

2. These joints include immovable synchondroses united by hyaline cartilage (epiphyseal plates between diaphyses and epiphyses) and slightly movable symphyses united by fibrocartilage (pubic symphysis).

Synovial Joints (p. 232)

1. Synovial joints contain a space between bones called the synovial cavity. All synovial joints are diarthroses.

2. Other characteristics of synovial joints are the presence of articular cartilage and an articular capsule, made up of a fibrous capsule and a synovial membrane.

3. The synovial membrane secretes synovial fluid, which forms a thin, viscous film over the surfaces within the articular capsule.

4. Many synovial joints also contain accessory ligaments (extracapsular and intracapsular) and articular discs (menisci).

5. Synovial joints contain an extensive nerve and blood supply. The nerves convey information about pain, joint movements, and the degree of stretch at a joint. Blood vessels penetrate the articular capsule and ligaments.

6. Bursae are saclike structures, similar in structure to joint capsules, that alleviate friction in joints such as the shoulder and knee joints.

7. Tendon sheaths are tubelike bursae that wrap around tendons where there is considerable friction.

8. Subtypes of synovial joints include planar, hinge, pivot, condyloid, saddle, and ball-and-socket.

9. In a planar joint the articulating surfaces are flat, and the bones glide back and forth and side to side (nonaxial); examples are joints between carpals and tarsals.

10. In a hinge joint, the convex surface of one bone fits into the concave surface of another, and motion is angular around one axis (monaxial); examples are the elbow, knee, and ankle joints.

11. In a pivot joint, a round or pointed surface of one bone fits into a ring formed by another bone and a ligament, and movement is rotation (monaxial); examples are the atlanto-axial and radioulnar joints.

12. In a condyloid joint, an oval projection of one bone fits into an oval cavity of another, and motion is angular around two axes (biaxial); examples include the wrist joint and metacarpophalangeal joints of the second through fifth digits.

13. In a saddle joint, the articular surface of one bone is shaped like a saddle and the other bone fits into the "saddle" like a sitting rider; motion is angular around two axes (biaxial). An example is the carpometacarpal joint between the trapezium and the metacarpal of the thumb.

14. In a ball-and-socket joint, the ball-shaped surface of one bone fits into the cuplike depression of another; motion is angular and rotational around three axes and all directions in between (multiaxial). Examples include the shoulder and hip joints.

15. Table 9.1 on page 237 summarizes the structural and functional categories of joints.

Types of Movements at Synovial Joints (p. 236)

1. In a gliding movement, the nearly flat surfaces of bones move back and forth and from side to side. Gliding movements occur at planar joints.

2. In angular movements, a change in the angle between bones occurs. Examples are flexion–extension, lateral flexion, hyperextension, and abduction–adduction. Circumduction refers to flexion, abduction, extension, and adduction in succession. Angular movements occur at hinge, condyloid, saddle, and ball-and-socket joints.

3. In rotation, a bone moves around its own longitudinal axis. Rotation can occur at pivot and ball-and-socket joints.

4. Special movements occur at specific synovial joints. Examples are elevation–depression, protraction–retraction, inversion–eversion, dorsiflexion–plantar flexion, supination–pronation, and opposition.

5. Table 9.2 on page 242 summarizes the various types of movements at synovial joints.

Selected Joints of the Body (p. 243)

1. A summary of selected joints of the body, including articular components, structural and functional classifications, and movements, is presented in Tables 9.3 and 9.4 on pages 243 and 244.

2. The temporomandibular joint (TMJ) is between the condyle of the mandible and mandibular fossa and articular tubercle of the temporal bone (Exhibit 9.1, page 245).

3. The shoulder (humeroscapular or glenohumeral) joint is between the head of the humerus and glenoid cavity of the scapula (Exhibit 9.2 on page 246).

4. The elbow joint is between the trochlea of the humerus, the trochlear notch of the ulna, and the head of the radius (Exhibit 9.3 on page 249).

5. The hip (coxal) joint is between the head of the femur and acetabulum of the hip bone (Exhibit 9.4 on page 250).

6. The knee (tibiofemoral) joint is between the patella and patellar

surface of the femur; the lateral condyle of the femur, the lateral meniscus, and the lateral condyle of the tibia; and the medial condyle of the femur, the medial meniscus, and the medial condyle of the tibia (Exhibit 9.5 on page 253).

7. The ankle (talocrural) joint is formed by the distal end of the tibia and its medial malleolus with the talus and the lateral malleolus of the fibula.

Factors Affecting Contact and Range of Motion at Synovial Joints (p. 259)

1. The ways that articular surfaces of synovial joints contact one another determines the type of movement possible.

2. Factors that contribute to keeping the surfaces in contact and affect range of motion are structure or shape of the articulating bones, strength and tension of the ligaments, arrangement and tension of the muscles, apposition of soft parts, hormones, and disuse.

Aging and Joints (p. 259)

1. With aging, a decrease in synovial fluid, thinning of articular cartilage, and decreased flexibility of ligaments occur.

2. Most individuals experience some degeneration in the knee, elbow, hip, and shoulder joints due to the aging process.

Arthroplasty (p. 259)

1. Arthroplasty refers to the surgical replacement of joints.

2. The most commonly replaced joints are the hips, knees, and shoulders.

 SELF-QUIZ QUESTIONS

Choose the one best answer to the following questions.

1. Choose the pair of terms that is most closely associated or matched:
   a. wrist joint–pronation
   b. intertarsal joints–inversion
   c. elbow joint–hyperextension
   d. ankle joint–eversion
   e. interphalangeal joint–circumduction.

2. Which of the following is a ball-and-socket joint?
   a. temporomandibular joint
   b. the knee joint
   c. the shoulder joint
   d. the elbow joint
   e. Both b and c are correct.

3. Which of the following structures is not associated with the knee joint?
   a. glenoid labrum
   b. patellar ligament
   c. infrapatellar bursa
   d. cruciate ligaments
   e. tibial collateral ligament

4. The lambdoid suture is an example of a(n):
   a. synarthrosis
   b. amphiarthrosis
   c. diarthrosis
   d. fibrous joint
   e. Both a and d are correct.

5. Pads of fibrocartilage that extend into the space between articulating bones in the knee are called:
   a. ligaments.
   b. bursae.
   c. articular capsules.
   d. menisci.
   e. gomphoses.

6. The joint at which the atlas rotates around the axis is an example of a:
   a. hinge joint.
   b. ball-and-socket joint.
   c. gliding joint.
   d. saddle joint.
   e. pivot joint.

7. The inner layer of the articular capsule is the:
   a. synovial membrane.
   b. fibrous capsule.
   c. bursa.
   d. meniscus.
   e. articular cartilage.

8. A broad flat disc of fibrocartilage joins the bones at the:
   a. knee joint.
   b. shoulder joint.
   c. pubic symphysis.
   d. coronal suture.
   e. atlanto-occipital joint.

Complete the following:

9. The epiphyseal plate is an example of the structural joint classification known as a _____ because _____ joins the epiphysis and diaphysis of the growing bone.

10. When the fibers of the fibrous capsule of an articular capsule are arranged in parallel bundles, the structure is called a(n) _____.

11. The temporomandibular joint is functionally classified as a(n) _____.

12. The ends of bones at synovial joints are covered with a protective layer of _____.

13. Sac-like structures that reduce friction between body parts at joints are called _____.

Are the following statements true or false?

14. Synovial fluid becomes more viscous as joint activity increases.

15. Tipping the head backward at the atlanto-occipital joint is an example of hyperextension.

16. Raising the arm to point straight ahead is an example of abduction.

17. The nerves that supply a joint are different from those that supply the skeletal muscles that move that joint.

18. A shoulder dislocation is an injury to the acromioclavicular joint.

Matching

19. Match the following:
   ___ (a) distal tibiofibular joint
   ___ (b) pubic symphysis
   ___ (c) coronal suture
   ___ (d) tooth in alveolar socket
   ___ (e) atlanto-axial joint
   ___ (f) intercarpal joints
   ___ (g) elbow joint
   ___ (h) carpometacarpal joint of thumb
   ___ (i) epiphyseal plate

   (1) synovial, saddle
   (2) gomphosis
   (3) syndesmosis
   (4) synovial, pivot
   (5) cartilaginous (fibrocartilage)
   (6) cartilaginous (hyaline cartilage)
   (7) fibrous joint, synarthrosis
   (8) synovial, gliding
   (9) synovial, hinge

20. Match the following:

___ **(a)** movement of a body part anteriorly, horizontally to the ground

___ **(b)** horizontal movement of an anteriorly projected body part back into the anatomical position

___ **(c)** movement of the soles medially at the intertarsal joints

___ **(d)** movement of the soles laterally

___ **(e)** action that occurs when you stand on your heels

___ **(f)** position of foot when the heel is on the floor and rest of the foot is raised

___ **(g)** movement of the forearm to turn the palm anteriorly

___ **(h)** movement of the forearm to turn the palm posteriorly

(1) pronation
(2) plantar flexion
(3) eversion
(4) retraction
(5) inversion
(6) protraction
(7) dorsiflexion
(8) supination

CRITICAL THINKING QUESTIONS

1. Burt and Al have been golf partners for 50 years. Burt's golf game improved by 5 points this spring and he credits the hip replacement he had last year. Al's knee has been bothering him for years but when he asked his orthopedist about a new knee joint he was told "it's not that simple." Al told Burt "one joint's like any other joint" and wants a second opinion. What will the new doctor say?

2. After your second Human Anatomy exam, you dropped to one knee, raised your arm over your head with one hand clenched into a fist, pumped your arm up and down, bent your head back, looked straight up, and yelled "*Yes!*" Use the proper terms to describe the movements at the various joints.

3. Lars was just getting the hang of bodysurfing during his first trip to the shore when he got caught in the break of a wave. While he was being rolled in the surf, he felt his shoulder "pop." When Lars finally got back to his towel he was out of breath, in pain, and his arm was hanging at an odd angle. What's the prognosis for the rest of Lars's vacation at the shore?

4. Arthur was in a serious motorcycle accident in which he broke his tibia, fibula, and patella. His knee was twisted sideways in the process. His bones have since healed, but he still experiences considerable pain in the knee joint and says it feels like "there's stuff in there." He also has trouble with stability on the joint while walking. To what specific anatomical problems would you attribute Arthur's joint difficulties?

5. Chuck has gone to the chiropractor because his back hurts. The chiropractor tells him, "You know, your pelvis is out of alignment." This news came shortly after his orthopedist told him he had slightly bowed legs. On top of all this, he has severe osteoarthritis in his ankle joint. Do you think all of these things could be related? Why or why not?

ANSWERS TO FIGURE QUESTIONS

9.1 Functionally, sutures are classified as synarthroses because they are immovable; syndesmoses are classified as amphiarthroses because they are slightly movable.

9.2 The structural difference between a synchondrosis and a symphysis is the type of cartilage that holds the joint together: hyaline cartilage in a synchondrosis and fibrocartilage in a symphysis.

9.3 Functionally, synovial joints are diarthroses, freely movable joints.

9.4 Condyloid and saddle joints are biaxial joints.

9.5 Gliding movements occur at planar joints.

9.6 Two examples of flexion that do not occur in the sagittal plane are flexion of the thumb and lateral flexion of the trunk.

9.7 When you adduct your arm or leg, you bring it closer to the midline of the body, thus "adding" it to the trunk.

9.8 Circumduction involves flexion, abduction, extension, and adduction in continuous sequence.

9.9 The anterior surface of a bone or limb rotates toward the midline in medial rotation, and away from the midline in lateral rotation.

9.10 Bringing the arms forward until the elbows touch is an example of protraction.

9.11 The lateral ligament prevents displacement of the mandible.

9.12 The shoulder joint is the most freely movable joint in the body because of the looseness of its articular capsule and the shallowness of the glenoid cavity in relation to the size of the head of the humerus.

9.13 A hinge joint permits flexion and extension.

9.14 Tension in three ligaments—iliofemoral, pubofemoral, and ischiofemoral—limits the degree of extension at the hip joint.

9.15 Contraction of the quadriceps femoris muscle causes extension at the knee joint.

9.16 Sprains of the lateral ankle are more common than those of the medial ankle because the lateral ligament is weaker than the deltoid (medial) ligament.

9.17 The purpose of arthroplasty is to relieve joint pain and permit greater range of motion.

MUSCULAR TISSUE | 10

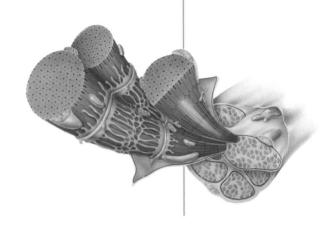

INTRODUCTION   Although bones provide leverage and form the framework of the body, they cannot move body parts by themselves. Motion results from the alternating contraction and relaxation of muscles, which constitute 40–50% of total body weight. Your muscular strength reflects the prime function of muscle—changing chemical energy into mechanical energy to generate force, perform work, and produce movement. In addition, muscular tissue stabilizes the body's position, regulates organ volume, generates heat, and propels fluids and food through various body systems. The scientific study of muscles is known as **myology** (mī-OL-ō-jē; *myo-=muscle; -logy=*study of).

Can you recognize whose work this may be? What is the intent of this image? Is it merely a chart displaying muscles that are to be memorized? Or is it a reference plate for painters and sculptors?

www.wiley.com/college/apcentral

OVERVIEW OF MUSCULAR TISSUE

• Correlate the three types of muscular tissue with their functions and special properties.

Types of Muscular Tissue

There are three types of muscular tissue: skeletal, cardiac, and smooth (see the comparison in Table 3.4 on pages 85–86). Although the three types of muscular tissue share some properties, they differ from one another in their microscopic anatomy, location, and control by the nervous and endocrine systems.

Skeletal muscle tissue is so-named because the function of most skeletal muscles is to move bones of the skeleton. (A few skeletal muscles attach to structures other than bone, such as the skin or even other skeletal muscles.) Skeletal muscle tissue is termed **striated** because alternating light and dark bands (**striations)** are visible when the tissue is examined under a microscope (see Figure 10.4). Skeletal muscle tissue works primarily in a **voluntary** manner; that is, its activity can be consciously (voluntarily) controlled.

Cardiac muscle tissue is found only in the heart, where it forms most of the heart wall. Like skeletal muscle, cardiac muscle is **striated,** but its action is **involuntary**—its alternating contraction and relaxation cannot be consciously controlled. The heart beats because it has a pacemaker that initiates each contraction; this built-in (intrinsic) rhythm is called **autorhythmicity.** Several hormones and neurotransmitters adjust heart rate by speeding or slowing the pacemaker.

Smooth muscle tissue is located in the walls of hollow internal structures, such as blood vessels, airways, and most organs in the abdominopelvic cavity. It is also attached to hair follicles in the skin. Smooth muscle tissue gets its name from the fact that, under a microscope, it appears **nonstriated** or **smooth.** The action of smooth muscle is usually **involuntary,** and, like cardiac muscle, some smooth muscle tissue has autorhythmicity. Both cardiac muscle and smooth muscle are regulated by neurons that are part of the autonomic (involuntary) division of the nervous system and by hormones released by endocrine glands.

Functions of Muscular Tissue

Through sustained contraction or alternating contraction and relaxation, muscular tissue has four key functions: producing body movements, stabilizing body positions, storing and moving substances within the body, and generating heat.

1. *Producing body movements.* Total body movements such as walking and running, and localized movements such as grasping a pencil or nodding the head, rely on the integrated functioning of bones, joints, and skeletal muscles.

2. *Stabilizing body positions.* Skeletal muscle contractions stabilize joints and help maintain body positions, such as standing or sitting. Postural muscles contract continuously when a person is awake; for example, sustained contractions in neck muscles hold the head upright.

3. *Storing and moving substances within the body.* Sustained contractions of ringlike bands of smooth muscles called *sphincters* may prevent outflow of the contents of a hollow organ. Temporary storage of food in the stomach or urine in the urinary bladder is possible because smooth muscle sphincters close off the outlets of these organs. Cardiac muscle contractions pump blood through the body's blood vessels. Contraction and relaxation of smooth muscle in the walls of blood vessels help adjust their diameter and thus regulate the rate of blood flow. Smooth muscle contractions also move food and substances such as bile and enzymes through the gastrointestinal tract, push gametes (sperm and oocytes) through the reproductive systems, and propel urine through the urinary system. Skeletal muscle contractions promote the flow of lymph and aid the return of blood to the heart.

4. *Producing heat.* As muscular tissue contracts, it also produces heat. Much of the heat released by muscle is used to maintain normal body temperature. Involuntary contractions of skeletal muscle, known as shivering, can dramatically increase the rate of heat production.

Properties of Muscular Tissue

Muscular tissue has four special properties that enable it to function and contribute to homeostasis:

1. **Electrical excitability,** a property of both muscle cells and neurons, is the ability to respond to certain stimuli by producing electrical signals—for example, **action potentials (impulses).** The action potentials propagate (travel) along the plasma membrane due to the presence of specific ion channels. The stimuli that trigger action potentials may be electrical signals arising in the muscle tissue itself, such as occurs in the heart's pacemaker, or chemical stimuli, such as neurotransmitters released by neurons, hormones distributed by the blood, or even local changes in pH.

2. **Contractility** is the ability of muscular tissue to contract forcefully when stimulated by an action potential. When muscle contracts, it generates tension (force of contraction) while pulling on its attachment points. In an **isometric contraction** (*iso-*=equal; *-metric*=measure or length), the muscle develops tension but does not shorten. An example is holding a book in an outstretched hand. If the tension generated is great enough to overcome the resistance of the object to being moved, the muscle shortens and movement occurs. In an **isotonic contraction** (*-tonic*=tension), the tension developed by the muscle remains almost constant while the muscle shortens. An example is lifting a book off a table.

3. **Extensibility** is the ability of muscular tissue to stretch without being damaged. Extensibility allows a muscle to contract forcefully even if it is already stretched. Normally, smooth muscle is subject to the greatest amount of stretching. For example, each time the stomach fills with food, the

muscle in the wall is stretched. Cardiac muscle also is stretched each time the heart fills with blood.

4. **Elasticity** is the ability of muscular tissue to return to its original length and shape after contraction or extension.

Skeletal muscle is the focus of much of this chapter. Cardiac muscle and smooth muscle are described briefly here. Cardiac muscle is discussed in more detail in Chapter 14 (the heart), and smooth muscle is included in Chapter 20 (the autonomic nervous system), as well as in discussions of the various organs containing smooth muscle.

CHECKPOINT

1. What features distinguish the three types of muscular tissue?
2. Summarize the functions of muscular tissue.
3. Describe the properties of muscular tissue.

SKELETAL MUSCLE TISSUE

OBJECTIVES

- Explain the relationship of connective tissue components, blood vessels, and nerves to skeletal muscles.
- Describe the microscopic anatomy of a skeletal muscle fiber.
- Explain how a skeletal muscle fiber contracts and relaxes.

Each skeletal muscle is a separate organ composed of hundreds to thousands of cells. The cells are called **muscle fibers** because of their elongated shapes. Connective tissues surround muscle fibers and whole muscles, and blood vessels and nerves penetrate skeletal muscles to exert their effects on individual muscle fibers (Figure 10.1). To understand skeletal muscle contraction, you first need to learn about its gross and microscopic anatomy.

Connective Tissue Components

Connective tissue surrounds and protects muscular tissue. A **fascia** (FASH-ē-a=bandage) is a sheet or broad band of fibrous connective tissue that supports and surrounds organs of the body. The **superficial fascia** (or **subcutaneous layer**) separates muscle from skin (see Figure 11.23 on page 365). It is composed of areolar connective tissue and adipose tissue and provides a pathway for nerves, blood vessels, and lymphatic vessels to enter and exit muscles. The adipose tissue of superficial fascia stores most of the body's triglycerides, serves as an insulating layer that reduces heat loss, and protects muscles from physical trauma. **Deep fascia** is dense irregular connective tissue that lines the body wall and limbs and holds muscles with similar functions together (see Figure 11.23 on page 365). Deep fascia allows free movement of muscles, carries nerves and blood and lymphatic vessels, and fills spaces between muscles.

Three layers of connective tissue extend from the deep fascia to further protect and strengthen skeletal muscle (Figure 10.1). The outermost layer, which encircles the entire muscle, is the **epimysium** (ep-i-MĪZ-ē-um; *epi-*=upon). **Perimysium** (per-i-MĪZ-ē-um; *peri-*=around) surrounds groups of 10 to 100 or more individual muscle fibers, separating them into bundles

called **fascicles** (FAS-i-kuls=little bundles). Many fascicles are large enough to be seen with the naked eye. They give a cut of meat its characteristic "grain," and if you tear a piece of meat, it rips apart along the fascicles. Both epimysium and perimysium are dense irregular connective tissue. Penetrating the interior of each fascicle and separating individual muscle fibers from one another is **endomysium** (en'-dō-MĪZ-ē-um; *endo-*=within), a thin sheath of areolar connective tissue.

Epimysium, perimysium, and endomysium are all continuous with the connective tissue that attaches skeletal muscle to other structures, such as bone or another muscle. All three connective tissue layers may extend beyond the muscle fibers to form a **tendon**—a cord of dense regular connective tissue composed of parallel bundles of collagen fibers that attaches a muscle to the periosteum of a bone. An example is the calcaneal (Achilles) tendon of the gastrocnemius (calf) muscle, which attaches the muscle to the calcaneus (see Figure 11.24c on page 369). When the connective tissue elements extend as a broad, flat layer, the tendon is called an **aponeurosis** (*apo-*= from; *neur-*=a sinew). An example of an aponeurosis is the epicranial aponeurosis on top of the skull between the occipital and frontal bellies of the occipitofrontalis muscle (see Figure 11.4c on page 300).

Certain tendons, especially those of the wrist and ankle, are enclosed by tubes of fibrous connective tissue called **tendon (synovial) sheaths,** which are similar in structure to bursae. The inner layer of a tendon sheath, the *visceral layer,* is attached to the surface of the tendon. The outer layer is known as the *parietal layer* and is attached to bone (see Figure 11.20a on page 350). Between the layers is a cavity that contains a film of synovial fluid. Tendon sheaths reduce friction as tendons slide back and forth.

GANGLION CYST

A **ganglion (synovial) cyst** is a swelling, usually on the dorsum of the wrist and hand, that appears periodically, especially following activities that involve repetitive hand motions. It results from an accumulation of fluid that has leaked from a tendon sheath or joint. The cyst contains fluid within fibrous connective tissue and is an extension of a tendon sheath that encloses a long extensor tendon in the wrist. Although they are usually not painful, ganglion cysts do cause some discomfort when they are inadvertently hit or when the wrist is flexed or extended. As a rule they are harmless, generally requiring no treatment and having little effect on activities. ∎

Microscopic Anatomy of a Skeletal Muscle Fiber

The most important components of a skeletal muscle are the muscle fibers themselves. During embryonic development, each skeletal muscle fiber arises from the fusion of a hundred or more small mesodermal cells called *myoblasts* (Figure 10.2a on page 269). Hence, each mature skeletal muscle fiber is a single cell with a hundred or more nuclei. Once fusion has occurred, the muscle fiber loses its ability to undergo mitosis. Thus, the number of skeletal muscle fibers is set before birth, and most of these cells last a lifetime. The dramatic muscle growth that occurs after birth occurs mainly by enlargement of existing muscle fibers, called **hypertrophy** (hī-PER-trō-fē; *hyper-*=above or excessive; *-trophy*=nourishment), rather than by **hyperplasia** (hī-per-PLĀ

-zē-a; *-plasis*=molding), an increase in the number of fibers. During childhood, human growth hormone and other hormones stimulate an increase in the size of skeletal muscle fibers. The hormone testosterone (from the testes in males and in small amounts from other tissues in females) promotes further

Figure 10.1 Organization of skeletal muscle and its connective tissue coverings.

A skeletal muscle consists of individual muscle fibers (cells) bundled into fascicles and surrounded by three connective tissue layers that are extensions of the deep fascia.

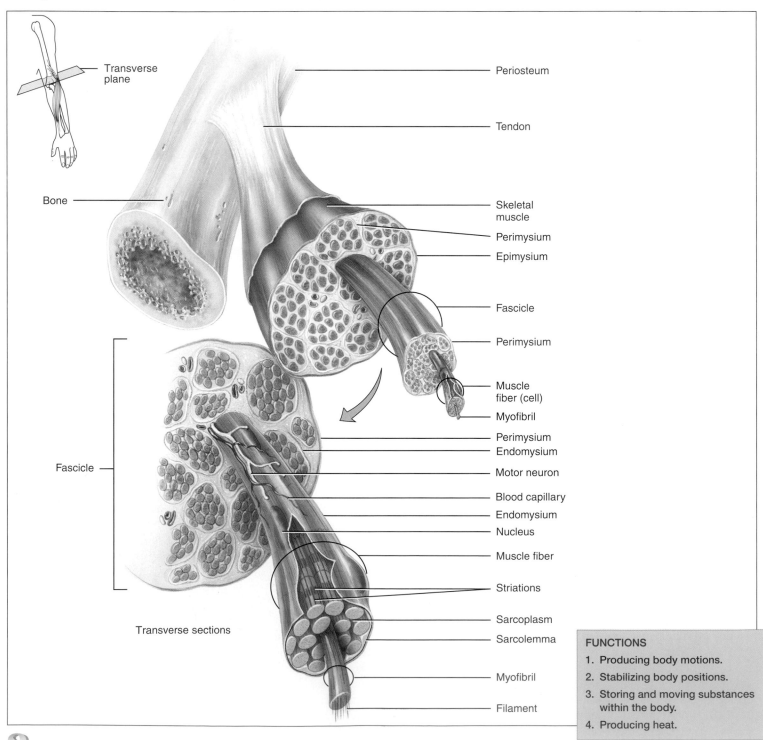

Transverse plane

Bone

Fascicle

Transverse sections

Periosteum

Tendon

Skeletal muscle

Perimysium

Epimysium

Fascicle

Perimysium

Muscle fiber (cell)

Myofibril

Perimysium

Endomysium

Motor neuron

Blood capillary

Endomysium

Nucleus

Muscle fiber

Striations

Sarcoplasm

Sarcolemma

Myofibril

Filament

FUNCTIONS

1. Producing body motions.
2. Stabilizing body positions.
3. Storing and moving substances within the body.
4. Producing heat.

Which connective tissue coat surrounds groups of muscle fibers, separating them into fascicles?

enlargement of muscle fibers. A few myoblasts do persist in mature skeletal muscle as *satellite cells* (Figure 10.2a). These cells retain the capacity to fuse with one another or with damaged muscle fibers to regenerate functional muscle fibers. The number of new skeletal muscle fibers formed, however, is not enough to compensate for significant skeletal muscle damage or degeneration. In such cases, skeletal muscle tissue undergoes **fibrosis,** the replacement of muscle fibers by fibrous scar tissue. For this reason, skeletal muscle tissue can regenerate only to a limited extent.

Figure 10.2 Microscopic organization of skeletal muscle. The sarcolemma of the muscle fiber encloses sarcoplasm and myofibrils, and sarcoplasmic reticulum (SR) wraps around each myofibril. Thousands of T tubules filled with extracellular fluid invaginate from the sarcolemma toward the center of the muscle fiber. A T tubule and the two terminal cisterns of the SR on either side of it form a triad. A photomicrograph of skeletal muscle tissue is shown in Table 3.4A on page 85.

The contractile elements of muscle fibers are the myofibrils, which contain overlapping thick and thin filaments.

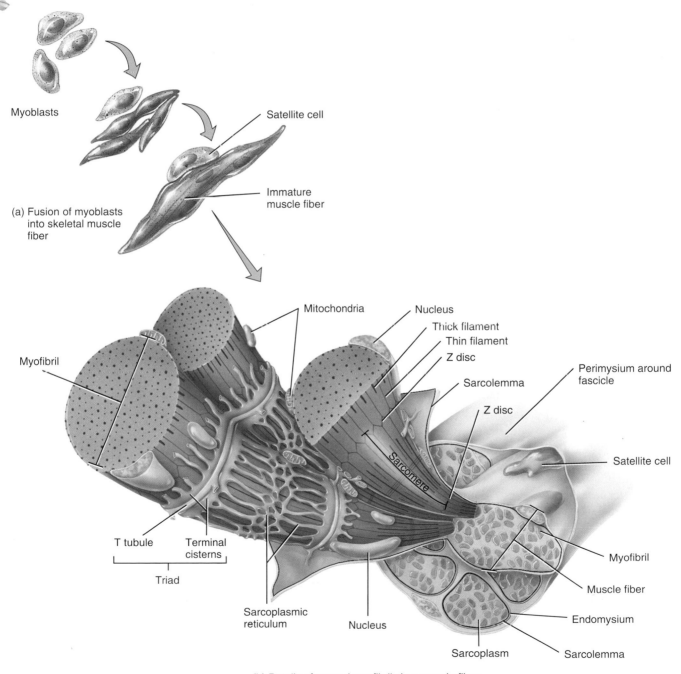

(a) Fusion of myoblasts into skeletal muscle fiber

(b) Details of several myofibrils in a muscle fiber

Which structure shown here releases calcium ions to trigger muscle contraction?

Mature muscle fibers range from 10 to 100 μm* in diameter. Typical muscle fiber length is about 10 cm (4 in.) in humans, although some are up to 30 cm (12 in.) long.

Sarcolemma, T Tubules, and Sarcoplasm

The multiple nuclei of a skeletal muscle fiber are located just beneath the **sarcolemma** (*sarc-*=flesh; *-lemma*=sheath), the plasma membrane of a muscle fiber (Figure 10.2b). Thousands of tiny invaginations of the sarcolemma, called **T (transverse) tubules,** tunnel from the surface in toward the center of each muscle fiber. T tubules are open to the outside of the fiber and thus are filled with interstitial fluid. Muscle action potentials propagate along the sarcolemma and through the T tubules, quickly spreading throughout the muscle fiber. This arrangement ensures that all the superficial and deep parts of the muscle fiber become excited by an action potential virtually simultaneously.

Within the sarcolemma is the **sarcoplasm,** the cytoplasm of a muscle fiber. Sarcoplasm includes a substantial amount of glycogen, which can be split into glucose that is used for synthesis of ATP. In addition, the sarcoplasm contains a red-colored protein called **myoglobin** (MĪ-o-glōb-in). This protein, found only in muscle, binds oxygen molecules that diffuse into muscle fibers from interstitial fluid. Myoglobin releases oxygen when mitochondria need it for ATP production. The mitochondria lie in rows throughout the muscle fiber, strategically close to the muscle proteins that use ATP during contraction.

Myofibrils and Sarcoplasmic Reticulum

At high magnification the sarcoplasm appears stuffed with little threads. These small structures are the contractile elements of skeletal muscle, the **myofibrils** (Figure 10.2b). Myofibrils are about 2 μm in diameter and extend the entire length of the muscle fiber. Their prominent striations make the whole muscle fiber look striped.

A fluid-filled system of membranous sacs called the **sarcoplasmic reticulum** (sar′-kō-PLAZ-mik re-tik-ū-lum) or **SR** encircles each myofibril (Figure 10.2b). This elaborate system is similar to smooth endoplasmic reticulum in nonmuscle cells. Dilated end sacs of the sarcoplasmic reticulum called **terminal cisterns** (=reservoirs) butt against the T tubules from both sides. One T tubule and the two terminal cisterns on either side of it form a **triad** (*tri-*=three). In a relaxed muscle fiber, the sarcoplasmic reticulum stores calcium ions (Ca^{2+}). Release of Ca^{2+} from the terminal cisterns into the sarcoplasm triggers muscle contraction.

MUSCULAR ATROPHY AND HYPERTROPHY

Muscular atrophy (a-trō-fē; *a-*=without, *-trophy*=nourishment) is a wasting away of muscles. Individual muscle fibers decrease in size as a result of progressive loss of myofibrils. The atrophy that occurs if muscles are not used is termed *disuse atrophy.* Bedridden individuals and people with casts experience disuse atrophy because the flow of nerve impulses to inactive

skeletal muscle is greatly reduced. If the nerve supply to a muscle is disrupted or cut, the muscle undergoes *denervation atrophy.* Over a period of from 6 months to 2 years, the muscle shrinks to about one-fourth its original size, and the muscle fibers are replaced by fibrous connective tissue. The transition to connective tissue, when complete, cannot be reversed.

As noted previously **muscular hypertrophy** is an increase in the size of muscle fibers owing to increased production of myofibrils, mitochondria, sarcoplasmic reticulum, and so forth. It results from very forceful, repetitive muscular activity, such as strength training. Because hypertrophied muscles contain more myofibrils, they are capable of more forceful contractions. ■

Filaments and the Sarcomere

Within myofibrils are even smaller structures called **filaments,** which are only 1–2 μm long (Figure 10.2b). Two of the filaments, thin and thick filaments, are only 1–2 μm long and are directly involved in the contractile process. The diameter of the *thin filaments* is about 8 nm,† whereas that of the *thick filaments* is about 16 nm. Overall, there are two thin filaments for every thick filament. The filaments inside a myofibril do not extend the entire length of a muscle fiber. Instead, they are arranged in basic functional units called **sarcomeres** (*-mere*=part) (Figure 10.3a). Narrow, plate-shaped regions of dense material called **Z discs** separate one sarcomere from the next.

The thick and thin filaments overlap one another to a greater or lesser extent, depending on whether the muscle is contracted, relaxed, or stretched. The pattern of their overlap, consisting of a variety of zones and bands (Figure 10.3b), creates the striations that can be seen both in single myofibrils and in whole muscle fibers. The darker middle part of the sarcomere is the **A band,** which extends the entire length of the thick filaments (Figure 10.3b). A narrow **H zone** in the center of each A band contains thick filaments but no thin filaments. Toward each end of the A band is a *zone of overlap,* where the thick and thin filaments lie side by side. The **I band** is a lighter, less dense area that contains thin filaments but no thick filaments (Figure 10.3b). A Z disc passes through the center of each I band. Supporting proteins that hold the thick filaments together at the center of the H zone form the **M line,** so-named because it is at the *middle* of the sarcomere. Figure 10.4 shows the relations of the zones, bands, and lines as seen in a transmission electron micrograph.

EXERCISE-INDUCED MUSCLE DAMAGE

Comparison of electron micrographs of muscle tissue taken from athletes before and after intense exercise reveal considerable exercise-induced muscle damage, including torn sarcolemmas in some muscle fibers, damaged myofibrils, and disrupted Z discs. Microscopic muscle damage after exercise is also indicated by increases in blood levels of proteins, such as myoglobin and the enzyme creatine kinase, that are normally confined within

---

*One micrometer (μm) is 10^{-6} meter (1/25,000 in.).

†One nanometer (nm) is 10^{-9} meter (0.001 μm).

Figure 10.3 The arrangement of filaments within a sarcomere. A sarcomere extends from one Z disc to the next.

Myofibrils contain two types of contractile filaments: thick filaments and thin filaments.

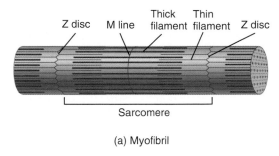

(a) Myofibril

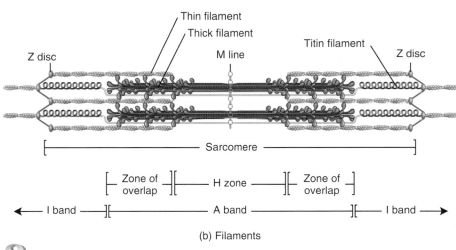

(b) Filaments

Among the following, which is smallest: muscle fiber, thick filament, or myofibril? Which is largest?

Figure 10.4 Transmission electron micrograph showing the characteristic zones and bands of a sarcomere.

The striations of skeletal muscle are alternating darker A bands and lighter I bands.

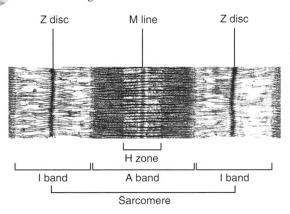

 21,600x

What is the ratio between thick and thin filaments in skeletal muscle?

muscle fibers. From 12 to 48 hours after a period of strenuous exercise, skeletal muscles often become sore. Such **delayed onset muscle soreness (DOMS)** is accompanied by stiffness, tenderness, and swelling. Although the causes of DOMS are not completely understood, microscopic muscle damage appears to be a major factor. ■

Muscle Proteins

Myofibrils are built from three kinds of proteins: (1) contractile proteins, which generate force during contraction; (2) regulatory proteins, which help switch the contraction process on and off; and (3) structural proteins, which keep the thick and thin filaments in the proper alignment, give the myofibril elasticity and extensibility, and link the myofibrils to the sarcolemma and extracellular matrix.

The two *contractile proteins* in muscle are myosin and actin, which are the main components of thick and thin filaments, respectively. **Myosin** functions as a motor protein in all three types of muscle tissue. Motor proteins push or pull their cargo to achieve movement by converting the chemical energy in ATP to mechanical energy of motion or force production. About 300 molecules of myosin form a single thick filament.

Each myosin molecule is shaped like two golf clubs twisted together (Figure 10.5a). The *myosin tail* (golf club handles) points toward the M line in the center of the sarcomere. Tails of neighboring myosin molecules lie parallel to one another, forming the shaft of the thick filament. The two projections of each myosin molecule (golf club heads) are called *myosin heads* or *crossbridges*. The heads project outward from the shaft in a spiraling fashion, each extending toward one of the six thin filaments that surround the thick filament.

Thin filaments extend from anchoring points within the Z discs (see Figure 10.3b). Their main component is the protein **actin.** Individual actin molecules join to form an actin filament that is twisted into a helix (Figure 10.5b). On each actin molecule is a myosin-binding site, where a myosin head can attach. Smaller amounts of two *regulatory proteins*—**tropomyosin** and **troponin**—are also part of the thin filament. In relaxed muscle, myosin is blocked from binding to actin because tropomyosin covers the myosin-binding site on actin. The tropomyosin strand, in turn, is held in place by troponin.

Besides contractile and regulatory proteins, muscle contains about a dozen *structural proteins*, which contribute to the alignment, stability, elasticity, and extensibility of myofibrils. Several key structural proteins are titin, myomesin, and dystrophin. *Titin* (*titan*=gigantic) is the third most plentiful protein in skeletal muscle (after actin and myosin). This molecule's name reflects its huge size. With a molecular weight of about 3 million daltons, titin is 50 times larger than an average-sized protein. Each titin molecule spans half a sarcomere, from a Z disc to an M line (see Figure 10.3b), a distance of 1 to 1.2 μm in relaxed muscle. Titin anchors a thick filament to both a Z disc and the M line, thereby helping stabilize the position of the thick filament. Because it can stretch to at least four times its resting length and then spring back unharmed, titin accounts for much of the elasticity and extensibility of myofibrils. Titin probably helps the sarcomeres return to their resting length after a muscle has contracted or been stretched, may help prevent overextension of sarcomeres, and maintains the central location of the A bands.

Molecules of the protein *myomesin* form the M line. The M line proteins bind to titin and connect adjacent thick filaments to one another. Myomesin holds the thick filaments in register at the M line. *Dystrophin* is a cytoskeletal protein that links thin filaments of the sarcomere to integral membrane proteins of the sarcolemma. In turn, the membrane proteins attach to proteins in the connective tissue matrix that surrounds muscle fibers. Hence, dystrophin and its associated proteins are thought to reinforce the sarcolemma and help transmit the tension generated by the sarcomeres to the tendons.

Nerve and Blood Supply

Skeletal muscles are well supplied with nerves and blood vessels (see Figure 11.18c on page 340). Generally, an artery and one or two veins accompany each nerve that penetrates a skeletal muscle. The neurons (nerve cells) that stimulate skeletal muscle fibers to contract are called **motor neurons.** A motor neuron has a threadlike extension, called an **axon,** from the brain or spinal cord to a group of skeletal muscle fibers. A muscle fiber contracts in response to one or more action potentials propagating along its sarcolemma and through its T tubule system. Muscle action potentials arise at the **neuromuscular junction (NMJ),** the synapse between a motor neuron and a skeletal muscle fiber (Figure 10.6a). A **synapse** is a region where communication occurs between two neurons, or between a neuron and a target cell—for example, between a motor neuron and a muscle fiber. At most synapses a small gap, called the **synaptic cleft,** separates the two cells. Because the cells do not physically touch, the action potential from one cell cannot "jump the gap" to directly excite the next cell. Instead, the first cell communicates with the second indirectly, by releasing a chemical called a **neurotransmitter.**

At a neuromuscular junction, the motor neuron axon terminal divides into a cluster of **synaptic end bulbs** (Figure 10.6a, b). Suspended in the cytosol within each bulb are hundreds of membrane-enclosed sacs called **synaptic vesicles.** Inside each synaptic vesicle are thousands of molecules of **acetylcholine** (as'-ē-til-KŌ-lēn), abbreviated **ACh,** the neurotransmitter released at the NMJ. The region of the sarcolemma that is adjacent to the synaptic end bulbs is called the **motor end plate.** It contains 30–40 million acetylcholine receptors, which are integral transmembrane proteins that bind specifically to ACh. A neuromuscular junction thus includes all the synaptic

Figure 10.5 Structure of thick and thin filaments. (a) The thick filament contains about 300 myosin molecules, one of which is pictured. The myosin tails form the shaft of the thick filament, whereas the myosin heads project outward toward the surrounding thin filaments. (b) Thin filaments contain actin, troponin, and tropomyosin.

🔑 **Contractile proteins (myosin and actin) generate force during contraction, whereas regulatory proteins (troponin and tropomyosin) help switch contraction on and off.**

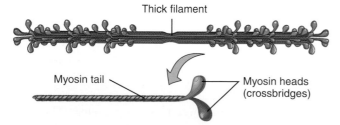

Thick filament

Myosin tail

Myosin heads (crossbridges)

(a) One thick filament (above) and a myosin molecule (below)

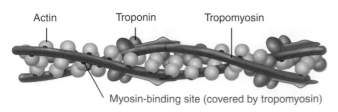

Actin   Troponin   Tropomyosin

Myosin-binding site (covered by tropomyosin)

(b) Portion of a thin filament

❓ **Which proteins connect to the Z disc? Which proteins are present in the A band? In the I band?**

Figure 10.6 Structure of the neuromuscular junction (NMJ), the synapse between a motor neuron and a skeletal muscle fiber.

Synaptic end bulbs at the tips of axon terminals contain synaptic vesicles filled with acetylcholine.

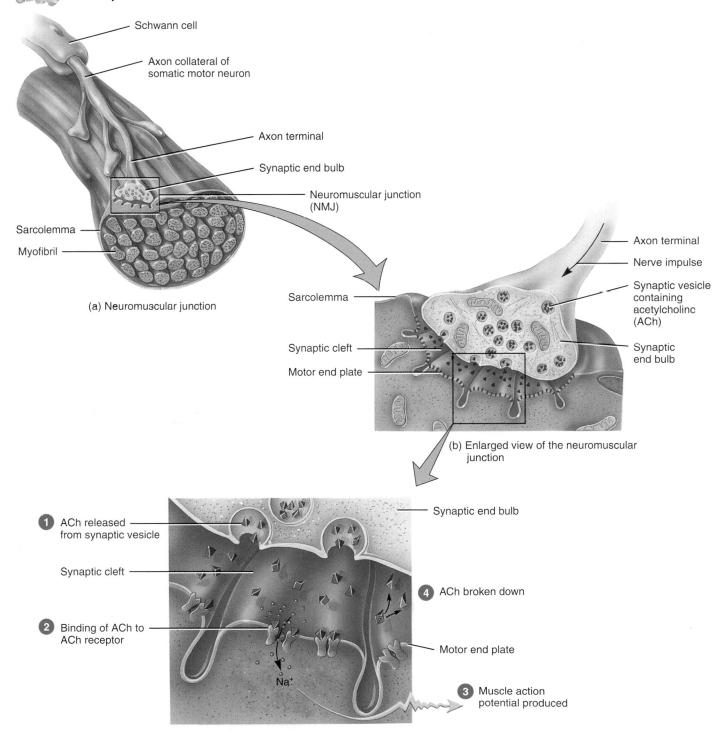

(a) Neuromuscular junction

(b) Enlarged view of the neuromuscular junction

(c) Binding of acetylcholine to ACh receptors in the motor end plate

continues

Figure 10.6 (continued)

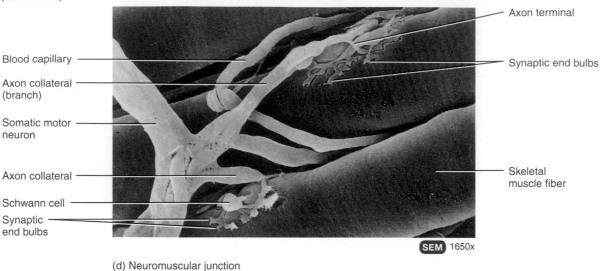

Blood capillary

Axon collateral
(branch)

Somatic motor
neuron

Axon collateral

Schwann cell

Synaptic
end bulbs

Axon terminal

Synaptic end bulbs

Skeletal
muscle fiber

SEM 1650x

(d) Neuromuscular junction

What portion of the sarcolemma contains acetylcholine receptors?

end bulbs on one side of the synaptic cleft, plus the motor end plate of the muscle fiber on the other side.

Even though each skeletal muscle fiber has only a single neuromuscular junction, the axon of a motor neuron branches out and forms neuromuscular junctions with many different muscle fibers. A motor neuron plus all the skeletal muscle fibers it stimulates is called a **motor unit.** A single motor neuron makes contact with an average of 150 muscle fibers, and all muscle fibers in one motor unit contract in unison. Muscles that control precise movements consist of many small motor units. For example, muscles of the larynx (voice box) that control voice production have as few as two or three muscle fibers per motor unit, and muscles controlling eye movements may have 10–20 muscle fibers per motor unit. In contrast, some motor units in skeletal muscles responsible for large-scale and powerful movements, such as the biceps brachii muscle in the arm and the gastrocnemius muscle in the leg, may have 2000–3000 muscle fibers each. The total strength of a muscle contraction depends, in part, on how large its motor units are and how many motor units are activated at the same time.

Microscopic blood vessels called capillaries are plentiful in muscle tissue; each muscle fiber is in close contact with one or more capillaries (see Figure 10.1). Capillary blood brings oxygen and nutrients to the muscle fibers and removes heat and the waste products of muscle metabolism. Especially during contraction, a muscle fiber synthesizes and uses considerable ATP (adenosine triphosphate); these reactions require oxygen, glucose, fatty acids, and other substances that are supplied in the blood.

Contraction and Relaxation of Skeletal Muscle Fibers

Arrival of the nerve impulse (nerve action potential) at the synaptic end bulbs causes synaptic vesicles to liberate ACh. The

ACh diffuses across the synaptic cleft between the motor neuron and the motor end plate (Figure 10.6c). Binding of ACh to its receptor in the sarcolemma allows small cations, most importantly Na^+, to flow across the membrane. The inflow of Na^+ triggers a muscle action potential. The muscle action potential propagates along the sarcolemma through the T tubule system and to the sarcoplasmic reticulum (SR), where it causes the release of calcium ions (Ca^{2+}) into the sarcoplasm of the muscle fiber. In the presence of Ca^{2+} and ATP, the skeletal muscle shortens because the thick and thin filaments slide past each other. In this so-called **sliding filament mechanism,** the myosin heads attach to the thin filaments at both ends of a sarcomere, progressively pulling the thin filaments toward the M line (Figure 10.7). As a result, the thin filaments slide inward and meet at the center of a sarcomere. They may even move so far inward that their ends overlap (Figure 10.7c). As the thin filaments slide inward, the Z discs come closer together, and the sarcomere shortens. Note that the lengths of the individual thick and thin filaments do not change; it is the amount of overlap between them that changes. Shortening of the individual sarcomeres causes shortening of the whole muscle fiber, and ultimately shortening of the entire muscle.

The effect of ACh binding lasts only briefly because the neurotransmitter is rapidly broken down by an enzyme in the synaptic cleft called **acetylcholinesterase (AChE).** AChE breaks down ACh into products that cannot activate the ACh receptor. When action potentials cease being transmitted by the motor neuron, ACh release stops and AChE rapidly breaks down the ACh already present in the synaptic cleft. This ends the generation of muscle action potentials, and the Ca^{2+} moves from the sarcoplasm of the muscle fiber back into the membrane of the sarcoplasmic reticulum.

Because skeletal muscle fibers often are very long cells, the NMJ usually is located near the midpoint of the fiber. Muscle action potentials arise at the NMJ and then propagate toward both ends of the fiber. This arrangement permits nearly simultaneous activation (and thus contraction) of all parts of the fiber.

Several plant products and drugs selectively block certain events at the NMJ. *Botulinum toxin*, produced by the bacterium *Clostridium botulinum*, blocks exocytosis of synaptic vesicles at the NMJ. As a result, ACh is not released, and muscle contraction does not occur. The bacteria proliferate in improperly canned foods and their toxin is one of the most lethal chemicals known. A tiny amount can cause death by paralyzing skeletal muscles. Breathing stops due to paralysis of respiratory muscles, including the diaphragm. Yet, it is also the first bacterial toxin to be used as a medicine (Botox). Injections of Botox into the affected muscles can help patients who have strabismus (crossed eyes) or blepharospasm (uncontrollable blinking). It is also used to alleviate chronic back pain due to muscle spasms in the lumbar region, and as a cosmetic treatment to relax muscles that cause facial wrinkles.

The plant derivative *curare*, a poison used by South American Indians on arrows and blowgun darts, causes muscle paralysis by binding to and blocking ACh receptors. Curare-like drugs are often used during surgery to relax skeletal muscles. A family of chemicals called *anticholinesterase agents* have the property of slowing the enzymatic activity of acetylcholinesterase, thus slowing removal of ACh from the synaptic cleft. At low doses, these agents can strengthen weak muscle contractions. One example is neostigmine, which is used to treat patients with myasthenia gravis (see page 283). Neostigmine is also used as an antidote for curare poisoning and to terminate the effects of curare after surgery.

RIGOR MORTIS

After death, cellular membranes start to become leaky. Calcium ions leak out of the sarcoplasmic reticulum into the sarcoplasm and allow myosin heads to bind to actin. ATP synthesis has ceased, however, so the crossbridges cannot detach from actin. The resulting condition, in which muscles are in a state of rigidity (cannot contract or stretch), is called **rigor mortis** (rigidity of death). Rigor mortis begins 3–4 hours after death and lasts about 24 hours; then it disappears as proteolytic enzymes from lysosomes digest the crossbridges. ■

Muscle Tone

In a skeletal muscle, a small number of motor units are involuntarily activated to produce a sustained contraction of their muscle fibers while the majority of the motor units are not activated and their muscle fibers remain relaxed. This process results in **muscle tone** (*tonos*=tension), which keeps muscles firm. In order to sustain muscle tone, small groups of motor units become active and then become inactive, in a constantly shifting pattern. Although muscle tone keeps skeletal muscles firm, it does not result in a contraction strong enough to produce movement. For example, when the muscles in the back of the neck are in normal tonic contraction, they keep the head upright and prevent it from slumping forward on the chest, but they do not generate enough force to draw the head backward into hyperextension. Muscle tone is important in smooth muscle such as that found in the gastrointestinal tract, where the walls of the digestive organs maintain a steady pressure on their contents. The tone of smooth muscles surrounding the walls of blood vessels plays a crucial role in maintaining a steady blood pressure.

Figure 10.7 Sliding filament mechanism of muscle contraction, as it occurs in two adjacent sarcomeres.

During muscle contractions, thin filaments move toward the M line of each sarcomere.

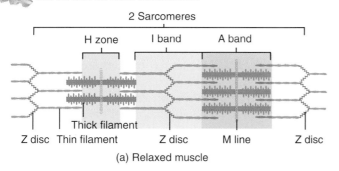

(a) Relaxed muscle

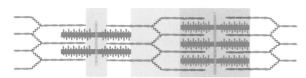

(b) Partially contracted muscle

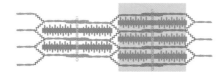

(c) Maximally contracted muscle

What happens to the I band and H zone during contraction? Do the lengths of the thick and thin filaments change?

HYPOTONIA AND HYPERTONIA

Hypotonia (*hypo-*=below) refers to decreased or lost muscle tone. Such muscles are said to be **flaccid** (FLAK-sid or FLAS-sid). Flaccid muscles are loose and appear flattened rather than rounded; the affected limbs are hyperextended. Certain disorders of the nervous system and electrolyte disturbances (especially sodium, calcium, and, to a lesser extent, magnesium) may result in **flaccid paralysis,** which is characterized by loss of muscle tone, loss or reduction of tendon reflexes, and atrophy (wasting away) and degeneration of muscles.

Hypertonia (*hyper-*=above) refers to increased muscle tone and is expressed in two ways: spasticity or rigidity. **Spasticity** (spas-TISS-i-tē) is characterized by increased muscle tone (stiffness) associated with an increase in tendon reflexes and pathological reflexes (such as the Babinski sign, in which the great toe extends with or without fanning of the other toes in response to stroking the outer margin of the sole). Certain disorders of the nervous system and electrolyte disturbances such as those previously noted may result in **spastic paralysis,** partial paralysis in which the muscles exhibit spasticity. **Rigidity** refers to increased muscle tone in which reflexes are not affected. ■

CHECKPOINT

4. Describe the types of connective tissue that cover skeletal muscles.
5. Describe the components of a sarcomere. How do thin and thick filaments differ structurally?
6. What proteins comprise skeletal muscle tissue and what are their functions?
7. Describe the nerve supply to a skeletal muscle fiber.
8. Why is a rich blood supply so important to muscle contraction?
9. Why is muscle tone important?

TYPES OF SKELETAL MUSCLE FIBERS

OBJECTIVE

● Compare the structure and function of the three types of skeletal muscle fibers.

Skeletal muscle fibers vary structurally in their content of myoglobin, the red protein that binds oxygen in muscle fibers. Those with a high myoglobin content are called **red muscle fibers,** while those that have a low myoglobin content are called **white muscle fibers.** Red muscle fibers also contain more mitochondria and are supplied by more blood capillaries than white muscle fibers.

Skeletal muscle fibers also contract and relax at different speeds. A fiber is categorized as either slow or fast depending on how rapidly the ATPase in its myosin heads hydrolyzes ATP. In addition, skeletal muscle fibers vary in which metabolic reactions they use to generate ATP and in how quickly they fatigue. Based on these structural and functional characteristics, skeletal muscle fibers are classified into three main types: (1) slow oxidative fibers, (2) fast oxidative-glycolytic fibers, and (3) fast glycolytic fibers.

Slow Oxidative Fibers

Slow oxidative (SO) fibers are smallest in diameter and thus are the least powerful type of muscle fibers. They appear dark red because they contain large amounts of myoglobin and many blood capillaries. Because they have many large mitochondria, SO fibers generate ATP mainly by aerobic cellular respiration, which is why they are called oxidative fibers. These fibers are said to be "slow" because they use ATP at a slow rate. As a result, SO fibers have a slow speed of contraction. However, slow fibers are very resistant to fatigue and are capable of pro-

longed, sustained contractions for many hours. These fibers are adapted for maintaining posture and for aerobic, endurance-type activities such as running a marathon.

Fast Oxidative-Glycolytic Fibers

Fast oxidative-glycolytic (FOG) fibers are intermediate in diameter between the other two types of fibers. Like slow oxidative fibers, they contain large amounts of myoglobin and many blood capillaries. Thus, they have a dark red appearance. FOG fibers can generate considerable ATP by aerobic cellular respiration, which gives them a moderately high resistance to fatigue. Because their intracellular glycogen level is high, they also generate ATP by anaerobic (oxygen free) glycolysis. FOG fibers are "fast" because they use ATP at a fast rate, which makes their speed of contraction faster than SO fibers. FOG fibers contribute to activities such as walking and sprinting.

Fast Glycolytic Fibers

Fast glycolytic (FG) fibers are largest in diameter and contain the most myofibrils. Hence, they can generate the most powerful contractions. FG fibers have low myoglobin content, relatively few blood capillaries, few mitochondria, and appear white in color. They contain large amounts of glycogen and generate ATP mainly by glycolysis. Due to their large size and their ability to use ATP at a fast rate, FG fibers contract strongly and quickly. These fast-twitch fibers are adapted for intense anaerobic movements of short duration, such as weight lifting or throwing a ball, but they fatigue quickly. Strength training programs that engage a person in activities requiring great strength for short times produce increases in the size, strength, and glycogen content of fast glycolytic fibers. The FG fibers of a weight-lifter may be 50% larger than those of a sedentary person or endurance athlete. The increase in size is due to increased synthesis of muscle proteins. The overall result is muscle enlargement due to hypertrophy of the FG fibers.

Most skeletal muscles are a mixture of all three types of skeletal muscle fibers. The proportions vary somewhat, depending on the action of the muscle, the person's training regimen, and genetic factors. For example, the continually active postural muscles of the neck, back, and legs have a high proportion of SO fibers. Muscles of the shoulders and arms, in contrast, are not constantly active but are used briefly now and then to produce large amounts of tension, such as in lifting and throwing. These muscles have a high proportion of FG fibers. Leg muscles, which not only support the body but are also used for walking and running, have large numbers of both SO and FOG fibers.

The skeletal muscle fibers of any given motor unit are all of the same type. However, the different motor units in a muscle are recruited in a specific order, depending on need. For example, if weak contractions suffice to perform a task, only SO motor units are activated. If more force is needed, the motor units of FOG fibers are also recruited. Finally, if maximal force is required, motor units of FG fibers are also called into action.

TABLE 10.1 CHARACTERISTICS OF THE THREE TYPES OF SKELETAL MUSCLE FIBERS			
	SLOW OXIDATIVE (SO) FIBERS	**FAST OXIDATIVE- GLYCOLYTIC (FOG) FIBERS**	**FAST GLYCOLYTIC (FG) FIBERS**
Structural Characteristic			
Fiber diameter	Smallest.	Intermediate.	Largest.
Myoglobin content	Large amount.	Large amount.	Small amount.
Mitochondria	Many.	Many.	Few.
Capillaries	Many.	Many.	Few.
Color	Red.	Red-pink.	White (pale).
Functional Characteristic			
Capacity for generating ATP and method used	High capacity, by aerobic (oxygen-requiring) cellular respiration.	Intermediate capacity, by both aerobic (oxygen-requiring) cellular respiration and anaerobic cellular respiration (glycolysis).	Low capacity, by anaerobic cellular respiration (glycolysis).
Rate of ATP use	Slow.	Fast.	Fast.
Contraction velocity	Slow.	Fast.	Fast.
Location where fibers are abundant	Postural muscles such as those of the neck.	Leg muscles.	Arm muscles.
Primary functions of fibers	Maintaining posture and aerobic endurance activities.	Walking, sprinting.	Rapid, intense movements of short duration.

Slow oxidative fiber

Fast glycolytic fiber

Fast oxidative-glycolytic fiber

LM 440x

Transverse section of three types of skeletal muscle fibers

Activation of various motor units is controlled by the brain and spinal cord.

Table 10.1 summarizes the characteristics of the three types of skeletal muscle fibers.

EXERCISE AND SKELETAL MUSCLE TISSUE

OBJECTIVE

- Describe the effects of exercise on different types of skeletal muscle fibers.

The relative ratio of fast glycolytic (FG) and slow oxidative (SO) in each muscle is genetically determined and helps account for individual differences in physical performance. For example, people with a higher proportion of fast glycolytic (FG) fibers often excel in activities that require periods of intense activity, such as weight lifting or sprinting. People with higher percentages of slow oxidative (SO) fibers are better at activities that require endurance, such as long-distance running.

Although the total number of skeletal muscle fibers usually does not increase, the characteristics of those present can change to some extent. Various types of exercises can induce changes in the fibers in a skeletal muscle. Endurance-type (aerobic) exercises, such as running or swimming, cause a gradual transformation of some fast glycolytic (FG) fibers into fast oxidative-glycolytic (FOG) fibers. The transformed muscle fibers show slight increases in diameter, number of mitochondria, blood supply, and strength. Endurance exercises also result in cardiovascular and respiratory changes that cause skeletal muscles to receive better supplies of oxygen and nutrients, but do not increase muscle mass. By contrast, exercises that require great strength for short periods produce an increase in the size and strength of fast glycolytic (FG) fibers. The increase in size is due to increased synthesis of thick and thin filaments. The overall result is muscle enlargement (hypertrophy), as evidenced by the bulging muscles of body builders.

ANABOLIC STEROIDS

The illegal use of **anabolic steroids** by athletes has received widespread attention: These steroid hormones, similar to testosterone, are taken to increase muscle size and thus strength during athletic contests. The large doses needed to produce an effect, however, have damaging, sometimes even devastating side effects, including liver cancer, kidney damage, increased risk of heart disease, stunted growth, wide mood swings, and increased irritability and aggression. Additionally, females who take anabolic steroids may experience atrophy of the breasts and uterus, menstrual irregularities, sterility, facial hair growth, and deepening of the voice. Males may experience diminished testosterone secretion, atrophy of the testes, and baldness. ■

CHECKPOINT

10. What is the basis for classifying skeletal muscle fibers into three types?

CARDIAC MUSCLE TISSUE

OBJECTIVE

● Describe the main structural and functional characteristics of cardiac muscle tissue.

The principal tissue in the heart wall is **cardiac muscle tissue.** Although it is striated like skeletal muscle, its activity cannot be controlled voluntarily. Also, certain cardiac muscle fibers (as well as some smooth muscle fibers and nerve cells in the brain and spinal cord) display **autorhythmicity,** the ability to repeatedly generate spontaneous action potentials. In the heart, these action potentials cause alternating contraction and relaxation of the heart muscle fibers. Compared with skeletal muscle fibers, cardiac muscle fibers are shorter in length and less circular in transverse section (Figure 10.8a). They also exhibit branching, which gives individual cardiac muscle fibers a "stair-step" appearance. A typical cardiac muscle fiber is 50–100 μm long and has a diameter of about 14 μm. Usually one centrally located nucleus is present, although an occasional cell may have two nuclei. The ends of cardiac muscle fibers connect to neighboring fibers by irregular transverse thickenings of the sarcolemma referred to **intercalated discs** (in-TER-kā-lāt-ed; *intercalat-*=to insert between). The discs contain **desmosomes,** which hold the fibers together, and **gap junctions,** which allow muscle action potentials to conduct from one muscle fiber to its neighbors.

Mitochondria are larger and more numerous in cardiac muscle fibers than in skeletal muscle fibers. Cardiac muscle fibers have the same arrangement of actin and myosin, and the same bands, zones, and Z discs, as skeletal muscle fibers (Figure 10.8b). The transverse (T) tubules of cardiac muscle are wider but less abundant than those of skeletal muscle; there is one T tubule per sarcomere, located at the Z disc. The sarcoplasmic reticulum of cardiac muscle fibers is somewhat smaller than the SR of skeletal muscle fibers.

REGENERATION OF HEART CELLS

The heart of a heart attack survivor often has regions of infarcted (dead) cardiac muscle tissue that typically are replaced with noncontractile fibrous scar tissue over time. Our inability to repair damage from a heart attack has been attributed to a lack of stem cells in cardiac muscle and to the absence of mitosis in mature cardiac muscle fibers. A recent study of heart transplant recipients by American and Italian scientists, however, provides evidence for replacement of heart cells. The researchers studied men who had received a heart from a female, and then looked for the presence of a Y chromosome in heart cells. (All female cells except gametes have two X chromosomes and lack the Y chromosome.) Several years after the transplant surgery, between 7% and 16% of the heart cells in the transplanted tissue, including cardiac muscle fibers and endothelial cells in coronary arterioles and capillaries, had been replaced by the recipient's own cells, as evidenced by the presence of a Y chromosome. The study also revealed cells with some of the characteristics of stem cells in both transplanted hearts and control hearts. Evidently, stem cells can migrate from the blood into the heart and differentiate into functional muscle and endothelial cells. The hope is that researchers can learn how to "turn on" such regeneration of heart cells to treat people with heart failure or cardiomyopathy (diseased heart). ■

Under normal resting conditions, cardiac muscle tissue contracts and relaxes about 75 times per minute. This continuous, rhythmic activity is a major functional difference between cardiac and skeletal muscle tissue. Another difference is the source of stimulation. Skeletal muscle tissue contracts only when stimulated by acetylcholine released by an action potential in a motor neuron. In contrast, cardiac muscle tissue can contract without extrinsic (outside) nervous or hormonal stimulation. Its source of stimulation is a conducting network of specialized cardiac muscle fibers within the heart. Stimulation from the body's nervous system merely causes the conducting fibers to increase or decrease their rate of discharge. Cardiac muscle tissue remains contracted 10 to 15 times longer than skeletal muscle tissue. Cardiac muscle tissue also allows time for the heart chambers to relax and fill with blood between beats. This permits the heart rate to increase significantly but prevents the heart from undergoing *tetanus* (sustained contraction), which would cause blood flow to cease. Like skeletal muscle, cardiac muscle fibers can undergo hypertrophy in response to increased workload. Hence, many athletes have enlarged hearts.

CHECKPOINT

11. How do cardiac and skeletal muscle tissue differ structurally and functionally?

SMOOTH MUSCLE TISSUE

OBJECTIVE

● Describe the main structural and functional characteristics of smooth muscle tissue.

Like cardiac muscle tissue, **smooth muscle tissue** is usually activated involuntarily. Of the two types of smooth muscle tissue, the more common type is **visceral (single-unit) smooth**

Figure 10.8 **Histology of cardiac muscle.** A photomicrograph of cardiac muscle tissue is shown in Table 3.4B on page 85.

Muscle fibers of the atria form one functional network, whereas muscle fibers of the ventricles form a second functional network.

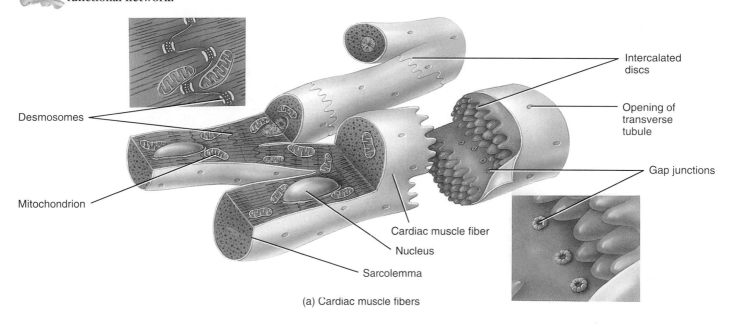

Desmosomes

Mitochondrion

Intercalated discs

Opening of transverse tubule

Gap junctions

Cardiac muscle fiber

Nucleus

Sarcolemma

(a) Cardiac muscle fibers

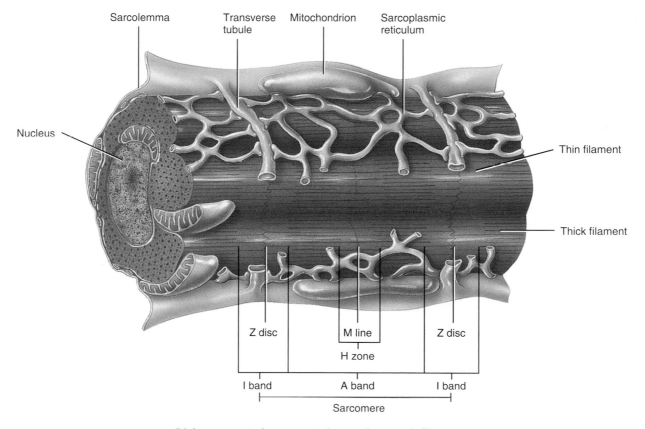

Sarcolemma

Transverse tubule

Mitochondrion

Sarcoplasmic reticulum

Nucleus

Thin filament

Thick filament

Z disc

M line

Z disc

H zone

I band

A band

I band

Sarcomere

(b) Arrangement of components in a cardiac muscle fiber

What are the functions of intercalated discs in cardiac muscle fibers?

muscle tissue (Figure 10.9). It is found in wraparound sheets that form part of the walls of small arteries and veins and of hollow viscera such as the stomach, intestines, uterus, and urinary bladder. Like cardiac muscle, visceral smooth muscle is autorhythmic. Because the fibers connect to one another by gap junctions, muscle action potentials spread throughout the network. For example, when a neurotransmitter, hormone, or autorhythmic signal stimulates one fiber, the muscle action potential spreads to neighboring fibers, which then contract in unison, as a single unit.

The second kind of smooth muscle tissue is **multiunit smooth muscle tissue** (Figure 10.9b). It consists of individual fibers, each with its own motor neuron terminals and with few gap junctions between neighboring fibers. Whereas stimulation of one visceral muscle fiber causes contraction of many adjacent fibers, stimulation of one multiunit fiber causes contraction of that fiber only. The walls of large arteries, the airways to the lungs, the arrector pili muscles that attach to hair follicles, the muscles of the iris that adjust pupil diameter, and the ciliary body that adjusts focus of the lens in the eye all contain multiunit smooth muscle tissue.

Smooth muscle fibers are considerably smaller than skeletal muscle fibers. A single relaxed smooth muscle fiber is 30–200 μm long, thickest in the middle (3–8 μm), and tapered at each end (Figure 10.9c). Within each fiber is a single, oval, centrally located nucleus. The sarcoplasm of smooth muscle fibers contains both thick filaments and thin filaments, in ratios between about 1:10

and 1:15, but they are not arranged in orderly sarcomeres as in striated muscle. Smooth muscle fibers also contain **intermediate filaments.** These filaments contain the protein desmin and appear to have a structural rather than a contractile role. Because the various filaments have no regular pattern of overlap, smooth muscle fibers do not exhibit striations—thus the name *smooth*. Smooth muscle fibers also lack transverse tubules and have little sarcoplasmic reticulum for storage of Ca^{2+}.

In smooth muscle fibers, intermediate filaments attach to structures called **dense bodies,** which are functionally similar to Z discs in striated muscle fibers. Some dense bodies are dispersed throughout the sarcoplasm; others are attached to the sarcolemma. Bundles of intermediate filaments stretch from one dense body to another (Figure 10.9c). During contraction, the sliding filament mechanism involving thick and thin filaments generates tension that is transmitted to intermediate filaments. These, in turn, pull on the dense bodies attached to the sarcolemma, causing a shortening of the muscle fiber (Figure 10.9c). When a smooth muscle fiber contracts, it turns like a corkscrew; it rotates in the opposite direction as it relaxes.

Although the principles of contraction are similar in all three types of muscle tissue, smooth muscle tissue exhibits some important physiological differences. Compared with contraction in a skeletal muscle fiber, contraction in a smooth muscle fiber starts more slowly and lasts much longer. Moreover, smooth muscle can both shorten and stretch to a greater extent than other muscle types.

Figure 10.9   Histology of smooth muscle tissue. In (a), one autonomic motor neuron synapses with several visceral smooth muscle fibers, and action potentials spread to neighboring fibers through gap junctions. In (b), three autonomic motor neurons synapse with individual multiunit smooth muscle fibers. Stimulation of one multiunit fiber causes contraction of that fiber only. A photomicrograph of smooth muscle tissue is shown in Table 3.4C on page 86.

Smooth muscle fibers have thick and thin filaments but no transverse tubules and little sarcoplasmic reticulum.

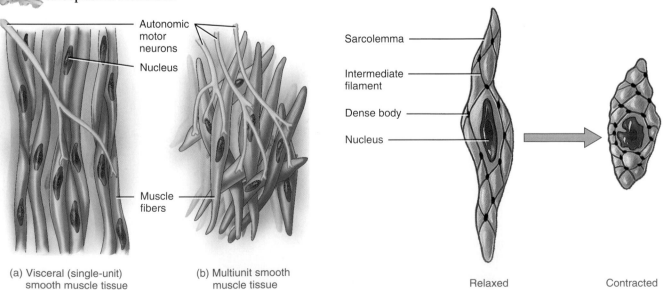

Autonomic motor neurons

Nucleus

Muscle fibers

(a) Visceral (single-unit) smooth muscle tissue

(b) Multiunit smooth muscle tissue

Sarcolemma

Intermediate filament

Dense body

Nucleus

Relaxed   Contracted

(c) Details of a smooth muscle fiber

? **Which type of smooth muscle is more like cardiac muscle than skeletal muscle, with respect to both its structure and function?**

An increase in the concentration of Ca^{2+} in smooth muscle cytosol initiates contraction, just as in striated muscle. There is far less sarcoplasmic reticulum (the reservoir for Ca^{2+} in striated muscle) in smooth muscle than in skeletal muscle. Calcium ions flow into smooth muscle sarcoplasm from both the interstitial fluid and sarcoplasmic reticulum, but because there are no transverse tubules in smooth muscle fibers, it takes longer for Ca^{2+} to reach the filaments in the center of the fiber and trigger the contractile process. This accounts, in part, for the slow onset and prolonged contraction of smooth muscle.

Calcium ions also move out of the muscle fiber slowly, which delays relaxation. The prolonged presence of Ca^{2+} in the cytosol provides for **smooth muscle tone,** a state of continued partial contraction. Smooth muscle tissue can thus sustain long-term tone, which is important in the gastrointestinal tract, where the walls maintain a steady pressure on the contents of the tract, and in the walls of blood vessels called arterioles, which maintain a steady pressure on blood.

Most smooth muscle fibers contract or relax in response to action potentials from the autonomic nervous system. In addi-

TABLE 10.2 SUMMARY OF THE MAJOR FEATURES OF THE THREE TYPES OF MUSCULAR TISSUE			
CHARACTERISTIC	**SKELETAL MUSCLE**	**CARDIAC MUSCLE**	**SMOOTH MUSCLE**
Microscopic appearance and features	Long cylindrical fiber with many peripherally located nuclei; striated.	Branched cylindrical fiber with one centrally located nucleus; intercalated discs join neighboring fibers; striated.	Fiber is thickest in the middle, tapered at each end, and has one centrally positioned nucleus; not striated.
Location	Attached primarily to bones by tendons.	Heart.	Walls of hollow viscera, airways, blood vessels, iris and ciliary body of eye, arrector pili of hair follicles.
Fiber diameter	Very large (10–100 μm).	Large (10–20 μm).	Small (5–10 μm).
Connective tissue components	Endomysium, perimysium, and epimysium.	Endomysium.	Endomysium.
Fiber length	30–40 cm.	50–100 μm.	20–500 μm.
Contractile proteins organized into sarcomeres	Yes.	Yes.	No.
Sarcoplasmic reticulum	Abundant.	Some.	Scanty.
Transverse tubules present	Yes, aligned with each A–I band junction.	Yes, aligned with each Z disc.	No.
Junctions between fibers	None.	Intercalated discs contain gap junctions and desmosomes.	Gap junctions in visceral smooth muscle; none in multiunit smooth muscle.
Autorhythmicity	No.	Yes.	Yes, in visceral smooth muscle.
Source of Ca^{2+} for contraction	Sarcoplasmic reticulum.	Sarcoplasmic reticulum and interstitial fluid.	Sarcoplasmic reticulum and interstitial fluid.
Speed of contraction	Fast.	Moderate.	Slow.
Nervous control	Voluntary.	Involuntary.	Involuntary.
Capacity for regeneration	Limited, via satellite cells.	Limited, under certain conditions.	Considerable, via pericytes (compared with other muscle tissues, but limited compared with epithelium).

tion, many smooth muscle fibers contract or relax in response to stretching, hormones, or local changes in pH, oxygen and carbon dioxide levels, temperature, and ion concentrations.

Unlike striated muscle fibers, smooth muscle fibers can stretch considerably and still maintain their contractile function. When smooth muscle fibers are stretched, they initially contract, developing increased tension. Within a minute or so, the tension decreases. This phenomenon, termed the **stress–relaxation response,** allows smooth muscle to undergo great changes in length while still retaining the ability to contract effectively. Even though smooth muscle in the walls of blood vessels and hollow organs such as the stomach, intestines, and urinary bladder can stretch, the pressure on the contents within them changes very little. After the organ empties, however, the smooth muscle rebounds, and the wall of the organ retains its firmness.

Smooth muscle tissue, like skeletal and cardiac muscle tissue, can undergo hypertrophy. In addition, certain smooth muscle fibers, such as those in the uterus, retain their capacity for division and thus can grow by hyperplasia. Also, new smooth muscle fibers can arise from cells called *pericytes,* stem cells found in association with blood capillaries and small veins. Smooth muscle fibers can also proliferate in certain pathological conditions, such as occur in the development of atherosclerosis (see page 445). Although smooth muscle tissue has considerably greater powers of regeneration than skeletal muscle or cardiac muscle, such powers are limited when compared with tissues such as epithelium.

Table 10.2 on page 281 summarizes the major characteristics of the three types of muscular tissue.

CHECKPOINT

12. How do visceral and multiunit smooth muscle differ?
13. Compare the properties of skeletal and smooth muscle.

DEVELOPMENT OF MUSCLES

OBJECTIVE

● Describe the development of muscles.

Except for the muscles of the iris of the eyes and the arrector pili muscles attached to hairs, all muscles of the body are derived from **mesoderm.** Recall from Chapter 4 that by about the seventeenth day after fertilization, the mesoderm adjacent to the notochord and neural tube forms paired longitudinal columns of paraxial mesoderm (see Figure 4.9b on page 105). The paraxial mesoderm soon undergoes segmentation into paired, cube-shaped structures called **somites.** (See Figure 10.10a.) The first pair of somites appears on the twentieth day of embryonic development. By the end of the fifth week, 42 to 44 pairs of somites are formed. The number of somites that develop over a given period can be correlated to the approximate age of the embryo.

With the exception of the skeletal muscles of the head and limbs, *skeletal muscles* develop from the **mesoderm of somites.** Because there are very few somites in the head region of the embryo, most of the skeletal muscles there develop from the **general mesoderm** in the head region. The skeletal muscles of the limbs develop from masses of general mesoderm around developing bones in embryonic limb buds (origins of future limbs; see Figure 8.18 on page 225).

As you learned in Chapter 4, the cells of a somite differentiate into three regions: (1) a **myotome,** which forms the skeletal muscles of the head, neck, and limbs; (2) a **dermatome,** which forms the connective tissues, including the dermis of the skin; and (3) a **sclerotome,** which gives rise to the vertebrae (Figure 10.10b).

Cardiac muscle develops from **mesodermal cells** that migrate to and envelop the developing heart while it is still in the form of primitive heart tubes (see in Figure 14.12b on page 443).

Figure 10.10 Location and structure of somites, key structures in the development of the muscular system.

🔑 **Most muscles are derived from mesoderm.**

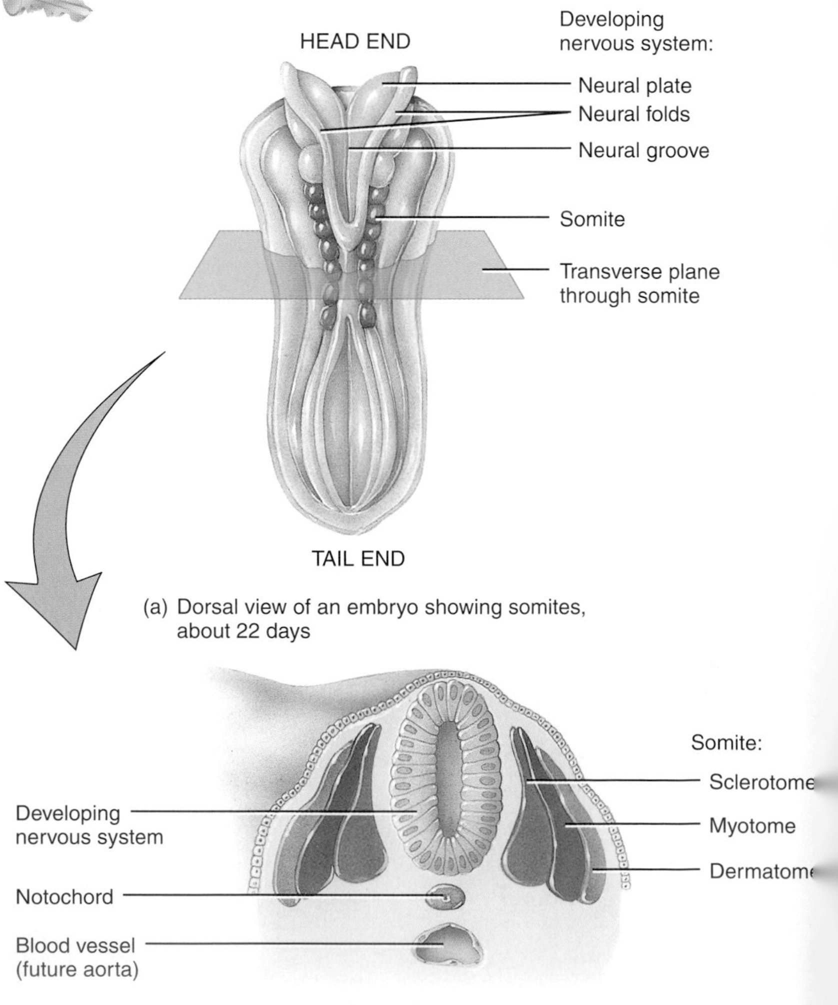

(a) Dorsal view of an embryo showing somites, about 22 days

(b) Transverse section through a somite

? Which part of a somite differentiates into skeletal muscle?

Smooth muscle develops from **mesodermal cells** that migrate to and envelop the developing gastrointestinal tract and viscera.

14. From which type of embryonic tissue do most muscles develop?

AGING AND MUSCULAR TISSUE

● Explain how aging affects skeletal muscle.

Beginning at about 30 years of age, humans undergo a slow, progressive loss of skeletal muscle mass that is replaced largely by fibrous connective tissue and adipose tissue. In part, this decline is due to increasing inactivity. Accompanying the loss of muscle mass is a decrease in maximal strength, a slowing of muscle reflexes, and a loss of flexibility. In some muscles, a selective loss of muscle fibers of a given type may occur. With aging, the relative number of slow oxidative fibers appears to increase. This could be due to either atrophy of the other fiber types or their conversion into slow oxidative fibers. Whether this is an effect of aging itself or mainly reflects the more limited physical activity of older people is still an unresolved question. Nevertheless, aerobic activities and strength training programs are effective in older people and can slow or even reverse the age-associated decline in muscular performance.

15. Why does muscle strength decrease with aging?

APPLICATIONS TO HEALTH

Skeletal muscle function may be abnormal due to disease or damage of any of the components of a motor unit: somatic motor neuron, neuromuscular junctions, or muscle fibers. The term **neuromuscular disease** encompasses problems at all three sites, whereas the term **myopathy** (mī-OP-a-thē; *-pathy*=disease) signifies a disease or disorder of the skeletal muscle tissue itself.

MYASTHENIA GRAVIS

Myasthenia gravis (mī-as-THĒ-nē-a GRAV-is; *myo-*=muscle; *-disthesis*=sensation) is an autoimmune disease that causes chronic, progressive damage of the neuromuscular junction. In people with myasthenia gravis, the immune system inappropriately produces antibodies that bind to and block some ACh receptors, thereby decreasing the number of functional ACh receptors at the motor end plates of skeletal muscles (see Figure 10.6c). Because 75% of patients with myasthenia gravis have hyperplasia or tumors of the thymus, it is possible that thymic abnormalities cause the disorder. As the disease progresses, more ACh receptors are lost. Thus, muscles become increasingly weaker, fatigue more easily, and may eventually cease to function.

Myasthenia gravis occurs in about 1 in 10,000 people and is more common in women, who typically are ages 20 to 40 at onset, than in men, who usually are ages 50 to 60 at onset. The muscles of the face and neck are most often affected. Initial symptoms include weakness of the eye muscles, which may produce double vision, and difficulty in swallowing. Later, the person has difficulty chewing and talking. Eventually the muscles of the limbs may become involved. Death may result from paralysis of the respiratory muscles, but often the disorder does not progress to this stage.

Anticholinesterase drugs such as pyridostigmine (Mestinon) or neostigmine are the first line of treatment for myasthenia gravis. They act as inhibitors of acetylcholinesterase, the enzyme that breaks down ACh. Thus, the inhibitors raise the level of ACh that is available to bind with still-functional receptors. More recently, steroid drugs, such as prednisone, have been used with success to reduce antibody levels. Another treatment is plasmapheresis, a procedure that removes the antibodies from the blood. Often, surgical removal of the thymus (thymectomy) is helpful.

MUSCULAR DYSTROPHY

The term **muscular dystrophy** refers to a group of inherited muscle-destroying diseases that cause progressive degeneration of skeletal muscle fibers. The most common form of muscular dystrophy is *Duchenne muscular dystrophy* (doo-SHĀN) (*DMD*). Because the mutated gene is on the X chromosome, which males have only one of, DMD strikes boys almost exclusively. Worldwide, about 1 in every 3500 male babies—21,000 in all—are born with DMD each year. The disorder usually becomes apparent between the ages of 2 and 5, when parents notice the child falls often and has difficulty running, jumping, and hopping. By age 12 most boys with DMD are unable to walk. Respiratory or cardiac failure usually causes death between the ages of 20 and 30.

In DMD, the gene that codes for the protein dystrophin is mutated, and little or no dystrophin is present. Without the reinforcing effect of dystrophin, the sarcolemma tears easily during muscle contraction. Because their plasma membrane is damaged, muscle fibers slowly rupture and die. The dystrophin gene was discovered in 1987, and by 1990 the first attempts were made to treat DMD patients with gene therapy. The muscles of three boys with DMD were injected with myoblasts bearing functional dystrophin genes, but only a few muscle fibers gained the ability to produce dystrophin. Similar clinical trials with additional patients have also failed. An alternate approach to the problem is to find a way to induce muscle fibers to produce the protein utrophin, which is similar to dystrophin. Experiments with dystrophin-deficient mice suggest this approach may work.

FIBROMYALGIA

Fibromyalgia (fī-brō-mī-AL-jē-a; *algia*=painful condition) is a painful, nonarticular rheumatic disorder that usually appears between the ages of 25 and 50. An estimated 3 million people in the United States suffer from fibromyalgia, which is 15 times more common in women than in men. The disorder affects the fibrous connective tissue components of muscles, tendons, and ligaments. A striking sign is pain that results from gentle pressure at specific "tender points." Even without pressure, there is pain, tenderness, and stiffness of muscles, tendons, and surrounding soft tissues. Besides muscle pain, those with fibromyalgia report severe fatigue, poor sleep, headaches, depression, and inability to carry out their daily activities. Treatment consists of stress reduction, regular exercise, application of heat, gentle massage, physical therapy, medication for pain, and a low-dose antidepressant to help improve sleep.

ABNORMAL CONTRACTIONS OF SKELETAL MUSCLE

One kind of abnormal muscular contraction is a **spasm,** a sudden involuntary contraction of a single muscle in a large group of muscles. A painful spasmodic contraction is known as a **cramp.** A **tic** is a spasmodic involuntary twitching by muscles that are ordinarily under voluntary control. Twitching of the eyelid and facial muscles are examples of tics. A **tremor** is a rhythmic, involuntary, purposeless contraction that produces a quivering or shaking movement. A **fasciculation** (fa-sik-ū-LĀ-shun) is an involuntary, brief twitch of an entire motor unit that is visible under the skin; it occurs irregularly and is not associated with movement of the affected muscle. Fasciculations may be seen in multiple sclerosis (see page 549) or in amyotrophic lateral sclerosis (Lou Gehrig's disease; see page 664). A **fibrillation** (fi-bri-LĀ-shun) is a spontaneous contraction of a single muscle fiber that is not visible under the skin but can be recorded by electromyography. Fibrillations may signal destruction of motor neurons.

KEY MEDICAL TERMS ASSOCIATED WITH MUSCULAR TISSUE

Muscle strain Tearing of a muscle because of forceful impact, accompanied by bleeding and severe pain. Also known as a **Charley horse** or pulled muscle. It often occurs in contact sports and typically affects the quadriceps femoris muscle on the anterior surface of the thigh. The condition is treated by RICE therapy: rest (R), ice immediately after the injury (I), compression via a supportive wrap (C), and elevation of the limb (E).

Myalgia (mī-AL-jē-a; *-algia*=painful condition) Pain in or associated with muscles.

Myoma (mī-Ō-ma; *-oma*=tumor) A tumor consisting of muscular tissue.

Myomalacia (mī-Ō-ma-LĀ-shē-a; *-malacia*=soft) Pathological softening of muscular tissue.

Myositis (mī′-ō-SĪ-tis; *-itis*=inflammation of) Inflammation of muscle fibers (cells).

Myotonia (mī′-ō-TŌ-nē-a; *-tonia*=tension) Increased muscular excitability and contractility, with decreased power of relaxation; tonic spasm of the muscle.

Volkmann's contracture (FŌLK-manz kon-TRAK-tur; *contra-*= against) Permanent shortening (contracture) of a muscle due to replacement of destroyed muscle fibers by fibrous connective tissue, which lacks extensibility. Destruction of muscle fibers may occur from interference with circulation caused by a tight bandage, a piece of elastic, or a cast.

 STUDY OUTLINE

Introduction (p. 265)

1. Muscles constitute 40–50% of total body weight.
2. The prime function of muscle is changing chemical energy into mechanical energy to perform work.

Overview of Muscular Tissue (p. 266)

1. The three types of muscular tissue are skeletal, cardiac, and smooth. Skeletal muscle tissue is primarily attached to bones; it is striated and under voluntary control. Cardiac muscle tissue forms the wall of the heart; it is striated and involuntary. Smooth muscle tissue is located primarily in internal organs; it is nonstriated (smooth) and involuntary.
2. Through contraction and relaxation, muscular tissue performs four important functions: producing body movements, stabilizing body positions, storing and moving substances within the body, and producing heat.

3. Four special properties of muscular tissues are electrical excitability, the property of responding to stimuli by producing action potentials; contractility, the ability to generate tension to do work; extensibility, the ability to be extended (stretched); and elasticity, the ability to return to original shape after contraction or extension.

4. An isotonic contraction occurs when a muscle shortens and moves a constant load; tension remains almost constant. An isometric contraction occurs without shortening of the muscle; tension increases greatly.

Skeletal Muscle Tissue (p. 267)

1. Connective tissues surrounding skeletal muscles are the epimysium, covering the entire muscle; perimysium, covering fascicles; and the endomysium, covering muscle fibers. Superficial fascia separates muscle from skin.

2. Tendons and aponeuroses are extensions of connective tissue beyond muscle fibers that attach the muscle to bone or to other muscle.

3. Each skeletal muscle fiber has 100 or more nuclei because it arises from fusion of many myoblasts. Satellite cells are myoblasts that persist after birth. The sarcolemma is a muscle fiber's plasma membrane; it surrounds the sarcoplasm. T tubules are invaginations of the sarcolemma.

4. Each fiber contains myofibrils, the contractile elements of skeletal muscle. Sarcoplasmic reticulum surrounds each myofibril. Within a myofibril are thin and thick filaments arranged in compartments called sarcomeres.

5. The overlapping of thick and thin filaments produces striations; darker A bands alternate with lighter I bands.

6. Myofibrils are built from three types of proteins: contractile, regulatory, and structural. The contractile proteins are myosin (thick filament) and actin (thin filament). Regulatory proteins are tropomyosin and troponin (both are part of the thin filament). Structural proteins include titin (links Z disc to M line and stabilizes thick filament).

7. Projecting myosin heads (crossbridges) contain actin-binding and ATP-binding sites and are the motor proteins that power muscle contraction.

8. Skeletal muscles are well supplied with nerves and blood vessels. Generally, an artery and one or two veins accompany each nerve that penetrates a skeletal muscle. Blood capillaries bring in oxygen and nutrients and remove heat and waste products of muscle metabolism.

9. Motor neurons provide the nerve impulses that stimulate skeletal muscle to contract.

10. The neuromuscular junction (NMJ) is the synapse between a somatic motor neuron and a skeletal muscle fiber. The NMJ includes the axon terminals and synaptic end bulbs of a motor neuron, plus the adjacent motor end plate of the muscle fiber sarcolemma.

11. When an action potential reaches the end bulbs of a somatic motor neuron, it triggers exocytosis of the synaptic vesicles. ACh diffuses across the synaptic cleft and binds to ACh receptors, initiating a muscle action potential. Acetylcholinesterase then quickly destroys ACh.

12. Muscle contraction occurs because myosin heads attach to and "walk" along the thin filaments at both ends of a sarcomere, progressively pulling the thin filaments toward the center of a sarcomere. As the thin filaments slide inward, the Z discs come closer together, and the sarcomere shortens.

Types of Skeletal Muscle Fibers (p. 276)

1. On the basis of their structure and function, skeletal muscle fibers are classified as slow oxidative (SO), fast oxidative-glycolytic (FOG), and fast glycolytic (FG) fibers.

2. Most skeletal muscles contain a mixture of all three fiber types. Their proportions vary with the typical action of the muscle.

3. The motor units of a muscle are recruited in the following order: first SO fibers, then FOG fibers, and finally FG fibers.

4. Table 10.1 on page 277 summarizes the three types of skeletal muscle fibers.

Exercise and Skeletal Muscle Tissue (p. 277)

1. Various types of exercises can induce changes in the fibers in a

skeletal muscle. Endurance-type (aerobic) exercises cause a gradual transformation of some fast glycolytic (FG) fibers into fast oxidative-glycolytic (FOG) fibers.

2. Exercises that require great strength for short periods produce an increase in the size and strength of fast-glycolytic (FG) fibers. The increase in size is due to increased synthesis of thick and thin filaments.

Cardiac Muscle Tissue (p. 278)

1. This muscle tissue is found only in the heart. It is striated and involuntary.

2. Cardiac muscle fibers are branching cylinders and usually contain a single centrally located nucleus.

3. Compared to skeletal muscle tissue, cardiac muscle tissue has more sarcoplasm, more mitochondria, less well-developed sarcoplasmic reticulum, and wider transverse tubules located at Z discs rather than at A–I band junctions. Filaments are not arranged in discrete myofibrils.

4. Cardiac muscle fibers branch and are connected end to end via desmosomes.

5. Intercalated discs provide strength and aid in conduction of muscle action potentials by way of gap junctions located in the discs.

6. Unlike skeletal muscle tissue, cardiac muscle tissue contracts and relaxes rapidly, continuously, and rhythmically.

7. Cardiac muscle tissue can contract without extrinsic stimulation and can remain contracted longer than skeletal muscle tissue.

Smooth Muscle Tissue (p. 278)

1. Smooth muscle tissue is nonstriated and involuntary.

2. Smooth muscle fibers contain intermediate filaments and dense bodies that function as Z discs.

3. Visceral (single-unit) smooth muscle is found in the walls of viscera and small blood vessels. The fibers are arranged in a network.

4. Multiunit smooth muscle is found in large blood vessels, arrector pili muscles, and the iris of the eye. The fibers operate independently rather than in unison.

5. The duration of contraction and relaxation of smooth muscle is longer than in skeletal muscle.

6. Smooth muscle fibers contract in response to nerve impulses, hormones, and local factors.

7. Smooth muscle fibers can stretch considerably without developing tension.

8. Table 10.2 on page 281 summarizes the principal characteristics of the three types of muscle tissue.

Development of Muscles (p. 282)

1. With few exceptions, muscles develop from mesoderm.

2. Skeletal muscles of the head and limbs develop from general mesoderm. Other skeletal muscles develop from the mesoderm of somites. Cardiac muscle and smooth muscle develop from mesodermal cells that migrate during the development process to the heart and to the gastrointestinal tract and viscera, respectively.

Aging and Muscular Tissue (p. 283)

1. Beginning at about 30 years of age, there is a slow, progressive loss of skeletal muscle, which is replaced by fibrous connective tissue and fat.

2. Aging also results in a decrease in muscle strength, slower muscle reflexes, and loss of flexibility, which can be compensated for to an extent by increased physical activity.

SELF-QUIZ QUESTIONS

Choose the one best answer to the following questions.

1. The specialized region of the sarcolemma at the neuromuscular junction is called the:
   a. synapse.   b. synaptic cleft.   c. motor end plate.
   d. cisterna.   e. motor unit.

2. The ability of muscular tissue to respond to a stimulus by producing action potentials is referred to as:
   a. contractility.   b. excitability.   c. elasticity.
   d. extensibility.   e. conductivity.

3. The function of titin in muscle cells is to:
   a. contract in response to increases in calcium ion levels.
   b. block the myosin binding sites on actin molecules.
   c. block the actin binding sites on myosin molecules.
   d. stabilize the position of thick filaments.
   e. provide energy to fuel the shortening of the sarcomere.

4. The most fatigue-resistant skeletal muscle fibers are the:
   a. slow oxidative fibers.   b. fast oxidative-glycolytic fibers.
   c. fast glycolytic fibers.   d. satellite cells.
   e. All types of fibers are equally fatigue-resistant.

5. The skeletal muscle fibers with the largest diameters that generate the most powerful contractions due to their high myofibril content are the:
   a. slow oxidative fibers.   b. fast oxidative-glycolytic fibers.
   c. fast glycolytic fibers.   d. satellite cells.
   e. All types of fibers produce equally strong contractions.

6. A motor unit is defined as a:
   a. nerve and a muscle.
   b. single motor neuron and a single muscle fiber.
   c. motor neuron and all the muscle fibers it stimulates.
   d. single muscle fiber and the nerves that innervate it.
   e. muscle and the motor and sensory nerves that innervate it.

7. Arrange the following from largest to smallest: (1) myofibril, (2) filament, (3) muscle fiber, (4) fascicle:
   a. 1, 3, 2, 4   b. 3, 4 ,2, 1   c. 3, 1, 4, 2
   d. 4, 3, 1, 2   e. 4, 2, 3, 1

Complete the following.

8. The _____ separates muscle from skin.

9. Connective tissue that separates one muscle fiber from another is the _____.

10. Plate-shaped regions of dense material separating one sarcomere from the next are called _____.

11. The two regulatory proteins in thin filaments are _____ and _____.

12. To describe a sarcomere during contraction, complete each statement with one of the following: L (lengthens), S (shortens), or U (is unchanged in length).
   a. The sarcomere _____.
   b. Each thick myofilament (A band) _____.
   c. Each thin myofilament _____.
   d. The I band _____.
   e. The H zone _____.

13. The muscle tissues that exhibit autorhythmicity are _____ and _____.

Are the following statements true or false?

14. Enlargement of existing muscle fibers is called hyperplasia.

15. In adult life, growth of skeletal muscle is due to an increase in the number of muscle cells rather than an increase in the size of existing cells.

16. Elasticity is the property of muscular tissue that allows it to be stretched.

17. Stimulation of one visceral (single-unit) smooth muscle fiber causes contraction of several adjacent fibers; stimulation of a multiunit fiber causes contraction of that fiber only.

18. Thick filaments are made mostly of actin.

Matching

19. Match the following:
   ___ (a) involuntary muscle found in blood vessels and the intestines
   ___ (b) involuntary striated muscle
   ___ (c) striated voluntary muscle attached to bones
   ___ (d) spindle-shaped cells
   ___ (e) intercalated discs
   ___ (f) no transverse tubules
   ___ (g) several nuclei, peripherally arranged

   (1) cardiac
   (2) skeletal
   (3) smooth

20. Match the following:
   ___ (a) functional units of myofibrils
   ___ (b) red protein that binds oxygen in the sarcoplasm
   ___ (c) releases acetylcholine into the synaptic cleft at the neuromuscular junction
   ___ (d) invaginations of the plasma membrane of muscle cells
   ___ (e) where the acetylcholine receptors are located at the neuromuscular junction
   ___ (f) stores and releases calcium ions during muscle relaxation and contraction

   (1) sarcolemma
   (2) T tubules
   (3) myoglobin
   (4) sarcoplasmic reticulum
   (5) sarcomeres
   (6) motor neuron

CRITICAL THINKING QUESTIONS

1. A marathon runner and a weight-lifter were discussing muscle types while working out. Both were convinced that their own muscle types were the best. How does the leg muscle of the runner compare to the arm of the lifter?

2. Bill tore some ligaments in his knee while skiing. He was in a toe-to-thigh cast for 6 weeks. When the cast was removed, the newly healed leg was noticeably thinner than the uncasted leg. What happened to his leg?

3. The newspaper reported several cases of botulism poisoning following a fund-raiser potluck dinner for the local clinic. The cause appeared to be a potato salad that contained the toxin of the soil bacteria *Clostridium botulinum*. This toxin blocks the release of acetylcholine. What would you expect the major effect of botulism poisoning to be, and why?

4. Research is underway to grow new cardiac muscle cells for ailing hearts. Skeletal muscle transplants have been tried, but they don't work as well as cardiac muscle. Both skeletal and cardiac muscle are striated, so why does cardiac muscle support rhythmic contraction while skeletal muscle does not?

5. Your study partner says that the contractility, extensibility, and elasticity of muscle tissue can be explained by the "stretchiness" of the contractile proteins in muscle cells. "After all," he says, "why else would they be called 'contractile' proteins?" How should you respond to your study partner? Explain.

ANSWERS TO FIGURE QUESTIONS

10.1 Perimysium bundles groups of muscle fibers into fascicles.

10.2 The sarcoplasmic reticulum releases calcium ions to trigger muscle contraction.

10.3 Size, from smallest to largest: thick filament, myofibril, muscle fiber.

10.4 There are two thin filaments for each thick filament in skeletal muscle.

10.5 Actin and titin connect to the Z disc. A bands contain myosin, actin, troponin, tropomysin, and titin; I bands contain actin, troponin, tropomysin, and titin.

10.6 The portion of the sarcolemma that contains acetylcholine receptors is the motor end plate.

10.7 During muscle contraction, the I bands and H zones of the sarcomere disappear. The lengths of the thin and thick filaments do not change.

10.8 The intercalated discs contain desmosomes that hold the cardiac muscle fibers together and gap junctions that enable action potentials to be transmitted from one muscle fiber to another.

10.9 Visceral smooth muscle and cardiac muscle are similar in that both contain gap junctions, which allow action potentials to spread from one cell to its neighbors.

10.10 As its name suggests, the myotome of a somite differentiates into skeletal muscle.

11 THE MUSCULAR SYSTEM

INTRODUCTION Together, the voluntarily controlled muscles of your body comprise the **muscular system.** Almost all of the 700 individual skeletal muscles that make up the muscular system, such as, the biceps brachii muscle, include both skeletal muscle tissue and connective tissue. The function of most muscles is to produce movements of body parts. A few muscles function mainly to stabilize bones so that other skeletal muscles can execute a movement more effectively. This chapter presents many of the major muscles in the body. Most of them are found on both the right and left sides. We'll identify the attachment sites and innervation—the nerve or nerves that stimulate it to contract—of each muscle described. Developing a working knowledge of these key aspects of skeletal muscle anatomy will enable you to understand how normal movements occur. This knowledge is especially crucial for professionals, such as those in the allied health and physical rehabilitation fields, who work with patients whose normal patterns of movement and physical mobility have been disrupted by physical trauma, surgery, or muscular paralysis.

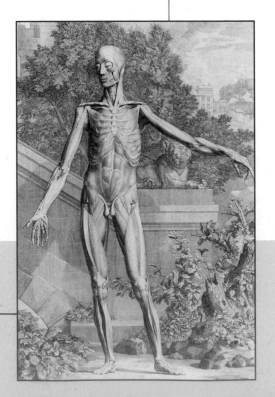

Have you ever pulled a hamstring or experienced a shin splint as a result of an athletic activity? Understanding the relation of muscles to movement can explain how these injuries occur.

www.wiley.com/college/apcentral

HOW SKELETAL MUSCLES PRODUCE MOVEMENTS

Muscle Attachment Sites: Origin and Insertion

OBJECTIVES

- Describe the relationship between bones and skeletal muscles in producing body movements.
- Define lever and fulcrum, and compare the three types of levers based on location of the fulcrum, effort, and load.
- Identify the types of fascicle arrangements in a skeletal muscle, and relate the arrangements to strength of contraction and range of motion.
- Explain how the prime mover, antagonist, synergist, and fixator in a muscle group work together to produce movement.

As we have already mentioned, not all skeletal muscles produce movements. Skeletal muscles that do produce movements do so by exerting force on tendons, which in turn pull on bones or other structures (such as skin). Most muscles cross at least one joint and are usually attached to articulating bones that form the joint (Figure 11.1a).

When a skeletal muscle contracts, it pulls one of the articulating bones toward the other. The two articulating bones usually do not move equally in response to contraction. One bone remains stationary or near its original position, either because other muscles stabilize that bone by contracting and pulling it in the opposite direction or because its structure makes it less movable. Ordinarily, the attachment of a muscle's tendon to the stationary bone is called the **origin**; the attachment of the muscle's other tendon to the movable bone is called the **insertion**. A good analogy is a spring on a door. In this example, the part of the spring attached to the frame is the origin; the part attached to the door represents the insertion. A useful rule of thumb is that the origin is usually proximal and the insertion distal, espe-

Figure 11.1 Relationship of skeletal muscles to bones. (a) Muscles are attached to bones by tendons known as the origin and the insertion. (b) Skeletal muscles produce movements by pulling on bones. Bones serve as levers, and joints act as fulcrums for the levers. Here the lever-fulcrum principle is illustrated by the movement of the forearm. Note where the load (resistance) and effort are applied in this example.

In the limbs, the origin of a muscle is usually proximal and the insertion is usually distal.

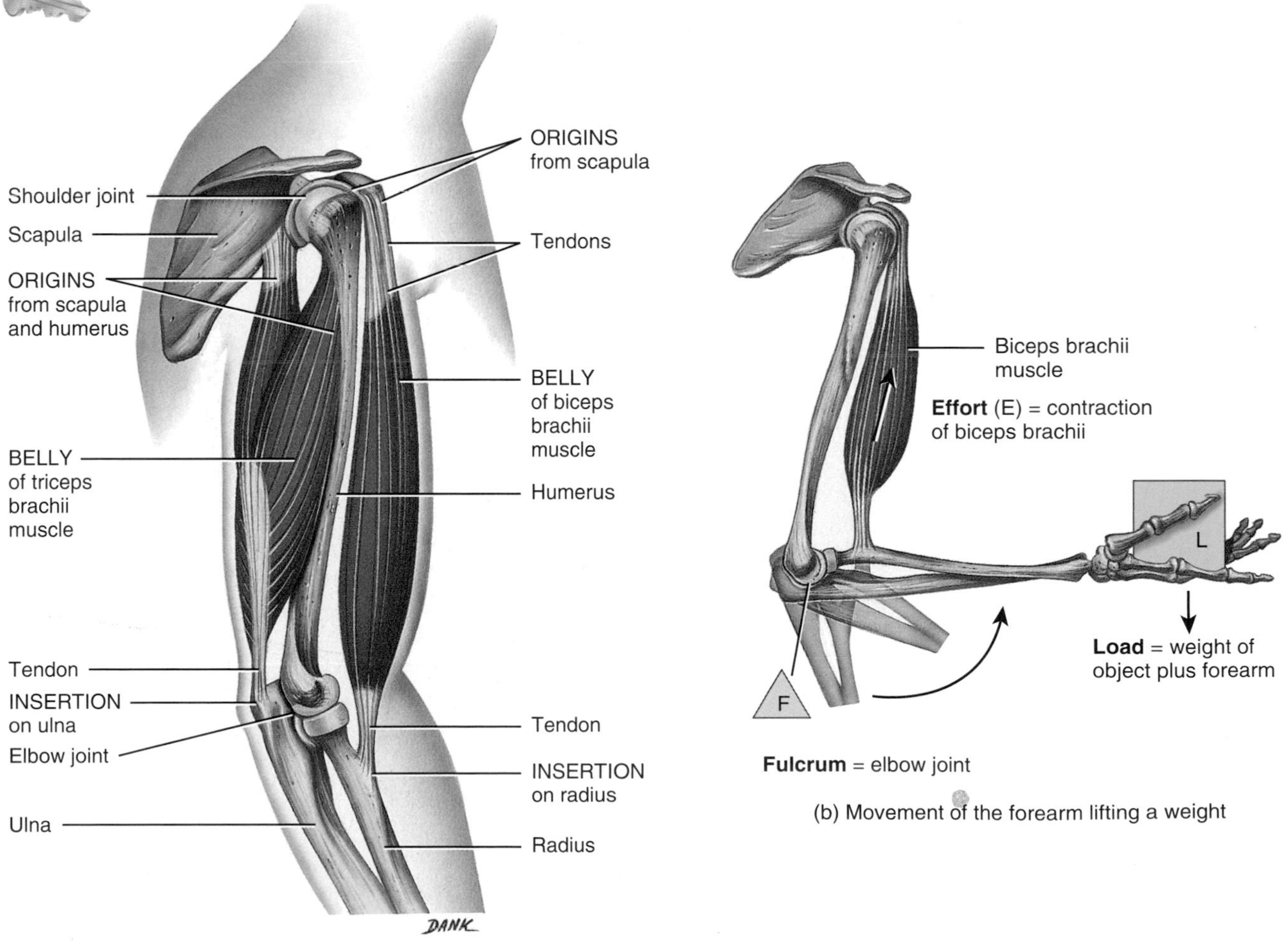

(a) Origin and insertion of a skeletal muscle

(b) Movement of the forearm lifting a weight

Where is the belly of the muscle that extends the forearm located?

cially in the limbs; the insertion is usually pulled toward the origin. The fleshy portion of the muscle between the tendons is called the **belly** (*gaster*). The **actions** of a muscle are the main movements that occur when the muscle contracts. In our spring example, this would be the closing of the door.

Muscles that move a body part often do not cover the moving part. Figure 11.1b shows that although one of the functions of the biceps brachii muscle is to move the forearm, the belly of the muscle lies over the humerus, not over the forearm. You will also see that muscles that cross two joints, such as the rectus femoris and sartorius of the thigh, have more complex actions than muscles that cross only one joint.

TENOSYNOVITIS

Tenosynovitis (ten′-ō-sin-ō-VĪ-tis) is an inflammation of the tendons, tendon sheaths, and synovial membranes surrounding certain joints. The tendons most often affected are at the wrists, shoulders, elbows (resulting in a condition called *tennis elbow*), finger joints (resulting in a condition called *trigger finger*), ankles, and feet. The affected sheaths sometimes become visibly swollen because of fluid accumulation. Tenderness and pain are frequently associated with movement of the body part. The condition often follows trauma, strain, or excessive exercise. Tenosynovitis of the dorsum of the foot may be caused by tying shoelaces too tightly. Also, gymnasts are prone to developing the condition as a result of chronic, repetitive, and maximum hyperextension at the wrists. Other repetitive movements involving activities such as typing, haircutting, carpentry, and assembly line work can also result in tenosynovitis. ■

Lever Systems and Leverage

In producing movement, bones act as levers, and joints function as the fulcrums of these levers. A **lever** is a rigid structure that can move around a fixed point called a **fulcrum,** ⚖. A lever is acted on at two different points by two different forces: the **effort** (E), which causes movement, and the **load** Ⓛ or **resistance,** which opposes movement. The effort is the force exerted by muscular contraction, whereas the load is typically the weight of the body part that is moved. Motion occurs when the effort applied to the bone at the insertion exceeds the load. Consider the biceps brachii flexing the forearm at the elbow as an object is lifted (Figure 11.1b). When the forearm is raised, the elbow is the fulcrum. The weight of the forearm plus the weight of the object in the hand is the load. The force of contraction of the biceps brachii pulling the forearm up is the effort.

Levers produce trade-offs between effort and the speed and range of motion. A lever operates at a *mechanical advantage*—has **leverage**—when a smaller effort can move a heavier load. Here the trade-off is that the effort must move a greater distance (must have a longer range of motion) and faster than the load. Recall from Chapter 9 that range of motion refers to the range,

measured in degrees of a circle, through which the bones of a joint can be moved. The lever formed by the mandible at the temporomandibular joints (fulcrums) and the effort provided by contraction of the jaw muscles produce a high mechanical advantage that crushes food. In contrast, a lever operates at a *mechanical disadvantage* when a larger effort moves a lighter load. In this case the trade-off is that the effort must move more slowly and for a shorter distance than the load. The lever formed by the humerus at the shoulder joint (fulcrum) and the effort provided by the back and shoulder muscles produces a mechanical "disadvantage" that enables a major-league pitcher to hurl a baseball at nearly 100 miles per hour!

The positions of the effort, load, and fulcrum on the lever determine whether the lever operates at a mechanical advantage or disadvantage. When the load is close to the fulcrum and the effort is applied farther away, the lever operates at a mechanical advantage. When you chew food, the load (the food) is positioned close to the fulcrums (your temporomandibular joints) while your jaw muscles exert effort farther out from the joints. By contrast, when the effort is applied close to the fulcrum and the load is farther away, the lever operates at a mechanical disadvantage. When a pitcher throws a baseball, the back and shoulder muscles apply intense effort very close to the fulcrum (the shoulder joint) while the lighter load (the ball) is propelled at the far end of the lever (the arm bone).

Levers are categorized into three types according to the positions of the fulcrum, the effort, and the load:

1. The fulcrum is between the effort and the load in **first-class levers** (Figure 11.2a). (Think E*F*L.) Scissors and seesaws are examples of first-class levers. A first-class lever can produce either a mechanical advantage or disadvantage depending on whether the effort or the load is closer to the fulcrum. (Think of an adult and a child on a seesaw.) As we've seen in the preceding examples, if the effort (child) is farther from the fulcrum than the load (adult), a heavy load can be moved, but not very far or fast. If the effort is closer to the fulcrum than the load, only a lighter load can be moved, but it moves far and fast. There are few first-class levers in the body. One example is the lever formed by the head resting on the vertebral column (Figure 11.2a). When the head is raised, the contraction of the posterior neck muscles provides the effort (E), the joint between the atlas and the occipital bone (atlanto-occipital joint) forms the fulcrum (F), and the weight of the anterior portion of the skull is the load.

2. The load is between the fulcrum and the effort in **second-class levers** (Figure 11.2b). (Think F*L*E.) They operate like a wheelbarrow. Second-class levers always produce a mechanical advantage because the load is always closer to the fulcrum than the effort. This arrangement sacrifices speed and range of motion for force; this type of lever produces the most force. Most authorities believe that there are no second-class levers in the body.

3. The effort is between the fulcrum and the load in **third-class levers** (Figure 11.2c). (Think F*E*L.) These levers operate like a pair of forceps and are the most common

Figure 11.2   Types of levers.

Levers are divided into three types based on the placement of the fulcrum, effort, and load (resistance).

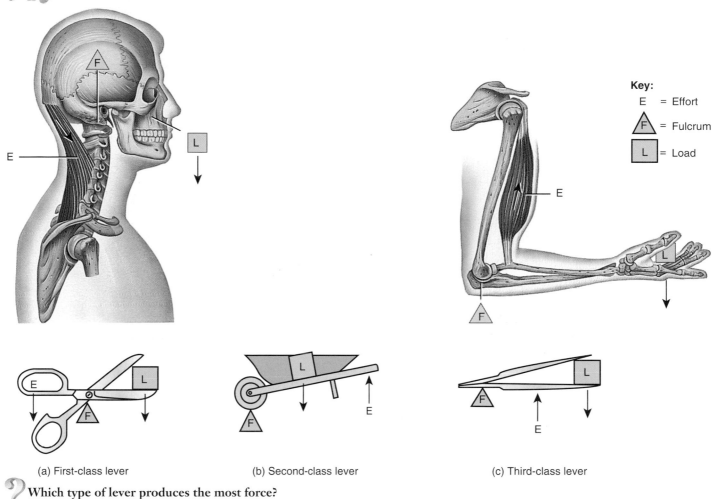

Key:
E = Effort
F = Fulcrum
L = Load

(a) First-class lever   (b) Second-class lever   (c) Third-class lever

Which type of lever produces the most force?

levers in the body. Third-class levers always produce a mechanical disadvantage because the effort is always closer to the fulcrum than the load. In the body, this arrangement favors speed and range of motion over force. The elbow joint, the biceps brachii muscle, and the bones of the arm and forearm are one example of a third-class lever (Figure 11.1b). As we have seen, in flexing the forearm at the elbow, the elbow joint is the fulcrum (F), the contraction of the biceps brachii muscle provides the effort (E), and the weight of the hand and forearm is the load (L). Another example of the action of a third-class lever is adduction of the thigh, in which the hip joint is the fulcrum, the contraction of the adductor muscles is the effort, and the thigh is the load.

Effects of Fascicle Arrangement

Recall from Chapter 10 that the skeletal muscle fibers (cells) within a muscle are arranged in bundles known as **fascicles.**

Within a fascicle, all muscle fibers are parallel to one another. The fascicles, however, may form one of five patterns with respect to the tendons: parallel, fusiform (shaped like a cigar), circular, triangular, or pennate (shaped like a feather) (Table 11.1).

Fascicular arrangement affects a muscle's power and range of motion. As a muscle fiber contracts, it shortens to about 70% of its resting length. Thus, the longer the fibers in a muscle, the greater the range of motion it can produce. However, the power of a muscle depends not on length but on its total cross-sectional area, because a short fiber can contract as forcefully as a long one. Fascicular arrangement often represents a compromise between power and range of motion. Pennate muscles, for instance, have a large number of fascicles distributed over their tendons, giving them greater power but a smaller range of motion. Parallel muscles, in contrast, have comparatively few fascicles that extend the length of the muscle, so they have a greater range of motion but less power.

TABLE 11.1 ARRANGEMENT OF FASCICLES

PARALLEL	FUSIFORM
Fascicles parallel to longitudinal axis of muscle; terminate at either end in flat tendons.	Fascicles nearly parallel to longitudinal axis of muscle; terminate in flat tendons; muscle tapers toward tendons, where diameter is less than at belly.

Example: Stylohyoid muscle (see Figure 11.8)	*Example:* Digastric muscle (see Figure 11.8)

CIRCULAR	TRIANGULAR
Fascicles in concentric circular arrangements form sphincter muscles that enclose an orifice (opening).	Fascicles spread over broad area converge at thick central tendon; gives muscle a triangular appearance.

Example: Orbicularis oculi muscle (see Figure 11.4)	*Example:* Pectoralis major muscle (see Figure 11.3a)

PENNATE

Short fascicles in relation to total muscle length; tendon extends nearly entire length of muscle.

Unipennate	**Bipennate**	**Multipennate**
Fascicles are arranged on only one side of tendon.	Fascicles are arranged on both sides of centrally positioned tendons.	Fascicles attach obliquely from many directions to several tendons.

Example: Extensor digitorum longus muscle (see Figure 11.24b)	*Example:* Rectus femoris muscle (see Figure 11.22a)	*Example:* Deltoid muscle (see Figure 11.12a)

INTRAMUSCULAR INJECTIONS

An **intramuscular (IM) injection** penetrates the skin and subcutaneous tissue to enter the muscle itself. Intramuscular injections are preferred when prompt absorption is desired, when larger doses than can be given subcutaneously are indicated, or when the drug is too irritating to give subcutaneously. The common sites for intramuscular injections include the gluteus medius muscle of the buttock (see Figure 11.3b), lateral side of the thigh in the midportion of the vastus lateralis muscle (see Figure 11.3a), and the deltoid muscle of the shoulder (see Figure 11.3b). Muscles in these areas, especially the gluteal muscles in the buttock, are fairly thick, and absorption is promoted by their extensive blood supply. To avoid injury, intramuscular injections are given deep within the muscle, away from major nerves and blood vessels. In terms of speed of delivery, intramuscular injections are faster than oral medications, but slower than intravenous infusions. ■

Coordination Within Muscle Groups

Movements often are the result of several skeletal muscles acting as a group rather than acting alone. Most skeletal muscles are arranged in opposing (antagonistic) pairs at joints—that is, flexors–extensors, abductors–adductors, and so on. Within opposing pairs, one muscle, called the **prime mover** or **agonist** (=leader), contracts to cause an action while the other muscle, the **antagonist** (*anti-*=against), stretches and yields to the effects of the prime mover. In the process of flexing the forearm at the elbow, for instance, the biceps brachii is the prime mover, and the triceps brachii is the antagonist (see Figure 11.1). The antagonist and prime mover are usually located on opposite sides of the bone or joint, as is the case in this example.

With an opposing pair of muscles, the roles of the prime mover and antagonist can switch for different movements. For example, while extending the forearm at the elbow (i.e., lowering the load shown in Figure 11.1), the triceps brachii becomes the prime mover, and the biceps brachii is the antagonist. If a prime mover and its antagonist contract at the same time with equal force, there will be no movement.

Sometimes a prime mover crosses other joints before it reaches the joint at which its primary action occurs. The biceps brachii, for example, spans both the shoulder and elbow joints, with primary action on the forearm. To prevent unwanted movements at intermediate joints or to otherwise aid the movement of the prime mover, muscles called **synergists** (SIN-er-gists; *syn-*=together; *-ergon*=work) contract and stabilize the intermediate joints. As an example, muscles that flex the fingers (prime movers) cross the intercarpal and radiocarpal joints (intermediate joints). If movement at these intermediate joints was unrestrained, you would not be able to flex your fingers without flexing the wrist at the same time. Synergistic contraction of the wrist extensor muscles stabilizes the wrist joint and prevents unwanted movement, while the flexor muscles of the fingers contract to bring about the primary action, efficient flexion of the fingers. Synergists are usually located close to the prime mover.

Some muscles in a group also act as **fixators**, which stabilize the origin of the prime mover so that the prime mover can act more efficiently. Fixators steady the proximal end of a limb while movements occur at the distal end. For example, the scapula in the pectoral (shoulder) girdle is a freely movable bone that serves as the origin for several muscles that move the arm. When the arm muscles contract, the scapula must be held steady. In abduction of the arm, the deltoid muscle serves as the prime mover, whereas fixators (pectoralis minor, trapezius, subclavius, serratus anterior muscles, and others) hold the scapula firmly against the back of the chest (see Figure 11.16). The insertion of the deltoid muscle pulls on the humerus to abduct the arm. Under different conditions—that is, for different movements—and at different times, many muscles may act as prime movers, antagonists, synergists, or fixators.

In the limbs, a **compartment** is a group of skeletal muscles, along with their associated blood vessels and nerves, that have a common function. In the upper limbs, for example, flexor compartment muscles are anterior, whereas extensor compartment muscles are posterior.

BENEFITS OF STRETCHING

The overall goal of **stretching** is to achieve normal range of motion of joints and mobility of soft tissues surrounding the joints. For most individuals, the best stretching routine involves *static stretching*, that is, slow sustained stretching that holds a muscle in a lengthened position. The muscles should be stretched to the point of slight discomfort (not pain) and held for about 15–30 seconds. Stretching should be done after warming up to increase the range of motion most effectively. Among the benefits of stretching are the following:

1. *Improved physical performance.* A flexible joint has the ability to move through a greater range of motion, which improves performance.

2. *Decreased risk of injury.* Stretching decreases resistance in various soft tissues so there is less likelihood of exceeding maximum tissue extensibility during an activity (i.e., injuring the soft tissues).

3. *Reduced muscle soreness.* Stretching can reduce some of the muscle soreness that results after exercise.

4. *Improved posture.* Poor posture results from improper position of various parts of the body and the effects of gravity over a number of years. Stretching can help realign soft tissues to improve and maintain good posture. ■

CHECKPOINT

1. Using the terms origin, insertion, and belly in your discussion, describe how skeletal muscles produce body movements by pulling on bones.
2. Describe the three types of levers, and give an example of a first- and third-class lever found in the body.
3. Describe the various arrangements of fascicles.
4. Define the roles of the prime mover (agonist), antagonist, synergist, and fixator in producing various movements of the upper limb.

HOW SKELETAL MUSCLES ARE NAMED

OBJECTIVE

• Explain seven features used in naming skeletal muscles.

The names of most of the skeletal muscles contain combinations of the word roots of their distinctive features. This works two ways. You can remember the names of muscles by learning the terms that refer to muscle features, such as the pattern of the muscle's fascicles; the size, shape, action, number of origins, and location of the muscle; and the sites of origin and insertion of the muscle. Familiarity with the names of the muscles, then, will give you clues to their features. Study Table 11.2 to become familiar with the terms used in muscle names.

CHECKPOINT

5. Select 10 muscles in Figure 11.3 and identify the features on which their names are based. (Hint: *Use the prefix, suffix, and root of each muscle's name as a guide.*)

TABLE 11.2 CHARACTERISTICS USED TO NAME MUSCLES

NAME	MEANING	EXAMPLE	FIGURE
DIRECTION: Orientation of muscle fascicles relative to the body's midline.			
Rectus	Parallel to midline	Rectus abdominis	11.12b
Transverse	Perpendicular to midline	Transversus abdominis	11.12b
Oblique	Diagonal to midline	External oblique	11.12a
SIZE: Relative size of the muscle.			
Maximus	Largest	Gluteus maximus	11.3b
Minimus	Smallest	Gluteus minimus	11.22c
Longus	Long	Adductor longus	11.22a
Brevis	Short	Adductor brevis	11.22b
Latissimus	Widest	Latissimus dorsi	11.17b
Longissimus	Longest	Longissimus capitis	11.21a
Magnus	Large	Adductor magnus	11.22b
Major	Larger	Pectoralis major	11.12a
Minor	Smaller	Pectoralis minor	11.16a
Vastus	Huge	Vastus lateralis	11.22a
SHAPE: Relative shape of the muscle.			
Deltoid	Triangular	Deltoid	11.12a
Trapezius	Trapezoid	Trapezius	11.3b
Serratus	Saw-toothed	Serratus anterior	11.16b
Rhomboid	Diamond-shaped	Rhomboid major	11.16d
Orbicularis	Circular	Orbicularis oculi	11.4a
Pectinate	Comblike	Pectineus	11.22a
Piriformis	Pear-shaped	Piriformis	11.22c
Platys	Flat	Platysma	11.4c
Quadratus	Square, four-sided	Quadratus femoris	11.22c
Gracilis	Slender	Gracilis	11.22a
ACTION: Principal action of the muscle.			
Flexor	Decreases a joint angle	Flexor carpi radialis	11.19a
Extensor	Increases a joint angle	Extensor carpi ulnaris	11.19c
Abductor	Moves a bone away from the midline	Abductor pollicis longus	11.19c
Adductor	Moves a bone closer to the midline	Adductor longus	11.22a
Levator	Raises or elevates a body part	Levator scapulae	11.16d
Depressor	Lowers or depresses a body part	Depressor labii inferioris	11.4b
Supinator	Turns palm anteriorly	Supinator	11.19b
Pronator	Turns palm posteriorly	Pronator teres	11.19a
Sphincter	Decreases the size of an opening	External anal sphincter	11.14
Tensor	Makes a body part rigid	Tensor fasciae latae	11.22a
Rotator	Rotates a bone around its longitudinal axis	Rotatores	11.21b
NUMBER OF ORIGINS: Number of tendons of origin.			
Biceps	Two origins	Biceps brachii	11.18a
Triceps	Three origins	Triceps brachii	11.18b
Quadriceps	Four origins	Quadriceps femoris	11.22a
LOCATION: Structure near which a muscle is found. *Example:* Temporalis, a muscle near the temporal bone.			11.4c
ORIGIN AND INSERTION: Sites where muscle originates and inserts. *Example:* Sternocleidomastoid, originating on the sternum and clavicle and Inserting on mastoid process of temporal bone.			11.3a

Figure 11.3 Principal superficial skeletal muscles.

Most movements require several skeletal muscles acting in groups rather than individually.

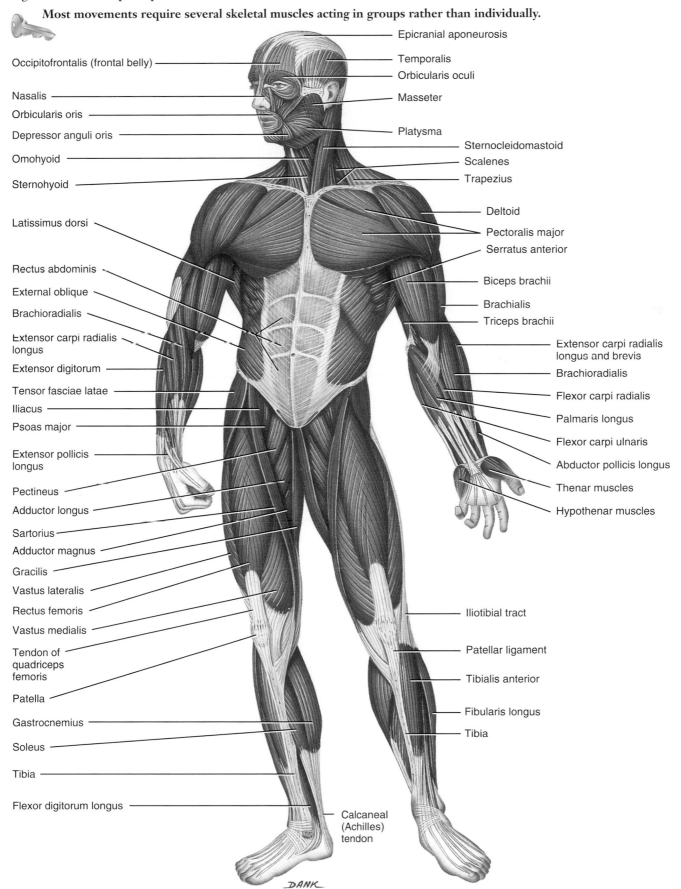

Epicranial aponeurosis

Occipitofrontalis (frontal belly)

Temporalis

Orbicularis oculi

Nasalis

Masseter

Orbicularis oris

Depressor anguli oris

Platysma

Omohyoid

Sternocleidomastoid

Scalenes

Sternohyoid

Trapezius

Latissimus dorsi

Deltoid

Pectoralis major

Serratus anterior

Rectus abdominis

Biceps brachii

External oblique

Brachialis

Brachioradialis

Triceps brachii

Extensor carpi radialis longus

Extensor digitorum

Extensor carpi radialis longus and brevis

Tensor fasciae latae

Brachioradialis

Iliacus

Flexor carpi radialis

Psoas major

Palmaris longus

Extensor pollicis longus

Flexor carpi ulnaris

Abductor pollicis longus

Pectineus

Thenar muscles

Adductor longus

Hypothenar muscles

Sartorius

Adductor magnus

Gracilis

Vastus lateralis

Rectus femoris

Vastus medialis

Iliotibial tract

Tendon of quadriceps femoris

Patellar ligament

Patella

Tibialis anterior

Gastrocnemius

Fibularis longus

Soleus

Tibia

Tibia

Flexor digitorum longus

Calcaneal (Achilles) tendon

DANK

(a) Anterior view

continues

Figure 11.3   **(continued)**

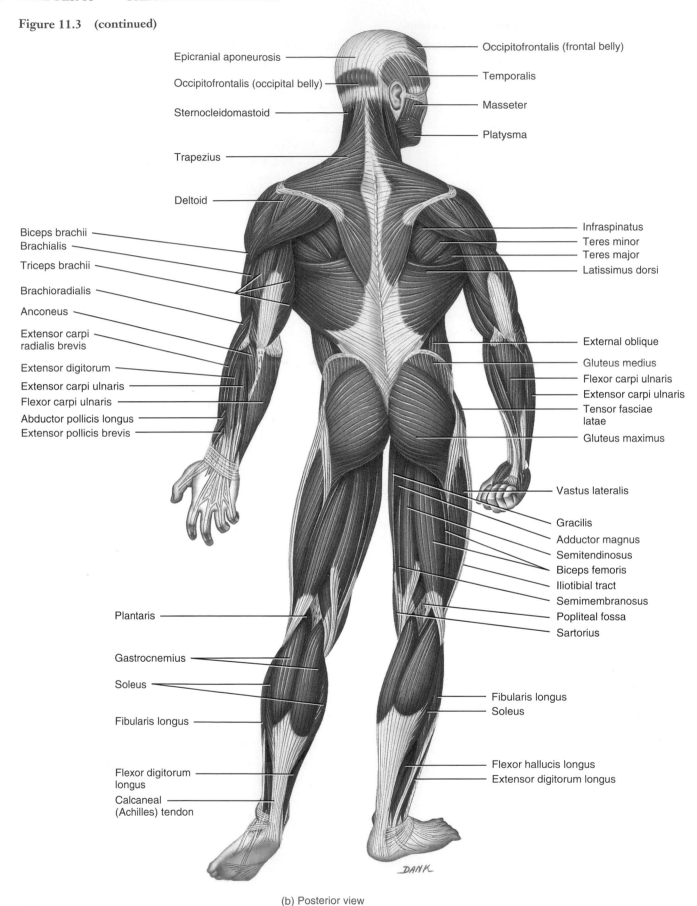

Epicranial aponeurosis

Occipitofrontalis (occipital belly)

Sternocleidomastoid

Trapezius

Deltoid

Biceps brachii
Brachialis
Triceps brachii

Brachioradialis

Anconeus

Extensor carpi
radialis brevis

Extensor digitorum

Extensor carpi ulnaris

Flexor carpi ulnaris

Abductor pollicis longus

Extensor pollicis brevis

Plantaris

Gastrocnemius

Soleus

Fibularis longus

Flexor digitorum
longus

Calcaneal
(Achilles) tendon

Occipitofrontalis (frontal belly)

Temporalis

Masseter

Platysma

Infraspinatus
Teres minor
Teres major
Latissimus dorsi

External oblique

Gluteus medius
Flexor carpi ulnaris
Extensor carpi ulnaris
Tensor fasciae
latae
Gluteus maximus

Vastus lateralis

Gracilis
Adductor magnus
Semitendinosus
Biceps femoris
Iliotibial tract
Semimembranosus
Popliteal fossa
Sartorius

Fibularis longus
Soleus

Flexor hallucis longus
Extensor digitorum longus

DANK

(b) Posterior view

Give an example of a muscle named for each of the following characteristics: direction of fibers, shape, action, size, origin and insertion, location, and number of tendons of origin.

PRINCIPAL SKELETAL MUSCLES

Exhibits 11.1 through 11.22 will assist you in learning the names of the principal skeletal muscles in various regions of the body. The muscles in the exhibits are divided into groups according to the part of the body on which they act. As you study groups of muscles in the exhibits, refer to Figure 11.3 on pages 295-296 to see how each group is related to the others.

The exhibits contain the following elements:

- *Objective.* This statement describes what you should learn from the exhibit.

- *Overview.* These paragraphs provide a general introduction to the muscles under consideration and emphasize how the muscles are organized within various regions. The discussion also highlights any distinguishing features of the muscles.

- *Muscle names.* The word roots indicate how the muscles are named. As noted previously, once you have mastered the naming of the muscles, you can more easily understand their actions.

- *Origins, insertions, and actions.* You are also given the origin, insertion, and actions of each muscle.

- *Relating muscles to movements.* These exercises will help you organize the muscles in the body region under consideration according to the actions they produce.

- *Innervation.* This section lists the nerve or nerves that cause contraction of each muscle. In general, cranial nerves, which arise from the lower parts of the brain, serve muscles in the head region. By contrast, spinal nerves, which arise from the spinal cord within the vertebral column, innervate muscles in the rest of the body. Cranial nerves are designated by both a name and a Roman numeral—for example, the facial (VII) nerve. Spinal nerves are numbered in groups according to the part of the spinal cord from which they arise: C = cervical (neck region), T = thoracic (chest region), L = lumbar (lower back region), and S = sacral (buttocks region). An example is T1, the first thoracic spinal nerve.

- *Questions.* These knowledge checkpoints relate specifically to information in each exhibit, and take the form of review, critical thinking, and/or application questions.

- *Clinical Applications.* Selected exhibits include clinical applications, which, like those in the text, explore the clinical, professional, or everyday relevance of a particular muscle or its function through descriptions of disorders or clinical procedures.

- *Figures.* The figures in the exhibits may present superficial and deep, anterior and posterior, or medial and lateral views to show each muscle's position as clearly as possible. The muscle names in all capital letters are specifically referred to in the tabular part of the exhibit.

The following is a list of the exhibits and accompanying figures that describe the principal skeletal muscles:

EXHIBIT 11.1   MUSCLES OF FACIAL EXPRESSION (Figure 11.4)

OBJECTIVE

● Describe the origin, insertion, action, and innervation of the muscles of facial expression.

The muscles of facial expression provide humans with the ability to express a wide variety of emotions. The muscles themselves lie within the layers of superficial fascia. They usually originate in the fascia or bones of the skull and insert into the skin. Because of their insertions, the muscles of facial expression move the skin rather than a joint when they contract.

Among the noteworthy muscles in this group are those surrounding the orifices (openings) of the head such as the eyes, nose, and mouth. These muscles function as *sphincters* (SFINGK-ters), which close the orifices, and *dilators,* which open the orifices. For example, the **orbicularis oculi** muscle closes the eye, whereas the **levator palpebrae superioris** muscle opens the eye. The **occipitofrontalis** is an unusual muscle in this group because it is made up of two parts: an anterior part called the **frontal belly,** which is superficial to the frontal bone, and a posterior part called the **occipital belly,** which is superficial to the occipital bone. The two muscular portions are held together by a strong aponeurosis (sheetlike tendon), the **epicranial aponeurosis** (ep-i-KRĀ-nē-al ap-ō-noo-RŌ-sis) or **galea aponeurotica** (GĀ-lē-a ap-ō-noo'-RŌ-ti-ka), which covers the superior and lateral surfaces of the skull. The **buccinator** muscle forms the major muscular portion of the cheek. The duct of the parotid gland (a salivary gland) pierces the buccinator muscle to reach the oral cavity. The buccinator muscle is so-named because it compresses the cheeks (*bucc-*=cheek) during blowing — for example, when a musician plays a wind instrument such as a trumpet. It functions in whistling, blowing, and sucking and assists in chewing.

Innervation

● All muscles of facial expression except levator palpebrae superioris: facial (VII) nerve

● Levator palpebrae superioris: oculomotor (III) nerve

BELL'S PALSY

Bell's palsy, also known as **facial paralysis,** is a unilateral paralysis of the muscles of facial expression. It is due to damage or

MUSCLE	ORIGIN	INSERTION	ACTION
SCALP MUSCLES **Occipitofontalis** (ok'-sip'-i-tō-frun-TĀ-lis)			
Frontal belly	Epicranial aponeurosis.	Skin superior to supraorbital margin.	Draws scalp anteriorly, raises eyebrows, and wrinkles skin of forehead horizontally.
Occipital belly (*occipit-*=backof the head)	Occipital bone and mastoid process of temporal bone.	Epicranial aponeurosis.	Draws scalp posteriorly.
MOUTH MUSCLES **Orbicularis oris** (or-bi'-kū-LAR-is OR-is; *orb-*=circular; *oris*=of the mouth)	Muscle fibers surrounding opening of mouth.	Skin at corner of mouth.	Closes and protrudes lips, compresses lips against teeth, and shapes lips during speech.
Zygomaticus major (zī-gō-MA-ti-kus; *zygomatic*=cheek bone; *major*=greater)	Zygomatic bone.	Skin at angle of mouth and orbicularis oris.	Draws angle of mouth superiorly and laterally, as in smiling or laughing.
Zygomaticus minor (*minor*=lesser)	Zygomatic bone.	Upper lip.	Raises (elevates) upper lip, exposing maxillary teeth.
Levator labii superioris (le-VĀ-tor LĀ-bē-ī soo-per'-ē-OR-is; *levator*=raises or elevates; *labii*=lip; *superioris*=upper)	Superior to infraorbital foramen of maxilla.	Skin at angle of mouth and orbicularis oris.	Raises upper lip.
Depressor labii inferioris (de-PRE-sor LĀ-bē-ī in-fer'-ē-OR-is; *depressor*=depresses or lowers; *inferioris*=lower)	Mandible.	Skin of lower lip.	Depresses (lowers) lower lip.
Depressor anguli oris (*angul*=angle or corner)	Mandible.	Angle of mouth.	Draws angle of mouth laterally and inferiorly, as in opening mouth.
Levator anguli oris	Inferior to infraorbital foramen.	Skin of lower lip and orbicularis oris.	Draws angle of mouth laterally and superiorly.
Buccinator (BUK-si-nā'-tor; *bucc-*=cheek)	Alveolar processes of maxilla and mandible and pterygomandibular raphe (fibrous band extending from the pterygoid process to the mandible).	Orbicularis oris.	Presses cheeks as in whistling, blowing, and sucking; draws corner of mouth laterally; and assists in mastication (chewing) by keeping food between the teeth (and not between teeth and cheeks).
Risorius (ri-ZOR-ē-us; *risor*=laughter)	Fascia over parotid (salivary) gland.	Skin at angle of mouth.	Draws angle of mouth laterally, as in tenseness.
Mentalis (men-TĀ-lis; *ment-*=the chin)	Mandible.	Skin of chin.	Elevates and protrudes lower lip and pulls skin of chin up, as in pouting.

continues on page 300

disease of the facial (VII) nerve. Although the cause is unknown, inflammation of the facial nerve or infection by the herpes simplex virus have been suggested. The paralysis causes the entire side of the face to droop in severe cases. The person cannot wrinkle the forehead, close the eye, or pucker the lips on the affected side. Drooling and difficulty in swallowing also occur. Eighty percent of patients recover completely within a few weeks to a few months. For others, paralysis is permanent. The symptoms of Bell's palsy mimic those of a stroke. ■

Relating Muscles to Movements

Arrange the muscles in this exhibit into two groups: (1) those that act on the mouth and (2) those that act on the eyes.

■ CHECKPOINT

1. Why do the muscles of facial expression move the skin rather than a joint?

Figure 11.4 **Muscles of facial expression.** (See Tortora, *A Photographic Atlas of the Human Body*, Figures 5.2 through 5.4.)

When they contract, muscles of facial expression move the skin rather than a joint.

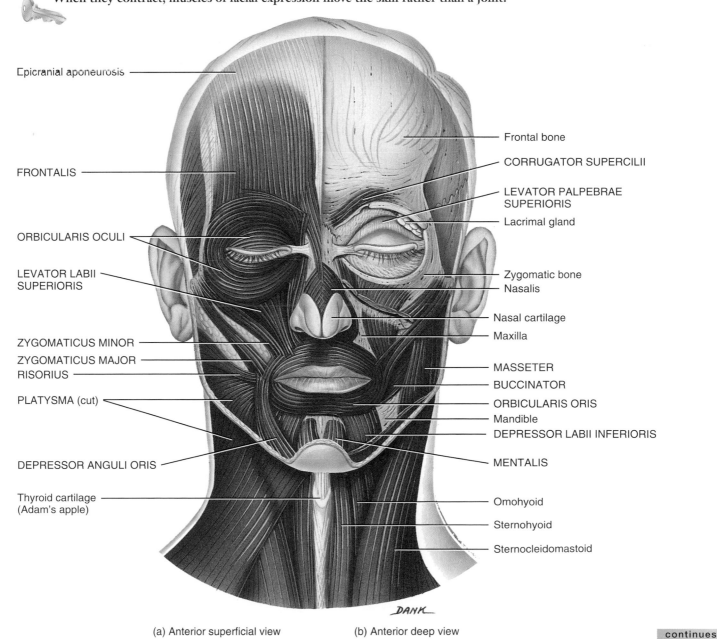

Epicranial aponeurosis

Frontal bone

CORRUGATOR SUPERCILII

FRONTALIS

LEVATOR PALPEBRAE SUPERIORIS

Lacrimal gland

ORBICULARIS OCULI

LEVATOR LABII SUPERIORIS

Zygomatic bone

Nasalis

Nasal cartilage

Maxilla

ZYGOMATICUS MINOR

ZYGOMATICUS MAJOR

RISORIUS

MASSETER

BUCCINATOR

PLATYSMA (cut)

ORBICULARIS ORIS

Mandible

DEPRESSOR LABII INFERIORIS

DEPRESSOR ANGULI ORIS

MENTALIS

Thyroid cartilage (Adam's apple)

Omohyoid

Sternohyoid

Sternocleidomastoid

DANK

(a) Anterior superficial view   (b) Anterior deep view

continues

EXHIBIT 11.1   MUSCLES OF FACIAL EXPRESSION (Figure 11.4)

CONTINUED

MUSCLE	ORIGIN	INSERTION	ACTION
NECK MUSCLE			
Platysma (pla-TIZ-ma; *platy*=flat, broad)	Fascia over deltoid and pectoralis major muscles.	Mandible, muscle around angle of mouth, and skin of lower face.	Draws outer part of lower lip inferiorly and posteriorly as in pouting; depresses mandible.
ORBIT AND EYEBROW MUSCLES			
Orbicularis oculi (or-bi'-kū-LAR-is OK-ū-lī; *oculi*=of the eye)	Medial wall of orbit.	Circular path around orbit.	Closes eye.
Corrugator supercilii (KOR-a-gā'-tor soo-per-SI-lē-ī; *corrugat*=wrinkle; *supercilii*=of the eyebrow)	Medial end of superciliary arch of frontal bone.	Skin of eyebrow.	Draws eyebrow inferiorly and wrinkles skin of forehead vertically as in frowning.
Levator palpebrae superioris (le-VĀ-tor PAL-pe-brē soo-per'-ē-OR-is; *palpebrae*=eyelids) (see also Figure 11.5a)	Roof of orbit (lesser wing of sphenoid bone).	Skin of upper eyelid.	Elevates upper eyelid (opens eye).

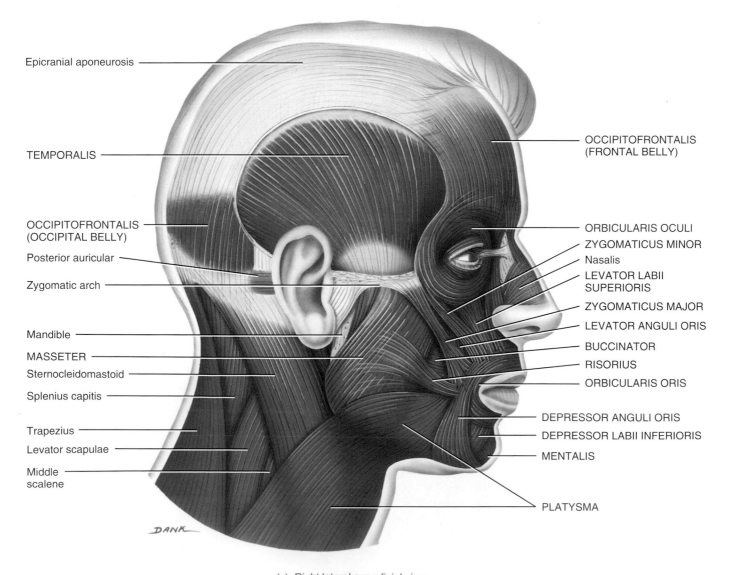

(c) Right lateral superficial view

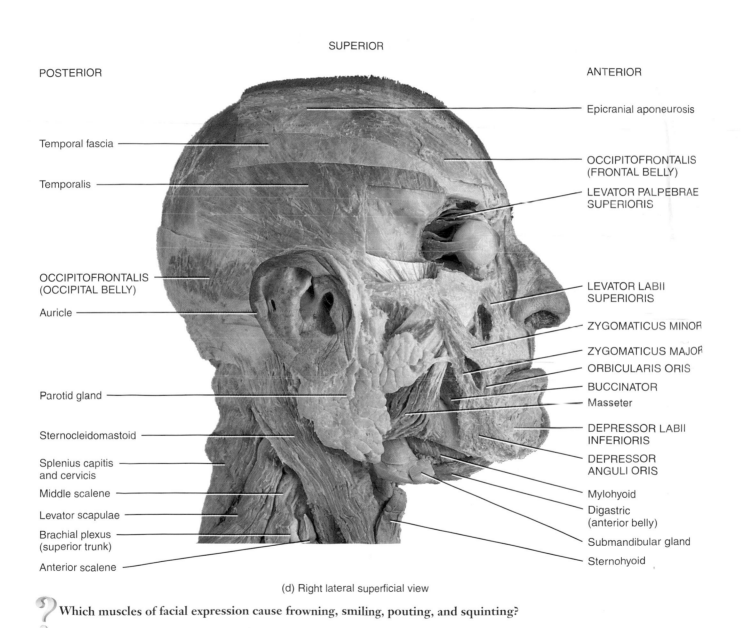

POSTERIOR

SUPERIOR

ANTERIOR

Temporal fascia

Temporalis

OCCIPITOFRONTALIS
(OCCIPITAL BELLY)

Auricle

Parotid gland

Sternocleidomastoid

Splenius capitis
and cervicis

Middle scalene

Levator scapulae

Brachial plexus
(superior trunk)

Anterior scalene

Epicranial aponeurosis

OCCIPITOFRONTALIS
(FRONTAL BELLY)

LEVATOR PALPEBRAE
SUPERIORIS

LEVATOR LABII
SUPERIORIS

ZYGOMATICUS MINOR

ZYGOMATICUS MAJOR

ORBICULARIS ORIS

BUCCINATOR

Masseter

DEPRESSOR LABII
INFERIORIS

DEPRESSOR
ANGULI ORIS

Mylohyoid

Digastric
(anterior belly)

Submandibular gland

Sternohyoid

(d) Right lateral superficial view

Which muscles of facial expression cause frowning, smiling, pouting, and squinting?

OBJECTIVE

- Describe the origin, insertion, action, and innervation of the extrinsic eye muscles.

Muscles that move the eyeballs are called **extrinsic eye muscles** because they originate outside the eyeballs (in the orbit) and insert on the outer surface of the sclera ("white of the eye"). The extrinsic eye muscles are some of the fastest contracting and most precisely controlled skeletal muscles in the body.

Three pairs of extrinsic eye muscles control movements of the eyeballs: (1) superior and inferior recti, (2) lateral and medial recti, and (3) superior and inferior oblique. The four recti muscles (superior, inferior, lateral, and medial) arise from a tendinous ring in the orbit and insert into the sclera of the eye. As their names imply, the superior and inferior recti move the eyeballs superiorly and inferiorly, respectively; the lateral and medial recti move the eyeballs laterally and medially, respectively.

The actions of the oblique muscles cannot be deduced from their names. The **superior oblique** muscle originates posteriorly near the tendinous ring, then passes anteriorly, and ends in a round tendon. The tendon extends through a pulleylike loop called the *trochlea* (=pulley) in the anterior and medial part of the roof of the orbit. Finally, the tendon turns and inserts on the posterolateral aspect of the eyeballs. Accordingly, the superior oblique muscle moves the eyeballs inferiorly and laterally. The **inferior oblique** muscle originates on the maxilla at the anteromedial aspect of the floor of the orbit. It then passes posteriorly and laterally and inserts on the posterolateral aspect of the eyeballs. Because of this arrangement, the inferior oblique muscle moves the eyeballs superiorly and laterally.

Innervation

- Superior rectus, inferior rectus, medial rectus, and inferior oblique: oculomotor (III) nerve
- Lateral rectus: abducens (VI) nerve
- Superior oblique: trochlear (IV) nerve

STRABISMUS

Strabismus is a condition in which the two eyes are not properly aligned. This can be hereditary or it can be due to birth injuries, poor attachments of the muscles, problems with the brain's control center, or localized disease. Strabismus can be constant or intermittent. In strabismus, each eye sends an image to a different area of the brain. Because the brain usually ignores the messages sent by one of the eyes, the ignored eye becomes weaker, hence "lazy eye" or *amblyopia,* develops. *External strabismus* results when a lesion in the oculomotor (III) cranial nerve causes the eyeball to

MUSCLE	ORIGIN	INSERTION	ACTION
SUPERIOR RECTUS (*rectus*=fascicles parallel to midline)	Common tendinous ring (attached to orbit around optic foramen).	Superior and central part of eyeball.	Moves eyeball superiorly (elevation) and medially (adduction), and rotates it medially.
INFERIOR RECTUS	Same as above.	Inferior and central part of eyeball.	Moves eyeball inferiorly (depression) and medially (adduction), and rotates it laterally.
LATERAL RECTUS	Same as above.	Lateral side of eyeball.	Moves eyeball laterally (abduction).
MEDIAL RECTUS	Same as above.	Medial side of eyeball.	Moves eyeball medially (adduction).
SUPERIOR OBLIQUE (*oblique*=fascicles diagonal to midline)	Sphenoid bone, superior and medial to the tendinous ring in the orbit.	Eyeball between superior and lateral recti. The muscle inserts into the superior and lateral surfaces of the eyeball via a tendon that passes through the trochlea.	Moves eyeball inferiorly (depression) and laterally (abduction), rotates it medially.
INFERIOR OBLIQUE	Maxilla in floor of orbit.	Eyeball between inferior and lateral recti.	Moves eyeball superiorly (elevation) and laterally (abduction), and rotates it laterally.

move laterally when at rest, and results in an inability to move the eyeball medially and inferiorly. A lesion in the abducens (VI) cranial nerve results in *internal strabismus,* a condition in which the eyeball moves medially when at rest and cannot move laterally.

Treatment options for strabismus depend on the specific type of problem an individual has, and the treatments are often used in conjunction with one another. Treatment modalities include surgery, visual therapy (retraining the brain's control center), and orthoptics (eye muscle training to straighten the eyes). ■

Relating Muscles to Movements

Arrange the muscles in this exhibit according to their actions on the eyeballs: (1) elevation, (2) depression, (3) abduction, (4) adduction, (5) medial rotation, and (6) lateral rotation. The same muscle may be mentioned more than once.

■ CHECKPOINT

1. Which muscles contract and relax in each eye as you gaze to your left without moving your head?

Figure 11.5 **Extrinsic muscles of the eyeball.**

The extrinsic muscles of the eyeball are among the fastest contracting and most precisely controlled skeletal muscles in the body.

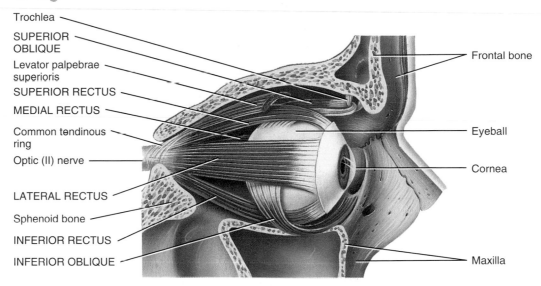

Trochlea
SUPERIOR OBLIQUE
Levator palpebrae superioris
SUPERIOR RECTUS
MEDIAL RECTUS
Common tendinous ring
Optic (II) nerve
LATERAL RECTUS
Sphenoid bone
INFERIOR RECTUS
INFERIOR OBLIQUE

Frontal bone
Eyeball
Cornea
Maxilla

(a) Lateral view of right eyeball

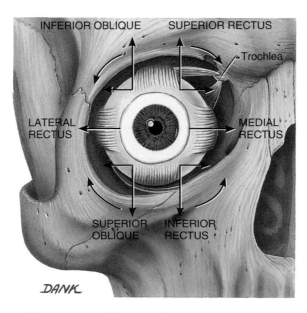

INFERIOR OBLIQUE   SUPERIOR RECTUS
Trochlea
LATERAL RECTUS   MEDIAL RECTUS
SUPERIOR OBLIQUE   INFERIOR RECTUS

DANK

(b) Movements of right eyeball in response to contraction of extrinsic muscles

continues

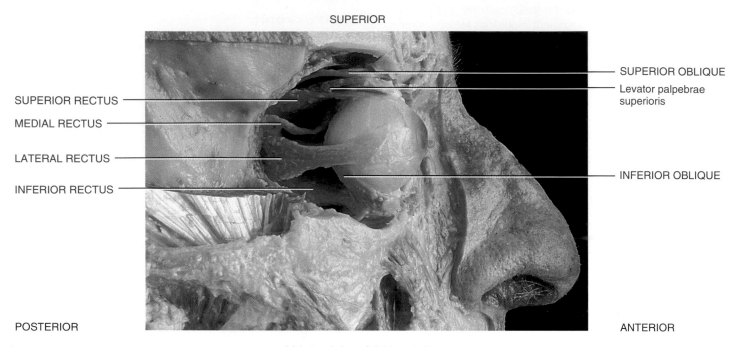

(c) Lateral view of right eyeball

How does the inferior oblique muscle move the eyeball superiorly and laterally?

EXHIBIT 11.3 MUSCLES THAT MOVE THE MANDIBLE (LOWER JAW BONE) (Figure 11.6)

OBJECTIVE

● Describe the origin, insertion, action, and innervation of the muscles that move the mandible.

The muscles that move the mandible (lower jaw bone) at the temporo-mandibular joint (TMJ) are known as the muscles of mastication because they are involved in chewing (mastication). Of the four pairs of muscles involved in mastication, three are powerful closers of the jaw and account for the strength of the bite: **masseter, temporalis,** and **medial pterygoid.** Of these, the masseter is the strongest muscle of mastication. The medial and **lateral pterygoid** muscles assist in mastication by moving the mandible from side to side to help grind food. Additionally, these muscles protrude the mandible.

Innervation

• Masseter, temporalis, medial pterygoid, lateral pterygoid: mandibular division of the trigeminal (V) nerve

Relating Muscles to Movements

Arrange the muscles in this exhibit according to their actions on the mandible: (1) elevation, (2) depression, (3) retraction, (4) protraction, and (5) side-to-side movement. The same muscle may be mentioned more than once.

■ **CHECKPOINT**

1. What would happen if you lost tone in the masseter and temporalis muscles?

MUSCLE	ORIGIN	INSERTION	ACTION
MASSETER (MA-se-ter=a chewer) (see Figure 11.4c)	Maxilla and zygomatic arch.	Angle and ramus of mandible.	Elevates mandible, as in closing mouth, and retracts (draws back) mandible.
TEMPORALIS (tem'-pō-RĀ-lis; tempor-=time or temples)	Temporal bone.	Coronoid process and ramus of mandible.	Elevates and retracts mandible.
MEDIAL PTERYGOID (TER-i-goyd; medial=closer to midline; pterygoid=like a wing)	Medial surface of lateral portion of pterygoid process of sphenoid bone; maxilla.	Angle and ramus of mandible.	Elevates and protracts (protrudes) mandible and moves mandible from side to side.
LATERAL PTERYGOID (TER-i-goyd; lateral=farther from midline)	Greater wing and lateral surface of lateral portion of pterygoid process of sphenoid bone.	Condyle of mandible; temporomandibular joint (TMJ).	Protracts mandible, depresses mandible as in opening mouth, and moves mandible from side to side.

continues

EXHIBIT 11.3   **MUSCLES THAT MOVE THE MANDIBLE (LOWER JAW BONE)** (Figure 11.6)

Figure 11.6   Muscles that move the mandible (lower jaw bone).

The muscles that move the mandible are also known as muscles of mastication.

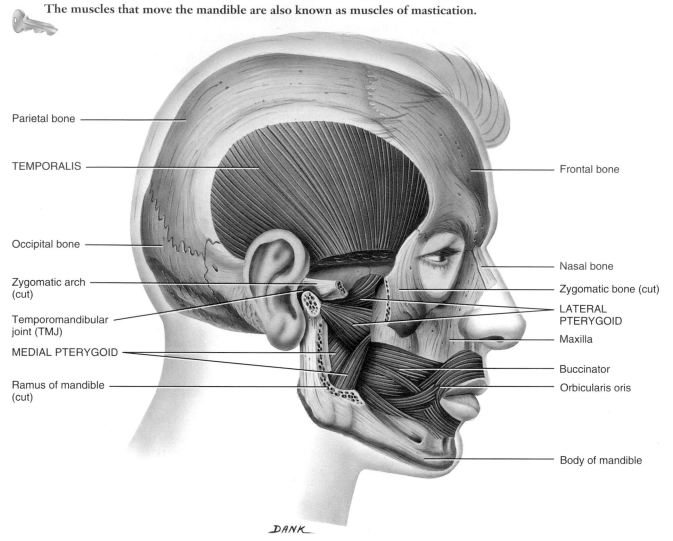

Right lateral superficial view

? Which is the strongest muscle of mastication?

OBJECTIVE

- Describe the origin, insertion, action and innervation of the extrinsic muscles of the tongue.

The tongue is a highly mobile structure that is vital to digestive functions such as mastication, perception of taste, and deglutition (swallowing). It is also important in speech. The tongue's mobility is greatly aided by its suspension from the mandible, styloid process of the temporal bone, and hyoid bone.

The tongue is divided into lateral halves by a median fibrous septum. The septum extends throughout the length of the tongue. Inferiorly, the septum attaches to the hyoid bone. Muscles of the tongue are of two principal types: extrinsic and intrinsic. **Extrinsic tongue muscles** originate outside the tongue and insert into it. They move the entire tongue in various directions, such as anteriorly, posteriorly, and laterally. **Intrinsic tongue muscles** originate and insert within the tongue. These muscles alter the shape of the tongue rather than moving the entire tongue. The extrinsic and intrinsic muscles of the tongue insert into both lateral halves of the tongue.

When you study the extrinsic tongue muscles, you will notice that all of their names end in *glossus,* meaning tongue. You will notice that the actions of the muscles are obvious, considering the positions of the mandible, styloid process, hyoid bone, and soft palate, which serve as origins for these muscles. For example, the **genioglossus** (origin: the mandible) pulls the tongue downward and forward, the **styloglossus** (origin: the styloid process) pulls the tongue upward and backward, the **hyoglossus** (origin: the hyoid bone) pulls the tongue downward and flattens it, and the **palatoglossus** (origin: the soft palate) raises the back portion of the tongue.

Innervation

- Genioglossus, styloglossus, and hyoglossus: hypoglossal (XII) nerve

- Palatoglossus: the pharyngeal plexus, which contains axons from both the vagus (X) nerve and the accessory (XI) nerve

INTUBATION DURING ANESTHESIA

When general anesthesia is administered during surgery, a total relaxation of the muscles results. Once one of the various types of drugs for anesthesia have been given (especially the paralytic agents), the patient's airway must be protected and the lungs ventilated because the muscles involved with respiration are among those paralyzed. Paralysis of the genioglossus muscle causes the tongue to fall posteriorly, which may obstruct the airway to the lungs. To avoid this, the mandible is either manually thrust forward and held in place (known as the "sniffing position"), or a tube is inserted from the lips through the laryngopharynx (inferior portion of the throat) into the trachea (endotracheal intubation). People can also be intubated nasally. ■

Relating Muscles to Movements

Arrange the muscles in this exhibit according to the following actions on the tongue: (1) depression, (2) elevation, (3) protraction, and (4) retraction. The same muscle may be mentioned more than once.

■ CHECKPOINT

1. When your physician says, "open your mouth, stick out your tongue and say *ahh*" so she can examine the inside of your mouth for possible signs of infection, which muscles do you contract?

MUSCLE	ORIGIN	INSERTION	ACTION
GENIOGLOSSUS (jē′-nē-ō-GLOS-us; *genio-*=chin; *glossus*=tongue)	Mandible.	Undersurface of tongue and hyoid bone.	Depresses tongue and thrusts it anteriorly (protraction).
STYLOGLOSSUS (stī′-lō-GLOS-us; *stylo-*=stake or pole; styloid process of temporal bone)	Styloid process of temporal bone.	Side and undersurface of tongue.	Elevates tongue and draws it posteriorly (retraction).
PALATOGLOSSUS (pal′-a-tō-GLOS-us; *palate-*=the roof of the mouth or palate)	Anterior surface of soft palate.	Side of tongue.	Elevates posterior portion of tongue and draws soft palate down on tongue.
HYOGLOSSUS (hī′-ō-GLOS-us)	Greater horn and body of hyoid bone.	Side of tongue.	Depresses tongue and draws down its sides.

continues

Figure 11.7   Muscles that move the tongue.

The extrinsic and intrinsic muscles of the tongue are arranged in both lateral halves of the tongue.

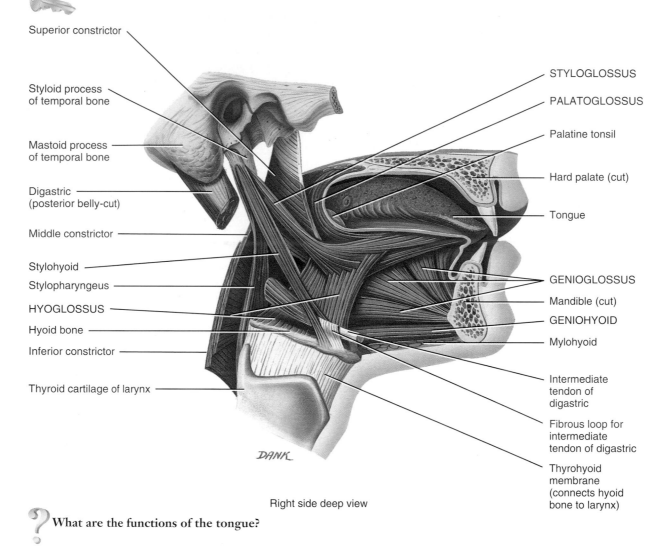

Superior constrictor

Styloid process
of temporal bone

Mastoid process
of temporal bone

Digastric
(posterior belly-cut)

Middle constrictor

Stylohyoid

Stylopharyngeus

HYOGLOSSUS

Hyoid bone

Inferior constrictor

Thyroid cartilage of larynx

DANK

STYLOGLOSSUS

PALATOGLOSSUS

Palatine tonsil

Hard palate (cut)

Tongue

GENIOGLOSSUS

Mandible (cut)

GENIOHYOID

Mylohyoid

Intermediate
tendon of
digastric

Fibrous loop for
intermediate
tendon of digastric

Thyrohyoid
membrane
(connects hyoid
bone to larynx)

Right side deep view

? **What are the functions of the tongue?**

EXHIBIT 11.5 MUSCLES OF THE ANTERIOR NECK (Figure 11.8)

OBJECTIVE

● Describe the origin, insertion, action, and innervation of the muscles of the anterior neck.

Two groups of muscles are associated with the anterior aspect of the neck: (1) the **suprahyoid muscles,** so-called because they are located superior to the hyoid bone, and (2) the **infrahyoid muscles,** so-named because they lie inferior to the hyoid bone. Acting with the infrahyoid muscles, the suprahyoid muscles stabilize the hyoid bone. Thus, the hyoid bone serves as a firm base upon which the tongue can move.

As a group, the suprahyoid muscles elevate the hyoid bone, floor of the oral cavity, and tongue during swallowing. As its name suggests, the **digastric** muscle has two bellies, anterior and posterior, united by an intermediate tendon that is held in position by a fibrous loop (see Figure 11.7). This muscle elevates the hyoid bone and larynx (voice box) during swallowing and speech and depresses the mandible. The **stylohyoid** muscle elevates and draws the hyoid bone posteriorly, thus elongating the floor of the oral cavity during swallowing. The **mylohyoid** muscle elevates the hyoid bone and helps press the tongue against the roof of the oral cavity during swallowing to move food from the oral cavity into the throat. The **geniohyoid** muscle (see Figure 11.7) elevates and draws the hyoid bone anteriorly to shorten the floor of the oral cavity and to widen the throat to receive food that is being swallowed. It also depresses the mandible.

The infrahyoid muscles are sometimes called "strap" muscles because of their ribbonlike appearance. Most of the infrahyoid muscles depress the hyoid bone, and some move the larynx during swallowing and speech. The **omohyoid** muscle, like the digastric muscle, is composed of two bellies connected by an intermediate tendon. In this case, however, the two bellies are referred to as

superior and inferior, rather than anterior and posterior. Together, the omohyoid, **sternohyoid,** and **thyrohyoid** muscles depress the hyoid bone. In addition, the **sternothyroid** muscle depresses the thyroid cartilage (Adam's apple) of the larynx, whereas the thyrohyoid muscle elevates the thyroid cartilage. The actions are necessary during the production of low and high tones respectively, during phonation.

Innervation

- Anterior belly of the digastric and the mylohyoid muscles: mandibular division of the trigeminal (V) nerve
- Posterior belly of the digastric and the stylohyoid: facial (VII) nerve
- Geniohyoid musle: first cervical spinal nerve
- Omohyoid, sternohyoid, and sternothyroid: branches of ansa cervicalis (spinal nerves C1–C3)
- Thyrohyoid: branches of ansa cervicalis (spinal nerves C1–C2) and descending hypoglossal (XII) nerve

Relating Muscles to Movements

Arrange the muscles in this exhibit according to the following actions on the hyoid bone: (1) elevating it, (2) drawing it anteriorly, (3) drawing it posteriorly, and (4) depressing it; and on the thyroid cartilage: (1) elevating it and (2) depressing it. The same muscle may be mentioned more than once.

■ CHECKPOINT

1. Which tongue, facial, and mandibular muscles do you use for chewing?

MUSCLE	ORIGIN	INSERTION	ACTION
SUPRAHYOID MUSCLES			
Digastric (dī′-GAS-trik; *di-*=two; *gastr-*=belly)	Anterior belly from inner side of inferior border of mandible; posterior belly from temporal bone.	Body of hyoid bone via an intermediate tendon.	Elevates hyoid bone and depresses mandible, as in opening the mouth.
Stylohyoid (stī′-lō-HĪ-oid; *stylo-*=stake or pole, styloid process of temporal bone; *hyo-*=U-shaped, pertaining to hyoid bone)	Styloid process of temporal bone.	Body of hyoid bone.	Elevates hyoid bone and draws it posteriorly.
Mylohyoid (mī′-lō-HĪ-oid) (*mylo-*=mill)	Inner surface of mandible.	Body of hyoid bone.	Elevates hyoid bone and floor of mouth and depresses mandible.
Geniohyoid (jē′-nē-ō-HĪ-oid; *genio-*=chin) (see also Figure 11.7)	Inner surface of mandible.	Body of hyoid bone.	Elevates hyoid bone, draws hyoid bone and tongue anteriorly, and depresses mandible.
INFRAHYOID MUSCLES			
Omohyoid (ō-mō-HĪ-oid; *omo-*=relationship to the shoulder)	Superior border of scapula and superior transverse ligament.	Body of hyoid bone.	Depresses hyoid bone.
Sternohyoid (ster′-nō-HĪ-oid; *sterno-*=sternum)	Medial end of clavicle and manubrium of sternum.	Body of hyoid bone.	Depresses hyoid bone.
Sternothyroid (ster′-nō-THĪ-roid; *thyro-*=thyroid gland)	Manubrium of sternum.	Thyroid cartilage of larynx.	Depresses thyroid cartilage of larynx.
Thyrohyoid (thī-rō-HĪ-oid)	Thyroid cartilage of larynx.	Greater horn of hyoid bone.	Elevates thyroid cartilage and depresses hyoid bone.

continues

EXHIBIT 11.5   MUSCLES OF THE ANTERIOR NECK (Figure 11.8)

Figure 11.8   Muscles of the floor of the oral cavity and front of the neck.

The suprahyoid muscles elevate the hyoid bone, the floor of the oral cavity, and the tongue during swallowing.

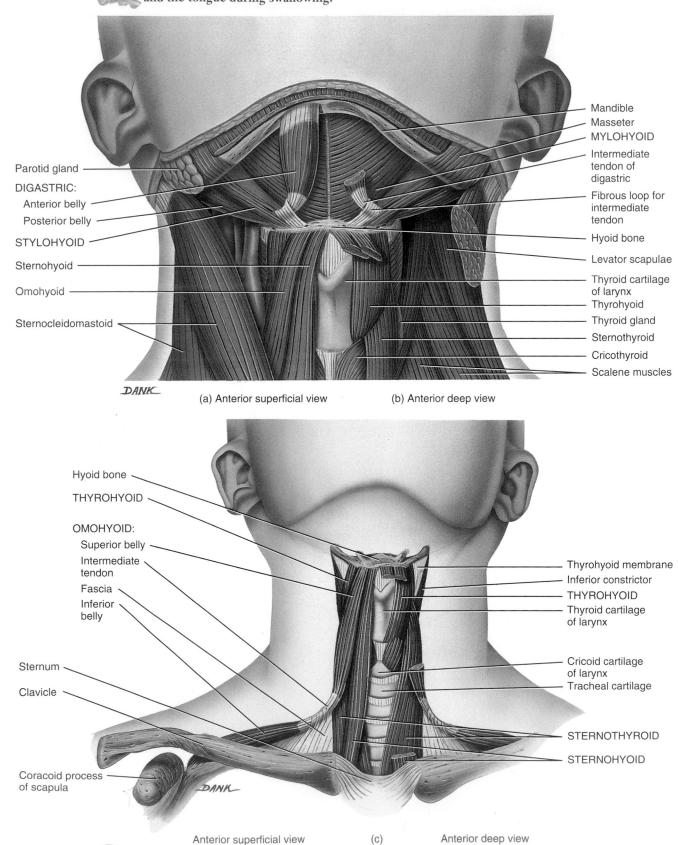

Mandible
Masseter
MYLOHYOID
Intermediate tendon of digastric
Fibrous loop for intermediate tendon
Hyoid bone
Levator scapulae
Thyroid cartilage of larynx
Thyrohyoid
Thyroid gland
Sternothyroid
Cricothyroid
Scalene muscles

Parotid gland
DIGASTRIC:
  Anterior belly
  Posterior belly
STYLOHYOID
Sternohyoid
Omohyoid
Sternocleidomastoid

DANK

(a) Anterior superficial view   (b) Anterior deep view

Hyoid bone
THYROHYOID
OMOHYOID:
  Superior belly
  Intermediate tendon
  Fascia
  Inferior belly
Sternum
Clavicle
Coracoid process of scapula

DANK

Thyrohyoid membrane
Inferior constrictor
THYROHYOID
Thyroid cartilage of larynx
Cricoid cartilage of larynx
Tracheal cartilage
STERNOTHYROID
STERNOHYOID

Anterior superficial view   (c)   Anterior deep view

What is the combined action of the suprahyoid and infrahyoid muscles?

EXHIBIT 11.6   MUSCLES THAT MOVE THE HEAD (Figure 11.9)

OBJECTIVE

• Describe the origin, insertion, action, and innervation of the muscles that move the head.

The head is attached to the vertebral column at the atlanto-occipital joints formed by the atlas and occipital bone. Balance and movement of the head on the vertebral column involves the action of several neck muscles. For example, acting together (bilaterally), contraction of the two **sternocleidomastoid** muscles flexes the cervical portion of the vertebral column and extends the head. Acting singly (unilaterally), each sternocledomastioid muscle laterally flexes and rotates the head. Bilateral contraction of the **semispinalis capitis, splenius capitis,** and **longissimus capitis** muscles extends the head. However, when these same muscles contract unilaterally, their actions are quite different, involving primarily rotation of the head.

The sternocleidomastoid muscle is an important landmark that divides the neck into two major triangles: anterior and posterior. The triangles are important anatomically and surgically because of the structures that lie within their boundaries.

The **anterior triangle** is bordered superiorly by the mandible, inferiorly by the sternum, medially by the cervical midline, and laterally by the anterior border of the sternocleidomastoid muscle. The anterior triangle is subdivided into an unpaired submental triangle and three paired triangles: submandibular, carotid, and muscular. The anterior triangle contains submental, submandibular, and deep cervical lymph nodes; the submandibular salivary gland and a portion of the parotid salivary gland; the facial artery and vein; carotid arteries and internal jugular vein; and the following cranial nerves: glossopharyngeal (IX), vagus (X), accessory (XI), and hypoglossal (XII).

The **posterior triangle** is bordered inferiorly by the clavicle, anteriorly by the posterior border of the sternocleidomastoid muscle, and posteriorly by the anterior border of the trapezius muscle. The posterior triangle is subdivided into two triangles, occipital and supraclavicular (omoclavicular), by the inferior belly of the omohyoid muscle. The posterior triangle contains part of the subclavian artery, external jugular vein, cervical lymph nodes, brachial plexus, and the accessory (XI) nerve.

Innervation

• Sternocleidomastoid: accessory (XI) nerve

• All capitis muscles: cervical spinal nerves

Relating Muscles to Movements

Arrange the muscles in this exhibit according to the following actions on the head: (1) flexion, (2) lateral flexion, (3) extension, (4) rotation to side opposite contracting muscle, and (5) rotation to same side as contracting muscle. The same muscle may be mentioned more than once.

▪ CHECKPOINT

1. What muscles do you contract to signify "yes" and "no"?

MUSCLE	ORIGIN	INSERTION	ACTION
STERNOCLEIDOMASTOID (ster'-nō-klī'-dō-MAS-toid; *sterno-*=breastbone; *cleido-*=clavicle; *mastoid*=mastoid process of temporal bone)	Sternum and clavicle.	Mastoid process of temporal bone.	Acting together (bilaterally), flex cervical portion of vertebral column, extend head, and elevate sternum during forced inhalation; acting singly (unilaterally), laterally flex and rotate head to side opposite contracting muscle.
SEMISPINALIS CAPITIS (se'-mē-spi-NĀ-lis KAP-i-tis; *semi-*=half; *spine*=spinous process; *capit-*=head) (see Figure 11.21a)	Transverse processes of first six or seven thoracic vertebrae and seventh cervical vertebra, and articular processes of fourth, fifth, and sixth cervical vertebrae.	Occipital bone between superior and inferior nuchal lines.	Acting together, extend head; acting singly, rotate head to side opposite contracting muscle.
SPLENIUS CAPITIS (SPLĒ-nē-us KAP-i-tis; *splenion-*=bandage) (see Figure 11.21a)	Ligamentum nuchae and spinous processes of seventh cervical vertebra and first three or four thoracic vertebrae.	Occipital bone and mastoid process of temporal bone.	Acting together, extend head; acting singly, laterally flex and rotate head to same side as contracting muscle.
LONGISSIMUS CAPITIS (lon-JIS-i-mus KAP-i-tis; *longissimus*=longest) (see Figure 11.21a)	Transverse processes of upper four thoracic vertebrae and articular processes of last four cervical vertebrae.	Mastoid process of temporal bone.	Acting together, extend head; acting singly, laterally flex and rotate head to same side as contracting muscle.

continues

EXHIBIT 11.6   MUSCLES THAT MOVE THE HEAD (Figure 11.9)

Figure 11.9   Triangles of the neck.

The sternocleidomastoid muscle divides the neck into two principal triangles: anterior and posterior.

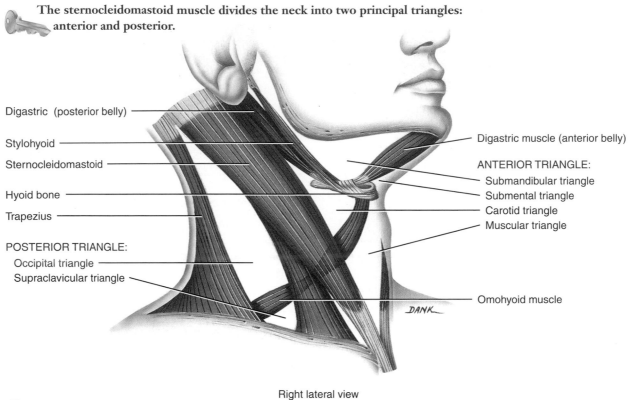

Digastric (posterior belly)

Stylohyoid

Sternocleidomastoid

Hyoid bone

Trapezius

POSTERIOR TRIANGLE:
  Occipital triangle
  Supraclavicular triangle

Digastric muscle (anterior belly)

ANTERIOR TRIANGLE:
  Submandibular triangle
  Submental triangle
  Carotid triangle
  Muscular triangle

Omohyoid muscle

DANK

Right lateral view

 Why are triangles important?

EXHIBIT 11.7   MUSCLES OF THE LARYNX (VOICE BOX) (Figure 11.10)

● Describe the origin, insertion, action and innervation of the muscles of the larynx.

The muscles of the larynx (voice box), like those of the eyeballs and tongue, are grouped into **extrinsic muscles of the larynx** and **intrinsic muscles of the larynx.** The extrinsic muscles of the larynx, which are associated with the anterior aspect of the neck, are called **infrahyoid muscles** because they lie inferior to the hyoid bone. Refer to Exhibit 11.5 and Figure 11.8 for a description of these muscles.

Whereas the extrinsic muscles move the larynx as a whole, the intrinsic muscles of the larynx only move parts of the larynx. Based on their actions, the intrinsic muscles may be grouped into three functional sets according to their actions. The first set includes the **cricothyroid** and **thyroarytenoid** muscles, which regulate the tension of the vocal folds (true vocal cords). The second set varies the size of the rima glottidis (space between the vocal folds). These muscles include the **lateral cricoarytenoid,** which brings the vocal folds together (adduction), thus closing the rima glottidis, and the **posterior cricoarytenoid,** which moves the vocal folds apart

(abduction), thus opening the rima glottidis. The **transverse arytenoid** closes the posterior portion of the rima glottidis. The last intrinsic muscle functions as a sphincter to control the size of the inlet of the larynx, which is the opening anteriorly from the pharynx (throat) into the larynx. This muscle is the **oblique arytenoid.**

Innervation

• The intrinsic muscles are innervated by the recurrent laryngeal branch of the vagus (X) nerve.

Relating Muscles to Movements

Arrange the intrinsic muscles of the larynx in this exhibit according to the following actions on the vocal cords: (1) increasing tension on the vocal cords; (2) moving the vocal cords apart, and (3) moving the vocal cords together. The same muscle may be mentioned more than once.

■ **CHECKPOINT**

1. How are the intrinsic muscles of the larynx grouped functionally?

MUSCLE	ORIGIN	INSERTION	ACTION
EXTRINSIC MUSCLES of the larynx (Infrahyoid muscles) **Omohyoid** **Sternohyoid** **Sternothyroid** **Thyrohyoid**	See Exhibit 11.5		
INTRINSIC MUSCLES of the larynx **Cricothyroid** (kri-kō-THĪ-roid; *crico-*=cricoid cartilage of larynx)	Anterior and lateral portion of cricoid cartilage of larynx.	Anterior border of thyroid cartilage of larynx and posterior part of inferior border of thyroid cartilage of larynx.	Elongates and places tension on vocal folds.
THYROARYTENOID (thī'-rō-ar'-i-TĒ-noid; *-arytaina*=shaped like a jug)	Inferior portion of thyroid cartilage of larynx and middle of cricothyroid ligament.	Base and anterior surface of arytenoid cartilage of larynx.	Shortens and relaxes vocal folds.
LATERAL CRICOARYTENOID (kri'-kō-ar'-i-TĒ-noid)	Superior border of cricoid cartilage of larynx.	Anterior surface of arytenoid cartilage of larynx.	Brings vocal folds together (adduction), thus closing the rima glottidis.
POSTERIOR CRICOARYTENOID	Posterior surface of cricoid cartilage of larynx.	Posterior surface of arytenoid cartilage of larynx.	Moves the vocal folds apart (abduction), thus opening the rima glottidis.
TRANSVERSE AND OBLIQUE ARYTENOID (ar'-i-TĒ-noid)	Posterior surface and lateral border of one arytenoid cartilage of larynx.	Corresponding parts of opposite arytenoid cartilage of larynx.	The transverse arytenoid closes the posterior portion of the rima glottidis; the oblique arytenoid regulates the size of the inlet of the larynx.

continues

EXHIBIT 11.7   MUSCLES OF THE LARYNX (VOICE BOX) (Figure 11.10)

CONTINUED

Figure 11.10   Muscles of the larynx (voice box). (See Tortora, *A Photographic Atlas of the Human Body*, Figure 5.6.)

Intrinsic muscles of the larynx adjust the tension of the vocal folds and open or close the rima glottidis.

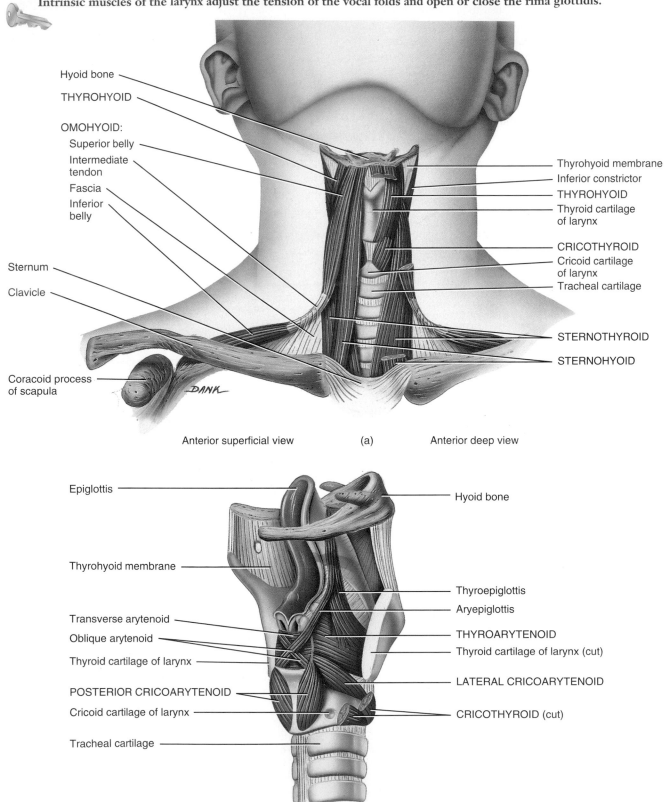

Anterior superficial view   (a)   Anterior deep view

(b) Right posterolateral view

How do the extrinsic and intrinsic muscles of the larynx differ functionally?

EXHIBIT 11.8   MUSCLES OF THE PHARYNX (THROAT) (Figure 11.11)

OBJECTIVE

● Describe the origin, insertion, action, and innervation of the muscles of the pharynx.

The pharynx (throat) is a somewhat funnel-shaped tube posterior to the nasal and oral cavities. It is a common chamber for the respiratory and digestive systems, opening anteriorly into the larynx and posteriorly into the esophagus.

The muscles of the pharynx are arranged in two layers, an outer circular layer and an inner longitudinal layer. The **circular layer** is composed of three constrictor muscles, each overlapping the muscle above it, an arrangement that resembles stacked flowerpots. Their names are the inferior, middle, and superior constrictor muscles. The **longitudinal layer** is composed of three muscles that descend from the styloid process of the temporal bone, auditory (eustachian) tube, and soft palate. Their names are the stylopharyngeus, salpingopharyngeus, and palatopharyngeus, respectively.

The **inferior constrictor** muscle is the thickest of the constrictor muscles. Its inferior fibers are continuous with the musculature of the esophagus, whereas its superior fibers overlap the middle constrictor. The **middle constrictor** is fan-shaped and smaller than the inferior constrictor, and it overlaps the superior constrictor. The **superior constrictor** is quadrilateral and thinner than the other constrictors. As a group, the constrictor muscles constrict the pharynx during swallowing. The sequential contraction of these muscles moves food and drink from the mouth into the esophagus.

The **stylopharyngeus** muscle is a long, thin muscle that enters the pharynx between the middle and superior constrictors and inserts along with the palatopharyngeus on the thyroid cartilage. The **salpingopharyngeus** muscle is a thin muscle that descends in the lateral wall of the pharynx and also inserts along with the palatopharyngeus muscle. The **palatopharyngeus** muscle also descends in the lateral wall of the pharynx and inserts along with the stylopharyngeus. All three muscles of the longitudinal layer elevate the pharynx and larynx during swallowing and speaking. Elevation of the pharynx widens it to receive food and liquids, and elevation of the larynx causes a structure called the epiglottis to close over the rima glottidis (space between the vocal folds) and seal the respiratory passageway. Food and drink are further kept out of the respiratory tract by the suprahyoid muscles of the larynx, which elevate the hyoid bone. Additionally, the respiratory passageway is sealed by the action of the intrinsic muscles of the larynx, which bring the vocal folds together to close off the rima glottidis. After swallowing, the infrahyoid muscles of the larynx depress the hyoid bone and larynx.

Innervation

All muscles of the pharynx except the stylopharyngeus are innervated by the pharyngeal plexus, branches of the vagus (X) nerve. The stylopharyngeus is innervated by the glossopharyngeal (IX) nerve.

Relating Muscles to Movements

Arrange the muscles in this exhibit according to the following actions on the pharynx: (1) constriction and (2) elevation, and according to the following action on the larynx: elevation. The same muscle may be mentioned more than once.

■ CHECKPOINT

1. What happens when the pharynx and larynx are elevated?

MUSCLE	ORIGIN	INSERTION	ACTION
CIRCULAR LAYER			
Inferior constrictor (*inferior*=below; *constrictor*=decreases diameter of a lumen)	Cricoid and thyroid cartilages of larynx.	Posterior median raphe (slender band of collage fibers) of pharynx.	Constricts inferior portion of pharynx to propel food and drink into esophagus.
Middle constrictor	Greater and lesser horns of hyoid bone and stylohyoid ligament.	Posterior median raphe of pharynx.	Constricts middle portion of pharynx to propel food and drink into esophagus.
Superior constrictor (*superior*=above)	Pterygoid process of sphenoid, pterygomandibular raphe, and medial surface of mandible.	Posterior median raphe of pharynx.	Constricts superior portion of pharynx to propel food and drink into esophagus.
LONGITUDINAL LAYER			
Stylopharyngeus (stī′-lō-far-IN-jē-us; *stylo-*= stake or pole; styloid process of temporal bone; *pharyngo-*=pharynx (see also Figure 11.7)	Styloid process of temporal bone.	Thyroid cartilage with the palatopharyngeus.	Elevates larynx and pharynx.
Salpingopharyngeus (sal-pin′-gō-far-IN-jē-us; *salping-*=pertaining to the auditory or uterine tube)	Inferior portion of auditory (eustachian) tube.	Thyroid cartilage with the palatopharyngeus.	Elevates larynx and pharynx and opens orifice of auditory (eustachian) tube.
Palatopharyngeus (pal′-a-tō-far-IN-jē-us; *palato-*=palate)	Soft palate.	Thyroid cartilage with the stylopharyngeus.	Elevates larynx and pharynx and helps close nasopharynx during swallowing.

continues

315

EXHIBIT 11.8   MUSCLES OF THE PHARYNX (THROAT) (Figure 11.11)

CONTINUED

Figure 11.11   Muscles of the pharynx.

The muscles of the pharynx assist in swallowing and in closing off the respiratory passageway.

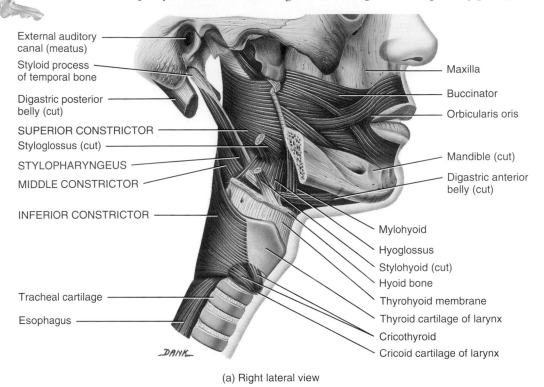

External auditory canal (meatus)
Styloid process of temporal bone
Digastric posterior belly (cut)
SUPERIOR CONSTRICTOR
Styloglossus (cut)
STYLOPHARYNGEUS
MIDDLE CONSTRICTOR
INFERIOR CONSTRICTOR
Tracheal cartilage
Esophagus

Maxilla
Buccinator
Orbicularis oris
Mandible (cut)
Digastric anterior belly (cut)
Mylohyoid
Hyoglossus
Stylohyoid (cut)
Hyoid bone
Thyrohyoid membrane
Thyroid cartilage of larynx
Cricothyroid
Cricoid cartilage of larynx

(a) Right lateral view

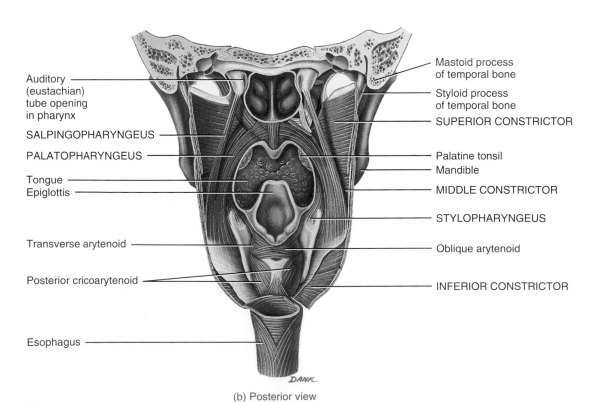

Auditory (eustachian) tube opening in pharynx
SALPINGOPHARYNGEUS
PALATOPHARYNGEUS
Tongue
Epiglottis
Transverse arytenoid
Posterior cricoarytenoid
Esophagus

Mastoid process of temporal bone
Styloid process of temporal bone
SUPERIOR CONSTRICTOR
Palatine tonsil
Mandible
MIDDLE CONSTRICTOR
STYLOPHARYNGEUS
Oblique arytenoid
INFERIOR CONSTRICTOR

(b) Posterior view

What are the antagonistic roles of the longitudinal muscles of the pharynx and the infrahyoid muscles?

EXHIBIT 11.9   MUSCLES THAT ACT ON THE ABDOMINAL WALL (Figure 11.12)

OBJECTIVE

● Describe the origin, insertion, action, and innervation of the muscles that act on the abdominal wall.

The anterolateral abdominal wall is composed of skin, fascia, and four pairs of muscles: the external oblique, internal oblique, transversus abdominis, and rectus abdominis. The first three muscles are arranged from superficial to deep. The **external oblique** is the superficial muscle. Its fascicles extend inferiorly and medially. The **internal oblique** is the intermediate flat muscle. Its fascicles extend at right angles to those of the external oblique. The **transversus abdominis** is the deep muscle, with most of its fascicles directed transversely around the abdominal wall. Together, the external oblique, internal oblique, and transversus abdominis form three layers of muscle around the abdomen. In each layer, the muscle fascicles extend in a different direction. This is a structural arrangement that affords considerable protection to the abdominal viscera, especially when the muscles have good tone.

The **rectus abdominis** muscle is a long muscle that extends the entire length of the anterior abdominal wall, from the pubic crest and pubic symphysis to the cartilages of ribs 5–7 and the xiphoid process of the sternum. The anterior surface of the muscle is interrupted by three transverse fibrous bands of tissue called **tendinous intersections,** believed to be remnants of septa that separated myotomes during embryological development (see Figure 10.10 on page 282).

As a group, the muscles of the anterolateral abdominal wall help contain and protect the abdominal viscera; flex, laterally flex, and rotate the vertebral column at the intervertebral joints; compress the abdomen during forced exhalation; and produce the force required for defecation, urination, and childbirth.

The aponeuroses of the external oblique, internal oblique, and transversus abdominis muscles form the **rectus sheaths,** which enclose the rectus abdominis muscles. The sheaths meet at the midline to form the **linea alba** (=white line), a tough, fibrous band that extends from the xiphoid process of the sternum to the pubic symphysis. In the latter stages of pregnancy, the linea alba stretches to increase the distance between the rectus abdominis muscles. The inferior free border of the external oblique aponeurosis forms the **inguinal ligament,** which runs from the anterior superior iliac spine to the pubic tubercle (see Figure 11.22a). Just superior to the medial end of the inguinal ligament is a triangular slit in the aponeurosis referred to as the **superficial inguinal ring,** the outer opening of the **inguinal canal** (see Figure 27.2 on page 838). The canal contains the spermatic cord and ilioinguinal nerve in males, and the round ligament of the uterus and ilioinguinal nerve in females.

The posterior abdominal wall is formed by the lumbar vertebrae, parts of the ilia of the hip bones, psoas major and iliacus muscles (described in Exhibit 11.19), and quadratus lumborum muscle. Whereas the anterolateral abdominal wall can contract and distend, the posterior abdominal wall is bulky and stable by comparison.

Innervation

- Rectus abdominis: thoracic spinal nerves T7–T12

- External oblique: thoracic spinal nerves T7–T12 and the iliohypogastric nerve

- Internal oblique and transversus abdominis: thoracic spinal nerves T8–T12, the iliohypogastric nerve, and the ilioinguinal nerve

- Quadratus lumborum: thoracic spinal nerve T12 and lumbar spinal nerves L1–L3 or L1–L4

 INGUINAL HERNIA

A hernia is a lump that can be seen or felt through the skin's surface. The inguinal region is a weak area in the abdominal wall. It is often the site of an **inguinal hernia,** a rupture or separation of a portion of the inguinal area of the abdominal wall resulting in the protrusion of a part of the small intestine. Hernia is much more common in males than in females because the inguinal canals in males are larger to accommodate the spermatic cord and ilioinguinal nerve. Treatment of hernias most often involves surgery. The organ that protrudes is "tucked" back into the abdominal cavity and the defect in the abdominal muscle is repaired. In addition, a mesh is often applied to reinforce the area of weakness. ■

Relating Muscles to Movements

Arrange the muscles in this exhibit according to the following actions on the vertebral column: (1) flexion, (2) lateral flexion, (3) extension, and (4) rotation. The same muscle may be mentioned more than once.

▬ **CHECKPOINT**

1. Which muscles do you contract when you "suck in your tummy," thereby compressing the anterior abdominal wall?

continues

EXHIBIT 11.9 MUSCLES THAT ACT ON THE ABDOMINAL WALL (Figure 11.12)

CONTINUED

MUSCLE	ORIGIN	INSERTION	ACTION
RECTUS ABDOMINIS (REK-tus ab-DOM-in-is; *rectus-*=fascicles parallel to midline; *abdomin*=abdomen)	Pubic crest and pubic symphysis.	Cartilage of fifth to seventh ribs and xiphoid process.	Flexes vertebral column,* especially lumbar portion, and compresses abdomen to aid in defecation, urination, forced exhalation, and childbirth.
EXTERNAL OBLIQUE (ō-BLĒK; *external*=closer to surface; *oblique*=fascicles diagonal to midline)	Inferior eight ribs.	Iliac crest and linea alba.	Acting together (bilaterally), compress abdomen and flex vertebral column; acting singly (unilaterally), laterally flex vertebral column, especially lumbar portion, and rotate vertebral column.
INTERNAL OBLIQUE (ō-BLEK; *internal*=farther from surface)	Iliac crest, inguinal ligament, and thoracolumbar fascia.	Cartilage of last three or four ribs and linea alba.	Acting together, compress abdomen and flex vertebral column; acting singly, laterally flex vertebral column, especially lumbar portion, and rotate vertebral column.
TRANSVERSUS ABDOMINIS (tranz-VER-sus ab-DOM-in-is; *transverse*=fascicles perpendicular to midline)	Iliac crest, inguinal ligament, lumbar fascia, and cartilages of inferior six ribs.	Xiphoid process, linea alba, and pubis.	Compresses abdomen.
QUADRATUS LUMBORUM (kwod-RĀ-tus lum-BOR-um; *quad-*=four; *lumbo-*=lumbar region) (see Figure 11.13)	Iliac crest and iliolumbar ligament.	Inferior border of twelfth rib and first four lumbar vertebrae.	Acting together, pull twelfth ribs inferiorly during forced exhalation, fix twelfth ribs to prevent their elevation during deep inhalation, and help extend lumbar portion of vertebral column; acting singly, laterally flex vertebral column, especially lumbar portion.

* The attachments of a muscle (origin and insertion) may be reversed, depending on the action. For example, when the rectus abdominis muscles act from below, they flex the vertebral column. However, when they act from above, they flex the pelvis on the vertebral column.

Figure 11.12 Muscles of the male anterolateral abdominal wall.

The anterolateral abdominal muscles protect the abdominal viscera, move the vertebral column, and assist in forced expiration, defecation, urination, and childbirth.

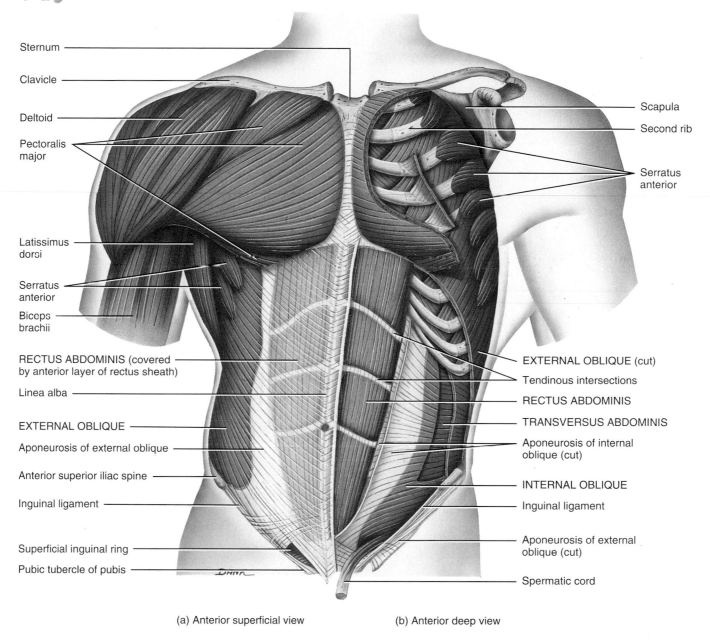

Sternum

Clavicle

Deltoid

Pectoralis major

Latissimus dorsi

Serratus anterior

Biceps brachii

RECTUS ABDOMINIS (covered by anterior layer of rectus sheath)

Linea alba

EXTERNAL OBLIQUE

Aponeurosis of external oblique

Anterior superior iliac spine

Inguinal ligament

Superficial inguinal ring

Pubic tubercle of pubis

Scapula

Second rib

Serratus anterior

EXTERNAL OBLIQUE (cut)

Tendinous intersections

RECTUS ABDOMINIS

TRANSVERSUS ABDOMINIS

Aponeurosis of internal oblique (cut)

INTERNAL OBLIQUE

Inguinal ligament

Aponeurosis of external oblique (cut)

Spermatic cord

(a) Anterior superficial view

(b) Anterior deep view

continues

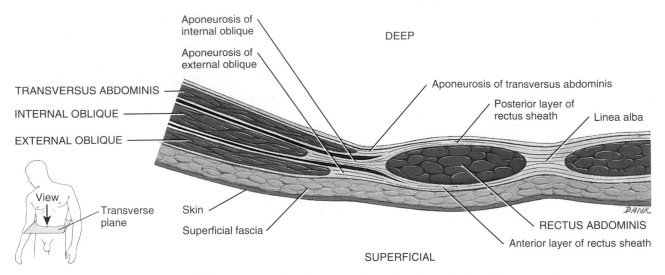

(c) Transverse section of anterior abdominal wall superior to umbilicus (navel)

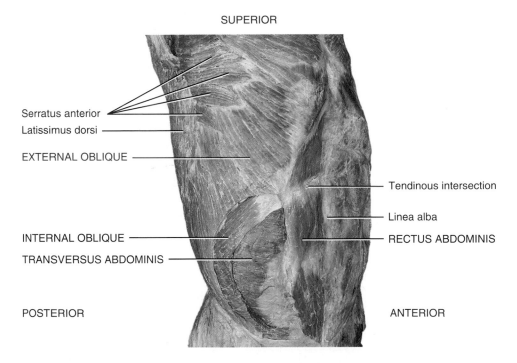

(d) Right anterolateral view

Which abdominal muscle aids in urination?

EXHIBIT 11.10   MUSCLES USED IN BREATHING (Figure 11.13)

OBJECTIVE

● Describe the origin, insertion, action, and innervation of the muscles used in breathing.

The muscles described here alter the size of the thoracic cavity so that breathing can occur. Inhalation (breathing in) occurs when the thoracic cavity increases in size, and exhalation (breathing out) occurs when the thoracic cavity decreases in size.

The dome-shaped **diaphragm** is the most important muscle that powers breathing. It also separates the thoracic and abdominal cavities. The diaphragm has a convex superior surface that forms the floor of the thoracic cavity (Figure 11.13c) and a concave, inferior surface that forms the roof of the abdominal cavity (Figure 11.13d). The **peripheral muscular portion** of the diaphragm originates on the xiphoid process of the sternum, the inferior six ribs and their costal cartilages, and the lumbar vertebrae and their intervertebral discs and the twelfth rib (Figure 11.13d). From their various origins, the fibers of the muscular portion converge and insert into the **central tendon,** a strong aponeurosis located near the center of the muscle (Figure 11.13c, d). The central tendon fuses with the inferior surface of the pericardium (covering of the heart) and the pleurae (coverings of the lungs).

The diaphragm has three major openings through which various structures pass between the thorax and abdomen. These structures include the aorta, along with the thoracic duct and azygos vein, which pass through the **aortic hiatus;** the esophagus with accompanying vagus (X) cranial nerves, which pass through the **esophageal hiatus;** and the inferior vena cava, which passes through the **caval opening (foramen for the vena cava).** In a condition called a hiatus hernia, the stomach protrudes superiorly through the esophageal hiatus.

Movements of the diaphragm also help return venous blood passing through the abdomen to the heart. Together with the anterolateral abdominal muscles, the diaphragm helps to increase intra-abdominal pressure to evacuate the pelvic contents during defecation, urination, and childbirth. This mechanism is further assisted when you take a deep breath and close the rima glottidis (the space between vocal folds). The trapped air in the respiratory system prevents the diaphragm from elevating. The increase in intra-abdominal pressure will also help support the vertebral column and prevent flexion during weightlifting. This greatly assists the back muscles in lifting a heavy weight.

Other muscles involved in breathing are called **intercostal muscles.** They span the intercostal spaces, the spaces between ribs. These muscles are arranged in three layers. The 11 **external intercostal muscles** occupy the superficial layer, and their fibers run obliquely inferiorly and anteriorly from the rib above to the rib below. They elevate the ribs during inhalation to help expand the thoracic cavity. The 11 **internal intercostal muscles** occupy the intermediate layer of the intercostal spaces. The fibers of these muscles run obliquely inferiorly and posteriorly from the inferior border of the rib above to the superior border of the rib below. They draw adjacent ribs together during forced exhalation to help decrease the size of the thoracic cavity. The deepest muscle layer is made up of the **innermost intercostal muscles.** These poorly developed muscles (not illustrated) extend in the same direction as the internal intercostals and may have the same role.

As you will see in Chapter 24, the diaphragm and external intercostal muscles are used during quiet inhalation and exhalation. However, during deep, forceful inhalation (during exercise or playing a wind instrument), the sternocleidomastoid, scalene, and pectoralis minor muscles are also used; during deep, forceful exhalation, the external oblique, internal oblique, transversus abdominis, rectus abdominis, and internal intercostals are also used.

Innervation

● Diaphragm: phrenic nerve, which contains axons from cervical spinal nerves C3–C5

● External and internal intercostals: thoracic spinal nerves T2–T12

Relating Muscles to Movements

Arrange the muscles in this exhibit according to the following actions on the size of the thorax: (1) increase in vertical dimension, (2) increase in lateral and anteroposterior dimensions, and (3) decrease in lateral and anteroposterior dimensions.

■ **CHECKPOINT**

1. What are the names of the three openings in the diaphragm, and which structures pass through each?

MUSCLE	ORIGIN	INSERTION	ACTION
DIAPHRAGM (DĪ-a-fram; *dia-*=across; *-phragm*=wall)	Xiphoid process of the sternum, costal cartilages and adjacent portions of the inferior six ribs, lumbar vertebrae and their intervertebral discs, and the twelfth rib.	Central tendon.	Contraction of the diaphragm causes it to flatten and increases the vertical dimension of the thoracic cavity, resulting in inhalation; relaxation of the diaphragm causes it to move superiorly and decreases the vertical dimension of the thoracic cavity, resulting in exhalation.
EXTERNAL INTERCOSTALS (in'-ter-KOS-tals; *external*=closer to surface; *inter-*=between; *costa-*=rib)	Inferior border of rib above.	Superior border of rib below.	Contraction elevates the ribs and increases the anteroposterior and lateral dimensions of thoracic cavity, resulting in inhalation; the relaxation depresses the ribs and decreases the anteroposterior and lateral dimensions of the thoracic cavity, resulting in exhalation.
INTERNAL INTERCOSTALS (in'-ter-KOS-tals; *internal*=farther from surface)	Superior border of rib below.	Inferior border of rib above.	Contraction draws adjacent ribs together to further decrease the anteroposterior and lateral dimensions of the thoracic cavity during forced exhalation.

continues

EXHIBIT 11.10   MUSCLES USED IN BREATHING (Figure 11.13)

Figure 11.13   Muscles used in breathing, as seen in a male.

Openings in the diaphragm permit the passage of the aorta, esophagus, and inferior vena cava.

Clavicle

INTERNAL INTERCOSTALS

EXTERNAL INTERCOSTALS

Pectoralis minor (cut)

Ribs

External oblique (cut)

Rectus abdominis (cut)

Transversus abdominis and aponeurosis

Rectus abdominis (covered by anterior layer of rectus sheath [cut])

Linea alba

Internal oblique

Aponeurosis of internal oblique

Anterior superior iliac spine

Inguinal ligament

Spermatic cord

Ribs

EXTERNAL INTERCOSTALS

INTERNAL INTERCOSTALS

Sternum

Central tendon

DIAPHRAGM

Arcuate ligament of diaphragm

QUADRATUS LUMBORUM

Transversus abdominis

Fourth lumbar vertebra

Iliac crest

Sacrum

Pubis

Pubic symphysis

DANK

(a) Anterior superficial view   (b) Anterior deep view

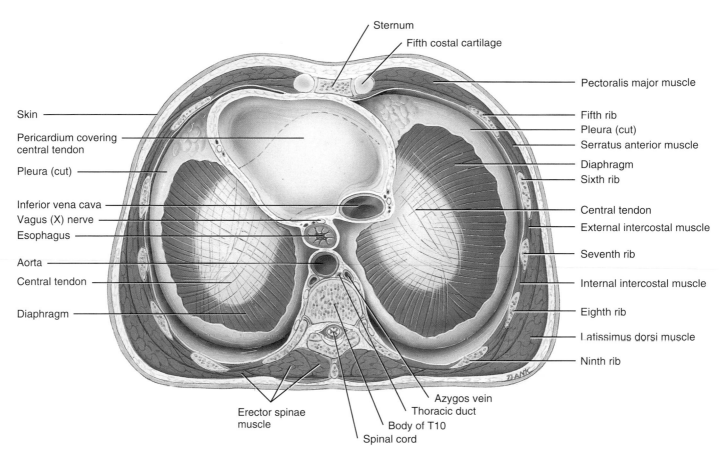

Sternum

Fifth costal cartilage

Pectoralis major muscle

Skin

Pericardium covering
central tendon

Pleura (cut)

Inferior vena cava

Vagus (X) nerve

Esophagus

Aorta

Central tendon

Diaphragm

Fifth rib

Pleura (cut)

Serratus anterior muscle

Diaphragm

Sixth rib

Central tendon

External intercostal muscle

Seventh rib

Internal intercostal muscle

Eighth rib

Latissimus dorsi muscle

Ninth rib

Erector spinae
muscle

Azygos vein

Thoracic duct

Body of T10

Spinal cord

(c) Superior view of diaphragm

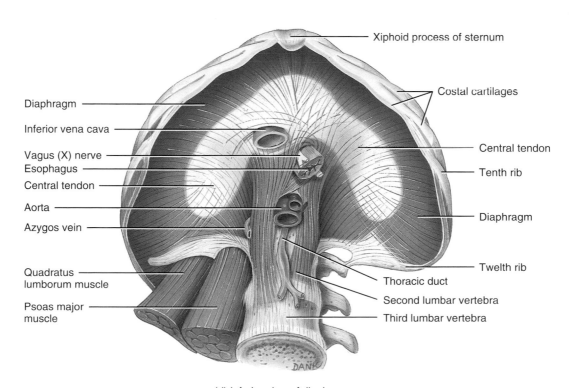

Xiphoid process of sternum

Diaphragm

Inferior vena cava

Vagus (X) nerve

Esophagus

Central tendon

Aorta

Azygos vein

Quadratus
lumborum muscle

Psoas major
muscle

Costal cartilages

Central tendon

Tenth rib

Diaphragm

Twelth rib

Thoracic duct

Second lumbar vertebra

Third lumbar vertebra

(d) Inferior view of diaphragm

? **Which muscle associated with breathing is innervated by the phrenic nerve?**

323

EXHIBIT 11.11   MUSCLES OF THE PELVIC FLOOR (Figure 11.14)

OBJECTIVE

● Describe the origin, insertion, action, and innervation of the muscles of the pelvic floor.

The muscles of the pelvic floor are the levator ani and coccygeus. Together with the fascia covering their internal and external surfaces, these muscles are referred to as the **pelvic diaphragm,** which stretches from the pubis anteriorly to the coccyx posteriorly, and from one lateral wall of the pelvis to the other. This arrangement gives the pelvic diaphragm the appearance of a funnel suspended from its attachments. The anal canal and urethra pierce the pelvic diaphragm in both sexes, and the vagina goes through it in females.

The two components of the **levator ani** muscle are the **pubococcygeus** and **iliococcygeus.** Figure 11.14 shows these muscles in the female and Figure 11.15 illustrates them in the male. The levator ani is the largest and most important muscle of the pelvic floor. It supports the pelvic viscera and resists the inferior thrust that accompanies increases in intra-abdominal pressure during functions such as forced exhalation, coughing, vomiting, urination, and

defecation. The muscle also functions as a sphincter at the anorectal junction, urethra, and vagina. In addition to assisting the levator ani, the **coccygeus** pulls the coccyx anteriorly after it has been pushed posteriorly during defecation or childbirth.

Innervation

- Pubococcygeus and iliococcygeus: sacral spinal nerves S2–S4
- Coccygeus: sacral spinal nerves S4–S5

INJURY OF LEVATOR ANI AND STRESS INCONTINENCE

During childbirth, the levator ani muscle supports the head of the fetus, and the muscle may be injured during a difficult childbirth or traumatized during an *episiotomy* (a cut made with surgical scissors to prevent or direct tearing of the perineum during the birth of a

MUSCLE	ORIGIN	INSERTION	ACTION
LEVATOR ANI (le-VĀ-tor Ā-nē; *levator*=raises; *ani*=anus)	This muscle is divisible into two parts, the pubococcygeus muscle and the iliococcygeus muscle.		
PUBOCOCCYGEUS (pū´-bō-kok-SIJ-ē-us; *pubo-*=pubis; *coccygeus*=coccyx)	Pubis.	Coccyx, urethra, anal canal, perineal body of the perineum (a wedge-shaped mass of fibrous tissue in the center of the perineum), and anococcygeal raphe (narrow fibrous band that extends from anus to coccyx).	Supports and maintains position of pelvic viscera; resists increase in intra-abdominal pressure during forced exhalation, coughing, vomiting, urination, and defecation; constricts anus, urethra, and vagina; and supports fetal head during childbirth.
ILIOCOCCYGEUS (il´-ē-ō-kok-SIJ-ē-us; *ilio-*=ilium)	Ischial spine.	Coccyx.	As above.
COCCYGEUS (kok-SIJ-ē-us)	Ischial spine.	Lower sacrum and upper coccyx.	Supports and maintains position of pelvic viscera; resists increase in intra-abdominal pressure during forced exhalation, coughing, vomiting, urination, and defecation; and pulls coccyx anteriorly following defecation or childbirth.

baby). The consequence of such injury may be urinary stress incontinence, leakage of urine whenever intra-abdominal pressure is increased — for example, during coughing. One way to treat urinary stress incontinence is to strengthen and tighten the muscles that support the pelvic viscera. This is accomplished by *Kegel exercises,* the alternate contraction and relaxation of muscles of the pelvic floor. To find the right muscles, the person imagines that she is urinating and then stops in midstream. Next, hold for a count of three, then relax for a count of three. This should be done 5–10 times each hour — sitting, standing, and lying down. Kegel exercises are also encouraged during pregnancy to strengthen the muscles for delivery. ■

Relating Muscles to Movements

Arrange the muscles in this exhibit according to the following actions: (1) supporting and maintaining the position of the pelvic viscera, and resisting an increase in intra-abdominal pressure; (2) constriction of the anus, urethra, and vagina; (3) expulsion of urine and semen, (4) erection of the clitoris and penis. The same muscle may be mentioned more than once.

■ **CHECKPOINT**

1. Which muscles are strengthened by Kegel exercises?

Figure 11.14   Muscles of the pelvic floor, as seen in the female perineum.

 The pelvic diaphragm supports the pelvic viscera.

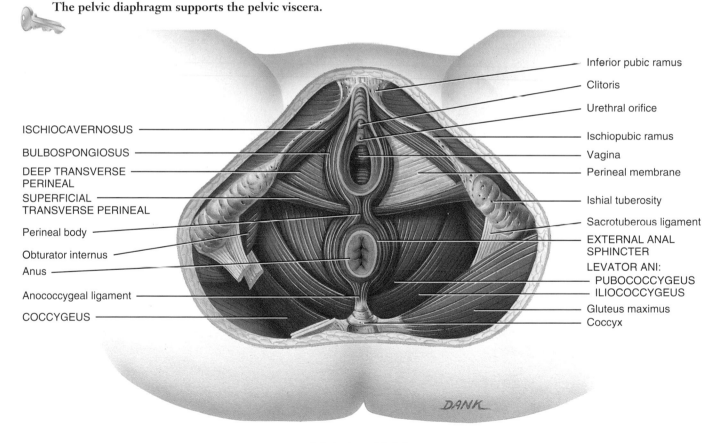

ISCHIOCAVERNOSUS

BULBOSPONGIOSUS

DEEP TRANSVERSE PERINEAL

SUPERFICIAL TRANSVERSE PERINEAL

Perineal body

Obturator internus

Anus

Anococcygeal ligament

COCCYGEUS

Inferior pubic ramus

Clitoris

Urethral orifice

Ischiopubic ramus

Vagina

Perineal membrane

Ishial tuberosity

Sacrotuberous ligament

EXTERNAL ANAL SPHINCTER

LEVATOR ANI:
PUBOCOCCYGEUS
ILIOCOCCYGEUS

Gluteus maximus

Coccyx

Inferior superficial view

 What are the borders of the pelvic diaphragm?

EXHIBIT 11.12   MUSCLES OF THE PERINEUM (Figures 11.14 and 11.15)

• Describe the origin, insertion, action, and innervation of the muscles of the perineum.

The **perineum** is the region of the trunk inferior to the pelvic diaphragm. It is a diamond-shaped area that extends from the pubic symphysis anteriorly, to the coccyx posteriorly, and to the ischial tuberosities laterally. The female and the male perineums may be compared in Figures 11.14 and 11.15, respectively. A transverse line drawn between the ischial tuberosities divides the perineum into an anterior **urogenital triangle** that contains the external genitals and a posterior **anal triangle** that contains the anus (see Figure 27.22 on page 864). Several perineal muscles insert into the perineal body of the perineum (described on page 324). Clinically, the perineum is very important to physicians who care for women during pregnancy and who treat disorders related to the female genital tract, urogenital organs, and the anorectal region.

The muscles of the perineum are arranged in two layers: **superficial** and **deep**. The muscles of the superficial layer are the **superficial transverse perineal muscle,** the **bulbospongiosus,** and the **ischiocavernosus.** The deep muscles of the perineum are the **deep transverse perineal muscle** and the **external urethral sphincter.** The deep muscles of the perineum assist in urination and ejaculation in males and urination in females. The **external anal sphincter** closely adheres to the skin around the margin of the anus and keeps the anal canal and anus closed except during defecation.

Innervation

• All muscles of the perineum except external anal sphincter: perineal branch of the pudendal nerve of the sacral plexus (shown in Exhibit 18.4 on page 575).

• External anal sphincter: sacral spinal nerve S4 and the inferior rectal branch of the pudendal nerve.

Relating Muscles to Movements

Arrange the muscles in this exhibit according to the following actions: (1) expulsion of urine and semen, (2) erection of the clitoris and penis, (3) closing the anal orifice, and (4) constricting the vaginal orifice. The same muscle may be mentioned more than once.

■ CHECKPOINT

1. What are the borders and contents of the urogenital triangle and the anal triangle?

MUSCLE	ORIGIN	INSERTION	ACTION
SUPERFICIAL PERINEAL MUSCLES			
Superficial transverse perineal (per-i-NĒ-al; *superficial*=closer to surface; *transverse*=across; *perineus*=perineum)	Ischial tuberosity.	Perineal body of perineum.	Helps stabilize perineal body of perineum.
Bulbospongiosus (bul′-bō-spon′-jē-Ō-sus; *bulb*=a bulb; *spongio-*=sponge)	Perineal body of perineum.	Perineal membrane of deep muscles of perineum, corpus spongiosum of penis, and deep fascia on dorsum of penis in male; pubic arch and root and dorsum of clitoris in female.	Helps expel urine during urination, helps propel semen along urethra, assists in erection of the penis in male; constricts vaginal orifice and assists in erection of clitoris in female.
Ischiocavernosus (is′-kē-ō-ka′-ver-NŌ-sus; *ischio-*=the hip)	Ischial tuberosity and ischial and pubic rami.	Corpus cavernosum of penis in male and clitoris in female.	Maintains erection of penis in male and clitoris in female.
DEEP PERINEAL MUSCLES			
Deep transverse perineal (*deep*=farther from surface)	Ischial rami.	Perineal body of perineum.	Helps expel last drops of urine and semen in male and urine in female.
External urethral sphincter (ū-RĒ-thral SFINGK-ter)	Ischial and pubic rami.	Median raphe in male and vaginal wall in female.	Helps expel last drops of urine and semen in male and urine in female.
External anal sphincter (Ā-nal)	Anococcygeal ligament.	Perineal body of perineum.	Keeps anal canal and anus closed.

Figure 11.15 Muscles of the male perineum.

The urogenital diaphragm assists in urination in females and males, ejaculation in males, and helps strengthen the pelvic floor.

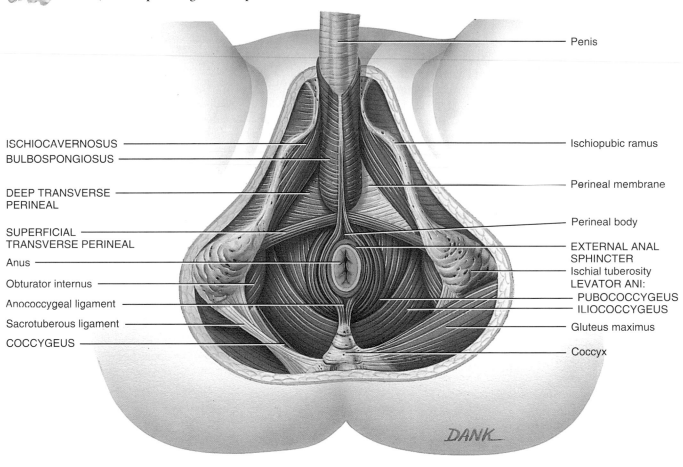

ISCHIOCAVERNOSUS

BULBOSPONGIOSUS

DEEP TRANSVERSE PERINEAL

SUPERFICIAL TRANSVERSE PERINEAL

Anus

Obturator internus

Anococcygeal ligament

Sacrotuberous ligament

COCCYGEUS

Penis

Ischiopubic ramus

Perineal membrane

Perineal body

EXTERNAL ANAL SPHINCTER

Ischial tuberosity

LEVATOR ANI:
PUBOCOCCYGEUS
ILIOCOCCYGEUS

Gluteus maximus

Coccyx

Inferior superficial view

? What are the borders of the perineum?

EXHIBIT 11.13 MUSCLES THAT MOVE THE PECTORAL (SHOULDER) GIRDLE (Figure 11.16)

OBJECTIVE

● Describe the origin, insertion, action, and innervation of the muscles that move the pectoral girdle.

The main action of the muscles that move the pectoral girdle is to stabilize the scapula so it can function as a steady origin for most of the muscles that move the humerus. Because scapular movements usually accompany humeral movements in the same direction, the muscles also move the scapula to increase the range of motion of the humerus. For example, it would not be possible to abduct the humerus past the horizontal if the scapula did not move with the humerus. During abduction, the scapula follows the humerus by rotating upward.

Muscles that move the pectoral girdle can be classified into two groups based on their location in the thorax: **anterior** and **posterior thoracic muscles.** The anterior thoracic muscles are the subclavius, pectoralis minor, and serratus anterior. The **subclavius** is a small, cylindrical muscle under the clavicle that extends from the clavicle to the first rib. It steadies the clavicle during movements of the pectoral girdle. The **pectoralis minor** is a thin, flat, triangular muscle that is deep to the pectoralis major. Besides its role in movements of the scapula, the pectoralis minor muscle assists in forced inhalation. The **serratus anterior** is a large, flat, fan-shaped muscle between the ribs and scapula. It is named because of the saw-toothed appearance of its origins on the ribs.

The posterior thoracic muscles are the trapezius, levator scapulae, rhomboid major, and rhomboid minor. The **trapezius** is a large, flat, triangular sheet of muscle extending from the skull and vertebral column medially to the pectoral girdle laterally. It is the most superficial back muscle and covers the posterior neck region and superior portion of the trunk. The two trapezius muscles form a trapezium (diamond-shaped quadrangle)—hence its name. The **levator scapulae** is a narrow, elongated muscle in the posterior portion of the neck. It is deep to the sternocleidomastoid and trapezoid muscles. As its name suggests, one of its actions is to elevate the scapula. The **rhomboid major** and **rhomboid minor** lie deep to the trapezius and are not always distinct from each other. They appear as parallel bands that pass inferiolaterally from the vertebrae to the scapula. They are named based on their shape—that is, a rhomboid (an oblique parallelogram). The rhomboid major is about two times wider than the rhomboid minor. Both muscles are used when forcibly lowering the raised upper limbs, as in driving a stake with a sledgehammer.

To understand the actions of muscles that move the scapula, it is first helpful to describe the various movements of the scapula:

• **Elevation:** superior movement of the scapula, such as shrugging the shoulders or lifting a weight over the head.

• **Depression:** inferior movement of the scapula, as in doing a "pull-up."

• **Abduction (protraction):** movement of the scapula laterally and anteriorly, as in doing a "push-up" or punching.

• **Adduction (retraction):** movement of the scapula medially and posteriorly, as in pulling the oars in a rowboat.

• **Upward rotation:** movement of the inferior angle of the scapula laterally so that the glenoid cavity is moved upward. This movement is required to abduct the humerus past the horizontal.

• **Downward rotation:** movement of the inferior angle of the scapula medially so that the glenoid cavity is moved downward. This movement is seen when a gymnast on parallel bars supports the weight of the body on the hands.

Innervation

Muscles that move the shoulder receive their innervation mainly from nerves that emerge from the cervical and brachial plexuses (shown in Exhibits 18.1 and 18.2 on pages 566 and 568).

• Subclavius: subclavian nerve

• Pectoralis minor: medial pectoral nerve

• Serratus anterior: long thoracic nerve

• Trapezius: accessory (XI) nerve and cervical spinal nerves C3–C5

• Levator scapulae: dorsal scapular nerve and cervical spinal nerves C3–C5

• Rhomboid major and rhomboid minor: dorsal scapular nerve

Relating Muscles to Movements

Arrange the muscles in this exhibit according to the following actions on the scapula: (1) depression, (2) elevation, (3) abduction, (4) adduction, (5) upward rotation, and (6) downward rotation. The same muscle may be mentioned more than once.

■ **CHECKPOINT**

1. What muscles in this exhibit are used to raise your shoulders, lower your shoulders, join your hands behind your back, and join your hands in front of your chest?

MUSCLE	ORIGIN	INSERTION	ACTION
ANTERIOR THORACIC MUSCLES			
Subclavius (sub-KLĀ-vē-us; *sub-*=under; *clavius*=clavicle)	First rib.	Clavicle.	Depresses and moves clavicle anteriorly and helps stabilize pectoral girdle.
Pectoralis minor (pek'-tō-RĀ-lis; *pector-*=the breast, chest, thorax; *minor*=lesser)	Second through fifth, third through fifth, or second through fourth ribs.	Coracoid process of scapula.	Abducts scapula and rotates it downward; elevates third through fifth ribs during forced inhalation when scapula is fixed.
Serratus anterior (ser-Ā-tus; *serratus*=saw-toothed; *anterior*=front)	Superior eight or nine ribs.	Vertebral border and inferior angle of scapula.	Abducts scapula and rotates it upward; elevates ribs when scapula is stabilized; known as "boxer's muscle" because it is important in horizontal arm movements such as punching and pushing.
POSTERIOR THORACIC MUSCLES			
Trapezius (tra-PĒ-zē-us; *trapezi-*=trapezoid-shaped)	Superior nuchal line of occipital bone, ligamentum nuchae, and spines of seventh cervical and all thoracic vertebrae.	Clavicle and acromion and spine of scapula.	Superior fibers elevate scapula and can help extend head; middle fibers adduct scapula; inferior fibers depress scapula; superior and inferior fibers together rotate scapula upward; stabilizes scapula.
Levator scapulae (le-VĀ-tor SKA-pū-lē; *levator*=raises; *scapulae*=of the scapula)	Superior four or five cervical vertebrae.	Superior vertebral border of scapula.	Elevates scapula and rotates it downward.
Rhomboid major (rom-BOYD; *rhomboid*=rhomboid or diamond-shaped)	Spines of second to fifth thoracic vertebrae.	Vertebral border of scapula inferior to spine.	Elevates and adducts scapula and rotates it downward; stabilizes scapula.
Rhomboid minor	Spines of seventh cervical and first thoracic vertebrae.	Vertebral border of scapula superior to spine.	Elevates and adducts scapula and rotates it downward; stabilizes scapula.

continues

Figure 11.16 **Muscles that move the pectoral (shoulder) girdle.** (See Tortora, *A Photographic Atlas of the Human Body*, Figure 5.8.)

Muscles that move the pectoral girdle originate on the axial skeleton and insert on the clavicle or scapula.

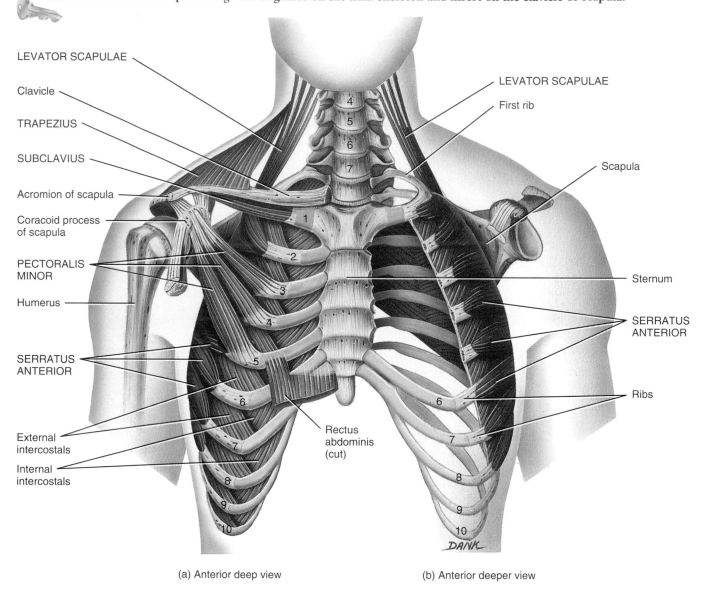

(a) Anterior deep view (b) Anterior deeper view

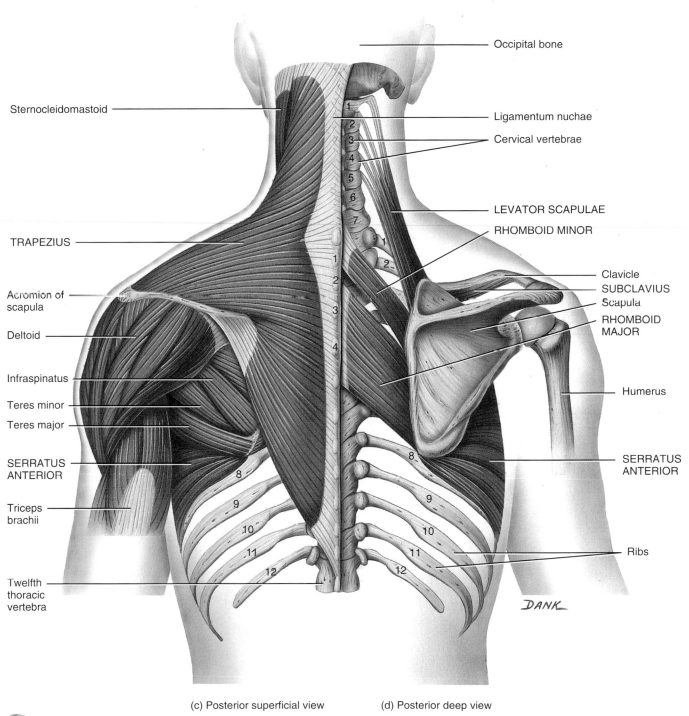

Sternocleidomastoid

Occipital bone

Ligamentum nuchae

Cervical vertebrae

1
2
3
4
5
6
7

LEVATOR SCAPULAE

RHOMBOID MINOR

TRAPEZIUS

1
2
3
4

1
2

Clavicle
SUBCLAVIUS
Scapula
RHOMBOID MAJOR

Acromion of scapula

Deltoid

Infraspinatus

Teres minor

Teres major

SERRATUS ANTERIOR

8
9
10
11
12

8
9
10
11
12

Humerus

SERRATUS ANTERIOR

Triceps brachii

Ribs

Twelfth thoracic vertebra

DANK

(c) Posterior superficial view   (d) Posterior deep view

What is the main action of the muscles that move the pectoral girdle?

EXHIBIT 11.14 MUSCLES THAT MOVE THE HUMERUS (ARM BONE) (Figure 11.17)

OBJECTIVE

- Describe the origin, insertion, action, and innervation of the muscles that move the humerus.

Of the nine muscles that cross the shoulder joint, all except the pectoralis major and latissimus dorsi originate on the scapula. The pectoralis major and latissimus dorsi thus are called **axial muscles** because they originate on the axial skeleton. The remaining seven muscles, the **scapular muscles,** arise from the scapula.

Of the two axial muscles that move the humerus, the **pectoralis major** is a large, thick, fan-shaped muscle that covers the superior part of the thorax. It has two origins: a smaller clavicular head and a larger sternocostal head. The **latissimus dorsi** is a broad, triangular muscle located on the inferior part of the back. It is commonly called the "swimmer's muscle"; because its many actions are used while swimming, many competitive swimmers have well-developed "lats."

Among the scapular muscles, the **deltoid** is a thick, powerful shoulder muscle that covers the shoulder joint and forms the rounded contour of the shoulder. This muscle is a frequent site of intramuscular injections. As you study the deltoid, note that its fascicles originate from three different points and that each group of fascicles moves the humerus differently. The **subscapularis** is a large triangular muscle that fills the subscapular fossa of the scapula and forms part of the posterior wall of the axilla. The **supraspinatus** is a rounded muscle, named for its location in the supraspinous fossa of the scapula. It lies deep to the trapezius. The **infraspinatus** is a triangular muscle, also named for its location in the infraspinous fossa of the scapula. The **teres major** is a thick, flattened muscle inferior to the teres minor that also helps form part of the posterior wall of the axilla. The **teres minor** is a cylindrical, elongated muscle, often inseparable from the infraspinatus, which lies along its superior border. The **coracobrachialis** is an elongated, narrow muscle in the arm.

The strength and stability of the shoulder joint are not provided by the shape of the articulating bones or its ligaments. Instead, four deep muscles of the shoulder—subscapularis, supraspinatus, infraspinatus, and teres minor—strengthen and stabilize the shoulder joint. These muscles join the scapula to the humerus. Their flat tendons fuse together to form the **rotator (musculotendinous) cuff,** a nearly complete circle of tendons around the shoulder joint, like the cuff on a shirtsleeve. The supraspinatus muscle is especially predisposed to wear and tear because of its location between the head of the humerus and acromion of the scapula, which compresses its tendon during shoulder movements, especially abduction of the arm.

Innervation

Nerves that emerge from the brachial plexus (shown in Exhibit 18.2 on page 568) innervate muscles that move the arm.

- Pectoralis major: medial and lateral pectoral nerves
- Latissimus dorsi: thoracodorsal nerve
- Deltoid and teres minor: axillary nerve
- Subscapularis: upper and lower subscapular nerve
- Supraspinatus and infraspinatus: suprascapular nerve
- Teres major: lower subscapular nerve
- Coracobrachialis: musculocutaneous nerve

 IMPINGEMENT SYNDROME

One of the most common causes of shoulder pain and dysfunction in athletes is known as **impingement syndrome.** Another is compartment syndrome, discussed on page 376. The repetitive movement of the arm over the head that is common in baseball, overhead racquet sports, lifting weights over the head, spiking a volleyball, and swimming puts these athletes at risk for developing impingement syndrome. It may also be caused by a direct blow or stretch injury. Continual pinching of the supraspinatus tendon as a result of overhead motions causes it to become inflamed and results in pain. If movement is continued despite the pain, the tendon may degenerate near the attachment to the humerus and ultimately may tear away from the bone (rotator cuff injury). Treatment consists of resting the injured tendons, strengthening the shoulder through exercise, and surgery if the injury is particularly severe. ■

Relating Muscles to Movements

Arrange the muscles in this exhibit according to the following actions on the humerus at the shoulder joint: (1) flexion, (2) extension, (3) abduction, (4) adduction, (5) medial rotation, and (6) lateral rotation. The same muscle may be mentioned more than once.

■ CHECKPOINT

1. Why are two muscles that cross the shoulder joint called axial muscles, and the seven others called scapular muscles?

MUSCLE	ORIGIN	INSERTION	ACTION
AXIAL MUSCLES that move the humerus			
Pectoralis major (pek′-tō-RĀ-lis; *pector-*=chest; *major*=larger) (see also Figure 11.12a)	Clavicle (clavicular head), sternum, and costal cartilages of second to sixth ribs and sometimes first to seventh ribs (sternocostal head).	Greater tubercle and lateral lip of the intertubercular sulcus of humerus.	As a whole, adducts and medially rotates arm at shoulder joint; clavicular head flexes arm, and sternocostal head extends (against resistance) the flexed arm to side of trunk at shoulder joint.
Latissimus dorsi (la-TIS-i-mus DOR-sī; *latissimus*=widest; *dorsi*=of the back)	Spines of inferior six thoracic vertebrae, lumbar vertebrae, crests of sacrum and ilium, inferior four ribs.	Intertubercular sulcus of humerus.	Extends, adducts, and medially rotates arm at shoulder joint; draws arm inferiorly and posteriorly.
SCAPULAR MUSCLES that move the humerus			
Deltoid (DEL-toyd=triangularly shaped)	Acromial extremity of clavicle (anterior fibers), acromion of scapula (lateral fibers), and spine of scapula (posterior fibers).	Deltoid tuberosity of humerus.	Lateral fibers abduct arm at shoulder joint; anterior fibers flex and medially rotate arm at shoulder joint; posterior fibers extend and laterally rotate arm at shoulder joint.
Subscapularis (sub-scap′-ū-LĀ-ris; *sub-*=below; *scapularis*=scapula)	Subscapular fossa of scapula.	Lesser tubercle of humerus.	Medially rotates arm at shoulder joint.
Supraspinatus (soo-pra-spī-NĀ-tus; *supra-*=above; *spina-*=spine [of the scapula])	Supraspinous fossa of scapula.	Greater tubercle of humerus.	Assists deltoid muscle in abducting arm at shoulder joint.
Infraspinatus (in′-fra-spī-NĀ-tus; *infra-*=below)	Infraspinous fossa of scapula.	Greater tubercle of humerus.	Laterally rotates and adducts arm at shoulder joint.
Teres major (TE-rēz; *teres*=long and round)	Inferior angle of scapula.	Medial lip of intertubercular sulcus of humerus.	Extends arm at shoulder joint and assists in adduction and medial rotation of arm at shoulder joint.
Teres minor	Inferior lateral border of scapula.	Greater tubercle of humerus.	Laterally rotates, extends, and adducts arm at shoulder joint.
Coracobrachialis (kor′-a-kō-brā-kē-Ā′-lis; *coraco-*=coracoid process [of the scapula]; *brachi-*=arm)	Coracoid process of scapula.	Middle of medial surface of shaft of humerus.	Flexes and adducts arm at shoulder joint.

continues

EXHIBIT 11.14   MUSCLES THAT MOVE THE HUMERUS (ARM BONE) (Figure 11.17)

Figure 11.17   Muscles that move the humerus (arm bone). (See Tortora, *A Photographic Atlas of the Human Body*, Figure 5.10.)

The strength and stability of the shoulder joint are provided by the tendons that form the rotator cuff.

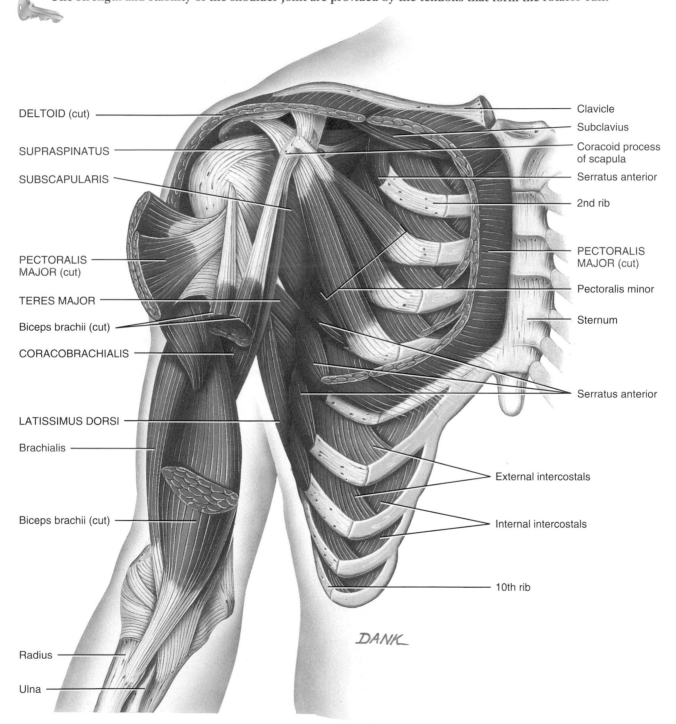

DELTOID (cut)

SUPRASPINATUS

SUBSCAPULARIS

PECTORALIS
MAJOR (cut)

TERES MAJOR

Biceps brachii (cut)

CORACOBRACHIALIS

LATISSIMUS DORSI

Brachialis

Biceps brachii (cut)

Radius

Ulna

Clavicle

Subclavius

Coracoid process
of scapula

Serratus anterior

2nd rib

PECTORALIS
MAJOR (cut)

Pectoralis minor

Sternum

Serratus anterior

External intercostals

Internal intercostals

10th rib

DANK

(a) Anterior deep view (the intact pectoralis major muscle is shown in figure 11.12a)

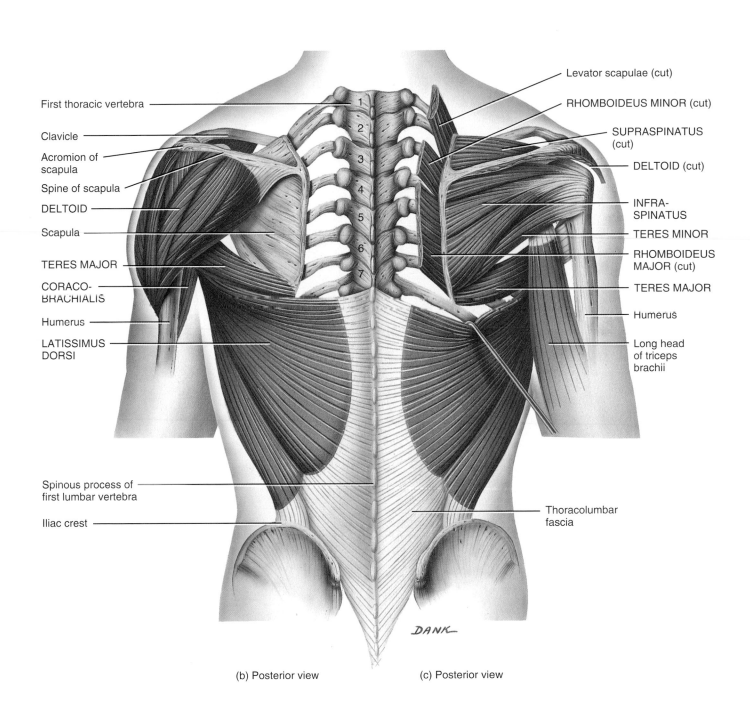

First thoracic vertebra

Clavicle

Acromion of scapula

Spine of scapula

DELTOID

Scapula

TERES MAJOR

CORACO-BRACHIALIS

Humerus

LATISSIMUS DORSI

Spinous process of first lumbar vertebra

Iliac crest

Levator scapulae (cut)

RHOMBOIDEUS MINOR (cut)

SUPRASPINATUS (cut)

DELTOID (cut)

INFRA-SPINATUS

TERES MINOR

RHOMBOIDEUS MAJOR (cut)

TERES MAJOR

Humerus

Long head of triceps brachii

Thoracolumbar fascia

1
2
3
4
5
6
7

DANK

(b) Posterior view   (c) Posterior view

continues

SUPERIOR

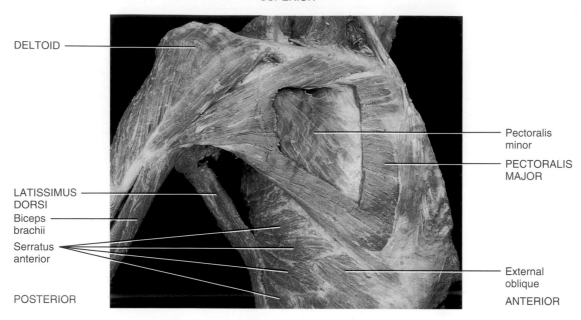

DELTOID

Pectoralis minor

PECTORALIS MAJOR

LATISSIMUS DORSI

Biceps brachii

Serratus anterior

External oblique

POSTERIOR

ANTERIOR

(d) Right lateral superficial view

SUPERIOR

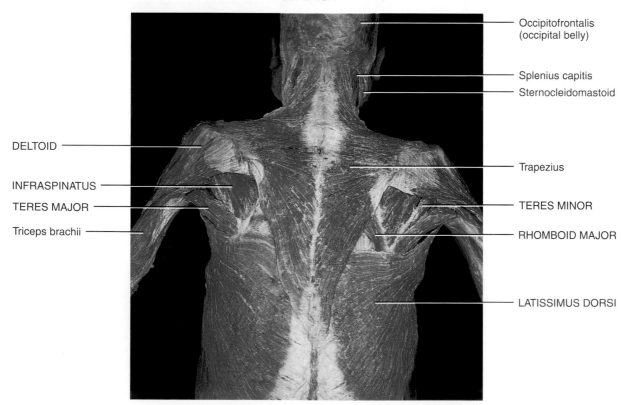

Occipitofrontalis (occipital belly)

Splenius capitis

Sternocleidomastoid

DELTOID

Trapezius

INFRASPINATUS

TERES MAJOR

TERES MINOR

Triceps brachii

RHOMBOID MAJOR

LATISSIMUS DORSI

(e) Posterior view

? Which tendons make up the rotator cuff?

● Describe the origin, insertion, action, and innervation of the muscles that move the radius and ulna.

Most of the muscles that move the radius and ulna (forearm bones) cause flexion and extension at the elbow, which is a hinge joint. The biceps brachii, brachialis, and brachioradialis muscles are the flexor muscles. The extensor muscles are the triceps brachii and the anconeus.

The **biceps brachii** is the large muscle located on the anterior surface of the arm. As indicated by its name, it has two heads of origin (long and short), both from the scapula. The muscle spans both the shoulder and elbow joints. In addition to its role in flexing the forearm at the elbow joint, it also supinates the forearm at the radioulnar joints and flexes the arm at the shoulder joint. The **brachialis** is deep to the biceps brachii muscle. It is the most powerful flexor of the forearm at the elbow joint. For this reason, it is called the "workhorse" of the elbow flexors. The **brachioradialis** flexes the forearm at the elbow joint, especially when a quick movement is required or when a weight is lifted slowly during flexion of the forearm.

The **triceps brachii** is the large muscle located on the posterior surface of the arm. It is the more powerful of the extensors of the forearm at the elbow joint. As its name implies, it has three heads of origin, one from the scapula (long head) and two from the humerus (lateral and medial heads). The long head crosses the shoulder joint; the other heads do not. The **anconeus** is a small muscle located on the lateral part of the posterior aspect of the elbow that assists the triceps brachii in extending the forearm at the elbow joint.

Some muscles that move the radius and ulna are involved in pronation and supination at the radioulnar joints. The pronators, as suggested by their names, are the **pronator teres** and **pronator quadratus** muscles. The supinator of the forearm is aptly named

the **supinator** muscle. You use the powerful action of the supinator when you twist a corkscrew or turn a screw with a screwdriver.

In the limbs, functionally related skeletal muscles and their associated blood vessels and nerves are grouped together by fascia into regions called **compartments.** In the arm, the biceps brachii, brachialis, and coracobrachialis muscles comprise the **anterior (flexor) compartment.** The triceps brachii muscle forms the **posterior (extensor) compartment.**

Innervation

Muscles that move the radius and ulna are innervated by nerves derived from the brachial plexus (shown in Exhibit 18.2 on page 568).

- Biceps brachii: musculocutaneous nerve.
- Brachialis: musculocutaneous and radial nerves.
- Brachioradialis, triceps brachii, and anconeus: radial nerve
- Pronator teres and pronator quadratus: median nerve
- Supinator: deep radial nerve

Relating Muscles to Movements

Arrange the muscles in this exhibit according to the following actions on the elbow joint: (1) flexion and (2) extension; the following actions on the forearm at the radioulnar joints: (1) supination and (2) pronation; and the following actions on the humerus at the shoulder joint: (1) flexion and (2) extension. The same muscle may be mentioned more than once.

■ **CHECKPOINT**

1. Which muscles are in the anterior and posterior compartments of the arm?

continues

CONTINUED

MUSCLE	ORIGIN	INSERTION	ACTION
FOREARM FLEXORS			
Biceps brachii (BĪ-ceps BRĀ-kē-ī; *biceps*=two heads [of origin]; *brachii*=arm)	Long head originates from tubercle above glenoid cavity of scapula (supraglenoid tubercle); short head originates from coracoid process of scapula.	Radial tuberosity of radius and bicipital aponeurosis.*	Flexes forearm at elbow joint, supinates forearm at radioulnar joints, and flexes arm at shoulder joint.
Brachialis (brā-kē-Ā-lis)	Distal, anterior surface of humerus.	Ulnar tuberosity and coronoid process of ulna.	Flexes forearm at elbow joint.
Brachioradialis (bra′-kē-ō-rā-dē-Ā-lis; *radi*=radius) (see Figure 11.19a)	Lateral border of distal end of humerus.	Superior to styloid process of radius.	Flexes forearm at elbow joint; supinates and pronates forearm at radioulnar joints to neutral position.
FOREARM EXTENSORS			
Triceps brachii (TRĪ-ceps BRĀ-kē-ī; *triceps*=three heads [of origin])	Long head: infraglenoid tubercle, a projection inferior to glenoid cavity of scapula. Lateral head: lateral and posterior surface of humerus superior to radial groove. Medial head: entire posterior surface of humerus inferior to a groove for the radial nerve.	Olecranon of ulna.	Extends forearm at elbow joint and extends arm at shoulder joint.
Anconeus (an-KŌ-nē-us; *ancon*=the elbow) (see Figure 11.19c)	Lateral epicondyle of humerus.	Olecranon and superior portion of shaft of ulna.	Extends forearm at elbow joint.
FOREARM PRONATORS			
Pronator teres (PRŌ-nā-tor TE-rēz; *pronator*=turns palm posteriorly) (see Figure 11.19a)	Medial epicondyle of humerus and coronoid process of ulna.	Midlateral surface of radius.	Pronates forearm at radioulnar joints and weakly flexes forearm at elbow joint.
Pronator quadratus (PRŌ-nā-tor kwod-RĀ-tus; *quadratus*=square, four-sided) (see Figure 11.19a)	Distal portion of shaft of ulna.	Distal portion of shaft of radius.	Pronates forearm at radioulnar joints.
FOREARM SUPINATOR			
Supinator (SOO-pi-nā-tor; *supinator*=turns palm anteriorly) (see Figure 11.19b)	Lateral epicondyle of humerus and ridge near radial notch of ulna (supinator crest).	Lateral surface of proximal one-third of radius.	Supinates forearm at radioulnar joints.

*The **bicipital aponeurosis** is a broad aponeurosis from the tendon of insertion of the biceps brachii muscle that descends medially across the brachial artery and fuses with deep fascia over the forearm flexor muscles.

Figure 11.18   Muscles that move the radius and ulna (forearm bones).

Whereas the anterior arm muscles flex the forearm, the posterior arm muscles extend it.

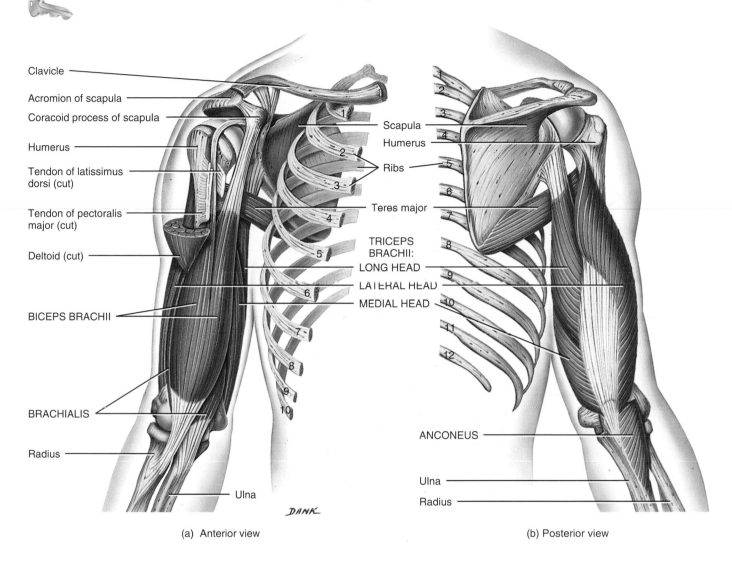

Clavicle

Acromion of scapula

Coracoid process of scapula

Humerus

Tendon of latissimus dorsi (cut)

Tendon of pectoralis major (cut)

Deltoid (cut)

BICEPS BRACHII

BRACHIALIS

Radius

Ulna

Scapula

Humerus

Ribs

Teres major

TRICEPS BRACHII:

LONG HEAD

LATERAL HEAD

MEDIAL HEAD

ANCONEUS

Ulna

Radius

DANK

(a) Anterior view

(b) Posterior view

continues

CONTINUED

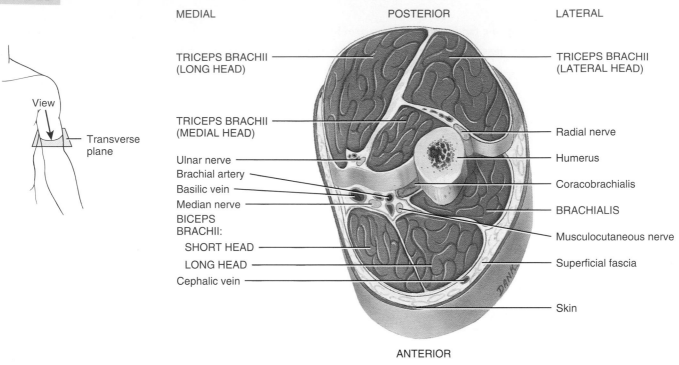

(c) Superior view of transverse section of arm

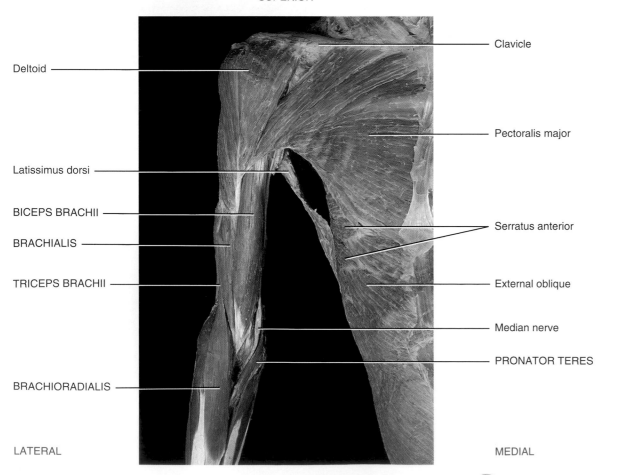

(d) Anterior superficial view

Which muscles are the most powerful flexor and the most powerful extensor of the forearm?

● Describe the origin, insertion, action, and innervation of the muscles that move the wrist, hand, thumb, and fingers.

Muscles of the forearm that move the wrist, hand, thumb, and fingers are many and varied. Those in this group that act on the digits are known as **extrinsic hand muscles** because they originate *outside* the hand and insert within it. As you will see, the names for the muscles that move the wrist, hand, and digits give some indication of their origin, insertion, or action. Based on location and function, the muscles of the forearm are divided into two groups: (1) anterior compartment muscles and (2) posterior compartment muscles. The **anterior (flexor) compartment** muscles of the forearm originate on the humerus, typically insert on the carpals, metacarpals, and phalanges, and function as flexors. The bellies of these muscles form the bulk of the forearm. The **posterior (extensor) compartment** muscles of the forearm originate on the humerus, insert on the metacarpals and phalanges, and function as extensors. Within each compartment, the muscles are grouped as superficial or deep.

The **superficial anterior compartment** muscles are arranged in the following order from lateral to medial: **flexor carpi radialis, palmaris longus** (absent in about 10% of the population), and **flexor carpi ulnaris** (the ulnar nerve and artery are just lateral to the tendon of this muscle at the wrist). The **flexor digitorum superficialis** muscle is deep to the other three muscles and is the largest superficial muscle in the forearm.

The **deep anterior compartment** muscles are arranged in the following order from lateral to medial: **flexor pollicis longus** (the only flexor of the distal phalanx of the thumb) and **flexor digitorum profundus** (ends in four tendons that insert into the distal phalanges of the fingers).

The **superficial posterior compartment** muscles are arranged in the following order from lateral to medial: **extensor carpi radialis longus, extensor carpi radialis brevis, extensor digitorum** (occupies most of the posterior surface of the forearm and divides into four tendons that insert into the middle and distal phalanges of the fingers), **extensor digiti minimi** (a slender muscle generally connected to the extensor digitorum), and the **extensor carpi ulnaris.**

The deep posterior compartment muscles are arranged in the following order from lateral to medial: abductor pollicis longus, extensor pollicis brevis, extensor pollicis longus, and extensor indicis.

The tendons of the muscles of the forearm that attach to the wrist or continue into the hand, along with blood vessels and nerves, are held close to bones by strong fasciae. The tendons are also surrounded by tendon sheaths. At the wrist, the deep fascia is thickened into fibrous bands called **retinacula** (*retinacul*=a holdfast). The **flexor retinaculum** is located over the palmar surface of the carpal bones. The long flexor tendons of the digits and wrist and the median nerve pass through the flexor retinaculum. The **extensor retinaculum** is located over the dorsal surface of the carpal bones. The extensor tendons of the wrist and digits pass through it.

Innervation

Nerves derived from the brachial plexus (shown in Exhibit 18.2 on page 568) innervate muscles that move the wrist, hand, thumb, and fingers.

- Flexor carpi radialis, palmaris longus, flexor digitorum superficialis, and flexor pollicis longus: median nerve
- Flexor carpi ulnaris: ulnar nerve
- Flexor digitorum profundus: median and ulnar nerves
- Extensor carpi radialis longus, extensor carpi radialis brevis, and extensor digitorum: radial nerve
- Extensor digiti minimi, extensor carpi ulnaris, and all of the deep muscles of the posterior compartment: deep radial nerve

Relating Muscles to Movements

Arrange the muscles in this exhibit according to the following actions on the wrist joint: (1) flexion, (2) extension, (3) abduction, and (4) adduction; the following actions on the fingers at the metacarpophalangeal joints: (1) flexion and (2) extension; the following actions on the fingers at the interphalangeal joints: (1) flexion and (2) extension; the following actions on the thumb at the carpometacarpal, metacarpophalangeal, and interphalangeal joints: (1) extension and (2) abduction; and the following action on the thumb at the interphalangeal joint: flexion. The same muscle may be mentioned more than once.

■ **CHECKPOINT**

1. Which muscles and actions of the wrist, hand, and digits are used when writing?

continues

CONTINUED

MUSCLE	ORIGIN	INSERTION	ACTION
SUPERFICIAL ANTERIOR (flexor) compartment of the forearm			
Flexor carpi radialis (FLEK-sor KAR-pē rā′-dē-Ā-lis; *flexor*=decreases angle at joint; *carpi*=of the wrist; *radi-*=radius)	Medial epicondyle of humerus.	Second and third metacarpals.	Flexes and abducts hand (radial deviation) at wrist joint.
Palmaris longus (pal-MA-ris LON-gus; *palma*=palm; *longus*=long)	Medial epicondyle of humerus.	Flexor retinaculum and palmar aponeurosis (deep fascia in center of palm).	Weakly flexes hand at wrist joint.
Flexor carpi ulnaris (FLEK-sor KAR-pē ul-NAR-is; *ulnar-*=ulna)	Medial epicondyle of humerus and superior posterior border of ulna.	Pisiform, hamate, and base of fifth metacarpal.	Flexes and adducts hand (ulnar deviation) at wrist joint.
Flexor digitorum superficialis (FLEK-sor di-ji-TOR-um soo′-per-fish′-ē-Ā-lis; *digit*=finger or toe; *superficialis*=closer to surface)	Medial epicondyle of humerus, coronoid process of ulna, and a ridge along lateral margin of anterior surface (anterior oblique line) of radius.	Middle phalanx of each finger.*	Flexes middle phalanx of each finger at proximal interphalangeal joint, proximal phalanx of each finger at metacarpophalangeal joint, and hand at wrist joint.
DEEP ANTERIOR (flexor) compartment of the forearm			
Flexor pollicis longus (FLEK-sor POL-li-sis LON-gus; *pollic-*=thumb)	Anterior surface of radius and interosseous membrane (sheet of fibrous tissue that holds shafts of ulna and radius together).	Base of distal phalanx of thumb.	Flexes distal phalanx of thumb at interphalangeal joint.
Flexor digitorum profundus (FLEK-sor di′-ji-TOR-um prō-FUN-dus; *profundus*=deep)	Anterior medial surface of body of ulna.	Base of distal phalanx of each finger.	Flexes distal and middle phalanges of each finger at interphalangeal joints, proximal phalanx of each finger at metacarpophalangeal joint, and hand at wrist joint.
SUPERFICIAL POSTERIOR (extensor) compartment of the forearm			
Extensor carpi radialis longus (eks-TEN-sor KAR-pē rā′-dē-Ā-lis LON-gus; *extensor*=increases angle at joint)	Lateral supracondylar ridge of humerus.	Second metacarpal.	Extends and adducts hand at wrist joint.
Extensor carpi radialis brevis (eks-TEN-sor KAR-pē rā′-dē-Ā-lis BREV-is; *brevis*=short)	Lateral epicondyle of humerus.	Third metacarpal.	Extends and abducts hand at wrist joint.

MUSCLE	ORIGIN	INSERTION	ACTION
Extensor digitorum (eks-TEN-sor di′-ji-TOR-um)	Lateral epicondyle of humerus.	Distal and middle phalanges of each finger.	Extends distal and middle phalanges of each finger at interphalangeal joints, proximal phalanx of each finger at metacarpophalangeal joint, and hand at wrist joint.
Extensor digiti minimi (eks-TEN-sor DIJ-i-tē MIN-i-mē; *digit*=finger or toe; *minimi*=smallest)	Lateral epicondyle of humerus.	Tendon of extensor digitorum on fifth phalanx.	Extends proximal phalanx of little finger at metacarpophalangeal joint and hand at wrist joint.
Extensor carpi ulnaris (eks-TEN-sor KAR-pē ul-NAR-is)	Lateral epicondyle of humerus and posterior border of ulna.	Fifth metacarpal.	Extends and abducts hand at wrist joint.
DEEP POSTERIOR (extensor) compartment of the forearm			
Abductor pollicis longus (ab-DUK-tor POL-li-sis LON-gus; *abductor*=moves part away from midline)	Posterior surface of middle of radius and ulna and interosseous membrane.	First metacarpal.	Abducts and extends thumb at carpometacarpal joint and abducts hand at wrist joint.
Extensor pollicis brevis (eks-TEN-sor POL-li-sis BREV-is)	Posterior surface of middle of radius and interosseous membrane.	Base of proximal phalanx of thumb.	Extends proximal phalanx of thumb at metacarpophalangeal joint, first metacarpal of thumb at carpometacarpal joint, and hand at wrist joint.
Extensor pollicis longus (eks-TEN-sor POL-li-sis LON-gus)	Posterior surface of middle of ulna and interosseous membrane.	Base of distal phalanx of thumb.	Extends distal phalanx of thumb at interphalangeal joint, first metacarpal of thumb at carpometacarpal joint, and abducts hand at wrist joint.
Extensor indicis (eks-TEN-sor IN-di-kis; *indicis*=index)	Posterior surface of ulna.	Tendon of extensor digitorum of index finger.	Extends distal and middle phalanges of index finger at interphalangeal joints, proximal phalanx of index finger at metacarpophalangeal joint, and hand at wrist joint.

*Reminder: The thumb or pollex is the first digit and has two phalanges: proximal and distal. The remaining digits, the fingers, are numbered 2–5, and each has three phalanges: proximal, middle, and distal.

continues

CONTINUED

Figure 11.19   Muscles that move the wrist, hand, and digits. (See Tortora, *A Photographic Atlas of the Human Body*, Figures 5.12 and 5.13.) The anterior compartment muscles function as flexors, and the posterior compartment muscles function as extensors.

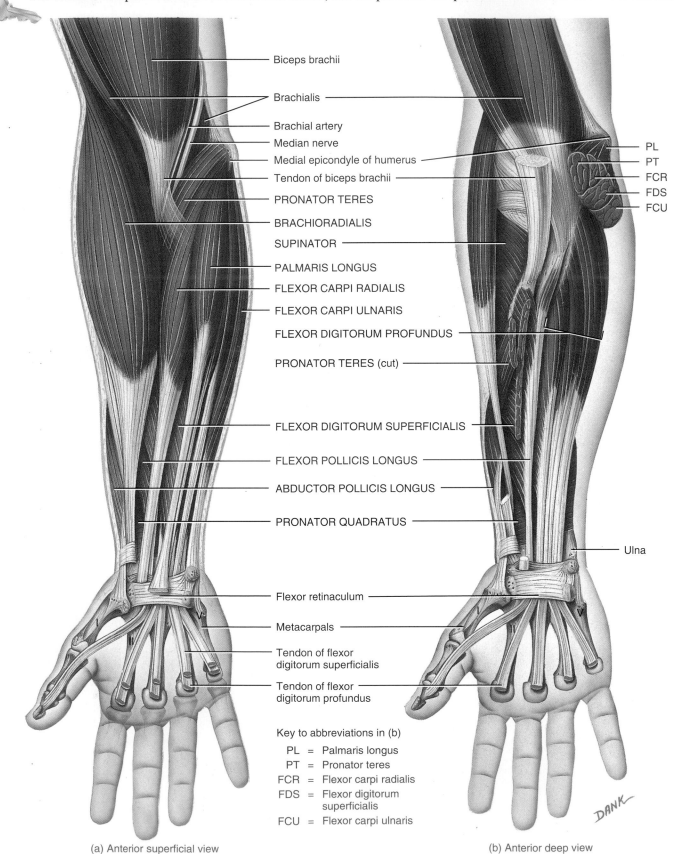

Biceps brachii

Brachialis

Brachial artery

Median nerve

Medial epicondyle of humerus

Tendon of biceps brachii

PRONATOR TERES

BRACHIORADIALIS

SUPINATOR

PALMARIS LONGUS

FLEXOR CARPI RADIALIS

FLEXOR CARPI ULNARIS

FLEXOR DIGITORUM PROFUNDUS

PRONATOR TERES (cut)

FLEXOR DIGITORUM SUPERFICIALIS

FLEXOR POLLICIS LONGUS

ABDUCTOR POLLICIS LONGUS

PRONATOR QUADRATUS

Flexor retinaculum

Metacarpals

Tendon of flexor digitorum superficialis

Tendon of flexor digitorum profundus

PL
PT
FCR
FDS
FCU

Ulna

Key to abbreviations in (b)

PL = Palmaris longus
PT = Pronator teres
FCR = Flexor carpi radialis
FDS = Flexor digitorum superficialis
FCU = Flexor carpi ulnaris

(a) Anterior superficial view

(b) Anterior deep view

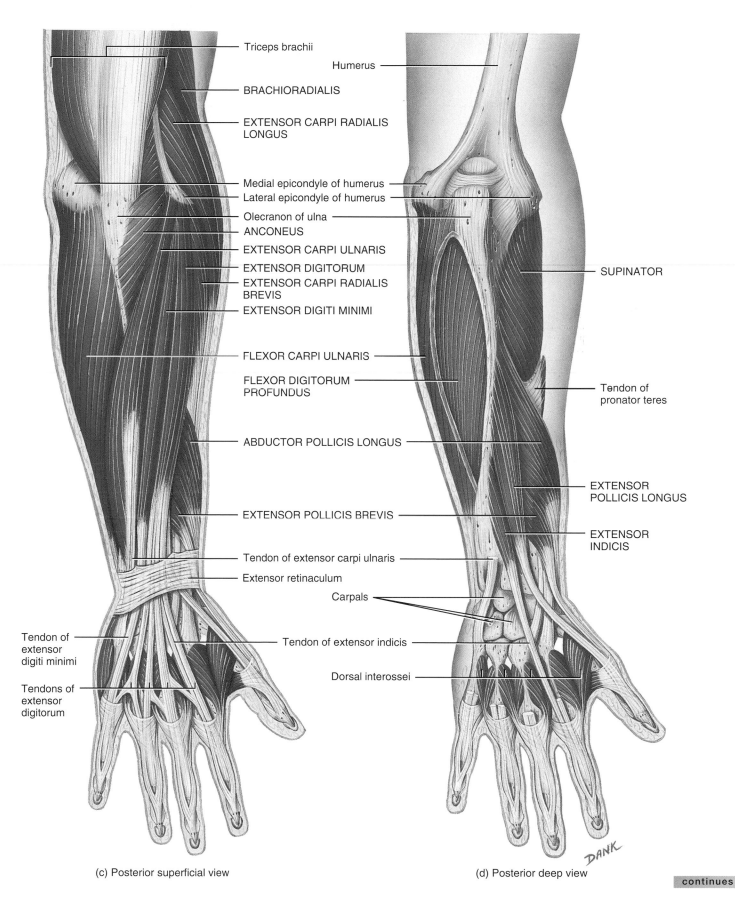

Triceps brachii

Humerus

BRACHIORADIALIS

EXTENSOR CARPI RADIALIS
LONGUS

Medial epicondyle of humerus

Lateral epicondyle of humerus

Olecranon of ulna

ANCONEUS

EXTENSOR CARPI ULNARIS

EXTENSOR DIGITORUM

EXTENSOR CARPI RADIALIS
BREVIS

EXTENSOR DIGITI MINIMI

SUPINATOR

FLEXOR CARPI ULNARIS

FLEXOR DIGITORUM
PROFUNDUS

Tendon of
pronator teres

ABDUCTOR POLLICIS LONGUS

EXTENSOR
POLLICIS LONGUS

EXTENSOR POLLICIS BREVIS

EXTENSOR
INDICIS

Tendon of extensor carpi ulnaris

Extensor retinaculum

Carpals

Tendon of
extensor
digiti minimi

Tendon of extensor indicis

Dorsal interossei

Tendons of
extensor
digitorum

(c) Posterior superficial view

(d) Posterior deep view

DANK

continues

345

CONTINUED

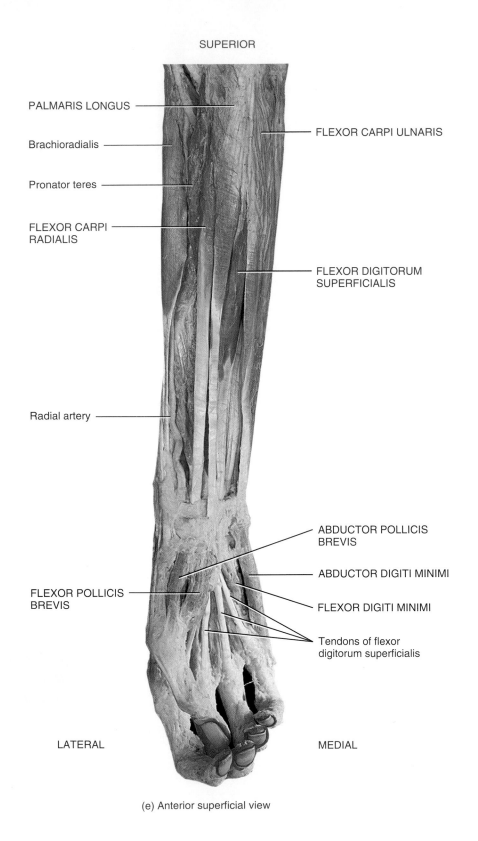

SUPERIOR

PALMARIS LONGUS

Brachioradialis

Pronator teres

FLEXOR CARPI RADIALIS

Radial artery

FLEXOR POLLICIS BREVIS

FLEXOR CARPI ULNARIS

FLEXOR DIGITORUM SUPERFICIALIS

ABDUCTOR POLLICIS BREVIS

ABDUCTOR DIGITI MINIMI

FLEXOR DIGITI MINIMI

Tendons of flexor digitorum superficialis

LATERAL

MEDIAL

(e) Anterior superficial view

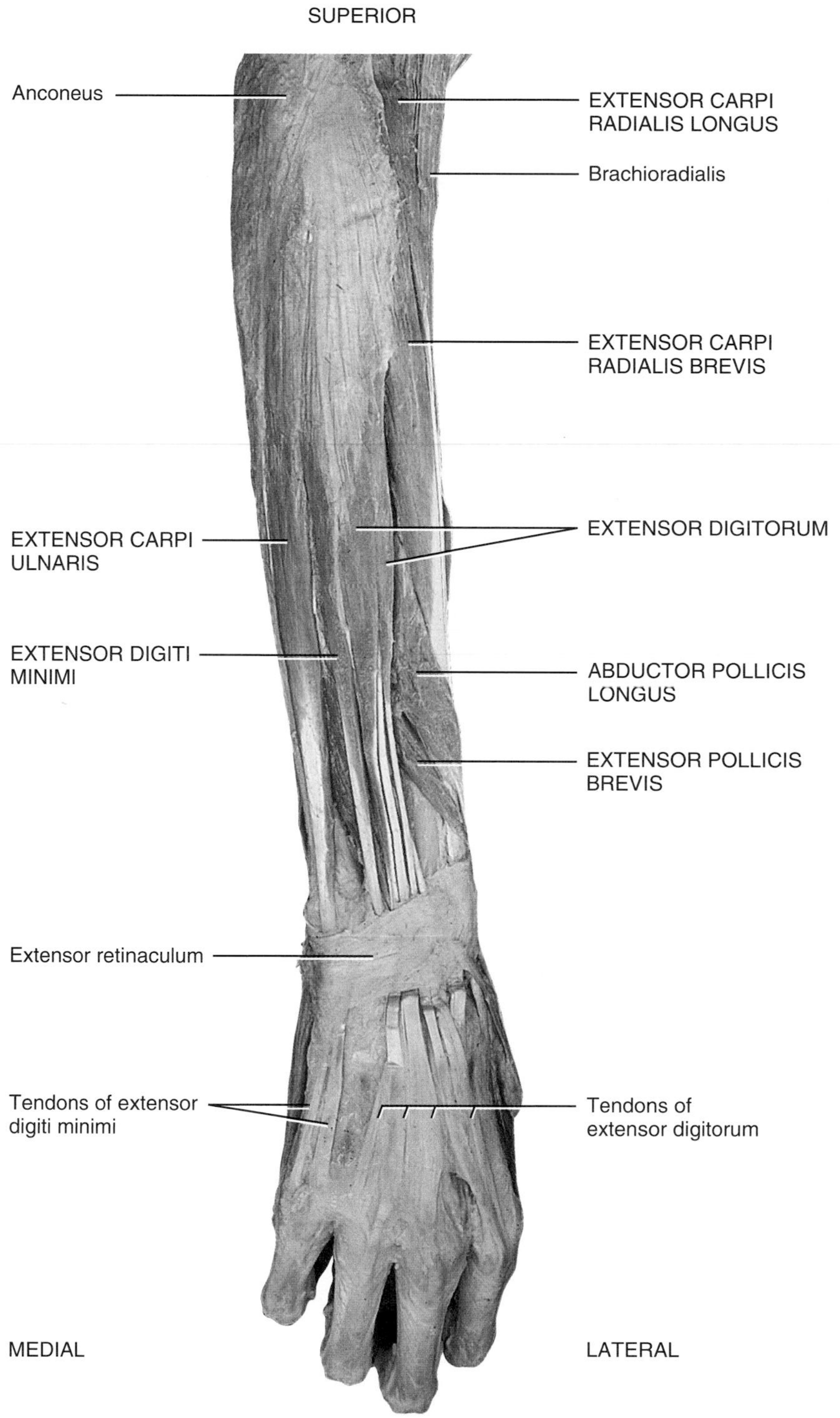

SUPERIOR

Anconeus

EXTENSOR CARPI RADIALIS LONGUS

Brachioradialis

EXTENSOR CARPI RADIALIS BREVIS

EXTENSOR CARPI ULNARIS

EXTENSOR DIGITORUM

EXTENSOR DIGITI MINIMI

ABDUCTOR POLLICIS LONGUS

EXTENSOR POLLICIS BREVIS

Extensor retinaculum

Tendons of extensor digiti minimi

Tendons of extensor digitorum

MEDIAL

LATERAL

(f) Posterior superficial view

? **What structures pass through the flexor retinaculum?**

347

EXHIBIT 11.17   INTRINSIC MUSCLES OF THE HAND (Figure 11.20)

OBJECTIVE

● Describe the origin, insertion, action, and innervation of the intrinsic muscles of the hand.

Several of the muscles discussed in Exhibit 11.16 move the digits in various ways and are known as extrinsic muscles. They produce the powerful but crude movements of the digits. The **intrinsic muscles** in the palm produce the weak but intricate and precise movements of the digits that characterize the human hand. The muscles in this group are so named because their origins and insertions are *within* the hand.

The intrinsic muscles of the hand are divided into three groups: (1) **thenar,** (2) **hypothenar,** and (3) **intermediate.** The four thenar muscles act on the thumb and form the **thenar eminence,** the lateral rounded contour on the palm that is also called the ball of the thumb. The thenar muscles include the abductor policis brevis, opponens pollicis, flexor pollicis brevis, and adductor pollicis. The **abductor pollicis brevis** is a thin, short, relatively broad superficial muscle on the lateral side of the thenar eminence. The **opponens pollicis** is a small, triangular muscle that is deep to the abductor pollicis brevis muscle. The **flexor pollicis brevis** is a short, wide muscle that is medial to the abductor pollicis brevis muscle. The **adductor pollicis** is fan-shaped and has two heads (oblique and transverse) separated by a gap through which the radial artery passes.

The three hypothenar muscles act on the little finger and form the **hypothenar eminence,** the medial rounded contour on the palm that is also called the ball of the little finger. The hypothenar muscles are the abductor digiti minimi, flexor digiti minimi brevis, and opponens digiti minimi. The **abductor digiti minimi** is a short, wide muscle and is the most superficial of the hypothenar muscles. It is a powerful muscle that plays an important role in grasping an object with outspread fingers. The **flexor digiti minimi brevis** muscle is also short and wide and is lateral to the abductor digiti minimi muscle. The **opponens digiti minimi** muscle is triangular and deep to the other two hypothenar muscles.

The 11 intermediate (midpalmar) muscles act on all the digits except the thumb. The intermediate muscles include the lumbricals, palmar interossei, and dorsal interossei. The **lumbricals,** as their name indicates, are worm-shaped. They originate from and insert into the tendons of other muscles (flexor digitorum profundus and extensor digitorum). The **palmar interossei** are the smaller and most superficial of the interossei muscles. The **dorsal interossei** are the deep interossei muscles. Both sets of interossei muscles are located between the metacarpals and are important in abduction, adduction, flexion, and extension of the fingers, and in movements in skilled activities such as writing, typing, and playing a piano.

The functional importance of the hand is readily apparent when one considers that certain hand injuries can result in permanent disability. Most of the dexterity of the hand depends on movements of the thumb. The general activities of the hand are free motion, power grip (forcible movement of the fingers and thumb against the palm, as in squeezing), precision handling (a change in position of a handled object that requires exact control of finger and thumb positions, as in winding a watch or threading a needle), and pinch (compression between the thumb and index finger or between the thumb and first two fingers).

Movements of the thumb are very important in the precise activities of the hand, and they are defined in different planes from comparable movements of other digits because the thumb is positioned at a right angle to the other digits. The five principal movements of the thumb are illustrated in Figure 11.20f and include *flexion* (movement of the thumb medially across the palm), *extension* (movement of the thumb laterally away from the palm), *adduction* (movement of the thumb in an anteroposterior plane toward the palm), and *opposition* (movement of the thumb across the palm so that the tip of the thumb meets the tip of a finger). Opposition is the single most distinctive digital movement that gives humans and other primates the ability to grasp and manipulate objects precisely.

Innervation

Nerves derived from the brachial plexus (shown in Exhibit 18.2 on page 568) innervate intrinsic muscles of the hand.

● Abductor pollicis brevis and opponens pollicis: median nerve

● Adductor pollicis, abductor digiti minimi, flexor digiti minimi brevis, opponens digiti minimi, dorsal interossei, and palmar interossei: ulnar nerve

● Flexor pollicis brevis and lumbricals: median and ulnar nerves

CARPAL TUNNEL SYNDROME

The **carpal tunnel** is a narrow passageway formed anteriorly by the flexor retinaculum and posteriorly by the carpal bones. Through this tunnel pass the median nerve, the most superficial structure, and the long flexor tendons for the digits. Structures within the carpal tunnel, especially the median nerve, are vulnerable to compression, and the resulting condition is called **carpal tunnel syndrome.** Compression of the median nerve leads to sensory changes over the lateral side of the hand, and muscle weakness in the thenar eminence. This results in pain, numbness, and tingling of the fingers. The condition may be caused by inflammation of the digital tendon sheaths, fluid retention, excessive exercise, infection, trauma, and/or repetitive activities that involve flexion of the wrist, such as keyboarding, cutting hair, and playing a piano. ■

Relating Muscles to Movements

Arrange the muscles in this exhibit according to the following actions on the thumb at the carpometacarpal and metaphalangeal joints: (1) abduction, (2) adduction, (3) flexion, and (4) opposition; and the following actions on the fingers at the metacarpophalangeal and interphalangeal joints: (1) abduction, (2) adduction, (3) flexion, and (4) extension. The same muscle may be mentioned more than once.

■ CHECKPOINT

1. How do the actions of the extrinsic and intrinsic muscles of the hand differ?

MUSCLE	ORIGIN	INSERTION	ACTION
THENAR (lateral aspect of palm)			
Abductor pollicis brevis (ab-DUK-tor POL-li-sis BREV-is; *abductor*=moves part away from middle; *pollic-*= the thumb; *brevis*= short)	Flexor retinaculum, scaphoid, and trapezium.	Lateral side of proximal phalanx of thumb.	Abducts thumb at carpometacarpal joint.
Opponens pollicis (op-PŌ-nenz POL-li-sis; *opponens*=opposes)	Flexor retinaculum and trapezium.	Lateral side of first metacarpal (thumb).	Moves thumb across palm to meet little finger (opposition) at the carpometacarpal joint.
Flexor pollicis brevis (FLEK-sor POL-li-sis BREV-is; *flexor*=decreases angle at joint)	Flexor retinaculum, trapezium, capitate, and trapezoid.	Lateral side of proximal phalanx of thumb.	Flexes thumb at carpometacarpal and metacarpophalangeal joints.
Adductor pollicis (ad-DUK-tor POL-li-sis; *adductor*=moves part toward midline)	Oblique head: capitate and second and third metacarpals; transverse head: third metacarpal.	Medial side of proximal phalanx of thumb by a tendon containing a sesamoid bone.	Adducts thumb at carpometacarpal and metacarpophalangeal joints.
HYPOTHENAR (medial aspect of palm)			
Abductor digiti minimi (ab-DUK-tor DIJ-i-tē MIN-i-mē; *digit*=finger or toe; *minimi*=little)	Pisiform and tendon of flexor carpi ulnaris.	Medial side of proximal phalanx of little finger.	Abducts and flexes little finger at metacarpophalangeal joint.
Flexor digiti minimi brevis (FLEK-sor DIJ-i-tē MIN-i-mē BREV-is)	Flexor retinaculum and hamate.	Medial side of proximal phalanx of little finger.	Flexes little finger at carpometacarpal and metacarpophalangeal joints.
Opponens digiti minimi (op-PŌ-nenz DIJ-i-tē MIN-i-mē)	Flexor retinaculum and hamate.	Medial side of fifth metacarpal (little finger).	Moves little finger across palm to meet thumb (opposition) at the carpometacarpal joint.
INTERMEDIATE (midpalmar)			
Lumbricals (LUM-bri-kals; *lumbric-*=earthworm) (four muscles)	Lateral sides of tendons and flexor digitorum profundus of each finger.	Lateral sides of tendons of extensor digitorum on proximal phalanges of each finger.	Flex each finger at metacarpophalangeal joints and extend each finger at interphalangeal joints.
Palmar interossei (in'-ter-OS-ē-ī; *palmar*=palm; *inter-*=between; *ossei*= bones) (three muscles)	Sides of shafts of metacarpals of all digits (except the middle one).	Sides of bases of proximal phalanges of all digits (except the middle one).	Adduct each finger at metacarpophalangeal joints; flex each finger at metacarpophalangeal joints.
Dorsal interossei (in'-ter-OS-ē-ī; *dorsal*=back surface) (four muscles)	Adjacent sides of metacarpals.	Proximal phalanx of each finger.	Abduct fingers 2–4 at metacarpophalangeal joints; flex fingers 2–4 at metacarpophalangeal joints; and extend each finger at interphalangeal joints.

continues

EXHIBIT 11.17   INTRINSIC MUSCLES OF THE HAND (Figure 11.20)

CONTINUED

Figure 11.20   Intrinsic muscles of the hand. (See Tortora, *A Photographic Atlas of the Human Body,* 2e, Figures 5.14 and 5.15.)

The intrinsic muscles of the hand produce the intricate and precise movements of the digits that characterize the human hand.

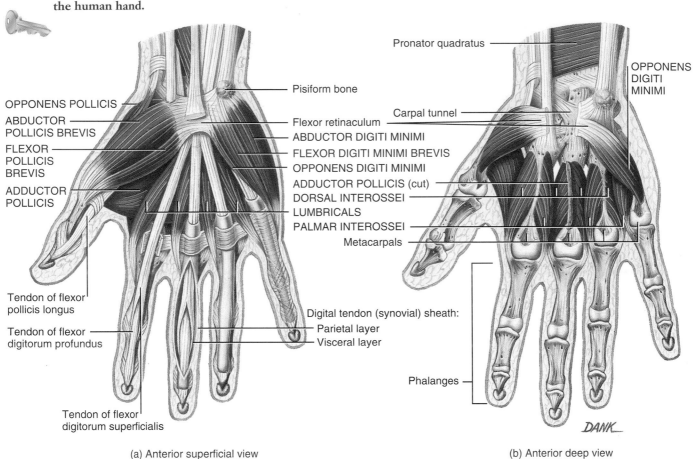

OPPONENS POLLICIS
ABDUCTOR POLLICIS BREVIS
FLEXOR POLLICIS BREVIS
ADDUCTOR POLLICIS
Tendon of flexor pollicis longus
Tendon of flexor digitorum profundus
Tendon of flexor digitorum superficialis

Pisiform bone
Flexor retinaculum
ABDUCTOR DIGITI MINIMI
FLEXOR DIGITI MINIMI BREVIS
OPPONENS DIGITI MINIMI
ADDUCTOR POLLICIS (cut)
DORSAL INTEROSSEI
LUMBRICALS
PALMAR INTEROSSEI
Metacarpals
Digital tendon (synovial) sheath:
Parietal layer
Visceral layer

Pronator quadratus
OPPONENS DIGITI MINIMI
Carpal tunnel
Phalanges

(a) Anterior superficial view

(b) Anterior deep view

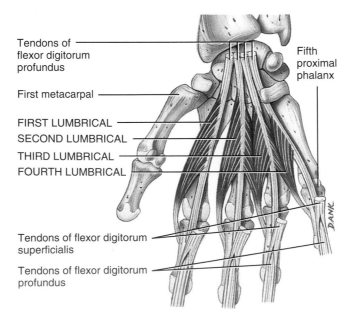

Tendons of flexor digitorum profundus
First metacarpal
FIRST LUMBRICAL
SECOND LUMBRICAL
THIRD LUMBRICAL
FOURTH LUMBRICAL
Tendons of flexor digitorum superficialis
Tendons of flexor digitorum profundus

Fifth proximal phalanx

(c) Anterior deep view showing lumbricals

350

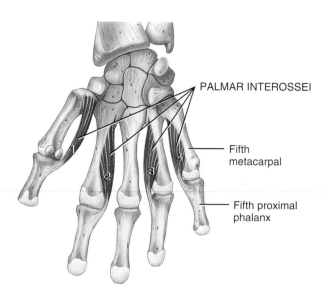

PALMAR INTEROSSEI

Fifth
metacarpal

Fifth proximal
phalanx

(d) Anterior deep view of palmar interossei

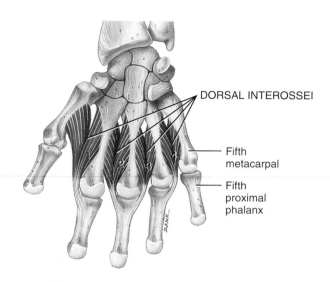

DORSAL INTEROSSEI

Fifth
metacarpal

Fifth
proximal
phalanx

(e) Anterior deep view of dorsal interossei

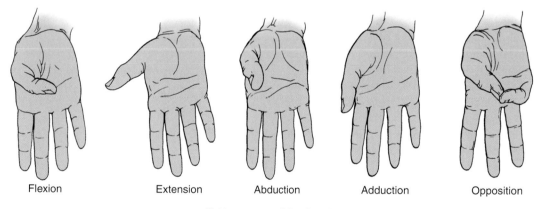

| Flexion | Extension | Abduction | Adduction | Opposition |

(f) Movements of the thumb

Muscles of the thenar eminence act on which digit?

EXHIBIT 11.18   MUSCLES THAT MOVE THE VERTEBRAL COLUMN (BACKBONE) (Figure 11.21)

OBJECTIVE

● Describe the origin, insertion, action, and innervation of the muscles that move the vertebral column.

The muscles that move the vertebral column (backbone) are quite complex because they have multiple origins and insertions and there is considerable overlap among them. One way to group the muscles is on the basis of the general direction of the muscle bundles and their approximate lengths. For example, the splenius muscles arise from the midline and extend laterally and superiorly to their insertions (Figure 11.21a). The erector spinae muscle group (consisting of the iliocostalis, longissimus, and spinalis muscles) arises from either the midline or more laterally but usually runs almost longitudinally, with neither a significant lateral nor medial direction as it is traced superiorly. The muscles of the transversospinalis group (semispinalis, rotatores, multifidus) arise laterally but extend toward the midline as they are traced superiorly. Deep to these three muscle groups are small segmental muscles that extend between spinous processes or transverse processes of vertebrae. Because the scalene muscles assist in moving the vertebral column, they are also included in this exhibit. Note in Exhibit 11.9 that the rectus abdominis, external oblique, internal oblique, and quadratus lumborum muscles also play a role in moving the vertebral column.

The bandage-like **splenius** muscles are attached to the sides and back of the neck. The two muscles in this group are named on the basis of their superior attachments (insertions): **splenius capitis** (head region) and **splenius cervicis** (cervical region). They extend the head and laterally flex and rotate the head.

The **erector spinae** is the largest muscle mass of the back, forming a prominent bulge on either side of the vertebral column. It is the chief extensor of the vertebral column. It is also important in controlling flexion, lateral flexion, and rotation of the vertebral column and in maintaining the lumbar curve, because the main mass of the muscle is in the lumbar region. As noted above, it consists of three groups: iliocostalis (laterally placed), longissimus (intermediately placed), and spinalis (medially placed). These groups, in turn, consist of a series of overlapping muscles, and the muscles within the groups are named according to the regions of the body with which they are associated. The **iliocostalis group** consists of three muscles: the **iliocostalis cervicis** (cervical region), **iliocostalis thoracis** (thoracic region), and **iliocostalis lumborum** (lumbar region). The **longissimus group** resembles a herringbone and consists of three muscles: the **longissimus capitis** (head region), **longissimus cervicis** (cervical region), and **longissimus thoracis** (thoracic region). The **spinalis group** also consists of three muscles: the **spinalis capitis, spinalis cervicis,** and **spinalis thoracis.**

The **transversospinales** are so named because their fibers run from the transverse processes to the spinous processes of the vertebrae. The semispinalis muscles in this group are also named according to the region of the body with which they are associated: **semispinalis capitis** (head region), **semispinalis cervicis** (cervical region), and **semispinalis thoracis** (thoracic region). These muscles extend the vertebral column and rotate the head. The **multifidus** muscle in this group, as its name implies, is segmented into several bundles. It extends and laterally flexes the vertebral column and

rotates the head. The **rotatores** muscles of this group are short and are found along the entire length of the vertebral column. They extend and rotate the vertebral column.

Within the **segmental** muscle group (Figure 11.21b), the **interspinales** and **intertransversarii** muscles unite the spinous and transverse processes of consecutive vertebrae. They function primarily in stabilizing the vertebral column during its movements.

Within the **scalene** group (Figure 11.21c), the **anterior scalene** muscle is anterior to the middle scalene muscle, the **middle scalene** muscle is intermediate in placement and is the longest and largest of the scalene muscles, and the **posterior scalene** muscle is posterior to the middle scalene muscle and is the smallest of the scalene muscles. These muscles flex, laterally flex, and rotate the head and assist in deep inhalation.

Innervation

- Splenius capitis: middle cervical nerves
- Splenius cervicis: inferior cervical nerves
- Iliocostalis cervicis and semispinalis capitis: cervical and thoracic nerves
- Iliocostalis thoracis: thoracic nerves
- Iliocostalis lumborum: lumbar nerves
- Longissimus capitis: middle and inferior cervical nerves
- Longissimus cervicis, longissimus thoracis, all spinalis muscles, multifidus, rotatores, and interspinales: cervical, thoracic, and lumbar nerves
- Semispinalis cervicis and semispinalis thoracis: cervical and thoracic nerves
- Intertransversarii: cervical, thoracic and lumbar spinal nerves
- Anterior scalene: cervical spinal nerves C5–C6
- Middle scalene: cervical spinal nerves C3–C8
- Posterior scalene: cervical spinal nerves C6–C8.

BACK INJURIES AND HEAVY LIFTING

Next to headaches, medical experts note that back problems are the most common medical complaint that lead people to seek treatment. Second only to the common cold as the greatest cause of lost workdays, back injuries cost U.S. industry $10–14 billion in workers' compensation costs and about 100 million lost workdays annually.

The four factors associated with increased risk of back injury are amount of force, repetition, posture, and stress applied to the backbone. Poor physical condition, poor posture, lack of exercise, and excessive body weight contribute to the number and severity of sprains and strains. Back pain caused by a muscle strain or ligament sprain will normally heal within a short time and may never cause further problems. However, if ligaments and muscles are weak, discs in the lower back can become weakened and may herniate (rupture) with excessive lifting or a sudden fall. After years of back abuse, or with aging, the discs may simply wear out and cause chronic pain. Degeneration of the spine due to aging is often misdiagnosed as a sprain or strain.

Full flexion at the waist, as in touching your toes, overstretches the erector spinae muscles. Muscles that are overstretched cannot contract effectively. Straightening up from such a position is therefore initiated by the hamstring muscles on the back of the thigh and the gluteus maximus muscles of the buttocks. The erector spinae muscles join in as the degree of flexion decreases. Improperly lifting a heavy weight, however, can strain the erector spinae muscles. The result can be painful muscle spasms, tearing of tendons and ligaments of the lower back, and herniating of intervertebral discs. The lumbar muscles are adapted for maintaining posture, not for lifting. This is why it is important to bend at the knees and use the powerful extensor muscles of the thighs and buttocks while lifting a heavy load. ■

Relating Muscles to Movements

Arrange the muscles in this exhibit according to the following actions on the head at the atlanto-occipital and intervertebral joints: (1) flexion, (2) extension, (3) lateral flexion, (4) rotation to same side as contracting muscle, and (5) rotation to opposite side as contracting muscle; the following actions on the vertebral column at the intervertebral joints: (1) flexion, (2) extension, (3) lateral flexion, (4) rotation, and (5) stabilization; and the following action on the ribs: elevation during deep inhalation. The same muscle may be mentioned more than once.

■ **CHECKPOINT**

1. What is the largest muscle group of the back?

MUSCLE	ORIGIN	INSERTION	ACTION
SPLENIUS (SPLĒ-nē-us)			
Splenius capitis (KAP-i-tis; *splenium*=bandage; *capit-*=head)	Ligamentum nuchae and spinous processes of seventh cervical vertebra and first three or four thoracic vertebrae.	Occipital bone and mastoid process of temporal bone.	Acting together (bilaterally), extend head; acting singly (unilaterally), laterally flex and rotate head to same side as contracting muscle.
Splenius cervicis (SER-vi-kis; *cervic-*=neck)	Spinous processes of third through sixth thoracic vertebrae.	Transverse processes of first two or four cervical vertebrae.	Acting together, extend head; acting singly, laterally flex and rotate head to same side as contracting muscle.
ERECTOR SPINAE (e-REK-tor SPI-nē) Consists of iliocostalis muscles (lateral), longissimus muscles (intermediate), and spinalis muscles (medial).			
ILIOCOSTALIS GROUP (Lateral)			
Iliocostalis cervicis (il'-ē-ō-kos-TAL-is SER-vi-kis; *ilio-*=flank; *costa-*=rib)	Superior six ribs.	Transverse processes of fourth to sixth cervical vertebrae.	Acting together, muscles of each region (cervical, thoracic, and lumbar) extend and maintain erect posture of vertebral column of their respective regions; acting singly, laterally flex vertebral column of their respective regions.
Iliocostalis thoracis (il'-ē-ō-kos-TAL-is thō-RĀ-sis; *thorac-*=chest)	Inferior six ribs.	Superior six ribs.	
Iliocostalis lumborum (il'-ē-ō-kos-TĀL-is lum-BOR-um)	Iliac crest.	Inferior six ribs.	
LONGISSIMUS GROUP (Intermediate)			
Longissimus capitis (lon-JIS-i-mus KAP-i-tis; *longissimus*=longest)	Transverse processes of superior four thoracic vertebrae and articular processes of inferior four cervical vertebrae.	Mastoid process of temporal bone.	Acting together, both longissimus capitis muscles extend head; acting singly, rotate head to same side as contracting muscle. Acting together, longissimus cervicis and both longissimus thoracis muscles extend vertebral column of their respective regions; acting singly, laterally flex vertebral column of their respective regions.
Longissimus cervicis (lon-JIS-i-mus SER-vi-kis)	Transverse processes of fourth and fifth thoracic vertebrae.	Transverse processes of second to sixth cervical vertebrae.	
Longissimus thoracis (lon-JIS-i-mus thō-RĀ-sis)	Transverse processes of lumbar vertebrae.	Transverse processes of all thoracic and superior lumbar vertebrae and ninth and tenth ribs.	

continues

EXHIBIT 11.18   MUSCLES THAT MOVE THE VERTEBRAL COLUMN (BACKBONE) (Figure 11.21)

CONTINUED

MUSCLE	ORIGIN	INSERTION	ACTION
SPINALIS GROUP (Medial)			
Spinalis capitis (spi-NĀ-lis KAP-i-tis; *spinal*=vertebral column)	Arises with semispinalis capitis.	Occipital bone.	Acting together, muscles of each region (head, cervical, and thoracic) extend vertebral column of their respective regions.
Spinalis cervicis (spi-NĀ-lis SER-vi-kis)	Ligamentum nuchae and spinous process of seventh cervical vertebra.	Spinous process of axis.	
Spinalis thoracis (spi-NĀ-lis thō-RĀ-sis)	Spinous processes of superior lumbar and inferior thoracic vertebrae.	Spinous processes of superior thoracic vertebrae.	
TRANSVERSOSPINALES (trans-ver-sō-spi-NĀ-lēz)			
Semispinalis capitis (sem′-ē-spi-NĀ-lis KAP-i-tis; *semi-*=partially one-half)	Transverse processes of first six or seven thoracic vertebrae and seventh cervical vertebra, and articular processes of fourth, fifth, and sixth cervical vertebrae.	Occipital bone.	Acting together, extend head; acting singly, rotate head to side opposite contracting muscle.
Semispinalis cervicis (SER-vi-kis)	Transverse processes of superior five or six thoracic vertebrae.	Spinous processes of first to fifth cervical vertebrae.	Acting together, both semispinalis cervicis and both semispinalis thoracis muscles extend vertebral column of their respective regions; acting singly, rotate head to side opposite contracting muscle.
Semispinalis thoracis (thō-RĀ-sis)	Transverse processes of sixth to tenth thoracic vertebrae.	Spinous processes of superior four thoracic and last two cervical vertebrae.	
Multifidus (mul-TIF-i-dus; *multi*=many; *fid-*=segmented)	Sacrum, ilium, transverse processes of lumbar, thoracic, and inferior four cervical vertebrae.	Spinous process of a more superior vertebra.	Acting together, extend vertebral column; acting singly, laterally flex vertebral column and rotate head to side opposite contracting muscle.
Rotatores (rō-ta-TŌ-rēz; singular is *rotatore*; *rotator*=to rotate)	Transverse processes of all vertebrae.	Spinous process of vertebra superior to the one of origin.	Acting together, extend vertebral column; acting singly, rotate vertebral column to side opposite contracting muscle.
SEGMENTAL (seg-MEN-tal)			
Interspinales (in-ter-spī-NĀ-lēz; *inter-*=between)	Superior surface of all spinous processes.	Inferior surface of spinous process of vertebra superior to the one of origin.	Acting together, extend vertebral column; acting singly, stabilize vertebral column during movement.
Intertransversarii (in′-ter-trans-vers-AR-ē-ī; singular is *intertransversarius*)	Transverse processes of all vertebrae.	Transverse process of vertebra superior to the one of origin.	Acting together, extend vertebral column; acting singly, laterally flex vertebral column and stabilize it during movements.
SCALENES (SKĀ-lēnz)			
Anterior scalene (SKĀ-lēn; *anterior*=front; *scalene*=uneven)	Transverse processes of third through sixth cervical vertebrae.	First rib.	Acting together, right and left anterior scalene and middle scalene muscles flex head and elevate first ribs during deep inhalation; acting singly, laterally flex head and rotate head to side opposite contracting muscle.
Middle scalene	Transverse processes of inferior six cervical vertebrae.	First rib.	
Posterior scalene	Transverse processes of fourth through sixth cervical vertebrae.	Second rib.	Acting together, flex head and elevate second ribs during deep inhalation; acting singly, laterally flex head and rotate head to side opposite contracting muscle.

Figure 11.21   **Muscles that move the vertebral column (backbone).**

The erector spinae group (iliocostalis, longissimus, and spinalis muscles) is the largest muscular mass of the body and is the chief extensor of the vertebral column.

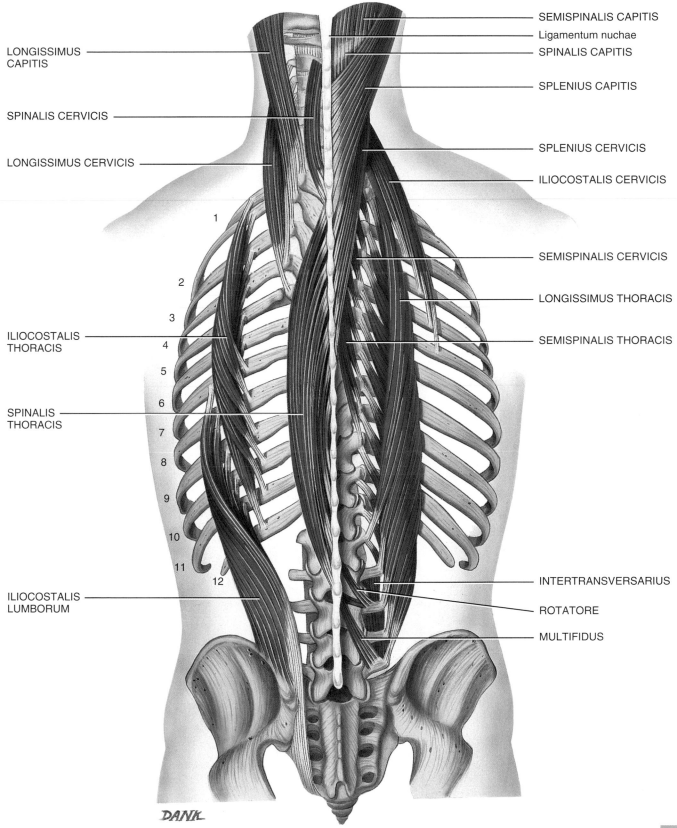

LONGISSIMUS CAPITIS

SPINALIS CERVICIS

LONGISSIMUS CERVICIS

ILIOCOSTALIS THORACIS

SPINALIS THORACIS

ILIOCOSTALIS LUMBORUM

SEMISPINALIS CAPITIS

Ligamentum nuchae

SPINALIS CAPITIS

SPLENIUS CAPITIS

SPLENIUS CERVICIS

ILIOCOSTALIS CERVICIS

SEMISPINALIS CERVICIS

LONGISSIMUS THORACIS

SEMISPINALIS THORACIS

INTERTRANSVERSARIUS

ROTATORE

MULTIFIDUS

DANK

(a) Posterior view

continues

EXHIBIT 11.18   MUSCLES THAT MOVE THE VERTEBRAL COLUMN (BACKBONE) (Figure 11.21)

CONTINUED

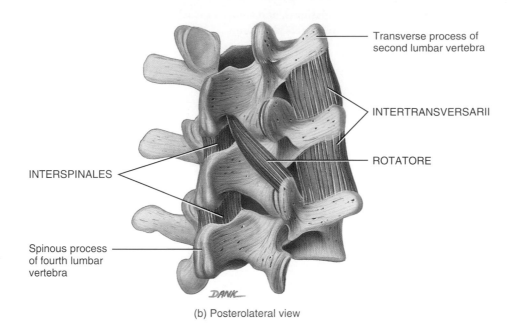

Transverse process of
second lumbar vertebra

INTERTRANSVERSARII

ROTATORE

INTERSPINALES

Spinous process
of fourth lumbar
vertebra

(b) Posterolateral view

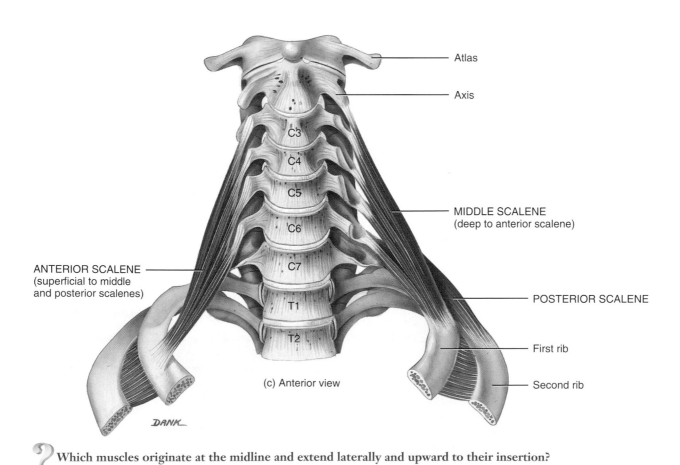

Atlas

Axis

C3

C4

C5

C6

C7

T1

T2

MIDDLE SCALENE
(deep to anterior scalene)

ANTERIOR SCALENE
(superficial to middle
and posterior scalenes)

POSTERIOR SCALENE

First rib

Second rib

(c) Anterior view

? Which muscles originate at the midline and extend laterally and upward to their insertion?

EXHIBIT 11.19   MUSCLES THAT MOVE THE FEMUR (THIGH BONE) (Figure 11.22)

OBJECTIVE

● Describe the origin, insertion, action, and innervation of the muscles that move the femur.

As you will see, muscles of the lower limbs are larger and more powerful than those of the upper limbs because lower limb muscles function in stability, locomotion, and maintenance of posture. Upper limb muscles are characterized by versatility of movement. In addition, muscles of the lower limbs often cross two joints and act equally on both.

The majority of muscles that move the femur originate on the pelvic girdle and insert on the femur. The **psoas major** and **iliacus** muscles have a common insertion (lesser trochanter of the femur) and together, they are known as the **iliopsoas** (il′-ē-ō-SŌ-as) muscle. Sometimes people "cheat" when doing situps by using their iliopsoas muscle instead of the abdominal muscles. This can potentially produce back pain due to the origin of the iliopsoas on the backbone. There are three gluteal muscles: gluteus maximus, gluteus medius, and gluteus minimus. The **gluteus maximus** is the largest and heaviest of the three muscles and is one of the largest muscles in the body. It is the chief extensor of the femur. The **gluteus medius** is mostly deep to the gluteus maximus and is a powerful abductor of the femur at the hip joint. It is a common site for an intramuscular injection. The **gluteus minimus** is the smallest of the gluteal muscles and lies deep to the gluteus medius.

The **tensor fasciae latae** muscle is located on the lateral surface of the thigh. The *fascia lata* is a layer of deep fascia, composed of dense connective tissue, that encircles the entire thigh. It is well developed laterally where, together with the tendons of the tensor fasciae and gluteus maximus muscles, it forms a structure called the **iliotibial tract.** The tract inserts into the lateral condyle of the tibia.

The **piriformis, obturator internus, obturator externus, superior gemellus, inferior gemellus,** and **quadratus femoris** muscles are all deep to the gluteus maximus muscle and function as lateral rotators of the femur at the hip joint.

Three muscles on the medial aspect of the thigh are the **adductor longus, adductor brevis,** and **adductor magnus.** They originate on the pubic bone and insert on the femur. All three muscles adduct, flex, and medially rotate the femur at the hip joint. The **pectineus** muscle also adducts and flexes the femur at the hip joint.

Technically, the adductor muscles and pectineus muscles are components of the medial compartment of the thigh and could be included in Exhibit 11.20. However, they are included here because they act on the femur.

At the junction between the trunk and lower limb is a space called the **femoral triangle.** The base is formed superiorly by the inguinal ligament, medially by the lateral border of the adductor longus muscle, and laterally by the medial border of the sartorius muscle. The apex is formed by the crossing of the adductor longus by the sartorius muscles (Figure 11.22a). The contents of the femoral triangle, from lateral to medial, are the femoral nerve and its branches, the femoral artery and several of its branches, the femoral vein and its proximal tributaries, and the deep inguinal lymph nodes.

Innervation

Nerves derived from the lumbar and sacral plexuses (shown in Exhibits 18.3 and 18.4 on pages 572 and 575) innervate the muscles that move the femur.

• Psoas major: lumbar nerves L2–L3

• Iliacus and pectineus: femoral nerve

• Gluteus maximus: inferior gluteal nerve

• Gluteus medius and minimus and tensor fasciae latae: superior gluteal nerve

• Piriformis: sacral spinal nerves S1 or S2, mainly S1

• Obturator internus and superior gemellus: nerve to obturator internus

• Obturator externus, adductor longus, and adductor brevis: obturator nerve

• Inferior gemellus and quadratus femoris: nerve to quadratus femoris

• Adductor magnus: obturator and sciatic nerves

GROIN STRAIN (ADDUCTOR MUSCLE RUPTURE)

The five major muscles of the inner thigh function to move the legs medially. This muscle group is important in activities such as sprinting, hurdling, and horseback riding. A rupture or tear of one or more of these muscles can cause a **groin strain.** Groin strains most often occur during sprinting or twisting, or from kicking a solid, perhaps stationary object. Symptoms of a groin strain may be sudden or may not surface until the day after the injury, and include pain, swelling, bruising, or inability to contract the muscles. As with most muscle strain injuries, treatment involves RICE therapy, which stands for *R*est, *I*ce, *C*ompression, and *E*levation. Immediately apply ice, and rest and elevate the injured part. Then apply an elastic bandage, if possible, to compress the injured tissue. ■

Relating Muscles to Movements

Arrange the muscles in this exhibit according to the following actions on the thigh at the hip joint: (1) flexion, (2) extension, (3) abduction, (4) adduction, (5) medial rotation, and (6) lateral rotation. The same muscle may be mentioned more than once.

■ CHECKPOINT

1. What forms the iliotibial tract?

continues

EXHIBIT 11.19   MUSCLES THAT MOVE THE FEMUR (THIGH BONE) (Figure 11.22)

CONTINUED

MUSCLE	ORIGIN	INSERTION	ACTION
ILIOPSOAS			
Psoas major (SŌ-as; *psoa*=a muscle of the loin)	Transverse processes and bodies of lumbar vertebrae.	With iliacus into lesser trochanter of femur.	Psoas major and iliacus muscles acting together flex thigh at hip joint, rotate thigh laterally, and flex trunk on the hip as in sitting up from the supine position.
Iliacus (il'-ē-a-cus; *iliac-*=ilium)	Iliac fossa and sacrum.	With psoas major into lesser trochanter of femur.	
GLUTEUS MAXIMUS (GLOO-tē-us MAK-si-mus; *glute-*=rump or buttock; *maximus*=largest)	Iliac crest, sacrum, coccyx, and aponeurosis of sacrospinalis.	Iliotibial tract of fascia lata and lateral part of linea aspera (gluteal tuberosity) under greater trochanter of femur.	Extends thigh at hip joint and laterally rotates thigh.
GLUTEUS MEDIUS (GLOO-tē-us MĒ-dē-us; *medi-*=middle)	Ilium.	Greater trochanter of femur.	Abducts thigh at hip joint and medially rotates thigh.
GLUTEUS MINIMUS (GLOO-tē-us MIN-i-mus; *minim-*=smallest)	Ilium.	Greater trochanter of femur.	Abducts thigh at hip joint and medially rotates thigh.
TENSOR FASCIAE LATAE (TEN-sor FA-shē-ē LĀ-tē; *tensor*=makes tense; *fasciae*=of the band; *lat-*=wide)	Iliac crest.	Tibia by way of the iliotibial tract.	Flexes and abducts thigh at hip joint.
PIRIFORMIS (pir-i-FOR-mis; *piri-*=pear; *form-*=shape)	Anterior sacrum.	Superior border of greater trochanter of femur.	Laterally rotates and abducts thigh at hip joint.
OBTURATOR INTERNUS (OB-too-rā'-tor in-TER-nus; *obturator*=obturator foramen; *intern-*=inside)	Inner surface of obturator foramen, pubis, and ischium.	Medial surface of greater trochanter of femur.	Laterally rotates and abducts thigh at hip joint.
OBTURATOR EXTERNUS (ex-TER-nus; *extern-*=outside)	Outer surface of obturator membrane.	Deep depression inferior to greater trochanter (trochanteric fossa) of femur.	Laterally rotates and abducts thigh at hip joint.
SUPERIOR GEMELLUS (jem-EL-lus; *superior*=above; *gemell-*=twins)	Ischial spine.	Medial surface of greater trochanter of femur.	Laterally rotates and abducts thigh at hip joint.
INFERIOR GEMELLUS (*inferior*=below)	Ischial tuberosity.	Medial surface of greater trochanter of femur.	Laterally rotates and abducts thigh at hip joint.
QUADRATUS FEMORIS (kwod-RĀ-tus FEM-or-is; *quad-*=square, four-sided; *femoris*=femur)	Ischial tuberosity.	Elevation superior to mid-portion of intertrochanteric crest (quadrate tubercle) on posterior femur.	Laterally rotates and stabilizes hip joint.
ADDUCTOR LONGUS (LONG-us; *adductor*=moves part closer to midline; *longus*=long)	Pubic crest and pubic symphysis.	Linea aspera of femur.	Adducts and flexes thigh at hip joint and medially rotates thigh.
ADDUCTOR BREVIS (BREV-is; *brevis*=short)	Inferior ramus of pubis.	Superior half of linea aspera of femur.	Adducts and flexes thigh at hip joint and medially rotates thigh.
ADDUCTOR MAGNUS (MAG-nus; *magnus*=large)	Inferior ramus of pubis and ischium to ischial tuberosity.	Linea aspera of femur.	Adducts thigh at hip joint and medially rotates thigh; anterior part flexes thigh at hip joint, and posterior part extends thigh at hip joint.
PECTINEUS (pek-TIN-ē-us; *pectin-*=a comb)	Superior ramus of pubis.	Pectineal line of femur, between lesser trochanter and linea aspera.	Flexes and adducts thigh at hip joint.

Figure 11.22 **Muscles that move the femur (thigh bone).**

Most muscles that move the femur originate on the pelvic (hip) girdle and insert on the femur.

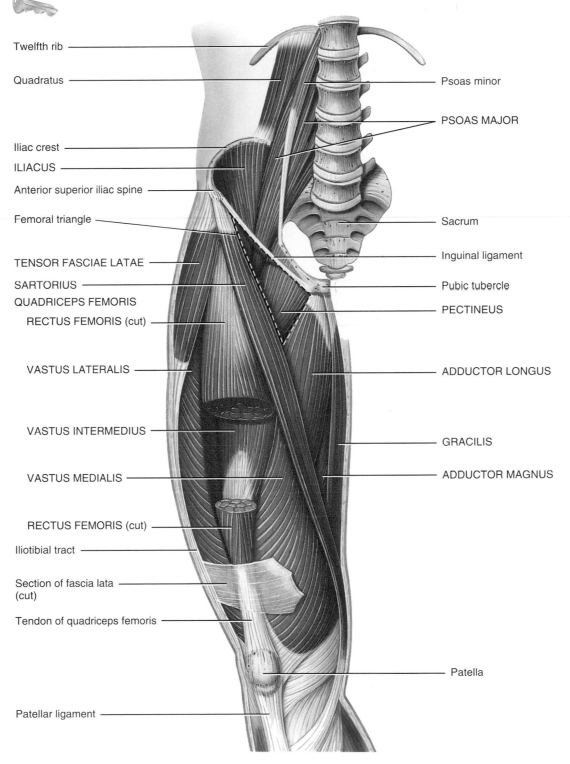

Twelfth rib

Quadratus

Iliac crest

ILIACUS

Anterior superior iliac spine

Femoral triangle

TENSOR FASCIAE LATAE

SARTORIUS

QUADRICEPS FEMORIS

RECTUS FEMORIS (cut)

VASTUS LATERALIS

VASTUS INTERMEDIUS

VASTUS MEDIALIS

RECTUS FEMORIS (cut)

Iliotibial tract

Section of fascia lata (cut)

Tendon of quadriceps femoris

Patellar ligament

Psoas minor

PSOAS MAJOR

Sacrum

Inguinal ligament

Pubic tubercle

PECTINEUS

ADDUCTOR LONGUS

GRACILIS

ADDUCTOR MAGNUS

Patella

(a) Anterior superficial view (the femoral triangle is indicated by a dashed line)

continues

EXHIBIT 11.19   MUSCLES THAT MOVE THE FEMUR (THIGH BONE) (Figure 11.22)

CONTINUED

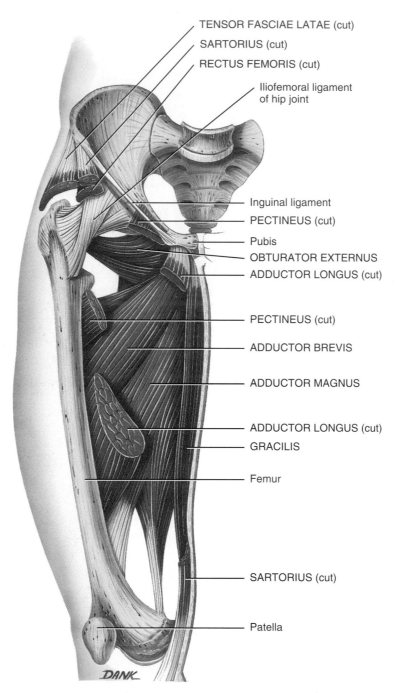

TENSOR FASCIAE LATAE (cut)

SARTORIUS (cut)

RECTUS FEMORIS (cut)

Iliofemoral ligament
of hip joint

Inguinal ligament

PECTINEUS (cut)

Pubis

OBTURATOR EXTERNUS

ADDUCTOR LONGUS (cut)

PECTINEUS (cut)

ADDUCTOR BREVIS

ADDUCTOR MAGNUS

ADDUCTOR LONGUS (cut)

GRACILIS

Femur

SARTORIUS (cut)

Patella

(b) Anterior deep view (femur rotated laterally)

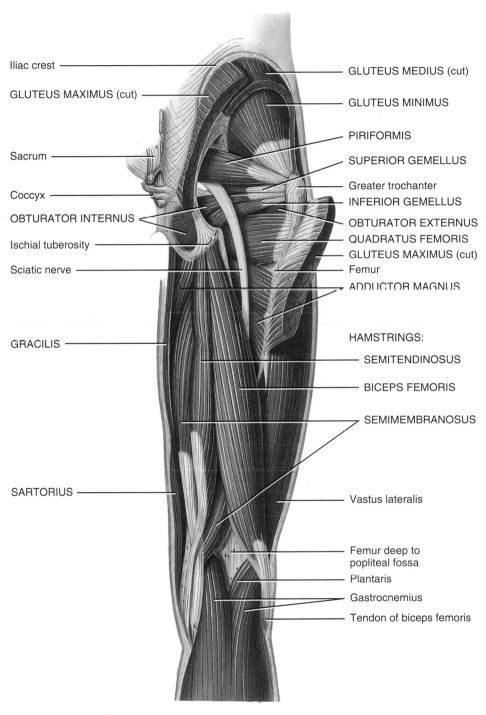

Iliac crest

GLUTEUS MAXIMUS (cut)

Sacrum

Coccyx

OBTURATOR INTERNUS

Ischial tuberosity

Sciatic nerve

GRACILIS

SARTORIUS

GLUTEUS MEDIUS (cut)

GLUTEUS MINIMUS

PIRIFORMIS

SUPERIOR GEMELLUS

Greater trochanter

INFERIOR GEMELLUS

OBTURATOR EXTERNUS

QUADRATUS FEMORIS

GLUTEUS MAXIMUS (cut)

Femur

ADDUCTOR MAGNUS

HAMSTRINGS:

SEMITENDINOSUS

BICEPS FEMORIS

SEMIMEMBRANOSUS

Vastus lateralis

Femur deep to popliteal fossa

Plantaris

Gastrocnemius

Tendon of biceps femoris

(c) Posterior superficial view

continues

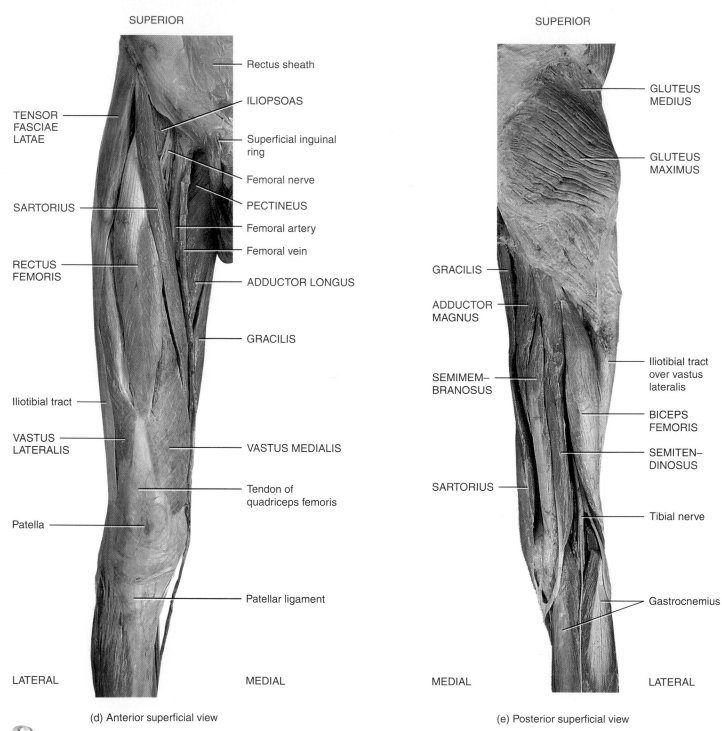

SUPERIOR

Rectus sheath

ILIOPSOAS

Superficial inguinal ring

Femoral nerve

PECTINEUS

Femoral artery

Femoral vein

ADDUCTOR LONGUS

GRACILIS

VASTUS MEDIALIS

Tendon of quadriceps femoris

Patellar ligament

TENSOR FASCIAE LATAE

SARTORIUS

RECTUS FEMORIS

Iliotibial tract

VASTUS LATERALIS

Patella

LATERAL

MEDIAL

(d) Anterior superficial view

SUPERIOR

GLUTEUS MEDIUS

GLUTEUS MAXIMUS

GRACILIS

ADDUCTOR MAGNUS

SEMIMEM– BRANOSUS

SARTORIUS

Iliotibial tract over vastus lateralis

BICEPS FEMORIS

SEMITEN– DINOSUS

Tibial nerve

Gastrocnemius

MEDIAL

LATERAL

(e) Posterior superficial view

? **What are the principal differences between the muscles of the upper and lower limbs?**

OBJECTIVE

● Describe the origin, insertion, action, and innervation of the muscles that act on the femur, tibia, and fibula.

Deep fascia separate the muscles that act on the femur (thigh bone) and tibia and fibula (leg bones) into medial, anterior, and posterior compartments. The muscles of the **medial (adductor) compartment of the thigh** adduct the femur at the hip joint. (See the adductor magnus, adductor longus, adductor brevis, and pectineus, which are components of the medial compartment, in Exhibit 11.19.) The **gracilis,** the other muscle in the medial compartment, not only adducts the thigh, but also flexes the leg at the knee joint. For this reason, it is discussed here. The gracilis is a long, straplike muscle on the medial aspect of the thigh and knee.

The muscles of the **anterior (extensor) compartment of the thigh** extend the leg (and flex the thigh). This compartment contains the quadriceps femoris and sartorius muscles. The **quadriceps femoris** muscle is the biggest muscle in the body, covering most of the anterior surface and sides of the thigh. The muscle is actually a composite muscle, usually described as four separate muscles: (1) **rectus femoris,** on the anterior aspect of the thigh; (2) **vastus lateralis,** on the lateral aspect of the thigh; (3) **vastus medialis,** on the medial aspect of the thigh; and (4) **vastus intermedius,** located deep to the rectus femoris between the vastus lateralis and vastus medialis. The common tendon for the four muscles is known as the **quadriceps tendon,** which inserts into the patella. The tendon continues below the patella as the **patellar ligament,** which attaches to the tibial tuberosity. The quadriceps femoris muscle is the great extensor muscle of the leg. The **sartorius** is a long, narrow muscle that forms a band across the thigh from the ilium of the hip bone to the medial side of the tibia. The various movements it produces help effect the cross-legged sitting position in which the heel of one limb is placed on the knee of the opposite limb. It is known as the tailor's muscle because tailors often assume this cross-legged sitting position. (Because the major action of the sartorius muscle is to move the thigh rather than the leg, it could have been included in Exhibit 11.19.)

The muscles of the **posterior (flexor) compartment of the thigh** flex the leg (and extend the thigh). This compartment is composed of three muscles collectively called the **hamstrings:** (1) **biceps femoris,** (2) **semitendinosus,** and (3) **semimembranosus.** The hamstrings are so-named because butchers hung hams for smoking by their tendons, which are long and stringlike in the popliteal area. Because the hamstrings span two joints (hip and knee), they are both extensors of the thigh and flexors of the leg. The **popliteal fossa** is a diamond-shaped space on the posterior aspect of the knee bordered laterally by the tendons of the biceps femoris muscle and medially by the tendons of the semitendinosus and semimembranosus muscles (see Figure 12.15b on page 399).

Innervation

The obturator and femoral nerves, derived from the lumbar plexus (shown in Exhibit 18.3 on page 000), and the sciatic nerve, derived from the sacral plexus (shown in Exhibit 18.4 on page 575), innervate muscles that act on the femur, tibia, and fibula.

- Gracilis: obturator nerve
- Quadriceps femoris and sartorius: femoral nerve
- Biceps femoris: tibial and common peroneal nerves from the sciatic nerve
- Semimembranosus and semitendinosus: tibial nerve from the sciatic nerve

PULLED HAMSTRINGS

A strain or partial tear of the proximal hamstring muscles is referred to **as pulled hamstrings** or **hamstring strains.** Like groin strains (see Exhibit 11.19), they are common sports injuries in individuals who run very hard and/or are required to perform quick starts and stops. Sometimes the violent muscular exertion required to perform a feat tears off part of the tendinous origins of the hamstrings, especially the biceps femoris, from the ischial tuberosity. This is usually accompanied by a contusion (bruising), tearing of some of the muscle fibers, and rupture of blood vessels, producing a hematoma (collection of blood) and pain. Adequate training with good balance between the quadriceps femoris and hamstings and stretching exercises before running or competing are important in preventing this injury. ■

Relating Muscles to Movements

Arrange the muscles in this exhibit according to the following actions on the thigh at the hip joint: (1) abduction, (2) adduction, (3) lateral rotation (4) flexion, and (5) extension; and according to the following actions on the leg at the knee joint: (1) flexion and (2) extension. The same muscle may be mentioned more than once.

CHECKPOINT

1. Which muscles are part of the medial, anterior, and posterior compartments of the thigh?

continues

CONTINUED

MUSCLE	ORIGIN	INSERTION	ACTION
MEDIAL (adductor) compartment of the thigh			
Adductor magnus (MAG-nus)			
Adductor longus (LONG-us)	See Exhibit 11.19		
Adductor brevis (BREV-is)			
Pectineus (pek-TIN-ē-us)			
Gracilis (gra-SIL-is; *gracilis*=slender)	Body and inferior ramus of pubis.	Medial surface of body of tibia.	Adducts thigh at hip joint, medially rotates thigh, and flexes leg at knee joint.
ANTERIOR (extensor) compartment of the thigh **Quadriceps femoris** (KWOD-ri-ceps FEM-or-is; *quadriceps*=four heads [of origin]; *femoris*=femur)			
Rectus femoris (REK-tus FEM-or-is; *rectus*=fascicles parallel to midline)	Anterior inferior iliac spine.		
Vastus lateralis (VAS-tus lat′-e-RĀ-lis; *vast*=huge; *lateralis*=lateral)	Greater trochanter and linea aspera of femur.	Patella via quadriceps tendon and then tibial tuberosity via patellar ligament.	All four heads extend leg at knee joint; rectus femoris muscle acting alone also flexes thigh at hip joint.
Vastus medialis (VAS-tus mē′-dē-Ā-lis; *medialis*=medial)	Linea aspera of femur.		
Vastus intermedius (VAS-tus in′-ter-MĒ-dē-us; *intermedius*=middle)	Anterior and lateral surfaces of body of femur.		
SARTORIUS (sar-TOR-ē-us; *sartor*=tailor; longest muscle in body)	Anterior superior iliac spine.	Medial surface of body of tibia.	Flexes leg at knee joint; flexes, abducts, and laterally rotates thigh at hip joint.
POSTERIOR (flexor) compartment of the thigh **Hamstrings** A collective designation for three separate muscles.			
Biceps femoris (BĪ-ceps FEM-or-is; *biceps*=two heads of origin)	Long head arises from ischial tuberosity; short head arises from linea aspera of femur.	Head of fibula and lateral condyle of tibia.	Flexes leg at knee joint and extends thigh at hip joint.
Semitendinosus (sem′-ē-ten-di-NŌ-sus; *semi-*=half; *tendo*=tendon)	Ischial tuberosity.	Proximal part of medial surface shaft of tibia.	Flexes leg at knee joint and extends thigh at hip joint.
Semimembranosus (sem′-ē-mem-bra-NŌ-sus; *membran-*=membrane)	Ischial tuberosity.	Medial condyle of tibia.	Flexes leg at knee joint and extends thigh at hip joint.

Figure 11.23 **Muscles that act on the femur (thigh bone) and tibia and fibula (leg bones).**

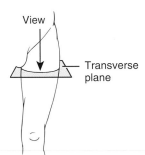

Muscles that act on the leg originate in the hip and thigh and are separated into compartments by deep fascia.

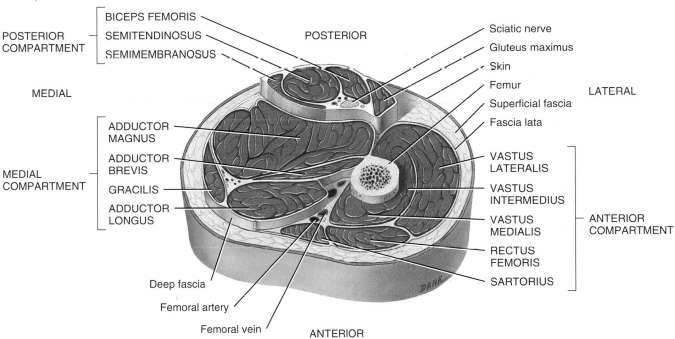

Superior view of transverse section of thigh

Which muscles constitute the quadriceps femoris and hamstring muscles?

continues

EXHIBIT 11.21   MUSCLES THAT MOVE THE FOOT AND TOES (Figure 11.24)

OBJECTIVE

● Describe the origin, insertion, action, and innervation of the muscles that move the foot and toes.

Muscles that move the foot and toes are located in the leg. The muscles of the leg, like those of the thigh, are divided by deep fascia into three compartments: anterior, lateral, and posterior. The **anterior compartment of the leg** consists of muscles that dorsiflex the foot. In a situation analogous to the wrist, the tendons of the muscles of the anterior compartment are held firmly to the ankle by thickenings of deep fascia called the **superior extensor retinaculum** (*transverse ligament of the ankle*) and **inferior extensor retinaculum** (*cruciate ligament of the ankle*).

Within the anterior compartment, the **tibialis anterior** is a long, thick muscle against the lateral surface of the tibia, where it is easy to palpate. The **extensor hallucis longus** is a thin muscle between and partly deep to the tibialis anterior and **extensor digitorum longus** muscles. This featherlike muscle is lateral to the tibialis anterior muscle, where it can be easily palpated. The **fibularis (peroneus) tertius** muscle is part of the extensor digitorum longus, with which it shares a common origin.

The lateral (fibular) compartment of the leg contains two muscles that plantar flex and evert the foot: the **fibularis (peroneus) longus** and **fibularis (peroneus) brevis.**

The **posterior compartment of the leg** consists of muscles in superficial and deep groups. The superficial muscles share a common tendon of insertion, the **calcaneal (Achilles) tendon,** the strongest tendon of the body. It inserts into the calcaneal bone of the ankle. The superficial and most of the deep muscles plantar flex the foot at the ankle joint. The superficial muscles of the posterior compartment are the gastrocnemius, soleus, and plantaris — the so-called calf muscles. The large size of these muscles is directly related to the characteristic upright stance of humans. The **gastrocnemius** is the most superficial muscle and forms the prominence of the calf. The **soleus,** which lies deep to the gastrocnemius, is broad and flat. It derives its name from its resemblance to a flat fish (sole). The **plantaris** is a small muscle that may be absent; sometimes there are two of them in each leg. It runs obliquely between the gastrocnemius and soleus muscles.

The deep muscles of the posterior compartment are the popliteus, tibialis posterior, flexor digitorum longus, and flexor hallucis longus. The **popliteus** is a triangular muscle that forms the floor of the popliteal fossa. The **tibialis posterior** is the deepest muscle in the posterior compartment. It lies between the flexor digitorum longus and flexor hallucis longus muscles. The **flexor digitorum longus** is smaller than the **flexor hallucis longus,** even though the former flexes four toes, whereas the latter flexes only the great toe at the interphalangeal joint.

Innervation

Nerves derived from the sacral plexuses (shown in Exhibit 18.4 on page 575) innervate muscles that move the foot and toes.

- Muscles of the anterior compartment: deep fibular (peroneal) nerve
- Muscles of the lateral compartment: superficial fibular (peroneal) nerve
- Muscles of the posterior compartment: tibial nerve

Shin Splint Syndrome

Shin splint syndrome, or simply **shin splints,** refers to pain or soreness along the tibia, specifically the medial, distal two-thirds. It may be caused by tendinitis of the anterior compartment muscles, especially the tibialis anterior muscle, inflammation of the periosteum (periostitis) around the tibia, or stress fractures of the tibia. The tendinitis usually occurs when poorly conditioned runners run on hard or banked surfaces with poorly supportive running shoes. The condition may also occur with vigorous activity of the legs following a period of relative inactivity. The muscles in the anterior compartment (mainly the tibialis anterior) can be strengthened to balance the stronger posterior compartment muscles. ■

Relating Muscles to Movements

Arrange the muscles in this exhibit according to the following actions on the foot at the ankle joint: (1) dorsiflexion and (2) plantar flexion; according to the following actions on the foot at the intertarsal joints: (1) inversion and (2) eversion; and according to the following actions on the toes at the metatarsophalangeal and interphalangeal joints: (1) flexion and (2) extension. The same muscle may be mentioned more than once.

■ **CHECKPOINT**

1. What are the superior extensor retinaculum and inferior extensor retinaculum?

MUSCLE	ORIGIN	INSERTION	ACTION
ANTERIOR compartment of the leg			
Tibialis anterior (tib′-ē-Ā-lis=tibia; *anterior*=front)	Lateral condyle and body of tibia and interosseous membrane (sheet of fibrous tissue that holds shafts of tibia and fibula together).	First metatarsal and first (medial) cuneiform.	Dorsiflexes foot at ankle joint and inverts foot at intertarsal joints.
Extensor hallucis longus (HAL-ū-sis LON-gus; *extensor*=increases angle at joint; *halluc-*=hallux or great toe; *longus*=long)	Anterior surface of fibula and interosseous membrane.	Distal phalanx of great toe.	Dorsiflexes foot at ankle joint and extends proximal phalanx of great toe at metatarsophalangeal joint.
Extensor digitorum longus (di-ji-TOR-um)	Lateral condyle of tibia, anterior surface of fibula, and interosseous membrane.	Middle and distal phalanges of toes 2–5.*	Dorsiflexes foot at ankle joint and extends distal and middle phalanges of each toe at interphalangeal joints and proximal phalanx of each toe at metatarsophalangeal joint.
Fibularis (Peroneus) tertius (fib-ū-LĀR-is TER-shus; *peron-*=fibula; *tertius*=third)	Distal third of fibula and interosseous membrane.	Base of fifth metatarsal.	Dorsiflexes foot at ankle joint and everts foot at intertarsal joints.
LATERAL (fibular) compartment of the leg			
Fibularis (peroneus) longus (fib-ū-LĀR-is LON-gus)	Head and body of fibula and lateral condyle of tibia.	First metatarsal and first cuneiform.	Plantar flexes foot at ankle joint and everts foot at intertarsal joints.
Fibularis (peroneus) brevis (fib-ū-LĀR-is BREV-is; *brevis*=short)	Body of fibula.	Base of fifth metatarsal.	Plantar flexes foot at ankle joint and everts foot at intertarsal joints.
SUPERFICIAL POSTERIOR compartment of the leg			
Gastrocnemius (gas′-trok-NĒ-mē-us; *gastro-*=belly; *cnem-*=leg)	Lateral and medial condyles of femur and capsule of knee.	Calcaneus by way of calcaneal (Achilles) tendon.	Plantar flexes foot at ankle joint and flexes leg at knee joint.
Soleus (SŌ-lē-us; *sole*=a type of flat fish)	Head of fibula and medial border of tibia.	Calcaneus by way of calcaneal (Achilles) tendon.	Plantar flexes foot at ankle joint.
Plantaris (plan-TĀR-is; *plantar-*=sole of foot)	Femur superior to lateral condyle.	Calcaneus by way of calcaneal (Achilles) tendon.	Plantar flexes foot at ankle joint and flexes leg at knee joint.
DEEP POSTERIOR compartment of the leg			
Popliteus (pop-LIT-ē-us; *poplit-*=the back of the knee)	Lateral condyle of femur.	Proximal tibia.	Flexes leg at knee joint and medially rotates tibia to unlock the extended knee.
Tibialis posterior (tib′-ē-Ā-lis; *posterior*=back)	Tibia, fibula, and interosseous membrane.	Second, third, and fourth metatarsals; navicular; all three cuneiforms; and cuboid.	Plantar flexes foot at ankle joint and inverts foot at intertarsal joints.
Flexor digitorum longus (di′-ji-TOR-um LON-gus; *digit*=finger or toe)	Posterior surface of tibia.	Distal phalanges of toes 2–5.	Plantar flexes foot at ankle joint; flexes distal and middle phalanges of each toe at interphalangeal joints and proximal phalanx of each toe at metatarsophalangeal joint.
Flexor hallucis longus (HAL-ū-sis LON-gus; *flexor*=decreases angle at point	Inferior two-thirds of fibula.	Distal phalanx of great toe.	Plantar flexes foot at ankle joint; flexes distal phalanx of great toe at interphalangeal joint and proximal phalanx of great toe at metatarso-phalangeal joint.

*Reminder: The great toe or hallux is the first toe and has two phalanges: proximal and distal. The remaining toes are numbered 2–5, and each has three phalanges: proximal, middle, and distal.

continues

EXHIBIT 11.21   MUSCLES THAT MOVE THE FOOT AND TOES (Figure 11.24)

Figure 11.24   Muscles that move the foot and toes.

The superficial muscles of the posterior compartment share a common tendon of insertion, the calcaneal (Achilles) tendon, that inserts into the calcaneal bone of the ankle.

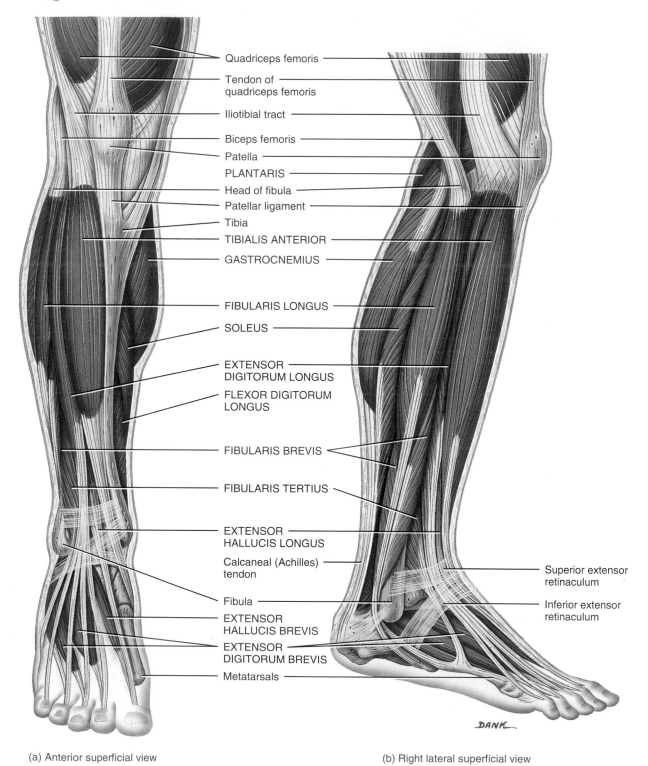

Quadriceps femoris
Tendon of quadriceps femoris
Iliotibial tract
Biceps femoris
Patella
PLANTARIS
Head of fibula
Patellar ligament
Tibia
TIBIALIS ANTERIOR
GASTROCNEMIUS
FIBULARIS LONGUS
SOLEUS
EXTENSOR DIGITORUM LONGUS
FLEXOR DIGITORUM LONGUS
FIBULARIS BREVIS
FIBULARIS TERTIUS
EXTENSOR HALLUCIS LONGUS
Calcaneal (Achilles) tendon
Fibula
EXTENSOR HALLUCIS BREVIS
EXTENSOR DIGITORUM BREVIS
Metatarsals

Superior extensor retinaculum
Inferior extensor retinaculum

DANK

(a) Anterior superficial view

(b) Right lateral superficial view

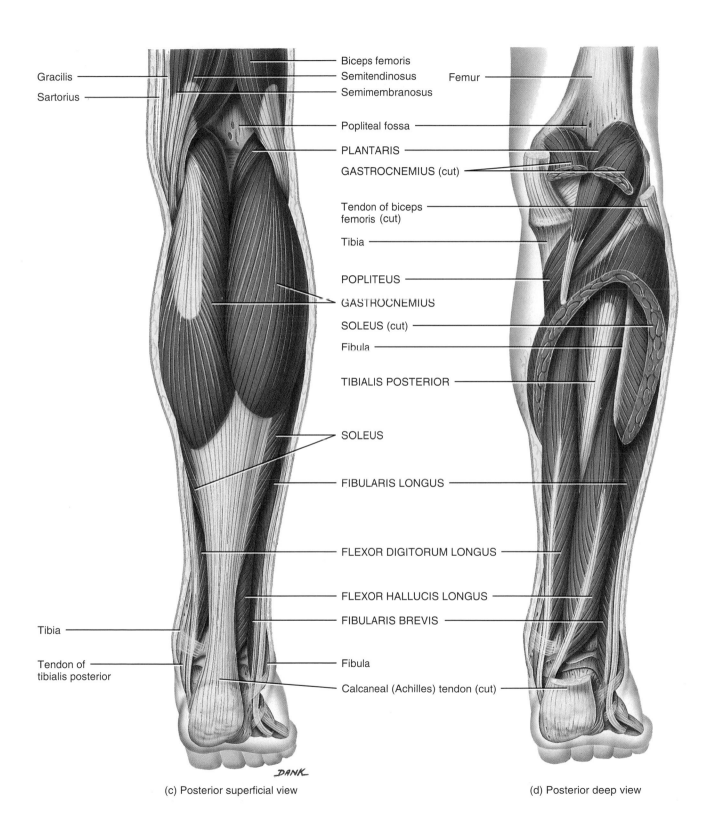

Gracilis

Sartorius

Biceps femoris

Semitendinosus

Semimembranosus

Femur

Popliteal fossa

PLANTARIS

GASTROCNEMIUS (cut)

Tendon of biceps
femoris (cut)

Tibia

POPLITEUS

GASTROCNEMIUS

SOLEUS (cut)

Fibula

TIBIALIS POSTERIOR

SOLEUS

FIBULARIS LONGUS

FLEXOR DIGITORUM LONGUS

FLEXOR HALLUCIS LONGUS

FIBULARIS BREVIS

Tibia

Tendon of
tibialis posterior

Fibula

Calcaneal (Achilles) tendon (cut)

(c) Posterior superficial view

(d) Posterior deep view

DANK

continues

EXHIBIT 11.21   MUSCLES THAT MOVE THE FOOT AND TOES (Figure 11.24)

CONTINUED

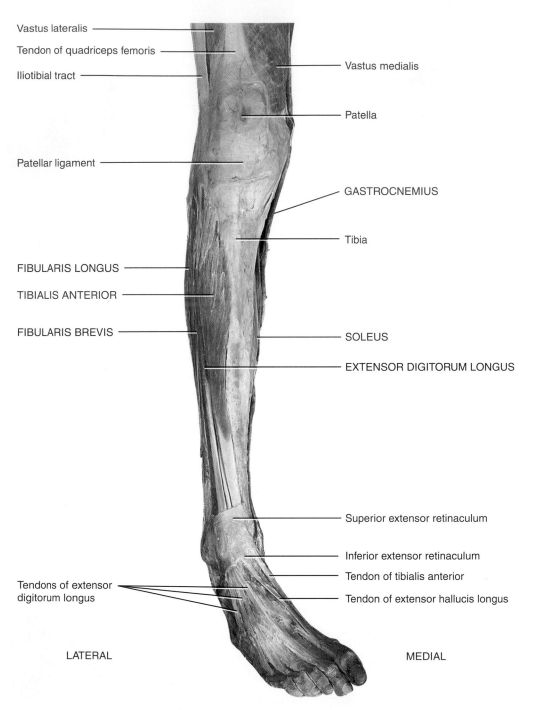

Vastus lateralis

Tendon of quadriceps femoris

Iliotibial tract

Patellar ligament

FIBULARIS LONGUS

TIBIALIS ANTERIOR

FIBULARIS BREVIS

Tendons of extensor
digitorum longus

LATERAL

Vastus medialis

Patella

GASTROCNEMIUS

Tibia

SOLEUS

EXTENSOR DIGITORUM LONGUS

Superior extensor retinaculum

Inferior extensor retinaculum

Tendon of tibialis anterior

Tendon of extensor hallucis longus

MEDIAL

(e) Anterior superficial view

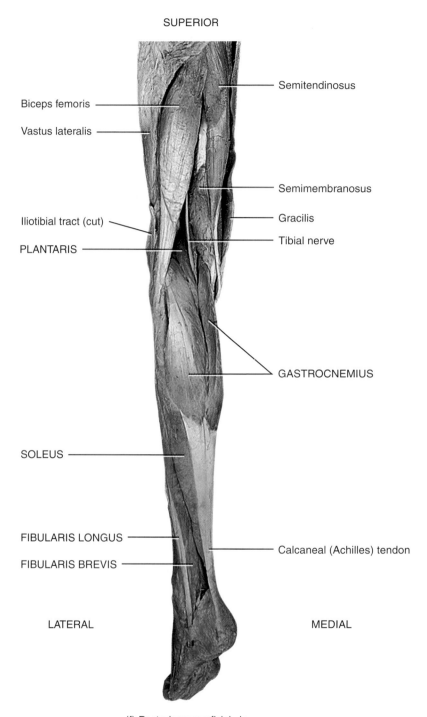

SUPERIOR

Semitendinosus

Biceps femoris

Vastus lateralis

Semimembranosus

Gracilis

Iliotibial tract (cut)

Tibial nerve

PLANTARIS

GASTROCNEMIUS

SOLEUS

FIBULARIS LONGUS

Calcaneal (Achilles) tendon

FIBULARIS BREVIS

LATERAL

MEDIAL

(f) Posterior superficial view

What structures firmly hold the tendons of the anterior compartment muscles to the ankle?

EXHIBIT 11.22   INTRINSIC MUSCLES OF THE FOOT (Figure 11.25)

OBJECTIVE

● Describe the origin, insertion, action, and innervation of the intrinsic muscles of the foot.

The muscles in this exhibit are termed **intrinsic muscles** because they originate and insert *within* the foot. Whereas the muscles of the hand are specialized for precise and intricate movements, those of the foot are limited to support and locomotion. The deep fascia of the foot forms the **plantar aponeurosis (fascia)** that extends from the calcaneus bone to the phalanges of the toes. The aponeurosis supports the longitudinal arch of the foot and encloses the flexor tendons of the foot.

The intrinsic muscles of the foot are divided into two groups: **dorsal** and **plantar.** There is only one dorsal muscle, the **extensor digitorum brevis,** a four-part muscle deep to the tendons of the extensor digitorum longus muscle, which extends toes 2–5 at the metatarsophalangeal joints.

The plantar muscles are arranged in four layers. The most superficial layer is called the first layer. Three muscles are in the first layer. The **abductor hallucis,** which lies along the medial border of the sole, is comparable to the abductor pollicis brevis in the hand, and abducts the great toe at the metatarsophalangeal joint. The **flexor digitorum brevis,** which lies in the middle of the sole, flexes toes 2–5 at the interphalangeal and metatarsophalangeal joints. The **abductor digiti minimi,** which lies along the lateral border of the sole, is comparable to the same muscle in the hand, and abducts the little toe.

The second layer consists of the **quadratus plantae,** a rectangular muscle that arises by two heads and flexes toes 2–5 at the metatarsophalangeal joints, and the **lumbricals,** four small muscles that are similar to the lumbricals in the hands. They flex the proximal phalanges and extend the distal phalanges of toes 2–5.

Three muscles comprise the third layer. The **flexor hallucis brevis,** which lies adjacent to the plantar surface of the metatarsal of the great toe, is comparable to the same muscle in the hand, and flexes the great toe. The **adductor hallucis,** which has an oblique and transverse head like the adductor pollicis in the hand, adducts the great toe. The **flexor digiti minimi brevis,** which lies superficial to the metatarsal of the little toe, is comparable to the same muscle in the hand, and flexes the little toe.

The fourth layer is the deepest and consists of two muscle groups. The **dorsal interossei** are four muscles that abduct toes 2–4, flex the proximal phalanges, and extend the distal phalanges. The three **plantar interossei** abduct toes 3–5, flex the proximal phalanges, and extend the distal phalanges. The interossei of the feet are similar to those of the hand. However, their actions are rela-

tive to the midline of the second digit rather than the third digit as in the hand.

Innervation

Nerves derived from the sacral plexus (shown in Exhibit 18.4 on page 575) innervate intrinsic muscles of the foot.

- Extensor digitorum brevis: deep fibular (peroneal) nerve
- Abductor hallucis and flexor digitorum brevis: medial plantar nerve
- Abductor digiti minimi, quadratus plantae, adductor hallucis, flexor digiti minimi brevis, and dorsal and plantar interossei: lateral plantar nerve
- Lumbricals: medial and lateral plantar nerves

 PLANTAR FASCIITIS

Plantar fasciitis (fas-ē-Ī-tis) or **painful heel syndrome** is an inflammatory reaction due to chronic irritation of the plantar aponeurosis (fascia) at its origin on the calcaneus (heel bone). The aponeurosis becomes less elastic with age. This condition is also related to weight-bearing activities (walking, jogging, lifting heavy objects), improperly constructed or ill-fitting shoes, excess weight (puts pressure on the feet), and poor biomechanics (flat feet, high arches, and abnormalities in gait may cause uneven distribution of weight on the feet). Plantar fasciitis is the most common cause of heel pain in runners and arises in response to the repeated impact of running. Treatments include ice, deep heat, stretching exercises, weight loss, prosthetics (such as shoe inserts or heel lifts), steroid injections, and surgery. ■

Relating Muscles to Movements

Arrange the muscles in this exhibit according to the following actions on the great toe at the metatarsophalangeal joint: (1) flexion, (2) extension, (3) abduction, and (4) adduction; and according to the following actions on toes 2–5 at the metatarsophalangeal and interphalangeal joints: (1) flexion, (2) extension, (3) abduction, and (4) adduction. The same muscle may be mentioned more than once.

▦ **CHECKPOINT**

1. How do the intrinsic muscles of the hand and foot differ in function?

MUSCLE	ORIGIN	INSERTION	ACTION
DORSAL			
Extensor digitorum brevis (*extensor*=increases angle at joint; *digit*=finger or toe; *brevis*=short) (see Figure 11.24a, b)	Calcaneus and inferior extensor retinaculum.	Tendons of extensor digitorum longus on toes 2–4 and proximal phalanx of great toe.*	Extensor hallucis brevis extends great toe at metatarsophalangeal joint and extensor digitorum brevis extends toes 2–4 at interphalangeal joints.
PLANTAR **First Layer (most superficial)**			
Abductor hallucis (*abductor*= moves part away from midline; *hallucis*=hallux or great toe)	Calcaneus, plantar aponeurosis, and flexor retinaculum.	Medial side of proximal phalanx of great toe with the tendon of the flexor hallucis brevis.	Abducts and flexes great toe at metatarsophalangeal joint.
Flexor digitorum brevis (*flexor*=decreases angle at joint)	Calcaneus and plantar aponeurosis.	Sides of middle phalanx of toes 2–5.	Flexes toes 2–5 at proximal interphalangeal and metatarsophalangeal joints.
Abductor digiti minimi (*minimi*=smallest)	Calcaneus and plantar aponeurosis.	Lateral side of proximal phalanx of little toe with the tendon of the flexor digiti minimi brevis.	Abducts and flexes little toe at metatarsophalangeal joint.
Second layer			
Quadratus plantae (kwod-RĀ-tus; *quad-*=square, four-sided; *planta*=the sole)	Calcaneus.	Tendon of flexor digitorum longus.	Assists flexor digitorum longus to flex toes 2–5 at interphalangeal and metatarsophalangeal joints.
Lumbricals (LUM-bri-kals; *lumbric-*=earthworm)	Tendons of flexor digitorum longus.	Tendons of extensor digitorum longus on proximal phalanges of toes 2–5.	Extend toes 2–5 at interphalangeal joints and flex toes 2–5 at metatarsophalangeal joints.
Third Layer			
Flexor hallucis brevis	Cuboid and third (lateral) cuneiform.	Medial and lateral sides of proximal phalanx of great toe via a tendon containing a sesamoid bone.	Flexes great toe at metatarsophalangeal joint.
Adductor hallucis	Metatarsals 2–4, ligaments of 3–5 metatarsophalangeal joints, and tendon of peroneus longus.	Lateral side of proximal phalanx of great toe.	Adducts and flexes great toe at metatarsophalangeal joint.
Flexor digiti minimi brevis	Metatarsal 5 and tendon of peroneus longus.	Lateral side of proximal phalanx of little toe.	Flexes little toe at metatarsophalangeal joint.
Fourth Layer (deepest)			
Dorsal interossei (in-ter-OS-ē-ī) (not illustrated)	Adjacent side of metatarsals.	Proximal phalanges: both sides of toe 2 and lateral side of toes 3 and 4.	Abduct and flex toes 2–4 at metatarsophalangeal joints and extend toes at interphalangeal joints.
Plantar interossei	Metatarsals 3–5.	Medial side of proximal phalanges of toes 3–5.	Adduct and flex proximal metatarsophalangeal joints and extend toes at interphalangeal joints.

*The tendon that inserts into the proximal phalanx of the great toe, together with its belly, is often described as a separate muscle, the extensor hallucis brevis.

continues

EXHIBIT 11.22   **INTRINSIC MUSCLES OF THE FOOT** (Figure 11.25)

Figure 11.25   Intrinsic muscles of the foot.

Whereas the muscles of the hand are specialized for precise and intricate movements, those of the foot are limited to support and movement.

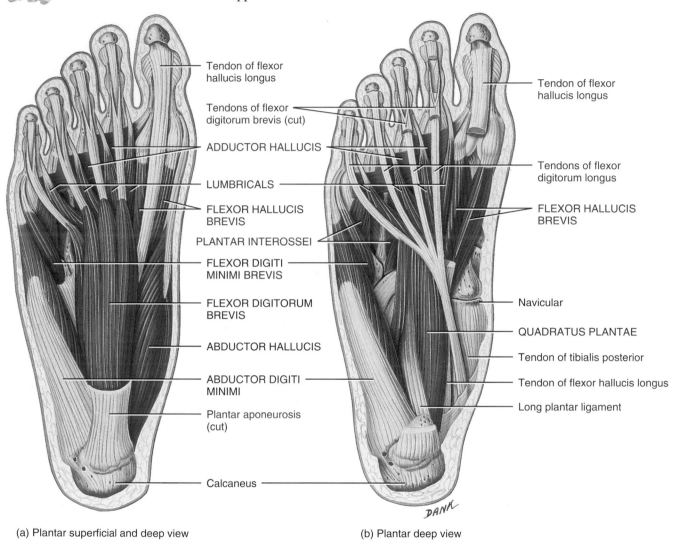

Tendon of flexor
hallucis longus

Tendons of flexor
digitorum brevis (cut)

ADDUCTOR HALLUCIS

LUMBRICALS

FLEXOR HALLUCIS
BREVIS

PLANTAR INTEROSSEI

FLEXOR DIGITI
MINIMI BREVIS

FLEXOR DIGITORUM
BREVIS

ABDUCTOR HALLUCIS

ABDUCTOR DIGITI
MINIMI

Plantar aponeurosis
(cut)

Calcaneus

Tendon of flexor
hallucis longus

Tendons of flexor
digitorum longus

FLEXOR HALLUCIS
BREVIS

Navicular

QUADRATUS PLANTAE

Tendon of tibialis posterior

Tendon of flexor hallucis longus

Long plantar ligament

DANK

(a) Plantar superficial and deep view

(b) Plantar deep view

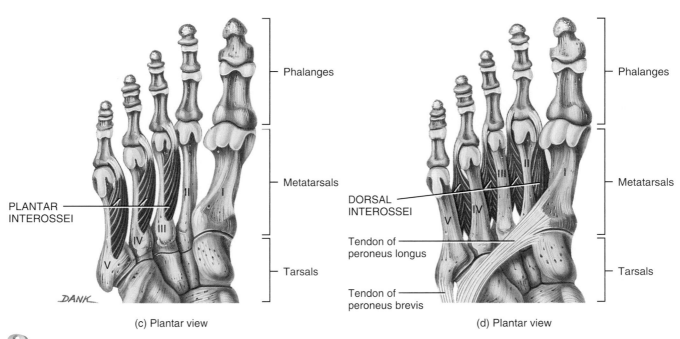

PLANTAR
INTEROSSEI

Phalanges

Metatarsals

Tarsals

(c) Plantar view

DORSAL
INTEROSSEI

Tendon of
peroneus longus

Tendon of
peroneus brevis

Phalanges

Metatarsals

Tarsals

(d) Plantar view

What structure supports the longitudinal arch and encloses the flexor tendons of the foot?

APPLICATIONS TO HEALTH

RUNNING INJURIES

Nearly 70% of those who jog or run sustain some type of running-related injury. Although most such injuries are minor, some are quite serious. Moreover, untreated or inappropriately treated minor injuries may become chronic. Among runners, common sites of injury include the ankle, knee, calcaneal (Achilles) tendon, hip, groin, foot, and back. Of these, the knee often is the most severely injured area.

Running injuries are frequently related to faulty training techniques. This may involve improper (or lack of) warm-up routines, running too much, or running too soon after an injury. Or it might involve extended running on hard and/or uneven surfaces. Poorly constructed or worn-out running shoes can also contribute to injury, as can any biomechanical problem (such as a fallen arch) that is aggravated by running.

Most sports injuries should be treated initially with RICE therapy, which stands for *Rest, Ice, Compression,* and *Elevation.* Immediately apply ice, and rest and elevate the injured part. Then apply an elastic bandage, if possible, to compress the injured tissue. Continue using RICE for 2 to 3 days, and resist the temptation to apply heat, which may worsen the swelling. Follow-up treatment may include alternating moist heat and ice massage to enhance blood flow in the injured area. Sometimes it is helpful to take nonsteroidal antiinflammatory drugs (NSAIDs) or to have local injections of corticosteroids. During the recovery period, it is important to keep active using an alternative fitness program that does not worsen the original injury. This activity should be determined in consultation with a physician.

Finally, careful exercise is needed to rehabilitate the injured area itself.

COMPARTMENT SYNDROME

As noted earlier in this chapter, skeletal muscles in the limbs are organized into units called *compartments*. In a disorder called **compartment syndrome,** some external or internal pressure constricts the structures within a compartment, resulting in damaged blood vessels and subsequent reduction of the blood supply (ischemia) to the structures within the compartment. Common causes of compartment syndrome include crushing and penetrating injuries, contusion (damage to subcutaneous tissues without the skin being broken), muscle strain (overstretching of a muscle), or an improperly fitted cast. The pressure increase in the compartment can have serious consequences, such as hemorrhage, tissue injury, and edema (buildup of interstitial fluid). Because the fasciae that enclose the compartments are very strong, accumulated blood and interstitial fluid cannot escape, and the increased pressure can literally choke off the blood flow and deprive nearby muscles and nerves of oxygen. One treatment option is **fasciotomy** (fash-ē-OT-ō-mē), a surgical procedure in which muscle fascia is cut to relieve the pressure. Without intervention, nerves can suffer damage, and muscles can develop scar tissue that results in permanent shortening of the muscles, a condition called *contracture.* If left untreated, tissues may die (necrosis), and the limb may no longer be able to function. Once the syndrome has reached this stage, amputation may be the only treatment option.

STUDY OUTLINE

How Skeletal Muscles Produce Movements (p. 289)

1. Skeletal muscles that produce movement do so by pulling on bones.
2. The attachment to the more stationary bone is the origin; the attachment to the more movable bone is the insertion.
3. Bones serve as levers, and joints serve as fulcrums. Two different forces act on the lever: load (resistance) and effort.
4. Levers are categorized into three types—first-class, second-class, and third-class (most common)—according to the positions of the fulcrum, the effort, and the load on the lever.
5. Fascicular arrangements include parallel, fusiform, circular, triangular, and pennate. Fascicular arrangement affects a muscle's power and range of motion.
6. A prime mover produces the desired action; an antagonist produces an opposite action. Synergists assist a prime mover by reducing unnecessary movement. Fixators stabilize the origin of a prime mover so that it can act more efficiently.

How Skeletal Muscles Are Named (p. 293)

1. Distinctive features of different skeletal muscles include direction of muscle fascicles; size, shape, action, number of origins (or heads), and location of the muscle; and sites of origin and insertion of the muscle.
2. Most skeletal muscles are named based on combinations of these features.

Principal Skeletal Muscles (p. 297)

1. Muscles of facial expression move the skin rather than a joint when they contract, and they permit us to express a wide variety of emotions.
2. The extrinsic muscles that move the eyeballs are among the fastest contracting and most precisely controlled skeletal muscles in the body. They permit us to elevate, depress, abduct, adduct, and medially and laterally rotate the eyeballs.
3. Muscles that move the mandible (lower jaw) are also known as the muscles of mastication because they are involved in chewing.

4. The extrinsic muscles that move the tongue are important in chewing, swallowing, and speech.

5. Muscles of the floor of the anterior neck, called suprahyoid muscles, are located above the hyoid bone. They elevate the hyoid bone, oral cavity, and tongue during swallowing.

6. Muscles that move the head alter its position and help balance the head on the vertebral column.

7. Most muscles of the larynx (voice box) depress the hyoid bone and larynx during swallowing and speech and move the internal portion of the larynx during speech.

8. Muscles of the pharynx arc arranged into a circular layer, which functions in swallowing, and a longitudinal layer, which functions in swallowing and speaking.

9. Muscles that act on the abdominal wall help contain and protect the abdominal viscera, move the vertebral column, compress the abdomen, and produce the force required for defecation, urination, vomiting, and childbirth.

10. Muscles used in breathing alter the size of the thoracic cavity so that ventilation can occur and assist in venous return of blood to the heart.

11. Muscles of the pelvic floor support the pelvic viscera, resist the thrust that accompanies increases in intraabdominal pressure, and function as sphincters at the anorectal junction, urethra, and vagina.

12. Muscles of the perineum assist in urination, erection of the penis and clitoris, ejaculation, and defecation.

13. Muscles that move the pectoral (shoulder) girdle stabilize the scapula so it can function as a stable point of origin for most of the muscles that move the humerus.

14. Muscles that move the humerus (arm bone) originate for the most part on the scapula (scapular muscles); the remaining muscles originate on the axial skeleton (axial muscles).

15. Muscles that move the radius and ulna (forearm bones) are involved in flexion and extension at the elbow joint and are organized into flexor and extensor compartments.

16. Muscles that move the wrist, hand, thumb, and fingers are many and varied; those muscles that act on the digits are called extrinsic muscles.

17. The intrinsic muscles of the hand are important in skilled activities and provide humans with the ability to grasp and manipulate objects precisely.

18. Muscles that move the vertebral column are quite complex because they have multiple origins and insertions and because there is considerable overlap among them.

19. Muscles that move the femur (thigh bone) originate for the most part on the pelvic girdle and insert on the femur; these muscles are larger and more powerful than comparable muscles in the upper limb.

20. Muscles that move the femur (thigh bone) and tibia and fibula (leg bones) are separated into medial (adductor), anterior (extensor), and posterior (flexor) compartments.

21. Muscles that move the foot and toes are divided into anterior, lateral, and posterior compartments.

22. Intrinsic muscles of the foot, unlike those of the hand, are limited to the functions of support and locomotion.

SELF-QUIZ QUESTIONS

Choose the one best answer to the following questions.

1. Which of these statements is *false* when you hyperextend your head as if to look at the sky?
   a. The weight of the face and jaw serves as the load (L).
   b. The posterior neck muscles provide the E (effort).
   c. The F (fulcrum) is the atlanto-occipital joint.
   d. This is an example of a second-class lever.
   e. All statements are false.

2. The thenar muscles act on the:
   a. wrist.   b. thumb.   c. little finger.
   d. all fingers except the thumb.   e. the great toe.

3. During exhalation the diaphragm:
   a. contracts and flattens.
   b. relaxes and flattens.
   c. contracts and forms a dome.
   d. relaxes and forms a dome.
   e. does not contract or relax.

4. Choose the muscle(s) that depress(es) the mandible:
   a. lateral ptcrygoid   b. digastric
   c. medial pterygoid   d. Both a and b are correct.
   e. All three muscles depress the mandible.

5. The action of the pectoralis major muscle is to:
   a. abduct the arm and rotate the arm laterally.
   b. flex, adduct, and rotate that arm medially.
   c. adduct the arm and rotate the arm laterally.
   d. abduct the arm and rotate the arm medially.
   e. abduct and raise the arm.

6. Which of the following is part of the medial compartment of muscles that act on the femur?
   a. semitendinosus   b. biceps femoris   c. rectus femoris
   d. gluteus medius   e. gracilis

7. Muscles of which group share a common origin on the ischium and act to extend the thigh and flex the leg?
   a. gluteal muscles   b. quadriceps muscles
   c. adductor muscles   d. hamstring muscles
   e. peroneal muscles

Complete the following.

8. A muscle that contracts to cause the desired action is called the _____. Muscles that assist or cooperate with the muscle that causes the desired action are known as _____.

9. The attachment of a muscle's tendon to the stationary bone is called the _____, and the attachment of the muscle's other tendon to the movable bone is called the _____.

10. In an anatomical lever system, _____ act as levers, _____ act as fulcrums, and the effort is provided by _____.

11. The muscles of the posterior compartment of muscles also known as the hamstrings are the _____, the _____, and the _____.

12. The sternocleidomastoid and longissimus capitis muscles insert on the _____ of the temporal bone.

13. The diaphragm inserts onto the _____ tendon.

14. Three posterior thoracic muscles that adduct the scapula are the _____, the _____, and the _____.

15. The diamond-shaped area that extends from the pubic symphysis anteriorly, to the coccyx posteriorly, and to the ischial tuberosities laterally is the _____.

16. A circular muscle that decreases the size of an opening, such as the anus, is known as a _____.

17. The pectineus, psoas major, adductor longus, and iliacus muscles have one action in common: _____ of the thigh.

18. The quadriceps femoris is the main muscle mass on the _____ surface of the thigh. The name of this muscle mass indicates it has _____ heads of origin. All converge to insert on the _____ (bone) by means of the patellar ligament.

19. The three muscles that utilize the calcaneal tendon to insert on the calcaneus are the _____, the _____, and the _____.

Are the following statements true or false?

20. The flexor retinaculum is located on the anterior aspect of the wrist and serves to hold the tendons, nerves, and blood vessels close to the bones of the wrist.

21. The action of the biceps brachii on the forearm at the elbow joint illustrates the structure and action of a first-class lever system.

22. The muscles that plantar flex the foot at the ankle joint are located in the anterior compartment of the leg.

Matching

23. Match each of these muscles with the clue(s) given by the parts of its name.

___ (a) transversus abdominis   (1) action
___ (b) gluteus minimus   (2) direction of fibers
___ (c) biceps brachii   (3) location
___ (d) sternocleidomastoid   (4) number of tendons of origin
___ (e) adductor magnus   (5) sites of origin and/or insertion
___ (f) tibialis anterior
___ (g) rhomboid major   (6) size or shape

24. Match the muscles with the actions:

___ (a) presses cheeks against teeth and lips   (1) latissimus dorsi
___ (b) moves eyeball inferiorly and laterally, rotates eyeball medially   (2) quadriceps femoris
                                                       (3) superior oblique
                                                       (4) buccinator
___ (c) elevate and adduct scapula   (5) trapezius
___ (d) extend leg at knee joint   (6) external intercostals
___ (e) extends the forearm at the shoulder joint   (7) triceps brachii
___ (f) extends, adducts, and medially rotates arm at shoulder joint   (8) deltoid
                                                       (9) external oblique
___ (g) flexes and rotates vertebral column   (10) gluteus maximus
___ (h) extends and laterally rotates thigh at hip joint
___ (i) abducts arm at shoulder joint
___ (j) elevates ribs during inspiration

25. Match the following muscles with the appropriate nerves:

___ (a) most of the muscles of facial expression   (1) hypoglossal (XII) nerve
___ (b) muscles that move the mandible   (2) thoracic nerves 7–12
___ (c) most muscles that move the tongue   (3) facial (VII) nerve
___ (d) most muscles of the pharynx   (4) tibial nerve
___ (e) muscles that move the head   (5) femoral nerve
___ (f) muscles of the anterolateral abdominal wall   (6) trigeminal (V) nerve, mandibular division
___ (g) diaphragm   (7) deep peroneal nerve
___ (h) quadriceps femoris muscles   (8) accessory (XI) and cervical spinal nerves
___ (i) muscles of the anterior compartment of the leg (9) superficial peroneal nerve
___ (j) muscles of the lateral compartment of the leg (10) phrenic nerve
___ (k) muscles of the posterior compartment of the leg   (11) vagus (X) nerve

CRITICAL THINKING QUESTIONS

1. Three-year-old Ming likes to form her tongue into a cylinder shape and use it as a straw when she drinks her milk. Name the muscles that Ming uses to protrude her lips and tongue and to suck up the milk.

2. Baby Eddie weighed 13 lb 4 oz at birth and he's been gaining ever since. His mom has noticed a sharp pain between her shoulder blades when she picks up Eddie to put him in his high chair (again). Which muscle may Eddie's mother have strained?

3. A group of dental students was working out while singing "we might, we might, we might improve our bite!" Which muscles work to elevate and depress the jaw?

4. Perry was leading his fantasy baseball league until his best pitcher went on the disabled list with a shoulder injury. What is the most likely injury to this player? What specific muscles are likely to be involved?

5. Wyman has been doing a lot of heavy lifting lately. He has suddenly noticed a bulge on the anterior aspect of his torso down near his groin. What has probably happened to him? Do you think he needs to see a doctor? Why or why not?

ANSWERS TO FIGURE QUESTIONS

11.1   The belly of the muscle that extends the forearm, the triceps brachii, is located posterior to the humerus.

11.2   Second-class levers produce the most force.

11.3   For muscles named after their various characteristics, here are possible correct responses (for others, see Table 11.2): direction of fibers: external oblique; shape: deltoid; action: extensor digitorum; size: gluteus maximus; origin and insertion: sternocleidomastoid; location: tibialis anterior; number of tendons of origin: biceps brachii.

11.4   The frontal belly of the occipitofrontalis muscle is involved in frowning; the zygomaticus major muscle contracts when you smile; the platysma muscle contributes to pouting; the orbicularis oculi muscle contributes to squinting.

11.5   The inferior oblique muscle moves the eyeball superiorly and laterally because it originates at the anteromedial aspect of the floor of the orbit and inserts on the posterolateral aspect of the eyeball.

11.6   The masseter is the strongest of the chewing muscles (muscles of mastication).

11.7   Functions of the tongue include chewing, perception of taste, swallowing, and speech.

11.8   The suprahyoid and infrahyoid muscles stabilize the hyoid bone to assist in tongue movements.

11.9   The triangles in the neck formed by the sternocleidomastoid muscles are important anatomically and surgically because of the structures that lie within their boundaries.

11.10   The extrinsic muscles move the larynx as a whole, while the intrinsic muscles move parts of the larynx.

11.11   The longitudinal muscles elevate the larynx, whereas the infrahyoid muscles depress the larynx.

11.12   The rectus abdominis muscle aids in urination.

11.13   The diaphragm is innervated by the phrenic nerve.

11.14   The borders of the pelvic diaphragm are the pubic symphysis anteriorly, the coccyx posteriorly, and the walls of the pelvis laterally.

11.15   The borders of the perineum are the pubic symphysis anteriorly, the coccyx posteriorly, and the ischial tuberosities laterally.

11.16   The main action of the muscles that move the pectoral girdle is to stabilize the scapula to assist in movements of the humerus.

11.17   The rotator cuff consists of the flat tendons of the subscapularis, supraspinatus, infraspinatus, and teres minor muscles that form a nearly complete circle around the shoulder joint.

11.18   The brachialis is the most powerful forearm flexor; the triceps brachii is the most powerful forearm extensor.

11.19   Flexor tendons of the digits and wrist and the median nerve pass through the flexor retinaculum.

11.20   Muscles of the thenar eminence act on the thumb.

11.21   The splenius muscles arise from the midline and extend laterally and superiorly to their insertion.

11.22   Upper limb muscles exhibit diversity of movement; lower limb muscles function in stability, locomotion, and maintenance of posture. Moreover, lower limb muscles usually cross two joints and act equally on both.

11.23   The quadriceps femoris consists of the rectus femoris, vastus lateralis, vastus medialis, and vastus intermedius; the hamstrings consists of the biceps femoris, semitendinosus, and semimembranosus.

11.24   The superior and inferior extensor retinacula firmly hold the tendons of the anterior compartment muscles to the ankle.

11.25   The plantar aponeurosis supports the longitudinal arch and encloses the flexor tendons of the foot.

12 | SURFACE ANATOMY

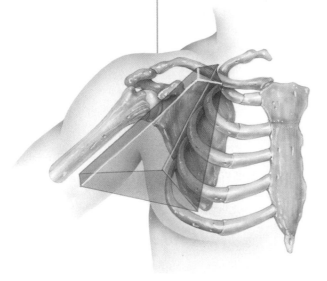

INTRODUCTION In Chapter 1 we introduced several branches of anatomy and noted their relationship to our understanding of the body's structure. In this chapter we will take a closer look at **surface anatomy**. Recall that surface anatomy is the study of the anatomical landmarks on the exterior of the body. A knowledge of surface anatomy will not only help you to identify structures on the body's exterior, but it will also assist you in locating the positions of various internal structures.

The study of surface anatomy involves two related, although distinct, activities: visualization and palpation. **Visualization** involves looking in a very selective and purposeful manner at a specific part of the body. **Palpation** (pal-PĀ-shun) means using the sense of touch to determine the location of an internal part of the body through the skin. Like visualization, palpation is performed selectively and purposefully, and it supplements information already gained through other methods, including visualization. Because of the varying depth of different

With the knowledge you have gained from studying the skeletal and muscular systems, can you identify any surface anatomy landmarks in this painting?

www.wiley.com/college/apcentral

structures from the surface of the body and due to differences in the thickness of the subcutaneous layer over different parts of the body between females (thicker) and males (thinner), palpations may range from light, to moderate, to deep.

Knowledge of surface anatomy has many applications, both anatomically and clinically. From an anatomical viewpoint, a knowledge of surface anatomy can provide valuable information regarding the location of structures such as bones, muscles, blood vessels, nerves, lymph nodes, and internal organs. Clinically, a knowledge of surface anatomy is the basis for conducting a physical examination and performing certain diagnostic tests. Health care professionals use their understanding of surface anatomy to learn where to take the pulse, measure blood pressure, draw blood, stop bleeding, insert needles and tubes, make surgical incisions, reduce (straighten) fractured bones, and listen to sounds

made by the heart, lungs, and intestines. Clinicians also draw on a knowledge of surface anatomy to assess the status of lymph nodes and to identify the presence of tumors or other unusual masses in the body.

Recall that there are five principal regions of the body: (1) head, (2) neck, (3) trunk, (4) upper limbs, and (5) lower limbs. Using these regions as our guide, we will first study the surface anatomy of the head and then discuss the other regions, concluding with the lower limbs. Because living bodies are best suited for studying surface anatomy, you will find it helpful to visualize and/or palpate the various structures described on your own body as you study each region. Following a brief introduction to a region of the body, you will be presented with a list of the prominent structures to locate. This directed approach will help to organize your learning efforts. Labeled photographs illustrate most of the structures listed in each region.

SURFACE ANATOMY OF THE HEAD

OBJECTIVE

- Describe the surface anatomy features of the head.

The **head** (*cephalic region*, or *caput*) contains the brain and sense organs (eyes, ears, nose, and tongue) and is divided into the cranium and face. The **cranium** (*skull*, or *brain case*) is that portion of the head that surrounds and protects the brain; the **face** is the anterior portion of the head.

Regions of the Cranium and Face

The cranium and face are divided into several regions, which are described here and illustrated in Figure 12.1.

- **Frontal region:** forms the front of the skull and includes the frontal bone.
- **Parietal region:** forms the crown of the skull and includes the parietal bones.
- **Temporal region:** forms the side of the skull and includes the temporal bones.
- **Occipital region:** forms the base of the skull and includes the occipital bone.
- **Orbital (ocular) region:** includes the eyeball, eyebrow, and eyelid.
- **Infraorbital region:** the region inferior to the orbit.
- **Zygomatic region:** the region inferolateral to the orbit that includes the zygomatic bone.
- **Nasal region:** region of the nose.
- **Oral region:** region of the mouth.
- **Mental region:** anterior portion of the mandible.
- **Buccal region** (BUK-al): region of the cheek.
- **Auricular region** (aw-RIK-ū-lar): region of the external ear.

In addition to locating the various regions of the head, it is also possible to visualize and/or palpate certain structures within each region. We will examine various bone and muscle structures of the head and then look at specific areas such as the eyes, ears, and nose.

Many of the bony landmarks of the head and muscles of facial expression are labeled in Figure 12.2 on page 383. As you read about each feature, refer to the figure, and try to visualize and/or palpate each feature on your own head or on the head of a study partner.

Bony Landmarks of the Head

Several bony structures of the skull can be detected through palpation, including the following:

- **Sagittal suture.** Move the fingers from side to side over the superior aspect of the scalp in order to palpate this suture.
- **Coronal** and **lambdoid sutures.** Use the same side-to-side technique to palpate these structures located on the frontal and occipital regions of the skull.
- **External occipital protuberance.** This is the most prominent bony landmark on the occipital region of the skull.
- **Orbit.** Deep to the eyebrow, the superior aspect of the orbit, the **supraorbital margin** of the frontal bone, can be palpated. Actually, the entire circumference of the orbit can be palpated.
- **Nasal bones.** These can be palpated in the nasal region between the orbits on either side of the midline.
- **Mandible.** The **ramus** (vertical portion), **body** (horizontal portion), and **angle** (area where the ramus meets the body) of the mandible can be easily palpated at the mental and buccal regions of the head.
- **Zygomatic arch.** This can be palpated in the zygomatic region.

Figure 12.1 Principal regions of the cranium and face.

The head consists of an anterior face and a cranium that surrounds the brain.

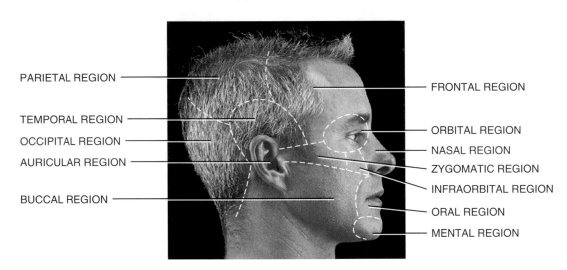

(a) Anterior view

FRONTAL REGION

TEMPORAL REGION

ORBITAL REGION

NASAL REGION

ZYGOMATIC REGION

AURICULAR REGION

INFRAORBITAL REGION

BUCCAL REGION

ORAL REGION

MENTAL REGION

(b) Right lateral view

PARIETAL REGION

FRONTAL REGION

TEMPORAL REGION

ORBITAL REGION

OCCIPITAL REGION

NASAL REGION

AURICULAR REGION

ZYGOMATIC REGION

INFRAORBITAL REGION

BUCCAL REGION

ORAL REGION

MENTAL REGION

Which regions are named for bones that are deep to them?

Muscles of Facial Expression

Several muscles of facial expression can be palpated while they are contracting:

- **Occipitofrontalis.** By raising and lowering the eyebrows, it is possible to palpate the frontal belly and the occipital belly in the frontal and occipital regions, respectively, as they alternately contract and the scalp moves forward and backward.

- **Orbicularis oculi.** By closing the eyes and placing the fin-

gers on the eyelids, the muscle can be palpated in the orbital region by tightly squeezing the eyes shut.

- **Corrugator supercilii.** By frowning, this muscle can be felt above the nose near the medial end of the eyebrow.

- **Zygomaticus major.** By smiling, this muscle can be palpated between the corner of the mouth and zygomatic arch.

- **Depressor labii inferioris.** This muscle can be palpated in the mental region between the lower lip and chin when moving the lower lip inferiorly to expose the lower teeth.

Figure 12.2 Surface anatomy of the head.

Several muscles of facial expression can be palpated while they are contracting.

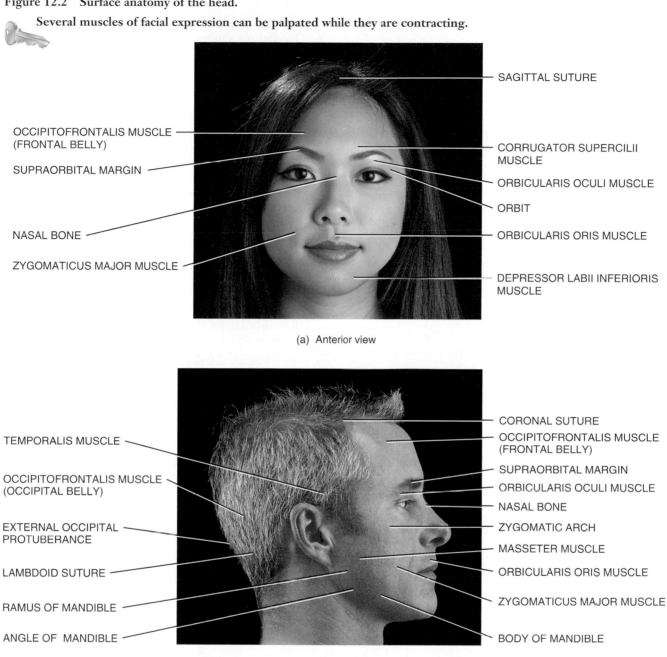

OCCIPITOFRONTALIS MUSCLE (FRONTAL BELLY)

SUPRAORBITAL MARGIN

NASAL BONE

ZYGOMATICUS MAJOR MUSCLE

SAGITTAL SUTURE

CORRUGATOR SUPERCILII MUSCLE

ORBICULARIS OCULI MUSCLE

ORBIT

ORBICULARIS ORIS MUSCLE

DEPRESSOR LABII INFERIORIS MUSCLE

(a) Anterior view

TEMPORALIS MUSCLE

OCCIPITOFRONTALIS MUSCLE (OCCIPITAL BELLY)

EXTERNAL OCCIPITAL PROTUBERANCE

LAMBDOID SUTURE

RAMUS OF MANDIBLE

ANGLE OF MANDIBLE

CORONAL SUTURE

OCCIPITOFRONTALIS MUSCLE (FRONTAL BELLY)

SUPRAORBITAL MARGIN

ORBICULARIS OCULI MUSCLE

NASAL BONE

ZYGOMATIC ARCH

MASSETER MUSCLE

ORBICULARIS ORIS MUSCLE

ZYGOMATICUS MAJOR MUSCLE

BODY OF MANDIBLE

(b) Right lateral view

When palpating the sagittal suture, what sheetlike tendon is also palpated deep to the skin?

- **Orbicularis oris.** By closing the lips tightly, this muscle can be palpated in the oral region around the margin of the lips.
- **Temporalis.** By alternately clenching the teeth and then opening the mouth, it is possible to palpate this muscle in the temporal region just superior to the zygomatic arch.
- **Masseter.** Again, by alternately clenching the teeth and then opening the mouth, this muscle can be palpated just superior and anterior to the angle of the mandible.

CHECKPOINT

1. What is surface anatomy? Why are visualization and palpation important in learning surface anatomy?
2. What are some of the applications of a knowledge of surface anatomy?
3. List the regions of the head and name one structure in each.

Surface Features of the Eyes

Now we will examine the surface features of the eyes, illustrated in Figure 12.3.

- **Iris** (=rainbow). Circular pigmented muscular structure behind the cornea (transparent covering over the iris).
- **Pupil.** Opening in the center of the iris through which light travels.
- **Sclera** (*skleros*=hard). "White" of the eye; a coat of fibrous tissue that covers the entire eyeball except for the cornea.
- **Conjunctiva** (kon-junk-TĪ-va). Membrane that covers the exposed surface of the eyeball and lines the eyelids.
- **Eyelids (palpebrae).** Folds of skin and muscle lined on their inner aspect by conjunctiva.
- **Palpebral fissure.** Space between the eyelids when they are open; when the eye is closed, it lies just inferior to the level of the pupil.
- **Medial commissure** (KOM-i-syūr). Medial (near the nose) site of the union of the upper and lower eyelids.
- **Lateral commissure.** Lateral (away from the nose) site of the union of the upper and lower eyelids.
- **Lacrimal** (*lacrim*=tears) **caruncle.** Fleshy, yellowish projection of the medial commissure that contains modified sudoriferous (sweat) and sebaceous (oil) glands.

- **Eyelashes.** Hairs on the margin of the eyelids, usually arranged in two or three rows.
- **Eyebrows (supercilia).** Several rows of hairs superior to the upper eyelids.

Surface Features of the Ears

Next, locate the surface features of the ears in Figure 12.4.

- **Auricle** (AW-ri-kul). Shell-shaped portion of the external ear; also called the *pinna*. It funnels sound waves into the external auditory canal. Just posterior to it, the **mastoid process** of the temporal bone can be palpated.
- **Tragus** (TRĀ-gus; *tragos*=goat; hair that grows on the tragus resembles the beard of a goat). Cartilaginous projection anterior to the external auditory canal. Just anterior to the tragus and posterior to the neck of the mandible you can palpate the **superficial temporal artery** and feel the pulse in the vessel. Inferior to this point, it is possible to feel the **temporomandibular joint (TMJ).** As you open and close your jaw, you can feel the movement of the mandibular condyle.
- **Antitragus.** Cartilaginous projection opposite the tragus.
- **Concha** (KON-ka). Hollow of the auricle.
- **Helix.** Superior and posterior free margin of the auricle.

Figure 12.3 Surface anatomy of the right eye.

The eyeball is protected by the eyebrow, eyelashes, and eyelids.

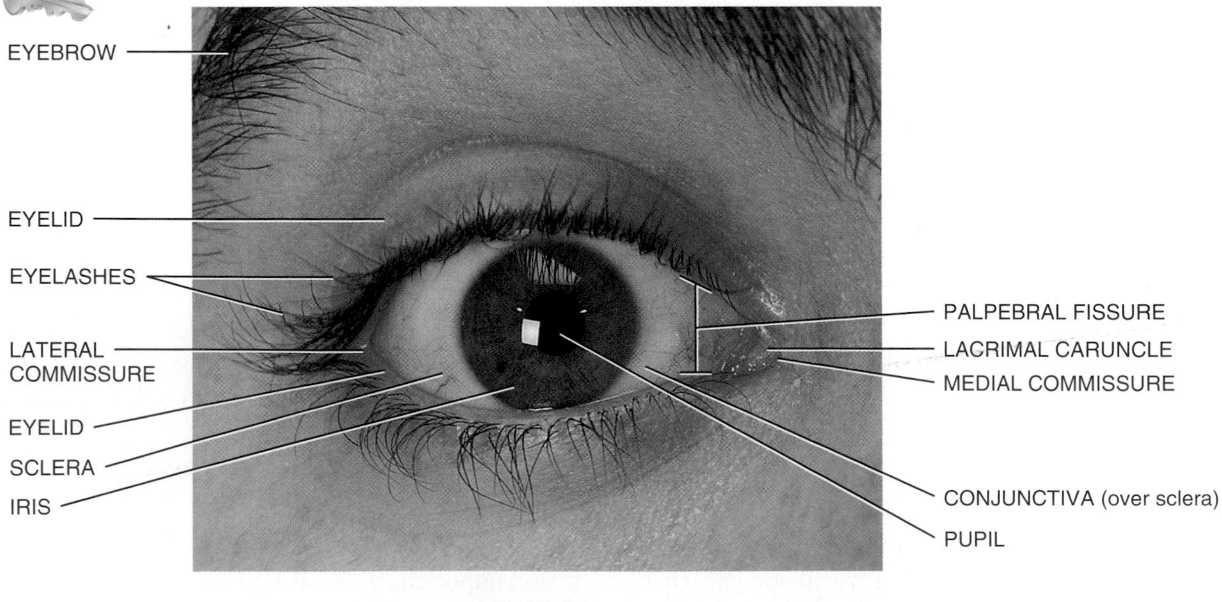

Anterior view

? Through which part of the eye does light enter?

Figure 12.4   Surface anatomy of the right ear.

The most conspicuous surface feature of the ear is the auricle.

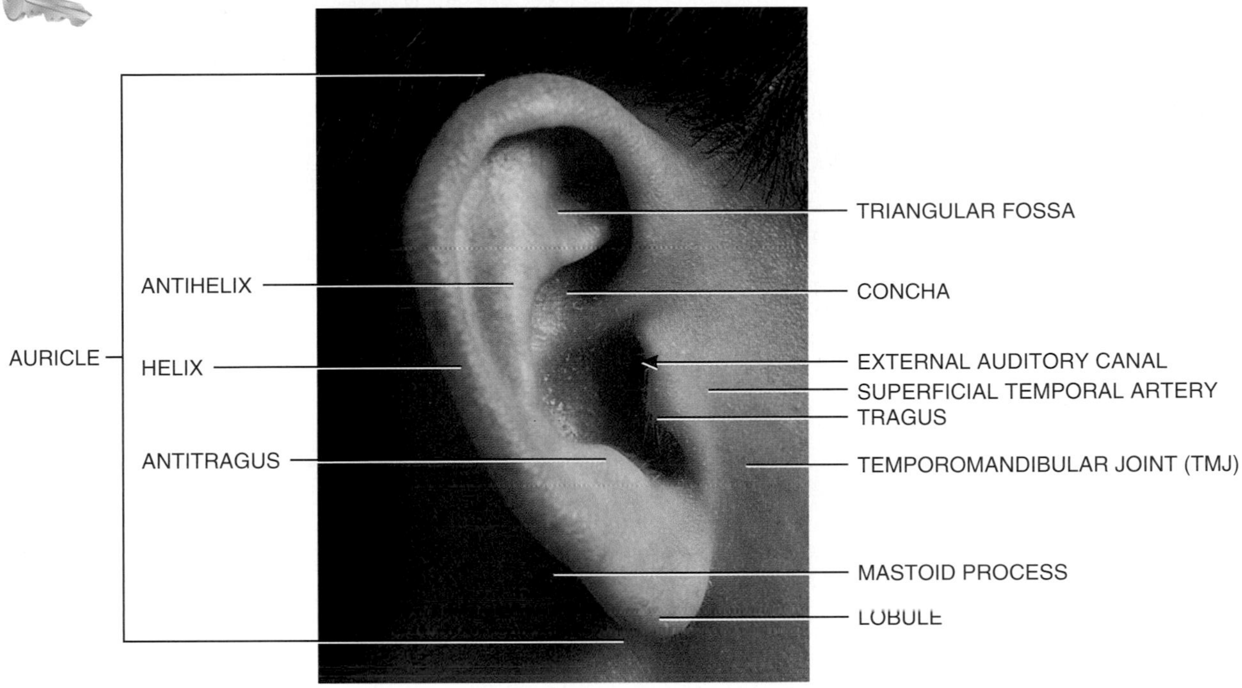

Right lateral view

Which part of the temporal bone can be palpated just posterior to the auricle?

- **Antihelix.** Semicircular ridge superior and posterior to the tragus.
- **Triangular fossa.** Depression in the superior portion of the antihelix.
- **Lobule.** Inferior portion of the auricle; it does not contain cartilage. Commonly referred to as the earlobe.
- **External auditory canal (meatus).** Canal about 3 cm (1 in.) long extending from the external ear to the eardrum (tympanic membrane). It contains ceruminous glands that secrete cerumen (earwax). The **condylar process** of the mandible can be palpated by placing your little finger in the canal and opening and closing your mouth.

Surface Features of the Nose and Mouth

To complete the discussion of the surface anatomy of the head, locate the following features of the nose and mouth (Figure 12.5):

- **Root.** Superior attachment of the nose at the forehead, between the eyes.

- **Apex.** Tip of the nose.
- **Dorsum nasi.** Rounded anterior border connecting the root and apex; in profile, it may be straight, convex, concave, or wavy.
- **External naris.** External opening into the nose.
- **Ala.** Convex flared portion of the inferior lateral surfaces of the nose.
- **Bridge.** Superior portion of the dorsum nasi, superficial to the nasal bones.
- **Philtrum** (FIL-trum=depression on upper lip). The vertical groove on the upper lip that extends along the midline to the inferior portion of the nose.
- **Lips (labia).** Superior and inferior fleshy borders of the oral cavity.

CHECKPOINT

4. List and define five surface features of the eyes, ears, and nose.
5. How would you locate the superficial temporal artery, temporomandibular joint, zygomatic arch, mastoid process, and condylar process of the mandible?

Figure 12.5 **Surface anatomy of the nose and lips.**

The external nares permit air to move into and out of the nose during breathing.

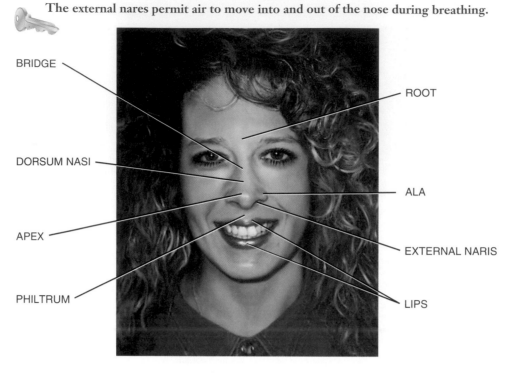

BRIDGE

ROOT

DORSUM NASI

ALA

APEX

EXTERNAL NARIS

PHILTRUM

LIPS

Anterior view

What is the name of the anterior border of the nose that connects the root and apex?

SURFACE ANATOMY OF THE NECK

OBJECTIVE

● Describe the surface anatomy features of the neck.

The **neck** (*cervical*) is the region of the body that connects the head to the trunk and is divided into an *anterior cervical region*, two *lateral cervical regions*, and a *posterior cervical region* (*nucha*).

The following are the major surface features of the neck, most of which are illustrated in Figure 12.6:

● **Thyroid cartilage (Adam's apple).** The largest of the cartilages that compose the larynx (voice box). It is the most prominent structure in the midline of the anterior cervical region.

● **Hyoid bone.** Located just superior to the thyroid cartilage. It is the first structure palpated in the midline inferior to the chin.

● **Cricoid cartilage.** A laryngeal cartilage located just inferior to the thyroid cartilage, it attaches the larynx to the trachea (windpipe). After you pass your finger over the cricoid cartilage moving inferiorly, your fingertip sinks in. The cricoid cartilage is used as a landmark for performing a tracheotomy, which is described on page 741.

● **Thyroid gland.** Two-lobed gland with one lobe just inferior to the larynx with one lobe on either side.

● **Sternocleidomastoid muscle.** Forms the major portion of the lateral aspect of the neck. If you rotate your head to either side, you can palpate the muscle from its origin on the sternum and clavicle to its insertion on the mastoid process of the temporal bone.

● **Common carotid artery.** Lies just deep to the sternocleidomastoid muscle along its anterior border.

● **Internal jugular vein.** Located lateral to the common carotid artery.

● **Subclavian artery.** Located just lateral to the inferior portion of the sternocleidomastoid muscle. Pressure on this artery can stop bleeding in the upper limb because this vessel supplies blood to the entire limb.

● **External carotid artery.** Located superior to the larynx, just anterior to the sternocleidomastoid muscle, this artery is the site of the carotid (neck) pulse.

● **External jugular vein.** Located superficial to the sternocleidomastoid muscle, this vessel is readily seen if you are angry or if your collar is too tight.

● **Trapezius muscle.** Extends inferiorly and laterally from the base of the skull and occupies a portion of the lateral cervical region. A "stiff neck" is frequently associated with inflammation of this muscle.

● **Vertebral spines.** The spinous processes of the cervical vertebrae may be felt along the midline of the posterior region of the neck. Especially prominent at the base of the neck is the spinous process of the seventh cervical vertebra,

Figure 12.6   Surface anatomy of the neck.

The anatomical subdivisions of the neck are the anterior cervical region, lateral cervical regions, and posterior cervical region.

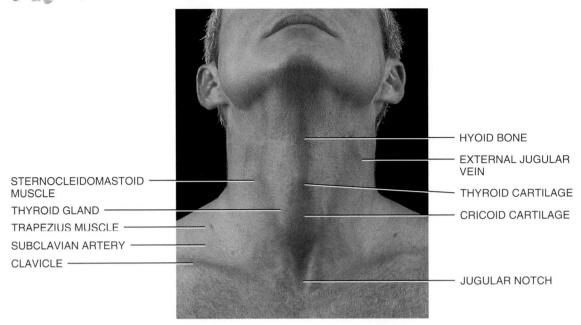

HYOID BONE

EXTERNAL JUGULAR VEIN

STERNOCLEIDOMASTOID MUSCLE

THYROID CARTILAGE

THYROID GLAND

CRICOID CARTILAGE

TRAPEZIUS MUSCLE

SUBCLAVIAN ARTERY

CLAVICLE

JUGULAR NOTCH

(a) Anterior view of surface anatomy features

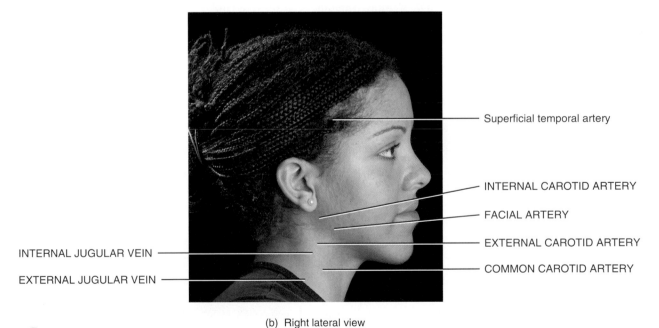

Superficial temporal artery

INTERNAL CAROTID ARTERY

FACIAL ARTERY

INTERNAL JUGULAR VEIN

EXTERNAL CAROTID ARTERY

COMMON CAROTID ARTERY

EXTERNAL JUGULAR VEIN

(b) Right lateral view

Which muscle divides the neck into anterior and posterior triangles?

called the vertebra prominens (see Figure 12.7). The sternocleidomastoid muscle is not only a landmark for several arteries and veins, it is also the muscle that divides the neck into two major triangles: anterior and posterior (see Figure 11.9 on page 312). The triangles are important because of the structures that lie within their boundaries (see Exhibit 11.6 on page 311).

CHECKPOINT

6. What is the neck?
7. How would you palpate the hyoid bone, cricoid cartilage, thyroid gland, common carotid artery, external carotid artery, and external jugular vein?

SURFACE ANATOMY OF THE TRUNK

• Describe the surface anatomy of the various features of the trunk—back, chest, abdomen, and pelvis.

The **trunk** is divided into the back, chest, abdomen, and pelvis. We will consider each region separately.

Surface Features of the Back

Among the prominent features of the **back,** or *dorsum*, are several superficial bones and muscles (Figure 12.7).

• **Vertebral spines.** The spinous processes of vertebrae, especially the thoracic and lumbar vertebrae, are quite prominent when the vertebral column is flexed.

• **Scapulae.** These easily identifiable surface landmarks on the back lie between ribs 2–7. In fact, it is also possible to palpate some ribs on the back. Depending on how lean a person is, it might be possible to palpate various parts of the scapula, such as the **vertebral border, axillary border, inferior angle, spine,** and **acromion.** The **spinous process of T3** is at about the same level as the spine of the scapula, and the **spinous process of T7** is approximately opposite the inferior angle of the scapula.

• **Latissimus dorsi muscle.** A broad, flat, triangular muscle of the lumbar region that extends superiorly to the axilla.

• **Erector spinae (sacrospinalis) muscle.** Located on either side of the vertebral column between the twelfth ribs and iliac crests.

• **Infraspinatus muscle.** Located inferior to the spine of the scapula.

• **Trapezius muscle.** Extends from the cervical and thoracic vertebrae to the spine and acromion of the scapula and the lateral end of the clavicle. It also occupies a portion of the lateral region of the neck and forms the posterior border of the posterior triangle of the neck.

• **Teres major muscle.** Located inferior to the infraspinatus muscle; together with the latissimus dorsi muscle it forms the inferior border of the posterior axillary fold (posterior wall of the axilla).

• **Posterior axillary fold.** Formed by the latissimus dorsi and teres major muscles, the posterior axillary fold can be palpated between the fingers and thumb at the posterior aspect of the axilla; forms the posterior wall of the axilla.

• **Triangle of auscultation** (aw-skul-TĀ-shun; *ausculto-*=listening). A triangular region of the back just medial to the inferior part of the scapula, where the rib cage is not covered by superficial muscles. It is bounded by the latissimus dorsi and trapezius muscles and vertebral border of the scapula.

Figure 12.7 Surface anatomy of the back.

The posterior boundary of the axilla, the posterior axillary fold, is formed mainly by the latissimus dorsi and teres major muscles.

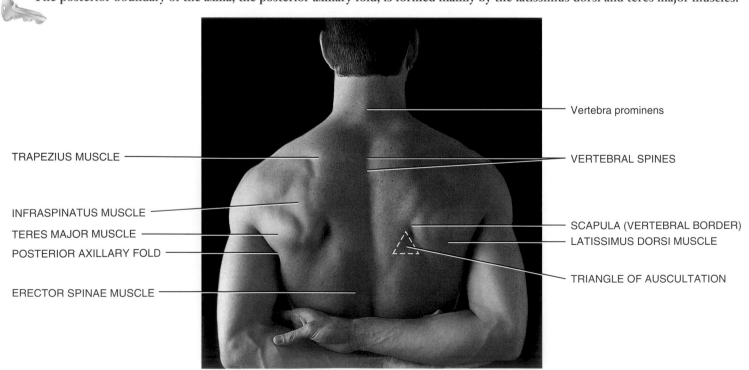

TRAPEZIUS MUSCLE

INFRASPINATUS MUSCLE

TERES MAJOR MUSCLE

POSTERIOR AXILLARY FOLD

ERECTOR SPINAE MUSCLE

Vertebra prominens

VERTEBRAL SPINES

SCAPULA (VERTEBRAL BORDER)

LATISSIMUS DORSI MUSCLE

TRIANGLE OF AUSCULTATION

Posterior view

What is the clinical significance of the triangle of auscultation?

TRIANGLE OF AUSCULTATION AND LUNG SOUNDS

The triangle of auscultation is a landmark of clinical significance because in this area respiratory sounds can be heard clearly through a stethoscope pressed against the skin. If a patient folds the arms across the chest and bends forward, the lung sounds can be heard clearly in the intercostal space between ribs 6 and 7. ∎

CHECKPOINT

8. What are the divisions of the trunk?
9. Describe the location of the scapulae relative to the ribs and vertebrae.
10. What is the triangle of auscultation?

Surface Features of the Chest

The surface anatomy of the **chest,** or *thorax,* includes several distinguishing surface landmarks (Figure 12.8), as well as some surface markings used to identify the location of the heart and lungs within the chest.

- **Clavicle.** Visible at the junction of the neck and thorax. Inferior to the clavicle, especially where it articulates with the manubrium (superior portion) of the sternum, the **first rib** can be palpated. In a depression superior to the medial end of the clavicle, just lateral to the sternocleidomastoid muscle, the **trunks of the brachial plexus** can be palpated (see Figure 18.9a on page 426).

- **Suprasternal (jugular) notch of sternum.** A depression on the superior border of the manubrium of the sternum between the medial ends of the clavicles; the **trachea** (windpipe) can be palpated posterior to the notch.

- **Manubrium of sternum.** Superior portion of the sternum at the same levels as the bodies of the third and fourth thoracic vertebrae and anterior to the arch of the aorta.

- **Sternal angle of sternum.** Formed at the junction between the manubrium and body of the sternum, located about 4 cm (1.5 in.) inferior to the suprasternal notch. It is palpable deep to the skin and locates the costal cartilage of the second rib. It is the most reliable landmark of the chest and the starting point from which ribs are counted. At or inferior to the sternal angle and slightly to the right, the trachea divides into right and left primary bronchi.

- **Body of sternum.** Midportion of the sternum anterior to the heart and the vertebral bodies of T5–T8.

- **Xiphoid process of sternum.** Inferior portion of the sternum, medial to the seventh costal cartilages. The joint between the xiphoid process and body of the sternum is called the **xiphisternal joint.** The heart lies on the diaphragm deep to this joint.

- **Costal margin.** Inferior edges of the costal cartilages of ribs 7 through 10. The first costal cartilage lies inferior to the medial end of the clavicle; the seventh costal cartilage is the most inferior costal cartilage to articulate directly with the sternum; the tenth costal cartilage forms the most inferior part of the costal margin when viewed anteriorly. At the superior end of the costal margin is the xiphisternal joint.

- **Serratus anterior muscle.** Inferior and lateral to the pectoralis major muscle.

- **Ribs.** Twelve pairs of ribs help to form the bony cage of the thoracic cavity. Depending on body leanness, they may or may not be visible. One site for listening to the heartbeat in adults is the left fifth intercostal space, just medial to the left midclavicular line.

Figure 12.8 Surface anatomy of the chest.

 Thoracic viscera are protected by the sternum, ribs, and thoracic vertebrae, which form the skeleton of the thorax.

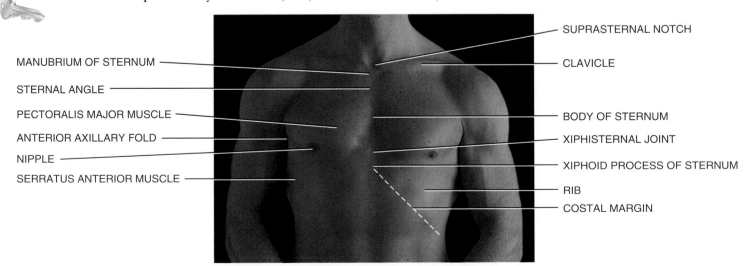

MANUBRIUM OF STERNUM

STERNAL ANGLE

PECTORALIS MAJOR MUSCLE

ANTERIOR AXILLARY FOLD

NIPPLE

SERRATUS ANTERIOR MUSCLE

SUPRASTERNAL NOTCH

CLAVICLE

BODY OF STERNUM

XIPHISTERNAL JOINT

XIPHOID PROCESS OF STERNUM

RIB

COSTAL MARGIN

Anterior view of surface features of the chest

What is the most reliable landmark of the chest?

- **Mammary glands.** Accessory organs of the female reproductive system located inside the breasts. These glands overlie the pectoralis major muscle (two-thirds) and serratus anterior muscle (one-third). After puberty, the mammary glands enlarge to their hemispherical shape, and in young adult females they extend from the second through the sixth ribs and from the lateral margin of the sternum to the midaxillary line. The **midaxillary line** is a vertical line that extends downward on the lateral thoracic wall from the center of the axilla.

- **Nipples.** Superficial to the fourth intercostal space or fifth rib, about 10 cm (4 in.) from the midline in males and most females. The position of the nipples in females is variable depending on the size and pendulousness of the breasts. The right dome of the diaphragm is just inferior to the right nipple, the left dome is about 2–3 cm (1 in.) inferior to the left nipple, and the central tendon is at the level of the xiphisternal joint.

- **Anterior axillary fold.** Formed by the lateral border of the pectoralis major muscle; can be palpated between the fingers and thumb; forms the anterior wall of the axilla.

- **Pectoralis major muscle.** Principal upper chest muscle; in the male, the inferior border of the muscle forms a curved line leading to the anterior wall of the axilla and serves as a guide to the fifth rib. In the female, the inferior border is mostly covered by the breast.

Although structures within the chest are almost totally hidden by the sternum, ribs, and thoracic vertebrae, it is important to indicate the surface markings of the heart and lungs. The projection (shape of an organ on the surface of the body) of the heart on the anterior surface of the chest is indicated by four points as follows (see Figure 14.1c on page 426). The *inferior left point* is the apex (inferior, pointed end) of the heart, which projects downward and to the left and can be palpated in the fifth intercostal space, about 9 cm (3.5 in.) to the left of the midline. The *inferior right point* is at the lower border of the costal cartilage of the right sixth rib, about 3 cm to the right of the midline. The *superior right point* is located at the superior border of the costal cartilage of the right third rib, about 3 cm to the right of the midline. The *superior left point* is located at the inferior border of the costal cartilage of the left second rib, about 3 cm to the left of the midline. If you connect the four points, you can determine the heart's location and size (about the size of a closed fist).

The lungs almost totally fill the thorax. The *apex* (superior, pointed end) of the lungs lies just superior to the medial third of the clavicles and is the only area that can be palpated (see Figure 24.9a on page 745). The anterior, lateral, and posterior surfaces of the lungs lie against the ribs. The *base* (inferior, broad end) of the lungs is concave and fits over the convex area of the diaphragm. It extends from the sixth costal cartilage anteriorly to the spinous process of the tenth thoracic vertebra posteriorly. (The base of the right lung is usually slightly higher than that of the left due to the position of the liver below it.) The covering around the lungs is called the *pleura*. At the base of each lung, the pleura extends about 5 cm below the base from the sixth costal cartilage anteriorly to the twelfth rib posteriorly. Thus, the lungs do not completely fill the pleural cavity in this area. A needle is typically inserted into this space in the pleural cavity to withdraw fluid from the chest.

CHECKPOINT

11. Describe the location of the sternum relative to the ribs and vertebrae.
12. What is the significance of the xiphisternal joint as a landmark?
13. Where would you palpate the first rib and brachial plexus?

Figure 12.9 Surface anatomy of the abdomen and pelvis.

🔑 **The linea alba is a frequent site for an abdominal incision because cutting through it severs no muscles and few blood vessels and nerves.**

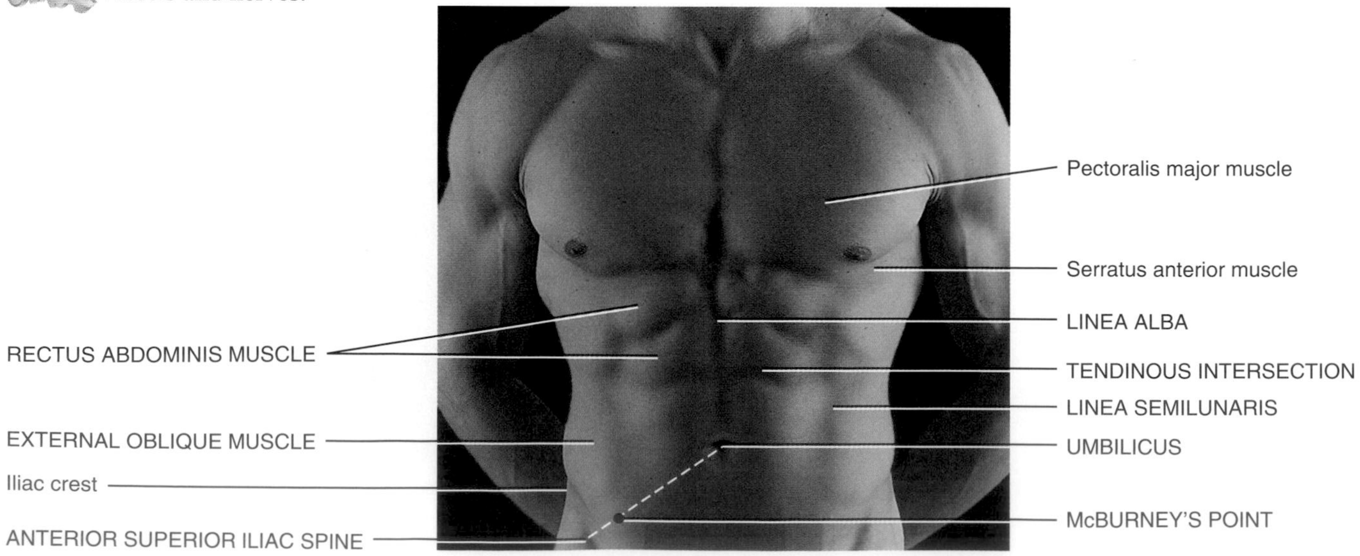

RECTUS ABDOMINIS MUSCLE

EXTERNAL OBLIQUE MUSCLE

Iliac crest

ANTERIOR SUPERIOR ILIAC SPINE

Pectoralis major muscle

Serratus anterior muscle

LINEA ALBA

TENDINOUS INTERSECTION

LINEA SEMILUNARIS

UMBILICUS

McBURNEY'S POINT

(a) Anterior view of abdomen

14. Why is the midaxillary line an important anatomical landmark?
15. How would you indicate the surface markings of the heart?
16. Describe the position of the lungs in the thorax.
17. Why do the lungs not completely fill the pleural cavity? Why is this important clinically?

Surface Features of the Abdomen and Pelvis

Following are some of the prominent surface anatomy features of the **abdomen** and **pelvis** (Figure 12.9):

- **Umbilicus.** Also called the *navel*; it marks the site of attachment of the umbilical cord to the fetus. It is level with the intervertebral disc between the bodies of vertebrae L3 and L4. The **abdominal aorta,** which branches into the right and left common iliac arteries anterior to the body of verte-

bra L4, can be easily palpated through the upper part of the anterior abdominal wall just to the left of the midline. The **inferior vena cava** lies to the right of the abdominal aorta and is wider; it arises anterior to the body of vertebra L5.

- **External oblique muscle.** Located inferior to the serratus anterior muscle. The aponeurosis of the muscle on its inferior border is the **inguinal ligament,** a structure along which hernias frequently occur.

- **Rectus abdominis muscles.** Located just lateral to the midline of the abdomen. They can be seen by raising the shoulders while in the supine position without using the arms.

- **Linea alba.** Flat, tendinous raphe forming a furrow along the midline between the rectus abdominis muscles. The furrow extends from the xiphoid process to the pubic symphysis. It is broad superior to the umbilicus and narrow

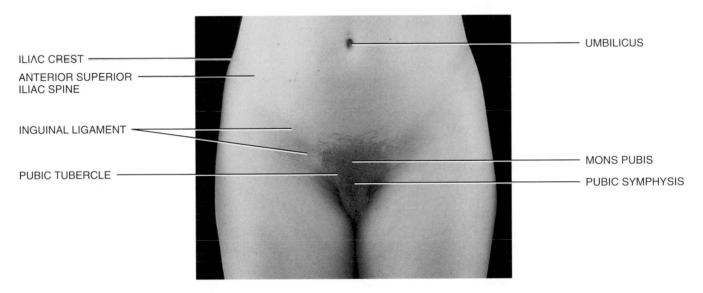

(b) Anterior view of pelvis

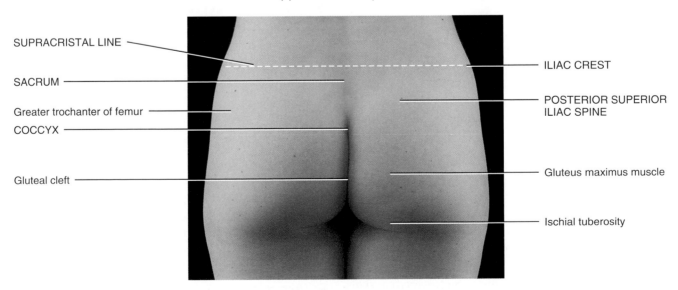

(c) Posterior view of pelvis

 What is the clinical significance of McBurney's point?

inferior to it. The linea alba is a frequently selected site for abdominal surgery because an incision through it severs no muscles and only a few blood vessels and nerves.

- **Tendinous intersection.** A fibrous band that runs transversely or obliquely across the rectus abdominis muscles. One intersection is at the level of the umbilicus, one is at the level of the xiphoid process, and one is midway between the two. The rectus abdominis muscles thus contain four prominent bulges and three tendinous intersections.

- **Linea semilunaris.** The lateral edge of the rectus abdominis muscle can be seen as a line that crosses the costal margin at the top of the ninth costal cartilage.

- **McBurney's point.** A clinically important site (see next section) located two-thirds of the way down an imaginary line drawn between the umbilicus and anterior superior iliac spine.

MCBURNEY'S POINT AND THE APPENDIX

McBurney's point is an important landmark related to the appendix. When the appendix must be removed in an operation called an **appendectomy,** an oblique incision is made through McBurney's point. Moreover, pressure of the finger on McBurney's point produces pain in acute **appendicitis,** inflammation of the appendix, aiding in diagnosis. ∎

- **Iliac crest.** Superior margin of the ilium of the hip bone. It forms the outline of the superior border of the buttock. When you rest your hands on your hips, they rest on the iliac crests. A horizontal line drawn across the highest point of each iliac crest is called the **supracristal line,** which intersects the spinous process of the fourth lumbar vertebra. This vertebra is a landmark for performing a spinal tap (see page 554).

- **Anterior superior iliac spine.** The anterior end of the iliac crest that lies at the upper lateral end of the fold of the groin.

- **Posterior superior iliac spine.** The posterior end of the iliac crest, indicated by a dimple in the skin that coincides with the middle of the sacroiliac joint, where the hip bone attaches to the sacrum.

- **Pubic tubercle.** Projection on the superior border of the pubis of the hip bone. Attached to it is the medial end of the **inguinal ligament,** the inferior free edge of the aponeurosis of the external oblique muscle that forms the **inguinal canal.** The lateral end of the ligament is attached to the anterior superior iliac spine. The spermatic cord in males and the round ligament of the uterus in females pass through the inguinal canal.

- **Pubic symphysis.** Anterior joint of the hip bones; palpated as a firm resistance in the midline at the inferior portion of the anterior abdominal wall.

- **Mons pubis.** An elevation of adipose tissue covered by skin and pubic hair that is anterior to the pubic symphysis.

- **Sacrum.** The fused spinous processes of the sacrum, called

the **median sacral crest,** can be palpated beneath the skin superior to the **gluteal cleft,** a depression along the midline that separates the buttocks (and is part of the discussion on the buttocks on page 398).

- **Coccyx.** The inferior surface of the tip of the coccyx can be palpated in the gluteal cleft, about 2.5 cm (1 in.) posterior to the anus.

Chapter 1 discussed the division of the abdomen and pelvis into nine regions (see Figure 1.9 on page 16). The nine-region designation subdivided the abdomen and pelvis by drawing two vertical lines (left and right clavicular), an upper horizontal line (subcostal), and a lower horizontal line (transtubercular). Take time to examine Figure 1.9 carefully, so that you can see which organs or parts of organs lie within each region.

In Chapter 11, as part of your study of skeletal muscles, you were introduced to the **perineum,** the diamond-shaped region medial to the thigh and buttocks. It is bounded by the pubic symphysis anteriorly, the ischial tuberosities laterally, and the coccyx posteriorly. A transverse line drawn between the ischial tuberosities divides the perineum into an anterior *urogenital triangle* that contains the external genitals, and a posterior *anal triangle* that contains the anus. Details concerning the functions of the perineum are provided in Chapter 11. At this point keep in mind that the musculature of the perineum constitutes the floor of the pelvic cavity.

CHECKPOINT

18. Why are the linea alba, linea semilunaris, McBurney's point, and supracristal line important landmarks?
19. What is the inguinal ligament, and why is it important?
20. How would you palpate the sacrum and coccyx?
21. What is the perineum?
22. Define the buttock. What is the gluteal cleft? Gluteal fold?
23. Why is the greater trochanter of the femur important?

SURFACE ANATOMY OF THE UPPER LIMB (EXTREMITY)

OBJECTIVE

- Describe the surface anatomy of the various regions of the upper limb—armpit, shoulder, arm, elbow, forearm, wrist, and hand.

The **upper limb** consists of the shoulder, armpit, arm, elbow, forearm, wrist, and hand.

Surface Features of the Shoulder

The **shoulder,** or *acromial region,* is located at the lateral aspect of the clavicle, where the clavicle joins the scapula and the scapula joins the humerus. The region presents several conspicuous surface features (Figure 12.10).

- **Acromioclavicular joint.** A slight elevation at the lateral end of the clavicle. It is the joint between the acromion of the scapula and the clavicle.

Figure 12.10 Surface anatomy of the shoulder.

 The deltoid muscle gives the shoulder its rounded prominence.

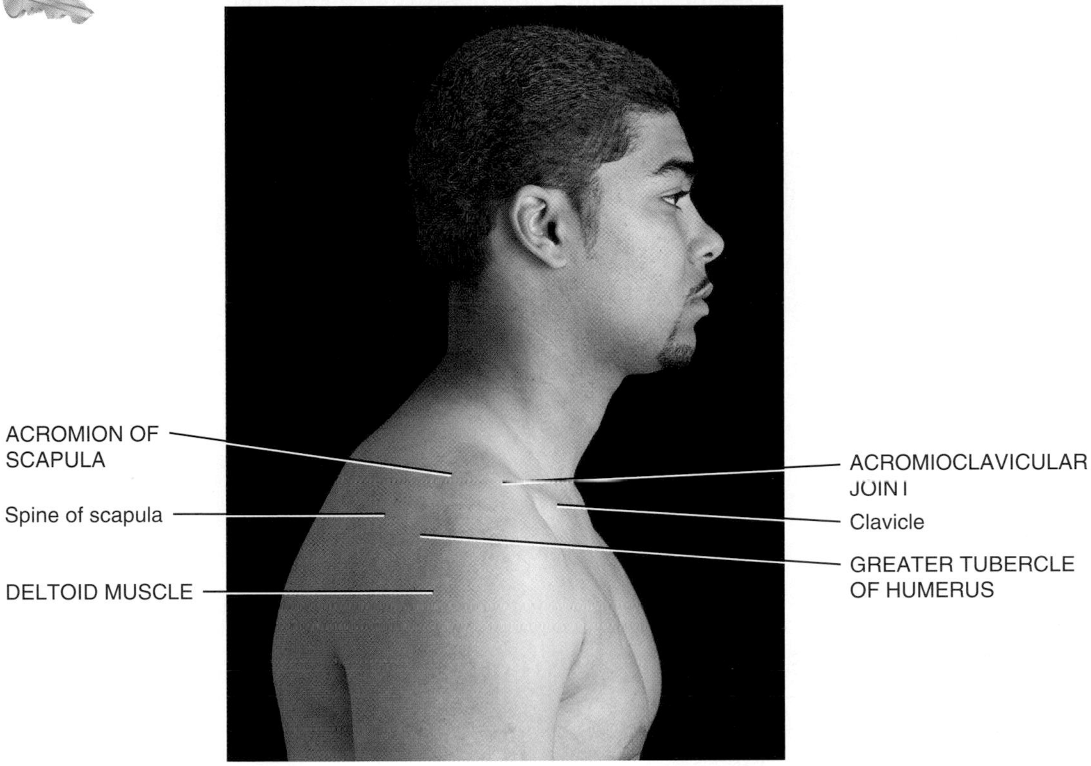

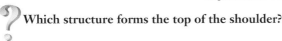

ACROMION OF SCAPULA

Spine of scapula

DELTOID MUSCLE

ACROMIOCLAVICULAR JOINT

Clavicle

GREATER TUBERCLE OF HUMERUS

Right Lateral View

? **Which structure forms the top of the shoulder?**

- **Acromion.** The expanded lateral end of the spine of the scapula that forms the top of the shoulder. It can be palpated about 2.5 cm (1 in.) distal to the acromioclavicular joint.

- **Humerus.** The **greater tubercle** of the humerus may be palpated on the superior aspect of the shoulder. It is the most laterally palpable bony structure.

- **Deltoid muscle.** Triangular muscle that forms the rounded prominence of the shoulder.

 **SHOULDER INTRAMUSCULAR INJECTION**

The *deltoid muscle* is a frequent site for an intramuscular injection. To avoid injury to major blood vessels and nerves, the injection is given in the midportion of the muscle about 2–3 finger-widths inferior to the acromion of the scapula and lateral to the axilla. ■

Surface Features of the Armpit

The armpit, or **axilla,** is a pyramid-shaped area at the junction of the arm and the chest that enables blood vessels and nerves

to pass between the neck and the upper limbs (Figure 12.11 a and b). The **apex** of the axilla is surrounded by the clavicle, scapula, and first rib. The **base** of the axilla is formed by the concave skin and fascia that extends from the arm to the chest wall. It contains hair, and deep to the base the axillary lymph nodes can be palpated. The **anterior wall** of the axilla is composed mainly of the pectoralis major muscle (anterior axillary fold; see also Figure 12.8). The **posterior wall** of the axilla is formed mainly by the teres major and latissimus dorsi muscles (posterior axillary fold; see also Figure 12.7). The **medial wall** of the axilla is formed by ribs 1–4 and their corresponding intercostal muscles, plus the overlying serratus anterior muscle. Finally, the **lateral wall** of the axilla is formed by the coracobrachialis muscle and the superior portion of the shaft of the humerus. Passing through the axilla are the axillary artery and vein, branches of the brachial plexus, and axillary lymph nodes.

CHECKPOINT

24. What are the major divisions of the upper limb?

25. What three bones are present at the shoulder? How would you palpate each?

26. What is the clinical significance of the deltoid muscle?

27. Define the axilla. What are its borders? Why is it important?

Figure 12.11   Surface anatomy of the axilla, arm, and elbow. The location of the muscles that form the walls of the axilla are shown in Figure 11.17a on page 334.

 The biceps brachii and triceps brachii muscles form the bulk of the musculature of the arm.

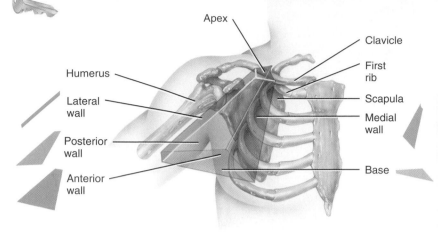

(a) Location and parts of the axilla

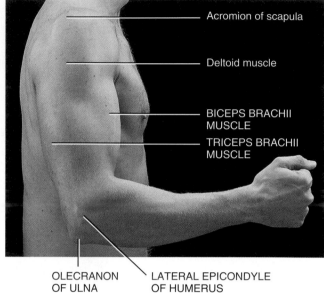

(c) Right lateral view of arm

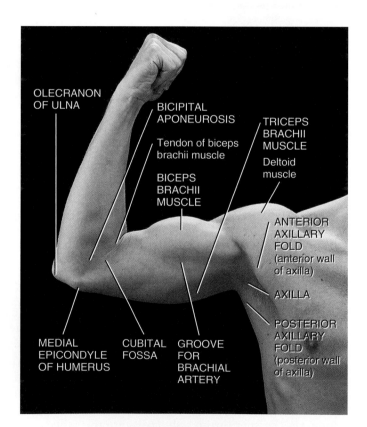

(b) Medial view of arm

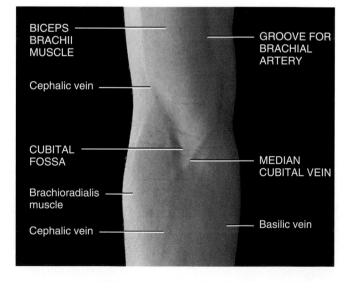

(d) Anterior view of cubital fossa

Which blood vessel in the cubital fossa is frequently used to withdraw blood?

Surface Features of the Arm and Elbow

The **arm,** or *brachium*, is the region between the shoulder and elbow. The **elbow,** or *cubitus*, is the region where the arm and forearm join. The arm and elbow present several surface anatomy features (Figure 12.11b-d).

- **Humerus.** This arm bone may be palpated along its entire length, especially near the elbow (see descriptions of the medial and lateral epicondyles that follow).

- **Biceps brachii muscle.** Forms the bulk of the anterior surface of the arm. On the medial side of the muscle is a groove that contains the **brachial artery.**

- **Triceps brachii muscle.** Forms the bulk of the posterior surface of the arm.

- **Medial epicondyle.** Medial projection of the humerus near the elbow.

- **Lateral epicondyle.** Lateral projection of the humerus near the elbow.

- **Olecranon.** Projection of the proximal end of the ulna between and slightly superior to the epicondyles when the forearm is extended; it forms the elbow.

- **Ulnar nerve.** Can be palpated in a groove posterior to the medial epicondyle. The "funny bone" is the region where the ulnar nerve rests against the medial epicondyle. Hitting the nerve at this point produces a sharp pain along the medial side of the forearm.

- **Cubital fossa.** Triangular space in the anterior region of the elbow bounded proximally by an imaginary line between the humeral epicondyles, laterally by the medial border of the brachioradialis muscle, and medially by the lateral border of the pronator teres muscle; contains the tendon of the biceps brachii muscle, the median cubital vein, brachial artery and its terminal branches (radial and ulnar arteries), and parts of the median and radial nerves.

- **Median cubital vein.** Crosses the cubital fossa obliquely and connects the lateral cephalic with the medial basilic veins of the arm. The median cubital vein is frequently used to withdraw blood from a vein for diagnostic purposes or to introduce substances into blood, such as medications, contrast media for radiographic procedures, nutrients, and blood cells and/or plasma for transfusion.

- **Brachial artery.** Continuation of the axillary artery that passes posterior to the coracobrachialis muscle and then medial to the biceps brachii muscle. It enters the middle of the cubital fossa and passes deep to the bicipital aponeurosis, which separates it from the median cubital vein.

- **Bicipital aponeurosis.** An aponeurosis that inserts the biceps brachii muscle into the deep fascia in the medial aspect of the forearm. It can be felt when the muscle contracts.

BLOOD PRESSURE

Blood pressure is usually measured in the brachial artery, when the cuff of a *sphygmomanometer* (blood pressure instrument) is wrapped around the arm and a stethoscope is placed over the brachial artery in the cubital fossa. Pulse can also be detected in the artery in the cubital fossa. However, blood pressure can be measured at any artery where you can obstruct blood flow. This becomes important in case the brachial artery cannot be utilized. In such situations, the radial or popliteal arteries might be used to obtain a blood pressure reading. ■

CHECKPOINT

28. In which arteries of the upper limb can a pulse be detected?
29. What is the "funny bone"?
30. Define the borders and state the significance of the cubital fossa.
31. What is the clinical importance of the median cubital vein?
32. What artery is normally used to measure blood pressure?

Surface Features of the Forearm and Wrist

The **forearm,** or *antebrachium*, is the region between the elbow and wrist. The **wrist,** or carpus, is between the forearm and palm. Following are some prominent surface anatomy features of the forearm and wrist (Figure 12.12).

- **Ulna.** The medial bone of the forearm. It can be palpated along its entire length from the olecranon (see Figure 12.11a, b) to the **styloid process,** a projection on the distal end of the bone at the medial side of the wrist. The **head of the ulna** is a conspicuous enlargement just proximal to the styloid process.

- **Radius.** When the forearm is rotated, the distal half of the radius can be palpated; the proximal half is covered by muscles. The **styloid process** of the radius is a projection on the distal end of the bone at the lateral side of the wrist.

- **Muscles.** Because of their close proximity, it is difficult to identify individual muscles of the forearm. Instead, it is much easier to identify tendons as they approach the wrist and then trace them proximally to the following muscles:

Figure 12.12 Surface anatomy of the forearm and wrist.

Muscles of the forearm are most easily identified by locating their tendons near the wrist and tracing them proximally.

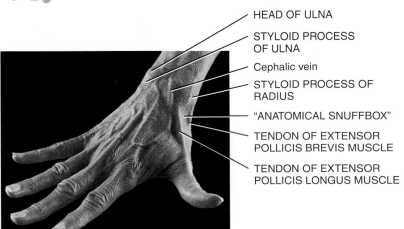

HEAD OF ULNA

STYLOID PROCESS OF ULNA

Cephalic vein

STYLOID PROCESS OF RADIUS

"ANATOMICAL SNUFFBOX"

TENDON OF EXTENSOR POLLICIS BREVIS MUSCLE

TENDON OF EXTENSOR POLLICIS LONGUS MUSCLE

(a) Dorsum of wrist

continues

**Figure 12.12
(continued)**

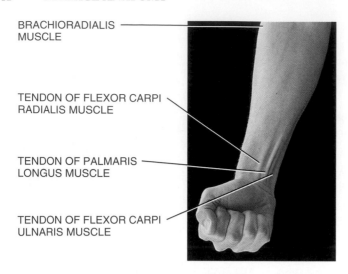

BRACHIORADIALIS
MUSCLE

TENDON OF FLEXOR CARPI
RADIALIS MUSCLE

TENDON OF PALMARIS
LONGUS MUSCLE

TENDON OF FLEXOR CARPI
ULNARIS MUSCLE

(b) Anterior aspect of forearm
and wrist

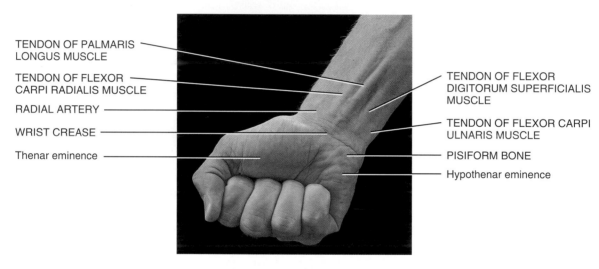

TENDON OF PALMARIS
LONGUS MUSCLE

TENDON OF FLEXOR
CARPI RADIALIS MUSCLE

RADIAL ARTERY

WRIST CREASE

Thenar eminence

TENDON OF FLEXOR
DIGITORUM SUPERFICIALIS
MUSCLE

TENDON OF FLEXOR CARPI
ULNARIS MUSCLE

PISIFORM BONE

Hypothenar eminence

(c) Anterior aspect of wrist

 Which blood vessel is frequently used to take a pulse?

Brachioradialis muscle. Located at the superior and lateral aspect of the forearm.

Flexor carpi radialis muscle. The tendon of this muscle is on the lateral side of the forearm about 1 cm medial to the styloid process of the radius.

Palmaris longus muscle. The tendon of this muscle is medial to the flexor carpi radialis tendon and can be seen if the wrist is slightly flexed and the base of the thumb and little finger are drawn together.

Flexor digitorum superficialis muscle. The tendon of this muscle is medial to the palmaris longus tendon and can be palpated by flexing the fingers at the metacarpophalangeal and proximal interphalangeal joints.

Flexor carpi ulnaris muscle. The tendon of this muscle is on the medial aspect of the forearm.

• **Radial artery.** Located on the lateral aspect of the wrist

between the flexor carpi radialis tendon and styloid process of the radius. It is frequently used to take a pulse.

• **Pisiform bone.** Medial bone of the proximal row of carpals that can be palpated as a projection distal and anterior to the styloid process of the ulna.

• **"Anatomical snuffbox."** A triangular depression between tendons of extensor pollicis brevis and extensor pollicis longus muscles. It derives its name from a habit of previous centuries in which a person would take a pinch of snuff (powdered tobacco or scented powder) and place it in the depression before sniffing it into the nose. The styloid process of the radius, the base of the first metacarpal, trapezium, scaphoid, and radial artery can all be palpated in the depression.

• **Wrist creases.** Three more or less constant lines on the anterior aspect of the wrist (named *proximal, middle,* and *distal*) where the skin is firmly attached to underlying deep fascia.

Surface Features of the Hand

The **hand**, or *manus*, is the region from the wrist to the termination of the upper limb; it has several conspicuous surface features (Figure 12.13).

- **Knuckles.** Commonly refers to the dorsal aspect of the distal ends of metacarpals II–V (or 2–5), but also includes the dorsal aspects of the metacarpophalangeal and interphalangeal joints.

- **Dorsal venous network of the hand (dorsal venous arch).** Superficial veins on the dorsum of the hand that drain blood into the cephalic vein. It can be displayed by compressing the blood vessels at the wrist for a few moments as the hand is opened and closed.

- **Tendon of extensor digiti minimi muscle.** This can be seen on the dorsum of the hand in line with the phalanx of the little finger.

- **Tendons of extensor digitorum muscle.** These can be seen on the dorsum of the hand in line with the phalanges of the ring, middle, and index fingers.

- **Tendon of the extensor pollicis brevis muscle.** This tendon (described previously) is in line with the phalanx of the thumb (see Figure 12.12a).

- **Thenar eminence.** Larger, rounded contour on the lateral aspect of the palm formed by muscles that move the thumb. Also called the ball of the thumb.

- **Hypothenar eminence.** Smaller, rounded contour on the medial aspect of the palm formed by muscles that move the little finger. Also called the ball of the little finger.

- **Palmar flexion creases.** Skin creases on the palm.

- **Digital flexion creases.** Skin creases on the anterior surface of the fingers.

Figure 12.13 Surface anatomy of the hand.

Several tendons on the dorsum of the hand can be identified by their alignment with the phalanges of the digits.

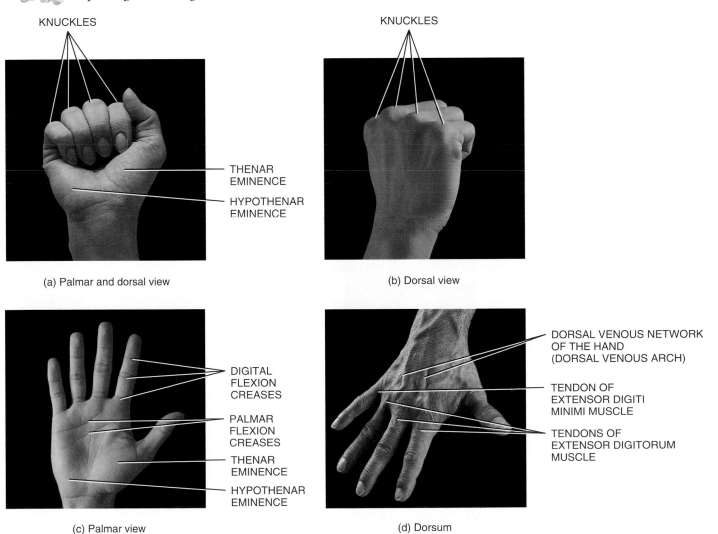

(a) Palmar and dorsal view

(b) Dorsal view

(c) Palmar view

(d) Dorsum

Muscles that form the thenar eminence move which digit?

33. How would you palpate the ulna, styloid process of the ulna, head of the ulna, and styloid process of the radius?
34. Explain the easiest way to identify the muscles of the forearm.
35. What is the "anatomical snuffbox"?
36. What are the knuckles?
37. Why are the thenar and hypothenar eminences important?

SURFACE ANATOMY OF THE LOWER LIMB (EXTREMITY)

• Describe the surface features of the various regions of the lower limb.

The **lower limb** consists of the buttock, thigh, knee, leg, ankle, and foot.

Surface Features of the Buttock

The **buttock,** or *gluteal region*, is formed mainly by the gluteus maximus muscle. The outline of the superior border of the buttock is formed by the iliac crests (see Figure 12.9c). Following are some surface features of the buttock (Figure 12.14).

• **Gluteus maximus muscle.** Forms the major portion of the prominence of the buttocks. The sciatic nerve is deep to this muscle.

• **Gluteus medius muscle.** Superior and lateral to the gluteus maximus muscle.

GLUTEAL INTRAMUSCULAR INJECTION

Another common site for an intramuscular injection is the *gluteus medius muscle.* In order to give this injection, the buttock is divided into quadrants and the upper outer quadrant is used as an injection site. The iliac crest serves as the landmark for this quadrant. This site is chosen because the gluteus medius muscle in this area is quite thick, and there is less chance of injury to the sciatic nerve or major blood vessels. ∎

• **Gluteal (natal) cleft.** Depression along the midline that separates the left and right buttocks.

• **Gluteal fold.** Inferior limit of the buttock that roughly corresponds to the inferior margin of the gluteus maximus muscle.

• **Ischial tuberosity.** Just superior to the medial side of the gluteal fold, the tuberosity bears the weight of the body when seated.

• **Greater trochanter.** A projection of the proximal end of the femur on the lateral side of the thigh. It is about 20 cm (8 in.) inferior to the highest point of the iliac crest.

38. What are the divisions of the lower limb?
39. Define the buttock. What is the gluteal cleft? Gluteal fold?
40. Why is the greater trochanter of the femur important?

Figure 12.14 Surface anatomy of the buttock.

The gluteus medius muscle is a frequent site for an intramuscular injection.

GREATER TROCHANTER

Iliac crest

GLUTEUS MEDIUS MUSCLE

GLUTEUS MAXIMUS MUSCLE

GLUTEAL (NATAL) CLEFT

ISCHIAL TUBEROSITY
GLUTEAL FOLD

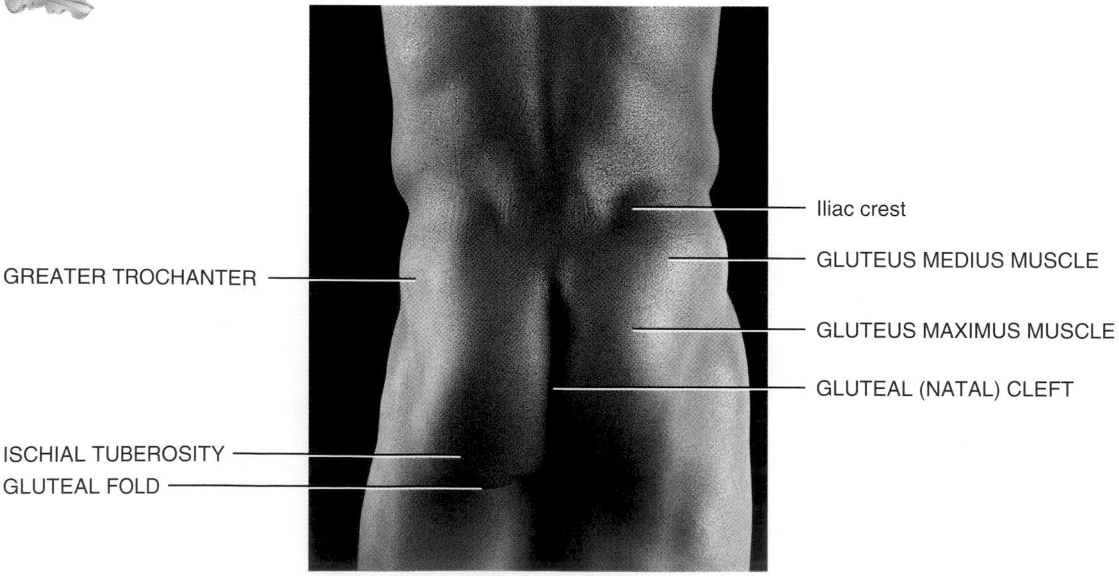

Posterior view

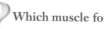Which muscle forms the bulk of the buttock?

Surface Features of the Thigh and Knee

The **thigh**, or *femoral region*, is the region from the hip to the knee. The **knee**, or *genu*, is the region where the thigh and leg join. Several muscles are clearly visible in the thigh. Following are several surface features of the thigh and knee (Figure 12.15).

- **Sartorius muscle.** Superficial anterior muscle that can be traced from the lateral aspect of the thigh to the medial aspect of the knee.

- **Quadriceps femoris muscle.** Three of the four components of the muscle can be seen: **rectus femoris** at the midpoint of the anterior aspect of the thigh; **vastus medialis** at the anteromedial aspect of the thigh; and **vastus lateralis** at the anterolateral aspect of the thigh. The fourth component, the vastus intermedius, is deep to the rectus femoris (see Figure 11.22a on page 359).

THIGH INTRAMUSCULAR INJECTION

The *vastus lateralis muscle* of the quadriceps femoris group is another site that may be used for an intramuscular injection. This site is located at a point midway between the greater trochanter (see Figure 12.14) of the femur and the patella (knee cap). Injection in this area reduces the chance of injury to major blood vessels and nerves. ■

Figure 12.15 Surface anatomy of the thigh and knee.

The quadriceps femoris and hamstrings form the bulk of the musculature of the thigh.

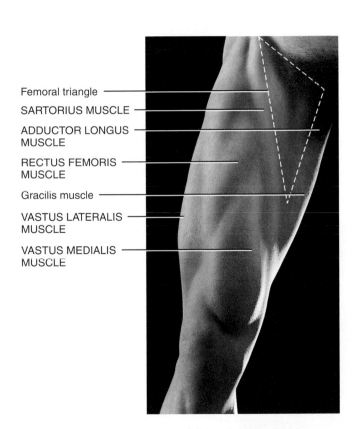

Femoral triangle
SARTORIUS MUSCLE
ADDUCTOR LONGUS MUSCLE
RECTUS FEMORIS MUSCLE
Gracilis muscle
VASTUS LATERALIS MUSCLE
VASTUS MEDIALIS MUSCLE

(a) Anterior view of thigh

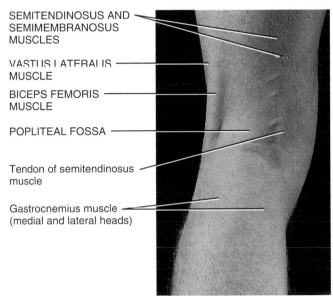

SEMITENDINOSUS AND SEMIMEMBRANOSUS MUSCLES
VASTUS LATERALIS MUSCLE
BICEPS FEMORIS MUSCLE
POPLITEAL FOSSA
Tendon of semitendinosus muscle
Gastrocnemius muscle (medial and lateral heads)

(b) Posterior view of popliteal fossa

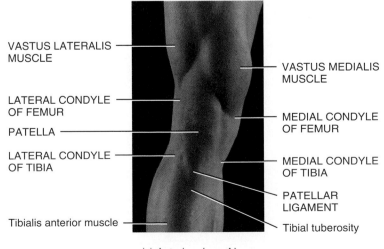

VASTUS LATERALIS MUSCLE
LATERAL CONDYLE OF FEMUR
PATELLA
LATERAL CONDYLE OF TIBIA
Tibialis anterior muscle
VASTUS MEDIALIS MUSCLE
MEDIAL CONDYLE OF FEMUR
MEDIAL CONDYLE OF TIBIA
PATELLAR LIGAMENT
Tibial tuberosity

(c) Anterior view of knee

Which muscles form the borders of the popliteal fossa?

- **Adductor longus muscle.** Located at the superior aspect of the medial thigh. It is the most superficial of the three adductor muscles (adductor magnus, adductor brevis, and adductor longus; see Figure 11.22b on page 360).

- **Femoral triangle.** A large space formed by the inguinal ligament superiorly, the sartorius muscle laterally, and the adductor longus muscle medially. The triangle contains the femoral artery, vein, and nerve and deep inguinal lymph nodes. The triangle is an important arterial pressure point in cases of severe hemorrhage of the lower limb. Hernias frequently occur in this area.

- **Hamstring muscles.** Superficial, posterior thigh muscles located below the gluteal folds. They are the **biceps femoris,** which lies more laterally as it passes inferiorly to the knee, and the **semitendinosus** and **semimembranosus,** which lie medially as they pass inferiorly to the knee. The tendons of the hamstring muscles can be palpated laterally and medially on the posterior aspect of the knee.

The knee presents several distinguishing surface features.

- **Patella.** Also called the kneecap, this large sesamoid bone is located within the tendon of the quadriceps femoris muscle on the anterior surface of the knee along the midline.

- **Patellar ligament.** Continuation of the quadriceps femoris tendon inferior to the patella. Infrapatellar fat pads cushion it on both sides.

- **Medial condyle of femur.** Medial projection on the distal end of the femur.

- **Medial condyle of tibia.** Medial projection on the proximal end of the tibia.

- **Lateral condyle of femur.** Lateral projection on the distal end of the femur.

- **Lateral condyle of tibia.** Lateral projection on the proximal end of the tibia. All four condyles can be palpated just inferior to the patella on either side of the patellar ligament.

- **Popliteal fossa** (pop-LIT-ē-al). A diamond-shaped area on the posterior aspect of the knee that is clearly visible when the knee is flexed. The fossa is bordered superolaterally by the biceps femoris muscle, superomedially by the semimembranosus and semitendinosus muscles, and inferolaterally and inferomedially by the lateral and medial heads of the gastrocnemius muscle, respectively. The **head of the fibula** can easily be palpated on the lateral side of the popliteal fossa. The fossa also contains the popliteal artery and vein. It is sometimes possible to detect a pulse in the popliteal artery.

Surface Features of the Leg, Ankle, and Foot

The **leg,** or *crus,* is the region between the knee and ankle. The **ankle,** or *tarsus,* is between the leg and foot. The **foot** is the region from the ankle to the termination of the lower limb. Following are several surface anatomy features of the leg, ankle, and foot (Figure 12.16).

Figure 12.16 Surface anatomy of the leg, ankle, and foot.

The calcaneal tendon is a common tendon for the gastrocnemius and soleus (calf muscles) and inserts into the calcaneus (heel bone).

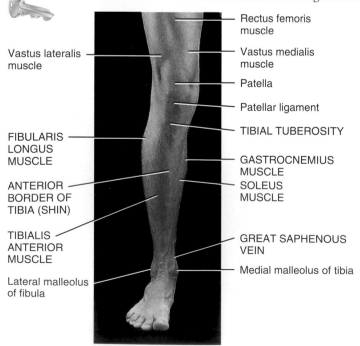

Vastus lateralis muscle

Rectus femoris muscle

Vastus medialis muscle

Patella

Patellar ligament

TIBIAL TUBEROSITY

FIBULARIS LONGUS MUSCLE

GASTROCNEMIUS MUSCLE

ANTERIOR BORDER OF TIBIA (SHIN)

SOLEUS MUSCLE

TIBIALIS ANTERIOR MUSCLE

GREAT SAPHENOUS VEIN

Medial malleolus of tibia

Lateral malleolus of fibula

(a) Anterior view of leg, ankle, and foot

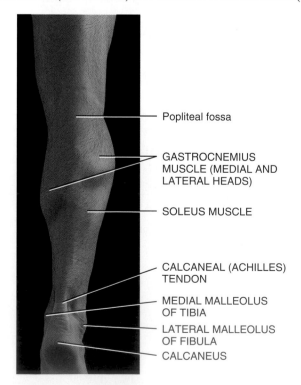

Popliteal fossa

GASTROCNEMIUS MUSCLE (MEDIAL AND LATERAL HEADS)

SOLEUS MUSCLE

CALCANEAL (ACHILLES) TENDON

MEDIAL MALLEOLUS OF TIBIA

LATERAL MALLEOLUS OF FIBULA

CALCANEUS

(b) Posterior view of leg and ankle

- **Tibial tuberosity.** Bony prominence on the superior, anterior surface of the tibia into which the patellar ligament inserts.

- **Tibialis anterior muscle.** Lies against the lateral surface of the tibia, where it is easy to palpate, particularly when the foot is dorsiflexed. The distal tendon of the muscle can be traced to its insertion on the medial cuneiform and base of the first metatarsal.

- **Tibia.** The medial surface and anterior border (shin) of the tibia are subcutaneous and can be palpated throughout the length of the bone.

- **Fibularis (peroneus) longus muscle.** A superficial lateral muscle that overlies the fibula.

- **Gastrocnemius muscle.** Forms the bulk of the midportion and superior portion of the posterior aspect of the leg. The medial and lateral heads can be seen clearly in a person standing on tiptoe.

- **Soleus muscle.** Located deep to the gastrocnemius muscle; it and the gastrocnemius together are referred to as the calf muscles.

- **Calcaneal (Achilles) tendon.** Prominent tendon of the gastrocnemius and soleus muscles on the posterior aspect of the ankle; inserts into the **calcaneus** (heel) bone of the foot.

- **Lateral malleolus of fibula.** Projection of the distal end of the fibula that forms the lateral prominence of the ankle. The head of the fibula, at the proximal end of the bone, lies at the same level as the tibial tuberosity.

- **Medial malleolus of tibia.** Projection of the distal end of the tibia that forms the medial prominence of the ankle.

- **Dorsal venous arch.** Superficial veins on the dorsum of the foot that unite to form the small and great saphenous veins. The great saphenous vein is the longest vein of the body, extending from the foot to the groin.

- **Tendons of extensor digitorum longus muscle.** Visible in line with phalanges II–V (2–5).

- **Tendon of extensor hallucis longus muscle.** Visible in line with phalanx I (great toe). Pulsations in the dorsalis pedis artery may be felt in most people just lateral to this tendon where the blood vessel passes over the navicular and cuneiform bones of the tarsus.

CHECKPOINT

41. What are the divisions of the lower limb?

42. What are the major anterior and posterior muscles of the thigh?

43. Define the femoral triangle. Why is it important?

44. What are the borders of the popliteal fossa? What structures pass through it?

45. Name the calf muscles. Where do they insert?

Now that you have studied some of the surface features of the principal regions of the body, you can appreciate their importance in locating the positions of many internal structures, especially bones, joints, and muscles. In later chapters you will learn to relate surface features to various blood vessels, lymph nodes, nerves, and organs in the chest, abdomen, and pelvis. Once you have visualized and palpated as many of these structures as possible on your own body, you will gain a better appreciation of the application of surface anatomy to your anatomical and clinical studies.

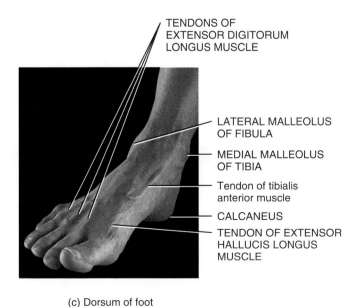

TENDONS OF EXTENSOR DIGITORUM LONGUS MUSCLE

LATERAL MALLEOLUS OF FIBULA

MEDIAL MALLEOLUS OF TIBIA

Tendon of tibialis anterior muscle

CALCANEUS

TENDON OF EXTENSOR HALLUCIS LONGUS MUSCLE

(c) Dorsum of foot

MEDIAL MALLEOLUS OF TIBIA

LATERAL MALLEOLUS OF FIBULA

DORSAL VENOUS ARCH

TENDONS OF EXTENSOR DIGITORUM LONGUS MUSCLE

TENDON OF EXTENSOR HALLUCIS LONGUS MUSCLE

(d) Dorsum of foot

 Which artery is just lateral to the tendon of the extensor digitorum muscle and is a site where a pulse may be detected?

STUDY OUTLINE

Introduction (p. 380)

1. Surface anatomy is the study of anatomical landmarks on the exterior of the body.
2. Surface features may be noted by visualization and/or palpation.
3. Knowledge of surface anatomy has many anatomical and clinical applications.
4. The regions of the body include the head, neck, trunk, upper limbs, and lower limbs.

Surface Anatomy of the Head (p. 381)

1. The head is divided into a cranium and face, each of which is divided into distinct regions.
2. Many muscles of facial expression and skull bones are easily palpated.
3. The eyes, ears, and nose present numerous readily identifiable surface features.

Surface Anatomy of the Neck (p. 386)

1. The neck connects the head to the trunk.
2. It contains several important arteries and veins.
3. The neck is divisible into several triangles by specific muscles and bones.

Surface Anatomy of the Trunk (p. 388)

1. The trunk is divided into the back, chest, abdomen, and pelvis.
2. The back presents several muscles and superficial bones that serve as landmarks.
3. The triangle of auscultation is a region of the back not covered by superficial muscles where respiratory sounds can be heard clearly.
4. The skeleton of the chest protects the organs within and provides several surface landmarks for identifying the position of the heart and lungs.
5. The abdomen and pelvis contain several important landmarks, such as the linea alba, linea semilunaris, McBurney's point, and supracristal line.
6. The perineum forms the floor of the pelvis.

Surface Anatomy of the Upper Limb (Extremity) (p. 392)

1. The upper limb consists of the shoulder, armpit, arm, elbow, forearm, wrist, and hand.
2. The prominence of the shoulder is formed by the deltoid muscle, a frequent site of an intramuscular injection.
3. The axilla is a pyramid-shaped area at the junction of the arm and chest that contains blood vessels and nerves that pass between the neck and upper limbs.
4. The arm contains several blood vessels used for withdrawing blood, introducing fluids, taking the pulse, and measuring blood pressure.
5. Several muscles of the forearm are more easily identifiable by their tendons. The radial artery is a principal blood vessel for detecting a pulse.
6. The hand contains the origin of the cephalic vein and extensor tendons on the dorsum, and the thenar and hypothenar eminences on the palm.

Surface Anatomy of the Lower Limb (Extremity) (p. 398)

1. The lower limb consists of the buttock, thigh, knee, leg, ankle, and foot.
2. The buttock is formed mainly by the gluteus maximus muscle.
3. The major muscles of the thigh are the anterior quadriceps femoris and the posterior hamstrings. An important landmark in the thigh is the femoral triangle.
4. The popliteal fossa is a diamond-shaped area on the posterior aspect of the knee.
5. The gastrocnemius muscle forms the bulk of the midportion of the posterior aspect of the leg and together with the soleus muscle forms the calf of the leg.
6. The foot contains the origin of the great saphenous vein, the largest vein in the body.

Q SELF-QUIZ QUESTIONS

Choose the one best answer to the following questions.

1. Which of the following is true of the sternal angle?
   a. It is palpated to locate the costal cartilage of the second rib.
   b. It is the junction between the manubrium and body of the sternum.
   c. It is external to the point at which the trachea branches into the right and left primary bronchi.
   d. Only a and b are correct.
   e. a, b, and c are all correct.

2. Which of the following terms refers to the region of the external ear?
   a. temporal   b. oral   c. mental   d. auricular   e. buccal

3. The apex of what structure can be palpated just superior to the medial third of the clavicle?
   a. heart   b. lung   c. larynx   d. diaphragm
   e. hyoid bone

4. The muscle that can be palpated just superior and anterior to the angle of the mandible by alternately clenching the teeth and then opening the mouth is the:
   a. masseter.   b. sternocleidomastoid.   c. temporalis.
   d. zygomaticus major.   e. orbicularis oris.

5. The mastoid process can be palpated just posterior to the:
   a. lateral commissure.   b. tragus.   c. auricle.
   d. dorsum nasi.   e. cricoid cartilage.

6. The pulse is commonly taken in the neck just anterior to the sternocleidomastoid muscle from a blood vessel called the:

a. external jugular vein.
b. subclavian artery.
c. internal jugular vein.
d. common carotid artery.
e. external carotid artery.

7. The femoral triangle is bounded by the adductor longus, the sartorius, and the:

a. psoas major. b. iliac crest. c. ischial tuberosity.
d. inguinal ligament. e. semitendinosus.

Complete the following.

8. The head is divided into the _____ region and the _____ region.

9. Use the directional terms (anterior, posterior, superior, inferior, proximal, distal, medial, lateral) to complete the following:

a. The mental region is _____ to the oral region.
b. The orbital region is _____ to the frontal region.
c. The antebrachium is _____ to the brachium.
d. The sternal angle is _____ to the xiphoid process.
e. The tragus of the ear is _____ to the external auditory canal.
f. The iris of the eye is _____ to the medial commissure.
g. The lobule is the _____ portion of the auricle.

10. The triangle of auscultation is bounded by the vertebral border of the scapula, the _____ muscle, and the _____ muscle.

11. The heart lies on the diaphragm deep to the _____ joint.

Are the following statements true or false?

12. The ulnar nerve lies in a groove posterior to the medial epicondyle.

13. All of the following surface features are located on the lower limbs: tibial tuberosity, calcaneal (Achilles) tendon, lateral malleolus.

14. McBurney's point is located at the halfway point along a line between the umbilicus and the anterior superior iliac spine.

15. A landmark for performing a tracheostomy is the cricoid cartilage.

16. A vein of the upper limb often used for intravenous therapy is located in the popliteal fossa.

Matching

17. Match the following:

___ **(a)** frequent site for intramuscular injections in the upper limb

___ **(b)** a "stiff neck" results from an inflammation of this muscle

___ **(c)** divides neck into anterior and posterior (lateral) triangles

___ **(d)** contributes to the anterior axillary fold

___ **(e)** occupies most of the posterior surface of the arm

___ **(f)** tendon is visible on the anterior surface of the wrist when making a fist

___ **(g)** tendon forms one border of the "anatomical snuffbox"

___ **(h)** frequently used as a site for insulin injection

(1) palmaris longus
(2) triceps brachii
(3) trapezius
(4) deltoid
(5) vastus lateralis
(6) extensor pollicis longus
(7) sternocleidomastoid
(8) pectoralis major

18. Match the following terms with their descriptions:

___ **(a)** prominent posterior projection in the midline at the base of the neck

___ **(b)** most prominent structure in the midline of the anterior cervical region

___ **(c)** superior border of the buttock

___ **(d)** forms the tip of the shoulder

___ **(e)** located deep to the dimple that coincides with the middle of the sacroiliac joint

___ **(f)** bears most of the weight of a person when sitting

___ **(g)** attachment point for calcaneal (Achilles) tendon

___ **(h)** protuberance at the distal end of the ulna

(1) thyroid cartilage
(2) iliac crest
(3) ischial tuberosity
(4) acromion
(5) vertebra prominens
(6) calcaneus
(7) styloid process
(8) posterior superior iliac spine

CRITICAL THINKING QUESTIONS

1. Fred, the football fanatic, ran into the goal post and landed on his back. He told the coach that he felt all right but when he reached behind his head, he felt a bump on the back of his skull and another at the base of his neck. The team physician examined Fred and told him that the bumps were nothing to worry about; that they had always been there. What were these bumps?

2. Great Aunt Edith heard you're bringing your new boyfriend, the one with an anatomically correct heart tattooed on his bicep (he can even make it beat) and "I brake for ambulances" bumper sticker, to Thanksgiving dinner. She told your mother that she's having "palpations." Your new boyfriend, a first-year medical student, said she probably meant "palpitations." Is there a difference?

3. After Great Aunt Edith meets your new boyfriend, she decides he's not so bad after all, and says to him, "Why did you tattoo that heart on your arm? Why didn't you put it on your chest where it belongs?" Your boyfriend thinks this is a great idea. Describe the projection points on the surface of the chest that the artist should use to be anatomically correct.

4. Randy has been told that he needs to receive a large dose of an antibiotic to treat his infection. When he hears that this will be administered by a gluteal intramuscular injection, he says he isn't worried because, "They're injecting me there because it's so padded with fat, I won't feel it." If the nurse injected him in the fatty part of his buttocks, would the injection have been administered correctly? What muscle should the nurse be aiming for, and what is the major landmark for locating the muscle? What are the risks of not injecting the correct muscle?

5. An ad for a new telephoto camera appeared in a medical journal. It states "you'll see the pores on your neighbor's ala, the sweat on his philtrum, the hairs in his external nares, the studs in his helix, the ring in his supercilia and the color of his sclera." Is this a great camera? Explain.

ANSWERS TO FIGURE QUESTIONS

12.1 The frontal, parietal, temporal, occipital, zygomatic, and nasal regions are all named for bones that are deep to them.

12.2 When palpating the sagittal suture, the sheetlike epicranial aponeurosis (galea aponeurotica) is also palpated deep to the skin.

12.3 Light enters the eye through the pupil.

12.4 The mastoid process of the temporal bone can be palpated just posterior to the auricle.

12.5 The anterior border of the nose that connects the root and apex is the dorsum nasi.

12.6 The sternocleidomastoid muscle divides the neck into anterior and posterior triangles.

12.7 With the aid of a stethoscope, respiratory sounds can be heard clearly in the triangle of auscultation.

12.8 The sternal angle is the most reliable landmark of the chest.

12.9 Pain produced by pressure of the finger on McBurney's point indicates acute appendicitis; the incision for an appendectomy is made through it.

12.10 The acromion of the scapula forms the top of the shoulder.

12.11 Blood is usually withdrawn from the median cubital vein; this vein is also used to administer medication.

12.12 A pulse is frequently taken by placing pressure on the radial artery in the wrist.

12.13 Muscles that form the thenar eminence move the thumb.

12.14 The gluteus maximus muscle forms the bulk of the buttock.

12.15 The popliteal fossa is bordered superolaterally by the biceps femoris, superomedially by the semimembranosus and semitendinosus, and inferolaterally and inferomedially by the lateral and medial heads of the gastrocnemius.

12.16 A pulse may be felt in the dorsalis pedis artery, located just lateral to the tendon of the extensor digitorum muscle.

THE CARDIOVASCULAR SYSTEM: BLOOD

13

INTRODUCTION The **cardiovascular system** (*cardio-*=heart; *vascular*=blood vessels) consists of three interrelated components: blood, the heart, and blood vessels. The focus of this chapter is blood; the next two chapters will examine the heart and blood vessels, respectively. Blood transports various substances, helps regulate several life processes, and affords protection against disease. For all of its similarities in origin, composition, and functions, blood is as unique from one person to another as are skin, bone, and hair. Health-care professionals routinely examine and analyze its differences through various blood tests when trying to determine the cause of different diseases. Despite these differences, blood is the most easily and widely shared of human tissues, saving many thousands of lives every year through blood transfusions. The branch of science concerned with the study of blood, blood-forming tissues, and the disorders associated with them is **hematology** (hēm-a-TOL-ō-jē; *hema-* or *hemato-*=blood; *-logy*=study of).

Can you guess what is being practiced in this illustration? Can you speculate as to the culture of the people in this image?

www.wiley.com/college/apcentral

Because of their common origins, the development of blood in the embryo and fetus is considered in Chapter 15 with the development of the blood vessels.

Most cells of a multicellular organism cannot move around to obtain oxygen and nutrients and get rid of carbon dioxide and other wastes. Instead, these needs are met by two fluids: blood and interstitial fluid. **Blood** is a connective tissue composed of a liquid portion called plasma (the extracellular matrix) and a cellular portion consisting of various cells and cell fragments. **Interstitial fluid** is the fluid that bathes body cells. Oxygen brought into the lungs and nutrients brought into the gastrointestinal tract are transported by the blood to the cells of the body. The oxygen and nutrients diffuse from the blood into the interstitial fluid and then into body cells. Carbon dioxide and other wastes move in the reverse direction from the body cells into the inter-

stitial fluid and then into the blood. Blood then transports the wastes to various organs—the lungs, kidneys, skin, and digestive system—for elimination from the body.

In order for blood to reach all the cells, it must be moved throughout the body. The *heart* is the pump that circulates the blood. *Blood vessels* convey blood from the heart to body cells and from body cells back to the heart. The largest blood vessels, which take blood away from the heart, are called *arteries*. Arteries branch into smaller vessels called *arterioles*. As an arteriole enters a tissue, it divides into numerous microscopic vessels called *capillaries*. Substances exchanged between the blood and interstitial fluid pass through the thin walls of capillaries. Before leaving a tissue, capillaries unite to form small vessels called *venules*, which in turn merge to form progressively larger vessels called *veins*. Veins convey blood back to the heart.

FUNCTIONS OF BLOOD

OBJECTIVE

● List and describe the functions of blood.

Blood, a liquid connective tissue, has three general functions:

1. *Transportation.* Blood transports oxygen from the lungs to the cells of the body and carbon dioxide from the body cells to the lungs for exhalation. It carries nutrients from the gastrointestinal tract to body cells and hormones from endocrine glands to other body cells. Blood also transports heat and waste products to the lungs, kidneys, and skin for elimination from the body.

2. *Regulation.* Circulating blood helps maintain homeostasis in all body fluids. Blood helps regulate pH through buffers. It also helps adjust body temperature through the heat-absorbing and coolant properties of the water in blood plasma and its variable rate of flow through the skin, where excess heat can be lost from the blood to the environment. Blood osmotic pressure also influences the water content of cells, mainly through interactions of dissolved ions and proteins.

3. *Protection.* Blood can clot, which protects against its excessive loss from the cardiovascular system after an injury. In addition, white blood cells protect against disease by carrying on phagocytosis. Several types of blood proteins, including antibodies, interferons, and complement, help protect against disease in a variety of ways.

CHECKPOINT

1. What substances does blood transport?

PHYSICAL CHARACTERISTICS OF BLOOD

OBJECTIVE

● List the principal physical characteristics of blood.

Blood is denser and more viscous (sticky) than water, which is part of the reason it flows more slowly than water. The temperature of blood is about 38°C (100.4°F), which is slightly higher

than normal body temperature, and it has a slightly alkaline pH ranging from 7.35 to 7.45. Blood constitutes about 8% of the total body weight. The blood volume is 5–6 liters (1.5 gal) in an average-sized adult male and 4–5 liters (1.2 gal) in an average-sized adult female.

WITHDRAWING BLOOD

Blood samples for laboratory testing may be obtained in several ways. The most frequently used procedure is **venipuncture,** withdrawal of blood from a vein using a hypodermic needle and syringe. A tourniquet is wrapped around the arm above the venipuncture site to cause the blood to accumulate in the vein. This increased blood volume causes the vein to stand out. Opening and closing the fist causes it to stand out even more, making the venipuncture more successful. A common site for venipuncture is the median cubital vein anterior to the elbow (see Figure 15.12b on page 489). Another method of drawing blood is through a **finger** or **heel stick.** This procedure is frequently used by diabetic patients to monitor their daily blood sugar as well as for drawing blood from infants and children. In an **arterial stick,** blood is withdrawn from an artery; this procedure is used when it is important to know the level of oxygen in oxygenated blood. ■

CHECKPOINT

2. How much does the blood of a 150-pound person weigh?

COMPONENTS OF BLOOD

OBJECTIVE

● Describe the principal components of blood.

Whole blood is composed of (1) blood plasma, a watery liquid that contains dissolved substances, and (2) formed elements, which are cells and cell fragments. If a sample of blood is centrifuged (spun) in a small glass tube, the cells sink to the bottom of the tube while the lighter-weight plasma forms a layer on top (Figure 13.1a). Blood is about 45% formed elements and about 55% blood plasma. Normally, more than 99% of the formed elements are red-colored red blood cells (RBCs). Pale or colorless

Figure 13.1 Components of blood in a normal adult.

Blood is a connective tissue that consists of blood plasma (liquid) plus formed elements (red blood cells, white blood cells, and platelets).

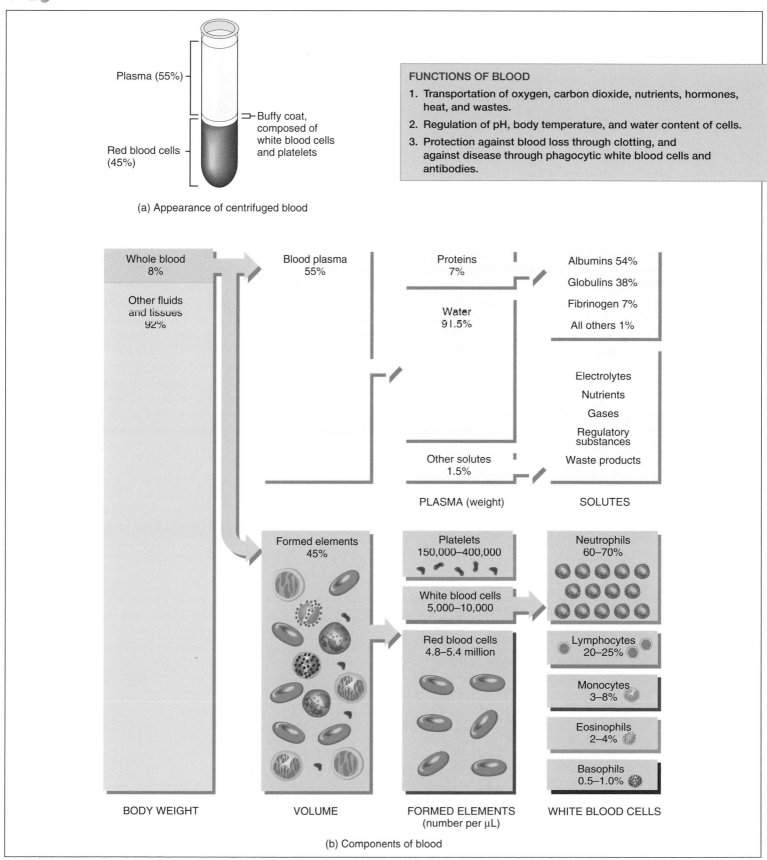

Plasma (55%)

Buffy coat, composed of white blood cells and platelets

Red blood cells (45%)

FUNCTIONS OF BLOOD

1. Transportation of oxygen, carbon dioxide, nutrients, hormones, heat, and wastes.
2. Regulation of pH, body temperature, and water content of cells.
3. Protection against blood loss through clotting, and against disease through phagocytic white blood cells and antibodies.

(a) Appearance of centrifuged blood

Whole blood 8%

Other fluids and tissues 92%

Blood plasma 55%

Proteins 7%

Water 91.5%

Other solutes 1.5%

Albumins 54%

Globulins 38%

Fibrinogen 7%

All others 1%

Electrolytes

Nutrients

Gases

Regulatory substances

Waste products

PLASMA (weight)

SOLUTES

Formed elements 45%

Platelets 150,000–400,000

White blood cells 5,000–10,000

Red blood cells 4.8–5.4 million

Neutrophils 60–70%

Lymphocytes 20–25%

Monocytes 3–8%

Eosinophils 2–4%

Basophils 0.5–1.0%

BODY WEIGHT

VOLUME

FORMED ELEMENTS (number per μL)

WHITE BLOOD CELLS

(b) Components of blood

What is the approximate volume of blood in your body?

white blood cells (WBCs) and platelets occupy less than 1% of total blood volume. They form a very thin layer, called the *buffy coat*, between the packed RBCs and blood plasma in centrifuged blood. Figure 13.1b shows the composition of blood plasma and the numbers of the various types of formed elements in blood.

Blood Plasma

When the formed elements are removed from blood, the straw-colored liquid called **blood plasma** (or simply **plasma**) is left. Plasma is about 91.5% water and 8.5% solutes, most of which (7% by weight) are proteins. Some of the proteins in plasma are also found elsewhere in the body, but those confined to blood are called **plasma proteins.** Hepatocytes (liver cells) synthesize most of the plasma proteins, which include the **albumins** (54% of plasma proteins), **globulins** (38%), and **fibrinogen** (7%). Certain blood cells develop into cells that produce gamma globulins, an important type of globulin. These plasma proteins are also called **antibodies** or **immunoglobulins** because they are produced during certain immune responses. Foreign substances (antigens) such as bacteria and viruses stimulate production of antibodies. An antibody binds specifically to the antigen that stimulated its production and this disables the invading antigen.

Besides proteins, other solutes in plasma include electrolytes, nutrients, regulatory substances such as enzymes and hormones, gases, and waste products such as urea, uric acid, creatinine, ammonia, and bilirubin.

Table 13.1 describes the chemical composition of blood plasma.

Formed Elements

The **formed elements** of the blood include **red blood cells (RBCs), white blood cells (WBCs),** and **platelets** (Figure 13.2). Although RBCs and WBCs are living cells, platelets are cell fragments. Unlike RBCs and platelets, which perform limited roles, WBCs have a number of more general functions. There are several distinct types of WBCs—neutrophils, lymphocytes, monocytes, eosinophils, and basophils—each having a unique microscopic appearance. The roles of each type of WBC are discussed later in this chapter.

The percentage of total blood volume occupied by RBCs is called the **hematocrit** (he-MAT-ō-krit). For example, a hematocrit of 40 means that 40% of the volume of blood is composed of RBCs. The normal range of hematocrit for adult females is about 38–46% (average = 42); for adult males it is about 40–54% (average = 47). The hormone testosterone, which is present in much higher concentration in males than in females, stimulates synthesis of a hormone called **erythropoietin** (e-rith′-rō-POY-e-tin) or **EPO** by the kidneys. This hormone, which stimulates production of RBCs, contributes to higher hematocrits in males. Lower values in women during their reproductive years may be due to excessive loss of blood during menstruation. A significant drop in hematocrit indicates anemia, a lower-than-normal number of RBCs. In *polycythemia* the percentage of RBCs is abnormally high, and the hematocrit may be 65% or higher. Polycythemia may be caused by conditions such as an unregulated increase in RBC production, tissue hypoxia, dehydration, or blood doping by athletes.

TABLE 13.1	SUBSTANCES IN BLOOD PLASMA
CONSTITUENT (%)	**DESCRIPTION**
Water (91.5)	Liquid portion of blood. Acts as solvent and suspending medium for components of blood, absorbs, transports, and releases heat.
Plasma proteins (7.0)	Exert colloid osmotic pressure, which helps maintain water balance between blood and tissues and regulates blood volume.
Albumins	Smallest and most numerous plasma proteins; produced by liver. Function as transport proteins for several steroid hormones and for fatty acids.
Globulins	Produced by liver and by plasma cells, which develop from B lymphocytes. Antibodies (immunoglobulins) help attack viruses and bacteria. Alpha and beta globulins transport iron, lipids, and fat-soluble vitamins.
Fibrinogen	Produced by liver. Plays essential role in blood clotting.
Other solutes (1.5)	
Electrolytes	Inorganic salts. Positively charged ions (cations) include Na^+, K^+, Ca^{2+}, Mg^{2+}; negatively charged ions (anions) include Cl^-, HPO_4^{2-}, SO_4^{2-}, and HCO_3^-. Help maintain osmotic pressure and play essential roles in the function of cells.
Nutrients	Products of digestion pass into blood for distribution to all body cells. Include amino acids (from proteins), glucose (from carbohydrates), fatty acids and glycerol (from triglycerides), vitamins, and minerals.
Gases	Oxygen (O_2), carbon dioxide (CO_2), and nitrogen (N_2). Whereas more O_2 is associated with hemoglobin inside red blood cells, more CO_2 is dissolved in plasma. N_2 is present but has no known function in the body.
Regulatory substances	Enzymes, produced by body cells, catalyze chemical reactions. Hormones, produced by endocrine glands, regulate metabolism, growth, and development. Vitamins are cofactors for enzymatic reactions.
Waste products	Most are breakdown products of protein metabolism and are carried by blood to organs of excretion. Include urea, uric acid, creatine, creatinine, bilirubin, and ammonia.

Figure 13.2 Scanning electron micrograph and photomicrograph of the formed elements of blood.

The formed elements of blood are red blood cells (RBCs), white blood cells (WBCs), and platelets.

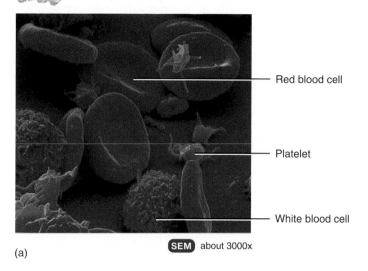

Red blood cell

Platelet

White blood cell

(a) SEM about 3000x

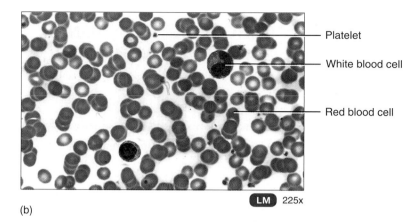

Platelet

White blood cell

Red blood cell

(b) LM 225x

Which formed elements of the blood are cell fragments?

INDUCED POLYCYTHEMIA IN ATHLETES

Delivery of oxygen to muscles is a limiting factor in muscular activity. As a result, increasing the oxygen-carrying capacity of the blood enhances athletic performance, especially in endurance events. Because red blood cells are the main transport vehicle for oxygen, athletes have tried several means of increasing their hematocrit, a practice called **induced polycythemia,** to gain a competitive edge. Training at higher altitudes produces a natural increase in hematocrit because the lower amount of oxygen in inhaled air triggers release of more EPO. In past years, some athletes performed *blood doping* to elevate their hematocrit. In this procedure, blood cells are removed from the body and stored for a month or so, during which time the hematocrit returns to normal. Then, a few days before an athletic event, the RBCs are reinjected to increase hematocrit. More recently, athletes have enhanced their RBC production by injecting *Epoetin alfa* (*Procrit* or *Epogen*), a drug that is used to treat anemia. Epoetin alfa is identical to naturally occurring EPO because it is manufactured by recombinant DNA technology using the human gene that codes for erythropoietin. Practices that increase hematocrit are dangerous, however, because they increase the workload of the heart. The elevated numbers of RBCs raise the viscosity of the blood, which increases the resistance to blood flow and makes the blood more difficult for the heart to pump. Increased viscosity also contributes to high blood pressure and increased risk of stroke. During the 1980s, at least 15 competitive cyclists died from heart attacks or strokes linked to suspected use of Epoetin alfa. Although the International Olympics Committee prohibits use of Epoetin alfa, enforcement is difficult because the drug is undetectable. ■

CHECKPOINT

3. What are some functions of blood plasma proteins?
4. What is the significance of lower-than-normal or higher-than-normal hematocrit?

FORMATION OF BLOOD CELLS

OBJECTIVE

● Explain the origin of blood cells.

Although some lymphocytes have a lifetime measured in years, most formed elements of the blood are continually dying and being replaced within hours, days, or weeks. Negative feedback systems regulate the total number of RBCs and platelets in circulation, and their numbers normally remain steady. The abundance of the different types of WBCs, however, varies in response to challenges by invading pathogens and other foreign antigens.

The process by which the formed elements of blood develop is called **hemopoiesis** (hē-mō-poy-Ē-sis; -*poiesis*=making) or *hematopoiesis*. As you will learn in Chapter 15, before birth, hemopoiesis first occurs in the yolk sac of an embryo and later in the liver, spleen, thymus, and lymph nodes of a fetus. In the last three months before birth, red bone marrow becomes the primary site of hemopoiesis and continues to be the main source of blood cells after birth and throughout life.

Red bone marrow is a highly vascularized connective tissue located in the microscopic spaces between trabeculae of spongy bone tissue. It is present chiefly in bones of the axial skeleton, pectoral and pelvic girdles, and the proximal epiphyses of the humerus and femur. About 0.05–0.1% of red bone marrow cells are derived from mesenchymal cells called **pluripotent stem cells** (ploo-RIP-ō-tent; *pluri-*=several). Pluripotent stem cells are cells that have the capacity to develop into several different types of cells (Figure 13.3).

BONE MARROW EXAMINATION

Sometimes a sample of red bone marrow must be obtained in order to diagnose certain blood disorders, such as leukemia and severe anemias. **Bone marrow examination** may involve *bone marrow aspiration* (withdrawal of a small amount of red bone marrow with a fine needle and syringe) or a *bone marrow biopsy* (removal of a core of red bone marrow with a larger needle). Both types of samples are usually taken from the iliac crest of the hip bone, although samples are sometimes aspirated from the sternum. In young children, bone marrow samples are taken from a vertebra or tibia (shin bone). The tissue or cell sample is then sent to a pathology lab for analysis. Specifically, the lab technicians look for signs of neoplastic (cancer) cells or other diseased cells to assist in diagnosis. ■

Figure 13.3   Origin, development, and structure of blood cells. Some of the generations of some cell lines have been omitted.

Blood cell production is called hemopoiesis and occurs mainly in red bone marrow after birth.

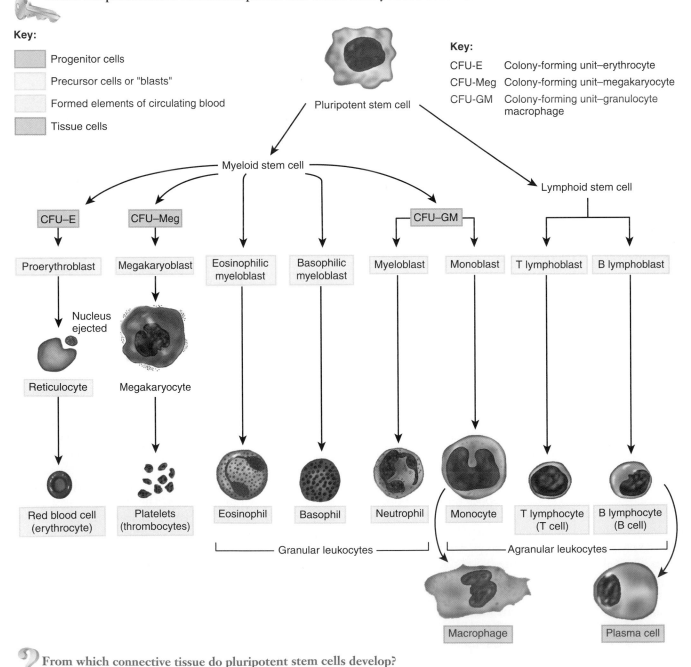

Key:

▢ Progenitor cells

▢ Precursor cells or "blasts"

▢ Formed elements of circulating blood

▢ Tissue cells

Key:

CFU-E   Colony-forming unit–erythrocyte

CFU-Meg   Colony-forming unit–megakaryocyte

CFU-GM   Colony-forming unit–granulocyte macrophage

From which connective tissue do pluripotent stem cells develop?

Stem cells in red bone marrow reproduce themselves, proliferate, and differentiate into cells that give rise to blood cells, macrophages, reticular cells, mast cells, and adipocytes. Some of the stem cells can also form osteoblasts, chondroblasts, and muscle cells, and may some day be used as a source of bone, cartilage, and muscular tissue for tissue and organ replacement. The reticular cells produce reticular fibers, which form the stroma (framework) that supports red bone marrow cells. Blood from nutrient and metaphyseal arteries (see Figure 6.4 on page 150) enters a bone and passes into the enlarged and leaky capillaries, called *sinuses*, that surround red bone marrow cells and fibers. After blood cells form, they enter the sinuses and other blood vessels and leave the bone through nutrient and periosteal veins (see Figure 6.4). With the exception of lymphocytes, formed elements do not divide once they leave red bone marrow.

Pluripotent stem cells produce two further types of stem cells, called *myeloid stem cells* and *lymphoid stem cells.* Myeloid stem cells begin and complete their development in red bone marrow and give rise to red blood cells, platelets, monocytes, neutrophils, eosinophils, and basophils. Lymphoid stem cells begin their development in red bone marrow but complete it in lymphatic tissues; they give rise to lymphocytes. Although stem cells have distinctive cell identity markers in their plasma membranes, they cannot be distinguished histologically and resemble lymphocytes.

During hemopoiesis, the myeloid stem cells differentiate into **progenitor cells** (prō-JEN-i-tor). Progenitor cells are no longer capable of reproducing themselves and are committed to giving rise to more specific elements of blood. Some progenitor cells are known as *colony-forming units (CFUs).* After the CFU designation is an abbreviation that designates the mature elements in blood that they will produce: CFU-E ultimately produces erythrocytes (red blood cells), CFU-Meg produces megakaryocytes, the source of platelets, and CFU-GM ultimately produces granulocytes (specifically, neutrophils) and monocytes. Progenitor cells, like stem cells, resemble lymphocytes and cannot be distinguished by their microscopic appearance alone. Other myeloid stem cells develop directly into cells called precursor cells (described next). Lymphoid stem cells differentiate into T lymphoblasts and B lymphoblasts, which ultimately develop into T lymphocytes (T cells) and B lymphocytes (B cells), respectively.

In the next generation, the cells are called **precursor cells,** also known as **blasts.** Over several cell divisions they develop into the actual formed elements of blood. For example, monoblasts develop into monocytes, eosinophilic myeloblasts develop into eosinophils, and so on. Precursor cells have recognizable microscopic appearances.

Several hormones called **hemopoietic growth factors** regulate the differentiation and proliferation of particular progenitor cells. **Erythropoietin** or **EPO** increases the number of red blood cell precursors. The main producers of EPO are cells in the kidneys that lie between the kidney tubules (peritubular interstitial cells). With renal failure, EPO release slows and RBC production is inadequate. **Thrombopoietin** (throm′-bō-POY-ē-tin) or **TPO** is a hormone produced by the liver that stimulates the formation of platelets (thrombocytes) from megakaryocytes. Several different **cytokines,** small glycoproteins produced by red bone marrow cells, leukocytes, macrophages, fibroblasts, and endothelial cells, regulate development of different blood cell types. They typically act as local hormones (autocrines or paracrines). Cytokines stimulate proliferation of progenitor cells in red bone marrow and regulate the activities of cells involved in nonspecific defenses (such as phagocytes) and immune responses (such as B cells and T cells). Two important families of cytokines that stimulate the formation of white blood cell formation are **colony-stimulating factors (CSFs)** and **interleukins.**

MEDICAL USES OF HEMOPOIETIC GROWTH FACTORS

Hemopoietic growth factors made available through recombinant DNA technology hold tremendous potential for medical uses when a person's natural ability to form new blood cells is diminished or defective. Recombinant erythropoietin (Epoetin alfa) is very effective in treating the diminished red blood cell production that accompanies end-stage kidney disease. Granulocyte-macrophage colony-stimulating factor and granulocyte CSF are given to stimulate white blood cell formation in cancer patients who are undergoing chemotherapy, which kills some of their red bone marrow cells as well as the cancer cells because both cell types are undergoing mitosis. Thrombopoietin shows great promise for preventing platelet depletion during chemotherapy. CSFs and thrombopoietin also improve the outcome of patients who receive bone marrow transplants. Hemopoietic growth factors are also used to treat thrombocytopenia in neonates, other clotting disorders, and various types of anemia. Research on these medications is ongoing and shows a great deal of promise. ■

CHECKPOINT

5. Which hemopoietic growth factors regulate differentiation and proliferation of CFU-E and formation of platelets from megakaryocytes?

RED BLOOD CELLS

OBJECTIVE

● Describe the structure, functions, life cycle, and production of red blood cells.

Red blood cells (RBCs) or **erythrocytes** (e-RITH-rō-sīts; *erythro-*=red; *-cyte*=cell) contain the oxygen-carrying protein **hemoglobin,** which is a pigment that gives whole blood its red color. A healthy adult male has about 5.4 million red blood cells per microliter (μL) of blood*, and a healthy adult female has about 4.8 million. (One drop of blood is about 50 μL.) To maintain normal numbers of RBCs, new mature cells must enter the

---

*1 μL = 1 mm^3 = 10^{-6} liter.

circulation at the astonishing rate of at least 2 million per second, a pace that balances the equally high rate of RBC destruction.

RBC Anatomy

RBCs are biconcave discs with a diameter of 7–8 μm (Figure 13.4a). Mature red blood cells have a simple structure. Their plasma membrane is both strong and flexible, which allows them to deform without rupturing as they squeeze through narrow capillaries. As you will see later, certain glycolipids in the plasma membrane of RBCs are antigens that account for the various blood groups such as the ABO and Rh groups. RBCs lack a nucleus and other organelles and can neither reproduce nor carry on extensive metabolic activities. The cytosol of RBCs contains hemoglobin molecules, which were synthesized before loss of the nucleus during RBC production and which constitute about 33% of the cell's weight.

RBC Functions

Red blood cells are highly specialized for their oxygen transport function. Because mature RBCs have no nucleus, all their internal space is available for oxygen transport. Moreover, RBCs lack mitochondria and generate ATP anaerobically (without oxygen). Hence, they do not use up any of the oxygen they transport. Even the shape of an RBC facilitates its function. A biconcave disc has a much greater surface area for its volume than, say, a sphere or a cube. This shape provides a large surface area for the diffusion of gas molecules into and out of the RBC.

Figure 13.4 The shapes of a red blood cell (RBC) and a hemoglobin molecule. In (b), note that each of the four polypeptide chains of a hemoglobin molecule (blue) has one heme group (gold), which contains an iron ion (Fe^{2+}), shown in red.

The iron portion of a heme group binds oxygen for transport by hemoglobin.

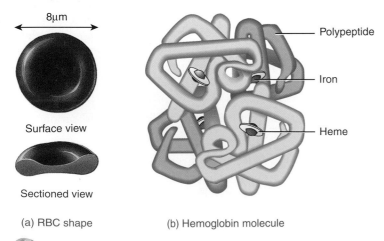

8μm

Surface view

Sectioned view

(a) RBC shape

Polypeptide

Iron

Heme

(b) Hemoglobin molecule

? How many molecules of O_2 can one hemoglobin molecule transport?

Each RBC contains about 280 million hemoglobin molecules; each hemoglobin molecule can carry up to four oxygen molecules. A hemoglobin molecule consists of a protein called **globin,** composed of four polypeptide chains (two alpha and two beta chains), plus four nonprotein pigments called **hemes** (Figure 13.4b). One ringlike heme binds to each polypeptide chain. At the center of the heme ring is an iron ion (Fe^{2+}) that can combine reversibly with one oxygen molecule. The oxygen picked up in the lungs is transported bound to the iron of the heme group. As blood flows through tissue capillaries, the iron-oxygen reaction reverses. Hemoglobin releases oxygen, which diffuses first into the interstitial fluid and then into cells.

Hemoglobin also transports about 13% of the total carbon dioxide, a waste product of metabolism. Blood flowing through tissue capillaries picks up carbon dioxide, some of which combines with amino acids in the globin part of hemoglobin. As blood flows through the lungs, the carbon dioxide is released from hemoglobin and then exhaled.

In addition to its key role in transporting oxygen and carbon dioxide, hemoglobin also plays a role in regulation of blood flow and blood pressure. The gaseous hormone **nitric oxide (NO),** produced by the endothelial cells that line blood vessels, binds to hemoglobin. Under some circumstances, hemoglobin releases NO. The released NO causes *vasodilation*, an increase in blood vessel diameter that occurs when the smooth muscle in the vessel wall relaxes. Vasodilation improves blood flow and enhances oxygen delivery to cells near the site of NO release.

IRON OVERLOAD

In cases of **iron overload,** the amount of iron present in the body builds up. Because we have no method for eliminating excess iron, any condition that increases dietary iron absorption can cause iron overload. At some point, the plasma protein that transports iron (transferrin) becomes saturated with iron ions, and the free iron level rises. Common consequences of iron overload are diseases of the liver, heart, pancreatic islets, and gonads. Iron overload also allows certain iron-dependent microbes to flourish. Such microbes normally are not pathogenic, but they multiply rapidly and can cause lethal effects in a short time when free iron is present. ■

RBC Life Cycle

Red blood cells live only about 120 days because of the wear and tear their plasma membranes undergo as they squeeze through blood capillaries. Without a nucleus and other organelles, RBCs cannot synthesize new components to replace damaged ones. The plasma membrane becomes more fragile with age, and the cells are more likely to burst, especially as they squeeze through narrow channels in the spleen. Ruptured red blood cells are removed from circulation and destroyed by fixed phagocytic macrophages in the spleen and liver, and the breakdown products are recycled.

Erythropoiesis: Production of RBCs

Erythropoiesis (e-rith′-rō-poy-Ē-sis), the production of RBCs, starts in the red bone marrow with a precursor cell called a **proerythroblast** (see Figure 13.3). The proerythroblast divides several times, producing cells that begin to synthesize hemoglobin. Ultimately, a cell near the end of the developmental sequence ejects its nucleus and becomes a **reticulocyte** (re-TIK-ū-lō-sīt). Loss of the nucleus causes the center of the cell to indent, producing a distinctive biconcave shape. Reticulocytes, which are composed of about 34% hemoglobin and retain some mitochondria, ribosomes, and endoplasmic reticulum, pass from red bone marrow into the bloodstream by squeezing through holes in the plasma membrane of the endothelial cells of blood capillaries. Reticulocytes usually develop into erythrocytes, or mature red blood cells, within 1–2 days after their release from red bone marrow.

Normally, erythropoiesis and red blood cell destruction proceed at roughly the same pace. If the oxygen-carrying capacity of the blood falls because erythropoiesis is not keeping up with RBC destruction, RBC production is increased. Cellular oxygen deficiency, called **hypoxia** (hī-POKS-ē-a), may occur if not enough oxygen enters the blood. For example, the reduced oxygen content of air at high altitudes results in a reduced level of oxygen in the blood. Oxygen delivery may also fall due to anemia, which has many causes: lack of iron, lack of certain amino acids, and lack of vitamin B12 are but a few (see page 418). Circulatory problems that reduce blood flow to tissues may also reduce oxygen delivery. Whatever the cause, hypoxia stimulates the kidneys to step up the release of erythropoietin. This hormone circulates through the blood to the red bone marrow, where it speeds the development of proerythroblasts into reticulocytes. When the number of circulating RBCs increases, more oxygen can be delivered to body tissues.

Premature newborns often exhibit anemia, due in part to inadequate production of erythropoietin. During the first weeks after birth, the liver, not the kidneys, produces most EPO. Because the liver is less sensitive than the kidneys to hypoxia, newborns have a smaller EPO response to anemia than do adults.

RETICULOCYTE COUNT

The rate of erythropoiesis is measured by a **reticulocyte count.** Normally, a little less than 1% of the oldest RBCs are replaced by newcomer reticulocytes on any given day. It then takes 1 to 2 days for the reticulocytes to lose the last vestiges of endoplasmic reticulum and become mature RBCs. Thus, reticulocytes account for about 0.5–1.5% of all RBCs in a normal blood sample. A low "retic" count in a person who is anemic might indicate a shortage of erythropoietin or an inability of the red bone marrow to respond to EPO, perhaps because of a nutritional deficiency or leukemia. A high "retic" count might indicate a good red bone marrow response to previous blood loss or to iron therapy in someone who had been iron deficient. It could also point to illegal use of Epoetin alfa by an athlete. ■

CHECKPOINT

6. Describe the size, microscopic appearance, and functions of RBCs.
7. Define erythropoiesis. Relate erythropoiesis to hematocrit. What factors accelerate and slow erythropoiesis?

Blood Group Systems

More than 100 kinds of antigens have been detected on the surface of red blood cells. These antigens are genetically determined and are referred to as **isoantigens** or **agglutinogens** (ag′-loo-TIN-ō-jens). Many of these antigens appear in characteristic patterns, a fact that enables scientists or health care professionals to identify a person's blood as belonging to one or more blood grouping systems; at least 14 are currently recognized. Each system is characterized by the presence or absence of specific antigens on the surface of a red blood cell's plasma membrane. The two major blood groups distinguished on the basis of these antigens are called the ABO and Rh blood groups.

The *ABO blood grouping system* is based on two antigens, symbolized as *A* and *B*. Individuals whose erythrocytes manufacture only antigen *A* are said to have blood type A. Those who manufacture only antigen *B* are type B. Individuals who manufacture both *A* and *B* are type AB. Those who manufacture neither are type O.

The *Rh blood grouping system* is so named because it was first worked out using the blood of the *Rh*esus monkey. Individuals whose erythrocytes have the Rh antigens (D antigens) are designated Rh^+. Those who lack Rh antigens are designated Rh^-.

As just noted, the presence or absence of certain antigens on red blood cells is the basis for classifying blood into several different groups. Such information is very important when a transfusion is given. A transfusion is the transfer of whole blood or blood components (red blood cells or blood plasma, for example) into the bloodstream. A transfusion may be given to treat low blood volume, anemia, or a low platelet count. However, in an incompatible blood transfusion, one person's red blood cell antigens may be recognized as foreign by antibodies in their blood plasma when transfused into someone with a different blood type. In this case, the transfused cells are attacked by antibodies and undergo hemolysis (burst), releasing hemoglobin into the plasma; the liberated hemoglobin may cause kidney damage.

WHITE BLOOD CELLS

OBJECTIVE

● Describe the structure, functions, and production of white blood cells.

WBC Anatomy and Types

Unlike red blood cells, white blood cells, or **leukocytes** (LOO-kō-sīts; *leuko-*=white), have a nucleus and do not contain hemoglobin (Figure 13.5). WBCs are classified as either granular or agranular, depending on whether they contain conspicuous cytoplasmic vesicles (originally called granules) that are made visible by staining. *Granular leukocytes* include neutrophils, eosinophils, and basophils; *agranular leukocytes* include lympho-cytes and monocytes. As shown in Figure 13.3, monocytes and granular leukocytes develop from myeloid stem cells. In con-trast, lymphocytes develop from lymphoid stem cells.

Granular Leukocytes

After staining, each of the three types of granular leukocytes dis-play conspicuous granules with distinctive coloration that can be recognized under a light microscope. The large, uniform-sized granules within an **eosinophil** (ē-ō-SIN-ō-fil) are *eosinophilic* (=eosin-loving)—they stain red-orange with acidic dyes (Figure 13.5a). The granules usually do not cover or obscure the nucleus, which most often has two or three lobes connected by either a thin strand or a thick strand of nuclear material. The round, variable-sized granules of a **basophil** (BĀ-sō-fil) are *basophilic* (=basic loving)—they stain blue-purple with basic dyes (Figure 13.5b). The granules commonly obscure the nucleus, which has two lobes. The granules of a **neutrophil** (NOO-trō-fil) are smaller, evenly distributed, and pale lilac in color (Figure 13.5c); the nucleus has two to five lobes, con-nected by very thin strands of chromatin. As the cells age, the number of nuclear lobes increases. Because older neutrophils thus have several differently shaped nuclear lobes, they are often called *polymorphonuclear leukocytes (PMNs)*, polymorphs, or "polys." Younger neutrophils are often called *bands* because their nucleus is more rod-shaped.

Agranular Leukocytes

Even though so-called agranular leukocytes possess cytoplasmic granules, the granules are not visible under a light microscope because of their small size and poor staining qualities.

The nucleus of a **lymphocyte** (LIM-fō-sīt) stains dark and is round or slightly indented. The cytoplasm stains sky blue and forms a rim around the nucleus. The larger the cell, the more cytoplasm is visible. Lymphocytes are classified as large or small based on cell diameter: 6–9 μm in small lymphocytes and 10–14 μm in large lymphocytes (Figure 13.5d). (Although the functional significance of the size difference between small and large lymphocytes is unclear, the distinction is still useful clini-cally because an increase in the number of large lymphocytes has diagnostic significance in acute viral infections and in some immunodeficiency diseases.)

Monocytes (MON-ō-sīts) are 12–20 μm in diameter (Figure 13.5e). The nucleus of a monocyte is usually kidney-shaped or horseshoe-shaped, and the cytoplasm is blue-gray and has a foamy appearance. This color and appearance are due to very fine *azurophilic granules* (az'-ū-rō-FIL-ik; *azur*=blue; *philos*=loving), which are lysosomes. The blood is merely a con-duit for monocytes, which migrate from the blood into the tis-sues, where they enlarge and differentiate into **macrophages** (=large eaters). Some become **fixed macrophages,** which means they reside in a particular tissue; examples are alveolar macrophages in the lungs, macrophages in the spleen, or stellate reticuloendothelial (Kupffer) cells in the liver. Others become **wandering macrophages,** which roam the tissues and gather at sites of infection or inflammation.

White blood cells and other nucleated body cells have pro-teins, called *major histocompatibility (MHC) antigens*, protruding from their plasma membranes into the extracellular fluid. These cell identity markers are unique for each person (except identical twins). Although RBCs possess blood group antigens, they lack the MHC antigens.

WBC Functions

In a healthy body, some WBCs, especially lymphocytes, can live for several months or years, but most live only a few days. During a period of infection, phagocytic WBCs may live only a few hours. WBCs are far less numerous than red blood cells, about 5000–10,000 cells per μL of blood. RBCs therefore out-number white blood cells by about 700:1. **Leukocytosis** (loo'-kō-sī-TŌ-sis), an increase in the number of WBCs (above 10,000/μL), is a normal, protective response to stresses such as invading microbes, strenuous exercise, anesthesia, and surgery. An abnormally low level of white blood cells (below 5000/μL) is termed **leukopenia** (loo-kō-PĒ-nē-a). It is never beneficial and may be caused by radiation, shock, and certain chemotherapeu-tic agents.

The skin and mucous membranes of the body are continu-ously exposed to microbes and their toxins. Some of these microbes can invade deeper tissues to cause disease. Once pathogens enter the body, the general function of white blood cells is to combat them by phagocytosis or immune responses. To accomplish these tasks, many WBCs leave the bloodstream

Figure 13.5 Structure of white blood cells.

🔑 **White blood cells are distinguished from one another by the shape of their nuclei and the staining properties of their cytoplasmic granules.**

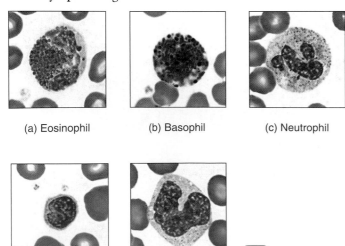

(a) Eosinophil   (b) Basophil   (c) Neutrophil

LM all 1100x

(d) Lymphocyte   (e) Monocyte

❓ **Which WBCs are called granular leukocytes? Why?**

and collect at points of pathogen invasion or inflammation. Once granulocytes and monocytes leave the bloodstream to fight injury or infection, they never return to it. Lymphocytes, on the other hand, continually recirculate—from blood to interstitial spaces of tissues to lymphatic fluid and back to blood. Only 2% of the total lymphocyte population is circulating in the blood at any given time; the rest are in lymphatic fluid and organs such as skin, lungs, lymph nodes, and spleen.

WBCs leave the bloodstream by a process termed **emigration** (em′-i-GRĀ-shun; *e-*=out; *migra-*=wander), formerly called *diapedesis*, in which they roll along the endothelium, stick to it, and then squeeze between endothelial cells. The precise signals that stimulate emigration through a particular blood vessel vary for the different types of WBCs.

Neutrophils and macrophages are active in **phagocytosis;** they can ingest bacteria and dispose of dead matter. Several different chemicals released by microbes and inflamed tissues attract phagocytes, a phenomenon called **chemotaxis.**

Among WBCs, neutrophils respond most quickly to tissue destruction by bacteria. After engulfing a pathogen during phagocytosis, a neutrophil unleashes several destructive chemicals to destroy the ingested pathogen. These chemicals include the enzyme **lysozyme,** which destroys certain bacteria, and **strong oxidants,** such as the superoxide anion (O_2^-), hydrogen peroxide (H_2O_2), and the hypochlorite anion (OCl^-), which is similar to household bleach. Neutrophils also contain **defensins,** proteins that exhibit a broad range of antibiotic activity against bacteria and fungi. Within a neutrophil, granules containing defensins merge with phagosomes containing microbes. Defensins form peptide "spears" that poke holes in microbe membranes; the resulting loss of cellular contents kills the invader.

Monocytes take longer to reach a site of infection than do neutrophils, but they arrive in large numbers and destroy more microbes. Upon arrival they enlarge and differentiate into wandering macrophages, which clean up cellular debris and microbes by phagocytosis following an infection.

Eosinophils leave the capillaries and enter tissue fluid. They are believed to release enzymes, such as histaminase, that combat the effects of histamine and other mediators of inflammation in allergic reactions. Eosinophils also phagocytize antigen–antibody complexes and are effective against certain parasitic worms. A high eosinophil count often indicates an allergic condition or a parasitic infection.

At sites of inflammation, basophils leave capillaries, enter tissues, and release heparin, histamine, and serotonin. These substances intensify the inflammatory reaction and are involved in hypersensitivity (allergic) reactions. Basophils are similar in function to mast cells, connective tissue cells that originate from pluripotent stem cells in red bone marrow. Like basophils, mast cells liberate mediators in inflammation, including heparin, histamine, and proteases. Mast cells are widely dispersed in the body, particularly in connective tissues of the skin and mucous membranes of the respiratory and gastrointestinal tracts.

Lymphocytes are the major soldiers in immune system battles (described in detail in Chapter 16). Three main types of lymphocytes are B cells, T cells, and natural killer cells. B cells are particularly effective in destroying bacteria and inactivating their toxins. T cells attack viruses, fungi, transplanted cells, cancer cells, and some bacteria. Immune responses carried out by B cells and T cells help combat infection and provide protection against some diseases. T cells also are responsible for transfusion reactions, allergies, and the rejection of transplanted organs. Natural killer cells attack a wide variety of infectious microbes and certain spontaneously arising tumor cells.

An increase in the number of circulating WBCs usually indicates inflammation or infection. A physician may order a **differential white blood cell count** to detect infection or inflammation, determine the effects of possible poisoning by chemicals or drugs, monitor blood disorders (such as leukemia) and the effects of chemotherapy, or detect allergic reactions and parasitic infections. Because each type of white blood cell plays a different role, determining the *percentage* of each type in the blood assists in diagnosing the condition.

Table 13.2 lists the significance of both elevated and depressed WBC counts.

CHECKPOINT

8. Explain the importance of emigration, chemotaxis, and phagocytosis in fighting bacterial invaders.
9. Distinguish between leukocytosis and leukopenia.
10. What is a differential white blood cell count?
11. What functions are performed by B cells, T cells, and natural killer cells?

TABLE 13.2	SIGNIFICANCE OF HIGH AND LOW WHITE BLOOD CELL COUNTS	
WBC TYPE	**HIGH COUNT MAY INDICATE**	**LOW COUNT MAY INDICATE**
Neutrophils	Bacterial infection, burns, stress, inflammation.	Radiation exposure, drug toxicity, vitamin B12 deficiency, and systemic lupus erythematosus (SLE).
Lymphocytes	Viral infections, some leukemias.	Prolonged illness, immunosuppression, and treatment with cortisol.
Monocytes	Viral or fungal infections, tuberculosis, some leukemias, other chronic diseases.	Bone marrow suppression, treatment with cortisol.
Eosinophils	Allergic reactions, parasitic infections, autoimmune diseases.	Drug toxicity, stress.
Basophils	Allergic reactions, leukemias, cancers, hypothyroidism.	Pregnancy, ovulation, stress, and hyperthyroidism.

PLATELETS

• Describe the structure, function, and origin of platelets.

Besides the immature cell types that develop into erythrocytes and leukocytes, hemopoietic stem cells also differentiate into cells that produce platelets. Under the influence of the hormone **thrombopoietin,** myeloid stem cells develop into megakaryocyte-colony-forming cells that in turn develop into precursor cells called megakaryoblasts (see Figure 13.3). Megakaryoblasts transform into megakaryocytes, huge cells that splinter into 2000–3000 fragments. Each fragment, enclosed by a piece of the cell membrane, is a **platelet (thrombocyte).** Platelets break off from the megakaryocytes in red bone marrow and then enter the blood circulation. Between 150,000 and 400,000 platelets are present in each μL of blood. They are disc-shaped, 2–4 μm in diameter, and exhibit many vesicles but no nucleus. Platelets help stop blood loss from damaged blood vessels by coming together to form a platelet plug that fills the gap in the blood vessel wall. Their vesicles also contain chemicals that, once released, promote blood clotting. Platelets can initiate a series of chemical reactions that culminates in the formation of a network of insoluble protein threads called *fibrin*. A **blood clot** is a gel-like mass that consists of fibrin threads, platelets, and any blood cells trapped in the fibrin (Figure 13.6). The blood clot not only provides a seal in the damaged area of a blood vessel to prevent blood loss, but also

Figure 13.6 Scanning electron micrograph (SEM) of a portion of a blood clot, showing a platelet and red blood cells trapped by fibrin threads.

A blood clot is a gel that contains formed elements of the blood entangled in fibrin threads.

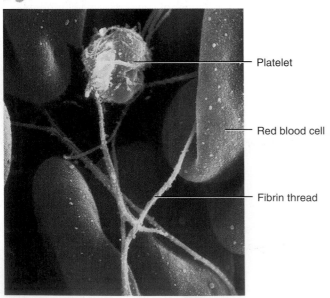

Platelet

Red blood cell

Fibrin thread

SEM 15,000x

What are the two functions of a blood clot?

pulls the edges of the damaged vessel together to help heal the damage. Platelets have a short life span, normally just 5–9 days. Aged and dead platelets are removed from the circulation by fixed macrophages in the spleen and liver.

Table 13.3 summarizes the formed elements in blood.

COMPLETE BLOOD COUNT

A **complete blood count (CBC)** is a valuable test that screens for anemia and various infections. Usually included are counts of RBCs, WBCs, and platelets per μL of whole blood; hematocrit; and differential white blood cell count. The amount of hemoglobin in grams per mL of blood also is determined. Normal hemoglobin ranges are: infants, 14–20 g/100 mL of blood; adult females, 12–16 g/100 mL of blood; and adult males, 13.5–18 g/100 mL of blood. ■

12. Compare RBCs, WBCs, and platelets with respect to size, number per μL, and life span.

STEM CELL TRANSPLANTS FROM BONE MARROW AND CORD-BLOOD

• Explain the importance of bone marrow transplants and stem cell transplants.

A **bone marrow transplant** is the replacement of cancerous or abnormal red bone marrow with healthy red bone marrow in order to establish normal blood cell counts. In patients with cancer or certain genetic diseases, the defective red bone marrow is destroyed by high doses of chemotherapy and whole body radiation just before the transplant takes place. These treatments kill the cancer cells and destroy the patient's immune system in order to decrease the chance of transplant rejection.

Healthy red bone marrow for transplanting may be supplied by a donor or by the patient when the underlying disease is inactive. The marrow from a donor is usually removed from the iliac crest of the hip bone under general anesthesia with a syringe and is then injected into the recipient's vein, much like a blood transfusion. The injected marrow migrates to the recipient's red bone marrow cavities and the stem cells in the marrow multiply. If all goes well, the recipient's red bone marrow is replaced entirely by healthy, noncancerous cells.

Bone marrow transplants have been used to treat aplastic anemia, certain types of leukemia, severe combined immunodeficiency disease (SCID), Hodgkin's disease, non-Hodgkin's lymphoma, multiple myeloma, thalassemia, sickle-cell disease, breast cancer, ovarian cancer, testicular cancer, and hemolytic anemia. However, there are some drawbacks. Since the recipient's white blood cells have been completely destroyed by chemotherapy and radiation, the patient is extremely vulnerable to infection. (It takes about 2–3 weeks for transplanted bone marrow to produce enough white blood cells to protect against

TABLE 13.3   SUMMARY OF FORMED ELEMENTS IN BLOOD

NAME AND APPEARANCE	NUMBER	CHARACTERICTICS*	FUNCTIONS
Red Blood Cells (RBCs) or Erythrocytes	4.8 million/μL in females; 5.4 million/μL in males.	7–8 μm diameter, biconcave discs, without nuclei; live for about 120 days.	Hemoglobin within RBCs transports most of the oxygen and part of the carbon dioxide in the blood.
White Blood Cells (WBCs) or Leukocytes	5000–10,000/μL.	Most live for a few hours to a few days.†	Combat pathogens and other foreign substances that enter the body.
Granular Leukocytes *Neutrophils*	60–70% of all WBCs.	10–12 μm diameter; nucleus has 2–5 lobes connected by thin strands of chromatin; cytoplasm has very fine, pale lilac granules.	Phagocytosis. Destruction of bacteria with lysozyme, defensins, and strong oxidants, such as superoxide anion, hydrogen peroxide, and hypochlorite anion.
Eosinophils	2–4% of all WBCs.	10–12 μm diameter; nucleus has 2 or 3 lobes; large, red-orange granules fill the cytoplasm.	Combat the effects of histamine in allergic reactions, phagocytize antigen–antibody complexes, and destroy certain parasitic worms.
Basophils	0.5–1% of all WBCs.	8–10 μm diameter; nucleus has 2 lobes; large cytoplasmic granules appear deep blue-purple.	Liberate heparin, histamine, and serotonin in allergic reactions that intensify the overall inflammatory response.
Agranular Leukocytes *Lymphocytes (T cells, B cells, and natural killer cells)*	20–25% of all WBCs.	Small lymphocytes are 6–9 μm in diameter; large lymphocytes are 10–14 μm in diameter; nucleus is round or slightly indented; cytoplasm forms a rim around the nucleus that looks sky blue; the larger the cell, the more cytoplasm is visible.	Mediate immune responses, including antigen–antibody reactions. B cells develop into plasma cells, which secrete antibodies. T cells attack invading viruses, cancer cells, and transplanted tissue cells. Natural killer cells attack a wide variety of infectious microbes and certain spontaneously arising tumor cells.
Monocytes	3–8% of all WBCs.	12–20 μm in diameter; nucleus is kidney shaped or horseshoe shaped; cytoplasm is blue-gray and has foamy appearance.	Phagocytosis (after transforming into fixed or wandering macrophages).
Platelets (thrombocytes)	150,000–400,000/μL.	2–4μm diameter cell fragments that live for 5–9 days; contain many vesicles but no nucleus.	Form platelet plug in hemostasis; release chemicals that promote vascular spasm and blood clotting.

*Colors are those seen when using Wright's stain.

†Some lymphocytes, called T and B memory cells, can live for many years once they are established.

Mnemonic for ranking WBCs from most to least numerous: "*N*ever *L*et *M*onkeys *E*at *B*ananas."

Neutrophils, **L**ymphocytes, **M**onocytes, **E**osinophils, **B**asophils

infection.) Another problem is that transplanted red bone marrow may produce T cells that attack the recipient's tissues, a reaction called *graft-versus-host disease.* Moreover, any of the recipient's T cells that survived the chemotherapy and radiation can attack donor transplant cells. Another drawback is that patients must take immunosuppressive drugs for life. Because these drugs reduce the level of immune system activity, they increase the risk of infection. Immunosuppressive drugs also have side effects such as fever, muscle aches, headache, nausea, fatigue, depression, high blood pressure, and kidney and liver damage.

A more recent advance for obtaining stem cells involves a **cord-blood transplant.** Recall from Chapter 4 that the connection between the placenta and embryo (and later the fetus) is the umbilical cord. The placenta contains stem cells, which may be obtained from the umbilical cord shortly after birth. The stem cells are removed from the cord with a syringe and then frozen. Stem cells from the cord have several advantages over those obtained from red bone marrow.

1. They are easily collected following permission of the newborn's parents.

2. They are more abundant than stem cells in red bone marrow.

3. They are less likely to cause graft-versus-host disease, so the match between donor and recipient does not have to be as close as in a bone marrow transplant. This provides a larger number of potential donors.

4. They are less likely to transmit infections.

5. They can be stored indefinitely in cord-blood banks.

CHECKPOINT

13. What are the similarities and differences between cord-blood transplants and bone marrow transplants?

APPLICATIONS TO HEALTH

ANEMIA

Anemia is a condition in which the oxygen-carrying capacity of blood is reduced. All of the many types of anemia are characterized by reduced numbers of RBCs or a decreased amount of hemoglobin in the blood. The person feels fatigued and is intolerant of cold, both of which are related to lack of oxygen needed for ATP and heat production. Also, the skin appears pale, due to the low content of red-colored hemoglobin circulating in skin blood vessels. Among the most important causes and types of anemia are the following:

- *Inadequate absorption of iron, excessive loss of iron, increased iron requirement, or insufficient intake of iron* causes **iron-deficiency anemia,** the most common type of anemia. Women are at greater risk for iron-deficiency anemia due to menstrual blood losses and increased iron demands of the growing fetus during pregnancy. Gastrointestinal losses, such as occurs with malignancy or ulceration, also contribute to this type of anemia.

- *Insufficient hemopoiesis* resulting from an inability of the stomach to produce intrinsic factor, which is needed for absorption of vitamin B12 in the small intestine, causes **pernicious anemia.**

- *Excessive loss of RBCs* through bleeding, resulting from large wounds, stomach ulcers, or especially heavy menstruation, leads to **hemorrhagic anemia.**

- *RBC plasma membranes rupture prematurely* in **hemolytic anemia.** The released hemoglobin pours into the plasma and may damage the filtering units (glomeruli) in the kidneys. The condition may result from inherited defects such as abnormal red blood cell enzymes, or from outside agents such as parasites, toxins, or antibodies from incompatible transfused blood.

- *Deficient synthesis of hemoglobin* occurs in **thalassemia** (thal′-a-SĒ-mē-a), a group of hereditary hemolytic anemias. The RBCs are small (microcytic), pale (hypochromic), and short-lived. Thalassemia occurs primarily in populations from countries bordering the Mediterranean Sea.

- *Destruction of red bone marrow* results in **aplastic anemia.** Toxins, gamma radiation, and certain medications that inhibit enzymes needed for hemopoiesis are causes.

SICKLE-CELL DISEASE

The RBCs of a person with **sickle-cell disease (SCD)** contain Hb-S, an abnormal type of hemoglobin. When Hb-S gives up oxygen to the interstitial fluid, it forms long, stiff, rodlike structures that bend the erythrocyte into a sickle shape (Figure 13.7).

Figure 13.7 Red blood cells from a patient with sickle-cell disease.

Sickle-cell disease is due to an abnormal kind of hemoglobin.

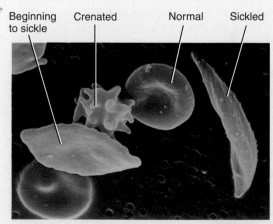

Red blood cells

A person with one defective gene for sickle-cell disease is more resistant to what other disease?

The sickled cells rupture easily. Even though erythropoiesis is stimulated by the loss of the cells, it cannot keep pace with hemolysis. People with sickle-cell disease always have some degree of anemia and mild jaundice and many experience joint or bone pain, breathlessness, rapid heart rate, abdominal pain, fever, and fatigue as a result of tissue damage caused by prolonged recovery oxygen uptake (oxygen debt). Any activity that reduces the amount of oxygen in the blood, such as vigorous exercise, may produce a *sickle-cell crisis* (worsening of the anemia, pain in the abdomen and long bones of the limbs, fever, and shortness of breath).

Sickle-cell disease is inherited. People with two sickle-cell genes have severe anemia, whereas those with only one defective gene have minor problems. Sickle-cell genes are found primarily among populations, or descendants of populations, that live in the malaria belt around the world, including parts of Mediterranean Europe, sub-Saharan Africa, and tropical Asia. The gene responsible for the tendency of the RBCs to sickle also alters the permeability of the plasma membranes of sickled cells, causing potassium ions to leak out. Low levels of potassium kill the malaria parasites that infect sickled cells. Because of this effect, a person with one normal gene and one sickle-cell gene has higher-than-average resistance to malaria. The possession of a single sickle-cell gene thus confers a survival advantage.

Treatment of SCD consists of administration of analgesics to relieve pain, fluid therapy to maintain hydration, oxygen to reduce the stimulus for the crisis, antibiotics to counter infections, and blood transfusions. People who suffer SCD have normal fetal hemoglobin (Hb-F), a slightly different form of hemoglobin that predominates at birth and is present in small amounts thereafter. In some patients with sickle-cell disease, a drug called hydroxyurea promotes transcription of the normal Hb-F gene, elevates the level of Hb-F, and reduces the chance that the RBCs will sickle. Unfortunately, this drug also has toxic effects on the bone marrow, so its safety for long-term use is questionable.

HEMOPHILIA

Hemophilia (hē-mō-FIL-ē-a; *-philia*=loving) is an inherited deficiency of clotting in which bleeding may occur spontaneously or after only minor trauma. It is the oldest known hereditary bleeding disorder; descriptions of the disease are found as early as the second century AD. Hemophilia usually affects males and is sometimes referred to as "the royal disease" because many descendants of Queen Victoria, beginning with her son, were affected by the disease. Different types of hemophilia are due to deficiencies of different blood clotting chemicals and exhibit varying degrees of severity, ranging from mild to severe bleeding tendencies. Hemophilia is characterized by spontaneous or traumatic subcutaneous and intramuscular hemorrhaging, nosebleeds, blood in the urine, and hemorrhages in joints that produce pain and tissue damage. Treatment involves transfusions of fresh blood plasma or concentrates of the deficient clotting chemicals to relieve the tendency to bleed. Another treatment is the drug desmopressin (DDAVP), which can boost the levels of clotting factors.

DISSEMINATED INTRAVASCULAR CLOTTING

Disseminated intravascular clotting (DIC) is a disorder of clotting characterized by simultaneous and unregulated blood clotting and hemorrhage throughout the body. Common causes of DIC are infections, hypoxia, low blood flow rates, trauma, tumors, hypotension, and hemolysis. DIC may be severe and life threatening. The clots decrease blood flow and eventually cause ischemia, infarction, and necrosis, resulting in multisystem organ dysfunction. A peculiar feature of DIC is that the patient often begins to bleed despite forming clots. Because so many clotting factors are removed by the widespread clotting, too few remain to allow normal clotting of the remaining blood.

LEUKEMIA

The term **leukemia** (loo-KĒ-mē-a; *leuko-*=white) refers to a group of red bone marrow cancers in which abnormal white blood cells multiply uncontrollably. The accumulation of the cancerous white blood cells in red bone marrow interferes with the production of red blood cells, white blood cells, and platelets. As a result the oxygen-carrying capacity of the blood is reduced, an individual is more susceptible to infection, and blood clotting is abnormal. In most leukemias, the cancerous white blood cells spread to the lymph nodes, liver, and spleen, causing them to enlarge. All leukemias produce the usual symptoms of anemia (fatigue, intolerance to cold, and pale skin). In addition, weight loss, fever, night sweats, excessive bleeding, and recurrent infections may also occur.

In general, leukemias are classified as acute (symptoms develop rapidly) and chronic (symptoms may take years to develop). Whereas adults may have either type, children usually have the acute type. Based on the speed of onset and the specific cells involved, following are several forms of leukemia:

1. **Acute lymphoblastic leukemia (ALL).** The abnormal type of white blood cell involved is a lymphoblast (immature lymphocte). See Figure 13.3.

2. **Acute myelocytic leukemia (AML).** The abnormal type of white blood cell involved is a myeloid stem cell. See Figure 13.3.

3. **Chronic lymphocytic leukemia (CLL).** The abnormal white blood cell involved is a mature lymphocyte. See Figure 13.3.

4. **Chronic myelocytic leukemia (CML).** The abnormal type of white blood cell is a myeloid stem cell. See Figure 13.3.

The cause of most types of leukemia is unknown. However, certain risk factors have been implicated. These include exposure to radiation or chemotherapy for other cancers, genetics (some genetic disorders such as Down's syndrome), environmental factors (smoking and benzene), and microbes such as the human T cell leukemia-lymphoma virus-1 (HTLV-1) and the Epstein-Barr virus.

Treatment options include chemotherapy, radiation, stem cell transplantation, interferon, antibodies, and blood transfusions.

KEY MEDICAL TERMS ASSOCIATED WITH BLOOD

Acute normovolemic hemodilution (nor-mō-vō-LĒ-mik hē-mō-di-LOO-shun) Removal of blood immediately before surgery and its replacement with a cell-free solution to maintain sufficient blood volume for adequate circulation. At the end of surgery, once bleeding has been controlled, the collected blood is returned to the body.

Autologous preoperative transfusion (aw-TOL-o-gus trans-FŪ-zhun; *auto-*=self) Donating one's own blood; can be done up to 6 weeks before elective surgery. Also called *predonation*. This procedure eliminates the risk of incompatibility and blood-borne disease.

Blood bank A facility that collects and stores a supply of blood for future use by the donor or others. Because blood banks have additional and diverse functions (immunohematology reference work, continuing medical education, bone and tissue storage, and clinical consultation), they are more appropriately referred to as *centers of transfusion medicine.*

Cyanosis (sī-a-NŌ-sis; *cyano-*=blue) Slightly bluish/dark-purple skin discoloration, most easily seen in the nail beds and mucous membranes, due to an increased quantity of reduced hemoglobin (hemoglobin not combined with oxygen) in systemic blood.

Embolus (EM-bō-lus=plug) A blood clot, bubble of air or fat from broken bones, mass of bacteria, or other foreign material transported by the blood.

Gamma globulin (GLOB-ū-lin) Solution of immunoglobulins from blood consisting of antibodies that react with specific pathogens, such as viruses. It is prepared by injecting the specific virus into animals, removing blood from the animals after antibodies have accumulated, isolating the antibodies, and injecting them into a human to provide short-term immunity.

Hemochromatosis (hē-mō-krō-ma-TŌ-sis; *chroma*=color) Disorder of iron metabolism characterized by excessive absorption of ingested iron and excess deposits of iron in tissues (especially the liver, heart, pituitary gland, gonads, and pancreas) that result in bronze discoloration of the skin, cirrhosis, diabetes mellitus, and bone and joint abnormalities.

Hemorrhage (HEM-or-ij; *rhegnynai*=bursting forth) Loss of a large amount of blood; can be either internal (from blood vessels into tissues) or external (from blood vessels directly to the surface of the body).

Jaundice (*jaund-*=yellow) An abnormal yellowish discoloration of the sclerae (white of the eyes), skin, and mucous membranes due to excess bilirubin (yellow-orange pigment) in the blood. The three main categories of jaundice are *prehepatic jaundice*, due to excess production of bilirubin; *hepatic jaundice*, due to abnormal bilirubin processing by the liver caused by congenital liver disease, cirrhosis (scar tissue formation) of the liver, or hepatitis (liver inflammation); and *extrahepatic jaundice*, due to blockage of bile drainage by gallstones or cancer of the bowel or pancreas.

Multiple myeloma (mī-e-LŌ-ma) Malignant disorder of plasma cells in red bone marrow; symptoms (pain, osteoporosis, hypercalcemia, thrombocytopenia, kidney damage) are caused by the growing tumor cell mass or antibodies produced by malignant cells.

Phlebotomist (fle-BOT-ō-mist; *phlebo-*=vein; *-tom*=cut) A technician who specializes in withdrawing blood.

Septicemia (sep′-ti-SĒ-mē-a; *septic-*=decay; *-emia*=condition of blood) Toxins or disease-causing bacteria in the blood. Also called "blood poisoning."

Thrombocytopenia (throm′-bō-sī′-tō-PĒ-nē-a; *-penia*=poverty) Very low platelet count that results in a tendency to bleed from capillaries.

Thrombus (THROM-bus=clot) A clot in the cardiovascular system formed in an unbroken blood vessel (usually a vein). The clot consists of a network of insoluble fibrin threads in which the formed elements of blood are trapped.

Transfusion (trans-FŪ-zhun) Transfer of whole blood, blood components (red blood cells only or plasma only), or red bone marrow directly into the bloodstream. Transfusions are given to increase the oxygen-carrying capacity of the blood, improve immunity, assist blood clotting, and restore blood volume.

Venesection (vē-ne-SEK-shun; *ven-*=vein) Opening of a vein for withdrawal of blood. Although **phlebotomy** (fle-BOT-ō-me) is a synonym for venesection, in clinical practice phlebotomy refers to therapeutic bloodletting, such as the removal of some blood to lower its viscosity in a patient with polycythemia.

Whole blood Blood containing all formed elements, plasma, and plasma solutes in natural concentrations.

STUDY OUTLINE

Introduction (p. 405)

1. The cardiovascular system consists of the blood, heart, and blood vessels.
2. Blood is a connective tissue composed of blood plasma (liquid portion) and formed elements (cells and cell fragments).

Functions of Blood (p. 406)

1. Blood transports oxygen, carbon dioxide, nutrients, wastes, and hormones.
2. It helps regulate pH, body temperature, and water content of cells.
3. It provides protection through clotting and by combating toxins and microbes, a function of certain phagocytic white blood cells or specialized plasma proteins.

Physical Characteristics of Blood (p. 406)

1. Physical characteristics of blood include a viscosity greater than that of water; a temperature of 38°C (100.4°F); and a pH of 7.35–7.45.
2. Blood constitutes about 8% of body weight, and its volume is 4–6 liters in adults.

Components of Blood (p. 406)

1. Blood consists of about 55% blood plasma and about 45% formed elements.
2. The hematocrit is the percentage of total blood volume occupied by red blood cells.
3. Blood plasma consists of 91.5% water and 8.5% solutes.
4. Principal solutes include proteins (albumins, globulins, fibrinogen), nutrients, vitamins, hormones, respiratory gases, electrolytes, and waste products.
5. The formed elements in blood include red blood cells (erythrocytes), white blood cells (leukocytes), and platelets.

Formation of Blood Cells (p. 409)

1. Hemopoiesis is the formation of blood cells from hemopoietic stem cells in red bone marrow.
2. Myeloid stem cells form RBCs, platelets, granulocytes, and monocytes. Lymphoid stem cells give rise to lymphocytes.
3. Several hemopoietic growth factors stimulate differentiation and proliferation of the various blood cells.

Red Blood Cells (p. 411)

1. Mature RBCs are biconcave discs that lack nuclei and contain hemoglobin.
2. The function of the hemoglobin in red blood cells is to transport oxygen and some carbon dioxide.

3. RBCs live about 120 days. A healthy male has about 5.4 million RBCs/μL of blood; a healthy female, about 4.8 million/μL.
4. After phagocytosis of aged RBCs by macrophages, hemoglobin is recycled.
5. RBC formation, called erythropoiesis, occurs in adult red bone marrow of certain bones. It is stimulated by hypoxia, which stimulates the release of erythropoietin by the kidneys.
6. A reticulocyte count is a diagnostic test that indicates the rate of erythropoiesis.

White Blood Cells (p. 413)

1. WBCs are nucleated cells. The two principal types are granulocytes (neutrophils, eosinophils, and basophils) and agranulocytes (lymphocytes and monocytes).
2. The general function of WBCs is to combat inflammation and infection. Neutrophils and macrophages (which develop from monocytes) act by phagocytosis.
3. Eosinophils combat the effects of histamine in allergic reactions, phagocytize antigen–antibody complexes, and combat parasitic worms; basophils liberate heparin, histamine, and serotonin in allergic reactions that intensify the inflammatory response.
4. B lymphocytes, in response to the presence of foreign substances called antigens, differentiate into plasma cells that produce antibodies. Antibodies attach to the antigens and render them harmless. This antigen–antibody response combats infection and provides immunity. T lymphocytes destroy foreign invaders directly. Natural killer cells attack infectious microbes and tumor cells.
5. Except for lymphocytes, which may live for years, WBCs usually live for only a few hours or a few days. Normal blood contains 5000–10,000 WBCs/μL.

Platelets (p. 416)

1. Platelets (thrombocytes) are disc-shaped structures without nuclei.
2. They are fragments derived from megakaryocytes and are involved in clotting.
3. Normal blood contains 150,000–400,000 platelets/μL.

Stem Cell Transplants from Bone Marrow and Cord-Blood (p. 416)

1. Bone marrow transplants involve removal of marrow as a source of stem cells from the iliac crest.
2. In a cord-blood transplant, stem cells from the placenta are removed from the umbilical cord.
3. Cord-blood transplants have several advantages over bone marrow transplants.

SELF-QUIZ QUESTIONS

Choose the one best answer to the following questions.

1. Which of the following is *not* a site of hemopoiesis in the adult body?
   a. sternum   b. liver   c. vertebrae
   d. head of humerus   e. parietal bones

2. Which of the following statements is *false*?
   a. Blood is more viscous than water.
   b. Blood normally has a pH of 6.5–7.0.
   c. The adult male body normally contains about 5–6 liters of blood.
   d. Normal blood temperature is 38 degrees C.
   e. Blood constitutes about 8% of total body weight.

3. The normal red blood cell count in healthy adult males is:
   a. 5.4 million RBCs/μL.   b. 2 million RBCs/μL.
   c. 0.5 million RBCs/μL.   d. 250,000 RBCs/μL.
   e. 8,000 RBCs/μL.

4. The *buffy coat* of a centrifuged sample of blood contains mostly:
   a. erythrocytes.   b. platelets.   c. blood plasma.
   d. white blood cells.   e. gamma globulins.

5. The predominant type of blood plasma protein is:
   a. albumins.   b. fibrinogen.   c. immunoglobulins.
   d. erythropoietin.   e. colony-stimulating factors.

6. Thrombopoietin is a hormone that stimulates the formation of:
   a. albumins.   b. B lymphocytes.   c. platelets.
   d. fibrinogen.   e. erythrocytes.

7. The structure of erythrocytes makes them highly specialized for:
   a. antibody production.   b. hormone production.
   c. phagocytosis.   d. clotting.   e. oxygen transport.

8. A gaseous hormone that can bind to hemoglobin and be released to cause localized vasodilation is:
   a. erythropoietin.   b. nitric oxide.   c. thrombopoietin.
   d. fibrinogen.   e. lysozyme.

9. Place the following cells in the correct order for the production of red blood cells: (1) proerythroblasts, (2) CFU-E cells, (3) myeloid stem cells, (4) erythroblasts, (5) pluripotent stem cells, (6) reticulocytes.
   a. 5, 3, 2, 1, 4, 6   b. 3, 2, 5, 6, 1, 4   c. 4, 5, 3, 6, 1, 2
   d. 5, 2, 1, 3, 4, 6   e. 2, 3, 5, 1, 4, 6

10. Which of the following best describes the structure of erythrocytes?
    a. irregular fragments of larger cells
    b. biconcave discs that lack nuclei
    c. spherical cells with lobed nuclei and large granules
    d. spherical, slightly indented cells with large, round nuclei
    e. large, irregularly shaped cells with kidney-bean-shaped nuclei

11. Platelets function in the:
    a. transport of carbon dioxide.   b. production of vitamin K.
    c. utilization of calcium and phosphorus.
    d. destruction of bacteria.   e. clotting of blood.

12. Place the following cells in the correct order for the process of neutrophil production: (1) myeloblasts, (2) pluripotent stem cells, (3) CFU-GM cells, (4) myeloid stem cells.
    a. 2, 3, 4, 1   b. 2, 4, 3, 1   c. 4, 2, 1, 3
    d. 3, 2, 4, 1   e. 3, 1, 2, 4

Complete the following.

13. A test to determine the percentage of each type of white blood cell is known as a _____ count.

14. Blood is a connective tissue that consists of about _____ % extracellular material and about _____ % formed elements.

15. The process by which white blood cells squeeze through capillaries to reach an injured or infected body tissue is called _____.

16. The process of red blood cell formation is called _____.

17. The rate of red blood cell production is measured by doing a _____ count.

18. Two families of cytokines that stimulate white blood cell formation are the _____ and the _____.

19. The center of a heme ring is a(n) _____ ion.

Are the following statements true or false?

20. Nucleated cells (including white blood cells) have surface MHC antigens.

21. Red blood cells have lobed nuclei and abundant mitochondria.

22. The three kinds of granular leukocytes are neutrophils, monocytes, and eosinophils.

Matching

23. Match the following terms with their description (answers are used more than once):
    ___ **(a)** constitute the largest percentage (about 60%–70%) of leukocytes
    ___ **(b)** make up 20%–25% of leukocytes
    ___ **(c)** are involved in immunity; form plasma cells for antibody production
    ___ **(d)** are involved in allergic response; release serotonin, heparin, and histamine
    ___ **(e)** are involved in allergic response; release antihistamines
    ___ **(f)** develop into wandering macrophages that clean up sites of infection
    ___ **(g)** are important in phagocytosis (two answers)
    ___ **(h)** contain large reddish granules in the cytoplasm
    ___ **(i)** contain large deep blue or purple granules in cytoplasm
    ___ **(j)** cytoplasm is blue-gray and foamy in appearance

    (1) basophils
    (2) eosinophils
    (3) lymphocytes
    (4) monocytes
    (5) neutrophils

CRITICAL THINKING QUESTIONS

1. A professional bicyclist finished second in a grueling cross-country bike race. Although he appeared to be in great physical condition, he suffered a heart attack a few hours after the race. The press speculated that his condition was caused by blood doping. Explain.

2. Maddy has been sniffling and sneezing her way through Human Anatomy class all semester. While checking a smear of her own blood, she noticed that her blood had a lot more of the bluish-black granular cells than her lab partner's blood. Maddy's eyes are watery and itchy so she asked her instructor to confirm her finding. What are these cells and why does Maddy have a higher number of them than usual?

3. Gus suffers from chronic renal (kidney) failure and undergoes dialysis regularly. One of his associated problems is chronic anemia, which the doctor says is directly related to the renal failure. What is likely to be the connection between the two, and how might it be treated?

4. Millie was at her grandmother's house and found a very old bottle of a tonic that promised to relieve the symptoms of "iron-poor blood." What do you think the symptoms of "iron-poor blood" would be, and how would enriching the iron content of blood relieve these symptoms?

5. Raoul plans on having some elective surgery performed in about a month, after the semester is over. His doctor suggested that he donate some of his own blood now. Raoul's not sure about this; he figures he's going to need all the blood he has for the surgery. Why should he give up some of his blood now?

ANSWERS TO FIGURE QUESTIONS

13.1 Blood volume is about 6 liters in males and 4–5 liters in females, representing about 8% of body weight.

13.2 Platelets are cell fragments.

13.3 Pluripotent stem cells develop from mesenchyme.

13.4 One hemoglobin molecule can transport four O_2 molecules—one bound to each heme group.

13.5 Neutrophils, eosinophils, and basophils are called granulocytes because all have cytoplasmic granules that are visible through a light microscope when stained.

13.6 The two functions of a blood clot are to seal a damaged blood vessel and pull the edges of the vessel together to prevent blood loss.

13.7 A person with one defective gene for sickle-cell disease is more resistant to malaria.

14

THE CARDIOVASCULAR SYSTEM: THE HEART

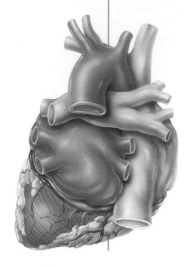

INTRODUCTION   In the last chapter we examined the composition and functions of blood. For blood to reach body cells and exchange materials with them, it must be constantly pumped by the heart through the body's blood vessels. The heart beats about 100,000 times every day, which adds up to about 35 million beats in a year. The left side of the heart pumps blood through an estimated 100,000 km (60,000 mi) of blood vessels. The right side of the heart pumps blood through the lungs, enabling blood to pick up oxygen and unload carbon dioxide. Even while you are sleeping, your heart pumps 30 times its own weight each minute, which amounts to about 5 liters (5.3 qt) to the lungs and the same volume to the rest of the body. At this rate, the heart pumps more than 14,000 liters (3,600 gal) of blood in a day, or 10 million liters (2.6 million gal) in a year. You don't spend all your time sleeping, however, and your heart pumps more vigorously when you are active. Thus, the actual blood volume the heart pumps in a single day is much larger.

The scientific study of the normal heart and the diseases associated with it is **cardiology** (kar′-dē-OL-ō-jē; *cardio-*=heart; *-logy*=study of). This chapter explores the design of the heart and the unique properties that permit it to pump for a lifetime without rest.

**Can you recognize whose work this is?
Is this a modern or historical composition?**

www.wiley.com/college/apcentral

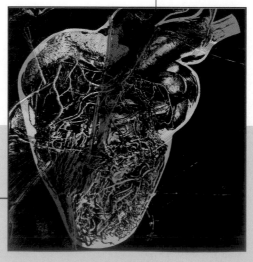

LOCATION AND SURFACE PROJECTION OF THE HEART

OBJECTIVE

● Describe the location of the heart, and trace its outline on the surface of the chest.

For all its might, the cone-shaped heart is relatively small, roughly the same size as a closed fist—about 12 cm (5 in.) long, 9 cm (3.5 in.) wide at its broadest point, and 6 cm (2.5 in.) thick. Its mass averages 250 g (8 oz) in adult females and 300 g (10 oz) in adult males. The heart rests on the diaphragm, near the midline of the thoracic cavity in the **mediastinum** (mē-dē-a-STĪ-num), a mass of tissue that extends from the sternum to the vertebral column and between the coverings (pleurae) of the lungs (Figure 14.1a, b).

About two-thirds of the mass of the heart lies to the left of the body's midline. The position of the heart in the mediastinum is more readily appreciated by examining its ends, surfaces, and borders (Figure 14.1b). Visualize the heart as a cone lying on its side. The pointed end of the heart is the **apex,** which is directed anteriorly, inferiorly, and to the left. The broad portion of the heart opposite the apex is the **base,** which is directed posteriorly, superiorly, and to the right. The major blood vessels of the heart enter and exit at the base. In addition to the apex and base, the heart has several surfaces and borders (margins) that are useful

Figure 14.1 Position of the heart and associated structures in the mediastinum (dashed outline), and the points of the heart that correspond to its surface projection. (See Tortora, *A Photographic Atlas of the Human Body,* 2e, Figures 6.5 and 6.6.)

🔑 **The heart is located in the mediastinum; two-thirds of its mass is to the left of the midline.**

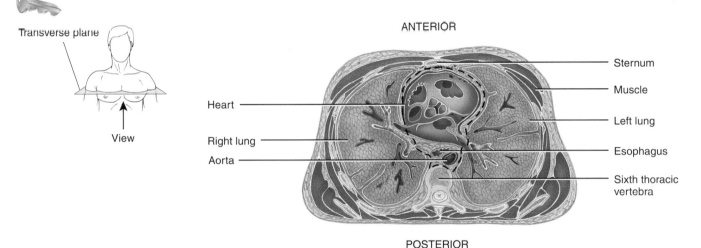

(a) Inferior view of transverse section of thoracic cavity showing the heart in the mediastinum

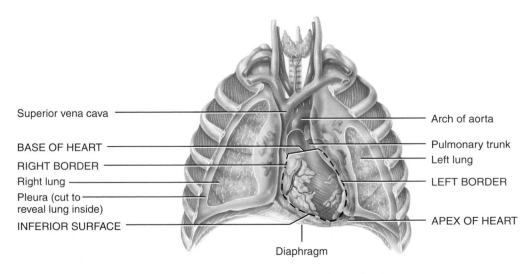

(b) Anterior view of the heart in the mediastinum

Figure 14.1 (continued)

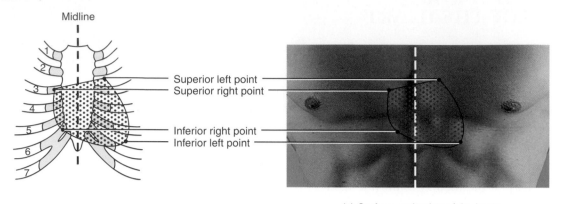

(c) Surface projection of the heart

? **What is the mediastinum?**

in determining its surface projection (described shortly). The **anterior surface** is deep to the sternum and ribs. The **inferior surface** is the portion of the heart that rests mostly on the diaphragm and is found between the apex and right border. The **right border** faces the right lung and extends from the inferior surface to the base; the **left border,** also called the *pulmonary border,* faces the left lung and extends from the base to the apex.

Determining an organ's **surface projection** means outlining its dimensions with respect to landmarks on the surface of the body. This practice is useful when conducting diagnostic procedures (for example, a lumbar puncture), auscultation (for example, listening to heart and lung sounds), and anatomical studies. We can project the heart on the anterior surface of the chest by locating the following landmarks (Figure 14.1c): The **superior right point** is located at the superior border of the third right costal cartilage, about 3 cm (1 in.) to the right of the midline. The **superior left point** is located at the inferior border of the second left costal cartilage, about 3 cm to the left of the midline. A line connecting these two points corresponds to the base of the heart. The **inferior left point** is located at the apex of the heart in the fifth left intercostal space, about 9 cm (3.5 in.) to the left of the midline. A line connecting the superior and inferior left points corresponds to the left border of the heart. The **inferior right point** is located at the superior border of the sixth right costal cartilage, about 3 cm to the right of the midline. A line connecting the inferior left and right points corresponds to the inferior surface of the heart, and a line connecting the inferior and superior right points corresponds to the right border of the heart. When all four points are connected, they form an outline that roughly reveals the size and shape of the heart.

⚕ CARDIOPULMONARY RESUSCITATION

Because the heart lies between two rigid structures—the vertebral column and the sternum (Figure 14.1a)—external pressure (compression) on the chest can be used to force blood out of the

heart and into the circulation. In cases in which the heart suddenly stops beating, **cardiopulmonary resuscitation (CPR)**— properly applied cardiac compressions, performed together with artificial ventilation of the lungs via mouth-to-mouth respiration—saves lives. CPR keeps oxygenated blood circulating until the heart can be restarted.

In a 2000 study conducted in Seattle, researchers found that chest compressions alone are at least as effective as, if not better than, traditional CPR with lung ventilation. This is good news because it is easier for an emergency dispatcher to give instructions limited to chest compressions to frightened, nonmedical bystanders. In addition, as public fear of contracting contagious diseases such as hepatitis, HIV, and TB continues to rise, people are much more likely to perform chest compressions alone than treatment involving mouth-to-mouth rescue breathing. ■

CHECKPOINT

1. Describe the position of the heart in the mediastinum by defining its apex, base, anterior and posterior surfaces, and right and left borders.
2. Explain the location of the superior right point, superior left point, inferior left point, and inferior right point. Why are these points significant?

STRUCTURE AND FUNCTION OF THE HEART

OBJECTIVES

● Describe the structure of the pericardium and the heart wall.
● Discuss the external and internal anatomy of the chambers of the heart.
● Describe the structure and function of the valves of the heart.

Pericardium

The membrane that surrounds and protects the heart is the **pericardium** (*peri-*=around). It confines the heart to its position in the mediastinum, while allowing sufficient freedom of movement for vigorous and rapid contraction. The pericardium consists of two principal portions: the fibrous pericardium and the serous pericardium (Figure 14.2a, b). The superficial **fibrous pericardium** is a tough, inelastic, dense irregular connective tissue. It resembles a bag that rests on and attaches to the diaphragm; its open end is fused to the connective tissues of the blood vessels entering and leaving the heart. The fibrous pericardium prevents overstretching of the heart, provides protection, and anchors the heart in the mediastinum.

The deeper **serous pericardium** is a thinner, more delicate membrane that forms a double layer around the heart (Figure 14.2a, b). The outer **parietal layer** of the serous pericardium is fused to the fibrous pericardium. The inner **visceral layer** of the serous pericardium, also called the **epicardium** (*epi-*=on top of), adheres tightly to the surface of the heart. Between the parietal and visceral layers of the serous pericardium is a thin film of serous fluid. This fluid, known as **pericardial fluid,** is a slippery secretion of the pericardial cells that reduces friction between the membranes as the heart moves. The space that contains the few milliliters of pericardial fluid is called the **pericardial cavity.**

PERICARDITIS AND CARDIAC TAMPONADE

Inflammation of the pericardium is known as **pericarditis.** If production of pericardial fluid diminishes, painful rubbing together of the parietal and visceral serous pericardial layers may result. A buildup of pericardial fluid (which may also occur in pericarditis) or extensive bleeding into the pericardium are life-threatening conditions. Because the pericardium cannot stretch, the buildup of fluid or blood compresses the heart. This com-

Figure 14.2   Pericardium and heart wall.

The pericardium is a triple-layered sac that surrounds and protects the heart.

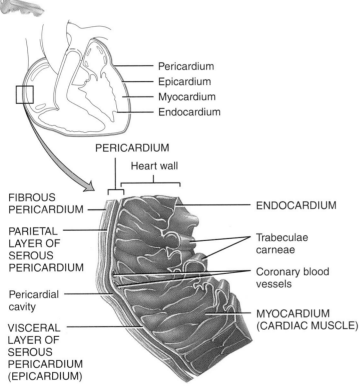

Pericardium
Epicardium
Myocardium
Endocardium

PERICARDIUM

Heart wall

FIBROUS PERICARDIUM

PARIETAL LAYER OF SEROUS PERICARDIUM

Pericardial cavity

VISCERAL LAYER OF SEROUS PERICARDIUM (EPICARDIUM)

ENDOCARDIUM

Trabeculae carneae

Coronary blood vessels

MYOCARDIUM (CARDIAC MUSCLE)

(a) Portion of pericardium and right atrial heart wall showing the divisions of the pericardium and layers of the heart wall

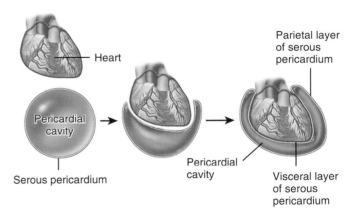

Heart

Parietal layer of serous pericardium

Pericardial cavity

Serous pericardium

Pericardial cavity

Visceral layer of serous pericardium

(b) Simplified relationship of the serous pericardium to the heart

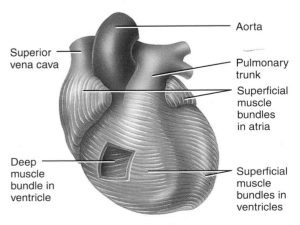

Aorta

Superior vena cava

Pulmonary trunk

Superficial muscle bundles in atria

Deep muscle bundle in ventricle

Superficial muscle bundles in ventricles

(c) Cardiac muscle bundles of the myocardium

 Which layer is both a part of the pericardium and a part of the heart wall?

pression, known as **cardiac tamponade** (tam′-pon-ĀD), restricts ventricular filling, reduces cardiac output, decreases venous return to the heart, and decreases blood pressure. Trauma is the most common cause of cardiac tamponade. Other causes include cancer, surgery, and infections. ■

Layers of the Heart Wall

The wall of the heart consists of three layers (Figure 14.2a): the epicardium (external layer), the myocardium (middle layer), and the endocardium (inner layer). The outermost **epicardium,** which as you just learned is also called the *visceral layer of the serous pericardium,* is the thin, transparent outer layer of the wall. It is composed of mesothelium and delicate connective tissue that imparts a smooth, slippery texture to the outermost surface of the heart. The middle **myocardium** (*myo-*=muscle), which is composed of cardiac muscle tissue, makes up the bulk of the heart and is responsible for its pumping action (see Figure 10.8 on page 279). Although it is striated like skeletal muscle, cardiac muscle is involuntary like smooth muscle. The cardiac muscle fibers swirl diagonally around the heart in interlacing bundles (Figure 14.2c). The innermost **endocardium** (*endo-*=within) is a thin layer of endothelium overlying a thin layer of connective tissue. It provides a smooth lining for the chambers of the heart and covers the valves of the heart. The endocardium is continuous with the endothelial lining of the large blood vessels attached to the heart.

Chambers of the Heart

The heart contains four chambers. The two upper chambers are the **atria** (=entry halls or chambers), and the two lower chambers are the **ventricles** (=little bellies). On the anterior surface of each atrium is a wrinkled pouchlike structure called an **auricle** (OR-i-kul; *auri-*=ear), so named because of its resemblance to a dog's ear (Figure 14.3). Each auricle slightly increases the capacity of an atrium so that it can hold a greater volume of blood. Also on the surface of the heart are a series of grooves, called **sulci** (SUL-sē), that contain coronary blood vessels and a variable amount of fat. Each sulcus (*SUL-kus;* singular) marks the external boundary between two chambers of the heart. The deep **coronary sulcus** (*coron-*=resembling a crown) encircles most of the heart and marks the boundary between the superior atria and inferior ventricles. The **anterior interventricular sulcus** is a shallow groove on the anterior surface of the heart that marks the boundary between the right and left ventricles. This sulcus continues around to the posterior surface of the heart as the **posterior interventricular sulcus,** which marks the boundary between the ventricles on the posterior aspect of the heart (Figure 14.3c).

Right Atrium

The **right atrium** forms the right border of the heart (see Figure 14.1b). It receives blood from three veins: *superior vena cava,*

Figure 14.3 Structure of the heart: surface features.

Sulci are grooves that contain blood vessels and fat and mark the boundaries between the various chambers.

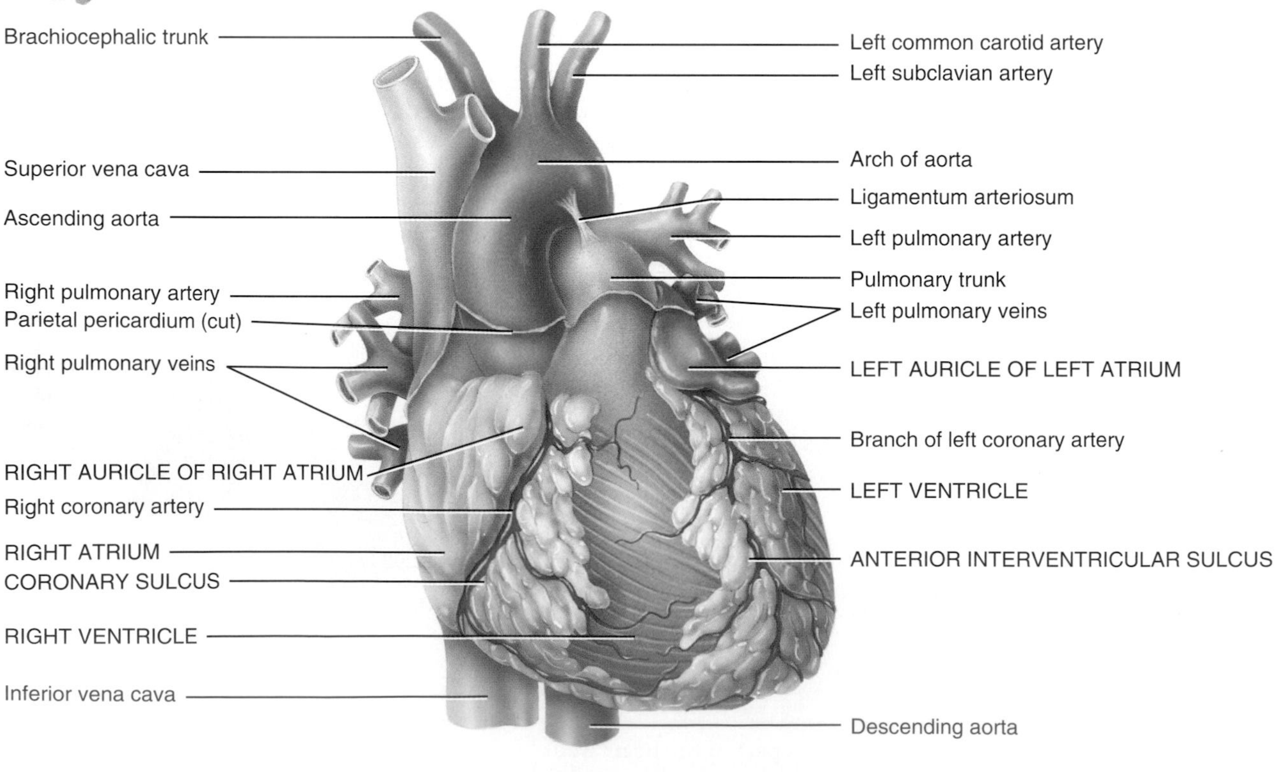

(a) Anterior external view showing surface features

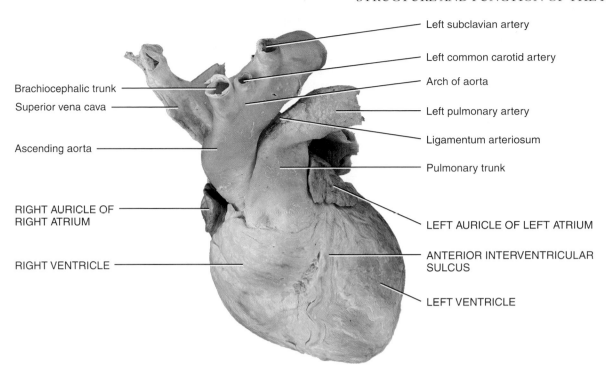

Left subclavian artery

Left common carotid artery

Arch of aorta

Brachiocephalic trunk

Superior vena cava

Left pulmonary artery

Ligamentum arteriosum

Ascending aorta

Pulmonary trunk

RIGHT AURICLE OF
RIGHT ATRIUM

LEFT AURICLE OF LEFT ATRIUM

ANTERIOR INTERVENTRICULAR
SULCUS

RIGHT VENTRICLE

LEFT VENTRICLE

(b) Anterior external view showing surface features

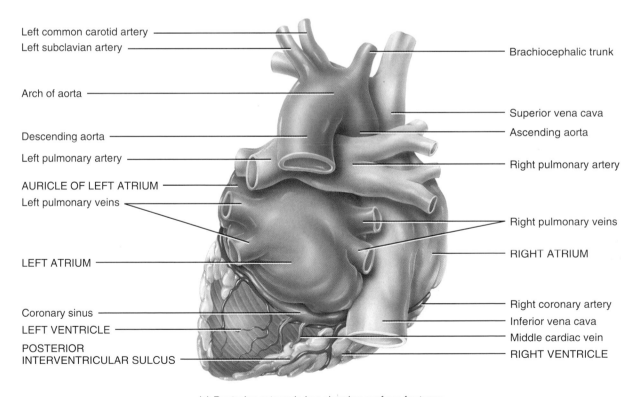

Left common carotid artery

Left subclavian artery

Brachiocephalic trunk

Arch of aorta

Superior vena cava

Ascending aorta

Descending aorta

Left pulmonary artery

Right pulmonary artery

AURICLE OF LEFT ATRIUM

Left pulmonary veins

Right pulmonary veins

LEFT ATRIUM

RIGHT ATRIUM

Coronary sinus

Right coronary artery

LEFT VENTRICLE

Inferior vena cava

POSTERIOR
INTERVENTRICULAR SULCUS

Middle cardiac vein

RIGHT VENTRICLE

(c) Posterior external view showing surface features

The coronary sulcus forms a boundary between which chambers of the heart?

inferior vena cava, and *coronary sinus* (Figure 14.4a). The anterior and posterior walls within the right atrium differ considerably. Whereas the posterior wall is smooth, the anterior wall is rough due to the presence of muscular ridges called **pectinate muscles** (*pectin*=comb), which also extend into the auricle (Figure 14.4b). Between the right atrium and left atrium is a thin partition called the **interatrial septum** (*inter-*=between; *septum*=a dividing wall or partition). A prominent feature of this septum is an oval depression called the **fossa ovalis,** which is the remnant of the *foramen ovale,* an opening in the interatrial septum of the fetal heart that directs blood from the right to left atrium in order to bypass the nonfunctioning fetal lungs. The foreamen ovale normally closes soon after birth (see Figure 15.17 on page 503). Blood passes from the right atrium into the right ventricle through a valve called the **tricuspid valve** (trī-KUS-pid; *tri-*=three; *cuspid*=point) because it consists of three leaflets or cusps (Figure 14.4a). The valves of the heart are composed of dense connective tissue covered by endocardium.

Right Ventricle

The **right ventricle** forms most of the anterior surface of the heart (see Figure 14.3a). The inside of the right ventricle contains a series of ridges formed by raised bundles of cardiac muscle fibers called **trabeculae carneae** (tra-BEK-ū-lē KAR-nē-ē; *trabeculae*=little beams; *carneae*=fleshy) (Figure 14.4). Some of the trabeculae carneae contain part of the conduction system of the heart (described on page 436). The cusps of the tricuspid valve are connected to tendonlike cords, the **chordae tendineae** (KOR-dē ten-DIN-ē-ē; *chord-*=cord; *tend-*=tendon), which, in turn, are connected to cone-shaped trabeculae carneae called **papillary muscles** (*papill-*=nipple). The right ventricle is separated from the left ventricle by a partition called the **interventricular septum.** Blood passes from the right ventricle through the **pulmonary valve** into a large vessel called the *pulmonary trunk,* which divides into right and left *pulmonary arteries.*

Left Atrium

The **left atrium** forms most of the base of the heart (see Figure 14.1b). It receives blood from the lungs through four *pulmonary veins.* Like the right atrium, the inside of the left atrium has a smooth posterior wall. Because the ridged pectinate muscles are confined to the auricle of the left atrium, the anterior wall of the left atrium is smooth. Blood passes from the left atrium into the left ventricle through the **bicuspid (mitral) valve** (*bi-*=two), which has two cusps. The term mitral refers to its resemblance to a bishop's miter (hat), which is two-sided.

Figure 14.4 Structure of the heart: internal anatomy.

The thickness of the four chambers varies according to their functions.

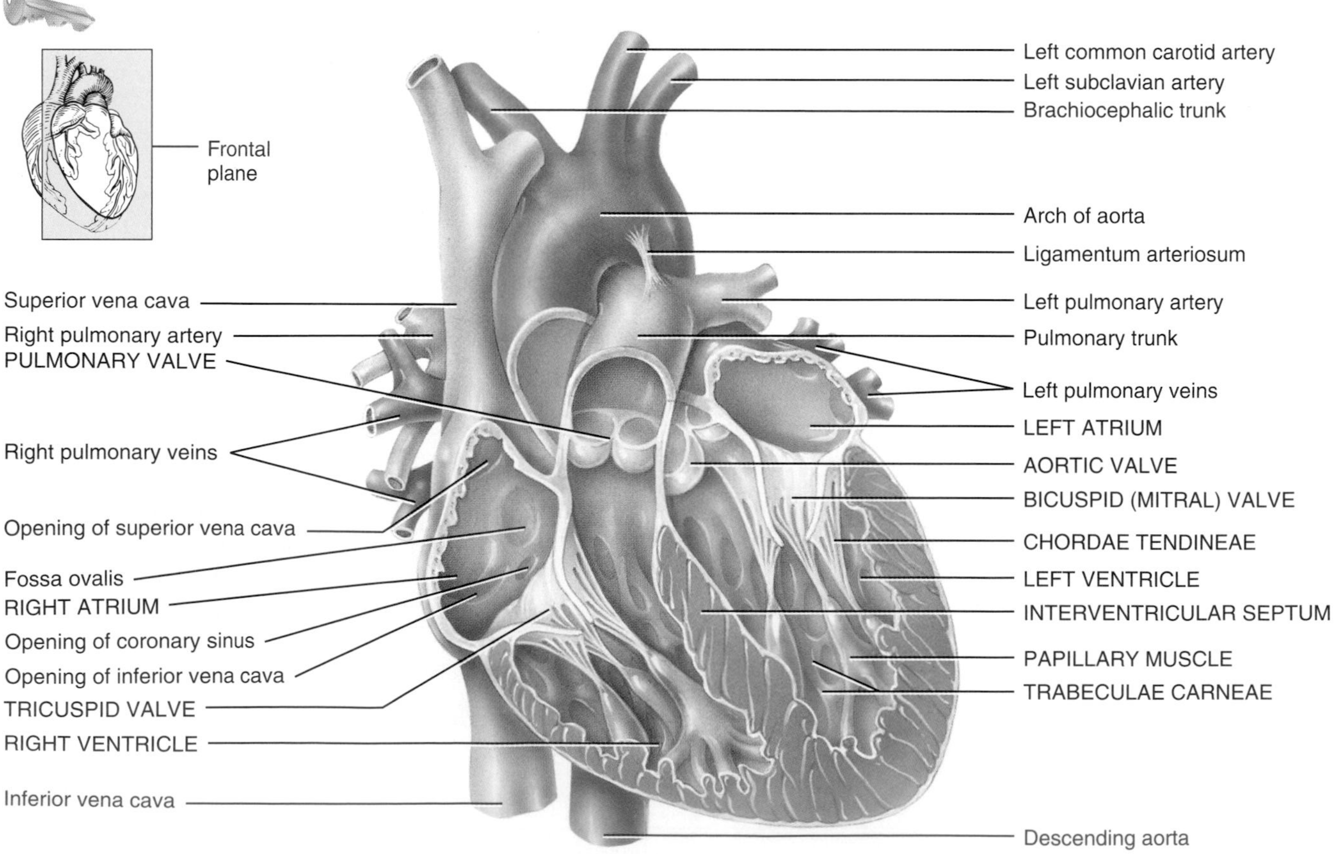

Frontal plane

Left common carotid artery
Left subclavian artery
Brachiocephalic trunk
Arch of aorta
Ligamentum arteriosum
Left pulmonary artery
Pulmonary trunk
Left pulmonary veins
LEFT ATRIUM
AORTIC VALVE
BICUSPID (MITRAL) VALVE
CHORDAE TENDINEAE
LEFT VENTRICLE
INTERVENTRICULAR SEPTUM
PAPILLARY MUSCLE
TRABECULAE CARNEAE
Descending aorta

Superior vena cava
Right pulmonary artery
PULMONARY VALVE
Right pulmonary veins
Opening of superior vena cava
Fossa ovalis
RIGHT ATRIUM
Opening of coronary sinus
Opening of inferior vena cava
TRICUSPID VALVE
RIGHT VENTRICLE
Inferior vena cava

(a) Anterior view of frontal section showing internal anatomy

Left Ventricle

The **left ventricle** forms the apex of the heart (see Figure 14.1b). Like the right ventricle, the left ventricle contains trabeculae carneae and has chordae tendinae that anchor the cusps of the bicuspid valve to papillary muscles. Blood passes from the left ventricle through the **aortic valve** into the largest artery of the body, the *ascending aorta* (*aorte*=to suspend, because the aorta once was believed to lift up the heart). Some of the blood in the aorta flows into the *coronary arteries*, which branch from the ascending aorta and carry blood to the heart wall; the remainder of the blood passes into the *arch of the aorta* and *descending aorta*. Branches of the arch of the aorta and descending aorta carry blood throughout the body.

Between the left pulmonary artery and arch of the aorta is a structure called the **ligamentum arteriosum** (Figure 14.4a). It is the remnant of a blood vessel in fetal circulation (*ductus arteriosus*) that allows most blood to bypass the nonfunctional fetal lungs (see Figure 15.17a on page 503).

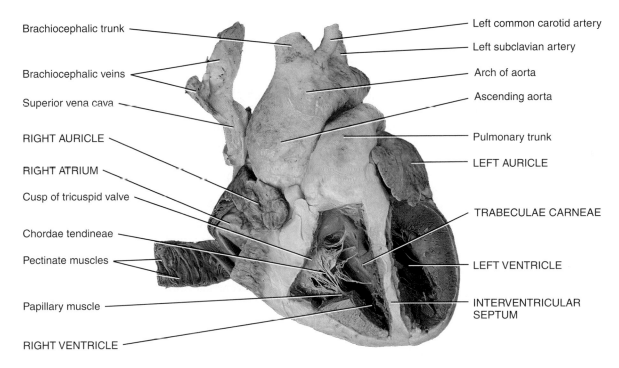

Brachiocephalic trunk

Brachiocephalic veins

Superior vena cava

RIGHT AURICLE

RIGHT ATRIUM

Cusp of tricuspid valve

Chordae tendineae

Pectinate muscles

Papillary muscle

RIGHT VENTRICLE

Left common carotid artery

Left subclavian artery

Arch of aorta

Ascending aorta

Pulmonary trunk

LEFT AURICLE

TRABECULAE CARNEAE

LEFT VENTRICLE

INTERVENTRICULAR SEPTUM

(b) Anterior view of partially sectioned heart showing internal anatomy

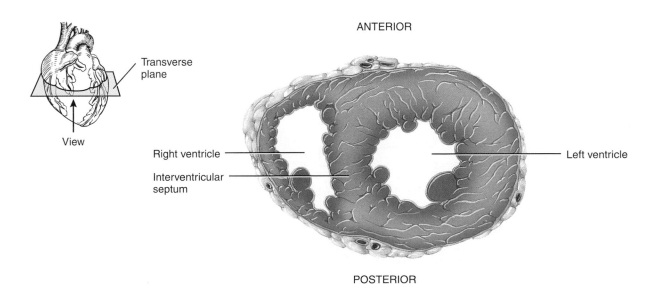

Transverse plane

View

ANTERIOR

Right ventricle

Interventricular septum

Left ventricle

POSTERIOR

(c) Inferior view of transverse section showing differences in thickness of ventricular walls

 Which chamber has the thickest wall?

Myocardial Thickness and Function

The thickness of the myocardium of the four chambers varies according to function. The atria are thin-walled because they deliver blood into the adjacent ventricles; since the ventricles pump blood greater distances, their walls are thicker (Figure 14.4a). Although the right and left ventricles act as two separate pumps that simultaneously eject equal volumes of blood, the right side has a much smaller workload. It pumps blood a short distance to the lungs and there is low resistance to blood flow. The left ventricle pumps blood great distances to all other parts of the body and the resistance to blood flow is higher. Thus, the left ventricle works harder than the right ventricle to maintain the same rate of blood flow. The anatomy of the two ventricles confirms this functional difference: The muscular wall of the left ventricle is considerably thicker than that of the right ventricle (Figure 14.4c). Note also that the perimeter of the lumen (space) of the left ventricle is circular, whereas that of the right ventricle is crescent-shaped.

Fibrous Skeleton of the Heart

In addition to cardiac muscle tissue, the heart wall contains dense connective tissue that forms the **fibrous skeleton of the heart.** The skeleton forms the foundation to which the heart valves attach, serves as a point of insertion for cardiac muscle bundles (see Figure 14.2c), prevents overstretching of the valves as blood passes through them, and acts as an electrical insulator that prevents the direct spread of action potentials from the atria to the ventricles.

Essentially, the fibrous skeleton consists of dense connective tissue rings that surround the valves of the heart, fuse with one another, and merge with the interventricular septum. The components of the fibrous skeleton of the heart are as follows (Figure 14.5):

1. Four **fibrous rings** that support the four valves of the heart and are fused to each other: **right atrioventricular fibrous ring, left atrioventricular fibrous ring, pulmonary fibrous ring,** and **aortic fibrous ring.**

2. **Right fibrous trigone** (TRĪ-gōn), a relatively large triangular mass formed by the fusion of the fibrous connective tissue of the left atrioventricular, aortic, and right atrioventricular fibrous rings.

3. **Left fibrous trigone,** a smaller mass formed by the fusion of the fibrous connective tissue of the left atrioventricular and aortic fibrous rings.

4. **Conus tendon,** a mass formed by the fusion of the fibrous connective tissue of the pulmonary and aortic fibrous rings.

Heart Valves

As each chamber of the heart contracts, it pushes a volume of blood into a ventricle or out of the heart into an artery. Valves open and close in response to pressure changes as the heart contracts and relaxes. Each of the four valves helps to ensure one-way flow of blood by opening to let blood through and closing to prevent its backflow.

Atrioventricular Valves

Because they are located between an atrium and a ventricle, the tricuspid and bicuspid valves are termed **atrioventricular (AV) valves.** When an AV valve is open, the pointed ends of the cusps (leaflets) project into the ventricle. Blood moves from the atria into the ventricles through open AV valves when atrial pressure is higher than ventricular pressure (Figure 14.6a, c). At this time, the papillary muscles are relaxed, and the chordae tendineae are slack. When the ventricles contract, the pressure of the blood drives the cusps upward until their edges meet and close the opening (Figure 14.6b, d). At the same time, the papil-

Figure 14.5 Fibrous skeleton of the heart.

 The fibrous skeleton provides a base for the attachment of heart valves, prevents overstretching of the valves, serves as a point of insertion for cardiac muscle bundles, and prevents the direct spread of action potentials from the atria to the ventricles.

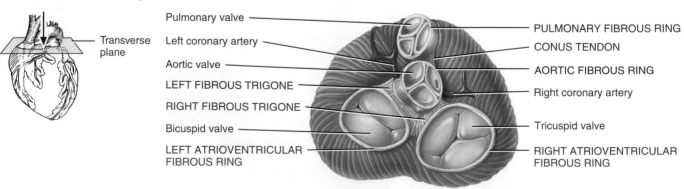

Superior view (the atria have been removed)

? **Which component of the fibrous skeleton supports the heart valves?**

Figure 14.6   Valves of the heart.

Heart valves prevent the backflow of blood.

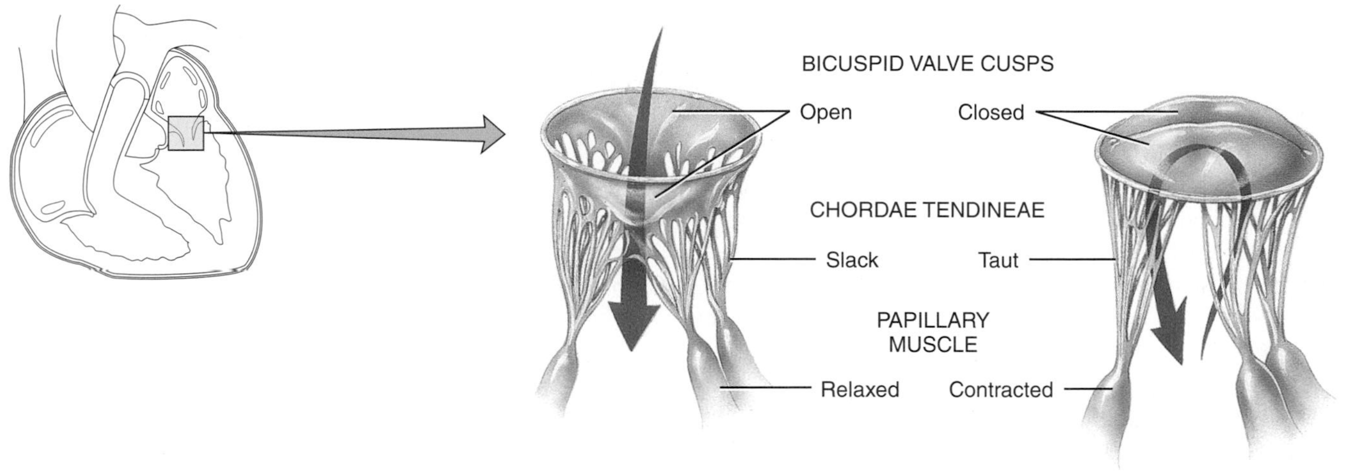

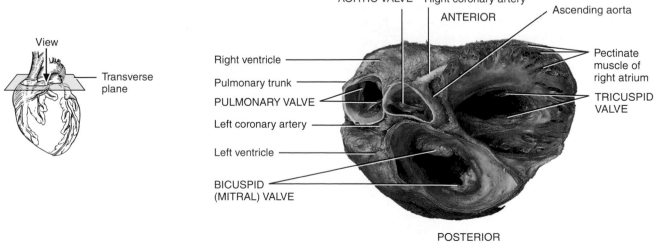

BICUSPID VALVE CUSPS

Open   Closed

CHORDAE TENDINEAE

Slack   Taut

PAPILLARY
MUSCLE

Relaxed   Contracted

(a) Bicuspid valve open

(b) Bicuspid valve closed

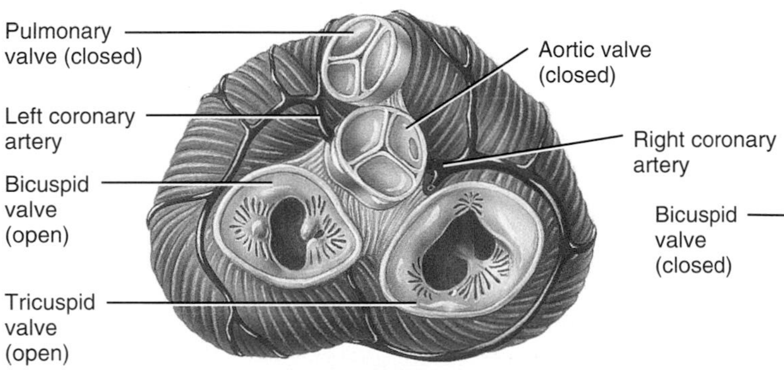

ANTERIOR

Pulmonary
valve (closed)

Aortic valve
(closed)

Left coronary
artery

Right coronary
artery

Bicuspid
valve
(open)

Tricuspid
valve
(open)

POSTERIOR

(c) Superior view with atria removed:
pulmonary and aortic valves closed,
bicuspid and tricuspid valves open.

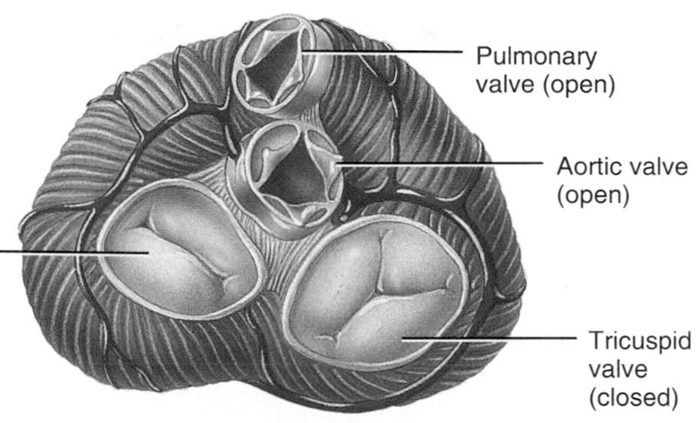

ANTERIOR

Pulmonary
valve (open)

Aortic valve
(open)

Bicuspid
valve
(closed)

Tricuspid
valve
(closed)

POSTERIOR

(d) Superior view with atria removed:
pulmonary and aortic valves open,
bicuspid and tricuspid valves closed.

AORTIC VALVE   Right coronary artery

ANTERIOR

Ascending aorta

View

Transverse
plane

Right ventricle

Pulmonary trunk

PULMONARY VALVE

Left coronary artery

Left ventricle

BICUSPID
(MITRAL) VALVE

Pectinate
muscle of
right atrium

TRICUSPID
VALVE

POSTERIOR

(e) Superior view of atrioventricular and semilunar valves

 How do papillary muscles prevent AV valve cusps from everting or swinging upward into the atria?

lary muscles are also contracting, which pulls on and tightens the chordae tendineae, preventing the valve cusps from everting (opening in the opposite direction into the atria due to the high ventricular pressure). If the AV valves or chordae tendineae are damaged, blood may regurgitate (flow back) into the atria when the ventricles contract.

Semilunar Valves

The two **semilunar (SL) valves** (*semi*=half; *lunar*=moon-shaped), the pulmonary and aortic valves, allow ejection of blood from the heart into arteries, but prevent backflow of blood into the ventricles. Both SL valves consist of three crescent-shaped cusps (Figure 14.6c), each attached to the arterial wall by its convex outer margin. The free borders of the cusps curve outward and project into the lumen of the artery. When the ventricles contract, pressure builds up within them. The semilunar valves open when pressure in the ventricles exceeds the pressure in the arteries, permitting ejection of blood from the ventricles into the pulmonary trunk and aorta (Figure 14.6d). As the ventricles relax, blood starts to flow back toward

the heart. This backflowing blood fills the valve cusps, which causes the semilunar valves to close tightly (Figure 14.6c).

HEART VALVE DISORDERS

When heart valves operate normally, they open fully and close completely at the proper times. Failure of a heart valve to open fully is known as **stenosis** (=a narrowing) whereas failure of a valve to close completely is termed **insufficiency** or **incompetence**. In **mitral stenosis**, scar formation or a congenital defect causes narrowing of the mitral valve. **Mitral insufficiency** allows backflow of blood from the left ventricle into the left atrium. One cause is **mitral valve prolapse (MVP)**, in which one or both cusps of the mitral valve protrude into the left atrium during ventricular contraction. Mitral valve prolapse is one of the most common valvular disorders, affecting as much as 30% of the population. It is more prevalent in women than in men. Although a small volume of blood may flow back into the left atrium during ventricular contraction, mitral valve prolapse

Figure 14.7 Systemic and pulmonary circulations. Throughout this book, blood vessels that carry oxygenated blood (which looks bright red) are colored red, whereas those that carry deoxygenated blood (which looks dark red) are colored blue.

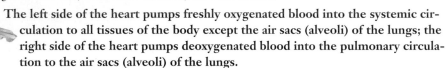

The left side of the heart pumps freshly oxygenated blood into the systemic circulation to all tissues of the body except the air sacs (alveoli) of the lungs; the right side of the heart pumps deoxygenated blood into the pulmonary circulation to the air sacs (alveoli) of the lungs.

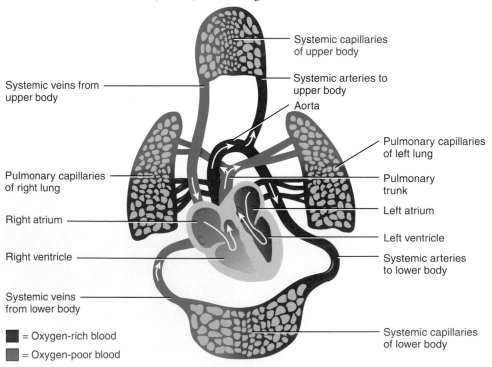

(a) Systemic and pulmonary circulations

does not always pose a serious threat. Mitral insufficiency also can be due to a damaged mitral valve or ruptured chordae tendineae. In **aortic stenosis** the aortic valve is narrowed, and in **aortic insufficiency** there is backflow of blood from the aorta into the left ventricle.

Certain infectious diseases can damage or destroy the heart valves. One example is **rheumatic fever,** an acute systemic inflammatory disease that usually occurs after a streptococcal infection of the throat. The bacteria trigger an immune response in which antibodies that are produced to destroy the bacteria attack and inflame the connective tissues in joints, heart valves, and other organs. Even though rheumatic fever may weaken the entire heart wall, most often it damages the bicuspid (mitral) and aortic valves.

If daily activities are affected by symptoms and if a heart valve cannot be repaired surgically, then the valve must be replaced. Tissue valves may be provided by human donors or pigs; sometimes, mechanical replacements are used. In any case, valve replacement involves open heart surgery. The aortic valve is the most commonly replaced heart valve. ■

3. Define the following external features of the heart: auricle, coronary sulcus, anterior interventricular sulcus, and posterior interventricular sulcus.
4. Describe the characteristic internal features of each chamber of the heart.
5. List the blood vessels that deliver blood to or receive ejected blood from each chamber of the heart, and name the valve that blood passes through on its way to the next heart chamber or blood vessel.
6. Describe the relationship between wall thickness and function for each heart chamber.
7. How does the fibrous skeleton of the heart assist the operation of heart valves?
8. What causes the heart valves to open and to close?

CIRCULATION OF BLOOD

● Describe the flow of blood through the chambers of the heart and through the systemic and pulmonary circulations.
● Discuss the coronary circulation.

Systemic and Pulmonary Circulations

In postnatal (after birth) circulation, the heart pumps blood into two circuits—the **systemic circulation** and the **pulmonary circulation** (*pulmon-*=lung) with each beat. As you will see later, the two circuits are arranged in series so that the output of one becomes the input of the other, as would happen if you attached two garden hoses together. The left side of the heart, which receives bright red freshly oxygenated (oxygen-rich) blood from the lungs, is the pump for the systemic circulation. The left ventricle ejects blood into the *aorta*, which branches into (Figure 14.7a on page 434) progressively smaller *systemic arteries* that carry the blood to all organs throughout the body—except for the air sacs (alveoli) of the lungs, which are supplied by the pulmonary circulation. In systemic tissues, arteries give rise to smaller-diameter *arterioles*, which finally lead into extensive beds of *systemic capillaries*. Exchange of nutrients and gases occurs across the thin capillary walls: In the tissues, blood unloads O_2 (oxygen) and picks up CO_2 (carbon dioxide). In most cases, blood flows through only one capillary and then enters a *systemic venule*. Venules carry deoxygenated (oxygen-poor) blood away from tissues and merge to form larger *systemic veins*, and ultimately the blood flows back to the right atrium.

The right side of the heart is the pump for the pulmonary circulation; it receives all the dark red deoxygenated blood returning from the systemic circulation. Blood ejected from the right ventricle flows into the *pulmonary trunk*, which branches into *pulmonary arteries* that carry blood to the right and left lungs. In pulmonary capillaries, blood unloads CO_2, which is exhaled, and picks up O_2. The freshly oxygenated blood then flows into pulmonary veins and returns to the left atrium. The flowchart in Figure 14.7b shows the route of blood flow through the chambers and valves of the heart and the pulmonary and systemic circulations.

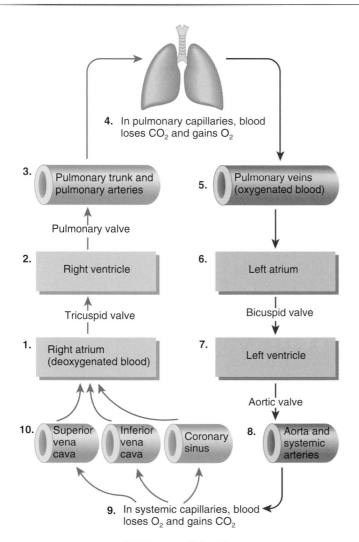

4. In pulmonary capillaries, blood loses CO_2 and gains O_2

3. Pulmonary trunk and pulmonary arteries

Pulmonary valve

2. Right ventricle

Tricuspid valve

1. Right atrium (deoxygenated blood)

10. Superior vena cava / Inferior vena cava / Coronary sinus

9. In systemic capillaries, blood loses O_2 and gains CO_2

5. Pulmonary veins (oxygenated blood)

6. Left atrium

Bicuspid valve

7. Left ventricle

Aortic valve

8. Aorta and systemic arteries

(b) Diagram of blood flow

In part (b), which numbers constitute the pulmonary circulation? Which represent the systemic circulation?

Coronary Circulation

Nutrients could not possibly diffuse from blood in the chambers of the heart through all the layers of cells that make up the heart tissue. For this reason, the wall of the heart has its own blood supply. The flow of blood through the many vessels that pierce the myocardium is called the **coronary (cardiac) circulation** because the arteries of the heart encircle it like a crown encircles the head (*corona*=crown). While it is contracting, the heart receives little oxygenated blood by way of the **coronary arteries,** which branch from the ascending aorta (Figure 14.8a). When the heart relaxes, however, the high pressure of blood in the aorta propels blood through the coronary arteries, into capillaries, and then into **coronary veins** (Figure 14.8b).

Coronary Arteries

Two coronary arteries, the right and left coronary arteries, branch from the ascending aorta and supply oxygenated blood to the myocardium (Figure 14.8a). The **left coronary artery** passes inferior to the left auricle and divides into the anterior interventricular and circumflex branches. The **anterior interventricular branch** or *left anterior descending (LAD) artery* is in the anterior interventricular sulcus and supplies oxygenated blood to the walls of both ventricles. The **circumflex branch** lies in the coronary sulcus and distributes oxygenated blood to the walls of the left ventricle and left atrium.

The **right coronary artery** supplies small branches (*atrial branches*) to the right atrium. It continues inferior to the right auricle and ultimately divides into the posterior interventricular and marginal branches. The **posterior interventricular branch** follows the posterior interventricular sulcus and supplies the walls of the two ventricles with oxygenated blood. The **marginal branch** in the coronary sulcus transports oxygenated blood to the myocardium of the right ventricle.

Most parts of the body receive blood from branches of more than one artery, and where two or more arteries supply the same region, they usually connect. These connections, called **anastomoses** (a-nas′-tō-MŌ-sēs), provide alternate routes for blood to reach a particular organ or tissue. The myocardium contains many anastomoses that connect branches of a given coronary artery or extend between branches of different coronary arteries. They provide detours for arterial blood if a main route becomes obstructed. Thus, the heart muscle may receive sufficient oxygen even if one of its coronary arteries is partially blocked.

Coronary Veins

After blood passes through the arteries of the coronary circulation, where it delivers oxygen and nutrients to the heart muscle, it passes into capillaries, where it collects carbon dioxide and wastes, and then into veins. Most of the deoxygenated blood from the myocardium drains into a large vascular sinus in the coronary sulcus on the posterior surface of the heart, called the **coronary sinus** (Figure 14.8b), which empties into the right atrium. (A *vascular sinus* is a thin-walled vein that has no smooth muscle to alter its diameter.) The principal tributaries carrying blood into the coronary sinus include the following:

- **Great cardiac vein** in the anterior interventricular sulcus, which drains the areas of the heart supplied by the left coronary artery (left and right ventricles and left atrium).
- **Middle cardiac vein** in the posterior interventricular sulcus, which drains the areas supplied by the posterior interventricular branch of the right coronary artery (left and right ventricles).
- **Small cardiac vein** in the coronary sulcus, which drains the right atrium and right ventricle.
- **Anterior cardiac veins,** which drain the right ventricle and open directly into the right atrium.

🩺 REPERFUSION DAMAGE

When blockage of a coronary artery deprives the heart muscle of oxygen, **reperfusion,** reestablishing the blood flow, may damage the tissue further. This surprising effect is due to the formation of oxygen **free radicals** from the reintroduced oxygen. Free radicals are electrically charged molecules that have an unpaired electron. Such molecules are unstable and highly reactive. They cause chain reactions that lead to cellular damage and death. To counter the effects of oxygen free radicals, body cells produce enzymes that convert free radicals to less reactive substances. Two such enzymes are *superoxide dismutase* and *catalase.* In addition, some nutrients such as vitamin E, vitamin C, beta-carotene, and selenium are antioxidants, which remove oxygen free radicals. Drugs that lessen reperfusion damage after a heart attack or stroke are currently being developed. ■

CHECKPOINT

9. In correct sequence, list the heart chambers, heart valves, and blood vessels encountered by a drop of blood as it flows out of the right atrium until it reaches the aorta.
10. Which arteries deliver oxygenated blood to the myocardium of the left and right ventricles?

THE CARDIAC CONDUCTION SYSTEM

OBJECTIVES

- Explain the structural and functional features of the conduction system of the heart.
- Explain the meaning of an electrocardiogram and its diagnostic importance.

During embryonic development, about 1% of the cardiac muscle fibers become *autorhythmic cells* (*auto-*=self), that is, cells that repeatedly and rhythmically generate action potentials. Autorhythmic cells act as a **pacemaker,** setting the rhythm for the contraction of the entire heart, and they form the **conduction system,** the route that delivers action potentials throughout the heart muscle. The conduction system assures that cardiac chambers are stimulated to contract in a coordinated

Figure 14.8 **Coronary (cardiac) circulation.** The views of the heart from the anterior aspect in (a) and (b) are drawn as if the heart were transparent to reveal blood vessels on the posterior aspect. (See Tortora, *A Photographic Atlas of the Human Body*, 2e, Figure 6.9.)

The right and left coronary arteries deliver blood to the heart; the coronary veins drain blood from the heart into the coronary sinus.

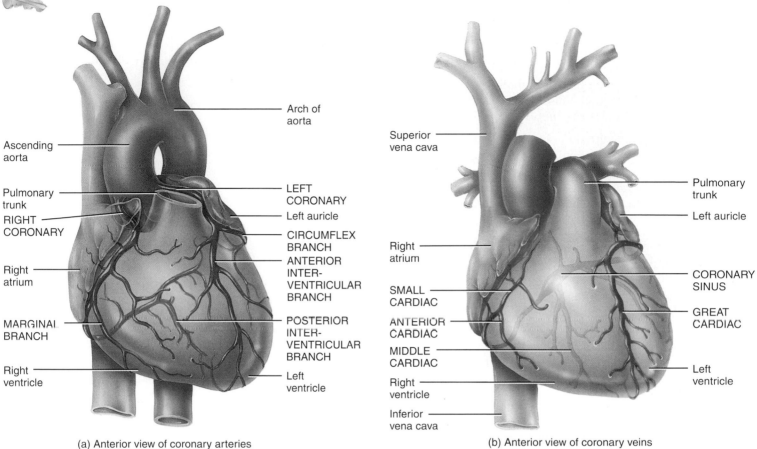

(a) Anterior view of coronary arteries

(b) Anterior view of coronary veins

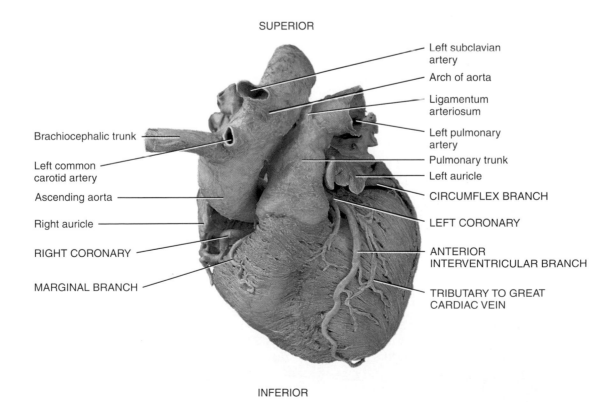

(c) Anterior view

Which coronary blood vessel delivers oxygenated blood to the left atrium and left ventricle?

manner, which makes the heart an effective pump. Cardiac action potentials propagate through the following components of the conduction system (Figure 14.9):

❶ Normally, cardiac excitation begins in the **sinoatrial (SA) node,** located in the right atrial wall just inferior to the opening of the superior vena cava. Each action potential from the SA node propagates throughout both atria via gap junctions in the intercalated discs of atrial fibers. With the arrival of the action potential, the atria contract.

❷ By propagating along atrial muscle fibers, the action potential reaches the **atrioventricular (AV) node,** located in the septum between the two atria, just anterior to the opening of the coronary sinus. At the AV node, the action potential slows considerably, providing time for the atria to empty their blood into the ventricles.

❸ From the AV node, the action potential enters the **atrioventricular (AV) bundle** (also known as the **bundle of His**), the only site where action potentials can conduct from the atria to the ventricles. (Elsewhere, the fibrous skeleton of the heart electrically insulates the atria from the ventricles.)

❹ After conducting along the AV bundle, the action potential then enters both the **right** and **left bundle branches** that course through the interventricular septum toward the apex of the heart.

❺ Finally, the large-diameter **Purkinje fibers** rapidly conduct the action potential, first to the apex of the ventricular myocardium, pushing the blood upward to the semilunar valves. About 0.20 sec (200 milliseconds) after the atria contract, the ventricles contract.

The SA node initiates action potentials 90 to 100 times per minute, faster than any other region of the conducting system. Thus, the SA node sets the rhythm for contraction of the heart—it is the *pacemaker* of the heart. Various hormones and neurotransmitters can speed or slow pacing of the heart by SA node fibers. In a person at rest, for example, acetylcholine released by the parasympathetic division of the ANS typically slows SA node pacing to about 75 action potentials per minute, causing 75 heartbeats per minute. If the SA node becomes diseased or damaged, the slower AV node fibers can become the pacemaker. With pacing by the AV node, however, heart rate is slower, only 40 to 60 beats/min. If the activity of both nodes is suppressed, the heartbeat may still be maintained by the AV bundle, a bundle branch, or conduction myofibers. These fibers generate action potentials very slowly, about 20 to 35 times per minute. At such a low heart rate, blood flow to the brain is inadequate. When this condition occurs, normal heart rhythm can

Figure 14.9   The conduction system of the heart and a normal electrocardiogram. The route of action potentials through the numbered components of the conduction system is described in the text.

🔑 **The conduction system ensures that cardiac chambers contract in a coordinated manner.**

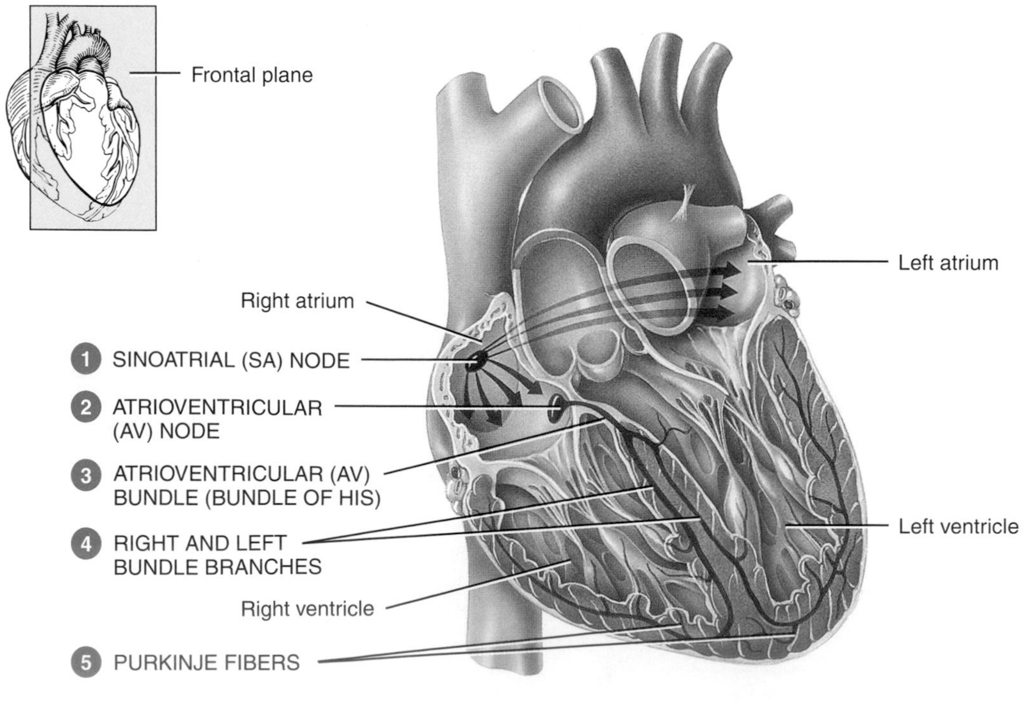

Frontal plane

Right atrium

❶ SINOATRIAL (SA) NODE

❷ ATRIOVENTRICULAR (AV) NODE

❸ ATRIOVENTRICULAR (AV) BUNDLE (BUNDLE OF HIS)

❹ RIGHT AND LEFT BUNDLE BRANCHES

Right ventricle

❺ PURKINJE FIBERS

Left atrium

Left ventricle

(a) Anterior view of frontal section

be restored and maintained by surgically implanting an **artificial pacemaker,** a device that sends out small electrical currents to stimulate ventricular contractions of the heart to maintain adequate cardiac output. Many of the newer pacemakers, called activity-adjusted pacemakers, automatically speed up the heartbeat during exercise.

Sometimes, a site other than the SA node develops abnormal self-excitability and becomes the pacemaker. Such a site is called an *ectopic pacemaker* (ek-TOP-ik; *ectop-*=displaced). An ectopic pacemaker may operate only occasionally, producing an irregular heartbeat, or it may pace the heart for some period of time. Triggers of ectopic activity include caffeine and nicotine, electrolyte imbalances, hypoxia, and toxic reactions to drugs such as digitalis.

Transmission of action potentials through the conduction system generates an electric current that can be detected on the body's surface. A recording of the electrical changes that accompany the heartbeat is called an **electrocardiogram** (e-lek′-trō-KAR-dē-ō-gram), which is abbreviated as either *ECG* or *EKG*. The action potentials are graphed as a series of up-and-down waves during an ECG. Three clearly recognizable waves normally accompany each cardiac cycle (Figure 14.10b). The first, called the **P wave,** is the spread of an action potential from the SA node through the two atria. A fraction of a second after the P wave begins, the atria contract. The second wave, called

the **QRS wave,** is the spread of the action potential through the ventricles. Shortly after the QRS wave begins, the ventricles contract. The third wave, the **T wave,** indicates ventricular relaxation. There is no wave to show atrial relaxation because the stronger QRS wave masks this event.

Variations in the size and duration of deflection waves of an ECG are useful in diagnosing abnormal cardiac rhythms and conduction patterns and in following the course of recovery from a heart attack. It can also detect the presence of a living fetus.

Sometimes it is necessary to evaluate the heart's response to the stress of physical exercise. Such a test is called a **stress electrocardiogram,** or **stress test.** Although narrowed coronary arteries may carry adequate oxygenated blood while a person is at rest, during strenuous exercise they will be unable to meet the heart's increased need for oxygen, creating changes that can be noted on an electrocardiogram.

CHECKPOINT

11. What are autorhythmic cells?
12. Trace an action potential through the conduction system of the heart.
13. What is an electrocardiogram? What is its diagnostic significance?

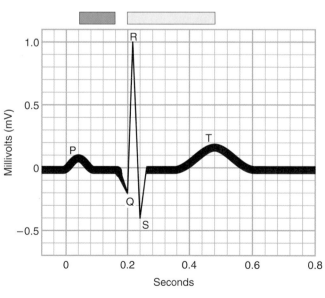

Key:

🟪 Atrial contraction
⬜ Ventricular contraction

(b) Waves associated with a normal electrocardiogram of a single heartbeat

❓ **Which component of the conduction system provides the only electrical connection between the atria and the ventricles?**

CARDIAC CYCLE (HEARTBEAT)

• Describe the phases associated with a cardiac cycle.

The **cardiac cycle** comprises all the events associated with one heartbeat. In a normal cardiac cycle, the two atria contract while the two ventricles relax. Then while the two ventricles contract, the two atria relax. **Systole** (SIS-tō-lē=contraction) refers to the phase of contraction of a chamber of the heart; **diastole** (dī′-AS-tō-lē=dilation or expansion) is the phase of relaxation. For the purposes of our discussion, we will divide the cardiac cycle into the following phases illustrated in Figure 14.10:

❶ **Relaxation period.** At the end of a cardiac cycle when the ventricles start to relax, all four chambers are in diastole. This is the beginning of the *relaxation period.* As the ventricles relax, pressure within the chambers drops, and blood starts to flow from the pulmonary trunk and aorta back toward the ventricles. As this blood becomes trapped in the semilunar cusps, the semilunar valves close. As the ventricles continue to relax, the space inside expands, and the pressure falls. When ventricular pressure drops below atrial pressure, the AV valves open and ventricular filling begins. The major part of ventricular filling (75%) occurs just after the AV valves open. This occurs *without atrial systole.*

❷ **Atrial systole (contraction).** Atrial systole marks the end of the relaxation period and accounts for the remaining 25% of the blood that fills the ventricles. Throughout the period of ventricular filling, the AV valves are still open and the semilunar valves are still closed.

❸ **Ventricular systole (contraction).** Ventricular contraction pushes blood up against the AV valves, forcing them shut. For a very brief period, all four valves are closed again. As

ventricular contraction continues, pressure inside the chambers rises sharply. When left ventricular pressure rises above the pressure in the arteries, both semilunar valves open, and ejection of blood from the heart begins. This lasts until the ventricles start to relax. Then, the semilunar valves close and another relaxation period begins.

14. What is a cardiac cycle?
15. Describe the major events of the phases of a cardiac cycle.

HEART SOUNDS

Auscultation (aws-kul-TĀ-shun; *ausculta-*=listening) is the act of listening to sounds within the body, and it is usually done with a stethoscope. The sound of the heartbeat comes primarily from blood turbulence caused by the closing of the heart valves. Smoothly flowing blood is silent. Compare the sounds made by whitewater rapids or a waterfall to the silence of a smoothly flowing river. During each cardiac cycle, there are four **heart sounds,** but in a normal heart only the first and second heart sounds (S1 and S2) are loud enough to be heard by listening through a stethoscope.

The first sound (S1), which can be described as a **lubb** sound, is louder and a bit longer than the second sound. S1 is caused by blood turbulence associated with closure of the AV valves soon after ventricular systole begins. The second sound (S2), which is shorter and not as loud as the first, can be described as a **dupp** sound. S2 is caused by blood turbulence associated with closure of the SL valves at the beginning of ventricular diastole. Although S1 and S2 are due to blood turbulence associated with the closure of valves, they are best heard at the surface of the chest in locations that are slightly different from the locations of the valves (Figure 14.11). Normally not loud enough to be heard, S3 is due to blood turbulence during rapid ventricular filling, and S4 is due to blood turbulence during atrial systole.

HEART MURMURS

Heart sounds provide valuable information about the mechanical operation of the heart. A **heart murmur** is an abnormal sound consisting of a clicking, rushing, or gurgling noise that is heard before, between, or after the normal heart sounds, or that may mask the normal heart sounds. Although some heart murmurs are "innocent," meaning they are not associated with a significant heart problem, most often a murmur indicates a valve disorder. When a heart valve exhibits stenosis, the heart murmur is heard while the valve should be fully open but is not. For example, mitral stenosis produces a murmur during the relaxation period, between S2 and the next S1. An incompetent heart valve, by contrast, causes a murmur to appear when the valve should be fully closed but is not. So, a murmur due to mitral incompetence occurs during ventricular systole, between S1 and S2. Heart murmurs in children are extremely common and usually do not represent a health condition. Murmurs are most frequently discovered in children between the ages of two and

Figure 14.10 The cardiac cycle (heartbeat).

🔑 A cardiac cycle comprises all the events associated with a single heartbeat.

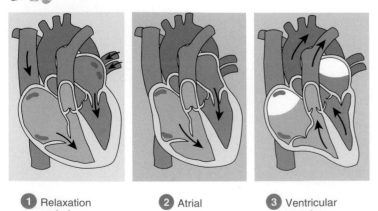

❶ Relaxation period ❷ Atrial systole ❸ Ventricular systole

❓ **What is the status of the four heart valves during ventricular filling?**

Figure 14.11 Location of valves (purple) and auscultation sites (red) for heart sounds.

Listening to sounds within the body is called auscultation; it is usually done with a stethoscope.

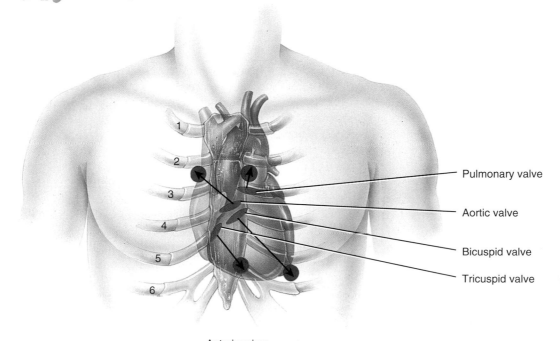

Pulmonary valve

Aortic valve

Bicuspid valve

Tricuspid valve

Anterior view

Which heart sound is related to blood turbulence associated with closure of the AV valves?

four. These types of heart murmurs are referred to as *innocent* or *functional heart murmurs*; they often subside or disappear with growth. ■

CHECKPOINT

16. Where are the two heart sounds best heard on the surface of the chest?

EXERCISE AND THE HEART

OBJECTIVE

● Explain the relationship between exercise and the heart.

A person's fitness, regardless of level, can be improved at any age with regular exercise. Of the various types of exercise, some are more effective than others for improving the health of the cardiovascular system. **Aerobics,** any activity that works large body muscles for at least 20 minutes, elevates cardiac output and accelerates metabolic rate. Three to five such sessions a week are usually recommended for improving the health of the cardiovascular system. Brisk walking, running, bicycling, cross-country skiing, and swimming are examples of aerobic activities.

Sustained exercise increases the oxygen demand of the muscles. Whether the demand is met depends primarily on the adequacy of cardiac output and proper functioning of the respiratory system. After several weeks of training, a healthy person increases maximal cardiac output, thereby increasing the maximal rate of oxygen delivery to the tissues. Oxygen delivery also

rises because hemoglobin level increases and skeletal muscles develop more capillary networks in response to long-term training.

During strenuous activity, a well-trained athlete can achieve a cardiac output double that of a sedentary person, in part because training causes hypertrophy (enlargement) of the heart. Even though the heart of a well-trained athlete is larger, *resting* cardiac output is about the same as in a healthy untrained person, because stroke volume is increased while heart rate is decreased. The resting heart rate of a trained athlete often is only 40–60 beats per minute. Regular exercise also helps to reduce blood pressure, anxiety, and depression; control weight; and increase the body's ability to dissolve blood clots by increasing fibrinolytic activity.

HELP FOR FAILING HEARTS

As the heart fails, a person has decreasing ability to exercise or even to move around. A variety of surgical techniques and medical devices exist to aid a failing heart. For some patients, even a 10% increase in the volume of blood ejected from the ventricles can mean the difference between being bedridden and having limited mobility. **Heart transplants** are common today and produce good results, but the availability of donor hearts is very limited. There are 50 potential candidates for each of the 2500 donor hearts available each year in the United States. In the procedure, most of the diseased heart is removed, but the posterior walls of the atria are retained, and the ventricles of the donor heart are then attached to the remaining areas of the recipient's heart. Another approach is to use **cardiac assist devices** and surgical procedures that augment heart function without removing the heart. Table 14.1 describes several of these.

Finally, scientists continue to develop and refine **artificial hearts,** mechanical devices that completely replace the functions of the natural heart. During the 1980s several patients received a Jarvik-7 artificial heart, which used an external power source to drive an internal pump by compressed air. In 1990, the U.S. Food and Drug Administration (FDA) banned use of the device because persistent problems with blood clotting caused strokes, and the chest tube for compressed air led to infections. More than a decade later, in July 2001, the first person received a fully self-contained artificial heart called the AbioCor Implantable Replacement Heart. Made of titanium, plastic, and epoxy, the 2-pound AbioCor heart is powered by a battery pack worn outside but with no wires penetrating through the skin. It alternately pumps blood from the left and then the right side of the heart. Because a permanent opening through the chest is not needed, the risk of infection is much lower than with the Jarvik-7. Without the artificial heart, the first recipient was given little chance to survive more than a month due to his congestive heart failure, kidney disease, and diabetes. After surgery, he lived for 151 days (almost 5 months), recovering enough to give several interviews and to enjoy a fishing trip. However, internal bleeding and organ failure unrelated to the AbioCor heart caused his death. ■

CHECKPOINT

17. What are some of the cardiovascular benefits of regular exercise?

DEVELOPMENT OF THE HEART

OBJECTIVE

● Describe the development of the heart.

The cardiovascular system is the first system to form and function in an embryo and the heart is the first functional organ. This is necessary because the rapidly growing embryo needs an efficient way to obtain oxygen and nutrients and get rid of waste. Recall that oxygen and nutrients in the mother's intervillous spaces diffuse into embryonic chorionic villi and wastes diffuse in the opposite direction as early as 21 days after fertilization (see Figure 4.10 on page 107). Blood vessels in the chorionic villi connect to the embryonic heart by way of the umbilical arteries and umbilical vein (see Figure 4.10c on page 107).

As we trace the development of the heart, keep in mind that many congenital (present at birth) disorders of the heart develop during embryonic life. Such disorders, which are described on pages 447–448, are responsible for almost half of all deaths from birth defects.

The *heart* begins its development from splanchnic **mesoderm** on day 18 or 19 following fertilization. In the head end of the embryo, the heart develops from a group of mesodermal cells called the **cardiogenic area** (kar-dē-ō-JEN-ik; *cardio-*= heart; *-genic*=producing) (Figure 14.12a). In response to induction signals from the underlying endoderm, the mesoderm in the cardiogenic area forms a pair of elongated strands called **cardiogenic cords.** Shortly after, these cords develop a hollow center and then become known as **endocardial tubes** (Figure 14.12b). With lateral folding of the embryo, the paired endocardial tubes approach each other and fuse into a single tube called

TABLE 14.1	CARDIAC ASSIST DEVICES AND PROCEDURES
DEVICE	**DESCRIPTION**
Intra-aortic balloon pump (IABP)	A 40-mL polyurethane balloon mounted on a catheter is inserted into an artery in the groin and threaded into the thoracic aorta. An external pump inflates the balloon with gas at the beginning of ventricular diastole. As the balloon inflates, it pushes blood both backward toward the heart, which improves coronary blood flow, and forward toward peripheral tissues. The balloon then is rapidly deflated just before the next ventricular systole, making it easier for the left ventricle to eject blood. Because the balloon is inflated between heartbeats, this technique is called intra-aortic balloon counterpulsation.
Hemopump	This propeller-like pump is threaded through an artery in the groin and then into the left ventricle. There, the blades of the pump whirl at about 25,000 revolutions per minute, pulling blood out of the left ventricle and pushing it into the aorta.
Left ventricular assist device (LVAD)	The LVAD is a completely portable assist device. It is implanted within the abdomen and powered by a battery pack worn in a shoulder holster. The LVAD is connected to the patient's weakened left ventricle and pumps blood into the aorta. The pumping rate increases automatically during exercise.
Cardiomyoplasty	A large piece of the patient's own skeletal muscle (left latissimus dorsi) is partially freed from its connective tissue attachments and wrapped around the heart, leaving the blood and nerve supply intact. An implanted pacemaker stimulates the skeletal muscle's motor neurons to cause contraction 10–20 times per minute, in synchrony with some of the heartbeats.
Skeletal muscle assist device	A piece of the patient's own skeletal muscle is used to fashion a pouch that is inserted between the heart and the aorta, functioning as a booster heart. A pacemaker stimulates the muscle's motor neurons to elicit contraction.

the **primitive heart tube** on day 21 following fertilization (Figure 14.12c).

On the twenty-second day, the primitive heart tube develops into five distinct regions and begins to pump blood. From tail end to head end (and the direction of blood flow) they are the (1) **sinus venosus,** (2) **atrium,** (3) **ventricle,** (4) **bulbus cordis,** and (5) **truncus arteriosus.** The sinus venosus initially receives blood from all the veins in the embryo; contractions of the heart begin in this region and follow sequentially in the other regions. Thus, at this stage, the heart consists of a series of unpaired regions. The fates of the regions are as follows:

1. The sinus venosus develops into part of the *right atrium, coronary sinus,* and *sinoatrial (SA) node.*
2. The atrium develops into part of the *right atrium* and the *left atrium.*
3. The ventricle gives rise to the *left ventricle.*
4. The bulbus cordis develops into the *right ventricle.*
5. The truncus arteriosus gives rise to the *ascending aorta* and *pulmonary trunk.*

On day 23, the primitive heart tube elongates. Because the bulbus cordis and ventricle grow more rapidly than other parts of the tube and because the atrial and venous ends of the tube are confined by the pericardium, the tube begins to loop and fold. At first, the primitive heart tube assumes a U-shape; later it becomes S-shaped (Figures 14.12e). As a result of these movements, which are completed by day 28, the atria and ventricles of the future heart are reoriented to assume their final adult positions. The remainder of heart development consists of reconstruction of the chambers and the formation of septa and valves to form a four-chambered heart.

On about day 28, thickenings of mesoderm of the inner lining of the heart wall, called **endocardial cushions,** appear (Figure 14.13). They grow toward each other, fuse, and divide the single **atrioventricular canal** (region between atria and ventricles) into smaller, separate left and right atrioventricular canals. Also, the *interatrial septum* begins its growth toward the fused endocardial cushions. Ultimately, the interatrial septum and endocardial cushions unite and an opening in the septum, the **foramen ovale,** develops. The interatrial septum divides the

Figure 14.12 Development of the heart. Arrows within the structures indicate the direction of blood flow.

The heart begins its development from mesoderm on day 18 or 19 following fertilization.

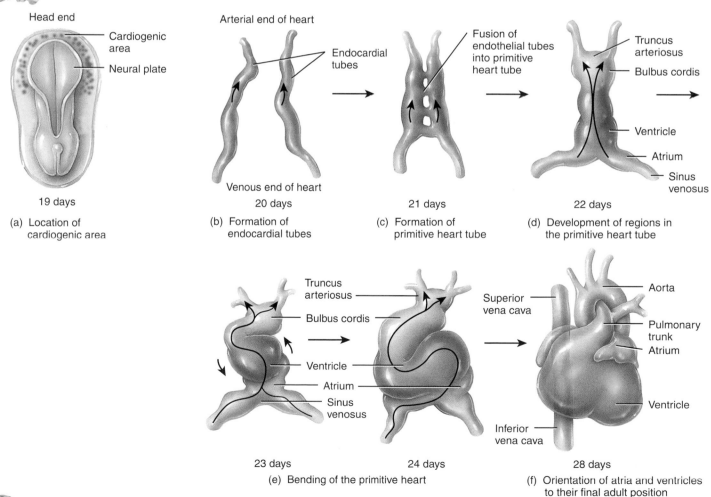

(a) Location of cardiogenic area — 19 days

(b) Formation of endocardial tubes — 20 days

(c) Formation of primitive heart tube — 21 days

(d) Development of regions in the primitive heart tube — 22 days

(e) Bending of the primitive heart — 23 days, 24 days

(f) Orientation of atria and ventricles to their final adult position — 28 days

When during embryonic development does the primitive heart begin to contract?

Figure 14.13   Partitioning of the heart into four chambers.

Partitioning of the heart begins on about the 28th day after fertilization.

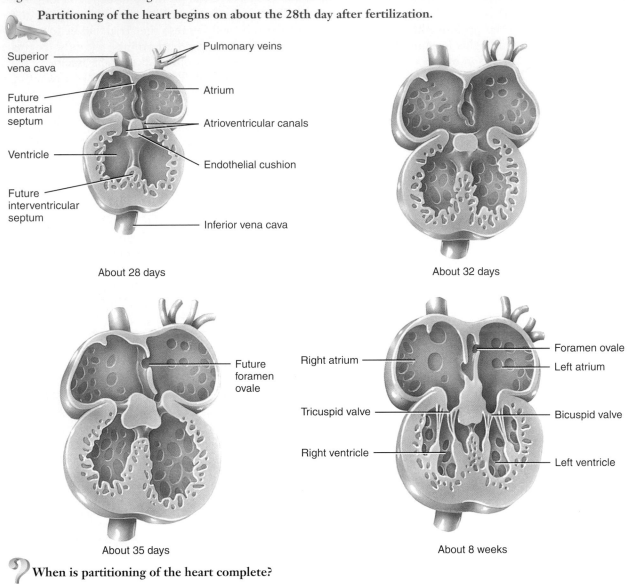

When is partitioning of the heart complete?

atrial region into a *right atrium* and a *left atrium*. Before birth, the foramen ovale allows most blood entering the right atrium to pass into the left atrium. After birth, it normally closes so that the interatrial septum is a complete partition. The remnant of the foramen ovale is the fossa ovalis (see Figure 14.4a). Formation of the *interventricular septum* partitions the ventricular region into a *right ventricle* and a *left ventricle*. Partitioning of the atrioventricular canal, atrial region, and ventricular region is

basically complete by the end of the fifth week. The *atrioventricular valves* form between the fifth and eighth weeks. The *semilunar valves* form between the fifth and ninth weeks.

CHECKPOINT

18. Why is the cardiovascular system the first of all the systems to develop?
19. From which tissue does the heart develop?

APPLICATIONS TO HEALTH

CORONARY ARTERY DISEASE

Coronary artery disease (CAD) is a serious medical problem that annually affects about 7 million people. Responsible for nearly

three quarters of a million deaths in the United States each year, it is the leading cause of death for both men and women. CAD is defined as the effects of the accumulation of atherosclerotic plaques (described shortly) in coronary arteries that lead to a reduction in blood flow to the myocardium. Some individuals have no signs or symptoms; others experience angina pectoris (chest pain), and still others suffer heart attacks.

Risk Factors for CAD

People who possess combinations of certain risk factors are more likely to develop CAD. *Risk factors* (characteristics, symptoms, or signs present in a disease-free person that are statistically associated with a greater chance of developing a disease) include smoking, high blood pressure, diabetes, high cholesterol levels, obesity, "type A" personality, sedentary lifestyle, and family history of CAD. Most of these are modifiable; that is, they can be altered by changing diet and other habits or can be controlled by taking medications. However, other risk factors are unmodifiable—that is, beyond our control—including genetic predisposition (family history of CAD at an early age), age, and gender. (For example, adult males are more likely than adult females to develop CAD; after age 70 the risks are roughly equal.) Smoking is undoubtedly the number-one risk factor in all CAD-associated diseases (ischemic heart disease, strokes, etc.). Smoking roughly doubles the risk of morbidity and mortality.

Development of Atherosclerotic Plaques

Although the following discussion of atherosclerosis applies to coronary arteries, the process can also occur in arteries outside the heart. Thickening of the walls of arteries and loss of elasticity are the main characteristics of a group of diseases referred to as **arteriosclerosis** (ar-tē-rē-ō-skle-RŌ-sis; *sclero-*=hardening). One form of arteriosclerosis is **atherosclerosis** (ath-er-ō-skle-RŌ-sis), a progressive disease characterized by the formation in the walls of large and medium-sized arteries of lesions called **atherosclerotic plaques** (Figure 14.14).

To understand how atherosclerotic plaques develop, you will need to know about molecules produced by the liver and small intestine called **lipoproteins.** These spherical particles consist of an inner core of triglycerides and other lipids and an outer shell of proteins, phospholipids, and cholesterol. Most lipids, including cholesterol, do not dissolve in water and must be made water-soluble in order to be transported in the blood. This is accomplished by combining them with lipoproteins. Two major lipoproteins are called **low-density lipoproteins (LDLs)** and **high-density lipoproteins (HDLs).** LDLs transport cholesterol from the liver to body cells for use in cell membrane repair and the production of steroid hormones and bile salts. However, excessive amounts of LDLs promote atherosclerosis, so the cholesterol in these particles is known as "bad cholesterol." HDLs, on the other hand, remove excess cholesterol from body cells and transport it to the liver for elimination. Because HDLs decrease blood cholesterol level, the cholesterol in HDLs is known as "good cholesterol." Basically, you want your LDL to be low and your HDL to be high.

It has recently been learned that inflammation, a defensive response of the body to tissue damage, plays a key role in the development of atherosclerotic plaques. As a result of tissue damage, blood vessels dilate and increase their permeability, and phagocytes, including macrophages, appear in large numbers. The formation of atherosclerotic plaques begins when excess LDLs from the blood accumulate in the inner layer of an artery wall (layer closest to the bloodstream) and the lipids and proteins in the LDLs undergo oxidation and the proteins also bind to sugars (glycation). In response, endothelial and smooth muscle cells of the artery secrete substances that attract monocytes from the blood and convert them into macrophages. The macrophages then ingest and become so filled with the oxidized LDL particles that they have a foamy appearance when viewed microscopically

Figure 14.14 Photomicrographs of a transverse section of (a) a normal artery and (b) one partially obstructed by an atherosclerotic plaque.

Inflammation plays a key role in the development of atherosclerotic plaques.

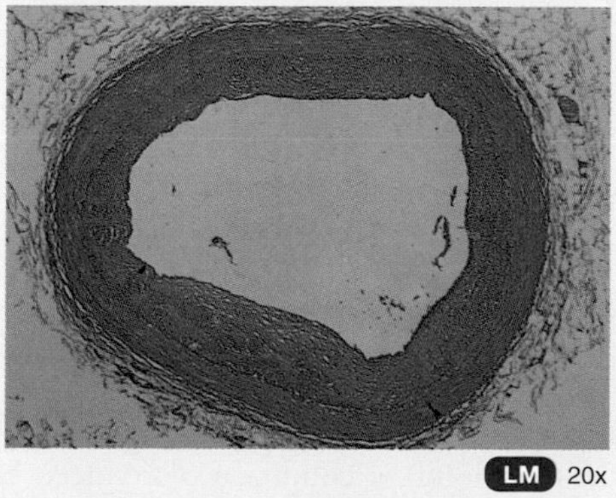

LM 20x

(a) Normal artery

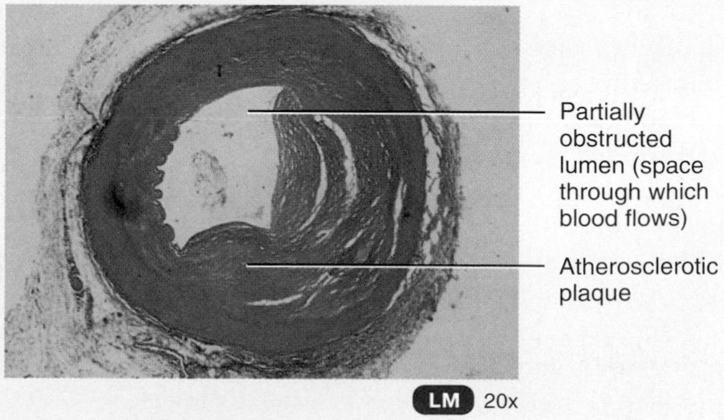

LM 20x

Partially obstructed lumen (space through which blood flows)

Atherosclerotic plaque

(b) Obstructed artery

? **What is the role of HDL?**

(**foam cells**). T cells (lymphocytes) follow monocytes into the inner lining of an artery and there release chemicals that intensify the inflammatory response. Together, the foam cells, macrophages, and T cells form a **fatty streak,** the beginning of an atherosclerotic plaque.

In most inflammatory responses, macrophages release chemicals that promote healing following fatty streak formation. However, macrophages and endothelial cells secrete chemicals that cause smooth muscle cells of the middle layer of an artery to migrate to the top of the atherosclerotic plaque, forming a cap over it and thus walling it off from the blood.

Because most atherosclerotic plaques expand away from the bloodstream rather than into it, blood can flow through an artery with relative ease, often for decades. Only about 15% of heart attacks occur when plaque in a coronary artery expands into the bloodstream and restricts blood flow. Most heart attacks occur when the cap over the plaque breaks open in response to chemicals produced by foam cells. In addition, T cells induce foam cells to produce tissue factor (TF), a chemical that begins the cascade of reactions that results in blood clot formation. If the clot in a coronary artery is large enough, it can significantly decrease or stop the flow of blood and result in a heart attack.

In addition to LDL, several other factors are implicated in the development of atherosclerosis. For example, in diabetes, blood sugar level is high and the excess glucose can bind to proteins in LDL (glycation), thus enhancing inflammation. The carbon monoxide in cigarette smoke may speed up the oxidation of lipids and proteins in LDL and trigger inflammation. Many cases of hypertension (high blood pressure) are associated with the hormone angiotensin II, which also enhances inflammation. Certain microbes, such as the herpes virus and *Chlamydia pneumoniae* (the causative agent of many respiratory infections), can also trigger the inflammatory response. Despite all of the factors that may trigger inflammation, HDL may reduce inflammation by transporting enzymes that can break down oxidized lipids.

Diagnosis of CAD

Cardiac catheterization (kath′-e-ter-i-ZĀ-shun) is an invasive procedure used to visualize the heart's coronary arteries, chambers, valves, and great vessels. It may also be used to measure pressure in the heart and blood vessels; to assess function, cardiac output, and diastolic properties of the left ventricle; to measure the flow of blood through the heart and blood vessels, the oxygen content of blood, and the status of heart valves and conduction system; and to identify the exact location of septal and valvular defects. The basic procedure involves inserting a long, flexible, radiopaque **catheter** (plastic tube) into a peripheral vein (for right heart catheterization) or a peripheral artery (for left heart catheterization) and guiding it under fluoroscopy (x-ray observation).

Coronary angiography (an′-jē-OG-ra-fē) is another invasive procedure in which a cardiac catheter is used to inject a radiopaque contrast medium into blood vessels or heart chambers. The procedure may be used to visualize coronary arteries, the aorta, pulmonary blood vessels, and the ventricles to assess

structural abnormalities in blood vessels (such as atherosclerotic plaques and emboli), ventricular volume, wall thickness, and wall motion. The procedure can also be used to inject clot-dissolving drugs, such as streptokinase or tissue plasminogen activator (t-PA), into a coronary artery to dissolve an obstructing thrombus.

Treatment of CAD

Treatment options for CAD include drugs (antihypertensives, nitroglycerine, beta blockers, and cholesterol-lowering and clot-dissolving agents) and various surgical and nonsurgical procedures designed to increase the blood supply to the heart.

Coronary artery bypass grafting (CABG) is a surgical procedure in which a blood vessel from another part of the body is attached ("grafted") to a coronary artery to bypass an area of blockage. A piece of the grafted blood vessel is sutured between the aorta and the unblocked portion of the coronary artery (Figure 14.15a).

A nonsurgical procedure used to treat CAD is termed **percutaneous transluminal coronary angioplasty (PTCA)** (*percutaneous*=through the skin; *trans-*=across; *lumen*=an opening or channel in a tube; *angio-*=blood vessel; *-plasty*=to mold or to shape). In this procedure, a balloon catheter (plastic tube) is inserted into an artery of an arm or leg and gently guided into a coronary artery (Figure 14.15b). Then, while dye is released, angiograms (x-rays of blood vessels) are taken to locate the plaques. Next, the catheter is advanced to the point of obstruction, and a balloonlike device is inflated with air to squash the plaque against the blood vessel wall. Because about 30–50% of PTCA-opened arteries fail due to restenosis (renarrowing) within six months after the procedure is done, a special device called a stent may be inserted via a catheter. A **stent** is a stainless steel device, resembling a spring coil, that is permanently placed in an artery to keep the artery patent (open), permitting blood to circulate (Figure 14.15c). Restenosis may be due to damage from the procedure itself, for PTCA may damage the coronary artery wall, leading to platelet activation, proliferation of smooth muscle fibers, and plaque formation.

One area of current research involves cooling the body's core temperature during procedures such as coronary artery bypass grafting (CABG) and during other vascular disease processes or procedures. For example, there have been some promising results from the application of cold therapy during a cerebral vascular accident (CVA or stroke). This research stemmed from observations of people who had suffered a hypothermic incident (such as cold water drowning) and recovered with relatively minimal neurologic deficits.

MYOCARDIAL ISCHEMIA AND INFARCTION

Partial obstruction of blood flow in the coronary arteries may cause **myocardial ischemia** (is-KĒ-mē-a; *ische-*=to obstruct; *-emia*=in the blood), a condition of reduced blood flow to the myocardium. Usually, ischemia causes **hypoxia** (reduced oxygen supply), which may weaken cells without killing them. **Angina pectoris** (an-JĪ-na, or AN-ji-na, PEK-to-ris), literally meaning

Figure 14.15 Procedures for reestablishing blood flow in occluded coronary arteries.

Treatment options for CAD include drugs and various nonsurgical and surgical procedures.

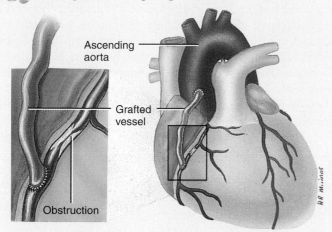

(a) Coronary artery bypass grafting (CABG)

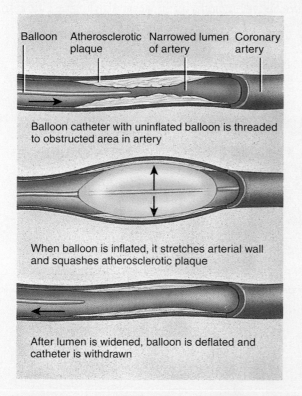

Balloon catheter with uninflated balloon is threaded to obstructed area in artery

When balloon is inflated, it stretches arterial wall and squashes atherosclerotic plaque

After lumen is widened, balloon is deflated and catheter is withdrawn

(b) Percutaneous transluminal coronary angioplasty (PTCA)

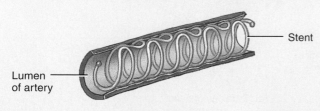

(c) Stent in an artery

Which diagnostic procedure for CAD is used to visualize coronary blood vessels?

"strangled chest," is a severe pain that usually accompanies myocardial ischemia. Typically, sufferers describe it as a tightness or squeezing sensation, as though the chest were in a vise. Angina pectoris often occurs during exertion, when the heart demands more oxygen, and then disappears with rest. The pain associated with angina pectoris is often referred to the neck, chin, or down the left arm to the elbow. In some people, ischemic episodes occur without producing pain. This condition is known as **silent myocardial ischemia** and is particularly dangerous because the person has no forewarning of an impending heart attack.

A complete obstruction to blood flow in a coronary artery may result in a **myocardial infarction** (in-FARK-shun), or **MI**, commonly called a heart attack. *Infarction* means the death of an area of tissue because of interrupted blood supply. Because the heart tissue distal to the obstruction dies and is replaced by noncontractile scar tissue, the heart muscle loses some of its strength. The aftereffects depend partly on the size and location of the infarcted (dead) area. Besides killing normal heart tissue, an infarction may disrupt the conduction system of the heart and cause sudden death by triggering ventricular fibrillation. Treatment for a myocardial infarction may involve injection of a thrombolytic (clot-dissolving) agent such as streptokinase or t-PA, plus heparin (an anticoagulant), or performing coronary angioplasty or coronary artery bypass grafting. Fortunately, heart muscle can remain alive in a resting person if it receives as little as 10–15% of its normal blood supply. Furthermore, the extensive anastomoses of coronary blood vessels are a significant factor in helping some people survive myocardial infarctions.

Recent studies have shown that women and men experience MIs in different ways. Women and men with acute coronary syndromes (such as an MI) have different clinical profiles, presentation, and outcomes. These differences may reflect pathophysiological and anatomical differences between women and men. Women and men also have different risk factors and symptoms. Women have significantly higher rates of dyspnea (painful or labored breathing), nausea, and epigastric pain, but less diaphoresis (profuse perspiration) than men. After MI, younger women have higher rates of death during hospitalization than men of the same age (this difference does not appear in older women and men). The younger the age of the female patient, the higher the risk of death. With the rise in awareness by the scientific research community that disease affects different populations in different ways, medical studies are now including a wider cross-section of the general population.

CONGENITAL HEART DEFECTS

A defect that is present at birth, and usually before, is called a **congenital defect.** Many such defects are not serious and may go unnoticed for a lifetime. Others are life-threatening and must be surgically repaired. Among the several congenital defects that affect the heart are the following (Figure 14.16):

- **Coarctation** (kō′-ark-TĀ-shun) **of the aorta.** In this condition, a segment of the aorta is too narrow, and thus the flow of oxygenated blood to the body is reduced, the left ventricle is forced to pump harder, and high blood pressure develops. Coarctation is usually repaired surgically by removing the area of obstruction. Surgical interventions that are done in childhood may require revisions in adulthood. Another surgical procedure is balloon dilatation, insertion and inflation of a device in the aorta to stretch the vessel. A metal coil (stent) can be inserted and left in place to hold the vessel open.

- **Patent ductus arteriosus (PDA).** In some babies, the ductus arteriosus, a temporary blood vessel between the aorta and the pulmonary trunk, remains open rather than closing shortly after birth. As a result, aortic blood flows into the lower-pressure pulmonary trunk, thus increasing the pulmonary trunk blood pressure and overworking both ventricles. In uncomplicated PDA, medication can be used to facilitate the closure of the defect. In more severe cases, surgical intervention may be required.

- **Septal defect.** A septal defect is an opening in the septum that separates the interior of the heart into left and right sides. In an **interatrialseptal defect** the fetal foramen ovale between the two atria fails to close after birth. An **interventricularseptal defect** is caused by incomplete development of the interven-

tricular septum. In such cases, oxygenated blood flows directly from the left ventricle into the right ventricle, where it mixes with deoxygenated blood. The condition is treated surgically.

- **Tetralogy of Fallot** (tet-RAL-ō-jē of fal-Ō). This condition is a combination of four developmental defects: an interventricular septal defect, an aorta that emerges from both ventricles instead of from the left ventricle only, a stenosed pulmonary valve, and an enlarged right ventricle. There is a decreased flow of blood to the lungs and mixing of blood from both sides of the heart. This causes cyanosis, the bluish discoloration most easily seen in nail beds and mucous membranes when the level of deoxygenated hemoglobin is high. For this reason, tetralogy of Fallot is one of the conditions that cause a "blue baby." Despite the apparent complexity of this condition, it can usually be completely repaired via surgery.

ARRHYTHMIAS

The usual rhythm of heartbeats, established by the SA node, is **normal sinus rhythm.** An **arrhythmia** (a-RITH-mē-a), or **dysrhythmia,** is an irregularity in heart rhythm resulting from a

Figure 14.16 Congenital heart defects.

A congenital defect is one that is present at birth, and usually before.

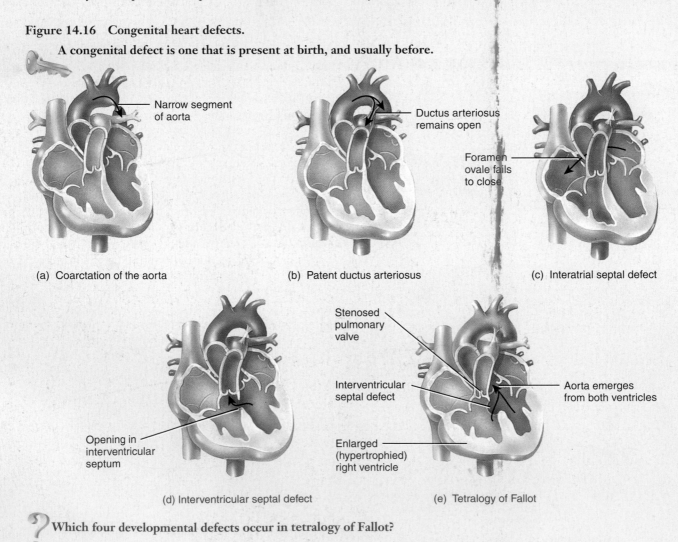

(a) Coarctation of the aorta   Narrow segment of aorta

(b) Patent ductus arteriosus   Ductus arteriosus remains open

(c) Interatrial septal defect   Foramen ovale fails to close

(d) Interventricular septal defect   Opening in interventricular septum

(e) Tetralogy of Fallot   Stenosed pulmonary valve   Interventricular septal defect   Aorta emerges from both ventricles   Enlarged (hypertrophied) right ventricle

Which four developmental defects occur in tetralogy of Fallot?

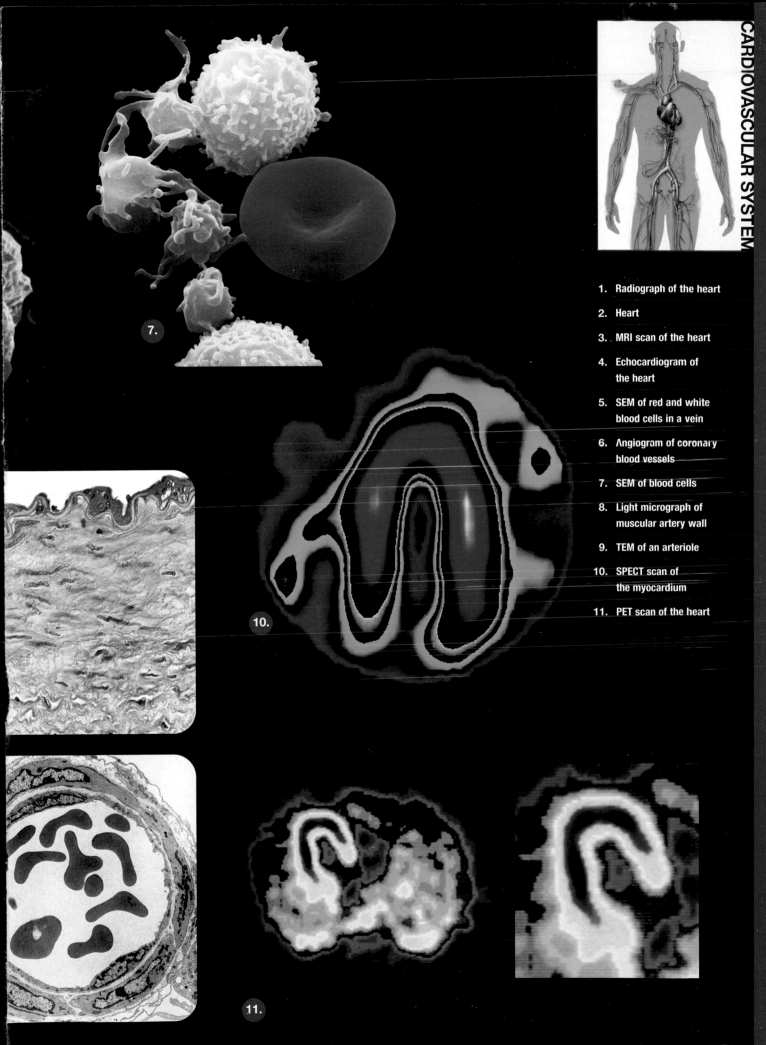

1. Radiograph of the heart

2. Heart

3. MRI scan of the heart

4. Echocardiogram of the heart

5. SEM of red and white blood cells in a vein

6. Angiogram of coronary blood vessels

7. SEM of blood cells

8. Light micrograph of muscular artery wall

9. TEM of an arteriole

10. SPECT scan of the myocardium

11. PET scan of the heart

CARDIOVASCULAR SYSTEM

1. **Radiograph of the heart.** Radiograph of the heart in anterior view.

2. **Heart.** Heart seen in anterior view of frontal section. Note the lungs on either side of the heart.

3. **MRI scan of the heart.** Magnetic Resonance Image (MRI) scan of the heart in frontal section.

4. **Echocardiogram of the heart.** Echocardiogram of the heart with an EKG tracing.

5. **SEM of red and white blood cells in a vein.** Scanning Electron Micrograph (SEM) of red and white blood cells in a vein (saphenous).

6. **Angiogram of coronary blood vessels.** Angiogram of coronary blood vessels.

7. **SEM of blood cells.** Scanning Electron Micrograph (SEM) of blood cells.

8. **Light micrograph of muscular artery wall.** Light micrograph of the wall of a muscular artery.

9. **TEM of an arteriole.** Transmission Electron Micrograph (TEM) of an arteriole.

10. **SPECT scan of the myocardium.** Single Photon Emission Computed Tomography (SPECT) scan of the myocardium in transverse section.

11. **PET scan of the heart.** Positron Emission Tomography (PET) scan of the heart.

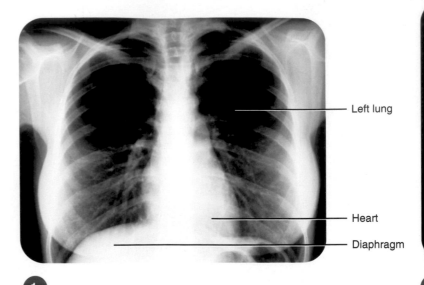

Left lung

Heart

Diaphragm

1.

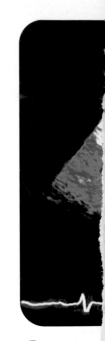

3.

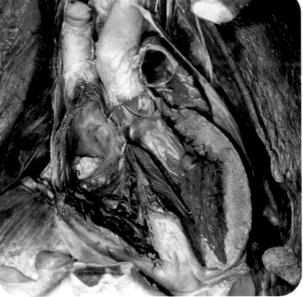

2.

4.

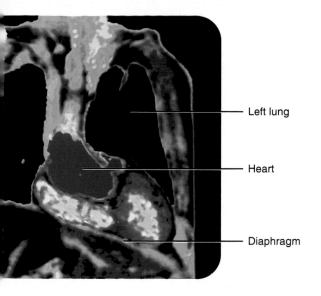

Left lung

Heart

Diaphragm

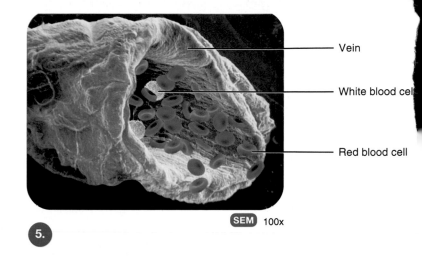

Vein

White blood cell

Red blood cell

SEM 100x

5.

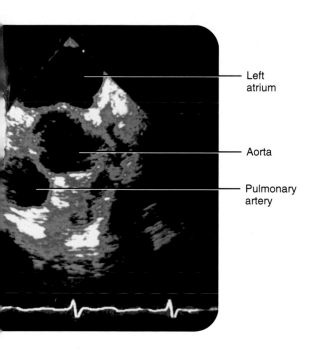

Left atrium

Aorta

Pulmonary artery

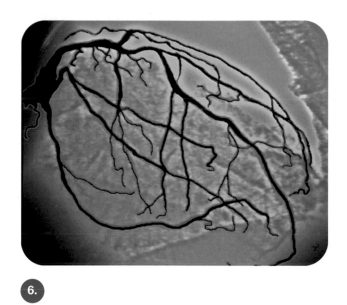

6.

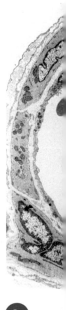

8.

9.

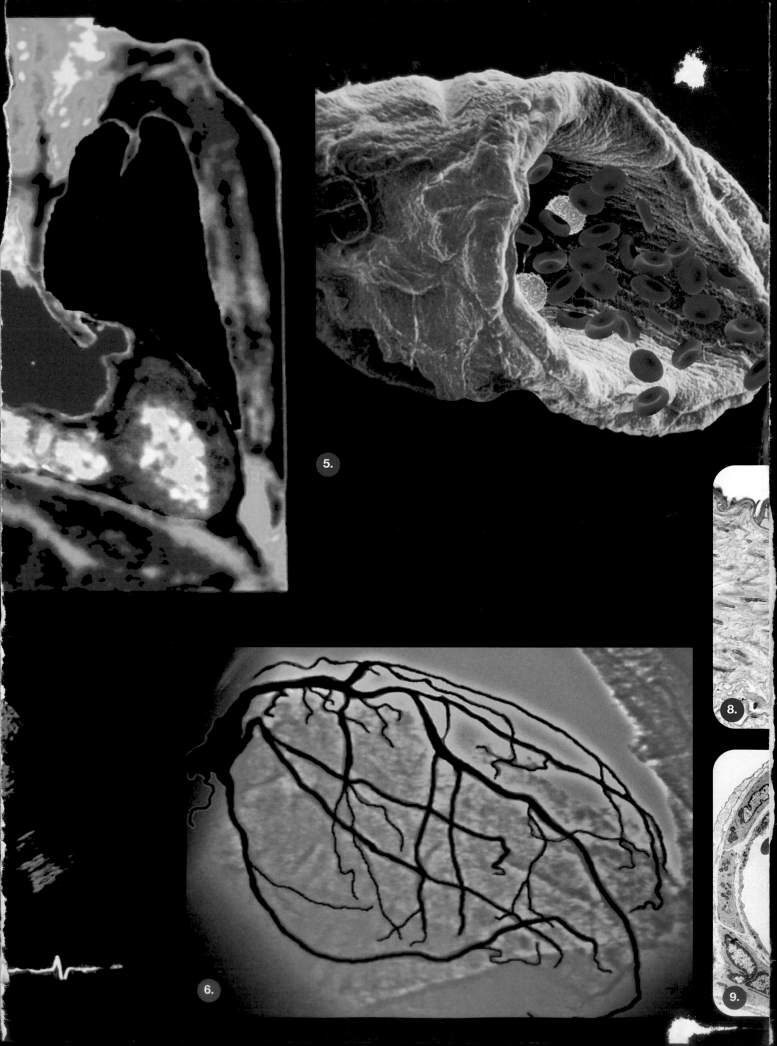

5.

6.

8.

9.

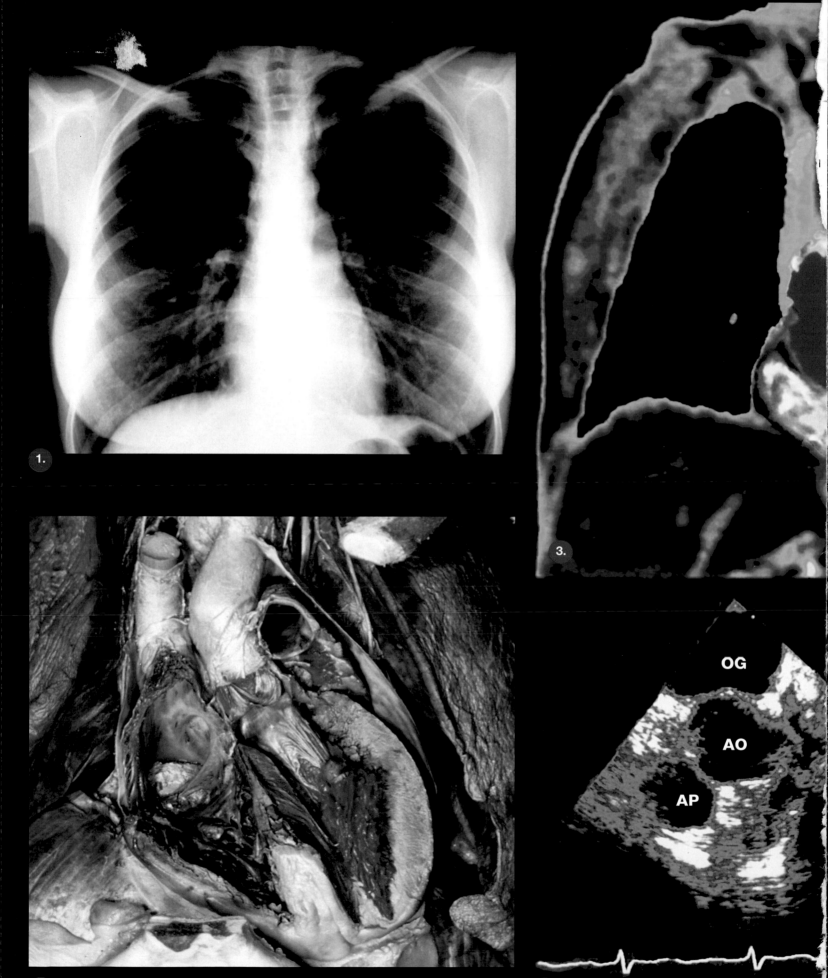

1.

2.

3.

4. OG

AO

AP

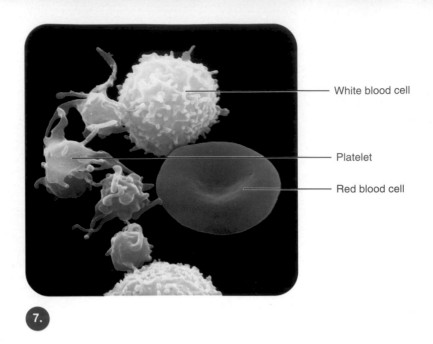

White blood cell

Platelet

Red blood cell

7.

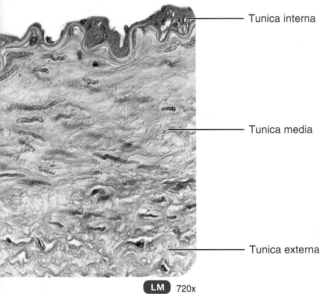

Tunica interna

Tunica media

Tunica externa

LM 720x

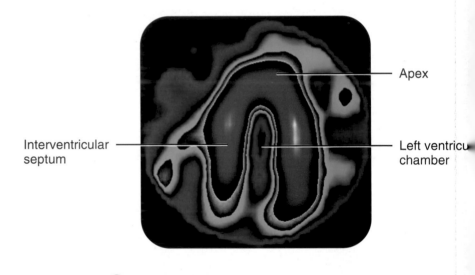

Apex

Interventricular septum

Left ventricu... chamber

10.

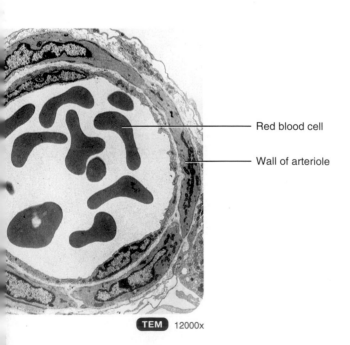

Red blood cell

Wall of arteriole

TEM 12000x

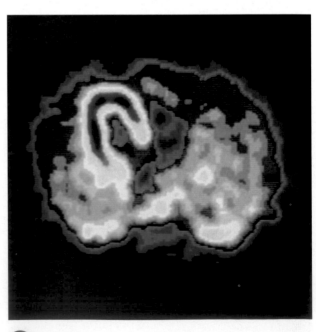

11.

defect in the conduction system of the heart. Arrhythmias can be caused by caffeine, nicotine, alcohol, and certain other drugs; anxiety; hyperthyroidism; potassium deficiency; and certain heart diseases.

Heart Block

One serious arrhythmia is **heart block,** in which propagation of action potentials through the conduction system is either slowed or blocked at some point. The most common site of blockage is the atrioventricular node, a condition called **atrioventricular (AV) block.** In *first-degree AV block,* the P-Q interval is prolonged, usually because conduction through the AV node is slower than normal. In *second-degree AV block,* some of the action potentials from the SA node are not conducted through the AV node. The result is "dropped" beats because excitation doesn't always reach the ventricles. In *third-degree (complete) AV block,* no SA node action potentials get through the AV node. Autorhythmic fibers in the atria and ventricles pace the upper and lower chambers separately. With complete AV block, the ventricular contraction rate is less than 40 beats/min.

Flutter and Fibrillation

Two other abnormal rhythms are **flutter** and **fibrillation. Atrial flutter** consists of rapid atrial contractions (240–360 beats/min) accompanied by a second-degree AV block. Flutter may result from rheumatic heart disease, coronary artery disease, or certain congenital heart diseases. In **atrial fibrillation,** contraction of atrial fibers is asynchronous (not in unison) so that atrial pumping ceases altogether. Atrial fibrillation may occur as a result of myocardial infarction, acute and chronic rheumatic heart disease, hypertension, and hyperthyroidism. In an otherwise strong heart, atrial fibrillation reduces the pumping effectiveness of the heart by 20–30%.

Ventricular fibrillation is the most deadly arrhythmia, in which contractions of ventricular muscle fibers are completely asynchronous. Ventricular pumping stops, blood ejection ceases, and circulatory failure and death occur unless the situation is quickly corrected. A procedure known as **cardioversion** (kar′-dē-ō-VER-zhun), also called **defibrillation,** in which a strong, brief electrical current is passed through the heart, often can stop ventricular fibrillation. The electric shock is applied by a device called a **defibrillator** via large paddle-shaped electrodes pressed against the skin of the chest. The shock is syncronized to the QRS complex. More than 250,000 people in the United States suffer from sudden cardiac arrest (SCA) every year. Fewer

than 5 percent of those who experience an SCA "in the field" (outside a health care facility) survive, because of a lack of access to a defibrillator. In response, **automated external defibrillators (AEDs)** have been developed. These devices allow rapid defibrillation by virtually any bystander. AEDs are becoming more and more visible in airports, shopping malls, and other public places. Patients who face a high risk of dying from heart rhythm disorders) now can receive an **automatic implantable cardioverter defibrillator (AICD),** an implanted device that monitors their heart rhythm and delivers a small shock directly to the heart when a life-threatening rhythm disturbance occurs. Thousands of patients around the world have AICDs, including Dick Cheney, Vice President of the United States, who received a combination pacemaker-defibrillator in 2001.

Ventricular Premature Contraction

Another form of arrhythmia arises when an *ectopic focus,* a region of the heart other than the conduction system, becomes more excitable than normal and causes an occasional abnormal action potential to occur. As a wave of depolarization spreads outward from the ectopic focus, it causes a **ventricular premature contraction.** The contraction occurs early in diastole before the SA node is normally scheduled to discharge its action potential. Ventricular premature contractions may be relatively benign and may be caused by emotional stress, excessive intake of stimulants such as caffeine or nicotine, and lack of sleep. In other cases, the premature beats may reflect an underlying pathology.

CONGESTIVE HEART FAILURE

In **congestive heart failure (CHF),** the heart is a failing pump. Causes of CHF include coronary artery disease, congenital defects, long-term high blood pressure, myocardial infarctions (regions of dead heart tissue due to a previous heart attack), and valve disorders. As the pump becomes less effective, more blood remains in the ventricles at the end of each cycle. Often, one side of the heart starts to fail before the other. If the left ventricle fails first, it can't pump out all the blood it receives. As a result, blood backs up in the lungs and causes *pulmonary edema,* fluid accumulation in the lungs that can suffocate an untreated person. If the right ventricle fails first, blood backs up in the systemic veins and, over time, the kidneys cause an increase in blood volume. In this case, the resulting *peripheral edema* usually is most noticeable in the feet and ankles.

KEY MEDICAL TERMS ASSOCIATED WITH THE HEART

Asystole (ā-SIS-tō-lē; *a-*=without) Failure of the myocardium to contract.

Cardiac arrest (KAR-dē-ak a-REST) A clinical term meaning cessation of an effective heartbeat. The heart may be completely stopped or in ventricular fibrillation.

Cardiomegaly (kar′-dē-ō-MEG-a-lē; *mega*=large) Heart enlargement.

Cardiac rehabilitation (rē-ha-bil-i-TĀ-shun) A supervised program of progressive exercise, psychological support, education, and training to enable a patient to resume normal activities following a myocardial infarction.

Cardiomyopathy (kar′-dē-ō-mī-OP-a-thē; *myo*=muscle; *-pathos*= disease) A progressive disorder in which ventricular structure or function is impaired. In **dilated cardiomyopathy,** the ventricles enlarge (stretch) and become weaker and reduce the heart's pumping action. In **hypertrophic cardiomyopathy,** the ventricular walls thicken and the pumping efficiency of the ventricle is reduced.

Cor pulmonale (CP) (kor pul-mōn-ALE; *cor-*=heart; *pulmon-*= lung) A term referring to right ventricular hypertrophy from disorders that bring about hypertension (high blood pressure) in the pulmonary circulation.

Echocardiography (ek-ō-kar-dē-OG-ra-fē; *cardio*=heart; *-grapho*= to record) The use of ultrasound to study the structure of the heart and its motions and the presence of pericardial fluid.

Electrocardiography (e-LEK-trō-kar-dē-og′-ra-fē) (ECG or EKG). A method of recording electrical currents generated by the heart muscle by means of an electrocardiograph.

Electrophysiological testing (ē-lek′-trō-fiz′-ē-OL-ō-ji-kal) A procedure in which a catheter with an electrode is passed through blood vessels and introduced into the heart. It is used to detect the exact locations of abnormal electrical conduction pathways. Once an abnormal pathway is located, it can be destroyed by sending a current through the electrode, a procedure called *radio frequency ablation.*

Palpitation (pal′-pi-TĀ-shun) A fluttering of the heart or an abnormal rate or rhythm of the heart.

Paroxysmal tachycardia (par′-ok-SIZ-mal tak′-e-KAR-dē-a; *tachy-*=swift or fast) A period of rapid heartbeats that begins and ends suddenly.

Radionuclide imaging (rā′-dē-ō-NOO-klīd; *-radio*=x-ray or gamma ray) A procedure in which a radioactively labeled substance (radionuclide) is introduced into a vein and distributed throughout the body. Gamma cameras provide an image that provides information about organ structure and function. One type of radionuclide imaging, called **thallium scanning,** is used to look at the activity of the heart during exercise. It reveals areas of heart muscle with a poor blood supply.

Sudden cardiac death The unexpected cessation of circulation and breathing due to an underlying heart disease such as ischemia, myocardial infarction, or a disturbance in cardiac rhythm.

 ## STUDY OUTLINE

Location and Surface Projection of the Heart (p. 425)

1. The heart is located in the mediastinum; about two-thirds of its mass is to the left of the midline.
2. The heart is basically a cone lying on its side; it consists of an apex, a base, anterior and inferior surfaces, and right and left borders.
3. Four points are used to project the heart's location to the surface of the chest.

Structure and Function of the Heart (p. 426)

1. The pericardium is the membrane that surrounds and protects the heart; it consists of an outer fibrous layer and an inner serous pericardium, which is composed of a parietal layer and a visceral layer.
2. Between the parietal and visceral layers of the serous pericardium is the pericardial cavity, a potential space filled with a few milliliters of pericardial fluid that reduces friction between the two membranes.
3. The wall of the heart has three layers: epicardium (visceral layer of the serous pericardium), myocardium, and endocardium.
4. The epicardium consists of mesothelium and connective tissue, the myocardium is composed of cardiac muscle tissue, and the endocardium consists of endothelium and connective tissue.
5. The heart chambers include two superior chambers, the right and left atria, and two inferior chambers, the right and left ventricles.
6. External features of the heart include the auricles (flaps on each atrium that increase their volume), the coronary sulcus between the atria and ventricles, and the anterior and posterior sulci

between the ventricles on the anterior and posterior surfaces of the heart, respectively.

7. The right atrium receives blood from the superior vena cava, inferior vena cava, and coronary sinus. It is separated from the left atrium by the interatrial septum, which contains the fossa ovalis. Blood exits the right atrium through the tricuspid valve.
8. The right ventricle receives blood from the right atrium. It is separated from the left ventricle by the interventricular septum and pumps blood to the lungs through the pulmonary valve and pulmonary trunk.
9. Oxygenated blood enters the left atrium from the pulmonary veins and exits through the bicuspid (mitral) valve.
10. The left ventricle pumps oxygenated blood into the systemic circulation through the aortic valve and aorta.
11. The thickness of the myocardium of the four chambers varies according to the chamber's function. The left ventricle has the thickest wall because of its high workload.
12. The fibrous skeleton of the heart is dense connective tissue that surrounds and supports the valves of the heart.
13. Heart valves prevent the backflow of blood within the heart.
14. The atrioventricular (AV) valves, which lie between atria and ventricles, are the tricuspid valve on the right side of the heart and the bicuspid (mitral) valve on the left. The chordae tendineae and papillary muscles stabilize the flaps of the AV valves and stop blood from backing into the atria.
15. Each of the two arteries that leaves the heart has a semilunar valve (aortic and pulmonary).

Circulation of Blood (p. 435)

1. The left side of the heart is the pump for the systemic circulation, the circulation of blood throughout the body except for the air sacs of the lungs. The left ventricle ejects blood into the aorta, and blood then flows into systemic arteries, arterioles, capillaries, venules, and veins, which carry it back to the right atrium.
2. The right side of the heart is the pump for pulmonary circulation, the circulation of blood through the lungs. The right ventricle ejects blood into the pulmonary trunk, and blood then flows into pulmonary arteries, pulmonary capillaries, and pulmonary veins, which carry it back to the left atrium.
3. The flow of blood through the heart is called the coronary (cardiac) circulation.
4. The principal arteries of the coronary circulation are the left and right coronary arteries; the principal veins are the cardiac vein and the coronary sinus.

The Cardiac Conduction System (p. 436)

1. Autorhythmic cells form the conduction system; these are cardiac muscle fibers that spontaneously generate action potentials.
2. Components of the conduction system are the sinoatrial (SA) node (pacemaker), atrioventricular (AV) node, atrioventricular (AV) bundle (bundle of His), bundle branches, and Purkinje fibers.
3. The record of electrical changes during the course of cardiac cycles is called an electrocardiogram (ECG).
4. A normal ECG consists of a P wave (atrial depolarization), a QRS complex (onset of ventricular depolarization), and a T wave (ventricular repolarization).

Cardiac Cycle (Heartbeat) (p. 440)

1. A cardiac cycle consists of the systole (contraction) and diastole (relaxation) of both atria, plus the systole and diastole of both ventricles.
2. The phases of the cardiac cycle are (a) relaxation period, (b) ventricular filling, and (c) ventricular systole.

Heart Sounds (p. 440)

1. S1, the first heart sound (lubb), is caused by blood turbulence associated with the closing of the atrioventricular valves.
2. S2, the second sound (dupp), is caused by blood turbulence associated with the closing of semilunar valves.

Exercise and the Heart (p. 441)

1. Sustained exercise increases oxygen demand on muscles.
2. Among the benefits of aerobic exercise are increased cardiac output, decreased blood pressure, weight control, and increased fibrinolytic activity.

Development of the Heart (p. 442)

1. The heart develops from mesoderm.
2. The endothelial tubes develop into the four-chambered heart and great vessels of the heart.

SELF-QUIZ QUESTIONS

Choose the one best answer to the following questions.

1. On a surface projection, the apex of the heart is found at the:
   a. superior right point.　　b. superior left point.
   c. inferior right point.　　d. inferior left point.
   e. a line connecting the inferior right and left points.

2. The layer of the heart wall that is composed of cardiac muscle is the:
   a. endocardium.　　b. myocardium.
   c. epicardium.　　d. pericardium.
   e. a, b, and c are all correct.

3. Which of the following is the correct route of blood through the heart from the systemic circulation to the pulmonary circulation and back to the systemic circulation?
   a. right atrium, tricuspid valve, right ventricle, pulmonary valve, left atrium, mitral valve, left ventricle, aortic valve
   b. left atrium, triscuspid valve, left ventricle, pulmonary valve, right atrium, mitral valve, right ventricle, aortic valve
   c. left atrium, pulmonary valve, right atrium, tricuspid valve, left ventricle, aortic valve, right ventricle, mitral valve
   d. left ventricle, mitral valve, left atrium, pulmonary valve, right ventricle, tricuspid valve, right atrium, aortic valve
   e. right atrium, mitral valve, right ventricle, pulmonary valve, left atrium, tricuspid valve, left ventricle, aortic valve

4. Which of the following vessels carries the most highly oxygenated blood?
   a. superior vena cava　　b. coronary sinus　　c. inferior vena cava
   d. pulmonary veins　　e. pulmonary arteries

5. Blood in the superior vena cava passes next into the:
   a. inferior vena cava.　　b. right atrium.　　c. left atrium.
   d. coronary sinus.　　e. Both a and d are correct.

6. Which of the following sequences correctly represents the conduction of an impulse through the heart?
   a. SA node, AV node, AV bundle, bundle branches
   b. SA node, AV bundle, AV node, bundle branches
   c. AV node, SA node, AV bundle, bundle branches
   d. SA node, bundle branches, AV node, AV bundle
   e. AV node, AV bundle, SA node, bundle branches

Complete the following.

7. The membrane surrounding and protecting the heart is the _____.
8. The first two vessels to branch from the ascending aorta are the _____ and the _____.
9. The first heart sound is created by turbulence of blood at the closing of the _____ valves.
10. The branch of the left coronary artery that lies in the coronary sulcus and distributes oxygenated blood to the walls of the left atrium and ventricle is the _____.

11. During the period of ventricular filling, the AV valves are in the _____ position, and the semilunar valves are in the _____ position.

12. The term _____ refers to the contraction of a heart chamber, and the word _____ refers to the relaxation of a heart chamber.

13. The heart chamber with the thickest wall is the _____.

14. The _____ point of the heart is located in the fifth left intercostal space, about 9 cm to the left of the midline.

Are the following statements true or false?

15. Trabeculae carneae are ridges located on the inner surface of the right ventricle.

16. The right side of the heart is the pump for the pulmonary circulation.

17. Blood passes into the aorta from the right ventricle.

18. Blood in the coronary sinus flows into the right atrium.

19. Most of the ventricular filling occurs during atrial systole.

20. The Purkinje fibers distribute action potentials first to the apex of the ventricles, then superiorly to the remainder of the ventricles.

21. Chordae tendineae attach AV valve cusps to pectinate muscles.

22. The visceral layer of the serous pericardium is also called the endocardium.

Matching

23. Match the following (answers may be used more than once):

___ (a) also called the mitral valve

___ (b) prevents backflow of blood into right atrium

___ (c) prevents backflow of blood into right ventricle

___ (d) prevents backflow of blood into left ventricle

___ (e) located between the right atrium and right ventricle

___ (f) located between the left atrium and left ventricle

(1) aortic valve
(2) pulmonary valve
(3) bicuspid valve
(4) tricuspid valve

CRITICAL THINKING QUESTIONS

1. Why don't the AV valves flip open backwards when the ventricles contract?

2. Mr. Williams was diagnosed with blockages in his anterior interventricular branch or left anterior descending (LAD) and circumflex branch of the left coronary artery. What regions of the heart may be affected by these blockages?

3. The heart is constantly beating, which means it's constantly moving. Why doesn't the heart beat its way out of position? Why doesn't the muscle contraction pull the fibers apart?

4. Aleesha's three-year-old daughter has a sore throat. Her pediatrician wants Aleesha to bring her daughter into the office so she can be tested for a streptococcal infection. Aleesha wonders why this is necessary, since children get so many sore throats, and they usually recover quickly. What would you tell her?

5. Franz has been diagnosed with aortic valve stenosis. What does this mean, and how do you think it will affect his heart function?

ANSWERS TO FIGURE QUESTIONS

14.1 The mediastinum is the mass of tissue that extends from the sternum to the vertebral column between the pleurae of the lungs.

14.2 The visceral layer of the serous pericardium (epicardium) is both a part of the pericardium and a part of the heart wall.

14.3 The coronary sulcus forms a boundary between the superior atria and inferior ventricles.

14.4 The left ventricle is the chamber of the heart that has the thickest wall.

14.5 The four fibrous rings support the heart valves.

14.6 The papillary muscles contract, which pulls on the chordae tendineae and prevents valve cusps from everting and letting blood flow back into the atria.

14.7 Numbers 2 (right ventricle) through 6 in part b depict the pulmonary circulation, whereas numbers 7 (left ventricle) through 10 and 1 (right atrium) depict the systemic circulation.

14.8 The circumflex branch delivers oxygenated blood to the walls of the left atrium and left ventricle.

14.9 The only electrical connection between the atria and the ventricles is the atrioventricular bundle.

14.10 The atrioventricular valves are open and the semilunar valves are closed during ventricular filling.

14.11 The first heart sound (S1), or lubb, is associated with the closure of AV valves.

14.12 The heart begins to contract by the twenty-second day of gestation.

14.13 Partitioning of the heart is complete by the end of the fifth week.

14.14 HDL removes excess cholesterol from body cells and transports it to the liver for elimination.

14.15 Coronary angiography is used to visualize many blood vessels.

14.16 Tetralogy of Fallot involves an interventricular septal defect, an aorta that emerges from both ventricles, a stenosed pulmonary valve, and an enlarged right ventricle.

THE CARDIOVASCULAR SYSTEM: BLOOD VESSELS

15

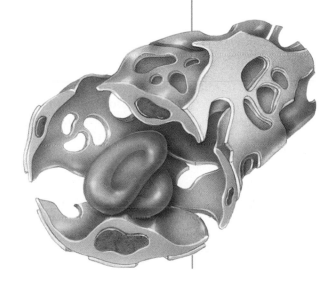

INTRODUCTION   Blood vessels form a system of tubes that carries blood away from the heart, transports it to the tissues of the body, and then returns it to the heart. This chapter focuses on the structure and functions of the various types of blood vessels and on the vessels that constitute the major circulatory routes.

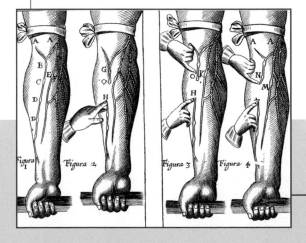

Can you determine what is being demonstrated in this image?

www.wiley.com/college/apcentral

ANATOMY OF BLOOD VESSELS

• Contrast the structure and function of arteries, arterioles, capillaries, venules, and veins.

The five main types of blood vessels are arteries, arterioles, capillaries, venules, and veins. **Arteries** (AR-ter-ēz) carry blood *away from the heart* to other organs. Large, elastic arteries leave the heart and divide into medium-sized, muscular arteries that branch out into the various regions of the body. Medium-sized arteries then divide into small arteries, which, in turn, divide into still smaller arteries called **arterioles** (ar-TER-ē-ōls). As the arterioles enter a tissue, they branch into myriad tiny vessels called **capillaries** (KAP-i-lar′-ēz=hairlike). The thin walls of capillaries allow exchange of substances between the blood and body tissues. Groups of capillaries within a tissue reunite to form small veins called **venules** (VEN-ūls). These, in turn, merge to form progressively larger blood vessels called veins. **Veins** (VĀNZ) are the blood vessels that convey blood from the tissues *back to the heart.* Because blood vessels require oxygen (O_2) and nutrients just like other tissues of the body, larger blood vessels are served by their own blood vessels, called **vasa vasorum** (literally, vasculature of vessels), located within their walls.

Arteries

Because **arteries** (*ar-*=air; *ter-*=to carry) were found empty at death, in ancient times they were thought to contain only air. The wall of an artery has three coats, or tunics: (1) tunica interna, (2) tunica media, and (3) tunica externa (Figure 15.1). The innermost coat, the **tunica interna (intima),** is composed of a lining of simple squamous epithelium called *endothelium*, a *basement membrane*, and a layer of elastic tissue called the *internal elastic lamina*. The endothelium is a continuous layer of cells that line the inner surface of the entire cardiovascular system (the heart and all blood vessels). Normally, endothelium is the only tissue that makes contact with blood. The tunica interna is closest to the **lumen,** the hollow center through which blood flows. The middle coat, or **tunica media,** is usually the thickest layer; it consists of elastic fibers and smooth muscle fibers (cells), arranged in rings around the lumen. Due to their plentiful elastic fibers, arteries normally have high *compliance*, which means that their walls stretch easily or expand without tearing in response to a small increase in pressure. The outer coat, the **tunica externa,** is composed principally of elastic and collagen fibers. In muscular arteries (described shortly), an *external elastic lamina* composed of elastic tissue separates the tunica externa from the tunica media.

Sympathetic fibers of the autonomic nervous system innervate the smooth muscle of blood vessels. An increase in sympathetic stimulation typically stimulates the smooth muscle to contract, squeezing the vessel wall and narrowing the lumen. Such a decrease in the diameter of the lumen of a blood vessel is called **vasoconstriction.** In contrast, when sympathetic stimulation decreases, or in the presence of certain chemicals (such as nitric oxide, H^+, and lactic acid), smooth muscle fibers relax. The resulting increase in lumen diameter is called **vasodilation.**

Additionally, when an artery or arteriole is damaged, its smooth muscle contracts, producing vascular spasm of the vessel. Such a vasospasm limits blood flow through the damaged vessel and helps reduce blood loss if the vessel is small.

Elastic Arteries

The largest-diameter arteries, termed **elastic arteries** because the tunica media contains a high proportion of elastic fibers, have walls that are relatively thin in proportion to their overall diameter. Elastic arteries perform an important function: They help propel blood onward while the ventricles are relaxing. As blood is ejected from the heart into elastic arteries, their highly elastic walls stretch, accommodating the surge of blood. By stretching, the elastic fibers momentarily store mechanical energy, functioning as a **pressure reservoir.** Then, the elastic fibers recoil and convert stored (potential) energy in the vessel into kinetic energy of the blood. Thus, blood continues to move through the arteries even while the ventricles are relaxed. Because they conduct blood from the heart to medium-sized, more-muscular arteries, elastic arteries also are called *conducting arteries*. The aorta and the brachiocephalic, common carotid, subclavian, vertebral, pulmonary, and common iliac arteries are elastic arteries (see Figure 15.6).

Muscular Arteries

Medium-sized arteries are called **muscular arteries** because their tunica media contains more smooth muscle and fewer elastic fibers than elastic arteries. Thus, muscular arteries are capable of greater vasoconstriction and vasodilation to adjust the rate of blood flow. The large amount of smooth muscle makes the walls of muscular arteries relatively thick. Muscular arteries also are called *distributing arteries* because they distribute blood to various parts of the body. Examples include the brachial artery in the arm and radial artery in the forearm (see Figure 15.6).

Arterioles

An **arteriole** (=small artery) is a very small (almost microscopic) artery that delivers blood to capillaries (Figure 15.2 on page 456). Arterioles near the arteries from which they branch have a tunica interna like that of arteries, a tunica media composed of smooth muscle and very few elastic fibers, and a tunica externa composed mostly of elastic and collagen fibers. In the smallest-diameter arterioles, which are closest to capillaries, the tunics consist of little more than a ring of endothelial cells surrounded by a few scattered smooth muscle fibers.

Arterioles play a key role in regulating blood flow from arteries into capillaries by regulating *resistance*, the opposition to blood flow. In a blood vessel, resistance is due mainly to friction between blood and the inner walls of blood vessels. When the blood vessel diameter is smaller, the friction is greater. Because contraction and relaxation of the smooth muscle in arteriole walls can change their diameter, arterioles are known as *resistance vessels*. Contraction of arteriolar smooth muscle causes vasoconstriction, which increases resistance and decreases blood flow into capillaries supplied by that arteriole. By contrast, relaxation of arteriolar smooth muscle causes vasodilation, which decreases resistance and increases blood flow into capillaries.

Figure 15.1 Comparative structure of blood vessels. The relative size of the capillary in (c) is enlarged.

Arteries carry blood from the heart to tissues; veins carry blood from tissues to the heart.

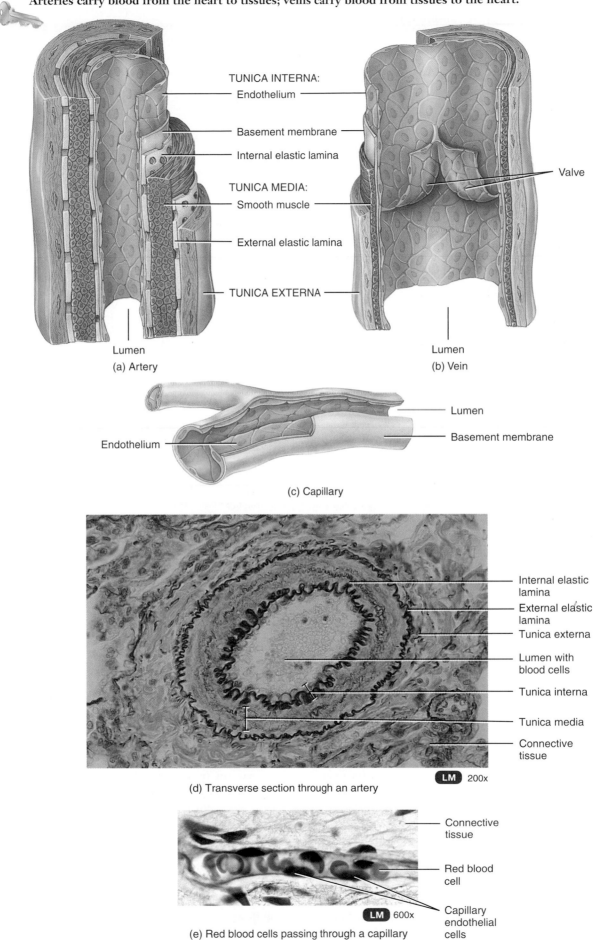

TUNICA INTERNA:
Endothelium

Basement membrane

Internal elastic lamina

TUNICA MEDIA:
Smooth muscle

External elastic lamina

TUNICA EXTERNA

Valve

Lumen
(a) Artery

Lumen
(b) Vein

Lumen

Endothelium

Basement membrane

(c) Capillary

Internal elastic lamina
External elastic lamina
Tunica externa
Lumen with blood cells
Tunica interna
Tunica media
Connective tissue

LM 200x
(d) Transverse section through an artery

Connective tissue
Red blood cell
Capillary endothelial cells

LM 600x
(e) Red blood cells passing through a capillary

Which vessel—the femoral artery or the femoral vein—has a thicker wall? Which has a wider lumen?

Figure 15.2 Arteriole, capillaries, and venules.

🔑 Arterioles regulate blood flow into capillaries, where nutrients, gases, and wastes are exchanged between the blood and interstitial fluid.

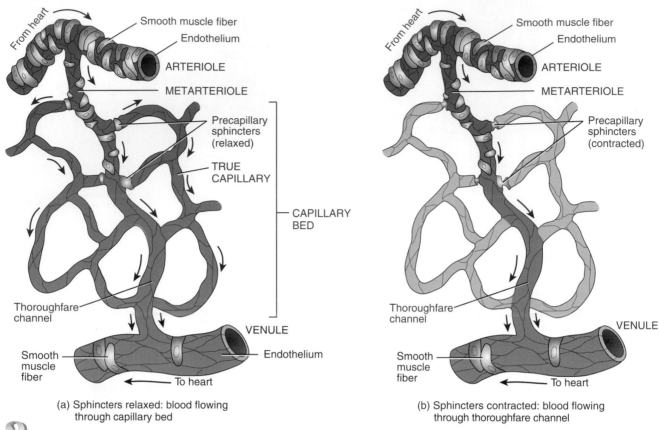

(a) Sphincters relaxed: blood flowing through capillary bed

(b) Sphincters contracted: blood flowing through thoroughfare channel

❓ Why do metabolically active tissues have extensive capillary networks?

A change in diameter of arterioles can also significantly affect blood pressure. For example, nicotine causes vasoconstriction in arterioles, which increases blood pressure.

Capillaries

Capillaries are microscopic vessels that connect arterioles to venules (Figure 15.2). The flow of blood from arterioles to venules through capillaries is called the **microcirculation.** Capillaries are found near almost every cell in the body, but their number varies with the metabolic activity of the tissue they serve. Body tissues with high metabolic requirements, such as muscles, the liver, the kidneys, and the nervous system, use more O₂ and nutrients and thus have extensive capillary networks. Tissues with lower metabolic requirements, such as tendons and ligaments, contain fewer capillaries. Capillaries are absent in a few tissues, for instance, all covering and lining epithelia, the cornea and lens of the eye, and cartilage.

Capillaries are known as *exchange vessels* because their prime function is the exchange of nutrients and wastes between the blood and tissue cells through the interstitial fluid. The structure of capillaries is well suited to this function. Capillary walls are composed of only a single layer of endothelial cells and a basement membrane (see Figure 15.1e). They have no tunica media or tunica externa. Thus, a substance in the blood must

pass through just one cell layer to reach the interstitial fluid and tissue cells. Exchange of materials occurs only through the walls of capillaries and the beginning of venules; the walls of arteries, arterioles, most venules, and veins present too thick a barrier. Capillaries form extensive branching networks that increase the surface area available for rapid exchange of materials. In most tissues, blood flows through only a small part of the capillary network when metabolic needs are low. However, when a tissue is active, such as contracting muscle, the entire capillary network fills with blood.

A **metarteriole** (*met-*=beyond) is a vessel that emerges from an arteriole and supplies a group of 10–100 capillaries that constitute a **capillary bed** (Figure 15.2a). The proximal end of a metarteriole is surrounded by scattered smooth muscle fibers whose contraction and relaxation help regulate blood flow through the capillary bed. The distal end of a metarteriole, which empties into a venule, has no smooth muscle fibers and is called a **thoroughfare channel.** Blood flowing through a thoroughfare channel bypasses the capillary bed.

True capillaries emerge from arterioles or metarterioles. At their sites of origin, a ring of smooth muscle fibers called a **precapillary sphincter** controls the flow of blood into a true capillary. When the precapillary sphincters are relaxed (open), blood flows into the capillary bed (Figure 15.2a); when precapillary sphincters contract (close or partially close), blood flow through the capillary bed ceases or decreases (Figure 15.2b). Typically, blood flows intermittently through a capillary bed due to alternating contraction and relaxation of the smooth muscle of metarterioles and the precapillary sphincters. This intermittent contraction and relaxation, which may occur 5 to 10 times per minute, is called **vasomotion.** In part, vasomotion is due to chemicals released by the endothelial cells; nitric oxide is one example. At any given time, blood flows through only about 25% of a capillary bed.

The body contains three different types of capillaries: continuous capillaries, fenestrated capillaries, and sinusoids (Figure 15.3). Many capillaries are **continuous capillaries,** in which the plasma membranes of endothelial cells form a continuous tube that is interrupted only by **intercellular clefts,** gaps between neighboring endothelial cells (Figure 15.3a). Continuous capillaries are found in skeletal and smooth muscle, connective tissues, and the lungs.

Other capillaries of the body are **fenestrated capillaries** (*fenestr-*=window). The plasma membranes of the endothelial cells in these capillaries have many **fenestrations,** small pores (holes) ranging from 70 to 100 nm in diameter (Figure 15.3b). Fenestrated capillaries are found in the kidneys, villi of the small intestine, choroid plexuses of the ventricles in the brain, ciliary processes of the eyes, and endocrine glands.

Sinusoids are wider and more winding than other capillaries. Their endothelial cells may have unusually large fenestrations. In addition to having an incomplete or absent basement membrane (Figure 15.3c), sinusoids have very large intercellular clefts that allow proteins and in some cases even blood cells to pass from a tissue into the bloodstream. For example, newly formed blood cells enter the bloodstream through the sinusoids of red bone marrow. In addition, sinusoids contain specialized lining cells that are adapted to the function of the tissue. Sinusoids in the liver, for example, contain phagocytic cells that remove bacteria and other debris from the blood. The spleen, anterior pituitary, and parathyroid glands also have sinusoids.

Usually blood passes from the heart and then in sequence through arteries, arterioles, capillaries, venules, and veins and then back to the heart. In some parts of the body, however, blood passes from one capillary network into another through a vein called a portal vein. Such a circulation of blood is called a **portal system.** The name of the portal system gives the name of the second capillary location. As you will see later, there are portal systems associated with the liver (hepatic portal circulation) and the pituitary gland (hypophyseal portal system).

Venules

When several capillaries unite, they form small veins called **venules** (=little veins). Venules collect blood from capillaries and drain into veins. The smallest venules, those closest to the capillaries, consist of a tunica interna of endothelium and a tunica media that has only a few scattered smooth muscle fibers and fibroblasts. Like capillaries, the walls of the smallest venules are very porous and are the site where many phagocytic white blood cells emigrate from the bloodstream into an inflamed or infected tissue. As venules become larger and converge to form veins, they contain the tunica externa characteristic of veins.

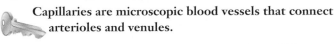

Figure 15.3 Types of capillaries (shown in transverse sections).

Capillaries are microscopic blood vessels that connect arterioles and venules.

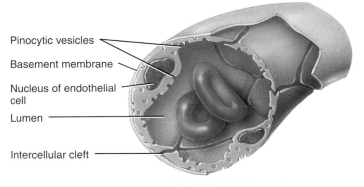

Pinocytic vesicles
Basement membrane
Nucleus of endothelial cell
Lumen
Intercellular cleft

(a) Continuous capillary formed by endothelial cells

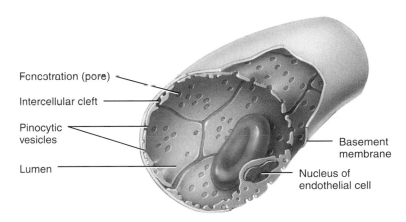

Fenestration (pore)
Intercellular cleft
Pinocytic vesicles
Lumen
Basement membrane
Nucleus of endothelial cell

(b) Fenestrated capillary

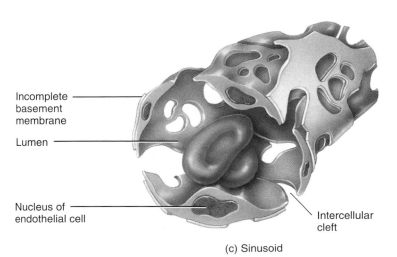

Incomplete basement membrane
Lumen
Nucleus of endothelial cell
Intercellular cleft

(c) Sinusoid

How do materials cross capillary walls?

Veins

Although **veins** are composed of essentially the same three coats as arteries, the relative thicknesses of the layers are different. The tunica interna of veins is thinner than that of arteries; the tunica media of veins is much thinner than in arteries, with relatively little smooth muscle and elastic fibers. The tunica externa of veins is the thickest layer and consists of collagen and elastic fibers; the tunica externa of the inferior vena cava also contains longitudinal fibers of smooth muscle. Veins lack the external or internal elastic laminae found in arteries (see Figure 15.1b). Still, veins are distensible enough to adapt to variations in the volume and pressure of blood passing through them, although they are not designed to withstand high pressure. Furthermore, the lumen of a vein is larger than that of a comparable artery, and veins often appear collapsed (flattened) when sectioned.

The pumping action of the heart is a major factor in moving venous blood back to the heart. The contraction of skeletal muscles in the lower limbs also helps boost venous return to the heart. Valves in veins, described shortly, also assume a key function in returning venous blood to the heart. The average blood pressure in veins is considerably lower than in arteries. The difference in pressure can be noticed when blood flows from a cut vessel. Blood leaves a cut vein in an even, slow flow but spurts rapidly from a cut artery. Most of the structural differences between arteries and veins reflect this pressure difference. For example, the walls of veins are not as strong as those of arteries. Many veins feature **valves** (see Figure 15.1b), which are needed because venous blood pressure is so low. When you stand, the pressure pushing blood up the veins in your lower limbs is barely enough to overcome the force of gravity pulling it back down.

Each valve is composed of two or more thin folds of tunica interna that form flaplike cusps; the cusps project into the lumen of the veins pointing toward the heart (Figure 15.4). Veins pass between groups of skeletal muscles, and when these muscles contract, venous pressure is increased; the valve opens as the blood pushes the cusps against the wall of the vein in its journey through the valve toward the heart. When the muscles relax, blood would tend to move back toward the feet due to the force of gravity, but is stopped from doing so as the cusps come together and close the lumen of the vein. In this way, valves prevent backflow of blood and aid in moving venous blood in one direction only—toward the heart.

A **vascular (venous) sinus** is a vein with a thin endothelial wall that has no smooth muscle to alter its diameter. In a vascular sinus, the surrounding dense connective tissue replaces the tunica media and tunica externa in providing support. For example, dural venous sinuses, which are supported by the dura mater, convey deoxygenated blood from the brain to the heart. Another example of a vascular sinus is the coronary sinus of the heart (see Figure 14.3c on page 429).

VARICOSE VEINS

Leaky venous valves can cause veins to become dilated and twisted in appearance, a condition called **varicose veins** or **varices** (VAR-i-sēz; *varic-*=a swollen vein). The singular is *varix* (VAR-iks). The condition may occur in the veins of almost any

Figure 15.4 Role of skeletal muscle contractions and venous valves in returning blood to the heart. (a) When skeletal muscles contract, the proximal valve opens, and blood is forced toward the heart. (b) Sections of a venous valve.

 Venous return depends on the pumping action of the heart, skeletal muscle contractions, and valves in veins.

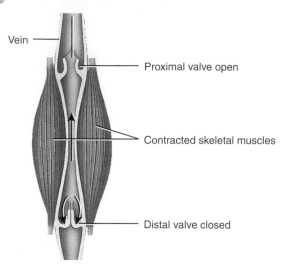

Vein

Proximal valve open

Contracted skeletal muscles

Distal valve closed

(a) Diagram of contracted skeletal muscles

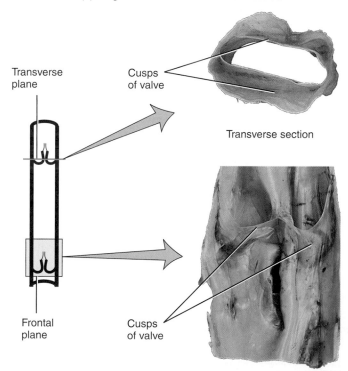

Transverse plane

Cusps of valve

Transverse section

Frontal plane

Cusps of valve

Longitudinally cut

(b) Photograph of a valve in a vein

 Why are valves more important in arm veins and leg veins than in neck veins?

body part, but it is most common in the esophagus and in superficial veins of the lower limbs. The valvular defect may be congenital or may result from mechanical stress (prolonged standing or pregnancy) or aging. The leaking venous valves allow the backflow of blood, which causes pooling of blood. This in turn creates pressure that distends the vein and allows fluid to leak into surrounding tissue. As a result, the affected vein and the tissue around it may become inflamed and painfully tender. Veins close to the surface of the legs, especially the saphenous vein, are highly susceptible to varicosities, whereas deeper veins are not as vulnerable because surrounding skeletal muscles prevent their walls from stretching excessively. Varicosed veins in the anal canal are referred to as *hemorrhoids.* Esophageal varices result from dilated veins in the walls of the lower part of the esophagus and sometimes the upper part of the stomach. Bleeding esophageal varices are life-threatening and are usually a result of chronic liver disease. ∎

Anastomoses

Most tissues of the body receive blood from more than one artery. The union of the branches of two or more arteries supplying the same body region is called an **anastomosis** (a-nas-tō-MŌ-sis‾ connecting; plural is *anastomoses*). Anastomoses between arteries provide alternate routes for blood to reach a tissue or organ. If blood flow stops momentarily when normal movements compress a vessel, or if a vessel is blocked by disease, injury, or surgery, then circulation to a part of the body is not necessarily stopped. The alternate route of blood flow to a body part through an anastomosis is known as **collateral circulation.** Anastomoses may also occur between veins and between arterioles and venules. Arteries that do not anastomose are known as **end arteries.** Obstruction of an end artery interrupts the blood supply to a whole segment of an organ, producing necrosis (death) of that segment. Alternate blood routes may also be provided by nonanastomosing vessels that supply the same region of the body.

Blood Distribution

The largest portion of your blood volume at rest—about 64%—is in systemic veins and venules. Systemic arteries hold about 13% of the blood volume, systemic capillaries hold about 7%, pulmonary blood vessels hold about 9%, and the heart holds about 7%. Because systemic veins and venules contain a large percentage of the blood volume, they function as **blood reservoirs** from which blood can be diverted quickly if the need arises. For example, when there is increased muscular activity, the cardiovascular center in the brain stem sends more sympathetic impulses to veins. The result is *venoconstriction*, constriction of veins, which reduces the volume of blood in reservoirs and allows a greater blood volume to flow to skeletal muscles, where it is needed most. A similar mechanism operates in cases of hemorrhage, when blood volume and pressure decrease; in this case, venoconstriction helps counteract the drop in blood pressure. Among the principal blood reservoirs are the veins of the abdominal organs (especially the liver and spleen) and the veins of the skin.

1. Discuss the importance of elastic fibers and smooth muscle in the tunica media of arteries.
2. Distinguish between elastic and muscular arteries in terms of location, histology, and function.
3. Describe the structural features of capillaries that allow exchange of materials between blood and body cells.
4. What are the main structural and functional differences between arteries and veins?
5. Describe the relationship between anastomoses and collateral circulation.

CIRCULATORY ROUTES

• Describe and compare the major routes that blood takes through various regions of the body.

Arteries, arterioles, capillaries, venules, and veins are organized into routes that deliver blood throughout the body. We can now look at the basic routes the blood takes as it is transported through its vessels.

Figure 15.5 shows the **circulatory routes** for blood flow. The routes are parallel; that is, in most cases a portion of the cardiac output flows separately to each tissue of the body. Thus each organ receives its own supply of freshly oxygenated blood. The two basic postnatal (after birth) routes for blood flow are the systemic and pulmonary circulations. The **systemic circulation** includes all the arteries and arterioles that carry oxygenated blood from the left ventricle to systemic capillaries, plus the veins and venules that carry deoxygenated blood returning to the right atrium after flowing through body organs. Blood leaving the aorta and flowing through the systemic arteries is a bright red color. As it moves through capillaries, it loses some of its oxygen and picks up carbon dioxide, so that blood in systemic veins is a dark red color.

Some subdivisions of the systemic circulation are the **coronary (cardiac) circulation** (see Figure 14.8 on page 437), which supplies the myocardium of the heart; **cerebral circulation,** which supplies the brain (see Figure 15.7c); and the **hepatic portal circulation,** which extends from the gastrointestinal tract to the liver (see Figure 15.15). The nutrient arteries to the lungs, such as the bronchial arteries, also are part of the systemic circulation.

When blood returns to the heart from the systemic route, it is pumped out of the right ventricle through the **pulmonary circulation** to the lungs (see Figure 15.16). In capillaries of the air sacs (alveoli) of the lungs, the blood loses some of its carbon dioxide and takes on oxygen. Bright red again, it returns to the left atrium of the heart and reenters the systemic circulation as it is pumped out by the left ventricle.

Another major route—the **fetal circulation**—exists only in the fetus and contains special structures that allow the developing fetus to exchange materials with its mother (see Figure 15.17).

Systemic Circulation

The systemic circulation carries oxygen and nutrients to body tissues and removes carbon dioxide and other wastes and heat from the tissues. All systemic arteries branch from the aorta. Deoxygenated blood returns to the heart through the systemic

Figure 15.5 Circulatory routes. Large black arrows indicate the systemic circulation (detailed in Exhibits 15.1–15.12), small black arrows the pulmonary circulation (detailed in Figure 15.16), and red arrows the hepatic portal circulation (detailed in Figure 15.15). Refer to Figure 14.8 on page 437 for details of the coronary circulation, and to Figure 15.17 for details of the fetal circulation.

Blood vessels are organized into various routes that deliver blood to tissues of the body.

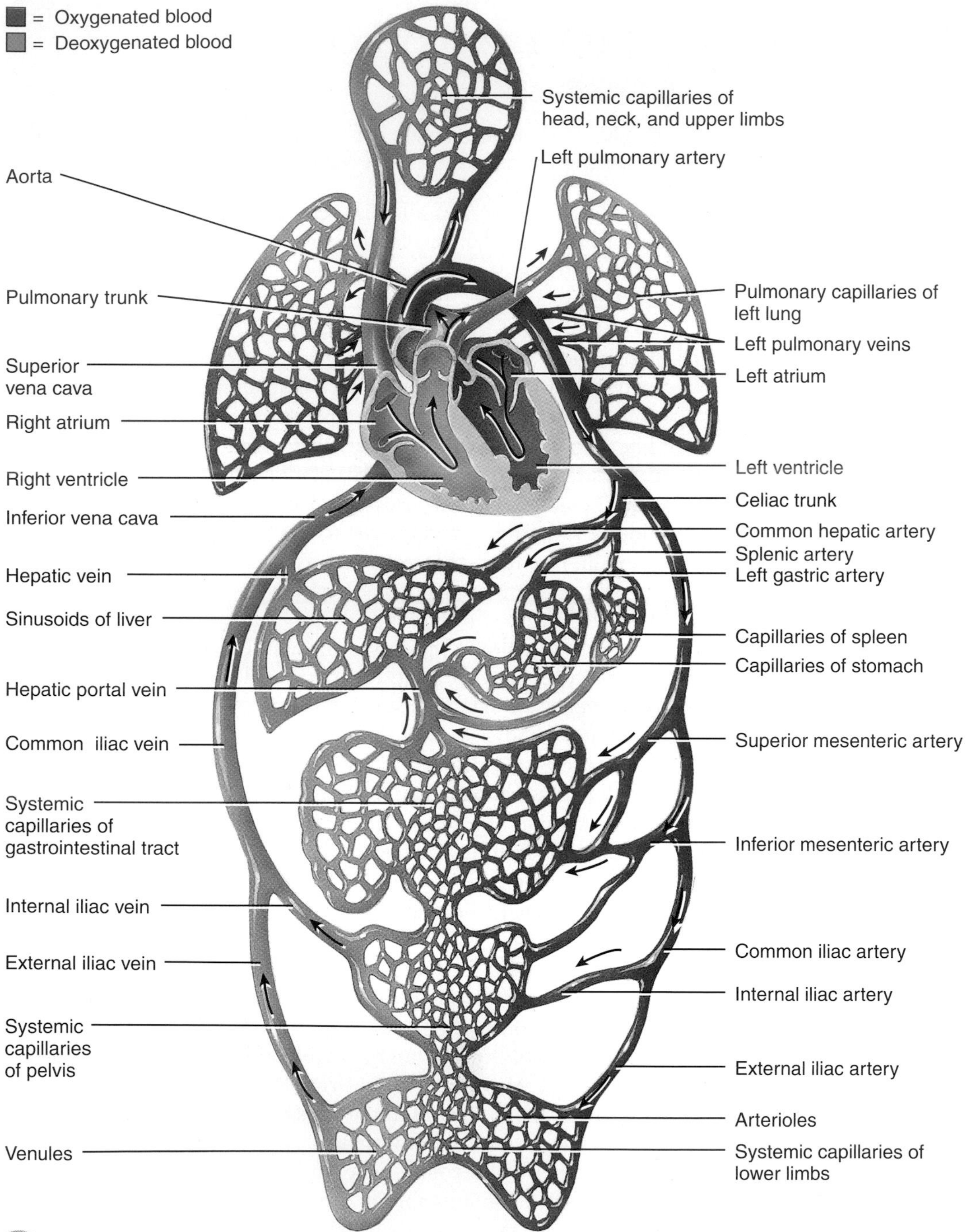

■ = Oxygenated blood
■ = Deoxygenated blood

What are the two main circulatory routes?

names. The blood vessels are organized in the exhibits according to regions of the body. Figure 15.6a shows an overview of the major arteries, whereas Figure 15.10 shows an overview of the major veins. As you study the various blood vessels in the exhibits, refer to these two figures to see the relationships of the blood vessels under consideration to other regions of the body. Each of the exhibits contains the following information:

- **An overview.** This information provides a general orientation to the blood vessels under consideration, with emphasis on how the blood vessels are organized into various regions and on distinguishing and/or interesting features about the blood vessels.

- **Blood vessel names.** Students often have difficulty with the pronunciations and meanings of blood vessels' names. To learn them more easily, study the phonetic pronunciations and word derivations that indicate how blood vessels get their names.

- **Region supplied or drained.** For each artery listed, there is a description of the parts of the body that receive blood from the vessel. For each vein listed, there is a description of the parts of the body that are drained by the vessel.

- **Illustrations and photographs.** The figures that accompany the exhibits contain several elements. There is an illustration of the blood vessels under consideration. In many cases, flow diagrams are provided to indicate the patterns of blood distribution or drainage. Cadaver photographs are also included in selected exhibits to provide more realistic views of the blood vessels.

veins. All the veins of the systemic circulation drain into the **superior vena cava, inferior vena cava,** or **coronary sinus,** which in turn empty into the right atrium.

The principal arteries and veins of the systemic circulation are described and illustrated in Exhibits 15.1–15.12 and Figures 15.6–15.14 to assist you in learning their

EXHIBIT 15.1   THE AORTA AND ITS BRANCHES (Figure 15.6)

OBJECTIVES

- Identify the four principal divisions of the aorta.
- Locate the major arterial branches arising from each division.

The **aorta** (=to lift up) is the largest artery of the body, with a diameter of 2–3 cm (about 1 in.). Its four principal divisions are the ascending aorta, arch of the aorta, thoracic aorta, and abdominal aorta. The portion of the aorta that emerges from the left ventricle posterior to the pulmonary trunk is the **ascending aorta.** The beginning of the aorta contains the aortic valve (see Figure 14.4a on page 430). The ascending aorta gives off two coronary artery branches that supply the myocardium of the heart. Then the ascending aorta turns to the left, forming the **arch of the aorta,** which descends and ends at the level of the intervertebral disc between the fourth and fifth thoracic vertebrae. As the aorta continues to descend, it lies close to the vertebral bodies, passes through the aortic hiatus of the diaphragm, and divides at the level of the fourth lumbar vertebra into two **common iliac arteries,** which carry blood to the lower limbs. The section of the aorta between the arch of the aorta and the diaphragm is called the **thoracic aorta;** the section between the diaphragm and the common iliac arteries is the **abdominal aorta.** Each division of the aorta gives off arteries that branch into distributing arteries that lead to various organs. Within the organs, the arteries divide into arterioles and then into capillaries that service the systemic tissues (all tissues except the alveoli of the lungs).

■ CHECKPOINT

1. What general regions do each of the four principal divisions of the aorta supply?

DIVISION AND BRANCHES	REGION SUPPLIED
ASCENDING AORTA	
Right and left coronary arteries	Heart.
ARCH OF THE AORTA	
Brachiocephalic trunk (brā′-kē-ō-se-FAL-ik)	
Right common carotid artery (ka-ROT-id)	Right side of head and neck.
Right subclavian artery (sub-KLĀ-vē-an)	Right upper limb.
Left common carotid artery	Left side of head and neck.
Left subclavian artery	Left upper limb.
THORACIC AORTA (*thorac-*=chest)	
Pericardial arteries (per-i-KAR-dē-al)	Pericardium.
Bronchial arteries (BRONG-kē-al)	Bronchi of lungs.
Esophageal arteries (e-sof′-a-JĒ-al)	Esophagus.
Mediastinal arteries (mē′-dē-as-TĪ-nal)	Structures in mediastinum.
Posterior intercostal arteries (in′-ter-KOS-tal)	Intercostal and chest muscles.
Subcostal arteries (sub-KOS-tal)	Same as posterior intercostals.
Superior phrenic arteries (FREN-ik)	Superior and posterior surfaces of diaphragm.
ABDOMINAL AORTA	
Inferior phrenic arteries (FREN-ik)	Inferior surface of diaphragm.
Celiac trunk (SĒ-lē-ak)	
Common hepatic artery (he-PAT-ik)	Liver.
Left gastric artery (GAS-trik)	Stomach and esophagus.
Splenic artery (SPLĒN-ik)	Spleen, pancreas, and stomach.
Superior mesenteric artery (MES-en-ter′-ik)	Small intestine, cecum, ascending and transverse colons, and pancreas.
Suprarenal arteries (soo′-pra-RĒ-nal)	Adrenal (suprarenal) glands.
Renal arteries (RĒ-nal)	Kidneys.
Gonadal arteries (gō-NAD-al)	
Testicular arteries (tes-TIK-ū-lar)	Testes (male).
Ovarian arteries (ō-VAR-ē-an)	Ovaries (female).
Inferior mesenteric artery	Transverse, descending, and sigmoid colons; rectum.
Common iliac arteries (IL-ē-ak)	
External iliac arteries	Lower limbs.
Internal iliac arteries	Uterus (female), prostate (male), muscles of buttocks, and urinary bladder.

continues

EXHIBIT 15.1   THE AORTA AND ITS BRANCHES (Figure 15.6)

CONTINUED

Figure 15.6   Aorta and its principal branches.

All systemic arteries branch from the aorta.

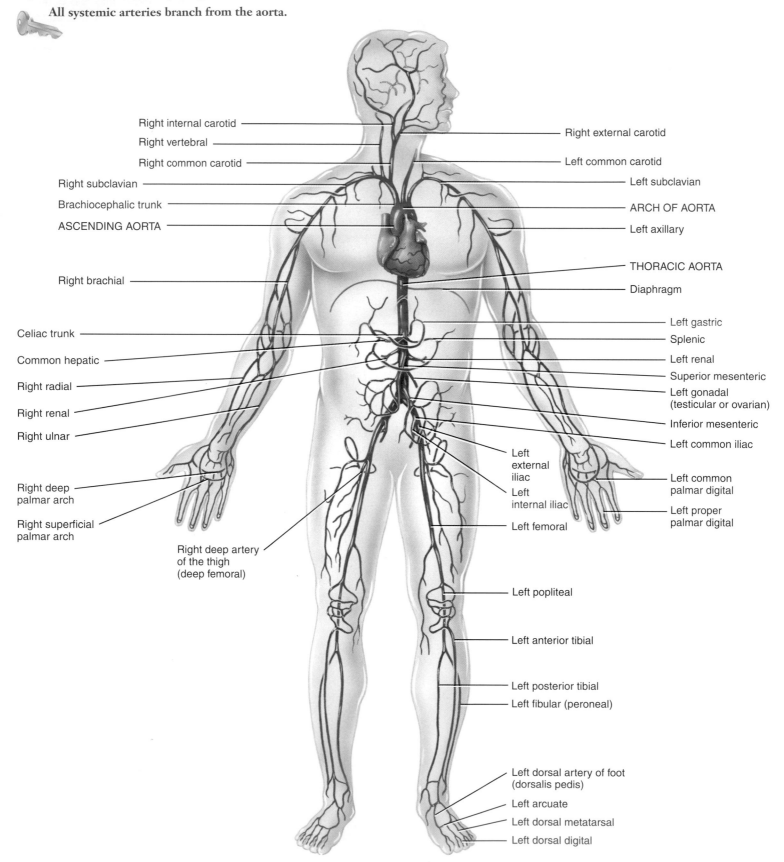

Right internal carotid

Right vertebral

Right common carotid

Right subclavian

Brachiocephalic trunk

ASCENDING AORTA

Right brachial

Celiac trunk

Common hepatic

Right radial

Right renal

Right ulnar

Right deep palmar arch

Right superficial palmar arch

Right deep artery of the thigh (deep femoral)

Right external carotid

Left common carotid

Left subclavian

ARCH OF AORTA

Left axillary

THORACIC AORTA

Diaphragm

Left gastric

Splenic

Left renal

Superior mesenteric

Left gonadal (testicular or ovarian)

Inferior mesenteric

Left common iliac

Left external iliac

Left internal iliac

Left femoral

Left common palmar digital

Left proper palmar digital

Left popliteal

Left anterior tibial

Left posterior tibial

Left fibular (peroneal)

Left dorsal artery of foot (dorsalis pedis)

Left arcuate

Left dorsal metatarsal

Left dorsal digital

(a) Overall anterior view of the principal branches of the aorta

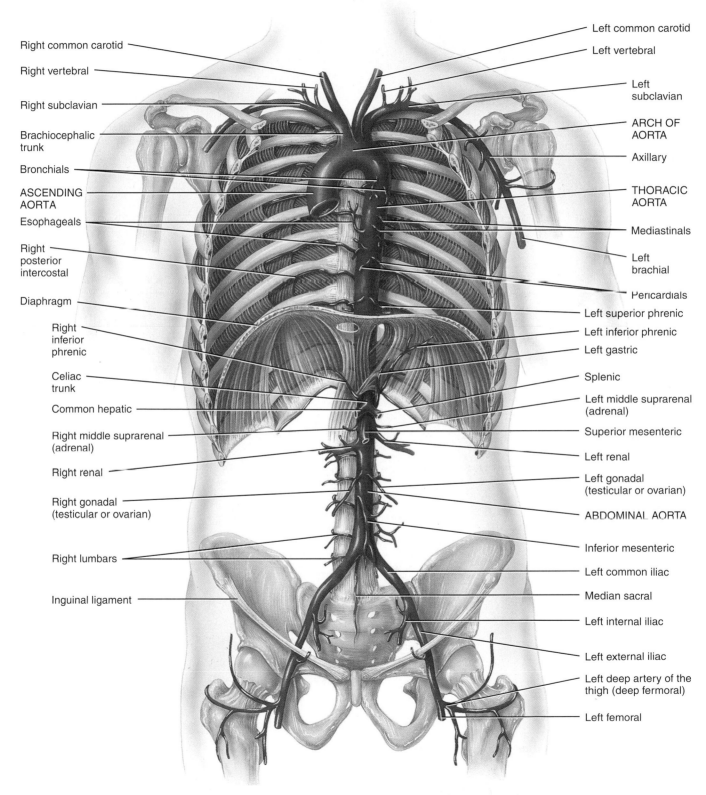

Right common carotid

Right vertebral

Right subclavian

Brachiocephalic
trunk

Bronchials

ASCENDING
AORTA

Esophageals

Right
posterior
intercostal

Diaphragm

Right
inferior
phrenic

Celiac
trunk

Common hepatic

Right middle suprarenal
(adrenal)

Right renal

Right gonadal
(testicular or ovarian)

Right lumbars

Inguinal ligament

Left common carotid

Left vertebral

Left
subclavian

ARCH OF
AORTA

Axillary

THORACIC
AORTA

Mediastinals

Left
brachial

Pericardials

Left superior phrenic

Left inferior phrenic

Left gastric

Splenic

Left middle suprarenal
(adrenal)

Superior mesenteric

Left renal

Left gonadal
(testicular or ovarian)

ABDOMINAL AORTA

Inferior mesenteric

Left common iliac

Median sacral

Left internal iliac

Left external iliac

Left deep artery of the
thigh (deep fermoral)

Left femoral

(b) Detailed anterior view of the principal branches of the aorta

continues

EXHIBIT 15.1   THE AORTA AND ITS BRANCHES (Figure 15.6)

SUPERIOR

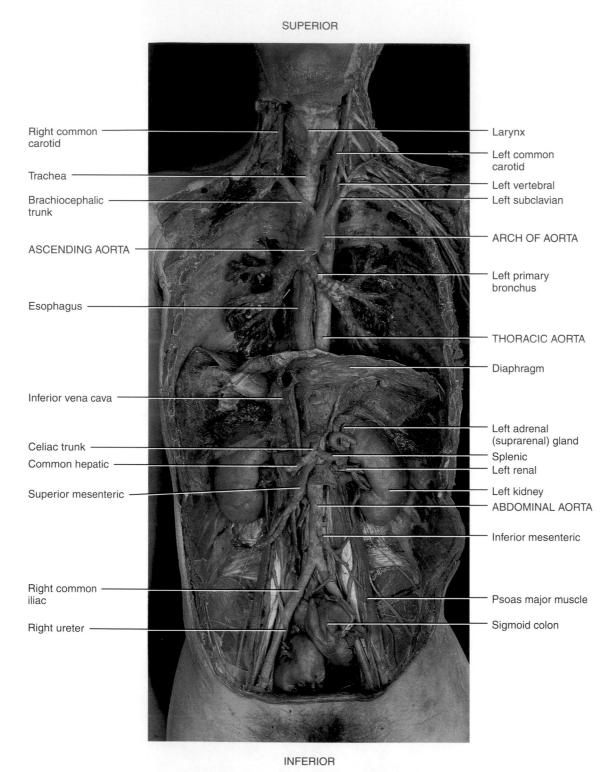

Right common carotid

Trachea

Brachiocephalic trunk

ASCENDING AORTA

Esophagus

Inferior vena cava

Celiac trunk

Common hepatic

Superior mesenteric

Right common iliac

Right ureter

Larynx

Left common carotid

Left vertebral

Left subclavian

ARCH OF AORTA

Left primary bronchus

THORACIC AORTA

Diaphragm

Left adrenal (suprarenal) gland

Splenic

Left renal

Left kidney

ABDOMINAL AORTA

Inferior mesenteric

Psoas major muscle

Sigmoid colon

INFERIOR

(c) Anterior view of the principal branches of the aorta

? **What are the four principal subdivisions of the aorta?**

EXHIBIT 15.2 ASCENDING AORTA (See Figure 14.8 on page 437)

OBJECTIVE

• Identify the two primary arterial branches of the ascending aorta.

The **ascending aorta** is about 5 cm (2 in.) in length and begins at the aortic valve. It is directed superiorly, slightly anteriorly, and to the right. It ends at the level of the sternal angle, where it becomes the arch of the aorta. The beginning of the ascending aorta is posterior to the pulmonary trunk and right auricle; the right pulmonary artery is posterior to it. At its origin, the ascending aorta contains three dilations called *aortic sinuses.* Two of these, the right and left sinuses, give rise to the right and left coronary arteries, respectively.

The right and left **coronary arteries** (*coron-*=crown) arise from the ascending aorta just superior to the aortic semilunar valve. They form a crownlike ring around the heart, giving off branches to the atrial and ventricular myocardium. The **posterior interventricular branch** (in-ter-ven-TRIK-ū-lar; *inter-*=between) of the right coronary artery supplies both ventricles, and the **marginal branch** supplies the right ventricle. **The anterior interventricular branch,** also known as the **left anterior descending (LAD) branch,** of the left coronary artery supplies both ventricles, and the **circumflex branch** (SER-kum-flex; *circum-*=around; *-flex*=to bend) supplies the left atrium and left ventricle.

CHECKPOINT

1. Which branches of the coronary arteries supply the left ventricle? Why does the left ventricle have such an extensive arterial blood supply?

SCHEME OF DISTRIBUTION

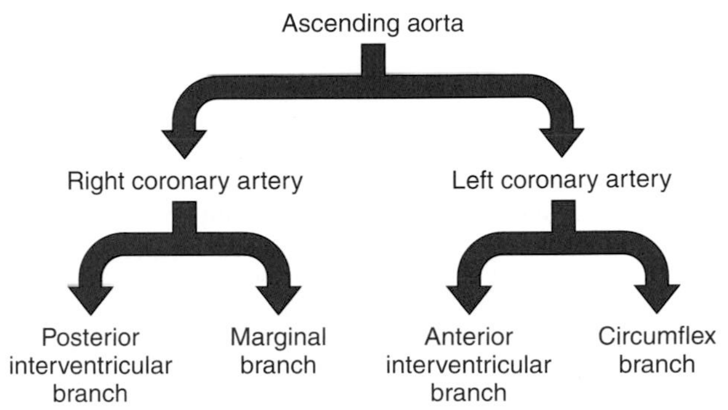

Ascending aorta

Right coronary artery | Left coronary artery

Posterior interventricular branch | Marginal branch | Anterior interventricular branch | Circumflex branch

465

EXHIBIT 15.3   THE ARCH OF THE AORTA (Figure 15.7)

OBJECTIVE

- Identify the three principal arteries that branch from the arch of the aorta.

The **arch of the aorta** is 4–5 cm (almost 2 in.) in length and is the continuation of the ascending aorta. It emerges from the pericardium posterior to the sternum at the level of the sternal angle. The arch is directed superiorly and posteriorly to the left and then inferiorly; it ends at the intervertebral disc between the fourth and fifth thoracic vertebrae, where it becomes the thoracic aorta. Three major arteries branch from the superior aspect of the arch of the aorta: the brachiocephalic trunk, the left common carotid, and the left subclavian. The first and largest branch from the arch of the aorta is the **brachiocephalic trunk** (brā′-kē-ō-se-FAL-ik; *brachio-*= arm; *-cephalic*=head). It extends superiorly, bending slightly to the right, and divides at the right sternoclavicular joint to form the right subclavian artery and right common carotid artery. The second

BRANCH	DESCRIPTION AND REGION SUPPLIED
BRACHIOCEPHALIC TRUNK	The **brachiocephalic trunk** divides to form the right subclavian artery and right common carotid artery (Figure 15.7a).
Right subclavian artery (sub-KLĀ-vē-an)	The **right subclavian artery** extends from the brachiocephalic trunk to the first rib and then passes into the armpit (axilla). The general distribution of the artery is to the brain and spinal cord, neck, shoulder, thoracic viscera and wall, and scapular muscles.
Internal thoracic or mammary *artery* (thor-AS-ik; *thorac-*=chest)	The **internal thoracic artery** arises from the first part of the subclavian artery and descends posterior to the costal cartilages of the superior six ribs. It terminates at the sixth intercostal space. It supplies the anterior thoracic wall and structures in the mediastinum. In coronary artery bypass grafting, if only a single vessel is obstructed, the internal thoracic artery (usually the left) is used to create the bypass. The upper end of the artery is left attached to the subclavian artery and the cut end is connected to the coronary artery at a point distal to the blockage. The lower end of the internal thoracic artery is tied off. Artery grafts are preferred over vein grafts because arteries can withstand the greater pressure of blood flowing through coronary arteries and are less likely to become obstructed over time.
Vertebral artery (VER-te-bral)	Before passing into the axilla, the right subclavian artery gives off a major branch to the brain called the **right vertebral artery** (Figure 15.7b). The right vertebral artery passes through the foramina of the transverse processes of the sixth through first cervical vertebrae and enters the skull through the foramen magnum to reach the inferior surface of the brain. Here it unites with the left vertebral artery to form the **basilar** (BAS-i-lar) **artery.** The vertebral artery supplies the posterior portion of the brain with blood. The basilar artery passes along the midline of the anterior aspect of the brain stem. It gives off several branches (posterior cerebral and cerebellar arteries) that supply the cerebellum and pons of the brain and the inner ear.
Axillary artery (AK-sil-ār-ē=armpit)	Continuation of the right subclavian artery into the axilla is called the **axillary artery.** (Note that the right subclavian artery, which passes deep to the clavicle, is a good example of the practice of giving the same vessel different names as it passes through different regions.) Its general distribution is the shoulder, thoracic and scapular muscles, and humerus.
Brachial artery (BRĀ-kē-al=arm)	The **brachial artery** is the continuation of the axillary artery into the arm. The brachial artery provides the main blood supply to the arm and is superficial and palpable along its course. It begins at the tendon of the teres major muscle and ends just distal to the bend of the elbow. At first, the brachial artery is medial to the humerus, but as it descends it gradually curves laterally and passes through the cubital fossa, a triangular depression anterior to the elbow where you can easily detect the pulse of the brachial artery and listen to the various sounds when taking a person's blood pressure. Just distal to the bend in the elbow, the brachial artery divides into the radial artery and ulnar artery. Blood pressure is usually measured in the brachial artery. In order to control hemorrhage, the best place to compress the brachial artery is near the middle of the arm.
Radial artery (RĀ-dē-al=radius)	The **radial artery** is the smaller branch and is a direct continuation of the brachial artery. It passes along the lateral (radial) aspect of the forearm and then through the wrist and hand, supplying these structures with blood. At the wrist, the radial artery contacts the distal end of the radius, where it is covered only by fascia and skin. Because of its superficial location at this point, it is a common site for measuring radial pulse.
Ulnar artery (UL-nar=ulna)	The **ulnar artery,** the larger branch of the brachial artery, passes along the medial (ulnar) aspect of the forearm and then into the wrist and hand, supplying these structures with blood. In the palm, branches of the radial and ulnar arteries anastomose to form the superficial palmar arch and the deep palmar arch.

branch from the arch of the aorta is the **left common carotid artery** (ka-ROT-id), which divides into the same branches with the same names as the right common carotid artery. The third branch from the arch of the aorta is the **left subclavian artery** (sub-KLĀ-vē-an), which distributes blood to the left vertebral artery and vessels of the left upper limb. Arteries branching from the left subclavian artery are similar in distribution and name to those branching from the right subclavian artery. The following description focuses on the principal arteries originating from the brachiocephalic trunk.

■ CHECKPOINT

1. What general regions are supplied by the arteries that arise from the arch of the aorta?

BRANCH	DESCRIPTION AND REGION SUPPLIED
Superficial palmar arch (*palma*=palm)	The **superficial palmar arch** is formed mainly by the ulnar artery, with a contribution from a branch of the radial artery. The arch is superficial to the long flexor tendons of the fingers and extends across the palm at the bases of the metacarpals. It gives rise to **common palmar digital arteries,** which supply the palm. Each divides into a pair of **proper palmar digital arteries,** which supply the fingers.
Deep palmar arch	Mainly the radial artery forms the **deep palmar arch,** with a contribution from a branch of the ulnar artery. The arch is deep to the long flexor tendons of the fingers and extends across the palm, just distal to the bases of the metacarpals. Arising from the deep palmar arch are **palmar metacarpal arteries,** which supply the palm and anastomose with the common palmar digital arteries of the superficial palmar arch.
Right common carotid artery	The **right common carotid artery** begins at the bifurcation (division into two branches) of the brachiocephalic trunk, posterior to the right sternoclavicular joint, and passes superiorly in the neck to supply structures in the head (Figure 15.7b). At the superior border of the larynx (voice box), it divides into the right external and right internal carotid arteries. Pulse may be detected in the common carotid artery, just lateral to the larynx. It is convenient to detect a carotid pulse when exercising or when administering cardiopulmonary resuscitation.
External carotid artery	The **external carotid artery** begins at the superior border of the larynx and terminates near the temporomandibular joint in the parotid gland, where it divides into two branches: the superficial temporal and maxillary arteries. The carotid pulse can be detected in the external carotid artery just anterior to the sternocleidomastoid muscle at the superior border of the larynx. The general distribution of the external carotid artery is to structures *external* to the skull.
Internal carotid artery	The **internal carotid artery** has no branches in the neck and supplies structures *internal* to the skull. It enters the cranial cavity through the carotid foramen in the temporal bone. The internal carotid artery supplies blood to the eyeball and other orbital structures, ear, most of the cerebrum of the brain, pituitary gland, and external nose. The terminal branches of the internal carotid artery are the **anterior cerebral artery,** which supplies most of the medial surface of the cerebrum and deep masses of gray matter within the cerebrum, and the **middle cerebral artery,** which supplies most of the lateral surface of the cerebrum (Figure 15.7c). Inside the cranium, anastomoses of the left and right internal carotid arteries along with the basilar artery form an arrangement of blood vessels at the base of the brain near the hypophyseal fossa called the **cerebral arterial circle (circle of Willis).** From this circle (Figure 15.7c) arise arteries supplying most of the brain. Essentially, the cerebral arterial circle is formed by the union of the **anterior cerebral arteries** (branches of internal carotids) and **posterior cerebral arteries** (branches of basilar artery). The posterior cerebral arteries supply the inferolateral surface of the temporal lobe and lateral and medial surfaces of the occipital lobe of the cerebrum, deep masses of gray matter within the cerebrum, and midbrain. The posterior cerebral arteries are connected with the internal carotid arteries by the **posterior communicating arteries.** The **anterior communicating artery** connects the anterior cerebral arteries. The **internal carotid arteries** are also considered part of the cerebral arterial circle. The functions of the cerebral arterial circle are to equalize blood pressure to the brain and provide alternate routes for blood flow to the brain, should the arteries become damaged.
LEFT COMMON CAROTID ARTERY	See description in the Introduction above.
LEFT SUBCLAVIAN ARTERY	See description in the Introduction above.

continues

EXHIBIT 15.3   THE ARCH OF THE AORTA (Figure 15.7)

CONTINUED

SCHEME OF DISTRIBUTION

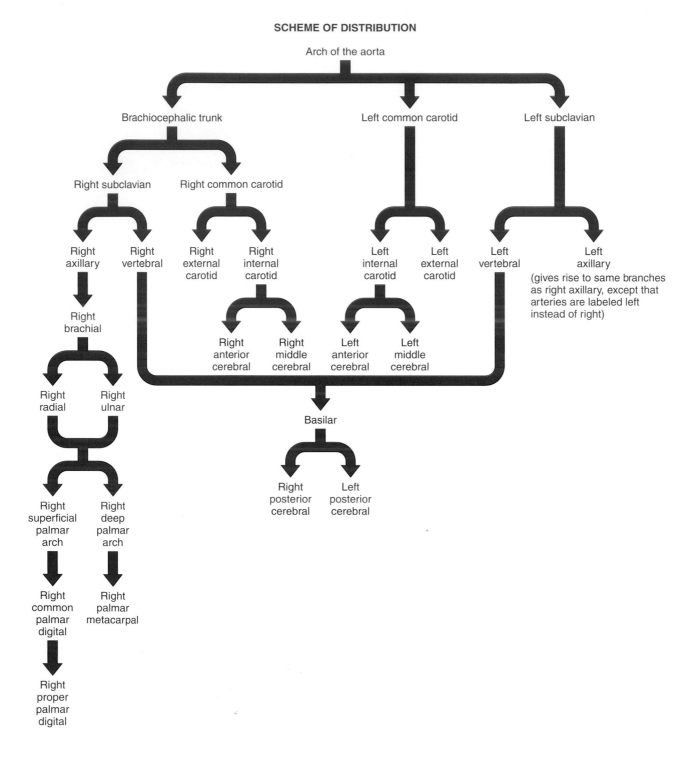

Figure 15.7 Arch of the aorta and its branches. Note in (c) the arteries that constitute the cerebral arterial circle (circle of Willis). (See Tortora, *A Photographic Atlas of the Human Body*, 2e, Figures 6.11 and 6.12.)

The arch of the aorta ends at the level of the intervertebral disc between the fourth and fifth thoracic vertebrae.

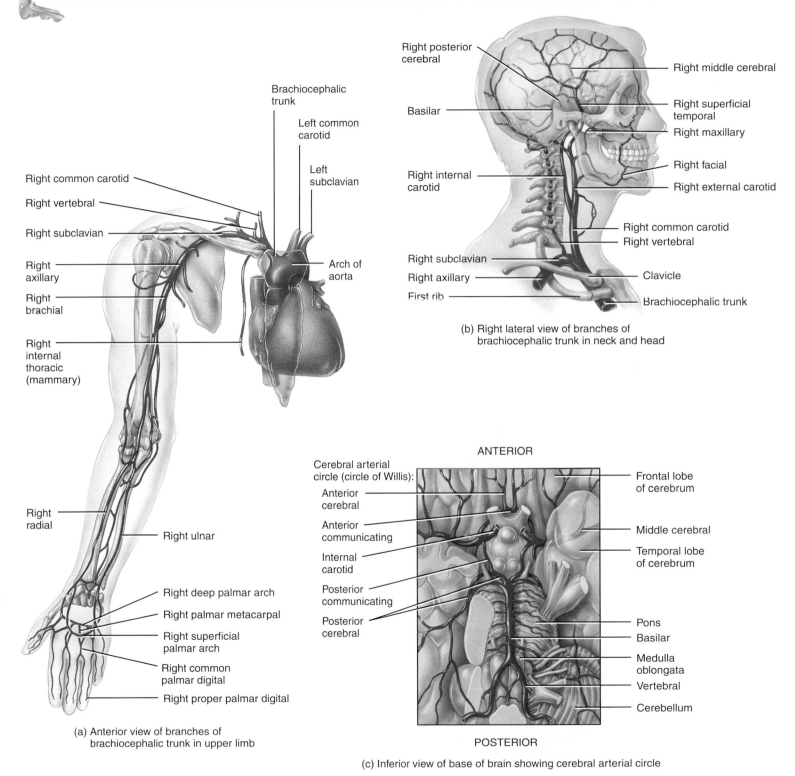

(a) Anterior view of branches of brachiocephalic trunk in upper limb

(b) Right lateral view of branches of brachiocephalic trunk in neck and head

(c) Inferior view of base of brain showing cerebral arterial circle

continues

EXHIBIT 15.3 **THE ARCH OF THE AORTA** (Figure 15.7)

CONTINUED

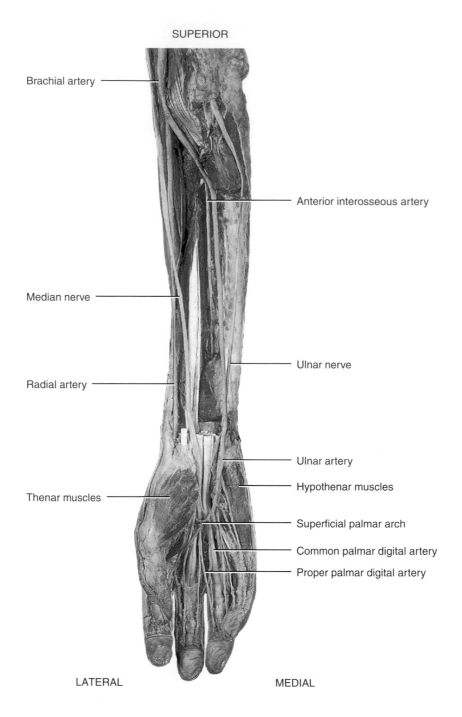

SUPERIOR

Brachial artery

Anterior interosseous artery

Median nerve

Ulnar nerve

Radial artery

Ulnar artery

Hypothenar muscles

Thenar muscles

Superficial palmar arch

Common palmar digital artery

Proper palmar digital artery

LATERAL   MEDIAL

Anterior view

(d) Anterior view of arteries of upper limb

**? What are the three major branches of the arch of the aorta,
in order of their origination?**

EXHIBIT 15.4 THORACIC AORTA (See Figure 15.6b)

OBJECTIVE

● Identify the visceral and parietal branches of the thoracic aorta.

The **thoracic aorta** is about 20 cm (8 in.) long and is a continuation of the arch of the aorta. It begins at the level of the intervertebral disc between the fourth and fifth thoracic vertebrae, where it lies to the left of the vertebral column. As it descends, it moves closer to the midline and extends through an opening in the diaphragm (aortic hiatus), which is located anterior to the vertebral column at the level of the intervertebral disc between the twelfth thoracic and first lumbar vertebrae.

Along its course, the thoracic aorta sends off numerous small arteries, **visceral branches** to viscera and **parietal branches** to body wall structures.

CHECKPOINT

1. What general regions do the visceral and parietal branches of the thoracic aorta supply?

BRANCH	DESCRIPTION AND REGION SUPPLIED
VISCERAL	
Pericardial arteries (per′-i-KAR-dē-al; *peri-*=around; *cardia-*=heart)	Two or three tiny **pericardial arteries** supply blood to the pericardium.
Bronchial arteries (BRONG-kē-al=windpipe)	One right and two left **bronchial arteries** supply the bronchial tubes, pleurae, bronchial lymph nodes, and esophagus. (Whereas the right bronchial artery arises from the third posterior intercostal artery, the two left bronchial arteries arise from the thoracic aorta.)
Esophageal arteries (e-sof′-a-JĒ-al; *eso-*=to carry; *phage-*=food)	Four or five **esophageal arteries** supply the esophagus.
Mediastinal arteries (mē′-dē-as-TĪ-nal)	Numerous small **mediastinal arteries** supply blood to structures in the mediastinum.
PARIETAL	
Posterior intercostal arteries (in′-ter-KOS-tal; *inter-*=between; *costa*=rib)	Nine pairs of **posterior intercostal arteries** supply the intercostal, pectoralis major and minor, and serratus anterior muscles; overlying subcutaneous tissue and skin; mammary glands; and vertebrae, meninges, and spinal cord.
Subcostal arteries (sub-KOS-tal; *sub-*=under)	The left and right **subcostal arteries** have a distribution similar to that of the posterior intercostals.
Superior phrenic arteries (FREN-ik=pertaining to the diaphragm)	Small **superior phrenic arteries** supply the superior and posterior surfaces of the diaphragm.

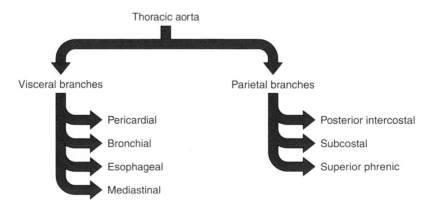

SCHEME OF DISTRIBUTION

EXHIBIT 15.5 ABDOMINAL AORTA (Figure 15.8)

OBJECTIVE

• Identify the visceral and parietal branches of the abdominal aorta.

The **abdominal aorta** is the continuation of the thoracic aorta (see Figure 15.6b). It begins at the aortic hiatus in the diaphragm and ends at about the level of the fourth lumbar vertebra, where it divides into the right and left common iliac arteries. The abdominal aorta lies anterior to the vertebral column.

As with the thoracic aorta, the abdominal aorta gives off visceral and parietal branches. The unpaired visceral branches arise from the anterior surface of the aorta and include the **celiac trunk** and the **superior mesenteric** and **inferior mesenteric arteries**

BRANCH	DESCRIPTION AND REGION SUPPLIED
UNPAIRED VISCERAL BRANCHES	

Celiac trunk (SĒ-lē-ak)	The **celiac trunk (artery)** is the first visceral branch from the aorta inferior to the diaphragm, at about the level of the twelfth thoracic vertebra (Figure 15.8a). Almost immediately, the celiac trunk divides into three branches: the left gastric, splenic, and common hepatic arteries (Figure 15.8a).
	
	1. The **left gastric artery** (GAS-trik=stomach) is the smallest of the three branches. It passes superiorly to the left toward the esophagus and then turns to follow the lesser curvature of the stomach. It supplies the stomach and esophagus.
	
	2. The **splenic artery** (SPLĒN-ik=spleen) is the largest branch of the celiac trunk. It arises from the left side of the celiac trunk distal to the left gastric artery, and passes horizontally to the left along the pancreas. Before reaching the spleen, it gives rise to three arteries:
	
	  • **Pancreatic artery** (pan-krē-AT-ik), which supplies the pancreas.
	  • **Left gastroepiploic artery** (gas′-trō-ep′-i-PLŌ-ik; *epiplo-*=omentum), which supplies the stomach and greater omentum.
	  • **Short gastric arteries,** which supply the stomach.
	
	3. The **common hepatic artery** (he-PAT-ik=liver) is intermediate in size between the left gastric and splenic arteries. Unlike the other two branches of the celiac trunk, the common hepatic artery arises from the right side. It gives rise to three arteries:
	
	  • **Proper hepatic artery,** which supplies the liver, gallbladder, and stomach.
	  • **Right gastric artery,** which supplies the stomach.
	  • **Gastroduodenal artery** (gas′-trō-doo′-ō-DĒ-nal), which supplies the stomach, duodenum of the small intestine, pancreas, and greater omentum.

Superior mesenteric artery (MES-en-ter′-ik; *meso-*=middle; *enteric*=pertaining to the intestines)	The **superior mesenteric artery** (Figure 15.8b) arises from the anterior surface of the abdominal aorta about 1 cm inferior to the celiac trunk at the level of the first lumbar vertebra. It extends inferiorly and anteriorly and between the layers of mesentery, which is a portion of the peritoneum that attaches the small intestine to the posterior abdominal wall. It anastomoses extensively and has five branches: 1. The **inferior pancreaticoduodenal artery** (pan′-krē-at′-i-kō-doo′-ō-DĒ-nal) supplies the pancreas and duodenum. 2. The **jejunal** (je-JOO-nal) and **ileal arteries** (IL-ē-al) supply the jejunum and ileum of the small intestine, respectively. 3. The **ileocolic artery** (il′-ē-ō-KŌL-ik) supplies the ileum and ascending colon of the large intestine. 4. The **right colic artery** (KŌL-ik) supplies the ascending colon. 5. The **middle colic artery** supplies the transverse colon of the large intestine.
Inferior mesenteric artery	The **inferior mesenteric artery** (Figure 15.8c) arises from the anterior aspect of the abdominal aorta at the level of the third lumbar vertebra and then passes inferiorly to the left of the aorta. It anastomoses extensively and has three branches: 1. The **left colic artery** supplies the transverse colon and descending colon of the large intestine. 2. The **sigmoid arteries** (SIG-moyd) supply the descending colon and sigmoid colon of the large intestine. 3. The **superior rectal artery** (REK-tal) supplies the rectum of the large intestine.

(see Figure 15.6b). The paired visceral branches arise from the lateral surfaces of the aorta and include the **suprarenal, renal,** and **gonadal arteries.** The unpaired parietal branch is the **median sacral artery.** The paired parietal branches arise from the posterolateral surfaces of the aorta and include the **inferior phrenic** and **lumbar arteries.**

■ CHECKPOINT

1. Name the paired visceral and parietal branches and the unpaired visceral and parietal branches of the abdominal aorta, and indicate the general regions they supply.

BRANCH	DESCRIPTION AND REGION SUPPLIED
PAIRED VISCERAL BRANCHES	
Suprarenal arteries (soo′-pra-RĒ-nal; *supra-*=above; *ren-*=kidney)	Although there are three pairs of **suprarenal (adrenal) arteries** that supply the adrenal (suprarenal) glands (superior, middle, and inferior), only the middle pair originates directly from the abdominal aorta (see Figure 15.6). The middle suprarenal arteries arise at the level of the first lumbar vertebra at or superior to the renal arteries. The superior suprarenal arteries arise from the inferior phrenic artery, and the inferior suprarenal arteries originate from the renal arteries.
Renal arteries (RĒ-nal=pertaining to the kidney)	The right and left **renal arteries** usually arise from the lateral aspects of the abdominal aorta at the superior border of the second lumbar vertebra, about 1 cm inferior to the superior mesenteric artery (see Figure 15.6). The right renal artery, which is longer than the left, arises slightly lower than the left and passes posterior to the right renal vein and inferior vena cava. The left renal artery is posterior to the left renal vein and is crossed by the inferior mesenteric vein. The renal arteries carry blood to the kidneys, adrenal (suprarenal) glands, and ureters. Their distribution within the kidneys is discussed in Chapter 26.
Gonadal (gō-NAD-al; *gon-*=seed) [**testicular** (test-TIK-ū-lar) or **ovarian arteries** (ō-VAR-ē-an)]	The **gonadal arteries** arise from the abdominal aorta at the level of the second lumbar vertebra just inferior to the renal arteries (see Figure 15.6). In males, the gonadal arteries are specifically referred to as the **testicular arteries.** They pass through the inguinal canal and supply the testes, epididymis, and ureters. In females, the gonadal arteries are called the **ovarian arteries.** They are much shorter than the testicular arteries and supply the ovaries, uterine (fallopian) tubes, and ureters.
UNPAIRED PARIETAL BRANCH	
Median sacral artery (SĀ-kral=pertaining to the sacrum)	The **median sacral artery** arises from the posterior surface of the abdominal aorta about 1 cm superior to the bifurcation of the aorta into the right and left common iliac arteries (see Figure 15.6). The median sacral artery supplies the sacrum and coccyx.
PAIRED PARIETAL BRANCHES	
Inferior phrenic arteries (FREN-ik=pertaining to the diaphragm)	The **inferior phrenic arteries** are the first paired branches of the abdominal aorta, immediately superior to the origin of the celiac trunk (see Figure 15.6). (They may also arise from the renal arteries.) The inferior phrenic arteries are distributed to the inferior surface of the diaphragm and adrenal (suprarenal) glands.
Lumbar arteries (LUM-bar=pertaining to the loin)	The four pairs of **lumbar arteries** arise from the posterolateral surface of the abdominal aorta (see Figure 15.6). They supply the lumbar vertebrae, spinal cord and its meninges, and the muscles and skin of the lumbar region of the back.

continues

EXHIBIT 15.5 ABDOMINAL AORTA (Figure 15.8)

CONTINUED

SCHEME OF DISTRIBUTION

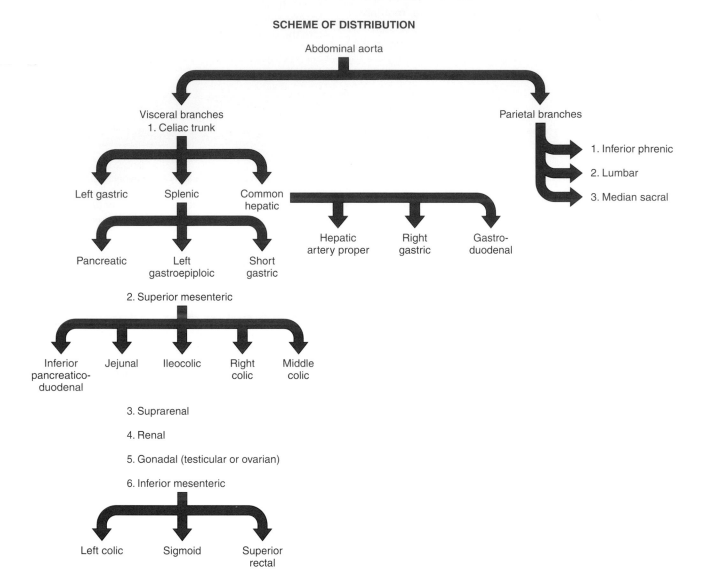

Figure 15.8 Abdominal aorta and principal branches. (See Tortora, *A Photographic Atlas of the Human Body*, 2e, Figure 6.14.)
The abdominal aorta is the continuation of the thoracic aorta.

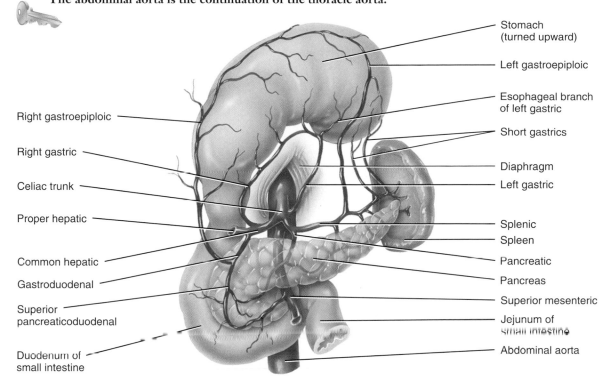

(a) Anterior view of celiac trunk and its branches

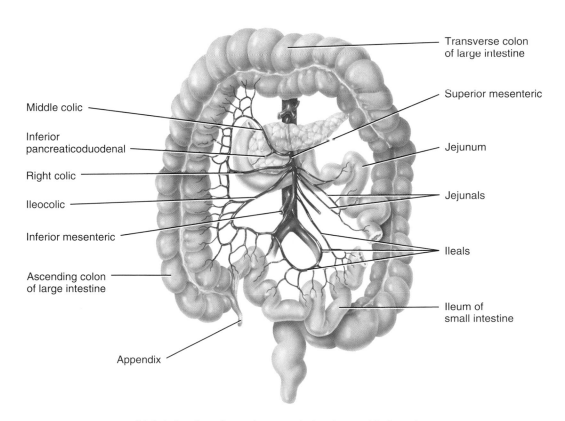

(b) Anterior view of superior mesenteric artery and its branches

continues

EXHIBIT 15.5   ABDOMINAL AORTA (Figure 15.8)

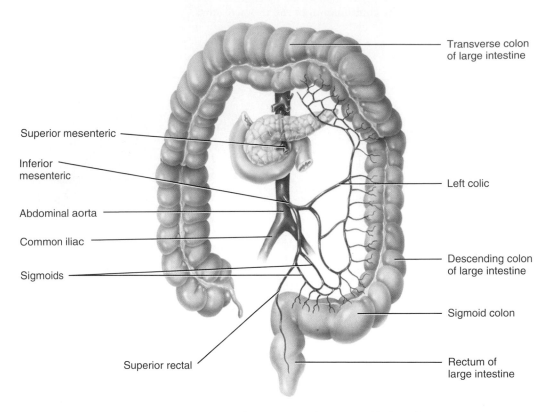

Transverse colon
of large intestine

Superior mesenteric

Inferior
mesenteric

Abdominal aorta

Common iliac

Sigmoids

Left colic

Descending colon
of large intestine

Sigmoid colon

Superior rectal

Rectum of
large intestine

(c) Anterior view of inferior mesenteric artery and its branches

SUPERIOR

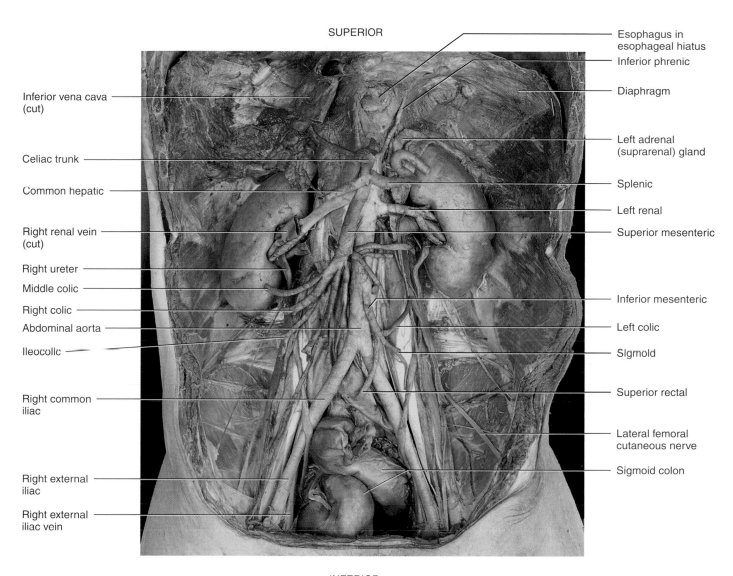

Inferior vena cava (cut)

Celiac trunk

Common hepatic

Right renal vein (cut)

Right ureter

Middle colic

Right colic

Abdominal aorta

Ileocolic

Right common iliac

Right external iliac

Right external iliac vein

Esophagus in esophageal hiatus

Inferior phrenic

Diaphragm

Left adrenal (suprarenal) gland

Splenic

Left renal

Superior mesenteric

Inferior mesenteric

Left colic

Sigmoid

Superior rectal

Lateral femoral cutaneous nerve

Sigmoid colon

INFERIOR

(d) Anterior view of arteries of the abdomen and pelvis

 Where does the abdominal aorta begin?

EXHIBIT 15.6   ARTERIES OF THE PELVIS AND LOWER LIMBS (Figure 15.9)

OBJECTIVE

● Identify the two major branches of the common iliac arteries.

The abdominal aorta ends by dividing into the right and left **common iliac arteries.** These, in turn, divide into the **internal** and **external iliac arteries.** In sequence, the external iliac become the **femoral arteries** in the thighs, the **popliteal arteries** posterior to the knee, and the **anterior** and **posterior tibial arteries** in the legs.

■ CHECKPOINT

1. What general regions do the internal and external iliac arteries supply?

BRANCH	DESCRIPTION AND REGION SUPPLIED
Common iliac arteries (IL-ē-ak=pertaining to the ilium)	At about the level of the fourth lumbar vertebra, the abdominal aorta divides into the right and left **common iliac arteries,** the terminal branches of the abdominal aorta. Each passes inferiorly about 5 cm (2 in.) and gives rise to two branches: internal iliac and external iliac arteries. The general distribution of the common iliac arteries is to the pelvis, external genitals, and lower limbs.
Internal iliac arteries	The **internal iliac (hypogastric) arteries** are the primary arteries of the pelvis. They begin at the bifurcation (division into two branches) of the common iliac arteries anterior to the sacroiliac joint at the level of the lumbosacral intervertebral disc. They pass posteromedially as they descend in the pelvis and divide into anterior and posterior divisions. The general distribution of the internal iliac arteries is to the pelvis, buttocks, external genitals, and thigh.
External iliac arteries	The **external iliac arteries** are larger than the internal iliac arteries. Like the internal iliac arteries, they begin at the bifurcation of the common iliac arteries. They descend along the medial border of the psoas major muscles following the pelvic brim, pass posterior to the midportion of the inguinal ligaments, and become the femoral arteries. The general distribution of the external iliac arteries is to the lower limbs. Specifically, branches of the external iliac arteries supply the muscles of the anterior abdominal wall, the cremaster muscle in males and the round ligament of the uterus in females, and the lower limbs.
Femoral arteries (FEM-o-ral=pertaining to the thigh)	The **femoral arteries** descend along the anteromedial aspects of the thighs to the junction of the middle and lower third of the thighs. Here they pass through an opening in the tendon of the adductor magnus muscle, where they emerge posterior to the femurs as the popliteal arteries. A pulse may be felt in the femoral artery just inferior to the midpoint of the inguinal ligament. Recall from Chapters 11 and 12 that the femoral artery, along with the femoral vein and nerve and deep inguinal lymph nodes, are all located in the *femoral triangle* (see Figure 12.15a on page 399). The general distribution of the femoral arteries is to the lower abdominal wall, groin, external genitals, and muscles of the thigh. A major branch of the femoral artery, the **deep artery of the thigh (deep femoral),** supplies most of the muscles of the thigh: quadriceps femoris, adductors, and hamstrings. Recall that in cardiac catheterization, a catheter is inserted through a blood vessel and advanced into the major vessels and heart chamber. A catheter often contains a measuring instrument or other device at its tip. To reach the left side of the heart, the catheter is inserted in the femoral artery and passed into the aorta to the coronary arteries or heart chamber.
Popliteal arteries (pop′-li-TĒ-al=posterior surface of the knee)	The **popliteal arteries** are the continuation of the femoral arteries through the popliteal fossa (space behind the knee). They descend to the inferior border of the popliteus muscles, where they divide into the anterior and posterior tibial arteries. A pulse may be detected in the popliteal arteries. In addition to supplying the adductor magnus and hamstring muscles and the skin on the posterior aspect of the legs, branches of the popliteal arteries also supply the gastrocnemius, soleus, and plantaris muscles of the calf, knee joint, femur, patella, and fibula.
Anterior tibial arteries (TIB-ē-al=pertaining to the shin bone)	The **anterior tibial arteries** descend from the bifurcation of the popliteal arteries. They are smaller than the posterior tibial arteries. The anterior tibial arteries descend through the anterior muscular compartment of the leg. They pass through the interosseous membrane that connects the tibia and fibula, lateral to the tibia. The anterior tibial arteries supply the knee joints, anterior compartment muscles of the legs, skin over the anterior aspects of the legs, and ankle joints. At the ankles, the anterior tibial arteries become the **dorsal arteries of the foot (dorsalis pedis arteries),** also arteries from which a pulse may be detected. A pulse in this artery may be taken to evaluate the peripheral vascular system. The dorsal arteries of the foot supply the muscles, skin, and joints on the dorsal aspects of the feet. On the dorsum of the feet, the dorsalis pedis arteries give off a transverse branch at the first (medial) cuneiform bone called the **arcuate arteries** (*arcuat-*=bowed) that run laterally over the bases of the metatarsals. From the arcuate arteries branch **dorsal metatarsal arteries,** which supply the feet. The dorsal metatarsal arteries terminate by dividing into the **dorsal digital arteries,** which supply the toes.
Posterior tibial arteries	The **posterior tibial arteries,** the direct continuations of the popliteal arteries, descend from the bifurcation of the popliteal arteries. They pass down the posterior muscular compartment of the legs posterior to the medial malleolus of the tibia. They terminate by dividing into the medial and lateral plantar arteries. Their general distribution is to the muscles, bones, and joints of the leg and foot. Major branches of the posterior tibial arteries are the **fibular (peroneal) arteries,** which supply the fibularis, soleus, tibialis posterior, and flexor hallucis muscles. They also supply the fibula, tarsus, and lateral aspect of the heel. The bifurcation of the posterior tibial arteries into the medial and lateral plantar arteries occurs deep to the flexor retinaculum on the medial side of the feet. The **medial plantar arteries** (PLAN-tar=sole) supply the abductor hallucis and flexor digitorum brevis muscles and the toes. The **lateral plantar arteries** unite with a branch of the dorsal arteries of the foot to form the **plantar arch.** The arch begins at the base of the fifth metatarsal and extends medially across the metacarpals. As the arch crosses the foot, it gives off **plantar metatarsal arteries,** which supply the feet. These terminate by dividing into **plantar digital arteries,** which supply the toes.

SCHEME OF DISTRIBUTION

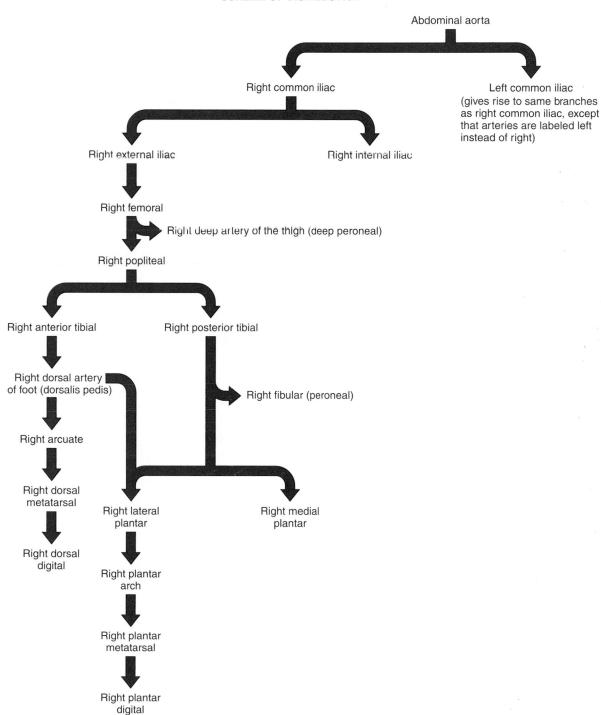

continues

EXHIBIT 15.6   ARTERIES OF THE PELVIS AND LOWER LIMBS (See Figure 15.9)

CONTINUED

Figure 15.9   Arteries of the pelvis and right lower limb.

The internal iliac arteries carry most of the blood supply to the pelvic viscera and wall.

Abdominal aorta

Right common iliac

Left common iliac

Right internal iliac

Right external iliac

Right deep artery of the thigh (deep femoral)

Right femoral

Right popliteal

Right anterior tibial

Right posterior tibial

Right fibular (peroneal)

Right dorsal artery of foot (dorsalis pedis)

Right lateral plantar

Right medial plantar

Right arcuate

Right dorsal metatarsal

Right plantar arch

Right plantar metatarsal

Right dorsal digital

Right plantar digital

(a) Anterior view

(b) Posterior view

SUPERIOR

External iliac vein

External iliac artery

Inguinal ligament

Tensor fasciae
latae muscle

Femoral nerve

Deep artery of
the thigh (deep femoral)

Lateral circumflex vein

Femoral vein

Femoral artery

Muscular branch

Iliotibial tract

Vastus lateralis
muscle

Sartorius muscle

Rectus femoris
muscle

LATERAL

Right ureter

Urinary bladder

Ductus (vas) deferens

Spermatic cord

Pectineus muscle

Inguinal lymph node

Scrotum

Adductor longus muscle

Gracilis muscle

MEDIAL

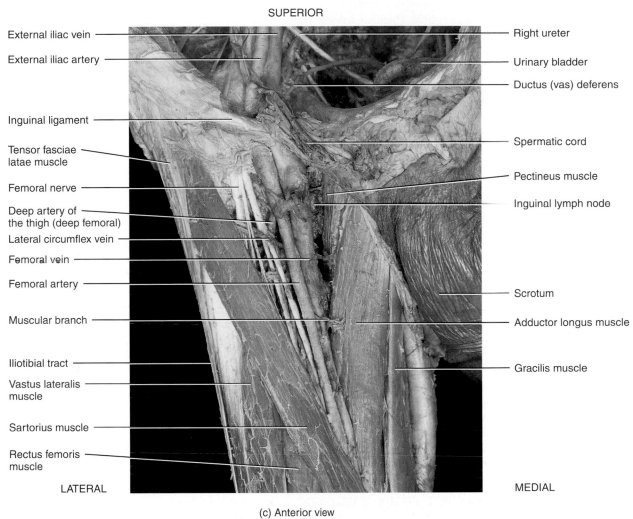

(c) Anterior view

? At what point does the abdominal aorta divide into the common iliac arteries?

481

EXHIBIT 15.7   VEINS OF THE SYSTEMIC CIRCULATION (Figure 15.10)

- Identify the three systemic veins that return deoxygenated blood to the heart.

Whereas arteries distribute blood to various parts of the body, veins drain blood away from them. For the most part, arteries are deep, whereas veins may be superficial or deep. Superficial veins are located just beneath the skin and can be seen easily. Because there are no large superficial arteries, the names of superficial veins do not correspond to those of arteries. Superficial veins are clinically important as sites for withdrawing blood or giving injections. Deep veins generally travel alongside arteries and usually bear the same name. Arteries usually follow definite pathways; veins are more difficult to follow because they connect in irregular networks in which many tributaries merge to form a large vein. Also, the number and spatial distribution of veins are more variable than arteries. Although only one systemic artery, the aorta, takes oxygenated blood away from the heart (left ventricle), three systemic veins, the **coronary sinus, superior vena cava,** and **inferior vena cava,** return deoxygenated blood to the heart (right atrium). The coronary sinus receives blood from the cardiac veins; the superior vena cava receives blood from other veins superior to the diaphragm, except the air sacs (alveoli) of the lungs; the inferior vena cava receives blood from veins inferior to the diaphragm.

■ CHECKPOINT

1. What are the three tributaries of the coronary sinus?

VEIN	DESCRIPTION AND REGION DRAINED
CORONARY SINUS (KOR-ō-nar-ē; *corona*=crown)	The **coronary sinus** is the main vein of the heart; it receives almost all venous blood from the myocardium. It is located in the coronary sulcus (see Figure 14.3c on page 429) and opens into the right atrium between the orifice of the inferior vena cava and the tricuspid valve. It is a wide venous channel into which three veins drain. It receives the **great cardiac vein** (in the anterior interventricular sulcus) into its left end, and the **middle cardiac vein** (in the posterior interventricular sulcus) and the **small cardiac vein** into its right end. Several **anterior cardiac veins** drain directly into the right atrium.
SUPERIOR VENA CAVA (SVC) (VĒ-na CĀ-va; *vena*=vein; *cava*=cavelike)	The **superior vena cava** is about 7.5 cm (3 in.) long and 2 cm (1 in.) in diameter and empties its blood into the superior part of the right atrium. It begins posterior to the right first costal cartilage by the union of the right and left brachiocephalic veins and ends at the level of the right third costal cartilage, where it enters the right atrium. The SVC drains the head, neck, chest, and upper limbs.
INFERIOR VENA CAVA (IVC)	The **inferior vena cava** is the largest vein in the body, about 3.5 cm (1.4 in.) in diameter. It begins anterior to the fifth lumbar vertebra by the union of the common iliac veins, ascends behind the peritoneum to the right of the midline, pierces the caval opening of the diaphragm at the level of the eighth thoracic vertebra, and enters the inferior part of the right atrium. The IVC drains the abdomen, pelvis, and lower limbs. The inferior vena cava is commonly compressed during the later stages of pregnancy by the enlarging uterus, producing edema of the ankles and feet and temporary varicose veins.

Figure 15.10   Principal veins.

Deoxygenated blood returns to the heart via the superior and inferior venae cavae and the coronary sinus.

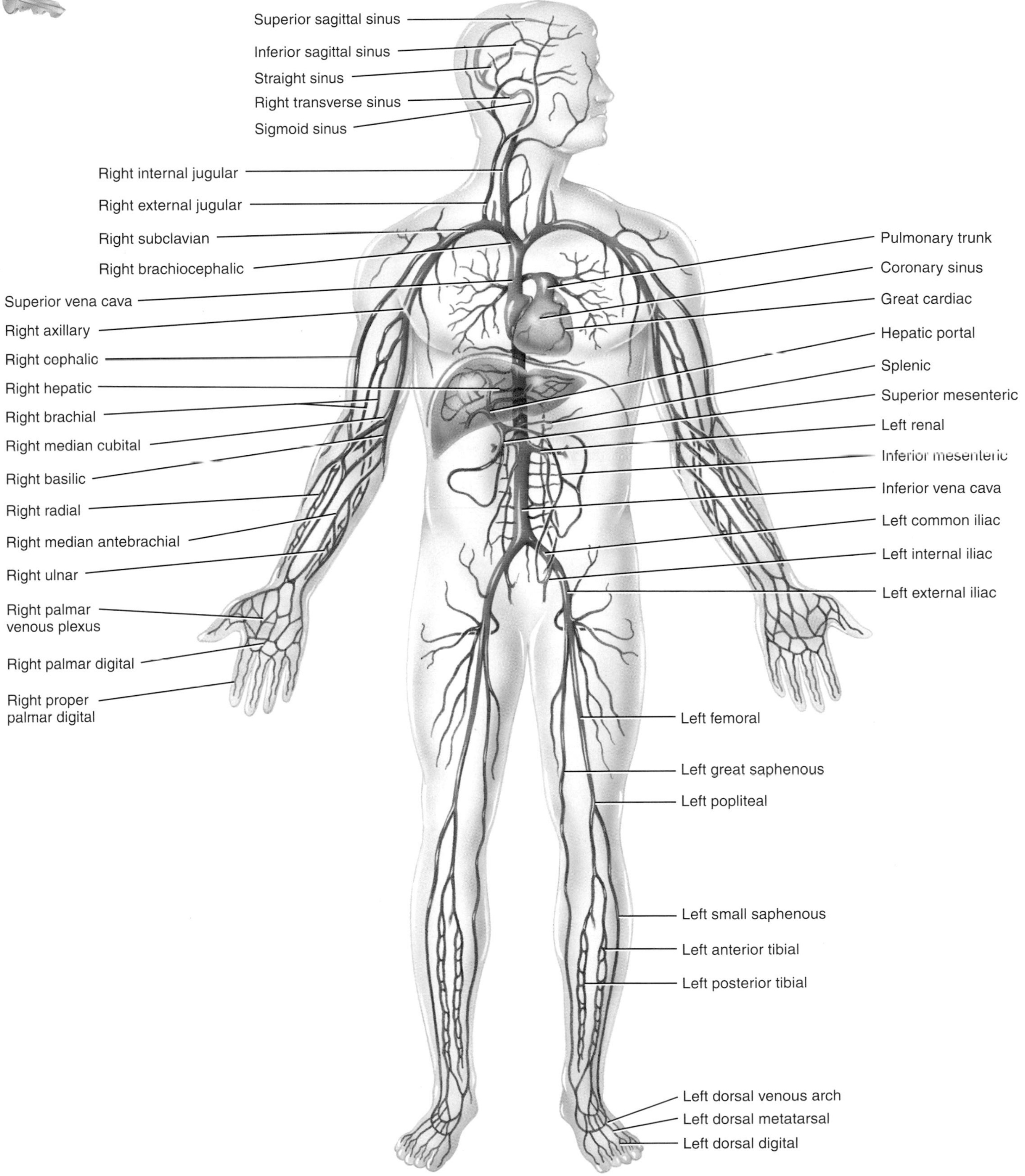

Superior sagittal sinus

Inferior sagittal sinus

Straight sinus

Right transverse sinus

Sigmoid sinus

Right internal jugular

Right external jugular

Right subclavian

Right brachiocephalic

Superior vena cava

Right axillary

Right cophalic

Right hepatic

Right brachial

Right median cubital

Right basilic

Right radial

Right median antebrachial

Right ulnar

Right palmar
venous plexus

Right palmar digital

Right proper
palmar digital

Pulmonary trunk

Coronary sinus

Great cardiac

Hepatic portal

Splenic

Superior mesenteric

Left renal

Inferior mesenteric

Inferior vena cava

Left common iliac

Left internal iliac

Left external iliac

Left femoral

Left great saphenous

Left popliteal

Left small saphenous

Left anterior tibial

Left posterior tibial

Left dorsal venous arch

Left dorsal metatarsal

Left dorsal digital

Overall anterior view of the principal veins

Which general regions of the body are drained by the superior vena cava and the inferior vena cava?

EXHIBIT 15.8   VEINS OF THE HEAD AND NECK (Figure 15.11)

OBJECTIVE

● Identify the three major veins that drain blood from the head.

Most blood draining from the head passes into three pairs of veins: the **internal jugular, external jugular,** and **vertebral veins.** Within the brain, all veins drain into dural venous sinuses and then into the internal jugular veins. **Dural venous sinuses** are endothelial-lined venous channels between layers of the cranial dura mater.

■ **CHECKPOINT**

1. Which general areas are drained by the internal jugular, external jugular, and vertebral veins?

VEIN	DESCRIPTION AND REGION DRAINED
INTERNAL JUGULAR VEINS (JUG-ū-lar; *jugular*=throat)	The flow of blood from the dural venous sinuses into the internal jugular veins is as follows (Figure 15.11): The **superior sagittal sinus** (SAJ-i-tal=straight) begins at the frontal bone, where it receives a vein from the nasal cavity, and passes posteriorly to the occipital bone. Along its course, it receives blood from the superior, medial, and lateral aspects of the cerebral hemispheres, meninges, and cranial bones. The superior sagittal sinus usually turns to the right and drains into the right transverse sinus. The **inferior sagittal sinus** is much smaller than the superior sagittal sinus; it begins posterior to the attachment of the falx cerebri and receives the great cerebral vein to become the straight sinus. The great cerebral vein drains the deeper parts of the brain. Along its course the inferior sagittal sinus also receives tributaries from the superior and medial aspects of the cerebral hemispheres.
	The **straight sinus** runs in the tentorium cerebelli and is formed by the union of the inferior sagittal sinus and the great cerebral vein. The straight sinus also receives blood from the cerebellum and usually drains into the left transverse sinus.
	The **transverse sinuses** begin near the occipital bone, pass laterally and anteriorly, and become the sigmoid sinuses near the temporal bone. The transverse sinuses receive blood from the cerebrum, cerebellum, and cranial bones.
	The **sigmoid sinuses** (SIG-moyd=S-shaped) are located along the temporal bone. They pass through the jugular foramina, where they terminate in the internal jugular veins. The sigmoid sinuses drain the transverse sinuses.
	The **cavernous sinuses** (KAV-er-nus=cavelike) are located on either side of the sphenoid bone. They receive blood from the ophthalmic veins from the orbits, and from the cerebral veins from the cerebral hemispheres. They ultimately empty into the transverse sinuses and internal jugular veins. The cavernous sinuses are unique because they have nerves and a major blood vessel passing through them on their way to the orbit and face. The oculomotor (III) nerve, trochlear (IV) nerve, and ophthalmic and maxillary branches of the trigeminal (V) nerve, as well as the internal carotid arteries pass through the cavernous sinuses.
	The right and left **internal jugular veins** pass inferiorly on either side of the neck lateral to the internal carotid and common carotid arteries. They then unite with the subclavian veins posterior to the clavicles at the sternoclavicular joints to form the right and left **brachiocephalic veins** (brā'-kē-ō-se-FAL-ik; *brachio-*=arm; *cephal-*=head). From here blood flows into the superior vena cava. The general structures drained by the internal jugular veins are the brain (through the dural venous sinuses), face, and neck.
EXTERNAL JUGULAR VEINS	The right and left **external jugular veins** begin in the parotid glands near the angle of the mandible. They are superficial veins that descend through the neck across the sternocleido-mastoid muscles. They terminate at a point opposite the middle of the clavicle, where they empty into the subclavian veins. The general structures drained by the external jugular veins are external to the cranium, such as the scalp and superficial and deep regions of the face. When venous pressure rises, for example, during heavy coughing or straining or in cases of heart failure, the external jugular veins become very prominent along the side of the neck.
VERTEBRAL VEINS (VER-te-bral; *vertebra*=vertebrae)	The right and left **vertebral veins** originate inferior to the occipital condyles. They descend through successive transverse foramina of the first six cervical vertebrae and emerge from the foramina of the sixth cervical vertebra to enter the brachiocephalic veins in the root of the neck. The vertebral veins drain deep structures in the neck such as the cervical vertebrae, cervical spinal cord, and some neck muscles.

SCHEME OF DRAINAGE

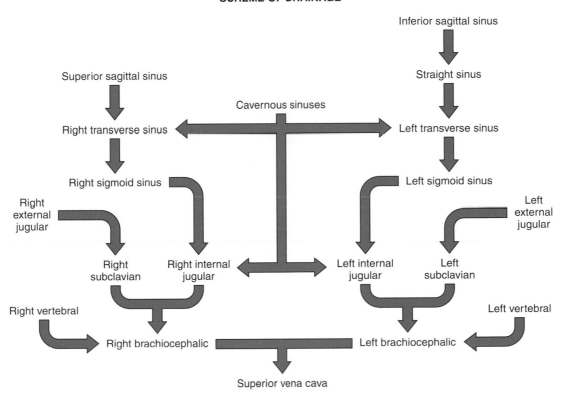

EXHIBIT 15.8   VEINS OF THE HEAD AND NECK (Figure 15.11)

CONTINUED

Figure 15.11   Principal veins of the head and neck.

Blood draining from the head passes into the internal jugular, external jugular, and vertebral veins.

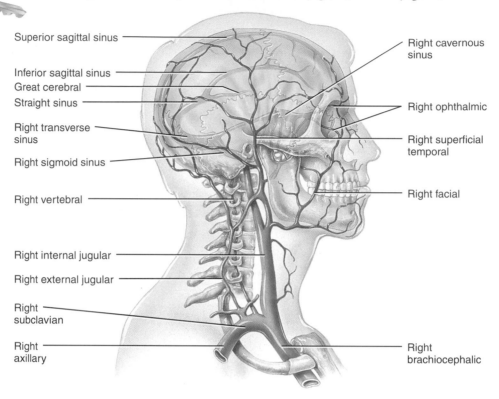

Superior sagittal sinus

Right cavernous sinus

Inferior sagittal sinus

Great cerebral

Straight sinus

Right ophthalmic

Right transverse sinus

Right superficial temporal

Right sigmoid sinus

Right vertebral

Right facial

Right internal jugular

Right external jugular

Right subclavian

Right axillary

Right brachiocephalic

(a) Right lateral view

SUPERIOR

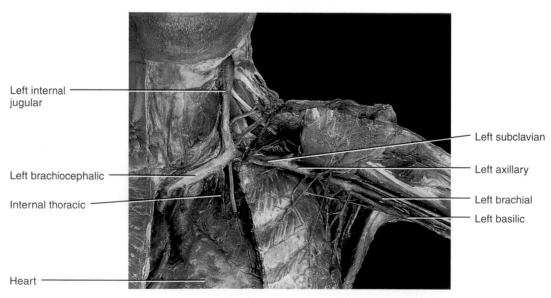

Left internal jugular

Left brachiocephalic

Internal thoracic

Heart

Left subclavian

Left axillary

Left brachial

Left basilic

INFERIOR

(b) Anterior view

Into which veins in the neck does all venous blood in the brain drain?

EXHIBIT 15.9   VEINS OF THE UPPER LIMBS (Figure 15.12)

OBJECTIVE

● Identify the principal veins that drain the upper limbs.

Both superficial and deep veins return blood from the upper limbs to the heart. **Superficial veins** are located just deep to the skin and are often visible. They anastomose extensively with one another and with deep veins, and they do not accompany arteries. Superficial veins are larger than deep veins and return most of the blood from the upper limbs. **Deep veins** are located deep in the body. They usually accompany arteries and have the same names as the corresponding arteries. Both superficial and deep veins have valves, but they are more numerous in the deep veins.

■ **CHECKPOINT**

1. Where do the cephalic, basilic, median antebrachial, radial, and ulnar veins originate?

VEIN	DESCRIPTION AND REGION DRAINED
SUPERFICIAL	
Cephalic veins (se-FAL-ik=pertaining to the head)	The principal superficial veins that drain the upper limbs are the cephalic and basilic veins. They originate in the hand and convey blood from the smaller superficial veins into the axillary veins. The **cephalic veins** begin on the lateral aspect of the **dorsal venous networks of the hands (dorsal venous arches),** networks of veins on the dorsum of the hands formed by the **dorsal metacarpal veins** (Figure 15.12a). These veins, in turn, drain the **dorsal digital veins,** which pass along the sides of the fingers. Following their formation from the dorsal venous networks of the hands, the cephalic veins arch around the radial side of the forearms to the anterior surface and ascend through the entire limbs along the anterolateral surface. The cephalic veins end where they join the axillary veins, just inferior to the clavicles. **Accessory cephalic veins** originate either from a venous plexus on the dorsum of the forearms or from the medial aspects of the dorsal venous networks of the hands and unite with the cephalic veins just inferior to the elbow. The cephalic veins drain blood from the lateral aspect of the upper limbs.
Basilic veins (ba-SIL-ik=royal, of prime importance)	The **basilic veins** begin on the medial aspects of the dorsal venous networks of the hands and ascend along the posteromedial surface of the forearm and anteromedial surface of the arm (Figure 15.12b). They drain blood from the medial aspects of the upper limbs. Anterior to the elbow, the basilic veins are connected to the cephalic veins by the **median cubital veins** (*cubital*=pertaining to the elbow), which drain the forearm. If veins must be punctured for an injection, transfusion, or removal of a blood sample, the medial cubital veins are preferred. After receiving the median cubital veins, the basilic veins continue ascending until they reach the middle of the arm. There they penetrate the tissues deeply and run alongside the brachial arteries until they join the brachial veins. As the basilic and brachial veins merge in the axillary area, they form the axillary veins.
Median antebrachial veins (an′-tē-BRĀ-kē-al; *ante-*=before, in front of; *brachi-*=arm)	The **median antebrachial veins (median veins of the forearm)** begin in the **palmar venous plexuses,** networks of veins on the palms. The plexuses drain the **palmar digital veins** in the fingers. The median antebrachial veins ascend anteriorly in the forearms to join the basilic or median cubital veins, sometimes both. They drain the palms and forearms.
DEEP	
Radial veins (RĀ-dē-al=pertaining to the radius)	The paired **radial veins** begin at the **deep palmar venous arches** (Figure 15.12c). These arches drain the **palmar metacarpal veins** in the palms. The radial veins drain the lateral aspects of the forearms and pass alongside the radial arteries. Just inferior to the elbow joint, the radial veins unite with the ulnar veins to form the brachial veins.
Ulnar veins (UL-nar=pertaining to the ulna)	The paired **ulnar veins,** which are larger than the radial veins, begin at the **superficial palmar venous arches.** These arches drain the **common palmar digital veins** and the **proper palmar digital veins** in the fingers. The ulnar veins drain the medial aspect of the forearms, pass alongside the ulnar arteries, and join with the radial veins to form the brachial veins.
Brachial veins (BRĀ-kē-al; *brachi-*=arm)	The paired **brachial veins** accompany the brachial arteries. They drain the forearms, elbow joints, arms, and humerus. They pass superiorly and join with the basilic veins to form the axillary veins.
Axillary veins (AK-sil-ār-ē; *axilla*=armpit)	The **axillary veins** ascend to the outer borders of the first ribs, where they become the subclavian veins. The axillary veins receive tributaries that correspond to the branches of the axillary arteries. The axillary veins drain the arms, axillas, and superolateral chest wall.
Subclavian veins (sub-KLĀ-vē-an; *sub-*=under; *clavian*=pertaining to the clavicle)	The **subclavian veins** are continuations of the axillary veins that terminate at the sternal end of the clavicles, where they unite with the internal jugular veins to form the brachiocephalic veins. The subclavian veins drain the arms, neck, and thoracic wall. The thoracic duct of the lymphatic system delivers lymph into the junction between the left subclavian vein and left internal jugular veins. The right lymphatic duct delivers lymph into the junction between the right subclavian and right internal jugular veins (see Figure 16.3 on page 515). In a procedure called *central line placement,* the right subclavian vein is frequently used to administer nutrients and medication and measure venous pressure.

continues

EXHIBIT 15.9   VEINS OF THE UPPER LIMBS (Figure 15.12)

SCHEME OF DRAINAGE

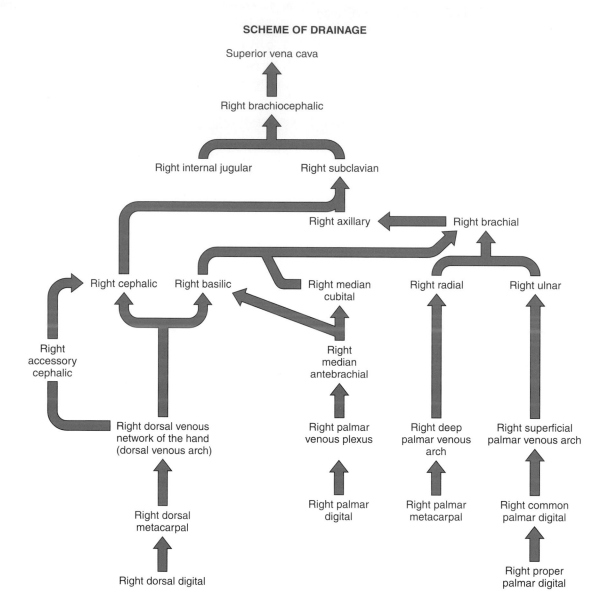

Figure 15.12 **Principal veins of the right upper limb.** (See Tortora, *A Photographic Atlas of the Human Body*, 2e, Figure 6.20.)

Deep veins usually accompany arteries that have similar names.

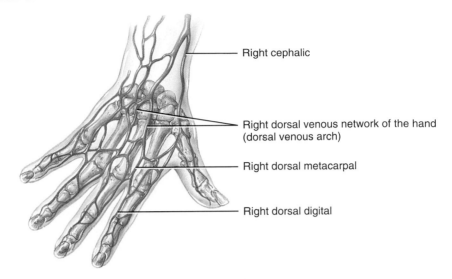

Right cephalic

Right dorsal venous network of the hand (dorsal venous arch)

Right dorsal metacarpal

Right dorsal digital

(a) Posterior view of superficial veins of the hand

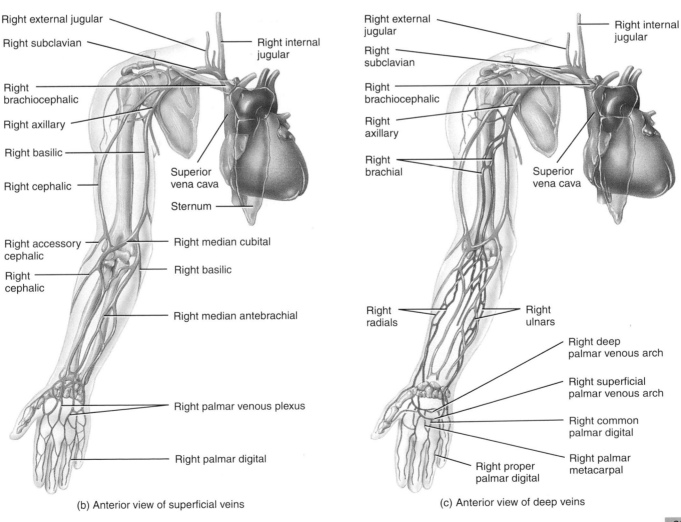

Right external jugular

Right subclavian

Right brachiocephalic

Right axillary

Right basilic

Right cephalic

Right internal jugular

Superior vena cava

Sternum

Right accessory cephalic

Right median cubital

Right cephalic

Right basilic

Right median antebrachial

Right palmar venous plexus

Right palmar digital

(b) Anterior view of superficial veins

Right external jugular

Right subclavian

Right brachiocephalic

Right axillary

Right brachial

Right internal jugular

Superior vena cava

Right radials

Right ulnars

Right deep palmar venous arch

Right superficial palmar venous arch

Right common palmar digital

Right palmar metacarpal

Right proper palmar digital

(c) Anterior view of deep veins

continues

EXHIBIT 15.9 **VEINS OF THE UPPER LIMBS** (Figure 15.12)

SUPERIOR

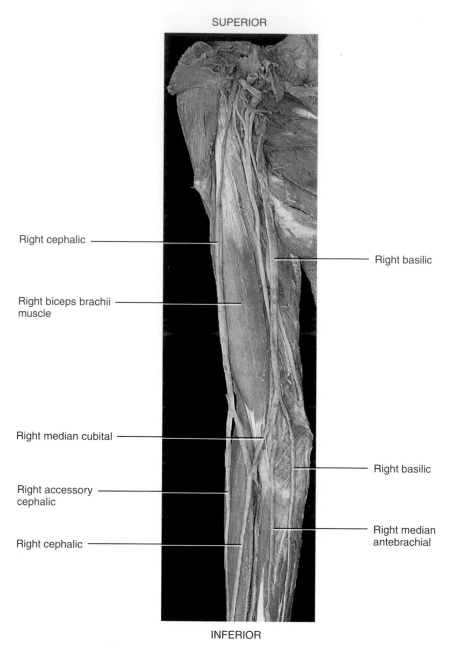

Right cephalic

Right basilic

Right biceps brachii
muscle

Right median cubital

Right basilic

Right accessory
cephalic

Right median
antebrachial

Right cephalic

INFERIOR

(d) Anterior view of superficial veins of arm and forearm

From which vein in the upper limb is a blood sample often taken?

EXHIBIT 15.10   VEINS OF THE THORAX (Figure 15.13)

OBJECTIVE

● Identify the components of the azygos system of veins.

Although the brachiocephalic veins drain some portions of the thorax, most thoracic structures are drained by a network of veins, called the **azygos system,** that runs on either side of the vertebral column. The system consists of three veins — the **azygos, hemiazygos,** and **accessory hemiazygos veins** — that show considerable variation in origin, course, tributaries, anastomoses, and termination. Ultimately they empty into the superior vena cava.

■ CHECKPOINT

1. What is the importance of the azygos system relative to the inferior vena cava?

VEIN	DESCRIPTION AND REGION DRAINED
BRACHIOCEPHALIC VEIN (brā-kē-ō-se-FAL-ik; *brachio-*=arm; *cephalic*=pertaining to the head)	The right and left **brachiocephalic veins,** formed by the union of the subclavian and internal jugular veins, drain blood from the head, neck, upper limbs, mammary glands, and superior thorax. The brachiocephalic veins unite to form the superior vena cava. Because the superior vena cava is to the right of the body's midline, the left brachiocephalic vein is longer than the right. The right brachiocephalic vein is anterior and to the right of the brachiocephalic trunk. The left brachiocephalic vein is anterior to the brachiocephalic trunk, the left common carotid and left subclavian arteries, the trachea, the left vagus (x) nerve, and the phrenic nerve.
AZYGOS SYSTEM (az-Ī-gos=unpaired)	The **azygos system,** besides collecting blood from the thorax and abdominal wall, may serve as a bypass for the inferior vena cava that drains blood from the lower body. Several small veins directly link the azygos system with the inferior vena cava. Large veins that drain the lower limbs and abdomen conduct blood into the azygos system. If the inferior vena cava or hepatic portal vein becomes obstructed, the azygos system can return blood from the lower body to the superior vena cava.
Azygos vein	The **azygos vein** is anterior to the vertebral column, slightly to the right of the midline. It usually begins at the junction of the right ascending lumbar and right subcostal veins near the diaphragm. At the level of the fourth thoracic vertebra, it arches over the root of the right lung to end in the superior vena cava. Generally, the azygos vein drains the right side of the thoracic wall, thoracic viscera, and abdominal wall. Specifically, the azygos vein receives blood from most of the **right posterior intercostal, hemiazygos, accessory hemiazygos, esophageal, mediastinal, pericardial,** and **bronchial veins.**
Hemiazygos vein (HEM-ē-az-Ī-gos; *hemi-*=half)	The **hemiazygos vein** is anterior to the vertebral column and slightly to the left of the midline. It often begins at the junction of the left ascending lumbar and left subcostal veins. It terminates by joining the azygos vein at about the level of the ninth thoracic vertebra. Generally, the hemiazygos vein drains the left side of the thoracic wall, thoracic viscera, and abdominal wall. Specifically, the hemiazygos vein receives blood from the ninth through eleventh **left posterior intercostal, esophageal, mediastinal,** and sometimes the **accessory hemiazygos veins.**
Accessory hemiazygos vein	The **accessory hemiazygos vein** is also anterior to the vertebral column and to the left of the midline. It begins at the fourth or fifth intercostal space and descends from the fifth to the eighth thoracic vertebra or ends in the hemiazygos vein. It terminates by joining the azygos vein at about the level of the eighth thoracic vertebra. The accessory hemiazygos vein drains the left side of the thoracic wall. It receives blood from the fourth through eighth **left posterior intercostal veins** (the first through third left posterior intercostal veins open into the left brachiocephalic vein), **left bronchial,** and **mediastinal veins.**

SCHEME OF DRAINAGE

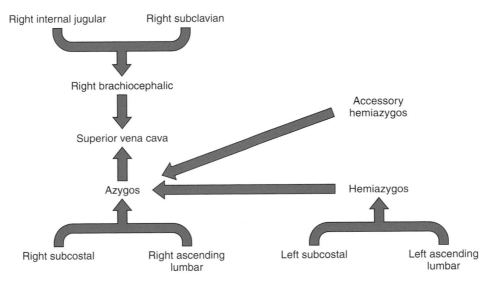

continues

EXHIBIT 15.10   VEINS OF THE THORAX (Figure 15.13)

Figure 15.13   Principal veins of the thorax, abdomen, and pelvis.

Most thoracic structures are drained by the azygos system of veins.

Right internal jugular

Right external jugular

Right brachiocephalic

Superior vena cava

Right posterior intercostal

Azygos

Mediastinals

Bronchial

Pericardial

Diaphragm

Hepatics

Right suprarenal

Right subcostal

Right renal

Right ascending lumbar

Right gonadal (testicular or ovarian)

Right lumbar

Right common iliac

Right internal iliac

Right external iliac

Left internal jugular

Left external jugular

Left subclavian

Left brachiocephalic

Left superior intercostal

Left axillary

Left cephalic

Left posterior intercostal

Left brachial

Accessory hemiazygos

Left basilic

Esophageals

Left inferior phrenics

Hemiazygos

Left suprarenal

Left renal

Left ascending lumbar

Left gonadal (testicular or ovarian)

Inferior vena cava

Left common iliac

Inguinal ligament

Middle sacral

Left internal iliac

Left external iliac

Left femoral

(a) Anterior view

SUPERIOR

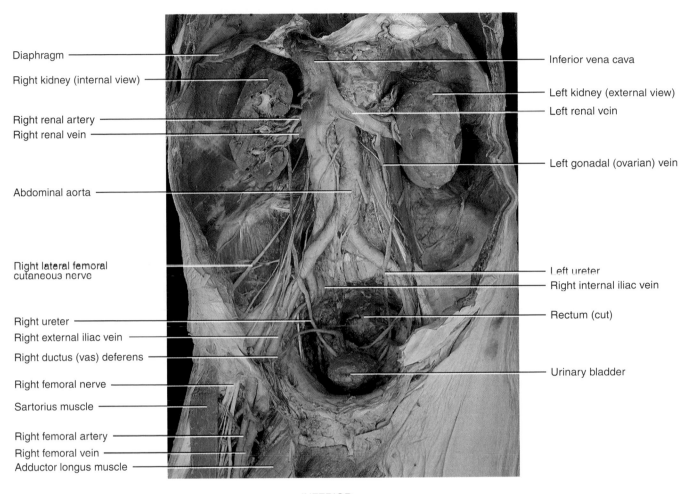

Diaphragm

Right kidney (internal view)

Right renal artery
Right renal vein

Abdominal aorta

Right lateral femoral
cutaneous nerve

Right ureter
Right external iliac vein
Right ductus (vas) deferens

Right femoral nerve

Sartorius muscle

Right femoral artery
Right femoral vein
Adductor longus muscle

Inferior vena cava

Left kidney (external view)
Left renal vein

Left gonadal (ovarian) vein

Left ureter
Right internal iliac vein

Rectum (cut)

Urinary bladder

INFERIOR

(b) Anterior view

Which vein returns blood from the abdominopelvic viscera to the heart?

493

EXHIBIT 15.11   VEINS OF THE ABDOMEN AND PELVIS

OBJECTIVE

● Identify the principal veins that drain the abdomen and pelvis.

Blood from the abdominal and pelvic viscera and abdominal wall returns to the heart via the **inferior vena cava.** Many small veins enter the inferior vena cava. Most carry return flow from parietal branches of the abdominal aorta, and their names correspond to the names of the arteries.

The inferior vena cava does not receive veins directly from the gastrointestinal tract, spleen, pancreas, and gallbladder. These organs pass their blood into a common vein, the **hepatic portal vein,** which delivers the blood to the liver. The superior mesenteric

and splenic veins unite to form the hepatic portal vein (see Figure 15.15). This special flow of venous blood, called the **hepatic portal circulation,** will be described shortly. After passing through the liver for processing, blood drains into the hepatic veins, which empty into the inferior vena cava.

■ **CHECKPOINT**

1. What structures do the lumbar, gonadal, renal, suprarenal, inferior phrenic, and hepatic veins drain?

VEIN	DESCRIPTION AND REGION DRAINED
Inferior Vena Cava (VĒ-na CĀ-va; *vena*=vein; *cava*=cavelike)	The two common iliac veins that drain the lower limbs, pelvis, and abdomen unite to form the **inferior vena cava.** The inferior vena cava extends superiorly through the abdomen and thorax to the right atrium.
Common Iliac Veins (IL-ē-ak=pertaining to the ilium)	The **common iliac veins** are formed by the union of the internal and external iliac veins anterior to the sacroiliac joint and represent the distal continuation of the inferior vena cava at their bifurcation. The right common iliac vein is much shorter than the left and is also more vertical. Generally, the common iliac veins drain the pelvis, external genitals, and lower limbs.
Internal Iliac Veins	The **internal iliac veins** begin near the superior portion of the greater sciatic notch and run medial to their corresponding arteries. Generally, the veins drain the thigh, buttocks, external genitals, and pelvis.
External Iliac Veins	The **external iliac veins** are companions of the internal iliac arteries and begin at the inguinal ligaments as continuations of the femoral veins. They end anterior to the sacroiliac joint where they join with the internal iliac veins to form the common iliac veins. The external iliac veins drain the lower limbs, cremaster muscle in males, and the abdominal wall.
Lumbar Veins (LUM-bar=pertaining to the loin)	A series of parallel **lumbar veins,** usually four on each side, drain blood from both sides of the posterior abdominal wall, vertebral canal, spinal cord, and meninges. The lumbar veins run horizontally with the lumbar arteries. The lumbar veins connect at right angles with the right and left **ascending lumbar veins,** which form the origin of the corresponding azygos or hemiazygos vein. The lumbar veins drain blood into the ascending lumbars and then run to the inferior vena cava, where they release the remainder of the flow.
Gonadal Veins (gō-NAD-al; *gono*=seed) [**testicular** (tes-TIK-ū-lar) or **ovarian** (ō-VAR-ē-an)]	The **gonadal veins** ascend with the gonadal arteries along the posterior abdominal wall. In the male, the gonadal veins are called the testicular veins. The **testicular veins** drain the testes (the left testicular vein empties into the left renal vein, and the right testicular vein drains into the inferior vena cava). In the female, the gonadal veins are called the ovarian veins. The **ovarian veins** drain the ovaries. The left ovarian vein empties into the left renal vein, and the right ovarian vein drains into the inferior vena cava.
Renal Veins (RĒ-nal=pertaining to the kidney)	The **renal veins** are large and pass anterior to the renal arteries. The left renal vein is longer than the right renal vein and passes anterior to the abdominal aorta. It receives the left testicular (or ovarian), left inferior phrenic, and usually left suprarenal veins. The right renal vein empties into the inferior vena cava posterior to the duodenum. The renal veins drain the kidneys.
Suprarenal Veins (soo'-pra-RĒ-nal; *supra-*=above)	The **suprarenal veins** drain the adrenal (suprarenal) glands (the left suprarenal vein empties into the left renal vein, and the right suprarenal vein empties into the inferior vena cava).
Inferior Phrenic Veins (FREN-ik=pertaining to diaphragm)	The **inferior phrenic veins** drain the diaphragm (the left inferior phrenic vein usually sends one tributary to the left suprarenal vein, which empties into the left renal vein, and another tributary that empties into the inferior vena cava; the right inferior phrenic vein empties into the inferior vena cava).
Hepatic Veins (he-PAT-ik=pertaining to the liver)	The **hepatic veins** drain the liver.

SCHEME OF DRAINAGE

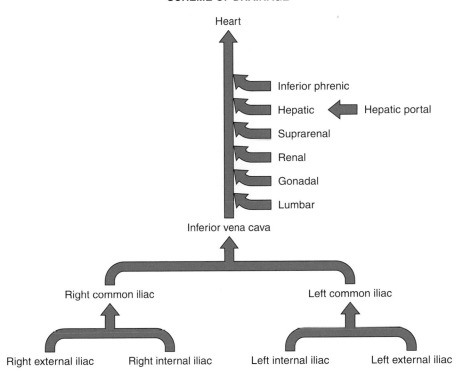

EXHIBIT 15.12   VEINS OF THE LOWER LIMBS (Figure 15.14)

OBJECTIVE

- Identify the principal superficial and deep veins that drain the lower limbs.

As with the upper limbs, blood from the lower limbs is drained by both **superficial** and **deep veins.** The superficial veins often anastomose with each other and with deep veins along their length.

Deep veins, for the most part, have the same names as corresponding arteries. All veins of the lower limbs have valves, which are more numerous than in veins of the upper limbs.

■ CHECKPOINT

1. Why are the great saphenous veins clinically important?

VEIN	DESCRIPTION AND REGION DRAINED
SUPERFICIAL VEINS	
Great saphenous veins (sa-FĒ-nus; *saphen-*=clearly visible)	The **great (long) saphenous veins,** the longest veins in the body, ascend from the foot to the groin in the subcutaneous layer. They begin at the medial end of the dorsal venous arches of the foot. The **dorsal venous arches** (VĒ-nus) are networks of veins on the dorsum of the foot formed by the **dorsal digital veins,** which collect blood from the toes, and then unite in pairs to form the **dorsal metatarsal veins,** which parallel the metatarsals. As the dorsal metatarsal veins approach the foot, they combine to form the dorsal venous arches. The great saphenous veins pass anterior to the medial malleolus of the tibia and then superiorly along the medial aspect of the leg and thigh just deep to the skin. They receive tributaries from superficial tissues and connect with the deep veins as well. They empty into the femoral veins at the groin. Generally, the great saphenous veins drain mainly the medial side of the leg and thigh, the groin, external genitals, and abdominal wall. Along their length, the great saphenous veins have from 10 to 20 valves, with more located in the leg than the thigh. These veins are more likely to be subject to varicosities than other veins in the lower limbs because they must support a long column of blood and are not well supported by skeletal muscles. The great saphenous veins are often used for prolonged administration of intravenous fluids. This is particularly important in very young children and in patients of any age who are in shock and whose veins are collapsed. In coronary artery bypass grafting, if multiple blood vessels need to be grafted, sections of the great saphenous vein are used along with at least one artery as a graft. After the great saphenous vein is removed and divided into sections, the sections are used to bypass the blockages. The vein grafts are reversed so that the valves do not obstruct the flow of blood.
Small saphenous veins	The **small (short) saphenous veins** begin at the lateral aspect of the dorsal venous arches of the foot. They pass posterior to the lateral malleolus of the fibula and ascend deep to the skin along the posterior aspect of the leg. They empty into the popliteal veins in the popliteal fossa, posterior to the knee. Along their length, the small saphenous veins have from 9 to 12 valves. The small saphenous veins drain the foot and posterior aspect of the leg. They may communicate with the great saphenous veins in the proximal thigh.

VEIN	DESCRIPTION AND REGION DRAINED
DEEP VEINS	
Posterior tibial veins (TIB-ē-al)	The **plantar digital veins** on the plantar surfaces of the toes unite to form the **plantar metatarsal veins,** which parallel the metatarsals. They unite to form the **deep plantar venous arches.** From each arch emerges the **medial** and **lateral plantar veins.** The medial and lateral plantar veins, posterior to the medial malleolus of the tibia, form the paired **posterior tibial veins,** which sometimes merge into a single vessel. They accompany the posterior tibial artery through the leg. They ascend deep to the muscles in the posterior aspect of the leg and drain the foot and posterior compartment muscles. About two-thirds of the way up the leg, the posterior tibial veins drain blood from the **fibular (peroneal) veins,** which drain the lateral and posterior leg muscles. The posterior tibial veins unite with the anterior tibial veins just inferior to the popliteal fossa to form the popliteal veins.
Anterior tibial veins	The paired **anterior tibial veins** arise in the dorsal venous arch and accompany the anterior tibial artery. They ascend in the interosseous membrane between the tibia and fibula and unite with the posterior tibial veins to form the popliteal vein. The anterior tibial veins drain the ankle joint, knee joint, tibiofibular joint, and anterior portion of the leg.
Popliteal veins (pop′-li-TĒ-al=pertaining to the hollow behind knee)	The anterior and posterior tibial veins unite to form the **popliteal veins.** The popliteal veins also receive blood from the small saphenous veins and tributaries that correspond to branches of the popliteal artery. The popliteal veins drain the knee joint and the skin, muscles, and bones of portions of the calf and thigh around the knee joint.
Femoral veins (FEM-o-ral)	The **femoral veins** accompany the femoral arteries and are the continuations of the popliteal veins just superior to the knee. The femoral veins extend up the posterior surface of the thighs and drain the muscles of the thighs, femurs, external genitals, and superficial lymph nodes. The largest tributaries of the femoral veins are the **deep veins of the thigh (deep femoral veins).** Just before penetrating the abdominal wall, the femoral veins receive the deep veins and the great saphenous veins. The veins formed from this union penetrate the body wall and enter the pelvic cavity. Here they are known as the **external iliac veins.** In order to take blood samples or pressure recordings from the right side of the heart, a catheter is inserted into the femoral vein as it passes through the femoral triangle. The catheter passes through the external and common iliac veins and inferior vena cava into the right atrium.

continues

EXHIBIT 15.12   VEINS OF THE LOWER LIMBS (Figure 15.14)

CONTINUED

SCHEME OF DRAINAGE

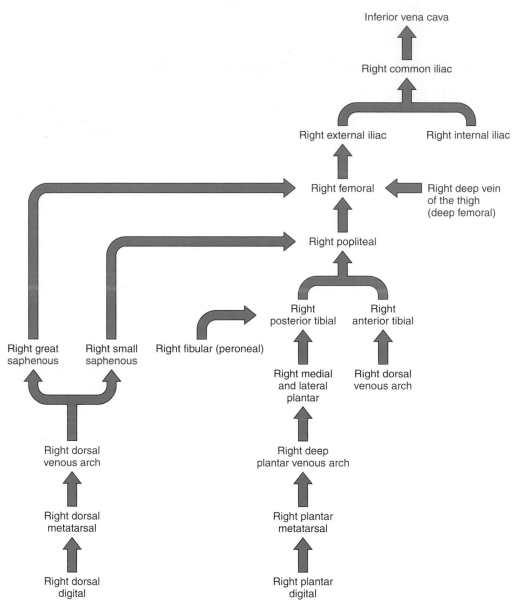

Figure 15.14 **Principal veins of the pelvis and lower limbs.** (See Tortora, *A Photographic Atlas of the Human Body*, 2e, Figure 6.22.)

Deep veins usually bear the names of their companion arteries.

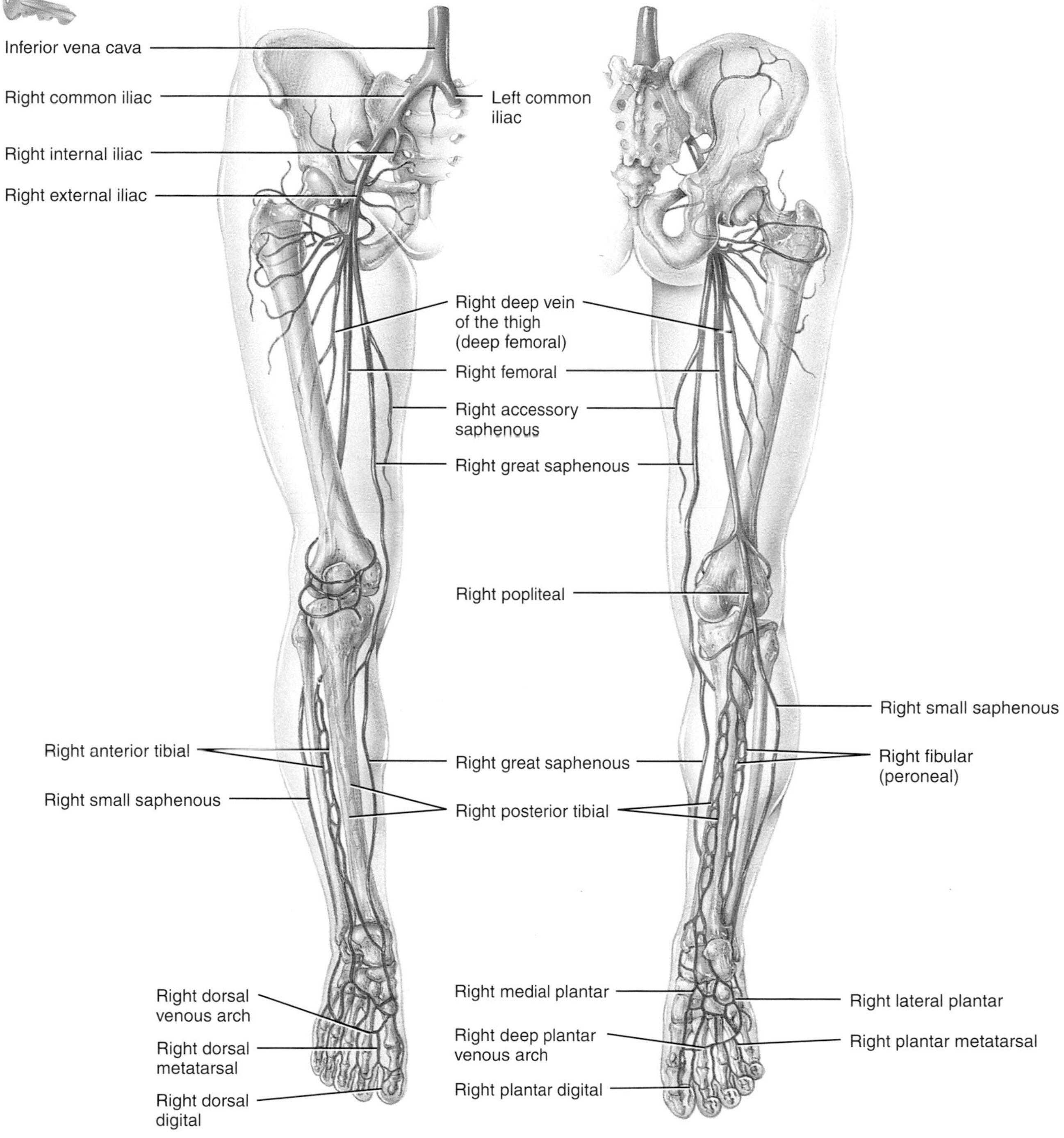

Inferior vena cava

Right common iliac

Left common iliac

Right internal iliac

Right external iliac

Right deep vein of the thigh (deep femoral)

Right femoral

Right accessory saphenous

Right great saphenous

Right popliteal

Right small saphenous

Right anterior tibial

Right great saphenous

Right fibular (peroneal)

Right small saphenous

Right posterior tibial

Right dorsal venous arch

Right medial plantar

Right lateral plantar

Right dorsal metatarsal

Right deep plantar venous arch

Right plantar metatarsal

Right dorsal digital

Right plantar digital

(a) Anterior view

(b) Posterior view

Which veins of the lower limb are superficial?

The Hepatic Portal Circulation

We noted earlier that the two principal circulatory routes are the systemic and pulmonary circulations. Before looking at the pulmonary circulation, we will examine the hepatic portal circulation, a subdivision of the systemic circulation. The **hepatic portal circulation** (*hepat-*=liver) detours venous blood from the gastrointestinal organs and spleen through the liver before it returns to the heart (Figure 15.15). A *portal system*, as noted earlier, carries blood between two capillary networks, from one location in the body to another, without passing through the heart. In this case, blood flows from capillaries of the gastrointestinal tract to sinusoids (capillaries) of the liver. After a meal, hepatic portal blood is rich in substances absorbed from the gastrointestinal tract. The liver stores some of them and modifies others before they pass into the general circulation. For example, the liver converts glucose into glycogen for storage, thereby helping to maintain homeostasis of blood glucose during a fast; the liver also detoxifies harmful substances such as alcohol that have been absorbed from the gastrointestinal tract and destroys bacteria by phagocytosis.

The **hepatic portal vein** is formed by the union of the superior mesenteric and splenic veins. The **superior mesenteric vein** drains blood from the small intestine and portions of

Figure 15.15 Hepatic portal circulation. A schematic diagram of blood flow through the liver, including arterial circulation, is shown in (b); deoxygenated blood is indicated in blue, oxygenated blood in red.

🔑 **The hepatic portal circulation delivers venous blood from the organs of the gastrointestinal tract and spleen to the liver.**

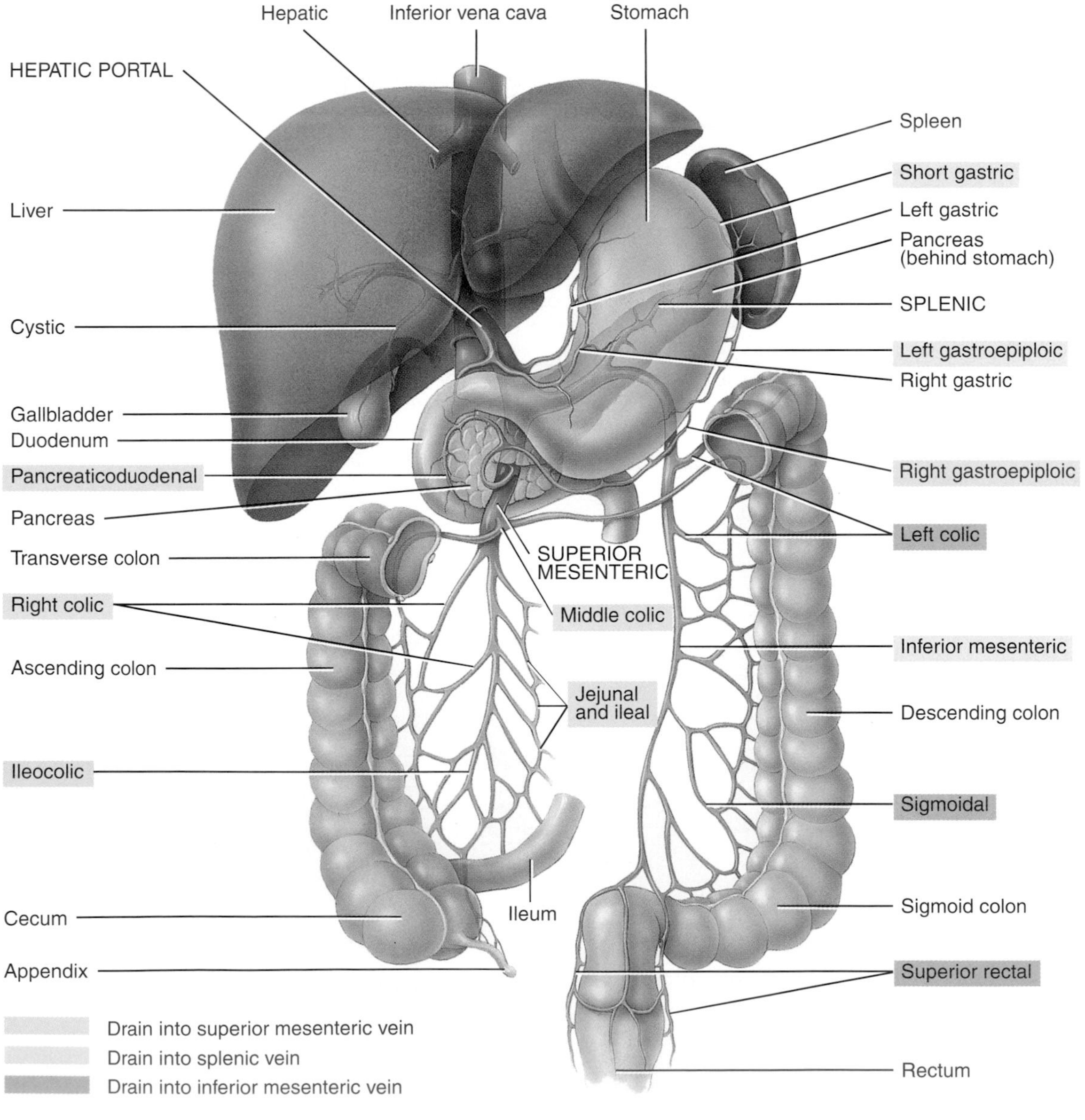

Drain into superior mesenteric vein
Drain into splenic vein
Drain into inferior mesenteric vein

(a) Anterior view of veins draining into the hepatic portal vein

the large intestine, stomach, and pancreas through the *jejunal, ileal, ileocolic, right colic, middle colic, pancreaticoduodenal,* and *right gastroepiploic veins*. The **splenic vein** drains blood from the stomach, pancreas, and portions of the large intestine through the *short gastric, left gastroepiploic, pancreatic,* and *inferior mesenteric veins*. The inferior mesenteric vein, which passes into the splenic vein, drains portions of the large intestine through the superior *rectal, sigmoidal,* and *left colic veins*. The *right* and *left gastric veins*, which open directly into the hepatic portal vein, drain the stomach. The *cystic vein*, which also opens into the hepatic portal vein, drains the gallbladder.

At the same time that the liver receives nutrient-rich deoxygenated blood via the hepatic portal system, it also receives oxygenated blood via the proper hepatic artery, a branch of the celiac trunk. Ultimately, all blood leaves the liver through the **hepatic veins,** which drain into the inferior vena cava.

The Pulmonary Circulation

The **pulmonary circulation** (*pulmo-*=lung) carries deoxygenated blood from the right ventricle to the air sacs (alveoli) within the lungs and returns oxygenated blood from the air sacs to the left atrium (Figure 15.16). The **pulmonary trunk** emerges from the right ventricle and passes superiorly, posteriorly, and to the left. It then divides into two branches: the **right pulmonary artery** to the right lung and the **left pulmonary artery** to the left lung. After birth, the pulmonary arteries are the only arteries that carry deoxygenated blood. On entering the lungs, the branches divide and subdivide until finally they form capillaries around the air sacs within the lungs. CO_2 passes from the blood into the air sacs and is exhaled. Inhaled O_2 passes from the air

within the lungs into the blood. The pulmonary capillaries unite to form venules and eventually **pulmonary veins,** which exit the lungs and transport the oxygenated blood to the left atrium. Two left and two right pulmonary veins enter the left atrium. After birth, the pulmonary veins are the only veins that carry oxygenated blood. Contractions of the left ventricle then eject the oxygenated blood into the systemic circulation.

The pulmonary and systemic circulations are different in two important ways. First, blood in the pulmonary circulation need not be pumped as far as blood in the systemic circulation. Second, compared to systemic arteries, pulmonary arteries have larger diameters, thinner walls, and less elastic tissue; as a result, the resistance to pulmonary blood flow is very low, which means that less pressure is needed to move blood through the lungs.

Because resistance in the pulmonary circulation is low, normal *pulmonary* capillary hydrostatic pressure, the principal force that moves fluid out of capillaries into interstitial fluid, is only 10 mmHg. This compares to the average *systemic* capillary pressure of about 25 mmHg. The relatively low capillary hydrostatic pressure tends to prevent pulmonary edema. However, if capillary blood pressure in the lungs increases (due to increased left atrial pressure as may occur in mitral valve stenosis) or capillary permeability increases (as may occur from bacterial toxins), edema may develop. Pulmonary edema reduces the rate of diffusion of oxygen and carbon dioxide and thus slows the exchange of respiratory gases in the lungs.

The Fetal Circulation

The circulatory system of a fetus, called the **fetal circulation,** exists only in the fetus and contains special structures that allow

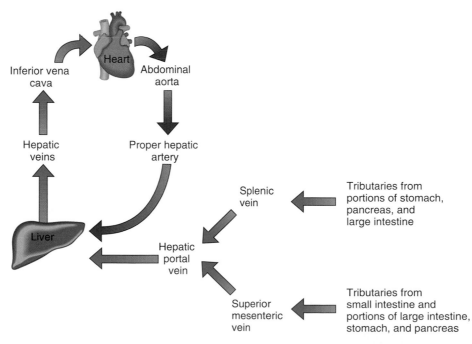

(b) Scheme of principal blood vessels of hepatic portal circulation and arterial supply and venous drainage of liver

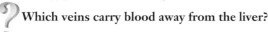
Which veins carry blood away from the liver?

Figure 15.16   Pulmonary circulation.

The pulmonary circulation brings deoxygenated blood from the right ventricle to the lungs and returns oxygenated blood from the lungs to the left atrium.

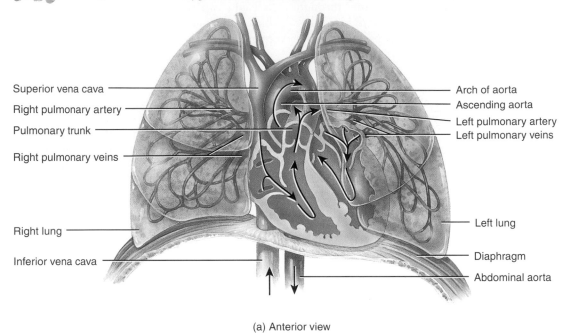

Superior vena cava

Right pulmonary artery

Pulmonary trunk

Right pulmonary veins

Right lung

Inferior vena cava

Arch of aorta

Ascending aorta

Left pulmonary artery

Left pulmonary veins

Left lung

Diaphragm

Abdominal aorta

(a) Anterior view

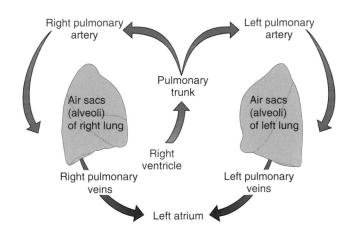

Right pulmonary artery

Left pulmonary artery

Pulmonary trunk

Air sacs (alveoli) of right lung

Air sacs (alveoli) of left lung

Right ventricle

Right pulmonary veins

Left pulmonary veins

Left atrium

(b) Scheme of pulmonary circulation

After birth, which are the only arteries that carry deoxygenated blood?

the developing fetus to exchange materials with its mother (Figure 15.17a). It differs from the postnatal (after birth) circulation shown in Figure 15.17b because the lungs, kidneys, and gastrointestinal organs do not begin to function until birth. The fetus obtains O_2 and nutrients from and eliminates CO_2 and other wastes into the maternal blood.

The exchange of materials between fetal and maternal circulations occurs through the **placenta** (pla-SEN-ta), which forms inside the mother's uterus and attaches to the umbilicus (navel) of the fetus by the **umbilical cord** (um-BIL-i-kal). The placenta communicates with the mother's cardiovascular system through many small blood vessels that emerge from the uterine wall. The umbilical cord contains blood vessels that branch into

capillaries in the placenta. Wastes from the fetal blood diffuse out of the capillaries, into spaces containing maternal blood (intervillous spaces) in the placenta, and finally into the mother's uterine veins. Nutrients travel the opposite route—from the maternal blood vessels to the intervillous spaces to the fetal capillaries. Normally, there is no direct mixing of maternal and fetal blood because all exchanges occur by diffusion through capillary walls.

Blood passes from the fetus to the placenta via two **umbilical arteries** (Figure 15.17a, c). These branches of the internal iliac (hypogastric) arteries are within the umbilical cord. At the placenta, fetal blood picks up O_2 and nutrients and eliminates CO_2 and wastes. The oxygenated blood returns from the pla-

Figure 15.17 Fetal circulation and changes at birth. The gold boxes between parts (a) and (b) describe the fate of certain fetal structures once postnatal circulation is established.

🔑 **The lungs and gastrointestinal organs do not begin to function until birth.**

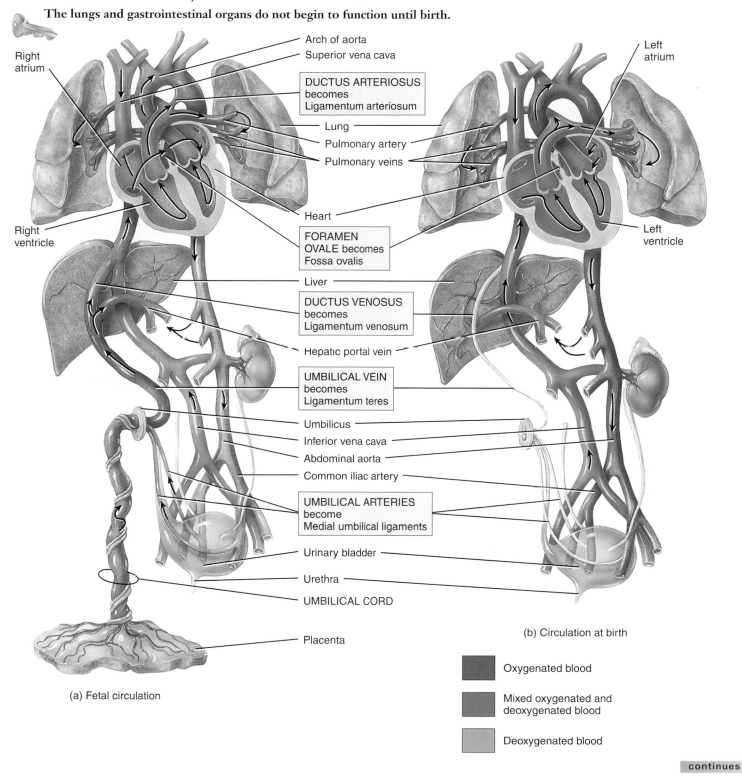

(a) Fetal circulation

(b) Circulation at birth

■ Oxygenated blood

■ Mixed oxygenated and deoxygenated blood

■ Deoxygenated blood

continues

centa via a single **umbilical vein.** This vein ascends to the liver of the fetus, where it divides into two branches. Whereas some blood flows through the branch that joins the hepatic portal vein and enters the liver, most of the blood flows into the second branch, the **ductus venosus** (DUK-tus ve-NŌ-sus), which drains into the inferior vena cava.

Figure 15.17 (continued)

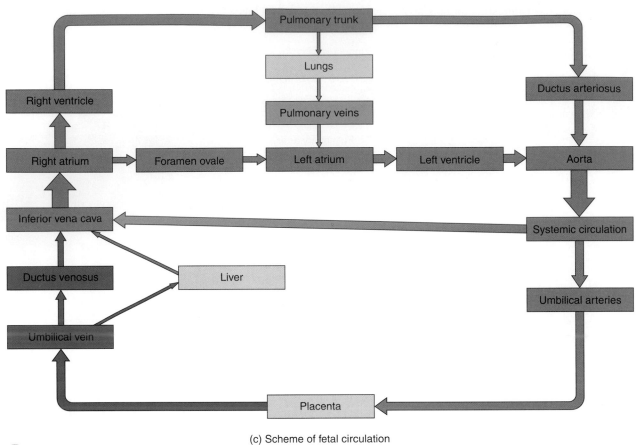

(c) Scheme of fetal circulation

Through which structure does the exchange of materials between mother and fetus occur?

Deoxygenated blood returning from lower body regions of the fetus mingles with oxygenated blood from the ductus venosus in the inferior vena cava. This mixed blood then enters the right atrium. Deoxygenated blood returning from upper body regions of the fetus enters the superior vena cava and passes into the right atrium.

Most of the fetal blood does not pass from the right ventricle to the lungs, as it does in postnatal circulation, because an opening called the **foramen ovale** (fō-RĀ-men ō-VAL-ē) exists in the septum between the right and left atria. About one-third of the blood that enters the right atrium passes through the foramen ovale into the left atrium and joins the systemic circulation. The blood that does pass into the right ventricle is pumped into the pulmonary trunk, but little of this blood reaches the nonfunctioning fetal lungs. Instead, most is sent through the **ductus arteriosus** (ar-tē-rē-Ō-sus), a vessel that connects the pulmonary trunk with the aorta, so that most blood bypasses the fetal lungs. The blood in the aorta is carried to all fetal tissues through the systemic circulation. When the common iliac arteries branch into the external and internal iliacs, part of the blood flows into the internal iliacs, into the umbilical arteries, and back to the placenta for another exchange of materials.

After birth, when pulmonary (lung), renal, and digestive functions begin, the following vascular changes occur (Figure 15.17b):

1. When the umbilical cord is tied off, blood no longer flows through the umbilical arteries, they fill with connective tissue, and the distal portions of the umbilical arteries become fibrous cords called the **medial umbilical ligaments.** Although the arteries are closed functionally only a few minutes after birth, complete obliteration of the lumens may take 2 to 3 months.

2. The umbilical vein collapses but remains as the **ligamentum teres (round ligament),** a structure that attaches the umbilicus to the liver.

3. The ductus venosus collapses but remains as the **ligamentum venosum,** a fibrous cord on the inferior surface of the liver.

4. The placenta is expelled as the **"afterbirth."**

5. The foramen ovale normally closes shortly after birth to become the **fossa ovalis,** a depression in the interatrial septum. When an infant takes its first breath, the lungs expand and blood flow to the lungs increases. Blood returning from

the lungs to the heart increases pressure in the left atrium. This closes the foramen ovale by pushing the valve that guards it against the interatrial septum. Permanent closure occurs in about a year.

6. The ductus arteriosus closes by vasoconstriction almost immediately after birth and becomes the **ligamentum arteriosum.** Complete anatomical obliteration of the lumen takes 1 to 3 months.

CHECKPOINT

6. What is the purpose of systemic circulation?
7. Prepare a diagram to show the hepatic portal circulation. Why is this route important?
8. Prepare a diagram to show the route of the pulmonary circulation.
9. Discuss the anatomy and physiology of the fetal circulation. Indicate the function of the umbilical arteries, umbilical vein, ductus venosus, foramen ovale, and ductus arteriosus.

DEVELOPMENT OF BLOOD VESSELS AND BLOOD

OBJECTIVE

• Describe the development of blood vessels and blood.

The development of blood cells and the formation of blood vessels begins outside the embryo as early as fifteen to sixteen days in the **mesoderm** of the wall of the yolk sac, chorion, and connecting stalk. About two days later, blood vessels form within the embryo. The early formation of the cardiovascular system is linked to the small amount of yolk in the ovum and yolk sac. As the embryo develops rapidly during the third week, there is a greater need to develop a cardiovascular system to supply sufficient nutrients to the embryo and remove wastes from it.

Blood vessels and blood cells develop from the same precursor cell, called a **hemangioblast** (hē-MAN-jē-ō-blast; *hema-*=blood; *-blast*=immature stage). Once mesenchyme develops into hemangioblasts, they can give rise to cells that produce blood vessels or cells that produce blood cells.

Blood vessels develop from **angioblasts** derived from hemangioblasts (Figure 15.18). Angioblasts aggregate to form isolated masses and cords throughout the embryonic disc called **blood islands** (Figure 15.18). Spaces soon appear in the islands and become the lumens of the blood vessels. Some of the angioblasts immediately around the spaces give rise to the *endothelial lining of the blood vessels.* Angioblasts around the endothelium form the *tunics* (interna, media, and externa) of the larger blood vessels. Growth and fusion of blood islands form an extensive network of blood vessels throughout the embryo. By continuous branching, blood vessels outside the embryo connect with those inside the embryo, linking the embryo with the placenta.

Blood cells develop from **pluripotent stem cells** derived from hemangioblasts. This development occurs in the walls of blood vessels in the yolk sac, chorion, and allantois at about three weeks after fertilization. Blood formation in the embryo

Figure 15.18 Development of blood vessels and blood cells from blood islands.

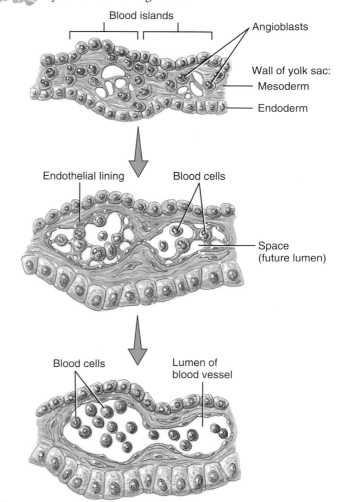

Blood vessel development begins in the embryo on about day seventeen or eighteen.

Blood islands — Angioblasts

Wall of yolk sac:
— Mesoderm
— Endoderm

Endothelial lining — Blood cells

Space (future lumen)

Blood cells — Lumen of blood vessel

From which germ cell layer are blood vessels and blood derived?

itself begins at about the fifth week in the liver and the twelfth week in the spleen, red bone marrow, and thymus.

CHECKPOINT

10. What are the sites of blood cell production outside the embryo and within the embryo?

AGING AND THE CARDIOVASCULAR SYSTEM

OBJECTIVE

• Explain the effects of aging on the cardiovascular system.

General changes in the cardiovascular system associated with aging include decreased compliance of the aorta, reduction in cardiac muscle fiber size, progressive loss of cardiac muscular strength, reduced cardiac output, a decline in maximum heart rate, and an increase in systolic blood pressure. Total blood cholesterol tends to increase with age, as does low-density lipoprotein (LDL); high-density lipoprotein (HDL) tends to decrease. There is an increase in the incidence of coronary artery disease (CAD), the major cause of heart disease and death in older Americans. Congestive heart failure, a set of symptoms associated with impaired pumping of the heart, is also prevalent in older individuals. Changes in blood vessels that serve brain tissue—for example, atherosclerosis—reduce nourishment to the brain and result in the malfunction or death of brain cells. By age 80, cerebral blood flow is 20% less and renal blood flow is 50% less than in the same person at age 30.

CHECKPOINT

11. How does aging affect the heart?

APPLICATIONS TO HEALTH

HYPERTENSION

About 50 million Americans have hypertension, or persistently high blood pressure. It is the most common disorder affecting the heart and blood vessels and is the major cause of heart failure, kidney disease, and stroke. In May 2003, the Joint National Committee on Prevention, Detection, Evaluation, and Treatment of High Blood Pressure published new guidelines for hypertension because clinical studies have linked what were once considered fairly low blood pressure readings to an increased risk of cardiovascular disease. The new guidelines are as follows:

Category	Systolic (mmHg)	Diastolic (mmHg)
Normal	Less than 120 *and*	Less than 80
Prehypertension	120–139 *or*	80–89
Stage 1 hypertension	140–159 *or*	90–99
Stage 2 hypertension	Greater than 160 *or*	Greater than 100

Under the new guidelines, the normal classification was previously considered optimal; prehypertension now includes many more individuals previously classified as normal or high-normal; stage 1 hypertension is the same as in previous guidelines; and stage 2 hypertension now combines the previous stage 2 and stage 3 categories since treatment options are the same for the former stages 2 and 3.

Types and Causes of Hypertension

Between 90% and 95% of all cases of hypertension are **primary hypertension,** which is a persistently elevated blood pressure that cannot be attributed to any identifiable cause. The remaining 5–10% of cases are **secondary hypertension,** which has an identifiable underlying cause.

Damaging Effects of Untreated Hypertension

High blood pressure is known as the "silent killer" because it can cause considerable damage to the blood vessels, heart, brain, and kidneys before it causes pain or other noticeable symptoms. It is a major risk factor for the number-one (heart disease) and number-three (stroke) causes of death in the United States. In blood vessels, hypertension causes thickening of the tunica media, accelerates development of atherosclerosis and coronary artery disease, and increases systemic vascular resistance. In the heart, hypertension leads to myocardial hypertrophy that is accompanied by muscle damage and fibrosis (a buildup of collagen fibers between the muscle fibers). As a result, the left ventricle enlarges, weakens, and dilates. Because arteries in the brain are usually less protected by surrounding tissues than are the major arteries in other parts of the body, prolonged hypertension can eventually cause them to rupture, and a stroke occurs due to a brain hemorrhage. Hypertension also damages kidney arterioles, causing them to thicken, which narrows the lumen; because the blood supply to the kidneys is thereby reduced, the kidneys secrete more renin, which elevates the blood pressure even more.

Lifestyle Changes to Reduce Hypertension

Although several categories of drugs (described next) can reduce elevated blood pressure, the following lifestyle changes are also effective in managing hypertension:

- **Lose weight.** This is the best treatment for high blood pressure short of using drugs. Loss of even a few pounds helps reduce blood pressure in overweight hypertensive individuals.

- **Limit alcohol intake.** Drinking in moderation may lower the risk of coronary heart disease, mainly among males over 45 and females over 55. Moderation is defined as no more than one 12-oz beer per day for females and no more than two 12-oz beers per day for males.

- **Exercise.** Becoming more physically fit by engaging in moderate activity (such as brisk walking) several times a week for 30 to 45 minutes can lower systolic blood pressure by about 10 mmHg.

- **Reduce intake of sodium (salt).** Roughly half the people with hypertension are "salt sensitive." For them, a high-salt diet appears to promote hypertension, and a low-salt diet can lower their blood pressure.

- **Maintain recommended dietary intake of potassium, calcium, and magnesium.** Higher levels of potassium, calcium, and magnesium in the diet are associated with a lower risk of hypertension.

- **Don't smoke.** Smoking has devastating effects on the heart and can augment the damaging effects of high blood pressure by promoting vasoconstriction.

● *Manage stress.* Various meditation and biofeedback techniques help some people reduce high blood pressure. These methods may work by decreasing the daily release of epinephrine and norepinephrine by the adrenal medulla.

Drug Treatment of Hypertension

Many people are successfully treated with *diuretics*, agents that decrease blood pressure by decreasing blood volume through increased elimination of water and salt in the urine. *ACE* *(angiotensin converting enzyme) inhibitors* promote vasodilation and decrease the liberation of aldosterone. *Beta blockers* reduce blood pressure by decreasing heart rate and contractility. *Vasodilators* relax the smooth muscle in arterial walls, causing vasodilation and lowering blood pressure. An important category of vasodilators are the *calcium channel blockers*, which slow the inflow of Ca^{2+} into vascular smooth muscle cells. They reduce the heart's workload by slowing Ca^{2+} entry into pacemaker cells and regular myocardial fibers, thereby decreasing heart rate and the force of myocardial contraction.

KEY MEDICAL TERMS ASSOCIATED WITH BLOOD VESSELS

Aneurysm (AN-ū-rizm) A thin, weakened section of the wall of an artery or a vein that bulges outward, forming a balloon-like sac. Common causes are atherosclerosis, syphilis, congenital blood vessel defects, and trauma. If untreated, the aneurysm enlarges and the blood vessel wall becomes so thin that it bursts. The result is massive hemorrhage with shock, severe pain, stroke, or death. Treatment may involve surgery in which the weakened area of the blood vessel is removed and replaced with a graft of synthetic material.

Angiogenesis (an'-jē-ō-JEN-e-sis; *angio-*=vessel; *-genesis*=production) Formation of new blood vessels.

Angiography (an'-jē-OG-rā-fē; *angio-*=vessel; *-grapho*=to write) A diagnostic procedure in which a radiopaque dye is injected through a catheter that has been introduced into a blood vessel and guided to the blood vessel to be examined. The dye flows into the appropriate blood vessel, making abnormalities such as blockages visible on x-rays. Angiography is used to study the blood vessels of the heart (*coronary angiography*), aorta and its branches (*aortography*), and blood vessels in the legs (*femoral angiography*).

Arteritis (ar'-te-RĪ-tis; *-itis*=inflammation of) Inflammation of an artery, probably due to an autoimmune response.

Carotid endarterectomy (ka-ROT-id end'-ar-ter-EK-tō-mē) The removal of atherosclerotic plaque from the carotid artery to restore greater blood flow to the brain.

Claudication (klaw'-di-KĀ-shun) Pain and lameness or limping caused by defective circulation of the blood in the vessels of the limbs.

Deep vein thrombosis The presence of a thrombus (blood clot) in a deep vein of the lower limbs. It may lead to (1) pulmonary embolism, if the thrombus dislodges and then lodges within the pulmonary arterial blood flow, and (2) postphlebitic syndrome, which consists of edema, pain, and skin changes due to destruction of venous valves.

Doppler ultrasound scanning Imaging technique commonly used to measure blood flow. A transducer is placed on the skin and an image is displayed on a monitor that provides the exact position and severity of a blockage.

Hypotension (hī-pō-TEN-shun) Low blood pressure; most commonly used to describe an acute drop in blood pressure, as occurs during excessive blood loss.

Normotensive (nor'-mō-TEN-siv) Characterized by normal blood pressure.

Occlusion (ō-KLOO-shun) The closure or obstruction of the lumen of a structure such as a blood vessel. An example is an atherosclerotic plaque in an artery.

Orthostatic hypotension (or'-thō-STAT-ik; *ortho-*=straight; *-static*=causing to stand) An excessive lowering of systemic blood pressure when a person assumes an erect or semierect posture; it is usually a sign of a disease. May be caused by excessive fluid loss, certain drugs, and cardiovascular or neurogenic factors. Also called **postural hypotension**.

Phlebitis (fle-BĪ-tis; *phleb-*=vein) Inflammation of a vein, often in a leg.

Shock Failure of the cardiovascular system to deliver enough oxygen and nutrients to meet cellular metabolic needs as a result of inadequate blood flow to tissues. It may be caused by decreased blood volume, poor heart function, inappropriate vasodilation, or obstruction of blood flow.

Syncope (SIN-kō-pē=cutting short) Fainting. A sudden, temporary loss of consciousness followed by spontaneous recovery. It is usually due to decreased blood flow to the brain.

Thrombectomy (throm-BEK-tō-mē; *thrombo-*=clot) An operation to remove a blood clot from a blood vessel.

Thrombophlebitis (throm'-bō-fle-BĪ-tis) Inflammation of a vein involving clot formation. Superficial thrombophlebitis occurs in veins under the skin, especially in the calf.

Venipuncture (VEN-i-punk-chur; *vena-*=vein) The puncture of a vein, usually to withdraw blood for analysis or introduce a solution, for example, an antibiotic. The median cubital vein is frequently used.

White coat (office) hypertension A stress-induced syndrome found in patients who have elevated blood pressure when being examined by health-care personnel, but otherwise have normal blood pressure.

STUDY OUTLINE

Anatomy of Blood Vessels (p. 454)

1. Arteries carry blood away from the heart. The wall of an artery consists of a tunica interna, a tunica media (which maintains elasticity and contractility), and a tunica externa.

2. Large arteries are termed elastic (conducting) arteries, and medium-sized arteries are called muscular (distributing) arteries.

3. Many arteries anastomose, which means the distal ends of two or more vessels unite. An alternate blood route from an anastomosis is called collateral circulation. Arteries that do not anastomose are called end arteries.

4. Arterioles are small arteries that deliver blood to capillaries.

5. Through constriction and dilation, arterioles assume key roles in regulating blood flow from arteries into capillaries and in altering arterial blood pressure.

6. Capillaries are microscopic blood vessels through which materials are exchanged between blood and tissue cells; some capillaries are continuous, whereas others are fenestrated.

7. Capillaries branch to form an extensive network throughout a tissue. This network increases the surface area available for the exchange of materials between the blood and the body tissue; it also allows rapid exchange of large quantities of materials.

8. Precapillary sphincters regulate blood flow through capillaries.

9. Microscopic blood vessels in the liver are called sinusoids.

10. Venules are small vessels that form from the merging capillaries; venules merge to form veins.

11. Veins consist of the same three tunics as arteries but have a thinner tunica interna and tunica media. The lumen of a vein is also larger than that of a comparable artery.

12. Veins contain valves to prevent backflow of blood.

13. Weak valves can lead to varicose veins.

14. Vascular (venous) sinuses are veins with very thin walls.

15. Systemic veins are collectively called blood reservoirs because they hold a large volume of blood. If the need arises, this blood can be shifted into other blood vessels through vasoconstriction of veins.

16. The principal blood reservoirs are the veins of the abdominal organs (liver and spleen) and skin.

Circulatory Routes (p. 459)

1. The two basic postnatal circulatory routes are the systemic and pulmonary circulations.

2. Among the subdivisions of the systemic circulation are the coronary (cardiac) and the hepatic portal circulations.

3. Fetal circulation exists only in the fetus.

4. The systemic circulation carries oxygenated blood from the left ventricle through the aorta to all parts of the body (including some lung tissue, but *not* the air sacs of the lungs) and returns the deoxygenated blood to the right atrium.

5. The aorta is divided into the ascending aorta, the arch of the aorta, and the descending aorta. Each section gives off arteries that branch to supply the whole body.

6. Blood returns to the heart through the systemic veins. All veins of the systemic circulation drain into the superior or inferior venae cavae or the coronary sinus; these, in turn, empty into the right atrium.

7. The principal blood vessels of the systemic circulation may be reviewed in Exhibits 15.1–15.12.

8. The hepatic portal circulation detours venous blood from the gastrointestinal organs and spleen and directs it into the hepatic portal vein of the liver before it is returned to the heart. It enables the liver to utilize nutrients and detoxify harmful substances in the blood.

9. The pulmonary circulation takes deoxygenated blood from the right ventricle to the alveoli within the lungs and returns oxygenated blood from the alveoli to the left atrium. It allows blood to be oxygenated for systemic circulation.

10. The fetal circulation involves the exchange of materials between fetus and mother.

11. The fetus derives O_2 and nutrients and eliminates CO_2 and wastes through the maternal blood supply via the placenta.

12. At birth, when pulmonary (lung), digestive, and liver functions begin, the special structures of fetal circulation are no longer needed.

Development of Blood Vessels and Blood (p. 505)

1. Blood vessels develop from mesenchyme (hemangioblasts → angioblasts → blood islands) in mesoderm called blood islands.

2. Blood cells also develop from mesenchyme (hemangioblasts → pluripotent stem cells).

Aging and the Cardiovascular System (p. 505)

1. General changes associated with aging include reduced elasticity of blood vessels, reduction in cardiac muscle size, reduced cardiac output, and increased systolic blood pressure.

2. The incidence of coronary artery disease (CAD), congestive heart failure (CHF), and atherosclerosis increases with age.

SELF-QUIZ QUESTIONS

Choose the one best answer to the following questions.

1. Which statement best describes arteries?
   a. All carry oxygenated blood to the heart.
   b. All contain valves to prevent the backflow of blood.
   c. All carry blood away from the heart.
   d. Only large arteries are lined with endothelium.
   e. All branch from the descending aorta.

2. Which statement is *true* of veins?
   a. Their tunica interna is thicker than in arteries.
   b. Their tunica externa is thicker than in arteries.
   c. Most veins in the limbs have valves.
   d. They always carry deoxygenated blood.
   e. All ultimately empty into the inferior vena cava.

3. Put the following vessels in the correct order to trace the route of a drop of blood moving from the right side of the heart to

the left side of the heart. (1) pulmonary vein, (2) pulmonary artery, (3) arterioles, (4) venules, (5) capillaries.

a. 1,4,5,3,2   b. 1,3,5,4,2   c. 2,4,5,3,1
d. 2,5,4,1,3   e. 2,3,5,4,1

4. Vessels that function as a pressure reservoir to ensure blood flow while the ventricles are relaxed are the:

a. venules.   b. elastic arteries.   c. capillaries.
d. sinusoids.   e. aorta.

5. In fetal circulation, blood bypasses pulmonary circulation by passing through the:

a. ductus venosus.   b. ductus arteriosus.   c. fossa ovalis.
d. All of the above.   e. b and c.

6. *All* of the following are elastic arteries *except* the:

a. aorta.   b. brachiocephalic.   c. common carotid.
d. vertebral.   e. femoral.

7. Metarterioles lead into:

a. the aorta.   b. vasa recta.   c. capillary beds.
d. elastic arteries.   e. Both b and c are correct.

8. The hepatic portal vein is formed by the union of the:

a. left and right common iliac veins.
b. superior and inferior mesenteric veins.
c. superior mesenteric vein and splenic vein.
d. inferior mesenteric vein and splenic vein.
e. right and left hepatic veins.

9. The cerebral arterial circle (circle of Willis) forms anastomoses between the:

a. middle cerebral arteries.
b. internal and external jugular veins.
c. internal and external carotid arteries.
d. internal carotid arteries and the basilar artery.
e. middle cerebral arteries and the internal carotid arteries.

10. *All* of the following vessels branch directly from the aorta *except* the:

a. right subclavian artery.   b. left subclavian artery.
c. left common carotid artery.   d. brachiocephalic artery.
e. right coronary artery.

11. Blood in the internal jugular veins has just come from the:

a. brachiocephalic veins.   b. external jugular veins.
c. sinuses of the brain.   d. superior vena cava.
e. vertebral veins.

Complete the following.

12. The three branches of the celiac trunk are the _____, the _____, and the _____.

13. The walls of large blood vessels receive their oxygen and nutrients from blood in vessels known as _____.

14. The layer of an artery wall that contains endothelium is the _____.

15. The two common iliac veins unite to form the _____.

16. Blood from gastrointestinal organs and the spleen is transported to the liver via the _____ vein.

17. The union of the distal branches of two or more arteries supplying the same body region is called a(n) _____.

18. The tunica _____ of artery walls is composed of smooth muscle.

19. A decrease in the size of the lumen of a blood vessel due to contraction of smooth muscle is called _____.

20. Three types of capillaries include _____ capillaries found in skeletal and smooth muscle; _____ capillaries found in the kidneys and choroid plexuses of the brain ventricles; and _____, which are wide capillaries seen in red bone marrow and the liver.

21. The _____ veins are the only veins that carry oxygenated blood after birth.

Are the following statements true or false?

22. At rest, the largest portion of blood is located in the capillaries.

23. The external jugular veins join with the internal jugular veins to form the superior vena cava.

24. Arterioles play a key role in regulating blood pressure because of their role in regulating resistance to flow.

25. The median cubital veins connect the basilic and cephalic veins anterior to the elbow.

26. An increase in stimulation by sympathetic nerves typically stimulates vasodilation of blood vessels.

Matching

27. Match the following:

___ **(a)** passes through cervical transverse foramina

___ **(b)** continues as left axillary artery

___ **(c)** gives rise to the left common carotid artery

___ **(d)** is a branch of the thoracic aorta

___ **(e)** abdominal vessel that branches to supply blood to the pancreas, small intestine and part of the large intestine

___ **(f)** becomes the dorsalis pedis artery at the ankle

___ **(g)** receives venous blood from the myocardium

___ **(h)** is located in the arm

___ **(i)** is the longest vein in the body

___ **(j)** is completely located in the leg

___ **(k)** is completely located in the head

(1) anterior tibial artery
(2) esophageal artery
(3) cephalic vein
(4) sigmoid sinus
(5) coronary sinus
(6) posterior tibial vein
(7) left subclavian artery
(8) arch of aorta
(9) great saphenous vein
(10) vertebral artery
(11) superior mesenteric artery

CRITICAL THINKING QUESTIONS

1. Which structures present in the fetal circulation are absent in the adult circulatory system? Why do these changes occur?

2. Mike spent the week before the exam lying on a beach studying the back of his eyelids. "What was that question about sinuses doing on the circulatory test?" he complained. "We did that back with bones!" See if you can enlighten Mike.

3. You've read about varicose veins. Why aren't there varicose arteries?

4. Enrique is scheduled for a medical procedure called a cardiac catheterization. During this procedure, a tube will be inserted into

Enrique's femoral artery and snaked through the blood vessels to reach the heart. If the ultimate target is the pulmonary arteries, what vessels will the tube need to pass through?

5. Gina is 45 years old and works as an executive secretary to her college's biology department. She is 5′3″ tall, weighs 163 pounds, and has smoked since she was 15 years old. Her doctor measured her blood pressure at "132 over 86." What do you think the doctor will be advising Gina to do, and why?

ANSWERS TO FIGURE QUESTIONS

15.1 The femoral artery has the thicker wall; the femoral vein has the wider lumen.

15.2 Metabolically active tissues have extensive capillary networks because they use O_2 and produce wastes more rapidly than inactive tissues.

15.3 Materials cross capillary walls through intercellular clefts and fenestrations, via transcytosis in pinocytic vesicles, and through the plasma membranes of endothelial cells.

15.4 Valves are more important in arm veins and leg veins than in neck veins because gravity tends to cause pooling of blood in the veins of the limbs when you are standing. The valves prevent backflow into the limbs so the blood proceeds toward the right atrium after each heartbeat. When you are erect, gravity aids the flow of blood in neck veins back toward the heart.

15.5 The two main circulatory routes are the systemic and pulmonary circulations.

15.6 The four principal subdivisions of the aorta are the ascending aorta, arch of the aorta, thoracic aorta, and abdominal aorta.

15.7 The three major branches of the arch of the aorta, in order of their origination, are the brachiocephalic trunk, left common carotid artery, and left subclavian artery.

15.8 The abdominal aorta begins at the aortic hiatus in the diaphragm.

15.9 The abdominal aorta divides into the common iliac arteries at about the level of L4.

15.10 The superior vena cava drains regions above the diaphragm, and the inferior vena cava drains regions below the diaphragm.

15.11 All venous blood in the brain drains into the internal jugular veins.

15.12 The median cubital vein of the upper limb is often used for withdrawing blood.

15.13 The inferior vena cava returns blood from abdominopelvic viscera to the heart.

15.14 The superficial veins of the lower limbs are the dorsal venous arch and the great saphenous and small saphenous veins.

15.15 The hepatic veins carry blood away from the liver.

15.16 After birth, the only arteries that carry deoxygenated blood are the pulmonary arteries.

15.17 Exchange of materials between mother and fetus occurs across the placenta.

15.18 Blood vessels and blood cells are derived from mesoderm.

THE LYMPHATIC AND IMMUNE SYSTEM

16

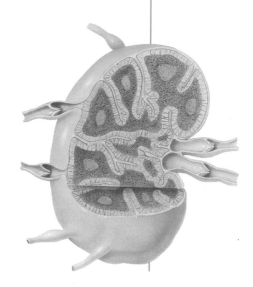

INTRODUCTION Maintaining physical health requires continuous combat against harmful agents in our internal and external environment. Despite constant exposure to a variety of **pathogens** (PATH-o-jens), disease-producing microbes such as bacteria and viruses, most people remain healthy. The body surface also endures cuts and bumps, exposure to ultraviolet rays in sunlight, chemical toxins, and minor burns with an array of defensive ploys. **Resistance** is the ability to ward off damage or disease through our defenses. Lack of resistance is termed **susceptibility.**

The two general types of resistance are (1) nonspecific resistance or innate defenses and (2) specific resistance or immunity. Nonspecific resistance is present at birth and includes defense mechanisms that provide *immediate* but *general* protection against invasion by a wide range of pathogens. Mechanical and chemical barriers of the skin and mucous membranes provide the first line of defense in nonspecific resistance.

If you were asked to construct an anatomical model, which methods and materials would you use? From what material is this model constructed? Which human structures are being demonstrated?

www.wiley.com/college/apcentral

511

The acidity of gastric juice in the stomach, for example, kills many bacteria in food. The second line of defense in nonspecific resistance consists of antimicrobial proteins (interferons and complement), phagocytes (mostly neutrophils and macrophages) and natural killer cells, inflammation, and fever. Specific resistance (immunity) develops in response to contact with a particular invader. It occurs more slowly than nonspecific resistance mechanisms and involves activation of specific lymphocytes that can combat a specific invader. The body system responsible for specific resistance (and some aspects of nonspecific resistance) is the lymphatic and immune system.

LYMPHATIC AND IMMUNE SYSTEM STRUCTURE AND FUNCTIONS

OBJECTIVES

- Describe the components and major functions of the lymphatic and immune system.
- Describe the organization of lymphatic vessels.
- Describe the formation and flow of lymph.
- Compare the structure and functions of the primary and secondary lymphatic organs and tissues.

The **lymphatic and immune system** (lim-FAT-ik) consists of a fluid called lymph, vessels called lymphatic vessels to transport the lymph, a number of structures and organs containing lymphatic tissue, and red bone marrow, where stem cells develop into the various types of blood cells, including lymphocytes (Figure 16.1). The lymphatic and immune system assists in circulating body fluids and helps defend the body against disease-causing agents. Most components of blood plasma filter through blood capillary walls to form interstitial fluid. After interstitial fluid passes into lymphatic vessels, it is referred to as **lymph** (LIMF=clear fluid). Therefore, interstitial fluid and lymph are very similar; the major difference between the two is location. Whereas interstitial fluid is found between cells, lymph is located within lymphatic vessels and lymphatic tissue.

Lymphatic tissue is a specialized form of reticular connective tissue (see Table 3.3C on page 77) that contains large numbers of lymphocytes. Recall from Chapter 13 that lymphocytes are agranular white blood cells. Two types of lymphocytes participate in immune responses: *B cells* and *T cells*.

Functions of the Lymphatic and Immune System

The lymphatic and immune system has three primary functions:

1. *Draining excess interstitial fluid.* Lymphatic vessels drain excess interstitial fluid from tissue spaces and return it to the blood.
2. *Transporting dietary lipids.* Lymphatic vessels transport the lipids and lipid-soluble vitamins (A, D, E, and K) absorbed by the gastrointestinal tract to the blood.
3. *Carrying out immune responses.* Lymphatic tissue initiates highly specific responses directed against particular microbes or abnormal cells. T and B cells (lymphocytes), aided by macrophages, recognize foreign cells, microbes, toxins, and cancer cells and respond to them in two basic ways: (1) T cells destroy the intruders by causing them to rupture or by releasing cytotoxic (cell-killing) substances. (2) B cells differentiate into plasma cells that protect us against disease by producing antibodies, proteins that combine with and cause destruction of specific foreign substances.

Lymphatic Vessels and Lymph Circulation

Lymphatic vessels begin as **lymphatic capillaries.** These tiny vessels are closed at one end and located in the spaces between cells (Figure 16.2 on page 514). Just as blood capillaries converge to form venules and then veins, lymphatic capillaries unite to form larger **lymphatic vessels** (see Figure 16.1), which resemble veins in structure but have thinner walls and more valves. At intervals along the lymphatic vessels, lymph flows through **lymph nodes,** encapsulated masses of B cells and T cells. In the skin, lymphatic vessels lie in the subcutaneous tissue and generally follow the same route as veins; lymphatic vessels of the viscera generally follow arteries, forming plexuses (networks) around them. Tissues that lack lymphatic capillaries include avascular tissues (such as cartilage, the epidermis, and the cornea of the eye), the central nervous system, portions of the spleen, and red bone marrow.

Lymphatic Capillaries

Lymphatic capillaries are slightly larger in diameter than blood capillaries and have a unique structure that permits interstitial fluid to flow into them but not out. The ends of endothelial cells that make up the wall of a lymphatic capillary overlap (Figure 16.2b). When pressure is greater in the interstitial fluid than in lymph, the cells separate slightly, like the opening of a one-way swinging door, and interstitial fluid enters the lymphatic capillary. When pressure is greater inside the lymphatic capillary, the cells adhere more closely, and lymph cannot escape back into interstitial fluid. Attached to the lymphatic capillaries are *anchoring filaments*, which contain elastic fibers. They extend out from the lymphatic capillary, attaching lymphatic endothelial cells to surrounding tissues. When excess interstitial fluid accumulates and causes tissue swelling, the anchoring filaments are pulled, making the openings between cells even larger so that more fluid can flow into the lymphatic capillary.

In the small intestine, specialized lymphatic capillaries called **lacteals** (LAK-tē-als; *lact-*=milky) carry dietary lipids into lymphatic vessels and ultimately into the blood. The presence of these lipids causes the lymph draining from the small intestine to appear creamy white; such lymph is referred to as **chyle** (KĪL=juice). Elsewhere, lymph is a clear, pale-yellow fluid.

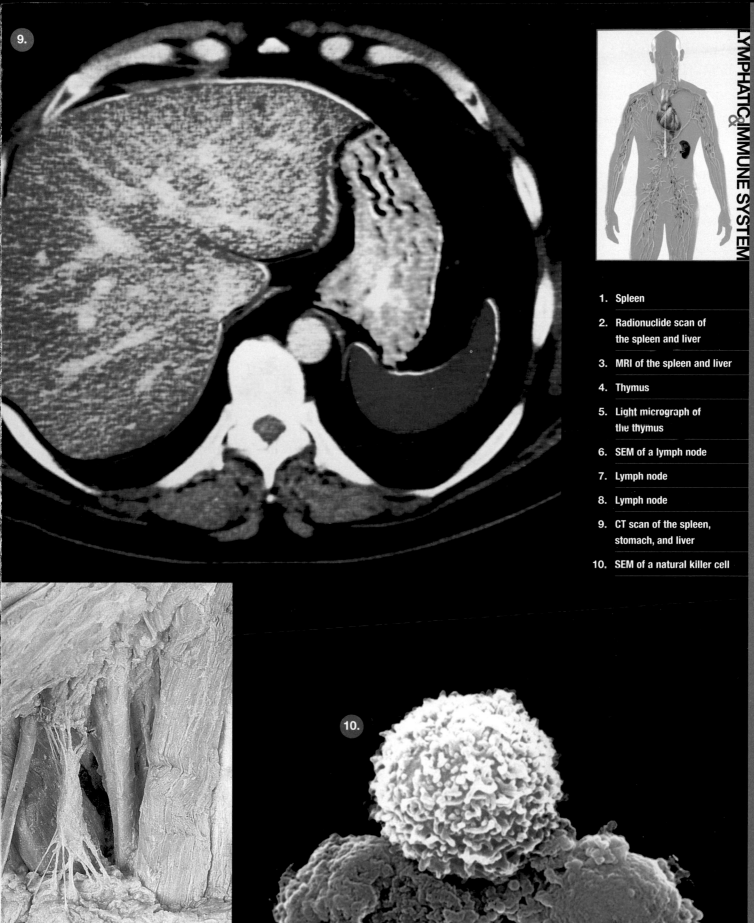

9.

1. Spleen

2. Radionuclide scan of the spleen and liver

3. MRI of the spleen and liver

4. Thymus

5. Light micrograph of the thymus

6. SEM of a lymph node

7. Lymph node

8. Lymph node

9. CT scan of the spleen, stomach, and liver

10. SEM of a natural killer cell

10.

LYMPHATIC AND IMMUNE SYSTEM

1. **Spleen.** Spleen in anterior view.

2. **Radionuclide scan of the spleen and liver.** Radionuclide scan of the spleen and liver in anterior view.

3. **MRI of the spleen and liver.** Magnetic Resonance Image (MRI) of the spleen and liver in transverse section.

4. **Thymus.** Thymus in anterior view.

5. **Light micrograph of the thymus.** Light micrograph of a portion of the thymus showing the cortex and medulla of the thymic lobules.

6. **SEM of a lymph node.** Scanning Electron Micrograph (SEM) of a lymph node.

7. **Lymph node.** Light micrograph of a lymph node with numerous germinal centers.

8. **Lymph node.** Lymph node in anterior view.

9. **CT scan of the spleen, stomach, and liver.** Computed Tomography (CT) scan of the spleen, stomach, and liver in transverse section.

10. **SEM of a natural killer cell.** Scanning Electron Micrograph (SEM) of a natural killer cell attacking two cancer cells.

WILEY

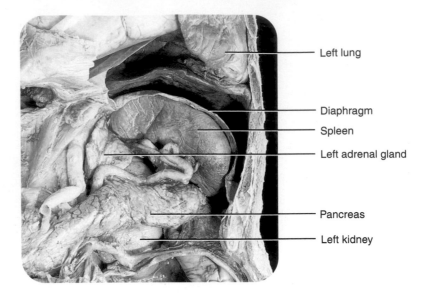

Left lung — Diaphragm — Spleen — Left adrenal gland — Pancreas — Left kidney

1.

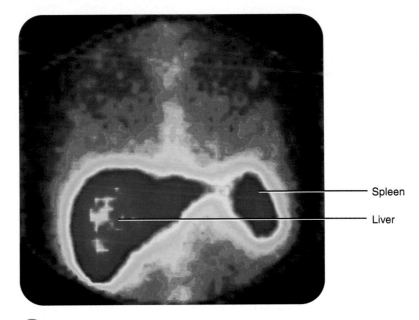

Spleen — Liver

2.

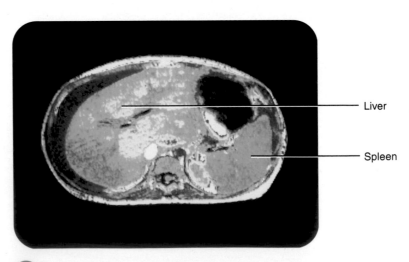

Liver — Spleen

3.

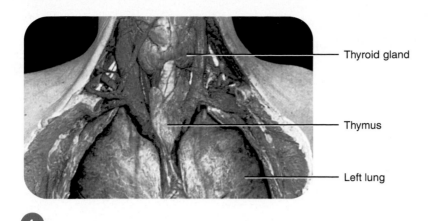

Thyroid gland

Thymus

Left lung

4.

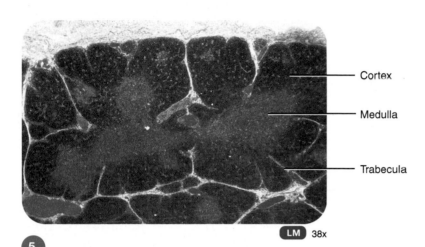

Cortex

Medulla

Trabecula

LM 38x

5.

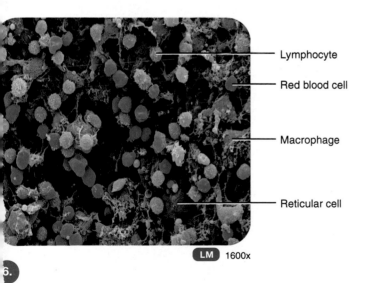

Lymphocyte

Red blood cell

Macrophage

Reticular cell

LM 1600x

6.

8.

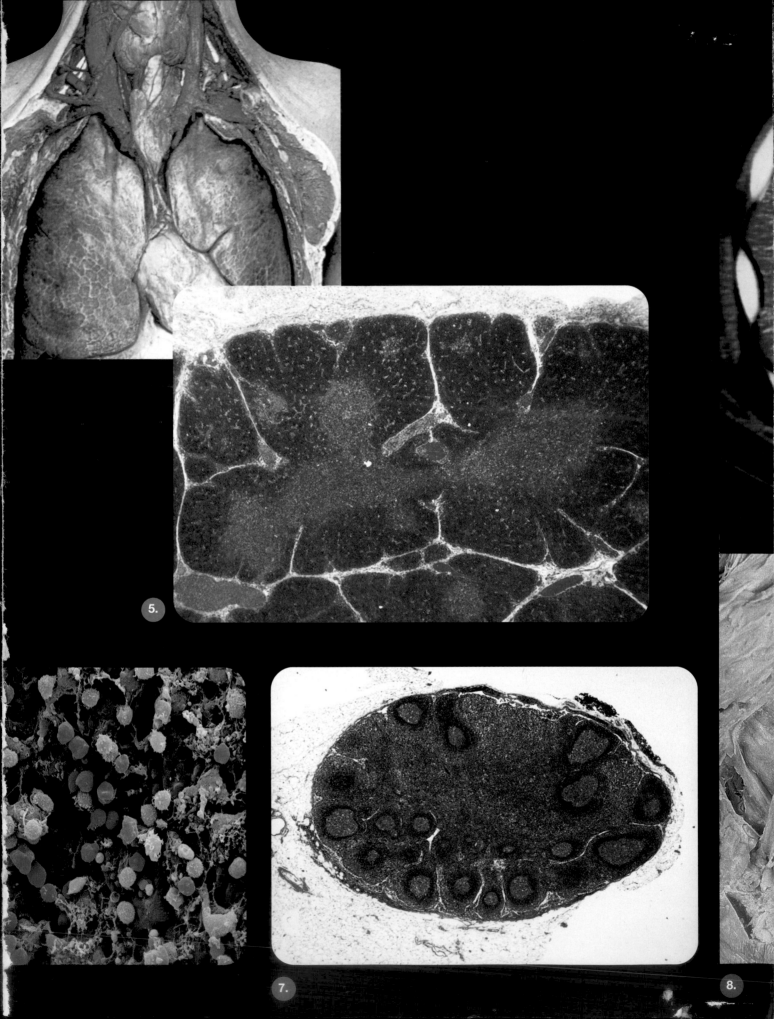

5.

7.

8.

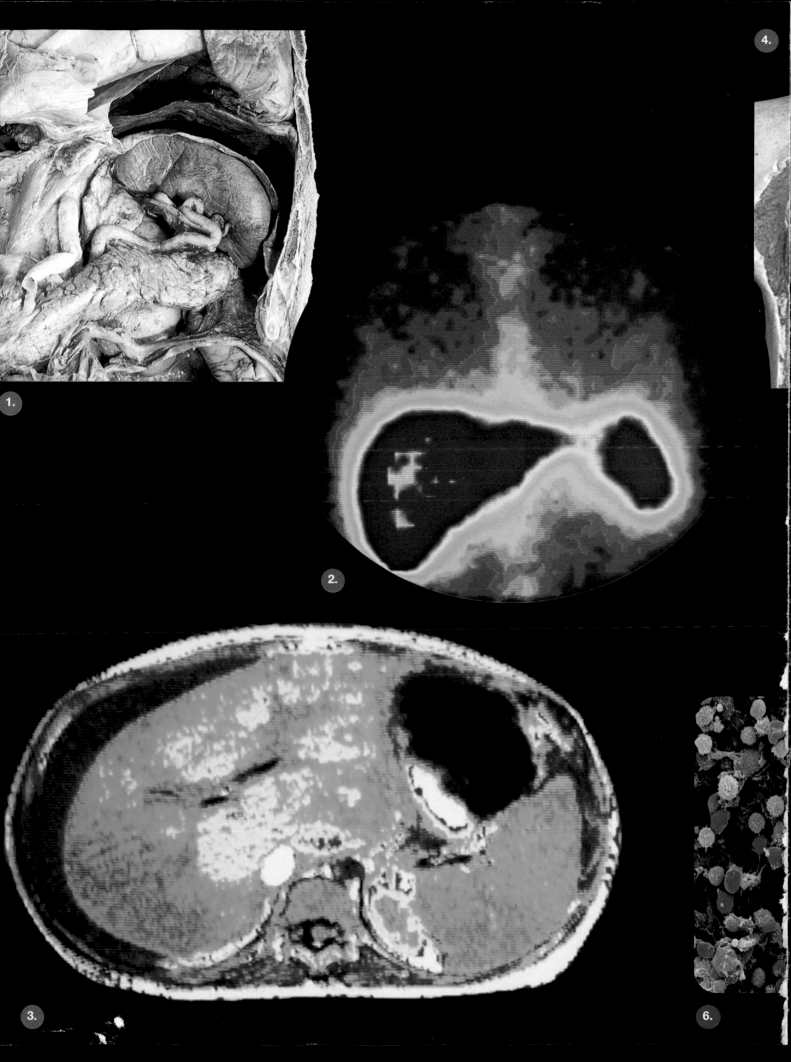

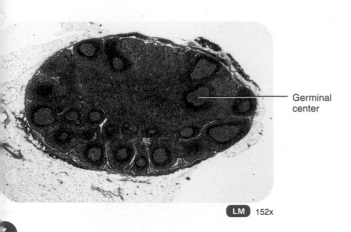

Germinal center

LM 152x

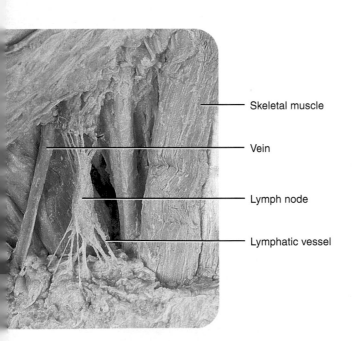

Skeletal muscle

Vein

Lymph node

Lymphatic vessel

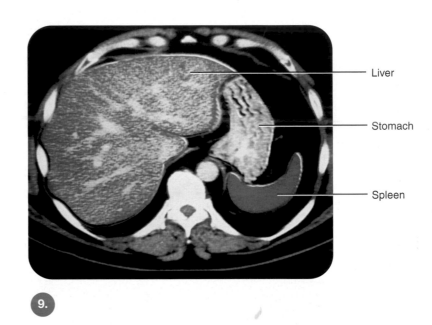

Liver

Stomach

Spleen

9.

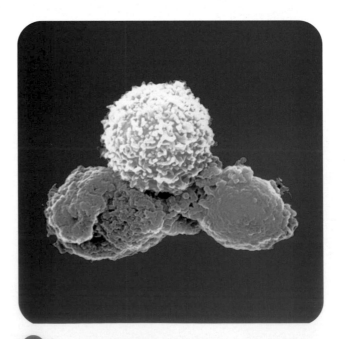

10.

Figure 16.1 Components of the lymphatic system.

🔑 The lymphatic and immune system consists of lymph, lymphatic vessels, lymphatic tissues, and red bone marrow.

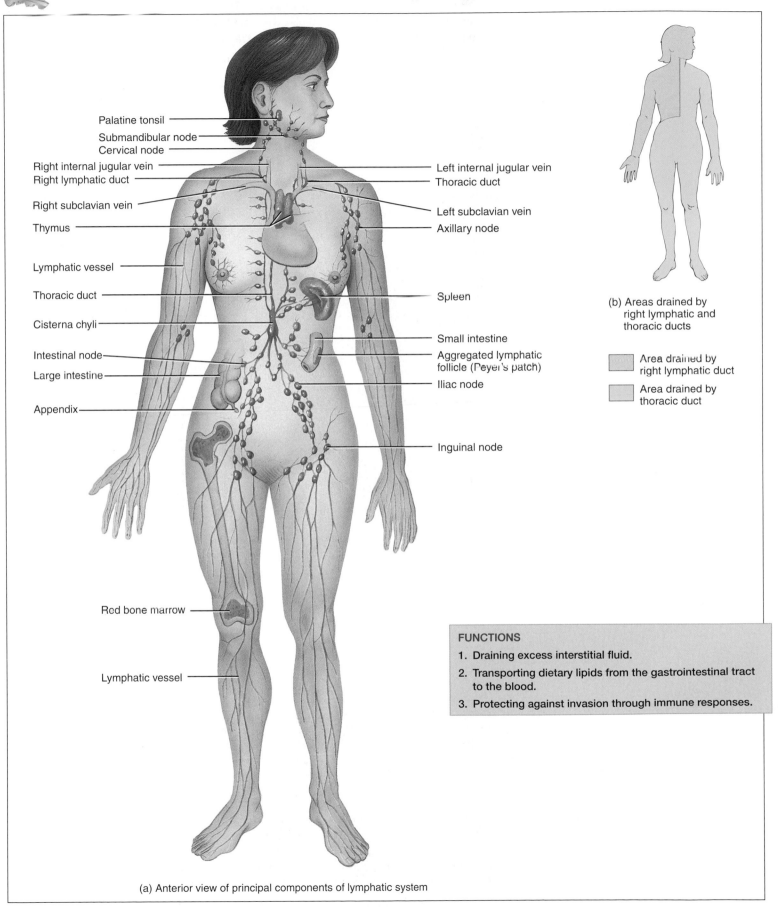

Palatine tonsil
Submandibular node
Cervical node
Right internal jugular vein
Right lymphatic duct
Right subclavian vein
Thymus
Lymphatic vessel
Thoracic duct
Cisterna chyli
Intestinal node
Large intestine
Appendix
Red bone marrow
Lymphatic vessel

Left internal jugular vein
Thoracic duct
Left subclavian vein
Axillary node
Spleen
Small intestine
Aggregated lymphatic follicle (Peyer's patch)
Iliac node
Inguinal node

(b) Areas drained by right lymphatic and thoracic ducts

☐ Area drained by right lymphatic duct

☐ Area drained by thoracic duct

FUNCTIONS

1. Draining excess interstitial fluid.
2. Transporting dietary lipids from the gastrointestinal tract to the blood.
3. Protecting against invasion through immune responses.

(a) Anterior view of principal components of lymphatic system

❓ **What tissue contains stem cells that develop into lymphocytes?**

Figure 16.2   Lymphatic capillaries.

Lymphatic capillaries are found throughout the body except in avascular tissues, the central nervous system, portions of the spleen, and red bone marrow.

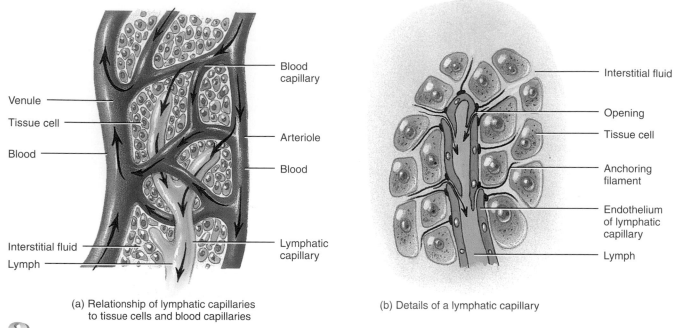

(a) Relationship of lymphatic capillaries to tissue cells and blood capillaries

(b) Details of a lymphatic capillary

Is lymph more similar to blood plasma or to interstitial fluid? Why?

Lymph Trunks and Ducts

Lymph passes from lymphatic capillaries into lymphatic vessels and then through lymph nodes. Lymphatic vessels exiting lymph nodes pass lymph either toward another node within the same group or on to another group of nodes. From the most proximal group of each chain of nodes, the exiting vessels unite to form **lymph trunks.** The principal trunks are the **lumbar, intestinal, bronchomediastinal, subclavian,** and **jugular trunks** (Figure 16.3). Lymph passes from lymph trunks into two main channels, the thoracic duct and the right lymphatic duct, and then drains into venous blood.

The **thoracic (left lymphatic) duct** is about 38–45 cm (15–18 in.) long and begins as a dilation called the **cisterna chyli** (sis-TER-na KĪ-lē; *cisterna*=cavity or reservoir) anterior to the second lumbar vertebra. The thoracic duct is the main duct for return of lymph to blood. It receives lymph from the left side of the head, neck, and chest, the left upper limb, and the entire body inferior to the ribs (see Figure 16.1b). The thoracic duct drains lymph into venous blood at the junction of the left internal jugular and left subclavian veins.

The cisterna chyli receives lymph from the right and left lumbar trunks and from the intestinal trunk. The lumbar trunks drain lymph from the lower limbs, the wall and viscera of the pelvis, the kidneys, the adrenal glands, and the deep lymphatic vessels that drain lymph from most of the abdominal wall. The intestinal trunk drains lymph from the stomach, intestines, pancreas, spleen, and part of the liver.

In the neck, the thoracic duct also receives lymph from the left jugular, left subclavian, and left bronchomediastinal trunks. The left jugular trunk drains lymph from the left side of the head and neck, and the left subclavian trunk drains lymph from the left upper limb. The left bronchomediastinal trunk drains lymph from the left side of the deeper parts of the anterior thoracic wall, the superior part of the anterior abdominal wall, the anterior part of the diaphragm, the left lung, and the left side of the heart.

The **right lymphatic duct** (Figure 16.3) is about 1.2 cm (0.5 in.) long and drains lymph from the upper-right side of the body into venous blood at the junction of the right internal jugular and right subclavian veins (see Figure 16.1b). Three lymphatic trunks drain into the right lymphatic duct: (1) The right jugular trunk drains the right side of the head and neck. (2) The right subclavian trunk drains the right upper limb. (3) The right bronchomediastinal trunk drains the right side of the thorax, the right lung, the right side of the heart, and part of the liver.

Formation and Flow of Lymph

Most components of blood plasma filter freely through the capillary walls to form interstitial fluid. More fluid filters out of blood capillaries, however, than returns to them by reabsorption. The excess filtered fluid—about 3 liters per day—drains into lymphatic vessels and becomes lymph. Because most blood plasma proteins are too large to leave blood vessels, interstitial

Figure 16.3   Routes for drainage of lymph from lymph trunks into the thoracic and right lymphatic ducts. The arrows indicate the direction of lymph flow.

🔑 **All lymph returns to the bloodstream through the thoracic (left) lymphatic duct and right lymphatic duct.**

Right internal jugular vein

RIGHT JUGULAR TRUNK

RIGHT SUBCLAVIAN TRUNK

RIGHT LYMPHATIC DUCT

Right subclavian vein

Right brachiocephalic vein

RIGHT BRONCHO MEDIASTINAL TRUNK

Superior vena cava

Rib

Intercostal muscle

Azygos vein

CISTERNA CHYLI

RIGHT LUMBAR TRUNK

Inferior vena cava

Esophagus

Trachea

Left internal jugular vein

LEFT JUGULAR TRUNK

LEFT SUBCLAVIAN TRUNK

THORACIC (LEFT LYMPHATIC) DUCT

Left brachiocephalic vein

Left subclavian vein

First rib

LEFT BRONCHO-MEDIASTINAL TRUNK

Accessory hemiazygos vein

THORACIC (LEFT LYMPHATIC) DUCT

Hemiazygos vein

LEFT LUMBAR TRUNK

INTESTINAL TRUNK

(a) Overall anterior view

Right jugular trunk

Right lymphatic duct

Right subclavian trunk

Left jugular trunk

Left subclavian trunk

Thoracic (left lymphatic) duct

Right broncho-mediastinal trunk

Left broncho-mediastinal trunk

(b) Detailed anterior view

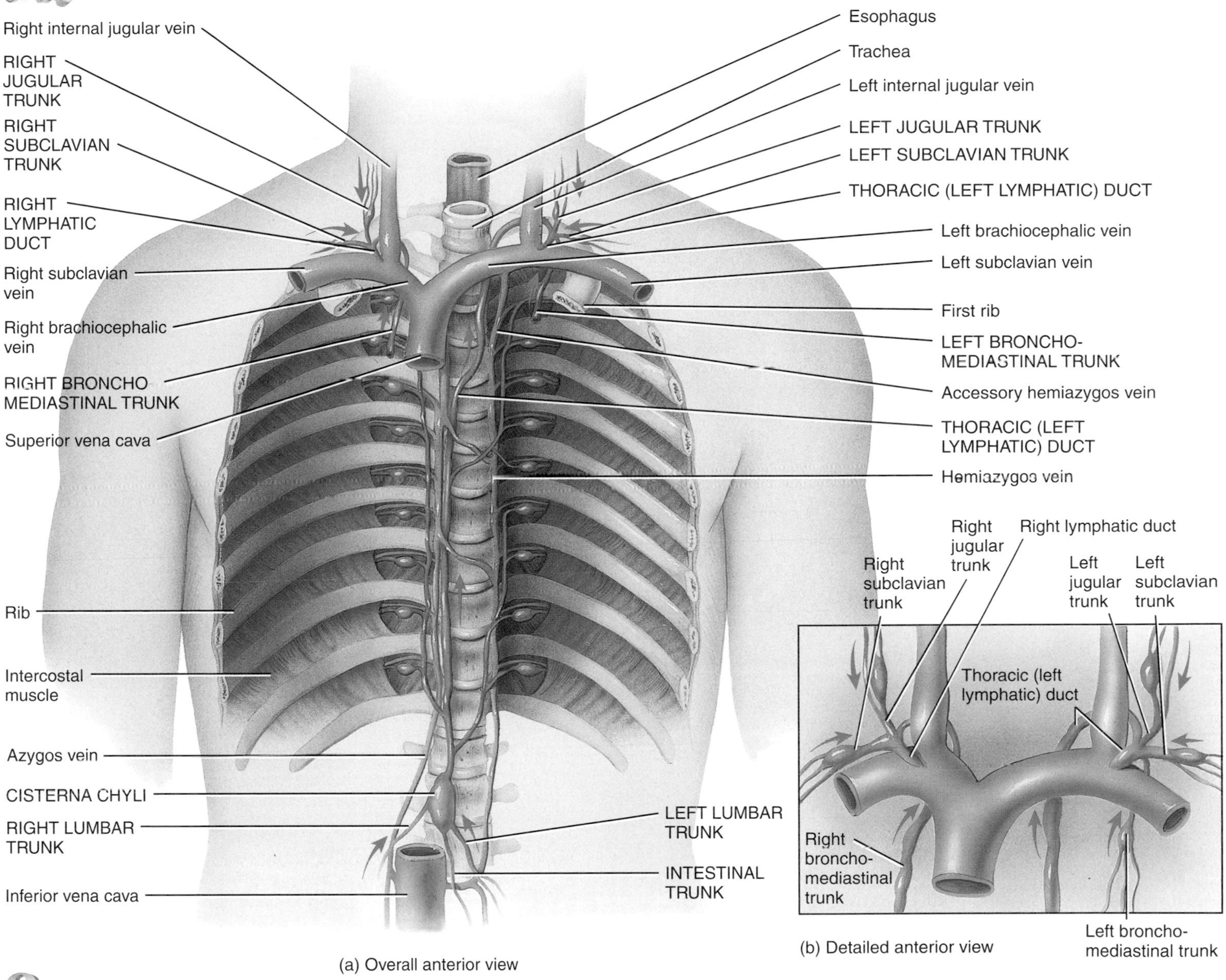

❓ **Which lymphatic vessels empty into the cisterna chyli, and which duct receives lymph from the cisterna chyli?**

fluid contains only a small amount of protein. Proteins that do leave blood plasma cannot return to the blood directly by diffusion because the concentration gradient (high level of proteins inside blood capillaries, low level outside) opposes such movement. The proteins can, however, move readily through the more permeable lymphatic capillaries into lymph. Thus, an important function of lymphatic vessels is to return lost plasma proteins to the bloodstream.

Like veins, lymphatic vessels contain valves, which ensure the one-way movement of lymph. As noted previously, lymph drains into venous blood through the right lymphatic duct and the thoracic duct at the junction of the internal jugular and subclavian veins (Figure 16.3). Thus, the sequence of fluid flow is blood capillaries (blood) → interstitial spaces (interstitial fluid) → lymphatic capillaries (lymph) → lymphatic vessels (lymph) → lymphatic ducts (lymph) → junction of internal jugular and subclavian veins (blood). Figure 16.4 illustrates this sequence, as well as the relationship of the lymphatic and cardiovascular systems.

Figure 16.4 Schematic diagram showing the relationship of the lymphatic system to the cardiovascular system. Arrows show direction of flow of lymph and blood.

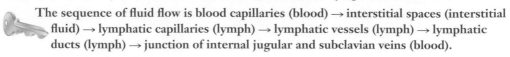

The sequence of fluid flow is blood capillaries (blood) → interstitial spaces (interstitial fluid) → lymphatic capillaries (lymph) → lymphatic vessels (lymph) → lymphatic ducts (lymph) → junction of internal jugular and subclavian veins (blood).

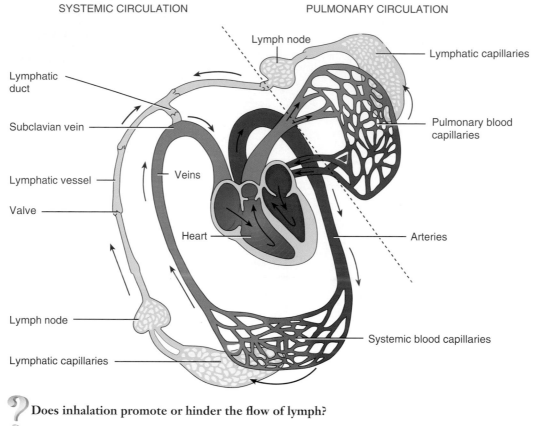

Does inhalation promote or hinder the flow of lymph?

EDEMA AND LYMPH FLOW

Edema (*edemas*=swelling), an excessive accumulation of interstitial fluid in tissue spaces, may be caused by an obstruction to lymph flow, such as an infected lymph node or a blocked lymphatic vessel. Edema may also result from increased capillary blood pressure, which causes excess interstitial fluid to form faster than it can pass into lymphatic vessels or be reabsorbed back into the capillaries. ∎

Lymphatic Organs and Tissues

Lymphatic organs and tissues, which are widely distributed throughout the body, are classified into two groups based on their functions. **Primary lymphatic organs** are the sites where stem cells divide and become *immunocompetent*, that is, capable of mounting an immune response. The primary lymphatic organs are the red bone marrow (in flat bones and the epiphyses of long bones of adults) and the thymus. Pluripotent stem cells in red bone marrow give rise to mature, immunocompetent B cells and to pre-T cells (immature T cells) which migrate to and become immunocompetent T cells in the thymus. The **secondary lymphatic organs** and **tissues** are the sites where most

immune responses occur. They include lymph nodes, the spleen, and lymphatic nodules (follicles). The thymus, lymph nodes, and spleen are considered organs because each is surrounded by a connective tissue capsule; lymphatic nodules, in contrast, are not organs because they lack such a capsule.

Thymus

The **thymus** is a bilobed organ located in the mediastinum between the sternum and the aorta (Figure 16.5a). An enveloping layer of connective tissue holds the two lobes closely together, but a connective tissue **capsule** encloses each lobe separately. Extensions of the capsule, called **trabeculae** (tra-BEK-ū-lē=little beams), penetrate inward and divide each lobe into **lobules** (Figure 16.5b).

Each thymic lobule consists of a deeply staining outer cortex and a lighter-staining central medulla (Figure 16.5b). The **cortex** is composed of large numbers of T cells and scattered dendritic cells, epithelial cells, and macrophages. Immature T cells migrate from red bone marrow to the cortex of the thymus, where they proliferate and begin to mature. **Dendritic cells** (*dendr-*=a tree), so-named because they have long, branched projections that resemble the dendrites of a neuron, assist the

Figure 16.5 Thymus. (See Tortora, *A Photographic Atlas of the Human Body,* 2e, Figure 7.2a.)

 The bilobed thymus is largest at puberty and then atrophies with age.

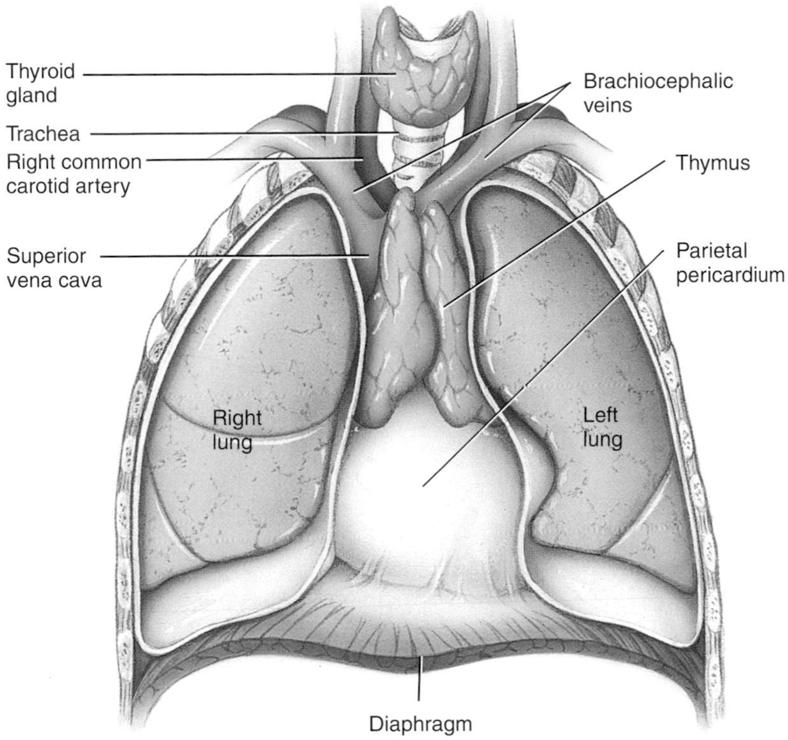

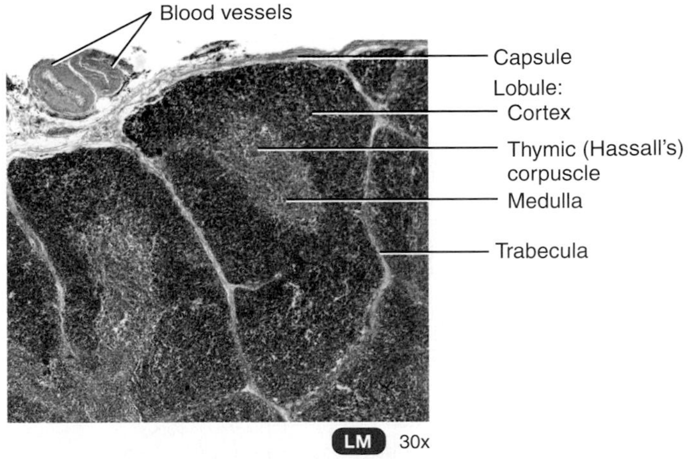

Blood vessels

Capsule
Lobule:
Cortex
Thymic (Hassall's) corpuscle
Medulla
Trabecula

LM 30x

(b) Thymic lobules

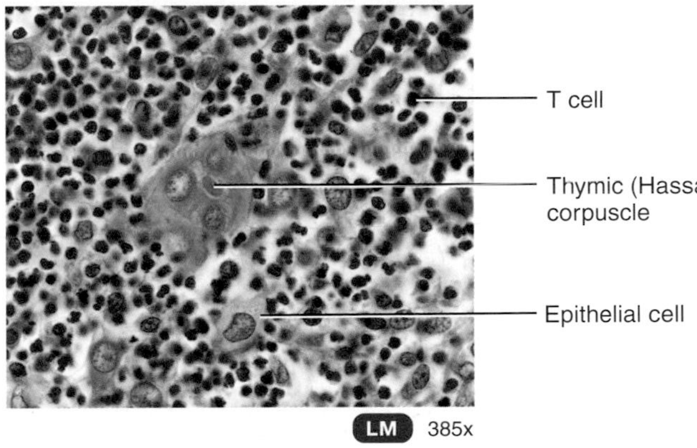

Thyroid gland
Trachea
Right common carotid artery
Superior vena cava

Brachiocephalic veins
Thymus
Parietal pericardium

Right lung
Left lung

Diaphragm

(a) Thymus of adolescent

T cell

Thymic (Hassall's) corpuscle

Epithelial cell

LM 385x

(c) Details of the thymic medulla

Which types of lymphocytes mature in the thymus?

maturation process. As you will see shortly, dendritic cells in other parts of the body, such as lymph nodes, play another key role in immune responses. Each of the specialized **epithelial cells** in the cortex has several long processes that surround and serve as a framework for as many as 50 T cells. These epithelial cells help pre-T cells mature into T cells. Additionally, the epithelial cells produce thymic hormones that are thought to aid in the maturation of T cells. Only about 2% of developing T cells survive in the cortex. The remaining cells die via **apoptosis** (programmed cell death). Thymic macrophages help clear out the debris of dead and dying cells. The surviving T cells enter the medulla.

The **medulla** consists of widely scattered, more mature T cells, epithelial cells, dendritic cells, and macrophages (Figure 16.5c). Some of the epithelial cells become arranged into concentric layers of flat cells that degenerate and become filled with keratohyalin granules and keratin. These clusters are called **thymic (Hassall's) corpuscles.** Although their role is uncertain, they may serve as sites of T cell death in the medulla. T cells that leave the thymus via the blood are carried to lymph nodes, the spleen, and other lymphatic tissues where they colonize parts of these organs and tissues.

In infants, the thymus is large, with a mass of about 70 g (2.3 oz). After puberty, adipose and areolar connective tissue begin to replace the thymic tissue. By the time a person reaches maturity, the gland has atrophied considerably, and in old age it may weigh only 3 g (0.1 oz). Before the thymus atrophies, it populates the secondary lymphatic organs and tissues with T cells. However, some T cells continue to proliferate in the thymus throughout an individual's lifetime.

Lymph Nodes

Located along lymphatic vessels are about 600 bean-shaped **lymph nodes.** They are scattered throughout the body, both superficially and deep, and usually occur in groups (see Figure 16.1). Large groups of lymph nodes are present near the mammary glands and in the axillae and groin. Later in the chapter, the principal groups of lymph nodes in various regions of the body will be presented in a series of Exhibits, starting on page 522.

Lymph nodes are 1–25 mm (0.04–1 in.) long and, like the thymus, are covered by a capsule of dense connective tissue that extends into the node (Figure 16.6). The capsular extensions, called **trabeculae,** divide the node into compartments, provide

Figure 16.6 Structure of a lymph node. Arrows indicate the direction of lymph flow through a lymph node.

Lymph nodes are present throughout the body, usually clustered in groups.

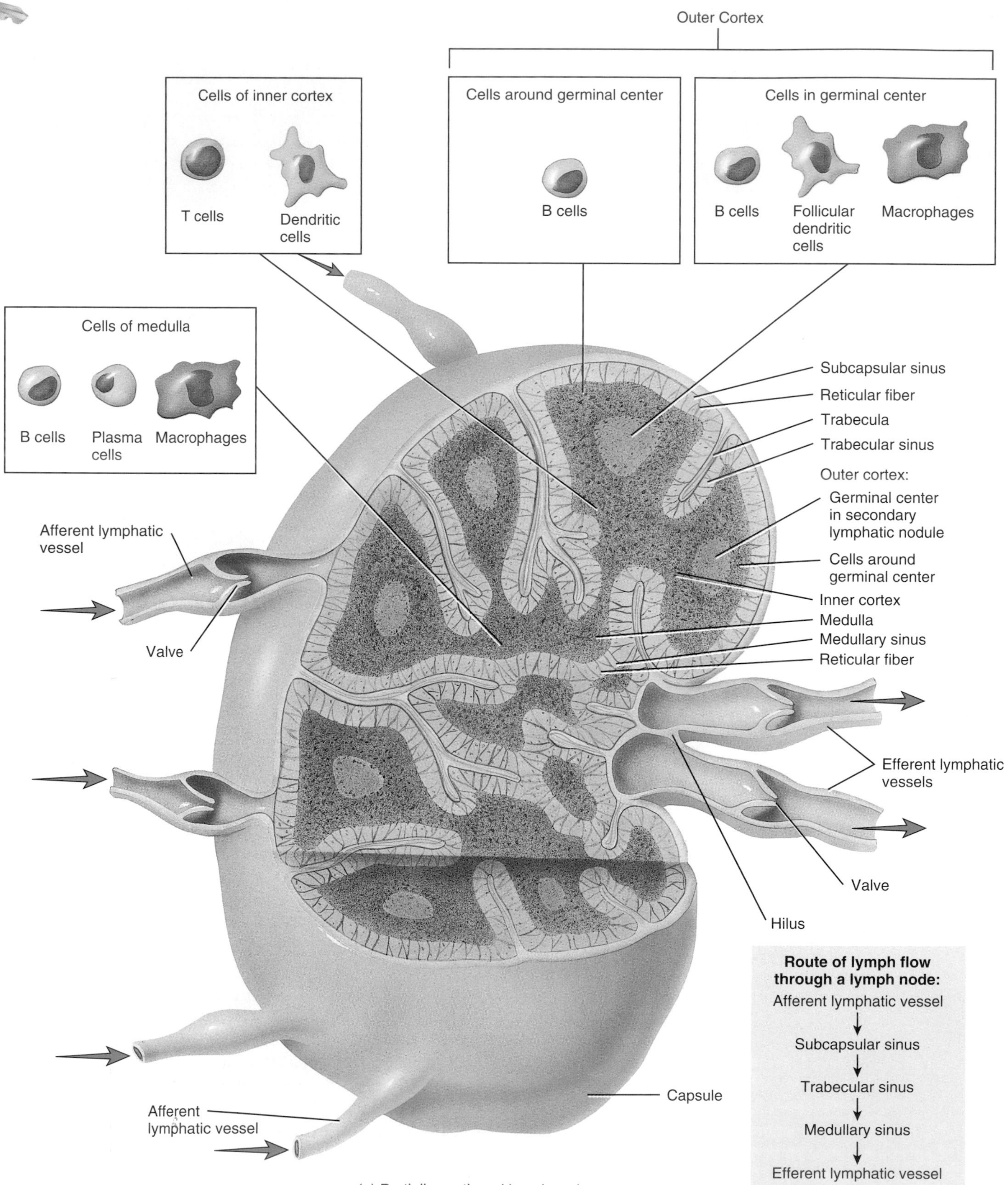

Outer Cortex

Cells of inner cortex

T cells
Dendritic cells

Cells around germinal center

B cells

Cells in germinal center

B cells
Follicular dendritic cells
Macrophages

Cells of medulla

B cells
Plasma cells
Macrophages

Afferent lymphatic vessel

Valve

Subcapsular sinus
Reticular fiber
Trabecula
Trabecular sinus
Outer cortex:
Germinal center in secondary lymphatic nodule
Cells around germinal center
Inner cortex
Medulla
Medullary sinus
Reticular fiber

Efferent lymphatic vessels

Valve

Hilus

Capsule

Afferent lymphatic vessel

Route of lymph flow through a lymph node:

Afferent lymphatic vessel
↓
Subcapsular sinus
↓
Trabecular sinus
↓
Medullary sinus
↓
Efferent lymphatic vessel

(a) Partially sectioned lymph node showing overall structure

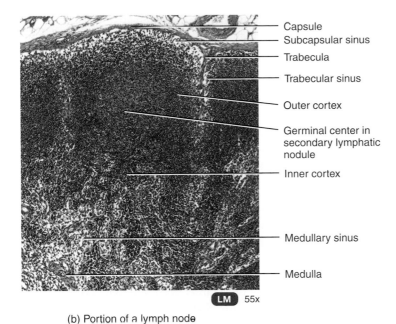

- Capsule
- Subcapsular sinus
- Trabecula
- Trabecular sinus
- Outer cortex
- Germinal center in secondary lymphatic nodule
- Inner cortex
- Medullary sinus
- Medulla

LM 55x

(b) Portion of a lymph node

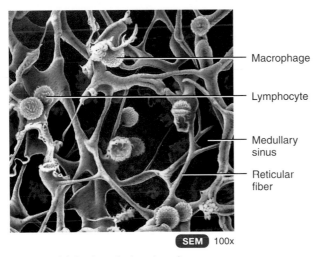

- Macrophage
- Lymphocyte
- Medullary sinus
- Reticular fiber

SEM 100x

(c) Portion of a lymph node

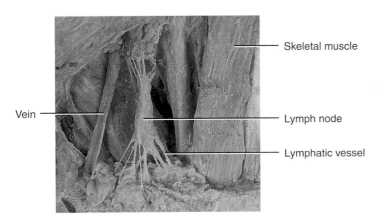

- Skeletal muscle
- Vein
- Lymph node
- Lymphatic vessel

(d) Anterior view of a lymph node

? **What happens to foreign substances in the lymph when they enter a lymph node?**

support, and provide a route for blood vessels into the interior of a node. Internal to the capsule is a supporting network of reticular fibers and fibroblasts. The capsule, trabeculae, reticular fibers, and fibroblasts constitute the *stroma*, or framework, of a lymph node.

The *parenchyma* of a lymph node is divided into a superficial cortex and a deep medulla. Within the **outer cortex** are eggshaped aggregates of B cells called **lymphatic nodules** (follicles). A lymphatic nodule consisting chiefly of B cells is called a *primary lymphatic nodule*. Most lymphatic nodules in the outer cortex are *secondary lymphatic nodules* (Figure 16.6), which form in response to an antigenic challenge and are sites of plasma cell and memory B cell formation. After B cells in a primary lymphatic nodule recognize an antigen, the primary lymphatic nodule develops into a secondary lymphatic nodule. The center of a secondary lymphatic nodule contains a region of light-staining cells called a *germinal center*. In the germinal center are B cells, follicular dendritic cells (a special type of dendritic cell), and macrophages. When follicular dendritic cells "present" an antigen, B cells proliferate and develop into antibody-producing plasma cells or develop into memory B cells. Memory B cells persist after an immune response and "remember" having encountered a specific antigen. B cells that do not develop properly undergo apoptosis and are destroyed by macrophages. The region of a secondary lymphatic nodule surrounding the germinal center is composed of dense accumulations of B cells that have migrated away from their site of origin within the nodule.

The **inner cortex**, also called the *paracortex*, does not contain lymphatic nodules. It consists mainly of T cells and dendritic cells that enter a lymph node from other tissues. The dendritic cells present antigens to T cells, causing their proliferation. The newly formed T cells then migrate from the lymph node to areas of the body where there is antigenic activity.

The **medulla** of a lymph node contains B cells, antibody-producing plasma cells that have migrated out of the cortex into the medulla, and macrophages. The various cells are embedded in a network of reticular fibers and reticular cells.

As you have already learned, lymph flows through a lymph node in one direction only (Figure 16.6a). It enters through **afferent lymphatic vessels** (*afferent*=to carry toward), which penetrate the convex surface of the node at several points. The afferent vessels contain valves that open toward the center of the node so that the lymph is directed *inward*. Within the node, lymph enters **sinuses**, a series of irregular channels that contain branching reticular fibers, lymphocytes, and macrophages. From the afferent lymphatic vessels, lymph flows into the **subcapsular sinus,** immediately beneath the capsule. From there the lymph flows through **trabecular sinuses,** which extend through the cortex parallel to the trabeculae, and into **medullary sinuses,** which extend through the medulla. The medullary sinuses drain into one or two **efferent lymphatic vessels** (*efferent*=to carry away), which are wider than afferent vessels and fewer in number. They contain valves that open away from the center of the lymph node to convey lymph, antibodies secreted by plasma cells, and activated T cells *out* of the lymph node. Efferent lymphatic vessels emerge from one side of the lymph node at a slight depression called a **hilus** (HĪ-lus). Blood vessels also enter and leave the node at the hilus.

Lymph nodes function as a type of filter. As lymph enters one end of a lymph node, foreign substances are trapped by the reticular fibers within the sinuses of the node. Then macrophages destroy some foreign substances by phagocytosis while lymphocytes destroy others by immune responses. The filtered lymph then leaves the other end of the lymph node.

METASTASIS THROUGH LYMPHATIC VESSELS

Metastasis (me-TAS-ta-sis; *meta-*=beyond; *stasis*=to stand), the spread of a disease from one part of the body to another, can occur via lymphatic vessels. All malignant tumors eventually metastasize. Cancer cells may travel in the blood or lymph and establish new tumors where they lodge. When metastasis occurs via lymphatic vessels, secondary tumor sites can be predicted according to the direction of lymph flow from the primary tumor site. Cancerous lymph nodes feel enlarged, firm, non-tender, and fixed to underlying structures. By contrast, most lymph nodes that are enlarged due to an infection are softer, moveable, and very tender. ■

Spleen

The oval **spleen** is the largest single mass of lymphatic tissue in the body, measuring about 12 cm (5 in.) in length (Figure 16.7a). It is located in the left hypochondriac region between the stomach and diaphragm. The superior surface of the spleen is smooth and convex and conforms to the concave surface of the diaphragm. Neighboring organs make indentations in the visceral surface of the spleen—the *gastric impression* (stomach), the *renal impression* (left kidney), and the *colic impression* (left flexure of colon). Like lymph nodes, the spleen has a hilus. Through it pass the splenic artery, splenic vein, and efferent lymphatic vessels.

A capsule of dense connective tissue surrounds the spleen and is covered, in turn, by a serous membrane, the visceral peritoneum. Trabeculae extend inward from the capsule. The capsule plus trabeculae, reticular fibers, and fibroblasts constitute the stroma of the spleen; the parenchyma of the spleen consists of two different kinds of tissue called white pulp and red pulp (Figure 16.7c, d). **White pulp** is lymphatic tissue, consisting mostly of lymphocytes and macrophages arranged around branches of the splenic artery called **central arteries.** The **red pulp** consists of blood-filled **venous sinuses** and cords of splenic tissue called **splenic (Billroth's) cords.** Splenic cords consist of red blood cells, macrophages, lymphocytes, plasma cells, and granulocytes. Veins are closely associated with the red pulp.

Blood flowing into the spleen through the splenic artery enters the central arteries of the white pulp. Within the white pulp, B cells and T cells carry out immune functions, similar to lymph nodes, while spleen macrophages destroy blood-borne pathogens by phagocytosis. Within the red pulp, the spleen performs three functions related to blood cells: (1) removal by macrophages of ruptured, worn out, or defective blood cells and platelets; (2) storage of platelets, up to one-third of the body's supply; and (3) production of blood cells (hemopoiesis) during fetal life.

RUPTURED SPLEEN

The spleen is one of the organs most often damaged in cases of abdominal trauma. Severe blows over the inferior left chest or superior abdomen can fracture the protecting ribs. Such crushing injury may **rupture the spleen,** which causes severe intraperitoneal hemorrhage and shock. Prompt removal of the spleen, called a **splenectomy,** is needed to prevent death due to bleeding. Other structures, particularly red bone marrow and the liver, can take over some functions normally carried out by the spleen. Immune functions, however, decrease in the absence of a spleen. The spleen's absence also places the patient at higher risk for **sepsis** (a blood infection) due to loss of the filtering and phagocytic functions of the spleen. To reduce the risk of sepsis, patients who have undergone a splenectomy take prophylactic (preventive) antibiotics before any invasive procedures. ■

Lymphatic Nodules

Lymphatic nodules are egg-shaped masses of lymphatic tissue that are not surrounded by a capsule. Because they are scattered throughout the lamina propria (connective tissue) of mucous membranes lining the gastrointestinal, urinary, and reproductive tracts and the respiratory airways, lymphatic nodules in these areas are also referred to as **mucosa-associated lymphatic tissue (MALT).**

Although many lymphatic nodules are small and solitary, some occur in multiple large aggregations in specific parts of the body. Among these are the tonsils in the pharyngeal region and the aggregated lymphatic follicles (Peyer's patches) in the ileum of the small intestine. Aggregations of lymphatic nodules also occur in the appendix. Usually there are five **tonsils,** which form a ring at the junction of the oral cavity and oropharynx and at the junction of the nasal cavity and nasopharynx (see Figure 24.2b on page 734). The tonsils are strategically positioned to participate in immune responses against inhaled or ingested foreign substances. The single **pharyngeal tonsil** (fa-RIN-jē-al) or **adenoid** is embedded in the posterior wall of the nasopharynx. The two **palatine tonsils** (PAL-a-tīn) lie at the posterior region of the oral cavity, one on either side; these are the tonsils commonly removed in a tonsillectomy. The paired **lingual tonsils** (LIN-gwal), located at the base of the tongue, may also require removal during a tonsillectomy.

CHECKPOINT

1. How are interstitial fluid and lymph similar, and how do they differ?
2. How do lymphatic vessels differ in structure from veins?
3. Construct a diagram that shows the route of lymph circulation.
4. What is the role of the thymus in immunity?
5. What functions do lymph nodes serve?
6. Describe the functions of the spleen and tonsils.

Figure 16.7   Structure of the spleen.

The spleen is the largest single mass of lymphatic tissue in the body.

SUPERIOR

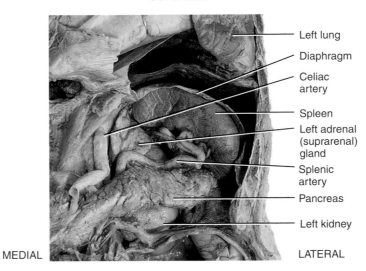

Left lung

Diaphragm

Celiac artery

Spleen

Left adrenal (suprarenal) gland

Splenic artery

Pancreas

Left kidney

MEDIAL

LATERAL

(a) Anterior view of a portion of the abdominal cavity

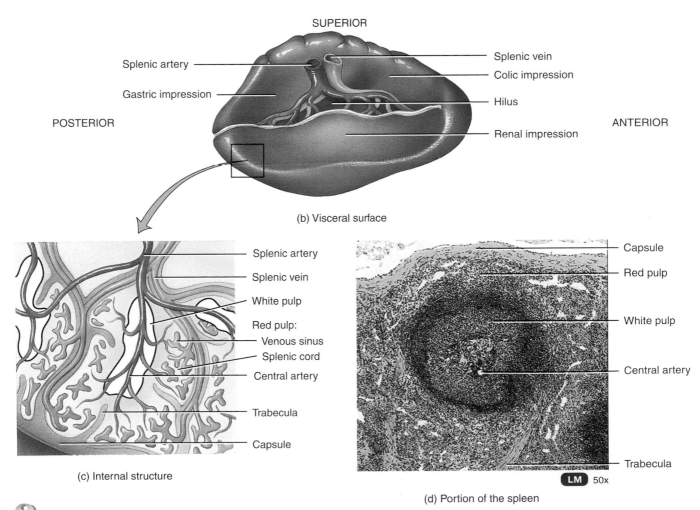

SUPERIOR

Splenic artery

Gastric impression

POSTERIOR

Splenic vein

Colic impression

Hilus

Renal impression

ANTERIOR

(b) Visceral surface

Splenic artery

Splenic vein

White pulp

Red pulp:
Venous sinus
Splenic cord
Central artery

Trabecula

Capsule

(c) Internal structure

Capsule

Red pulp

White pulp

Central artery

Trabecula

LM 50x

(d) Portion of the spleen

After birth, what are the main functions of the spleen?

EXHIBIT 16.1   PRINCIPAL LYMPH NODES OF THE HEAD AND NECK (Figure 16.8)

LYMPH NODES OF THE HEAD	LOCATION	DRAINAGE
OCCIPITAL NODES (ok-SIP-ī-tal)	Near trapezius and semispinalis capitis muscles.	Occipital portion of scalp and upper neck.
RETROAURICULAR NODES (re'-trō-aw-RIK-ū-lar=behind auricle)	Posterior to ear.	Skin of ear and posterior parietal region of scalp.
PRE-AURICULAR NODES (*pre-*=before)	Anterior to ear.	Auricle of ear and temporal region of scalp.
PAROTID NODES (pa-ROT-id; *para-*=beside; *ot-*=ear)	Embedded in and inferior to parotid gland.	Root of nose, eyelids, anterior temporal region, external auditory meatus, tympanic cavity, nasopharynx, and posterior portions of nasal cavity.
FACIAL NODES	Consist of three groups: infraorbital, buccal, and mandibular.	
Infraorbital nodes (in'-fra-OR-bi-tal; *infra-*=below; *-orbital*=orbit)	Inferior to the orbit.	Eyelids and conjunctiva.
Buccal nodes (BUK-al; *bucca-*=cheek)	At angle of mouth.	Skin and mucous membrane of nose and cheek.
Mandibular nodes (man-DIB-ū-lar; *mand-*=to chew)	Over mandible.	Skin and mucous membrane of nose and cheek.

LYMPH NODES OF THE NECK	LOCATION	DRAINAGE
SUBMANDIBULAR NODES	Along inferior border of mandible.	Chin, lips, nose, nasal cavity, cheeks, gums, inferior surface of palate, and anterior portion of tongue.
SUBMENTAL NODES (sub-MEN-tal; *sub-*=beneath; *-ment*=chin)	Between diagastric muscles.	Chin, lower lip, cheeks, tip of tongue, and floor of mouth.
SUPERFICIAL CERVICAL NODES (SER-vi-kul; *cervic-*=neck)	Along external jugular vein.	Inferior part of ear and parotid region.
DEEP CERVICAL NODES	Largest group of nodes in neck, consisting of numerous large nodes forming a chain extending from base of skull to root of neck. They are arbitrarily divided into superior deep cervical nodes and inferior deep cervical nodes.	
Superior deep cervical nodes	Deep to sternocleidomastoid muscle.	Posterior head and neck, auricle, tongue, larynx, esophagus, thyroid gland, nasopharynx, nasal cavity, palate, and tonsils.
Inferior deep cervical nodes	Near subclavian vein.	Posterior scalp and neck, superficial pectoral region, and part of arm.

PRINCIPAL GROUPS OF LYMPH NODES

OBJECTIVE

● Identify the locations and drainage regions of the principal groups of lymph nodes.

Exhibits 16.1–16.5 (pages 522–529) provide a listing of the principal groups of lymph nodes, by region, and the general areas that they drain.

text continues on page 528

Figure 16.8   Principal lymph nodes of the head and neck.

The facial lymph nodes include the infraorbital, buccal, and mandibular lymph nodes.

PRE-AURICULAR
LYMPH NODE

RETROAURICULAR
LYMPH NODE

OCCIPITAL
LYMPH NODE

SUPERFICIAL CERVICAL
LYMPH NODE

DEEP CERVICAL
LYMPH NODE

Orbicularis oculi muscle

INFRAORBITAL
LYMPH NODE

PAROTID LYMPH NODE

Parotid salivary gland

BUCCAL LYMPH NODE

MANDIBULAR LYMPH NODE

Submandibular salivary gland

SUBMENTAL LYMPH NODE

SUBMANDIBULAR
LYMPH NODE

Sternocleidomastoid muscle

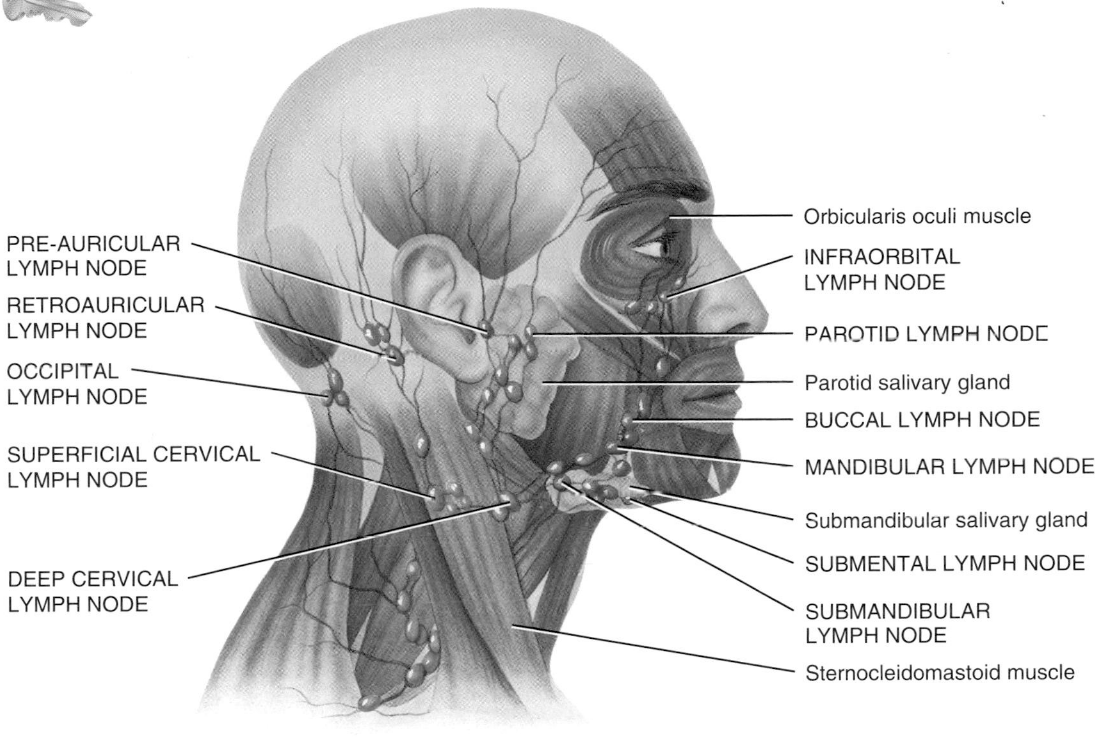

Lateral view of lymph nodes of the head and neck

Which is the largest group of lymph nodes in the neck?

EXHIBIT 16.2   PRINCIPAL LYMPH NODES OF THE UPPER LIMBS (Figure 16.9)

LYMPH NODES	LOCATION	DRAINAGE
SUPRATROCHLEAR NODES (soo-pra-TROK-lē-ar; *supra-*=above; *-trochlea*=pulley)	Superior to medial epicondyle of humerus.	Medial fingers, palm, and forearm.
DELTOPECTORAL NODES (del′-tō-PEK-tō-ral; *-pectus*=breast)	Inferior to clavicle.	Lymphatic vessels on radial side of upper limb.
AXILLARY NODES (AK-sil-ār-ē; *axil-*=armpit)*	Most deep lymph nodes of the upper limbs are in the axilla and are called the axillary nodes. They are large.	
Lateral nodes	Medial and posterior aspects of axillary artery.	Most of entire upper limb.
Pectoral (anterior) nodes	Along inferior border of the pectoralis minor muscle.	Skin and muscles of anterior and lateral thoracic walls and central and lateral portions of mammary gland.
Subscapular (posterior) nodes	Along subscapular artery.	Skin and muscles of posterior part of neck and thoracic wall.
Central (intermediate) nodes	Base of axilla embedded in adipose tissue.	Lateral, pectoral (anterior), and subscapular (posterior) nodes.
Subclavicular (apical) nodes	Posterior and superior to pectoralis minor muscle.	Deltopectoral nodes.

*Because infection or malignancy of the upper limbs may cause tenderness and swelling in the axilla, the axillary nodes, especially the lateral group, are clinically important in that they filter lymph from much of the upper limb. More than 75% of the lymph drainage of the breast is to the pectoral (anterior) group of axillary lymph nodes. In breast cancer, it is possible that cancer cells leave the breast and lodge in the pectoral nodes. From here, metastasis may develop in other axillary nodes. Most of the remaining lymphatic drainage is to the sternal (parasternal) lymph nodes (see Figure 16.12 also).

Figure 16.9   Principal lymph nodes of the upper limbs.
In (b) the direction of drainage is indicated by arrows.

🔑 **Most of the lymph drainage of the breast is to the pectoral group of axillary lymph nodes.**

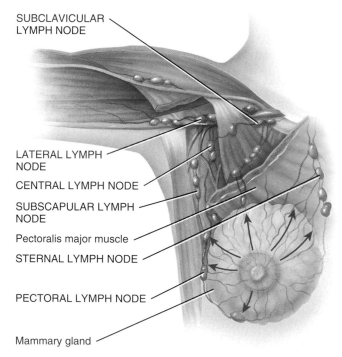

SUBCLAVICULAR LYMPH NODE

LATERAL LYMPH NODE

CENTRAL LYMPH NODE

SUBSCAPULAR LYMPH NODE

Pectoralis major muscle

STERNAL LYMPH NODE

PECTORAL LYMPH NODE

Mammary gland

(b) Anterior view of mostly axillary lymph nodes

❓ **Which lymph nodes drain most of the upper limb?**

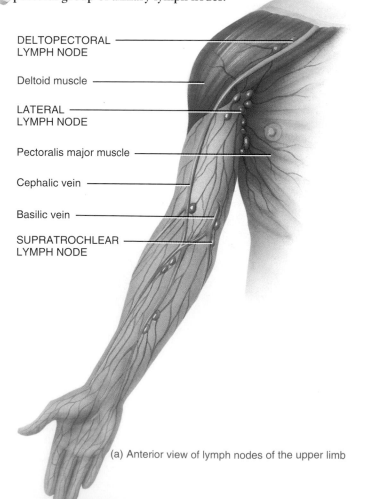

DELTOPECTORAL LYMPH NODE

Deltoid muscle

LATERAL LYMPH NODE

Pectoralis major muscle

Cephalic vein

Basilic vein

SUPRATROCHLEAR LYMPH NODE

(a) Anterior view of lymph nodes of the upper limb

EXHIBIT 16.3   PRINCIPAL LYMPH NODES OF THE LOWER LIMBS (Figure 16.10)

LYMPH NODES	LOCATION	DRAINAGE
POPLITEAL NODES (pop-LIT-ē-al; *poples-*=ham of the knee)	In adipose tissue in popliteal fossa. (Not shown)	Knee and portions of leg and foot, especially heel.
SUPERFICIAL INGUINAL NODES (ING-gwi-nal; *inguen-*=groin)	Parallel to saphenous vein.	Anterior and lateral abdominal wall to level of umbilicus, gluteal region, external genitals, perineal region, and entire superficial lymphatics of lower limb.
DEEP INGUINAL NODES	Medial to femoral vein.	Deep lymphatics of lower limb, penis, and clitoris.

Figure 16.10   Principal lymph nodes of the lower limbs.

The inguinal lymph nodes drain the lymphatics of the lower limbs.

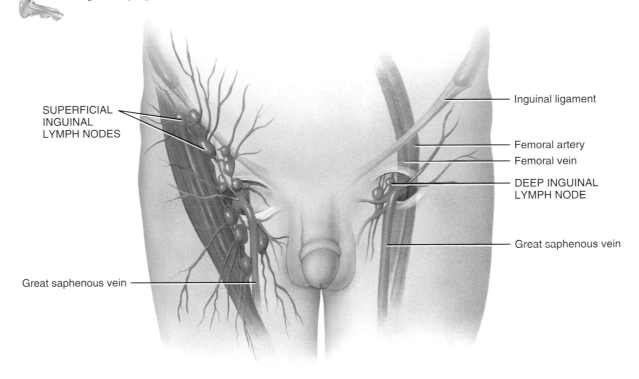

Anterior view of inguinal lymph nodes

? Which lymph nodes are parallel to the saphenous vein?

EXHIBIT 16.4 PRINCIPAL LYMPH NODES OF THE ABDOMEN AND PELVIS (Figure 16.11)

PARIETAL LYMPH NODES	LOCATION	DRAINAGE
Parietal nodes are located retroperitoneally (behind the parietal peritoneum) and in close association with larger blood vessels.		
EXTERNAL ILIAC NODES (IL-ē-ak=ilium)	Along external iliac vessels.	Deep lymphatics of abdominal wall inferior to umbilicus, adductor region of thigh, urinary bladder, prostate, ductus (vas) deferens, seminal vesicles, prostatic and membranous urethra, uterine (fallopian) tubes, uterus, and vagina.
COMMON ILIAC NODES	Along course of common iliac vessels.	Pelvic viscera.
INTERNAL ILIAC NODES	Near internal iliac artery.	Pelvic viscera, perineum, gluteal region, and posterior surface of thigh.
SACRAL NODES (SĀ-krum=holy bone)	In hollow of sacrum.	Rectum, prostate, and posterior pelvic wall.
LUMBAR NODES (LUM-bar; *lumb-*=loin)	From aortic bifurcation to diaphragm; arranged around aorta and designated as *right lateral aortic nodes, left lateral aortic nodes, preaortic nodes* and *retroaortic nodes.*	Efferents from testes, ovaries, uterine (fallopian) tubes, uterus, kidneys, adrenal (suprarenal) glands, abdominal surface of diaphragm, and lateral abdominal wall.

Figure 16.11 Principal lymph nodes of the abdomen and pelvis.

🔑 The parietal lymph nodes are retroperitoneal and in close association with larger blood vessels.

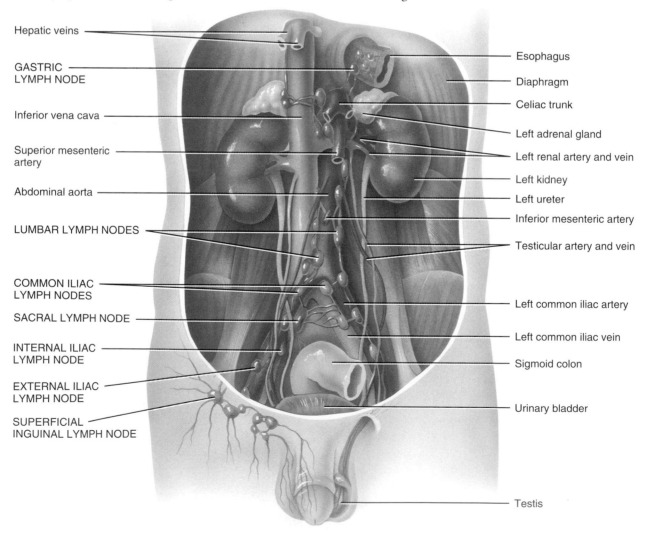

(a) Anterior view of abdominal and pelvic lymph nodes

Visceral nodes are found in association with visceral arteries.

VISCERAL LYMPH NODES	LOCATION	DRAINAGE
CELIAC NODES (SĒ-lē-ak; *koilia-*=belly)	Consist of three groups: gastric, hepatic, and pancreaticosplenic.	
Gastric nodes	Along lesser curvature of stomach.	Lesser curvature of stomach; inferior, anterior, and posterior aspects of stomach; esophagus.
Hepatic nodes	Along hepatic artery. (Not shown)	Stomach, duodenum, liver, gallbladder, and pancreas.
Pancreaticosplenic nodes (pan-krē-at'-i-kō-SPLĒN-ik)	Along splenic artery. (Not shown)	Stomach, spleen, and pancreas.
SUPERIOR MESENTERIC NODES (MES-en-ter'-ik; *meso-*=middle; *-enteric*=intestines)	Consist of three groups: mesenteric, ileocolic, and transverse mesocolic.	
Mesenteric nodes	Along superior mesenteric artery.	Jejunum and all parts of ileum, except for terminal portion.
Ileocolic nodes (il'-ē-ō-KŌL-ik)	Along ileocolic artery.	Terminal portion of ileum, appendix, cecum, and ascending colon.
Transverse mesocolic nodes	Between layers of transverse mesocolon.	Descending iliac and sigmoid parts of colon.
INFERIOR MESENTERIC NODES	Near left colic, sigmoid, and superior rectal arteries.	Descending, iliac, and sigmoid parts of colon; superior part of rectum; superior anal canal.

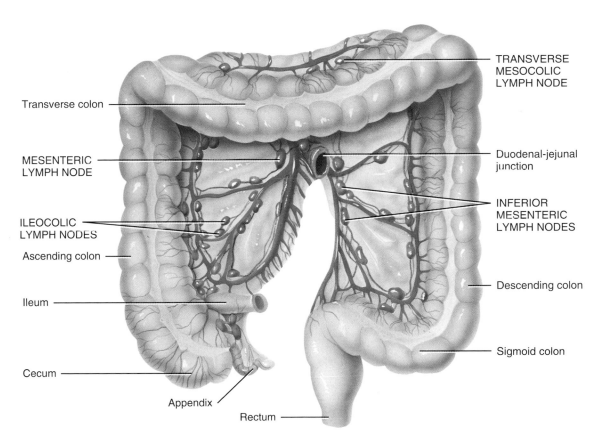

(b) Anterior view of superior and inferior mesenteric lymph nodes

What are the three groups of celiac lymph nodes?

527

EXHIBIT 16.5   PRINCIPAL LYMPH NODES OF THE THORAX (Figure 16.12)

PARIETAL LYMPH NODES	LOCATION	DRAINAGE
Parietal nodes drain the wall of the thorax.		
STERNAL (PARASTERNAL) NODES	Alongside internal thoracic artery.	Central and lateral parts of mammary gland, deeper structures of anterior abdominal wall superior to umbilicus, diaphragmatic surface of liver, and deeper parts of anterior portion of thoracic wall.
INTERCOSTAL NODES (in′-ter-KOS-tal; *inter-*=between; *costa*=rib)	Near heads of ribs at posterior parts of intercostal spaces.	Posterolateral aspect of thoracic wall.
PHRENIC (DIAPHRAGMATIC) NODES (FREN-ik=diaphragm)	On thoracic aspect of diaphragm and divisible into three sets called anterior phrenic, middle phrenic, and posterior phrenic.	
Anterior phrenic nodes	Posterior to base of xiphoid process.	Convex surface of liver, diaphragm, and anterior abdominal wall.
Middle phrenic nodes	Close to phrenic nerves where they pierce diaphragm.	Medial part of diaphragm and convex surface of liver.
Posterior phrenic nodes	Posterior surface of diaphragm near aorta.	Posterior part of diaphragm.

VISCERAL LYMPH NODES	LOCATION	DRAINAGE
Visceral nodes drain the viscera in the thorax.		
ANTERIOR MEDIASTINAL NODES (mē′-dē-as-TĪ-nal; *media-*=middle; *-stinum*=partition)	Anterior part of superior mediastinum anterior to arch of aorta.	Thymus and pericardium.
POSTERIOR MEDIASTINAL NODES	Posterior to pericardium.	Esophagus, posterior aspect of the pericardium, diaphragm, and convex surface of liver.
TRACHEOBRONCHIAL NODES	Are divided into five groups: tracheal, superior and inferior tracheobronchial, bronchopulmonary, and pulmonary nodes.	
Tracheal nodes	Either side of trachea.	Trachea and upper esophagus.
Superior tracheobronchial nodes (trā′-kē-ō-BRONG-kē-al)	Between trachea and bronchi.	Trachea and bronchi.
Inferior tracheobronchial nodes	Between bronchi.	Trachea and bronchi.
Bronchopulmonary nodes (brong-kō-PUL-mō-nar-ē)	In hilus of each lung.	Lungs and bronchi.
Pulmonary nodes	Within lungs on larger bronchial tube branches.	Lungs and bronchi.

DEVELOPMENT OF LYMPHATIC TISSUES

OBJECTIVE

● Describe the development of lymphatic tissues.

Lymphatic tissues begin to develop by the end of the fifth week of embryonic life. *Lymphatic vessels* develop from **lymph sacs** that arise from developing veins, which are derived from **mesoderm.**

The first lymph sacs to appear are the paired **jugular lymph sacs** at the junction of the internal jugular and subclavian veins (Figure 16.13 on page 530). From the jugular lymph sacs, lymphatic capillary plexuses spread to the thorax, upper limbs, neck, and head. Some of the plexuses enlarge and form lymphatic vessels in their respective regions. Each jugular lymph sac retains at least one connection with its jugular vein, the left one developing into the superior portion of the thoracic duct (left lymphatic duct).

The next lymph sac to appear is the unpaired **retroperitoneal lymph sac** at the root of the mesentery of the intestine. It develops from the primitive vena cava and mesonephric (primitive kidney) veins. Capillary plexuses and lymphatic vessels spread from the retroperitoneal lymph sac to the abdominal

Figure 16.12   Principal lymph nodes of the thorax.

 The parietal lymph nodes drain the thoracic wall, while the visceral lymph nodes drain the viscera of the thorax.

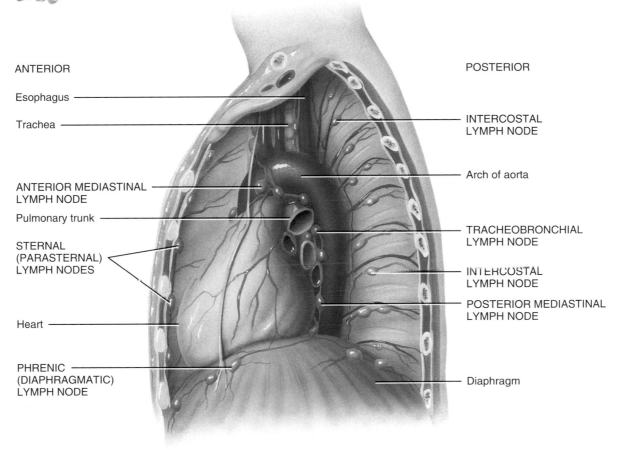

ANTERIOR

Esophagus

Trachea

ANTERIOR MEDIASTINAL LYMPH NODE

Pulmonary trunk

STERNAL (PARASTERNAL) LYMPH NODES

Heart

PHRENIC (DIAPHRAGMATIC) LYMPH NODE

POSTERIOR

INTERCOSTAL LYMPH NODE

Arch of aorta

TRACHEOBRONCHIAL LYMPH NODE

INTERCOSTAL LYMPH NODE

POSTERIOR MEDIASTINAL LYMPH NODE

Diaphragm

View from left side showing thoracic lymph nodes

Which group of lymph nodes drains most of the diaphragm?

viscera and diaphragm. The sac establishes connections with the cisterna chyli but loses its connections with neighboring veins.

At about the time the retroperitoneal lymph sac is developing, another lymph sac, the **cisterna chyli,** develops inferior to the diaphragm on the posterior abdominal wall. It gives rise to the inferior portion of the *thoracic duct* and the *cisterna chyli* of the thoracic duct. Like the retroperitoneal lymph sac, the cisterna chyli also loses its connections with surrounding veins.

The last of the lymph sacs, the paired **posterior lymph sacs,** develop from the iliac veins. The posterior lymph sacs produce capillary plexuses and lymphatic vessels of the abdominal wall, pelvic region, and lower limbs. The posterior lymph sacs join the cisterna chyli and lose their connections with adjacent veins.

With the exception of the anterior part of the sac from which the cisterna chyli develops, all lymph sacs become invaded by **mesenchymal cells** and are converted into groups of *lymph nodes.*

The *spleen* develops from **mesenchymal cells** between layers of the dorsal mesentery of the stomach. The *thymus* arises as an outgrowth of the **third pharyngeal pouch** (see Figure 23.8 on page 722).

Figure 16.13 Development of lymphatic tissue.

The lymphatic system is derived from mesoderm.

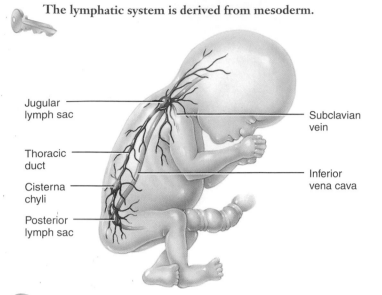

Jugular lymph sac

Subclavian vein

Thoracic duct

Inferior vena cava

Cisterna chyli

Posterior lymph sac

When do lymphatic tissues begin to develop?

CHECKPOINT

7. Name the four lymph sacs from which lymphatic vessels develop.

AGING AND THE LYMPHATIC AND IMMUNE SYSTEM

OBJECTIVE

• Describe the effects of aging on the lymphatic and immune system.

With advancing age, elderly individuals become more susceptible to all types of infections and malignancies. Their response to vaccines is decreased, and they tend to produce more autoantibodies (antibodies against their body's own molecules). In addition, the immune system exhibits lowered levels of function. For example, T cells become less responsive to antigens (foreign substances), and fewer T cells respond to infections. This may result from age-related atrophy of the thymus or decreased production of thymic hormones. Because the T cell population decreases with age, B cells are also less responsive. Consequently, antibody levels do not increase as rapidly in response to a challenge by an antigen, resulting in increased susceptibility to various infections. It is for this key reason that elderly individuals are encouraged to get influenza (flu) vaccinations each year.

CHECKPOINT

8. What changes occur in the lymphatic and immune system with advancing age?

APPLICATIONS TO HEALTH

AIDS: ACQUIRED IMMUNODEFICIENCY SYNDROME

Acquired immunodeficiency syndrome (AIDS) is a condition in which a person experiences a telltale assortment of infections due to the progressive destruction of immune system cells by the **human immunodeficiency virus (HIV).** AIDS represents the end stage of infection by HIV. A person who is infected with HIV may be symptom-free for many years, even while the virus is actively attacking the immune system. In the two decades after the first five cases were reported in 1981, 22 million people died of AIDS. Worldwide, 35 to 40 million people are currently infected with HIV.

HIV Transmission

Because HIV is present in the blood and some body fluids, it is most effectively transmitted (spread from one person to another) by practices that involve the exchange of blood or body fluids. HIV is transmitted in semen or vaginal fluid during unprotected (without a condom) anal, vaginal, or oral sex. HIV also is transmitted by direct blood-to-blood contact, such as occurs in intravenous drug users who share hypodermic needles or health-care professionals who may be accidentally stuck by HIV-contaminated hypodermic needles. In addition, HIV can

be transmitted from an HIV-infected mother to her baby at birth or during breast-feeding.

The chance of transmitting or of being infected by HIV during vaginal or anal intercourse can be greatly reduced—although not eliminated—by the use of latex condoms. Public health programs aimed at encouraging drug users not to share needles have proved effective at checking the increase in new HIV infections in this population. Also, giving certain drugs to pregnant HIV-infected women greatly reduces the risk of transmission of the virus to their babies.

HIV is a very fragile virus; it cannot survive for long outside the human body. The virus is not transmitted by insect bites. One cannot become infected by casual physical contact with an HIV-infected person, such as by hugging or sharing household items. The virus can be eliminated from personal care items and medical equipment by exposing them to heat (135 °F for 10 minutes) or by cleaning them with common disinfectants such as hydrogen peroxide, rubbing alcohol, household bleach, or germicidal cleansers such as Betadine or Hibiclens. Standard dishwashing and clothes washing also kills HIV.

HIV: Structure and Infection

HIV consists of an inner core of ribonucleic acid (RNA) covered by a protein coat (capsid) surrounded by an outer layer, the envelope, composed of a lipid bilayer penetrated by glycoproteins (Figure 16.14). Glycoproteins assist HIV's binding to and entering host cells. Outside a living host cell, a virus is unable to replicate. However, when the virus infects and enters a host cell, its RNA uses the host cell's resources to make thousands of copies of the virus. New viruses eventually leave and then infect other cells.

HIV mainly damages T cells. Over 10 billion viral copies may be made each day. The viruses bud so rapidly from an infected cell's plasma membrane that the cell ruptures and dies. In most HIV-infected people, T cells are initially replaced as fast as they are destroyed. After several years, however, the body's ability to replace T cells is slowly exhausted, and the number of T cells in circulation gradually declines.

Signs, Symptoms, and Diagnosis of HIV Infection

Soon after being infected with HIV, most people experience a brief flu-like illness. Common signs and symptoms are fever, fatigue, rash, headache, joint pain, sore throat, and swollen lymph nodes. In addition, about 50% of infected people have night sweats. After three to four weeks, plasma cells begin secreting antibodies to components of the HIV protein coat. These antibodies are detectable in blood plasma and form the basis for some of the screening tests for HIV. When people test "HIV-positive," it usually means they have antibodies to HIV antigens in their bloodstream. If a recent HIV infection is suspected but the antibody test is negative, laboratory tests based on detection of HIV's RNA or its coat protein in blood plasma can confirm the presence of HIV.

Figure 16.14 Human immunodeficiency virus (HIV), the causative agent of AIDS. HIV consists of an inner core of RNA covered by a protein coat (capsid) surrounded by an outer layer, the envelope, composed of a lipid bilayer studded with glycoproteins.

HIV is most effectively transmitted by practices that involve the exchange of body fluids.

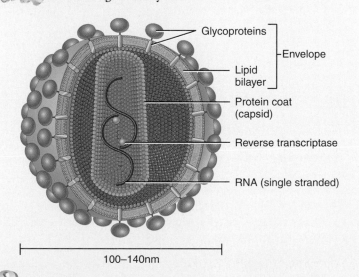

100–140nm

Which cells of the immune system are attacked by HIV?

Progression to AIDS

After a period of 2 to 10 years, the virus destroys enough helper T cells that most infected people begin to experience symptoms of immunodeficiency. HIV-infected people commonly have enlarged lymph nodes and experience persistent fatigue, involuntary weight loss, night sweats, skin rashes, diarrhea, and various lesions of the mouth and gums. In addition, the virus may begin to infect neurons in the brain, affecting the person's memory and producing visual disturbances.

As the immune system slowly collapses, an HIV-infected person becomes susceptible to a host of *opportunistic infections.* These are diseases caused by microorganisms that are normally held in check but now proliferate because of the defective immune system. AIDS is diagnosed when the helper T cell count drops below 200 cells per microliter (=cubic millimeter) of blood or when opportunistic infections arise, whichever occurs first. In time, opportunistic infections usually are the cause of death.

Treatment of HIV Infection

At present, infection with HIV cannot be cured. Vaccines designed to block new HIV infections and to reduce the viral load (the number of copies of HIV RNA in a microliter of blood plasma) in those who are already infected are in clinical trials. Meanwhile, two categories of drugs have proved successful in extending the life of many HIV-infected people:

1. **Reverse transcriptase inhibitors** interfere with the action of reverse transcriptase, the enzyme that the virus uses to convert its RNA into a DNA copy. Among the drugs in this category are zidovudine (ZDV, previously called AZT), didanosine (ddI), and stavudine (d4T). Trizivir, approved in 2000 for treatment of HIV infection, combines three reverse transcriptase inhibitors in one pill.

2. **Protease inhibitors** interfere with the action of protease, a viral enzyme that cuts proteins into pieces to assemble the coat of newly produced HIV particles. Drugs in this category include nelfinavir, saquinavir, ritonavir, and indinavir.

In 1996, physicians treating HIV-infected patients widely adopted *highly active antiretroviral therapy (HAART)*—a combination of two differently acting reverse transcriptase inhibitors and one protease inhibitor. Most HIV-infected individuals receiving HAART experience a drastic reduction in viral load and an increase in the number of T cells in their blood. Not only does HAART delay the progression of HIV infection to AIDS, but many people with AIDS have seen the remission or disappearance of opportunistic infections and an apparent return to health. Unfortunately, HAART is very costly (exceeding $10,000 per year), the dosing schedule is grueling, and not all people can tolerate the toxic side effects of these drugs. Although HIV may virtually disappear from the blood with drug treatment (and thus a blood test may be "negative" for HIV), the virus typically still lurks in various lymphatic tissues. In such cases, the infected person can still transmit the virus to another person.

ALLERGIC REACTIONS

A person who is overly reactive to a substance that is tolerated by most other people is said to be **allergic.** Whenever an allergic reaction takes place, some tissue injury occurs. The antigens that induce an allergic reaction are termed **allergens.** Common allergens include certain foods (milk, peanuts, shellfish, eggs), antibiotics (penicillin, tetracycline), vaccines (pertussis, typhoid), venoms (honeybee, wasp, snake), cosmetics, chemicals in plants such as poison ivy, pollens, dust, molds, iodine-containing dyes used in certain x-ray procedures, and even microbes.

Allergic reactions typically occur within a few minutes after a person who was previously sensitized to an allergen is reexposed to it. In response to certain allergens, some people produce IgE antibodies that bind to the surface of mast cells and basophils. The next time the same allergen enters the body, it attaches to the IgE antibodies already present. In response, the mast cells and basophils release histamine, prostaglandins, and other chemicals. Collectively, these chemicals cause vasodilation, increased blood capillary permeability, increased smooth muscle contraction in the airways of the lungs, and increased mucus secretion. As a result, a person may experience inflammatory responses, difficulty in breathing through the narrowed airways, and a runny nose from excess mucus secretion. In **anaphylactic shock,** which may occur in a susceptible individual who has just received a triggering drug or been stung by a wasp, wheezing and shortness of breath as airways constrict are usually accompanied by shock due to vasodilation and fluid loss from blood. Injecting epinephrine to dilate the airways and strengthen the heartbeat usually is effective in this life-threatening emergency.

Delayed hypersensitivity reactions usually appear 12 to 72 hours after exposure to an allergen. These reactions occur when allergens are taken up by antigen-presenting cells (such as Langerhans cells in the skin) that migrate to lymph nodes and present the allergen to T cells, which then divide many times. Some of the new T cells return to the site of allergen entry into the body, where they produce an interferon that activates macrophages, and other chemicals that stimulate an inflammatory response. Bacteria such as *Mycobacterium tuberculosis* trigger this type of cell-mediated immune response, as does *poison ivy* toxin.

INFECTIOUS MONONUCLEOSIS

Infectious mononucleosis or "mono" is a contagious disease caused by the *Epstein-Barr virus (EBV).* It occurs mainly in children and young adults, and more often in females than in males. The virus commonly enters the body through intimate oral contact such as kissing, which accounts for its common name, the "kissing disease." EBV then multiplies in lymphatic tissues and spreads into the blood, where it infects and multiplies in B cells, the primary host cells. Because of this infection, the B cells become enlarged and abnormal in appearance so that they resemble monocytes, the primary reason for the term **mononucleosis.** Besides an elevated white blood cell count with an abnormally high percentage of lymphocytes, signs and symptoms include fatigue, headache, dizziness, sore throat, enlarged and tender lymph nodes, and fever. There is no cure for infectious mononucleosis, but the disease usually runs its course in a few weeks.

LYMPHOMAS

Lymphomas (lim-FŌ-mas; *lymph-*=clear water; *-oma*=tumor) are cancers of the lymphatic organs, especially the lymph nodes. Most have no known cause. The two main types of lymphomas are Hodgkin's disease and non-Hodgkin's lymphoma.

Hodgkin's disease (HD) is characterized by painless, nontender enlargement of one or more lymph nodes, most commonly in the neck, chest, and axillae (armpits). If the disease has metastasized from these sites, fevers, night sweats, weight loss, and bone pain also occur. HD primarily affects individuals between ages 15 and 35 and those over 60; it is more common in males. If diagnosed early, HD has a 90–95% cure rate.

Non-Hodgkin's lymphoma (NHL), which is more common than HD, occurs in all age groups. NHL may start the same way as HD but may also include an enlarged spleen, anemia, and general malaise. Up to half of all individuals with NHL are cured or survive for a lengthy period. Treatment options for both HD and NHL include radiation therapy, chemotherapy, and bone marrow transplantation.

KEY MEDICAL TERMS ASSOCIATED WITH THE LYMPHATIC AND IMMUNE SYSTEM

Adenitis (ad-e-NĪ-tis; *aden-*=gland; *-itis*=inflammation of) Enlarged, tender, and inflamed lymph nodes resulting from an infection.

Allograft (AL-ō-graft; *allo-*=other) A transplant between genetically distinct individuals of the same species. Skin transplants from other people and blood transfusions are allografts.

Autograft (AW-tō-graft; *auto-*=self) A transplant in which one's own tissue is grafted to another part of the body (such as skin grafts for burn treatment or plastic surgery).

Autoimmune disease (aw-tō-i-MŪN) A disease in which the immune system fails to recognize self-antigens and attacks the person's own cells. Examples are rheumatoid arthritis (RA), systemic lupus erythematosus (SLE), rheumatic fever, hemolytic and pernicious anemias, Addison's disease, Graves' disease, insulin-dependent diabetes mellitus, myasthenia gravis, multiple sclerosis (MS), and ulcerative colitis. Also called autoimmunity.

Chronic fatigue syndrome (CFS) A disorder, usually occurring in young adults and primarily in females, characterized by

(1) extreme fatigue that impairs normal activities for at least 6 months and (2) the absence of other known diseases (cancer, infections, drug abuse, toxicity, or psychiatric disorders) that might produce similar symptoms.

Gamma globulin (GLOB-ū-lin) Suspension of immunoglobulins from blood consisting of antibodies that react with a specific pathogen. It is prepared by injecting the pathogen into animals, removing blood from the animals after antibodies have been produced, isolating the antibodies, and injecting them into a human to provide short-term immunity.

Hypersplenism (hī′-per-SPLĒN-izm; *hyper-*=over) Abnormal splenic activity due to splenic enlargement and associated with an increased rate of destruction of normal blood cells.

Lymphadenopathy (lim-fad′-e-NOP-a-thē; *lymph-*=clear fluid; *-pathy*=disease) Enlarged, sometimes tender lymph nodes as a response to infection; also called **swollen glands.**

Lymphedema (lim′-fe-DĒ-ma; *edema*=swelling) Accumulation of lymph in lymphatic vessels, causing painless swelling of a limb.

Splenomegaly (splē′-nō-MEG-a-lē; *mega-*=large) Enlarged spleen.

Systemic lupus erythematosus (er-e′-thēm-a-TŌ-sus) [*SLE*, or *lupus* (*lupus*=wolf)] An autoimmune, noncontagious, inflammatory disease of connective tissue, occurring mostly in young women. In SLE, damage to blood vessel walls results in the release of chemicals that mediate inflammation. Symptoms of SLE include joint pain, slight fever, fatigue, oral ulcers, weight loss, enlarged lymph nodes and spleen, photosensitivity, rapid loss of large amounts of scalp hair, and sometimes an eruption across the bridge of the nose and cheeks called a "butterfly rash."

Tonsillectomy (ton′-si-LEK-tō-mē; *-ectomy*=excision) Removal of a tonsil.

Transplantation (tranz-plan-TĀ-shun; *trans-*=across; *-planto*=to plant) The transfer of living cells, tissues, or organs from a donor to a recipient or from one part of the body to another part of the same body in order to restore a lost function. Donor cells, tissues, or organs may come from a living person or someone who has recently died. Normally, the immune system attacks and destroys transplanted tissues (graft rejection) and for this reason donated tissues should match the recipient's tissues as closely as possible. The success of a transplant depends on *major histocompatibility (MHC) antigens* on the surfaces of white blood cells and other body cells (except for red blood cells). MHC antigens are unique for each person; the better the match between donor and recipient MHC antigens, the greater the probability that graft rejection will be avoided. Even if MHC antigens are closely matched, it is necessary to use immunosuppressive drugs, which suppress the activity of the immune system, to prevent rejection.

Xenograft (ZEN-ō-graft; *xeno-*=strange or foreign) A transplant between animals of different species. Xenografts from porcine (pig) or bovine (cow) tissue may be used in a human as a physiological dressing for severe burns. Other xenografts include pig heart valves and baboon hearts.

 STUDY OUTLINE

Introduction (p. 511)

1. The ability to ward off disease is called resistance. Lack of resistance is called susceptibility.

2. Nonspecific resistance refers to a wide variety of body responses against a wide range of pathogens; specific resistance or immunity involves activation of specific lymphocytes to combat a particular foreign substance.

Lymphatic and Immune System Structure and Functions (p. 512)

1. The lymphatic system carries out immune responses and consists of lymph, lymphatic vessels, and structures and organs that contain lymphatic tissue (specialized reticular tissue containing many lymphocytes).

2. The lymphatic system drains interstitial fluid, transports dietary lipids, and protects against pathogens through immune responses.

3. Lymphatic vessels begin as lymph capillaries with one closed end in tissue spaces between cells.

4. Interstitial fluid drains into lymphatic capillaries, thus forming lymph.

5. Lymph capillaries merge to form larger vessels, called lymphatic vessels, which convey lymph into and out of structures called lymph nodes.

6. The route of lymph flow is from lymph capillaries to lymphatic vessels to lymph trunks to the thoracic duct or right lymphatic duct to the subclavian veins.

7. Lymph flows as a result of skeletal muscle contractions and respiratory movements. It is also aided by valves in lymphatic vessels.

8. The primary lymphatic organs are red bone marrow and the thymus. Secondary lymphatic organs are lymph nodes, spleen, and lymphatic nodules.

9. The thymus lies between the sternum and the large blood vessels above the heart. It is the site of T cell maturation.

10. Lymph nodes are encapsulated, oval structures located along lymphatic vessels.

11. Lymph enters nodes through afferent lymphatic vessels, is filtered, and exits through efferent lymphatic vessels.

12. Lymph nodes are the site of proliferation of plasma cells and T cells.

13. The spleen is the largest single mass of lymphatic tissue in the

body. It is a site of B cell proliferation into plasma cells and phagocytosis of bacteria and worn-out red blood cells.

14. Lymphatic nodules are scattered throughout the mucosa of the gastrointestinal, respiratory, urinary, and reproductive tracts. This lymphatic tissue is termed mucosa-associated lymphoid tissue (MALT).

Principal Groups of Lymph Nodes (p. 522)

1. Lymph nodes are scattered throughout the body in superficial and deep groups.
2. The principal groups of lymph nodes are found in the head and neck, upper limbs, lower limbs, abdomen and pelvis, and thorax.

Development of Lymphatic Tissues (p. 528)

1. Lymphatic vessels develop from lymph sacs, which arise from developing veins. Thus, they are derived from mesoderm.
2. Lymph nodes develop from lymph sacs that become invaded by mesenchymal cells.

Aging and the Lymphatic and Immune System (p. 530)

1. With advancing age, individuals become more susceptible to infections and malignancies, respond less well to vaccines, and produce more autoantibodies.
2. Immune responses also diminish with age.

 SELF-QUIZ QUESTIONS

Choose the one best answer to the following questions.

1. Put the following items in the correct order for the pathway of lymph from the stomach to the blood. (1) thoracic duct, (2) lymphatic vessels and lymph nodes, (3) lymph capillaries, (4) interstitial fluid, (5) junction of left internal jugular and left subclavian veins.
   a. 2,3,4,5,1   b. 4,3,2,1,5   c. 1,5,3,2,4   d. 4,2,1,5,3
   e. 3,2,4,1,5

2. Which of the following is *not* a secondary lymphatic tissue or organ?
   a. thymus   b. spleen   c. liver
   d. pharyngeal tonsil   e. axillary lymph nodes

3. Which of the following is an important function of the lymphatic system?
   a. controlling body temperature by evaporation of sweat
   b. manufacturing all white blood cells
   c. transporting fluids out to, and back from, the body tissues
   d. returning fluid and proteins to the cardiovascular system
   e. All of the above are correct.

4. The spleen:
   a. serves as storage site for blood platelets.
   b. is an organ in which phagocytosis of aged erythrocytes takes place.
   c. is an organ in which phagocytosis of aged or damaged platelets occurs.
   d. is a site of blood formation in the fetus.
   e. All of the above are correct.

5. Plasma cells are a differentiated form of:
   a. T lymphocyte.   b. B lymphocyte.   c. natural killer cell.
   d. dendritic cell.   e. endothelial cell.

6. Which of the following statements about HIV transmission is true?
   a. Standard dishwashing and clothes washing kills HIV.
   b. The transmission of HIV requires the transfer of, or direct contact with, infected body fluids.
   c. Organ transplants and artificial insemination are routes for HIV transmission.
   d. HIV may be present in semen, blood, and breast milk.
   e. All of the above are correct.

7. The thoracic duct drains lymph into the venous blood at the junction of the:
   a. left brachiocephalic vein and the superior vena cava.
   b. left internal jugular vein and the left subclavian vein.
   c. right internal jugular vein and the right subclavian vein.
   d. left internal and external jugular veins.
   e. superior vena cava and right atrium.

8. Which of the following statements is true about the flow of lymph? (1) It is aided by a pressure difference—toward the thoracic region. (2) It is maintained primarily by the milking action of muscles. (3) It is possible because of valves in lymphatic vessels.
   a.1 only   b. 2 only   c. 3 only   d. Both 2 and 3 are correct.
   e. 1, 2, and 3 are all correct.

Complete the following.

9. Specialized lymphatic capillaries in the small intestine are called _____.

10. Lymphatic vessels are similar in structure to _____ but have thinner walls and more valves.

11. The primary lymphatic organs, so-called because they produce the lymphocytes for the immune system, are the _____ and the _____.

12. The tonsils include the _____ tonsil in the posterior wall of the nasopharynx; the _____ in the posterior region of the oral cavity; and the _____ tonsils at the base of the tongue.

13. Disease-producing microbes are collectively known as _____.

14. Lymphatic nodules in lamina propria of mucous membranes are collectively referred to as _____.

Are the following statements true or false?

15. B cells are located in the medulla of a lymph node.
16. Lymph is conveyed into a node at several points by afferent lymphatic vessels.
17. The main difference between lymph and interstitial fluid is location.
18. There are no lymphatic capillaries in cartilage.
19. Lymphatic capillaries are slightly smaller in diameter and less porous than systemic capillaries.

Matching

20. Match the following lymph nodes with their descriptions:

___ **(a)** facial nodes

___ **(b)** axillary nodes

___ **(c)** popliteal nodes

___ **(d)** common iliac nodes

___ **(e)** celiac nodes

___ **(f)** phrenic nodes

___ **(g)** sacral nodes

___ **(h)** deep cervical nodes

(1) drain the knee, leg, and heel

(2) three groups of nodes that drain the stomach, liver, pancreas, and spleen

(3) largest group of nodes in the neck

(4) three groups of nodes that drain the mucous membranes of the cheeks, eyelids, and nose

(5) anterior to sacrum; drain the rectum, prostate gland, and pelvic wall

(6) three groups of nodes that drain parts of the liver, diaphragm, and abdominal wall

(7) deep nodes; drain the upper limbs, skin and muscles of the thorax, and part of the neck

(8) arranged along blood vessels of the same name; drain the pelvic viscera

---

CRITICAL THINKING QUESTIONS

1. Jamal got a splinter in his right heel while he was playing beach volleyball. He neglected to clean it properly and it became infected. The first aid station warned him to have the wound taken care of before the infection spreads to his blood. How can an infection travel from his foot to his cardiovascular system?

2. Four-year old Kelsey had a history of repeated throat infections, averaging 10 per year. She breathed loudly through her mouth and even snored. Following the pediatrician's recommendation, she had an operation to remove the troublesome organs. When she returned to preschool, she told the other kids about all the ice cream she got to eat following her "tonsil-X-me." Where are these organs located, and what is their function?

3. After several years of struggling with a chronic autoimmune disease, Kelly's doctor recommended a splenectomy. Since then, every time she goes to the dentist, she is advised to take antibiotics for a period of time prior to her appointment. Why would the doctor have recommended removal of such an important organ, and why would the dentist recommend the antibiotics?

4. Nan was ordering her dinner at an Italian restaurant. She informed the waiter that she was "highly allergic to eggplant" and that it was extremely important that her meal be eggplant-free. Dinner arrived, and shortly after she started eating her grilled vegetables, she started having trouble breathing and talking. There were no visible pieces of eggplant on her plate, so she announced to her dinner companion that the vegetables had been cooked on the same grill as the eggplant for other diners. How would her difficulty breathing lead her to this conclusion? How does being allergic to eggplant relate to difficulty breathing? If the difficulty breathing continues to worsen, what treatment might be required? Explain why such treatment would help.

5. An infection with a tropical parasite may block a lymphatic vessel. What would be the effect of blockage of the left subclavian trunk?

ANSWERS TO FIGURE QUESTIONS

16.1 Red bone marrow contains stem cells that develop into lymphocytes.

16.2 Lymph is more similar to interstitial fluid than to blood plasma because the protein content of lymph is low.

16.3 The left and right lumbar trunks and the intestinal trunk empty into the cisterna chyli, which then drains into the thoracic duct.

16.4 Inhalation promotes the movement of lymph from abdominal lymphatic vessels toward the thoracic region.

16.5 T cells mature in the thymus.

16.6 When foreign substances enter a lymph node, they may be phagocytized by macrophages or attacked by lymphocytes that mount immune responses.

16.7 After birth, white pulp of the spleen functions in immunity, and red pulp of the spleen removes worn-out blood cells and stores platelets.

16.8 The deep cervical nodes are the largest group of lymph nodes in the neck.

16.9 The lateral lymph nodes drain most of the upper limb.

16.10 The superficial inguinal lymph nodes are parallel to the saphenous vein.

16.11 The three groups of celiac lymph nodes are the gastric, hepatic, and pancreaticosplenic nodes.

16.12 The phrenic lymph nodes drain most of the diaphragm.

16.13 Lymphatic tissues begin to develop by the end of the fifth week of gestation.

16.14 HIV attacks T cells.

17

NERVOUS TISSUE

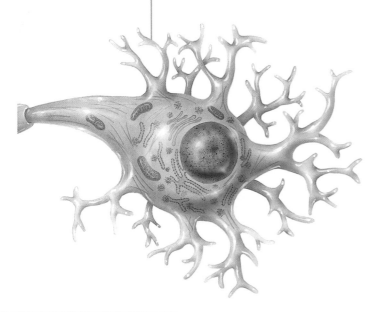

INTRODUCTION   With a mass of only 2 kg (4.5 lb), about 3% of total body weight, the nervous system is one of the smallest yet most complex of the 11 body systems. The two main subdivisions of the nervous system are the **central nervous system (CNS),** which consists of the brain and spinal cord, and the **peripheral** (pe-RIF-er-al) **nervous system (PNS),** which includes all nervous tissue outside the CNS.

Together, the nervous system and the endocrine system share responsibility for maintaining homeostasis, which means keeping the body's internal environment within normal limits. The objective is the same but the two systems achieve that objective differently. The nervous system regulates body activities by responding rapidly using nerve impulses (action potentials); the endocrine system responds more slowly, though no less effectively, by releasing hormones.

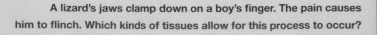
A lizard's jaws clamp down on a boy's finger. The pain causes him to flinch. Which kinds of tissues allow for this process to occur?

www.wiley.com/college/apcentral

The nervous system is also responsible for our perceptions, behaviors, and memories, and initiates all voluntary movements. Because the nervous system is quite complex, we will consider different aspects of its structure and function in several related chapters. This chapter focuses on the organization of the nervous system and the properties of the cells that make up nervous tissue—neurons (nerve cells) and neuroglia (cells that support the activities of neurons). In chapters that follow, we will examine the structure and functions of the spinal cord and spinal nerves (Chapter 18), and of the brain and cranial nerves (Chapter 19). Then we will discuss the autonomic nervous system, the part of the nervous system that operates without voluntary control (Chapter 20). Next, we examine the somatic senses—touch, pressure, warmth, cold, pain, and others—and the sensory and motor pathways to understand how nerve impulses pass into the spinal cord and brain or from the spinal cord and brain to muscles and glands (Chapter 21). Our exploration concludes with a discussion of the special senses: smell, taste, vision, hearing, and equilibrium (Chapter 22).

The branch of medical science that deals with the normal functioning and disorders of the nervous system is **neurology** (noo-ROL-ō-jē; *neuro-*=nerve or nervous system; *-logy*=study of). **A neurologist** is a physician who specializes in the diagnosis and treatment of disorders of the neuromuscular system.

OVERVIEW OF THE NERVOUS SYSTEM

OBJECTIVES

- List the structures and basic functions of the nervous system.
- Describe the organization of the nervous system.

Structures of the Nervous System

The **nervous system** is a complex, highly organized network of billions of neurons and even more neuroglia. The structures that make up the nervous system include the brain, cranial nerves and their branches, the spinal cord, spinal nerves and their branches, ganglia, enteric plexuses, and sensory receptors (Figure 17.1). The skull encloses the **brain,** which contains

Figure 17.1   Major structures of the nervous system.

The nervous system includes the brain, cranial nerves, spinal cord, spinal nerves, ganglia, enteric plexuses, and sensory receptors.

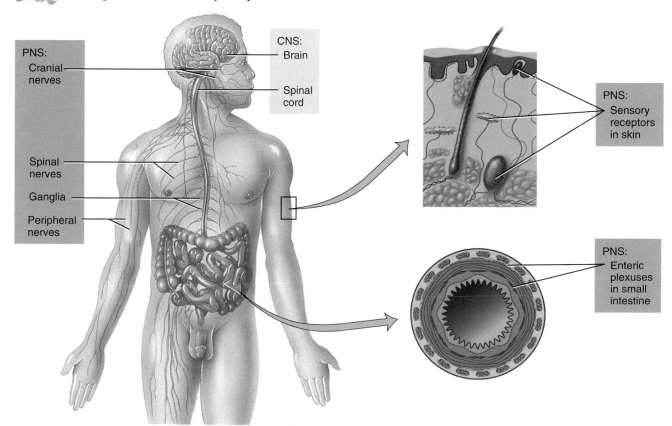

What is the total number of cranial and spinal nerves in the human body?

about 100 billion (10^{11}) neurons. Twelve pairs (right and left) of **cranial nerves,** numbered I through XII, emerge from the base of the brain. A **nerve** is a bundle of hundreds to thousands of axons plus associated connective tissue and blood vessels that lies outside the brain and spinal cord. Each nerve follows a defined path and serves a specific region of the body. For example, cranial nerve I carries signals for the sense of smell from the nose to the brain.

The **spinal cord** connects to the brain through the foramen magnum of the skull and is encircled by the bones of the vertebral column. It contains about 100 million neurons. Thirty-one pairs of **spinal nerves** emerge from the spinal cord, each serving a specific region on the right or left side of the body. **Ganglia** (GANG-lē-a=swelling or knot) are small masses of nervous tissue, containing primarily cell bodies of neurons, that are located outside the brain and spinal cord. Ganglia are closely associated with cranial and spinal nerves. In the walls of organs of the gastrointestinal tract, extensive networks of neurons, called **enteric plexuses,** help regulate the digestive system. The term **sensory receptor** is used for both a dendrite of a sensory neuron (described shortly) and a separate, specialized cell that monitors changes in the internal or external environment, such as a photoreceptor in the retina of the eye.

Functions of the Nervous System

The nervous system carries out a complex array of tasks, such as sensing various smells, producing speech, remembering, providing signals that control body movements, and regulating the operation of internal organs. These diverse activities can be grouped into three basic functions: sensory, integrative, and motor.

- *Sensory function.* Sensory receptors *detect* internal stimuli, such as an increase in blood acidity, and external stimuli, such as a raindrop landing on your arm. The neurons that carry sensory information from cranial and spinal nerves into the brain and spinal cord are called **sensory** or **afferent neurons** (AF-er-ent NOO-ronz; *af-*=toward; *-ferrent*= carried).

- *Integrative function.* The nervous system *integrates* (processes) sensory information by analyzing and storing some of it and by making decisions for appropriate responses. Many of the neurons that participate in integration are **interneurons,** whose axons extend only for a short distance and contact nearby neurons in the brain, spinal cord, or a ganglion. Interneurons comprise the vast majority of neurons in the body.

- *Motor function.* The nervous system's motor function involves *responding* to integration decisions. The neurons that serve this function are called **motor** or **efferent neurons** (EF-er-ent; *ef-*=away from). Motor neurons carry information from the brain toward the spinal cord or out of the brain and spinal cord into cranial or spinal nerves. The cells and organs contacted by motor neurons in cranial and spinal nerves are termed **effectors.** Muscle fibers and glandular cells are examples of effectors.

Organization of the Nervous System

The CNS (brain and spinal cord) integrates and correlates many different kinds of incoming sensory information. The CNS is also the source of thoughts, emotions, and memories. Most nerve impulses that stimulate muscles to contract and glands to secrete originate in the CNS. Components of the peripheral nervous system (PNS) include cranial nerves and their branches, spinal nerves and their branches, ganglia, and sensory receptors. The PNS may be subdivided further into a **somatic nervous system (SNS)** (*somat-*=body), an **autonomic nervous system (ANS)** (*auto-*=self; *-nomic*=law), and an **enteric nervous system (ENS)** (*enter-*=intestines) (Figure 17.2). The SNS consists of (1) sensory neurons that convey information from somatic receptors in the head, body wall, and limbs and from receptors for the special senses of vision, hearing, taste, and smell to the CNS and (2) motor neurons that conduct impulses from the CNS to *skeletal muscles* only. Because these motor responses can be consciously controlled, the action of this part of the PNS is considered *voluntary.*

The ANS consists of (1) sensory neurons that convey information from autonomic sensory receptors, located primarily in visceral organs such as the stomach and lungs, to the CNS, and (2) motor neurons that conduct nerve impulses from the CNS to *smooth muscle, cardiac muscle,* and *glands.* Because its motor responses are not normally under conscious control, the action of the ANS is considered *involuntary.* The motor part of the ANS consists of two branches, the **sympathetic division** and the **parasympathetic division.** With a few exceptions, effectors are innervated by both divisions, and usually the two divisions have opposing actions. For example, sympathetic neurons increase heart rate, whereas parasympathetic neurons slow it down. In general, the sympathetic division helps support exercise or emergency actions, so-called "fight-or-flight" responses, whereas the parasympathetic division takes care of "rest-and-digest" activities.

The operation of the ENS, the "brain of the gut," is involuntary. Once considered part of the ANS, the ENS consists of approximately 100 million neurons in enteric plexuses that extend the entire length of the gastrointestinal (GI) tract. Many of the neurons of the enteric plexuses function independently of the ANS and CNS to some extent, although they also communicate with the CNS via sympathetic and parasympathetic neurons. Sensory neurons of the ENS monitor chemical changes within the GI tract and the stretching of its walls. Enteric motor neurons govern contraction of GI tract smooth muscle to propel food through the digestive system, secretions of the GI tract organs such as acid from the stomach, and activity of GI tract endocrine cells, which secrete hormones.

CHECKPOINT

1. What are the components of the CNS and PNS?
2. What kinds of problems would result from damage to each of the following: sensory neurons, interneurons, and motor neurons?
3. Describe the components and functions of the SNS, ANS, and ENS, and indicate which of these subdivisions of the PNS control voluntary actions and which regulate involuntary actions.

Figure 17.2 Organization of the nervous system. The subdivisions of the PNS are the somatic nervous system (SNS), the autonomic nervous system (ANS), and the enteric nervous system (ENS).

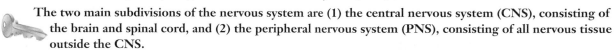

The two main subdivisions of the nervous system are (1) the central nervous system (CNS), consisting of the brain and spinal cord, and (2) the peripheral nervous system (PNS), consisting of all nervous tissue outside the CNS.

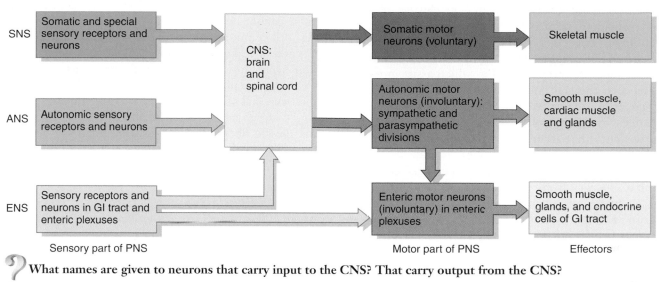

? **What names are given to neurons that carry input to the CNS? That carry output from the CNS?**

HISTOLOGY OF NERVOUS TISSUE

OBJECTIVES

- Compare the histological characteristics and functions of neurons and neuroglia.
- Distinguish between gray matter and white matter.

Nervous tissue consists of two types of cells: neurons and neuroglia. Neurons provide most of the unique functions of the nervous system, such as sensing, thinking, remembering, controlling muscle activity, and regulating glandular secretions. Neuroglia support, nourish, and protect neurons, and maintain the interstitial fluid that bathes them.

Neurons

Neurons or **nerve cells** possess **excitability,** the ability to respond to a stimulus and convert it into a nerve impulse (a *stimulus* is any change in the environment that is strong enough to initiate a nerve impulse). Very simply, a **nerve impulse (action potential)** is an electrical signal that propagates (travels) along the surface of the membrane of a neuron. It begins and travels due to the movement of ions (such as sodium and potassium) between interstitial fluid and the inside of a neuron through specific ion channels in the plasma membrane of a neuron. Once begun, a nerve impulse travels rapidly and at a constant strength.

Some neurons are tiny and propagate impulses over a short distance (less than 1 mm) within the CNS. Others are the longest cells in your body. Motor neurons that cause muscles to wiggle your toes, for example, extend from the lumbar region of

your spinal cord (just above waist level) to your foot. Some sensory neurons are even longer. Those that allow you to feel the position of your wiggling toes stretch all the way from your foot to the lower portion of your brain. Nerve impulses travel these great distances at speeds ranging from 0.5 to 130 meters per second (1 to 280 mi/hr).

NEUROTOXINS AND LOCAL ANESTHETICS

Certain shellfish and other creatures contain **neurotoxins** (noo-rō-TOK-sins), substances that produce their poisonous effects by acting on the nervous system. One particularly lethal neurotoxin is tetrodotoxin (TTX), present in the viscera of Japanese pufferfish. TTX effectively blocks production of nerve impulses by inserting itself into Na^+ channels so they cannot open.

Local anesthetics are drugs that block pain and other somatic sensations. Examples include procaine (Novocaine) and lidocaine, which may be used to produce anesthesia in the skin during suturing of a gash, in the mouth during dental work, or in the lower body during childbirth. Like TTX, these drugs act by blocking nerve impulses so pain signals do not reach the CNS. Smaller-diameter axons, such as those carrying pain signals, are more sensitive than larger-diameter axons to low doses of anesthetic drugs.

In addition, localized cooling of a nerve produces an anesthetic effect. It decreases impulse conduction because axons propagate impulses at lower speeds when cooled. Pain resulting from injured tissue can be reduced by the application of ice because propagation of the pain sensations along axons is partially blocked. ■

Parts of a Neuron

Most neurons have three parts: (1) a cell body, (2) dendrites, and (3) an axon (Figure 17.3). The **cell body** contains a nucleus surrounded by cytoplasm that includes typical cellular organelles such as lysosomes, mitochondria, and a Golgi complex. Neuronal cell bodies also contain prominent clusters of rough endoplasmic reticulum, termed *Nissl bodies*, which are the sites of protein synthesis. The cytoskeleton includes both *neurofibrils*, composed of bundles of intermediate filaments that provide the cell shape and support, and *microtubules*, which assist in moving materials between the cell body and axon. Many neurons also contain *lipofuscin*, a pigment that occurs as clumps of yellowish brown granules in the cytoplasm. Lipofuscin is a product of neuronal lysosomes that accumulate in a neuron, but does not seem to cause harm to the neuron as it ages.

Nerve fiber is a general term for any neuronal process or extension that emerges from the cell body of a neuron. Most neurons have two kinds of processes: multiple dendrites and a single axon. **Dendrites** (=little trees) are the receiving or input portions of a neuron. They usually are short, tapering, and highly branched. In many neurons the dendrites form a tree-shaped array of processes extending from the cell body. Their cytoplasm contains Nissl bodies, mitochondria, and other organelles.

The single **axon** (=axis) of a neuron carries nerve impulses toward another neuron, a muscle fiber, or a gland cell. An axon is a long, thin, cylindrical projection that often joins the cell body at a cone-shaped elevation called the **axon hillock** (=small hill). The part of the axon closest to the axon hillock is the **initial segment.** In most neurons, impulses arise at the junction of the axon hillock and the initial segment, an area called the **trigger zone,** and then travel along the axon. An axon contains mitochondria, microtubules, and neurofibrils. Because rough endoplasmic reticulum is not present, protein synthesis does not occur in the axon. The cytoplasm of an axon, called **axoplasm,** is surrounded by a plasma membrane known as the **axolemma** (*lemma*=sheath or husk). Along the length of an axon, side branches called **axon collaterals** may branch off, typically at a right angle to the axon. The axon and its collaterals end by dividing into many fine processes called **axon terminals.**

The site of communication between two neurons or between a neuron and an effector cell is called a **synapse** (SIN-aps). The term **presynaptic neuron** refers to a nerve cell that carries a nerve impulse toward a synapse. It is the cell that sends a signal. A **postsynaptic cell** is a nerve cell or effector (muscle or gland) that carries a nerve impulse away from a synapse. It is the cell that receives the signal. The tips of some axon terminals swell into bulb-shaped structures called **synaptic end bulbs,** whereas others exhibit a string of swollen bumps called **varicosities.** Both synaptic end bulbs and varicosities contain many tiny membrane-enclosed sacs called **synaptic vesicles** that store a chemical called a neurotransmitter. A **neurotransmitter** is a molecule released from a synaptic vesicle that excites or inhibits postsynaptic neurons, muscle fibers, or gland cells. Neurons were long thought to liberate just one type of neurotransmitter at all synaptic end bulbs. We now know that many neurons contain two or even three neurotransmitters.

The synapse between a motor neuron and a muscle fiber is called a **neuromuscular junction** (see Figure 10.6 on pages 273–274), whereas the synapse between a neuron and a glandular cell is called a **neuroglandular junction.** Recall from Chapter 10 that most synapses contain a small gap between cells called a **synaptic cleft** (see Figure 10.6c) and that a nerve impulse cannot jump the gap to excite the next cell. Instead, the first cell communicates with the second cell indirectly by releasing a neurotransmitter. For example, at a neuromuscular junction, synaptic vesicles release the neurotransmitter acetylcholine (ACh) (see Figure 10.6c). The ACh is picked up by ACh receptors (integral transmembrane proteins) on muscle fibers, thus triggering an action potential in the muscle fibers. This causes the muscle fibers to contract. Once ACh has accomplished its task, it is rapidly broken down by an enzyme in the synaptic cleft called acetylcholinesterase (AChE).

Neurotransmitters

About 100 substances are either known or suspected neurotransmitters. **Acetylcholine (ACh)** is released by many PNS neurons and by some CNS neurons. ACh is an excitatory neurotransmitter at the neuromuscular junction. It is also known to be an inhibitory neurotransmitter at other synapses. For example, parasympathetic neurons slow heart rate by releasing ACh at inhibitory synapses.

Several amino acids are neurotransmitters in the CNS. **Glutamate** and **aspartate** have powerful excitatory effects. Two other amino acids, **gamma aminobutyric acid** (GAM-ma am-i-nō-bū-TIR-ik) **(GABA)** and **glycine,** are important inhibitory neurotransmitters. Antianxiety drugs such as diazepam (Valium) enhance the action of GABA.

Some neurotransmitters are modified amino acids. These include norepinephrine, dopamine, and serotonin. **Norepinephrine (NE)** plays roles in arousal (awakening from deep sleep), dreaming, and regulating mood. Brain neurons containing the neurotransmitter **dopamine (DA)** are active during emotional responses, addictive behaviors, and pleasurable experiences. In addition, dopamine-releasing neurons help regulate skeletal muscle tone and some aspects of movement due to contraction of skeletal muscles. One form of schizophrenia is due to accumulation of excess dopamine. **Serotonin** is thought to be involved in sensory perception, temperature regulation, control of mood, appetite, and the onset of sleep.

Neurotransmitters consisting of amino acids linked by peptide bonds are called **neuropeptides** (noor-ō-PEP-tīds). The **endorphins** (en-DOR-fins) are neuropeptides that are the body's natural painkillers. Acupuncture may produce analgesia (loss of pain sensation) by increasing the release of endorphins. They have also been linked to improved memory and learning and to feelings of pleasure or euphoria.

An important newcomer to the ranks of recognized neurotransmitters is the simple gas **nitric oxide (NO),** which is different from all previously known neurotransmitters because it is not synthesized in advance and packaged into synaptic vesicles. Rather, it is formed on demand, diffuses out of cells that produce it and into neighboring cells, and acts immediately. Some research suggests that NO plays a role in learning and memory.

Figure 17.3 Structure of a typical neuron. Arrows indicate the direction of information flow: dendrites → cell body → axon → axon terminals. The axon is actually much longer than shown.

The basic parts of a neuron are several dendrites, a cell body, and an axon.

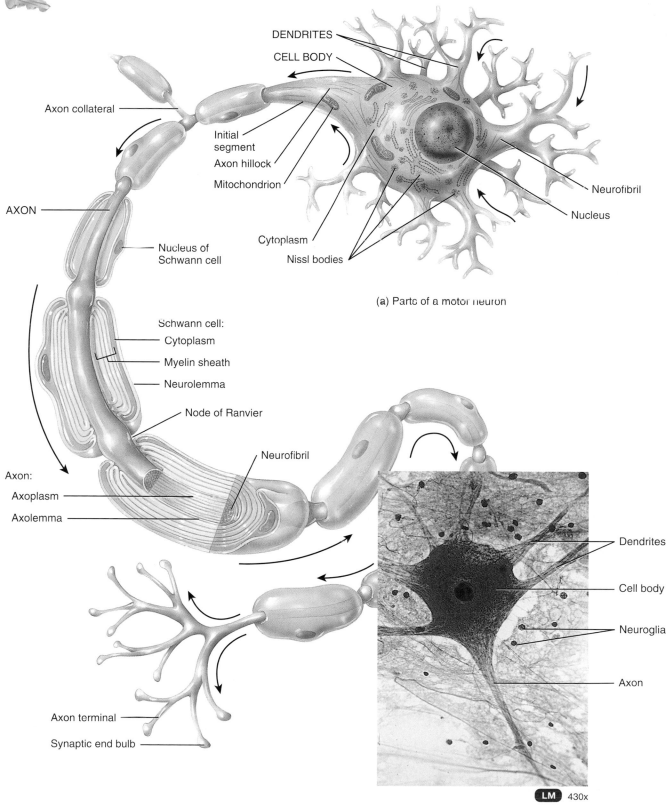

(a) Parts of a motor neuron

(b) Motor neuron

LM 430x

What roles do the dendrites, cell body, and axon play in communication of nerve impulses?

MODIFYING NEUROTRANSMITTERS

Substances naturally present in the body as well as drugs and toxins can **modify the effects of neurotransmitters** in several ways. Cocaine produces euphoria—intensely pleasurable feelings—by blocking reuptake of dopamine. This action allows dopamine to linger longer in synaptic clefts, producing excessive stimulation of certain brain regions. Loss of dopamine-releasing neurons results in Parkinson's disease. Isoproterenol (Isuprel) can be used to dilate the airways during an asthma attack because it binds to and activates receptors for norepinephrine. Zyprexa, a drug prescribed for schizophrenia, is effective because it binds to and blocks receptors for serotonin and dopamine.

Antidepressants are drugs that relieve the symptoms of depression. There are two major groups of antidepressants—older "tricyclic" antidepressants and the newer "SSRIs" (selective serotonin reuptake inhibitors). They both operate by altering the way in which certain neurotransmitters work in our brains. In depression, some of the neurotransmitter systems, particularly those involving serotonin and noradrenaline, don't seem to work properly. It is believed that antidepressants work by increasing the activity of these chemicals in our brains. Tri-cyclics are just as effective as SSRIs but, on the whole, the SSRIs have fewer side-effects. A major advantage of the SSRIs is that an overdose is not dangerous, whereas tricyclic overdose is extremely serious and has a high morbidity rate. ■

Structural Diversity in Neurons

Neurons display great diversity in size and shape. For example, their cell bodies range in diameter from 5 micrometers (μm) (slightly smaller than a red blood cell) up to 135 μm (barely large enough to see with the unaided eye). The pattern of dendritic branching is varied and distinctive for neurons in different parts of the nervous system. A few small neurons lack an axon, and many others have very short axons. As noted earlier in the chapter, the longest axons are almost as long as a person is tall, extending from the toes to the lowest part of the brain.

Both structural and functional features are used to classify the various neurons in the body. Structurally, neurons are classified according to the number of processes extending from the cell body (Figure 17.4).

- **Multipolar neurons** usually have several dendrites and one axon (see also Figure 17.3). Most neurons in the brain and spinal cord are of this type.

- **Bipolar neurons** have one main dendrite and one axon. They are found in the retina of the eye, in the inner ear, and in the olfactory area of the brain.

- **Unipolar neurons** are sensory neurons that begin in the embryo as bipolar neurons. During development, the axon and dendrite fuse into a single process that divides into two branches a short distance from the cell body. Both branches have the characteristic structure and function of an axon. They are long, cylindrical processes that propagate action potentials. However, the axon branch that extends into the periphery has dendrites at its distal tip, whereas the axon branch that extends into the CNS ends in synaptic end bulbs. The dendrites monitor a sensory stimulus such as touch or stretching. The trigger zone for nerve impulses in a unipolar neuron is at the junction of the dendrites and axon (Figure 17.4c). The impulses then propagate toward the synaptic end bulbs.

Figure 17.4 Structural classification of neurons. Breaks indicate that axons are longer than shown.

A multipolar neuron has many processes extending from the cell body, whereas a bipolar neuron has two and a unipolar neuron has one.

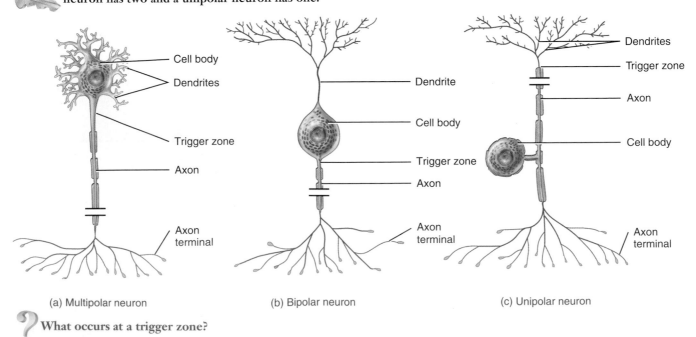

(a) Multipolar neuron   (b) Bipolar neuron   (c) Unipolar neuron

What occurs at a trigger zone?

Some neurons are named for the histologist who first described them or for an aspect of their shape or appearance; examples include **Purkinje cells** (pur-KIN-jē) in the cerebellum (Figure 17.5a) and **pyramidal cells** (pi-RAM-i-dal), found in the cerebral cortex of the brain, which have pyramid-shaped cell bodies (Figure 17.5b). Often, a distinctive pattern of dendritic branching allows identification of a particular type of neuron in the CNS.

Neuroglia

Neuroglia (noo-RŌG-lē-a; -*glia*=glue) or **glia** constitute about half the volume of the CNS. Their name derives from the idea of early histologists that they were the "glue" that held nervous tissue together. We now know that neuroglia are not merely passive bystanders but rather active participants in the function of nervous tissue. Generally, neuroglia are smaller than neurons, and they are 5 to 50 times more numerous. In contrast to neurons, glia do not generate or propagate action potentials, and they have the ability to multiply and divide in the mature nervous system. In cases of injury or disease, neuroglia multiply to fill in the spaces formerly occupied by neurons. Brain tumors derived from glia, called **gliomas,** tend to be highly malignant and rapidly growing.

Neuroglia of the CNS

Neuroglia of the CNS can be distinguished on the basis of size, cytoplasmic processes, and intracellular organization into four types: astrocytes, oligodendrocytes, microglia, and ependymal cells (Figure 17.6).

ASTROCYTES (AS-trō-sīts; *astro-*=star; -*cyte*=cell). These are star-shaped cells with many processes and are the largest and most numerous of the neuroglia. There are two types of astrocytes. *Protoplasmic astrocytes* have many short branching processes and are found in gray matter (described shortly). *Fibrous astrocytes* have many long unbranched processes and are located mainly in white matter (also described shortly). The processes of astrocytes make contact with blood capillaries, neurons, and the pia mater (a thin membrane around the brain and spinal cord).

Following are the functions of astrocytes: (1) Astrocytes contain microfilaments that provide them with considerable strength, which enables them to support neurons. (2) Since neurons of the CNS must be isolated from various potentially harmful substances in blood, the endothelial cells of CNS blood capillaries have very selective permeability characteristics. In effect, the endothelial cells create a *blood–brain barrier,* which restricts the movement of substances between the blood and interstitial fluid of the CNS. Processes of astrocytes wrapped around blood capillaries secrete chemicals that maintain the unique permeability characteristics of the endothelial cells. Details of the blood–brain barrier are discussed in Chapter 19. (3) In the embryo, astrocytes secrete chemicals that appear to regulate the growth, migration, and interconnection between neurons in the brain. (4) Astrocytes help to maintain the appropriate chemical environment for the generation of nerve impulses. For example, they regulate the concentration of

Figure 17.5 Two examples of CNS neurons. Arrows indicate the direction of information flow.

The dendritic branching pattern often is distinctive for a particular type of neuron.

(a) Purkinje cell (b) Pyramidal cell

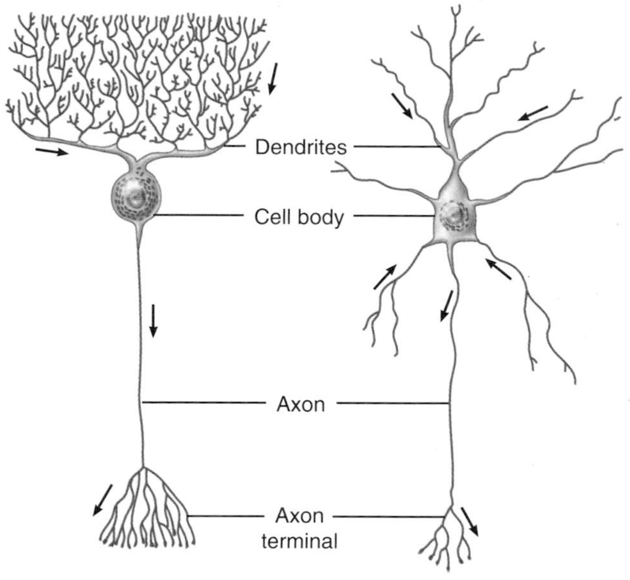

? Why are pyramidal cells so named?

important ions such as K^+; take up excess neurotransmitters; and serve as a conduit for the passage of nutrients and other substances between blood capillaries and neurons.

OLIGODENDROCYTES (OL-i-gō-den′-drō-sīts; *oligo-*=few; *dendro-*=tree). These resemble astrocytes, but are smaller and contain fewer processes. Oligodendrocyte processes are responsible for forming and maintaining the myelin sheath around CNS axons. As you will see shortly, the myelin sheath is a lipid and protein covering around some axons that insulates the axon and increases the speed of nerve impulse conduction. Such axons are said to be myelinated. Oligodendrocyte processes that make contact with cell bodies of neurons are thought to influence the chemistry of the interstitial fluid of neurons.

MICROGLIA (mī-KROG-lē-a; *micro-*=small). These neuroglia are small cells with slender processes that give off numerous spinelike projections. Unlike other neuroglial cells, which develop from the neural tube, microglia originate in red bone marrow and migrate into the CNS as it develops. Microglia function as phagocytes. Like tissue macrophages, they remove cellular debris formed during normal development of the nervous system and phagocytize microbes and damaged nervous tissue.

EPENDYMAL CELLS (ep-EN-de-mal; *epen-*=above; *dym-*=garment). Ependymal cells are cuboidal to columnar cells arranged in a single layer that possess microvilli and cilia. These cells line the ventricles of the brain and central canal of the

Figure 17.6 **Neuroglia of the central nervous system (CNS).**

Neuroglia of the CNS are distinguished on the basis of size, cytoplasmic processes, and intracellular organization.

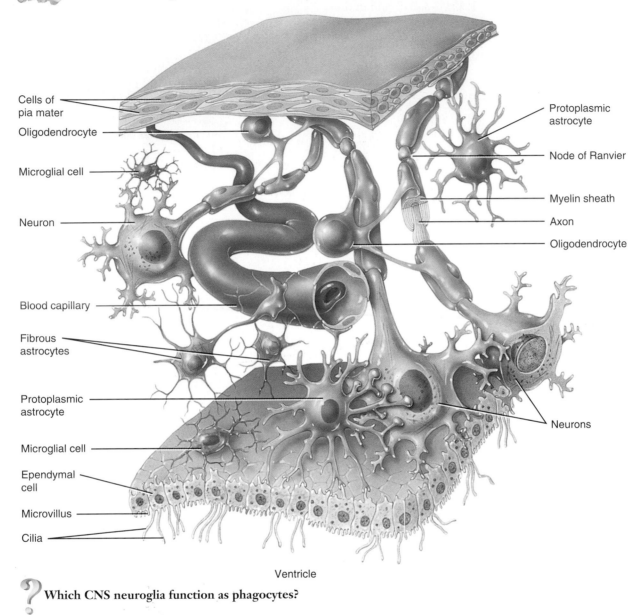

Cells of pia mater

Oligodendrocyte

Microglial cell

Neuron

Blood capillary

Fibrous astrocytes

Protoplasmic astrocyte

Microglial cell

Ependymal cell

Microvillus

Cilia

Protoplasmic astrocyte

Node of Ranvier

Myelin sheath

Axon

Oligodendrocyte

Neurons

Ventricle

? **Which CNS neuroglia function as phagocytes?**

spinal cord (spaces filled with cerebrospinal fluid). Functionally, ependymal cells produce, possibly monitor, and assist in the circulation of cerebrospinal fluid. They also form the blood–cerebrospinal fluid barrier, which is discussed in Chapter 19.

Neuroglia of the PNS

Neuroglia of the PNS completely surround axons and cell bodies. The two types of glial cells in the PNS are Schwann cells and satellite cells (Figure 17.7).

SCHWANN CELLS (SCHVON). These flat cells encircle PNS axons. Like oligodendrocytes, they form the myelin sheath around axons. However, whereas a single oligodendrocyte

myelinates several axons, each Schwann cell myelinates a single axon (Figure 17.7a; see also Figure 17.8a, c). A single Schwann cell can also enclose as many as 20 or more unmyelinated axons (axons that lack a myelin sheath) (Figure 17.7b). Schwann cells participate in axon regeneration, which is more easily accomplished in the PNS than in the CNS.

SATELLITE CELLS (SAT-i-līt). These flat cells surround the cell bodies of neurons of PNS ganglia (Figure 17.7c). (Recall that ganglia are collections of neuronal cell bodies outside the CNS.) Besides providing structural support, satellite cells regulate the exchange of materials between neuronal cell bodies and interstitial fluid.

Figure 17.7 Neuroglia of the peripheral nervous system (PNS).

Neuroglia of the PNS completely surround axons and cell bodies of neurons.

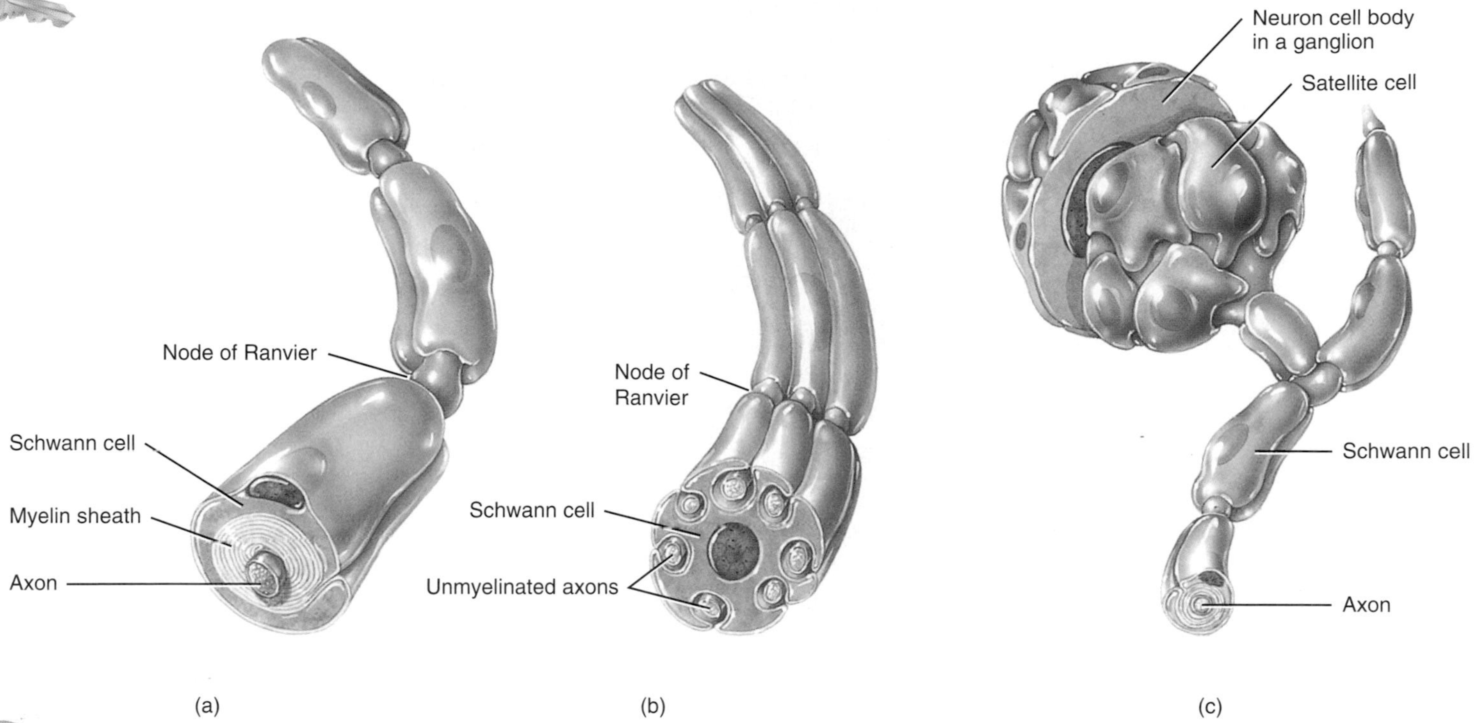

Neuron cell body in a ganglion

Satellite cell

Node of Ranvier

Schwann cell

Myelin sheath

Axon

Node of Ranvier

Schwann cell

Unmyelinated axons

Schwann cell

Axon

(a)   (b)   (c)

How do Schwann cells and oligodendrocytes differ with respect to number of axons myelinated?

Myelination

Axons that are surrounded by a multilayered lipid and protein covering, called the **myelin sheath,** are said to be **myelinated** (Figure 17.8a). The sheath electrically insulates the axon of a neuron and increases the speed of nerve impulse conduction. Axons without such a covering are said to be **unmyelinated** (Figure 17.8b).

Two types of neuroglia produce myelin sheaths: Schwann cells (in the PNS) and oligodendrocytes (in the CNS). In the PNS, Schwann cells begin to form myelin sheaths around axons during fetal development. Each Schwann cell wraps about 1 millimeter (1 mm=0.04 in.) of a single axon's length by spiraling many times around the axon (Figure 17.8a). Eventually, multiple layers of glial plasma membrane surround the axon, with the Schwann cell's cytoplasm and nucleus forming the outermost layer. The inner portion, consisting of up to 100 layers of Schwann cell membrane, is the myelin sheath. The outer nucleated cytoplasmic layer of the Schwann cell, which encloses myelinated or unmyelinated axons, is the **neurolemma (sheath of Schwann).** A neurolemma is found only around axons in the PNS. When an axon is injured, the neurolemma aids regeneration by forming a regeneration tube that guides and stimulates regrowth of the axon. Gaps in the myelin sheath, called **nodes of Ranvier** (RON-vē-ā), appear at intervals along the axon (see Figure 17.3). Each Schwann cell wraps one axon segment between two nodes.

In the CNS, an oligodendrocyte myelinates parts of several axons. Each oligodendrocyte puts forth about 15 broad, flat processes that spiral around CNS axons, forming a myelin sheath. A neurolemma is not present, however, because the oligodendrocyte cell body and nucleus do not envelop the axon. Nodes of Ranvier are present, but they are fewer in number. Axons in the CNS display little regrowth after injury. This is thought to be due, in part, to the absence of a neurolemma, and in part to an inhibitory influence on axon regrowth exerted by the oligodendrocytes.

The amount of myelin increases from birth to maturity, and its presence greatly increases the speed of nerve impulse conduction. An infant's responses to stimuli are neither as rapid nor as coordinated as those of an older child or an adult, in part because myelination is still in progress during infancy.

DEMYELINATION

Demyelination (dē-mī-e-li-NĀ-shun) refers to the loss or destruction of myelin sheaths around axons. It may result from disorders such as multiple sclerosis (see page 549) or Tay-Sachs disease (see page 37), or from medical treatments such as radiation therapy and chemotherapy. Any single episode of demyelination may cause deterioration of affected nerves. ■

Figure 17.8 Myelinated and unmyelinated axons.

Axons of neurons surrounded by a myelin sheath produced by Schwann cells in the PNS and by oligodendrocytes in the CNS are said to be myelinated.

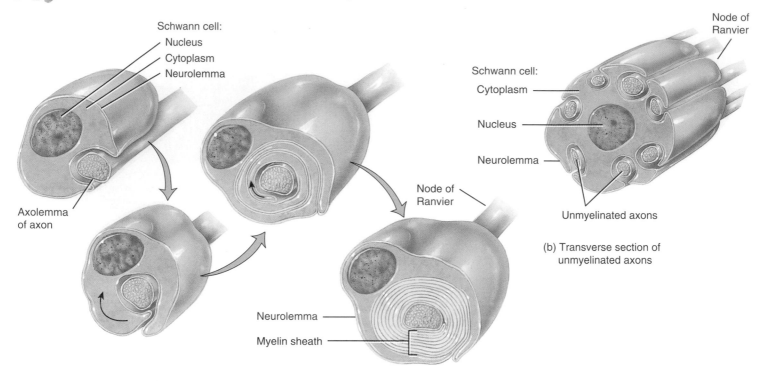

(a) Transverse sections of stages in the formation of a myelin sheath

(b) Transverse section of unmyelinated axons

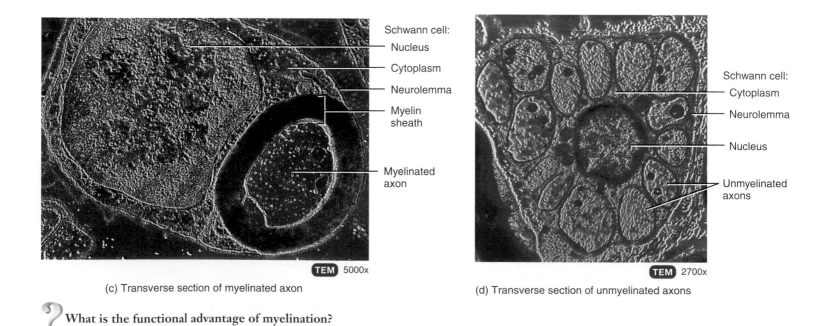

(c) Transverse section of myelinated axon

(d) Transverse section of unmyelinated axons

What is the functional advantage of myelination?

Gray and White Matter

In a freshly dissected section of the brain or spinal cord, some regions look white and glistening, whereas others appear gray (Figure 17.9). The **white matter** is aggregations of myelinated and unmyelinated axons of many neurons. The whitish color of myelin gives white matter its name. The **gray matter** of the nervous system contains neuronal cell bodies, dendrites, unmyelinated axons, axon terminals, and neuroglia. It appears grayish, rather than white, because the Nissl bodies impart a gray color and there is little or no myelin in these areas. Blood vessels are present in both white and gray matter.

Figure 17.9   Distribution of gray and white matter in the spinal cord and brain.

White matter consists of myelinated and unmyelinated axons of many neurons. Gray matter consists of neuron cell bodies, dendrites, axon terminals, bundles of unmyelinated axons, and neuroglia.

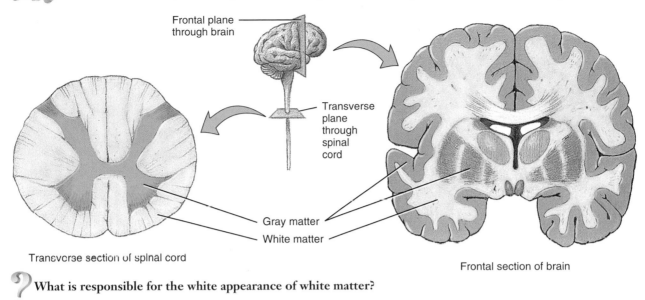

Frontal plane through brain

Transverse plane through spinal cord

Gray matter

White matter

Transverse section of spinal cord

Frontal section of brain

What is responsible for the white appearance of white matter?

In the spinal cord, the white matter surrounds an inner core of gray matter shaped like a butterfly or the letter H; in the brain, a thin shell of gray matter covers the surface of the largest portions of the brain, the cerebrum and cerebellum (Figure 17.9). When used to describe nervous tissue, a **nucleus** is a cluster of neuronal cell bodies within the CNS. (Recall that the term *ganglion* refers to a similar arrangement within the PNS.) Many nuclei of gray matter also lie deep within the brain. The arrangements of gray and white matter in the spinal cord and brain are discussed more extensively in Chapters 18 and 19, respectively.

CHECKPOINT

4. Describe the parts of a neuron and the functions of each.
5. Give several examples of the structural diversity of neurons.
6. What is a neurolemma, and why is it important?
7. With reference to the central nervous system, what is a nucleus?

NEURAL CIRCUITS

OBJECTIVE

● Identify the structure and function of the various types of neural circuits in the nervous system.

The CNS contains billions of neurons organized into complicated networks called **neural circuits,** each of which is a functional group of neurons that processes a specific kind of information. In a **simple series circuit,** a presynaptic neuron stimulates a single postsynaptic neuron. The second neuron then stimulates another, and so on. However, most neural circuits are more complex.

A single presynaptic neuron may synapse with several postsynaptic neurons. Such an arrangement, called **divergence,** permits one presynaptic neuron to influence several postsynaptic neurons (or several muscle fibers or gland cells) at the same time. In a **diverging circuit,** the nerve impulse from a single presynaptic neuron causes the stimulation of increasing numbers of cells along the circuit (Figure 17.10a). For example, a small number of neurons in the brain that govern a particular body movement stimulate a much larger number of neurons in the spinal cord. Sensory signals also spread into diverging circuits and are often relayed to several regions of the brain. This arrangement has the effect of amplifying the signal.

In another arrangement, called **convergence,** several presynaptic neurons synapse with a single postsynaptic neuron. This arrangement permits more effective stimulation or inhibition of the postsynaptic neuron. In a **converging circuit** (Figure 17.10b), the postsynaptic neuron receives nerve impulses from several different sources. For example, a single motor neuron that synapses with skeletal muscle fibers at neuromuscular junctions receives input from several pathways that originate in different brain regions.

Some circuits are constructed so that once the presynaptic cell is stimulated, it will cause the postsynaptic cell to transmit a series of nerve impulses. One such circuit is called a **reverberating circuit** (Figure 17.10c). In this pattern, the incoming impulse stimulates the first neuron, which stimulates the second, which stimulates the third, and so on. Branches from later neurons synapse with earlier ones. This arrangement sends impulses back through the circuit again and again. The output signal may last from a few seconds to many hours, depending on the number of synapses and the arrangement of neurons in the circuit. Inhibitory neurons may turn off a reverberating circuit after a period of time. Among the body responses thought to be the result of output signals from reverberating circuits are breathing, coordinated muscular activities, waking up, sleeping (when reverberation stops), and short-term memory.

Figure 17.10 Examples of neural circuits.

🔑 A neural circuit is a functional group of neurons that processes a specific kind of information.

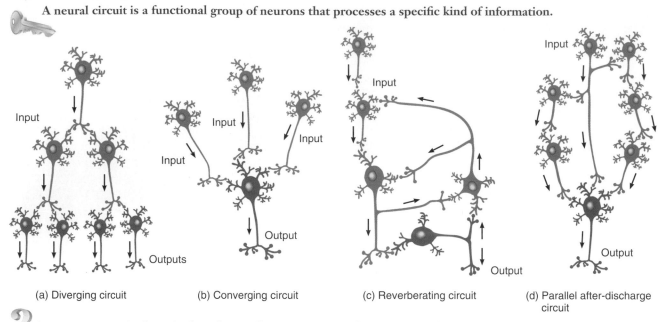

(a) Diverging circuit   (b) Converging circuit   (c) Reverberating circuit   (d) Parallel after-discharge circuit

❓ **A motor neuron in the spinal cord typically receives input from neurons that originate in several different regions of the brain. Is this an example of convergence or divergence?**

A fourth type of circuit is the **parallel after-discharge circuit** (Figure 17.10d). In this circuit, a single presynaptic cell stimulates a group of neurons, each of which synapses with a common postsynaptic cell. Parallel after-discharge circuits may be involved in precise activities such as mathematical calculations.

CHECKPOINT

8. What is a neural circuit?
9. What are the functions of diverging, converging, reverberating, and parallel after-discharge circuits?

REGENERATION AND NEUROGENESIS

OBJECTIVE

• Define regeneration and neurogenesis.

Throughout your life, your nervous system exhibits **plasticity,** the capability to change based on experience. At the level of individual neurons, the changes that can occur include the sprouting of new dendrites, synthesis of new proteins, and changes in synaptic contacts with other neurons. Undoubtedly, both chemical and electrical signals drive the changes that occur. Despite plasticity, mammalian neurons have very limited powers of **regeneration,** the capability to replicate or repair themselves. In the PNS, damage to dendrites and myelinated axons may be repaired if the cell bodies remain intact and if the myelinating Schwann cells remain active. In the CNS, little or no repair of

damage to neurons occurs. Even when the cell body remains intact, a severed axon cannot be repaired or regrown.

Neurogenesis—the birth of new neurons from undifferentiated stem cells—occurs regularly in some animals. For example, new neurons appear and disappear every year in some songbirds. Until relatively recently, the dogma in humans and other primates was "no new neurons" in the adult brain. In 1992, researchers discovered that **epidermal growth factor (EGF)** stimulated cells taken from the brains of adult mice to proliferate into both neurons and astrocytes. Previously, EGF was known to trigger mitosis in a variety of nonneuronal cells and to promote wound healing and tissue regeneration. In 1998 scientists discovered that significant numbers of new neurons do arise from ependymal stem cells in the adult human hippocampus, an area of the brain that is crucial for learning.

The nearly complete lack of neurogenesis in other regions of the brain and spinal cord seems to result from two factors: (1) inhibitory influences from neuroglia, particularly oligodendrocytes, and (2) absence of growth-stimulating cues that were present during fetal development. Thus, injury of the brain or spinal cord usually is permanent. Nevertheless, ongoing research seeks ways to improve the environment for existing spinal cord axons to bridge the injury gap. Scientists also are trying to find ways to stimulate dormant stem cells to replace neurons lost through damage or disease and to develop tissue-cultured neurons that can be used for transplantation purposes.

CHECKPOINT

10. What factors contribute to the absence of neurogenesis in most parts of the brain?

APPLICATIONS TO HEALTH

MULTIPLE SCLEROSIS

Multiple sclerosis (MS) is a disease that causes a progressive destruction of myelin sheaths of neurons in the CNS. It afflicts about 350,000 people in the United States and 2 million people worldwide. It usually appears between the ages of 20 and 40, affecting females twice as often as males. MS is most common in whites, less common in blacks, and rare in Asians. MS is an autoimmune disease—the body's own immune system spearheads the attack. The condition's name describes the anatomical pathology: In *multiple* regions the myelin sheaths deteriorate to *scleroses*, which are hardened scars or plaques. Magnetic resonance imaging (MRI) studies reveal numerous plaques in the white matter of the brain and spinal cord. The destruction of myelin sheaths slows and then short-circuits propagation of nerve impulses.

The most common form of the condition is relapsing–remitting MS, which usually appears in early adulthood. The first symptoms may include a feeling of heaviness or weakness in the muscles, abnormal sensations, or double vision. An attack is followed by a period of remission during which the symptoms temporarily disappear. One attack follows another over the years, usually every year or two. The result is a progressive loss of function interspersed with remission periods, during which symptoms abate.

Although the cause of MS is unclear, both genetic susceptibility and exposure to some environmental factor (perhaps a herpes virus) appear to contribute. Many patients with relapsing–remitting MS are treated with injections of beta interferon. This treatment lengthens the time between relapses, decreases the severity of relapses, and slows formation of new lesions in some cases. Unfortunately, not all MS patients can tolerate beta interferon, and therapy becomes less effective as the disease progresses.

EPILEPSY

Epilepsy is characterized by short, recurrent attacks of motor, sensory, or psychological malfunction, although it almost never affects intelligence. The attacks, called *epileptic seizures*, afflict about 1% of the world's population. They are initiated by abnormal, synchronous electrical discharges from millions of neurons in the brain, perhaps resulting from abnormal reverberating circuits. The discharges stimulate many of the neurons to send nerve impulses over their conduction pathways. As a result, lights, noise, or smells may be sensed when the eyes, ears, and nose have not been stimulated. Moreover, the skeletal muscles of a person having a seizure may contract involuntarily. *Partial seizures* (petit mal) begin in a small focus on one side of the brain and produce milder symptoms, whereas *generalized seizures* (grand mal) involve larger areas on both sides of the brain and loss of consciousness.

Epilepsy has many causes, including brain damage at birth (the most common cause); metabolic disturbances (hypoglycemia, hypocalcemia, uremia, hypoxia); infections (encephalitis or meningitis); toxins (alcohol, tranquilizers, hallucinogens); vascular disturbances (hemorrhage, hypotension); head injuries; and tumors and abscesses of the brain. Seizures associated with fever are most common in children under the age of two. However, most epileptic seizures have no demonstrable cause.

Epileptic seizures often can be eliminated or alleviated by antiepileptic drugs, such as phenytoin, carbamazepine, and valproate sodium. An implantable device that stimulates the vagus (X) nerve also has produced dramatic results in reducing seizures in some patients whose epilepsy was not well-controlled by drugs. In very severe cases, surgical intervention may be an option.

KEY MEDICAL TERMS ASSOCIATED WITH NERVOUS TISSUE

Guillain-Barré Syndrome (GBS) (GĒ-an ba-RĀ) An acute demyelinating disorder in which macrophages strip myelin from axons in the PNS. It is the most common cause of acute paralysis in North America and Europe and may result from the immune system's response to a bacterial infection. Most patients recover completely or partially, but about 15% remain paralyzed.

Neuropathy (noo-ROP-a-thē; *neuro-*=a nerve; *-pathy*=disease) Any disorder that affects the nervous system but particularly a disorder of a cranial or spinal nerve. An example is *facial neuropathy* (Bell's palsy), a disorder of the facial (VII) nerve.

Neuroblastoma (noor-ō-blas-TŌ-ma) A malignant tumor that consists of immature nerve cells (neuroblasts); occurs most commonly in the abdomen and most frequently in the adrenal glands. Although rare, it is the most common tumor in infants.

Rabies (RĀ-bēz; *rabi-*=mad, raving) A fatal disease caused by a virus that reaches the CNS via an axon. It is usually transmitted by the bite of a dog or other meat-eating animal. The symptoms are excitement, aggressiveness, and madness, followed by paralysis and death.

STUDY OUTLINE

Overview of the Nervous System (p. 537)

1. The central nervous system (CNS) consists of the brain and spinal cord. The peripheral nervous system (PNS) consists of all nervous tissue outside the CNS.

2. Structures that make up the nervous system are the brain, 12 pairs of cranial nerves and their branches, the spinal cord, 31 pairs of spinal nerves and their branches, ganglia, enteric plexuses, and sensory receptors.

3. The nervous system helps maintain homeostasis and integrates all body activities by sensing changes (sensory function), interpreting them (integrative function), and reacting to them (motor function).

4. Sensory (afferent) neurons carry sensory information from cranial and spinal nerves into the brain and spinal cord or from a lower to a higher level in the spinal cord and brain. Interneurons have short axons that contact nearby neurons in the brain, spinal cord, or a ganglion. Motor (efferent) neurons carry information from the brain toward the spinal cord or out of the brain and spinal cord into cranial or spinal nerves.

5. Components of the PNS include the somatic nervous system (SNS), autonomic nervous system (ANS), and enteric nervous system (ENS).

6. The SNS consists of neurons that conduct impulses from somatic and special sense receptors to the CNS and motor neurons from the CNS to skeletal muscles.

7. The ANS contains sensory neurons from visceral organs and motor neurons that convey impulses from the CNS to smooth muscle tissue, cardiac muscle tissue, and glands.

8. The ENS consists of neurons in enteric plexuses in the gastrointestinal (GI) tract that function independently of the ANS and CNS to some extent. The ENS monitors sensory changes in and controls operation of the GI tract.

Histology of Nervous Tissue (p. 539)

1. Nervous tissue consists of neurons (nerve cells) and neuroglia. Neurons have the property of electrical excitability and are responsible for most unique functions of the nervous system: sensing, thinking, remembering, controlling muscle activity, and regulating glandular secretions.

2. Most neurons have three parts. The dendrites are the main receiving or input region. Integration occurs in the cell body, which includes typical cellular organelles. The output part typically is a single axon, which propagates nerve impulses toward another neuron, a muscle fiber, or a gland cell.

3. Synapses are the site of functional contact between two excitable cells. Axon terminals contain synaptic vesicles filled with neurotransmitter molecules.

4. On the basis of their structure, neurons are classified as multipolar, bipolar, or unipolar.

5. Neuroglia support, nurture, and protect neurons and maintain the interstitial fluid that bathes them. Neuroglia in the CNS include astrocytes, oligodendrocytes, microglia, and ependymal cells. Neuroglia in the PNS include Schwann cells and satellite cells.

6. Two types of neuroglia produce myelin sheaths: oligodendrocytes myelinate axons in the CNS, and Schwann cells myelinate axons in the PNS.

7. White matter consists of aggregates of myelinated processes, whereas gray matter contains cell bodies, dendrites, and axon terminals of neurons, unmyelinated axons, and neuroglia.

8. In the spinal cord, gray matter forms an H-shaped inner core that is surrounded by white matter. In the brain, a thin, superficial shell of gray matter covers the cerebral and cerebellar hemispheres.

Neural Circuits (p. 547)

1. Neurons in the central nervous system are organized into networks called neural circuits.

2. Neural circuits include simple series, diverging, converging, reverberating (oscillatory), and parallel after-discharge circuits.

Regeneration and Neurogenesis (p. 548)

1. The nervous system exhibits plasticity (the capability to change based on experience), but it has very limited powers of regeneration (the capability to replicate or repair damaged neurons).

2. Neurogenesis, the birth of new neurons from undifferentiated stem cells, is normally very limited in adult humans. Repair of damaged axons does not occur in most regions of the CNS.

SELF-QUIZ QUESTIONS

Choose the one best answer to the following questions.

1. Which of the following statements is *true*?
   a. A neuron usually has many axons that are connected to other neurons.
   b. Most unipolar neurons are found in the central nervous system.
   c. Multipolar neurons possess numerous dendrites and only one axon.
   d. Sensory neurons are usually bipolar.
   e. All of the above are true.

2. The myelin sheath of peripheral nervous system neurons is produced by:
   a. Schwann cells.   b. astrocytes.   c. oligodendrocytes.
   d. microglia.   e. ependymal cells.

3. Endorphins are neuropeptides that:
   a. stimulate skeletal muscle contraction.
   b. increase heart rate.
   c. stimulate vasoconstriction.
   d. reduce pain sensations.
   e. regulate body temperature.

4. The point of contact between a nerve fiber and a muscle is called the:
   a. axon hillock. b. node of Ranvier. c. trigger zone.
   d. neuromuscular junction. e. ganglion.

5. Which of the following is *true* for multipolar neurons?
   a. They are primarily sensory neurons.
   b. They have several dendrites and one axon.
   c. They have no nucleus.
   d. They are common in the brain and spinal cord.
   e. Both b and d are correct.

6. Neuroglia that are important components of the blood–brain barrier are the:
   a. astrocytes. b. oligodendrocytes. c. microglia.
   d. ependymal cells. e. Schwann cells.

7. Neuroglia that are phagocytic are the:
   a. astrocytes. b. oligodendrocytes. c. microglia.
   d. ependymal cells. e. Schwann cells.

8. Which of the following best describes a diverging circuit?
   a. Several presynaptic neurons stimulate one postsynaptic neuron.
   b. One presynaptic neuron stimulates several postsynaptic neurons.
   c. Neuron #1 stimulates Neuron #2, which stimulates Neuron #3. Neurons #2 and #3 stimulate Neuron #1.
   d. One presynaptic neuron stimulates one postsynaptic neuron.
   e. None of these fits the definition of a diverging circuit.

Complete the following.

9. Cerebrospinal fluid is produced by neuroglia known as _____.

10. The peripheral nervous system is subdivided into the _____ nervous system, the _____ nervous system, and the _____ nervous system.

11. A bundle of nerve fibers located in the central nervous system is known as a(n) _____.

12. Cells and organs contacted by motor neurons in cranial and spinal nerves are termed _____.

13. Two amino acids that are important inhibitory neurotransmitters are _____ and _____.

14. Sacs in the axon terminals that store neurotransmitters are called _____.

15. The myelin sheath of axons in the central nervous system is produced by cells called _____.

Are the following statements true or false?

16. A unipolar neuron originates as a bipolar neuron, the axon and dendrite of which fuse close to the cell body.

17. Enteric motor neurons control secretions from organs in the gastrointestinal tract.

18. Motor responses of the autonomic nervous system are under subconscious control.

19. White matter refers to aggregations of myelinated processes from many axons.

20. The nucleus of a neuron is located in the axon terminal.

21. Skeletal muscles are effectors in the autonomic nervous system.

Matching

22. Match the following:
    ___ (a) connects an axon to the cell body
    ___ (b) yellowish pigment that increases with age; appears to be a byproduct of lysosomes
    ___ (c) plasma membrane of the axon
    ___ (d) long, thin fibrils composed of bundles of intermediate filaments
    ___ (e) orderly arrangement of rough endoplasmic reticulum, site of protein synthesis
    ___ (f) conducts nerve impulses toward cell body
    ___ (g) conducts nerve impulses away from cell body; has synaptic end bulbs that contain neurotransmitters
    ___ (h) fine filaments that are branching ends of axons and axon collaterals

    (1) axon
    (2) axon terminal
    (3) axon hillock
    (4) Nissl bodies
    (5) dendrite
    (6) lipofuscin
    (7) axolemma
    (8) neurofibrils

23. Match the following (answers are used more than once):
    ___ (a) association neurons
    ___ (b) unipolar neurons
    ___ (c) afferent neurons
    ___ (d) efferent neurons
    ___ (e) most central nervous system neurons are of this type
    ___ (f) these neurons conduct impulses away from the central nervous system
    ___ (g) these neurons conduct impulses toward the central nervous system

    (1) sensory neurons
    (2) motor neurons
    (3) interneurons

CRITICAL THINKING QUESTIONS

1. Varudhini was reviewing her anatomy notes when she heard a catchy advertising jingle on the radio. Hours later, on her way to class, the jingle was still running through her mind. Varudhini was suddenly reminded of a neuronal circuit from her anatomy notes. Name the type of circuit and explain its uses in the body.

2. Miranda said she was so nervous before giving her speech in Communications class that she "could feel the sparks jump from nerve to nerve." You laughed, but said, "Sparks don't jump between neurons." Miranda replied that she had learned in another class that the nervous system controlled the body "electri-

cally" and used "circuits," so the sparks must fly! How would you respond to Miranda?

3. When asked to define "gray matter," Jen answered that it's white matter from a really old person. How would you define gray matter?

4. The buzzing of the alarm clock woke Mohamed. He yawned, stretched, and started to salivate as he smelled the coffee brewing. He could feel his stomach rumble. List the divisions of the nervous system involved in each of these actions.

5. Althea is in research and development for a large pharmaceutical company. She has just been pleasantly surprised by the results of her recent experiments on a new drug to treat a viral infection of the brain. Her secretary exclaims, "Great! With my shares of stock in the company, I'll be rich!" Althea told her secretary, "Don't buy that new Porsche just yet. What works in the Petri dish in the lab may not work in a real person. First of all, the drug has to be able to reach the infection." The lab assistant was perplexed, and said, "Well, why not? Just inject it into a vein, and it will get there." What would your response be?

 ANSWERS TO FIGURE QUESTIONS

17.1 The total number of cranial and spinal nerves in your body is $(12 \times 2) + (31 \times 2) = 86$.

17.2 Sensory or afferent neurons carry input to the CNS; motor or efferent neurons carry output from the CNS.

17.3 Dendrites receive (motor neurons or interneurons) or generate (sensory neurons) inputs; the cell body also receives input signals; the axon conducts nerve impulses (action potentials) and transmits the message to another neuron or effector cell by releasing a neurotransmitter at its synaptic end bulbs.

17.4 Nerve impulses arise at the trigger zone.

17.5 The cell body of a pyramidal cell is shaped like a pyramid.

17.6 Microglia function as phagocytes in the central nervous system.

17.7 One Schwann cell myelinates a single axon, whereas one oligodendrocyte myelinates several axons.

17.8 Myelination increases the speed of nerve impulse conduction.

17.9 Myelin makes white matter look shiny and white, giving white matter its name.

17.10 A motor neuron receiving input from several other neurons is an example of convergence.

THE SPINAL CORD AND THE SPINAL NERVES

18

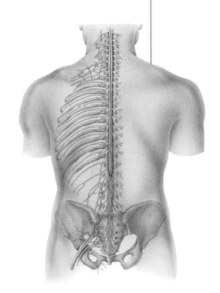

INTRODUCTION Together, the spinal cord and spinal nerves contain neural circuits that mediate some of your quickest reactions to environmental changes. If you pick up something hot, for example, the grasping muscles may relax and you may drop it even before the sensation of extreme heat or pain reaches your conscious perception. This is an example of a **spinal cord reflex**—a quick, automatic response to certain kinds of stimuli that involves neurons only in the spinal nerves and spinal cord. In addition to processing reflexes, the spinal cord is the site for integrating excitation or inhibition of neurons that arises locally or is triggered by nerve impulses from the peripheral nervous system (the periphery) and the brain. Moreover, the spinal cord is the highway traveled by sensory nerve impulses headed for the brain, and by motor nerve impulses destined for spinal nerves. Keep in mind that the spinal cord is continuous with the brain and that together they constitute the central nervous system (CNS).

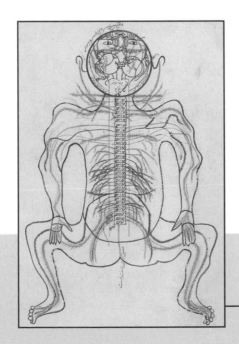

Islamic anatomists believed that God was responsible for creating the human body. It has been theorized that the position of the head (upward, toward the heavens) in this image suggests this belief.

www.wiley.com/college/apcentral

SPINAL CORD ANATOMY

• Describe the protective structures and the gross anatomical features of the spinal cord.

Protective Structures

Two types of connective tissue coverings—bony vertebrae and tough, connective tissue meninges—plus a cushion of cerebrospinal fluid produced in the brain surround and protect the delicate nervous tissue of the spinal cord (and the brain).

Vertebral Column

The spinal cord is located within the vertebral canal of the vertebral column. The vertebral foramina of all the vertebrae, stacked one on top of the other, form the canal. The surrounding vertebrae provide a sturdy shelter for the enclosed spinal cord (see Figure 18.1b). The vertebral ligaments, meninges, and cerebrospinal fluid provide additional protection.

VERTEBRAL CANAL AND SPINAL CORD INJURIES

The **size of the vertebral canal** varies in different regions of the vertebral column, a significant factor in spinal cord injuries. For example, although minor dislocations and/or fractures are common in the cervical region, the relatively large size of the vertebral canal often prevents damage to the spinal cord. The same type of injuries to the thoracic region, where the vertebral canal is relatively smaller, can result in severe spinal cord injury.   ■

Meninges

The **meninges** (me-NIN-jēz) are three connective tissue coverings that encircle the spinal cord and brain. The **spinal meninges** surround the spinal cord (Figure 18.1a) and are continuous with the **cranial meninges,** which encircle the brain (shown in Figure 19.4a on 588). The most superficial of the three spinal meninges is the **dura mater** (DOO-ra MĀ-ter=tough mother), which is composed of dense, irregular connective tissue. It forms a sac from the level of the foramen magnum in the occipital bone, where it is continuous with the dura mater of the brain, to the second sacral vertebra. The spinal cord is also protected by a cushion of fat and connective tissue located in the **epidural space,** a space between the dura mater and the wall of the vertebral canal (Figure 18.1b).

The middle **meninx** (MĒ-ninks, singular form of meninges) is an avascular covering called the **arachnoid mater** (a-RAK-noyd; *arachn-*=spider; *-oid*=similar to) because of its spider's web arrangement of delicate collagen fibers and some elastic fibers. It is deep to the dura mater and is continuous with the arachnoid mater of the brain. Between the dura mater and the arachnoid mater is a thin **subdural space,** which contains interstitial fluid.

The innermost meninx is the **pia mater** (PĒ-a MĀ-ter; *pia*=delicate), a thin transparent connective tissue layer that adheres to the surface of the spinal cord and brain. It consists of interlacing bundles of collagen fibers and some fine elastic fibers. Within the pia mater are many blood vessels that supply oxygen and nutrients to the spinal cord. Between the arachnoid mater and the pia mater is the **subarachnoid space,** which contains cerebrospinal fluid.

SPINAL TAP

In a **spinal tap (lumbar puncture),** a local anesthetic is given, and a long needle is inserted into the subarachnoid space. The procedure is used to withdraw cerebrospinal fluid (CSF) for diagnostic purposes; to introduce antibiotics, contrast media for myelography, or anesthetics, for example, during childbirth; to administer chemotherapy; to measure CSF pressure; and/or to evaluate the effects of treatment for diseases such as meningitis. In adults, a spinal tap is normally performed between the third and fourth or fourth and fifth lumbar vertebrae. Because this region is inferior to the lowest portion of the spinal cord, it provides relatively safe access. (A line drawn across the highest points of the iliac crests, called the *supracristal line,* passes through the spinous process of the fourth lumbar vertebra.)   ■

All three spinal meninges cover the spinal nerves up to the point of exit from the spinal column through the intervertebral foramina. Triangular-shaped membranous extensions of the pia mater suspend the spinal cord in the middle of its dural sheath. These extensions, called **denticulate ligaments** (den-TIK-ū-lāt=small tooth), are thickenings of the pia mater. They project laterally and fuse with the arachnoid mater and inner surface of the dura mater between the anterior and posterior nerve roots of spinal nerves on either side (Figure 18.1b). Extending all along the length of the spinal cord, the denticulate ligaments protect the spinal cord against sudden displacement that could result in shock.

External Anatomy of the Spinal Cord

The **spinal cord,** although roughly cylindrical, is flattened slightly in its anterior-posterior dimension. In adults, it extends from the medulla oblongata, the inferior part of the brain, to the superior border of the second lumbar vertebra (Figure 18.2 on pages 556–557). In newborn infants, it extends to the third or fourth lumbar vertebra. During early childhood, both the spinal cord and the vertebral column grow longer as part of overall body growth. Around age 4 or 5, however, elongation of the spinal cord stops. Because the vertebral column continues to elongate, the spinal cord does not extend the entire length of the vertebral column in adults. The length of the adult spinal cord ranges from 42 to 45 cm (16–18 in.). Its diameter is about 2 cm (0.75 in.) in the midthoracic region, somewhat larger in the lower cervical and midlumbar regions, and smallest at the inferior tip.

When the spinal cord is viewed externally, two conspicuous enlargements can be seen. The superior enlargement, the **cervical enlargement,** extends from the fourth cervical vertebra to

Figure 18.1 **Gross anatomy of the spinal cord.** The spinal meninges are evident in both views.

Meninges are connective tissue coverings that surround the spinal cord and brain.

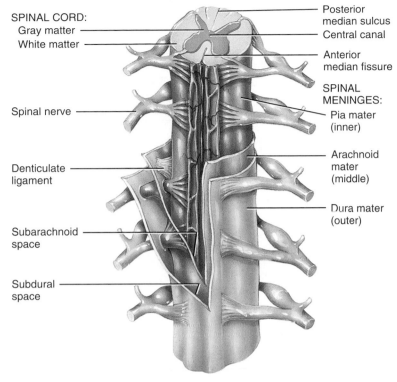

SPINAL CORD:
Gray matter
White matter

Spinal nerve

Denticulate
ligament

Subarachnoid
space

Subdural
space

Posterior
median sulcus
Central canal
Anterior
median fissure

SPINAL
MENINGES:
Pia mater
(inner)

Arachnoid
mater
(middle)

Dura mater
(outer)

(a) Anterior view and transverse section through spinal cord

View

Transverse plane

POSTERIOR

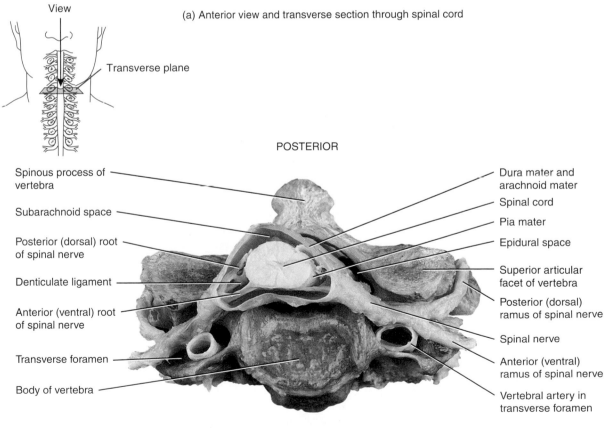

Spinous process of
vertebra

Subarachnoid space

Posterior (dorsal) root
of spinal nerve

Denticulate ligament

Anterior (ventral) root
of spinal nerve

Transverse foramen

Body of vertebra

Dura mater and
arachnoid mater

Spinal cord

Pia mater

Epidural space

Superior articular
facet of vertebra

Posterior (dorsal)
ramus of spinal nerve

Spinal nerve

Anterior (ventral)
ramus of spinal nerve

Vertebral artery in
transverse foramen

ANTERIOR

(b) Transverse section of the spinal cord within a cervical vertebra

 What are the superior and inferior boundaries of the spinal dura mater?

Figure 18.2 **External anatomy of the spinal cord and the spinal nerves.**

The spinal cord extends from the medulla oblongata of the brain to the superior border of the second lumbar vertebra.

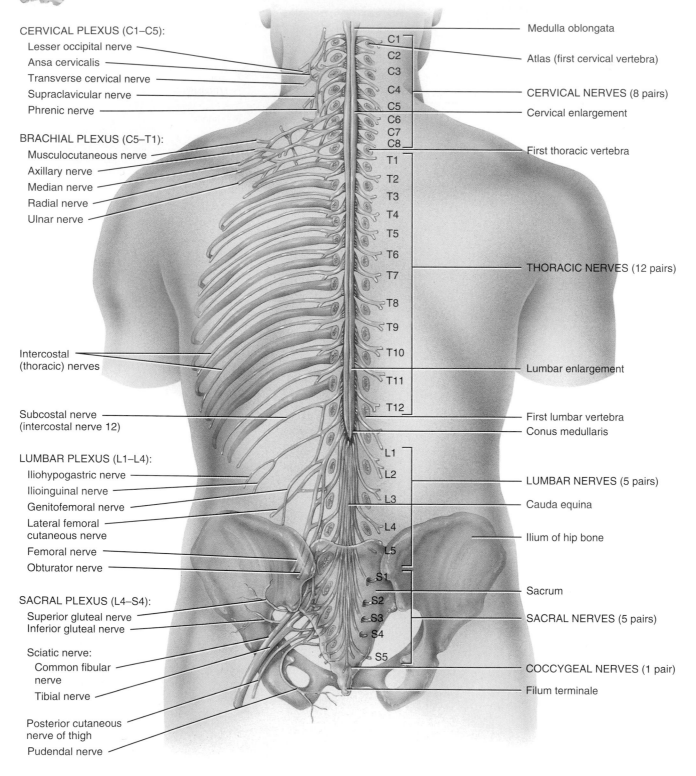

CERVICAL PLEXUS (C1–C5):
Lesser occipital nerve
Ansa cervicalis
Transverse cervical nerve
Supraclavicular nerve
Phrenic nerve

BRACHIAL PLEXUS (C5–T1):
Musculocutaneous nerve
Axillary nerve
Median nerve
Radial nerve
Ulnar nerve

Intercostal (thoracic) nerves

Subcostal nerve (intercostal nerve 12)

LUMBAR PLEXUS (L1–L4):
Iliohypogastric nerve
Ilioinguinal nerve
Genitofemoral nerve
Lateral femoral cutaneous nerve
Femoral nerve
Obturator nerve

SACRAL PLEXUS (L4–S4):
Superior gluteal nerve
Inferior gluteal nerve
Sciatic nerve:
  Common fibular nerve
  Tibial nerve

Posterior cutaneous nerve of thigh
Pudendal nerve

C1
C2
C3
C4
C5
C6
C7
C8
T1
T2
T3
T4
T5
T6
T7
T8
T9
T10
T11
T12
L1
L2
L3
L4
L5
S1
S2
S3
S4
S5

Medulla oblongata
Atlas (first cervical vertebra)
CERVICAL NERVES (8 pairs)
Cervical enlargement
First thoracic vertebra

THORACIC NERVES (12 pairs)

Lumbar enlargement

First lumbar vertebra
Conus medullaris

LUMBAR NERVES (5 pairs)
Cauda equina

Ilium of hip bone

Sacrum

SACRAL NERVES (5 pairs)

COCCYGEAL NERVES (1 pair)
Filum terminale

(a) Posterior view of entire spinal cord and portions of spinal nerves

SUPERIOR

Fourth ventricle

Glossopharyngeal (IX)
and vagus (X) nerves

Accessory (XI) nerve

Gracile fasciculus

Cuneate fasciculus

Cerebellum of brain (cut)

Occipital bone (cut)

Posterior median sulcus

Vertebral artery

Denticulate ligament

Dura mater
and arachnoid

Posterior (dorsal) rootlets
of spinal nerve

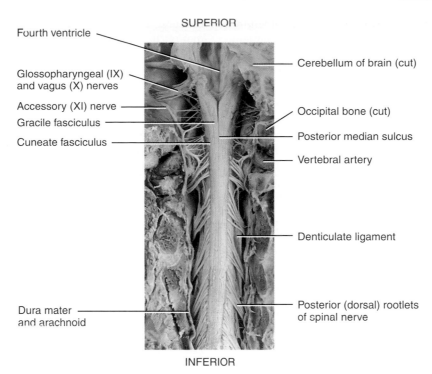

INFERIOR

(b) Posterior view of cervical region of spinal cord

SUPERIOR

Conus medullaris

Dura mater
and arachnoid

Posterior (dorsal)
rami of spinal nerves

Cauda equina

Conus medullaris

Sacrum

Gluteus maximus

Filum terminale

Right coccygeal
nerve

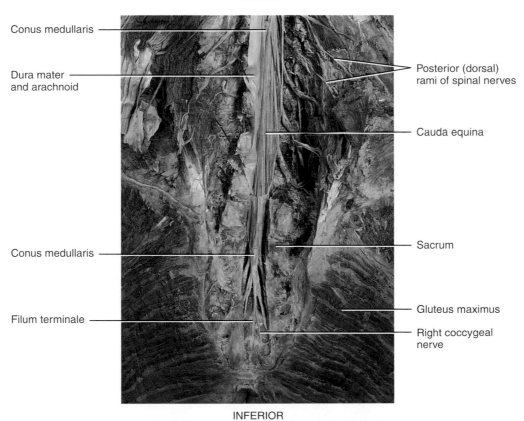

INFERIOR

(c) Posterior view of inferior portion of spinal cord

 What portion of the spinal cord connects with sensory and motor nerves of the upper limbs?

the first thoracic vertebra. Nerves to and from the upper limbs arise from the cervical enlargement. The inferior enlargement, called the **lumbar enlargement,** extends from the ninth to the twelfth thoracic vertebra. Nerves to and from the lower limbs arise from the lumbar enlargement.

Inferior to the lumbar enlargement, the spinal cord tapers to a conical portion referred to as the **conus medullaris** (KŌ-nus med-ū-LAR-is; *conus*=cone), which ends at the level of the intervertebral disc between the first and second lumbar vertebrae in adults (Figure 18.2c). Arising from the conus medullaris is the **filum terminale** (FĪ-lum ter-mi-NAL-ē=terminal filament), an extension of the pia mater that extends inferiorly and anchors the spinal cord to the coccyx (Figure 18.2c).

Nerves that arise from most of the spinal cord do not leave the vertebral column at the same level as they exit from the cord because the spinal cord is shorter than the vertebral column. The roots (points of attachment of spinal nerves to the spinal cord) of most of the spinal nerves, especially in the inferior portion of the spinal cord, angle inferiorly in the vertebral canal from the end of the spinal cord like wisps of hair. Appropriately, the roots of these nerves are collectively named the **cauda equina** (KAW-da ē-KWĪ-na), meaning "horse's tail" (Figure 18.2a).

Spinal cord organization appears to be segmented because the 31 pairs of spinal nerves emerge at regular intervals from intervertebral foramina (Figure 18.2). Indeed, each pair of spinal nerves is said to arise from a *spinal segment*. Within the spinal cord, however, there is no obvious segmentation of the white matter or the gray matter. The naming of spinal nerves and spinal segments is based on their location. There are 8 pairs of *cervical nerves* (represented as C1–C8), 12 pairs of *thoracic nerves* (T1–T12), 5 pairs of *lumbar nerves* (L1–L5), 5 pairs of *sacral nerves* (S1–S5), and 1 pair of *coccygeal nerves* (Col).

Spinal nerves are the paths of communication between the spinal cord and the nerves innervating specific regions of the body. Two bundles of axons, called **roots,** connect each spinal nerve to a segment of the cord (see Figure 18.3a). The **posterior (dorsal) root** contains only sensory axons, which conduct nerve impulses from sensory receptors in the skin, muscles, and internal organs into the central nervous system. Each posterior root also has a swelling, the **posterior (dorsal) root ganglion,** which contains the cell bodies of sensory neurons. The **anterior (ventral) root** contains axons of motor neurons, which conduct nerve impulses from the CNS to effector organs and cells. The spinal nerve roots pass through the intervertebral foramina as they enter or exit the spinal cord.

SPINAL NERVE ROOT DAMAGE

Spinal nerve roots exit from the vertebral canal through intervertebral foramina. The most common cause of **spinal nerve root damage** is a herniated (slipped) intervertebral disc. Damage to vertebrae as a result of osteoporosis, cancer, or injury can also damage spinal nerve roots. Another cause of damage is osteoarthritis. Symptoms include pain, muscle weakness, and loss of feeling. Rest, physical therapy, pain medications, and epidural injections are the most widely used conservative treatments. It is recommended that 6 to 12 weeks of conservative therapy be attempted first. If the pain continues, is intense, or is impairing normal functioning, surgery is often the next step. ∎

Internal Anatomy of the Spinal Cord

Two grooves penetrate the white matter of the spinal cord and divide it into right and left sides (Figure 18.3). The **anterior median fissure** is a deep, wide groove on the anterior (ventral) side. The **posterior median sulcus** is a shallower, narrow furrow on the posterior (dorsal) side. The gray matter of the spinal cord is shaped like the letter H or a butterfly and is surrounded by white matter. The gray matter consists primarily of the cell bodies of neurons, neuroglia, unmyelinated axons, and the dendrites of interneurons and motor neurons. The white matter consists of bundles of myelinated and unmyelinated axons of sensory neurons, interneurons, and motor neurons. The **gray commissure** (KOM-mi-shur) forms the crossbar of the H. In the center of the gray commissure is a small space called the **central canal;** it extends the entire length of the spinal cord and contains cerebrospinal fluid. At its superior end, the central canal is continuous with the fourth ventricle (a space that also contains cerebrospinal fluid) in the medulla oblongata of the brain. Anterior to the gray commissure is the **anterior (ventral) white commissure,** which connects the white matter of the right and left sides of the spinal cord.

In the gray matter of the spinal cord and brain, clusters of neuronal cell bodies form functional groups called **nuclei.** *Sensory nuclei* receive input from receptors via sensory neurons, whereas *motor nuclei* provide output to effector tissues via motor neurons. The gray matter on each side of the spinal cord is subdivided into regions called **horns.** The **anterior (ventral) gray horns** contain cell bodies of somatic motor neurons and motor nuclei, which provide nerve impulses for contraction of skeletal muscles. The **posterior (dorsal) gray horns** contain somatic and autonomic sensory nuclei. Between the anterior and posterior gray horns are the **lateral gray horns,** which are present only in the thoracic, upper lumbar, and sacral segments of the spinal cord. The lateral horns contain the cell bodies of autonomic motor neurons that regulate activity of smooth muscle, cardiac muscle, and glands.

The white matter, like the gray matter, is organized into regions. The anterior and posterior gray horns divide the white matter on each side into three broad areas called **columns:** (1) **anterior (ventral) white columns,** (2) **posterior (dorsal) white columns,** and (3) **lateral white columns.** Each column, in turn, contains distinct bundles of axons having a common origin or destination and carrying similar information. These bundles, which may extend long distances up or down the spinal cord, are called **tracts. Sensory (ascending) tracts** consist of axons that conduct nerve impulses toward the brain. Tracts consisting of axons that carry nerve impulses down the spinal cord are called **motor (descending) tracts.** Sensory and motor tracts of the spinal cord are continuous with sensory and motor tracts in the brain.

Figure 18.3 **Internal anatomy of the spinal cord: the organization of gray matter and white matter.**
For simplicity, dendrites are not shown in this and several other illustrations of transverse sections of the spinal cord. Blue and red arrows in (a) indicate the direction of nerve impulse propagation.

In the spinal cord, white matter surrounds the gray matter.

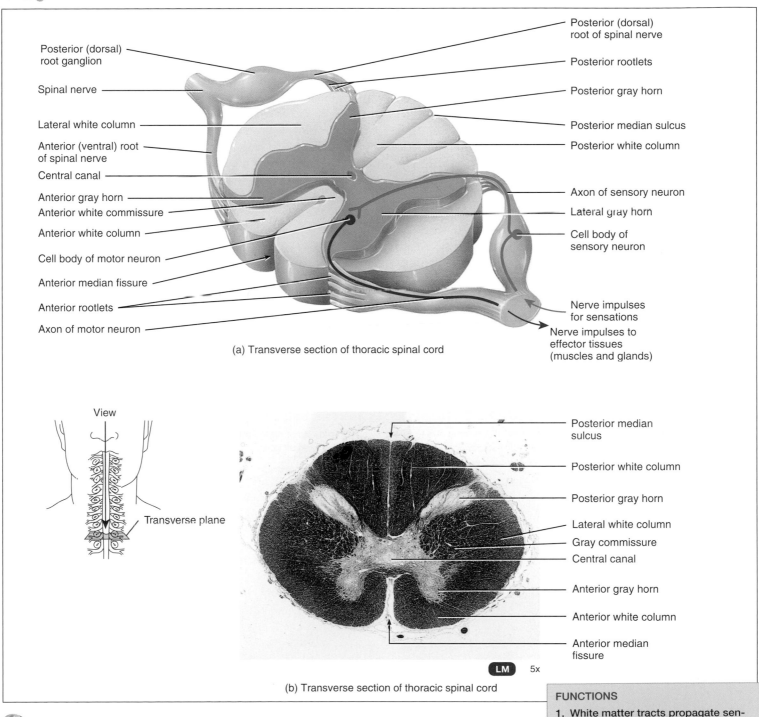

Posterior (dorsal) root ganglion

Spinal nerve

Lateral white column

Anterior (ventral) root of spinal nerve

Central canal

Anterior gray horn

Anterior white commissure

Anterior white column

Cell body of motor neuron

Anterior median fissure

Anterior rootlets

Axon of motor neuron

Posterior (dorsal) root of spinal nerve

Posterior rootlets

Posterior gray horn

Posterior median sulcus

Posterior white column

Axon of sensory neuron

Lateral gray horn

Cell body of sensory neuron

Nerve impulses for sensations

Nerve impulses to effector tissues (muscles and glands)

(a) Transverse section of thoracic spinal cord

View

Transverse plane

Posterior median sulcus

Posterior white column

Posterior gray horn

Lateral white column

Gray commissure

Central canal

Anterior gray horn

Anterior white column

Anterior median fissure

LM 5x

(b) Transverse section of thoracic spinal cord

What is the difference between a horn and a column in the spinal cord?

FUNCTIONS

1. White matter tracts propagate sensory impulses from receptors to the brain and motor impulses from the brain to effectors.

2. Gray matter receives and integrates incoming and outgoing information.

The various spinal cord segments vary in size, shape, relative amounts of gray and white matter, and distribution and shape of gray matter. These features are summarized in Table 18.1.

CHECKPOINT

1. Where are the spinal meninges located? Where are the epidural, subdural, and subarachnoid spaces located?
2. What are the cervical and lumbar enlargements?
3. Define conus medullaris, filum terminale, and cauda equina.
4. What is a spinal segment? How is the spinal cord partially divided into right and left sides?
5. Define each of the following terms: gray commissure, central canal, anterior gray horn, lateral gray horn, posterior gray horn, anterior white column, lateral white column, posterior white column, ascending tract, descending tract.

SPINAL CORD FUNCTIONS

OBJECTIVES

- Describe the functions of the major sensory and motor tracts of the spinal cord.
- Describe the functional components of a reflex arc and the ways reflexes maintain homeostasis.

The spinal cord has two principal functions in maintaining homeostasis: nerve impulse propagation and integration of information. The *white matter tracts* in the spinal cord are highways for nerve impulse propagation. Along these tracts, sensory impulses from receptors flow toward the brain, and motor impulses flow from the brain toward skeletal muscles and other effector tissues. The *gray matter* of the spinal cord receives and integrates incoming and outgoing information.

TABLE 18.1   COMPARISON OF VARIOUS SPINAL CORD SEGMENTS

SEGMENT	DISTINGUISHING CHARACTERISTICS
Cervical 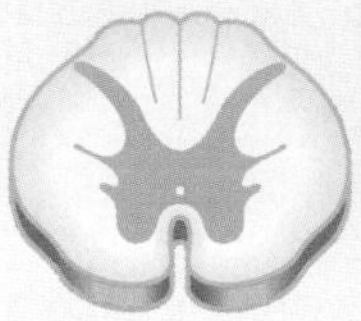 (Segment C1) 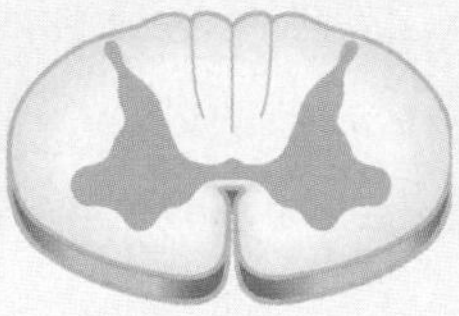 (Segment C8)	Relatively large diameter, relatively large amounts of white matter, oval in shape; in upper cervical segments (C1–C6), posterior gray horn is large, but anterior gray horn is relatively small; in lower cervical segments (C6 and below), posterior gray horns are enlarged and anterior gray horns are well developed.
Thoracic 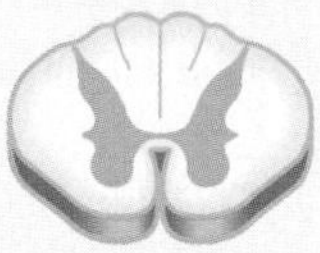 (Segment T2)	Small diameter is due to relatively small amounts of gray matter; except for first thoracic segment, anterior and posterior gray horns are relatively small; a small lateral gray horn is present.
Lumbar 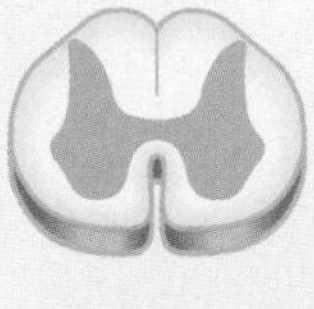 (Segment L4)	Nearly circular; very large anterior and posterior gray horns; relatively less white matter than cervical segments.
Sacral  (Segment S3)	Relatively small, but with relatively large amounts of gray matter; relatively small amounts of white matter; anterior and posterior gray horns are large and thick.
Coccygeal	Resemble lower sacral spinal segments, but much smaller.

Sensory and Motor Tracts

The first way the spinal cord promotes homeostasis is by conducting nerve impulses along tracts. Often, the name of a tract indicates its position in the white matter and where it begins and ends. For example, the anterior spinothalamic tract is located in the *anterior* white column; it begins in the *spinal cord* and ends in the *thalamus* (a region of the brain). Notice that the location of the axon terminals comes last in the name. This regularity in naming allows you to determine the direction of information flow along any tract named according to this convention. Thus, because the anterior spinothalamic tract conveys nerve impulses from the spinal cord toward the brain, it is a sensory (ascending) tract. Figure 18.4 highlights the major sensory and motor tracts in the spinal cord. These tracts are described in detail in Chapter 21 and summarized in Tables 21.3 and 21.4 on pages 662 and 665.

Nerve impulses from sensory receptors propagate up the spinal cord to the brain along two main routes on each side: the spinothalamic tracts and the posterior columns. The **lateral** and **anterior spinothalamic tracts** convey nerve impulses for sensing pain, warmth, coolness, itching, tickling, deep pressure, and a crude, poorly localized sense of touch. The right and left **posterior columns** carry nerve impulses for several kinds of sensations. These include (1) proprioception, awareness of the positions and movements of muscles, tendons, and joints;

(2) discriminative touch, the ability to feel exactly what part of the body is touched; (3) two-point discrimination, the ability to distinguish the touching of two different points on the skin, even though they are close together; (4) light pressure sensations; and (5) vibration sensations.

The sensory systems keep the CNS informed of changes in the external and internal environments. Responses to this information are brought about by motor systems, which enable us to move about and change our physical relationship to the world around us. As sensory information is conveyed to the CNS, it becomes part of a large pool of sensory input. Each piece of incoming information is integrated with all the other information arriving from activated sensory neurons.

Through the activity of interneurons, integration occurs in several regions of the spinal cord and brain. As a result, motor impulses to make a muscle contract or a gland secrete can be initiated at several levels. Most regulation of involuntary activities of smooth muscle, cardiac muscle, and glands by the autonomic nervous system (ANS) originates in the brain stem (the lower part of the brain that is continuous with the spinal cord) and in a nearby brain region called the hypothalamus.

The cerebral cortex (superficial gray matter of the cerebrum) plays a major role in controlling precise, voluntary muscular movements. Other brain regions integrate automatic movements, such as arm swinging during walking. Motor output to skeletal muscles travels down the spinal cord in two types of

Figure 18.4 **The locations of selected sensory and motor tracts, shown in a transverse section of the spinal cord.** Sensory tracts are indicated on one half and motor tracts on the other half of the cord, but in fact all tracts are present on both sides.

The name of a tract often indicates its location in the white matter and where it begins and ends.

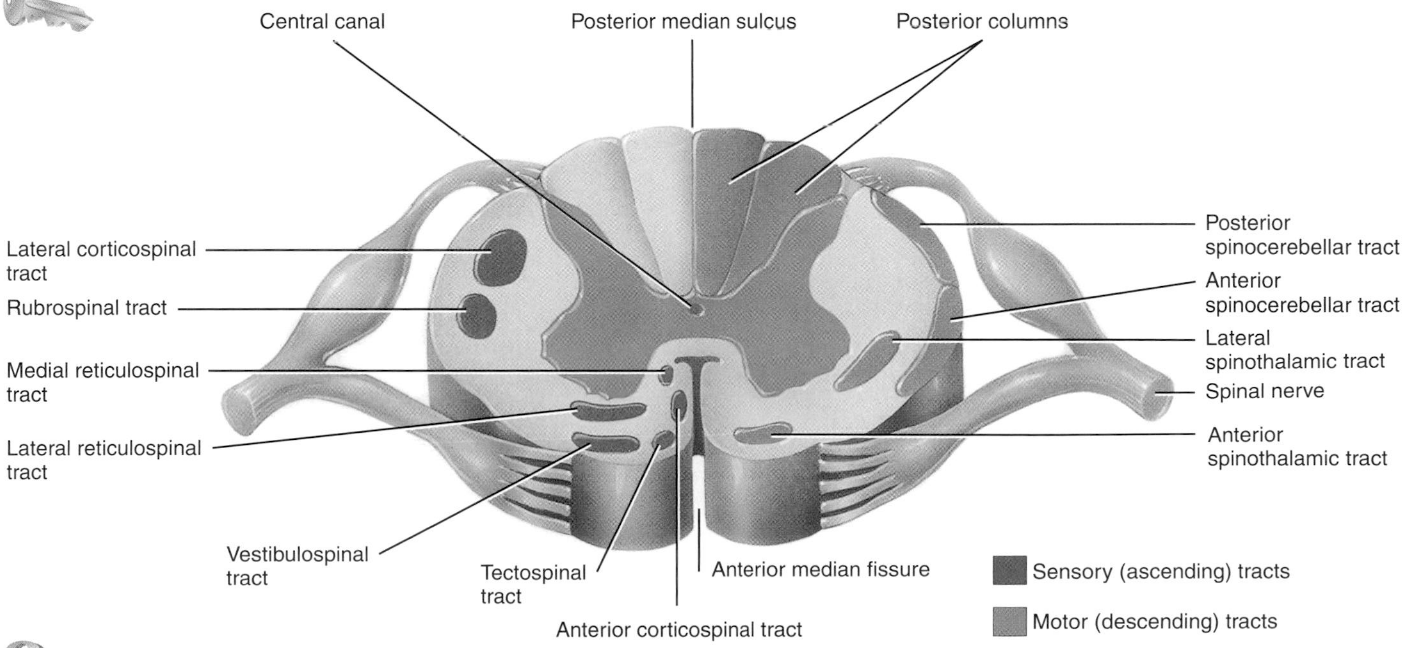

Based on its name, what are the position in the spinal cord, origin, and destination of the anterior corticospinal tract? Is this a sensory or a motor tract?

descending pathways: direct and indirect. The **direct pathways** include the *lateral corticospinal, anterior corticospinal,* and *corticobulbar tracts.* They convey nerve impulses that originate in the cerebral cortex and are destined to cause precise, *voluntary* movements of skeletal muscles. **Indirect pathways** include the *rubrospinal, tectospinal,* and *vestibulospinal tracts.* They convey nerve impulses from the brain stem and other parts of the brain that govern *automatic movements* and help coordinate body movements with visual stimuli. Indirect pathways also maintain skeletal muscle tone, maintain contraction of postural muscles, and play a major role in equilibrium by regulating muscle tone in response to movements of the head.

Reflexes and Reflex Arcs

The second way the spinal cord promotes homeostasis is by serving as an integrating center for some reflexes. A **reflex** is a fast, involuntary, unplanned sequence of actions that occurs in response to a particular stimulus. Some reflexes are inborn, such as pulling your hand away from a hot surface before you even feel that it is hot. Other reflexes are learned or acquired. For instance, you learn many reflexes while acquiring driving expertise. Slamming on the brakes in an emergency is one example. When integration takes place in the spinal cord gray matter, the reflex is a **spinal reflex.** An example is the familiar patellar reflex (knee jerk). By contrast, if integration occurs in the brain stem rather than the spinal cord, the reflex is a **cranial reflex.** An example is the tracking movements of your eyes as you read this sentence. You are probably most aware of **somatic reflexes,** which involve contraction of skeletal muscles. Equally important, however, are the **autonomic (visceral) reflexes,** which generally are not consciously perceived. They involve responses of smooth muscle, cardiac muscle, and glands. As you will see in Chapter 20, body functions such as heart rate, digestion, urination, and defecation are controlled by the autonomic nervous system through autonomic reflexes.

Nerve impulses propagating into, through, and out of the CNS follow specific pathways, depending on the kind of information, its origin, and its destination. The pathway followed by nerve impulses that produce a reflex is a **reflex arc (reflex circuit).** Using the **patellar reflex** (knee jerk) as an example, the basic components of a reflex arc are as follows (Figure 18.5):

① Sensory receptor. The distal end of a sensory neuron (dendrite) or an associated sensory structure serves as a sensory receptor. Sensory receptors respond to a specific type of *stimulus* (a change in the internal or external environment) by generating one or more nerve impulses. In the patellar reflex, sensory receptors known as *muscle spindles* detect slight stretching of the quadriceps femoris muscle (anterior thigh) when the patellar (knee cap) ligament is tapped with a reflex hammer.

② Sensory neuron. The nerve impulses conduct from the sensory receptor along the axon of a sensory neuron to its axon terminals, which are located in the CNS gray matter. Axon branches of the sensory neuron also relay nerve impulses to the brain, allowing conscious awareness that the reflex has occurred.

③ Integrating center. One or more regions of gray matter in the CNS act as an integrating center. In the simplest type of reflex, such as the patellar reflex, the integrating center is a single synapse between a sensory neuron and a motor neuron. In other types of reflexes, the integrating center includes one or more interneurons.

④ Motor neuron. Impulses triggered by the integrating center pass out of the spinal cord (or brain stem, in the case of a cranial reflex) along a motor neuron to the part of the body that will respond. In the patellar reflex, the axon of the motor neuron extends to the quadriceps femoris muscle.

⑤ Effector. The part of the body that responds to the motor nerve impulse, such as a muscle or gland, is the effector. The patellar reflex is a somatic reflex because its effector is a skeletal muscle, the quadriceps femoris muscle, which contracts and thereby relieves the stretching that initiated the reflex. In sum, the patellar reflex causes extension of the knee by contraction of the quadriceps femoris muscle in response to tapping the patellar ligament.

REFLEXES AND DIAGNOSIS

Reflexes are often used for diagnosing disorders of the nervous system and locating injured tissue. If a reflex ceases to function or functions abnormally, the physician may suspect that the damage lies somewhere along a particular conduction pathway. Many somatic reflexes can be tested simply by tapping or stroking the body. Among the somatic reflexes of clinical significance are the following:

1. **Patellar reflex (knee jerk).** This reflex, as just noted, involves extension of the knee joint by contraction of the quadriceps femoris muscle in response to tapping the patellar ligament (Figure 18.5). This reflex is blocked by damage to the sensory or motor nerves supplying the muscle or to the integrating centers in the second, third, or fourth lumbar segments of the spinal cord. It is often absent in people with chronic diabetes mellitus or neurosyphilis, which cause degeneration of nerves. It is exaggerated in disease or injury involving certain motor tracts descending from the higher centers of the brain to the spinal cord.

2. **Achilles reflex (ankle jerk).** This stretch reflex involves extension (plantar flexion) of the foot by contraction of the gastrocnemius and soleus muscles in response to tapping the calcaneal (Achilles) tendon. Absence of the Achilles reflex indicates damage to the nerves supplying the posterior leg muscles or to neurons in the lumbosacral region of the spinal cord. This reflex may also disappear in people with chronic diabetes, neurosyphilis, alcoholism, and subarachnoid hemorrhages. An exaggerated Achilles reflex indicates cervical cord compression or a lesion of the motor tracts of the first or second sacral segments of the cord.

3. **Babinski sign.** This reflex results from gentle stroking of the lateral outer margin of the sole. The great toe extends, with or without fanning of the other toes. This phenome-

Figure 18.5 Patellar reflex showing the general components of a reflex arc.
The arrows show the direction of nerve impulse propagation.

🔑 **Reflexes are fast involuntary responses to particular stimuli.**

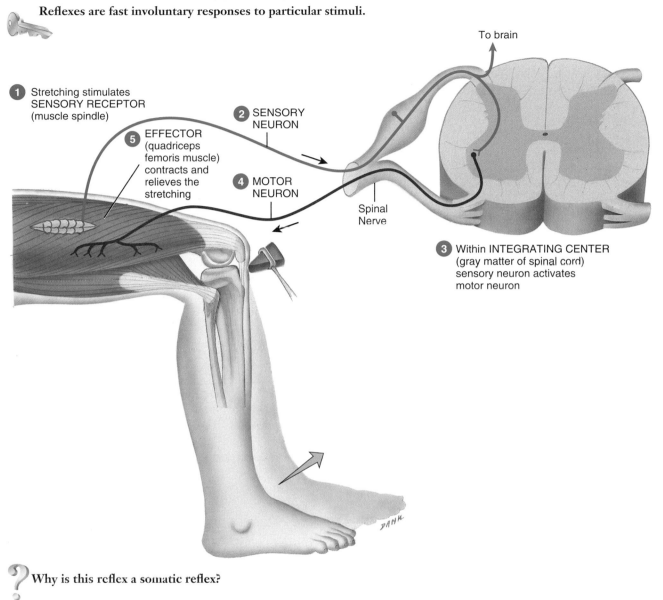

1 Stretching stimulates SENSORY RECEPTOR (muscle spindle)

2 SENSORY NEURON

5 EFFECTOR (quadriceps femoris muscle) contracts and relieves the stretching

4 MOTOR NEURON

Spinal Nerve

To brain

3 Within INTEGRATING CENTER (gray matter of spinal cord) sensory neuron activates motor neuron

❓ **Why is this reflex a somatic reflex?**

non occurs in normal children under 1 ½ years of age and is due to incomplete myelination of fibers in the corticospinal tract. A positive Babinski sign after age 1 ½ is abnormal and indicates an interruption of the corticospinal tract as the result of a lesion of the tract, usually in the upper portion. The normal response after 1 ½ years of age is the **plantar flexion reflex,** or **negative Babinski**—a curling under of all the toes, accompanied by a slight turning in and flexion of the anterior part of the foot.

4. **Abdominal reflex.** This reflex involves contraction of the muscles that compress the abdominal wall in response to stroking the side of the abdomen. The response is an abdominal muscle contraction that causes a lateral movement of the umbilicus to the side opposite the stimulus. Absence of this reflex is associated with lesions of the corti-

cospinal tracts. It may also be absent because of lesions of the peripheral nerves, lesions of integrating centers in the thoracic part of the cord, and multiple sclerosis.

Most autonomic reflexes are not practical diagnostic tools because it is difficult to stimulate visceral receptors, which are deep inside the body. An exception is the pupillary light reflex, in which the pupils of both eyes decrease in diameter when either eye is exposed to light. Because the reflex arc includes synapses in lower parts of the brain, the **absence of a normal pupillary light reflex** may indicate brain damage or injury. ■

CHECKPOINT

6. What are the functions of the anterior spinothalamic tract and the posterior columns?

7. Describe the components of the patellar reflex arc.

SPINAL NERVES

- Describe the components, connective tissue coverings, and branching of a spinal nerve.
- Define a plexus, and identify the distribution of nerves of the cervical, brachial, lumbar, and sacral plexuses.
- Describe the clinical significance of dermatomes.

Spinal nerves and the nerves that branch from them are part of the peripheral nervous system (PNS). They connect the CNS to sensory receptors, muscles, and glands in all parts of the body. The 31 pairs of spinal nerves are named and numbered according to the region and level of the vertebral column from which they emerge (see Figure 18.2). The first cervical pair emerges between the atlas (first cervical vertebra) and the occipital bone. All other spinal nerves emerge from the vertebral column through the intervertebral foramina between adjoining vertebrae.

However, not all spinal cord segments are aligned with their corresponding vertebrae. Recall that the spinal cord ends near the level of the superior border of the second lumbar vertebra, and that the roots of the lower lumbar, sacral, and coccygeal nerves descend at an angle to reach their respective foramina before emerging from the vertebral column. This arrangement constitutes the cauda equina (see Figure 18.2).

As noted earlier, a typical **spinal nerve** has two connections to the cord: a posterior root and an anterior root (see Figure 18.3a). The posterior and anterior roots unite to form a spinal nerve at the intervertebral foramen. Because the posterior root contains sensory axons and the anterior root contains motor axons, the spinal nerves are considered **mixed nerves.**

Connective Tissue Coverings of Spinal Nerves

Each cranial nerve and spinal nerve contains layers of protective connective tissue coverings (Figure 18.6). Individual axons, whether myelinated or unmyelinated, are wrapped in **endoneurium** (en'-dō-NOO-rē-um; *endo-*=within or inner). Groups of axons with their endoneurium are arranged in bundles called **fascicles** (not to be confused with the muscle fiber bundles of the same name). Each fascicle is wrapped in **perineurium** (per'-i-NOO-rē-um; *peri-*=around). The superficial covering over the entire nerve is the **epineurium** (ep'-i-NOO-rē-um; *epi-*=over). Extensions of the epineurium also occur between fascicles, a situation that does not occur in skeletal muscle. The dura mater of the spinal meninges fuses with the epineurium as the nerve passes through the intervertebral fora-

Figure 18.6 Organization and connective tissue coverings of a spinal nerve. (Part b: © Dr. Richard Kessel and Dr. Randy Kardon/Visuals Unlimited. *Tissues and Organs: A Text-Altas of Scanning Electron Microscopy.* © 1979 by W. H. Freeman and Company. Reprinted by permission.)

 Three layers of connective tissue wrappings protect axons: endoneurium surrounds individual axons, perineurium surrounds bundles of axons (fascicles), and epineurium surrounds an entire nerve.

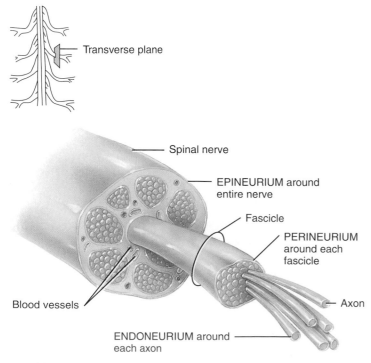

(a) Transverse section showing the coverings of a spinal nerve

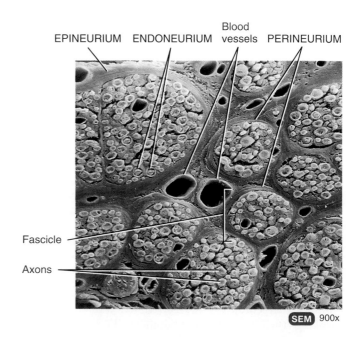

(b) Transverse section of 12 nerve fascicles

? Why are all spinal nerves classified as mixed nerves?

men. Note the presence of many blood vessels, which nourish nerves, within the perineurium and epineurium (Figure 18.6b). You may recall from Chapter 9 that the connective tissue coverings of skeletal muscles—endomysium, perimysium, and epimysium—are similar in organization to those of nerves.

Distribution of Spinal Nerves

Branches

A short distance after passing through its intervertebral foramen, a spinal nerve divides into several branches (Figure 18.7). These branches are known as **rami** (RĀ-mī=branches). The **posterior (dorsal) ramus** (RĀ-mus; singular form) serves the deep muscles and skin of the dorsal surface of the trunk. The **anterior (ventral) ramus** serves the muscles and structures of the upper and lower limbs and the skin of the lateral and ventral surfaces of the trunk. In addition to posterior and anterior rami, spinal nerves also give off a **meningeal branch.** This branch reenters the vertebral canal through the intervertebral foramen and supplies the vertebrae, vertebral ligaments, blood vessels of the spinal cord, and meninges. Other branches of a spinal nerve

are the **rami communicantes** (kō-mū-ni-KAN-tēz), components of the autonomic nervous system whose structure and function are discussed in Chapter 20.

Plexuses

The anterior rami of spinal nerves, except for thoracic nerves T2–T12, do not go directly to the body structures they supply. Instead, they form networks on both the left and right sides of the body by joining with various numbers of axons from anterior rami of adjacent nerves. Such a network of axons is called a **plexus** (=braid or network). The principal plexuses are the **cervical plexus, brachial plexus, lumbar plexus,** and **sacral plexus.** A smaller **coccygeal plexus** is also present. Refer to Figure 18.2 to see their relationships to one another. Emerging from the plexuses are nerves bearing names that are often descriptive of the general regions they serve or the course they take. Each of the nerves, in turn, may have several branches named for the specific structures they innervate.

Exhibits 18.1–18.4 summarize the principal plexuses. The anterior rami of spinal nerves T2–T12, called intercostal nerves, will be discussed next.

Figure 18.7 Branches of a typical spinal nerve, shown in transverse section through the thoracic portion of the spinal cord. (See also Figure 18.1b.)

The branches of a spinal nerve are the posterior ramus, the anterior ramus, the meningeal branch, and the rami communicantes.

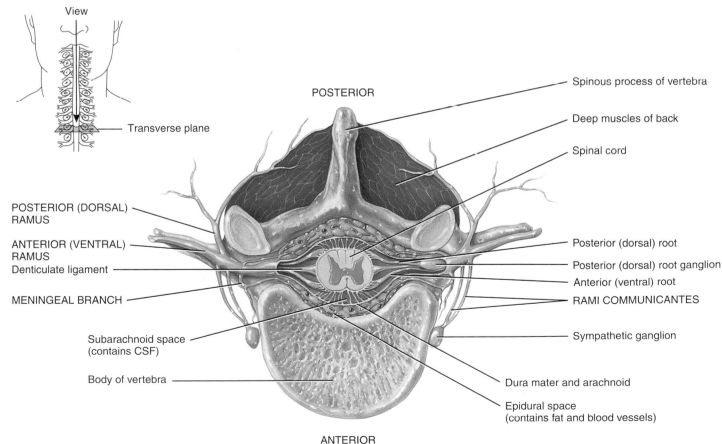

Which spinal nerve branches serve the upper and lower limbs?

EXHIBIT 18.1   CERVICAL PLEXUS (Figure 18.8)

OBJECTIVE

• Describe the origin and distribution of the cervical plexus.

The **cervical plexus** (SER-vi-kul) is formed by the anterior (ventral) rami of the first four cervical nerves (C1–C4), with contributions from C5. There is one on each side of the neck alongside the first four cervical vertebrae. The roots of the plexus indicated in Figure 18.8 are the anterior rami.

The cervical plexus supplies the skin and muscles of the head, neck, and superior part of the shoulders and chest. The phrenic nerves arise from the cervical plexuses and supply motor fibers to the diaphragm. Branches of the cervical plexus also run parallel to the accessory (XI) nerve and hypoglossal (XII) nerve.

INJURIES TO THE PHRENIC NERVES

Complete severing of the spinal cord above the origin of the phrenic nerves (C3, C4, and C5) causes respiratory arrest. Breathing stops because the phrenic nerves no longer send impulses to the diaphragm. ■

■ CHECKPOINT

1. Which nerve that arises from the cervical plexus causes contraction of the diaphragm?

NERVE	ORIGIN	DISTRIBUTION
SUPERFICIAL (SENSORY) BRANCHES		
Lesser occipital	C2	Skin of scalp posterior and superior to ear.
Great auricular (aw-RIK-ū-lar)	C2–C3	Skin anterior, inferior, and over ear, and over parotid glands.
Transverse cervical	C2–C3	Skin over anterior aspect of neck.
Supraclavicular	C3–C4	Skin over superior portion of chest and shoulder.
DEEP (LARGELY MOTOR) BRANCHES		
Ansa cervicalis (AN-sa ser-vi-KAL-is)		This nerve divides into superior and inferior roots.
Superior root	C1	Infrahyoid and geniohyoid muscles of neck.
Inferior root	C2–C3	Infrahyoid muscles of neck.
Phrenic (FREN-ik)	C3–C5	Diaphragm.
Segmental branches	C1–C5	Prevertebral (deep) muscles of neck, levator scapulae, and middle scalene muscles.

Figure 18.8 Cervical plexus in anterior view. (See Tortora, *A Photographic Atlas of the Human Body*, 2e, Figure 8.7.)

The cervical plexus supplies the skin and muscles of the head, neck, superior portion of the shoulders and chest, and diaphragm.

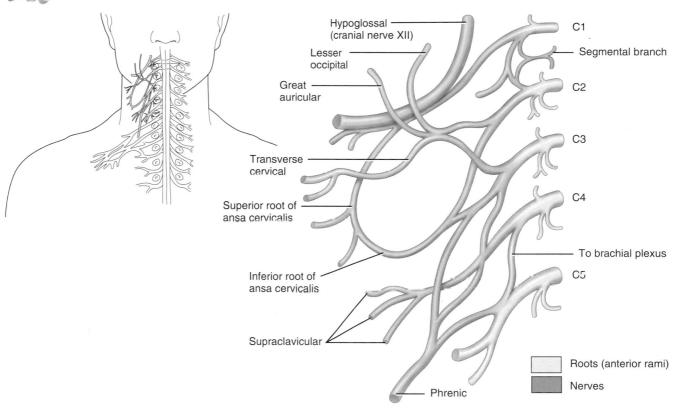

Origin of cervical plexus

Why does complete severing of the spinal cord at level C2 cause respiratory arrest?

EXHIBIT 18.2   BRACHIAL PLEXUS (Figures 18.9 and 18.10)

OBJECTIVE

● Describe the origin, distribution, and effects of damage to the brachial plexus.

The anterior (ventral) rami of spinal nerves C5–C8 and T1 form the **brachial plexus** (BRĀ-kē-al), which extends inferiorly and laterally on either side of the last four cervical and first thoracic vertebrae (Figure 18.9a). It passes above the first rib posterior to the clavicle and then enters the axilla.

The brachial plexus provides the entire nerve supply of the shoulders and upper limbs (Figure 18.9b). Five important nerves arise from the brachial plexus: (1) The **axillary nerve** supplies the deltoid and teres minor muscles. (2) The **musculocutaneous nerve** supplies the flexors of the arm. (3) The **radial nerve** supplies the muscles on the posterior aspect of the arm and forearm. (4) The **median nerve** supplies most of the muscles of the anterior forearm and some of the muscles of the hand. (5) The **ulnar nerve** supplies

the anteromedial muscles of the forearm and most of the muscles of the hand.

INJURIES TO NERVES EMERGING FROM THE BRACHIAL PLEXUS

Injury to the superior roots of the brachial plexus (C5–C6) may result from forceful pulling away of the head from the shoulder, as might occur from a heavy fall on the shoulder or excessive stretching of an infant's head during childbirth. The presentation of this injury is characterized by an upper limb in which the shoulder is adducted, the arm is medially rotated, the elbow is extended, the forearm is pronated, and the wrist is flexed (Figure 18.10a). This condition is called **Erb-Duchene palsy** or **waiter's tip position.** There is loss of sensation along the lateral side of the arm.

NERVE	ORIGIN	DISTRIBUTION
DORSAL SCAPULAR (SKAP-ū-lar)	C5	Levator scapulae, rhomboid major, and rhomboid minor muscles.
LONG THORACIC (thor-RAS-ik)	C5–C7	Serratus anterior muscle.
SUBCLAVIAN (sub-KLĀ-vē-an)	C5–C6	Subclavius muscle.
SUPRASCAPULAR	C5–C6	Supraspinatus and infraspinatus muscles.
MUSCULOCUTANEOUS (mus′-kū-lo-kū-TĀN-ē-us)	C5–C7	Coracobrachialis, biceps brachii, and brachialis muscles.
LATERAL PECTORAL (PEK-to-ral)	C5–C7	Pectoralis major muscle.
UPPER SUBSCAPULAR	C5–C6	Subscapularis muscle.
THORACODORSAL (tho-RĀ-kō-dor-sal)	C6–C8	Latissimus dorsi muscle.
LOWER SUBSCAPULAR	C5–C6	Subscapularis and teres major muscles.
AXILLARY (AK-si-lar-ē)	C5–C6	Deltoid and teres minor muscles; skin over deltoid and superior posterior aspect of arm.
MEDIAN	C5–T1	Flexors of forearm, except flexor carpi ulnaris and some muscles of the hand (lateral palm); skin of lateral two-thirds of palm of hand and fingers.
RADIAL	C5–C8, T1	Triceps brachii and other extensor muscles of arm and extensor muscles of forearm; skin of posterior arm and forearm, lateral two-thirds of dorsum of hand, and fingers over proximal and middle phalanges.
MEDIAL PECTORAL	C8–T1	Pectoralis major and pectoralis minor muscles.
MEDIAL CUTANEOUS NERVE OF ARM (kū′-TĀN-ē-us)	C8–T1	Skin of medial and posterior aspects of distal third of arm.
MEDIAL CUTANEOUS NERVE OF FOREARM	C8–T1	Skin of medial and posterior aspects of forearm.
ULNAR	C8–T1	Flexor carpi ulnaris, flexor digitorum profundus, and most muscles of the hand; skin of medial side of hand, little finger, and medial half of ring finger.

Radial (and axillary) **nerve injury** can be caused by improperly administered intramuscular injections into the deltoid muscle. The radial nerve may also be injured when a cast is applied too tightly around the mid-humerus. Radial nerve injury is indicated by **wrist-drop,** the inability to extend the wrist and fingers (Figure 18.10b). Sensory loss is minimal due to the overlap of sensory innervation by adjacent nerves.

Median nerve injury may result in **median nerve palsy,** which is indicated by numbness, tingling, and pain in the palm and fingers. There is also inability to pronate the forearm and flex the proximal interphalangeal joints of all digits and the distal interphalangeal joints of the second and third digits (Figure 18.10c). In addition, wrist flexion is weak and is accompanied by adduction, and thumb movements are weak.

Ulnar nerve injury may result in **ulnar nerve palsy,** which is indicated by an inability to abduct or adduct the fingers, atrophy of the interosseus muscles of the hand, hyperextension of the metacarpophalangeal joints, and flexion of the interphalangeal joints, a condition called **clawhand** (Figure 18.10d). There is loss of sensation in the little finger.

Long thoracic nerve injury results in paralysis of the serratus anterior muscle. The medial border of the scapula protrudes, giving it the appearance of a wing. When the arm is raised, the vertebral border and inferior angle of the scapula pull away from the thoracic wall and protrude outward, a condition called **winged scapula** (Figure 18.10e). The arm cannot be abducted beyond the horizontal position. ■

■ **CHECKPOINT**

1. Injury of which nerve could cause paralysis of the serratus anterior muscle?

Figure 18.9   **Brachial plexus in anterior view.** (See Tortora, *A Photographic Atlas of the Human Body*, 2e, Figure 8.9.)

The brachial plexus supplies the shoulders and upper limbs.

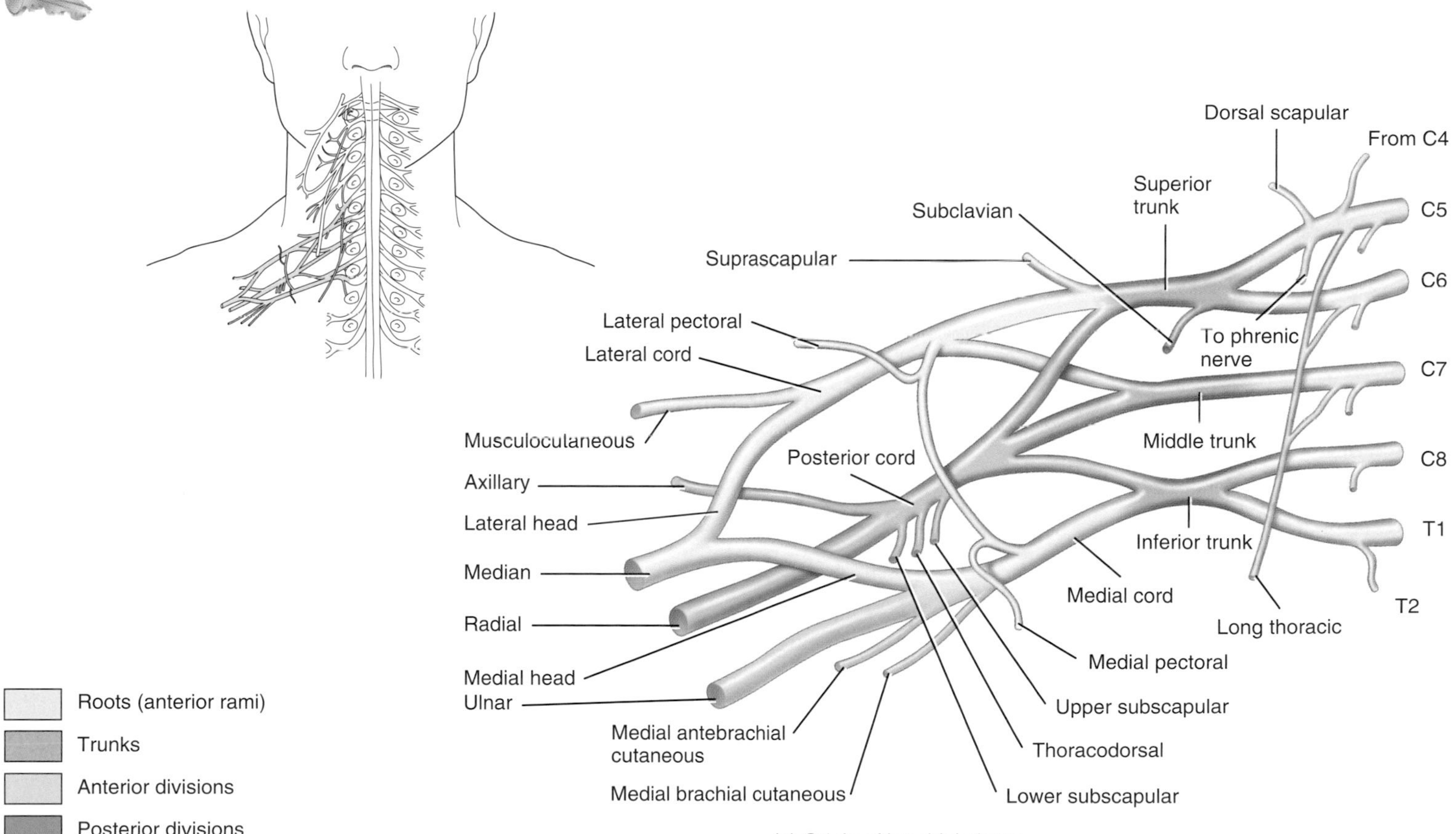

(a) Origin of brachial plexus

Roots (anterior rami)

Trunks

Anterior divisions

Posterior divisions

MNEMONIC for subunits of the brachial plexus:
Risk **T**akers **D**on't **C**autiously **B**ehave.
Roots, **T**runks, **D**ivisions, **C**ords, **B**ranches

continues

569

EXHIBIT 18.2   BRACHIAL PLEXUS (Figures 18.9 and 18.10)

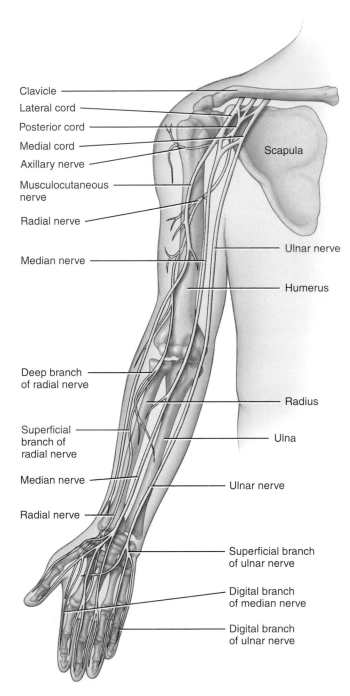

Clavicle

Lateral cord

Posterior cord

Medial cord

Axillary nerve

Musculocutaneous nerve

Radial nerve

Median nerve

Deep branch of radial nerve

Superficial branch of radial nerve

Median nerve

Radial nerve

Scapula

Ulnar nerve

Humerus

Radius

Ulna

Ulnar nerve

Superficial branch of ulnar nerve

Digital branch of median nerve

Digital branch of ulnar nerve

(b) Distribution of nerves from the brachial plexus

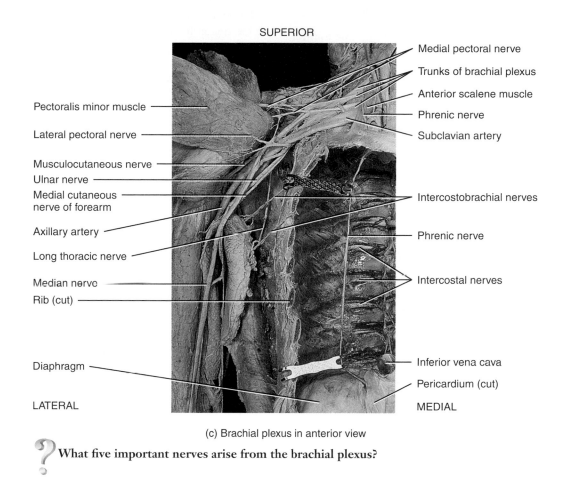

SUPERIOR

Pectoralis minor muscle

Lateral pectoral nerve

Musculocutaneous nerve
Ulnar nerve
Medial cutaneous
nerve of forearm

Axillary artery

Long thoracic nerve

Median nerve
Rib (cut)

Diaphragm

LATERAL

Medial pectoral nerve
Trunks of brachial plexus
Anterior scalene muscle
Phrenic nerve
Subclavian artery

Intercostobrachial nerves

Phrenic nerve

Intercostal nerves

Inferior vena cava
Pericardium (cut)

MEDIAL

(c) Brachial plexus in anterior view

What five important nerves arise from the brachial plexus?

Figure 18.10 Injuries to the brachial plexus.

Injuries to the brachial plexus affect the sensations and movements of the upper limbs.

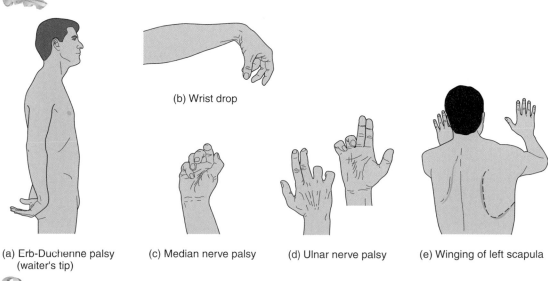

(b) Wrist drop

(a) Erb-Duchenne palsy
(waiter's tip)

(c) Median nerve palsy

(d) Ulnar nerve palsy

(e) Winging of left scapula

Injury to which nerve of the brachial plexus affects sensations on the palm and fingers?

EXHIBIT 18.3 LUMBAR PLEXUS (Figure 18.11)

OBJECTIVE

● Describe the origin and distribution of the lumbar plexus.

Anterior (ventral) rami of spinal nerves L1–L4 form the **lumbar plexus** (LUM-bar) (Figure 18.11). Unlike the brachial plexus, there is no intricate intermingling of fibers in the lumbar plexus. On either side of the first four lumbar vertebrae, the lumbar plexus passes obliquely outward, posterior to the psoas major muscle and anterior to the quadratus lumborum muscle. It then gives rise to its peripheral nerves.

The lumbar plexus supplies the anterolateral abdominal wall, external genitals, and part of the lower limbs.

NERVE	ORIGIN	DISTRIBUTION
ILIOHYPOGASTRIC (Il'-ē-ō-hī-pō-GAS-trik)	L1	Muscles of anterolateral abdominal wall; skin of inferior abdomen and buttocks.
ILIOINGUINAL (Il'-ē-ō-IN-gwi-nal)	L1	Muscles of anterolateral abdominal wall; skin of superior medial aspect of thigh, root of penis and scrotum in male, and labia majora and mons pubis in female.
GENITOFEMORAL (jen'-i-tō-FEM-or-al)	L1–L2	Cremaster muscle; skin over middle anterior surface of thigh, scrotum in male, and labia majora in female.
LATERAL CUTANEOUS NERVE OF THIGH	L2–L3	Skin over lateral, anterior, and posterior aspects of thigh.
FEMORAL	L2–L4	Flexor muscles of thigh and extensor muscles of leg; skin over anterior and medial aspect of thigh and medial side of leg and foot.
OBTURATOR (OB-too-rā-tor)	L2–L4	Adductor muscles of leg; skin over medial aspect of thigh.

Figure 18.11   Lumbar plexus in anterior view.

The lumbar plexus supplies the anterolateral abdominal wall, external genitals, and part of the lower limbs.

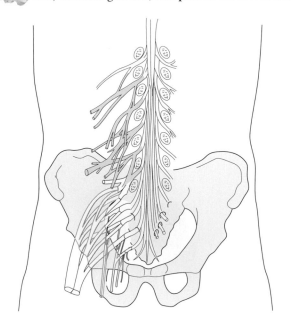

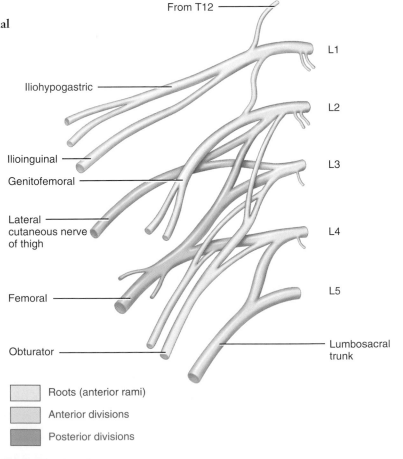

From T12

L1

Iliohypogastric

L2

Ilioinguinal

Genitofemoral

L3

Lateral cutaneous nerve of thigh

L4

L5

Femoral

Obturator

Lumbosacral trunk

☐ Roots (anterior rami)

☐ Anterior divisions

☐ Posterior divisions

(a) Origin of lumbar plexus

LUMBAR PLEXUS INJURIES

The largest nerve arising from the lumbar plexus is the femoral nerve. **Femoral nerve injury,** which can occur in stab or gunshot wounds, is indicated by an inability to extend the leg and by loss of sensation in the skin over the anteromedial aspect of the thigh.

Obturator nerve injury, a common complication of childbirth, results in paralysis of the adductor muscles of the leg and loss of sensation over the medial aspect of the thigh. ■

■ CHECKPOINT

1. Injury of which nerve is indicated by inability to extend the leg?

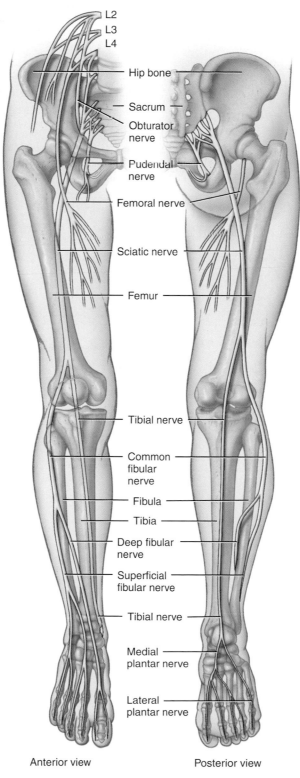

L2
L3
L4

Hip bone

Sacrum

Obturator nerve

Pudendal nerve

Femoral nerve

Sciatic nerve

Femur

Tibial nerve

Common fibular nerve

Fibula

Tibia

Deep fibular nerve

Superficial fibular nerve

Tibial nerve

Medial plantar nerve

Lateral plantar nerve

Anterior view   Posterior view

(b) Distribution of nerves from the lumbar and sacral plexuses

continues

EXHIBIT 18.3   **LUMBAR PLEXUS** (Figure 18.11)

CONTINUED

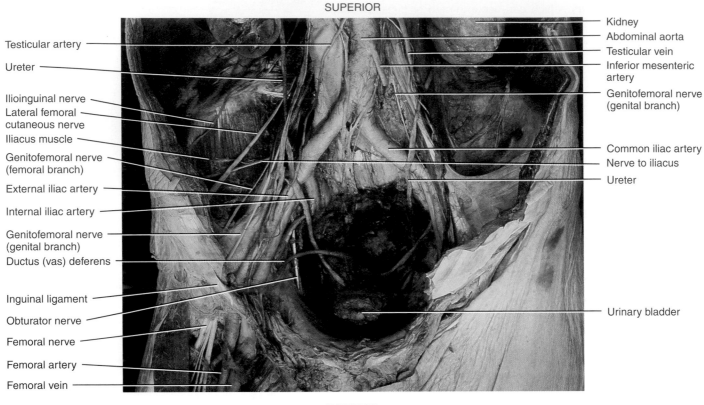

Testicular artery

Ureter

Ilioinguinal nerve

Lateral femoral cutaneous nerve

Iliacus muscle

Genitofemoral nerve (femoral branch)

External iliac artery

Internal iliac artery

Genitofemoral nerve (genital branch)

Ductus (vas) deferens

Inguinal ligament

Obturator nerve

Femoral nerve

Femoral artery

Femoral vein

SUPERIOR

Kidney

Abdominal aorta

Testicular vein

Inferior mesenteric artery

Genitofemoral nerve (genital branch)

Common iliac artery

Nerve to iliacus

Ureter

Urinary bladder

INFERIOR

(c) Anterior view

 What are the signs of femoral nerve injury?

EXHIBIT 18.4   SACRAL AND COCCYGEAL PLEXUSES (Figure 18.12)

• Describe the origin and distribution of the sacral plexus.

The anterior (ventral) rami of spinal nerves L4–L5 and S1–S4 form the **sacral plexus** (SĀ-kral) (Figure 18.12). This plexus is situated largely anterior to the sacrum. The sacral plexus supplies the buttocks, perineum, and lower limbs. The largest nerve in the body — the sciatic nerve — arises from the sacral plexus.

The anterior (ventral) rami of spinal nerves S4–S5 and the coccygeal nerves form a small **coccygeal plexus,** which supplies a small area of skin in the coccygeal region.

SCIATIC NERVE INJURY

The most common form of back pain is caused by compression or irritation of the sciatic nerve, the longest nerve in the human body. Injury to the sciatic nerve and its branches results in **sciatica,** pain that may extend from the buttock down the posterior and lateral aspect of the leg and the lateral aspect of the foot. The sciatic nerve may be injured because of a herniated (slipped) disc, dislocated hip, osteoarthritis of the lumbosacral spine, pressure from the uterus during pregnancy, inflammation, irritation, or an improperly administered gluteal intramuscular injection.

In the majority of sciatic nerve injuries, the common fibular portion is the most affected, frequently from fractures of the fibula or by pressure from casts or splints. Damage to the common fibular nerve causes the foot to be plantar flexed, a condition called **footdrop,** and inverted, a condition called **equinovarus.** There is also loss of function along the anterolateral aspects of the leg and dorsum of the foot and toes. Injury to the tibial portion of the sciatic nerve results in dorsiflexion of the foot plus eversion, a condition called **calcaneovalgus.** Loss of sensation on the sole also occurs. Treatments for sciatica are similar to those outlined earlier for a herniated disc — rest, pain medications, exercises, ice or heat, and massage. ■

■ **CHECKPOINT**

1. Injury of which nerve causes footdrop?

NERVE	ORIGIN	DISTRIBUTION
SUPERIOR GLUTEAL (GLOO-tē-al)	L4–L5 and S1	Gluteus minimus and gluteus medius muscles and tensor fasciae latae.
INFERIOR GLUTEAL	L5–S2	Gluteus maximus muscle.
NERVE TO PIRIFORMIS (pir-i-FORM-is)	S1–S2	Piriformis muscle.
NERVE TO QUADRATUS FEMORIS (qwad-RĀ-tus FEM-or-is) **AND INFERIOR GEMELLUS** (jem-EL-us)	L4–L5 and S1	Quadratus femoris and inferior gemellus muscles.
NERVE TO OBTURATOR INTERNUS (OB-too-rā′-tor in-TER-nus) **AND SUPERIOR GEMELLUS**	L5–S2	Obturator internus and superior gemellus muscles.
PERFORATING CUTANEOUS (kū′-TĀ-ne-us)	S2–S3	Skin over inferior medial aspect of buttock.
POSTERIOR CUTANEOUS NERVE OF THIGH	S1–S3	Skin over anal region, inferior lateral aspect of buttock, superior posterior aspect of thigh, superior part of calf, scrotum in male, and labia majora in female.
SCIATIC (sī-AT-ik)	L4–S3	Actually two nerves — tibial and common fibular — bound together by common sheath of connective tissue. It splits into its two divisions, usually at the knee. (See below for distributions.) As sciatic nerve descends through the thigh, it sends branches to hamstring muscles and adductor magnus.
Tibial (TIB-ē-al)	L4–S3	Gastrocnemius, plantaris, soleus, popliteus, tibialis posterior, flexor digitorum longus, and flexor hallucis longus muscles. Branches of tibial nerve in foot are medial plantar nerve and lateral plantar nerve (see Figure 18.11b).
Medial plantar (PLAN-tar)		Abductor hallucis, flexor digitorum brevis, and flexor hallucis brevis muscles; skin over medial two-thirds of plantar surface of foot.
Lateral plantar		Remaining muscles of foot not supplied by medial plantar nerve; skin over lateral third of plantar surface of foot.
Common fibular (FIB-ū-lar)	L4–S2	Divides into a superficial fibular and a deep fibular branch (see Figure 18.11b).
Superficial fibular		Fibularis longus and fibularis brevis muscles; skin over distal third of anterior aspect of leg and dorsum of foot.
Deep fibular		Tibialis anterior, extensor hallucis longus, fibularis tertius, and extensor digitorum longus and brevis muscles; skin on adjacent sides of great and second toes.
PUDENDAL (pū-DEN-dal)	S2–S4	Muscles of perineum; skin of penis and scrotum in male and clitoris, labia majora, labia minora, and vagina in female.

continues

EXHIBIT 18.4   SACRAL AND COCCYGEAL PLEXUSES (Figure 18.12)

CONTINUED

Figure 18.12   Sacral plexus in anterior view. The distribution of the nerves of the sacral plexus is shown in Figure 18.11b. (See Tortora, *A Photographic Atlas of the Human Body*, 2e, Figure 8.11.)

🔑 **The sacral plexus supplies the buttocks, perineum, and lower limbs.**

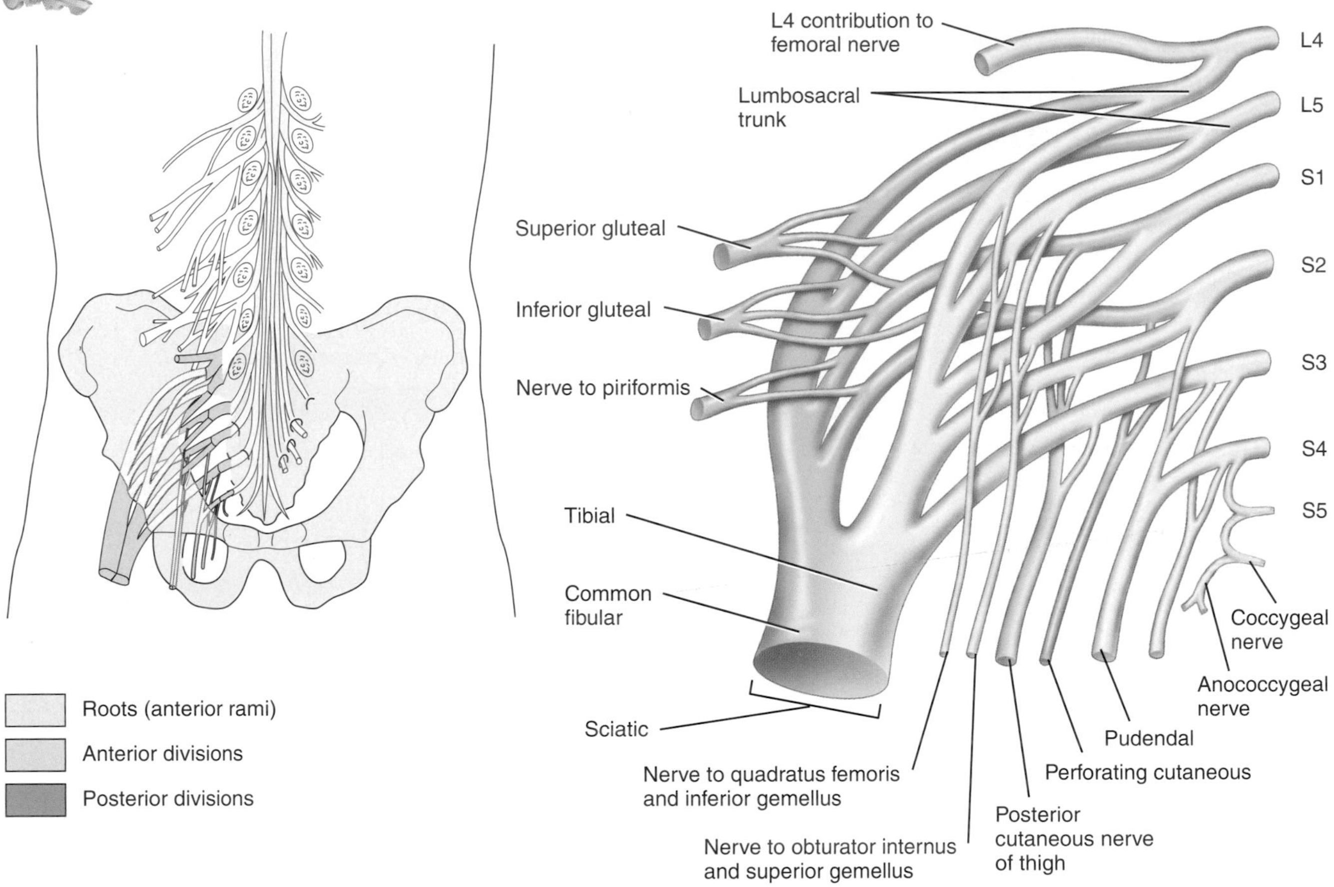

L4 contribution to femoral nerve

Lumbosacral trunk

Superior gluteal

Inferior gluteal

Nerve to piriformis

Tibial

Common fibular

Sciatic

Nerve to quadratus femoris and inferior gemellus

Nerve to obturator internus and superior gemellus

Posterior cutaneous nerve of thigh

Perforating cutaneous

Pudendal

Anococcygeal nerve

Coccygeal nerve

L4
L5
S1
S2
S3
S4
S5

Roots (anterior rami)

Anterior divisions

Posterior divisions

Origin of sacral plexus

❓ **What is the origin of the sacral plexus?**

Intercostal Nerves

The anterior rami of spinal nerves T2–T12 do not enter into the formation of plexuses and are known as **intercostal** or **thoracic nerves.** These nerves directly innervate the structures they supply in the intercostal spaces. After leaving its intervertebral foramen, the anterior ramus of nerve T2 innervates the intercostal muscles of the second intercostal space and supplies the skin of the axilla and posteromedial aspect of the arm. Nerves T3–T6 extend along the costal grooves of the ribs and then to the intercostal muscles and skin of the anterior and lateral chest wall. Nerves T7–T12 supply the intercostal muscles and abdominal muscles, and the overlying skin. The posterior rami of the intercostal nerves supply the deep back muscles and skin of the posterior aspect of the thorax.

Dermatomes

The skin over the entire body is supplied by somatic sensory neurons that carry nerve impulses from the skin into the spinal cord and brain stem. Likewise, somatic motor neurons that carry impulses out of the spinal cord innervate the underlying skeletal muscles. Each spinal nerve contains sensory neurons that serve a specific, predictable segment of the body. The trigeminal (V) nerve serves most of the skin of the face and scalp. The area of the skin that provides sensory input to the CNS via one pair of spinal nerves or the trigeminal (V) nerve is called a **dermatome** (*derma-*=skin; *-tome*=thin segment) (Figure 18.13). The nerve supply in adjacent dermatomes overlaps somewhat. Knowing which spinal cord segments supply each dermatome makes it possible to locate damaged regions of the spinal cord. If the skin in a particular region is stimulated but the sensation is not perceived, the nerves supplying that dermatome are probably damaged. In regions where the overlap is considerable, little loss of sensation may result if only one of the nerves supplying the dermatome is damaged. Information about the innervation patterns of spinal nerves can also be used therapeutically. Cutting posterior roots or infusing local anesthetics can block pain either permanently or transiently. Because dermatomes overlap, deliberate production of a region of complete anesthesia may require that at least three adjacent spinal nerves be cut or blocked by an anesthetic drug.

CHECKPOINT

8. How are spinal nerves named and numbered? Why are all spinal nerves classified as mixed nerves?
9. Describe how a spinal nerve is connected with the spinal cord.
10. Describe the branches and innervations of a typical spinal nerve.
11. Describe the principal plexuses and the regions they supply.

Figure 18.13 Distribution of dermatomes.

A dermatome is an area of skin that provides sensory input via the posterior roots of one pair of spinal nerves or via the trigeminal (V) nerve.

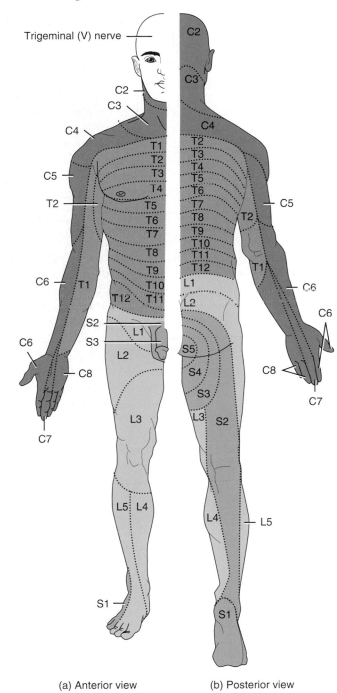

(a) Anterior view   (b) Posterior view

Which is the only spinal nerve that does not have a corresponding dermatome?

APPLICATIONS TO HEALTH

SPINAL CORD DISORDERS

The spinal cord can be damaged in several ways. Outcomes range from little or no long-term neurological deficits to severe deficits and even death.

Traumatic Injuries

Most spinal cord injuries are due to trauma as a result of factors such as automobile accidents, falls, contact sports, diving, and acts of violence. The effects of the injury depend on the extent of direct trauma to the spinal cord or compression of the cord by fractured or displaced vertebrae or blood clots. Although any segment of the spinal cord may be involved, most common sites of injury are in the cervical, lower thoracic, and upper lumbar regions. Depending on the location and extent of spinal cord damage, paralysis may occur. **Monoplegia** (*mono-*=one; *-plegia*=blow or strike) is paralysis of one limb only. **Diplegia** (*di-*=two) is paralysis of both upper limbs or both lower limbs. **Paraplegia** (*para-*=beyond) is paralysis of both lower limbs. **Hemiplegia** (*hemi-*=half) is paralysis of the upper limb, trunk, and lower limb on one side of the body, and **quadriplegia** (*quad-*=four) is paralysis of all four limbs.

    Complete transection (tran-SEK-shun; *trans-*=across; *-section*=a cut) of the spinal cord means that the cord is severed from one side to the other, thus cutting all sensory and motor tracts. It results in a loss of all sensations and voluntary movement *below* the level of the transection. A person will have permanent loss of all sensations in dermatomes below the injury because ascending nerve impulses cannot propagate past the transection to reach the brain. At the same time, all voluntary muscle contractions will be lost below the transection because nerve impulses descending from the brain also cannot pass. The extent of paralysis of skeletal muscles depends on the level of injury. The following list outlines which muscle functions may be *retained* at progressively lower levels of spinal cord transection.

- C1–C3: no function maintained from the neck down; ventilator needed for breathing
- C4–C5: diaphragm, which allows breathing
- C6–C7: some arm and chest muscles, which allows feeding, some dressing, and propelling wheelchair
- T1–T3: intact arm function
- T4–T9: control of trunk above the umbilicus
- T10–L1: most thigh muscles, which allows walking with long leg braces
- L1–L2: most leg muscles, which allows walking with short leg braces

    Hemisection is a partial transection of the cord on either the right or the left side. After hemisection, three main symptoms, known together as *Brown-Séquard syndrome* (sa-KAR), occur below the level of the injury: (1) Damage of the posterior column causes loss of proprioception and fine touch sensations on the ipsilateral (same) side as the injury. (2) Damage of the lateral corticospinal tract causes ipsilateral paralysis. (3) Damage of the spinothalamic tracts causes loss of pain and temperature sensations on the contralateral (opposite) side.

    Following complete transection, and to varying degrees after hemisection, spinal shock occurs. **Spinal shock** is an immediate response to spinal cord injury characterized by temporary **areflexia** (a′-rē-FLEK-sē-a), loss of reflex function. The areflexia occurs in parts of the body served by spinal nerves below the level of the injury. Signs of acute spinal shock include slow heart rate, low blood pressure, flaccid paralysis of skeletal muscles, loss of somatic sensations, and urinary bladder dysfunction. Spinal shock may begin within 1 hour after injury and may last from several minutes to several months, after which reflex activity gradually returns.

    In many cases of traumatic injury of the spinal cord, the patient may have an improved outcome if an anti-inflammatory corticosteroid drug called methylprednisolone is given within 8 hours of the injury. This is because the degree of neurologic deficit is greatest immediately following traumatic injury as a result of edema (collection of fluid within tissues) as the immune system responds to the injury.

Spinal Cord Compression

Although the spinal cord is normally protected by the vertebral column, certain disorders may put pressure on it and disrupt its normal functions. Spinal cord compression may result from fractured vertebrae, herniated intervertebral discs, tumors, osteoporosis, or infections. If the source of the compression is determined before neural tissue is destroyed, spinal cord function usually returns to normal. Depending on the location and degree of compression, symptoms include pain, weakness or paralysis and either decreased or complete loss of sensation below the level of the injury.

Degenerative Diseases

A number of degenerative diseases affect the functions of the spinal cord. One of these is multiple sclerosis, the details of which were presented in Chapter 17 (see page 549). Another progressive degenerative disease is amyotrophic lateral sclerosis (Lou Gehrig's disease), which affects motor neurons of the brain and spinal cord and results in muscle weakness and atrophy. Details are presented in Chapter 21 (page 664).

SHINGLES

Shingles is an acute infection of the peripheral nervous system caused by herpes zoster (HER-pēz ZOS-ter), the virus that also causes chickenpox. After a person recovers from chickenpox, the virus retreats to a posterior root ganglion. If the virus is reactivated, the immune system usually prevents it from spreading.

From time to time, however, the reactivated virus overcomes a weakened immune system, leaves the ganglion, and travels down sensory neurons of the skin. The result is pain, discoloration of the skin, and a characteristic line of skin blisters. The line of blisters marks the distribution (dermatome) of the particular cutaneous sensory nerve belonging to the infected posterior root ganglion.

POLIOMYELITIS

Poliomyelitis, or simply **polio,** is caused by a virus called *poliovirus.* The onset of the disease is marked by fever, severe headache, a stiff neck and back, deep muscle pain and weakness, and loss of certain somatic reflexes. In its most serious form, the virus produces paralysis by destroying cell bodies of motor neurons, specifically those in the anterior horns of the spinal cord and in the nuclei of the cranial nerves. Polio can cause death from respiratory or heart failure if the virus invades neurons in vital centers that control breathing and heart functions in the brain stem. Even though polio vaccines have virtually eradicated polio in the United States, outbreaks of polio continue through-

out the world. Due to international travel, polio could easily be reintroduced into North America if individuals are not vaccinated appropriately.

Several decades after suffering a severe attack of polio and following their recovery from it, some individuals develop a condition called **post-polio syndrome.** This neurological disorder is characterized by progressive muscle weakness, extreme fatigue, loss of function, and pain, especially in muscles and joints. Post-polio syndrome seems to involve a slow degeneration of motor neurons that innervate muscle fibers. Triggering factors appear to be a fall, a minor accident, surgery, or prolonged bed rest. Possible causes include overuse of surviving motor neurons over time, smaller motor neurons because of the initial infection by the virus, reactivation of dormant polio viral particles, immune-mediated responses, hormone deficiencies, and environmental toxins. Treatment consists of muscle-strengthening exercises, administration of pyridostigmine to enhance the action of acetylcholine in stimulating muscle contraction, and administration of nerve growth factors to stimulate both nerve and muscle growth.

KEY MEDICAL TERMS ASSOCIATED WITH THE SPINAL CORD AND SPINAL NERVES

Epidural block Injection of an anesthetic drug into the epidural space, the space between the dura mater and the vertebral column, in order to cause a temporary loss of sensation. Such injections in the lower lumbar region are used to control pain during childbirth.

Meningitis (men-in-JĪ-tis; *-itis*=inflammation) Inflammation of the meninges.

Myelitis (mī-e-LĪ-tis; *myel-*=spinal cord) Inflammation of the spinal cord.

Nerve block Loss of sensation in a region due to injection of a local anesthetic; an example is local dental anesthesia.

Neuralgia (noo-RAL-jē-a; *neur-*=nerve; *-algia*=pain) Attacks of pain along the entire course or a branch of a sensory nerve.

Neuritis (*neur-*=nerve; *-itis*=inflammation) Inflammation of one or several nerves that may result from irritation to the nerve produced by direct blows, bone fractures, contusions,

or penetrating injuries. Additional causes include infections, vitamin deficiency (usually thiamine), and poisons such as carbon monoxide, carbon tetrachloride, heavy metals, and some drugs.

Paresthesia (par-es-THĒ-zē-a; *par-*=departure from normal; *-esthesia*=sensation) An abnormal sensation such as burning, pricking, tickling, or tingling resulting from a disorder of a sensory nerve.

Myelography (mī-e-LOG-ra-fē, *myel-*=marrow, *-graph*=to write) A procedure in which a CT scan or x-ray image of the spinal cord is taken after injection of a radiopaque dye (contrast medium) to diagnose abnormalities such as tumors and herniated intervertebral discs. MRI has largely replaced myelography because the former shows greater detail, is safer, and is simpler.

STUDY OUTLINE

Spinal Cord Anatomy (p. 554)

1. The spinal cord is protected by the vertebral column, the meninges, cerebrospinal fluid, and denticulate ligaments.

2. The three meninges are coverings that run continuously around the spinal cord and brain. They are the dura mater, arachnoid mater, and pia mater.

3. The spinal cord begins as a continuation of the medulla oblongata and ends at about the second lumbar vertebra in an adult.

4. The spinal cord contains cervical and lumbar enlargements that serve as points of origin for nerves to the limbs.

5. The tapered inferior portion of the spinal cord is the conus medullaris, from which arise the filum terminale and cauda equina.

6. Spinal nerves connect to each segment of the spinal cord by two roots. The posterior or dorsal root contains sensory axons, and the anterior or ventral root contains motor neuron axons.

7. The anterior median fissure and the posterior median sulcus partially divide the spinal cord into right and left sides.

8. The gray matter in the spinal cord is divided into horns, and the white matter into columns. In the center of the spinal cord is the central canal, which runs the length of the spinal cord and is filled with cerebrospinal fluid.

9. Parts of the spinal cord observed in transverse section are the gray commissure; central canal; anterior, posterior, and lateral gray horns; and anterior, posterior, and lateral white columns, which contain ascending and descending tracts. Each part has specific functions.

10. The spinal cord conveys sensory and motor information by way of ascending and descending tracts, respectively.

Spinal Cord Functions (p. 560)

1. The white matter tracts in the spinal cord are highways for nerve impulse propagation. Along these tracts, sensory impulses flow toward the brain, and motor impulses flow from the brain toward skeletal muscles and other effector tissues.

2. Sensory impulses travel along two main routes in the white matter of the spinal cord: the spinothalamic tracts and the posterior columns.

3. Motor impulses propagate along two main routes in the white matter of the spinal cord: direct pathways and indirect pathways.

4. A second major function of the spinal cord is to serve as an integrating center for spinal reflexes. This integration occurs in the gray matter.

5. A reflex is a fast, predictable sequence of involuntary actions, such as muscle contractions or glandular secretions, which occurs in response to certain changes in the environment.

6. Reflexes may be spinal or cranial and somatic or autonomic (visceral).

7. The components of a reflex arc are sensory receptor, sensory neuron, integrating center, motor neuron, and effector.

Spinal Nerves (p. 564)

1. The 31 pairs of spinal nerves are named and numbered according to the region and level of the spinal cord from which they emerge.

2. There are 8 pairs of cervical, 12 pairs of thoracic, 5 pairs of lumbar, 5 pairs of sacral, and 1 pair of coccygeal nerves.

3. Spinal nerves typically are connected with the spinal cord by a posterior root and an anterior root. All spinal nerves contain both sensory and motor axons (are mixed nerves).

4. Three connective tissue coverings associated with spinal nerves are the endoneurium, perineurium, and epineurium.

5. Branches of a spinal nerve include the posterior ramus, anterior ramus, meningeal branch, and rami communicantes.

6. The anterior rami of spinal nerves, except for T2–T12, form networks of nerves called plexuses.

7. Emerging from the plexuses are nerves bearing names that typically describe the general regions they supply or the route they follow.

8. Nerves of the cervical plexus supply the skin and muscles of the head, neck, and upper part of the shoulders; they connect with some cranial nerves and innervate the diaphragm.

9. Nerves of the brachial plexus supply the upper limbs and several neck and shoulder muscles.

10. Nerves of the lumbar plexus supply the anterolateral abdominal wall, external genitals, and part of the lower limbs.

11. Nerves of the sacral plexus supply the buttocks, perineum, and part of the lower limbs.

12. Nerves of the coccygeal plexus supply the skin of the coccygeal region.

13. Anterior rami of nerves T2–T12 do not form plexuses and are called intercostal (thoracic) nerves. They are distributed directly to the structures they supply in intercostal spaces.

14. Sensory neurons within spinal nerves and cranial nerve V (trigeminal nerve) serve specific, constant segments of the skin called dermatomes.

15. Knowledge of dermatomes helps a physician determine which segment of the spinal cord or which spinal nerve is damaged.

 SELF-QUIZ QUESTIONS

Choose the one best answer to the following questions.

1. Choose the *true* statement:
   a. Tracts are located only in the spinal cord and not in the brain.
   b. Tracts appear white because they consist of bundles of unmyelinated nerve fibers.
   c. Ascending tracts are all motor tracts.
   d. Ascending tracts are in the white matter; descending tracts are in the gray matter.
   e. None of the above is true.

2. Which of the following lists the meninges and spaces in the correct order from most superficial to deepest?
   a. pia mater, subarachnoid space, arachnoid mater, subdural space, dura mater, epidural space
   b. dura mater, epidural space, arachnoid mater, subdural space, pia mater, subarachnoid space
   c. arachnoid mater, subarachnoid space, epidural space, dura mater, subdural space, pia mater
   d. subdural space, dura mater, epidural space, arachnoid mater, subarachnoid space, pia mater
   e. epidural space, dura mater, subdural space, arachnoid mater, subarachnoid space, pia mater

3. The posterior (dorsal) root ganglion contains:
   a. cell bodies of motor neurons.
   b. cell bodies of sensory neurons.   c. cranial nerve axons.
   d. synapses.   e. All of the above are correct.

4. Which sequence best represents the course of a nerve impulse over a reflex arc?
   a. receptor, integrating center, sensory neuron, motor neuron, effector
   b. effector, sensory neuron, integrating center, motor neuron, receptor

c. receptor, sensory neuron, integrating center, motor neuron, effector

d. receptor, motor neuron, integrating center, sensory neuron, effector

e. effector, integrating center, sensory neuron, motor neuron, receptor

5. The spinothalamic tracts convey sensory information to the brain regarding:

a. pain. b. two-point discrimination. c. temperature.
d. proprioception. e. Both a and c are correct.

6. The extension of the pia mater that anchors the spinal cord to the coccyx is the:

a. cauda equina. b. denticulate ligament. c. conus medullaris.
d. filum terminale. e. dorsal root.

7. Which of the following is *true* regarding the number of spinal nerves?

a. There are 7 pairs of cervical nerves, 12 pairs of thoracic nerves, 4 pairs of lumbar nerves, 1 pair of sacral nerves, and 1 pair of coccygeal nerves.

b. There are 8 pairs of cervical nerves, 12 pairs of thoracic nerves, 5 pairs of lumbar nerves, 5 pairs of sacral nerves, and 5 pairs of coccygeal nerves.

c. There are 7 pairs of cervical nerves, 12 pairs of thoracic nerves, 4 pairs of lumbar nerves, 5 pairs of sacral nerves, and 1 pair of coccygeal nerves.

d. There are 7 pairs of cervical nerves, 14 pairs of thoracic nerves, 4 pairs of lumbar nerves, 5 pairs of sacral nerves, and 1 pair of coccygeal nerves.

e. There are 8 pairs of cervical nerves, 12 pairs of thoracic nerves, 5 pairs of lumbar nerves, 5 pairs of sacral nerves, and 1 pair of coccygeal nerves.

8. What fills the central canal of the spinal cord?

a. blood b. bone c. cerebrospinal fluid
d. axon bundles e. lymph

Complete the following.

9. The spinal cord extends from the _____ of the brain to the level of the _____ of the vertebral column.

10. Lateral extensions of the pia mater that help to protect the spinal cord against displacement are called _____.

11. The effectors in somatic reflexes are _____, while the effectors for autonomic reflexes are _____, _____, and _____.

12. During a spinal tap, the needle is inserted between the _____ and _____ vertebrae into the _____ space.

13. Cell bodies of motor neurons leading to skeletal muscles are located in the _____ gray horns of the spinal cord, while cell bodies of neurons supplying smooth muscle, cardiac muscle, or glands lie in the _____ gray horns of the spinal cord.

14. The area of skin that provides sensory input to the central nervous system via one pair of spinal nerves of the trigeminal nerve is called a(n) _____.

Are the following statements true or false?

15. Cell bodies of motor neurons are located in the posterior (dorsal) root ganglia.

16. The lateral spinothalamic tract is a motor tract.

17. The tapering inferior end of the spinal cord at the level of the first two lumbar vertebrae is called the conus medullaris.

18. The main functions of the spinal cord are to propagate nerve impulses and to integrate sensory input and motor output.

19. The rubrospinal and vestibulospinal tracts are examples of indirect motor pathways.

20. Individual axons are protected by a connective tissue wrapping known as the epineurium.

Matching

21. Match the following:

___ (a) provides the entire nerve supply for the shoulder and upper limbs

___ (b) gives rise to the phrenic nerve that supplies the diaphragm

___ (c) gives rise to the median, radial, and axillary nerves

___ (d) not a plexus at all, but rather segmentally arranged nerves

___ (e) supplies fibers to the scalp, neck, and part of the shoulder and chest

___ (f) supplies fibers to the femoral nerve that innervates the flexor muscles of the thigh and the extensor muscles of the leg

___ (g) gives rise to the largest nerve in the body, the sciatic nerve, that supplies the posterior aspect of the thigh and the leg

(1) brachial plexus
(2) cervical plexus
(3) sacral plexus
(4) lumbar plexus
(5) intercostal nerves

22. Match the following:

___ (a) the joining together of the anterior rami of adjacent nerves

___ (b) spinal nerve branches that serve the deep muscles and skin of the posterior surface of the trunk

___ (c) spinal nerve branches that serve the muscles and structures of the upper and lower limbs and the lateral and ventral trunk

___ (d) area of the spinal cord from which nerves to the upper limbs arise

___ (e) area of the spinal cord from which nerves to the lower limbs arise

___ (f) contains motor neuron axons and conducts impulses from the spinal cord to the periphery

___ (g) contains sensory nerve fibers and conducts impulses from the periphery into the spinal cord

(1) cervical enlargement
(2) lumbar enlargement
(3) posterior root
(4) anterior root
(5) posterior ramus
(6) anterior ramus
(7) plexus

CRITICAL THINKING QUESTIONS

1. A high school senior dove headfirst into a murky pond. Unfortunately, his head hit a submerged log and he is now paralyzed. Can you deduce the location of the injury? What is the likelihood of recovery from his injury?

2. The spinal cord is covered in protective layers and enclosed by the vertebral column. How is it able to send and receive messages from the periphery of the body?

3. Why doesn't the spinal cord "creep" up toward the head every time you bend over? Why doesn't it get all twisted out of position when you exercise?

4. Nischal slipped on the ice and broke his "tailbone." However, he did no damage to his spinal cord. How is it possible that he broke part of his vertebral column without damaging the spinal cord?

5. Not long after Nischal fell on the ice, he began experiencing pain in his lower back and down his leg, as well as some numbness in his foot. When he walks, he has trouble controlling the flexion of his foot. He told his friend that "the doctor said something about compression of the lumbar vertebrae." What has probably happened to Nischal?

ANSWERS TO FIGURE QUESTIONS

18.1 The superior boundary of the spinal dura mater is the foramen magnum of the occipital bone. The inferior boundary is the second sacral vertebra.

18.2 The cervical enlargement of the spinal cord connects with sensory and motor nerves of the upper limbs.

18.3 A horn is an area of gray matter, and a column is a region of white matter in the spinal cord.

18.4 The anterior corticospinal tract is located on the anterior side of the spinal cord, originates in the cortex of the cerebrum, and ends in the spinal cord. It contains descending axons and thus is a motor tract.

18.5 It is a somatic reflex because the effector is a skeletal muscle.

18.6 All spinal nerves are mixed (have sensory and motor components) because the posterior root containing sensory axons and the anterior root containing motor axons unite to form the spinal nerve.

18.7 The anterior rami serve the upper and lower limbs.

18.8 Severing the spinal cord at level C2 causes respiratory arrest because it prevents descending nerve impulses from reaching the phrenic nerve, which stimulates contraction of the diaphragm, the main muscle needed for breathing.

18.9 The axillary, musculocutaneous, radial, median, and ulnar nerves are five important nerves that arise from the brachial plexus.

18.10 Injury to the median nerve affects sensations on the palm and fingers.

18.11 Signs of femoral nerve injury include inability to extend the leg and loss of sensation in the skin over the anterolateral aspect of the thigh.

18.12 The origin of the sacral plexus is the anterior rami of spinal nerves L4–L5 and S1–S4.

18.13 The only spinal nerve without a corresponding dermatome is C1.

THE BRAIN AND THE CRANIAL NERVES

19

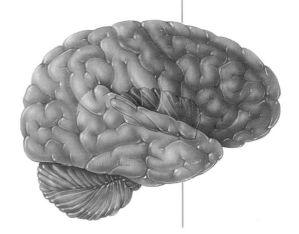

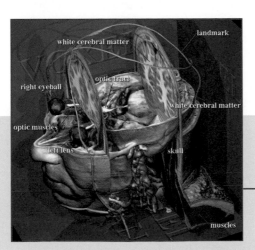

INTRODUCTION Solving an equation, feeling hungry, laughing—the neural processes needed for each of these activities occur in different regions of the **brain,** that portion of the central nervous system contained within your cranium. About 100 billion neurons and 10–50 trillion neuroglia make up the brain, which has a mass of about 1300 g (almost 3 lb) in adults. On average, each neuron forms 1000 synapses with other neurons. The total number of synapses, about a thousand trillion (10^{15}), is larger than the number of stars in the galaxy.

The brain is the center for registering sensations, correlating them with one another and with stored information, making decisions, and taking actions. It also is the center for the intellect, emotions, behavior, and memory. As you will see, different regions of the brain are specialized for different functions, yet they must work together to accomplish certain tasks. This chapter explores how the brain is protected and nourished, what functions occur in the major regions of the brain, and how it is connected to the spinal cord and the 12 pairs of cranial nerves.

What is your first reaction when looking at this image? Can you guess how such an image was generated? Is this a painting, a photograph, or a virtual cadaver?

www.wiley.com/college/apcentral

OVERVIEW OF BRAIN ORGANIZATION AND BLOOD SUPPLY

OBJECTIVES

● Identify the major parts of the brain.
● Describe how the brain is protected.
● Describe the blood supply of the brain.

Knowledge of the embryological development of the brain is necessary to understand the terminology used for the principal parts of the adult brain. The development of the brain is introduced briefly here, and discussed in more detail at the end of the chapter.

Recall from Chapter 4 that the brain and spinal cord develop from ectoderm arranged in a tubular structure called the neural tube (see also Figure 19.26). The anterior part of the neural tube expands, and constrictions appear that create three regions called primary brain vesicles: prosencephalon, mesencephalon, and rhombencephalon (see Figure 19.27). The mesencephalon gives rise to the midbrain and aqueduct of the midbrain (cerebral aqueduct). Both the prosencephalon and rhombencephalon subdivide further, forming secondary brain vesicles. The prosencephalon gives rise to the telencephalon and diencephalon, and the rhombencephalon develops into the metencephalon and myelencephalon. The telencephalon develops into the cerebrum and lateral ventricles. The diencephalon forms the thalamus, hypothalamus, epithalamus, subthalamus, and the third ventricle. The metencephalon becomes the pons, cerebellum, and upper part of the fourth ventricle. Finally, the myelencephalon forms the medulla oblongata and lower part of the fourth ventricle.

These relationships are summarized in Table 19.1.

Major Parts of the Brain

The adult brain consists of four major parts: brain stem, cerebellum, diencephalon, and cerebrum (Figure 19.1). The **brain stem** is continuous with the spinal cord and consists of the medulla oblongata, pons, and midbrain. Posterior to the brain stem is the **cerebellum** (ser'-e-BEL-um=little brain). Superior to the brain stem is the **diencephalon** (dī'-en-SEF-a-lon; *di-*=through; *-encephalon*=brain), which as noted previously consists of the thalamus and hypothalamus and includes the epithalamus and the subthalamus. Supported on the diencephalon and brain stem is the **cerebrum** (se-RĒ-brum=brain), the largest part of the brain.

Protective Coverings of the Brain

The cranium (see Figure 7.4 on page 170) and the cranial meninges surround and protect the brain. The **cranial meninges** (me-NIN-jēz) are continuous with the spinal meninges, have the same basic structure, and bear the same names: the outer **dura mater,** the middle **arachnoid mater,** and the inner **pia mater** (Figure 19.2 on page 586). Note that the cranial dura mater has two layers; the spinal dura mater has only one. Blood vessels that enter brain tissue pass along the surface of the brain, and as they penetrate inward, they are sheathed by a loose-fitting sleeve of pia mater. Three extensions of the dura mater separate parts of the brain. (1) The **falx cerebri** (FALKS CER-e-brē; *falx*= sickle-shaped) separates the two hemispheres (sides) of the cerebrum. (2) The **falx cerebelli** (cer-e-BEL-ī) separates the two hemispheres of the cerebellum. (3) The **tentorium cerebelli** (ten-TŌ-rē-um=tent) separates the cerebrum from the cerebellum.

TABLE 19.1   DEVELOPMENT OF THE BRAIN

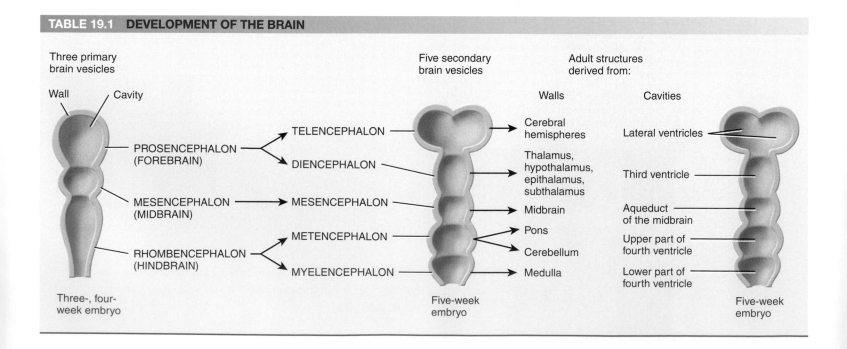

Figure 19.1 **The brain.** The pituitary gland is discussed with the endocrine system in Chapter 23. (See Tortora, *A Photographic Atlas of the Human Body*, 2e, Figures 8.12, 8.13, and 8.15.)

The four principal parts of the brain are the brain stem, cerebellum, diencephalon, and cerebrum.

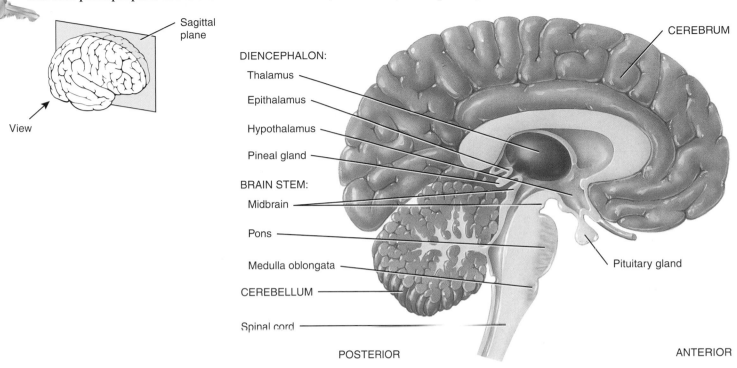

(a) Sagittal section, medial view

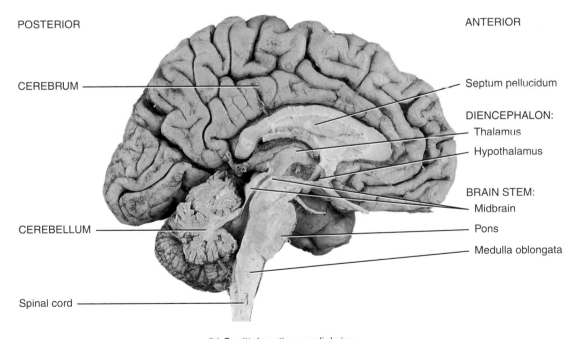

(b) Sagittal section, medial view

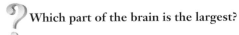 **Which part of the brain is the largest?**

Figure 19.2 **The protective coverings of the brain.**

Cranial bones and the cranial meninges protect the brain.

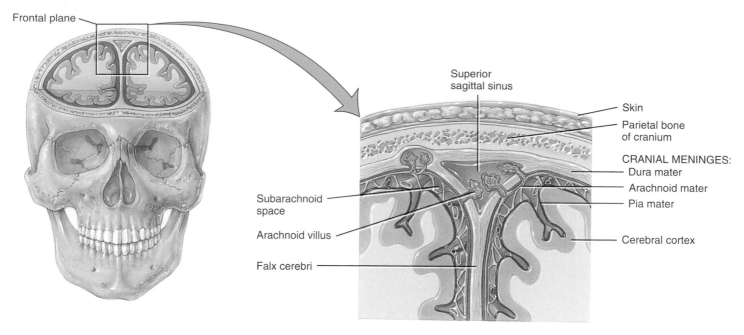

Frontal plane

Superior sagittal sinus

Skin

Parietal bone of cranium

CRANIAL MENINGES:
Dura mater
Arachnoid mater
Pia mater

Subarachnoid space

Arachnoid villus

Cerebral cortex

Falx cerebri

Frontal section through skull showing the cranial meninges

What are the three layers of the cranial meninges, from superficial to deep?

SUBDURAL HEMORRHAGE

A **subdural hemorrhage** results from tearing of the superior cerebral veins where they enter the superior sagittal sinus (see Figure 15.11a on page 486). The cause is typically a head blow on the front or back of the skull. As a result of blood vessel damage, bleeding occurs into the subdural space (space between the dura mater and arachnoid mater) and a hematoma (blood clot) forms. Following a head injury, bleeding may occur within minutes (*acute subdural hemorrhage*) or take several weeks or months (*chronic subdural hemorrhage*). Symptoms of the acute type are drowsiness, confusion, and coma. The chronic type is characterized by headaches, gradually developing confusion and drowsiness, and visual problems. Both types may additionally include nausea and vomiting, muscular weakness or paralysis, and seizures. Subdural hemorrhage is a potentially life-threatening condition that requires immediate medical attention. ■

Brain Blood Flow and the Blood–Brain Barrier

Blood flows to the brain mainly via the internal carotid and vertebral arteries (see Figure 15.7b on page 469); the veins that return blood from the head to the heart are shown in Figure 15.11a (page 486).

In an adult, the brain represents only 2% of total body weight, but consumes about 20% of the oxygen and glucose used even at rest. Neurons synthesize ATP almost exclusively from glucose via reactions that use oxygen. When activity of neurons and neuroglia increases in a region of the brain, blood flow to that area also increases. Even a brief slowing of brain blood flow may cause unconsciousness. Typically, an interruption in blood flow for 1 or 2 minutes impairs neuronal function, and total deprivation of oxygen for about 4 minutes causes permanent injury. Because virtually no glucose is stored in the brain, the supply of glucose also must be continuous. If blood entering the brain has a low level of glucose, mental confusion, dizziness, convulsions, and loss of consciousness may occur.

The existence of a **blood–brain barrier (BBB)** protects brain cells from harmful substances and pathogens by preventing passage of many substances from blood into brain tissue. The blood–brain barrier consists mainly of tight junctions (see Figure 3.1a on page 59) that seal together the endothelial cells of brain capillaries and partly of a thick basement membrane around the capillaries. The processes of many astrocytes, which as you learned in Chapter 17 are one type of neuroglia, press up against the capillaries and secrete chemicals that maintain the permeability characteristics of the tight junctions. A few water-soluble substances, such as glucose, cross the BBB by active transport. Other substances, such as creatinine, urea, and most ions, cross the BBB very slowly. Still other substances—proteins and most antibiotic drugs—do not pass at all from the blood into brain tissue. However, lipid-soluble substances, such as oxygen, carbon dioxide, alcohol, and most anesthetic agents, access brain tissue freely. Trauma, certain toxins, and inflammation can cause a breakdown of the blood–brain barrier.

BREACHING THE BLOOD–BRAIN BARRIER

We have seen how the BBB prevents the passage of potentially harmful substances into brain tissue. But another consequence of the BBB's efficient protection is that it also prevents the passage of certain drugs that could be therapeutic for brain cancer or other CNS disorders. Researchers are exploring ways to move drugs past the BBB. In one method, the drug is injected in a concentrated sugar solution. The high osmotic pressure of the sugar solution causes the endothelial cells of the capillaries to shrink, which opens gaps between their tight junctions and makes the BBB more leaky. As a result, the drug can enter the brain tissue. ■

CHECKPOINT

1. Compare the sizes and locations of the cerebrum and cerebellum.
2. Describe the locations of the cranial meninges.
3. Explain the blood supply to the brain and the importance of the blood–brain barrier.

CEREBROSPINAL FLUID

OBJECTIVE

● Explain the formation and circulation of cerebrospinal fluid.

Cerebrospinal fluid (CSF) is a clear, colorless liquid that protects the brain and spinal cord against chemical and physical injuries. It also carries oxygen, glucose, and other needed chemicals from the blood to neurons and neuroglia. CSF continuously circulates through cavities in the brain and spinal cord and around the brain and spinal cord in the subarachnoid space (between the arachnoid mater and pia mater).

Formation of CSF in the Ventricles

Figure 19.3 shows the four CSF-filled cavities within the brain, which are called **ventricles** (VEN-tri-kuls=little cavities). A **lateral ventricle** is located in each hemisphere of the cerebrum. Anteriorly, the lateral ventricles are separated by a thin membrane, the **septum pellucidum** (SEP-tum pe-LOO-si-dum; *pellucid*=transparent). The **third ventricle** is a narrow cavity along the midline superior to the hypothalamus and between the right and left halves of the thalamus. The **fourth ventricle** lies between the brain stem and the cerebellum.

The total volume of CSF is 80 to 150 mL (3 to 5 oz) in an adult. CSF contains glucose, proteins, lactic acid, urea, cations (Na^+, K^+, Ca^{2+}, Mg^{2+}), and anions (Cl^- and $HCO3^-$); it also contains some white blood cells. The CSF contributes to homeostasis in three main ways:

1. ***Mechanical protection.*** CSF serves as a shock-absorbing medium that protects the delicate tissues of the brain and spinal cord from jolts that would otherwise cause them to hit the bony walls of the cranial and vertebral cavities. The fluid also buoys the brain so that it "floats" in the cranial cavity.

Figure 19.3   Locations of ventricles within a "transparent" brain. One interventricular foramen on each side connects a lateral ventricle to the third ventricle, and the aqueduct of the midbrain connects the third ventricle to the fourth ventricle.

Ventricles are cavities within the brain that are filled with cerebrospinal fluid.

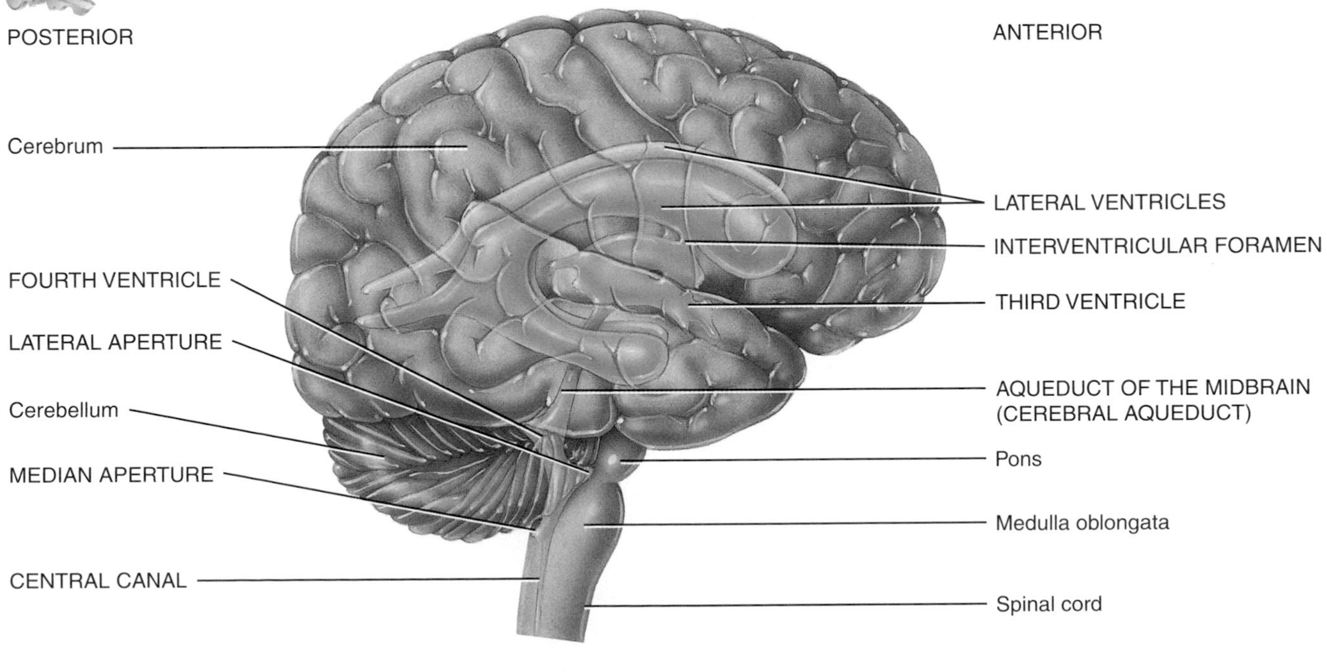

POSTERIOR

ANTERIOR

Cerebrum

LATERAL VENTRICLES

INTERVENTRICULAR FORAMEN

FOURTH VENTRICLE

THIRD VENTRICLE

LATERAL APERTURE

Cerebellum

AQUEDUCT OF THE MIDBRAIN (CEREBRAL AQUEDUCT)

MEDIAN APERTURE

Pons

Medulla oblongata

CENTRAL CANAL

Spinal cord

Right lateral view of brain

Which brain region is anterior to the fourth ventricle? Which is posterior to it?

Figure 19.4 Pathways of circulating cerebrospinal fluid. (See Tortora, *A Photographic Atlas of the Human Body*, 2e, Figure 8.13.)

CSF is formed by ependymal cells that cover the choroid plexuses of the ventricles.

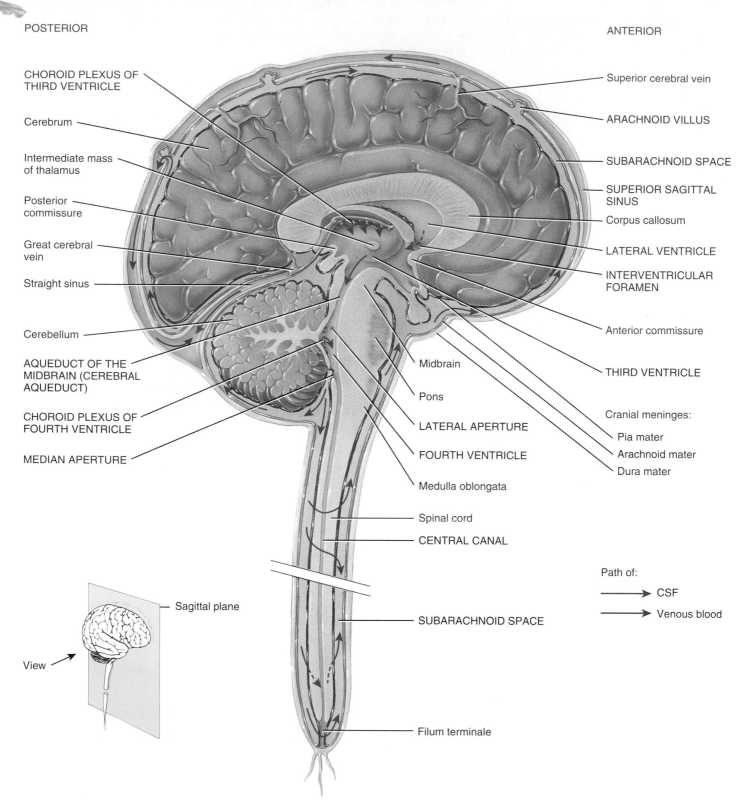

POSTERIOR

ANTERIOR

CHOROID PLEXUS OF THIRD VENTRICLE

Cerebrum

Intermediate mass of thalamus

Posterior commissure

Great cerebral vein

Straight sinus

Cerebellum

AQUEDUCT OF THE MIDBRAIN (CEREBRAL AQUEDUCT)

CHOROID PLEXUS OF FOURTH VENTRICLE

MEDIAN APERTURE

Superior cerebral vein

ARACHNOID VILLUS

SUBARACHNOID SPACE

SUPERIOR SAGITTAL SINUS

Corpus callosum

LATERAL VENTRICLE

INTERVENTRICULAR FORAMEN

Anterior commissure

THIRD VENTRICLE

Cranial meninges:

Pia mater

Arachnoid mater

Dura mater

Midbrain

Pons

LATERAL APERTURE

FOURTH VENTRICLE

Medulla oblongata

Spinal cord

CENTRAL CANAL

Sagittal plane

View

SUBARACHNOID SPACE

Path of:

→ CSF

→ Venous blood

Filum terminale

(a) Sagittal section of brain and spinal cord

2. *Chemical protection.* CSF provides an optimal chemical environment for accurate neuronal signaling. Even slight changes in the ionic composition of CSF within the brain can seriously disrupt production of action potentials.

3. *Circulation.* CSF is a medium for exchange of nutrients and waste products between the blood and nervous tissue.

The sites of CSF production are the **choroid plexuses** (KŌ-royd=membrane like), which are networks of capillaries (microscopic blood vessels) in the walls of the ventricles. The capillaries are covered by ependymal cells that form cerebrospinal fluid from blood plasma by filtration and secretion. Because the ependymal cells are joined by tight junctions, materials entering CSF from choroid capillaries cannot leak between these cells; instead, they must pass through the ependymal cells. This **blood–cerebrospinal fluid barrier** permits certain substances to enter the CSF but excludes others, protecting the brain and spinal cord from potentially harmful blood-borne substances. Whereas the blood–brain barrier is formed mainly by tight junctions of brain capillary endothelial cells, the blood cerebrospinal fluid barrier is formed by tight junctions of ependymal cells.

Circulation of CSF

The CSF formed in the choroid plexuses of each lateral ventricle flows into the third ventricle through two narrow, oval openings, the **interventricular foramina** (Figure 19.4a on page 588). More CSF is added by the choroid plexus in the roof of the third ventricle. The fluid then flows through the **aqueduct of the midbrain (cerebral aqueduct),** which passes through the midbrain, into the fourth ventricle. The choroid plexus of the fourth ventricle contributes more fluid. CSF enters the subarachnoid space through three openings in the roof of the fourth ventricle: a **median aperture** and the paired **lateral apertures,** one on each side. CSF then circulates in the central canal of the spinal cord and in the subarachnoid space around the surface of the brain and spinal cord.

CSF is gradually reabsorbed into the blood through **arachnoid villi,** fingerlike extensions of the arachnoid that project into the dural venous sinuses, especially the **superior sagittal sinus** (see Figure 19.2). (A sinus is like a vein but has thinner walls.) Normally, CSF is reabsorbed as rapidly as it is formed by the choroid plexuses, at a rate of about 20 mL/hr (480 mL/day). Because the rates of formation and reabsorption are the same, the pressure of CSF normally is constant.

HYDROCEPHALUS

Abnormalities in the brain—tumors, inflammation, or developmental malformations—can interfere with the drainage of CSF from the ventricles into the subarachnoid space. When excess

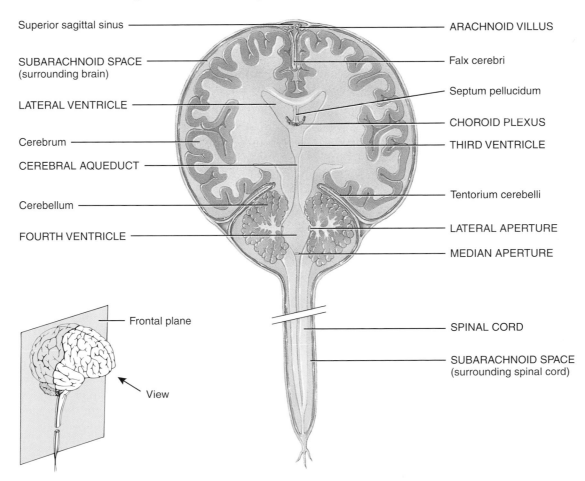

(b) Frontal section of brain and spinal cord

 **Where is CSF reabsorbed?**

CSF accumulates in the ventricles, the CSF pressure rises. Elevated CSF pressure causes a condition called **hydrocephalus** (hī′-drō-SEF-a-lus; *hydro-*=water; *cephal-*=head). In a baby whose fontanels have not yet closed, the head bulges due to the increased pressure. If the condition persists, the fluid buildup compresses and damages the delicate nervous tissue. Hydrocephalus is relieved by draining the excess CSF. A neurosurgeon may implant a drain line, called a shunt, into the lateral ventricle to divert CSF into the superior vena cava or abdominal cavity. In adults, hydrocephalus may occur after head injury, meningitis, or subarachnoid hemorrhage. This condition can quickly become life-threatening and requires immediate intervention since the skull bones have already fused. ■

CHECKPOINT

4. What structures are the sites of CSF production, and where are they located?
5. What is the difference between the blood–brain barrier and the blood–cerebrospinal fluid barrier?

THE BRAIN STEM

OBJECTIVE

● Describe the structures and functions of the brain stem.

The brain stem is the part of the brain between the spinal cord and the diencephalon; it consists of the (1) medulla oblongata, (2) pons, and (3) midbrain. Extending through the brain stem is the reticular formation, a netlike region of interspersed gray and white matter.

Medulla Oblongata

The **medulla oblongata** (me-DOOL-la ob′-long-GA-ta), or more simply the **medulla,** is a continuation of the superior part of the spinal cord; it forms the inferior part of the brain stem (Figure 19.5; also see Figure 19.1). The medulla begins at the foramen magnum and extends to the inferior border of the pons, a distance of about 3 cm (1.2 in.).

Within the medulla's white matter are all sensory (ascending) and motor (descending) tracts extending between the spinal cord and other parts of the brain. Some of the white matter forms bulges on the anterior aspect of the medulla. These protrusions are the **pyramids** (Figure 19.6 on page 592; see also Figure 19.5), formed by the largest motor tracts that pass from the cerebrum to the spinal cord. Just superior to the junction of the medulla with the spinal cord, 90% of the axons in the left pyramid cross to the right side, and 90% of the axons in the right pyramid cross to the left side. This crossing is called the **decussation of pyramids** (dē′-ka-SĀ-shun; *decuss-*=crossing) and explains why one side of the brain controls movements on the opposite side of the body.

The medulla also contains several nuclei, masses of gray matter where neurons form synapses with one another. Several of these nuclei control vital body functions. The **cardiovascular center** regulates the rate and force of the heartbeat and the diameter of blood vessels. The **medullary rhythmicity area** of

the respiratory center adjusts the basic rhythm of breathing. Other nuclei in the medulla control reflexes for vomiting, coughing, and sneezing.

Just lateral to each pyramid is an oval-shaped swelling called an **olive** (see Figures 19.5 and 19.6). Within the olive is the **inferior olivary nucleus.** Neurons here relay impulses from proprioceptors (monitoring joint and muscle positions) to the cerebellum.

Nuclei associated with sensations of touch, conscious proprioception, pressure, and vibration are located in the posterior part of the medulla. These nuclei are the right and left **gracile nucleus** (GRAS-il=slender) and **cuneate nucleus** (KŪ-nē-āt=wedge). Many ascending sensory axons form synapses in these nuclei, and postsynaptic neurons then relay the sensory information to the thalamus on the opposite side of the brain (see Figure 21.4a on page 659). The axons ascend to the thalamus in a band of white matter called the **medial lemniscus** (lem-NIS-kus=ribbon), which extends through the medulla, pons, and midbrain (see Figure 19.7b).

Finally, the medulla contains nuclei associated with the following five pairs of cranial nerves (see Figure 19.5).

1. **Vestibulocochlear (VIII) nerves.** Several nuclei in the medulla receive sensory input from and provide motor output to the cochlea of the internal ear via the cochlear branches of the vestibulocochlear nerves. These nerves convey impulses related to hearing.

2. **Glossopharyngeal (IX) nerves.** Nuclei in the medulla relay sensory and motor impulses related to taste, swallowing, and salivation via the glossopharyngeal nerves.

3. **Vagus (X) nerves.** Nuclei in the medulla receive sensory impulses from and provide motor impulses to the pharynx and larynx and many thoracic and abdominal viscera via the vagus nerves.

4. **Accessory (XI) nerves (cranial portion).** Nuclei in the medulla are the origin for nerve impulses that control swallowing via the cranial portion of the accessory nerves.

5. **Hypoglossal (XII) nerves.** Nuclei in the medulla are the origin for nerve impulses that control tongue movements during speech and swallowing via the hypoglossal nerves.

INJURY OF THE MEDULLA

Given the many vital activities controlled by the medulla, it is not surprising that a hard blow to the back of the head or upper neck can be fatal. Damage to the medullary rhythmicity area, which controls the basic rhythm of respiration, is particularly serious and can rapidly lead to death. Symptoms of nonfatal injury to the medulla may include cranial nerve malfunctions on the same side of the body as the injury, paralysis and loss of sensation on the opposite side of the body, and irregularities in breathing or heart rhythm. ■

Pons

The **pons** (=bridge) lies directly superior to the medulla and anterior to the cerebellum and is about 2.5 cm (1 in.) long (see Figures 19.1 and 19.5). Like the medulla, the pons consists of both nuclei and tracts. As its name implies, the pons is a bridge that connects parts of the brain with one another. These connections are provided by bundles of axons. Some axons of the pons connect the right and left sides of the cerebellum. Others are part of ascending sensory tracts and descending motor tracts.

Several **pontine nuclei** (PON-tīn) are the sites at which signals for voluntary movements that originate in the cerebral cortex are relayed into the cerebellum. Other nuclei located in the pons are the **pneumotaxic area** (noo-mō-TAK-sik) and the **apneustic area** (ap-NOO-stik), shown in Figure 24.14 on page

753. Together with the medullary rhythmicity area, the pneumotaxic and apneustic areas help control breathing.

The pons also contains nuclei associated with the following four pairs of cranial nerves (see Figure 19.5):

1. **Trigeminal (V) nerves.** Nuclei in the pons receive sensory impulses for somatic sensations from the head and face and provide motor impulses that govern chewing via the trigeminal nerves.

2. **Abducens (VI) nerves.** Nuclei in the pons provide motor

Figure 19.5 Medulla oblongata in relation to the rest of the brain stem. (See Tortora, *A Photographic Atlas of the Human Body*, 2e, Figure 8.19.)

The brain stem consists of the medulla oblongata, pons, and midbrain.

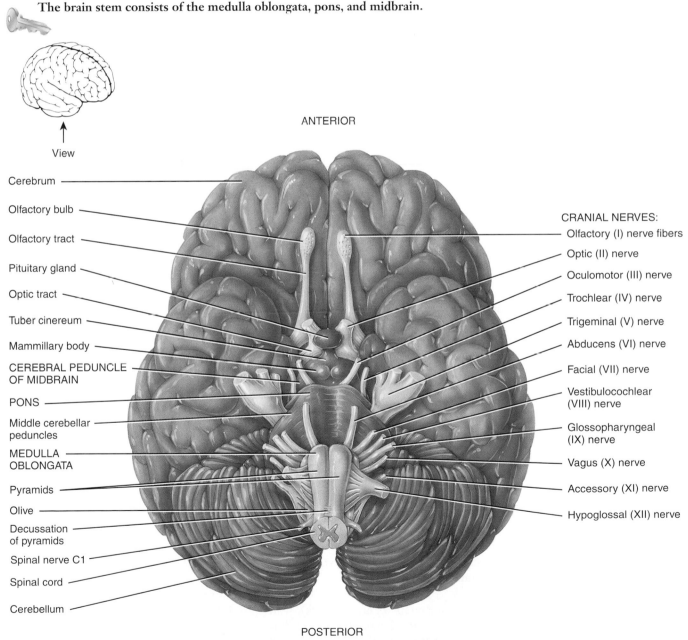

What part of the brain stem contains the pyramids? The cerebral peduncles? Literally means "bridge"?

Figure 19.6 Internal anatomy of the medulla oblongata.

The pyramids of the medulla contain the largest motor tracts that run from the cerebrum to the spinal cord.

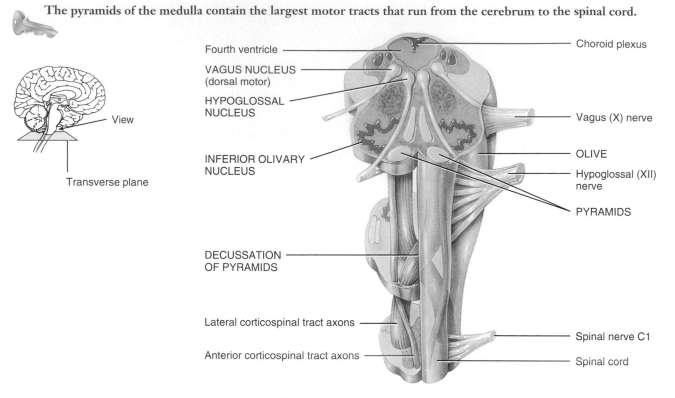

Fourth ventricle

VAGUS NUCLEUS
(dorsal motor)

HYPOGLOSSAL
NUCLEUS

INFERIOR OLIVARY
NUCLEUS

DECUSSATION
OF PYRAMIDS

Lateral corticospinal tract axons

Anterior corticospinal tract axons

Choroid plexus

Vagus (X) nerve

OLIVE

Hypoglossal (XII)
nerve

PYRAMIDS

Spinal nerve C1

Spinal cord

View

Transverse plane

Transverse section and anterior surface of medulla oblongata

What does decussation mean? What is the functional consequence of decussation of the pyramids?

impulses that control eyeball movement via the abducens nerves.

3. **Facial (VII) nerves.** Nuclei in the pons receive sensory impulses for taste and provide motor impulses to regulate secretion of saliva and tears and contraction of muscles of facial expression via the facial nerves.

4. **Vestibulocochlear (VIII) nerves.** Nuclei in the pons receive sensory impulses from and provide motor impulses to the vestibular apparatus via the vestibular branches of the vestibulocochlear nerves. These nerves convey impulses related to balance and equilibrium.

Midbrain

The **midbrain** or **mesencephalon** extends from the pons to the diencephalon (see Figures 19.1 and 19.5) and is about 2.5 cm (1 in.) long. The cerebral aqueduct passes through the midbrain, connecting the third ventricle above with the fourth ventricle below. Like the medulla and the pons, the midbrain contains both tracts and nuclei.

The anterior part of the midbrain contains a pair of tracts called **cerebral peduncles** (pe-DUNK-kuls or PĒ-dung-kuls=little feet; see Figures 19.5 and 19.7b). They contain axons of corticospinal, corticopontine, and corticobulbar motor neurons, which conduct nerve impulses from the cerebrum to the spinal cord, medulla, and pons. The cerebral peduncles also

contain axons of sensory neurons that extend from the medulla to the thalamus.

The posterior part of the midbrain, called the **tectum** (TEK-tum;=roof), contains four rounded elevations (Figure 19.7a). The two superior elevations are known as the **superior colliculi** (kō-LIK-ū-lī=little hills, singular is *colliculus*). These nuclei serve as reflex centers for certain visual activities. Through neural circuits from the retina of the eye to the superior colliculi to the extrinsic eye muscles, visual stimuli elicit eye movements for tracking moving images (such as a moving car) and scanning stationary images (as you are doing to read this sentence). Other superior colliculi reflexes are the accommodation reflex (adjusts shape of lens for close versus far vision) and reflexes that govern movements of the eyes, head, and neck in response to visual stimuli. The two inferior elevations, the **inferior colliculi,** are part of the auditory pathway, relaying impulses from the receptors for hearing in the ear to the thalamus. These two nuclei also are reflex centers for the startle reflex, sudden movements of the head and body that occur when you are surprised by a loud noise.

The midbrain contains several nuclei, including the left and right **substantia nigra** (sub-STAN-shē-a=substance; NĪ-gra= black), which are large, darkly pigmented nuclei. Neurons that release dopamine extend from the substantia nigra to the basal ganglia and help control subconscious muscle activities (Figure 19.7b). Loss of these neurons is associated with Parkinson's dis-

Figure 19.7 Midbrain.

The midbrain connects the pons to the diencephalon.

View

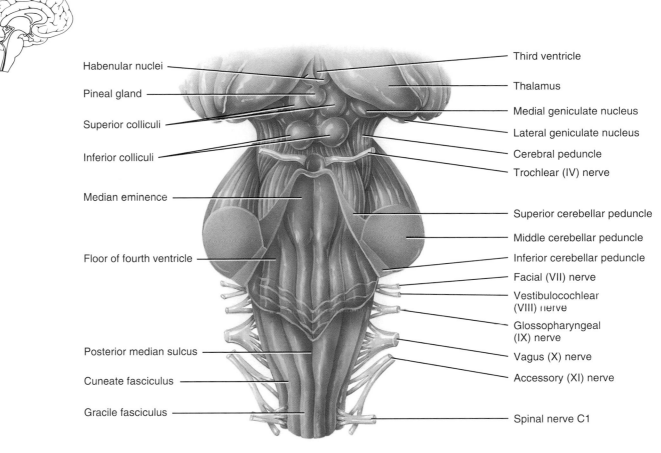

Habenular nuclei

Pineal gland

Superior colliculi

Inferior colliculi

Median eminence

Floor of fourth ventricle

Posterior median sulcus

Cuneate fasciculus

Gracile fasciculus

Third ventricle

Thalamus

Medial geniculate nucleus

Lateral geniculate nucleus

Cerebral peduncle

Trochlear (IV) nerve

Superior cerebellar peduncle

Middle cerebellar peduncle

Inferior cerebellar peduncle

Facial (VII) nerve

Vestibulocochlear (VIII) nerve

Glossopharyngeal (IX) nerve

Vagus (X) nerve

Accessory (XI) nerve

Spinal nerve C1

(a) Posterior view of midbrain in relation to brain stem

POSTERIOR

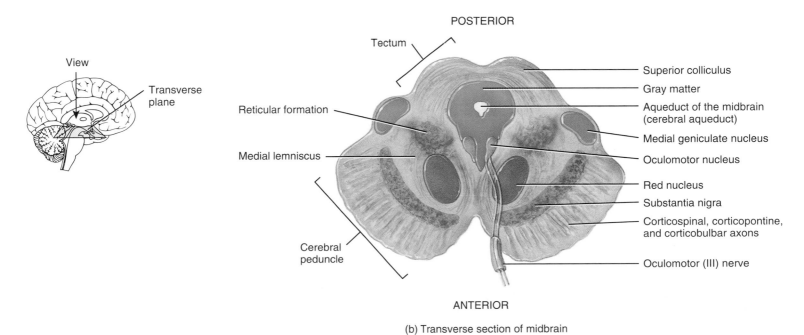

View

Transverse plane

Tectum

Reticular formation

Medial lemniscus

Cerebral peduncle

Superior colliculus

Gray matter

Aqueduct of the midbrain (cerebral aqueduct)

Medial geniculate nucleus

Oculomotor nucleus

Red nucleus

Substantia nigra

Corticospinal, corticopontine, and corticobulbar axons

Oculomotor (III) nerve

ANTERIOR

(b) Transverse section of midbrain

continues

Figure 19.7 (continued)

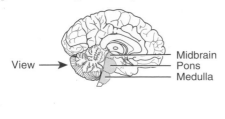

View → Midbrain
Pons
Medulla

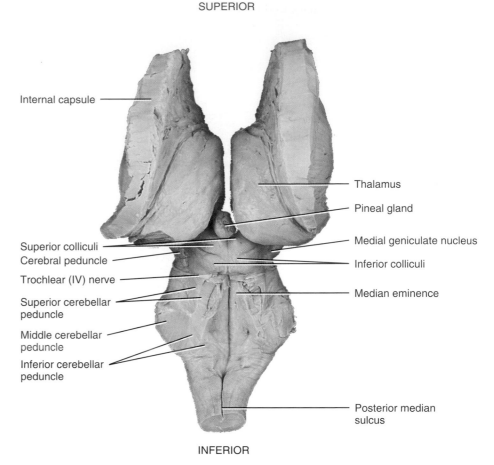

SUPERIOR

Internal capsule

Thalamus

Pineal gland

Superior colliculi

Cerebral peduncle

Trochlear (IV) nerve

Superior cerebellar
peduncle

Middle cerebellar
peduncle

Inferior cerebellar
peduncle

Medial geniculate nucleus

Inferior colliculi

Median eminence

Posterior median
sulcus

INFERIOR

(c) Posterior view of midbrain in relation to brain stem

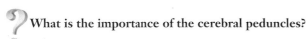

What is the importance of the cerebral peduncles?

ease (see page 667). Also present are the left and right **red nuclei** (Figure 19.7b), which look reddish due to their rich blood supply and an iron-containing pigment in their neuronal cell bodies. Axons from the cerebellum and cerebral cortex form synapses in the red nuclei, which function with the cerebellum to coordinate muscular movements.

Finally, nuclei in the midbrain are associated with two pairs of cranial nerves (see Figure 19.5):

1. **Oculomotor (III) nerves.** Nuclei in the midbrain provide motor impulses that control movements of the eyeball, constriction of the pupil, and changes in shape of the lens via the oculomotor nerves.

2. **Trochlear (IV) nerves.** Nuclei in the midbrain provide motor impulses that control movements of the eyeball via the trochlear nerves.

Reticular Formation

In addition to the well-defined nuclei already described, much of the brain stem consists of small clusters of neuronal cell bodies (gray matter) interspersed among small bundles of myelinated axons (white matter). The broad region where white matter and gray matter exhibit a netlike arrangement is known

as the **reticular formation** (*ret-*=net; Figure 19.7b). It extends from the upper part of the spinal cord, throughout the brain stem, and into the lower part of the diencephalon. Neurons within the reticular formation have both ascending (sensory) and descending (motor) functions. Part of the reticular formation, called the **reticular activating system (RAS),** consists of sensory axons that project to the cerebral cortex. The RAS helps maintain consciousness and is active during awakening from sleep. For example, we awaken to the sound of an alarm clock, to a flash of lightning, or to a painful pinch because of RAS activity that arouses the cerebral cortex. The reticular formation's main descending function is to help regulate muscle tone, which is the slight degree of contraction in normal resting muscles.

The functions of the brain stem are summarized in Table 19.2 on page 606.

CHECKPOINT

6. Where are the medulla, pons, and midbrain located relative to one another?

7. Define decussation of pyramids. Why is it important?

8. What body functions are governed by nuclei in the brain stem?

9. What are two important functions of the reticular formation?

THE CEREBELLUM

OBJECTIVE

- Describe the structure and functions of the cerebellum.

The **cerebellum,** the second-largest part of the brain, occupies the inferior and posterior aspects of the cranial cavity. It accounts for about a tenth of the brain mass yet contains nearly half of the neurons in the brain. The cerebellum is posterior to the medulla and pons and inferior to the posterior portion of the cerebrum (see Figure 19.1). A deep groove known as the **transverse fissure,** and the **tentorium cerebelli,** which supports the posterior part of the cerebrum, separate the cerebellum from the cerebrum (see Figure 19.4b). The tentorium cerebelli is a tent-like fold of the dura mater that is attached to the temporal and occipital bones.

In superior or inferior views, the shape of the cerebellum resembles a butterfly. The central constricted area is the **vermis** (=worm), and the lateral "wings" or lobes are the **cerebellar hemispheres** (Figure 19.8a, b). Each hemisphere consists of lobes separated by deep and distinct fissures. The **anterior lobe** and **posterior lobe** govern subconscious aspects of skeletal muscle movements. The **flocculonodular lobe** (flok-ū-lō-NOD-ū-lar; *flocculo-*=wool-like tuft) on the inferior surface contributes to equilibrium and balance.

The superficial layer of the cerebellum, called the **cerebellar cortex,** consists of gray matter in a series of slender, parallel ridges called **folia** (=leaves). Deep to the gray matter are tracts of white matter called **arbor vitae** (=tree of life) that resemble branches of a tree. Even deeper, within the white matter, are the **cerebellar nuclei,** regions of gray matter that give rise to axons carrying impulses from the cerebellum to other brain centers and to the spinal cord.

Three paired **cerebellar peduncles** (pc-DUNG-kls) attach the cerebellum to the brain stem (see also Figure 19.7c). These bundles of white matter consist of axons that conduct impulses between the cerebellum and other parts of the brain. The **inferior cerebellar peduncles** carry sensory information from the vestibular apparatus of the inner ear and from proprioceptors throughout the body into the cerebellum; their axons extend from the inferior olivary nucleus of the medulla and from the spinocerebellar tracts of the spinal cord into the cerebellum. The **middle cerebellar peduncles** are the largest peduncles; their axons carry commands for voluntary movements (those that originate in motor areas of the cerebral cortex) from the pontine nuclei into the cerebellum. The **superior cerebellar peduncles** contain axons that extend from the cerebellum to the red nuclei of the midbrain and to several nuclei of the thalamus.

A main function of the cerebellum is to evaluate how well movements initiated by motor areas in the cerebrum are actually being carried out. When movements initiated by the cerebral motor areas are not being carried out correctly, the cerebellum detects the discrepancies. It then sends feedback signals to motor areas of the cerebral cortex, via its connections to the red nucleus and thalamus. The feedback signals help correct the errors, smooth the movements, and coordinate complex sequences of skeletal muscle contractions. Aside from this coordination of skilled movements, the cerebellum is the main brain

Figure 19.8 Cerebellum. (See Tortora, *A Photographic Atlas of the Human Body*, 2e, Figure 8.25.)

The cerebellum coordinates skilled movements and regulates posture and balance.

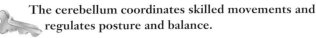
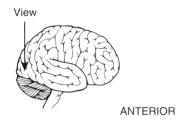
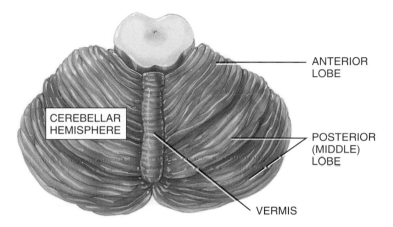

(a) Superior view

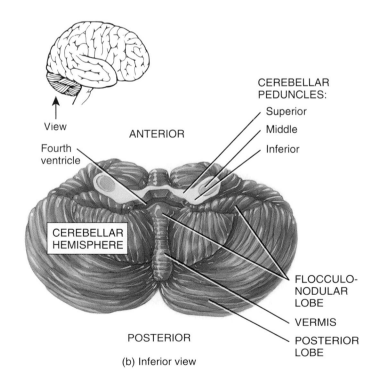

(b) Inferior view

continues

Figure 19.8 (continued)

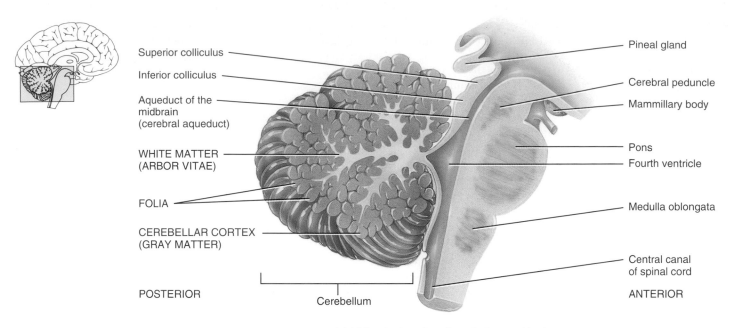

Superior colliculus

Inferior colliculus

Aqueduct of the midbrain (cerebral aqueduct)

WHITE MATTER (ARBOR VITAE)

FOLIA

CEREBELLAR CORTEX (GRAY MATTER)

POSTERIOR

Cerebellum

Pineal gland

Cerebral peduncle

Mammillary body

Pons

Fourth ventricle

Medulla oblongata

Central canal of spinal cord

ANTERIOR

(c) Midsagittal section of cerebellum and brain stem

? **Which fiber tracts carry information into and out of the cerebellum?**

region that regulates posture and balance. These aspects of cerebellar function make possible all skilled muscular activities, from catching a baseball to dancing to speaking.

The presence of reciprocal connections between the cerebellum and association areas of the cerebral cortex suggest that the cerebellum may also have nonmotor functions such as cognition (acquisition of knowledge) and language processing. This view is supported by imaging studies like MRI and PET.

 **ATAXIA**

Damage to the cerebellum through trauma or disease disrupts muscle coordination, a condition called **ataxia** (*a-*=without; *-taxia*=order). Blindfolded people with ataxia cannot touch the tip of their nose with a finger because they cannot coordinate movement with their sense of where a body part is located. Another sign of ataxia is a changed speech pattern due to uncoordinated speech muscles. Cerebellar damage may also result in staggering or abnormal walking movements. People who consume too much alcohol show signs of ataxia because alcohol inhibits activity of the cerebellum. ■

The functions of the cerebellum are summarized in Table 19.2 on page 606.

CHECKPOINT

10. Describe the location and principal parts of the cerebellum.
11. Where do the axons of each of the three pairs of cerebellar peduncles begin and end? What are their functions?

THE DIENCEPHALON

OBJECTIVE

● Describe the components and functions of the diencephalon.

The **diencephalon** extends from the brain stem to the cerebrum and surrounds the third ventricle; it includes the thalamus, hypothalamus, epithalamus, and subthalamus.

Thalamus

The **thalamus** (THAL-a-mus=inner chamber), which measures about 3 cm (1.2 in.) in length and makes up 80% of the diencephalon, consists of paired oval masses of gray matter organized into nuclei with interspersed tracts of white matter (Figure 19.9). A bridge of gray matter called the **intermediate mass (interthalamic adhesion)** joins the right and left halves of the thalamus in about 70% of human brains.

The thalamus is the major relay station for most sensory impulses that reach the primary sensory areas of the cerebral cortex from the spinal cord, brain stem, and midbrain. Although crude perception of painful, thermal, and pressure sensations arises at the level of the thalamus, precise localization of these sensations depends on nerve impulses arriving at the cerebral cortex.

The thalamus contributes to motor functions by transmitting information from the cerebellum and basal ganglia to the primary motor area of the cerebral cortex. It also relays nerve impulses between different areas of the cerebrum. The thalamus also plays a role in the regulation of autonomic activities and the maintenance of consciousness. Axons that connect the thalamus and cerebral cortex pass through the **internal capsule,** a thick band of white matter lateral to the thalamus (see Figure 19.13b).

A vertical Y-shaped sheet of white matter called the **internal medullary lamina** divides the gray matter of the right and left sides of the thalamus (Figure 19.9c). It consists of myelinated axons that enter and leave the various thalamic nuclei.

Based on their positions and functions, there are seven major groups of nuclei on each side of the thalamus (Figure 19.9c, d).

1. The **anterior nucleus** connects to the hypothalamus and limbic system (described on page 604). It functions in emotions, regulation of alertness, and memory.

2. The **medial nuclei** connect to the cerebral cortex, limbic system, and basal ganglia. They function in emotions, learning, memory, awareness, and cognition (thinking and knowing).

3. Nuclei in the **lateral group** connect to the superior colliculi, limbic system, and cortex in all lobes of the cerebrum. The **lateral dorsal nucleus** functions in the expression of emotions. The **lateral posterior nucleus** and **pulvinar nucleus** help integrate sensory information.

4. Five nuclei are part of the **ventral group.** The **ventral anterior nucleus** contributes to motor functions, possibly movement planning. The **ventral lateral nucleus** connects to the cerebellum and motor parts of the cerebral cortex. Its

Figure 19.9 **Thalamus.** Note the position of the thalamus in (a), the lateral view, and (b), the medial view. Various thalamic nuclei in (c) and (d) are correlated by color to the cortical regions to which they project in (a) and (b).

🔑 **The thalamus is the principal relay station for sensory impulses that reach the cerebral cortex from other parts of the brain and the spinal cord.**

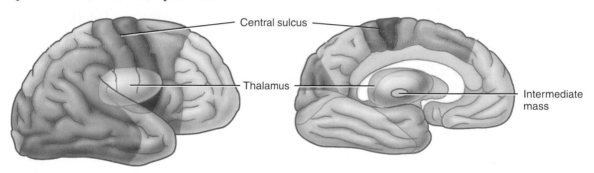

(a) Lateral view of right cerebral hemisphere

(b) Medial view of left cerebral hemisphere

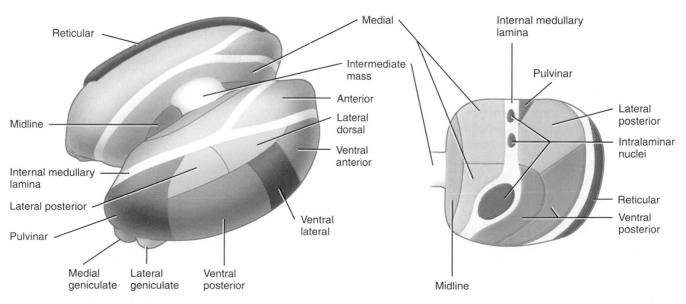

(c) Superolateral view of thalamus showing locations of thalamic nuclei (reticular nucleus is shown on the left side only; all other nuclei are shown on the right side)

(d) Transverse section of right side of thalamus showing locations of thalamic nuclei

continues

Figure 19.9 (continued)

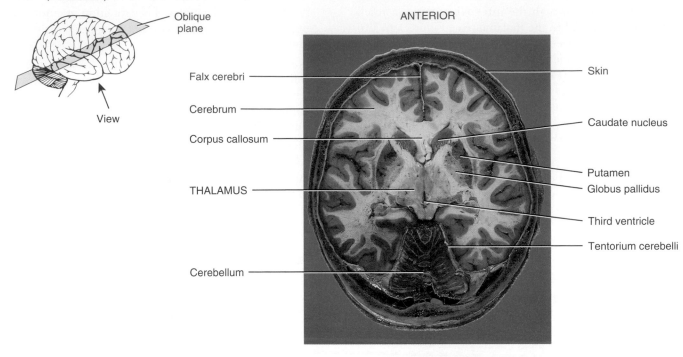

(e) Oblique section of brain

What structure connects the right and left halves of the thalamus?

neurons are active during movements on the opposite side of the body. The **ventral posterior nucleus** relays impulses for somatic sensations such as touch, pressure, proprioception, vibration, heat, cold, and pain from the face and body to the cerebral cortex. The **lateral geniculate nucleus** (je-NIK-ū-lat=bent like a knee) relays visual impulses for sight from the retina to the primary visual area of the cerebral cortex. The **medial geniculate nucleus** relays auditory impulses for hearing from the ear to the primary auditory area of the cerebral cortex.

5. **Intralaminar nuclei** lie within the internal medullary lamina and make connections with the reticular formation, cerebellum, basal ganglia, and wide areas of the cerebral cortex. They function in pain perception, integration of sensory and motor information, and arousal (activation of the cerebral cortex from the brain stem reticular formation).

6. The **midline nucleus** forms a thin band adjacent to the third ventricle and has a presumed function in memory and olfaction.

7. The **reticular nucleus** surrounds the lateral aspect of the thalamus, next to the internal capsule. This nucleus monitors, filters, and integrates activities of other thalamic nuclei.

Hypothalamus

The **hypothalamus** (*hypo-*=under) is a small part of the diencephalon located inferior to the thalamus. It is composed of a dozen or so nuclei in four major regions:

1. The **mammillary region** (*mammill-*=nipple-shaped), adjacent to the midbrain, is the most posterior part of the hypothalamus. It includes the mammillary bodies and posterior hypothalamic nuclei (Figure 19.10). The **mammillary bodies** are two small, rounded projections that serve as relay stations for reflexes related to the sense of smell (see also Figure 19.5).

2. The **tuberal region,** the widest part of the hypothalamus, includes the dorsomedial, ventromedial, and arcuate nuclei, plus the stalklike **infundibulum** (in-fun-DIB-ū-lum=funnel), which connects the pituitary gland to the hypothalamus (Figure 19.10). The **median eminence** is a slightly raised region that encircles the infundibulum.

3. The **supraoptic region** (*supra-*=above; *-optic*=eye) lies superior to the optic chiasm (point of crossing of optic nerves) and contains the paraventricular nucleus, supraoptic nucleus, anterior hypothalamic nucleus, and suprachiasmatic nucleus (Figure 19.10). Axons from the paraventricular and supraoptic nuclei form the hypothalamohypophyseal tract, which extends through the infundibulum to the posterior lobe of the pituitary.

4. The **preoptic region** anterior to the supraoptic region is usually considered part of the hypothalamus because it participates with the hypothalamus in regulating certain autonomic activities. The preoptic region contains the medial and lateral preoptic nuclei (Figure 19.10).

The hypothalamus controls many body activities and is one of the major regulators of homeostasis. Sensory impulses related

Figure 19.10 Hypothalamus. Selected portions of the hypothalamus and a three-dimensional representation of hypothalamic nuclei are shown (after Netter).

🔑 **The hypothalamus controls many body activities and is an important regulator of homeostasis.**

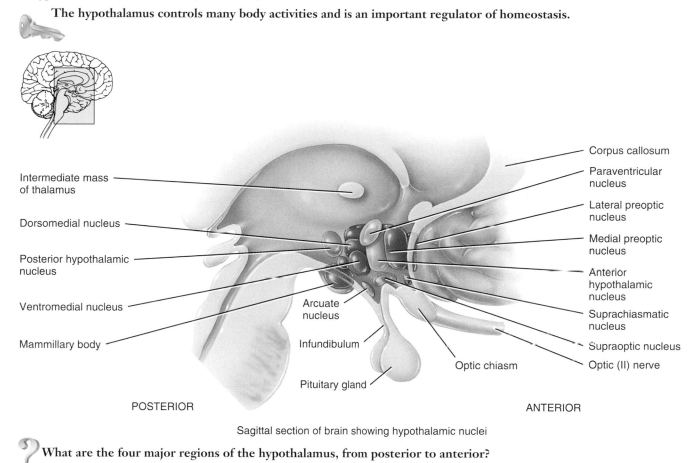

Sagittal section of brain showing hypothalamic nuclei

❓ **What are the four major regions of the hypothalamus, from posterior to anterior?**

to both somatic and visceral senses arrive at the hypothalamus, as do impulses from receptors for vision, taste, and smell. Other receptors within the hypothalamus itself continually monitor osmotic pressure, glucose level, certain hormone concentrations, and the temperature of blood. The hypothalamus has several very important connections with the pituitary gland and produces a variety of hormones, which are described in more detail in Chapter 23. Whereas some functions can be attributed to specific hypothalamic nuclei, others are not so precisely localized. Important functions of the hypothalamus include the following:

1. ***Control of the ANS.*** The hypothalamus controls and integrates activities of the autonomic nervous system, which regulates contraction of smooth and cardiac muscle and the secretions of many glands. Axons extend from the hypothalamus to sympathetic and parasympathetic nuclei in the brain stem and spinal cord. Through the ANS, the hypothalamus is a major regulator of visceral activities, including regulation of heart rate, movement of food through the gastrointestinal tract, and contraction of the urinary bladder.

2. ***Production of hormones.*** The hypothalamus produces several hormones and has two types of important connections with the pituitary gland, an endocrine gland located inferior to the hypothalamus (see Figure 19.1). First, hypothalamic

hormones are released into capillary networks in the median eminence. The bloodstream carries these hormones directly to the anterior lobe of the pituitary, where they stimulate or inhibit secretion of anterior pituitary hormones. Second, axons extend from the paraventricular and supraoptic nuclei through the infundibulum into the posterior lobe of the pituitary. The cell bodies of these neurons make one of two hormones (oxytocin or antidiuretic hormone). Their axons transport the hormones to the posterior pituitary, where they are released.

3. ***Regulation of emotional and behavioral patterns.*** Together with the limbic system (described shortly), the hypothalamus participates in expressions of rage, aggression, pain, and pleasure, and the behavioral patterns related to sexual arousal.

4. ***Regulation of eating and drinking.*** The hypothalamus regulates food intake through the arcuate and paraventricular nuclei. It also contains a thirst center. When certain cells in the hypothalamus are stimulated by rising osmotic pressure of the extracellular fluid, they cause the sensation of thirst. The intake of water by drinking restores the osmotic pressure to normal, removing the stimulation and relieving the thirst.

5. ***Control of body temperature.*** If the temperature of blood

flowing through the hypothalamus is above normal, the hypothalamus directs the autonomic nervous system to stimulate activities that promote heat loss. When blood temperature is below normal, the hypothalamus generates impulses that promote heat production and retention.

6. ***Regulation of circadian rhythms and states of consciousness.*** The suprachiasmatic nucleus establishes patterns of awakening and sleep that occur on a circadian (daily) schedule. This nucleus receives input from the eyes (retina) and sends output to other hypothalamic nuclei, the reticular formation, and the pineal gland.

Epithalamus

The **epithalamus** (*epi-*=above), a small region superior and posterior to the thalamus, consists of the pineal gland and habenular nuclei. The **pineal gland** (PĪN-ē-al=pinecone-like) is about the size of a small pea and protrudes from the posterior midline of the third ventricle (see Figure 19.1). The pineal gland is part of the endocrine system because it secretes the hormone **melatonin.** As more melatonin is liberated during darkness than in light, this hormone is thought to promote sleepiness. Melatonin also appears to contribute to the setting of the body's biological clock. The **habenular nuclei** (ha-BEN-ū-lar), shown in Figure 19.7a, are involved in olfaction, especially emotional responses to odors such as a loved one's cologne or Mom's chocolate chip cookies baking in the oven.

Subthalamus

The **subthalamus,** a small area immediately below the thalamus, includes tracts and the paired **subthalamic nuclei,** which connect to motor areas of the cerebrum. Parts of two pairs of midbrain nuclei—the red nucleus and the substantia nigra—also extend into the subthalamus. The subthalamic nuclei, red nuclei, and substantia nigra work together with the basal ganglia, cerebellum, and cerebrum in the control of body movements.

The functions of the four parts of the diencephalon are summarized in Table 19.2 on page 606.

Circumventricular Organs

Parts of the diencephalon, called **circumventricular** (ser'-kum-ven-TRIK-ū-lar) **organs (CVOs)** because they lie in the walls of the third and fourth ventricles, can monitor chemical changes in the blood because they lack a blood–brain barrier. CVOs include part of the hypothalamus, the pineal gland, the pituitary gland, and a few other nearby structures. Functionally, these regions coordinate homeostatic activities of the endocrine and nervous systems, such as the regulation of blood pressure, fluid balance, hunger, and thirst. CVOs are also thought to be the sites of entry into the brain of HIV, the virus that causes AIDS. Once in the brain, HIV may cause dementia (irreversible deterioration of mental state) and other neurological disorders.

CHECKPOINT

12. Why is the thalamus considered a "relay station" in the brain?
13. Why is the hypothalamus considered part of both the nervous system and the endocrine system?

THE CEREBRUM

OBJECTIVES

● Describe the cortex, convolutions, fissures, and sulci of the cerebrum.
● List and locate the lobes of the cerebrum.
● Describe the nuclei that comprise the basal ganglia.
● List the structures and describe the functions of the limbic system.

The cerebrum is the "seat of intelligence." It provides us with the ability to read, write, and speak; to make calculations and compose music; and to remember the past, plan for the future, and imagine things that have never existed before.

The right and left halves of the cerebrum, called **cerebral hemispheres,** are separated by the falx cerebri. The hemispheres consist of an outer rim of gray matter, an internal region of cerebral white matter, and gray matter nuclei deep within the white matter. The outer rim of gray matter is the **cerebral cortex** (*cortex*=rind or bark) (Figure 19.11a). Although only 2–4 mm (0.08–0.16 in.) thick, the cerebral cortex contains billions of neurons. Deep to the cerebral cortex lies the cerebral white matter.

During embryonic development, when brain size increases rapidly, the gray matter of the cortex enlarges much faster than the deeper white matter. As a result, the cortical region rolls and folds upon itself. The folds are called **gyri** (JĪ-rī=circles) or **convolutions** (Figure 19.11a, b). The deepest grooves between folds are known as **fissures;** the shallower grooves between folds are termed **sulci** (SUL-sī=grooves). (The singular terms are *gyrus* and *sulcus.*) The most prominent fissure, the **longitudinal fissure,** separates the cerebrum into right and left halves called **cerebral hemispheres.** The hemispheres are connected internally by the **corpus callosum** (kal-LŌ-sum; *corpus*=body; *callosum*=hard), a broad band of white matter containing axons that extend between the hemispheres (see Figure 19.12).

Lobes of the Cerebrum

Each cerebral hemisphere can be further subdivided into four lobes. The lobes are named after the bones that cover them: frontal, parietal, temporal, and occipital lobes (see Figure 19.11a, b). The **central sulcus** (SUL-kus) separates the **frontal lobe** from the **parietal lobe.** A major gyrus, the **precentral gyrus**—located immediately anterior to the central sulcus—contains the primary motor area of the cerebral cortex. Another major gyrus, the **postcentral gyrus,** which is located immediately posterior to the central sulcus, contains the primary somatosensory area of the cerebral cortex. The **lateral cerebral sulcus (fissure)** separates the **frontal lobe** from the **temporal lobe.** The **parieto-occipital sulcus** separates the **parietal lobe** from the **occipital lobe.** A fifth part of the cerebrum, the **insula,** cannot be seen at the surface of the brain because it lies within the lateral cerebral sulcus, deep to the parietal, frontal, and temporal lobes (Figure 19.11b).

Figure 19.11   Cerebrum. Because the insula cannot be seen externally, it has been projected to the surface in (b).

The cerebrum is the "seat of intelligence"; it provides us with the ability to read, write, and speak; to make calculations and compose music; to remember the past and plan for the future; and to create.

ANTERIOR

Frontal lobe

Longitudinal fissure

Precentral gyrus

Central sulcus

Postcentral gyrus

Parietal lobe

Occipital lobe

Gyrus

Sulcus

Cerebral cortex

Cerebral white matter

Fissure

Left hemisphere

Right hemisphere

POSTERIOR

Details of a gyrus, sulcus, and fissure

(a) Superior view

Postcentral gyrus

Parietal lobe

Parieto-occipital sulcus

Occipital lobe

Transverse fissure

Cerebellum

Central sulcus

Precentral gyrus

Frontal lobe

Insula (projected to surface)

Lateral cerebral sulcus

Temporal lobe

(b) Right lateral view

POSTERIOR

ANTERIOR

Central sulcus

Postcentral gyrus

Parietal lobe

Precentral gyrus

Frontal lobe

Insula

Occipital lobe

Temporal lobe (cut)

Cerebellum

Medulla oblongata

Spinal cord

(c) Right lateral view with temporal lobe cut away

During development, does the gray matter or white matter enlarge more rapidly?
What are the brain folds, shallow grooves, and deep grooves called?

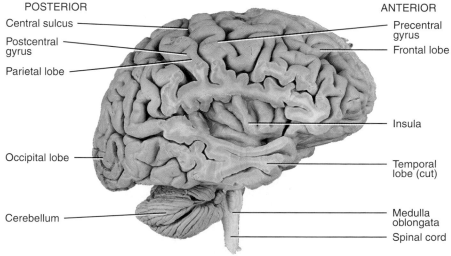

Cerebral White Matter

The **cerebral white matter** consists of both myelinated and unmyelinated axons in three types of tracts (Figure 19.12 and see also Figure 19.4a):

1. **Association tracts** contain axons that conduct nerve impulses between gyri in the same hemisphere.

2. **Commissural tracts** contain axons that conduct nerve impulses from gyri in one cerebral hemisphere to corresponding gyri in the other cerebral hemisphere. Three important groups of commissural tracts are the **corpus callosum** (the largest fiber bundle in the brain, containing about 300 million fibers), **anterior commissure**, and **posterior commissure.**

3. **Projection tracts** contain axons that conduct nerve impulses from the cerebrum to lower parts of the CNS (thalamus, brainstem, or spinal cord) or from lower parts of the CNS to the cerebrum. An example is the **internal capsule,** a thick band of white matter that contains both ascending and descending axons (see Figure 19.13b).

Basal Ganglia

Deep within each cerebral hemisphere are three nuclei (masses of gray matter) that are collectively termed the **basal ganglia** (Figure 19.13). (Recall that "ganglion" usually means a collection of neuronal cells bodies *outside* the CNS. The name here is the one exception to that general rule. An alternative term used in some textbooks—*basal nuclei*—is not used by most neuroscientists. It can be confused with the names of other brain regions, such as the nucleus basalis, which deteriorates in people who suffer from Alzheimer's disease.)

Two of the basal ganglia are side-by-side, just lateral to the thalamus. They are the **globus pallidus** (GLŌ-bus PAL-i-dus; *globus*=ball; *pallidus*=pale), which is closer to the thalamus, and the **putamen** (pū-TĀ-men=shell), which is closer to the cerebral cortex. Together, the globus pallidus and putamen are referred to as the **lentiform nucleus** (LEN-ti-form=shaped like a lens.) The third of the basal ganglia is the **caudate nucleus** (KAW-dāt; *caud-*=tail), which has a large "head" connected to a smaller "tail" by a long comma-shaped "body." Together, the lentiform and caudate nuclei are known as the **corpus striatum** (strī-Ā-tum; *corpus*=body; *striatum*=striated). The term corpus striatum refers to the striated (striped) appearance of the internal capsule as it passes among the basal ganglia. Nearby structures that are functionally linked to the basal ganglia are the *substantia nigra* of the midbrain (see Figure 19.7b) and the *subthalamic nuclei* of the diencephalon (see Figure 19.13b). Axons from the substantia nigra terminate in the caudate nucleus and putamen. The subthalamic nuclei interconnect with the globus pallidus.

The basal ganglia receive input from the cerebral cortex and provide output to motor parts of the cortex via the medial and ventral group nuclei of the thalamus. In addition, the nuclei of the basal ganglia have extensive connections with one another. A major function of the basal ganglia is to help regulate initiation and termination of movements. Activity of neurons in the putamen precedes or anticipates body movements; activity of neurons in the caudate nucleus occurs prior to eye movements. The globus pallidus helps regulate the muscle tone required for specific body movements. The basal ganglia also control subconscious contractions of skeletal muscles. Examples include automatic arm swings while walking and laughing in response to a joke.

Figure 19.12   Organization of fibers into white matter tracts of the left cerebral hemisphere.

🔑   **Association tracts, commissural tracts, and projection tracts form white matter tracts in the cerebral hemispheres.**

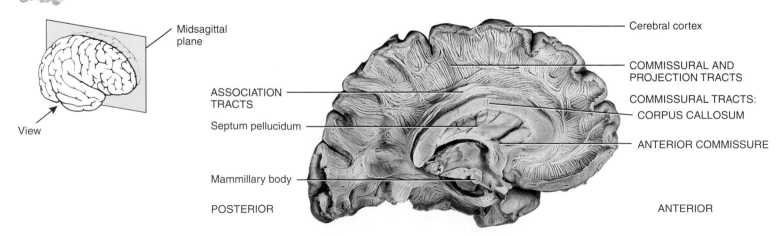

Midsagittal plane

View

Cerebral cortex

COMMISSURAL AND PROJECTION TRACTS

COMMISSURAL TRACTS: CORPUS CALLOSUM

ASSOCIATION TRACTS

Septum pellucidum

ANTERIOR COMMISSURE

Mammillary body

POSTERIOR

ANTERIOR

Medial view of tracts revealed by removing gray matter from a midsagittal section

❓   **Which fibers carry impulses between gyri of the same hemisphere? Between gyri in opposite hemispheres? Between the cerebrum and thalamus, brain stem, and spinal cord?**

Figure 19.13 Basal ganglia. In (a) and (b), the basal ganglia have been projected to the surface and are shown in dark blue. (See Tortora, *A Photographic Atlas of the Human Body*, 2e, Figure 8.22.)

The basal ganglia control automatic movements of skeletal muscles and muscle tone.

Lateral ventricle — Body of caudate nucleus

— Frontal lobe of cerebrum

Thalamus — — Putamen

Tail of caudate nucleus — — Head of caudate nucleus

Occipital lobe of cerebrum —

POSTERIOR ANTERIOR

(a) Lateral view of right side of brain

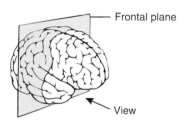

Frontal plane

View

Longitudinal fissure — — Cerebrum

Septum pellucidum — — Corpus callosum

Internal capsule — — Lateral ventricle

Insula — — Caudate nucleus

Putamen — Corpus striatum

Thalamus — — Globus pallidus

Subthalamic nucleus — — Third ventricle

Hypothalamus — — Optic tract

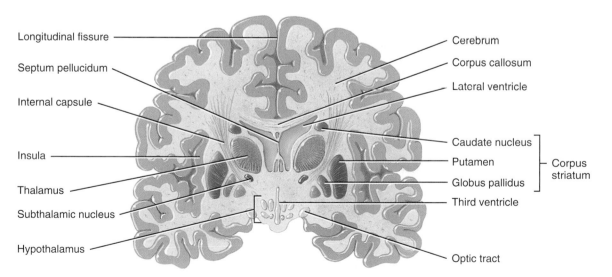

(b) Anterior view of frontal section

continues

Figure 19.13 (continued)

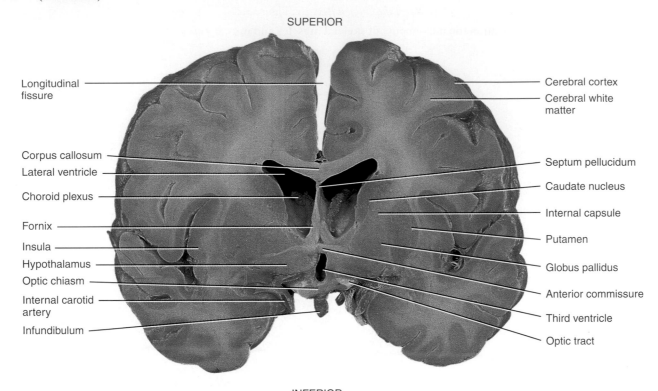

(c) Anterior view of frontal section

? Where are the basal ganglia located relative to the thalamus?

DAMAGE TO THE BASAL GANGLIA

Damage to the basal ganglia results in uncontrollable shaking (tremor), muscular rigidity (stiffness), and involuntary muscle movements. Movement disruptions such as these are a hallmark of disorders like Parkinson's disease (see page 667). In this disorder, neurons that extend from the substantia nigra to the putamen and caudate nucleus degenerate, causing the disruptions. ■

The basal ganglia have other roles in addition to influencing motor function. They help initiate and terminate some cognitive processes, such as attention, memory, and planning, and may act with the limbic system to regulate emotional behaviors. Some psychiatric disorders, such as obsessive-compulsive disorder, schizophrenia, and chronic anxiety, are thought to involve dysfunction of circuits between the basal ganglia and the limbic system.

The Limbic System

Encircling the upper part of the brain stem and the corpus callosum is a ring of structures on the inner border of the cerebrum and floor of the diencephalon that constitutes the **limbic system** (*limbic*=border). The main components of the limbic system are as follows (Figure 19.14):

1. The so-called **limbic lobe** is a rim of cerebral cortex on the medial surface of each hemisphere. It includes the **cingulate gyrus** (*cingul-*=belt), which lies above the corpus callosum, and the **parahippocampal gyrus** (par′-a-hip-ō-KAM-pal), which is in the temporal lobe below. The **hippocampus** (=seahorse) is a portion of the parahippocampal gyrus that extends into the floor of the lateral ventricle.

2. The **dentate gyrus** (*dentate*=toothed) lies between the hippocampus and parahippocampal gyrus.

3. The **amygdala** (*amygda-*=almond-shaped) is composed of several groups of neurons located close to the tail of the caudate nucleus.

4. The **septal nuclei** are located within the septal area formed by the regions under the corpus callosum and the paraterminal gyrus (a cerebral gyrus).

5. The **mammillary bodies** of the hypothalamus are two round masses close to the midline near the cerebral peduncles.

6. Two nuclei of the thalamus, the **anterior nucleus** and the **medial nucleus,** participate in limbic circuits.

7. The **olfactory bulbs** are flattened bodies of the olfactory pathway that rest on the cribriform plate.

8. The **fornix, stria terminalis, stria medullaris, medial forebrain bundle,** and **mammillothalamic tract** (mam-i-lō-tha-LAM-ik) are linked by bundles of interconnecting myelinated axons.

Figure 19.14 Components of the limbic system and surrounding structures.

 The limbic system governs emotional aspects of behavior.

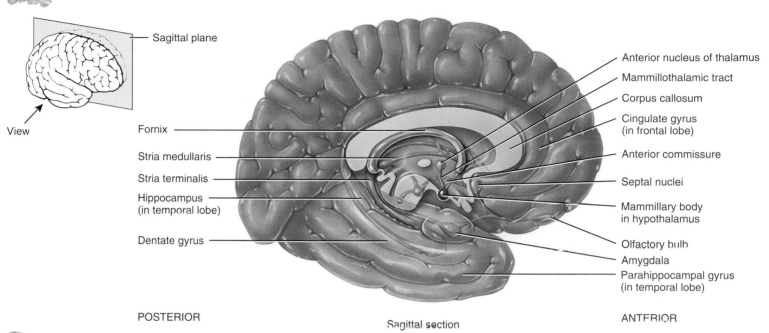

Sagittal plane

View

Fornix

Stria medullaris

Stria terminalis

Hippocampus
(in temporal lobe)

Dentate gyrus

Anterior nucleus of thalamus

Mammillothalamic tract

Corpus callosum

Cingulate gyrus
(in frontal lobe)

Anterior commissure

Septal nuclei

Mammillary body
in hypothalamus

Olfactory bulb

Amygdala

Parahippocampal gyrus
(in temporal lobe)

POSTERIOR

Sagittal section

ANTERIOR

Which part of the limbic system functions with the cerebrum in memory?

The limbic system is sometimes called the "emotional brain" because it plays a primary role in a range of emotions, including pain, pleasure, docility, affection, and anger. It also is involved in olfaction (smell) and memory. Experiments have shown that when different areas of animals' limbic system are stimulated, the animals' reactions indicate that they are experiencing intense pain or extreme pleasure. Stimulation of other limbic system areas in animals produces tameness and signs of affection. Stimulation of a cat's amygdala or certain nuclei of the hypothalamus produces a behavioral pattern called rage—the cat extends its claws, raises its tail, opens its eyes wide, hisses, and spits. By contrast, removal of the amygdala produces an animal that lacks fear and aggression. Likewise, a person whose amygdala is damaged fails to recognize fearful expressions in others or to express fear in situations where this emotion would normally be appropriate.

The hippocampus, together with other parts of the cerebrum, functions in memory. People with damage to certain limbic system structures forget recent events and cannot commit anything to memory.

The functions of the cerebrum are summarized in Table 19.2.

 **BRAIN INJURIES**

Brain injuries are commonly associated with head trauma and result in part from displacement and distortion of neuronal tissue at the moment of impact. Additional tissue damage occurs when normal blood flow is restored after a period of ischemia (reduced blood flow). The sudden increase in oxygen level pro-

duces large numbers of oxygen free radicals (charged oxygen molecules with an unpaired electron). Brain cells recovering from the effects of a stroke or cardiac arrest also release free radicals. Free radicals cause damage by disrupting cellular DNA and enzymes and by altering plasma membrane permeability. Brain injuries can also result from hypoxia.

Various degrees of brain injury are described by specific terms. A **concussion** is an injury characterized by an abrupt, but temporary, loss of consciousness (from seconds to hours), disturbances of vision, and problems with equilibrium. It is caused by a blow to the head or the sudden stopping of a moving head and is the most common brain injury. A concussion produces no obvious bruising of the brain. Signs of a concussion are headache, drowsiness, nausea and/or vomiting, lack of concentration, confusion, or post-traumatic amnesia (memory loss).

A **contusion** is bruising of the brain due to trauma, and includes the leakage of blood from microscopic vessels. It is usually associated with a concussion. In a contusion, the pia mater may be torn, allowing blood to enter the subarachnoid space. The area most commonly affected is the frontal lobe. A contusion usually results in an immediate loss of consciousness (generally lasting no longer than 5 minutes), loss of reflexes, transient cessation of respiration, and decreased blood pressure. Vital signs typically stabilize in a few seconds.

A **laceration** is a tear of the brain, usually from a skull fracture or a gunshot wound. A laceration results in rupture of large blood vessels, with bleeding into the brain and subarachnoid space. Consequences include cerebral hematoma (localized pool of blood, usually clotted, that swells against the brain tissue), edema, and increased intracranial pressure. If the blood clot is small enough, it may pose no major threat and may be absorbed.

TABLE 19.2 SUMMARY OF FUNCTIONS OF PRINCIPAL PARTS OF THE BRAIN

PART	FUNCTION	PART	FUNCTION
Brain stem 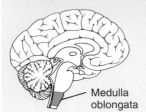	*Medulla oblongata:* Relays motor and sensory impulses between other parts of the brain and the spinal cord. Reticular formation (also in pons, midbrain, and diencephalon) functions in consciousness and arousal. Vital centers regulate heartbeat, breathing (together with pons), and blood vessel diameter. Other centers coordinate swallowing, vomiting, coughing, sneezing, and hiccupping. Contains nuclei of origin for cranial nerves VIII, IX, X, XI, and XII.	**Diencephalon** 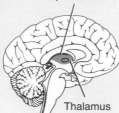	*Epithalamus:* Consists of pineal gland, which secretes melatonin, and the habenular nuclei. *Thalamus:* Relays almost all sensory input to the cerebral cortex. Provides crude perception of touch, pressure, pain, and temperature. Includes nuclei involved in movement planning and control.
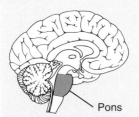	*Pons:* Relays impulses from one side of the cerebellum to the other and between the medulla and midbrain. Contains nuclei of origin for cranial nerves, V, VI, VII, and VIII. Pneumotaxic area and apneustic area, together with the medulla, help control breathing.	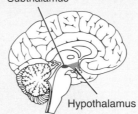	*Subthalamus:* Contains the subthalamic nuclei and portions of the red nucleus and the substantia nigra. These regions communicate with the basal ganglia to help control body movements. *Hypothalamus:* Controls and integrates activities of the autonomic nervous system and pituitary gland. Regulates emotional and behavioral patterns and circadian rhythms. Controls body temperature and regulates eating and drinking behavior. Helps maintain the waking state and establishes patterns of sleep. Produces the hormones oxytocin and antidiuretic hormone (ADH).
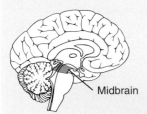	*Midbrain:* Relays motor impulses from the cerebral cortex to the pons and sensory impulses from the spinal cord to the thalamus. Superior colliculi coordinate movements of the eyeballs in response to visual and other stimuli, and the inferior colliculi coordinate movements of the head and trunk in response to auditory stimuli. Most of substantia nigra and red nucleus contribute to control of movement. Contains nuclei of origin for cranial nerves III and IV.		
Cerebellum 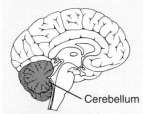	Compares intended movements with what is actually happening to smooth and coordinate complex, skilled movements. Regulates posture and balance. May have a role in cognition and language processing.	**Cerebrum** 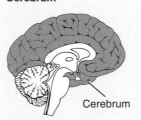	Sensory areas interpret sensory impulses, motor areas control muscular movement, and association areas function in emotional and intellectual processes. Basal ganglia coordinate gross, automatic muscle movements and regulate muscle tone. Limbic system functions in emotional aspects of behavior related to survival.

If the blood clot is large it may require surgical evacuation. Swelling infringes on the limited space that the brain occupies in the cranial cavity. Swelling causes excruciating headaches. Brain tissue can become necrotic due to the swelling; if the swelling is severe enough, the brain can herniate through the foramen magnum, resulting in death. ■

14. How are the lobes of the cerebrum separated from one another? What is the insula?
15. How is cerebral white matter organized?
16. What are the functions of the basal ganglia?
17. What system is the amygdala part of and what is its function?

FUNCTIONAL ORGANIZATION OF THE CEREBRAL CORTEX

OBJECTIVE

- Describe the locations and functions of the sensory, association, and motor areas of the cerebral cortex.

Specific types of sensory, motor, and integrative signals are processed in certain cerebral regions (Figure 19.15). Generally, **sensory areas** receive and interpret sensory impulses, **motor areas** initiate movements, and **association areas** deal with more complex integrative functions such as memory, emotions, reasoning, will, judgment, personality traits, and intelligence.

Sensory Areas

Sensory impulses arrive mainly in the posterior half of both cerebral hemispheres, in regions behind the central sulci. In the cortex, primary sensory areas have the most direct connections with peripheral sensory receptors.

Secondary sensory areas and sensory association areas often are adjacent to the primary areas. They usually receive input both from the primary areas and from other brain regions. Secondary sensory areas and sensory association areas integrate sensory experiences to generate meaningful patterns of recognition and awareness. Whereas a person with damage in the *primary* visual area would be blind in at least part of his visual field, a person with damage to a visual *association* area might see normally yet be unable to recognize a friend.

Figure 19.15   Functional areas of the cerebrum. Broca's speech area and Wernicke's area are in the left cerebral hemisphere of most people; they are shown here to indicate their relative locations. The numbers, still used today, are from K. Brodmann's map of the cerebral cortex, first published in 1909.

Particular areas of the cerebral cortex process sensory, motor, and integrative signals.

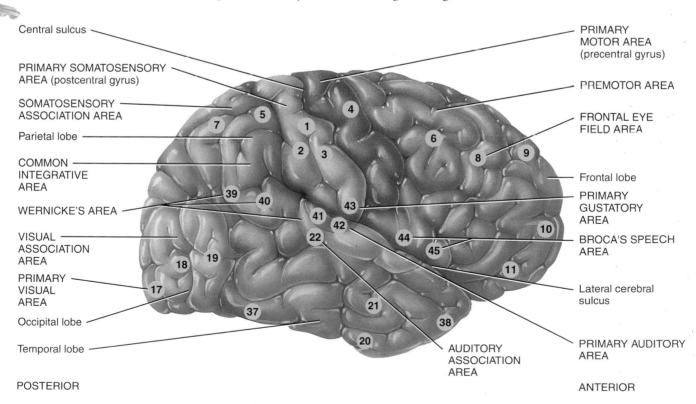

Lateral view of right cerebral hemisphere

What area(s) of the cerebrum integrate(s) interpretation of visual, auditory, and somatic sensations? Translates thoughts into speech? Controls skilled muscular movements? Interprets sensations related to taste? Interprets pitch and rhythm? Interprets shape, color, and movement of objects? Controls voluntary scanning movements of the eyes?

The following are some important sensory areas (Figure 19.15):

- The **primary somatosensory area** (areas 1, 2, and 3) is located directly posterior to the central sulcus of each cerebral hemisphere in the postcentral gyrus of each parietal lobe. It extends from the lateral cerebral sulcus, along the lateral surface of the parietal lobe to the longitudinal fissure, and then along the medial surface of the parietal lobe within the longitudinal fissure.

  The primary somatosensory area receives nerve impulses for touch, proprioception (joint and muscle position), pain, itching, tickle, and thermal sensations. A "map" of the entire body is present in the primary somatosensory area: Each point within the area receives impulses from a specific part of the body (see Figure 21.5a on page 661). The size of the cortical area receiving impulses from a particular part of the body depends on the number of receptors present there rather than on the size of the body part. For example, a larger region of the somatosensory area receives impulses from the lips and fingertips than from the thorax or hip. The major function of the primary somatosensory area is to pinpoint the points where sensations originate, so that you know exactly where on your body to swat at that mosquito.

- The **primary visual area** (area 17), located at the posterior tip of the occipital lobe mainly on the medial surface (next to the longitudinal fissure), receives impulses that convey information for vision. Axons of neurons with cell bodies in the eye form the optic (II) nerve, which ends in the lateral geniculate nucleus of the thalamus. From the thalamus, neurons project to the primary visual area, carrying information concerning shape, color, and movement of visual stimuli.

- The **primary auditory area** (areas 41 and 42), located in the superior part of the temporal lobe near the lateral cerebral sulcus, interprets the basic characteristics of sound such as pitch and rhythm.

- The **primary gustatory area** (area 43), located at the base of the postcentral gyrus superior to the lateral cerebral sulcus in the parietal cortex, receives impulses for taste.

- The **primary olfactory area** (area 28), located in the temporal lobe on the medial aspect (and thus not visible in Figure 19.15), receives impulses for smell.

Motor Areas

Motor output from the cerebral cortex flows mainly from the anterior part of each hemisphere. Among the most important motor areas are the following (Figure 19.15):

- The **primary motor area** (area 4) is located in the precentral gyrus of the frontal lobe. Each region in the primary motor area controls voluntary contractions of specific muscles or groups of muscles (see Figure 21.5b on page 661). Electrical stimulation of any point in the primary motor area causes contraction of specific skeletal muscle fibers on the opposite side of the body. As is true for the primary somatosensory area, body parts do not "map" to the primary motor area in proportion to their size. More cortical area is devoted to those muscles involved in skilled, complex, or delicate movements. For instance, the cortical region devoted to muscles that move the fingers is much larger than the region for muscles that move the toes.

- **Broca's** (BRŌ-kaz) **speech area** (areas 44 and 45) is located in the frontal lobe close to the lateral cerebral sulcus. Speaking and understanding language are complex activities that involve several sensory, association, and motor areas of the cortex. In 97% of the population, these language areas are localized in the *left* hemisphere. The planning and production of speech occurs in the *left* frontal lobe in most people. From Broca's speech area, nerve impulses pass to the premotor regions that control the muscles of the larynx, pharynx, and mouth. The impulses from the premotor area result in specific, coordinated muscle contractions. Simultaneously, impulses propagate from Broca's speech area to the primary motor area. From here, impulses also control the breathing muscles to regulate the proper flow of air past the vocal cords. The coordinated contractions of your speech and breathing muscles enable you to speak your thoughts. People who suffer a cerebrovascular accident (CVA) or stroke in this area can still have clear thoughts, but are unable to form words (nonfluent aphasia; see clinical application).

Association Areas

The association areas of the cerebrum consist of some motor and sensory areas, plus large areas on the lateral surfaces of the occipital, parietal, and temporal lobes and on the frontal lobes anterior to the motor areas. Association areas are connected with one another by association tracts and include the following (Figure 19.15):

- The **somatosensory association area** (areas 5 and 7) is just posterior to and receives input from the primary somatosensory area, as well as from the thalamus and other parts of the brain. Its role is to integrate and interpret sensations. This area permits you to determine the exact shape and texture of an object without looking at it, to determine the orientation of one object with respect to another as they are felt, and to sense the relationship of one body part to another. Another role of the somatosensory association area is the storage of memories of past sensory experiences, allowing you to compare current sensations with previous experiences.

- The **visual association area** (areas 18 and 19), located in the occipital lobe, receives sensory impulses from the primary visual area and the thalamus. It relates present and past visual experiences and is essential for recognizing and evaluating what is seen.

- The **auditory association area** (area 22), located inferior and posterior to the primary auditory area in the temporal cortex, allows you to discern whether a sound is speech, music, or noise.

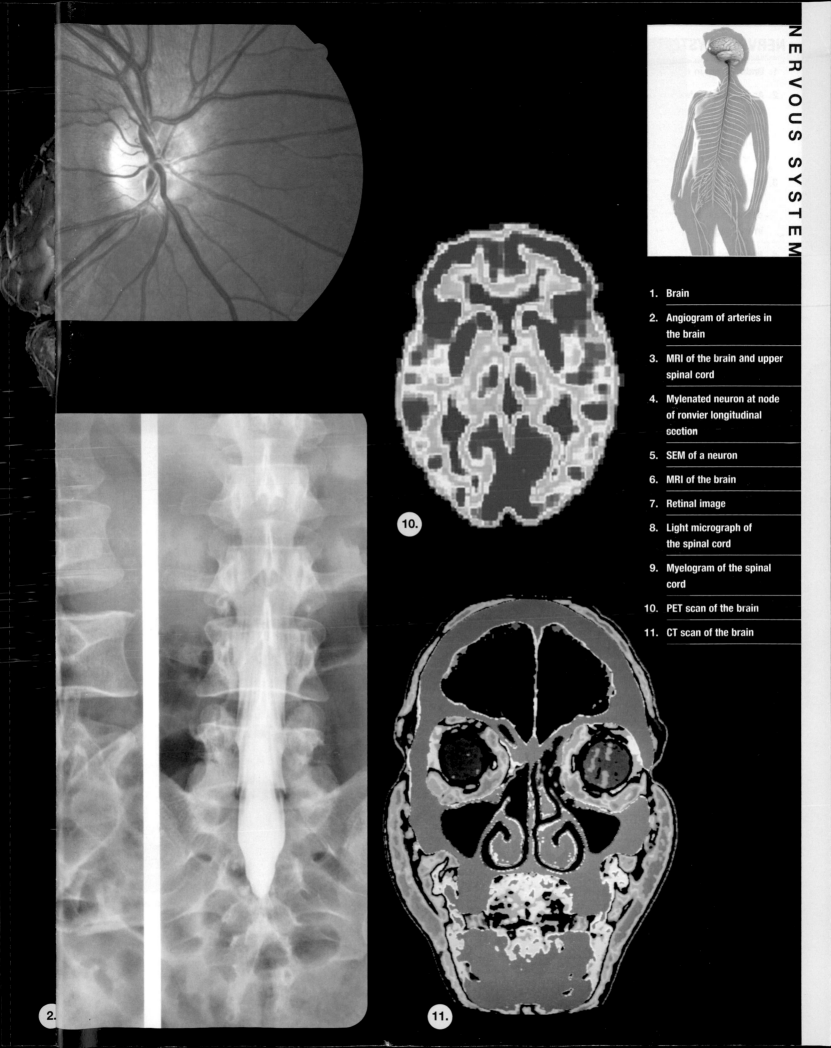

1. Brain

2. Angiogram of arteries in the brain

3. MRI of the brain and upper spinal cord

4. Mylenated neuron at node of ronvier longitudinal section

5. SEM of a neuron

6. MRI of the brain

7. Retinal image

8. Light micrograph of the spinal cord

9. Myelogram of the spinal cord

10. PET scan of the brain

11. CT scan of the brain

10.

2.

11.

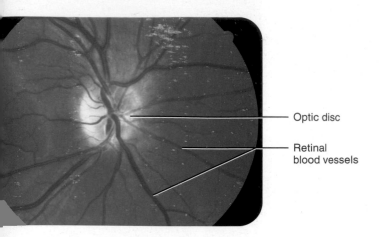

Optic disc

Retinal
blood vessels

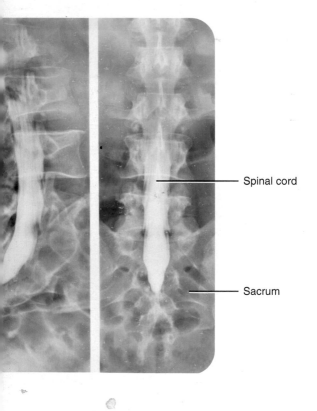

Spinal cord

Sacrum

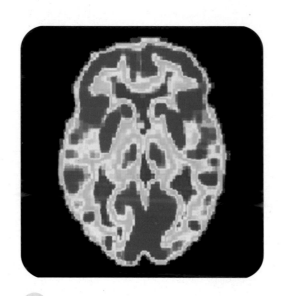

10.

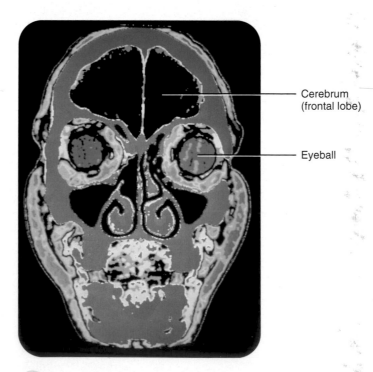

Cerebrum
(frontal lobe)

Eyeball

11.

- **Wernicke's (posterior language) area** (VER-ni-kēz) (area 22, and possibly areas 39 and 40), a broad region in the temporal and parietal lobes, interprets the meaning of speech by recognizing spoken words. It is active as you translate words into thoughts. The regions in the *right* hemisphere that correspond to Broca's and Wernicke's areas in the left hemisphere also contribute to verbal communication by adding emotional content, such as anger or joy, to spoken words. Unlike those who have CVAs in Broca's area, people who suffer strokes in Wernicke's area can still speak, but cannot arrange words in a coherent fashion (fluent aphasia, or "word salad"; see clinical application).

- The **common integrative area** (areas 5, 7, 39, and 40) is bordered by the somatosensory, visual, and auditory association areas. It receives nerve impulses from these areas and from the primary gustatory area, the primary olfactory area, the thalamus, and parts of the brain stem. This area integrates sensory interpretations from the association areas and impulses from other areas, allowing the formation of thoughts based on a variety of sensory inputs. It then transmits signals to other parts of the brain for the appropriate response to the sensory signals it has interpreted.

- The **premotor area** (area 6) is a motor association area that is immediately anterior to the primary motor area. Neurons in this area communicate with the primary motor cortex, the sensory association areas in the parietal lobe, the basal ganglia, and the thalamus. The premotor area deals with learned motor activities of a complex and sequential nature. It generates nerve impulses that cause specific groups of muscles to contract in a specific sequence, as when you write your name. The premotor area also serves as a memory bank for such movements.

- The **frontal eye field area** (area 8) in the frontal cortex is sometimes included in the premotor area. It controls voluntary scanning movements of the eyes—like those you just used in reading this sentence.

APHASIA

Much of what we know about language areas comes from studies of patients with language or speech disturbances that have resulted from brain damage. Broca's speech area, the auditory association area, and other language areas are located in the left cerebral hemisphere of most people, regardless of whether they are left-handed or right-handed. Injury to language areas of the cerebral cortex results in **aphasia** (a-FĀ-zē-a; *a-*=without; *-phasia*=speech), an inability to use or comprehend words. Damage to Broca's speech area results in *nonfluent aphasia*, an inability to properly articulate or form words; people with nonfluent aphasia know what they wish to say but cannot speak. Damage to the common integrative area or auditory association area (areas 39 and 22) results in *fluent aphasia*, characterized by faulty understanding of spoken or written words. A person experiencing this type of aphasia may fluently produce strings of words that have no meaning ("word salad"). The underlying deficit may be **word deafness** (an inability to understand spoken words), **word blindness** (an inability to understand written words), or both. ■

Hemispheric Lateralization

Although the brain is almost symmetrical on its right and left sides, subtle anatomical differences between the two hemispheres exist. For example, in about two-thirds of the population, the planum temporale, a region of the temporal lobe that includes Wernicke's area, is 50% larger on the left side than on the right side. This asymmetry appears in the human fetus at about 30 weeks of gestation. Moreover, although the two hemispheres share performance of many functions, each hemisphere also specializes in performing certain unique functions. This functional asymmetry is termed **hemispheric lateralization.**

The most obvious example of hemispheric lateralization is this: The left hemisphere receives somatic sensory signals from and controls muscles on the right side of the body, whereas the right hemisphere receives sensory signals from and controls the left side of the body. In most people the left hemisphere is more important for reasoning, numerical and scientific skills, spoken and written language, and the ability to use and understand sign language. Patients with damage in the left hemisphere, for example, often exhibit aphasia. Conversely, the right hemisphere is more specialized for musical and artistic awareness; spatial and pattern perception; recognition of faces and emotional content of language; discrimination of different smells; and generating mental images of sight, sound, touch, taste, and smell to compare relationships among them. Patients with damage in the right hemisphere regions that correspond to Broca's and Wernicke's areas in the left hemisphere speak in a monotonous voice, having lost the ability to impart emotional inflection to what they say.

Despite some dramatic differences in functions of the two hemispheres, there is considerable variation from one person to another. Also, lateralization seems less pronounced in females than in males, both for language (left hemisphere) and for visual and spatial skills (right hemisphere). For instance, females are less likely than males to suffer aphasia after damage to the left hemisphere. A possibly related observation is that the anterior commissure is 12% larger and the corpus callosum has a broader posterior portion in females. Recall that both the anterior commissure and the corpus callosum are commissural tracts that provide communication between the two hemispheres.

Table 19.3 summarizes some of the distinctive functions that are more likely to exhibit hemispheric lateralization.

Brain Waves

At any instant, brain neurons are generating millions of nerve impulses (action potentials). Taken together, these electrical signals are called **brain waves.** Brain waves generated by neurons close to the brain surface, mainly neurons in the cerebral cortex, can be detected by sensors called electrodes placed on the forehead and scalp. A record of such waves is called an **electroencephalogram** (e-lek'-trō-en-SEF-a-lō-gram; *electro-*=

TABLE 19.3 FUNCTIONAL DIFFERENCES BETWEEN THE TWO CEREBRAL HEMISPHERES	
LEFT HEMISPHERE	**RIGHT HEMISPHERE**
Receives somatic sensory signals from and controls muscles on right side of body. Also receives images from right half of visual field.	Receives somatic sensory signals from and controls muscles on left side of body. Also receives images from left half of visual field.
Reasoning.	Musical and artistic awareness.
Numerical and scientific skills.	Space and pattern perception.
Ability to use and understand sign language.	Recognition of faces and emotional content of facial expressions.
Spoken and written language.	Generating emotional content of language.
	Generating mental images to compare spatial relationships.
	Identifying and discriminating among odors.

electricity; -*gram*=recording) or **EEG.** Electroencephalograms are useful both in studying normal brain functions, such as changes that occur during sleep, and in diagnosing a variety of brain disorders, such as epilepsy, tumors, metabolic abnormalities, sites of trauma, and degenerative diseases. The EEG is also utilized to determine if "life" is present, that is, to establish or confirm that death or brain death has occurred.

CHECKPOINT

18. Describe the cortex, convolutions, fissures, and sulci of the cerebrum.
19. List and locate the lobes of the cerebrum. How are they separated from one another? What is the insula?
20. Describe the organization of cerebral white matter and indicate the function of each major group of fibers.
21. Give the name and function of each of the nuclei that form basal ganglia, and describe the effects of basal ganglia damage.
22. Define the limbic system and list several of its functions.
23. Compare the functions of the sensory, motor, and association areas of the cerebral cortex.
24. What is hemispheric lateralization?
25. What is the diagnostic value of an EEG?

CRANIAL NERVES

OBJECTIVE

● Identify the cranial nerves by name, number, and type, and give the function of each.

The 12 pairs of **cranial nerves** are so-named because they pass through various foramina in the bones of the cranium. Like the 31 pairs of spinal nerves, they are part of the peripheral nervous system (PNS). Each cranial nerve has both a number, designated by a roman numeral, and a name (see Figure 19.5). The numbers indicate the order, from anterior to posterior, in which the nerves arise from the brain. The names designate a nerve's distribution or function.

Cranial nerves emerge from the nose (cranial nerve I), the eyes (cranial nerve II), the inner ear (cranial nerve VIII), the brain stem (cranial nerves III–XII), and the spinal cord (part of

cranial nerve XI). Two cranial nerves (cranial nerves I and II) contain only sensory axons and thus are called **sensory nerves.** The rest are **mixed nerves** because they contain axons of both sensory and motor neurons. The cell bodies of sensory neurons are located in ganglia outside the brain, whereas the cell bodies of motor neurons lie in nuclei within the brain. Cranial nerves III, IV, VI, XI, and XII are mainly motor. Whereas a few of their axons are sensory axons from muscle proprioceptors, most of their axons are motor neurons that innervate skeletal muscles. Cranial nerves III, VII, IX, and X include both somatic and autonomic motor axons. The somatic axons innervate skeletal muscles whereas the autonomic axons, which are part of the parasympathetic division, innervate glands, smooth muscle, and cardiac muscle. Although the cranial nerves are mentioned singly in the following descriptions of their type, location, and function, remember that they are paired structures.

Olfactory (I) Nerve

The **olfactory (I) nerve** (ol-FAK-tō-rē; *olfact-*=to smell) is entirely sensory; it contains axons that conduct nerve impulses for the sense of smell, called olfaction (Figure 19.16). The olfactory epithelium occupies the superior part of the nasal cavity, covering the inferior surface of the cribriform plate and extending down along the superior nasal concha. The olfactory receptors are within the olfactory epithelium. These bipolar neurons have a single odor-sensitive dendrite projecting from one side of the cell body and an unmyelinated axon extending from the other side. Bundles of axons of olfactory receptors extend through about 20 olfactory foramina in the cribriform plate of the ethmoid bone on each side of the nose. These 40 or so bundles of axons collectively form the right and left olfactory nerves.

Olfactory nerves end in the brain in paired masses of gray matter called the **olfactory bulbs,** two extensions of the brain that rest on the cribriform plate. Within the olfactory bulbs, the axon terminals of olfactory receptors form synapses with the dendrites and cell bodies of the next neurons in the olfactory pathway. The axons of these neurons make up the **olfactory tracts,** which extend posteriorly from the olfactory bulbs (see Figure 19.5). Axons in the olfactory tracts end in the primary olfactory area in the temporal lobe of the cerebral cortex.

Figure 19.16 Olfactory (I) nerve.

🔑 The olfactory epithelium is located on the inferior surface of the cribriform plate and superior nasal conchae.

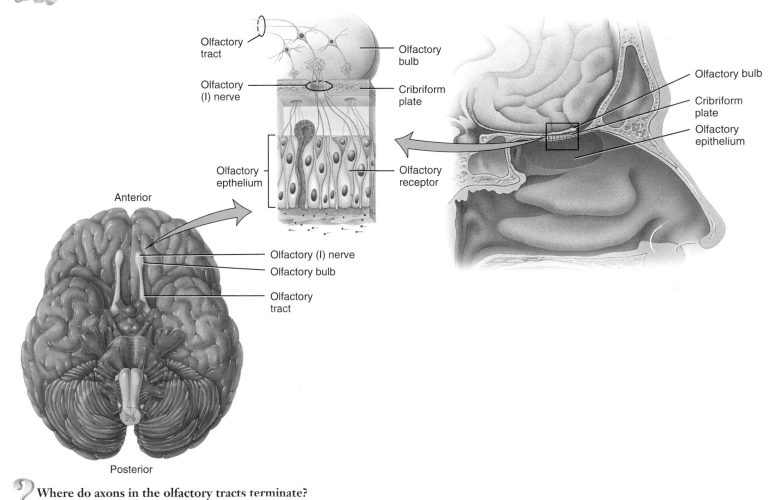

❓ **Where do axons in the olfactory tracts terminate?**

Optic (II) Nerve

The **optic (II) nerve** (OP-tik; *opti-*=the eye, vision) is entirely sensory; it contains axons that conduct nerve impulses for vision (Figure 19.17). In the retina, rods and cones initiate visual signals and relay them to bipolar cells, which pass the signals on to ganglion cells. Axons of all the ganglion cells in the retina of each eye join to form an optic nerve, which passes through the optic foramen. About 10 mm (0.4 in.) posterior to the eyeball, the two optic nerves merge to form the **optic chiasm** (KĪ-azm=a crossover, as in the letter X). Within the chiasm, axons from the medial half of each eye cross to the opposite side; axons from the lateral half remain on the same side. Posterior to the chiasm, the regrouped axons, some from each eye, form the **optic tracts.** Most axons in the optic tracts end in the lateral geniculate nucleus of the thalamus. There they synapse with neurons whose axons extend to the primary visual area in the occipital lobe of the cerebral cortex (area 17 in Figure 19.15). A few axons pass through the optic chiasm and then extend to the superior colliculi of the midbrain. They synapse with motor neurons that control the extrinsic and intrinsic eye muscles.

Oculomotor (III) Nerve

The **oculomotor (III) nerve** (ok′-ū-lō-MŌ-tor; *oculo-*=eye; *-motor*=a mover) is a mixed but mainly motor cranial nerve. Its motor nucleus is in the ventral part of the midbrain (Figure 19.18a on page 613). The oculomotor nerve extends anteriorly and divides into superior and inferior branches, both of which pass through the superior orbital fissure into the orbit. Axons in the superior branch innervate the superior rectus (an extrinsic eyeball muscle) and the levator palpebrae superioris (the muscle of the upper eyelid). Axons in the inferior branch supply the medial rectus, inferior rectus, and inferior oblique muscles—all extrinsic eyeball muscles. These somatic motor neurons control movements of the eyeball and upper eyelid.

The inferior branch of the oculomotor nerve also provides parasympathetic innervation to intrinsic eyeball muscles, which consist of smooth muscle. They include the ciliary muscle of the eyeball and the circular muscles (sphincter pupillae) of the iris. Parasympathetic impulses propagate from the oculomotor nucleus in the midbrain to the **ciliary ganglion,** a relay center of the autonomic nervous system. From the ciliary ganglion,

Figure 19.17 Optic (II) nerve.

In sequence, visual signals are relayed from rods and cones to bipolar cells to ganglion cells.

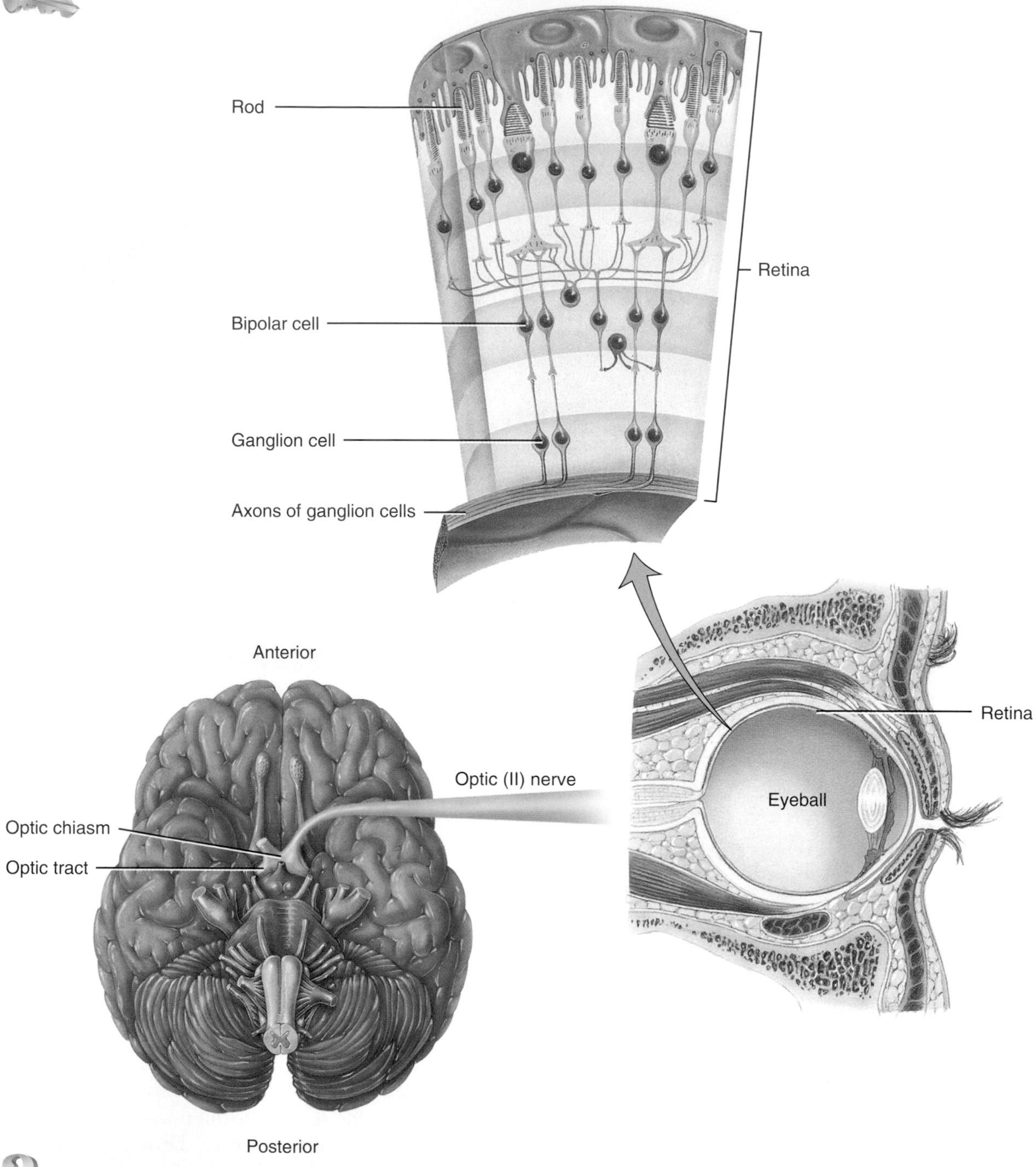

Rod

Retina

Bipolar cell

Ganglion cell

Axons of ganglion cells

Anterior

Optic (II) nerve

Optic chiasm

Optic tract

Retina

Eyeball

Posterior

? Where do most axons in the optic tracts terminate?

parasympathetic axons extend to the ciliary muscle, which adjusts the lens for near vision. Other parasympathetic axons stimulate the circular muscles of the iris to contract when bright light stimulates the eye, causing a decrease in the size of the pupil (constriction).

The sensory portion of the oculomotor nerve consists of afferent axons extending from proprioceptors in the extrinsic eyeball muscles supplied by the nerve to the midbrain. These axons convey nerve impulses for **proprioception,** the nonvisual perception of the movements and position of the body.

Figure 19.18 Oculomotor (III), trochlear (IV), and abducens (VI) nerves.

The oculomotor nerve has the widest distribution among extrinsic eye muscles.

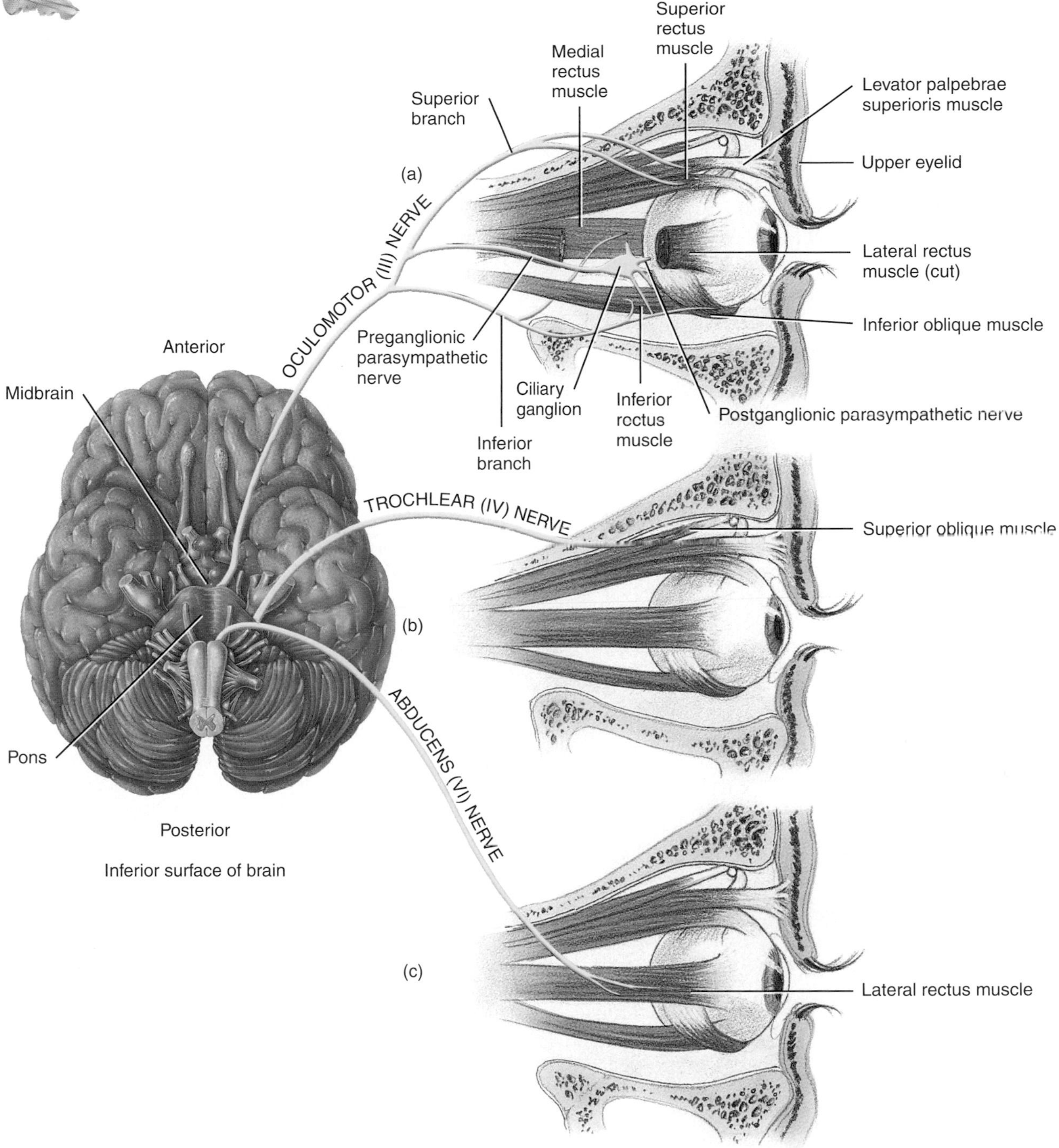

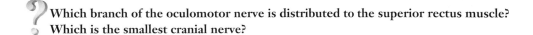

Which branch of the oculomotor nerve is distributed to the superior rectus muscle?
Which is the smallest cranial nerve?

Trochlear (IV) Nerve

The **trochlear (IV) nerve** (TRŌK-lē-ar; *trochle-*=a pulley) is a mixed but mainly motor cranial nerve. It is the smallest of the 12 cranial nerves and is the only one that arises from the posterior aspect of the brain stem.

The motor portion originates in a nucleus in the midbrain, and axons from the nucleus pass through the superior orbital fissure of the orbit (Figure 19.18b). These somatic motor axons innervate the superior oblique muscle of the eyeball, another extrinsic eyeball muscle that controls movement of the eyeball.

The sensory portion of the trochlear nerve consists of axons that extend from proprioceptors in the superior oblique muscle to a nucleus of the nerve in the midbrain. Like those of the oculomotor nerve, these axons convey nerve impulses for proprioception.

Trigeminal (V) Nerve

The **trigeminal (V) nerve** (trī-JEM-i-nal=triple, for its three branches) is a mixed cranial nerve and the largest of the cranial nerves. The trigeminal nerve emerges from two roots on the ventrolateral surface of the pons. The large sensory root has a swelling called the **trigeminal ganglion,** which is located in a fossa on the inner surface of the petrous portion of the temporal bone. The ganglion contains cell bodies of most of the primary sensory neurons. The smaller motor root originates in a nucleus in the pons.

As indicated by its name, the trigeminal nerve has three branches: ophthalmic, maxillary, and mandibular (Figure 19.19). The **ophthalmic nerve** (of-THAL-mik; *ophthalm-*=the eye), the smallest branch, passes the orbit via the superior orbital fissure. The **maxillary nerve** (*maxilla*=upper jaw bone) is intermediate in size between the ophthalmic and mandibular nerves and passes through the foramen rotundum. The **mandibular nerve** (*mandibula*=lower jaw bone), the largest branch, passes through the foramen ovale.

Figure 19.19 Trigeminal (V) nerve.

The three branches of the trigeminal nerve leave the cranium through the superior orbital fissure, foramen rotundum, and foramen ovale.

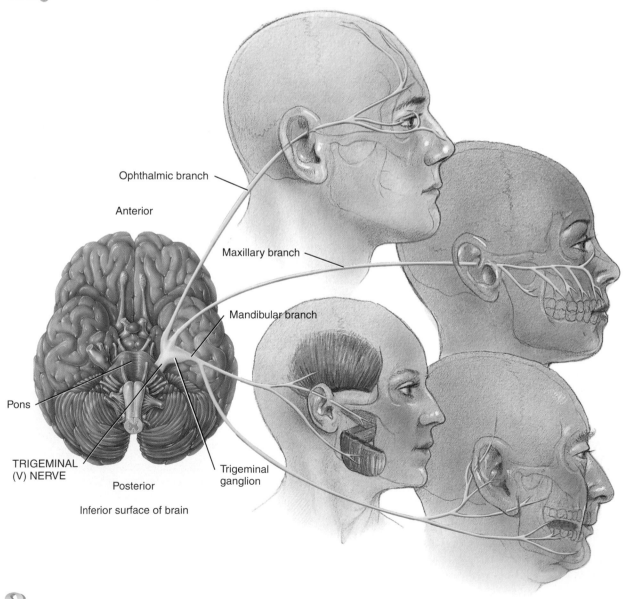

Ophthalmic branch

Anterior

Maxillary branch

Mandibular branch

Pons

TRIGEMINAL (V) NERVE

Trigeminal ganglion

Posterior

Inferior surface of brain

How does the trigeminal nerve compare in size to the other cranial nerves?

Sensory axons in the trigeminal nerve carry nerve impulses for touch, pain, and thermal sensations. The ophthalmic nerve contains sensory axons from the skin over the upper eyelid, eyeball, lacrimal glands, upper part of the nasal cavity, side of the nose, forehead, and anterior half of the scalp. The maxillary nerve includes sensory axons from the mucosa of the nose, palate, part of the pharynx, upper teeth, upper lip, and lower eyelid. The mandibular nerve contains sensory axons from the anterior two-thirds of the tongue (not taste), cheek and mucosa deep to it, lower teeth, skin over the mandible and side of the head anterior to the ear, and mucosa of the floor of the mouth. The sensory axons from the three branches enter the semilunar ganglion and terminate in nuclei in the pons. The trigeminal nerve also contains sensory fibers from proprioceptors located in the muscles of mastication.

Somatic motor axons of the trigeminal nerve are part of the mandibular nerve and supply muscles of mastication (masseter, temporalis, medial pterygoid, lateral pterygoid, anterior belly of digastric, and mylohyoid muscles). These motor neurons control chewing movements.

DENTAL ANESTHESIA

The inferior alveolar nerve, a branch of the mandibular nerve, supplies all the teeth in one-half of the mandible; it is often anesthetized in dental procedures. The same procedure will anesthetize the lower lip because the mental nerve is a branch of the inferior alveolar nerve. Because the lingual nerve runs very close to the inferior alveolar nerve near the mental foramen, it too is often anesthetized at the same time. For anesthesia to the upper teeth, the superior alveolar nerve endings, which are branches of the maxillary nerve, are blocked by inserting the needle beneath the mucous membrane. The anesthetic solution is then infiltrated slowly throughout the area of the roots of the teeth to be treated. ■

Abducens (VI) Nerve

The **abducens (VI) nerve** (ab-DOO-senz; *ab-*=away; *-ducens*= to lead) is a mixed but mainly motor cranial nerve that originates from a nucleus in the pons (see Figure 19.18c). Somatic motor axons extend from the nucleus to the lateral rectus muscle of the eyeball, an extrinsic eyeball muscle, through the superior orbital fissure of the orbit. The abducens nerve is so named because nerve impulses cause abduction of the eyeball (lateral rotation). The sensory axons extend from proprioceptors in the lateral rectus muscle to the pons.

Facial (VII) Nerve

The **facial (VII) nerve** (FĀ-shal=face) is a mixed cranial nerve. Its sensory axons extend from the taste buds of the anterior two-thirds of the tongue through the **geniculate ganglion** (je-NIK-ū-lāt), a cluster of cell bodies of sensory neurons that lie beside the facial nerve, and end in the pons (Figure 19.20).

The sensory portion of the facial nerve also contains axons from proprioceptors in muscles of the face and scalp.

Axons of somatic motor neurons arise from a nucleus in the pons, enter the petrous portion of the temporal bone, and innervate facial, scalp, and neck muscles. Nerve impulses propagating along these axons cause contraction of the muscles of facial expression plus the stylohyoid muscle and the posterior belly of the digastric muscle.

Axons of parasympathetic neurons that are part of the facial nerve end in two parasympathetic ganglia: the **pterygopalatine** (ter'-i-gō-PAL-a-tīn) **ganglion** and the **submandibular ganglion.** From the two ganglia, other parasympathetic axons extend to lacrimal glands (which secrete tears), nasal glands, palatine glands, and saliva-producing sublingual and submandibular glands.

Vestibulocochlear (VIII) Nerve

The **vestibulocochlear (VIII) nerve** (vest-tib-ū-lō-KOK-lē-ar; *ves-tibulo-*=small cavity; *-cochlear*=a spiral, snail-like) was formerly known as the **acoustic** or **auditory nerve.** It is a mixed but mainly sensory cranial nerve and has two branches, the vestibular branch and the cochlear branch (Figure 19.21). The **vestibular branch** carries impulses for equilibrium whereas the **cochlear branch** carries impulses for hearing.

Sensory axons in the vestibular branch extend from the semicircular canals, the saccule, and the utricle of the inner ear to the **vestibular ganglion,** where their cell bodies are located (see Figure 22.13b on page 689), and end in vestibular nuclei in the pons. Some sensory axons also enter the cerebellum via the inferior cerebellar peduncle. Axons of motor neurons in the vestibular branch project from the pons to hair cells of the semicircular canals, saccule, and utricle.

Sensory axons in the cochlear branch arise in the spiral organ (organ of Corti) in the cochlea of the internal ear. The cell bodies of cochlear branch sensory neurons are located in the **spiral ganglion** of the cochlea (see Figure 22.13b). From there, axons extend to nuclei in the medulla oblongata. Axons of motor neurons in the cochlear branch project from the pons to hair cells of the spiral organ.

Glossopharyngeal (IX) Nerve

The **glossopharyngeal (IX) nerve** (glos'-ō-fa-RIN-jē-al; *glosso-*=tongue; *-pharyngeal*= throat) is a mixed cranial nerve. Sensory axons of the glossopharyngeal nerve arise from taste buds and somatic sensory receptors on the posterior one-third of the tongue, from proprioceptors in swallowing muscles supplied by the motor portion, from baroreceptors (stretch receptors) in the carotid sinus, and from chemoreceptors in the carotid body near the carotid arteries (Figure 19.22 on page 617). The cell bodies of these sensory neurons are located in the superior and inferior ganglia. From the ganglia, sensory axons pass through the jugular foramen and end in the medulla.

Axons of motor neurons in the glossopharyngeal nerve arise in nuclei of the medulla and exit the skull through the jugular

Figure 19.20   Facial (VII) nerve.

The facial nerve causes contraction of the muscles of facial expression.

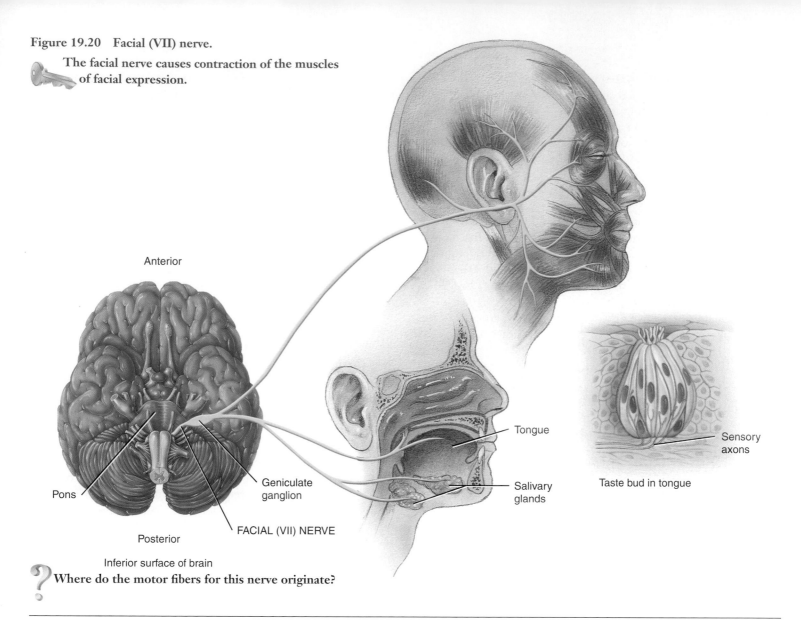

Anterior

Pons

Posterior

Inferior surface of brain

Geniculate ganglion

FACIAL (VII) NERVE

Tongue

Salivary glands

Sensory axons

Taste bud in tongue

? Where do the motor fibers for this nerve originate?

Figure 19.21   Vestibulocochlear (VIII) nerve.

The vestibular branch of the vestibulocochlear nerve carries impulses for equilibrium, while the cochlear branch carries impulses for hearing.

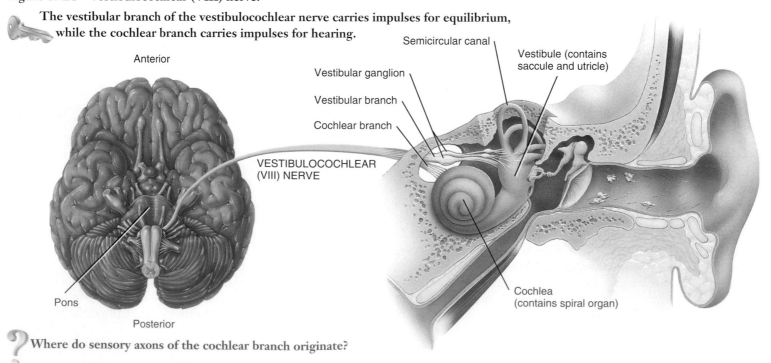

Anterior

Pons

Posterior

Vestibular ganglion

Vestibular branch

Cochlear branch

VESTIBULOCOCHLEAR (VIII) NERVE

Semicircular canal

Vestibule (contains saccule and utricle)

Cochlea (contains spiral organ)

? Where do sensory axons of the cochlear branch originate?

Figure 19.22 Glossopharyngeal (IX) nerve.

Sensory fibers of the glossopharyngeal nerve supply the taste buds.

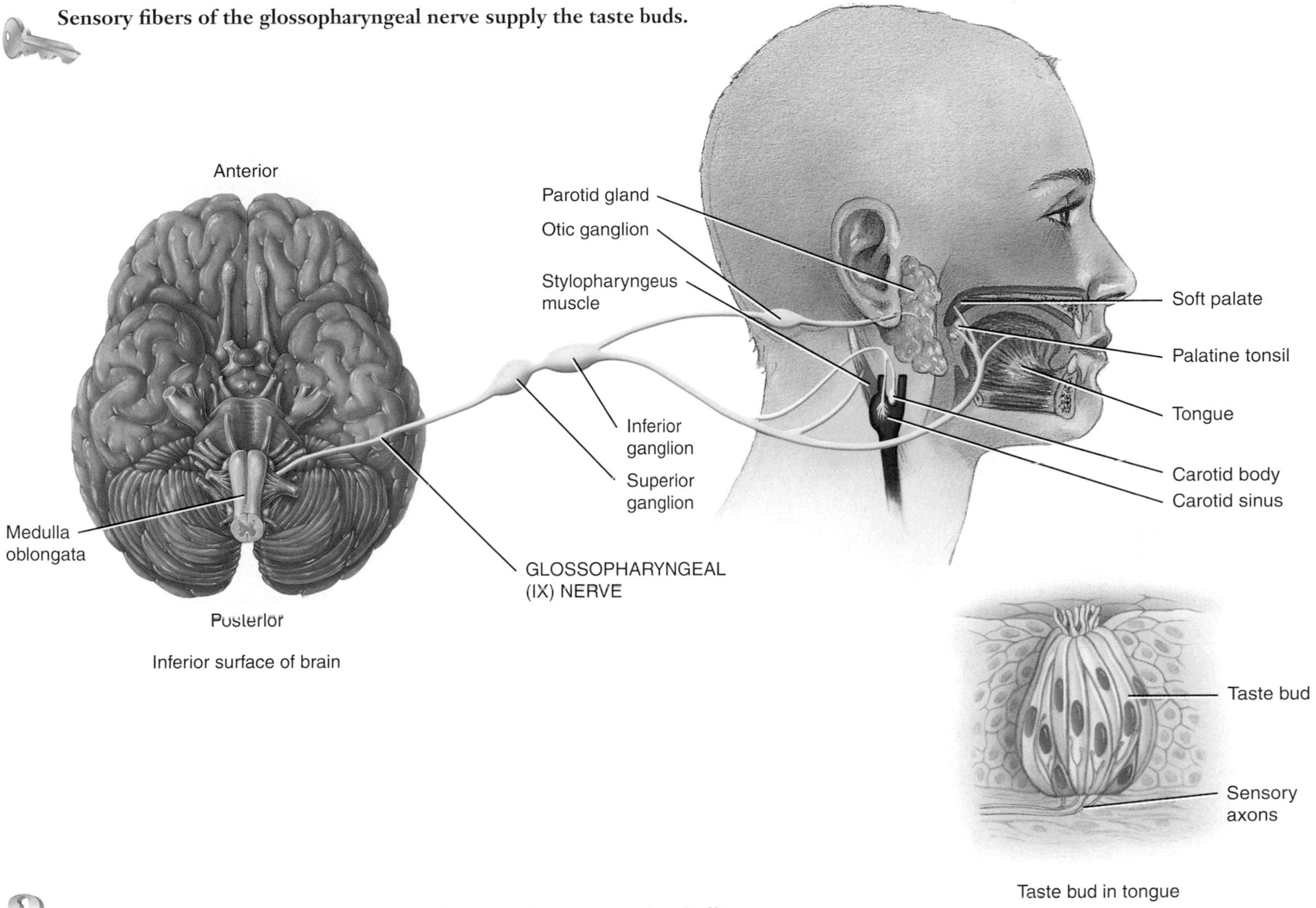

Inferior surface of brain

Taste bud in tongue

Through which foramen does the glossopharyngeal nerve exit the skull?

foramen. Somatic motor neurons innervate the stylopharyngeus muscle, and autonomic motor neurons (parasympathetic) stimulate the parotid gland to secrete saliva. Some of the cell bodies of parasympathetic motor neurons are located in the **otic ganglion.**

Vagus (X) Nerve

The **vagus (X) nerve** (VĀ-gus=vagrant or wandering) is a mixed cranial nerve that is distributed from the head and neck into the thorax and abdomen (Figure 19.23). The nerve derives its name from its wide distribution. In the neck, it lies medial and posterior to the internal jugular vein and common carotid artery.

Sensory axons in the vagus nerve arise from the skin of the external ear, a few taste buds in the epiglottis and pharynx, and proprioceptors in muscles of the neck and throat. Also, sensory axons come from baroreceptors (stretch receptors) in the carotid sinus and chemoreceptors in the carotid body near the carotid arteries and in the aortic bodies near the arch of the aorta,

and visceral sensory receptors in most organs of the thoracic and abdominal cavities. These axons pass through the jugular foramen and end in the medulla and pons.

Axons of autonomic motor neurons (parasympathetic) in the vagus nerve originate in nuclei of the medulla and end in the lungs and heart. Vagal parasympathetic axons also supply glands of the gastrointestinal (GI) tract and smooth muscle of the respiratory passageways, esophagus, stomach, gallbladder, small intestine, and most of the large intestine (see Figure 20.3 on page 637).

Accessory (XI) Nerve

The **accessory (XI) nerve** (ak-SES-ō-rē=assisting) is a mixed cranial nerve. It differs from all other cranial nerves because it originates from *both* the brain stem and the spinal cord (Figure 19.24 on page 619). The **cranial root** is motor and arises from nuclei in the medulla oblongata, passes through the jugular foramen, and supplies the voluntary muscles of the pharynx, larynx, and soft palate that are used in swallowing. The **spinal root** is mixed but mainly motor. Its motor axons arise in the anterior

Figure 19.23 Vagus (X) nerve.

The vagus nerve is widely distributed in the head, neck, thorax, and abdomen.

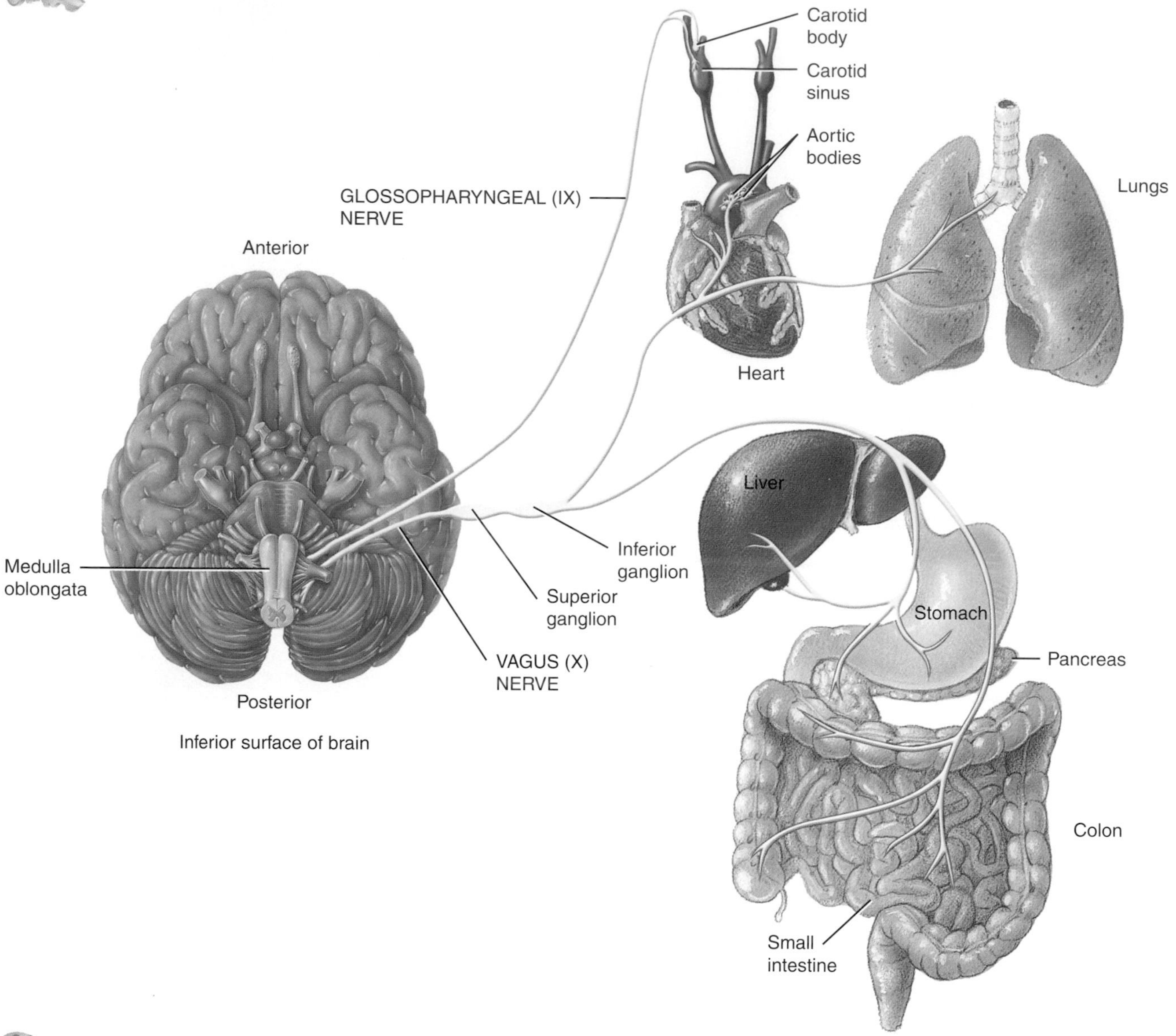

Inferior surface of brain

Where is the vagus nerve located in the neck?

gray horn of the first five segments of the cervical portion of the spinal cord. The axons from the segments come together, pass through the foramen magnum, and then exit through the jugular foramen along with axons in the cranial root. The spinal root conveys motor impulses to the sternocleidomastoid and trapezius muscles to coordinate head movements. Sensory axons in the spinal root originate from proprioceptors in the muscles supplied by its motor neurons and end in the medulla oblongata.

Hypoglossal (XII) Nerve

The **hypoglossal (XII) nerve** (hī′-pō-GLOS-al; *hypo-*=below; *-glossal*=tongue) is a mixed cranial nerve. The sensory portion of the hypoglossal nerve consists of axons that originate from proprioceptors in the tongue muscles and end in the medulla

oblongata (Figure 19.25). The sensory fibers conduct nerve impulses for proprioception. The somatic motor axons originate in a nucleus in the medulla oblongata, pass through the hypoglossal canal, and supply the muscles of the tongue. These axons conduct nerve impulses for speech and swallowing.

Table 19.4 on pages 620–623 presents a summary of cranial nerves, including clinical applications related to their dysfunction.

CHECKPOINT

26. How are cranial nerves named and numbered?
27. What is the difference between a mixed cranial nerve and a sensory cranial nerve?
28. What sort of test could reveal damage to each of the 12 cranial nerves?

Figure 19.24 Accessory (XI) nerve.

The accessory nerve exits the cranium through the jugular foramen.

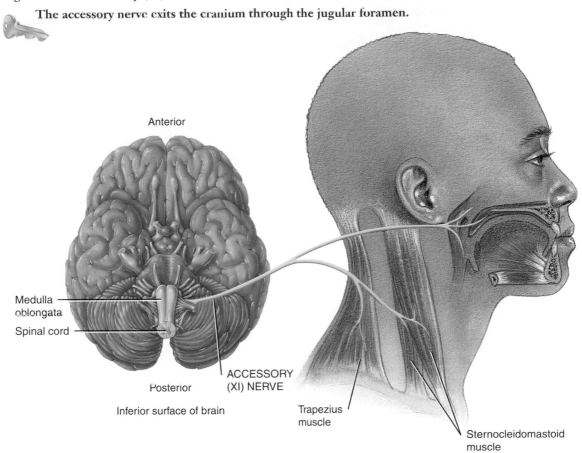

Anterior

Medulla oblongata

Spinal cord

ACCESSORY (XI) NERVE

Posterior

Inferior surface of brain

Trapezius muscle

Sternocleidomastoid muscle

How does the accessory nerve differ from the other cranial nerves?

Figure 19.25 Hypoglossal (XII) nerve.

The hypoglossal nerve exits the cranium through the hypoglossal canal.

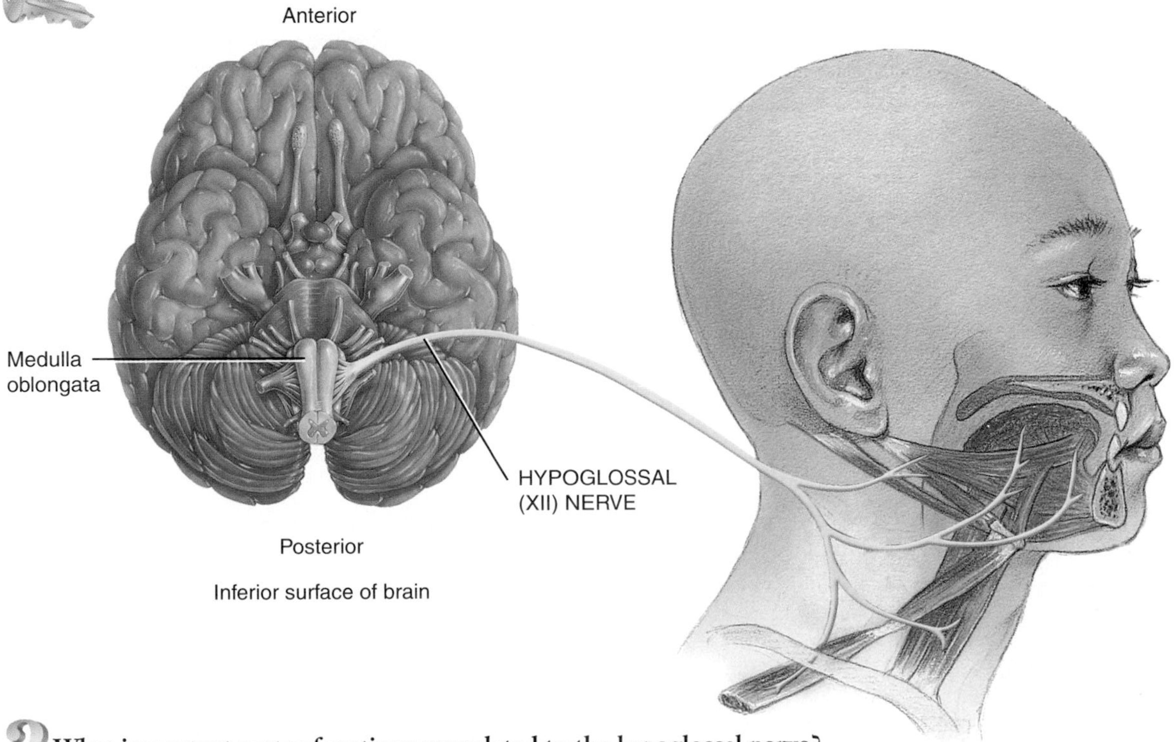

Anterior

Medulla oblongata

HYPOGLOSSAL (XII) NERVE

Posterior

Inferior surface of brain

What important motor functions are related to the hypoglossal nerve?

619

TABLE 19.4   SUMMARY OF CRANIAL NERVES*

NUMBER AND NAME	TYPE AND LOCATION	FUNCTION AND CLINICAL APPLICATION
Olfactory (I) nerve 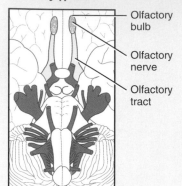 Olfactory bulb / Olfactory nerve / Olfactory tract	**Sensory** Arises in olfactory mucosa, passes through foramina in the cribriform plate of the ethmoid bone, and ends in the olfactory bulb. The olfactory tract extends via two pathways to olfactory areas of cerebral cortex.	*Function:* Smell. *Clinical application:* Loss of the sense of smell, called *anosmia* (an-OZ-mē-a), may result from head injuries in which the cribriform plate of the ethmoid bone is fractured, or from lesions along the olfactory pathway.
Optic (II) nerve 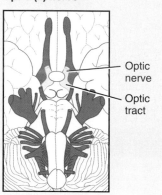 Optic nerve / Optic tract	**Sensory** Arises in the retina of the eye, passes through the optic foramen, forms the optic chiasm and then the optic tracts, and terminates in the lateral geniculate nuclei of the thalamus. From the thalamus, axons extend to the primary visual area (area 17) of the cerebral cortex.	*Function:* Vision *Clinical application:* Fractures in the orbit, damage along the visual pathway, or diseases of the nervous system may result in visual field defects and loss of visual acuity. Blindness due to a defect in or loss of one or both eyes is called *anopia*.
Oculomotor (III) nerve 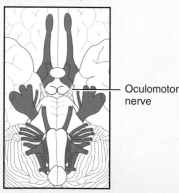 Oculomotor nerve	**Mixed (mainly motor)** *Sensory portion:* Consists of axons from proprioceptors in eyeball muscles that pass through the superior orbital fissure and terminate in the midbrain. *Motor portion:* Originates in the midbrain and passes through the superior orbital fissure. Axons of somatic motor neurons innervate the levator palpebrae superioris muscle of the upper eyelid and four extrinsic eyeball muscles (superior rectus, medial rectus, inferior rectus, and inferior oblique). Parasympathetic axons innervate the ciliary muscle of the eyeball and the circular muscles (sphincter pupillae) of the iris.	*Sensory function:* Proprioception. *Somatic motor function:* Movement of upper eyelid and eyeball. *Autonomic motor function (parasympathetic):* Accommodation of lens for near vision and constriction of pupil. *Clinical application:* Nerve damage causes strabismus (a deviation of the eye in which both eyes do not fix on the same object), *ptosis* (drooping) of the upper eyelid, dilation of the pupil, movement of the eyeball downward and outward on the damaged side, loss of accommodation for near vision, or *diplopia* (double vision).

*MNEMONIC for cranial nerves:

Oh Olfactory	**Oh** Optic	**Oh** Oculomotor	**To** Trochlear	**Touch** Trigeminal	**And** Abducens	**Feel** Facial
Very Vestibulocochlear		**Green** Glossopharyngeal		**Vegetables** Vagus	**AH!** Accessory	Hypoglossal

NUMBER AND NAME	TYPE AND LOCATION	FUNCTION AND CLINICAL APPLICATION
Trochlear (IV) nerve 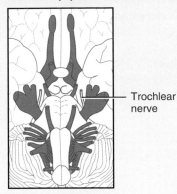 Trochlear nerve	**Mixed (mainly motor)** *Sensory portion*: Consists of axons from proprioceptors in the superior oblique muscles, which pass through the superior orbital fissure and terminate in the midbrain. *Motor portion*: Originates in the midbrain and passes through the superior orbital fissure. Innervates the superior oblique muscle, an extrinsic muscle.	*Sensory function*: Proprioception. *Somatic motor function*: Movement of the eyeball. *Clinical application*: In trochlear nerve paralysis, diplopia and strabismus occur.
Trigeminal (V) nerve 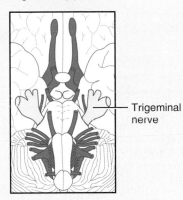 Trigeminal nerve	**Mixed** *Sensory portion*: Consists of three branches, all of which end in the pons. (1) The **ophthalamic nerve** (*ophthalm-*=the eye) contains axons from the skin over the upper eyelid, eyeball, lacrimal glands, nasal cavity, side of nose, forehead, and anterior half of scalp that pass through superior orbital fissure. (2) The **maxillary nerve** (*maxilla*=upper jaw bone) contains axons from the mucosa of the nose, palate, parts of the pharynx, upper teeth, upper lip, and lower eyelid that pass through the foramen rotundum. (3) The **mandibular nerve** (*mandibula*=lower jaw bone) contains axons from the anterior two-thirds of the tongue (somatic sensory axons but not axons for the special sense of taste), the lower teeth, skin over mandible, cheek and mucosa deep to it, and side of head in front of ear that pass through the foramen ovale. *Motor portion*: Is part of the mandibular branch, which originates in the pons, passes through the foramen ovale, and innervates muscles of mastication (masseter, temporalis, medial pterygoid, lateral pterygoid, anterior belly of digastric, and mylohyoid muscles).	*Sensory function*: Conveys impulses for touch, pain, and temperature sensations and proprioception. *Somatic motor function*: Chewing. *Clinical application*: *Neuralgia* (pain) of one or more branches of the trigeminal nerve is called *trigeminal neuralgia* (tic douloureux). Injury of the mandibular nerve may cause paralysis of the chewing muscles and a loss of the sensations of touch, temperature, and proprioception in the lower part of the face. Dentists apply anesthetic drugs to branches of the maxillary nerve for anesthesia of upper teeth and to branches of the mandibular nerve for anesthesia of lower teeth.
Abducens (VI) nerve  Abducens nerve	**Mixed (mainly motor)** *Sensory portion*: Consists of axons from proprioceptors in the lateral rectus muscle, which pass through the superior orbital fissure and end in the pons. *Motor portion*: Originates in the pons, passes through the superior orbital fissure, and innervates the lateral rectus muscle, an extrinsic eyeball muscle.	*Sensory function*: Proprioception. *Somatic motor function*: Movement of the eyeball. *Clinical application*: With damage to this nerve, the affected eyeball cannot move laterally beyond the midpoint, and the eye usually is directed medially.

continues

TABLE 19.4   SUMMARY OF CRANIAL NERVES (c o n t i n u e d)

NUMBER AND NAME	TYPE AND LOCATION	FUNCTION AND CLINICAL APPLICATION
Facial (VII) nerve 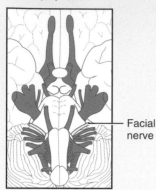 Facial nerve	**Mixed** *Sensory portion*: Arises from taste buds on the anterior two-thirds of the tongue, passes through the stylomastoid foramen and geniculate ganglion (located beside the facial nerve), and ends in the pons. From there, axons extend to the thalamus, and then to the gustatory areas of the cerebral cortex. Also contains axons from proprioceptors in muscles of the face and scalp. *Motor portion*: Originates in the pons and passes through the stylomastoid foramen. Axons of somatic motor neurons innervate facial, scalp, and neck muscles. Parasympathetic axons innervate lacrimal, sublingual, submandibular, nasal, and palatine glands.	*Sensory function*: Proprioception and taste. *Somatic motor function*: Facial expression. *Autonomic motor function (parasympathetic)*: Secretion of saliva and tears. *Clinical application*: Damage due to a viral infection (shingles) or a bacterial infection (Lyme disease) produces Bell's palsy (paralysis of the facial muscles), loss of taste, decreased salivation, and loss of ability to close the eyes, even during sleep.
Vestibulocochlear (VIII) nerve  Vestibulocochlear nerve	**Mixed (mainly sensory)** *Vestibular nerve, sensory portion*: Arises in the semicircular canals, saccule, and utricle and forms the vestibular ganglion. Axons end in the pons and cerebellum. *Vestibular nerve, motor portion*: Originates in the pons and terminates on hair cells of the semicircular canals, saccule, and utricle. *Cochlear nerve, sensory portion*: Arises in the spiral organ (organ of Corti), forms the spiral ganglion, passes through nuclei in the medulla, and ends in the thalamus. Axons synapse with thalamic neurons that relay impulses to the primary auditory area (areas 41 and 42) of the cerebral cortex. *Cochlear nerve, motor portion*: Originates in the pons and terminates on hair cells of the spiral organ.	*Vestibular branch sensory function*: Conveys impulses related to equilibrium. *Vestibular branch motor function*: Adjusts sensitivity of hair cells. *Cochlear branch sensory function:* Conveys impulses for hearing. *Cochlear branch motor function*: Modifies function of hair cells by altering their response to sound waves. *Clinical application*: Injury to the vestibular branch may cause *vertigo,* an illusory feeling that one's own body or the environment is rotating, *ataxia* (muscular incoordination), or *nystagmus* (involuntary rapid movement of the eyeball). Injury to the cochlear branch may cause tinnitus (ringing in the ears) or deafness.
Glossopharyngeal (IX) nerve  Glossopharyngeal nerve	**Mixed** *Sensory portion*: Consists of axons from taste buds and somatic sensory receptors on posterior one-third of the tongue, from proprioceptors in swallowing muscles supplied by the motor portion, and from stretch receptors in carotid sinus and chemoreceptors in carotid body near the carotid arteries. Axons pass through the jugular foramen and end in the medulla. *Motor portion*: Originates in the medulla and passes through the jugular foramen. Axons of somatic motor neurons innervate the stylopharyngeus muscle, a muscle of the pharynx that elevates the larynx during swallowing. Parasympathetic axons innervate the parotid (salivary) gland.	*Sensory function*: Taste and somatic sensations (touch, pain, temperature) from posterior third of tongue; proprioception in swallowing muscles; monitoring of blood pressure; monitoring of O_2 and CO_2 in blood for regulation of breathing rate and depth. *Somatic motor function*: Elevates the pharynx during swallowing and speech. *Autonomic motor function (parasympathetic)*: Stimulates secretion of saliva. *Clinical application*: Injury causes difficulty in swallowing, reduced secretion of saliva, loss of sensation in the throat, and loss of taste sensation.

NUMBER AND NAME	TYPE AND LOCATION	FUNCTION AND CLINICAL APPLICATION
Vagus (X) nerve  Vagus nerve	**Mixed** *Sensory portion*: Consists of axons from a small number of taste buds in the epiglottis and pharynx, proprioceptors in muscles of the neck and throat, stretch receptors and chemoreceptors in the carotid sinus and carotid body near the carotid arteries, chemoreceptors in the aortic bodies near the arch of the aorta, and visceral sensory receptors in most organs of the thoracic and abdominal cavities. Axons pass through the jugular foramen and end in the medulla and pons. *Motor portion*: Originates in medulla and passes through the jugular foramen. Axons of somatic motor neurons innervate skeletal muscles in the throat and neck. Parasympathetic axons innervate smooth muscle in the airways, esophagus, stomach, small intestine, most of large intestine, and gallbladder; cardiac muscle in the heart; and glands of the gastrointestinal (GI) tract.	*Sensory function*: Taste and somatic sensations (touch, pain, temperature, and proprioception) from epiglottis and pharynx; monitoring of blood pressure; monitoring of O_2 and CO_2 in blood for regulation of breathing rate and depth; sensations from visceral organs in thorax and abdomen. *Somatic motor function*: Swallowing, coughing, and voice production. *Autonomic motor function (parasympathetic)*: Smooth muscle contraction and relaxation in organs of the GI tract; slowing of the heart rate; secretion of digestive fluids. *Clinical application*: Injury interrupts sensations from many organs in the thoracic and abdominal cavities, interferes with swallowing, paralyzes vocal cords, and causes heart rate to increase.
Accessory (XI) nerve 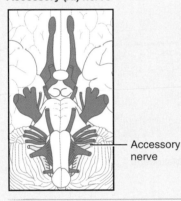 Accessory nerve	**Mixed (mainly motor)** *Sensory portion*: Consists of axons from proprioceptors in muscles of the pharynx, larynx, and soft palate that pass through the jugular foramen and end in the medulla. *Motor portion*: Consists of a cranial root and a spinal root. *Cranial root* arises in the medulla, passes through the jugular foramen, and supplies muscles of the pharynx, larynx, and soft palate. *Spinal root* originates in the anterior gray horn of the first five cervical segments of the spinal cord, passes through the jugular foramen, and supplies the sternocleidomastoid and trapezius muscles.	*Sensory function*: Proprioception. *Somatic motor function*: Cranial root mediates swallowing movements; spinal root mediates movement of head and shoulders. *Clinical application*: If nerves are damaged, the sternocleidomastoid and trapezius muscles become paralyzed, with a resulting inability to raise the shoulders and difficulty in turning the head.
Hypoglossal (XII) nerve 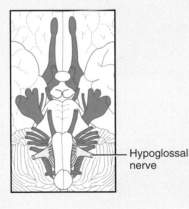 Hypoglossal nerve	**Mixed (mainly motor)** *Sensory portion*: Consists of axons from proprioceptors in tongue muscles that pass through the hypoglossal canal and end in the medulla. *Motor portion*: Originates in the medulla, passes through the hypoglossal canal, and supplies muscles of the tongue.	*Sensory function*: Proprioception. *Motor function*: Movement of tongue during speech and swallowing. *Clinical application*: Injury results in difficulty in chewing, speaking, and swallowing. The tongue, when protruded, curls toward the affected side, and the affected side atrophies.

DEVELOPMENT OF THE NERVOUS SYSTEM

● Describe how the parts of the brain develop.

As you learned in Chapter 4, development of the nervous system begins in the third week of gestation with a thickening of the **ectoderm** called the **neural plate** (Figure 19.26). The plate folds inward and forms a longitudinal groove, the **neural groove**. The raised edges of the neural plate are called **neural folds**. As development continues, the neural folds increase in height and meet to form a tube called the **neural tube.**

Three layers of cells differentiate from the wall that encloses the neural tube. The outer or **marginal layer** cells develop into the *white matter* of the nervous system. The middle or **mantle layer** cells develop into the *gray matter.* The inner or **ependymal layer** cells eventually form the *lining of the central canal of the spinal cord and ventricles of the brain.*

The **neural crest** is a mass of tissue between the neural tube and the skin ectoderm (Figure 19.26b). It differentiates and eventually forms the *posterior (dorsal) root ganglia of spinal nerves, spinal nerves, ganglia of cranial nerves, cranial nerves, ganglia of the autonomic nervous system, adrenal medulla,* and *meninges.*

Figure 19.26 Origin of the nervous system. (a) Dorsal view of an embryo in which the neural folds have partially united, forming the early neural tube. (b) Transverse sections through the embryo showing the formation of the neural tube.

The nervous system begins developing in the third week from a thickening of ectoderm called the neural plate.

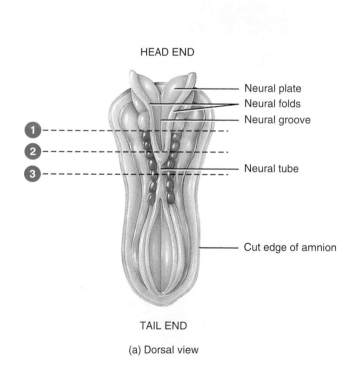

HEAD END

Neural plate
Neural folds
Neural groove

Neural tube

Cut edge of amnion

TAIL END

(a) Dorsal view

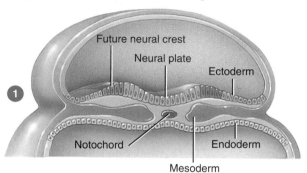

Future neural crest
Neural plate
Ectoderm

Notochord
Endoderm
Mesoderm

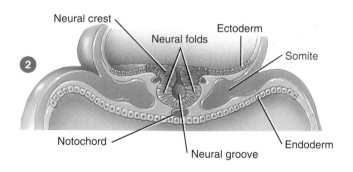

Neural crest
Ectoderm
Neural folds
Somite

Notochord
Neural groove
Endoderm

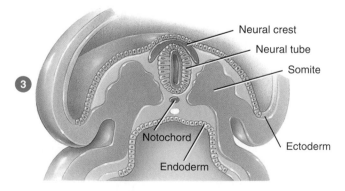

Neural crest
Neural tube
Somite

Notochord
Endoderm
Ectoderm

(b) Transverse sections

What is the origin of the gray matter of the nervous system?

As you learned earlier in this chapter on page 584, during the third to fourth week of embryonic development, the anterior part of the neural tube develops into three enlarged areas called **primary brain vesicles** that are named for their relative positions. These are the **prosencephalon** (prōs'-en-SEF-a-lon; *pros-*=before) or forebrain, **mesencephalon** (mes'-en-SEF-a-lon; *mes-*=middle) or midbrain, and **rhombencephalon** (rom'-ben-SEF-a-lon; *rhomb-*=behind) or hindbrain (Figure 19.27a; see also Table 19.1). During the fifth week of development, **secondary brain vesicles** begin to develop. The prosencephalon develops into two secondary brain vesicles called the **telencephalon** (tel'-en-SEF-a-lon; *tel-*=distant) and the **diencephalon** (dī-en-SEF-a-

lon; *di-*=through) (Figure 19.27b). The rhombencephalon also develops into two secondary brain vesicles called the **metencephalon** (met'-en-SEF-a-lon; *met-*=after) and the **myelencephalon** (mī-el-en-SEF-a-lon; *myel-*=marrow). The area of the neural tube inferior to the myelencephalon gives rise to the *spinal cord*.

The brain vesicles continue to develop as follows (Figure 19.27c, d; see also Table 19.1).

- The telencephalon develops into the *cerebral hemispheres*, including the *basal ganglia*, and houses the paired *lateral ventricles*.

Figure 19.27 Development of the brain and spinal cord.

The various parts of the brain develop from the primary brain vesicles.

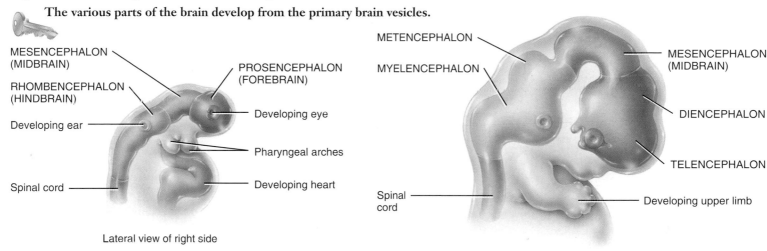

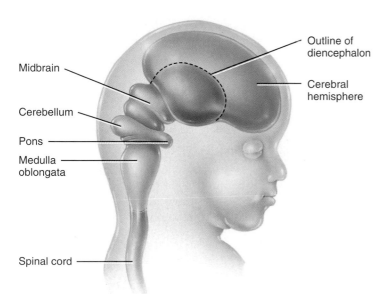

(a) Three-four week embryo showing primary brain vesicles

(b) Seven-week embryo showing secondary brain vesicles

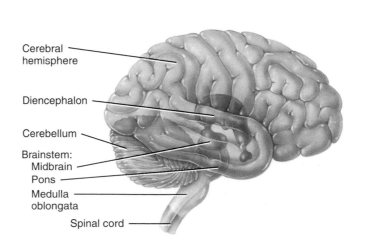

(c) Eleven-week fetus showing expanding cerebral hemispheres overgrowing the diencephalon

(d) Brain at birth (the diencephalon and superior portion of the brain stem have been projected to the surface)

 Which primary brain vesicle does not develop into a secondary brain vesicle?

- The diencephalon develops into the *epithalamus, thalamus, subthalamus, hypothalamus,* and *pineal gland* and houses the third ventricle.

- The mesencephalon develops into the *midbrain,* which surrounds the *aqueduct of the midbrain (cerebral aqueduct).*

- The metencephalon becomes the *pons* and *cerebellum* and houses the upper part of the *fourth ventricle.*

- The myelencephalon develops into the *medulla oblongata* and houses the remainder of the *fourth ventricle.*

Two neural tube defects—spina bifida (see page 197) and anencephaly (absence of the skull and cerebral hemispheres, discussed on page 106)—are associated with low levels of folic acid (folate), one of the B vitamins, in the first few weeks of development. Many foods, especially grain products such as cereals and bread, are now fortified with folic acid; however, the incidence of both disorders is greatly reduced when women who are or may become pregnant take folic acid supplements.

CHECKPOINT

29. What parts of the brain develop from each primary brain vesicle?

AGING AND THE NERVOUS SYSTEM

OBJECTIVE

- Describe the effects of aging on the nervous system.

The brain grows rapidly during the first few years of life. Growth is due mainly to an increase in the size of neurons already present, the proliferation and growth of neuroglia, the development of dendritic branches and synaptic contacts, and continuing myelination of axons. From early adulthood onward, brain mass declines. By the time a person reaches age 80, the brain weighs about 7% less than it did in young adulthood. Although the number of neurons present does not decrease very much, the number of synaptic contacts declines. Associated with the decrease in brain mass is a decreased capacity for sending nerve impulses to and from the brain. As a result, processing of information diminishes, conduction velocity decreases, voluntary motor movements slow down, and reflex times increase.

CHECKPOINT

30. How is brain mass related to age?

APPLICATIONS TO HEALTH

CEREBROVASCULAR ACCIDENT

The most common brain disorder is a **cerebrovascular accident (CVA),** also called a **stroke** or **brain attack.** CVAs affect 500,000 people a year in the United States and represent the third leading cause of death, behind heart attacks and cancer. A CVA is characterized by abrupt onset of persistent neurological symptoms, such as paralysis or loss of sensation, that arise from destruction of brain tissue. Common causes of CVAs are intracerebral hemorrhage (from a blood vessel in the pia mater or brain), emboli (blood clots), and atherosclerosis (formation of cholesterol-containing plaques that block blood flow) of the cerebral arteries.

Among the risk factors implicated in CVAs are high blood pressure, high blood cholesterol, heart disease, narrowed carotid arteries, transient ischemic attacks (TIAs; discussed next), diabetes, smoking, obesity, and excessive alcohol intake.

A clot-dissolving drug called tissue plasminogen activator (t-PA) is now being used to open up blocked blood vessels in the brain. The drug is most effective when administered within three hours of the onset of the CVA, however, and is helpful only for CVAs due to a blood clot. Use of t-PA can decrease the permanent disability associated with these types of CVAs by 50%. New studies show that "cold therapy" might be successful in limiting the amount of residual damage from a CVA. These "cooling" therapies developed from knowledge obtained following examination of cold water drowning victims. States of hypothermia seem to trigger a survival response in which the body requires less oxygen; this principle has been showing promise in the treatment of stroke victims. Some commercial companies now provide "CVA survival kits," which include cooling blankets that can be kept in the home.

TRANSIENT ISCHEMIC ATTACK

A **transient ischemic attack (TIA)** is an episode of temporary cerebral dysfunction caused by impaired blood flow to the brain. Symptoms include dizziness, weakness, numbness, or paralysis in a limb or in one side of the body; drooping of one side of the face; headache; slurred speech or difficulty understanding speech; and a partial loss of vision or double vision. Sometimes nausea or vomiting also occurs. The onset of symptoms is sudden and reaches maximum intensity almost immediately. A TIA usually persists for 5 to 10 minutes and only rarely lasts as long as 24 hours. It leaves no permanent neurological deficits. The causes of the impaired blood flow that lead to TIAs include blood clots, atherosclerosis, and certain blood disorders. About one-third of patients who experience a TIA will have a CVA eventually. Therapy for TIAs includes drugs such as aspirin, which blocks the aggregation of blood platelets, and anticoagulants; cerebral artery bypass grafting; and carotid endarterectomy (removal of the cholesterol-containing plaques and inner lining of an artery).

ALZHEIMER'S DISEASE

Alzheimer's disease (ALTZ-hī-mers) or **AD** is a disabling senile dementia, the loss of reasoning and ability to care for oneself, that afflicts about 11% of the population over age 65. In the United States, there are about 4 million people who suffer from AD; it claims over 100,000 lives a year. Thus, AD is the fourth leading cause of death among the elderly, after heart disease, cancer, and stroke. The cause of most AD cases is still unknown, but evidence suggests it is due to a combination of genetic factors, environmental or lifestyle factors, and the aging process. Mutations in three different genes (coding for presenilin-1, presenilin-2, and amyloid precursor protein) lead to early-onset forms of AD in afflicted families but account for less than 1% of all cases. An environmental risk factor for developing AD is a history of head injury. A similar dementia occurs in boxers, probably caused by repeated blows to the head.

Individuals with AD initially have trouble remembering recent events. They then become confused and forgetful, often repeating questions or getting lost while traveling to familiar places. Disorientation grows and memories of past events disappear, and episodes of paranoia, hallucination, or violent changes in mood may occur. As their minds continue to deteriorate, they lose their ability to read, write, talk, eat, or walk. The disease culminates in dementia. A person with AD usually dies of some complication that afflicts bedridden patients, such as pneumonia.

At autopsy, brains of AD victims show three distinct structural abnormalities.

1. *Loss of neurons that liberate acetylcholine.* A major center of neurons that liberate ACh is the nucleus basalis, which is below the globus pallidus. Axons of these neurons project widely throughout the cerebral cortex and limbic system. Their destruction is a hallmark of Alzheimer's disease.

2. *Beta-amyloid plaques*, clusters of abnormal proteins deposited outside neurons.

3. *Neurofibrillary tangles*, abnormal bundles of protein filaments inside neurons in affected brain regions.

Drugs that inhibit acetylcholinesterase (AChE), the enzyme that inactivates ACh, improve alertness and behavior in about 5% of AD patients. Tacrine, the first anticholinesterase inhibitor approved for treatment of AD in the United States, has significant side effects and requires dosing four times a day. Donepezil, approved in 1998, is less toxic to the liver and has the advantage of once-a-day dosing. Some evidence suggests that vitamin E (an antioxidant), estrogens, ibuprofen, and ginkgo biloba extract may have slight beneficial effects in AD patients.

BRAIN TUMORS

A **brain tumor** is an abnormal growth of tissue in the brain and may be malignant or benign. Unlike most other tumors in the body, malignant and benign tumors may be equally serious, compressing adjacent tissues and causing a buildup of pressure in the skull. The most common malignant tumors are secondary tumors that metastasize from other cancers in the body, such as those in the lungs, breasts, skin (malignant melanoma), blood (leukemia), and lymphatic organs (lymphoma). Most primary brain tumors (those that originate within the brain) are gliomas, which develop in neuroglia. The symptoms of a brain tumor depend on its size, location, and rate of growth. Among the symptoms are headache, poor balance and coordination, dizziness, double vision, slurred speech, nausea and vomiting, fever, abnormal pulse and breathing rates, personality changes, numbness and weakness of the limbs, and seizures. Treatment options for brain tumors vary with their size, location, and type and may include surgery, radiation therapy, or chemotherapy alone or in combination. Unfortunately, chemotherapeutic agents do not readily cross the blood–brain barrier.

KEY MEDICAL TERMS ASSOCIATED WITH THE BRAIN AND THE CRANIAL NERVES

Agnosia (ag-NŌ-zē-a; *a-*=without; *-gnosia*=knowledge) Inability to recognize the significance of sensory stimuli such as sounds, sights, smells, tastes, and touch.

Apraxia (a-PRAK-sē-a; *-praxia*=coordinated) Inability to carry out purposeful movements in the absence of paralysis.

Arousal (a-ROW-zal) Awakening from sleep, a response due to stimulation of the reticular activating system (RAS).

Coma A state of deep unconsciousness from which a person cannot be aroused due to damage to the RAS or other parts of the brain. A comatose patient lies in a sleeplike state with eyes closed. In the lightest stages of coma, brain stem and spinal cord reflexes persist, but in the deepest stages, even these reflexes are lost. If respiratory and cardiovascular controls are lost, the patient dies.

Consciousness (KON-shus-nes) A state of wakefulness in which an individual is fully alert, aware, and oriented, partly as a result of feedback between the cerebral cortex and reticular activating system.

Delirium (de-LIR-ē-um=off the track) A transient disorder of abnormal cognition and disordered attention accompanied by disturbances of the sleep–wake cycle and psychomotor behavior (hyperactivity or hypoactivity of movements and speech). Also called **acute confusional state (ACS).**

Dementia (de-MEN-shē-a; *de-*=away from; *-mentia*=mind) Permanent or progressive general loss of intellectual abilities, including impairment of memory, judgment and abstract thinking, and changes in personality.

Encephalitis (en′-sef-a-LĪ-tis) An acute inflammation of the brain caused by either a direct attack by any of several viruses or an allergic reaction to any of the many viruses that are normally harmless to the central nervous system. If the virus affects the spinal cord as well, the condition is called **encephalomyelitis.**

Encephalopathy (en-sef′-a-LOP-a-thē; *encephalo*=brain, *-pathos*= disease) Any disorder of the brain.

Learning The ability to acquire new knowledge or skills through instruction or experience.

Lethargy (LETH-ar-jē) A condition of functional sluggishness.

Memory The process by which knowledge acquired through learning is retained over time.

Microcephaly (mī-krō-SEF-a-lē; *micro-*=small; *-cephal*=head) A congenital condition that involves the development of a small brain and skull and frequently results in mental retardation.

Reye's (RĪZ) syndrome Occurs after a viral infection, particularly chickenpox or influenza, most often in children or teens who have taken aspirin; characterized by vomiting and brain dysfunction (disorientation, lethargy, and personality changes) that may progress to coma and death.

Sleep A state of altered consciousness or partial unconsciousness from which an individual can be aroused by many different stimuli.

Stupor (STOO-por) Unresponsiveness from which a patient can be aroused only briefly and only by vigorous and repeated stimulation.

STUDY OUTLINE

Overview of Brain Organization and Blood Supply (p. 584)

1. The major parts of the brain are the brain stem, cerebellum, diencephalon, and cerebrum.
2. The brain is protected by cranial bones and the cranial meninges.
3. The cranial meninges are continuous with the spinal meninges. From superficial to deep they are the dura mater, arachnoid mater, and pia mater.
4. Blood flow to the brain is mainly via the internal carotid and vertebral arteries.
5. Any interruption of the oxygen or glucose supply to the brain can result in weakening of, permanent damage to, or death of brain cells.
6. The blood–brain barrier (BBB) causes different substances to move between the blood and the brain tissue at different rates and prevents the movement of some substances from blood into the brain.

Cerebrospinal Fluid (p. 587)

1. Cerebrospinal fluid (CSF) is formed in the choroid plexuses and circulates through the lateral ventricles, third ventricle, fourth ventricle, subarachnoid space, and central canal. Most of the fluid is absorbed into the blood across the arachnoid villi of the superior sagittal blood sinus.
2. Cerebrospinal fluid provides mechanical protection, chemical protection, and circulation of nutrients.

The Brain Stem (p. 590)

1. The medulla oblongata is continuous with the superior part of the spinal cord and contains both motor and sensory tracts. It contains nuclei that are reflex centers for regulation of heart rate, respiratory rate, vasoconstriction, swallowing, coughing, vomiting, and sneezing. It also contains nuclei associated with cranial nerves VIII through XII.
2. The pons is superior to the medulla. It connects the spinal cord with the brain and links parts of the brain with one another by way of tracts. Pontine nuclei relay nerve impulses related to voluntary skeletal movements from the cerebral cortex to the cerebellum. The pons contains the pneumotaxic and apneustic centers, which help control breathing. It contains nuclei associated with cranial nerves V–VII and the vestibular branch of cranial nerve VIII.
3. The midbrain connects the pons and diencephalon and surrounds the cerebral aqueduct. It conveys motor impulses from the cerebrum to the cerebellum and spinal cord, sends sensory impulses from the spinal cord to the thalamus, and regulates auditory and

visual reflexes. It also contains nuclei associated with cranial nerves III and IV.
4. A large part of the brain stem consists of small areas of gray matter and white matter called the reticular formation, which helps maintain consciousness, causes awakening from sleep, and contributes to regulating muscle tone.

The Cerebellum (p. 595)

1. The cerebellum occupies the inferior and posterior aspects of the cranial cavity. It consists of two lateral hemispheres and a medial, constricted vermis.
2. The cerebellum connects to the brain stem by three pairs of cerebellar peduncles.
3. The cerebellum coordinates contractions of skeletal muscles and maintains normal muscle tone, posture, and balance.

The Diencephalon (p. 596)

1. The diencephalon surrounds the third ventricle and consists of the thalamus, hypothalamus, epithalamus, and subthalamus.
2. The thalamus is superior to the midbrain and contains nuclei that serve as relay stations for all sensory impulses to the cerebral cortex. It also allows crude appreciation of pain, temperature, and pressure and mediates some motor activities.
3. The hypothalamus is inferior to the thalamus. It controls and integrates the autonomic nervous system, connects the nervous and endocrine systems, functions in rage and aggression, controls body temperature, regulates food and fluid intake, and establishes circadian rhythms.
4. The epithalamus consists of the pineal gland and the habenular nuclei. The pineal gland secretes melatonin, which is thought to promote sleep and to help set the body's biological clock.
5. The subthalamus connects to motor areas of the cerebrum.
6. Circumventricular organs (CVOs) can monitor chemical changes in the blood because they lack the blood–brain barrier.

The Cerebrum (p. 600)

1. The cerebrum is the largest part of the brain. Its cortex contains gyri (convolutions), fissures, and sulci.
2. The cerebral hemispheres are divided into four lobes: frontal, parietal, temporal, and occipital.
3. The white matter of the cerebrum is deep to the cortex and consists of myelinated and unmyelinated axons extending to other regions as association, commissural, and projection fibers.
4. The basal ganglia are several groups of nuclei in each cerebral

hemisphere. They help control large, automatic movements of skeletal muscles and help regulate muscle tone.

5. The limbic system encircles the upper part of the brain stem and the corpus callosum. It functions in emotional aspects of behavior and memory.

6. Table 19.2 on page 606 summarizes the functions of various parts of the brain.

Functional Organization of the Cerebral Cortex (p. 607)

1. The sensory areas of the cerebral cortex allow perception of sensory impulses. The motor areas are the regions that govern muscular movement. The association areas are concerned with more complex integrative functions.

2. The primary somatosensory area (areas 1, 2, and 3) receives nerve impulses from somatic sensory receptors for touch, proprioception, pain, and temperature. Each point within the area receives impulses from a specific part of the face or body.

3. The primary visual area (area 17) receives impulses that convey visual information. The primary auditory area (areas 41 and 42) interprets the basic characteristics of sound such as pitch and rhythm. The primary gustatory area (area 43) receives impulses for taste. The primary olfactory area (area 28) receives impulses for smell.

4. Motor areas include the primary motor area (area 4), which controls voluntary contractions of specific muscles or groups of muscles, and Broca's speech area (areas 44 and 45), which controls production of speech.

5. The somatosensory association area (areas 5 and 7) permits you to determine the exact shape and texture of an object without looking at it and to sense the relationship of one body part to another. The visual association area (areas 18 and 19) relates present to past visual experiences and is essential for recognizing and evaluating what is seen. The auditory association area (area 22) deals with the meanings of sounds.

6. Wernicke's area (area 22 and possibly 39 and 40) interprets the meaning of speech by translating words into thoughts. The common integrative area (areas 5, 7, 39, and 40) integrates sensory interpretations from the association areas and impulses from other areas, allowing thoughts based on sensory inputs.

7. The premotor area (area 6) generates nerve impulses that cause specific groups of muscles to contract in specific sequences. The frontal eye field area (area 8) controls voluntary scanning movements of the eyes.

8. Subtle anatomical differences exist between the two hemispheres, and each has unique functions. Each hemisphere receives sensory signals from and controls the opposite side of the body. The left hemisphere is more important for language, numerical and scientific skills, and reasoning. The right hemisphere is more important for musical and artistic awareness, spatial and pattern perception, recognition of faces, emotional content of language, identifying odors, and generating mental images of sight, sound, touch, taste, and smell.

9. Brain waves generated by the cerebral cortex are recorded from the surface of the head in an electroencephalogram (EEG). The EEG may be used to diagnose epilepsy, infections, and tumors.

Cranial Nerves (p. 610)

1. Twelve pairs of cranial nerves originate from the nose, eyes, inner ear, brain stem, and spinal cord.

2. They are named primarily based on their distribution and are numbered I–XII in order of attachment to the brain. Table 19.4 on pages 620–623 summarizes the types, locations, functions, and disorders of the cranial nerves.

Development of the Nervous System (p. 624)

1. The development of the nervous system begins with a thickening of a region of the ectoderm called the neural plate.

2. During embryological development, primary brain vesicles form from the neural tube and serve as forerunners of various parts of the brain.

3. The telencephalon forms the cerebrum, the diencephalon develops into the thalamus and hypothalamus, the mesencephalon develops into the midbrain, the metencephalon develops into the pons and cerebellum, and the myelencephalon forms the medulla.

Aging and the Nervous System (p. 626)

1. The brain grows rapidly during the first few years of life.

2. Age-related effects involve loss of brain mass and decreased capacity for sending nerve impulses.

Q SELF-QUIZ QUESTIONS

Choose the one best answer to the following questions.

1. The diencephalon consists of the:
   a. insula and hypothalamus.
   b. pons and hypothalamus.
   c. hypothalamus and midbrain.
   d. thalamus and hypothalamus.
   e. thalamus and midbrain.

2. Blood flows into the brain mainly via the:
   a. superior sagittal sinus.
   b. internal and external jugular veins.
   c. internal carotid and vertebral arteries.
   d. anterior and posterior communicating arteries.
   e. external carotid and subclavian arteries.

3. The blood–cerebrospinal fluid barrier is formed primarily by the tight junctions between:
   a. microglia.
   b. oligodendrocytes.
   c. neurons.
   d. endothelial cells.
   e. ependymal cells.

4. Which of the following is not located in the cerebrum?
   a. globus pallidus
   b. corpora quadrigemina
   c. corpus striatum
   d. caudate nucleus
   e. lentiform nucleus

5. Which of the following does not contribute to the formation of a blood–brain barrier?
   a. tight junctions between the endothelial cells
   b. the continuous, thick basement membrane of the capillary endothelium
   c. a layer of oligodendrocytes
   d. processes of astrocytes that contact the capillary walls
   e. All of the above are correct.

6. Damage to the occipital lobe of the cerebrum would most likely cause:

    a. loss of hearing. b. loss of vision. c. loss of ability to smell.
    d. paralysis.   e. loss of muscle sense (proprioception).

7. The medullary rhythmicity center in the medulla oblongata controls:

    a. heart rate.
    b. the basic rhythm of breathing.
    c. blood pressure.
    d. the pattern of hormone secretion throughout the lifespan.
    e. swinging the arms while walking.

8. Which of the following is *not* a function of the hypothalamus?

    a. control of conscious skeletal movements
    b. regulation of hormone secretion from the pituitary gland
    c. regulation of body temperature
    d. regulation of water intake
    e. establishment of sleep/wake cycles

Complete the following.

9. The structures responsible for the production of cerebrospinal fluid are the _____.

10. The ventricle located between the right and left halves of the thalamus is the _____.

11. Two areas of the pons that help control breathing are the _____ center and the _____ center.

12. A group of sensory axons that projects from the brain stem to the cerebral cortex that helps maintain consciousness is known as the _____.

13. White matter fibers that transmit nerve impulses between gyri in the same hemisphere are called _____ fibers.

14. The outer layer of the cerebrum is called the _____. It is composed of _____ matter, which means that it contains mainly neuron cell bodies. The surface of the cerebrum is a series of tightly packed ridges called _____, with shallow grooves between them called _____.

15. Write "w" for white matter or "g" for gray matter:

    (a) corpus callosum: _____
    (b) olive: _____
    (c) cerebral cortex: _____
    (d) corpora quadrigemina: _____
    (e) arbor vitae: _____
    (f) caudate and lenticular nuclei: _____

16. Number the following in the correct order for the circulation of CSF, from its production to its reabsorption:

    (a) arachnoid villi: _____
    (b) median and lateral apertures: _____
    (c) aqueduct of the midbrain: _____
    (d) lateral ventricle: _____
    (e) third ventricle: _____
    (f) fourth ventricle: _____
    (g) interventricular foramen: _____
    (h) subarachnoid space: _____
    (i) choroid plexuses: _____
    (j) superior sagittal sinus: _____

17. The most posterior part of the hypothalamus, which serves as a relay station for reflexes related to the sense of smell, is a pair of small, rounded projections called the _____.

18. The pineal gland secretes the hormone _____, which is thought to promote sleepiness and to help set the body's biological clock.

Are the following statements true or false?

19. Circumventricular organs coordinate homeostatic activities of the endocrine and nervous systems.

20. The limbic system functions in the control of behavior.

21. The three parts of the brain stem are the medulla oblongata, pons, and cerebellum.

22. The basal ganglia contain the supraoptic nucleus and the geniculate nuclei.

Matching

23. Match the following (answers may be used more than once):

    __ (a) has two main parts, separated by a longitudinal fissure, but joined internally by the corpus callosum

    __ (b) responsible for coordination of skilled movements and regulation of posture and balance

    __ (c) cranial nerves III–IV attach to this brain part

    __ (d) cranial nerves V–VIII attach to this brain part

    __ (e) cranial nerves VIII–XII attach to this brain part

    __ (f) regulates food and fluid intake and body temperature

    __ (g) all sensations are relayed through here

    __ (h) centers for control of heart rate and respiration are located here

    __ (i) constitutes four-fifths of the diencephalon

    __ (j) connects to the pituitary gland via the infundibulum

    __ (k) part of the brain stem; contains tracts that connect the cerebellum, midbrain, and medulla oblongata

    __ (l) surrounds the cerebral aqueduct, and contains nuclei that serve as reflex centers for head and eye movements

    (1) hypothalamus
    (2) medulla
    (3) midbrain
    (4) pons
    (5) thalamus
    (6) cerebellum
    (7) cerebrum

24. Match the cranial nerve and the description:

    __ (a) accommodation of lens for near vision
    __ (b) control of secretion of digestive fluids; slow heart rate
    __ (c) chewing
    __ (d) facial expression, secretion of saliva, and tears
    __ (e) movement of tongue during speech and swallowing
    __ (f) taste; sensations from tongue; monitor blood pressure and blood oxygen levels
    __ (g) smell
    __ (h) movements of head and shoulders
    __ (i) movement of eyeball via lateral rectus muscle
    __ (j) vision
    __ (k) equilibrium
    __ (l) movement of eyeball via superior oblique muscle

    (1) olfactory
    (2) optic
    (3) oculomotor
    (4) trochlear
    (5) trigeminal
    (6) abducens
    (7) facial
    (8) vestibulo-cochlear
    (9) glosso-pharyngeal
    (10) vagus
    (11) accessory
    (12) hypoglossal

CRITICAL THINKING QUESTIONS

1. An elderly relative suffered a stroke and now has difficulty moving her right arm and also has speech problems. What areas of the brain were damaged by the stroke?

2. Wolfgang partied a little too hard one night and passed out drunk in his bathroom at home. He awoke with a lump on his head from hitting the sink, but he thought he was okay. However, when Tony came over a day later, he said, "Man, are you still drunk? You're not making any sense, and you look like you're going to pass out again!" Before Wolfgang could answer, he vomited. What do you think—is Wolfgang still drunk? Why, or why not?

3. Dr. M. D. Hatter has developed a drug that inhibits the activity of the amygdala of the limbic system. Do you think this is good or a bad thing? Explain your answer.

4. Alicia just figured out that her Human Anatomy class actually starts at 9:00 A.M. and not at 9:15 A.M., which has been her arrival time since the beginning of the term. One of the other students remarks that Alicia's "gray matter is pretty thin." Should Alicia thank him?

5. Dwayne's first trip to the dentist after a 10-year absence resulted in extensive dental work. He received numbing injections of anesthetic in several locations during the session. While having lunch right after the appointment, soup dribbles out of Dwayne's mouth because he has no feeling in his left upper lip, right lower lip, and tip of his tongue. What happened to Dwayne?

ANSWERS TO FIGURE QUESTIONS

19.1 The largest part of the brain is the cerebrum.

19.2 From superficial to deep, the three cranial meninges are the dura mater, arachnoid mater, and pia mater.

19.3 The brain stem is anterior to the fourth ventricle, and the cerebellum is posterior to it.

19.4 Cerebrospinal fluid is reabsorbed by the arachnoid villi that project into the dural venous sinuses.

19.5 The medulla oblongata contains the pyramids; the midbrain contains the cerebral peduncles; pons means "bridge."

19.6 Decussation means crossing to the opposite side. The functional consequence of decussation of the pyramids is that one side of the cerebrum controls muscles on the opposite side of the body.

19.7 The cerebral peduncles are the main sites where tracts extend and nerve impulses are conducted between the superior parts of the brain and the inferior parts of the brain and spinal cord.

19.8 The cerebellar peduncles contain the axons that carry information into and out of the cerebellum.

19.9 The intermediate mass connects the right and left halves of the thalamus.

19.10 From posterior to anterior, the four major regions of the hypothalamus are the mammillary, tuberal, supraoptic, and preoptic regions.

19.11 The gray matter enlarges more rapidly during development, in the process producing convolutions or gyri (folds), sulci (shallow grooves), and fissures (deep grooves).

19.12 Association tracts connect gyri of the same hemisphere; commissural tracts connect gyri in opposite hemispheres; projection tracts connect the cerebrum with the thalamus, brain stem, and spinal cord.

19.13 The basal ganglia are lateral, superior, and inferior to the thalamus.

19.14 The hippocampus is the component of the limbic system that functions with the cerebrum in memory.

19.15 The common integrative area integrates interpretation of visual, auditory, and somatic sensations; the motor speech area translates thoughts into speech; the premotor area controls skilled muscular movements; the gustatory areas interpret sensations related to taste; the auditory areas interpret pitch and rhythm; the visual areas interpret shape, color, and movement of objects; the frontal eye field area controls voluntary scanning movements of the eyes.

19.16 Axons in the olfactory tracts terminate in the primary olfactory area in the temporal lobe of the cerebral cortex.

19.17 Most axons in the optic tracts terminate in the primary visual area in the occipital lobe of the cerebral cortex.

19.18 The superior branch of the oculomotor nerve is distributed to the superior rectus muscle; the trochlear nerve is the smallest cranial nerve.

19.19 The trigeminal nerve is the largest cranial nerve.

19.20 Motor axons of the facial nerve originate in the pons.

19.21 Sensory axons of the cochlear branch originate in the spiral organ, the organ of hearing.

19.22 The glossopharyngeal nerve exits the skull through the jugular foramen.

19.23 The vagus nerve is located medial and posterior to the internal jugular vein and common carotid artery in the neck.

19.24 The accessory nerve is the only cranial nerve that originates from both the brain and spinal cord.

19.25 Two important motor functions of the hypoglossal nerve are speech and swallowing.

19.26 The gray matter of the nervous system derives from the mantle layer cells of the neural tube.

19.27 The mesencephalon does not develop into a secondary brain vesicle.

20 | THE AUTONOMIC NERVOUS SYSTEM

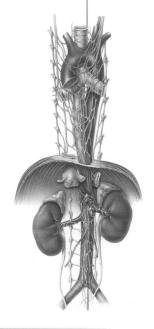

INTRODUCTION   As you learned in Chapter 17, the **autonomic nervous system (ANS)** is a division of the peripheral nervous system that consists of (1) *autonomic sensory neurons* in visceral organs and blood vessels that convey information into (2) *integrating centers* in the central nervous system (CNS), and (3) *autonomic motor neurons* that propagate from the CNS to various effector tissues, thereby regulating the activity of smooth muscle, cardiac muscle, and many glands. Functionally, the ANS usually operates without conscious control. The system was originally named *autonomic* because it was thought to operate autonomously or in a self-governing manner, without control by the CNS. However, centers in the hypothalamus and brain stem do regulate ANS reflexes.

In this chapter, we compare structural and functional features of the autonomic nervous system with those of the somatic nervous system, which was introduced in Chapter 17. Then we discuss the anatomy of the ANS and compare the organization and actions of its two major parts, the sympathetic and parasympathetic divisions.

Examine this image closely. What do you see? Can you identify any of the organs? Can you guess which processes are being depicted?

www.wiley.com/college/apcentral

COMPARISON OF SOMATIC AND AUTONOMIC NERVOUS SYSTEMS

OBJECTIVE

- Outline the structural and functional differences between the somatic and autonomic nervous systems.

As you learned in Chapter 17, the somatic nervous system includes both sensory and motor neurons. Sensory neurons convey input from receptors for somatic senses (pain, thermal, tactile, and proprioceptive sensations; see Chapter 21) and from receptors for the special senses (vision, hearing, taste, smell, and equilibrium; see Chapter 22). All these sensations normally are consciously perceived. In turn, somatic motor neurons innervate skeletal muscle—the effector tissue of the somatic nervous system—and produce voluntary movements. When a somatic motor neuron stimulates a skeletal muscle, the muscle contracts; the effect always is excitation. If somatic motor neurons cease to stimulate a muscle, the result is a paralyzed, limp muscle that has no muscle tone. In addition, even though we are generally not conscious of breathing, the muscles that generate respiratory movements are skeletal muscles controlled by somatic motor neurons. If the respiratory motor neurons become inactive, breathing stops. A few skeletal muscles, such as those in the middle ear, are controlled by reflexes and cannot be contracted voluntarily.

The main input to the ANS comes from **autonomic sensory neurons.** Mostly, these neurons are associated with interoceptors (receptors inside the body), such as chemoreceptors that monitor blood CO_2 level, and mechanoreceptors that detect the degree of stretch in the walls of organs or blood vessels. These sensory signals are not consciously perceived most of the time, although intense activation may produce conscious sensations. Two examples of perceived visceral sensations are sensations of pain from damaged viscera and angina pectoris (chest pain) from inadequate blood flow to the heart. Some sensations monitored by somatic sensory (Chapter 21) and special sensory neurons (Chapter 22) also influence the ANS. For example, pain can produce dramatic changes in some autonomic activities.

Autonomic motor neurons regulate visceral activities by either increasing (exciting) or decreasing (inhibiting) ongoing activities in their effector tissues, which are cardiac muscle, smooth muscle, and glands. Changes in the diameter of the pupils, dilation and constriction of blood vessels, and adjustment of the rate and force of the heartbeat are examples of autonomic motor responses. Unlike skeletal muscle, tissues innervated by the ANS often function to some extent even if their nerve supply is damaged. For example, the heart continues to beat when it is removed for transplantation. Single-unit smooth muscle, like that found in the lining of the gastrointestinal tract, contracts rhythmically on its own, and glands produce some secretions in the absence of ANS control.

Most autonomic responses cannot be consciously altered or suppressed to any great degree. You probably cannot voluntarily slow your heartbeat to half its normal rate. For this reason, some autonomic responses are the basis for polygraph ("lie detector") tests. Nevertheless, practitioners of yoga or other techniques of meditation may learn how to modulate at least some of their autonomic activities through long practice. (Biofeedback, in which monitoring devices display information about a body function such as heart rate or blood pressure, enhances the ability to learn such conscious control.) Signals from the general somatic and special senses, acting via the limbic system, also influence responses of autonomic motor neurons. For example, seeing a bike about to hit you, hearing the squealing brakes of a nearby car as you cross the street, or being grabbed from behind by an attacker would increase the rate and force of your heartbeat.

Recall from Chapter 10 that the axon of a single, myelinated somatic motor neuron extends from the CNS all the way to the skeletal muscle fibers in its motor unit (Figure 20.1a). By contrast, most autonomic motor pathways consist of two motor neurons *in series*, one following the other (Figure 20.1b). The first neuron has its cell body in the CNS; its myelinated axon extends from the CNS to an **autonomic ganglion.** (Recall that a ganglion is a collection of neuronal cell bodies outside the CNS.) The cell body of the second neuron is also in that autonomic ganglion; its unmyelinated axon extends directly from the ganglion to the effector (smooth muscle, cardiac muscle, or a gland). In some autonomic pathways, the first motor neuron extends to the adrenal medullae (inner portions of the adrenal glands) rather than an autonomic ganglion. Whereas all somatic motor neurons release only acetylcholine (ACh) as their neurotransmitter, autonomic motor neurons release either ACh or norepinephrine (NE).

The output (motor) part of the ANS has two divisions: the **sympathetic division** and the **parasympathetic division.** Most organs have **dual innervation:** They receive impulses from both sympathetic and parasympathetic neurons. In general, nerve impulses from one division of the ANS stimulate the organ to increase its activity (excitation), whereas impulses from the other division decrease the organ's activity (inhibition). For example, an increased rate of nerve impulses from the sympathetic division increases heart rate, whereas an increased rate of nerve impulses from the parasympathetic division decreases heart rate. Table 20.1 on page 635 summarizes the similarities and differences between the somatic and autonomic nervous systems.

1. Why is the autonomic nervous system so-named?
2. What are the main input and output components of the autonomic nervous system?

Figure 20.1   Motor neuron pathways in the (a) somatic nervous system and (b) autonomic nervous system (ANS). Note that autonomic motor neurons release either acetylcholine (ACh) or norepinephrine (NE); somatic motor neurons release ACh.

🔑 Somatic nervous system stimulation always excites its effectors (skeletal muscle fibers); stimulation by the autonomic nervous system either excites or inhibits visceral effectors.

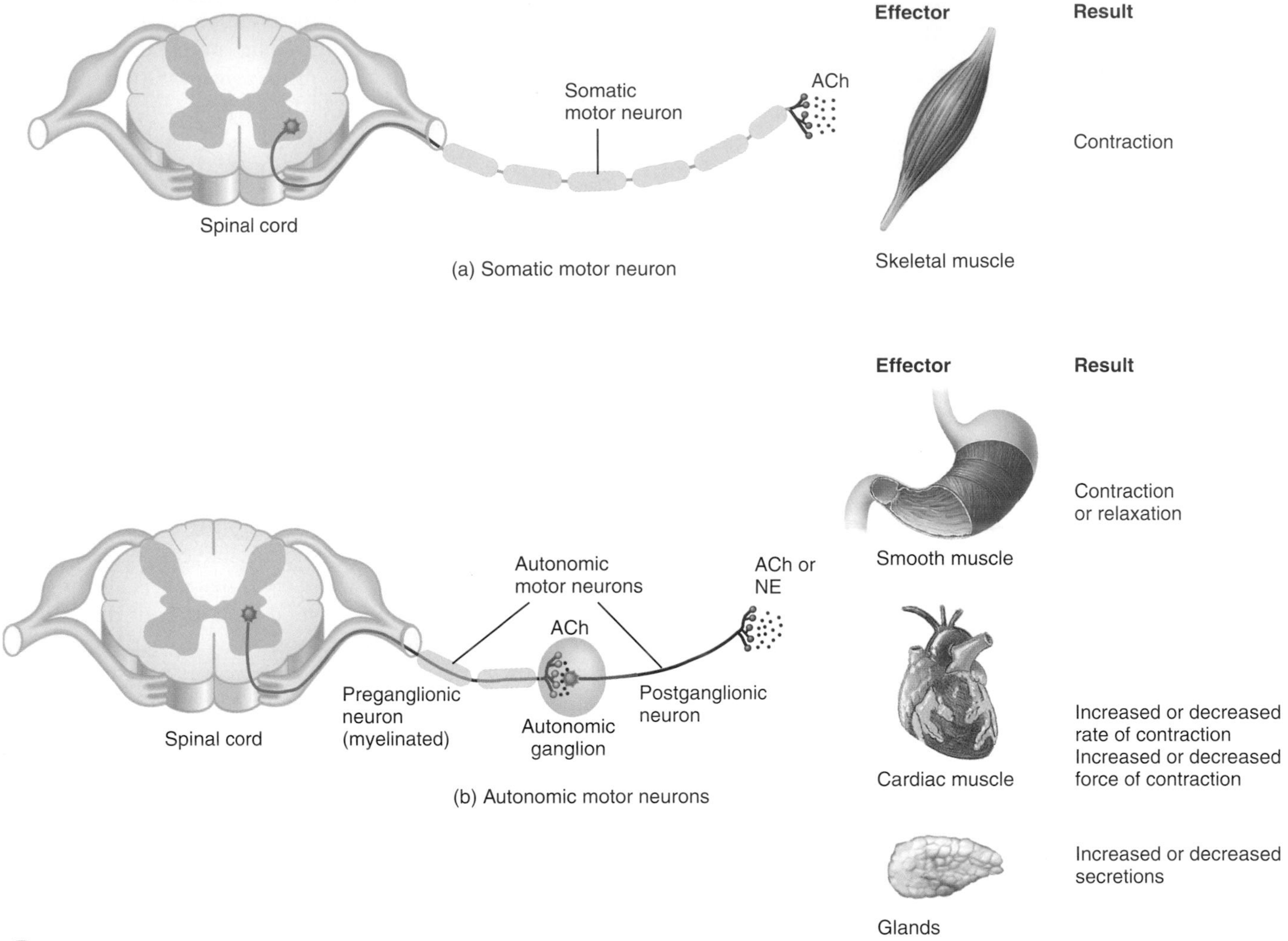

Effector   **Result**

ACh

Somatic
motor neuron

Contraction

Spinal cord

(a) Somatic motor neuron

Skeletal muscle

Effector   **Result**

Contraction
or relaxation

Smooth muscle

Autonomic
motor neurons

ACh or
NE

ACh

Preganglionic
neuron
(myelinated)

Postganglionic
neuron

Spinal cord

Autonomic
ganglion

(b) Autonomic motor neurons

Increased or decreased
rate of contraction
Increased or decreased
force of contraction

Cardiac muscle

Increased or decreased
secretions

Glands

 **What does dual innervation mean?**

ANATOMY OF AUTONOMIC MOTOR PATHWAYS

OBJECTIVES

• Describe preganglionic and postganglionic neurons of the autonomic nervous system.
• Compare the anatomical components of the sympathetic and parasympathetic divisions of the autonomic nervous system.

Anatomical Components of an Autonomic Motor Pathway

The first of the two motor neurons in any autonomic motor pathway is called a **preganglionic neuron** (Figure 20.1b). Its cell body is in the brain or spinal cord, and its axon exits the CNS as part of a cranial or spinal nerve. The axon of a pregan-

glionic neuron is a small-diameter, myelinated type B fiber that usually extends to an autonomic ganglion. There it synapses with a **postganglionic neuron,** the second neuron in the autonomic motor pathway (Figure 20.1b). Notice that the postganglionic neuron lies entirely outside the CNS. Its cell body and dendrites are located in an autonomic ganglion, where it forms one or more synapses with preganglionic neurons. The axon of a postganglionic neuron is a small-diameter, unmyelinated type C fiber that terminates in a visceral effector. Thus, preganglionic neurons convey nerve impulses from the CNS to autonomic ganglia, and postganglionic neurons relay the impulses from autonomic ganglia to visceral effectors.

Preganglionic Neurons

In the sympathetic division, the preganglionic neurons have their cell bodies in the lateral horns of the gray matter in the 12

TABLE 20.1 COMPARING THE SOMATIC AND AUTONOMIC NERVOUS SYSTEMS

	SOMATIC NERVOUS SYSTEM	AUTONOMIC NERVOUS SYSTEM
Sensory input	Special senses and somatic senses.	Mainly from interoceptors; some from special senses and somatic senses.
Control of motor output	Voluntary control from cerebral cortex, with contributions from basal ganglia, cerebellum, brain stem, and spinal cord.	Involuntary control from limbic system, hypothalamus, brain stem, and spinal cord; limited control from cerebral cortex.
Motor neuron pathway	One-neuron pathway: Somatic motor neurons extending from CNS synapse directly with effector.	Usually two-neuron pathway: Preganglionic neurons extending from CNS synapse with postganglionic neurons in an autonomic ganglion, and postganglionic neurons extending from ganglion synapse with a visceral effector. Also, preganglionic neurons may extend from CNS to synapse with cells of adrenal medullae.
Neurotransmitters and hormones	All somatic motor neurons release ACh.	All preganglionic axons release acetylcholine (ACh); most sympathetic postganglionic neurons release norepinephrine (NE), those to most sweat glands release ACh; all parasympathetic postganglionic neurons release ACh; adrenal medullae release epinephrine and NE.
Effectors	Skeletal muscle.	Smooth muscle, cardiac muscle, and glands.
Responses	Contraction of skeletal muscle.	Contraction or relaxation of smooth muscle; increased or decreased rate and force of contraction of cardiac muscle; increased or decreased secretions of glands.

thoracic segments and the first two lumbar segments of the spinal cord (Figure 20.2). For this reason, the sympathetic division is also called the **thoracolumbar division** (thōr´-a-kō-LUM-bar), and the axons of the sympathetic preganglionic neurons are known as the **thoracolumbar outflow.**

Cell bodies of preganglionic neurons of the parasympathetic division are located in the nuclei of four cranial nerves in the brain stem (III, VII, IX, and X) and in the lateral gray horns of the second through fourth sacral segments of the spinal cord (see Figure 20.3). Hence, the parasympathetic division is also known as the **craniosacral division** (krā´-nē-ō-SĀ-kral), and the axons of the parasympathetic preganglionic neurons are referred to as the **craniosacral outflow.**

Autonomic Ganglia

The autonomic ganglia may be divided into three general groups: Two of the groups are components of the sympathetic division, and one group is a component of the parasympathetic division.

SYMPATHETIC GANGLIA The sympathetic ganglia are the sites of synapses between sympathetic preganglionic and postganglionic neurons. The two groups of sympathetic ganglia are sympathetic trunk ganglia and prevertebral ganglia. **Sympathetic trunk ganglia** (also called *vertebral chain ganglia* or *paravertebral ganglia*) lie in a vertical row on either side of the vertebral column. These ganglia extend from the base of the skull to the coccyx (Figure 20.2). Because the sympathetic trunk ganglia are near the spinal cord, most sympathetic preganglionic axons are short. Postganglionic axons from sympathetic trunk

ganglia mostly innervate organs above the diaphragm. Examples of sympathetic trunk ganglia are the **superior, middle,** and **inferior cervical ganglia** (Figure 20.2).

The second group of sympathetic ganglia, the **prevertebral** (*collateral*) **ganglia,** lies anterior to the vertebral column and close to the large abdominal arteries. In general, postganglionic axons from prevertebral ganglia innervate organs below the diaphragm. There are three major prevertebral ganglia: (1) The **celiac ganglion** (SĒ-lē-ak) is on either side of the celiac artery just inferior to the diaphragm. (2) The **superior mesenteric ganglion** is near the beginning of the superior mesenteric artery in the upper abdomen. (3) The **inferior mesenteric ganglion** is near the beginning of the inferior mesenteric artery in the middle of the abdomen (Figure 20.2; see also Figure 20.4).

PARASYMPATHETIC GANGLIA Preganglionic axons of the parasympathetic division synapse with postganglionic neurons in **terminal** (*intramural*) **ganglia.** Most of these ganglia are located close to or actually within the wall of a visceral organ. Because the axons of parasympathetic preganglionic neurons extend from the CNS to a terminal ganglion in an innervated organ, they are longer than most of the axons of sympathetic preganglionic neurons. Examples of terminal ganglia include the ciliary ganglion, pterygopalatine ganglion, submandibular ganglion, and otic ganglion (Figure 20.3 on page 637).

Autonomic Plexuses

In the thorax, abdomen, and pelvis, axons of both sympathetic and parasympathetic neurons form tangled networks called **autonomic plexuses,** many of which lie along major arteries.

Figure 20.2   Structure of the sympathetic division of the autonomic nervous system. Solid lines represent preganglionic axons; dashed lines represent postganglionic axons. Although the innervated structures are shown only for one side of the body for diagrammatic purposes, the sympathetic division actually innervates tissues and organs on both sides.

Cell bodies of sympathetic preganglionic neurons are located in the lateral horns of gray matter in the 12 thoracic and first two lumbar segments of the spinal cord.

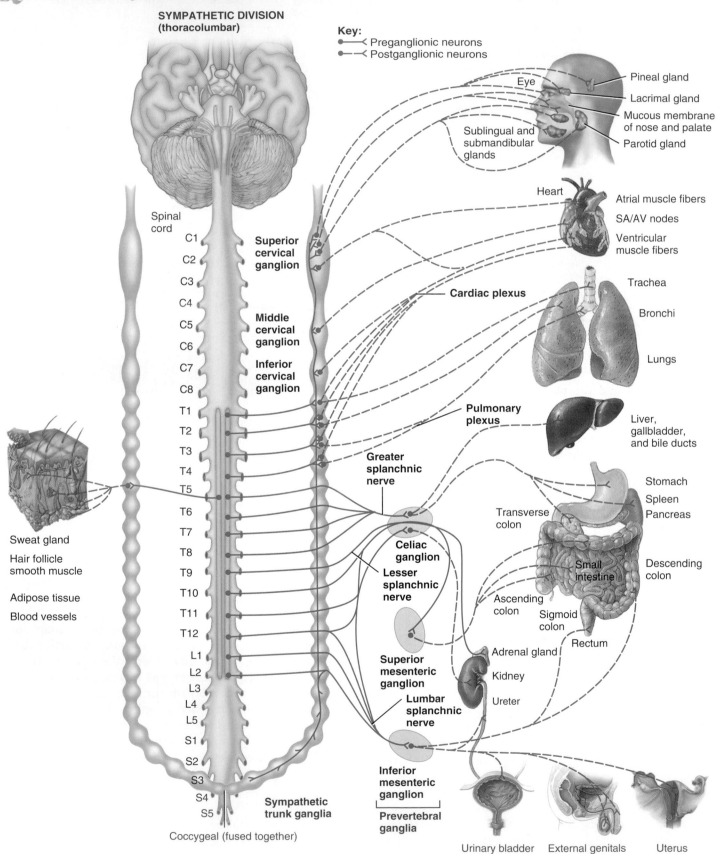

Which division, sympathetic or parasympathetic, has longer preganglionic axons? Why?

Figure 20.3 **Structure of the parasympathetic division of the autonomic nervous system.** Solid lines represent preganglionic axons; dashed lines represent postganglionic axons. Although the innervated structures are shown only for one side of the body for diagrammatic purposes, the parasympathetic division actually innervates tissues and organs on both sides.

🗝 Cell bodies of parasympathetic preganglionic neurons are located in brain stem nuclei and in the lateral horns of gray matter in the second through fourth sacral segments of the spinal cord.

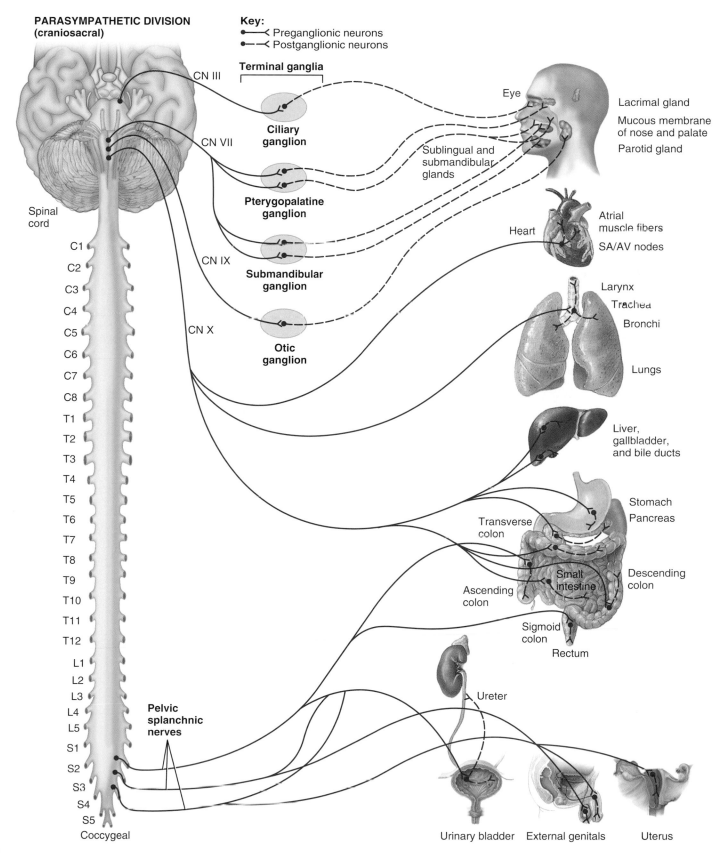

PARASYMPATHETIC DIVISION
(craniosacral)

Key:
━━━◁ Preganglionic neurons
━ ─ ─◁ Postganglionic neurons

❓ Which ganglia are associated with the parasympathetic division? Sympathetic division?

The autonomic plexuses also may contain sympathetic ganglia and axons of autonomic sensory neurons. The major plexuses in the thorax are the **cardiac plexus,** which supplies the heart, and the **pulmonary plexus,** which supplies the bronchial tree (Figure 20.4; see also Figure 20.2).

The abdomen and pelvis also contain major autonomic plexuses and often are named after the artery along which they are distributed (Figure 20.4). The **celiac** (*solar*) **plexus** is the largest autonomic plexus and surrounds the celiac and superior mesenteric arteries. It contains two large celiac ganglia and a dense network of autonomic axons and is distributed to the liver, gallbladder, stomach, pancreas, spleen, kidneys, medulla (inner region) of the adrenal gland, testes, and ovaries. The **superior mesenteric plexus** contains the superior mesenteric ganglion and supplies the small and large intestine. The **inferior mesenteric plexus** contains the inferior mesenteric ganglion, which innervates the large intestine. The **hypogastric plexus** supplies pelvic viscera. The **renal plexuses** contain the renal ganglion and supply the renal arteries within the kidneys and the ureters.

Postganglionic Neurons

SYMPATHETIC POSTGANGLIONIC NEURONS Once axons of sympathetic preganglionic neurons pass to sympathetic trunk ganglia, they may connect with postganglionic neurons in one of the following ways (Figure 20.5):

❶ An axon may synapse with postganglionic neurons in the first ganglion it reaches.

❷ An axon may ascend or descend to a higher or lower ganglion before synapsing with postganglionic neurons.

❸ An axon may continue, without synapsing, through the sympathetic trunk ganglion to end at a prevertebral ganglion and synapse with postganglionic neurons there.

Figure 20.4 Autonomic plexuses in the thorax, abdomen, and pelvis.

An autonomic plexus is a network of sympathetic and parasympathetic axons that sometimes also includes autonomic sensory axons and sympathetic ganglia.

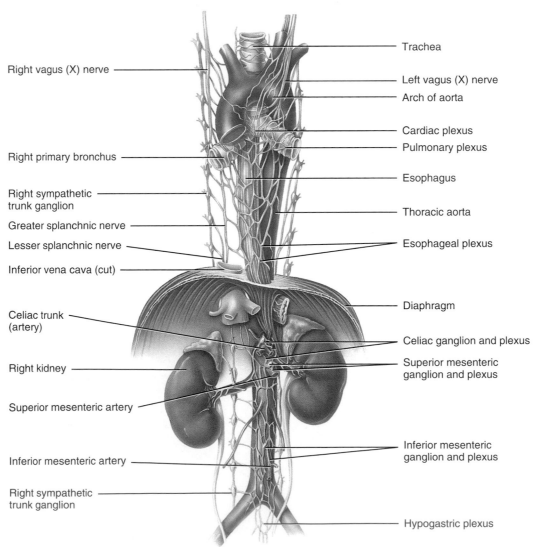

Right vagus (X) nerve

Right primary bronchus

Right sympathetic trunk ganglion

Greater splanchnic nerve

Lesser splanchnic nerve

Inferior vena cava (cut)

Celiac trunk (artery)

Right kidney

Superior mesenteric artery

Inferior mesenteric artery

Right sympathetic trunk ganglion

Trachea

Left vagus (X) nerve

Arch of aorta

Cardiac plexus

Pulmonary plexus

Esophagus

Thoracic aorta

Esophageal plexus

Diaphragm

Celiac ganglion and plexus

Superior mesenteric ganglion and plexus

Inferior mesenteric ganglion and plexus

Hypogastric plexus

Which is the largest autonomic plexus?

Figure 20.5 Types of connections between ganglia and postganglionic neurons in the sympathetic division of the ANS. Numbers correspond to descriptions in the text. Also illustrated are the gray and white rami communicantes.

🔑 Sympathetic ganglia lie in two chains on either side of the vertebral column (sympathetic trunk ganglia) and near large abdominal arteries anterior to the vertebral column (prevertebral ganglia).

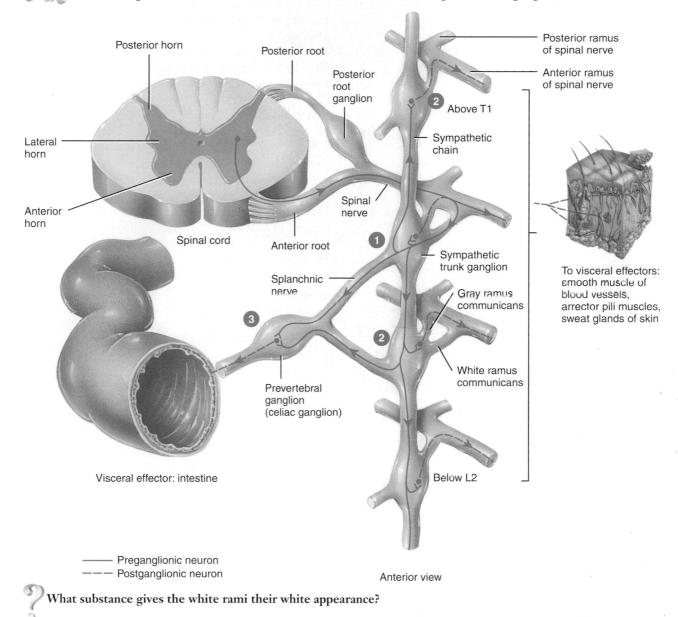

Preganglionic neuron

--- Postganglionic neuron

Anterior view

❓ **What substance gives the white rami their white appearance?**

In addition, some preganglionic sympathetic axons extend to and terminate in the adrenal medullae.

A single sympathetic preganglionic fiber has many axon collaterals (branches) and may synapse with 20 or more postganglionic neurons. This pattern of projection is an example of divergence (see Chapter 17) and helps explain why many sympathetic responses affect almost the entire body simultaneously. After exiting their ganglia, the postganglionic axons typically terminate in several visceral effectors (see Figure 20.2).

PARASYMPATHETIC POSTGANGLIONIC NEURONS Axons of preganglionic neurons of the parasympathetic division pass to terminal ganglia near or within a visceral effector (see

Figure 20.3). In the ganglion, the presynaptic neuron usually synapses with only four or five postsynaptic neurons, all of which supply a single visceral effector. Thus, parasympathetic responses can be localized to a single effector. With this background in mind, we can now examine some specific structural features of the sympathetic and parasympathetic divisions of the ANS.

Structure of the Sympathetic Division

Cell bodies of sympathetic preganglionic neurons are part of the lateral horns of all thoracic segments and of the first two lumbar segments of the spinal cord (see Figure 20.2). The preganglionic axons leave the spinal cord through the anterior root of a spinal nerve along with the somatic motor neurons at the same segmental level. After exiting through the intervertebral foramina, the myelinated preganglionic sympathetic axons enter a short pathway called a **white ramus** before passing to the nearest sympathetic trunk ganglion on the same side (Figure 20.5). Collectively, the white rami are called the **white rami communicantes** (kō-mū-ni-KAN-tēz; singular is *ramus communicans*). The "white" in their name indicates that they contain myelinated axons. Only the thoracic and first two or three lumbar nerves have white rami communicantes. The white rami communicantes connect the anterior ramus of the spinal nerve with the ganglia of the sympathetic trunk. The paired sympathetic trunk ganglia are arranged anterior and lateral to the vertebral column, one on either side. Typically, there are 3 cervical, 11 or 12 thoracic, 4 or 5 lumbar, 4 or 5 sacral sympathetic trunk ganglia, and 1 coccygeal ganglion. The right and left coccygeal ganglia are fused together and usually lie at the midline. Although the sympathetic trunk ganglia extend interiorly from the neck, chest, and abdomen to the coccyx, they receive preganglionic axons only from the thoracic and lumbar segments of the spinal cord (see Figure 20.2).

The cervical portion of each sympathetic trunk ganglion is located in the neck and is subdivided into superior, middle, and inferior ganglia (see Figure 20.2). Postganglionic neurons leaving the **superior cervical ganglion** serve the head and heart. They are distributed to sweat glands, smooth muscle of the eye, blood vessels of the face, lacrimal glands, nasal mucosa, salivary glands (submandibular, sublingual, and parotid), and the heart. Gray rami communicantes (described shortly) from the superior cervical ganglion also pass to the upper two to four cervical spinal nerves. Postganglionic neurons leaving the **middle cervical ganglion** and **inferior cervical ganglion** innervate the heart.

The thoracic portion of each sympathetic trunk ganglion lies anterior to the necks of the corresponding ribs. This region of the sympathetic trunk receives most of the sympathetic preganglionic axons, and its postganglionic neurons innervate the heart, lungs, bronchi, and other thoracic viscera. In the skin, these neurons also innervate sweat glands, blood vessels, and arrector pili muscles of hair follicles.

The lumbar portion of each sympathetic trunk ganglion lies lateral to the corresponding lumbar vertebrae. The sacral region of the sympathetic trunk ganglion lies in the pelvic cavity on the medial side of the sacral foramina. Unmyelinated postganglionic axons from the lumbar and sacral sympathetic trunk ganglia enter a short pathway called a **gray ramus** and then merge with a spinal nerve or join the hypogastric plexus via direct visceral branches. The **gray rami communicantes** are structures containing the postganglionic axons that connect the ganglia of the various portions of the sympathetic trunk ganglion to spinal nerves (Figure 20.5). The axons are unmyelinated. Gray rami communicantes outnumber the white rami because there is a gray ramus leading to each of the 31 pairs of spinal nerves.

As preganglionic axons extend from a white ramus communicans into the sympathetic trunk ganglion, they give off several axon collaterals (branches). These collaterals terminate and synapse in several ways. Some synapse in the first ganglion at the level of entry. Others pass up or down the sympathetic trunk for a variable distance to form the **sympathetic chains,** the fibers on which the ganglia are strung (Figure 20.5). Many postganglionic axons rejoin the spinal nerves through gray rami and supply peripheral visceral effectors such as sweat glands, smooth muscle in blood vessels, and arrector pili muscles of hair follicles.

Some preganglionic axons pass through the sympathetic trunk without terminating in it. Beyond the trunk, they form nerves known as **splanchnic nerves** (SPLANK-nik; see Figure 20.2), which extend to and terminate in the outlying prevertebral ganglia. Splanchnic nerves from the thoracic area terminate in the **celiac ganglion,** where the preganglionic neurons synapse with postganglionic cell bodies. Preganglionic axons from the fifth through ninth or tenth thoracic ganglia (T5–T9 or T10) form the **greater splanchnic nerve,** which pierces the diaphragm and enters the celiac ganglion of the celiac plexus. From there, postganglionic neurons extend to the stomach, spleen, liver, kidney, and small intestine. Preganglionic axons from the tenth and eleventh thoracic ganglia (T10–T11) form the **lesser splanchnic nerve,** which pierces the diaphragm and passes through the celiac plexus to enter the superior mesenteric ganglion of the superior mesenteric plexus. Postganglionic neurons from the superior mesenteric ganglion innervate the small intestine and colon. The **lowest splanchnic nerve,** which is not always present, is formed by preganglionic axons from the twelfth thoracic ganglia (T12) or a branch of the lesser splanchnic nerve. It passes through the diaphragm, and enters the renal plexus near the kidney. Postganglionic neurons from the renal plexus supply kidney arterioles and the ureter.

Preganglionic axons that form the **lumbar splanchnic nerve** from the first through fourth lumbar ganglia (L1–L4) enter the inferior mesenteric plexus and terminate in the inferior mesenteric ganglion, where they synapse with postganglionic neurons. Axons of postganglionic neurons extend through the hypogastric plexus and supply the distal colon and rectum, urinary bladder, and genital organs. Postganglionic axons leaving the prevertebral ganglia follow the course of various arteries to abdominal and pelvic visceral effectors.

Sympathetic preganglionic neurons also extend to the adrenal medullae. Developmentally, the adrenal medullae and sympathetic ganglia are derived from the same tissue, the neural crest (see Figure 19.26 on page 624). The adrenal medullae are modified sympathetic ganglia, and their cells are similar to sympathetic postganglionic neurons. Rather than extending to another organ, however, these cells release hormones into the blood. Upon stimulation by sympathetic preganglionic neurons, the adrenal medullae release a mixture of hormones—about 80% **epinephrine**, 20% **norepinephrine**, and a trace amount of **dopamine.**

HORNER'S SYNDROME

In **Horner's syndrome,** the sympathetic innervation to one side of the face is lost due to an inherited mutation, an injury, or a disease that affects sympathetic outflow through the superior cervical ganglion. Symptoms occur on the affected side and include ptosis (drooping of the upper eyelid), miosis (constricted pupil), flushing of the face, and anhidrosis (lack of sweating). ■

Structure of the Parasympathetic Division

Cell bodies of parasympathetic preganglionic neurons are found in nuclei in the brain stem and in the lateral horns of the second through fourth sacral segments of the spinal cord (see Figure 20.3). Their axons emerge as part of a cranial nerve or as part of the anterior root of a spinal nerve. The **cranial parasympathetic outflow** consists of preganglionic axons that extend from the brain stem in four cranial nerves. The **sacral parasympathetic outflow** consists of preganglionic axons in anterior roots of the second through fourth sacral nerves. The preganglionic axons of both the cranial and sacral outflows end in terminal ganglia, where they synapse with postganglionic neurons.

The cranial outflow has five components: four pairs of ganglia and the plexuses associated with the vagus (X) nerve. The four pairs of cranial parasympathetic ganglia innervate structures in the head and are located close to the organs they innervate (see Figure 20.3).

1. The **ciliary ganglia** lie lateral to each optic (II) nerve near the posterior aspect of the orbit. Preganglionic axons pass with the oculomotor (III) nerves to the ciliary ganglia. Postganglionic axons from the ganglia innervate smooth muscle fibers in the eyeball.

2. The **pterygopalatine ganglia** (ter′-i-gō-PAL-a-tīn) are located lateral to the sphenopalatine foramen, between the sphenoid and palatine bones. They receive preganglionic axons from the facial (VII) nerve and send postganglionic axons to the nasal mucosa, palate, pharynx, and lacrimal glands.

3. The **submandibular ganglia** are found near the ducts of the submandibular salivary glands. They receive preganglionic axons from the facial nerves and send postganglionic axons to the submandibular and sublingual salivary glands.

4. The **otic ganglia** are situated just inferior to each foramen ovale. They receive preganglionic axons from the glossopharyngeal (IX) nerves and send postganglionic axons to the parotid salivary glands.

Preganglionic axons that leave the brain as part of the vagus (X) nerves carry nearly 80% of the total craniosacral outflow. Vagal axons extend to many terminal ganglia in the thorax and abdomen. Because the terminal ganglia are close to or in the walls of their visceral effectors, postganglionic parasympathetic axons are very short. As the vagus nerve passes through the thorax, it sends axons to the heart and the airways of the lungs. In the abdomen, it supplies the liver, gallbladder, stomach, pancreas, small intestine, and part of the large intestine.

The sacral parasympathetic outflow consists of preganglionic axons from the anterior roots of the second through fourth sacral nerves (S2–S4), which form the **pelvic splanchnic nerves** (see Figure 20.3). These nerves synapse with parasympathetic postganglionic neurons located in terminal ganglia in the

walls of the innervated viscera. From the ganglia, parasympathetic postganglionic axons innervate smooth muscle and glands in the walls of the colon, ureters, urinary bladder, and reproductive organs.

3. Why is the sympathetic division called the thoracolumbar division even though its ganglia extend from the cervical to the sacral region?

4. Prepare a list of the organs served by each sympathetic and parasympathetic ganglion.

5. Describe the locations of sympathetic trunk ganglia, prevertebral ganglia, and terminal ganglia. Which types of autonomic neurons synapse in each type of ganglion?

6. Why may the sympathetic division produce simultaneous effects throughout the body, whereas parasympathetic effects typically are localized to specific organs?

ANS NEUROTRANSMITTERS AND RECEPTORS

● Describe the neurotransmitters and receptors involved in autonomic responses.

Autonomic neurons are classified based on the neurotransmitter they produce and release. The receptors for the neurotransmitters are integral membrane proteins that are located in the plasma membrane of the postsynaptic neuron or effector cell.

Cholinergic Neurons and Receptors

Cholinergic neurons (kō′-lin-ER-jik) release the neurotransmitter **acetylcholine** (as′-ē-til-KŌ-lēn) or **ACh**. In the ANS, the cholinergic neurons include (1) all sympathetic and parasympathetic preganglionic neurons, (2) sympathetic postganglionic neurons that innervate most sweat glands, and (3) all parasympathetic postganglionic neurons (Figure 20.6).

ACh is stored in synaptic vesicles and released by exocytosis. It then diffuses across the synaptic cleft and binds with specific **cholinergic receptors,** integral membrane proteins in the *postsynaptic* plasma membrane. The two types of cholinergic receptors, both of which bind ACh, are nicotinic receptors and muscarinic receptors. **Nicotinic receptors** are present in the plasma membranes of dendrites and cell bodies of both sympathetic and parasympathetic postganglionic neurons (Figure 20.6a, b) and in the motor end plate at the neuromuscular junction. They are so-named because nicotine mimics the action of ACh by binding to these receptors. (Nicotine, a natural substance in tobacco leaves, is not normally present in the bodies of nonsmokers). **Muscarinic receptors** are present in the plasma membranes of all effectors (smooth muscle, cardiac muscle, and glands) innervated by parasympathetic postganglionic axons. In addition, most sweat glands, which receive their innervation from cholinergic sympathetic postganglionic

Figure 20.6 Cholinergic neurons (aqua) and adrenergic neurons (orange) in the sympathetic and parasympathetic divisions. Cholinergic neurons release acetylcholine, whereas adrenergic neurons release norepinephrine. Cholinergic and adrenergic receptors all are integral membrane proteins located in the plasma membrane of a postsynaptic neuron or an effector cell.

🔑 **Most sympathetic postganglionic neurons are adrenergic; other autonomic neurons are cholinergic.**

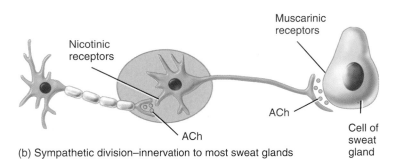

(a) Sympathetic division–innervation to most effector tissues

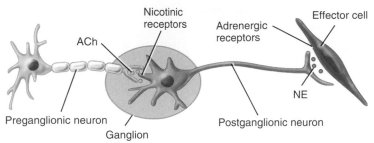

(b) Sympathetic division–innervation to most sweat glands

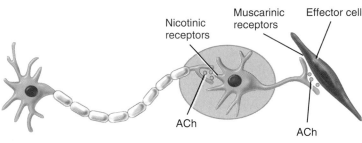

(c) Parasympathetic division

❓ **Which neurons are cholinergic and possess nicotinic ACh receptors? What type of receptors for ACh do the effector tissues innervated by these neurons possess?**

neurons, possess muscarinic receptors (see Figure 20.6). These receptors are so-named because a mushroom poison called muscarine mimics the actions of ACh by binding to muscarinic receptors.

Activation of nicotinic receptors by ACh causes depolarization and thus excitation of the postsynaptic cell—which can be a postganglionic neuron, an autonomic effector, or a skeletal muscle fiber. Activation of muscarinic receptors by ACh sometimes causes depolarization (excitation) and sometimes causes hyperpolarization (inhibition), depending on which particular cell bears the muscarinic receptors. For example, binding of ACh to muscarinic receptors inhibits (relaxes) smooth muscle sphincters in the gastrointestinal tract. By contrast, ACh excites smooth muscle fibers in the circular muscles of the iris of the eye, causing them to contract. Because acetylcholine is quickly inactivated by the enzyme **acetylcholinesterase (AChE),** effects triggered by cholinergic neurons are brief.

Adrenergic Neurons and Receptors

In the ANS, **adrenergic neurons** (ad′-ren-ER-jik) release **norepinephrine** (nor′-ep-i-NEF-rin) or **NE,** also known as **noradrenalin** (Figure 20.6a). Most sympathetic postganglionic neurons are adrenergic. Like ACh, NE is synthesized and stored in synaptic vesicles and released by exocytosis. Molecules of NE diffuse across the synaptic cleft and bind to specific adrenergic receptors on the postsynaptic membrane, causing either excitation or inhibition of the effector cell.

Adrenergic receptors bind both NE and epinephrine, a hormone with actions similar to NE. As noted previously, NE is released as a neurotransmitter by sympathetic postganglionic neurons. Both epinephrine and NE are released as hormones into the blood by the adrenal medullae. The two main types of adrenergic receptors are **alpha (α) receptors** and **beta (β) receptors,** which are found on visceral effectors innervated by most sympathetic postganglionic axons. These receptors are further classified into subtypes—α_1, α_2, β_1, β_2, and β_3—based on the specific responses they elicit and by their selective binding of the drugs that activate or block them. Although there are some exceptions, activation of α_1 and β_1 receptors generally produces excitation, whereas activation of α_2 and β_2 receptors causes inhibition of effector tissues. β_3 receptors are present only on cells of brown adipose tissue, where their activation causes thermogenesis (heat production). Cells of most effectors contain either α or β receptors; some visceral effector cells contain both. NE stimulates alpha receptors more strongly than beta receptors, whereas epinephrine is a potent stimulator of both alpha and beta receptors.

The activity of NE at a synapse is terminated either when the NE is taken up by the axon that released it or when the NE is enzymatically inactivated by either **catechol-O-methyltransferase (COMT)** or **monoamine oxidase (MAO).** Compared to ACh, NE lingers in the synaptic cleft for a longer time. Thus, effects triggered by adrenergic neurons typically are longer lasting than those triggered by cholinergic neurons.

DRUGS AND RECEPTOR SELECTIVITY

A large variety of drugs and natural products can selectively activate or block specific cholinergic or adrenergic receptors. An **agonist** is a substance that binds to and activates a receptor, in the process mimicking the effect of a natural neurotransmitter or hormone. Phenylephrine, an adrenergic agonist at α_1 receptors, is a common ingredient in cold and sinus medications. Because it constricts blood vessels in the nasal mucosa, phenylephrine reduces production of mucus, thus relieving nasal congestion. An **antagonist** is a substance that binds to and blocks a receptor, thereby preventing a natural neurotransmitter or hormone from exerting its effect. For example, atropine, which blocks muscarinic ACh receptors, dilates the pupils, reduces glandular secretions, and relaxes smooth muscle in the gastrointestinal tract. It is used to dilate the pupils during eye examinations, in the treatment of smooth muscle disorders such as iritis and intestinal hypermotility, and as an antidote for chemical warfare agents that inactivate AChE.

Propranolol (Inderal) often is prescribed for patients with hypertension (high blood pressure). It is a nonselective beta blocker, meaning it binds to all types of beta receptors and prevents their activation by epinephrine and norepinephrine. The desired effects of propranolol are due to its *blockade* of β_1 receptors—namely, decreased heart rate and force of contraction and a consequent decrease in blood pressure. Undesired effects due to blockade of β_2 receptors may include hypoglycemia (low blood glucose), resulting from decreased glycogen breakdown and decreased gluconeogenesis (the conversion of a noncarbohydrate into glucose in the liver), and mild bronchoconstriction (narrowing of the airways). If these side effects pose a threat to the patient, a selective β_1 blocker such as metoprolol (Lopressor) can be prescribed. ■

CHECKPOINT

7. Why are cholinergic and adrenergic neurons so-named?
8. What neurotransmitters and hormones bind to adrenergic receptors?
9. What do the terms *agonist* and *antagonist* mean?

FUNCTIONS OF THE ANS

OBJECTIVE

● Describe the major responses of the body to stimulation by the sympathetic and parasympathetic divisions of the ANS.

As noted earlier, most body organs receive innervation from both divisions of the ANS, which typically work in opposition to one another. The balance between sympathetic and parasympathetic activity is regulated by the hypothalamus. Typically, the hypothalamus turns up sympathetic activity at the same time it turns down parasympathetic activity, and vice versa. As you learned in the last section of this chapter, the two divisions affect body organs differently because of the different neurotransmit-

ters released by their postganglionic neurons and the different adrenergic and cholinergic receptors on the cells of their effector organs. A few structures receive only sympathetic innervation—sweat glands, arrector pili muscles attached to hair follicles in the skin, the kidneys, the spleen, most blood vessels, and the adrenal medullae (see Figure 20.2). In these structures there is no opposition from the parasympathetic division; increases and decreases in sympathetic activity are responsible for the changes.

Sympathetic Responses

During physical or emotional stress, the sympathetic division dominates the parasympathetic division. High sympathetic activity favors body functions that can support vigorous physical activity and rapid production of ATP. At the same time, the sympathetic division decreases body functions that favor the storage of energy. Physical exertion and a variety of emotions—such as fear, embarrassment, or rage—stimulate the sympathetic division. Visualizing body changes that occur during "E situations" such as exercise, emergency, excitement, and embarrassment will help you remember most of the sympathetic responses. Activation of the sympathetic division and release of hormones by the adrenal medullae set in motion a series of physiological responses collectively called the **fight-or-flight response**, which includes the following effects:

1. The pupils of the eyes dilate.
2. Heart rate, force of heart contraction, and blood pressure increase.
3. The airways dilate, allowing faster movement of air into and out of the lungs.
4. Blood vessels that supply organs involved in exercise or fighting off danger—skeletal muscles, cardiac muscle, liver, and adipose tissue—dilate, allowing greater blood flow through these tissues.
5. Liver cells perform glycogenolysis (breakdown of glycogen to glucose), and adipose tissue cells perform lipolysis (breakdown of triglycerides to fatty acids and glycerol).
6. Release of glucose by the liver increases blood glucose level.
7. Processes that are not essential for meeting the stressful situation are inhibited. For example, the blood vessels that supply the kidneys and gastrointestinal tract constrict, which decreases blood flow through these tissues. The result is a slowing of urine formation and digestive activities, which are not essential during exercise.

The effects of sympathetic stimulation are longer lasting and more widespread than the effects of parasympathetic stimulation for three reasons: (1) Sympathetic postganglionic axons diverge more extensively; as a result, many tissues are activated simultaneously. (2) AChE quickly inactivates ACh, whereas NE lingers in the synaptic cleft for a longer period. (3) Epinephrine and NE secreted as hormones into the blood from the adrenal medulla intensify and prolong the responses caused by NE released as a neurotransmitter from sympathetic postganglionic axons. These blood-borne hormones circulate throughout the body, affecting all tissues that have α and β receptors. In time, blood-borne NE and epinephrine are inactivated by enzymatic destruction in the liver.

Parasympathetic Responses

In contrast to the "fight-or-flight" activities of the sympathetic division, the parasympathetic division enhances "rest-and-digest" activities. Parasympathetic responses support body functions that conserve and restore body energy during times of rest and recovery. In the quiet intervals between periods of exercise, parasympathetic impulses to the digestive glands and the smooth muscle of the gastrointestinal tract predominate over sympathetic impulses. This allows energy-supplying food to be digested and absorbed. At the same time, parasympathetic responses decrease body functions that support physical activity.

The acronym "SLUDD" can be helpful in remembering five parasympathetic responses. It stands for salivation (S), lacrimation (L), urination (U), digestion (D), and defecation (D). All of these activities are stimulated mainly by the parasympathetic division. Besides the increasing SLUDD responses, other important parasympathetic responses are "three

decreases": decreased heart rate, decreased diameter of airways (bronchoconstriction), and decreased diameter (constriction) of the pupils.

Table 20.2 compares the structural and functional features of the sympathetic and parasympathetic divisions of the ANS. Table 20.3 lists the responses of glands, cardiac muscle, and smooth muscle to stimulation by the sympathetic and parasympathetic divisions of the ANS.

CHECKPOINT

10. Define cholinergic and adrenergic neurons and receptors.
11. What are some examples of the antagonistic effects of the sympathetic and parasympathetic divisions of the autonomic nervous system?
12. What happens during the fight-or-flight response?
13. Why is the parasympathetic division of the ANS called an energy conservation/restoration system?
14. Using Table 20.3 as a reference, describe the sympathetic response in a frightening situation for each of the following body parts: hair follicles, iris of eye, lungs, spleen, adrenal medullae, urinary bladder, stomach, intestines, gallbladder, liver, heart, arterioles of the abdominal viscera, and arterioles of skeletal muscles.

TABLE 20.2	COMPARISON OF SYMPATHETIC AND PARASYMPATHETIC DIVISIONS OF THE ANS	
	SYMPATHETIC	**PARASYMPATHETIC**
Distribution	Wide regions of the body: skin, sweat glands, arrector pili muscles of hair follicles, adipose tissue, smooth muscle of blood vessels.	Limited mainly to head and to viscera of thorax, abdomen, and pelvis; some blood vessels.
Location of preganglionic neuron cell bodies and site of outflow	Cell bodies of preganglionic neurons are located in lateral gray horns of spinal cord segments T1–L2. Axons of preganglionic neurons constitute thoracolumbar outflow.	Cell bodies of preganglionic neurons are located in the nuclei of cranial nerves III, VII, IX, and X and the lateral gray horns of spinal cord segments S2–S4. Axons of preganglionic neurons constitute craniosacral outflow.
Associated ganglia	Two types: sympathetic trunk ganglia and prevertebral ganglia.	One type: terminal ganglia.
Ganglia locations	Close to CNS and distant from visceral effectors.	Typically near or within wall of visceral effectors.
Axon length and divergence	Preganglionic neurons with short axons synapse with many postganglionic neurons with long axons that pass to many visceral effectors.	Preganglionic neurons with long axons usually synapse with four to five postganglionic neurons with short axons that pass to a single visceral effector.
Rami communicantes	Both present; white rami communicantes contain myelinated preganglionic axons, and gray rami communicantes contain unmyelinated postganglionic axons.	Neither present.
Neurotransmitters	Preganglionic neurons release acetylcholine (ACh), which is excitatory and stimulates postganglionic neurons; most postganglionic neurons release norepinephrine (NE); postganglionic neurons that innervate most sweat glands and some blood vessels in skeletal muscle release ACh.	Preganglionic neurons release acetylcholine (ACh), which is excitatory and stimulates postganglionic neurons; postganglionic neurons release ACh.
Physiological effects	Fight-or-flight responses.	Rest-and-digest activities.

TABLE 20.3 EFFECTS OF SYMPATHETIC AND PARASYMPATHETIC DIVISIONS OF THE ANS

VISCERAL EFFECTOR	EFFECT OF SYMPATHETIC STIMULATION (α OR β ADRENERGIC RECEPTORS, EXCEPT AS NOTED)*	EFFECT OF PARASYMPATHETIC STIMULATION (MUSCARINIC ACh RECEPTORS)
Glands		
Adrenal medullae	Secretion of epinephrine and NE (niocotinic ACh receptors).	No known effect.
Lacrimal (tear)	Slight secretion of tears (α).	Secretion of tears.
Pancreas	Inhibits secretion of digestive enzymes and the hormone insulin (α_2); promotes secretion of the hormone glucagon (β_2).	Secretion of digestive enzymes and the hormone insulin.
Posterior pituitary	Secretion of antidiuretic hormone (ADH) (β_1).	No known effect.
Pineal	Increases synthesis and release of melatonin (β).	No known effect.
Sweat	Increases sweating in most body regions (muscarinic ACh receptors); sweating on palms and soles (α_1).	No known effect.
Adipose tissue[†]	Lipolysis (breakdown of triglycerides into fatty acids and glycerol) (β_1); release of fatty acids into blood (β_1 and β_3).	No known effect.
Liver[†]	Glycogenolysis (conversion of glycogen into glucose); glyconeogenesis (conversion of noncarbohydrates into glucose); decreased bile secretion (α_1 and β_2).	Glycogen synthesis; increased bile secretion.
Kidney, juxtaglomerular cells[†]	Secretion of renin (β_1).	No known effect.
Cardiac (heart) muscle	Increased heart rate and force of atrial and ventricular contractions (β_1).	Decreased heart rate; decreased force of atrial contraction.
Smooth muscle		
Iris, radial muscle	Contraction → dilation of pupil (α_1).	No known effect.
Iris, circular muscle	No known effect.	Contraction → constriction of pupil.
Ciliary muscle of eye	Relaxation for distant vision (β_2).	Contraction for close vision.
Lungs, bronchial muscle	Relaxation → airway dilation (β_2).	Contraction → airway constriction.
Gallbladder and ducts	Relaxation (β_2).	Contraction → increased release of bile into small intestine.
Stomach and intestines	Decreased motility and tone (α_1, α_2, β_2); contraction of sphincters (α_1).	Increased motility and tone; relaxation of sphincters.
Spleen	Contraction and discharge of stored blood into general circulation (α_1).	No known effect.
Ureter	Increases motility (α_1).	Increases motility (?).
Urinary bladder	Relaxation of muscular wall (β_2); contraction of sphincter (α_1).	Contraction of muscular wall; relaxation of sphincter.
Uterus	Inhibits contraction in nonpregnant woman (β_2); promotes contraction in pregnant woman (α_1).	Minimal effect.
Sex organs	In males: contraction of smooth muscle of ductus (vas) deferens, seminal vesicle, prostate → ejaculation of semen (α_1).	Vasodilation; erection of clitoris (females) and penis (males).
Hair follicles, arrector pili	Contraction → erection of hairs (α_1).	No known effect.
Vascular smooth muscle		
Salivary gland arterioles	Vasoconstriction, which decreases K^+ and water secretion (α_1)..	Vasodilation, which increases K^+ and water secretion.
Gastric gland arterioles	Vasoconstriction, which inhibits secretion of gastric juice (α_1).	Secretion of gastric juice.
Intestinal gland arterioles	Vasoconstriction, which inhibits secretion of intestinal juice (α_1).	Secretion of intestinal juice.
Coronary (heart) arterioles	Relaxation → vasodilation (β_2); contraction → vasoconstriction (α_1, α_2).	Contraction → vasoconstriction.
Skin and mucosal arterioles	Contraction → vasoconstriction (α_1).	Vasodilation, which may not be physiologically significant.
Skeletal muscle arterioles	Contraction → vasoconstriction (α_1); relaxation → vasodilation (β_2).	No known effect.
Abdominal viscera arterioles	Contraction → vasoconstriction (α_1, β_2).	No known effect.
Brain arterioles	Slight contraction → vasoconstriction (α_1).	No known effect.
Kidney arterioles	Constriction of blood vessels → decreased urine volume (α_1).	No known effect.
Systemic veins	Contraction → constriction (α_1); relaxation → dilation (β_2).	No known effect.

*Subcategories of α and β receptors are listed if known.

[†]Grouped with glands because they release substances into the blood.

INTEGRATION AND CONTROL OF AUTONOMIC FUNCTIONS

- Describe the components of an autonomic reflex.
- Explain the relationship of the hypothalamus to the ANS.

Autonomic Reflexes

Autonomic reflexes are responses that occur when nerve impulses pass over an autonomic reflex arc. These reflexes play a key role in regulating controlled conditions in the body, such as *blood pressure*, by adjusting heart rate, force of ventricular contraction, and blood vessel diameter; *digestion*, by adjusting the motility (movement) and muscle tone of the gastrointestinal tract; and *defecation* and *urination*, by regulating the opening and closing of sphincters.

The components of an autonomic reflex arc are as follows:

1. *Receptor.* Like the receptor in a somatic reflex arc (see Figure 18.5 on page 563), the receptor in an autonomic reflex arc is the distal end of a sensory neuron, which responds to a stimulus and produces a change that will ultimately trigger nerve impulses. Autonomic sensory receptors are mostly associated with interoceptors.

2. *Sensory neuron.* Conducts nerve impulses from receptors to the CNS.

3. *Integrating center.* Interneurons within the CNS relay signals from sensory neurons to motor neurons. The main integrating centers for most autonomic reflexes are located in the hypothalamus and brain stem. Some autonomic reflexes, such as those for urination and defecation, have integrating centers in the spinal cord.

4. *Motor neurons.* Nerve impulses triggered by the integrating center propagate out of the CNS along motor neurons to an effector. In an autonomic reflex arc, two motor neurons connect the CNS to an effector: The preganglionic neuron conducts motor impulses from the CNS to an autonomic ganglion, and the postganglionic neuron conducts motor impulses from an autonomic ganglion to an effector (see Figure 20.1).

5. *Effector.* In an autonomic reflex arc, the effectors are smooth muscle, cardiac muscle, and glands, and the reflex is called an autonomic reflex.

Autonomic Control by Higher Centers

Normally, we are not aware of muscular contractions of the digestive organs, heartbeat, changes in the diameter of blood vessels, and pupil dilation and constriction because the integrating centers for these autonomic responses are in the spinal cord or the lower regions of the brain. Somatic or autonomic sensory neurons deliver input to these centers, and autonomic motor neurons provide output that adjusts activity in the visceral effector, usually without our conscious perception.

The hypothalamus is the major control and integration center of the ANS. The hypothalamus receives sensory input related to visceral functions, olfaction (smell), and gustation (taste), as well as input related to changes in temperature, and levels of various substances in blood. In addition, it receives input relating to emotions from the limbic system. Output from the hypothalamus influences autonomic centers in the brain stem (such as the cardiovascular, salivation, swallowing, and vomiting centers) and in the spinal cord (such as the defecation and urination reflex centers in the sacral spinal cord).

Anatomically, the hypothalamus is connected to both the sympathetic and parasympathetic divisions of the ANS by axons of neurons whose dendrites and cell bodies are in various hypothalamic nuclei. The axons form tracts from the hypothalamus to sympathetic and parasympathetic nuclei in the brain stem and spinal cord through relays in the reticular formation. The posterior and lateral parts of the hypothalamus control the sympathetic division. As you might expect, stimulation of the posterior or lateral hypothalamus produces an increase in heart rate and force of contraction, a rise in blood pressure due to constriction of blood vessels, an increase in body temperature, dilation of the pupils, and inhibition of the gastrointestinal tract. In contrast, stimulation of the anterior and medial parts of the hypothalamus, which control the parasympathetic division, results in a decrease in heart rate, lowering of blood pressure, constriction of the pupils, and increased secretion and motility of the gastrointestinal tract.

15. Give three examples of activities in the body that are controlled by autonomic reflexes.
16. How does an autonomic reflex arc differ from a somatic reflex arc?

APPLICATIONS TO HEALTH

RAYNAUD'S DISEASE

In **Raynaud's disease** (rā-NOZ) the digits (fingers and toes) become ischemic (lack blood) after exposure to cold or with emotional stress. The condition is due to excessive sympathetic stimulation of smooth muscle in the arterioles of the digits and a heightened response to stimuli that cause vasoconstriction. When arterioles in the digits vasoconstrict in response to sympathetic stimulation, blood flow is greatly diminished. As a result, the digits may blanch (look white due to blockage of blood flow) or become cyanotic (look blue due to deoxygenated blood in capillaries). In extreme cases, the digits may become necrotic from lack of oxygen and nutrients. With rewarming after cold exposure, the arterioles may dilate, causing the fingers and toes to look red. Many patients with Raynaud's disease have low blood pressure. Some have increased numbers of alpha-adrenergic receptors. The disease is most common in young women and occurs more often in cold climates. Patients with Raynaud's disease should avoid exposure to cold, wear warm clothing, and keep the hands and feet warm. Drugs used to treat the condition include nifedipine, a calcium channel blocker that relaxes vascular smooth muscle, and prazosin, which relaxes smooth muscle by blocking alpha receptors. Smoking and the use of alcohol or illicit drugs can exacerbate the symptoms of this disease.

AUTONOMIC DYSREFLEXIA

Autonomic dysreflexia (dis′-rē-FLEK-sē-a) is an exaggerated response of the sympathetic division of the ANS that occurs in about 85% of individuals with spinal cord injury at or above the level of T6. The condition is seen after recovery from spinal shock (see page 578) and occurs due to interruption of the con-

trol of ANS neurons by higher centers. When certain sensory impulses, such as those resulting from stretching of a full urinary bladder, are unable to ascend the spinal cord, mass stimulation of the sympathetic nerves inferior to the level of injury occurs. Other triggers include stimulation of pain receptors and the visceral contractions resulting from sexual stimulation, labor/delivery, and bowel stimulation. Among the effects of increased sympathetic activity is severe vasoconstriction, which elevates blood pressure. In response, the cardiovascular center in the medulla oblongata (1) increases parasympathetic output via the vagus (X) nerve, which decreases heart rate, and (2) decreases sympathetic output, which causes dilation of blood vessels superior to the level of the injury.

Autonomic dysreflexia is characterized by a pounding headache; hypertension; flushed, warm skin with profuse sweating above the injury level; pale, cold, and dry skin below the injury level; and anxiety. It is an emergency condition that requires immediate intervention. The first approach is to quickly identify the problematic stimulus and remove it. If this does not relieve the symptoms, an antihypertensive drug such as clonidine or nitroglycerine can be administered. If left untreated, autonomic dysreflexia can cause seizures, stroke, or heart attack.

KEY MEDICAL TERMS ASSOCIATED WITH THE AUTONOMIC NERVOUS SYSTEM

Autonomic nerve neuropathy (noo-ROP-a-thē) If a neuropathy (specifically a disorder of a cranial or spinal nerve) affects one or more autonomic nerves, there can be multiple effects on the autonomic nervous system that interfere with reflexes. These include fainting and low blood pressure when standing (orthostatic hypotension) due to decreased sympathetic control of the cardiovascular system, constipation, urinary incontinence, and impotence. This type of neuropathy is often caused by long-term diabetes mellitus and is known as *diabetic neuropathy*.

Biofeedback A technique in which an individual is provided with information regarding an autonomic response such as heart rate, blood pressure, or skin temperature. Various electronic monitoring devices provide visual or auditory signals about the autonomic responses. By concentrating on positive thoughts, individuals learn to alter autonomic responses. For example, biofeedback has been used to decrease heart rate and blood pressure and increase skin temperature in order to decrease the severity of migraine headaches.

Dysautonomia (dis-aw-tō-NŌ-mē-a; *dys-*=difficult; *autonomia*=self-governing) An inherited disorder in which the autonomic nervous system functions abnormally, resulting in reduced tear gland secretions, poor vasomotor control, motor incoordination, skin blotching, absence of pain sensa-

tion, difficulty in swallowing, hyporeflexia, excessive vomiting, and emotional instability.

Hyperhydrosis (hi′-per-hī-DRŌ-sis; *hyper-*=above or too much; *hidrosis*=sweat; *-osis*=condition) Excessive or profuse sweating due to intense stimulation of sweat glands.

Mass reflex In cases of severe spinal cord injury above the level of the sixth thoracic vertebra, stimulation of the skin or overfilling of a visceral organ (such as the urinary bladder or colon) below the level of the injury results in intense activation of autonomic and somatic output from the spinal cord as reflex activity returns. The exaggerated response occurs because there is no inhibitory input from the brain. The mass reflex consists of flexor spasms of the lower limbs, evacuation of the urinary bladder and colon, and profuse sweating below the level of the lesion.

Megacolon (*mega-*=big) An abnormally large colon. In congenital megacolon, parasympathetic nerves to the distal segment of the colon do not develop properly. Loss of motor function in the segment causes massive dilation of the normal proximal colon. The condition results in extreme constipation, abdominal distention, and occasionally, vomiting. Surgical removal of the affected segment of the colon corrects the disorder.

STUDY OUTLINE

Comparison of Somatic and Autonomic Nervous Systems (p. 633)

1. The somatic nervous system operates under conscious control; the ANS usually operates without conscious control.

2. Sensory input to the somatic nervous system is mainly from the special senses and somatic senses; sensory input to the ANS is primarily from interoceptors, with some contributions from the special senses and somatic senses.

3. The axons of somatic motor neurons extend from the CNS and synapse directly with an effector. Autonomic motor pathways consist of two motor neurons in series. The axon of the first motor neuron extends from the CNS and synapses in a ganglion with the second motor neuron; the second neuron synapses with an effector.

4. The output (motor) portion of the ANS has two divisions: sympathetic and parasympathetic. Most body organs receive dual innervation; usually one ANS division causes excitation and the other causes inhibition.

5. Somatic nervous system effectors are skeletal muscles; ANS effectors include cardiac muscle, smooth muscle, and glands.

6. Table 20.1 on page 635 compares the somatic and autonomic nervous systems.

Anatomy of Autonomic Motor Pathways (p. 634)

1. Preganglionic neurons are myelinated; postganglionic neurons are unmyelinated.

2. The cell bodies of sympathetic preganglionic neurons are in the lateral gray horns of the 12 thoracic and the first two or three lumbar segments of the spinal cord; the cell bodies of parasympathetic preganglionic neurons are in four cranial nerve nuclei (III, VII, IX, and X) in the brain stem and lateral gray horns of the second through fourth sacral segments of the spinal cord.

3. Autonomic ganglia are classified as sympathetic trunk ganglia (on both sides of the vertebral column), prevertebral ganglia (anterior to the vertebral column), and terminal ganglia (near or inside visceral effectors).

4. Sympathetic preganglionic neurons synapse with postganglionic neurons in ganglia of the sympathetic trunk ganglia or in prevertebral ganglia; parasympathetic preganglionic neurons synapse with postganglionic neurons in terminal ganglia.

ANS Neurotransmitters and Receptors (p. 641)

1. Cholinergic neurons release acetylcholine, which binds to nicotinic or muscarinic cholinergic receptors.

2. In the ANS, the cholinergic neurons include all sympathetic and parasympathetic preganglionic neurons, all parasympathetic postganglionic neurons, and sympathetic postganglionic neurons that innervate most sweat glands.

3. In the ANS, adrenergic neurons release norepinephrine. Both epinephrine and norepinephrine bind to alpha and beta adrenergic receptors.

4. Most sympathetic postganglionic neurons are adrenergic.

5. An agonist is a substance that binds to and activates a receptor, mimicking the effect of a natural neurotransmitter or hormone. An antagonist is a substance that binds to and blocks a receptor, thereby preventing a natural neurotransmitter or hormone from exerting its effect.

Functions of the ANS (p. 643)

1. The sympathetic division favors body functions that can support vigorous physical activity and rapid production of ATP in a series of physiological responses called the fight-or-flight response; the parasympathetic division regulates activities that conserve and restore body energy.

2. The effects of sympathetic stimulation are longer-lasting and more widespread than the effects of parasympathetic stimulation.

3. Table 20.2 on page 644 compares structural and functional features of the sympathetic and parasympathetic divisions.

4. Table 20.3 on page 645 lists the effects of sympathetic and parasympathetic stimulation on effectors throughout the body.

Integration and Control of Autonomic Functions (p. 646)

1. An autonomic reflex adjusts the activities of smooth muscle, cardiac muscle, and glands.

2. An autonomic reflex arc consists of a receptor, a sensory neuron, an integrating center, two autonomic motor neurons, and a visceral effector.

3. The hypothalamus is the major control and integration center of the ANS. It is connected to both the sympathetic and the parasympathetic divisions.

SELF-QUIZ QUESTIONS

Choose the one best answer to the following questions.

1. Which of the following statements is *true* for the parasympathetic division of the autonomic nervous system?
   a. Ganglia are close to the central nervous system and distant from the effector.
   b. It forms the craniosacral outflow.
   c. It supplies nerves to blood vessels, sweat glands, and adrenal (suprarenal) glands.
   d. It has some of its nerve fibers passing through paravertebral ganglia.
   e. All of the above statements are true.

2. Which of the following cranial nerve nuclei is *not* a site of cell bodies of preganglionic neurons of the parasympathetic division?
   a. III   b. V   c. VII   d. IX   e. X

3. Which of the following statements about gray rami is *false?*
   a. They carry preganglionic neurons from the anterior rami of spinal nerves to trunk ganglia.
   b. They contain only sympathetic nerve fibers.
   c. They carry impulses from sympathetic trunk ganglia to spinal nerves.
   d. They are located at all levels of the vertebral column.
   e. None of the above is false.

4. Which of the following is *not* a sympathetic ganglion?
   a. ciliary   b. celiac   c. inferior cervical
   d. superior mesenteric   e. All of the above are sympathetic ganglia.

5. The ANS provides the chief nervous control in which of these activities?
   a. following a moving object with the eyes
   b. moving a hand reflexively from a hot object
   c. typing   d. digesting food   e. surfing the Internet

6. White rami communicantes connect:
   a. preganglionic and postganglionic parasympathetic fibers.
   b. preganglionic and postganglionic sympathetic fibers.
   c. sympathetic and parasympathetic ganglia.
   d. sympathetic trunk and prevertebral ganglia.
   e. the anterior ramus of a spinal nerve with sympathetic trunk ganglia.

7. Cholinergic fibers are thought to include:
   a. all preganglionic axons.
   b. all postganglionic parasympathetic axons.
   c. a few postganglionic sympathetic axons.
   d. all axons of somatic motor neurons.
   e. All of the above are correct.

8. Which of the following statements would *not* be included in a description of the sympathetic trunks?
   a. The trunks lie anterior and lateral to the spinal cord.
   b. Each trunk contains 24 ganglia.
   c. Each trunk receives preganglionic fibers from the thoracic and sacral regions of the spinal cord.
   d. The ganglia of the trunks are called paravertebral ganglia.
   e. White rami direct nerve fibers into the trunk; gray rami direct nerve fibers out of the trunk.

9. Which of the following could *not* be the result of the binding of acetylcholine to nicotinic receptors?
   a. contraction of a skeletal muscle
   b. stimulation of a postganglionic sympathetic neuron
   c. stimulation of a postganglionic parasympathetic neuron
   d. increase in heart rate
   e. All of the above are possible.

Complete the following.

10. _____ nerves are extensions of preganglionic sympathetic axons that pass through sympathetic trunk ganglia and extend to prevertebral ganglia.

11. Organs such as the heart or stomach that receive impulses from both divisions of the ANS are said to have _____ innervation.

12. Norepinephrine is enzymatically inactivated by COMT or _____.

13. The two types of adrenergic receptors are _____ receptors and _____ receptors.

14. The two types of cholinergic receptors are _____ receptors and _____ receptors.

15. The endocrine gland stimulated directly by sympathetic preganglionic neurons is the _____.

16. Nearly 80% of the craniosacral outflow of the parasympathetic division is conducted via preganglionic axons leaving the brain with the _____ nerve.

Are the following statements true or false?

17. In the fight-or-flight response, blood glucose levels increase.

18. A major difference between autonomic ganglia and posterior root ganglia is that only autonomic ganglia contain synapses.

19. The ganglia that make up the sympathetic trunks are also called prevertebral or collateral ganglia.

20. Postganglionic neurons begin in the spinal cord and extend to effectors.

21. An event occurring during the fight-or-flight response is increased urine production due to blood being shunted to the kidneys.

22. All sympathetic and parasympathetic preganglionic neurons release the neurotransmitter norepinephrine.

Matching

23. Match the following plexuses with their descriptions (answers may be used more than once):
    ___ **(a)** plexus located mostly posterior to each lung
    ___ **(b)** the largest autonomic plexus; consists of 2 large ganglia and a network of fibers that surround the celiac and superior mesenteric arteries
    ___ **(c)** plexus located anterior to the fifth lumbar vertebra; supplies pelvic viscera
    ___ **(d)** plexus containing the inferior mesenteric ganglion; supplies the large intestine
    ___ **(e)** plexus at the base of the heart; surrounds the large blood vessels emerging from the heart
    ___ **(f)** plexus containing the superior mesenteric ganglion; supplies the small and large intestine

    (1) applies to the sympathetic division of the ANS
    (2) applies to the parasympathetic division of the ANS
    (3) applies to both divisions of the ANS

24. Match the following (answers may be used more than once):
    ___ **(a)** thoracolumbar outflow
    ___ **(b)** has long preganglionic fibers and short postganglionic fibers
    ___ **(c)** has short preganglionic fibers and long postganglionic fibers
    ___ **(d)** sends some preganglionic fibers through cranial nerves
    ___ **(e)** has some preganglionic fibers synapsing in sympathetic trunk ganglia
    ___ **(f)** has more widespread and longer lasting effect in the body
    ___ **(g)** has some fibers running in gray rami to supply sweat glands, hair follicle muscles, and blood vessels
    ___ **(h)** has fibers in white rami (connecting spinal nerve with sympathetic trunk ganglia)
    ___ **(i)** contains fibers that supply viscera with motor impulses
    ___ **(j)** celiac and superior mesenteric ganglia are sites of postganglionic neuron cell bodies

    (1) cardiac plexus
    (2) pulmonary plexus
    (3) superior mesenteric plexus
    (4) inferior mesenteric plexus
    (5) celiac (solar) plexus
    (6) hypogastric plexus

CRITICAL THINKING QUESTIONS

1. Skydiving, hang gliding, and bungee jumping can all give you a great rush (or get you killed). How do these activities cause this rush?

2. "The quickest way to a man's heart is through his stomach," Sophia's grandmother is fond of saying. Trace the pathway that an impulse would follow from a full stomach to a happy heart.

3. Mando has been living in a refugee camp for two years. The guards are hostile to the refugees, and they regularly fire off rounds from their machine guns at night and frequently make threatening remarks and gestures to the women in the camp. Every time a guard comes near, Mando feels her heart race and her mouth go dry. Physiologically, what is happening to Mando, and what do you think some of the long-term physiological effects might be?

4. The autonomic and the enteric divisions of the nervous system both control the digestive system. How do they compare?

5. While reviewing the chapter on the brain, your study partner says, "The hypothalamus can't be very important. It's so small." Would you agree with your friend? Why or why not?

ANSWERS TO FIGURE QUESTIONS

20.1 Dual innervation means that a body organ innervated by the ANS receives both sympathetic and parasympathetic neurons.

20.2 Most parasympathetic preganglionic axons are longer than most sympathetic preganglionic axons because most parasympathetic ganglia are in the walls of visceral organs, whereas most sympathetic ganglia are close to the spinal cord in the sympathetic trunk.

20.3 Terminal ganglia are associated with the parasympathetic division; sympathetic trunk and prevertebral ganglia are associated with the sympathetic division.

20.4 The celiac plexus is the largest autonomic plexus.

20.5 White rami look white due to the presence of myelin.

20.6 Cholinergic neurons with nicotinic ACh receptors include sympathetic postganglionic neurons innervating sweat glands and all parasympathetic postganglionic neurons. The effectors innervated by these cholinergic neurons possess muscarinic receptors.

SOMATIC SENSES

21

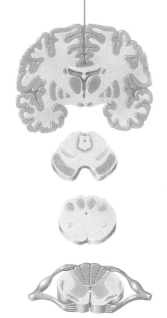

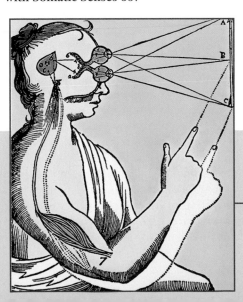

INTRODUCTION Consider what would happen if you could not feel the pain of a hot pot handle or an inflamed appendix, or if you could not see an oncoming car, hear a baby's cry, smell smoke, taste spoiled milk or food, or maintain your balance on a flight of stairs. In short, if you could not "sense" your environment and make the necessary adjustments, you could not survive very well on your own.

The previous four chapters described the organization of the nervous system. Now we will see how certain parts cooperate to carry out its three basic functions: (1) receiving sensory input; (2) integrating the sensory input, that is, processing and interpreting the input, deciding a course of action, and storing information; and (3) transmitting motor impulses that result in a response (muscular contraction or glandular secretion). In this chapter we will explore the nature and types of sensations, the pathways that convey somatic sensory input from the body to the brain, and the pathways that carry motor commands from the brain to effectors. Chapter 22 deals with the special senses of smell, taste, vision, hearing, and equilibrium.

Take a good look around you — at this book, your classmates, and your professor. Have you ever wondered how we see an object that is before us? Or how a mental intention is carried out by a physical action?

www.wiley.com/college/apcentral

OVERVIEW OF SENSATIONS

- Define a sensation and describe the conditions necessary for a sensation to occur.
- Describe the different ways to classify sensory receptors.

Definition of Sensations

A **sensation** is the awareness of changes in the external or internal conditions of the body. For a sensation to occur, four conditions must be satisfied:

1. A *stimulus*, or change in the environment, capable of activating certain sensory neurons, must occur.
2. A *sensory receptor* must convert the stimulus to nerve impulses.
3. The nerve impulses must be *conducted* along a neural pathway from the sensory receptor to the brain.
4. A region of the brain must receive and *integrate* the nerve impulses, producing a sensation.

A stimulus that activates a sensory receptor may be in the form of light, heat, pressure, mechanical energy, or chemical energy. A sensory receptor responds to a stimulus by altering its membrane's permeability to small ions. In most types of sensory receptors, the resulting flow of ions across the membrane produces a change that triggers one or more nerve impulses. The impulses are then conducted along the sensory neuron toward the CNS.

Characteristics of Sensations

A **perception** is the conscious awareness and interpretation of a sensation. Perceptions are integrated in the cerebral cortex. You seem to see with your eyes, hear with your ears, and feel pain in an injured part of your body. This is because sensory impulses from each part of the body arrive in a specific region of the cerebral cortex, which interprets the sensation as coming from the stimulated sensory receptors.

The distinct quality that makes one sensation different from others is its **sensory modality** (mō-DAL-i-tē). Based on the receptor stimulated, a sensory neuron carries information for one sensory modality only. Neurons relaying impulses for touch, for example, do not also transmit impulses for pain. The specialization of sensory neurons enables nerve impulses from the eyes to be perceived as sight, and those from the ears to be perceived as sounds.

A characteristic of most sensory receptors is **adaptation,** a decrease in sensation during a prolonged stimulus. Adaptation is caused in part by a decrease in the responsiveness of sensory receptors. As a result of adaptation, the perception of a sensation decreases or disappears even though the stimulus persists. For example, when you first step into a hot shower, the water may feel very hot, but soon the sensation decreases to one of comfortable warmth even though the stimulus (the high temperature of the water) does not change. Receptors vary in how quickly they adapt.

Classification of Sensations

The senses can be grouped into two classes: general senses and special senses.

1. The **general senses** include both **somatic senses** (*somat-* = of the body) and **visceral senses.** Somatic senses are tactile sensations (touch, pressure, and vibration); thermal sensations (warm and cold); pain sensations; and proprioceptive sensations, which allow perception of both the static (nonmoving) positions of limbs and body parts (joint and muscle position sense) and movements of the limbs and head. Visceral senses provide information about conditions within internal organs.
2. The **special senses** include smell, taste, vision, hearing, and equilibrium (balance).

In addition to introducing the mechanism of sensation, this chapter deals with the somatic senses; the special senses will be covered in Chapter 22.

Types of Sensory Receptors

Several structural and functional characteristics of sensory receptors can be used to group them into different classes. On a microscopic level, sensory receptors may be (1) free nerve endings of sensory neurons, (2) encapsulated nerve endings of sensory neurons, or (3) separate cells that synapse with sensory neurons (see Figure 21.1)

Free nerve endings are bare dendrites; when viewed under a light microscope they lack structural specialization. Receptors for pain, thermal, tickle, itch, and some touch sensations are free nerve endings. Receptors for other somatic sensations and visceral sensations, such as touch, pressure, and vibration, are **encapsulated nerve endings.** Their dendrites are enclosed in a connective tissue capsule that has a distinctive microscopic structure—for example, lamellated (pacinian) corpuscles (see Figure 21.1). The different types of capsules enhance the sensitivity or specificity of the receptor. Sensory receptors for some special senses are specialized, **separate cells** that synapse with sensory neurons. These include hair cells for hearing and equilibrium in the inner ear and gustatory receptor cells in taste buds (see Figure 22.2c). Photoreceptors, sensory receptors for vision located in the retina, are also specialized, separate cells.

Another way to group sensory receptors is based on the location of the receptors and the origin of the stimuli that activate them. **Exteroceptors** (EKS-ter-ō-sep′-tors) are located at or near the external surface of the body; they are sensitive to stimuli originating outside the body and provide information about the external environment. The sensations of hearing, vision, smell, taste, touch, pressure, vibration, temperature, and pain are conveyed by exteroceptors. **Interoceptors** (IN-ter-ō-sep′-tors) are located in blood vessels, visceral organs, muscles, and the nervous system and monitor conditions in the *internal* environment. The nerve impulses produced by interoceptors usually are not consciously perceived; occasionally, however, activation of interoceptors by strong stimuli may be felt as pain or pressure. **Proprioceptors** (PRŌ-prē-ō-sep′-tors; *proprio-* = one's own) are located in muscles, tendons, joints, and the inner ear. They provide information about body position, muscle length and tension, and the position and movement of our joints.

A third way to group sensory receptors is according to the type of stimulus they detect. Most stimuli are in the form of electromagnetic energy, such as light or heat; mechanical energy, such as sound waves or pressure changes; or chemical energy, such as in a molecule of glucose.

Table 21.1 summarizes the classification of sensory receptors and lists different types of sensory receptors according to the type of stimulus they detect.

CHECKPOINT

1. Distinguish between sensation and perception.
2. Define sensory modality and adaptation.
3. What events are needed for a sensation to occur?
4. Describe the classification of sensory receptors.

SOMATIC SENSATIONS

OBJECTIVES

- Describe the location and function of the receptors for tactile, thermal, and pain sensations.
- Identify the receptors for proprioception and describe their functions.

Somatic sensations arise from stimulation of sensory receptors embedded in the skin or subcutaneous layer; in mucous membranes of the mouth, vagina, and anus; in muscles, tendons, and joints; and in the inner ear. The sensory receptors for somatic sensations are distributed unevenly—some parts of the body surface are densely populated with receptors whereas other parts contain only a few. The areas with the highest density of somatic sensory receptors are the tip of the tongue, the lips, and the fingertips. Somatic sensations that arise from stimulating the skin surface are **cutaneous sensations** (kū-TĀ-nē-us; *cutane-=* skin). Somatic sensations are of four sensory modalities: tactile, thermal, pain, and proprioceptive.

Tactile Sensations

The **tactile sensations** (TAK-tĭl; *tact-=*touch) include touch, pressure, vibration, itch, and tickle. Several types of encapsulated mechanoreceptors mediate sensations of touch, pressure, and vibration. Other touch sensations, as well as itch and tickle sensations, are detected by free nerve endings. Tactile receptors in the skin or subcutaneous layer include corpuscles of touch, hair root plexuses, type I and II cutaneous mechanoreceptors, lamellated corpuscles, and free nerve endings (Figure 21.1).

Touch

Sensations of **touch** generally result from stimulation of tactile receptors in the skin or subcutaneous layer. **Crude touch** is the ability to perceive that something has contacted the skin, even though its exact location, shape, size, or texture cannot be determined. **Fine touch** provides specific information about a touch sensation, such as exactly what point on the body is touched plus the shape, size, and texture of the source of stimulation.

TABLE 21.1 CLASSIFICATION OF SENSORY RECEPTORS	
BASIS OF CLASSIFICATION	**DESCRIPTION**
Microscopic features	
Free nerve endings	Bare dendrites associated with pain, thermal, tickle, itch, and some touch sensations.
Encapsulated nerve endings	Dendrites enclosed in a connective tissue capsule, such as a corpuscle of touch.
Separate cells	Receptor cells that synapse with sensory neurons; located in the retina of the eye (photoreceptors), inner ear (hair cells), and taste buds of the tongue (gustatory receptor cells).
Receptor location and activating stimuli	
Exteroceptors	Located at or near body surface; sensitive to stimuli originating outside body; provide information about external environment; convey visual, smell, taste, touch, pressure, vibration, thermal, and pain sensations.
Interoceptors	Located in blood vessels, visceral organs, and nervous system; provide information about internal environment; impulses produced usually are not consciously perceived but occasionally may be felt as pain or pressure.
Proprioceptors	Located in muscles, tendons, joints, and inner ear; provide information about body position, muscle length and tension, position and motion of joints, and equilibrium (balance).
Type of stimulus detected	
Mechanoreceptors	Detect mechanical pressure; provide sensations of touch, pressure, vibration, proprioception, and hearing and equilibrium; also monitor stretching of blood vessels and internal organs.
Thermoreceptors	Detect changes in temperature.
Nociceptors	Respond to stimuli resulting from physical or chemical damage to tissue.
Photoreceptors	Detect light that strikes the retina of the eye.
Chemoreceptors	Detect chemicals in mouth (taste), nose (smell), and body fluids.
Osmoreceptors	Sense the osmotic pressure of body fluids.

Figure 21.1 Structure and location of sensory receptors in the skin and subcutaneous layer.

The somatic sensations of touch, pressure, vibration, warmth, cold, and pain arise from sensory receptors in the skin, subcutaneous layer, and mucous membranes.

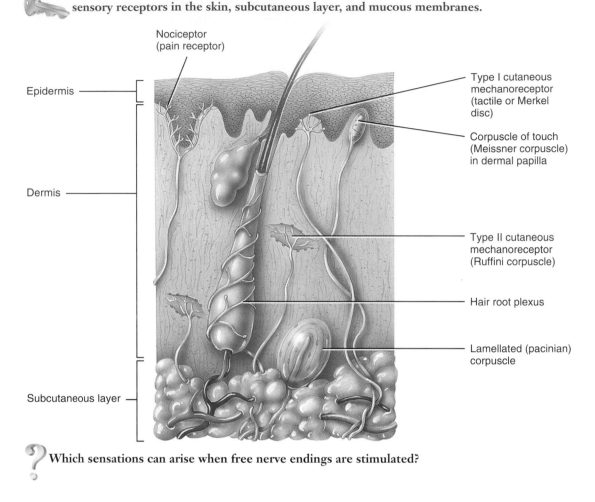

Which sensations can arise when free nerve endings are stimulated?

There are two types of rapidly adapting touch receptors. **Corpuscles of touch,** or **Meissner corpuscles** (MĪS-ner), are receptors for fine touch that are located in the dermal papillae of hairless skin. Each corpuscle is an egg-shaped mass of dendrites enclosed by a capsule of connective tissue. Because corpuscles of touch are rapidly adapting receptors, they generate nerve impulses mainly at the onset of a touch. They are abundant in the hands, eyelids, tip of the tongue, lips, nipples, soles, clitoris, and tip of the penis. **Hair root plexuses** are rapidly adapting touch receptors found in hairy skin; they consist of free nerve endings wrapped around hair follicles. Hair root plexuses detect movements on the skin surface that disturb hairs. For example, a flea landing on a hair causes movement of the hair shaft that stimulates the free nerve endings.

There also are two types of slowly adapting touch receptors. **Type I cutaneous mechanoreceptors,** also known as **Merkel discs,** function in fine touch. Merkel discs are saucer-shaped, flattened free nerve endings that contact Merkel cells of the stratum basale (see Figure 5.2d on page 123). These mechanoreceptors are plentiful in the fingertips, hands, lips, and external genitalia. **Type II cutaneous mechanoreceptors,** or **Ruffini corpuscles,** are elongated, encapsulated receptors located deep in the dermis, and in ligaments and tendons. Present in the

hands and abundant on the soles, they are most sensitive to stretching that occurs as digits or limbs are moved.

Pressure and Vibration

Pressure is a sustained sensation that is felt over a larger area than is touch. It occurs with deformation of deeper tissues. Receptors that contribute to sensations of pressure include corpuscles of touch, type I mechanoreceptors, and lamellated corpuscles. **Lamellated,** or **pacinian, corpuscles** (pa-SIN-ē-an) are large oval structures composed of a multilayered connective tissue capsule that encloses a dendrite. Like corpuscles of touch, lamellated corpuscles adapt rapidly. They are widely distributed in the body: in the dermis and subcutaneous layer; in submucosal tissues that underlie mucous and serous membranes; around joints, tendons, and muscles; in the periosteum; and in the mammary glands, external genitalia, and certain viscera, such as the pancreas and urinary bladder.

Sensations of **vibration** result from rapidly repetitive sensory signals from tactile receptors. The receptors for vibration sensations are corpuscles of touch and lamellated corpuscles. Whereas corpuscles of touch can detect lower-frequency vibrations, lamellated corpuscles detect higher-frequency vibrations.

Itch and Tickle

The **itch** sensation results from stimulation of free nerve endings by certain chemicals, such as bradykinin, often because of a local inflammatory response. Free nerve endings and lamellated corpuscles are thought to mediate the **tickle** sensation. This intriguing sensation typically arises only when someone else touches you, not when you touch yourself. The explanation of this puzzle seems to lie in the impulses that conduct to and from the cerebellum when you are moving your fingers and touching yourself that don't occur when someone else is tickling you.

PHANTOM LIMB SENSATION

Patients who have had a limb amputated may still experience sensations such as itching, pressure, tingling, or pain as if the limb were still there. This phenomenon is called **phantom limb sensation.** One explanation for phantom limb sensation is that the cerebral cortex interprets impulses arising in the proximal portions of sensory neurons that previously carried impulses from the limb as coming from the nonexistent (phantom) limb. Another explanation for phantom limb sensation is that the brain itself contains networks of neurons that generate sensations of body awareness. In this view, neurons in the brain that previously received sensory impulses from the missing limb are still active, giving rise to false sensory perceptions. Phantom limb pain can be very distressing to an amputee. Many report that the pain is severe or extremely intense, and that it often does not respond to traditional pain medication therapy. In such cases, alternative treatments may include electrical nerve stimulation, acupuncture, and biofeedback. ■

Thermal Sensations

Thermoreceptors are free nerve endings. Two distinct **thermal sensations**—coldness and warmth—are mediated by different receptors. **Cold receptors** are located in the stratum basale of the epidermis. Temperatures between 10° and 40 °C (50–105 °F) activate cold receptors. **Warm receptors** are located in the dermis and are activated by temperatures between 32° and 48 °C (90–118 °F). Cold and warm receptors both adapt rapidly at the onset of a stimulus but continue to generate impulses at a lower frequency throughout a prolonged stimulus. Temperatures below 10 °C and above 48 °C stimulate mainly pain receptors rather than thermoreceptors.

Pain Sensations

Pain is indispensable for survival. It serves a protective function by signaling the presence of noxious, tissue-damaging conditions. From a medical standpoint, the subjective description of pain along with an indication of where it occurs may help pinpoint the underlying cause of disease.

Nociceptors (nō-sī-SEP-tors; *noci*=harmful), the receptors for pain, are free nerve endings found in every tissue of the body except the brain (Figure 21.1). Intense thermal, mechanical, or chemical stimuli can activate nociceptors. Tissue irritation or injury releases chemicals such as prostaglandins, kinins, and potas-

sium ions (K$^+$) that stimulate nociceptors. Pain may persist even after a pain-producing stimulus is removed because pain-mediating chemicals linger, and because nociceptors exhibit very little adaptation. Conditions that elicit pain include excessive distention (stretching) of a structure, prolonged muscular contractions, muscle spasms, or ischemia (inadequate blood flow to an organ).

Types of Pain

There are two types of pain: fast and slow. The perception of **fast pain** occurs very rapidly, usually within 0.1 second after a stimulus is applied. This type of pain is also known as acute, sharp, or pricking pain. The pain felt from a needle puncture or knife cut to the skin are examples of fast pain. Fast pain is not felt in deeper tissues of the body. The perception of **slow pain,** by contrast, begins a second or more after a stimulus is applied. It then gradually increases in intensity over a period of several seconds or minutes. This type of pain, which may be excruciating, is also referred to as chronic, burning, aching, or throbbing pain. Slow pain can occur both in the skin and in deeper tissues or internal organs. An example is the pain associated with a toothache.

Pain that arises from stimulation of receptors in the skin is called **superficial somatic pain,** whereas stimulation of receptors in skeletal muscles, joints, tendons, and fascia causes **deep somatic pain. Visceral pain** results from stimulation of nociceptors in visceral organs.

Localization of Pain

Fast pain is very precisely localized to the stimulated area. For example, if someone pricks you with a pin, you know exactly which part of your body was stimulated. Slow somatic pain also is well localized but more diffuse (involves large areas); it usually appears to come from a larger area of the skin. In some instances of visceral slow pain, the pain is felt in the affected area. If the pleural membranes around the lungs are inflamed, for example, you experience chest pain.

In many instances of visceral pain, the pain is felt in or just deep to the skin that overlies the stimulated organ, or in a surface area far from the stimulated organ. This phenomenon is called **referred pain.** Figure 21.2 shows skin regions to which visceral pain may be referred. In general, the visceral organ involved and the area to which the pain is referred are served by the same segment of the spinal cord. For example, sensory fibers from the heart, the skin over the heart, and the skin along the medial aspect of the left arm enter spinal cord segments T1 to T5. Thus, the pain of a heart attack typically is felt in the skin over the heart and along the left arm.

ANALGESIA: RELIEF FROM PAIN

Pain sensations sometimes occur out of proportion to minor damage, persist chronically due to an injury, or even appear for no obvious reason rather than warning of actual or impending damage. In such cases, **analgesia** (an-al-JĒ-zē-a; *an-*=without; *-algesia*=pain) or pain relief is needed. Analgesic drugs such as aspirin and ibuprofen (for example, Advil or Motrin) block the

Figure 21.2 **Distribution of referred pain.** The colored portions of the diagrams indicate skin areas to which visceral pain is referred.

Nociceptors are present in almost every tissue of the body.

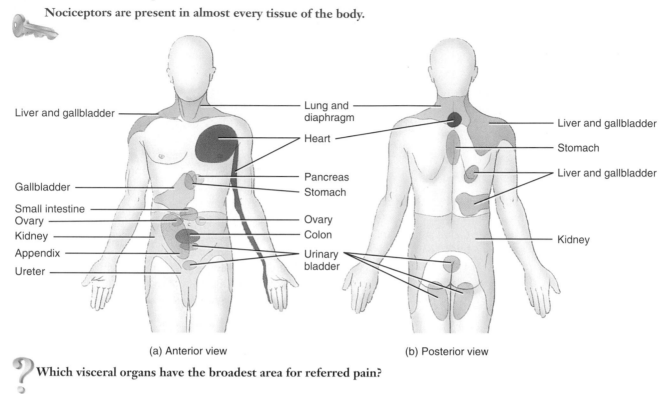

(a) Anterior view   (b) Posterior view

Which visceral organs have the broadest area for referred pain?

formation of prostaglandins, which stimulate nociceptors. Local anesthetics, such as Novocaine, provide short-term pain relief by blocking conduction of nerve impulses along the axons of first-order pain neurons. Morphine and other opiate drugs alter the quality of pain perception in the brain; pain is still sensed, but it is no longer perceived as being so noxious. Many pain clinics use anticonvulsant and antidepressant medications to treat those suffering from chronic pain. ■

Proprioceptive Sensations

Proprioceptive sensations allow us to know where our head and limbs are located and how they are moving even if we are not looking at them, so that we can walk, type, or dress without using our eyes. **Kinesthesia** (kin′-es-THĔ-zē-a; *kin-*=motion; *-esthesia*=perception) is the perception of body movements. Proprioceptive sensations arise in receptors termed **proprioceptors.** Those proprioceptors imbedded in muscles (especially postural muscles) and tendons inform us of the degree to which muscles are contracted, the amount of tension on tendons, and the positions of joints. Hair cells of the vestibular apparatus in the inner ear are proprioceptors that monitor the orientation of the head relative to the ground and head position during movements. They provide information for maintaining balance and equilibrium (described in Chapter 22). Because proprioceptors adapt slowly and only slightly, the brain continually receives nerve impulses related to the position of different body parts and makes adjustments to ensure coordination.

Proprioceptive sensations also allow us to estimate the weight of objects and determine the muscular effort necessary to perform a task. For example, as you pick up a bag you quickly realize whether it contains popcorn or books, and you then exert the correct amount of effort needed to lift it. Here we discuss three types of proprioceptors: muscle spindles within skeletal muscles, tendon organs within tendons, and joint kinesthetic receptors within synovial joint capsules.

Muscle Spindles

Each **muscle spindle** consists of several slowly adapting sensory nerve endings that wrap around 3 to 10 specialized muscle fibers, called **intrafusal muscle fibers** (*intrafusal*=within a spindle). A connective tissue capsule encloses the sensory nerve endings and intrafusal fibers and anchors the spindle to endomysium and perimysium (Figure 21.3). Muscle spindles are interspersed among ordinary skeletal muscle fibers and aligned parallel to them. In muscles that produce finely controlled movements, such as movements of the fingers or eyes, muscle spindles are plentiful. Muscles involved in coarser but more forceful movements, such as the quadriceps femoris and hamstrings muscles of the thigh, have fewer muscle spindles. The only skeletal muscles that lack spindles are the tiny muscles of the middle ear.

The main function of muscle spindles is to measure *muscle length*—how much a muscle is being stretched. Either sudden or prolonged stretching of the central areas of the intrafusal muscle fibers stimulates the sensory nerve endings. The resulting nerve impulses propagate into the CNS. Information from muscle spindles arrives quickly at the somatic sensory areas of the cerebral cortex, which allows conscious perception of limb positions and movements. At the same time, impulses from

Figure 21.3 Two types of proprioceptors: a muscle spindle and a tendon organ. In muscle spindles, which monitor changes in skeletal muscle length, sensory nerve endings wrap around the central portion of intrafusal muscle fibers. In tendon organs, which monitor the force of muscle contraction, sensory nerve endings are activated by increasing tension on a tendon.

Proprioceptors provide information about body position and movement.

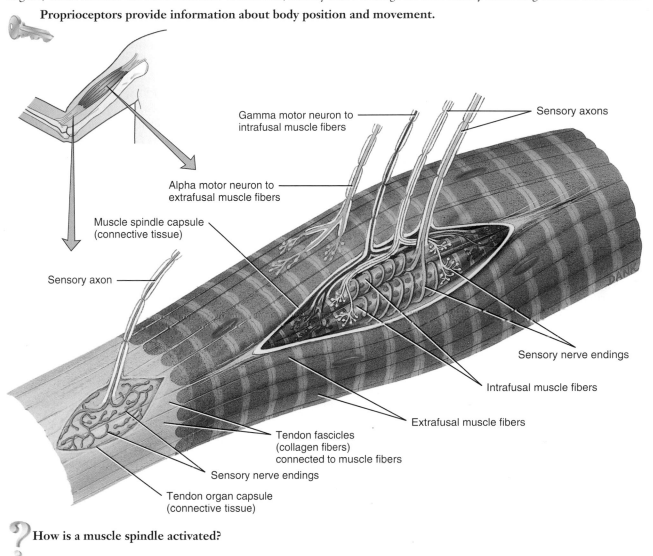

? **How is a muscle spindle activated?**

muscle spindles also pass to the cerebellum, where the input is used to coordinate muscle contractions.

In addition to their sensory nerve endings near the middle of intrafusal fibers, muscle spindles contain motor neurons called **gamma motor neurons.** The brain regulates sensitivity of muscle spindles through pathways to these motor neurons, which terminate near both ends of the intrafusal fibers. As the two ends of an intrafusal fiber contract as a result of stimulation by the gamma motor neurons, they exert a "pull" on the sensory nerve endings in the middle. As the frequency of impulses in its gamma motor neurons increases, a muscle spindle becomes more sensitive to stretching of its midregion.

Surrounding muscle spindles are ordinary skeletal muscle fibers, called **extrafusal muscle fibers** (*extrafusal*=outside a spindle), which are innervated by **alpha motor neurons.** The cell bodies of both gamma and alpha motor neurons are located in the anterior gray horn of the spinal cord (or in the brain stem for muscles in the head). During stretching of a muscle, impulses in muscle spindle sensory axons propagate into the spinal cord and brain stem and activate alpha motor neurons that innervate extrafusal muscle fibers in the same muscle. In this way, activation of its muscle spindles causes contraction of an entire skeletal muscle, which relieves the stretching.

Tendon Organs

Tendon organs are located at the junction of a tendon and a muscle and protect tendons and their associated muscles from damage due to excessive tension. (When a muscle contracts, it exerts a force that tends to pull the points of attachment at either end toward each other. This force is the *muscle tension.*) Each **tendon organ** consists of a thin capsule of connective tissue that encloses a few tendon fascicles (bundles of collagen fibers) (Figure 21.3). One or more sensory nerve endings penetrate the capsule, winding among the collagen fibers. When tension is applied to a muscle, the tendon organs generate nerve impulses that propagate into the CNS, providing information about changes in muscle tension. Tendon reflexes decrease muscle tension by causing relaxation.

Joint Kinesthetic Receptors

Several types of **joint kinesthetic receptors** are present within and around the articular capsules of synovial joints (see Chapter 9). Free nerve endings and type II cutaneous mechanoreceptors (Ruffini corpuscles) in the capsules of joints respond to pressure. Small lamellated (pacinian) corpuscles in the connective tissue outside articular capsules respond to acceleration and deceleration in the movement of joints. Articular ligaments contain receptors similar to tendon organs that adjust the adjacent muscles when excessive strain is placed on the joint.

Table 21.2 summarizes the somatic sensory receptors and the sensations they convey.

CHECKPOINT

5. Which somatic sensory receptors are encapsulated?
6. Which somatic sensory receptors adapt slowly, and which adapt rapidly?
7. Which somatic sensory receptors mediate fine touch sensations?
8. How does fast pain differ from slow pain?
9. What is referred pain, and how is it useful in diagnosing internal disorders?
10. What aspects of muscle function are monitored by muscle spindles and tendon organs?

SOMATIC SENSORY PATHWAYS

OBJECTIVE

● Describe the neuronal components and functions of the posterior column–medial lemniscus pathway, the anterolateral pathway, and the spinocerebellar pathway.

Somatic sensory pathways relay information from the somatic sensory receptors just described to the primary somatosensory area in the cerebral cortex and to the cerebellum. The pathways to the cerebral cortex consist of thousands of sets of three neurons: first-order neurons, second-order neurons, and third-order neurons.

1. **First-order neurons** conduct impulses from somatic receptors into the brain stem or spinal cord. From the face, mouth, teeth, and eyes, somatic sensory impulses propagate along *cranial nerves* into the brain stem. From the neck, body, and posterior aspect of the head, somatic sensory impulses propagate along *spinal nerves* into the spinal cord.

2. **Second-order neurons** conduct impulses from the brain stem and spinal cord to the thalamus or cerebellum. Axons of second-order neurons *decussate* (cross over to the opposite side) in the brain stem or spinal cord before ascending to

TABLE 21.2 SUMMARY OF RECEPTORS FOR SOMATIC SENSATIONS

RECEPTOR TYPE	RECEPTOR STRUCTURE AND LOCATION	SENSATIONS
Tactile receptors		
Corpuscles of touch (Meissner corpuscles)	Capsule surrounds mass of dendrites in dermal papillae of hairless skin.	Fine touch, pressure, and slow vibrations.
Hair root plexuses	Free nerve endings wrapped around hair follicles in skin.	Touch.
Type I cutaneous mechanoreceptors (tactile or Merkel discs)	Saucer-shaped free nerve endings make contact with Merkel cells in epidermis.	Touch and pressure.
Type II cutaneous mechanoreceptors (Ruffini corpuscles)	Elongated capsule surrounds dendrites deep in dermis and in ligaments and tendons.	Stretching of skin.
Lamellated (pacinian) corpuscles	Oval, layered capsule surrounds dendrites; present in dermis and subcutaneous layer, submucosal tissues, joints, periosteum, and some viscera.	Pressure, fast vibrations, and tickling.
Itch and tickle receptors	Free nerve endings and lamellated corpuscles in skin and mucous membranes.	Itching and tickling.
Thermoreceptors		
Warm receptors and cold receptors	Free nerve endings in skin and mucous membranes of mouth, vagina, and anus.	Warmth or cold.
Pain receptors		
Nociceptors	Free nerve endings in every tissue of the body except the brain.	Pain.
Proprioceptors		
Muscle spindles	Sensory nerve endings wrap around central area of encapsulated intrafusal muscle fibers within most skeletal muscles.	Muscle length.
Tendon organs	Capsule encloses collagen fibers and sensory nerve endings at junction of tendon and muscle.	Muscle tension.
Joint kinesthetic receptors	Lamellated corpuscles, Ruffini corpuscles, tendon organs, and free nerve endings.	Joint position and movement.

the ventral posterior nucleus of the thalamus. Thus, all somatic sensory information from one side of the body reaches the thalamus on the opposite side.

3. **Third-order neurons** conduct impulses from the thalamus to the primary somatosensory area of the cortex on the same side.

Somatic sensory impulses entering the spinal cord ascend to the cerebral cortex via two general pathways: (1) the posterior column–medial lemniscus pathway and (2) the anterolateral (spinothalamic) pathways. Somatic sensory impulses entering the spinal cord reach the cerebellum via the spinocerebellar tracts.

Posterior Column–Medial Lemniscus Pathway to the Cortex

Nerve impulses for conscious proprioception and most tactile sensations ascend to the cerebral cortex along the **posterior column–medial lemniscus pathway** (Figure 21.4a). The name of the pathway comes from the names of two white-matter tracts that convey the impulses: the posterior column of the spinal cord and the medial lemniscus of the brain stem.

First-order neurons extend from sensory receptors in the trunk and limbs into the spinal cord and ascend to the medulla oblongata on the same side of the body. The cell bodies of these

Figure 21.4 Somatic sensory pathways.

Nerve impulses are conducted along sets of first-order, second-order, and third-order neurons to the primary somatosensory area (postcentral gyrus) of the cerebral cortex.

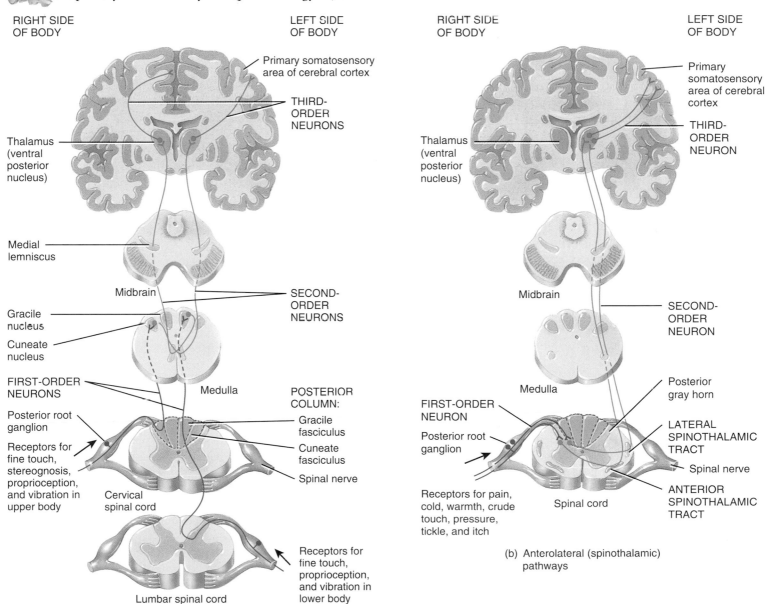

(a) Posterior column-medial lemniscus pathway

(b) Anterolateral (spinothalamic) pathways

What types of sensory deficits could be produced by damage to the right lateral spinothalamic tract?

first-order neurons are in the posterior (dorsal) root ganglia of spinal nerves. In the spinal cord, their axons form the **posterior (dorsal) columns,** which consist of two parts: the **gracile fasciculus** (GRAS-īl fa-SIK-ū-lus) and the **cuneate fasciculus** (KŪ-nē-āt). (See Table 21.3 on page 662.) The axon terminals synapse with second-order neurons whose cell bodies are located in the gracile nucleus or the cuneate nucleus of the medulla. Impulses from the neck, upper limbs, and upper chest propagate along axons in the cuneate fasciculus and arrive at the cuneate nucleus. Impulses from the trunk and lower limbs propagate along axons in the gracile fasciculus and arrive at the gracile nucleus. Impulses for somatic sensations from the head arrive in the brain stem via sensory axons that are part of all cranial nerves except the olfactory (I) nerve and optic (II) nerve. Most somatic sensory axons carrying impulses from the face are part of the trigeminal (V) nerve.

The axons of the second-order neurons cross to the opposite side of the medulla and enter the **medial lemniscus** (lem-NIS-kus=ribbon), a thin ribbonlike projection tract that extends from the medulla to the ventral posterior nucleus of the thalamus. In the thalamus, the axon terminals of second-order neurons synapse with third-order neurons, which project their axons to the primary somatosensory area of the cerebral cortex.

Impulses conducted along the posterior column–medial lemniscus pathway give rise to several highly evolved and refined sensations:

- **Fine touch** is the ability to recognize specific information about a touch sensation, such as what point on the body is touched plus the shape, size, and texture of the source of stimulation.

- **Stereognosis** is the ability to recognize the size, shape, and texture of an object by feeling it. Examples are reading Braille or identifying a paperclip by feeling it.

- **Proprioception** is the awareness of the precise position of body parts, and **kinesthesia** is the awareness of directions of movement. Proprioceptors also allow **weight discrimination,** the ability to assess the weight of an object.

- **Vibratory sensations** arise when rapidly fluctuating touch stimuli are present.

Anterolateral Pathways to the Cortex

Like the posterior column–medial lemniscus pathway just described, the **anterolateral** or **spinothalamic pathways** (spī-nō-tha-LAM-ik) are composed of three-neuron sets (Figure 21.4b). The first-order neurons connect a receptor of the neck, trunk, or limbs with the spinal cord. The cell bodies of the first-order neurons are in the posterior root ganglion. The axon terminals of the first-order neurons synapse with the second-order neurons, whose cell bodies are located in the posterior gray horn of the spinal cord.

The axons of the second-order neurons cross to the opposite side of the spinal cord. Then, they pass upward to the brain stem in either the **lateral spinothalamic tract** or the **anterior spinothalamic tract.** The lateral spinothalamic tract conveys sensory impulses for pain and temperature, whereas the

anterior spinothalamic tract conveys impulses for tickle, itch, crude touch, pressure, and vibrations. The axons of the second-order neurons end in the ventral posterior nucleus of the thalamus, where they synapse with the third-order neurons. The axons of the third-order neurons project to the primary somatosensory area on the same side of the cerebral cortex as the thalamus.

Mapping the Primary Somatosensory Area

Specific areas of the cerebral cortex receive somatic sensory input from particular parts of the body and other areas of the cerebral cortex provide output instructions for movement of particular parts of the body. The somatic sensory map and the somatic motor map relate body parts to these cortical areas.

Precise localization of somatic sensations occurs when nerve impulses arrive at the **primary somatosensory area** (areas 1, 2, and 3 in Figure 19.15 on page 607), which occupies the postcentral gyri of the parietal lobes of the cerebral cortex. Each region in this area receives sensory input from a different part of the body. Figure 21.5a maps the destination of somatic sensory signals from different parts of the left side of the body in the somatosensory area of the right cerebral hemisphere. The left cerebral hemisphere has a similar primary somatosensory area that receives sensory input from the right side of the body.

Note that some parts of the body—chiefly the lips, face, tongue, and thumb—provide input to large regions in the somatosensory area. Other parts of the body, such as the trunk and lower limbs, project to much smaller cortical regions. The relative sizes of these regions in the somatosensory area are proportional to the number of specialized sensory receptors within the corresponding part of the body. For example, there are many sensory receptors in the skin of the lips but few in the skin of the trunk. The size of the cortical region that represents a body part may expand or shrink somewhat, depending on the quantity of sensory impulses received from that body part. For example, people who learn to read Braille eventually have a larger cortical region in the somatosensory area to represent the fingertips.

Somatic Sensory Pathways to the Cerebellum

Two tracts in the spinal cord—the **posterior spinocerebellar tract** (spī-nō-ser-e-BEL-ar) and the **anterior spinocerebellar tract**—are the major routes whereby proprioceptive impulses reach the cerebellum. Although not consciously perceived, sensory impulses conveyed to the cerebellum along these two pathways are critical for posture, balance, and coordination of skilled movements. Table 21.3 on page 662 summarizes the major somatic sensory tracts in the spinal cord and pathways in the brain.

 **SYPHILIS**

Syphilis is a sexually transmitted disease caused by the bacterium *Treponema pallidum.* It is easily treated with penicillin. If the infection is not treated, the third stage of syphilis typically

Figure 21.5   Somatic sensory and somatic motor maps in the cerebral cortex. (a) Primary somatosensory area (postcentral gyrus) and (b) primary motor area (precentral gyrus) of the right cerebral hemisphere. The left hemisphere has similar representation. (After Penfield and Rasmussen.)

 Each point on the body surface maps to a specific region in both the primary somatosensory area and the primary motor area.

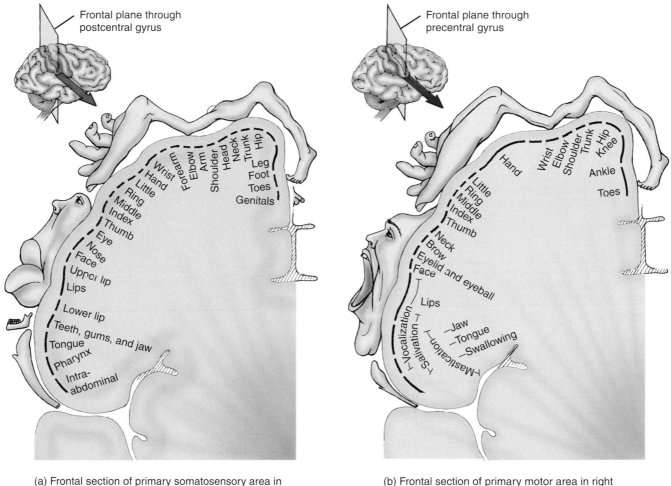

(a) Frontal section of primary somatosensory area in right cerebral hemisphere

(b) Frontal section of primary motor area in right cerebral hemisphere

 How do the somatosensory and motor representations compare for the hand, and what does this difference imply?

causes debilitating neurological symptoms. A common outcome is progressive degeneration of the posterior portions of the spinal cord, including the posterior columns, posterior spinocerebellar tracts, and posterior roots. Somatic sensations are lost, and the person's gait becomes uncoordinated and jerky because proprioceptive impulses fail to reach the cerebellum. ∎

CHECKPOINT

11. What are three differences between the posterior column–medial lemniscus pathway and the anterolateral pathways?

12. Which body parts have the largest representation in the primary somatosensory area?

13. What type of sensory information is carried in the spinocerebellar tracts, and what is its usefulness?

SOMATIC MOTOR PATHWAYS

OBJECTIVES

• Identify the locations and functions of the different types of neurons in the somatic motor pathways.

• Compare the locations and functions of the direct and indirect motor pathways.

• Explain how the basal ganglia and cerebellum contribute to movements.

Neural circuits in the brain and spinal cord orchestrate all voluntary and involuntary movements. Ultimately, all excitatory and inhibitory signals that control movement converge on the alpha motor neurons that extend out of the brain stem and spinal cord to innervate skeletal muscles in the head and body.

TRACT AND LOCATION	FUNCTIONS AND PATHWAYS
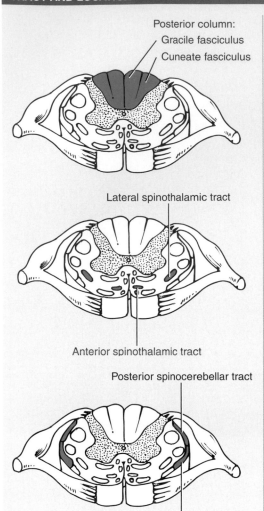 Posterior column: Gracile fasciculus, Cuneate fasciculus	***Posterior column*** Conveys nerve impulses for the sensations of fine touch, stereognosis, conscious proprioception, kinesthesia, weight discrimination, and vibration. Axons of first-order neurons from one side of the body form the posterior column on the same side and end in the medulla, where they synapse with dendrites and cell bodies of second-order neurons. Axons of second-order neurons decussate, enter the medial lemniscus on the opposite side, and extend to the thalamus. Third-order neurons transmit nerve impulses from the thalamus to the primary somatosensory cortex on the side opposite the side of stimulation.
Lateral spinothalamic tract / Anterior spinothalamic tract	***Lateral spinothalamic*** Conveys nerve impulses for pain and thermal sensations. Axons of first-order neurons from one side of the body synapse with dendrites and cell bodies of second-order neurons in the posterior gray horn on the same side. Axons of second-order neurons decussate, enter the lateral spinothalamic tract on the opposite side, and extend to the thalamus. Third-order neurons transmit nerve impulses from the thalamus to the primary somatosensory cortex on the side opposite the site of stimulation.
	Anterior spinothalamic Conveys nerve impulses for itch, tickle, pressure, vibrations, and crude, poorly localized touch sensations. Axons of first-order neurons from one side of the body synapse with dendrites and cell bodies of second-order neurons in the posterior gray horn on the same side. Axons of second-order neurons decussate, enter the anterior spinothalamic tract on the opposite side, and extend to the thalamus. Third-order neurons transmit nerve impulses from the thalamus to the primary somatosensory cortex on the side opposite the site of stimulation.
Posterior spinocerebellar tract / Anterior spinocerebellar tract	***Anterior and posterior spinocerebellar*** Convey nerve impulses from proprioceptors in the trunk and lower limb of one side of the body to the same side of the cerebellum. The proprioceptive input informs the cerebellum of actual movements, allowing it to coordinate, smooth, and refine skilled movements and maintain posture and balance.

These neurons, also known as **lower motor neurons (LMNs),** have their cell bodies in the brain stem and spinal cord. Their axons extend from the motor nuclei of cranial nerves to skeletal muscles of the face and head and from all levels of the spinal cord to skeletal muscles of the limbs and trunk. Only lower motor neurons provide output from the CNS to skeletal muscle fibers. For this reason, they are also called the *final common pathway.*

Several groups of neurons participate in control of movement by providing input to lower motor neurons.

1. *Local circuit neurons.* Input arrives at lower motor neurons from nearby interneurons called **local circuit neurons.** These neurons are located close to the lower motor neuron cell bodies in the brain stem and spinal cord. Local circuit neurons receive input from somatic sensory receptors, such as nociceptors and muscle spindles, as well as from higher centers in the brain. They help coordinate rhythmic activity in specific muscle groups, such as alternating flexion and extension of the lower limbs during walking.

2. *Upper motor neurons.* Both local circuit neurons and lower motor neurons receive input from **upper motor neurons**

(UMNs). Most upper motor neurons synapse with local circuit neurons, which in turn synapse with lower motor neurons. (A few upper motor neurons synapse directly with lower motor neurons.) UMNs from the cerebral cortex are essential for planning, initiating, and directing sequences of voluntary movements. Other UMNs originate in motor centers of the brain stem: the red nucleus, the vestibular nucleus, the superior colliculus, and the reticular formation. UMNs from the brain stem regulate muscle tone, control postural muscles, and help maintain balance and orientation of the head and body. Both the basal ganglia and cerebellum exert an influence on upper motor neurons.

3. *Basal ganglia neurons.* **Basal ganglia neurons** assist movement by providing input through the thalamus to upper motor neurons. Neural circuits interconnect the basal ganglia with motor areas of the cerebral cortex, thalamus, subthalamic nucleus, and substantia nigra. These circuits help initiate and terminate movements, suppress unwanted movements, and establish a normal level of muscle tone.

4. *Cerebellar neurons.* **Cerebellar neurons** also aid movement by controlling the activity of upper motor neurons

through the thalamus. Neural circuits interconnect the cerebellum with motor areas of the cerebral cortex and the brain stem. A prime function of the cerebellum is to monitor differences between intended movements and movements actually performed. Then, it issues commands to upper motor neurons to reduce errors in movement. The cerebellum thus coordinates body movements and helps maintain normal posture and balance.

PARALYSIS

Damage or disease of *lower* motor neurons produces **flaccid paralysis** of muscles on the same side of the body. There is neither voluntary nor reflex action of the innervated muscle fibers, muscle tone is decreased or lost, and the muscle remains limp or flaccid. Injury or disease of *upper* motor neurons in the cerebral cortex causes **spastic paralysis** of muscles on the opposite side of the body. In this condition muscle tone is increased, reflexes are exaggerated, and pathological reflexes such as the Babinski sign (see page 562) appear. ■

The axons of upper motor neurons extend from the brain to lower motor neurons via two types of somatic motor pathways—direct and indirect. **Direct motor pathways** provide input to lower motor neurons via axons that extend directly from the cerebral cortex. **Indirect motor pathways** provide input to lower motor neurons from motor centers in the brain stem. These brain stem centers, in turn, receive signals from neurons in the basal ganglia, cerebellum, and cerebral cortex. Direct and indirect pathways both govern generation of nerve impulses in the lower motor neurons, the neurons that stimulate contraction of skeletal muscles.

Before we examine these pathways we consider the role of the motor cortex in voluntary movement.

Mapping the Motor Areas

Control of body movements occurs via neural circuits in several regions of the brain. The **primary motor area** (area 4 in Figure 19.15 on page 607), located in the precentral gyrus of the frontal lobe (Figure 21.5b) of the cerebral cortex, is a major control region for planning and initiating voluntary movements. The adjacent **premotor area** (area 6) and even the primary somatosensory area in the postcentral gyrus also contribute axons to the descending motor pathways. As is true for somatic sensory representation in the somatosensory area, different muscles are represented unequally in the primary motor area. The cortical area devoted to a muscle is proportional to the number of motor units in that muscle. Muscles in the thumb, fingers, lips, tongue, and vocal cords have large representations, whereas the trunk has a much smaller representation. By comparing Figures 21.5a and b, you can see that somatosensory and somatic motor representations are similar but not identical for most parts of the body.

Direct Motor Pathways

Nerve impulses for voluntary movements propagate from the cerebral cortex to lower motor neurons via the direct motor pathways (Figure 21.6), also known as the *pyramidal pathways*.

Figure 21.6 Direct motor pathways in which signals initiated by the primary motor area in the right hemisphere control skeletal muscles on the left side of the body. Spinal cord tracts carrying impulses of direct motor pathways are the lateral corticospinal tract and anterior corticospinal tract.

> 🔑 **Direct pathways convey impulses that result in precise, voluntary movements.**

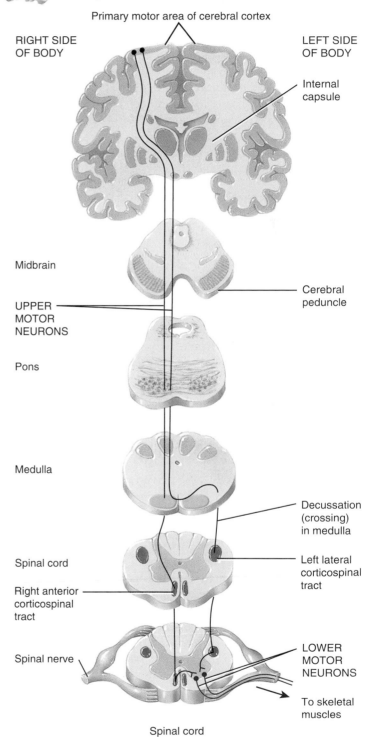

? **What other tracts (not shown in the figure) convey impulses that result in precise, voluntary movements?**

Areas of the cerebral cortex that contain large, pyramid-shaped cell bodies of upper motor neurons include not only the primary motor area in the precentral gyrus (area 4 in Figure 19.15 on page 607) but also the premotor area (area 6) and even the primary somatosensory area in the postcentral gyrus (areas 1, 2, and 3). Axons of these cortical UMNs descend through the internal capsule of the cerebrum. In the medulla oblongata, the axon bundles form the ventral bulges known as the pyramids.

About 90% of the axons of upper motor neurons *decussate* (cross over) to the *contralateral* (opposite) side in the medulla oblongata. The 10% that remain on the *ipsilateral* (same) side eventually decussate at the spinal cord levels where they synapse with an interneuron or lower motor neuron. Thus, the right cerebral cortex controls muscles on the left side of the body, and the left cerebral cortex controls muscles on the right side of the body.

Three tracts contain axons of upper motor neurons belonging to the direct motor pathways.

1. *Lateral corticospinal tracts.* Axons of UMNs that decussate in the medulla form the **lateral corticospinal tracts** (kor′-ti-kō-SPĪ-nal) in the right and left lateral white columns of the spinal cord (Figure 21.6 and Table 21.4). These motor neurons control muscles located in distal parts of the limbs. The distal muscles are responsible for precise, agile, and highly skilled movements of the limbs, hands, and feet. Examples include the movements needed to button a shirt or play the piano.

2. *Anterior corticospinal tracts.* Axons of cortical UMNs that do not decussate in the medulla form the **anterior corticospinal tracts** in the right and left anterior white columns (Figure 21.6 and Table 21.4). At each spinal cord level, some of these axons decussate via the anterior white commissure. Then they synapse with interneurons or lower motor neurons in the anterior gray horn. Axons of these lower motor neurons exit the cervical and upper thoracic segments of the cord in the anterior roots of spinal nerves. They terminate in skeletal muscles that control movements of the neck and part of the trunk, thus coordinating movements of the axial skeleton.

3. *Corticobulbar tracts.* Some axons of upper motor neurons that conduct impulses for the control of skeletal muscles in the head extend through the internal capsule to the midbrain, where they join the **corticobulbar tracts** (kor′-ti-kō-BUL-bar) in the right and left cerebral peduncles (see Table 21.4). Some of the axons in the corticobulbar tracts have decussated, whereas others have not. The axons terminate in the motor nuclei of nine pairs of cranial nerves in the pons and medulla oblongata: the oculomotor (III), trochlear (IV), trigeminal (V), abducens (VI), facial (VII), glossopharyngeal (IX), vagus (X), accessory (XI), and hypoglossal (XII). The lower motor neurons of cranial nerves convey impulses that control precise, voluntary movements of the eyes, tongue, and neck, plus chewing, facial expression, and speech.

Table 21.4 summarizes the functions and pathways of the tracts in the direct motor pathways.

AMYOTROPHIC LATERAL SCLEROSIS

Amyotrophic lateral sclerosis (ALS) (ā′-mī-ō-TROF-ik; *a-*= without; *myo-*=muscle; *trophic*=nourishment) is a progressive degenerative disease of unknown cause. The disease attacks motor areas of the cerebral cortex, axons of upper motor neurons in the lateral white columns (corticospinal and rubrospinal tracts), and lower motor neuron cell bodies. It causes progressive muscle weakness and atrophy. ALS often begins in sections of the spinal cord that serve the hands and arms but rapidly spreads to involve the whole body and face, without affecting intellect or sensations. Death typically occurs in 2 to 5 years. ALS is commonly known as *Lou Gehrig's disease* after the New York Yankees baseball player who died of it at age 37 in 1941. ∎

Indirect Motor Pathways

The **indirect motor pathways** or **extrapyramidal pathways** include all somatic motor tracts other than the corticospinal and corticobulbar tracts. Nerve impulses conducted along the indirect pathways follow complex, polysynaptic circuits that involve the motor cortex, basal ganglia, thalamus, cerebellum, reticular formation, and nuclei in the brain stem. Axons of upper motor neurons that carry nerve impulses from the indirect pathways descend from various nuclei of the brain stem into five major tracts of the spinal cord and terminate on local circuit neurons or lower motor neurons. These tracts are the **rubrospinal** (ROO-brō-spī-nal), **tectospinal** (TEK-tō-spī-nal), **vestibulospinal** (ves-TIB-ū-lō-spī-nal), **lateral reticulospinal** (re-TIK-ū-lō-spī-nal), and **medial reticulospinal tracts.**

Table 21.4 summarizes the functions and pathways of the tracts in the indirect motor pathways.

Roles of the Basal Ganglia

As noted previously, the basal ganglia and cerebellum influence movement through their effects on upper motor neurons. Two parts of the basal ganglia, the caudate nucleus and the putamen, receive input from sensory, association, and motor areas of the cerebral cortex and from the substantia nigra. Output from the basal ganglia comes from the globus pallidus and substantia nigra, which send signals to the motor cortex by way of the thalamus. (Figure 19.13b on page 603 shows these parts of the basal ganglia.) This circuit—from cortex to basal ganglia to thalamus to cortex—appears to function in initiating and terminating movements. Neurons in the putamen generate impulses just before body movements occur whereas neurons in the caudate nucleus generate impulses just before eye movements occur.

The basal ganglia also suppress unwanted movements by their inhibitory effects on the thalamus and superior colliculus. This function of the basal ganglia is revealed in certain types of basal ganglia damage, such as occurs in Huntington's disease. Abnormal, uncontrollable movements or tremors arise with loss of basal ganglia inhibition. The basal ganglia also influence muscle tone. The globus pallidus sends impulses that reduce muscle tone into the reticular formation. Damage or destruction of

TABLE 21.4 MAJOR SOMATIC MOTOR PATHWAYS IN THE BRAIN AND TRACTS IN THE MIDBRAIN AND SPINAL CORD

TRACT AND LOCATION	FUNCTIONS AND PATHWAYS
Direct (pyramidal) tracts 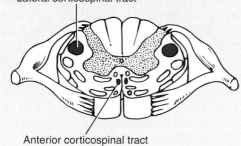 Lateral corticospinal tract Anterior corticospinal tract Spinal cord 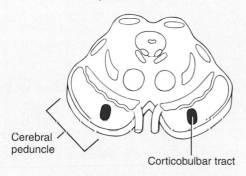 Cerebral peduncle Corticobulbar tract Midbrain of brain stem	*Lateral corticospinal* Conveys nerve impulses from the motor cortex to skeletal muscles on the opposite side of the body for precise, voluntary movements of the limbs, hands, and feet. Axons of upper motor neurons (UMNs) descend from the precentral and postcentral gyri of the cortex into the medulla. Here 90% decussate (cross over to the opposite side) and then enter the contralateral side of the spinal cord to form this tract. At their level of termination, these UMNs end in the anterior gray horn on the same side. They provide input to lower motor neurons, which innervate skeletal muscles. *Anterior corticospinal* Conveys nerve impulses from the motor cortex to skeletal muscles on opposite side of body for movements of the axial skeleton. Axons of UMNs descend from the cortex into the medulla. Here the 10% that do not decussate enter the spinal cord and form this tract. At their level of termination, these UMNs decussate and end in the anterior gray horn on the opposite side. They provide input to lower motor neurons, which innervate skeletal muscles. *Corticobulbar* Conveys nerve impulses from the motor cortex to skeletal muscles of the head and neck to coordinate precise, voluntary movements. Axons of UMNs descend from the cortex into the brain stem, where some decussate and others do not. They provide input to lower motor neurons in the nuclei of cranial nerves III, IV, V, VI, VII, IX, X, XI, and XII, which control voluntary movements of the eyes, tongues and neck; chewing; facial expression; and speech.
Indirect (extrapyramidal) tracts 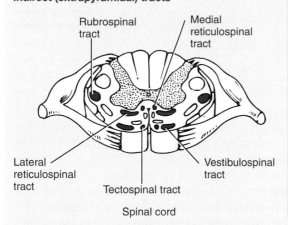 Rubrospinal tract Medial reticulospinal tract Lateral reticulospinal tract Tectospinal tract Vestibulospinal tract Spinal cord	*Rubrospinal* Conveys nerve impulses from the red nucleus (which receives input from the cerebral cortex and cerebellum) to contralateral skeletal muscles that govern precise movements of the distal parts of the limbs. *Tectospinal* Conveys nerve impulses from the superior colliculus to contralateral skeletal muscles that move the head and eyes in response to visual stimuli. *Vestibulospinal* Conveys nerve impulses from the vestibular nucleus (which receives input about head movements from the vestibular apparatus in the inner ear) to regulate ipsilateral muscle tone for maintaining balance in response to head movements. *Lateral reticulospinal* Conveys nerve impulses from the reticular formation to facilitate flexor reflexes, inhibit extensor reflexes, and decrease muscle tone in muscles of the axial skeleton and proximal parts of the limbs. *Medial reticulospinal* Conveys nerve impulses from the reticular formation to facilitate extensor reflexes, inhibit flexor reflexes, and increase muscle tone in muscles of the axial skeleton and proximal parts of the limbs.

some basal ganglia connections causes a generalized increase in muscle tone.

In addition to their motor functions, the basal ganglia also influence many aspects of cortical function, including sensory, limbic, cognitive, and linguistic functions.

DAMAGE TO THE BASAL GANGLIA

Damage to the basal ganglia results in uncontrollable, abnormal body movements (such as "pill rolling" with the fingertips), often accompanied by muscle rigidity and tremors (shaking) while at rest. In **Parkinson's disease (PD),** neurons that extend from the substantia nigra to the caudate nucleus and putamen degenerate. Loss of these dopamine-releasing neurons causes a generalized increase in muscle tone and stiffness of the face, arms, and legs. For more on Parkinson's disease, see the Applications to Health section at the end of the chapter. **Huntington's disease (HD)** is an hereditary disorder in which the caudate nucleus and putamen degenerate, with loss of neurons that normally release GABA or acetylcholine. A key sign of HD is **chorea** (KŌ-rē-a=a dance), in which rapid, jerky movements occur involuntarily and without purpose. Progressive mental deterioration also occurs. Symptoms of HD often do not appear until age 30 or 40. Death occurs 10 to 20 years after symptoms first appear. ■

Roles of the Cerebellum

In addition to maintaining proper posture and balance, the cerebellum is active in learning and performing rapid, coordinated, highly skilled movements such as hitting a golf ball, speaking, and swimming. Cerebellar function involves four activities.

1. The cerebellum *monitors intentions for movement* by receiving impulses from the motor cortex and basal ganglia via the pontine nuclei in the pons regarding what movements are planned.

2. The cerebellum *monitors actual movement* by receiving input from proprioceptors in joints and muscles that reveals what actually is happening. These nerve impulses travel in the anterior and posterior spinocerebellar tracts. Nerve impulses from the vestibular (equilibrium-sensing) apparatus in the inner ear and from the eyes also enter the cerebellum.

3. The cerebellum *compares the command signals* (intentions for movement) *with sensory information* (actual movement performed).

4. If there is a discrepancy between intended and actual movement, the cerebellum *sends out corrective signals* to upper motor neurons. This information travels via the thalamus to UMNs in the cerebral cortex but goes directly to UMNs in brain stem motor centers. As movements occur, the cerebellum continuously provides error corrections to upper motor neurons, which decreases errors and smoothes the motion. It also contributes over longer periods to the learning of new motor skills.

Skilled activities such as tennis or volleyball provide good examples of the contribution of the cerebellum to movement. To make a good serve or to block a spike, you must bring your racket or arms forward just far enough to make solid contact. How do you stop at exactly the right point? For example, in tennis before you even hit the ball, the cerebellum has sent nerve impulses to the cerebral cortex and basal ganglia informing them where your arm or racket swing must stop. In response to impulses from the cerebellum, the cortex and basal ganglia transmit motor impulses to opposing body muscles to stop the swing.

🩺 DAMAGE TO THE CEREBELLUM

A hallmark of cerebellar trauma or disease is **ataxia** (a-TAK-sē-a; *a-*=without; *-taxia*=coordination), an abnormal condition in which movements are jerky and uncoordinated. People with ataxia have a staggering gait and, if blindfolded, cannot touch the tip of their nose with a finger because they cannot coordinate movements with their sense of where a body part is located. Often, speech is slurred due to a lack of coordination of speech muscles. Cerebellar damage also produces **intention tremor,** a shaking that occurs during deliberate voluntary movement. Chronic alcoholics often suffer degeneration of the anterior part of the cerebellum, which controls the lower limbs. Such people have an impaired walking gait with relatively normal control of arm movements. ■

CHECKPOINT

14. Which parts of the body have the largest representation in the motor cortex? Which have the smallest?
15. Explain why the two main somatic motor pathways are named "direct" and "indirect."
16. Compare the effects of disease of the basal ganglia with damage to the cerebellum.

INTEGRATION OF SENSORY INPUT AND MOTOR OUTPUT

OBJECTIVE

● Explain how sensory input and motor output are integrated.

Sensory pathways provide the input that keeps the central nervous system informed of changes in the external and internal environment. Output from the CNS is then conveyed through motor pathways, which enable movement and glandular secretions to occur. As sensory information reaches the CNS, it becomes part of a large pool of sensory information. However, the CNS does not necessarily respond to every impulse. Rather, the incoming sensory information is **integrated;** that is, it is processed and interpreted and a course of action is taken.

The integration process occurs not just once but at many places along the pathways of the CNS and at both conscious and subconscious levels. It occurs within the spinal cord, brain stem, cerebellum, basal ganglia, and cerebral cortex. As a result, the output descending along a motor pathway that makes a muscle contract or a gland secrete can be modified and responded to at any of these levels. Motor portions of the cerebral cortex play the major role for initiating and controlling precise movements of muscles. The basal ganglia largely integrate semivoluntary, automatic movements such as walking, swimming, and laughing. The cerebellum assists the motor cortex and basal ganglia by making body movements smooth and coordinated and by contributing significantly to the maintenance of normal posture and balance.

CHECKPOINT

17. What does integration of sensory input mean?

APPLICATIONS TO HEALTH

SPINAL CORD INJURY

The spinal cord may be damaged by a tumor either within or adjacent to it, herniated intervertebral discs, blood clots, penetrating wounds caused by projectile fragments, or other trauma. In many cases of traumatic injury of the spinal cord, the patient has an improved outcome if an anti-inflammatory corticosteroid drug called methylprednisolone is given within 8 hours of the injury.

Depending on the location and extent of spinal cord damage, paralysis may occur. Since monoplegia, diplegia, paraplegia, hemiplegia, quadriplegia, complete transection, hemisection, and spinal shock have already been discussed in Chapter 18, please refer to page 578.

PARKINSON'S DISEASE

Parkinson's disease (PD) is a progressive disorder of the CNS that typically affects its victims around age 60. Neurons that extend from the substantial nigra to the putamen and caudate nucleus, where they release the neurotransmitter dopamine (DA), degenerate in PD. In the caudate nucleus of the basal ganglia are neurons that liberate the neurotransmitter acetylcholine (ACh). Although the level of ACh does not change as the level of DA declines, the imbalance of neurotransmitter activity—too little DA and too much ACh—is thought to cause most of the symptoms. The cause of PD is unknown, but toxic environmental chemicals, such as pesticides, herbicides, and carbon monoxide, are suspected contributing agents. Only 5% of PD patients have a family history of the disease.

In PD patients, involuntary skeletal muscle contractions often interfere with voluntary movement. For instance, the muscles of the upper limb may alternately contract and relax, causing the hand to shake. This shaking, called **tremor,** is the most common symptom of PD. Also, muscle tone may increase greatly, causing rigidity of the involved body part. Rigidity of

the facial muscles gives the face a masklike appearance. The expression is characterized by a wide-eyed, unblinking stare and a slightly open mouth with uncontrolled drooling.

Motor performance is also impaired by **bradykinesia** (*brady-*=slow), slowness of movements. Activities such as shaving, cutting food, and buttoning a blouse take longer and become increasingly more difficult as the disease progresses. Muscular movements also exhibit **hypokinesia** (*hypo-*=under), decreasing range of motion. For example, words are written smaller, letters are poorly formed, and eventually handwriting becomes illegible. Often, walking is impaired; steps become shorter and shuffling, and arm swing diminishes. Even speech may be affected.

Treatment of PD is directed toward increasing levels of DA and decreasing levels of ACh. Although people with PD do not manufacture enough dopamine, taking it orally is useless because DA cannot cross the blood–brain barrier. Even though symptoms are partially relieved by a drug developed in the 1960s called levodopa (L-dopa), a precursor of DA, the drug does not slow the progression of the disease. As more and more affected brain cells die, the drug becomes useless. Another drug called selegiline (deprenyl) is used to inhibit monoamine oxidase, an enzyme that degrades catecholamine neurotransmitters such as dopamine. This drug slows progression of PD and may be used together with levodopa.

For more than a decade, surgeons have sought to reverse the effects of Parkinson's disease by transplanting dopamine-rich fetal nerve tissue into the basal ganglia (usually the putamen) of patients with severe PD. Only a few postsurgical patients have shown any degree of improvement, such as less rigidity and improved quickness of motion. Another surgical technique that has produced improvement for some patients is *pallidotomy*, in which a part of the globus pallidus that generates tremors and produces muscle rigidity is destroyed.

KEY MEDICAL TERMS ASSOCIATED WITH SOMATIC SENSES

Acupuncture (ak-ū-PUNK-chur) The use of fine needles (lasers, ultrasound, or electricity) inserted into specific exterior body locations (acupoints) and manipulated to relieve pain and provide therapy for various conditions. The placement of needles may cause the release of neurotransmitters such as endorphins, painkillers that may inhibit pain pathways.

Cerebral palsy (CP) A motor disorder that results in the loss of muscle control and coordination; caused by damage of the motor areas of the brain during fetal life, birth, or infancy. Radiation during fetal life, temporary lack of oxygen during birth, and hydrocephalus during infancy may also cause cerebral palsy.

Dysarthria (dis-AR-thrē-a; *dys-*=difficult; *-arthro*=to articulate) Difficult or imperfect speech due to loss of muscular control as a result of disorders of motor pathways.

Pain threshold The smallest intensity of a painful stimulus at which a person perceives pain. All individuals have the same pain threshold.

Pain tolerance The greatest intensity of painful stimulation that a person is able to tolerate. Individuals vary in their tolerance to pain.

Synesthesia (sin-es-THĒ-zē-a; *syn-*=together; *-aisthesis*=sensation) A condition in which sensations of two or more modalities accompany one another. In some cases, a stimulus for one sensation is perceived as a stimulus for another, for example, a sound produces a sensation of color. In other cases, a stimulus from one part of the body is experienced as coming from a different part.

Overview of Sensations (p. 652)

1. Sensation is the awareness of changes in the external and internal conditions of the body.

2. For a sensation to occur, a stimulus must reach a sensory receptor, the stimulus must be converted to a nerve impulse, and the impulse conducted to the brain; finally, the impulse must be integrated by a region of the brain.

3. When stimulated, most sensory receptors ultimately produce one or more nerve impulses.

4. Sensory impulses from each part of the body arrive in specific regions of the cerebral cortex.

5. Modality is the distinct quality that makes one sensation different from others.

6. Adaptation is a decrease in sensation during a prolonged stimulus.

7. Two classes of senses are general senses, which include somatic senses and visceral senses, and special senses, which include the modalities of smell, taste, vision, hearing, and equilibrium (balance).

8. Types of sensory receptors are based on microscopic structure, location and origin of the stimuli that activate them, and type of stimulus they detect.

9. Table 21.1 on page 653 summarizes the classification of sensory receptors.

Somatic Sensations (p. 653)

1. Somatic sensations include tactile sensations (touch, pressure, vibration, itch, and tickle), thermal sensations (warmth and cold), pain, and proprioception.

2. Receptors for tactile, thermal, and pain sensations are located in the skin, subcutaneous layer, and mucous membranes of the mouth, vagina, and anus.

3. Receptors for proprioceptive sensations (position and movement of body parts) are located in muscles, tendons, joints, and the inner ear.

4. Receptors for touch are (a) hair root plexuses and corpuscles of touch (Meissner corpuscles), which are rapidly adapting, and (b) slowly adapting type I cutaneous mechanoreceptors (tactile or Merkel discs). Type II cutaneous mechanoreceptors (Ruffini corpuscles), which are slowly adapting, are sensitive to stretching. Receptors for pressure include corpuscles of touch, type I mechanoreceptors, and lamellated (pacinian) corpuscles. Receptors for vibration are corpuscles of touch and lamellated corpuscles. Itch receptors are free nerve endings, whereas both free nerve endings and lamellated corpuscles mediate the tickle sensation.

5. Thermoreceptors are free nerve endings. Cold receptors are located in the stratum basale of the epidermis; warm receptors are located in the dermis.

6. Pain receptors (nociceptors) are free nerve endings that are located in nearly every body tissue.

7. Proprioceptors include muscle spindles, tendon organs, joint kinesthetic receptors, and hair cells of the inner ear.

8. Table 21.2 on page 658 summarizes the somatic sensory receptors and the sensations they convey.

Somatic Sensory Pathways (p. 658)

1. Somatic sensory pathways from receptors to the cerebral cortex involve three-neuron sets: first-order, second-order, and third-order.

2. Axon collaterals (branches) of somatic sensory neurons simultaneously carry signals into the cerebellum and the reticular formation of the brain stem.

3. Impulses propagating along the posterior column–medial lemniscus pathway relay fine touch, stereognosis, proprioception, and vibratory sensations.

4. The neural pathway for pain and thermal sensations is the lateral spinothalamic tract.

5. The neural pathway for tickle, itch, crude touch, and pressure sensations is the anterior spinothalamic pathway.

6. The pathways to the cerebellum are the anterior and posterior spinocerebellar tracts, which transmit impulses for subconscious muscle and joint position sense from the trunk and lower limbs.

7. Table 21.3 on page 662 summarizes the major somatic sensory pathways.

8. Specific regions of the primary somstosensory area (postcentral gyrus) of the cerebral cortex receive somatic sensory input from different parts of the body.

Somatic Motor Pathways (p. 661)

1. All excitatory and inhibitory signals that control movement converge on the alpha motor neurons, also known as lower motor neurons (LMNs) or the final common pathway.

2. Several groups of neurons participate in control of movement by providing input to lower motor neurons: local circuit neurons, upper motor neurons, basal ganglia neurons, and cerebellar neurons.

3. The primary motor area (precentral gyrus) of the cortex is the major control region for planning and initiating voluntary movements.

4. The axons of upper motor neurons extend from the brain to lower motor neurons via direct and indirect motor pathways. The direct (pyramidal) pathways include the lateral and anterior corticospinal tracts and corticobulbar tracts. Indirect (extrapyramidal) pathways extend from several motor centers of the brain stem into the spinal cord.

5. Neurons of the basal ganglia assist movement by providing input to the upper motor neurons. They help initiate and terminate movements, suppress unwanted movements, and establish a normal level of muscle tone.

6. The cerebellum is active in learning and performing rapid, coordinated, highly skilled movements. It also contributes to the maintenance of balance and posture.

7. Table 21.4 on page 665 summarizes the major somatic motor pathways.

Integration of Sensory Input and Motor Output (p. 666)

1. Sensory input keeps the CNS informed of changes in the environment.

2. Incoming sensory information is integrated at many stations along the CNS at both conscious and subconscious levels.

3. A motor response makes a muscle contract or a gland secrete.

SELF-QUIZ QUESTIONS

Choose the one best answer to the following questions:

1. Which of the following is the receptor for pain?

    a. chemoreceptor   b. photoreceptor   c. nociceptor
    d. thermoreceptor   e. mechanoreceptor

2. Which of the following is *not* classified as a somatic sense?

    a. pain   b. joint position   c. touch
    d. temperature (thermal)   e. equilibrium

3. The two major routes whereby proprioceptive input reaches the cerebellum are the:

    a. anterior and posterior spinocerebellar tracts
    b. anterior and lateral spinothalamic tracts
    c. anterior and lateral corticospinal tracts
    d. direct and indirect pathways
    e. gracile fasciculus and cuneate to fasciculus tracts

4. Which of the following statements is *true* for muscle spindles?

    a. They are located within and around articular capsules of synovial joints.
    b. They are located at the junction of a tendon and a muscle.
    c. They respond to acceleration and deceleration of joint movement, and they monitor pressure at joints.
    d. They monitor changes in the length of a skeletal muscle.
    e. All of these statements are true.

5. Gamma motor neurons stimulate:

    a. pacinian corpuscles   b. Meissner corpuscles
    c. cutaneous mechanoreceptors   d. intrafusal muscle fibers
    e. extrafusal muscle fibers

6. The primary somatosensory area is located in the:

    a. precentral gyrus of the cerebral cortex
    b. postcentral gyrus of the cerebral cortex
    c. cerebellum   d. thalamus
    e. gray horns of the spinal cord

Complete the following:

7. A change in the environment capable of activating certain sensory neurons is called a(n) _____.

8. Sensory receptors may be classified by location into three groups: _____, _____ , and _____.

9. The receptors for touch, pressure, and vibration are all _____-receptors, based on the type of stimulus to which they respond.

10. Upper motor neurons of the indirect pathways begin in the _____ region of the brain and synapse with either lower motor neurons or _____ neurons.

11. The pain felt in the skin of the thorax and left arm during a heart attack is known as _____ pain.

12. The axons of sensory receptors in the trunk and limbs form the _____ of the spinal cord.

13. The ability to recognize size, shape, and texture of an object by feeling it is known as _____.

14. Second- and third-order neurons of the posterior column–medial lemniscus pathway synapse in the _____.

Are the following statements true or false?

15. Stimulation of pain receptors in skeletal muscles, joints, and tendons results in visceral pain.

16. A sensory pathway, from receptor to the cerebral cortex, involves three neurons.

17. The precentral gyrus of the cerebral cortex is the premotor area.

18. Cold receptors are located in the epidermis.

19. Third-order neurons conduct impulses from somatic receptors to the central nervous system.

Matching

20. Match the following (answers are used more than once):

    ___ **(a)** located in skeletal muscles
    ___ **(b)** located within and around articular capsules of synovial joints
    ___ **(c)** located at the junction of a tendon and a muscle
    ___ **(d)** help protect tendons and their associated muscles from damage due to excessive tension
    ___ **(e)** respond to acceleration and deceleration of joint movement; monitor pressure at joints
    ___ **(f)** monitor changes in the length of a skeletal muscle

    (1) tendon organ
    (2) muscle spindle
    (3) joint kinesthetic receptor

21. Match the following:

    ___ **(a)** sensitive to movement of hair shaft
    ___ **(b)** egg-shaped masses located in dermal papillae; plentiful in fingertips, palms, and soles; receptors for discriminative touch
    ___ **(c)** onion-shaped structures sensitive to pressure in skin, membranes, joints, and some viscera
    ___ **(d)** dendrites that contact epidermal cells in the stratum basale of skin; function in discriminative touch

    (1) corpuscles of touch (Meissner corpuscles)
    (2) tactile (Merkel) discs
    (3) lamellated (pacinian) corpuscles
    (4) hair root plexuses

CRITICAL THINKING QUESTIONS

1. When Sau Lan held the cup of hot cocoa in her hands, at first it felt comfortably warm and within minutes, she didn't notice the temperature at all. Absently she took a big gulp and almost choked when she felt the hot cocoa burn her mouth and throat. What happened—did the hot cocoa get hotter?

2. Jenny was scheduled for surgery. As a part of her preparation, she was to take a course of antibiotic treatment for 10 days prior to the surgery. Unfortunately, in a rare reaction to the drugs, the antibiotics destroyed her muscle spindles. How will this affect Jenny's life?

3. Stella was visiting a long-term care facility as a part of her observational experiences in her allied health program. There she saw a patient who was having great difficulty controlling his body movements. She was not allowed to look at the patient's charts, so she was trying to figure out all the possible nervous system problems he could have. Damage to what specific parts of the brain might result in such loss of control?

4. Very young children are usually given only spoons (no forks or knives) and cups with lids at the dinner table. Why?

5. Jon used to predict the weather by how much the bunion (abnormally swollen joint on big toe) on his left foot was bothering him. Last year, Jon's left foot was amputated due to complications from diabetes. But sometimes, he still thinks he feels that bunion. Explain Jon's weather toe.

ANSWERS TO FIGURE QUESTIONS

21.1 Pain, thermal sensations, tickle, and itch involve activation of different free nerve endings.

21.2 The kidneys have the broadest area for referred pain.

21.3 Muscle spindles are activated when the central areas of their intrafusal fibers are stretched.

21.4 Damage to the right lateral spinothalamic tract could result in loss of pain and thermal sensations on the left side of the body.

21.5 The hand has a larger representation in the motor area than in the somatosensory area, which implies greater precision in the hand's movement control than discriminative ability in its sensation.

21.6 The corticobulbar and rubrospinal tracts (see Table 21.4 on page 665) convey impulses that result in precise, voluntary movements.

SPECIAL SENSES 22

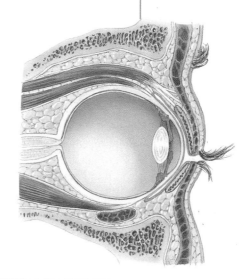

How many senses do we have? Reconcile this caricature of the senses with what you read in this chapter, along with the senses you learned about in Chapter 21.

INTRODUCTION   Recall from Chapter 21 that general senses include somatic senses (tactile, thermal, pain, and proprioceptive) and visceral sensations. Receptors for the general senses are scattered throughout the body and are relatively simple in structure. They are mostly modified dendrites of sensory neurons. Receptors for the special senses—smell, taste, vision, hearing, and equilibrium—are anatomically distinct from one another and are concentrated in specific locations in the head. They are usually embedded in the epithelial tissue within complex sensory organs such as the eyes and ears. Neural pathways for the special senses are more complex than those for the general senses.

In this chapter we examine the structure and function of the special sense organs, and the pathways involved in conveying information from them to the central nervous system. **Ophthalmology** (of-thal-MOL-ō-jē; *opthalmo-*=eye; *-logy*=study of) is the science that deals with the eye and its disorders. The other special senses are, in large part, the concern of **otorhinolaryngology** (ō′-tō-rī′-nō-lar-in-GOL-ō-jē; *oto-*=ear; *rhino-*=nose; *laryngo-*= larynx), the science that deals with ear, nose, pharynx (throat), and larynx and their disorders.

www.wiley.com/college/apcentral

OLFACTION: SENSE OF SMELL

• Describe the olfactory receptors and the neural pathway for olfaction.

Both smell and taste are chemical senses; the sensations arise from the interaction of molecules with smell or taste receptors. To be detected by either sense, the stimulating molecules must be dissolved. Because impulses for smell and taste propagate to the limbic system (and to higher cortical areas as well), certain odors and tastes can evoke strong emotional responses or a flood of memories.

Anatomy of Olfactory Receptors

The nose contains 10–100 million receptors for the sense of smell or **olfaction** (ol-FAK-shun; *olfact-*=smell). The total area of the olfactory epithelium is 5 cm^2 (a little less than 1 in.^{2}). It occupies the superior part of the nasal cavity, covering the inferior surface of the cribriform plate and extending along the superior nasal concha (Figure 22.1a). The olfactory epithelium consists of three kinds of cells: olfactory receptors, supporting cells, and basal cells (Figure 22.1b).

Olfactory receptors are the first-order neurons of the olfactory pathway. They are bipolar neurons; the knob-shaped dendrite forms the exposed tip, and the axon extends through the cribriform plate to the olfactory bulb in the brain. The parts of the olfactory receptor that respond to inhaled chemicals are the **olfactory hairs,** cilia that project from the dendrite. Chemicals that have an odor and can therefore stimulate the olfactory hairs are termed **odorants.** Olfactory receptors respond to the chemical stimulation of an odorant molecule by initiating the olfactory response.

Supporting cells are columnar epithelial cells of the mucous membrane lining the nose. They provide physical support, nourishment, and electrical insulation for the olfactory receptors, and they help detoxify chemicals that come in contact with the olfactory epithelium. **Basal cells** lie between the bases of the supporting cells and continually undergo cell division to produce new olfactory receptors, which live for only a month or so before being replaced. This process is remarkable when you consider that olfactory receptors are neurons, which are generally not replaced.

Within the connective tissue that supports the olfactory epithelium are **olfactory (Bowman's) glands,** which produce a serous secretion that is carried to the surface of the epithelium by ducts. The secretion moistens the surface of the olfactory epithelium and dissolves odorants. Both supporting cells of the nasal epithelium and olfactory glands are innervated by branches of the facial (VII) nerve, which can be stimulated by certain chemicals. Impulses in these nerves, in turn, stimulate the lacrimal glands in the eyes and nasal mucous glands. The result is tears and a runny nose after inhaling substances such as pepper, onion, or the vapors of household ammonia.

The Olfactory Pathway

On each side of the nose, bundles of the slender, unmyelinated axons of olfactory receptors extend through about 20 olfactory foramina in the cribriform plate of the ethmoid bone (Figure 22.1b). These 40 or so bundles of axons collectively form the right and left **olfactory (I) nerves.** The olfactory nerves terminate in the brain in paired masses of gray matter called the **olfactory bulbs,** which are located below the frontal lobes of the cerebrum and lateral to the crista galli of the ethmoid bone. Within the olfactory bulbs, the axon terminals of olfactory receptors—the first-order neurons—form synapses with the dendrites and cell bodies of second-order neurons in the olfactory pathway.

Axons of second-order olfactory bulb neurons extend posteriorly and form the **olfactory tract** (Figure 22.1b). These axons project to the lateral olfactory area, which is located at the inferior and medial surface of the temporal lobe. This cortical region is part of the limbic system and includes part of the amygdala (see Figure 19.14 on page 605). Because many olfactory tract axons end in the lateral olfactory area, it is probably the *primary olfactory area*, where conscious awareness of smells begins. Connections to other limbic system regions and to the hypothalamus account for our emotional and memory-evoked responses to odors. Examples include sexual excitement upon smelling a certain perfume, nausea upon smelling a food that once made you violently ill, or an odor-evoked memory of a childhood experience.

From the lateral olfactory area, pathways also extend to the frontal lobe, both directly and indirectly via the thalamus. An important region for odor identification and discrimination is the orbitofrontal area (area 11 in Figure 19.15 on page 607). People who suffer damage in this area have difficulty identifying different odors. Positron emission tomography (PET) studies suggest some degree of hemispheric lateralization: The orbitofrontal area of the *right* hemisphere exhibits greater activity during olfactory processing.

HYPOSMIA

Women often have a keener sense of smell than men do, especially at the time of ovulation. Smoking seriously impairs the sense of smell in the short term and may cause long-term damage to olfactory receptors. With aging the sense of smell deteriorates. **Hyposmia** (hī-POZ-mē-a; *osmi*=smell, odor), a reduced ability to smell, affects half of those over age 65 and 75% of those over age 80. Hyposmia also can be caused by neurological changes, such as a head injury, Alzheimer's disease, or Parkinson's disease; certain drugs, such as antihistamines, analgesics, or steroids; and the damaging effects of smoking. ■

1. How do basal cells contribute to olfaction?
2. What is the sequence of events from the binding of an odorant molecule to an olfactory hair to the arrival of a nerve impulse in an olfactory bulb?

Figure 22.1 Olfactory epithelium and olfactory receptors. (a) Location of olfactory epithelium in nasal cavity. (b) Anatomy of olfactory receptors, which are first-order neurons whose axons extend through the cribriform plate and terminate in the olfactory bulb. (c) Histology of the olfactory epithelium. (See Tortora, *A Photographic Atlas of the Human Body*, 2e, Figure 9.1b.)

🔑 **The olfactory epithelium consists of olfactory receptors, supporting cells, and basal cells.**

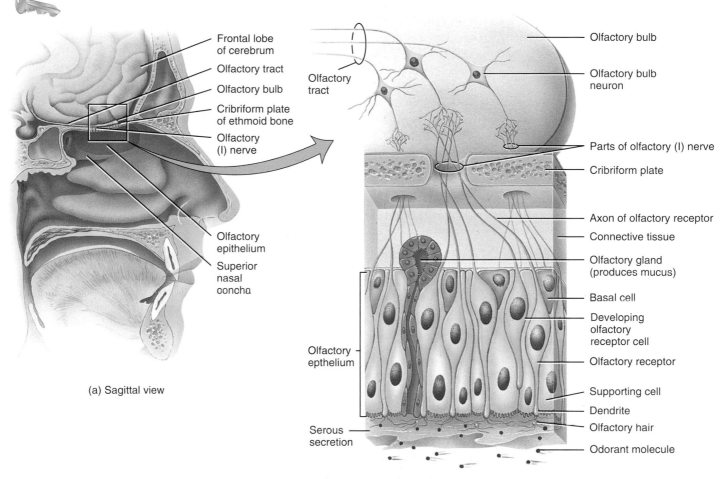

(a) Sagittal view

- Frontal lobe of cerebrum
- Olfactory tract
- Olfactory bulb
- Cribriform plate of ethmoid bone
- Olfactory (I) nerve
- Olfactory epithelium
- Superior nasal concha

- Olfactory tract
- Olfactory bulb
- Olfactory bulb neuron
- Parts of olfactory (I) nerve
- Cribriform plate
- Axon of olfactory receptor
- Connective tissue
- Olfactory gland (produces mucus)
- Basal cell
- Developing olfactory receptor cell
- Olfactory receptor
- Supporting cell
- Dendrite
- Olfactory hair
- Odorant molecule
- Olfactory epthelium
- Serous secretion

(b) Enlarged aspect of olfactory receptors

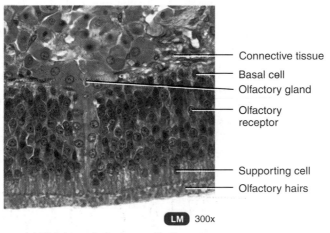

- Connective tissue
- Basal cell
- Olfactory gland
- Olfactory receptor
- Supporting cell
- Olfactory hairs

LM 300x

(c) Histology of olfactory epithelium

❓ **Which part of an olfactory receptor cell detects an odorant molecule?**

GUSTATION: SENSE OF TASTE

• Describe the gustatory receptors and the neural pathway for gustation.

Taste or **gustation** (gus-TĀ-shun; *gust-*=taste) is much simpler than olfaction in that only five primary tastes can be distinguished: *sour, sweet, bitter, salty,* and *umami* (ū-MAM-ē). The umami taste, recently described by Japanese scientists, is described as "meaty" or "savory."

Umami is believed to arise from taste receptors that are stimulated by monosodium glutamate (MSG), a substance naturally present in many foods and added to others as a flavor enhancer. All other flavors, such as chocolate, pepper, and coffee, are combinations of the five primary tastes, plus accompanying olfactory and tactile (touch) sensations. Odors from food can pass upward from the mouth into the nasal cavity, where they stimulate olfactory receptors. Because olfaction is much more sensitive than taste, a given concentration of a food substance may stimulate the olfactory system thousands of times more strongly than it stimulates the gustatory system. When people with colds or allergies complain that they cannot taste their food, they are reporting blockage of olfaction, not of taste.

Anatomy of Gustatory Receptors

The receptors for sensations of taste are located in the taste buds (Figure 22.2). The nearly 10,000 taste buds of a young adult are mainly on the tongue, but they are also found on the soft palate (posterior portion of roof of mouth), pharynx (throat), and epiglottis (cartilage lid over voice box). The number of taste buds declines with age. Each **taste bud** is an oval body consisting of three kinds of epithelial cells: supporting cells, gustatory receptor cells, and basal cells (Figure 22.2c). The **supporting cells** contain microvilli and surround about 50 **gustatory receptor cells.** A single, long microvillus, called a **gustatory hair,** projects from each gustatory receptor cell to the external surface through the **taste pore,** an opening in the taste bud. **Basal cells,** found at the periphery of the taste bud near the connective tissue layer, produce supporting cells. The supporting cells then develop into gustatory receptor cells, which have a life span of about 10 days. At their base, the gustatory receptor cells synapse with dendrites of the first-order sensory neurons that form the first part of the gustatory pathway. Each first-order neuron has many dendrites, which receive input from many gustatory receptor cells in numerous taste buds.

Taste buds are found in elevations on the tongue called **papillae** (pa-PIL-ē), which provide a rough texture to the upper surface of the tongue (Figure 22.2a, b). Three types of papillae contain taste buds.

1. About 12 very large, circular **vallate (circumvallate) papillae** (VAL-āt=wall-like) form an inverted V-shaped row at the back of the tongue. Each of these papillae houses 100–300 taste buds.

2. **Fungiform papillae** (FUN-ji-form=mushroomlike) are mushroom-shaped elevations scattered over the entire surface of the tongue that contain about five taste buds each.

3. **Foliate papillae** (FŌ-lē-āt=leaflike) are located in small trenches on the lateral margins of the tongue, but most of their taste buds degenerate in early childhood.

In addition, the entire surface of the tongue has **filiform papillae** (FIL-i-form=threadlike). These pointed, threadlike structures contain tactile receptors but no taste buds. They increase friction between the tongue and food, making it easier for the tongue to move food in the oral cavity.

Chemicals that stimulate gustatory receptor cells are known as **tastants.** Once a tastant is dissolved in saliva, it can make contact with the gustatory hairs, ultimately triggering nerve impulses in the first-order sensory neurons that synapse with gustatory receptor cells.

The Gustatory Pathway

Three cranial nerves contain axons of first-order gustatory neurons from taste buds. The facial (VII) nerve serves the anterior two-thirds of the tongue, the glossopharyngeal (IX) nerve serves the posterior one-third of the tongue, and the vagus (X) nerve serves the throat and epiglottis (see Figures 19.20, 19.22 and 19.23 on pages 616–618). From taste buds, impulses propagate along these cranial nerves to the medulla oblongata. From the medulla, some axons carrying taste signals project to the limbic system and the hypothalamus, whereas others project to the thalamus. Taste signals that project from the thalamus to the *primary gustatory area* in the parietal lobe of the cerebral cortex (see area 43 in Figure 19.15 on page 607) give rise to the conscious perception of taste.

TASTE AVERSION

Probably because of taste projections to the hypothalamus and limbic system, there is a strong link between taste and pleasant or unpleasant emotions. Sweet foods evoke reactions of pleasure, while bitter ones cause expressions of disgust even in newborn babies. This phenomenon is the basis for **taste aversion,** in which people and animals quickly learn to avoid a food if it upsets the digestive system. The advantage of avoiding foods that cause such illness is longer survival. However, the drugs and radiation treatments used to combat cancer often cause nausea and gastrointestinal upset regardless of the type of food consumed. Thus, cancer patients may lose their appetite because they develop taste aversions for most foods. ∎

3. How do olfactory receptors and gustatory receptor cells differ in structure and function?
4. Compare the olfactory and gustatory pathways.

Figure 22.2 **The relationship of gustatory receptor cells in taste buds to tongue papillae.**
(See Tortora, *A Photographic Atlas of the Human Body*, 2e, Figure 9.2.)

Gustatory receptor cells are located in taste buds.

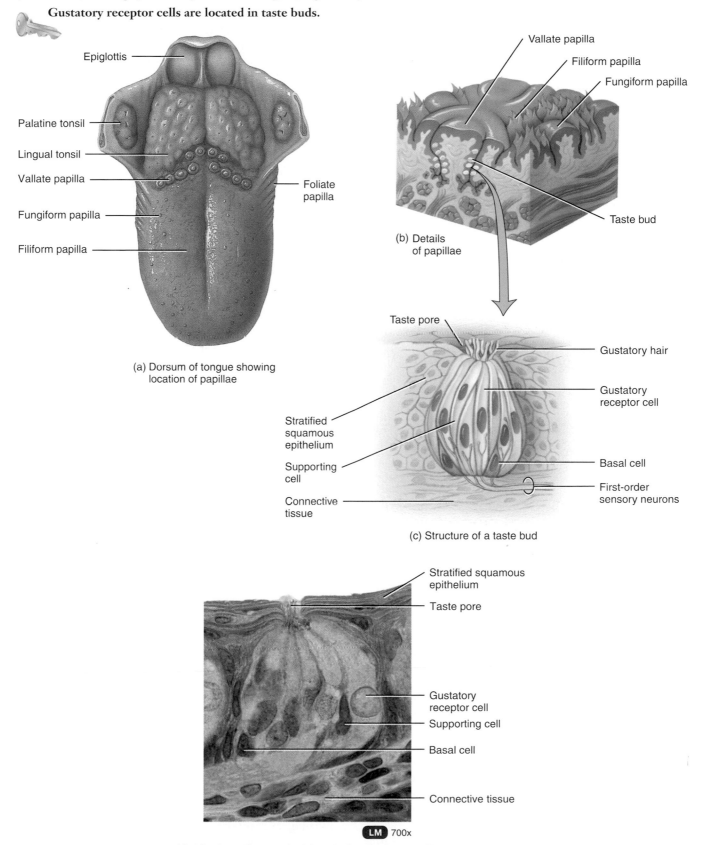

(a) Dorsum of tongue showing location of papillae

(b) Details of papillae

(c) Structure of a taste bud

(d) Histology of a taste bud from a circumvallate papilla

LM 700x

Beginning at the gustatory receptor cells, what structures form the gustatory pathway?

VISION

OBJECTIVES

• List and describe the accessory structures of the eye and the structural components of the eyeball.
• Describe the neural pathway for vision.

More than half the sensory receptors in the human body are located in the eyes, and a large part of the cerebral cortex is devoted to processing visual information. In this section of the chapter, we examine the accessory structures of the eye, the structure of the eyeball, the formation of visual images, the physiology of vision, and the visual pathway.

Accessory Structures of the Eye

The **accessory structures of the eye** include the eyelids, eyelashes, eyebrows, lacrimal (tearing) apparatus, and extrinsic eye muscles.

Eyelids

The upper and lower **eyelids,** or palpebrae (PAL-pe-brē), shade the eyes during sleep, protect the eyes from excessive light and foreign objects, and spread lubricating secretions over the eyeballs (Figure 22.3; see also Figure 12.3 on page 384). The upper eyelid is more movable than the lower and contains in its superior region the levator palpebrae superioris muscle. Sometimes a person may experience a twitch in an eyelid. It is an involuntary quivering similar to muscle twitches in the hand, forearm, leg, or foot. Twitches are almost always harmless and usually last for only a few seconds. They are often associated with stress and fatigue. The space between the upper and lower eyelids that exposes the eyeball is the **palpebral fissure.** Its angles are known as the **lateral commissure** (KOM-i-shur), which is narrower and closer to the temporal bone, and the **medial commissure,** which is broader and nearer the nasal bone. In the medial commissure is a small, reddish elevation, the **lacrimal caruncle** (KAR-ung-kul), which contains sebaceous (oil) glands and sudoriferous (sweat) glands. The whitish material that sometimes collects in the medial commissure comes from these glands.

From superficial to deep, each eyelid consists of epidermis, dermis, subcutaneous tissue, fibers of the orbicularis oculi muscle, a tarsal plate, tarsal glands, and conjunctiva (Figure 22.3). The **tarsal plate** is a thick fold of connective tissue that gives form and support to the eyelids. Embedded in each tarsal plate is a row of elongated modified sebaceous glands, known as tarsal or **Meibomian glands** (mī-BŌ-mē-an), that secrete a fluid that helps keep the eyelids from adhering to each other. Infection of the tarsal glands produces a relatively painless tumor or cyst on the eyelid called a **chalazion** (ka-LĀ-zē-on=small bump). The **conjunctiva** (kon′-junk-TĪ-va=to bind together) is a thin, protective mucous membrane composed of stratified columnar epithelium with numerous goblet cells that is supported by areolar connective tissue. The **palpebral conjunctiva** lines the inner aspect of the eyelids, and the **bulbar conjunctiva** passes from the eyelids onto the anterior surface of the eyeball. Dilation and congestion of the blood vessels of the bulbar conjunctiva due to local irritation or infection are the cause of **bloodshot eyes.**

Eyelashes and Eyebrows

The **eyelashes,** which project from the border of each eyelid, and the **eyebrows,** which arch transversely above the upper eyelids, help protect the eyeballs from foreign objects, perspiration, and the direct rays of the sun. Sebaceous glands at the base of the hair follicles of the eyelashes, called **sebaceous ciliary glands,** release a lubricating fluid into the follicles. Infection of these glands, usually by bacteria, cause a painful, pus-filled swelling called a **sty.**

The Lacrimal Apparatus

The **lacrimal apparatus** (*lacrim-*=tears) is a group of structures that produces and drains **lacrimal fluid** or **tears.** The **lacrimal glands,** each about the size and shape of an almond, secrete lacrimal fluid, which drains into 6–12 excretory lacrimal ducts that empty tears onto the surface of the conjunctiva of the upper lid (Figure 22.3b). From here the tears pass medially over the anterior surface of the eyeball to enter two small openings called **lacrimal puncta.** Tears then pass into two ducts, the **lacrimal canals,** which lead into the **lacrimal sac** and then into the **nasolacrimal duct.** This duct carries the lacrimal fluid into the nasal cavity just inferior to the inferior nasal concha. An infection of the lacrimal sacs is called **dacryocystitis** (*dacryo-*=lacrimal sac; *-itis*=inflammation of). It is usually caused by a bacterial infection and results in blockage in the nasolacrimal ducts.

Lacrimal fluid is a watery solution containing salts, some mucus, and **lysozyme,** a protective bactericidal enzyme. The fluid protects, cleans, lubricates, and moistens the eyeball. After being secreted, lacrimal fluid is spread medially over the surface of the eyeball by the blinking of the eyelids. Each gland produces about 1 mL of lacrimal fluid per day.

Normally, tears are cleared away as fast as they are produced, either by evaporation or by passing into the lacrimal canals and then into the nasal cavity. If an irritating substance makes contact with the conjunctiva, however, the lacrimal glands are stimulated to oversecrete, and tears accumulate (*watery eyes*). Lacrimation is a protective mechanism, as the tears dilute and wash away the irritating substance. Watery eyes also occur when an inflammation of the nasal mucosa, such as occurs with a cold, obstructs the nasolacrimal ducts and blocks drainage of tears. Humans are unique in expressing emotions, both happiness and sadness, by **crying.** In response to parasympathetic stimulation, the lacrimal glands produce excessive lacrimal fluid that may spill over the edges of the eyelids and even fill the nasal cavity with fluid. Thus, crying often produces a runny nose.

Extrinsic Eye Muscles

Six extrinsic eye muscles move each eye: the **superior rectus, inferior rectus, lateral rectus, medial rectus, superior oblique,** and **inferior oblique** (see Figures 22.3a and 22.4). They receive innervation from cranial nerves III, IV, or VI. In general, the motor units in these muscles are small. Some motor neurons serve only two or three muscle fibers—fewer than in any other part of the body except the larynx (voice box).

Figure 22.3 Accessory structures of the eye.

Accessory structures of the eye include the eyelids, eyelashes, eyebrows, lacrimal apparatus, and extrinsic eye muscles.

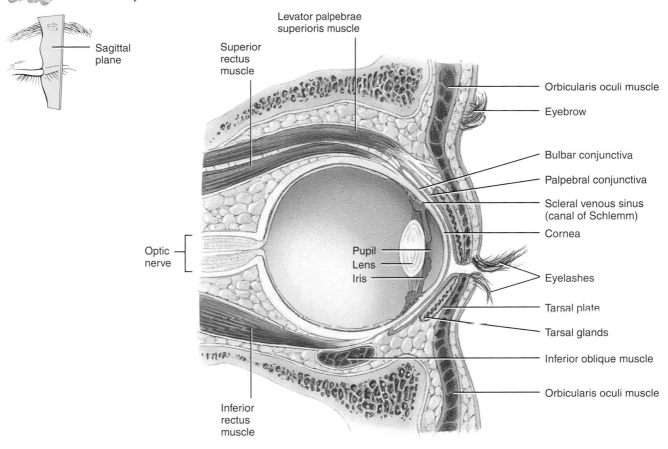

(a) Sagittal section of eye and its accessory structures

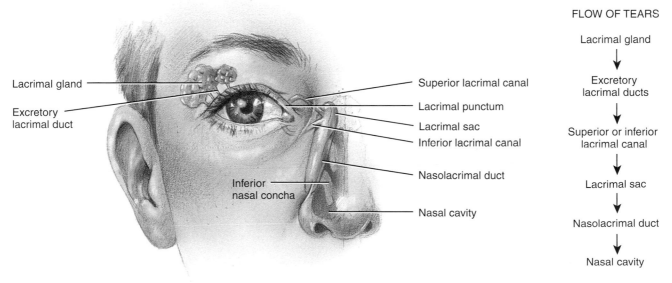

(b) Anterior view of the lacrimal apparatus

FLOW OF TEARS

Lacrimal gland
↓
Excretory lacrimal ducts
↓
Superior or inferior lacrimal canal
↓
Lacrimal sac
↓
Nasolacrimal duct
↓
Nasal cavity

? What is lacrimal fluid, and what are its functions?

Such small motor units permit smooth, precise, and rapid movement of the eyes. As indicated in Exhibit 11.2 on page 302, the extrinsic eye muscles move the eyeball laterally, medially, superiorly, and inferiorly. Looking to the right, for example, requires simultaneous contraction of the right lateral rectus and left medial rectus muscles and relaxation of the left lateral rectus and right medial rectus. The oblique muscles preserve rotational stability of the eyeball. Circuits in the brain stem and cerebellum coordinate and synchronize the movements of the eyes.

The surface anatomy of the accessory structures of the eye and the eyeball may be reviewed in Figure 12.3 on page 384.

Anatomy of the Eyeball

The adult **eyeball** measures about 2.5 cm (1 in.) in diameter. Of its total surface area, only the anterior one-sixth is exposed; the remainder is recessed and protected by the orbit, into which it fits. Anatomically, the wall of the eyeball consists of three layers: fibrous tunic, vascular tunic, and retina.

Fibrous Tunic

The **fibrous tunic,** the superficial coat of the eyeball, is avascular and consists of the anterior cornea and posterior sclera (Figure 22.4). The **cornea** (KOR-nē-a) is a transparent coat that covers the colored iris. Because it is curved, the cornea helps

Figure 22.4   Gross structure of the eyeball. (See Tortora, *A Photographic Atlas of the Human Body,* 2e, Figure 9.3.)

The wall of the eyeball consists of three layers: the fibrous tunic, the vascular tunic, and the retina.

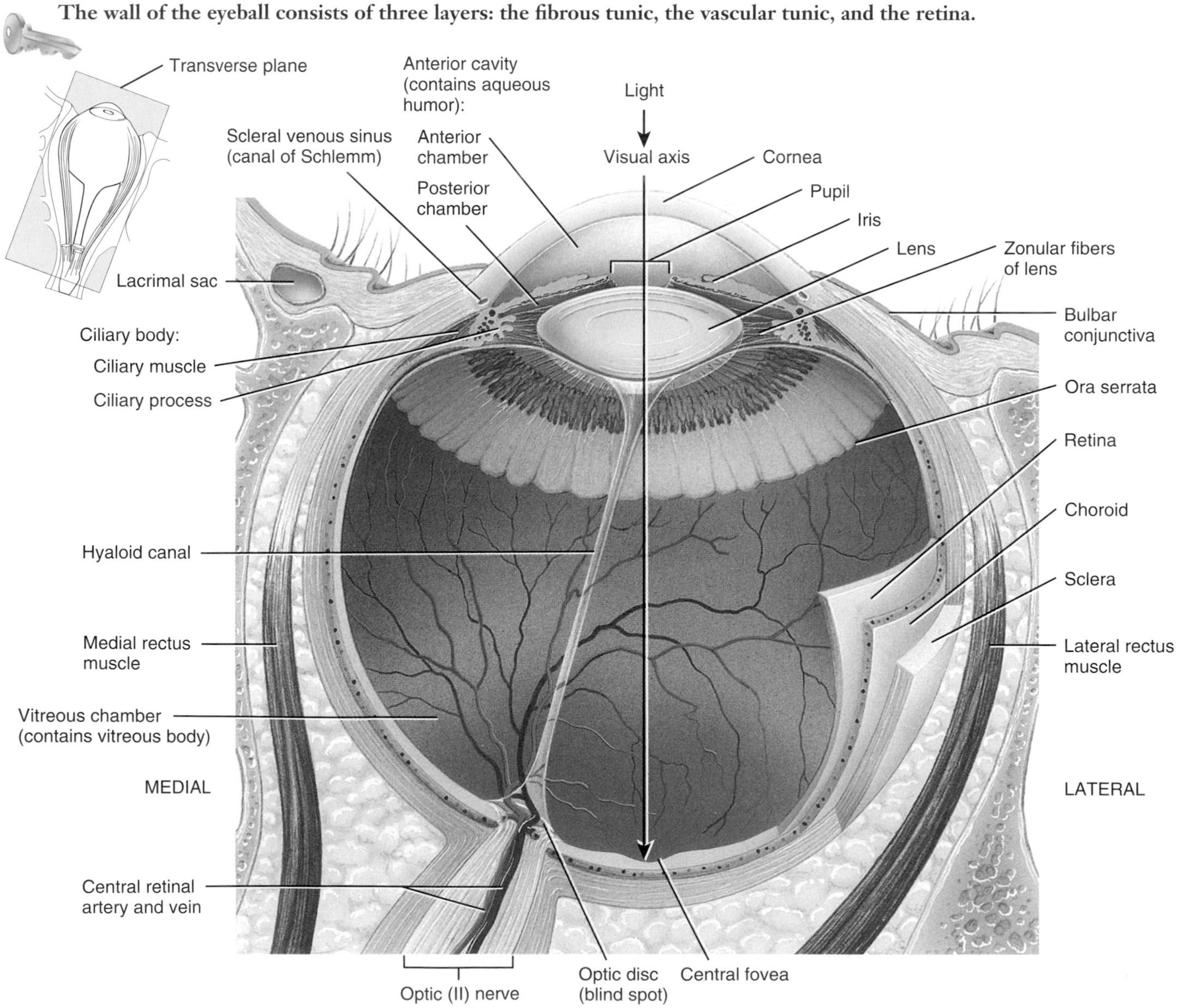

Superior view of transverse section of right eyeball

? What are the components of the fibrous tunic and vascular tunic?

focus light onto the retina. Its outer surface consists of nonkeratinized stratified squamous epithelium. The middle coat of the cornea consists of collagen fibers and fibroblasts, and the inner surface is simple squamous epithelium. Since the central part of the cornea receives oxygen from the outside air, contact lenses that are worn for long periods of time must be permeable to permit oxygen to pass through them. The **sclera** (SKLE-ra; *scler-*=hard), the "white" of the eye, is a layer of dense connective tissue made up mostly of collagen fibers and fibroblasts. The sclera covers the entire eyeball except the cornea; it gives shape to the eyeball, makes it more rigid, and protects its inner parts. At the junction of the sclera and cornea is an opening known as the **scleral venous sinus (canal of Schlemm).** A fluid called aqueous humor drains into this sinus (see Figure 22.8).

LASIK

An increasingly popular alternative to wearing glasses or contact lenses is refractive surgery to correct the curvature of the cornea for conditions such as farsightedness, nearsightedness, and astigmatism. The most common type of refractive surgery is **LASIK** (laser-assisted in-situ keratomileusis). After anesthetic drops are placed in the eye, a circular flap of tissue from the center of the cornea is cut. The flap is folded out of the way and the underlying layer of cornea is reshaped with a laser, one microscopic layer at a time. A computer assists the physician in removing very precise layers of the cornea. After the sculpting is complete, the corneal flap is repositioned over the treated area. A patch is placed over the eye overnight and the flap quickly reattaches to the rest of the cornea. ■

Vascular Tunic

The **vascular tunic** or **uvea** (Ū-vē-a) is the middle layer of the eyeball and has three parts: choroid, ciliary body, and iris (Figure 22.4). The highly vascularized **choroid** (KŌ-royd), which is the posterior portion of the vascular tunic, lines most of the internal surface of the sclera. It provides nutrients to the posterior surface of the retina.

In the anterior portion of the vascular tunic, the choroid becomes the **ciliary body** (SIL-ē-ar′-ē). It extends from the **ora serrata** (Ō-ra ser-RĀ-ta), the jagged anterior margin of the retina, to a point just posterior to the junction of the sclera and cornea. The ciliary body consists of the ciliary processes and the ciliary muscle. The **ciliary processes** are protrusions or folds on the internal surface of the ciliary body. They contain blood capillaries that secrete aqueous humor. Extending from the ciliary process are **zonular fibers (suspensory ligaments)** that attach to the lens. The **ciliary muscle** is a circular band of smooth muscle that alters the shape of the lens, adapting it for near or far vision. When the muscle contracts, the ciliary muscle is pulled anteriorly and inward toward the pupil, the tension in the zonular fibers is reduced, and the shape of the lens is altered into a more spherical shape.

The **iris,** the colored portion of the eyeball, is shaped like a flattened donut. It is suspended between the cornea and the lens and is attached at its outer margin to the ciliary processes. It consists of pigmented (melanin) epithelium and circular and radial smooth muscle fibers. The eyes are dark when the number of melanocytes is great but blue when the number of melanocytes is less. A principal function of the iris is to regulate the amount of light entering the vitreous chamber of the eyeball through the **pupil,** the hole in the center of the iris. The pupil appears black because, as you look through the lens, you see the heavily pigmented back of the eye (choroid and retina). However, if bright light is directed into the pupil, the reflected light is red because of the blood vessels on the surface of the retina. It is for this reason that a person's eyes appear red in a photograph when a bright light is directed into the pupil. Autonomic reflexes regulate pupil diameter in response to light levels (Figure 22.5). When bright light stimulates the eye, parasympathetic neurons stimulate the **circular muscles (sphincter pupillae)** of the iris to contract, causing a decrease in the size of the pupil (constriction). In dim light, sympathetic neurons stimulate the **radial muscles (dilator pupillae)** of the iris to contract, causing an increase in the pupil's size (dilation).

Retina

The third and innermost coat of the eyeball, the **retina,** lines the posterior three-quarters of the eyeball and is the beginning of the visual pathway (see Figure 22.4). An ophthalmoscope allows an observer to peer through the pupil, providing a magnified image of the retina and the blood vessels that course across the retina's anterior surface (Figure 22.6). The surface of the retina is the only place in the body where blood vessels can be viewed directly and examined for pathological changes, such as those that occur with hypertension or diabetes mellitus. Several landmarks are visible. The **optic disc** is the site where the optic (II) nerve exits the eyeball. Bundled together with the optic nerve are the **central retinal artery,** a branch of the ophthalmic artery, and the **central retinal vein** (see Figure 22.4). Branches of the central retinal artery fan out to nourish the anterior surface of the retina; the central retinal vein drains blood from the retina through the optic disc.

Figure 22.5   Responses of the pupil to light of varying brightness.

🔑 Contraction of the circular muscles causes constriction of the pupil; contraction of the radial muscles causes dilation of the pupil.

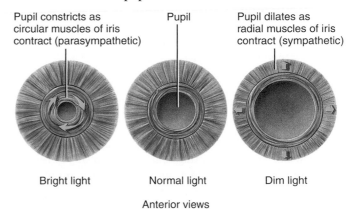

Pupil constricts as circular muscles of iris contract (parasympathetic)

Pupil

Pupil dilates as radial muscles of iris contract (sympathetic)

Bright light   Normal light   Dim light

Anterior views

❓ **Which division of the autonomic nervous system causes pupillary constriction? Which causes pupillary dilation?**

Figure 22.6 A normal retina, as seen through an ophthalmoscope.

Blood vessels in the retina can be viewed directly and examined for pathological changes.

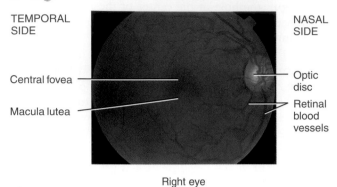

Right eye

Evidence of what diseases may be seen through an ophthalmoscope?

The optic part of the retina consists of a pigmented layer and a neural layer. The **pigmented layer** is a sheet of melanin-containing epithelial cells located between the choroid and the neural part of the retina. Melanin in the choroid and in the pigmented layer absorbs stray light rays, which prevents reflection and scattering of light within the eyeball. As a result, the image cast on the retina by the cornea and lens remains sharp and clear. Albinos lack melanin in all parts of the body, including the eye. They often need to wear sunglasses, even indoors, because even moderately bright light is perceived as bright glare due to light scattering.

The **neural layer** of the retina is a multilayered outgrowth of the brain that extensively processes visual data before sending nerve impulses into axons that form the optic nerve. Three distinct layers of retinal neurons—the **photoreceptor layer,** the **bipolar cell layer,** and the **ganglion cell layer**—are separated by two zones, the **outer** and **inner synaptic layers,** where synaptic contacts are made (Figure 22.7). Note that light passes through the ganglion and bipolar cell layers and both synaptic layers before it reaches the photoreceptor layer. Two other types of cells present in the bipolar cell layer of the retina are called **horizontal cells** and **amacrine cells.** These cells form laterally directed neural circuits that modify the signals being transmitted along the pathway from photoreceptors to bipolar cells to ganglion cells.

DETACHED RETINA

A **detached retina** may occur in trauma, such as a blow to the head, in various eye disorders, or age-related degeneration. The detachment occurs between the neural portion of the retina and the pigment epithelium. Fluid accumulates between these layers, forcing the thin, pliable retina to billow outward. The result is distorted vision and blindness in the corresponding field of vision. The retina may be reattached by laser surgery or cryosurgery (localized application of extreme cold). ■

Two types of photoreceptors are specialized to begin the process by which light rays are ultimately converted to nerve impulses: rods and cones. Each retina has about 6 million cones and 120 million rods. **Rods** allow us to see in dim light, such as moonlight. Because they do not provide color vision, in dim light we see only shades of gray. Brighter lights stimulate the **cones,** which produce color vision. Most of our visual experiences are mediated by the cone system, the loss of which produces legal blindness. In contrast, a person who loses rod vision mainly has difficulty seeing in dim light and thus should not drive at night. Rods and cones contain **photopigments,** colored proteins that absorb light and begin the process of nerve impulse generation.

COLOR BLINDNESS AND NIGHT BLINDNESS

Most forms of **color blindness,** an inherited inability to distinguish between certain colors, result from the absence or a deficiency of one or more of the cone photopigments. The most common type is *red-green color blindness*, in which a photopigment sensitive to orange-red light or green light is missing. As a result, the person cannot distinguish between red and green. Prolonged vitamin A deficiency and the consequent below-normal amount of the photopigment in cones may cause **night blindness** or **nyctalopia** (nik′-ta-LŌ-pē-a), an inability to see well at low light levels. ■

The **macula lutea** (MAK-ū-la LOO-tē-a; *macula*=a small, flat spot; *lute-*=yellowish) is in the exact center of the posterior portion of the retina, at the visual axis of the eye. The central fovea (see Figure 22.4), a small depression in the center of the macula lutea, contains only cones. In addition, the layers of bipolar and ganglion cells, which scatter light to some extent, do not cover the cones here; these layers are displaced to the periphery of the central fovea. As a result, the central fovea is the area of highest **visual acuity** or **resolution** (sharpness of vision). A main reason that you move your head and eyes while looking at something is to place images of interest on your central fovea—as you do to read the words in this sentence! Rods are absent in the central fovea and are more plentiful toward the periphery of the retina. Because rod vision is more sensitive than cone vision, you can see a faint object (such as a dim star) better if you gaze slightly to one side rather than looking directly at it.

From photoreceptors, information flows to bipolar cells and then to ganglion cells. The axons of ganglion cells extend posteriorly to the optic disc and exit the eyeball as the optic nerve. The optic disc is also called the blind spot. Because it contains no rods or cones, we cannot see an image that strikes the blind spot. Normally, you are not aware of having a blind spot, but you can easily demonstrate its presence. Cover your left eye and gaze directly at the cross below. Then increase or decrease the distance between the book and your eye. At some point the square will disappear as its image falls on the blind spot.

+   ■

Figure 22.7 Microscopic structure of the retina. The downward blue arrow at left indicates the direction of the signals passing through the neural portion of the retina. Eventually, nerve impulses arise in ganglion cells and propagate along their axons, which make up the optic (II) nerve.

In the retina, visual signals pass from photoreceptors to bipolar cells to ganglion cells.

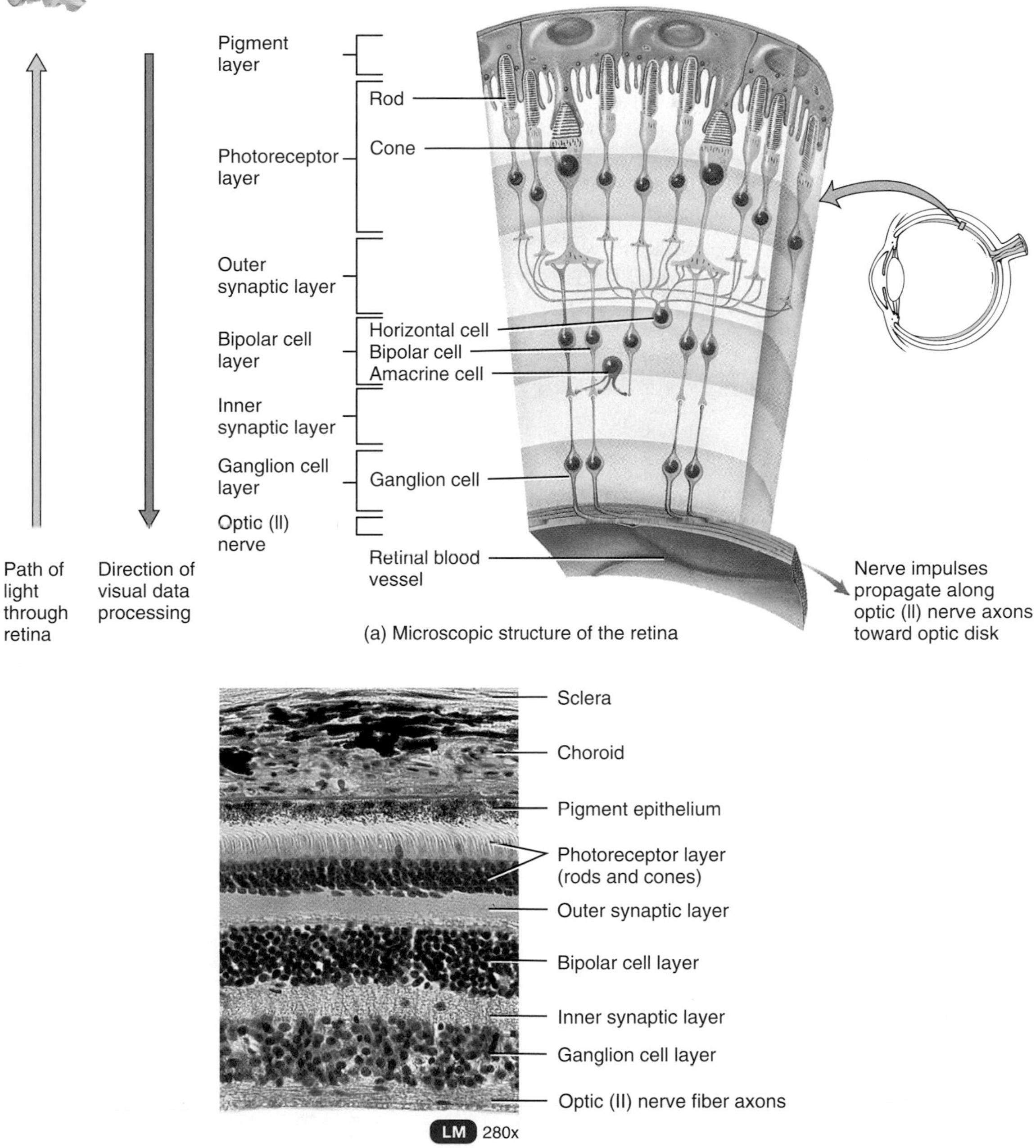

Pigment layer

Rod

Photoreceptor layer

Cone

Outer synaptic layer

Bipolar cell layer

Horizontal cell
Bipolar cell
Amacrine cell

Inner synaptic layer

Ganglion cell layer

Ganglion cell

Optic (II) nerve

Retinal blood vessel

Path of light through retina

Direction of visual data processing

Nerve impulses propagate along optic (II) nerve axons toward optic disk

(a) Microscopic structure of the retina

Sclera

Choroid

Pigment epithelium

Photoreceptor layer (rods and cones)

Outer synaptic layer

Bipolar cell layer

Inner synaptic layer

Ganglion cell layer

Optic (II) nerve fiber axons

LM 280x

(b) Histology of a portion of the retina

What are the two types of photoreceptors, and how do their functions differ?

Lens

Posterior to the pupil and iris, within the cavity of the eyeball, is the lens (see Figure 22.4). Proteins called *crystallins*, arranged like the layers of an onion, make up the lens, which normally is perfectly transparent and lacks blood vessels. It is enclosed by a clear connective tissue capsule and held in position by encircling zonular fibers, which attach to the ciliary processes. The lens helps focus images on the retina to facilitate clear vision.

PRESBYOPIA

With aging, the lens loses elasticity and thus its ability to curve to focus on objects that are close. Therefore, older people cannot read print at the same close range as can youngsters. This condition is called **presbyopia** (prez-bē-Ō-pē-a; *presby-* = old; *-opia* = pertaining to the eye or vision). By age 40 the near point of vision may have increased to 20 cm (8 in.), and at age 60 it may be as much as 80 cm (31 in.). Presbyopia usually begins in the mid-forties. At about that age, people who have not previously worn glasses begin to need them for reading. Those who already wear glasses typically start to need bifocals, lenses that can focus for both distant and close vision. ■

Interior of the Eyeball

The lens divides the interior of the eyeball into two cavities: the anterior cavity and the vitreous chamber. The **anterior cavity**— the space anterior to the lens—consists of two chambers. The **anterior chamber** lies between the cornea and the iris, whereas the **posterior chamber** lies behind the iris and in front of the zonular fibers and lens (Figure 22.8). Both chambers of the anterior cavity are filled with **aqueous humor** (*aqua*=water), a watery fluid that nourishes the lens and cornea. Aqueous humor continually filters out of blood capillaries in the ciliary processes and enters the posterior chamber. It then flows forward between the iris and the lens, through the pupil, and into the anterior chamber. From the anterior chamber, aqueous humor drains into the scleral venous sinus (canal of Schlemm) and then into

the blood. Normally, aqueous humor is completely replaced about every 90 minutes.

The pressure in the eye, called **intraocular pressure,** is produced mainly by the aqueous humor and partly by the vitreous body (described shortly); normally it is about 16 mmHg (millimeters of mercury). The intraocular pressure maintains the shape of the eyeball and prevents the eyeball from collapsing. Puncture wounds to the eyeball may cause the loss of aqueous humor and the vitreous body. This causes a decrease in intraocular pressure, a detached retina, and maybe even blindness.

The second, and larger, cavity of the eyeball is the **vitreous chamber,** which lies between the lens and the retina. Within the vitreous chamber is the **vitreous body,** a jellylike substance that contributes to intraocular pressure. It holds the retina flush against the choroid, so that the retina provides an even surface for the reception of clear images. Unlike the aqueous humor, the vitreous body does not undergo constant replacement. It is formed during embryonic life and is not replaced thereafter. The vitreous body also contains phagocytic cells that remove

Figure 22.8   The anterior and posterior chambers of the eye, seen in a section through the anterior portion of the eyeball at the junction of the cornea and sclera. Arrows indicate the direction of flow of aqueous humor.

The lens separates the posterior chamber of the anterior cavity from the vitreous chamber.

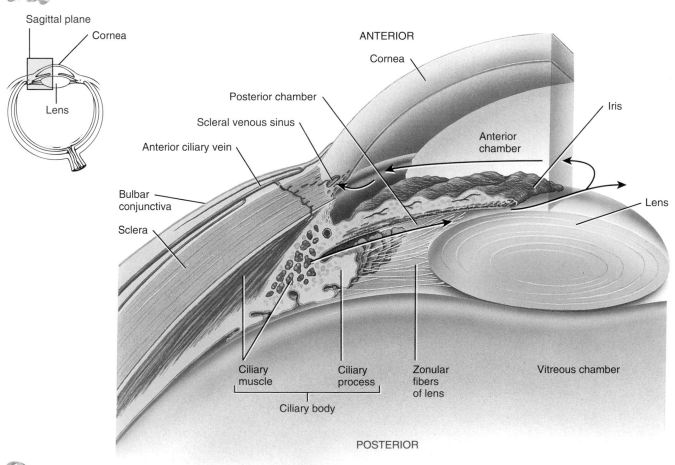

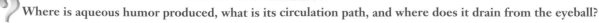

Where is aqueous humor produced, what is its circulation path, and where does it drain from the eyeball?

debris, keeping this part of the eye clear for unobstructed vision. Occasionally, collections of debris may cast a shadow on the retina and create the appearance of specks, hairs, and fine strings that dart in and out of the field of vision. These *vitreal floaters* result from changes in the vitreous body related to aging and from certain conditions such as nearsightedness and inflammation. Vitreal floaters are usually harmless and do not require treatment. The **hyaloid canal** (see Figure 22.4) is a narrow channel that runs through the vitreous body from the optic disc to the posterior aspect of the lens. In the fetus, this channel is occupied by the hyaloid artery (see Figure 22.17).

Table 22.1 summarizes the structures associated with the eyeball.

AGE-RELATED MACULAR DISEASE

Age-related macular disease (AMD), also known as *macular degeneration*, is a degenerative disorder of the retina and pigmented layer in persons 50 years of age and older. In AMD, abnormalities occur in the region of the macula lutea, which is ordinarily the area of most acute vision. Victims of advanced AMD retain their peripheral vision but lose the ability to see straight ahead. For instance, they cannot see facial features to identify a person in front of them. AMD is the leading cause of blindness in those over age 75, afflicting 13 million Americans, and is 2.5 times more common in pack-a-day smokers than in nonsmokers. Initially, a person may experience blurring and distortion at the center of the visual field. In "dry" AMD, central vision gradually diminishes because the pigmented layer atrophies and degenerates. There is no effective treatment. In about 10% of cases, dry AMD progresses to "wet" AMD, in which new blood vessels form in the choroid and leak plasma or blood under the retina. Vision loss can be slowed by using laser surgery to destroy leaking blood vessels. ∎

The Visual Pathway

After considerable processing of visual signals in the retina at synapses among the various types of neurons, the axons of retinal ganglion cells provide output from the retina to the brain. They exit the eyeball as the optic (II) nerve (see Figure 22.7).

TABLE 22.1	SUMMARY OF STRUCTURES OF THE EYEBALL		
STRUCTURE	**FUNCTION**	**STRUCTURE**	**FUNCTION**
Fibrous tunic Cornea / Sclera	*Cornea* Admits and refracts (bends) light. *Sclera* Provides shape and protects inner parts.	**Lens** Lens	Focuses light on the retina.
Vascular tunic Iris / Ciliary body / Choroid	*Iris* Regulates the amount of light that enters eyeball. *Ciliary body* Secretes aqueous humor and alters the shape of the lens for near or far vision (accommodation). *Choroid* Provides blood supply and absorbs scattered light.	**Anterior cavity** Anterior cavity	Contains aqueous humor that helps maintain the shape of the eyeball and supplies oxygen and nutrients to the lens and the cornea.
Retina Retina	Receives light and converts it into nerve impulses. Provides output to the brain via axons of ganglion cells, which form the optic (II) nerve.	**Vitreous chamber** Vitreous chamber	Contains the vitreous body that helps maintain the shape of the eyeball and keeps the retina attached to the choroid.

Processing of Visual Input in the Retina

Within the retina, certain features of visual input are enhanced while other features may be discarded. Input from several cells may either converge upon a smaller number of postsynaptic neurons or diverge to a larger number. On the whole, convergence predominates: 1 million ganglion cells receive input from about 126 million photoreceptor cells.

Chemicals (neurotransmitters) released by rods and cones induce changes in both bipolar cells and horizontal cells that lead to the generation of nerve impulses (see Figure 22.7). Amacrine cells synapse with ganglion cells and also transmit information to them. When bipolar, horizontal, or amacrine cells transmit signals to ganglion cells, the ganglion cells initiate nerve impulses.

Pathway in the Brain

The axons of the optic (II) nerve pass through the **optic chiasm** (kī-AZM=a crossover, as in the letter X), a crossing point of the optic nerves (Figure 22.9). Some fibers cross to the opposite side, whereas others remain uncrossed. After passing through the optic chiasm, the fibers, now part of the **optic tract,** enter the brain and terminate in the lateral geniculate nucleus of the thalamus. Here they synapse with neurons whose axons form the **optic radiations,** which project to the primary visual areas in the occipital lobes of the cerebral cortex (area 17 in Figure 19.15 on page 607).

CHECKPOINT

5. Describe the structure and importance of the eyelids, eyelashes, and eyebrows.
6. What is the function of the lacrimal apparatus?
7. What are the components of the fibrous tunic, vascular tunic, and retina, and what are their functions?
8. Describe the histology of the neural portion of the retina.
9. Describe the pathway for vision, beginning with light entering the eye and ending in the brain.

HEARING AND EQUILIBRIUM

OBJECTIVES

● Describe the anatomy of the structures in the three principal regions of the ear.
● List the principal events involved in hearing.
● Identify the receptor organs for equilibrium, and describe how they function.
● Describe the auditory and equilibrium pathways.

The ear is an engineering marvel because its sensory receptors can transduce sound vibrations with amplitudes as small as the diameter of an atom of gold (0.3 nm) into electrical signals 1000 times faster than photoreceptors can respond to light. Besides receptors for sound waves, the ear also contains receptors for equilibrium.

Figure 22.9 The visual pathway. Partial dissection of the brain in part (a) reveals the optic radiations (axons extending from the thalamus to the occipital lobe), along with the rest of the visual pathway, which is diagrammed in part (b).

 The optic chiasm is the crossing point of the optic nerves.

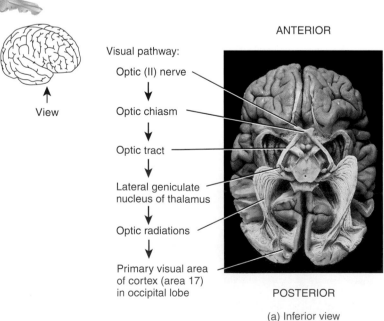

(a) Inferior view

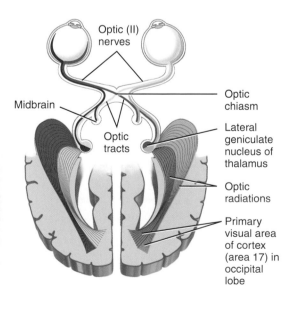

(b) Superior view of transverse section through eyeballs and brain

Where does the optic tract terminate?

Anatomy of the Ear

The **ear** is divided into three main regions: the external ear, which collects sound waves and channels them inward; the middle ear, which conveys sound vibrations to the oval window; and the internal ear, which houses the receptors for hearing and equilibrium.

External (Outer) Ear

The **external (outer) ear** consists of the auricle, external auditory canal, and eardrum (Figure 22.10). The **auricle (pinna)** is a flap of elastic cartilage shaped like the flared end of a trumpet and covered by skin. The rim of the auricle is the **helix;** the inferior portion is the **lobule.** Ligaments and muscles attach the auricle to the head. The **external auditory canal** (*audit-*= hearing) is a curved tube about 2.5 cm (1 in.) long that lies in the temporal bone and leads from the auricle to the eardrum. The **eardrum** or **tympanic membrane** (tim-PAN-ik; *tympan-*=a drum) is a thin, semitransparent partition between the external auditory canal and middle ear. The eardrum is covered by epi-

dermis and lined by simple cuboidal epithelium. Between the epithelial layers is connective tissue composed of collagen, elastic fibers, and fibroblasts. Tearing of the tympanic membrane is called a **perforated eardrum.** It may be due to pressure from a cotton swab, trauma, or a middle ear infection, and usually heals within a month. The eardrum may be examined directly by an **otoscope** (*oto-*=ear; *-skopeo*=to view), a viewing instrument that illuminates and magnifies the external ear and eardrum.

Near the exterior opening, the external auditory canal contains a few hairs and specialized sebaceous (oil) glands called **ceruminous glands** (se-ROO-mi-nus) that secrete earwax or **cerumen** (se-ROO-men). The combination of hairs and cerumen helps prevent dust and foreign objects from entering the ear. Cerumen usually dries up and falls out of the ear canal. Some people, however, produce a large amount of cerumen, which can become impacted and can muffle incoming sounds. The treatment for **impacted cerumen** is usually periodic ear irrigation or removal of wax with a blunt instrument by trained medical personnel.

Figure 22.10   Structure of the ear.

The ear has three principal regions: the external (outer) ear, the middle ear, and the internal (inner) ear.

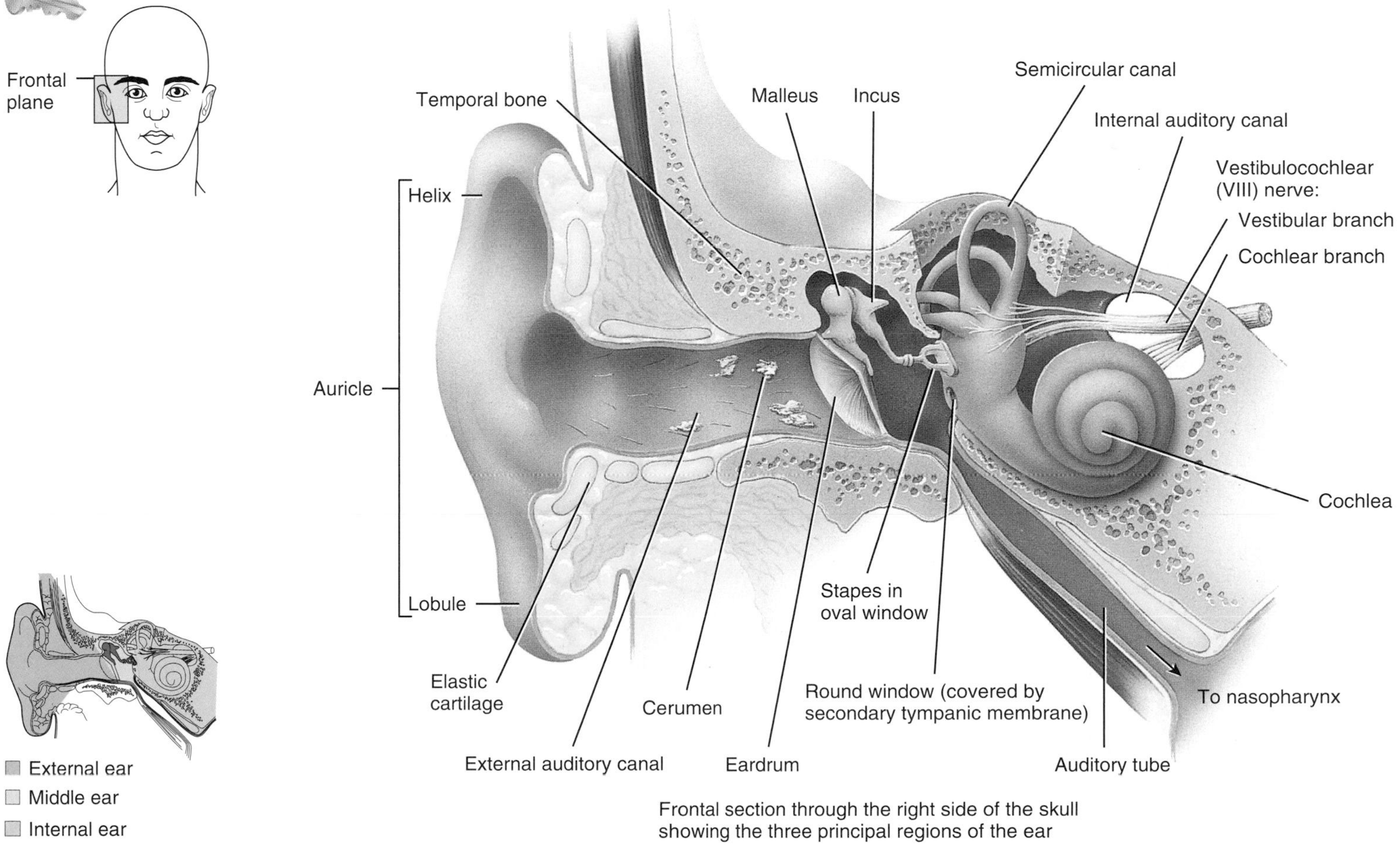

Frontal section through the right side of the skull showing the three principal regions of the ear

■ External ear
□ Middle ear
■ Internal ear

To which structure of the external ear does the malleus of the middle ear attach?

Middle Ear

The **middle ear** is a small, air-filled cavity in the temporal bone that is lined by epithelium (Figure 22.11). It is separated from the external ear by the eardrum and from the internal ear by a thin bony partition that contains two small membrane-covered openings: the oval window and the round window. Extending across the middle ear and attached to it by ligaments are the three smallest bones in the body, the **auditory ossicles** (OS-si-kuls), which are connected by synovial joints. The bones, named for their shapes, are the malleus, incus, and stapes—commonly called the hammer, anvil, and stirrup, respectively. The "handle" of the **malleus** (MAL-ē-us) attaches to the internal surface of the eardrum. The head of the malleus articulates with the body of the incus. The **incus** (ING-kus), the middle bone in the series, articulates with the head of the stapes. The base or footplate of the **stapes** (STĀ-pēz) fits into the **oval window.** Directly below the oval window is another opening, the **round window,** which is enclosed by a membrane called the **secondary tympanic membrane.**

Besides the ligaments, two tiny skeletal muscles also attach to the ossicles (Figure 22.11). The **tensor tympani muscle,** which is innervated by the mandibular branch of the trigeminal (V) nerve, limits movement and increases tension on the eardrum to prevent damage to the inner ear from loud noises. The **stapedius muscle,** which is innervated by the facial (VII) nerve, is the smallest of all skeletal muscles. By dampening large vibrations of the stapes due to loud noises, it protects the oval window, but it also decreases the sensitivity of hearing. For this reason, paralysis of the stapedius muscle is associated with **hyperacusia** (abnormally sensitive hearing). Because it takes a fraction of a second for the tensor tympani and stapedius muscles to contract, they can protect the inner ear from prolonged loud noises such as thunder, but not from brief ones such as a gunshot.

The anterior wall of the middle ear contains an opening that leads directly into the **auditory (pharyngotympanic) tube,** commonly known as the **eustachian tube.** The auditory tube, which consists of both bone and hyaline cartilage, connects the middle ear with the nasopharynx (upper portion of the throat). It is normally closed at its medial (pharyngeal) end. During swallowing and yawning, it opens, allowing air to enter or leave the middle ear until the pressure in the middle ear

Figure 22.11 The right middle ear containing the auditory ossicles. (See Tortora, *A Photographic Atlas of the Human Body*, 2e, Figure 3.14)

Common names for the malleus, incus, and stapes are the hammer, anvil, and stirrup, respectively.

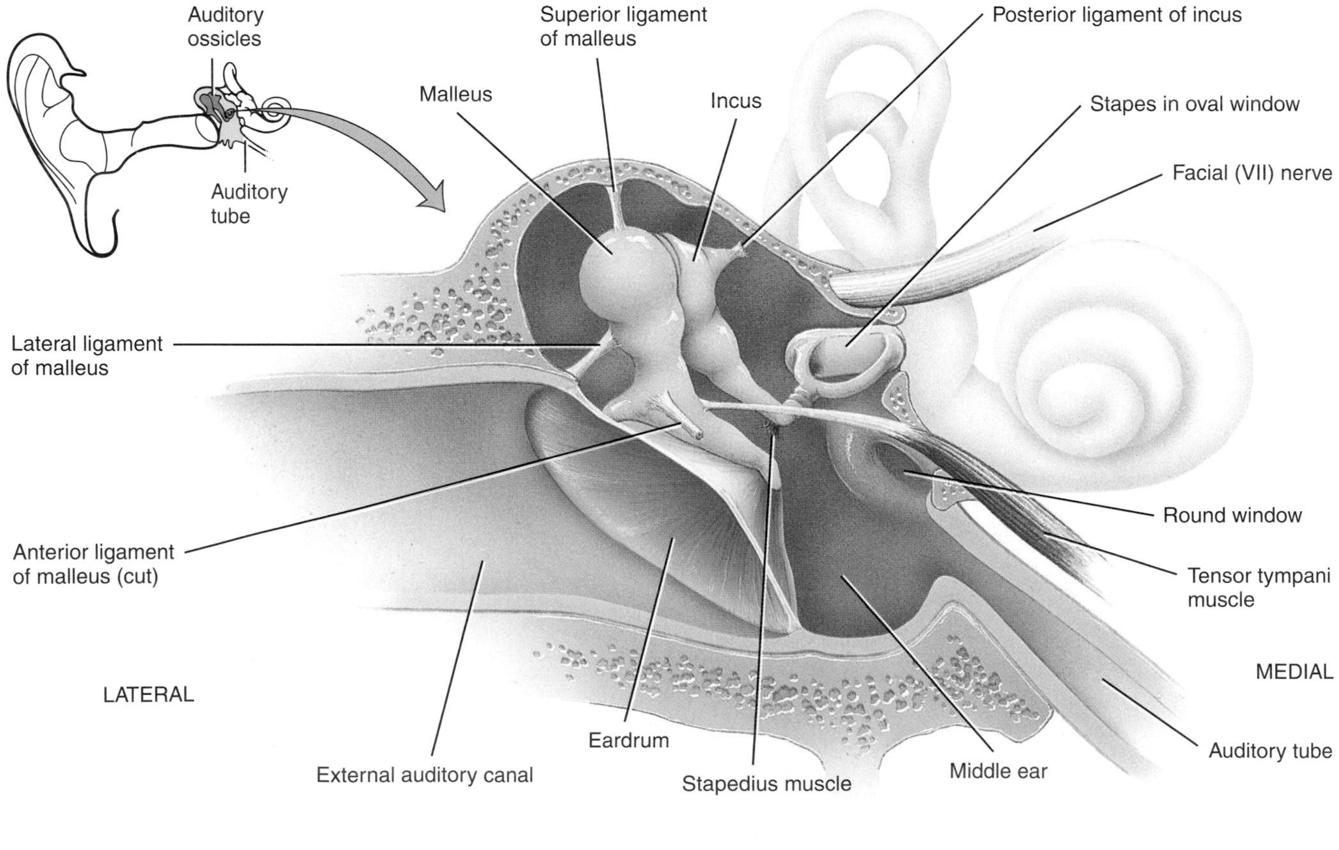

Frontal section showing location of auditory ossicles

What structures separate the middle ear from the internal ear?

equals the atmospheric pressure. When the pressures are balanced, the eardrum vibrates freely as sound waves strike it. If the pressure is not equalized, intense pain, hearing impairment, ringing in the ears, and vertigo could develop. The auditory tube also is a route whereby pathogens may travel from the nose and throat to the middle ear, causing the most common type of ear infection (see Otitis Media in the Applications to Health section at the end of this chapter).

Internal (Inner) Ear

The **internal (inner) ear** is also called the **labyrinth** (LAB-i-rinth) because of its complicated series of canals (Figure 22.12). Structurally, it consists of two main divisions: an outer bony labyrinth that encloses an inner membranous labyrinth. The **bony labyrinth** is a series of cavities in the temporal bone divided into three areas: (1) the semicircular canals and (2) the vestibule, both of which contain receptors for equilibrium, and (3) the cochlea, which contains receptors for hearing. The bony labyrinth is lined with periosteum and contains **perilymph.** This fluid, which is chemically similar to cerebrospinal fluid, surrounds the **membranous labyrinth,** a series of sacs and tubes inside the bony labyrinth and having the same general form. The membranous labyrinth is lined by epithelium and contains **endolymph.** The level of potassium ions (K^+) in endolymph is unusually high for an interstitial fluid, and potassium ions play a role in the generation of interstitial auditory signals (described shortly).

The **vestibule** (VES-ti-būl) is the oval central portion of the bony labyrinth. The membranous labyrinth in the vestibule consists of two sacs called the **utricle** (Ū-tri-kl=little bag) and the **saccule** (SAK-ūl=little sac), which are connected by a small duct. Projecting superiorly and posteriorly from the vestibule are the three bony **semicircular canals,** each of which lies at approximately right angles to the other two. Based on their positions, they are named the anterior, posterior, and lateral semicircular canals. The anterior and posterior semicircular canals are vertically oriented; the lateral one is horizontally oriented. At one end of each canal is a swollen enlargement called the **ampulla** (am-PUL-la=saclike duct). The portions of the membranous labyrinth that lie inside the bony semicircular canals are called the **semicircular ducts.** These structures communicate with the utricle of the vestibule.

Figure 22.12   The right internal ear. The outer, cream-colored area is part of the bony labyrinth; the inner, pink-colored area is the membranous labyrinth.

The bony labyrinth contains perilymph, and the membranous labyrinth contains endolymph.

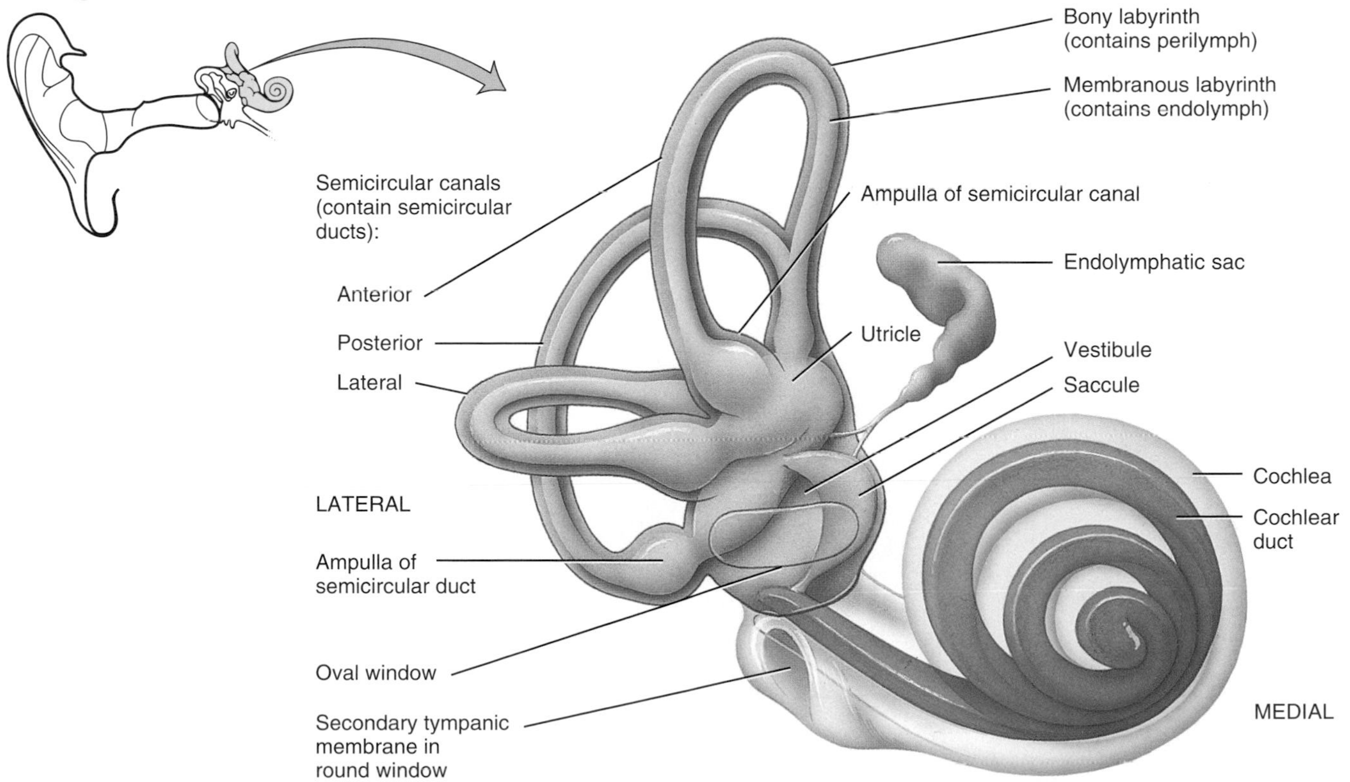

What are the names of the two sacs that lie in the vestibule?

The vestibular branch of the vestibulocochlear (VIII) nerve consists of *ampullary, utricular,* and *saccular nerves.* These nerves contain both first-order sensory neurons and motor neurons that synapse with receptors for equilibrium. The first-order sensory neurons carry sensory information from the receptors, and the motor neurons carry feedback signals to the receptors, apparently to modify their sensitivity. Cell bodies of the sensory neurons are located in the **vestibular ganglia** (see Figure 22.13b).

Anterior to the vestibule is the **cochlea** (KOK-lē-a=snail-shaped), a bony spiral canal (Figure 22.13a) that resembles a snail's shell and makes almost three turns around a central bony core called the **modiolus** (mō-DĪ-ō-lus; Figure 22.13b). Sections through the cochlea reveal that it is divided into three channels (Figure 22.13a–c). Together, the partitions that separate the channels are shaped like the letter Y. The stem of the Y is a bony shelf that protrudes into the canal; the wings of the Y are composed mainly of membranous labyrinth. The channel above the bony partition is the **scala vestibuli,** which ends at the oval window; the channel below is the **scala tympani,** which ends at the round window.

The scala vestibuli and scala tympani both contain perilymph and are completely separated, except for an opening at the apex of the cochlea, the **helicotrema** (hel-i-kō-TRĒ-ma; see Figure 22.14). The cochlea adjoins the wall of the vestibule, into which the scala vestibuli opens. The perilymph in the vestibule is continuous with that of the scala vestibuli. The third channel (between the wings of the Y) is the **cochlear duct** or **scala media.** The **vestibular membrane** separates the cochlear duct from the scala vestibuli, and the **basilar membrane** separates the cochlear duct from the scala tympani.

Resting on the basilar membrane is the **spiral organ** or **organ of Corti** (Figure 22.13c, d). The spiral organ is a coiled sheet of epithelial cells, including supporting cells and about 16,000 **hair cells,** which are the receptors for hearing. There are two groups of hair cells: The *inner hair cells* are arranged in a single row whereas the *outer hair cells* are arranged in three rows. At the apical tip of each hair cell is a **hair bundle,** consisting of 30–100 *stereocilia* that extend into the endolymph of the cochlear duct. Despite their name, stereocilia are actually long, hairlike microvilli arranged in several rows of graded height.

At their basal ends, inner and outer hair cells synapse both with first-order sensory neurons and with motor neurons from the cochlear branch of the vestibulocochlear (VIII) nerve. Cell bodies of the sensory neurons are located in the **spiral ganglion** (Figure 22.13b, c). Although outer hair cells outnumber them by 3 to 1, the inner hair cells synapse with 90–95% of the first-

Figure 22.13 Semicircular canals, vestibule, and cochlea of the right ear. Note that the cochlea makes almost three complete turns.

The three channels in the cochlea are the scala vestibuli, the scala tympani, and the cochlear duct.

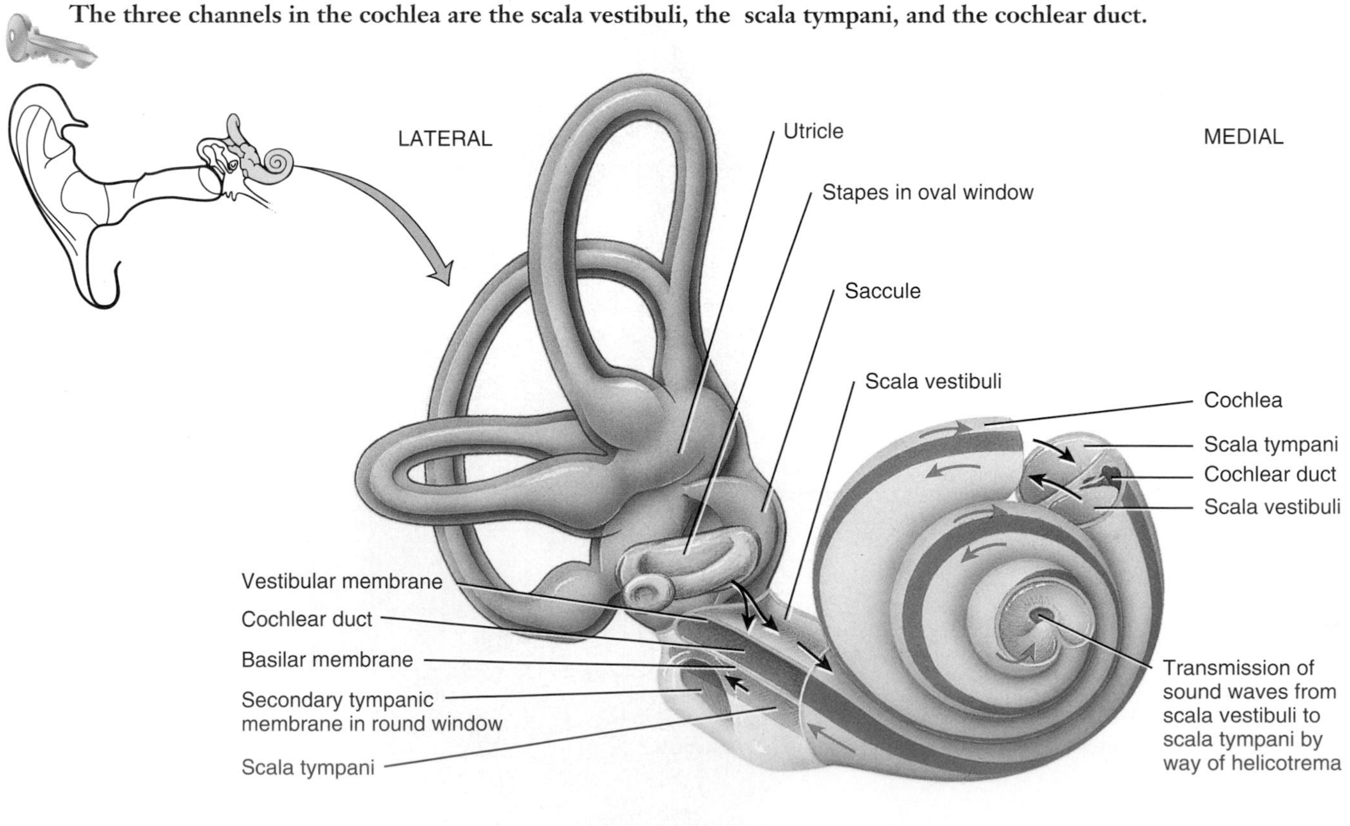

LATERAL

Utricle

Stapes in oval window

Saccule

Scala vestibuli

Cochlea

Scala tympani

Cochlear duct

Scala vestibuli

MEDIAL

Vestibular membrane

Cochlear duct

Basilar membrane

Secondary tympanic membrane in round window

Scala tympani

Transmission of sound waves from scala vestibuli to scala tympani by way of helicotrema

(a) Sections through the cochlea

order sensory neurons in the cochlear nerve that relay auditory information to the brain. By contrast, 90% of the motor neurons in the cochlear nerve synapse with outer hair cells. Whereas inner hair cells are the actual receptors that begin the process of nerve impulse generation that result in sound, outer hair cells amplify sound. Projecting over and in contact with the hair cells of the spiral organ is the **tectorial membrane** (*tector-*=covering), a flexible gelatinous membrane.

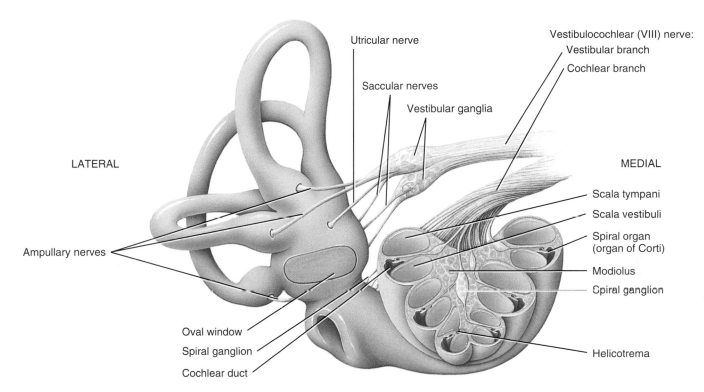

(b) Components of the vestibulocochlear nerve (cranial nerve VIII)

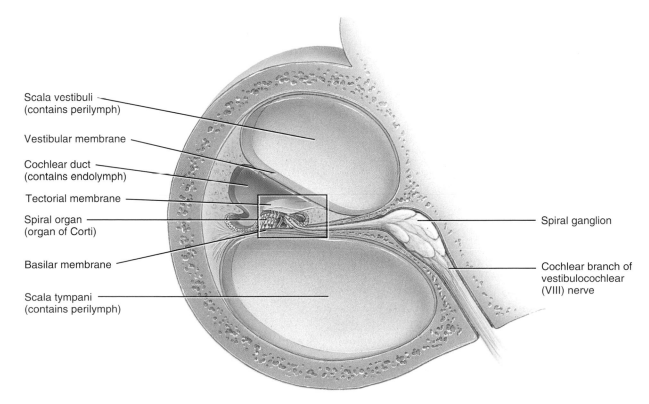

(c) Section through one turn of the cochlea

continues

Figure 22.13 (continued)

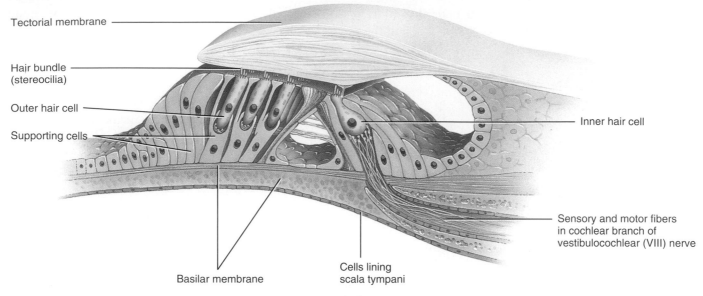

Tectorial membrane

Hair bundle (stereocilia)

Outer hair cell

Supporting cells

Inner hair cell

Sensory and motor fibers in cochlear branch of vestibulocochlear (VIII) nerve

Basilar membrane

Cells lining scala tympani

(d) Enlargement of spiral organ (organ of Corti)

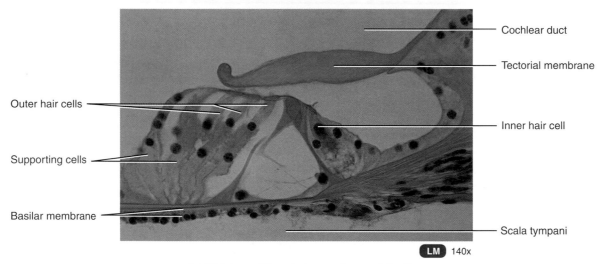

Cochlear duct

Tectorial membrane

Outer hair cells

Supporting cells

Inner hair cell

Basilar membrane

Scala tympani

LM 140x

(e) Histology of the spiral organ (organ of Corti)

What are the three subdivisions of the bony labyrinth?

LOUD SOUNDS AND HAIR CELL DAMAGE

Exposure to loud music and the engine roar of jet planes, revved-up motorcycles, lawn mowers, and vacuum cleaners damages hair cells of the cochlea. Because prolonged noise exposure causes hearing loss, employers in the United States must require workers to use hearing protectors when occupational noise levels are high. Continued exposure to high-intensity sounds is one cause of **deafness,** a significant or total hearing loss. The louder the sounds, the more rapid is the hearing loss. Deafness usually begins with loss of sensitivity for high-pitched sounds. If you are listening to music through head-phones and bystanders can hear it, the noise level is in the damaging range. Most people fail to notice their progressive hearing loss until destruction is extensive and they begin having difficulty understanding speech. Wearing earplugs while engaging in noisy activities can protect the sensitivity of your ears. ■

Mechanism of Hearing

Sound waves are a series of alternating high- and low-pressure regions traveling in the same direction through some medium (such as air). They originate from a vibrating object in much the same way that ripples arise and travel over the surface of a pond when you toss a stone into it.

The following events are involved in hearing (Figure 22.14):

1 The auricle directs sound waves into the external auditory canal.

2 When sound waves strike the eardrum, the alternating waves of high and low pressure in the air cause the eardrum to vibrate back and forth. The eardrum vibrates slowly in response to low-frequency (low-pitched) sounds and rapidly in response to high-frequency (high-pitched) sounds.

3 The central area of the eardrum connects to the malleus, which vibrates along with the eardrum. This vibration is transmitted from the malleus to the incus and then to the stapes.

4 As the stapes moves back and forth, it pushes the membrane of the oval window in and out. The oval window vibrates about 20 times more vigorously than the eardrum because the auditory ossicles efficiently transmit small vibrations spread over a large surface area (the eardrum) into larger vibrations of a smaller surface (the oval window).

5 The movement of the oval window sets up fluid pressure waves in the perilymph of the cochlea. As the oval window bulges inward, it pushes on the perilymph of the scala vestibuli.

6 Pressure waves are transmitted from the scala vestibuli to the scala tympani and eventually to the round window, causing it to bulge outward into the middle ear. (See **9** in the figure.)

7 As the pressure waves deform the walls of the scala vestibuli and scala tympani, they also push the vestibular membrane back and forth, creating pressure waves in the endolymph inside the cochlear duct.

8 The pressure waves in the endolymph cause the basilar membrane to vibrate, which moves the hair cells of the spiral organ against the tectorial membrane. Bending of the stereocilia ultimately leads to the generation of nerve impulses in first-order neurons in cochlear nerve fibers.

Besides its role in detecting sounds, the cochlea has the surprising ability to produce sounds. These usually inaudible sounds, called **otoacoustic emissions,** can be picked up by placing a sensitive microphone next to the eardrum. They are caused by vibrations of the outer hair cells that occur in response to sound waves and to signals from motor neurons. This vibratory behavior appears to change the stiffness of the tectorial membrane and is thought to enhance the movement of the basilar membrane, which amplifies the responses of the

Figure 22.14   Events in the stimulation of auditory receptors in the right ear. The numbers correspond to the events listed in the text. The cochlea has been uncoiled to more easily visualize the transmission of sound waves and their distortion of the vestibular and basilar membranes of the cochlear duct.

The function of hair cells of the spiral organ (organ of Corti) is to ultimately convert a mechanical vibration (stimulus) into an electrical signal (nerve impulse).

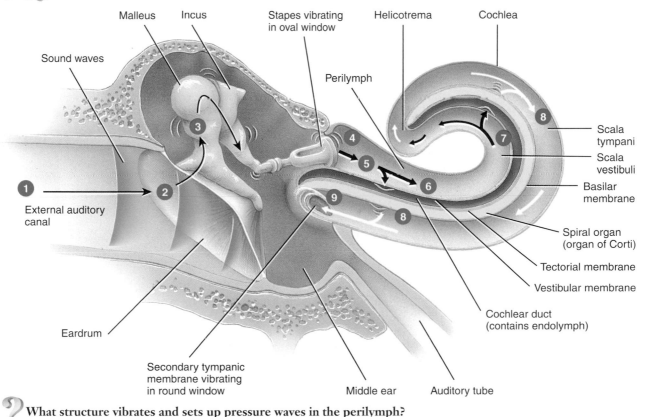

? **What structure vibrates and sets up pressure waves in the perilymph?**

inner hair cells. At the same time, the outer hair cell vibrations set up a traveling wave that goes back toward the stapes and leaves the ear as an otoacoustic emission. Detection of these inner ear–produced sounds is a fast, inexpensive, and noninvasive way to screen newborns for hearing defects. In deaf babies, otoacoustic emissions are not produced or are greatly reduced in size.

OTOSCLEROSIS

Otosclerosis (ō-tō-sklē-RŌ-sis; *oto-*=ear; *-sklerosis*=hardening) is a condition in which there is an overgrowth of spongy bone tissue over the oval window in the middle ear that immobilizes the stapes. This prevents the transmission of sound waves to the inner ear and leads to progressive hearing loss. Treatment may involve use of a hearing aid or a surgical procedure in which the stapes is replaced by a prosthetic device. ■

The Auditory Pathway

First-order sensory neurons in the cochlear branch of each vestibulocochlear (VIII) nerve terminate in the cochlear nuclei of the medulla oblongata on the same side. From there, axons carrying auditory signals project to the superior olivary nuclei in the pons on both sides. Slight differences in the timing of impulses arriving from the two ears at the olivary nuclei allow us to locate the source of a sound. From both the cochlear nuclei and the olivary nuclei, axons ascend to the inferior colliculus in the midbrain, and then to the medial geniculate body of the thalamus. From the thalamus, auditory signals project to the primary auditory area in the superior temporal gyrus of the cerebral cortex (Brodmann's areas 41 and 42 in Figure 19.15 on page 607). Because many auditory axons decussate (cross over) in the medulla while others remain on the same side, the right and left primary auditory areas receive nerve impulses from both ears.

COCHLEAR IMPLANTS

Cochlear implants are devices that translate sounds into electronic signals that can be interpreted by the brain. They are useful for people with deafness that is caused by damage to hair cells in the cochlea. Sound waves are picked up by a tiny microphone and converted into electrical signals. The signals then travel to electrodes implanted in the cochlea, where they trigger nerve impulses in sensory neurons in the cochlear branch of the vestibulocochlear nerve. These artificially induced nerve impulses propagate over their normal pathways to the brain. The sounds perceived are crude compared to normal hearing, but they provide a sense of rhythm and loudness; information about certain noises, such as those made by telephones and automobiles; and the pitch and cadence of speech. Some patients hear well enough with a cochlear implant to use the telephone. ■

Mechanism of Equilibrium

There are two types of **equilibrium** (balance). **Static equilibrium** refers to the maintenance of the position of the body (mainly the head) relative to the force of gravity. **Dynamic equilibrium** is the maintenance of body position (mainly the head) in response to sudden movements such as rotation, acceleration, and deceleration. Collectively, the receptor organs for equilibrium are called the **vestibular apparatus** (ves-TIB-ū-lar), which includes the saccule, utricle, and semicircular ducts.

Otolithic Organs: Saccule and Utricle

The walls of both the utricle and the saccule contain a small, thickened region called a **macula** (MAK-ū-la; Figure 22.15). The two *maculae* (plural), which are perpendicular to one another, are the receptors for static equilibrium, and contribute to some aspects of dynamic equilibrium as well. Their role in static equilibrium is to provide sensory information on the position of the head in space, essential to maintaining appropriate posture and balance. The maculae contribute to dynamic equilibrium by detecting linear acceleration and deceleration—for example, the sensations you feel while in an elevator or a car that is speeding up or slowing down.

The two maculae consist of two kinds of cells: **hair cells,** which are the sensory receptors, and **supporting cells.** Hair cells feature **hair bundles** that consist of 70 or more *stereocilia,* which are actually microvilli, plus one *kinocilium,* a conventional cilium anchored firmly to its basal body and extending beyond the longest stereocilia. Scattered among the hair cells are columnar supporting cells that probably secrete the thick, gelatinous, glycoprotein layer, called the **otolithic membrane,** that rests on the hair cells. A layer of dense calcium carbonate crystals, called **otoliths** (*oto*=ear; *-liths*=stones), extends over the entire surface of the otolithic membrane.

Because the otolithic membrane sits on top of the macula, when you tilt your head forward, the otolithic membrane and otoliths are pulled by gravity and slide downhill over the hair cells in the direction of the tilt, bending the hair bundles. The movement of the hair bundles initiates responses that ultimately lead to the generation of nerve impulses. The hair cells synapse with first-order sensory neurons in the vestibular branch of the vestibulocochlear (VIII) nerve.

Semicircular Ducts

The three semicircular ducts, together with the saccule and the utricle, function in dynamic equilibrium. The ducts lie at right angles to one another in three planes (Figure 22.16 on page 694). The two vertical ducts are the anterior and posterior semicircular ducts, and the horizontal one is the lateral semicircular duct (see also Figure 22.12). This positioning permits detection of rotational acceleration or deceleration. In the **ampulla,** the dilated portion of each duct, is a small elevation called the **crista.** Each crista contains a group of **hair cells** and **supporting cells** covered by a mass of gelatinous material called the **cupula** (KŪ-pū-la). When the head moves, the attached semicircular ducts and hair cells move with it. The endolymph within the ampulla, however, is not attached and lags behind due to its inertia. As the moving hair cells drag along the stationary fluid, the hair bundles bend. Bending of the hair bundles produces responses that lead to nerve impulses that pass along the ampullary nerve, a branch of the vestibular division of the vestibulocochlear (VIII) nerve.

Figure 22.15 Location and structure of receptors in the maculae of the right ear. Both first-order sensory neurons (blue) and motor neurons (red) synapse with the hair cells.

The movement of stereocilia initiates responses that ultimately lead to the generation of nerve impulses.

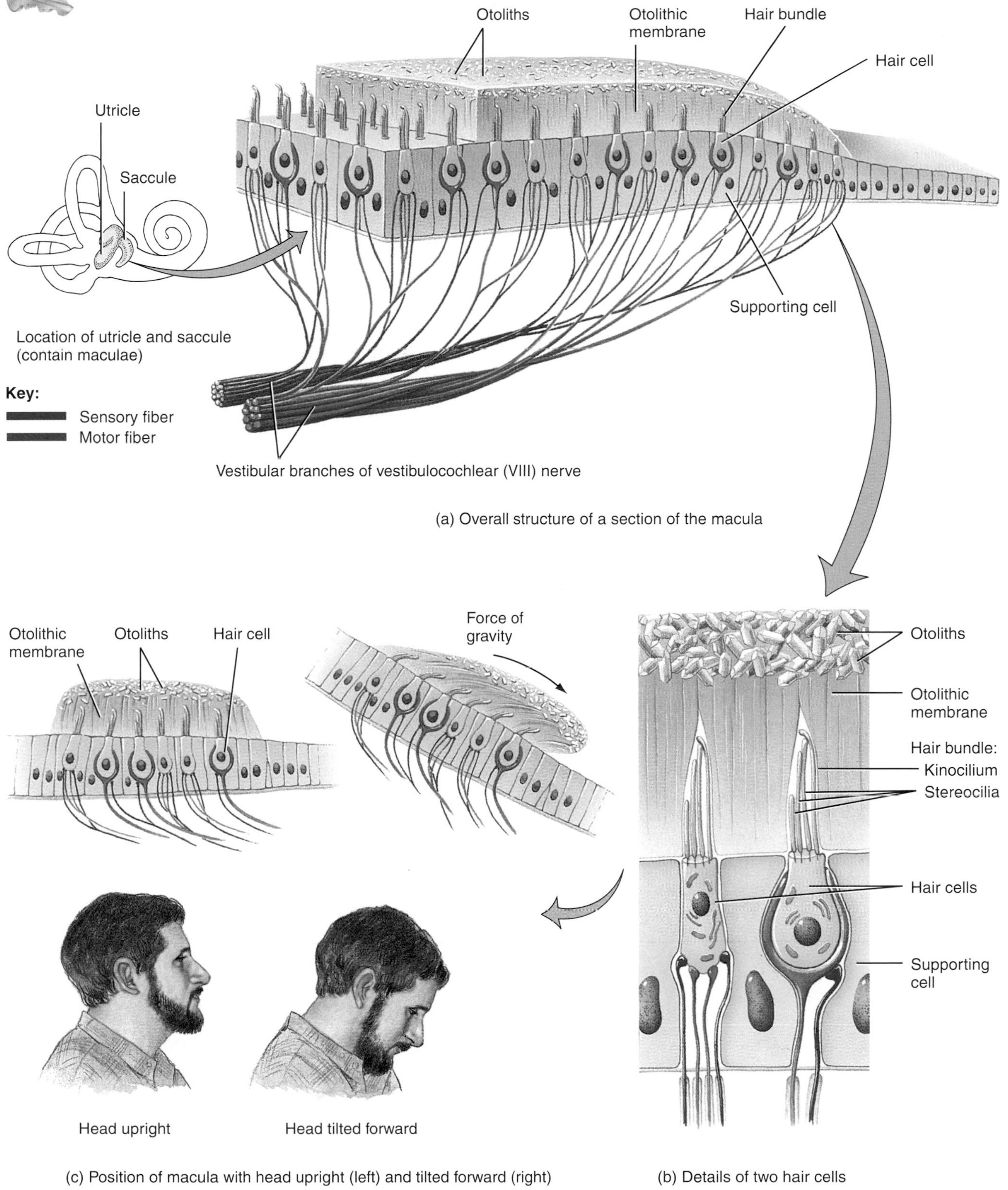

Otoliths

Otolithic membrane

Hair bundle

Hair cell

Utricle

Saccule

Supporting cell

Location of utricle and saccule (contain maculae)

Key:
▬ Sensory fiber
▬ Motor fiber

Vestibular branches of vestibulocochlear (VIII) nerve

(a) Overall structure of a section of the macula

Otolithic membrane   Otoliths   Hair cell

Force of gravity

Otoliths

Otolithic membrane

Hair bundle:
Kinocilium
Stereocilia

Hair cells

Supporting cell

Head upright

Head tilted forward

(c) Position of macula with head upright (left) and tilted forward (right)

(b) Details of two hair cells

With which type of equilibrium are the maculae mainly concerned?

Figure 22.16 Location and structure of the membranous semicircular ducts of the right ear. Both first-order sensory neurons (blue) and motor neurons (red) synapse with the hair cells. The ampullary nerves are branches of the vestibular division of the vestibulocochlear (VIII) nerve.

🔑 **The positions of the membranous semicircular ducts permit detection of rotational movements.**

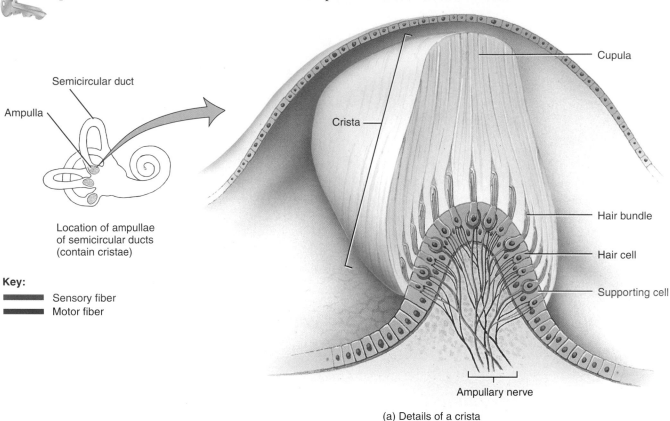

Semicircular duct

Ampulla

Location of ampullae
of semicircular ducts
(contain cristae)

Key:
━━━ Sensory fiber
━━━ Motor fiber

Cupula

Crista

Hair bundle

Hair cell

Supporting cell

Ampullary nerve

(a) Details of a crista

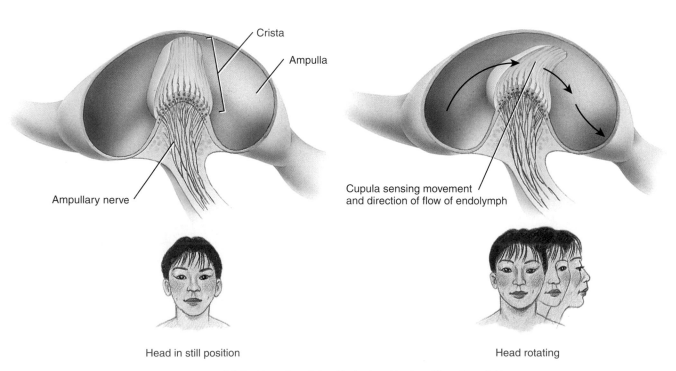

Crista

Ampulla

Ampullary nerve

Cupula sensing movement
and direction of flow of endolymph

Head in still position

Head rotating

(b) Position of a crista with the head in the still position (left)
and when the head rotates (right)

❓ **With which type of equilibrium are the membranous semicircular ducts, the utricle, and the saccule associated?**

Equilibrium Pathways

Most of the vestibular branch axons of the vestibulocochlear (VIII) nerve enter the brain stem and terminate in several vestibular nuclei in the medulla and pons. The remaining axons enter the cerebellum through the inferior cerebellar peduncle (see Figure 19.7a on page 593). Bidirectional pathways connect the vestibular nuclei and cerebellum.

Axons from all the vestibular nuclei extend to the nuclei of cranial nerves that control eye movements—oculomotor (III), trochlear (IV), and abducens (V). Other axons from vestibular nuclei extend to the nucleus of the accessory (XI) nerve, which helps control head and neck movements. In addition, axons from the lateral vestibular nucleus form the vestibulospinal tract, which conveys impulses to skeletal muscles that regulate muscle tone in response to head movements.

Various pathways among the vestibular nuclei, cerebellum, and cerebrum enable the cerebellum to play a key role in maintaining static and dynamic equilibrium. The cerebellum continuously receives updated sensory information from the utricle and saccule. It monitors this information and makes corrective adjustments. Essentially, in response to input from the utricle, saccule, and semicircular ducts, the cerebellum continuously sends nerve impulses to the motor areas of the cerebrum. This feedback allows correction of signals from the motor cortex to specific skeletal muscles to smooth movements and coordinate complex sequences of muscle contractions to help maintain equilibrium.

Table 22.2 summarizes the structures of the ear related to hearing and equilibrium.

CHECKPOINT

10. List the components of the external, middle, and internal ear and their functions.
11. Explain the mechanism of hearing.
12. Describe the auditory pathway.
13. Compare the function of the maculae in maintaining static equilibrium with the role of the cristae in maintaining dynamic equilibrium.
14. Describe the equilibrium pathways.
15. Describe the role of vestibular input to the cerebellum.

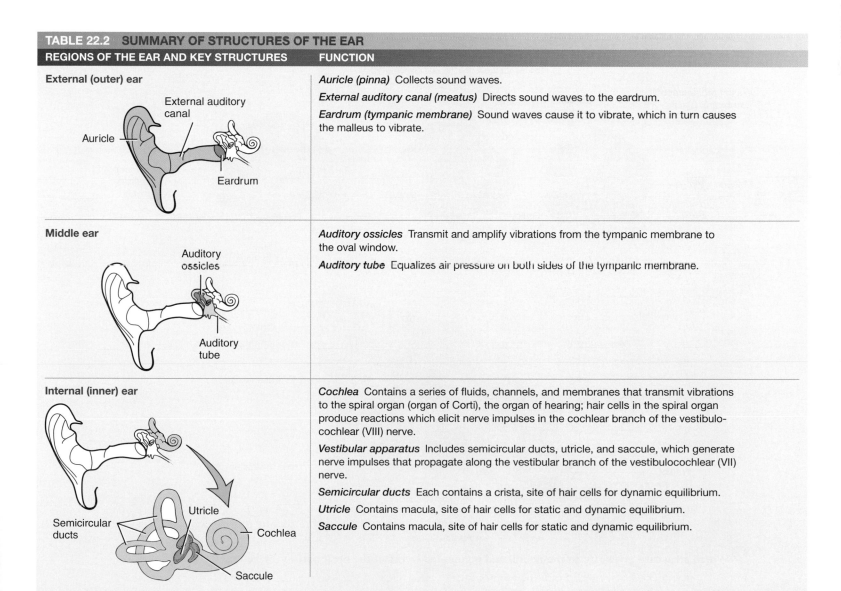

TABLE 22.2	SUMMARY OF STRUCTURES OF THE EAR
REGIONS OF THE EAR AND KEY STRUCTURES	**FUNCTION**
External (outer) ear	***Auricle (pinna)*** Collects sound waves.
	External auditory canal (meatus) Directs sound waves to the eardrum.
	Eardrum (tympanic membrane) Sound waves cause it to vibrate, which in turn causes the malleus to vibrate.
Middle ear	***Auditory ossicles*** Transmit and amplify vibrations from the tympanic membrane to the oval window.
	Auditory tube Equalizes air pressure on both sides of the tympanic membrane.
Internal (inner) ear	***Cochlea*** Contains a series of fluids, channels, and membranes that transmit vibrations to the spiral organ (organ of Corti), the organ of hearing; hair cells in the spiral organ produce reactions which elicit nerve impulses in the cochlear branch of the vestibulocochlear (VIII) nerve.
	Vestibular apparatus Includes semicircular ducts, utricle, and saccule, which generate nerve impulses that propagate along the vestibular branch of the vestibulocochlear (VII) nerve.
	Semicircular ducts Each contains a crista, site of hair cells for dynamic equilibrium.
	Utricle Contains macula, site of hair cells for static and dynamic equilibrium.
	Saccule Contains macula, site of hair cells for static and dynamic equilibrium.

DEVELOPMENT OF THE EYES AND EARS

• Describe the development of the eyes and the ears.

Development of the Eyes

The *eyes* begin to develop about 22 days after fertilization when the **ectoderm** of the lateral walls of the prosencephalon (fore-brain) bulges out to form a pair of shallow grooves called the **optic grooves** (Figure 22.17a). Within a few days, as the neural tube is closing, the optic grooves enlarge and grow toward the surface ectoderm and become known as the **optic vesicles** (Figure 22.17b). When the optic vesicles reach the surface ecto-derm, the surface ectoderm thickens to form the **lens placodes.** In addition, the distal portions of the optic vesicles invaginate (Figure 22.17c), forming the **optic cups;** they remain attached to the prosencephalon by narrow, hollow proximal structures called **optic stalks** (Figure 22.17d).

Figure 22.17   Development of the eyes.

🔑 **The eyes begin to develop about 22 days after fertilization from ectoderm of the prosencephalon.**

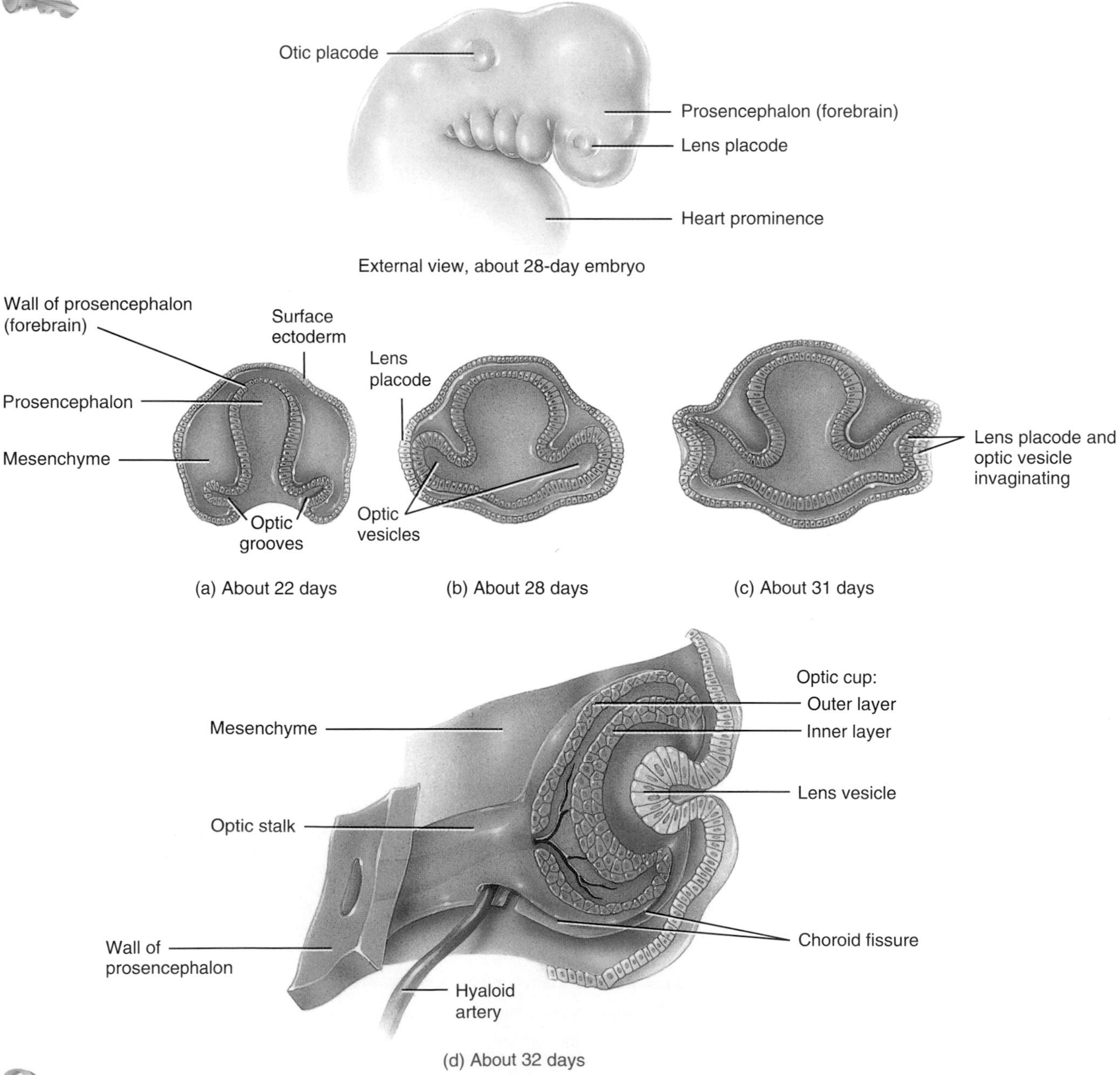

External view, about 28-day embryo

(a) About 22 days   (b) About 28 days   (c) About 31 days

(d) About 32 days

❓ **Which structure gives rise to the neural and pigmented layers of the optic part of the retina?**

The lens placodes also invaginate and develop into lens vesicles that sit in the optic cups. The lens vesicles eventually develop into the *lenses.* Blood is supplied to the developing lenses (and retina) by the hyaloid arteries. These arteries gain access to the developing eyes through a groove on the inferior surface of the optic cup and optic stalk called the **choroid fissure.** As the lenses mature, part of the hyaloid arteries that pass through the vitreous chamber degenerate; the remaining portions of the hyaloid arteries become the *central retinal arteries.*

The inner wall of the optic cup forms the *neural layer* of the optic part of the retina, while the outer layer forms the *pigmented layer* of the optic part of the retina. Axons from the neural layer grow through the optic stalk to the brain, converting the optic stalk to the *optic (II) nerve.* Although myelination of the optic nerves begins late in fetal life, it is not completed until the tenth week after birth.

The anterior portion of the optic cup forms the epithelium of the *ciliary body, iris,* and *circular and radial muscles* of the iris. The connective tissue of the ciliary body, *ciliary muscle,* and *zonular fibers* of the lens develop from **mesenchyme** around the anterior portion of the optic cup.

Mesenchyme surrounding the optic cup and optic stalk differentiates into an inner layer that gives rise to the *choroid* and an outer layer that develops into the *sclera* and part of the *cornea.* The remainder of the cornea is derived from surface **ectoderm.**

The *anterior chamber* develops from a cavity that forms in the **mesenchyme** between the iris and cornea; the *posterior chamber* develops from a cavity that forms in the **mesenchyme** between the iris and lens.

Some mesenchyme around the developing eye enters the optic cup through the choroid fissure. This mesenchyme occupies the space between the lens and retina and differentiates into a delicate network of fibers. Later the spaces between the fibers fill with a jellylike substance, thus forming the *vitreous body* in the vitreous chamber.

The *eyelids* form from surface **ectoderm** and **mesenchyme.** The upper and lower eyelids meet and fuse at about eight weeks of development and remain closed until about 26 weeks of development.

Development of the Ears

The first portion of the ear to develop is the *internal ear.* It begins to form about 22 days after fertilization as a thickening of the surface ectoderm, called **otic placodes** (Figure 22.18a), which appear on either side of the rhombencephalon (hindbrain). The otic placodes invaginate quickly (Figure 22.18b) to form the **otic pits** (Figure 22.18c). Next, the otic pits pinch off from the surface ectoderm to form the **otic vesicles** within the mesenchyme of the head (Figure 22.18d). During later development, the otic vesicles

Figure 22.18   Development of the ears.

   The first part of the ears to develop are the internal ears, which begin to form about 22 days after fertilization as thickenings of surface ectoderm.

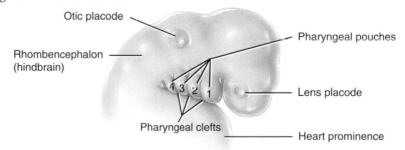

Otic placode
Rhombencephalon (hindbrain)
Pharyngeal pouches
Lens placode
Pharyngeal clefts
Heart prominence

External view, about 28-day embryo

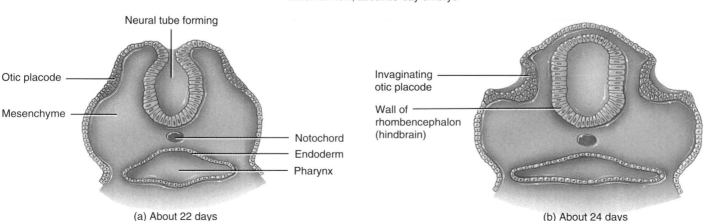

Neural tube forming
Otic placode
Mesenchyme
Notochord
Endoderm
Pharynx

(a) About 22 days

Invaginating otic placode
Wall of rhombencephalon (hindbrain)

(b) About 24 days

continues

Figure 22.18 (continued)

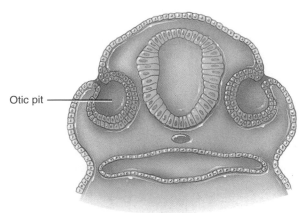

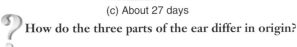

(c) About 27 days

How do the three parts of the ear differ in origin?

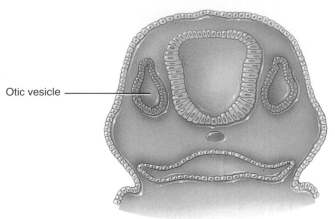

(d) About 32 days

will form the structures associated with the *membranous labyrinth* of the internal ear. **Mesenchyme** around the otic vesicles produces cartilage that later ossifies to form the bone associated with the *bony labyrinth* of the internal ear.

The *middle ear* develops from a structure called the first **pharyngeal (branchial) pouch,** an **endoderm**-lined outgrowth of the primitive pharynx (see the inset in Figure 22.18). The pharyngeal pouches were discussed in detail in Chapter 4 on page 109. The *auditory ossicles* develop from the first and second pharyngeal pouches.

The *external ear* develops from the first **pharyngeal cleft,** an **endoderm**-lined groove between the first and second pharyngeal pouches (see the inset in Figure 22.18). The pharyngeal clefts were discussed in detail in Chapter 4 on page 109.

CHECKPOINT

16. How do the origins of the eyes and ears differ?

AGING AND THE SPECIAL SENSES

OBJECTIVE

• Describe the age-related changes that occur in the eyes and ears.

Most people do not experience any problems with the senses of smell and taste until about age 50. This is due to a gradual loss of olfactory receptors and gustatory receptor cells coupled with their slower rate of replacement as we age.

Several age-related changes occur in the eyes. As noted earlier, the lens loses some of its elasticity and thus cannot change shape as easily, resulting in presbyopia (see page 681). Cataracts (loss of transparency of the lenses) also occur with aging (see page 699). In old age, the sclera ("white" of the eye) becomes thick and rigid and develops a yellowish or brownish coloration due to many years of exposure to ultraviolet light, wind, and dust. The sclera may also develop random splotches of pigment, especially in people with dark complexions. The iris fades or develops irregular pigment. The muscles that regulate the size of the pupil weaken with age and the pupils become smaller, react more slowly to light, and dilate move slowly in the dark. For these reasons, elderly people find that objects are not as bright, their eyes may adjust more slowly when going outdoors, and they have problems going from brightly lit to darkly lit places. Some diseases of the retina are more likely to occur in old age, including age-related macular disease (see page 683) and detached retina (see page 680). A disorder called glaucoma (see page 699) develops in the eyes of aging people as a result of the buildup of aqueous humor. The number of mucous cells in the conjunctiva and tear production may decrease with age, resulting in dry eyes. The eyelids lose their elasticity, becoming baggy and wrinkled. The amount of fat around the orbits may decrease, causing the eyeballs to sink into the orbits. Finally, as we age the sharpness of vision decreases, color and depth perception are reduced, and "vitreal floaters" increase.

By about age 60, around 25 percent of individuals experience a noticeable hearing loss, especially for higher-pitched sounds. The age-associated progressive loss of hearing in both ears is called **presbycusis** (pres′-bī-KŪ-sis; *presby-*=old; *-acou*= hearing; *-sis*=condition). It may be related to damaged and lost hair cells in the spiral organ or degeneration of the nerve pathway for hearing. Tinnitus (ringing in the ears) and vestibular imbalance also occur more frequently in the elderly.

CHECKPOINT

17. What changes in the eyes are related to the aging process, and how do they take place?

APPLICATIONS TO HEALTH

CATARACTS

A common cause of blindness is a loss of transparency of the lens known as a **cataract** (CAT-a-rakt=waterfall). The lens becomes cloudy (less transparent) due to changes in the structure of the lens proteins. Cataracts often occur with aging but may also be caused by injury, excessive exposure to ultraviolet rays, certain medications (such as long-term use of steroids), or complications of other diseases (for example, diabetes). People who smoke also have increased risk of developing cataracts. Fortunately, sight can usually be restored by surgical removal of the old lens and implantation of a new artificial one.

GLAUCOMA

Glaucoma (glaw-KŌ-ma) is the most common cause of blindness in the United States, afflicting about 2% of the population over age 40. Glaucoma is an abnormally high intraocular pressure due to a buildup of aqueous humor within the anterior cavity. The fluid compresses the lens into the vitreous body and puts pressure on the neurons of the retina. Persistent pressure results in a progression from mild visual impairment to irreversible destruction of neurons of the retina, damage to the optic nerve, and blindness. Glaucoma is painless, and the other eye compensates largely, so a person may experience considerable retinal damage and loss of vision before the condition is diagnosed. Because glaucoma occurs more often with advancing age, regular measurement of intraocular pressure is an increasingly important part of an eye exam as people grow older. Risk factors include race (blacks are more susceptible), increasing age, family history, and past eye injuries and disorders.

DEAFNESS

Deafness is significant or total hearing loss. **Sensorineural deafness** is caused by either impairment of hair cells in the cochlea or damage of the cochlear branch of the vestibulocochlear (VIII) nerve. This type of deafness may be caused by atherosclerosis, which reduces blood supply to the ears; by repeated exposure to loud noise, which destroys hair cells of the spiral organ; and/or by certain drugs such as aspirin and streptomycin. **Conduction deafness** is caused by impairment of the external and middle ear mechanisms for transmitting sounds to the cochlea. Causes of conduction deafness include otosclerosis, the deposition of new bone around the oval window; impacted cerumen; injury to the eardrum; and aging, which often results in thickening of the eardrum and stiffening of the joints of the auditory ossicles. A hearing test called *Weber's test* is used to distinguish between sensorineural and conduction deafness. In the test, the stem of a vibrating fork is held to the forehead. In people with normal hearing, the sound is heard equally in both ears. If the sound is heard best in the affected ear, the deafness is probably of the conduction type; if the sound is heard best in the normal ear, it is probably of the sensorineural type.

MÉNIÈRE'S DISEASE

Ménière's disease (men′-ē-ĀRZ) results from an increased amount of endolymph that enlarges the membranous labyrinth. Among the symptoms are fluctuating hearing loss (caused by distortion of the basilar membrane of the cochlea) and roaring tinnitus (ringing). Spinning or whirling vertigo (dizziness) is characteristic of Ménière's disease. Almost total destruction of hearing may occur over a period of years.

OTITIS MEDIA

Otitis media is an acute infection of the middle ear caused mainly by bacteria and associated with infections of the nose and throat. Symptoms include pain, malaise, fever, and a reddening and outward bulging of the eardrum, which may rupture unless prompt treatment is received. (This may involve draining pus from the middle ear.) Bacteria passing into the auditory tube from the nasopharynx are the primary cause of middle ear infections. Children are more susceptible than adults are to middle ear infections because their auditory tubes are almost horizontal, which decreases drainage. If otitis media occurs frequently, a surgical procedure called **tympanotomy** (tim′-pa-NOT-ō-mē; *tympano-*=drum; *-tome*=incision) is often employed. This consists of the insertion of a small tube into the eardrum to provide a pathway for the drainage of fluid from the middle ear.

MOTION SICKNESS

Motion sickness is nausea and vomiting brought on by repetitive angular, linear, or vertical motion. The cause is excessive stimulation of the vestibular apparatus by motion, usually incurred in travel by car, boat, or airplane. Nerve impulses pass from the internal ear to the vomiting center in the medulla. Visual stimuli and emotional factors such as fear or anxiety can also contribute to motion sickness. Susceptible people can take medication (for example, Dramamine) before traveling because prevention is more successful than treatment of symptoms once they have developed.

KEY MEDICAL TERMS ASSOCIATED WITH SPECIAL SENSES

Ageusia (a-GŪ-sē-a; *a-*=without; *-geusis*=taste) Loss of the sense of taste.

Amblyopia (am′-blē-Ō-pē-a; *ambly-*=dull or dim) Term used to describe the loss of vision in an otherwise normal eye that, because of muscle imbalance, cannot focus in synchrony with the other eye. Sometimes called "wandering eyeball" or a "lazy eye."

Anosmia (an-OZ-mē-a; *a-*=without; *osmi*=smell, odor) Total lack of the sense of smell.

Barotrauma (bar′-ō-TRAW-ma; *baros-*=weight) Damage or pain, mainly affecting the middle ear, as a result of pressure changes. It occurs when pressure on the outer side of the tympanic membrane is higher than on the inner side, for example when flying in an airplane or driving. Swallowing or holding your nose and exhaling with your mouth closed usually opens the auditory tubes, allowing air into the middle ear to equalize the pressure.

Blepharitis (blef-a-RĪ-tis; *blephar-*=eyelid; *-itis*=inflammation of) An inflammation of the eyelid.

Conjunctivitis (pinkeye) An inflammation of the conjunctiva; when caused by bacteria such as pneumococci, staphylococci, or *Hemophilus influenzae*, it is very contagious and more common in children. Conjunctivitis may also be caused by irritants, such as dust, smoke, or pollutants in the air, in which case it is not contagious.

Corneal abrasion (KOR-nē-al a-BRĀ-zhun) A scratch on the surface of the cornea, for example, from a speck of dirt or damaged contact lenses. Symptoms include pain, redness, watering, blurry vision, sensitivity to bright light, and frequent blinking.

Corneal transplant A procedure in which a defective cornea is removed and a donor cornea of similar diameter is sewn in. It is the most common and most successful transplant operation. Since the cornea is avascular, antibodies in the blood that might cause rejection do not enter the transplanted tissue, and rejection rarely occurs. The shortage of donor corneas has been partially overcome by the development of artificial corneas made of plastic.

Diabetic retinopathy (ret-i-NOP-a-thē; *retino-*=retina; *-pathos*=suffering) Degenerative disease of the retina due to diabetes mellitus, in which blood vessels in the retina are damaged or new ones grow and interfere with vision.

Exotropia (ek′-sō-TRŌ-pē-a; *ex-*=out; *-tropia*=turning) Turning outward of the eyes.

Keratitis (ker′-a-TĪ-tis; *kerat-*=cornea) An inflammation or infection of the cornea.

Miosis (mī-Ō-sis) Constriction of the pupil.

Mydriasis (mi-DRĪ-a-sis) Dilation of the pupil.

Nystagmus (nis-TAG-mus; *nystagm-*=nodding or drowsy) A rapid involuntary movement of the eyeballs, possibly caused by a disease of the central nervous system. It is associated with conditions that cause vertigo.

Otalgia (ō-TAL-jē-a; *oto-*=ear; *-algia*=pain) Earache.

Photophobia (fō′-tō-FŌ-bē-a; *photo-*=light; *-phobia*=fear) Abnormal visual intolerance to light.

Ptosis (TŌ-sis=fall) Falling or drooping of the eyelid (or slippage of any organ below its normal position).

Retinoblastoma (ret-i-nō-blas-TŌ-ma; *-oma*=tumor) A highly malignant tumor arising from immature retinal cells; it accounts for 2% of childhood cancers.

Scotoma (skō-TŌ-ma=darkness) An area of reduced or lost vision in the visual field.

Strabismus (stra-BIZ-mus; *strabismos*=squinting) Misalignment of the eyeballs so that the eyes do not move in unison when viewing an object; the affected eye turns either medially or laterally with respect to the normal eye and the result is double vision (diplopia). It may be caused by physical trauma, vascular injuries, or tumors of the extrinsic eye muscle or the oculomotor (III), trochlear (IV), or abducens (VI) cranial nerves.

Tinnitus (ti-NĪ-tus) A ringing, roaring, or clicking in the ears.

Tonometer (tō-NOM-e-ter; *tono-*=tension or pressure; *-metron*=measure) An instrument for measuring pressure, especially intraocular pressure.

Trachoma (tra-KŌ-ma) A serious form of conjunctivitis and the greatest single cause of blindness in the world. It is caused by the bacterium *Chlamydia trachomatis*. The disease produces an excessive growth of subconjunctival tissue and invasion of blood vessels into the cornea, which progresses until the entire cornea is opaque.

Vertigo (VER-ti-gō=dizziness) A sensation of spinning or movement in which the world seems to revolve or the person seems to revolve in space, often associated with nausea and in some cases, vomiting. It may be caused by arthritis of the neck or an infection of the vestibular apparatus.

STUDY OUTLINE

Olfaction: Sense of Smell (p. 672)

1. The receptors for olfaction, which are bipolar neurons, are in the nasal epithelium.
2. Axons of olfactory receptors form the olfactory (I) nerves, which convey nerve impulses to the olfactory bulbs, olfactory tracts, limbic system, and cerebral cortex (temporal and frontal lobes).

Gustation: Sensation of Taste (p. 674)

1. The receptors for gustation, the gustatory receptor cells, are located in taste buds.
2. Dissolved chemicals, called tastants, stimulate gustatory receptor cells.

3. Gustatory receptor cells trigger nerve impulses in cranial nerves VII, IX, and X. Taste signals then pass to the medulla oblongata, thalamus, and cerebral cortex (parietal lobe).

Vision (p. 676)

1. Accessory structures of the eyes include the eyebrows, eyelids, eyelashes, lacrimal apparatus, and extrinsic eye muscles.

2. The lacrimal apparatus consists of structures that produce and drain tears.

3. The eye is constructed of three layers: (a) fibrous tunic (sclera and cornea), (b) vascular tunic (choroid, ciliary body, and iris), and (c) retina.

4. The retina consists of a pigmented layer and a neural layer that includes a photoreceptor layer, bipolar cell layer, ganglion cell layer, horizontal cells, and amacrine cells.

5. The anterior cavity contains aqueous humor; the vitreous chamber contains the vitreous body.

6. Chemicals (neurotransmitters) released by rods and cones induce changes in bipolar cells and horizontal cells that ultimately lead to the generation of nerve impulses.

7. Impulses from ganglion cells are conveyed into the optic (II) nerve, through the optic chiasm and optic tract, to the thalamus. From the thalamus, impulses for vision propagate to the cerebral cortex (occipital lobe). Axon collaterals of retinal ganglion cells extend to the midbrain and hypothalamus.

Hearing and Equilibrium (p. 684)

1. The external (outer) ear consists of the auricle, external auditory canal, and eardrum (tympanic membrane).

2. The middle ear consists of the auditory (eustachian) tube, ossicles, oval window, and round window.

3. The internal (inner) ear consists of the bony labyrinth and membranous labyrinth. The internal ear contains the spiral organ (organ of Corti), the organ of hearing.

4. Sound waves enter the external auditory canal and strike the eardrum, causing it to vibrate. The vibrations pass through the ossicles, strike the oval window, set up waves in the perilymph, strike the vestibular membrane and scala tympani, increase pressure in the endolymph, vibrate the basilar membrane, and stimulate hair bundles on the spiral organ (organ of Corti).

5. Bending of stereocilia ultimately leads to generation of nerve impulses in first-order sensory neurons.

6. Sensory fibers in the cochlear branch of the vestibulocochlear (VIII) nerve terminate in the medulla oblongata. Auditory signals then pass to the inferior colliculus, thalamus, and temporal lobes of the cerebral cortex.

7. Static equilibrium is the orientation of the body relative to the pull of gravity. The maculae of the utricle and saccule are the sense organs of static equilibrium.

8. Dynamic equilibrium is the maintenance of body position in response to movement. The cristae in the membranous semicircular ducts are the principal sense organs of dynamic equilibrium.

9. Most vestibular branch axons of the vestibulocochlear (VIII) nerve enter the brain stem and terminate in the medulla and pons; other axons enter the cerebellum.

Development of the Eyes and Ears (p. 696)

1. The eyes begin their development about 22 days after fertilization from ectoderm of the lateral walls of the prosencephalon (forebrain).

2. The ears begin their development about 22 days after fertilization from a thickening of ectoderm on either side of the rhombencephalon (hindbrain). The sequence of development of the ear is internal ear, middle ear, and external ear.

Aging and the Special Senses (p. 698)

1. Most people do not experience problems with the senses of smell and taste until about age 50.

2. Among the age-related changes to the eyes are presbyopia, cataracts, difficulty adjusting to light, macular disease, glaucoma, dry eyes, and decreased sharpness of vision.

3. With age there is a progressive loss of hearing and tinnitus occurs more frequently.

 SELF-QUIZ QUESTIONS

Choose the one best answer to the following questions:

1. The lateral (primary) olfactory area is located in the:
   a. occipital lobe of the cerebrum.
   b. cerebellum.
   c. frontal lobe of the cerebrum.
   d. temporal lobe of the cerebrum.
   e. midbrain.

2. Which of the following sensory cells are also first-order sensory neurons?
   a. olfactory receptors
   b. gustatory receptor cells
   c. hair cells of the spiral organ (organ of Corti)
   d. hair cells of the otolithic organs
   e. All of the above are correct.

3. The shape of the lens is altered by the contraction and relaxation of the:
   a. superior rectus muscle.   b. ciliary muscle.
   c. circular muscles.   d. radial muscles.
   e. Both c and d are correct.

4. In the eye, only cones are present in the:
   a. lens.   b. bipolar cell layer.   c. central fovea.
   d. ganglion cell layer.   e. vitreous body.

5. The bony labyrinth of the inner ear:
   a. is lined with ceruminous glands.
   b. consists of semicircular canals, vestibule, and tympanic antrum.
   c. contains a fluid called perilymph.
   d. houses hearing and equilibrium receptors.
   e. Both c and d are correct.

6. The iris separates the:
   a. anterior and posterior chambers.
   b. anterior cavity and the vitreous chamber.
   c. pupil and the lens.
   d. posterior chamber and the vitreous chamber.
   e. sclera and the choroid.

7. Increased sympathetic nervous stimulation of the radial muscles of the eye causes:
   a. increased production of tears.
   b. the eyeballs to turn toward the midline.
   c. the lens to become flatter in shape.
   d. improved color vision.
   e. dilation of the pupil.

8. The optic tract extends from the:
   a. retina to the optic chiasm.
   b. optic chiasm to the occipital lobe.
   c. thalamus to the occipital lobe.
   d. macula lutea to the optic disc.
   e. pupil to the retina.

Complete the following:

9. The two skeletal muscles that attach to the ossicles of the ear are the _____ muscle and the _____ muscle.

10. Taste buds are located on elevated projections of the tongue called _____. The _____ type is scattered over the entire surface of the tongue.

11. Aqueous humor is produced by the _____ body, and is reabsorbed into the _____ sinus.

12. The cochlear duct is separated from the scala tympani by the _____, on which rests the _____ that contains the receptors for hearing.

13. The mucous membrane that lines the eyelids and covers the anterior surface of the eyeball is called the _____.

14. _____ equilibrium is the maintenance of body position in response to sudden movement, and _____ equilibrium is the maintenance of body position in response to gravity.

15. Number the following in the correct sequence for the conduction pathway for the sense of smell. (a) olfactory bulbs: _____; (b) olfactory receptors: _____; (c) olfactory (I) nerves: _____; (d) olfactory tract: _____; (e) cortical region of temporal lobe: _____

16. Number the following in the correct sequence for the pathway of sound waves. (a) external auditory canal: _____; (b) stapes: _____; (c) malleus: _____; (d) incus: _____; (e) oval window: _____; (f) eardrum (tympanic membrane): _____

17. Otoliths are calcium carbonate crystals that are essential components of the receptors for the sense of _____ found in the _____ and the _____.

Are the following statements true or false?

18. Dendrites of olfactory receptors collectively make up the olfactory (I) nerves.

19. Cones are most important for color vision, whereas rods are most important for seeing shades of gray in dim light.

20. The membranous labyrinth of the internal (inner) ear contains perilymph.

21. The primary function of the auditory (eustachian) tube is to conduct sound waves from the exterior to the tympanic membrane.

Matching

22. Match the following terms to their descriptions:
    ___ (a) "white of the eye"
    ___ (b) a clear structure, composed of protein layers arranged like an onion
    ___ (c) blind spot; area in which there are no cones or rods
    ___ (d) exact center of the posterior portion of the retina
    ___ (e) nonvascular, transparent, fibrous coat; most anterior eye structure
    ___ (f) layer containing neurons; if detached, causes blindness
    ___ (g) dark brown layer; prevents reflection of light rays; also nourishes eyeball because it is vascular
    ___ (h) a hole; appears black, like a circular doorway leading into a dark room
    ___ (i) regulates the amount of light entering the eye; colored part of the eye
    ___ (j) attaches to the lens by means of radially arranged fibers called the suspensory ligaments
    ___ (k) serrated margin of the retina
    ___ (l) located at the junction of sclera and cornea; drains aqueous humor

    (1) macula lutea
    (2) choroid
    (3) cornea
    (4) ciliary muscle
    (5) iris
    (6) lens
    (7) optic disc
    (8) ora serrata
    (9) pupil
    (10) retina
    (11) sclera
    (12) scleral venous sinus (canal of Schlemm)

23. Match the following terms to their descriptions:
    ___ (a) organs of dynamic equilibrium; located in the ampullae of the semicircular canals
    ___ (b) eardrum
    ___ (c) air-filled cavity of temporal bone
    ___ (d) opening between the middle and inner ear; receives base of stapes
    ___ (e) oval central portion of the bony labyrinth
    ___ (f) connects the middle ear to the nasopharynx, allowing for equalization of pressure on both sides of the eardrum
    ___ (g) organ for hearing
    ___ (h) ear bones: malleus, incus, and stapes
    ___ (i) receptor for static equilibrium; also contributes to some aspects of dynamic equilibrium; consists of hair cells and supporting cells
    ___ (j) opening into the middle ear; is enclosed by a membrane called the secondary tympanic membrane
    ___ (k) contains the spiral organ (organ of Corti)
    ___ (l) receptor organs for equilibrium; the saccule, utricle, and semicircular canals

    (1) macula
    (2) round window
    (3) oval window
    (4) tympanic membrane
    (5) cristae
    (6) auditory (eustachian) tube
    (7) spiral organ (organ of Corti)
    (8) auditory ossicles
    (9) middle ear
    (10) vestibular apparatus
    (11) cochlea
    (12) vestibule

CRITICAL THINKING QUESTIONS

1. Suestia was diagnosed with a brain tumor. Early symptoms had included false sensations of smell. How/why might this have occurred?

2. Deirdre refuses to cook anything with onions for a date because whenever she slices and dices them, her nose runs and her eyes water. "And how can you impress a date when you have puffy eyes and a runny nose?" Explain why Deirdre has such trouble with onions.

3. Lucas noticed that the colored ring in one of his mother's eyes was a different shape than the other eye. His mother explained that she had gotten poked in the eye with a stick when she was a little girl

about his age and that had caused the damage to her eye. What structure of the eye was injured by the stick? What effect would the damage have on her vision?

4. Reuben was on his first cruise and he felt miserable! The ship's doctor mentioned something about messages from his eyes confusing his ears (or the other way around) but Reuben was too sick to listen. He took his medication and went to bed. Explain the cause of Reuben's seasickness.

5. Why do the eyes in some photographs taken with a flash appear to be red?

ANSWERS TO FIGURE QUESTIONS

22.1 The olfactory hairs detect odorant molecules.

22.2 The gustatory pathway is as follows: Gustatory receptor cells → cranial nerves VII, IX, or X → medulla oblongata → either limbic system and hypothalamus, or thalamus → primary gustatory area in the parietal lobe of the cerebral cortex.

22.3 Lacrimal fluid, or tears, is a watery solution containing salts, some mucus, and lysozyme that protects, cleans, lubricates, and moistens the eyeball.

22.4 The fibrous tunic consists of the cornea and sclera; the vascular tunic consists of the choroid, ciliary body, and lens.

22.5 The parasympathetic division of the ANS causes pupillary constriction, whereas the sympathetic division causes pupillary dilation.

22.6 An ophthalmoscopic examination can reveal evidence of hypertension, diabetes mellitus, cataract, and macular degeneration.

22.7 The two types of photoreceptors are rods and cones. Rods provide black-and-white vision in dim light, whereas cones provide high visual acuity and color vision in bright light.

22.8 After its secretion by the ciliary process, aqueous humor flows into the posterior chamber, around the iris, into the anterior chamber, and out of the eyeball through the scleral venous sinus.

22.9 The optic tract terminates in the lateral geniculate nucleus of the thalamus.

22.10 The malleus of the middle ear is attached to the eardrum, considered part of the external ear.

22.11 The oval and round windows separate the middle ear from the internal ear.

22.12 The two sacs in the vestibule are the utricle and saccule.

22.13 The three subdivisions of the bony labyrinth are the semicircular canals, vestibule, and cochlea.

22.14 The oval window vibrates and sets up presure waves in the perilymph.

22.15 Maculae mainly contribute to static equilibrium by providing sensory information on the position of the head in space.

22.16 The semicircular ducts, utricle, and saccule are mainly associated with dynamic equilibrium.

22.17 The optic cup forms the neural and pigmented layers of the optic part of the retina.

22.18 The internal ear develops from surface ectoderm, the middle ear develops from pharyngeal pouches, and the external ear develops from a pharyngeal cleft.

23

THE ENDOCRINE SYSTEM

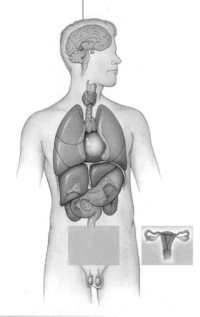

INTRODUCTION   Together, the nervous and endocrine systems coordinate functions of all body systems. As you learned in the last several chapters, the nervous system exerts its control through nerve impulses conducted along axons of neurons. At synapses, nerve impulses trigger the release of mediator (messenger) molecules called neurotransmitters. In contrast, the endocrine system releases mediator molecules called **hormones** (*hormon* = to excite or get moving) into interstitial fluid and then the bloodstream. The circulating blood delivers the hormones to virtually all the cells of the body; cells that recognize a particular hormone then respond. The science that deals with the structure and function of the endocrine glands and the diagnosis and treatment of disorders of the endocrine system is **endocrinology** (en′-dō-kri-NOL-ō-jē; *endo-* = within; *-crino* = to secrete; *-logy* = study of).

    The nervous and endocrine systems are both coordinated as an interlocking supersystem termed the neuroendocrine system. Certain parts of the nervous system stimulate or inhibit the release of hor-

Many people read their horoscopes regularly. Are you one of them? Do you know what sign you were born under? How much do you think the heavens influence our bodies? Is such a belief superstition or science?

www.wiley.com/college/apcentral

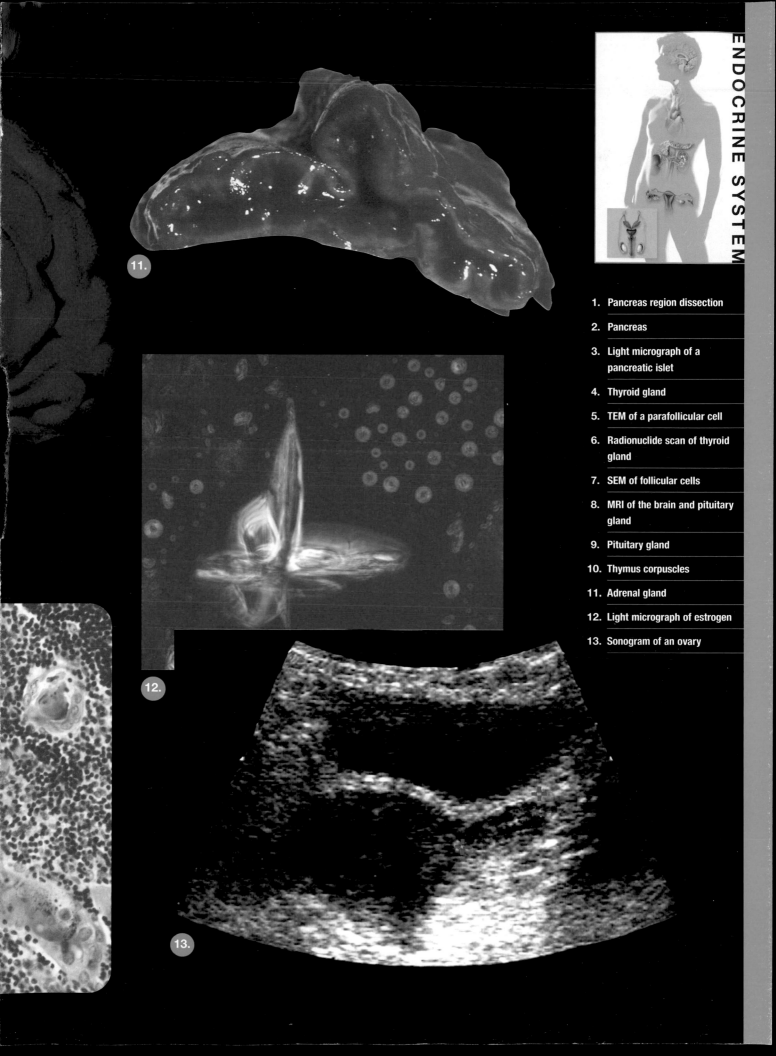

1. Pancreas region dissection

2. Pancreas

3. Light micrograph of a pancreatic islet

4. Thyroid gland

5. TEM of a parafollicular cell

6. Radionuclide scan of thyroid gland

7. SEM of follicular cells

8. MRI of the brain and pituitary gland

9. Pituitary gland

10. Thymus corpuscles

11. Adrenal gland

12. Light micrograph of estrogen

13. Sonogram of an ovary

ENDOCRINE SYSTEM

1. **Pancreas region dissection.** Pancreas region dissection.

2. **Pancreas.** Pancreas in anterior view.

3. **Light micrograph of a pancreatic islet.** Light micrograph of a pancreatic islet (islet of Langerhans).

4. **Thyroid gland.** Thyroid gland in anterior view.

5. **TEM of a parafollicular cell.** Transmission Electron Micrograph (TEM) of the thyroid gland showing a parafollicular cell and several follicular cells.

6. **Radionuclide scan of thyroid gland.** Radionuclide scan of the thyroid gland in anterior view.

7. **SEM of follicular cells.** Scanning Electron Micrograph (SEM) of follicular cells of the thyroid gland.

8. **MRI of the brain and pituitary gland.** Magnetic Resonance Image (MRI) of the brain and pituitary gland in sagittal section.

9. **Pituitary gland.** Pituitary gland in sagittal section.

10. **Thymus corpuscles.** Light micrograph of the thymus gland showing thymic (Hassal's) corpuscles.

11. **Adrenal gland.** Adrenal gland in frontal section showing adrenal cortex and adrenal medulla.

12. **Light micrograph of estrogen.** Light micrograph of crystals of estrogen.

13. **Sonogram of an ovary.** Sonogram of an ovary (between the red

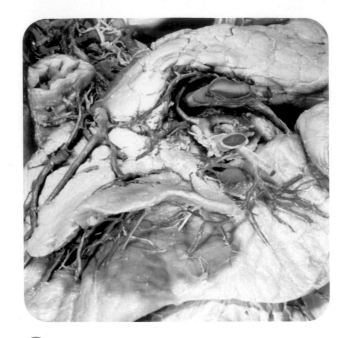

1.

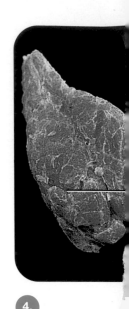

4.

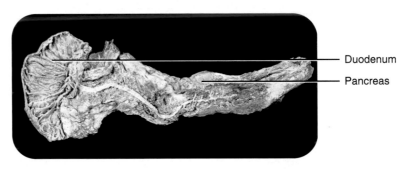

Duodenum

Pancreas

2.

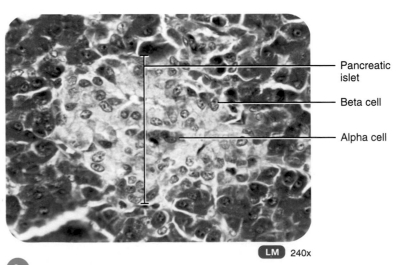

Pancreatic islet

Beta cell

Alpha cell

LM 240x

3.

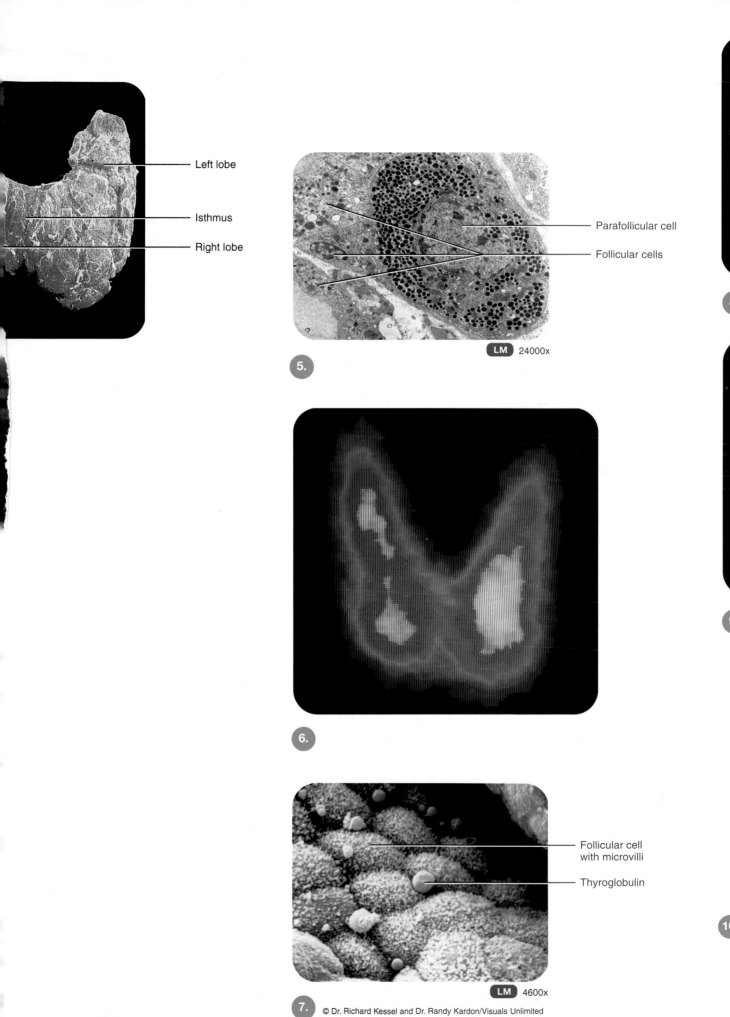

Left lobe

Isthmus

Right lobe

Parafollicular cell

Follicular cells

LM 24000x

5.

6.

Follicular cell
with microvilli

Thyroglobulin

LM 4600x

7.

© Dr. Richard Kessel and Dr. Randy Kardon/Visuals Unlimited

8.

9.

10.

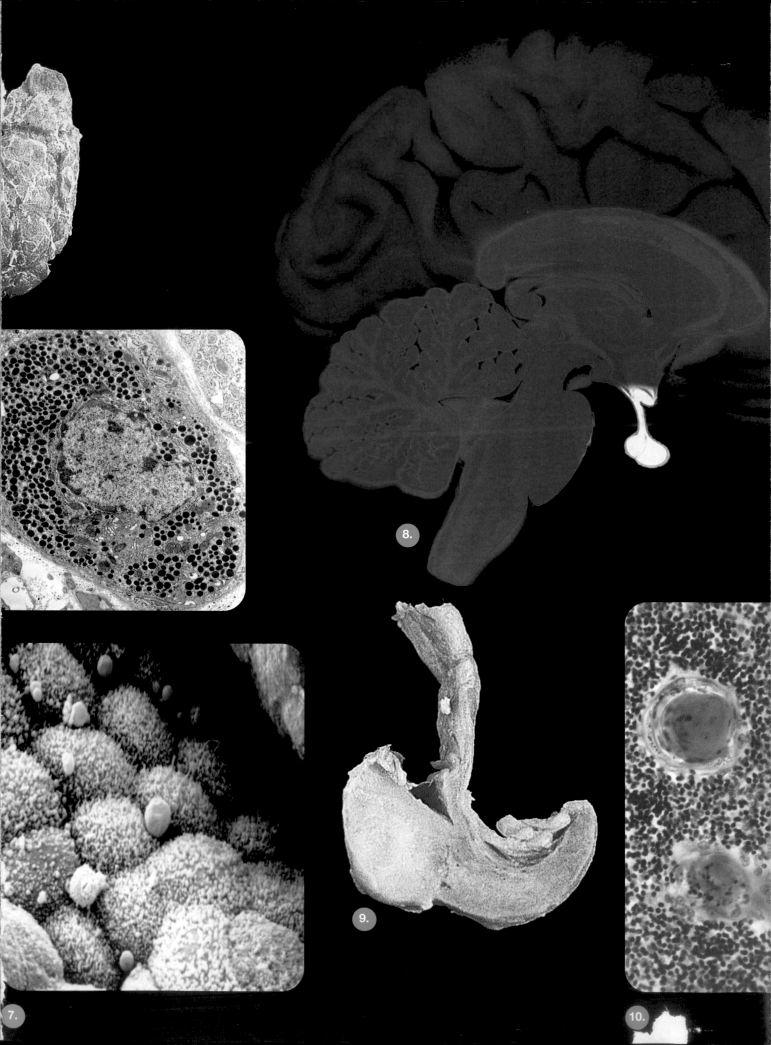

7.

8.

9.

10.

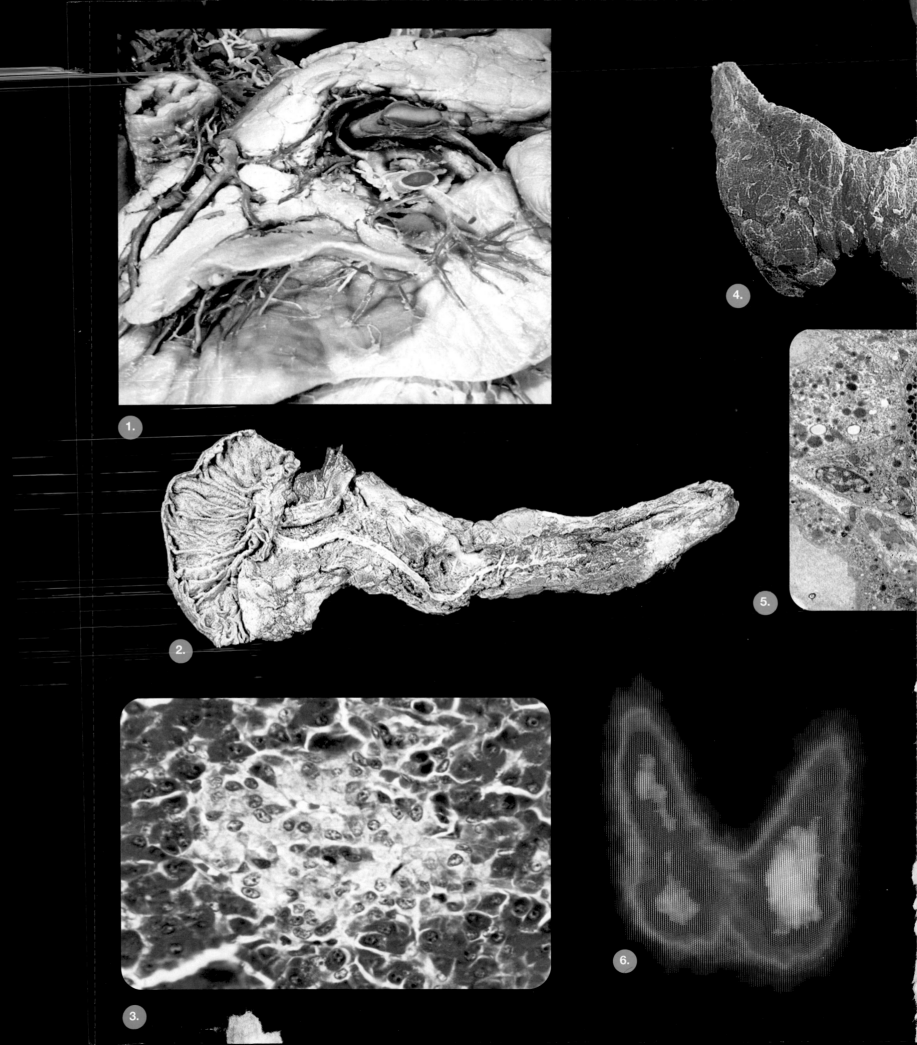

1.

2.

3.

4.

5.

6.

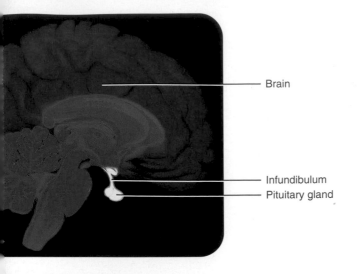

Brain

Infundibulum
Pituitary gland

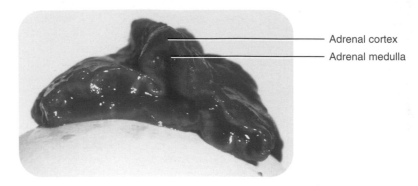

Adrenal cortex
Adrenal medulla

11.

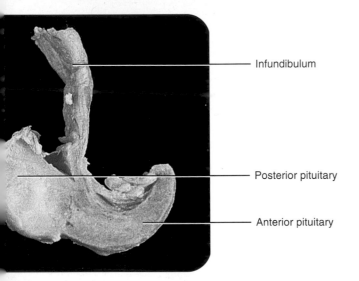

Infundibulum

Posterior pituitary

Anterior pituitary

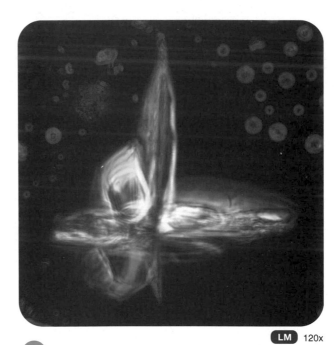

LM 120x

12.

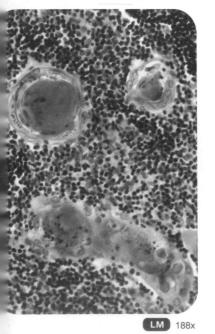

LM 188x

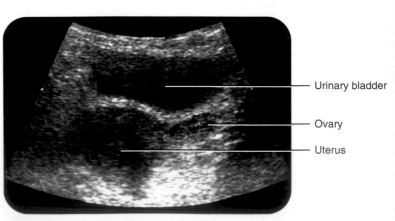

Urinary bladder

Ovary

Uterus

13.

mones, which in turn may promote or inhibit the generation of nerve impulses. The nervous system causes muscles to contract and glands to secrete either more or less of their product. The endocrine system not only helps regulate the activity of smooth muscle, cardiac muscle, and some glands; it affects virtually all other tissues as well. Hormones alter metabolism, regulate growth and development, and influence reproductive processes.

The nervous and endocrine systems respond to stimuli at different rates. Nerve impulses most often produce an effect within a few milliseconds; some hormones can act within seconds, whereas others may take several hours or more to cause a response. The effects of nervous system activation are generally briefer than the effects produced by the endocrine system. Table 23.1 compares the characteristics of the nervous and endocrine systems.

In this chapter we examine the principal endocrine glands and hormone-producing tissues, along with their roles in coordinating body activities.

ENDOCRINE GLANDS DEFINED

OBJECTIVE

• Distinguish between exocrine glands and endocrine glands.

Recall from Chapter 3 that the body contains two kinds of glands: exocrine glands and endocrine glands. **Exocrine glands** (*exo-*=outside) secrete their products into ducts that carry the secretions into body cavities, into the lumen of an organ, or to the outer surface of the body. Exocrine glands include sudoriferous (sweat), sebaceous (oil), mucous, and digestive glands. **Endocrine glands,** by contrast, secrete their products (hormones) into the interstitial fluid surrounding the secretory cells, rather than into ducts. From the interstitial fluid, hormones diffuse into capillaries and blood carries them throughout the body. Very small amounts of most hormones are required in most cases, so circulating levels typically are low.

The endocrine glands include the pituitary, thyroid, parathyroid, adrenal, and pineal glands (Figure 23.1). In addition, several other organs and tissues do not function exclusively as endocrine glands, but contain cells that secrete hormones. These include the hypothalamus, thymus, pancreas, ovaries, testes, kidneys, stomach, liver, small intestine, skin, heart, adipose tissue, and placenta. Taken together, all endocrine glands and hormone-secreting cells constitute the **endocrine system.**

CHECKPOINT

1. Which of the following are exclusively exocrine glands or endocrine glands: thymus, pancreas, placenta, pituitary, kidneys, sudoriferous glands, mucous glands, testes, ovaries?

HORMONES

OBJECTIVE

• Describe how hormones interact with target-cell receptors.

Hormones have powerful effects, even when present in very low concentrations. As a rule, most of the body's 50 or so hormones affect only a few types of cells. The reason that some cells respond to a particular hormone and others do not depends on whether the cells have hormone receptors.

Although a given hormone travels throughout the body in the blood, it affects only certain **target cells.** Hormones, like neurotransmitters, influence their target cells by chemically binding to specific protein or glycoprotein **receptors.** Only the target cells for a given hormone have receptors that bind and recognize that hormone. For example, thyroid-stimulating hormone (TSH) binds to receptors on cells of the thyroid gland, but it does not bind to cells of the ovaries because ovarian cells do not have TSH receptors.

TABLE 23.1	COMPARISON OF CONTROL BY THE NERVOUS AND ENDOCRINE SYSTEMS	
CHARACTERISTIC	**NERVOUS SYSTEM**	**ENDOCRINE SYSTEM**
Mediator molecules	Neurotransmitters released locally in response to nerve impulses.	Hormones delivered to tissues throughout the body by the blood.
Site of mediator action	Close to site of release, at a synapse; binds to receptors in postsynaptic membrane.	Far from site of release (usually); binds to receptors on or in target cells.
Types of target cells	Muscle (smooth, cardiac, and skeletal) cells, gland cells, other neurons.	Virtually all body cells.
Time to onset of action	Typically within milliseconds (thousandths of a second).	Seconds to hours or days.
Duration of action	Generally briefer (milliseconds).	Generally longer (seconds to days).

Figure 23.1 Location of many endocrine glands. Also shown are other organs that contain endocrine tissue, and associated structures.

Endocrine glands secrete hormones, which circulating blood delivers to target tissues.

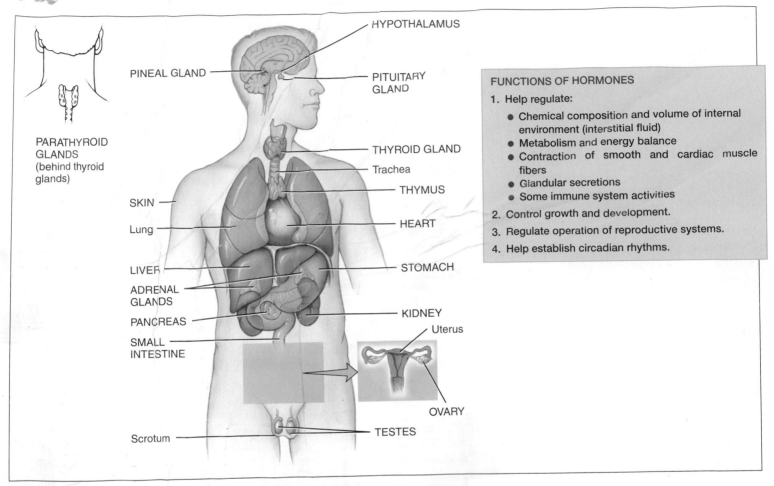

FUNCTIONS OF HORMONES

1. Help regulate:
   - Chemical composition and volume of internal environment (interstitial fluid)
   - Metabolism and energy balance
   - Contraction of smooth and cardiac muscle fibers
   - Glandular secretions
   - Some immune system activities
2. Control growth and development.
3. Regulate operation of reproductive systems.
4. Help establish circadian rhythms.

What is the basic difference between endocrine glands and exocrine glands?

Receptors, like other cellular proteins, are constantly being synthesized and broken down. Generally, a target cell has 2000–100,000 receptors for a particular hormone. When a hormone is present in excess, the number of target-cell receptors may decrease. This decreases the responsiveness of target cells to the hormone. In contrast, when a hormone (or neurotransmitter) is deficient, the number of receptors may increase to make the target tissue more sensitive.

BLOCKING HORMONE RECEPTORS

Synthetic hormones that **block the receptors** for certain naturally occurring hormones are available as drugs. For example, RU486 (mifepristone), which is used to induce abortion, binds to the receptors for progesterone (a female sex hormone) and prevents progesterone from exerting its normal effects. When RU486 is given to a pregnant woman, the uterine conditions needed for nurturing an embryo are not maintained, embryonic

development stops, and the embryo is sloughed off along with the uterine lining. This example illustrates an important endocrine principle: If a hormone is prevented from interacting with its receptors, the hormone cannot perform its normal functions. RU486 is often used in conjunction with one or two other drugs, methotrexate and misoprostol. ■

Most hormones are released in short bursts, with little or no secretion between bursts. When stimulation increases, an endocrine gland will release its hormone in more frequent bursts, thus elevating the hormone's overall blood concentration; in the absence of stimulation, bursts are minimal or inhibited, and the blood level of the hormone decreases. Regulation of hormone secretion normally maintains homeostasis and prevents overproduction or underproduction of any given hormone.

CHECKPOINT

2. Explain the relationship between target cell receptors and hormones.

HYPOTHALAMUS AND PITUITARY GLAND

OBJECTIVES

- Explain why the hypothalamus is classified as an endocrine gland.
- Describe the location, histology, hormones, and functions of the anterior and posterior pituitary.

For many years the **pituitary** (pi-TOO-i-tar-ē) **gland** or **hypophysis** (hī-POF-i-sis) was called the "master" endocrine gland because it secretes several hormones that control other endocrine glands. We now know that the pituitary gland itself has a master—the **hypothalamus.** This small region of the brain, inferior to the thalamus, is the major integrating link between the nervous and endocrine systems. It receives input from several other regions of the brain, including the limbic system, cerebral cortex, thalamus, and reticular activating system. It also receives sensory signals from internal organs and from the retina.

Painful, stressful, and emotional experiences all cause changes in hypothalamic activity. In turn, the hypothalamus controls the autonomic nervous system, regulating body temperature, thirst, hunger, sexual behavior, and defensive reactions such as fear and rage. Thus, not only is the hypothalamus an important regulatory center in the nervous system, it is also a crucial endocrine gland. Hormones secreted by the hypothalamus (described shortly) and the pituitary gland play important roles in the regulation of virtually all aspects of growth, development, metabolism, and homeostasis.

The pituitary gland is a pea-shaped structure measuring about 1–1.5 cm (0.5 in.) in diameter that lies in the hypophyscal fossa of the sphenoid bone and attaches to the hypothalamus by a stalk, the **infundibulum** (=a funnel; Figure 23.2a). The pituitary gland has two anatomically and functionally separate portions. The **anterior pituitary (anterior lobe)** accounts for about 75% of the total weight of the gland. It develops from an outgrowth of ectoderm called the hypophyseal pouch in the roof of the mouth (see Figure 23.8b). The anterior pituitary consists of two parts in an adult: The **pars distalis** is the larger bulbar portion, and the **pars tuberalis** forms a sheath around the infundibulum. The **posterior pituitary (posterior lobe)** also develops from an ectodermal outgrowth, this one called the neurohypophyseal bud (see Figure 23.8b). The posterior pituitary also consists of two parts: the **pars nervosa,** the larger bulbar portion, and the infundibulum. The posterior pituitary contains axons and axon terminals of more than 10,000 neurons whose cell bodies are located in the supraoptic and paraventricular nuclei of the hypothalamus (see Figure 19.10 on page 599). The axon terminals in the posterior pituitary gland are associated with specialized neuroglia called **pituicytes** (pi-TOO-i-sītz). These cells have a supporting role similar to that of astrocytes (see Chapter 17).

A third region, called the **pars intermedia,** atrophies during fetal development and ceases to exist as a separate lobe in adults (see Figure 23.8b). However, some of its cells migrate into adjacent parts of the anterior pituitary, where they persist.

Anterior Pituitary

The **anterior pituitary,** or **adenohypophysis** (ad′-e-nō-hī-POF-i-sis; *adeno-*=gland; *-hypophysis*=undergrowth), secretes hormones that regulate a wide range of bodily activities, from growth to reproduction. Release of anterior pituitary hormones is stimulated by **releasing hormones** and suppressed by **inhibiting hormones** from the hypothalamus (Figure 23.2b). These hypothalamic hormones are an important link between the nervous and endocrine systems.

Hypothalamic hormones reach the anterior pituitary through a portal system. As you learned in Chapter 13, blood usually passes from the heart through an artery to a capillary to a vein and back to the heart. In a portal system, blood flows from one capillary network into a portal vein, and then into a second capillary network without passing through the heart. The name of the portal system gives the location of the *second* capillary network. In the **hypophyseal portal system** (hī′-pō-FIZ-ē-al), blood flows from capillaries in the hypothalamus into portal veins that carry blood to capillaries of the anterior pituitary.

The **superior hypophyseal arteries,** branches of the internal carotid arteries, bring blood into the hypothalamus (Figure 23.2a). At the juncture of the median eminence of the hypothalamus and the infundibulum, these arteries divide into a capillary network called the **primary plexus of the hypophyseal portal system.** From the primary plexus, blood drains into the **hypophyseal portal veins** that pass down the surface of the infundibulum. In the anterior pituitary, the hypophyseal portal veins divide again and form another capillary network called the **secondary plexus of the hypophyseal portal system.**

Near the median eminence and above the optic chiasm are clusters of specialized neurons, called **neurosecretory cells.** They synthesize the hypothalamic releasing and inhibiting hormones in their cell bodies and package the hormones inside vesicles, which reach the axon terminals by axonal transport. When nerve impulses reach the axon terminals, they stimulate the vesicles to undergo exocytosis. The hormones then diffuse into the primary plexus of the hypophyseal portal system.

Quickly, the hypothalamic hormones flow with the blood through the portal veins and into the secondary plexus. This direct route permits hypothalamic hormones to act immediately on anterior pituitary cells, before the hormones are diluted or destroyed in the general circulation. Hormones secreted by anterior pituitary cells pass into the secondary plexus capillaries, which drain into the anterior hypophyseal veins. Now in the general circulation, anterior pituitary hormones travel to target tissues throughout the body.

The following list describes the seven major hormones secreted by five types of anterior pituitary cells:

1. **Human growth hormone (hGH),** or **somatotropin** (sō′-ma-tō-TRŌ-pin; *somato-*=body; *-tropin*=change), is secreted by cells called **somatotrophs.** Human growth hormone in turn stimulates several tissues to secrete **insulinlike growth factors,** hormones that stimulate general body growth and regulate various aspects of metabolism.

Figure 23.2 Hypothalamus and pituitary gland, and their blood supply. Figure 23.2b indicates that releasing and inhibiting hormones synthesized by hypothalamic neurons are transported within axons and released from the axon terminals. The hormones diffuse into capillaries of the primary plexus of the hypophyseal portal system and are carried by the hypophyseal portal veins to the secondary plexus of the hypophyseal portal system for distribution to target cells in the anterior pituitary.

Hypothalamic hormones are an important link between the nervous and endocrine systems.

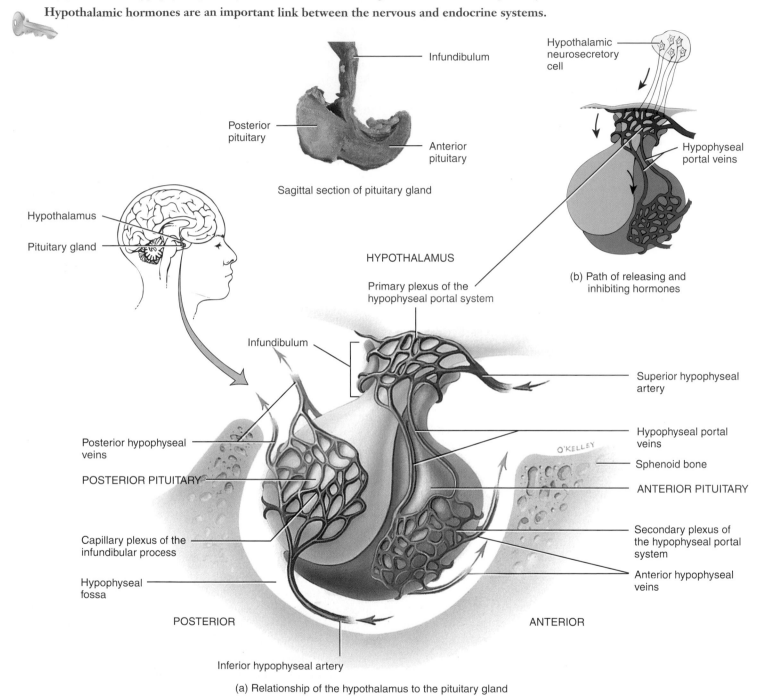

Infundibulum

Posterior pituitary

Anterior pituitary

Sagittal section of pituitary gland

Hypothalamic neurosecretory cell

Hypophyseal portal veins

(b) Path of releasing and inhibiting hormones

Hypothalamus

Pituitary gland

HYPOTHALAMUS

Primary plexus of the hypophyseal portal system

Infundibulum

Superior hypophyseal artery

Posterior hypophyseal veins

Hypophyseal portal veins

O'KELLEY

Sphenoid bone

POSTERIOR PITUITARY

ANTERIOR PITUITARY

Secondary plexus of the hypophyseal portal system

Capillary plexus of the infundibular process

Hypophyseal fossa

Anterior hypophyseal veins

POSTERIOR

ANTERIOR

Inferior hypophyseal artery

(a) Relationship of the hypothalamus to the pituitary gland

2. **Thyroid-stimulating hormone (TSH),** or **thyrotropin** (thī-rō-TRŌ-pin; *thyro-*=shield), which controls the secretions and other activities of the thyroid gland, is secreted by **thyrotrophs.**

3. **Follicle-stimulating hormone (FSH)** and **luteinizing hormone (LH)** (LOO-tē-in′-īz-ing) are secreted by **gonadotrophs** (*gonado-*=seed). FSH and LH both act on

the gonads: They stimulate secretion of estrogens and progesterone and the maturation of oocytes in the ovaries, and they stimulate secretion of testosterone and sperm production in the testes.

4. **Prolactin (PRL),** which initiates milk production in the mammary glands, is released by **lactotrophs** (*lacto-*=milk).

5. **Adrenocorticotropic hormone (ACTH),** or **corticotropin**

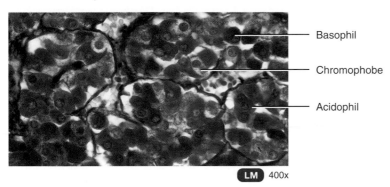

— Basophil

— Chromophobe

— Acidophil

LM 400x

(c) Histology of the anterior pituitary

? **What is the functional importance of the hypophyseal portal veins?**

(*cortico-*=rind or bark), which stimulates the adrenal cortex to secrete glucocorticoids, is synthesized by **corticotrophs.** Some corticotrophs also secrete **melanocyte-stimulating hormone (MSH).**

The five different types of cells of the anterior pituitary may be classified according to their staining reaction (Figure 23.2c). **Basophils** (thyrotrophs, gonadotrophs, and corticotrophs) make up about 10% of the cells, stain blue with basic dyes, and contain secretory granules. **Acidophils** (somatotrophs and lactotrophs) make up about 40% of the cells, stain red with acidic dyes, and also contain secretory granules. **Chromophobes** (KRŌ-mō-fōbs) make up about 50% of the cells and have little affinity for basic or acidic dyes. They have few or no secretory granules and do not secrete hormones. They are believed to be basophils or acidophils that have already released their granules.

Four of the anterior pituitary hormones influence the activity of another endocrine gland and are called **tropic hormones,** or **tropins.** These include TSH, ACTH, FSH, and LH. The two **gonadotropins,** FSH and LH, are hormones that specifically regulate the functions of the gonads (ovaries and testes).

The principal actions of the hormones of the anterior pituitary gland are summarized in Table 23.2.

TABLE 23.2   SUMMARY OF THE PRINCIPAL ACTIONS OF ANTERIOR PITUITARY HORMONES

HORMONE AND TARGET TISSUES	PRINCIPAL ACTIONS	HORMONE AND TARGET TISSUES	PRINCIPAL ACTIONS
Human growth hormone (hGH) or somatotropin Liver	Stimulates liver, muscle, cartilage, bone, and other tissues to synthesize and secrete insulinlike growth factors (IGFs); IGFs promote growth of body cells, protein synthesis, tissue repair, lipolysis, and elevation of blood glucose concentration.	**Luteinizing hormone (LH)** Ovaries Testes	In females, stimulates secretion of estrogens and progesterone, ovulation, and formation of corpus luteum. In males, stimulates interstitial cells in testes to develop and produce testosterone.
Thyroid-stimulating hormone (TSH) or thyrotropin Thyroid gland	Stimulates synthesis and secretion of thyroid hormones by thyroid gland.	**Prolactin (PRL)** Mammary glands	Together with other hormones, promotes milk secretion by the mammary glands.
Follicle-stimulating hormone (FSH) Ovaries Testes	In females, initiates development of oocytes and induces ovarian secretion of estrogens. In males, stimulates testes to produce sperm.	**Adrenocorticotropic hormone (ACTH) or corticotropin** Adrenal cortex	Stimulates secretion of glucocorticoids (mainly cortisol) by adrenal cortex.
		Melanocyte-stimulating hormone (MSH) Brain	Exact role in humans is unknown but may influence brain activity; when present in excess, can cause darkening of skin.

Posterior Pituitary

Although the **posterior pituitary**, or **neurohypophysis,** does not *synthesize* hormones, it does *store* and *release* two hormones. As noted earlier, it consists of pituicytes and axon terminals of hypothalamic neurosecretory cells. The cell bodies of the neurosecretory cells are in the paraventricular and supraoptic nuclei of the hypothalamus; their axons form the **hypothalamo-hypophyseal tract** (hī′-pō-thal′-a-mō-hī-pō-FIZ-ē-al), which begins in the hypothalamus and ends near blood capillaries in the posterior pituitary gland (Figure 23.3). Different neurosecretory cells produce two hormones: **oxytocin (OT;** ok′-sē-TŌ-sin; *oxytoc-*=quick birth) and **antidiuretic hormone (ADH),** also called **vasopressin** (vā-sō-PRES-in; *vaso-*=blood vessel; *-pressus*=to press).

After their production in the cell bodies of neurosecretory cells, oxytocin and antidiuretic hormone are packed into vesicles and transported to the axon terminals in the posterior pituitary gland. Nerve impulses that propagate along the axon and reach the axon terminals trigger exocytosis of these secretory vesicles.

During and after delivery of a baby, oxytocin has two target tissues: the mother's uterus and breasts. During delivery, oxytocin enhances contraction of smooth muscle cells in the wall of the uterus; after delivery, it stimulates milk ejection ("letdown") from the mammary glands in response to the mechanical stimulus provided by a suckling infant. The function of oxytocin in males and in nonpregnant females is unclear. Experiments with animals have suggested that it has actions within the brain that foster parental caretaking behavior toward young offspring. It may also be responsible, in part, for the feelings of sexual pleasure during and after intercourse.

An **antidiuretic** (an-tī-dī-Ū-ret-ik; *anti-*=against; *-diuretic* =increases urine production) is a substance that decreases urine production. ADH causes the kidneys to return more water to

Figure 23.3   Axons of hypothalamic neurosecretory cells form the hypothalamohypophyseal tract, which extends from the paraventricular and supraoptic nuclei to the posterior pituitary. Hormone molecules synthesized in the cell body of a neurosecretory cell are packaged into secretory vesicles that move down to the axon terminals. Nerve impulses trigger exocytosis of the vesicles and release of the hormone.

Oxytocin and antidiuretic hormone are synthesized in the hypothalamus and released into capillaries of the posterior pituitary.

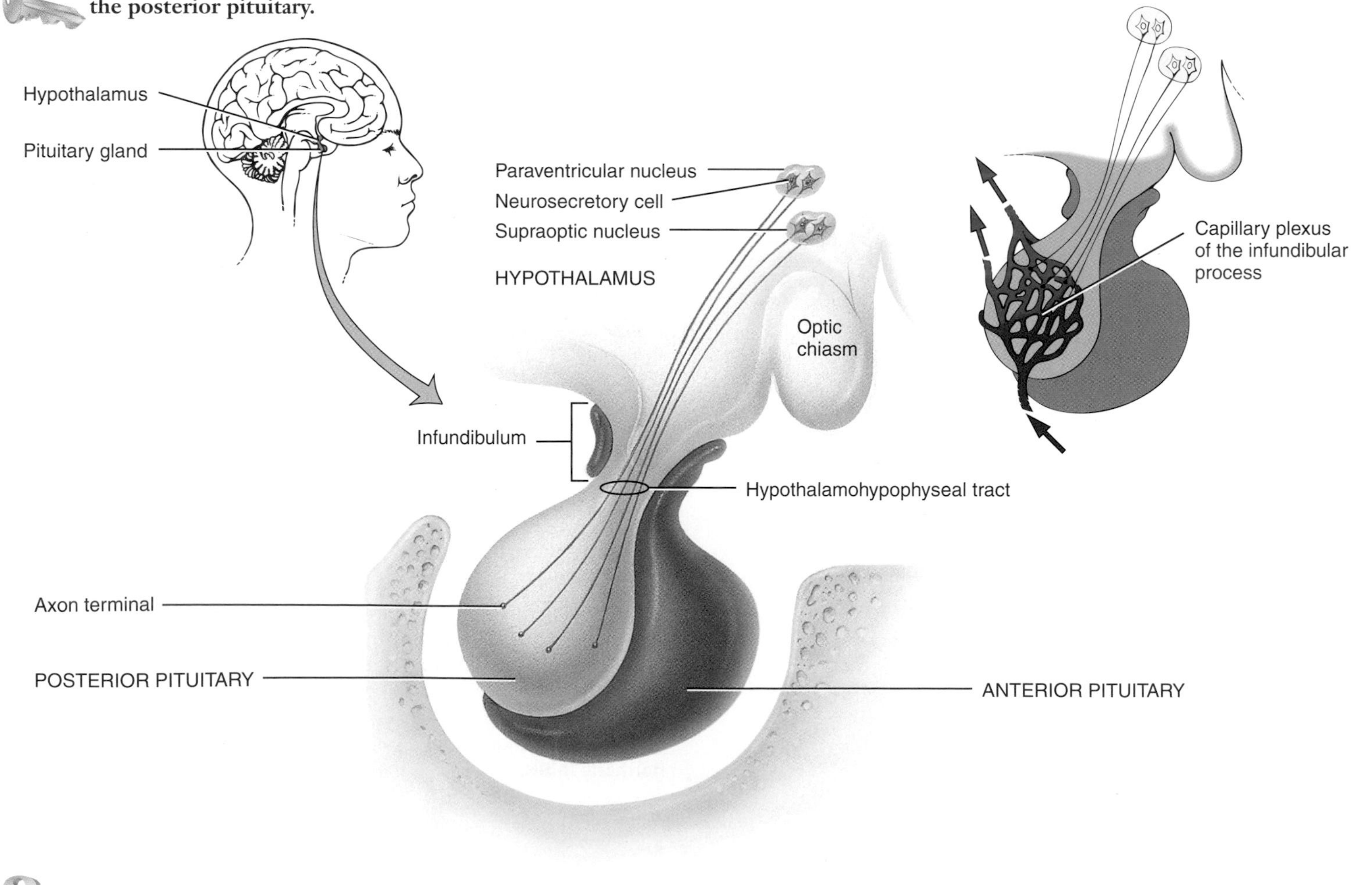

Functionally, how are the hypothalamohypophyseal tract and the hypophyseal portal veins similar? Structurally, how are they different?

the blood, thus decreasing urine volume. In the absence of ADH, urine output increases more than tenfold, from the normal 1–2 liters to about 20 liters a day. Drinking alcohol often causes frequent and copious urination because the alcohol inhibits secretion of ADH. ADH also decreases the water lost through sweating and causes constriction of arterioles, which increases blood pressure. The other name for this hormone, vasopressin, reflects its effect on blood pressure.

Blood is supplied to the posterior pituitary by the **inferior hypophyseal arteries** (see Figure 23.2a), which branch from the internal carotid arteries. In the posterior pituitary, the inferior hypophyseal arteries drain into the **capillary plexus of the infundibular process,** a capillary network that receives oxytocin and antidiuretic hormone secreted from the neurosecretory cells of the hypothalamus (see Figure 23.2a). From this plexus, hormones pass into the **posterior hypophyseal veins** for distribution to target cells in other tissues.

OXYTOCIN AND CHILDBIRTH

Years before oxytocin was discovered, it was common practice in midwifery to let a first-born twin nurse at the mother's breast to speed the birth of the second child. Now we know why this practice is helpful— it stimulates release of oxytocin. Even after a single birth, nursing promotes expulsion of the placenta (afterbirth) and helps the uterus regain its smaller size. Synthetic OT often is given to induce labor or to increase uterine tone and control hemorrhage just after giving birth. ■

A summary of posterior pituitary hormones is presented in Table 23.3.

CHECKPOINT

3. In what respect is the pituitary gland actually two glands?
4. Describe the structure and importance of the hypothalamo-hypophyseal tract.

THYROID GLAND

OBJECTIVE

● Describe the location, histology, hormones, and functions of the thyroid gland.

The butterfly-shaped **thyroid gland** is located just inferior to the larynx (voice box); the right and left **lateral lobes** lie on either side of the trachea (Figure 23.4a). Connecting the lobes is a mass of tissue called an **isthmus** (IS-mus) that lies anterior to the trachea. A small, pyramidal-shaped lobe sometimes extends upward from the isthmus. The gland usually weighs about 30 g (1 oz) and has a rich blood supply, receiving 80–120 mL of blood per minute.

Microscopic spherical sacs called **thyroid follicles** (Figure 23.4b) make up most of the thyroid gland. The wall of each follicle consists primarily of cells called **follicular cells,** most of which extend to the lumen (internal space) of the follicle. When the follicular cells are inactive, their shape is low cuboidal to squamous, but under the influence of TSH they become cuboidal or low columnar and actively secrete hormones. The follicular cells produce two hormones: **thyroxine** (thī-ROK-sēn), which is also called **tetraiodothyronine** (tet-ra-ī-ō-dō-THĪ-rō-nēn), or T_4, because it contains four atoms of iodine, and **triiodothyronine** (trī-ī′-ō-dō-THĪ-rō-nēn), or T_3, which

TABLE 23.3	SUMMARY OF POSTERIOR PITUITARY HORMONES	
HORMONE AND TARGET TISSUES	**CONTROL OF SECRETION**	**PRINCIPAL ACTIONS**
Oxytocin (OT) Uterus Mammary glands	Neurosecretory cells of hypothalamus secrete OT in response to uterine distention and stimulation of nipples.	Stimulates contraction of smooth muscle cells of uterus during childbirth; stimulates contraction of myoepithelial cells in mammary glands to cause milk ejection.
Antidiuretic hormone (ADH) or vasopressin Kidneys Sudoriferous (sweat) glands Arterioles	Neurosecretory cells of hypothalamus secrete ADH in response to elevated blood osmotic pressure, dehydration, loss of blood volume, pain, or stress; low blood osmotic pressure, high blood volume, and alcohol inhibit ADH secretion.	Conserves body water by decreasing urine volume; decreases water loss through perspiration; raises blood pressure by constricting arterioles.

Figure 23.4   Location, blood supply, and histology of the thyroid gland.

Thyroid hormones regulate (1) oxygen use and basal metabolic rate, (2) cellular metabolism, and (3) growth and development.

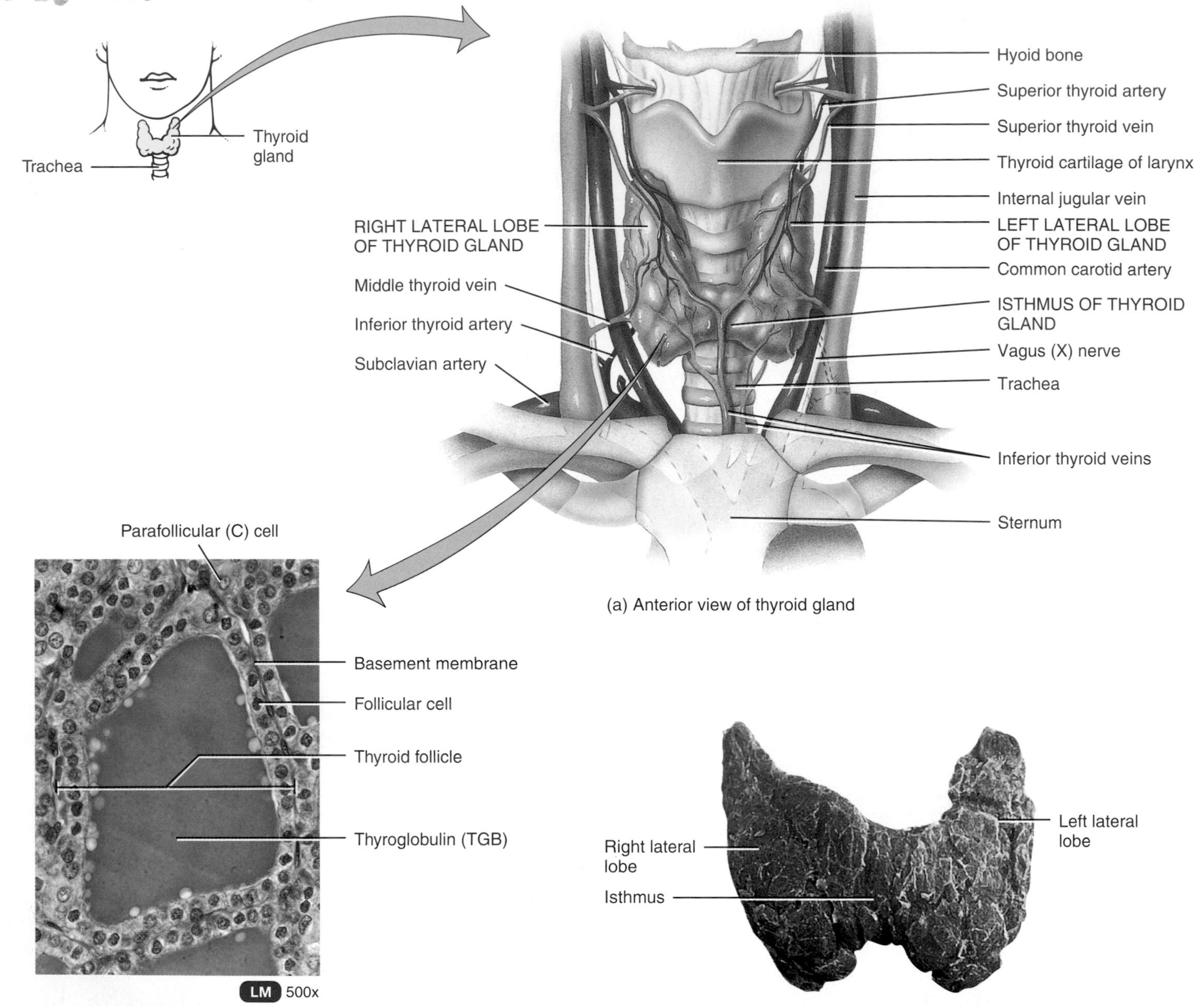

Trachea — Thyroid gland

Hyoid bone

Superior thyroid artery

Superior thyroid vein

Thyroid cartilage of larynx

Internal jugular vein

RIGHT LATERAL LOBE OF THYROID GLAND

LEFT LATERAL LOBE OF THYROID GLAND

Middle thyroid vein

Common carotid artery

Inferior thyroid artery

ISTHMUS OF THYROID GLAND

Subclavian artery

Vagus (X) nerve

Trachea

Inferior thyroid veins

Sternum

(a) Anterior view of thyroid gland

Parafollicular (C) cell

Basement membrane

Follicular cell

Thyroid follicle

Thyroglobulin (TGB)

LM 500x

(b) Thyroid follicles

Right lateral lobe

Left lateral lobe

Isthmus

(c) Anterior view of thyroid gland

Which cells secrete T_3 and T_4? Which secrete calcitonin? Which of these hormones are also called thyroid hormones?

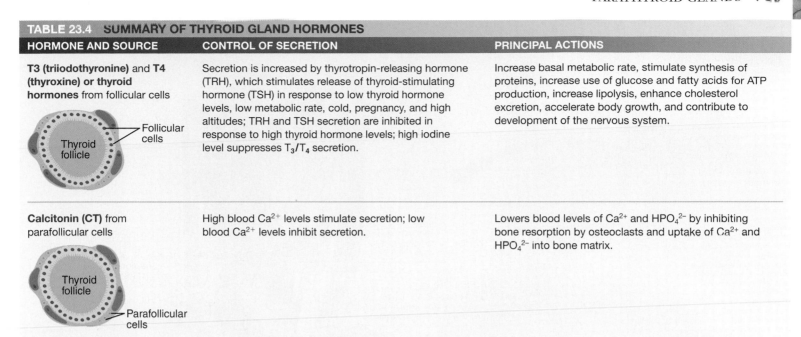

TABLE 23.4	SUMMARY OF THYROID GLAND HORMONES	
HORMONE AND SOURCE	**CONTROL OF SECRETION**	**PRINCIPAL ACTIONS**
T3 (triiodothyronine) and **T4 (thyroxine)** or **thyroid hormones** from follicular cells	Secretion is increased by thyrotropin-releasing hormone (TRH), which stimulates release of thyroid-stimulating hormone (TSH) in response to low thyroid hormone levels, low metabolic rate, cold, pregnancy, and high altitudes; TRH and TSH secretion are inhibited in response to high thyroid hormone levels; high iodine level suppresses T_3/T_4 secretion.	Increase basal metabolic rate, stimulate synthesis of proteins, increase use of glucose and fatty acids for ATP production, increase lipolysis, enhance cholesterol excretion, accelerate body growth, and contribute to development of the nervous system.
Calcitonin (CT) from parafollicular cells	High blood Ca^{2+} levels stimulate secretion; low blood Ca^{2+} levels inhibit secretion.	Lowers blood levels of Ca^{2+} and HPO_4^{2-} by inhibiting bone resorption by osteoclasts and uptake of Ca^{2+} and HPO_4^{2-} into bone matrix.

contains three atoms of iodine. T_3 and T_4 are also known as **thyroid hormones.** The thyroid hormones regulate (1) oxygen use and basal metabolic rate, (2) cellular metabolism, and (3) growth and development. A few cells called **parafollicular cells,** or **C cells,** may be embedded within a follicle or lie between follicles. They produce the hormone **calcitonin** (kal-si-TŌ-nin), which helps regulate calcium homeostasis.

The main blood supply of the thyroid gland is from the superior thyroid artery, a branch of the external carotid artery, and the inferior thyroid artery, a branch of the thyrocervical trunk from the subclavian artery. The thyroid gland is drained by the superior and middle thyroid veins, which pass into the internal jugular veins, and the inferior thyroid veins, which join the brachiocephalic veins or internal jugular veins (Figure 23.4a).

The nerve supply of the thyroid gland consists of postganglionic fibers from the superior and middle cervical sympathetic ganglia. Preganglionic fibers from the ganglia are derived from the second through seventh thoracic segments of the spinal cord.

A summary of thyroid gland hormones and their actions is presented in Table 23.4.

CHECKPOINT

5. Name the hormones produced by follicular and parafollicular cells and describe their actions.

PARATHYROID GLANDS

OBJECTIVE

● Describe the location, histology, hormones, and functions of the parathyroid glands.

Partially embedded in the posterior surface of the lateral lobes of the thyroid gland are several small, round masses of tissue called the **parathyroid glands** (*para-*=beside). Each parathyroid gland has a mass of about 40 mg (0.04 g). Usually, one superior and one inferior parathyroid gland are attached to each lateral thyroid lobe (Figure 23.5a).

Microscopically, the parathyroid glands contain two kinds of epithelial cells (Figure 23.5b). The more numerous cells are called **chief (principal) cells** and produce **parathyroid hormone (PTH),** or **parathormone.** The function of the other type of parathyroid epithelial cell, called an **oxyphil cell,** is unknown.

PTH increases the number and activity of osteoclasts (bone-destroying cells). The result is elevated bone resorption, which releases ionic calcium (Ca^{2+}) and phosphates (HPO_4^{2-}) into the blood. PTH also produces two changes in the kidneys: (1) It increases the rate at which the kidneys remove Ca^{2+} and magnesium (Mg^{2+}) from urine that is being formed and returns them to the blood, and (2) it inhibits the reabsorption of (HPO_4^{2-}) filtered by the kidneys, so that more of it is excreted in urine. More (HPO_4^{2-}) is lost in the urine than is gained from the bones. Overall, then, PTH decreases blood HPO_4^{2-} level and increases blood Ca^{2+} and Mg^{2+} levels. With respect to blood Ca^{2+} level, PTH and calcitonin are *antagonists;* that is, they have opposite actions. A third effect of PTH on the kidneys is to promote formation of the hormone **calcitriol,** which is the active form of vitamin D.

The parathyroids are abundantly supplied with blood from branches of the superior and inferior thyroid arteries. Blood is drained by the superior, middle, and inferior thyroid veins. The nerve supply of the parathyroids is derived from the thyroid branches of cervical sympathetic ganglia.

A summary of parathyroid hormone and its actions is presented in Table 23.5 on page 715.

Figure 23.5 Location, blood supply, and histology of the parathyroid glands.

The parathyroid glands, normally four in number, are embedded in the posterior surface of the thyroid gland.

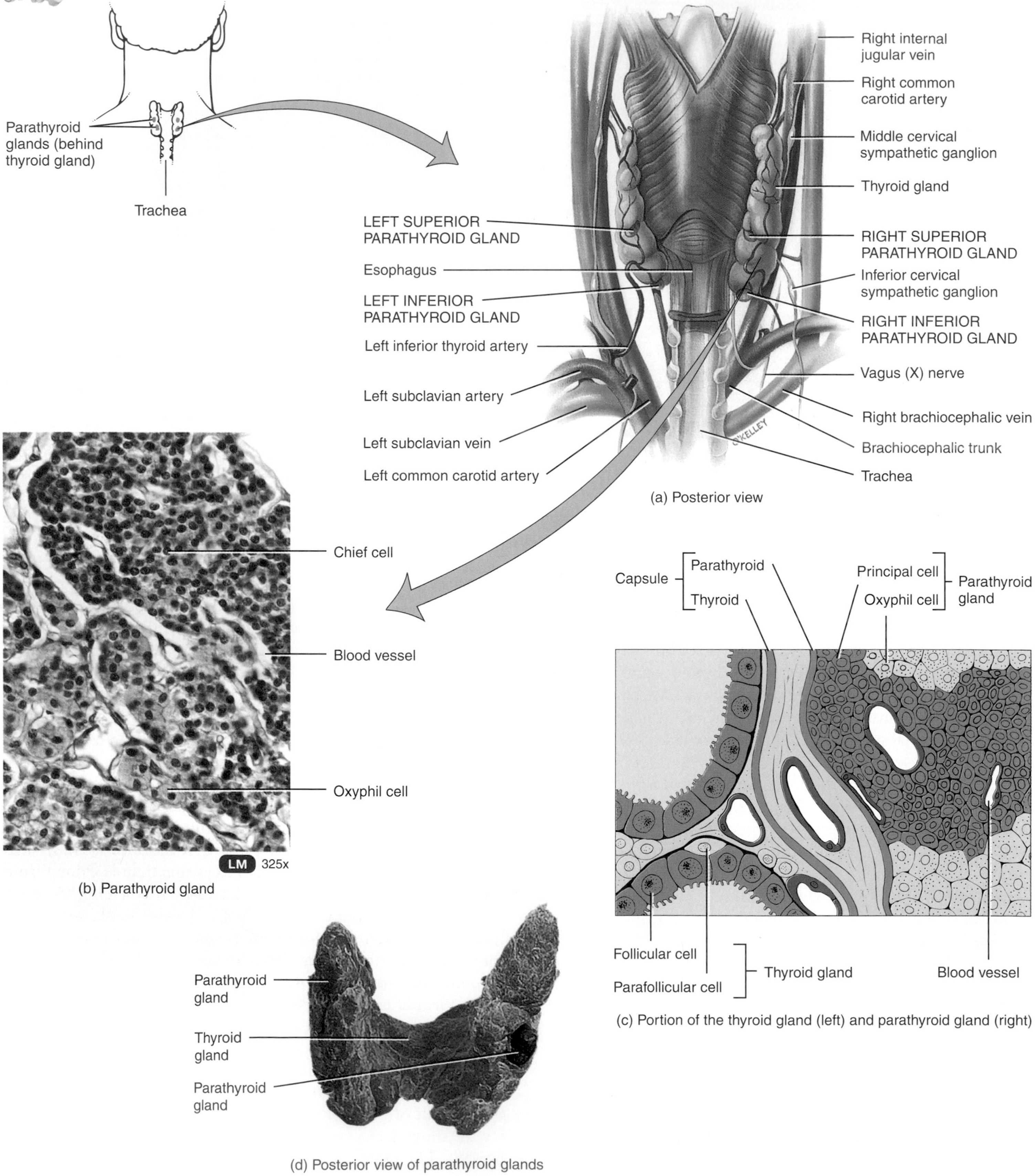

Parathyroid glands (behind thyroid gland)

Trachea

Right internal jugular vein

Right common carotid artery

Middle cervical sympathetic ganglion

Thyroid gland

LEFT SUPERIOR PARATHYROID GLAND

Esophagus

LEFT INFERIOR PARATHYROID GLAND

Left inferior thyroid artery

Left subclavian artery

Left subclavian vein

Left common carotid artery

RIGHT SUPERIOR PARATHYROID GLAND

Inferior cervical sympathetic ganglion

RIGHT INFERIOR PARATHYROID GLAND

Vagus (X) nerve

Right brachiocephalic vein

Brachiocephalic trunk

Trachea

(a) Posterior view

Chief cell

Blood vessel

Oxyphil cell

LM 325x

(b) Parathyroid gland

Capsule

Parathyroid

Thyroid

Principal cell

Oxyphil cell

Parathyroid gland

Follicular cell

Parafollicular cell

Thyroid gland

Blood vessel

(c) Portion of the thyroid gland (left) and parathyroid gland (right)

Parathyroid gland

Thyroid gland

Parathyroid gland

(d) Posterior view of parathyroid glands

What are the secretory products of (1) parafollicular cells of the thyroid gland and (2) chief cells of the parathyroid glands?

TABLE 23.5 SUMMARY OF PARATHYROID GLAND HORMONE

HORMONE AND SOURCE	CONTROL OF SECRETION	PRINCIPAL ACTIONS
Parathyroid hormone (PTH) from chief cells 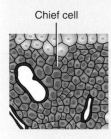 Chief cell	Low blood Ca^{2+} levels stimulate secretion. High blood Ca^{2+} levels inhibit secretion.	Increases blood Ca^{2+} and Mg^{2+} levels and decreases blood HPO_4^{2-} level; increases bone resorption by osteoclasts; increases Ca^{2+} reabsorption and phosphate excretion by kidneys; and promotes formation of calcitriol (active form of vitamin D), which increases rate of dietary Ca^{2+} and Mg^{2+} absorption.

CHECKPOINT

6. Compare the actions of PTH and calcitonin with respect to blood Ca^{2+} level.

ADRENAL GLANDS

OBJECTIVE

• Describe the location, histology, hormones, and functions of the adrenal glands.

The paired **adrenal glands,** or **suprarenal** (*supra-*=above; *renal*=kidney) **glands,** one of which lies superior to each kidney (Figure 23.6a, c), have a flattened pyramidal shape. In an adult, each adrenal gland is 3–5 cm in height, 2–3 cm in width, and a little less than 1 cm thick; it weighs 3.5–5 g, only half its weight at birth. During embryonic development, the adrenal glands differentiate into two structurally and functionally distinct regions: A large, peripherally located **adrenal cortex,** representing 80–90% of the gland by weight, develops from mesoderm; a small, centrally located **adrenal medulla** develops from ectoderm (Figure 23.6b). The adrenal cortex produces hormones that are essential for life. Complete loss of adrenocortical hormones leads to death due to dehydration and electrolyte imbalances in a few days to a week, unless hormone replacement therapy begins promptly. The adrenal medulla produces two hormones: norepinephrine and epinephrine. Covering the gland is a connective tissue capsule. The adrenal glands, like the thyroid gland, are highly vascularized.

Adrenal Cortex

The adrenal cortex is subdivided into three zones, each of which secretes different hormones (Figure 23.6d). The outer zone, just deep to the connective tissue capsule, is called the **zona glomerulosa** (*zona*=belt; *glomerul-*=little ball). Its cells, which are closely packed and arranged in spherical clusters and arched columns, secrete hormones called **mineralocorticoids** (min′-er-al-ō-KOR-ti-koyds) because they affect metabolism of the minerals sodium and potassium. The middle zone, or **zona fasciculata** (*fascicul-*=little bundle), is the widest of the three zones and consists of cells arranged in long, straight cords. The cells of the zona fasciculata secrete mainly **glucocorticoids** (gloo′-kō-KOR-ti-koyds), so named because they affect glucose metabolism. The cells of the inner zone, the **zona reticularis** (*reticul-*=network), are arranged in branching cords. They synthesize small amounts of weak **androgens** (*andro-*=a man), hormones that have masculinizing effects.

Adrenal Medulla

The inner region of the adrenal gland, the **adrenal medulla,** is a modified sympathetic ganglion of the autonomic nervous system (ANS). It develops from the same embryonic tissue as all other sympathetic ganglia, but its cells lack axons and form clusters around large blood vessels. Rather than releasing a neurotransmitter, the cells of the adrenal medulla secrete hormones. The hormone-producing cells, called **chromaffin cells** (KRŌ-maf-in; *chrom-*=color; *-affin*=affinity for; see Figure 23.6d), are innervated by sympathetic preganglionic neurons of the splanchnic nerve. Because the ANS controls the chromaffin cells directly, hormone release can occur very quickly.

The two principal hormones synthesized by the adrenal medulla are **epinephrine** and **norepinephrine (NE),** also called adrenaline and noradrenaline, respectively. Epinephrine constitutes about 80% of the total secretion of the gland. Both hormones are **sympathomimetic** (sim′-pa-thō-mi-MET-ik)— their effects mimic those brought about by the sympathetic division of the ANS. To a large extent, they are responsible for the fight-or-flight response. Like the glucocorticoids of the adrenal cortex, these hormones help resist stress. Unlike the hormones of the adrenal cortex, however, the medullary hormones are not essential for life.

The main arteries that supply the adrenal glands are the several superior suprarenal arteries arising from the inferior phrenic artery, the right and left middle suprarenal arteries from the aorta, and the inferior suprarenal arteries from the renal arteries. The suprarenal vein of the right adrenal gland drains

Figure 23.6 Location, blood supply, and histology of the adrenal (suprarenal) glands.

The adrenal cortex secretes hormones that are essential for life; the adrenal medulla secretes norepinephrine and epinephrine.

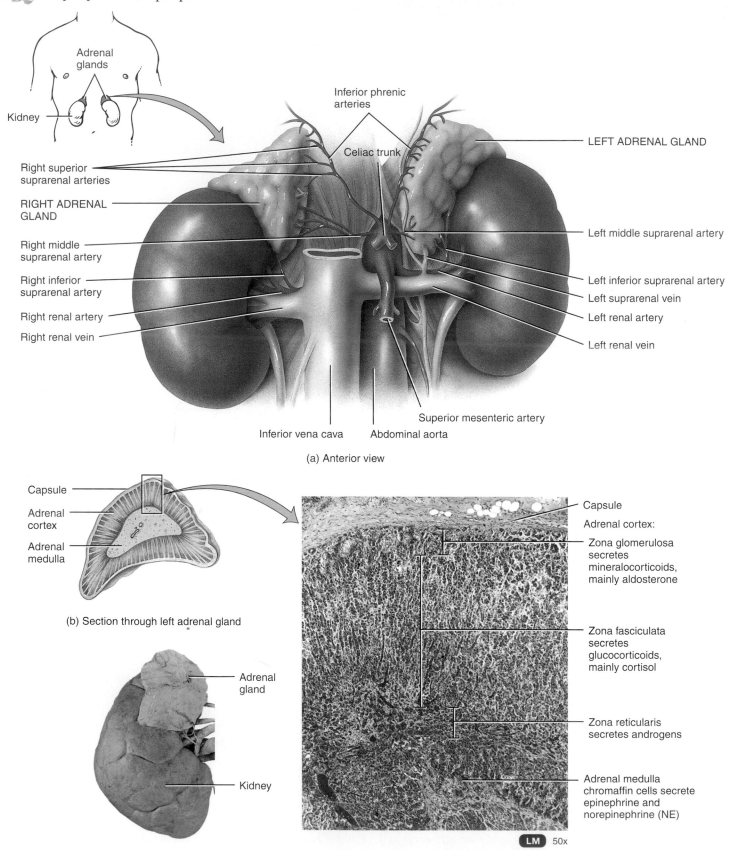

Adrenal glands

Kidney

Inferior phrenic arteries

Celiac trunk

LEFT ADRENAL GLAND

Right superior suprarenal arteries

RIGHT ADRENAL GLAND

Right middle suprarenal artery

Right inferior suprarenal artery

Right renal artery

Right renal vein

Left middle suprarenal artery

Left inferior suprarenal artery

Left suprarenal vein

Left renal artery

Left renal vein

Superior mesenteric artery

Inferior vena cava

Abdominal aorta

(a) Anterior view

Capsule

Adrenal cortex

Adrenal medulla

(b) Section through left adrenal gland

Adrenal gland

Kidney

(c) Anterior view of adrenal gland and kidney

Capsule

Adrenal cortex:

Zona glomerulosa secretes mineralocorticoids, mainly aldosterone

Zona fasciculata secretes glucocorticoids, mainly cortisol

Zona reticularis secretes androgens

Adrenal medulla chromaffin cells secrete epinephrine and norepinephrine (NE)

LM 50x

(d) Subdivisions of the adrenal gland

What is the position of the adrenal glands relative to the kidneys?

TABLE 23.6 SUMMARY OF ADRENAL GLAND HORMONES

HORMONE AND SOURCE	CONTROL OF SECRETION	PRINCIPAL ACTIONS
Adrenal cortex hormones		
Mineralocorticoids (mainly aldosterone) from zona glomerulosa cells	Increased blood K^+ level and angiotensin II stimulate secretion.	Increase blood levels of Na^+ and water and decrease blood level of K^+.
Glucocorticoids (mainly cortisol) from zona fasciculata cells	ACTH stimulates release; corticotropin-releasing hormone (CRH) promotes ACTH secretion in response to stress and low blood levels of glucocorticoids.	Increase protein breakdown (except in liver), stimulate gluconeogenesis and lipolysis, provide resistance to stress, dampen inflammation, and depress immune responses.
Androgens from zona reticularis cells	ACTH stimulates secretion.	Assist in early growth of axillary and pubic hair in both sexes; in females, contribute to libido and are source of estrogens after menopause.
— Adrenal cortex		
Adrenal medulla hormones		
Epinephrine and **norepinephrine** from chromaffin cells	Sympathetic preganglionic neurons release acetylcholine, which stimulates secretion.	Produce effects that enhance those of the sympathetic division of the autonomic nervous system (ANS) during stress.
— Adrenal medulla		

directly into the inferior vena cava, whereas the suprarenal vein of the left adrenal gland empties into the left renal vein (see Figure 23.6a).

The principal nerve supply to the adrenal glands is from preganglionic fibers from the thoracic splanchnic nerves, which pass through the celiac and associated sympathetic plexuses. These myelinated fibers innervate the secretory cells of the gland found in a region of the medulla.

A summary of adrenal gland hormones and their actions is presented in Table 23.6.

CHECKPOINT

7. Compare the location and histology of the adrenal cortex and the adrenal medulla.
8. Describe the relationship of the adrenal medulla to the autonomic nervous system.

PANCREAS

OBJECTIVE

- Describe the location, histology, hormones, and functions of the pancreas.

The **pancreas** (*pan-*=all; *-creas*=flesh) is both an endocrine gland and an exocrine gland. We discuss its endocrine functions here and its exocrine functions in Chapter 25 on the digestive system. The pancreas is a flattened organ that measures about 12.5–15 cm (4.5–6 in.) in length. It is located posterior and slightly inferior to the stomach and consists of a head, a body, and a tail (Figure 23.7a). Roughly 99% of the pancreatic cells

are arranged in clusters called **acini** (singular is *acinus*); these cells produce digestive enzymes, which flow into the gastrointestinal tract through a network of ducts. Scattered among the exocrine acini are 1–2 million tiny clusters of endocrine cells called **pancreatic islets** or **islets of Langerhans** (LAHNG-er-hanz; Figure 23.7b, c). Abundant capillaries serve both the exocrine and endocrine portions of the pancreas.

Each pancreatic islet contains four types of hormone-secreting cells:

1. **Alpha** or **A cells** constitute about 15% of pancreatic islet cells and secrete **glucagon** (GLOO-ka-gon).

2. **Beta** or **B cells** constitute about 80% of pancreatic islet cells and secrete **insulin** (IN-soo-lin).

3. **Delta** or **D cells** constitute about 5% of pancreatic islet cells and secrete **somatostatin** (identical to growth hormone–inhibiting hormone secreted by the hypothalamus).

4. **F cells** constitute the remainder of pancreatic islet cells and secrete **pancreatic polypeptide.**

The superior and inferior pancreaticoduodenal arteries and the splenic and superior mesenteric arteries supply blood to the pancreas (Figure 23.7a). The veins, in general, correspond to the arteries. Venous blood reaches the hepatic portal vein by means of the splenic and superior mesenteric veins.

The nerves to the pancreas are autonomic nerves derived from the celiac and superior mesenteric plexuses. Included are preganglionic vagal, postganglionic sympathetic, and sensory fibers. Parasympathetic vagal fibers are said to terminate at both acinar (exocrine) and islet (endocrine) cells. Although nervous

Figure 23.7 Location, blood supply, and histology of the pancreas.

Pancreatic hormones regulate blood glucose level.

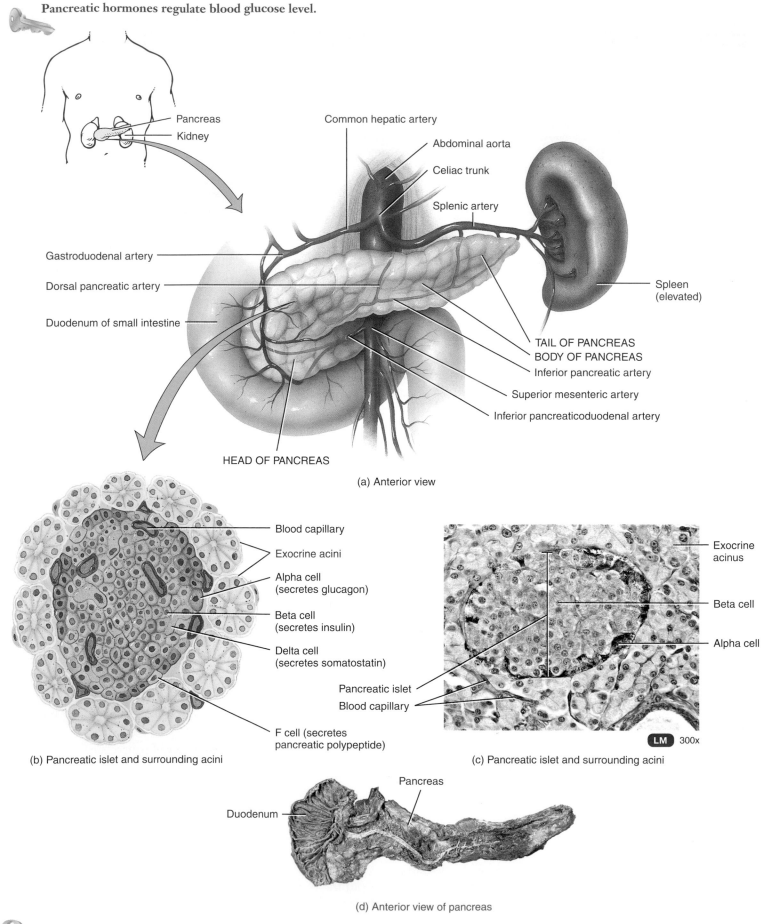

Pancreas
Kidney

Common hepatic artery
Abdominal aorta
Celiac trunk
Splenic artery

Gastroduodenal artery
Dorsal pancreatic artery
Duodenum of small intestine

Spleen (elevated)

TAIL OF PANCREAS
BODY OF PANCREAS
Inferior pancreatic artery
Superior mesenteric artery
Inferior pancreaticoduodenal artery

HEAD OF PANCREAS

(a) Anterior view

Blood capillary
Exocrine acini
Alpha cell (secretes glucagon)
Beta cell (secretes insulin)
Delta cell (secretes somatostatin)
F cell (secretes pancreatic polypeptide)

(b) Pancreatic islet and surrounding acini

Exocrine acinus
Beta cell
Alpha cell
Pancreatic islet
Blood capillary

LM 300x

(c) Pancreatic islet and surrounding acini

Pancreas
Duodenum

(d) Anterior view of pancreas

Is the pancreas an exocrine gland or an endocrine gland?

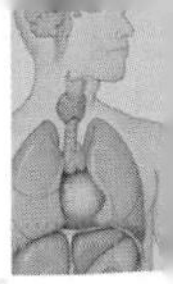

TABLE 23.7 SUMMARY OF PANCREATIC ISLET HORMONES

HORMONE AND SOURCE	CONTROL OF SECRETION	PRINCIPAL ACTIONS
Glucagon from alpha cells of pancreatic islets 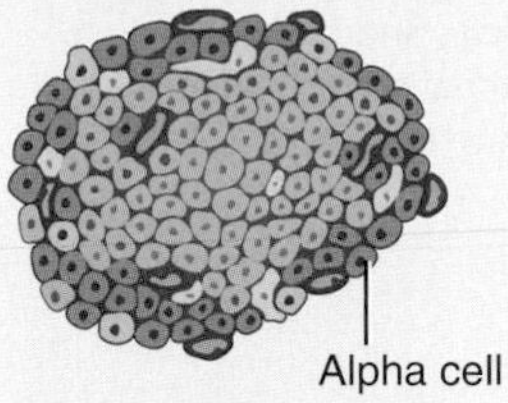 Alpha cell	Decreased blood level of glucose, exercise, and mainly protein meals stimulate secretion; somatostatin and insulin inhibit secretion.	Raises blood glucose level by accelerating breakdown of glycogen into glucose in liver (glycogenolysis), converting other nutrients into glucose in liver (gluconeogenesis), and releasing glucose into the blood.
Insulin from beta cells of pancreatic islets 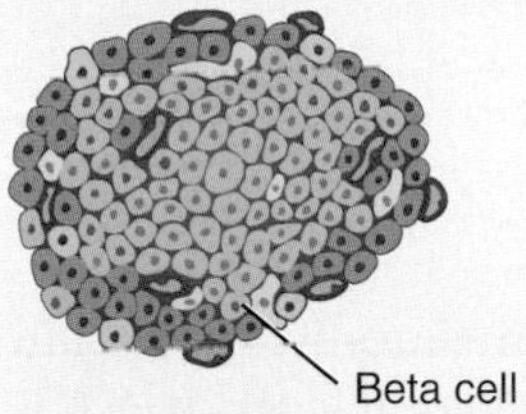 Beta cell	Increased blood level of glucose, acetylcholine (released by parasympathetic vagus nerve fibers), arginine and leucine (two amino acids), glucagon, GIP, hGH, and ACTH stimulate secretion; somatostatin inhibits secretion.	Lowers blood glucose level by accelerating transport of glucose into cells, converting glucose into glycogen (glycogenesis), and decreasing glycogenolysis and gluconeogenesis; also increases lipogenesis and stimulates protein synthesis.
Somatostatin from delta cells of pancreatic islets 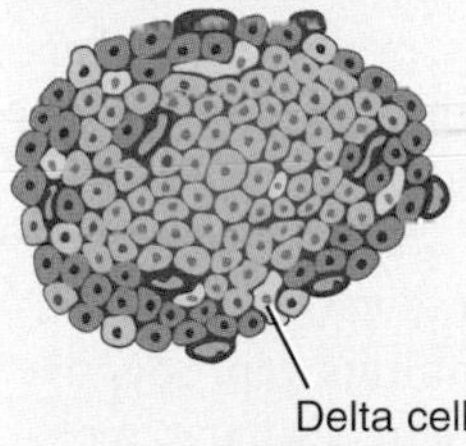 Delta cell	Pancreatic polypeptide inhibits secretion.	Inhibits secretion of insulin and glucagon and slows absorption of nutrients from the gastrointestinal tract.
Pancreatic polypeptide from F cells of pancreatic islets 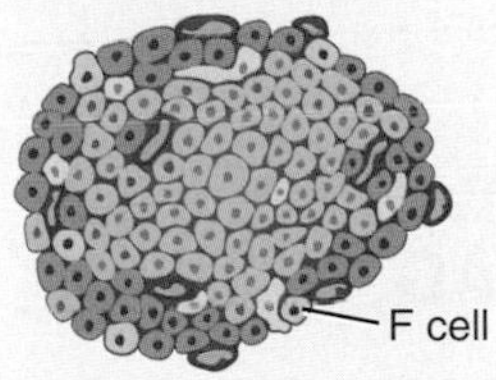 F cell	Meals containing protein, fasting, exercise, and acute hypoglycemia stimulate secretion; somatostatin and elevated blood glucose level inhibit secretion.	Inhibits somatostatin secretion, gallbladder contraction, and secretion of pancreatic digestive enzymes.

innervation is presumed to influence enzyme formation, pancreatic secretion is controlled largely by the hormones secretin and cholecystokinin (CCK) released by the small intestine. The sympathetic fibers that enter the islets (and also innervate blood vessels) are vasomotor and are accompanied by sensory fibers that transmit impulses, especially for pain.

A summary of pancreatic hormones and their actions is presented in Table 23.7.

CHECKPOINT

9. Why is the pancreas classified as both an exocrine gland and an endocrine gland?

OVARIES AND TESTES

OBJECTIVE

● Describe the location, hormones, and functions of the male and female gonads.

The female gonads, called the **ovaries,** are paired oval bodies located in the pelvic cavity. The ovaries produce female sex hormones called **estrogens** and **progesterone.** Along with the gonadotropic hormones of the pituitary gland, the sex hormones regulate the female reproductive cycle, maintain pregnancy, and prepare the mammary glands for lactation. These hormones are

TABLE 23.8	SUMMARY OF HORMONES OF THE OVARIES AND TESTES		
HORMONE	**PRINCIPAL ACTIONS**	**HORMONE**	**PRINCIPAL ACTIONS**
Ovarian hormones		**Testicular hormones**	
Estrogens and **progesterone** Ovaries	Together with gonadotropic hormones of the anterior pituitary, regulate the female reproductive cycle, regulate oogenesis, maintain pregnancy, prepare the mammary glands for lactation, and promote development and maintenance of female secondary sexual characteristics.	**Testosterone** Testes	Stimulates descent of testes before birth, regulates spermatogenesis, and promotes development and maintenance of male secondary sexual characteristics.
Relaxin	Increases flexibility of pubic symphysis during pregnancy and helps dilate uterine cervix during labor and delivery.	**Inhibin**	Inhibits secretion of FSH from anterior pituitary.
Inhibin	Inhibits secretion of FSH from anterior pituitary.		

also responsible for the development and maintenance of female secondary sexual characteristics. The ovaries also produce **inhibin,** a hormone that inhibits secretion of follicle-stimulating hormone (FSH). During pregnancy, the ovaries and placenta produce a hormone called **relaxin,** which increases the flexibility of the pubic symphysis during pregnancy and helps dilate the uterine cervix during labor and delivery. These actions help ease the baby's passage by enlarging the birth canal.

The male has two oval gonads, called **testes,** that produce **testosterone,** the primary androgen. Testosterone regulates production of sperm and stimulates the development and maintenance of male secondary sexual characteristics such as beard growth. The testes also produce inhibin, which inhibits secretion of FSH. The specific roles of gonadotropic hormones and sex hormones are discussed in Chapter 27.

Table 23.8 summarizes the hormones produced by the ovaries and testes and their principal actions.

CHECKPOINT

10. Explain why the ovaries and testes are considered endocrine glands.

PINEAL GLAND

OBJECTIVE

• Describe the location, histology, hormones, and functions of the pineal gland.

The **pineal gland** (PĪN-ē-al;=pinecone shape) is a small endocrine gland attached to the roof of the third ventricle of the brain at the midline (see Figure 23.1). It is part of the epithalamus, positioned between the two superior colliculi, and it weighs 0.1–0.2 g. The gland, which is covered by a capsule formed by the pia mater, consists of masses of neuroglia and secretory cells called **pinealocytes** (pin-ē-AL-ō-sīts). Sympathetic postganglionic fibers from the superior cervical ganglion terminate in the pineal gland.

Although many anatomical features of the pineal gland have been known for years, its physiological role is still unclear. One

hormone secreted by the pineal gland is **melatonin.** Melatonin contributes to the setting of the body's biological clock, which is controlled from the suprachiasmatic nucleus of the hypothalamus. During sleep, levels of melatonin in the bloodstream increase tenfold and then decline to a low level again before awakening. Small doses of melatonin given orally can induce sleep and reset daily rhythms, which might benefit workers whose shifts alternate between daylight and nighttime hours. Melatonin also is a potent antioxidant that may provide some protection against damaging oxygen free radicals. In animals with specific breeding seasons, melatonin inhibits reproductive functions outside the breeding season. What effect, if any, melatonin exerts on human reproductive function is still unclear.

The posterior cerebral artery supplies the pineal gland with blood, and the great cerebral vein drains it.

SEASONAL AFFECTIVE DISORDER AND JET LAG

Seasonal affective disorder (SAD) is a type of depression that afflicts some people during the winter months, when the length of days is short. It is thought to be due, in part, to overproduction of melatonin. Bright light therapy—repeated doses of several hours of artificial light—provides relief for some people. Three to six hours of exposure to bright light also appears to speed recovery from **jet lag,** the fatigue suffered by travelers who cross several time zones. ■

CHECKPOINT

11. What is the relationship between melatonin and sleep?

THYMUS

OBJECTIVE

• Describe the role of the thymus in immunity.

Because of its role in immunity, the details of the structure and functions of the **thymus** are discussed in Chapter 16, which examines the lymphatic system and immunity. In this chapter, only its hormonal role in immunity will be discussed.

As you learned in Chapter 13, lymphocytes are one type of white blood cell. There are two types of lymphocytes, called T cells and B cells, based on their specific roles in immunity. Hormones produced by the thymus, called **thymosin, thymic humoral factor (THF), thymic factor (TF),** and **thymopoietin,** promote the proliferation and maturation of T cells, which destroy foreign substances and microbes. There is also some evidence that thymic hormones may retard the aging process.

CHECKPOINT

12. Which thymic hormones assume a role in immunity?

OTHER ENDOCRINE TISSUES

OBJECTIVE

● List the hormones secreted by cells in other tissues and organs, and describe their functions.

As you learned at the beginning of this chapter, cells in organs other than those usually classified as endocrine glands have an endocrine function and secrete hormones. You learned about

several of these in this chapter: the hypothalamus, thymus, pancreas, ovaries, and testes. Other tissues with endocrine functions are summarized in Table 23.9.

CHECKPOINT

13. List the hormones secreted by the gastrointestinal tract, placenta, kidneys, skin, adipose tissue, and heart, and indicate their functions.

DEVELOPMENT OF THE ENDOCRINE SYSTEM

OBJECTIVE

● Describe the development of endocrine glands.

The development of the endocrine system is not as localized as the development of other systems because endocrine organs develop in widely separated parts of the embryo.

About three weeks after fertilization, the *pituitary gland (hypophysis)* begins to develop from two different regions of the **ectoderm.** The *posterior pituitary (neurohypophysis)* is derived from an outgrowth of ectoderm called the **neurohypophyseal bud,** located on the floor of the hypothalamus (Figure 23.8).

TABLE 23.9	SUMMARY OF HORMONES PRODUCED BY OTHER ORGANS AND TISSUES THAT CONTAIN ENDOCRINE CELLS
HORMONE	**PRINCIPAL ACTIONS**
Gastrointestinal tract	
Gastrin	Promotes secretion of gastric juice and increases movements of the stomach.
Glucose-dependent insulinotropic peptide (GIP)	Stimulates release of insulin by pancreatic beta cells.
Secretin	Stimulates secretion of pancreatic juice and bile.
Cholecystokinin (CCK)	Stimulates secretion of pancreatic juice, regulates release of bile from the gallbladder, and brings about a feeling of fullness after eating.
Placenta	
Human chorionic gonadotropin (hCG)	Stimulates the corpus luteum in the ovary to continue the production of estrogens and progesterone to maintain pregnancy.
Estrogens and **progesterone**	Maintain pregnancy and help prepare mammary glands to secrete milk.
Human chorionic somatomammotropin (hCS)	Stimulates the development of the mammary glands for lactation.
Kidneys	
Renin	Part of a sequence of reactions that raises blood pressure by bringing about vasoconstriction and secretion of aldosterone.
Erythropoietin (EPO)	Increases rate of red blood cell formation.
Calcitriol* (active form of vitamin D)	Aids in the absorption of dietary calcium and phosphorus.
Heart	
Atrial natriuretic peptide (ANP)	Decreases blood pressure.
Adipose tissue	
Leptin	Suppresses appetite and may increase the activity of FSH and LH.

*Synthesis begins in the skin, continues in the liver, and ends in the kidneys.

Figure 23.8 Development of the endocrine system.

Glands of the endocrine system develop from all three primary germ layers.

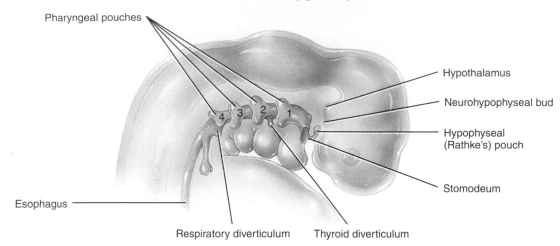

(a) Location of the neurohypophyseal bud, hypophyseal (Rathke's) pouch, thyroid diverticulum and pharyngeal pouches in a 28-day embryo

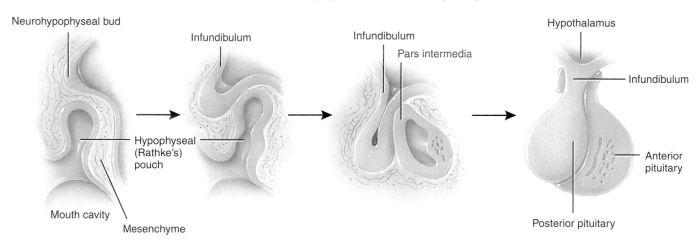

(b) Development of the pituitary gland between five and sixteen weeks

Which two endocrine glands develop from tissues with two different embryological origins?

The *infundibulum*, also an outgrowth of the neurohypophyseal bud, connects the posterior pituitary to the hypothalamus. The *anterior pituitary (adenohypophysis)* is derived from an outgrowth of ectoderm from the roof of the mouth called the **hypophyseal (Rathke's) pouch.** The pouch grows toward the neurohypophyseal bud and eventually loses its connection with the roof of the mouth.

The *thyroid gland* develops during the fourth week as a midventral outgrowth of **endoderm,** called the **thyroid diverticulum,** from the floor of the pharynx at the level of the second pair of pharyngeal pouches. The outgrowth projects inferiorly and differentiates into the right and left lateral lobes and the isthmus of the gland.

The *parathyroid glands* develop during the fourth week from **endoderm** as outgrowths from the third and fourth **pharyngeal pouches,** which help to form structures of the head and neck.

The adrenal cortex and adrenal medulla develop during the fifth week and have completely different embryological origins. The *adrenal cortex* is derived from intermediate **mesoderm** from the same region that produces the gonads. Endocrine tissues that secrete steroid hormones all are derived from mesoderm. The *adrenal medulla* is derived from **ectoderm** from **neural crest** cells that migrate to the superior portion of the kidney. Recall that neural crest cells also give rise to sympathetic ganglia and other structures of the nervous system (see Figure 19.26 on page 624).

The *pancreas* develops during the fifth through seventh weeks from two outgrowths of **endoderm** from the part of the **foregut** that later becomes the duodenum (see Figure 4.12c on page 110). The two outgrowths eventually fuse to form the pancreas. The origin of the ovaries and testes is discussed in the section on the reproductive system.

The *pineal gland* arises during the seventh week as an outgrowth between the thalamus and colliculi of the midbrain from **ectoderm** associated with the **diencephalon** (see Figure 19.27 on page 625).

The *thymus* arises during the fifth week from **endoderm** of the third pharyngeal pouches.

14. Compare the origins of the adrenal cortex and adrenal medulla.

AGING AND THE ENDOCRINE SYSTEM

OBJECTIVE

● Describe the effects of aging on the endocrine system.

Although some endocrine glands shrink as we get older, their performance may or may not be compromised. Production of human growth hormone by the anterior pituitary decreases, which is one cause of muscle atrophy as aging proceeds. The thyroid gland often decreases its output of thyroid hormones with age, causing a decrease in metabolic rate, an increase in body fat, and hypothyroidism, which is seen more often in older people. Due to less negative feedback (lower levels of thyroid hormones), thyroid-stimulating hormone level increases with age.

With aging, the blood level of PTH rises, perhaps due to inadequate dietary intake of calcium. In a study of older women who took 2,400 mg/day of supplemental calcium, blood levels of PTH were lower, as in younger women. Both calcitriol and calcitonin levels are lower in older persons. Together, the rise in PTH and the fall in calcitonin level heighten the age-related decrease in bone mass that leads to osteoporosis and increased risk of fractures.

The adrenal glands contain increasingly more fibrous tissue and thus produce less cortisol and aldosterone with advancing age. However, production of epinephrine and norepinephrine remains normal. The pancreas releases insulin more slowly with age, and receptor sensitivity to glucose declines. As a result, blood glucose levels in older people increase faster and return to normal more slowly than in younger individuals.

The thymus is largest in infancy. After puberty, its size begins to decrease, and thymic tissue is replaced by adipose and areolar connective tissue. In older adults, the thymus has atrophied significantly. However, it still produces new T cells for immune responses.

The ovaries decrease in size with age, and they no longer respond to gonadotropins. The resultant decreased output of estrogens leads to conditions such as osteoporosis, high blood cholesterol, and atherosclerosis. FSH and LH levels are high due to less negative feedback inhibition of estrogens. Although testosterone production by the testes decreases with age, the effects are not usually apparent until very old age, and many elderly males can still produce active sperm in normal numbers.

15. Which hormone is related to muscle atrophy associated with aging?

APPLICATIONS TO HEALTH

Disorders of the endocrine system often involve either **hyposecretion** (*hypo-*=too little or under), inadequate release of a hormone, or **hypersecretion** (*hyper-*=too much or above), excessive release of a hormone. In other cases, the problem is faulty hormone receptors, or an inadequate number of receptors. Yet another cause of endocrine system disorders is the "death," or complete shut-down, of the gland. Because hormones are distributed in the blood to target tissues throughout the body, problems associated with endocrine dysfunction may also be widespread.

PITUITARY GLAND DISORDERS

Pituitary Dwarfism, Giantism, and Acromegaly

Several disorders of the anterior pituitary involve human growth hormone (hGH). Hyposecretion of hGH during the growth years slows bone growth, and the epiphyseal plates close before normal height is reached. This condition is called **pituitary dwarfism.** Other organs of the body also fail to grow, and the body proportions are childlike. Treatment requires administration of hGH during childhood, before the epiphyseal plates close.

Hypersecretion of hGH during childhood causes **giantism,** an abnormal increase in the length of long bones. The person grows to be very tall, but body proportions are about normal. Figure 23.9a shows identical twins; one brother developed giantism due to a pituitary tumor. Hypersecretion of hGH during adulthood is referred to as **acromegaly** (ak′-rō-MEG-a-lē). Although hGH cannot produce further lengthening of the long bones because the epiphyseal plates are already closed, the bones of the hands, feet, cheeks, and jaws thicken and other tissues enlarge. In addition, the eyelids, lips, tongue, and nose enlarge, and the skin thickens and develops furrows, especially on the forehead and soles (Figure 23.9b).

DIABETES INSIPIDUS

The most common abnormality associated with dysfunction of the posterior pituitary is **diabetes insipidus** (dī-a-BĒ-tēs in-SIP-i-dus; *diabetes*=overflow; *insipidus*=tasteless) or **DI.** This disorder is due to defects in antidiuretic hormone (ADH) receptors or an inability to secrete ADH. Blood glucose levels are normal. *Neurogenic diabetes insipidus* results from hyposecretion of ADH, usually caused by a brain tumor, head trauma, or brain

Figure 23.9   Various endocrine disorders.

 Disorders of the endocrine system often involve hyposecretion or hypersecretion of hormones.

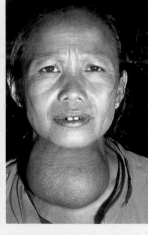

(b) Acromegaly (excess hGH during adulthood)

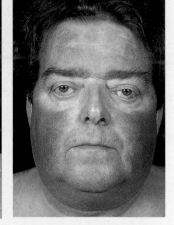

(c) Exophthalmos (excess thyroid hormones, as in Graves' disease)

(a) A 22-year old man with pituitary giantism shown beside his identical twin

(d) Goiter (enlargement of thyroid gland)

(e) Cushing's syndrome (excess glucocorticoids)

Which of the disorders is due to antibodies that mimic the action of TSH?

surgery that damages the posterior pituitary or the hypothalamus. In *nephrogenic diabetes insipidus*, the kidneys do not respond to ADH. The ADH receptors may be nonfunctional, or the kidneys may be damaged. A common symptom of both forms of DI is excretion of large volumes of urine, with resulting dehydration and thirst. Bed-wetting is common in afflicted children. Because so much water is lost in the urine, a person with DI may die of dehydration if deprived of water for only a day or so.

Treatment of neurogenic diabetes insipidus involves hormone replacement, usually for life. Either subcutaneous injection or nasal spray application of ADH analogs is effective. Treatment of nephrogenic diabetes insipidus is more complex and depends on the nature of the kidney dysfunction. Restriction of salt in the diet and, paradoxically, the use of certain diuretic drugs, are helpful.

THYROID GLAND DISORDERS

Thyroid gland disorders affect all major body systems and are among the most common endocrine disorders. **Congenital hypothyroidism,** hyposecretion of thyroid hormones that is present at birth, has devastating consequences if not treated promptly. Previously termed *cretinism,* this condition causes severe mental retardation. At birth, the baby typically is normal because lipid-soluble maternal thyroid hormones crossed the placenta during pregnancy and allowed normal development. Most states require testing of all newborns to ensure adequate thyroid function. If congenital hypothyroidism exists, oral thyroid hormone treatment must be started soon after birth and continued for life.

Hypothyroidism during the adult years produces **myxedema** (mix-e-DĒ-ma), which occurs about five times more often in females than in males. A hallmark of this disorder is edema (accumulation of interstitial fluid) that causes the facial tissues to swell and look puffy. A person with myxedema has a slow heart rate, low body temperature, sensitivity to cold, dry hair and skin, muscular weakness, general lethargy, and a tendency to gain weight easily. Because the brain has already reached maturity, mental retardation does not occur, but the person may be less alert. Oral thyroid hormones reduce the symptoms.

The most common form of hyperthyroidism is **Graves' disease,** which also occurs seven to ten times more often in females than in males, usually before age 40. Graves' disease is an autoimmune disorder in which the person produces antibodies that mimic the action of thyroid-stimulating hormone (TSH). The antibodies continually stimulate the thyroid gland to grow and produce thyroid hormones. A primary sign is an enlarged thyroid, which may be two to three times its normal size. Graves patients often have a peculiar edema behind the eyes, called **exophthalmos** (ek'-sof-THAL-mos), which causes the eyes to protrude (Figure 23.9c). Treatment may include surgically removing part or all of the thyroid gland (thyroidectomy), using radioactive iodine (^{131}I) to selectively destroy thyroid tissue, and/or using antithyroid drugs to block synthesis of thyroid hormones.

A **goiter** (GOY-ter; *guttur-*=throat) is simply an enlarged thyroid gland. It may be associated with hyperthyroidism, hypothyroidism, or **euthyroidism** (*eu-*=good), which means normal secretion of thyroid hormone. In some places in the world, dietary iodine intake is inadequate; the resultant low level of thyroid hormone in the blood stimulates secretion of TSH, which causes thyroid gland enlargement (Figure 23.9d).

PARATHYROID GLAND DISORDERS

Hypoparathyroidism—too little parathyroid hormone—leads to a deficiency of Ca^{2+}, which causes neurons and muscle fibers to depolarize and produce action potentials spontaneously. This leads to twitches, spasms, and **tetany** (maintained contraction) of skeletal muscle. The leading cause of hypoparathyroidism is accidental damage to the parathyroid glands or to their blood supply during thyroidectomy surgery.

Hyperparathyroidism, an elevated level of parathyroid hormone, most often is due to a tumor of one of the parathyroid glands. An elevated level of PTH causes excessive resorption of bone matrix, raising the blood levels of calcium and phosphate ions and causing bones to become soft and easily fractured. High blood calcium level promotes formation of kidney stones. Fatigue, personality changes, and lethargy are also seen in patients with hyperparathyroidism.

ADRENAL GLAND DISORDERS

Cushing's Syndrome

Hypersecretion of cortisol by the adrenal cortex produces **Cushing's syndrome** (Figure 23.9e). Causes include a tumor of the adrenal gland that secretes cortisol, or a tumor elsewhere that secretes adrenocorticotropic hormone (ACTH), which in turn stimulates excessive secretion of cortisol. The condition is characterized by breakdown of muscle proteins and redistribu-

tion of body fat, resulting in spindly arms and legs accompanied by a rounded "moon face," "buffalo hump" on the back, and pendulous (hanging) abdomen. Facial skin is flushed, and the skin covering the abdomen develops stretch marks. The person also bruises easily, and wound healing is poor. The elevated level of cortisol causes hyperglycemia, osteoporosis, weakness, hypertension, increased susceptibility to infection, decreased resistance to stress, and mood swings. People who need long-term glucocorticoid therapy—for instance, to prevent rejection of a transplanted organ—may develop a cushinoid appearance.

ADDISON'S DISEASE

Undersecretion of glucocorticoids and aldosterone causes **Addison's disease (chronic adrenocortical insufficiency).** The majority of cases are autoimmune disorders in which antibodies cause adrenal cortex destruction or block binding of ACTH to its receptors. Pathogens, such as the bacterium that causes tuberculosis, also may trigger adrenal cortex destruction. Symptoms, which typically do not appear until 90% of the adrenal cortex has been destroyed, include mental lethargy, anorexia, nausea and vomiting, weight loss, hypoglycemia, and muscular weakness. Loss of aldosterone leads to elevated potassium and decreased sodium in the blood, low blood pressure, dehydration, decreased cardiac output, arrhythmias, and even cardiac arrest. The skin may have a "bronzed" appearance that often is mistaken for a suntan. Such was true in the case of President John F. Kennedy, whose Addison's disease was known to only a few while he was alive. Treatment consists of replacing glucocorticoids and mineralocorticoids and increasing sodium in the diet.

Pheochromocytomas

Usually benign, tumors of the chromaffin cells of the adrenal medulla, called **pheochromocytomas** (fē-ō-krō'-mō-sī-TŌ-mas; *pheo-*=dusky; *chromo-*=color; *cyto-*=cell), cause hypersecretion of epinephrine and norepinephrine. The result is a prolonged version of the fight-or-flight response: rapid heart rate, high blood pressure, high levels of glucose in blood and urine, an elevated basal metabolic rate (BMR), flushed face, nervousness, sweating, and decreased gastrointestinal motility. Treatment is surgical removal of the tumors.

PANCREATIC ISLET DISORDERS

The most common endocrine disorder is **diabetes mellitus** (MEL-i-tus; *melli-*=honey sweetened), caused by an inability to produce or use insulin. Diabetes mellitus is the fourth leading cause of death by disease in the United States, primarily because of its damage to the cardiovascular system. Because insulin is unavailable to aid transport of glucose into body cells, blood glucose level is high and glucose "spills" into the urine (glucosuria). Hallmarks of diabetes mellitus are the three "polys": *polyuria,* excessive urine production due to an inability of the kidneys to reabsorb water; *polydipsia,* excessive thirst; and *polyphagia,* excessive eating.

Both genetic and environmental factors contribute to onset of the two types of diabetes mellitus—type 1 and type 2—but the exact mechanisms are still unknown. In **type 1 diabetes** insulin level is low because the person's immune system destroys the pancreatic beta cells. It is also called **insulin-dependent diabetes mellitus (IDDM)** because insulin injections are required to prevent death. Most commonly, IDDM develops in people younger than age 20, though it persists throughout life. By the time symptoms of IDDM arise, 80–90% of the islet beta cells have been destroyed. IDDM is most common in northern Europe, especially in Finland where nearly 1% of the population develops IDDM by 15 years of age. In the United States, IDDM is 1.5–2.0 times more common in whites than in African American or Asian populations.

The cellular metabolism of an untreated type 1 diabetic is similar to that of a starving person. Because insulin is not present to aid the entry of glucose into body cells, most cells use fatty acids to produce ATP. Stores of triglycerides in adipose tissue are catabolized to yield fatty acids and glycerol. The byproducts of fatty acids breakdown—organic acids called ketones or ketone bodies—accumulate. Buildup of ketones causes blood pH to fall, a condition known as **ketoacidosis.** Unless treated quickly, ketoacidosis can cause death.

The breakdown of stored triglycerides also causes weight loss. As lipids are transported by the blood from storage depots to cells, lipid particles are deposited on the walls of blood vessels, leading to atherosclerosis and a multitude of cardiovascular problems, including cerebrovascular insufficiency, ischemic heart disease, peripheral vascular disease, and gangrene. A major complication of diabetes is loss of vision due either to cataracts (excessive glucose attaches to lens proteins, causing cloudiness) or damage to blood vessels of the retina. Severe kidney problems also may result from damage to renal blood vessels.

Type 1 diabetes is treated through self-monitoring of blood glucose level (up to 7 times daily), regular meals containing 45–50% carbohydrates and less than 30% fats, exercise, and periodic insulin injections (up to 3 times a day). Several implantable pumps are available to provide insulin without the need for repeated injections. Because they lack a reliable glucose sensor, however, the person must self-monitor blood glucose level to determine insulin doses. It is also possible to successfully transplant a pancreas, but immunosuppressive drugs must then be taken for life. Most pancreatic transplants are done in people who also need a kidney transplant because of renal failure. Another promising approach under investigation is transplantation of isolated islets in semipermeable hollow tubes. The tubes allow glucose and insulin to enter and leave but prevent entry of immune system cells that might attack the islet cells.

Type 2 diabetes, also called **non-insulin-dependent diabetes mellitus (NIDDM),** is much more common than type 1, representing more than 90% of all cases. Type 2 diabetes most often occurs in obese people who are over age 35. Clinical symptoms are mild, and the high glucose levels in the blood often can be controlled by diet, exercise, and weight loss. Sometimes, a drug such as *glyburide* (DiaBeta) is used to stimulate secretion of insulin by pancreatic beta cells. Although some type 2 diabetics need insulin, many have a sufficient amount (or even a surplus) of insulin in the blood. For these people, diabetes arises not from a shortage of insulin but because target cells become less sensitive to it due to down-regulation of insulin receptors.

Hyperinsulinism most often results when a diabetic injects too much insulin. The main symptom is **hypoglycemia,** decreased blood glucose level, which occurs because the excess insulin stimulates too much uptake of glucose by body cells. The resulting hypoglycemia stimulates the secretion of epinephrine, glucagon, and human growth hormone. As a consequence, anxiety, sweating, tremor, increased heart rate, hunger, and weakness occur. When blood glucose falls, brain cells are deprived of the steady supply of glucose they need to function effectively. Severe hypoglycemia leads to mental disorientation, convulsions, unconsciousness, and shock. Shock due to an insulin overdose is termed **insulin shock.** Death can occur quickly unless blood glucose level is raised. From a clinical standpoint, a diabetic suffering from either a hyperglycemia or a hypoglycemia crisis can have very similar symptoms—mental changes, coma, seizures, and so on. It is important to quickly and correctly identify the cause of the underlying symptoms and treat them appropriately.

KEY MEDICAL TERMS ASSOCIATED WITH THE ENDOCRINE SYSTEM

Gynecomastia (gī-ne′-kō-MAS-tē-a; *gyneco-*=woman; *mast-*= breast) Excessive development of mammary glands in a male. Sometimes a tumor of the adrenal gland may secrete sufficient amounts of estrogens to cause the condition.

Hirsutism (HER-soo-tizm; *hirsut-*=shaggy) Presence of excessive bodily and facial hair in a male pattern, especially in women; may be caused by excess androgen production due to tumors or drugs.

Thyroid crisis (storm) A severe state of hyperthyroidism that can be life-threatening. It is characterized by high body temperature, rapid heart rate, high blood pressure, gastrointestinal symptoms (abdominal pain, vomiting, diarrhea), agitation, tremors, confusion, seizures, and possibly coma.

Virilism (VIR-il-izm; *virilis*=masculine) The presence of mature masculine characteristics in females or prepubescent males. In a female, virile characteristics include growth of a beard, development of a much deeper voice, occasionally the development of baldness, development of a masculine distribution of hair on the body and on the pubis, growth of the clitoris such that it may resemble a penis, atrophy of the breasts, infrequent or absent menstruation, and increased muscularity that produces a male-like physique. In prepubescent males the syndrome causes the same characteristics as in females, plus rapid development of the male sexual organs and emergence of sexual desires.

Virilizing adenoma (*aden*=gland; *oma*=tumor) Tumor of the adrenal gland that liberates excessive androgens, causing virilism (masculinization) in females. Occasionally, adrenal tumor cells liberate estrogens to the extent that a male patient develops gynecomastia. Such a tumor is called a **feminizing adenoma.**

STUDY OUTLINE

Introduction (p. 704)

1. The nervous system controls homeostasis through nerve impulses; the endocrine system uses hormones.
2. The nervous system causes muscles to contract and glands to secrete; the endocrine system affects virtually all body tissues.
3. Table 23.1 on page 705 compares the characteristics of the nervous and endocrine systems.

Endocrine Glands Defined (p. 705)

1. Exocrine (sudoriferous, sebaceous, and digestive) glands secrete their products through ducts into body cavities or onto body surfaces.
2. Endocrine glands secrete hormones into the blood.
3. The endocrine system consists of endocrine glands and several organs that contain endocrine tissue (see Figure 23.1 on page 706).
4. Hormones regulate the internal environment, metabolism, and energy balance.
5. They also help regulate muscular contraction, glandular secretion, and certain immune responses.
6. Hormones affect growth, development, and reproduction.

Hormones (p. 705)

1. The amount of hormone released is determined by the body's need for the hormone.
2. Cells that respond to the effects of hormones are called target cells.
3. The combination of hormone and receptor activates a chain of events in a target cell that produce the physiological effects of the hormone.

Hypothalamus and Pituitary Gland (p. 707)

1. The hypothalamus is the major integrating link between the nervous and endocrine systems.
2. The hypothalamus and pituitary gland regulate virtually all aspects of growth, development, and metabolism, and they also affect other body activities.
3. The pituitary gland is located in the hypophyseal fossa and is divided into the anterior pituitary (glandular portion), the posterior pituitary (nervous portion), and pars intermedia (avascular zone in between).
4. Hormones of the anterior pituitary are controlled by releasing or inhibiting hormones produced by the hypothalamus.
5. The blood supply to the anterior pituitary is from the superior hypophyseal arteries. It carries releasing and inhibiting hormones from the hypothalamus.

6. Histologically, the anterior pituitary consists of somatotrophs that produce human growth hormone (hGH); lactotrophs that produce prolactin (PRL); corticotrophs that secrete adrenocorticotropic hormone (ACTH) and melanocyte-stimulating hormone (MSH); thyrotrophs that secrete thyroid-stimulating hormone (TSH); and gonadotrophs that synthesize follicle-stimulating hormone (FSH) and luteinizing hormone (LH).
7. hGH stimulates body growth. TSH regulates thyroid gland activities. FSH and LH both regulate the activities of the ovaries and testes. PRL helps initiate milk secretion. MSH increases skin pigmentation. ACTH regulates the activities of the adrenal cortex.
8. The neural connection between the hypothalamus and posterior pituitary is via the supraopticohypophyseal tract.
9. Hormones made by the hypothalamus and stored in the posterior pituitary are oxytocin (OT), which stimulates contraction of the uterus and ejection of milk, and antidiuretic hormone (ADH), which stimulates water reabsorption by the kidneys and arteriole constriction.
10. The hormones of the anterior pituitary are summarized in Table 23.2 on page 709, and the posterior pituitary hormones are summarized in Table 23.3 on page 711.

Thyroid Gland (p. 711)

1. The thyroid gland is located inferior to the larynx.
2. Histologically, the thyroid gland consists of thyroid follicles composed of follicular cells, which secrete the thyroid hormones thyroxine (T4) and triiodothyronine (T3), and parafollicular cells, which secrete calcitonin (CT).
3. Thyroid hormones regulate the rate of metabolism, growth and development, and the reactivity of the nervous system.
4. Calcitonin (CT) lowers the blood level of calcium.
5. A summary of thyroid gland hormones and their actions is presented in Table 23.4 on page 713.

Parathyroid Glands (p. 713)

1. The parathyroid glands are embedded on the posterior surfaces of the lateral lobes of the thyroid gland.
2. The parathyroids consist of chief and oxyphil cells.
3. Parathyroid hormone (PTH) increases blood calcium level and decreases blood phosphate level.
4. A summary of parathyroid hormone and its actions is presented in Table 23.5 on page 715.

Adrenal Glands (p. 715)

1. The adrenal glands are located superior to the kidneys. They consist of an outer adrenal cortex and an inner adrenal medulla.

2. Histologically, the adrenal cortex is divided into a zona glomerulosa, zona fasciculata, and zona reticularis; the adrenal medulla consists of chromaffin cells and large blood vessels.

3. Cortical secretions include mineralocorticoids, glucocorticoids, and gonadocorticoids.

4. Medullary secretions are epinephrine and norepinephrine (NE), which produce effects similar to sympathetic responses. They are released during stress.

5. A summary of adrenal gland hormones and their actions is presented in Table 23.6 on page 717.

Pancreas (p. 717)

1. The pancreas is posterior and slightly inferior to the stomach.

2. Histologically, the pancreas consists of pancreatic islets, or islets of Langerhans (endocrine cells), and clusters of enzyme-producing cells (acini) (exocrine cells). The four types of cells in the endocrine portion are alpha, beta, delta, and F cells.

3. Alpha cells secrete glucagon, beta cells secrete insulin, delta cells secrete somatostatin, and F cells secrete pancreatic polypeptide.

4. Glucagon increases blood sugar level.

5. Insulin decreases blood sugar level.

6. A summary of pancreatic hormones and their actions is presented in Table 23.7 on page 719.

Ovaries and Testes (p. 719)

1. The ovaries are located in the pelvic cavity and produce sex hormones that function in the development and maintenance of female secondary sexual characteristics, the reproductive cycle, pregnancy, lactation, and normal reproductive functions.

2. The testes lie inside the scrotum and produce sex hormones that function in the development and maintenance of male secondary sexual characteristics and normal reproductive functions.

3. Table 23.8 on page 720 summarizes the hormones produced by the ovaries and testes and their principal actions.

Pineal Gland (p. 720)

1. The pineal gland is attached to the roof of the third ventricle.

2. Histologically, it consists of secretory cells called pinealocytes, neuroglial cells, and scattered postganglionic sympathetic fibers.

3. The pineal gland secrets melatonin, which contributes to setting the body's biological clock (which is controlled by the suprachiasmatic nucleus). During sleep, levels of melatonin in the bloodstream increase tenfold and then decline to a low level again before awakening.

Thymus (p. 720)

1. The thymus secretes several hormones related to immunity.

2. Thymosin, thymic humoral factor (THF), thymic factor (TF), and thymopoietin promote the maturation of T cells.

Other Endocrine Tissues (p. 721)

1. The gastrointestinal tract synthesizes several hormones, including gastrin, gastric inhibitory peptide (GIP), secretin, and cholecystokinin (CCK).

2. The placenta produces human chorionic gonadotropin (hCG), estrogens, progesterone, and human chorionic somatomammotropin (hCS).

3. The kidneys release erythropoietin.

4. The skin begins the synthesis of vitamin D.

5. The atria of the heart produce atrial natriuretic peptide (ANP).

6. Adipose tissue produces leptin.

7. A summary of hormones secreted by other endocrine tissues is included in Table 23.9 on page 721.

Development of the Endocrine System (p. 721)

1. The development of the endocrine system is not as localized as in other systems because endocrine organs develop in widely separated parts of the embryo.

2. The pituitary gland, adrenal medulla, and pineal gland develop from ectoderm; the adrenal cortex develops from mesoderm; and the thyroid gland, parathyroid glands, pancreas, and thymus develop from endoderm.

Aging and the Endocrine System (p. 723)

1. Although some endocrine glands shrink as we get older, their performance may or may not be compromised.

2. Production of human growth hormone, thyroid hormones, cortisol, aldosterone, and estrogens decrease with advancing age.

3. With aging, the blood levels of TSH, LH, FSH, and PTH rise.

4. The pancreas releases insulin more slowly with age, and receptor sensitivity to glucose declines.

5. After puberty, thymus size begins to decrease, and thymic tissue is replaced by adipose and areolar connective tissue.

Q SELF-QUIZ QUESTIONS

Choose the one best answer to the following questions:

1. A generalized anti-inflammatory effect is most closely associated with:
   a. glucocorticoids. b. mineralocorticoids.
   c. parathyroid hormone (PTH). d. insulin. e. melatonin.

2. A chemical produced by the hypothalamus that causes the anterior pituitary to secrete a hormone is called a:
   a. gonadotropic hormone. b. tropic hormone.
   c. releasing hormone. d. target hormone.
   e. neurotransmitter.

3. Which of the following is *not* released by the adenohypophysis?
   a. ADH b. hGH c. TSH d. FSH e. ACTH

4. The thyroid gland is located:
   a. anterior to the sternum.
   b. in the abdominal cavity, inferior to the liver.
   c. in the hypophyseal fossa of the sphenoid bone.
   d. inferior to the larynx with lobes on either side of the trachea.
   e. in the roof of the third ventricle of the brain.

5. The primary mineralocorticoid is:
   a. ACTH. b. antidiuretic hormone. c. cortisol.
   d. aldosterone. e. epinephrine.

6. Which one of these glands is called the "emergency gland" and helps the body meet sudden stress?

   a. pituitary gland   b. pancreas   c. thyroid gland
   d. thymus   e. adrenal (suprarenal) glands

Complete the following:

7. Place numbers in the blanks to indicate the correct order of the vessels that supply blood to the anterior pituitary. (a) hypophyseal portal veins: _____, (b) primary plexus: _____, (c) superior hypophyseal arteries: _____, (d) secondary plexus: _____

8. Cell bodies of neurons whose axons make up the posterior pituitary are located in the _____.

9. Thyroid stimulating hormone (TSH) controls the secretion of hormones by follicular cells of the thyroid gland. Therefore TSH is a _____ hormone and the follicular cells are the _____ cells.

10. Glucagon, produced by _____ cells of the pancreatic (islets of Langerhans) causes blood sugar level to _____.

11. Insulin-like growth factors are secreted by tissues in response to _____, which is produced by the anterior pituitary.

12. The superior thyroid artery is a branch of the _____ artery.

13. Increased production of calcitriol, the active form of vitamin D, by the kidney is an effect of _____.

Are the following statements true or false?

14. Increasing the number of receptors on a target cell decreases the cell's sensitivity to a hormone.

15. The secretory portion of the posterior pituitary is the axons of neurosecretory cells; the secretory portion of the anterior pituitary is glandular epithelium.

16. Pituicytes are the hormone-secreting cells of the posterior pituitary.

17. Oxytocin stimulates milk production.

18. Chromaffin cells are the principal secreting cells of the pineal gland.

19. The outer region of the adrenal (suprarenal) gland is called the adrenal cortex.

Matching

20. Match the following glands with the hormones they produce:
    ___ **(a)** adrenal cortex   **(1)** ACTH
    ___ **(b)** adrenal medulla   **(2)** inhibin
    ___ **(c)** anterior pituitary   **(3)** epinephrine
    ___ **(d)** pineal gland   **(4)** glucagon
    ___ **(e)** posterior pituitary   **(5)** melatonin
    ___ **(f)** testes   **(6)** cortisol
    ___ **(g)** pancreas   **(7)** oxytocin

21. Match the following hormone-secreting cells to the hormones they secrete:
    ___ **(a)** insulin   **(1)** corticotrophs
    ___ **(b)** glucagon   **(2)** somatotrophs
    ___ **(c)** calcitonin   **(3)** thyrotrophs
    ___ **(d)** TSH   **(4)** gonadotrophs
    ___ **(e)** hGH   **(5)** beta cells of pancreatic islets
    ___ **(f)** testosterone   (islets of Langerhans)
    ___ **(g)** ACTH   **(6)** alpha cells of pancreatic islets
    ___ **(h)** progesterone   (islets of Langerhans)
    ___ **(i)** thyroxine and   **(7)** follicular cells of the thyroid
        triiodothyronine   gland
    ___ **(j)** FSH and LH   **(8)** parafollicular cells of the
                                     thyroid gland
                                 (9) interstitial cells of the testis
                                (10) corpus luteum of the ovary

CRITICAL THINKING QUESTIONS

1. You've won a trip to beautiful Tropicanaland, a 12-hour time difference from where you live. Your coworkers gave you a bottle of melatonin, a bottle of melanocyte stimulating hormone, and a very bright flashlight as a bon voyage present. You'll be arriving at 8 P.M. Tropicanaland time, which is 8 A.M. your time. How can you adjust to Tropicanaland time most quickly?

2. Amadu, who has just arrived in the United States from Africa, has what appears to be a tumor in his neck. His doctor says it is not a tumor but a goiter. After some blood work, the doctor determines that a diet rich in seafoods and iodized salt should be sufficient to solve the problem. What is a goiter, and how is a change of diet going to help Amadu?

3. Lester's son isn't growing as tall as he had hoped, although he is not abnormally short. "He's not going to be any taller than me," said Lester. "He'll never make it in the NBA and be able to

support me in my old age!" Lester thinks that maybe he should find a doctor who will treat his son with human growth hormone. Besides stimulating growth of tissues, human growth hormone raises blood glucose. Do you think this treatment is a good idea (physiologically speaking) in the long-term? Explain your answer.

4. For several years, military pilots were given radiation treatments in their nasal cavity to reduce sinus problems that interfered with flying. Years later, some of these former pilots began to exhibit problems with their pituitary gland hormones. Can you propose an explanation for this relationship?

5. A patient was found to have markedly elevated blood sugar. Tests showed that his insulin level was actually a bit elevated. How can someone with elevated insulin have elevated blood sugar?

ANSWERS TO FIGURE QUESTIONS

23.1 Secretions of endocrine glands diffuse into interstitial fluid and then into the blood; exocrine secretions flow into ducts that lead into body cavities or to the body surface.

23.2 The hypophyseal portal veins carry blood from the median eminence of the hypothalamus, where hypothalamic releasing and inhibiting hormones are secreted, to the anterior pituitary, where these hormones act.

23.3 Functionally, both the hypothalamohypophyseal tract and the hypophyseal portal veins carry hypothalamic hormones to the pituitary gland. Structurally, the tract is composed of axons of neurons that extend from the hypothalamus to the posterior pituitary, whereas the portal veins are blood vessels that extend to the anterior pituitary.

23.4 Follicular cells secrete T_3 and T_4, also known as thyroid hormones. Parafollicular cells secrete calcitonin.

23.5 Parafollicular cells of the thyroid gland secrete calcitonin; chief cells of the parathyroid gland secrete PTH.

23.6 The adrenal glands are superior to the kidneys.

23.7 The pancreas is both an endocrine and an exocrine gland.

23.8 The pituitary gland and the adrenal glands both include tissues having two different embryological origins.

23.9 In Graves' disease, antibodies are produced that mimic the action of TSH.

THE RESPIRATORY SYSTEM

24

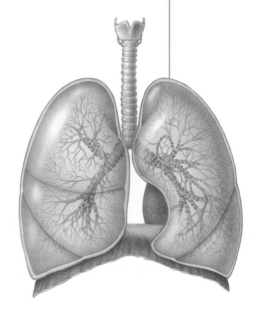

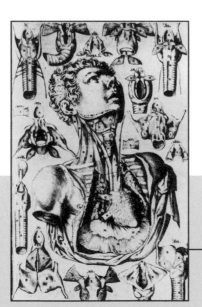

INTRODUCTION Cells continually use oxygen (O_2) for the metabolic reactions that release energy from nutrient molecules and produce ATP. At the same time, these reactions release carbon dioxide (CO_2). Because an excessive amount of CO_2 produces acidity that can be toxic to cells, excess CO_2 must be eliminated quickly and efficiently. The two systems that cooperate to supply O_2 to and eliminate CO_2 from body cells are the cardiovascular and respiratory systems. The respiratory system is responsible for gas exchange—intake of O_2 and elimination of CO_2—whereas the cardiovascular system transports blood containing the gases between the lungs and body cells. Failure of either system disrupts homeostasis by causing rapid death of cells from oxygen starvation and buildup of waste products. In addition to functioning in gas exchange, the respiratory system also participates in regulating blood pH, contains receptors for the sense of smell, filters inhaled air, produces sounds, and rids the body of small amounts of water and heat in exhaled air.

Can you identify the organs depicted in this image?

www.wiley.com/college/apcentral

RESPIRATORY SYSTEM ANATOMY

OBJECTIVES

- Describe the anatomy and histology of the nose, pharynx, larynx, trachea, bronchi, and lungs.
- Identify the functions of each respiratory system structure.

The **respiratory system** consists of the nose, pharynx (throat), larynx (voice box), trachea (windpipe), bronchi, and lungs (Figure 24.1). *Structurally*, the respiratory system consists of two parts: (1) The **upper respiratory system** includes the nose, pharynx, and associated structures. (2) The **lower respiratory system** includes the larynx, trachea, bronchi, and lungs. *Functionally*, the respiratory system also consists of two parts: (1) The **conducting portion** consists of a series of interconnecting cavities and tubes both outside and within the lungs—the nose, pharynx, larynx, trachea, bronchi, bronchioles, and terminal bronchioles—that filter, warm, and moisten air and conduct it into the lungs. (2) The **respiratory portion** consists of tissues within the lungs where gas exchange occurs—the respiratory bronchioles, alveolar ducts, alveolar sacs, and alveoli, the main sites of gas exchange between air and blood. The volume of the conducting portion in an adult is about 150 mL; that of the respiratory portion is 5 to 6 liters.

The branch of medicine that deals with the diagnosis and treatment of diseases of the ears, nose, and throat is called **otorhinolaryngology** (ō'-tō-rī'-nō-lar'-in-GOL-ō-jē; *oto-*=ear; *rhino-*=nose; *laryngo-*=voice box; *-logy*=study of). A **pulmonologist** (pul-mō-NOL-ō-gist; *pulmo-*=lung) is a specialist in the diagnosis and treatment of diseases of the lungs.

Nose

The **nose** can be divided into external and internal portions. The *external nose* consists of a supporting framework of bone and hyaline cartilage covered with muscle and skin and lined by a mucous membrane. The frontal bone, nasal bones, and maxillae form the bony framework of the external nose (Figure 24.2a on page 734). The cartilaginous framework of the external nose consists of the **septal cartilage,** which forms the anterior portion of the nasal septum; the **lateral nasal cartilages** inferior to the nasal bones; and the **alar cartilages,** which form a portion of the walls of the nostrils. Because it consists of pliable hyaline cartilage, the cartilaginous framework of the external nose is somewhat flexible. On the undersurface of the external nose are two openings called the **external nares** (NĀ-rez; singular is *naris*) or **nostrils.** The surface anatomy of the nose is shown in Figure 12.5 on page 386.

The interior structures of the external nose have three functions: (1) warming, moistening, and filtering incoming air; (2) detecting olfactory (smell) stimuli; and (3) modifying speech vibrations as they pass through the large, hollow resonating chambers. *Resonance* refers to prolonging, amplifying, or modifying a sound by vibration.

The *internal nose* is a large cavity in the anterior aspect of the skull that lies inferior to the nasal bone and superior to the mouth; it also includes muscle and mucous membrane. Anteriorly, the internal nose merges with the external nose, and

posteriorly it communicates with the pharynx through two openings called the **internal nares** or **choanae** (kō-Ā-nē) (Figure 24.2b). Ducts from the paranasal sinuses (frontal, sphenoidal, maxillary, and ethmoidal paranasal sinuses) and the nasolacrimal ducts also open into the internal nose (see Figure 7.13 on page 184). The lateral walls of the internal nose are formed by the ethmoid, maxillae, lacrimal, palatine, and inferior nasal conchae bones (see Figure 7.9 on page 178); the ethmoid also forms the roof. The palatine bones and palatine processes of the maxillae, which together constitute the hard palate, form the floor of the internal nose.

The space within the internal nose is called the **nasal cavity.** The anterior portion of the nasal cavity, just inside the nostrils, is called the **vestibule** and is surrounded by cartilage; the superior part of the nasal cavity is surrounded by bone. A vertical partition, the **nasal septum,** divides the nasal cavity into right and left sides. The anterior portion of the septum consists primarily of hyaline cartilage; the remainder is formed by the vomer, perpendicular plate of the ethmoid, maxillae, and palatine bones (see Figure 7.11 on page 181).

When air enters the nostrils, it passes first through the vestibule, which is lined by skin containing coarse hairs that filter out large dust particles. Three shelves formed by projections of the superior, middle, and inferior nasal conchae extend out of each lateral wall of the cavity. The conchae, almost reaching the septum, subdivide each side of the nasal cavity into a series of groovelike passageways—the **superior, middle,** and **inferior meatuses** (mē-Ā-tus-ēz=openings or passages). Mucous membrane lines the cavity and its shelves. The arrangement of conchae and meatuses increases surface area in the internal nose and prevents dehydration by acting as a baffle that traps water droplets during exhalation.

The olfactory receptors lie in the membrane lining the superior nasal conchae and adjacent septum. This region is called the **olfactory epithelium.** Inferior to the olfactory epithelium, the mucous membrane contains capillaries and pseudostratified ciliated columnar epithelium with many goblet cells. As inhaled air whirls around the conchae and meatuses, it is warmed by blood circulating in the abundant capillaries. Mucus secreted by the goblet cells moistens the air and traps dust particles. Drainage from the nasolacrimal ducts and perhaps secretions from the paranasal sinuses also help moisten the air. The cilia move the mucus and trapped dust particles toward the pharynx, at which point they can be removed (i.e., swallowed or spit out) from the respiratory tract.

NASAL POLYPS

Nasal polyps (*poly-*=many; *-pous*=foot) are outgrowths of the mucous membrane of the nose that resemble peeled, seedless grapes. They typically form around the openings of the paranasal sinuses. Nasal polyps form in response to inflammation and, unlike polyps in the colon or urinary bladder, they do not suggest an increased risk of cancer. They are most common in people who suffer from chronic sinus infections or environmental allergies, both of which result in increased tissue inflammation. Treatment involves the use of corticosteroids in nasal sprays or oral tablets, or surgery. ■

Figure 24.1   Structures of the respiratory system.

The upper respiratory system includes the nose, pharynx, and associated structures; the lower respiratory system includes the larynx, trachea, bronchi, and lungs.

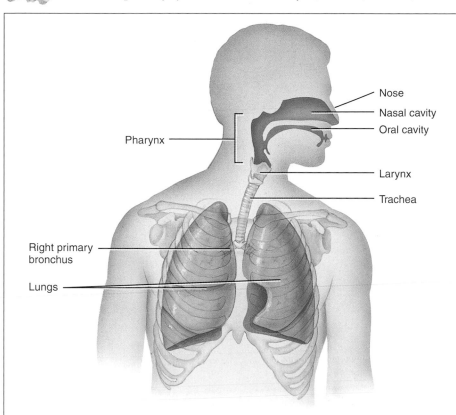

FUNCTIONS

1. Provides for gas exchange — intake of O_2 for delivery to body cells and elimination of CO_2 produced by body cells.
2. Helps regulate blood pH.
3. Contains receptors for the sense of smell, filters inspired air, produces vocal sounds (phonation), and excretes small amounts of water and heat.

Nose
Nasal cavity
Oral cavity
Pharynx
Larynx
Trachea
Right primary bronchus
Lungs

(a) Anterior view showing organs of respiration

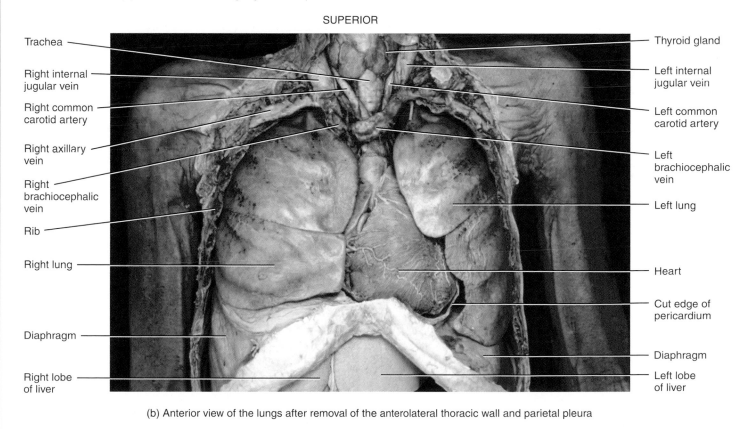

SUPERIOR

Trachea
Right internal jugular vein
Right common carotid artery
Right axillary vein
Right brachiocephalic vein
Rib
Right lung
Diaphragm
Right lobe of liver

Thyroid gland
Left internal jugular vein
Left common carotid artery
Left brachiocephalic vein
Left lung
Heart
Cut edge of pericardium
Diaphragm
Left lobe of liver

(b) Anterior view of the lungs after removal of the anterolateral thoracic wall and parietal pleura

Which structures are part of the conducting portion of the respiratory system?

Figure 24.2   Respiratory structures in the head and neck. (See Tortora, *A Photographic Atlas of the Human Body*, 2e, Figure 11.2.)

As air passes through the nose, it is warmed, filtered, and moistened and olfaction occurs.

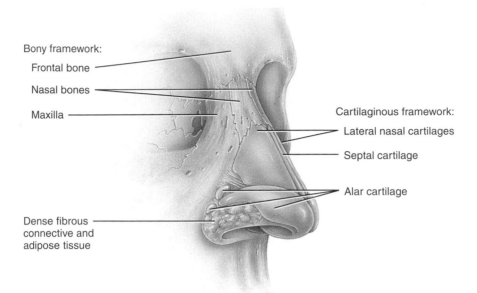

Bony framework:
Frontal bone
Nasal bones
Maxilla

Cartilaginous framework:
Lateral nasal cartilages
Septal cartilage
Alar cartilage

Dense fibrous connective and adipose tissue

(a) Anterolateral view of external portion of nose showing cartilaginous and bony framework

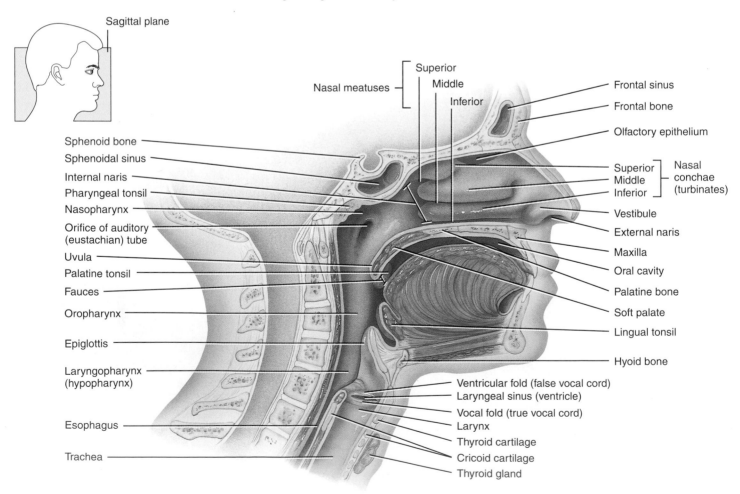

Sagittal plane

Nasal meatuses
Superior
Middle
Inferior

Frontal sinus
Frontal bone
Olfactory epithelium

Sphenoid bone
Sphenoidal sinus
Internal naris
Pharyngeal tonsil
Nasopharynx
Orifice of auditory (eustachian) tube
Uvula
Palatine tonsil
Fauces
Oropharynx
Epiglottis
Laryngopharynx (hypopharynx)
Esophagus
Trachea

Superior
Middle
Inferior
Nasal conchae (turbinates)

Vestibule
External naris
Maxilla
Oral cavity
Palatine bone
Soft palate
Lingual tonsil
Hyoid bone
Ventricular fold (false vocal cord)
Laryngeal sinus (ventricle)
Vocal fold (true vocal cord)
Larynx
Thyroid cartilage
Cricoid cartilage
Thyroid gland

(b) Sagittal section of the left side of the head and neck showing the location of respiratory structures

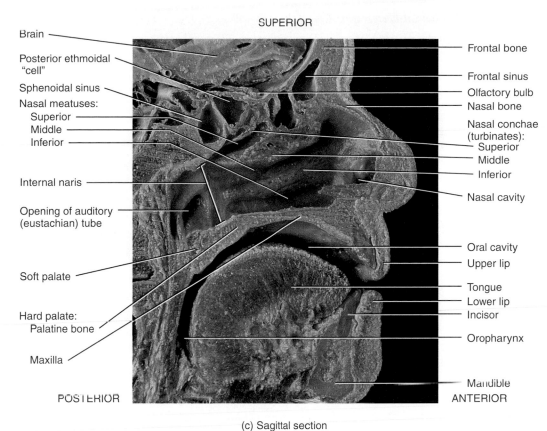

SUPERIOR

Brain
Posterior ethmoidal "cell"
Sphenoidal sinus
Nasal meatuses:
    Superior
    Middle
    Inferior
Internal naris
Opening of auditory (eustachian) tube
Soft palate
Hard palate:
    Palatine bone
Maxilla

Frontal bone
Frontal sinus
Olfactory bulb
Nasal bone
Nasal conchae (turbinates):
    Superior
    Middle
    Inferior
Nasal cavity
Oral cavity
Upper lip
Tongue
Lower lip
Incisor
Oropharynx
Mandible

POSTERIOR   ANTERIOR

(c) Sagittal section

What is the path taken by air molecules into and through the nose?

The arterial supply to the nasal cavity is principally from the sphenopalatine branch of the maxillary artery. The remainder is supplied by the ophthalmic artery. The veins of the nasal cavity drain into the sphenopalatine vein, the facial vein, and the ophthalmic vein.

The nerve supply of the nasal cavity consists of olfactory cells in the olfactory epithelium associated with the olfactory (I) nerve and the nerves of general sensation. These nerves are branches of the ophthalmic and maxillary divisions of the trigeminal (V) nerve.

RHINOPLASTY

Rhinoplasty (RĪ-nō-plas'-tē; *-plasty*=to mold or to shape), commonly called a "nose job," is a surgical procedure in which the structure of the external nose is altered. Although rhinoplasty is often done for cosmetic reasons, it is sometimes performed to repair a fractured nose or a deviated nasal septum. In the procedure, both local and general anesthetics are given. Instruments are then inserted through the nostrils, the nasal cartilage is reshaped, and the nasal bones are fractured and repositioned to achieve the desired shape. An internal packing and splint are inserted to keep the nose in the desired position as it heals. ■

CHECKPOINT

1. What functions do the respiratory and cardiovascular systems have in common?
2. What structural and functional features are different in the upper and lower respiratory systems?
3. How do the structure and functions of the external nose and the internal nose compare?

Pharynx

The **pharynx** (FAIR-inks), or throat, is a funnel-shaped tube about 13 cm (5 in.) long that starts at the internal nares and extends to the level of the cricoid cartilage, the most inferior cartilage of the larynx (voice box) (Figure 24.3). The pharynx lies just posterior to the nasal and oral cavities, superior to the larynx, and just anterior to the cervical vertebrae. Its wall is composed of skeletal muscles and is lined with a mucous membrane. The pharynx functions as a passageway for air and food, provides a resonating chamber for speech sounds, and houses the tonsils, which participate in immunological reactions against foreign invaders.

The pharynx can be divided into three anatomical regions: (1) nasopharynx, (2) oropharynx, and (3) laryngopharynx. (See the lower orientation diagram in Figure 24.3.) The muscles of the entire pharynx are arranged in two layers, an outer circular layer and an inner longitudinal layer.

Figure 24.3 **Pharynx.** (See Tortora, *A Photographic Atlas of the Human Body, 2e,* Figure 11.4.)

The three subdivisions of the pharynx are the (1) nasopharynx, (2) oropharynx, and (3) laryngopharynx.

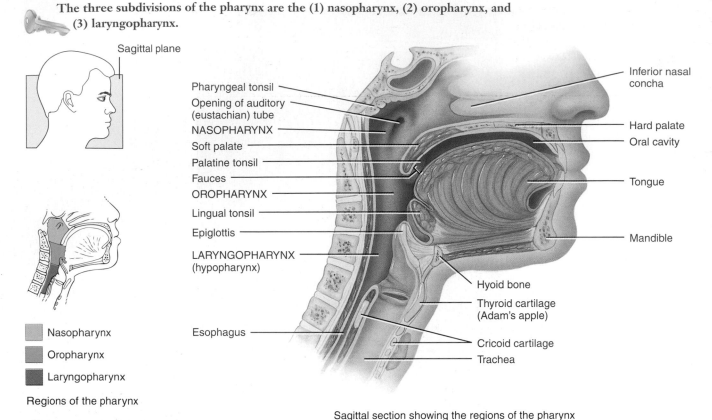

Sagittal plane

Pharyngeal tonsil
Opening of auditory (eustachian) tube
NASOPHARYNX
Soft palate
Palatine tonsil
Fauces
OROPHARYNX
Lingual tonsil
Epiglottis
LARYNGOPHARYNX (hypopharynx)

Esophagus

Inferior nasal concha
Hard palate
Oral cavity
Tongue
Mandible
Hyoid bone
Thyroid cartilage (Adam's apple)
Cricoid cartilage
Trachea

Nasopharynx
Oropharynx
Laryngopharynx

Regions of the pharynx

Sagittal section showing the regions of the pharynx

? **What are the superior and inferior borders of the pharynx?**

The superior portion of the pharynx, called the **nasopharynx,** lies posterior to the nasal cavity and extends to the plane of the soft palate. There are five openings in its wall: two internal nares, two openings that lead into the auditory (pharyngotympanic) tubes (commonly known as the eustachian tubes), and the opening into the oropharynx. The posterior wall also contains the **pharyngeal tonsil.** Through the internal nares, the nasopharynx receives air from the nasal cavity and receives packages of dust-laden mucus. The nasopharynx is lined with pseudostratified ciliated columnar epithelium, and the cilia move the mucus down toward the most inferior part of the pharynx. The nasopharynx also exchanges small amounts of air with the auditory (eustachian) tubes to equalize air pressure between the pharynx and the middle ear.

The intermediate portion of the pharynx, the **oropharynx,** lies posterior to the oral cavity and extends from the soft palate inferiorly to the level of the hyoid bone. It has only one opening, the **fauces** (FAW-sēz=throat), the opening from the mouth. This portion of the pharynx has both respiratory and digestive functions because it is a common passageway for air, food, and drink. Because the oropharynx is subject to abrasion by food particles, it is lined with nonkeratinized stratified squamous epithelium. Two pairs of tonsils, the **palatine** and **lingual tonsils,** are found in the oropharynx.

The inferior portion of the pharynx, the **laryngopharynx** (la-rin'-gō-FAIR-inks), or **hypopharynx,** begins at the level of

the hyoid bone. It opens into the esophagus (food tube) posteriorly and the larynx (voice box) anteriorly. Like the oropharynx, the laryngopharynx is both a respiratory and a digestive pathway and is lined by nonkeratinized stratified squamous epithelium.

The arterial supply of the pharynx includes the ascending pharyngeal artery, the ascending palatine branch of the facial artery, the descending palatine and pharyngeal branches of the maxillary artery, and the muscular branches of the superior thyroid artery. The veins of the pharynx drain into the pterygoid plexus and the internal jugular veins.

Most of the muscles of the pharynx are innervated by the pharyngeal plexus. This plexus is formed by the pharyngeal branches of the glossopharyngeal (IX), vagus (X), and cranial portion of the accessory (XI) nerves, and the superior cervical sympathetic ganglion.

Larynx

The **larynx** (LAIR-inks), or voice box, is a short passageway that connects the laryngopharynx with the trachea. It lies in the midline of the neck anterior to the fourth through sixth cervical vertebrae (C4–C6).

The wall of the larynx is composed of nine pieces of cartilage (Figure 24.4). Three occur singly (thyroid cartilage, epiglottis, and cricoid cartilage), and three occur in pairs (arytenoid, cuneiform, and corniculate cartilages). Of the paired

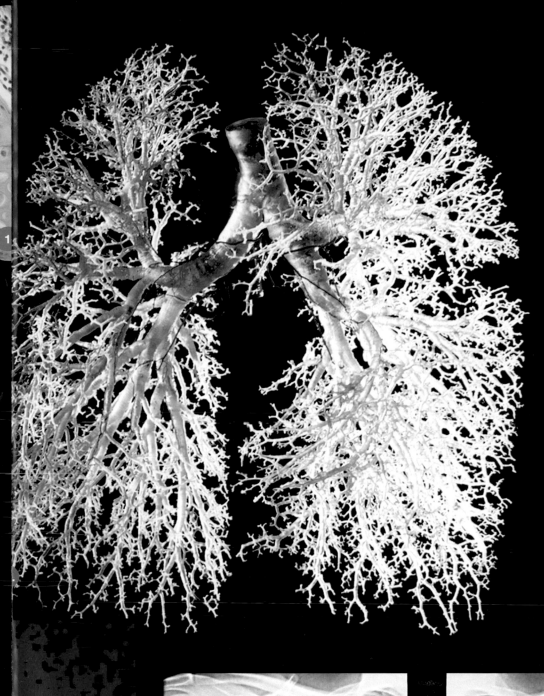

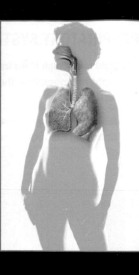

1. **Light micrograph of the tracheal wall**

2. **SEM of cells lining a bronchus**

3. **Lungs and heart**

4. **Endoscopic view of the larynx**

5. **Magnetic resonance angiogram of the pulmonary arteries**

6. **SEM of alveoli**

7. **Bronchograph of the bronchial tree**

8. **Bronchogram**

9. **Radionuclide scan of the lungs**

10. **Resin "corrosion" cast of the bronchial tree**

11. **X-ray of normal lungs**

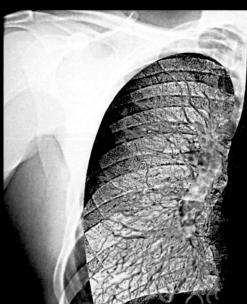

4.

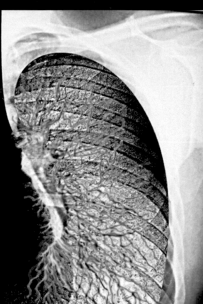

11.

eoli

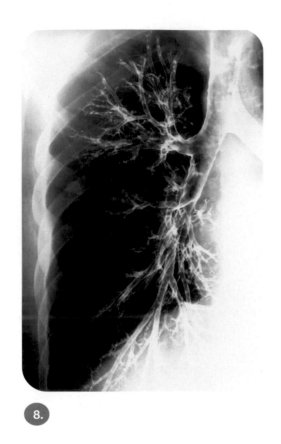

8.

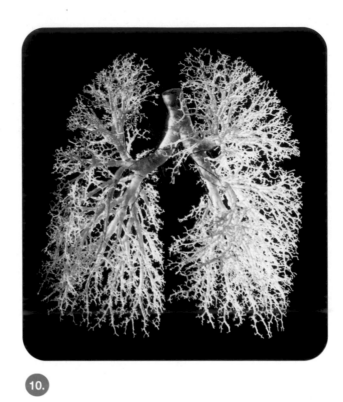

10.

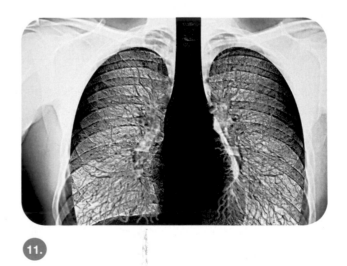

ary

ry

9.

11.

Figure 24.4   Larynx. (See Tortora, *A Photographic Atlas of the Human Body,* 2e, Figures 11.5 and 11.6.)

The larynx is composed of nine pieces of cartilage.

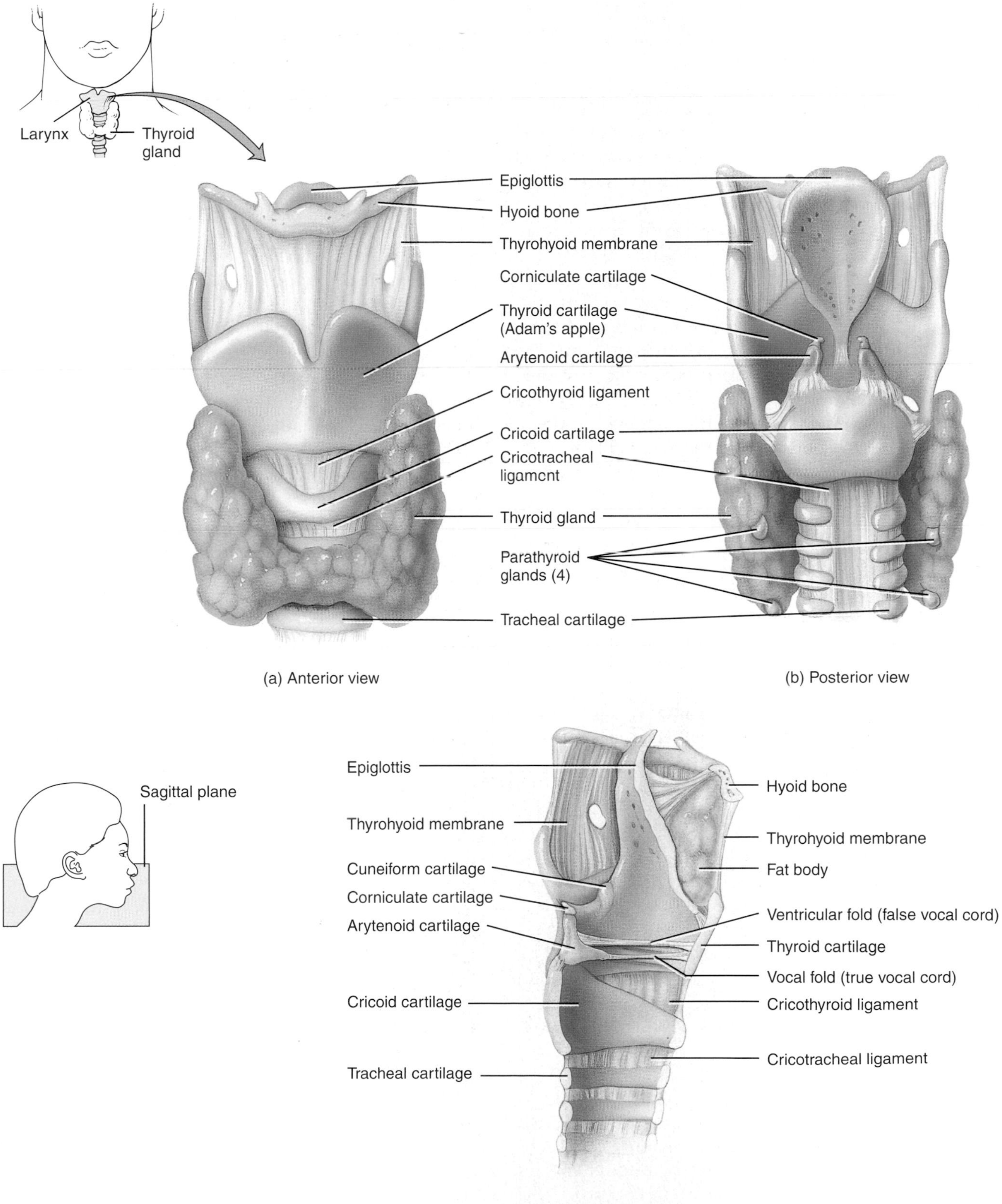

Larynx   Thyroid gland

Epiglottis

Hyoid bone

Thyrohyoid membrane

Corniculate cartilage

Thyroid cartilage (Adam's apple)

Arytenoid cartilage

Cricothyroid ligament

Cricoid cartilage

Cricotracheal ligament

Thyroid gland

Parathyroid glands (4)

Tracheal cartilage

(a) Anterior view

(b) Posterior view

Sagittal plane

Epiglottis

Hyoid bone

Thyrohyoid membrane

Thyrohyoid membrane

Cuneiform cartilage

Fat body

Corniculate cartilage

Ventricular fold (false vocal cord)

Arytenoid cartilage

Thyroid cartilage

Vocal fold (true vocal cord)

Cricoid cartilage

Cricothyroid ligament

Cricotracheal ligament

Tracheal cartilage

(c) Sagittal section

How does the epiglottis prevent aspiration of foods and liquids?

cartilages, the arytenoid cartilages are the most important because they influence the positions and tensions of the vocal folds (true vocal cords). Whereas the extrinsic muscles of the larynx connect the cartilages to other structures in the throat, the intrinsic muscles connect the cartilages to each other (see Figure 11.10 on page 314).

The **thyroid cartilage (Adam's apple)** consists of two fused plates of hyaline cartilage that form the anterior wall of the larynx and give it a triangular shape. It is usually larger in males than in females due to the influence of male sex hormones on its growth during puberty. The ligament that connects the thyroid cartilage to the hyoid bone is called the **thyrohyoid membrane.**

The **epiglottis** (*epi-*=over; *glottis*=tongue) is a large, leaf-shaped piece of elastic cartilage that is covered with epithelium (see Figure 24.2). The "stem" of the epiglottis is the tapered inferior portion that is attached to the anterior rim of the thyroid cartilage and hyoid bone. The broad superior "leaf" portion of the epiglottis is unattached and is free to move up and down like a trap door. During swallowing, the pharynx and larynx rise. Elevation of the pharynx widens it to receive food or drink; elevation of the larynx causes the epiglottis to move down and form a lid over the glottis, closing it off. The **glottis** consists of a pair of folds of mucous membrane, the vocal folds in the larynx, and the space between them called the **rima glottidis** (RĪ-ma GLOT-ti-dis). The closing of the larynx in this way during swallowing routes liquids and foods into the esophagus and keeps them out of the larynx and airways inferior to it. When small particles of dust, smoke, food, or liquids pass into the larynx, a cough reflex occurs, usually expelling the material.

The **cricoid cartilage** (KRĪ-koyd=ringlike) is a ring of hyaline cartilage that forms the inferior wall of the larynx. It is attached to the first ring of cartilage of the trachea by the **crico-tracheal ligament.** The thyroid cartilage is connected to the cricoid cartilage by the **cricothyroid ligament.** The cricoid cartilage is the landmark for making an emergency airway (a tracheotomy; see page 741).

The paired **arytenoid cartilages** (ar′-i-TĒ-noyd=ladle-like) are triangular pieces of mostly hyaline cartilage located at the posterior, superior border of the cricoid cartilage. They attach to the vocal folds and intrinsic pharyngeal muscles. Supported by the arytenoid cartilages, the intrinsic pharyngeal muscles contract and thus move the vocal folds.

The paired **corniculate cartilages** (kor-NIK-ū-lāt=shaped like a small horn), horn-shaped pieces of elastic cartilage, are located at the apex of each arytenoid cartilage. The paired **cuneiform cartilages** (KŪ-nē-i-form=wedge-shaped), club-shaped elastic cartilages anterior to the corniculate cartilages, support the vocal folds and lateral aspects of the epiglottis.

The lining of the larynx superior to the vocal folds is nonkeratinized stratified squamous epithelium. The lining of the larynx inferior to the vocal folds is pseudostratified ciliated columnar epithelium consisting of ciliated columnar cells, goblet cells, and basal cells. The mucus secreted by these cells helps trap dust not removed in the upper passages. In contrast to the action of the cilia in the upper respiratory tract, which move mucus and trapped particles *down* toward the pharynx, the cilia in the lower respiratory tract move them *up* toward the pharynx.

The Structures of Voice Production

The mucous membrane of the larynx forms two pairs of folds (Figure 24.4c): a superior pair called the **ventricular folds (false vocal cords)** and an inferior pair called simply the **vocal folds (true vocal cords).** The space between the ventricular folds is known as the **rima vestibuli. The laryngeal sinus (ventricle)** is a lateral expansion of the middle portion of the laryngeal cavity; it is bordered superiorly by the ventricular folds and inferiorly by the vocal folds (see Figure 24.2b).

When the ventricular folds are brought together, they function in holding the breath against pressure in the thoracic cavity, such as might occur when a person strains to lift a heavy object. Deep to the mucous membrane of the vocal folds, which is lined by nonkeratinized stratified squamous epithelium, are bands of elastic ligaments stretched between pieces of rigid cartilage like the strings on a guitar. Intrinsic laryngeal muscles attach to both the rigid cartilage and the vocal folds. When the muscles contract, they pull the elastic ligaments tight and stretch the vocal folds out into the airways so that the rima glottidis is narrowed. If air is directed against the vocal folds, they vibrate and produce sounds (phonation) and set up sound waves in the column of air in the pharynx, nose, and mouth. The greater the pressure of air, the louder the sound.

When the intrinsic muscles of the larynx contract, they pull on the arytenoid cartilages, which causes them to pivot. Contraction of the posterior cricoarytenoid muscles, for example, moves the vocal folds apart (abduction), thereby opening the rima glottidis (Figure 24.5). By contrast, contraction of the lateral cricoarytenoid muscles moves the vocal folds together (adduction), thereby closing the rima glottidis (Figure 24.5b). Other intrinsic muscles can elongate (and place tension on) or shorten (and relax) the vocal folds.

Pitch is controlled by the tension on the vocal folds. If they are pulled taut by the muscles, they vibrate more rapidly, and a higher pitch results. Decreasing the muscular tension on the vocal folds produces lower-pitch sounds. Due to the influence of androgens (male sex hormones), vocal folds are usually thicker and longer in males than in females, and therefore they vibrate more slowly. Thus, men's voices generally have a lower range of pitch than women's.

Sound originates from the vibration of the vocal folds, but other structures are necessary for converting the sound into recognizable speech. The pharynx, mouth, nasal cavity, and paranasal sinuses all act as resonating chambers that give the voice its human and individual quality. We produce the vowel sounds by constricting and relaxing the muscles in the wall of the pharynx. Muscles of the face, tongue, and lips help us enunciate words.

Whispering is accomplished by closing all but the posterior portion of the rima glottidis. Because the vocal folds do not vibrate during whispering, there is no pitch to this form of speech. However, we can still produce intelligible speech while whispering by changing the shape of the oral cavity as we enunciate. As the size of the oral cavity changes, its resonance qualities change, which imparts a vowel-like pitch to the air as it rushes toward the lips.

Figure 24.5 Movement of the vocal folds.

The glottis consists of a pair of folds of mucous membrane, the vocal folds in the larynx, and the space between them (the rima glottidis).

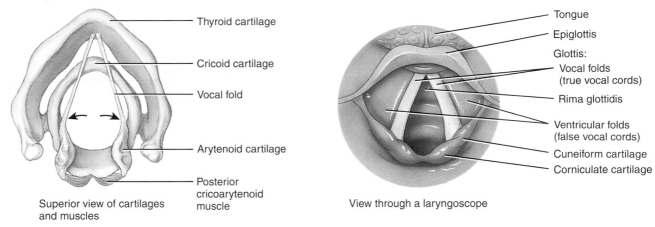

Thyroid cartilage

Cricoid cartilage

Vocal fold

Arytenoid cartilage

Posterior cricoarytenoid muscle

Superior view of cartilages and muscles

Tongue

Epiglottis

Glottis:
Vocal folds (true vocal cords)

Rima glottidis

Ventricular folds (false vocal cords)

Cuneiform cartilage

Corniculate cartilage

View through a laryngoscope

(a) Movement of vocal folds apart (abduction)

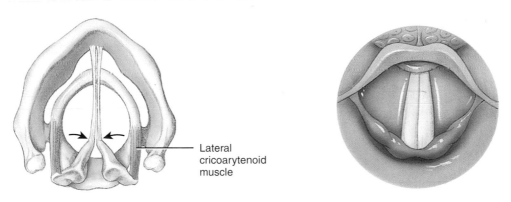

Lateral cricoarytenoid muscle

(b) Movement of vocal folds together (adduction)

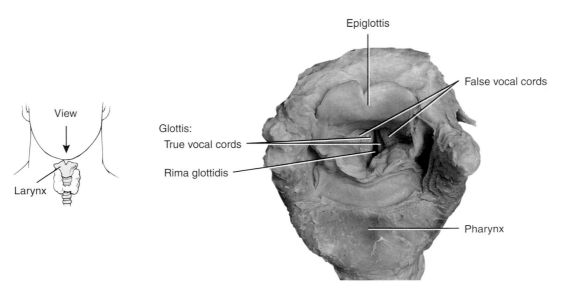

Epiglottis

False vocal cords

View

Larynx

Glottis:
True vocal cords

Rima glottidis

Pharynx

(c) Superior view

 What is the main function of the vocal folds?

LARYNGITIS AND CANCER OF THE LARYNX

Laryngitis is an inflammation of the larynx that is most often caused by a respiratory infection or irritants such as cigarette smoke. Inflammation of the vocal folds causes hoarseness or loss of voice by interfering with the contraction of the folds or by causing them to swell to the point where they cannot vibrate freely. Many long-term smokers acquire a permanent hoarseness from the damage done by chronic inflammation. **Cancer of the larynx** is found almost exclusively in individuals who smoke or chew tobacco products. The condition is characterized by hoarseness, pain on swallowing, or pain radiating to an ear. Treatment consists of radiation therapy and/or surgery. ■

The arteries of the larynx are the superior and inferior laryngeal arteries. The superior and inferior laryngeal veins accompany the arteries. The superior laryngeal vein empties into the superior thyroid vein, and the inferior laryngeal vein empties into the inferior thyroid vein.

The nerves of the larynx are the superior and recurrent (inferior) laryngeal branches of the vagus (X) nerves.

Trachea

The **trachea** (TRĀ-kē-a=sturdy), or windpipe, is a tubular passageway for air that is about 12 cm (5 in.) long and 2.5 cm (1 in.) in diameter. It is located anterior to the esophagus (Figure 24.6) and extends from the larynx to the superior border of the fifth thoracic vertebra (T5), where it divides into the right and left primary bronchi (see Figure 24.7).

The layers of the tracheal wall, from deep to superficial, are the (1) mucosa, (2) submucosa, (3) hyaline cartilage, and (4) adventitia, which is composed of areolar connective tissue. The mucosa of the trachea consists of an epithelial layer of pseudostratified ciliated columnar epithelium and an underlying layer of lamina propria that contains elastic and reticular fibers. The epithelium consists of ciliated columnar cells and goblet cells that reach the luminal surface, plus basal cells that do not (see Table 3.1E on page 65). The epithelium provides the same protection against dust as the membrane lining the nasal cavity and larynx. The submucosa consists of areolar connective tissue that contains seromucous glands and their ducts.

The 16–20 incomplete, horizontal rings of hyaline cartilage resemble the letter C and are stacked one on top of another. They may be felt through the skin inferior to the larynx. The open part of each C-shaped cartilage ring faces the esophagus (Figure 24.6), an arrangement that accommodates slight expansion of the esophagus into the trachea during swallowing. Transverse smooth muscle fibers, called the **trachealis muscle,** and elastic connective tissue stabilize the open ends of the cartilage rings. The solid C-shaped cartilage rings provide a semirigid support so that the tracheal wall does not collapse inward (especially during inhalation) and obstruct the air passageway. The adventitia of the trachea consists of areolar connective tissue that joins the trachea to surrounding tissues.

Figure 24.6   Location of the trachea in relation to the esophagus.

> The trachea is anterior to the esophagus and extends from the larynx to the superior border of the fifth thoracic vertebra.

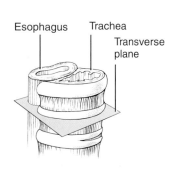

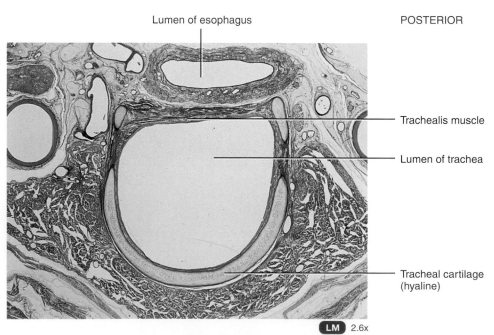

Transverse section of the trachea and esophagus

LM 2.6x

What is the benefit of not having cartilage between the trachea and the esophagus?

TRACHEOTOMY AND INTUBATION

Several conditions may block airflow by obstructing the trachea. For example, the rings of cartilage that support the trachea may collapse due to a crushing injury to the chest, inflammation of the mucous membrane may cause it to swell so much that the airway closes, vomit or a foreign object may be aspirated into it, or a cancerous tumor may protrude into the airway. Two methods are used to reestablish airflow past a tracheal obstruction. If the obstruction is superior to the level of the larynx, a **tracheotomy** (trā-kē-O-tō-mē; *-tome*=cutting), an operation to make an opening into the trachea, may be performed. In this procedure, also called a *tracheostomy*, a skin incision is followed by a short longitudinal incision into the trachea inferior to the cricoid cartilage. The patient can then breathe through a metal or plastic tracheal tube inserted through the incision. The second method used to restore or protect the airway is **intubation**, in which a tube is inserted into the mouth or nose and passed inferiorly through the larynx and trachea. The firm wall of the tube pushes aside any flexible obstruction, and the lumen of

the tube provides a passageway for air; any mucus clogging the trachea can be suctioned out through the tube. ■

The arteries of the trachea are branches of the inferior thyroid, internal thoracic, and bronchial arteries. The veins of the trachea terminate in the inferior thyroid veins.

The smooth muscle and glands of the trachea are innervated parasympathetically via the vagus (X) nerves directly and by their recurrent laryngeal branches. Sympathetic innervation is through branches from the sympathetic trunk and its ganglia.

Bronchi

At the superior border of the fifth thoracic vertebra, the trachea divides into a **right primary bronchus** (BRON-kus=windpipe), which goes into the right lung, and a **left primary bronchus,** which goes into the left lung (Figure 24.7). The right primary bronchus is more vertical, shorter, and wider than the left. As a result, an aspirated object is more likely to enter and lodge in the right primary bronchus than the left. Like the trachea,

Figure 24.7 Branching of airways from the trachea: the bronchial tree.

The bronchial tree begins at the trachea and ends at the terminal bronchioles.

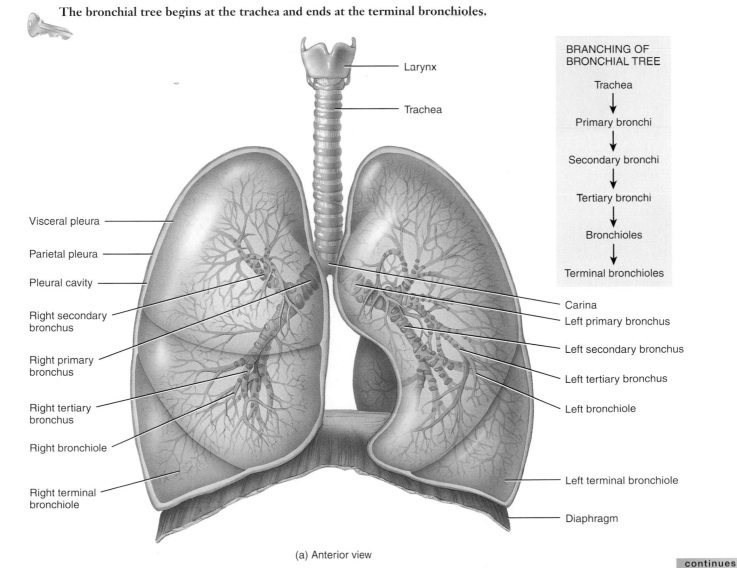

(a) Anterior view

continues

Figure 24.7 (continued)

SUPERIOR

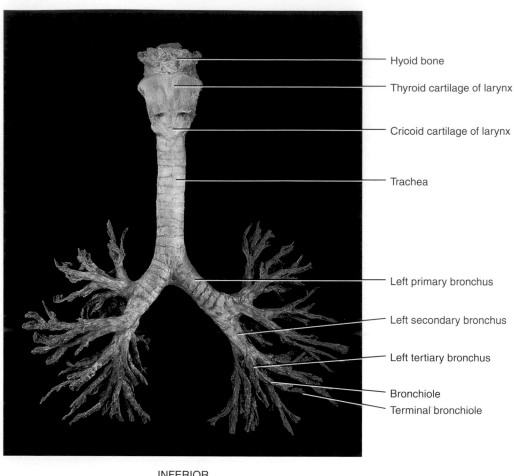

Hyoid bone

Thyroid cartilage of larynx

Cricoid cartilage of larynx

Trachea

Left primary bronchus

Left secondary bronchus

Left tertiary bronchus

Bronchiole
Terminal bronchiole

INFERIOR

(b) Anterior view

the primary bronchi (BRON-kē) contain incomplete rings of cartilage and are lined by pseudostratified ciliated columnar epithelium.

At the point where the trachea divides into right and left primary bronchi is an internal ridge called the **carina** (ka-RĪ-na=keel of a boat). It is formed by a posterior and somewhat inferior projection of the last tracheal cartilage. The mucous membrane of the carina is one of the most sensitive areas of the entire larynx and trachea for triggering a cough reflex. Widening and distortion of the carina is a serious sign because it usually indicates a carcinoma of the lymph nodes around the region where the trachea divides.

On entering the lungs, the primary bronchi divide to form smaller bronchi—the **secondary (lobar) bronchi,** one for each lobe of the lung. (The right lung has three lobes; the left lung has two.) The secondary bronchi continue to branch, forming still smaller bronchi, called **tertiary (segmental) bronchi,** that divide into **bronchioles.** Bronchioles, in turn, branch repeatedly, and the smallest ones branch into even smaller tubes called **terminal bronchioles.** This extensive branching from the

trachea resembles an inverted tree and is commonly referred to as the **bronchial tree.**

As the branching becomes more extensive in the bronchial tree, several structural changes may be noted. First, the mucous membrane in the bronchial tree changes from pseudostratified ciliated columnar epithelium in the primary bronchi, secondary bronchi, and tertiary bronchi to ciliated simple columnar epithelium with some goblet cells in larger bronchioles, to mostly ciliated simple cuboidal epithelium with no goblet cells in smaller bronchioles, to mostly nonciliated simple cuboidal epithelium in terminal bronchioles. (In regions where simple nonciliated cuboidal epithelium is present, inhaled particles are removed by macrophages.) Second, plates of cartilage gradually replace the incomplete rings of cartilage in primary bronchi and finally disappear in the distal bronchioles. Third, as the amount of cartilage decreases, the amount of smooth muscle increases. Smooth muscle encircles the lumen in spiral bands. Because there is no supporting cartilage, however, muscle spasms can close off the airways. This is what happens during an asthma attack, and it can be a life-threatening situation.

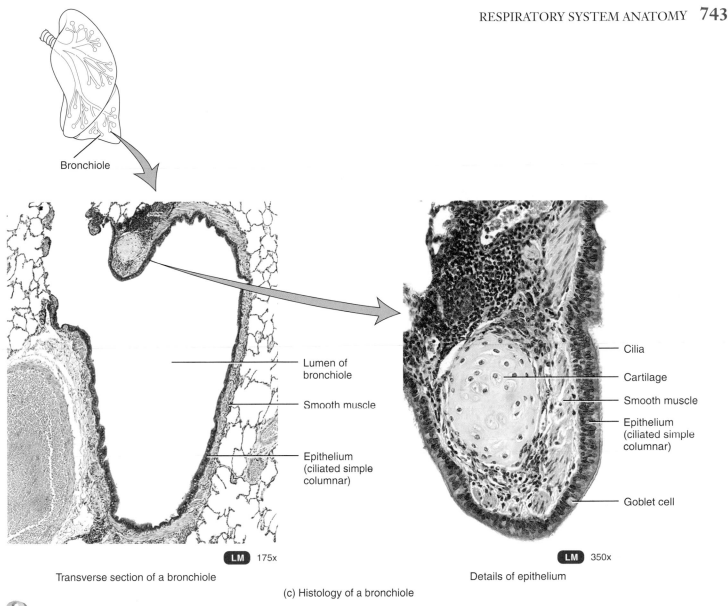

Bronchiole

Lumen of bronchiole

Smooth muscle

Epithelium (ciliated simple columnar)

LM 175x

Transverse section of a bronchiole

Cilia

Cartilage

Smooth muscle

Epithelium (ciliated simple columnar)

Goblet cell

LM 350x

Details of epithelium

(c) Histology of a bronchiole

? How many lobes and secondary bronchi are present in each lung?

During exercise, activity in the sympathetic division of the autonomic nervous system (ANS) increases and causes the adrenal medullae to release the hormones epinephrine and norepinephrine, both of which cause relaxation of smooth muscle in the bronchioles, which dilates the airways. The result is improved lung ventilation because air reaches the alveoli more quickly. The parasympathetic division of the ANS and mediators of allergic reactions such as histamine cause contraction of bronchiolar smooth muscle and result in constriction of distal bronchioles.

NEBULIZATION

Many respiratory disorders are treated by means of **nebulization** (neb-ū-li-ZĀ-shun). This procedure consists of administering medication into the respiratory tract in the form of droplets that are suspended in air. The patient inhales the medication as a fine mist. Nebulization therapy can be used with many different types of drugs, such as chemicals that relax the smooth muscle of the airways, chemicals that reduce the thickness of mucus, and antibiotics. ■

The blood supply to the bronchi is via the left bronchial and right bronchial arteries. The veins that drain the bronchi are the right bronchial vein, which enters the azygos vein, and the left bronchial vein, which empties into the hemiazygos vein or the left superior intercostal vein.

CHECKPOINT

4. List the roles of the three anatomical regions of the pharynx in respiration.
5. How does the larynx function in respiration and voice production?
6. Describe the location, structure, and function of the trachea.
7. What is the structure and function of the bronchial tree?

Lungs

The **lungs** (=lightweights, because they float) are paired cone-shaped organs in the thoracic cavity. They are separated from each other by the heart and other structures in the mediastinum, which separates the thoracic cavity into two anatomically distinct chambers. Because of this separation, if trauma causes one lung to collapse, the other may remain expanded. Two layers of serous membrane, collectively called the **pleural membrane** (PLOOR-al; *pleur-*=side), enclose and protect each lung. The superficial layer lines the wall of the thoracic cavity and is called the **parietal pleura;** the deep layer, the **visceral pleura,** is attached to the lungs themselves (Figure 24.8). Between the visceral and parietal pleurae is a small space, the **pleural cavity,** which contains a small amount of lubricating fluid secreted by the two layers. This fluid reduces friction between the membranes, allowing them to slide easily over one another during breathing. Pleural fluid also causes the pleurae to adhere to one another just as a film of water causes two glass slides to stick together, a phenomenon called surface tension. Separate pleural cavities surround the left and right lungs. Inflammation of the pleural membrane, called **pleurisy** or **pleuritis,** may in its early stages cause pain due to friction between the parietal and visceral layers of the pleura. If the inflammation persists, excess fluid accumulates in the pleural space, a condition known as **pleural effusion.**

PNEUMOTHORAX AND HEMOTHORAX

In certain conditions, the pleural cavities may fill with air (**pneumothorax;** *pneumo-*=air or breath), blood (**hemothorax),** or pus. Air in the pleural cavities, most commonly introduced in a surgical opening of the chest or as a result of a stab or gunshot wound, may cause the lungs to collapse. This collapse of a part of a lung, or rarely an entire lung, is called **atelectasis** (at'-e-LEK-ta-sis; *ateles-*=incomplete; *-ectasis*=expansion). The goal of treatment is the evacuation of air (or blood) from the pleural space, which allows the lung to reinflate. A small pneumothorax may resolve on its own, but it is often necessary to insert a chest tube to assist in evacuation. ■

Figure 24.8   Relationship of the pleural membranes to the lungs. The arrow in the inset indicates the direction from which the lungs are viewed (superior).

The parietal pleura lines the thoracic cavity, whereas the visceral pleura covers the lungs.

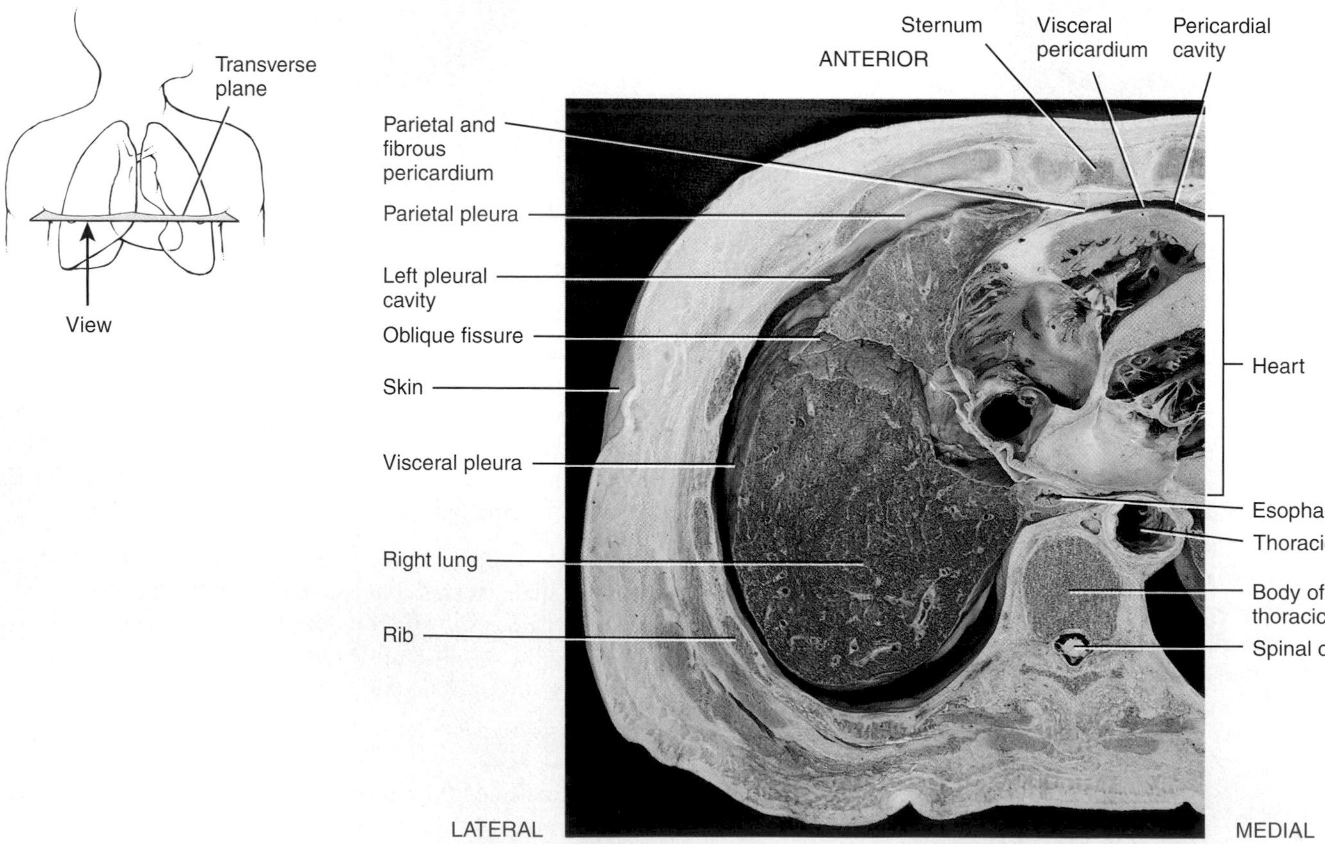

Transverse plane

View

ANTERIOR

Sternum

Visceral pericardium

Pericardial cavity

Parietal and fibrous pericardium

Parietal pleura

Left pleural cavity

Oblique fissure

Skin

Visceral pleura

Right lung

Rib

Heart

Esophagus

Thoracic aorta

Body of fifth thoracic vertebra

Spinal cord

LATERAL

MEDIAL

Inferior view of a transverse section through the thoracic cavity showing the pleural cavity and pleural membranes

What type of membrane is the pleural membrane?

The lungs extend from the diaphragm to just slightly superior to the clavicles and lie against the ribs anteriorly and posteriorly (Figure 24.9a). The broad inferior portion of the lung, the **base,** is concave and fits over the convex area of the diaphragm.

The narrow superior portion of the lung is the **apex.** The surface of the lung lying against the ribs, the **costal surface,** matches the rounded curvature of the ribs. The **mediastinal (medial) surface** of each lung contains a region, the **hilus,**

Figure 24.9   Surface anatomy of the lungs.   (See Tortora, *A Photographic Atlas of the Human Body*, Figures 11.12 and 11.14.)

The oblique fissure divides the left lung into two lobes. The oblique and horizontal fissures divide the right lung into three lobes.

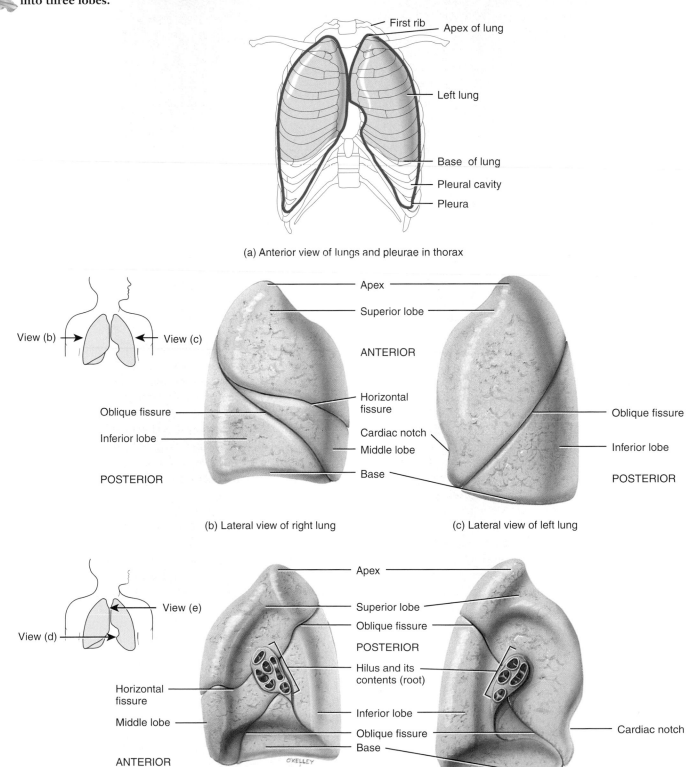

(a) Anterior view of lungs and pleurae in thorax

(b) Lateral view of right lung

(c) Lateral view of left lung

(d) Medial view of right lung

(e) Medial view of left lung

continues

Figure 24.9 (continued)

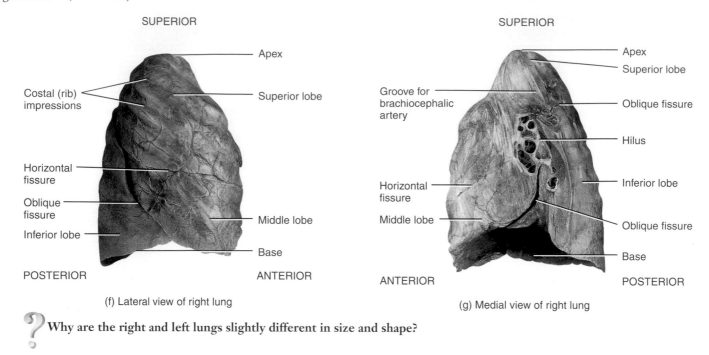

(f) Lateral view of right lung

(g) Medial view of right lung

❓ Why are the right and left lungs slightly different in size and shape?

through which bronchi, pulmonary blood vessels, lymphatic vessels, and nerves enter and exit (Figure 24.9e). These structures are held together by the pleura and connective tissue and constitute the **root** of the lung. Medially, the left lung also contains a concavity, the **cardiac notch,** in which the heart lies. Due to the space occupied by the heart, the left lung is about 10% smaller than the right lung. The right lung is thicker and broader, but it is also somewhat shorter than the left lung because the diaphragm is higher on the right side to accommodate the liver, which lies inferior to it.

The lungs almost fill the thorax (Figure 24.9a). The apex of the lungs lies superior to the medial third of the clavicles and is the only area that can be palpated. The anterior, lateral, and posterior surfaces of the lungs lie against the ribs. The base of the lungs extends from the sixth costal cartilage anteriorly to the spinous process of the tenth thoracic vertebra posteriorly. The pleura extends about 5 cm below the base from the sixth costal cartilage anteriorly to the twelfth rib posteriorly. Thus, the lungs do not completely fill the pleural cavity in this area. Removal of excessive fluid in the pleural cavity can be accomplished without injuring lung tissue by inserting a needle posteriorly through the seventh intercostal space, a procedure termed **thoracentesis** (thor′-a-sen-TĒ-sis; -*centesis*=puncture). The needle is passed along the superior border of the lower rib to avoid damage to the intercostal nerves and blood vessels. Inferior to the seventh intercostal space there is danger of penetrating the diaphragm.

Lobes, Fissures, and Lobules

One or two fissures divide each lung into lobes (Figure 24.9b–e). Both lungs have an **oblique fissure,** which extends inferiorly and anteriorly; the right lung also has a **horizontal fissure.** The oblique fissure in the left lung separates the **superior lobe** from

the **inferior lobe.** In the right lung, the superior part of the oblique fissure separates the superior lobe from the inferior lobe; the inferior part of the oblique fissure separates the inferior lobe from the **middle lobe,** which is bordered superiorly by the horizontal fissure.

Each lobe receives its own secondary (lobar) bronchus. Thus, the right primary bronchus gives rise to three secondary (lobar) bronchi called the **superior, middle,** and **inferior secondary (lobar) bronchi,** whereas the left primary bronchus gives rise to **superior** and **inferior secondary (lobar) bronchi.** Within the lung, the secondary bronchi give rise to the **tertiary (segmental) bronchi;** there are 10 tertiary bronchi in each lung. The segment of lung tissue that each tertiary bronchus supplies is called a **bronchopulmonary segment** (Figure 24.10). Bronchial and pulmonary disorders (such as tumors or abscesses) that are localized in a particular bronchopulmonary segment may be surgically removed without seriously disrupting the surrounding lung tissue.

Each bronchopulmonary segment of the lungs has many small compartments called **lobules;** each lobule is wrapped in elastic connective tissue and contains a lymphatic vessel, an arteriole, a venule, and a branch from a terminal bronchiole (Figure 24.11a on page 748). Terminal bronchioles subdivide into microscopic branches called **respiratory bronchioles** (Figure 24.11b). As the respiratory bronchioles penetrate more deeply into the lungs, the epithelial lining changes from simple cuboidal to simple squamous. Respiratory bronchioles, in turn, subdivide into several (2–11) **alveolar ducts.** The respiratory passages from the trachea to the alveolar ducts contain about 25 orders of branching; that is, branching occurs about 25 times—from the trachea into primary bronchi (first-order branching) into secondary bronchi (second-order branching) and so on down to the alveolar ducts.

Figure 24.10 Bronchopulmonary segments of the lungs. The bronchial branches are shown in the center of the figure. The bronchopulmonary segments within the lungs are numbered and named for convenience.

There are 10 tertiary (segmental) bronchi in each lung; each is composed of smaller compartments called lobules.

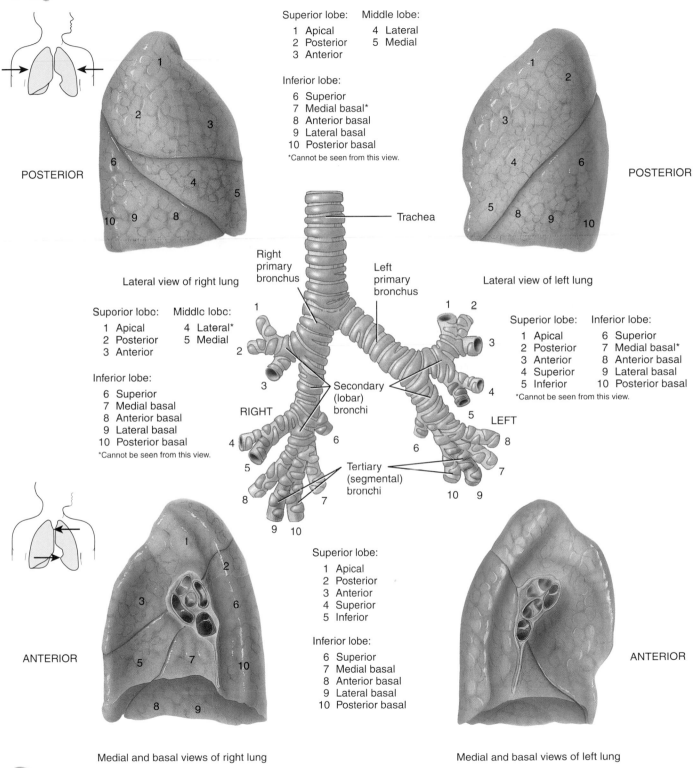

Superior lobe:
1 Apical
2 Posterior
3 Anterior

Middle lobe:
4 Lateral
5 Medial

Inferior lobe:
6 Superior
7 Medial basal*
8 Anterior basal
9 Lateral basal
10 Posterior basal
*Cannot be seen from this view.

POSTERIOR

Lateral view of right lung

POSTERIOR

Lateral view of left lung

Trachea

Right primary bronchus

Left primary bronchus

Secondary (lobar) bronchi

RIGHT

LEFT

Tertiary (segmental) bronchi

Suporior lobo:
1 Apical
2 Posterior
3 Anterior

Middlc lobc:
4 Lateral*
5 Medial

Inferior lobe:
6 Superior
7 Medial basal
8 Anterior basal
9 Lateral basal
10 Posterior basal
*Cannot be seen from this view.

Superior lobe:
1 Apical
2 Posterior
3 Anterior
4 Superior
5 Inferior

Inferior lobe:
6 Superior
7 Medial basal*
8 Anterior basal
9 Lateral basal
10 Posterior basal
*Cannot be seen from this view.

Superior lobe:
1 Apical
2 Posterior
3 Anterior
4 Superior
5 Inferior

Inferior lobe:
6 Superior
7 Medial basal
8 Anterior basal
9 Lateral basal
10 Posterior basal

ANTERIOR

Medial and basal views of right lung

ANTERIOR

Medial and basal views of left lung

 Which bronchi supply a bronchopulmonary segment?

Figure 24.11 Microscopic anatomy of a lobule of the lungs.

🔑 **Alveolar sacs consist of two or more alveoli that share a common opening.**

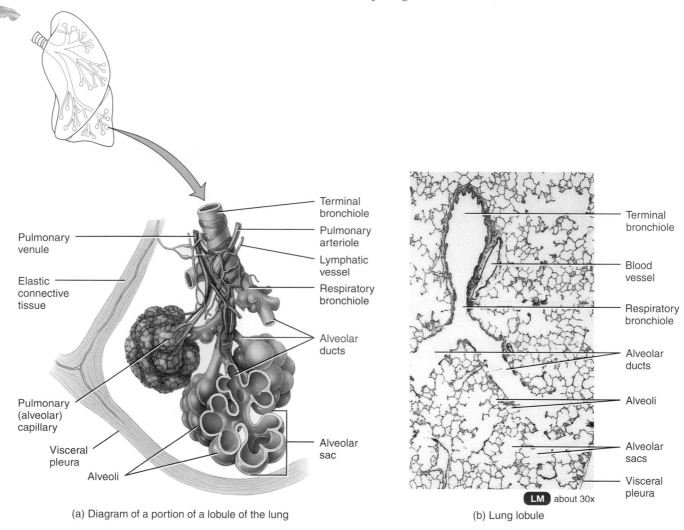

(a) Diagram of a portion of a lobule of the lung

(b) Lung lobule

LM about 30x

Labels on diagram (a): Terminal bronchiole; Pulmonary arteriole; Lymphatic vessel; Respiratory bronchiole; Alveolar ducts; Alveolar sac; Pulmonary venule; Elastic connective tissue; Pulmonary (alveolar) capillary; Visceral pleura; Alveoli

Labels on micrograph (b): Terminal bronchiole; Blood vessel; Respiratory bronchiole; Alveolar ducts; Alveoli; Alveolar sacs; Visceral pleura

Alveoli

Around the circumference of the alveolar ducts are numerous alveoli and alveolar sacs. An **alveolus** (al-VĒ-ō-lus) is a cup-shaped outpouching lined by simple squamous epithelium and supported by a thin elastic basement membrane; an **alveolar sac** consists of two or more alveoli that share a common opening (Figure 24.11a, b). The walls of alveoli consist of two types of alveolar epithelial cells (Figure 24.12 on page 750). **Type I alveolar cells,** the predominant cells, are simple squamous epithelial cells that form a nearly continuous lining of the alveolar wall. **Type II alveolar cells,** also called **septal cells,** are fewer in number and are found between type I alveolar cells. The thin type I alveolar cells are the main sites of gas exchange. Type II alveolar cells, which are rounded or cuboidal epithelial cells with free surfaces containing microvilli, secrete alveolar fluid. This fluid keeps the surface between the cells and the air moist. Included in the alveolar fluid is **surfactant** (sur-FAK-tant), a complex mixture of phospholipids and lipoproteins. Surfactant

lowers the surface tension of alveolar fluid, which reduces the tendency of alveoli to collapse (described later).

Associated with the alveolar wall are **alveolar macrophages (dust cells),** wandering phagocytes that remove fine dust particles and other debris in the alveolar spaces. Also present are fibroblasts that produce reticular and elastic fibers. Underlying the layer of type I alveolar cells is an elastic basement membrane. On the outer surface of the alveoli, the pulmonary arterioles and venules disperse into a network of blood capillaries (see Figure 24.11a) that consist of a single layer of endothelial cells and basement membrane.

The exchange of O_2 and CO_2 between the air spaces in the lungs and the blood takes place by diffusion across the alveolar and capillary walls, which together form the **respiratory membrane.** Extending from the alveolar air space to blood plasma, the respiratory membrane consists of four layers (see Figure 24.12b):

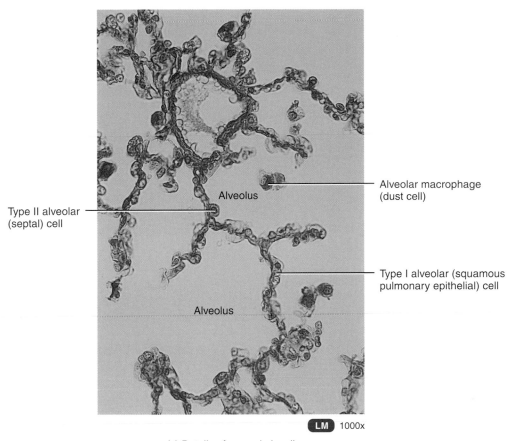

Type II alveolar (septal) cell

Alveolus

Alveolar macrophage (dust cell)

Type I alveolar (squamous pulmonary epithelial) cell

Alveolus

LM 1000x

(c) Details of several alveoli

What types of cells make up the wall of an alveolus?

1. A layer of type I and type II alveolar cells and associated alveolar macrophages that constitutes the **alveolar wall.**

2. An **epithelial basement membrane** underlying the alveolar wall.

3. A **capillary basement membrane** that is often fused to the epithelial basement membrane.

4. The **endothelial cells** of the capillary.

Despite having several layers, the respiratory membrane is very thin—only 0.5 μm thick, about one-sixteenth the diameter of a red blood cell. This thinness allows rapid diffusion of gases. It has been estimated that the lungs contain 300 million alveoli, providing an immense surface area of 70 m^2 (750 ft^2)—about the size of a handball court—for the exchange of gases.

Blood Supply to the Lungs

The lungs receive blood via two sets of arteries: pulmonary arteries and bronchial arteries. Deoxygenated blood passes through the pulmonary trunk, which divides into a left pulmonary artery that enters the left lung and a right pulmonary artery that enters the right lung. Return of the oxygenated blood to the heart occurs by way of the four pulmonary veins, which drain into the left atrium (see Figure 15.16 on page 502). A unique feature of pulmonary blood vessels is their constriction in response to localized hypoxia (low O$_2$ level). In all other body tissues, hypoxia causes dilation of blood vessels, which serves to increase blood flow. In the lungs, however, blood vessels constrict in response to hypoxia, diverting oxygenated blood from poorly ventilated areas to well-ventilated regions of the lungs.

Bronchial arteries, which branch from the aorta, deliver oxygenated blood to the lungs. This blood mainly passes through the walls of the bronchi and bronchioles. Connections do exist between branches of the bronchial arteries and branches of the pulmonary arteries; most blood returns to the heart via pulmonary veins. Some blood, however, drains into bronchial veins, branches of the azygos system, and returns to the heart via the superior vena cava.

The nerve supply of the lungs is derived from the pulmonary plexus, located anterior and posterior to the roots of the lungs. The pulmonary plexus is formed by branches of the vagus (X) nerves and sympathetic trunks. Motor parasympathetic fibers arise from the dorsal nucleus of the vagus (X) nerve, whereas motor sympathetic fibers are postganglionic fibers of the second to fifth thoracic paravertebral ganglia of the sympathetic trunk.

Figure 24.12 Structural components of an alveolus. The respiratory membrane consists of a layer of type I and type II alveolar cells, an epithelial basement membrane, a capillary basement membrane, and the capillary endothelium.

 The exchange of respiratory gases occurs by diffusion across the respiratory membrane.

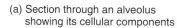

Monocyte

Reticular fiber

Elastic fiber

Type II alveolar (septal) cell

Respiratory membrane

Alveolus

Type I alveolar cell

Alveolar macrophage

Red blood cell in pulmonary capillary

Diffusion of O_2

Diffusion of CO_2

Alveolus

Red blood cell

Capillary endothelium

Capillary basement membrane

Epithelial basement membrane

Type I alveolar cell

Interstitial space

Alveolar fluid with surfactant

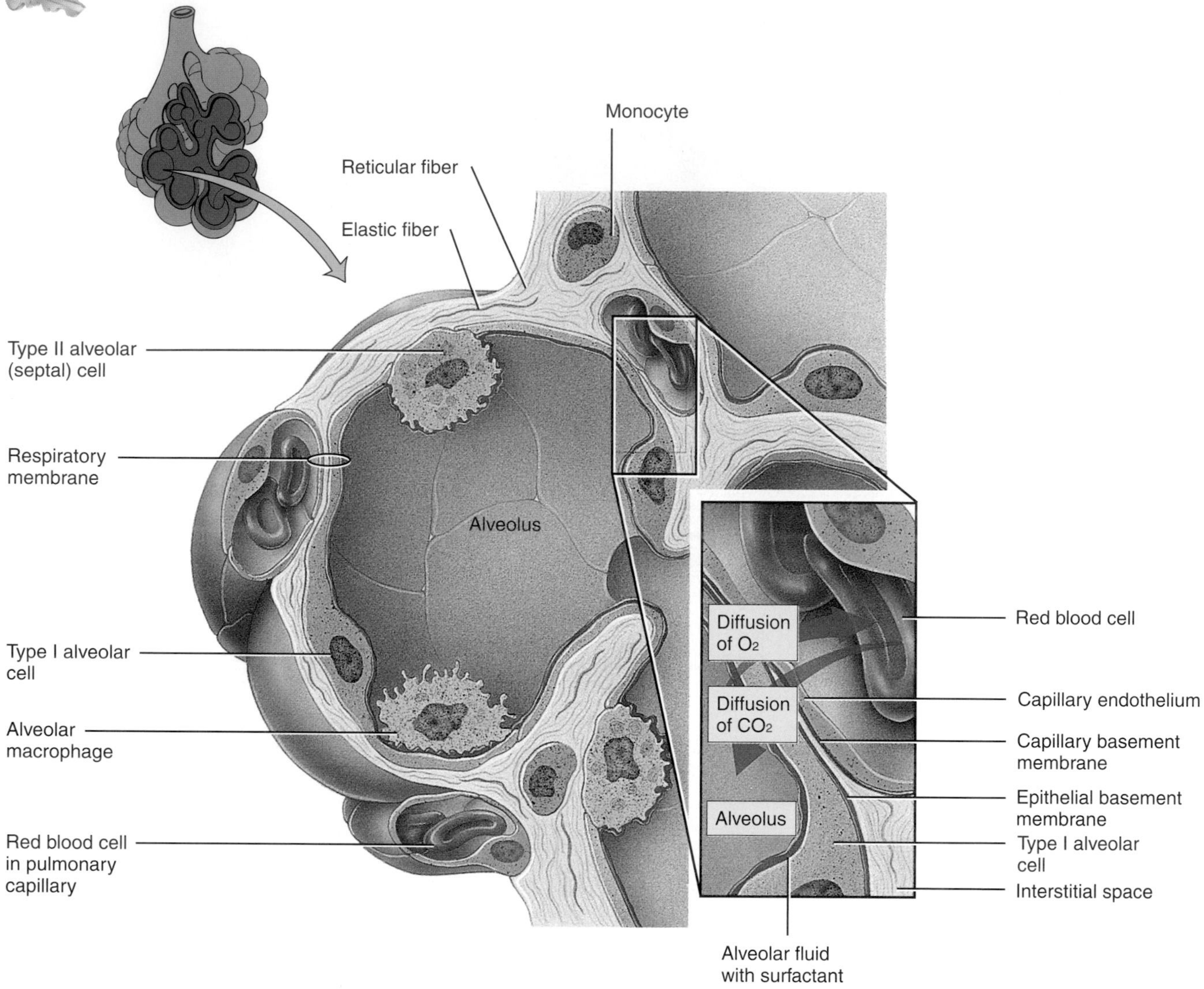

(a) Section through an alveolus showing its cellular components

(b) Details of respiratory membrane

How thick is the respiratory membrane?

CHECKPOINT

8. Where are the lungs located in relation to the clavicles, ribs, and heart? Where can the lungs be palpated?
9. Distinguish the parietal pleura from the visceral pleura.
10. Define each of the following parts of a lung: base, apex, costal surface, medial surface, hilus, root, cardiac notch, lobe, and lobule.
11. What is a bronchopulmonary segment?
12. Describe the histology and function of the respiratory membrane.

MECHANICS OF PULMONARY VENTILATION (BREATHING)

OBJECTIVES

• Distinguish among pulmonary ventilation, external respiration, and internal respiration.
• Describe how inspiration and expiration occur.

Respiration is the exchange of gases between the atmosphere, blood, and body cells. It takes place in three basic steps:

1. **Pulmonary ventilation.** The first process, pulmonary (*pulmo*=lung) ventilation, or breathing, is the inhalation (inflow) and exhalation (outflow) of air between the atmosphere and the air spaces of the lungs.

2. **External (pulmonary) respiration.** This is the exchange of gases between the air spaces of the lungs and blood in pulmonary capillaries across the respiratory membrane. The blood gains O_2 and loses CO_2.

3. **Internal (tissue) respiration.** The is the exchange of gases between blood in systemic capillaries and tissue cells. The blood loses O_2 and gains CO_2.

The flow of air between the atmosphere and lungs occurs for the same reason that blood flows through the body: A pressure gradient (difference) exists. Air moves into the lungs when the pressure inside the lungs is less than the air pressure in the atmosphere. Air moves out of the lungs when the pressure inside the lungs is greater than the pressure in the atmosphere.

Inhalation

Breathing in is called **inhalation (inspiration).** Just before each inhalation, the air pressure inside the lungs is equal to the pressure of the atmosphere, which at sea level is about 760 millimeters of mercury (mmHg), or 1 atmosphere (atm). For air to flow into the lungs, the pressure inside the alveoli must become lower than the atmospheric pressure. This condition is achieved by increasing the volume of the lungs.

For inhalation to occur, the lungs must expand. This increases lung volume and thus decreases the pressure in the lungs below atmospheric pressure. The first step in expanding the alveoli of the lungs during normal quiet breathing involves contraction of the principal muscles of inhalation—the diaphragm and/or external intercostals (Figure 24.13).

The diaphragm, the most important muscle of inhalation, is a dome-shaped skeletal muscle that forms the floor of the thoracic cavity. It is innervated by fibers of the phrenic nerves, which emerge from both sides of the spinal cord at cervical levels 3, 4, and 5. Contraction of the diaphragm causes it to flatten, lowering its dome. This increases the vertical diameter of the thoracic cavity and accounts for the movement of about 75% of the air that enters the lungs during normal quiet inhalation. The distance the diaphragm moves during inspiration ranges from 1 cm (0.4 in.) during normal quiet breathing up to about 10 cm (4 in.) during strenuous exercise. Advanced pregnancy, excessive obesity, or confining abdominal clothing can prevent a complete descent of the diaphragm. At the same time the diaphragm contracts, the external intercostals contract. These skeletal muscles run obliquely downward and forward between adjacent ribs, and when these muscles contract, the ribs are pulled superiorly and the sternum is pushed anteriorly. This increases the anteroposterior and lateral diameters of the thoracic cavity.

As the diaphragm and internal intercostals contract and the overall size of the thoracic cavity increases, the walls of the lungs are pulled outward. The parietal and visceral pleurae normally adhere strongly to each other because of the below-atmospheric pressure between them and because of the surface tension created by their moist adjoining surfaces. As the thoracic cavity expands, the parietal pleura lining the cavity is pulled outward in all directions, and the visceral pleura and lungs are pulled along with it, increasing the volume of the lungs.

When the volume of the lungs increases, alveolar pressure decreases from 760 to 758 mmHg. A pressure gradient is thus established between the atmosphere and the alveoli. Air rushes from the atmosphere into the lungs due to a gas pressure difference, and inhalation takes place. Air continues to move into the lungs as long as the pressure difference exists.

During deep, forceful inhalations, accessory muscles of inspiration also participate in increasing the size of the thoracic cavity (Figure 24.13a). The muscles are so-named because they make little, if any, contribution during normal quiet inhalation, but during exercise or forced ventilation they may contract vigorously. The accessory muscles of inhalation include the sternocleidomastoid muscles, which elevate the sternum; the scalene muscles, which elevate the first two ribs; and the pectoralis minor muscles, which elevate the third through fifth ribs.

Exhalation

Breathing out, called **exhalation (expiration),** is also achieved by a pressure gradient, but in this case the gradient is reversed: The pressure in the lungs is greater than the pressure of the atmosphere. Normal exhalation during quiet breathing depends on two factors: (1) the recoil of elastic fibers that were stretched during inspiration and (2) the inward pull of surface tension due to the film of alveolar fluid.

Exhalation starts when the inspiratory muscles relax. As the external intercostals relax, the ribs move inferiorly; as the diaphragm relaxes, its dome moves superiorly owing to its elasticity. These movements decrease the vertical, anteroposterior, and lateral diameters of the thoracic cavity. Also, surface tension exerts an inward pull between the parietal and visceral pleurae and the elastic basement membranes of the alveoli and elastic fibers in bronchioles and alveolar ducts recoil. As a result, lung volume decreases and the alveolar pressure increases to 762 mmHg. Air then flows from the area of higher pressure in the alveoli to the area of lower pressure in the atmosphere.

During labored breathing and when air movement out of the lungs is impeded, muscles of exhalation—abdominal and internal intercostals—contract. Contraction of the abdominal muscles moves the inferior ribs inferiorly and compresses the abdominal viscera, thus forcing the diaphragm superiorly. Contraction of the internal intercostals, which extend inferiorly and posteriorly between adjacent ribs, also pulls the ribs inferiorly.

Respirations also provide humans with methods for expressing emotions such as laughing, sighing, and sobbing. Moreover, respiratory air can be used to expel foreign matter from the lower air passages through actions such as sneezing and coughing. Respiratory movements can also be modified and controlled when you talk or sing. Some of the modified respiratory movements that express emotion or clear the airways

Figure 24.13   Muscles of inhalation and exhalation and their actions. The pectoralis minor muscle (not shown here) is illustrated in Figure 11.13a on page 322.

During deep, labored breathing, accessory muscles of inhalation (sternocleidomastoids, scalenes, and pectoralis minors) participate.

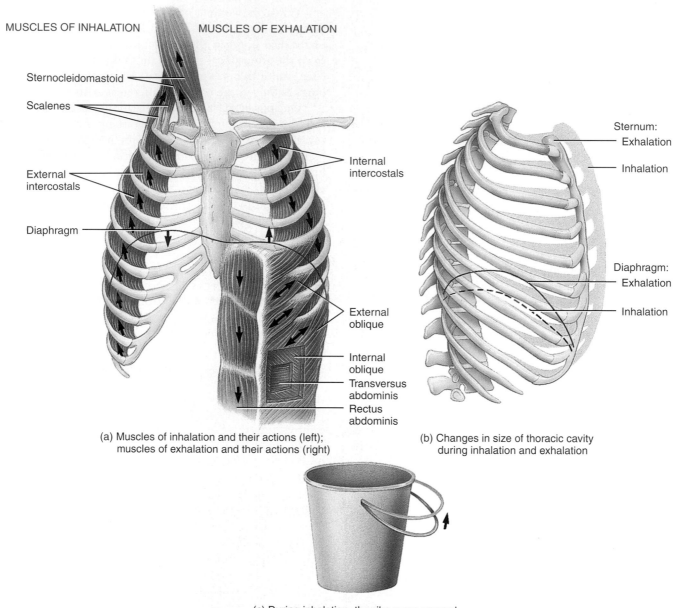

MUSCLES OF INHALATION   MUSCLES OF EXHALATION

Sternocleidomastoid

Scalenes

External intercostals

Diaphragm

Internal intercostals

External oblique

Internal oblique

Transversus abdominis

Rectus abdominis

Sternum: Exhalation — Inhalation

Diaphragm: Exhalation — Inhalation

(a) Muscles of inhalation and their actions (left); muscles of exhalation and their actions (right)

(b) Changes in size of thoracic cavity during inhalation and exhalation

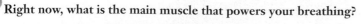

(c) During inhalation, the ribs move upward and outward like the handle on a bucket

Right now, what is the main muscle that powers your breathing?

are listed in Table 24.1. All these movements are reflexes, but some of them can be initiated voluntarily.

13. What are the basic differences among pulmonary ventilation, external respiration, and internal respiration?
14. Compare what happens during quiet ventilation and forceful ventilation.
15. List and describe the various types of modified respiratory movements.

REGULATION OF RESPIRATION

• Describe the various factors that regulate the rate and depth of respiration.

Although breathing can be controlled voluntarily for short periods, the nervous system usually controls respirations automatically to meet the body's demand without conscious effort.

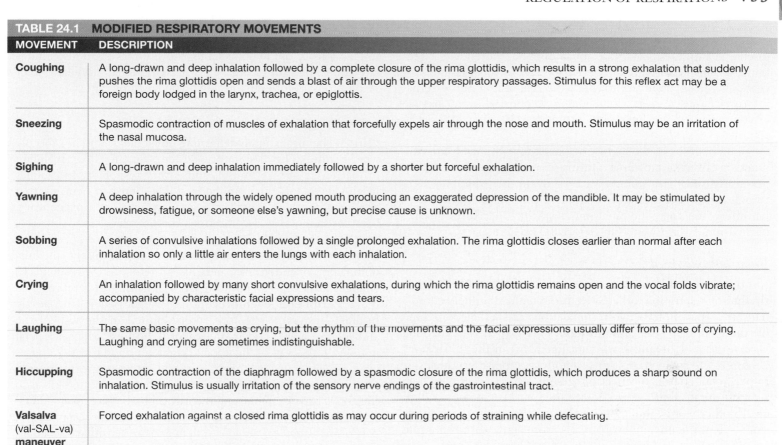

TABLE 24.1 MODIFIED RESPIRATORY MOVEMENTS

MOVEMENT	DESCRIPTION
Coughing	A long-drawn and deep inhalation followed by a complete closure of the rima glottidis, which results in a strong exhalation that suddenly pushes the rima glottidis open and sends a blast of air through the upper respiratory passages. Stimulus for this reflex act may be a foreign body lodged in the larynx, trachea, or epiglottis.
Sneezing	Spasmodic contraction of muscles of exhalation that forcefully expels air through the nose and mouth. Stimulus may be an irritation of the nasal mucosa.
Sighing	A long-drawn and deep inhalation immediately followed by a shorter but forceful exhalation.
Yawning	A deep inhalation through the widely opened mouth producing an exaggerated depression of the mandible. It may be stimulated by drowsiness, fatigue, or someone else's yawning, but precise cause is unknown.
Sobbing	A series of convulsive inhalations followed by a single prolonged exhalation. The rima glottidis closes earlier than normal after each inhalation so only a little air enters the lungs with each inhalation.
Crying	An inhalation followed by many short convulsive exhalations, during which the rima glottidis remains open and the vocal folds vibrate; accompanied by characteristic facial expressions and tears.
Laughing	The same basic movements as crying, but the rhythm of the movements and the facial expressions usually differ from those of crying. Laughing and crying are sometimes indistinguishable.
Hiccupping	Spasmodic contraction of the diaphragm followed by a spasmodic closure of the rima glottidis, which produces a sharp sound on inhalation. Stimulus is usually irritation of the sensory nerve endings of the gastrointestinal tract.
Valsalva (val-SAL-va) maneuver	Forced exhalation against a closed rima glottidis as may occur during periods of straining while defecating.

Role of the Respiratory Center

The size of the thorax is altered by the action of the respiratory muscles, which contract and relax as a result of nerve impulses transmitted to them from centers in the brain. The area from which nerve impulses are sent to respiratory muscles consists of clusters of neurons located bilaterally in the medulla oblongata and pons of the brain stem. This area, called the **respiratory center,** consists of a widely dispersed group of neurons functionally divided into three areas: (1) the medullary rhythmicity area in the medulla oblongata; (2) the pneumotaxic area in the pons; and (3) the apneustic area, also in the pons (Figure 24.14).

Medullary Rhythmicity Area

The function of the **medullary rhythmicity area** (rith-MIS-i-tē) is to control the basic rhythm of respiration, which in the normal resting state is about 2 seconds of inhalation and 3 seconds of exhalation. Within the medullary rhythmicity area are both inspiratory and expiratory neurons that constitute inspiratory and expiratory areas, respectively. We first consider the role of the inspiratory neurons in respiration.

Nerve impulses generated in the **inspiratory area** establish the basic rhythm of breathing. While the inspiratory area is active, it generates nerve impulses for about 2 seconds. The impulses propagate to the external intercostal muscles via intercostal nerves and to the diaphragm via the phrenic nerves. When the nerve impulses reach the diaphragm and external intercostal muscles, the muscles contract and inhalation occurs.

Figure 24.14   Location of the respiratory center in the brain.

> The respiratory center is composed of neurons in the medullary rhythmicity area in the medulla oblongata plus the pneumotaxic and apneustic areas in the pons.

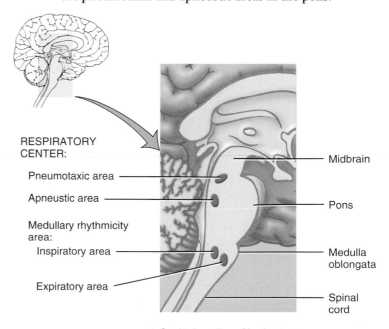

RESPIRATORY CENTER:
Pneumotaxic area
Apneustic area
Medullary rhythmicity area:
Inspiratory area
Expiratory area

Midbrain
Pons
Medulla oblongata
Spinal cord

Sagittal section of brain stem

Which area contains neurons that are active and then inactive in a repeating cycle?

Even when all incoming nerve connections to the inspiratory area are cut or blocked, neurons in this area still rhythmically discharge impulses that cause inhalation. At the end of 2 seconds, the inspiratory area becomes inactive and nerve impulses cease. With no impulses arriving, the diaphragm and external intercostal muscles relax for about 3 seconds, allowing passive elastic recoil of the lungs and thoracic wall. Then, the cycle repeats.

The neurons of the **expiratory area** remain inactive during quiet breathing. However, during forceful breathing nerve impulses from the inspiratory area activate the expiratory area. Impulses from the expiratory area cause contraction of the internal intercostal and abdominal muscles, which decreases the size of the thoracic cavity and causes forceful exhalation.

Pneumotaxic Area

Although the medullary rhythmicity area controls the basic rhythm of respiration, other sites in the brain stem help coordinate the transition between inhalation and exhalation. One of these sites is the **pneumotaxic area** (noo-mō-TAK-sik; *pneumo-*=air or breath; *-taxic*=arrangement) in the upper pons (Figure 24.14), which transmits inhibitory impulses to the inspiratory area. The major effect of these nerve impulses is to help turn off the inspiratory area before the lungs become too full of air. In other words, the impulses shorten the duration of inhalation. When the pneumotaxic area is more active, breathing rate is more rapid.

Apneustic Area

Another part of the brain stem that coordinates the transition between inhalation and exhalation is the **apneustic area** (ap-NOO-stik) in the lower pons (Figure 24.14). This area sends stimulatory impulses to the inspiratory area that activate it and prolong inhalation. The result is a long, deep inhalation. When the pneumotaxic area is active, it overrides the signals from the apneustic area.

Regulation of the Respiratory Center

Although the basic rhythm of respiration is set and coordinated by the inspiratory area, the rhythm can be modified in response to inputs from other brain regions, receptors in the peripheral nervous system, and other factors.

Cortical Influences on Respiration

Because the cerebral cortex has connections with the respiratory center, we can voluntarily alter our pattern of breathing. We can even refuse to breathe at all for a short time. Voluntary control is protective because it enables us to prevent water or irritating gases from entering the lungs. The ability to not breathe, however, is limited by the buildup of CO_2 and H^+ in the body. When CO_2 and H^+ concentrations increase to a certain level, the inspiratory area is strongly stimulated, nerve impulses are sent along the phrenic and intercostal nerves to inspiratory muscles, and breathing resumes, whether you want it to or not. It is impossible for people to kill themselves by voluntarily holding

their breath. Even if you hold your breath long enough that you faint, breathing resumes when consciousness is lost. Nerve impulses from the hypothalamus and limbic system also stimulate the respiratory center, allowing emotional stimuli to alter respirations as, for example, when you laugh or cry.

Chemoreceptor Regulation of Respiration

Certain chemical stimuli determine how quickly and how deeply we breathe. The respiratory system functions to maintain proper levels of CO_2 and O_2 and is very responsive to changes in the levels of either in body fluids. Sensory neurons that are responsive to chemicals are termed **chemoreceptors.** Chemoreceptors in two locations monitor levels of CO_2, H^+, and O_2 and provide input to the respiratory center (Figure 24.15). **Central chemoreceptors** are located in the medulla oblongata in the *central* nervous system. They respond to changes in H^+ or CO_2 concentration, or both, in cerebrospinal fluid. **Peripheral chemoreceptors** are located in the **aortic bodies,** clusters of chemoreceptors located in the wall of the arch of the aorta, and in the **carotid bodies,** which are oval nodules in the wall of the left and right common carotid arteries where they divide into the internal and external carotid arteries. (The chemoreceptors of the aortic bodies are located close to the aortic baroreceptors, and the carotid bodies are located close to the carotid sinus baroreceptors.) These chemoreceptors are part of the *peripheral* nervous system and are sensitive to changes in O_2, H^+, and CO_2 in the blood. Axons of sensory neurons from the aortic bodies are part of the vagus (X) nerves, whereas those from the carotid bodies are part of the right and left glossopharyngeal (IX) nerves.

If there is even a slight increase in CO_2, central and peripheral chemoreceptors are stimulated. The chemoreceptors send nerve impulses to the brain that cause the inspiratory area to become highly active, and the rate of respiration increases. This allows the body to expel more CO_2 until the CO_2 is lowered to normal. If arterial CO_2 is lower than normal, the chemoreceptors are not stimulated, and stimulatory impulses are not sent to the inspiratory area. Consequently, the rate of respiration decreases until CO_2 accumulates and the CO_2 level rises to normal.

The Inflation Reflex

Located in the walls of bronchi and bronchioles are stretch-sensitive receptors called **baroreceptors** or **stretch receptors.** When these receptors become stretched during overinflation of the lungs, nerve impulses are sent along the vagus (X) nerves to the inspiratory and apneustic areas. In response, the inspiratory area is inhibited, and the apneustic area is inhibited from activating the inspiratory area. As a result, expiration begins. As air leaves the lungs during expiration, the lungs deflate and the stretch receptors are no longer stimulated. Thus, the inspiratory and apneustic areas are no longer inhibited, and a new inspiration begins. Some evidence suggests that this reflex, referred to as the **inflation (Hering-Breuer) reflex,** is mainly a protective mechanism for preventing excessive inflation of the lungs rather than a key component in the normal regulation of respiration.

Figure 24.15   Locations of peripheral chemoreceptors.

Chemoreceptors are sensory neurons that respond to changes in the levels of certain chemicals in the body.

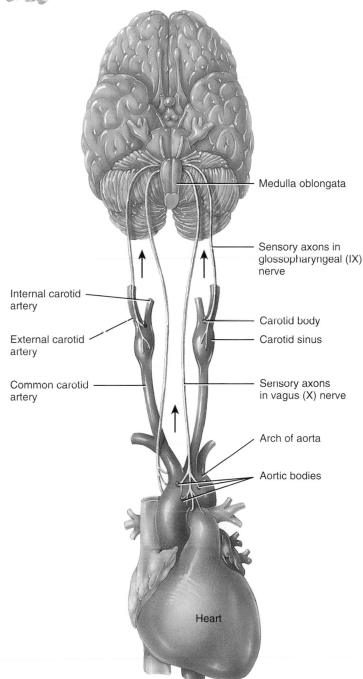

Medulla oblongata

Sensory axons in glossopharyngeal (IX) nerve

Internal carotid artery

Carotid body

External carotid artery

Carotid sinus

Common carotid artery

Sensory axons in vagus (X) nerve

Arch of aorta

Aortic bodies

Heart

Which chemicals stimulate peripheral chemoreceptors?

CHECKPOINT

16. How does the medullary rhythmicity area function in regulating respiration? How are the apneustic and pneumotaxic areas related to the control of respiration?

17. Explain how each of the following modifies respiration: cerebral cortex, inflation reflex, CO_2 levels, and O_2 levels.

EXERCISE AND THE RESPIRATORY SYSTEM

OBJECTIVE

● Describe the effects of exercise on the respiratory system.

The respiratory and cardiovascular systems make adjustments in response to both the intensity and duration of exercise. The effects of exercise on the heart are discussed in Chapter 14. Here we focus on how exercise affects the respiratory system.

The heart pumps the same amount of blood to the lungs as to all the rest of the body. Thus, as cardiac output rises, the rate of blood flow through the lungs also increases. As blood flows more rapidly through the lungs, it picks up more O_2. In addition, the rate at which O_2 diffuses from alveolar air into the blood increases during maximal exercise because blood flows through a larger percentage of the pulmonary capillaries, providing a greater surface area for the diffusion of O_2 into blood.

When muscles contract during exercise, they consume large amounts of O_2 and produce large amounts of CO_2. During vigorous exercise, O_2 consumption and pulmonary ventilation both increase dramatically. At the onset of exercise, an abrupt increase in pulmonary ventilation is due to neural changes that send excitatory impulses to the inspiratory area in the medulla oblongata. The more gradual increase in ventilation during moderate exercise is due to chemical and physical changes in the bloodstream. During moderate exercise, it is mostly an increase in the depth of ventilation (bigger breaths) rather than an increase in the breathing rate. When exercise is more strenuous, the frequency of breathing also increases.

At the end of an exercise session, an abrupt decrease in pulmonary ventilation is followed by a more gradual decline to the resting level. The initial decrease is due mainly to changes in neural factors when movement stops or slows, whereas the more gradual phase reflects the slower return of blood chemistry levels and temperature to the resting state.

THE EFFECT OF SMOKING ON RESPIRATORY EFFICIENCY

Smoking may cause a person to become easily "winded" during even moderate exercise because several factors decrease respiratory efficiency in smokers: (1) Nicotine constricts terminal bronchioles, which decreases airflow into and out of the lungs. (2) Carbon monoxide in smoke binds to hemoglobin and reduces its oxygen-carrying capability. (3) Irritants in smoke cause increased mucus secretion by the mucosa of the bronchial tree and swelling of the mucosal lining, both of which impede airflow into and out of the lungs. (4) Irritants in smoke also inhibit the movement of cilia and destroy cilia in the lining of the respiratory system. Thus, excess mucus and foreign debris are not easily removed, which further adds to the difficulty in breathing. (5) With time, smoking leads to destruction of elastic fibers in the lungs and is the prime cause of emphysema (described on page 758). These changes cause collapse of small bronchioles and trapping of air in alveoli at the end of exhalation. The result is less efficient gas exchange. ■

18. How does exercise affect the inspiratory area?
19. Describe the changes in pulmonary ventilation caused by a brisk walk in the park.

DEVELOPMENT OF THE RESPIRATORY SYSTEM

OBJECTIVE

• Describe the development of the respiratory system.

The development of the mouth and pharynx are discussed in Chapter 25. Here we consider the development of the remainder of the respiratory system.

At about four weeks of development, the respiratory system begins as an outgrowth of the foregut (precursor of some digestive organs) just anterior to the pharynx. This outgrowth is called the **respiratory diverticulum** (Figure 24.16; see also Figure 23.8a). The **endoderm** lining the respiratory diverticulum gives rise to the epithelium and glands of the trachea, bronchi, and alveoli. **Splanchnic mesoderm** (see Figure 4.9d on page 105) surrounding the respiratory diverticulum gives rise to the connective tissue, cartilage, and smooth muscle of these structures.

The epithelial lining of the *larynx* develops from the **endoderm** of the respiratory diverticulum; the cartilages and muscles originate from the **fourth** and **sixth pharyngeal arches** (see Figure 4.13 on page 111).

As the respiratory diverticulum elongates, its distal end enlarges to form a globular **tracheal bud,** which gives rise to the *trachea.* Soon after, the tracheal bud divides into **bronchial buds,** which branch repeatedly and develop with the *bronchi.* By 24 weeks, 17 orders of branches have formed and *respiratory bronchioles* have developed.

During weeks 6 to 16, all major elements of the *lungs* have formed, except for those involved in gaseous exchange (respiratory bronchioles, alveolar ducts, and alveoli). Since respiration is not possible at this stage, fetuses born during this time cannot survive.

During weeks 16 to 26, lung tissue becomes highly vascular and respiratory bronchioles, alveolar ducts, and some primitive alveoli develop. Although it is possible for a fetus born near the end of this period to survive if given intensive care, death frequently occurs due to the immaturity of the respiratory and other systems.

From 26 weeks to birth, many more primitive alveoli develop; they consist of type I alveolar cells (main sites of gaseous exchange) and type II surfactant-producing cells. Blood capillaries also establish close contact with the primitive alveoli. Recall that surfactant is necessary to lower surface tension of alveolar fluid and thus reduce the tendency of alveoli to collapse on exhalation. Although surfactant production begins by 20 weeks, it is present in only small quantities. Amounts sufficient to permit survival of a premature (preterm) infant are not produced until 26 to 28 weeks gestation. Infants born before 26–28

Figure 24.16   Development of the bronchial tubes and lungs.

 The respiratory system develops from endoderm and mesoderm.

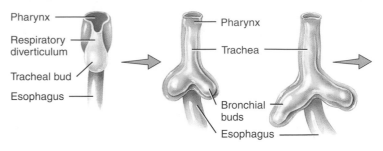

Fourth week

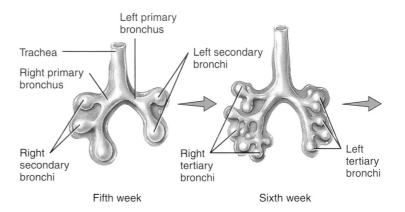

Fifth week   Sixth week

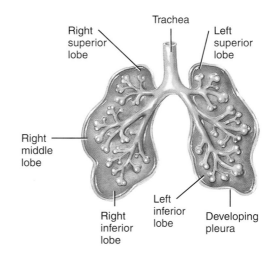

Eighth week

When does the respiratory system begin to develop in an embryo?

weeks are severely at risk of **respiratory distress syndrome (RDS),** in which the alveoli collapse during exhalation and must be reinflated during inhalation. The condition is treated by employing respirators that force air into the lungs and by administering surfactant.

At about 30 weeks, mature alveoli develop. However, it is estimated that only about one-sixth of the full complement of alveoli develop before birth; the remainder develop after birth during the first eight years.

As the lungs develop, they acquire their *pleural sacs*. The *visceral pleura* develops from **splanchnic mesoderm** and the *parietal pleura* develops from **somatic mesoderm** (see Figure 4.9d on page 105). The space between the pleural layers is the *pleural cavity*.

During development, breathing movements of the fetus cause the aspiration of fluid into the lungs. The fluid is a mixture of amniotic fluid, mucus from the bronchial glands, and surfactant. At birth, the lungs are about half-filled with fluid. When breathing begins at birth, most of the fluid is rapidly reabsorbed by blood and lymph capillaries and a small amount is expelled through the nose and mouth during delivery.

20. What structures develop from the respiratory diverticulum?
21. Which respiratory structures develop from endoderm? From mesoderm?
22. How many weeks old must a fetus be for it to survive as a preterm infant? Why?

AGING AND THE RESPIRATORY SYSTEM

● Describe the effects of aging on the respiratory system.

With advancing age, the airways and tissues of the respiratory tract, including the alveoli, become less elastic and more rigid; the chest wall becomes more rigid as well. The result is a decrease in lung capacity. In fact, vital capacity (the maximum amount of air that can be expired after maximal inhalation) can decrease as much as 35% by age 70. A decrease in blood level of O_2, decreased activity of alveolar macrophages, and diminished ciliary action of the epithelium lining the respiratory tract also occur. Owing to all these age-related factors, elderly people are more susceptible to pneumonia, bronchitis, emphysema, and other pulmonary disorders. Age-related changes in the structure and functions of the lung can also contribute to an older person's reduced ability to perform vigorous exercises, such as running.

23. What accounts for the decrease in lung capacity with aging?

APPLICATIONS TO HEALTH

ASTHMA

Asthma (AZ-ma=panting) is a disorder characterized by chronic airway inflammation, airway hypersensitivity to a variety of stimuli, and airway obstruction that is at least partially reversible, either spontaneously or with treatment. It affects 3–5% of the U.S. population and is more common in children than in adults. Airway obstruction may be due to smooth muscle spasms in the walls of smaller bronchi and bronchioles, edema of the mucosa of the airways, increased mucus secretion, and/or damage to the epithelium of the airway.

Individuals with asthma typically react to concentrations of agents too low to cause symptoms in people without asthma. Sometimes the trigger is an allergen such as pollen, house dust mites, molds, or a particular food. Other common triggers of asthma attacks are emotional upset, aspirin, sulfiting agents (used in wine and beer and to keep greens fresh in salad bars), exercise, and breathing cold air or cigarette smoke. In the early phase (acute) response, smooth muscle spasm is accompanied by excessive secretion of mucus that may clog the bronchi and bronchioles and worsen the attack. The late phase (chronic) response is characterized by inflammation, fibrosis, edema, and necrosis (death) of bronchial epithelial cells. A host of mediator chemicals, including leukotrienes, prostaglandins, thromboxane, platelet-activating factor, and histamine, take part.

Symptoms include difficult breathing, coughing, wheezing, chest tightness, tachycardia, fatigue, moist skin, and anxiety. An acute attack is treated by giving an inhaled beta$_2$-adrenergic agonist (albuterol) to help relax smooth muscle in the bronchioles and open up the airways. However, long-term therapy of asthma strives to suppress the underlying inflammation. The anti-inflammatory drugs that are used most often are inhaled corticosteroids (glucocorticoids), cromolyn sodium (Intal), and leukotriene blockers (Accolate).

CHRONIC OBSTRUCTIVE PULMONARY DISEASE

Chronic obstructive pulmonary disease (COPD) is a type of respiratory disorder characterized by chronic and recurrent obstruction of airflow, which increases airway resistance. COPD affects about 30 million Americans and is the fourth leading cause of death behind heart disease, cancer, and cerebrovascular disease. The principal types of COPD are emphysema and chronic bronchitis. In most cases, COPD is preventable because its most common cause is cigarette smoking or breathing secondhand smoke. Other causes include air pollution, pulmonary infection, occupational exposure to dusts and gases, and genetic factors. Because men, on average, have more years of exposure to cigarette smoke than women, they are twice as likely as women to suffer from COPD; still, the incidence of COPD in women has risen sixfold in the past 50 years, a reflection of increased smoking among women.

EMPHYSEMA

Emphysema (em′-fi-SĒ-ma′=blown up or full of air) is a disorder characterized by destruction of the walls of the alveoli, producing abnormally large air spaces that remain filled with air during exhalation. With less surface area for gas exchange, O_2 diffusion across the damaged respiratory membrane is reduced. Blood O_2 level is somewhat lowered, and any mild exercise that raises the O_2 requirements of the cells leaves the patient breathless. As increasing numbers of alveolar walls are damaged, lung elastic recoil decreases due to loss of elastic fibers, and an increasing amount of air becomes trapped in the lungs at the end of exhalation. Over several years, added exertion during inhalation increases the size of the chest cage, resulting in a "barrel chest."

Emphysema is generally caused by a long-term irritation; cigarette smoke, air pollution, and occupational exposure to industrial dust are the most common irritants. Some destruction of alveolar sacs may be caused by an enzyme imbalance. Treatment consists of cessation of smoking, removal of other environmental irritants, exercise training under careful medical supervision, breathing exercises, use of bronchodilators, and oxygen therapy.

CHRONIC BRONCHITIS

Chronic bronchitis is a disorder characterized by excessive secretion of bronchial mucus accompanied by a productive cough (sputum is raised) that lasts for at least three months of the year for two successive years. Cigarette smoking is the leading cause of chronic bronchitis. Inhaled irritants lead to chronic inflammation with an increase in the size and number of mucous glands and goblet cells in the airway epithelium. The thickened and excessive mucus narrows the airway and impairs ciliary function. Thus, inhaled pathogens become embedded in airway secretions and multiply rapidly. Besides a productive cough, symptoms of chronic bronchitis are shortness of breath, wheezing, cyanosis, and pulmonary hypertension. Treatment for chronic bronchitis is similar to that for emphysema.

LUNG CANCER

In the United States **lung cancer** is the leading cause of cancer death in both males and females, accounting for 160,000 deaths annually. At the time of diagnosis, lung cancer is usually well advanced, with distant metastases present in about 55% of patients, and regional lymph node involvement in an additional 25%. Most people with lung cancer die within a year of the initial diagnosis; the overall survival rate is only 10–15%. Cigarette smoke is the most common cause of lung cancer. Roughly 85% of lung cancer cases are related to smoking, and the disease is 10 to 30 times more common in smokers than nonsmokers. Exposure to secondhand smoke is also associated with lung cancer and heart disease. In the United States, secondhand smoke causes an estimated 4000 deaths a year from lung cancer, and nearly 40,000 deaths a year from heart disease. Other causes of lung cancer are ionizing radiation and inhaled irritants, such as asbestos and radon gas. Emphysema is a common precursor to the development of lung cancer.

The most common type of lung cancer, **bronchogenic carcinoma,** starts in the epithelium of the bronchial tubes. Bronchogenic tumors are named based on where they arise. For example, *adenocarcinomas* develop in peripheral areas of the lungs from bronchial glands and alveolar cells, *squamous cell carcinomas* develop from the epithelium of larger bronchial tubes, and *small (oat) cell carcinomas* develop from epithelial cells in primary bronchi near the hilus of the lungs and tend to involve the mediastinum early on. Depending on the type of bronchogenic tumors, they may be aggressive, locally invasive, and undergo widespread metastasis. The tumors begin as epithelial lesions that grow to form masses that obstruct the bronchial tubes or invade adjacent lung tissue. Bronchogenic carcinomas metastasize to lymph nodes, the brain, bones, liver, and other organs.

Symptoms of lung cancer are related to the location of the tumor. These may include a chronic cough, spitting blood from the respiratory tract, wheezing, shortness of breath, chest pain, hoarseness, difficulty swallowing, weight loss, anorexia, fatigue, bone pain, confusion, problems with balance, headache, anemia, thrombocytopenia, and jaundice.

Treatment consists of partial or complete surgical removal of a diseased lung (pulmonectomy), radiation therapy, and chemotherapy.

PNEUMONIA

Pneumonia is an acute infection or inflammation of the alveoli. It is the most common infectious cause of death in the United States, where an estimated 4 million cases occur annually. When certain microbes enter the lungs of susceptible individuals, they release damaging toxins, stimulating inflammation and immune responses that have damaging side effects. The toxins and immune response damage alveoli and bronchial mucous membranes; inflammation and edema cause the alveoli to fill with fluid, interfering with ventilation and gas exchange.

The most common cause of pneumonia is the pneumococcal bacterium *Streptococcus pneumoniae,* but other microbes may also cause pneumonia. Those who are most susceptible to pneumonia are the elderly, infants, immunocompromised individuals (AIDS or cancer patients, or those taking immunosuppressive drugs), cigarette smokers, and individuals with an obstructive lung disease. Most cases of pneumonia are preceded by an upper respiratory infection that often is viral. Individuals then develop fever, chills, productive or dry cough, malaise, chest pain, and sometimes dyspnea (difficult breathing) and hemoptysis (spitting blood).

Treatment may involve antibiotics, bronchodilators, oxygen therapy, increased fluid intake, and chest physiotherapy (percussion, vibration, and postural drainage).

TUBERCULOSIS

The bacterium *Mycobacterium tuberculosis* produces an infectious, communicable disease called **tuberculosis (TB)** that most often affects the lungs and the pleurae but may involve other parts of the body. Once the bacteria are inside the lungs, they multiply and cause inflammation, which stimulates neutrophils and macrophages to migrate to the area and engulf the bacteria to prevent their spread. If the immune system is not impaired, the bacteria remain dormant for life, but impaired immunity may enable the bacteria to escape into blood and lymph to infect other organs. In many people, symptoms—fatigue, weight loss, lethargy, anorexia, a low-grade fever, night sweats, cough, dyspnea, chest pain, and hemoptysis—do not develop until the disease is advanced.

During the past several years, the incidence of TB in the United States has risen dramatically. Perhaps the single most important factor related to this increase is the spread of the human immunodeficiency virus (HIV). People infected with HIV are much more likely to develop tuberculosis because their immune systems are impaired. Among the other factors that have contributed to the increased number of cases are homelessness, increased drug abuse, increased immigration from countries with a high prevalence of tuberculosis, increased crowding in housing among the poor, and airborne transmission of tuberculosis in prisons and shelters. In addition, recent outbreaks of tuberculosis involving multi-drug-resistant strains of *Mycobacterium tuberculosis* have occurred because patients fail to complete their antibiotic and other treatment regimens. TB is treated with the medication isoniazid.

CORYZA AND INFLUENZA

Hundreds of viruses can cause **coryza** (ko-RĪ-za) or the **common cold,** but a group of viruses called *rhinoviruses* is responsible for about 40% of all colds in adults. Typical symptoms include sneezing, excessive nasal secretion, dry cough, and congestion. The uncomplicated common cold is not usually accompanied by a fever. Complications include sinusitis, asthma, bronchitis, ear infections, and laryngitis. Recent investigations suggest an association between emotional stress and the common cold. The higher the stress level, the greater the frequency and duration of colds.

Influenza (flu) is also caused by a virus. Its symptoms include chills, fever (usually higher than 101 °F=39 °C), headache, and muscular aches. Influenza can become life-threatening and may develop into pneumonia. It is important to recognize that influenza is a respiratory disease, not a gastrointestinal (GI) disease. Many people mistakenly report having "the flu" when they are suffering from a GI illness.

PULMONARY EDEMA

Pulmonary edema is an abnormal accumulation of fluid in the interstitial spaces and alveoli of the lungs. The edema may arise from increased permeability of the pulmonary capillaries (pulmonary origin) or increased pressure in the pulmonary capillaries (cardiac origin); the latter cause may coincide with congestive heart failure. The most common symptom is dyspnea. Others include wheezing, tachypnea (rapid breathing rate), restlessness, a feeling of suffocation, cyanosis, pallor (paleness), diaphoresis (excessive perspiration), and pulmonary hypertension. Treatment consists of administering oxygen, drugs that dilate the bronchioles and lower blood pressure, diuretics to rid the body of excess fluid, and drugs that correct acid–base imbalance; suctioning of airways; and mechanical ventilation. One of the recent culprits for causing pulmonary edema was the "phen-phen" diet pills.

CYSTIC FIBROSIS

Cystic fibrosis (CF) is an inherited disease of secretory epithelia that affects the airways, liver, pancreas, small intestine, and sweat glands. It is the most common lethal genetic disease in whites: 5% of the population are thought to be genetic carriers. The cause of cystic fibrosis is a genetic mutation affecting a transporter protein that carries chloride ions across the plasma membranes of many epithelial cells. Because dysfunction of sweat glands causes perspiration to contain excessive sodium chloride (salt), measurement of the excess chloride is one index for diagnosing CF. The mutation also disrupts the normal functioning of several organs by causing ducts within them to become obstructed by thick mucus secretions that do not drain easily from the passageways. Buildup of these secretions leads to inflammation and replacement of injured cells with connective tissue that further blocks the ducts. Clogging and infection of the airways leads to difficulty in breathing and eventual destruction of lung tissue. Lung disease accounts for most deaths from CF. Obstruction of small bile ducts in the liver interferes with digestion and disrupts liver function; clogging of pancreatic ducts prevents digestive enzymes from reaching the small intestine. Because pancreatic juice contains the main fat-digesting enzyme, the person fails to absorb fats or fat-soluble vitamins and thus suffers from vitamin A, D, and K deficiency diseases. With respect to the reproductive systems, blockage of the ductus (vas) deferens leads to infertility in males; the formation of dense mucus plugs in the vagina restricts the entry of sperm into the uterus and can lead to infertility in females.

A child suffering from cystic fibrosis is given pancreatic extract and large doses of vitamins A, D, and K. The recommended diet is high in calories, fats, and proteins, with vitamin supplementation and liberal use of salt. One of the newest treatments for CF is heart-lung transplants.

ACUTE RESPIRATORY DISTRESS SYNDROME

Acute respiratory distress syndrome (ARDS) is a form of respiratory failure characterized by excessive leakiness of the respiratory membranes and severe hypoxia. Situations that can cause ARDS include near drowning, aspiration of acidic gastric juice, drug reactions, inhalation of an irritating gas such as ammonia, allergic reactions, various lung infections such as pneumonia or tuberculosis, and pulmonary hypertension. ARDS strikes about 250,000 people a year in the United States, and about half of them die despite intensive medical care.

ASBESTOS-RELATED DISEASES

Asbestos-related diseases are serious lung disorders that develop as a result of inhaling asbestos particles decades earlier.

When asbestos particles are inhaled, they penetrate lung tissue. In response, white blood cells attempt to destroy them by phagocytosis. However, the fibers usually destroy the white blood cells and scarring of lung tissue may follow. Asbestos-related diseases include **asbestosis** (widespread scarring of lung tissue), **diffuse pleural thickening** (thickening of the pleurae), and **mesothelioma** (cancer of the pleurae or, less commonly, the peritoneum).

KEY MEDICAL TERMS ASSOCIATED WITH THE RESPIRATORY SYSTEM

Abdominal thrust (Heimlich) maneuver (HĪM-lik ma-NOO-ver) First-aid procedure designed to clear the airways of obstructing objects. It is performed by applying a quick upward thrust between the navel and costal margin that causes sudden elevation of the diaphragm and forceful, rapid expulsion of air from the lungs; this action forces air out the trachea to eject the obstructing object. The Heimlich maneuver is also used to expel water from the lungs of near-drowning victims before resuscitation is begun.

Apnea (AP-nē-a; *a*=without; *pnoia*=air or breath) Absence of ventilatory movements.

Asphyxia (as-FIK-sē-a; *sphyxia*=pulse) Oxygen starvation due to low atmospheric oxygen or interference with ventilation, external respiration, or internal respiration.

Aspiration (as′-pi-RĀ-shun) Inhalation of a foreign substance such as water, food, or a foreign body into the bronchial tree; also, the drawing of a substance in or out by suction.

Bronchiectasis (bron′-kē-EK-ta-sis; *-ektasis*=stretching) A chronic dilation of the bronchi or bronchioles resulting from damage to the bronchial wall, for example, from respiratory infections.

Bronchography (bron-KOG-ra-fē) An imaging technique used to visualize the bronchial tree using x-rays. After an opaque contrast medium is inhaled through an intratracheal catheter, radiographs of the chest in various positions are taken, and the developed film, a **bronchogram** (BRON-kō-gram), provides a picture of the bronchial tree.

Bronchoscopy (bron-KOS-kō-pē) Visual examination of the bronchi through a **bronchoscope,** an illuminated, flexible tubular instrument that is passed through the mouth (or nose), larynx, and trachea into the bronchi. The examiner can view the interior of the trachea and bronchi to biopsy a tumor, clear an obstructing object or secretions from an airway, take cultures or smears for microscopic examination, stop bleeding, or deliver drugs.

Cheyne-Stokes respiration (CHĀN STŌKS res′-pi-RĀ-shun) A repeated cycle of irregular breathing that begins with shallow breaths that increase in depth and rapidity and then decrease and cease altogether for 15–20 seconds. Cheyne-Stokes is normal in infants; it is also often seen just before death from pulmonary, cerebral, cardiac, or kidney disease.

Dyspnea (DISP-nē-a; *dys-*=painful, difficult) Painful or labored breathing.

Epistaxis (ep′-i-STAK-sis) Loss of blood from the nose due to trauma, infection, allergy, malignant growths, or bleeding disorders. It can be arrested by cautery with silver nitrate, electrocautery, or firm packing. Also called **nosebleed.**

Hemoptysis (hē-MOP-ti-sis; *hemo-*=blood; *-ptysis*=spit) Spitting of blood from the respiratory tract.

Hyperventilation (*hyper-*=above) Rapid and deep breathing.

Hypoventilation (*hypo-*=below) Slow and shallow breathing.

Mechanical ventilation The use of an automatically cycling device (ventilator or respirator) to assist breathing. A plastic tube is inserted into the nose or mouth and the tube is attached to a device that forces air into the lungs. Exhalation occurs passively due to the elastic recoil of the lungs.

Pulmonary embolism (EM-bō-lizm) The blockage of a pulmonary artery or its branches by a blood clot that travels to the lungs, usually from a vein in a leg or the pelvis.

Rales (RĀLS) Sounds sometimes heard in the lungs that resemble bubbling or rattling. Rales are to the lungs what murmurs are to the heart. Different types are due to the presence of an abnormal type or amount of fluid or mucus within the bronchi or alveoli, or to bronchoconstriction that causes turbulent airflow.

Respirator (RES-pi-rā′-tor) An apparatus fitted to a mask over the nose and mouth, or hooked directly to an endotracheal or tracheotomy tube, that is used to assist or support ventilation or to provide nebulized medication to the air passages.

Respiratory failure A condition in which the respiratory system either cannot supply sufficient O_2 to maintain metabolism or cannot eliminate enough CO_2 to prevent respiratory acidosis (a lower-than-normal pH in interstitial fluid).

Rhinitis (rī-NĪ-tis; *rhin-*=nose) Chronic or acute inflammation of the mucous membrane of the nose due to viruses, bacteria, or irritants. Excessive mucus production produces a runny nose, nasal congestion, and postnasal drip.

Sleep apnea (AP-nē-a; *a-*=without; *-pnea*=breath) A disorder in which a person repeatedly stops breathing for 10 or more seconds while sleeping. Most often, it occurs because loss of muscle tone in pharyngeal muscles allows the airway to collapse.

Sputum (SPŪ-tum=to spit) Mucus and other fluids from the air passages that is expectorated (expelled by coughing).

Strep throat Inflammation of the pharynx caused by the bacterium *Streptococcus pyogenes*. It may also involve the tonsils and middle ear.

Sudden infant death syndrome (SIDS) Death of infants between the ages of 1 week and 12 months thought to be due to hypoxia while sleeping in a prone position (on the stomach) and the rebreathing of exhaled air trapped in a depression of the mattress. It is now recommended that normal newborns be placed on their backs for sleeping: "back to sleep."

Tachypnea (tak′-ip-NĒ-a; *tachy-*=rapid; *-pnea*=breath) Rapid breathing rate.

Wheeze (HWĒZ) A whistling, squeaking, or musical high-pitched sound during breathing resulting from a partially obstructed airway.

 STUDY OUTLINE

Respiratory System Anatomy (p. 732)

1. The respiratory system consists of the nose, pharynx, larynx, trachea, bronchi, and lungs. They act with the cardiovascular system to supply oxygen (O_2) to and remove carbon dioxide (CO_2) from the blood.

2. The external portion of the nose is made of cartilage and skin and is lined with a mucous membrane. Openings to the exterior are the external nares.

3. The internal portion of the nose communicates with the paranasal sinuses and nasopharynx through the internal nares.

4. The nasal cavity is divided by a septum. The anterior portion of the cavity is called the vestibule. The nose warms, moistens, and filters air and functions in olfaction and speech.

5. The pharynx (throat) is a muscular tube lined by a mucous membrane. The anatomic regions are the nasopharynx, oropharynx, and laryngopharynx.

6. The nasopharynx functions in respiration. The oropharynx and laryngopharynx function both in digestion and in respiration.

7. The larynx (voice box) is a passageway that connects the pharynx with the trachea. It contains the thyroid cartilage (Adam's apple); the epiglottis, which prevents food from entering the larynx; the cricoid cartilage, which connects the larynx and trachea; and the paired arytenoid, corniculate, and cuneiform cartilages.

8. The larynx contains vocal folds, which produce sound as they vibrate. Taut folds produce high pitches, and relaxed ones produce low pitches.

9. The trachea (windpipe) extends from the larynx to the primary bronchi. It is composed of C-shaped rings of cartilage and smooth muscle and is lined with pseudostratified ciliated columnar epithelium.

10. The bronchial tree consists of the trachea, primary bronchi, secondary bronchi, tertiary bronchi, bronchioles, and terminal bronchioles. Walls of bronchi contain rings of cartilage; walls of bronchioles contain increasingly smaller plates of cartilage and increasing amounts of smooth muscle.

11. Lungs are paired organs in the thoracic cavity enclosed by the pleural membrane. The parietal pleura is the superficial layer that lines the thoracic cavity; the visceral pleura is the deep layer that covers the lungs.

12. The right lung has three lobes separated by two fissures; the left lung has two lobes separated by one fissure, along with a depression, the cardiac notch.

13. Secondary bronchi give rise to branches called segmental bronchi, which supply segments of lung tissue called bronchopulmonary segments.

14. Each bronchopulmonary segment consists of lobules, which con-

tain lymphatics, arterioles, venules, terminal bronchioles, respiratory bronchioles, alveolar ducts, alveolar sacs, and alveoli.

15. Alveolar walls consist of type I alveolar cells, type II alveolar cells, and associated alveolar macrophages.

16. Gas exchange occurs across the respiratory membranes.

Mechanics of Pulmonary Ventilation (p. 750)

1. Pulmonary ventilation, or breathing, consists of inhalation and exhalation.

2. Inhalation occurs when alveolar pressure falls below atmospheric pressure. Contraction of the diaphragm and external intercostals increases the size of the thorax, thereby decreasing the intrapleural pressure so that the lungs expand. Expansion of the lungs decreases alveolar pressure so that air moves down a pressure gradient from the atmosphere into the lungs.

3. During forceful inhalation, accessory muscles of inhalation (sternocleidomastoids, scalenes, and pectoralis minors) are also used.

4. Exhalation occurs when alveolar pressure is higher than atmospheric pressure. Relaxation of the diaphragm and external intercostals results in elastic recoil of the chest wall and lungs, which increases intrapleural pressure; lung volume decreases and alveolar pressure increases, so air moves from the lungs to the atmosphere.

5. Forceful exhalation involves contraction of the internal intercostal and abdominal muscles.

Regulation of Respiration (p. 752)

1. The respiratory center consists of a medullary rhythmicity area, a pneumotaxic area, and an apneustic area.

2. The inspiratory area sets the basic rhythm of respiration.

3. The pneumotaxic and apneustic areas coordinate the transition between inspiration and expiration.

4. Respiration may be modified by a number of factors, including cortical influences; the inflation reflex; and chemical stimuli, such as O_2, CO_2, and H^+ levels.

Exercise and the Respiratory System (p. 755)

1. The rate and depth of ventilation change in response to both the intensity and duration of exercise.

2. An increase in pulmonary blood flow and O_2-diffusing capacity occurs during exercise.

3. The abrupt increase in ventilation at the start of exercise is due to neural changes that send excitatory impulses to the inspiratory area in the medulla oblongata. The more gradual increase in ventilation during moderate exercise is due to chemical and physical changes in the bloodstream.

Development of the Respiratory System (p. 756)

1. The respiratory system begins as an outgrowth of endoderm called the respiratory diverticulum.
2. Smooth muscle, cartilage, and connective tissue of the bronchial tubes and pleural sacs develop from mesoderm.

Aging and the Respiratory System (p. 757)

1. Aging results in decreased vital capacity, decreased blood level of O_2, and diminished alveolar macrophage activity.
2. Elderly people are more susceptible to pneumonia, emphysema, bronchitis, and other pulmonary disorders.

Q SELF-QUIZ QUESTIONS

Choose the one best answer to the following questions:

1. Which of the following is a part of the cartilaginous framework of the nose?
   a. arytenoid cartilage   b. cricoid cartilage   c. alar cartilage
   d. cartilage of the carina   e. cuneiform cartilage

2. Which is correct order, from superficial to deep, of the following five structures? (1) parietal pleura, (2) visceral pleura, (3) pleural cavity, (4) lung, (5) wall of thoracic cavity.
   a. 5,1,3,2,4   b. 5,3,1,2,4   c. 4,2,3,1,5
   d. 5,1,2,4,3   e. 5,2,3,1,4

3. The nasopharynx extends from the:
   a. exterior of the body to the nasal conchae.
   b. oropharynx to the laryngopharynx.
   c. nasal cavity to the soft palate.
   d. soft palate to the hyoid bone.
   e. mouth to the esophagus.

4. The trachea extends from the:
   a. laryngopharynx to the larynx.
   b. epiglottis to the bronchi.
   c. hyoid bone to the fourth cervical vertebra.
   d. fourth to sixth cervical vertebrae.
   e. larynx to the fifth thoracic vertebra.

5. Pseudostratified ciliated columnar epithelium is found in the:
   a. visceral pleura.   b. alveoli.   c. oropharynx.
   d. trachea.   e. Both c and d are correct.

6. Which of the following is *not* a component of the conducting portion of the respiratory system?
   a. terminal bronchiole   b. larynx   c. bronchus
   d. trachea   e. respiratory bronchiole

7. Which of the following structures is the smallest in diameter?
   a. left primary bronchus   b. respiratory bronchioles
   c. secondary bronchi   d. alveolar ducts
   e. right primary bronchus

8. Inhalation (inspiration) occurs when:
   a. the phrenic nerve stimulates contraction of the diaphragm, which then flattens, thus increasing the volume of the thoracic cavity.
   b. the diaphragm relaxes and flattens, thus increasing the volume of the thoracic cavity.
   c. the phrenic nerve stimulates contraction of the diaphragm, which then forms a dome, thus decreasing the volume of the thoracic cavity.
   d. the vagus (X) nerve stimulates contraction of the diaphragm, which then flattens, thus increasing the volume of the thoracic cavity.
   e. the vagus (X) nerve stimulates contraction of the diaphragm, which then forms a dome, thus decreasing the volume of the thoracic cavity.

Complete the following:

9. C-shaped cartilage rings support the _____.
10. The flap-like structure that prevents aspiration of solid and liquid substances into the trachea is called the _____.
11. The space between the vocal folds is called the rima _____, whereas the space between the ventricular folds is called the rima _____.
12. A section of lung tissue supplied by a tertiary bronchus is called a _____.
13. Place numbers in the blanks to indicate the route through which air passes as it enters a lobule enroute to alveoli. (a) alveolar ducts: _____; (b) respiratory bronchiole: _____; (c) terminal bronchiole: _____.
14. Collapse of all or part of a lung is called _____.
15. The component of alveolar fluid that reduces the tendency for alveoli to collapse is _____.

Are the following statements true or false?

16. Bronchioles have more smooth muscle and less cartilage than bronchi.
17. The pharynx, larynx, and trachea are parts of the upper respiratory system.
18. The trachea is posterior to the esophagus.
19. The branches of the aorta that carry blood to the lungs are the pulmonary arteries.

Matching

20. Match the following terms to their descriptions:
    ___ (a) nasal conchae and meatuses are located here
    ___ (b) contains pharyngeal tonsil
    ___ (c) palatine tonsils are located here
    ___ (d) this structure leads directly into the esophagus
    ___ (e) cricoid, epiglottis, and thyroid cartilages are here
    ___ (f) vocal cords here enable voice production
    ___ (g) both a respiratory and digestive pathway (two answers)
    ___ (h) fauces opens into this structure

    (1) nasal cavity
    (2) oropharynx
    (3) laryngopharynx
    (4) larynx
    (5) nasopharynx

21. Match the following:
    ___ (a) cuboidal epithelium with microvilli that secrete alveolar fluid
    ___ (b) found in the medulla oblongata; sensitive to H^+ and CO_2 levels in the cerebrospinal fluid
    ___ (c) simple squamous epithelial cells forming the lining of the alveolar wall
    ___ (d) located in the walls of bronchi and bronchioles; sensitive to stretch
    ___ (e) make up the capillary wall of the respiratory membrane
    ___ (f) found in the wall of the aortic arch; sensitive to levels of O_2, CO_2, and H^+ in blood
    ___ (g) phagocytic cells

    (1) type I alveolar cells
    (2) type II alveolar cells
    (3) alveolar macrophages
    (4) endothelial cells
    (5) central chemoreceptors
    (6) peripheral chemoreceptors
    (7) baroreceptors

22. Match the following terms to their descriptions:
    ___ (a) thicker, broader, and shorter
    ___ (b) has a cardiac notch
    ___ (c) has only two secondary bronchi
    ___ (d) has a horizontal fissure
    ___ (e) has a primary bronchus that is shorter, wider, and more vertical

    (1) right lung
    (2) left lung

CRITICAL THINKING QUESTIONS

1. Your friend Hedge wants to pierce his nose to go along with the 6 earrings in his ear. He thinks a ring through the center would be awesome but wonders if there's a difference between piercing the center versus the side of the nose. Is there?

2. Suzanne is traveling high into the Andes Mountains of South America to visit some ancient Incan ruins. Though she is in excellent physical condition, she finds herself breathing rapidly. What is happening to Suzanne and why/how?

3. Dawson was riding his new mountain bike through the woods near his home when he crashed into some brush. He felt a sharp pain in his thorax and when he rolled over, a large stick was impaled in his side. Now he's having trouble breathing. Why?

4. Gretchen, who is nine years old, is upset because after finally persuading her daddy to play soccer with her, he had to sit down and rest after only 15 minutes. She indignantly berated him, saying, "You know, Daddy, if you'd stop smoking, you could play longer." Why is it so hard for Gretchen's father to catch his breath? Be specific.

5. LaTasha is losing her patience with her little sister, LaTonya. "I'm going to hold my breath 'til I turn blue and die and then you're gonna get it!" screams LaTonya. LaTasha is not too worried. Why not?

ANSWERS TO FIGURE QUESTIONS

24.1 The conducting portion of the respiratory system includes the nose, pharynx, larynx, trachea, bronchi, and bronchioles (except the respiratory bronchioles).

24.2 The path of air is external nares → vestibule → nasal cavity → internal nares.

24.3 The superior border of the pharynx is the internal nares; the inferior border of the pharynx is the cricoid cartilage.

24.4 During swallowing, the epiglottis closes over the rima glottidis, the entrance to the trachea, to prevent aspiration of food and liquids into the lungs.

24.5 The main function of the vocal folds is voice production.

24.6 Because the tissues between the esophagus and trachea are soft, the esophagus can bulge and press against the trachea during swallowing.

24.7 The left lung has two lobes and secondary bronchi; the right lung has three of each.

24.8 The pleural membrane is a serous membrane.

24.9 Because two-thirds of the heart lies to the left of the midline, the left lung contains a cardiac notch to accommodate the presence of the heart. The right lung is shorter than the left because the diaphragm is higher on the right side to accommodate the liver.

24.10 Tertiary bronchi supply a broncopulmonary segment.

24.11 The wall of an alveolus is made up of type I alveolar cells, type II alveolar cells, and associated alveolar macrophages.

24.12 The respiratory membrane averages 0.5 μm in thickness.

24.13 If you are at rest while reading, your diaphragm is responsible for about 75% of each inhalation.

24.14 The medullary inspiratory area contains autorhythmic neurons that have cycles of activity/inactivity.

24.15 Peripheral chemoreceptors are responsive to changes in the partial pressures of oxygen and carbon dioxide, and concentrations of H^+ in the blood.

24.16 The respiratory system begins to develop about 4 weeks after fertilization.

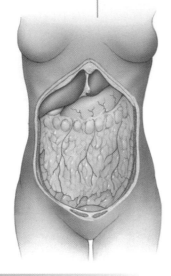

25

THE DIGESTIVE SYSTEM

INTRODUCTION Food contains a variety of nutrients—molecules needed for building new body tissues, repairing damaged tissues, and sustaining needed chemical reactions. Food is also vital for life because it is the source of the energy that drives the chemical reactions occurring in every cell of the body. As consumed, however, most food cannot be used as a source of cellular energy. It must first be broken down into molecules small enough to cross the plasma membranes of cells, a process known as **digestion.** The passage of these smaller molecules through cells into the blood and lymph is termed **absorption.** The organs that perform these functions—collectively called the **digestive system**—are the focus of this chapter.

The medical specialty that deals with the structure, function, diagnosis, and treatment of diseases of the stomach and intestines is called **gastroenterology** (gas′-trō-en′-ter-OL-ō-jē; *gastro-*=stomach; *enter-*=intestines; *-ology*=study of). The medical specialty that deals with the diagnosis and treatment of disorders of the rectum and anus is called **proctology** (prok-TOL-ō-jē; *proct-*=rectum).

How many body organs can you identify in this illustration?

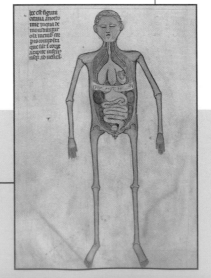

OVERVIEW OF THE DIGESTIVE SYSTEM

- Identify the organs of the digestive system.
- Describe the basic processes performed by the digestive system.

The digestive system (Figure 25.1) is composed of two groups of organs: the gastrointestinal (GI) tract and the accessory digestive organs. The **gastrointestinal (GI) tract,** or **alimentary canal** (*alimentary*=nourishment), is a continuous tube that extends from the mouth to the anus through the ventral body cavity. Organs of the gastrointestinal tract include the mouth, pharynx, esophagus, stomach, small intestine, and large intestine. The length of the GI tract taken from a cadaver is about 9 m (30 ft). In a living person it is shorter because the muscles along the walls of GI tract organs are in a state of *tonus* (sustained contraction). The **accessory digestive organs** include the teeth, tongue, salivary glands, liver, gallbladder, and pancreas. Teeth aid in the physical breakdown of food, and the tongue assists in chewing and swallowing. The other accessory digestive organs never come into direct contact with food. They produce or store secretions that flow into the GI tract through ducts and aid in the chemical breakdown of food.

The GI tract contains food from the time it is eaten until it is digested and absorbed or eliminated. Muscular contractions in the wall of the GI tract physically break down the food by churning it, and propel the food along the tract, from the esophagus to the anus. The contractions also help to dissolve foods by mixing them with fluids secreted into the tract. Enzymes secreted by accessory digestive organs and cells that line the tract break down the food chemically.

Overall, the digestive system performs six basic functions:

1. **Ingestion.** This process involves taking foods and liquids into the mouth (eating).

2. **Secretion.** Each day, cells within the walls of the GI tract and accessory digestive organs secrete a total of about 7 liters of water, acid, buffers, and enzymes into the lumen (interior space) of the tract.

3. **Mixing and propulsion.** Alternating contraction and relaxation of smooth muscle in the walls of the GI tract mix food and secretions and propel them toward the anus. This capability of the GI tract to mix and move material along its length is termed **motility.**

4. **Digestion.** Mechanical and chemical processes break down ingested food into small molecules. In **mechanical digestion** the teeth cut and grind food before it is swallowed, and then smooth muscles of the stomach and small intestine churn the food. As a result, food molecules become dissolved and thoroughly mixed with digestive enzymes. **Chemical digestion** is the breakdown of the large carbohydrate, lipid, protein, and nucleic acid molecules in food into smaller molecules. Digestive enzymes produced by the salivary glands, tongue, stomach, pancreas, and small intestine catalyze these catabolic reactions. A few substances in food can be absorbed without chemical digestion, including amino acids, cholesterol, glucose, vitamins, minerals, and water.

5. **Absorption.** The secreted fluids and the small molecules and ions that are products of digestion enter the epithelial cells lining the lumen of the GI tract. The absorbed substances pass into interstitial fluid and then into blood or lymph and circulate to cells throughout the body.

6. **Defecation.** Wastes, indigestible substances, bacteria, cells sloughed from the lining of the GI tract, and digested materials that were not absorbed leave the body through the anus in a process called **defecation.** The eliminated material is called **feces.**

1. Which components of the digestive system are GI tract organs and which are accessory digestive organs?
2. Which organs of the digestive system come in contact with food, and what are some of their digestive functions?

LAYERS OF THE GI TRACT

- Describe the layers that form the wall of the gastrointestinal tract.

The wall of the GI tract, from the lower esophagus to the anal canal, has the same basic, four-layered arrangement of tissues. The four layers of the tract, from deep to superficial, are the mucosa, submucosa, muscularis, and serosa (Figure 25.2 on page 767).

Mucosa

The **mucosa,** or inner lining of the GI tract, is a mucous membrane. It is composed of (1) a layer of epithelium in direct contact with the contents of the GI tract, (2) areolar connective tissue, and (3) a thin layer of smooth muscle (muscularis mucosae).

1. The **epithelium** in the mouth, pharynx, esophagus, and anal canal is mainly nonkeratinized stratified squamous epithelium that serves a protective function. Simple columnar epithelium, which functions in secretion and absorption, lines the stomach and intestines. Neighboring simple columnar epithelial cells are firmly sealed to one another by tight junctions that restrict leakage between the cells. The rate of renewal of GI tract epithelial cells is rapid: Every 5–7 days they slough off and are replaced by new cells. Located among the absorptive epithelial cells are exocrine cells that secrete mucus and fluid into the lumen of the tract, and several types of endocrine cells, collectively called **enteroendocrine cells,** that secrete hormones into the bloodstream.

Figure 25.1 Organs of the digestive system.

The organs of the gastrointestinal (GI) tract are the mouth, pharynx, esophagus, stomach, small intestine, and large intestine. Accessory digestive organs include the teeth, tongue, salivary glands, liver, gallbladder, and pancreas.

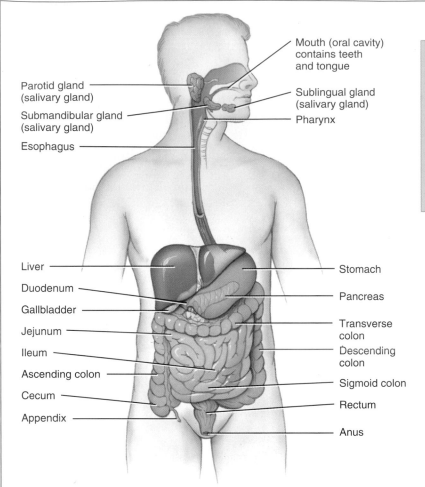

Mouth (oral cavity) contains teeth and tongue

Parotid gland (salivary gland)

Sublingual gland (salivary gland)

Submandibular gland (salivary gland)

Pharynx

Esophagus

Liver

Stomach

Duodenum

Pancreas

Gallbladder

Jejunum

Transverse colon

Ileum

Descending colon

Ascending colon

Sigmoid colon

Cecum

Rectum

Appendix

Anus

FUNCTIONS

1. *Ingestion.* Taking food into the mouth.
2. *Secretion.* Release of water, acid, buffers, and enzymes into the lumen of the GI tract.
3. *Mixing and propulsion.* Churning and propulsion of food through the GI tract.
4. *Digestion.* Mechanical and chemical breakdown of food.
5. *Absorption.* Passage of digested products from the GI tract into the blood and lymph.
6. *Defecation.* The elimination of feces from the GI tract.

(a) Right lateral view of head and neck and anterior view of trunk

SUPERIOR

Transverse colon

Diaphragm

Greater omentum

Peritoneum

Jejunum

Ascending colon

Ileum

Cecum

Appendix

Descending colon

Sigmoid colon

(b) Anterior view

Which structures of the digestive system secrete digestive enzymes?

Figure 25.2 Layers of the gastrointestinal tract.

The four layers of the gastrointestinal tract, from deep to superficial, are the mucosa, submucosa, muscularis, and serosa.

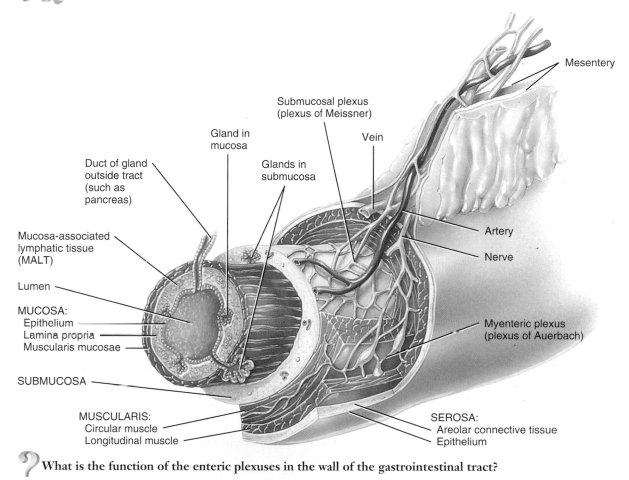

What is the function of the enteric plexuses in the wall of the gastrointestinal tract?

2. The **lamina propria** (*lamina* = thin, flat plate; *propria* = one's own) is an areolar connective tissue layer containing many blood and lymphatic vessels that carry the nutrients absorbed by the GI tract to the other tissues of the body. This layer supports the epithelium and binds it to the muscularis mucosae (discussed next). The lamina propria also contains most of the cells of the **mucosa-associated lymphoid tissue (MALT).** These prominent lymphatic nodules contain immune system cells that protect against disease. MALT is present all along the GI tract, especially in the tonsils, small intestine, appendix, and large intestine, and it contains about as many immune cells as are present in all the rest of the body. The lymphocytes and macrophages in MALT mount immune responses against microbes, such as bacteria, that may penetrate the epithelium.

3. A thin layer of smooth muscle fibers called the **muscularis mucosae** causes the mucous membrane of the stomach and small intestine to form many small folds, increasing the surface area for digestion and absorption. Movements of the muscularis mucosae ensure that all absorptive cells are fully exposed to the contents of the GI tract.

Submucosa

The **submucosa** consists of areolar connective tissue that binds the mucosa to the third layer, the muscularis. It is highly vascular and contains the **submucosal plexus,** or *plexus of Meissner*, a portion of the **enteric nervous system (ENS).** The ENS is the "brain of the gut" and consists of approximately 100 million neurons in two enteric plexuses that extend the entire length of the GI tract. The submucosal plexus contains sensory and motor enteric neurons, plus parasympathetic and sympathetic postganglionic fibers that innervate the mucosa and submucosa. It regulates movements of the mucosa and vasoconstriction of blood vessels. Because it also innervates secretory cells of mucosal glands, the submucosal plexus is important in controlling secretions of the GI tract. The submucosa may also contain glands and lymphatic tissue.

Muscularis

The **muscularis** of the mouth, pharynx, and superior and middle parts of the esophagus contains *skeletal muscle* that produces voluntary swallowing. Skeletal muscle also forms the external anal sphincter, which permits voluntary control of defecation. Throughout the rest of the tract, the muscularis consists of *smooth muscle* that is generally found in two sheets: an inner sheet of circular fibers and an outer sheet of longitudinal fibers. Involuntary contractions of the smooth muscles assist in the mechanical breakdown of food, mix it with digestive secretions, and propel it along the tract. The muscularis also contains the second plexus of the enteric nervous system—the **myenteric plexus** (*my-*=muscle), or *plexus of Auerbach*, which contains enteric neurons, parasympathetic ganglia and postganglionic fibers, and sympathetic postganglionic fibers that innervate the muscularis. This plexus mostly controls GI tract motility (movement), in particular the frequency and strength of the contractions of the muscularis.

Serosa

Those portions of the GI tract that are suspended in the abdominopelvic cavity have a superficial layer called the **serosa**. As its name implies, the serosa is a serous membrane composed of areolar connective tissue and simple squamous epithelium. As we will see shortly, the esophagus, which passes through the mediastinum, has a superficial layer called the *adventitia* com-

posed of areolar connective tissue. Inferior to the diaphragm, the serosa is also called the visceral peritoneum; it forms a portion of the peritoneum, which we examine in detail next.

CHECKPOINT

3. Where along the GI tract is the muscularis composed of skeletal muscle? Is control of this skeletal muscle voluntary or involuntary?
4. What two plexuses form the enteric nervous system, and where are they located?

PERITONEUM

OBJECTIVE

• Describe the peritoneum and its folds.

The **peritoneum** (per′-i-tō-NĒ-um; *peri-*=around) is the largest serous membrane of the body; it consists of a layer of simple squamous epithelium (mesothelium) with an underlying supporting layer of connective tissue. The peritoneum is divided into the **parietal peritoneum,** which lines the wall of the abdominopelvic cavity, and the **visceral peritoneum** or serosa,

Figure 25.3 Relationship of the peritoneal folds to each other and to organs of the digestive system. The size of the peritoneal cavity in (a) is exaggerated for emphasis.

The peritoneum is the largest serous membrane in the body.

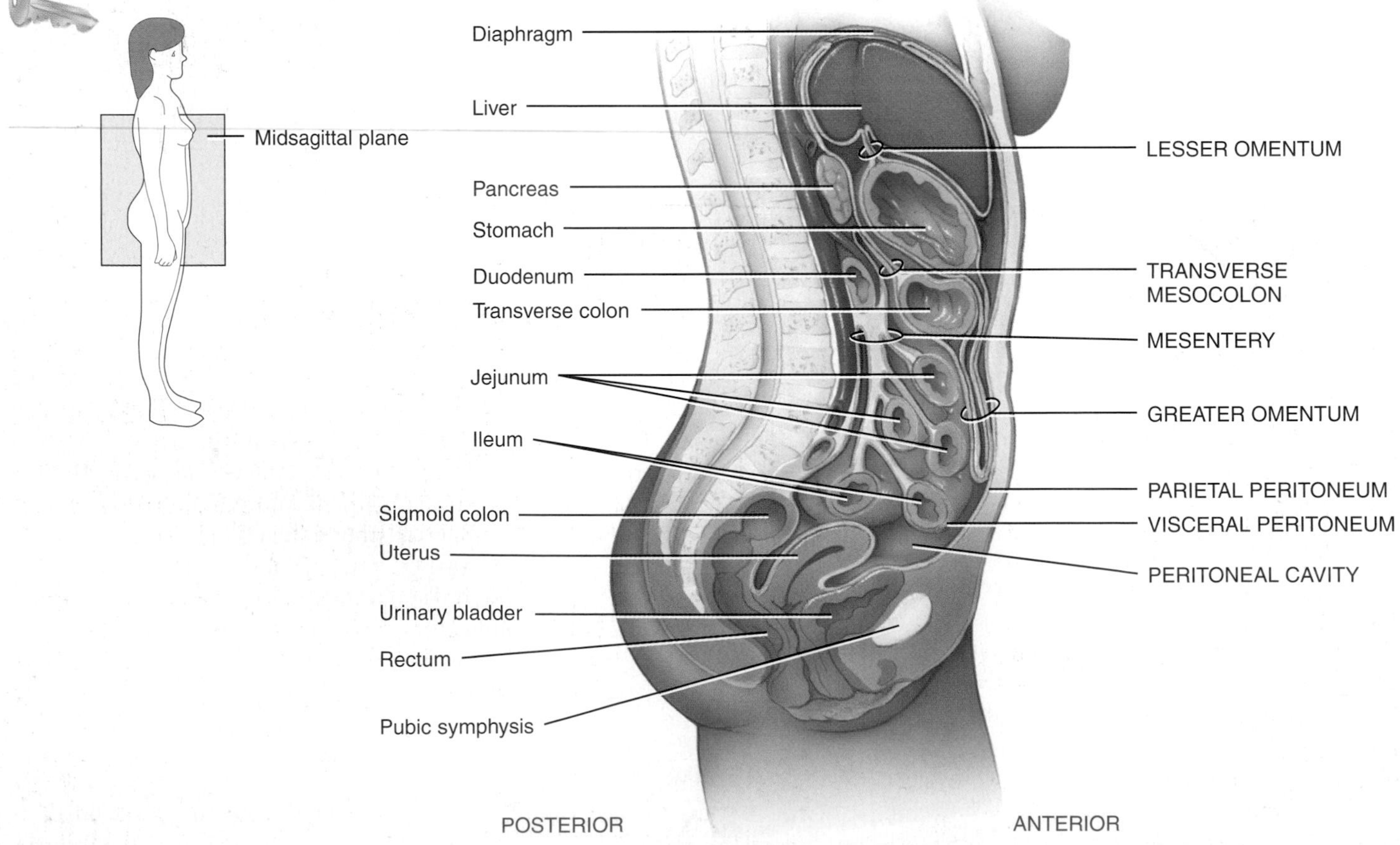

(a) Midsagittal section showing the peritoneal folds

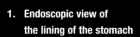

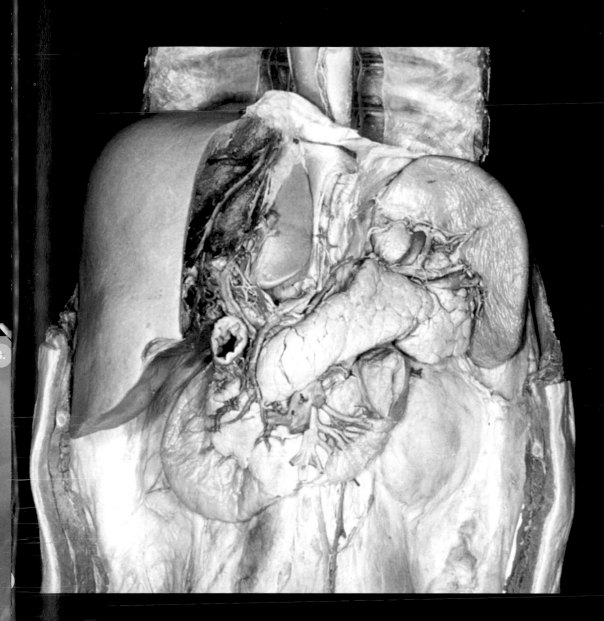

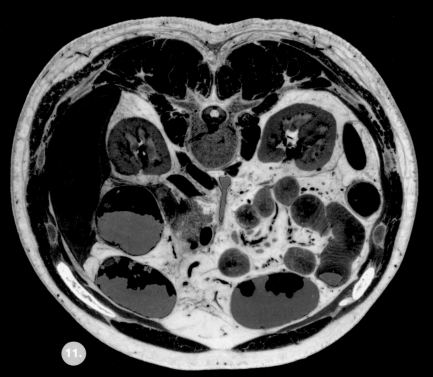

1. Endoscopic view of the lining of the stomach

2. Sonogram of the gallbladder

3. ERCP of the liver, gallbladder, and pancreas

4. Light micrograph of the stomach wall

5. Radiograph of the stomach

6. Endoscopic view of the lining of the esophagus

7. Endoscopic view of the lining of the colon

8. SEM of villi of the small intestine

9. Radiograph of the large Intestine

10. Liver, gallbladder, pancreas, duodenum, and spleen

11. Transverse section through the abdomen

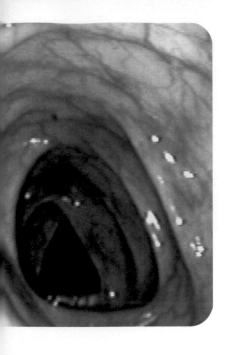

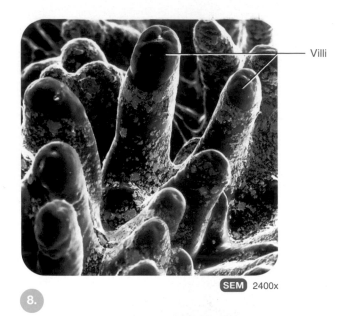

Villi

SEM 2400x

8.

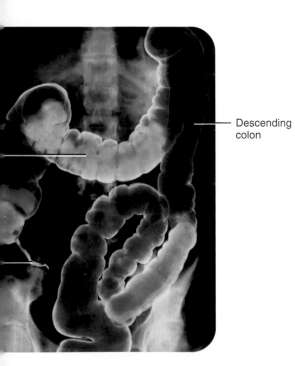

Descending colon

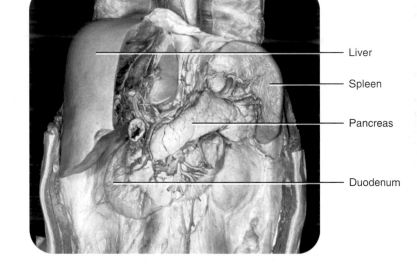

Liver

Spleen

Pancreas

Duodenum

10.

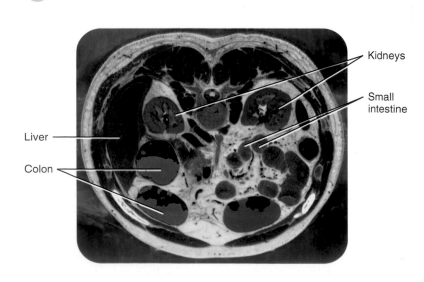

Kidneys

Small intestine

Liver

Colon

11.

which as you just learned covers some of the organs in the cavity (Figure 25.3a). The slim space between the parietal and visceral portions of the peritoneum, called the **peritoneal cavity,** contains serous fluid. In certain diseases, the peritoneal cavity may become distended by the accumulation of several liters of fluid, a condition called **ascites** (a-SĪ-tēz).

As we will see, some organs lie against the posterior abdominal wall and are covered by peritoneum only on their anterior surfaces. Such organs, including the kidneys and pancreas, are said to be **retroperitoneal** (*retro-*=behind).

Unlike the pericardium and pleurae, which smoothly cover the heart and lungs, the peritoneum contains large folds that weave between the viscera. The folds bind the organs to each other and to the walls of the abdominal cavity. They also contain blood vessels, lymphatic vessels, and nerves that supply the abdominal organs.

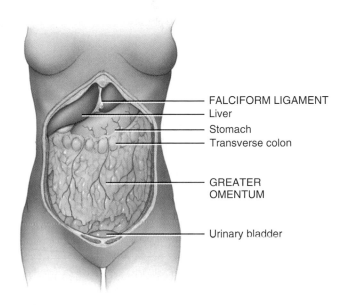

(b) Anterior view

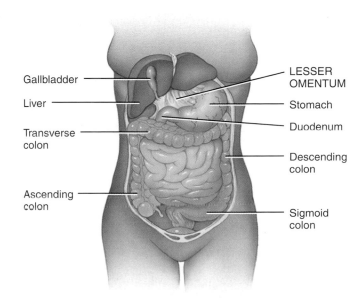

(c) Lesser omentum, anterior view
(liver and gallbladder lifted)

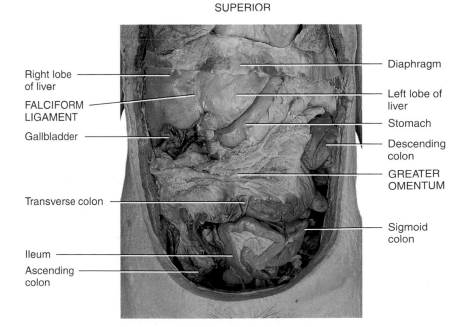

(d) Anterior view (greater omentum
lifted and small intestine reflected
to right side)

(e) Anterior view

 Which peritoneal fold binds the small intestine to the posterior abdominal wall?

The **greater omentum** (ō-MEN-tum=fat skin), the largest peritoneal fold, drapes over the transverse colon and coils of the small intestine like a "fatty apron" (Figure 25.3a, d). Because the greater omentum is a double sheet that folds back upon itself, it is a four-layered structure. From attachments along the stomach and duodenum, the greater omentum extends inferiorly and anteriorly to the small intestine, then turns and extends superiorly, where it attaches to the transverse colon. The greater omentum normally contains a considerable amount of adipose tissue, which can greatly expand with weight gain, giving rise to the characteristic "beer belly" seen in some overweight individuals. The many lymph nodes of the greater omentum contribute macrophages and antibody-producing plasma cells that help combat and contain infections of the GI tract. The greater omentum has considerable mobility and moves around the peritoneal cavity in response to changes in posture and movements of the viscera. It is capable of wrapping itself around an inflamed organ (such as the appendix) and isolating it from other organs.

The **falciform ligament** (FAL-si-form; *falc-*=sickle-shaped) attaches the liver to the anterior abdominal wall and diaphragm (Figure 25.3b). The liver is the only digestive organ that is attached to the anterior abdominal wall.

The **lesser omentum,** which arises as two folds in the serosa of the stomach and duodenum, suspends the stomach and duodenum from the liver (Figure 25.3a, c). It contains some lymph nodes.

Another fold of the peritoneum, called the **mesentery** (MEZ-en-ter′-ē; *mes-*=middle), is fan-shaped and binds the small intestine to the posterior abdominal wall (Figure 25.3a, d). It extends from the posterior abdominal wall to wrap around the small intestine and then returns to its origin, forming a double-layered structure. Between the two layers are blood vessels, lymphatic vessels, and lymph nodes.

A fold of peritoneum called the **mesocolon** (mez′-ō-KŌ-lon) binds the transverse and sigmoid colons of the large intestine to the posterior abdominal wall (Figure 25.3a); it also carries blood and lymphatic vessels to the intestines. The mesentery and mesocolon hold the intestines loosely in place, allowing for a great amount of movement as muscular contractions mix and move the contents along the GI tract.

PERITONITIS

Peritonitis is an acute inflammation of the peritoneum. A common cause of the condition is contamination of the peritoneum by infectious microbes, which can result from accidental or surgical wounds in the abdominal wall, or from perforation or rupture of abdominal organs. If, for example, bacteria gain access to the peritoneal cavity through an intestinal perforation or rupture of the appendix, they can produce an acute, life-threatening form of peritonitis. A less serious (but still painful) form of peritonitis can result from the rubbing together of inflamed peritoneal surfaces. Peritonitis is of particularly grave concern to those who rely on peritoneal dialysis. ■

5. Describe the locations of the visceral peritoneum and parietal peritoneum.
6. Describe the attachment sites and functions of the mesentery, mesocolon, falciform ligament, lesser omentum, and greater omentum.

MOUTH

OBJECTIVES

● Identify the locations of the salivary glands, and describe the functions of their secretions.
● Describe the structure and functions of the tongue.
● Identify the parts of a typical tooth, and compare deciduous and permanent dentitions.

The **mouth,** also referred to as the **oral** or **buccal cavity** (BUK-al; *bucca*=cheeks), is formed by the cheeks, hard and soft palates, and tongue (Figure 25.4). Forming the lateral walls of the oral cavity are the **cheeks**—muscular structures covered externally by skin and internally by nonkeratinized stratified squamous epithelium. The anterior portions of the cheeks end at the lips.

The **lips** or **labia** (=fleshy borders), fleshy folds surrounding the opening of the mouth, are covered externally by skin and internally by a mucous membrane. There is a transition zone where the two kinds of covering tissue meet. This portion of the lips is nonkeratinized, and the color of the blood in the underlying blood vessels is visible through the transparent surface layer. The inner surface of each lip is attached to its corresponding gum by a midline fold of mucous membrane called the **labial frenulum** (LĀ-bē-al FREN-ū-lum; *frenulum*=small bridle).

The orbicularis oris muscle and connective tissue lie between the skin and the mucous membrane of the oral cavity. During chewing, contraction of the buccinator muscles in the cheeks and orbicularis oris muscle in the lips help keep food between the upper and lower teeth. These muscles also assist in speech.

The **vestibule** (=entrance to a canal) of the oral cavity is a space bounded externally by the cheeks and lips and internally by the gums and teeth. The **oral cavity proper** is a space that extends from the gums and teeth to the **fauces** (FAW-sēs=passages), the opening between the oral cavity and the pharynx or throat.

The **hard palate,** the anterior portion of the roof of the mouth, is formed by the maxillae and palatine bones. It is covered by mucous membrane and forms a bony partition between the oral and nasal cavities. The **soft palate,** which forms the posterior portion of the roof of the mouth, is an arch-shaped muscular partition between the oropharynx and nasopharynx that is lined by mucous membrane.

Hanging from the free border of the soft palate is a conical muscular process called the **uvula** (Ū-vū-la=little grape). During swallowing, the soft palate and uvula are drawn superiorly, closing off the nasopharynx and preventing swallowed foods and liquids from entering the nasal cavity. Lateral to the base of the uvula are two muscular folds that run down the lateral sides of the soft palate: Anteriorly, the **palatoglossal arch** extends to the side of the base of the tongue; posteriorly, the **palatopharyngeal arch** (PAL-a-tō-fa-rin′-jē-al) extends to the side of the pharynx. The palatine tonsils are situated between the arches, and the lingual tonsils are situated at the base of the tongue. At the posterior border of the soft palate, the mouth opens into the oropharynx through the fauces (Figure 25.4).

Salivary Glands

A **salivary gland** releases a secretion called *saliva* into the oral cavity. Ordinarily, just enough saliva is secreted to keep the mucous membranes of the mouth and pharynx moist and to cleanse the mouth and teeth. When food enters the mouth, however, secretion of saliva increases to lubricate, dissolve, and begin the chemical breakdown of the food.

The mucous membrane of the mouth and tongue contains many small salivary glands that open directly, or indirectly via short ducts, to the oral cavity. These glands, which include *labial, buccal,* and *palatal glands* in the lips, cheeks, and palate, respectively, and *lingual glands* in the tongue, all make a small contribution to saliva.

However, most saliva is secreted by the **major salivary glands,** which lie beyond the oral mucosa; their secretions empty into ducts that lead to the oral cavity. The three pairs of major salivary glands are the parotid, submandibular, and sublingual glands (Figure 25.5a). The **parotid glands** (*par-*=near; *ot-*=ear) are located inferior and anterior to the ears, between the skin and the masseter muscle. Each secretes saliva into the oral cavity via a **parotid (Stensen's) duct** that pierces the buccinator muscle to open into the vestibule opposite the second maxillary (upper) molar tooth. The **submandibular glands** are found beneath the base of the tongue in the posterior part of the floor of the mouth. Their ducts, the **submandibular (Wharton's) ducts,** run under the mucosa on either side of the midline of the floor of the mouth and enter the oral cavity proper lateral to the lingual frenulum. The **sublingual glands** are superior to the submandibular glands. Their ducts, the **lesser sublingual (Rivinus') ducts,** open into the floor of the mouth in the oral cavity proper.

The parotid gland receives its blood supply from branches of the external carotid artery and is drained by tributaries of the external jugular vein. The submandibular gland is supplied by branches of the facial artery and drained by tributaries of the facial vein. The sublingual gland is supplied by the sublingual branch of the lingual artery and the submental branch of the

Figure 25.4   Structures of the mouth (oral cavity).

The mouth is formed by the cheeks, hard and soft palates, and tongue.

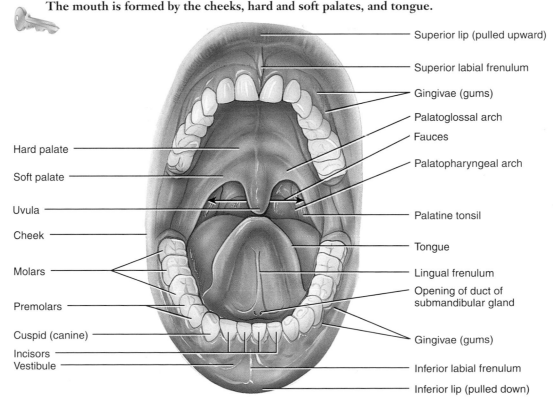

Superior lip (pulled upward)

Superior labial frenulum

Gingivae (gums)

Palatoglossal arch

Fauces

Palatopharyngeal arch

Hard palate

Soft palate

Uvula

Cheek

Molars

Premolars

Cuspid (canine)

Incisors

Vestibule

Palatine tonsil

Tongue

Lingual frenulum

Opening of duct of submandibular gland

Gingivae (gums)

Inferior labial frenulum

Inferior lip (pulled down)

Anterior view

What is the function of the uvula?

Figure 25.5 **The three major salivary glands—parotid, sublingual, and submandibular.** The submandibular glands, shown in the light micrograph in (b), consist mostly of serous acini (serous-fluid-secreting portions of gland) and a few mucous acini (mucus-secreting portions of gland); the parotid glands consist of serous acini only; and the sublingual glands consist of mostly mucous acini and a few serous acini. (See Tortora, *A Photographic Atlas of the Human Body*, 2e, Figure 12.6a.)

Saliva lubricates and dissolves foods and begins the chemical breakdown of carbohydrates and lipids.

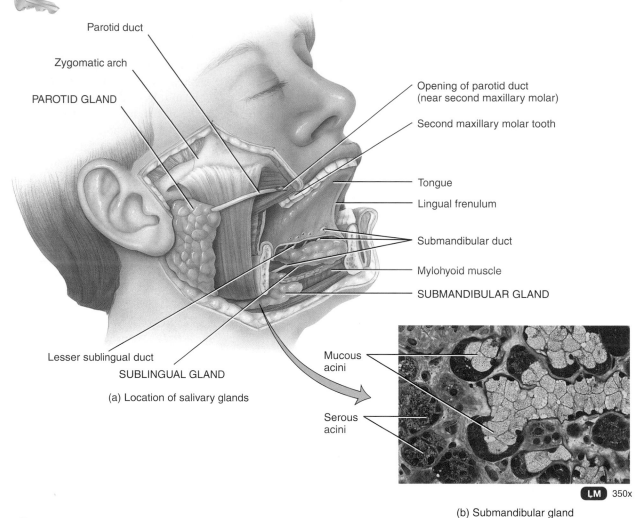

(a) Location of salivary glands

(b) Submandibular gland

LM 350x

The ducts of which salivary glands empty on either side of the lingual frenulum?

facial artery and is drained by tributaries of the sublingual and submental veins.

The salivary glands receive both sympathetic and parasympathetic innervation. The sympathetic fibers form plexuses on the blood vessels that supply the glands and initiate vasoconstriction, which decreases the production of saliva. The parotid gland receives sympathetic fibers from the plexus on the external carotid artery, whereas the submandibular and sublingual glands receive sympathetic fibers that contribute to the sympathetic plexus and accompany the facial artery to the glands. The parasympathetic fibers of the glands produce vasodilation and thus increase the production of saliva.

The fluids secreted by the buccal glands, minor salivary glands, and the three pairs of major salivary glands constitute

saliva. Amounts of saliva secreted daily vary considerably but range from 1000 to 1500 mL (1 to 1.6 qt). Chemically, saliva is 99.5% water and 0.5% solutes and has a slightly acidic pH (6.35 to 6.85). The solute portion includes mucus, an enzyme that destroys bacteria (lysozyme), the digestive enzymes salivary amylase and lingual lipase, and traces of salts, proteins, and other organic compounds. **Salivary amylase** initiates the breakdown of starch in the mouth. **Lingual lipase** and mucus are secreted by **lingual glands** on the dorsum of the tongue. This enzyme, which is active in the stomach, can digest as much as 30% of dietary triglycerides (fats) into simpler fatty acids and monoglycerides.

Secretion of saliva, or **salivation** (sal-i-VĀ-shun), is controlled by the nervous system. Normally, parasympathetic stimulation promotes continuous secretion of a moderate amount of saliva, which keeps the mucous membranes moist and lubricates the movements of the tongue and lips during speech. The saliva is then swallowed and helps moisten the esophagus. Eventually, most components of saliva are reabsorbed, which prevents fluid loss. Sympathetic stimulation dominates during stress, resulting in dryness of the mouth. During dehydration, the salivary glands stop secreting saliva to conserve water; the resulting dryness in the mouth contributes to the sensation of thirst. Drinking will then not only restore the homeostasis of body water but also moisten the mouth.

The feel and taste of food also are potent stimulators of salivary gland secretions. Chemicals in the food stimulate receptors in taste buds on the tongue, and impulses are conveyed from the taste buds to two salivary nuclei in the brain stem (**superior** and **inferior salivatory nuclei**). Returning parasympathetic impulses in fibers of the facial (VII) and glossopharyngeal (IX) nerves stimulate the secretion of saliva. Saliva continues to be heavily secreted for some time after food is swallowed; this flow of saliva washes out the mouth and dilutes and buffers the remnants of irritating chemicals. The smell, sight, sound, or thought of food may also stimulate secretion of saliva.

MUMPS

Although any of the salivary glands may be the target of a nasopharyngeal infection, the mumps virus (myxovirus) typically attacks the parotid glands. **Mumps** is an inflammation and enlargement of the parotid glands accompanied by moderate fever, malaise (general discomfort), and extreme pain in the throat, especially when swallowing sour foods or acidic juices. Swelling occurs on one or both sides of the face, just anterior to the ramus of the mandible. In about 30% of males past puberty, the testes may also become inflamed; sterility rarely occurs because testicular involvement is usually unilateral (one testis only). Since a vaccine became available for mumps in 1967, the incidence of the disease has declined. ■

Tongue

The **tongue** is an accessory digestive organ composed of skeletal muscle covered with mucous membrane. Together with its associated muscles, it forms the floor of the oral cavity. The tongue is divided into symmetrical lateral halves by a median septum that extends its entire length, and it is attached inferiorly to the hyoid bone, styloid process of the temporal bone, and mandible. Each half of the tongue consists of an identical complement of extrinsic and intrinsic muscles.

The **extrinsic muscles** of the tongue, which originate outside the tongue (attach to bones in the area) and insert into connective tissues in the tongue, include the hyoglossus, genioglossus, and styloglossus muscles (see Figure 11.7 on page 308). The extrinsic muscles move the tongue from side to side and in and out to maneuver food for chewing, shape the food into a rounded mass, and force the food to the back of the

mouth for swallowing. They also form the floor of the mouth and hold the tongue in position, and assist in speech. The **intrinsic muscles** originate in and insert into connective tissue within the tongue and alter the shape and size of the tongue for speech and swallowing. The intrinsic muscles include the longitudinalis superior, longitudinalis inferior, transversus linguae, and verticalis linguae muscles. The **lingual frenulum** (*lingua* =the tongue), a fold of mucous membrane in the midline of the undersurface of the tongue, is attached to the floor of the mouth and aids in limiting the movement of the tongue posteriorly (see Figures 25.4 and 25.5). If a lingual frenulum is abnormally short or rigid—a condition called **ankyloglossia** (ang′-kē-lō-GLOSS-ē-a)—the person is said to be "tongue-tied" because of the resulting impairment in eating and speaking.

The dorsum (upper surface) and lateral surfaces of the tongue are covered with **papillae** (pa-PIL-ē=nipple-shaped projections), projections of the lamina propria covered with keratinized epithelium (see Figure 22.2a, b on page 675). Many papillae contain taste buds, the receptors for gustation (taste). As their name implies, **fungiform papillae** (FUN-ji-form=mushroomlike) are mushroomlike elevations distributed among the filiform papillae that are more numerous near the tip of the tongue. They appear as red dots on the surface of the tongue, and most of them contain taste buds. **Vallate (circumvallate) papillae** (VAL-āt−wall-like) are arranged in an inverted V shape on the posterior surface of the tongue; all of them contain taste buds. **Foliate papillae** (FŌ-lē-āt=leaflike) are located in small trenches on the lateral margins of the tongue, but most of their taste buds degenerate in early childhood. **Filiform papillae** (FIL-i-form=threadlike) are pointed, threadlike projections distributed in parallel rows over the anterior two-thirds of the tongue. Although filiform papillae lack taste buds, they contain receptors for touch and increase friction between the tongue and food, making it easier for the tongue to move food in the oral cavity.

Teeth

The **teeth,** or **dentes** (Figure 25.6), are accessory digestive organs located in sockets of the alveolar processes of the mandible and maxillae. The alveolar processes are covered by the **gingivae** (jin-JI-vē), or gums, which extend slightly into each socket to form the **gingival sulcus.** The sockets are lined by the **periodontal ligament** or **membrane** (*odont-*=tooth), which consists of dense fibrous connective tissue and is attached to the socket walls and cemental surface of the roots. The periodontal ligament anchors the teeth in position and acts as a shock absorber during chewing.

A typical tooth consists of three principal regions. The **crown** is the visible portion above the level of the gums. One to three **roots** are embedded in each socket. The **neck** is the constricted junction of the crown and root near the gum line.

Teeth are composed primarily of **dentin,** a calcified connective tissue that gives the tooth its basic shape and rigidity. It is harder than bone because of its higher content (70%) of calcium salts. The dentin encloses a cavity. The enlarged part of the cavity, the **pulp cavity,** lies within the crown and is filled with

Figure 25.6 **A typical tooth and surrounding structures.**

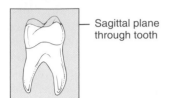 Teeth are anchored in sockets of the alveolar processes of the mandible and maxillae.

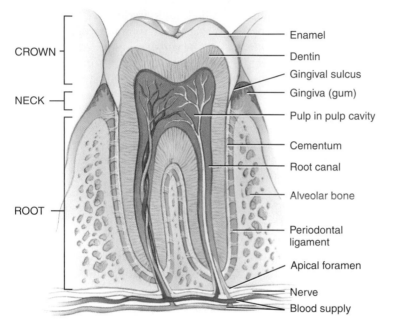

Sagittal section of a mandibular (lower) molar

What type of tissue is the main component of teeth?

pulp, a connective tissue containing blood vessels, nerves, and lymphatic vessels. Narrow extensions of the pulp cavity, called **root canals,** run through the root(s) of the tooth. Each root canal has an opening at its base, the **apical foramen,** through which blood vessels, lymphatic vessels, and nerves pass.

ROOT CANAL THERAPY

Root canal therapy refers to a procedure, accomplished in several phases, in which all traces of pulp tissue are removed from the pulp cavity and root canals of a badly diseased tooth. After a hole is made in the tooth, the root canals are filed out and irrigated to remove bacteria. Then the canals are treated with medication and sealed tightly. The damaged crown is then repaired. ∎

The dentin of the crown is covered by **enamel** that consists primarily of calcium phosphate and calcium carbonate. Enamel, the hardest substance in the body and the richest in calcium salts (about 95%), protects the tooth from the wear of chewing. It is

also a barrier against acids that easily dissolve the dentin. The dentin of the root is covered by **cementum,** also a bonelike substance, which attaches the root to the periodontal ligament.

The arteries that supply blood to the teeth are distributed to the pulp cavity and surrounding periodontal ligament. These include the superior alveolar branches of the maxillary artery (anterior and posterior) and the incisive and dental branches of the inferior alveolar artery.

The teeth receive sensory fibers from branches of the maxillary and mandibular divisions of the trigeminal (V) nerve—the maxillary teeth from branches of the maxillary division and the mandibular teeth from branches of the mandibular division.

The branch of dentistry that is concerned with the prevention, diagnosis, and treatment of diseases that affect the pulp, root, periodontal ligament, and alveolar bone is known as **endodontics** (en′-dō-DON-tiks; *endo-*=within). **Orthodontics** (or′-thō-DON-tiks; *ortho-*=straight) is a branch of dentistry that is concerned with the prevention and correction of abnormally aligned teeth, whereas **periodontics** (per′-ē-ō-DON-tiks) is a branch of dentistry concerned with the treatment of abnormal conditions of the tissues immediately surrounding the teeth.

Humans have two **dentitions,** or sets of teeth: deciduous and permanent. The first of these—the **deciduous teeth** (*decidu-*=falling out), also called **primary teeth, milk teeth,** or **baby teeth**—begin to erupt (emerge) at about 6 months of age, and one pair of teeth appears at about each month thereafter, until all 20 are present (Figure 25.7a, c). The incisors, which are closest to the midline, are chisel-shaped and adapted for cutting into food. They are referred to as either **central** or **lateral incisors** on the basis of their position. Next to the incisors, moving posteriorly, are the **cuspids (canines),** which have a pointed surface called a cusp. Cuspids are used to tear and shred food. Incisors and cuspids have only one root apiece. Posterior to them lie the **first** and **second molars,** which have four cusps. Maxillary (upper) molars have three roots; mandibular (lower) molars have two roots. The molars crush and grind food.

All the deciduous teeth are lost—generally between the ages of 6 and 12 years—and are replaced by the **permanent (secondary) teeth** (Figure 25.7b, d). The permanent dentition contains 32 teeth that erupt between age 6 and adulthood. The pattern resembles the deciduous dentition, with the following exceptions. The deciduous molars are replaced by the **first** and **second premolars (bicuspids),** which have two cusps and one root (upper first bicuspids have two roots) and are used for crushing and grinding. The permanent molars, which erupt into the mouth posterior to the bicuspids, do not replace any deciduous teeth and erupt as the jaw grows to accommodate them— the **first molars** at age 6, the **second molars** at age 12, and the **third molars (wisdom teeth)** after age 17.

Often the human jaw does not have enough room posterior to the second molars to accommodate the eruption of the third molars. In this case, the third molars remain embedded in the alveolar bone and are said to be "impacted." They often cause pressure and pain and must be removed surgically. In some people, third molars may be dwarfed in size or may not develop at all.

Table 25.1 on page 776 lists terms used to describe the surface orientation of teeth.

Figure 25.7 Dentitions and times of eruptions (indicated in parentheses). A designated letter (deciduous teeth) or number (permanent teeth) uniquely identifies each tooth. Deciduous teeth begin to erupt at 6 months of age, and one pair of teeth appears about each month thereafter, until all 20 are present.

🔑 **There are 20 teeth in a complete deciduous set and 32 teeth in a complete permanent set.**

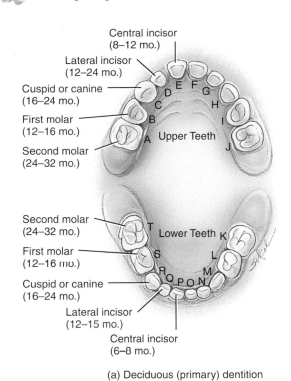

Central incisor (8–12 mo.)
Lateral incisor (12–24 mo.)
Cuspid or canine (16–24 mo.)
First molar (12–16 mo.)
Second molar (24–32 mo.)

Upper Teeth

Second molar (24–32 mo.)
First molar (12–16 mo.)
Cuspid or canine (16–24 mo.)
Lateral incisor (12–15 mo.)
Central incisor (6–8 mo.)

Lower Teeth

(a) Deciduous (primary) dentition

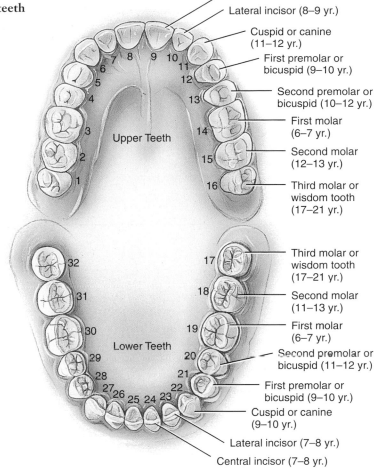

Central incisor (7–8 yr.)
Lateral incisor (8–9 yr.)
Cuspid or canine (11–12 yr.)
First premolar or bicuspid (9–10 yr.)
Second premolar or bicuspid (10–12 yr.)
First molar (6–7 yr.)
Second molar (12–13 yr.)
Third molar or wisdom tooth (17–21 yr.)

Upper Teeth

Third molar or wisdom tooth (17–21 yr.)
Second molar (11–13 yr.)
First molar (6–7 yr.)
Second premolar or bicuspid (11–12 yr.)
First premolar or bicuspid (9–10 yr.)
Cuspid or canine (9–10 yr.)
Lateral incisor (7–8 yr.)
Central incisor (7–8 yr.)

Lower Teeth

(b) Permanent (secondary) dentition

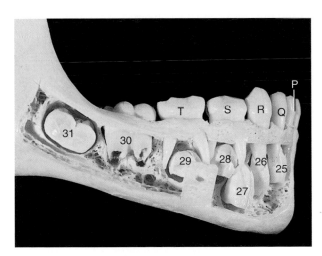

Right lateral view

Deciduous teeth:
P - Central incisor
Q - Lateral incisor
R - Cuspid (canine)
S - First molar (bicuspid)
T - Second molar

Permanent teeth:
25 - Central incisor
26 - Lateral incisor
27 - Cuspid (canine)
28 - First premolar (bicuspid)
29 - Second premolar
30 - First molar
31 - Second molar

(c) Mandible of a six year old child showing erupted deciduous teeth and unerupted permanent teeth

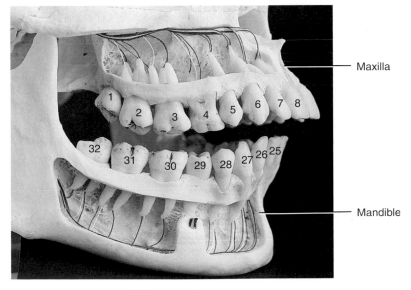

Maxilla

Mandible

Right lateral view

8, 25 - Central incisor
7, 26 - Lateral incisor
6, 27 - Cuspid (canine)
5, 28 - First premolar (bicuspid)

4, 29 - Second premolar
3, 30 - First molar
2, 31 - Second molar
1, 32 - Third molar (wisdom tooth)

(d) Mandible (and maxilla) showing permanent teeth and blood and nerve supply to them

❓ **Which permanent teeth do not replace any deciduous teeth?**

TABLE 25.1   SURFACE ORIENTATION OF TEETH

TOOTH SURFACE	DESCRIPTION
Labial	Contacting the lips.
Buccal	Contacting or facing the cheeks.
Lingual	Facing the tongue (teeth of the mandible only).
Palatal	Facing the palate (teeth of the maxillae only).
Mesial	Anterior or medial side relative to dental arch.
Distal	Posterior or lateral side relative to dental arch.
Occlusal	The biting surface.

CHECKPOINT

7. What structures form the mouth (oral cavity)?
8. How are the major salivary glands distinguished on the basis of location and structure?
9. How do the extrinsic and intrinsic muscles of the tongue differ in function?
10. Contrast the functions of incisors, cuspids, premolars, and molars.

PHARYNX

OBJECTIVE

• Describe the structure and function of the pharynx.

Through chewing, or **mastication** (mas′-ti-KĀ-shun; *masticare*= to chew), the tongue manipulates food, the teeth grind it, and the food is mixed with saliva. As a result, the food is reduced to a soft, flexible mass called a **bolus** (*bolos*=lump) that is easily swallowed. When food is first swallowed, it passes from the mouth into the pharynx.

The **pharynx** (*pharynx*=throat) is a funnel-shaped tube that extends from the internal nares to the esophagus posteriorly and the larynx anteriorly (see Figure 24.3 on page 736). The pharynx is composed of skeletal muscle and lined by mucous membrane. Whereas the nasopharynx functions only in respiration, the oropharynx and laryngopharynx have both digestive and respiratory functions. Swallowing, or **deglutition** (dē-gloo-TISH-un), is a mechanism that moves food from the mouth to the stomach. It is helped by saliva and mucus and involves the mouth, pharynx, and esophagus. Food that is swallowed passes from the mouth into the oropharynx and laryngopharynx before passing into the esophagus. Muscular contractions of the oropharynx and laryngopharynx help propel food into the esophagus and then into the stomach.

CHECKPOINT

11. What is a bolus? How is it formed?
12. Where does the pharynx begin and end?

ESOPHAGUS

OBJECTIVE

• Describe the location, anatomy, histology, and function of the esophagus.

The **esophagus** (e-SOF-a-gus=eating gullet) is a collapsible muscular tube that lies posterior to the trachea. It is about 25 cm (10 in.) long. It begins at the inferior end of the laryngopharynx, passes through the mediastinum anterior to the vertebral column, pierces the diaphragm through an opening called the **esophageal hiatus,** and ends in the superior portion of the stomach (see Figure 25.1). Sometimes, a portion of the stomach protrudes above the diaphragm through the esophageal hiatus. This condition, called a hiatal hernia, will be described later in the chapter.

The arteries of the esophagus are derived from the arteries along its length: inferior thyroid, thoracic aorta, intercostal arteries, phrenic, and left gastric arteries. It is drained by the adjacent veins. Innervation of the esophagus is by recurrent laryngeal nerves, the cervical sympathetic chain, and vagus (X) nerves.

Histology

The **mucosa** of the esophagus consists of nonkeratinized stratified squamous epithelium, lamina propria (areolar connective tissue), and a muscularis muscosae (smooth muscle) (Figure 25.8). Near the stomach, the mucosa of the esophagus also contains mucous glands. The stratified squamous epithelium associated with the lips, mouth, tongue, oropharynx, laryngopharynx, and esophagus affords considerable protection against abrasion and wear-and-tear from food particles that are chewed, mixed with secretions, and swallowed. The **submucosa** contains areolar connective tissue, blood vessels, and mucous glands. The **muscularis** of the superior third of the esophagus is skeletal muscle, the intermediate third is skeletal and smooth muscle, and the inferior third is smooth muscle. The superficial layer of the esophagus is known as the **adventitia** (ad-ven-TISH-a), rather than serosa as in the stomach, because the areolar connective tissue of this layer is not covered by mesothelium and because the connective tissue merges with the connective tissue of surrounding structures of the mediastinum, through which it passes. The adventitia attaches the esophagus to surrounding structures.

Functions

The esophagus secretes mucus and transports food into the stomach. It does not produce digestive enzymes, and it does not carry on absorption. The passage of food from the laryngopharynx into the esophagus is regulated at the entrance to the esophagus by a sphincter (a circular band or ring of muscle that is normally contracted) called the **upper esophageal sphincter** (e-sof′-a-JĒ-al). It consists of the cricopharyngeus muscle attached to the cricoid cartilage. The elevation of the larynx causes the sphincter to relax, allowing the bolus to enter the esophagus. This sphincter also relaxes during exhalation.

Figure 25.8 Histology of the esophagus. A higher magnification of nonkeratinized, stratified squamous epithelium is shown in Table 3.1f on page 65. (See Tortora, *A Photographic Atlas of the Human Body*, Figure 12.8a.)

The esophagus secretes mucus and transports food to the stomach.

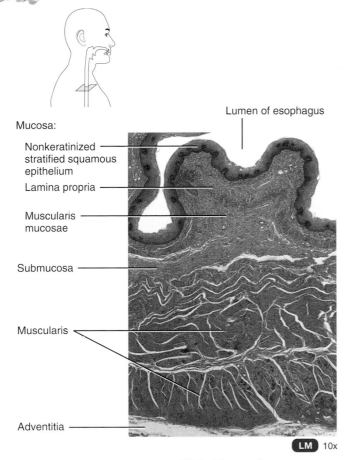

Wall of the esophagus

In which layers of the esophagus are the glands that secrete lubricating mucus located?

Food is pushed through the esophagus by a progression of involuntary coordinated contractions and relaxations of the circular and longitudinal layers of the muscularis called **peristalsis** (per′-i-STAL-sis; *stalsis*=constriction) (Figure 25.9). Peristalsis occurs in other tubular structures, including other portions of the GI tract and the ureters, bile ducts, and uterine tubes; in the esophagus it is controlled by the medulla oblongata. In the section of the esophagus lying just superior to the bolus, the circular muscle fibers contract, constricting the esophageal wall and squeezing the bolus toward the stomach. Meanwhile, longitudinal fibers inferior to the bolus also contract, which shortens this inferior section and pushes its walls outward so it can receive the bolus. The contractions are repeated in a wave that pushes the food toward the stomach. Mucus secreted by esophageal glands lubricates the bolus and reduces friction.

Just superior to the level of the diaphragm, the esophagus narrows slightly. This narrowing is a physiological sphincter in the inferior part of the esophagus known as the **lower**

Figure 25.9 Peristalsis during deglutition (swallowing).

Peristalsis consists of progressive, wavelike contractions of the muscularis.

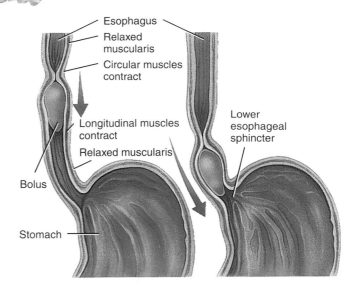

Anterior view of frontal sections of peristalsis in esophagus

Does peristalsis "push" or "pull" food along the gastrointestinal tract?

esophageal (gastroesophageal or **cardiac) sphincter.** (A *physiological sphincter* is a section of a tubular structure, in this case the esophagus, that functions like a sphincter even though no sphincter muscle is actually present.) The lower esophageal sphincter relaxes during swallowing and thus allows the bolus to pass from the esophagus into the stomach.

GASTROESOPHAGEAL REFLUX DISEASE

If the lower esophageal sphincter fails to close adequately after food has entered the stomach, the stomach contents can reflux (back up) into the inferior portion of the esophagus. This condition is known as **gastroesophageal reflux disease (GERD)**. Hydrochloric acid (HCl) from the stomach contents can irritate the esophageal wall, resulting in a burning sensation called **heartburn** because it is experienced in a region very near the heart, even though it is unrelated to any cardiac problem. Drinking alcohol and smoking can cause the sphincter to relax, worsening the problem. The symptoms of GERD often can be controlled by avoiding foods that strongly stimulate stomach acid secretion (caffeinated coffee, chocolate, tomatoes, fatty foods, peppermint, spearmint, and onions). Other acid-reducing strategies include taking over-the-counter histamine-2 (H2) blockers such as Tagamet HB or Pepcid AC 30–60 minutes before eating, and neutralizing already secreted acid with antacids such as Tums or Maalox. Symptoms are less likely to occur if food is eaten in smaller amounts and if the person does not lie down immediately after a meal. GERD may erode the esophagus and may be associated with cancer of the esophagus. ∎

CHECKPOINT

13. Describe the location and histology of the esophagus.
14. What is the role of the esophagus in digestion?
15. Compare the operation of the upper and lower esophageal sphincters. Which one is a physiological sphincter?

STOMACH

OBJECTIVE

● Describe the location, anatomy, histology, and functions of the stomach.

The **stomach** is a typically J-shaped enlargement of the GI tract directly inferior to the diaphragm in the epigastric, umbilical, and left hypochondriac regions of the abdomen (see Figure 1.9 on page 16). The stomach connects the esophagus to the duodenum, the first part of the small intestine (Figure 25.10). Because a meal can be eaten much more quickly than the intestines can digest and absorb it, the stomach functions as a mixing vat and holding reservoir. At appropriate intervals after food is ingested, the stomach forces a small quantity of material into the first portion of the small intestine. The position and size of the stomach vary continually; the diaphragm pushes it inferiorly with each inspiration and pulls it superiorly with each expiration. Empty, it is about the size of a large sausage, but it is the most distensible portion of the GI tract and can accommodate a large quantity of food. In the stomach, the digestion of starch that began in the mouth continues, digestion of proteins and triglycerides begins, the semisolid bolus is converted to a liquid, and certain substances are absorbed.

Anatomy

The stomach has four main regions: the cardia, fundus, body, and pylorus (Figure 25.10). The **cardia** (CAR-dē-a) surrounds the superior opening of the stomach. The rounded portion superior and to the left of the cardia is the **fundus** (FUN-dus). Inferior to the fundus is the large central portion of the stomach, called the **body.** The region of the stomach that connects to the duodenum is the **pylorus** (pī-LOR-us; *pyl-*=gate; *-orus*=guard); it has two parts, the **pyloric antrum** (AN-trum=cave), which connects to the body of the stomach, and the **pyloric canal,** which leads into the duodenum. When the stomach is empty, the mucosa lies in large folds, called **rugae** (ROO-gē=wrinkles), that can be seen with the unaided eye. The pylorus communicates with the duodenum of the small intestine via the **pyloric sphincter.** The concave medial border of the stomach is called the **lesser curvature,** and the convex lateral border is called the **greater curvature.**

The arterial supply of the stomach is derived from the celiac trunk. The right and left gastric arteries form an anastomosing arch along the lesser curvature, and the right and left gastroepiploic arteries form a similar arch on the greater curvature. Short gastric arteries supply the fundus. The veins of the same name accompany the arteries and drain, directly or indirectly, into the hepatic portal vein.

The vagus (X) nerves convey parasympathetic fibers to the stomach. These fibers form synapses within the submucosal plexus (plexus of Meissner) in the submucosa and the myenteric plexus (plexus of Auerbach) in the muscularis. The sympathetic nerves arise from the celiac ganglia, and the nerves reach the stomach along the branches of the celiac artery.

PYLOROSPASM AND PYLORIC STENOSIS

Two abnormalities of the pyloric sphincter can occur in infants. In **pylorospasm** (pī-LOR-ō-spazm) the muscle fibers of the sphincter fail to relax normally, so food does not pass easily from the stomach to the small intestine, the stomach becomes overly full, and the infant vomits often to relieve the pressure. Pylorospasm is treated by drugs that relax the muscle fibers of the sphincter. **Pyloric stenosis** (ste-NŌ-sis) is a narrowing of the pyloric sphincter that must be corrected surgically. The hallmark symptom is *projectile vomiting*—the spraying of liquid vomitus some distance from the infant. ∎

Histology

The stomach wall is composed of the same four basic layers as the rest of the GI tract, with certain modifications (Figure 25.11a on page 780). The surface of the **mucosa** is a layer of simple columnar epithelial cells called **surface mucous cells.** The mucosa contains a **lamina propria** (areolar connective tissue) and a **muscularis mucosae** (smooth muscle). Epithelial cells extend down into the lamina propria, where they form columns of secretory cells called **gastric glands** that line many narrow channels called **gastric pits.** Secretions from several gastric glands flow into each gastric pit and then into the lumen of the stomach.

The gastric glands contain three types of *exocrine gland cells* that secrete their products into the stomach lumen: mucous neck cells, chief cells, and parietal cells. Both mucous surface cells and **mucous neck cells** secrete mucus (Figure 25.11b). **Parietal cells** produce intrinsic factor (needed for absorption of vitamin B_{12}) and hydrochloric acid. The **chief (zymogenic) cells** secrete pepsinogen and gastric lipase. The secretions of the mucous, parietal, and chief cells form **gastric juice,** which totals 2000–3000 mL (roughly 2–3 qt.) per day. In addition, gastric glands include a type of enteroendocrine cell, the **G cell,** which is located mainly in the pyloric antrum and secretes the hormone gastrin into the bloodstream. Gastrin stimulates growth of the gastric glands and secretion of large amounts of gastric juice. It also strengthens contraction of the lower esophageal sphincter, increases motility of the stomach, and relaxes the pyloric and ileocecal sphincters (described later).

ZOLLINGER-ELLISON SYNDROME

Zollinger-Ellison Syndrome (ZOL-in-jer EL-i-son) is a syndrome in which individuals produce too much hydrochloric acid. It is caused by a tumor in the duodenum, bile ducts, or

Figure 25.10 External and internal anatomy of the stomach.

The four regions of the stomach are the cardia, fundus, body, and pylorus.

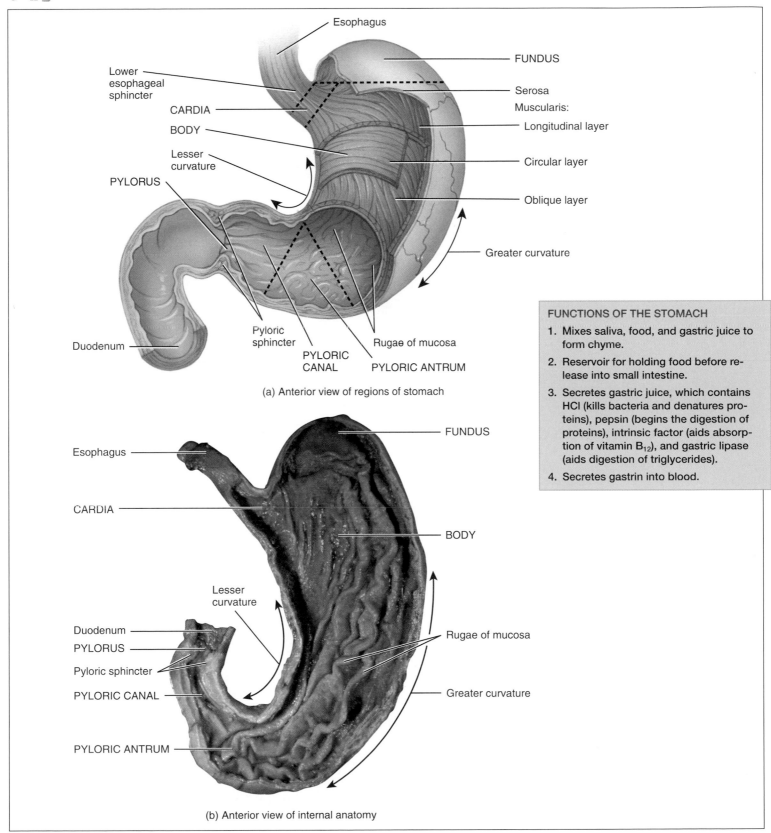

(a) Anterior view of regions of stomach

FUNCTIONS OF THE STOMACH

1. Mixes saliva, food, and gastric juice to form chyme.

2. Reservoir for holding food before release into small intestine.

3. Secretes gastric juice, which contains HCl (kills bacteria and denatures proteins), pepsin (begins the digestion of proteins), intrinsic factor (aids absorption of vitamin B_{12}), and gastric lipase (aids digestion of triglycerides).

4. Secretes gastrin into blood.

(b) Anterior view of internal anatomy

After a very large meal, does your stomach have rugae?

Figure 25.11 **Histology of the stomach.**

The muscularis of the stomach has three layers of smooth muscle tissue.

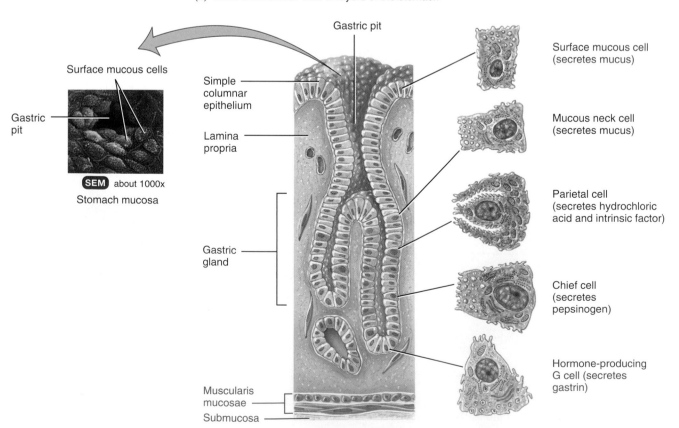

Lumen of stomach

Gastric pits

Simple columnar epithelium

Lamina propria

Gastric gland

Lymphatic nodule

Muscularis mucosae

Lymphatic vessel

Venule

Arteriole

Oblique layer of muscle

Circular layer of muscle

Myenteric plexus

Longitudinal layer of muscle

MUCOSA

SUBMUCOSA

MUSCULARIS

SEROSA

(a) Three dimensional view of layers of the stomach

Gastric pit

Surface mucous cells

Gastric pit

SEM about 1000x

Stomach mucosa

Simple columnar epithelium

Lamina propria

Gastric gland

Muscularis mucosae

Submucosa

Surface mucous cell (secretes mucus)

Mucous neck cell (secretes mucus)

Parietal cell (secretes hydrochloric acid and intrinsic factor)

Chief cell (secretes pepsinogen)

Hormone-producing G cell (secretes gastrin)

(b) Sectional view of the stomach mucosa showing gastric glands and cell types

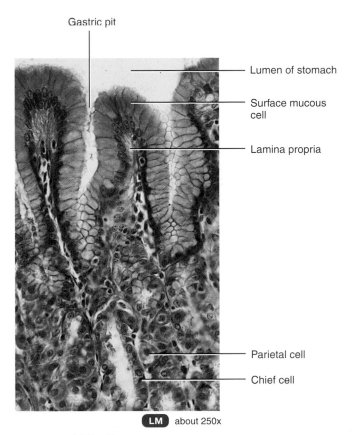

Gastric pit

Lumen of stomach

Surface mucous cell

Lamina propria

Parietal cell

Chief cell

LM about 250x

(c) Fundic mucosa

What types of cells are found in gastric glands, and what does each secrete?

Functions

Several minutes after food enters the stomach, gentle, rippling, peristaltic movements called **mixing waves** pass over the stomach every 15–25 seconds. These waves macerate food, mix it with secretions of the gastric glands, and reduce it to a soupy liquid called **chyme** (KĪM=juice). Few mixing waves are observed in the fundus, which primarily has a storage function. As digestion proceeds, more vigorous mixing waves begin at the body of the stomach and intensify as they reach the pylorus. The pyloric sphincter normally remains almost, but not completely, closed; as food reaches the pylorus, each mixing wave forces several milliliters of chyme into the duodenum through the pyloric sphincter. Most of the chyme is forced back into the body of the stomach, where mixing continues. The next wave pushes the chyme forward again and forces a little more into the duodenum.

The enzymatic digestion of proteins begins in the stomach. In the adult, this is achieved mainly through the enzyme **pepsin,** secreted by chief cells in an inactive form called *pepsinogen.* Pepsin breaks certain peptide bonds between the amino acids making up proteins. Thus, a protein chain of many amino acids is broken down into smaller fragments called **peptides.** Pepsin also brings about the clumping and digestion of milk proteins. Another enzyme of the stomach is **gastric lipase.** Gastric lipase splits the short-chain triglycerides (fats) in butterfat molecules found in milk. The enzyme has a limited role in the adult stomach. To digest fats, adults rely almost exclusively on the lingual lipase secreted by salivary glands in the mouth, and **pancreatic lipase,** an enzyme secreted by the pancreas into the small intestine.

Within 2–4 hours after eating a meal, the stomach has emptied its contents into the duodenum. Foods rich in carbohydrate spend the least time in the stomach; high-protein foods remain somewhat longer; and emptying is slowest after a fat-laden meal containing large amounts of triglycerides.

The stomach wall is impermeable to the passage of most materials into the blood; most substances are not absorbed until they reach the small intestine. However, the stomach does participate in the absorption of some water, electrolytes, certain drugs (especially aspirin), and alcohol.

VOMITING

Vomiting, or *emesis,* is the forcible expulsion of the contents of the upper GI tract (stomach and sometimes duodenum) through the mouth. The strongest stimuli for vomiting are irritation and distension of the stomach; other stimuli include unpleasant sights, general anesthesia, dizziness, and certain drugs, such as morphine and derivatives of digitalis. Nerve impulses are transmitted to the vomiting center in the medulla oblongata, and returning impulses propagate to the upper GI tract organs, diaphragm, and abdominal muscles. Vomiting basically involves squeezing the stomach between the diaphragm and abdominal muscles and expelling the contents through open esophageal sphincters. Prolonged vomiting, especially in infants and elderly people, can be serious because the loss of acidic gastric juice can lead to alkalosis (higher than normal blood pH), and dehydration. ■

pancreas that produces gastrin. Gastrin, in turn, stimulates the secretion of gastric juice. Treatment consists of drugs that decrease acid secretion, and surgery. ■

Three additional layers lie deep to the mucosa. The **submucosa** of the stomach is composed of areolar connective tissue. The **muscularis** has three layers of smooth muscle (rather than the two found in the small and large intestines): an outer longitudinal layer, a middle circular layer, and an inner oblique layer. The oblique layer is limited primarily to the body of the stomach. The **serosa** is composed of simple squamous epithelium (mesothelium) and areolar connective tissue and the portion covering the stomach is part of the visceral peritoneum. At the lesser curvature, the visceral peritoneum extends superiorly to the liver as the lesser omentum. At the greater curvature, the visceral peritoneum continues inferiorly as the greater omentum and drapes over the intestines.

CHECKPOINT

16. Describe the location and anatomical features of the stomach.
17. Compare the epithelium of the esophagus with that of the stomach. How is each adapted to the function of the organ?
18. Describe the importance of rugae, surface mucous cells, mucous neck cells, chief cells, parietal cells, and G cells in the stomach.

19. How does mechanical digestion occur in the stomach?
20. What are the functions of gastric lipase and lingual lipase in the stomach?
21. Describe the role of the stomach in absorption.

PANCREAS

● Describe the location, anatomy, histology, and function of the pancreas.

From the stomach, chyme passes into the small intestine. Because chemical digestion in the small intestine depends on activities of the pancreas, liver, and gallbladder, we first consider the activities of these accessory digestive organs and their contributions to digestion in the small intestine.

Anatomy

The **pancreas** (*pan-*=all; *-creas*=flesh), a retroperitoneal gland that is about 12–15 cm (5–6 in.) long and 2.5 cm (1 in.) thick, lies posterior to the greater curvature of the stomach (see Figure 25.1). The pancreas consists of a head, a body, and a tail and is usually connected to the duodenum by two ducts (Figure 25.12). The **head** is the expanded portion of the organ near the curve of the duodenum; superior to and to the left of the head are the central **body** and the tapering **tail.**

Pancreatic secretions pass from the secreting cells into small ducts that ultimately unite to form two larger ducts that convey the secretions into the small intestine. The larger of the two ducts is called the **pancreatic duct (duct of Wirsung).** In most people, the pancreatic duct joins the common bile duct from the liver and gallbladder and enters the duodenum as a common duct called the **hepatopancreatic ampulla (ampulla of Vater).** The ampulla opens onto an elevation of the duodenal mucosa, the **major duodenal papilla,** that lies about 10 cm (4 in.)

Figure 25.12   Relation of the pancreas to the liver, gallbladder, and duodenum. The inset shows details of the common bile duct and pancreatic duct forming the hepatopancreatic ampulla (ampulla of Vater) and emptying into the duodenum.

🔑 **Pancreatic enzymes digest starches (polysaccharides), proteins, triglycerides, and nucleic acids.**

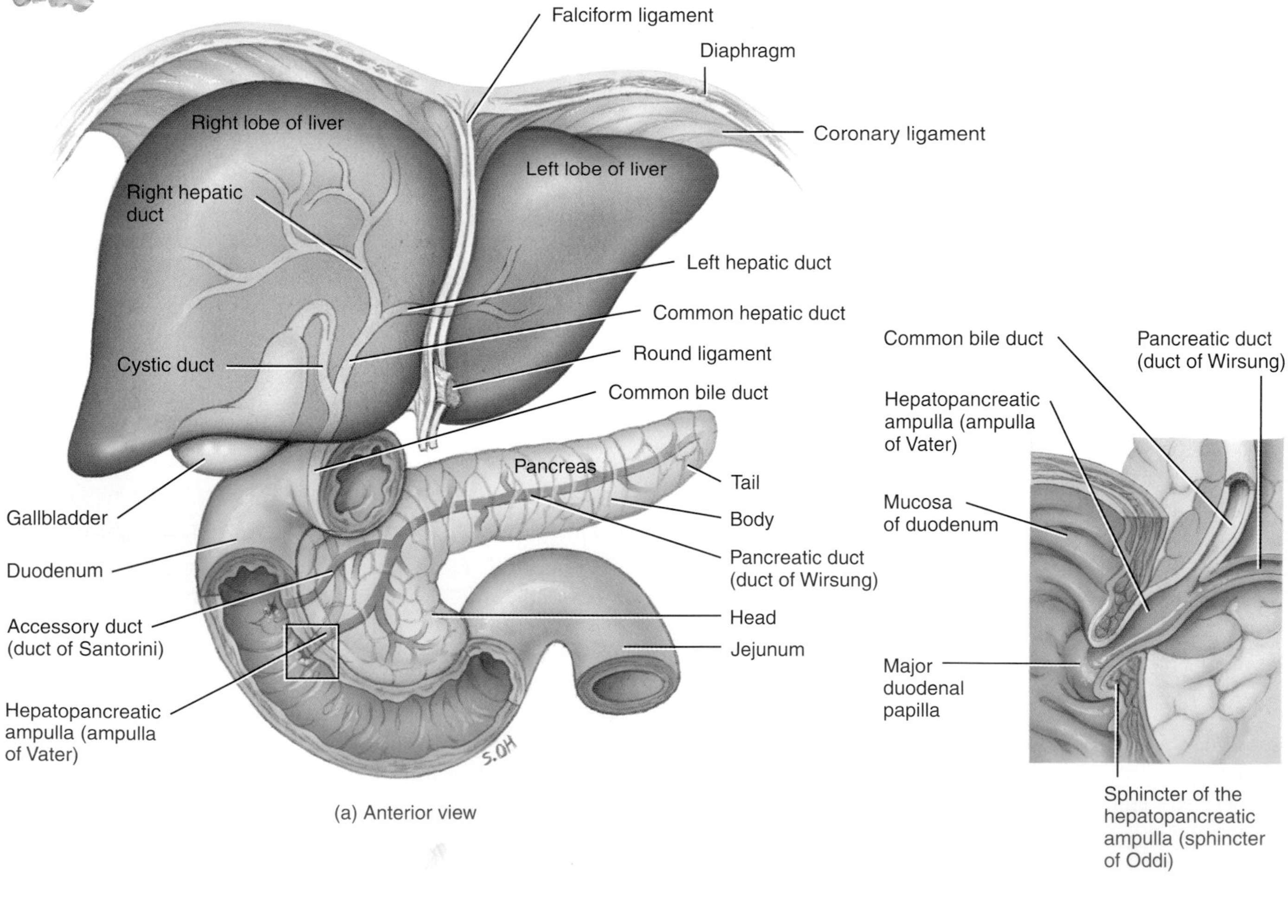

(a) Anterior view

inferior to the pyloric sphincter of the stomach. The smaller of the two ducts, the **accessory duct (duct of Santorini),** leads from the pancreas and empties into the duodenum about 2.5 cm (1 in.) superior to the hepatopancreatic ampulla.

The arterial supply of the pancreas is from the superior and inferior pancreaticoduodenal arteries and from the splenic and superior mesenteric arteries. The veins, in general, correspond to the arteries. Venous blood reaches the hepatic portal vein by means of the splenic and superior mesenteric veins.

The nerves to the pancreas are autonomic nerves derived from the celiac and superior mesenteric plexuses. Included are preganglionic vagal, postganglionic sympathetic, and sensory fibers. Parasympathetic vagal fibers are said to terminate at both acinar (exocrine) and islet (endocrine) cells. Although the innervation is presumed to influence enzyme formation, pancreatic secretion is controlled largely by the hormones secretin and cholecystokinin (CCK) released by the small intestine. The sympathetic fibers enter the islets and also end on blood vessels; these fibers are vasomotor and accompanied by sensory fibers, especially for pain.

Histology

The pancreas is made up of small clusters of glandular epithelial cells, about 99% of which are arranged in clusters called **acini** (AS-i-nē) and constitute the *exocrine* portion of the organ (see Figure 23.7b, c on page 718). The cells within acini secrete a mixture of fluid and digestive enzymes called **pancreatic juice.** The remaining 1% of the cells are organized into clusters called **pancreatic islets (islets of Langerhans),** the *endocrine* portion

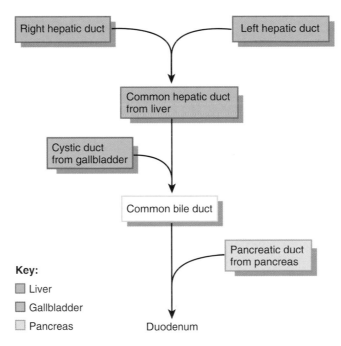

Key:
- Liver
- Gallbladder
- Pancreas

(b) Ducts carrying bile from liver and gallbladder and pancreatic juice from pancreas to the duodenum

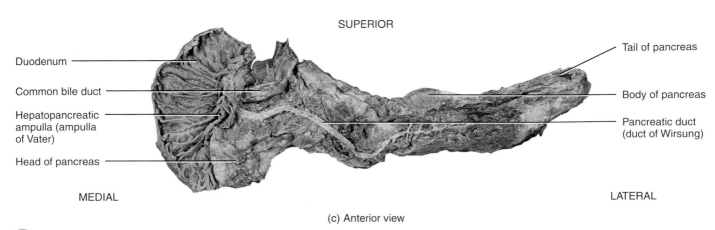

(c) Anterior view

? **What type of fluid is found in the pancreatic duct? The common bile duct? The hepatopancreatic ampulla?**

of the pancreas. These cells secrete the hormones glucagon, insulin, somatostatin, and pancreatic polypeptide. The functions of these hormones are discussed in Table 23.7 on page 719.

Functions

Each day the pancreas produces 1200–1500 mL (about 1.2–1.5 qt) of **pancreatic juice,** a clear, colorless liquid consisting mostly of water, some salts, sodium bicarbonate, and several enzymes. The sodium bicarbonate gives pancreatic juice a slightly alkaline pH (7.1–8.2) that buffers acidic gastric juice in chyme, stops the action of pepsin from the stomach, and creates the proper pH for the action of digestive enzymes in the small intestine. The enzymes in pancreatic juice include a carbohydrate-digesting enzyme called **pancreatic amylase;** several protein-digesting enzymes called **trypsin** (TRIP-sin), **chymotrypsin** (kī′-mō-TRIP-sin), **carboxypeptidase** (kar-bok′-sē-PEP-ti-dās), and **elastase** (ē-LAS-tās); the principal triglyceride-digesting enzyme in adults, called **pancreatic lipase;** and nucleic acid–digesting enzymes called **ribonuclease** and **deoxyribonuclease.**

🩺 PANCREATITIS AND PANCREATIC CANCER

Inflammation of the pancreas, as may occur in association with alcohol abuse or chronic gallstones, is called **pancreatitis** (pan′-krē-a-TĪ-tis). In a more severe condition known as **acute pancreatitis,** which is associated with heavy alcohol intake or biliary tract obstruction, the pancreatic cells may release either trypsin instead of trypsinogen or insufficient amounts of trypsin inhibitor, and the trypsin begins to digest the pancreatic cells. Patients with acute pancreatitis usually respond to treatment, but recurrent attacks are the rule. In some people pancreatitis is *idiopathic,* meaning that the cause is unknown. Other causes of pancreatitis include cystic fibrosis, high levels of calcium in the blood (hypercalcemia), high levels of blood fats (hyperlipidemia or hypertriglyceridemia), some drugs, and certain autoimmune conditions. However, in roughly 70 percent of adults with pancreatitis, the cause is alcoholism. Often the first episode happens between ages 30 and 40.

Pancreatic cancer usually affects people over 50 years of age and occurs more frequently in males. Typically, there are few symptoms until the disorder reaches an advanced stage and often not until it has metastasized to other parts of the body such as the lymph nodes, liver, or lungs. The disease is nearly always fatal and is the fourth most common cause of death from cancer in the United States. Pancreatic cancer has been linked to fatty foods, high alcohol consumption, genetic factors, smoking, and chronic pancreatitis. ■

CHECKPOINT

22. Describe the duct system connecting the pancreas to the duodenum.
23. What are pancreatic acini? Contrast their functions with those of the pancreatic islets (islets of Langerhans).
24. Describe the composition and functions of pancreatic juice.

LIVER AND GALLBLADDER

OBJECTIVE

● Describe the location, anatomy, histology, and functions of the liver and gallbladder.

The **liver** is the largest internal organ and heaviest gland of the body, weighing about 1.4 kg (about 3 lb) in an average adult. Of the organs of the body, it is second in size only to the skin. The liver is inferior to the diaphragm and occupies most of the right hypochondriac and part of the epigastric regions of the abdominopelvic cavity (see Figure 1.9 on page 16).

The **gallbladder** (*gall-*=bile) is a pear-shaped sac that is located on the inferior surface of the liver. It is 7–10 cm (3–4 in.) long and part of it typically hangs below the anterior inferior margin of the liver (Figure 25.12).

Anatomy

The liver is almost completely covered by visceral peritoneum and *is* completely covered by a capsule composed of dense irregular connective tissue that lies deep to the peritoneum. The liver is divided into two principal lobes—a large **right lobe** and a smaller **left lobe**—by the **falciform ligament,** a fold of the parietal peritoneum (Figures 25.12 and 25.13). The right lobe is considered by many anatomists to include an inferior **quadrate lobe** and a posterior **caudate lobe;** however, on the basis of internal morphology (primarily the distribution of blood vessels), the quadrate and caudate lobes more appropriately belong to the left lobe. The falciform ligament extends from the undersurface of the diaphragm between the two principal lobes of the liver to the superior surface of the liver, helping to suspend the liver in the abdominal cavity. The free border of the falciform ligament is the **ligamentum teres (round ligament),** a fibrous cord that is a remnant of the umbilical vein of the fetus; it extends from the liver to the umbilicus. The right and left **coronary ligaments** are narrow extensions of the parietal peritoneum that suspend the liver from the diaphragm.

The parts of the gallbladder are the broad **fundus,** which projects downward beyond the inferior border of the liver; the central portion, called the **body;** and a tapered portion called the **neck.** The body and neck project superiorly.

Histology

The lobes of the liver are made up of many functional units called **lobules** (Figure 25.14 on page 786). A lobule consists of specialized epithelial cells, called **hepatocytes** (*hepat-*=liver; -*cytes*=cells), arranged in irregular, branching, interconnected plates around a **central vein.** Instead of capillaries, the liver has larger, endothelium-lined spaces called **sinusoids,** through which blood passes. Also present in the sinusoids are fixed phagocytes called **stellate reticuloendothelial (Kupffer) cells,** which destroy worn-out leukocytes and red blood cells, bacteria, and other foreign matter in the venous blood draining from the gastrointestinal tract.

Bile, a yellow, brownish, or olive-green liquid, is partially an excretory product and partially a digestive secretion secreted

Figure 25.13 External anatomy of the liver. The anterior view is illustrated in Figure 25.12a.

🔑 **The two principal lobes of the liver, the right and left lobes, are separated by the falciform ligament.**

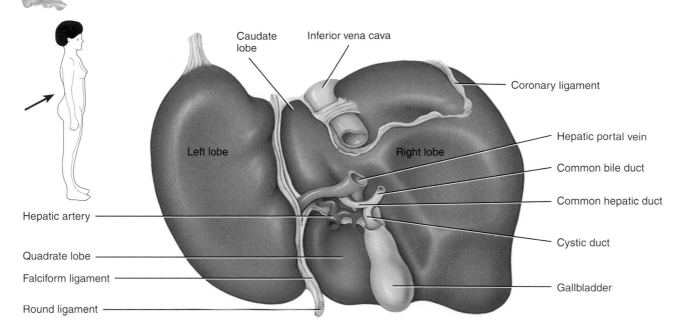

(a) Posteroinferior surface of liver

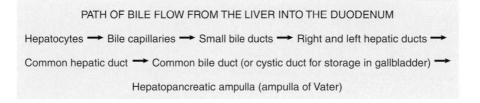

PATH OF BILE FLOW FROM THE LIVER INTO THE DUODENUM

Hepatocytes ➝ Bile capillaries ➝ Small bile ducts ➝ Right and left hepatic ducts ➝

Common hepatic duct ➝ Common bile duct (or cystic duct for storage in gallbladder) ➝

Hepatopancreatic ampulla (ampulla of Vater)

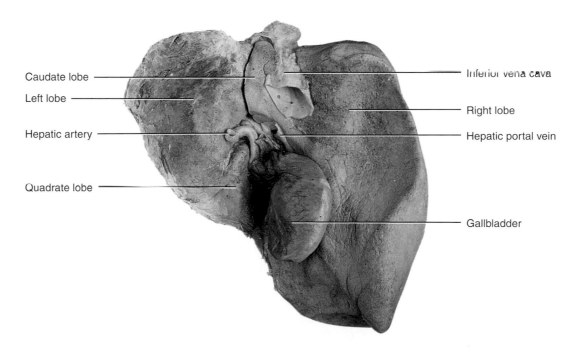

(b) Posteroinferior surface of liver

❓ **Within which abdominopelvic region (see Figure 1.9a on page 16) could you palpate (feel) most of the liver to decide if it is enlarged?**

Figure 25.14 **Histology of a lobule, the functional unit of the liver.**

A lobule consists of hepatocytes arranged around a central vein.

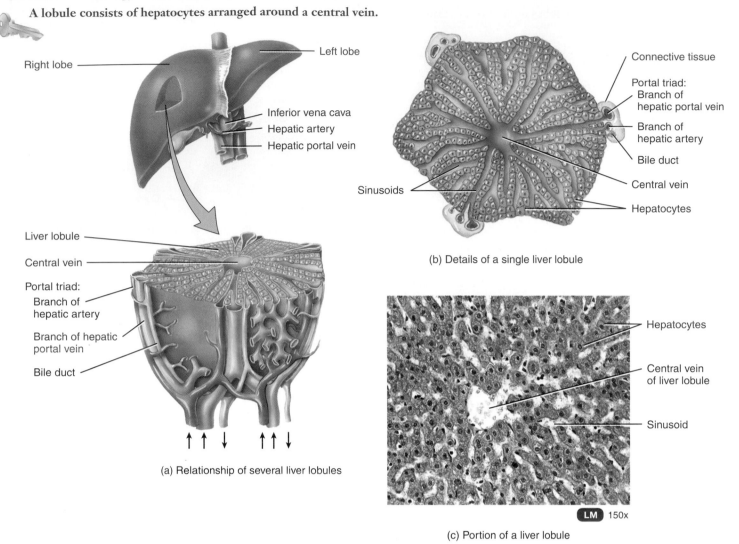

Right lobe

Left lobe

Inferior vena cava

Hepatic artery

Hepatic portal vein

Liver lobule

Central vein

Portal triad:
Branch of hepatic artery

Branch of hepatic portal vein

Bile duct

(a) Relationship of several liver lobules

Connective tissue

Portal triad:
Branch of hepatic portal vein

Branch of hepatic artery

Bile duct

Central vein

Hepatocytes

Sinusoids

(b) Details of a single liver lobule

Hepatocytes

Central vein of liver lobule

Sinusoid

LM 150x

(c) Portion of a liver lobule

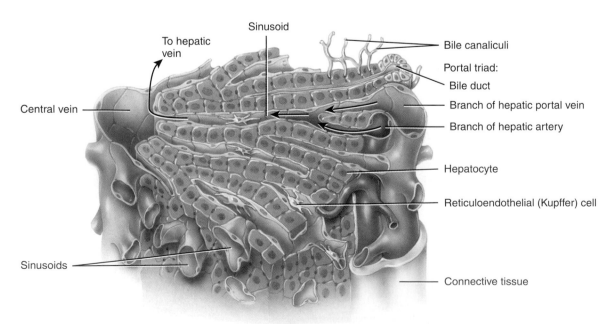

Sinusoid

To hepatic vein

Bile canaliculi

Portal triad:

Bile duct

Branch of hepatic portal vein

Branch of hepatic artery

Central vein

Hepatocyte

Reticuloendothelial (Kupffer) cell

Sinusoids

Connective tissue

(d) Details of a portion of a liver lobule

Which type of liver cell is phagocytic?

by hepatocytes. Bile enters **bile canaliculi** (kan′-a-LIK-ū-lī = small canals), which are narrow intercellular canals that empty into small **bile ductules.** The ductules pass bile into **bile ducts** at the periphery of the lobules of the liver. The bile ducts merge and eventually form the larger **right** and **left hepatic ducts,** which unite and exit the liver as the **common hepatic duct** (see Figure 25.12). Farther on, the common hepatic duct joins the **cystic duct** (*cystic*=bladder) from the gallbladder to form the **common bile duct.** Bile enters the cystic duct and is temporarily stored in the gallbladder.

The mucosa of the gallbladder consists of simple columnar epithelium arranged in rugae resembling those of the stomach. The wall of the gallbladder lacks a submucosa. The middle, muscular coat consists of smooth muscle fibers; the contraction of these fibers ejects the contents of the gallbladder into the **cystic duct.** The gallbladder's outer coat is the visceral peritoneum. The functions of the gallbladder are to store and concentrate bile (up to tenfold) until it is needed in the small intestine.

 JAUNDICE

Jaundice (JAWN-dis=yellow) is a yellowish coloration of the sclerae (white of the eyes), skin, and mucous membranes due to a buildup in the body of a yellow compound called bilirubin. Bilirubin, which is formed from the breakdown of the heme pigment in aged red blood cells, is transported to the liver, where it is normally processed and eventually excreted into bile. The three main categories of jaundice are (1) *prehepatic jaundice,* due to excess production of bilirubin; (2) *hepatic jaundice,* due to congenital liver disease, cirrhosis of the liver, or hepatitis; and (3) *extrahepatic jaundice,* due to blockage of bile drainage by gallstones or cancer of the bowel or the pancreas.

Because the liver of a newborn functions poorly for the first week or so, many babies experience a mild form of jaundice called *neonatal (physiological) jaundice* that disappears as the liver matures. Usually, it is treated by exposing the infant to blue light, which converts bilirubin into substances the kidneys can excrete. ■

Blood and Nerve Supply

The liver receives blood from two sources (Figure 25.15). From the hepatic artery it obtains oxygenated blood, and from the hepatic portal vein it receives deoxygenated blood containing newly absorbed nutrients, drugs, and possibly microbes and toxins from the gastrointestinal tract. Branches of both the hepatic artery and the hepatic portal vein carry blood into liver sinusoids, where oxygen, most of the nutrients, and certain toxic substances are taken up by the hepatocytes. Products manufactured by the hepatocytes and nutrients needed by other cells are secreted back into the blood, which then drains into the central vein and eventually passes into a hepatic vein. Because blood from the gastrointestinal tract passes through the liver as part of the hepatic portal circulation, the liver is often a site for metastasis of cancer that originates in the GI tract. Branches of

the hepatic portal vein, hepatic artery, and bile duct typically accompany one another in their distribution through the liver. Collectively, these three structures are called a **portal triad** (see Figure 25.14).

The nerve supply to the liver consists of parasympathetic innervation from the vagus (X) nerves and sympathetic innervation from the greater splanchnic nerves through the celiac ganglia.

The gallbladder is supplied by the cystic artery, which usually arises from the right hepatic artery. The cystic veins drain the gallbladder. The nerves to the gallbladder include branches from the celiac plexus and the vagus (X) nerve.

Functions

Hepatocytes continuously secrete 800–1000 mL (about 1 qt) of bile per day. Bile salts, which are sodium salts and potassium salts of bile acids (mostly cholic acid and chenodeoxycholic acid), play roles in (1) **emulsification,** the breakdown of large lipid globules into a suspension of droplets about 1 μm in diameter, and (2) the absorption of digested lipids.

Figure 25.15 Hepatic blood flow: sources, path through the liver, and return to the heart.

The liver receives oxygenated blood via the hepatic artery and nutrient-rich deoxygenated blood via the hepatic portal vein.

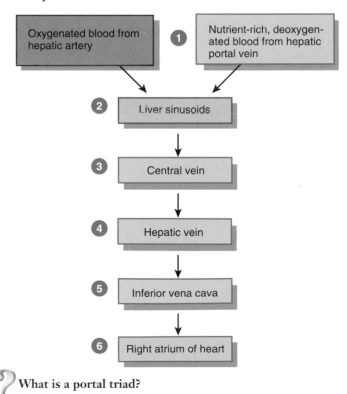

What is a portal triad?

GALLSTONES

If bile contains insufficient bile salts, lecithin, or excessive cholesterol, the cholesterol may crystallize to form **gallstones.** As they grow in size and number, gallstones may cause minimal, intermittent, or complete obstruction to the flow of bile from the gallbladder into the duodenum. Treatment consists of using gallstone-dissolving drugs, lithotripsy (shock-wave therapy), or surgery. For people with recurrent gallstones or for whom drugs or lithotripsy is not indicated, *cholecystectomy*—the removal of the gallbladder and its contents—is necessary. More than half a million cholecystectomies are performed each year in the United States. This surgery used to require a rather large incision in the right upper quadrant. Now, the gallbladder is often removed using a less invasive laparoscopic method, which results in faster recovery time. ■

Between meals, bile flows into the gallbladder for storage because the **sphincter of the hepatopancreatic ampulla** (**sphincter of Oddi**; Figure 25.12) closes off the entrance to the duodenum. After a meal, several neural and hormonal stimuli promote the production and release of bile. Parasympathetic impulses along the vagus (X) nerve fibers can stimulate the liver to increase bile production to more than twice the baseline rate. Fatty acids and amino acids in chyme entering the duodenum stimulate some duodenal enteroendocrine cells to secrete the hormone cholecystokinin (CCK) into the blood. CCK causes contraction of the walls of the gallbladder, which squeezes stored bile out of the gallbladder into the cystic duct and through the common bile duct. CCK also causes relaxation of the sphincter of the hepatopancreatic ampulla, which allows bile to flow into the duodenum.

The liver performs many other vital functions:

- *Carbohydrate metabolism.* The liver is especially important in maintaining a normal blood glucose level. When blood glucose is low, the liver can break down glycogen to glucose and release glucose into the bloodstream. The liver can also convert certain amino acids and lactic acid to glucose, and it can convert other sugars, such as fructose and galactose, into glucose. When blood glucose is high, as occurs just after eating a meal, the liver converts glucose to glycogen and triglycerides for storage.

- *Lipid metabolism.* Hepatocytes store some triglycerides; break down fatty acids to generate ATP; synthesize lipoproteins, which transport fatty acids, triglycerides, and cholesterol to and from body cells; synthesize cholesterol; and use cholesterol to make bile salts.

- *Protein metabolism.* Hepatocytes *deaminate* [remove the amino group (NH_2) from] amino acids so that the amino acids can be used for ATP production or converted to carbohydrates or fats. The resulting toxic ammonia (NH_3) is then converted into the much less toxic urea, which is excreted in urine. Hepatocytes also synthesize most plasma proteins, such as alpha and beta globulins, albumin, prothrombin, and fibrinogen.

- *Processing of drugs and hormones.* The liver can detoxify substances such as alcohol or secrete drugs such as penicillin, erythromycin, and sulfonamides into bile. It can also inactivate hormones such as thyroid hormones, estrogens, and aldosterone.

- *Excretion of bilirubin.* As previously noted, bilirubin, derived from the heme of aged red blood cells, is absorbed by the liver from the blood and secreted into bile. Most of the bilirubin in bile is metabolized in the small intestine by bacteria and eliminated in feces.

- *Synthesis of bile salts.* Bile salts are used in the small intestine for the emulsification and absorption of lipids, cholesterol, phospholipids, and lipoproteins.

- *Storage.* In addition to glycogen, the liver is a prime storage site for certain vitamins (A, B_{12}, D, E, and K) and minerals (iron and copper), which are released from the liver when needed elsewhere in the body.

- *Phagocytosis.* The stellate reticuloendothelial (Kupffer) cells of the liver phagocytize aged red blood cells and white blood cells and some bacteria.

- *Activation of vitamin D.* The skin, liver, and kidneys participate in synthesizing the active form of vitamin D.

LIVER BIOPSY

In a **liver biopsy** (BĪ-op-sē; *bios-*=life; *-opsis*=vision) a sample of living liver tissue is removed to diagnose a number of disorders, such as cancer, cirrhosis, hepatitis, and cysts. The puncture needle is commonly inserted through the tenth intercostal space after the patient exhales as much as possible while holding his or her breath. This minimizes the possibility of damaging the lung and contaminating the pleural cavity. ■

CHECKPOINT

25. Draw and label a diagram of a liver lobule.
26. Describe the pathways of blood flow into, through, and out of the liver.
27. How are the liver and gallbladder connected to the duodenum?
28. Describe the functions of the liver and gallbladder.

SMALL INTESTINE

OBJECTIVE

- Describe the location, anatomy, histology, and function of the small intestine.

The major events of digestion and absorption occur in a long tube called the **small intestine.** Because almost all digestion and absorption of nutrients occur in the small intestine, its structure is specially adapted for this function. Its length alone provides a large surface area for digestion and absorption, and that area is further increased by circular folds, villi, and microvilli. The small intestine begins at the pyloric sphincter of the stomach, coils through the central and inferior part of the abdominal cavity, and eventually opens into the large intestine. It averages 2.5 cm (1 in.) in diameter; its length is about 3 m (10 ft.) in a living person and about 6.5 m (21 ft.) in a cadaver due to the loss of smooth muscle tone after death.

Anatomy

The small intestine is divided into three regions (Figure 25.16). The **duodenum** (doo-ō′-DĒ-num), the shortest region, is retroperitoneal. *Duodenum* means "12"; it is so named because it is about as long as the width of 12 fingers. It starts at the pyloric sphincter of the stomach and extends about 25 cm (10 in.) until it merges with the next section, called the jejunum. The **jejunum** (jē-JOO-num) is about 1 m (3 ft.) long and extends to the ileum. *Jejunum* means "empty," which is how it is found at death. The final and longest region of the small intestine, the **ileum** (IL-ē-um=twisted), measures about 2 m (6 ft.) and joins the large intestine at the **ileocecal** (il′-ē-ō-SĒ-kal) **sphincter.**

Projections called **circular folds** (or **plicae circulares**) are permanent ridges in the mucosa about 10 mm (0.4 in.) high (Figure 25.16b). The circular folds begin near the proximal portion of the duodenum and end at about the midportion of the ileum; some extend all the way around the circumference of the intestine, and others extend only part of the way around. They enhance absorption by increasing surface area and causing the chyme to spiral, rather than move in a straight line, as it passes through the small intestine.

The arterial blood supply of the small intestine is from the superior mesenteric artery and the gastroduodenal artery, which arises from the hepatic artery of the celiac trunk. Blood is returned by way of the superior mesenteric vein, which, with the splenic vein, forms the hepatic portal vein.

The nerves to the small intestine are supplied by the superior mesenteric plexus. The branches of the plexus contain postganglionic sympathetic fibers, preganglionic parasympathetic fibers, and sensory fibers. The sensory fibers are components of the vagus (X) nerves and spinal nerves. In the wall of the small intestine are two autonomic plexuses: the myenteric plexus between the muscular layers and the submucosal plexus in the submucosa. The nerve fibers are derived chiefly from the sympathetic division of the autonomic nervous system, but some of them originate from the vagus (X) nerve.

Histology

Although the wall of the small intestine is composed of the same four coats that make up most of the GI tract, special features of both the mucosa and the submucosa facilitate the processes of digestion and absorption. The mucosa forms a series of fingerlike **villi** (=tufts of hair), projections that are 0.5–1 mm long (Figures 25.17 and 25.18 on page 791). The large number of villi (20–40 per square millimeter) vastly increases the surface area of the epithelium available for absorption and digestion and gives the intestinal mucosa a velvety appearance. Each *villus* (singular form) has a core of lamina propria (areolar connective tissue); embedded in this connective tissue are an arteriole, a

Figure 25.16 Regions of the small intestine. See also Figure 25.1b.

 Most digestion and absorption occur in the small intestine.

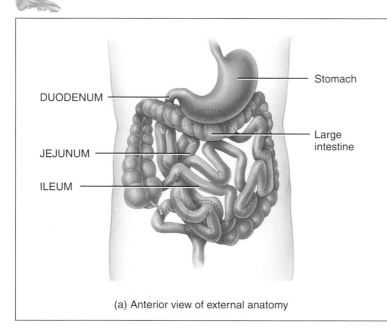

(a) Anterior view of external anatomy

Labels: DUODENUM, JEJUNUM, ILEUM, Stomach, Large intestine

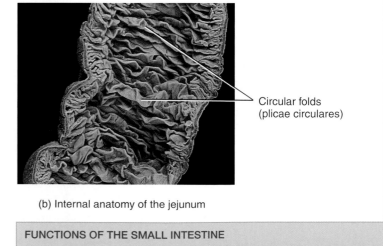

(b) Internal anatomy of the jejunum

Label: Circular folds (plicae circulares)

FUNCTIONS OF THE SMALL INTESTINE

1. Segmentations mix chyme with digestive juices and bring food into contact with the mucosa for absorption; peristalsis propels chyme through the small intestine.
2. Completes the digestion of carbohydrates (starches), proteins, and lipids; begins and completes the digestion of nucleic acids.
3. Absorbs about 90% of nutrients and water.

Which portion of the small intestine is the longest?

Figure 25.17   Histology of the small intestine.

Circular folds, villi, and microvilli increase the surface area of the small intestine for digestion and absorption.

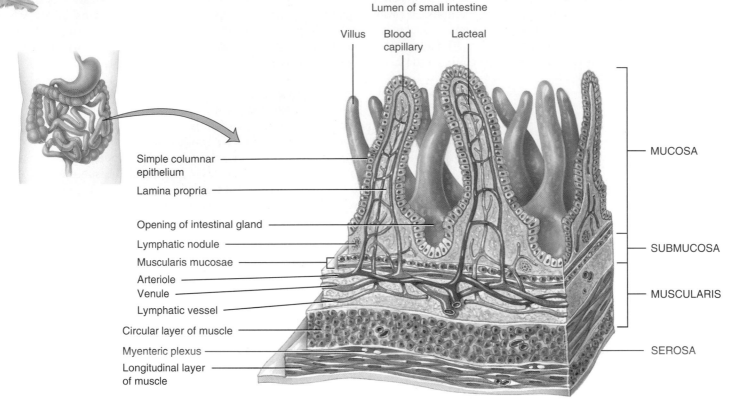

(a) Three-dimensional view of layers of the small intestine showing villi

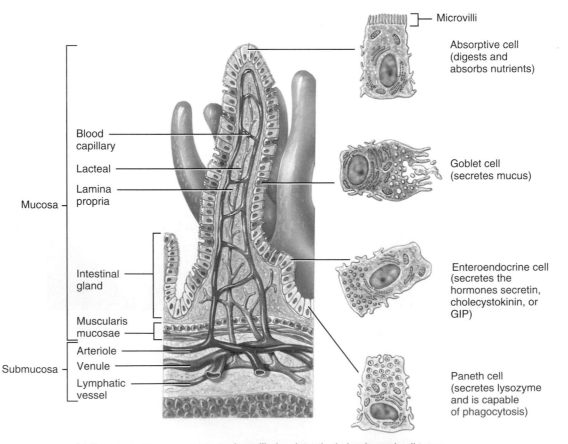

(b) Enlarged villus showing lacteal, capillaries, intestinal glands, and cell types

What is the functional significance of the blood capillary network and lacteal in the center of each villus?

Figure 25.18 Histology of the duodenum and ileum.

Microvilli greatly increase the surface area of the small intestine for digestion and absorption.

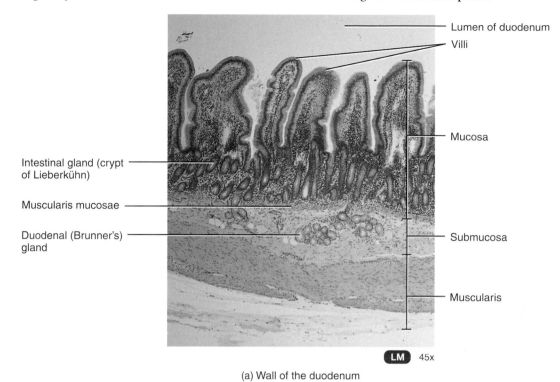

Intestinal gland (crypt of Lieberkühn)

Muscularis mucosae

Duodenal (Brunner's) gland

Lumen of duodenum

Villi

Mucosa

Submucosa

Muscularis

LM 45x

(a) Wall of the duodenum

Villi

Lumen of duodenum

Brush border

Simple columnar epithelium

Goblet cell

Absorptive cell

Lamina propria

Intestinal glands (crypts of Lieberkühn)

Muscularis mucosae

Duodenal (Brunner's) gland in submucosa

Duodenum

LM 160x

(b) Three villi from the duodenum of the small intestine

continues

Figure 25.18 (continued)

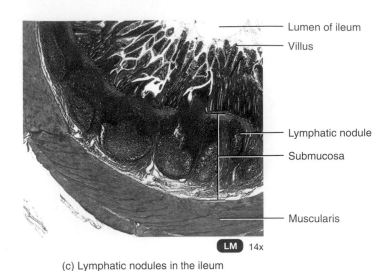

(c) Lymphatic nodules in the ileum

LM 14x

(d) Several microvilli from the duodenum

TEM 46,800x

What is the function of the fluid secreted by duodenal (Brunner's) glands?

venule, a blood capillary network, and a **lacteal** (LAK-tē-al= milky), which is a lymphatic capillary. Nutrients absorbed by the epithelial cells covering the villus pass through the wall of a capillary or a lacteal to enter blood or lymph, respectively.

The epithelium of the mucosa consists of simple columnar epithelium that contains four types of cells: absorptive cells, goblet cells, enteroendocrine cells, and Paneth cells (Figure 25.17b). The apical (free) membrane of **absorptive cells** features **microvilli** (mī′-krō-VIL-ī; *micro-*=small); each microvillus is a 1 μm-long cylindrical projection of the plasma membrane that contains a bundle of 20–30 actin filaments. When viewed through a light microscope, the microvilli are too small to be seen individually; instead they form a fuzzy line, called the **brush border,** extending into the lumen of the small intestine (Figure 25.18b, d). There are an estimated 200 million microvilli per square millimeter of small intestine. Because the microvilli greatly increase the surface area of the plasma membrane, larger amounts of digested nutrients can diffuse into absorptive cells in a given period of time. The brush border also contains several brush-border enzymes that have digestive functions (discussed shortly).

The mucosa contains many deep crevices lined with glandular epithelium. Cells lining the crevices form the **intestinal glands (crypts of Lieberkühn)** and secrete intestinal juice. Many of the epithelial cells in the mucosa are **goblet cells,** which secrete mucus. **Paneth cells,** found in the deepest parts of the intestinal glands, secrete the bactericidial enzyme lysozyme, and are also capable of phagocytosis. They may have a role in regulating the microbial population in the intestines. Three types of enteroendocrine cells, also in the deepest part of the intestinal glands, secrete hormones: secretin (by S cells), cholecystokinin (by CCK cells), and glucose-dependent insulinotropic peptide (by K cells). The lamina propria of the

small intestine has an abundance of mucosa-associated lymphoid tissue (MALT; see Chapter 16). **Solitary lymphatic nodules** are most numerous in the distal part of the ileum; groups of lymphatic nodules referred to as **aggregated lymphatic follicles (Peyer's patches)** are also numerous in the ileum. The muscularis mucosae consists of smooth muscle. The submucosa of the duodenum contains **duodenal (Brunner's) glands** (Figure 25.18a), which secrete an alkaline mucus that helps neutralize gastric acid in the chyme.

The **muscularis** of the small intestine consists of two layers of smooth muscle. The outer, thinner layer contains longitudinal fibers; the inner, thicker layer contains circular fibers. Except for a major portion of the duodenum, the serosa (or visceral peritoneum) completely surrounds the small intestine.

Functions

Chyme entering the small intestine contains partially digested carbohydrates, proteins, and lipids (mostly triglycerides). The completion of the digestion of carbohydrates, proteins, and lipids is the result of the collective action of pancreatic juice, bile, and intestinal juice in the small intestine.

Intestinal juice is a clear yellow fluid secreted in amounts of 1 to 2 liters (about 1 to 2 quarts) each day. It has a pH of 7.6, which is slightly alkaline, and contains water and mucus. Together, pancreatic and intestinal juice provide a vehicle for the absorption of substances from chyme as they come in contact with the villi.

The absorptive epithelial cells synthesize several digestive enzymes, called **brush-border enzymes,** and insert them in the plasma membrane of the microvilli. Thus, some enzymatic digestion occurs at the surface of the epithelial cells that line the villi; in other parts of the GI tract enzymatic digestion occurs in the lumen exclusively. Among the brush-border enzymes are four carbohydrate-digesting enzymes called **α-dextrinase, maltase, sucrase,** and **lactase;** protein-digesting enzymes called **peptidases** (**aminopeptidase** and **dipeptidase**); and two types of nucleotide-digesting enzymes, **nucleosidases** and **phosphatases.** Also, as cells slough off into the lumen of the small intestine, they break apart and release enzymes that help digest nutrients in the chyme.

LACTOSE INTOLERANCE

In some people the mucosal cells of the small intestine fail to produce enough lactase, which is essential for the digestion of lactose. This results in a condition called **lactose intolerance,** in which undigested lactose in chyme retains fluid in the feces, and bacterial fermentation of lactose results in the production of gases. Symptoms of lactose intolerance include diarrhea, gas, bloating, and abdominal cramps after consumption of milk and other dairy products. The severity of symptoms varies from relatively minor to sufficiently serious to require medical attention. The *hydrogen breath test* is often used to aid in diagnosis of lactose intolerance. The patient consumes a high-lactose drink, and is then asked to breathe into balloon-like bags over the next several hours. Very little hydrogen can be detected in the breath of a normal person. Hydrogen and other gases are produced when undigested lactose in the colon is fermented by bacteria; the hydrogen is absorbed from the intestines and carried through the bloodstream to the lungs, where it is exhaled. Persons with lactose intolerance can take dietary supplements to aid in the digestion of lactose. ■

The two types of movements of the small intestine—segmentations and a type of peristalsis called migrating motility complexes—are governed mainly by the myenteric plexus. **Segmentations** are a localized, mixing type of contraction that occurs in portions of the intestine distended by a large volume of chyme. Segmentations mix chyme with the digestive juices and bring the particles of food into contact with the mucosa for absorption; they do not push the intestinal contents along the tract. A segmentation starts with the contractions of circular muscle fibers in a portion of the small intestine, an action that constricts the intestine into segments. Next, muscle fibers that encircle the middle of each segment also contract, dividing each segment again. Finally, the fibers that first contracted relax, and each small segment unites with an adjoining small segment so that large segments are formed again. As this sequence of events repeats, the chyme sloshes back and forth. Segmentations occur most rapidly in the duodenum, about 12 times per minute, and progressively decrease to about 8 times per minute in the ileum. This movement is similar to alternately squeezing the middle and then the ends of a capped tube of toothpaste.

After most of a meal has been absorbed, which lessens distention of the wall of the small intestine, segmentation stops and peristalsis begins. The type of peristalsis that occurs in the small intestine, termed a **migrating motility complex (MMC),**

begins in the lower portion of the stomach and pushes chyme forward along a short stretch of small intestine before dying out. The MMC slowly migrates down the small intestine, reaching the end of the ileum in 90–120 minutes. Then another MMC begins in the stomach. Altogether, chyme remains in the small intestine for 3–5 hours.

All the chemical and mechanical phases of digestion from the mouth through the small intestine are directed toward changing food into forms that can pass through the epithelial cells lining the mucosa into the underlying blood and lymphatic vessels. These forms are monosaccharides (glucose, fructose, and galactose) from carbohydrates; single amino acids, dipeptides, and tripeptides from proteins; fatty acids, glycerol, and monoglycerides from lipids; and pentoses and nitrogenous bases from nucleic acids. Passage of these digested nutrients from the gastrointestinal tract into the blood or lymph is called **absorption.** Absorption occurs by diffusion, facilitated diffusion, osmosis, and active transport.

About 90% of all absorption of nutrients takes place in the small intestine. The other 10% occurs in the stomach and large intestine. Any undigested or unabsorbed material left in the small intestine passes on to the large intestine.

ABSORPTION OF ALCOHOL

The intoxicating and incapacitating effects of alcohol depend on the blood alcohol level. Because it is lipid soluble, alcohol begins to be absorbed in the stomach. However, the surface area available for absorption is much greater in the small intestine than in the stomach, so when alcohol passes into the duodenum, it is absorbed more rapidly. The longer that alcohol remains in the stomach, the more slowly blood alcohol level rises. Because fatty acids in chyme slow gastric emptying, blood alcohol level will rise more slowly when fat-rich foods, such as pizza, hamburgers, or nachos, are consumed with alcoholic beverages. Also, the enzyme alcohol dehydrogenase, which is present in gastric mucosa cells of the stomach, breaks down some of the alcohol to acetaldehyde, which is not intoxicating. When the rate of gastric emptying is slower, proportionally more alcohol will be absorbed and converted to acetaldehyde in the stomach, and thus less alcohol will reach the bloodstream. Given identical consumption of alcohol, females often develop higher blood alcohol levels (and therefore experience greater intoxication) than males of comparable size because the activity of gastric alcohol dehydrogenase is up to 60% lower in females than in males. Asian males may also have lower levels of this gastric enzyme. ■

CHECKPOINT

29. Name and describe the regions of the small intestine.
30. In what ways are the mucosa and submucosa of the small intestine adapted for digestion and absorption?
31. Describe the types of movement in the small intestine.
32. What is absorption? In what form are the products of carbohydrate, protein, and lipid digestion absorbed?

LARGE INTESTINE

OBJECTIVE

- Describe the anatomy, histology, and functions of the large intestine.

The large intestine is the terminal portion of the GI tract and is divided into four principal regions. As chyme moves through the large intestine, bacteria act on it and water, ions, and vitamins are absorbed. As a result, feces are formed and then eliminated from the body.

Anatomy

The **large intestine,** which is about 1.5 m (5 ft) long and 6.5 cm (2.5 in.) in diameter, extends from the ileum to the anus and is attached to the posterior abdominal wall by its **mesocolon,** a double layer of peritoneum. Structurally, the four principal regions of the large intestine are the cecum, colon, rectum, and anal canal (Figure 25.19a).

The opening from the ileum into the large intestine is guarded by a fold of mucous membrane called the **ileocecal sphincter** or **valve,** which allows materials from the small intestine to pass into the large intestine. Hanging inferior to the ileocecal valve is the **cecum,** a small pouch about 6 cm (2.4 in.) long. Attached to the cecum is a twisted, coiled tube, measuring about 8 cm (3 in.) in length, called the **appendix** or **vermiform appendix** (*vermiform*=worm-shaped; *appendix*=appendage). The mesentery of the appendix, called the **mesoappendix,** attaches the appendix to the inferior part of the mesentery of the ileum.

The open end of the cecum merges with a long tube called the **colon** (=food passage), which is divided into ascending, transverse, descending, and sigmoid portions. Both the ascending and descending colon are retroperitoneal, whereas the transverse and sigmoid colon are not. The **ascending colon** ascends on the right side of the abdomen, reaches the inferior surface of the liver, and turns abruptly to the left to form the **right colic (hepatic) flexure.** The colon continues across the abdomen to the left side as the **transverse colon.** It curves beneath the inferior end of the spleen on the left side as the **left colic (splenic) flexure** and passes inferiorly to the level of the iliac crest as the **descending colon.** The **sigmoid colon** (*sigm-*=S-shaped) begins near the left iliac crest, projects medially to the midline, and terminates as the rectum at about the level of the third sacral vertebra.

The **rectum,** the last 20 cm (8 in.) of the GI tract, lies anterior to the sacrum and coccyx. The terminal 2–3 cm (1 in.) of the rectum is called the **anal canal** (Figure 25.19b). The mucous membrane of the anal canal is arranged in longitudinal folds called **anal columns** that contain a network of arteries and veins. The opening of the anal canal to the exterior, called the **anus,** is guarded by an internal sphincter of smooth muscle (involuntary) and an external sphincter of skeletal muscle (voluntary). Normally the anus is closed except during the elimination of feces.

APPENDICITIS

Inflammation of the appendix, termed **appendicitis,** is preceded by obstruction of the lumen of the appendix by chyme, inflammation, a foreign body, a carcinoma of the cecum, stenosis, or kinking of the organ. It is characterized by high fever, elevated white cell count, and a neutrophil count higher than 75%. The infection that follows may result in edema and ischemia and may progress to gangrene and perforation within 24 to 36 hours. Typically, appendicitis begins with referred pain in the umbilical region of the abdomen, followed by anorexia (loss of appetite for food), nausea, and vomiting. After several hours the pain localizes in the right lower quadrant (RLQ) and is continuous, dull or severe, and intensified by coughing, sneezing, or body movements. Early appendectomy (removal of the appendix) is recommended because it is safer to operate than to risk rupture, peritonitis, and gangrene. Although it required major abdominal surgery in the past, today appendectomies are usually performed laparoscopically. ■

The arterial supply of the cecum and colon is derived from branches of the superior mesenteric and inferior mesenteric arteries. The venous return is by way of the superior and inferior mesenteric veins ultimately to the hepatic portal vein and into the liver. The arterial supply of the rectum and anal canal is derived from the superior, middle, and inferior rectal arteries. The rectal veins correspond to the rectal arteries.

The nerves to the large intestine consist of sympathetic, parasympathetic, and sensory components. The sympathetic innervation is derived from the celiac, superior, and inferior mesenteric ganglia and superior and inferior mesenteric plexuses. The fibers reach the viscera by way of the thoracic and lumbar splanchnic nerves. The parasympathetic innervation is derived from the vagus (X) and pelvic splanchnic nerves.

Histology

The wall of the large intestine differs from that of the small intestine in several respects. No villi or permanent circular folds are found in the **mucosa,** which consists of simple columnar epithelium, lamina propria (areolar connective tissue), and muscularis mucosae (smooth muscle) (Figure 25.20 on pages 796-797). The epithelium contains mostly absorptive and goblet cells (Figure 25.20b and c). The absorptive cells function primarily in water absorption, whereas the goblet cells secrete mucus that lubricates the passage of the colonic contents. Both absorptive and goblet cells are located in long, straight, tubular intestinal glands that extend the full thickness of the mucosa. Solitary lymphatic nodules are also found in the mucosa. The **submucosa** of the large intestine is similar to that found in the rest of the GI tract. The **muscularis** consists of an external layer of longitudinal smooth muscle and an internal layer of circular smooth muscle. Unlike other parts of the GI tract, portions of the longitudinal muscles are thickened, forming three conspicuous longitudinal bands called **teniae coli** (TĒ-nē-ē KŌ-lī; *taenia*=flat band), that run most of the length of the large intestine (see Figure 25.19a).

Figure 25.19   Anatomy of the large intestine.

The regions of the large intestine are the cecum, colon, rectum, and anal canal.

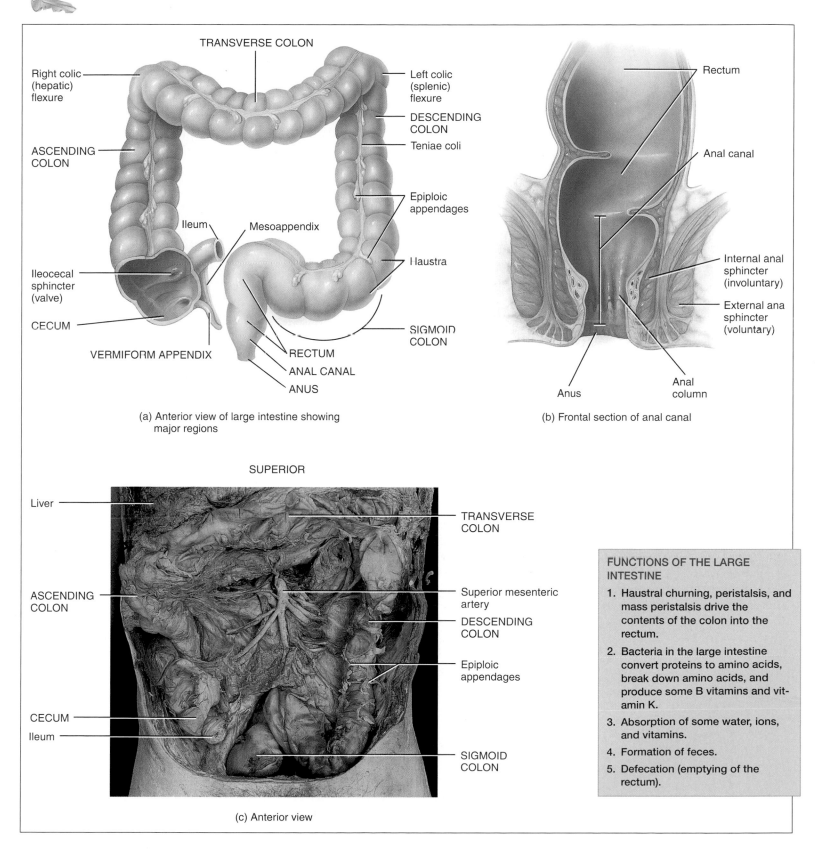

Right colic (hepatic) flexure

TRANSVERSE COLON

Left colic (splenic) flexure

ASCENDING COLON

DESCENDING COLON

Teniae coli

Ileum

Mesoappendix

Epiploic appendages

Ileocecal sphincter (valve)

Haustra

CECUM

SIGMOID COLON

VERMIFORM APPENDIX

RECTUM

ANAL CANAL

ANUS

(a) Anterior view of large intestine showing major regions

Rectum

Anal canal

Internal anal sphincter (involuntary)

External anal sphincter (voluntary)

Anus

Anal column

(b) Frontal section of anal canal

SUPERIOR

Liver

TRANSVERSE COLON

ASCENDING COLON

Superior mesenteric artery

DESCENDING COLON

Epiploic appendages

CECUM

Ileum

SIGMOID COLON

(c) Anterior view

FUNCTIONS OF THE LARGE INTESTINE

1. Haustral churning, peristalsis, and mass peristalsis drive the contents of the colon into the rectum.

2. Bacteria in the large intestine convert proteins to amino acids, break down amino acids, and produce some B vitamins and vitamin K.

3. Absorption of some water, ions, and vitamins.

4. Formation of feces.

5. Defecation (emptying of the rectum).

Which portions of the colon are retroperitoneal?

Figure 25.20 Histology of the large intestine.

Intestinal glands formed by simple columnar epithelial cells and goblet cells extend the full thickness of the mucosa.

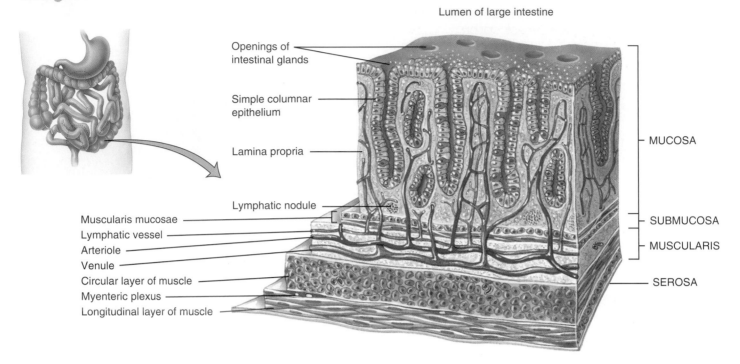

Lumen of large intestine

Openings of intestinal glands

Simple columnar epithelium

Lamina propria

MUCOSA

Lymphatic nodule

Muscularis mucosae
Lymphatic vessel
Arteriole
Venule
Circular layer of muscle
Myenteric plexus
Longitudinal layer of muscle

SUBMUCOSA

MUSCULARIS

SEROSA

(a) Three-dimensional view of layers of the large intestine

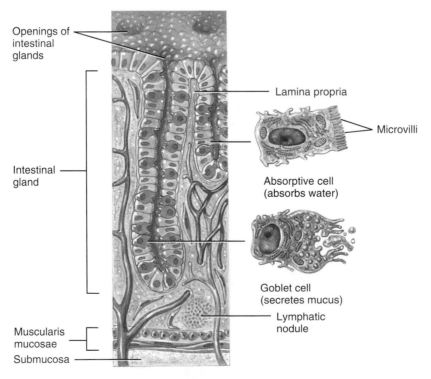

Openings of intestinal glands

Lamina propria

Microvilli

Intestinal gland

Absorptive cell (absorbs water)

Goblet cell (secretes mucus)

Muscularis mucosae

Lymphatic nodule

Submucosa

(b) Sectional view of intestinal glands and cell types

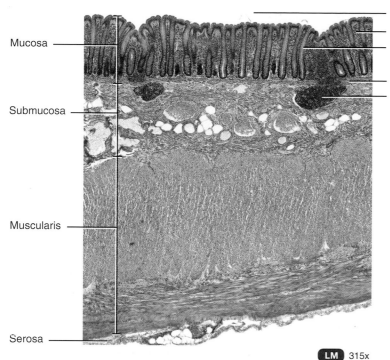

Mucosa

Submucosa

Muscularis

Serosa

Lumen of large intestine
Lamina propria
Intestinal gland

Muscularis mucosae
Lymphatic nodule

LM 315x

(c) Portion of the wall of the large intestine

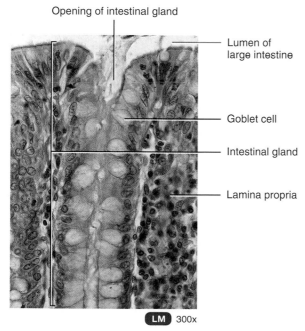

Opening of intestinal gland

Lumen of
large intestine

Goblet cell

Intestinal gland

Lamina propria

LM 300x

(d) Details of mucosa of large intestine

What is the function of the goblet cells of the large intestine?

The teniae coli are separated by portions of the wall with less longitudinal muscle or none at all. Tonic contractions of the bands gather the colon into a series of pouches called **haustra** (HAWS-tra=shaped like pouches), which give the colon a puckered appearance. A single layer of circular smooth muscle lies between the teniae coli. The **serosa** of the large intestine is part of the visceral peritoneum. Small pouches of visceral peritoneum filled with fat are attached to teniae coli and are called **epiploic appendages.**

POLYPS IN THE COLON

Polyps in the colon are generally slow-developing benign growths that arise from the mucosa of the large intestine. Often, they do not cause symptoms. If symptoms do occur, they include diarrhea, blood in the feces, and mucus discharged from the anus. The polyps are removed by colonoscopy (see page 799) or surgery because some of them may become cancerous. ■

Functions

The passage of chyme from the ileum into the cecum is regulated by the action of the ileocecal sphincter. Normally, the valve remains partially closed, and the passage of chyme into the cecum is a slow process. Immediately after a meal, ileal peristalsis intensifies, the sphincter relaxes, and chyme is forced from the ileum into the cecum. As food passes through the ileocecal sphincter, it fills the cecum and accumulates in the ascending colon, and movements of the colon begin.

One movement characteristic of the large intestine is **haustral churning.** In this process, the haustra remain relaxed and distended while they fill up. When the distension reaches a certain point, the wall contracts and squeezes the contents into the next haustrum. **Peristalsis** also occurs, although at a slower rate (3 to 12 contractions per minute) than in other portions of the tract. A final type of movement is **mass peristalsis,** a strong peristaltic wave that begins at about the middle of the transverse colon and quickly drives the colonic contents into the rectum. Mass peristalsis usually takes place three or four times a day, during or immediately after a meal.

The final stage of digestion occurs in the colon through the activity of bacteria that inhabit the lumen. Mucus is secreted by the glands of the large intestine, but no enzymes are secreted. Chyme is prepared for elimination by the action of bacteria, which ferment any remaining carbohydrates and release hydrogen, carbon dioxide, and methane gases. These gases contribute to flatus (gas) in the colon, termed *flatulence* when it is excessive. Bacteria also convert any remaining proteins to amino acids and break down the amino acids into simpler substances: indole, skatole, hydrogen sulfide, and fatty acids. Some of the indole and skatole is eliminated in the feces and contributes to their odor; the rest is absorbed and transported to the liver, where these compounds are converted to less toxic compounds and excreted in the urine. Bacteria also decompose bilirubin to simpler pigments, including stercobilin, which give feces their brown color. Several vitamins needed for normal metabolism, including some B vitamins and vitamin K, are bacterial products that are absorbed in the colon.

By the time chyme has remained in the large intestine 3–10 hours, it has become solid or semisolid as a result of water absorption and is now called **feces.** Chemically, feces consist of water, inorganic salts, sloughed-off epithelial cells from the mucosa of the gastrointestinal tract, bacteria, products of bacterial decomposition, unabsorbed digested materials, and indigestible parts of food.

Although 90% of all water absorption occurs in the small intestine, the large intestine absorbs enough to make it an important organ in maintaining the body's water balance. Of the 0.5–1.0 liter of water that enters the large intestine, all but about 100–200 mL is absorbed via osmosis. The large intestine also absorbs electrolytes, including sodium and chloride, and some vitamins.

OCCULT BLOOD

The term **occult blood** refers to blood that is hidden; it is not detectable by the human eye. The main diagnostic value of occult blood testing is to screen for colorectal cancer. Two substances frequently examined for occult blood are feces and urine. Several types of products are available for at-home testing for hidden blood in feces. The tests are based on color changes when reagents are added to feces. The presence of occult blood in urine may be detected at home by using dip-and-read reagent strips. ■

Mass peristaltic movements push fecal material from the sigmoid colon into the rectum. The resulting distention of the rectal wall stimulates stretch receptors, which initiates a **defecation reflex** that empties the rectum. The defecation reflex occurs as follows: In response to distention of the rectal wall, the receptors send sensory nerve impulses to the sacral spinal cord. Motor impulses from the cord travel along parasympathetic nerves back to the descending colon, sigmoid colon, rectum, and anus. The resulting contraction of the longitudinal rectal muscles shortens the rectum, thereby increasing the pressure within it. This pressure, along with voluntary contractions of the diaphragm and abdominal muscles, plus parasympathetic stimulation, opens the internal sphincter.

The external sphincter is voluntarily controlled. If it is voluntarily relaxed, defecation occurs and the feces are expelled through the anus; if it is voluntarily constricted, defecation can be postponed. Voluntary contractions of the diaphragm and abdominal muscles aid defecation by increasing the pressure within the abdomen, which pushes the walls of the sigmoid colon and rectum inward. If defecation does not occur, the feces back up into the sigmoid colon until the next wave of mass peristalsis again stimulates the stretch receptors, further creating the urge to defecate. In infants, the defecation reflex causes automatic emptying of the rectum because voluntary control of the external anal sphincter has not yet developed.

Diarrhea (dī-a-RĒ-a; *dia-*=through; *rrhea*=flow) is an increase in the frequency, volume, and fluid content of the feces caused by increased motility of and decreased absorption by the intestines. When chyme passes too quickly through the small intestine and feces pass too quickly through the large intestine, there is not enough time for absorption. Frequent diarrhea can result in dehydration and electrolyte imbalances. Excessive motility may be caused by lactose intolerance, stress, or microbes that irritate the gastrointestinal mucosa.

Constipation (kon-sti-PĀ-shun; *con-*=together; *stip-*=to press) refers to infrequent or difficult defecation caused by decreased motility of the intestines. Because the feces remain in the colon for prolonged periods of time, excessive water absorption occurs, and the feces become dry and hard. Constipation may be caused by poor habits (delaying defecation), spasms of the colon, insufficient fiber in the diet, inadequate fluid intake, lack of exercise, emotional stress, and certain drugs. A common treatment is a mild laxative, such as milk of magnesia, which induces defecation. However, many physicians maintain that laxatives are habit-forming, and that adding fiber to the diet, increasing the amount of exercise, and increasing fluid intake are safer ways of controlling this common problem.

Dietary fiber consists of indigestible plant carbohydrates—such as cellulose, lignin, and pectin—found in fruits, vegetables, grains, and beans. **Insoluble fiber,** which does not dissolve in water, includes the woody or structural parts of plants such as the skins of fruits and vegetables and the bran coating around wheat and corn kernels. Insoluble fiber passes through the GI tract largely unchanged and speeds up the passage of material through the tract. **Soluble fiber** dissolves in water and forms a gel, which slows the passage of material through the tract; it is found in abundance in beans, oats, barley, broccoli, prunes, apples, and citrus fruits.

People who choose a fiber-rich diet may reduce their risk of developing obesity, diabetes, atherosclerosis, gallstones, hemorrhoids, diverticulitis, appendicitis, and colorectal cancer. Soluble fiber also may help lower blood cholesterol. The liver normally converts cholesterol to bile salts, which are released into the small intestine to help fat digestion. Having accomplished their task, the bile salts are reabsorbed by the small intestine and recycled back to the liver. Since soluble fiber binds to bile salts to prevent their reabsorption, the liver makes more bile salts to replace those lost in feces. Thus, the liver uses more cholesterol to make more bile salts and blood cholesterol level is lowered.

COLONOSCOPY

Colonoscopy (kō-lon-OS-kō-pē; *-skopes*=to view) is the visual examination of the lining of the colon using an elongated, flexible, fiberoptic endoscope called a *colonoscope*. It is used to detect disorders such as polyps, cancer, and diverticulosis, to take tissue samples, and to remove small polyps. Most tumors of the large intestine occur in the rectum. ■

A summary of the digestive organs and their functions is presented in Table 25.2.

33. What are the principal regions of the large intestine?
34. How does the muscularis of the large intestine differ from that of the rest of the gastrointestinal tract? What are haustra?
35. Describe the mechanical movements that occur in the large intestine.
36. Define defecation. How does it occur?
37. Explain the activities of the large intestine that change its contents into feces.

DEVELOPMENT OF THE DIGESTIVE SYSTEM

OBJECTIVE

● Describe the development of the digestive system.

During the fourth week of development, the cells of the **endoderm** form a cavity called the **primitive gut,** the forerunner of the gastrointestinal tract. See Figure 4.12b on page 110. Soon afterward the mesoderm forms and splits into two layers (somatic and splanchnic), as shown in Figure 4.9d on page 105. The splanchnic mesoderm associates with the endoderm of the primitive gut; as a result, the primitive gut has a double-layered wall. The **endodermal layer** gives rise to the *epithelial lining* and *glands* of most of the gastrointestinal tract; the **mesodermal layer** produces the *smooth muscle* and *connective tissue* of the tract.

The primitive gut elongates and differentiates into an anterior **foregut,** an intermediate **midgut,** and a posterior **hindgut** (see Figure 4.12c on page 110). Until the fifth week of development, the midgut opens into the yolk sac; after that time, the yolk sac constricts and detaches from the midgut, and the

ORGANS	FUNCTIONS
TABLE 25.2	**SUMMARY OF ORGANS OF THE DIGESTIVE SYSTEM AND THEIR FUNCTIONS**
Tongue	Maneuvers food for mastication, shapes food into a bolus, maneuvers food for deglutition, detects sensations for taste, and initiates digestion of triglycerides.
Salivary glands	Saliva produced by these glands softens, moistens, and dissolves foods; cleanses mouth and teeth; initiates the digestion of starch.
Teeth	Cut, tear, and pulverize food to reduce solids to smaller particles for swallowing.
Pancreas	Pancreatic juice buffers acidic gastric juice in chyme, stops the action of pepsin from the stomach, creates the proper pH for digestion in the small intestine, and participates in the digestion of carbohydrates, proteins, triglycerides, and nucleic acids.
Liver	Produces bile, which is required for the emulsification and absorption of lipids in the small intestine.
Gallbladder	Stores and concentrates bile and releases it into the small intestine.
Mouth	See the functions of the tongue, salivary glands, and teeth, all of which are in the mouth. Additionally, the lips and cheeks keep food between the teeth during mastication, and buccal glands lining the mouth produce saliva.
Pharynx	Receives a bolus from the oral cavity and passes it into the esophagus.
Esophagus	Receives a bolus from the pharynx and moves it into the stomach. This requires relaxation of the upper esophageal sphincter and secretion of mucus.
Stomach	Mixing waves combine saliva, food, and gastric juice, which activates pepsin, initiates protein digestion, kills microbes in food, helps absorb vitamin B12, contracts the lower esophageal sphincter, increases stomach motility, relaxes the pyloric sphincter, and moves chyme into the small intestine.
Small intestine	Segmentation mixes chyme with digestive juices; peristalsis propels chyme toward the ileocecal sphincter; digestive secretions from the small intestine, pancreas, and liver complete the digestion of carbohydrates, proteins, lipids, and nucleic acids; circular folds, villi, and microvilli help absorb about 90% of digested nutrients.
Large intestine	Haustral churning, peristalsis, and mass peristalsis drive the colonic contents into the rectum; bacteria produce some B vitamins and vitamin K; absorption of some water, ions, and vitamins occurs; defecation.

midgut seals. In the region of the foregut, a depression consisting of ectoderm, the **stomodeum** (stō-mō-DĒ-um), appears (see Figure 4.12d on page 110). This develops into the *oral cavity*. The **oropharyngeal membrane** is a depression of fused ectoderm and endoderm on the surface of the embryo that separates the foregut from the stomodeum. The membrane ruptures during the fourth week of development, so that the foregut is continuous with the outside of the embryo through the oral cavity. Another depression consisting of ectoderm, the **proctodeum** (prok-tō-DĒ-um), forms in the hindgut and goes on to develop into the *anus*. See Figure 4.12d on page 110. The **cloacal membrane** (klō-Ā-kul) is a fused membrane of ectoderm and endoderm that separates the hindgut from the proctodeum. After the cloacal membrane ruptures during the seventh week, the hindgut forms a continuous tube from mouth to anus.

The foregut develops into the *pharynx, esophagus, stomach,* and *part of the duodenum*. The midgut is transformed into the *remainder of the duodenum*, the *jejunum*, the *ileum*, and *portions of the large intestine* (cecum, appendix, ascending colon, and most of the transverse colon). The hindgut develops into the *remainder of the large intestine*, except for a portion of the anal canal that is derived from the proctodeum.

As development progresses, the endoderm at various places along the foregut develops into hollow buds that grow into the mesoderm. These buds will develop into the *salivary glands, liver, gallbladder,* and *pancreas*. Each of these organs retains a connection with the gastrointestinal tract through ducts.

CHECKPOINT

38. What structures develop from the foregut, midgut, and hindgut?

AGING AND THE DIGESTIVE SYSTEM

OBJECTIVE

- Describe the effects of aging on the digestive system.

Overall changes of the digestive system associated with aging include decreased secretory mechanisms, decreased motility of the digestive organs, loss of strength and tone of the muscular tissue and its supporting structures, changes in neurosensory feedback regarding enzyme and hormone release, and diminished response to pain and internal sensations. In the upper portion of the GI tract, common changes include reduced sensitivity to mouth irritations and sores, loss of taste, periodontal disease, difficulty in swallowing, hiatal hernia, gastritis, and peptic ulcer disease. Changes in the small intestine may include duodenal ulcers, appendicitis, malabsorption, and maldigestion. Other pathologies that increase in incidence with age are gallbladder problems, jaundice, cirrhosis, and acute pancreatitis. Large intestinal changes such as constipation, hemorrhoids, and diverticular disease may also occur. Cancer of the colon or rectum is quite common, as are bowel obstructions and impactions.

CHECKPOINT

39. What overall changes in the digestive system appear with age?

 # APPLICATIONS TO HEALTH

DENTAL CARIES

Dental caries, or tooth decay, involve a gradual demineralization (softening) of the enamel and dentin. If untreated, microorganisms may invade the pulp, causing inflammation and infection, with subsequent death of the pulp and abscess of the alveolar bone surrounding the root's apex. Such teeth are treated by root canal therapy.

Dental caries begin when bacteria, acting on sugars, produce acids that demineralize the enamel. **Dextran,** a sticky polysaccharide produced from sucrose, causes the bacteria to stick to the teeth. Masses of bacterial cells, dextran, and other debris adhering to teeth constitute **dental plaque.** Saliva cannot reach the tooth surface to buffer the acid because of the plaque covering the teeth. Brushing the teeth immediately after eating removes the plaque from flat surfaces before the bacteria can produce acids. Dentists also recommend that the plaque between the teeth be removed every 24 hours with dental floss.

PERIODONTAL DISEASE

Periodontal disease is a collective term for a variety of conditions characterized by inflammation and degeneration of the gingivae, alveolar bone, periodontal ligament, and cementum. One such condition is called **pyorrhea,** the initial symptoms of which are enlargement and inflammation of the soft tissue and bleeding of the gums. Without treatment, the soft tissue may deteriorate and the alveolar bone may be resorbed, causing loosening of the teeth and recession of the gums. Periodontal diseases are often caused by poor oral hygiene; by local irritants, such as bacteria, impacted food, and cigarette smoke; or by a poor "bite."

PEPTIC ULCER DISEASE

In the United States, 5–10% of the population develops **peptic ulcer disease (PUD).** An **ulcer** is a craterlike lesion in a membrane; ulcers that develop in areas of the GI tract exposed to acidic gastric juice are called **peptic ulcers.** The most common complication of peptic ulcers is bleeding, which can lead to anemia if enough blood is lost. In acute cases, peptic ulcers can lead to shock and death. Three distinct causes of PUD are recognized: (1) the bacterium *Helicobacter pylori;* (2) nonsteroidal anti-inflammatory drugs (NSAIDs) such as aspirin; and (3) hypersecretion of HCl, as occurs in Zollinger-Ellison syndrome, a gastrin-producing tumor usually of the pancreas (see page 778).

Helicobacter pylori (previously named *Campylobacter pylori*) is the most frequent cause of PUD. The bacterium produces an enzyme called urease, which splits urea into ammonia and carbon dioxide. While shielding the bacterium from the acidity of the stomach, the ammonia also damages the protective mucous layer of the stomach and the underlying gastric cells. *H. pylori* also produces catalase, an enzyme that may protect the microbe from phagocytosis by neutrophils, plus several adhesion proteins that allow the bacterium to attach itself to gastric cells.

Several therapeutic approaches are helpful in the treatment of PUD. Because cigarette smoke, alcohol, caffeine, and NSAIDs can impair mucosal defensive mechanisms, in the process increasing mucosal susceptibility to the damaging effects of HCl, these substances should be avoided. In cases associated with *H. pylori*, treatment with an antibiotic drug often resolves the problem. Oral antacids such as Tums or Maalox can help temporarily by buffering gastric acid. When hypersecretion of HCl is the cause of PUD, H_2 blockers (such as Tagamet) or proton pump inhibitors such as omeprazole (Prilosec), which block secretion of H^+ from parietal cells, may be used.

DIVERTICULAR DISEASE

In **diverticular disease,** saclike outpouchings of the wall of the colon, called **diverticula,** occur in places where the muscularis has weakened; the diverticula may become inflamed. Development of diverticula is known as **diverticulosis.** Many people who develop diverticulosis have no symptoms and experience no complications. Of those people known to have diverticulosis, 10–25% eventually develop an inflammation known as **diverticulitis.** This condition may be characterized by pain, either constipation or increased frequency of defecation, nausea, vomiting, and low-grade fever. Because diets low in fiber contribute to development of diverticulitis, patients who change to high-fiber diets show marked relief of symptoms. In severe cases, affected portions of the colon may require surgical removal. If diverticula rupture, the release of bacteria into the abdominal cavity can cause peritonitis.

COLORECTAL CANCER

Colorectal cancer is among the deadliest of malignancies, ranking second to lung cancer in males and third after lung cancer and breast cancer in females. Genetics plays a very important role; an inherited predisposition contributes to more than half of all cases of colorectal cancer. Intake of alcohol and diets high in animal fat and protein are associated with increased risk of colorectal cancer, whereas dietary fiber, aspirin, calcium, and selenium may be protective. Signs and symptoms of colorectal cancer include diarrhea, constipation, cramping, abdominal pain, and rectal bleeding, either visible or occult. Precancerous growths on the mucosal surface, called **polyps,** also increase the risk of developing colorectal cancer. Screening includes testing for blood in the feces, digital rectal examination, sigmoidoscopy, colonoscopy, and barium enema. Tumors may be removed endoscopically or surgically.

HEPATITIS

Hepatitis is an inflammation of the liver that can be caused by viruses, drugs, and chemicals, including alcohol. Clinically, several types of viral hepatitis are recognized. **Hepatitis A (infectious hepatitis)** is caused by the hepatitis A virus and is spread via fecal contamination of objects such as food, clothing, toys, and eating utensils (fecal-oral route). It is generally a mild disease of children and young adults characterized by loss of appetite, malaise, nausea, diarrhea, fever, and chills. Eventually, jaundice appears. This type of hepatitis does not cause lasting liver damage. Most people recover in 4 to 6 weeks.

Hepatitis B is caused by the hepatitis B virus and is spread primarily by sexual contact, contaminated syringes, or infected transfusion equipment. It can also be spread via saliva and tears. Hepatitis B virus can be present (without symptoms) for years or even a lifetime, and it can produce cirrhosis and possibly cancer of the liver. Individuals who harbor the active hepatitis B virus also become carriers. Vaccines produced through recombinant DNA technology (for example, Recombivax HB) are available to prevent hepatitis B infection.

Hepatitis C, caused by the hepatitis C virus, is clinically similar to hepatitis B. Hepatitis C can cause cirrhosis and possibly liver cancer. In developed nations, donated blood is screened for the presence of hepatitis B and C.

Hepatitis D is caused by the hepatitis D virus. It is transmitted like hepatitis B and, in fact, a person must be infected with hepatitis B in order to contract hepatitis D. Hepatitis D results in severe liver damage and has a higher fatality rate than infection with hepatitis B virus alone.

Hepatitis E is caused by the hepatitis E virus and is spread like hepatitis A. Although it does not cause chronic liver disease, hepatitis E virus is responsible for a very high mortality rate in pregnant women.

OBESITY

Obesity—body weight more than 20% above some desirable standard due to an excessive accumulation of adipose tissue—affects one-third of the adult population in the United States. (An athlete may be *overweight* due to higher-than-normal amounts of muscle tissue without being *obese.*) Even moderate obesity is hazardous to health; it is implicated as a risk factor in cardiovascular disease, hypertension, pulmonary disease, non-insulin-dependent diabetes mellitus, arthritis, certain cancers (breast, uterus, and colon), varicose veins, and gallbladder disease.

In a few cases, obesity may result from trauma of or tumors in the food-regulating centers in the hypothalamus. In most cases of obesity, no specific cause can be identified. Contributing factors include genetic factors, eating habits taught early in life, overeating to relieve tension, and social customs. Studies indicate that some obese people burn fewer calories during diges-tion and absorption of a meal. Additionally, after they lose weight, formerly obese people require about 15% fewer calories to maintain normal body weight than people who have never been obese. Although the hormone leptin suppresses appetite and produces satiety in experimental animals, it is not deficient in most obese people.

KEY MEDICAL TERMS ASSOCIATED WITH THE DIGESTIVE SYSTEM

Achalasia (ak′-a-LĀ-zē-a; *a-*=without; *chalasis*=relaxation) A condition, caused by malfunction of the myenteric plexus, in which the lower esophageal sphincter fails to relax normally as food approaches. A whole meal may become lodged in the esophagus and enter the stomach very slowly. Distension of the esophagus results in chest pain that is often confused with pain originating from the heart.

Anal fissure (FISH-ur; *findo*=to cleave) A longitudinal tear in the mucosa of the anal canal frequently caused by injury from passing hard feces. Anal fissures may cause pain and bleeding, usually during or shortly after a bowel movement. Treatment consists of using stool softeners and lubricating suppositories and, in severe cases, surgery.

Anorexia nervosa A chronic disorder characterized by self-induced weight loss, negative perception of body image, and physiological changes that result from nutritional depletion. Patients have a fixation on weight control and often abuse laxatives, which worsens their fluid and electrolyte imbalances and nutrient deficiencies. The disorder is found predominantly in young, single females; it may be inherited. Individuals may become emaciated and may ultimately die of starvation or one of its complications.

Barrett's esophagus A pathological change in the epithelium of the esophagus from nonkeratinized stratified squamous epithelium to columnar epithelium so that the lining resembles that of the stomach or small intestine. This abnormal development of tissue is due to long-term exposure of the esophagus to stomach acid and increases the risk of developing cancer of the esophagus.

Borborygmus (bor′-bō-RIG-mus) A rumbling noise caused by the propulsion of gas through the intestines.

Bulimia (*bu-*=ox; *limia*=hunger or **binge-purge syndrome**) A disorder that typically affects young, single, middle-class, white females, characterized by overeating at least twice a week followed by purging by self-induced vomiting, strict dieting or fasting, vigorous exercise, or use of laxatives or diuretics; it occurs in response to fears of being overweight or to stress, depression, and physiological disorders such as hypothalamic tumors.

Canker sore (KANG-ker) Painful ulcer on the mucous membrane of the mouth that affects females more often than males, usually between ages 10 and 40; may be an autoimmune reaction or a food allergy.

Celiac disease (SĒ-lē-ak; *koilia*=hollow) A common malabsorption disorder caused by sensitivity to gluten, a protein found in some grains such as wheat, rye, barley, and oats. The gluten damages intestinal villi and reduces the length of the microvilli. Symptoms include foul-smelling diarrhea, bloating, weight loss, abdominal pain, and anemia.

Cholecystitis (kō′-lē-sis-TĪ-tis; *chole-*=bile; *cyst-*=bladder; *-itis*=inflammation of) An inflammation of the gallbladder, caused in some cases by an autoimmune reaction; other cases are caused by obstruction of the cystic duct by bile stones.

Cirrhosis Distorted or scarred liver as a result of chronic inflammation due to hepatitis, chemicals that destroy hepatocytes, parasites that infect the liver, or alcoholism; the hepatocytes are replaced by fibrous or adipose connective tissue. Symptoms include jaundice, edema in the legs, uncontrolled bleeding, and increased sensitivity to drugs.

Colitis (ko-LĪ-tis) Inflammation of the mucosa of the colon and rectum in which absorption of water and salts is reduced, producing watery, bloody feces and, in severe cases, dehydration and salt depletion. Spasms of the irritated muscularis produce cramps. It is thought to be an autoimmune condition.

Colostomy (kō-LOS-tō-mē; *-stomy*=provide an opening) The diversion of feces through an opening in the colon, creating a surgical "stoma" (artificial opening) that is made in the exterior of the abdominal wall. This opening serves as a substitute anus through which feces are eliminated into a bag worn on the abdomen.

Dysphagia (dis-FĀ-jē-a; *dys-*=abnormal; *phagia*=to eat) Difficulty in swallowing that may be caused by inflammation, paralysis, obstruction, or trauma.

Flatus (FLĀ-tus) Air (gas) in the stomach or intestine, usually expelled through the anus. If the gas is expelled through the mouth, it is called **eructation** or **belching** (burping). Flatus may result from gas released during the breakdown of foods in the stomach or from swallowing air or gas-containing substances such as carbonated drinks.

Food poisoning A sudden illness caused by ingesting food or drink contaminated by an infectious microbe (bacterium, virus, or protozoan) or a toxin (poison). The most common cause of food poisoning is the toxin produced by the bacterium *Staphylococcus aureus*. Most types of food poisoning cause diarrhea and/or vomiting, often associated with abdominal pain.

Gastroenteritis (gas'-trō-en-ter-Ī-tis; *gastro-*=stomach; *enteron*=intestine; *-itis*=inflammation) Inflammation of the lining of the stomach and intestine (especially the small intestine). It is usually caused by a viral or bacterial infection that may be acquired by contaminated food or water or by people in close contact. Symptoms include diarrhea, vomiting, fever, loss of appetite, cramps, and abdominal discomfort.

Gastroscopy (gas-TROS-kō-pē; *-scopy*=to view with a lighted instrument) Endoscopic examination of the stomach in which the examiner can view the interior of the stomach directly to evaluate an ulcer, tumor, inflammation, or source of bleeding.

Halitosis (hal'-i-TŌ-sis; *halitus-*=breath; *-osis*=condition) A foul odor from the mouth; also called **bad breath.**

Heartburn A burning sensation in a region near the heart due to irritation of the mucosa of the esophagus from hydrochloric acid in stomach contents. It is caused by failure of the lower esophageal sphincter to close properly, so that the stomach contents enter the inferior esophagus. It is not related to any cardiac problem.

Hemorrhoids (HEM-ō-royds; *hemi*=blood; *rhoia*=flow) Varicosed (enlarged and inflamed) superior rectal veins. Hemorrhoids develop when the veins are put under pressure and become engorged with blood. If the pressure continues, the wall of the vein stretches. Such a distended vessel oozes blood; bleeding or itching is usually the first sign that a hemorrhoid has developed. Stretching of a vein also favors clot formation, further aggravating swelling and pain. Hemorrhoids may be caused by constipation, which may be brought on by low-fiber diets. Also, repeated straining during defecation forces blood down into the rectal veins, increasing pressure in these veins and possibly causing hemorrhoids. Also called **piles.**

Hernia (HER-nē-a) Protrusion of all or part of an organ through a membrane or cavity wall, usually the abdominal cavity. *Hiatal hernia* is the protrusion of a part of the stomach into the thoracic cavity through the esophageal hiatus. *Inguinal hernia* is the protrusion of the hernial sac into the inguinal opening; it may contain a portion of the bowel in an advanced stage and may extend into the scrotal compartment in males, causing strangulation of the herniated part.

Indigestion (in-di-JES-chun) A nonspecific term used to describe many symptoms associated with abdominal distress, especially after eating. Symptoms include discomfort or a feeling of dullness in the upper abdomen, nausea, and a sensation of bloating, often relieved by belching. Among the factors related to indigestion are reactions to drugs, alcohol, and supplements; overproduction of stomach acid; infections; food sensitivity; stress; and psychological factors such as depression and anxiety. Also called **dyspepsia.**

Inflammatory bowel disease (in-FLAM-a-tō'-rē BOW-el) Inflammation of the gastrointestinal tract that exists in two forms. (1) **Crohn's disease** is an inflammation of any part of the gastrointestinal tract in which the inflammation extends from the mucosa through the submucosa, muscularis, and serosa. (2) **Ulcerative colitis** is an inflammation of the mucosa of the colon and rectum, usually accompanied by rectal bleeding. Curiously, cigarette smoking increases one's risk for Crohn's disease but decreases the risk of ulcerative colitis.

Irritable bowel syndrome (IBS) Disease of the entire gastrointestinal tract in which a person reacts to stress by developing symptoms (such as cramping and abdominal pain) associated with alternating patterns of diarrhea and constipation. Excessive amounts of mucus may appear in feces; other symptoms include flatulence, nausea, and loss of appetite. The condition is also known as **irritable colon** or **spastic colitis.**

Laparoscopy (lap'-a-ROS-kō-pē; *laparo-*=flank, *-skopy*=to view) Examination of the contents of the peritoneum with a laparoscope (a type of endoscope) passed through the abdominal wall. Frequently used to examine female reproductive organs.

Malabsorption (mal-ab-SORP-shun; *mal-*=bad) A number of disorders in which nutrients from food are not absorbed properly. It may be due to disorders that result in the inadequate breakdown of food during digestion (due to inadequate digestive enzymes or juices), damage to the lining of the small intestine (surgery, infections, and drugs like neomycin and alcohol), and impairment of motility. Symptoms may include diarrhea, weight loss, weakness, vitamin deficiencies, and bone demineralization.

Malocclusion (mal'-ō-KLOO-zhun; *mal-*=bad; *occlusion*=to fit together) Condition in which the surfaces of the maxillary (upper) and mandibular (lower) teeth fit together poorly.

Nausea (NAW-sē-a; *nausia*=seasickness) Discomfort characterized by a loss of appetite and the sensation of impending vomiting. Its causes include local irritation of the gastrointestinal tract, a systemic disease, brain disease or injury, overexertion, or the effects of medication or drug overdosage.

Pruritus ani (proo-RĪ-tus; *pruire*=to itch) Anal itching that may be caused by pinworms (in children), infections, dermatitis, hemorrhoids, or dry, aging skin.

Traveler's diarrhea Infectious disease of the gastrointestinal tract that results in loose, urgent bowel movements, cramping, abdominal pain, malaise, nausea, and occasionally fever and dehydration. It is acquired through ingestion of food or water contaminated with fecal material typically containing bacteria (especially *Escherichia coli*); viruses or protozoan parasites are a less common cause.

Xerostomia (zē-rō-STŌ-mē-a; *xeros-*=dry; *-stoma*=mouth) Dryness of the mouth due to decreased or inhibited salivary secretion.

STUDY OUTLINE

Introduction (p. 764)

1. The breakdown of larger food molecules into smaller molecules is called digestion.
2. The passage of these smaller molecules into blood and lymph is termed absorption.

Overview of the Digestive System (p. 765)

1. The organs that collectively perform digestion and absorption constitute the digestive system and are usually divided into two main groups: the gastrointestinal (GI) tract and accessory digestive organs.
2. The GI is a continuous tube extending from the mouth to the anus.
3. The accessory digestive organs include the teeth, tongue, salivary glands, liver, gallbladder, and pancreas.
4. Digestion includes six basic processes: ingestion, secretion, mixing and propulsion, mechanical and chemical digestion, absorption, and defecation.
5. Mechanical digestion consists of mastication and movements of the gastrointestinal tract that aid chemical digestion.
6. Chemical digestion is a series of hydrolysis reactions that break down large carbohydrates, lipids, proteins, and nucleic acids in foods into smaller molecules that are usable by body cells.

Layers of the GI Tract (p. 765)

1. The basic arrangement of layers in most of the gastrointestinal tract, from deep to superficial, is the mucosa, submucosa, muscularis, and serosa.
2. Associated with the lamina propria of the mucosa are extensive patches of lymphatic tissue called mucosa-associated lymphoid tissue (MALT).

Peritoneum (p. 768)

1. The peritoneum is the largest serous membrane of the body; it lines the wall of the abdominal cavity and covers some abdominal organs.
2. Folds of the peritoneum include the mesentery, mesocolon, falciform ligament, lesser omentum, and greater omentum.

Mouth (p. 770)

1. The mouth is formed by the cheeks, hard and soft palates, lips, and tongue.
2. The vestibule is the space bounded externally by the cheeks and lips and internally by the teeth and gums.
3. The oral cavity proper extends from the vestibule to the fauces.
4. The tongue, together with its associated muscles, forms the floor of the oral cavity. It is composed of skeletal muscle covered with mucous membrane.
5. The upper surface and sides of the tongue are covered with papillae, some of which contain taste buds.
6. The major portion of saliva is secreted by the salivary glands, which lie outside the mouth and pour their contents into ducts that empty into the oral cavity.

7. There are three pairs of salivary glands: parotid, submandibular (submaxillary), and sublingual glands.
8. Saliva lubricates food and starts the chemical digestion of carbohydrates.
9. Salivation is controlled by the nervous system.
10. The teeth (dentes) project into the mouth and are adapted for mechanical digestion.
11. A typical tooth consists of three principal regions: crown, root, and neck.
12. Teeth are composed primarily of dentin and are covered by enamel, the hardest substance in the body.
13. There are two dentitions: deciduous and permanent.
14. Through mastication, food is mixed with saliva and shaped into a soft, flexible mass called a bolus.
15. Salivary amylase begins the digestion of starches, and lingual lipase acts on triglycerides.

Pharynx (p. 776)

1. Deglutition, or swallowing, moves a bolus from the mouth to the stomach.
2. Muscular contractions of the oropharynx and laryngopharynx propel a bolus into the esophagus.

Esophagus (p. 776)

1. The esophagus is a collapsible, muscular tube that connects the pharynx to the stomach.
2. It passes a bolus into the stomach by peristalsis.
3. It contains an upper and a lower esophageal sphincter.

Stomach (p. 778)

1. The stomach connects the esophagus to the duodenum.
2. The principal anatomic regions of the stomach are the cardia, fundus, body, and pylorus.
3. Adaptations of the stomach for digestion include rugae; glands that produce mucus, hydrochloric acid, pepsin, gastric lipase, and intrinsic factor; and a three-layered muscularis.
4. Mechanical digestion consists of mixing waves.
5. Chemical digestion consists mostly of the conversion of proteins into peptides by pepsin.
6. The stomach wall is impermeable to most substances.
7. Among the substances the stomach can absorb are water, certain ions, drugs, and alcohol.

Pancreas (p. 782)

1. The pancreas consists of a head, a body, and a tail and is connected to the duodenum via the pancreatic duct and accessory duct.
2. Endocrine pancreatic islets (islets of Langerhans) secrete hormones, and exocrine acini secrete pancreatic juice.
3. Pancreatic juice contains enzymes that digest starch (pancreatic amylase), proteins (trypsin, chymotrypsin, carboxypeptidase, and elastase), triglycerides (pancreatic lipase), and nucleic acids (ribonuclease and deoxyribonuclease).

Liver and Gallbladder (p. 784)

1. The liver has left and right lobes; the right lobe includes a quadrate and caudate lobe. The gallbladder is a sac located in a depression on the posterior surface of the liver that stores and concentrates bile.

2. The lobes of the liver are made up of lobules that contain hepatocytes (liver cells), sinusoids, stellate reticuloendothelial (Kupffer's) cells, and a central vein.

3. Hepatocytes produce bile that is carried by a duct system to the gallbladder for concentration and temporary storage. Cholecystokinin (CCK) stimulates ejection of bile into the common bile duct.

4. Bile's contribution to digestion is the emulsification of dietary lipids.

5. The liver also functions in carbohydrate, lipid, and protein metabolism; processing of drugs and hormones; excretion of bilirubin; synthesis of bile salts; storage of vitamins and minerals; phagocytosis; and activation of vitamin D.

6. Bile secretion is regulated by neural and hormonal mechanisms.

Small Intestine (p. 788)

1. The small intestine extends from the pyloric sphincter to the ileocecal sphincter.

2. It is divided into duodenum, jejunum, and ileum.

3. Its glands secrete fluid and mucus, and the circular folds, villi, and microvilli of its wall provide a large surface area for digestion and absorption.

4. Brush-border enzymes digest α-dextrins, maltose, sucrose, lactose, peptides, and nucleotides at the surface of mucosal epithelial cells.

5. Pancreatic and intestinal brush-border enzymes break down carbohydrates, proteins, and nucleic acids.

6. Mechanical digestion in the small intestine involves segmentation and migrating motility complexes.

7. Absorption is the passage of digested nutrients from the gastrointestinal tract into blood or lymph.

8. The absorbed nutrients include monosaccharides, amino acids, fatty acids, monoglycerides, pentoses, and nitrogenous bases.

Large Intestine (p. 794)

1. The large intestine extends from the ileocecal sphincter to the anus.

2. Its regions include the cecum, colon, rectum, and anal canal.

3. The mucosa contains many goblet cells, and the muscularis consists of teniae coli and haustra.

4. Mechanical movements of the large intestine include haustral churning, peristalsis, and mass peristalsis.

5. The last stages of chemical digestion occur in the large intestine through bacterial action. Substances are further broken down, and some vitamins are synthesized.

6. The large intestine absorbs water, electrolytes, and vitamins.

7. Feces consist of water, inorganic salts, epithelial cells, bacteria, and undigested foods.

8. The elimination of feces from the rectum is called defecation.

9. Defecation is a reflex action aided by voluntary contractions of the diaphragm and abdominal muscles and relaxation of the external anal sphincter.

Development of the Digestive System (p. 799)

1. The endoderm of the primitive gut forms the epithelium and glands of most of the gastrointestinal tract.

2. The mesoderm of the primitive gut forms the smooth muscle and connective tissue of the gastrointestinal tract.

Aging and the Digestive System (p. 800)

1. General changes include decreased secretory mechanisms, decreased motility, and loss of tone.

2. Specific changes may include loss of taste, pyorrhea, hernias, peptic ulcer disease, constipation, hemorrhoids, and diverticular diseases.

Q SELF-QUIZ QUESTIONS

Choose the one best answer to the following questions.

1. The cells of the gastric glands that produce secretions directly involved in chemical digestion are:
   a. mucous neck cells.   b. parietal cells.   c. chief cells.
   d. G cells.   e. goblet cells.

2. Which anatomical region of the stomach is closest to the esophagus?
   a. body   b. pylorus   c. fundus
   d. cardia   e. pyloric sphincter

3. *All* of the following are considered accessory organs of the digestive system *except* the:
   a. liver.   b. stomach.   c. pancreas.
   d. teeth.   e. salivary glands.

4. Which of the following digestive juices contains enzymes that digest carbohydrates, lipids, and proteins?
   a. pancreatic juice   b. bile   c. saliva   d. gastric juice
   e. none of the above (because no one digestive juice contains enzymes for digesting all three classes of foods)

5. The type of tissue that lines the stomach and intestines is:
   a. stratified squamous epithelium.
   b. pseudostratified ciliated columnar epithelium.
   c. simple columnar epithelium.
   d. simple cuboidal epithelium.
   e. endothelium.

6. The roots of a tooth are covered by a tissue that is harder and denser than bone and is called:
   a. gingivae.   b. cementum.   c. dentin.
   d. periodontal ligament.   e. enamel.

7. The parotid glands are located:
   a. inferior and anterior to the ears.
   b. at the base of the tongue.
   c. inferior to the liver.
   d. in the gastric mucosa.
   e. in the folds of the peritoneum.

8. The region of the stomach that connects to the small intestine is the:
   a. fundus.   b. body.   c. cardia.   d. rugae.   e. pylorus.

Complete the following.

9. The largest peritoneal fold, which drapes over the transverse colon and small intestine, is the _____.

10. Place numbers in the blanks to indicate the correct order for the flow of bile. (a) bile canaliculi: _____; (b) common bile duct: _____; (c) common hepatic duct: _____; (d) right and left hepatic ducts: _____; (e) hepatopancreatic ampulla (ampulla of Vater) and duodenum: _____.

11. The scientific name for swallowing is _____.

12. The process by which contents of the gastrointestinal tract are propelled forward by coordinated contractions and relaxations of the circular and longitudinal muscles of the muscularis is called _____.

13. First and second premolars and first, second, and third molars are characteristic of the _____ dentition.

14. The _____ is an organ of the digestive system that assumes a role in phagocytosis, manufacture of plasma proteins, detoxification, and interconversions of nutrients.

15. The duct that conveys bile to and from the gallbladder is the _____ duct.

16. Parietal cells of the stomach secrete _____ and _____, and the chief cells secrete _____ and _____.

17. The two large lobes of the liver are the right and left lobes, and the two smaller lobes are the posterior _____ lobe and the inferior _____ lobe.

18. The three protein-digesting enzymes found in pancreatic juice are _____, _____, and _____.

Are the following statements true or false?

19. The submucosal plexus (plexus of Meissner) is important in controlling motility of the GI tract whereas the myenteric plexus (plexus of Auerbach) controls secretions of the GI tract.

20. The gallbladder produces bile.

21. The largest lobe of the liver is the right lobe.

22. Acini are clusters of cells that make up the exocrine portions of the pancreas.

23. The sinusoids of the liver receive blood from two sources: oxygenated blood from branches of the hepatic artery, and deoxygenated blood from branches of the hepatic portal vein.

24. The duodenum is the most distal region of the small intestine.

Matching

25. Match the following structures and descriptions:

    ___ **(a)** secrete lysozyme in intestinal juice

    ___ **(b)** site of maltase, sucrase, and lactase in the small intestine

    ___ **(c)** the pouches seen in the large intestine

    ___ **(d)** folds of the mucosa of the stomach and gallbladder

    ___ **(e)** binds the small intestine to the posterior abdominal wall

    ___ **(f)** produce bile

    ___ **(g)** secrete saliva

    ___ **(h)** secretes both enzymes and hormones

    (1) mesentery
    (2) parotid glands
    (3) Paneth cells
    (4) hepatocytes
    (5) pancreas
    (6) brush border
    (7) rugae
    (8) haustra

CRITICAL THINKING QUESTIONS

1. If you could leave French fried potatoes in your mouth long enough after chewing them, they would start to taste sweeter and maybe even a little tart. Why?

2. When Zelda turned 50, her doctor said, "Well, it's time to schedule your first colonoscopy!" What is a colonoscopy? Describe the specific anatomical characteristics the doctor would be examining with the fiber optics.

3. The small intestine is the primary site of digestion and absorption in the gastrointestinal tract. What structural modifications are unique to the small intestine and what are their functions?

4. Krystal and her friend were giggling over their milk and fries at her birthday party. In mid-giggle, milk started coming out of Krystal's nose. How did this happen?

5. Four-year-old Billy was lying down with his head resting on his mother's abdomen. He started laughing and said, "Mommy, your tummy sure is making a lot of funny noises!" What (specifically) is Billy hearing?

ANSWERS TO FIGURE QUESTIONS

25.1 Digestive enzymes are secreted by the salivary glands, tongue, stomach, pancreas, and small intestine.

25.2 The enteric plexuses help regulate secretions and motility of the gastrointestinal tract.

25.3 Mesentery binds the small intestine to the posterior abdominal wall.

25.4 The uvula helps prevent foods and liquids from entering the nasal cavity during swallowing.

25.5 The ducts of the submandibular glands empty on either side of the lingual frenulum.

25.6 The main component of teeth is a type of connective tissue called dentin.

25.7 The first, second, and third molars do not replace any deciduous teeth.

25.8 The esophageal mucosa and submucosa contain mucus-secreting glands that lubricate the gastrointestinal tract to allow food to pass through more easily.

25.9 Food is pushed along the gastrointestinal tract by contraction of smooth muscle behind the bolus and relaxation of smooth muscle in front of it.

25.10 After a large meal, the rugae stretch and disappear as the stomach fills; as the stomach empties, the rugae reappear.

25.11 In the gastric glands of the stomach, surface mucous cells and mucous neck cells secrete mucus; chief cells secrete pepsinogen and gastric lipase; parietal cells secrete HCl and intrinsic factor; and G cells secrete gastrin.

25.12 The pancreatic duct contains pancreatic juice (fluid and digestive enzymes); the common bile duct contains bile; the hepatopancreatic ampulla contains pancreatic juice and bile.

25.13 The epigastric region, which contains most of the liver, can be palpated in a clinical examination to check for enlargement.

25.14 The phagocytic cell in the liver is the stellate reticuloendothelial (Kupffer) cell.

25.15 A portal triad is a unit composed of branches of the hepatic portal vein, hepatic artery, and a bile duct.

25.16 The ileum is the longest part of the small intestine.

25.17 Nutrients being absorbed enter the blood via the capillaries or the lymph via the lacteals in the center of each villus.

25.18 The fluid secreted by duodenal glands—alkaline mucus—neutralizes gastric acid and protects the mucosal lining of the duodenum.

25.19 The ascending and descending portions of the colon are retroperitoneal.

25.20 The goblet cells of the large intestine secrete mucus to lubricate the colonic contents.

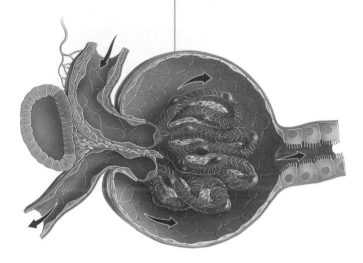

26

THE URINARY SYSTEM

INTRODUCTION   The **urinary system** consists of two kidneys, two ureters, one urinary bladder, and a single urethra (Figure 26.1). The kidneys do the major work of the urinary system, as the other parts of the system are primarily passageways and storage areas. In filtering blood and forming urine, the kidneys perform several functions:

1. *Regulation of blood volume and composition.* The kidneys regulate the composition and volume of the blood and remove wastes from the blood. In the process, urine is formed. They also excrete selected amounts of various wastes including excess H^+, which helps control blood pH.

2. *Regulation of blood pressure.* The kidneys help regulate blood pressure by secreting the enzyme renin, which activates the renin–angiotensin pathway. This results in an increase in blood pressure.

3. *Contribution to metabolism.* The kidneys contribute to metabolism by (1) synthesizing new glucose molecules (gluconeogenesis) during

When you look at anatomical images that are horizontal slices, are you able to identify relevant structures? Or are you confused by the unusual perspective? Transverse sectional anatomy is not easy to learn. Are you up to the challenge?

www.wiley.com/college/apcentral

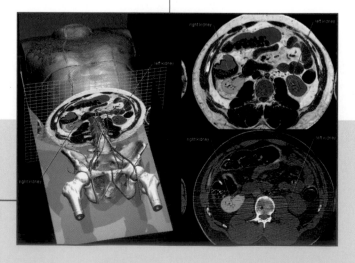

Figure 26.1   Organs of the urinary system, shown in relation to the surrounding structures in a female.
(See Tortora, *A Photographic Atlas of the Human Body*, 2e, Figures 13.3 and 13.4.)

> Urine formed by the kidneys passes first into the ureters, then to the urinary bladder for storage, and finally through the urethra for elimination from the body.

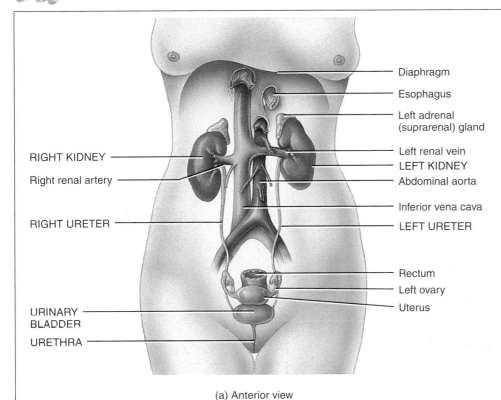

Diaphragm
Esophagus
Left adrenal (suprarenal) gland
Left renal vein
RIGHT KIDNEY
LEFT KIDNEY
Right renal artery
Abdominal aorta
Inferior vena cava
RIGHT URETER
LEFT URETER
Rectum
Left ovary
Uterus
URINARY BLADDER
URETHRA

FUNCTIONS OF THE URINARY SYSTEM

1. The kidneys regulate blood volume and composition, help regulate blood pressure, synthesize glucose, release erythropoietin, and participate in vitamin D synthesis.

2. The ureters transport urine from the kidneys to the urinary bladder.

3. The urinary bladder stores urine.

4. The urethra eliminates urine from the body.

(a) Anterior view

SUPERIOR

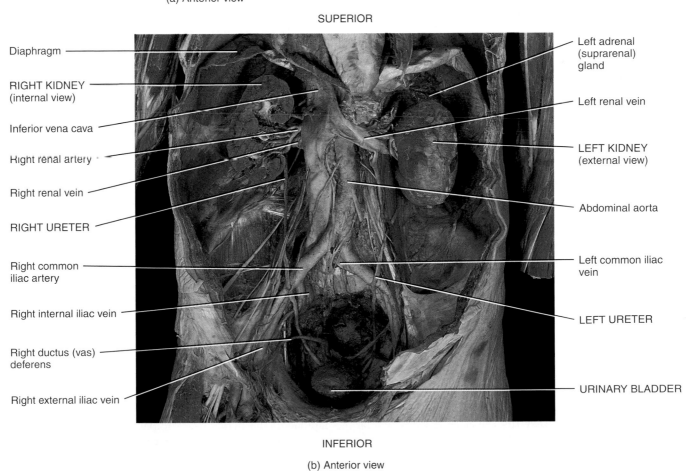

Diaphragm
RIGHT KIDNEY (internal view)
Inferior vena cava
Right renal artery
Right renal vein
RIGHT URETER
Right common iliac artery
Right internal iliac vein
Right ductus (vas) deferens
Right external iliac vein

Left adrenal (suprarenal) gland
Left renal vein
LEFT KIDNEY (external view)
Abdominal aorta
Left common iliac vein
LEFT URETER
URINARY BLADDER

INFERIOR

(b) Anterior view

? Which organs constitute the urinary system?

periods of fasting or starvation, (2) secreting erythropoietin, a hormone that stimulates red blood cell production, and (3) participating in synthesis of vitamin D.

4. *Transportation, storage, and elimination of urine.* Urine from each kidney is transported through its ureter and is stored in the urinary bladder until it is eliminated from the body through the urethra.

Nephrology (nef-ROL-ō-jē; *nephr-*=kidney; *-ology*=study of) is the scientific study of the anatomy, physiology, and pathology of the kidneys. The branch of medicine that deals with the male and female urinary systems and the male reproductive system is **urology** (ū-ROL-ō-jē; *uro-*=urine). A physician who specializes in this branch of medicine is called a **urologist** (ū-ROL-ō-jist).

ANATOMY AND HISTOLOGY OF THE KIDNEYS

OBJECTIVES

- Describe the external and internal gross anatomical features of the kidneys.
- Trace the path of blood flow through the kidneys.
- Describe the structure of renal corpuscles and renal tubules.

The paired **kidneys** are reddish, kidney-bean-shaped organs located just above the waist between the peritoneum and the posterior wall of the abdomen. Because their position is posterior to the peritoneum of the abdominal cavity, they are said to be **retroperitoneal** (re′-trō-per-i-tō-NĒ-al; *retro-*=behind) organs (Figure 26.2). The kidneys are located between the levels of the last thoracic and third lumbar vertebrae, a position where they are partially protected by the eleventh and twelfth pairs of ribs. The right kidney is slightly lower than the left (see Figure 26.1) because the liver occupies considerable space on the right side superior to the kidney.

External Anatomy of the Kidneys

A typical kidney in an adult is 10–12 cm (4–5 in.) long, 5–7 cm (2–3 in.) wide, and 3 cm (1 in.) thick—about the size of a bar of bath soap—and has a mass of 125–170 g (4.5–5 oz). The concave medial border of each kidney faces the vertebral column (see Figure 26.1). Near the center of the concave border is a deep vertical fissure called the **renal hilus** (RĒ-nal; *ren-*=kidney) (see Figure 26.3), through which the ureter emerges from the kidney along with blood vessels, lymphatic vessels, and nerves.

Three layers of tissue surround each kidney (Figure 26.2). The deep layer, the **renal capsule,** is a smooth, transparent sheet of dense irregular connective tissue that is continuous with the outer coat of the ureter. It serves as a barrier against trauma and helps maintain the shape of the kidney. The middle layer, the **adipose capsule,** is a mass of fatty tissue surrounding the renal capsule. It also protects the kidney from trauma and holds it firmly in place within the abdominal cavity. The superficial layer, the **renal fascia,** is another thin layer of dense irregular connective tissue that anchors the kidney to the surrounding structures and to the abdominal wall. On the anterior surface of the kidneys, the renal fascia is deep to the peritoneum.

NEPHROPTOSIS (FLOATING KIDNEY)

Nephroptosis (nef′-rōp-TŌ-sis; *ptosis*=falling), or **floating kidney,** is an inferior displacement or dropping of the kidney. It occurs when the kidney slips from its normal position because it is not securely held in place by adjacent organs or its covering of fat. Nephroptosis develops most often in very thin people whose adipose capsule or renal fascia is deficient. It is dangerous because the ureter may kink and block urine flow. The resulting backup of urine puts pressure on the kidney, which damages the tissue. Twisting of the ureter also causes pain. Nephroptosis is very common, with about one in four people having some degree of weakening of the fibrous bands that hold the kidney in place; it is 10 times more common in females than males. Because it happens during life it is very easy to distinguish from congenital anomalies. ■

Internal Anatomy of the Kidneys

A frontal section through the kidney reveals two distinct regions: a superficial, smooth-textured reddish area called the **renal cortex** (*cortex*=rind or bark) and a deep, reddish-brown inner region called the **renal medulla** (*medulla*=inner portion) (Figure 26.3 on page 812). The medulla consists of 8 to 18 cone-shaped **renal pyramids.** The base (wider end) of each pyramid faces the renal cortex, and its apex (narrower end), called a **renal papilla,** points toward the renal hilus. The renal cortex is the smooth-textured area extending from the renal capsule to the bases of the renal pyramids and into the spaces between them. It is divided into an outer *cortical zone* and an inner *juxtamedullary zone.* Those portions of the renal cortex that extend between renal pyramids are called **renal columns. A renal lobe** consists of a renal pyramid, its overlying area of renal cortex, and one-half of each adjacent renal column.

Together, the renal cortex and renal pyramids of the renal medulla constitute the **parenchyma** (functional portion) of the kidney. Within the parenchyma are the functional units of the kidney—about 1 million microscopic structures called **nephrons** (NEF-rons). Urine formed by the nephrons drains into large **papillary ducts,** which extend through the renal papillae of the pyramids. The papillary ducts drain into cuplike structures called **minor** and **major calyces** (KĀ-li-sēz=cups; singular is *calyx*). Each kidney has 8 to 18 minor calyces and 2 to 3 major calyces. A minor calyx receives urine from the papillary

Figure 26.2   Position and coverings of the kidneys.

The kidneys are surrounded by a renal capsule, adipose capsule, and renal fascia.

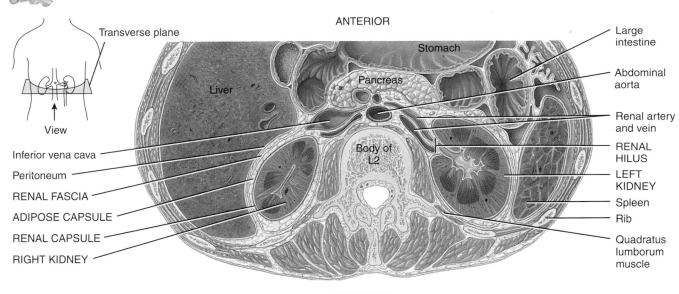

(a) Inferior view of transverse section of abdomen (L2)

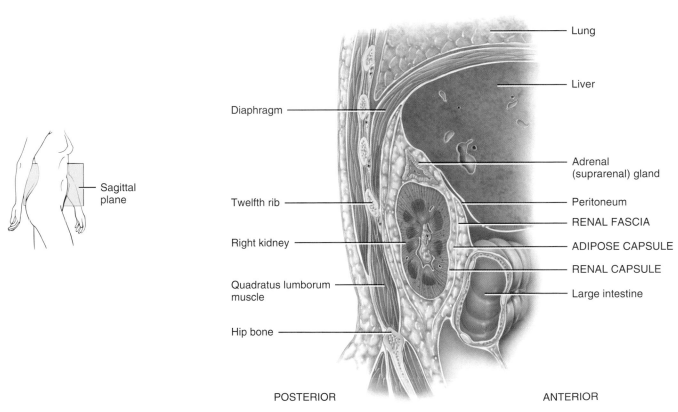

(b) Sagittal section through the right kidney

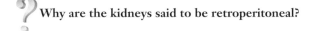
Why are the kidneys said to be retroperitoneal?

ducts of one renal papilla and delivers it to a major calyx. From the major calyces, urine drains into a single large cavity called the **renal pelvis** (*pelv-*=basin) and then out through the ureter to the urinary bladder.

The hilus expands into a cavity within the kidney called the **renal sinus,** which contains part of the renal pelvis, the calyces, and branches of the renal blood vessels and nerves. Adipose tissue helps stabilize the position of these structures in the renal sinus.

Figure 26.3 Internal anatomy of the kidneys.

The two main regions of the kidney parenchyma are the renal cortex and the renal pyramids in the renal medulla.

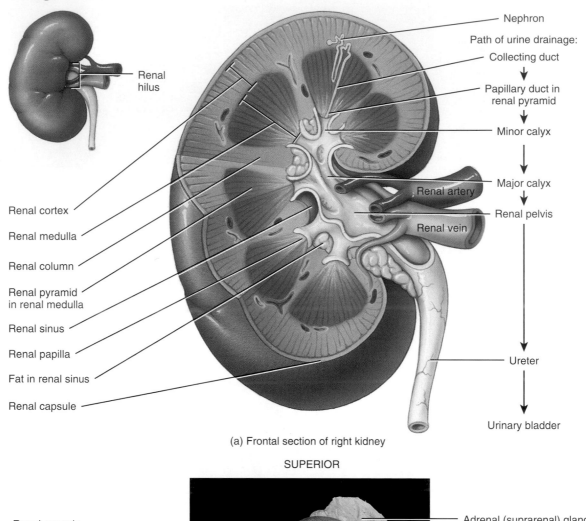

Renal hilus

Nephron

Path of urine drainage:

Collecting duct
↓
Papillary duct in renal pyramid
↓
Minor calyx
↓
Major calyx
↓
Renal pelvis
↓

Renal artery

Renal vein

Renal cortex

Renal medulla

Renal column

Renal pyramid in renal medulla

Renal sinus

Renal papilla

Fat in renal sinus

Renal capsule

Ureter
↓
Urinary bladder

(a) Frontal section of right kidney

SUPERIOR

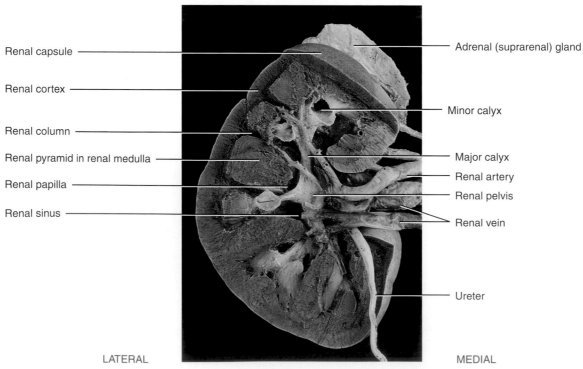

Renal capsule

Renal cortex

Renal column

Renal pyramid in renal medulla

Renal papilla

Renal sinus

Adrenal (suprarenal) gland

Minor calyx

Major calyx

Renal artery

Renal pelvis

Renal vein

Ureter

LATERAL

MEDIAL

(b) Frontal section of right kidney

What structures pass through the renal hilus?

Blood and Nerve Supply of the Kidneys

Because the kidneys remove wastes from the blood and regulate its volume and ionic composition, it is not surprising that they are abundantly supplied with blood vessels. Although the kidneys constitute less than 0.5% of total body mass, they receive 20–25% of the resting cardiac output via the right and left **renal arteries** (Figure 26.4). In adults, renal blood flow is about 1700 mL per minute.

Within the kidney, the renal artery divides into several **segmental arteries,** which supply different segments (areas) of the kidney. Each segmental artery gives off several branches that enter the parenchyma and pass through the renal columns between the lobes of the kidneys as the **interlobar arteries.** At the bases of the renal pyramids, the interlobar arteries arch between the renal medulla and cortex; here they are known as the **arcuate arteries** (AR-kū-āt=shaped like a bow), because they arch over the bases of the renal pyramids. Divisions of the arcuate arteries produce a series of **interlobular arteries.** These arteries are so-named because they pass between lobules of the kidney (a **renal lobule** is a group of nephrons that open into branches of the same collecting duct; see Figure 26.5).

Figure 26.4   Blood supply of the kidneys. In the cast of the kidney shown in (c), the arteries are red, the veins are blue, and the urine-draining structures are yellow.

The renal arteries deliver 20–25% of the resting cardiac output to the kidneys.

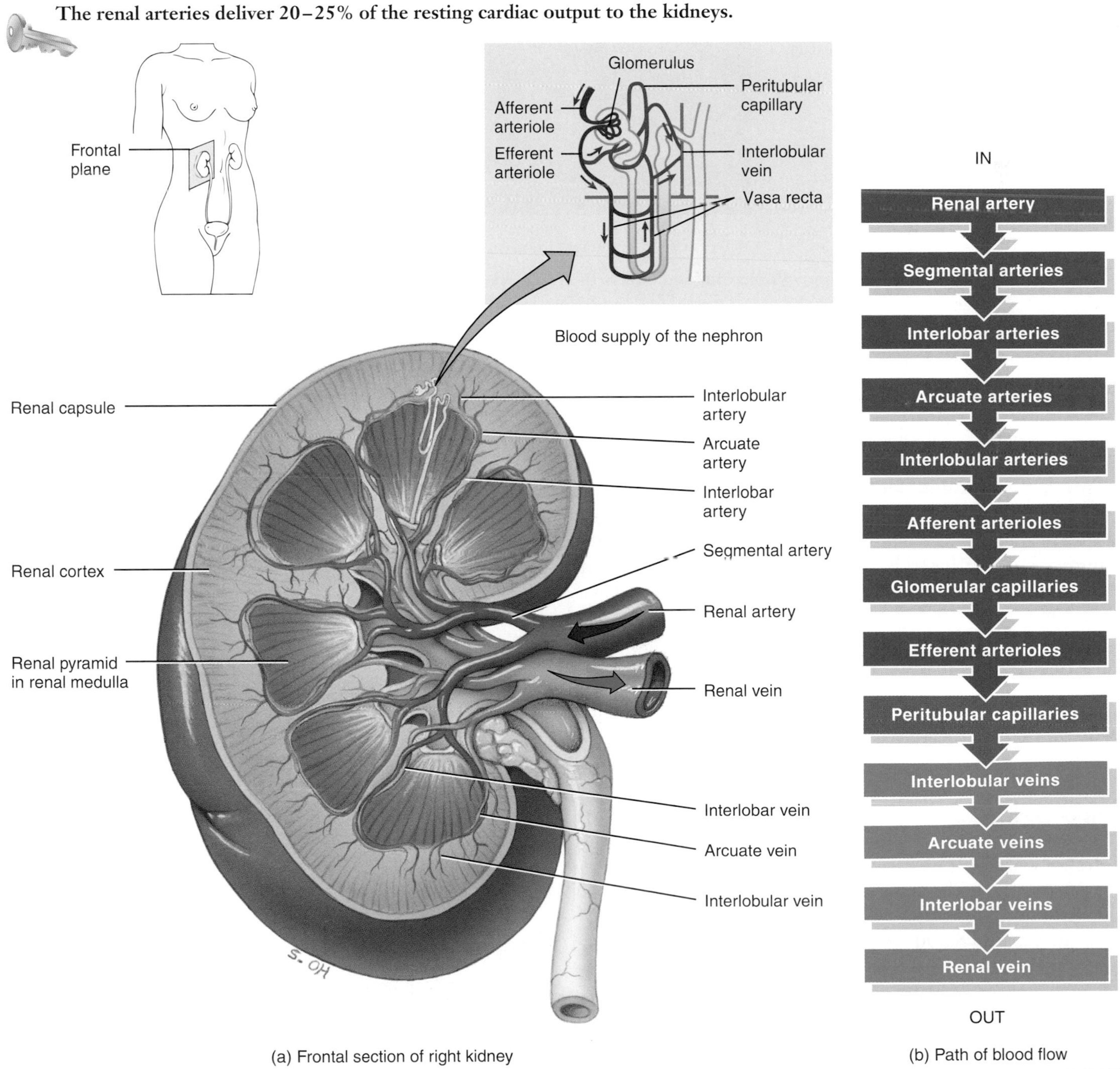

Blood supply of the nephron

(a) Frontal section of right kidney

(b) Path of blood flow

continues

Figure 26.4 (continued)

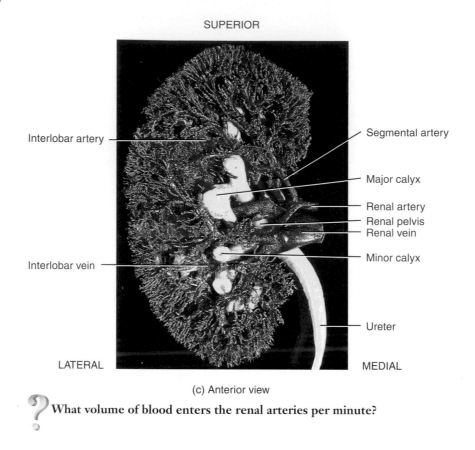

SUPERIOR

Interlobar artery

Segmental artery

Major calyx

Renal artery
Renal pelvis
Renal vein

Minor calyx

Interlobar vein

Ureter

LATERAL

MEDIAL

(c) Anterior view

What volume of blood enters the renal arteries per minute?

Interlobular arteries enter the renal cortex and give off branches called **afferent arterioles** (*af-*=toward; *-ferrent*=to carry).

Each nephron receives one afferent arteriole, which divides into a tangled, ball-shaped capillary network called the **glomerulus** (glō-MER-ū-lus=little ball; plural is *glomeruli*). The glomerular capillaries then reunite to form an **efferent arteriole** (*ef-*=out) that carries blood out of the glomerulus. Glomerular capillaries are unique among capillaries in the body because they are positioned between two arterioles, rather than between an arteriole and a venule. Because they are capillary networks and they also play an important role in urine formation, the glomeruli are considered part of both the cardiovascular and the urinary systems.

The efferent arterioles divide to form the **peritubular capillaries** (*peri-*=around), which surround tubular parts of the nephron in the renal cortex. Extending from some efferent arterioles are long loop-shaped capillaries called **vasa recta** (VĀ-sa REK-ta; *vasa*=vessels; *recta*=straight) that supply tubular portions of the nephron in the renal medulla (see Figure 26.5b).

The peritubular capillaries eventually reunite to form **peritubular venules** and then **interlobular veins,** which also receive blood from the vasa recta. Then the blood drains through the **arcuate veins** to the **interlobar veins** running between the renal pyramids. Blood leaves the kidney through a single **renal vein** that exits at the renal hilus and carries venous blood to the inferior vena cava.

Most renal nerves originate in the *celiac ganglion* (see Figure 20.4 on page 638) and pass through the *renal plexus* into the kidneys along with the renal arteries. Renal nerves are part of the sympathetic division of the autonomic nervous system. Most are vasomotor nerves that regulate the flow of blood through the kidney by causing vasodilation or vasoconstriction of renal arterioles.

KIDNEY TRANSPLANT

A **kidney transplant** is the transfer of a kidney from a living donor or a cadaver to a recipient whose kidney(s) no longer function. In the procedure, the donor kidney is placed in the pelvis of the recipient through an abdominal incision. The renal artery and vein of the transplanted kidney are attached to the renal artery and vein of the recipient. The ureter of the transplanted kidney is then attached to the urinary bladder. During a kidney transplant, the patient receives only one donor kidney, since only one kidney is needed to maintain sufficient renal function. The diseased kidneys are usually left in place. As with all organ transplants, kidney transplant patients must be ever vigilant for signs of infection or organ rejection. The transplant patient will take immunosuppressive drugs for the rest of his or her life to avoid rejection of the "foreign" organ. ■

The Nephron

Parts of a Nephron

Nephrons are the functional units of the kidneys. Each nephron (Figure 26.5) consists of two parts: a **renal corpuscle** (KOR-pus-sul=tiny body), where blood plasma is filtered, and a **renal** tubule into which the filtered fluid passes. The two components of a renal corpuscle are the **glomerulus** (capillary network) and the **glomerular (Bowman's) capsule,** a double-walled epithelial cup that surrounds the glomerular capillaries. In the order that fluid passes through them, the renal tubule consists of a (1) **proximal convoluted tubule,** (2) **loop of Henle (nephron loop),** and (3) **distal convoluted tubule.** *Proximal* denotes the part of the tubule attached to the glomerular capsule, and *distal* denotes the part that is farther away. *Convoluted* means the tubule is tightly coiled rather than straight. The renal corpuscle and both convoluted tubules lie within the renal cortex, whereas the loop of Henle extends into the renal medulla, makes a hairpin turn, and then returns to the renal cortex.

The distal convoluted tubules of several nephrons empty into a single **collecting duct.** Collecting ducts then unite and converge until eventually there are only several hundred large **papillary ducts,** which drain into the minor calyces. The collecting ducts and papillary ducts extend from the renal cortex through the renal medulla to the renal pelvis. Although one kidney has about 1 million nephrons, it has a much smaller number of collecting ducts and even fewer papillary ducts.

In a nephron, the loop of Henle connects the proximal and distal convoluted tubules. The first part of the loop of Henle dips into the renal medulla, where it is called the **descending limb of the loop of Henle** (Figure 26.5). It then makes that hairpin turn and returns to the renal cortex as the **ascending**

Figure 26.5   The structure of nephrons (colored gold) and associated blood vessels. (a) A cortical nephron. (b) A juxtamedullary nephron.

Nephrons are the functional units of the kidneys.

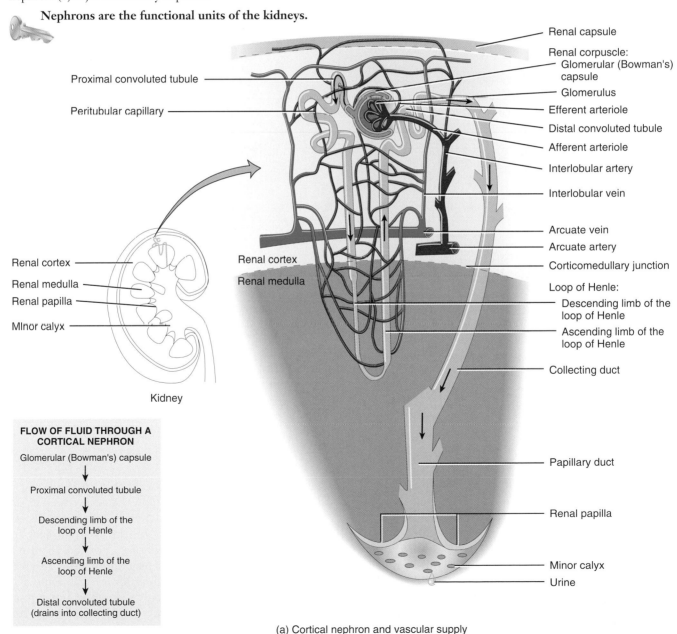

FLOW OF FLUID THROUGH A CORTICAL NEPHRON

Glomerular (Bowman's) capsule
↓
Proximal convoluted tubule
↓
Descending limb of the loop of Henle
↓
Ascending limb of the loop of Henle
↓
Distal convoluted tubule (drains into collecting duct)

(a) Cortical nephron and vascular supply

continues

Figure 26.5 (continued)

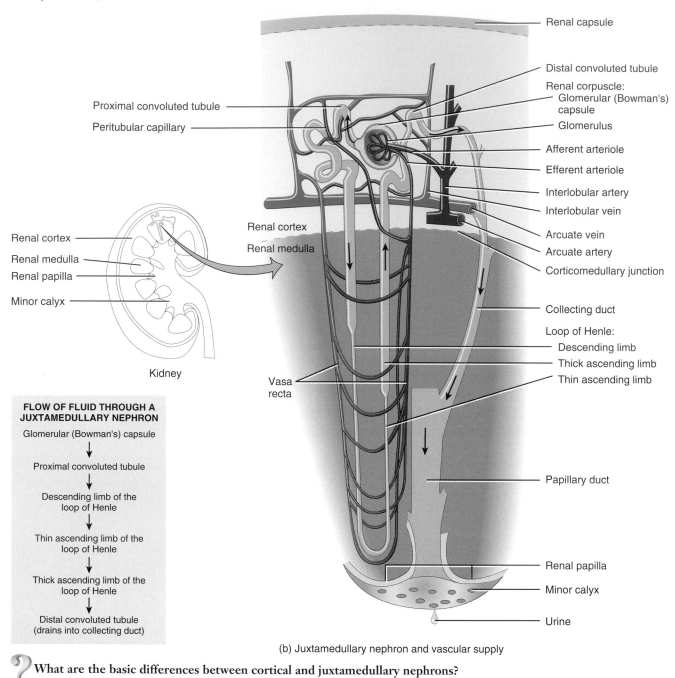

FLOW OF FLUID THROUGH A
JUXTAMEDULLARY NEPHRON

Glomerular (Bowman's) capsule
↓
Proximal convoluted tubule
↓
Descending limb of the
loop of Henle
↓
Thin ascending limb of the
loop of Henle
↓
Thick ascending limb of the
loop of Henle
↓
Distal convoluted tubule
(drains into collecting duct)

(b) Juxtamedullary nephron and vascular supply

? **What are the basic differences between cortical and juxtamedullary nephrons?**

limb of the loop of Henle. About 80–85% of the nephrons are **cortical nephrons.** Their renal corpuscles lie in the outer portion of the renal cortex, and they have *short* loops of Henle that lie mainly in the cortex and penetrate only into the outer region of the renal medulla (Figure 26.5a). The short loops of Henle receive their blood supply from peritubular capillaries that arise from efferent arterioles. The other 15–20% of the nephrons are **juxtamedullary nephrons** (*juxta-*=near to). Their renal corpuscles lie deep in the cortex, close to the medulla, and they have *long* loops of Henle that extend into the deepest region of the medulla (Figure 26.5b). Long loops of Henle receive their blood supply from peritubular capillaries and from the vasa recta that arise from efferent arterioles. In addition, the ascending limb of the loop of Henle of juxtamedullary nephrons consists of two portions: a **thin ascending limb** followed by a **thick ascending limb** (Figure 26.5b). The lumen of the thin ascending limb is the same as in other areas of the renal tubule; it is only the epithelium that is thinner. Nephrons with long loops of Henle enable the kidneys to excrete very dilute or very concentrated urine.

Histology of the Nephron and Collecting Duct

A single layer of epithelial cells forms the entire wall of the glomerular capsule, renal tubule, and ducts. Each part, however, has distinctive histological features that reflect its particular functions. In the order that fluid flows through them, the parts are the glomerular capsule, the renal tubule, and the collecting duct.

GLOMERULAR CAPSULE The glomerular (Bowman's) capsule consists of visceral and parietal layers (Figure 26.6a). The visceral layer consists of modified simple squamous epithelial cells called **podocytes** (PŌ-dō-sīts; *podo-*=foot; *-cytes*=cells). The many footlike projections of these cells (pedicels) wrap around the single layer of endothelial cells of the glomerular capillaries and form the inner wall of the capsule. The parietal layer of the glomerular capsule consists of simple squamous

Figure 26.6   Histology of a renal corpuscle.

A renal corpuscle consists of a glomerular (Bowman's) capsule and a glomerulus.

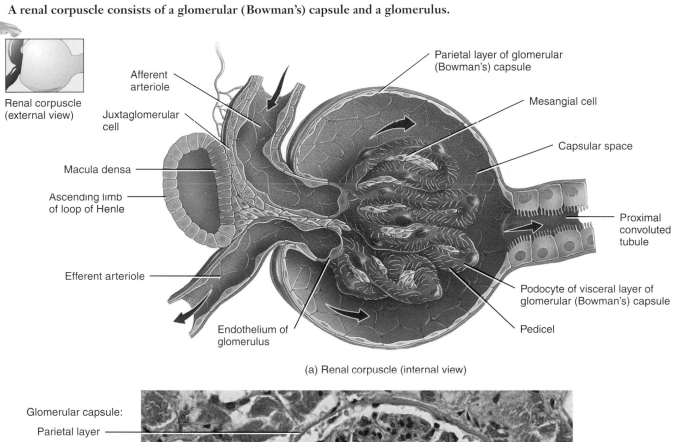

(a) Renal corpuscle (internal view)

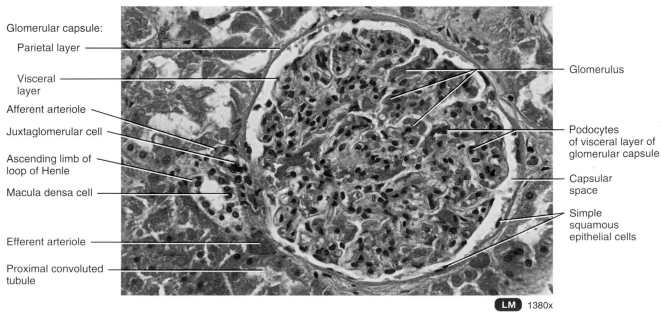

(b) Renal corpuscle

Is the photomicrograph in (b) from a section through the renal cortex or renal medulla? How can you tell?

epithelium and forms the outer wall of the capsule. Fluid filtered from the glomerular capillaries enters the **capsular (Bowman's) space,** the space between the two layers of the glomerular capsule. Think of the relationship between the glomerulus and glomerular capsule in the following way. The glomerulus is a fist punched into a limp balloon (the glomerular capsule) until the fist is covered by two layers of balloon (visceral and parietal layers) with a space in between, the capsular space.

RENAL TUBULE AND COLLECTING DUCT Table 26.1 illustrates the histology of the cells that form the renal tubule and collecting duct. In the proximal convoluted tubule, the cells are simple cuboidal epithelial cells with a prominent brush border of microvilli on their apical surface (surface facing the lumen). These microvilli, like those of the small intestine, increase the surface area for reabsorption and secretion. The descending limb of the loop of Henle and the first part of the ascending limb of the loop of Henle (the thin ascending limb) are composed of simple squamous epithelium. (Recall that cortical or short-loop nephrons lack the thin ascending limb.) The thick ascending limb of the loop of Henle is composed of simple cuboidal to low columnar epithelium.

In each nephron, the final part of the ascending limb of the loop of Henle makes contact with the afferent arteriole serving that renal corpuscle (Figure 26.6a). Because the columnar tubule cells in this region are crowded together, they are known as the **macula densa** (*macula*=spot; *densa*=dense). Alongside the macula densa, the wall of the afferent arteriole (and sometimes the efferent arteriole) contains modified smooth muscle fibers called **juxtaglomerular (JG) cells.** Together with the macula densa, they constitute the **juxtaglomerular apparatus (JGA).** The JGA helps regulate blood pressure within the kidneys. The distal convoluted tubule (DCT) begins a short distance past the macula densa. In the last part of the DCT and continuing into the collecting ducts, two different types of cells are present. Most are **principal cells,** which have receptors for both antidiuretic hormone (ADH) and aldosterone, two hormones that regulate their functions. A smaller number are **intercalated cells,** which play a role in the homeostasis of blood pH. The collecting ducts drain into large papillary ducts, which are lined by simple columnar epithelium.

NUMBER OF NEPHRONS

The **number of nephrons** is constant from birth. Any increase in kidney size is due solely to the growth of individual nephrons. If nephrons are injured or become diseased, new ones do not form. Signs of kidney dysfunction usually do not become apparent until function declines to less than 25% of normal because the remaining functional nephrons adapt to handle a larger-than-normal load. Surgical removal of one kidney, for example, stimulates hypertrophy (enlargement) of the remaining kidney, which eventually is able to filter blood at 80% of the rate of two normal kidneys. ■

CHECKPOINT

1. Describe the location of the kidneys. Why are they said to be retroperitoneal?
2. Which branch of the autonomic nervous system innervates renal blood vessels?
3. How do cortical nephrons and juxtamedullary nephrons differ structurally?
4. Describe the histology of the various portions of a nephron and collecting duct.
5. Describe the structure of the juxtaglomerular apparatus (JGA).

TABLE 26.1	HISTOLOGICAL FEATURES OF THE RENAL TUBULE AND COLLECTING DUCT	
REGION AND HISTOLOGY		**DESCRIPTION**
Proximal convoluted tubule (PCT) Microvilli, Mitochondrion, Apical surface		Simple cuboidal epithelial cells with prominent brush borders of microvilli.
Loop of Henle: descending limb and thin ascending limb		Simple squamous epithelial cells.
Loop of Henle: thick ascending limb		Simple cuboidal to low columnar epithelial cells.
Most of distal convoluted tubule (DCT)		Simple cuboidal epithelial cells.
Last part of DCT and all of collecting duct (CD) Intercalated cell, Principal cell		Simple cuboidal epithelium consisting of principal cells and intercalated cells.

FUNCTIONS OF NEPHRONS

- Identify the three basic tasks performed by nephrons and collecting ducts, and indicate where each task occurs.
- Describe the filtration membrane.

To produce urine, nephrons and collecting ducts perform three basic processes—glomerular filtration, tubular secretion, and tubular reabsorption (Figure 26.7):

1 *Glomerular filtration.* In the first step of urine production, water and most solutes in blood plasma move across the wall of glomerular capillaries into the glomerular capsule and then into the renal tubule.

2 *Tubular reabsorption.* As filtered fluid flows along the renal tubule and through the collecting duct, tubule cells reabsorb about 99% of the filtered water and many useful solutes. The water and solutes return to the blood as it flows through the peritubular capillaries and vasa recta. Note that the term *reabsorption* refers to the return of filtered water and solutes to the bloodstream. The term *absorption*, by contrast, means entry of new substances into the body, as occurs in the gastrointestinal tract.

3 *Tubular secretion.* As fluid flows along the tubule and through the collecting duct, the tubule and duct cells secrete other materials, such as wastes, drugs, and excess ions, into the fluid. Notice that tubular secretion *removes* a substance from the blood. In other instances of secretion—such as secretion of hormones—cells release substances into interstitial fluid and blood.

Solutes in the fluid that drains into the renal pelvis remain in the urine and are excreted. The rate of urinary excretion of any solute is equal to its rate of glomerular filtration, plus its rate of secretion, minus its rate of reabsorption.

By filtering, reabsorbing, and secreting, nephrons help maintain homeostasis of the blood's volume and composition. The situation is somewhat analogous to a recycling center: Garbage trucks dump refuse into an input hopper, where the smaller refuse passes onto a conveyor belt (glomerular filtration of plasma). As the conveyor belt carries the garbage along, workers remove useful items, such as aluminum cans, plastics, and glass containers (reabsorption). Other workers place additional garbage left at the center and larger items onto the conveyor belt (secretion). At the end of the belt, all remaining garbage falls into a truck for transport to the landfill (excretion of wastes in urine).

Glomerular Filtration

The fluid that enters the capsular space is called the **glomerular filtrate.** On average, the daily volume of glomerular filtrate in adults is 150 liters in females and 180 liters in males, a volume that represents about 65 times the entire blood plasma volume. More than 99% of the glomerular filtrate returns to the bloodstream via tubular reabsorption, however, so only 1–2 liters (about 1–2 qt) are excreted as urine.

Together, endothelial cells of glomerular capillaries and podocytes, which completely encircle the capillaries, form a leaky barrier referred to as the **filtration membrane** or **endothelial–capsular membrane.** This sandwichlike assembly permits filtration of water and small solutes but prevents

Figure 26.7 Relation of a nephron's structure to its three basic functions: glomerular filtration, tubular reabsorption, and tubular secretion. Secreted substances remain in the urine and are subsequently excreted by the body.

Glomerular filtration occurs in the renal corpuscle, whereas tubular reabsorption and tubular secretion occur all along the renal tubule and collecting duct.

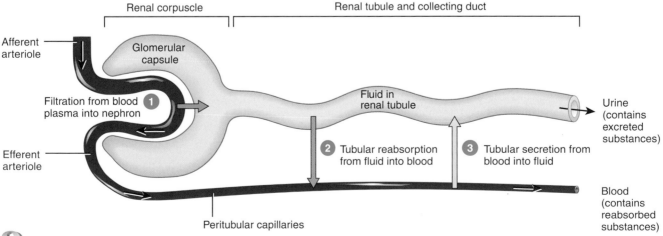

When cells of the renal tubules secrete the drug penicillin, is the drug being added to or removed from the bloodstream?

filtration of most plasma proteins, blood cells, and platelets. Filtered substances move from the bloodstream through three barriers—a glomerular endothelial cell, the basal lamina, and a filtration slit formed by a podocyte (Figure 26.8):

1 Glomerular endothelial cells are quite leaky because they have large **fenestrations** (pores) that are 70–100 nm (0.07–0.1 μm) in diameter. This size permits all solutes in blood plasma to exit glomerular capillaries but prevents filtration of blood cells and platelets. Located among the glomerular capillaries and in the cleft between afferent and efferent arterioles are **mesangial cells** (*mes-*=in the middle; *-angi*=blood vessel), contractile cells that help regulate glomerular filtration (see Figure 26.6a).

2 The **basal lamina,** a layer of material between the endothelium and the podocytes, consists of minute fibers in a glycoprotein matrix; it prevents filtration of larger plasma proteins.

3 Extending from each podocyte are thousands of footlike processes termed **pedicels** (PED-i-sels=little feet) that wrap around glomerular capillaries. The spaces between pedicels are the **filtration slits.** A thin membrane, the **slit membrane,** extends across each filtration slit; it permits the passage of molecules having a diameter smaller than 6–7 nm (0.006–0.007 μm), including water, glucose, vitamins, amino acids, very small plasma proteins, ammonia, urea, and ions. Because the most plentiful plasma protein—albumin—has a diameter of 7.1 nm, less than 1% of it passes the slit membrane.

The principle of *filtration*—the use of pressure to force fluids and solutes through a membrane—is the same in glomerular capillaries as in capillaries elsewhere in the body. However, the volume of fluid filtered by the renal corpuscle is much larger than in other capillaries of the body for three reasons:

1. Glomerular capillaries present a large surface area for filtration because they are long and extensive. The mesangial cells regulate how much of this surface area is available for filtration. When mesangial cells are relaxed, surface area is maximal, and glomerular filtration is very high. Contraction of mesangial cells reduces the available surface area, and glomerular filtration decreases.

2. The filtration membrane is thin and porous. Despite having several layers, the thickness of the filtration membrane is only 0.1 μm. Glomerular capillaries also are about 50 times leakier than capillaries in most other tissues, mainly because of their large fenestrations.

3. Glomerular capillary blood pressure is high. Because the efferent arteriole is smaller in diameter than the afferent arteriole, resistance to the outflow of blood from the glomerulus is high. As a result, blood pressure in glomerular capillaries is considerably higher than in capillaries elsewhere in the body, and a higher pressure produces more filtrate.

OLIGURIA AND ANURIA

Conditions that greatly reduce blood pressure, such as severe hemorrhage, may cause glomerular blood pressure to fall so low that net filtration pressure drops despite constriction of efferent arterioles. Then, glomerular filtration slows, or even stops entirely. The result is **oliguria** (*olig-*=scanty; *-uria*=urine production), a daily urine output between 50 and 250 mL, or **anuria,** a daily urine output of less than 50 mL. Obstructions, such as a kidney stone that blocks a ureter or an enlarged prostate that blocks the urethra in a male, can also decrease net filtration pressure and thereby reduce urine output. ■

Tubular Reabsorption

The normal rate of glomerular filtration is so high that the volume of fluid entering the proximal convoluted tubules in half an hour is greater than the total blood plasma volume. Reabsorption is the second function of the nephron and collecting duct. Epithelial cells all along the renal tubule and duct carry out reabsorption, but proximal convoluted tubule cells make the largest contribution. Solutes that are reabsorbed by both active and passive processes include glucose, amino acids, urea, and ions such as Na^+ (sodium), K^+ (potassium), Ca^{2+} (calcium), Cl^- (chloride), HCO_3^- (bicarbonate), and HPO_4^{2-} (phosphate). Cells located more distally fine-tune the reabsorption processes to maintain the appropriate concentrations of water and selected ions. Most small proteins and peptides that pass through the filter also are reabsorbed, usually via bulk-phase endocytosis.

DIURETICS

Diuretics are substances that slow renal reabsorption of water and thereby cause *diuresis*, an elevated urine flow rate, which in turn reduces blood volume. Diuretic drugs often are prescribed to treat *hypertension* (high blood pressure), because lowering blood volume usually reduces blood pressure. Naturally occurring diuretics include *caffeine* in coffee, tea, and sodas and *alcohol* in beer, wine, and mixed drinks. Some diuretics are potassium sparing (they leave potassium in the body), while others are not. People who are taking certain diuretics must closely monitor their potassium levels and may have to take potassium supplements. ■

Tubular Secretion

The third function of nephrons and collecting ducts is tubular secretion, the transfer of materials from the blood and tubule cells into tubular fluid. Secreted substances include H^+, K^+, ammonium ions (NH_4^+), creatinine, and certain drugs such as penicillin. Tubular secretion has two important outcomes: The secretion of H^+ helps control blood pH, and the secretion of other substances helps eliminate them from the body.

Figure 26.8 The filtration (endothelial-capsular) membrane. The size of the endothelial fenestrations and filtration slits in (a) have been exaggerated for emphasis. (Dr. Richard K. Kessel and Dr. Randy H. Kardon, *Tissues and Organs: A Text-Atlas of Scanning Electron Microscopy.* © 1979 by W. H. Freeman and Company. Reprinted by permission.)

During glomerular filtration, water and solutes pass from blood plasma into the capsular space.

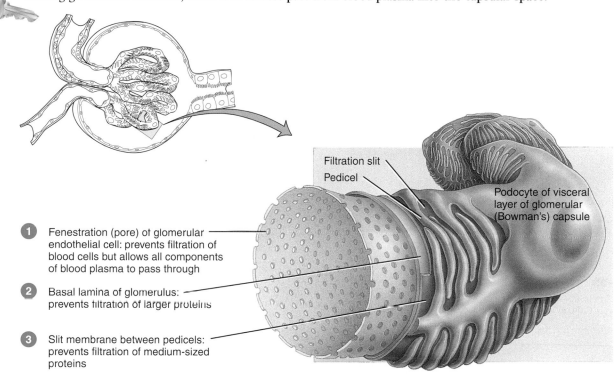

1 Fenestration (pore) of glomerular endothelial cell: prevents filtration of blood cells but allows all components of blood plasma to pass through

2 Basal lamina of glomerulus: prevents filtration of larger proteins

3 Slit membrane between pedicels: prevents filtration of medium-sized proteins

Filtration slit
Pedicel
Podocyte of visceral layer of glomerular (Bowman's) capsule

(a) Details of filtration membrane

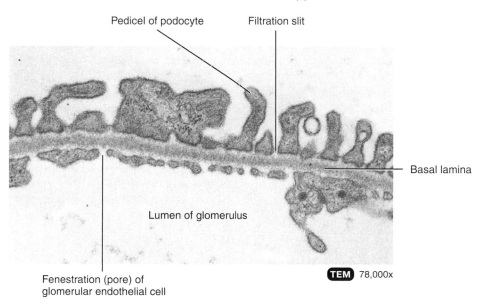

Pedicel of podocyte
Filtration slit
Basal lamina
Lumen of glomerulus
Fenestration (pore) of glomerular endothelial cell

TEM 78,000x

(b) Filtration membrane

Which part of the filtration membrane prevents red blood cells from entering the capsular space?

CHECKPOINT

6. What are the functions of glomerular filtration, tubular reabsorption, and tubular secretion?

7. Describe the factors that allow a considerably greater filtration through glomerular capillaries than through capillaries elsewhere in the body.

URINE TRANSPORTATION, STORAGE, AND ELIMINATION

OBJECTIVE

• Describe the anatomy, histology, and functions of the ureters, urinary bladder, and urethra.

Urine drains through papillary ducts into the minor calyces, which join to become major calyces that unite to form the renal pelvis (see Figure 26.3). From the renal pelvis, urine first drains into the ureters and then into the urinary bladder; urine is then discharged from the body through the single urethra (see Figure 26.1).

Ureters

Each of the two **ureters** (Ū-re-ters) transports urine from the renal pelvis of one kidney to the urinary bladder. Peristaltic contractions of the muscular walls of the ureters push urine toward the urinary bladder, but hydrostatic pressure and gravity also contribute. Peristaltic waves that pass from the renal pelvis to the urinary bladder vary in frequency from one to five per minute, depending on how fast urine is being formed.

The ureters are 25–30 cm (10–12 in.) long and are thick-walled, narrow tubes that vary in diameter from 1 mm to 10 mm along their course between the renal pelvis and the urinary bladder. Like the kidneys, the ureters are retroperitoneal. At the base of the urinary bladder the ureters curve medially and pass obliquely through the wall of the posterior aspect (Figure 26.9).

Even though there is no anatomical valve at the opening of each ureter into the urinary bladder, there is a physiological one that is quite effective. As the urinary bladder fills with urine, pressure within it compresses the oblique openings into the ureters and prevents the backflow of urine. When this physiological valve is not operating properly, it is possible for microbes to travel up the ureters from the urinary bladder to infect one or both kidneys.

Three coats of tissue form the wall of the ureters (Figure 26.10 on page 824). The deepest coat, or **mucosa,** is a mucous membrane with **transitional epithelium** (see Table 3.1I on page 67) and an underlying **lamina propria** of areolar connective tissue with considerable collagen, elastic fibers, and lymphatic tissue. Transitional epithelium is able to stretch—a marked advantage for any organ that must accommodate a variable volume of fluid. Mucus secreted by the mucosa prevents the cells of the walls of the ureter from coming in contact with urine, which is important because the solute concentration and pH of urine may differ drastically from the cytosol of the cells of the walls of the ureter. Throughout most of the length of the ureters, the intermediate coat, the **muscularis,** is composed of inner longitudinal and outer circular layers of smooth muscle fibers, an arrangement opposite that of the gastrointestinal tract, which contains inner circular and outer longitudinal layers; the muscularis of the distal third of the ureters also contains a third, outer layer of longitudinal muscle fibers. Peristalsis is the major function of the muscularis. The superficial coat of the ureters is the **adventitia,** a layer of areolar connective tissue containing blood vessels, lymphatic vessels, and nerves that serve the muscularis

and mucosa. The adventitia blends in with surrounding connective tissue and anchors the ureters in place.

The arterial supply of the ureters is from the renal, testicular or ovarian, common iliac, and inferior vesical arteries (arising from the internal iliac artery, a trunk with the internal pudendal and superior gluteal arteries, or a branch of the internal pudendal artery). The veins terminate in the corresponding trunks.

The ureters are innervated by the renal plexuses, which are supplied by sympathetic and parasympathetic fibers from the lesser and lowest splanchnic nerves.

Urinary Bladder

The **urinary bladder** is a hollow, distensible muscular organ situated in the pelvic cavity posterior to the pubic symphysis. In males, it is directly anterior to the rectum; in females it is anterior to the vagina and inferior to the uterus (see Figure 26.12). It is held in position by folds of the peritoneum. The shape of the urinary bladder depends on how much urine it contains. Empty, it is collapsed; when slightly distended it becomes spherical; as urine volume increases it becomes pear-shaped and rises into the abdominal cavity. Urinary bladder capacity averages 700–800 mL; it is smaller in females because the uterus occupies the space just superior to the urinary bladder.

In the floor of the urinary bladder is a small triangular area called the **trigone** (TRĪ-gōn=triangle). The two posterior corners of the trigone contain the two ureteral openings, whereas the opening into the urethra, the **internal urethral orifice,** lies in the anterior corner (see Figure 26.9). Because its mucosa is firmly bound to the muscularis, the trigone has a smooth appearance.

Three coats make up the wall of the urinary bladder (Figure 26.11 on page 824). The deepest is the **mucosa,** a mucous membrane composed of **transitional epithelium** and an underlying **lamina propria** similar to that of the ureters. Rugae (the folds in the mucosa) are also present. Surrounding the mucosa is the intermediate **muscularis,** also called the **detrusor muscle** (de-TROO-ser=to push down), which consists of three layers of smooth muscle fibers: the inner longitudinal, middle circular, and outer longitudinal layers. Around the opening to the urethra the circular fibers form an **internal urethral sphincter** (see Figure 26.9a); inferior to it is the **external urethral sphincter,** which is composed of skeletal muscle. The most superficial coat of the urinary bladder on the posterior and inferior surfaces is the **adventitia,** a layer of areolar connective tissue that is continuous with that of the ureters. Over the superior surface of the urinary bladder is the **serosa,** a layer of peritoneum.

Discharge of urine from the urinary bladder, called **micturition** (mik′-too-RISH-un; *mictur-*=urinate), is also known as *urination* or *voiding.* Micturition occurs via a combination of involuntary and voluntary muscle contractions. When the volume of urine in the urinary bladder exceeds 200–400 mL, pressure within the urinary bladder increases considerably, and stretch receptors in its wall transmit nerve impulses into the spinal cord. These impulses propagate to the **micturition center** in sacral spinal cord segments S2 and S3 and trigger a spinal

Figure 26.9   Ureters, urinary bladder, and urethra (shown in a female).

   Urine is stored in the urinary bladder before being expelled by micturition.

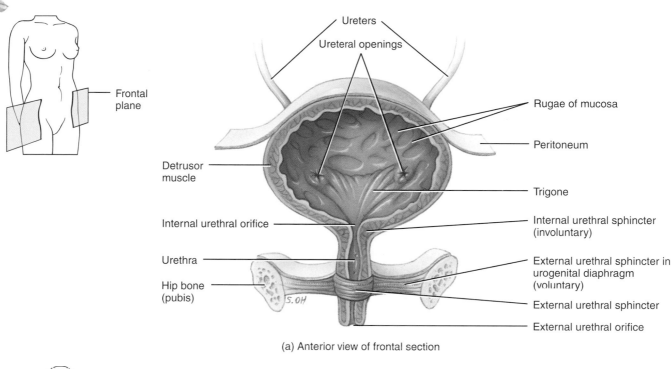

(a) Anterior view of frontal section

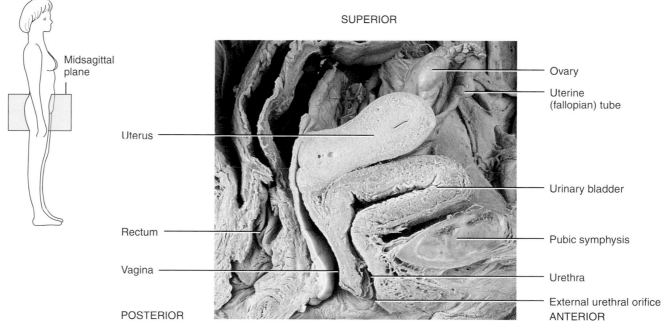

(b) Midsagittal section

What is a lack of voluntary control over micturition called?

reflex called the **micturition reflex.** In this reflex arc, parasympathetic impulses from the micturition center propagate to the urinary bladder wall and internal urethral sphincter. The nerve impulses cause *contraction* of the detrusor muscle and *relaxation* of the internal urethral sphincter muscle. Simultaneously, the micturition center inhibits somatic motor neurons that innervate skeletal muscle in the external urethral sphincter. Upon contraction of the urinary bladder wall and relaxation of the sphincters, urination takes place. Urinary bladder filling causes a sensation of fullness that initiates a conscious desire to urinate before the micturition reflex actually occurs. Although emptying of the urinary bladder is a reflex, in early childhood we learn to initiate it and stop it voluntarily. Through learned control of the external urethral sphincter muscle and certain muscles of the pelvic floor, the cerebral cortex can initiate micturition or delay its occurrence for a limited period of time.

   The arteries of the urinary bladder are the superior vesical (arises from the umbilical artery), the middle vesical (arises from

Figure 26.10 Histology of the ureter.

Three coats of tissue form the wall of the ureters: mucosa, muscularis, and adventitia.

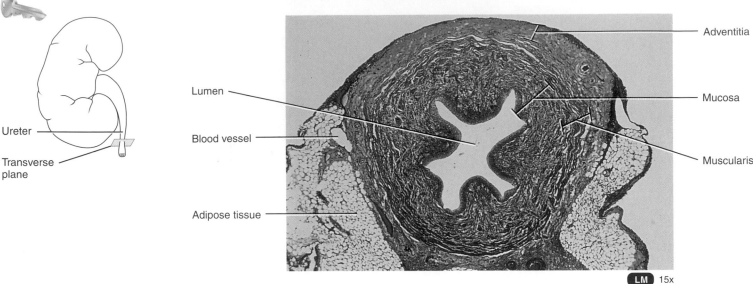

Transverse section of ureter

How does the muscularis of most of the ureters differ from that of the gastrointestinal tract?

the umbilical artery or a branch of the superior vesical), and the inferior vesical (arises from the internal iliac artery, a trunk with the internal pudendal and superior gluteal arteries, or a branch of the internal pudendal artery). The veins from the urinary bladder pass to the internal iliac trunk.

The nerves are derived partly from the hypogastric sympathetic plexus and partly from the second and third sacral nerves (pelvic splanchnic nerve).

CYSTOSCOPY

Cystoscopy (sis-TOS-kō-pē; *cysto-*=bladder; *-skopy*=to examine) is a very important procedure for direct examination of the mucosa of the urethra and urinary bladder and prostate in males. In the procedure, a *cystoscope* (a flexible narrow tube with a light) is inserted into the urethra to examine the structures through which it passes. With special attachments, tissue samples can be removed for examination (biopsy) and small stones can also be removed. Cystoscopy is useful for evaluating urinary bladder problems such as cancer and infections. It can also evaluate the degree of obstruction resulting from an enlarged prostate. ■

Urethra

The **urethra** is a small tube leading from the internal urethral orifice in the floor of the urinary bladder to the exterior of the body (see Figure 26.9a). In both males and females, the urethra is the terminal portion of the urinary system and the passageway for discharging urine from the body; in males it discharges semen as well.

In females, the urethra lies directly posterior to the pubic symphysis, is directed obliquely inferiorly and anteriorly, and has a length of 4 cm (1.5 in.) (Figure 26.12a). The opening of

Figure 26.11 Histology of the urinary bladder.

Discharge of urine from the urinary bladder is a combination of voluntary and involuntary muscular contractions called micturition.

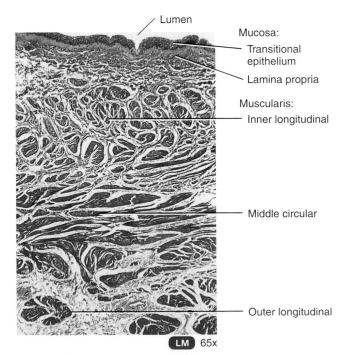

Transverse section of urinary bladder

What is the trigone?

Figure 26.12   Comparison between female and male urethras.

The urethra carries urine from the urinary bladder to the exterior.

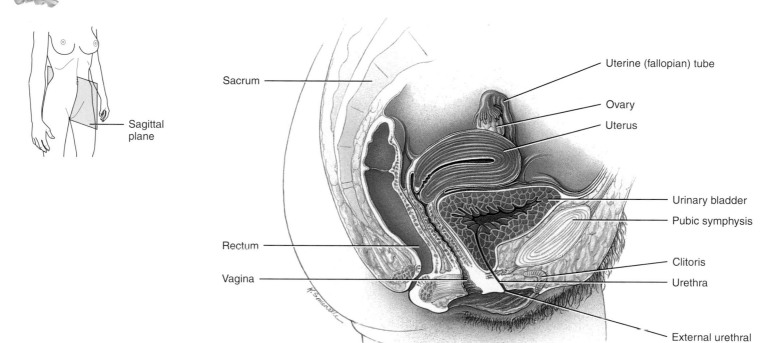

Sacrum

Uterine (fallopian) tube

Ovary

Uterus

Urinary bladder

Pubic symphysis

Rectum

Vagina

Clitoris

Urethra

Sagittal plane

External urethral orifice

(a) Sagittal section

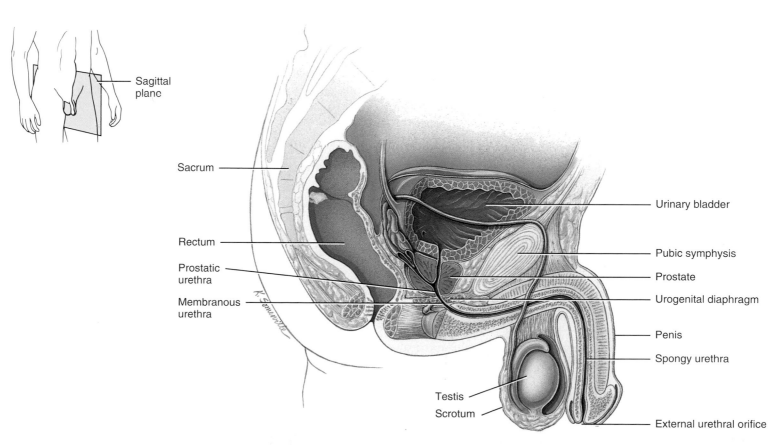

Sagittal plane

Sacrum

Rectum

Prostatic urethra

Membranous urethra

Urinary bladder

Pubic symphysis

Prostate

Urogenital diaphragm

Penis

Spongy urethra

Testis

Scrotum

External urethral orifice

(b) Sagittal section

Through which three main structures does the male urethra pass?

the urethra to the exterior, the **external urethral orifice,** is located between the clitoris and the vaginal opening. The wall of the female urethra consists of a deep **mucosa** and a superficial **muscularis.** The mucosa is a mucous membrane composed of **epithelium** and **lamina propria** (areolar connective tissue with elastic fibers and a plexus of veins). The muscularis consists of circularly arranged smooth muscle fibers and is continuous with that of the urinary bladder. Near the urinary bladder, the mucosa contains transitional epithelium that is continuous with that of the urinary bladder; near the external urethral orifice the epithelium is nonkeratinized stratified squamous epithelium. Between these areas, the mucosa contains stratified columnar or pseudostratified columnar epithelium.

In males, the urethra also extends from the internal urethral orifice to the exterior, but its length and passage through the body are considerably different than in females (Figure 26.12b). The male urethra first passes through the prostate, then through the urogenital diaphragm and finally through the penis, a distance of about 20 cm (8 in.).

The male urethra, which also consists of a deep **mucosa** and a superficial **muscularis,** is subdivided into three anatomical regions: (1) The **prostatic urethra** passes through the prostate; (2) the **membranous urethra,** the shortest portion, passes through the urogenital diaphragm; and (3) the **spongy urethra,** the longest portion, passes through the penis. The mucosa of the prostatic urethra is continuous with that of the urinary bladder and consists of transitional epithelium that becomes stratified columnar or pseudostratified columnar epithelium more distally. The mucosa of the membranous urethra contains stratified columnar or pseudostratified columnar epithelium. The epithelium of the spongy urethra is stratified columnar or pseudostratified columnar epithelium, except near the external urethral orifice, which is nonkeratinized stratified squamous epithelium. The **lamina propria** of the male urethra is areolar connective tissue with elastic fibers and a plexus of veins.

The muscularis of the prostatic urethra is composed of wisps of mostly circular smooth muscle fibers superficial to the lamina propria; these circular fibers help form the internal urethral sphincter of the urinary bladder. The muscularis of the membranous urethra consists of circularly arranged skeletal muscle fibers of the urogenital diaphragm that help form the external urethral sphincter of the urinary bladder.

Several glands and other structures associated with reproduction deliver their contents into the male urethra. The prostatic urethra contains the openings of (1) ducts that transport secretions from the **prostate** and (2) the **seminal vesicles** and **ductus (vas) deferens,** which deliver sperm into the urethra and provide secretions that both neutralize the acidity of the female reproductive tract and contribute to sperm motility and viability. The openings of the ducts of the **bulbourethral (Cowper's) glands** empty into the spongy urethra. They deliver an alkaline substance prior to ejaculation that neutralizes the acidity of the urethra. The glands also secrete mucus, which lubricates the end of the penis during sexual arousal. Throughout the urethra, but especially in the spongy urethra, the openings of the ducts of **urethral (Littré) glands** discharge mucus during sexual arousal and ejaculation.

URINARY INCONTINENCE

A lack of voluntary control over micturition is called **urinary incontinence.** In infants and children under 2–3 years old, incontinence is normal because neurons to the external urethral sphincter muscle are not completely developed; voiding occurs whenever the urinary bladder is sufficiently distended to stimulate the micturition reflex. Urinary incontinence also occurs in adults. There are four types of urinary incontinence—stress, urge, overflow, and functional. Of the more than 10 million U.S. adults who suffer from urinary incontinence, 1–2 million have **stress incontinence.** In this condition, physical stresses that increase abdominal pressure, such as coughing, sneezing, laughing, exercising, pregnancy, or simply walking, cause leakage of urine from the urinary bladder. Other causes of incontinence in adults are injury to the nerves controlling the urinary bladder, loss of bladder flexibility with age, disease or irritation of the bladder or urethra, damage to the external urethral sphincter, and certain drugs. And, compared with nonsmokers, those who smoke have twice the risk of developing incontinence. Choosing the right treatment option depends on correct diagnosis of the type of incontinence. Treatments include Kegel exercises, urinary bladder training, medication, and possibly even surgery. ■

CHECKPOINT

8. What forces help propel urine from the renal pelvis to the bladder?
9. What is micturition? Describe the micturition reflex.
10. Compare the location, length, and histology of the urethra in males and females.

DEVELOPMENT OF THE URINARY SYSTEM

OBJECTIVE

• Describe the development of the urinary system.

Starting in the third week of fetal development, a portion of the mesoderm along the posterior aspect of the embryo, the **intermediate mesoderm,** differentiates into the kidneys. The intermediate mesoderm is located in paired elevations called **urogenital ridges.** Three pairs of kidneys form within the intermediate mesoderm in succession: the pronephros, the mesonephros, and the metanephros (Figure 26.13). Only the last pair remains as the functional kidneys of the newborn.

The first kidney to form, the **pronephros** (prō-NEF-rōs; *pro-*=before; *-nephros*=kidney), is the most superior of the three and has an associated **pronephric duct.** This duct empties into the **cloaca,** the expanded terminal part of the hindgut, which functions as a common outlet for the urinary, digestive, and reproductive ducts. The pronephros begins to degenerate during the fourth week and is completely gone by the sixth week.

The second kidney, the **mesonephros** (mez′-ō-NEF-rōs; *meso-*=middle), replaces the pronephros. The retained portion

Figure 26.13 Development of the urinary system.

Three pairs of kidneys form within intermediate mesoderm in successive time periods: pronephros, mesonephros, and metanephros.

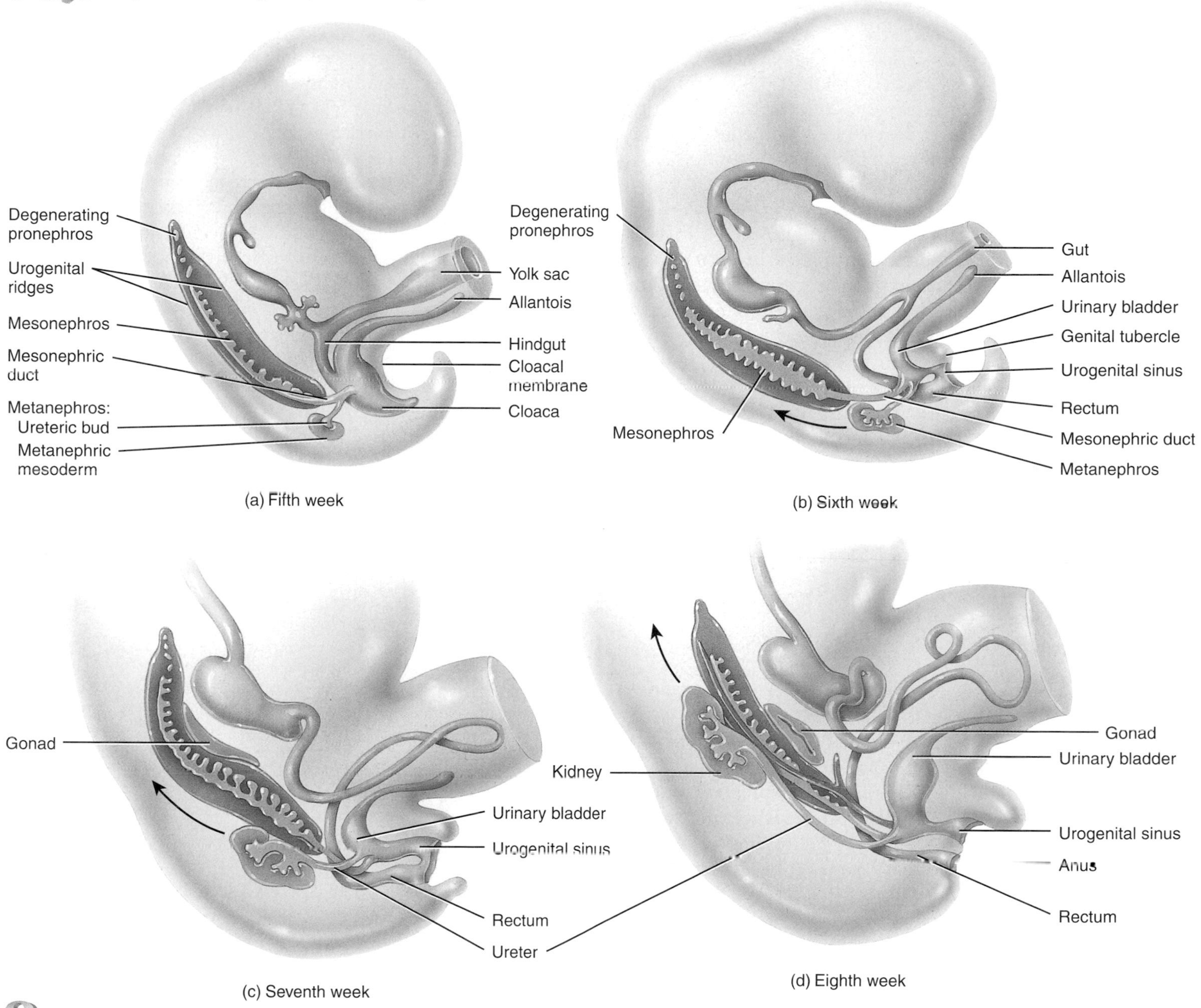

(a) Fifth week

Degenerating pronephros
Urogenital ridges
Mesonephros
Mesonephric duct
Metanephros:
Ureteric bud
Metanephric mesoderm

Yolk sac
Allantois
Hindgut
Cloacal membrane
Cloaca

(b) Sixth week

Degenerating pronephros
Mesonephros

Gut
Allantois
Urinary bladder
Genital tubercle
Urogenital sinus
Rectum
Mesonephric duct
Metanephros

(c) Seventh week

Gonad

Kidney
Urinary bladder
Urogenital sinus
Rectum
Ureter

(d) Eighth week

Gonad
Urinary bladder
Urogenital sinus
Anus
Rectum

? When do the kidneys begin to develop?

of the pronephric duct, which connects to the mesonephros, develops into the **mesonephric duct.** The mesonephros begins to degenerate by the sixth week and is almost gone by the eighth week.

At about the fifth week, a mesodermal outgrowth, called a **ureteric bud** (ū-rē-TER-ik), develops from the distal portion of the mesonephric duct near the cloaca. The **metanephros** (met-a-NEF-rōs; *meta-*=after), or ultimate kidney, develops from the ureteric bud and metanephric mesoderm. The ureteric bud

forms the *collecting ducts, calyces, renal pelvis,* and *ureter.* The **metanephric mesoderm** forms the *nephrons* of the kidneys. By the third month, the fetal kidneys begin excreting urine into the surrounding amniotic fluid; indeed, fetal urine makes up most of the amniotic fluid.

During development, the cloaca divides into a **urogenital sinus,** into which urinary and genital ducts empty, and a *rectum* that discharges into the anal canal. The *urinary bladder* develops from the urogenital sinus. In females, the *urethra* develops as a result of lengthening of the short duct that extends from the urinary bladder to the urogenital sinus. In males, the urethra is considerably longer and more complicated, but it is also derived from the urogenital sinus.

Although the metanephric kidneys form in the pelvis, they ascend to their ultimate destination in the abdomen. As they do so, they receive renal blood vessels. Although the inferior blood vessels usually degenerate as superior ones appear, sometimes the inferior vessels do not degenerate. Consequently, some individuals (about 30%) develop multiple renal vessels.

In a condition called **unilateral renal agenesis** (ā-JEN-e-sis; *a-*=without; *genesis*=production; *unilateral*=one side) only one kidney develops (usually the right) due to the absence of a ureteric bud. The condition occurs once in every 1000 newborn infants and usually affects males more than females. Other kidney abnormalities that occur during development are **malrotated kidneys** (the hilus faces anteriorly, posteriorly, or laterally instead of medially); **ectopic kidney** (one or both kidney may be in an abnormal position, usually inferior); and **horseshoe kidney** (the fusion of the two kidneys, usually inferiorly, into a single U-shaped kidney).

CHECKPOINT

11. Which type of embryonic tissue develops into nephrons?
12. Which tissue gives rise to collecting ducts, calyces, renal pelves, and ureters?

AGING AND THE URINARY SYSTEM

OBJECTIVE

● Describe the effects of aging on the urinary system.

With aging, the kidneys shrink in size, have a decreased blood flow, and filter less blood. The mass of the two kidneys decreases from an average of nearly 300 g in 20-year-olds to less than 200 g by age 80, a decrease of about one-third. Likewise, renal blood flow and filtration rates decline by 50% between ages 40 and 70. By age 80, about 40% of glomeruli are not functioning and thus filtration, reabsorption, and secretion decrease. Kidney diseases that become more common with age include acute and chronic kidney inflammations and renal calculi (kidney stones). Because the sensation of thirst diminishes with age, older individuals also are susceptible to dehydration. Urinary bladder changes that occur with aging include a reduction in size and capacity and weakening of the muscles. Urinary tract infections are more common among the elderly, as are polyuria (excessive urine production), nocturia (excessive urination at night), increased frequency of urination, dysuria (painful urination), urinary retention or incontinence, and hematuria (blood in the urine).

CHECKPOINT

13. To what extent do kidney mass and filtration rate decrease with age?

APPLICATIONS TO HEALTH

DIALYSIS

If a person's kidneys are so impaired by disease or injury that they are unable to function adequately, then blood must be cleansed artificially by **dialysis,** which utilizes the same methods as kidney filtration: the separation of large solutes from smaller ones through use of a selectively permeable membrane. The leading cause of renal failure is diabetes. One method of dialysis is the artificial kidney machine, which performs **hemodialysis** (*hemo-*=blood) because it directly filters the patient's blood. As blood flows through tubing made of selectively permeable dialysis membrane, waste products diffuse from the blood into a dialysis solution surrounding the membrane. The dialysis solution is continuously replaced to maintain favorable concentration gradients for diffusion of solutes into and out of the blood. After passing through the dialysis tubing, the cleansed blood flows back into the body. As a general rule, most affected people require 6–12 hours on dialysis each week (roughly every other day).

Continuous ambulatory peritoneal dialysis (CAPD) uses the peritoneal lining of the abdominal cavity as the dialysis membrane to filter the blood. The tip of a catheter is surgically placed in the patient's peritoneal cavity and connected to a sterile dialysis solution. The dialysis solution flows into the peritoneal cavity from a plastic container by gravity. The solution remains in the cavity until metabolic waste products, excess elec-

trolytes, and extracellular fluid diffuse into the dialysis solution. The solution is then drained from the cavity by gravity into a sterile bag that is discarded. The procedure is repeated several times each day.

URINALYSIS

An analysis of the volume and physical, chemical, and microscopic properties of urine, called a **urinalysis** (ū-ri-NAL-i-sis), reveals much information about the health of the body. The principal physical characteristics of normal urine are summarized in Table 26.2. Of the 1–2 liters (about 1–2 quarts) of urine eliminated per day by a normal adult, about 95% is water. The remaining 5% consists of solutes, among them urea, sodium, potassium, phosphate, and sulfate ions; creatinine; and uric acid. In addition, much smaller amounts of calcium, magnesium, and bicarbonate ions are also found in urine. If disease alters body metabolism or kidney function, traces of substances not normally present may appear in the urine, or normal constituents may appear in abnormal amounts. Table 26.3 lists several abnormal constituents in urine that may be detected as part of a urinalysis.

RENAL CALCULI

The crystals of salts present in urine occasionally precipitate and solidify into insoluble stones called **renal calculi** (*calculi*=

pebbles) or **kidney stones.** They commonly contain crystals of calcium oxalate, uric acid, or calcium phosphate. Conditions leading to calculus formation include the ingestion of excessive calcium, low water intake, abnormally alkaline or acidic urine, and overactivity of the parathyroid glands. When a stone lodges in a narrow passage, such as a ureter, the pain can be intense. **Shock-wave lithotripsy** (LITH-ō-trip′-sē; *litho-* = stone) offers an alternative to surgical removal of kidney stones. A device,

TABLE 26.2	PHYSICAL CHARACTERISTICS OF NORMAL URINE
CHARACTERISTIC	**DESCRIPTION**
Volume	One to two liters (about 1 to 2 quarts) in 24 hours but varies considerably.
Color	Yellow or amber, but varies with urine concentration and diet. Color is due to urochrome (pigment produced from breakdown of bile) and urobilin (from breakdown of hemoglobin). Concentrated urine is darker in color. Diet, medications, and certain diseases affect color.
Turbidity	Transparent when freshly voided, but becomes turbid (cloudy) after a while.
Odor	Mildly aromatic but becomes ammonia-like after a time. Some people inherit the ability to form methylmercaptan from digested asparagus, which gives urine a characteristic odor.
pH	Ranges between 4.6 and 8.0; average 6.0; varies considerably with diet. High-protein diets increase acidity; vegetarian diets increase alkalinity.
Specific gravity	Specific gravity (density) is the ratio of the weight of a volume of a substance to the weight of an equal volume of distilled water. Urine specific gravity ranges from 1.001 to 1.035. The higher the concentration of solutes, the higher the specific gravity.

TABLE 26.3	SUMMARY OF ABNORMAL CONSTITUENTS IN URINE
ABNORMAL CONSTITUENT	**COMMENTS**
Albumin	A normal constituent of blood plasma that usually appears in only very small amounts in urine because it is too large to be filtered. The presence of excessive albumin in the urine, **albuminuria** (al′-bū-mi-NOO-rē-a), indicates an increase in the permeability of filtering membranes due to injury or disease, increased blood pressure, or damage of kidney cells.
Glucose	**Glucosuria,** the presence of glucose in the urine, usually indicates diabetes mellitus.
Red blood cells (erythrocytes)	**Hematuria** (hēm-a-TOO-rē-a), the presence of hemoglobin from ruptured red blood cells in the urine, can occur with acute inflammation of the urinary organs as a result of disease or irritation from kidney stones, tumors, trauma, and kidney disease.
White blood cells (leukocytes)	The presence of white blood cells and other components of pus in the urine, referred to as **pyuria** (pī-Ū-rē-a), indicates infection in the kidneys or other urinary organs.
Ketone bodies	High levels of ketone bodies in the urine, called **ketonuria** (kē-tō-NOO-rē-a), may indicate diabetes mellitus, anorexia, starvation, or too little carbohydrate in the diet.
Bilirubin	When red blood cells are destroyed by macrophages, the globin portion of hemoglobin is split off and the heme is converted to biliverdin. Most of the biliverdin is converted to bilirubin. An above-normal level of bilirubin in urine is called **bilirubinuria** (bil′-ē-roo-bi-NOO-rē-a).
Urobilinogen	The presence of urobilinogen (breakdown product of hemoglobin) in urine is called **urobilinogenuria** (u′-rō-bi-lin′-ō-jē-NOO-rē-a). Traces are normal, but elevated urobilinogen may be due to hemolytic or pernicious anemia, infectious hepatitis, obstruction of bile ducts, jaundice, cirrhosis, congestive heart failure, or infectious mononucleosis.
Casts	**Casts** are tiny masses of material that have hardened and assumed the shape of the lumen of a tubule in which they formed. They are flushed out of the tubule when glomerular filtrate builds up behind them. Casts are named after the cells or substances that compose them or based on their appearance. For example, there are white blood cell casts, red blood cell casts, and epithelial cell casts (cells from the epithelial cells of the renal tubules).
Microbes	Normal urine is sterile. The number and type of bacteria vary with specific infections in the urinary tract. One of the most common is *E. coli.* The most common fungus to appear in urine is *Candida albicans,* a cause of vaginitis. The most frequent protozoan seen is *Trichomonas vaginalis,* a cause of vaginitis in females and urethritis in males.

called a *lithotripter*, delivers brief, high-intensity sound waves through a water-filled cushion. Over a period of 30 to 60 minutes, 1000 or more hydraulic shock waves pulverize the stone, creating fragments that are small enough to wash out in the urine.

URINARY TRACT INFECTIONS

The term **urinary tract infection (UTI)** is used to describe either an infection of a part of the urinary system or the presence of large numbers of microbes in urine. UTIs are more common in females due to the shorter length of the urethra. Symptoms include painful or burning urination, urgent and frequent urination, low back pain, and bed-wetting. UTIs include *urethritis* (inflammation of the urethra), *cystitis* (inflammation of the urinary bladder), and *pyelonephritis* (inflammation of the kidneys). If pyelonephritis becomes chronic, scar tissue can form in the kidneys and severely impair their function. Drinking cranberry juice can prevent the attachment of *E. coli* bacteria to the lining of the urinary bladder so that they are more readily flushed away during urination.

GLOMERULAR DISEASES

A variety of conditions may damage the kidney glomeruli, either directly or indirectly because of disease elsewhere in the body. Typically, the filtration membrane sustains damage, and its permeability increases.

Glomerulonephritis is an inflammation of the kidneys that involves the glomeruli. One of the most common causes is an allergic reaction to the toxins produced by streptococcal bacteria that have recently infected another part of the body, especially the throat. The glomeruli become so inflamed, swollen, and engorged with blood that the filtration membranes allow blood cells and plasma proteins to enter the filtrate. As a result, the urine contains many erythrocytes (hematuria) and a lot of protein. The glomeruli may be permanently damaged, leading to chronic renal failure.

Nephrotic syndrome is a condition characterized by *proteinuria* (protein in the urine) and *hyperlipidemia* (high blood levels of cholesterol, phospholipids, and triglycerides). The proteinuria is due to an increased permeability of the filtration membrane, which permits proteins, especially albumin, to escape from blood into urine. Loss of albumin results in *hypoalbuminemia* (low blood albumin level) once liver production of albumin fails to meet increased urinary losses. Edema, usually seen around the eyes, ankles, feet, and abdomen, occurs in nephrotic syndrome because loss of albumin from the blood decreases blood osmotic pressure. Nephrotic syndrome is associated with several glomerular diseases of unknown cause, as well as with systemic disorders such as diabetes mellitus, systemic lupus erythematosus (SLE), a variety of cancers, and AIDS.

RENAL FAILURE

Renal failure is a decrease or cessation of glomerular filtration. In **acute renal failure (ARF)**, the kidneys abruptly stop working entirely (or almost entirely). The main feature of ARF is the suppression of urine flow, usually characterized either by oliguria (daily urine output between 50 mL and 250 mL), or by anuria (daily urine output less than 50 mL). Causes include low blood volume (for example, due to hemorrhage), decreased cardiac output, damaged renal tubules, kidney stones, tissue damage caused by the dyes used to visualize blood vessels in angiograms, nonsteroidal anti-inflammatory drugs, and some antibiotic drugs. It is also common in people who suffer a devastating illness or overwhelming traumatic injury; in such cases it can be related to a more general organ failure known as Multiple Organ Dysfunction Syndrome (MODS).

Renal failure causes a multitude of problems. There is edema due to salt and water retention and acidosis due to an inability of the kidneys to excrete acidic substances. In the blood, urea builds up due to impaired renal excretion of metabolic waste products, and potassium level rises, which can lead to cardiac arrest. Often, there is anemia because the kidneys no longer produce enough erythropoietin for adequate red blood cell production. Because the kidneys are no longer able to convert vitamin D to calcitriol, which is needed for adequate calcium absorption from the small intestine, osteomalacia also may occur.

Chronic renal failure (CRF) refers to a progressive and usually irreversible decline in glomerular filtration rate (GFR). CRF may result from chronic glomerulonephritis, pyelonephritis, polycystic kidney disease, or traumatic loss of kidney tissue. CRF develops in three stages. In the first stage, *diminished renal reserve*, nephrons are destroyed until about 75% of the functioning nephrons are lost. At this stage, a person may have no signs or symptoms because the remaining nephrons enlarge and take over the function of those that have been lost. Once 75% of the nephrons are lost, the person enters the second stage, called *renal insufficiency*, characterized by a decrease in GFR and increased blood levels of nitrogen-containing wastes and creatinine. Also, the kidneys cannot effectively concentrate or dilute the urine. The final stage, called *end-stage renal failure*, occurs when about 90% of the nephrons have been lost. At this stage, GFR diminishes to 10–15% of normal, oliguria is present, and blood levels of nitrogen-containing wastes and creatinine increase further. People with end-stage renal failure need dialysis therapy and are possible candidates for a kidney transplant operation.

POLYCYSTIC KIDNEY DISEASE

Polycystic kidney disease (PKD) is one of the most common inherited disorders. In PKD, the kidney tubules become riddled with hundreds or thousands of cysts (fluid-filled cavities). In addition, inappropriate apoptosis (programmed cell death) of cells in noncystic tubules leads to progressive impairment of renal function and eventually to end-stage renal failure.

People with PKD also may have cysts and apoptosis in the liver, pancreas, spleen, and gonads; increased risk of cerebral aneurysms; heart valve defects; and diverticuli in the colon. Typically, symptoms are not noticed until adulthood, when patients may have back pain, urinary tract infections, blood in the urine, hypertension, and large abdominal masses. Using drugs to restore normal blood pressure, restricting protein and salt in the diet, and controlling urinary tract infections may slow progression to renal failure.

URINARY BLADDER CANCER

Each year, nearly 12,000 Americans die from **urinary bladder cancer.** It generally strikes people over 50 years of age and is three times more likely to develop in males than females. The disease is typically painless as it develops, but in most cases blood in the urine is a primary sign of the disease. Less often, people experience painful and/or frequent urination.

As long as the disease is identified early and treated promptly, the prognosis is favorable. Fortunately, about 75% of the urinary bladder cancers are confined to the epithelium of the urinary bladder and are easily removed by surgery. The lesions tend to be low grade, meaning that they have only a small potential for metastasis.

Urinary bladder cancer is frequently the result of a carcinogen. About half of all cases occur in people who smoke or have at some time smoked cigarettes. The cancer also tends to develop in people who are exposed to chemicals called aromatic amines. Workers in the leather, dye, rubber, and aluminum industries, as well as painters, are often exposed to these chemicals.

KEY MEDICAL TERMS ASSOCIATED WITH THE URINARY SYSTEM

Azotemia (az-ō-TĒ-mē-a; *azot-*=nitrogen; *-emia*=condition of blood) Presence of urea or other nitrogen-containing substances in the blood.

Cystocele (SIS-tō-sēl; *cysto-*=bladder; *-cele*=hernia or rupture) Hernia of the urinary bladder.

Diabetic kidney disease A disorder caused by diabetes mellitus in which glomeruli are damaged. The result is the leakage of proteins into the urine and a reduction in the ability of the kidney to remove water and waste.

Dysuria (dis-Ū-rē-a; *dys*=painful; *uria*=urine) Painful urination.

Enuresis (en′-ū-RĒ-sis=to void urine) Involuntary voiding of urine after the age at which voluntary control has typically been attained.

Hydronephrosis (hī′-drō-ne-FRŌ-sis; *hydro-*=water; *nephros*=kidney; *-osis*=condition) Swelling of the kidney due to dilation of the renal pelvis and calyces as a result of an obstruction to the flow of urine. It may be due to a congenital abnormality, a narrowing of the ureter, a kidney stone, or an enlarged prostate.

Intravenous pyelogram (in′-tra-VĒ-nus PĪ-e-lō-gram′; *intra-*=within; *veno-*=vein; *pyelo-*=pelvis of kidney; *-gram*=record), or *IVP* Radiograph (x-ray) of the kidneys, ureters, and urinary bladder after venous injection of a radiopaque contrast medium.

Nephropathy (ne-FROP-a-thē; *neph-*=kidney; *-pathos*=suffering) Any disease of the kidneys. Types include analgesic (from long-term and excessive use of drugs such as ibuoprofen), lead (from ingestion of lead-based paint), and solvent (from carbon tetrachloride and other solvents).

Nocturnal enuresis (nok-TUR-nal en′-ū-RĒ-sis) Discharge of urine during sleep, resulting in bed-wetting; occurs in about 15% of 5-year-old children and generally resolves spontaneously, afflicting only about 1% of adults. It may have a genetic basis, as bed-wetting occurs more often in identical twins than in fraternal twins and more often in children whose parents or siblings were bed-wetters. Possible causes include smaller-than-normal urinary bladder capacity, failure to awaken in response to a full urinary bladder, and above-normal production of urine at night. Also referred to as **nocturia.**

Polyuria (pol′-ē-Ū-rē-a; *poly-*=too much) Excessive urine formation. It may occur in conditions such as diabetes mellitus and glomerulonephritis.

Stricture (STRIK-chur) Narrowing of the lumen of a canal or hollow organ, as may occur in the ureter, urethra, or any other tubular structure in the body.

Uremia (ū-RĒ-mē-a; *emia*=condition of blood) Toxic levels of urea in the blood resulting from severe malfunction of the kidneys.

Urinary retention A failure to completely or normally void urine; may be due to an obstruction in the urethra or neck of the urinary bladder, to nervous contraction of the urethra, or to lack of urge to urinate. In men, an enlarged prostate may constrict the urethra and cause urinary retention. If urinary retention is prolonged, a catheter (slender rubber drainage tube) must be placed into the urethra to drain the urine.

STUDY OUTLINE

Introduction (p. 808)

1. The organs of the urinary system are the kidneys, ureters, urinary bladder, and urethra.
2. After the kidneys filter blood and return most water and many solutes to the bloodstream, the remaining water and solutes constitute urine.
3. The kidneys regulate the ionic composition, osmolarity, volume, and pH of the blood, as well as blood pressure.
4. The kidneys also perform gluconeogenesis, release calcitriol and erythropoietin, and excrete wastes and foreign substances.

Anatomy and Histology of the Kidneys (p. 810)

1. The kidneys are retroperitoneal organs attached to the posterior abdominal wall.
2. Three layers of tissue surround the kidneys: the renal capsule, adipose capsule, and renal fascia.
3. Internally, the kidneys consist of a renal cortex, a renal medulla, renal pyramids, renal papillae, renal columns, calyces, and a renal pelvis.
4. Blood flows into the kidney through the renal artery and successively into segmental, interlobar, arcuate, and interlobular arteries; afferent arterioles; glomerular capillaries; efferent arterioles; peritubular capillaries and vasa recta; and interlobular, arcuate, and interlobar veins before flowing out of the kidney through the renal vein.
5. Vasomotor nerves from the sympathetic division of the autonomic nervous system supply kidney blood vessels; they help regulate blood flow through the kidney.
6. The nephron is the functional unit of the kidneys. A nephron consists of a renal corpuscle (glomerulus, and glomerular or Bowman's capsule) and a renal tubule.
7. A renal tubule consists of a proximal convoluted tubule, a loop of Henle, and a distal convoluted tubule, which drains into a collecting duct (shared by several nephrons). The loop of Henle consists of a descending limb and an ascending limb.
8. A cortical nephron has a short loop that dips only into the superficial region of the renal medulla; a juxtamedullary nephron has a long loop of Henle that stretches through the renal medulla almost to the renal papilla.
9. The wall of the entire glomerular capsule, renal tubule, and ducts consists of a single layer of epithelial cells. The epithelium has distinctive histological features in different parts of the tubule. Table 26.1 on page 818 summarizes the histological features of the renal tubule and collecting duct.
10. The juxtaglomerular apparatus (JGA) consists of the juxtaglomerular cells of an afferent arteriole and the macula densa of the final portion of the ascending limb of the loop of Henle.

Functions of Nephrons (p. 819)

1. Fluid that enters the capsular space is glomerular filtrate.

2. The filtration (endothelial-capsular) membrane consists of the glomerular endothelium, basal lamina, and filtration slits between pedicels of podocytes.
3. Most substances in plasma easily pass through the glomerular filter. However, blood cells and most proteins normally are not filtered.
4. Glomerular filtrate amounts to up to 180 liters of fluid per day. This large amount of fluid is filtered because the filter is porous and thin, the glomerular capillaries are long, and the capillary blood pressure is high.
5. Tubular reabsorption is a selective process that reclaims materials from tubular fluid and returns them to the bloodstream. Reabsorbed substances include water, glucose, amino acids, urea, and ions, such as sodium, chloride, potassium, bicarbonate, and phosphate.
6. Some substances not needed by the body are removed from the blood and discharged into the urine via tubular secretion. Included are ions (K^+, H^+, and NH_4^+), urea, creatinine, and certain drugs.

Urine Transportation, Storage, and Elimination (p. 822)

1. The ureters are retroperitoneal and consist of a mucosa, muscularis, and adventitia. They transport urine from the renal pelvis to the urinary bladder, primarily via peristalsis.
2. The urinary bladder is located in the pelvic cavity posterior to the pubic symphysis; its function is to store urine prior to micturition.
3. The urinary bladder consists of a mucosa with rugae, a muscularis (detrusor muscle), and an adventitia (serosa over the superior surface).
4. The micturition reflex discharges urine from the urinary bladder via parasympathetic impulses that cause contraction of the detrusor muscle and relaxation of the internal urethral sphincter muscle, and via inhibition of impulses in somatic motor neurons to the external urethral sphincter.
5. The urethra is a tube leading from the floor of the urinary bladder to the exterior. Its anatomy and histology differ in females and males. In both sexes the urethra functions to discharge urine from the body; in males it discharges semen as well.

Development of the Urinary System (p. 826)

1. The kidneys develop from intermediate mesoderm.
2. The kidneys develop in the following sequence: pronephros, mesonephros, and metanephros. Only the metanephros remains and develops into a functional kidney.

Aging and the Urinary System (p. 828)

1. With aging, the kidneys shrink in size, have a decreased blood flow, and filter less blood.
2. Common problems related to aging include urinary tract infections, increased frequency of urination, urinary retention or incontinence, and renal calculi.

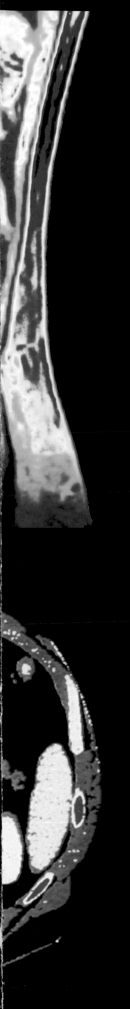

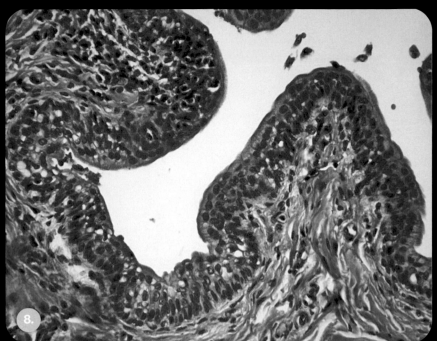

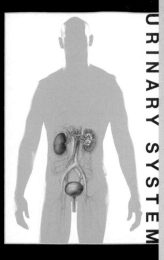

1. Kidney in frontal section

2. Kidneys, ureters, adrenal glands, and blood vessels

3. SEM of nephrons

4. SEM of renal corpuscles and renal tubules

5. Resin "corrosion" cast of the renal blood vessels

6. MRI of the kidneys

7. CT scan of the abdomen

8. Light micrograph of the urethra

9. Radiograph of the urinary system

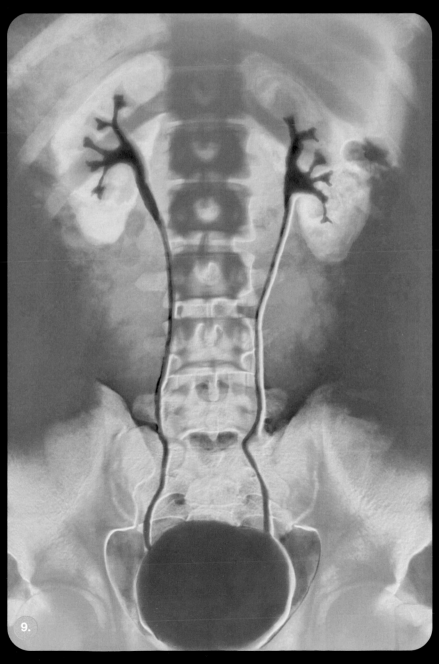

URINARY SYSTEM

1. **Kidney in frontal section.** Kidney in frontal section clearly showing the renal cortex and renal medulla.

2. **Kidneys, ureters, adrenal glands, and blood vessels.** Kidneys, ureters, adrenal glands, and blood vessels in anterior view.

3. **SEM of nephrons.** Scanning Electron Micrograph (SEM) of several nephrons.

4. **SEM of renal corpuscles and renal tubules.** Scanning Electron Micrograph (SEM) of several renal corpuscles and renal tubules.

5. **Resin "corrosion" cast of the renal blood vessels.** Resin "corrosion" cast of the renal blood vessels and duct system. The blood vessels were filled with latex and the surrounding tissue was dissolved with acid. The blood vessels are red and blue and the ureters are yellow.

6. **MRI of the kidneys.** Magnetic resonance image (MRI) of the kidneys in frontal section.

7. **CT scan of the abdomen.** Computed Tomography (CT) scan of a transverse section of the abdomen showing the kidneys in inferior view.

8. **Light micrograph of the urethra.** Light micrograph of the lining of the urethra.

9. **Radiograph of the urinary system.** Radiograph of the urinary system in anterior view following injection of a contrast medium into the blood, which is concentrated and excreted by the kidneys (intravenous pyelogram).

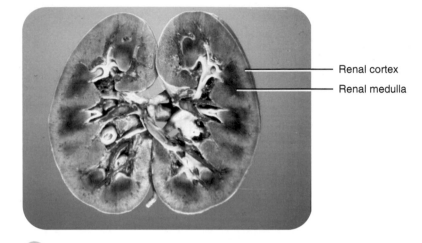

1.

Renal cortex
Renal medulla

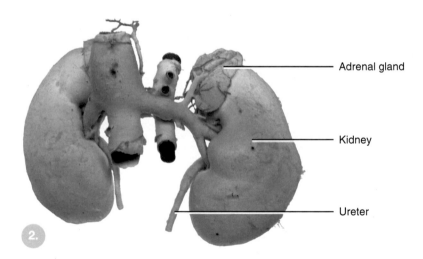

2.

Adrenal gland

Kidney

Ureter

SEM 460x

3.

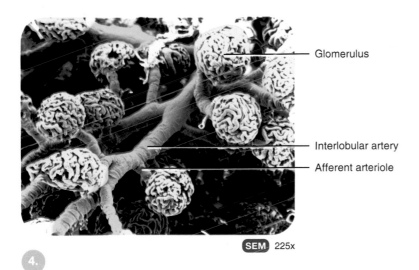

Glomerulus

Interlobular artery

Afferent arteriole

SEM 225x

4.

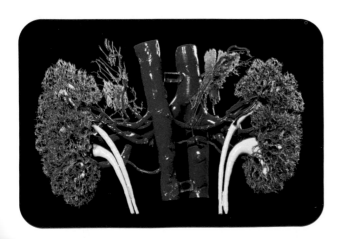

5.

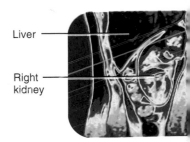

Liver

Right kidney

6.

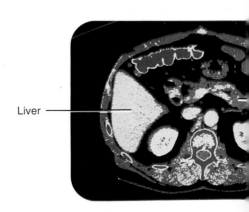

Liver

7.

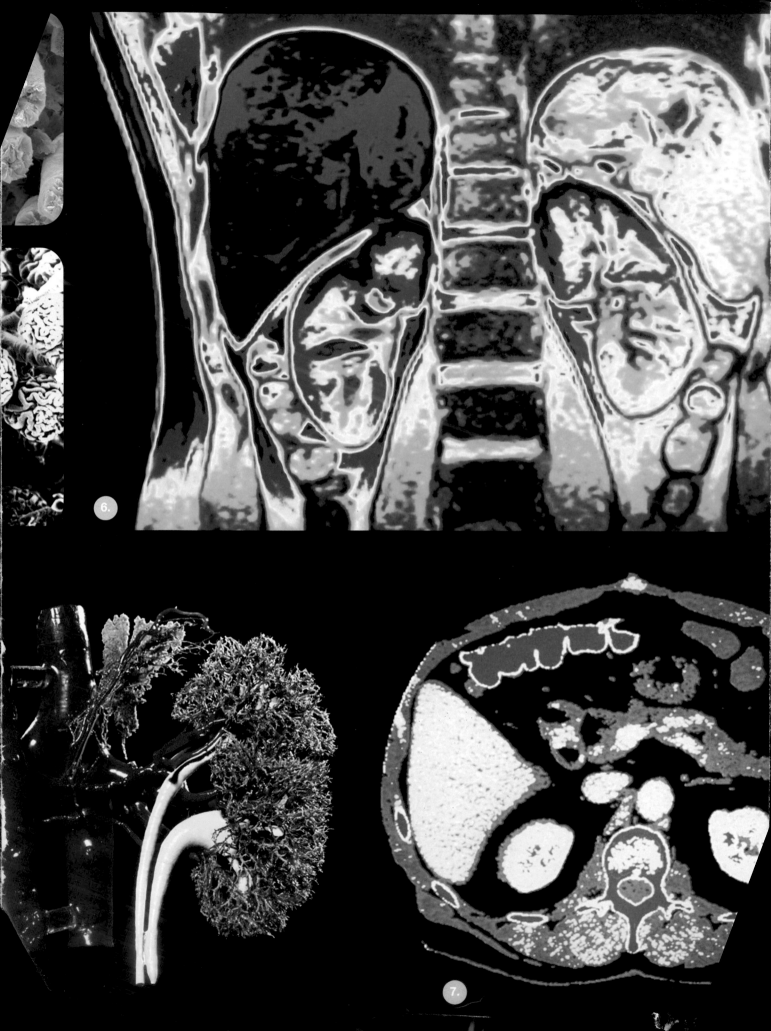

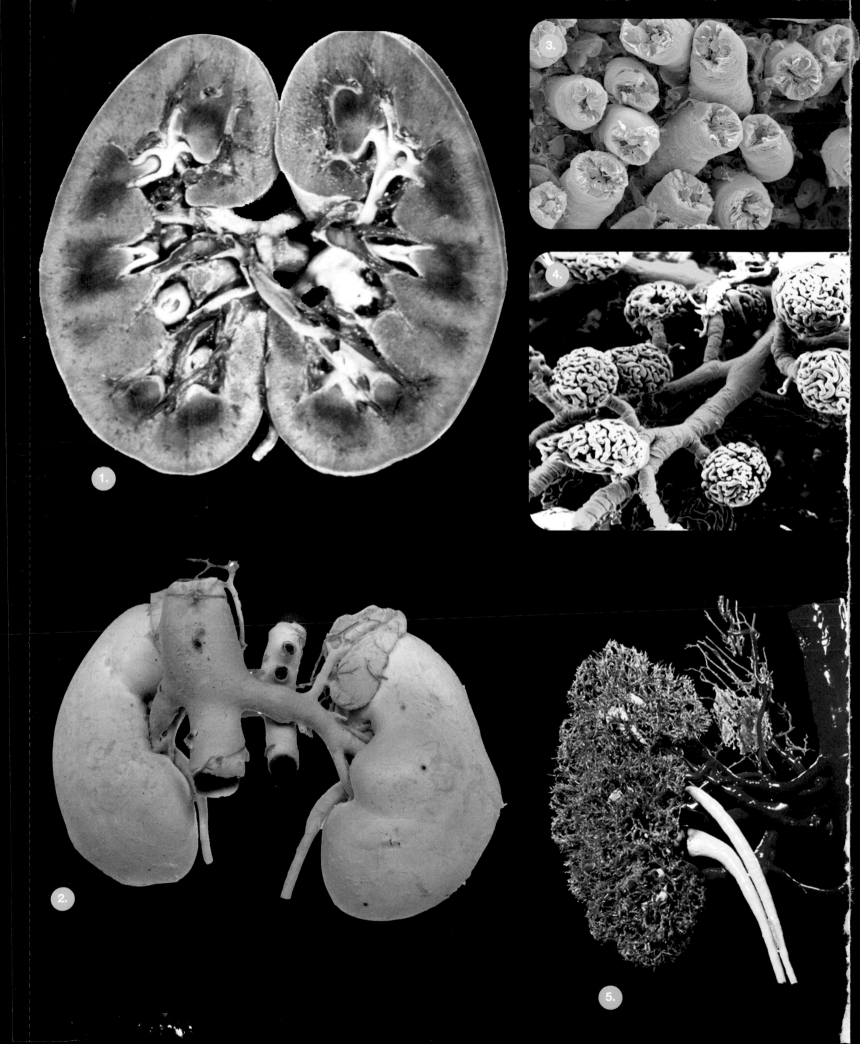

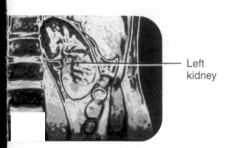

Left
kidney

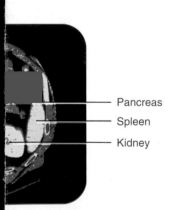

Pancreas

Spleen

Kidney

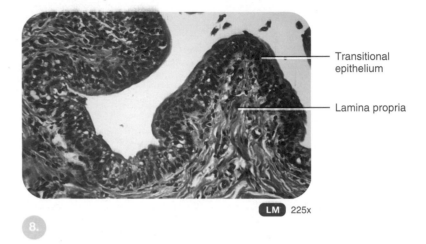

Transitional
epithelium

Lamina propria

LM 225x

8.

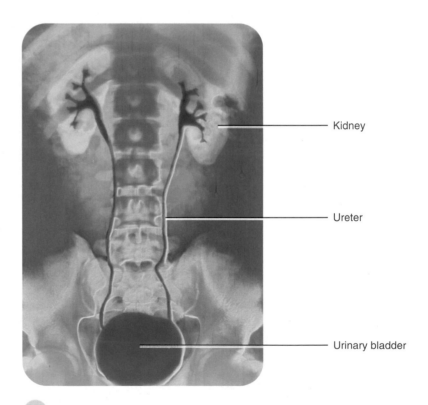

Kidney

Ureter

Urinary bladder

9.

SELF-QUIZ QUESTIONS

Choose the one best answer to the following questions.

1. Which of the following is *not* a function of the kidneys?
   a. participation in the formation of the active form of vitamin D
   b. regulation of the volume and composition of the blood
   c. removal of wastes from the blood in the form of urine
   d. production of white blood cells
   e. regulation of blood pressure by secretion of renin, which activates the renin–angiotensin pathway

2. Which of the following vessels split from the efferent arterioles?
   a. glomerular capillaries   b. peritubular capillaries
   c. afferent arterioles   d. vasa recta
   e. Both b and d are correct.

3. Urine leaving the distal convoluted tubule passes through various structures in which of the following sequences?
   a. collecting duct, renal hilus, calyx, ureter
   b. collecting duct, calyx, renal pelvis, ureter
   c. calyx, collecting duct, renal pelvis, ureter
   d. calyx, renal hilus, renal pelvis, ureter
   e. collecting duct, renal hilus, ureter, calyx

4. The trigone, a landmark in the urinary bladder, is a triangular area bounded by:
   a. the orifices of the ejaculatory ducts and the urethra.
   b. the internal urethral orifice and the inferior border of the detrusor muscle.
   c. the ureteral and the internal urethral orifices.
   d. the top of the fundus and the ureteral orifices.
   e. the major and minor calyces.

5. Fluid that filters into the glomerular capsule flows next into the:
   a. glomerulus.   b. efferent arteriole.
   c. proximal convoluted tubule.   d. distal convoluted tubule.
   e. loop of Henle.

6. The macula densa is located:
   a. between the glomerular capillaries and the glomerular capsule.
   b. between the renal papilla and the minor calyx.
   c. where the afferent arteriole branches from the interlobular artery.
   d. where the ascending limb of the loop of Henle makes contact with the afferent arteriole.
   e. at the bottom of the loop of Henle.

7. The detrusor muscle is:
   a. the middle layer of the wall of the urinary bladder.
   b. the muscle that controls the flow of urine from the renal pelvis into the ureter.
   c. another name for the muscularis of the ureter.
   d. a modification of the urogenital diaphragm muscle.
   e. Both a and d are correct.

Complete the following.

8. The apex of a renal pyramid, called the _____, points toward the interior of the kidney.

9. In the renal corpuscle, the _____ arteriole normally has a smaller diameter than the _____ arteriole.

10. The special simple squamous epithelial cells that have many projections and that are found in the visceral layer of the glomerular capsule are called _____.

11. The longest portion of the male urethra is the _____ urethra.

12. Most nephrons are cortical nephrons, but 15–20% are _____ nephrons.

13. The _____ carry urine from the kidney to the urinary bladder.

14. Place numbers in the blanks to arrange the following vessels in order. (a) arcuate arteries: _____; (b) interlobular arteries: _____; (c) renal arteries: _____; (d) peritubular capillaries and vasa recta: _____; (e) glomerular capillaries: _____; (f) efferent arteriole: _____; (g) afferent arteriole: _____; (h) interlobar arteries: _____; (i) segmental arteries: _____.

15. The process of emptying the urinary bladder is called _____.

Are the following statements true or false?

16. Signs of kidney dysfunction usually do not become apparent until more than 75% of normal kidney function is lost.

17. Most renal nerves arise from the celiac ganglion and are part of the sympathetic division of the autonomic nervous system.

18. The internal urethral sphincter is composed of voluntary skeletal muscle, whereas the external urethral sphincter is composed of involuntary smooth muscle.

19. The juxtaglomerular apparatus is part of the endothelial-capsular (filtration) membrane.

CRITICAL THINKING QUESTIONS

1. Imagine that a new super bug has emerged from a nuclear waste dump. It produces a toxin that blocks renal tubule function but leaves the glomerulus unaffected. Predict the effects of this toxin.

2. Little Caitlyn has a very sore throat. When her mother calls the pediatrician, she is told to bring the child in right away to be tested for "strep throat." Why is the doctor more concerned about a streptococcal sore throat than other types of sore throats?

3. Although urinary catheters come in one length only, the number of centimeters that must be inserted to release the urine differs significantly in males and females. Why? Why is volume of urine released from a full bladder different in males and females?

4. As Juan focused his microscope on his own urine sample during his Human Anatomy lab, he was concerned to see many cells in the field of view; his urine had appeared to be fluid. There's no evidence of blood or infection. What are these cells and where do they come from?

5. Craig's urine samples were repeatedly found to have high concentrations of albumin. The doctor is also tracking Craig's blood pressure. Are Craig's blood pressure and albuminuria related?

ANSWERS TO FIGURE QUESTIONS

26.1 The kidneys, ureters, urinary bladder, and urethra are the components of the urinary system.

26.2 The kidneys are retroperitoneal because they are posterior to the peritoneum of the abdominal cavity.

26.3 Blood vessels, lymphatic vessels, nerves, and a ureter pass through the renal hilus.

26.4 About 1200 mL of blood enters the renal arteries each minute.

26.5 Cortical nephrons have glomeruli in the superficial renal cortex, and their short loops of Henle penetrate only into the superficial renal medulla; juxtamedullary nephrons have glomeruli deep in the renal cortex, and their long loops of Henle extend through the renal medulla nearly to the renal papilla.

26.6 The section shown must be part of the renal cortex because there are no renal corpuscles in the renal medulla.

26.7 Penicillin secreted by the cells of the renal tubule is being removed from the bloodstream.

26.8 Endothelial fenestrations (pores) in glomerular capillaries are too small for red blood cells to pass through.

26.9 Lack of voluntary control over micturition is called urinary incontinence.

26.10 The muscularis of most of the ureters consists of an inner longitudinal layer and an outer circular layer, an arrangement opposite that of the gastrointestinal tract.

26.11 The trigone is a triangular area in the urinary bladder formed by the ureteral openings (posterior corners) and the internal urethral orifice (anterior corner).

26.12 The male urethra passes through the prostate, urogenital diaphragm, and penis.

26.13 The kidneys start to develop during the third week of gestation.

THE REPRODUCTIVE SYSTEMS

27

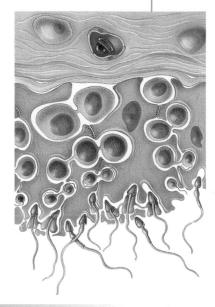

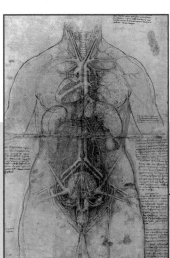

INTRODUCTION Sexual reproduction is the process by which organisms produce offspring through the union of germ cells called **gametes** (GAM-ēts=spouses). After the male gamete (sperm cell) unites with the female gamete (secondary oocyte)—an event called **fertilization**—the resulting cell contains one set of chromosomes from each parent. Males and females have anatomically distinct reproductive organs that are adapted for producing gametes, facilitating fertilization, and, in females, sustaining the growth of the embryo and fetus.

The male and female reproductive organs can be grouped by function. The **gonads**—testes in males and ovaries in females—produce gametes and secrete sex hormones. Various **ducts** then store and transport the gametes, and **accessory sex glands** produce substances that protect the gametes and facilitate their movement. Finally, **supporting structures,** such as the penis and the uterus, assist the delivery and joining of gametes and, in females, the growth of the fetus during pregnancy.

Do you recognize this image? Whose work is this? Which human organs are being depicted in this rendering?

www.wiley.com/college/apcentral

835

Gynecology (gī-ne-KOL-ō-jē; *gynec-*=woman; *-ology*=study of) is the specialized branch of medicine concerned with the diagnosis and treatment of diseases of the female reproductive system. As noted in Chapter 26, **urology** (ū-ROL-ō-jē) is the study of the urinary system. Urologists also diagnose and treat diseases and disorders of the male reproductive system.

MALE REPRODUCTIVE SYSTEM

● Describe the location, structure, and functions of the organs of the male reproductive system.
● Discuss the process of spermatogenesis in the testes.

The organs of the male reproductive system are the testes, a system of ducts (including the ductus deferens, ejaculatory ducts, and urethra), accessory sex glands (seminal vesicles, prostate, and bulbourethral gland), and several supporting structures, including the scrotum and the penis (Figure 27.1). The testes (male gonads) produce sperm and secrete hormones. A system of ducts transports and stores sperm, assists in their maturation, and conveys them to the exterior. Semen contains sperm plus the secretions provided by the accessory sex glands.

Scrotum

The **scrotum** (SKRŌ-tum=bag), the supporting structure for the testes, is a sac consisting of loose skin and superficial fascia that hangs from the root (attached portion) of the penis (Figure

Figure 27.1   Male organs of reproduction and surrounding structures.

🔑 Reproductive organs are adapted for producing new individuals and passing on genetic material from one generation to the next.

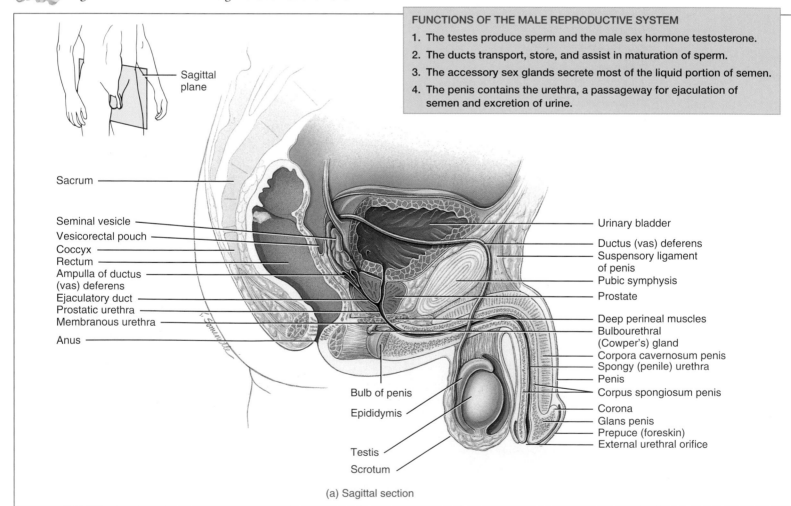

FUNCTIONS OF THE MALE REPRODUCTIVE SYSTEM

1. The testes produce sperm and the male sex hormone testosterone.
2. The ducts transport, store, and assist in maturation of sperm.
3. The accessory sex glands secrete most of the liquid portion of semen.
4. The penis contains the urethra, a passageway for ejaculation of semen and excretion of urine.

Sagittal plane

Sacrum

Seminal vesicle
Vesicorectal pouch
Coccyx
Rectum
Ampulla of ductus (vas) deferens
Ejaculatory duct
Prostatic urethra
Membranous urethra
Anus

Urinary bladder
Ductus (vas) deferens
Suspensory ligament of penis
Pubic symphysis
Prostate
Deep perineal muscles
Bulbourethral (Cowper's) gland
Corpora cavernosum penis
Spongy (penile) urethra
Penis
Corpus spongiosum penis
Corona
Glans penis
Prepuce (foreskin)
External urethral orifice

Bulb of penis
Epididymis

Testis
Scrotum

(a) Sagittal section

27.1a). As you will see later, the root of the penis is attached to the deep muscles of the perineum, and the ischium and pubis of the hip bone. Externally, the scrotum looks like a single pouch of skin separated into lateral portions by a median ridge called the **raphe** (RĀ-fē=seam); internally, the scrotal septum divides the scrotum into two sacs, each containing a single testis (Figure 27.2). The septum consists of superficial fascia and muscle tissue called the **dartos muscle** (DAR-tōs=skinned), which is composed of bundles of smooth muscle fibers. The dartos muscle is also found in the subcutaneous tissue of the scrotum and is directly continuous with the subcutaneous tissue of the abdominal wall. When it contracts, the dartos muscle causes wrinkling of the skin of the scrotum.

The location of the scrotum and the contraction of its muscle fibers regulate the temperature of the testes. A temperature about 2–3°C below core body temperature, which is required for normal sperm production, is maintained within the scrotum because it is outside the pelvic cavity. The **cremaster muscle** (krē-MAS-ter=suspender), a small band of skeletal muscle in the spermatic cord that is a continuation of the internal oblique muscle, elevates the testes upon exposure to cold (and during sexual arousal). This action moves the testes closer to the pelvic cavity, where they can absorb body heat. Exposure to warmth reverses the process. The dartos muscle also contracts in response to cold and relaxes in response to warmth.

The blood supply of the scrotum is derived from the internal pudendal branch of the internal iliac artery, the cremasteric branch of the inferior epigastric artery, and the external pudendal artery from the femoral artery. The scrotal veins follow the arteries.

The scrotal nerves are derived from the pudendal nerve, posterior cutaneous nerve of the thigh, and ilioinguinal nerves.

Testes

The **testes** (TES-tēz), or **testicles,** are paired oval glands in the scrotum measuring about 5 cm (2 in.) long and 2.5 cm (1 in.) in diameter (Figure 27.3 on page 839). Each **testis** (singular) weighs 10–15 grams. The testes develop near the kidneys, in the posterior portion of the abdomen, and they usually begin their descent into the scrotum through the inguinal canals (passageways in the anterior abdominal wall; see Figure 27.2) during the latter half of the seventh month of fetal development.

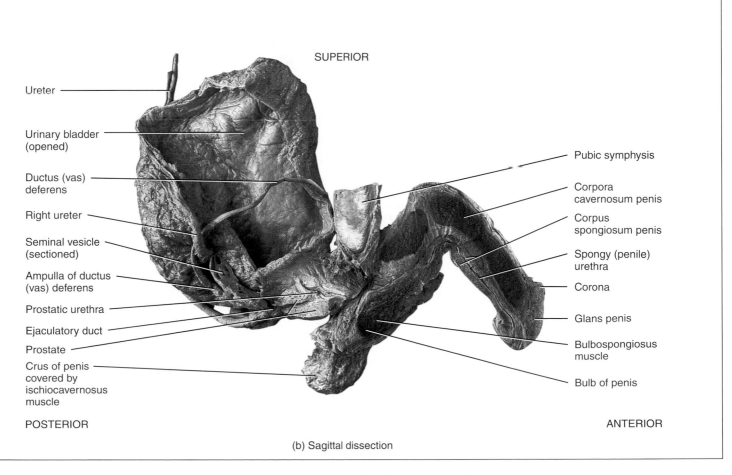

SUPERIOR

Ureter

Urinary bladder (opened)

Ductus (vas) deferens

Right ureter

Seminal vesicle (sectioned)

Ampulla of ductus (vas) deferens

Prostatic urethra

Ejaculatory duct

Prostate

Crus of penis covered by ischiocavernosus muscle

Pubic symphysis

Corpora cavernosum penis

Corpus spongiosum penis

Spongy (penile) urethra

Corona

Glans penis

Bulbospongiosus muscle

Bulb of penis

POSTERIOR

ANTERIOR

(b) Sagittal dissection

What are the groups of reproductive organs in males, and what are the functions of each group?

Figure 27.2   The scrotum, the supporting structure for the testes.

The scrotum consists of loose skin and superficial fascia and supports the testes.

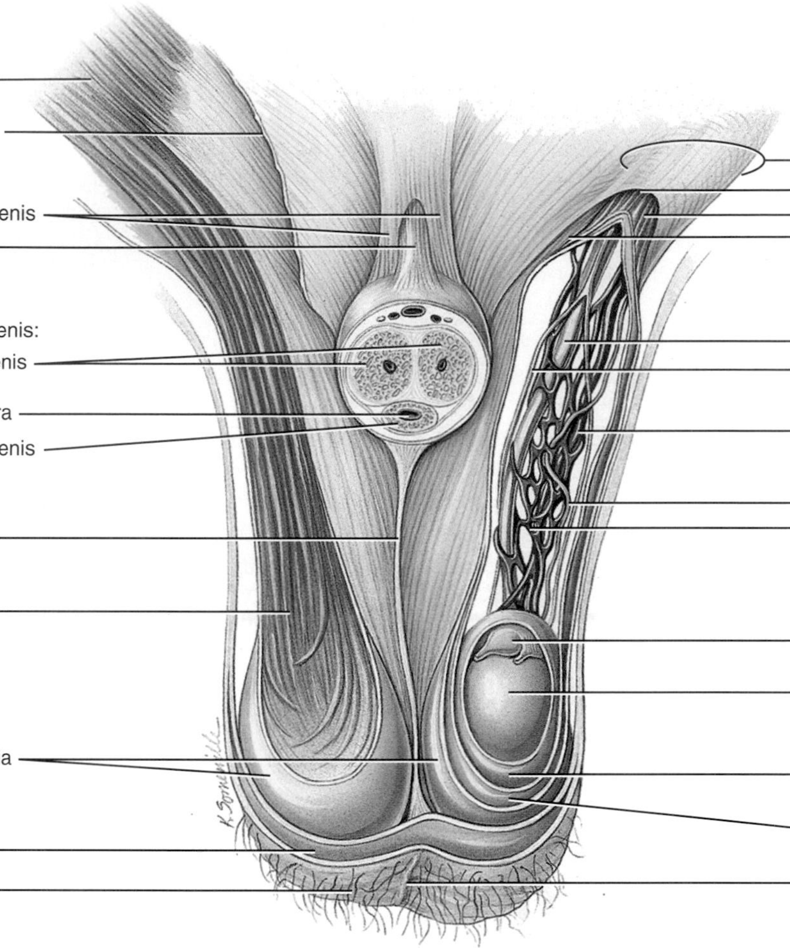

Internal oblique muscle

Aponeurosis of external oblique muscle (cut)

Fundiform ligament of penis

Suspensory ligament of penis

Transverse section of penis:

Corpora cavernosa penis

Spongy (penile) urethra

Corpus spongiosum penis

Scrotal septum

Cremaster muscle

External spermatic fascia

Dartos muscle

Skin of scrotum

Spermatic cord

Superficial inguinal ring

Cremaster muscle

Inguinal canal

Ductus (vas) deferens

Autonomic nerve

Testicular artery

Lymphatic vessel

Pampiniform plexus of testicular veins

Epididymis

Tunica albuginea of testis

Tunica vaginalis (peritoneum)

Internal spermatic fascia

Raphe

Anterior view of scrotum and testes and transverse section of penis

 **Which muscles help regulate the temperature of the testes?**

TESTICULAR INJURIES

Because of their location outside the abdomen, the testes are vulnerable to various injuries, and pain caused by striking the testes is frequently severe. Most **testicular injuries** are not serious due to the sponginess of testicular tissue and the flexibility of their location, which enable them to absorb considerable shock without sustaining permanent harm. Usually, if the pain and swelling from a testicular injury go away after about an hour, it can be assumed that there is no permanent damage. If not, medical help should be sought since infertility or loss of a testis may result.   ■

A serous membrane called the **tunica vaginalis** (*tunica*= sheath), which is derived from the peritoneum and forms during the descent of the testes, partially covers the testes. A collection of serous fluid in the tunica vaginalis is called a **hydrocele** (HĪ-drō-sēl; *hydro-*=water; *-kele*=hernia). It may be caused by injury

to the testes or inflammation of the epididymis. Usually, no treatment is required. Internal to the tunica vaginalis is a dense white fibrous capsule composed of dense irregular connective tissue, the **tunica albuginea** (al′-bū-JIN-ē-a; *albu-*= white); it extends inward, forming septa that divide each testis into a series of internal compartments called **lobules.** Each of the 200–300 lobules contains one to three tightly coiled tubules, the **seminiferous tubules** (*semin-*=seed; *fer-*=to carry), where sperm are produced (Figure 27.4 on page 840). The process by which the seminiferous tubules of the testes produce sperm is called **spermatogenesis** (sper′-ma-tō-JEN-e-sis; *genesis*=beginning process or production).

The seminiferous tubules contain two types of cells: **spermatogenic cells,** the sperm-forming cells, and **Sertoli cells,** which have several functions in supporting spermatogenesis (Figure 27.4). Starting at puberty, sperm production begins at the periphery of the seminiferous tubules in stem cells called

Figure 27.3 **Internal and external anatomy of a testis.**

The testes are the male gonads, which produce haploid sperm.

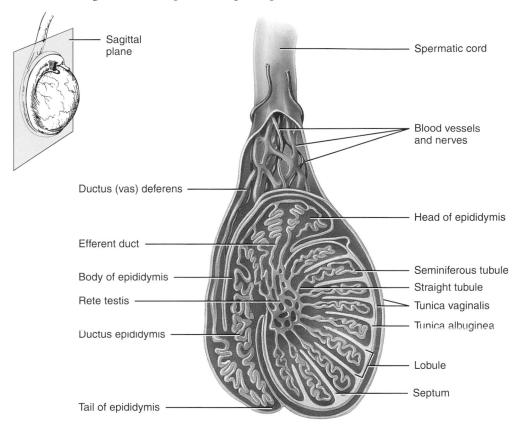

Sagittal plane

Spermatic cord

Blood vessels and nerves

Ductus (vas) deferens

Head of epididymis

Efferent duct

Body of epididymis

Seminiferous tubule

Straight tubule

Rete testis

Tunica vaginalis

Ductus epididymis

Tunica albuginea

Lobule

Septum

Tail of epididymis

(a) Sagittal section of a testis showing seminiferous tubules

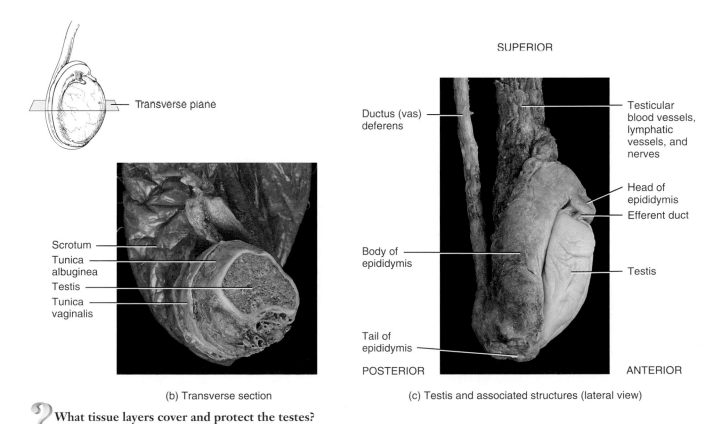

Transverse plane

SUPERIOR

Ductus (vas) deferens

Testicular blood vessels, lymphatic vessels, and nerves

Head of epididymis

Efferent duct

Body of epididymis

Testis

Scrotum

Tunica albuginea

Testis

Tunica vaginalis

Tail of epididymis

POSTERIOR

ANTERIOR

(b) Transverse section

(c) Testis and associated structures (lateral view)

What tissue layers cover and protect the testes?

Figure 27.4 **Microscopic anatomy of the seminiferous tubules and stages of sperm production (spermatogenesis).**
Arrows in (b) indicate the progression of spermatogenic cells from least mature to most mature. The (*n*) and (*2n*) refer to haploid and diploid chromosome number, respectively.

Spermatogenesis occurs in the seminiferous tubules of the testes.

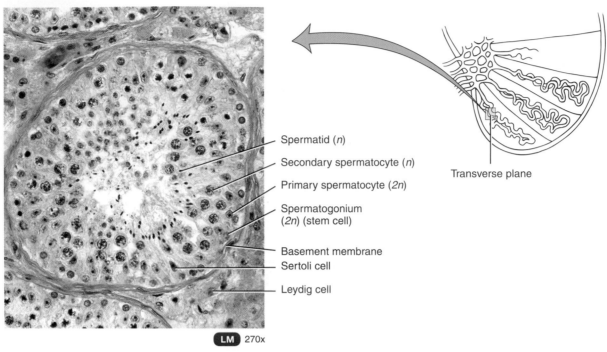

Spermatid (*n*)

Secondary spermatocyte (*n*)

Primary spermatocyte (*2n*)

Spermatogonium (*2n*) (stem cell)

Basement membrane

Sertoli cell

Leydig cell

Transverse plane

LM 270x

(a) Transverse section of several seminiferous tubules

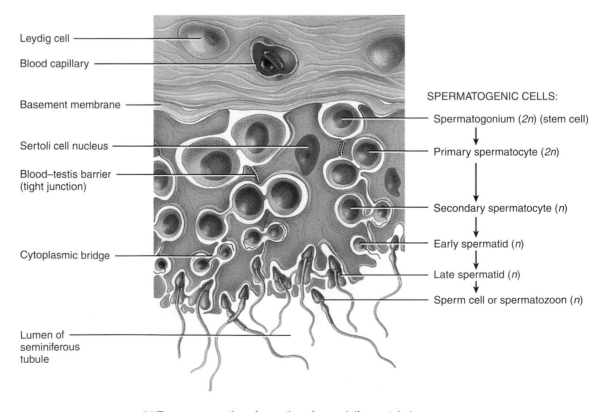

Leydig cell

Blood capillary

Basement membrane

Sertoli cell nucleus

Blood–testis barrier (tight junction)

Cytoplasmic bridge

Lumen of seminiferous tubule

SPERMATOGENIC CELLS:

Spermatogonium (*2n*) (stem cell)

Primary spermatocyte (*2n*)

Secondary spermatocyte (*n*)

Early spermatid (*n*)

Late spermatid (*n*)

Sperm cell or spermatozoon (*n*)

(b) Transverse section of a portion of a seminiferous tubule

Which cells produce testosterone?

spermatogonia (sper'-ma-tō-GŌ-nē-a; *-gonia*=offspring; singular is *spermatogonium*). These cells develop from **primordial germ cells** (*primordi-*=primitive or early form) that arise from the yolk sac endoderm and enter the testes during the fifth week of development. In the embryonic testes, the primordial germ cells differentiate into spermatogonia, which remain dormant during childhood and become active at puberty. Toward the lumen of the tubule are layers of progressively more mature cells. In order of advancing maturity, these are primary spermatocytes, secondary spermatocytes, spermatids, and sperm. When a **sperm cell,** or **spermatozoon** (sper'-ma-tō-ZŌ-on; *-zoon*=life), has nearly reached maturity, it is released into the lumen of the seminiferous tubule. (The plural terms are *sperm* and *spermatozoa*.)

Embedded among the spermatogenic cells in the tubules are large **Sertoli cells,** or *sustentacular cells* (sus'-ten-TAK-ū-lar), which extend from the basement membrane to the lumen of the tubule. Internal to the basement membrane and spermatogonia, tight junctions join neighboring Sertoli cells to one another. These junctions form an obstruction known as the **blood–testis barrier** because substances must first pass through the Sertoli cells before they can reach the developing sperm. By isolating the developing gametes from the blood, the blood–testis barrier prevents an immune response against the spermatogenic cell's surface antigens, which are recognized as "foreign" by the immune system. The blood–testis barrier does not include spermatogonia.

Sertoli cells support and protect developing spermatogenic cells in several ways. They nourish spermatocytes, spermatids, and sperm; phagocytize excess spermatid cytoplasm as development proceeds; and control movements of spermatogenic cells and the release of sperm into the lumen of the seminiferous tubule. They also produce fluid for sperm transport, secrete the hormone inhibin, which decreases the rate of spermatogenesis, and mediate the effects of testosterone and FSH (follicle-stimulating hormone).

In the spaces between adjacent seminiferous tubules are clusters of cells called **Leydig (interstitial) cells** (Figure 27.4). These cells secrete testosterone, the most important androgen (male sex hormone). Although androgens are hormones that promote development of masculine characteristics, they also have other functions, such as promoting libido (sexual desire) in both males and females.

CRYPTORCHIDISM

The condition in which the testes do not descend into the scrotum is called **cryptorchidism** (krip-TOR-ki-dizm; *crypt-*=hidden; *orchid*=testis); it occurs in about 3% of full-term infants and about 30% of premature infants. Untreated bilateral cryptorchidism results in sterility because the cells involved in the initial stages of spermatogenesis are destroyed by the higher temperature of the pelvic cavity. The chance of testicular cancer is 30–50 times greater in cryptorchid testes. The testes of about 80% of boys with cryptorchidism will descend spontaneously during the first year of life. When the testes remain unde-

scended, the condition can be corrected surgically, and should ideally be done before 18 months of age. ■

Before you read this section, please review the topic of reproductive cell devision in Chapter 2 on page 47. Also, pay particular attention to Figures 2.20 and 2.21 on pages 48 and 49, respectively.

In humans, spermatogenesis takes about 65–75 days. It begins in the spermatogonia, which contain the diploid (*2n*) chromosome number (Figure 27.5). Spermatogonia are *stem cells* because when they undergo mitosis, some cells remain near the basement membrane of the seminiferous tubule in an undifferentiated state to serve as a reservoir of cells for future mitosis and subsequent sperm production. The rest of the cells lose contact with the basement membrane, squeeze through the tight junctions of the blood–testis barrier, undergo developmental changes, and differentiate into **primary spermatocytes** (SPER-ma-tō-sītz'). Primary spermatocytes, like spermatogonia, are diploid (*2n*); that is, they have 46 chromosomes.

Each primary spermatocyte enlarges and then begins meiosis. Then, two nuclear divisions occur as part of meiosis (Figure

Figure 27.5 Events in spermatogenesis. Diploid cells (*2n*) have 46 chromosomes; haploid cells (*n*) have 23 chromosomes.

Spermiogenesis involves the maturation of spermatids into sperm.

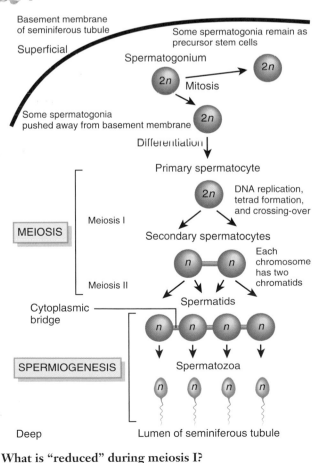

What is "reduced" during meiosis I?

27.5). In meiosis I, DNA replicates, homologous pairs of chromosomes line up at the metaphase plate, and crossing-over occurs. Then, the meiotic spindle forms and pulls one (duplicated) chromosome of each pair to an opposite pole of the dividing cell. This random assortment of maternally derived chromosomes and paternally derived chromosomes toward opposite poles is another reason for genetic variation among gametes. The two cells formed by meiosis I are called **secondary spermatocytes,** and each cell has 23 chromosomes—the haploid number. However, each chromosome within a secondary spermatocyte is made up of two chromatids (two copies of the DNA) still attached by a centromere.

In meiosis II, no replication of DNA occurs. The chromosomes line up in single file along the metaphase plate, and two chromatids of each chromosome separate from each other. The four haploid cells resulting from meiosis II are called **spermatids.** A single primary spermatocyte therefore produces four spermatids through two rounds of cell division (meiosis I and meiosis II).

A unique and very interesting process occurs during spermatogenesis. As the sperm cells proliferate following their production by spermatogonia, they fail to complete cytoplasmic separation (cytokinesis). The cells remain in contact by cytoplasmic bridges through their entire development (see Figures 27.4b and 27.5). This pattern of development most likely accounts for the synchronized production of sperm in any given area of a seminiferous tubule. It may have survival value in that half of the sperm contain an X chromosome and half contain a Y chromosome. The larger X chromosome may carry genes needed for spermatogenesis that are lacking on the smaller Y chromosome.

The final stage of spermatogenesis, **spermiogenesis** (sper′-mē-ō-JEN-e-sis), is the maturation of haploid spermatids into sperm. Because no cell division occurs in spermiogenesis, each spermatid develops into a single **sperm cell** (*sperma*=seed). During this process, spherical spermatids transform into elongated, slender sperm. An acrosome (described shortly) forms atop the nucleus, which condenses and elongates, a flagellum develops, and mitochondria multiply. Sertoli cells dispose of the excess cytoplasm that is sloughed off during this process. Finally, sperm are released from their connections to Sertoli cells, an event known as **spermiation.** Sperm then enter the lumen of the seminiferous tubule. Fluid secreted by Sertoli cells pushes sperm along their way, toward the ducts of the testes.

Sperm

Spermatogenesis produces about 300 million sperm per day. Once ejaculated, most do not survive more than 48 hours within the female reproductive tract. A sperm is about 60 μm long and contains several structures that are highly adapted for reaching and penetrating a secondary oocyte (Figure 27.6). The major parts of a sperm are the head and the tail. The flattened, pointed **head** of the sperm is about 4–5 μm long. It contains a **nucleus** that has highly condensed haploid chromosomes (23). Covering the anterior two-thirds of the nucleus is the **acrosome** (*acro-*=

Figure 27.6 Parts of sperm cell (spermatozoon).

About 300 million sperm mature each day.

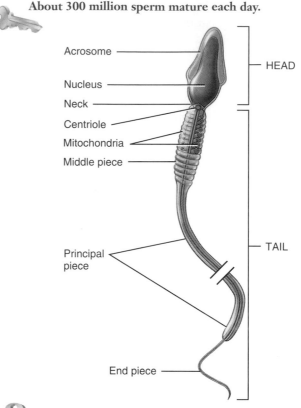

Acrosome
Nucleus
Neck
Centriole
Mitochondria
Middle piece
Principal piece
End piece
HEAD
TAIL

What are the functions of each part of a sperm cell?

atop; *-some*=body), a caplike vesicle filled with enzymes that help a sperm to penetrate a secondary oocyte to bring about fertilization. Among the enzymes are hyaluronidase and proteases. The **tail** of a sperm is subdivided into four parts: neck, middle piece, principal piece, and end piece. The **neck** is the constricted region just behind the head that contains centrioles. The centrioles form the microtubules that comprise the remainder of the tail. The **middle piece** contains mitochondria arranged in a spiral, which provide the energy (ATP) for locomotion of sperm to the site of fertilization and for sperm metabolism. The **principal piece** is the longest portion of the tail and the **end piece** is the terminal, tapering portion of the tail.

CHECKPOINT

1. Describe the function of the scrotum in protecting the testes from temperature fluctuations.
2. Describe the internal structure of a testis. Where are sperm cells produced? What are the functions of Sertoli cells and Leydig cells?
3. Describe the principal events of spermatogenesis.
4. Identify the parts of a sperm cell, and list the functions of each.

Reproductive System Ducts in Males

Ducts of the Testis

Pressure generated by the fluid secreted by Sertoli cells pushes sperm and fluid along the lumen of seminiferous tubules and then into a series of very short ducts called **straight tubules.** The straight tubules lead to a network of ducts in the testis called the **rete testis** (RĒ-tē=network) (see Figure 27.3a). From the rete testis, sperm move into a series of coiled **efferent ducts** in the epididymis that empty into a single tube called the **ductus epididymis.**

Epididymis

The **epididymis** (ep′-i-DID-i-mis; *epi-*=over or above; *-didymis* =testis) is a comma-shaped organ about 4 cm (1.5 in.) long that lies along the posterior border of each testis (see Figure 27.3a). The plural is *epididymides* (ep′-i-did-ĪM-i-dēs). Each epididymis consists mostly of the tightly coiled **ductus epididymis.** The larger, superior portion of the epididymis, the **head,** is where the efferent ducts from the testis join the ductus epididymis. The **body** is the narrow midportion of the epididymis, and the **tail** is the smaller, inferior portion. At its distal end, the tail of the epididymis continues as the ductus (vas) deferens (discussed shortly).

The ductus epididymis would measure about 6 m (20 ft) in length if it were straightened out. It is lined with pseudostratified columnar epithelium and encircled by a layer of smooth muscle (Figure 27.7). The free surfaces of the columnar cells contain **stereocilia,** long, branching microvilli that increase surface area for the reabsorption of degenerated sperm. Connective tissue around the muscle layer attaches the loops of the ductus epididymis and carries blood vessels and nerves.

Functionally, the ductus epididymis is the site where sperm mature, that is, they acquire motility and the ability to fertilize an ovum. This occurs over a 10- to 14-day period. The ductus epididymis also stores sperm and helps propel them by peristaltic contraction of its smooth muscle into the ductus (vas) deferens. Sperm may remain in storage in the ductus epididymis for a month or more.

Ductus Deferens

Within the tail of the epididymis, the ductus epididymis becomes less convoluted, and its diameter increases. Beyond this point, the duct is referred to as the **ductus deferens** or **vas deferens** (see Figure 27.3a). The ductus deferens, which is about 45 cm (18 in.) long, ascends along the posterior border of the epididymis, passes through the inguinal canal (see Figure 27.2), and enters the pelvic cavity; there it loops over the ureter and passes over the side and down the posterior surface of the urinary bladder (see Figure 27.1a). The dilated terminal portion of the ductus deferens is known as the **ampulla** (am-POOL-la=little jar) (see Figure 27.9). The mucosa of the ductus deferens consists of pseudostratified columnar epithelium and lamina propria (areolar connective tissue). The muscularis is composed of three layers; the inner and outer layers are longitudinal, and the middle layer is circular (Figure 27.8 on page 845).

Figure 27.7 Histology of the ductus epididymis.

Stereocilia increase the surface area for the reabsorption of degenerated sperm.

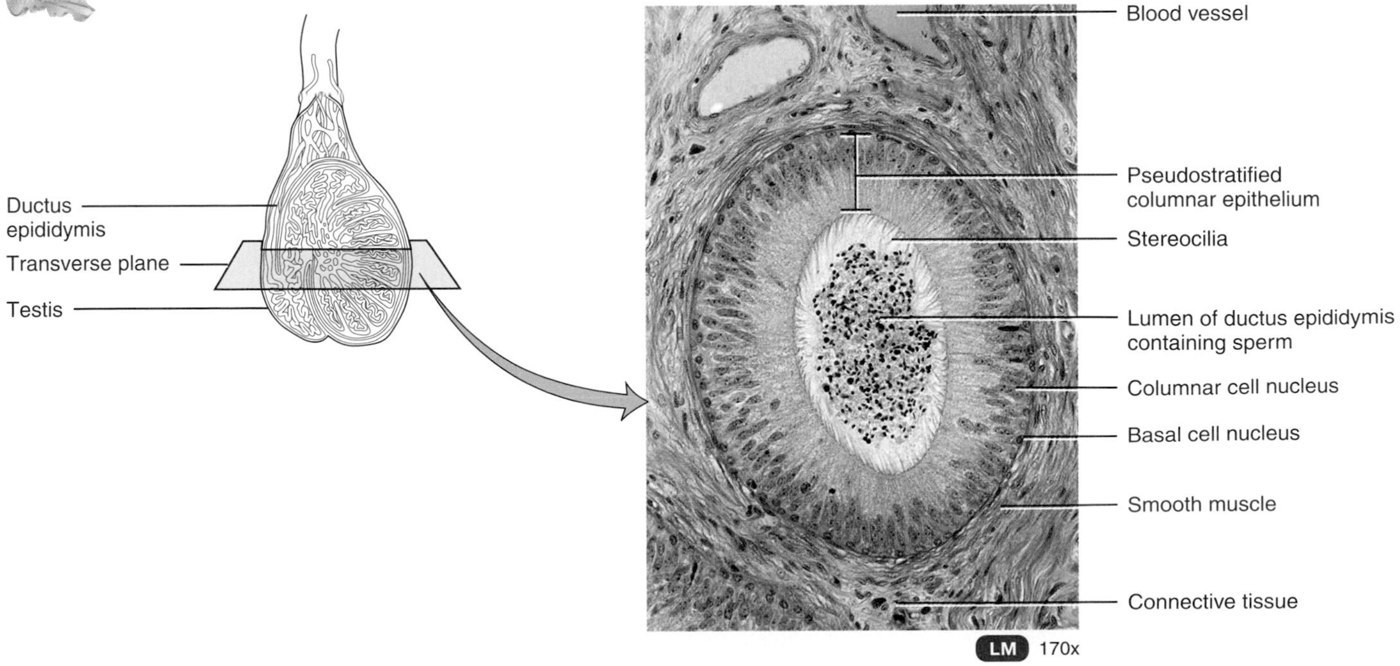

(a) Transverse section of the ductus epididymis

LM 170x

continues

Figure 27.7 (continued)

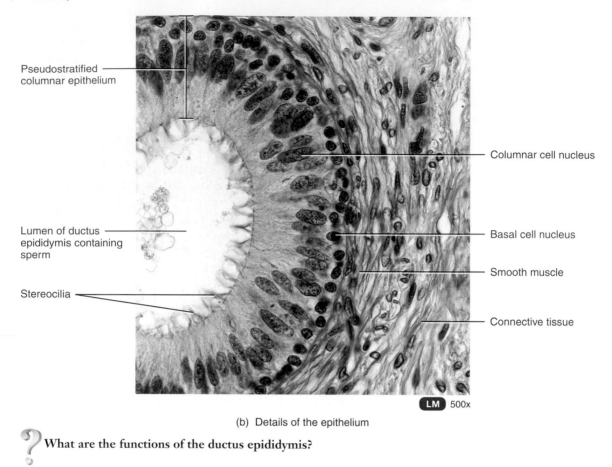

Pseudostratified columnar epithelium

Columnar cell nucleus

Lumen of ductus epididymis containing sperm

Basal cell nucleus

Smooth muscle

Stereocilia

Connective tissue

LM 500x

(b) Details of the epithelium

What are the functions of the ductus epididymis?

Functionally, the ductus deferens stores sperm; they can remain viable here for up to several months. The ductus deferens also conveys sperm from the epididymis toward the urethra by peristaltic contractions of its muscular coat. Sperm that are not ejaculated are ultimately reabsorbed.

VASECTOMY

The principal method for sterilization of males is a **vasectomy** (vas-EK-tō-mē; *-ectomy*=cut out), in which a portion of each ductus deferens is removed. An incision is made in the posterior side of the scrotum, the ducts are located and cut, each is tied in two places, and the portion between the ties is removed. Although sperm production continues in the testes, sperm can no longer reach the exterior. The sperm degenerate and are destroyed by phagocytosis. Because the blood vessels are not cut, testosterone levels in the blood remain normal, so vasectomy has no effect on sexual desire and performance. If done correctly, it is close to 100% effective. The procedure can be reversed, but the chance of regaining fertility is only 30–40%. ■

Ejaculatory Ducts

Each **ejaculatory duct** (e-JAK-ū-la-tō′-rē; *ejacul-*=to expel) is about 2 cm (1 in.) long and is formed by the union of the duct from the seminal vesicle and the ampulla of the ductus (vas) deferens (see Figure 27.9 on page 846). The ejaculatory ducts form just above the base (superior portion) of the prostate and pass inferiorly and anteriorly through the prostate. They terminate in the prostatic urethra, where they eject sperm and seminal vesicle secretions just before the release of semen from the urethra to the exterior.

Urethra

In males, the **urethra** is the shared terminal duct of the reproductive and urinary systems; it serves as a passageway for both semen and urine. About 20 cm (8 in.) long, it passes through the prostate, the urogenital diaphragm, and the penis, and is subdivided into three parts (see Figures 27.1a and 27.9). The **prostatic urethra** is 2–3 cm (1 in.) long and passes through the prostate. As this duct continues inferiorly, it passes through the deep perineal muscles (see Figure 11.15 on page 327), where it is known as the **membranous urethra.** The membranous urethra is about 1 cm (0.5 in.) in length. As this duct passes through the corpus spongiosum of the penis, it is known as the **spongy (penile) urethra,** which is about 15–20 cm (6–8 in.) long. The spongy urethra ends at the **external urethral orifice.** The histology of the male urethra may be reviewed on page 826 of Chapter 26.

Figure 27.8 Histology of the ductus (vas) deferens.

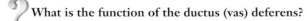

The ductus (vas) deferens enters the pelvic cavity through the inguinal canal.

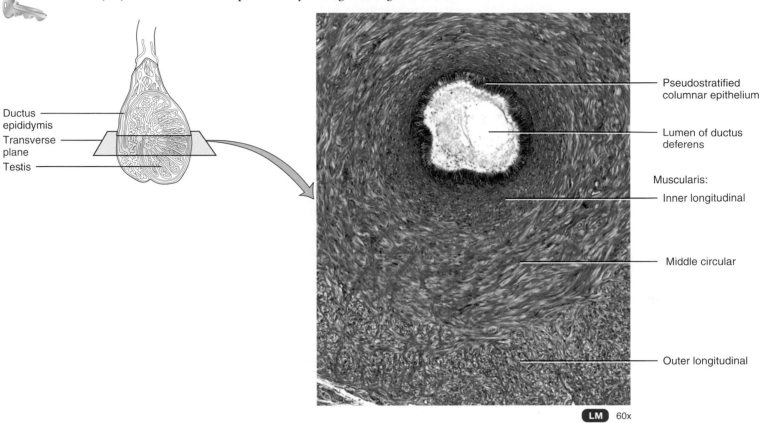

Ductus epididymis

Transverse plane

Testis

Pseudostratified columnar epithelium

Lumen of ductus deferens

Muscularis:

Inner longitudinal

Middle circular

Outer longitudinal

LM 60x

(a) Transverse section of the ductus deferens

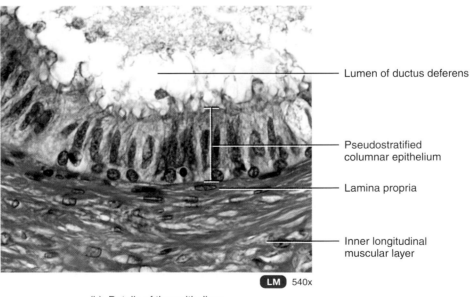

Lumen of ductus deferens

Pseudostratified columnar epithelium

Lamina propria

Inner longitudinal muscular layer

LM 540x

(b) Details of the epithelium

What is the function of the ductus (vas) deferens?

Figure 27.9 Locations of several accessory reproductive glands in males. The prostate, urethra, and penis have been sectioned to show internal details.

 The male urethra has three subdivisions: the prostatic, membranous, and spongy (penile) urethra.

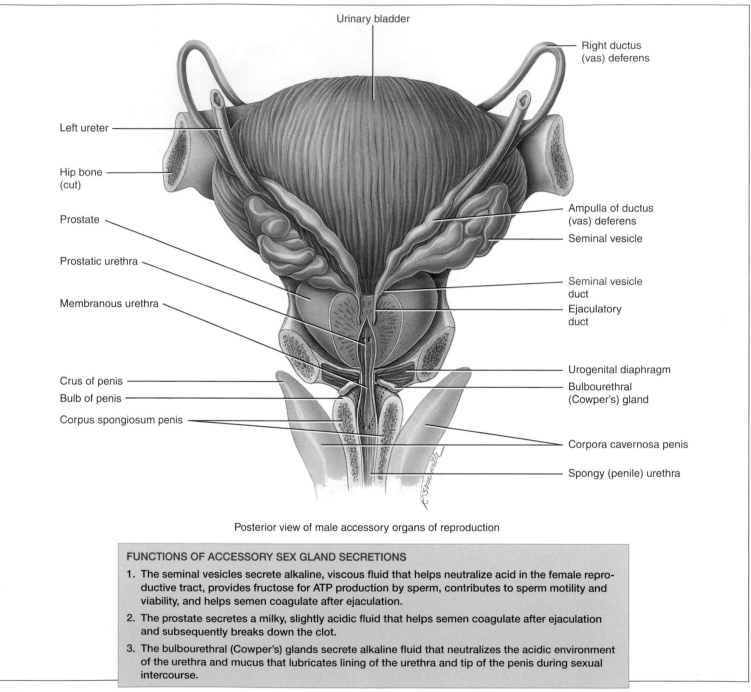

Urinary bladder

Right ductus (vas) deferens

Left ureter

Hip bone (cut)

Prostate

Prostatic urethra

Membranous urethra

Crus of penis

Bulb of penis

Corpus spongiosum penis

Ampulla of ductus (vas) deferens

Seminal vesicle

Seminal vesicle duct

Ejaculatory duct

Urogenital diaphragm

Bulbourethral (Cowper's) gland

Corpora cavernosa penis

Spongy (penile) urethra

Posterior view of male accessory organs of reproduction

FUNCTIONS OF ACCESSORY SEX GLAND SECRETIONS

1. The seminal vesicles secrete alkaline, viscous fluid that helps neutralize acid in the female reproductive tract, provides fructose for ATP production by sperm, contributes to sperm motility and viability, and helps semen coagulate after ejaculation.

2. The prostate secretes a milky, slightly acidic fluid that helps semen coagulate after ejaculation and subsequently breaks down the clot.

3. The bulbourethral (Cowper's) glands secrete alkaline fluid that neutralizes the acidic environment of the urethra and mucus that lubricates lining of the urethra and tip of the penis during sexual intercourse.

What accessory sex gland contributes the majority of the seminal fluid?

Spermatic Cord

The **spermatic cord** is a supporting structure of the male reproductive system that ascends out of the scrotum (see Figure 27.2). It consists of the ductus (vas) deferens as it ascends through the scrotum, the testicular artery, autonomic nerves, veins that drain the testes and carry testosterone into the circulation (the pampiniform plexus), lymphatic vessels, and the cremaster muscle. The term **varicocele** (VAR-i-kō-sēl; *varico-*= varicose; *-kele*=hernia) refers to a swelling in the scrotum due to varicosities in the veins that drain the testes. It usually disappears when the person lies down, and typically does not require treatment. The spermatic cord and ilioinguinal nerve pass through the **inguinal canal** (IN-gwin-al=groin), an oblique passageway in the anterior abdominal wall just superior and parallel to the medial half of the inguinal ligament. The canal, which is about 4–5 cm (about 2 in.) long, originates at the **deep (abdominal) inguinal ring,** a slitlike opening in the aponeurosis of the transversus abdominis muscle; the canal ends at the **superficial (subcutaneous) inguinal ring** (see Figure 27.2), a somewhat triangular opening in the aponeurosis of the external oblique muscle. In females, the round ligament of the uterus and ilioinguinal nerve pass through the inguinal canal.

INGUINAL HERNIAS

Because the inguinal region is a weak area in the abdominal wall, it is often the site of an **inguinal hernia**—a rupture or separation of a portion of the inguinal area of the abdominal wall resulting in the protrusion of a part of the small intestine. In an *indirect inguinal hernia*, a part of the small intestine protrudes through the deep inguinal ring and enters the scrotum. In a *direct inguinal hernia*, a portion of the small intestine pushes into the posterior wall of the inguinal canal, usually causing a localized bulging in the wall of the canal. Inguinal hernias are much more common in males than in females because the larger inguinal canals in males represent larger weak points in the abdominal wall. ■

CHECKPOINT

5. Which ducts transport sperm within the testes?
6. Describe the location, structure, and functions of the ductus epididymis, ductus (vas) deferens, and ejaculatory duct.
7. Give the locations of the three subdivisions of the male urethra.
8. Trace the course of sperm through the system of ducts from the seminiferous tubules through the urethra.
9. List the structures within the spermatic cord.

Accessory Sex Glands in Males

Whereas the ducts of the male reproductive system store and transport sperm cells, the **accessory sex glands** secrete most of the liquid portion of semen. The accessory sex glands are the seminal vesicles, the prostate, and the bulbourethral glands.

Seminal Vesicles

The paired **seminal vesicles** (VES-i-kuls) or **seminal glands** are convoluted pouchlike structures, about 5 cm (2 in.) in length, lying posterior to and at the base of the urinary bladder anterior to the rectum (Figure 27.9). They secrete an alkaline, viscous fluid that contains fructose (a monosaccharide sugar), prostaglandins, and clotting proteins (discussed shortly) unlike those found in blood. The alkaline nature of the fluid helps to neutralize the acidic environment of the male urethra and female reproductive tract that otherwise would inactivate and kill sperm. The fructose is used for ATP production by sperm. Prostaglandins contribute to sperm motility and viability and may also stimulate muscular contractions within the female reproductive tract. The clotting proteins help semen coagulate after ejaculation. Fluid secreted by the seminal vesicles normally constitutes about 60% of the volume of semen.

Prostate

The **prostate** (PROS-tāt) is a single, doughnut-shaped gland about the size of a golf ball. It measures about 4 cm (1.6 in.) from side to side, about 3 cm (1.2 in.) from top to bottom, and about 2 cm (0.8 in.) from front to back. It is inferior to the urinary bladder and surrounds the prostatic urethra (Figure 27.9). The prostate slowly increases in size from birth to puberty, and then it expands rapidly. The size attained by age 30 typically remains stable until about age 45, when further enlargement may occur.

The prostate secretes a milky, slightly acidic fluid (pH about 6.5) that contains several important substances: (1) *Citric acid* in prostatic fluid is used by sperm for ATP production via the Krebs cycle. (2) Several *proteolytic enzymes*, such as *prostate-specific antigen (PSA)*, pepsinogen, lysozyme, amylase, and hyaluronidase, eventually break down the clotting proteins from the seminal vesicles. (3) Acid phosphatase is secreted by the prostate, but its function is unknown. Secretions of the prostate enter the prostatic urethra through many prostatic ducts. Prostatic secretions make up about 25% of the volume of semen and contribute to sperm motility and viability.

PROSTATITIS

Prostatitis is a common group of disorders that can have very disabling effects, including inflammation, swelling, and often pain. Prostatitis is not generally considered a sexually transmitted disease. Symptoms, when present, can include any of the following: fever, chills, urinary frequency, frequent urination at night, difficulty urinating, burning or painful urination, perineal (referring to the perineum, the area between the scrotum and the anus) and low-back pain, joint or muscle pain, tender or swollen prostate, blood in the urine, or painful ejaculation. However, often there are no symptoms. ■

Bulbourethral Glands

The paired **bulbourethral glands** (bul'-bō-ū-RĒ-thral), or **Cowper's glands,** each about the size of a pea, lie inferior to the prostate on either side of the membranous urethra within the urogenital diaphragm; their ducts open into the spongy urethra (Figure 27.9). During sexual arousal, the bulbourethral glands secrete an alkaline substance that protects the passing sperm by neutralizing acids from urine in the urethra. At the same time, they secrete mucus that lubricates the end of the penis and the lining of the urethra, thereby decreasing the number of sperm damaged during ejaculation.

Semen

Semen (=seed) is a mixture of sperm and **seminal fluid,** a liquid that consists of the secretions of the seminiferous tubules, seminal vesicles, prostate, and bulbourethral glands. The volume of semen in a typical ejaculation is 2.5–5 mL, with a sperm count (concentration) of 50–150 million sperm/mL. A male whose sperm count falls below 20 million/mL is likely to be infertile. The very large number is required for successful fertilization because only a tiny fraction ever reach the secondary oocyte.

Despite the slight acidity of prostatic fluid, semen has a slightly alkaline pH of 7.2–7.7 due to the higher pH and large volume of fluid from the seminal vesicles. The prostatic secretion gives semen a milky appearance, and fluids from the seminal vesicles and bulbourethral glands give it a sticky consistency. Seminal fluid provides sperm with a transportation medium and nutrients, and it neutralizes the hostile acidic environment of the male's urethra and the female's vagina. Semen also contains an antibiotic, *seminalplasmin*, that can destroy certain bacteria. Seminalplasmin may help to control the abundance of bacteria both in semen and in the lower portion of the female reproductive tract.

Once ejaculated, liquid semen coagulates within 5 minutes due to the presence of clotting proteins from the seminal vesicles. The functional role of semen coagulation is not known, but the proteins involved are different from those that cause blood coagulation. After about 10–20 minutes, semen reliquefies because prostate-specific antigen (PSA) and other proteolytic enzymes produced by the prostate break down the clot. Abnormal or delayed liquefaction of clotted semen may cause complete or partial immobilization of sperm, thereby inhibiting their movement through the cervix of the uterus. The presence of blood in semen is called **hemospermia** (hē-mō-SPER-mē-a; *hemo-*=blood; *-sperma*=seed). In most cases, it is caused by inflammation of the blood vessels lining the seminal vesicles; it is usually treated with antibiotics.

Penis

The **penis** is the male reproductive organ that contains the urethra and is a passageway for the ejaculation of semen and the excretion of urine (Figure 27.10). It is cylindrical in shape and consists of a root, body, and glans penis. The **root of the penis** is the attached portion (proximal portion). It consists of the **bulb of the penis,** the expanded portion of the base of the corpus spongiosum penis (described shortly), and the **crura of the penis** (singular is *crus*=resembling a leg), the two separated and tapered portions of the corpora cavernosa penis (Figure 27.10a). The bulb of the penis is attached to the inferior surface of the deep muscles of the perineum and is enclosed by the bulbospongiosus muscle. Each crus of the penis is attached to the ischial and inferior pubic rami and is surrounded by the ischiocavernosus muscle (see Figure 27.1b and Figure 11.15 on page 327). Contraction of these skeletal muscles aids ejaculation (discussed shortly).

The **body of the penis** is composed of three cylindrical masses of tissue, each surrounded by fibrous tissue called the **tunica albuginea.** The two dorsolateral masses are called the **corpora cavernosa penis** (*corpora*=main bodies; *cavernosa*=hollow). The smaller midventral mass, the **corpus spongiosum penis,** contains the spongy urethra and functions in keeping the spongy urethra open during ejaculation. Fascia and skin enclose all three masses, which consist of erectile tissue. *Erectile tissue* is composed of numerous blood sinuses (vascular spaces) lined by

endothelial cells and surrounded by smooth muscle and elastic connective tissue.

The distal end of the corpus spongiosum penis is a slightly enlarged, acorn-shaped region called the **glans penis;** its margin is the **corona.** The distal urethra enlarges within the glans penis and forms a terminal slitlike opening, the **external urethral orifice.** Covering the glans in an uncircumcised penis is the loosely fitting **prepuce** (PRĒ-poos), or **foreskin.** The weight of the penis is supported by two ligaments that are continuous with the fascia of the penis. (1) The **fundiform ligament** arises from the inferior part of the linea alba. (2) The **suspensory ligament of the penis** arises from the pubic symphysis.

CIRCUMCISION

Circumcision (=to cut around) is a surgical procedure in which the entire prepuce or a portion of it is removed. It is usually performed just after delivery, 3 to 4 days after birth, or on the eighth day as part of a Jewish religious rite. Although most health-care professionals find no medical justification for circumcision, some feel that it has benefits, such as a lower risk of urinary tract infections, protection against penile cancer, and possibly a lower risk for sexually transmitted diseases. Indeed, studies in several African villages have found lower rates of HIV infection among circumcised men. ■

Upon sexual stimulation (visual, tactile, auditory, olfactory, or imagined), parasympathetic fibers from the lumbar and sacral portions of the spinal cord initiate and maintain an **erection,** the enlargement and stiffening of the penis. The parasympathetic fibers release and cause local production of nitric oxide (NO). The NO causes smooth muscle in the walls of arterioles supplying erectile tissue to relax, allowing the vessels to dilate. This, in turn, causes large amounts of blood to enter the erectile tissue. NO also causes the smooth muscle within the erectile tissue to relax, resulting in widening of the blood sinuses. The combination of increased blood flow and widening of the blood sinuses results in an erection. Expansion of the blood sinuses also compresses the veins that drain the penis; the slowing of blood outflow helps to maintain the erection. The penis returns to its flaccid (relaxed) state when the arterioles constrict and when the smooth muscle within erectile tissue contracts, making the blood sinuses smaller. This relieves pressure on the veins and allows blood to drain from the penis.

The term **priapism** (PRĪ-a-pizm) refers to a persistent and usually painful erection of the corpora cavernosa of the penis that does not involve sexual desire or excitement. The condition may last up to several hours and is accompanied by pain and tenderness. It results from abnormalities of blood vessels and nerves, usually in response to medication used to produce erections in males who otherwise cannot attain them. Other causes include a spinal cord disorder, leukemia, sickle-cell disease, or a pelvic tumor.

Ejaculation (ē-jak-ū-LĀ-shun; *ejectus*=to throw out), the powerful release of semen from the urethra to the exterior, is a sympathetic reflex. As part of the reflex, the smooth muscle sphincter at the base of the urinary bladder closes. Therefore, urine is not expelled during ejaculation, and semen does not nor-

Figure 27.10 **Internal structure of the penis.** The inset in (b) shows details of the skin and fasciae. (See Tortora, *A Photographic Atlas of the Human Body*, 2e, Figure 14.6.)

The penis contains the urethra, a pathway for the ejaculation of semen and the excretion of urine.

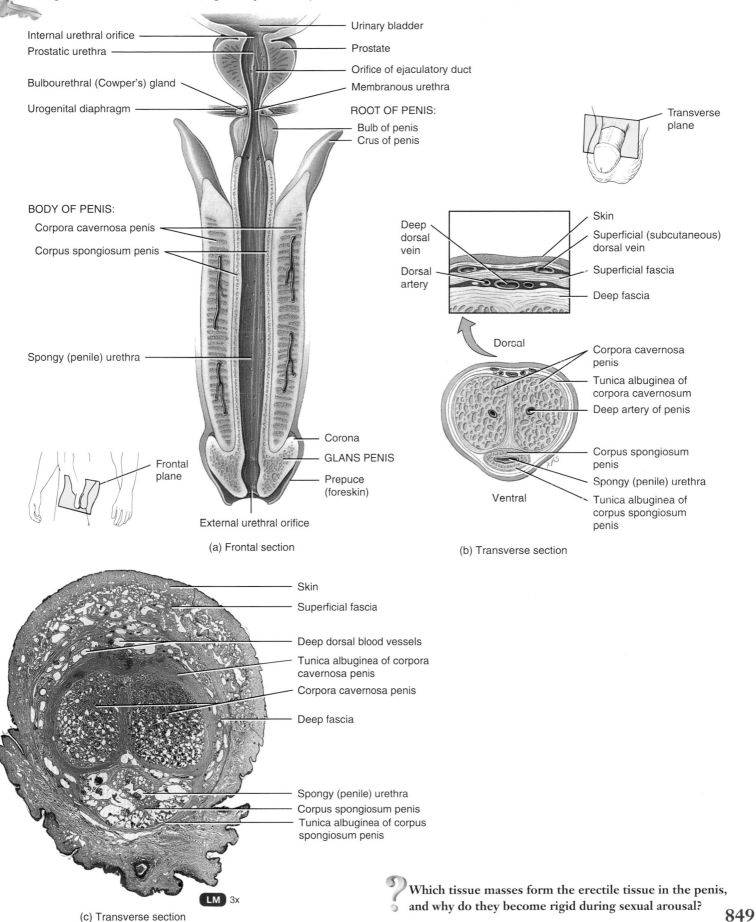

Internal urethral orifice

Prostatic urethra

Bulbourethral (Cowper's) gland

Urogenital diaphragm

ROOT OF PENIS:

BODY OF PENIS:

Corpora cavernosa penis

Corpus spongiosum penis

Spongy (penile) urethra

Frontal plane

Urinary bladder

Prostate

Orifice of ejaculatory duct

Membranous urethra

Bulb of penis
Crus of penis

Corona

GLANS PENIS

Prepuce (foreskin)

External urethral orifice

(a) Frontal section

Transverse plane

Deep dorsal vein

Dorsal artery

Skin

Superficial (subcutaneous) dorsal vein

Superficial fascia

Deep fascia

Dorsal

Corpora cavernosa penis

Tunica albuginea of corpora cavernosum

Deep artery of penis

Corpus spongiosum penis

Spongy (penile) urethra

Tunica albuginea of corpus spongiosum penis

Ventral

(b) Transverse section

Skin

Superficial fascia

Deep dorsal blood vessels

Tunica albuginea of corpora cavernosa penis

Corpora cavernosa penis

Deep fascia

Spongy (penile) urethra

Corpus spongiosum penis

Tunica albuginea of corpus spongiosum penis

LM 3x

(c) Transverse section

Which tissue masses form the erectile tissue in the penis, and why do they become rigid during sexual arousal?

mally enter the urinary bladder. Even before ejaculation occurs, peristaltic contractions in the ampulla of the ductus deferens, seminal vesicles, ejaculatory ducts, and prostate propel semen into the penile portion of the urethra (spongy urethra). Typically, this leads to **emission** (ē-MISH-un), the discharge of a small volume of semen before ejaculation. Emission may also occur during sleep (nocturnal emission). The musculature of the penis (bulbospongiosus, ischiocavernosus, and superficial transverse perineus muscles), which is supplied by the pudendal nerve, also contracts at ejaculation (see Figure 11.15 on page 327).

PREMATURE EJACULATION

A **premature ejaculation** is ejaculation that occurs too early, for example, during foreplay or upon or shortly after penetration. It is usually caused by anxiety, other psychological causes, or an unusually sensitive foreskin or glans penis. For most males, premature ejaculation can be overcome by various techniques (such as squeezing the penis between the glans penis and shaft as ejaculation approaches), behavioral therapy, or medication. ■

The penis has a very rich blood supply from the internal pudendal artery and the femoral artery. The veins drain into corresponding vessels.

The sensory nerves to the penis are branches from the pudendal and ilioinguinal nerves. The corpora have a parasympathetic and a sympathetic supply. As noted previously, parasympathetic stimulation causes the blood vessels to dilate, increasing the flow of blood into the erectile tissue and trapping blood within the penis to maintain the erection. At ejaculation, sympathetic stimulation causes the smooth muscle located in the walls of the ducts and accessory sex glands of the reproductive tract to contract and propel the semen along its course.

Figure 27.11 Organs of reproduction and surrounding structures in females.

The organs of reproduction in females include the ovaries, uterine (fallopian) tubes, uterus, vagina, vulva, and mammary glands.

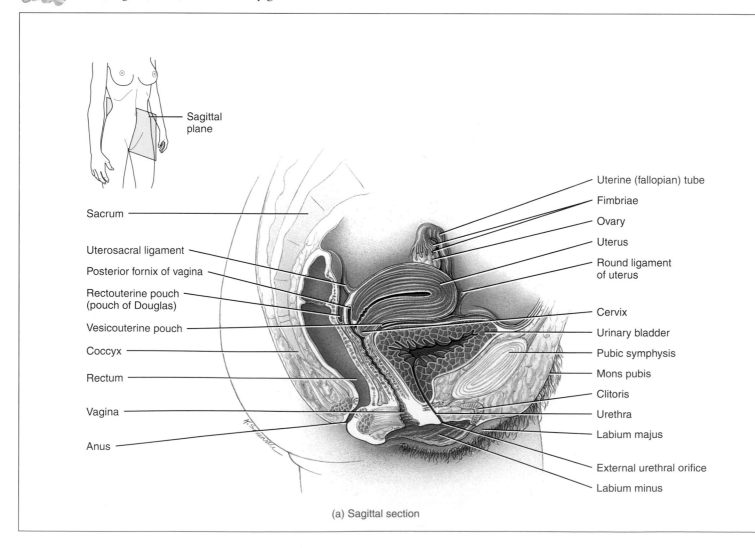

(a) Sagittal section

10. Briefly explain the locations and functions of the seminal vesicles, the prostate, and the bulbourethral (Cowper's) glands.
11. What is semen? What is its function?
12. Explain the physiological processes involved in erection and ejaculation.

FEMALE REPRODUCTIVE SYSTEM

OBJECTIVES

- Describe the location, structure, and functions of the organs of the female reproductive system.
- Discuss the process of oogenesis in the ovaries.

The organs of reproduction in females (Figure 27.11) include the ovaries, which produce secondary oocytes and hormones, such as progesterone and estrogens (the female sex hormones), inhibin, and relaxin; the uterine (fallopian) tubes, or oviducts, which transport secondary oocytes and fertilized ova to the uterus; the uterus, in which embryonic and fetal development occur; the vagina; and external organs that constitute the vulva, or pudendum. The mammary glands also are considered part of the female reproductive system.

FUNCTIONS OF THE FEMALE REPRODUCTIVE SYSTEM

1. The ovaries produce secondary oocytes and hormones, including progesterone and estrogens (female sex hormones), inhibin, and relaxin.
2. The uterine tubes transport a secondary oocyte to the uterus and normally are the sites where fertilization occurs.
3. The uterus is the site of implantation of a fertilized ovum, development of the fetus during pregnancy, and labor.
4. The vagina receives the penis during sexual intercourse and is a passageway for childbirth.
5. The mammary glands synthesize, secrete, and eject milk for nourishment of the newborn.

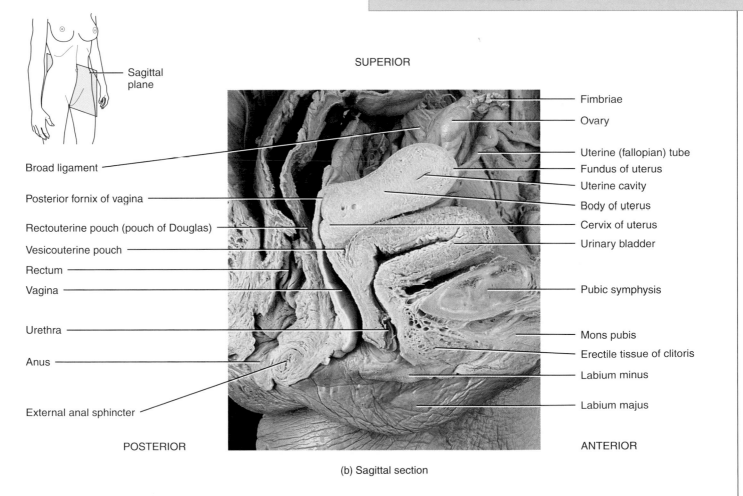

Sagittal plane

SUPERIOR

Broad ligament

Posterior fornix of vagina

Rectouterine pouch (pouch of Douglas)

Vesicouterine pouch

Rectum

Vagina

Urethra

Anus

External anal sphincter

POSTERIOR

Fimbriae

Ovary

Uterine (fallopian) tube

Fundus of uterus

Uterine cavity

Body of uterus

Cervix of uterus

Urinary bladder

Pubic symphysis

Mons pubis

Erectile tissue of clitoris

Labium minus

Labium majus

ANTERIOR

(b) Sagittal section

Which structures in males are homologous to the ovaries, the clitoris, the paraurethral glands, and the greater vestibular glands?

Ovaries

The **ovaries** (=egg receptacles), which are the female gonads, are paired glands that resemble unshelled almonds in size and shape; they are homologous to the testes. (Here *homologous* means that two organs have the same embryonic origin.) The ovaries, one on either side of the uterus, descend to the brim of the superior portion of the pelvic cavity during the third month of development. A series of ligaments holds them in position (Figure 27.12). The **broad ligament** of the uterus (see also Figure 27.11b), which is itself part of the parietal peritoneum, attaches to the ovaries by a double-layered fold of peritoneum called the **mesovarium.** The **ovarian ligament** anchors the ovaries to the uterus, and the **suspensory ligament** attaches them to the pelvic wall. Each ovary contains a **hilus,** the point of entrance and exit for blood vessels and nerves along which the mesovarium is attached.

Histology of the Ovaries

Each ovary consists of the following parts (Figure 27.13):

* The **germinal epithelium** (*germen*=sprout or bud) is a layer of simple epithelium (low cuboidal or squamous) that covers the surface of the ovary. It is continuous with the mesothelium that covers the mesovarium. The term germi-nal epithelium is a misnomer because it does not give rise to ova, although at one time people believed that it did. Now we know that the progenitors of ova arise from the endoderm of the yolk sac and migrate to the ovaries during embryonic development.

* The **tunica albuginea** is a whitish capsule of dense, irregular connective tissue immediately deep to the germinal epithelium.

* The **ovarian cortex** is a region just deep to the tunica albuginea. It consists of ovarian follicles (described shortly) surrounded by dense irregular connective tissue that contains scattered smooth muscle cells.

* The **ovarian medulla** is deep to the ovarian cortex. The border between the cortex and medulla is indistinct, but the medulla consists of more loosely arranged connective tissue and contains blood vessels, lymphatic vessels, and nerves.

* **Ovarian follicles** (*folliculus*=little bag) are located in the ovarian cortex and consist of **oocytes** in various stages of development, plus the cells surrounding them. When the surrounding cells form a single layer, they are called **follicular cells;** later in development, when they form several layers, they are referred to as **granulosa cells.** The surrounding cells nourish the developing oocyte and begin to secrete estrogens as the follicle grows larger.

Figure 27.12 Relative positions of the ovaries, the uterus, and the ligaments that support them. (See Tortora, *A Photographic Atlas of the Human Body*, 2e, Figure 14.9.)

Ligaments holding the ovaries in position include the mesovarium, the ovarian ligament, and the suspensory ligament.

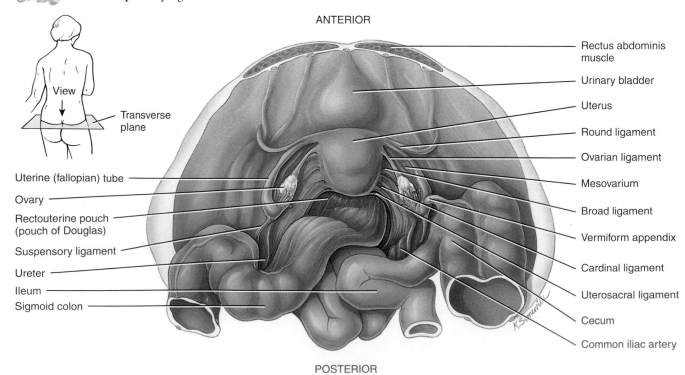

Superior view of transverse section

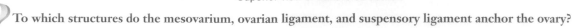

To which structures do the mesovarium, ovarian ligament, and suspensory ligament anchor the ovary?

Figure 27.13 Histology of the ovary. The arrows in (a) indicate the sequence of developmental stages that occur as part of the maturation of an ovum during the ovarian cycle.

🔑 **The ovaries are the female gonads; they produce haploid oocytes.**

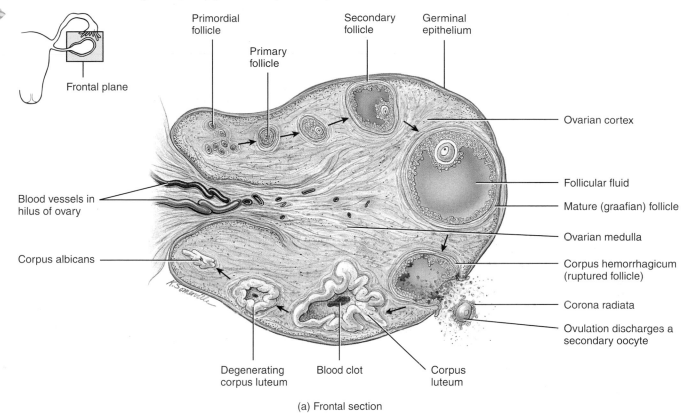

(a) Frontal section

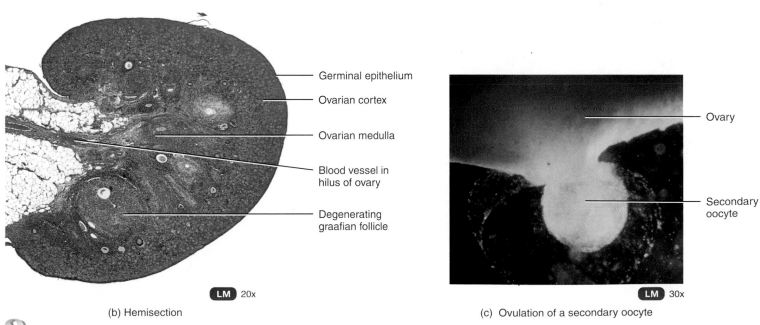

LM 20x

(b) Hemisection

LM 30x

(c) Ovulation of a secondary oocyte

❓ **What structures in the ovary contain endocrine tissue, and what hormones do they secrete?**

- A **mature (graafian) follicle** is a large, fluid-filled follicle that is ready to rupture and expel its secondary oocyte, a process known as **ovulation.**

- A **corpus luteum** (=yellow body) contains the remnants of a mature follicle after ovulation. The corpus luteum produces progesterone, estrogens, relaxin, and inhibin until it degenerates into fibrous scar tissue called the **corpus albicans** (=white body).

The ovarian blood supply is furnished by the ovarian arteries, which anastomose with branches of the uterine arteries. The ovaries are drained by the ovarian veins. On the right side they drain into the inferior vena cava, and on the left side they drain into the renal vein.

Sympathetic and parasympathetic nerve fibers to the ovaries terminate on the blood vessels and enter the ovaries.

Oogenesis and Follicular Development

The formation of gametes in the ovaries is termed **oogenesis** (ō-ō-JEN-e-sis; *oo-*=egg). Whereas spermatogenesis begins in males at puberty, oogenesis begins in females before they are even born. Oogenesis occurs in essentially the same manner as spermatogenesis: meiosis (see Chapter 2) takes place and the resulting germ cells undergo maturation.

During early fetal development, primordial (primitive) germ cells migrate from the endoderm of the yolk sac to the ovaries. There, germ cells differentiate within the ovaries into **oogonia** (ō'-ō-GŌ-nē-a; singular is *oogonium*). Oogonia are diploid (2*n*) stem cells that divide mitotically to produce millions of germ cells. Even before birth, most of these germ cells degenerate in a process known as **atresia** (a-TRĒ-zē-a). A few, however, develop into larger cells called **primary oocytes** (Ō-ō-sītz) that enter prophase of meiosis I during fetal development but do not complete that phase until after puberty. At birth, 200,000 to 2,000,000 oogonia and primary oocytes remain in each ovary. Of these, about 40,000 remain at puberty, and around 400 will mature and ovulate during a woman's reproductive lifetime. The rest undergo atresia.

At the same time that oogenesis is occurring, the follicle cells surrounding the oocyte are also undergoing developmental changes. In the beginning, a single layer of follicular cells surrounds each primary oocyte, and the entire structure is called a **primordial follicle** (Figure 27.14a). Although the stimulating

Figure 27.14 Ovarian follicles. (a) Primordial and primary follicles in the ovarian cortex. (b) A secondary follicle.

🔑 **As an ovarian follicle enlarges, follicular fluid accumulates in a cavity called the antrum.**

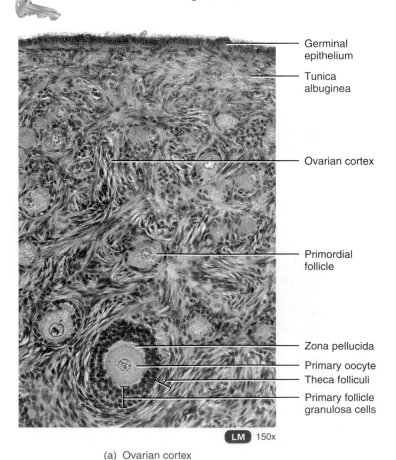

(a) Ovarian cortex

(b) Secondary follicle

? **What happens to most ovarian follicles?**

mechanism is unclear, a few primordial follicles periodically start to grow, even during childhood. They become **primary follicles,** which are surrounded first by one layer of cuboidal follicular cells and then by six to seven layers of cuboidal and low columnar cells called granulosa cells. As a follicle grows, it forms a clear glycoprotein layer, called the **zona pellucida** (pe-LOO-si-da), between the primary oocyte and the granulosa cells. The innermost layer of granulosa cells becomes firmly attached to the zona pellucida and is called the **corona radiata** (*corona*=crown; *radiata*=radiation) (Figure 27.14b).

The outermost granulosa cells rest on a basement membrane. Encircling the basement membrane is a region called the **theca folliculi.** Although the region of granulosa cells is avascular, many blood capillaries are located in the theca folliculi. As a primary follicle continues to grow, the theca differentiates into two layers: (1) the **theca interna,** a highly vascularized internal layer of cuboidal secretory cells, fibroblasts, and bundles of collagen fibers and (2) the **theca externa,** an outer layer of connective tissue cells, collagen fibers, and smooth muscle fibers. The granulosa cells begin to secrete follicular fluid, which builds up in a cavity called the **antrum** in the center of the follicle, now termed a **secondary follicle.**

Each month after puberty, gonadotropins secreted by the anterior pituitary stimulate the resumption of oogenesis (Figure 27.15). Meiosis I resumes in several secondary follicles, although only one will eventually reach the maturity needed for ovulation. The diploid primary oocyte completes meiosis I, producing two haploid cells of unequal size—both with 23 chromosomes (n) consisting of two chromatids each. The smaller cell produced by meiosis I, called the **first polar body,** is essentially a packet of discarded nuclear material. The larger cell, known as the **secondary oocyte,** receives most of the cytoplasm. Once a secondary oocyte is formed, it begins meiosis II but then stops in metaphase. The follicle in which these events are taking place—the **mature (graafian) follicle**—soon ruptures and releases its secondary oocyte. It takes 90 days or longer for a primary follicle to develop into a secondary follicle and then a mature follicle ready to release its secondary oocyte.

At ovulation, the secondary oocyte is expelled into the pelvic cavity together with the first polar body and corona radiata. Normally these cells are swept into the uterine tube. If fertilization does not occur, the cells degenerate. If sperm are present in the uterine tube and one penetrates the secondary oocyte, however, meiosis II resumes. The secondary oocyte splits into two haploid (n) cells, again of unequal size. The larger cell is the **ovum,** or mature egg; the smaller one is the **second polar body.** The nuclei of the sperm cell and the ovum then unite, forming a diploid ($2n$) **zygote.** If the first polar body undergoes another division to produce two polar bodies, then the primary oocyte ultimately gives rise to three haploid (n) polar bodies, which all degenerate, and a single haploid (n) ovum. Thus, one primary oocyte gives rise to a single gamete (an ovum). By contrast, recall that in males one primary spermatocyte produces four gametes (sperm).

Figure 27.15 Oogenesis. Diploid cells ($2n$) have 46 chromosomes; haploid cells (n) have 23 chromosomes.

In an oocyte, meiosis II is completed only if fertilization occurs.

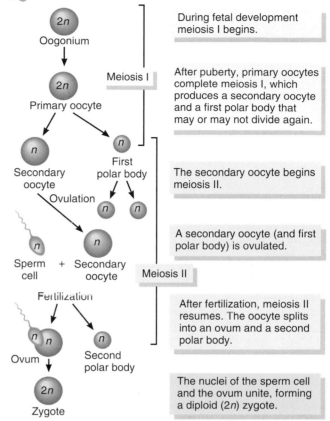

During fetal development meiosis I begins.

After puberty, primary oocytes complete meiosis I, which produces a secondary oocyte and a first polar body that may or may not divide again.

The secondary oocyte begins meiosis II.

A secondary oocyte (and first polar body) is ovulated.

After fertilization, meiosis II resumes. The oocyte splits into an ovum and a second polar body.

The nuclei of the sperm cell and the ovum unite, forming a diploid ($2n$) zygote.

How does the age of a primary oocyte in a female compare with the age of a primary spermatocyte in a male?

OVARIAN CYSTS

An **ovarian cyst** is a fluid-filled sac in or on an ovary. Such cysts are relatively common, are usually noncancerous, and frequently disappear on their own. Cancerous cysts are more likely to occur in women over 40. Ovarian cysts may cause pain, pressure, a dull ache, or fullness in the abdomen; pain during sexual intercourse; delayed, painful, or irregular menstrual periods; abrupt onset of sharp pain in the lower abdomen; and/or vaginal bleeding. Most ovarian cysts require no treatment, but larger ones (more than 5 cm or 2 in.) may be removed surgically. ■

Table 27.1 summarizes the events of oogenesis and follicular development.

13. How are the ovaries held in position in the pelvic cavity?
14. Describe the microscopic structure and functions of an ovary.
15. Describe the principal events of oogenesis.

TABLE 27.1 SUMMARY OF OOGENESIS AND FOLLICULAR DEVELOPMENT

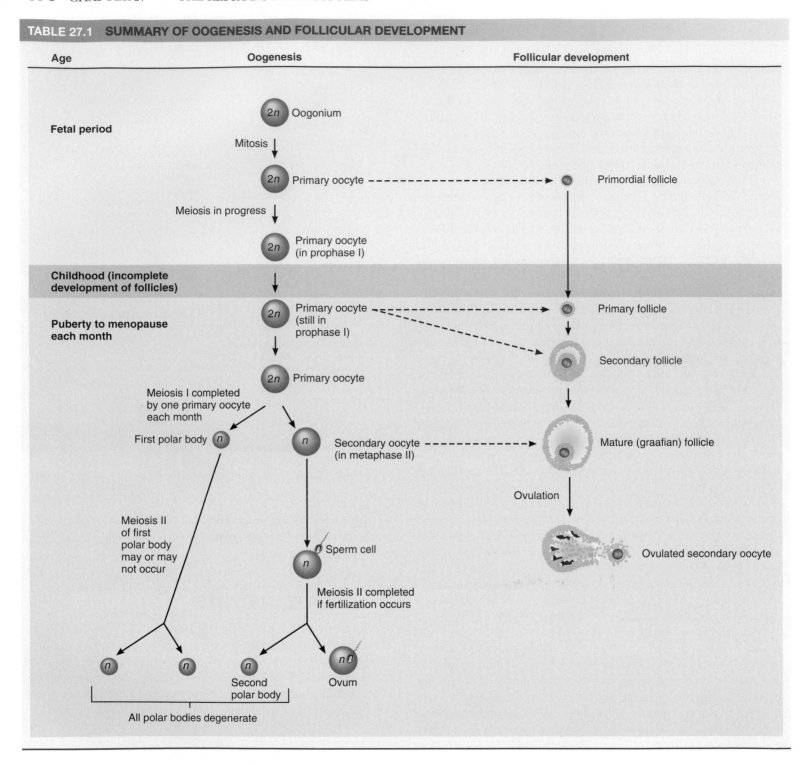

Age	Oogenesis	Follicular development

Uterine Tubes

Females have two **uterine (fallopian) tubes,** or **oviducts,** that extend laterally from the uterus (Figure 27.16). The tubes, which measure about 10 cm (4 in.) long and lie between the folds of the broad ligaments of the uterus, transport secondary oocytes and fertilized ova from the ovaries to the uterus. The funnel-shaped portion of each tube, called the **infundibulum,** is close to the ovary but is open to the pelvic cavity. It ends in a fringe of fingerlike projections called **fimbriae** (FIM-brē-ē= fringe), one of which is attached to the lateral end of the ovary. From the infundibulum, the uterine tube extends medially and eventually inferiorly and attaches to the superior lateral angle of the uterus. The **ampulla** of the uterine tube is the widest, longest portion, making up about the lateral two-thirds of its length. The **isthmus** of the uterine tube is the more medial, short, narrow, thick-walled portion that joins the uterus.

Figure 27.16   Relationship of the uterine (fallopian) tubes to the ovaries, uterus, and associated structures. In the left side of the drawing the uterine tube and uterus have been sectioned to show internal structures.

After ovulation, a secondary oocyte and its corona radiata move from the pelvic cavity into the infundibulum of the uterine tube. The uterus is the site of menstruation, implantation of a fertilized ovum, development of the fetus, and labor.

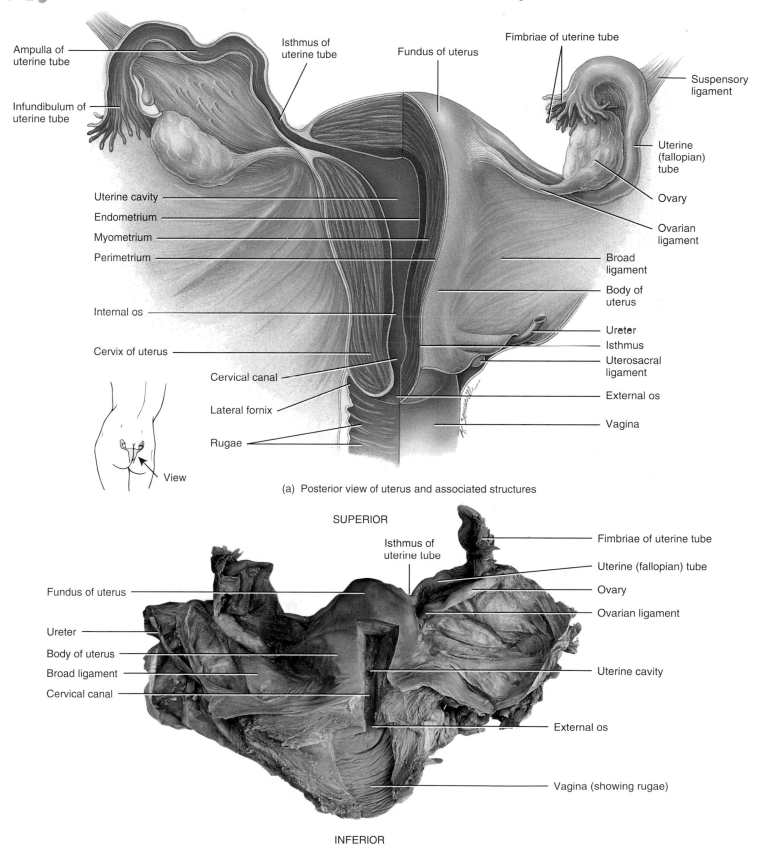

(a) Posterior view of uterus and associated structures

(b) Posterior view of uterus and associated structures

Where does fertilization usually occur?

Histologically, the uterine tubes are composed of three layers: mucosa, muscularis, and serosa (Figure 27.17a). The mucosa consists of epithelium and lamina propria (areolar connective tissue). The epithelium contains ciliated simple columnar cells, which function as a ciliary "conveyor belt" to help move a fertilized ovum (or secondary oocyte) along the tube toward the uterus, and nonciliated (peg) cells which have microvilli and secrete a fluid that provides nutrition for the ovum (Figure 27.17b). The middle layer, the muscularis, is composed of an inner, thick, circular ring of smooth muscle and an outer, thin, region of longitudinal smooth muscle. Peristaltic contractions of the muscularis and the ciliary action of the mucosa help move the oocyte or fertilized ovum toward the uterus. The outer layer of the uterine tubes is a serous membrane, the serosa.

Local currents produced by movements of the fimbriae, which surround the ovary during ovulation, sweep the ovulated secondary oocyte from the pelvic cavity into the uterine tube. A sperm cell usually encounters and fertilizes a secondary oocyte in the ampulla of the uterine tube, although fertilization in the pelvic cavity is not uncommon. Fertilization can occur at any time up to about 24 hours after ovulation. Some hours after fertilization, the nuclear materials of the haploid ovum and sperm unite. The diploid fertilized ovum is now called a **zygote** and begins to undergo cell divisions while moving toward the uterus. It arrives at the uterus 6 to 7 days after ovulation.

The uterine tubes are supplied by branches of the uterine (see Figure 27.19) and ovarian (see Figure 15.6b on page 463) arteries. Venous return is via the uterine veins.

The uterine tubes are supplied with sympathetic and parasympathetic nerve fibers from the hypogastric plexus and the pelvic splanchnic nerves. The fibers are distributed to the muscular coat of the tubes and their blood vessels.

Uterus

The **uterus** (womb) serves as part of the pathway for sperm deposited in the vagina to reach the uterine tubes (see Figure 27.16). It is also the site of implantation of a fertilized ovum, development of the fetus during pregnancy, and labor. During reproductive cycles when implantation does not occur, the uterus is the source of menstrual flow.

Situated between the urinary bladder and the rectum, the uterus is the size and shape of an inverted pear (see Figure 27.11). In females who have never been pregnant, it is about 7.5 cm (3 in.) long, 5 cm (2 in.) wide, and 2.5 cm (1 in.) thick. The uterus is larger in females who have recently been pregnant, and smaller (atrophied) when sex hormone levels are low, as occurs after menopause.

Anatomical subdivisions of the uterus include (see Figure 27.16): (1) a domeshaped portion superior to the uterine tubes

Figure 27.17   Histology of the uterine (fallopian) tube.

Peristaltic contractions of the muscularis and ciliary action of the mucosa of the uterine tube help move the oocyte or fertilized ovum toward the uterus.

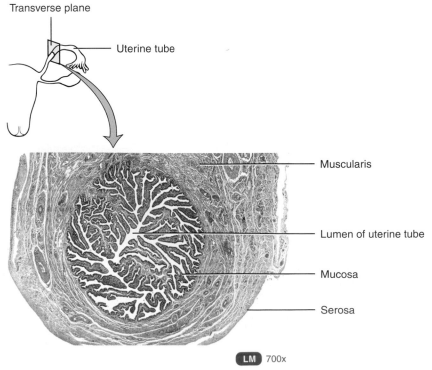

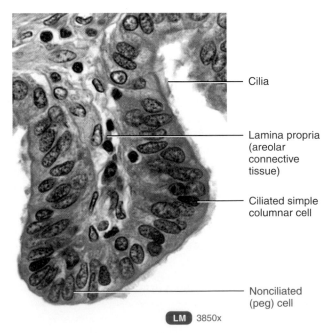

(a) Transverse section through uterine (fallopian) tube

(b) Details of epithelium

 What types of cells line the uterine tubes?

called the **fundus,** (2) a tapering central portion called the **body,** and (3) an inferior narrow portion called the **cervix** that opens into the vagina. Between the body of the uterus and the cervix is the **isthmus** (IS-mus), a constricted region about 1 cm (0.5 in.) long. The interior of the body of the uterus is called the **uterine cavity,** and the interior of the narrow cervix is called the **cervical canal.** The cervical canal opens into the uterine cavity at the **internal os** (*os*=mouthlike opening) and into the vagina at the **external os.**

Normally, the body of the uterus projects anteriorly and superiorly over the urinary bladder in a position called **anteflexion.** The cervix projects inferiorly and posteriorly and enters the anterior wall of the vagina at nearly a right angle (see Figure 27.11). Several ligaments that are either extensions of the parietal peritoneum or fibromuscular cords maintain the position of the uterus (see Figure 27.12). The paired **broad ligaments** are double folds of peritoneum attaching the uterus to either side of the pelvic cavity. The paired **uterosacral ligaments,** also peritoneal extensions, lie on either side of the rectum and connect the uterus to the sacrum. The **cardinal (lateral cervical) ligaments** are located inferior to the bases of the broad ligaments and extend from the pelvic wall to the cervix and vagina. The **round ligaments** are bands of fibrous connective tissue between the layers of the broad ligament; they extend from a point on the uterus just inferior to the uterine tubes to a portion of the labia majora of the external genitalia. Although the ligaments normally maintain the anteflexed position of the uterus, they also afford the uterine body enough movement such that the uterus may become malpositioned. A posterior tilting of the uterus is called **retroflexion** (*retro-*=backward or behind). It is a harmless variation of the normal position of the uterus. There is often no cause for the condition, but it may occur after childbirth or because of an ovarian cyst.

UTERINE PROLAPSE

A condition called **uterine prolapse** (*prolapse*=falling down or downward displacement) may result from weakening of supporting ligaments and pelvic musculature associated with age or disease, traumatic vaginal delivery, chronic straining from coughing or difficult bowel movements, or pelvic tumors. The prolapse may be characterized as *first degree (mild)*, in which the cervix remains within the vagina; *second degree (marked)*, in which the cervix protrudes to the exterior through the vagina; and *third degree (complete)*, in which the entire uterus is outside the vagina. Depending on the degree of prolapse, treatment may involve pelvic exercises, dieting if a patient is overweight, a stool softener to minimize straining during defecation, pessary therapy (placement of a rubber device around the uterine cervix that helps prop up the uterus), or surgery. ■

Histologically, the uterus consists of three layers of tissue: the perimetrium, myometrium, and endometrium (Figure 27.18). The outer layer—the **perimetrium** (*peri-*=around; *-metrium*=uterus) or serosa—is part of the parietal peritoneum; it is composed of simple squamous epithelium and a thin layer of areolar connective tissue. Laterally, it becomes the broad

Figure 27.18 Histology of the uterus.

🔑 The three layers of the uterus from superficial to deep are the perimetrium (serosa), the myometrium, and the endometrium.

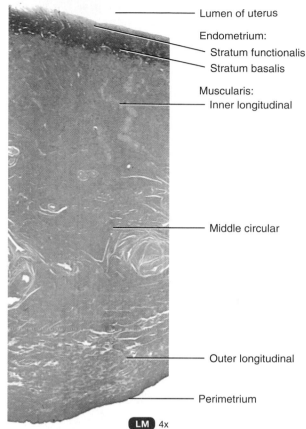

Lumen of uterus
Endometrium:
— Stratum functionalis
— Stratum basalis
Muscularis:
— Inner longitudinal
— Middle circular
— Outer longitudinal
— Perimetrium

LM 4x

(a) Transverse section through the uterus

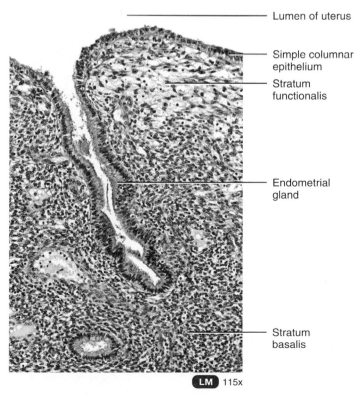

Lumen of uterus
Simple columnar epithelium
Stratum functionalis
Endometrial gland
Stratum basalis

LM 115x

(b) Details of endometrium

❓ **What structural features of the endometrium and myometrium contribute to their functions?**

859

ligament. Anteriorly, it covers the urinary bladder and forms a shallow pouch, the **vesicouterine pouch** (ves'-i-kō-Ū-ter-in; *vesico-*=bladder; see Figure 27.11). Posteriorly, it covers the rectum and forms a deep pouch, the **rectouterine pouch** (rek-tō-Ū-ter-in; *recto-*=rectum) or *pouch of Douglas*—the most inferior point in the pelvic cavity.

The middle layer of the uterus, the **myometrium** (*myo-*= muscle), consists of three layers of smooth muscle fibers that are thickest in the fundus and thinnest in the cervix. The thicker middle layer is circular, whereas the inner and outer layers are longitudinal or oblique. During labor and childbirth, coordinated contractions of the myometrium in response to stimulation by the hormone oxytocin from the posterior pituitary help expel the fetus from the uterus.

The inner layer of the uterus, the **endometrium** (*endo-*= within), is highly vascularized and has three components: (1) An innermost layer of simple columnar epithelium (ciliated and secretory cells) lines the lumen. (2) An underlying endometrial stroma is a very thick region of lamina propria (areolar connective tissue). (3) Endometrial (uterine) glands develop as invaginations of the luminal epithelium and extend almost to the myometrium. The endometrium is divided into two layers. The **stratum functionalis** (=functional layer) lines the uterine cavity and sloughs off during menstruation as a result of declining levels of progesterone from the ovaries. The deeper layer, the **stratum basalis** (=basal layer), is permanent and gives rise to a new stratum functionalis after each menstruation.

Branches of the internal iliac artery called **uterine arteries** (Figure 27.19) supply blood to the uterus. Uterine arteries give off branches called **arcuate arteries** (=shaped like a bow) that are arranged in a circular fashion in the myometrium. These arteries branch into **radial arteries** that penetrate deeply into the myometrium. Just before the branches enter the endometrium, they divide into two kinds of arterioles: **Straight arterioles** supply the stratum basalis with the materials needed to regenerate the stratum functionalis; **spiral arterioles** supply the stratum functionalis and change markedly during the menstrual cycle. Blood leaving the uterus is drained by the **uterine veins** into the internal iliac veins. The extensive blood supply of the uterus is essential to support regrowth of a new stratum functionalis after menstruation, implantation of a fertilized ovum, and development of the placenta.

The secretory cells of the mucosa of the cervix produce a secretion called **cervical mucus,** a mixture of water, glycoproteins, lipids, enzymes, and inorganic salts. During their reproductive years, females secrete 20–60 mL of cervical mucus per day. Cervical mucus is more hospitable to sperm at or near the time of ovulation because it is then less viscous and more alkaline (pH 8.5). At other times, viscous mucus forms a cervical plug that physically impedes sperm penetration. Cervical mucus supplements the energy needs of sperm, and both the cervix and cervical mucus protect sperm from phagocytes and the hostile environment of the vagina and uterus. They may also play a role in **capacitation**—a functional change that sperm undergo in

Figure 27.19   Blood supply of the uterus. The inset shows histological details of the blood vessels of the endometrium.

Straight arterioles supply the materials needed for regeneration of the stratum functionalis after menstruation.

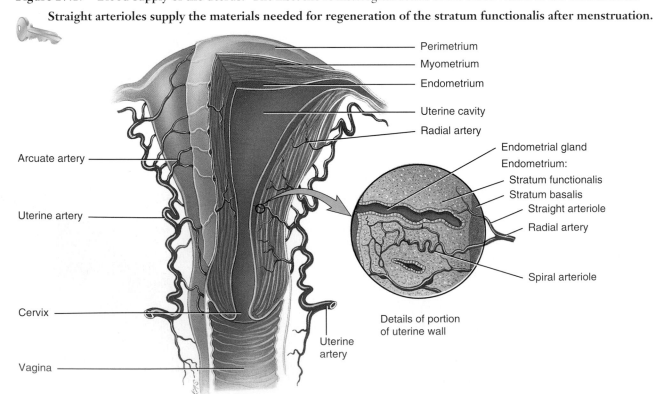

Perimetrium
Myometrium
Endometrium
Uterine cavity
Radial artery

Arcuate artery

Uterine artery

Cervix

Vagina

Uterine artery

Endometrial gland
Endometrium:
Stratum functionalis
Stratum basalis
Straight arteriole
Radial artery

Spiral arteriole

Details of portion of uterine wall

Anterior view with left side of uterus partially sectioned

What is the functional significance of the stratum basalis layer of the endometrium?

the female reproductive tract before they are able to fertilize a secondary oocyte.

HYSTERECTOMY

Hysterectomy (hiss-ter-EK-tō-mē; *hyster-*=uterus), the surgical removal of the uterus, is the most common gynecological operation. It may be indicated in conditions such as endometriosis, pelvic inflammatory disease, recurrent ovarian cysts, excessive uterine bleeding, or cancer of the cervix, uterus, or ovaries. In a *partial (subtotal) hysterectomy*, the body of the uterus is removed but the cervix is left in place. A *complete hysterectomy* is the removal of both the body and cervix of the uterus. A *radical hysterectomy* includes removal of the body and cervix of the uterus, uterine tubes, possibly the ovaries, the superior portion of the vagina, pelvic lymph nodes, and supporting structures, such as ligaments. A hysterectomy can be performed either through an incision in the abdominal wall, or through the vagina. ■

CHECKPOINT

16. Where are the uterine tubes located and what is their function?
17. What are the principal parts of the uterus? Where are they located in relation to one another?
18. Describe the arrangement of ligaments that hold the uterus in its normal position.
19. Describe the histology of the uterus.
20. Why is an abundant blood supply important to the uterus?

Vagina

The **vagina** (=sheath) is a tubular, 10-cm (4-in.) long fibromuscular canal lined with mucous membrane that extends from the exterior of the body to the uterine cervix (see Figures 27.11 and 27.16). It is the receptacle for the penis during sexual intercourse, the outlet for menstrual flow, and the passageway for childbirth. Situated between the urinary bladder and the rectum, the vagina is directed superiorly and posteriorly, where it attaches to the uterus. A recess called the **fornix** (=arch or vault) surrounds the vaginal attachment to the cervix. When properly inserted, a contraceptive diaphragm rests on the fornix, covering the cervix.

The mucosa of the vagina is continuous with that of the uterus. Histologically, it consists of nonkeratinized stratified squamous epithelium and lamina propria (Figure 27.20b)

Figure 27.20   Histology of the vagina.

The muscularis of the vagina can stretch considerably to accommodate the penis during sexual intercourse and an infant during birth.

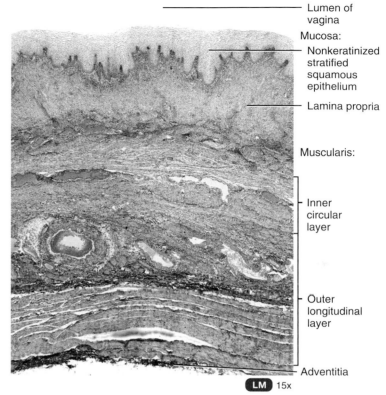

(a) Transverse section through the vaginal wall

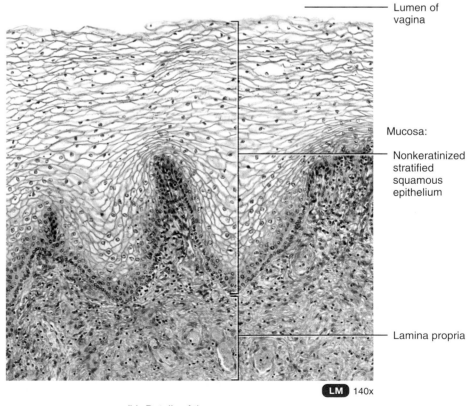

(b) Details of the mucosa

What are the functions of the vagina?

(areolar connective tissue) that lies in a series of transverse folds called **rugae** (ROO-gē). See Figure 27.16. Dendritic cells in the mucosa are antigen-presenting cells. Unfortunately, they also participate in the transmission of viruses—for example, HIV (the virus that causes AIDS)—to a female during intercourse with an infected male. The mucosa of the vagina contains large stores of glycogen, the decomposition of which produces organic acids. The resulting acidic environment retards microbial growth, but it is also harmful to sperm. Alkaline components of semen, mainly from the seminal vesicles, raise the pH of fluid in the vagina and increase viability of the sperm.

The muscularis is composed of an outer longitudinal layer and an inner circular layer of smooth muscle (Figure 27.20a) that can stretch considerably to accommodate the penis during sexual intercourse and an infant during birth.

The adventitia, the superficial layer of the vagina, consists of areolar connective tissue (Figure 27.20a). It anchors the vagina to adjacent organs such as the urethra and urinary bladder anteriorly, and the rectum and anal canal posteriorly.

A thin fold of vascularized mucous membrane, called the **hymen** (=membrane), forms a border around and partially closes the inferior end of the vaginal opening to the exterior, the **vaginal orifice** (see Figure 27.21). Sometimes the hymen completely covers the orifice, a condition called **imperforate hymen** (im-PER-fō-rāt). Surgery may be needed to open the orifice and permit the discharge of menstrual flow.

ABNORMAL VAGINAL DISCHARGE AND BLEEDING

A small amount of vaginal discharge, mostly mucus produced by the vagina and cervix of the uterus, is normal. Also, the amount and appearance of a normal vaginal discharge may vary under different circumstances. During ovulation more mucus is produced, and it is thinner than during other times in the menstrual cycle. Pregnancy, the use of oral contraceptives, and sexual arousal also affect the quantity and appearance of normal vaginal discharge. An **abnormal vaginal discharge** is one that is heavier or thicker than usual; puslike; white and clumpy; grayish, greenish, yellowish, or blood-tinged; foul-smelling; or accompanied by stinging, itching, burning, a rash, or soreness. Abnormal vaginal discharges may be caused by infections, chemical irritation (spermicides, diaphragm, latex, vaginal lubricants, harsh soaps, douches), tumors, or radiation therapy to the pelvis.

Abnormal vaginal bleeding refers to bleeding from the vagina before puberty, after menopause, or during the reproductive years in which menstrual periods are too light or heavy, last too long, occur too frequently, or are irregular. The condition may be caused by inflammation or infection, cancer of the uterus or vagina, sexual abuse, injuries, oral contraceptives, ectopic pregnancy, blood clotting disorders, endometriosis, noncancerous uterine growths, or thyroid disorders. Abnormal uterine bleeding can also occur following pregnancy or an abortion. ■

Vulva

The term **vulva** (VUL-va=to wrap around), or **pudendum** (pū-DEN-dum), refers to the external genitals of the female (Figure 27.21). The following are the components of the vulva:

- Anterior to the vaginal and urethral openings is the **mons pubis** (MONZ PŪ-bis; *mons*=mountain), an elevation of adipose tissue covered by skin and coarse pubic hair that cushions the pubic symphysis.

- From the mons pubis, two longitudinal folds of skin, the **labia majora** (LĀ-bē-a ma-JŌ-ra; *labia*=lips; *majora*= larger), extend inferiorly and posteriorly. The singular term is *labium majus*. The labia majora are covered by pubic hair and contain an abundance of adipose tissue, sebaceous (oil) glands, and apocrine sudoriferous (sweat) glands. They are homologous to the scrotum.

- Medial to the labia majora are two smaller folds of skin called the **labia minora** (mī-NŌ-ra; *minora*=smaller). The singular term is *labium minus*. Unlike the labia majora, the labia minora are devoid of pubic hair and fat and have few sudoriferous glands, but they do contain many sebaceous glands. The labia minora are homologous to the spongy (penile) urethra.

- The **clitoris** (KLI-to-ris) is a small cylindrical mass of erectile tissue and nerves located at the anterior junction of the labia minora. A layer of skin called the **prepuce of the clitoris** is formed at the point where the labia minora unite and covers the body of the clitoris. The exposed portion of the clitoris is the glans. The clitoris is homologous to the glans penis in males. Like the male structure, it is capable of enlargement upon tactile stimulation and has a role in sexual excitement in the female.

- The region between the labia minora is the **vestibule.** Within the vestibule are the hymen (if still present), the vaginal orifice, the external urethral orifice, and the openings of the ducts of several glands. The vestibule is homologous to the membranous urethra of males. The **vaginal orifice,** the opening of the vagina to the exterior, occupies the greater portion of the vestibule and is bordered by the hymen. Anterior to the vaginal orifice and posterior to the clitoris is the **external urethral orifice,** the opening of the urethra to the exterior. On either side of the external urethral orifice are the openings of the ducts of the **paraurethral (Skene's) glands.** These mucus-secreting glands are embedded in the wall of the urethra. The paraurethral glands are homologous to the prostate. On either side of the vaginal orifice itself are the **greater vestibular (Bartholin's) glands** (see Figure 27.22), which open by ducts into a groove between the hymen and labia minora. They produce a small quantity of mucus during sexual arousal and intercourse that adds to cervical mucus and also provides lubrication. The greater vestibular glands are homologous to the bulbourethral glands in males. Several **lesser vestibular glands** also open into the vestibule.

- The **bulb of the vestibule** (see Figure 27.22) consists of two elongated masses of erectile tissue just deep to the labia

Figure 27.21   Components of the vulva (pudendum). (See Tortora, *A Photographic Atlas of the Human Body*, 2e, Figure 14.7.)

The vulva refers to the external genitals of the female.

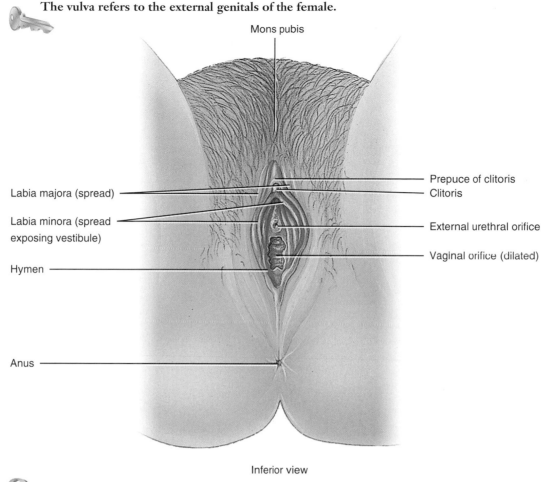

Mons pubis

Prepuce of clitoris

Labia majora (spread)

Clitoris

Labia minora (spread exposing vestibule)

External urethral orifice

Vaginal orifice (dilated)

Hymen

Anus

Inferior view

What surface structures are anterior to the vaginal opening? Lateral to it?

on either side of the vaginal orifice. The bulb of the vestibule becomes engorged with blood during sexual arousal, narrowing the vaginal orifice and placing pressure on the penis during intercourse. The bulb of the vestibule is homologous to the corpus spongiosum and bulb of the penis in males.

Table 27.2 summarizes the homologous structures of the female and male reproductive systems.

Perineum

The **perineum** (per'-i-NĒ-um) is the diamond-shaped area medial to the thighs and buttocks of both males and females (Figure 27.22). It contains the external genitals and anus. The perineum is bounded anteriorly by the pubic symphysis, laterally by the ischial tuberosities, and posteriorly by the coccyx. A transverse line drawn between the ischial tuberosities divides the perineum into an anterior **urogenital triangle** (ū'-rō-JEN-i-tal) that contains the external genitalia and a posterior **anal triangle** that contains the anus.

TABLE 27.2	SUMMARY OF HOMOLOGOUS STRUCTURES OF THE FEMALE AND MALE REPRODUCTIVE SYSTEMS
FEMALE STRUCTURES	**MALE STRUCTURES**
Ovaries	Testes
Ovum	Sperm cell
Labia majora	Scrotum
Labia minora	Spongy (penile) urethra
Vestibule	Membranous urethra
Bulb of vestibule	Corpus spongiosum penis and bulb of penis
Clitoris	Glans penis
Paraurethral glands	Prostate
Greater vestibular glands	Bulbourethral (Cowper's) glands

Figure 27.22 **Perineum of a female.** (Figure 11.14 on page 325 shows the perineum of a male.)

The perineum is a diamond-shaped area medial to the thighs and buttocks that contains the external genitals and anus.

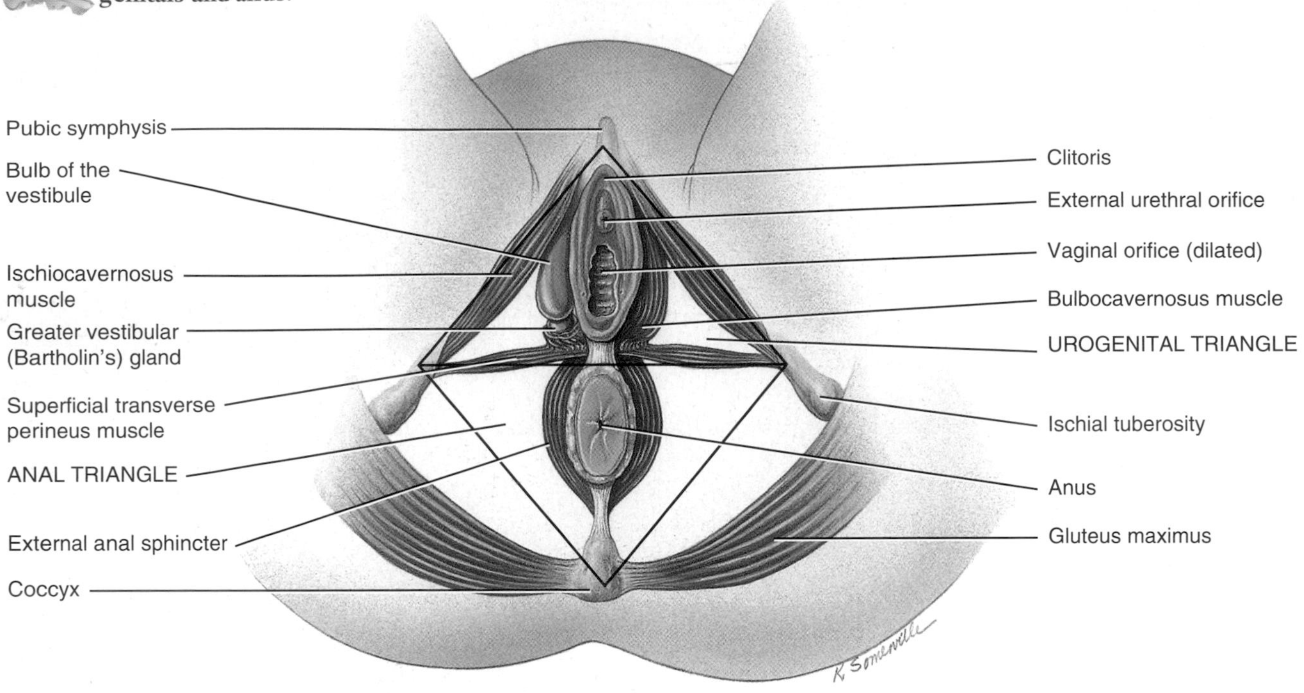

Pubic symphysis

Bulb of the vestibule

Ischiocavernosus muscle

Greater vestibular (Bartholin's) gland

Superficial transverse perineus muscle

ANAL TRIANGLE

External anal sphincter

Coccyx

Clitoris

External urethral orifice

Vaginal orifice (dilated)

Bulbocavernosus muscle

UROGENITAL TRIANGLE

Ischial tuberosity

Anus

Gluteus maximus

Inferior view

Why is the anterior portion of the perineum called the urogenital triangle?

EPISIOTOMY

During childbirth, the emerging fetus stretches the perineal region. To prevent excessive stretching and even tearing of this region, a physician sometimes performs an **episiotomy** (e-piz-ē-OT-ō-mē; *episi-*=vulva or pubic region; *-otomy*=incision), a perineal cut made with surgical scissors. The cut may be made along the midline of the perineum or at an approximately 45-degree angle. The cut enlarges the vaginal opening to make more room for the fetus to pass. In effect, a straight, more easily sutured cut is substituted for a jagged tear. The incision is closed in layers with a suture that is absorbed within a few weeks, so that stitches do not have to be removed. ■

Mammary Glands

The two **mammary glands** (*mamma*=breast) are modified sudoriferous (sweat) glands that produce milk. They lie over the pectoralis major and serratus anterior muscles and are attached to them by a layer of deep fascia composed of dense irregular connective tissue (Figure 27.23).

Each breast has one pigmented projection, the **nipple,** that has a series of closely spaced openings of ducts called **lactiferous ducts,** where milk emerges. The circular pigmented area of skin surrounding the nipple is called the **areola** (a-RĒ-ō-la= small space); it appears rough because it contains modified sebaceous (oil) glands. Strands of connective tissue called the **suspensory ligaments of the breast (Cooper's ligaments)** run between the skin and deep fascia and support the breast. These ligaments become looser with age or with excessive strain, as occurs in long-term jogging or high-impact aerobics. Wearing a supportive bra slows the appearance of "Cooper's droop."

Internally, the mammary gland consists of 15 to 20 lobes, or compartments, separated by a variable amount of adipose tissue. In each lobe are several smaller compartments called **lobules,** composed of grapelike clusters of milk-secreting glands termed **alveoli** (=small cavities) embedded in connective tissue (Figure 27.24 on page 866). Surrounding the alveoli are **myoepithelial cells,** the contraction of which helps propel milk toward the nipples. When milk is being produced, it passes from the alveoli into a series of **secondary tubules** and then into the **mammary ducts.** Near the nipple, the mammary ducts expand to form sinuses called **lactiferous sinuses** (*lact-*=milk), where some milk may be stored before draining into a lactiferous duct. Each lactiferous duct typically carries milk from one of the lobes to the exterior.

The essential functions of the mammary glands are the synthesis, secretion, and ejection of milk; these functions, called **lactation,** are associated with pregnancy and childbirth. Milk production is stimulated largely by the hormone prolactin from the anterior pituitary, with contributions from progesterone and estrogens. The ejection of milk is stimulated by oxytocin, which is released from the posterior pituitary in response to the sucking of an infant on the mother's nipple (suckling).

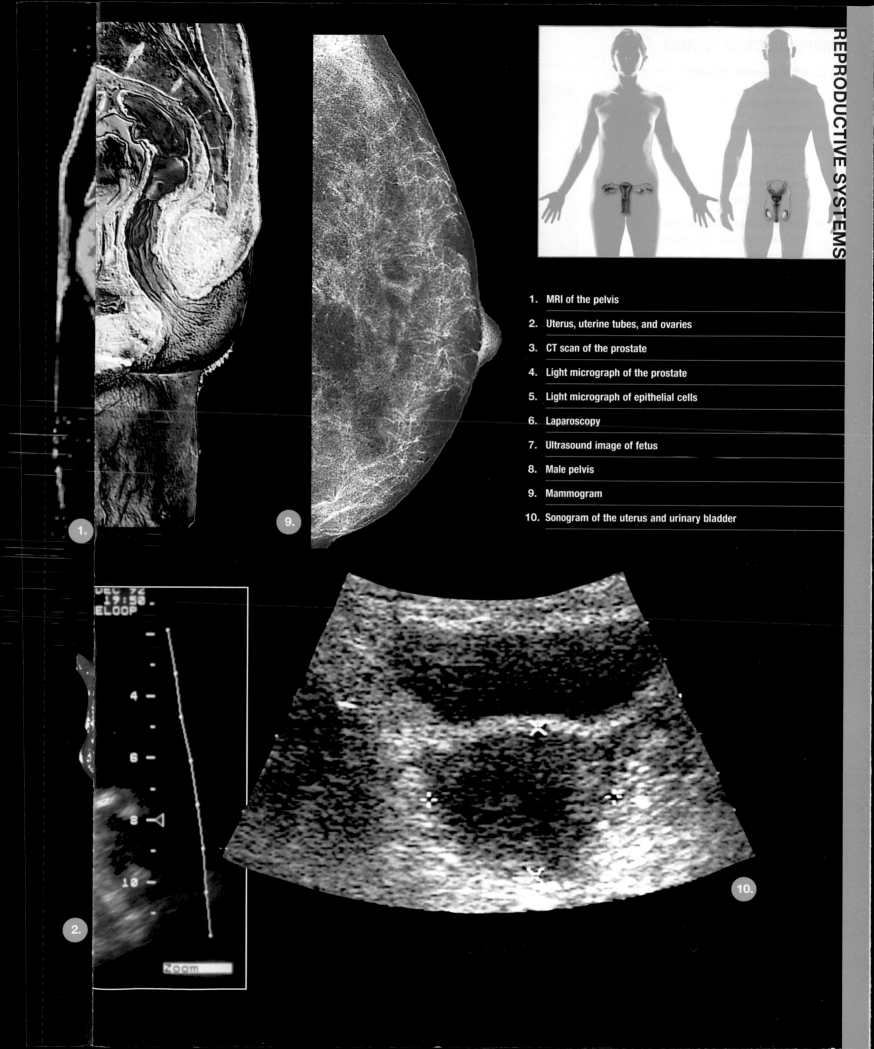

1. MRI of the pelvis

2. Uterus, uterine tubes, and ovaries

3. CT scan of the prostate

4. Light micrograph of the prostate

5. Light micrograph of epithelial cells

6. Laparoscopy

7. Ultrasound image of fetus

8. Male pelvis

9. Mammogram

10. Sonogram of the uterus and urinary bladder

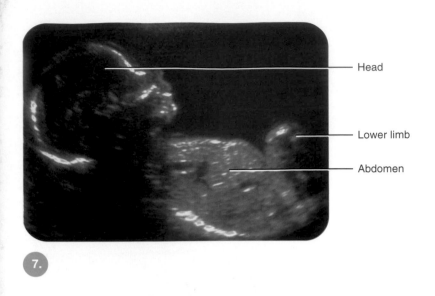

Head

Lower limb

Abdomen

7.

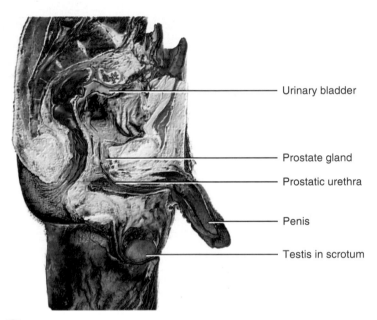

Urinary bladder

Prostate gland

Prostatic urethra

Penis

Testis in scrotum

8.

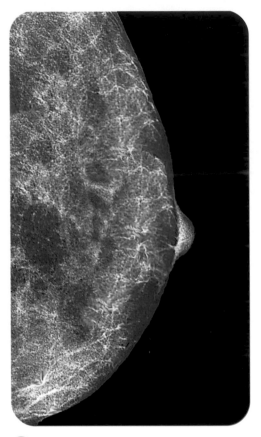

9.

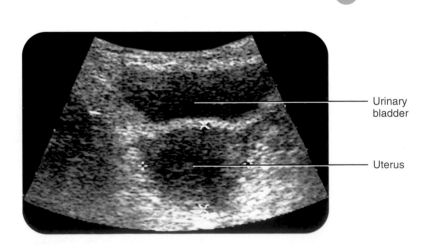

Urinary bladder

Uterus

10.

Figure 27.23 Mammary glands.

The mammary glands function in the synthesis, secretion, and ejection of milk (lactation).

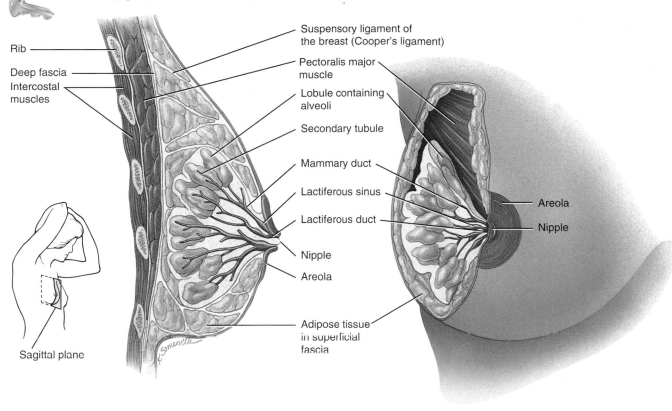

Rib

Deep fascia

Intercostal
muscles

Suspensory ligament of
the breast (Cooper's ligament)

Pectoralis major
muscle

Lobule containing
alveoli

Secondary tubule

Mammary duct

Lactiferous sinus

Lactiferous duct

Nipple

Areola

Adipose tissue
in superficial
fascia

Areola

Nipple

Sagittal plane

(a) Sagittal section

(b) Anterior view, partially sectioned

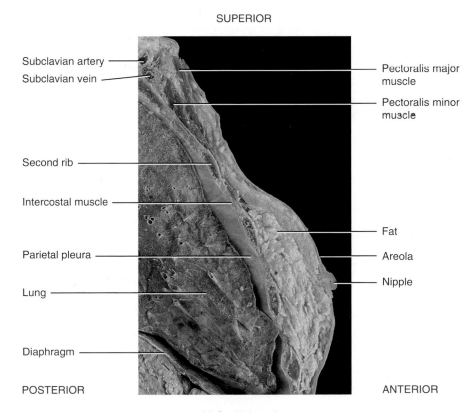

SUPERIOR

Subclavian artery

Subclavian vein

Pectoralis major
muscle

Pectoralis minor
muscle

Second rib

Intercostal muscle

Fat

Parietal pleura

Areola

Nipple

Lung

Diaphragm

POSTERIOR

ANTERIOR

(c) Sagittal section

What hormones regulate the synthesis and ejection of milk?

Figure 27.24   Histology of the mammary glands.

Contraction of myoepithelial cells helps propel milk toward the nipples.

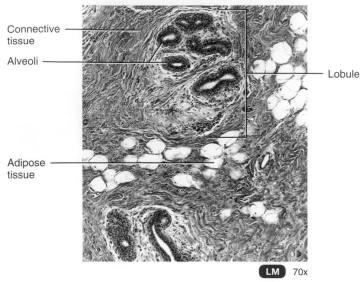

Connective tissue

Alveoli

Lobule

Adipose tissue

LM 70x

(a) Section of a nonlactating (inactive) mammary gland

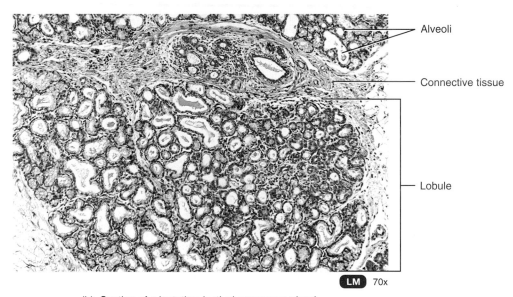

Alveoli

Connective tissue

Lobule

LM 70x

(b) Section of a lactating (active) mammary gland

In which portion of the mammary gland are alveoli located?

FIBROCYSTIC DISEASE OF THE BREASTS

The breasts of females are highly susceptible to cysts and tumors. In **fibrocystic disease,** the most common cause of breast lumps in females, one or more cysts (fluid-filled sacs) and thickening of alveoli (clusters of milk-secreting cells) develop. The condition, which occurs mainly in females between the ages of 30 and 50, is probably due to a relative excess of estrogens or a deficiency of progesterone in the postovulatory (luteal) phase of the reproductive cycle (discussed shortly). Fibrocystic disease usually causes one or both breasts to become lumpy, swollen, and tender a week or so before menstruation begins. ∎

CHECKPOINT

21. What is the function of the vagina?
22. Describe the histology of the vagina.
23. List the parts of the vulva, and explain the functions of each.
24. Describe the structure of the mammary glands and explain how they are supported.
25. What is the route of milk from the alveoli of the mammary gland to the nipple?

FEMALE REPRODUCTIVE CYCLE

• Describe the major events of the female reproductive cycle.

During their reproductive years, nonpregnant females normally exhibit cyclical changes in the ovaries and uterus. Each cycle takes about a month and involves both oogenesis and preparation of the uterus to receive a fertilized ovum. Hormones secreted by the hypothalamus, anterior pituitary, and ovaries control the main events. The **ovarian cycle** is a series of events in the ovaries that occur during and after the maturation of an oocyte. Steroid hormones released by the ovaries control the **uterine (menstrual) cycle,** a concurrent series of changes in the endometrium of the uterus to prepare it for the arrival and development of a fertilized ovum. If fertilization does not occur,

levels of ovarian hormones decrease, which causes the stratum functionalis of the endometrium to slough off. The general term **female reproductive cycle** encompasses the ovarian and uterine cycles, the hormonal changes that regulate them, and the related cyclical changes in the breasts and cervix.

Gonadotropin-releasing hormone (GnRH) secreted by the hypothalamus controls the events of the female reproductive cycle (Figure 27.25). GnRH stimulates the release of **follicle-stimulating hormone (FSH)** and **luteinizing hormone (LH)** from the anterior pituitary. FSH, in turn, initiates follicular growth and the secretion of estrogens by the growing ovarian follicles. LH stimulates the further development of follicles and their full secretion of estrogens. At midcycle, LH triggers ovulation and then promotes formation of the corpus luteum (the reason for the name luteinizing hormone). Stimulated by LH, the corpus luteum produces and secretes estrogens, progesterone, relaxin, and inhibin.

Figure 27.25 The female reproductive cycle. Events in the ovarian and uterine cycles and the release of anterior pituitary gland hormones are correlated with the sequence of the cycle's four phases. In the cycle shown, fertilization and implantation have not occurred.

The length of the female reproductive cycle typically is 24–36 days; the preovulatory phase is more variable in length than the other phases.

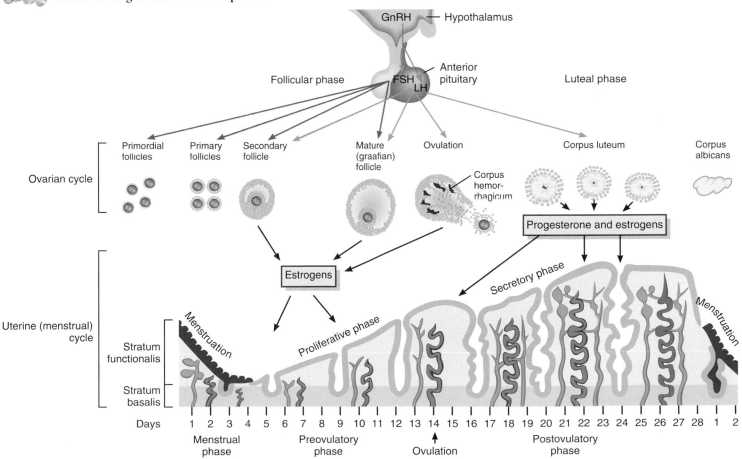

Hormonal regulation of changes in the ovary and uterus

Which hormones are responsible for the proliferative phase of endometrial growth, for ovulation, for growth of the corpus luteum, and for the surge of LH at midcycle?

The duration of the female reproductive cycle typically ranges from 24 to 35 days. For this discussion, we assume a duration of 28 days and divide it into four phases: the menstrual phase, the preovulatory phase, ovulation, and the postovulatory phase (Figure 27.25).

Menstrual Phase

The **menstrual phase** (MEN-stroo-al), also called **menstruation** (men′-stroo-Ā-shun) or **menses** (=month), lasts for roughly the first 5 days of the cycle. (By convention, the first day of menstruation is day one of a new cycle.)

Events in the Ovaries

Beginning about day 25 of the previous cycle and continuing during the menstrual phase, 20 or so small secondary follicles, some in each ovary, begin to enlarge. Follicular fluid, secreted by the granulosa cells and filtered from blood in the capillaries of the theca folliculi, accumulates in the enlarging antrum (space that forms within the follicle) while the oocyte remains near the edge of the follicle (see Figure 27.14b).

Events in the Uterus

Menstrual flow from the uterus consists of 50–150 mL of blood, tissue fluid, mucus, and epithelial cells shed from the endometrium. This discharge occurs because the declining level of ovarian hormones, especially progesterone, stimulates release of prostaglandins that cause the uterine spiral arterioles to constrict. As a result, the cells they supply become oxygen-deprived and start to die. Eventually, the entire stratum functionalis sloughs off. At this time the endometrium is very thin, about 2–5 mm, because only the stratum basalis remains. The menstrual flow passes from the uterine cavity through the cervix and vagina to the exterior.

Preovulatory Phase

The **preovulatory phase** is the time between the end of menstruation and ovulation. The preovulatory phase of the cycle is more variable in length than the other phases and accounts for most of the difference when cycles are shorter or longer than 28 days. It lasts from days 6 to 13 in a 28-day cycle.

Events in the Ovaries

Under the influence of FSH, the group of about 20 secondary follicles continues to grow and begins to secrete estrogens and inhibin. By about day 6, a single follicle in one of the two ovaries has outgrown all the others to become the **dominant follicle**. Estrogens and inhibin secreted by the dominant follicle decrease the secretion of FSH, which causes other, less well-developed follicles to stop growing and undergo atresia.

Normally, the one dominant follicle becomes the **mature (graafian) follicle,** which continues to enlarge until it is more than 20 mm in diameter and ready for ovulation (see Figure 27.13). This follicle forms a blisterlike bulge due to the swelling antrum on the surface of the ovary. During the final maturation process, the dominant follicle continues to increase its estrogen production under the influence of an increasing level of LH. Fraternal (nonidentical) twins or triplets result when two or three secondary follicles become codominant and later are ovulated and fertilized at about the same time. Although estrogens are the main ovarian hormones before ovulation, small amounts of progesterone are produced by the mature follicle a day or two before ovulation.

The menstrual and preovulatory phases together are called the **follicular phase** (fō-LIK-ū-lar) of the ovarian cycle because ovarian follicles are growing and developing.

Events in the Uterus

Estrogens liberated into the blood by growing ovarian follicles stimulate the repair of the endometrium; cells of the stratum basalis undergo mitosis and produce a new stratum functionalis. As the endometrium thickens, the short, straight endometrial glands develop, and the arterioles coil and lengthen as they penetrate the stratum functionalis. The thickness of the endometrium approximately doubles, to about 4–10 mm. The preovulatory phase is also referred to as the **proliferative phase** of the uterine cycle because the endometrium is proliferating.

Ovulation

Ovulation, the rupture of the mature follicle and the release of the secondary oocyte into the pelvic cavity, usually occurs on day 14 in a 28-day cycle. During ovulation, the secondary oocyte remains surrounded by its zona pellucida and corona radiata. Development of a secondary follicle into a fully mature follicle generally takes a total of about 20 days (spanning the last 6 days of the previous cycle and the first 14 days of the current cycle). During this time the primary oocyte completes meiosis I to become a secondary oocyte; the secondary oocyte then begins meiosis II but halts in metaphase until it is fertilized. An over-the-counter home test that detects a rising level of LH can be used to predict ovulation a day in advance for purposes of planning (or avoiding) pregnancy.

Postovulatory Phase

The **postovulatory phase** of the female reproductive cycle is the time between ovulation and onset of the next menses. In duration, it is the most constant part of the female reproductive cycle. It lasts for 14 days in a 28-day cycle, from day 15 to day 28 (see Figure 27.25).

Events in the Ovaries

After ovulation, the mature follicle collapses, and the basement membrane between the granulosa cells and theca interna breaks down. Once a blood clot forms from minor bleeding of the ruptured follicle, the follicle becomes the **corpus hemorrhagicum** (*hemo-*=blood; *rrhagic-*=bursting forth) (see Figure 27.13). Theca interna cells mix with the granulosa cells as they all become transformed into corpus luteum cells under the influence of LH. Stimulated by LH, the corpus luteum secretes progesterone, estrogens, relaxin, and inhibin. The luteal cells also absorb the blood clot. This phase is also called the **luteal phase** of the ovarian cycle.

Later events in an ovary that has ovulated an oocyte depend on whether the oocyte is fertilized. If the oocyte is *not fertilized*, the corpus luteum has a life span of only 2 weeks. At the end of this time period, its secretory activity declines, and it degenerates into a corpus albicans (see Figure 27.13). As the levels of progesterone, estrogens, and inhibin decrease, release of GnRH, FSH, and LH rise due to loss of negative feedback suppression by the ovarian hormones. Follicular growth resumes, and a new ovarian cycle begins.

If the secondary oocyte *is fertilized* and begins to divide, the corpus luteum persists past its normal 2-week life span. It is "rescued" from degeneration by **human chorionic gonadotropin** (kō-rē-ON-ik) (**hCG**). This hormone is produced by the chorion of the embryo beginning about 8 days after fertilization. Like LH, hCG stimulates the secretory activity of the corpus luteum. The presence of hCG in maternal blood or urine is an indicator of pregnancy and is the hormone detected by home pregnancy tests.

Events in the Uterus

Progesterone and estrogens produced by the corpus luteum promote growth and coiling of the endometrial glands, vascularization of the superficial endometrium, and thickening of the endometrium to 12–18 mm (0.48–0.72 in.) in preparation for the arrival of a fertilized ovum. Because of the secretory activity of the endometrial glands, which begin to secrete glycogen, this period is referred to as the **secretory phase** of the uterine cycle. These preparatory changes peak about one week after ovulation, at the time a fertilized ovum might arrive in the uterus. If fertilization does not occur, the level of progesterone declines due to degeneration of the corpus luteum. A decrease in progesterone level causes menstruation.

FEMALE ATHLETE TRIAD: DISORDERED EATING, AMENORRHEA, AND PREMATURE OSTEOPOROSIS

The female reproductive cycle can be disrupted by many factors, including weight loss, low body weight, disordered eating, and vigorous physical activity. The observation that three conditions—disordered eating, amenorrhea, and osteoporosis—often occur together in female athletes led researchers to coin the term **female athlete triad** to encompass these three conditions.

Many athletes experience intense pressure from coaches, parents, peers, and themselves to lose weight to improve performance. Hence, they may develop disordered eating behaviors and engage in other harmful weight-loss practices in a struggle to maintain a very low body weight. **Amenorrhea** (ā-men'-ō-RĒ-a; *a-*=without; *men-*=month; *-rrhea*=a flow) is the absence of menstruation. The most common causes of amenorrhea are pregnancy and menopause. In female athletes, amenorrhea results from reduced secretion of gonadotropin-releasing hor-

mone, which decreases the release of LH and FSH. As a result, ovarian follicles fail to develop, ovulation does not occur, synthesis of estrogens and progesterone wanes, and monthly menstrual bleeding ceases. Most cases of the female athlete triad occur in young women whose percentage of body fat is very low. Low levels of the hormone leptin, secreted by adipose cells, may be a contributing factor.

Because estrogens help bones retain calcium and other minerals, chronically low levels of estrogens are associated with loss of bone mineral density. The female athlete triad causes "old bones in young women." In one study, amenorrheic runners in their twenties had low bone mineral densities, similar to those of postmenopausal women 50 to 70 years old. Whereas short periods of amenorrhea in young athletes may cause no lasting harm, long-term cessation of the reproductive cycle may be accompanied by a loss of bone mass; adolescent athletes may fail to achieve an adequate bone mass before they even reach adulthood. Both situations can lead to premature osteoporosis and irreversible bone damage. ■

CHECKPOINT

26. Describe the function of each of the following hormones in the female reproductive cycle: GnRH, FSH, LH, estrogens, progesterone, and inhibin.
27. Briefly outline the major events of each phase of the female reproductive cycle, and correlate them with the events of the ovarian and uterine cycles.

BIRTH CONTROL METHODS

OBJECTIVE

● Explain the differences among the various types of birth control methods and compare their effectiveness.

Birth control refers to restricting the number of children by various methods designed to control fertility and prevent conception. No single, ideal method of birth control exists. The only method of preventing pregnancy that is 100% reliable is total **abstinence,** the avoidance of sexual intercourse. Several other methods are available; each has advantages and disadvantages. These include surgical sterilization, hormonal methods, intrauterine devices, spermicides, barrier methods, and periodic abstinence. Table 27.3 provides the failure rates for various methods of birth control. We will also discuss induced abortion, the intentional termination of pregnancy.

Surgical Sterilization

Sterilization is a procedure that renders an individual incapable of reproduction. The most common means of sterilization of males is vasectomy, discussed earlier in the chapter on page 844. Sterilization in females most often is achieved by performing a **tubal ligation** (lī-GĀ-shun) in which both uterine tubes are tied closed and then cut. This can be achieved in a few different ways. "Clips" or "clamps" can be placed on the fallopian tubes,

TABLE 27.3 FAILURE RATES OF SEVERAL BIRTH CONTROL METHODS		
	FAILURE RATES*	
METHOD	**PERFECT USE[†]**	**TYPICAL USE**
None	85%	85%
Complete abstinence	0%	0%
Surgical sterilization		
Vasectomy	0.10%	0.15%
Tubal ligation	0.5%	0.5%
Hormonal methods		
Oral contraceptives	0.1%	3%[‡]
Norplant	0.3%	0.3%
Depo-provera	0.05%	0.05%
Lunelle	0.1%	3%[‡]
Skin patch	0.1%	3%[‡]
Intrauterine device		
Copper T 380A	0.6%	0.8%
Spermicides	6%	26%[‡]
Barrier methods		
Male condom	3%	14%[‡]
Vaginal pouch	5%	21%
Diaphragm	6%	20%[‡]
Periodic abstinence		
Rhythm	9%	25%[‡]
Sympto-thermal	2%	20%[‡]

* Defined as percentage of women having an unintended pregnancy during the first year of use.

[†] Failure rate when the method is used correctly and consistently.

[‡] Includes couples who forgot to use the method.

the tubes can be tied and/or cut, and sometimes they are cauterized. In any case the result is that the secondary oocyte cannot pass through the uterine tubes, and sperm cannot reach the oocyte. Tubal ligation reduces the risk of pelvic inflammatory disease in women who are exposed to sexually transmitted infections; it may also reduce the risk of ovarian cancer.

Hormonal Methods

Aside from total abstinence or surgical sterilization, hormonal methods are the most effective means of birth control. Used by 50 million women worldwide, **oral contraceptives** ("the pill") contain various mixtures of synthetic estrogens and progestins (progestins are chemicals with actions similar to those of progesterone). They prevent pregnancy mainly by negative feedback inhibition of anterior pituitary secretion of the gonadotropins FSH and LH. The low levels of FSH and LH usually prevent development of a dominant follicle. As a result, estrogen level does not rise, the midcycle LH surge does not occur, and ovulation is not triggered. Thus, there is no secondary oocyte available for fertilization. Even if ovulation does occur, as it does in some cases, oral contraceptives also alter cervical mucus so that it is more hostile to sperm, and block implantation in the uterus. If taken properly, the pill is close to 100% effective.

Among the noncontraceptive benefits of oral contraceptives are regulation of the length of menstrual cycles and decreased menstrual flow (and therefore decreased risk of anemia). The pill also provides protection against endometrial and ovarian cancers and reduces the risk of endometriosis. However, oral contraceptives may not be advised for women with a history of blood clotting disorders, cerebral blood vessel damage, migraine headaches, hypertension, liver malfunction, or heart disease. Women who take the pill and smoke face far higher odds of having a heart attack or stroke than do nonsmoking pill users. Smokers should quit smoking or use an alternative method of birth control.

Oral contraceptives also may be used for **emergency contraception (EC),** the so-called "morning-after pill." The relatively high levels of estrogens and progestin in EC pills provide negative feedback inhibition of FSH and LH secretion. Loss of the stimulating effects of these gonadotropic hormones causes the ovaries to cease secretion of their own estrogens and progesterone. In turn, declining levels of estrogens and progesterone induce shedding of the uterine lining, thereby blocking implantation. When two pills are taken within 72 hours after unprotected intercourse, and another two pills are taken 12 hours later, the chance of pregnancy is reduced by 75%.

Other hormonal methods of contraception are also available. **Norplant** consists of six slender hormone-containing capsules that are surgically implanted under the skin of the arm using local anesthesia. They slowly and continually release a progestin, which inhibits ovulation and thickens the cervical mucus. The effects last for 5 years, and Norplant is about as reliable as sterilization. Removing the Norplant capsules restores fertility. **Depo-provera,** which is given as an intramuscular injection once every 3 months, contains progestin, a hormone similar to progesterone, that prevents maturation of the ovum and causes changes in the uterine lining that make pregnancy less likely. **Lunelle** is a once-a-month intramuscular injection. It contains estrogens and progestin and acts like an oral contraceptive. It is designed for females who have trouble remembering to take their pills every day. **Birth control skin patches** contain estrogens and progestin and are placed on the skin once a week for three weeks. Each week the patch is removed and a new one is placed on a different area of the skin. During the fourth week no patch is used so that menstruation can occur. The **vaginal ring** is a doughnut-shaped ring that fits in the vagina and releases either a progestin alone or a progestin and an estrogen. It is worn for 3 weeks and removed for 1 week to allow menstruation to occur.

Intrauterine Devices

An **intrauterine device (IUD)** is a small object made of plastic, copper, or stainless steel that is inserted into the cavity of the uterus. IUDs cause changes in the uterine lining that prevent implantation of a fertilized ovum. The IUD most commonly used in the United States today is the Copper T 380A, which is approved for up to 10 years of use and has long-term effectiveness comparable to that of tubal ligation. Some women cannot use IUDs because of expulsion, bleeding, or discomfort.

Spermicides

Various foams, creams, jellies, suppositories, and douches that contain sperm-killing agents, or **spermicides,** make the vagina and cervix unfavorable for sperm survival and are available without prescription. The most widely used spermicide is non-oxynol-9, which kills sperm by disrupting their plasma membranes. It also inactivates the AIDS virus and decreases the incidence of gonorrhea (described on page 877). A spermicide is more effective when used with a barrier method such as a diaphragm or a condom.

Barrier Methods

Barrier methods are designed to prevent sperm from gaining access to the uterine cavity and uterine tubes. In addition to preventing pregnancy, certain barrier methods (condom and vaginal pouch) may also provide some protection against sexually transmitted diseases (STDs) such as AIDS. In contrast, oral contraceptives and IUDs confer no such protection. Among the barrier methods are use of a condom, a vaginal pouch, or a diaphragm.

A **condom** is a nonporous, latex covering placed over the penis that prevents deposition of sperm in the female reproductive tract. A **vaginal pouch,** sometimes called a female condom, is made of two flexible rings connected by a polyurethane sheath. One ring lies inside the sheath and is inserted to fit over the cervix; the other ring remains outside the vagina and covers the female external genitals. A **diaphragm** is a rubber, dome-shaped structure that fits over the cervix and is used in conjunction with a spermicide. It can be inserted up to 6 hours before intercourse. The diaphragm stops most sperm from passing into the cervix and the spermicide kills most sperm that do get by. Although diaphragm use does decrease the risk of some STDs, it does not fully protect against HIV infection.

Periodic Abstinence

A couple can use their knowledge of the physiological changes that occur during the female reproductive cycle to decide either to abstain from intercourse on those days when pregnancy is a likely result, or to plan intercourse on those days if they wish to conceive a child. In females with normal and regular menstrual cycles, these physiological events help to predict the day on which ovulation is likely to occur.

The first physiologically based method, developed in the 1930s, is known as the **rhythm method.** It involves abstaining from sexual activity on the days that ovulation is likely to occur in each reproductive cycle. During this time (3 days before ovulation, the day of ovulation, and 3 days after ovulation) the couple abstains from intercourse. The effectiveness of the rhythm method for birth control is poor in many women due to the irregularity of the female reproductive cycle.

Another system is the **sympto-thermal method,** in which couples are instructed to know and understand certain signs of fertility. The signs of ovulation include increased basal body temperature; the production of abundant clear, stretchy cervical mucus; and pain associated with ovulation (mittelschmerz). If a couple abstains from sexual intercourse when the signs of ovulation are present and for 3 days afterward, the chance of pregnancy is decreased. A big problem with this method is that fertilization is very likely if intercourse occurs one or two days *before* ovulation.

Abortion

Abortion refers to the premature expulsion of the products of conception from the uterus, usually before the twentieth week of pregnancy. An abortion may be spontaneous (naturally occurring; also called a miscarriage) or induced (intentionally performed). Induced abortions may be performed by vacuum aspiration (suction), infusion of a saline solution, or surgical evacuation (scraping).

Certain drugs, most notably RU 486, can induce a so-called nonsurgical abortion. **RU 486 (mifepristone)** is an antiprogestin; it blocks the action of progesterone by binding to and blocking progesterone receptors. Progesterone prepares the uterine endometrium for implantation and then maintains the uterine lining after implantation. If the level of progesterone falls during pregnancy or if the action of the hormone is blocked, menstruation occurs, and the embryo sloughs off along with the uterine lining. Within 12 hours after taking RU 486, the endometrium starts to degenerate, and within 72 hours, it begins to slough off. A form of prostaglandin E (misoprostol), which stimulates uterine contractions, is given after RU 486 to aid in expulsion of the endometrium. RU 486 can be taken up to 5 weeks after conception. One side effect of the drug is uterine bleeding.

CHECKPOINT

28. How do oral contraceptives work to reduce the likelihood of pregnancy?
29. Why do some methods of birth control protect against sexually transmitted diseases, whereas others do not?

DEVELOPMENT OF THE REPRODUCTIVE SYSTEMS

OBJECTIVE

• Describe the development of the male and female reproductive systems.

The *gonads* develop from the **intermediate mesoderm.** During the fifth week of development, they appear as bulges that protrude into the ventral body cavity (Figure 27.26). Adjacent to the gonads are the **mesonephric (Wolffian) ducts,** which eventually develop into structures of the reproductive system in males. A second pair of ducts, the **paramesonephric (Müllerian) ducts,** develop lateral to the mesonephric ducts and eventually form structures of the reproductive system in females. Both sets of ducts empty into the urogenital sinus. An early embryo has the potential to follow either the male or the female pattern of development because it contains both sets of ducts and primitive gonads that can differentiate into either testes or ovaries.

Figure 27.26 Development of the internal reproductive systems.

The gonads develop from intermediate mesoderm.

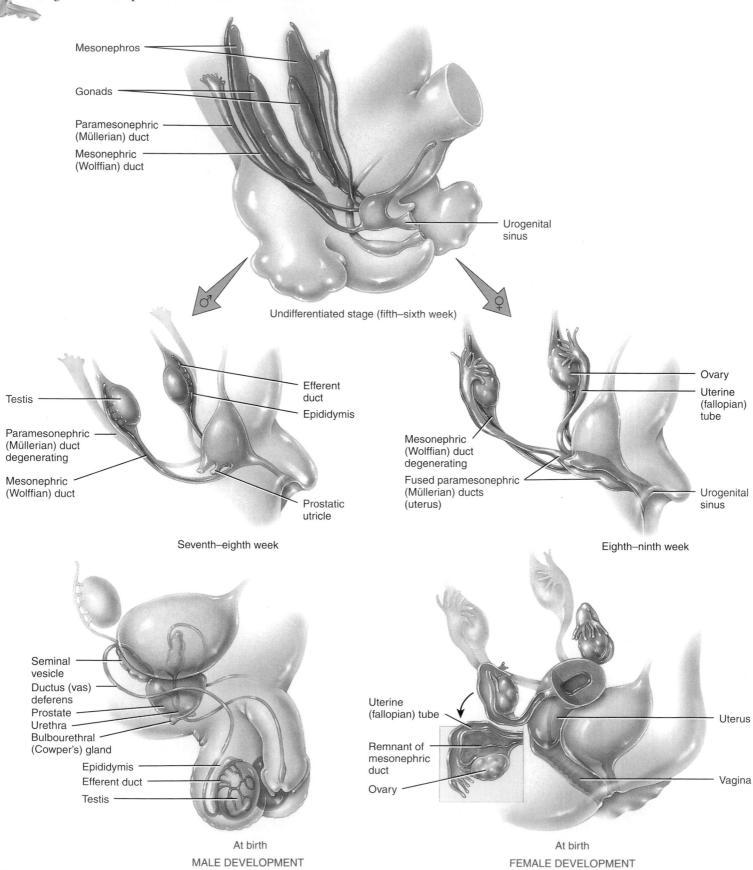

Mesonephros

Gonads

Paramesonephric (Müllerian) duct

Mesonephric (Wolffian) duct

Urogenital sinus

Undifferentiated stage (fifth–sixth week)

♂

♀

Testis

Efferent duct

Epididymis

Paramesonephric (Müllerian) duct degenerating

Mesonephric (Wolffian) duct

Prostatic utricle

Seventh–eighth week

Ovary

Uterine (fallopian) tube

Mesonephric (Wolffian) duct degenerating

Fused paramesonephric (Müllerian) ducts (uterus)

Urogenital sinus

Eighth–ninth week

Seminal vesicle

Ductus (vas) deferens

Prostate

Urethra

Bulbourethral (Cowper's) gland

Epididymis

Efferent duct

Testis

At birth
MALE DEVELOPMENT

Uterine (fallopian) tube

Remnant of mesonephric duct

Ovary

Uterus

Vagina

At birth
FEMALE DEVELOPMENT

Which gene is responsible for the development of the gonads into testes?

Cells of a male embryo have one X chromosome and one Y chromosome. The male pattern of development is initiated by a Y chromosome "master switch" gene named **SRY,** which stands for *Sex-determining Region* of the *Y* chromosome. When the *SRY* gene is expressed during development, its protein product causes the primitive Sertoli cells to begin to differentiate in the gonadal tissues during the seventh week. The developing Sertoli cells secrete a hormone called **Müllerian-inhibiting substance (MIS),** which causes apoptosis of cells within the paramesonephric (Müllerian) ducts. As a result, those cells do not contribute any functional structures to the male reproductive system. Stimulated by human chorionic gonadotropin (hCG), primitive Leydig cells in the gonadal tissue begin to secrete the androgen **testosterone** during the eighth week. Testosterone then stimulates development of the mesonephric duct on each side into the *epididymis, ductus (vas) deferens, ejaculatory duct,* and *seminal vesicle.* The *testes* connect to the mesonephric duct through a series of tubules that eventually become the *seminiferous tubules.* The *prostate* and *bulbourethral glands* are **endodermal** outgrowths of the urethra.

Cells of a female embryo have two X chromosomes and no Y chromosome. Because *SRY* is absent, the gonads develop into *ovaries,* and because MIS is not produced, the paramesonephric ducts flourish. The distal ends of the paramesonephric ducts fuse to form the *uterus* and *vagina* whereas the unfused proximal portions become the *uterine (fallopian) tubes.* The mesonephric ducts degenerate without contributing any functional structures to the female reproductive system because testosterone is absent. The *greater* and *lesser vestibular glands* develop from **endodermal** outgrowths of the vestibule.

The *external genitals* of both male and female embryos (penis and scrotum in males and clitoris, labia, and vaginal orifice in females) also remain undifferentiated until about the eighth week. Before differentiation, all embryos have an elevated midline swelling called the **genital tubercle** (Figure 27.27). The tubercle consists of the **urethral groove** (opening into the urogenital sinus), paired **urethral folds,** and paired **labioscrotal swellings.**

In male embryos, some testosterone is converted to a second androgen called **dihydrotestosterone (DHT).** DHT stimulates development of the urethra, prostate, and external genitals (scrotum and penis). Part of the genital tubercle elongates and develops into a penis. Fusion of the urethral folds forms the *spongy (penile) urethra* and leaves an opening to the exterior only at the distal end of the penis, the *external urethral orifice.* The labioscrotal swellings develop into the *scrotum.* In the absence of DHT, the genital tubercle gives rise to the *clitoris* in female embryos. The urethral folds remain open as the *labia*

Figure 27.27 Development of the external genitals.

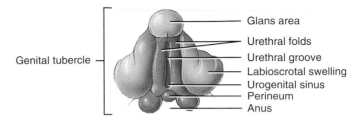

The external genitals of male and female embryos remain undifferentiated until about the eighth week.

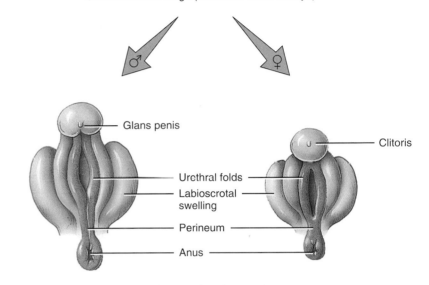

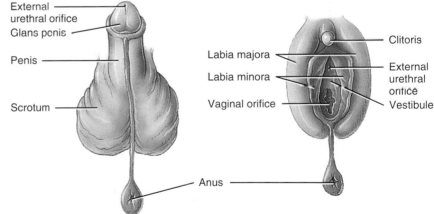

MALE DEVELOPMENT   FEMALE DEVELOPMENT

Which hormone is responsible for the differentiation of the external genitals?

minora, and the labioscrotal swellings become the *labia majora.* The urethral groove becomes the *vestibule.* After birth, androgen levels decline because hCG is no longer present to stimulate secretion of testosterone.

DEFICIENCY OF 5 ALPHA-REDUCTASE

A rare genetic mutation leads to a **deficiency of 5 alpha-reductase,** the enzyme that converts testosterone to dihydrotestosterone. At birth such a baby looks like a female externally because of the absence of dihydrotestosterone during development. At puberty, however, testosterone level rises, masculine characteristics start to appear, and the breasts fail to develop. Internal examination reveals testes and structures that normally develop from the mesonephric duct (epididymis, ductus deferens, seminal vesicle, and ejaculatory duct) rather than ovaries and a uterus. ■

CHECKPOINT

30. Describe the role of hormones in differentiation of the gonads, the mesonephric ducts, the paramesonephric ducts, and the external genitals.

AGING AND THE REPRODUCTIVE SYSTEMS

OBJECTIVE

• Describe the effects of aging on the reproductive systems.

During the first decade of life, the reproductive system is in a juvenile state. At about age 10, hormone-directed changes start to occur in both sexes. **Puberty** (PŪ-ber-tē=a ripe age) is the period when secondary sexual characteristics begin to develop and the potential for sexual reproduction is reached. The onset of puberty is marked by pulses or bursts of LH and FSH secretion, each triggered by a pulse of GnRH. Most pulses occur during sleep. As puberty advances, the hormone pulses occur during the day as well as at night. The pulses increase in frequency during a three- to four-year period until the adult pattern is established. The stimuli that cause the GnRH pulses are still unclear, but a role for the hormone leptin is starting to unfold. Just before puberty, leptin levels rise in proportion to adipose tissue mass. Interestingly, leptin receptors are present in both the hypothalamus and anterior pituitary. Mice that lack a functional leptin gene from birth are sterile and remain in a prepubertal state. Giving leptin to such mice elicits secretion of gonadotropins, and they become fertile. Leptin may signal the hypothalamus that long-term energy stores (triglycerides in adipose tissue) are adequate for reproductive functions to begin.

In females, the reproductive cycle normally occurs once each month from **menarche** (me-NAR-kē), the first menses, to **menopause,** the permanent cessation of menses. Thus, the female reproductive system has a time-limited span of fertility between menarche and menopause. For the first 1 to 2 years after menarche, ovulation only occurs in about 10% of the cycles and the luteal phase is short. Gradually, the percentage of ovulatory cycles increases, and the luteal phase reaches its normal duration of 14 days. With age, fertility declines. Between the ages of 40 and 50 the pool of remaining ovarian follicles becomes exhausted. As a result, the ovaries become less responsive to hormonal stimulation. The production of estrogens declines, despite copious secretion of FSH and LH by the anterior pituitary. Many women experience hot flashes and heavy sweating, which coincide with bursts of GnRH release. Other symptoms of menopause are headache, hair loss, muscular pains, vaginal dryness, insomnia, depression, weight gain, and mood swings. Some atrophy of the ovaries, uterine tubes, uterus, vagina, external genitalia, and breasts occurs in postmenopausal women. Due to loss of estrogens, most women experience a decline in bone mineral density after menopause. Sexual desire (libido) does not show a parallel decline; it may be maintained by adrenal sex steroids. The risk of having uterine cancer peaks at about 65 years of age, but cervical cancer is more common in younger women.

In males, declining reproductive function is much more subtle than in females. Healthy men often retain reproductive capacity into their eighties or nineties. At about age 55 a decline in testosterone synthesis leads to reduced muscle strength, fewer viable sperm, and decreased sexual desire. Although sperm production decreases 50–70% between ages 60 and 80, abundant sperm may still be present even in old age.

Enlargement of the prostate to two to four times its normal size occurs in approximately one-third of all males over age 60. This condition, called **benign prostatic hyperplasia (BPH),** decreases the size of the prostatic urethra and is characterized by frequent urination, nocturia (bed-wetting), hesitancy in urination, decreased force of urinary stream, postvoiding dribbling, and a sensation of incomplete emptying.

CHECKPOINT

31. What changes occur in males and females at puberty?
32. What do the terms menarche and menopause mean?

APPLICATIONS TO HEALTH

REPRODUCTIVE SYSTEM DISORDERS IN MALES

Testicular Cancer

Testicular cancer is the most common cancer in males between the ages of 20 and 35. More than 95% of testicular cancers arise from spermatogenic cells within the seminiferous tubules. An early sign of testicular cancer is a mass in the testis, often associated with a sensation of testicular heaviness or a dull ache in the lower abdomen; pain usually does not occur. To increase the chance for early detection of a testicular cancer, all males should perform regular self-examinations of the testes. The examination should be done starting in the teen years and once each month thereafter. After a warm bath or shower (when the scrotal skin is loose and relaxed) each testicle is examined as follows. The testicle is grasped and gently rolled between the index finger and thumb, feeling for lumps, swelling, hardness, or other changes. If a lump or other change is detected, a physician should be consulted as soon as possible.

Prostate Disorders

Because the prostate surrounds part of the urethra, any infection, enlargement, or tumor in it can obstruct the flow of urine. Acute and chronic infections of the prostate are common in postpubescent males, often in association with inflammation of the urethra. In acute prostatitis, the prostate becomes swollen and tender. **Chronic prostatitis** is one of the most common chronic infections in men of the middle and later years. On examination, the prostate feels enlarged, soft, and very tender, and its surface outline is irregular.

Prostate cancer is the leading cause of death from cancer in men in the United States, having surpassed lung cancer in 1991. Each year it is diagnosed in almost 200,000 U.S. men and causes nearly 40,000 deaths. The amount of PSA (prostate-specific antigen), which is produced only by prostate epithelial cells, increases with enlargement of the prostate and may indicate infection, benign enlargement, or prostate cancer. A blood test can measure the level of PSA in the blood. Males over the age of 40 should have an annual examination of the prostate gland. In a **digital rectal exam,** a physician palpates the gland through the rectum with the fingers (digits). Many physicians also recommend an annual PSA test for males over age 50. Treatment for prostate cancer may involve surgery, cryotherapy, radiation, hormonal therapy, and chemotherapy. Because many prostate cancers grow very slowly, some urologists recommend "watchful waiting" before treating small tumors in men over age 70.

Erectile Dysfunction

Erectile dysfunction (ED), previously termed *impotence*, is the consistent inability of an adult male to ejaculate or to attain or hold an erection long enough for sexual intercourse. Many cases of impotence are caused by insufficient release of nitric oxide, which relaxes the smooth muscle of the penile arterioles and erectile tissue. The drug *Viagra* (sildenafil) enhances smooth muscle relaxation by nitric oxide (NO) in the penis. Other causes of erectile dysfunction include diabetes mellitus, physical abnormalities of the penis, systemic disorders such as syphilis, vascular disturbances (arterial or venous obstructions), neurological disorders, surgery, testosterone deficiency, and drugs (alcohol, antidepressants, antihistamines, antihypertensives, narcotics, nicotine, and tranquilizers). Psychological factors such as anxiety or depression, fear of causing pregnancy, fear of sexually transmitted diseases, religious inhibitions, and emotional immaturity may also cause ED.

REPRODUCTIVE SYSTEM DISORDERS IN FEMALES

Premenstrual Syndrome and Premenstrual Dysphoric Disorder

Premenstrual syndrome (PMS) is a cyclical disorder of severe physical and emotional distress. It appears during the postovulatory (luteal) phase of the female reproductive cycle and dramatically disappears when menstruation begins. The signs and symptoms are highly variable from one woman to another. They may include edema, weight gain, breast swelling and tenderness, abdominal distension, backache, joint pain, constipation, skin eruptions, fatigue and lethargy, greater need for sleep, depression or anxiety, irritability, mood swings, headache, poor coordination and clumsiness, and cravings for sweet or salty foods. The cause of PMS is unknown. For some women, getting regular exercise; avoiding caffeine, salt, and alcohol; and eating a diet that is high in complex carbohydrates and lean proteins can bring considerable relief.

Premenstrual dysphoric disorder (PMDD) is a more severe syndrome in which PMS-like signs and symptoms do not resolve after the onset of menstruation. Clinical research studies have found that suppression of the reproductive cycle by a drug that interferes with GnRH (leuprolide) decreases symptoms significantly. Because symptoms reappear when estradiol or progesterone is given together with leuprolide, researchers propose that PMDD is caused by abnormal responses to normal levels of these ovarian hormones. *SSRIs* (selective serotonin receptor inhibitors) have shown promise in treating both PMS and PMDD.

Endometriosis

Endometriosis (en-dō-mē-trē-Ō-sis; *endo-*=within; *metri-*= uterus; *osis*=condition) is characterized by the growth of endometrial tissue outside the uterus. The tissue enters the pelvic cavity via the open uterine tubes and may be found in any of several sites—on the ovaries, the rectouterine pouch, the outer surface of the uterus, the sigmoid colon, pelvic and abdominal lymph nodes, the cervix, the abdominal wall, the kidneys, and the urinary bladder. Endometrial tissue responds to hormonal fluctuations, whether it is inside or outside the uterus. With each reproductive cycle, the tissue proliferates and then breaks down and bleeds. When this occurs outside the uterus, it can cause inflammation, pain, scarring, and infertility. Symptoms include premenstrual pain or unusually severe menstrual pain.

Breast Cancer

One in eight women in the United States faces the prospect of **breast cancer.** After lung cancer, it is the second-leading cause of death from cancer in U.S. women. Breast cancer can occur in males but is rare. In females, breast cancer is seldom seen before age 30; its incidence rises rapidly after menopause. An estimated 5% of the 180,000 cases diagnosed each year in the United States, particularly those that arise in younger women, stem from inherited genetic mutations (changes in the DNA). Researchers have now identified two genes that increase susceptibility to breast cancer: *BRCA1 (breast cancer 1)* and *BRCA2.* Mutation of *BRCA1* also confers a high risk for ovarian cancer. In addition, mutations of the *p53* gene increase the risk of breast cancer in both males and females, and mutations of the androgen receptor gene are associated with the occurrence of breast cancer in some males. Because breast cancer generally is not painful until it becomes quite advanced, any lump, no matter how small, should be reported to a physician at once. Early detection—by breast self-examination and mammograms—is the best way to increase the chance of survival.

The most effective technique for detecting tumors less than 1 cm (0.4 in.) in diameter is **mammography** (mam-OG-ra-fē; *-graphy*=to record), a type of radiography using very sensitive x-ray film. The image of the breast, called a **mammogram** (see Table 1.4 on page 18), is best obtained by compressing the breasts, one at a time, using flat plates. A supplementary procedure for evaluating breast abnormalities is **ultrasound.** Although ultrasound cannot detect tumors smaller than 1 cm in diameter, it can be used to determine whether a lump is a benign, fluid-filled cyst or a solid (and therefore possibly malignant) tumor.

Among the factors that increase the risk of developing breast cancer are (1) a family history of breast cancer, especially in a mother or sister; (2) nulliparity (never having borne a child) or having a first child after age 35; (3) previous cancer in one breast; (4) exposure to ionizing radiation, such as x-rays; (5) excessive alcohol intake; and (6) cigarette smoking.

The American Cancer Society recommends the following steps to help diagnose breast cancer as early as possible:

- All women over 20 should develop the habit of monthly breast self-examination.

- A physician should examine the breasts every 3 years when a woman is between the ages of 20 and 40, and every year after age 40.

- A mammogram should be taken in women between the ages of 35 and 39, to be used later for comparison (baseline mammogram).

- Women with no symptoms should have a mammogram every year or two between ages 40 and 49, and every year after age 50.

- Women of any age with a history of breast cancer, a strong family history of the disease, or other risk factors should consult a physician to determine a schedule for mammography.

Treatment for breast cancer may involve hormone therapy, chemotherapy, radiation therapy, **lumpectomy** (removal of the tumor and the immediate surrounding tissue), a modified or radical mastectomy, or a combination of these approaches. A **radical mastectomy** (*mast-*=breast) involves removal of the affected breast along with the underlying pectoral muscles and the axillary lymph nodes. (Lymph nodes are removed because metastasis of cancerous cells usually occurs through lymphatic or blood vessels). Radiation treatment and chemotherapy may follow the surgery to ensure the destruction of any stray cancer cells. Several types of chemotherapeutic drugs are used to decrease the risk of relapse or disease progression. *Nolvadex (tamoxifen)* is an antagonist to estrogens that binds to and blocks receptors for estrogens, thus decreasing the stimulating effect of estrogens on breast cancer cells. Tamoxifen has been used for 20 years, and greatly reduces the risk of cancer recurrence. *Herceptin*, a monoclonal antibody drug, targets an antigen on the surface of breast cancer cells. It is effective in causing regression of tumors and retarding progression of the disease. The early data from clinical trials of two new drugs, *Femara* and *Amimidex*, show relapse rates that are lower than those for tamoxifen. These drugs are inhibitors of aromatase, the enzyme needed for the final step in synthesis of estrogens. Finally, two drugs—tamoxifen and *Evista (raloxifene)*—are being marketed for breast cancer *prevention.* Interestingly, raloxifene blocks estrogen receptors in the breasts and uterus but activates estrogen receptors in bone. Thus, it can be used to treat osteoporosis without increasing a woman's risk of breast or endometrial (uterine) cancer.

Ovarian Cancer

Even though **ovarian cancer** is the sixth most common form of cancer in females, it is the leading cause of death from all gynecological malignancies (excluding breast cancer) because it is difficult to detect before it metastasizes (spreads) beyond the ovaries. Risk factors associated with ovarian cancer include age (usually over age 50); race (whites are at highest risk); family history of ovarian cancer; more than 40 years of active ovulation; nulliparity or first pregnancy after age 30; a high-fat, low-fiber, vitamin A-deficient diet; and prolonged exposure to asbestos or talc. Early ovarian cancer has no symptoms or only mild ones associated with other common problems, such as abdominal discomfort, heartburn, nausea, loss of appetite, bloating, and flatulence. Later-stage signs and symptoms include an enlarged abdomen, abdominal and/or pelvic pain, persistent gastrointestinal disturbances, urinary complications, menstrual irregularities, and heavy menstrual bleeding.

Cervical Cancer

Cervical cancer, carcinoma of the cervix of the uterus, starts with **cervical dysplasia** (dis-PLĀ-sē-a), a change in the shape, growth, and number of cervical cells. The cells may either return to normal or progress to cancer. In most cases, cervical cancer may be detected in its earliest stages by a Pap test (see page 69). Some evidence links cervical cancer to the virus that causes genital warts, human papillomavirus (HPV). Increased risk is associated with having a large number of sexual partners, having first intercourse at a young age, and smoking cigarettes.

Vulvovaginal Candidiasis

Candida albicans is a yeastlike fungus that commonly grows on mucous membranes of the gastrointestinal and genitourinary tracts. The organism is responsible for **vulvovaginal candidiasis** (vul-vō-VAJ-i-nal can-di-DĪ-a-sis), the most common form of **vaginitis** (vaj′-i-NĪ-tis), inflammation of the vagina. Candidiasis is characterized by severe itching; a thick, yellow, cheesy discharge; a yeasty odor; and pain. The disorder, experienced at least once by about 75% of females, is usually a result of proliferation of the fungus following antibiotic therapy for another condition. Predisposing conditions include the use of oral contraceptives or cortisone-like medications, pregnancy, and diabetes.

SEXUALLY TRANSMITTED DISEASES

A **sexually transmitted disease (STD)** is one that is spread by sexual contact. In most developed countries of the world, such as those of Western Europe, Japan, Australia, and New Zealand, the incidence of STDs has declined markedly during the past 25 years. In the United States, by contrast, STDs have been rising to near-epidemic proportions; they currently affect more than 65 million people. AIDS and hepatitis B, which are sexually transmitted diseases that also may be contracted in other ways, are discussed in Chapters 16 and 25, respectively.

Chlamydia

Chlamydia (kla-MID-ē-a) is a sexually transmitted disease caused by the bacterium *Chlamydia trachomatis* (*chlamy-*=cloak). This unusual bacterium cannot reproduce outside body cells; it "cloaks" itself inside cells, where it divides. At present, chlamydia is the most prevalent sexually transmitted disease in the United States. In most cases, the initial infection is asymptomatic and thus difficult to recognize clinically. In males, urethritis is the principal result, causing a clear discharge, burning on urination, frequent urination, and painful urination. Without treatment, the epididymides may also become inflamed, leading to sterility. In 70% of females with chlamydia, symptoms are absent, but chlamydia is the leading cause of pelvic inflammatory disease. Moreover, the uterine tubes may also become inflamed, which increases the risk of ectopic pregnancy (implantation of a fertilized ovum outside the uterus) and infertility due to the formation of scar tissue in the tubes.

Gonorrhea

Gonorrhea (gon-ō-RĒ-a) or **"the clap"** is caused by the bacterium *Neisseria gonorrhoeae*. In the United States, 1–2 million new cases of gonorrhea appear each year, most among individuals aged 15–29 years. Discharges from infected mucus membranes are the source of transmission of the bacteria either during sexual contact or during the passage of a newborn through the birth canal. The infection site can be in the mouth and throat after oral-genital contact, in the vagina and penis after genital intercourse, or in the rectum after recto-genital contact.

Males usually experience urethritis with profuse pus drainage and painful urination. The prostate and epididymis may also become infected. In females, infection typically occurs in the vagina, often with a discharge of pus. Both infected males and females may harbor the disease without any symptoms, however, until it has progressed to a more advanced stage; about 5–10% of males and 50% of females are asymptomatic. In females, the infection and consequent inflammation can proceed from the vagina into the uterus, uterine tubes, and pelvic cavity. An estimated 50,000 to 80,000 women in the United States are made infertile by gonorrhea every year as a result of scar tissue formation that closes the uterine tubes. If bacteria in the birth canal are transmitted to the eyes of a newborn, blindness can result. Administration of a 1% silver nitrate solution in the infant's eyes prevents infection.

Syphilis

Syphilis, caused by the bacterium *Treponema pallidum*, is transmitted through sexual contact or exchange of blood, or through the placenta to a fetus. The disease progresses through several stages. During the *primary stage*, the chief sign is a painless open sore, called a **chancre** (SHANG-ker), at the point of contact. The chancre heals within 1 to 5 weeks. From 6 to 24 weeks later, signs and symptoms such as a skin rash, fever, and aches in the joints and muscles usher in the *secondary state*, which is systemic—the infection spreads to all major body systems. When signs of organ degeneration appear, the disease is said to be in the *tertiary stage*. If the nervous system is involved, the tertiary stage is called **neurosyphilis.** As motor areas become extensively damaged, victims may be unable to control urine and bowel movements. Eventually they may become bedridden and unable even to feed themselves. In addition, damage to the cerebral cortex produces memory loss and personality changes that range from irritability to hallucinations.

Genital Herpes

Genital herpes is an incurable STD. Type II herpes simplex virus (HSV-2) causes genital infections, producing painful blisters on the prepuce, glans penis, and penile shaft in males and on the vulva or sometimes high up in the vagina in females. The blisters disappear and reappear in most patients, but the virus itself remains in the body. A related virus, type I herpes simplex virus (HSV-1), causes cold sores on the mouth and lips. Infected individuals typically experience recurrences of symptoms several times a year.

Genital Warts

Warts are an infectious disease caused by viruses. *Human papillomavirus (HPV)* causes **genital warts,** which is commonly transmitted sexually. Nearly one million people a year develop genital warts in the United States. Patients with a history of genital warts may be at increased risk for cancers of the cervix, vagina, anus, vulva, and penis. There is no cure for genital warts.

KEY MEDICAL TERMS ASSOCIATED WITH THE REPRODUCTIVE SYSTEMS

Castration (kas-TRĀ-shun=to prune) Removal, inactivation, or destruction of the gonads; commonly used in reference to removal of the testes only.

Colposcopy (kol-POS-kō-pē; *colpo-*=vagina; *-scopy*=to view) Visual inspection of the vagina and cervix of the uterus using a culposcope, an instrument that has a magnifying lens (between 5–50x) and a light. The procedure generally takes place after an unusual Pap smear.

Culdoscopy (kul-DOS-kō-pē; *cul-*=cul-de-sac; *-scopy*=to view) A procedure in which a culdoscope (endoscope) is inserted through the posterior wall of the vagina to view the rectouterine pouch in the pelvic cavity.

Dysmenorrhea (dis′-men-or-Ē-a; *dys-*=difficult or painful) Pain associated with menstruation; the term is usually reserved to describe menstrual symptoms that are severe enough to prevent a woman from functioning normally for one or more days each month. Some cases are caused by uterine tumors, ovarian cysts, pelvic inflammatory disease, or intrauterine devices.

Dyspareunia (dis-pa-ROO-nē-a; *dys-*=difficult; *para*=beside; *-enue*=bed) Pain during sexual intercourse. It may occur in the genital area or in the pelvic cavity, and may be due to inadequate lubrication, inflammation, infection, an improperly fitting diaphragm or cervical cap, endometriosis, pelvic inflammatory disease, pelvic tumors, or weakened uterine ligaments.

Endocervical curettage (kū-re-TAHZH; *curette*=scraper) A procedure in which the cervix is dilated and the endometrium of the uterus is scraped with a spoon-shaped instrument called a curette; commonly called a D and C (dilation and curettage).

Fibroids (FĪ-broyds; *fibro-*=fiber; *-eidos*=resemblance) Noncancerous tumors in the myometrium of the uterus composed of muscular and fibrous tissue. Their growth appears to be related to high levels of estrogens. They do not occur before puberty and usually stop growing after menopause. Symptoms include abnormal menstrual bleeding, and pain or pressure in the pelvic area.

Hermaphroditism (her-MAF-rō-dīt-izm) The presence of both ovarian and testicular tissue in one individual.

Hypospadias (hī′-pō-SPĀ-dē-as; *hypo-*=below) A common congenital abnormality in which the urethral opening is displaced. In males, the displaced opening may be on the underside of the penis, at the penoscrotal junction, between the scrotal folds, or in the perineum; in females, the urethra opens into the vagina. The problem can be corrected surgically.

Leukorrhea (loo′-kō-RĒ-a; *leuko-*=white) A whitish (nonbloody) vaginal discharge containing mucus and pus cells that may occur at any age and affects most women at some time.

Menorrhagia (men-ō-RA-jē-a; *meno-*=menstruation; *-rhage*=to burst forth) Excessively prolonged or profuse menstrual period. May be due to a disturbance in hormonal regulation of the menstrual cycle, pelvic infection, medications (anticoagulants), fibroids (noncancerous uterine tumors composed of muscle and fibrous tissue), endometriosis, or intrauterine devices.

Oophorectomy (ō′-of-ō-REK-tō-mē; *oophor-*=bearing eggs) Removal of the ovaries.

Orchitis (or-KĪ-tis; *orchi-*=testes; *-itis*=inflammation) Inflammation of the testes, for example, as a result of the mumps virus or a bacterial infection.

Ovarian cyst The most common form of ovarian tumor, in which a fluid-filled follicle or corpus luteum persists and continues growing.

Pelvic inflammatory disease (PID) A collective term for any extensive bacterial infection of the pelvic organs, especially the uterus, uterine tubes, or ovaries, which is characterized by pelvic soreness, lower back pain, abdominal pain, and urethritis. Often the early symptoms of PID occur just after menstruation. As infection spreads, fever may develop, along with painful abscesses of the reproductive organs.

Polycystic ovary syndrome (pol-ē-SIS-tik) or ***PCOS*** A disorder that usually develops during puberty and is characterized by enlarged ovaries with many fluid-filled sacs (cysts) and a tendency to have high levels of male hormones (androgens). Symptoms include failure to menstruate, unpredictable menstrual periods, an unusual amount of facial or body hair, high blood sugar, obesity, and increased risk of cardiovascular disease.

Salpingectomy (sal′-pin-JEK-tō-mē; *salpingo*=tube) Removal of a uterine (fallopian) tube.

Smegma (SMEG-ma) The secretion, consisting principally of desquamated epithelial cells, found chiefly around the external genitalia and especially under the foreskin of the male.

Introduction (p. 835)

1. Reproduction is the process by which new individuals of a species are produced and the genetic material is passed from generation to generation.

2. The organs of reproduction are grouped as gonads (produce gametes), ducts (transport and store gametes), accessory sex glands (produce materials that support gametes), and supporting structures (have various roles in reproduction).

Male Reproductive System (p. 836)

1. The male structures of reproduction include the testes, ductus epididymis, ductus (vas) deferens, ejaculatory duct, urethra, seminal vesicles, prostate, bulbourethral (Cowper's) glands, and penis.

2. The scrotum is a sac that hangs from the root of the penis and consists of loose skin and superficial fascia; it supports the testes.

3. The temperature of the testes is regulated by contraction of the cremaster muscle and dartos muscle, which either elevates them and brings them closer to the pelvic cavity or relaxes and moves them farther from the pelvic cavity.

4. The testes are paired oval glands (gonads) in the scrotum containing seminiferous tubules, in which sperm cells are made; Sertoli cells (sustentacular cells), which nourish sperm cells and secrete inhibin; and Leydig (interstitial) cells, which produce the male sex hormone testosterone.

5. The testes descend into the scrotum through the inguinal canals during the seventh month of fetal development. Failure of the testes to descend is called cryptorchidism.

6. Secondary oocytes and sperm, both of which are called gametes, are produced in the gonads.

7. Spermatogenesis, which occurs in the testes, is the process whereby immature spermatogonia develop into mature sperm. The spermatogenesis sequence, which includes meiosis I, meiosis II, and spermiogenesis, results in the formation of four haploid sperm (spermatozoa) from each primary spermatocyte.

8. The principal parts of a mature sperm are a head and a tail. The function of sperm is to fertilize a secondary oocyte.

9. The duct system of the testes includes the seminiferous tubules, straight tubules, and rete testis. Sperm flow out of the testes through the efferent ducts.

10. The ductus epididymis is the site of sperm maturation and storage.

11. The ductus (vas) deferens stores sperm and propels them toward the urethra during ejaculation.

12. Each ejaculatory duct, formed by the union of the duct from the seminal vesicle and ductus (vas) deferens, is the passageway for ejection of sperm and secretions of the seminal vesicles into the first portion of the urethra, the prostatic urethra.

13. The urethra in males is subdivided into three portions: the prostatic, membranous, and spongy (penile) urethra.

14. The seminal vesicles secrete an alkaline, viscous fluid that constitutes about 60% of the volume of semen and contributes to sperm viability.

15. The prostate secretes a slightly acidic fluid that constitutes about 25% of the volume of semen and contributes to sperm motility.

16. The bulbourethral (Cowper's) glands secrete mucus for lubrication and an alkaline substance that neutralizes acid.

17. Semen is a mixture of sperm and seminal fluid; it provides the fluid in which sperm are transported, supplies nutrients, and neutralizes the acidity of the male urethra and the vagina.

18. The penis consists of a root, a body, and a glans penis.

19. Engorgement of the penile blood sinuses under the influence of sexual excitation is called erection.

Female Reproductive System (p. 851)

1. The female organs of reproduction include the ovaries (gonads), uterine (fallopian) tubes or oviducts, uterus, vagina, and vulva.

2. The mammary glands are considered part of the reproductive system in females.

3. The ovaries, the female gonads, are located in the superior portion of the pelvic cavity, lateral to the uterus.

4. Ovaries produce secondary oocytes, discharge secondary oocytes (ovulation), and secrete estrogens, progesterone, relaxin, and inhibin.

5. Oogenesis (production of haploid secondary oocytes) begins in the ovaries. The oogenesis sequence includes meiosis I and meiosis II, which goes to completion only after an ovulated secondary oocyte is fertilized by a sperm cell.

6. The uterine (fallopian) tubes transport secondary oocytes from the ovaries to the uterus and are the normal sites of fertilization. Ciliated cells and peristaltic contractions help move a secondary oocyte or fertilized ovum toward the uterus.

7. The uterus is an organ the size and shape of an inverted pear that functions in menstruation, implantation of a fertilized ovum, development of a fetus during pregnancy, and labor. It is also part of the pathway for sperm to reach the uterine tubes to fertilize a secondary oocyte. Normally, the uterus is held in position by a series of ligaments.

8. Histologically, the layers of the uterus are an outer perimetrium (serosa), a middle myometrium, and an inner endometrium.

9. The vagina is a passageway for sperm and the menstrual flow, the receptacle of the penis during sexual intercourse, and the inferior portion of the birth canal. It is capable of considerable distension.

10. The vulva, a collective term for the external genitals of the female, consists of the mons pubis, labia majora, labia minora, clitoris, vestibule, vaginal and urethral orifices, hymen, bulb of the vestibule, and the paraurethral (Skene's), greater vestibular (Bartholin's), and lesser vestibular glands.

11. The perineum is a diamond-shaped area at the inferior end of the trunk medial to the thighs and buttocks.

12. The mammary glands are modified sweat glands lying superficial to the pectoralis major muscles. Their function is to synthesize, secrete, and eject milk (lactation).

13. Mammary gland development depends on estrogens and progesterone.

14. Milk production is stimulated by prolactin, estrogens, and progesterone; milk ejection is stimulated by oxytocin.

Female Reproductive Cycle (p. 867)

1. The function of the ovarian cycle is to develop a secondary oocyte, whereas that of the uterine (menstrual) cycle is to prepare the endometrium each month to receive a fertilized egg. The female reproductive cycle includes both the ovarian and uterine cycles.

2. The female reproductive cycle is controlled by GnRH from the hypothalamus, which stimulates the release of FSH and LH by the anterior pituitary gland.

3. FSH stimulates development of secondary follicles and initiates secretion of estrogens by the follicles. LH stimulates further development of the follicles, secretion of estrogens by follicular cells, ovulation, formation of the corpus luteum, and the secretion of progesterone and estrogens by the corpus luteum.

4. During the menstrual phase, the stratum functionalis of the endometrium is shed, discharging blood, tissue fluid, mucus, and epithelial cells.

5. During the preovulatory phase, a group of follicles in the ovaries begin to undergo final maturation. One follicle outgrows the others and becomes dominant while the others degenerate. At the same time, endometrial repair occurs in the uterus. Estrogens are the dominant ovarian hormones during the preovulatory phase.

6. Ovulation is the rupture of the dominant mature (graafian) follicle and the release of a secondary oocyte into the pelvic cavity. It is brought about by a surge of LH. Signs and symptoms of ovulation include increased basal body temperature; clear, stretchy cervical mucus; changes in the uterine cervix; and ovarian pain.

7. During the postovulatory phase, both progesterone and estrogens are secreted in large quantity by the corpus luteum of the ovary, and the uterine endometrium thickens in readiness for implantation.

8. If fertilization and implantation do not occur, the corpus luteum degenerates, and the resulting low level of progesterone allows discharge of the endometrium followed by the initiation of another reproductive cycle.

9. If fertilization and implantation occur, the corpus luteum is maintained by placental hCG, and the corpus luteum and later the placenta secrete progesterone and estrogens to support pregnancy and breast development for lactation.

Birth Control Methods (p. 869)

1. Methods include surgical sterilization (vasectomy, tubal ligation), hormonal methods, intrauterine devices, spermatocides, barrier methods (condom, vaginal pouch, diaphragm), periodic abstinence (rhythm and sympto-thermal methods), and induced abortion. See Table 27.3 on page 870 for failure rates for these methods.

2. Contraceptive pills of the combination type contain estrogens and progestins in concentrations that decrease the secretion of FSH and LH and thereby inhibit development of ovarian follicles and ovulation.

3. An abortion is the premature expulsion from the uterus of the products of conception; it may be spontaneous or induced. RU 486 can induce abortion by blocking the action of progesterone.

Development of the Reproductive Systems (p. 871)

1. The gonads develop from intermediate mesoderm. In the presence of the *SRY* gene, the gonads begin to differentiate into testes during the seventh week. The gonads differentiate into ovaries when the *SRY* gene is absent.

2. In males, testosterone stimulates development of each mesonephric duct into an epididymis, ductus (vas) deferens, ejaculatory duct, and seminal vesicle, and Müllerian-inhibiting substance (MIS) causes the paramesonephric duct cells to die. In females, testosterone and MIS are absent; the paramesonephric ducts develop into the uterine tubes, uterus, and vagina and the mesonephric ducts degenerate.

3. The external genitals develop from the genital tubercle and are stimulated to develop into typical male structures by the hormone dihydrotestosterone (DHT). The external genitals develop into female structures when DHT is not produced, the normal situation in female embryos.

Aging and the Reproductive Systems (p. 874)

1. Puberty is the period when secondary sex characteristics begin to develop and the potential for sexual reproduction is reached.

2. Onset of puberty is marked by pulses or bursts of LH and FSH secretion, each triggered by a pulse of GnRH. The hormone leptin, released by adipose tissue, may signal the hypothalamus that long-term energy stores (triglycerides in adipose tissue) are adequate for reproductive functions to begin.

3. In females, the reproductive cycle normally occurs once each month from menarche, the first menses, to menopause, the permanent cessation of menses.

4. Between the ages of 40 and 50, the pool of remaining ovarian follicles becomes exhausted and levels of progesterone and estrogens decline. Most women experience a decline in bone mineral density after menopause, together with some atrophy of the ovaries, uterine tubes, uterus, vagina, external genitalia, and breasts. Uterine and breast cancer increase in incidence with age.

5. In older males, decreased levels of testosterone are associated with decreased muscle strength, waning sexual desire, and fewer viable sperm; prostate disorders are common.

Q SELF-QUIZ QUESTIONS

Choose the one best answer to the following questions.

1. The secondary sex gland(s) that contribute(s) the most volume to the semen is/are the:
   a. seminal vesicles.
   b. bulbourethral (Cowper's) glands.
   c. paraurethral (Skene's) glands.
   d. greater vestibular (Bartholin's) glands.
   e. prostate.

2. The dense irregular connective tissue that covers the testes and extends inward to subdivide the testes into lobules is the:

   a. tunica albuginea.   b. tunica vaginalis.   c. raphe.
   d. rete testis.   e. epididymis.

3. Sperm are produced in the:
   a. prostate.   b. seminal vesicles.
   c. seminiferous tubules.
   d. Sertoli cells.   e. ductus (vas) deferens.

4. Which of the following produce(s) a secretion that helps maintain the motility and viability of sperm?
   a. prostate   b. penis
   c. bulbourethral (Cowper's) glands   d. ejaculatory duct
   e. All of the above.

5. Testosterone is secreted by:
   a. Sertoli (sustentacular) cells.   b. spermatogonia.
   c. spermatids.   d. Leydig (interstitial) cells.
   e. All of these are correct.

6. In the postovulatory phase of the menstrual cycle, LH stimulates the corpus luteum to secrete hormones including:
   a. prolactin.   b. FSH.   c. hCG.
   d. progesterone.   e. Both c and d are correct.

Complete the following.

7. Place numbers in the blanks to arrange the following structures in the correct sequence for the pathway of sperm. (a) ejaculatory duct: _____; (b) testis: _____; (c) urethra: _____; (d) ductus (vas) deferens: _____; (e) epididymis: _____.

8. Branches of the radial arteries called _____ supply the stratum functionalis of the endometrium.

9. Fill in the blanks with the male homologues for each of the following. (a) labia majora: _____; (b) clitoris: _____; (c) paraurethral (Skene's) glands: _____; (d) greater vestibular (Bartholin's) glands: _____.

10. The layer of endometrium that is not shed during menstruation is the stratum _____.

11. The union of the ampulla of the ductus deferens and the duct from the seminal vesicle forms the _____.

12. If fertilization and implantation do not occur, the corpus luteum degenerates and becomes the _____.

13. The male accessory gland that is located inferior to the urinary bladder and that surrounds the urethra is the _____.

14. The ovary is attached to the uterus by the _____ ligament.

Are the following statements true or false?

15. The mass of tissue that surrounds the penile urethra is called the corpus spongiosum.

16. The female urethra is located posterior to the vagina.

17. Fertilization of a secondary oocyte by a sperm cell normally occurs in the uterus.

18. During the preovulatory phase of the menstrual cycle, levels of estrogens increase and the endometrium thickens.

19. The acrosome is the portion of a sperm that contains the genetic material.

20. The uterine tubes are open to the pelvic cavity.

Matching

21. Match the following terms to their descriptions:
    ___ (a) 200–300 of these per testis
    ___ (b) tightly coiled tubes (1–3 per lobule); composed of cells that develop into sperm
    ___ (c) cells located between developing sperm cells; form the blood–testis barrier and provide nourishment
    ___ (d) cells located between seminiferous tubules; secrete testosterone
    ___ (e) highly coiled duct in which sperm are stored for maturation
    ___ (f) ejects sperm and fluid into the urethra just before ejaculation
    ___ (g) enters the pelvic cavity through the inguinal canal

    (1) Leydig (interstitial) cells
    (2) ejaculatory duct
    (3) Sertoli (sustentacular) cells
    (4) seminiferous tubules
    (5) ductus (vas) deferens
    (6) ductus epididymis
    (7) lobule

 ## CRITICAL THINKING QUESTIONS

1. Esther nearly died from peritonitis (inflammation of the peritoneum) that her doctor said had spread from an infection in her reproductive tract. How was this possible?

2. Occasionally, someone feels that he or she has been born the wrong gender and undergoes a "sex-change" or "gender reassignment" process involving hormone treatment and surgery. However, a born male can never truly become a fully biological female or a born female become a fully biological male. Why not?

3. Thirty-nine-year-old Meg has been advised to have a hysterectomy due to medical problems. She is worried that the procedure will cause menopause. Is this a valid concern?

4. Darby has been doing peritoneal dialysis for kidney failure at home for several years. However, the weight of the dialysis fluid in his peritoneal cavity has caused him to develop inguinal hernias repeatedly. Darby is tired of this, and in consultation with his doctor, decides to have the inguinal canal surgically closed off. As a result, Darby will also have to have his testes removed. Why?

5. Phil has promised his wife that he will get a vasectomy after the birth of their next child. However, he's concerned about possible effects on his virility. How would you respond to Phil's concerns?

ANSWERS TO FIGURE QUESTIONS

27.1 The gonads (testes) produce gametes (sperm) and hormones; the ducts transport, store, receive gametes; the accessory sex glands secrete materials that help transport and protect gametes; and supporting structures, such as the penis, convey semen to the exterior.

27.2 The cremaster and dartos muscles help regulate the temperature of the testes.

27.3 The tunica vaginalis and tunica albuginea are tissue layers that cover and protect the testes.

27.4 The Leydig cells of the testes secrete testosterone.

27.5 During meiosis I, the number of chromosomes in each cell is reduced by half.

27.6 The sperm head contains the nucleus with highly condensed haploid chromosomes and an acrosome that contains enzymes for penetration of a secondary oocyte; the neck contains centrioles that produce microtubules for the rest of the tail; the midpiece contains mitochondria for ATP production for locomotion and metabolism; the principal and end pieces of the tail provide motility.

27.7 The functions of the ductus epididymis are sperm maturation, sperm storage, and propulsion of sperm into the ductus (vas) deferens.

27.8 The ductus deferens stores sperm and conveys sperm toward the urethra.

27.9 Seminal vesicles are the accessory sex glands that contribute the largest volume to seminal fluid.

27.10 Two corpora cavernosa penis and one corpus spongiosum penis contain blood sinuses that fill with blood that cannot flow out of the penis as quickly as it flows in. The trapped blood engorges and stiffens the tissue, producing an erection. The corpus spongiosum penis keeps the spongy urethra open so that ejaculation can occur.

27.11 The testes are homologous to the ovaries; the glans penis is homologous to the clitoris; the prostate is homologous to the paraurethral glands; and the bulbourethral gland is homologous to the greater vestibular glands.

27.12 The mesovarium anchors the ovary to the broad ligament of the uterus and the uterine tube; the ovarian ligament anchors it to the uterus; the suspensory ligament anchors it to the pelvic wall.

27.13 Ovarian follicles secrete estrogens; the corpus luteum secretes progesterone, estrogens, relaxin, and inhibin.

27.14 Most ovarian follicles undergo atresia (degeneration).

27.15 Primary oocytes are present in the ovary at birth, so they are as old as the woman is. In males, primary spermatocytes are continually being formed from stem cells (spermatogonia) and thus are only a few days old.

27.16 Fertilization most often occurs in the ampulla of the uterine tube.

27.17 Ciliated simple columnar epithelial cells and nonciliated (peg) cells with microvilli line the uterine tubes.

27.18 The endometrium is a highly vascular, secretory epithelium that provides the oxygen and nutrients needed to sustain a fertilized egg; the myometrium is a thick smooth muscle layer that supports the uterine wall during pregnancy and contracts to expel the fetus at birth.

27.19 The stratum basalis of the endometrium provides cells to replace those that shed (the stratum functionalis) during each menstruation.

27.20 The vagina receives the penis during sexual intercourse, serves as the outlet for menstrual flow, and is the passageway for childbirth.

27.21 Anterior to the vaginal opening are the mons pubis, clitoris, and prepuce. Lateral to the vaginal opening are the labia minora and labia majora.

27.22 The anterior portion of the perineum is called the urogenital triangle because its borders form a triangle that encloses the urethral (uro-) and vaginal (-genital) orifices.

27.23 Prolactin, estrogens, and progesterone are the hormones that regulate the synthesis of milk. Oxytocin regulates the ejection of milk.

27.24 Alveoli are located in lobules of the mammary glands.

27.25 The hormones responsible for the proliferative phase of endometrial growth are estrogens; for ovulation, LH; for growth of the corpus luteum, LH; and for the midcycle surge of LH, estrogens.

27.26 The *SRY* gene on the Y chromosome is responsible for the development of the gonads into testes.

27.27 The presence of dihydrotestosterone (DHT) stimulates differentiation of the external genitals in males; its absence allows differentiation of the external genitals in females.

APPENDIX A
MEASUREMENTS

U.S. CUSTOMARY SYSTEM

PARAMETER	UNIT	RELATION TO OTHER U.S. UNITS	SI (METRIC) EQUIVALENT
Length	inch	1/12 foot	2.54 centimeters
	foot	12 inches	0.305 meter
	yard	36 inches	0.914 meters
	mile	5,280 feet	1.609 kilometers
Mass	grain	1/1000 pound	64.799 milligrams
	dram	1/16 ounce	1.772 grams
	ounce	16 drams	28.350 grams
	pound	16 ounces	453.6 grams
	ton	2,000 pounds	907.18 kilograms
Volume (Liquid)	ounce	1/16 pint	29.574 milliliters
	pint	16 ounces	0.473 liter
	quart	2 pints	0.946 liter
	gallon	4 quarts	3.785 liters
Volume (Dry)	pint	1/2 quart	0.551 liter
	quart	2 pints	1.101 liters
	peck	8 quarts	8.810 liters
	bushel	4 pecks	35.239 liters

INTERNATIONAL SYSTEM (SI)

BASE UNITS

UNIT	QUANTITY	SYMBOL
meter	length	m
kilogram	mass	kg
second	time	s
liter	volume	l
mole	amount of matter	mol

PREFIXES

PREFIX	MULTIPLIER	SYMBOL
tera-	$10^{12} = 1,000,000,000,000$	T
giga-	$10^9 = 1,000,000,000$	G
mega-	$10^6 = 1,000,000$	M
kilo-	$10^3 = 1,000$	k
hecto-	$10^2 = 100$	h
deca-	$10^1 = 10$	da
deci-	$10^{-1} = 0.1$	d
centi-	$10^{-2} = 0.01$	c
milli-	$10^{-3} = 0.001$	m
micro-	$10^{-6} = 0.000,001$	μ
nano-	$10^{-9} = 0.000,000,001$	n
pico-	$10^{-12} = 0.000,000,000,001$	p

TEMPERATURE CONVERSION

FAHRENHEIT (F) TO CELSIUS (C)

$$°C = (°F - 32) \div 1.8$$

CELSIUS (C) TO FAHRENHEIT (F)

$$°F = (°C \times 1.8) + 32$$

U.S TO SI (METRIC) CONVERSION

WHEN YOU KNOW	MULTIPLY BY	TO FIND
inches	2.54	centimeters
feet	30.48	centimeters
yards	0.91	meters
miles	1.61	kilometers
ounces	28.35	grams
pounds	0.45	kilograms
tons	0.91	metric tons
fluid ounces	29.57	milliliters
pints	0.47	liters
quarts	0.95	liters
gallons	3.79	liters

SI (METRIC) TO U.S. CONVERSION

WHEN YOU KNOW	MULTIPLY BY	TO FIND
millimeters	0.04	inches
centimeters	0.39	inches
meters	3.28	feet
kilometers	0.62	miles
liters	1.06	quarts
cubic meters	35.32	cubic feet
grams	0.035	ounces
kilograms	2.21	pounds

APPENDIX B
ANSWERS

ANSWERS TO SELF-QUIZ QUESTIONS

Chapter 1
1. (e) 2. (e) 3. (b) 4. (d) 5. (d) 6. (c) 7. epithelia, connective, muscular, nervous 8. peritoneum 9. epigastric 10. parasagittal 11. organs 12. ipsi 13. inferior, lateral (or superior, also) 14. F 15. T 16. T 17. F 18. T 19. 1(d), 2(e), 3(g), 4(h), 5(a), 6(b), 7(c) 20. 1(c), 2(g), 3(a), 4(e), 5(d), 6(h), 7(f), 8(b)

Chapter 2
1. (a) 2. (a) 3. (a) 4. (e) 5. (d) 6. (c) 7. (c) 8. (b) 9. fitration 10. meiosis 11. receptor-mediated endocytosis 12. (phospho)lipids 13. ligands 14. interstitial fluid 15. osmosis 16. F 17. F 18. F 19. 1(j), 2(l), 3(c), 4(k), 5(e), 6(d), 7(h), 8(g), 9(a), 10(i), 11(b), 12(f) 20. 1(j), 2(a), 3(l), 4(h), 5(c), 6(g), 7(b), 8(f), 9(k), 10(e), 11(i), 12(d)

Chapter 3
1. (b) 2. (d) 3. (e) 4. (b) 5. (d) 6. (a) 7. (a) 8. (a) 9. cardiac 10. exocrine 11. gap junction 12. collagen, elastic, reticular 13. serous 14. lamina propria 15. T 16. T 17. T 18. T 19. 1(e), 2(f), 3(h), 4(a), 5(c), 6(g), 7(b), 8(d) 20. 1(d), 2(e), 3(a), 4(h), 5(b), 6(g), 7(c), 8(f), 9(j), 10(i)

Chapter 4
1. (d) 2. (b) 3. (a) 4. (c) 5. (b) 6. (c) 7. syngamy 8. ectoderm, endoderm, mesoderm 9. gastrulation 10. amniotic fluid 11. angiogenesis 12. decidua 13. human chorionic gonadotropin (hCG) 14. F 15. F 16. F 17. F 18. F 19. F 20. (a)2, (b)1, (c)3, (d)3, (e)3, (f)1, (g)2 21. 1(d), 2(h), 3(b), 4(j), 5(a), 6(e), 7(c), 8(f), 9(i), 10(g) 22. 1(d), 2(a), 3(f), 4(c), 5(g), 6(b), 7(e)

Chapter 5
1. (c) 2. (b) 3. (a) 4. (d) 5. (a) 6. (d) 7. ceruminous 8. epidermis, dermis 9. sudoriferous 10. F 11. F 12. F 13. T 14. T 15. F 16. 1(b), 2(d), 3(e), 4(a), 5(f), 6(c) 17. 1(c), 2(e), 3(b), 4(a), 5(d), 6(f)

Chapter 6
1. (e) 2. (d) 3. (b) 4. (e) 5. (d) 6. dense irregular fibrous connective tissue 7. endochondral 8. calcification 9. lacunae 10. yellow bone marrow 11. osteoclasts 12. appositional 13. T 14. F 15. F 16. F 17. T 18. 1(a), 2(e), 3(c) 4(b), 5(d)

Chapter 7
1. (c) 2. (d) 3. (b) 4. (a) 5. (d) 6. (a) 7. (e) 8. (a) 9. (e) 10. (e) 11. 1(i), 2(j), 3(r), 4(o,k), 5(p), 6(n,a), 7(q), 8(d), 9(k), 10(g,m), 11(d,f,b), 12. atlas, axis 13. long, short, flat 14. Wormian 15. hyoid 16. T 17. T 18. T 19. F 20. F

Chapter 8
1. (c) 2. (c) 3. (d) 4. (e) 5. (c) 6. (b) 7. (a) 8. (b) 9. (e) 10. (b) 11. ischium 12. patella 13. posterior, femur 14. false 15. head, femur 16. F 17. T 18. T 19. F 20. 1(f), 2(h,p), 3(a,k), 4(e,n), 5(b,j,m), 6(i,o), 7(c,q), 8(d), 9(g), 10(l)

Chapter 9
1. (b) 2. (c) 3. (a) 4. (e) 5. (d) 6. (e) 7. (a) 8. (c) 9. cartilaginous joint (synchondrosis), hyaline cartilage 10. ligament 11. diarthrosis 12. hyaline cartilage (articular cartilage) 13. bursae 14. F 15. T 16. F 17. T 18. F 19. 1(h), 2(d), 3(a), 4(e), 5(b), 6(i), 7(c), 8(f), 9(g) 19. 1(c), 2(e), 3(b), 4(d), 5(a) 20. 1(h), 2(f), 3(d), 4(b), 5(c), 6(a), 7(e), 8(g)

Chapter 10
1. (c) 2. (b) 3. (d) 4. (a) 5. (c) 6. (c) 7. (d) 8. superficial fascia 9. endomysium 10. Z lines 11. troponin, tropomyosin 12. S, U, U, S, S 13. smooth and cardiac 14. F 15. F 16. F 17. T 18. F 19. 1(b) and (e), 2(c) and (g), 3(a), (d), and (f), 20. 1(e), 2(d), 3(b), 4(f), 5(a), 6(c)

Chapter 11
1. (d) 2. (b) 3. (d) 4. (e) 5. (b) 6. (e) 7. (d) 8. prime mover (agonist), synergists 9. origin, insertion 10. bones, joints, muscle contraction 11. semitendinosus, semimembranosus, biceps femoris 12. mastoid process 13. central 14. trapezius, rhomboideus major, rhomboideus minor 15. perineum 16. sphincter 17. flexion 18. anterior, four, tibia 19. gastrocnemius, soleus, plantaris 20. T 21. F 22. F 23. (a)2 and 3, (b)3 and 6, (c)4 and 5, (d)5, (e)1 and 6, (f)3, (g)6 24. 1(f), 2(d), 3(b), 4(a), 5(c), 6(j), 7(e), 8(i), 9(g), 10(h) 25. (a)3, (b)6, (c)1, (d)11, (e)8, (f)2, (g)10, (h)5, (i)7, (j)9, (k)4

Chapter 12
1. (e) 2. (d) 3. (b) 4. (a) 5. (c) 6. (e) 7. (d) 8. cranial, facial 9. (a) inferior (b) inferior (c) distal (d) superior (e) anterior (f) lateral (g) inferior 10. latissimus dorsi, trapezius 11. xiphisternal 12. T 13. T 14. T 15. T 16. F 17. 1(f), 2(e), 3(b), 4(a), 5(h), 6(g), 7(c), 8(d) 18. 1(b), 2(c), 3(f), 4(d), 5(a), 6(g), 7(h), 8(e)

Chapter 13
1. (b) 2. (b) 3. (a) 4. (d) 5. (a) 6. (c) 7. (e) 8. (b) 9. (a) 10. (b) 11. (e) 12. (b) 13. differential white blood cell 14. 45%, 55% 15. emigration 16. erythropoiesis 17. reticulocyte 18. colony-stimulating factors, interleukins 19. iron 20. T 21. F 22. F 23. (a)5, (b)3, (c)3, (d)1, (e)2, (f)4, (g)4 and 5, (h)2, (i)1, (j)4

Chapter 14
1. (d) 2. (b) 3. (a) 4. (d) 5. (b) 6. (a) 7. pericardium 8. right and left coronary arteries 9. atrioventricular 10. circumflex artery 11. open, closed 12. systole, diastole 13. left ventricle 14. inferior left 15. F 16. T 17. F 18. T 19. F 20. T 21. F 22. F 23. 1(d), 2(c), 3(a,f), 4(b,e)

Chapter 15
1. (c) 2. (c) 3. (e) 4. (b) 5. (e) 6. (e) 7. (c) 8. (c) 9. (d) 10. (a) 11. (c) 12. common hepatic artery, left gastric artery, splenic artery 13. vasa recta 14. tunica intima 15. inferior vena cava 16. hepatic portal vein 17. anastomosis 18. media 19. vasoconstriction 20. continuous,

fenestrated, sinusoids 21. pulmonary 22. F 23. F 24. T 25. T 26. F 27. 1(f), 2(d), 3(h), 4(k), 5(g), 6(j) 7(b), 8(c), 9(i), 10(a), 11(e)

Chapter 16

1. (b) 2. (a) 3. (d) 4. (e) 5. (b) 6. (e) 7. (b) 8. (e) 9. lacteals 10. veins 11. red bone marrow, thymus 12. pharyngeal, palatine, lingual 13. pathogens 14. MALT (mucosa associated lymphatic tissue) 15. T 16. T 17. T 18. T 19. F 20. (a)4, (b)7, (c)1, (d)8, (e)2, (f)6, (g)5, (h)3

Chapter 17

1. (c) 2. (a) 3. (d) 4. (d) 5. (e) 6. (a) 7. (c) 8. (b) 9. ependymal cells 10. somatic, autonomic, enteric 11. tract 12. effectors 13. GABA and glycine 14. synaptic vesicles 15. oligodendrocytes 16. T 17. T, 18. T 19. T 20. F 21. F 22. (a)3, (b)6, (c)7, (d)8. (e)4, (f)5, (g)1, (h)2 23. (a)3, (b)1, (c)1, (d)2, (e)3, (f)2, (g)1

Chapter 18

1. (e) 2. (e) 3. (b) 4. (c) 5. (e) 6. (d) 7. (e) 8. (c) 9. medulla oblongata, second lumbar vertebra 10. denticulate ligament 11. skeletal muscles, cardiac muscle, smooth muscle, glands 12. 4(th), 5(th) [or third and fourth], subarachnoid 13. anterior, lateral 14. dermatome 15. F 16. F 17. T 18. T 19. T 20. F 21. (a)1, (b)2, (c)1, (d)5, (e)2, (f)4, (g)3 22. (a)7, (b)5, (c)6, (d)1, (e)2, (f)4, (g)3

Chapter 19

1. (d) 2. (c) 3. (e) 4. (b) 5. (c) 6. (b) 7. (b) 8. (A) 9. choroid plexuses 10. third ventricle 11. apneustic, pneumotaxic 12. reticular activating system 13. association 14. cerebral cortex, gray, gyri, sulci 15. (a)w, (b)g, (c)g, (d)g, (e)w, (f)g 16. (a)9, (b)7, (c)5, (d)2, (e)4, (f)6, (g)3, (h)8, (i)1, (j)10 19. (a)1, (b)6, (c)3, (d)4, (e)2, (f)1, (g)5, (h)2, (i)5, (j)1, (k)4 17. mamillary bodies 18. melatonin 19. T 20. T 21. F 22. F 23. 1(f,j), 2(e,h), 3(c,l), 4(d,k), 5(g), 6(b), 7(a) 24. 1(g), 2(j), 3(i), 4(a), 5(c), 6(l), 7(d), 8(k), 9(f), 10(b), 11(h), 12(e)

Chapter 20

1. (b) 2. (b) 3. (a) 4. (a) 5. (d) 6. (e) 7. (e) 8. (c) 9. (d) 10. splanchnic 11. dual 12. MAO 13. alpha, beta 14. nicotinic, muscarinic 15. adrenal medulla 16. vagus 17. T 18. T 19. F 20. F 21. F 22. F 23. (a)2, (b)5, (c)6, (d)4, (e)1, (f)3 24. (a)1, (b)2, (c)1, (d)2, (e)1, (f)1, (g)1, (h)1, (i)3, (j)1

Chapter 21

1. (c) 2. (e) 3. (a) 4. (d) 5. (d) 6. (b) 7. stimulus 8. exteroceptors, interoceptors (visceroceptors), proprioceptors 9. mechano- 10. brain stem, association 11. referred 12. spinocerebellar tracts 13. stereognosis

14. thalamus 15. F 16. T 17. F 18. T 19. F 20. (a)4, (b)1, (c)3, (d)2 21. (a)4, (b)3, (c)1, (d)2, (e)6, (f)5

Chapter 22

1. (d) 2. (a) 3. (b) 4. (c) 5. (e) 6. (a) 7. (e) 8. (b) 9. stapedius, tensor tympani 10. papillae, fungiform 11. ciliary, scleral venous 12. basilar membrane, organ of Corti 13. conjunctiva 14. dynamic, static 15. (a)3, (b)1, (c)2, (d)4, (e)5 16. (a)1, (b)5, (c)3, (d)4, (e)6, (f)2 17. static equilibrium, utricle, saccule 18. F 19. T 20. F 21. F 22. (a)11, (b)6, (c)7, (d)1, (e)3, (f)10, (g)2, (h)9, (i)5, (j)4, (k)8, (l)12 23. (a)5, (b)4, (c)9, (d)3, (e)12, (f)6, (g)7, (h)8, (i)1, (j)2, (k)11, (l)10

Chapter 23

1. (a) 2. (c) 3. (a) 4. (d) 5. (d) 6. (e) 7. (a)3, (b)2, (c)1, (d)4 8. hypothalamus 9. tropic, target 10. alpha, increase 11. human growth hormone 12. external carotid artery 13. parathyroid hormone 14. F 15. T 16. F 17. F 18. F 19. T 20. 1(c), 2(f), 3(b), 4(g), 5(d), 6(a), 7(e) 21. (a)5, (b)6, (c)8, (d)3, (e)2, (f)9, (g)1, (h)10, (i)7, (j)4

Chapter 24

1. (c) 2. (a) 3. (c) 4. (e) 5. (d) 6. (e) 7. (d) 8. (a) 9. trachea 10. epiglottis 11. glottidis, vestibuli 12. lobule 13. (a)3, (b)2, (c)1 14. atelectasis 15. surfactant 16. T 17. T 18. F 19. F 20. (a)1, (b)5, (c)2, (d)3, (e)4, (f)4, (g)2 and 3, (h)2 21. 1(c), 2(a), 3(g), 4(e), 5(b), 6(f), 7(d), 22. (a)1, (b)2, (c)2, (d)1, (e)1

Chapter 25

1. (c) 2. (d) 3. (b) 4. (a) 5. (c) 6. (b) 7. (a) 8. (e) 9. greater omentum 10. (a)1, (b)4, (c)3, (d)2, (e)5 11. deglutition 12. peristalsis 13. permanent 14. liver 15. cystic 16. hydrochloric acid, intrinsic factor, pepsinogen, gastric lipase 17. caudate, quadrate 18. trypsin, chymotrypsin, carboxypeptidase 19. F 20. F 21. T 22. T 23. T 24. F 25. 1(e), 2(g), 3(a), 4(f), 5(h), 6(b), 7(d), 8(c)

Chapter 26

1. (d) 2. (e) 3. (b) 4. (c) 5. (c) 6. (d) 7. (a) 8. renal papilla 9. efferent, afferent 10. podocytes 11. spongy 12. juxtaglomerular 13. ureters 14. (a)4, (b)5, (c)1, (d)9, (e)7, (f)8, (g)6, (h)3, (i)2 15. micturition 16. T 17. T 18. F 19. F

Chapter 27

1. (a) 2. (a) 3. (c) 4. (a) 5. (d) 6. (d) 7. (a)4, (b)1, (c)5, (d)3, (e)2 8. spiral arterioles 9. scrotum, penis, prostate gland, bulbourethral (Cowper's) glands 10. basalis 11. ejaculatory duct 12. corpus albicans 13. prostate 14. ovarian 15. T 16. F 17. F 18. T 19. F 20. T 21. (a)7, (b)4, (c)3, (d)1, (e) 6, (f)2, (g)5

ANSWERS TO CRITICAL THINKING QUESTIONS

Chapter 1

1. The kidneys are located behind the parietal peritoneum. To view the posterior surface of the kidney, a scope would penetrate the posterior body wall but would not enter the peritoneal cavity.

2. Taylor's arm will be extended lateral to the trunk, hand inferior and distal to the elbow, palm facing anteriorly, thumb lateral, and radius lateral to ulna.

3. The alien would have 2 tails, 4 arms, 2 legs, and a mouth where its navel is usually located.

4. Lung, diaphragm, stomach, large intestine, small intestine; possibly part of pancreas, ovary or uterine tube, kidney.

5. The peritoneum, the largest serous membrane in the body, covers most organs in the abdominal cavity. Therefore, an infection in this structure can spread to any or all organs in the cavity.

Chapter 2

1. An increase in neutrophils indicates an increased demand for phagocytosis, most likely due to bacterial infection.

2. Maternal inheritance is due to the DNA found in extranuclear organelles such as the mitochondria. The sperm contributes only nuclear DNA. Mitochondria are inherited only from the mother.

3. Water would move out of the red blood cells, where the concentration of water is higher, causing them to shrink.

4. Arsenic will poison the mitochondrial enzymes involved in cellular respiration. This will halt production of ATP by the mitochondria. Cells with a high metabolic rate such as muscle cells will be particularly affected.

5. Disruption of microtubules would halt cell division. Microtubules are also involved in cell motility, transportation, and movement of cilia, all of which would be affected.

Chapter 3

1. The drug will make it easier for the cilia on the epithelial cells to move the mucus, in which microbes are trapped, away from the lungs. Coughing up the thinned mucus should also be easier.

2. Some tissues found in a chicken leg would include: epithelium in the skin, areolar and adipose tissues in the subcutaneous layer under the skin, skeletal muscle (the meat), blood, hyaline cartilage, and bone.

3. Janelle's kidneys may drop out of position from lack of supporting fat. There would also be less fat for padding in joints and buttocks, and eyes may appear sunken.

4. The tissue is keratinized stratified squamous epithelium, which is avascular.

5. Bacteria would need a mechanism to pass through or between keratin-filled cells of the epidermis and to combat any phagocytic cells present. They would also need to survive any enzymes produced by cells the bacteria might encounter.

Chapter 4

1. Identical (monozygotic) twins develop from the same fertilized ovum or zygote. They are essentially the same person genetically and therefore must be the same gender. These must be fraternal (dizygotic) twins.

2. All tissues and organs form from the embryo's three primary germ layers. The ectoderm, the outermost germ layer, develops into the nervous system and the epithelial layer of the skin. This early embryological connection may sometimes result in disorders showing signs in both tissues.

3. Josefina is in false labor. In true labor, the contractions are regular and the pain may localize in the back. The pain may be intensified by walking. True labor is indicated by the "show" of bloody mucus and cervical dilation.

4. Certain microorganisms can infect the placenta itself to infect the embryo/fetus (transplacental transmission). Risk is greatest during the first trimester, a crucial time in the differentiation of cells.

5. Pregnancy increases appetite to accommodate the increase in nutritional demands. Increasing pressure on the urinary bladder increases frequency of urination. As the uterus increases in size, abdominal contents are pushed upward, and the stomach and its contents may push up into the esophagus (heartburn). Elena may need to prop herself up higher at night to help prevent heartburn and make breathing easier.

Chapter 5

1. Chronic exposure to UV light causes damage to elastin and collagen, promoting wrinkles. The suspicious growth may be skin cancer caused by UV photodamage to skin cells.

2. Dilation of the blood vessels in the skin increases blood flow, which means more body heat will reach the surface of the skin and be radiated away. Therefore, although they will feel warmer superficially, body temperature will decrease. They should wear their coats!

3. Felicity's plan is not wise. Synthesis of Vitamin D is dependent on exposure to sunlight. Temperature control is dependent on evaporation of sweat to cool the body.

4. The surface layer of the skin is keratinized stratified squamous epithelium. The keratinocytes are dead and filled with the protective protein keratin. Lipids from lamellar granules and sebum also function in waterproofing.

5. The higher the humidity, the less evaporation can occur, which means the less cooling can occur. Therefore people perceive the same temperature as hotter at a higher humidity.

Chapter 6

1. Lynda's diet lacks several nutrients necessary for bone health, including essential vitamins (such as A, D), minerals (such as calcium), and proteins. Her lack of exercise will weaken bone strength. Her age and smoking habit may result in a lack of estrogens causing demineralization of bone.

2. Without a skeleton, the skeletal muscles could not provide movement, the internal organs would lack support and protection, mineral storage (such as calcium and phosphorus) would be inadequate, and blood could not be produced in the red bone marrow.

3. At Aunt Edith's age, the production of several hormones necessary for bone remodeling (such as estrogens and human growth hormone) would be decreased. She may have loss of bone mass, brittleness, and possibly osteoporosis. Increased susceptibility to fractures results in damage to the vertebrae and loss of height.

4. Exercise causes mechanical stress on bones but since there is no gravity in space, the pull of gravity on bones is missing. The lack of stress results in demineralization of bone and weakness.

5. Because babies have fontanels (soft spots), the cranial bones tend to be more movable than after the sutures close. As the baby passed through the birth canal (and possibly because of the use of forceps to assist the delivery), the bones could move into a more "cone-shaped" position. Likewise, laying the baby on its back all the time flattens the occipital bone. In addition, infant bones tend to be more flexible than older bones, which lose moisture and become more brittle over time.

Chapter 7

1. Fontanels, the soft spots between cranial bones, are fibrous connective tissue membrane–filled spaces. They allow the infant's head to be molded during its passage through the birth canal and allow for brain and skull growth in the infant.

2. The pituitary gland lies under the brain as it sits in the sella turcica of the sphenoid bone. Thus, the gland is heavily protected and difficult to reach from any direction. In addition, there are important nerves, such as the optic nerve, and blood vessels passing nearby.

3. Infants are born with a single concave curve. Adults have four curves in their vertebral column at the cervical, thoracic, lumbar, and sacral regions.

4. The occipital bone protects key areas of the cerebrum, but also the brain stem, which joins the spinal cord at the foramen magnum. Damage to the spinal cord and/or the vital centers in the medulla oblongata controlling breathing, heart activity, and blood pressure could be fatal.

5. The x-rays were taken to view the paranasal sinuses—the "fuzzy looking holes." John probably had sinusitis due to infected paranasal sinuses.

Chapter 8

1. There are 14 phalanges in each hand: 2 bones in the thumb and 3 in each of the other fingers. Farmer Ramsey has lost 5 phalanges on his left hand so he has 9 remaining on his left and 14 remaining on his right for a total of 23.

2. The talus receives all of the body's weight. Half that weight is normally transmitted to the calcaneus but the weight distribution will be shifted on the hallux by dancing. Problems may appear in the talocrural (ankle) joint and hallux. Rose also has a deviated hallux or bunion that may have been caused by tight ballet shoes. The dancing did strengthen the foot muscles and correct the flat feet.

3. Bone growth occurs in response to mechanical stress. Therefore, any activity that causes the deltoid muscle to pull on the deltoid

tuberosity would increase its growth. Such activity might include paddling the large kayaks frequently over time and/or against turbulent waters.

4. Grandmother Amelia has probably fractured the neck of her femur. This is a common fracture in the elderly and is often called a "broken hip." The sacrum does indeed articulate with the ilium, and Grandpa Jeremiah could have osteoarthritis in that joint, which could be quite painful. There are, however, other lower back issues (e.g., herniated discs, sciatica, etc.) that could be equally painful.

5. Derrick has patellofemoral syndrome (runner's knee). Running daily in the same direction on a banked track has stressed the downhill knee. The patella is tracking laterally to its proper position resulting in pain after exercise.

Chapter 9

1. Structurally the hip joint is a simple ball-and-socket synovial joint composed of the head of the femur (ball) and the acetabulum of the hip bone (socket). The knee is really three joints: patellofemoral, lateral tibiofemoral, and medial tibiofemoral. It is a combination gliding and hinge joint.

2. Flexion at knee, extension at hip, extension of vertebral column except for hyperextension at neck, abduction at shoulder, flexion of fingers, adduction of thumb, flexion/extension at elbow, and depression of mandible.

3. No more body surfing for Lars. He has a dislocated shoulder. The head of the humerus was displaced from the glenoid cavity causing tearing of the supporting ligaments and tendons (rotator cuff) of the shoulder joint.

4. Most likely the cartilage of the menisci were torn in the accident. Pieces of cartilage (and possibly bone) may be in the joint cavity. In addition, it is possible that the articular cartilage of the tibia (and possibly the femur) was damaged. Damage to the menisci and ligaments associated with the knee would contribute to stability problems.

5. There are many possibilities in Chuck's situation. The problems could be all unrelated, and the back pain could be completely related to vertebral column issues. On the opposite end of the spectrum, it is possible that a long-term ankle problem caused Chuck to walk differently in order to favor the damaged ankle, thus causing the bowing of the legs as the muscles pull the leg in a slightly different direction. From there, the change in gait could cause the muscles that attach to the pelvis to pull more forcefully on points that they normally would not, which could pull the pelvic bones slightly out of alignment in relation to the vertebral column, thus causing pain at the sacroiliac joint.

Chapter 10

1. The runner's leg muscles will contain a higher amount of slow fibers—rich in myoglobin, mitochondria, and blood. The weightlifter's arm muscles will contain abundant fast fibers- high in glycogen, with less myoglobin and blood capillaries than the slow fibers. The slow fibers appear red while the fast fibers are white in color.

2. Bill's muscles in the casted leg were not being used, so they decreased in size due to loss of myofibrils. Bill's leg shows disuse atrophy.

3. Acetylcholine (ACh) is the neurotransmitter used to "bridge the gap" at the neuromuscular junction. If ACh release is blocked, the neuron cannot send a signal to the muscle and the muscle will not contract.

4. Some cardiac muscle cells are autorhythmic and contraction is started intrinsically. Gap junctions connect the cardiac cells, spreading the impulse through the fiber network so it contracts as a unit. Cardiac muscle has a long refractory period so it will exhibit tetanus like skeletal muscle does.

5. First, it is the term "extensibility" (not "contractility") that refers to stretching of a muscle fiber. In addition, the contractile proteins never change length during muscle contraction or during stretching of a muscle fiber—they only change position relative to each other.

Chapter 11

1. The orbicularis oris protrudes the lips. The genioglossus protracts the tongue. The buccinator aids in sucking.

2. One likely possibility is the rhomboid major muscle.

3. Masseter, temporalis, medial pterygoid, and lateral pterygoid.

4. The most likely injury is a rotator cuff injury. The rotator cuff includes tendons of the subscapularis, supraspinatus, infraspinatus, and teres minor muscles. The most common injury is to the supraspinatus muscle and/or tendon (impingement syndrome).

5. Wyman probably has an inguinal hernia. He should see a doctor, since constriction of the intestines pushing out through the opening can cause serious damage.

Chapter 12

1. The spine of C7 forms a prominence along the midline at the base of the neck. The spine of T1 forms a second prominence slightly inferior to C7.

2. Palpation is touching or examining using the hands. Palpitation is a forceful beating of the heart that can often be felt by the patient.

3. The projection (shape of an organ on the surface of the body) of the heart on the anterior surface of the chest is indicated by four points as follows: The *inferior left point* is the apex (inferior, pointed end) of the heart, which projects downward and to the left and can be palpated in the fifth intercostal space, about 9 cm (3.5 in.) to the left of the midline. The *inferior right point* is at the lower border of the costal cartilage of the right sixth rib, about 3 cm to the right of the midline. The *superior right point* is located at the superior border of the costal cartilage of the right third rib, about 3 cm to the right of the midline. The *superior left point* is located at the inferior border of the costal cartilage of the left second rib, about 3 cm to the left of the midline. If you connect the four points, you can determine the heart's location and size (about the size of a closed fist).

4. The *gluteus medius muscle* is the site for an intramuscular injection, not the gluteus maximus, which Randy seems to think. In order to give this injection, the buttock is divided into quadrants and the upper outer quadrant is used as an injection site, where there is less chance of injury to the sciatic nerve or major blood vessels.

5. The skin superior to the buttocks has a dimple or depression on each side of the spine, due to the attachment of the skin and fascia to bone.

Chapter 13

1. Blood doping is used by some athletes to improve performance but the practice strains the heart. A blood test would reveal polycythemia, an increased number of red blood cells.

2. Basophils appear as bluish-black granular cells in a stained blood smear. An elevated number of basophils may suggest an allergic response such as "hay fever."

3. As the kidney fails, it secretes less erythropoietin, which results in a lower rate of red blood cell production. Recombinant erythropoietin can be administered to boost red blood cell production.

4. Iron is an important element found in the heme portion of hemoglobin. If levels of iron are low, then hemoglobin production is low. Low levels of hemoglobin reduce the oxygen-carrying capacity of blood, and a person with low hemoglobin levels would demonstrate symptoms of anemia. Adding iron to the diet increases one of the raw materials needed for adequate hemoglobin production.

5. Raoul's doctor suggested autologous preoperative transfusion (predonation). If Raoul does need a transfusion during surgery it is safer to use his own blood. Possible problems with matching blood types and blood-borne disease is eliminated.

Chapter 14

1. The cusps of the valves "catch" the blood like the fabric of the parachute "catches" the air. The cusps are anchored to the papillary muscles of the ventricle by the chordae tendineae, just as the fabric of the parachute is anchored by the lines, so the cusps do not flip open backward.

2. The anterior ventricular branch is located in the anterior interventricular sulcus and supplies both ventricles. The circumflex branch is located in the coronary sulcus and supplies the left ventricle and the left atrium.

3. The pericardium surrounds the heart. The fibrous pericardium anchors the heart to the diaphragm, sternum, and major blood vessels. The intercalated discs contain desmosomes, which hold the cardiac muscle fibers together.

4. Some people develop an immune response to streptococcal infections; this can result in rheumatic fever, which can damage the valves of the heart, especially the mitral valve. It would be important to pinpoint the cause of the sore throat to ensure that antibiotic treatment is started if the infection is streptococcal to reduce the possibility of heart damage.

5. Aortic valve stenosis is a narrowing and/or stiffening of the valve, which makes the valve harder to open. Therefore, the heart must pump harder to open the valve sufficiently to pump out an amount of blood equal to that of a normal heart. If the heart is unable to pump strongly enough, more blood can remain in the left ventricle following ventricular systole than is normal, and the system backs up from there.

Chapter 15

1. The foramen ovale and ductus arteriosus close to establish the separate pulmonary and systemic circulations. The umbilical artery and veins close since the placenta is no longer functioning. The ductus venosus closes so that the liver is no longer bypassed.

2. A vascular sinus is a type of vein that lacks smooth muscle in its tunica media. The tunica media and externa are replaced by dense connective tissue as in the intracranial sinuses.

3. Varicose veins are caused by faulty valves. The weak valves allow pooling of blood in the veins. Valves are found only in veins and not in arteries.

4. The catheter would snake from the femoral artery to the external iliac artery to the common iliac artery to the abdominal aorta to the arch of the aorta to the left ventricle.

5. Gina's doctor will either say her blood pressure is too high or is "borderline high." The doctor should encourage Gina to stop smoking and to lose some weight, since both smoking and being overweight are known risk factors for high blood pressure. The doctor should also suggest regularly monitoring her blood pressure to make sure that this one reading was not unusual for Gina. If Gina's blood pressure stays high and she is making no progress with the weight loss and smoking cessation, she may require medication to control her blood pressure.

Chapter 16

1. The route is from lymph capillaries to lymphatic vessels to the popliteal nodes to the superficial inguinal nodes to the right lumbar trunk to the cisterna chyli to the thoracic duct to the junction of the left internal jugular and left subclavian veins.

2. The two palatine tonsils are in the lateral, posterior oral cavity. The two lingual tonsils are at the base of the tongue. The single pharyngeal tonsil is in the posterior nasopharynx. The five tonsils function in the immune response to inhaled or ingested foreign invaders.

3. The spleen is an important secondary lymphatic organ. By removing the organ, the doctor hopes to reduce the level of the immune response causing Kelly's symptoms. However, because the immune response would be reduced, taking antibiotics prior to invasive procedures, such as dental work, reduce the risk of serious infection.

4. Nan recognizes the difficulty breathing as a part of her anaphylactic allergic reaction to the eggplant. Release of histamine from mast cells causes her airways to constrict, thus reducing the flow of air. In very serious reactions, the airway can narrow so much that an affected individual cannot take in sufficient oxygen. In such cases, epinephrine should be administered; this will dilate the airways.

5. The left subclavian trunk drains lymph from the left upper limb. Blockage would cause a buildup of lymph and interstitial fluid in the upper limb causes it to swell due to edema.

Chapter 17

1. The jingle that won't stop is similar to a reverberating circuit in that a signal will be repeated over and over again. Reverberating circuits function in breathing, memory, waking, and coordinated muscle activity.

2. At synapses, chemicals called neurotransmitters are released from the end of the presynaptic axon when it is stimulated. The neurotransmitters cross the synapse and bind to the postsynaptic neuron, thus transmitting the "message."

3. Gray matter appears gray in color due to the absence of myelin. It is composed of cell bodies, dendrites, and unmyelinated axons. White matter's color is due to the presence of myelin.

4. The somatic, afferent, peripheral division would detect sound and smell. The somatic, efferent, peripheral division sends messages to the skeletal muscles for stretching and yawning. Stomach rumbling and salivation are controlled by the autonomic, efferent, peripheral division. The enteric, efferent, peripheral division also controls GI tract organs.

5. The drug may not be able to pass the blood–brain barrier to reach the affected nervous tissue.

Chapter 18

1. The senior damaged the cervical region of his spinal cord. Recovery is unlikely.

2. The spinal cord is connected to the periphery by the spinal nerves. The nerves exit through the intervertebral foramina. Spaces are maintained between the vertebrae by the intervertebral discs. The dura mater of the spinal cord fuses with the epineurium. The nerves are connected to the cord by the posterior and anterior roots.

3. The spinal cord is anchored in place by the filum terminale and denticulate ligaments.

4. The spinal cord only extends to the level of the superior border of the second lumbar vertebra—well above the level of the coccygeal vertebrae.

5. The compression of the vertebrae may cause herniation of intervertebral discs, which then can affect spinal nerves. Nischal may have damage to one or more spinal nerves. In this case, damage to the sciatic nerve is a likely possibility.

Chapter 19

1. Movement of the right upper limb is controlled by the left hemisphere's primary motor area, located in the post central gyrus. Speech is controlled by Broca's area in the left hemisphere's frontal lobe just superior to the lateral cerebral sulcus.

2. While it is conceivable that he could still be drunk, the signs suggest Wolfgang may have suffered a subdural hemorrhage, which is very dangerous.

3. The amygdala is the center for fear, rage, and aggression. While it might seem useful to be "fearless" or to be nonaggressive, this could prove dangerous, if not fatal, in some circumstances in which fear might prevent someone from doing risky things or aggression might be an appropriate response.

4. The gray matter is located on the outer surface of the cerebrum. Since the cerebrum functions in intelligence, thought, mathematical ability, creativity, and so on, Alicia was being insulted when she was told that her gray matter was thin.

5. The dentist has injected anesthetic into the inferior alveolar nerve, a branch of the mandibular nerve that numbs the lower teeth and the lower lip. The tongue is numbed by blocking the lingual nerve. The upper teeth and lip are numbed by injecting the superior alveolar nerve, a branch of the maxillary branch.

Chapter 20

1. The exciting and potentially dangerous activities activate the sympathetic nervous system, resulting in a "fight-or-flight" reaction. The sympathetic nervous system stimulates the release of epinephrine (adrenalin) and norepinephrine from the adrenal medulla. These hormones prolong the fight-or-flight response.

2. The autonomic sensory neurons detect stretch in the stomach and send impulses to the brain. The hypothalamus sends impulses through the parasympathetic division, along the vagus nerve, to the heart resulting in decreased heart rate and force of contraction.

3. Mando's fight-or-flight response is continually being stimulated, which means, among other effects, that her heart rate is often high, her blood glucose may become chronically high, and her adrenal glands are being continually stimulated. Such chronically altered physiological conditions can lead to other systemic problems. (See the endocrine system chapter for possibilities.)

4. The autonomic division controls the digestive system and many other organs including the heart, lungs, and eyes. Most organs are controlled by dual innervation with the sympathetic and parasympathetic divisions causing opposite effects. The sympathetic division results in widespread effects over the entire body while the parasympathetic effects are more localized. The enteric system controls only the digestive system and may function independently of the ANS.

5. The hypothalamus controls many aspects of behavior and physiology, including those regulated by the autonomic nervous system. Though it is a physically small area of the brain, its effects are wide-reaching.

Chapter 21

1. The warm (thermal) receptors in her hands were at first activated by the warmth of the cup, then they adapted to the stimulus as she continued to hold the cup. The high temperature of the hot cocoa stimulated the thermal receptors in her mouth as well as pain receptors.

2. Muscle spindles monitor the change in length of skeletal muscles to the brain with proprioceptive information about body position. Without this input, Jenny would find it difficult to know where her body parts are without actually seeing them. If the spindle function does not return, Jenny may need to learn to move her arms, legs, facial muscles, etc., all over again in a different way.

3. Many brain areas are involved in controlling skeletal muscle activity, including the basal ganglia of the cerebrum, the premotor and primary motor areas of the cerebrum, the cerebellum, etc. In addition, sensory spinal and brain tracts are necessary for providing input and motor tracts for transmitting the "orders" for movement.

4. In an infant or very young child, the myelination of the CNS is not complete. The corticospinal tracts, which control fine, voluntary motor movement, are not fully myelinated until the child is about 2 years old. An infant would not be able to manipulate a knife or fork safely due to lack of complete motor control.

5. Jon's perception of feeling in his amputated foot is called phantom limb sensation. Impulses from the remaining proximal section of the sensory neuron are perceived by the brain as still coming from the amputated foot.

Chapter 22

1. The pressure from the tumor may be stimulating Suestia's olfactory areas in the temporal lobe, which would make her think she is smelling something that is not there.

2. Within the connective tissue that supports the olfactory epithelium are olfactory (Bowman's) glands, which produce a serous secretion that is carried to the surface of the epithelium by ducts. The secretion moistens the surface of the olfactory epithelium and dissolves odorants. Both supporting cells of the nasal epithelium and olfactory glands are innervated by branches of the facial (VII) nerve, which can be stimulated by certain chemicals. Impulses in these nerves, in turn, stimulate the lacrimal glands in the eyes and nasal mucous glands. The result is tears and a runny nose after inhaling substances such as onion.

3. The stick damaged the iris (the colored ring) which is part of the vascular tunic. The iris controls the diameter of the pupil but does not directly affect the focusing of light rays on the retina. If permanent damage was done to the cornea or ciliary muscles, then focusing may be affected.

4. The endolymph in the saccule, utricle, and membranous semicircular ducts moves in response to the up and down motion of the ship on the ocean. The vestibular apparatus is stimulated and sends impulses to the medulla, pons, and cerebellum. The eyes also send feedback to the cerebellum to help maintain balance. Excessive stimulation and conflicting messages (eyes say the ship is still, ears say it's moving) cause motion sickness.

5. Pigment in the iris and choroid are usually sufficient to absorb light entering the eye. A flash photograph captures the picture before the iris has time to constrict in response to the high intensity of light. The picture captures the reflection of the light back through the pupil (and iris if it is lightly pigmented). The eyes appear red due to the vascular tunic. Albinos have no pigment in their eyes so the "red eye" effect is more pronounced.

Chapter 23

1. Melatonin could be taken to help induce sleep during the evening at Tropicanaland. For the quickest adjustment, exposure to very bright light in the morning would help reset the body's clock. Sunlight would be better for this than a bright flashlight. The MSH may help with getting a tan but won't help reset the body's clock.

2. Amadu has a goiter, which is an increase in the size of the thyroid gland. In this case the thyroid gland has increased in size because Amadu's diet lacked sufficient iodine to produce normal levels of thyroid hormones. Because levels of the thyroid hormones were low, TSH levels increased in an attempt to stimulate the thyroid gland to produce more thyroid hormones, but without the iodine,

this was not possible. Adding iodine back into the diet should provide the necessary raw materials, and the gland will not be stimulated to grow more.

3. Because blood glucose might be chronically elevated by such treatment, this could make the pancreas work harder to produce insulin to lower the levels, which could have an adverse effect on the pancreas.

4. The pituitary gland is located in the hypophyseal fossa in the base of the sphenoid bone. The sphenoid is posterior to the ethmoid bone, which makes up a major portion of the walls and roof of the nasal cavity. Due to its proximity to the nasal cavity, the pituitary gland could have been affected by the radiation.

5. All hormones, including insulin, need an adequate number of functioning receptors to perform their function. The problem in this patient is a lack of functioning receptors, not a problem with insulin level.

Chapter 24

1. The septum is composed of hyaline cartilage covered with mucous membrane. The lateral wall of the nose is skin and muscle, lined with mucous membrane. The upper ear is skin covering elastic cartilage.

2. Lower oxygen levels at the higher altitudes means lower oxygen levels in Suzanne's blood and cerebrospinal fluid. Central and peripheral chemoreceptors will detect changes in blood gas levels and proportions and signal the medullary rhythmicity center to increase the rate of breathing.

3. The stick has penetrated the pleural cavity and perhaps the visceral pleura and the lung. Air in the pleural cavity (pneumothorax) has caused the lung to collapse (atelectasis). Only one lung collapsed since the lungs are surrounded by two separate cavities.

4. Several factors decrease respiratory efficiency in smokers: (1) Nicotine constricts terminal bronchioles, which decreases airflow into and out of the lungs. (2) Carbon monoxide in smoke binds to hemoglobin and reduces its oxygen-carrying capability. (3) Irritants in smoke cause increased mucus secretion by the mucosa of the bronchial tree and swelling of the mucosal lining, both of which impede airflow into and out of the lungs. (4) Irritants in smoke also inhibit the movement of cilia and destroy cilia in the lining of the respiratory system. Thus, excess mucus and foreign debris are not easily removed, which further adds to the difficulty in breathing. (5) With time, smoking leads to destruction of elastic fibers in the lungs and is the prime cause of emphysema. These changes cause collapse of small bronchioles and trapping of air in alveoli at the end of exhalation. The result is less efficient gas exchange.

5. LaTonya's cerebral cortex can allow her to voluntarily hold her breath for a short time. Increasing levels of carbon dioxide and H^+ will stimulate the inspiratory area and normal breathing will resume despite LaTonya's desire to get her sister in trouble.

Chapter 25

1. Saliva contains salivary amylase, which will break down the starch of the potatoes into sugar (sweet), and lingual lipase, which will break down the fat from the frying into fatty acids (tart) and monoglycerides.

2. The doctor will be looking at the mucosa of the regions of the large intestine to check for smoothness (vs. polyp growth), normal-appearing simple columnar epithelium, normal-appearing haustra, etc.

3. Modifications to increase surface area include overall length, circular folds, villi, and microvilli (brush border). The circular folds also enhance absorption by imparting a spiraling motion to the chyme.

4. During swallowing, the soft palate and uvula close off the nasopharynx. If the nervous system sends conflicting signals (such as breathe, swallow, and giggle), the palate may be in the wrong position to block food from entering the nasal cavity.

5. Billy is hearing the normal bowel sounds produced as a result of the movement of intestinal contents via peristalsis, haustral churning, and mass peristalsis. In addition, gas is being produced as bacteria digest undigested materials, which contributes to the sounds as the bubbles move through the intestinal contents.

Chapter 26

1. The glomerulus would continue to filter the blood, producing filtrate containing water, glucose, ions, and other compounds. The toxin would block the normal reabsorption of 99% of the filtrate by the renal tubules. The infected person would rapidly become dehydrated from lost water and would also lose ions, glucose, and essential vitamins and nutrients. The infected person would most likely soon die.

2. Streptococcal infections may trigger an adverse immune reaction that results in glomerulonephritis, destruction of the glomerular capillaries and tubules of the nephron. Early antibiotic treatment can help prevent such problems from occurring.

3. In females the urethra is about 4 cm long. In males, the urethra is about 15–20 cm long, including its passage through the penis, the urogenital diaphragm, and prostate. The female urinary bladder holds less urine than the male bladder due to the presence of the uterus in females.

4. Urine samples contain an abundance of epithelial cells that are shed from the mucosal lining of the urinary system.

5. Possibly. If Craig's blood pressure is chronically very high—or even occassionally very high—proteins that would not normally pass the glomerular–capsular membrane may be forced through. The chronic situation can be very serious and requires medical attention to reduce the blood pressure.

Chapter 27

1. The infection could pass through the vagina to the uterus through the uterine tubes, which are open to the pelvic cavity.

2. The presence or absence of the SRY gene on the Y chromosome determines the development of the urogenital structures before birth. The presence of SRY and secretion of testosterone by the fetal testes results in a male. Absence of SRY results in a female. Structures can be altered by surgery and hormones somewhat but the female will still lack receptors and organs necessary for sperm production and the male will lack receptors and organs necessary for the production of ova. The chromosome composition cannot be changed.

3. A hysterectomy will not cause menopause. A hysterectomy is the removal of the uterus, which does not produce hormones. The ovaries are left intact and continue to produce estrogens and progesterone.

4. The testicular artery and veins that drain the testes pass through the inguinal ring; if their blood supply is cut off by the closure of the inguinal canal, the testes would have no source of nourishment or waste removal.

5. A vasectomy cuts only the vas (ductus) deferens and leaves the testis untouched. Male secondary sex characteristics and libido (sex drive) are maintained by androgens including testosterone. The secretion of these hormones by the testis is not interrupted by the vasectomy.

GLOSSARY

PRONUNCIATION KEY

1. The most strongly accented syllable appears in capital letters, for example, bilateral (bī-LAT-er-al) and diagnosis (dī-ag-NŌ-sis).

2. If there is a secondary accent, it is noted by a prime ('), for example, constitution (kon'-sti-TOO-shun) and physiology (fiz'-ē-OL-ō-jē). Any additional secondary accents are also noted by a prime, for example, decarboxylation (dē'-kar-bok'-si-LĀ-shun).

3. Vowels marked by a line above the letter are pronounced with the long sound, as in the following common words:

   ā as in māke   ō as in pōle
   ē as in bē   ū as in cūte
   ī as in īvy

4. Vowels not marked by a line above the letter are pronounced with the short sound, as in the following words:

   a as in above or at   o as in not
   e as in bet   u as in bud
   i as in sip

5. Other vowel sounds are indicated as follows:

   oy as in oil
   oo as in root

6. Consonant sounds are pronounced as in the following words:

   b as in bat   m as in mother
   ch as in chair   n as in no
   d as in dog   p as in pick
   f as in father   r as in rib
   g as in get   s as in so
   h as in hat   t as in tea
   j as in jump   v as in very
   k as in can   w as in welcome
   ks as in tax   z as in zero
   kw as in quit   zh as in lesion
   l as in let

A

Abdomen (ab-DŌ-men or AB-dō-men) The area between the diaphragm and pelvis.

Abdominal (ab-DŌM-i-nal) **cavity** Superior portion of the abdominopelvic cavity that contains the stomach, spleen, liver, gallbladder, most of the small intestine, and part of the large intestine.

Abdominal thrust maneuver A first-aid procedure for choking. Employs a quick, upward thrust against the diaphragm that forces air out of the lungs with sufficient force to eject any lodged material. Also called the **Heimlich** (HĪM-lik) **maneuver.**

Abdominopelvic (ab-dom'-i-nō-PEL-vic) **cavity** Inferior component of the ventral body cavity that is subdivided into a superior abdominal cavity and an inferior pelvic cavity.

Abduction (ab-DUK-shun) Movement away from the midline of the body.

Abortion (a-BOR-shun) The premature loss (spontaneous) or removal (induced) of the embryo or nonviable fetus; miscarriage due to a failure in the normal process of developing or maturing.

Abscess (AB-ses) A localized collection of pus and liquefied tissue in a cavity.

Absorption (ab-SORP-shun) Intake of fluids or other substances by cells of the skin or mucous membranes; the passage of digested foods from the gastrointestinal tract into blood or lymph.

Accessory duct A duct of the pancreas that empties into the duodenum about 2.5 cm (1 in.) superior to the ampulla of Vater (hepatopancreatic ampulla). Also called the **duct of Santorini** (san'-tō-RĒ-nē).

Acetabulum (as'-e-TAB-ū-lum) The rounded cavity on the external surface of the hip bone that receives the head of the femur.

Acetylcholine (as'-e-til-KŌ-lēn) **(ACh)** A neurotransmitter liberated by many peripheral nervous system neurons and some central nervous system neurons. It is excitatory at neuromuscular junctions but inhibitory at some other synapses (for example, it slows heart rate).

Achalasia (ak'-a-LĀ-zē-a) A condition, caused by malfunction of the myenteric plexus, in which the lower esophageal sphincter fails to relax normally as food approaches. A whole meal may become lodged in the esophagus and enter the stomach very slowly. Distension of the esophagus results in chest pain that is often confused with pain originating from the heart.

Achilles tendon See **Calcaneal tendon.**

Acini (AS-i-nē) Groups of cells in the pancreas that secrete digestive enzymes.

Acoustic (a-KOOS-tik) Pertaining to sound or the sense of hearing.

Acquired immunodeficiency syndrome (AIDS) A fatal disease caused by the human immunodeficiency virus (HIV). Characterized by a positive HIV-antibody test, low helper T cell count, and certain indicator diseases (for example Kaposi's sarcoma, pneumocystis carinii pneumonia, tuberculosis, fungal diseases). Other symptoms include fever or night sweats, coughing, sore throat, fatigue, body aches, weight loss, and enlarged lymph nodes.

Acrosome (AK-rō-sōm) A lysosomelike organelle in the head of a sperm cell containing enzymes that facilitate the penetration of a sperm cell into a secondary oocyte.

Actin (AK-tin) A contractile protein that is part of thin filaments in muscle fibers.

Action potential An electrical signal that propagates along the membrane of a neuron or muscle fiber (cell); a rapid change in membrane potential that involves a depolarization followed by a repolarization. Also called a **nerve action potential** or **nerve impulse** as it relates to a neuron, and a **muscle action potential** as it relates to a muscle fiber.

Activation (ak′-ti-VA-shun) **energy** The minimum amount of energy required for a chemical reaction to occur.

Active transport The movement of substances across cell membranes against a concentration gradient, requiring the expenditure of cellular energy (ATP).

Acute (a-KŪT) Having rapid onset, severe symptoms, and a short course; not chronic.

Adaptation (ad′-ap-TA-shun) The adjustment of the pupil of the eye to changes in light intensity. The property by which a sensory neuron relays a decreased frequency of action potentials from a receptor, even though the strength of the stimulus remains constant; the decrease in perception of a sensation over time while the stimulus is still present.

Adduction (ad-DUK-shun) Movement toward the midline of the body.

Adenoids (AD-e-noyds) The pharyngeal tonsils.

Adenosine triphosphate (a-DEN-ō-sēn trī-FOS-fāt) **(ATP)** The main energy currency in living cells; used to transfer the chemical energy needed for metabolic reactions. ATP consists of the purine base *adenine* and the five-carbon sugar *ribose*, to which are added, in linear array, three *phosphate* groups.

Adhesion (ad-HE-zhun) Abnormal joining of parts to each other.

Adipocyte (AD-i-pō-sīt) Fat cell, derived from a fibroblast.

Adipose (AD-i-pōz) **tissue** Tissue composed of adipocytes specialized for triglyceride storage and present in the form of soft pads between various organs for support, protection, and insulation.

Adrenal cortex (a-DRE-nal KOR-teks) The outer portion of an adrenal gland, divided into three zones; the zona glomerulosa secretes mineralocorticoids, the zona fasciculata secretes glucocorticoids, and the zona reticularis secretes androgens.

Adrenal glands Two glands located superior to each kidney. Also called the **suprarenal** (soo′-pra-RE-nal) **glands.**

Adrenal medulla (me-DUL-a) The inner part of an adrenal gland, consisting of cells that secrete epinephrine, norepinephrine, and a small amount of dopamine in response to stimulation by sympathetic preganglionic neurons.

Adrenergic (ad′-ren-ER-jik) **neuron** A neuron that releases epinephrine (adrenaline) or norepinephrine (noradrenaline) as its neurotransmitter.

Adrenocorticotropic (ad-rē′-nō-kor-ti-kō-TRŌP-ik) **hormone (ACTH)** A hormone produced by the anterior pituitary that influences the production and secretion of certain hormones of the adrenal cortex.

Adventitia (ad-ven-TISH-a) The outermost covering of a structure or organ.

Aerobic (air-Ō-bik) Requiring molecular oxygen.

Afferent arteriole (AF-er-ent ar-TE-rē-ōl) A blood vessel of a kidney that divides into the capillary network called a glomerulus; there is one afferent arteriole for each glomerulus.

Agglutination (a-gloo′-ti-NA-shun) Clumping of microorganisms or blood cells, typically due to an antigen–antibody reaction.

Aggregated lymphatic follicles Clusters of lymph nodules that are most numerous in the ileum. Also called **Peyer's** (PI-erz) **patches.**

Albinism (AL-bin-izm) Abnormal, nonpathological, partial, or total absence of pigment in skin, hair, and eyes.

Aldosterone (al-DOS-ter-ōn) A mineralocorticoid produced by the adrenal cortex that promotes sodium and water reabsorption by the kidneys and potassium excretion in urine.

Allantois (a-LAN-tō-is) A small, vascularized outpouching of the yolk sac that serves as an early site for blood formation and development of the urinary bladder.

Alleles (a-LELZ) Alternate forms of a single gene that control the same inherited trait (such as type A blood) and are located at the same position on homologous chromosomes.

Allergen (AL-er-jen) An antigen that evokes a hypersensitivity reaction.

Alopecia (al′-ō-PE-shē-a) The partial or complete lack of hair as a result of factors such as genetics, aging, endocrine disorders, chemotherapy, and skin diseases.

Alpha (AL-fa) **cell** A type of cell in the pancreatic islets (islets of Langerhans) in the pancreas that secretes the hormone glucagon. Also termed an **A cell.**

Alpha receptor A type of receptor for norepinephrine and epinephrine; present on visceral effectors innervated by sympathetic postganglionic neurons.

Alveolar-capillary (al-VE-ō-lar) **membrane** Structure in the lungs consisting of the alveolar wall and basement membrane and a capillary endothelium and basement membrane through which the diffusion of respiratory gases occurs. Also called the **respiratory membrane.**

Alveolar duct Branch of a respiratory bronchiole around which alveoli and alveolar sacs are arranged.

Alveolar macrophage (MAK-rō-fāj) Highly phagocytic cell found in the alveolar walls of the lungs. Also called a **dust cell.**

Alveolar sac A cluster of alveoli that share a common opening.

Alveolus (al-VE-ō-lus) A small hollow or cavity; an air sac in the lungs; milk-secreting portion of a mammary gland. *Plural is* **alveoli** (al-VE-ol-ī).

Alzheimer's (ALTZ-hī-merz) **disease (AD)** Disabling neurological disorder characterized by dysfunction and death of specific cerebral neurons, resulting in widespread intellectual impairment, personality changes, and fluctuations in alertness.

Amenorrhea (ā-men-ō-RE-a) Absence of menstruation.

Amnesia (am-NE-zē-a) A lack or loss of memory.

Amnion (AM-nē-on) A thin, protective fetal membrane that develops from the epiblast; holds the fetus suspended in amniotic fluid. Also called the "**bag of waters.**"

Amniotic (am′-nē-OT-ik) **fluid** Fluid in the amniotic cavity, the space between the developing embryo (or fetus) and amnion; the fluid is initially produced as a filtrate from maternal blood and later includes fetal urine. It functions as a shock absorber, helps regulate fetal body temperature, and helps prevent desiccation.

Amphiarthrosis (am′-fē-ar-THRŌ-sis) A slightly movable joint, in which the articulating bony surfaces are separated by fibrous connective tissue or fibrocartilage to which both are attached; types are syndesmosis and symphysis.

Ampulla (am-PUL-la) A saclike dilation of a canal or duct.

Ampulla of Vater *See* **Hepatopancreatic ampulla.**

Anabolism (a-NAB-ō-lizm) Synthetic, energy-requiring reactions whereby small molecules are built up into larger ones.

Anaerobic (an-ar-Ō-bik) Not requiring oxygen.

Anal (Ā-nal) **canal** The last 2 or 3 cm (1 in.) of the rectum; opens to the exterior through the anus.

Anal column A longitudinal fold in the mucous membrane of the anal canal that contains a network of arteries and veins.

Anal triangle The subdivision of the female or male perineum that contains the anus.

Analgesia (an-al-JĒ-zē-a) Pain relief; absence of the sensation of pain.

Anaphase (AN-a-fāz) The third stage of mitosis in which the chromatids that have separated at the centromeres move to opposite poles of the cell.

Anaphylaxis (an′-a-fi-LAK-sis) A hypersensitivity (allergic) reaction in which IgE antibodies attach to mast cells and basophils, causing them to produce mediators of anaphylaxis (histamine, leukotrienes, kinins, and prostaglandins) that bring about increased blood permeability, increased smooth muscle contraction, and increased mucus production. Examples are hay fever, hives, and anaphylactic shock.

Anastomosis (a-nas-tō-MŌ-sis) An end-to-end union or joining of blood vessels, lymphatic vessels, or nerves.

Anatomic dead space Spaces of the nose, pharynx, larynx, trachea, bronchi, and bronchioles totaling about 150 mL of the 500 mL in a quiet breath (tidal volume); air in the anatomic dead space does not reach the alveoli to participate in gas exchange.

Anatomical (an′-a-TOM-i-kal) **position** A position of the body universally used in anatomical descriptions in which the body is erect, the head is level, the eyes face forward, the upper limbs are at the sides, the palms face forward, and the feet are flat on the floor.

Anatomy (a-NAT-ō-mē) The structure or study of structure of the body and the relation of its parts to each other.

Androgens (AN-drō-jenz) Masculinizing sex hormones produced by the testes in males and the adrenal cortex in both sexes; responsible for libido (sexual desire); the two main androgens are testosterone and dihydrotestosterone.

Anemia (a-NĒ-mē-a) Condition of the blood in which the number of functional red blood cells or their hemoglobin content is below normal.

Anesthesia (an′-es-THĒ-zē-a) A total or partial loss of feeling or sensation; may be general or local.

Aneurysm (AN-ū-rizm) A saclike enlargement of a blood vessel caused by a weakening of its wall.

Angina pectoris (an-JI-na *or* AN-ji-na PEK-tō-ris) A pain in the chest related to reduced coronary circulation due to coronary artery disease (CAD) or spasms of vascular smooth muscle in coronary arteries.

Angiogenesis (an-jē-ō-JEN-e-sis) The formation of blood vessels in the extraembryonic mesoderm of the yolk sac, connecting stalk, and chorion at the beginning of the third week of development.

Ankylosis (ang′-ki-LŌ-sis) Severe or complete loss of movement at a joint as the result of a disease process.

Anomaly (a-NOM-a-lē) An abnormality that may be a developmental (congenital) defect; a variant from the usual standard.

Anoxia (an-OK-sē-a) Deficiency of oxygen.

Antagonist (an-TAG-ō-nist) A muscle that has an action opposite that of the prime mover (agonist) and yields to the movement of the prime mover.

Antagonistic (an-tag-ō-NIST-ik) **effect** A hormonal interaction in which the effect of one hormone on a target cell is opposed by another hormone. For example, calcitonin (CT) lowers blood calcium level, whereas parathyroid hormone (PTH) raises it.

Anterior (an-TĒR-ē-or) Nearer to or at the front of the body. Equivalent to **ventral** in bipeds.

Anterior pituitary Anterior lobe of the pituitary gland. Also called the **adenohypophysis** (ad′-e-nō-hī-POF-i-sis).

Anterior root The structure composed of axons of motor (efferent) neurons that emerges from the anterior aspect of the spinal cord and extends laterally to join a posterior root, forming a spinal nerve. Also called a **ventral root.**

Anterolateral (an′-ter-ō-LAT-er-al) **pathway** Sensory pathway that conveys information related to pain, temperature, crude touch, pressure, tickle, and itch.

Antibody (AN-ti-bod′-ē) A protein produced by plasma cells in response to a specific antigen; the antibody combines with that antigen to neutralize, inhibit, or destroy it. Also called an **immunoglobulin** (im-ū-nō-GLOB-ū-lin) or **Ig.**

Anticoagulant (an-tī-cō-AG-ū-lant) A substance that can delay, suppress, or prevent the clotting of blood.

Antidiuretic (an′-ti-dī-ū-RET-ik) Substance that inhibits urine formation.

Antidiuretic hormone (ADH) Hormone produced by neurosecretory cells in the paraventricular and supraoptic nuclei of the hypothalamus that stimulates water reabsorption from kidney tubule cells into the blood and vasoconstriction of arterioles. Also called **vasopressin** (vāz-ō-PRES-in).

Antigen (AN-ti-jen) A substance that has immunogenicity (the ability to provoke an immune response) and reactivity (the ability to react with the antibodies or cells that result from the immune response); contraction of *anti*body *gen*erator. Also termed a **complete antigen.**

Antigen-presenting cell (APC) Special class of migratory cell that processes and presents antigens to T cells during an immune response; APCs include macrophages, B cells, and dendritic cells, which are present in the skin, mucous membranes, and lymph nodes.

Antrum (AN-trum) Any nearly closed cavity or chamber, especially one within a bone, such as a sinus.

Anulus fibrosus (AN-ū-lus fī-BRŌ-sus) A ring of fibrous tissue and fibrocartilage that encircles the pulpy substance (nucleus pulposus) of an intervertebral disc.

Anuria (an-Ū-rē-a) Absence of urine formation or daily urine output of less than 50 mL.

Anus (Ā-nus) The distal end and outlet of the rectum.

Aorta (ā-OR-ta) The main systemic trunk of the arterial system of the body that emerges from the left ventricle.

Aortic (ā-OR-tik) **body** Cluster of chemoreceptors on or near the arch of the aorta that respond to changes in blood levels of oxygen, carbon dioxide, and hydrogen ions (H^+).

Aortic reflex A reflex that helps maintain normal systemic blood pressure; initiated by baroreceptors in the wall of the ascending aorta and arch of the aorta. Nerve impulses from aortic baroreceptors reach the cardiovascular center via sensory axons of the vagus (X) nerves.

Aperture (AP-er-chur) An opening or orifice.

Apex (Ā-peks) The pointed end of a conical structure, such as the apex of the heart.

Aphasia (a-FĀ-zē-a) Loss of ability to express oneself properly through speech or loss of verbal comprehension.

Apnea (AP-nē-a) Temporary cessation of breathing.

Apneustic (ap-NOO-stik) **area** A part of the respiratory center in the pons that sends stimulatory nerve impulses to the inspiratory area that activate and prolong inhalation and inhibit exhalation.

Apocrine (AP-ō-krin) **gland** A type of gland in which the secretory products gather at the free end of the secreting cell and are pinched off, along with some of the cytoplasm, to become the secretion, as in mammary glands.

Aponeurosis (ap′-ō-noo-RŌ-sis) A sheetlike tendon joining one muscle with another or with bone.

Apoptosis (ap′-ō-TŌ-sis *or* ap′-ōp-TŌ-sis) Programmed cell death; a normal type of cell death that removes unneeded cells during embryological development, regulates the number of cells in tissues, and eliminates many potentially dangerous cells such as cancer cells. During apoptosis, the DNA fragments, the nucleus condenses, mitochondria cease to function, and the cytoplasm shrinks, but the plasma membrane remains intact. Phagocytes engulf and digest the apoptotic cells, and an inflammatory response does not occur.

Appositional (ap′-ō-ZISH-o-nal) **growth** Growth due to surface deposition of material, as in the growth in diameter of cartilage and bone. Also called **exogenous** (eks-OJ-e-nus) **growth.**

Aqueous humor (AK-wē-us HŪ-mer) The watery fluid, similar in composition to cerebrospinal fluid, that fills the anterior cavity of the eye.

Arachnoid (a-RAK-noyd) **mater** The middle of the three meninges (coverings) of the brain and spinal cord. Also termed the **arachnoid.**

Arachnoid villus (VIL-us) Berrylike tuft of the arachnoid mater that protrudes into the superior sagittal sinus and through which cerebrospinal fluid is reabsorbed into the bloodstream.

Arbor vitae (AR-bor VĪ-tē) The white matter tracts of the cerebellum, which have a treelike appearance when seen in midsagittal section.

Arch of the aorta The most superior portion of the aorta, lying between the ascending and descending segments of the aorta.

Areola (a-RĒ-ō-la) Any tiny space in a tissue. The pigmented ring around the nipple of the breast.

Arm The part of the upper limb from the shoulder to the elbow.

Arousal (a-ROW-zal) Awakening from sleep, a response due to stimulation of the reticular activating system (RAS).

Arrector pili (a-REK-tor PI-lē) Smooth muscles attached to hairs; contraction pulls the hairs into a vertical position, resulting in "goose bumps."

Arrhythmia (a-RITH-mē-a) An irregular heart rhythm. Also called a **dysrhythmia.**

Arteriole (ar-TĒ-rē-ōl) A small, almost microscopic, artery that delivers blood to a capillary.

Arteriosclerosis (ar-tē-rē-ō-skle-RŌ-sis) Group of diseases characterized by thickening of the walls of arteries and loss of elasticity.

Artery (AR-ter-ē) A blood vessel that carries blood away from the heart.

Arthritis (ar-THRI-tis) Inflammation of a joint.

Arthrology (ar-THROL-ō-jē) The study or description of joints.

Arthroplasty (AR-thrō-plas′-tē) Surgical replacement of joints, for example, the hip and knee joints.

Arthroscopy (ar-THROS-co-pē) A procedure for examining the interior of a joint, usually the knee, by inserting an arthroscope into a small incision; used to determine extent of damage, remove torn cartilage, repair cruciate ligaments, and obtain samples for analysis.

Arthrosis (ar-THRŌ-sis) A joint or articulation.

Articular (ar-TIK-ū-lar) **capsule** Sleevelike structure around a synovial joint composed of a fibrous capsule and a synovial membrane.

Articular cartilage (KAR-ti-lij) Hyaline cartilage attached to articular bone surfaces.

Articular disc Fibrocartilage pad between articular surfaces of bones of some synovial joints. Also called a **meniscus** (men-IS-kus).

Articulation (ar-tik′-ū-LĀ-shun) A joint; a point of contact between bones, cartilage and bones, or teeth and bones.

Arytenoid (ar′-i-TĒ-noyd) **cartilages** A pair of small, pyramidal cartilages of the larynx that attach to the vocal folds and intrinsic pharyngeal muscles and can move the vocal folds.

Ascending colon (KŌ-lon) The part of the large intestine that passes superiorly from the cecum to the inferior border of the liver, where it bends at the right colic (hepatic) flexure to become the transverse colon.

Ascites (as-SĪ-tēz) Abnormal accumulation of serous fluid in the peritoneal cavity.

Aseptic (ā-SEP-tik) Free from any infectious or septic material.

Association areas Large cortical regions on the lateral surfaces of the occipital, parietal, and temporal lobes and on the frontal lobes anterior to the motor areas connected by many motor and sensory axons to other parts of the cortex. The association areas are concerned with motor patterns, memory, concepts of word-hearing and word-seeing, reasoning, will, judgment, and personality traits.

Asthma (AZ-ma) Usually allergic reaction characterized by smooth muscle spasms in bronchi resulting in wheezing and difficult breathing. Also called **bronchial asthma.**

Astigmatism (a-STIG-ma-tizm) An irregularity of the lens or cornea of the eye causing the image to be out of focus and producing faulty vision.

Astrocyte (AS-trō-sīt) A neuroglial cell having a star shape that participates in brain development and the metabolism of neurotransmitters, helps form the blood–brain barrier, helps maintain the proper balance of K⁺ for generation of nerve impulses, and provides a link between neurons and blood vessels.

Ataxia (a-TAK-sē-a) A lack of muscular coordination, lack of precision.

Atherosclerotic (ath′-er-ō-skle-RO-tic) **plaque** (PLAK) A lesion that results from accumulated cholesterol and smooth muscle fibers (cells) of the tunica media of an artery; may become obstructive.

Atom Unit of matter that makes up a chemical element; consists of a nucleus (containing positively charged protons and uncharged neutrons) and negatively charged electrons that orbit the nucleus.

Atresia (a TRĒ-zē-a) Degeneration and reabsorption of an ovarian follicle before it fully matures and ruptures; abnormal closure of a passage, or absence of a normal body opening.

Atrial fibrillation (Ā-trē-al fib-ri-LĀ-shun) Asynchronous contraction of cardiac muscle fibers in the atria that results in the cessation of atrial pumping.

Atrial natriuretic (na′-trē-ū-RET-ik) **peptide (ANP)** Peptide hormone, produced by the atria of the heart in response to stretching, that inhibits aldosterone production and thus lowers blood pressure; causes natriuresis, increased urinary excretion of sodium.

Atrioventricular (AV) (ā′-trē-ō-ven-TRIK-ū-lar) **bundle** The part of the conduction system of the heart that begins at the atrioventricular (AV) node, passes through the cardiac skeleton separating the atria and the ventricles, then extends a short distance down the interventricular septum before splitting into right and left bundle branches. Also called the **bundle of His** (HISS).

Atrioventricular (AV) node The part of the conduction system of the heart made up of a compact mass of conducting cells located in the septum between the two atria.

Atrioventricular (AV) valve A heart valve made up of membranous flaps or cusps that allows blood to flow in one direction only, from an atrium into a ventricle.

Atrium (Ā-trē-um) A superior chamber of the heart.

Atrophy (AT-rō-fē) Wasting away or decrease in size of a part, due to a failure, abnormality of nutrition, or lack of use.

Auditory ossicle (AW-di-tō-rē OS-si-kul) One of the three small bones of the middle ear called the **malleus, incus,** and **stapes.**

Auditory tube The tube that connects the middle ear with the nose and nasopharynx region of the throat. Also called the **eustachian** (ū-STĀ-shun *or* ū-STĀ-kē-an) **tube** or **pharyngotympanic tube.**

Auscultation (aws-kul-TĀ-shun) Examination by listening to sounds in the body.

Autoimmunity An immunological response against a person's own tissues.

Autolysis (aw-TOL-i-sis) Self-destruction of cells by their own lysosomal digestive enzymes after death or in a pathological process.

Autonomic ganglion (aw′-tō-NOM-ik GANG-lē-on) A cluster of cell bodies of sympathetic or parasympathetic neurons located outside the central nervous system.

Autonomic nervous system (ANS) Visceral sensory (afferent) and visceral motor (efferent) neurons. Autonomic motor neurons, both sympathetic and parasympathetic, conduct nerve impulses from the central nervous system to smooth muscle, cardiac muscle, and glands. So named because this part of the nervous system was thought to be self-governing or spontaneous.

Autonomic plexus (PLEK-sus) A network of sympathetic and parasympathetic axons; examples are the cardiac, celiac, and pelvic plexuses, which are located in the thorax, abdomen, and pelvis, respectively.

Autophagy (aw-TOF-a-jē) Process by which worn-out organelles are digested within lysosomes.

Autopsy (AW-top-sē) The examination of the body after death.

Autorhythmic cells Cardiac or smooth muscle fibers that are self-excitable (generate impulses without an external stimulus); act as the heart's pacemaker and conduct the pacing impulse through the conduction system of the heart; self-excitable neurons in the central nervous system, as in the inspiratory area of the brain stem.

Autosome (AW-tō-sōm) Any chromosome other than the X and Y chromosomes (sex chromosomes).

Axilla (ak-SIL-a) The small hollow beneath the arm where it joins the body at the shoulders. Also called the **armpit.**

Axon (AK-son) The usually single, long process of a nerve cell that propagates a nerve impulse toward the axon terminals.

Axon terminal Terminal branch of an axon where synaptic vesicles undergo exocytosis to release neurotransmitter molecules.

Azygos (AZ-ī-gos) An anatomical structure that is not paired; occurring singly.

B

B cell A lymphocyte that can develop into a clone of antibody-producing plasma cells or memory cells when properly stimulated by a specific antigen.

Babinski (ba-BIN-skē) **sign** Extension of the great toe, with or without fanning of the other toes, in response to stimulation of the outer margin of the sole; normal up to 18 months of age and indicative of damage to descending motor pathways such as the corticospinal tracts after that.

Back The posterior part of the body; the dorsum.

Ball-and-socket joint A synovial joint in which the rounded surface of one bone moves within a cup-shaped depression or socket of another bone, as in the shoulder or hip joint. Also called a **spheroid** (SFĒ-royd) **joint.**

Baroreceptor (bar′-ō-re-SEP-tor) Neuron capable of responding to changes in blood, air, or fluid pressure. Also called a **pressoreceptor.**

Basal ganglia (GANG-glē-a) Paired clusters of gray matter deep in each cerebral hemisphere including the globus pallidus, putamen, and caudate nucleus. Together, the caudate nucleus and putamen are known as the **corpus striatum.** Nearby structures that are functionally linked to the basal ganglia are the substantia nigra of the midbrain and the subthalamic nuclei of the diencephalon.

Basement membrane Thin, extracellular layer between epithelium and connective tissue consisting of a basal lamina and a reticular lamina.

Basilar (BĀS-i-lar) **membrane** A membrane in the cochlea of the internal ear that separates the cochlear duct from the scala tympani and on which the spiral organ (organ of Corti) rests.

Basophil (BĀ-sō-fil) A type of white blood cell characterized by a pale nucleus and large granules that stain blue-purple with basic dyes.

Belly The abdomen. The gaster or prominent, fleshy part of a skeletal muscle.

Beta (BĀ-ta) **cell** A type of cell in the pancreatic islets (islets of Langerhans) in the pancreas that secretes the hormone insulin.

Beta receptor A type of adrenergic receptor for epinephrine and norepinephrine; found on visceral effectors innervated by sympathetic postganglionic neurons.

Bicuspid (bī-KUS-pid) **valve** Atrioventricular (AV) valve on the left side of the heart. Also called the **mitral valve.**

Bilateral (bī-LAT-er-al) Pertaining to two sides of the body.

Bile (BĪL) A secretion of the liver consisting of water, bile salts, bile pigments, cholesterol, lecithin, and several ions; it emulsifies lipids prior to their digestion.

Bilirubin (bil-ē-ROO-bin) An orange pigment that is one of the end products of hemoglobin breakdown in the hepatocytes and is excreted as a waste material in bile.

Blastocele (BLAS-tō-sēl) The fluid-filled cavity within the blastocyst.

Blastocyst (BLAS-tō-sist) In the development of an embryo, a hollow ball of cells that consists of a blastocele (the internal cavity), trophoblast (outer cells), and inner cell mass.

Blastomere (BLAS-tō-mēr) One of the cells resulting from the cleavage of a fertilized ovum.

Blastula (BLAS-tyū-la) An early stage in the development of a zygote.

Blind spot Area in the retina at the end of the optic (II) nerve in which there are no photoreceptors.

Blood The fluid that circulates through the heart, arteries, capillaries, and veins and that constitutes the chief means of transport within the body.

Blood–brain barrier (BBB) A barrier consisting of specialized brain capillaries and astrocytes that prevents the passage of materials from the blood to the cerebrospinal fluid and brain.

Blood island Isolated mass of mesoderm derived from angioblasts and from which blood vessels develop.

Blood pressure (BP) Force exerted by blood against the walls of blood vessels due to contraction of the heart and influenced by the elasticity of the vessel walls; clinically, a measure of the pressure in arteries during ventricular systole and ventricular diastole. *See also* **mean arterial blood pressure.**

Blood reservoir (REZ-er-vwar) Systemic veins that contain large amounts of blood that can be moved quickly to parts of the body requiring the blood.

Blood–testis barrier (BTB) A barrier formed by Sertoli cells that prevents an immune response against antigens produced by spermatogenic cells by isolating the cells from the blood.

Body cavity A space within the body that contains various internal organs.

Bolus (BŌ-lus) A soft, rounded mass, usually food, that is swallowed.

Bony labyrinth (LAB-i-rinth) A series of cavities within the petrous portion of the temporal bone forming the vestibule, cochlea, and semicircular canals of the inner ear.

Bowman's capsule *See* **Glomerular capsule.**

Brachial plexus (BRĀ-kē-al PLEK-sus) A network of nerve axons of the ventral rami of spinal nerves C5, C6, C7, C8, and T1. The nerves that emerge from the brachial plexus supply the upper limb.

Bradycardia (brād′-i-KAR-dē-a) A slow resting heart or pulse rate (under 50 beats per minute).

Brain The part of the central nervous system contained within the cranial cavity.

Brain stem The portion of the brain immediately superior to the spinal cord, made up of the medulla oblongata, pons, and midbrain.

Brain waves Electrical signals that can be recorded from the skin of the head due to electrical activity of brain neurons.

Broad ligament A double fold of parietal peritoneum attaching the uterus to the side of the pelvic cavity.

Broca's (BRŌ-kaz) **area** Motor area of the brain in the frontal lobe that translates thoughts into speech. Also called the **motor speech area.**

Bronchi (BRONG-kē) Branches of the respiratory passageway including primary bronchi (the two divisions of the trachea), secondary or lobar bronchi (divisions of the primary bronchi that are distributed to the lobes of the lung), and tertiary or segmental bronchi (divisions of the secondary bronchi that are distributed to bronchopulmonary segments of the lung). *Singular is* **bronchus.**

Bronchial tree The trachea, bronchi, and their branching structures up to and including the terminal bronchioles.

Bronchiole (BRONG-kē-ōl) Branch of a tertiary bronchus further dividing into terminal bronchioles (distributed to lobules of the lung), which divide into respiratory bronchioles (distributed to alveolar sacs).

Bronchitis (brong-KĪ-tis) Inflammation of the mucous membrane of the bronchial tree; characterized by hypertrophy and hyperplasia of seromucous glands and goblet cells that line the bronchi which results in a productive cough.

Bronchopulmonary (brong′-kō-PUL-mō-ner-ē) **segment** One of the smaller divisions of a lobe of a lung supplied by its own branches of a bronchus.

Brunner's gland *See* **Duodenal gland.**

Buccal (BUK-al) Pertaining to the cheek or mouth.

Bulb of penis Expanded portion of the base of the corpus spongiosum penis.

Bulbourethral (bul′-bō-ū-RĒ-thral) **gland** One of a pair of glands located inferior to the prostate on either side of the urethra that secretes an alkaline fluid into the cavernous urethra. Also called a **Cowper's** (KOW-perz) **gland.**

Bulimia (boo-LIM-e-a *or* boo-LĒ-mē-a) A disorder characterized by overeating at least twice a week followed by purging by self-induced vomiting, strict dieting or fasting, vigorous exercise, or use of laxatives or diuretics. Also called **binge–purge syndrome.**

Bulk-phase endocytosis A process by which most body cells can ingest membrane-surrounded droplets of interstitial fluid.

Bundle branch One of the two branches of the atrioventricular (AV) bundle made up of specialized muscle fibers (cells) that transmit electrical impulses to the ventricles.

Bundle of His *See* **Atrioventricular (AV) bundle.**

Bursa (BUR-sa) A sac or pouch of synovial fluid located at friction points, especially about joints.

Bursitis (bur-SĪ-tis) Inflammation of a bursa.

Buttocks (BUT-oks) The two fleshy masses on the posterior aspect of the inferior trunk, formed by the gluteal muscles.

C

Calcaneal (kal-KĀ-nē-al) **tendon** The tendon of the soleus, gastrocnemius, and plantaris muscles at the back of the heel. Also called the **Achilles** (a-KIL-ēz) **tendon.**

Calcification (kal-si-fi-KĀ-shun) Deposition of mineral salts, primarily hydroxyapatite, in a framework formed by collagen fibers in which the tissue hardens. Also called **mineralization** (min′-e-ral-i-ZĀ-shun).

Calcitonin (kal-si-TŌ-nin) **(CT)** A hormone produced by the parafollicular cells of the thyroid gland that can lower the amount of blood calcium and phosphates by inhibiting bone resorption (breakdown of bone matrix) and by accelerating uptake of calcium and phosphates into bone matrix.

Calculus (KAL-kū-lus) A stone, or insoluble mass of crystallized salts or other material, formed within the body, as in the gallbladder, kidney, or urinary bladder.

Callus (KAL-lus) A growth of new bone tissue in and around a fractured area, ultimately replaced by mature bone. An acquired, localized thickening.

Calyx (KĀL-iks) Any cuplike division of the kidney pelvis. *Plural is* **calyces** (KĀ-li-sēz).

Canal (ka-NAL) A narrow tube, channel, or passageway.

Canaliculus (kan′-a-LIK-ū-lus) A small channel or canal, as in bones, where they connect lacunae. *Plural is* **canaliculi** (kan′-a-LIK-ū-lī).

Canal of Schlemm *See* **Scleral venous sinus.**

Cancellous (KAN-sel-us) Having a reticular or latticework structure, as in spongy tissue of bone.

Capacitation (ka′-pas-i-TĀ-shun) The functional changes that sperm undergo in the female reproductive tract that allow them to fertilize a secondary oocyte.

Capillary (KAP-i-lar′-ē) A microscopic blood vessel located between an arteriole and venule through which materials are exchanged between blood and interstitial fluid.

Carcinogen (kar-SIN-ō-jen) A chemical substance or radiation that causes cancer.

Cardiac (KAR-dē-ak) **arrest** Cessation of an effective heartbeat in which the heart is completely stopped or in ventricular fibrillation.

Cardiac cycle A complete heartbeat consisting of systole (contraction) and diastole (relaxation) of both atria plus systole and diastole of both ventricles.

Cardiac muscle Striated muscle fibers (cells) that form the wall of the heart; stimulated by an intrinsic conduction system and autonomic motor neurons.

Cardiac notch An angular notch in the anterior border of the left lung into which part of the heart fits.

Cardinal ligament A ligament of the uterus, extending laterally from the cervix and vagina as a continuation of the broad ligament.

Cardiogenic area (kar-dē-ō-JEN-ik) A group of mesodermal cells in the head end of an embryo that gives rise to the heart.

Cardiology (kar-dē-OL-ō-jē) The study of the heart and diseases associated with it.

Cardiovascular (kar-dē-ō-VAS-kū-lar) **center** Groups of neurons scattered within the medulla oblongata that regulate heart rate, force of contraction, and blood vessel diameter.

Carotene (KAR-o-tēn) Antioxidant precursor of vitamin A, which is needed for synthesis of photopigments; yellow-orange pigment present in the stratum corneum of the epidermis. Accounts for the yellowish coloration of skin. Also termed **beta-carotene.**

Carotid (ka-ROT-id) **body** Cluster of chemoreceptors on or near the carotid sinus that respond to changes in blood levels of oxygen, carbon dioxide, and hydrogen ions.

Carotid sinus A dilated region of the internal carotid artery just superior to where it branches from the common carotid artery; it contains baroreceptors that monitor blood pressure.

Carpal bones The eight bones of the wrist. Also called **carpals.**

Carpus (KAR-pus) A collective term for the eight bones of the wrist.

Cartilage (KAR-ti-lij) A type of connective tissue consisting of chondrocytes in lacunae embedded in a dense network of collagen and elastic fibers and a matrix of chondroitin sulfate.

Cartilaginous (kar′-tī-LAJ-i-nus) **joint** A joint without a synovial (joint) cavity where the articulating bones are held tightly together by cartilage, allowing little or no movement.

Cast A small mass of hardened material formed within a cavity in the body and then discharged from the body; can originate in different areas and can be composed of various materials.

Catabolism (ka-TAB-ō-lizm) Chemical reactions that break down complex organic compounds into simple ones, with the net release of energy.

Cataract (KAT-a-rakt) Loss of transparency of the lens of the eye or its capsule or both.

Cauda equina (KAW-da ē-KWĪ-na) A tail-like array of roots of spinal nerves at the inferior end of the spinal cord.

Caudal (KAW-dal) Pertaining to any tail-like structure; inferior in position.

Cecum (SĒ-kum) A blind pouch at the proximal end of the large intestine that attaches to the ileum.

Celiac plexus (PLEK-sus) A large mass of autonomic ganglia and axons located at the level of the superior part of the first lumbar vertebra. Also called the **solar plexus.**

Cell The basic structural and functional unit of all organisms; the smallest structure capable of performing all the activities vital to life.

Cell cycle Growth and division of a single cell into two identical cells; consists of interphase and cell division.

Cell division Process by which a cell reproduces itself that consists of a nuclear division (mitosis) and a cytoplasmic division (cytokinesis); types include somatic and reproductive cell division.

Cell junction Point of contact between plasma membrane of tissue cells.

Cementum (se-MEN-tum) Calcified tissue covering the root of a tooth.

Center of ossification (os'-i-fi-KĀ-shun) An area in the cartilage model of a future bone where the cartilage cells hypertrophy and then secrete enzymes that result in the calcification of their matrix, resulting in the death of the cartilage cells, followed by the invasion of the area by osteoblasts that then lay down bone.

Central canal A microscopic tube running the length of the spinal cord in the gray commissure. A circular channel running longitudinally in the center of an osteon (haversian system) of mature compact bone, containing blood and lymphatic vessels and nerves. Also called a **haversian** (ha-VER-shan) **canal.**

Central fovea (FŌ-vē-a) A depression in the center of the macula lutea of the retina, containing cones only and lacking blood vessels; the area of highest visual acuity (sharpness of vision).

Central nervous system (CNS) That portion of the nervous system that consists of the brain and spinal cord.

Centrioles (SEN-trē-ōlz) Paired, cylindrical structures of a centrosome, each consisting of a ring of microtubules and arranged at right angles to each other.

Centromere (SEN-trō-mēr) The constricted portion of a chromosome where the two chromatids are joined; serves as the point of attachment for the microtubules that pull chromatids during anaphase of cell division.

Centrosome (SEN-trō-sōm) A dense network of small protein fibers near the nucleus of a cell, containing a pair of centrioles and pericentriolar material.

Cephalic (se-FAL-ik) Pertaining to the head; superior in position.

Cerebellar peduncle (ser-e-BEL-ar pe-DUNG-kul) A bundle of nerve axons connecting the cerebellum with the brain stem.

Cerebellum (ser-e-BEL-um) The part of the brain lying posterior to the medulla oblongata and pons; governs balance and coordinates skilled movements.

Cerebral aqueduct (SER-ē-bral AK-we-dukt) A channel through the midbrain connecting the third and fourth ventricles and containing cerebrospinal fluid. Also termed the **aqueduct of Sylvius.**

Cerebral arterial circle A ring of arteries forming an anastomosis at the base of the brain between the internal carotid and basilar arteries and arteries supplying the cerebral cortex. Also called the **circle of Willis.**

Cerebral cortex The surface of the cerebral hemispheres, 2–4 mm thick, consisting of gray matter; arranged in six layers of neuronal cell bodies in most areas.

Cerebral peduncle One of a pair of nerve axon bundles located on the anterior surface of the midbrain, conducting nerve impulses between the pons and the cerebral hemispheres.

Cerebrospinal (se-rē'-brō-SPĪ-nal) **fluid (CSF)** A fluid produced by ependymal cells that cover choroid plexuses in the ventricles of the brain; the fluid circulates in the ventricles, the central canal, and the subarachnoid space around the brain and spinal cord.

Cerebrovascular (se rē'-brō-VAS-kū-lar) **accident (CVA)** Destruction of brain tissue (infarction) resulting from obstruction or rupture of blood vessels that supply the brain. Also called a **stroke** or **brain attack.**

Cerebrum (SER-e-brum *or* se-RĒ-brum) The two hemispheres of the forebrain (derived from the telencephalon), making up the largest part of the brain.

Cerumen (se-ROO-men) Waxlike secretion produced by ceruminous glands in the external auditory meatus (ear canal). Also termed **ear wax.**

Ceruminous (se-ROO-mi-nus) **gland** A modified sudoriferous (sweat) gland in the external auditory meatus that secretes cerumen (ear wax).

Cervical ganglion (SER-vi-kul GANG-glē-on) A cluster of cell bodies of postganglionic sympathetic neurons located in the neck, near the vertebral column.

Cervical plexus (PLEK-sus) A network formed by nerve axons from the ventral rami of the first four cervical nerves and receiving gray rami communicantes from the superior cervical ganglion.

Cervix (SER-viks) Neck; any constricted portion of an organ, such as the inferior cylindrical part of the uterus.

Chemoreceptor (kē-mō-rē-SEP-tor) Sensory receptor that detects the presence of a specific chemical.

Chiasm (KĪ-azm) A crossing; especially the crossing of axons in the optic (II) nerve.

Chief cell The secreting cell of a gastric gland that produces pepsinogen, the precursor of the enzyme pepsin, and the enzyme gastric lipase. Also called a **zymogenic** (zī'-mō-JEN-ik) **cell.** Cell in the parathyroid glands that secretes parathyroid hormone (PTH). Also called a **principal cell.**

Chiropractic (kī-rō-PRAK-tik) A system of treating disease by using one's hands to manipulate body parts, mostly the vertebral column.

Cholecystectomy (kō'-lē-sis-TEK-tō-mē) Surgical removal of the gallbladder.

Cholecystitis (kō'-lē-sis-TĪ-tis) Inflammation of the gallbladder.

Cholesterol (kō-LES-te-rol) Classified as a lipid, the most abundant steroid in animal tissues; located in cell membranes and used for the synthesis of steroid hormones and bile salts.

Cholinergic (kō'-lin-ER-jik) **neuron** A neuron that liberates acetylcholine as its neurotransmitter.

Chondrocyte (KON-drō-sīt) Cell of mature cartilage.

Chondroitin (kon-DROY-tin) **sulfate** An amorphous matrix material found outside connective tissue cells.

Chordae tendineae (KOR-dē TEN-di-nē-ē) Tendonlike, fibrous cords that connect atrioventricular valves of the heart with papillary muscles.

Chorion (KŌ-rē-on) The most superficial fetal membrane that becomes the principal embryonic portion of the placenta; serves a protective and nutritive function.

Chorionic villi (kō-rē-ON-ik VIL-lī) Fingerlike projections of the chorion that grow into the decidua basalis of the endometrium and contain fetal blood vessels.

Chorionic villi sampling (CVS) The removal of a sample of chorionic villus tissue by means of a catheter to analyze the tissue for prenatal genetic defects.

Choroid (KŌ-royd) One of the vascular coats of the eyeball.

Choroid plexus (PLEK-sus) A network of capillaries located in the roof of each of the four ventricles of the brain; ependymal cells around choroid plexuses produce cerebrospinal fluid.

Chromaffin (KRŌ-maf-in) **cell** Cell that has an affinity for chrome salts, due in part to the presence of the precursors of the neurotransmitter epinephrine; found, among other places, in the adrenal medulla.

Chromatid (KRŌ-ma-tid) One of a pair of identical connected nucleoprotein strands that are joined at the centromere and separate during cell division, each becoming a chromosome of one of the two daughter cells.

Chromatin (KRŌ-ma-tin) The threadlike mass of genetic material, consisting of DNA and histone proteins, that is present in the nucleus of a nondividing or interphase cell.

Chromatolysis (krō-ma-TOL-i-sis) The breakdown of Nissl bodies into finely granular masses in the cell body of a neuron whose axon has been damaged.

Chromosome (KRŌ-mō-sōm) One of the small, threadlike structures in the nucleus of a cell, normally 46 in a human diploid cell, that bears the genetic material; composed of DNA and proteins (histones) that form a delicate chromatin thread during interphase; becomes packaged into compact rodlike structures that are visible under the light microscope during cell division.

Chronic (KRON-ik) Long term or frequently recurring; applied to a disease that is not acute.

Chronic obstructive pulmonary disease (COPD) A disease, such as bronchitis or emphysema, in which there is some degree of obstruction of airways and consequent increase in airway resistance.

Chyle (KĪL) The milky-appearing fluid found in the lacteals of the small intestine after absorption of lipids in food.

Chyme (KĪM) The semifluid mixture of partly digested food and digestive secretions found in the stomach and small intestine during digestion of a meal.

Ciliary (SIL-ē-ar'-ē) **body** One of the three parts of the vascular tunic of the eyeball, the others being the choroid and the iris; includes the ciliary muscle and the ciliary processes.

Ciliary ganglion (GANG-glē-on) A very small parasympathetic ganglion whose preganglionic axons come from the oculomotor (III) nerve and whose postganglionic axons carry nerve impulses to the ciliary muscle and the sphincter muscle of the iris.

Cilium (SIL-ē-um) A hair or hairlike process projecting from a cell that may be used to move the entire cell or to move substances along the surface of the cell. *Plural is* **cilia.**

Circle of Willis *See* **Cerebral arterial circle.**

Circular folds Permanent, deep, transverse folds in the mucosa and submucosa of the small intestine that increase the surface area for absorption. Also called **plicae circulares** (PLĪ-kē SER-kū-lar-ēs).

Circumduction (ser'-kum-DUK-shun) A movement at a synovial joint in which the distal end of a bone moves in a circle while the proximal end remains relatively stable.

Cirrhosis (si-RŌ-sis) A liver disorder in which the parenchymal cells are destroyed and replaced by connective tissue.

Cisterna chyli (sis-TER-na KĪ-lē) The origin of the thoracic duct.

Cleavage The rapid mitotic divisions following the fertilization of a secondary oocyte, resulting in an increased number of progressively smaller cells, called blastomeres.

Climacteric (klī-mak-TER-ik) Cessation of the reproductive function in the female or diminution of testicular activity in the male.

Climax The peak period or moments of greatest intensity during sexual excitement.

Clitoris (KLI-to-ris) An erectile organ of the female, located at the anterior junction of the labia minora, that is homologous to the male penis.

Clone (KLŌN) A population of identical cells.

Coarctation (kō'-ark-TĀ-shun) **of the aorta** A congenital heart defect in which a segment of the aorta is too narrow. As a result, the flow of oxygenated blood to the body is reduced, the left ventricle is forced to pump harder, and high blood pressure develops.

Coccyx (KOK-six) The fused bones at the inferior end of the vertebral column.

Cochlea (KOK-lē-a) A winding, cone-shaped tube forming a portion of the inner ear and containing the spiral organ (organ of Corti).

Cochlear duct The membranous cochlea consisting of a spirally arranged tube enclosed in the bony cochlea and lying along its outer wall. Also called the **scala media** (SCA-la MĒ-dē-a).

Coitus (KŌ-i-tus) Sexual intercourse.

Collagen (KOL-a-jen) A protein that is the main organic constituent of connective tissue.

Collateral circulation The alternate route taken by blood through an anastomosis.

Colliculus (ko-LIK-ū-lus) A small elevation.

Colon The portion of the large intestine consisting of ascending, transverse, descending, and sigmoid portions.

Colony-stimulating factor (CSF) One of a group of molecules that stimulates development of white blood cells. Examples are macrophage CSF and granulocyte CSF.

Colostrum (kō-LOS-trum) A thin, cloudy fluid secreted by the mammary glands a few days prior to or after delivery before true milk is produced.

Column (KOL-um) Group of white matter tracts in the spinal cord.

Common bile duct A tube formed by the union of the common hepatic duct and the cystic duct that empties bile into the duodenum at the hepatopancreatic ampulla (ampulla of Vater).

Compact (dense) bone tissue Bone tissue that contains few spaces between osteons (haversian systems); forms the external portion of all bones and the bulk of the diaphysis (shaft) of long bones; is found immediately deep to the periosteum and external to spongy bone.

Concha (KONG-ka) A scroll-like bone found in the skull. *Plural is* **conchae** (KONG-kē).

Concussion (kon-KUSH-un) Traumatic injury to the brain that produces no visible bruising but may result in abrupt, temporary loss of consciousness.

Conduction system A group of autorhythmic cardiac muscle fibers that generates and distributes electrical impulses to stimulate coordinated contraction of the heart chambers; includes the sinoatrial (SA) node, the atrioventricular (AV) node, the atrioventricular (AV) bundle, the right and left bundle branches, and the Purkinje fibers.

Conductivity (kon'-duk-TIV-i-tē) The ability of a cell to propagate (conduct) action potentials along its plasma membrane; characteristic of neurons and muscle fibers (cells).

Condyloid (KON-di-loyd) **joint** A synovial joint structured so that an oval-shaped condyle of one bone fits into an elliptical cavity of another bone, permitting side-to-side and back-and-forth movements, such as the joint at the wrist between the radius and carpals. Also called an **ellipsoidal** (ē-lip-SOYD-al) **joint.**

Cone (KŌN) The type of photoreceptor in the retina that is specialized for highly acute color vision in bright light.

Congenital (kon-JEN-i-tal) Present at the time of birth.

Conjunctiva (kon'-junk-TĪ-va) The delicate membrane covering the eyeball and lining the eyes.

Connective tissue One of the most abundant of the four basic tissue types in the body, performing the functions of binding and supporting; consists of relatively few cells in a generous matrix (the ground substance and fibers between the cells).

Consciousness (KON-shus-nes) A state of wakefulness in which an individual is fully alert, aware, and oriented, partly as a result of feedback between the cerebral cortex and reticular activating system.

Continuous conduction (kon-DUK-shun) Propagation of an action potential (nerve impulse) in a step-by-step depolarization of each adjacent area of an axon membrane.

Contraception (kon′-tra-SEP-shun) The prevention of fertilization or impregnation without destroying fertility.

Contractility (kon′-trak-TIL-i-tē) The ability of cells or parts of cells to actively generate force to undergo shortening for movements. Muscle fibers (cells) exhibit a high degree of contractility.

Contralateral (kon′-tra-LAT-er-al) On the opposite side; affecting the opposite side of the body.

Conus medullaris (KŌ-nus med-ū-LAR-is) The tapered portion of the spinal cord inferior to the lumbar enlargement.

Convergence (con-VER-jens) A synaptic arrangement in which the synaptic end bulbs of several presynaptic neurons terminate on one postsynaptic neuron. The medial movement of the two eyeballs so that both are directed toward a near object being viewed in order to produce a single image.

Convulsion (con-VUL-shun) Violent, involuntary contractions or spasms of an entire group of muscles.

Cornea (KOR-nē-a) The nonvascular, transparent fibrous coat through which the iris of the eye can be seen.

Corona (kō-RŌ-na) Margin of the glans penis.

Corona radiata The innermost layer of granulosa cells that is firmly attached to the zona pellucida around a secondary oocyte.

Coronary artery disease (CAD) A condition such as atherosclerosis that causes narrowing of coronary arteries so that blood flow to the heart is reduced. The result is **coronary heart disease (CHD),** in which the heart muscle receives inadequate blood flow due to an interruption of its blood supply.

Coronary circulation The pathway followed by the blood from the ascending aorta through the blood vessels supplying the heart and returning it to the right atrium. Also called **cardiac circulation.**

Coronary sinus (SĪ-nus) A wide venous channel on the posterior surface of the heart that collects the blood from the coronary circulation and returns it to the right atrium.

Corpus (KOR-pus) The principal part of any organ; any mass or body.

Corpus albicans (KOR-pus AL-bi-kanz) A white fibrous patch in the ovary that forms after the corpus luteum regresses.

Corpus callosum (kal-LŌ-sum) The great commissure of the brain between the cerebral hemispheres.

Corpuscle of touch The sensory receptor for the sensation of touch; found in the dermal papillae, especially in palms and soles. Also called a **Meissner** (MĪZ-ner) **corpuscle.**

Corpus luteum (LOO-tē-um) A yellowish body in the ovary formed when a follicle has discharged its secondary oocyte; secretes estrogens, progesterone, relaxin, and inhibin.

Corpus striatum (strī-Ā-tum) An area in the interior of each cerebral hemisphere composed of the caudate and putamen of the basal ganglia and white matter of the internal capsule, arranged in a striated manner.

Cortex (KOR-teks) An outer layer of an organ. The convoluted layer of gray matter covering each cerebral hemisphere.

Costal (KOS-tal) Pertaining to a rib.

Costal cartilage (KAR-ti-lij) Hyaline cartilage that attaches a rib to the sternum.

Cramp A spasmodic, usually painful contraction of a muscle.

Cranial (KRĀ-ne-al) **cavity** A subdivision of the dorsal body cavity formed by the cranial bones and containing the brain.

Cranial nerve One of 12 pairs of nerves that leave the brain; pass through foramina in the skull; and supply sensory and motor neurons to the head, neck, part of the trunk, and viscera of the thorax and abdomen. Each is designated by a Roman numeral and a name.

Craniosacral (krā-nē-ō-SĀ-kral) **outflow** The axons of parasympathetic preganglionic neurons, which have their cell bodies located in nuclei in the brain stem and in the lateral gray matter of the sacral portion of the spinal cord.

Cranium (KRĀ-nē-um) The skeleton of the skull that protects the brain and the organs of sight, hearing, and balance; includes the frontal, parietal, temporal, occipital, sphenoid, and ethmoid bones.

Crista (KRIS-ta) A crest or ridged structure. A small elevation in the ampulla of each semicircular duct that contains receptors for dynamic equilibrium.

Crossing-over The exchange of a portion of one chromatid with another during meiosis. It permits an exchange of genes among chromatids and is one factor that results in genetic variation of progeny.

Crus (KRUS) **of penis** Separated, tapered portion of the corpora cavernosa penis. *Plural is* **crura** (KROO-ra).

Crypt of Lieberkühn *See* **Intestinal gland.**

Cryptorchidism (krip-TOR-ki-dizm) The condition of undescended testes.

Cuneate (KŪ-nē-āt) **nucleus** A group of neurons in the inferior part of the medulla oblongata in which axons of the cuneate fasciculus terminate.

Cupula (KUP-ū-la) A mass of gelatinous material covering the hair cells of a crista; a sensory receptor in the ampulla of a semicircular canal stimulated when the head moves.

Cushing's syndrome Condition caused by a hypersecretion of glucocorticoids characterized by spindly legs, "moon face," "buffalo hump," pendulous abdomen, flushed facial skin, poor wound healing, hyperglycemia, osteoporosis, hypertension, and increased susceptibility to disease.

Cutaneous (kū-TĀ-nē-us) Pertaining to the skin.

Cyanosis (sī-a-NŌ-sis) A blue or dark purple discoloration, most easily seen in nail beds and mucous membranes, that results from an increased concentration of deoxygenated (reduced) hemoglobin (more than 5 gm/dL).

Cyst (SIST) A sac with a distinct connective tissue wall, containing a fluid or other material.

Cystic (SIS-tik) **duct** The duct that carries bile from the gallbladder to the common bile duct.

Cystitis (sis-TĪ-tis) Inflammation of the urinary bladder.

Cytokinesis (sī-tō-ki-NĒ-sis) Distribution of the cytoplasm into two separate cells during cell division; coordinated with nuclear division (mitosis).

Cytolysis (sī-TOL-i-sis) The rupture of living cells in which the contents leak out.

Cytoplasm (SĪ-tō-plazm) Cytosol plus all organelles except the nucleus.

Cytoskeleton Complex internal structure of cytoplasm consisting of microfilaments, microtubules, and intermediate filaments.

Cytosol (SĪ-tō-sol) Semifluid portion of cytoplasm in which organelles and inclusions are suspended and solutes are dissolved. Also called **intracellular fluid.**

D

Dartos (DAR-tōs) The contractile tissue deep to the skin of the scrotum.

Decidua (dē-SID-ū-a) That portion of the endometrium of the uterus (all but the deepest layer) that is modified during pregnancy and shed after childbirth.

Deciduous (dē-SID-ū-us) Falling off or being shed seasonally or at a particular stage of development. In the body, referring to the first set of teeth.

Decussation (dē-ku-SĀ-shun) A crossing-over to the opposite (contralateral) side; an example is the crossing of 90% of the axons in the large motor tracts to opposite sides in the medullary pyramids.

Deep Away from the surface of the body or an organ.

Deep fascia (FASH-ē-a) A sheet of connective tissue wrapped around a muscle to hold it in place.

Deep inguinal (IN-gwi-nal) **ring** A slitlike opening in the aponeurosis of the transversus abdominis muscle that represents the origin of the inguinal canal.

Deep-venous thrombosis (DVT) The presence of a thrombus in a vein, usually a deep vein of the lower limbs.

Defecation (def-e-KĀ-shun) The discharge of feces from the rectum.

Degeneration (dē-jen′-er-Ā-shun) A change from a higher to a lower state; a breakdown in structure.

Deglutition (dē-gloo-TISH-un) The act of swallowing.

Dehydration (dē-hī-DRĀ-shun) Excessive loss of water from the body or its parts.

Delta cell A cell in the pancreatic islets (islets of Langerhans) in the pancreas that secretes somatostatin. Also termed a **D cell.**

Demineralization (de-min′-er-al-i-ZĀ-shun) Loss of calcium and phosphorus from bones.

Dendrite (DEN-drīt) A neuronal process that carries electrical signals, usually graded potentials, toward the cell body.

Dendritic (den-DRIT-ik) **cell** One type of antigen-presenting cell with long branchlike projections that commonly is present in mucosal linings such as the vagina, in the skin (Langerhans cells in the epidermis), and in lymph nodes (follicular dendritic cells).

Dens (DENZ) Tooth.

Dental caries (KA-rēz) Gradual demineralization of the enamel and dentin of a tooth that may invade the pulp and alveolar bone. Also called **tooth decay.**

Denticulate (den-TIK-ū-lāt) Finely toothed or serrated; characterized by a series of small, pointed projections.

Dentin (DEN-tin) The bony tissues of a tooth enclosing the pulp cavity.

Dentition (den-TI-shun) The eruption of teeth. The number, shape, and arrangement of teeth.

Deoxyribonucleic (dē-ok′-sē-rī′-bō-noo-KLĒ-ik) **acid (DNA)** A nucleic acid constructed of nucleotides consisting of one of four bases (adenine, cytosine, guanine, or thymine), deoxyribose, and a phosphate group; encoded in the nucleotides is genetic information.

Depression (de-PRESH-un) Movement in which a part of the body moves inferiorly.

Dermal papilla (pa-PILL-a) Fingerlike projection of the papillary region of the dermis that may contain blood capillaries or corpuscles of touch (Meissner corpuscles).

Dermatology (der-ma-TOL-ō-jē) The medical specialty dealing with diseases of the skin.

Dermatome (DER-ma-tōm) The cutaneous area developed from one embryonic spinal cord segment and receiving most of its sensory innervation from one spinal nerve. An instrument for incising the skin or cutting thin transplants of skin.

Dermis (DER-mis) A layer of dense irregular connective tissue lying deep to the epidermis.

Descending colon (KŌ-lon) The part of the large intestine descending from the left colic (splenic) flexure to the level of the left iliac crest.

Detritus (de-TRĪ-tus) Particulate matter produced by or remaining after the wearing away or disintegration of a substance or tissue; scales, crusts, or loosened skin.

Detrusor (de-TROO-ser) **muscle** Smooth muscle that forms the wall of the urinary bladder.

Developmental biology The study of development from the fertilized egg to the adult form.

Diagnosis (dī-ag-NŌ-sis) Distinguishing one disease from another or determining the nature of a disease from signs and symptoms by inspection, palpation, laboratory tests, and other means.

Dialysis (dī-AL-i-sis) The removal of waste products from blood by diffusion through a selectively permeable membrane.

Diaphragm (DĪ-a-fram) Any partition that separates one area from another, especially the dome-shaped skeletal muscle between the thoracic and abdominal cavities. Also a dome-shaped device that is placed over the cervix, usually with a spermicide, to prevent conception.

Diaphysis (dī-AF-i-sis) The shaft of a long bone.

Diarrhea (dī-a-RE-a) Frequent defecation of liquid feces caused by increased motility of the intestines.

Diarthrosis (dī-ar-THRŌ-sis) A freely movable joint; types are gliding, hinge, pivot, condyloid, saddle, and ball-and-socket.

Diastole (dī-AS-tō-lē) In the cardiac cycle, the phase of relaxation or dilation of the heart muscle, especially of the ventricles.

Diastolic (dī-as-TOL-ik) **blood pressure** The force exerted by blood on arterial walls during ventricular relaxation; the lowest blood pressure measured in the large arteries, normally about 80 mmHg in a young adult.

Diencephalon (dī′-en-SEF-a-lon) A part of the brain consisting of the thalamus, hypothalamus, epithalamus, and subthalamus.

Diffusion (dif-Ū-zhun) A passive process in which there is a net or greater movement of molecules or ions from a region of high concentration to a region of low concentration until equilibrium is reached.

Digestion (dī-JES-chun) The mechanical and chemical breakdown of food to simple molecules that can be absorbed and used by body cells.

Dilate (DĪ-lāt) To expand or swell.

Diploid (DIP-loyd) Having the number of chromosomes characteristically found in the somatic cells of an organism; having two haploid sets of chromosomes, one each from the mother and father. Symbolized $2n$.

Direct motor pathways Collections of upper motor neurons with cell bodies in the motor cortex that project axons into the spinal cord, where they synapse with lower motor neurons or interneurons in the anterior horns. Also called the **pyramidal pathways.**

Disease Any change from a state of health.

Dislocation (dis-lō-KA-shun) Displacement of a bone from a joint with tearing of ligaments, tendons, and articular capsules. Also called **luxation** (luks-Ā-shun).

Dissect (di-SEKT) To separate tissues and parts of a cadaver or an organ for anatomical study.

Distal (DIS-tal) Farther from the attachment of a limb to the trunk; farther from the point of origin or attachment.

Diuretic (dī-ū-RET-ik) A chemical that increases urine volume by decreasing reabsorption of water, usually by inhibiting sodium reabsorption.

Divergence (dī-VER-jens) A synaptic arrangement in which the synaptic end bulbs of one presynaptic neuron terminate on several postsynaptic neurons.

Diverticulum (dī-ver-TIK-ū-lum) A sac or pouch in the wall of a canal or organ, especially in the colon.

Dorsal body cavity Cavity near the dorsal (posterior) surface of the body that consists of a cranial cavity and vertebral canal.

Dorsal ramus (RĀ-mus) A branch of a spinal nerve containing motor and sensory axons supplying the muscles, skin, and bones of the posterior part of the head, neck, and trunk.

Dorsiflexion (dor′-si-FLEK-shun) Bending the foot in the direction of the dorsum (upper surface).

Down-regulation Phenomenon in which there is a decrease in the number of receptors in response to an excess of a hormone or neurotransmitter.

Duct of Santorini *See* **Accessory duct.**

Duct of Wirsung *See* **Pancreatic duct.**

Ductus arteriosus (DUK-tus ar-tē-rē-O-sus) A small vessel connecting the pulmonary trunk with the aorta; found only in the fetus.

Ductus (vas) deferens (DEF-er-ens) The duct that carries sperm from the epididymis to the ejaculatory duct. Also called the **seminal duct.**

Ductus epididymis (ep′-i-DID-i-mis) A tightly coiled tube inside the epididymis, distinguished into a head, body, and tail, in which sperm undergo maturation.

Ductus venosus (ve-NŌ-sus) A small vessel in the fetus that helps the circulation bypass the liver.

Duodenal (doo-ō-DĒ-nal) **gland** Gland in the submucosa of the duodenum that secretes an alkaline mucus to protect the lining of the small intestine from the action of enzymes and to help neutralize the acid in chyme. Also called **Brunner's** (BRUN-erz) **gland.**

Duodenal papilla (pa-PILL-a) An elevation on the duodenal mucosa that receives the hepatopancreatic ampulla (ampulla of Vater).

Duodenum (doo′-ō-DĒ-num *or* doo-OD-e-num) The first 25 cm (10 in.) of the small intestine, which connects the stomach and the ileum.

Dura mater (DOO-ra MĀ-ter) The outermost of the three meninges (coverings) of the brain and spinal cord.

Dynamic equilibrium (ē-kwi-LIB-rē-um) The maintenance of body position, mainly the head, in response to sudden movements such as rotation.

Dysfunction (dis-FUNK-shun) Absence of completely normal function.

Dysmenorrhea (dis′-men-ō-RĒ-a) Painful menstruation.

Dysplasia (dis-PLĀ-zē-a) Change in the size, shape, and organization of cells due to chronic irritation or inflammation; may either revert to normal if stress is removed or progress to neoplasia.

Dyspnea (DISP-nē-a) Shortness of breath.

Dystrophia (dis-TRO-fē-a) Progressive weakening of a muscle.

E

Eardrum A thin, semitransparent partition of fibrous connective tissue between the external auditory meatus and the middle ear. Also called the **tympanic membrane.**

Ectoderm The primary germ layer that gives rise to the nervous system and the epidermis of skin and its derivatives.

Ectopic (ek-TOP-ik) Out of the normal location, as in ectopic pregnancy.

Edema (e-DĒ-ma) An abnormal accumulation of interstitial fluid.

Effector (e-FEK-tor) An organ of the body, either a muscle or a gland, that is innervated by somatic or autonomic motor neurons.

Efferent arteriole (EF-er-ent ar-TĒ-rē-ōl) A vessel of the renal vascular system that carries blood from a glomerulus to a peritubular capillary.

Efferent (EF-er-ent) **ducts** A series of coiled tubes that transport sperm from the rete testis to the epididymis.

Ejaculation (e-jak-ū-LĀ-shun) The reflex ejection or expulsion of semen from the penis.

Ejaculatory (e-JAK-ū-la-tō-rē) **duct** A tube that transports sperm from the ductus (vas) deferens to the prostatic urethra.

Elasticity (e-las-TIS-i-tē) The ability of tissue to return to its original shape after contraction or extension.

Electrocardiogram (e-lek′-trō-KAR-dē-ō-gram) (**ECG** or **EKG**) A recording of the electrical changes that accompany the cardiac cycle that can be detected at the surface of the body; may be resting, stress, or ambulatory.

Elevation (el-e-VĀ-shun) Movement in which a part of the body moves superiorly.

Embolism (EM-bō-lizm) Obstruction or closure of a vessel by an embolus.

Embolus (EM-bō-lus) A blood clot, bubble of air or fat from broken bones, mass of bacteria, or other debris or foreign material transported by the blood.

Embryo (EM-brē-ō) The young of any organism in an early stage of development; in humans, the developing organism from fertilization to the end of the eighth week of development.

Embryology (em′-brē-OL-ō-jē) The study of development from the fertilized egg to the end of the eighth week of development.

Emesis (EM-e-sis) Vomiting.

Emigration (em′-e-GRĀ-shun) Process whereby white blood cells (WBCs) leave the bloodstream by rolling along the endothelium, sticking to it, and squeezing between the endothelial cells. Adhesion molecules help WBCs stick to the endothelium. Also known as **migration** or **extravasation.**

Emission (ē-MISH-un) Propulsion of sperm into the urethra due to peristaltic contractions of the ducts of the testes, epididymides, and ductus (vas) deferens as a result of sympathetic stimulation.

Emphysema (em-fi-SĒ-ma) A lung disorder in which alveolar walls disintegrate, producing abnormally large air spaces and loss of elasticity in the lungs; typically caused by exposure to cigarette smoke.

Emulsification (ē-mul′-si-fi-KĀ-shun) The dispersion of large lipid globules into smaller, uniformly distributed particles in the presence of bile.

Enamel (e-NAM-el) The hard, white substance covering the crown of a tooth.

Endocardium (en-dō-KAR-dē-um) The layer of the heart wall, composed of endothelium and smooth muscle, that lines the inside of the heart and covers the valves and tendons that hold the valves open.

Endochondral ossification (en′-dō-KON-dral os′-i-fi-KĀ-shun) The replacement of cartilage by bone. Also called **intracartilaginous** (in′-tra-kar′-ti-LAJ-i-nus) **ossification.**

Endocrine (EN-dō-krin) **gland** A gland that secretes hormones into interstitial fluid and then the blood; a ductless gland.

Endocrinology (en′-dō-kri-NOL-ō-jē) The science concerned with the structure and functions of endocrine glands and the diagnosis and treatment of disorders of the endocrine system.

Endocytosis (en′-dō-sī-TŌ-sis) The uptake into a cell of large molecules and particles in which a segment of plasma membrane surrounds the substance, encloses it, and brings it in; includes phagocytosis, pinocytosis, and receptor-mediated endocytosis.

Endoderm (EN-dō-derm) A primary germ layer of the developing embryo; gives rise to the gastrointestinal tract, urinary bladder, urethra, and respiratory tract.

Endodontics (en′-dō-DON-tiks) The branch of dentistry concerned with the prevention, diagnosis, and treatment of diseases that affect the pulp, root, periodontal ligament, and alveolar bone.

Endogenous (en-DOJ-e-nus) Growing from or beginning within the organism.

Endolymph (EN-dō-limf′) The fluid within the membranous labyrinth of the internal ear.

Endometriosis (en′-dō-MĒ-trē-ō′-sis) The growth of endometrial tissue outside the uterus.

Endometrium (en′-dō-MĒ-trē-um) The mucous membrane lining the uterus.

Endomysium (en′-dō-MĪZ-ē-um) Invagination of the perimysium separating each individual muscle fiber (cell).

Endoneurium (en′-dō-NOO-rē-um) Connective tissue wrapping around individual nerve axons (cells).

Endoplasmic reticulum (en′-do-PLAZ-mik re-TIK-ū-lum) **(ER)** A network of channels running through the cytoplasm of a cell that serves in intracellular transportation, support, storage, synthesis, and packaging of molecules. Portions of ER where ribosomes are attached to the outer surface are called **rough ER**; portions that have no ribosomes are called **smooth ER.**

End organ of Ruffini *See* **Type II cutaneous mechanoreceptor.**

Endosteum (en-DOS-tē-um) The membrane that lines the medullary (marrow) cavity of bones, consisting of osteogenic cells and scattered osteoclasts.

Endothelial-capsular (en-dō-THĒ-lē-al) **membrane** A filtration membrane in a nephron of a kidney consisting of the endothelium and basement membrane of the glomerulus and the epithelium of the visceral layer of the glomerular (Bowman's) capsule.

Endothelium (en′-dō-THĒ-lē-um) The layer of simple squamous epithelium that lines the cavities of the heart, blood vessels, and lymphatic vessels.

Enteric (EN-ter-ik) **nervous system** The part of the nervous system that is embedded in the submucosa and muscularis of the gastrointestinal (GI) tract; governs motility and secretions of the GI tract.

Enteroendocrine (en-ter-ō-EN-dō-krin) **cell** A cell of the mucosa of the gastrointestinal tract that secretes a hormone that governs function of the GI tract; hormones secreted include gastrin, cholecystokinin, glucose-dependent insulinotropic peptide (GIP), and secretin.

Enzyme (EN-zīm) A substance that accelerates chemical reactions; an organic catalyst, usually a protein.

Eosinophil (ē′-ō-SIN-ō-fil) A type of white blood cell characterized by granules that stain red or pink with acid dyes.

Ependymal (e-PEN-de-mal) **cells** Neuroglial cells that cover choroid plexuses and produce cerebrospinal fluid (CSF); they also line the ventricles of the brain and probably assist in the circulation of CSF.

Epicardium (ep′-i-KAR-dē-um) The thin outer layer of the heart wall, composed of serous tissue and mesothelium. Also called the **visceral pericardium.**

Epidemiology (ep′-i-dē-mē-OL-ō-jē) Study of the occurrence and distribution of diseases and disorders in human populations.

Epidermis (ep-i-DERM-is) The superficial, thinner layer of skin, composed of keratinized stratified squamous epithelium.

Epididymis (ep′-i-DID-i-mis) A comma-shaped organ that lies along the posterior border of the testis and contains the ductus epididymis, in which sperm undergo maturation. *Plural is* **epididymides** (ep′-i-di-DIM-i-dēz).

Epidural (ep′-i-DOO-ral) **space** A space between the spinal dura mater and the vertebral canal, containing areolar connective tissue and a plexus of veins.

Epiglottis (ep′-i-GLOT-is) A large, leaf-shaped piece of cartilage lying on top of the larynx, attached to the thyroid cartilage; its unattached portion is free to move up and down to cover the glottis (vocal folds and rima glottidis) during swallowing.

Epimysium (ep′-i-MĪZ-ē-um) Fibrous connective tissue around muscles.

Epinephrine (ep-ē-NEF-rin) Hormone secreted by the adrenal medulla that produces actions similar to those that result from sympathetic stimulation. Also called **adrenaline** (a-DREN-a-lin).

Epineurium (ep′-i-NOO-rē-um) The superficial connective tissue covering around an entire nerve.

Epiphyseal (ep′-i-FIZ-ē-al) **line** The remnant of the epiphyseal plate in the metaphysis of a long bone.

Epiphyseal plate The hyaline cartilage plate in the metaphysis of a long bone; site of lengthwise growth of long bones.

Epiphysis (ē-PIF-i-sis) The end of a long bone, usually larger in diameter than the shaft (diaphysis).

Epiphysis cerebri (se-RĒ-brē) Pineal gland.

Episiotomy (e-piz′-ē-OT-ō-mē) A cut made with surgical scissors to avoid tearing of the perineum at the end of the second stage of labor.

Epistaxis (ep′-i-STAK-sis) Loss of blood from the nose due to trauma, infection, allergy, neoplasm, and bleeding disorders. Also called **nosebleed.**

Epithalamus (ep′-i-THAL-a-mus) Part of the diencephalon superior and posterior to the thalamus, comprising the pineal gland and associated structures.

Epithelial (ep′-i-THĒ-lē-al) **tissue** The tissue that forms innermost and outermost surfaces of body structures and forms glands.

Eponychium (ep′-o-NIK-ē-um) Narrow band of stratum corneum at the proximal border of a nail that extends from the margin of the nail wall. Also called the **cuticle.**

Erectile dysfunction Failure to maintain an erection long enough for sexual intercourse. Also known as **impotence** (IM-pō-tens).

Erection (ē-REK-shun) The enlarged and stiff state of the penis or clitoris resulting from the engorgement of the spongy erectile tissue with blood.

Eructation (e-ruk′-TĀ-shun) The forceful expulsion of gas from the stomach. Also called **belching.**

Erythema (er′-i-THĒ-ma) Skin redness usually caused by dilation of the capillaries.

Erythrocyte (e-RITH-rō-sīt) A mature red blood cell.

Erythropoietin (e-rith′-rō-POY-e-tin) A hormone released by the juxtaglomerular cells of the kidneys that stimulates red blood cell production.

Esophagus (e-SOF-a-gus) The hollow muscular tube that connects the pharynx and the stomach.

Estrogens (ES-tro-jenz) Feminizing sex hormones produced by the ovaries; govern development of oocytes, maintenance of female reproductive structures, and appearance of secondary sex characteristics; also affect fluid and electrolyte balance, and protein anabolism. Examples are β-estradiol, estrone, and estriol.

Etiology (ē′-tē-OL-ō-jē) The study of the causes of disease, including theories of the origin and organisms (if any) involved.

Eupnea (ŪP-nē-a) Normal quiet breathing.

Eustachian tube *See* **Auditory tube.**

Eversion (ē-VER-zhun) The movement of the sole laterally at the ankle joint or of an atrioventricular valve into an atrium during ventricular contraction.

Exacerbation (eg-zas′-er-BĀ-shun) An increase in the severity of symptoms or of a disease.

Excitability (ek-sīt′-a-BIL-i-tē) The ability of muscle fibers to receive and respond to stimuli; the ability of neurons to respond to stimuli and generate nerve impulses.

Excrement (EKS-kre-ment) Material eliminated from the body as waste, especially fecal matter.

Excretion (eks-KRĒ-shun) The process of eliminating waste products from the body; also the products excreted.

Exocrine (EK-sō-krin) **gland** A gland that secretes its products into ducts that carry the secretions into body cavities, into the lumen of an organ, or to the outer surface of the body.

Exocytosis (ex′-ō-sī-TŌ-sis) A process in which membrane-enclosed secretory vesicles form inside the cell, fuse with the plasma membrane, and release their contents into the interstitial fluid; achieves secretion of materials from a cell.

Exogenous (ex-SOJ-e-nus) Originating outside an organ or part.

Exhalation (eks-ha-LĀ-shun) Breathing out; expelling air from the lungs into the atmosphere. Also called **expiration.**

Extensibility (ek-sten′-si-BIL-i-tē) The ability of muscle tissue to stretch when it is pulled.

Extension (ek-STEN-shun) An increase in the angle between two bones; restoring a body part to its anatomical position after flexion.

External Located on or near the surface.

External auditory (AW-di-tōr-ē) **canal** or **meatus** (mē-Ā-tus) A curved tube in the temporal bone that leads to the middle ear.

External ear The outer ear, consisting of the pinna, external auditory canal, and tympanic membrane (eardrum).

External nares (NĀ-rez) The openings into the nasal cavity on the exterior of the body. Also called the **nostrils.**

External respiration The exchange of respiratory gases between the lungs and blood. Also called **pulmonary respiration.**

Exteroceptor (eks′-ter-ō-SEP-tor) A sensory receptor adapted for the reception of stimuli from outside the body.

Extracellular fluid (ECF) Fluid outside body cells, such as interstitial fluid and plasma.

Extracellular matrix (MĀ-triks) The ground substance and fibers between cells in a connective tissue.

Extravasation (eks-trav-a-SĀ-shun) The escape of fluid, especially blood, lymph, or serum, from a vessel into the tissues.

Extrinsic (eks-TRIN-sik) Of external origin.

Exudate (EKS-oo-dāt) Escaping fluid or semifluid material that oozes from a space and that may contain serum, pus, and cellular debris.

Eyebrow The hairy ridge superior to the eye.

F

F cell A cell in the pancreatic islets (islets of Langerhans) that secretes pancreatic polypeptide.

Face The anterior aspect of the head.

Falciform ligament (FAL-si-form LIG-a-ment) A sheet of parietal peritoneum between the two principal lobes of the liver. The ligamentum teres, or remnant of the umbilical vein, lies within its fold.

Falx cerebelli (FALKS ser′-e-BEL-lē) A small triangular process of the dura mater attached to the occipital bone in the posterior cranial fossa and projecting inward between the two cerebellar hemispheres.

Falx cerebri (FALKS SER-e-brē) A fold of the dura mater extending deep into the longitudinal fissure between the two cerebral hemispheres.

Fascia (FASH-ē-a) A fibrous membrane covering, supporting, and separating muscles.

Fascicle (FAS-i-kul) A small bundle or cluster, especially of nerve or muscle fibers (cells). Also called a **fasciculus** (fa-SIK-ū-lus). *Plural is* **fasciculi** (fa-SIK-yoo-lī).

Fasciculation (fa-sik′-ū-LĀ-shun) Abnormal, spontaneous twitch of all skeletal muscle fibers in one motor unit that is visible at the skin surface; not associated with movement of the affected muscle; present in progressive diseases of motor neurons, for example, poliomyelitis.

Fauces (FAW-sēz) The opening from the mouth into the pharynx.

Feces (FĒ-sēz) Material discharged from the rectum and made up of bacteria, excretions, and food residue. Also called **stool.**

Female reproductive cycle General term for the ovarian and uterine cycles, the hormonal changes that accompany them, and cyclic changes in the breasts and cervix; includes changes in the endometrium of a nonpregnant female that prepares the lining of the uterus to receive a fertilized ovum. Less correctly termed the **menstrual cycle.**

Fertilization (fer′-ti-li-ZĀ-shun) Penetration of a secondary oocyte by a sperm cell, meiotic division of secondary oocyte to form an ovum, and subsequent union of the nuclei of the gametes.

Fetal circulation The cardiovascular system of the fetus, including the placenta and special blood vessels involved in the exchange of materials between fetus and mother.

Fetus (FĒ-tus) In humans, the developing organism *in utero* from the beginning of the third month to birth.

Fever An elevation in body temperature above the normal temperature of 37°C (98.6°F) due to a resetting of the hypothalamic thermostat.

Fibroblast (FĪ-brō-blast) A large, flat cell that secretes most of the extracellular matrix of areolar and dense connective tissues.

Fibrous (FĪ-brus) **joint** A joint that allows little or no movement, such as a suture or a syndesmosis.

Fibrous tunic (TOO-nik) The superficial coat of the eyeball, made up of the posterior sclera and the anterior cornea.

Fight-or-flight response The effects produced upon stimulation of the sympathetic division of the autonomic nervous system.

Filiform papilla (FIL-i-form pa-PIL-a) One of the conical projections that are distributed in parallel rows over the anterior two-thirds of the tongue and lack taste buds.

Filtrate (fil-TRĀT) The fluid produced when blood is filtered by the endothelial-capsular membrane.

Filtration (fil-TRĀ-shun) The flow of a liquid through a filter (or membrane that acts like a filter) due to a hydrostatic pressure; occurs in capillaries due to blood pressure.

Filum terminale (FĪ-lum ter-mi-NAL-ē) Non-nervous fibrous tissue of the spinal cord that extends inferiorly from the conus medullaris to the coccyx.

Fimbriae (FIM-brē-ē) Fingerlike structures, especially the lateral ends of the uterine (Fallopian) tubes.

Fissure (FISH-ur) A groove, fold, or slit that may be normal or abnormal.

Fistula (FIS-tū-la) An abnormal passage between two organs or between an organ cavity and the outside.

Fixator A muscle that stabilizes the origin of the prime mover so that the prime mover can act more efficiently.

Fixed macrophage (MAK-rō-fāj) Stationary phagocytic cell found in the liver, lungs, brain, spleen, lymph nodes, subcutaneous tissue, and red bone marrow. Also called a **histiocyte** (HIS-tē-ō-sīt).

Flaccid (FLAS-sid) Relaxed, flabby, or soft; lacking muscle tone.

Flagellum (fla-JEL-um) A hairlike, motile process on the extremity of a bacterium, protozoan, or sperm cell. *Plural is* **flagella** (fla-JEL-a).

Flatus (FLĀ-tus) Gas in the stomach or intestines; commonly used to denote expulsion of gas through the anus.

Flexion (FLEK-shun) Movement in which there is a decrease in the angle between two bones.

Follicle (FOL-i-kul) A small secretory sac or cavity; the group of cells that contains a developing oocyte in the ovaries.

Follicle-stimulating hormone (FSH) Hormone secreted by the anterior pituitary; it initiates development of ova and stimulates the ovaries to secrete estrogens in females, and initiates sperm production in males.

Fontanel (fon′-ta-NEL) A fibrous connective tissue membrane-filled space where bone formation is not yet complete, especially between the cranial bones of an infant's skull.

Foot The terminal part of the lower limb, from the ankle to the toes.

Foramen (fō-RĀ-men) A passage or opening; a communication between two cavities of an organ, or a hole in a bone for passage of vessels or nerves. *Plural is* **foramina** (fō-RAM-i-na).

Foramen ovale (fō-RĀ-men ō-VAL-ē) An opening in the fetal heart in the septum between the right and left atria. A hole in the greater wing of the sphenoid bone that transmits the mandibular branch of the trigeminal (V) nerve.

Forearm (FOR-arm) The part of the upper limb between the elbow and the wrist.

Fornix (FOR-niks) An arch or fold; a tract in the brain made up of association fibers, connecting the hippocampus with the mammillary bodies; a recess around the cervix of the uterus where it protrudes into the vagina.

Fossa (FOS-a) A furrow or shallow depression.

Fourth ventricle (VEN-tri-kul) A cavity filled with cerebrospinal fluid within the brain lying between the cerebellum and the medulla oblongata and pons.

Fracture (FRAK-choor) Any break in a bone.

Frenulum (FREN-ū-lum) A small fold of mucous membrane that connects two parts and limits movement.

Frontal plane A plane at a right angle to a midsagittal plane that divides the body or organs into anterior and posterior portions. Also called a **coronal** (kō-RŌ-nal) **plane.**

Fundus (FUN-dus) The part of a hollow organ farthest from the opening.

Fungiform papilla (FUN-ji-form pa-PIL-a) A mushroomlike elevation on the upper surface of the tongue appearing as a red dot; most contain taste buds.

Furuncle (FYUR-ung-kul) A boil; painful nodule caused by bacterial infection and inflammation of a hair follicle or sebaceous (oil) gland.

G

Gallbladder A small pouch, located inferior to the liver, that stores bile and empties by means of the cystic duct.

Gallstone A solid mass, usually containing cholesterol, in the gallbladder or a bile-containing duct; formed anywhere between bile canaliculi in the liver and the hepatopancreatic ampulla (ampulla of Vater), where bile enters the duodenum. Also called a **biliary calculus.**

Gamete (GAM-ēt) A male or female reproductive cell; a sperm cell or secondary oocyte.

Ganglion (GANG-glē-on) Usually, a group of neuronal cell bodies lying outside the central nervous system (CNS). *Plural is* **ganglia** (GANG-glē-a).

Gastric (GAS-trik) **glands** Glands in the mucosa of the stomach composed of cells that empty their secretions into narrow channels called gastric pits. Types of cells are chief cells (secrete pepsinogen), parietal cells (secrete hydrochloric acid and intrinsic factor), surface mucous and mucous neck cells (secrete mucus), and G cells (secrete gastrin).

Gastroenterology (gas′-trō-en′-ter-OL-ō-jē) The medical specialty that deals with the structure, function, diagnosis, and treatment of diseases of the stomach and intestines.

Gastrointestinal (gas-trō-in-TES-ti-nal) **(GI) tract** A continuous tube running through the ventral body cavity extending from the mouth to the anus. Also called the **alimentary** (al′-i-MEN-tar-ē) **canal.**

Gastrulation (gas′-troo-LĀ-shun) The migration of groups of cells from the epiblast that transform a bilaminar embryonic disc into a trilaminar embryonic disc with three primary germ layers; transformation of the blastula into the gastrula.

Gene (JĒN) Biological unit of heredity; a segment of DNA located in a definite position on a particular chromosome; a sequence of DNA that codes for a particular mRNA, rRNA, or tRNA.

Genetic engineering The manufacture and manipulation of genetic material.

Genetics The study of genes and heredity.

Genitalia (jen′-i-TĀ-lē-a) Reproductive organs.

Genome (JĒ-nōm) The complete set of genes of an organism.

Genotype (JĒ-nō-tīp) The genetic makeup of an individual; the combination of alleles present at one or more chromosomal locations, as distinguished from the appearance, or phenotype, that results from those alleles.

Geriatrics (jer′-ē-AT-riks) The branch of medicine devoted to the medical problems and care of elderly persons.

Gestation (jes-TĀ-shun) The period of development from fertilization to birth.

Gingivae (jin-JI-vē) Gums. They cover the alveolar processes of the mandible and maxilla and extend slightly into each socket.

Gland Specialized epithelial cell or cells that secrete substances; may be exocrine or endocrine.

Glans penis (glanz PĒ-nis) The slightly enlarged region at the distal end of the penis.

Glaucoma (glaw-KŌ-ma) An eye disorder in which there is increased intraocular pressure due to an excess of aqueous humor.

Gliding joint A synovial joint having articulating surfaces that are usually flat, permitting only side-to-side and back-and-forth movements, as between carpal bones, tarsal bones, and the scapula and clavicle. Also called an **arthrodial** (ar-THRŌ-dē-al) **joint.**

Glomerular (glō-MER-ū-lar) **capsule** A double-walled globe at the proximal end of a nephron that encloses the glomerular capillaries. Also called **Bowman's** (BŌ-manz) **capsule.**

Glomerular filtrate (glō-MER-ū-lar FIL-trāt) The fluid produced when blood is filtered by the filtration membrane in the glomeruli of the kidneys.

Glomerular filtration The first step in urine formation in which substances in blood pass through the filtration membrane and the filtrate enters the proximal convoluted tubule of a nephron.

Glomerulus (glō-MER-ū-lus) A rounded mass of nerves or blood vessels, especially the microscopic tuft of capillaries that is surrounded by the glomerular (Bowman's) capsule of each kidney tubule. *Plural is* **glomeruli.**

Glottis (GLOT-is) The vocal folds (true vocal cords) in the larynx plus the space between them (rima glottidis).

Glucagon (GLOO-ka-gon) A hormone produced by the alpha cells of the pancreatic islets (islets of Langerhans) that increases blood glucose level.

Glucocorticoids (gloo-kō-KOR-ti-koyds) Hormones secreted by the cortex of the adrenal gland, especially cortisol, that influence glucose metabolism.

Glucose (GLOO-kōs) A hexose (six-carbon sugar), $C_6H_{12}O_6$, that is a major energy source for the production of ATP by body cells.

Glucosuria (gloo′-kō-SOO-rē-a) The presence of glucose in the urine; may be temporary or pathological. Also called **glycosuria.**

Glycogen (GLI-kō-jen) A highly branched polymer of glucose containing thousands of subunits; functions as a compact store of glucose molecules in liver and muscle fibers (cells).

Goblet cell A goblet-shaped unicellular gland that secretes mucus; present in epithelium of the airways and intestines.

Goiter (GOY-ter) An enlarged thyroid gland.

Golgi (GOL-jē) **complex** An organelle in the cytoplasm of cells consisting of four to six flattened sacs (cisternae), stacked on one another, with expanded areas at their ends; functions in processing, sorting, packaging, and delivering proteins and lipids to the plasma membrane, lysosomes, and secretory vesicles.

Golgi tendon organ *See* **Tendon organ.**

Gomphosis (gom-FŌ-sis) A fibrous joint in which a cone-shaped peg fits into a socket.

Gonad (GŌ-nad) A gland that produces gametes and hormones; the ovary in the female and the testis in the male.

Gonadotropic hormone Anterior pituitary hormone that affects the gonads.

Gout (GOWT) Hereditary condition associated with excessive uric acid in the blood; the acid crystallizes and deposits in joints, kidneys, and soft tissue.

Graafian follicle *See* **Vesicular ovarian follicle.**

Gracile (GRAS-īl) **nucleus** A group of nerve cells in the inferior part of the medulla oblongata in which axons of the gracile fasciculus terminate.

Gray commissure (KOM-i-shur) A narrow strip of gray matter connecting the two lateral gray masses within the spinal cord.

Gray matter Areas in the central nervous system and ganglia containing neuronal cell bodies, dendrites, unmyelinated axons, axon terminals, and neuroglia; Nissl bodies impart a gray color and there is little or no myelin in gray matter.

Gray ramus communicans (RĀ-mus kō-MŪ-ni-kans) A short nerve containing axons of sympathetic postganglionic neurons; the cell bodies of the neurons are in a sympathetic chain ganglion, and the unmyelinated axons extend via the gray ramus to a spinal nerve and then to the periphery to supply smooth muscle in blood vessels, arrector pili muscles, and sweat glands. *Plural is* **rami communicantes** (RĀ-mē kō-mū-ni-KAN-tēz).

Greater omentum (ō-MEN-tum) A large fold in the serosa of the stomach that hangs down like an apron anterior to the intestines.

Greater vestibular (ves-TIB-ū-lar) **glands** A pair of glands on either side of the vaginal orifice that open by a duct into the space between the hymen and the labia minora. Also called **Bartholin's** (BAR-to-linz) **glands.**

Groin (GROYN) The depression between the thigh and the trunk; the inguinal region.

Gross anatomy The branch of anatomy that deals with structures that can be studied without using a microscope. Also called **macroscopic anatomy.**

Growth An increase in size due to an increase in (1) the number of cells, (2) the size of existing cells as internal components increase in size, or (3) the size of intercellular substances.

Gustatory (GUS-ta-tō′-rē) Pertaining to taste.

Gynecology (gī′-ne-KOL-ō-jē) The branch of medicine dealing with the study and treatment of disorders of the female reproductive system.

Gynecomastia (gīn′-e-kō-MAS-tē-a) Excessive growth (benign) of the male mammary glands due to secretion of estrogens by an adrenal gland tumor (feminizing adenoma).

Gyrus (JĪ-rus) One of the folds of the cerebral cortex of the brain. *Plural is* **gyri** (JĪ-rī). Also called a **convolution.**

H

Hair A threadlike structure produced by hair follicles that develops in the dermis. Also called a **pilus** (PĪ-lus).

Hair follicle (FOL-li-kul) Structure, composed of epithelium and surrounding the root of a hair, from which hair develops.

Hair root plexus (PLEK-sus) A network of dendrites arranged around the root of a hair as free or naked nerve endings that are stimulated when a hair shaft is moved.

Hand The terminal portion of an upper limb, including the carpus, metacarpus, and phalanges.

Haploid (HAP-loyd) Having half the number of chromosomes characteristically found in the somatic cells of an organism; characteristic of mature gametes. Symbolized *n*.

Hard palate (PAL-at) The anterior portion of the roof of the mouth, formed by the maxillae and palatine bones and lined by mucous membrane.

Haustra (HAWS-tra) A series of pouches that characterize the colon; caused by tonic contractions of the teniae coli. *Singular is* **haustrum.**

Haversian canal *See* **Central canal.**

Haversian system *See* **Osteon.**

Head The superior part of a human, cephalic to the neck. The superior or proximal part of a structure.

Heart A hollow muscular organ lying slightly to the left of the midline of the chest that pumps the blood through the cardiovascular system.

Heart block An arrhythmia (dysrhythmia) of the heart in which the atria and ventricles contract independently because of a blocking of electrical impulses through the heart at some point in the conduction system.

Heart murmur (MER-mer) An abnormal sound that consists of a flow noise that is heard before, between, or after the normal heart sounds, or that may mask normal heart sounds.

Hemangioblast (hē-MAN-jē-ō-blast) A precursor mesodermal cell that develops into blood and blood vessels.

Hematocrit (hē-MAT-ō-krit) **(Hct)** The percentage of blood made up of red blood cells. Usually measured by centrifuging a blood sample in a graduated tube and then reading the volume of red blood cells and dividing it by the total volume of blood in the sample.

Hematology (hē′-ma-TOL-ō-jē) The study of blood.

Hematoma (hē-ma-TŌ-ma) A tumor or swelling filled with blood.

Hemiplegia (hem-i-PLĒ-jē-a) Paralysis of the upper limb, trunk, and lower limb on one side of the body.

Hemoglobin (hē′-mō-GLŌ-bin) **(Hb)** A substance in red blood cells consisting of the protein globin and the iron-containing red pigment heme that transports most of the oxygen and some carbon dioxide in blood.

Hemolysis (hē-MOL-i-sis) The escape of hemoglobin from the interior of a red blood cell into the surrounding medium; results from disruption of the cell membrane by toxins or drugs, freezing or thawing, or hypotonic solutions.

Hemolytic disease of the newborn A hemolytic anemia of a newborn child that results from the destruction of the infant's erythrocytes (red blood cells) by antibodies produced by the mother; usually the antibodies are due to an Rh blood type incompatibility. Also called **erythroblastosis fetalis** (e-rith′-rō-blas-TŌ-sis fe-TAL-is).

Hemophilia (hē′-mō-FIL-ē-a) A hereditary blood disorder where there is a deficient production of certain factors involved in blood clotting, resulting in excessive bleeding into joints, deep tissues, and elsewhere.

Hemopoiesis (hē-mō-poy-Ē-sis) Blood cell production, which occurs in red bone marrow after birth. Also called **hematopoiesis** (hem′-a-tō-poy-Ē-sis).

Hemorrhage (HEM-or-rij) Bleeding; the escape of blood from blood vessels, especially when the loss is profuse.

Hemorrhoids (HEM-ō-royds) Dilated or varicosed blood vessels (usually veins) in the anal region. Also called **piles.**

Hepatic (he-PAT-ik) Refers to the liver.

Hepatic duct A duct that receives bile from the bile capillaries. Small hepatic ducts merge to form the larger right and left hepatic ducts that unite to leave the liver as the common hepatic duct.

Hepatic portal circulation The flow of blood from the gastrointestinal organs to the liver before returning to the heart.

Hepatocyte (he-PAT-ō-cyte) A liver cell.

Hepatopancreatic (hep′-a-tō-pan′-krē-A-tik) **ampulla** A small, raised area in the duodenum where the combined common bile duct and main pancreatic duct empty into the duodenum. Also called the **ampulla of Vater** (VA-ter).

Hernia (HER-nē-a) The protrusion or projection of an organ or part of an organ through a membrane or cavity wall, usually the abdominal cavity.

Herniated (HER-nē-ā′-ted) **disc** A rupture of an intervertebral disc so that the nucleus pulposus protrudes into the vertebral cavity. Also called a **slipped disc.**

Heterocrine (HET-er-ō-krin) **gland** A gland, such as the pancreas, that is both an exocrine and an endocrine gland.

Hiatus (hī-Ā-tus) An opening; a foramen.

Hilus (HĪ-lus) An area, depression, or pit where blood vessels and nerves enter or leave an organ. Also called a **hilum.**

Hinge joint A synovial joint in which a convex surface of one bone fits into a concave surface of another bone, such as the elbow, knee, ankle, and interphalangeal joints. Also called a **ginglymus** (JIN-gli-mus) **joint.**

Hirsutism (HER-soot-izm) An excessive growth of hair in females and children, with a distribution similar to that in adult males, due to the conversion of vellus hairs into large terminal hairs in response to higher-than-normal levels of androgens.

Histamine (HISS-ta-mēn) Substance found in many cells, especially mast cells, basophils, and platelets, that is released when the cells are injured; results in vasodilation, increased permeability of blood vessels, and constriction of bronchioles.

Histology (hiss-TOL-ō-jē) Microscopic study of the structure of tissues.

Holocrine (HŌL-ō-krin) **gland** A type of gland in which entire secretory cells, along with their accumulated secretions, make up the secretory product of the gland, as in the sebaceous (oil) glands.

Homeostasis (hō′-mē-ō-STĀ-sis) The condition in which the body's internal environment remains relatively constant, within physiological limits.

Homologous (hō-MOL-ō-gus) Correspondence of two organs in structure, position, and origin.

Homologous chromosomes Two chromosomes that belong to a pair. Also called homologs.

Hormone (HOR-mōn) A secretion of endocrine cells that alters the physiological activity of target cells of the body.

Horn An area of gray matter (anterior, lateral, or posterior) in the spinal cord.

Human chorionic gonadotropin (kō-rē-ON-ik gō-nad-ō-TRŌ-pin) **(hCG)** A hormone produced by the developing placenta that maintains the corpus luteum.

Human chorionic somatomammotropin (sō-mat-ō-mam-ō-TRŌ-pin) **(hCS)** Hormone produced by the chorion of the placenta that stimulates breast tissue for lactation, enhances body growth, and regulates metabolism. Also called **human placental lactogen (hPL).**

Human growth hormone (hGH) Hormone secreted by the anterior pituitary that stimulates growth of body tissues, especially skeletal and muscular tissues. Also known as **somatotropin** and **somatotropic hormone (STH).**

Hyaluronic (hī′-a-loo-RON-ik) **acid** A viscous, amorphous extracellular material that binds cells together, lubricates joints, and maintains the shape of the eyeballs.

Hymen (HĪ-men) A thin fold of vascularized mucous membrane at the vaginal orifice.

Hyperextension (hī′-per-ek-STEN-shun) Continuation of extension beyond the anatomical position, as in bending the head backward.

Hyperplasia (hī′-per-PLĀ-zē-a) An abnormal increase in the number of normal cells in a tissue or organ, increasing its size.

Hypersecretion (hī′-per-se-KRĒ-shun) Overactivity of glands resulting in excessive secretion.

Hypersensitivity (hī′-per-sen-si-TI-vi-tē) Overreaction to an allergen that results in pathological changes in tissues. Also called **allergy.**

Hypertension (hī′-per-TEN-shun) High blood pressure.

Hyperthermia (hī′-per-THERM-ē-a) An elevated body temperature.

Hypertonia (hī′-per-TŌ-nē-a) Increased muscle tone that is expressed as spasticity or rigidity.

Hypertonic (hī′-per-TON-ik) Solution that causes cells to shrink due to loss of water by osmosis.

Hypertrophy (hī-PER-trō-fē) An excessive enlargement or overgrowth of tissue without cell division.

Hyperventilation (hī′-per-ven-ti-LĀ-shun) A rate of inhalation and exhalation higher than that required to maintain a normal partial pressure of carbon dioxide in the blood.

Hyponychium (hī′-pō-NIK-ē-um) Free edge of the fingernail.

Hypophyseal fossa (hī′-pō-FIZ-ē-al FOS-a) A depression on the superior surface of the sphenoid bone that houses the pituitary gland.

Hypophyseal (hī′-pō-FIZ-ē-al) **pouch** An outgrowth of ectoderm from the roof of the mouth from which the anterior pituitary develops.

Hypophysis (hī-POF-i-sis) Pituitary gland.

Hyposecretion (hī′-pō-se-KRĒ-shun) Underactivity of glands resulting in diminished secretion.

Hypothalamohypophyseal (hī-pō-tha-lam′-ō-hī-pō-FIZ-ē-al) **tract** A bundle of axons containing secretory vesicles filled with oxytocin or antidiuretic hormone that extend from the hypothalamus to the posterior pituitary.

Hypothalamus (hī′-pō-THAL-a-mus) A portion of the diencephalon, lying beneath the thalamus and forming the floor and part of the wall of the third ventricle.

Hypothermia (hī′-pō-THER-mē-a) Lowering of body temperature below 35°C (95°F); in surgical procedures, it refers to deliberate cooling of the body to slow down metabolism and reduce oxygen needs of tissues.

Hypotonia (hī′-pō-TŌ-nē-a) Decreased or lost muscle tone in which muscles appear flaccid.

Hypotonic (hī′-pō TON-ik) Solution that causes cells to swell and perhaps rupture due to gain of water by osmosis.

Hypoventilation (hī-pō-ven-ti-LĀ-shun) A rate of inhalation and exhalation lower than that required to maintain a normal partial pressure of carbon dioxide in plasma.

Hypoxia (hī-POKS-ē-a) Lack of adequate oxygen at the tissue level.

Hysterectomy (hiss-te-REK-tō-mē) The surgical removal of the uterus.

I

Ileocecal (il-ē-ō-SĒ-kal) **sphincter** A fold of mucous membrane that guards the opening from the ileum into the large intestine. Also called the **ileocecal valve.**

Ileum (IL-ē-um) The terminal part of the small intestine.

Immunity (im-Ū-ni-tē) The state of being resistant to injury, particularly by poisons, foreign proteins, and invading pathogens.

Immunoglobulin (im-ū-nō-GLOB-ū-lin) **(Ig)** An antibody synthesized by plasma cells derived from B lymphocytes in response to the introduction of an antigen. Immunoglobulins are divided into five kinds (IgG, IgM, IgA, IgD, IgE).

Immunology (im′-ū-NOL-ō-jē) The study of the responses of the body when challenged by antigens.

Imperforate (im-PER-fō-rāt) Abnormally closed.

Implantation (im-plan-TĀ-shun) The insertion of a tissue or a part into the body. The attachment of the blastocyst to the stratum basalis of the endometrium about 6 days after fertilization.

Incontinence (in-KON-ti-nens) Inability to retain urine, semen, or feces through loss of sphincter control.

Indirect motor pathways Motor tracts that convey information from the brain down the spinal cord for automatic movements, coordination of body movements with visual stimuli, skeletal muscle tone and posture, and balance. Also known as **extrapyramidal pathways.**

Induction (in-DUK-shun) The process by which one tissue (inducting tissue) stimulates the development of an adjacent unspecialized tissue (responding tissue) into a specialized one.

Infarction (in-FARK-shun) A localized area of necrotic tissue, produced by inadequate oxygenation of the tissue.

Infection (in-FEK-shun) Invasion and multiplication of microorganisms in body tissues, which may be inapparent or characterized by cellular injury.

Inferior (in-FĒR-ē-or) Away from the head or toward the lower part of a structure. Also called **caudad** (KAW-dad).

Inferior vena cava (VĒ-na CĀ-va) **(IVC)** Large vein that collects blood from parts of the body inferior to the heart and returns it to the right atrium.

Infertility Inability to conceive or to cause conception. Also called **sterility.**

Inflammation (in′-fla-MĀ-shun) Localized, protective response to tissue injury designed to destroy, dilute, or wall off the infecting agent or injured tissue; characterized by redness, pain, heat, swelling, and sometimes loss of function.

Infundibulum (in′-fun-DIB-ū-lum) The stalklike structure that attaches the pituitary gland to the hypothalamus of the brain. The funnel-shaped, open, distal end of the uterine (Fallopian) tube.

Ingestion (in-JES-chun) The taking in of food, liquids, or drugs, by mouth.

Inguinal (IN-gwi-nal) Pertaining to the groin.

Inguinal canal An oblique passageway in the anterior abdominal wall just superior and parallel to the medial half of the inguinal ligament that transmits the spermatic cord and ilioinguinal nerve in the male and round ligament of the uterus and ilioinguinal nerve in the female.

Inhalation (in-ha-LĀ-shun) The act of drawing air into the lungs. Also termed **inspiration.**

Inheritance The acquisition of body traits by transmission of genetic information from parents to offspring.

Inhibin A hormone secreted by the gonads that inhibits release of follicle-stimulating hormone (FSH) by the anterior pituitary.

Inhibiting hormone Hormone secreted by the hypothalamus that can suppress secretion of hormones by the anterior pituitary.

Inner cell mass A region of cells of a blastocyst that differentiates into the three primary germ layers—ectoderm, mesoderm, and endoderm—from which all tissues and organs develop; also called an **embryoblast.**

Insertion (in-SER-shun) The attachment of a muscle tendon to a movable bone or the end opposite the origin.

Insula (IN-soo-la) A triangular area of the cerebral cortex that lies deep within the lateral cerebral fissue, under the parietal, frontal, and temporal lobes.

Insulin (IN-soo-lin) A hormone produced by the beta cells of a pancreatic islet (islet of Langerhans) that decreases the blood glucose level.

Integrins (IN-te-grinz) A family of transmembrane glycoproteins in plasma membranes that function in cell adhesion; they are present in hemidesmosomes, which anchor cells to a basement membrane, and they mediate adhesion of neutrophils to endothelial cells during emigration.

Integumentary (in-teg′-ū-MEN-tar-e) Relating to the skin.

Intercalated (in-TER-ka-lāt-ed) **disc** An irregular transverse thickening of sarcolemma that contains desmosomes, which hold cardiac muscle fibers (cells) together, and gap junctions, which aid in conduction of muscle action potentials from one fiber to the next.

Intercostal (in′-ter-KOS-tal) **nerve** A nerve supplying a muscle located between the ribs.

Intermediate Between two structures, one of which is medial and one of which is lateral.

Intermediate filament Protein filament, ranging from 8 to 12 nm in diameter, that may provide structural reinforcement, hold organelles in place, and give shape to a cell.

Internal Away from the surface of the body.

Internal capsule A large tract of projection fibers lateral to the thalamus that is the major connection between the cerebral cortex and the brain stem and spinal cord; contains axons of sensory neurons carrying auditory, visual, and somatic sensory signals to the cerebral cortex plus axons of motor neurons descending from the cerebral cortex to the thalamus, subthalamus, brain stem, and spinal cord.

Internal ear The inner ear or labyrinth, lying inside the temporal bone, containing the organs of hearing and balance.

Internal nares (NĀ-rez) The two openings posterior to the nasal cavities opening into the nasopharynx. Also called the **choanae** (kō-Ā-nē).

Internal respiration The exchange of respiratory gases between blood and body cells. Also called **tissue respiration.**

Interneurons (in′-ter-NOO-ronz) Neurons whose axons extend only for a short distance and contact nearby neurons in the brain, spinal cord, or a ganglion; they comprise the vast majority of neurons in the body.

Interoceptor (in′-ter-ō-SEP-tor) Sensory receptor located in blood vessels and viscera that provides information about the body's internal environment.

Interphase (IN-ter-fāz) The period of the cell cycle between cell divisions, consisting of the G_1-(gap or growth) phase, when the cell is engaged in growth, metabolism, and production of substances required for division; S-(synthesis) phase, during which chromosomes are replicated; and G_2-phase.

Interstitial cell of Leydig *See* **Interstitial endocrinocyte.**

Interstitial (in′-ter-STISH-al) **endocrinocyte** A cell that is located in the connective tissue between seminiferous tubules in a mature testis that secretes testosterone. Also called an **interstitial cell of Leydig** (LĪ-dig).

Interstitial (in′-ter-STISH-al) **fluid** The portion of extracellular fluid that fills the microscopic spaces between the cells of tissues; the internal environment of the body. Also called **intercellular** or **tissue fluid.**

Interstitial growth Growth from within, as in the growth of cartilage. Also called **endogenous** (en-DOJ-e-nus) **growth.**

Interventricular (in′-ter-ven-TRIK-ū-lar) **foramen** A narrow, oval opening through which the lateral ventricles of the brain communicate with the third ventricle. Also called the **foramen of Monro.**

Intervertebral (in′-ter-VER-te-bral) **disc** A pad of fibrocartilage located between the bodies of two vertebrae.

Intestinal gland A gland that opens onto the surface of the intestinal mucosa and secretes digestive enzymes. Also called a **crypt of Lieberkühn** (LĒ-ber-kūn).

Intracellular (in′-tra-SEL-yū-lar) **fluid (ICF)** Fluid located within cells.

Intrafusal (in′-tra-FŪ-sal) **fibers** Three to ten specialized muscle fibers (cells), partially enclosed in a spindle-shaped connective tissue capsule, that make up a muscle spindle.

Intramembranous ossification (in′-tra-MEM-bra-nus os′-i-fi-KĀ-shun) The method of bone formation in which the bone is formed directly in membranous tissue.

Intraocular (in′-tra-OK-ū-lar) **pressure (IOP)** Pressure in the eyeball, produced mainly by aqueous humor.

Intrinsic (in-TRIN-sik) Of internal origin.

Intrinsic factor (IF) A glycoprotein, synthesized and secreted by the parietal cells of the gastric mucosa, that facilitates vitamin B_{12} absorption in the small intestine.

In utero (Ū-ter-ō) Within the uterus.

Invagination (in-vaj′-i-NĀ-shun) The pushing of the wall of a cavity into the cavity itself.

Inversion (in-VER-zhun) The movement of the sole medially at the ankle joint.

In vitro (VĒ-trō) Literally, in glass; outside the living body and in an artificial environment such as a laboratory test tube.

In vivo (VĒ-vō) In the living body.

Ipsilateral (ip′-si-LAT-er-al) On the same side, affecting the same side of the body.

Iris The colored portion of the vascular tunic of the eyeball seen through the cornea that contains circular and radial smooth muscle; the hole in the center of the iris is the pupil.

Irritable bowel syndrome (IBS) Disease of the entire gastrointestinal tract in which a person reacts to stress by developing symptoms (such as cramping and abdominal pain) associated with alternating patterns of diarrhea and constipation. Excessive amounts of mucus may appear in feces, and other symptoms include flatulence, nausea, and loss of appetite. Also known as **irritable colon** or **spastic colitis.**

Ischemia (is-KĒ-mē-a) A lack of sufficient blood to a body part due to obstruction or constriction of a blood vessel.

Islet of Langerhans *See* **Pancreatic islet.**

Isotonic (ī′-sō-TON-ik) Having equal tension or tone. A solution having the same concentration of impermeable solutes as cytosol.

Isthmus (IS-mus) A narrow strip of tissue or narrow passage connecting two larger parts.

J

Jaundice (JAWN-dis) A condition characterized by yellowness of the skin, the white of the eyes, mucous membranes, and body fluids because of a buildup of bilirubin.

Jejunum (je-JOO-num) The middle part of the small intestine.

Joint kinesthetic (kin′-es-THET-ik) **receptor** A proprioceptive receptor located in a joint, stimulated by joint movement.

Juxtaglomerular (juks-ta-glō-MER-ū-lar) **apparatus (JGA)** Consists of the macula densa (cells of the distal convoluted tubule adjacent to the afferent and efferent arteriole) and juxtaglomerular cells (modified cells of the afferent and sometimes efferent arteriole); secretes renin when blood pressure starts to fall.

K

Keratin (KER-a-tin) An insoluble protein found in the hair, nails, and other keratinized tissues of the epidermis.

Keratinocyte (ke-RAT-in′-ō-sīt) The most numerous of the epidermal cells; produces keratin.

Kidney (KID-nē) One of the paired reddish organs located in the lumbar region that regulates the composition, volume, and pressure of blood and produces urine.

Kidney stone A solid mass, usually consisting of calcium oxalate, uric acid, or calcium phosphate crystals, that may form in any portion of the urinary tract. Also called a **renal calculus.**

Kinesiology (ki-nē′-sē-OL-ō-jē) The study of the movement of body parts.

Kinesthesia (kin-es-THĒ-zē-a) The perception of the extent and direction of movement of body parts; this sense is possible due to nerve impulses generated by proprioceptors.

Kinetochore (ki-NET-ō-kor) Protein complex attached to the outside of a centromere to which kinetochore microtubules attach.

Kupffer's cell *See* **Stellate reticuloendothelial cell.**

Kyphosis (kī-FŌ-sis) An exaggeration of the thoracic curve of the vertebral column, resulting in a "round-shouldered" appearance. Also called **hunchback.**

L

Labial frenulum (LĀ-bē-al FREN-ū-lum) A medial fold of mucous membrane between the inner surface of the lip and the gums.

Labia majora (LĀ-bē-a ma-JŌ-ra) Two longitudinal folds of skin extending downward and backward from the mons pubis of the female.

Labia minora (min-OR-a) Two small folds of mucous membrane lying medial to the labia majora of the female.

Labium (LĀ-bē-um) A lip. A liplike structure. *Plural is* **labia** (LA-bē-a).

Labor The process of giving birth in which a fetus is expelled from the uterus through the vagina.

Labyrinth (LAB-i-rinth) Intricate communicating passageway, especially in the internal ear.

Lacrimal canal A duct, one on each eyelid, beginning at the punctum at the medial margin of an eyelid and conveying tears medially into the nasolacrimal sac.

Lacrimal gland Secretory cells, located at the superior anterolateral portion of each orbit, that secrete tears into excretory ducts that open onto the surface of the conjunctiva.

Lacrimal sac The superior expanded portion of the nasolacrimal duct that receives the tears from a lacrimal canal.

Lactation (lak-TĀ-shun) The secretion and ejection of milk by the mammary glands.

Lacteal (LAK tē-al) One of many lymphatic vessels in villi of the intestines that absorb triglycerides and other lipids from digested food.

Lacuna (la-KOO-na) A small, hollow space, such as that found in bones in which the osteocytes lie. *Plural is* **lacunae** (la-KOO-nē).

Lambdoid (lam-DOYD) **suture** The joint in the skull between the parietal bones and the occipital bone; sometimes contains sutural (Wormian) bones.

Lamellae (la-MEL-ē) Concentric rings of hard, calcified extracellular matrix found in compact bone.

Lamellated corpuscle Oval-shaped pressure receptor located in the dermis or subcutaneous tissue and consisting of concentric layers of connective tissue wrapped around the dendrites of a sensory neuron. Also called a **Pacinian** (pa-SIN-ē-an) **corpuscle.**

Lamina (LAM-i-na) A thin, flat layer or membrane, as the flattened part of either side of the arch of a vertebra. *Plural is* **laminae** (LAM-i-nē).

Lamina propria (PRŌ-prē-a) The connective tissue layer of a mucosa.

Langerhans (LANG-er-hans) **cell** Epidermal dendritic cell that functions as an antigen-presenting cell (APC) during an immune response.

Lanugo (la-NOO-gō) Fine downy hairs that cover the fetus.

Large intestine The portion of the gastrointestinal tract extending from the ileum of the small intestine to the anus, divided structurally into the cecum, colon, rectum, and anal canal.

Laryngopharynx (la-rin′-gō-FAR-inks) The inferior portion of the pharynx, extending downward from the level of the hyoid bone that divides posteriorly into the esophagus and anteriorly into the larynx. Also called the **hypopharynx.**

Laryngotracheal (la-rin′-gō-TRĀ-ke-al) **bud** An outgrowth of endoderm of the foregut from which the respiratory system develops.

Larynx (LAR-inks) The voice box, a short passageway that connects the pharynx with the trachea.

Lateral (LAT-er-al) Farther from the midline of the body or a structure.

Lateral ventricle (VEN-tri-kul) A cavity within a cerebral hemisphere that communicates with the lateral ventricle in the other cerebral hemisphere and with the third ventricle by way of the interventricular foramen.

Leg The part of the lower limb between the knee and the ankle.

Lens A transparent organ constructed of proteins (crystallins) lying posterior to the pupil and iris of the eyeball and anterior to the vitreous body.

Lesion (LĒ-zhun) Any localized, abnormal change in a body tissue.

Lesser omentum (ō-MEN-tum) A fold of the peritoneum that extends from the liver to the lesser curvature of the stomach and the first part of the duodenum.

Lesser vestibular (ves-TIB-ū-lar) **gland** One of the paired mucus-secreting glands with ducts that open on either side of the urethral orifice in the vestibule of the female.

Leukemia (loo-KĒ-mē-a) A malignant disease of the blood-forming tissues characterized by either uncontrolled production and accumulation of immature leukocytes in which many cells fail to reach maturity (acute) or an accumulation of mature leukocytes in the blood because they do not die at the end of their normal life span (chronic).

Leukocyte (LOO-kō-sīt) A white blood cell.

Leydig (LĪ-dig) **cell** A type of cell that secretes testosterone; located in the connective tissue between seminiferous tubules in a mature testis. Also known as **interstitial cell of Leydig** or **interstitial endocrinocyte.**

Libido (li-BĒ-dō) Sexual desire.

Ligament (LIG-a-ment) Dense regular connective tissue that attaches bone to bone.

Ligand (LĪ-gand) A chemical substance that binds to a specific receptor.

Limbic system A part of the forebrain, sometimes termed the visceral brain, concerned with various aspects of emotion and behavior; includes the limbic lobe, dentate gyrus, amygdala, septal nuclei, mammillary bodies, anterior thalamic nucleus, olfactory bulbs, and bundles of myelinated axons.

Lingual frenulum (LIN-gwal FREN-ū-lum) A fold of mucous membrane that connects the tongue to the floor of the mouth.

Lipase An enzyme that splits fatty acids from triglycerides and phospholipids.

Lipid (LIP-id) An organic compound composed of carbon, hydrogen, and oxygen that is usually insoluble in water, but soluble in alcohol, ether, and chloroform; examples include triglycerides (fats and oils), phospholipids, steroids, and eicosanoids.

Lipid bilayer Arrangement of phospholipid, glycolipid, and cholesterol molecules in two parallel sheets in which the hydrophilic "heads" face outward and the hydrophobic "tails" face inward; found in cellular membranes.

Lipoprotein (lip′-ō-PRŌ-tēn) One of several types of particles containing lipids (cholesterol and triglycerides) and proteins that make it water soluble for transport in the blood; high levels of **low-density lipoproteins (LDLs)** are associated with increased risk of atherosclerosis, whereas high levels of **high-density lipoproteins (HDLs)** are associated with decreased risk of atherosclerosis.

Liver Large organ under the diaphragm that occupies most of the right hypochondriac region and part of the epigastric region. Functionally, it produces bile and synthesizes most plasma proteins; interconverts nutrients; detoxifies substances; stores glycogen, iron, and vitamins; carries on phagocytosis of worn-out blood cells and bacteria; and helps synthesize the active form of vitamin D.

Locus coeruleus (LŌ-kus sē-ROO-lē-us) A group of neurons in the brain stem where norepinephrine (NE) is concentrated.

Long-term potentiation (LTP) Prolonged, enhanced synaptic transmission that occurs at certain synapses within the hippocampus of the brain; believed to underlie some aspects of memory.

Lordosis (lor-DŌ-sis) An exaggeration of the lumbar curve of the vertebral column. Also called **swayback.**

Lower limb The appendage attached at the pelvic (hip) girdle, consisting of the thigh, knee, leg, ankle, foot, and toes. Also called the **lower extremity.**

Lumbar (LUM-bar) Region of the back and side between the ribs and pelvis; loin.

Lumbar plexus (PLEK-sus) A network formed by the anterior (ventral) branches of spinal nerves L1 through L4.

Lumen (LOO-men) The space within an artery, vein, intestine, renal tubule, or other tubular structure.

Lungs Main organs of respiration that lie on either side of the heart in the thoracic cavity.

Lunula (LOO-noo-la) The moon-shaped white area at the base of a nail.

Luteinizing (LOO-tē-in′-īz-ing) **hormone (LH)** A hormone secreted by the anterior pituitary that stimulates ovulation, stimulates progesterone secretion by the corpus luteum, and readies the mammary glands for milk secretion in females; stimulates testosterone secretion by the testes in males.

Lymph (LIMF) Fluid confined in lymphatic vessels and flowing through the lymphatic system until it is returned to the blood.

Lymph node An oval or bean-shaped structure located along lymphatic vessels.

Lymphatic (lim-FAT-ik) **capillary** Closed-ended microscopic lymphatic vessel that begins in spaces between cells and converges with other lymphatic capillaries to form lymphatic vessels.

Lymphatic tissue A specialized form of reticular tissue that contains large numbers of lymphocytes.

Lymphatic vessel A large vessel that collects lymph from lymphatic capillaries and converges with other lymphatic vessels to form the thoracic and right lymphatic ducts.

Lymphocyte (LIM-fō-sīt) A type of white blood cell that helps carry out cell-mediated and antibody-mediated immune responses; found in blood and in lymphatic tissues.

Lysosome (LĪ-sō-sōm) An organelle in the cytoplasm of a cell, enclosed by a single membrane and containing powerful digestive enzymes.

Lysozyme (LĪ-sō-zīm) A bactericidal enzyme found in tears, saliva, and perspiration.

M

Macrophage (MAK-rō-fāj) Phagocytic cell derived from a monocyte; may be fixed or wandering.

Macula (MAK-ū-la) A discolored spot or a colored area. A small, thickened region on the wall of the utricle and saccule that contains receptors for static equilibrium.

Macula lutea (LOO-tē-a) The yellow spot in the center of the retina.

Major histocompatibility (MHC) antigens Surface proteins on white blood cells and other nucleated cells that are unique for each person (except for identical siblings); used to type tissues and help prevent rejection of transplanted tissues. Also known as **human leukocyte antigens (HLA).**

Malaise (ma-LĀYZ) Discomfort, uneasiness, and indisposition, often indicative of infection.

Malignant (ma-LIG-nant) Referring to diseases that tend to become worse and cause death, especially the invasion and spreading of cancer.

Mammary (MAM-ar-ē) **gland** Modified sudoriferous (sweat) gland of the female that produces milk for the nourishment of the young.

Mammillary (MAM-i-ler-ē) **bodies** Two small rounded bodies on the inferior aspect of the hypothalamus that are involved in reflexes related to the sense of smell.

Marrow (MAR-ō) Soft, spongelike material in the cavities of bone. Red bone marrow produces blood cells; yellow bone marrow contains adipose tissue that stores triglycerides.

Mast cell A cell found in areolar connective tissue that releases histamine, a dilator of small blood vessels, during inflammation.

Mastication (mas′-ti-KĀ-shun) Chewing.

Mature follicle A large, fluid-filled follicle containing a secondary oocyte and surrounding granulosa cells that secrete estrogens. Also called a **graafian** (GRAF-ē-an) **follicle.**

Meatus (mē-Ā-tus) A passage or opening, especially the external portion of a canal.

Mechanoreceptor (me-KAN-ō-rē-sep-tor) Sensory receptor that detects mechanical deformation of the receptor itself or adjacent cells; stimuli so detected include those related to touch, pressure, vibration, proprioception, hearing, equilibrium, and blood pressure.

Medial (MĒ-dē-al) Nearer the midline of the body or a structure.

Medial lemniscus (lem-NIS-kus) A white matter tract that originates in the gracile and cuneate nuclei of the medulla oblongata and extends to the thalamus on the same side; sensory axons in this tract conduct nerve impulses for the sensations of proprioception, fine touch, vibration, hearing, and equilibrium.

Median aperture (AP-er-choor) One of the three openings in the roof of the fourth ventricle through which cerebrospinal fluid enters the subarachnoid space of the brain and cord. Also called the **foramen of Magendie.**

Median plane A vertical plane dividing the body into right and left halves. Situated in the middle.

Mediastinum (mē′-dē-as-TĪ-num) The broad, median partition between the pleurae of the lungs, that extends from the sternum to the vertebral column in the thoracic cavity.

Medulla (me-DUL-la) An inner layer of an organ, such as the medulla of the kidneys.

Medulla oblongata (me-DUL-la ob′-long-GA-ta) The most inferior part of the brain stem. Also termed the **medulla.**

Medullary (MED-ū-lar′-ē) **cavity** The space within the diaphysis of a bone that contains yellow bone marrow. Also called the **marrow cavity.**

Medullary rhythmicity (rith-MIS-i-tē) **area** The neurons of the respiratory center in the medulla oblongata that control the basic rhythm of respiration.

Meibomian gland *See* **Tarsal gland.**

Meiosis (mī-Ō-sis) A type of cell division that occurs during production of gametes, involving two successive nuclear divisions that result in cells with the haploid *(n)* number of chromosomes.

Meissner's corpuscle *See* **Corpuscle of touch.**

Melanin (MEL-a-nin) A dark black, brown, or yellow pigment found in some parts of the body such as the skin, hair, and pigmented layer of the retina.

Melanocyte (MEL-a-nō-sīt′) A pigmented cell, located between or beneath cells of the deepest layer of the epidermis, that synthesizes melanin.

Melanocyte-stimulating hormone (MSH) A hormone secreted by the anterior pituitary that stimulates the dispersion of melanin granules in melanocytes in amphibians; continued administration produces darkening of skin in humans.

Melatonin (mel-a-TŌN-in) A hormone secreted by the pineal gland that helps set the timing of the body's biological clock.

Membrane A thin, flexible sheet of tissue composed of an epithelial layer and an underlying connective tissue layer, as in an epithelial membrane, or of arcolar connective tissue only, as in a synovial membrane.

Membranous labyrinth (mem-BRA-nus LAB-i-rinth) The part of the labyrinth of the internal ear that is located inside the bony labyrinth and separated from it by the perilymph; made up of the semicircular ducts, the saccule and utricle, and the cochlear duct.

Memory The ability to recall thoughts; commonly classifed as short-term (activated) and long-term.

Menarche (me-NAR-kē) The first menses (menstrual flow) and beginning of ovarian and uterine cycles.

Meninges (me-NIN-jēz) Three membranes covering the brain and spinal cord, called the dura mater, arachnoid mater, and pia mater. *Singular is* **meninx** (MEN-inks).

Menopause (MEN-ō-pawz) The termination of the menstrual cycles.

Menstrual (MEN-stru-al) **cycle** A series of changes in the endometrium of a nonpregnant female that prepares the lining of the uterus to receive a fertilized ovum.

Menstruation (men′-stroo-Ā-shun) Periodic discharge of blood, tissue fluid, mucus, and epithelial cells that usually lasts for 5 days; caused by a sudden reduction in estrogens and progesterone. Also called the **menstrual phase** or **menses.**

Merkel (MER-kel) **cell** Type of cell in the epidermis of hairless skin that makes contact with a tactile (Merkel) disc, which functions in touch.

Merocrine (MER-ō-krin) **gland** Gland made up of secretory cells that remain intact throughout the process of formation and discharge of the secretory product, as in the salivary and pancreatic glands.

Mesenchyme (MEZ-en-kīm) An embryonic connective tissue from which all other connective tissues arise.

Mesentery (MEZ-en-ter′-ē) A fold of peritoneum attaching the small intestine to the posterior abdominal wall.

Mesocolon (mez′-ō-KŌ-lon) A fold of peritoneum attaching the colon to the posterior abdominal wall.

Mesoderm The middle primary germ layer that gives rise to connective tissues, blood and blood vessels, and muscles.

Mesothelium (mez′-ō-THĒ-lē-um) The layer of simple squamous epithelium that lines serous membranes.

Mesovarium (mez′-ō-VAR-ē-um) A short fold of peritoneum that attaches an ovary to the broad ligament of the uterus.

Metabolism (me-TAB-ō-lizm) All the biochemical reactions that occur within an organism, including the synthetic (anabolic) reactions and decomposition (catabolic) reactions.

Metacarpus (met′-a-KAR-pus) A collective term for the five bones that make up the palm.

Metaphase (MET-a-phāz) The second stage of mitosis, in which chromatid pairs line up on the metaphase plate of the cell.

Metaphysis (me-TAF-i-sis) Region of a long bone between the diaphysis and epiphysis that contains the epiphyseal plate in a growing bone.

Metarteriole (met′-ar-TĒ-rē-ōl) A blood vessel that emerges from an arteriole, traverses a capillary network, and empties into a venule.

Metastasis (me-TAS-ta-sis) The spread of cancer to surrounding tissues (local) or to other body sites (distant).

Metatarsus (met′-a-TAR-sus) A collective term for the five bones located in the foot between the tarsals and the phalanges.

Microfilament (mī-krō-FIL-a-ment) Rodlike protein filament about 6 nm in diameter; constitutes contractile units in muscle fibers (cells) and provides support, shape, and movement in nonmuscle cells.

Microglia (mī-krō-GLĒ-a) Neuroglial cells that carry on phagocytosis.

Microtubule (mī-krō-TOO-būl′) Cylindrical protein filament, from 18 to 30 nm in diameter, consisting of the protein tubulin; provides support, structure, and transportation.

Microvilli (mī′-krō-VIL-ē) Microscopic, fingerlike projections of the plasma membranes of cells that increase surface area for absorption, especially in the small intestine and proximal convoluted tubules of the kidneys.

Micturition (mik′-choo-RISH-un) The act of expelling urine from the urinary bladder. Also called **urination** (ū-ri-NĀ-shun).

Midbrain The part of the brain between the pons and the diencephalon. Also called the **mesencephalon** (mes′-en-SEF-a-lon).

Middle ear A small, epithelial-lined cavity hollowed out of the temporal bone, separated from the external ear by the eardrum and from the internal ear by a thin bony partition containing the oval and round windows; extending across the middle ear are the three auditory ossicles. Also called the **tympanic** (tim-PAN-ik) **cavity.**

Midline An imaginary vertical line that divides the body into equal left and right sides.

Midsagittal plane A vertical plane through the midline of the body that divides the body or organs into *equal* right and left sides. Also called a **median plane.**

Mineralocorticoids (min′-er-al-ō-KOR-ti-koyds) A group of hormones of the adrenal cortex that help regulate sodium and potassium balance.

Mitochondrion (mī′-tō-KON-drē-on) A double-membraned organelle that plays a central role in the production of ATP; known as the "powerhouse" of the cell.

Mitosis (mī-TŌ-sis) The orderly division of the nucleus of a cell that ensures that each new nucleus has the same number and kind of chromosomes as the original nucleus. The process includes the replication of chromosomes and the distribution of the two sets of chromosomes into two separate and equal nuclei.

Mitotic spindle Collective term for a football-shaped assembly of microtubules (nonkinetochore, kinetochore, and aster) that is responsible for the movement of chromosomes during cell division.

Modality (mō-DAL-i-tē) Any of the specific sensory entities, such as vision, smell, taste, or touch.

Modiolus (mō-DĪ-ō′-lus) The central pillar or column of the cochlea.

Monocyte (MON-ō-sit′) The largest type of white blood cell, characterized by agranular cytoplasm.

Monounsaturated fat A fatty acid that contains one double covalent bond between its carbon atoms; it is not completely saturated with hydrogen atoms. Plentiful in triglycerides of olive and peanut oils.

Mons pubis (MONZ PŪ-bis) The rounded, fatty prominence over the pubic symphysis, covered by coarse pubic hair.

Morphology (mor-FOL-ō-jē) The study of the form and structure of things.

Morula (MOR-ū-la) A solid sphere of cells produced by successive cleavages of a fertilized ovum about four days after fertilization.

Mōtor area The region of the cerebral cortex that governs muscular movement, particularly the precentral gyrus of the frontal lobe.

Motor end plate Region of the sarcolemma of a muscle fiber (cell) that includes acetylcholine (ACh) receptors, which bind ACh released by synaptic end bulbs of somatic motor neurons.

Motor neurons (NOO-ronz) Neurons that conduct impulses from the brain toward the spinal cord or out of the brain and spinal cord into cranial or spinal nerves to effectors that may be either muscles or glands. Also called **efferent neurons.**

Motor unit A motor neuron together with the muscle fibers (cells) it stimulates.

Mucin (MŪ-sin) A protein found in mucus.

Mucosa-associated lymphatic tissue (MALT) Lymphatic nodules scattered throughout the lamina propria (connective tissue) of mucous membranes lining the gastrointestinal tract, respiratory airways, urinary tract, and reproductive tract.

Mucous (MŪ-kus) **cell** A unicellular gland that secretes mucus. Two types are mucous neck cells and surface mucous cells in the stomach.

Mucous membrane A membrane that lines a body cavity that opens to the exterior. Also called the **mucosa** (mū-KŌ-sa).

Mucus The thick fluid secretion of goblet cells, mucous cells, mucous glands, and mucous membranes.

Muscarinic (mus′-ka-RIN-ik) **receptor** Receptor for the neurotransmitter acetylcholine found on all effectors innervated by parasympathetic postganglionic axons and on sweat glands innervated by cholinergic sympathetic postganglionic axons; so named because

muscarine activates these receptors but does not activate nicotinic receptors for acetylcholine.

Muscle An organ composed of one of three types of muscle tissue (skeletal, cardiac, or smooth), specialized for contraction to produce voluntary or involuntary movement of parts of the body.

Muscle action potential A stimulating impulse that propagates along the sarcolemma and transverse tubules; in skeletal muscle, it is generated by acetylcholine, which increases the permeability of the sarcolemma to cations, especially sodium ions (Na^+).

Muscle fatigue (fa-TĒG) Inability of a muscle to maintain its strength of contraction or tension; may be related to insufficient oxygen, depletion of glycogen, and/or lactic acid buildup.

Muscle spindle An encapsulated proprioceptor in a skeletal muscle, consisting of specialized intrafusal muscle fibers and nerve endings; stimulated by changes in length or tension of muscle fibers.

Muscle tone A sustained, partial contraction of portions of a skeletal or smooth muscle in response to activation of stretch receptors or a baseline level of action potentials in the innervating motor neurons.

Muscular dystrophies (DIS-trō-fēz′) Inherited muscle-destroying diseases, characterized by degeneration of muscle fibers (cells), which causes progressive atrophy of the skeletal muscle.

Muscularis (MUS-kū-la′-ris) A muscular layer (coat or tunic) of an organ.

Muscularis mucosae (mū-KŌ-sē) A thin layer of smooth muscle fibers that underlie the lamina propria of the mucosa of the gastrointestinal tract.

Muscular tissue A tissue specialized to produce motion in response to muscle action potentials by its qualities of contractility, extensibility, elasticity, and excitability; types include skeletal, cardiac, and smooth.

Mutation (mū-TĀ-shun) Any change in the sequence of bases in a DNA molecule resulting in a permanent alteration in some inheritable trait.

Myasthenia (mī-as-THĒ-nē-a) **gravis** Weakness and fatigue of skeletal muscles caused by antibodies directed against acetylcholine receptors.

Myelin (MĪ-e-lin) **sheath** Multilayered lipid and protein covering, formed by Schwann cells and oligodendrocytes, around axons of many peripheral and central nervous system neurons.

Myenteric plexus A network of autonomic axons and postganglionic cell bodies located in the muscularis of the gastrointestinal tract. Also called the **plexus of Auerbach** (OW-er-bak).

Myocardial infarction (mī′-ō-KAR-dē-al in-FARK-shun) **(MI)** Gross necrosis of myocardial tissue due to interrupted blood supply. Also called a **heart attack.**

Myocardium (mī′-ō-KAR-dē-um) The middle layer of the heart wall, made up of cardiac muscle tissue, lying between the epicardium and the endocardium and constituting the bulk of the heart.

Myofibril (mī-ō-FĪ-bril) A threadlike structure, extending longitudinally through a muscle fiber (cell) consisting mainly of thick filaments (myosin) and thin filaments (actin, troponin, and tropomyosin).

Myoglobin (mī-ō-GLŌ-bin) The oxygen-binding, iron-containing protein present in the sarcoplasm of muscle fibers (cells); contributes the red color to muscle.

Myogram (MĪ-ō-gram) The record or tracing produced by a myograph, an apparatus that measures and records the force of muscular contractions.

Myology (mī-OL-ō-jē) The study of muscles.

Myometrium (mī′-ō-MĒ-trē-um) The smooth muscle layer of the uterus.

Myopathy (mī-OP-a-thē) Any abnormal condition or disease of muscle tissue.

Myopia (mī-Ō-pē-a) Defect in vision in which objects can be seen distinctly only when very close to the eyes; nearsightedness.

Myosin (MĪ-ō-sin) The contractile protein that makes up the thick filaments of muscle fibers.

Myotome (MĪ-ō-tōm) A group of muscles innervated by the motor neurons of a single spinal segment. In an embryo, the portion of a somite that develops into some skeletal muscles.

N

Nail A hard plate, composed largely of keratin, that develops from the epidermis of the skin to form a protective covering on the dorsal surface of the distal phalanges of the fingers and toes.

Nail matrix (MĀ-triks) The part of the nail beneath the body and root from which the nail is produced.

Nasal (NĀ-zal) **cavity** A mucosa-lined cavity on either side of the nasal septum that opens onto the face at the external nares and into the nasopharynx at the internal nares.

Nasal septum (SEP-tum) A vertical partition composed of bone (perpendicular plate of ethmoid and vomer) and cartilage, covered with a mucous membrane, separating the nasal cavity into left and right sides.

Nasolacrimal (nā′-zō-LAK-ri-mal) **duct** A canal that transports the lacrimal secretion (tears) from the nasolacrimal sac into the nose.

Nasopharynx (nā′-zō-FAR-inks) The superior portion of the pharynx, lying posterior to the nose and extending inferiorly to the soft palate.

Neck The part of the body connecting the head and the trunk. A constricted portion of an organ, such as the neck of the femur or uterus.

Necrosis (ne-KRŌ-sis) A pathological type of cell death that results from disease, injury, or lack of blood supply in which many adjacent cells swell, burst, and spill their contents into the interstitial fluid, triggering an inflammatory response.

Neonatal (nē-ō-NĀ-tal) Pertaining to the first four weeks after birth.

Neoplasm (NĒ-ō-plazm) A new growth that may be benign or malignant.

Nephron (NEF-ron) The functional unit of the kidney.

Nerve A cordlike bundle of neuronal axons and/or dendrites and associated connective tissue coursing together outside the central nervous system.

Nerve fiber General term for any process (axon or dendrite) projecting from the cell body of a neuron.

Nerve impulse A wave of depolarization and repolarization that self-propagates along the plasma membrane of a neuron; also called a **nerve action potential.**

Nervous tissue Tissue containing neurons that initiate and conduct nerve impulses to coordinate homeostasis, and neuroglia that provide support and nourishment to neurons.

Neuralgia (noo-RAL-jē-a) Attacks of pain along the entire course or branch of a peripheral sensory nerve.

Neural plate A thickening of ectoderm, induced by the notochord, that forms early in the third week of development and represents the beginning of the development of the nervous system.

Neural tube defect (NTD) A developmental abnormality in which the neural tube does not close properly. Examples are spina bifida and anencephaly.

Neuritis (noo-RĪ-tis) Inflammation of one or more nerves.

Neurofibral node *See* **Node of Ranvier.**

Neuroglia (noo-RŌG-lē-a) Cells of the nervous system that perform various supportive functions. The neuroglia of the central nervous system are the astrocytes, oligodendrocytes, microglia, and ependy-mal cells; neuroglia of the peripheral nervous system include Schwann cells and satellite cells. Also called **glial** (GLĒ-al) **cells.**

Neurohypophyseal (noo′-rō-hī′-pō-FIZ-ē-al) **bud** An outgrowth of ectoderm located on the floor of the hypothalamus that gives rise to the posterior pituitary.

Neurolemma (noo-rō-LEM-ma) The peripheral, nucleated cytoplasmic layer of the Schwann cell. Also called **sheath of Schwann** (SCHVON).

Neurology (noo-ROL-ō-jē) The study of the normal functioning and disorders of the nervous system.

Neuromuscular (noo-rō-MUS-kū-lar) **junction** A synapse between the axon terminals of a motor neuron and the sarcolemma of a muscle fiber (cell).

Neuron (NOO-ron) A nerve cell, consisting of a cell body, dendrites, and an axon.

Neurosecretory (noo-rō-SĒC-re-tō-rē) **cell** A neuron that secretes a hypothalamic releasing hormone or inhibiting hormone into blood capillaries of the hypothalmus; a neuron that secretes oxytocin or antidiuretic hormone into blood capillaries of the posterior pituitary.

Neurotransmitter One of a variety of molecules within axon terminals that are released into the synaptic cleft in response to a nerve impulse, and that change the membrane potential of the postsynaptic neuron.

Neurulation (noor-oo-LĀ-shun) The process by which the neural plate, neural folds, and neural tube develop.

Neutrophil (NOO-trō-fil) A type of white blood cell characterized by granules that stain pale lilac with a combination of acidic and basic dyes.

Nicotinic (nik′-ō-TIN-ik) **receptor** Receptor for the neurotransmitter acetylcholine found on both sympathetic and parasympathetic postganglionic neurons and on skeletal muscle in the motor end plate; so named because nicotine activates these receptors but does not activate muscarinic receptors for acetylcholine.

Nipple A pigmented, wrinkled projection on the surface of the breast that is the location of the openings of the lactiferous ducts for milk release.

Nociceptor (nō′-sē-SEP-tor) A free (naked) nerve ending that detects painful stimuli.

Node of Ranvier (ron-vē-Ā) A space, along a myelinated axon, between the individual Schwann cells that form the myelin sheath and the neurolemma. Also called a **neurofibral node.**

Norepinephrine (nor′-ep-ē-NEF-rin) **(NE)** A hormone secreted by the adrenal medulla that produces actions similar to those that result from sympathetic stimulation. Also called **noradrenaline** (nor-a-DREN-a-lin).

Notochord (NŌ-tō-cord) A flexible rod of mesodermal tissue that lies where the future vertebral column will develop and plays a role in induction.

Nuclear medicine The branch of medicine concerned with the use of radioisotopes in the diagnosis and therapy of disease.

Nucleic (noo-KLĒ-ic) **acid** An organic compound that is a long polymer of nucleotides, with each nucleotide containing a pentose sugar, a phosphate group, and one of four possible nitrogenous bases (adenine, cytosine, guanine, and thymine or uracil).

Nucleolus (noo-KLĒ-ō-lus) Spherical body within a cell nucleus composed of protein, DNA, and RNA that is the site of the assembly of small and large ribosomal subunits.

Nucleosome (NOO-klē-ō-sōm) Structural subunit of a chromosome consisting of histones and DNA.

Nucleus (NOO-klē-us) A spherical or oval organelle of a cell that contains the hereditary factors of the cell, called genes. A cluster of unmyelinated nerve cell bodies in the central nervous system. The central part of an atom made up of protons and neutrons.

Nucleus pulposus (pul-PŌ-sus) A soft, pulpy, highly elastic substance in the center of an intervertebral disc; a remnant of the notochord.

Nutrient A chemical substance in food that provides energy, forms new body components, or assists in various body functions.

O

Obesity (ō-BĒS-i-tē) Body weight more than 20% above a desirable standard due to excessive accumulation of fat.

Oblique (ō-BLĒK) **plane** A plane that passes through the body or an organ at an angle between the transverse plane and either the midsagittal, parasagittal, or frontal plane.

Obstetrics (ob-STET-riks) The specialized branch of medicine that deals with pregnancy, labor, and the period of time immediately after delivery (about 6 weeks).

Olfactory (ōl-FAK-tō-rē) Pertaining to smell.

Olfactory bulb A mass of gray matter containing cell bodies of neurons that form synapses with neurons of the olfactory (I) nerve, lying inferior to the frontal lobe of the cerebrum on either side of the crista galli of the ethmoid bone.

Olfactory receptor A bipolar neuron with its cell body lying between supporting cells located in the mucous membrane lining the superior portion of each nasal cavity; transduces odors into neural signals.

Olfactory tract A bundle of axons that extends from the olfactory bulb posteriorly to olfactory regions of the cerebral cortex.

Oligodendrocyte (ol′-i-gō-DEN-drō-sīt) A neuroglial cell that supports neurons and produces a myelin sheath around axons of neurons of the central nervous system.

Oligospermia (ol′-i-gō-SPER-mē-a) A deficiency of sperm cells in the semen.

Oliguria (ol′-i-GŪ-rē-a) Daily urinary output usually less than 250 ml.

Olive A prominent oval mass on each lateral surface of the superior part of the medulla oblongata.

Oncogenes (ONG-kō-jēnz) Cancer-causing genes; they derive from normal genes, termed proto-oncogenes, that encode proteins involved in cell growth or cell regulation but have the ability to transform a normal cell into a cancerous cell when they are mutated or inappropriately activated. One example is *p53*.

Oncology (ong-KOL-ō-jē) The study of tumors.

Oogenesis (ō′-ō-JEN-e-sis) Formation and development of female gametes (oocytes).

Oophorectomy (ō′-of-ō-REK-tō-me) Surgical removal of the ovaries.

Ophthalmic (of-THAL-mik) Pertaining to the eye.

Ophthalmologist (of′-thal-MOL-ō-jist) A physician who specializes in the diagnosis and treatment of eye disorders using drugs, surgery, and corrective lenses.

Ophthalmology (of′-thal-MOL-ō-jē) The study of the structure, function, and diseases of the eye.

Optic (OP-tik) Refers to the eye, vision, or properties of light.

Optic chiasm (KĪ-azm) A crossing point of the two branches of the optic (II) nerve, anterior to the pituitary gland. Also called **optic chiasma.**

Optic disc A small area of the retina containing openings through which the axons of the ganglion cells emerge as the optic (II) nerve. Also called the **blind spot.**

Optician (op-TISH-an) A technician who fits, adjusts, and dispenses corrective lenses on prescription of an ophthalmologist or optometrist.

Optic tract A bundle of axons that carry nerve impulses from the retina of the eye between the optic chiasm and the thalamus.

Optometrist (op-TOM-e-trist) Specialist with a doctorate degree in optometry who is licensed to examine and test the eyes and treat visual defects by prescribing corrective lenses.

Ora serrata (Ō-ra ser-RĀ-ta) The irregular margin of the retina lying internal and slightly posterior to the junction of the choroid and ciliary body.

Orbit (OR-bit) The bony, pyramidal-shaped cavity of the skull that holds the eyeball.

Organ A structure composed of two or more different kinds of tissues with a specific function and usually a recognizable shape.

Organelle (or-gan-EL) A permanent structure within a cell with characteristic morphology that is specialized to serve a specific function in cellular activities.

Organism (OR-ga-nizm) A total living form; one individual.

Organogenesis (or′-ga-nō-JEN-e-sis) The formation of body organs and systems. By the end of the eighth week of development, all major body systems have begun to develop.

Orgasm (OR-gazm) Sensory and motor events involved in ejaculation for the male and involuntary contraction of the perineal muscles in the female at the climax of sexual intercourse.

Orifice (OR-i-fis) Any aperture or opening.

Origin (OR-i-jin) The attachment of a muscle tendon to a stationary bone or the end opposite the insertion.

Oropharynx (or′-ō-FAR-inks) The intermediate portion of the pharynx, lying posterior to the mouth and extending from the soft palate to the hyoid bone.

Orthopedics (or′-thō-PĒ-diks) The branch of medicine that deals with the preservation and restoration of the skeletal system, articulations, and associated structures.

Osmoreceptor (oz′-mō-re-CEP-tor) Receptor in the hypothalamus that is sensitive to changes in blood osmolarity and, in response to high osmolarity (low water concentration), stimulates synthesis and release of antidiuretic hormone (ADH).

Osmosis (os-MŌ-sis) The net movement of water molecules through a selectively permeable membrane from an area of higher water concentration to an area of lower water concentration until equilibrium is reached.

Osseous (OS-ē-us) Bony.

Ossicle (OS-si-kul) One of the small bones of the middle ear (malleus, incus, stapes).

Ossification (os′-i-fi-KĀ-shun) Formation of bone. Also called **osteogenesis.**

Osteoblast (OS-tē-ō-blast) Cell formed from an osteogenic cell that participates in bone formation by secreting some organic components and inorganic salts.

Osteoclast (OS-tē-ō-clast′) A large, multinuclear cell that resorbs (destroys) bone matrix.

Osteocyte (OS-tē-ō-sīt′) A mature bone cell that maintains the daily activities of bone tissue.

Osteogenic (os′-tē-ō-JEN-i-ik) **cell** Stem cell derived from mesenchyme that has mitotic potential and the ability to differentiate into an osteoblast.

Osteogenic layer The inner layer of the periosteum that contains cells responsible for forming new bone during growth and repair.

Osteology (os′-tē-OL-ō-jē) The study of bones.

Osteon (OS-tē-on) The basic unit of structure in adult compact bone, consisting of a central (haversian) canal with its concentrically arranged lamellae, lacunae, osteocytes, and canaliculi. Also called a **haversian** (ha-VER-shan) **system.**

Osteoporosis (os′-tē-ō-pō-RŌ-sis) Age-related disorder characterized by decreased bone mass and increased susceptibility to fractures, often as a result of decreased levels of estrogens.

Otic (Ō-tik) Pertaining to the ear.

Otolith (Ō-tō-lith) A particle of calcium carbonate embedded in the otolithic membrane that functions in maintaining static equilibrium.

Otolithic (ō-tō-LITH-ik) **membrane** Thick, gelatinous, glycoprotein layer located directly over hair cells of the macula in the saccule and utricle of the internal ear.

Otorhinolaryngology (ō′-tō-rī-nō-lar′-in-GOL-ō-jē) The branch of medicine that deals with the diagnosis and treatment of diseases of the ears, nose, and throat.

Oval window A small, membrane-covered opening between the middle ear and inner ear into which the footplate of the stapes fits.

Ovarian (ō-VAR-ē-an) **cycle** A monthly series of events in the ovary associated with the maturation of a secondary oocyte.

Ovarian follicle (FOL-i-kul) A general name for oocytes (immature ova) in any stage of development, along with their surrounding epithelial cells.

Ovarian ligament (LIG-a-ment) A rounded cord of connective tissue that attaches the ovary to the uterus.

Ovary (Ō-var-ē) Female gonad that produces oocytes and the estrogens, progesterone, inhibin, and relaxin hormones.

Ovulation (ov-ū-LĀ-shun) The rupture of a mature ovarian (Graafian) follicle with discharge of a secondary oocyte into the pelvic cavity.

Ovum (Ō-vum) The female reproductive or germ cell; an egg cell; arises through completion of meiosis in a secondary oocyte after penetration by a sperm.

Oxyhemoglobin (ok′-sē-HĒ-mō-glō-bin) **(Hb–O₂)** Hemoglobin combined with oxygen.

Oxytocin (ok′-sē-TŌ-sin) **(OT)** A hormone secreted by neurosecretory cells in the paraventricular and supraoptic nuclei of the hypothalamus that stimulates contraction of smooth muscle in the pregnant uterus and myoepithelial cells around the ducts of mammary glands.

P

P wave The deflection wave of an electrocardiogram that signifies atrial depolarization.

Pacinian corpuscle *See* **Lamellated corpuscle.**

Palate (PAL-at) The horizontal structure separating the oral and the nasal cavities; the roof of the mouth.

Palpate (PAL-pāt) To examine by touch; to feel.

Pancreas (PAN-krē-as) A soft, oblong organ lying along the greater curvature of the stomach and connected by a duct to the duodenum. It is both an exocrine gland (secreting pancreatic juice) and an endocrine gland (secreting insulin, glucagon, somatostatin, and pancreatic polypeptide).

Pancreatic (pan′-krē-AT-ik) **duct** A single large tube that unites with the common bile duct from the liver and gallbladder and drains pancreatic juice into the duodenum at the hepatopancreatic ampulla (ampulla of Vater). Also called the **duct of Wirsung.**

Pancreatic islet A cluster of endocrine gland cells in the pancreas that secretes insulin, glucagon, somatostatin, and pancreatic polypeptide. Also called an **islet of Langerhans** (LANG-er-hanz).

Papanicolaou (pap′-a-NIK-ō-la-oo) **test** A cytological staining test for the detection and diagnosis of premalignant and malignant conditions of the female genital tract. Cells scraped from the epithelium of the cervix of the uterus are examined microscopically. Also called a **Pap test** or **Pap smear.**

Papilla (pa-PIL-a) A small nipple-shaped projection or elevation.

Paralysis (pa-RAL-a-sis) Loss or impairment of motor function due to a lesion of nervous or muscular origin.

Paranasal sinus (par′-a-NĀ-zal SĪ-nus) A mucus-lined air cavity in a skull bone that communicates with the nasal cavity. Paranasal sinuses are located in the frontal, maxillary, ethmoid, and sphenoid bones.

Paraplegia (par-a-PLĒ-jē-a) Paralysis of both lower limbs.

Parasagittal plane (par-a-SAJ-i-tal) A vertical plane that does not pass through the midline and that divides the body or organs into *unequal* left and right portions.

Parasympathetic (par′-a-sim-pa-THET-ik) **division** One of the two subdivisions of the autonomic nervous system, having cell bodies of preganglionic neurons in nuclei in the brain stem and in the lateral gray horn of the sacral portion of the spinal cord; primarily concerned with activities that conserve and restore body energy.

Parathyroid (par′-a-THĪ-royd) **gland** One of usually four small endocrine glands embedded in the posterior surfaces of the lateral lobes of the thyroid gland.

Parathyroid hormone (PTH) A hormone secreted by the chief (principal) cells of the parathyroid glands that increases blood calcium level and decreases blood phosphate level.

Paraurethral (par′-a-ū-RĒ-thral) **gland** Gland embedded in the wall of the urethra whose duct opens on either side of the urethral orifice and secretes mucus. Also called **Skene's** (SKĒNZ) **gland.**

Parenchyma (par-EN-ki-ma) The functional parts of any organ, as opposed to tissue that forms its stroma or framework.

Parietal (pa-RĪ-e-tal) Pertaining to or forming the outer wall of a body cavity.

Parietal cell A type of secretory cell in gastric glands that produces hydrochloric acid and intrinsic factor. Also called an **oxyntic cell.**

Parietal pleura (PLOO-ra) The outer layer of the serous pleural membrane that encloses and protects the lungs; the layer that is attached to the wall of the pleural cavity.

Parkinson's disease (PD) Progressive degeneration of the basal ganglia and substantia nigra of the cerebrum resulting in decreased production of dopamine (DA) that leads to tremor, slowing of voluntary movements, and muscle weakness.

Parotid (pa-ROT-id) **gland** One of the paired salivary glands located inferior and anterior to the ears and connected to the oral cavity via a duct (Stensen's) that opens into the inside of the cheek opposite the maxillary (upper) second molar tooth.

Pars intermedia A small avascular zone between the anterior and posterior pituitary glands.

Parturition (par′-too-RISH-un) Act of giving birth to young; childbirth, delivery.

Patent ductus arteriosus A congenital heart defect in which the ductus arteriosus remains open. As a result, aortic blood flows into the lower-pressure pulmonary trunk, increasing pulmonary trunk pressure and overworking both ventricles.

Pathogen (PATH-ō-jen) A disease-producing microbe.

Pathological (path′-ō-LOJ-i-kal) **anatomy** The study of structural changes caused by disease.

Pectinate (PEK-ti-nāt) **muscles** Projecting muscle bundles of the anterior atrial walls and the lining of the auricles.

Pectoral (PEK-tō-ral) Pertaining to the chest or breast.

Pediatrician (pē′-dē-a-TRISH-un) A physician who specializes in the care and treatment of children.

Pedicel (PED-i-sel) Footlike structure, as on podocytes of a glomerulus.

Pelvic (PEL-vik) **cavity** Inferior portion of the abdominopelvic cavity that contains the urinary bladder, sigmoid colon, rectum, and internal female and male reproductive structures.

Pelvic splanchnic (PEL-vik SPLANGK-nik) **nerves** Consist of preganglionic parasympathetic axons from the levels of S2, S3, and S4 that supply the urinary bladder, reproductive organs, and the descending and sigmoid colon and rectum.

Pelvis The basinlike structure formed by the two hip bones, the sacrum, and the coccyx. The expanded, proximal portion of the ureter, lying within the kidney and into which the major calyces open.

Penis (PĒ-nis) The organ of urination and copulation in males; used to deposit semen into the female vagina.

Pepsin Protein-digesting enzyme secreted by chief cells of the stomach in the inactive form pepsinogen, which is converted to active pepsin by hydrochloric acid.

Peptic ulcer An ulcer that develops in areas of the gastrointestinal tract exposed to hydrochloric acid; classified as a gastric ulcer if in the lesser curvature of the stomach and as a duodenal ulcer if in the first part of the duodenum.

Percussion (per-KUSH-un) The act of striking (percussing) an underlying part of the body with short, sharp blows as an aid in diagnosing the part by the quality of the sound produced.

Perforating canal A minute passageway by means of which blood vessels and nerves from the periosteum penetrate into compact bone. Also called **Volkmann's** (FŌLK-manz) **canal.**

Pericardial (per′-i-KAR-dē-al) **cavity** Small potential space between the visceral and parietal layers of the serous pericardium that contains pericardial fluid.

Pericardium (per′-i-KAR-dē-um) A loose-fitting membrane that encloses the heart, consisting of a superficial fibrous layer and a deep serous layer.

Perichondrium (per′-i-KON-drē-um) The membrane that covers cartilage.

Perilymph (PER-i-limf) The fluid contained between the bony and membranous labyrinths of the inner ear.

Perimetrium (per′-i-MĒ-trē-um) The serosa of the uterus.

Perimysium (per′-i-MĪZ-ē-um) Invagination of the epimysium that divides muscles into bundles.

Perineum (per′-i-NĒ-um) The pelvic floor; the space between the anus and the scrotum in the male and between the anus and the vulva in the female.

Perineurium (per′-i-NOO-rē-um) Connective tissue wrapping around fascicles in a nerve.

Periodontal (per-ē-ō-DON-tal) **disease** A collective term for conditions characterized by degeneration of gingivae, alveolar bone, periodontal ligament, and cementum.

Periodontal ligament The periosteum lining the alveoli (sockets) for the teeth in the alveolar processes of the mandible and maxillae.

Periosteum (per′-ē-OS-tē-um) The membrane that covers bone and consists of connective tissue, osteogenic cells, and osteoblasts; is essential for bone growth, repair, and nutrition.

Peripheral (pe-RIF-er-al) Located on the outer part or a surface of the body.

Peripheral nervous system (PNS) The part of the nervous system that lies outside the central nervous system, consisting of nerves and ganglia.

Peristalsis (per′-i-STAL-sis) Successive muscular contractions along the wall of a hollow muscular structure.

Peritoneum (per′-i-tō-NĒ-um) The largest serous membrane of the body that lines the abdominal cavity and covers the viscera.

Peritonitis (per′-i-tō-NĪ-tis) Inflammation of the peritoneum.

Peroxisome (per-OK-si-sōm) Organelle similar in structure to a lysosome that contains enzymes that use molecular oxygen to oxidize various organic compounds; such reactions produce hydrogen peroxide; abundant in liver cells.

Perspiration Sweat; produced by sudoriferous (sweat) glands and containing water, salts, urea, uric acid, amino acids, ammonia, sugar, lactic acid, and ascorbic acid. Helps maintain body temperature and eliminate wastes.

Peyer's patches *See* **Aggregated lymphatic follicles.**

pH A measure of the concentration of hydrogen ions (H^+) in a solution. The pH scale extends from 0 to 14, with a value of 7 expressing neutrality, values lower than 7 expressing increasing acidity, and values higher than 7 expressing increasing alkalinity.

Phagocytosis (fag′-ō-sī-TŌ-sis) The process by which phagocytes ingest and destroy microbes, cell debris, and other foreign matter.

Phalanx (FĀ-lanks) The bone of a finger or toe. *Plural is* **phalanges** (fa-LAN-jēz).

Pharmacology (far′-ma-KOL-ō-jē) The science of the effects and uses of drugs in the treatment of disease.

Pharynx (FAR-inks) The throat; a tube that starts at the internal nares and runs partway down the neck, where it opens into the esophagus posteriorly and the larynx anteriorly.

Phenotype (FĒ-nō-tīp) The observable expression of genotype; physical characteristics of an organism determined by genetic makeup and influenced by interaction between genes and internal and external environmental factors.

Phlebitis (fle-BĪ-tis) Inflammation of a vein, usually in a lower limb.

Photopigment A substance that can absorb light and undergo structural changes that can lead to the development of a receptor potential. An example is rhodopsin. In the eye, also called **visual pigment.**

Photoreceptor Receptor that detects light shining on the retina of the eye.

Physiology (fiz′-ē-OL-ō-jē) Science that deals with the functions of an organism or its parts.

Pia mater (PĪ-a-MĀ-ter *or* PĒ-a MA-ter) The innermost of the three meninges (coverings) of the brain and spinal cord.

Pineal (PĪN-ē-al) **gland** A cone-shaped gland located in the roof of the third ventricle that secretes melatonin. Also called the **epiphysis cerebri** (ē-PIF-i-sis se-RĒ-brē).

Pinealocyte (pin-ē-AL-ō-sīt) Secretory cell of the pineal gland that releases melatonin.

Pinna (PIN-na) The projecting part of the external ear composed of elastic cartilage and covered by skin and shaped like the flared end of a trumpet. Also called the **auricle** (OR-i-kul).

Pituicyte (pi-TOO-i-sīt) Supporting cell of the posterior pituitary.

Pituitary (pi-TOO-i-tār-ē) **gland** A small endocrine gland occupying the hypophyseal fossa of the sphenoid bone and attached to the hypothalamus by the infundibulum. Also called the **hypophysis** (hī-POF-i-sis).

Pivot joint A synovial joint in which a rounded, pointed, or conical surface of one bone articulates with a ring formed partly by another bone and partly by a ligament, as in the joint between the atlas and axis and between the proximal ends of the radius and ulna. Also called a **trochoid** (TRŌ-koyd) **joint.**

Placenta (pla-SEN-ta) The special structure through which the exchange of materials between fetal and maternal circulations occurs. Also called the **afterbirth.**

Plantar flexion (PLAN-tar FLEK-shun) Bending the foot in the direction of the plantar surface (sole).

Plaque (PLAK) A layer of dense proteins on the inside of a plasma membrane in adherens junctions and desmosomes. A mass of bacterial cells, dextran (polysaccharide), and other debris that adheres to teeth (dental plaque). *See also* **Atherosclerotic plaque.**

Plasma (PLAZ-ma) The extracellular fluid found in blood vessels; blood minus the formed elements.

Plasma cell Cell that develops from a B cell (lymphocyte) and produces antibodies.

Plasma (cell) membrane Outer, limiting membrane that separates the cell's internal parts from extracellular fluid or the external environment.

Platelet (PLĀT-let) A fragment of cytoplasm enclosed in a cell membrane and lacking a nucleus; found in the circulating blood; plays a role in hemostasis. Also called a **thrombocyte** (THROM-bō-sīt).

Platelet plug Aggregation of platelets (thrombocytes) at a site where a blood vessel is damaged that helps stop or slow blood loss.

Pleura (PLOOR-a) The serous membrane that covers the lungs and lines the walls of the chest and the diaphragm.

Pleural cavity Small potential space between the visceral and parietal pleurae.

Plexus (PLEK-sus) A network of nerves, veins, or lymphatic vessels.

Plexus of Auerbach *See* **Myenteric plexus.**

Plexus of Meissner *See* **Submucosal plexus.**

Pluripotent stem cell Immature stem cell in red bone marrow that gives rise to precursors of all the different mature blood cells.

Pneumotaxic (noo-mō-TAK-sik) **area** A part of the respiratory center in the pons that continually sends inhibitory nerve impulses to the inspiratory area, limiting inhalation and facilitating exhalation.

Podiatry (pō-DĪ-a-trē) The diagnosis and treatment of foot disorders.

Polar body The smaller cell resulting from the unequal division of primary and secondary oocytes during meiosis. The polar body has no function and degenerates.

Polycythemia (pol'-ē-sī-THĒ-mē-a) Disorder characterized by an above-normal hematocrit (above 55%) in which hypertension, thrombosis, and hemorrhage can occur.

Polyunsaturated fat A fatty acid that contains more than one double covalent bond between its carbon atoms; abundant in triglycerides of corn oil, safflower oil, and cottonseed oil.

Polyuria (pol'-ē-Ū-rē-a) An excessive production of urine.

Pons (PONZ) The part of the brain stem that forms a "bridge" between the medulla oblongata and the midbrain, anterior to the cerebellum.

Portal system The circulation of blood from one capillary network into another through a vein.

Postcentral gyrus Gyrus of cerebral cortex located immediately posterior to the central sulcus; contains the primary somatosensory area.

Posterior (pos-TĒR-ē-or) Nearer to or at the back of the body. Equivalent to **dorsal** in bipeds.

Posterior column–medial lemniscus pathways Sensory pathways that carry information related to proprioception, fine touch, two-point discrimination, pressure, and vibration. First-order neurons project from the spinal cord to the ipsilateral medulla in the posterior columns (gracile fasciculus and cuneate fasciculus). Second-order neurons project from the medulla to the contralateral thalamus in the medial lemniscus. Third-order neurons project from the thalamus to the somatosensory cortex (postcentral gyrus) on the same side.

Posterior pituitary Posterior lobe of the pituitary gland. Also called the **neurohypophysis** (noo-rō-hī-POF-i-sis).

Posterior root The structure composed of sensory axons lying between a spinal nerve and the dorsolateral aspect of the spinal cord. Also called the **dorsal (sensory) root.**

Posterior root ganglion (GANG-glē-on) A group of cell bodies of sensory neurons and their supporting cells located along the posterior root of a spinal nerve. Also called a **dorsal (sensory) root ganglion.**

Postganglionic neuron (pōst'-gang-lē-ON-ik NOO-ron) The second autonomic motor neuron in an autonomic pathway, having its cell body and dendrites located in an autonomic ganglion and its unmyelinated axon ending at cardiac muscle, smooth muscle, or a gland.

Postsynaptic (pōst-sin-AP-tik) **neuron** The nerve cell that is activated by the release of a neurotransmitter from another neuron and carries nerve impulses away from the synapse.

Pouch of Douglas *See* **Rectouterine pouch.**

Precapillary sphincter (SFINGK-ter) A ring of smooth muscle fibers (cells) at the site of origin of true capillaries that regulate blood flow into true capillaries.

Precentral gyrus Gyrus of cerebral cortex located immediately anterior to the central sulcus; contains the primary motor area.

Preganglionic (pre'-gang-lē-ON-ik) **neuron** The first autonomic motor neuron in an autonomic pathway, with its cell body and dendrites in the brain or spinal cord and its myelinated axon ending at an autonomic ganglion, where it synapses with a postganglionic neuron.

Pregnancy Sequence of events that normally includes fertilization, implantation, embryonic growth, and fetal growth and terminates in birth.

Premenstrual syndrome (PMS) Severe physical and emotional stress ocurring late in the postovulatory phase of the menstrual cycle and sometimes overlapping with menstruation.

Prepuce (PRĒ-poos) The loose-fitting skin covering the glans of the penis and clitoris. Also called the **foreskin.**

Presbyopia (prez-bē-Ō-pē-a) A loss of elasticity of the lens of the eye due to advancing age with resulting inability to focus clearly on near objects.

Presynaptic (prē-sin-AP-tik) **neuron** A neuron that propagates nerve impulses toward a synapse.

Prevertebral ganglion (prē-VER-te-bral GANG-glē-on) A cluster of cell bodies of postganglionic sympathetic neurons anterior to the spinal column and close to large abdominal arteries. Also called a **collateral ganglion.**

Primary germ layer One of three layers of embryonic tissue, called ectoderm, mesoderm, and endoderm, that give rise to all tissues and organs of the body.

Primary motor area A region of the cerebral cortex in the precentral gyrus of the frontal lobe of the cerebrum that controls specific muscles or groups of muscles.

Primary somatosensory area A region of the cerebral cortex posterior to the central sulcus in the postcentral gyrus of the parietal lobe of the cerebrum that localizes exactly the points of the body where somatic sensations originate.

Prime mover The muscle directly responsible for producing a desired motion. Also called an **agonist** (AG-ō-nist).

Primitive gut Embryonic structure formed from the dorsal part of the yolk sac that gives rise to most of the gastrointestinal tract.

Primordial (prī-MOR-dē-al) Existing first; especially primordial egg cells in the ovary.

Principal cell Cell type in the distal convoluted tubules and collecting ducts of the kidneys that is stimulated by aldosterone and antidiuretic hormone.

Proctology (prok-TOL-ō-jē) The branch of medicine concerned with the rectum and its disorders.

Progeny (PROJ-e-nē) Offspring or descendants.

Progesterone (prō-JES-te-rōn) A female sex hormone produced by the ovaries that helps prepare the endometrium of the uterus for implantation of a fertilized ovum and the mammary glands for milk secretion.

Prognosis (prog-NŌ-sis) A forecast of the probable results of a disorder; the outlook for recovery.

Prolactin (prō-LAK-tin) **(PRL)** A hormone secreted by the anterior pituitary that initiates and maintains milk secretion by the mammary glands.

Prolapse (PRŌ-laps) A dropping or falling down of an organ, especially the uterus or rectum.

Proliferation (prō-lif'-er-Ā-shun) Rapid and repeated reproduction of new parts, especially cells.

Pronation (prō-NĀ-shun) A movement of the forearm in which the palm is turned posteriorly.

Prophase (PRŌ-fāz) The first stage of mitosis during which chromatid pairs are formed and aggregate around the metaphase plate of the cell.

Proprioception (prō-prē-ō-SEP-shun) The perception of the position of body parts, especially the limbs, independent of vision; this sense is possible due to nerve impulses generated by proprioceptors.

Proprioceptor (prō′-prē-ō-SEP-tor) A receptor located in muscles, tendons, joints, or the internal ear (muscle spindles, tendon organs, joint kinesthetic receptors, and hair cells of the vestibular apparatus) that provides information about body position and movements.

Prostaglandin (pros′-ta-GLAN-din) **(PG)** A membrane-associated lipid; released in small quantities and acts as a local hormone.

Prostate (PROS-tāt) A doughnut-shaped gland inferior to the urinary bladder that surrounds the superior portion of the male urethra and secretes a slightly acidic solution that contributes to sperm motility and viability.

Proteasome (PRŌ-tē-a-sōm) Tiny cellular organelle in cytosol and nucleus containing proteases that destroy unneeded, damaged, or faulty proteins.

Protein An organic compound consisting of carbon, hydrogen, oxygen, nitrogen, and sometimes sulfur and phosphorus; synthesized on ribosomes and made up of amino acids linked by peptide bonds.

Prothrombin (prō-THROM-bin) An inactive blood-clotting factor synthesized by the liver, released into the blood, and converted to active thrombin in the process of blood clotting by the activated enzyme prothrombinase.

Proto-oncogene (prō′-tō-ONG-kō-jēn) Gene responsible for some aspect of normal growth and development; it may transform into an oncogene, a gene capable of causing cancer.

Protraction (prō-TRAK-shun) The movement of the mandible or shoulder girdle forward on a plane parallel with the ground.

Proximal (PROK-si-mal) Nearer the attachment of a limb to the trunk; nearer to the point of origin or attachment.

Pseudopods (SOO-dō-pods) Temporary protrusions of the leading edge of a migrating cell; cellular projections that surround a particle undergoing phagocytosis.

Pterygopalatine ganglion (ter′-i-gō-PAL-a-tīn GANG-glē-on) A cluster of cell bodies of parasympathetic postganglionic neurons ending at the lacrimal and nasal glands.

Ptosis (TŌ-sis) Drooping, as of the eyelid or the kidney.

Puberty (PŪ-ber-tē) The time of life during which the secondary sex characteristics begin to appear and the capability for sexual reproduction is possible; usually occurs between the ages of 10 and 17.

Pubic symphysis A slightly movable cartilaginous joint between the anterior surfaces of the hip bones.

Puerperium (pū′-er-PER-ē-um) The period immediately after childbirth, usually 4–6 weeks.

Pulmonary (PUL-mo-ner′-ē) Concerning or affected by the lungs.

Pulmonary circulation The flow of deoxygenated blood from the right ventricle to the lungs and the return of oxygenated blood from the lungs to the left atrium.

Pulmonary edema (e-DĒ-ma) An abnormal accumulation of interstitial fluid in the tissue spaces and alveoli of the lungs due to increased pulmonary capillary permeability or increased pulmonary capillary pressure.

Pulmonary embolism (EM-bō-lizm) **(PE)** The presence of a blood clot or a foreign substance in a pulmonary arterial blood vessel that obstructs circulation to lung tissue.

Pulmonary ventilation The inflow (inhalation) and outflow (exhalation) of air between the atmosphere and the lungs. Also called **breathing**.

Pulp cavity A cavity within the crown and neck of a tooth, which is filled with pulp, a connective tissue containing blood vessels, nerves, and lymphatic vessels.

Pulse (PULS) The rhythmic expansion and elastic recoil of a systemic artery after each contraction of the left ventricle.

Pupil The hole in the center of the iris, the area through which light enters the posterior cavity of the eyeball.

Purkinje (pur-KIN-jē) **fiber** Muscle fiber (cell) in the ventricular tissue of the heart specialized for conducting an action potential to the myocardium; part of the conduction system of the heart.

Pus The liquid product of inflammation containing leukocytes or their remains and debris of dead cells.

Pyloric (pī-LOR-ik) **sphincter** A thickened ring of smooth muscle through which the pylorus of the stomach communicates with the duodenum. Also called the **pyloric valve.**

Pyogenesis (pī′-ō-JEN-e-sis) Formation of pus.

Pyorrhea (pī-ō-RĒ-a) A discharge or flow of pus, especially in the alveoli (sockets) and the tissues of the gums.

Pyramid (PIR-a-mid) A pointed or cone-shaped structure. One of two roughly triangular structures on the anterior aspect of the medulla oblongata composed of the largest motor tracts that run from the cerebral cortex to the spinal cord. A triangular structure in the renal medulla.

Pyramidal (pi-RAM-i-dal) **tracts (pathways).** *See* **Direct motor pathways**.

Pyuria (pī-Ū-rē-a) The presence of leukocytes and other components of pus in urine.

Q

QRS wave The deflection waves of an electrocardiogram that represent onset of ventricular depolarization.

Quadrant (KWOD-rant) One of four parts.

Quadriplegia (kwod′-ri-PLĒ-jē-a) Paralysis of four limbs: two upper and two lower.

R

Radiographic (rā′-dē-ō-GRAF-ic) **anatomy** Diagnostic branch of anatomy that includes the use of x rays.

Rami communicantes (RĀ-mē kō-mū-ni-KAN-tēz) Branches of a spinal nerve. *Singular is* **ramus communicans** (RĀ-mus kō-MŪ-ni-kans).

Rathke's pouch *See* **Hypophyseal pouch.**

Receptor A specialized cell or a distal portion of a neuron that responds to a specific sensory modality, such as touch, pressure, cold, light, or sound, and converts it to an electrical signal (generator or receptor potential). A specific molecule or cluster of molecules that recognizes and binds a particular ligand.

Receptor-mediated endocytosis A highly selective process whereby cells take up specific ligands, which usually are large molecules or particles, by enveloping them within a sac of plasma membrane. Ligands are eventually broken down by enzymes in lysosomes.

Recombinant DNA Synthetic DNA, formed by joining a fragment of DNA from one source to a portion of DNA from another.

Rectouterine pouch A pocket formed by the parietal peritoneum as it moves posteriorly from the surface of the uterus and is reflected onto the rectum; the most inferior point in the pelvic cavity. Also called the **pouch** or **cul de sac of Douglas.**

Rectum (REK-tum) The last 20 cm (8 in.) of the gastrointestinal tract, from the sigmoid colon to the anus.

Recumbent (re-KUM-bent) Lying down.

Red bone marrow A highly vascularized connective tissue located in microscopic spaces between trabeculae of spongy bone tissue.

Red nucleus A cluster of cell bodies in the midbrain, occupying a large part of the tectum from which axons extend into the rubroreticular and rubrospinal tracts.

Red pulp That portion of the spleen that consists of venous sinuses filled with blood and thin plates of splenic tissue called splenic (Billroth's) cords.

Referred pain Pain that is felt at a site remote from the place of origin.

Reflex Fast response to a change (stimulus) in the internal or external environment that attempts to restore homeostasis.

Reflex arc The most basic conduction pathway through the nervous system, connecting a receptor and an effector and consisting of a receptor, a sensory neuron, an integrating center in the central nervous system, a motor neuron, and an effector.

Regional anatomy The division of anatomy dealing with a specific region of the body, such as the head, neck, chest, or abdomen.

Regurgitation (rē-gur′-ji-TĀ-shun) Return of solids or fluids to the mouth from the stomach; backward flow of blood through incompletely closed heart valves.

Relaxin (RLX) A female hormone produced by the ovaries and placenta that increases flexibility of the pubic symphysis and helps dilate the uterine cervix to ease delivery of a baby.

Releasing hormone Hormone secreted by the hypothalamus that can stimulate secretion of hormones of the anterior pituitary.

Remodeling Replacement of old bone by new bone tissue.

Renal (RĒ-nal) Pertaining to the kidneys.

Renal corpuscle (KOR-pus-l) A glomerular (Bowman's) capsule and its enclosed glomerulus.

Renal pelvis A cavity in the center of the kidney formed by the expanded, proximal portion of the ureter, lying within the kidney, and into which the major calyces open.

Renal pyramid A triangular structure in the renal medulla containing the straight segments of renal tubules and the vasa recta.

Reproduction (rē-prō-DUK-shun) The formation of new cells for growth, repair, or replacement; the production of a new individual.

Reproductive cell division Type of cell division in which gametes (sperm and oocytes) are produced; consists of meiosis and cytokinesis.

Respiration (res-pi-RĀ-shun) Overall exchange of gases between the atmosphere, blood, and body cells consisting of pulmonary ventilation, external respiration, and internal respiration.

Respiratory center Neurons in the pons and medulla oblongata of the brain stem that regulate the rate and depth of pulmonary ventilation.

Retention (rē-TEN-shun) A failure to void urine due to obstruction, nervous contraction of the urethra, or absence of sensation of desire to urinate.

Rete (RĒ-tē) **testis** The network of ducts in the testes.

Reticular (re-TIK-ū-lar) **activating system (RAS)** A portion of the reticular formation that has many ascending connections with the cerebral cortex; when this area of the brain stem is active, nerve impulses pass to the thalamus and widespread areas of the cerebral cortex, resulting in generalized alertness or arousal from sleep.

Reticular formation A network of small groups of neuronal cell bodies scattered among bundles of axons (mixed gray and white matter) beginning in the medulla oblongata and extending superiorly through the central part of the brain stem.

Reticulocyte (re-TIK-ū-lō-sīt) An immature red blood cell.

Reticulum (re-TIK-ū-lum) A network.

Retina (RET-i-na) The deep coat of the posterior portion of the eyeball consisting of nervous tissue (where the process of vision begins) and a pigmented layer of epithelial cells that contact the choroid.

Retinaculum (ret-i-NAK-ū-lum) A thickening of deep fascia that holds structures in place, for example, the superior and inferior retinacula of the ankle.

Retraction (rē-TRAK-shun) The movement of a protracted part of the body posteriorly on a plane parallel to the ground, as in pulling the lower jaw back in line with the upper jaw.

Retrograde degeneration (RE-trō-grād dē-jen-er-Ā-shun) Changes that occur in the proximal portion of a damaged axon only as far as the first node of Ranvier; similar to changes that occur during Wallerian degeneration.

Retroperitoneal (re′-trō-per-i-tō-NĒ-al) External to the peritoneal lining of the abdominal cavity.

Rh factor An inherited antigen on the surface of red blood cells in Rh⁺ individuals; not present in Rh⁻ individuals.

Rhinology (rī-NOL-ō-jē) The study of the nose and its disorders.

Ribonucleic (rī-bō-noo-KLĒ-ik) **acid (RNA)** A single-stranded nucleic acid made up of nucleotides, each consisting of a nitrogenous base (adenine, cytosine, guanine, or uracil), ribose, and a phosphate group; three types are messenger RNA (mRNA), transfer RNA (tRNA), and ribosomal RNA (rRNA), each of which has a specific role during protein synthesis.

Ribosome (RĪ-bō-sōm) An organelle in the cytoplasm of cells, composed of a small subunit and a large subunit that contain ribosomal RNA and ribosomal proteins; the site of protein synthesis.

Right lymphatic (lim-FAT-ik) **duct** A vessel of the lymphatic system that drains lymph from the upper right side of the body and empties it into the right subclavian vein.

Rigidity (ri-JID-i-tē) Hypertonia characterized by increased muscle tone, but reflexes are not affected.

Rigor mortis State of partial contraction of muscles after death due to lack of ATP; myosin heads (cross bridges) remain attached to actin, thus preventing relaxation.

Rod One of two types of photoreceptor in the retina of the eye; specialized for vision in dim light.

Root canal A narrow extension of the pulp cavity lying within the root of a tooth.

Root of penis Attached portion of penis that consists of the bulb and crura.

Rotation (rō-TĀ-shun) Moving a bone around its own axis, with no other movement.

Round ligament (LIG-a-ment) A band of fibrous connective tissue enclosed between the folds of the broad ligament of the uterus, emerging from the uterus just inferior to the uterine tube, extending laterally along the pelvic wall and through the deep inguinal ring to end in the labia majora.

Round window A small opening between the middle and internal ear, directly inferior to the oval window, covered by the secondary tympanic membrane.

Rugae (ROO-gē) Large folds in the mucosa of an empty hollow organ, such as the stomach and vagina.

S

Saccule (SAK-ūl) The inferior and smaller of the two chambers in the membranous labyrinth inside the vestibule of the internal ear containing a receptor organ for static equilibrium.

Sacral hiatus (hī-Ā-tus) Inferior entrance to the vertebral canal formed when the laminae of the fifth sacral vertebra (and sometimes fourth) fail to meet.

Sacral plexus (SĀ-kral PLEK-sus) A network formed by the ventral branches of spinal nerves L4 through S3.

Sacral promontory (PROM-on-tor′-ē) The superior surface of the body of the first sacral vertebra that projects anteriorly into the pelvic cavity; a line from the sacral promontory to the superior border of the pubic symphysis divides the abdominal and pelvic cavities.

Saddle joint A synovial joint in which the articular surface of one bone is saddle-shaped and the articular surface of the other bone is shaped like the legs of the rider sitting in the saddle, as in the joint between the trapezium and the metacarpal of the thumb.

Sagittal (SAJ-i-tal) **plane** A plane that divides the body or organs into left and right portions. Such a plane may be **midsagittal (median),** in which the divisions are equal, or **parasagittal,** in which the divisions are unequal.

Saliva (sa-LĪ-va) A clear, alkaline, somewhat viscous secretion produced mostly by the three pairs of salivary glands; contains various salts, mucin, lysozyme, salivary amylase, and lingual lipase (produced by glands in the tongue).

Salivary amylase (SAL-i-ver-ē AM-i-lās) An enzyme in saliva that initiates the chemical breakdown of starch.

Salivary gland One of three pairs of glands that lie external to the mouth and pour their secretory product (saliva) into ducts that empty into the oral cavity; the parotid, submandibular, and sublingual glands.

Sarcolemma (sar′-kō-LEM-ma) The cell membrane of a muscle fiber (cell), especially of a skeletal muscle fiber.

Sarcomere (SAR-kō-mēr) A contractile unit in a striated muscle fiber (cell) extending from one Z disc to the next Z disc.

Sarcoplasm (SAR-kō-plazm) The cytoplasm of a muscle fiber (cell).

Sarcoplasmic reticulum (sar′-kō-PLAZ-mik re-TIK-ū-lum) A network of saccules and tubes surrounding myofibrils of a muscle fiber (cell), comparable to endoplasmic reticulum; functions to reabsorb calcium ions during relaxation and to release them to cause contraction.

Satellite cell (SAT-i-līt) Flat neuroglial cells that surround cell bodies of peripheral nervous system ganglia to provide structural support and regulate the exchange of material between a neuronal cell body and interstitial fluid.

Saturated fat A fatty acid that contains only single bonds (no double bonds) between its carbon atoms; all carbon atoms are bonded to the maximum number of hydrogen atoms; prevalent in triglycerides of animal products such as meat, milk, milk products, and eggs.

Scala tympani (SKA-la TIM-pan-ē) The inferior spiral-shaped channel of the bony cochlea, filled with perilymph.

Scala vestibuli (ves-TIB-ū-lē) The superior spiral-shaped channel of the bony cochlea, filled with perilymph.

Schwann (SCHVON) **cell** A neuroglial cell of the peripheral nervous system that forms the myelin sheath and neurolemma around a nerve axon by wrapping around the axon in a jelly-roll fashion.

Sciatica (sī-AT-i-ka) Inflammation and pain along the sciatic nerve; felt along the posterior aspect of the thigh extending down the inside of the leg.

Sclera (SKLE-ra) The white coat of fibrous tissue that forms the superficial protective covering over the eyeball except in the most anterior portion; the posterior portion of the fibrous tunic.

Scleral venous sinus A circular venous sinus located at the junction of the sclera and the cornea through which aqueous humor drains from the anterior chamber of the eyeball into the blood. Also called the **canal of Schlemm** (SHLEM).

Sclerosis (skle-RŌ-sis) A hardening with loss of elasticity of tissues.

Scoliosis (skō′-lē-Ō-sis) An abnormal lateral curvature from the normal vertical line of the backbone.

Scrotum (SKRŌ-tum) A skin-covered pouch that contains the testes and their accessory structures.

Sebaceous (se-BĀ-shus) **gland** An exocrine gland in the dermis of the skin, almost always associated with a hair follicle, that secretes sebum. Also called an **oil gland.**

Sebum (SĒ-bum) Secretion of sebaceous (oil) glands.

Secondary sex characteristic A characteristic of the male or female body that develops at puberty under the influence of sex hormones but is not directly involved in sexual reproduction; examples are distribution of body hair, voice pitch, body shape, and muscle development.

Secretion (se-KRĒ-shun) Production and release from a cell or a gland of a physiologically active substance.

Selective permeability (per′-mē-a-BIL-i-tē) The property of a membrane by which it permits the passage of certain substances but restricts the passage of others.

Semen (SĒ-men) A fluid discharged at ejaculation by a male that consists of a mixture of sperm and the secretions of the seminiferous tubules, seminal vesicles, prostate, and bulbourethral (Cowper's) glands.

Semicircular canals Three bony channels (anterior, posterior, lateral), filled with perilymph, in which lie the membranous semicircular canals filled with endolymph. They contain receptors for equilibrium.

Semicircular ducts The membranous semicircular canals filled with endolymph and floating in the perilymph of the bony semicircular canals; they contain cristae that are concerned with dynamic equilibrium.

Semilunar (sem′-ē-LOO-nar) **valve** A valve between the aorta or the pulmonary trunk and a ventricle of the heart.

Seminal vesicle (SEM-i-nal VES-i-kul) One of a pair of convoluted, pouchlike structures, lying posterior and inferior to the urinary bladder and anterior to the rectum, that secrete a component of semen into the ejaculatory ducts. Also termed **seminal gland.**

Seminiferous tubule (sem′-i-NI-fer-us TOO-būl) A tightly coiled duct, located in the testis, where sperm are produced.

Senescence (se-NES-ens) The process of growing old.

Sensation A state of awareness of external or internal conditions of the body.

Sensory area A region of the cerebral cortex concerned with the interpretation of sensory impulses.

Sensory neurons (NOO-ronz) Neurons that carry sensory information from cranial and spinal nerves into the brain and spinal cord or from a lower to a higher level in the spinal cord and brain. Also called **afferent neurons.**

Septal defect An opening in the atrial septum (atrial septal defect) because the foramen ovale fails to close, or the ventricular septum (ventricular septal defect) due to incomplete development of the ventricular septum.

Septum (SEP-tum) A wall dividing two cavities.

Serous (SĒR-us) **membrane** A membrane that lines a body cavity that does not open to the exterior. The external layer of an organ formed by a serous membrane. The membrane that lines the pleural, pericardial, and peritoneal cavities. Also called a **serosa** (se-RŌ-sa).

Sertoli (ser-TŌ-lē) **cell** A supporting cell in the seminiferous tubules that secretes fluid for supplying nutrients to sperm and the hormone inhibin, removes excess cytoplasm from spermatogenic cells, and mediates the effects of FSH and testosterone on spermatogenesis. Also called a **sustentacular** (sus′-ten-TAK-ū-lar) **cell.**

Serum Blood plasma minus its clotting proteins.

Sesamoid (SES-a-moyd) **bones** Small bones usually found in tendons.

Sex chromosomes The twenty-third pair of chromosomes, designated X and Y, which determine the genetic sex of an individual; in males, the pair is XY; in females, XX.

Sexual intercourse The insertion of the erect penis of a male into the vagina of a female. Also called **coitus** (KŌ-i-tus).

Sheath of Schwann *See* **Neurolemma.**

Shock Failure of the cardiovascular system to deliver adequate amounts of oxygen and nutrients to meet the metabolic needs of the body due to inadequate cardiac output. It is characterized by hypotension; clammy, cool, and pale skin; sweating; reduced urine formation; altered mental state; acidosis; tachycardia; weak, rapid pulse; and thirst. Types include hypovolemic, cardiogenic, vascular, and obstructive.

Shoulder joint A synovial joint where the humerus articulates with the scapula.

Sigmoid colon (SIG-moyd KŌ-lon) The S-shaped part of the large intestine that begins at the level of the left iliac crest, projects medially, and terminates at the rectum at about the level of the third sacral vertebra.

Sign Any objective evidence of disease that can be observed or measured such as a lesion, swelling, or fever.

Sinoatrial (si-nō-Ā-trē-al) **(SA) node** A small mass of cardiac muscle fibers (cells) located in the right atrium inferior to the opening of the superior vena cava that spontaneously depolarize and generate a cardiac action potential about 100 times per minute. Also called the **pacemaker.**

Sinus (SĪ-nus) A hollow in a bone (paranasal sinus) or other tissue; a channel for blood (vascular sinus); any cavity having a narrow opening.

Sinusoid (SĪ-nū-soyd) A large, thin-walled, and leaky type of capillary, having large intercellular clefts that may allow proteins and blood cells to pass from a tissue into the bloodstream; present in the liver, spleen, anterior pituitary, parathyroid glands, and red bone marrow.

Skeletal muscle An organ specialized for contraction, composed of striated muscle fibers (cells), supported by connective tissue, attached to a bone by a tendon or an aponeurosis, and stimulated by somatic motor neurons.

Skene's gland *See* **Paraurethral gland.**

Skin The external covering of the body that consists of a superficial, thinner epidermis (epithelial tissue) and a deep, thicker dermis (connective tissue) that is anchored to the subcutaneous layer.

Skull The skeleton of the head consisting of the cranial and facial bones.

Sleep A state of partial unconsciousness from which a person can be aroused; associated with a low level of activity in the reticular activating system.

Sliding-filament mechanism The explanation of how thick and thin filaments slide relative to one another during striated muscle contraction to decrease sarcomere length.

Small intestine A long tube of the gastrointestinal tract that begins at the pyloric sphincter of the stomach, coils through the central and inferior part of the abdominal cavity, and ends at the large intestine; divided into three segments: duodenum, jejunum, and ileum.

Smooth muscle A tissue specialized for contraction, composed of smooth muscle fibers (cells), located in the walls of hollow internal organs, and innervated by autonomic motor neurons.

Sodium-potassium ATPase An active transport pump located in the plasma membrane that transports sodium ions out of the cell and potassium ions into the cell at the expense of cellular ATP. It functions to keep the ionic concentrations of these ions at physiological levels. Also called the **sodium-potassium pump.**

Soft palate (PAL-at) The posterior portion of the roof of the mouth, extending from the palatine bones to the uvula. It is a muscular partition lined with mucous membrane.

Somatic (sō-MAT-ik) **cell division** Type of cell division in which a single starting cell duplicates itself to produce two identical cells; consists of mitosis and cytokinesis.

Somatic nervous system (SNS) The portion of the peripheral nervous system consisting of somatic sensory (afferent) neurons and somatic motor (efferent) neurons.

Somite (SŌ-mīt) Block of mesodermal cells in a developing embryo that is distinguished into a myotome (which forms most of the skeletal muscles), dermatome (which forms connective tissues), and sclerotome (which forms the vertebrae).

Spasm (SPAZM) A sudden, involuntary contraction of large groups of muscles.

Spasticity (spas-TIS-i-tē) Hypertonia characterized by increased muscle tone, increased tendon reflexes, and pathological reflexes (Babinski sign).

Spermatic (sper-MAT-ik) **cord** A supporting structure of the male reproductive system, extending from a testis to the deep inguinal ring, that includes the ductus (vas) deferens, arteries, veins, lymphatic vessels, nerves, cremaster muscle, and connective tissue.

Spermatogenesis (sper′-ma-tō-JEN-e-sis) The formation and development of sperm in the seminiferous tubules of the testes.

Sperm cell A mature male gamete. Also termed **spermatozoon** (sper′-ma-tō-ZŌ-on).

Spermiogenesis (sper′-mē-ō-JEN-e-sis) The maturation of spermatids into sperm.

Sphincter (SFINGK-ter) A circular muscle that constricts an opening.

Sphincter of Oddi *See* **Sphincter of the hepatopancreatic ampulla.**

Sphincter of the hepatopancreatic ampulla A circular muscle at the opening of the common bile and main pancreatic ducts in the duodenum. Also called the **sphincter of Oddi** (OD-ē).

Spinal (SPĪ-nal) **cord** A mass of nerve tissue located in the vertebral canal from which 31 pairs of spinal nerves originate.

Spinal nerve One of the 31 pairs of nerves that originate on the spinal cord from posterior and anterior roots.

Spinal shock A period from several days to several weeks following transection of the spinal cord that is characterized by the abolition of all reflex activity.

Spinothalamic (spī-nō-tha-LAM-ik) **tracts** Sensory (ascending) tracts that convey information up the spinal cord to the thalamus for sensations of pain, temperature, crude touch, and deep pressure.

Spinous (SPĪ-nus) **process** A sharp or thornlike process or projection. Also called a **spine.** A sharp ridge running diagonally across the posterior surface of the scapula.

Spiral organ The organ of hearing, consisting of supporting cells and hair cells that rest on the basilar membrane and extend into the endolymph of the cochlear duct. Also called the **organ of Corti** (KOR-tē).

Splanchnic (SPLANK-nik) Pertaining to the viscera.

Spleen (SPLĒN) Large mass of lymphatic tissue between the fundus of the stomach and the diaphragm that functions in formation of blood cells during early fetal development, phagocytosis of ruptured blood cells, and proliferation of B cells during immune responses.

Spongy (cancellous) bone tissue Bone tissue that consists of an irregular latticework of thin plates of bone called trabeculae; spaces between trabeculae of some bones are filled with red bone marrow; found inside short, flat, and irregular bones and in the epiphyses (ends) of long bones.

Sprain Forcible wrenching or twisting of a joint with partial rupture or other injury to its attachments without dislocation.

Squamous (SKWĀ-mus) Flat or scalelike.

Starvation (star-VĀ-shun) The loss of energy stores in the form of glycogen, triglycerides, and proteins due to inadequate intake of nutrients or inability to digest, absorb, or metabolize ingested nutrients.

Stasis (STĀ-sis) Stagnation or halt of normal flow of fluids, as blood or urine, or of the intestinal contents.

Static equilibrium (ē-kwi-LIB-rē-um) The maintenance of posture in response to changes in the orientation of the body, mainly the head, relative to the ground.

Stellate reticuloendothelial (STEL-āt re-tik′-ū-lō-en′-dō-THĒ-lē-al) **cell** Phagocytic cell bordering a sinusoid of the liver. Also called a **Kupffer** (KOOP-fer) **cell.**

Stem cell An unspecialized cell that has the ability to divide for indefinite periods and give rise to a specialized cell.

Stenosis (sten-Ō-sis) An abnormal narrowing or constriction of a duct or opening.

Stereocilia (ste′-rē-ō-SIL-ē-a) Groups of extremely long, slender, nonmotile microvilli projecting from epithelial cells lining the epididymis.

Sterile (STE-ril) Free from any living microorganisms. Unable to conceive or produce offspring.

Sterilization (ster′-i-li-ZĀ-shun) Elimination of all living microorganisms. Any procedure that renders an individual incapable of reproduction (for example, castration, vasectomy, hysterectomy, or oophorectomy).

Stimulus Any stress that changes a controlled condition; any change in the internal or external environment that excites a sensory receptor, a neuron, or a muscle fiber.

Stomach The J-shaped enlargement of the gastrointestinal tract directly inferior to the diaphragm in the epigastric, umbilical, and left hypochondriac regions of the abdomen, between the esophagus and small intestine.

Straight tubule (TOO-būl) A duct in a testis leading from a convoluted seminiferous tubule to the rete testis.

Stratum (STRĀ-tum) A layer.

Stratum basalis (ba-SAL-is) The layer of the endometrium next to the myometrium that is maintained during menstruation and gestation and produces a new stratum functionalis following menstruation or parturition.

Stratum functionalis (funk′-shun-AL-is) The layer of the endometrium next to the uterine cavity that is shed during menstruation and that forms the maternal portion of the placenta during gestation.

Stretch receptor Receptor in the walls of blood vessels, airways, or organs that monitors the amount of stretching. Also termed **baroreceptor.**

Stroma (STRŌ-ma) The tissue that forms the ground substance, foundation, or framework of an organ, as opposed to its functional parts (parenchyma).

Subarachnoid (sub′-a-RAK-noyd) **space** A space between the arachnoid mater and the pia mater that surrounds the brain and spinal cord and through which cerebrospinal fluid circulates.

Subcutaneous (sub′-kū-TĀ-nē-us) Beneath the skin. Also called **hypodermic** (hi-pō-DER-mik).

Subcutaneous layer A continuous sheet of areolar connective tissue and adipose tissue between the dermis of the skin and the deep fascia of the muscles. Also called the **superficial fascia** (FASH-ē-a).

Subdural (sub-DOO-ral) **space** A space between the dura mater and the arachnoid mater of the brain and spinal cord that contains a small amount of fluid.

Sublingual (sub-LING-gwal) **gland** One of a pair of salivary glands situated in the floor of the mouth deep to the mucous membrane and to the side of the lingual frenulum, with a duct (Rivinus') that opens into the floor of the mouth.

Submandibular (sub′-man-DIB-ū-lar) **gland** One of a pair of salivary glands found inferior to the base of the tongue deep to the mucous membrane in the posterior part of the floor of the mouth, posterior to the sublingual glands, with a duct (Wharton's) situated to the side of the lingual frenulum. Also called the **submaxillary** (sub′-MAK-si-ler-ē) **gland.**

Submucosa (sub-mū-KŌ-sa) A layer of connective tissue located deep to a mucous membrane, as in the gastrointestinal tract or the urinary bladder; the submucosa connects the mucosa to the muscularis layer.

Submucosal plexus A network of autonomic nerve fibers located in the superficial part of the submucous layer of the small intestine. Also called the **plexus of Meissner** (MĪZ-ner).

Subserous fascia (sub-SE-rus FASH-ē-a) A layer of connective tissue internal to the deep fascia, lying between the deep fascia and the serous membrane that lines the body cavities.

Substrate A molecule upon which an enzyme acts.

Subthalamus (sub-THAL-a-mus) Part of the diencephalon inferior to the thalamus; the substantia nigra and red nucleus extend from the midbrain into the subthalamus.

Subthreshold stimulus A stimulus of such weak intensity that it cannot initiate an action potential (nerve impulse).

Sudoriferous (soo′-dor-IF-er-us) **gland** An apocrine or eccrine exocrine gland in the dermis or subcutaneous layer that produces perspiration. Also called a **sweat gland.**

Sulcus (SUL-kus) A groove or depression between parts, especially between the convolutions of the brain. *Plural is* **sulci** (SUL-sī).

Superficial (soo′-per-FISH-al) Located on or near the surface of the body or an organ.

Superficial fascia (FASH-ē-a) A continuous sheet of fibrous connective tissue between the dermis of the skin and the deep fascia of the muscles. Also called **subcutaneous** (sub′-kū-TĀ-nē-us) **layer.**

Superficial inguinal (IN-gwi-nal) **ring** A triangular opening in the aponeurosis of the external oblique muscle that represents the termination of the inguinal canal.

Superior (soo-PĒR-ē-or) Toward the head or upper part of a structure. Also called **cephalad** (SEF-a-lad) or **craniad.**

Superior vena cava (VĒ-na CĀ-va) **(SVC)** Large vein that collects blood from parts of the body superior to the heart and returns it to the right atrium.

Supination (soo-pi-NĀ-shun) A movement of the forearm in which the palm is turned anteriorly.

Surface anatomy The study of the structures that can be identified from the outside of the body.

Surfactant (sur-FAK-tant) Complex mixture of phospholipids and lipoproteins, produced by type II alveolar (septal) cells in the lungs, that decreases surface tension.

Susceptibility (sus-sep′-ti-BIL-i-tē) Lack of resistance to the damaging effects of an agent such as a pathogen.

Suspensory ligament (sus-PEN-so-rē LIG-a-ment) A fold of peritoneum extending laterally from the surface of the ovary to the pelvic wall.

Sutural (SOO-cher-al) **bone** A small bone located within a suture between certain cranial bones. Also called **Wormian** (WER-mē-an) **bone.**

Suture (SOO-cher) An immovable fibrous joint that joins skull bones.

Sympathetic (sim′-pa-THET-ik) **division** One of the two subdivisions of the autonomic nervous system, having cell bodies of preganglionic neurons in the lateral gray columns of the thoracic segment and the first two or three lumbar segments of the spinal cord; primarily concerned with processes involving the expenditure of energy.

Sympathetic trunk ganglion (GANG-glē-on) A cluster of cell bodies of sympathetic postganglionic neurons lateral to the vertebral column, close to the body of a vertebra. These ganglia extend inferiorly through the neck, thorax, and abdomen to the coccyx on both sides of the vertebral column and are connected to one another to form a chain on each side of the vertebral column. Also called **sympathetic chain** or **vertebral chain ganglia.**

Symphysis (SIM-fi-sis) A line of union. A slightly movable cartilaginous joint such as the pubic symphysis.

Symptom (SIMP-tum) A subjective change in body function not apparent to an observer, such as pain or nausea, that indicates the presence of a disease or disorder of the body.

Synapse (SYN-aps) The functional junction between two neurons or between a neuron and an effector, such as a muscle or gland; may be electrical or chemical.

Synapsis (sin-AP-sis) The pairing of homologous chromosomes during prophase I of meiosis.

Synaptic (sin-AP-tik) **cleft** The narrow gap at a chemical synapse that separates the axon terminal of one neuron from another neuron or muscle fiber (cell) and across which a neurotransmitter diffuses to affect the postsynaptic cell.

Synaptic end bulb Expanded distal end of an axon terminal that contains synaptic vesicles. Also called a **synaptic knob.**

Synaptic vesicle Membrane-enclosed sac in a synaptic end bulb that stores neurotransmitters.

Synarthrosis (sin′-ar-THRŌ-sis) An immovable joint such as a suture, gomphosis, or synchondrosis.

Synchondrosis (sin′-kon-DRŌ-sis) A cartilaginous joint in which the connecting material is hyaline cartilage.

Syndesmosis (sin′-dez-MŌ-sis) A slightly movable joint in which articulating bones are united by fibrous connective tissue.

Syndrome (SIN-drōm) A group of signs and symptoms that occur together in a pattern that is characteristic of a particular disease or abnormal condition.

Synergist (SIN-er-jist) A muscle that assists the prime mover by reducing undesired action or unnecessary movement.

Synergistic (syn-er-JIS-tik) **effect** A hormonal interaction in which the effects of two or more hormones acting together is greater or more extensive than the sum of each hormone acting alone.

Synostosis (sin′-os-TŌ-sis) A joint in which the dense fibrous connective tissue that unites bones at a suture has been replaced by bone, resulting in a complete fusion across the suture line.

Synovial (si-NŌ-vē-al) **cavity** The space between the articulating bones of a synovial joint, filled with synovial fluid. Also called a **joint cavity.**

Synovial fluid Secretion of synovial membranes that lubricates joints and nourishes articular cartilage.

Synovial joint A fully movable or diarthrotic joint in which a synovial (joint) cavity is present between the two articulating bones.

Synovial membrane The deeper of the two layers of the articular capsule of a synovial joint, composed of areolar connective tissue that secretes synovial fluid into the synovial (joint) cavity.

System An association of organs that have a common function.

Systemic (sis-TEM-ik) Affecting the whole body; generalized.

Systemic anatomy The anatomic study of particular systems of the body, such as the skeletal, muscular, nervous, cardiovascular, or urinary systems.

Systemic circulation The routes through which oxygenated blood flows from the left ventricle through the aorta to all the organs of the body and deoxygenated blood returns to the right atrium.

Systole (SIS-tō-lē) In the cardiac cycle, the phase of contraction of the heart muscle, especially of the ventricles.

Systolic (sis-TOL-ik) **blood pressure** The force exerted by blood on arterial walls during ventricular contraction; the highest pressure measured in the large arteries, about 120 mmHg under normal conditions for a young adult.

T

T cell A lymphocyte that becomes immunocompetent in the thymus and can differentiate into a helper T cell or a cytotoxic T cell, both of which function in cell-mediated immunity.

T wave The deflection wave of an electrocardiogram that represents ventricular repolarization.

Tachycardia (tak′-i-KAR-dē-a) An abnormally rapid resting heartbeat or pulse rate (over 100 beats per minute).

Tactile (TAK-tīl) Pertaining to the sense of touch.

Tactile disc Modified epidermal cell in the stratum basale of hairless skin that functions as a cutaneous receptor for discriminative touch. Also called a **Merkel** (MER-kel) **disc.**

Target cell A cell whose activity is affected by a particular hormone.

Tarsal bones The seven bones of the ankle. Also called **tarsals.**

Tarsal gland Sebaceous (oil) gland that opens on the edge of each eyelid. Also called a **Meibomian** (mī-BŌ-mē-an) **gland.**

Tarsal plate A thin, elongated sheet of connective tissue, one in each eyelid, giving the eyelid form and support. The aponeurosis of the levator palpebrae superioris is attached to the tarsal plate of the superior eyelid.

Tarsus (TAR-sus) A collective term for the seven bones of the ankle.

Tectorial (tek-TŌ-rē-al) **membrane** A gelatinous membrane projecting over and in contact with the hair cells of the spiral organ (organ of Corti) in the cochlear duct.

Teeth (TĒTH) Accessory structures of digestion, composed of calcified connective tissue and embedded in bony sockets of the mandible and maxilla, that cut, shred, crush, and grind food. Also called **dentes** (DEN-tēz).

Telophase (TEL-ō-fāz) The final stage of mitosis.

Tendon (TEN-don) A white fibrous cord of dense regular connective tissue that attaches muscle to bone.

Tendon organ A proprioceptive receptor, sensitive to changes in muscle tension and force of contraction, found chiefly near the junctions of tendons and muscles. Also called a **Golgi** (GOL-jē) **tendon organ.**

Tendon reflex A polysynaptic, ipsilateral reflex that protects tendons and their associated muscles from damage that might be brought about by excessive tension. The receptors involved are called tendon organs (Golgi tendon organs).

Teniae coli (TĒ-nē-ē KŌ-lī) The three flat bands of thickened, longitudinal smooth muscle running the length of the large intestine, except in the rectum. *Singular is* **tenia coli.**

Tentorium cerebelli (ten-TŌ-rē-um ser′-e-BEL-ē) A transverse shelf of dura mater that forms a partition between the occipital lobe of the cerebral hemispheres and the cerebellum and that covers the cerebellum.

Teratogen (TER-a-tō-jen) Any agent or factor that causes physical defects in a developing embryo.

Terminal ganglion (TER-min-al GANG-glē-on) A cluster of cell bodies of parasympathetic postganglionic neurons either lying very close to the visceral effectors or located within the walls of the visceral effectors supplied by the postganglionic neurons.

Testis (TES-tis) Male gonad that produces sperm and the hormones testosterone and inhibin. Also called a **testicle.**

Testosterone (tes-TOS-te-rōn) A male sex hormone (androgen) secreted by interstitial endocrinocytes (Leydig cells) of a mature testis; needed for development of sperm; together with a second androgen termed **dihydrotestosterone (DHT),** controls the growth and development of male reproductive organs, secondary sex characteristics, and body growth.

Tetralogy of Fallot (tet-RAL-ō-jē of fal-Ō) A combination of four congenital heart defects: (1) constricted pulmonary semilunar valve, (2) interventricular septal opening, (3) emergence of the sorta from both ventricles instead of from the left only, and (4) enlarged right ventricle.

Thalamus (THAL-a-mus) A large, oval structure located bilaterally on either side of the third ventricle, consisting of two masses of gray matter organized into nuclei; main relay center for sensory impulses ascending to the cerebral cortex.

Thermoreceptor (THER-mō-rē-sep-tor) Sensory receptor that detects changes in temperature.

Thigh The portion of the lower limb between the hip and the knee.

Third ventricle (VEN-tri-kul) A slitlike cavity between the right and left halves of the thalamus and between the lateral ventricles of the brain.

Thoracic (thō-RAS-ik) **cavity** Superior portion of the ventral body cavity that contains two pleural cavities, the mediastinum, and the pericardial cavity.

Thoracic duct A lymphatic vessel that begins as a dilation called the cisterna chyli, receives lymph from the left side of the head, neck, and chest, left arm, and the entire body below the ribs, and empties into the junction between the internal jugular and left subclavian veins. Also called the **left lymphatic** (lim-FAT-ik) **duct.**

Thoracolumbar (thō′-ra-kō-LUM-bar) **outflow** The axons of sympathetic preganglionic neurons, which have their cell bodies in the lateral gray columns of the thoracic segments and first two or three lumbar segments of the spinal cord.

Thorax (THŌ-raks) The chest.

Thrombosis (throm-BŌ-sis) The formation of a clot in an unbroken blood vessel, usually a vein.

Thrombus A stationary clot formed in an unbroken blood vessel, usually a vein.

Thymus (THĪ-mus) A bilobed organ, located in the superior mediastinum posterior to the sternum and between the lungs, in which T cells develop immunocompetence.

Thyroid cartilage (THĪ-royd KAR-ti-lij) The largest single cartilage of the larynx, consisting of two fused plates that form the anterior wall of the larynx.

Thyroid follicle (FOL-i-kul) Spherical sac that forms the parenchyma of the thyroid gland and consists of follicular cells that produce thyroxine (T_4) and triiodothyronine (T_3).

Thyroid gland An endocrine gland with right and left lateral lobes on either side of the trachea connected by an isthmus; located anterior to the trachea just inferior to the cricoid cartilage; secretes thyroxine (T_4), triiodothyronine (T_3), and calcitonin.

Thyroid-stimulating hormone (TSH) A hormone secreted by the anterior pituitary that stimulates the synthesis and secretion of thyroxine (T_4) and triiodothyronine (T_3).

Thyroxine (thī-ROK-sēn) **(T_4)** A hormone secreted by the thyroid gland that regulates metabolism, growth and development, and the activity of the nervous system.

Tic Spasmodic, involuntary twitching of muscles that are normally under voluntary control.

Tissue A group of similar cells and their intercellular substance joined together to perform a specific function.

Tissue rejection Phenomenon by which the body recognizes the protein (HLA antigens) in transplanted tissues or organs as foreign and produces antibodies against them.

Tongue A large skeletal muscle covered by a mucous membrane located on the floor of the oral cavity.

Tonsil (TON-sil) An aggregation of large lymphatic nodules embedded in the mucous membrane of the throat.

Topical (TOP-i-kal) Applied to the surface rather than ingested or injected.

Torn cartilage A tearing of an articular disc (meniscus) in the knee.

Trabecula (tra-BEK-ū-la) Irregular latticework of thin plates of spongy bone. Fibrous cord of connective tissue serving as supporting fiber by forming a septum extending into an organ from its wall or capsule. *Plural is* **trabeculae** (tra-BEK-ū-lē).

Trabeculae carneae (KAR-nē-ē) Ridges and folds of the myocardium in the ventricles.

Trachea (TRĀ-kē-a) Tubular air passageway extending from the larynx to the fifth thoracic vertebra. Also called the **windpipe.**

Tract A bundle of nerve axons in the central nervous system.

Transplantation (tranz-plan-TĀ-shun) The transfer of living cells, tissues, or organs from a donor to a recipient or from one part of the body to another in order to restore a lost function.

Transverse colon (trans-VERS KŌ-lon) The portion of the large intestine extending across the abdomen from the right colic (hepatic) flexure to the left colic (splenic) flexure.

Transverse fissure (FISH-er) The deep cleft that separates the cerebrum from the cerebellum.

Transverse plane A plane that divides the body or organs into superior and inferior portions. Also called a **horizontal plane.**

Transverse tubules (TOO-būls) **(T tubules)** Small, cylindrical invaginations of the sarcolemma of striated muscle fibers (cells) that conduct muscle action potentials toward the center of the muscle fiber.

Trauma (TRAW-ma) An injury, either a physical wound or psychic disorder, caused by an external agent or force, such as a physical blow or emotional shock; the agent or force that causes the injury.

Tremor (TREM-or) Rhythmic, involuntary, purposeless contraction of opposing muscle groups.

Triad (TRĪ-ad) A complex of three units in a muscle fiber composed of a transverse tubule and the sarcoplasmic reticulum terminal cisterns on both sides of it.

Tricuspid (trī-KUS-pid) **valve** Atrioventricular (AV) valve on the right side of the heart.

Triglyceride (trī-GLI-cer-īd) A lipid formed from one molecule of glycerol and three molecules of fatty acids that may be either solid (fats) or liquid (oils) at room temperature; the body's most highly concentrated source of chemical potential energy. Found mainly within adipocytes. Also called a **neutral fat** or a **triacylglycerol.**

Trigone (TRĪ-gon) A triangular region at the base of the urinary bladder.

Triiodothyronine (trī-ī-ō-dō-THĪ-rō-nēn) **(T_3)** A hormone produced by the thyroid gland that regulates metabolism, growth and development, and the activity of the nervous system.

Trophoblast (TRŌF-ō-blast) The superficial covering of cells of the blastocyst.

Tropic (TRŌ-pik) **hormone** A hormone whose target is another endocrine gland.

Trunk The part of the body to which the upper and lower limbs are attached.

Tubal ligation (lī-GĀ-shun) A sterilization procedure in which the uterine (Fallopian) tubes are tied and cut.

Tubular reabsorption The movement of filtrate from renal tubules back into blood in response to the body's specific needs.

Tubular secretion The movement of substances in blood into renal tubular fluid in response to the body's specific needs.

Tumor suppressor gene A gene coding for a protein that normally inhibits cell division; loss or alteration of a tumor suppressor gene called *p53* is the most common genetic change in a wide variety of cancer cells.

Tunica albuginea (TOO-ni-ka al′-bū-JIN-ē-a) A dense white fibrous capsule covering a testis or deep to the surface of an ovary.

Tunica externa (eks-TER-na) The superficial coat of an artery or vein, composed mostly of elastic and collagen fibers. Also called the **adventitia.**

Tunica interna (in-TER-na) The deep coat of an artery or vein, consisting of a lining of endothelium, basement membrane, and internal elastic lamina. Also called the **tunica intima** (IN-ti-ma).

Tunica media (MĒ-dē-a) The intermediate coat of an artery or vein, composed of smooth muscle and elastic fibers.

Tympanic antrum (tim-PAN-ik AN-trum) An air space in the middle ear that leads into the mastoid air cells or sinus.

Tympanic (tim-PAN-ik) **membrane** A thin, semitransparent partition of fibrous connective tissue between the external auditory meatus and the middle ear. Also called the **eardrum.**

Type II cutaneous mechanoreceptor A sensory receptor embedded deeply in the dermis and deeper tissues that detects stretching of skin. Also called a **Ruffini corpuscle.**

U

Umbilical cord The long, ropelike structure containing the umbilical arteries and vein that connect the fetus to the placenta.

Umbilicus (um-BIL-i-kus *or* um-bil-Ī-kus) A small scar on the abdomen that marks the former attachment of the umbilical cord to the fetus. Also called the **navel.**

Upper limb The appendage attached at the shoulder girdle, consisting of the arm, forearm, wrist, hand, and fingers. Also called **upper extremity.**

Uremia (ū-RĒ-mē-a) Accumulation of toxic levels of urea and other nitrogenous waste products in the blood, usually resulting from severe kidney malfunction.

Ureter (Ū-rē-ter) One of two tubes that connect the kidney with the urinary bladder.

Urethra (ū-RĒ-thra) The duct from the urinary bladder to the exterior of the body that conveys urine in females and urine and semen in males.

Urinalysis (ū-ri-NAL-i-sis) An analysis of the volume and physical, chemical, and microscopic properties of urine.

Urinary (Ū-ri-ner-ē) **bladder** A hollow, muscular organ situated in the pelvic cavity posterior to the pubic symphysis; receives urine via two ureters and stores urine until it is excreted through the urethra.

Urine The fluid produced by the kidneys that contains wastes and excess materials; excreted from the body through the urethra.

Urogenital (ū′-rō-JEN-i-tal) **triangle** The region of the pelvic floor inferior to the pubic symphysis, bounded by the pubic symphysis and the ischial tuberosities, and containing the external genitalia.

Urology (ū-ROL-ō-jē) The specialized branch of medicine that deals with the structure, function, and diseases of the male and female urinary systems and the male reproductive system.

Uterine (Ū-ter-in) **tube** Duct that transports ova from the ovary to the uterus. Also called the **fallopian** (fal-LŌ-pē-an) **tube** or **oviduct.**

Uterosacral ligament (ū′-ter-ō-SĀ-kral LIG-a-ment) A fibrous band of tissue extending from the cervix of the uterus laterally to the sacrum.

Uterovesical (yū′-ter-ō-VES-i-kal) **pouch** A shallow pouch formed by the reflection of the peritoneum from the anterior surface of the uterus, at the junction of the cervix and the body, to the posterior surface of the urinary bladder.

Uterus (Ū-te-rus) The hollow, muscular organ in females that is the site of menstruation, implantation, development of the fetus, and labor. Also called the **womb.**

Utricle (Ū-tri-kul) The larger of the two divisions of the membranous labyrinth located inside the vestibule of the inner ear, containing a receptor organ for static equilibrium.

Uvea (Ū-vē-a) The three structures that together make up the vascular tunic of the eye.

Uvula (Ū-vū-la) A soft, fleshy mass, especially the V-shaped pendant part, descending from the soft palate.

V

Vagina (va-JĪ-na) A muscular, tubular organ that leads from the uterus to the vestibule, situated between the urinary bladder and the rectum of the female.

Vallate papilla (VAL-at pa-PIL-a) One of the circular projections that is arranged in an inverted V-shaped row at the back of the tongue; the largest of the elevations on the upper surface of the tongue containing taste buds. Also called **circumvallate papilla.**

Varicocele (VAR-i-kō-sēl) A twisted vein; especially, the accumulation of blood in the veins of the spermatic cord.

Varicose (VAR-i-kōs) Pertaining to an unnatural swelling, as in the case of a varicose vein.

Vas A vessel or duct.

Vasa recta (VĀ-sa REK-ta) Extensions of the efferent arteriole of a juxtamedullary nephron that run alongside the loop of the nephron (Henle) in the medullary region of the kidney.

Vasa vasorum (va-SŌ-rum) Blood vessels that supply nutrients to the larger arteries and veins.

Vascular (VAS-kū-lar) Pertaining to or containing many blood vessels.

Vascular (venous) sinus A vein with a thin endothelial wall that lacks a tunica media and externa and is supported by surrounding tissue.

Vascular spasm Contraction of the smooth muscle in the wall of a damaged blood vessel to prevent blood loss.

Vascular tunic (TOO-nik) The middle layer of the eyeball, composed of the choroid, ciliary body, and iris. Also called the **uvea** (Ū-ve-a).

Vasectomy (va-SEK-tō-mē) A means of sterilization of males in which a portion of each ductus (vas) deferens is removed.

Vasoconstriction (vāz-ō-kon-STRIK-shun) A decrease in the size of the lumen of a blood vessel caused by contraction of the smooth muscle in the wall of the vessel.

Vasodilation (vāz′-ō-DĪ-lā-shun) An increase in the size of the lumen of a blood vessel caused by relaxation of the smooth muscle in the wall of the vessel.

Vein A blood vessel that conveys blood from tissues back to the heart.

Vena cava (VĒ-na KĀ-va) One of two large veins that open into the right atrium, returning to the heart all of the deoxygenated blood from the systemic circulation except from the coronary circulation.

Ventral (VEN-tral) Pertaining to the anterior or front side of the body; opposite of dorsal.

Ventral body cavity Cavity near the ventral aspect of the body that contains viscera and consists of a superior thoracic cavity and an inferior abdominopelvic cavity.

Ventral ramus (RĀ-mus) The anterior branch of a spinal nerve, containing sensory and motor fibers to the muscles and skin of the anterior surface of the head, neck, trunk, and the limbs.

Ventricle (VEN-tri-kul) A cavity in the brain filled with cerebrospinal fluid. An inferior chamber of the heart.

Ventricular fibrillation (ven-TRIK-ū-lar fib-ri-LĀ-shun) Asynchronous ventricular contractions; unless reversed by defibrillation, results in heart failure.

Venule (VEN-ūl) A small vein that collects blood from capillaries and delivers it to a vein.

Vermiform appendix (VER-mi-form a-PEN-diks) A twisted, coiled tube attached to the cecum.

Vermilion (ver-MIL-yon) The area of the mouth where the skin on the outside meets the mucous membrane on the inside.

Vermis (VER-mis) The central constricted area of the cerebellum that separates the two cerebellar hemispheres.

Vertebral (VER-te-bral) **canal** A cavity within the vertebral column formed by the vertebral foramina of all the vertebrae and containing the spinal cord. Also called the **spinal canal.**

Vertebral column The 26 vertebrae of an adult and 33 vertebrae of a child; encloses and protects the spinal cord and serves as a point of attachment for the ribs and back muscles. Also called the **backbone, spine,** or **spinal column.**

Vesicle (VES-i-kul) A small bladder or sac containing liquid.

Vesicouterine (ves′-ik-ō-Ū-ter-in) **pouch** A shallow pouch formed by the reflection of the peritoneum from the anterior surface of the uterus, at the junction of the cervix and the body, to the posterior surface of the urinary bladder.

Vestibular (ves-TIB-ū-lar) **apparatus** Collective term for the organs of equilibrium, which includes the saccule, utricle, and semicircular ducts.

Vestibular membrane The membrane that separates the cochlear duct from the scala vestibuli.

Vestibule (VES-ti-būl) A small space or cavity at the beginning of a canal, especially the inner ear, larynx, mouth, nose, and vagina.

Villus (VIL-lus) A projection of the intestinal mucosal cells containing connective tissue, blood vessels, and a lymphatic vessel; functions in the absorption of the end products of digestion. *Plural is* **villi** (VIL-ī).

Viscera (VIS-er-a) The organs inside the ventral body cavity. *Singular is* **viscus** (VIS-kus).

Visceral (VIS-er-al) Pertaining to the organs or to the covering of an organ.

Visceral effectors (e-FEK-torz) Organs of the ventral body cavity that respond to neural stimulation, including cardiac muscle, smooth muscle, and glands.

Vital signs Signs necessary to life that include temperature (T), pulse (P), respiratory rate (RR), and blood pressure (BP).

Vitamin An organic molecule necessary in trace amounts that acts as a catalyst in normal metabolic processes in the body.

Vitreous (VIT-rē-us) **body** A soft, jellylike substance that fills the vitreous chamber of the eyeball, lying between the lens and the retina.

Vocal folds Pair of mucous membrane folds below the ventricular folds that function in voice production. Also called **true vocal cords.**

Volkmann's canal *See* **Perforating canal.**

Vulva (VUL-va) Collective designation for the external genitalia of the female. Also called the **pudendum** (poo-DEN-dum).

W

Wallerian (wal-LE-rē-an) **degeneration** Degeneration of the portion of the axon and myelin sheath of a neuron distal to the site of injury.

Wandering macrophage (MAK-rō-fāj) Phagocytic cell that develops from a monocyte, leaves the blood, and migrates to infected tissues.

White matter Aggregations or bundles of myelinated and unmyelinated axons located in the brain and spinal cord.

White pulp The regions of the spleen composed of lymphatic tissue, mostly B lymphocytes.

White ramus communicans (RĀ-mus kō-MŪ-ni-kans) The portion of a preganglionic sympathetic axon that branches from the anterior ramus of a spinal nerve to enter the nearest sympathetic trunk ganglion.

X

Xiphoid (ZĪ-foyd) Sword-shaped. The inferior portion of the sternum is the **xiphoid process.**

Y

Yolk sac An extraembryonic membrane composed of the exocoelomic membrane and hypoblast. It transfers nutrients to the embryo, is a source of blood cells, contains primordial germ cells that migrate into the gonads to form primitive germ cells, forms part of the gut, and helps prevent desiccation of the embryo.

Z

Zona fasciculata (ZŌ-na fa-sik′-ū-LA-ta) The middle zone of the adrenal cortex consisting of cells arranged in long, straight cords that secrete glucocorticoid hormones, mainly cortisol.

Zona glomerulosa (glo-mer′-ū-LŌ-sa) The outer zone of the adrenal cortex, directly under the connective tissue covering, consisting of cells arranged in arched loops or round balls that secrete mineralocorticoid hormones, mainly aldosterone.

Zona pellucida (pe-LOO-si-da) Clear glycoprotein layer between a secondary oocyte and the surrounding granulosa cells of the corona radiata.

Zona reticularis (ret-ik′-ū-LAR-is) The inner zone of the adrenal cortex, consisting of cords of branching cells that secrete sex hormones, chiefly androgens.

Zygote (ZĪ-got) The single cell resulting from the union of male and female gametes; the fertilized ovum.

CREDITS

PHOTO CREDITS

Chapter 1 Opener: Courtesy Yale University, Harvey Cushing/John Hay Whitney Medical Library. Page 3: John Wilson. Page 9 (top): Stephen A. Kieffer and E. Robert Heitzman, *An Atlas of Cross-Sectional Anatomy*. Harper & Row, New York, 1979. Page 9 (center): Lester V. Bergman/Project Masters, Inc. Page 9 (bottom): Martin Rotker. Page 14: Mark Nielsen. Page 18 (top left): Biophoto Associates/Photo Researchers. Page 18 (top and center): Breast Cancer Unit, Kings College Hospital, London/Photo Researchers. Page 18 (top right): Zephyr/Photo Researchers. Page 18 (bottom left): Cardio-Thoracic Centre, Freeman Hospital, Newcastle-Upon-Tyne/Photo Researchers. Page 18 (bottom and center): CNRI/Science Photo Library/Photo Researchers. Page 18 (bottom right): Science Photo Library/Photo Researchers. Page 19 (left): Scott Camazine/Photo Researchers. Page 19 (right): Simon Fraser/Photo Researchers. Page 20 (top left): Courtesy Andrew Joseph Tortora and Damaris Soler. Page 20 (top right): Howard Sochurek/Medical Images, Inc. Page 20 (bottom left): SIU/Visuals Unlimited. Page 20 (bottom right): Dept. of Nuclear Medicine, Charing Cross Hospital/Photo Researchers.

Chapter 2 Opener: John Reader/Photo Researchers. Page 30: Courtesy Abbott Laboratories. Page 33: Courtesy Kent McDonald, UC Berkeley Electron Microscope Laboratory. Page 34: D.W. Fawcett/Photo Researchers. Page 35: Biophoto Associates/Photo Researchers. Page 37: Courtesy Daniel S. Friend, Harvard Medical School. Pages 39 and 40: D.W. Fawcett/Photo Researchers. Page 45: Courtesy Michael Ross, University of Florida.

Chapter 3 Opener: Kazuyoshi Nomachi/Photo Researchers. Page 62: Biophoto Assoicates/Photo Researchers. Pages 63, 64, 65 (top), 66–68 (bottom), 75, 76, 77 (top), 78 (bottom), 79, 80 (top), 81, 85 and 86: Courtesy Michael Ross, University of Florida. Page 65 (bottom): Biophoto Associates/Photo Researchers. Page 68 (top): Lester V. Bergman/Project Masters, Inc. Page 77 (bottom): Courtesy Andrew J. Kuntzman. Page 78 (top): Ed Reschke. Page 80 (bottom): John Burbidge/Photo Researchers. Page 87: Ed Reschke.

Chapter 4 Opener: Courtesy National Library of Medicine. Page 96 (top): David Phillips/Photo Researchers. Page 96 (bottom): Myriam Wharman/Phototake. Page 108: Siu, Biomedical Comm./Custom Medical Stock Photo, Inc. Pages 113 (top left) and 114 (bottom): Photos provided courtesy of Kohei Shiota, Congenital Anomaly Research Center, Kyoto University, Graduate School of Medicine. Page 113 (top right, center left, center right, bottom): Courtesy National Museum of Health and Medicine, Armed Forces Institute of Pathology. Page 114 (top): Photo by Lennart Nilsson/Albert Bonniers Förlag AB, From *A Child is Born*, Dell Publishing Company. Reproduced with permission. Page 114 (bottom): Photo provided courtesy of Kohei Shiota, Congenital Anomaly Research Center, Kyoto University, Graduate School of Medicine.

Chapter 5 Opener: Sylvain Grandadam/Photo Researchers. Page 124: ©L.V. Bergman/Bergman Collection. Page 129: Science Photo Library/Photo Researchers. Page 131: Courtesy Michael Ross, University of Florida. Page 138 (left): Alain Dex/Photo Researchers. Page 138 (right): Biophoto Associates/Photo Researchers. Page 139 (top left): Sheila Terry/Science Photo Library/Photo Researchers. Page 139 (top and center, top right): St. Stephen's Hospital/Science Photo Library/Photo Researchers. Page 139 (bottom): Dr. P. Marazzi/Science Photo Library/Photo Researchers.

Chapter 6 Opener: National Library of Medicine/Photo Researchers. Page 146: Mark Nielsen. Page 154 (top): The Bergman Collection. Page 154 (bottom): Biophoto Associates/Photo Researchers. Page 159: P. Motta/Photo Researchers.

Chapter 7 Opener: From *Andreas Versalius of Brussels*, Dover Publications. Pages 169, 171, 173, 175, 177. 189, 191, 193 and 196: Mark Nielsen. Page 199: Center for Disease Control/Project Masters, Inc.

Chapter 8 Opener: The Royal Collection ©2003 Her Majesty Queen Elizabeth II. Pages 212 and 223: Mark Nielsen.

Chapter 9 Opener: ©Werner Forman Archive, Philip Goldman Collection, London/Art Resource. Page 233: Mark Nielsen. Pages 237–241: John Wilson White. Pages 248, 252, 255 and 256: Mark Nielsen. Page 260 (center): SIU/Visuals Unlimited. Page 260 (right): ©imagingbody.com.

Chapter 10 Opener: ©Art Resource. Page 271: Courtesy Denah Appelt and Clara Franzini-Armstrong. Page 274: Don Fawcett/Photo Researchers. Page 277: Biophoto Associates/Photo Researchers.

Chapter 11 Opener: Bernhard Siegfried Albinus, *Ecorché*, 1747, Bibl.Nationale, Paris/Art Resource. Pages 301, 304, 320, 336, 340, 346, 347, 362, 370 and 371: Mark Nielsen.

Chapter 12 Opener: Michaelangelo Buonarroti, *Creation of Adam*. Sistine Chapel, Vatican Palace, Vatican State/Art Resource. Pages 382–385, 387 (top), 388, 390, 391, 394–400: John Wiley & Sons. Page 386: Courtesy Lynne Marie Barghesi. Pages 383 (top), 387 (bottom) and 393: Andy Washnik.

Chapter 13 Opener: Courtesy National Library of Medicine. Page 409 (left): From Lennart Nilsson, *Our Body Victorious*, Boehringer Ingelheim International GmbH. Reproduced with permission. Pages 409 (right) and 414: Courtesy Michael Ross, University of Florida. Page 416: Lewin/Royal Free Hospital/Photo Researchers.

Chapter 14 Opener: Andy Warhol, *Human Heart*, 1979, The Andy Warhol Foundation, Inc./Art Resource/Artists' Rights Society. Page 426: John Wiley & Sons. Pages 429, 431, 433, 437: Mark Nielsen. Page 445 (left): ©Vu/Cabisco/Visuals Unlimited. Page 445 (right): W. Ober/Visuals Unlimited.

Chapter 15 Opener: Omikron/Photo Researchers. Page 455 (center): Dennis Strete. Page 455 (bottom): Courtesy Michael Ross, University of Florida. Pages 458, 464, 470, 477, 481, 486, 490 and 493: Mark Nielsen.

Chapter 16 Opener: Erich Lessing/Art Resource. Pages 517 and 521 (bottom): Courtesy Michael Ross, University of Florida. Page 519 (top): Leroy, Biocosmos/Photo Researchers. Page 519 (center): Lester Bergman & Associates. Pages 519 (bottom) and 521 (top): Mark Nielsen.

Chapter 17 Opener: Michelangelo Merisi Caravaggio, *Boy Bitten by a Lizard*, Private Collection, London/Art Resource. Page 541: Science VU/Visuals Unlimited. Page 546 (left): Dennis Kunkel/Phototake. Page 546 (right): Martin Rotker/Phototake.

Chapter 18 Opener: National Library of Medicine/Photo Researchers. Pages 555, 557, 571 and 574: Mark Nielsen. Page 559: Jean Claude Revy/Phototake. Page 564: ©Dr. Richard Kessel and Dr. Randy Kardon/Visuals Unlimited.

Chapter 19 Opener: Courtesy Institute of Mathematics and Computer Science in Medicine, University of Hamburg, Germany. Page 585: Mark Nielsen. Page 594: Mark Nielsen. Page 598: From Stephen A. Kieffer and E. Robert Heitzman, *An Atlas of Cross-Sectional Anatomy*, Harper and Row, Publishers, 1979. Reproduced with permission. Page 601: Mark Nielsen. Pages 602 and 604: N. Gluhbegovic and T.H. Williams, *The Human Brain: A Photographic Guide*, Harper and Row, Publishers, Inc. Hagerstown, MD, 1980.

Chapter 20 Opener: Courtesy National Library of Medicine.

Chapter 21 Opener: Dr. Jeremy Burgess/Photo Researchers.

Chapter 22 Opener: George Bernard/Photo Researchers. Pages 673, 675, 680 and 681: Courtesy Michael Ross. Page 684: N. Gluhbegovic and T.H. Williams, *The Human Brain: A Photographic Guide*, Harper and Row, Publishers, Inc. Hagerstown, MD, 1980. Page 690: John D. Cunningham/Visuals Unlimited.

Chapter 23 Opener: Giraudon/Art Resource. Pages 708, 712 (right), 714 (bottom): Mark Nielsen. Pages 709, 712 (left), 714 (center), 716 (right), 718 (right): Courtesy Michael Ross, University of Florida. Page 716 (left): Andrew J. Kuntzman. Page 718 (bottom): Courtesy James Sheetz, Department of Cell Biology, University of Alabama, Birmingham. Page 724 (left): From New England Journal of Medicine, February 18, 1999, vol. 340, No. 7, page 524. Photo provided courtesy of Robert Gagel, Department of Internal Medicine, University of Texas M.D. Anderson Cancer Center, Houston Texas. Page 724 (top and cen-

ter, top and right, bottom left): ©The Bergman Collection/Project Masters, Inc. Page 724 (bottom right): Biophoto Associates/Photo Researchers.

Chapter 24 Opener: Courtesy National Library of Medicine. Page 733: From J. W. Rohen, Ch. Yokochi, E. Luetjen, Drecoll, *Color Atlas of Anatomy*, 5e., Lippincott Williams & Wilkins Publishers. Pages 735, 739, 742, 744, 746: Mark Nielsen. Page 740: John Cunningham/Visuals Unlimited. Page 743: Courtesy Michael Ross, University of Florida. Pages 748 and 749: Biophoto Associates/Photo Researchers.

Chapter 25 Opener: Giraurdon/Art Resource. Page 766: From *A Stereoscopic Atlas of Human Anatomy*. Reproduced with permission of Robert A. Chase. Pages 769, 783, 785, 789 and 795: Mark Nielsen. Pages 772, 777, 792 and 797: Courtesy Michael Ross, University of Florida. Page 775 (left): Courtesy Dr. D. Gentry Steele. Page 775 (right): H. Hubatka/Maritus Gnbj/Phototake. Page 779: From Johannes W. Rohen, Chihiro Yokochi and Elke Lütjen-Drecoll, *Color Atlas of Anatomy*, F.K. Schattauer, Stuttgart/Germany, and by Lippincott, Williams and Wilkins, Baltimore, 5ed., 2002. Page 780: Hessler/Vu/Visuals Unlimited. Page 781: Ed Reschke. Page 786: Courtesy Michael Ross, University of Florida. Page 791 (top): Fred E. Hossler/Visuals Unlimited. Page 791 (bottom): G. W. Willis/Visuals Unlimited.

Chapter 26 Opener: Courtesy Karl Heinz Hoehne, Institute of Mathematics and Computer Science in Medicine (IMDM), University of Hamburg. Pages 809, 812, 814 and 823: Mark Nielsen. Page 816: Dennis Strete. Page 821: Courtesy Michael Ross, University of Florida. Page 824: Biophoto Associates/Photo Researchers.

Chapter 27 Opener: The Royal Collection ©2003 Her Majesty Queen Elizabeth II. Pages 837, 839, 851, 857 and 865: Mark Nielsen. Pages 840, 843–845, 849, 853 (left), 854 (left), 858–859, 861 and 866: Courtesy Michael Ross, University of Florida. Page 853 (right): Claude Edelmann/Photo Researchers. Page 854 (right): Biophoto Associates/Photo Researchers.

ART CREDITS

Chapter 1 1.1: Tomo Narashima. 1.2, 1.3: Molly Borman. 1.5: Kevin Somerville. 1.6, 1.7a: Imagineering. 1.7b, 1.8–1.10: Kevin Somerville. Table 1.2: Keith Kasnot.

Chapter 2 2.1, 2.2: Tomo Narashima. 2.3–2.6: Imagineering. 2.7–2.10: Tomo Narashima/Imagineering. 2.12: Imagineering. 2.13–2.15: Tomo Narashima/Imagineering. 2.16: Imagineering. 2.17, 2.18: Jared Schneidman Design. 2.19–2.22: Imagineering.

Chapter 3 3.1: Kevin Somerville. 3.2: Imagineering. 3.3, 3.4: Kevin Somerville. 3.5, 3.6: Imagineering. 3.7a-b: Kevin Somerville. 3.7c: Keith Kasnot/Kevin Somerville. 3.7d: Leonard Dank/Kevin Somerville. Table 3.1, Table 3.2, Table 3.3: Nadine Sokol/Kevin Somerville. Table 3.4, Table 3.5: Nadine Sokol.

Chapter 4 4.1–4.13, 4.15, Table 4.2: Kevin Somerville.

Chapter 5 5.1–5.3: Kevin Somerville. 5.4: Nadine Sokol. 5.5–5.8: Kevin Somerville.

Chapter 6 6.1: Leonard Dank. 6.2: Lauren Keswick. 6.3–6.6, 6.8: Kevin Somerville. 6.9: Leonard Dank. 6.10: Kevin Somerville.

Chapter 7 7.1–7.9: Leonard Dank. 7.9e: Sharon Ellis. 7.10–7.12: Leonard Dank. 7.13a-b: Kevin Somerville. 7.13c, 7.14, 7.15: Leonard Dank. 7.16: Imagineering/Leonard Dank. 7.17–7.22: Leonard Dank. 7.23: Kevin Somerville. 7.24, 7.25: Leonard Dank. 7.26: Imagineering. Table 7.1, Table 7.3: Nadine Sokol. Table 7.5: Leonard Dank.

Chapter 8 8.1: Leonard Dank. 8.2: Kevin Somerville. 8.3–8.5: Leonard Dank. 8.6: Molly Borman. 8.7–8.17: Leonard Dank. 8.18a: Kevin Somerville. 8.18b-e, Table 8.1: Leonard Dank.

Chapter 9 9.1–9.4, 9.11–9.15: Leonard Dank. 9.16: Imagineering. 9.17: Leonard Dank.

Chapter 10 10.1, 10.2: Kevin Somerville. 10.3, 10.5: Imagineering. 10.6: Kevin Somerville. 10.7: Imagineering. 10.8: Kevin Somerville. 10.9: Imagineering. 10.10: Kevin Somerville.

Chapter 11 11.1–11.20a-e: Leonard Dank. 11.20f: Kevin Somerville. 11.21–11.25: Leonard Dank. Table 11.1: Kevin Somerville.

Chapter 12 12.11a: Kevin Somerville.

Chapter 13 13.1, 13.3, 13.4, Table 13.3: Nadine Sokol.

Chapter 14 14.1, 14.2a: Kevin Somerville. 14.2b: Imagineering. 14.2c, 14.3–14.6: Kevin Somerville. 14.7a: Nadine Sokol. 14.7b: Imagineering. 14.8, 14.9, 14.11–14.13: Kevin Somerville. 14.15: Hilda Muinos. 14.16: Kevin Somerville.

Chapter 15 15.1: Kevin Somerville. 15.2: Imagineering. 15.3: Kevin Somerville. 15.4: Leonard Dank. 15.5–15.12, 15.14–15.18: Kevin Somerville.

Chapter 16 16.1: Molly Borman. 16.2: Sharon Ellis. 16.3: Molly Borman. 16.4: Imagineering. 16.5: Steve Oh. 16.6: Molly Borman. 16.7: Steve Oh. 16.8–16.12: Molly Borman. 16.13: Kevin Somerville. 16.14: Nadine Sokol/Imagineering.

Chapter 17 17.1: Kevin Somerville. 17.2: Jared Schneidman Design. 17.3: Kevin Somerville. 17.4, 17.5: Imagineering. 17.6–17.8: Kevin Somerville. 17.9: Sharon Ellis. 17.10: Imagineering.

Chapter 18 18.1–18.4: Kevin Somerville. 18.5: Leonard Dank. 18.6, 18.7: Kevin Somerville. 18.8, 18.9: Steve Oh. 18.10: Imagineering.

18.11, 18.12: Steve Oh. 18.13: Imagineering. Table 18.1: Jared Schneidman Design.

Chapter 19 19.1–19.8: Kevin Somerville. 19.9: Imagineering. 19.10–19.15: Kevin Somerville. 19.16: Kevin Somerville/Tomo Narashima. 19.17: Kevin Somerville/Lynn O'Kelley/Sharon Ellis. 19.18–19.20: Kevin Somerville/Sharon Ellis. 19.21: Kevin Somerville/Tomo Narashima. 19.22: Kevin Somerville/Sharon Ellis. 19.23: Kevin Somerville/Imagineering. 19.24, 19.25: Kevin Somerville/Sharon Ellis. 19.26, 19.27: Kevin Somerville. Table 19.1: Imagineering.

Chapter 20 20.1–20.3: Imagineering. 20.4, 20.5: Kevin Somerville. 20.6: Imagineering.

Chapter 21 21.1, 21.2: Kevin Somerville. 21.3: Leonard Dank. 21.4: Kevin Somerville. 21.5: Imagineering. 21.6, Table 21.3: Kevin Somerville.

Chapter 22 22.1: Tomo Narashima. 22.2: Molly Borman. 22.3: Sharon Ellis. 22.4: Tomo Narashima. 22.7: Lynn O'Kelley. 22.8: Tomo Narashima. 22.9: Imagineering. 22.10–22.14: Tomo Narashima. 22.15, 22.16: Tomo Narashima/Sharon Ellis. 22.16: Tomo Narashima/Sharon Ellis. 22.17, 22.18, Table 22.1: Kevin Somerville. Table 22.2: Imagineering.

Chapter 23 23.1: Steve Oh. 23.2, 23.3: Lynn O'Kelley. 23.4–23.7: Molly Borman. 23.8: Kevin Somerville. Table 23.2, Table 23.4: Nadine Sokol. Table 23.7: Imagineering.

Chapter 24 24.1: Molly Borman. 24.2a: Kevin Somerville. 24.2b, 24.3, 24.4: Molly Borman. 24.5: Steve Oh. 24.7: Molly Borman. 24.9: Lynn O'Kelley. 24.10: Imagineering. 24.11–24.13: Kevin Somerville. 24.14: Imagineering. 24.15, 24.16: Kevin Somerville.

Chapter 25 25.1: Steve Oh. 25.2: Kevin Somerville. 25.3: Steve Oh. 25.4–25.7, 25.9: Nadine Sokol. 25.10: Steve Oh. 25.11: Kevin Somerville. 25.12a: Steve Oh. 25.12b: Jared Schneidman Design. 25.13: Steve Oh. 25.14: Kevin Somerville. 25.15: Jared Schneidman Design. 25.16, 25.17: Kevin Somerville. 25.19: Molly Borman. 25.20: Kevin Somerville.

Chapter 26 26.1, 26.2: Kevin Somerville. 26.3, 26.4: Steve Oh. 26.5: Nadine Sokol/Imagineering. 26.6: Kevin Somerville. 26.8: Kevin Somerville/Imagineering. 26.9: Steve Oh. 26.12, 26.13: Kevin Somerville. Table 26.1: Imagineering.

Chapter 27 27.1–27.4: Kevin Somerville. 27.5: Jared Schneidman Design. 27.6–27.8: Imagineering. 27.9–27.13: Kevin Somerville. 27.15: Jared Schneidman Design. 27.16, 27.19, 27.21–27.23: Kevin Somerville. 27.25: Jared Schneidman Design. 27.26, 27.27: Kevin Somerville.

GATEFOLD PHOTO CREDITS

The Integumentary System Figure 1: Dr. Arthur Tucker/Photo Researchers. Figure 2: David Becker/Photo Researchers. Figures 3, 6, 7 and 9: Courtesy Michael Ross, University of Florida. Figures 4 and 8: Courtesy Roger C. Wagner, University of Delaware, Newark. Figure 5: Dee Breger/Photo Researchers. Figure 10: Science Vu/Visuals Unlimited.

The Skeletal System Figure 1: ISM/Phototake. Figure 2: Alfred Pasieka/Photo Researchers. Figure 3: Science Photo Library/Photo Researchers. Figure 4: Susan Leavines/Photo Researchers. Figure 5: CNRI/Phototake. Figure 6: SIU/Visuals Unlimited. Figure 7: S. Fraser/Photo Researchers. Figure 8: Carolina Biological Supply Co./Phototake.

The Muscular System Figure 1: Courtesy Department of Radiology, University Hospitals of Cleveland. Figures 2, 8 and 9: Courtesy Roger C. Wagner, University of Delaware, Newark. Figures 3 and 7: Dennis Kunkel/Phototake. Figure 4: Eric Grave/Phototake. Figure 5: ©Quest/Photo Researchers. Figure 6: John T. Hansen/Phototake. Figure 8: Courtesy Roger C. Wagner, University of Delaware, Newark.

The Cardiovascular System Figures 1 and 11: CNRI/Photo Researchers. Figure 2: L. Bassett/Visuals Unlimited. Researchers. Figure 3: Photo Researchers. Figure 4: CNRI/Phototake. Figure 5: Image Shop/Phototake. Figure 6: ©GJLP/Phototake. Figure 7: Juergen Bergen, Max-Planck Institute/Photo Researchers. Figures 8 and 9: Courtesy Michael Ross, University of Florida. Figure 10: ©ISM/Phototake.

The Lymphatic and Immune System Figures 1 and 8: Mark Nielsen. Figure 2: Jean-Perrin/CNRI/Photo Researchers. Figure 3: CNRI/Science Photo Library/Photo Researchers. Figure 4: From J.W. Rohen and C. Yokochi, *The Color Atlas of Anatomy*, 3e. Reprinted with permission of Lippincott, Wiliams & Wilkins. Figures 5 and 7: Gladden Willis/Visuals Unlimited. Figure 6: Science Photo Library/Photo Researchers. Figure 9: Alfred Pasieka/Science Photo Library/Photo Researchers. Figure 10: Phototake.

The Nervous System Figure 1: L. Bassett/Visuals Unlimited. Figures 2 and 11: CNRI/Phototake. Figure 3: Mehau Kulyk/Photo Researchers. Figure 4: CMEABG/UCBL1/Phototake. Figure 5: Dr. David Scott/Phototake. Figure 6: ©ISM/Phototake. Figure 7: Barbara Galati/Phototake. Figure 8: Karl E. Deckart/Phototake. Figure 9: Science Photo Library/Photo Researchers. Figure 10: Courtesy National Institute of Aging.

The Endocrine System Figure 1: L. Bassett/Visuals Unlimited. Figures 2, 4 and 9: Mark Nielsen. Figure 3: CNRI/Phototake. Figure 5: Courtesy Michael Ross, University of Florida. Figure 6: Chris Priest/Photo Researchers. Figure 7: ©Dr. Richard Kessel and Dr. Randy Kardon/Visuals Unlimited. Figure 8: ISM/Phototake. Figures 10 and 11: The Bergman Collection/Project Masters, Inc. Figure 12: Al Lamme/Phototake. Figure 13: ISM/Phototake.

The Respiratory System Figure 1: Courtesy Roger C. Wagner, University of Delaware, Newark. Figure 2: Courtesy Michael Ross, University of Florida. Figures 3 and 10: ©imagingbody.com. Figures 4 and 7: CNRI/Photo Researchers. Figure 5: ©ISM/Phototake. Figure 6: David Phillips/Visuals Unlimited. Figure 8: Gjlp/Photo Researchers. Figure 9: CNRI/Photo Researchers. Figure 11: BSIP/Phototake.

The Digestive System Figure 1: David M. Martin/Science Photo Library/Photo Researchers. Figure 2: ©GJLP/Phototake. Figure 3: ©PHT/Photo Researchers. Figure 4: CNRI/Phototake. Figure 5: Biophoto Associates/Photo Researchers. Figures 6 and 7: ©Camal/Phototake. Figure 8: David Phillips/Visuals Unlimited. Figure 9: CNRI/Photo Researchers. Figure 10: L. Bassett/Visuals Unlimited. Figure 11: Courtesy National Library of Medicine's Visible Human Project.

The Urinary System Figure 1: L. Bassett/Visuals Unlimited. Figures 2, 4 and 5: ©imagingbody.com. Figure 3: Dennis Kunkel/Phototake. Figure 6: Geoff Tompkinson/Photo Researchers. Figures 7 and 9: CNRI/Photo Researchers. Figure 8: Frederick Skvara/Visuals Unlimited.

The Reproductive System Figure 1: CNRI/Phototake. Figure 2: L. Bassett/Visuals Unlimited. Figures 3, 9 and 10: ISM/Phototake. Figure 4: Mike Abbey/Visuals Unlimited. Figure 5: E. Walker/Photo Researchers. Figure 6: Courtesy Dan Martin, M.D. Figure 7: Eurelios/Phototake. Figure 8: ©imagingbody.com.

Index

EPONYMS USED IN THIS TEXT

An **eponym** is a term that includes reference to a person's name; for example, you may be more familiar with *Achilles tendon* than you are with the technical, but correct term *calcaneal tendon*. Because eponyms remain in frequent use, this glossary has been prepared to indicate which currect terms have been used to replace eponyms in this book. In the body of the text eponyms are cited in parentheses, immediately following the correct terms where they are used for the first time in a chapter or later in the book. In addition, although eponyms are included in the index, they have been cross-referenced to their currect terminology.

EPONYM	CURRENT TERMINOLOGY
Achilles tendon	calcaneal tendon
Adam's apple	thyroid cartilage
ampulla of Vater (VA-ter)	hepatopancreatic ampulla
Bartholin's (BAR-tō-linz) gland	greater vestibular gland
Billroth's (BIL-rōtz) cord	splenic cord
Bowman's (BŌ-manz) capsule	glomerular capsule
Bowman's (BŌ-manz) gland	olfactory gland
Broca's (BRŌ-kaz) area	motor speech area
Brunner's (BRUN-erz) gland	duodenal gland
bundle of His (HISS)	atrioventricular (AV) bundle
canal of Schlemm (SHLEM)	scleral venous sinus
circle of Willis (WIL-is)	cerebral arterial circle
Cooper's (KOO-perz) ligament	suspensory ligament of the breast
Cowper's (KOW-perz) gland	bulbourethral gland
crypt of Lieberkühn (LĒ-ber-kūn)	intestinal gland
duct of Rivinus (re-VĒ-nus)	lesser sublingual duct
duct of Santorini (san'-tō-RĒ-nē)	accessory duct
duct of Wirsung (VĒR-sung)	pancreatic duct
end organ of Ruffini (roo-FĒ-nē)	type II cutaneous mechanoreceptor
Eustachian (ū-STĀ-kē-an)	auditory tube
Fallopian (fal-LŌ-pē-an) tube	uterine tube
gland of Littré (LĒ-tra)	urethral gland
gland of Zeis (ZĪS)	sebaceous ciliary gland
Golgi (GOL-jē) tendon organ	tendon organ
Graafian (GRAF-ē-an) follicle	mature ovarian follicle
Hassall's (HAS-alz) corpuscle	thymic corpuscle
Haversian (ha-VĒR-shun) canal	central canal
Haversian (ha-VĒR-shun) system	osteon
Heimlich (HĪM-lik) maneuver	abdomial thrust maneuver
interstitial cell of Leydig (LĪ-dig)	interstitial endocrinocyte
islet of Langerhans (LANG-er-hanz)	pancreatic islet
Kupffer's (KOOP-ferz) cell	stellate reticuloendothelial cell

EPONYM	CURRENT TERMINOLOGY
loop of Henle (HEN-lē)	loop of the nephron
Luschka's (LUSH-kaz) aperture	lateral aperture
Magendie's (ma-JEN-dēz) aperture	median aperture
Malpighian (mal-PIG-ē-an) corpuscle	splenic nodule
Meibomian (mī-BŌ-mē-an) gland	tarsal gland
Meissner's (MĪS-nerz) corpuscle	corpuscle of touch
Merkel's (MER-kelz) disc	tactile disc
Müller's (MIL-erz) duct	paramesonephric duct
Nissl (NIS-l) bodies	chromatophilic substances
organ of Corti (KOR-tē)	spiral organ
Pacinian (pa-SIN-ē-an) corpuscle	lamellated corpuscle
Peyer's (PĪ-erz) patch	aggregated lymphatic follicle
plexus of Auerbach (OW-er-bak)	myenteric plexus
plexus of Meissner (MĪS-ner)	submucosal plexus
pouch of Douglas	rectouterine pouch
Rathke's (rath-KĒZ) pouch	hypophyseal pouch
Sertoli (ser-TŌ-lē) cell	sustentacular cell
Sharpey's (SHAR-pēz) fiber	perforating fiber
Sheath of Schwann (SCHVON)	neurolemma
Skene's (SKĒNZ) gland	paraurethral gland
sphincter of Oddi (OD-dē)	sphincter of the hepatopancreatic ampulla
Stensen's (STEN-senz) duct	parotid duct
Volkmann's (FŌLK-manz) canal	perforating canal
Wernicke's (VER-ni-kēz) area	auditory association area
Wharton's (HWAR-tunz) duct	submandibular duct
Wharton's (HWAR-tunz) jelly	mucous connective tissue
Wormian (WER-mē-an) bone	sutural bone